Physiology in Health and Disease

Published on behalf of the American Physiological Society by Springer

Physiology in Health and Disease

This book series is published on behalf of the American Physiological Society (APS) by Springer. Access to APS books published with Springer is free to APS members.
APS publishes three book series in partnership with Springer: *Physiology in Health and Disease* (formerly *Clinical Physiology*), *Methods in Physiology*, and *Perspectives in Physiology* (formerly *People and Ideas*), as well as general titles.

More information about this series at http://www.springer.com/series/11780

Kirk L. Hamilton • Daniel C. Devor
Editors

Studies of Epithelial Transporters and Ion Channels

Ion Channels and Transporters of Epithelia in Health and Disease - Vol. 3

Second Edition

Springer

american
physiological
society

Editors
Kirk L. Hamilton
Department of Physiology, School of Biomedical Sciences
University of Otago
Dunedin, Otago, New Zealand

Daniel C. Devor
Department of Cell Biology
University of Pittsburgh
Pittsburgh, PA, USA

ISSN 2625-252X ISSN 2625-2538 (electronic)
Physiology in Health and Disease
ISBN 978-3-030-55453-8 ISBN 978-3-030-55454-5 (eBook)
https://doi.org/10.1007/978-3-030-55454-5

This Springer imprint is published by the registered company Springer Nature Switzerland AG.
The registered company address is: Gewerbestrasse 11, 6330 Cham, Switzerland

We dedicate this second edition to our families . . . Judy, Nathan, and Emma for KLH, and Cathy, Caitlin, Emily, and Daniel for DCD.

Preface to Second Edition—Volume 3

Our ultimate goal for the first edition of *Ion Channels and Transporters of Epithelia in Health and Disease* was to provide a comprehensive and authoritative volume that encapsulated the most recent research findings in the basic molecular physiology of epithelial ion channels and transporters of molecular diseases from the laboratory bench top to the bedside. Additionally, we envisioned that the book would be very exciting and useful to a range of readers from undergraduate and postgraduate students, to postdoctoral fellows, and to research and clinical scientists providing a wealth of up-to-date research information in the field of epithelial ion channels and transporters in health and disease. We firmly believe that the first edition fulfilled a niche that was crucially required. We have been informed that the first edition of the book has proven to be the best performing APS/Springer book based on downloaded chapters, to date. This is a direct testament to the world-class scientists and clinicians who contributed excellent chapters to that edition. Of course, there were many epithelial ion channels and transporters which were not included in the first edition, but certainly warranted inclusion.

With our second edition, we have superseded our original expectations by increasing the number of chapters from 29 in the first edition to a 3-volume second edition including 54 chapters; resulting in 25 new chapters. All of the original chapters have been expanded. Again, we were very fortunate to recruit "key" outstanding scientists and clinicians who contributed excellent chapters, some who were unable to commit to the first edition. In the end, the second edition has a total of 128 authors from 13 countries across four continents and both hemispheres. We truly believe that this book series represents a worldwide collaboration of outstanding international scientists and clinicians.

Volume 3: Studies of Epithelial Transporters and Ion Channels

This is the third of three volumes highlighting the importance of epithelial ion channels and transporters in the basic physiology and pathophysiology of human diseases. This volume has been expanded for the second edition. Volume 3 consists of 30 chapters, including 11 new chapters, and 2 original chapters with new authors, written by experts of ion transporter and ion channel families. Additionally, this volume contains chapters from experts in the pharmacology/pharmaceutical world who have contributed chapters on the most recent preclinical drug discovery efforts, culminating in what they have learned from clinical trials. Chapter topics include the Na^+/K^+-ATPase, $Na^+/K^+/2Cl^-$ cotransporter, Na^+-Cl^- cotransporter, NBCe 1 bicarbonate cotransporter, Na^+-glucose cotransporters (including GLUT transporters), Na^+/H^+ exchangers, amino acid transporters, Na^+-Ca^{2+} exchanger, urea transporters, H^+:K^+-ATPase, zinc transporters, SLC26 transporters, CLC-2 chloride channel, Cl^-/H^+ antiporter CLC-5, CFTR, TMEM16 proteins (anoctamins), ENaC, ROMK, and Kir4.1. In addition, there are chapters on KCa3.1 and BK channels, including on their pharmacology. There is a chapter on the KCNE regulation of KCNQ channels, as well as chapters on Orai channels, TRP channels, P2X receptors, polycystins (PC1 and PC2), and aquaporins. This volume provides vast background information about these transport proteins, structure and function of specific ion transporters and ion channels, and normal physiology and pathophysiology of these transport proteins in disease.

It is our intent that the second edition continues to be the comprehensive and authoritative work that captures the recent research on basic molecular physiology of epithelial ion channels and transporters of molecular diseases. We hope this new edition will be the "go-to" compendium that provides significant detailed research results about specific epithelial ion channels and transporters, and how these proteins play roles in molecular disease in epithelial tissues.

As stated in the preface of the first edition, the massive undertaking of a book of this enormity would certainly be an "Everest" of work. We want to sincerely thank all of our authors, and their families, who have spared time from their very busy work and non-work schedules to provide exciting and dynamic chapters, which provide depth of knowledge, informative description, and coverage of the basic physiology and pathophysiology of the topic of their individual chapters.

We want to, again, thank Dr. Dee Silverthorn who planted the "initial seed" that developed into the first edition; which stemmed from a Featured Topic session entitled "Ion Channels in Health and Disease" held during the Experimental Biology meetings in Boston in April 2013 (chaired by KLH). Then, based on the performance of that edition, Dee "twisted" our arms, with love, to attempt a second edition in 2017. We, once again, want to extend our huge thanks, gratitude, and appreciation to the members of the American Physiology Society Book Committee for their continued faith in us to pursue such a monumental second edition.

As with the first edition, this 3-volume second edition would not have been possible without the excellent partnership between the American Physiological Society and Springer Nature and the publishing team in Heidelberg, Germany. Many thanks to Markus Spaeth, Associate Editor (Life Science and Books), and Dr. Andrea Schlitzberger, Project Coordinator (Book Production Germany and Asia), who guided us on our second book publication journey never dreaming that this edition would be a 3-volume book bonanza.

We extend special thanks to Anand Venkatachalam (Project Coordinator, Books, Chennai, India) at SPi Global who answered unending questions during the production process. We thank his production team who assisted us through the many stages of the publication of the second edition. We also thank Nancey Biswas (Project Management, SPi Content Solution, Puducherry, India), Nedounsejiane Narmadha (Production General, SPi Technologies, Puducherry, India), and Mahalakshmi Rajendran (Project Manager, SPi Technologies, Chennai, India) at Spi Global for their assistance for overseeing the production of the chapters during the final print and online file stages of the second edition.

We want to thank our mentors Douglas C. Eaton and the late Dale J. Benos for KLH; Michael E. Duffey and Raymond A. Frizzell for DCD; and our colleagues who guided us over the years to be able to undertake this book project.

Finally, and most importantly, we want to thank our families: Judy, Nathan, and Emma for KLH, and Cathy, Caitlin, Emily, and Daniel for DCD for all your love and support during this 8-year journey.

We dedicate this second edition to our families.

Dunedin, New Zealand
Pittsburgh, PA
July 2020

Kirk L. Hamilton
Daniel C. Devor

Preface

Ion channels and transporters play critical roles both in the homeostasis of normal function of the human body and during the disease process. Indeed, as of 2005, 16% of all Food and Drug Administration-approved drugs targeted ion channel and transporters, highlighting their importance in the disease process. Further, the Human Genome Project provided a wealth of genetic information that has since been utilized, and will again in the future, to describe the molecular pathophysiology of many human diseases. Over the years, our understanding of the pathophysiology of many diseases has been realized. The next great "step" is a combined scientific effort in basic, clinical, and pharmaceutical sciences to advance treatments of molecular diseases.

A number of unique ion channels and transporters are located within epithelial tissues of various organs including the kidney, intestine, pancreas, and respiratory tract, and all play crucial roles in various transport processes responsible for maintaining homeostasis. Ultimately, understanding the fundamentals of ion channels and transporters, in terms of function, modeling, regulation, molecular biology, trafficking, structure, and pharmacology, will shed light on the importance of ion channels and transporters in the basic physiology and pathophysiology of human diseases.

This book contains chapters written by notable world-leading scientists and clinicians in their respective research fields. The book consists of four sections. The first part of the book is entitled "Basic Epithelial Ion Transport Principles and Function" (Chaps. 1, 2, 3, 4, 5, 6, 7 and 8) and spans the broad fundamentals of chloride, sodium, potassium, and bicarbonate transepithelial ion transport, the most recent developments in cell volume regulation, the mathematical modeling of these processes, the mechanisms by which these membrane proteins are correctly sorted to the apical and basolateral membranes, and protein folding of ion channels and transporters. The chapters in Part 1 provide the foundation of the molecular "participants" and epithelial cell models that play key roles in transepithelial ion transport function of epithelia detailed throughout the rest of this volume.

The second part is entitled "Epithelial Ion Channels and Transporters" and contains seventeen chapters (Chaps. 9, 10, 11, 12, 13, 14, 15, 16, 17, 18, 19, 20, 21, 22, 23, 24 and 25) in which authors have concentrated their discussion on a particular ion channel or transporter ranging from chloride channels to the Na^+/K^+-ATPase, for example. Generally, the authors have initially provided a broad perspective of the physiology/biology of a particular ion channel or transporter in epithelial tissues, followed by a focused in-depth discussion of the latest physiology, cell biology, and molecular biology of the ion channel/transporter and then finish their discussion on aspects of pathophysiology and disease.

It will be appreciated following the discussion of the various ion channels and transporters that many of these transport proteins are potential pharmacological targets for possible treatment of disease. Therefore, the third part is entitled "Pharmacology of Potassium Channels" that consists of two chapters (Chaps. 26 and 27) that provide the latest developments on the pharmacology of calcium-activated potassium channels and small-molecule pharmacology of inward rectified potassium channels. It should be noted, however, that pharmacological information about various ion channels and transporters is also provided in some of the chapters found within Part II of this volume.

Finally, the last part in this book is entitled "Diseases in Epithelia" and consists of two chapters (Chaps. 28 and 29). These chapters are designed to bridge the basic cellular models and epithelial transport functions discussed throughout this volume with a compelling clinical perspective: from bench to bedside. In these chapters, Dr. Whitcomb discusses the role of ion channels and transporters in pancreatic disease, while Dr. Ameen and her colleagues similarly provide insight into the secretory diarrheas.

Our utmost goal, with this book, was to provide a comprehensive and authoritative volume that encapsulates the most recent research findings in the basic physiology of ion channels and transporters of molecular diseases from the laboratory bench top to the bedside. Additionally, we hope that the book will be very exciting and useful to a range of readers from students to research scientists providing a wealth of up-to-date research information in the field of epithelial ion channels and transporters in health and disease.

The undertaking of a book of this scale would always be a "mountain" of work. We want to give our heartfelt thanks to all of our authors who have taken time from their very busy work and non-work schedules to provide excellent chapters, which provided depth of knowledge, informative description, and coverage of the basic physiology and pathophysiology of the topic of their particular chapters.

We want to thank Dr. Dee Silverthorn who planted the "seed" that developed into this volume; which stemmed from a Featured Topic session entitled "Ion Channels in Health and Disease" held during the Experimental Biology meetings in Boston in April 2013 (chaired by KLH). We thank the members of the American Physiology Society (APS) Book Committee who had faith in us to pursue such an exciting book.

As with any book, this volume would not have been possible without the excellent partnership between the APS and Springer-Verlag and the publishing team at Heidelberg, Germany (Britta Mueller, Springer Editor, and Jutta Lindenborn,

Project Coordinator). We wish to thank Portia Wong, our Developmental Editor at Springer+Business Media (San Mateo, CA) and her team who assisted with the early stages of the publishing process that greatly added to this contribution. Finally, special thanks to Shanthi Ramamoorthy (Production Editor, Books) and Ramya Prakash (Project Manager) of Publishing—Springer, SP1 Content Solutions—Spi Global and their production team who assisted us through the final stages of the publication of our book.

Finally, we want to thank our mentors Douglas C. Eaton and the late Dale J. Benos for KLH; Michael E. Duffey and Raymond A. Frizzell for DCD; and our colleagues who guided us over the years to be able to undertake this volume.

Dunedin, New Zealand — Kirk L. Hamilton
Pittsburgh, PA — Daniel C. Devor
June 2015

Contents

About the Editors

Kirk L. Hamilton was born in Baltimore, Maryland, in 1953. He gained his undergraduate (biology/chemistry) and M.Sc. (ecology) degrees from the University of Texas at Arlington. He obtained his Ph.D. at Utah State University under the tutelage of Dr. James A. Gessaman, where he studied incubation physiology of Barn owls. His first postdoctoral position was at the University of Texas Medical Branch in Galveston, Texas, under the mentorship of Dr. Douglas C. Eaton where he studied epithelial ion transport, specifically the epithelial sodium channel (ENaC). He then moved to the Department of Physiology at the University of Alabama, Birmingham, for additional postdoctoral training under the supervision of the late Dr. Dale J. Benos where he further studied ENaC and nonspecific cation channels. He took his first academic post in the Department of Biology at Xavier University of Louisiana in New Orleans (1990–1994). He then joined the Department of Physiology at the University of Otago in 1994, and he is currently an Associate Professor. He has focused his research on the molecular physiology and trafficking of potassium channels (specifically KCa3.1). He has published more than 60 papers and book chapters. His research work has been funded by the NIH, American Heart Association, Cystic Fibrosis Foundation, and Lottery Health Board New Zealand. Dr. Devor and he have been collaborators since 1999. When not working, he enjoys playing guitar (blues and jazz) and volleyball. Kirk is married to Judith Rodda, a recent Ph.D. graduate in spatial ecology. They have two children, Nathan (b. 1995) and Emma (b. 1998).

Daniel C. Devor was born in Vandercook Lake, Michigan, in 1961. His education took him through the Southampton College of Long Island University, where he studied marine biology, before entering SUNY Buffalo for his Ph.D., under the guidance of Dr. Michael E. Duffey. During this time, he studied the role of basolateral potassium channels in regulating transepithelial ion transport. He subsequently did his postdoctoral work at the University of Alabama, Birmingham, under the mentorship of Dr. Raymond A. Frizzell, where he studied both apical CFTR and basolateral KCa3.1

in intestinal and airway epithelia. He joined the University of Pittsburgh faculty in 1995 where he is currently a Professor of Cell Biology. During this time, he has continued to study the regulation, gating, and trafficking of KCa3.1 as well as the related family member, KCa2.3, publishing more than 50 papers on these topics. These studies have been funded by the NIH, Cystic Fibrosis Foundation, American Heart Association, and pharmaceutical industry. When not in the lab, he enjoys photography and growing exotic plants. Dan is married to Catherine Seluga, an elementary school teacher. They have 3 children, Caitlin (b. 1990), Emily (b. 1993), and Daniel (b. 1997).

Chapter 1
Na^+/K^+-ATPase Drives Most Asymmetric Transports and Modulates the Phenotype of Epithelial Cells

Isabel Larre, Marcelino Cereijido, Omar Paez, Liora Shoshani, and Arturo Ponce

Abstract Usually, the history of an enzyme is the narrative of the works to isolate and purify it, measure its molecular weight, determine its crystal configuration, measure its activity, and so on, along years of research. The history of the Na^+/K^+-ATPase is instead a tortuous road full of pitfalls, skirmishes with physical chemistry, thermodynamics, and even philosophy. Fortunately, it has a happy ending, because it was the first known molecule to produce vectorial movement of ions, at the expense of chemical energy, cyclically modifying its selectivity. Later on, its role has evolved to act as a self-adhesion molecule at cell–cell contacts, to act as a receptor of the hormone ouabain, whose main physiological role is to modulate cell contacts, to generate a Na^+ gradient that enables co- and anti-transporters to transport net amount of ions, sugars, amino acids, i.e., to act as secondary pumps. It is enough to say that one of its crucial properties, i.e., to be expressed in a polarized manner at the intercellular membrane of transporting epithelial cells involves the β-subunit of the pump, that happens to be an adhesion molecule which plays a crucial role in that polarization mechanism.

Keywords Active transport · Na^+/K^+-ATPase · Ouabain · Epithelia phenotype · Cell–cell contacts

I. Larre
Department of Clinical and Translational Sciences, Joan C. Edwards School of Medicine, Marshall University, Huntington, WV, USA
e-mail: perezm@marshall.edu

M. Cereijido · O. Paez · L. Shoshani · A. Ponce (✉)
Departamento de Fisiología, Biofísica y Neurociencias, Centro de Investigacion y Estudios Avanzados (Cinvestav), Mexico City, Mexico
e-mail: cereijido@fisio.cinvestav.mx; shoshani@cinvestav.mx; aponce@fisio.cinvestav.mx

K. L. Hamilton, D. C. Devor (eds.), *Studies of Epithelial Transporters and Ion Channels*, Physiology in Health and Disease,
https://doi.org/10.1007/978-3-030-55454-5_1

1.1 Introduction

As recently as 1807, Humphry Davy was the first person to differentiate Na^+ from K^+. An ironic situation indeed, because if he was an average-sized person, he must have had 3.72×10^{13} cells (Bianconi et al. 2013), each one discriminating those cations along their whole life. Nevertheless, science would take another century and a half to start learning how cells identify certain ion species and handle each one differently, but from the very beginning, it was clear that, in order to explain ion distributions and movements in biological systems, the known laws of physical chemistry would not suffice. No wonder the history of the Na^+/K^+-ATPase can be compared to a double-sided step ladder, one climbed by biologists that investigated its structure, subunits and intrinsic mechanisms and the other by accomplished physicists who denounced (in a first step) but helped to solve (in a second step) formidable theoretical obstacles in the road toward understanding active transport. In a given moment discrepancies were so sharp that led to fear that life would never be explained in terms of physical and chemical principles (Cereijido et al. 2003, 2004).

Thus, for a long while, the peculiar composition of the cytoplasm was attributed to an alleged membrane impermeability to Na^+ because, if this ion were not able to penetrate to neutralize negative charges bound to macromolecules inside the cell, some other cations would have to do it to satisfy the principle of electroneutrality; hence the high K^+ content in the cell was accounted for by the claimed membrane impermeability to Na^+. This alternative was disproved right after the Second World War, when radioisotopes became available for biological research, and it was discovered that tracer Na^+ added to the bathing solution readily penetrates and distributes throughout the cytoplasm. This rekindled the question more emphatically: If Na^+ can readily penetrate into the cytoplasm, why does it remain at a concentration much lower (~5 mM) than in the extracellular water (~140 mM)?

1.2 It All Started with Émile Du Bois Raymond

In the second half of the nineteenth century, Émile Du Bois Raymond observed that a frog skin (Fig. 1.1a, *green bar*) maintains an electrical potential difference between its inner and its outer sides (i.e., an asymmetry). Galeotti (1904) studied this electrical potential as a function of the ionic composition of the solutions and proposed that it would be accounted for if the flux of sodium from the outside toward the inside (*influx*) were larger than the *outflux* in the opposite direction (Fig. 1.1b, *black arrows*). His proposal was rejected on the basis that it would be in violation of the laws of thermodynamics. Thus, in a *Gedankenexperiment*, a frog skin mounted in a doughnut-shaped chamber containing saline solution, an asymmetric permeability would increase the concentration of Na^+ on the right-hand side of the epithelium (*small magenta dots*), and the accumulation of Na^+ will start the diffusion of this ion down its gradient (counterclockwise), thus establishing a

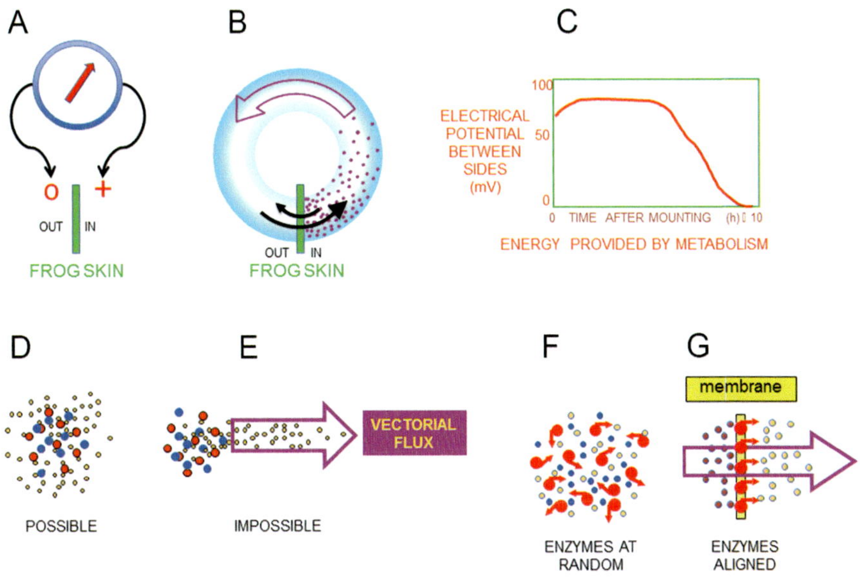

Fig. 1.1 Antecedents and roles of Na^+/K^+-ATPase. (**a**) Du Bois Raymond discovered that a frog skin can develop and maintain an electrical potential difference. (**b**) In 1905, Galeotti proposed that such potential would be accounted for if the frog skin were more permeable to Na^+ in one direction than in the opposite one. The proposal was easily refuted because it would originate a perpetuum mobile, i.e., Na^+ would rotate forever without a corresponding expenditure of free energy. (**c**) The electrical potential is not actually perpetual but lasts as long as the skin is alive, and energy is provided by metabolism. (**d–e**) Yet, according to the Curie principle, chemical reactions could not drive a unidirectional flux because this is a vectorial process. (**f**) Enzymes functioning as pumps are vectorial at the microscopic level, but when studied macroscopically, it becomes unnoticeable. (**g**). Pumps ordered as in a plasma membrane exhibit macroscopic vectoriality

perpetuum mobile (*magenta arrow*), that would in fact perform a work without the corresponding dissipation of energy. Yet biologists observed that a frog skin dissected and mounted between two chambers dies in a few hours and the electrical potential (Fig. 1.1c, *red curve*) (Ussing and Zerahn 1951) vanishes; in other words, far from being *perpetuum*, the potential lasts as long as the skin is alive. "Perhaps, the energy is afforded by metabolic energy" they proposed. Yet, given that metabolism is the sum of chemical reactions, physicists claimed that physiologists were now violating Curie's Principle: "Chemical reactions are scalar phenomena (again: a wrong assumption): they occur regardless of the orientation of the reacting molecules, and the resulting chemical products cannot diffuse in a particular direction" (Fig. 1.1d); therefore, chemical reactions would never originate a vectorial flux across a frog skin as in Fig. 1.1e (another incorrect supposition).

Biologists argued that each molecule (Fig. 1.1f, *red*) may be asymmetric, but since there are millions of molecules oriented at random, asymmetry cannot be observed at a macroscopic scale. However, if they were ordered in a membrane as suggested in Fig. 1.1g, the asymmetry would be recovered, and a macroscopic flux

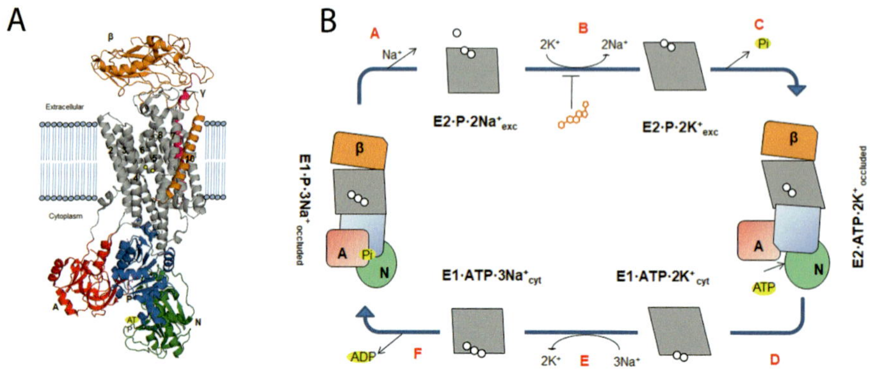

Fig. 1.2 (**a**) Ribbon model of the crystal structure of shark Na^+/K^+-ATPase ([Shinoda et al. 2009; PDB code: 2ZXE) indicating alpha subunit transmembrane domains (TM1–9, gray) and cytoplasmic domains N, P, and A in green, blue, and red, respectively. β-subunit is colored in orange. γ-subunit is colored in magenta. TM 1 and TM 9 are not visible from this projection. The two K^+ (yellow) are occluded in the crystal structure. In the N domain, the ATP binding site is also indicated. (**b**) Schematic depiction of the catalytic cycle of Na^+/K^+-ATPase. A structural rearrangement, especially in domain A (red), is suggested by the cartoons at both conformational states E1 and E2. For simplicity, the rest of the cycle stages are represented only with the TM region in gray. (1) Outward transport of three Na^+ ions (white circles) is coupled to the E1 to E1 transition. (2) Two K^+ (red circles) bind at binding sites oriented to the extracellular space. (3) Extracellularly bound K^+ activates dephosphorylation which in turn results in ion occlusion. Ouabain binds at this conformational state stopping the cycle. (4) ATP with a low affinity triggers the acceleration of inward transport of K^+ through the pore as the pump enters the E1 conformational transition with a low affinity for K^+. (5) Three Na^+ bind the intracellularly oriented sites. (6) Phosphorylation from ATP occurs and Na^+ are occluded again. Several transitional substates exist; nevertheless, they are not depicted for simplification

would take place. This made theoreticians happy, yet where was "the pump", i.e., the membrane molecule that would align and be responsible for the sided, asymmetrical movement of products? Around 1955–1959 Jens Christian Skou prepared an extract of crab tissue that contained an enzyme that splits molecules of ATP (hence deserving the name "ATPase") into ADP + Pi, provided the medium contains K^+ and Na^+ ions at concentrations that compare with those in the cell and in the surrounding extracellular space. Therefore, the enzyme was aptly named Na^+/K^+-ATPase. Interestingly, Skou (1957) was able to inhibit the ATP splitting activity of his extract by adding *ouabain*, a substance of vegetal origin that, a few years earlier was found to inhibit ion pumping when added from the outer, but not from the inner side of the cell membrane. By performing the K^+/Na^+ translocations cyclically Na^+/K^+-ATPase transfers those ions in a net amount toward the extracellular medium and toward the cytoplasm respectively, so it was justified to call it "pump." Today, thanks to high-resolution crystallography (Shinoda et al. 2009), the molecular structure of the Na^+/K^+-ATPase is known in great detail (Fig. 1.2a), and its cyclical reactions are described in Fig. 1.2b.

1.3 Polarized Distribution of Na^+/K^+-ATPase in Epithelial Cells

A unicellular organism in the ocean may consume nutrients and eliminate metabolic wastes without risking neither an exhaustion of nutrients nor pollution of the sea. In a metazoan, the ocean is represented by a narrow extracellular space that would be rapidly exhausted of nutrients and spoiled with wastes, were it not for a circulatory apparatus that moves fluids to and from tissues to transporting epithelia, where the true exchange with the external milieu takes place. Thanks to the circulation of blood, in spite of being so small, this intercellular "ocean" behaves as a constant and reliable reservoir (homeostasis). Figure 1.3a depicts the model developed by Koefoed-Johnsen and Ussing (1958) in which the pump, represented by Na^+/K^+-ATPase, is assumed to be located on the basal side of the epithelial cell, and this asymmetric distribution, together with the specific Na-permeability of the outer cell membrane (*Ocm*) and the specific K-permeability of the inner facing membrane (*Icm*) are responsible for net movement of Na^+. This model served as a blueprint to understand the net transport of important biological substances across most epithelia.

Notice that the Na^+/K^+-ATPase is *primarily* responsible for the pumping of Na^+ and K^+, but the concentration gradients that it generates result in the *secondary* transport of amino acids, sugars, and ions other than Na^+ and K^+. Since these transports occur in a net amount and can be inhibited with ouabain, for a while this was taken as a proof of the existence of glucose pumps, as well as other pumps for diverse amino acid species (Fig. 1.3b). Yet, eventually, it was demonstrated that carriers for sugars and for amino acids are not pumps, as they are not *directly* coupled to metabolism, but their affinity for sugars and amino acids drastically increases when loaded with Na^+. Today we are so used to talking about co- and

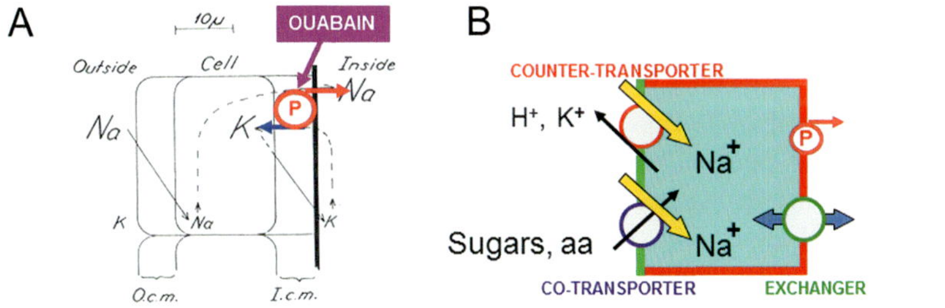

Fig. 1.3 Model proposed by Koefoed-Johnsen and Ussing (1958) to account for the net pond-to-blood transport of Na^+. (**a**) An outer cell membrane (O.c.m.) with a high selectivity for Na^+ and low selectivity for K^+ and an inner (I.c.m.) whose ion selectivity is high for K^+ and low for Na^+. A pump (P, red circle) located on the basal side takes Na^+ from the cytoplasm and transports it toward the blood (inner) side and takes K^+ from the inside and transports it toward the cytoplasm. This pump can be inhibited by ouabain added from the blood side. (**b**) The pump keeps the concentration of Na^+ in the cytoplasm at a level lower than in the extracellular fluid; this ion flows passively (yellow arrow) toward the cytoplasm, driving a cation in the opposite direction or an anion in the same direction. Furthermore, the penetration of Na^+ can also drive sugars and amino acids from the outer bathing solution to the cytoplasm

anti-transport that we forget it caused another famous, but easy to understand squirmish between biologists and physicists. Notice that co- and counter-transport assumed that the electrochemical potential difference for Na^+ can be used to move sugars and amino acids as depicted in Fig. 1.3b. Although we will not make a digression to explain how this important theoretical problem was solved, it is worth mentioning that for making a revolution in the Thermodynamics of Irreversible Processes, Lars Onsager was awarded the Nobel Prize in 1968. But since then, it was realized that, in principle, the flux of a substance (sugar, amino acid, chloride, etc.) can be moved by the energy arisen from any driving force present in the system. When Na^+/K^+-ATPase is inhibited with ouabain the concentration of sodium in the cytoplasm rises and its concentration gradient across the apical membrane vanishes. In summary, the Na^+/K^+-ATPase is the *primum movens*, responsible for the exchange of substances between metazoan and the environment across transporting epithelia, as well as for net exchange between the internal milieu and the cytoplasm.

1.3.1 Why Do Epithelial Cells Express Na^+/K^+-ATPase in a Polarized Manner?

Single cells, e.g., a leukocyte, a bacterium, or an amoeba, have Na^+/K^+-ATPase distributed at random all over the plasma membrane. The cells of transporting epithelia instead have Na^+/K^+-ATPase distributed in a peculiar way; in the model of Koefoed-Johnsen and Ussing (1958), for instance (Fig. 1.3a), they are assumed to be expressed mainly on the basal membrane. For a while, the apical/basolateral polarity of epithelial cells was erroneously attributed to the tight junction (TJ). This mistaken idea was based on the demonstration that the chelation of Ca^{2+} with EDTA or EGTA not only opens the TJs but destroys the apical/basolateral polarity, as well. However, TJs have no sorting mechanisms and, at most, maintain the apical/basolateral polarization that some molecular species have obtained by other mechanisms (e.g., lipids). This was clearly demonstrated by studies of Contreras et al. (1989), who found that MDCK (Madin–Darby canine kidney) cells adopt a spherical shape after harvesting and resuspension and a large fraction of plasma membrane is endocytosed along with a large amount of Na^+/K^+-ATPase units inserted on it, whereas those remaining at the surface lose their polarity and become randomly distributed. When cells are seeded again, and subject to the "calcium switch protocol" (Gonzalez-Mariscal et al. 1990; Contreras et al. 1992), TJ forms so quickly that catches Na^+/K^+-ATPase still randomized, but in spite of the presence of the barrier constituted by the already sealed TJ, the enzyme bypasses this structure and polarizes correctly. Hence, to explain the nonrandom distribution of molecules in the plasma membrane, it was necessary to study the intracellular routes and sorting compartments involved, such as the endoplasmic reticulum (ER) and the trans-Golgi network (TGN), and different carrier vesicles for delivery to their corresponding plasma membrane domains (Weisz and Rodriguez-Boulan 2009; Rodriguez-Boulan and Macara 2014). For each of these routes, polarity depends critically on the existence of specific signals (motifs) encoded within the membrane proteins themselves. The basolateral sorting signals are short peptide sequences most often found within the cytoplasmic domain of the protein. Some basolateral sorting signals resemble endocytic signals, e.g., variations of the canonical endocytic dileucine, YXXΦ, where Φ is any hydrophobic amino acid, and NPXY motifs. Other basolateral signals are unrelated to endocytic signals, e.g., the tyrosine motifs in the low-density lipoprotein receptor and the G-protein of the vesicular stomatitis virus (VSV). Early studies demonstrated that the Na^+/K^+-ATPase, comprising α and β subunits, is sorted in the TGN and delivered directly to the basolateral membrane without significant appearance at the apical surface in certain strains of MDCK cells (Caplan et al. 1986; Mays et al. 1995). Therefore, a basolateral signal was assumed to exist in the α-subunit of the Na^+/K^+-ATPase. The Na^+/K^+-ATPase and H^+/K^+-ATPase are highly homologous ion pumps, yet in LLC-PK1 cells, they are polarized to the basolateral and apical domain, respectively. The polarized expression of chimeric constructs of the α-subunit of the H^+/K^+-ATPase and the Na^+/K^+-ATPase in LLC-PK1 cells has been studied (Blostein et al. 1993). An apical sorting motif was identified within the fourth transmembrane domain of the α-subunit of the H^+/K^+-ATPase that is sufficient to redirect the Na^+/K^+-ATPase from the basolateral to the apical surface of these cells (Dunbar et al. 2000). However, it remains unclear whether basolateral sorting information exists in the fourth transmembrane domain of the α-subunit of the Na^+/K^+-ATPase; thus, a still unknown, noncanonical molecular signal seems to be involved in the basolateral targeting of Na^+/K^+-ATPase.

Clathrin plays a fundamental role in basolateral sorting. It interacts with endocytic or basolateral proteins through a variety of clathrin adaptors. It has been shown that the adaptor involved in basolateral protein sorting is the epithelial cell-specific AP-1B (adaptor protein 1B). Nevertheless, the correct polarization of Na^+/K^+-ATPase is not significantly affected by knocking down clathrin expression. It also remains correctly polarized in both the μ1B-deficient cell line LLC-PK1 (Ohka et al. 2001) and in MDCK cells in which μ1B expression had been suppressed via RNAi (Duffield et al. 2004). By taking advantage of the SNAP tag system to reveal the trafficking itinerary of the newly synthesized Na^+/K^+-ATPase, it was shown that basolateral delivery of the Na^+/K^+-ATPase is very fast and does not involve passage through recycling endosomes *en route* to the plasma membrane. Moreover, Na^+/K^+-ATPase trafficking is not regulated by the same small GTPases as other basolateral proteins (Farr et al. 2009). Some membrane proteins may achieve polarity by selective retention at the apical or basolateral surface.

Although less well understood, this polarity may reflect interactions with extracellular ligands or with intracellular scaffolds, such as cytoskeletal elements or arrays of PDZ domain-containing proteins (Zimmermann 2006). As described and discussed below, this is also the case of the epithelial Na^+/K^+-ATPase, which is retained at the lateral membrane domain due to the adhesion of its β1-subunits with those of neighboring cells. Readers interested in apical and basolateral sorting of transport proteins are directed to Chap. 5 of Volume 1.

The β-subunit functions as a molecular chaperone of the catalytic α-subunit. It facilitates the correct membrane integration and packing of the newly synthesized catalytic α-subunit, which is necessary for its protection against cellular degradation, acquisition of functional properties, and routing to the plasma membrane (Geering 2008). In addition to its chaperone function, β-subunits influence the transport properties of mature Na^+/K^+-ATPase. α-Subunits associated with different β-isoforms exhibit different apparent potassium affinities, and the β-structure influences the apparent sodium affinity of Na^+/K^+-ATPase (Geering 2001). During the catalytic cycle, there is a conformational rearrangement between α- and β-subunits (Dempski et al. 2005).

1.3.2 Cues Leading to a Model of Na^+/K^+-ATPase Polarity

As with most transporting epithelia, the monolayer of MDCK expresses Na^+/K^+-ATPase in a polarized manner toward the basolateral side (Cereijido et al. 1980). Figure 10.4a shows a monolayer of MDCK cells with nuclei stained with propidium iodide (*red*) and Na^+/K^+-ATPase (*green*). Contrary to the assumption of Koefoed-Johnsen and Ussing (Koefoed-Johnsen and Ussing 1958), the pump is not located on the *basal* domain of the plasma membrane, but only in the *lateral* domain (*first cue,* Fig. 1.4a). Although green lines in Fig. 1.4a, b appear as a single, undivided green line, upon treating the monolayer with EGTA to sequester Ca^{2+}, the apparent single green line splits into two (Fig. 1.4c), indicating that, in order to express Na^+/K^+-ATPase at a cell–cell contact, both neighboring cells have to contribute part of the enzyme (*second cue*). In order to express the Na^+/K^+-ATPase at a given lateral border, both contributing neighboring cells should be homotypic, for instance, MDCK/MDCK (dog/dog) but not heterotypic: MDCK/Ma104 (dog/monkey) (Shoshani et al. 2005), as depicted in Fig. 1.4d (*third cue*).

Figure 1.4e depicts the position and arrangement of Na^+/K^+-ATPase obtained by crystallography, showing the positioning of this trimer: α-subunit (*violet*), β-subunit (*green*), and γ-subunit (*red*). The β-subunit has the typical structure of a cell-attachment protein (Geering 2008). Karlish's group has confirmed that the β-subunit ectodomain contains an immunoglobulin-like structure that would be responsible for its adhesive properties (Bab-Dinitz et al. 2009). Accordingly, the β-subunit has a short cytoplasmic tail, a single transmembrane segment, and a long extracellular segment that is heavily glycosylated. We also showed that β1-subunit immobilized on Ni beads could specifically bind to the soluble extracellular domain of β1-subunits of the same animal species (dog; Padilla-Benavides et al. 2010). To test this property and to see if it works in MDCK cells, we transfected a gene coding for the β-subunit of the dog (remember that MDCK cells are also derived from a dog) into MDCK cells, and demonstrated that this considerably increases cell–cell adhesion (Fig. 1.4f). This constituted our *fourth cue*. We also examined this property by transfecting other cell types with dog β-subunit, that is, the same experiment as the one in Fig. 1.4d, except that this time the "other" cell (CHO cell from Chinese

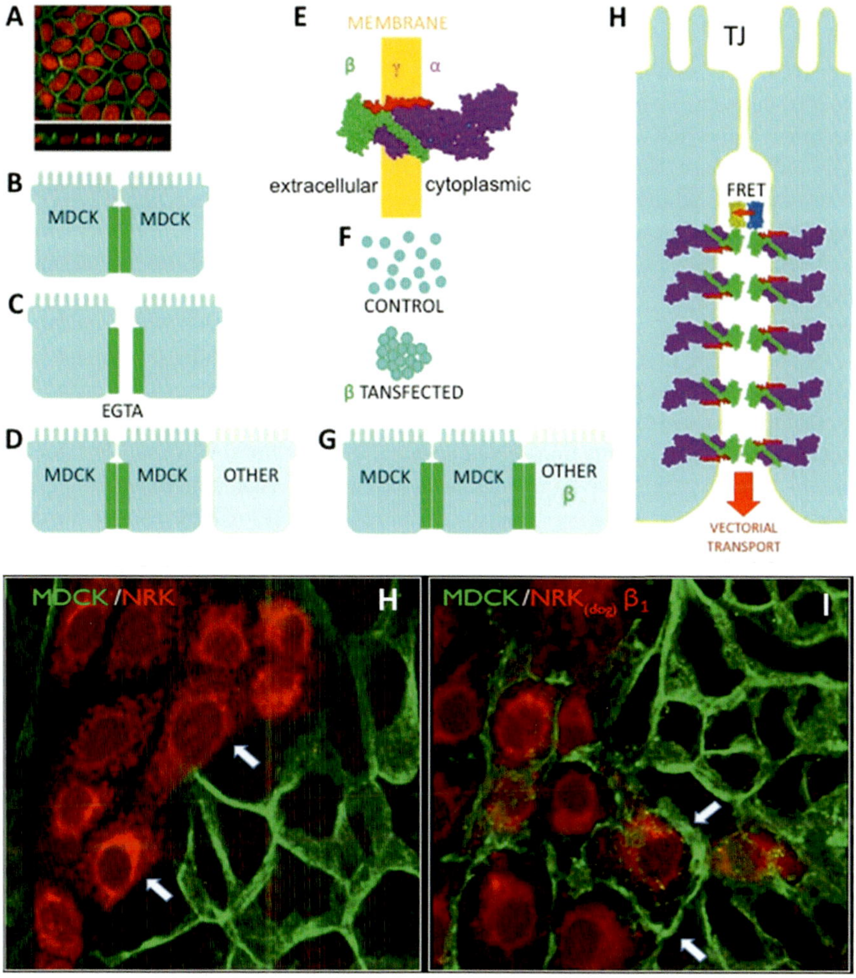

Fig. 1.4 Cues for the polarized distribution of Na^+/K^+-ATPase. (**a**) Monolayer of MDCK cells. Na^+/K^+-ATPase (green) and nuclei (red). The enzyme is expressed on the lateral membrane of the cells. (**b**, **c**). Although the image of the enzyme appears as a single green line, the use of EGTA splits the line, demonstrating that each cell contributes its own Na^+/K^+-ATPase. (**d**) In a monolayer prepared with a mixture of MDCK cells and epithelial cells of a different animal species (Chinese hamster ovary, CHO), the MDCK cell in the center only expresses its Na^+/K^+-ATPase on the side contacting another MDCK cell, but not on the side contacting the epithelial cell of a different animal species. Confocal image of a mixture of MDCK and NRK (normal rat kidney) cells is depicted in (**h**). Arrows indicate heterotypic borders lacking Na^+/K^+-ATPase. (**e**) The α-, β-, and γ-subunits of Na^+/K^+-ATPase as expressed in a cell membrane: most of the β-subunit (green) is exposed on the intercellular space (**f**) (above). Suspension of CHO cells (light blue) with a poor attachment to each other below CHO cells transfected with dog β-subunits tend to attach to each other and form aggregates, thereby demonstrating the self-adhesion tendency of this subunit. (**g**) A study essentially identical to the one described in (**d**), except that this time the third cell on the right is a CHO transfected with β-subunits of dog. MDCK cell at the center now expresses its β-subunits on both lateral sides. (**i**) A confocal image showing a mixture of MDCK and NRK cells transfected with dog β-subunit. Arrows indicate the presence of Na^+/K^+-ATPase at heterotypic borders. (**j**) Scheme showing the intercellular space of two neighbor MDCK cells in monolayer. Na^+/K^+-ATPase units

Hamster Ovary) is transfected with *dog* β-subunit. Notice that an MDCK and a CHO cell transfected with a dog β-subunit, offer the image of a single line, as in Fig. 10.4b. To complement the description of this cartoon, Fig. 1.4h, i show two types of mixed monolayers, one with MDCK/NRK cells (NRK, from rat kidney) and the other with NRK transfected with dog β-subunit, showing, again, that MDCK cells only expose Na^+/K^+-ATPase, when the neighboring cell exposes β-subunit from the same animal species. Finally, Fig. 1.4j, shows that the Na^+/K^+-ATPase expressed at the lateral border of the cell can only pump Na^+ into the intercellular space and, given that this space is sealed toward the outer side of the epithelium by the tight junctions (Fig. 1.4j), this ion can only flow vectorially toward the blood side.

Of course, the first question that arises is whether two matching β-subunits from different cells would get close enough to be able to span the intercellular space and interact as proposed. To answer this question, we prepared monolayers with a mixed population of MDCK cells transfected with a β-subunit fused to a cyan fluorescent protein (*blue*), with MDCK cells transfected with a β-subunit fused to yellow fluorescent protein, showing that in a fluorescence resonance energy transfer (FRET) analysis energy can be transferred from the first to the second cell type (Fig. 1.4j) (Padilla-Benavides et al. 2010). In other words, two β-subunits can interact directly at <10 nm, thereby anchoring the whole enzyme at the cell membrane facing the intercellular space (Padilla-Benavides et al. 2010). This basic mechanism seems to be strengthened through interactions with extracellular ligands and intracellular scaffolds such as cytoskeletal elements or arrays of PDZ domain-containing proteins. β_1-subunit has three conserved N-glycosylation sites. Remarkably, N-glycans stabilize the amino acid-mediated interactions between β1 subunits (Tokhtaeva et al. 2010). Moreover, intercellular adhesion can be regulated via N-glycan-mediated modulation of β1–β1 interactions. This modulation occurs because of cell-adhesion-dependent regulation of the level of the specific glycosyltransferases that control the degree of N-glycan branching of the Na^+/K^+-ATPase β_1 subunit (Tokhtaeva et al. 2011; Vagin et al. 2008). In this respect, several studies have shown that lateral targeting of the pump in transporting epithelial cells is related to the expression of the β_1 isoform, while apical targeting is related to expression of the β_2 isoform. Furthermore, this differential polarization is directly related to the variation of N-glycosylation between these two isoforms (Vagin et al. 2007a, b).

Fig. 1.4 (continued) are expressed at the lateral border, where they are anchored by the β-subunits. Notice that these pumps can only inject Na^+ into the intercellular space. Since this space is sealed toward the outer bathing solution by the tight junction (TJ), Na^+ can only diffuse toward the inner (blood) side. In summary, this cartoon depicts these two attributes whose mechanism of synthesis, assembly, and sealing, as well as the peculiar polarized distribution, are detailed elsewhere (Cereijido et al. 2008, 2012)

1.4 Structural Insights into the Na^+/K^+-ATPase Adhesion Mechanism

The crystal structure of the shark's Na^+/K^+-ATPase in the E2 state published by Shinoda and coworkers was the first resolving the atomic structure of the extracellular domain of the β subunit [PDB: 2ZXE) (Shinoda et al. 2009). The extracellular C-terminal domain of the protein folds into an Ig-like β-sheet sandwich as predicted in silico (Bab-Dinitz et al. 2009). Actually, the deletion of this C-terminal domain abolishes the β_1 adhesion capacity (unpublished observations). However, a large number of adhesion and non-adhesion proteins contain domains with an immunoglobulin-like topology [CATH database]. Structural alignments of the β_1 subunit extracellular domain against other well-studied cell adhesion molecules reveal no structural homolog of β subunits of any kind. Detailed inspection of the ectodomain structure uncovers several features distinctive to β subunit family members. Namely, its Ig-like fold has a unique topology given that its β-sheet sandwich is interrupted by a long α-helix secondary structure and has an atypical β-sheet disposition in relation to classical Ig folds. Also, the β subunit fold contains extensive loops and therefore its length is twice that of a typical Ig domain (unpublished observations). Furthermore, the β_1 subunit is structurally compromised with the catalytic α subunit in such a way that the C-terminal fold must be more rigid than the typical flexibility of whole adhesion domains such as in cadherins. Altogether, these observations suggest that the β subunit of the Na^+/K^+-ATPase must possess an adhesion mechanism that is particular to this family, as shown in Fig. 1.4f. The interaction between two β1 subunits of the same species (dog–dog or rat–rat) is more effective than the interaction between rat and dog β1 subunits (Padilla-Benavides et al. 2010). Based on this difference, Tokhtaeva et al. (2012) found that this difference in affinity between rat and dog is caused by the presence of different residues between the two species, all of them within 198–207 segment. Such a segment constitutes one of the characteristic protruding loops in the connecting β-strands B and C of the β1-subunit extracellular fold. The majority of the ectodomain surface-exposed residues located most distal from the membrane resides within loops interconnecting β-strands, some of which must be involved in the dimer interface in conjunction with segment 198–207. We have performed protein docking and molecular dynamics simulation to identify putative amino acids that participate in β1–β1 interaction. Our model suggests that Na^+/K^+-ATPase β1-subunit extracellular domain is a stable structure and that species-specific residues 198–207 interact with segment 223–229 on the associating β1-subunit protein [unpublished results].

1.5 Cardiac Steroids

Cardiac steroids (CS) became a medical tool since William Withering (1741–1799) observed that patients with a weak heart improved when drinking a herb tea of foxglove (*Digitalis purpurea*). He was emulated by European physicians using tisane of plant species from other regions, such as *Digitalis lanata* and West African plant (*Strophanthus gratus*) that led to coining of the name Strophanthin (ouabain). In 1953, Hans J. Schatzmann observed that ouabain inhibits active potassium and sodium transport in the erythrocyte membrane (Schatzmann 1953). Later, Jens Christian Skou prepared an extract from crab nerves that contained an ATPase that hydrolyzed ATP into ADP and Pi. Since the activity of this extract was enhanced when exposed to a certain combination of Na^+, $K^{+,}$ and Mg^{2+}, Skou hypothesized that the enzyme involved might be related to the active movement of sodium and potassium across the plasma membrane, and properly named it Na^+/K^+-ATPase. Furthermore, this enzyme was inhibited by ouabain, an observation that reinforced the suspicion that there must be an endogenous analog (Skou 1957).

The remarkable affinity of cardiac steroids for Na^+/K^+-ATPase in mammalian organisms stressed the possibility that there was an endogenous substance analogous to the one extracted from plants. Hamlyn et al. (1991) and Mathews et al. (1991) demonstrated the presence of a substance in plasma that could not be distinguished from ouabain of vegetal origin, a finding confirmed with more advanced methods such as 1H-NMR and mass ionization spectrometry. There are other analog substances in human blood (Kawamura et al. 1999; Komiyama et al. 2005; Takahashi 2000; Tymiak et al. 1993) and in other animal species, such as digoxin-like factor (Goto et al. 1990; Qazzaz et al. 2004) marinobufagenin (Bagrov et al. 1998); telocinobufagenin (Komiyama et al. 2005), 19-nobufalin (Lichtstein et al. 1993), and proscillaridin A (Hilton et al. 1996; Li et al. 1997). Endogenous ouabain is synthesized and secreted by the hypothalamus (Ludens et al. 1992; el-Masri et al. 2002) and the adrenocortical gland (Doris et al. 1996; Laredo et al. 1994; Qazzaz et al. 2004; Schreiber and Stepan 1986). The status of ouabain as an endogenous hormone was unambiguously recognized when it was demonstrated that it increases during exercise (Bauer et al. 2005), salty meals (Fedorova et al. 2005a, b, 2000), and pathological conditions such as arterial hypertension and myocardial infarction (Gottlieb et al. 1992; Manunta et al. 2009). Once ouabain was accepted as a hormone, it became of interest to find its physiological role. Our laboratory showed that ouabain binding to the Na^+/K^+-ATPase modulates epithelial cell adhesion (Larre et al. 2011a).

1.5.1 The Search for the Physiological Role of Hormone Ouabain

Less than a decade ago, the evidence that ouabain was a hormone was gaining recognition; therefore, investigators began to wonder what the physiological role of ouabain might be. We observed that MDCK cells exposed to high concentrations of ouabain (>1μM) do not show sign of damage, yet retrieve from the plasma membrane molecules involved in cell–cell and cell–substrate attachment and, as a consequence, those cells detach from each other as well as from the substrate (Contreras et al. 1999). This suggests that there is a mechanism that relates the occupancy of the pump (P) by ouabain to adhesion mechanisms (A). Accordingly, we called this mechanism P→A and put forward the working hypothesis that ouabain at nanomolar concentrations, i.e., within the hormonal range in mammalian plasma, may act on the same junctional structures without provoking irreversible damage (Larre et al. 2010). To explore the plausibility of this idea, we have studied the effect of ouabain on several types of cell contacts, as will be described next.

1.5.2 Ouabain Modulates the Tight Junction

The tight junction is the most apical adhesion structure of cell–cell contact. It seals the intercellular space between epithelial cells, thereby transforming the layer of cells into an effective permeability barrier separating biological compartments such as the intestinal lumen from blood or the interior of a nephron and blood (Cereijido et al. 1978, 1980). Several observations indicate that the effects of β1-subunit expression on tight junction formation may reflect the role of β-subunits in the maturation of Na^+/K^+-ATPase complexes and the consequent increase in Na^+/K^+-ATPase expression and activity (Rajasekaran et al. 2001; Cibrián-Uhalte et al. 2007). We have shown that high concentrations of ouabain (1 Mm) opens the TJ and causes a drastic drop of transepithelial resistance (TER), yet at 10–100 nM, it has the opposite effect, as it increases the degree of sealing of the TJ (Fig. 10.5a) and decreases the paracellular flux of neutral 3 kDa dextran (JDEX) (Larre et al. 2006, 2010; Rajasekaran et al. 2003; Rincon-Heredia et al. 2014). Figure 1.5b illustrates that ouabain modulates the permeability of the TJ by acting on the amount as well as the pattern of distribution of claudins of a given type independently.

1.5.3 Ouabain Modulates Adherens Junctions

Another prominent cell–cell contact is the adherens junction (AJ), and one of the scaffolding proteins of this junction is β-catenin, a key member of the Wnt signaling pathway, a family of highly conserved secreted signaling molecules that regulate

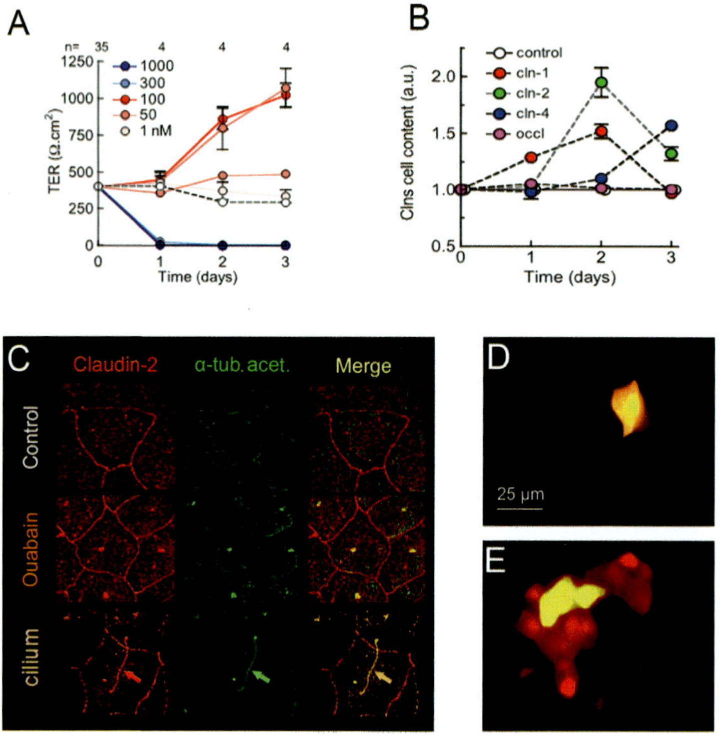

Fig. 1.5 Ouabain modulates several types of cell contacts. (**a**) Tight junctions (TJ): The hermeticity of TJs was gauged by measuring TER (transepithelial electrical resistance). At 1μM (blue circles) ouabain is toxic and opens the TJ, and TER drops to zero. (**b**) Ouabain modulates the expression of claudins individually; each one varies with its own kinetics. (**c**) The cilium, located at the center of the apical domain, freely waves in the fluid bathing the epithelium. In the case of monolayers of MDCK claudin-2 adopts the pattern of a chicken fence (red), because it belongs to the TJ. When monolayers are treated with 10 nM ouabain, claudin-2 is also expressed at the procilium and cilium of each cell, as evidenced by its colocalization with α-tubulin (green) that is a specific protein of the cilium. (**d**, **e**) Gap junctional communication (GJC): Tracer injection assays made on MDCK cells in monolayers that were either left as control (**d**) or treated with 10 nM ouabain for 60 min (**e**). In both cases, a single cell was injected with a mix of neurobiotin as tracer and dextran-fluorescein as a marker of injection. Neurobiotin diffusion to neighboring, GJ-coupled cells was revealed by streptavidin-rhodamine

cell-to-cell interactions during embryogenesis (Logan and Nusse 2004). During the activation of this pathway, a condition that can be achieved by changes in cell adhesion, β-catenin is translocated to the nucleus, where it modifies gene expression (Lien and Fuchs 2014). We observed that 10 nM ouabain, as well as 1μM, binding to Na^+/K^+-ATPase induces the translocation of β-catenin to the nucleus (Contreras et al. 2004). Liu and co-workers recently found evidence that Na^+/K^+-ATPase and E-cadherin are closely associated, indicating the pump might be a component of the adherens junction or that E-cadherin could be part of the signalosome of the Na^+/K^+-ATPase (Liu et al. 2011). Recent unpublished studies from our lab show that 10 nM ouabain enhances the localization of E-cadherin at the cell–cell borders and accelerates the arrival of E-cadherin on the cell–cell border in a calcium switch protocol. Therefore, our results suggest that Na^+/K^+-ATPase is a regulator of adherens junctions and a potential regulator of Wnt signaling and morphogenesis.

1.5.4 Ouabain Stimulates Ciliogenesis

The cilium is a slender protuberance with polarized expressed at the center of the apical domain of most eukaryotic cells (Colegio et al. 2003; Hou et al. 2006). Its development (ciliogenesis) is the last step of the process of apical/basolateral polarization. Therefore, we gauged polarity through the expression of a cilium in MDCK cells. Ciliogenesis occurs in quiescent cells that have reached confluence and is drastically accelerated by ouabain binding to Na^+/K^+-ATPase (Larre et al. 2010). MDCK cells express claudin-2 at the TJ, where it accounts for specific cation permeation (Rosenthal et al. 2010; Overgaard et al. 2011) (Fig. 1.5c, *above*). Surprisingly, we observed that this claudin is also localized at the cilium that, contrary to TJs, does not separate into two different fluid compartments, and therefore might not play a role in permeation (Larre et al. 2011b). Nevertheless, at the TJ claudin-2 is responsible for the specific permeation of cations, in particular Na^+ (Günzel and Fromm 2012), indicating that it has a sequence of amino acids or adopts a configuration that is able to sense the presence of this ion. Therefore, the possibility exists that claudin-2 in the cilium may act as a sensor of Na^+ concentration in the fluid bathing the apical domain (Fig. 1.5c, *below*).

1.5.5 Ouabain Modulates Cell–Cell Communication Through Gap Junctions

Gap junctions (GJ) are a type of cellular structure that allow adjacent cells to communicate through the exchange of ions and small molecules. We have shown (by dye transfer, as well as electric coupling assays) that incubation of mature monolayers of MDCK with 10 nM ouabain increases the intercellular communication through Gap Junctions (ICGJ). This response occurs within minutes and reaches a peak after 1 h (Fig. 1.5d, e). We have shown also, by silencing assays, that connexins 32 and 43 are both involved in these responses; however, no synthesis of new proteins is required, but rather by modulating the distribution of subunits previously synthesized. Furthermore, we have shown that Na^+/K^+-ATPase is the receptor that mediates this response and that csrc and Erk1/2 are involved in the signaling pathway (Ponce et al. 2014, 2016).

Therefore, ouabain should be listed among hormones that influence cell–cell communication, such as luteinizing hormones, which decrease the passage of cAMP from follicle cells to the oocyte (Dekel 2005); steroid hormones, which reduce cell communication in endometrial carcinoma (Saito et al. 2004); and FGF, which upregulates intercellular dye transfer in primary cultures of embryonic chick lens cells (Le and Musil 2001).

1.5.6 Ouabain Modulates the Epithelial Transporting Phenotype

Apical/basolateral polarity and TJs are the two fundamental features of the so-called "epithelial transporting phenotype" (Cereijido et al. 2003, 2008). Given that hormone ouabain modulates the TJ as well as apical/basolateral polarity as gauged by ciliogenesis, we may conclude that the binding of ouabain to the sodium pump regulates this phenotype. It is pertinent to remember that up to 85% of all human deaths are caused by the collapse of an epithelium (hepatic, renal, etc.) and most cancers in human adults start and occur in epithelia. Interestingly, ouabain helps develop transporting epithelia, that are the site where other adrenal corticoids regulate the homeostasis of metazoan.

1.5.7 Na^+/K^+-ATPase Is a Receptor of Cardiac Steroids

The primary function of the sodium pump is to transport sodium and potassium ions against the electrochemical gradient. Studies during the last two decades have revealed an additional role for the Na^+/K^+-ATPase as a signal transducer. Na^+/K^+-ATPase acts as a receptor of CS and forms a complex receptor regulating intracellular signaling pathways via, e.g., the inositol 1,4,5-triphosphate receptor and the Src kinase. Moreover, recently in vivo studies show the Na^+/K^+-ATPase is highly mobile and has in fact a "clustering or scaffolding function" (Liebmann et al. 2018).

The Na^+/K^+-ATPase and Src are enriched in caveolae of renal epithelial cells (Wang et al. 2004), and it has been suggested that they directly interact with each other through two domains. However, whether the Src kinase binds directly with a high affinity to the α1 subunit or regulates enzyme–substrate interaction as part of a multiprotein complex is not entirely clear. More importantly, ouabain binding to Na^+/K^+-ATPase regulates c-Src activity that increases protein tyrosine phosphorylation and consequently stimulates protein kinase cascades and phospholipase (Xie et al. 1999; Xie 2003; Tian et al. 2006).

The interaction between the Na^+/K^+-ATPase and Ins(1,4,5)P3R was first reported by Aperia's laboratory during the investigation of ouabain-induced calcium low-frequency oscillations in renal epithelial cells. This calcium oscillation activates and sends NF-κB to the nucleus (Aizman et al. 2001). Ouabain triggers the direct interaction between Na^+/K^+-ATPase and the Ins(1,4,5) P3R. Amino acid residues

LKK in the N-terminus of the catalytic α-subunit of Na^+/K^+-ATPase are essential for binding with the Ins(1,4,5)P3R N-terminus. This interaction is also supported by the scaffolding protein Ankyrin-B. The ouabain-Na^+/K^+-ATPase-Ins(1,4,5)P3R complex triggers slow Ca^{2+} oscillations that subsequently activate the NF-κB p65 and protects from apoptosis (Fontana et al. 2013).

Wild-type MDCK (W-MDCK) cells are sensitive to extremely low concentrations of ouabain (Kd in the order of 10^{-8} M). We also studied a strand of MDCK cells resistant to high concentrations of ouabain (up to 10μM) because, as mentioned above, their Na^+/K^+-ATPases have extremely low affinity to ouabain (R-MDCK cells) (Canessa et al. 1992; Soderberg et al. 1983; Cargnelli et al. 1983). Ten nanomolar ouabain cannot elicit the hormonal effect on the TJ, adherent junction, ciliogenesis nor gap junction in monolayers of R-MDCK cells indicating that hormone ouabain elicits its effect when binding to the same site used by toxic concentrations of ouabain to block the pump (Larre and Cereijido 2010).

1.6 Signaling Pathways

Na^+/K^+-ATPase is able to activate c-Src even in mutated pumps in which the pumping ability is suppressed (Tian et al. 2006). This activation of c-Src enables Na^+/K^+-ATPase to modulate signaling routes involving ERK1/2 and oxygen-reactive species (Xie 2003; Xie et al. 1999). In keeping with these characteristics, we have found that the modulation of cell contacts by ouabain involves all these molecules and signaling routes. The function and composition of TJ are regulated by the receptor complex of Na^+/K^+-ATPase-c-Src that activates signaling pathways. For example, while TER depends on c-Src and partially on ERK1/2, J_{DEX} depends on ERK1/2 but not on c-Src indicating that Na^+/K^+-ATPase can interact with other signal proteins and form a different receptor complex (Larre et al. 2010). The changes in the permeability of the TJ are accompanied by changes in the cell content and distribution patterns of specific claudins, each one regulated independently (Fig. 1.5b). Claudin-1 depends on c-Src and ERK1/2, but claudin-4 depends instead on an as yet unidentified signal molecule and ERK1/2. Binding of ouabain to the Na^+/K^+-ATPase accelerates ciliogenesis through the ERK1/2 pathway. At variance with the modulation of the TJ, that of adherens junctions and gap junctions completely depends on c-Src-ERK1/2 pathway. In turn, inhibitors c-Src, such as PP2, and ERK1/2, such as PD98059, inhibit GJ communication, as well as the addressing of β-catenin to the nucleus. Although all the effects attributed to hormone ouabain on the several cell contacts investigated are triggered by its binding to the Na^+/K^+-ATPase at the same moment, these effects are not simultaneous. Thus the enhancement of GJC is observed in minutes, the TJ effect is observed in hours, adherent junctions in 8–10 hr. and ciliogenesis in 1–3 days (Larre et al. 2010; Ponce et al. 2014, 2016).

1.7 Pathologies Related to Na^+/K^+-ATPase

In spite of the paramount importance of the Na^+/K^+-ATPase and the number of fundamental cellular functions in which it participates, there are not many known pathologies directly associated with its malfunctioning. In neural tissue, where Na^+/K^+-ATPase is abundant, there has been an association between the Na^+/K^+-ATPase α-subunit isoforms and some neurological diseases. For instance, mutations in the gene encoding the α2 isoform result in familial hemiplegic migraine type 2 (FHM2) (De Fusco et al. 2003), and mutations in the gene encoding the α3 subunit result in rapid-onset dystonia-parkinsonism (RDP) (Brashear et al. 2007). In epithelia, however, pathological conditions in which the Na^+/K^+-ATPase is involved are more related to adhesion and polarity rather than transport function. Na^+/K^+-ATPase β1-subunit is highly reduced in cells with deteriorated cell–cell adhesion, such as the cystic epithelia in polycystic kidney disease (Wilson et al. 1991), clear cell renal carcinoma cells (Rajasekaran et al. 1999), bladder cancer cells (Espineda et al. 2003), and poorly differentiated carcinoma cell lines derived from the colon, breast, kidney, and pancreas (Espineda et al. 2004) and the gastric carcinoma cell line, HGT-1 (Vagin et al. 2005). As mentioned previously in this chapter, intercellular adhesion mediated in part by Na/K-ATPase β1-subunits is essential for normal maturation of the epithelium, and so the loss of adhesion occurring in several forms of carcinoma may contribute to cancer metastasis. Alterations in the polarity of the Na^+/K^+-ATPase have been described in autosomal-dominant polycystic kidney disease (ADPKD) and were attributed to aberrant expression of the β2 isoforms that would target the Na^+/K^+-ATPase to the apical membrane, thus contributing to cyst formation (Wilson et al. 1991, 2000). As for the γ-subunits of the Na^+/K^+-ATPase, the isoform FXYD3 is downregulated in lung cancer (Okudela et al. 2009) but overexpressed in some carcinomas such as FXYD3 in pancreatic ductal adenocarcinoma (Kayed et al. 2006), gastric adenocarcinoma, and esophageal squamous cell carcinoma. Although it has been suggested that FXYD3 may affect cellular adhesion and migration through interactions with Na^+/K^+-ATPase this hypothesis is unclear (Zhu et al. 2010). In this regard, FXYD5 or dysadherin has been suggested as a cell–cell contact molecule, with its unusually long extracellular domain, and is also overexpressed in various tumors (Lubarski et al. 2011). Lastly, an abnormal modulation of Na^+/K^+-ATPase transport by FXYD1 resulting in atypical neuronal activity has been proposed as a direct contribution to Rett syndrome pathogenesis (Deng et al. 2007).

1.8 Horizons and Perspectives

Until a couple of decades ago, cell contacts were regarded as mere architectonic structures to keep cells together, lest they would disaggregate on deformation. Yet the observation that they are constituted by a precise arrangement of dozens of specific protein species and the structure, physiological properties as well as the supracellular architecture that they build up (e.g., a renal glomerulus, a hepatic lobule) are of paramount importance, led to the discovery that cell contacts play fundamental roles such as gene expression, cell cycling, proliferation, differentiation, migration, tissue architecture, cancer, metastasis, and homing. Their importance can hardly be overestimated given that the brain, the most complex object in the Universe, assembles itself depending on which cells contact which other cells, when and how (i.e., the type of synapses established). Therefore, the finding that ouabain is a hormone that modulates cell contacts is promising from the physiological as well as the pathological views. However, we have not yet systematically explored the physiological significance of the distribution in time of the different effects elicited by ouabain binding to the Na^+/K^+-ATPase complex receptor. Interestingly, hormone ouabain causes these effects acting on the same receptor site (the extracellular domain of the α-subunit of Na^+/K^+-ATPase) and activates similar signaling routes. Therefore, the Na^+/K^+-ATPase is likely to play an important role in phenomena depending on cell contacts, such as proliferation, differentiation, migration, metastasis, and invasion.

Acknowledgments We wish to acknowledge the efficient and pleasant assistance of E. el Oso, A. Castillo, E. Méndez. M.L. Roldan. This work was supported by The Sectoral Fund for Education Research, CONACYT Grant 285263. I Larre was a recipient of a Postdoctoral Research Fellowship from ICyTDF. O Paez was a recipient of a Doctoral Fellowship from CONACYT.

References

Aizman O, Uhlen P, Lal M, Brismar H, Aperia A (2001) Ouabain, a steroid hormone that signals with slow calcium oscillations. Proc Natl Acad Sci USA 98:13420–13424

Bab-Dinitz E, Albeck S, Peleg Y, Brumfeld V, Gottschalk KE, Karlish SJ (2009) A C-terminal lobe of the beta subunit of Na^+,K^+-ATPase and H^+,K^+-ATPase resembles cell adhesion molecules. Biochemistry 48:8684–8691

Bagrov AY, Fedorova OV, Dmitrieva RI, Howald WN, Hunter AP, Kuznetsova EA, Shpen VM (1998) Characterization of a urinary bufodienolide Na^+,K^+-ATPase inhibitor in patients after acute myocardial infarction. Hypertension 31:1097–1103

Bauer N, Müller-Ehmsen J, Krämer U, Hambarchian N, Zobel C, Schwinger RH, Neu H, Kirch U, Grünbaum EG, Schoner W (2005) Ouabain-like compound changes rapidly on physical exercise in humans and dogs: effects of beta-blockade and angiotensin-converting enzyme inhibition. Hypertension 45:1024–1028

Bianconi E, Piovesan A, Facchin F, Beraudi A, Casadei R, Frabetti F, Vitale L, Pelleri MC, Tassani S, Piva F, Perez-Amodio S, Strippoli P, Canaider S (2013) An estimation of the number of cells in the human body. Ann Hum Biol 40, 463:–471

Blostein R, Zhang R, Gottardi CJ, Caplan MJ (1993) Functional properties of an H,K-ATPase/Na, K-ATPase chimera. J Biol Chem 268:10654–10658

Brashear A, Dobyns WB, de Carvalho AP, Borg M, Frijns CJ, Gollamudi S, Green A, Guimaraes J, Haake BC, Klein C, Linazasoro G, Munchau A, Raymond D, Riley D, Saunders-Pullman R, Tijssen MA, Webb D, Zaremba J, Bressman SB, Ozelius LJ (2007) The phenotypic spectrum of rapid-onset dystonia-parkinsonism (RDP) and mutations in the ATP1A3 gene. Brain 130:828–835

Canessa CM, Horisberger JD, Louvard D, Rossier BC (1992) Mutation of a cysteine in the first transmembrane segment of Na,K-ATPase alpha subunit confers ouabain resistance. EMBO J 11:1681–1687

Caplan MJ, Anderson HC, Palade GE, Jamieson JD (1986) Intracellular sorting and polarized cell surface delivery of (Na^+,K^+)ATPase, an endogenous component of MDCK cell basolateral plasma membranes. Cell 46:623–631

Cargnelli G, Trevisi L, Soderberg K, Rossi B, Lazdunski M, Louvard D (1983) Characterization of ouabain resistant mutants of a canine kidney cell line, MDCK. J Biol Chem 258:12300–12307

Cereijido M, Robbins ES, Dolan WJ, Rotunno CA, Sabatini DD (1978) Polarized monolayers formed by epithelial cells on a permeable and translucent support. J Cell Biol 77:853–880

Cereijido M, Ehrenfeld J, Meza I, Martinez-Palomo A (1980) Structural and functional membrane polarity in cultured monolayers of MDCK cells. J Membr Biol 52:147–159

Cereijido M, Contreras RG, Shoshani L, García-Villegas MR (2003) Membrane targeting. Prog Biophys Mol Biol 81:81–115

Cereijido M, Contreras RG, Shoshani L (2004) Cell adhesion, polarity, and epithelia in the dawn of metazoans. Physiol Rev 84:1229–12262

Cereijido M, Contreras RG, Shoshani L, Flores-Benitez D, Larre I (2008) Tight junction and polarity interaction in the transporting epithelial phenotype. Biochim Biophys Acta 1778:770–793

Cereijido M, Contreras RG, Shoshani L, Larre I (2012) The Na^+-K^+-ATPase as self-adhesion molecule and hormone receptor. Am J Physiol Cell Physiol 302:C473–C481

Cibrián-Uhalte E, Langenbacher A, Shu X, Chen JN, Abdelilah-Seyfried S (2007) Involvement of zebrafish Na^+,K^+ ATPase in myocardial cell junction maintenance. J Cell Biol 176:223–230

Colegio OR, Van Itallie C, Rahner C, Anderson JM (2003) Claudin extracellular domains determine paracellular charge selectivity and resistance but not tight junction fibril architecture. Am J Physiol Cell Physiol 284:C1346–C1354

Contreras RG, Avila G, Gutierrez C, Bolívar JJ, González-Mariscal L, Darzon A, Beaty G, Rodriguez-Boulan E, Cereijido M (1989) Repolarization of Na^+-K^+ pumps during establishment of epithelial monolayers. Am J Physio Cell Physiol 257:C896–C905

Contreras RG, Miller JH, Zamora M, González-Mariscal L, Cereijido M (1992) Interaction of calcium with plasma membrane of epithelial (MDCK) cells during junction formation. Am J Physiol Cell Physiol 263:C313–C318

Contreras RG, Shoshani L, Flores-Maldonado C, Lázaro A, Cereijido M (1999) Relationship between Na^+,K^+-ATPase and cell attachment. J Cell Sci 112:4223–4232

Contreras RG, Flores-Maldonado C, Lázaro A, Shoshani L, Flores-Benitez D, Larré I, Cereijido M (2004) Ouabain binding to Na^+,K^+-ATPase relaxes cell attachment and sends a specific signal (NACos) to the nucleus. J Membr Biol 198:147–158

De Fusco M, Marconi R, Silvestri L, Atorino L, Rampoldi L, Morgante L, Ballabio A, Aridon P, Casari G (2003) Haploinsufficiency of ATP1A2 encoding the Na^+/K^+ pump alpha2 subunit associated with familial hemiplegic migraine type 2. Nat Genet 33:192–196

Dekel N (2005) Cellular, biochemical and molecular mechanisms regulating oocyte maturation. Mol Cell Endocrinol 234:19–25

Dempski RE, Friedrich T, Bamberg E (2005) The beta subunit of the Na^+/K^+-ATPase follows the conformational state of the holoenzyme. J Gen Physiol 125:505–520

Deng V, Matagne V, Banine F, Frerking M, Ohliger P, Budden S, Pevsner J, Dissen GA, Sherman LS, Ojeda SR (2007) FXYD1 is an MeCP2 target gene overexpressed in the brains of Rett syndrome patients and Mecp2-null mice. Hum Mol Genet 15:640–650

Doris PA, Hayward-Lester A, Bourne D, Stocco DM (1996) Ouabain production by cultured adrenal cells. Endocrinology 137:533–539

Duffield A, Fölsch H, Mellman I, Caplan MJ (2004) Sorting of H,K-ATPase beta-subunit in MDCK and LLC-PK cells is independent of mu 1B adaptin expression. Traffic 5:449–461

Dunbar LA, Aronson P, Caplan MJ (2000) A transmembrane segment determines the steady-state localization of an ion-transporting adenosine triphosphatase. J Cell Biol 148:769–778 el-Masri MA, Clark BJ, Qazzaz HM, Valdes R (2002) Human adrenal cells in culture produce both ouabain-like and dihydroouabain-like factors. Clin Chem 48:1720–1730

Espineda C, Seligson DB, James Ball W Jr, Rao J, Palotie A, Horvath S, Huang Y, Shi T, Rajasekaran AK (2003) Analysis of the Na,K-ATPase α- and β-subunit expression profiles of bladder cancer using tissue microarrays. Cancer 97:1859–1868

Espineda CE, Chang JH, Twiss J, Rajasekaran SA, Rajasekaran AK (2004) Repression of Na,K-ATPase beta1-subunit by the transcription factor snail in carcinoma. Mol Biol Cell 15:1364–1373

Farr GA, Hull M, Mellman I, Caplan MJ (2009) Membrane proteins follow multiple pathways to the basolateral cell surface in polarized epithelial cells. J Cell Biol 186:269–282

Fedorova OV, Lakatta EG, Bagrov A (2000) Endogenous Na,K pump ligands aredifferentially regulated during acute NaCl loading of Dahl rats. Circulation 102:3009–3014

Fedorova OV, Agalakova NI, Talan MI, Lakatta EG, Bagrov AY (2005a) Brainouabain stimulates peripheral marinobufagenin via angiotensin II signalling in NaCl-loaded Dahl-S rats. J Hypertens 23:1515–1523

Fedorova OV, Kolodkin NI, Agalakova NI, Namikas AR, Bzhelyansky A, St-Louis J, Lakatta EG, Bagrov AY (2005b) Antibody to marinobufagenin lowers blood pressure inpregnant rats on a high NaCl intake. J Hypertens 23:835–842

Fontana JM, Burlaka I, Khodus G, Brismar H, Aperia A (2013) Calcium oscillations triggered by cardiotonic steroids. FEBS J 280:5450–5455

Galeotti G (1904) Concerning the E.M.F. which is generated at the surface of animal membranes on contact with different electrolytes. Z Phys Chem 49:542–562

Geering K (2001) The functional role of beta subunits in oligomeric P-type ATPases. J Bioenerg Biomembr 33:425–438

Geering K (2008) Functional roles of Na,K-ATPase subunits. Curr Opin Nephrol Hypertens 17:526–532

Gonzalez-Mariscal L, Contreras RG, Bolívar JJ, Ponce A, Chávez De Ramirez B, Cereijido M (1990) Role of calcium in tight junction formation between epithelial cells. Am J Physiol Cell Physiol 259:C978–C986

Goto A, Ishiguro T, Yamada K, Ishii M, Yoshioka M, Eguchi C, Shimora M, Sugimoto T (1990) Isolation of a urinary digitalis-like factor indistinguishable from digoxin. Biochem Biophys Res Commun 173:1093–1101

Gottlieb SS, Rogowski AC, Weinberg M, Krichten CM, Hamilton BP, Hamlyn JM (1992) Elevated concentrations of endogenous ouabain in patients with congestive heart failure. Circulation 86:420–425

Günzel D, Fromm M (2012) Claudins and other tight junction proteins. Compr Physiol 2:1819–1852

Hamlyn JM, Blaustein MP, Bova S, DuCharme DW, Harris DW, Mandel F, Mathews WR, Ludens JH (1991) Identification and characterization of a ouabain-like compound from human plasma. Proc Natl Acad Sci USA 88:6259–6263

Hilton PJ, White RW, Lord GA, Garner GV, Gordon DB, Hilton MJ, Forni LG, McKinnon W, Ismail FM, Keenan M et al (1996) An inhibitor of the sodium pump obtained from human placenta. Lancet 348:303–305

Hou J, Gomes AS, Paul DL, Goodenough DA (2006) Study of claudin function by RNA interference. J Biol Chem 281:36117–36123

Kawamura A, Guo J, Itagaki Y, Bell C, Wang Y, Haupert GT Jr, Magil S, Gallagher RT, Berova N, Nakanishi K (1999) On the structure of endogenous ouabain. Proc Natl Acad Sci USA 96:6654–6659

Kayed H, Kleeff J, Kolb A, Ketterer K, Keleg S, Felix K, Giese T, Penzel R, Zentgraf H, Büchler MW, Korc M, Friess H (2006) FXYD3 is overexpressed in pancreatic ductal adenocarcinoma and influences pancreatic cancer cell growth. Int J Cancer 118:43–54

Koefoed-Johnsen V, Ussing HH (1958) The nature of the frog skin potential. Acta Physiol Scand 42 (3–4):298–308. https://doi.org/10.1111/j.1748-1716.1958.tb01563.x

Komiyama Y, Dong XH, Nishimura N, Masaki H, Yoshika M, Masuda M, Takahashi H (2005) A novel endogenous digitalis, telocinobufagin, exhibits elevated plasma levels in patients with terminal renal failure. Clin Biochem 38:36–45

Laredo J, Hamilton BP, Hamlyn JM (1994) Ouabain is secreted by bovine adrenocortical cells. Endocrinology 135:794–797

Larre I, Cereijido M (2010) Na,K-ATPase is the putative membrane receptor of hormone ouabain. Commun Integr Biol 3:625–628

Larre I, Contreras RG, Cereijido M (2011a) Ouabain modulates cell contacts as well as functions that depend on cell adhesion. Methods Mol Biol 763:155–168. https://doi.org/10.1007/978-1-61779-191-8_10

Larre I, Castillo A, Flores-Maldonado C, Contreras RG, Galvan I, Muñoz-Estrada J, Cereijido M (2011b) Ouabain modulates ciliogenesis in epithelial cells. Proc Natl Acad Sci USA 108 (51):20591–20596. https://doi.org/10.1073/pnas.1102617108

Larre I, Ponce A, Fiorentino R, Shoshani L, Contreras RG, Cereijido M (2006) Contacts and cooperation between cells depend on the hormone ouabain. Proc Natl Acad Sci USA 103:10911–10916

Larre I, Lazaro A, Contreras RG, Balda MS, Matter K, Flores-Maldonado C, Ponce A, Flores-Benitez D, Rincon-Heredia R, Padilla-Benavides T et al (2010) Ouabain modulates epithelial cell tight junction. Proc Natl Acad Sci USA 107:11387–11392

Le AC, Musil LS (2001) A novel role for FGF and extracellular signal-regulated kinase in gap junction-mediated intercellular communication in the lens. J Cell Biol 1154:197–216

Li SQ, Eim C, Schneider R, Sich B, Kirch U, Schoner W (1997) Proscillaridin A immunoreactivity: a new endogenous cardiac glycoside? Ann N Y Acad Sci 834:646–647

Lichtstein D, Gati I, Ovadia H (1993) Digitalis-like compounds in the toad *Bufo viridis*: interactions with plasma proteins. J Cardiovasc Pharmacol 22(Suppl 2):S102–S105

Liebmann T, Fritz N, Kruusmägi M, Westin L, Bernhem K, Bondar A, Aperia A, Brismar H (2018) Regulation of neuronal Na,K-ATPase by extracellular scaffolding proteins. Int J Mol Sci 19(8): e2214

Lien WH, Fuchs E (2014) Wnt some lose some: transcriptional governance of stem cells by Wnt/β-catenin signaling. Genes Dev 28:1517–1532

Liu L, Ivanov AV, Gable ME, Jolivel F, Morrill GA, Askari A (2011) Comparative properties of caveolar and noncaveolar preparations of kidney Na^+/K^+-ATPase. Biochemistry 50:8664–8673. https://doi.org/10.1021/bi2009008

Logan CY, Nusse R (2004) The Wnt signaling pathway in development and disease. Annu Rev Cell Dev Biol 20:781–810. https://doi.org/10.1146/annurev.cellbio.20.010403.113126

Lubarski I, Asher C, Garty H (2011) FXYD5 (dysadherin) regulates the paracellular permeability in cultured kidney collecting duct cells. Am J Physiol Renal Physiol 301:F1270–F1280. https://doi.org/10.1152/ajprenal.00142.2011

Ludens JH, Clark MA, Robinson FG, DuCharme DW (1992) Rat adrenal cortex is a source of a circulating ouabain like compound. Hypertension 19:721–724

Manunta P, Ferrandi M, Bianchi G, Hamlyn JM (2009) Endogenous ouabain in cardiovascular function and disease. J Hypertens 27:9–18

Mathews WR, DuCharme DW, Hamlyn JM, Harris DW, Mandel F, Clark MA, Ludens JH (1991) Mass spectral characterization of an endogenous digitalislike factor from human plasma. Hypertension 17:930–935

Mays RW, Siemers KA, Fritz BA, Lowe AW, van Meer G, Nelson WJ (1995) Hierarchy of mechanisms involved in generating Na/K-ATPase polarity in MDCK epithelial cells. J Cell Biol 130:1105–1115

Ohka S, Ohno H, Tohyama K, Nomoto A (2001) Basolateral sorting of human poliovirus receptor alpha involves an interaction with the mu1B subunit of the clathrin adaptor complex in polarized epithelial cells. Biochem Biophys Res Commun 287:941–948

Okudela K, Yazawa T, Ishii J, Woo T, Mitsui H, Bunai T, Sakaeda M, Shimoyamada H, Sato H, Tajiri M, Ogawa N, Masuda M, Sugimura H, Kitamura H (2009) Down-regulation of FXYD3

expression in human lung cancers: its mechanism and potential role in carcinogenesis. Am J Pathol 175:2646–2456
Overgaard CE, Daugherty BL, Mitchell LA, Koval M (2011) Claudins: control of barrier function and regulation in response to oxidant stress. Antioxid Redox Signal 15:1179–1193
Padilla-Benavides T, Roldán ML, Larre I, Flores-Benitez D, Villegas-Sepúlveda N, Contreras RG, Cereijido M, Shoshani L (2010) The polarized distribution of Na^+,K^+-ATPase: role of the interaction between β subunits. Mol Biol Cell 21:2217–2225
Ponce A, Larre I, Castillo A, García-Villegas R, Romero A, Flores-Maldonado C, Martinez-Rendón J, Contreras RG, Cereijido M (2014) Ouabain increases gap junctional communication in epithelial cells. Cell Physiol Biochem 34:2081–2090. https://doi.org/10.1159/000366403
Ponce A, Larre I, Castillo A, Flores-Maldonado C, Verdejo-Torres O, Contreras RG, Cereijido M (2016) Ouabain modulates the distribution of connexin 43 in epithelial cells. Cell Physiol Biochem 39:1329–1338. https://doi.org/10.1159/000447837
Qazzaz HM, Cao Z, Bolanowski DD, Clark BJ, Valdes R Jr (2004) De novo biosynthesis and radiolabeling of mammalian digitalis-like factors. Clin Chem 50:612–620
Rajasekaran SA, Ball WJ Jr, Bander NH, Liu H, Pardee JD, Rajasekaran AK (1999) Reduced expression of the beta-subunit of Na^+/K^+-ATPase in human clear-cell renal cell carcinoma. J Urol 162:574–580
Rajasekaran SA, Palmer LG, Quan K, Harper JF, Ball WJ Jr, Bander NH, Peralta Soler A, Rajasekaran AK (2001) Na,K-ATPase beta-subunit is required for epithelial polarization, suppression of invasion, and cell motility. Mol Biol Cell 12:279–295
Rajasekaran SA, Hu J, Gopal J, Gallemore R, Ryazantsev S, Bok D, Rajasekaran AK (2003) Na,K-ATPase inhibition alters tight junction structure and permeability in human retinal pigment epithelial cells. Am J Physiol Cell Physiol 284:C1497–C1507
Rincon-Heredia R, Flores-Benitez D, Flores-Maldonado C, Bonilla-Delgado J, García-Hernández V, Verdejo-Torres O, Castillo AM, Larré I, Poot-Hernández AC, Franco M, Gariglio P, Reyes JL, Contreras RG (2014) Ouabain induces endocytosis and degradation of tight junction proteins through ERK1/2-dependent pathways. Exp Cell Res 320:108–118
Rodriguez-Boulan E, Macara IG (2014) Organization and execution of the epithelial polarity programme. Nat Rev Mol Cell Biol 15:225–242
Rosenthal R, Milatz S, Krug SM, Oelrich B, Schulzke JD, Amasheh S, Günzel D, Fromm M (2010) Claudin-2, a component of the tight junction, forms a paracellular water channel. J Cell Sci 123 (Pt 11):1913–1921. https://doi.org/10.1242/jcs.060665
Saito T, Tanaka R, Wataba K, Kudo R, Yamasaki H (2004) Overexpression of estrogen receptor-alpha gene suppresses gap junctional intercellular communication in endometrial carcinoma cells. Oncogene 23:1109–1116
Schatzmann HJ (1953) Cardiac glycosides as inhibitors of active potassium and sodium transport by erythrocyte membrane. Helv Physiol Pharmacol Acta 11:346–354
Schreiber V, Stepan J (1986) Digoxin-like immunoreactivity of 19-NOR and 19-OHandrost-4-ene-3,17-dione. Physiol Bohemoslov 35:190–192
Shinoda T, Ogawa H, Cornelius F, Toyoshima C (2009) Crystal structure of the sodium-potassium pump at 2.4 Å resolution. Nature 459:446–450
Shoshani L, Contreras RG, Roldán ML, Moreno J, Lázaro A, Balda MS, Matter K, Cereijido M (2005) The polarized expression of Na^+,K^+-ATPase in epithelia depends on the association between beta-subunits located in neighboring cells. Mol Biol Cell 16:1071–1081
Skou JC (1957) The influence of some cations on an adenosine triphosphatase from peripheral nerves. Biochim Biophys Acta 23:394–401
Soderberg K, Rossi B, Lazdunski M, Louvard D (1983) Characterization of ouabain-resistant mutants of a canine kidney cell line, MDCK. J Biol Chem 258:12300–12307
Takahashi H (2000) Endogenous digitalislike factor: an update. Hypertens Res 23(Suppl):S1–S5
Tian J, Cai T, Yuan Z, Wang H, Liu L, Haas M, Maksimova E, Huang XY, Xie ZJ (2006) Binding of Src to Na^+/K^+-ATPase forms a functional signaling complex. Mol Biol Cell 17:317–326

Tokhtaeva E, Munson K, Sachs G, Vagin O (2010) N-glycan-dependent quality control of the Na, K-ATPase beta(2) subunit. Biochemistry 49(14):3116–3128. https://doi.org/10.1021/bi100115a

Tokhtaeva E, Sachs G, Souda P, Bassilian S, Whitelegge JP, Shoshani L, Vagin O (2011) Epithelial junctions depend on intercellular trans-interactions between the Na,K-ATPase β1 subunits. J Biol Chem 286:25801–25812

Tokhtaeva E, Sachs G, Sun H, Dada LA, Sznajder JI, Vagin O (2012) Identification of the amino acid region involved in the intercellular interaction between the β1subunits of Na^+/K^+-ATPase. J Cell Sci 125:1605–1616

Tymiak AA, Norman JA, Bolgar M, DiDonato GC, Lee H, Parker WL, Lo LC, Berova N, Nakanishi K, Haber E et al (1993) Physicochemical characterization of a ouabain isomer isolated from bovine hypothalamus. Proc Natl Acad Sci USA 90:8189–8193

Ussing HH, Zerahn K (1951) Active transport of sodium as the source of electric current in the short-circuited isolated frog skin. Acta Physiol Scand 23(2–3):110–127. https://doi.org/10.1111/j.1748-1716.1951.tb00800.x

Vagin O, Turdikulova S, Sachs G (2005) Recombinant addition of N-glycosylation sites to the basolateral Na,K-ATPase β1 subunit results in its clustering in caveolae and apical sorting in HGT-1 cells. J Biol Chem 280:43159–43167

Vagin O, Sachs G, Tokhtaeva E (2007a) The roles of the Na,K-ATPase beta 1 subunit in pump sorting and epithelial integrity. J Bioenerg Biomembr 39:367–372

Vagin O, Turdikulova S, Tokhtaeva E (2007b) Polarized membrane distribution of potassium-dependent ion pumps in epithelial cells: different roles of the N-glycans of their beta subunits. Cell Biochem Biophys 47:376–391

Vagin O, Tokhtaeva E, Yakubov I, Shevchenko E, Sachs G (2008) Inverse correlation between the extent of N-glycan branching and intercellular adhesion in epithelia. Contribution of the Na,K-ATPase beta1 subunit. J Biol Chem 283:2192–2202

Wang H, Haas M, Liang M, Cai T, Tian J, Li S, Xie Z (2004) Ouabain assembles signaling cascades through the caveolar Na^+/K^+-ATPase. J Biol Chem 279:17250–17259

Weisz OA, Rodriguez-Boulan E (2009) Apical trafficking in epithelial cells: signals, clusters and motors. J Cell Sci 122:4253–4266

Wilson PD, Sherwood AC, Palla K, Du J, Watson R, Norman JT (1991) Reversed polarity of Na^+-K^+-ATPase: mislocation to apical plasma membranes in polycystic kidney disease epithelia. Am J Physiol Renal Physiol 260:F420–F430

Wilson PD, Devuyst O, Li X, Gatti L, Falkenstein D, Robinson S, Fambrough D, Burrow CR (2000) Apical plasma membrane mispolarization of NaK-ATPase in polycystic kidney disease epithelia is associated with aberrant expression of the β2 isoform. Am J Pathol 156:253–268

Xie Z (2003) Molecular mechanisms of Na/K-ATPase-mediated signal transduction. Ann N YAcad Sci 986:497–503

Xie Z, Kometiani P, Liu J, Li J, Shapiro JI, Askari A (1999) Intracellular reactive oxygen species mediate the linkage of Na^+/K^+-ATPase to hypertrophy and its marker genes in cardiac myocytes. J Biol Chem 274:19323–19328

Zhu ZL, Zhao ZR, Zhang Y, Yang YH, Wang ZM, Cui DS, Wang MW, Kleeff J, Kayed H, Yan BY, Sun XF (2010) Expression and significance of FXYD-3 protein in gastric adenocarcinoma. Dis Markers 28:63–69

Zimmermann P (2006) PDZ domain-phosphoinositide interactions in cell-signaling. Verh K Acad Geneeskd Belg 68:271–286

Chapter 2
Na^+-K^+-$2Cl^-$ Cotransporter

Eric Delpire and Kenneth B. Gagnon

Abstract The conceptual breakthrough that the energy of the Na^+ gradient generated by the Na^+/K^+ ATPase (pump) could be used as the driving force for another membrane transport protein has led to the functional and molecular identification of multiple secondary active transporters. We have organized this chapter to address the expression, function, regulation, and evolutionary importance of the two isoforms of the electroneutral sodium–potassium–chloride cotransporter (NKCC). The combination of basolateral expression of the sodium–potassium pump and NKCC1 in various non-renal epithelial results in salt and water secretion, whereas basolateral expression of the pump with an apical expression of NKCC2 in the thick ascending limb of Henle of the kidney nephron results in salt and water reabsorption. NKCCs are regulated by phosphorylation of specific serine/threonine residues in their cytosolic amino-terminal domains, and the evolutionary conservation of these cotransporters from protists to humans confirms their vital role in cellular and whole-organism physiology.

Keywords Cl^- secretion · Na^+ reabsorption · Cell volume · Electroneutral cotransport · Loop diuretics

2.1 Introduction

Prokaryotic and eukaryotic cells have surrounded themselves with two leaflets of phospholipids forming a membrane or "oil" interface that isolates their biochemical reactions from the extracellular environment. However, complete isolation from the

---

E. Delpire (✉)
Department of Anesthesiology, Vanderbilt University School of Medicine, Nashville, TN, USA
e-mail: eric.delpire@vanderbilt.edu

K. B. Gagnon
Department of Medicine, University of Louisville School of Medicine, Louisville, KY, USA
e-mail: kenneth.gagnon@louisville.edu

K. L. Hamilton, D. C. Devor (eds.), *Studies of Epithelial Transporters and Ion Channels*, Physiology in Health and Disease,
https://doi.org/10.1007/978-3-030-55454-5_2

outside environment is incompatible with life, as the uptake of nutrients and elimination of waste are necessary steps of cell metabolism. By insertion of transmembrane proteins, cells have acquired the capability to move water and solutes across this lipid barrier and gained the capacity to modulate the flow of these substrates. Sugars, amino acids, organic compounds, and inorganic ions such as Na^+, K^+, Cl^-, Ca^{2+}, Mg^{2+}, and HCO_3^- are able to cross the lipid membrane (in conjunction with water) in and out of cells. In a multicellular organism, depending on its specific function within the organism, a cell will express a certain array of channels, pumps, cotransporters, and exchangers at its plasma membrane, allowing for a defined and regulated movement of solutes.

2.2 Ouabain-Insensitive Cation Pump?

In the early 1960s, physiologists believed that transport of Na^+ and K^+ across human red blood cell membranes was either energy-dependent, through the ATP-dependent cation (Na^+/K^+) pump, or a result of passive diffusion in the direction of the ions electrochemical gradient. The Na^+/K^+ ATPase exchanges K^+ outside for Na^+ inside and is inhibited by cardiac glycosides (e.g., ouabain). Identification of another transport pathway for Na^+ and K^+ which was "ouabain-insensitive" but still dependent on external Na^+ prompted investigators to search for another type of cation pump (Hoffman and Kregenow 1966). In fact, what they considered to be an active transport mechanism dependent on external Na^+ was actually the first evidence of a secondary active transport mechanism (i.e., Na-K-2Cl cotransporter) dependent on the electrochemical gradient of Na^+ generated by the "ouabain-sensitive" cation pump. This "sodium-gradient hypothesis" was first proposed by R.K. Crane and represents a conceptual breakthrough in ion transport (Crane 1965).

In Fig. 2.1, the cation exchange activity of the Na^+/K^+ ATPase maintains the intracellular Na^+ concentration significantly lower than the extracellular concentration while simultaneously maintaining the intracellular K^+ higher than the extracellular concentration. These concentration gradients provide the energy that secondary active transporters, such as the Na-K-2Cl cotransporter and the Na-glucose cotransporter, use to power the movement of other solutes against their own concentration gradients. In Box 2.1, the individual driving forces acting on Na^+, K^+, and Cl^- across the cell membrane can be added together to determine the net driving force and direction of the Na-K-2Cl cotransporter. Although accepted today, there was considerable resistance to the concept of a membrane transport protein using the downhill concentration gradient of one solute to move another solute uphill against its concentration gradient. Such a mechanism did not conform to the Second Law of Thermodynamics. Today, there are multiple Na^+-coupled cotransporters which not only move inorganic ions but also amino acids, neurotransmitters, and even carbohydrates. In the remainder of this chapter, we will focus on a subset of the inorganic ion transporters, the Na^+-coupled cation–chloride cotransporters, and their role in epithelial health and disease.

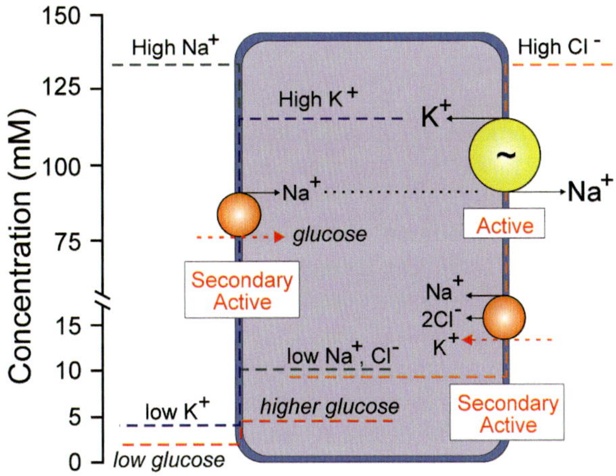

Fig. 2.1 The Na-K-2Cl cotransporter is a secondary active transport mechanism. Schematic representation of a cell with the Na^+/K^+ ATPase accumulating K^+ and extruding Na^+, thus creating energy in ionic concentration gradients that can be used by other transport mechanisms. As the very low [Na^+] inside cells constitutes a driving force for the inward movement of glucose (even if higher glucose inside), the low [Na^+] and [Cl^-] inside cells also facilitate the inward movement of ions through the Na-K-2Cl cotransporter

Box 2.1 Calculating the Electrochemical Driving Forces on the Na-K-2Cl Cotransporter

The electrochemical potential gradient for an ion, $\Delta\mu_{ion}$, is the sum of the chemical and electrical components: $\Delta\mu_{ion} = RT\ \ln([ion]_i/[ion]_o) + ZFE_m$. For a cotransport system that couples the movement of multiple substrates, the net free energy or overall chemical potential gradient is the sum of the gradient for each substrate: $\Delta\mu_{Na,K,Cl} = \Delta\mu_{Na} + \Delta\mu_K + 2\Delta\mu_{Cl}$ (because of two Cl ions transported on NKCC). Note that there is no electrical component as the cotransport of two cations with two anions is electroneutral.

Thus,

$$\Delta\mu_{Na,K,Cl} = RT\ \ln\left([Na]_i/[Na]_o\right) + RT\ \ln\left([K]_i/[K]_o\right) + 2\,RT\ \ln\left([Cl]_i/[Cl]_o\right)$$
$$\text{or} \quad = RT\ \ln\left([Na]_i[K]_i[Cl]_i^{\ 2}\right)/\left([Na]_o[K]_o[Cl]_o^{\ 2}\right)$$

We can see that the direction of transport will be dictated by the ratio of the products. For a ratio larger than 1, $\Delta\mu_{Na,K,Cl}$ will be positive, whereas for a ratio lower than 1, $\Delta\mu_{Na,K,Cl}$ will be negative. At 1, the chemical potential gradient will be 0 (since $\ln(1) = 0$), which represents thermodynamic equilibrium. If we consider external and internal ion concentrations of $[Na]_i = 10$ mM, $[Cl]_i = 30$ mM, $[K]_i = 100$ mM, $[Na]_o = 130$ mM, $[Cl]_o = 110$ mM, and $[K]_o = 3$, the ratio favors inward transport by 5:1.

2.3 Electrically Silent Plasma Membrane Cotransporters

In a set of companion papers in 1971, Floyd Kregenow described the two-phase response for duck erythrocytes incubated in nonhemolytic hypotonic and hypertonic media (Kregenow 1971a, b). He found that the instantaneous cell shrinkage induced by hypertonic shock was followed by a slower "volume regulatory phase" in which the cells swell back to their original size through a Na^+-dependent gain of K^+, Cl^-, and water (Kregenow 1971b). The accumulation of K^+ against its electrochemical gradient suggested an active transport mechanism; however, blockage of the Na^+/K^+ ATPase by ouabain had no effect on this volume controlling mechanism. Interestingly, regardless of the severity of the hypertonic challenge, there was no water recovery when external K^+ concentration was maintained at normal levels (i.e., 2.8 mM). However, when the external K^+ concentration was increased to 15 mM, the rate of cell water recovery was faster the greater the hypertonic challenge (see Fig. 2.2a). Although at first counter-intuitive, if we once again consider the individual driving forces acting on the Na-K-2Cl cotransporter (see Box 2.1), an increase in the external K^+ concentration actually reduces the resistance of the outwardly

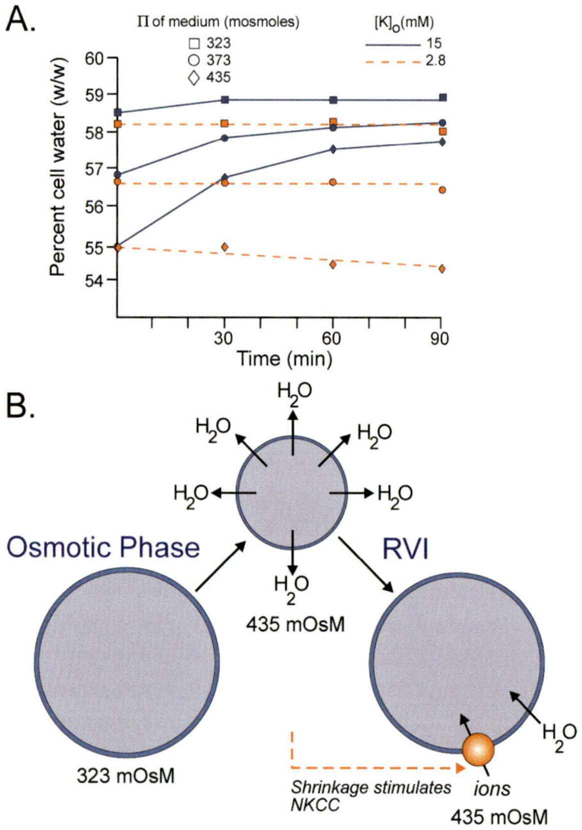

Fig. 2.2 The Na-K-2Cl cotransporter participates in cell volume regulation. (**a**) Data redrawn from Kregenow showing cell water in duck red blood cells recovering after an initial water loss due to the osmotic phase during a hypertonic stimulus (Kregenow 1971a, b). Note that the recovery is larger in cells exposed to larger osmotic shock and it occurs only when the driving force for Na-K-2Cl cotransport is increased by raising external K^+. (**b**) Schematic representation of a cell responding to a hypertonic stimulus. The cell loses water during the osmotic phase, leading to NKCC1 activation, uptake of ions, and obligatory water

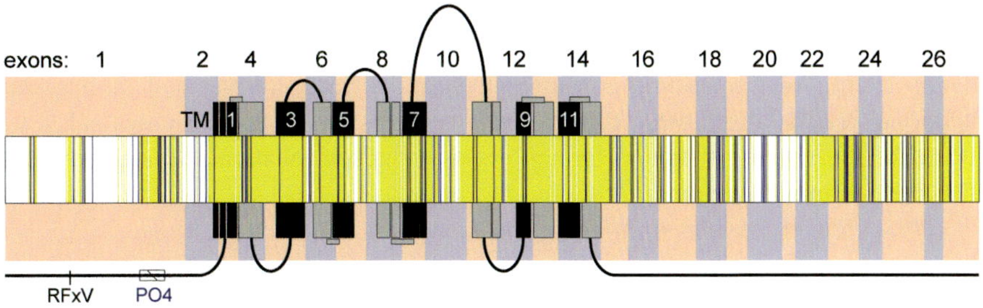

Fig. 2.3 Schematic representation of NKCC1 and NKCC2 proteins. Alignment of the full-length mouse NKCC1 and NKCC2 amino acid sequences illustrating identical (yellow), conserved (blue), and non-conserved (white) regions of the two proteins are superimposed on the topology of the cotransporters with odd-numbered transmembrane domains in black and even-numbered domains in gray. Also shown in the cytosolic amino-terminal domain are conserved SPAK/OSR1 binding sites (RFxV sequences) and a region of multiple key phosphorylation sites (PO4). The exons of the *SLC12A1* (NKCC2) and *SLC12A2* (NKCC1) genes are represented in the background by alternate pink (odd) and blue (even) boxes

directed K^+ gradient on the net activity of the Na-K-2Cl cotransporter. This reduced resistance allows the cotransporter, to work more efficiently (ratio favors inward transport ~26:1) in transporting ions (and cell water) to promote cell volume recovery (see Fig. 2.2b).

Although we have been identifying the Na^+-dependent K^+ and Cl^- transport mechanism described by Kregenow as the Na-K-2Cl cotransporter, it actually took another 6 years before Geck and coworkers proposed the existence of a "secondary-active" cotransport system for Na^+, K^+, and Cl^- in Ehrlich cells which exhibited the same "ouabain-insensitivity" and "volume control" identified by Kregenow. Thermodynamic experiments revealed a tight stoichiometric coupling of 1 Na^+, 1 K^+, and 2 Cl^- ions in every transport cycle (Geck et al. 1980).

In 1994, independent research teams identified cDNA sequences for two Na-K-2Cl cotransporter isoforms. The first isoform, termed NKCC1, isolated from fish (Xu et al. 1994) and mammals (Delpire et al. 1994), appears to be ubiquitously expressed and has been mapped to chromosome 5q23 in humans and chromosome 18 in mice. The second isoform, termed NKCC2, was isolated from the mammalian kidney (Gamba et al. 1994; Payne and Forbush 1994), is expressed exclusively in the apical membrane of the thick ascending limb of Henle, and has been mapped to chromosome 15 in humans and chromosome 2 in mice. In Fig. 2.3, amino acid sequences of the two mouse isoforms have been aligned to illustrate identical (yellow), conserved (blue) and non-conserved (white) regions of the two proteins. The predominance of yellow color throughout the protein illustrates the high (58% overall and 76% in the transmembrane core) degree of amino acid residue similarity between the two cotransporters. Alternating light pink and light blue boxes identify the exons of the cotransporters and the portion of the protein (NH_2-terminal, transmembrane, or COOH-terminal) encoded by each exon. There are three distinctive properties that functionally define Na-K-2Cl cotransport: (1) all three transported ions bind the cotransporter on the same side of the membrane; (2) transporter inhibition by loop diuretics (e.g., bumetanide and furosemide); and (3) electrically silent stoichiometry of ion translocation. In Sect. 2.5, we will discuss the functional regulation of these cotransporter proteins and the involvement of conserved kinase-binding sites and serine/threonine phosphorylation sites on the cytosolic NH_2-terminal domain.

2.4 NKCC1

The human *SLC12A2* gene encodes for a 1205 amino acid protein with 58% similarity with the second kidney-specific human isoform encoded by the *SLC12A1* gene (see below). NKCC1 has 12 alpha-helical membrane-spanning regions flanked by large cytosolic NH_2- and COOH-terminal domains. There are also several *N*-linked glycosylation sites on a large extracellular loop between transmembrane domain 7 and 8. RNA and protein expression studies have demonstrated that NKCC1 is expressed in various tissues including eye, stomach, heart, lung, brain, thymus, smooth and skeletal muscle, neurons, testis, colon and red blood cells (Delpire et al. 1994; Payne et al. 1995). This ubiquitous distribution suggests multiple physiological roles beyond salt reabsorption including Cl^- and fluid secretion (Cook and Young 1989; Liedkte 1992), acid secretion (Soybel et al. 1995), cell volume homeostasis (Palfrey and O'Donnell 1992), and possibly even cell division and proliferation (Panet et al. 1994, 2000).

2.4.1 *Electroneutrality, Stoichiometry, and Kinetic Properties*

Geck and coworkers first described the "electrically-silent" nature of NKCC1 transport in 1980 (Geck et al. 1980). In their study, pulse-response experiments revealed a furosemide-sensitive tight coupling of 1 Na^+, 1 K^+, and 2 Cl^- ions. As the two positive charges are nullified by the two negative charges, the cotransport of these ions neither affect the membrane potential nor does the membrane potential affect cotransport activity. In 1998, Lytle and coworkers (Lytle 1998) proposed a model of ordered binding and gliding symmetry where Na^+ binds first, followed by Cl^-, then K^+, and a second Cl^- and on the inside, the ions were released in the same order as binding: first on–first off (Fig. 2.4a). Through manipulation of the ionic composition both inside and outside the cell, they found that the tightly coupled ratio of 1 Na^+; 1 K^+; and 2 Cl^- was maintained regardless of the internal and external ion composition. They also observed two partial reaction cycles: Na^+/Na^+ exchange in cells with high $[Na^+]_i$, and K^+/K^+ exchange in cells with high $[K^+]_i$ (Lytle 1998). Although their model adequately explained the 1 Na^+, 1 K^+, 2 Cl^- stoichiometry of the cotransporter, their first on-first off gliding symmetry model does not explain

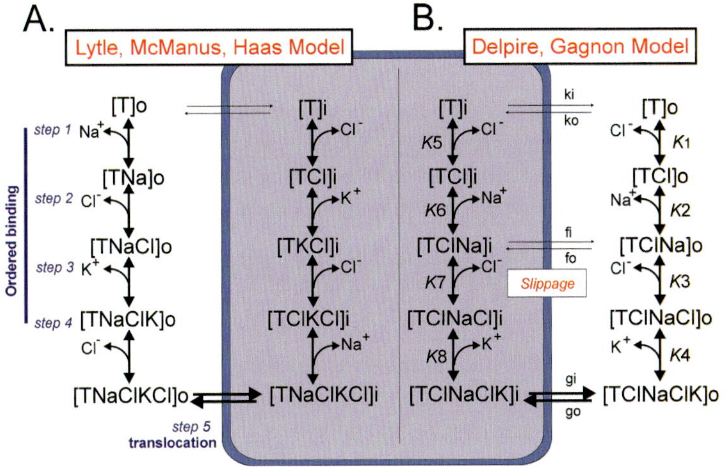

Fig. 2.4 Kinetic Models of NKCC1 cotransport. (**a**) Model proposed by Lytle et al. (1998). The cotransporter outside loads ions in an *ordered* fashion with Na$^+$ binding first, followed by Cl, K$^+$, then the second Cl$^-$ (steps 1–4). The fully loaded cotransporter then translocates (step 5) and releases ions on the other side on a first-on first-off basis. (**b**) Model proposed by Delpire and Gagnon (2011) where the cotransporter outside also loads ions in an *ordered* fashion but with Cl$^-$ binding first, followed by Na$^+$, a second Cl$^-$, then K$^+$. After translocation, the release occurs in a more traditional fashion. Binding rate constants for each reaction are indicated by labels above and below the directional arrows. Computer simulations performed by Delpire and Gagnon (2011) provided several sets of rate constants, depending upon the number of cotransporters inserted in the membrane. Note that translocation of the loaded cotransporter is much faster (thicker arrows) than the translocation of the empty cotransporter (thinner arrows). Kinetics allow for some partial reactions to occur (slippage step)

why Na$^+$/Na$^+$ exchange would be dependent on external K$^+$ or why K$^+$/K$^+$ exchange would require internal Na$^+$ but not external Na$^+$?

We and others have observed atypical NKCC transport stoichiometries in many cell types which cannot be resolved by first on–first off gliding symmetry (Canessa et al. 1986; Gagnon and Delpire 2010; Hall and Ellory 1985; Lytle 1998; Orlov et al. 1996; Russell 1983). In 2011, we proposed a different model for ion binding in a preferred order based on velocity equations for K$^+$ influx under both rapid equilibrium assumptions and combined equilibrium and steady-state assumptions. Our model has Cl$^-$ binding first, followed by a Na$^+$, the second Cl$^-$, and finally a K$^+$ ion (Fig. 2.4b). Upon binding, the fully loaded cotransporter undergoes a conformational change ("translocates") and releases the ions inside the cell in the reverse order to which they bound (i.e., K$^+$, Cl$^-$, Na$^+$, and Cl$^-$). In rapid equilibrium kinetics, the rate-limiting steps are the translocation steps, with the "fully-loaded" cotransporter translocating much faster than the "empty" cotransporter. In this model, some partial reactions are permitted (i.e., slippage), while preserving transport electroneutrality. Rapid translocation of fully loaded transporter and slippage steps can explain why under certain conditions, hypertonicity, for instance, the transporter unidirectionally moves more K$^+$ than Cl$^-$, and more Cl$^-$ than Na$^+$. In these conditions, the preferred mode of transport is K^+/K^+ exchange, followed by K-Cl cotransport (Na^+ is not released), and finally Na-K-2Cl cotransport (Delpire and Gagnon 2011).

In the Lytle study (Lytle 1998), they measured endogenous transport activity in duck red blood cells, whereas in our study (Delpire and Gagnon 2011), we measured transport activity of mouse NKCC1 heterologously overexpressed in *Xenopus laevis* oocytes. In our study, we altered the tonicity of the extracellular solutions, whereas, the Lytle study replaced transportable cations (Na^+ and K^+) with less-transportable cations (Li^+ and Rb^+), respectively. Our study used radiolabeled $^{22}Na^+$, $^{86}Rb^+$, and $^{36}Cl^-$ as tracers, whereas, the Haas study used a combination of radiolabeled tracers ($^{22}Na^+$, $^{86}Rb^+$, and $^{36}Cl^-$) and ion measurements from dilute perchloric acid extracts via air-acetylene flame spectrophotometry. Any combination of these differences could explain the two divergent models, and more experiments will be necessary to distinguish which better characterizes actual transport activity.

2.4.2 NKCC1 in Cl^- Secreting Epithelia

Contrary to the salt reabsorption observed with NKCC2 (see next section), the basolateral expression of NKCC1 in most non-renal epithelial cells results in salt secretion. In Fig. 2.5, a model of a prototypical epithelial cell illustrates how NKCC1

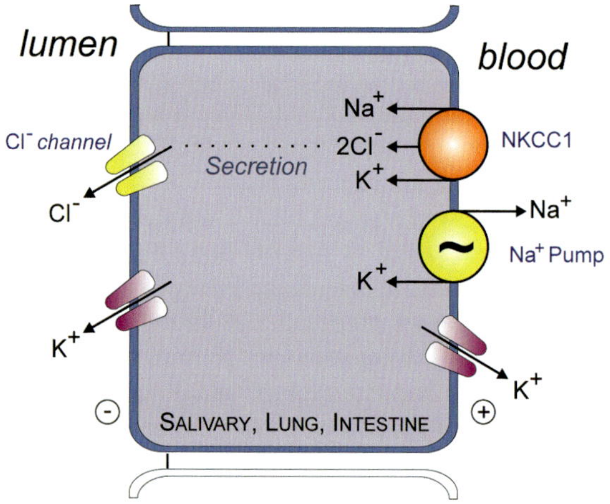

Fig. 2.5 Schematic representation of a prototypical Cl^- secreting non-renal epithelial cell. In this model, the Na^+/K^+ ATPase provides the energy for Cl^- transport from the basolateral (blood) side to the apical (luminal) side of the epithelium. Cl^- is secreted to the lumen at the apical membrane through a Cl^- channel (typically CFTR) and the Na-K-2Cl cotransporter on the basolateral membrane constitutes the entry pathway for Cl^- into the epithelial cell. Activation of the Cl^- channel (e.g., through cAMP) results in a drop in intracellular Cl^- which stimulates the NKCC1 function. The Na^+/K^+ ATPase recycles Na^+ ions entering through the cotransporter. Basolateral K^+ channels recycle K^+ ions entering through the cotransporter and the Na^+/K^+ ATPase

utilizes the Na^+ gradient generated by the basolateral expression of the Na^+/K^+ ATPase to import Na^+, K^+, and Cl^- into the cell. The resulting increase in intracellular $[Cl^-]$ above electrochemical equilibrium provides the driving force for Cl^- secretion into the lumen through apical Cl^- channels. In many epithelia, the apical Cl^- channel is the cAMP/Protein Kinase A (PKA)-regulated cystic fibrosis transmembrane conductance regulator (CFTR). CFTR is so named as it is the primary chloride channel mutated in patients with cystic fibrosis. Activation of CFTR by cAMP/PKA stimulation results in Cl^- secretion into the lumen thereby depleting the cells of intracellular Cl^-, which is then replenished through NKCC1 activity. This generic model is applicable to salivary, sweat, and lacrymal glands, as well as airway, gastric, and intestinal epithelial cells.

In 1989, Wiener and colleagues were the first to demonstrate that Cl^- secretion in colonic epithelial cells (Wiener and van Os 1989) follows the model of the shark rectal gland (i.e., driven by Na^+-dependent Cl^- transporter on the basolateral membrane and Cl^- channels on the apical membrane) (Silva et al. 1977). Using [3H]-bumetanide binding and loop diuretic-sensitive isotope uptake experiments, Weiner and colleagues showed that a Na-K-Cl cotransporter was present on the basolateral membrane of both surface and crypt cells of the rabbit distal colon epithelium. The physiological relevance of this transporter was highlighted in studies using bumetanide in *Ussing* chamber preparations. For example, in rat colon, cAMP stimulation by forskolin greatly increased short circuit current due to activation of apical Cl^- channels. The addition of bumetanide to the basolateral side completely abolishes this effect, confirming the importance of NKCC1 expression in the secretion process (Cheng 2012). A study in mouse intestine and human intestinal (HT29) cells showed that rosiglitazone and pioglitazone, two PPARγ agonists, significantly reduced forskolin-induced (cAMP) and carbachol-induced (Ca^{2+}) electrogenic Cl^- secretion. This decrease was associated with a significant reduction in CFTR, KCNQ1, and NKCC1 expression (Bajwa et al. 2009). Similarly, butyrate decreased NKCC1 expression in human colonic T84 epithelial cells, but not to a degree that affected forskolin or carbachol-mediated Cl^- secretion (Resta-Lenert et al. 2001). In contrast to the agents described above that decrease Cl^- secretion, there are also many factors that increase Cl^- and fluid secretion in the intestine. These include pathogens such as *Vibrio cholerae* that causes diarrhea predominantly by activating the net secretion of chloride ions by the colon (Das et al. 2018). The mechanism is through internalization of the cholera toxin, stimulation of cAMP production, and activation of CFTR (Das et al. 2018, see also Chap. 2 of Volume 2 of this book). Again, similar studies were performed using other Cl^- secreting epithelia, such as salivary, lacrimal, and airway epithelia.

2.4.3 NKCC1 in Kidney

Even though NKCC2 is the renal-specific cotransporter, NKCC1 is also present in several structures within the mammalian kidney. In the mouse, the cotransporter has been detected in smooth muscle cells of the afferent arteriole of the glomerulus, the extraglomerular mesangium, and the inner medullary collecting duct (Kaplan et al. 1996). In the rat, NKCC1 is also expressed in alpha intercalated cells of the outer medullary collecting duct (Ginns et al. 1996). These cells are acid-secreting cells and the cotransporter on the basolateral membrane might provide a pathway for Cl^- entry from the blood side, thus helping H^+ and Cl^- secretion on the apical side. Note that the mouse versus rat difference in expression of the cotransporter in the medulla may reflect significant physiological differences between the two species. In fact, a recent study demonstrated that in middle-aged mice, expression of NKCC1 and reduced age-related hearing loss is dependent upon aldosterone (Halonen et al. 2016). Expression of NKCC1 in renin-producing smooth muscle cells suggests a role for the cotransporter in tubuloglomerular feedback and/or renin release, both sensitive to luminal Cl^- concentrations (Kaplan et al. 1996). Basolateral expression of the cotransporter in the collecting duct epithelial cells may serve several roles. Physiological studies have shown that atrial natriuretic peptide (ANP) stimulates upregulation of Na-K-2Cl cotransporter expression and downregulation of the Na^+ channel and Na^+/K^+ ATPase, in effect, altering the inner medullary collecting duct epithelial cells from re-absorbing Na^+ to secreting Na^+ (Rocha and Kudo 1990; Sonnenberg et al. 1990; Zeidel et al. 1988). The capacity of NH_4^+ to displace K^+ at the binding site within the cotransporter suggests that basolateral expression of NKCC1 in the collecting duct epithelial cells may participate in H^+/NH_4^+ secretion similar to that observed in the stomach (Soybel et al. 1995). Finally, the "volume sensitivity" of the cotransporter initially identified by Kregenow (1971b) in duck red blood cells may have a significant role in collecting duct epithelial cells that are routinely exposed to severe osmotic challenges.

2.4.4 NKCC1 in Other Epithelia

The stria vascularis is a multilayered epithelium of the inner ear extending from Reissner's membrane to the spiral prominence. This unusual tissue has three primary cell types (marginal, intermediate, and basal cells) as well as intraepithelial capillaries. Basolateral expression of NKCC1 in the marginal cells, which directly face the endolymphatic compartment, does not participate in Cl^- secretion like in most other epithelia, but in K^+ secretion thereby generating a low-sodium, high-potassium cochlear endolymph (Tasaki and Spiropoulos 1959). On the apical membrane of these cells, a K^+ channel (KCNQ1) associated with a beta-subunit (KCNE1) participates in the apical secretion of the cation. Mutations in either of the subunits result in Jervell and Lange-Nielsen syndrome, a type of long QT syndrome, associated with severe, bilateral hearing loss (Tyson et al. 1997). Knockout of the KCNQ1 channel in mice (Casimiro et al. 2001) or of its beta-subunit (Vetter et al. 1996) results in sensorineural deafness. Knockout of the Na-K-2Cl cotransporter in mice similarly results in sensorineural deafness (Delpire et al. 1999; Flagella et al. 1999). This precise localization of the cotransporter in the stria vascularis also explains the

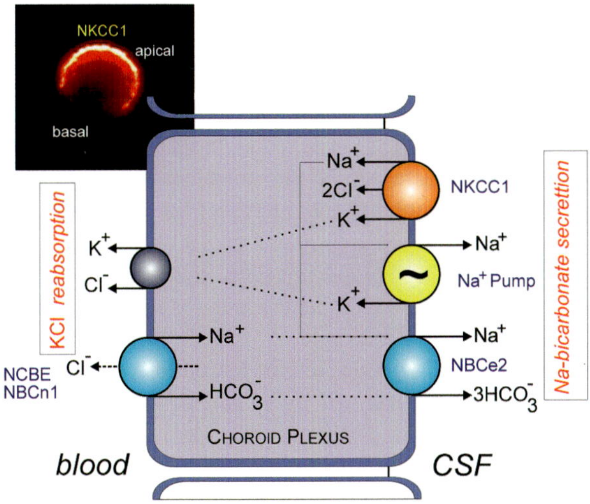

Fig. 2.6 Na-K-2Cl cotransport in choroid plexus. *Picture:* Single choroid plexus cell isolated with collagenase to keep apical versus basolateral polarity. After fixation, the cell was permeabilized and exposed to rabbit anti-NKCC1 polyclonal antibody followed by Cy3-conjugated anti-rabbit secondary antibody. *Model:* In these cuboidal epithelial cells, NKCC1 is located on the apical membrane alongside the Na^+/K^+ ATPase. The pump provides the energy for the secretion of Na-bicarbonate from the blood side to the luminal (cerebrospinal fluid (CSF)) side. Bicarbonate enters and exits the cell through distinct Na^+-coupled bicarbonate transporters. The pump also provides the driving force for K^+ reabsorption or movement of K^+ from CSF to blood. Expression of NKCC1 on the apical membrane may allow for the uncoupling of K^+ reabsorption from Na^+ secretion

ototoxicity caused by the use of large doses of loop diuretics (Ikeda et al. 1997; Rybak 1993).

Another epithelial tissue expressing NKCC1 is the four choroid plexuses of the brain. Interestingly, NKCC1 expression is localized on the apical rather than the basolateral membrane of the choroid plexus (CP). As illustrated in Fig. 2.6, choroid plexus epithelial cells participate in the secretion of cerebrospinal fluid. Na^+ movement through the Na^+/K^+ ATPase generates fluid movement. What is actually secreted in the ventricle is Na^+-bicarbonate. On the basolateral side, Na^+ enters through Na^+-coupled bicarbonate transporters, such as the Na^+-dependent Cl^-/HCO_3^- exchanger (NCBE) or the Na^+-HCO_3^- cotransporter (NBCn1) (Christensen et al. 2013). The location of NKCC1 on the apical membrane is counter to the movement of Na^+ as Na-K-2Cl cotransport is inward. It is also known that choroid plexus epithelial cells can participate in K^+ reabsorption. This is an important function, as the cerebrospinal fluid (CSF) K^+ content, and consequently, the K^+ concentration surrounding neurons need to be tightly regulated. The Na^+/K^+ ATPase can also serve as the primary mechanism for K^+ reabsorption. However, placing the cotransporter alongside the pump has the advantage of uncoupling the pump obligatory exchange of Na^+ with K^+. Indeed, the Na^+ that exits through the pump can now be reclaimed by the cotransporter (see Fig. 2.6).

Several studies were recently published that question the role of NKCC1 in choroid plexus. First, it was shown that following brain hemorrhage, an inflammatory response involving toll-4 receptor activation, leads to the activation of SPAK, stimulation of NKCC1 function, and hypersecretion of CSF (Karimy et al. 2017). Intracerebroventricular (ICV) injection of blood into 8-week-old Wistar rats showed TLR4 signaling activation, increased nuclear translocation of p65 (a subunit of the NFκB–p65 transcription complex), and increased numbers of activated CD68+ choroid plexus myeloid cells. This response could be prevented using an NFκB inhibitor. Some 48 h after ICV blood injection, the rate of CSF production was threefold greater than control and remained stimulated for an additional 7 days. ICV injection of acetazolamide, an inhibitor of carbonic anhydrase activity, did not significantly reduce this CSF hypersecretion. In contrast, the rate of secretion was reduced 80% by ICV injection of bumetanide, implicating the Na-K-2Cl cotransporter. The levels of phospho-NKCC1 and phospho-SPAK, the active forms of the cotransporter and its activating kinase, were increased by 6.8-fold and twofold, respectively. The use of an inhibitor that prevents interaction between the kinase and the cotransporter also reduced CSF secretion. These data clearly indicate that NKCC1 function is key to CSF production. How can an inwardly poised cotransporter expressed on the apical membrane of choroid plexus epithelial cells participate to CSF secretion?

One possibility was offered in a second recent paper that argued for outward water transport associated with outward movement of Na^+, K^+, and Cl^- ions through NKCC1 (Steffensen et al. 2018). They demonstrated that (1) water can move in association with NKCC1 activity even against an osmotic gradient, and (2) that the ion concentrations in the epithelial cells and CSF produced gradients that actually favored the outward instead of inward movement of ions. A third paper, however, challenged this second point, reporting ion concentrations in CP epithelial cells that are more consistent with traditional inward transport (Gregoriades et al. 2019). In a viewpoint published in the same issue of the *American Journal of Physiology Cell Physiology*, we identified possible weaknesses with each approach, arguing that the measured CP concentrations in the Steffensen paper were possibly affected by their assumptions of extracellular volume contamination. Any small changes in the amount of extracellular volume (rich in NaCl) associated with the tissue will affect the calculated intracellular ion concentrations. We also argued that isolation conditions and incubation of isolated cells in artificial bathing solutions for extended periods of time might have also affected the intracellular ion concentration in the Gregoriades study.

Interestingly, Gregoriades and coworkers, with evidence for inward Na-K-2Cl cotransport, did not argue against the role of NKCC1 in CSF secretion (Gregoriades et al. 2019). They propose that the activity of the cotransporter must be critical in maintaining CP epithelial cell volume and ion homeostasis to support CSF secretion. In agreement with the Karimy study (Karimy et al. 2017), Steffenson and coworkers also demonstrated that the application of bumetanide on the luminal side significantly reduced the level of CSF secretion in both in vitro and in vivo models (Steffensen et al. 2018). Thus, while NKCC1 clearly plays a role in the secretion of cerebrospinal fluid by the choroid plexus, the exact mechanism is still a matter of debate. Additional information can also be found in Chap. 10 of Volume 2 of this book.

2.4.5 NKCC1 in Non-epithelial Cells

Aside from being expressed in epithelial cells, NKCC1 is also found in many other cell types, such as myocytes, vascular smooth muscle cells, and neurons. In muscle, the cotransporter participates in conjunction with the Na^+/K^+ ATPase in the accumulation of K^+. This is a critical function as skeletal muscle cells store 70–75% of the body K^+ (Gosmanov et al. 2003). Furthermore, a high K^+ concentration is needed for proper contraction, as a significant decrease in plasma K^+ (and therefore muscle K^+) caused by high doses of bumetanide is known to produce musculoskeletal pain, cramping, and/or muscle weakness (Howard and Dunn 1997; Vaduganathan et al. 2013). As chloride conductances in muscle cells are critical in stabilizing the membrane potential at the resting level, NKCC1 might also be involved in this stabilization by accumulating Cl^- above electrochemical potential equilibrium (Aickin et al. 1989). In addition, NKCC1 activity is also critical in helping muscle cells maintain their water content (Gosmanov et al. 2003). In vasculature, NKCC1 is thought to contribute to vasoconstriction (Akar et al. 1999), as vascular smooth muscle tone is reduced in NKCC1 knockout mice (Meyer et al. 2002). In neurons, the cotransporter affects the level of intracellular Cl^- that ultimately determines the direction and strength of GABA- and/or glycine-mediated Cl^- currents. Thus, the cotransporter is a key determinant of inhibitory synaptic neurotransmission.

2.4.6 NKCC1 in Disease

Considering the wide-expression pattern of NKCC1 and its importance in the control of cell volume, the modulation of inhibitory synaptic transmission, and epithelial transport, it would be reasonable to expect disruptions in NKCC1 expression and/or function to have severe health consequences. Although we know that the multiple NKCC1-deficient models generated are viable, global and tissue-specific NKCC1 knockout mice do exhibit multiple and severe phenotypes that greatly affect their health (Delpire et al. 1999; Dixon et al. 1999; Flagella et al. 1999; Pace et al. 2000). In addition to the inner ear defects already mentioned that affects both balance and hearing, the mice exhibit intestinal obstruction (Flagella et al. 1999; Grubb et al. 2000), deficit in saliva secretion (Evans et al. 2000), decreased pain perception (Laird et al. 2004; Sung et al. 2000), and male infertility (Gagnon and Delpire 2013a; Pace et al. 2000). The defect in intestinal transit is most relevant as the mice transition from milk to solid food. Indeed, many mutant pups are lost during the peri-weaning period, confirming the severe consequence of loss of

NKCC1. The observed intestinal dysfunction could be a result of a deficit in fluid secretion, as the transporter is involved in Cl-mediated fluid secretion (Grubb et al. 2000). A second hypothesis is a deficit in peristalsis, as Na-K-2Cl cotransporter is expressed in interstitial cells of Cajal surrounding the myenteric plexus (Wouters et al. 2006; Zhu et al. 2016). Another possibility is a deficit in autonomic control of intestinal function, as the transporter is highly expressed in the fibers that provide sensory feedback to the central nervous system (Alvarez-Leefmans et al. 1988; Sung et al. 2000).

Analysis of genomic databases reveals that, as with all other genes, *SLC12A2* is not "immune" from mutations in the human population. The Exome Aggregation Consortium (ExAC) database (http://exac.broadinstitute.org) reports some 280 missense mutations in *SLC12A2*. It is highly likely that some of them result in loss of function of the mutated allele. It is also possible that a 50% reduction in NKCC1 expression results in no pathophysiological effect. However, combined with another mutation or particular circumstance, a reduction in NKCC1 expression might facilitate the development of severe health issues.

In 2010, the whole-genome exome sequencing of an 8-year-old female with multiple health issues identified an 11 bp deletion in exon 22 of *SLC12A2* resulting in a frameshift mutation in NKCC1 and premature truncation of the translated protein. At the time, the patient suffered from orthostatic hypotension, autonomic bladder dysfunction, small intestine dysmotility, dietary intolerance, decreased energy, and seizure episodes (Delpire et al. 2016). Now 16, she recently experienced a complete loss of intestinal and bladder function. In addition, pancreatic, thyroid, and parathyroid gland deficiencies are affecting her endocrine system. A credible explanation for the failure of all these systems is some type of dysautonomia or dysregulation of the autonomic nervous system.

Development of a mouse model replicating this 11 bp deletion resulted in no overt phenotype (Koumangoye et al. 2018). However, the assessment of any deficits in sensory feedback in this mouse still needs to be undertaken. One interesting aspect of the mutant transporter in epithelial cells is its mistrafficking away from the basolateral membrane due to the truncation of a portion of the carboxyl-terminal tail. We showed that in Madin–Darby Canine Kidney (MDCK) cells, the transporter is not trapped in the endoplasmic reticulum, but accumulates at the apical pole, both at the apical membrane co-localizing with Gp135 (podocalyxin) and in subapical vesicles (Koumangoye et al. 2018). The subapical vesicles were identified as Rab5-positive apical early endosomes, indicating retrieval of transporters units from the apical membrane (Koumangoye et al. 2019). Due to dimerization, some wild-type monomers are also found at the apical membrane. Serial deletions and mutagenesis of the C-terminal tail of NKCC1 identified a di-leucine motif responsible for this effect (Koumangoye et al. 2019). Certainly, either deletion of this motif or mutation of the two leucine residues to alanine residues resulted in behavior similar to that of the patient NKCC1 mutant transporter. Immunohistochemistry analysis of NKCC1 in the NKCC1 mutant mouse confirms the mislocalization of the transporter in the salivary gland and colonic epithelia (Koumangoye et al. 2018). Interestingly, some

Table 2.1 Physiological comparison of human NKCC1 mutations with NKCC1 KO mouse

Phenotype	16-year-old girl (DFX)	5-year-old boy (KO)	NKCC1 KO mouse
Sensorineural deafness	−	+	+
Developmental delays	−	+	−
Gastrointestinal deficits	+	+	+
Salivary gland (saliva) deficit	+	+	+
Lacrimal gland (tear) deficit		+	Undetermined
Respiratory distress	+	+	Undetermined
Dysmorphic features	−	+	−
Low blood pressure	+	+	+
Hypotonia	+	+	Undetermined
Decreased cerebral volume	+	−	+

of the transporters appear to still be expressed on the basolateral membrane, most likely enough to sustain saliva secretion.

A 5-year-old boy with sensorineural hearing loss, developmental delays, and a variety of other clinical presentations was shown to have a homozygous 22 kb deletion in *SLC12A2* resulting in exons 2–7 being absent and a premature truncation codon in exon 8. In this condition, if the RNA were stable, the gene would only produce a fragment (89%) of the cysotolic N-terminal tail of the cotransporter. Analysis of the patient's fibroblasts demonstrated a complete absence of NKCC1 protein expression (Macnamara et al. 2019). Table 2.1 illustrates striking similarities (i.e., gastrointestinal deficits, lack or reduced saliva and tear secretion, low blood pressure) between these two human cases of NKCC1 disruption with the known phenotypes of the NKCC1 knockout mouse.

2.5 NKCC2

In man, 99.8% of the filtered Na^+ is reabsorbed, with 15–25% of this reabsorption occurring in the thick ascending limb of Henle (TALH). The *SLC12A1* gene encodes for an 1100 amino acid protein that RNA and protein expression studies have demonstrated is kidney-specific, localizes to the TALH epithelial apical membrane, and serves to re-absorb sodium and chloride from the tubule lumen. However, as salt is reabsorbed along the TALH, the luminal concentration of Na^+ and Cl^- decreases approximately fivefold from medulla to cortex. Interestingly, alternative splicing of the exons encoding for transmembrane domain 2 has resulted in three variants of NKCC2. Variant F which has the lowest binding affinity ($K_m(Na^+) = 66.72 \pm 5.8$ and a $K_m(Cl^-) = 111.3 \pm 13.4$) is expressed in the medullar TALH where the Na^+ concentration is the highest. The mid- (variant A) and high-affinity (variant B) cotransporters are expressed in the cortical TALH where the Na^+ concentration is

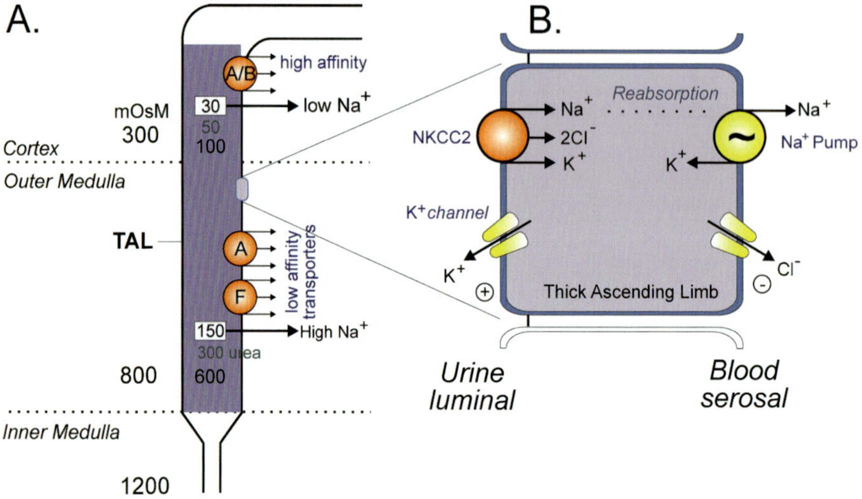

Fig. 2.7 NKCC2 function in the kidney. (**a**) Schematic representation of the thick ascending limb of Henle (TALH) showing its location in cortex and medulla. In the deep medulla, the osmolarity is elevated whereas, in the superficial cortex, the osmolarity is equivalent to blood. Therefore, there is a gradient of Na^+ in the TALH from the inner medulla to the cortex. Represented are three variants of the Na-K-2Cl cotransporter, NKCC2. In the medulla where the Na^+ concentration is high, the low (variant F) and mid (variant A) affinity cotransporters are expressed, whereas, in the cortex where the Na^+ is low, the high-affinity cotransporter (variant B) is expressed. (**b**) Schematic representation of a TALH epithelial cell with expression of NKCC2 and a K^+ channel (ROMK) at the urine or luminal membrane, the Na^+/K^+ ATPase, and Cl^- channels (CLCKA & B) at the blood or serosal membrane

lowest. Variant A has a $K_m(Na^+) = 16.45 \pm 1.9$ and a $K_m(Cl^-) = 44.65 \pm 3.87$) and Variant B has a $K_m(Na^+) = 20.65 \pm 2.4$ and a $K_m(Cl^-) = 8.95 \pm 1.3$) (Giménez et al. 2002). This expression pattern results in maximal reabsorption of luminal salt and water (see Fig. 2.7a). Similar to NKCC1, the driving force for apical NKCC2 cotransport activity is the basolateral expression of the Na^+/K^+ pump. However, now combined with Cl^- channels on the basolateral membrane, apical NKCC2 serves to absorb Na^+ and Cl^- into the blood. The K^+ ions transported by NKCC2 are recycled back to the lumen through ROMK, an apical K^+ channel (see Fig. 2.7b).

Along with NKCC2 in the TALH, Na^+ reabsorption in the distal kidney tubule is accomplished by the thiazide-sensitive Na-Cl cotransporter (NCC) and the amiloride-sensitive epithelial Na^+ channel (ENaC). Each of these transport mechanisms is a major target for pharmacological modification. Proper water and salt retention involves these transport mechanisms along with activation of the renin-angiotensin-aldosterone system, a secondary target for anti-hypertensive drugs. In Fig. 2.8, the inhibitory actions of furosemide, thiazide, and amiloride determine the amount of fluid reabsorbed into the blood and ultimately blood pressure. The juxtaglomerular cells in the macula densa (MD) match glomerular filtration rate (GFR) with tubular reabsorption through tubuloglomerular feedback (TGF) mechanism. Renin release into the blood from the MD converts angiotensinogen to

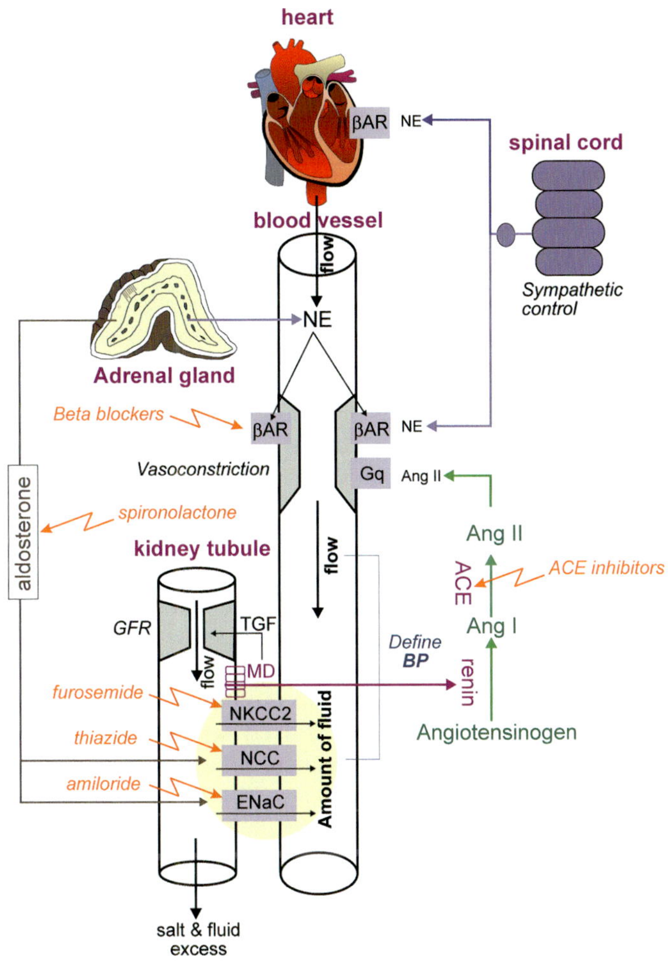

Fig. 2.8 Schematic representation of major targets of anti-hypertensive drugs. The Na^+ reabsorptive transport mechanisms (NKCC2, NCC, and ENaC) in kidney tubules constitute major drug targets. The renin–angiotensin–aldosterone system, which affects both blood vessel contraction and Na^+ transport, constitutes a second site of drug targets. Both angiotensin-converting enzyme (ACE) and aldosterone receptors are targets of anti-hypertensive drugs. By acting on adrenergic receptors of heart and blood vessels, beta-blockers decrease blood flow and contractility of vessels. There are two major sources of catecholamines: the adrenal gland, and postganglion sympathetic fibers. The flow of the urine, which is regulated by the glomerular filtration rate (GFR) and renin–angiotensin-mediated tubuloglomerular feedback (TGF) mechanism, also affects the rate of Na^+ reabsorption. *BO* blood pressure; *MD* macula densa; *GFR* glomerular filtration rate; *TGF* tubuloglomerular feedback; *ACE* angiotensin-converting enzyme; *Ang* angiotensin; *NE* norepinephrine; *βAR* beta adrenergic receptors

angiotensin I (Ang I) which is then converted to angiotensin II (Ang II) by angiotensin-converting-enzyme (ACE). Ang II increases the blood pressure through several mechanisms: (1) increase in sympathetic activity of the heart; (2) stimulation of aldosterone release from the adrenal gland which increases tubular Na^+, Cl^- reabsorption, K^+ excretion and water retention; (3) arteriolar vasoconstriction; and (4) stimulation of anti-diuretic hormone release from the pituitary which increases water absorption in the collecting duct. Beta-blockers acting on adrenergic receptors, ACE inhibitors preventing the conversion of Ang I to Ang II, spironolactone acting on mineralocorticoid (aldosterone) receptor and loop diuretics are used to regulate salt and water reabsorption in the kidney tubule (Fig. 2.8).

2.5.1 NKCC2 in Intestine

Euryhaline fish are capable of living in fresh, brackish, and even saltwater. In freshwater, these fish are hyposmotically challenged producing tremendous amounts of dilute urine to eliminate excess water. Conversely, in seawater, these fish are hyperosmotically challenged and must drink and absorb water to prevent dehydration (Lin et al. 2001; Seale et al. 2014). Although NKCC2 expression is considered kidney-specific in man, real-time PCR demonstrates that the cotransporter (variant B) is present in the intestine of the Mozambique tilapia and the Japanese seawater eel (Ando et al. 2014; Hiroi et al. 2008). Since only mucosal application of bumetanide on seawater eel intestine inhibited ion transport, expression of the variant B isoform of the cotransporter is likely located on the apical membrane. As such, similar to the mammalian kidney, NKCC2 variant B most likely serves as a key mechanism for water absorption and prevention of dehydration.

2.5.2 NKCC2 in Disease

Human mutations in NKCC2 which disrupt cotransporter function are responsible for Type I Bartter's syndrome, a rare inherited disorder characterized by prenatal onset, low potassium levels (hypokalemic), low blood pressure (hypotension), increased blood pH (alkalosis), excessive urination (polyuria), and excessive amounts of calcium in the urine (hypercalciuria) (Bartter et al. 1962). More than 70 mutations in the *SLC12A1* gene have been identified in patients suffering from Type I Bartter's syndrome (Simon et al. 1996). Both homozygous mutations (same mutation on both alleles) and compound heterozygous mutations (each allele has a different mutation) will cause this salt-wasting disorder to become clinically relevant. Symptoms of Type I Bartter's syndrome often appear before birth; however, patients can also manifest the disorder later in life as a result of residual NKCC2 function (Pressler et al. 2006).

2.6 NKCC Activity Is Regulated by Phosphorylation

The activity of the Na-K-2Cl cotransporter is increased by phosphorylation and decreased by dephosphorylation. Consensus exists as to the major phosphorylation event affecting the intrinsic transport activity of carriers already inserted in the plasma membrane. However, this consensus does not rule out the possibility that other phosphorylation events can affect the forward trafficking or retrograde recycling of transporters from the plasma membrane.

Original evidence that phosphorylation activates Na-K-Cl cotransport comes from studies performed by John Russell in 1988 in squid giant axon (Altamirano et al. 1988). The study showed that ATP depletion causes a marked decrease in Na-K-Cl cotransport activity. While this decrease could have been due to the collapse of the Na^+ gradient generated by the ATP-driven Na^+/K^+ ATPase, the fact that it was diminished by the addition of protein phosphatase inhibitors (such as orthovanadate or fluoride) indicated that instead, NKCC phosphorylation was likely affected by ATP depletion. This was later confirmed in studies performed in the shark rectal gland by Chris Lytle and Biff Forbush (Lytle and Forbush 1992a, b). They demonstrated that a specific residue in the cytosolic NH_2-terminal tail of the cotransporter acquired a phosphate residue upon cotransporter activation. In fact, we know now that multiple neighboring threonine and serine residues are phosphorylated upon cotransporter activation (Fig. 2.9). How phosphorylation of the NH_2-terminus leads to conformational changes in the protein core and consequently increased cotransport activity is currently unknown. Moreover, whether specificity exists in the multiple phosphorylation sites that were identified is also unknown.

Identification of the kinases involved in the activation of the cotransporter was made 10 years later (Dowd and Forbush 2003; Piechotta et al. 2003; Piechotta et al. 2002). There are still many details that are not resolved, mostly concerning the precise role of each kinase identified as regulators of the Na-K-2Cl cotransporter. We will start by addressing the current consensus. Upstream of the phosphorylation sites, in both NKCC1 and NKCC2 reside a short sequence that constitutes a binding site for serine/threonine kinases. The minimal amino acid sequence is an arginine residue, followed by a phenylalanine, any residue, then a valine or isoleucine (RFxV/I). This site is found once in the cytosolic amino-terminal domain of NKCC2 and twice in the amino-terminal domain of NKCC1. The kinases that bind to the RFxV/I site are members of a relatively large family of serine-threonine kinases: the mammalian Ste20p-like kinase family (Delpire 2009). SPAK (STE20/SPS1-related Proline/Alanine-rich Kinase, also known as PASK and STK39) and OSR1 (Oxidative stress response 1, also known as OxSR1) are the two members regulating the cotransporters. OSR1 is the original kinase, as it is found from protist to human, whereas SPAK originated from gene duplication late during vertebrate evolution (Gagnon and Delpire 2012). The two kinases possess an NH_2-terminal catalytic domain, followed by a COOH-terminal regulatory domain, whose last 90 residues form a protein fold that accommodates or binds RFxV/I peptides (Austin et al. 2014; Villa et al. 2007). Thus, constitutively active SPAK and OSR1 when

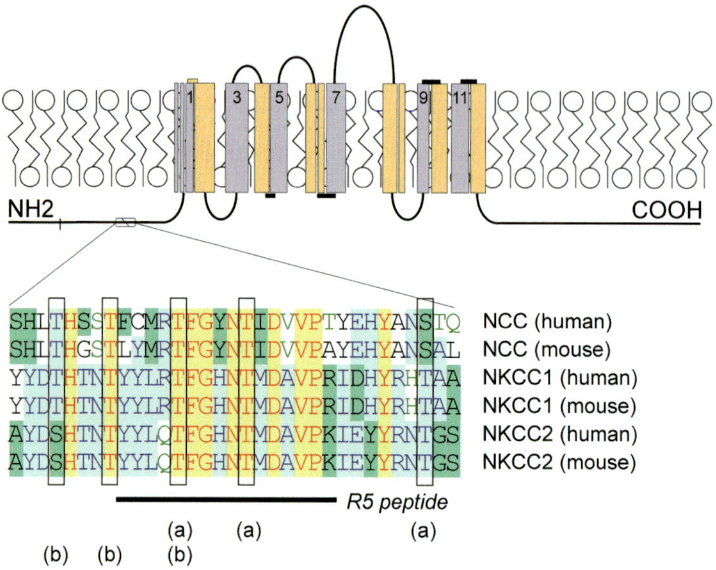

Fig. 2.9 Phosphorylation "hot spot" in Na^+-dependent cation–chloride cotransporters. Depicted is an amino acid alignment of human and mouse sequences for NCC, NKCC1, and NKCC2 of a small portion of the cytosolic NH_2-terminal tail of the cotransporters. Note the presence of five conserved serine/threonine residues. Highlighted by boxes are phosphorylation sites identified by mass spectrometry in two studies (**a**) (Darman and Forbush 2002) and (**b**) (Vitari et al. 2006). Horizontal bar delineates the peptide epitope of a phosphospecific antibody (anti-phospho NKCC1 R5 antibody) created by the Forbush Laboratory (Flemmer et al. 2002)

overexpressed in heterologous expression systems, like *Xenopus laevis* oocytes, significantly activate both Na-K-2Cl cotransporters. Interestingly, when overexpressed as native kinases in oocytes, they fail to activate the cotransporters and they themselves require activation by an upstream WNK kinase. The C-terminal domain of WNK kinases also contains RFxV/I sequences which makes them binding partners of SPAK/OSR1, thus facilitating their activation by WNK kinases (Fig. 2.10). A recent study (Piala et al. 2014) has indicated that WNK kinases also possess a portion of the binding domain of SPAK, which means that this upstream kinase can also bind to the cotransporters (Fig. 2.10). Figure 2.10 also highlights an adaptor protein (Cab39) which has been shown to facilitate Ste20 and WNK kinase activation, as well as the serum kinase SGK1 as a possible upstream regulator of WNK kinases. Finally, recent structural studies have uncovered the presence of a Cl^- ion within the crystal structure of the WNK1 catalytic domain, indicating a possible role for the anion in modulating kinase function (Piala et al. 2014).

As kinases that directly bind and phosphorylate NKCC1 and NKCC2, SPAK and OSR1 are likely to be essential components of cotransporter function. Thus, disruption of their expression should have significant physiological consequences. Global knockout mice were generated for both kinases. While the complete absence of OSR1 results in embryonic lethality, the absence of SPAK results in a viable mouse

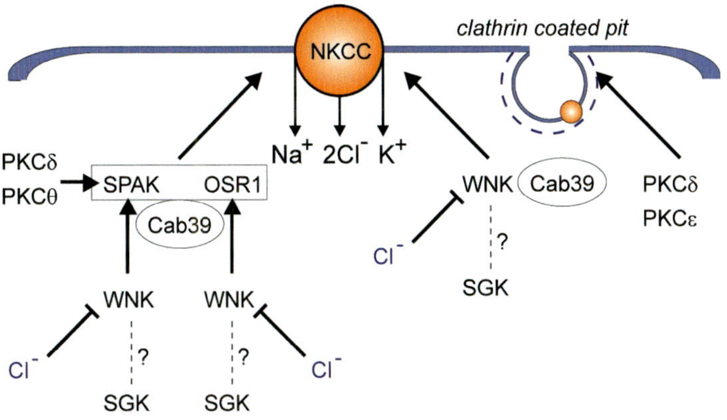

Fig. 2.10 Illustration of phosphoregulation of NKCC2 by WNK-SPAK pathway. Depiction of the signaling cascade leading to NKCC activation with two members of the Ste20p-like family of protein kinases (SPAK and OSR1) anchoring to the cytosolic NH_2-terminal tail and phosphorylating the cotransporter. Upstream of the Ste20p kinases are WNK kinases that can, through a similar binding mechanism (RFxV/I motif), anchor and activate SPAK and OSR1. In addition, an adaptor protein (Cab39) is depicted as a facilitator of kinase activation. Upstream regulatory mechanisms of WNK kinases, such as the serum glucocorticoid regulated kinase (SGK) and intracellular Cl^-, are also shown. WNK kinase also possesses a portion of the binding domain of SPAK, which means that this upstream kinase can bind directly to the cotransporter. Also depicted is the PKC-induced clathrin-dependent internalization of the cotransporter

with no overt phenotype (Delpire and Gagnon 2008). However, upon closer analysis, SPAK knockout mice demonstrate a renal phenotype that resembles the loss of NCC function in the distal convoluted tubule: salt-sensitive hypotension accompanied by electrolyte imbalance, such as hypokalemia, hypomagnesemia, and hypercalciuria (Grimm et al. 2012; McCormick et al. 2011; Yang et al. 2010). In contrast, the targeted deletion of OSR1 in the kidney led to a viable mouse with a phenotype resembling, in this case, the loss of NKCC2 function. Indeed, the mice displayed impaired Na^+ reabsorption in the TALH on a low-Na^+ diet with a significant reduction in NKCC2 phosphorylation, and a blunted response to furosemide (Lin et al. 2011). Thus, whereas these two kinases fulfill the same function in heterologous expression systems, they seem to have distinct functions in the kidney: OSR1 mainly regulating NKCC2, and SPAK mostly regulating NCC. In agreement with SPAK mostly regulating NCC, genetic variations in the human SPAK gene have been linked to hypertension (Wang et al. 2009). In other tissues, such as primary afferent sensory neurons, the two kinases seem to be complementing each other, as disruption of one kinase reduces NKCC1 function by half (Geng et al. 2009).

Mutations in WNK1 and WNK4 have been identified in the human population causing a disorder called Gordon syndrome or pseudohypoaldosteronism type II (PHAII). Features of the disorder are opposite to Gittleman's, indicating an increase instead of a decrease in NCC function. It is now recognized that these mutations prevent degradation of WNK1 or WNK4 by the cullin-3/Kelch-like

3 ubiquitinylation-mediated protein degradation pathway. In fact, mutations in both proteins also result in Gordon syndrome (Boyden et al. 2012). The consequence of all these mutations is an increased abundance of WNK proteins, leading to increased function of the Na-Cl cotransporter, and increased distal Na^+ reabsorption. This phenotype has been recapitulated in the mouse (Yang et al. 2007). Furthermore, disruption of the WNK4 gene in mice results in a mild Gitelman phenotype, consistent with WNK4 being a positive modulator of NCC function (Castañeda-Bueno et al. 2012; Ohta et al. 2008). Much less effort has been placed on determining whether manipulation of the WNK kinases also affects the function of the Na-K-2Cl cotransporters.

Aside from the major phosphorylation event that leads to stimulation of the cotransporters, there is evidence that Protein Kinase C (PKC) mediates the internalization of NKCC1 (Mykoniatis et al. 2010; Tang et al. 2010) (Fig. 2.10). Using a green fluorescent protein-tagged NKCC1 cotransporter expressed in MDCK cells, it was shown that phorbol ester activated PKC leads to internalization of NKCC1 through a clathrin-dependent endocytic pathway (Mykoniatis et al. 2010). Using small hairpin RNA (shRNA) delivered with adenovirus, PKCδ and PKCε were identified as the isotypes modulating cell surface expression of the cotransporter (Tang et al. 2010).

As discussed, in Sect. 2.4.2, activation of PKA through cAMP also leads to an increase in NKCC1 stimulation. However, the effect is indirect and mediated by the drop in the intracellular Cl^- concentration that follows the cAMP-induced opening of CFTR (Fig. 2.5). This indirect stimulation not only pertains to the cAMP stimulation of colonic NKCC1 in mammals but also to the cAMP/PKA mediated stimulation of NKCC1 in the shark rectal gland (Lytle and Forbush 1992a, b).

2.7 Gene Structure, Cotransporter Family, and Super Family

We mentioned in Sect. 2.3 that human NKCC1 and NKCC2 are the products of two distinct genes located on chromosomes 5 and 15, respectively. Analysis of animal genomes revealed that the *SLC12A2* gene originated first since NKCC1 is found in bacteria and protists (e.g., NCBI accession numbers: WP_008693040 and EFW46279), whereas *SLC12A1* originated from gene duplication 400–500 million years ago in early vertebrate evolution, as NKCC2 is found in cartilaginous fish (NCBI accession number: AAM749868). Examination of their genetic structure revealed that despite this very significant length of time since duplication, the exons have been well-conserved (Fig. 2.11). Indeed, 21 out of 27 exons are conserved (including 2 exons which have a single codon added). Therefore, there are only 6 exons which are somewhat different between the two genes: exons 1 and 2 for *SLC12A1* (NKCC2) are different from exon 1 in *SLC12A1* (NKCC1). Despite exon 3 of *SLC12A1* and exon 2 of *SLC12A2* starting differently, they end up with

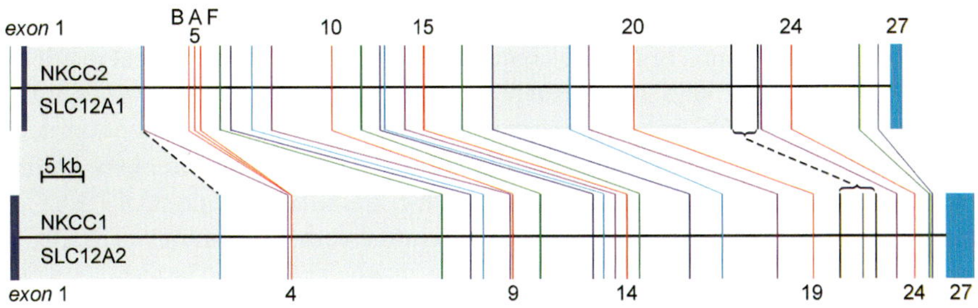

Fig. 2.11 Alignment of the human *SLC12A1* and *SLC12A2* genes. The gene structure with all 27 exons spanning 100 kb genomic DNA is displayed. The vast majority of exons are conserved with 21 out of 27 exons having identical sizes and coding the same protein fragments. Exons 1–3 which encode the NH_2-terminal tails of both cotransporters show the lowest degree of conservation. The 3′-untranslated region of the transcript, encoded by exon 27, is typically gene-specific, differs significantly between the two genes. Exon 5 in the *SLC12A1* gene is triplicated to give rise to the three splice variants: NKCC2F, NKCC2A, and NKCC2B. A short 48 bp alternatively spliced cassette in exon 21 is unique to the *SLC12A2* gene. Large genomic regions present between exons are depicted by a gray background

conserved sequences. Another difference can be seen in exons 21 and 22 for *SLC12A1* which corresponds to exons 20, 21, 22 in *SLC12A2* with very little conservation, if we exclude the terminal part of exon 22. Finally, exon 27 encodes the end of the open reading frame of both cotransporters, followed by gene-specific 3'untranslated region. This high degree of overall exon (Fig. 2.11) and protein (Fig. 2.3) conservation reflects the evolutionary pressure that was imposed to keep intact the vital function of the two cotransporters. In the end, there are very few functional differences between the two cotransporters, their strict stoichiometry with three (even four if we include the second Cl$^-$) distinct ions require tertiary, secondary, and primary structure integrity. To this note, we can add the surprising finding that while sequences of the 5′ end of the N-terminal tails are highly divergent, the cotransporters have managed to retain a SPAK/OSR1 binding motif.

Very little is known about transcriptional promoters for the two genes. While the promoter of *SLC12A2* NKCC1 contains features of housekeeping genes with a TATA-less promoter and several activator protein-2 (AP-2) and Specificity Protein 1 (SPS1) putative binding sites, the promoter of *SLC12A1*/NKCC2 only contains a TATA-box, and seems to be under the control of Hepatocyte Nuclear Factor (Igarashi et al. 1996). There is no indication that either transcript can be generated from alternative promoters. In contrast, as already mentioned, NKCC2 exists in three variants, due to alternative splicing of exon 5, whereas NKCC1 exists in two variants one with and one without exon 21 (Randall et al. 1997). Exon 5, as shown in Fig. 2.3, encodes for the second transmembrane domain and the sequence variations that affect the cotransporters binding affinity for ions (Giménez et al. 2002).

As mentioned earlier, the membrane targeting of the two NKCC isoforms is directly related to their different physiological roles. In an elegant set of experiments, Biff Forbush and coworkers generated chimeras of the two isoforms to identify what motifs within the two proteins mediate their basolateral (NKCC1) versus apical

(NKCC2) membrane expression (Carmosino et al. 2008). Transposition of the cytosolic amino-terminus of NKCC1 onto NKCC2 did not alter the apical distribution of the chimeric protein, whereas replacement of the cytosolic carboxy-terminus of NKCC2 onto NKCC1 altered the membrane targeting of the chimeric protein from basolateral to apical. Through a series of chimeras where regions of the carboxy-terminus of NKCC1 was replaced with corresponding regions of NKCC2, they identified a basolateral sorting motif within the carboxy-terminus of NKCC1 which is encoded by exon 21 that contains a di-leucine motif (Carmosino et al. 2008). Di-leucine motifs have been shown to mediate endocytosis and basolateral sorting of membrane receptors in epithelial cells (Hunziker and Fumey 1994). Both the alternative NKCC1 isoform that lacks exon 21 and the apically targeted NKCC2 lack this di-leucine motif. A recent paper, however, challenged the role of exon 21 in membrane targeting. Removal of exon 21 was indeed not sufficient to affect basolateral targeting of NKCC1 in MDCK cells (Koumangoye et al. 2019). In contrast, a far downstream C-terminal di-leucine motif was shown essential to basolateral membrane expression of NKCC1. Whether the alternatively spliced NKCC1 isoform, abundantly expressed in the brain, is localized to the apical membrane in choroid plexus epithelial cells due to the absence of this di-leucine motif is still unknown. However, the absence of this motif does explain the localization of NKCC2 on the apical membrane of the thick ascending limb of Henle (Carmosino et al. 2008).

The two Na-K-2Cl cotransporters are also closely related to the thiazide-sensitive Na-Cl cotransporter (NCC). Together, these cotransporters constitute the Na^+ transporting branch of the cation–chloride cotransporter family and share high homology to four Na^+-independent K-Cl cotransporters (KCC1-KCC4; (Gagnon and Delpire 2013a). All these transporters belong to the family of *SLC12A* cotransporters which itself shares ancestry with many other families of transport proteins which share the basic structure of an inverted repeat of five transmembrane domains (TM1–5 and TM6–10), followed by two additional transmembrane α-helical segments located either at their N- and/or C-termini (Wong et al. 2012). Because of high conservation at the structure level, the three-dimensional structural resolution of related transporters (e.g., a bacterium sodium-galactose symporter PBD#: 3dh4 and a bacterium glutamate/GABA antiporter PDB#: 4dji) allows for modeling of NKCC proteins. Using the Phyre2 protein fold recognition server (www.sbg.bio.ic.ac.uk/phyre2/), we drew a model of the first 10 transmembrane domains of NKCC1 (Fig. 2.12). We can see that specific helices run parallel to each other (TM1 and 6 in blue, TM2 and 7 in red, TM3 and 8 in green, TM4 and 9 in white). Considering the distance that separates LeuTAa, a bacterial homolog of a neurotransmitter sodium symporter, it is extraordinary that four out of six residues in TM9 and that seven out of ten residues in TM10, shown to promote LeuTAa dimerization (Yamashita et al. 2005), are conserved in NKCC1. In a 2012 study, Biff Forbush and coworkers identified several pore-lining residues within TM3 (green helix in Fig. 2.12) as functionally important in ion coordination (Tyr-383, Ala-379, Ala-375, Asn-376, Ile-368, and Gly-369), loop diuretic binding (Phe-372 and Ile-371), and complete bumetanide-insensitivity (Met-382) (Somasekharan et al. 2012).

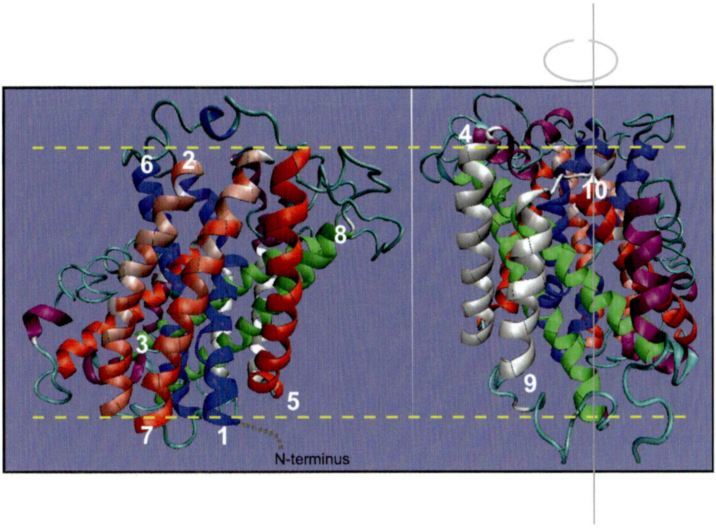

Fig. 2.12 Three-dimensional model of the transmembrane core of NKCC1. Three-dimensional resolution of two related transporter structures: a bacterium sodium-galactose symporter (PDB#: 3dh4) and a bacterium glutamate/GABA antiporter (PDB#: 4dji) were used to model the transmembrane core of NKCC1. Anti-parallel transmembrane helices from the first and second halves of the transmembrane core are identically colored. The structure is rotated 180 degrees (right) to visualize the helices visible from the back. The model was created using the Phyre2 software (see text)

Conserved identity within the transmembrane domains of the cation–chloride cotransporter family suggests that future homology studies may help understand which pore-lining residues define the specific characteristics of ion transport in both the Na^+-dependent and Na^+-independent cotransporters.

2.8 Summary

Over the past 50 years, physiologists have used functional, biochemical, and molecular biological experiments to go from the initial Na^+/K^+ Pump II hypothesis to the conceptual breakthrough of secondary active transport where the energy of the Na^+ gradient generated by the Na^+/K^+ ATPase is used by the Na-K-2Cl cotransporter in multiple organ systems. Expression of the older of the two isoforms, NKCC1, in various tissues suggests multiple physiological roles including Cl^- and fluid secretion, acid secretion, cell volume homeostasis, and possibly cell division and proliferation. The kidney-specific expression of three variants of NKCC2 to optimize salt reabsorption demonstrates its vital role in renal homeostasis and regulation of systemic blood pressure. The evolutionary conservation of these cotransporters from protists to humans confirms their vital role in cellular and whole-organism physiology.

References

Aickin CC, Betz WJ, Harris GL (1989) Intracellular chloride and the mechanism for its accumulation in rat lumbrical muscle. J Physiol 411:437–455

Akar F, Skinner E, Klein JD, Jena M, Paul RJ, O'Neill WC (1999) Vasoconstrictors and nitrovasodilators reciprocally regulate the Na^+-K^+-$2Cl^-$ cotransporter in rat aorta. Am J Phys 276:C1383–C1390

Altamirano AA, Breitwieser GE, Russel JM (1988) Vanadate and fluoride effects on Na-K-Cl cotransport in squid giant axon. Am J Phys 254:C582–C586

Alvarez-Leefmans FJ, Gamiño SM, Giraldez F, Nogueron I (1988) Intracellular chloride regulation in amphibian dorsal root ganglion neurons studied with ionselective microelectrodes. J Physiol Lond 406:225–246

Ando M, Wong MK, Takei Y (2014) Mechanisms of guanylin action on water and ion absorption at different regions of seawater eel intestine. Am J Physiol Regul Integr Comp Physiol 307:R653–R663

Austin T, Nannemann DP, Deluca SL, Meiler J, Delpire E (2014) In silico analysis and experimental verification of OSR1 kinase - peptide interaction. J Struct Biol 187:58–65

Bajwa PJ, Lee JW, Straus DS, Lytle C (2009) Activation of PPARγ by rosiglitazone attenuates intestinal Cl^- secretion. Am J Physiol Gastrointest Liver Physiol 297:G82–G97

Bartter FC, Pronove P, Gill JRJ, Maccardle RC (1962) Hyperplasia of the juxtaglomerular complex with hyperaldosteronism and hypokalemic alkalosis. A new syndrome. Am J Med 33:811–828

Boyden LM et al (2012) Mutations in kelch-like 3 and cullin 3 cause hypertension and electrolyte abnormalities. Nature 482:98–102

Canessa M, Brugnara C, Cusi D, Tosteson DC (1986) Modes of operation and variable stoichiometry of the furosemide-sensitive Na and K fluxes in human red cells. J Gen Physiol 87:113–142

Carmosino M, Giménez I, Caplan M, Forbush B (2008) Exon loss accounts for differential sorting of Na-K-Cl Cotransporters in polarized epithelial cells. Mol Biol Cell 19:4341–4351

Casimiro MC, Knollmann BC, Ebert SN, Vary JC Jr, Greene AE, Franz MR, Grinberg A, Huang SP, Pfeifer K (2001) Targeted disruption of the Kcnq1 gene produces a mouse model of Jervell and Lange-Nielsen syndrome. Proc Natl Acad Sci USA 98:2526–2531

Castañeda-Bueno M, Cervantes-Pérez LG, Vázquez N, Uribe N, Kantesaria S, Morla L, Bobadilla NA, Doucet A, Alessi DR, Gamba G (2012) Activation of the renal Na^+:Cl^- cotransporter by angiotensin II is a WNK4-dependent process. Proc Natl Acad Sci USA 109:7929–7934

Cheng SX (2012) Calcium-sensing receptor inhibits secretagogue-induced electrolyte secretion by intestine via the enteric nervous system. Am J Physiol Gasterointest Liver Physiol 303:G60–G70

Christensen HL, Nguyen AT, Pedersen FD, Damkier HH (2013) Na(+) dependent acid-base transporters in the choroid plexus; insights from slc4 and slc9 gene deletion studies. Front Physiol 4:1–10

Cook DI, Young JA (1989) Fluid and electrolyte secretion by salivary glands. In: Handbook of physiology. The gastrointestinal system. Salivary, pancreatic, gastric and hepatobiliary secretion. American Physiological Society, Bethesda, MD, pp 1–23

Crane RK (1965) Na^+-dependent transport in the intestine and other animal tissues. Fed Proc 24:1000–1006

Darman RB, Forbush B (2002) A regulatory locus of phosphorylation in the N terminus of the Na-K-Cl cotransporter, NKCC1. J Biol Chem 277:37542–37550

Das S, Jayaratne R, Barrett KE (2018) The role of ion transporters in the pathophysiology of infectious diarrhea. Cell Mol Gastroenterol Hepatol 6:33–45

Delpire E (2009) The mammalian family of Sterile20p-like protein kinases. Pflügers Arch 458:953–967

Delpire E, Gagnon KB (2008) SPAK and OSR1: STE20 kinases involved in the regulation of ion homoeostasis and volume control in mammalian cells. Biochem J 409:321–331

Delpire E, Gagnon KB (2011) Kinetics of hyperosmotically-stimulated Na-K-2Cl cotransporter in *Xenopus laevis* oocytes. Am J Physiol Cell Physiol 301:C1074–C1085

Delpire E, Rauchman MI, Beier DR, Hebert SC, Gullans SR (1994) Molecular cloning and chromosome localization of a putative basolateral Na-K-2Cl cotransporter from mouse inner medullary collecting duct (mIMCD-3) cells. J Biol Chem 269:25677–25683

Delpire E, Lu J, England R, Dull C, Thorne T (1999) Deafness and imbalance associated with inactivation of the secretory Na-K-2Cl co-transporter. Nat Genet 22:192–195

Delpire E, Wolfe L, Flores B, Koumangoye R, Schornak CC, Omer S, Pusey B, Lau C, Markello T, Adams DR (2016) A patient with multisystem dysfunction carries a truncation mutation in human SLC12A2, the gene encoding the Na-K-2Cl cotransporter, NKCC1. Cold Spring Harb Mol Case Studies 2:a001289

Dixon MJ, Gazzard J, Chaudhry SS, Sampson N, Schulte BA, Steel KP (1999) Mutation of the Na-K-Cl co-transporter gene Slc12a2 results in deafness in mice. Hum Mol Genet 8:1579–1584

Dowd BF, Forbush B (2003) PASK (Proline-alanine-rich STE20-related kinase), a regulatory kinase of the Na-K-Cl cotransporter (NKCC1). J Biol Chem 278:27347–27353

Evans RL, Park K, Turner RJ, Watson GE, Nguyen H-V, Dennett MR, Hand AR, Flagella M, Shull GE, Melvin JE (2000) Severe impairment of salivation in Na$^+$/K$^+$/2Cl$^-$ cotransporter (NKCC1)-deficient mice. J Biol Chem 275:26720–26726

Flagella M, Clarke LL, Miller ML, Erway LC, Giannella RA, Andringa A, Gawenis LR, Kramer J, Duffy JJ, Doetschman T, Lorenz JN, Yamoah EN, Cardell EL, Shull GE (1999) Mice lacking the basolateral Na-K-2Cl cotransporter have impaired epithelial chloride secretion and are profoundly deaf. J Biol Chem 274:26946–26955

Flemmer AW, Gimenez I, Dowd BF, Darman RB, Forbush B (2002) Activation of the Na-K-Cl cotransporter NKCC1 detected with a phospho-specific antibody. J Biol Chem 277:37551–37558

Gagnon KB, Delpire E (2010) Molecular determinants of hyperosmotically activated NKCC1-mediated K$^+$/K$^+$ exchange. J Physiol Lond 588:3385–3396

Gagnon KB, Delpire E (2012) Molecular physiology of SPAK and OSR1: two Ste20-related protein kinases regulating ion transport. Physiol Rev 92:1577–1617

Gagnon KB, Delpire E (2013a) Physiology of SLC12 transporters: lessons from inherited human genetic mutations and genetically-engineered mouse knockouts. Am J Physiol Cell Physiol 304: C693–C714

Gamba G, Miyanoshita A, Lombardi M, Lytton J, Lee W-S, Hediger M, Hebert SC (1994) Molecular cloning, primary structure, and characterization of two members of the mammalian electroneutral sodium-(potassium)-chloride cotransporter family expressed in kidney. J Biol Chem 269:17713–17722

Geck P, Pietrzyk C, Burckhardt B-C, Pfeiffer B, Heinz E (1980) Electrically silent cotransport of Na$^+$, K$^+$ and Cl$^-$ in Ehrlich cells. Biochim Biophys Acta 600:432–447

Geng Y, Hoke A, Delpire E (2009) The Ste20 kinases SPAK and OSR1 regulate NKCC1 function in sensory neurons. J Biol Chem 284:14020–14028

Giménez I, Isenring P, Forbush B (2002) Spatially distributed alternative splice variants of the renal Na-K-Cl cotransporter exhibit dramatically different affinities for the transported ions. J Biol Chem 277:8767–8770

Ginns SM, Knepper MA, Ecelbarger CA, Terris J, He X, Coleman RA, Wade JB (1996) Immunolocalization of the secretory isoform of Na-K-Cl cotransporter in rat renal intercalated cells. J Am Soc Nephrol 7:2533–2542

Gosmanov AR, Lindinger MI, Thomason DB (2003) Riding the tides: K+ concentration and volume regulation by muscle Na+-K+-2Cl^{-} cotransport activity. News Physiol Sci 18:196–200
Gregoriades JMC, Madaris A, Francisco J, Alvarez FJ, Alvarez-Leefmans FJ (2019) Genetic and pharmacologic inactivation of apical NKCC1 in choroid plexus epithelial cells reveals the physiological function of the cotransporter. Am J Physiol Cell Physiol. https://doi.org/10.1152/ajpcell.00026.2018
Grimm PR, Taneja TK, Liu J, Coleman R, Chen YY, Delpire E, Wade JB, Welling PA (2012) SPAK isoforms and OSR1 regulate sodium-chloride co-transporters in a nephron-specific manner. J Biol Chem 287:37673–37690
Grubb BR, Lee E, Pace AJ, Koller BH, Boucher RC (2000) Intestinal ion transport in NKCC1-deficient mice. Am J Physiol Gastrointest Liver Physiol 279:G707–G718
Hall AC, Ellory JC (1985) Measurement and stoichiometry of bumetanide-sensitive (2Na:1K:3Cl) cotransport in ferret red cells. J Membrane Biol 85:205–213
Halonen J, Hinton AS, Frisina RD, Ding B, Zhu X, Walton JP (2016) Long-term treatment with aldosterone slows the progression of age-related hearing loss. Hear Res 336:63–71
Hiroi J, Yasumasu S, McCormick SD, Hwang PP, Kaneko T (2008) Evidence for an apical Na-Cl cotransporter involved in ion uptake in a teleost fish. J Exp Biol 211:2584–2599
Hoffman JF, Kregenow FM (1966) The characterization of new energy dependent cation transport processes in red blood cells. Ann N Y Acad Sci 137:566–576
Howard PA, Dunn MI (1997) Severe musculoskeletal symptoms during continuous infusion of bumetanide. Chest 111:359–364
Hunziker W, Fumey C (1994) A di-leucine motif mediates endocytosis and basolateral sorting of macrophage IgG Fc receptors in MDCK cells. EMBO J 13:2963–2969
Igarashi P, Whyte DA, Li K, Nagami GT (1996) Cloning and kidney cell-specific activity of the promoter of the murine renal Na-K-Cl cotransporter gene. J Biol Chem 271:9666–9674
Ikeda K, Oshima T, Hidaka H, Takasaka T (1997) Molecular and clinical implications of loop diuretic ototoxicity. Hearing Res 107:1–8
Kaplan MR, Plotkin MD, Brown D, Hebert SC, Delpire E (1996) Expression of the mouse Na-K-2Cl cotransporter, mBSC2, in the terminal IMCD, the glomerular and extraglomerular mesangium and the glomerular afferent arteriole. J Clin Invest 98:723–730
Karimy JK, Zhang J, Kurland DB, Theriault BC, Duran D, Stokum JA, Furey CG, Zhou X, Mansuri MS, Montejo J, Vera A, DiLuna ML, Delpire E, Alper SL, Gunel M, Gerzanich V, Medzhitov R, Simard JM, Kahle KT (2017) Inflammation-dependent cerebrospinal fluid hypersecretion by the choroid plexus epithelium in posthemorrhagic hydrocephalus. Nat Med 23:997–1003
Koumangoye R, Omer S, Delpire E (2018) Mistargeting of a truncated Na-K-2Cl cotransporter in epithelial cells. Am J Physiol Cell Physiol 315:C258–C276
Koumangoye R, Omer S, Delpire E (2019) A dileucine motif in the C-terminal domain of NKCC1 targets the cotransporter to the plasma membrane. Am J Physiol Cell Physiol. https://doi.org/10.1152/ajpcell.00023.2019
Kregenow FM (1971a) The response of duck erythrocytes to nonhemolytic hypotonic media. Evidence for a volume-controlling mechanism. J Gen Physiol 58:372–395
Kregenow FM (1971b) The response of duck erythrocytes to hypertonic media. Further evidence for a volume-controlling mechanism. J Gen Physiol 58:396–411
Laird JM, Garcia-Nicas E, Delpire EJ, Cervero F (2004) Presynaptic inhibition and spinal pain processing in mice: a possible role of the NKCC1 cation-chloride co-transporter in hyperalgesia. Neurosci Lett 361:200–203
Liedkte CM (1992) Electrolyte transport in the epithelium of pulmonary segments of normal and cystic fibrosis lung. FASEB J 6:3076–3084
Lin LY, Weng CF, Hwang PP (2001) Regulation of drinking rate in euryhaline tilapia larvae (*Oreochromis mossambicus*) during salinity challenges. Physiol Biochem Zool 74:171–177
Lin SH, Yu IS, Jiang ST, Lin SW, Chu P, Chen A, Sytwu HK, Sohara E, Uchida S, Sasaki S, Yang SS (2011) Impaired phosphorylation of Na^{+}-K^{+}-2Cl^{-} cotransporter by oxidative stress-

responsive kinase-1 deficiency manifests hypotension and Bartter-like syndrome. Proc Natl Acad Sci USA 108:17538–17543
Lytle C (1998) A volume-sensitive protein kinase regulates the Na-K-2Cl cotransporter in duck red blood cells. Am J Physiol Cell Physiol 274:C1002–C1010
Lytle C, Forbush BI (1992a) Na-K-Cl cotransport in the shark rectal gland. II. Regulation in isolated tubules. Am J Physiol Cell Physiol 262:C1009–C10117
Lytle C, Forbush BI (1992b) The Na-K-Cl cotransport protein of shark rectal gland. II. Regulation by direct phosphorylation. J Biol Chem 267:25438–25443
Lytle C, McManus TJ, Haas M (1998) A model of Na-K-2Cl cotransport based on ordered ion binding and glide symmetry. Am J Phys 274:C299–C309
Macnamara EF, Koehler AE, D'Souza P, Estwick T, Lee P, Vezina G, Network MUD, Fauni H, Braddock SR, Torti E, Holt JM, Sharma P, Malicdan MCV, Tifft CJ (2019) Kilquist syndrome: a novel syndromic hearing loss disorder caused by homozygous deletion of SLC12A2. Hum Mut 40(5):532–538
McCormick JA, Mutig K, Nelson JH, Saritas T, Hoorn EJ, Yang C-L, Rogers S, Curry J, Delpire E, Bachmann S, Ellison DH (2011) A SPAK isoform switch modulates renal salt transport and blood pressure. Cell Metab 14:352–364
Meyer JW, Flagella M, Sutliff RL, Lorenz JN, Nieman ML, Weber CS, Paul RJ, Shull GE (2002) Decreased blood pressure and vascular smooth muscle tone in mice lacking basolateral Na(+)-K(+)-2Cl(−) cotransporter. Am J Physiol Heart Circ Physiol 283:H1846–H1855
Mykoniatis A, Shen L, Fedor-Chaiken M, Tang J, Tang X, Worrell RT, Delpire E, Turner JR, Matlin KS, Bouyer P, Matthews JB (2010) Phorbol 12-myristate 13-acetate-induced endocytosis of the Na-K-2Cl cotransporter in MDCK cells is associated with a clathrin-dependent pathway. Am J Physiol Cell Physiol 298:C85–C97
Ohta A, Rai T, Yui N, Chiga M, Yang SS, Lin SH, Sohara E, Sasaki S, Uchida S (2008) Targeted disruption of the Wnk4 gene decreases phosphorylation of Na-Cl cotransporter, increases Na excretion and lowers blood pressure. Hum Mol Genet 18:3978–3986
Orlov SN, Tremblay J, Hamet P (1996) Bumetanide-sensitive ion fluxes in vascular smooth muscle cells: lack of functional Na^+, K^+, 2 Cl^- cotransport. J Membr Biol 153:125–135
Pace AJ, Lee E, Athirakul K, Coffman TM, O'Brien DA, Koller BH (2000) Failure of spermatogenesis in mouse lines deficient in the Na^+-K^+-$2Cl^-$ cotransporter. J Clin Invest 105:441–450
Palfrey HC, O'Donnell ME (1992) Characteristics and regulation of the Na/K/2Cl cotransporter. Cell Physiol Biochem 2:293–307
Panet R, Markus M, Atlan H (1994) Bumetanide and furosemide inhibited vascular endothelial cell proliferation. J Cell Physiol 158:121–127
Panet R, Marcus M, Atlan H (2000) Overexpression of the Na^+/K^+/Cl^- cotransporter gene induces cell proliferation and phenotypic transformation in mouse fibroblasts. J Cell Physiol 182:109–118
Payne JA, Forbush BI (1994) Alternatively spliced isoforms of the putative renal Na-K-Cl cotransporter are differentially distributed within the rabbit kidney. Proc Natl Acad Sci USA 91:4544–4548
Payne JA, Xu J-C, Haas M, Lytle CY, Ward D, Forbush BI (1995) Primary structure, functional expression, and chromosome localization of the bumetanide sensitive Na-K-Cl cotransporter in human colon. J Biol Chem 270:17977–17985
Piala AT, Moon TM, Akella R, He H, Cobb MH, Goldsmith EJ (2014) Chloride sensing by WNK1 involves inhibition of autophosphorylation. Sci Signal 7:ra41
Piechotta K, Lu J, Delpire E (2002) Cation-chloride cotransporters interact with the stress-related kinases SPAK and OSR1. J Biol Chem 277:50812–50819
Piechotta K, Garbarini NJ, England R, Delpire E (2003) Characterization of the interaction of the stress kinase SPAK with the Na^+-K^+-$2Cl^-$ cotransporter in the nervous system: evidence for a scaffolding role of the kinase. J Biol Chem 278:52848–52856
Pressler CA, Heinzinger J, Jeck N, Waldegger P, Pechmann U, Reinalter S, Konrad M, Beetz R, Seyberth HW, Waldegger S (2006) Late-onset manifestation of antenatal Bartter syndrome as a

result of residual function of the mutated renal Na^+-K^+-$2Cl^-$ co-transporter. J Am Soc Nephrol 17:2136–2142

Randall J, Thorne T, Delpire E (1997) Partial cloning and characterization of *Slc12a2*: the gene encoding the secretory Na^+-K^+-$2Cl^-$ cotransporter. Am J Physiol Cell Physiol 273:C1267–C1277

Resta-Lenert S, Truong F, Barrett KE, Eckmann L (2001) Inhibition of epithelial chloride secretion by butyrate: role of reduced adenylyl cyclase expression and activity. Am J Physiol Cell Physiol 281:C1837–C1849

Rocha AS, Kudo LH (1990) Atrial peptide and cGMP effects on NaCl transport in inner medullary collecting duct. Am J Physiol (Renal Fluid Electrolyte Physiol) 259:F258–F268

Russell JM (1983) Cation-coupled chloride influx in squid axon. Role of potassium and stoichiometry of the transport process. J Gen Physiol 81:909–925

Rybak LP (1993) Ototoxicity of loop diuretics. Otolaryngol Clin N Am 26:829–844

Seale AP, Stagg JJ, Yamaguchi Y, Breves JP, Soma S, Watanabe S, Kaneko T, Cnaani A, Harpaz S, Lerner DT, Grau EG (2014) Effects of salinity and prolactin on gene transcript levels of ion transporters, ion pumps and prolactin receptors in Mozambique tilapia intestine. Gen Comp Endocrinol 206:146–154

Silva P, Stoff J, Field M, Fine L, Forrest JN, Epstein FH (1977) Mechanism of active chloride secretion by shark rectal gland: role of Na-K-ATPase in chloride transport. Am J Physiol (Renal Fluid Electrolyte Physiol) 233:F298–F306

Simon DB, Karet FE, Rodriquez-Soriano J, Hamdan JH, DiPietro A, Trachtman H, Sanjad SA, Lifton RP (1996) Genetic heterogeneity of Bartter's syndrome revealed by mutations in the K^+ channel, ROMK. Nat Genet 14:152–156

Somasekharan S, Tanis J, Forbush B (2012) Loop diuretic and ion-binding residues revealed by scanning mutagenesis of transmembrane helix 3 (TM3) of Na-K-Cl cotransporter (NKCC1). J Biol Chem 287:17308–17317

Sonnenberg H, Honrath U, Wilson DR (1990) In vivo microperfusion of inner medullary collecting duct in rats: effect of amiloride and ANF. Am J Physiol (Renal Fluid Electrolyte Physiol) 259: F222–F226

Soybel DI, Gullans SR, Maxwell F, Delpire E (1995) Role of basolateral Na-K-Cl cotransport in HCl secretion by amphibian gastric mucosa. Am J Physiol Cell Physiol 269:C242–C249

Steffensen AB, Oernbo EK, Stoica A, Gerkau NJ, Barbuskaite D, Tritsaris K, Rose CR, MacAulay N (2018) Cotransporter-mediated water transport underlying cerebrospinal fluid formation. Nat Commun 9:2167

Sung K-W, Kirby M, McDonald MP, Lovinger DM, Delpire E (2000) Abnormal $GABA_A$-receptor mediated currents in dorsal root ganglion neurons isolated from Na-K-2Cl cotransporter null mice. J Neurosci 20:7531–7538

Tang J, Bouyer P, Mykoniatis A, Buschmann M, Matlin KS, Matthews JB (2010) Activated PKCδ and PKCε inhibit epithelial chloride secretion response to cAMP via inducing internalization of the Na^+-K^+-$2Cl^-$ cotransporter NKCC1. J Biol Chem 285:34072–34085

Tasaki I, Spiropoulos CS (1959) Stria vascularis as source of endocochlear potential. J Neurophysiol 22:149–155

Tyson J, Tranebjaerg, L., Bellman, S., Wren, C., Taylor, J. F. N., Bathen, J., Aslaksen, B., Sorland, S. J., Lund, O., Malcolm, S., Pembrey, M., Bhattacharya, S., Bitner-Glindzicz, M. (1997) IsK and KvLQT1: mutation in either of the two subunits of the slow component of the delayed rectifier potassium channel can cause Jervell and Lange-Nielsen syndrome. Hum Mol Genet 61:2179–2185

Vaduganathan M, Allegretti AS, Manchette AM, Patel SS, Olson KR, Bazari H (2013) Intravenous moderate-dose bumetanide continuous infusion and severe musculoskeletal pain. Int J Cardiol 168:e29–e31

Vetter DE, Mann JR, Wangemann P, Liu J, McLaughlin KJ, Lesage F, Marcus DC, Lazdunski M, Heinemann SF, Barhanin J (1996) Inner ear defects induced by null mutation of the *isk* gene. Neuron 17:1251–1264

Villa F, Goebel J, Rafiqi FH, Deak M, Thastrup J, Alessi DR, van Aalten DMF (2007) Structural insights into the recognition of substrates and activators by the OSR1 kinase. EMBO Rep 8:839–845
Vitari AC, Thastrup J, Rafiqi FH, Deak M, Morrice NA, Karlsson HK, Alessi DR (2006) Functional interactions of the SPAK/OSR1 kinases with their upstream activator WNK1 and downstream substrate NKCC1. Biochem J 397:223–231
Wang Y et al (2009) Whole-genome association study identifies STK39 as a hypertension susceptibility gene. Proc Natl Acad Sci USA 106:226–231
Wiener H, van Os CH (1989) Rabbit distal colon epithelium: II. Characterization of (Na+,K+,Cl-)-cotransport and [3H]-bumetanide binding. J Membr Biol 110:163–174
Wong FH, Chen JS, Reddy V, Day JL, Shlykov MA, Wakabayashi ST, Saier MHJ (2012) The amino acid-polyamine-organocation superfamily. J Mol Microbiol Biotechnol 22:105–113
Wouters M, De Laet A, Ver Donck L, Delpire E, van Bogaert PP, Timmermans JP, de Kerchove d'Exaerde A, Smans K, Vanderwinden JM (2006) Subtractive hybridization unravels a role for the ion co-transporter NKCC1 in the murine intestinal pacemaker. Am J Physiol Gastrointest Liver Physiol 290:G1219–G1227
Xu J-C, Lytle C, Zhu TT, Payne JA, Benz EJ, Forbush BI (1994) Molecular cloning and functional expression of the bumetanide-sensitive Na-K-2Cl cotransporter. Proc Natl Acad Sci USA 91:2201–2205
Yamashita A, Singh SK, Kawate T, Jin Y, Gouaux E (2005) Crystal structure of a bacterial homologue of Na^+/Cl^--dependent neurotransmitter transporters. Nature 437:215–223
Yang SS, Morimoto T, Rai T, Chiga M, Sohara E, Ohno M, Uchida K, Lin SH, Moriguchi T, Shibuya H, Kondo Y, Sasaki S, Uchida S (2007) Molecular pathogenesis of pseudohypoaldosteronism type II: generation and analysis of a Wnk4(D561A/+) knockin mouse model. Cell Metab 5:331–344
Yang SS, Lo YF, Wu CC, Lin SW, Yeh CJ, Chu P, Sytwu HK, Uchida S, Sasaki S, Lin SH (2010) SPAK-knockout mice manifest Gitelman syndrome and impaired vasoconstriction. J Am Soc Nephrol 21:1868–1877
Zeidel ML, Kikeri D, Silva P, Burrowes M, Brenner BM (1988) Atrial natriuretic peptides inhibit conductive sodium uptake by rabbit inner medullary collecting duct cells. J Clin Invest 82:1067–1074
Zhu MH, Sung TS, Kurahashi M, O'Kane LE, O'Driscoll K, Koh SD, Sanders KM (2016) Na^+-K^+-Cl^- cotransporter (NKCC) maintains the chloride gradient to sustain pacemaker activity in interstitial cells of Cajal. Am J Physiol Gastrointest Liver Physiol 311:G1037–G1046

Selected Readings

Gagnon KB, Delpire E (2013b) Physiology of Slc12 transporters: lessons from inherited human genetic mutations and genetically-engineered mouse knockouts. Am J Physiol Cell Physiol 304: C693–C714
Gamba G (2009) The sodium-dependent chloride cotransporters. In: Alvarez-Leefmans FJ, Delpire E (eds) Physiology and pathology of chloride transporters and channels in the nervous system. Academic Press (Elsevier), London, pp 307–332
Gamba G, Garbarini N, Delpire E (2009) Regulation of cation-chloride cotransporters. In: Alvarez-Leefmans FJ, Delpire E (eds) Physiology and pathology of chloride transporters and channels in the nervous system. Academic Press (Elsevier), London, pp 357–382
Russell JM (2000) Sodium-potassium-chloride cotransport. Physiol Rev 80:212–276
Russell JM (2009) Sodium-coupled chloride cotransporters: Discovery and newly emerging concepts. In: Alvarez-Leefmans FJ, Delpire E (eds) Physiology and pathology of chloride transporters and channels in the nervous system. Academic Press (Elsevier), London, pp 17–26

Chapter 3
Thiazide-Sensitive NaCl Cotransporter

Arohan R. Subramanya

Abstract The thiazide-sensitive NaCl cotransporter (NCC, SLC12A3) is a member of the solute carrier 12 (SLC12) family of electroneutral cotransporters. NCC is expressed in the distal convoluted tubule (DCT) of the kidney, where it mediates the transport of sodium with chloride in 1:1 stoichiometry. NCC is responsible for reabsorbing approximately 5–10% of sodium from the glomerular filtrate. NCC plays a key role in determining the blood pressure set point and potassium balance. NCC is the major molecular target of thiazide-type diuretics, drugs commonly used as first-line agents in the treatment of essential hypertension and edema. Since its cloning decades ago, much has been learned about the physiological transport properties of NCC, its regulation, and its connections to human disease. In this chapter, I provide an overview of the biology of this important cotransporter.

Keywords NCC · Thiazide · WNK · SPAK · OSR1 · Distal convoluted tubule · Gitelman syndrome · Pseudohypoaldosteronism Type 2

3.1 Introduction

The thiazide-sensitive Na-Cl cotransporter (NCC, SLC12A3) is a member of the SLC12 family of cation–chloride cotransporters. It is the primary mediator of sodium transport in the distal convoluted tubule (DCT) of the kidney, a short but important nephron segment positioned between the Loop of Henle and collecting duct that regulates blood pressure and electrolyte balance. This chapter provides a summary of our current knowledge of NCC, extending from early studies of its transport characteristics and cloning, to recent work on its physiologic regulation in vivo, and its role in human disease.

A. R. Subramanya (✉)
Department of Medicine, Renal-Electrolyte Division, University of Pittsburgh School of Medicine, Pittsburgh, PA, USA
e-mail: ars129@pitt.edu

K. L. Hamilton, D. C. Devor (eds.), *Studies of Epithelial Transporters and Ion Channels*, Physiology in Health and Disease,
https://doi.org/10.1007/978-3-030-55454-5_3

3.2 Early Studies and Cloning of NCC

The first studies of thiazide-sensitive salt transport in the kidney were conducted in the 1950s. These studies were motivated by the search for new diuretic drugs that lacked the toxic effects of organic mercurial diuretics such as Mersalyl (Freeman et al. 1962). Proximal tubule carbonic anhydrase inhibitors, such as the sulfonamide acetazolamide, were among the first nontoxic diuretics identified, although it was immediately recognized that their ability to stimulate Na and water excretion waned over repeated uses. In addition, the diuretic effects of acetazolamide are almost always accompanied by clinically significant metabolic acidosis (Novello and Sprague 1957). In an effort to develop new sulfonamides with enhanced effects on sodium chloride excretion, investigators at Merck synthesized the first thiazide diuretic, chlorothiazide (Diuril), in 1957 (Novello and Sprague 1957). Although chlorothiazide inhibited carbonic anhydrase to some degree, it also stimulated chloride excretion in a manner similar to mercurial compounds, which act more distally. In the 1960s, experiments by Orloff and others suggested that chlorothiazide blocks NaCl reabsorption in the distal nephron (Earley and Orloff 1964). In the mid-1970s, Kunau et al. narrowed the site of action to the distal convolution (Kunau 1974), studies that were later verified by Constanzo et al. (Costanzo 1988; Costanzo and Windhager 1978). Ellison et al. established that thiazide diuretics block a coupled sodium chloride cotransport pathway in the DCT (Ellison et al. 1987). These important papers established that the natriuretic effects of thiazides were predominantly due to inhibition of a distal nephron salt cotransporter, not proximal tubule carbonic anhydrase.

Initial efforts to identify the molecular identity of the "thiazide receptor" from kidneys using an expression cloning approach were unsuccessful, possibly because the DCT is the shortest tubular segment of the nephron and NCC, therefore, comprises only a tiny fraction of total mRNA from a kidney extract. Thus, a radically different strategy had to be employed. In the 1980s, studies by Stokes (1984) and Renfro (1977) identified a salt cotransporter in the bladder of the winter flounder (*Pseudopleuronectes americanus*) whose activity was sensitive to thiazides. The transport properties were thought to be identical to the NCC in human DCT. Thus, in 1993, Gamba and Hebert astutely used mRNA from flounder bladder as starting material to derive the first NCC cDNA (Baggio et al. 1993). The flounder sequence was then used as a template to isolate mammalian NCC cDNAs via homology-based strategies (Gamba et al. 1994).

3.3 Primary Structure and Molecular Architecture

Analysis of the primary structure of NCC suggested a complex topology consisting of 12 transmembrane domains (TMs) flanked by large intracellular amino- and carboxy-termini (Baggio et al. 1993). It is believed that the N-terminal intracellular

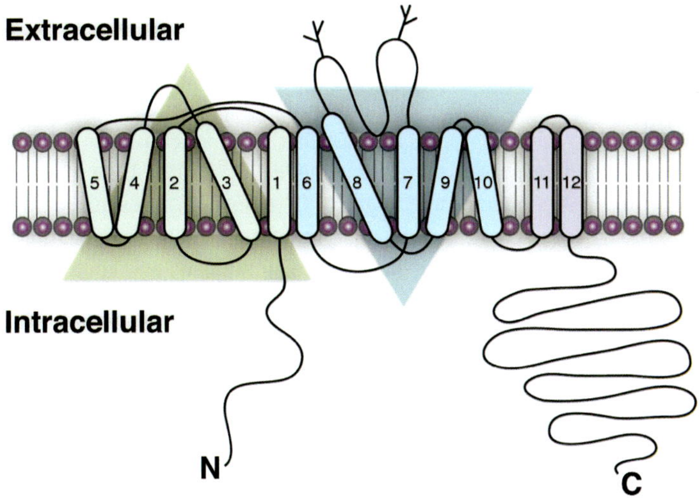

Fig. 3.1 Two-dimensional topology diagram of the thiazide-sensitive NaCl cotransporter (NCC), based on the 5 + 5 transmembrane inverted repeat common to members of the APC superfamily, and the solved Cryo-EM structure of NKCC1. The inverted "pseudosymmetric" folds are indicated with triangles. NCC has two large intracellular termini and a sizeable extracellular loop that contains N-glycosylation sites. The N-terminus is a regulatory domain that undergoes phosphorylation, while the structured C-terminus is important for homodimerization

domain is unstructured, while the C-terminus adopts a complex and compact fold. A large extracellular loop containing *N*-glycosylation sites was predicted to reside between TMs 7 and 8 (Fig. 3.1). In mammals, this extracellular loop contains two *N*-glycosylation sites, whereas, in teleosts, three sites are present (Gamba 2005). Only one of these asparagines, Asn-403 in flounder, is conserved among all NCCs. This general topology is consistent with that of the two other electroneutral sodium-coupled cotransporters within the SLC12 family, the bumetanide-sensitive Na^+-K^+-$2Cl^-$ cotransporters NKCC1 (SLC12A2) and NKCC2 (SLC12A1).

Initial clues into the transmembrane topology of NCC were extrapolated from the solved crystal structures of other sodium-coupled secondary transporters, such as the bacterial sodium-coupled leucine transporter LeuT (Krishnamurthy and Gouaux 2012). This topology was recently further supported when the structure of *Danio rerio* NKCC1 (*Dr*NKCC1) was solved in a partially inward open configuration by cryo-electron microscopy (Chew et al. 2019). Based on the fold of this cotransporter and other members of the amino acid-polyamine-cation (APC) transporter superfamily (of which the SLC12 cotransporters are a part of), the first ten TMs organize into two internally folded domains (Krishnamurthy et al. 2009). The two domains each consist of five helices that are oriented perpendicular to the membrane in opposite directions to each other in a "pseudosymmetric" manner (Fig. 3.1). The *Dr*NKCC1 fold suggests that the sites for sodium and chloride binding and translocation in NCC are located near TMs 1 and 6. Discussed in further detail below, phosphorylation of the intracellular amino terminus increases NCC activity. Since this intracellular domain is immediately contiguous with TM1, it is conceivable that phosphorylation of the NCC N-terminus may provoke structural alterations in the positioning of transmembrane domains within the solvent-accessible vestibule, thereby enhancing the efficiency of sodium chloride cotransport.

Previous studies by de Jong et al. indicate that NCC assembles into homodimers prior to plasma membrane delivery (de Jong et al. 2003). Like many proteins (Zerangue et al. 1999), macromolecular NCC assembly takes place in the ER, since high mannose immature ER-localized glycoforms of NCC readily form dimers (unpublished observations). Protein interaction studies suggest that the large and highly folded C-terminal domain (CTD) is essential for homodimer formation. These studies have been confirmed by the recently reported crystal structure of *Dr*NKCC1 (Chew et al. 2019) as well as the intracellular C-terminus of MaCCC, a bacterial cation–chloride cotransporter whose C-terminus shares some homology to NCC (Warmuth et al. 2009). The resolved structure of the CTD of one *Dr*NKCC1 subunit interacts with the transmembrane domain (TMD) of the opposing subunit in a domain-swap configuration (Chew et al. 2019). The connections between the TMDs and CTDs are mediated by a flexible helix of ~10 amino acids in length (termed the "scissor helix"). This structure wraps around a scissor helix from its associated dimeric partner, resulting in the domain-swap orientation. Conformational variations noted by cryo-electron microscopy revealed that the structure is therefore likely highly dynamic, undergoing swiveling and rocking motions that alter the orientations between the two swapped domains.

3.4 NCC Transport Characteristics

The initial functional characterization of flounder NCC confirmed that ion influx is saturable, consistent with a carrier-mediated mechanism. According to this study and others, the K_m values for Na^+ and Cl^- are approximately 30 and 15 in flounder NCC, respectively (Gamba et al. 1993; Vazquez et al. 2002). Mammalian (rat) NCC, in contrast, exhibits a much higher affinity for ions with Km values for Na^+ and Cl^- of 6 and 0.3, respectively (Vazquez et al. 2002). The mammalian NCC appears to be much more thiazide-sensitive, with an IC_{50} for metolazone of 0.3 μM, compared to the flounder NCC IC_{50} of 4.0 μM (Moreno et al. 2006; Vazquez et al. 2002).

Over a decade ago, studies carried out with NCC-NKCC2 chimeras revealed that, as expected, the hydrophobic 12 transmembrane domain region of NCC confers ion translocation and diuretic sensitivity (Tovar-Palacio et al. 2004). This is consistent with more recent structural information obtained from the solved structures of *Dr*NKCC1 and other APC superfamily members (Chew et al. 2019; Krishnamurthy and Gouaux 2012; Krishnamurthy et al. 2009; Singh et al. 2008). Moreno et al. (2006) took advantage of the functional differences between rat and flounder NCC to define sites of ion translocation and diuretic binding. Using several rat-flounder chimeric constructs, they concluded that transmembrane domains 1–7 contained residues that alter chloride affinity. This was recently supported by the solved structure of NKCC1, in which TMs 1, 3, 6, and 8 form the solvent-accessible vestibule for ion translocation (Chew et al. 2019). Within this vestibule, the binding site for sodium is identical to the classical "Na2" site contained in other APC transporters. Within human NCC, two adjacent serine residues at positions 467 and 468 appear to be critical. Molecular dynamics simulations of NKCC1 also indicate that Cl^- diffuses between TM1s and 6, indicating that the vestibule is important for the translocation of both ions. These studies also suggested that transport of Na^+ and Cl^- are likely cooperative as residency of Na^+ in the Na2 binding site enhances Cl^- residence time, and vice versa.

Moreno et al. also reported that swapping transmembrane domains 8–12 between species altered the thiazide binding affinity (Moreno et al. 2006). This indicates that the binding sites for chloride and thiazide diuretics are different. Further analysis of these transmembrane domains revealed that substituting residues within TM11 was sufficient to confer altered sensitivity to metolazone (Castaneda-Bueno et al. 2010). Mutating serine 575 in the rat cotransporter to the corresponding residue in flounder (a cysteine) was sufficient to mediate this effect. Because only cysteine substitution at this site was capable of altering thiazide sensitivity, it is possible that the flounder cotransporter contains an extra intramembranous disulfide bond that alters the NCC fold. Although it is unclear if rat serine 575 participates in a local binding site for metolazone, these findings implicate TM11 as an important structural component of NCC that defines its thiazide-sensitivity.

Mammalian NCC undergoes N-linked glycosylation on two asparagines harbored within the third extracellular loop, located between TM5 and TM6 (Hoover et al. 2003). N-glycosylation at these sites may play an important role in thiazide and chloride binding, since mutation of the glycosylation site asparagines to glutamine decreased the rat NCC K_m for chloride binding and IC_{50} for thiazide inhibition substantially, without affecting its affinity for sodium (Hoover et al. 2003). As mentioned above, the binding sites of chloride and thiazide diuretics are different (Moreno et al. 2006). Therefore, these findings suggest that N-linked glycosylation alters NCC conformation at multiple sites. One interpretation of the data is that glycosylation may introduce broad changes into the NCC tertiary structure, allowing it to achieve a mature conformation that can mediate thiazide-sensitive salt transport. This would be consistent with the well-appreciated role of N-glycosylation in the global folding of other polytopic membrane proteins, such as the cystic fibrosis transmembrane conductance regulator (CFTR) (Glozman et al. 2009).

3.5 NCC Biogenesis

3.5.1 NCC Processing in the ER

Given the complex topology of NCC, it must undergo considerable processing within the biosynthetic pathway before it can achieve a proper fold. Current data suggest that, similar to other polytopic membrane proteins (Vembar and Brodsky

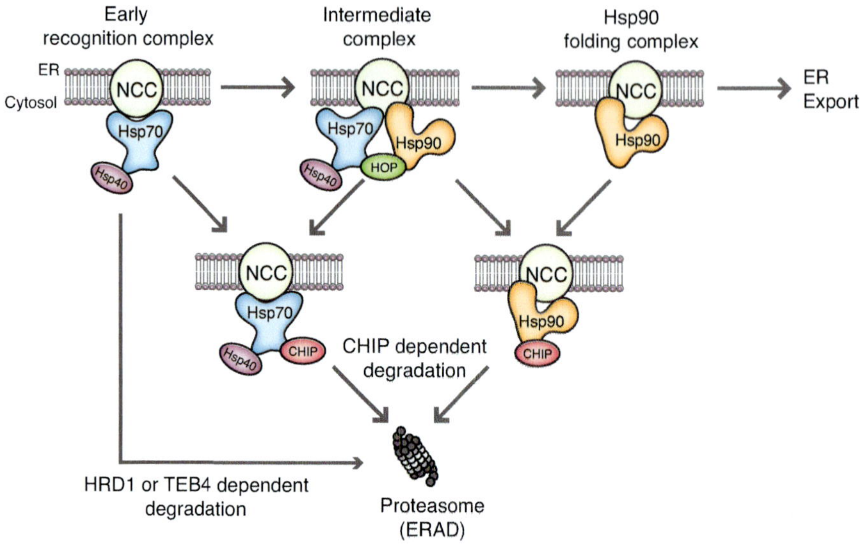

Fig. 3.2 Early steps in NCC biogenesis. NCC is recognized at an early stage of its biogenesis by an Hsp70/Hsp40 complex. Hsp70 recognizes misfolded regions of the cotransporter and attempts to fold it into a stable conformation. NCC will then be transferred from Hsp70 to Hsp90 via HOP. Cotransporters that achieve a stable Hsp90-bound conformational intermediate will then advance further along the biosynthetic pathway and out of the ER. At various stages of this pathway, the E3 ubiquitin ligases HRD, TEB4, and CHIP can ubiquitylate misfolded NCC conformers, targeting them for ER-associated proteasomal degradation (ERAD)

2008), this process is slow and/or inefficient. This is sensed by protein quality control mechanisms, which target suboptimally folded NCC conformers for ER-associated degradation (ERAD). Once the NCC polypeptide becomes integrated into the plasma membrane, molecular chaperones facilitate its proper folding. Heterologous expression studies in yeast and in mammalian cells have identified a host of cytoplasmic chaperone proteins are responsible for this process (Fig. 3.2). These molecular chaperones are exclusively located within the cytoplasm. ER luminal chaperones, in contrast, appear to be less important (Needham et al. 2011). The cytoplasmic Hsp70/Hsp40 chaperone complex participates in one of the earliest quality control steps during NCC processing in the ER (Needham et al. 2011). Within this complex, Hsp70 is critically important, while Hsp40 is bound to Hsp70 and NCC but is dispensable for NCC quality control. If Hsp70 is able to fold NCC into a stable intermediate conformation, it will then be transferred from Hsp70 to Hsp90 via the Hsp70/Hsp90 organizer cochaperone (HOP/Stip1) (Donnelly et al. 2013) (Fig. 3.2). Hsp90 will then engage NCC and attempt to fold it further. If the cotransporter is interacting with either of these chaperones and becomes terminally misfolded, it will be targeted for ubiquitylation and ERAD (Donnelly et al. 2013). Thus far, three E3 ubiquitin ligases have been identified that can mediate this process. Based on studies of NCC ERAD in yeast, suboptimal folding and aggregation of transmembrane domains are likely a common mishap, resulting in targeted degradation by HRD1 (Needham et al. 2011). Misfolded cytoplasmic regions of NCC, on the other hand, can be recognized by the E3 ligase Doa10/TEB4 (Needham et al. 2011), or by the C-terminus of Hsp70 interacting protein (CHIP/Stub1) (Donnelly et al. 2013) (Fig. 3.2). Following ubiquitylation, NCC is then extracted from the ER membrane and extruded into the cytosol by the AAA+ ATPase Cdc48 and degraded by the 26S proteasome (Needham et al. 2011). In each case, interaction with Hsp70 or Hsp90 appears to be a prerequisite prior to ubiquitylation and ERAD (Needham et al. 2011; Donnelly et al. 2013).

As NCC is being folded in the ER, it undergoes glycosylation on two asparagines harbored within its third extracellular loop (Hoover et al. 2003). Like other transmembrane glycoproteins such as CFTR (Ward and Kopito 1994), the N-linked sugars attached to NCC are processed in the ER into a high-mannosylated form. Upon transit to the Golgi, the mannoses are cleaved into mature shorter glycan chains. Proper glycosylation is essential for NCC to exit the biosynthetic pathway and reach the plasma membrane (Hoover et al. 2003).

3.5.2 Post-ER Processing and Endosomal Storage of NCC

Little is known about how later steps of NCC maturation are carried out from within the Golgi. Presumably, most NCC will be processed in the Golgi cisternae when it is already in its homodimeric form, as oligomerization events commonly take place in the ER (Geva and Schuldiner 2014). As it matures, the dimeric NCC passes through the Golgi cisternae, is packaged into vesicles at the trans-Golgi network (TGN), and is sent to the apical plasma membrane via an endosomal system. In some cases, however, a subpopulation of cotransporters can be sorted directly from the TGN to the lysosome for degradation. Protein interaction studies suggest that this process may require the adaptor protein complex AP-3 (Subramanya et al. 2009), or the lysosomal sorting receptor sortilin (Zhou et al. 2009). In heterologous expression systems, this alternative TGN-to-lysosome transport pathway appears to be greatly enhanced in cells overexpressing With-No-Lysine kinase 4 (WNK4, discussed in detail below).

Ultrastructural studies performed in rodent kidneys commonly demonstrate that NCC exists in apical and subapical populations in DCT cells (Lee et al. 2009; Sandberg et al. 2006). The subapical population resides within endosomes, which can be rapidly mobilized to the surface by hormones such as vasopressin and angiotensin II (Pedersen et al. 2010; Mutig et al. 2010; Sandberg et al. 2007). Conversely, physiological states such as acute hypertension can cause the NCC to quickly retract from the plasma membrane into endosomes and lysosomes (Lee et al. 2009). This process is linked to site-specific ubiquitylation of the cotransporter at the plasma membrane at several lysine residues localized within the cytoplasmic NCC CTD (Rosenbaek et al. 2017).

3.6 NCC Activity at the Plasma Membrane

3.6.1 NCC Phosphorylation

In order for NCC to be active, it must first be "switched on" by phosphorylation (Pacheco-Alvarez et al. 2006). These phosphoactivation events specifically occur on serines and threonines present in the intracellular NCC amino terminus. Figure 3.3a illustrates sites of phosphorylation that have been identified, either through phosphoproteomic studies or through analysis of homologous phosphorylation sites experimentally identified on other sodium-coupled cotransporters that are members of the SLC12 family, such as NKCC1 (Darman and Forbush 2002). To date, seven phosphoacceptor residues have been identified. According to the human NCC numeration, these amino acids correspond to threonines 46, 50, 55, and 60, and serines 73, 91, and 126 (Richardson et al. 2008; Rosenbaek et al. 2012). Phosphorylation at each of these sites enhances NCC activity, although some evidence suggests that threonine 60 may have a unique and dominant role in dictating NCC-mediated sodium transport (Pacheco-Alvarez et al. 2006) (Fig. 3.3a). Since all of these phosphosites are harbored within the NCC amino terminus, they are located immediately proximal to the first transmembrane domain. Thus, phosphorylation of the NCC N-terminus may somehow alter the positioning or structure of residues related to TM 1. Since structural data from other members of the APC superfamily indicate that TMs 1 and 6 help form the Na-Cl translocation pathway (Chew et al. 2019; Krishnamurthy et al. 2009), phosphorylation may enhance the efficiency of Na-Cl transport by altering the structure of the transporter vestibule.

Recent data also indicate a relationship between NCC phosphorylation status and expression at the plasma membrane. Immunolocalization and ultrastructural studies using phosphospecific antibodies to one or more of the aforementioned NCC phosphorylation sites indicate that phosphorylated NCC is tightly restricted to the DCT apical membrane, whereas the unphosphorylated form appears to be more diffusely spread in both plasma membrane and intracellular compartments (Yang et al. 2007a; Mutig et al. 2010; Sandberg et al. 2006; Lee et al. 2009). This suggests that phosphorylation enhances NCC accumulation at the plasma membrane. Consistent with this, phosphorylated forms of NCC are not internalized via clathrin-mediated endocytosis as rapidly as unphosphorylated forms of the cotransporter, and exhibit decreased ubiquitylation (Rosenbaek et al. 2014). Although the mechanism by which the phosphorylated form of NCC is relatively spared from endocytosis is unclear, it seems to be related to a change in its membrane complex properties. Blue native PAGE electrophoresis (BN-PAGE) studies suggest that upon phosphorylation, plasma membrane NCC shifts from a 400 kDa complex to heavier complexes of ~700 kDa in molecular mass (Lee et al. 2013). Included in the 700 kDa complex is γ-adducin, a membrane cytoskeletal protein that stimulates NCC (Lee et al. 2013; Dimke et al. 2011). Other data suggest that plasma membrane-localized NCC may be associated with other membrane proteins in heteromeric complexes, such as the epithelial sodium channel ENaC (Mistry et al. 2016). The incorporation of

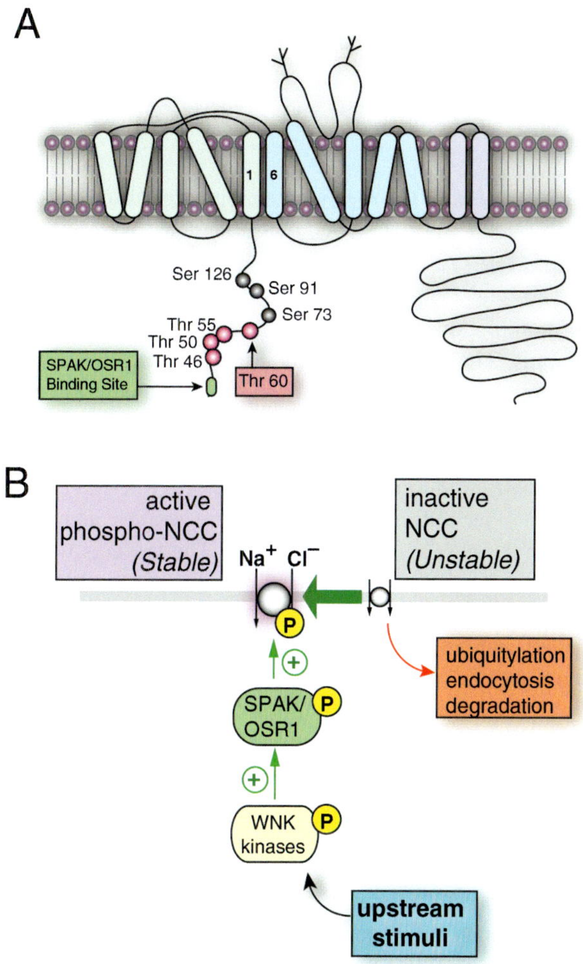

Fig. 3.3 Control of NCC activity by phosphorylation. (**a**) A regulatory locus of phosphorylation consistent of a cluster of serines and threonines is contained within the cytoplasmic amino terminus of NCC. The amino terminus is contiguous with the TM1, which likely forms a sodium binding site with TM6. A binding motif for the serine/threonine kinases SPAK and OSR1 is located at the extreme end of the N-terminus. SPAK/OSR1 phosphorylation sites are shown in pink. Threonine 60 is critical for NCC activity. Other known phosphoacceptor residues are shown in gray. The kinases that phosphorylate these residues are unknown, but some of them (including serine 73) are phosphorylated in concert with SPAK/OSR1 activation. This suggests that other kinases may cooperate with SPAK/OSR1 to activate NCC. (**b**) Current model for SPAK/OSR1 activation. SPAK and OSR1 are activated by WNK kinases and phosphorylate NCC at the plasma membrane. The WNKs are a convergent regulatory point for upstream stimuli, including low intracellular chloride levels and hormones that stimulate renal salt reabsorption. Phosphorylation increases NCC stability at the surface. Inactive nonphosphorylated forms of NCC do not transport NaCl and are ubiquitylated and endocytosed more rapidly from the plasma membrane

phosphorylated NCC into high molecular weight complexes at the plasma membrane may prevent the cotransporter from being incorporated into clathrin-coated vesicles.

3.6.2 *The WNK-SPAK/OSR1 Signaling Pathway*

3.6.2.1 SPAK and OSR1

Two structurally homologous serine–threonine kinases directly phosphorylate the NCC amino terminus. These kinases are the Ste20-like proline-alanine rich kinase (SPAK, encoded by the gene *STK39*), and oxidative stress-responsive kinase 1 (OSR1, encoded by *OXSR1*) (Richardson et al. 2008). The relationship between SPAK, OSR1, and sodium-coupled cation–chloride cotransporters was discovered initially in yeast two-hybrid studies of the secretory Na^{+}-K^{+}-$2Cl^{-}$ cotransporter NKCC1 (*SLC12A2*) (Piechotta et al. 2003). These studies identified SPAK as a binding partner with the NKCC1 amino terminus. The findings were then extended to OSR1, whose amino acid sequence is 67% identical to SPAK (Delpire and Gagnon 2006). Both SPAK and OSR1 bind to the NKCC1 N-terminus via an [R/K]FX[V/I] motif (where X can be any amino acid residue) (Piechotta et al. 2003). Later, these motifs were identified on NCC and the renal thick ascending limb Na^{+}-K^{+}-$2Cl^{-}$ cotransporter NKCC2 (*SLC12A1*) and were confirmed to be SPAK and OSR1 binding sites on those cotransporters as well (Gagnon et al. 2006; Richardson et al. 2008; Villa et al. 2007). Consistent with a broad role for SPAK and OSR1 in stimulating the activity of SLC12 cotransporters in the kidney, SPAK and OSR1 are highly expressed in the thick ascending limb of the Loop of Henle (TAL) and DCT (McCormick et al. 2011). Moreover, SPAK mutant knock-in mice that express a catalytically inactive form of SPAK that cannot be phosphorylated by upstream signals in its kinase activation loop (i.e., "T-loop"), exhibit decreased salt reabsorption and NKCC2 and NCC phosphorylation in both the TAL and DCT (Rafiqi et al. 2010). Presumably, this inactive form of SPAK acts as a dominant-negative, preventing the association of native SPAK and/or OSR1 with the N-terminal binding sites in NKCC2 and NCC. Many of the identified phosphorylation sites on NCC are SPAK/OSR1 phosphoacceptor residues. These sites correspond to amino acids 46, 50, 55, and 60 of the human NCC N-terminus (Richardson et al. 2008). Serine 73 may also be directly phosphorylated by SPAK, as SPAK KO mice exhibit decreased phosphorylation at this residue (Yang et al. 2010), this site was not modified in in vitro studies. Likewise, phosphorylation of serine 91 is diminished in cells lacking SPAK activity, although this residue was not a direct SPAK/OSR1 phosphorylation site. Thus, some of the phosphoacceptor residues in the NCC N-terminus may be targeted by other kinases that cooperate with SPAK and OSR1 to stimulate salt transport in the DCT.

Studies in SPAK and OSR1 knockout animals illustrate the importance of these kinases in renal salt transport. Although both of SPAK and OSR1 can activate NCC, their genetic ablation in the kidney results in disparate renal salt-wasting phenotypes. Mice deficient in SPAK exhibit salt wasting similar to a mild form of Gitelman syndrome (Yang et al. 2010; McCormick et al. 2011), suggesting that this kinase stimulates salt reabsorption in the DCT. These mice exhibit decreased NCC phosphorylation, but their mild salt wasting is due to the fact that they exhibit a dramatic *increase* in NKCC2 activity. In contrast, OSR1 knockout mice exhibit a more Bartter-like phenotype, due to a strong decrease in NKCC2 activity in the TAL (Lin et al. 2011). NCC activity is relatively unaffected.

Current evidence suggests that these different phenotypes are due to discrepancies in the expression and activity of SPAK and OSR1 in the TAL and DCT. In the TAL, SPAK activity appears to be low, due to the expression of short forms of the kinase (McCormick et al. 2011; Grimm et al. 2012). These short forms of SPAK are exclusively expressed in the TAL and are either truncated transcripts or proteolytic fragments that lack kinase activity (McCormick et al. 2008; Markadieu et al. 2014). These kinase-deficient species can suppress the kinase activity of full-length SPAK and OSR1, functioning in a dominant-interfering manner. Genetic deletion of SPAK eliminates the expression of both its long and short forms in the TAL; this effectively releases OSR1 from inhibition in the TAL, causing it to hyperphosphorylate NKCC2. Since only the full-length form of SPAK is expressed in the DCT, SPAK KO mice exhibit a net decrease in NCC phosphorylation, resulting in the Gitelman-like phenotype. In the absence of SPAK, OSR1 redistributes into an intracellular location, collecting in large puncta (McCormick et al. 2011; Grimm et al. 2012; Terker et al. 2015). These puncta, termed "WNK bodies," contain upregulated components of the WNK-SPAK/OSR1 pathway and are further discussed below (see section on WNK1) (Boyd-Shiwarski et al. 2018). More work is needed to further elucidate the relationship between SPAK, OSR1, and cation–chloride cotransporters in the kidney, but the current body of work suggests that SPAK is the dominant activator of NCC in the DCT, while OSR1 serves as the main activator of NKCC2 in the TAL.

3.6.2.2 WNK Kinases

In order for SPAK and OSR1 to phosphorylate and activate the sodium-coupled cation–chloride cotransporters, they must first be in an active state. SPAK and OSR1 activation occurs through phosphorylation of key residues in their kinase domains (Vitari et al. 2005, 2006; Thastrup et al. 2012). A family of proteins called With-No-Lysine (amino acid = [K]; WNK) kinases mediates this process. The mammalian WNK family consists of four genes, which encode large serine–threonine kinases found in many eukaryotes. All four WNK kinases share a similar domain organization, consisting of an N-terminal kinase domain, an autoinhibitory domain that suppresses intrinsic kinase activity and can cross-inhibit other WNKs, 2–3 coiled-coil domains which facilitate oligomerization, and multiple SPAK/OSR1 binding motifs (Richardson et al. 2008). They were named "With-No-Lysine" kinases after bioinformatic analysis of the kinase domain revealed that the catalytic lysine required for ATP binding and phosphoryl transfer is missing from its usual location in β strand 3 of the active site; instead, this lysine is positioned in β strand 2 (Xu et al. 2000; Min et al. 2004). Due to this unusual arrangement of residues within the kinase domain, chloride can bind directly to the catalytic site and block its ability to trigger autophosphorylation and kinase domain activation (Piala et al. 2014). Thus, the unique structure of WNK kinases allows them to act as intracellular chloride sensors.

The WNKs were identified as NCC regulators over a decade ago, through studies of a rare chloride-dependent thiazide-sensitive Mendelian blood pressure disorder called Familial Hyperkalemic Hypertension (FHHt, also known as Pseudohypoaldosteronism type 2 or Gordon Syndrome). In 2001, mutations in two WNK kinases, WNK1 and WNK4, were linked to FHHt (Wilson et al. 2001). Both of these kinases were expressed with NCC in the DCT (Wilson et al. 2001; Rinehart et al. 2005). While important roles for WNK3 have been established in the brain, and some reports indicate that it is also expressed in the kidney, WNK3 KO mice do not exhibit a renal phenotype, suggesting that it not be as important as WNK1 and WNK4 in renal tubular physiology (Mederle et al. 2013). WNK2, is a brain, heart, and intestine-expressed WNK kinase that is excluded from the kidney and therefore does not appear to be a major NCC regulator (Rinehart et al. 2011).

WNK1 is unique from WNK4 and WNK3 in that it contains multiple promoters and undergoes extensive alternative splicing in the kidney (O'Reilly et al. 2003; Delaloy et al. 2003). Thus, its renal expression pattern consists of multiple isoforms with disparate functions. These isoforms can be classified into two major functional groups: *L*ong kinase active isoforms ("L-WNK1") with intact serine–threonine kinase activity, or short *K*idney *S*pecific isoforms that are kinase-defective ("KS-WNK1"). KS-WNK1 isoforms are generated by an alternative promoter located between exons 4 and 5. This promoter is only active in the kidney and drives the high expression of KS-WNK1 in the DCT (O'Reilly et al. 2006; Vidal-Petiot et al. 2012; Liu et al. 2011).

Although mutations in WNK1 and WNK4 were the first to be implicated in the pathogenesis of FHHt, most FHHt affected harbor mutations in two other genes, the E3 ubiquitin ligase *Cullin 3* (CUL3) and its adaptor, *Kelch-like 3* (KLHL3) (Boyden et al. 2012; Louis-Dit-Picard et al. 2012). The CUL3-KLHL3 complex binds to WNK1 and WNK4 and facilitates their ubiquitylation and degradation (Shibata et al. 2013; Ohta et al. 2013).

3.6.2.2.1 WNK4

The first studies defining relationships between the WNKs and NCC were carried out in the *Xenopus laevis* oocyte expression system. This in vitro system had been used for the initial expression cloning studies that permitted the isolation of the cDNAs encoding NCC and other members of the cation–chloride cotransporter family. Thus, it was a natural extension to utilize the system to test the effects of WNK kinases on the NCC function. The general approach for these experiments was to coinject in vitro transcribed RNA encoding a WNK kinase along with NCC and measure NCC activity by performing $^{22}Na^+$ radioisotopic fluxes in the presence or absence of thiazide. These initial experiments indicated that overexpression of WNK4 suppresses NCC activity (Yang et al. 2003; Wilson et al. 2003). Later, this finding was also seen in mammalian overexpression systems and mouse distal convoluted tubule cell lines (Subramanya et al. 2009; Ko et al. 2012; Cai et al. 2006; Zhou et al. 2009). Follow-up studies revealed that this overexpression phenotype is due to the suppression of NCC trafficking from the TGN to the plasma membrane (Golbang et al. 2006; Subramanya et al. 2009). Instead of being trafficked from the TGN to the surface, NCC is routed directly to lysosomes for degradation via an intracellular route. This event was associated with enhanced interaction between NCC and the AP-3 adaptor protein complex, which sorts cargo to the lysosome from the TGN (Subramanya et al. 2009; Simpson et al. 1997; Dell'Angelica et al. 1997). In addition, WNK4 overexpression promotes NCC packaging into vesicles that contain the lysosomal sorting receptor sortilin (Zhou et al. 2009). Although some studies suggested that intact WNK4 kinase activity is required for this inhibitory effect on NCC forward trafficking (Wilson et al. 2003; Golbang et al. 2006; Cai et al. 2006), other studies have argued that the effect is mediated by the WNK4 C-terminus, and therefore, is a kinase-independent process (Yang et al. 2005). The MAP kinase ERK1/2 may mediate the inhibitory effect of WNK4 on NCC, as ERK1/2 activation is associated with decreased NCC plasma membrane expression, and WNK4 overexpression can activate the ERK1/2 signaling pathway (Zhou et al. 2012; Ko et al. 2007). However, these studies remain largely correlative to date, and conflicting experiments have implicated ERK1/2 as a stimulator of NCC endocytosis, a trafficking operation that is not affected by WNK4 overexpression (Golbang et al. 2006; Subramanya et al. 2009).

Although numerous laboratories have independently corroborated this "inhibitory" effect of WNK4 (Subramanya et al. 2009; Yang et al. 2003, 2005; Golbang et al. 2006; Wilson et al. 2003; Cai et al. 2006; Ko et al. 2012), caution should be exercised when interpreting these experiments. Importantly, nearly all of the in vitro studies looking at this aspect of WNK4-NCC regulation have been performed in heterologous expression systems. Overexpression of a WNK kinase can cause aberrant sequestration and mislocalization of other components of the signaling pathway, including other WNK kinases or SPAK and OSR1. Thus, the inhibitory phenotype seen in in vitro systems may not have an in vivo correlate. Indeed, transgenic mice overexpressing wild type WNK4 have provided conflicting results; one of the mouse models exhibited strong inhibition of NCC expression and DCT atrophy, while the other mouse actually exhibited *increased* NCC activity due to overstimulation of SPAK and OSR1 (Wakabayashi et al. 2013; Lalioti et al. 2006).

Consistent with this second mouse model, a number of studies have shown that WNK4 can activate NCC via SPAK/OSR1 (Vitari et al. 2005; Yang et al. 2007a). In contrast to the inhibitory effect of WNK4 described above, there is a general consensus that the stimulatory effect requires direct phosphorylation of SPAK/OSR1 by WNK4. This, in turn, results in downstream enhanced SPAK/OSR1 mediated phosphorylation of NCC. Moreover, several hormones that activate salt transport via NCC (discussed in detail below) appear to trigger NCC

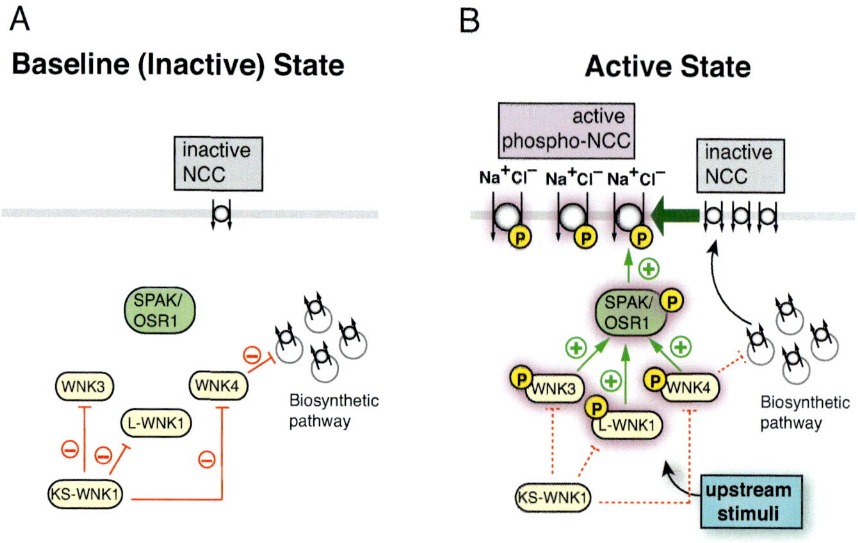

Fig. 3.4 WNK kinases. A model for WNK-mediated regulation of SPAK/OSR1 and NCC. (**a**) In the baseline unstimulated state, WNK4 may act as an inhibitor, suppressing NCC forward trafficking from endosomes derived from the biosynthetic pathway. WNK4, WNK3, and full-length WNK1 (L-WNK1) are all expressed in the DCT. The kinase-defective short isoform of WNK1, KS-WNK1, suppresses the kinase activity of WNK4, WNK3, and L-WNK1 at baseline. (**b**) When the DCT is challenged with a physiological stimulus, the kinase activity of WNK4, WNK3, and L-WNK1 increases relative to KS-WNK1. This converts these kinases into activators of SPAK and OSR1, which triggers NCC phosphorylation

phosphorylation by stimulating WNK4-mediated activation of SPAK/OSR1. Thus, many experts in the field view the stimulatory effect of WNK4 on NCC phosphorylation status as its primary and more physiologically relevant mode of operation (Takahashi et al. 2014; Richardson et al. 2008).

In summary, current data on the role of WNK4 in NCC regulation provide evidence on both sides of the fence: some studies support its role as an inhibitor of NCC trafficking to the plasma membrane, while others have shown that it can stimulate NCC activity via SPAK- and OSR1-mediated phosphorylation. So, how does one reconcile these two radically different effects into a unifying model? One view is that the kinase-inactive form of WNK4 may act as an inhibitor at baseline, holding NCC in an intracellular compartment via a mechanism that does not require intact kinase activity (Subramanya and Ellison 2014; McCormick et al. 2008) (Fig. 3.4a). Upon exposure to physiological factors that stimulate salt reabsorption in the DCT, WNK4 is converted into an active state. This results in WNK4 kinase domain activation and downstream SPAK and OSR1 phosphorylation, and loss of the inhibitory effect (Fig. 3.4b). This would increase NCC phosphorylation and the number of active NCC cotransporters expressed at the plasma membrane due to enhanced forward trafficking. Studies utilizing RNA interference in DCT-derived cell lines that endogenously express SPAK, WNK4, and NCC provide evidence in support of such a model (Ko et al. 2010, 2012).

3.6.2.2.2 WNK1

Unlike WNK4, early studies with L-WNK1 in the *Xenopus* oocyte expression system were carried out with the original L-WNK1 cDNA, which was derived from rat brain (Xu et al. 2000). Studies with this cDNA revealed no effect of L-WNK1 on NCC activity or surface expression when it was coexpressed with the cotransporter in oocytes. However, when WNK4, L-WNK1, and NCC were co-expressed together, L-WNK1 associated with WNK4 and reversed its inhibitory effect on NCC transport back to baseline levels (Yang et al. 2003). This effect was not seen when either the kinase defective KS-WNK1 isoform or a kinase-dead mutant of L-WNK1 was coexpressed with WNK4 and NCC, suggesting that intact L-WNK1 kinase activity is required to suppress WNK4-mediated inhibition (Yang et al. 2005; Subramanya et al. 2006). As expected, like the other WNKs, L-WNK1 can also stimulate SPAK/OSR1 activation by phosphorylating residues within the SPAK and OSR1 kinase domains (Vitari et al. 2005). A more recent study that used human L-WNK1 found a stimulatory effect on NCC in *Xenopus* oocytes, via SPAK/OSR1 activation (Chavez-Canales et al. 2014). Collectively, these studies indicate that L-WNK1 can activate SPAK and OSR1, and can also override the inhibitory effect of WNK4 commonly seen in heterologous expression systems (Fig. 3.4b).

As mentioned above, DCT-localized KS-WNK1 isoforms do not contain a functional kinase domain. They do, however, contain C-terminal coiled coil domains that are important for interactions with other WNK gene products, including L-WNK1 and WNK4 (Subramanya et al. 2006). Thus, KS-WNK1 appears to function as a scaffold for other WNK kinases. Consistent with this, a recent study showed that KS-WNK1 is required for WNK kinases to organize into large membraneless punctate foci in the DCT. These structures, termed "WNK bodies," have been visualized in the setting of NCC activation by hypokalemia, aldosterone, and in mouse models of FHHt (Boyd-Shiwarski et al. 2018). They have also been visualized in SPAK and OSR1 knockout mice, where genetic ablation of SPAK and OSR1 would be expected to result in upstream stimulation of WNK kinases to enhance NCC activity. Though the functional role of WNK bodies is being elucidated, this suggests that the formation of these structures is a signature of WNK pathway activation.

Currently, there are conflicting data regarding the role of KS-WNK1 in NCC regulation. KS-WNK1 contains an "autoinhibitory" region that suppresses L-WNK1 kinase activity in vitro (Subramanya et al. 2006), suggesting that it may function as an inhibitory scaffold for WNK kinases. Consistent with this, early studies in Xenopus oocytes suggested that KS-WNK1 blocks the ability of L-WNK1 to activate NCC (Subramanya et al. 2006). These findings were supported by two mouse models of KS-WNK1 inactivation (Hadchouel et al. 2010; Liu et al. 2011). However, more recent studies have argued that KS-WNK1 can function as an NCC activator (Argaiz et al. 2018), a finding that would be consistent with the role of KS-WNK1 in the formation of WNK bodies. Thus, though it is clear that KS-WNK1 is required for WNK body formation in the DCT, its precise role in NCC regulation remains unresolved and an active area of investigation.

3.6.2.2.3 WNK3

Comparative studies of the WNK kinases indicate that WNK3 is a potent NCC activator (Yang et al. 2007a; Rinehart et al. 2005) (Fig. 3.4b). In vitro, WNK3 stimulates NCC transport activity by enhancing its phosphorylation and surface expression (Rinehart et al. 2005; Yang et al. 2007a; Chavez-Canales et al. 2014). This effect can be blocked by KS-WNK1 (Yang et al. 2007a). Immunolocalization studies in kidney have identified WNK3 in several segments of the renal tubule, including the proximal tubule, Loop of Henle, DCT, and collecting duct (Rinehart et al. 2005). Based on the strong stimulatory effects of WNK3 on NCC activity, and its colocalization with NCC in the DCT, some have proposed that WNK3 may be a critically important regulator of thiazide-sensitive NaCl reabsorption. This hypothesis predicts that deleting WNK3 expression from the kidney should cause substantial salt wasting. Surprisingly, however, WNK3 knockout mice do not have an appreciable renal phenotype (Oi et al. 2012; Mederle et al. 2013). Although the reason for this is not entirely clear, compensation by L-WNK1 may play a role, since WNK1 mRNA is increased in WNK3 knockout mice relative to controls (Mederle et al. 2013). These findings suggest a minor role for WNK3 in the control of NCC activity in vivo.

3.6.3 NCC Dephosphorylation by Protein Phosphatases

Although much of the work on NCC phosphorylation has been centered on understanding how protein kinases stimulate NCC activity, a growing body of evidence indicates that protein phosphatases inhibit NCC. Some of these phosphatases likely directly remove phosphate groups from the NCC amino terminus, while others may influence NCC phosphorylation status by inhibiting the WNK-SPAK/OSR1 pathway.

Early studies of the sodium-coupled SLC12 cotransporters indicated that inhibitors of the ubiquitously expressed protein phosphatase 1 (PP1) could suppress NKCC1 activity through dephosphorylation (Darman and Forbush 2002; Lytle and Forbush 1996). Indeed, PP1 also appears to play an important role in NCC dephosphorylation, as one of its canonical inhibitors, PP1 inhibitor 1 (Ppp1r1a) is highly coexpressed with the cotransporter in the DCT (Picard et al. 2014). The deficiency of Ppp1r1a leads to a mild thiazide-sensitive salt-wasting phenotype. This suggests that the absence of Ppp1r1a leads to enhanced PP1 activity and NCC dephosphorylation. In 2010, Glover et al. identified PP4 as another phosphatase that is expressed in the DCT (Glover et al. 2010). The coexpression of PP4 with NCC reduces its phosphorylation at threonine 60 and therefore decreases its transport activity.

Protein phosphatase 3 (also known as calcineurin) was also recently identified as a potential negative regulator of NCC activity. The calcineurin inhibitor tacrolimus (FK506, "Prograf") commonly causes hypertension, hyperkalemia, hypercalciuria, and metabolic acidosis, despite normal glomerular filtration rate (GFR) (Hoorn et al. 2011). This constellation of findings is similar to the hypertension and hyperkalemia seen in FHHt (Hadchouel et al. 2006; Mayan et al. 2002). In 2012, Hoorn and colleagues showed that calcineurin inhibitors cause hypertension by activating thiazide-sensitive NaCl cotransport in the kidney. The mechanism appears to be directly related to an increase in NCC phosphorylation status. WNK4 abundance and SPAK/OSR1 activation were also increased. These findings suggest that tacrolimus may stimulate NCC activity through an indirect effect, possibly by preventing the dephosphorylation and inactivation of WNK kinases.

3.7 NCC Endocytosis

NCC is removed from the plasma membrane by dynamin- and clathrin-dependent endocytosis (Ko et al. 2010; Rosenbaek et al. 2014). Studies of overexpressed NCC in MDCK renal epithelial cells indicate that the cotransporter can be constitutively endocytosed, with one-third of the total NCC surface pool undergoing internalization within 10 min; this endocytic rate is compatible with other kidney tubule-expressed membrane proteins such as the epithelial sodium channel (ENaC) (Butterworth et al. 2005). Clathrin-dependent endocytic retrieval of NCC is associated with increased cotransporter ubiquitylation and routing to the lysosomal pathway (Rosenbaek et al. 2014). While some groups have reported that endocytosed NCC undergoes polyubiquitylation (Ko et al. 2010), others have reported that the endocytosed ubiquitylated cotransporter migrates at molecular mass similar to monomeric NCC, suggesting oligo- or monoubiquitylation (Rosenbaek et al. 2014). Endocytosis of NCC is dependent on the phosphorylation status of the protein, since phosphorylation at threonines 55, 60, and 73 stabilize NCC surface expression by slowing the rate of endocytosis (Rosenbaek et al. 2014; Yang et al. 2013). This, in turn, is associated with decreased NCC ubiquitylation. Ubiquitylation of the surface NCC pool is mediated by the aldosterone-regulated E3 ubiquitin ligase Nedd4-2 (Ronzaud et al. 2013; Arroyo et al. 2011), though other currently unidentified E3 ubiquitin ligases also probably participate in this process.

Treatment of cultured mDCT cells with the diacylglycerol analog 12-*O*-tetradecanoylphorbol-13-acetate (TPA) inhibits NCC by stimulating its endocytosis and ubiquitylation (Ko et al. 2010). Although TPA is classically used in the laboratory to activate protein kinase C (PKC), the effect appeared to be unrelated to this kinase, since PKC blockers did not reverse the inhibitory effect of TPA. Instead, the alternative TPA target Ras guanyl-releasing protein 1 (RasGRP1) was the primary mediator of NCC inhibition (Ko et al. 2007). The MAP kinase ERK1/2 may act downstream of RasGRP1 to trigger NCC endocytosis (Ko et al. 2010).

3.8 Mendelian Disorders of NCC Dysfunction

3.8.1 Gitelman Syndrome

In 1996, Simon et al. reported that mutations in the *SLC12A3* gene encoding NCC cause Gitelman syndrome, an autosomal recessive salt wasting disorder (Simon et al. 1996). Patients may be hypotensive, but often have normal blood pressures, since the urinary salt wasting is offset by hyperreninemia and upregulation of the renin–angiotensin system. Hypokalemia and hypomagnesemia, however, are always seen in the disorder. Other manifestations include hypocalciuria and metabolic alkalosis. Thus, the overall phenotype of Gitelman syndrome is similar to the constellation of findings seen in patients taking thiazide diuretics.

Since the initial linkage of NCC to Gitelman syndrome, hundreds of disease-causing mutations have been identified. These mutations are generally scattered throughout the NCC protein, with a slight clustering of lesions in the intracellular C-terminus. In vitro analyses indicate that mutations that cause the disorder generally promote NCC misfolding (Kunchaparty et al. 1999). These so-called "Class 2" mutations cause the NCC to be retained in the endoplasmic reticulum and targeted for ERAD (Ellison 2003). Consistent with this finding, a recent study found that Gitelman mutations localized to the NCC C-terminus cause the cotransporter to interact strongly with a molecular chaperone complex containing Hsp70, Hsp40, and the cochaperone/E3 ubiquitin ligase CHIP (Donnelly et al. 2013). On the other hand, some mutations that cause Gitelman syndrome do not affect the biosynthetic processing of NCC; cotransporters with these "Class 1" mutations appear to be processed normally and traffic to the plasma membrane without difficulty (Sabath et al. 2004; Riveira-Munoz et al. 2007; Ellison 2003).

In 2007, a study by Ji et al. suggested that mutations of NCC that are associated with Gitelman syndrome play an important role in determining susceptibility to essential hypertension (Ji et al. 2008). Based on the prevalence of Gitelman Syndrome, about 1% of the general population should be heterozygous for the disorder. On average, individuals in a large study population who were carriers of a loss of function NCC variant had a systolic blood pressure that was 6.3 mmHg lower throughout life, compared with individuals who had two wild type copies of NCC. Thus, the carrier state for Gitelman syndrome conferred lifelong protection from the development of hypertension (Acuna et al. 2011; Subramanya and Welling 2011). As the cost of next-generation sequencing technology continues to drop, one could imagine that screening for these mutations in normotensive individuals could provide information about the risk of developing hypertension over time.

3.8.2 Familial Hyperkalemic Hypertension

Familial Hyperkalemic Hypertension (FHHt) is a rare Mendelian syndrome featuring the unusual triad of chloride-dependent hypertension, severe hyperkalemia, and normal glomerular filtration rate. The disease was initially described by Paver and Pauline in the 1960s (Paver and Pauline 1964), and later by Gordon (Gordon 1986). For years, the disease was known as "Gordon Syndrome" or "Pseudohypoaldosteronism Type II," although more recently, some have elected to rename the syndrome "FHHt" (Yang et al. 2005; Hadchouel et al. 2006). The chloride-dependent hypertension and hyperkalemia can be cured by low doses of thiazide diuretics (Mayan et al. 2002). Those familiar with the syndrome immediately recognized that its constellation of findings is essentially a "mirror image" of Gitelman syndrome (Hadchouel et al. 2006). A natural conclusion, then, was that gain-of-function mutations of NCC should be linked to the disease. Candidate gene approaches, however, indicated that this was not the case, as targeted sequencing of the *SLC12A3* gene revealed no mutations (David Ellison, personal communication).

In 2001, Wilson et al. linked mutations in WNK1 and WNK4 to FHHt (Wilson et al. 2001). Although mutations in WNK kinases were the first to be implicated in the pathogenesis of FHHt, they only account for a minority of families affected by the disorder. As mentioned above, most FHHt affecteds harbor mutations in the genes encoding the E3 ubiquitin ligase CUL3 and its adaptor, KLHL3 (Boyden et al. 2012). These two proteins participate in a protein complex that degrades WNK1 and WNK4. Recent studies from several laboratories show that CUL3 covalently attaches ubiquitin molecules to WNK1 and WNK4, marking them for disposal (Shibata et al. 2013; Ohta et al. 2013). In order to do so, KLHL3 must be present, since it connects CUL3 to the WNKs.

An elaborate interrelationship between WNK1, WNK4, and the KLHL3/CUL3 complex underlies the pathogenesis of FHHt. FHHt-causing mutations in WNK4 are missense mutations that reduce WNK4 binding to KLHL3 (Shibata et al. 2013). Total WNK4 expression is, therefore, increased in patients who have these mutations. Disease-causing WNK4 mutants can phosphorylate SPAK and OSR1, and enhance NCC trafficking to the plasma membrane relative to the wild type protein (Yang et al. 2007b). Thus, increased mutant WNK4 protein expression causes NCC overactivation. FHHt mutations in WNK1 are intron deletions that do not alter the kinase's protein structure. Instead, these mutations increase total WNK1 mRNA and protein expression (Wilson et al. 2001). The increased protein expression probably overrides ubiquitination and degradation by the KLHL3/CUL3 complex, and the net effect of the mutation is to increase total L-WNK1 kinase activity (Vidal-Petiot et al. 2013). Thus, in FHHt caused by mutations in WNK1, increased L-WNK1 protein expression should suppress the inhibitory effect of WNK4 on NCC traffic while stimulating SPAK and OSR1, causing NCC to be overactive.

FHHt-associated mutations in KLHL3 reduce the binding of the CUL3–KLHL3 complex to WNK1 and WNK4 (Shibata et al. 2013; Ohta et al. 2013; Wakabayashi et al. 2013). Thus, mutations in KLHL3 diminish the amount of CUL3-mediated

WNK1 and WNK4 degradation. This increases WNK1 and WNK4 abundance, resulting in enhanced WNK-dependent signaling and NCC activation. Mutations in cullin-3 always cause skipping of an exon, resulting in an in-frame deletion (Boyden et al. 2012; Glover et al. 2014). Recent studies indicate that this specific mutation causes autocatalytic degradation of the KLHL3/CUL3 complex, reducing the number of available circulating E3 ubiquitin ligase complexes, thereby increasing WNK4 and WNK1 abundance (McCormick et al. 2014).

3.9 Physiologic Regulation of NCC

The discovery of the WNK-SPAK/OSR1 signaling pathway and the advent of new tools to study NCC function, including cell lines, genetically modified mice, and phosphospecific antibodies, has led to a number of fascinating insights into the underlying mechanisms that regulate NCC activity. This section summarizes the current understanding of the molecular bases of physiologic NCC regulation.

3.9.1 Regulation by Intracellular Chloride

SLC12 cation–chloride cotransporters are important regulators of intracellular chloride concentration (Gamba 2005). Studies examining the effect of chloride on NKCC1 activity found that intracellular Cl^- depletion increases NKCC1 activity (Flatman 2002). The stimulatory effect of chloride depletion on NKCC1 appears to be more than a consequence of generating a simple chemical gradient. Intracellular chloride depletion stimulates phosphorylation of the sodium-coupled SLC12 cotransporters at the same residues that are phosphorylated by SPAK and OSR1 (Richardson et al. 2008). Moreover, as described above, WNK kinases are chloride sensors that are activated by chloride depletion (Piala et al. 2014). Thus, the WNK-SPAK/OSR1 pathway represents an important intracellular signaling mechanism that responds to changes in intracellular Cl^- levels to activate NaCl influx via these electroneutral cotransporters. NCC is activated by intracellular chloride depletion in a manner identical to NKCC1 (Pacheco-Alvarez et al. 2006). In *Xenopus* oocytes and mammalian cells, preincubating cells expressing NCC with chloride-free solutions strongly activates NCC transport activity by stimulating amino-terminal phosphorylation at canonical SPAK/OSR1 target sites (Pacheco-Alvarez et al. 2006; Leaphart et al. 2008). Recently, the chloride sensing function of WNK4 was verified to be critically important in the in vivo regulation of NCC activity in the DCT (Chen et al. 2019).

3.9.2 Regulation by Extracellular Potassium

NCC activity is exquisitely sensitive to changes in the concentration of extracellular K^+. Hyperkalemia has been shown to be a potent NCC inhibitor, while hypokalemia activates NCC (Vallon et al. 2009; McDonough and Youn 2013; Rengarajan et al. 2014). Potassium predominantly affects NCC phosphorylation status. Within minutes of ingesting a high potassium load, rodents develop a natriuresis that is accompanied by rapid NCC dephosphorylation at the canonical SPAK/OSR1 phosphorylation sites (Sorensen et al. 2013). Conversely, hypokalemia appears to strongly trigger NCC phosphorylation at those same sites (Vallon et al. 2009). In both cases, the effects of potassium can occur independently of aldosterone action. The inverse relationship between potassium and NCC phosphorylation status is linear within the physiological range of plasma potassium levels (Terker et al. 2016). This indicates the DCT functions as a potassium sensor that controls luminal NaCl concentrations (and downstream Na^+-dependent collecting duct potassium secretion) via changes in NCC activity (Ellison and Terker 2015).

Extracellular potassium regulates NCC by altering the basolateral conductance of the DCT (Terker et al. 2015). This affects the chloride-dependent activity of the WNK-SPAK/OSR1 signaling pathway. Consider, for example, the physiological alterations seen during hyperkalemia. A high extracellular K^+ concentration in the blood causes the basolateral membrane of the DCT to become depolarized (Brown 1991). This decreased negativity of the cell membrane potential diminishes Cl^- exit via the basolateral DCT chloride channel ClC-Kb, raising the intracellular chloride concentration (Zhang et al. 2014). As described above, the corresponding increase in intracellular Cl^- suppresses WNK kinase autoactivation (Piala et al. 2014), resulting in reduced downstream SPAK/OSR1 and NCC activity. Supporting this concept, recent studies have shown that knockout mice lacking the major K^+ channel mediating basolateral K^+ conductance in the DCT, *KCNJ10* (Kir4.1), exhibit reduced NCC activity caused by depolarization of the DCT membrane potential and a corresponding increase in intracellular chloride, which inhibits SPAK and OSR1 (Zhang et al. 2014).

3.9.3 Luminal NaCl Delivery

NCC-mediated NaCl reabsorption in the DCT is dependent on the delivered load of NaCl (Khuri et al. 1975). The DCT responds to chronic increases in the delivery of NaCl with an increase in the capacity for NaCl cotransport, as well as marked ultrastructural changes in the DCT cell. These morphologic changes include an increase in the size of DCT cells, increased basolateral membrane surface area, and an increase in DCT mitochondrial mass (Ellison et al. 1989). Accompanying the functional and morphologic changes are an increase in basolateral a Na^+/K^+-ATPase activity, and an increase in thiazide-binding sites (Chen et al. 1990). It is not clear if these changes are due to transcriptional or post-translational effects on NCC expression, although they do appear to be a consequence of increased sodium entry into the DCT. Inhibition of NaCl entry into DCT cells with chronic thiazide treatment resulted in a loss of cell height, and polarity, and triggered apoptosis (Morsing et al. 1991). The cellular mechanisms whereby NaCl entry affects NCC transport function and DCT morphology are currently not known.

3.9.4 Hormonal Regulation of NCC

3.9.4.1 Angiotensin II

The peptide hormone angiotensin II (AngII) plays a critically important role in blood pressure homeostasis and tubular salt handling. Ang II has dual effects on NCC trafficking and phosphorylation. Infusion of AngII into animals causes the cotransporter to traffic from intracellular vesicles to the plasma membrane, while the treatment of rodents with angiotensin-converting enzyme (ACE) inhibitors redistributes NCC to intracellular vesicles (Sandberg et al. 2007). The surface-expressed cotransporter appears to be highly phosphorylated, based on immunolocalization studies and biochemical inquiry of the aforementioned canonical SPAK/OSR1 phosphorylation sites. The WNK-SPAK/OSR1 pathway mediates these effects, as the stimulatory effects of angiotensin II are associated with SPAK and OSR1 activation (Talati et al. 2010). Moreover, in vitro and in vivo experiments implicate WNK4 as being critically important (Castaneda-Bueno and Gamba 2012; San-Cristobal et al. 2009). Studies in the *Xenopus* oocyte system suggest that AngII converts WNK4 from its inhibitory state into a SPAK/OSR1-dependent NCC activator, and NCC expression and phosphorylation are unaffected by AngII infusion in WNK4 knockout mice. NCC stimulation could be seen in adrenalectomized animals, suggesting that the AngII effect occurred independently of aldosterone (van der Lubbe et al. 2011).

The classic view of the renin–angiotensin system is that it functions as a whole-body blood pressure and volume control mechanism that acts through the concerted effects of multiple organs (Guyton 1991). Renin released from the kidney triggers the conversion of liver-derived angiotensinogen to angiotensin I, which in turn is converted to AngII by lung-expressed ACE (Atlas 2007). However, it is clear that an intrarenal renin–angiotensin system also exists within the kidney, and that this system is critically important for the effects of AngII on renal hemodynamics (Navar and Rosivall 1984). This intrinsic system has important effects on the activity of NCC since the kidney-specific ablation of intrarenal ACE decreases NCC activity and phosphorylation (Gonzalez-Villalobos et al. 2013).

3.9.4.2 Adrenal Steroids

Current evidence indicates that adrenal steroid hormones stimulate NCC-mediated NaCl cotransport. The presence of both mineralocorticoid and glucocorticoid receptors in the DCT has been demonstrated by immunohistochemistry and by hormone binding experiments (Farman and Bonvalet 1983; Farman et al. 1991). Although mineralocorticoid receptors are expressed throughout the entire DCT, only the "late" portion of this segment (also known as the DCT2) expresses the steroid metabolizing enzyme 11-β-hydroxysteroid dehydrogenase 2 (11-βHSD2) (Velazquez et al. 1998; Bostanjoglo et al. 1998). This enzyme metabolizes cortisol to the inactive metabolite cortisone, thereby preventing circulating glucocorticoids from binding to mineralocorticoid receptors expressed in the DCT2. Because the mineralocorticoid receptors in the DCT2 can only be occupied by aldosterone, the DCT2 is highly sensitive to changes in circulating aldosterone levels.

Early studies by Chen and Fanestil (Chen et al. 1994a), Velazquez et al. (1995), and Kim et al. (1998) collectively demonstrated that aldosterone increases total NCC protein abundance without affects its mRNA expression, indicating that mineralocorticoids stimulate NCC activity through post-translational mechanisms. More recently, McDonough (Sandberg et al. 2006) showed that mineralocorticoid receptor blockade decreases NCC trafficking to the plasma membrane while simultaneously increasing its accumulation in lysosomes. Thus, one way in which aldosterone can increase NCC activity is by enhancing its expression at the plasma membrane. The molecular mechanism likely involves the serum and glucocorticoid regulated kinase, SGK1. The transcription of this serine–threonine kinase is upregulated in response to aldosterone binding to the mineralocorticoid receptor (Pearce 2001). The active form of SGK1 has been reported to regulate NCC trafficking via two mechanisms. First, in in vitro systems, SGK1 can reverse the inhibitory effect of overexpressed WNK4 on NCC delivery to the plasma membrane (Rozansky et al. 2009). Second, SGK1 phosphorylates and inactivates Nedd4-2, an E3 ubiquitin ligase which ubiquitylates and negatively regulates NCC surface expression (Arroyo et al. 2011). Although the first of these two findings has only been shown in the *Xenopus* oocyte expression system (Rozansky et al. 2009), the in vitro inhibitory effect of Nedd4–2 on NCC has been verified in mouse models (Arroyo et al. 2011; Ronzaud et al. 2013).

Aldosterone also appears to stimulate NCC transport activity by promoting NCC phosphorylation. Chiga et al. (2008) showed that NCC phosphorylation is dynamically regulated by adjustments in dietary NaCl intake and that this effect is at least partially dependent on the mineralocorticoid receptor (MR). Vallon et al. echoed these findings 1 year later (Vallon et al. 2009). In both reports, careful measurements of NCC abundance demonstrated enhanced NCC phosphorylation at all of its canonical SPAK/OSR1 phosphorylation residues. Consistent with this, aldosterone increased SPAK and OSR1 activation in a mineralocorticoid receptor-dependent manner (Chiga et al. 2008). A study by van der Lubbe and colleagues (van der Lubbe et al. 2012) implicated WNK4 as a mediator of the stimulatory effect of aldosterone on SPAK/OSR1 and NCC phosphorylation, though other WNK kinases expressed in the DCT may also be involved. Indeed, the E3 ubiquitin ligase Nedd4-2 binds to and ubiquitylates WNK1, and that this process is blocked by aldosterone, leading to an increase in WNK1 abundance in the DCT, providing a potential mechanism by which mineralocorticoids might augment NCC phosphorylation status (Roy et al. 2015). A more rapid and complex signal for aldosterone action involving cAMP, IP3, GPR30, and EGF Receptor-dependent networks was also recently described (Cheng et al. 2019).

Glucocorticoids such as dexamethasone increase thiazide-sensitive NaCl transport and increase the number of metolazone binding sites in the kidneys of adrenalectomized rodents (Chen et al. 1994a). This strongly suggests that glucocorticoids stimulate NCC-mediated salt transport in the DCT. The DCT1 may be the more relevant site of action of glucocorticoids, since, as mentioned above, the levels of active cortisol in the DCT2 should be relatively low due to its metabolism by 11β-HSD (Bostanjoglo et al. 1998). The WNK-SPAK/OSR1 pathway is expressed in both DCT1 and DCT2; thus, many of the same players involved in mineralocorticoid receptor-dependent regulation of NCC are likely involved. Although NCC overactivity may participate in the salt-sensitive hypertension seen in disorders of cortisol excess such as Cushing's syndrome, the role of glucocorticoids in the physiological regulation of Na^+ transport in the DCT remains understudied.

3.9.4.3 Gonadal Steroids and NCC

Gonadal steroid hormones may also influence NaCl cotransport in the DCT. Chen et al. (1994b) reported gender differences in the density of thiazide binding sites, and in the natriuretic response of thiazides in rodents. Specifically, female rats had higher levels of "thiazide receptors" in the renal cortex than males. The number of thiazide-binding sites decreased following ovariectomy, while levels in males rose following orchiectomy. Consistent with these results, Verlander et al. found that estradiol treatments increased NCC expression in the DCT (Verlander et al. 1998). Collectively, these results are consistent with the view that male sex hormones (e.g., testosterone) may downregulate NCC expression and salt transport, while estrogens increase NCC expression and salt transport in the DCT. The physiological relevance of these findings with respect to human hypertension is currently unclear, as no clear gender difference in the response of humans to thiazide diuretics has been established.

3.9.4.4 Vasopressin

Recent work indicates that vasopressin is a potent NCC activator (Gamba 2010). The vasopressin analog deamino-Cys-1, d-Arg-8 vasopressin (dDAVP) triggers an increase in NCC abundance, trafficking to the plasma membrane, and activation through SPAK/OSR1-mediated phosphorylation of its amino terminus (Saritas et al. 2013; Mutig et al. 2010; Pedersen et al. 2010). The effect probably requires cyclic

AMP and protein kinase A (PKA), well-established intermediaries of vasopressin-dependent signaling (Nedvetsky et al. 2009). The effect on NCC abundance may be a Nedd4–2-dependent process, since PKA can phosphorylate and inactivate Nedd4–2 through mechanisms similar to SGK1 (Snyder et al. 2004); however, this hypothesis is yet to be tested.

3.9.4.5 Insulin

One of the first studies suggesting a potential link between insulin and NCC was performed by Bickel et al., who noted that insulin-resistant obese Zucker rats retain salt and exhibit increased NCC abundance (Bickel et al. 2001). Subsequent studies confirmed that NCC is hyperphosphorylated in these animals, caused by an increase in total SPAK/OSR1 activity (Komers et al. 2012; (Chavez-Canales et al. 2013). Similar results were noted in another model of type 2 diabetes, the leptin receptor-deficient db/db mouse. Crossing db/db mice with SPAK/OSR1 catalytically inactive "T-loop" knock-in mice that cannot be activated by WNKs, the stimulatory effect of insulin on NCC was completely abrogated (Nishida et al. 2012). This indicates that insulin activates NCC via the WNK-SPAK/OSR1 pathway. Although it seems clear that SPAK/OSR1 is the key downstream mediator of NCC stimulation, there have been conflicting data with regards to which of the upstream WNK kinases mediate the effect. Data from Chavez-Canales et al. (2013) suggest that WNK3 is required for insulin to stimulate NCC activity, while studies by Sohara et al. (2011) and Takahashi et al. (2014) provide evidence that WNK4 is the dominant WNK kinase that mediates the effect. The phosphoinositol 3-kinase/-mTORC2-Akt pathway transduces the upstream signal from insulin receptors to the WNK-SPAK/OSR1 network (Nishida et al. 2012; Chavez-Canales et al. 2013).

3.10 Concluding Remarks

Although interest in the thiazide-sensitive cotransporter had only been modest for years, exciting genetic studies, the discovery of key NCC regulators, and the advent of tools to biochemically monitor NCC activation status have yielded an explosion of new insights into its role in physiology. Furthermore, the recent linkage of KLHL3 and CUL3 to FHHt, as well as the generation of new genetic models to study the WNK-SPAK/OSR1 pathway has opened up new and important questions and avenues of investigation. As the underlying molecular wiring that connects important blood pressure, potassium, and volume control systems to NCC in the DCT is elucidated further, perhaps new targets for the treatment of clinical disorders such as salt-sensitive hypertension, hyperkalemia, edema, and nephrolithiasis will be established.

Acknowledgments This work was supported by grants from the NIH (R01-DK098145 and R01-DK119252), and by the VA Pittsburgh Healthcare System Research Service.

References

Acuna R, Martinez-de-la-Maza L, Ponce-Coria J, Vazquez N, Ortal-Vite P, Pacheco-Alvarez D, Bobadilla NA, Gamba G (2011) Rare mutations in SLC12A1 and SLC12A3 protect against hypertension by reducing the activity of renal salt cotransporters. J Hypertens 29(3):475–483. https://doi.org/10.1097/HJH.0b013e328341d0fd

Argaiz ER, Chavez-Canales M, Ostrosky-Frid M, Rodriguez-Gama A, Vazquez N, Gonzalez-Rodriguez X, Garcia-Valdes J, Hadchouel J, Ellison D, Gamba G (2018) Kidney-specific WNK1 isoform (KS-WNK1) is a potent activator of WNK4 and NCC. Am J Physiol Renal Physiol 315(3):F734–F745. https://doi.org/10.1152/ajprenal.00145.2018

Arroyo JP, Lagnaz D, Ronzaud C, Vazquez N, Ko BS, Moddes L, Ruffieux-Daidie D, Hausel P, Koesters R, Yang B, Stokes JB, Hoover RS, Gamba G, Staub O (2011) Nedd4-2 modulates renal Na^+-Cl^- cotransporter via the aldosterone-SGK1-Nedd4-2 pathway. J Am Soc Nephrol 22 (9):1707–1719. https://doi.org/10.1681/ASN.2011020132

Atlas SA (2007) The renin-angiotensin aldosterone system: pathophysiological role and pharmacologic inhibition. JMCP 13(8 Suppl B):9–20

Baggio B, Bordin L, Clari G, Gambaro G, Moret V (1993) Functional correlation between the Ser/Thr-phosphorylation of band-3 and band-3-mediated transmembrane anion transport in human erythrocytes. Biochim Biophys Acta 1148(1):157–160

Bickel CA, Verbalis JG, Knepper MA, Ecelbarger CA (2001) Increased renal Na-K-ATPase, NCC, and beta-ENaC abundance in obese Zucker rats. Am J Physiol Renal Physiol 281(4):F639–F648

Bostanjoglo M, Reeves WB, Reilly RF, Velazquez H, Robertson N, Litwack G, Morsing P, Dorup J, Bachmann S, Ellison DH (1998) 11Beta-hydroxysteroid dehydrogenase, mineralocorticoid receptor, and thiazide-sensitive Na-Cl cotransporter expression by distal tubules. J Am Soc Nephrol 9(8):1347–1358

Boyden LM, Choi M, Choate KA, Nelson-Williams CJ, Farhi A, Toka HR, Tikhonova IR, Bjornson R, Mane SM, Colussi G, Lebel M, Gordon RD, Semmekrot BA, Poujol A, Valimaki MJ, De Ferrari ME, Sanjad SA, Gutkin M, Karet FE, Tucci JR, Stockigt JR, Keppler-Noreuil KM, Porter CC, Anand SK, Whiteford ML, Davis ID, Dewar SB, Bettinelli A, Fadrowski JJ, Belsha CW, Hunley TE, Nelson RD, Trachtman H, Cole TR, Pinsk M, Bockenhauer D, Shenoy M, Vaidyanathan P, Foreman JW, Rasoulpour M, Thameem F, Al-Shahrouri HZ, Radhakrishnan J, Gharavi AG, Goilav B, Lifton RP (2012) Mutations in kelch-like 3 and cullin 3 cause hypertension and electrolyte abnormalities. Nature 482(7383):98–102. https://doi.org/10.1038/nature10814

Boyd-Shiwarski CR, Shiwarski DJ, Roy A, Namboodiri HN, Nkashama LJ, Xie J, McClain KL, Marciszyn A, Kleyman TR, Tan RJ, Stolz DB, Puthenveedu MA, Huang CL, Subramanya AR (2018) Potassium-regulated distal tubule WNK bodies are kidney-specific WNK1 dependent. Mol Biol Cell 29(4):499–509. https://doi.org/10.1091/mbc.E17-08-0529

Brown RH Jr (1991) Molecular basis for hyperkalemic periodic paralysis. Int J Neurol 25-26:89–96

Butterworth MB, Edinger RS, Johnson JP, Frizzell RA (2005) Acute ENaC stimulation by cAMP in a kidney cell line is mediated by exocytic insertion from a recycling channel pool. J Gen Physiol 125(1):81–101. https://doi.org/10.1085/jgp.200409124

Cai H, Cebotaru V, Wang YH, Zhang XM, Cebotaru L, Guggino SE, Guggino WB (2006) WNK4 kinase regulates surface expression of the human sodium chloride cotransporter in mammalian cells. Kidney Int 69(12):2162–2170

Castaneda-Bueno M, Gamba G (2012) Mechanisms of sodium-chloride cotransporter modulation by angiotensin II. Curr Opin Nephrol Hypertens 21(5):516–522. https://doi.org/10.1097/MNH.0b013e32835571a4

Castaneda-Bueno M, Vazquez N, Bustos-Jaimes I, Hernandez D, Rodriguez-Lobato E, Pacheco-Alvarez D, Carino-Cortes R, Moreno E, Bobadilla NA, Gamba G (2010) A single residue in transmembrane domain 11 defines the different affinity for thiazides between the mammalian and flounder NaCl transporters. Am J Physiol Renal Physiol 299(5):F1111–F1119. https://doi.org/10.1152/ajprenal.00412.2010

Chavez-Canales M, Arroyo JP, Ko B, Vazquez N, Bautista R, Castaneda-Bueno M, Bobadilla NA, Hoover RS, Gamba G (2013) Insulin increases the functional activity of the renal NaCl cotransporter. J Hypertens 31(2):303–311. https://doi.org/10.1097/HJH.0b013e32835bbb83

Chavez-Canales M, Zhang C, Soukaseum C, Moreno E, Pacheco-Alvarez D, Vidal-Petiot E, Castaneda-Bueno M, Vazquez N, Rojas-Vega L, Meermeier NP, Rogers S, Jeunemaitre X, Yang CL, Ellison DH, Gamba G, Hadchouel J (2014) WNK-SPAK-NCC cascade revisited: WNK1 stimulates the activity of the Na-Cl cotransporter via SPAK, an effect antagonized by WNK4. Hypertension 64(5):1047–1053. https://doi.org/10.1161/HYPERTENSIONAHA.114.04036

Chen ZF, Vaughn DA, Beaumont K, Fanestil DD (1990) Effects of diuretic treatment and of dietary sodium on renal binding of 3H-metolazone. J Am Soc Nephrol 1(1):91–98

Chen Z, Vaughn DA, Blakely P, Fanestil DD (1994a) Adrenocortical steroids increase renal thiazide diuretic receptor density and response. J Am Soc Nephrol 5(6):1361–1368

Chen Z, Vaughn DA, Fanestil DD (1994b) Influence of gender on renal thiazide diuretic receptor density and response. J Am Soc Nephrol 5(4):1112–1119

Chen JC, Lo YF, Lin YW, Lin SH, Huang CL, Cheng CJ (2019) WNK4 kinase is a physiological intracellular chloride sensor. Proc Natl Acad Sci USA 116(10):4502–4507. https://doi.org/10.1073/pnas.1817220116

Cheng L, Poulsen SB, Wu Q, Esteva-Font C, Olesen ETB, Peng L, Olde B, Leeb-Lundberg LMF, Pisitkun T, Rieg T, Dimke H, Fenton RA (2019) Rapid aldosterone-mediated signaling in the DCT increases activity of the thiazide-sensitive NaCl cotransporter. J Am Soc Nephrol 30 (8):1454–1470. https://doi.org/10.1681/ASN.2018101025

Chew TA, Orlando BJ, Zhang J, Latorraca NR, Wang A, Hollingsworth SA, Chen DH, Dror RO, Liao M, Feng L (2019) Structure and mechanism of the cation-chloride cotransporter NKCC1. Nature 572(7770):488–492. https://doi.org/10.1038/s41586-019-1438-2

Chiga M, Rai T, Yang SS, Ohta A, Takizawa T, Sasaki S, Uchida S (2008) Dietary salt regulates the phosphorylation of OSR1/SPAK kinases and the sodium chloride cotransporter through aldosterone. Kidney Int 74(11):1403–1409. https://doi.org/10.1038/ki.2008.451

Costanzo LS (1988) Mechanism of action of thiazide diuretics. Semin Nephrol 8(3):234–241

Costanzo LS, Windhager EE (1978) Calcium and sodium transport by the distal convoluted tubule of the rat. Am J Phys 235(5):F492–F506

Darman RB, Forbush B (2002) A regulatory locus of phosphorylation in the N terminus of the Na-K-Cl cotransporter, NKCC1. J Biol Chem 277(40):37542–37550. https://doi.org/10.1074/jbc.M206293200

de Jong JC, Willems PH, Mooren FJ, van den Heuvel LP, Knoers NV, Bindels RJ (2003) The structural unit of the thiazide-sensitive NaCl cotransporter is a homodimer. J Biol Chem 278 (27):24302–24307. https://doi.org/10.1074/jbc.M303101200

Delaloy C, Lu J, Houot AM, Disse-Nicodeme S, Gasc JM, Corvol P, Jeunemaitre X (2003) Multiple promoters in the WNK1 gene: one controls expression of a kidney-specific kinase-defective isoform. Mol Cell Biol 23(24):9208–9221

Dell'Angelica EC, Ohno H, Ooi CE, Rabinovich E, Roche KW, Bonifacino JS (1997) AP-3: an adaptor-like protein complex with ubiquitous expression. EMBO J 16(5):917–928

Delpire E, Gagnon KB (2006) SPAK and OSR1, key kinases involved in the regulation of chloride transport. Acta Physiol (Oxf) 187(1–2):103–113

Dimke H, San-Cristobal P, de Graaf M, Lenders JW, Deinum J, Hoenderop JG, Bindels RJ (2011) γ-Adducin stimulates the thiazide-sensitive NaCl cotransporter. J Am Soc Nephrol 22 (3):508–517. https://doi.org/10.1681/ASN.2010060606
Donnelly BF, Needham PG, Snyder AC, Roy A, Khadem S, Brodsky JL, Subramanya AR (2013) Hsp70 and Hsp90 multichaperone complexes sequentially regulate thiazide-sensitive cotransporter endoplasmic reticulum-associated degradation and biogenesis. J Biol Chem 288 (18):13124–13135. https://doi.org/10.1074/jbc.M113.455394
Earley LE, Orloff J (1964) Thiazide diuretics. Annu Rev Med 15:149. https://doi.org/10.1146/Annurev.Me.15.020164.001053
Ellison DH (2003) The thiazide-sensitive Na-Cl cotransporter and human disease: reemergence of an old player. J Am Soc Nephrol 14(2):538–540
Ellison DH, Terker AS (2015) Why your mother was right: how potassium intake reduces blood pressure. Trans Am Clin Climatol Assoc 126:46–55
Ellison DH, Velazquez H, Wright FS (1987) Thiazide-sensitive sodium chloride cotransport in early distal tubule. Am J Phys 253(3 Pt 2):F546–F554
Ellison DH, Velazquez H, Wright FS (1989) Adaptation of the distal convoluted tubule of the rat. Structural and functional effects of dietary salt intake and chronic diuretic infusion. J Clin Invest 83(1):113–126. https://doi.org/10.1172/JCI113847
Farman N, Bonvalet JP (1983) Aldosterone binding in isolated tubules. III Autoradiography along the rat nephron. Am J Physiol 245(5 Pt 1):F606–F614
Farman N, Oblin ME, Lombes M, Delahaye F, Westphal HM, Bonvalet JP, Gasc JM (1991) Immunolocalization of gluco- and mineralocorticoid receptors in rabbit kidney. Am J Phys 260(2 Pt 1):C226–C233
Flatman PW (2002) Regulation of Na-K-2Cl cotransport by phosphorylation and protein-protein interactions. Biochim Biophys Acta 1566(1–2):140–151
Freeman RB, Maher JF, Schreiner GE, Mostofi FK (1962) Renal tubular necrosis due to nephrotoxicity of organic mercurial diuretics. Ann Intern Med 57:34–43
Gagnon KB, England R, Delpire E (2006) Characterization of SPAK and OSR1, regulatory kinases of the Na-K-2Cl cotransporter. Mol Cell Biol 26(2):689–698
Gamba G (2005) Molecular physiology and pathophysiology of electroneutral cation-chloride cotransporters. Physiol Rev 85(2):423–493. https://doi.org/10.1152/physrev.00011.2004
Gamba G (2010) Vasopressin regulates the renal Na^{+}-Cl^{-} cotransporter. Am J Physiol Renal Physiol 298(3):F500–F501. https://doi.org/10.1152/ajprenal.00723.2009
Gamba G, Saltzberg SN, Lombardi M, Miyanoshita A, Lytton J, Hediger MA, Brenner BM, Hebert SC (1993) Primary structure and functional expression of a cDNA encoding the thiazide-sensitive, electroneutral sodium-chloride cotransporter. Proc Natl Acad Sci USA 90 (7):2749–2753
Gamba G, Miyanoshita A, Lombardi M, Lytton J, Lee WS, Hediger MA, Hebert SC (1994) Molecular cloning, primary structure, and characterization of two members of the mammalian electroneutral sodium-(potassium)-chloride cotransporter family expressed in kidney. J Biol Chem 269(26):17713–17722
Geva Y, Schuldiner M (2014) The back and forth of cargo exit from the endoplasmic reticulum. Curr Biol 24(3):R130–R136. https://doi.org/10.1016/j.cub.2013.12.008
Glover M, Mercier Zuber A, Figg N, O'Shaughnessy KM (2010) The activity of the thiazide-sensitive Na(+)-Cl(−) cotransporter is regulated by protein phosphatase PP4. Can J Physiol Pharmacol 88(10):986–995. https://doi.org/10.1139/y10-080
Glover M, Ware JS, Henry A, Wolley M, Walsh R, Wain LV, Xu S, Van't Hoff WG, Tobin MD, Hall IP, Cook S, Gordon RD, Stowasser M, O'Shaughnessy KM (2014) Detection of mutations in KLHL3 and CUL3 in families with FHHt (familial hyperkalaemic hypertension or Gordon's syndrome). Clin Sci (Lond) 126(10):721–726. https://doi.org/10.1042/CS20130326
Glozman R, Okiyoneda T, Mulvihill CM, Rini JM, Barriere H, Lukacs GL (2009) N-glycans are direct determinants of CFTR folding and stability in secretory and endocytic membrane traffic. J Cell Biol 184(6):847–862. https://doi.org/10.1083/jcb.200808124

Golbang AP, Cope G, Hamad A, Murthy M, Liu CH, Cuthbert AW, O'Shaughnessy KM (2006) Regulation of the expression of the Na/Cl cotransporter (NCCT) by WNK4 and WNK1: evidence that accelerated dynamin-dependent endocytosis is not involved. Am J Physiol Renal Physiol. https://doi.org/10.1152/ajprenal.00468.2005

Gonzalez-Villalobos RA, Janjoulia T, Fletcher NK, Giani JF, Nguyen MT, Riquier-Brison AD, Seth DM, Fuchs S, Eladari D, Picard N, Bachmann S, Delpire E, Peti-Peterdi J, Navar LG, Bernstein KE, McDonough AA (2013) The absence of intrarenal ACE protects against hypertension. J Clin Invest 123(5):2011–2023. https://doi.org/10.1172/JCI65460

Gordon RD (1986) Syndrome of hypertension and hyperkalemia with normal glomerular filtration rate. Hypertension 8(2):93–102

Grimm PR, Taneja TK, Liu J, Coleman R, Chen YY, Delpire E, Wade JB, Welling PA (2012) SPAK isoforms and OSR1 regulate sodium-chloride co-transporters in a nephron-specific manner. J Biol Chem 287(45):37673–37690. https://doi.org/10.1074/jbc.M112.402800

Guyton AC (1991) Blood pressure control—special role of the kidneys and body fluids. Science 252(5014):1813–1816

Hadchouel J, Delaloy C, Faure S, Achard JM, Jeunemaitre X (2006) Familial hyperkalemic hypertension. J Am Soc Nephrol 17(1):208–217. https://doi.org/10.1681/ASN.2005030314

Hadchouel J, Soukaseum C, Busst C, Zhou XO, Baudrie V, Zurrer T, Cambillau M, Elghozi JL, Lifton RP, Loffing J, Jeunemaitre X (2010) Decreased ENaC expression compensates the increased NCC activity following inactivation of the kidney-specific isoform of WNK1 and prevents hypertension. Proc Natl Acad Sci USA 107(42):18109–18114. https://doi.org/10.1073/pnas.1006128107

Hoorn EJ, Walsh SB, McCormick JA, Furstenberg A, Yang CL, Roeschel T, Paliege A, Howie AJ, Conley J, Bachmann S, Unwin RJ, Ellison DH (2011) The calcineurin inhibitor tacrolimus activates the renal sodium chloride cotransporter to cause hypertension. Nat Med 17 (10):1304–1309. https://doi.org/10.1038/nm.2497

Hoover RS, Poch E, Monroy A, Vazquez N, Nishio T, Gamba G, Hebert SC (2003) N-Glycosylation at two sites critically alters thiazide binding and activity of the rat thiazide-sensitive Na(+):Cl(−) cotransporter. J Am Soc Nephrol 14(2):271–282

Ji W, Foo JN, O'Roak BJ, Zhao H, Larson MG, Simon DB, Newton-Cheh C, State MW, Levy D, Lifton RP (2008) Rare independent mutations in renal salt handling genes contribute to blood pressure variation. Nat Genet 40(5):592–599

Khuri RN, Strieder N, Wiederholt M, Giebisch G (1975) Effects of graded solute diuresis on renal tubular sodium transport in the rat. Am J Phys 228(4):1262–1268

Kim GH, Masilamani S, Turner R, Mitchell C, Wade JB, Knepper MA (1998) The thiazide-sensitive Na-Cl cotransporter is an aldosterone-induced protein. Proc Natl Acad Sci USA 95 (24):14552–14557

Ko B, Joshi LM, Cooke LL, Vazquez N, Musch MW, Hebert SC, Gamba G, Hoover RS (2007) Phorbol ester stimulation of RasGRP1 regulates the sodium-chloride cotransporter by a PKC-independent pathway. Proc Natl Acad Sci USA 104(50):20120–20125. https://doi.org/10.1073/pnas.0709506104

Ko B, Kamsteeg EJ, Cooke LL, Moddes LN, Deen PM, Hoover RS (2010) RasGRP1 stimulation enhances ubiquitination and endocytosis of the sodium-chloride cotransporter. Am J Physiol Renal Physiol 299(2):F300–F309. https://doi.org/10.1152/ajprenal.00441.2009

Ko B, Mistry AC, Hanson L, Mallick R, Cooke LL, Hack BK, Cunningham P, Hoover RS (2012) A new model of the distal convoluted tubule. Am J Physiol Renal Physiol 303(5):F700–F710. https://doi.org/10.1152/ajprenal.00139.2012

Komers R, Rogers S, Oyama TT, Xu B, Yang CL, McCormick J, Ellison DH (2012) Enhanced phosphorylation of Na(+)-Cl^{-} co-transporter in experimental metabolic syndrome: role of insulin. Clin Sci (Lond) 123(11):635–647. https://doi.org/10.1042/CS20120003

Krishnamurthy H, Gouaux E (2012) X-ray structures of LeuT in substrate-free outward-open and apo inward-open states. Nature 481(7382):469–474. https://doi.org/10.1038/nature10737

Krishnamurthy H, Piscitelli CL, Gouaux E (2009) Unlocking the molecular secrets of sodium-coupled transporters. Nature 459(7245):347–355. https://doi.org/10.1038/nature08143
Kunau R (1974) Nature of Chloruretic effect of thiazide diuretics. Kidney Int 6(6):A62–A63
Kunchaparty S, Palcso M, Berkman J, Velazquez H, Desir GV, Bernstein P, Reilly RF, Ellison DH (1999) Defective processing and expression of thiazide-sensitive Na-Cl cotransporter as a cause of Gitelman's syndrome. Am J Phys 277(4 Pt 2):F643–F649
Lalioti MD, Zhang J, Volkman HM, Kahle KT, Hoffmann KE, Toka HR, Nelson-Williams C, Ellison DH, Flavell R, Booth CJ, Lu Y, Geller DS, Lifton RP (2006) Wnk4 controls blood pressure and potassium homeostasis via regulation of mass and activity of the distal convoluted tubule. Nat Genet 38(10):1124–1132. https://doi.org/10.1038/ng1877
Leaphart CL, Dai S, Gribar SC, Richardson W, Ozolek J, Shi XH, Bruns JR, Branca M, Li J, Weisz OA, Sodhi C, Hackam DJ (2008) Interferon-gamma inhibits enterocyte migration by reversibly displacing connexin43 from lipid rafts. Am J Physiol Gastrointest Liver Physiol 295(3):G559–G569. https://doi.org/10.1152/ajpgi.90320.2008
Lee DH, Riquier AD, Yang LE, Leong PK, Maunsbach AB, McDonough AA (2009) Acute hypertension provokes acute trafficking of distal tubule Na-Cl cotransporter (NCC) to subapical cytoplasmic vesicles. Am J Physiol Renal Physiol 296(4):F810–F818. https://doi.org/10.1152/ajprenal.90606.2008
Lee DH, Maunsbach AB, Riquier-Brison AD, Nguyen MT, Fenton RA, Bachmann S, Yu AS, McDonough AA (2013) Effects of ACE inhibition and ANG II stimulation on renal Na-Cl cotransporter distribution, phosphorylation, and membrane complex properties. Am J Physiol Cell Physiol 304(2):C147–C163. https://doi.org/10.1152/ajpcell.00287.2012
Lin SH, Yu IS, Jiang ST, Lin SW, Chu P, Chen A, Sytwu HK, Sohara E, Uchida S, Sasaki S, Yang SS (2011) Impaired phosphorylation of Na(+)-K(+)-2Cl(−) cotransporter by oxidative stress-responsive kinase-1 deficiency manifests hypotension and Bartter-like syndrome. Proc Natl Acad Sci USA 108(42):17538–17543. https://doi.org/10.1073/pnas.1107452108
Liu Z, Xie J, Wu T, Truong T, Auchus RJ, Huang CL (2011) Downregulation of NCC and NKCC2 cotransporters by kidney-specific WNK1 revealed by gene disruption and transgenic mouse models. Hum Mol Genet 20(5):855–866. https://doi.org/10.1093/hmg/ddq525
Louis-Dit-Picard H, Barc J, Trujillano D, Miserey-Lenkei S, Bouatia-Naji N, Pylypenko O, Beaurain G, Bonnefond A, Sand O, Simian C, Vidal-Petiot E, Soukaseum C, Mandet C, Broux F, Chabre O, Delahousse M, Esnault V, Fiquet B, Houillier P, Bagnis CI, Koenig J, Konrad M, Landais P, Mourani C, Niaudet P, Probst V, Thauvin C, Unwin RJ, Soroka SD, Ehret G, Ossowski S, Caulfield M, International Consortium for Blood P, Bruneval P, Estivill X, Froguel P, Hadchouel J, Schott JJ, Jeunemaitre X (2012) KLHL3 mutations cause familial hyperkalemic hypertension by impairing ion transport in the distal nephron. Nat Genet 44 (4):456–460. https://doi.org/10.1038/ng.2218
Lytle C, Forbush B (1996) Regulatory phosphorylation of the secretory Na-K-Cl cotransporter: modulation by cytoplasmic Cl. Am J Phys 270(2 Pt 1):C437–C448
Markadieu N, Rios K, Spiller BW, McDonald WH, Welling PA, Delpire E (2014) Short forms of SPAK in the kidney are created by Dnpep-mediated proteolytic cleavage. J Biol Chem. https://doi.org/10.1074/jbc.M114.604009
Mayan H, Vered I, Mouallem M, Tzadok-Witkon M, Pauzner R, Farfel Z (2002) Pseudohypoaldosteronism type II: marked sensitivity to thiazides, hypercalciuria, normomagnesemia, and low bone mineral density. J Clin Endocrinol Metab 87(7):3248–3254
McCormick JA, Yang CL, Ellison DH (2008) WNK kinases and renal sodium transport in health and disease: an integrated view. Hypertension 51(3):588–596. https://doi.org/10.1161/HYPERTENSIONAHA.107.103788
McCormick JA, Mutig K, Nelson JH, Saritas T, Hoorn EJ, Yang CL, Rogers S, Curry J, Delpire E, Bachmann S, Ellison DH (2011) A SPAK isoform switch modulates renal salt transport and blood pressure. Cell Metab 14(3):352–364. https://doi.org/10.1016/j.cmet.2011.07.009
McCormick JA, Yang CL, Zhang C, Davidge B, Blankenstein KI, Terker AS, Yarbrough B, Meermeier NP, Park HJ, McCully B, West M, Borschewski A, Himmerkus N, Bleich M,

Bachmann S, Mutig K, Argaiz ER, Gamba G, Singer JD, Ellison DH (2014) Hyperkalemic hypertension-associated cullin 3 promotes WNK signaling by degrading KLHL3. J Clin Invest 124(11):4723–4736. https://doi.org/10.1172/JCI76126

McDonough AA, Youn JH (2013) Need to quickly excrete K^+? Turn off NCC. Kidney Int 83 (5):779–782. https://doi.org/10.1038/ki.2012.468

Mederle K, Mutig K, Paliege A, Carota I, Bachmann S, Castrop H, Oppermann M (2013) Loss of WNK3 is compensated for by the WNK1/SPAK axis in the kidney of the mouse. Am J Physiol Renal Physiol 304(9):F1198–F1209. https://doi.org/10.1152/ajprenal.00288.2012

Min X, Lee BH, Cobb MH, Goldsmith EJ (2004) Crystal structure of the kinase domain of WNK1, a kinase that causes a hereditary form of hypertension. Structure 12(7):1303–1311. https://doi.org/10.1016/j.str.2004.04.014

Mistry AC, Wynne BM, Yu L, Tomilin V, Yue Q, Zhou Y, Al-Khalili O, Mallick R, Cai H, Alli AA, Ko B, Mattheyses A, Bao HF, Pochynyuk O, Theilig F, Eaton DC, Hoover RS (2016) The sodium chloride cotransporter (NCC) and epithelial sodium channel (ENaC) associate. Biochem J 473(19):3237–3252. https://doi.org/10.1042/BCJ20160312

Moreno E, Cristobal PS, Rivera M, Vazquez N, Bobadilla NA, Gamba G (2006) Affinity-defining domains in the Na-Cl cotransporter: a different location for Cl^- and thiazide binding. J Biol Chem 281(25):17266–17275. https://doi.org/10.1074/jbc.M602614200

Morsing P, Velazquez H, Wright FS, Ellison DH (1991) Adaptation of distal convoluted tubule of rats. II Effects of chronic thiazide infusion. Am J Physiol 261(1 Pt 2):F137–F143

Mutig K, Saritas T, Uchida S, Kahl T, Borowski T, Paliege A, Bohlick A, Bleich M, Shan Q, Bachmann S (2010) Short-term stimulation of the thiazide-sensitive Na^+-Cl^- cotransporter by vasopressin involves phosphorylation and membrane translocation. Am J Physiol Renal Physiol 298(3):F502–F509. https://doi.org/10.1152/ajprenal.00476.2009

Navar LG, Rosivall L (1984) Contribution of the renin-angiotensin system to the control of intrarenal hemodynamics. Kidney Int 25(6):857–868

Nedvetsky PI, Tamma G, Beulshausen S, Valenti G, Rosenthal W, Klussmann E (2009) Regulation of aquaporin-2 trafficking. Handb Exp Pharmacol 190:133–157. https://doi.org/10.1007/978-3-540-79885-9_6

Needham PG, Mikoluk K, Dhakarwal P, Khadem S, Snyder AC, Subramanya AR, Brodsky JL (2011) The thiazide-sensitive NaCl cotransporter is targeted for chaperone-dependent endoplasmic reticulum-associated degradation. J Biol Chem 286(51):43611–43621. https://doi.org/10.1074/jbc.M111.288928

Nishida H, Sohara E, Nomura N, Chiga M, Alessi DR, Rai T, Sasaki S, Uchida S (2012) Phosphatidylinositol 3-kinase/Akt signaling pathway activates the WNK-OSR1/SPAK-NCC phosphorylation cascade in hyperinsulinemic db/db mice. Hypertension 60(4):981–990. https://doi.org/10.1161/HYPERTENSIONAHA.112.201509

Novello FC, Sprague JM (1957) Benzothiadiazine dioxides as novel diuretics. J Am Chem Soc 79 (8):2028–2029. https://doi.org/10.1021/Ja01565a079

O'Reilly M, Marshall E, Speirs HJ, Brown RW (2003) WNK1, a gene within a novel blood pressure control pathway, tissue-specifically generates radically different isoforms with and without a kinase domain. J Am Soc Nephrol 14(10):2447–2456

O'Reilly M, Marshall E, Macgillivray T, Mittal M, Xue W, Kenyon CJ, Brown RW (2006) Dietary electrolyte-driven responses in the renal WNK kinase pathway in vivo. J Am Soc Nephrol 17 (9):2402–2413

Ohta A, Schumacher FR, Mehellou Y, Johnson C, Knebel A, Macartney TJ, Wood NT, Alessi DR, Kurz T (2013) The CUL3-KLHL3 E3 ligase complex mutated in Gordon's hypertension syndrome interacts with and ubiquitylates WNK isoforms: disease causing mutations in KLHL3 and WNK4 disrupt interaction. Biochem J 451(1):111–122. https://doi.org/10.1042/BJ20121903

Oi K, Sohara E, Rai T, Misawa M, Chiga M, Alessi DR, Sasaki S, Uchida S (2012) A minor role of WNK3 in regulating phosphorylation of renal NKCC2 and NCC co-transporters in vivo. Biol Open 1(2):120–127. https://doi.org/10.1242/bio.2011048

Pacheco-Alvarez D, Cristobal PS, Meade P, Moreno E, Vazquez N, Munoz E, Diaz A, Juarez ME, Gimenez I, Gamba G (2006) The Na^+:Cl^- cotransporter is activated and phosphorylated at the amino-terminal domain upon intracellular chloride depletion. J Biol Chem 281 (39):28755–28763. https://doi.org/10.1074/jbc.M603773200
Paver WK, Pauline GJ (1964) Hypertension and hyperpotassaemia without renal disease in a young male. Med J Aust 2:305–306
Pearce D (2001) The role of SGK1 in hormone-regulated sodium transport. Trends Endocrinol Metab 12(8):341–347
Pedersen NB, Hofmeister MV, Rosenbaek LL, Nielsen J, Fenton RA (2010) Vasopressin induces phosphorylation of the thiazide-sensitive sodium chloride cotransporter in the distal convoluted tubule. Kidney Int 78(2):160–169. https://doi.org/10.1038/ki.2010.130
Piala AT, Moon TM, Akella R, He H, Cobb MH, Goldsmith EJ (2014) Chloride sensing by WNK1 involves inhibition of autophosphorylation. Sci Signal 7(324):ra41. https://doi.org/10.1126/scisignal.2005050
Picard N, Trompf K, Yang CL, Miller RL, Carrel M, Loffing-Cueni D, Fenton RA, Ellison DH, Loffing J (2014) Protein phosphatase 1 inhibitor-1 deficiency reduces phosphorylation of renal NaCl cotransporter and causes arterial hypotension. J Am Soc Nephrol 25(3):511–522. https://doi.org/10.1681/ASN.2012121202
Piechotta K, Garbarini N, England R, Delpire E (2003) Characterization of the interaction of the stress kinase SPAK with the Na^+-K^+-$2Cl^-$ cotransporter in the nervous system: evidence for a scaffolding role of the kinase. J Biol Chem 278(52):52848–52856
Rafiqi FH, Zuber AM, Glover M, Richardson C, Fleming S, Jovanovic S, Jovanovic A, O'Shaughnessy KM, Alessi DR (2010) Role of the WNK-activated SPAK kinase in regulating blood pressure. EMBO Mol Med 2(2):63–75. https://doi.org/10.1002/emmm.200900058
Renfro JL (1977) Interdependence of active Na^+ and Cl^- transport by the isolated urinary bladder of the teleost, *Pseudopleuronectes americanus*. J Exp Zool 199(3):383–390. https://doi.org/10.1002/jez.1401990311
Rengarajan S, Lee DH, Oh YT, Delpire E, Youn JH, McDonough AA (2014) Increasing plasma [K+] by intravenous potassium infusion reduces NCC phosphorylation and drives kaliuresis and natriuresis. Am J Physiol Renal Physiol 306(9):F1059–F1068. https://doi.org/10.1152/ajprenal.00015.2014
Richardson C, Rafiqi FH, Karlsson HK, Moleleki N, Vandewalle A, Campbell DG, Morrice NA, Alessi DR (2008) Activation of the thiazide-sensitive Na^+-Cl^- cotransporter by the WNK-regulated kinases SPAK and OSR1. J Cell Sci 121(Pt 5):675–684. https://doi.org/10.1242/jcs.025312
Rinehart J, Kahle KT, de Los HP, Vazquez N, Meade P, Wilson FH, Hebert SC, Gimenez I, Gamba G, Lifton RP (2005) WNK3 kinase is a positive regulator of NKCC2 and NCC, renal cation-Cl^- cotransporters required for normal blood pressure homeostasis. Proc Natl Acad Sci USA 102(46):16777–16782. https://doi.org/10.1073/pnas.0508303102
Rinehart J, Vazquez N, Kahle KT, Hodson CA, Ring AM, Gulcicek EE, Louvi A, Bobadilla NA, Gamba G, Lifton RP (2011) WNK2 kinase is a novel regulator of essential neuronal cation-chloride cotransporters. J Biol Chem 286(34):30171–30180. https://doi.org/10.1074/jbc.M111.222893
Riveira-Munoz E, Chang Q, Godefroid N, Hoenderop JG, Bindels RJ, Dahan K, Devuyst O, Belgian Network for Study of Gitelman S (2007) Transcriptional and functional analyses of SLC12A3 mutations: new clues for the pathogenesis of Gitelman syndrome. J Am Soc Nephrol 18(4):1271–1283. https://doi.org/10.1681/ASN.2006101095
Ronzaud C, Loffing-Cueni D, Hausel P, Debonneville A, Malsure SR, Fowler-Jaeger N, Boase NA, Perrier R, Maillard M, Yang B, Stokes JB, Koesters R, Kumar S, Hummler E, Loffing J, Staub O (2013) Renal tubular NEDD4-2 deficiency causes NCC-mediated salt-dependent hypertension. J Clin Invest 123(2):657–665. https://doi.org/10.1172/JCI61110

Rosenbaek LL, Assentoft M, Pedersen NB, MacAulay N, Fenton RA (2012) Characterization of a novel phosphorylation site in the sodium-chloride cotransporter, NCC. J Physiol 590 (Pt 23):6121–6139. https://doi.org/10.1113/jphysiol.2012.240986
Rosenbaek LL, Kortenoeven ML, Aroankins TS, Fenton RA (2014) Phosphorylation decreases ubiquitylation of the thiazide-sensitive cotransporter NCC and subsequent clathrin-mediated endocytosis. J Biol Chem 289(19):13347–13361. https://doi.org/10.1074/jbc.M113.543710
Rosenbaek LL, Rizzo F, Wu Q, Rojas-Vega L, Gamba G, MacAulay N, Staub O, Fenton RA (2017) The thiazide sensitive sodium chloride co-transporter NCC is modulated by site-specific ubiquitylation. Sci Rep 7(1):12981. https://doi.org/10.1038/s41598-017-12819-0
Roy A, Al-Qusairi L, Donnelly BF, Ronzaud C, Marciszyn AL, Gong F, Chang YP, Butterworth MB, Pastor-Soler NM, Hallows KR, Staub O, Subramanya AR (2015) Alternatively spliced proline-rich cassettes link WNK1 to aldosterone action. J Clin Invest 125(9):3433–3448. https://doi.org/10.1172/JCI75245
Rozansky DJ, Cornwall T, Subramanya AR, Rogers S, Yang YF, David LL, Zhu X, Yang CL, Ellison DH (2009) Aldosterone mediates activation of the thiazide-sensitive Na-Cl cotransporter through an SGK1 and WNK4 signaling pathway. J Clin Invest 119(9):2601–2612. https://doi.org/10.1172/JCI38323
Sabath E, Meade P, Berkman J, de los Heros P, Moreno E, Bobadilla NA, Vazquez N, Ellison DH, Gamba G (2004) Pathophysiology of functional mutations of the thiazide-sensitive Na-Cl cotransporter in Gitelman disease. Am J Physiol Renal Physiol 287(2):F195–F203. https://doi.org/10.1152/ajprenal.00044.2004
San-Cristobal P, Pacheco-Alvarez D, Richardson C, Ring AM, Vazquez N, Rafiqi FH, Chari D, Kahle KT, Leng Q, Bobadilla NA, Hebert SC, Alessi DR, Lifton RP, Gamba G (2009) Angiotensin II signaling increases activity of the renal Na-cl cotransporter through a WNK4-SPAK-dependent pathway. Proc Natl Acad Sci USA 106(11):4384–4389. https://doi.org/10.1073/pnas.0813238106
Sandberg MB, Maunsbach AB, McDonough AA (2006) Redistribution of distal tubule Na^+-Cl^- cotransporter (NCC) in response to a high-salt diet. Am J Physiol Renal Physiol 291(2):F503–F508. https://doi.org/10.1152/ajprenal.00482.2005
Sandberg MB, Riquier AD, Pihakaski-Maunsbach K, McDonough AA, Maunsbach AB (2007) ANG II provokes acute trafficking of distal tubule Na^+-Cl(−) cotransporter to apical membrane. Am J Physiol Renal Physiol 293(3):F662–F669. https://doi.org/10.1152/ajprenal.00064.2007
Saritas T, Borschewski A, McCormick JA, Paliege A, Dathe C, Uchida S, Terker A, Himmerkus N, Bleich M, Demaretz S, Laghmani K, Delpire E, Ellison DH, Bachmann S, Mutig K (2013) SPAK differentially mediates vasopressin effects on sodium cotransporters. J Am Soc Nephrol 24(3):407–418. https://doi.org/10.1681/ASN.2012040404
Shibata S, Zhang J, Puthumana J, Stone KL, Lifton RP (2013) Kelch-like 3 and Cullin 3 regulate electrolyte homeostasis via ubiquitination and degradation of WNK4. Proc Natl Acad Sci USA 110(19):7838–7843. https://doi.org/10.1073/pnas.1304592110
Simon DB, Nelson-Williams C, Bia MJ, Ellison D, Karet FE, Molina AM, Vaara I, Iwata F, Cushner HM, Koolen M, Gainza FJ, Gitleman HJ, Lifton RP (1996) Gitelman's variant of Bartter's syndrome, inherited hypokalaemic alkalosis, is caused by mutations in the thiazide-sensitive Na-Cl cotransporter. Nat Genet 12(1):24–30
Simpson F, Peden AA, Christopoulou L, Robinson MS (1997) Characterization of the adaptor-related protein complex, AP-3. J Cell Biol 137(4):835–845
Singh SK, Piscitelli CL, Yamashita A, Gouaux E (2008) A competitive inhibitor traps LeuT in an open-to-out conformation. Science 322(5908):1655–1661. https://doi.org/10.1126/science.1166777
Snyder PM, Olson DR, Kabra R, Zhou R, Steines JC (2004) cAMP and serum and glucocorticoid-inducible kinase (SGK) regulate the epithelial Na(+) channel through convergent phosphorylation of Nedd4-2. J Biol Chem 279(44):45753–45758. https://doi.org/10.1074/jbc.M407858200
Sohara E, Rai T, Yang SS, Ohta A, Naito S, Chiga M, Nomura N, Lin SH, Vandewalle A, Ohta E, Sasaki S, Uchida S (2011) Acute insulin stimulation induces phosphorylation of the Na-Cl

cotransporter in cultured distal mpkDCT cells and mouse kidney. PLoS One 6(8):e24277. https://doi.org/10.1371/journal.pone.0024277

Sorensen MV, Grossmann S, Roesinger M, Gresko N, Todkar AP, Barmettler G, Ziegler U, Odermatt A, Loffing-Cueni D, Loffing J (2013) Rapid dephosphorylation of the renal sodium chloride cotransporter in response to oral potassium intake in mice. Kidney Int 83(5):811–824. https://doi.org/10.1038/ki.2013.14

Stokes JB (1984) Sodium chloride absorption by the urinary bladder of the winter flounder. A thiazide-sensitive, electrically neutral transport system. J Clin Invest 74(1):7–16. https://doi.org/10.1172/JCI111420

Subramanya AR, Ellison DH (2014) Distal convoluted tubule. Clin J Am Soc Nephrol 9 (12):2147–2163. https://doi.org/10.2215/CJN.05920613

Subramanya AR, Welling PA (2011) Toward an understanding of hypertension resistance. Am J Physiol Renal Physiol 300(4):F838–F839. https://doi.org/10.1152/ajprenal.00078.2011

Subramanya AR, Yang CL, Zhu X, Ellison DH (2006) Dominant-negative regulation of WNK1 by its kidney-specific kinase-defective isoform. Am J Physiol Renal Physiol 290(3):F619–F624. https://doi.org/10.1152/ajprenal.00280.2005

Subramanya AR, Liu J, Ellison DH, Wade JB, Welling PA (2009) WNK4 diverts the thiazide-sensitive NaCl cotransporter to the lysosome and stimulates AP-3 interaction. J Biol Chem 284 (27):18471–18480. https://doi.org/10.1074/jbc.M109.008185

Takahashi D, Mori T, Nomura N, Khan MZ, Araki Y, Zeniya M, Sohara E, Rai T, Sasaki S, Uchida S (2014) WNK4 is the major WNK positively regulating NCC in the mouse kidney. Biosci Rep 34(3). https://doi.org/10.1042/BSR20140047

Talati G, Ohta A, Rai T, Sohara E, Naito S, Vandewalle A, Sasaki S, Uchida S (2010) Effect of angiotensin II on the WNK-OSR1/SPAK-NCC phosphorylation cascade in cultured mpkDCT cells and in vivo mouse kidney. Biochem Biophys Res Commun 393(4):844–848. https://doi.org/10.1016/j.bbrc.2010.02.096

Terker AS, Zhang C, McCormick JA, Lazelle RA, Zhang C, Meermeier NP, Siler DA, Park HJ, Fu Y, Cohen DM, Weinstein AM, Wang WH, Yang CL, Ellison DH (2015) Potassium modulates electrolyte balance and blood pressure through effects on distal cell voltage and chloride. Cell Metab 21(1):39–50. https://doi.org/10.1016/j.cmet.2014.12.006

Terker AS, Zhang C, Erspamer KJ, Gamba G, Yang CL, Ellison DH (2016) Unique chloride-sensing properties of WNK4 permit the distal nephron to modulate potassium homeostasis. Kidney Int 89(1):127–134. https://doi.org/10.1038/ki.2015.289

Thastrup JO, Rafiqi FH, Vitari AC, Pozo-Guisado E, Deak M, Mehellou Y, Alessi DR (2012) SPAK/OSR1 regulate NKCC1 and WNK activity: analysis of WNK isoform interactions and activation by T-loop trans-autophosphorylation. Biochem J 441(1):325–337. https://doi.org/10.1042/BJ20111879

Tovar-Palacio C, Bobadilla NA, Cortes P, Plata C, de los Heros P, Vazquez N, Gamba G (2004) Ion and diuretic specificity of chimeric proteins between apical Na(+)-K(+)-2Cl(−) and Na(+)-Cl(−) cotransporters. Am J Physiol Renal Physiol 287(3):F570–F577. https://doi.org/10.1152/ajprenal.00124.2004

Vallon V, Schroth J, Lang F, Kuhl D, Uchida S (2009) Expression and phosphorylation of the Na^+-Cl^- cotransporter NCC in vivo is regulated by dietary salt, potassium, and SGK1. Am J Physiol Renal Physiol 297(3):F704–F712. https://doi.org/10.1152/ajprenal.00030.2009

van der Lubbe N, Lim CH, Fenton RA, Meima ME, Jan Danser AH, Zietse R, Hoorn EJ (2011) Angiotensin II induces phosphorylation of the thiazide-sensitive sodium chloride cotransporter independent of aldosterone. Kidney Int 79(1):66–76. https://doi.org/10.1038/ki.2010.290

van der Lubbe N, Lim CH, Meima ME, van Veghel R, Rosenbaek LL, Mutig K, Danser AH, Fenton RA, Zietse R, Hoorn EJ (2012) Aldosterone does not require angiotensin II to activate NCC through a WNK4-SPAK-dependent pathway. Pflugers Arch 463(6):853–863. https://doi.org/10.1007/s00424-012-1104-0

Vazquez N, Monroy A, Dorantes E, Munoz-Clares RA, Gamba G (2002) Functional differences between flounder and rat thiazide-sensitive Na-Cl cotransporter. Am J Physiol Renal Physiol 282(4):F599–F607. https://doi.org/10.1152/ajprenal.00284.2001

Velazquez H, Bernstein PL, Bartiss A, Kunchaparty S, Ellison DH (1995) Thiazide-sensitive Nacl transport capacity of rat renal distal tubule is enhanced by adrenal-steroids. FASEB J 9(3): A586–A586

Velazquez H, Naray-Fejes-Toth A, Silva T, Andujar E, Reilly RF, Desir GV, Ellison DH (1998) Rabbit distal convoluted tubule coexpresses NaCl cotransporter and 11 beta-hydroxysteroid dehydrogenase II mRNA. Kidney Int 54(2):464–472. https://doi.org/10.1046/j.1523-1755.1998.00036.x

Vembar SS, Brodsky JL (2008) One step at a time: endoplasmic reticulum-associated degradation. Nat Rev Mol Cell Biol 9(12):944–957. https://doi.org/10.1038/nrm2546

Verlander JW, Tran TM, Zhang L, Kaplan MR, Hebert SC (1998) Estradiol enhances thiazide-sensitive NaCl cotransporter density in the apical plasma membrane of the distal convoluted tubule in ovariectomized rats. J Clin Invest 101(8):1661–1669. https://doi.org/10.1172/JCI601

Vidal-Petiot E, Cheval L, Faugeroux J, Malard T, Doucet A, Jeunemaitre X, Hadchouel J (2012) A new methodology for quantification of alternatively spliced exons reveals a highly tissue-specific expression pattern of WNK1 isoforms. PLoS One 7(5):e37751. https://doi.org/10.1371/journal.pone.0037751

Vidal-Petiot E, Elvira-Matelot E, Mutig K, Soukaseum C, Baudrie V, Wu S, Cheval L, Huc E, Cambillau M, Bachmann S, Doucet A, Jeunemaitre X, Hadchouel J (2013) WNK1-related familial hyperkalemic hypertension results from an increased expression of L-WNK1 specifically in the distal nephron. Proc Natl Acad Sci USA 110(35):14366–14371. https://doi.org/10.1073/pnas.1304230110

Villa F, Goebel J, Rafiqi FH, Deak M, Thastrup J, Alessi DR, van Aalten DM (2007) Structural insights into the recognition of substrates and activators by the OSR1 kinase. EMBO Rep 8 (9):839–845. https://doi.org/10.1038/sj.embor.7401048

Vitari AC, Deak M, Morrice NA, Alessi DR (2005) The WNK1 and WNK4 protein kinases that are mutated in Gordon's hypertension syndrome phosphorylate and activate SPAK and OSR1 protein kinases. Biochem J 391(Pt 1):17–24. https://doi.org/10.1042/BJ20051180

Vitari AC, Thastrup J, Rafiqi FH, Deak M, Morrice NA, Karlsson HK, Alessi DR (2006) Functional interactions of the SPAK/OSR1 kinases with their upstream activator WNK1 and downstream substrate NKCC1. Biochem J 397(1):223–231. https://doi.org/10.1042/BJ20060220

Wakabayashi M, Mori T, Isobe K, Sohara E, Susa K, Araki Y, Chiga M, Kikuchi E, Nomura N, Mori Y, Matsuo H, Murata T, Nomura S, Asano T, Kawaguchi H, Nonoyama S, Rai T, Sasaki S, Uchida S (2013) Impaired KLHL3-mediated ubiquitination of WNK4 causes human hypertension. Cell Rep 3(3):858–868. https://doi.org/10.1016/j.celrep.2013.02.024

Ward CL, Kopito RR (1994) Intracellular turnover of cystic fibrosis transmembrane conductance regulator. Inefficient processing and rapid degradation of wild-type and mutant proteins. J Biol Chem 269(41):25710–25718

Warmuth S, Zimmermann I, Dutzler R (2009) X-ray structure of the C-terminal domain of a prokaryotic cation-chloride cotransporter. Structure 17(4):538–546. https://doi.org/10.1016/j.str.2009.02.009

Wilson FH, Disse-Nicodeme S, Choate KA, Ishikawa K, Nelson-Williams C, Desitter I, Gunel M, Milford DV, Lipkin GW, Achard JM, Feely MP, Dussol B, Berland Y, Unwin RJ, Mayan H, Simon DB, Farfel Z, Jeunemaitre X, Lifton RP (2001) Human hypertension caused by mutations in WNK kinases. Science 293(5532):1107–1112. https://doi.org/10.1126/science.1062844

Wilson FH, Kahle KT, Sabath E, Lalioti MD, Rapson AK, Hoover RS, Hebert SC, Gamba G, Lifton RP (2003) Molecular pathogenesis of inherited hypertension with hyperkalemia: the Na-Cl cotransporter is inhibited by wild-type but not mutant WNK4. Proc Natl Acad Sci USA 100 (2):680–684. https://doi.org/10.1073/pnas.242735399

Xu B, English JM, Wilsbacher JL, Stippec S, Goldsmith EJ, Cobb MH (2000) WNK1, a novel mammalian serine/threonine protein kinase lacking the catalytic lysine in subdomain II. J Biol Chem 275(22):16795–16801

Yang CL, Angell J, Mitchell R, Ellison DH (2003) WNK kinases regulate thiazide-sensitive Na-Cl cotransport. J Clin Invest 111(7):1039–1045. https://doi.org/10.1172/JCI17443

Yang CL, Zhu X, Wang Z, Subramanya AR, Ellison DH (2005) Mechanisms of WNK1 and WNK4 interaction in the regulation of thiazide-sensitive NaCl cotransport. J Clin Invest 115 (5):1379–1387. https://doi.org/10.1172/JCI22452

Yang CL, Zhu X, Ellison DH (2007a) The thiazide-sensitive Na-Cl cotransporter is regulated by a WNK kinase signaling complex. J Clin Invest 117(11):3403–3411. https://doi.org/10.1172/JCI32033

Yang SS, Morimoto T, Rai T, Chiga M, Sohara E, Ohno M, Uchida K, Lin SH, Moriguchi T, Shibuya H, Kondo Y, Sasaki S, Uchida S (2007b) Molecular pathogenesis of pseudohypoaldosteronism type II: generation and analysis of a Wnk4(D561A/+) knockin mouse model. Cell Metab 5(5):331–344. https://doi.org/10.1016/j.cmet.2007.03.009

Yang SS, Lo YF, Wu CC, Lin SW, Yeh CJ, Chu P, Sytwu HK, Uchida S, Sasaki S, Lin SH (2010) SPAK-knockout mice manifest Gitelman syndrome and impaired vasoconstriction. J Am Soc Nephrol 21(11):1868–1877. https://doi.org/10.1681/ASN.2009121295

Yang SS, Fang YW, Tseng MH, Chu PY, Yu IS, Wu HC, Lin SW, Chau T, Uchida S, Sasaki S, Lin YF, Sytwu HK, Lin SH (2013) Phosphorylation regulates NCC stability and transporter activity in vivo. J Am Soc Nephrol 24(10):1587–1597. https://doi.org/10.1681/ASN.2012070742

Zerangue N, Schwappach B, Jan YN, Jan LY (1999) A new ER trafficking signal regulates the subunit stoichiometry of plasma membrane K(ATP) channels. Neuron 22(3):537–548

Zhang C, Wang L, Zhang J, Su XT, Lin DH, Scholl UI, Giebisch G, Lifton RP, Wang WH (2014) KCNJ10 determines the expression of the apical Na-Cl cotransporter (NCC) in the early distal convoluted tubule (DCT1). Proc Natl Acad Sci USA 111(32):11864–11869. https://doi.org/10.1073/pnas.1411705111

Zhou B, Zhuang J, Gu D, Wang H, Cebotaru L, Guggino WB, Cai H (2009) WNK4 enhances the degradation of NCC through a Sortilin-mediated lysosomal pathway. J Am Soc Nephrol. https://doi.org/10.1681/ASN.2008121275

Zhou B, Wang D, Feng X, Zhang Y, Wang Y, Zhuang J, Zhang X, Chen G, Delpire E, Gu D, Cai H (2012) WNK4 inhibits NCC protein expression through MAPK ERK1/2 signaling pathway. Am J Physiol Renal Physiol 302(5):F533–F539. https://doi.org/10.1152/ajprenal.00032.2011

Chapter 4
NBCe1: An Electrogenic Na^+ Bicarbonate Cotransporter, in Epithelia

Clayton T. Brady, Aleksandra Dugandžić, Mark D. Parker, and Michael F. Romero

Abstract The archetypal electrogenic Na^+-bicarbonate (HCO_3^-) cotransporter (NBCe1: SLC4A4) is one of five Na^+-coupled bicarbonate transporters encoded by the *SLC4* family of genes. NBCe1 cotransports 1 Na^+ with 2 HCO_3^- or, as an accumulation of evidence suggests, 1 Na^+ and with a single carbonate (CO_3^{2-}). NBCe1 action contributes to the buffering of blood plasma and intracellular pH (pH_i) and also supports transepithelial HCO_3^-/fluid secretion. The *SLC4A4* gene expresses three major NBCe1 isoforms: NBCe1-A (expressed in renal proximal tubular epithelial cells), NBCe1-B (expressed in a variety of epithelial and excitable cells), and NBCe1-C (expressed in neurons and glial cells). Aside from distribution differences, each isoform has a unique combination of amino- and carboxy-terminal sequences which dictates differences in baseline Na^+-$2HCO_3^-$ cotransport activity and regulation. In humans, recessive mutations in NBCe1 cause proximal renal tubular acidosis (pRTA) with ocular abnormalities: a devastating but rare disturbance of pH and fluid balance. This chapter will focus on the physiological and pathological roles of NBCe1-A and NBCe1-B in epithelial tissues.

Keywords Acid-base · Epithelia · Ion transport · Kidney · pH regulation

C. T. Brady · M. D. Parker
Physiology and Biophysics, Jacobs School of Medicine and Biomedical Sciences, The State University of New York, University at Buffalo, Buffalo, NY, USA
e-mail: cbrady@buffalo.edu; parker28@buffalo.edu

A. Dugandžić
Physiology, Croatian Institute for Brain Research, School of Medicine, University of Zagreb, Zagreb, Croatia
e-mail: aleksandra.dugandzic@mef.hr

M. F. Romero (✉)
Physiology and Biomedical Engineering, Nephrology and Hypertension, Mayo Clinic College of Medicine and Science, Rochester, MN, USA
e-mail: romero.michael@mayo.edu

K. L. Hamilton, D. C. Devor (eds.), *Studies of Epithelial Transporters and Ion Channels*, Physiology in Health and Disease,
https://doi.org/10.1007/978-3-030-55454-5_4

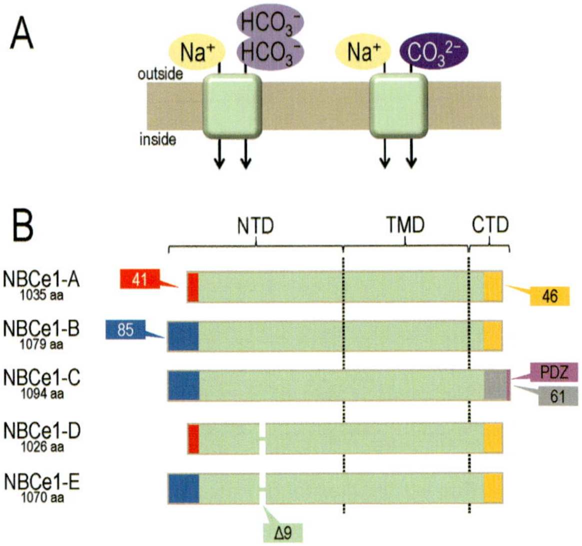

Fig. 4.1 NBCe1 and its isoforms. (**a**) Hypothetical modes of action of NBCe1: see Sect. 4.3.3. (**b**) NBCe1-A and NBCe1-D include an auto-stimulatory domain (ASD, 41-aa red region). All other variants include an auto-inhibitory domain (AID, 85-aa blue region): see Sect. 4.6.2. NBCe1-C terminates with an alternative CTD (61-aa gray region) that includes a PDZ-domain binding sequence (pink) rather than the general 46-aa CTD (gold). *NTD* N-terminal domain; *TMD* transmembrane domains; *CTD* C-terminal domain

4.1 Introduction

Most physiological processes are sensitive to pH and thus acid-base homeostasis is critical for health. In the face of a substantial daily acid load from diet and metabolism, the neutralization of acids in both blood plasma and intracellular compartments is crucial for maintaining pH within its narrow physiological range (plasma pH = 7.35–7.45). Bicarbonate (HCO_3^-), is the main physiological buffer and thus the transport of HCO_3^- across cell membranes is an essential element of acid-base homeostasis. The Solute Carrier 4 (*SLC4*) gene family encodes many of these HCO_3^- transporting membrane proteins including three Cl^-/HCO_3^- exchangers (anion exchangers, AEs) and five Na^+-coupled HCO_3^- transporters (NCBTs). Among the five NCBTs: NBCe1 (*SLC4A4*) and NBCe2 (*SLC4A5*) perform electrogenic Na^+-$2HCO_3^-$ cotransport (i.e., they carry net electrical charge across the membrane, Fig. 4.1a) (Romero et al. 1997; Sassani et al. 2002; Virkki et al. 2002), NBCn1 (*SLC4A7*) and NBCn2 (*SLC4A10*) perform electroneutral Na^+-HCO_3^- cotransport (Choi et al. 2000; Parker et al. 2008), and NDCBE (*SLC4A8*) performs electroneutral Na^+-driven Cl^-/HCO_3^- exchange (Grichtchenko et al. 2001). NBCe1 is the topic of this chapter, although the reader will find a discussion of the roles of the other NCBTs in support of pH regulation and fluid secretion in two others chapter of this book titled "Fundamentals of Bicarbonate Secretion in Epithelia" (Chap. 12 in Volume 1) and "Ion transport in the Choroid Plexus Epithelium" (Chap. 10 in Volume 2). A number of recently published review articles provide a broader consideration of the importance of SLC4 proteins (Parker and Boron 2013; Romero et al. 2013).

The electrogenic sodium bicarbonate cotransport activity mediated by NBCe1 was first described by Boron and Boulpaep in the basolateral membranes of the renal

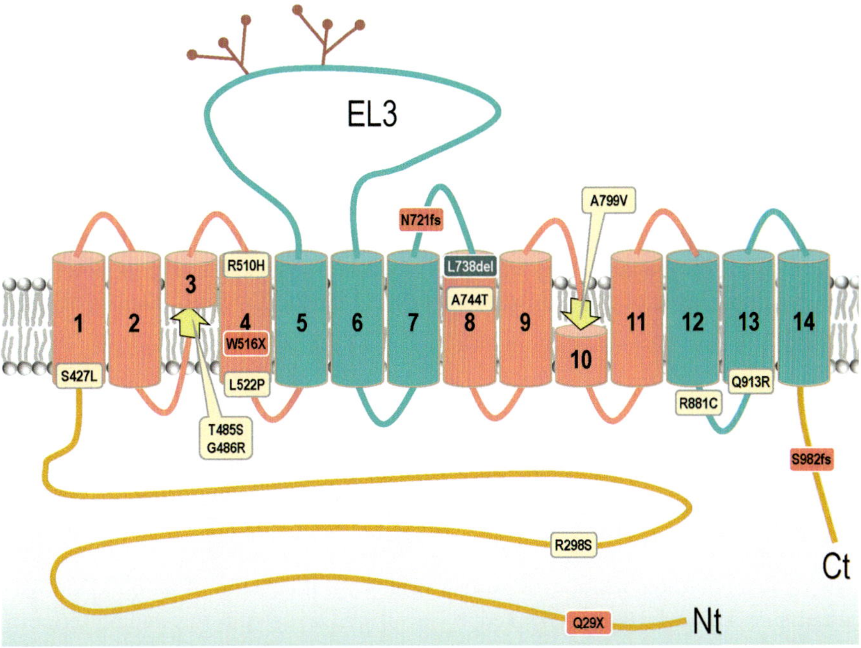

Fig. 4.2 Predicted membrane topology of NBCe1 transporter proteins showing human mutations. All isoforms have 14 transmembrane helices (1–14). Transmembrane spans 3 and 10 include short beta-sheets (yellow arrows). The N and C terminal domains of NBCe1 are located in the cell cytoplasm. Between TM5 and TM6 is a large extracellular glycosylated loop, a smaller non-glycosylated loop is located between TM7 and 8. Mutations in human NBCe1-A are illustrated (cream = missense, pink = nonsense/frameshift, grey = deletion). Based upon the structural model by Huynh et al. (2018). Blue TM = gate domain; pink TM = core domain; and yellow arrows = putative ion-coordination site

proximal tubule of the salamander (Boron and Boulpaep 1983). Now it is known that NBCe1 is expressed in a variety of epithelial and excitable cells throughout the vertebrate body where its function plays an important role in the physiology and pathophysiology of organs such as the brain, enamel organs, eyes, heart, kidneys, and both small and large intestines. NBCe1 is, however, not a singular entity. The *SLC4A4* gene, located within human chromosomal locus 4q13, includes 26 exons (Abuladze et al. 2000) that can be combined to produce five isoforms: NBCe1-A, -B, -C, -D, and -E (Fig. 4.1). All NBCe1 isoforms share a common transmembrane domain (TMD) that includes 14 transmembrane spans (TMs) with a large extracellular glycosylated loop between the fifth and sixth TMs (external loop 3, EL3: Fig. 4.2) and all NBCe1 isoforms have their amino- (N) and carboxy- (C-) terminal ends located in the cytoplasm of the cell. Sequence differences in these appendages confer differential basal activity and regulatory potential among isoforms (see Sect. 4.6.2).

The importance of NBCe1 is made clear when one considers the consequences of its dysfunction. Mutations in *SLC4A4* cause proximal renal tubular acidosis (pRTA;

OMIM 603345); a disease associated with acidic blood, progressive loss of vision, failure to thrive, and diverse neuromuscular defects. Nbce1-null mice are similarly afflicted with few surviving beyond the first few weeks of life (Gawenis et al. 2007).

This chapter provides an overview of our current understanding of the role of NBCe1 isoforms in support of the health and function of epithelial tissues and also brings together knowledge of the structure and molecular physiology of NBCe1 as well as the consequences of NBCe1 dysfunction in patients and animal models. The reader may be surprised to learn how much is still unknown about this important protein; even such fundamental properties as its substrates and transport stoichiometry remain subjects of active research and lively debate, 20 years after its original cloning.

4.2 Structure of NBCe1

4.2.1 General Features

NBCe1 is a polytopic membrane protein (Fig. 4.2). It consists of a ~60 kDa TMD comprising 14 TMs that is flanked by a ~50 kDa cytoplasmic amino-terminal domain (Nt) and a smaller ~10 kDa cytoplasmic carboxy-terminal domain (Ct). Two of the extracellular loops that connect TM helices—EL3 and EL4—are more prominent than the compactly folded extracellular surface, with EL3 being the longer of the two (Zhu et al. 2010, 2015). The predicted molecular weight of the NBCe1-A polypeptide is 116 kDa, however, on denaturing western blots of kidney lysate, NBCe1 migrates with an apparent molecular weight of ~130–145 kDa (Choi et al. 2003; Kao et al. 2008). This difference is explained by the addition of *N*-glycan chains at two of the three consensus glycosylation sites within EL3 (Choi et al. 2003). However, the presence of the glycan chains has no bearing on the intrinsic activity of NBCe1 as heterologously expressed in *Xenopus* oocytes (Choi et al. 2003).

In situ, NBCe1 is predominantly homodimeric with contact points between adjacent Nt domains and between TMD domains (Gill and Boron 2006a; Kao et al. 2008; Sergeev et al. 2012). Experiments performed on concatemeric NBCe1-A constructs, in which one monomer of each pair can be selectively disabled using a cysteine-reactive compound, indicate that individual monomers within a homodimer are independently capable of sodium-coupled bicarbonate transport (Kao et al. 2008). Interestingly, the cytoplasmic Nt domain of NBCe1 exerts a strong regulatory influence over the activity of the TMD; its presence is necessary for $Na^+/2HCO_3^-$ cotransport (Espiritu et al. 2006; McAlear et al. 2006) and Nt variations among NBCe1 isoforms can inhibit or stimulate $Na^+/2HCO_3^-$ cotransport (see Sect. 4.6.2). Less is known about the importance of the Ct domain although many studies indicate that the Ct includes important determinants for the trafficking and plasma-membrane stabilization of NBCe1 (Espiritu et al. 2006; Li et al. 2004, 2009; McAlear et al. 2006; Suzuki et al. 2010).

The 3D structure of an intact NBCe1 molecule has proven elusive, presumably due to the flexibility of the linkage between the Nt and TMD and the disordered nature of EL3 and the Ct. However, comparative studies of NBCe1 isoforms and structural studies of each NBCe1 domain in isolation promise significant progress in our understanding of how structure underlies NBCe1 function.

4.2.2 NBCe1 Isoforms

The five NBCe1 isoforms, NBCe1-A, -B, -C, -D, and -E, are a result of alternative promoter choice and alternate splicing of *SLC4A4* products. The first vertebrate NBCe1 clone was isolated from salamander kidney cDNA (Romero et al. 1997). This discovery was followed in short order by the cloning of its human (Burnham et al. 1997) and rat (Romero et al. 1998) orthologs, leading to a proliferation of acronyms—aNBC, rNBC, rkNBC, kNBC1, and hkNBCe1—all which are aliases for what is now commonly referred to as NBCe1-A.[1] Subsequently, NBCe1-B (aka pNBC, pNBC1, rpNBC, hhNBC, hcNBC, and rb1NBC)[2] was cloned from the human pancreas and heart (Abuladze et al. 1998; Choi et al. 1999). NBCe1-B is transcribed from an alternative promoter upstream from that which generates NBCe1-A, resulting in the first 41 amino acid residues (aa) of NBCe1-A being replaced by a different sequence of 85 aa in NBCe1-B (Fig. 4.1b) (Abuladze et al. 1998; Snead et al. 2011). In 2000, NBCe1-C (aka "rb2NBC") was cloned from rat brain (Bevensee et al. 2000). NBCe1-C is identical to NBCe1-B except that the 46 most C-terminal aa of NBCe1-B are replaced by a different sequence of 61 aa, that ends with a PDZ-binding sequence, due to an alternative splicing event (Fig. 4.1b). More recently, NBCe1-D and NBCe1-E transcripts have been cloned from the mouse reproductive tract cDNA (Liu et al. 2011). NBCe1-D and -E are identical to NBCe1-A and -B, respectively except for lack of a 9 aa stretch within their Nt domain (Fig. 4.1b). The functional consequences of this difference are unknown.

4.2.3 Structural Features of the N-Terminal Domain

The cytoplasmic Nt domain NBCe1 consists of a highly-structured core linked to the TMD via a glycine-rich linker. The NTDs from each monomer interlock with each

---

[1]aNBC is *Ambystoma* NBCe1-A (Romero et al. 1997); rNBC and rkNBC are rat NBCe1-A; kNBC1 is kidney NBCe1-A regardless of species; and hkNBCe1 is human kidney NBCe1-A.

[2]pNBC and pNBC1 are pancreas NBCe1-B; rpNBC is rat pancreas NBCe1-B; hhNBC is human heart NBCe1-B; hcNBC is human cardiac NBCe1-B; rb1NBC is rat brain NBCe1-B; while rb2NBC is rat brain NBCe1-C.

other, presumably stabilizing NBCe1 homodimers (Gill and Boron 2006a). The integrity of the Nt core is crucial for NBCe1 function, as highlighted by the deleterious effect of mutations within this region including a critical salt-bridge between R298 (human residue mutated in disease) and E91 (Chang et al. 2008). We now also appreciate that the unique 41-aa N-terminal sequence that precedes the Nt core in NBCe1-A constitutes a transport autostimulatory domain (ASD) whereas its 85-aa replacement in NBCe1-B or NBCe1-C (hereafter NBCe1-B/C) constitutes a transport autoinhibitory domain (AID), whose influence can be controlled by hormonal signaling cascades (see Sect. 4.6). Unfortunately, the ASD of NBCe1-A is not represented in early X-ray diffraction analyses of Nt crystals, presumably reflecting intrinsic disorder or a flexible linkage to the core (Gill and Boron 2006b), and the ASD was not included in constructs used to generate later crystals (Gill et al. 2013). Similarly, we have no structure of the AID of NBCe1-B/C. The mechanism of ASD and AID action remains unknown.

4.2.4 Structural Features of the Transmembrane Domain

Using cryo electron-microscopy (cryoEM), Huynh and coworkers determined the structure of the dimeric TMD of human NBCe1 (Huynh et al. 2018). They found that the TMD of each NBCe1 monomer is made up of 14 TMs, which, in common with many other membrane transport proteins (Bai et al. 2017), can be separated into two structurally similar inverted repeats. TMs 3 and 10 are unusual inasmuch as these spans are part β-sheet and part α-helix (see cartoon, Fig. 4.2). Their atomic model defines a "gate" and a "core" domain for each monomer (Fig. 4.3). TMs 5–7 and 12–14 make up the gate domain at the dimer interface (blue in Fig. 4.2; blue in Fig. 4.3), while TMs 1–4 and 8–11 constitute the core domain which is furthest from the dimer interface (pink in Fig. 4.2; pink in Fig. 4.3) and includes the putative ion-co-ordination site (yellow in Fig. 4.2; yellow in Fig. 4.3). The ion-coordination site is located at the most constricted region of the predicted ion-accessibility pathway and is formed by the antiparallel β-sheet formed at the crossover point between TM3 and TM10 (Huynh et al. 2018), but structural data has yet to reveal the precise location of Na^+ vs HCO_3^-/CO_3^{2-} coordinating sites. In support of the assignment of this region as the ion-coordinating site, several disease-causing mutations that compromise NBCe1 activity—T485S (TM3), G486R (TM3), and A799V (TM10)—cluster at this location (Kurtz 2014; Huynh et al. 2018). In the cryo-EM structure, this site is accessible only from the extracellular side, thus the structure represents an outward-open conformation. A comparative analysis of structures of other related SLC4 and Slc4-like proteins captured in other conformations indicates that the transport cycle involves the pivoting of the flanking core domains against the central, immobile gate domains: an action that exposes the ion-coordination site to the intracellular side (Thurtle-Schmidt and Stroud 2016).

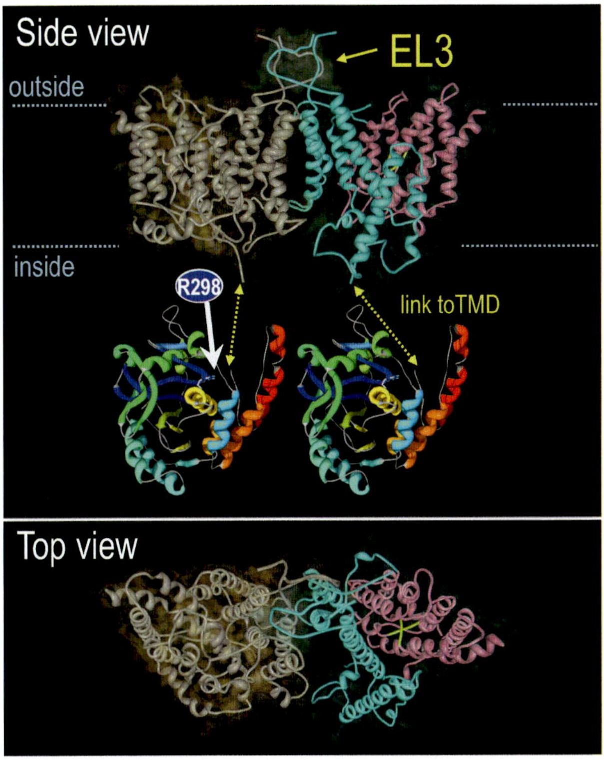

Fig. 4.3 Predicted structure of NBCe1 transporter proteins. Side view (top) and top view of NBCe1 from cryoEM (Huynh et al. 2018). The predicted *NTD* (showing R298) is based on human NBCe1 sequence mapped to the crystal coordinates of the NTD of AE1 (Chang et al. 2008). Note the two NTDs are indicated apart from one another to show details but are most likely in close proximity or even interacting. Residue R298, which forms a salt-bridge in this globular domain, and the connection to the TMD (yellow arrow) are indicated. Blue TM = gate domain; pink TM = core domain; yellow "X" (top view) = putative ion-coordination site

4.2.5 *Structural Features of the C-Terminal Domain*

The dynamic nature of the Ct domain did not permit cryoEM resolution of its structure along with the TMD; however the study of an isolated 20aa region of the Ct by circular dichroism indicates the presence of some α-helical content (Li et al. 2009).

4.3 Biophysics of NBCe1

4.3.1 *Electrogenicity*

Typical electrochemical gradients across an epithelial cell membrane favor HCO_3^- efflux. Therefore, HCO_3^- influx requires a transport mechanism that is energetically coupled to the inwardly directed Na^+ gradient: namely a Na^+-coupled HCO_3^- transporter. Because of the large constitutive Na^+ gradient generated by the Na^+/K^+-ATPase, an electroneutral Na^+-coupled HCO_3^- transporter such as NBCn1 is obliged to mediate influx under physiological conditions. On the other hand, the driving forces that influence HCO_3^- transport via an electrogenic NCBT such as NBCe1 are more malleable, being additionally influenced by membrane potential as well as by the stoichiometric ratio of Na^+ and HCO_3^- (or CO_3^{2-}) carried per transport cycle. The reversal potential for electrogenic Na^+: $\underline{n}HCO_3^-$ cotransport (where $\underline{n}$ = the number of HCO_3^- carried per Na^+; note that $CO_3^{2-} = \underline{2} \times HCO_3^-$)—the membrane potential at which equilibrium is achieved (and hence toward which the transport process drives V_m)—is described by the equation (Seki et al. 1993; Gross and Hopfer 1996):

$$E_{rev} = \frac{58}{(n-1)} \log\left(\frac{[Na^+]_i [HCO_3^-]_i^n}{[Na^+]_o [HCO_3^-]_o^n}\right).$$

In the majority of epithelial cells examined, NBCe1 performs Na^+:$2HCO_3^-$ cotransport (Seki et al. 1993; Lane et al. 2000; Gross et al. 2001; Yamaguchi and Ishikawa 2005; Shao et al. 2014). Assuming typical ionic values (e.g., $[Na^+]_i = 10$ mM, $[HCO_3^-]_i = 8$ mM, $[Na^+]_o = 140$ mM, $[HCO_3^-]_o = 24$ mM); when $n = 2$, $E_{rev} \sim -100$ mV. This value is far more negative than the typical epithelial membrane potential ($V_m \sim -60$ mV), and thus the influx of net negative charge (1 Na^+ coupled with $2HCO_3^-$), accompanied by a consequent rise in $[Na^+]_i$ and $[HCO_3^-]_i$, drives the system toward its electrochemical equilibrium (see Romero et al. 2010 for thermodynamic models). This is consistent with the physiological role of NBCe1 in moving HCO_3^- from the blood to replenish the HCO_3^- inside the cell. However, in renal proximal tubule cells, the opposite is true: NBCe1 moves HCO_3^- from the cell to replenish HCO_3^- in the blood plasma thereby buffering metabolically generated-acid (Boron and Boulpaep 1983). This inconsistency with the predicted action of a Na^+-$2HCO_3^-$ cotransporter, led to the hypothesis that n must be greater than 2 in these cells. If $n = 3$, $E_{rev} \sim -60$ mV and could be more positive if $[HCO_3^-]_o$ is low (Romero et al. 2010). This situation could support efflux. Several mechanisms have been proposed by which this change in stoichiometry—the presumed unveiling of an additional cryptic H^+ or HCO_3^--binding site—might be caused to happen. These include transporter phosphorylation and the influence of an unidentified binding partner, although no consensus has emerged (reviewed in Parker and Boron 2013). Some suggest that, with a more complete knowledge of the electrochemical gradients or other factors influencing renal NBCe1 action in situ, there may be no need to invoke a Na^+:$3HCO_3^-$ stoichiometry to explain the efflux mode in proximal tubule cells (Seki et al. 1993; Sciortino and Romero 1999; Zhu et al. 2013).

4.3.2 *Na^+ Dependence*

Before NBCe1 was cloned, early experiments on BSC-1 cells (monkey kidney epithelial cells) revealed an electrogenic Na^+-coupled HCO_3^- transport activity with apparent K_m for Na^+ of 20–40 mM at 28 mM HCO_3^-. This activity was not supported by Li^+ or by K^+ (Jentsch et al. 1985). These conclusions are in accord with studies of cloned NBCe1-A from various species expressed either in HEK-293 cells (Amlal et al. 1998) or *Xenopus* oocytes (Romero et al. 1998; Sciortino and Romero 1999; McAlear et al. 2006; Lee et al. 2013). These later experiments reveal that Li^+ can, to a small extent (5–10% activity of Na^+), substitute for Na^+ in NBCe1 transport cycles but, even then, only when present at a concentration that is several orders of magnitude greater than that which is likely to be encountered in vivo. Thus, NBCe1 is highly selective for Na^+ and Na^+ is likely to be the sole cation carried by NBCe1 in vivo.

4.3.3 *HCO_3^- Dependence*

Grichtchenko and colleagues examined the extracellular [HCO_3^-] dependence of NBCe1-A (salamander and rat) in *Xenopus* oocytes and determined an apparent K_m for external HCO_3^- of ~12 mM (Grichtchenko et al. 2000). A similar result was obtained in the pre-cloning era for the electrogenic Na^+-coupled HCO_3^- transport activity in BSC-1 cells (Jentsch et al. 1985). Other pre-cloning studies using BLMVs (basolateral membrane vesicles) prepared from the rabbit renal cortex indicated that SO_3^{2-} (or HSO_3^-, which might also be a significant component of SO_3^{2-} solutions at physiological pH) was also a substrate of NBCe1 (Soleimani and Aronson 1989). However, that observation has not been recapitulated in studies of cloned NBCe1-A expressed in *Xenopus* oocytes, indicating that the SO_3^{2-} transport activity may have been mediated by a different transporter in those vesicles (Grichtchenko et al. 2000; Lee et al. 2013). NBCe1-A does not support the electrogenic Na^+-coupled transport of Cl^-, NO_3^-, or oxalate^{2-}. (Lee et al. 2013). Thus, NBCe1 is highly selective for HCO_3^- equivalents and, moreover, HCO_3^- equivalents (i.e., HCO_3^-, CO_3^{2-}, or the $NaCO_3^-$ ion pair) are likely to be the sole anions carried by NBCe1 in vivo. Actually, the chemical form of "HCO_3^-" transported by the NBCe1 protein is still disputed, although recent evidence favors CO_3^{2-} over HCO_3^- (Lee et al. 2011a; Zhu et al. 2013). As CO_3^{2-} is relatively scarce at physiological pH (the pK_a for the dissociation of HCO_3^- into CO_3^{2-} is ~10.3), the K_m of NBCe1 for CO_3^{2-} is expected to be in the micromolar range (Moss et al. 2014). Interestingly there is some indirect evidence that NBCe1 can be made to perform electroneutral Na^+/NO_3^- cotransport, with NO_3^- structurally mimicking CO_3^{2-} (Zhu et al. 2013), although presumably with a substantially lower affinity.

4.3.4 Pharmacological Profile

There are currently no pharmaceuticals that deliberately target NBCe1 nor are there any drugs known to specifically target NBCe1 in a laboratory setting. However, relatively nonspecific transport inhibitors such as the stilbene derivative DIDS (Lu and Boron 2007), the nonsteroid anti-inflammatory drug tenidap (Ducoudret et al. 2001), the diBAC oxonol dye diBA(3)C4 (Liu et al. 2007), and S0859 (Ch'en et al. 2008), all inhibit NBCe1 with micromolar inhibitory constants in heterologous expression systems. The best characterized of these inhibitory interactions is that of DIDS, which covalently bonds with a lysine residue (K559 in NBCe1-A) that is located at the extracellular end of TM5 (Lu and Boron 2007), locking the transporter in an outward-facing conformation and blocking access to the ion-coordination site (Huynh et al. 2018). Notably, antibodies raised against extracellular epitopes in NBCe1 can exert stimulatory as well as inhibitory effects on NBCe1 action (De Giusti et al. 2011). A stimulator of NBCe1 activity would be a useful tool for countering the signs of proximal renal tubular acidosis (see Sect. 4.4.1) in cases where the mutation only partly inactivates NBCe1. Although the standard treatment for pRTA is alkali therapy (oral dosing of bicarbonate to correct blood pH), many signs of pRTA appear to be independent of acidemia, and therefore, require separate therapy (Salerno et al. 2019). The usefulness of NBCe1 inhibition in an epithelial setting is less obvious, although such action might be expected to counter secretory diarrhea (see Sect. 4.4.6). Beyond the epithelial realm, NBCe1 inhibition is a novel paradigm for treating heart disease (Fantinelli et al. 2014).

4.4 NBCe1 in Health and Disease

4.4.1 Systemic: Proximal Renal Tubular Acidosis (pRTA)

NBCe1 is expressed to some extent in almost every epithelial cell type that has been studied, as extensively reviewed by Parker and Boron (2013). Presumably, its presence and action contribute toward pH_i regulation in each of these cell types, but for this chapter, we will confine our considerations to those epithelial tissues for which NBCe1-A (predominantly renal) and NBCe1-B (predominantly non-renal) play a major and irreplaceable physiological role as disclosed by the pathology caused by NBCe1 dysfunction. Little is known of the expression or importance of NBCe1-C in epithelia and so we nominally exclude it from further consideration here. However, it should be noted that many PCR probes and antibodies do not discriminate between NBCe1-B and NBCe1-C so NBCe1-C could account for a subset of NBCe1 activity in some NBCe1-B expressing epithelia.

The importance of NBCe1 in overall health is underscored by the significant impairments observed in patients having autosomal recessive mutations in the *SLC4A4* gene (see Sect. 4.5). Mutations in NBCe1 result in a recessively, inherited form of proximal renal tubular acidosis (pRTA; OMIM 603345). Besides severe acidemia, the first patient described with inherited pRTA had an assortment of extrarenal manifestations including short stature, cataracts, band keratopathy, glaucoma, intellectual impairment, increased plasma amylase, and tooth defects (Igarashi et al. 1999). Since the first case, other reports of inherited pRTA associated with *SLC4A4* mutation have all included acidemia and at least one of the ocular abnormalities. Other features that are documented less frequently include growth defects, neurologic/neuromuscular signs (e.g., migraine, paralysis), intellectual impairment, dental abnormalities, calcification of the basal ganglia, elevated serum amylase, and hypokalemia (Table 4.1). Studies of human mutations and genetically modified mice suggest that some signs are related to loss of NBCe1-A while others are consequent to the loss of NBCe1-B (Sect. 4.5). The current understanding of NBCe1-related pathologies as they relate to epithelial function in particular organ systems are described in the rest of this section.

4.4.2 The Kidney: Acidemia

The kidneys receive 20% of cardiac output, filtering 180 L of blood plasma each day. That is to say: the kidneys filter 4320 mmoles (24 mM × 180 L, ~1 lb./day) of HCO_3^- each day, 99.9% of which must be reabsorbed (redeposited into the plasma) by the nephron in order to maintain normal plasma [HCO_3^-]. Eighty-five percent of renal HCO_3^- reabsorption is performed by proximal tubule (PT) epithelial cells. Furthermore, PT epithelia are able to produce "new" HCO_3^-, as a byproduct of ammoniagenesis, to replace the HCO_3^- that has been lost to the neutralization of the daily acid load. Both processes, reabsorption and generation of new HCO_3^-, require the activity of NBCe1-A in the basolateral membrane of PT epithelia (Boron and Boulpaep 1983; Abuladze et al. 1998; Maunsbach et al. 2000; Igarashi et al. 2001; Lee et al. 2018). Both mechanisms are described in Fig. 4.4.

Individuals with two mutant NBCe1 alleles have inherited pRTA and exhibit a severe acidemia due to metabolic acidosis (pH 7.04–7.27: normal range = 7.35–7.45) due to impaired HCO_3^- reabsorption ([HCO_3^-] 5–12 mM; normal range = 23–26-mM) and impaired ammoniagenesis in the PTs. NBCe1-A is the major renal isoform and is expressed in the earliest part of the proximal convoluted tubule (PCT aka the S1/S2 segments of the PT) (Schmitt et al. 1999; Fang et al. 2018). Mutations that specifically abrogate NBCe1-A expression are sufficient to cause acidemia as well as disturbed expression of ammoniagenic enzymes (Igarashi et al. 2001; Lee et al. 2018). Interestingly, Fang and colleagues have demonstrated the presence of a smaller quantity of NBCe1-B in the proximal straight tubules of mice (PST, S3 segment of PT), distal to the majority of NBCe1-A expression in the PCT, that is induced by acidosis (Fang et al. 2018). The physiological importance of NBCe1-B in the kidney remains to be demonstrated, although Nbce1b/c-null mice do not exhibit acidemia (Salerno et al. 2019): see Sect. 4.5.2.

Table 4.1 Disease signs in individuals with pRTA (Deda et al. 2001; Demirci et al. 2006; Dinour et al. 2004; Horita et al. 2005, 2018; Igarashi et al. 1999, 2001; Inatomi et al. 2004; Kari et al. 2014; Khan and Basamh 2018; Lo et al. 2011; Myers et al. 2016; Shiohara et al. 2000; Suzuki et al. 2008)

Mutation	Type	Effect	Age at study	Growth	Glaucoma	Cataract	Corneal opacity/ Band Keratopathy	Dental	Neurologic/ neuromuscular	Intellect	Basal Ganglion Calcium	Amylase	Hypo-kalemia	Original report
p.Q29X	Homozygous nonsense	Truncation	7, 24 years	Abnormal	x					Impaired				Igarashi et al. (2001)
p.R298S	Homozygous missense	Intracellular retention	16, 22 years	Abnormal	x	x	x			Impaired		High		Igarashi et al. (1999)
p.S427L	Homozygous missense	Intracellular retention Impaired ion transport	44 years	Abnormal	x	x	x	Abnormal						Dinour et al. (2004)
p.T485S	Homozygous missense	Impaired ion transport	2, 3 years	Abnormal		x	x							Horita et al. (2005)
p.G486R	Homozygous missense	Impaired ion transport	2 months, 7 years	Abnormal		x	x							Suzuki et al. (2008)
p.R510H	Homozygous missense	Intracellular retention	2, 12 years	Abnormal	x	x	x	Abnormal	x	Impaired		High		Igarashi et al. (1999), Shiohara et al. (2000)
p.W516X	Homozygous missense	Truncation	4, 7 years	Abnormal	x	x	x		x		x		x	Lo et al. (2011)
p.R510H and p.Q913R	Compound heterozygous	Intracellular retention Impaired ion transport	25 years	Abnormal	x	x	x	Abnormal	x		x			Myers et al. (2016)

p.L522P	Homozygous missense	Intracellular retention	27 years	Abnormal	x	x	x	Abnormal	x	Impaired					Demirci et al. (2006)
p. D721TfsX30 (2311A del)	Homozygous frameshift	Truncation	3, 12 years	Abnormal	x	x	x	Abnormal	x		x				Inatomi et al. (2004)
p.L738del	Homozygous deletion	Truncation	4 years	Abnormal			x			Impaired	x			x	Kari et al. (2014)
p.A744T	Homozygous missense	Unknown	3 years	Abnormal	x		x			Impaired					Khan and Basamh (2018)
p.A799V	Homozygous missense	Intracellular retention Impaired ion transport	4, 14 years	Abnormal	x	x	x		x	Impaired		High		x	Deda et al. (2001), Horita et al. (2005)
p.R881C	Homozygous missense	Intracellular retention	12 years	Abnormal	x	x				Impaired		High		x	Horita et al. (2005)
p. S982NFSX4	Homozygous frameshift	Intracellular retention	46, 50 years		x	x	x		x						Suzuki et al. (2010)
c.1076 + 3A>C and c. 1772 – 2A>T (splice regions)	Compound heterozygous	Decreased transcription	3, 7 years	Abnormal	x					Impaired				x	Horita et al. (2018)

Acidemia observed in all patients

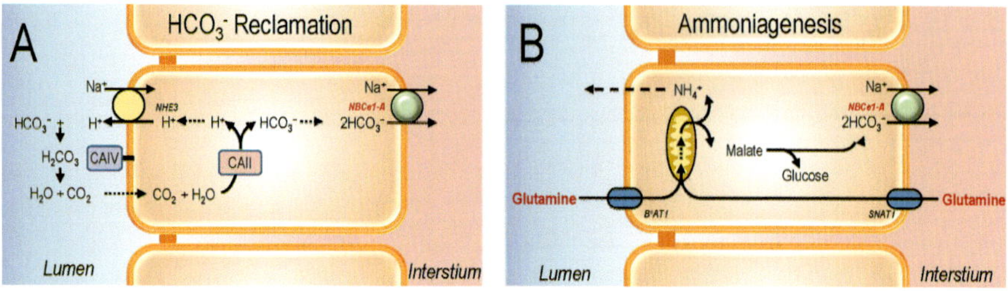

Fig. 4.4 Bicarbonate transport by proximal tubule (PT) cells. (**a**) The reabsorption of bicarbonate by the proximal tubule involves H^+ secretion at the apical cell membrane and transport of the bicarbonate into the interstitial space. Specifically, filtered bicarbonate combines with H^+ secreted in PT lumen by Na^+/H^+ exchanger type 3 (NHE3), which in turn is converted by carbonic anhydrase type IV (CAIV) to CO_2 and H_2O. Soluble CO_2 then moves into PT cells, in part via aquaporin 1, and then CO_2 is rehydrated to regenerate HCO_3^- and H^+ by the action of an intracellular CAII. HCO_3^- leaves the PT cell at the basolateral membrane via NBCe1-A. (**b**) This PT HCO_3^- absorption can be increased to ~90% of all renal HCO_3^- absorption through activation of PT ammoniagenesis. Ammoniagenesis (usually glutamine metabolism) generates both an H^+ for apical secretion as well as intracellular HCO_3^-, which is transported to the blood via NBCe1-A as above. The remaining filtered HCO_3^- is reabsorbed by the thick ascending limb (10%) and collecting duct (5%)

4.4.3 The Eye: Ocular Abnormalities

NBCe1, predominantly NBCe1-B, is expressed in numerous ocular tissues including the corneal endothelium, pigmented epithelium of the ciliary body, and lens epithelium (Bok et al. 2001). In the corneal endothelium, NBCe1-B is expressed in the basolateral membrane where it contributes to the vectorial ion movements that draw fluid out of the collagenous corneal stroma and into the aqueous humor (reviewed in Bonanno (2012), see Fig. 4.5). This action counters the effect of the oncotic pressure exerted by the stroma itself which would otherwise accumulate fluid, swell, and impair vision. Nbce1 deletion in mice results in the expected corneal edema (Lo et al. 2011; Salerno et al. 2019) although this is not a phenotype reported in humans with NBCe1 mutation, perhaps because other ocular pathologies complicate the assessment of corneal swelling. These other pathologies are band keratopathy (opacity of the cornea due to Ca^{2+} and Mg^{2+} deposits within the cornea), cataracts (opacity of the lens), and glaucoma (neuropathy of retina and optic nerve, as a result of increased intraocular pressure). However, not all individuals with pRTA exhibit all three signs (Table 4.1), which complicates the definitive assignment of a link between NBCe1 loss and pathology. Band keratopathy has been proposed to be a consequence of HCO_3^- accumulation in the corneal stroma, following loss of NBCe1 from the endothelia, and the consequent precipitation of divalent cation salts (Seki et al. 2013). Regarding cataracts, there is an established link between defective ion transport pathways in lens epithelial cells and decreased lens transparency, which could imply a role for NBCe1 (Delamere and Tamiya 2009). However, the link between NBCe1 defects and glaucoma is less straightforward. For example, NBCe1-B in the ciliary body is theoretically positioned to contribute toward the production

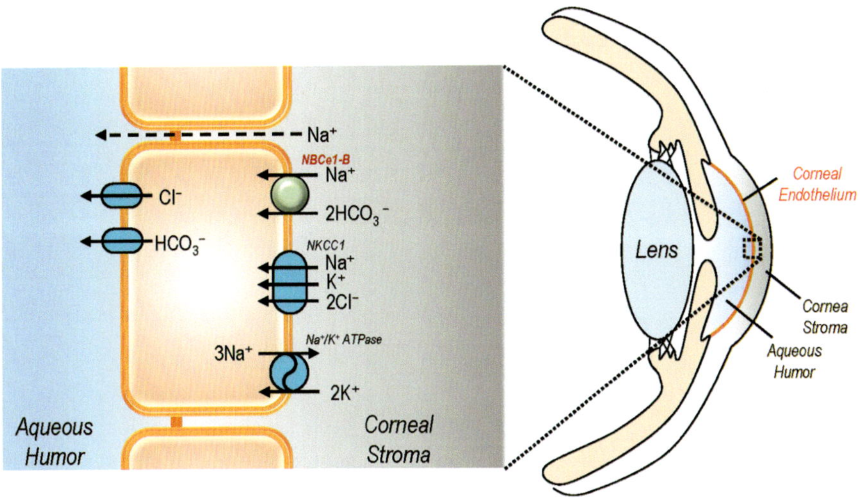

Fig. 4.5 Bicarbonate transport by corneal endothelial cells. The protein matrix of the corneal stroma draws fluid from the aqueous humor through the leaky tight junctions between corneal endothelial cells. NBCe1-B supports the transepithelial movement of anions (and therefore ultimately fluid) from the stroma back into the aqueous humor, countering the fluid accumulation that would otherwise distort and opacify the cornea

of aqueous humor, yet the phenotype seems consistent with *elevated* intraocular pressure. Furthermore, glaucoma/elevated intraocular pressure is the only ocular pathology exhibited by patients and mice with NBCe1-A-specific mutations (Igarashi et al. 2001; Romero et al. 2014). A link to the systemic effects of pRTA appears to be ruled out by glaucoma/elevated intraocular pressure and band keratopathy in individuals with reduced NBCe1 mRNA stability but normal blood pH (Patel et al. 2017). This leaves an NBCe1-A-specific role in aqueous humor drainage as a possible pathological mechanism. Yet, glaucoma is a clinical diagnosis referring to neuropathy of the retina and/or optic nerve, indicating that direct damage to these structures could also be part of the ocular pathophysiology in mice and humans.

4.4.4 The Enamel Organ: Hypomineralized Enamel

NBCe1-B is expressed in the ameloblasts of the enamel organs (Lacruz et al. 2010). In these cells it supports—either directly or indirectly—the secretion of HCO_3^- that neutralizes the acid generated by hydroxyapatite formation in the extracellular enamel-forming space, thereby promoting enamel mineralization (Jalali et al. 2014; Bronckers et al. 2016): see mechanism in Fig. 4.6. Some patients with NBCe1 mutations are noted as having poor dentition (Table 4.1). This pathology was originally thought to be secondary to acidemia or altered saliva chemistry. The

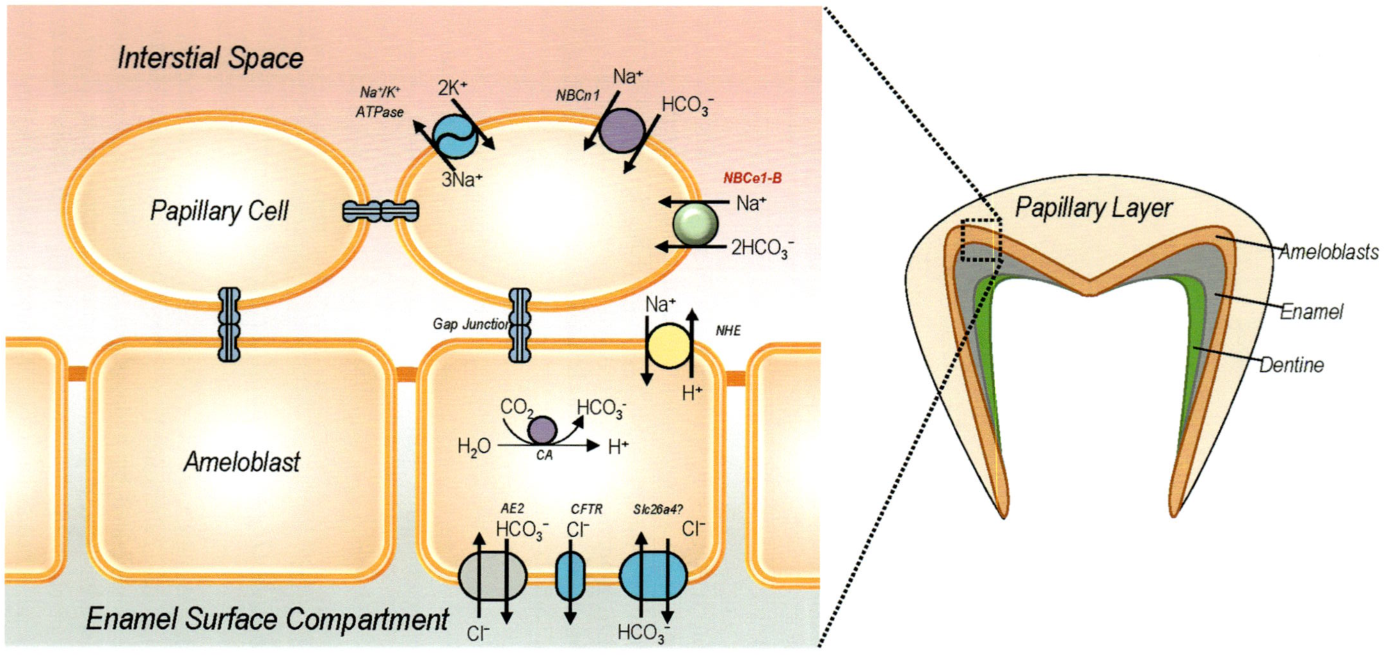

Fig. 4.6 Bicarbonate transport by ameloblasts. NBCe1-B, either in the papillary cells or in the basolateral membrane of the ameloblasts themselves (the two form a functional syncytium) supports the secretion of HCO_3^- into the enamel forming space. The neutralization of acid in this space promotes the maturation of the enamel rods that are formed as hydroxyapatite crystals form. AE2 = anion exchanger 2 (i.e., Slc4a2); CA = carbonic anhydrase; CFTR = cystic fibrosis transmembrane conductance regulator; NHE = Na^+/H^+ exchanger; Slc26a4 = pendrin

patient with the S427L recessive mutation was the first case whose parents and siblings had normal dentition (Dinour et al. 2004), and the L522P mutation was also associated with enamel hypoplasia (Demirci et al. 2006). The former case suggested that either improper enamel development could be directly associated with the NBCe1 mutation or secondary to academia. However, the discovery of hypomineralized enamel in pre-erupted teeth and in the teeth of Nbce1-B/C-null mice is most consistent with a local effect of NBCe1-B loss from the ameloblasts (Lacruz et al. 2010; Salerno et al. 2019).

4.4.5 *The Pancreas: Elevated Serum Amylase Levels*

In humans, NBCe1-B is abundantly expressed in the basolateral membrane of pancreatic duct cells (Marino et al. 1999; Satoh et al. 2003), where it is hypothesized to support the secretagogue-stimulated secretion of the HCO_3^--rich pancreatic juice (see mechanism in Fig. 4.7) that carries digestive enzymes to the duodenum. The alkalinity serves two purposes; it prevents premature activation of acid-sensitive pro-enzymes such as trypsinogen and neutralizes the acidic chyme that is released by the stomach, producing a pH-neutral environment in which digestion can proceed. Dysfunction of NBCe1-B and compromised secretion of the HCO_3^- rich fluid might permit premature activation of the secreted enzymes, leading to pancreatitis. Indeed, several individuals with pRTA exhibit signs of pancreatic dysfunction including elevated serum amylase and lipase levels (Table 4.1). NBCe1 is also reported to play a role in the endocrine portion of the pancreas in support of glucose-stimulated insulin secretion (Soyfoo et al. 2009), although the proposal that it does so by mediating HCO_3^- efflux to dispose of metabolic CO_2 is unusual and remains to be confirmed (Parker and Boron 2013). Recently, it has been reported that the loss of NBCe1 in pancreatic β-cells preserves β-cell function, while increased NBCe1 expression seems to be associated with type 2 diabetes (Brown et al. 2019).

4.4.6 *The Intestinal Tract: Blockage and Nutritional Deficits*

NBCe1-B is expressed in enterocytes and crypt cells throughout the digestive system (Jacob et al. 2000; Praetorius et al. 2001; Bartolo et al. 2009; Yu et al. 2009; Charoenphandhu et al. 2011; Jakab et al. 2011) where it contributes toward the basolateral import of HCO_3^-. This action is hypothesized to support the apical secretion of HCO_3^- that protects duodenal epithelia from acid damage, promotes intestinal fluid secretion, and accompanies the absorption of ions such as Na^+ and Cl^- (Gawenis et al. 2007; Yu et al. 2009): see Fig. 4.8). NBCe1 action also promotes H^+-coupled nutrient uptake by balancing enterocyte pH_i (Yu et al. 2016). The importance of NBCe1 action for these processes is underscored by several phenotypic observations. Notably, [a] NBCe1 has been genetically linked to the severity of

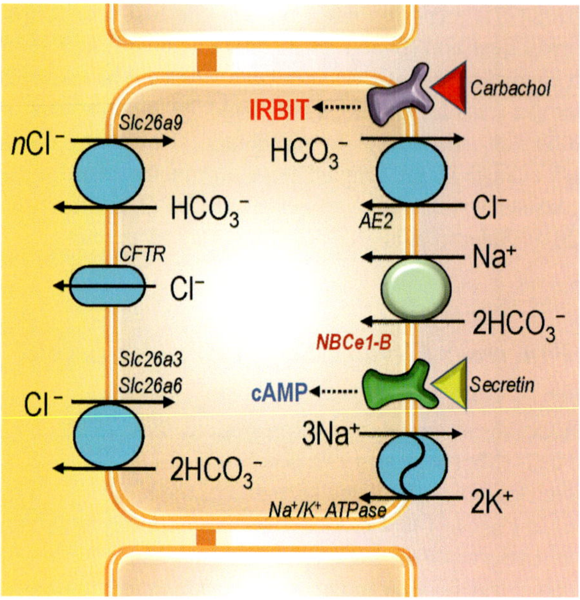

Fig. 4.7 Bicarbonate transport by pancreatic duct epithelia. Apical players are Slc26a3, Slc26a6, and Slc26a9 which can interact with CFTR (Chang et al. 2009; Ko et al. 2002, 2004). NBCe1-B is the major basolateral bicarbonate transporter, which is activated by IP_3 receptor-binding protein (IRBIT), thereby increasing bicarbonate secretion by pancreatic duct cells. At the basolateral membrane, carbachol and secretin both stimulate bicarbonate secretion by the pancreatic duct epithelium. Carbachol stimulation is mediated by PLC and IRBIT (see also Fig. 4.9). Secretin increases intracellular cAMP concentration which indirectly activates NBCe1-B (not shown) and therefore enhances bicarbonate secretion by pancreatic ducts. AE2 = anion exchanger 2

ileal obstruction in newborns with cystic fibrosis (Dorfman et al. 2009), [b] Failure to thrive in pRTA (Table 4.1), Nbce1-null mice (Gawenis et al. 2007) and the Nbce1-B/C mice (Salerno et al. 2019), and [c] Jejunal failure in Nbce1-null mice (Yu et al. 2016). In addition, colonic impaction is observed in Nbce1-null mice, although this is likely a phenotype exacerbated by elevated aldosterone in these mice, which activates the epithelial sodium channel ENaC resulting in intestinal hyperabsorption (Gawenis et al. 2007). Recent preliminary studies also indicate that in addition to intestinal epithelial transport, NBCe1 plays a role in pacing the intestinal, smooth muscle contraction mediated by the interstitial cells of Cajal (ICC) (Colmenares-Aguilar et al. 2019; Zhao et al. 2019).

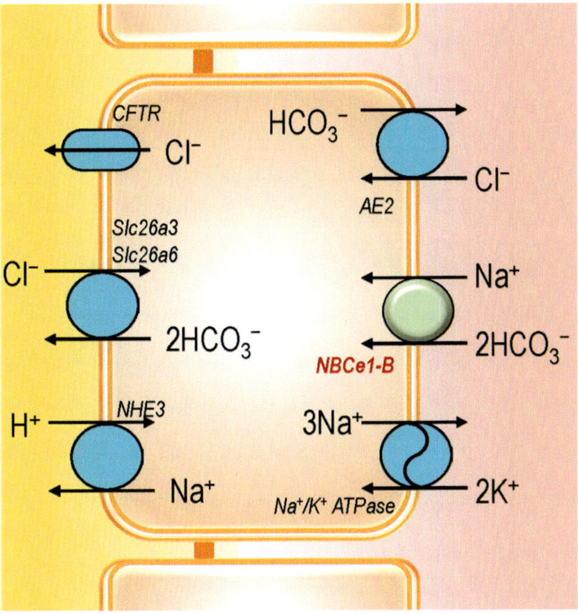

Fig. 4.8 Bicarbonate transport by intestinal epithelia. Bicarbonate from the blood and transported by basolateral NBCe1-B transporter is secreted from the cells via apical proteins (Slc26a3 and Slc26a6 which can interact with CFTR; Ko et al. 2002, 2004). AE2 = anion exchanger 2; NHE3 = Na^+/H^+ exchanger

4.5 The Genetic Basis of NBCe1-Linked Disease

4.5.1 Human Disease

Inherited pRTA is a devastating disease associated with acidemia, short stature, ocular abnormalities, and diverse other signs with variable penetrance and expressivity. To date, 16 autosomal recessive *SLC4A4* mutations have been documented in individuals with inherited pRTA. Of these mutations, 14 are homozygous and two are compound heterozygous mutations, all fall within coding exons, and all cause the partial or complete loss of functional expression of NBCe1 protein (summarized in Table 4.1). One further mutation, in the long terminal exon of *SLC4A4* that encodes the 3′-UTR, has been reported in six individuals with band keratopathy but no acidemia (Patel et al. 2017). The mutation creates an AU-rich destabilizing motif in the mRNA, reducing transcript lifetime. The apparent tissue specificity of the destabilizing mutant is hypothesized to be related either to the relative importance of transcript lifetime among cell-types, or the tissue-specific expression of factors that act upon the destabilizing motif (Patel et al. 2017).

Three of the pRTA-linked NBCe1 mutants exhibit unanticipated features: [a] T485S has been suggested to operate as an electroneutral Na^+:HCO_3^- cotransporter, and thus to be unable to support HCO_3^- efflux in the renal PT (Zhu et al. 2013) (see Sect. 4.3.1), [b] A799V is associated with a novel HCO_3^--independent, but DIDS-stimulated, Na^+ leak that has been proposed to contribute toward the expression of hyperkalemic periodic paralysis in the affected patient

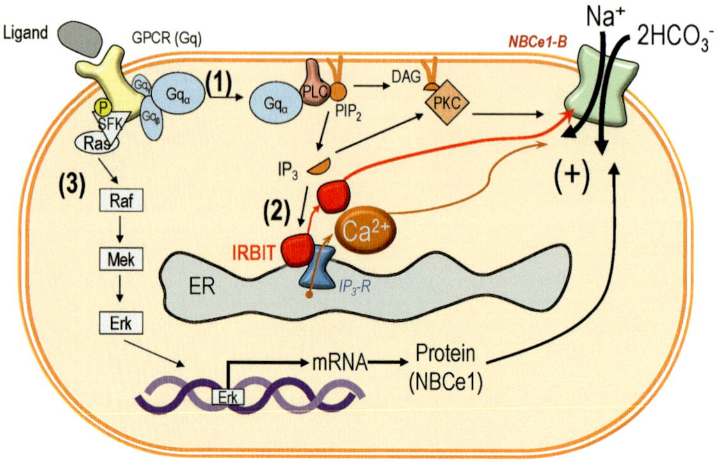

Fig. 4.9 Regulation of NBCe1. (1) Classic G_q signaling cascade: G_{qa} subunit activation of phospholipase C (PLC) results in cleavage of phosphatidylinositol 4,5-bisphosphate (PIP_2) into two intracellular second messengers, inositol 1,4,5 triphosphate (IP_3) and diacylglyerol (DAG), both of which activate PKC activity, subsequently modulating NBCe1 activity. PIP_2 also appears to have direct NBCe1 stimulatory effects. IP_3 binds the IP_3-Receptor (IP_3-R), releasing Ca^{2+} from the sarcoplasmic reticulum (SR), directly influencing NBCe1 as well as activating calcium-dependent kinases such as calmodulin kinase II (CaMKII; not shown). (2) IP_3 further displaces Inositol 1,4,5 trisphosphate receptor-binding protein released with inositol (IRBIT) from IP_3-R (see Fig. 4.10). (3) Mitogen-activated protein kinase (MAPK) cascade activation: muscarinic receptor 1 (MR1) and angiotensin II receptor-1 (AT1) agonism (both are G_{qa} coupled GPCRs) activate the MAPK cascade, resulting in transcriptional changes, upregulating NBCe1 expression and activity. Activation of the MAPK cascade is dependent on the SRC family kinases (SFK) and the GTP-associated protein RAS. ER = endoplasmic reticulum

(Parker et al. 2012), and [c] the Q913R mutant that is expressed in the individual with the compound heterozygous inheritance of R510H/Q913R alleles is associated with a novel HCO_3^--independent Cl^- leak that has been proposed to contribute toward the expression of diverse neuromuscular signs in the affected patient (Myers et al. 2016). Otherwise, there is little obvious genotype/phenotype correlation.

Two studies might be considered to distinguish the isolated local effects of NBCe1 mutation from the systemic effects of acidemia. The first is a pRTA case involving a mutation (p.Q29X) that is predicted to specifically affect the NBCe1-A variant, leaving NBCe1-B/C intact (Igarashi et al. 2001). This acidemic patient presented with glaucoma, short stature, and intellectual impairment. Thus, these signs are either direct results of non-renal NBCe1-A loss or, perhaps more likely given the sparse expression of NBCe1-A outside of the kidneys, secondary to acidemia. The second is a study of three families in which a point mutation in the 3′-UTR of NBCe1 creates a destabilizing motif in the mRNA. This mutation results in elevated intraocular pressure/glaucoma and band keratopathy in the absence of acidemia (Patel et al. 2017). Thus, these signs appear to be unrelated to the systemic

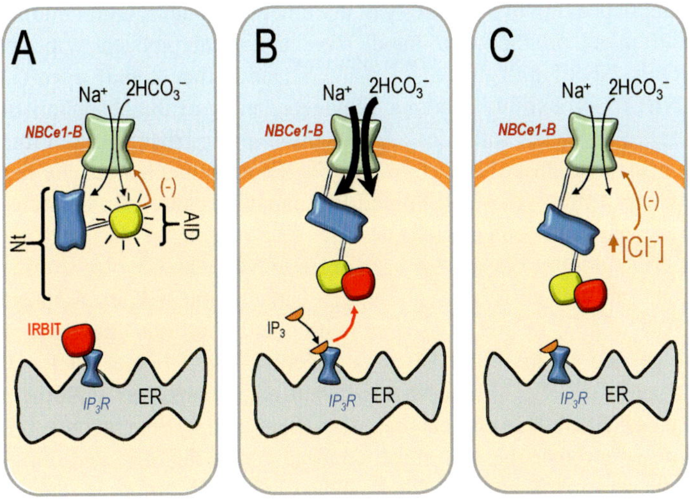

Fig. 4.10 Modulation of NBCe1-B activity by IRBIT and autoinhibition. (**a**, **b**) As described in Fig. 4.9 cleavage of PIP_2 results in the formation of second messenger IP_3; IP_3 competes with IRBIT for IP_3-R binding, causing dissociation of IRBIT from the IP_3-R allowing IRBIT to bind the auto-inhibitory domain (AID) of NBCe1-B. This relieves the autoinhibition of NBCe1-B, probably by preventing the inhibitory activities of WNK/SPAK kinases (not shown). (**c**) Recruitment by IRBIT of kinases such as Ca^{2+}/Calmodulin-dependent Kinase II and SPAK, and phosphatases such as protein phosphatase-1 and Calcineurin further modify the phosphorylation states of NBCe1-B. These changes affect the affinity of NBCe1-B to chloride (Cl^-) binding, which has an inhibitory effect on NBCe1-B. Thus, despite IRBIT activation of NBCe1-B activity, secondary IRBIT-dependent changes in Cl^- affinity can inhibit NBCe1-B activity, overall allowing NBCe1-B to respond to a wide range of intracellular Cl^- concentrations. IRBIT = IP_3 receptor binding protein

effects of acidemia. Further insights have been provided by the study of transgenic mouse models of *SLC4A4* mutations.

4.5.2 Mouse Models

Four mouse models of *SLC4A4* disruption have been studied: [a] “Nbce1-null,” a strain constructed to prevent the expression of all NBCe1 mRNA and protein variants (Gawenis et al. 2007), [b] “Nbce1$^{W516X/W516X}$,” a strain that carries the human W516X mutation, also preventing the expression of all NBCe1 variants (Lo et al. 2011), [c] “Nbce1a-null,” a strain in which only Nbce1a expression is disrupted leaving Nbce1b/c expression unaltered (Lee et al. 2018; Romero et al. 2014), and [d] “Nbce1b/c-null,” a strain in which only Nbce1b/c expression is disrupted leaving Nbce1a expression unaltered (Salerno et al. 2019).

Nbce1-null, Nbce1$^{W516X/W516X}$, and Nbce1b/c-null mice fail to thrive. These animals exhibit low body weight, reduced bone dimensions, and 100% mortality by 3 weeks (for mice lacking all NBCe1 variants) to 10 weeks (for Nbce1b/c-null mice), perhaps speaking to the scarcity of documented patients with inherited pRTA. Nbce1a-null mice, on the other hand, have a normal physical appearance and lifespan. Only Nbce1-null, Nbce1$^{W516X/W516X}$, and Nbce1a-null mice (i.e., those that lack NBCe1-A) exhibit acidemia. However, many of the characteristic phenotypes observed in Nbce1-null mice are also observed in Nbce1b/c-null mice. Thus, metabolic acidosis follows NBCe1-A loss, but the acidemia is not fully responsible for the developmental issues and limited lifespan. The cause of premature death of these animals has not been established.

Elevated intraocular pressure has been noted in Nbce1a-null mice, consistent with the expression of glaucoma in the individual carrying the NBCe1-A-specific mutation Q29X (Table 4.1) and consistent with the hypothesis that glaucoma is a phenotype related to loss of NBCe1-A (Sect. 4.4.3). Corneal edema has been noted in Nbce1$^{W516X/W516X}$ and Nbce1b/c-null mice, consistent with the proposed importance of NBCe1-B in corneal hydration (Sect. 4.4.3). Enamel defects are noted in Nbce1-null and Nbce1b/c-null mice, consistent with the proposed importance of NBCe1-B in enamel formation (Sect. 4.4.4). The exhibition of these traits in Nbce1b/c-null mice indicates that neither of these phenotypes is primarily due to acidemia, challenging the previous paradigm of alkali therapy as a panacea for all signs of inherited pRTA.

Interestingly, Nbce1-null mice have impacted intestines; a phenotype not observed in human patients with inherited pRTA. This was originally hypothesized to be due to defective fluid secretion due to the NBCe1-B/C loss from the intestines, together with ENaC-mediated hyperabsorption secondary to acidemia. However, this phenotype is not exhibited by Nbce1b/c-null mice suggesting that the latter may be a critical factor.

4.6 Regulation of NBCe1

4.6.1 General Comments

In the following subsections, we summarize our current understanding of NBCe1 regulation. Briefly, isoform-type, phosphorylation by protein kinase A (PKA) and protein kinase C (PKC), IP_3-receptor-binding protein released with IP3 (IRBIT), carbonic anhydrase (CA), Phosphatidylinositol 4,5-bisphosphate (PIP_2), and other small molecules all appear to have some regulatory influence. Although there is a wealth of studies regarding the regulation of epithelial HCO_3^- secretion and reabsorption, and NBCe1 includes numerous demonstrated phosphorylation sites (Feric et al. 2011), studies of how or if those signaling pathways act on NBCe1 are relatively few. Thus, the reader may glean additional information from studies of NBCe1 regulation in cardiomyocytes and heterologous systems, e.g., De Giusti et al. (2013); Perry et al. (2007); and Thornell and Bevensee (2015).

4.6.2 Autoregulation: NBCe1-A Versus NBCe1-B/C

Differences in Nt domain sequences between NBCe1-A versus NBCe1-B/C confer different intrinsic activities and different regulatory potential between these isoforms (McAlear et al. 2006). The first 41 aa of NBCe1-A constitute an ASD. An artificial NBCe1 clone that lacks the ASD has a twofold lower activity than full-length NBCe1-A (McAlear et al. 2006). NBCe1-B/C do not include this sequence; in its place is an 85 aa sequence that constitutes an AID (Fig. 4.2). An artificial clone that lacks the AID has a three- to fivefold greater activity than full-length NBCe1-B/C (Lee et al. 2011b; McAlear et al. 2006). Thus, NBCe1-A has a substantially greater activity than NBCe1-B/C, although the lower activity NBCe1-B/C variants can be stimulated by the AID-binding partner IRBIT (see Sect. 4.6.4). The basis for these functional differences is unknown as neither the ASD or AID are represented in current Nt structures (see Sect. 4.2.3).

4.6.3 Modulating NBCe1 to Control Renal HCO_3^- Reabsorption

The presence of the constitutively high-activity isoform NBCe1-A in proximal convoluted tubular epithelia reflects the requirement for these cells to transport large quantities of HCO_3^- into the blood (see Sect. 4.4.2). Whether NBCe1 requires specific modulation, such as phosphorylation by PKA, in order to achieve a state that favors mediation of HCO_3^- efflux is unresolved (see Sect. 4.3.1). Conditions that cause an increase in NBCe1 activity/plasma-membrane abundance include metabolic acidosis (which also induces NBCe1-B expression in the proximal straight tubule (Fang et al. 2018)), chronic hypercapnia, and hypovolemia (Kanaan et al. 2007; Ma et al. 2008; Preisig and Alpern 1988; de Seigneux et al. 2007). The responses to altered CO_2/HCO_3^- balance are dependent on the secretion of local angiotensin II and the subsequent agonism of luminal angiotensin AT_1 receptors (Fig. 4.9) (Zhou et al. 2007; Zhou and Boron 2008). Cholinergic agents also enhance NBCe1 activity via muscarinic M_1 receptors (Fig. 4.9). Pyk2 and Src family tyrosine kinases, as well as the MAP kinases (Fig. 4.9) all play roles in these processes (Ruiz et al. 1997; Espiritu et al. 2002; Robey et al. 2002). The means by which these pathways integrate upon NBCe1 are unknown, but clues are provided by heterologous studies: for example, the plasma membrane abundance of NBCe1-A is sensitive to the action of PKC (Ruiz et al. 1997; Perry et al. 2007) and PIP_2 directly activates NBCe1-A (Wu et al. 2009). Conditions that cause a decrease in NBCe1 activity include Na^+ loading and metabolic alkalosis (Amlal et al. 2001; Ma et al. 2008).

4.6.4 *Modulating NBCe1 to Control Intestinal HCO_3^- Secretion*

The presence of the low-activity, but stimulatable, isoform NBCe1-B in pancreatic and intestinal epithelia (see Sects. 4.4.5 and 4.4.6) reflects the requirement for these cells to increase secretion of HCO_3^- in response to the release of secretagogues such as secretin (see Chap. 12 in Volume 1: "Fundamentals of Bicarbonate Secretion in Epithelia"). The details are complex and are still being elucidated, but a key component of this system is IRBIT (Figs. 4.9 and 4.10). In short, following the agonism of G-protein coupled receptors, cAMP and Ca^{2+}-dependent signaling pathways converge upon IRBIT as well as residues with the NBCe1-B Nt domain allowing the binding of IRBIT to the AID of NBCe1-B. This interaction recruits numerous phosphatases and kinases to the vicinity of NBCe1-B, reveals a Cl^--sensing site, neutralizes the inhibitory effect of the AID, and increases NBCe1-B activity and plasma membrane abundance (Fig. 4.10c) (Shirakabe et al. 2006; Lee et al. 2011b; Ahuja et al. 2014; Ando et al. 2014; May et al. 2014; Vachel et al. 2018). The process is antagonized by the WNK/SPAK signaling pathway that retrieves NBCe1-B from the plasma membrane (Park et al. 2012).

4.7 Conclusion

It is evident that NBCe1 plays major physiological roles in the kidneys, eyes, teeth, and digestive system. The study of humans with recessive NBCe1 mutations and mouse models of NBCe1 dysfunction has greatly facilitated our understanding of the nature and importance of these processes and how they are impacted by NBCe1 mutation. Recent advances in our understanding of the atomic structure of NBCe1 structure provide the foundation for future studies that will determine the mechanism of action of NBCe1 as well as other SLC4 proteins and the reasons for the dysfunction of NBCe1 mutants, perhaps leading to the development of personalized treatments for these disorders.

Acknowledgments We thank our colleagues, collaborators, and past lab members who have contributed to our understanding of the importance of NBCe1. This work was supported by NIH EY017732 (MFR), EY028580 (MDP), F30-DK126330 (CTB), R25-DK101405 (MFR), the Cystic Fibrosis Foundation (Sindic-06F0, Romero-06G0), the Arlene and Robert Kogod Aging Center and the Mayo Foundation.

References

Abuladze N, Lee I, Newman D, Hwang J, Boorer K, Pushkin A, Kurtz I (1998) Molecular cloning, chromosomal localization, tissue distribution, and functional expression of the human pancreatic sodium bicarbonate cotransporter. J Biol Chem 273:17689–17695

Abuladze N, Song M, Pushkin A, Newman D, Lee I, Nicholas S, Kurtz I (2000) Structural organization of the human NBC1 gene: kNBC1 is transcribed from an alternative promoter in intron 3. Gene 251:109–122

Ahuja M, Jha A, Maléth J, Park S, Muallem S (2014) cAMP and Ca^{2+} signaling in secretory epithelia: crosstalk and synergism. Cell Calcium 55:385–393

Amlal H, Wang Z, Burnham C, Soleimani M (1998) Functional characterization of a cloned human kidney $Na^{+}:HCO_3^{-}$ cotransporter. J Biol Chem 273:16810–16815

Amlal H, Chen Q, Greeley T, Pavelic L, Soleimani M (2001) Coordinated down-regulation of NBC-1 and NHE-3 in sodium and bicarbonate loading. Kidney Int 60:1824–1836

Ando H, Kawaai K, Mikoshiba K (2014) IRBIT: a regulator of ion channels and ion transporters. Biochim Biophys Acta 1843:2195–2204

Bai X, Moraes TF, Reithmeier RAF (2017) Structural biology of solute carrier (SLC) membrane transport proteins. Mol Membr Biol 34:1–32

Bartolo RC, Harfoot N, Gill M, McLeod BJ, Butt AG (2009) Secretagogues stimulate electrogenic HCO_3- secretion in the ileum of the brushtail possum, Trichosurus vulpecula: evidence for the role of a $Na+/HCO_3-$ cotransporter. J Exp Biol 212:2645–2655. https://doi.org/10.1242/jeb.028928

Bevensee MO, Schmitt BM, Choi I, Romero MF, Boron WF (2000) An electrogenic $Na+-HCO_3-$ cotransporter (NBC) with a novel COOH-terminus, cloned from rat brain. Am J Phys Cell Phys 278:C1200–C1211

Bok D, Schibler MJ, Pushkin A, Sassani P, Abuladze N, Naser Z, Kurtz I (2001) Immunolocalization of electrogenic sodium-bicarbonate cotransporters pNBC1 and kNBC1 in the rat eye. Am J Physiol Renal Physiol 281:F920–F935

Bonanno JA (2012) Molecular mechanisms underlying the corneal endothelial pump. Exp Eye Res 95:2–7

Boron WF, Boulpaep EL (1983) Intracellular pH regulation in the renal proximal tubule of the salamander. Basolateral HCO_3^{-} transport. J Gen Physiol 81:53–94

Bronckers ALJJ, Lyaruu DM, Jalali R, DenBesten PK (2016) Buffering of protons released by mineral formation during amelogenesis in mice. Eur J Oral Sci 124:415–425

Brown M, Holmes H, Rakshit K, Javeed N, Romero MF, Matveyenko A (2019) NBCe1 is a novel regulator of pancreatic beta-cell function in diabetes. FASEB J 33:544.4

Burnham CE, Amlal H, Wang Z, Shull GE, Soleimani M (1997) Cloning and functional expression of a human kidney $Na+:HCO_3-$ cotransporter. J Biol Chem 272:19111–19114

Chang M-H, DiPiero J, Sönnichsen FD, Romero MF (2008) Entry to "formula tunnel" revealed by SLC4A4 human mutation and structural model. J Biol Chem 283:18402–18410

Chang M-H, Plata C, Sindic A, Ranatunga WK, Chen A-P, Zandi-Nejad K, Chan KW, Thompson J, Mount DB, Romero MF (2009) Slc26a9 is inhibited by the R-region of the cystic fibrosis transmembrane conductance regulator via the STAS domain. J Biol Chem 284:28306–28318

Charoenphandhu N, Laohapitakworn S, Kraidith K, Nakkrasae L-I, Jongwattanapisan P, Tharabenjasin P, Krishnamra N (2011) Electrogenic $Na+/HCO_3-$ co-transporter-1 is essential for the parathyroid hormone-stimulated intestinal HCO_3- secretion. Biochem Biophys Res Commun 409:775–779. https://doi.org/10.1016/j.bbrc.2011.05.087

Ch'en FF-T, Villafuerte FC, Swietach P, Cobden PM, Vaughan-Jones RD (2008) S0859, an N-cyanosulphonamide inhibitor of sodium-bicarbonate cotransport in the heart. Br J Pharmacol 153:972–982. https://doi.org/10.1038/sj.bjp.0707667

Choi I, Aalkjaer C, Boulpaep EL, Boron WF (2000) An electroneutral sodium/bicarbonate cotransporter NBCn1 and associated sodium channel. Nature 405:571–575

Choi I, Hu L, Rojas JD, Schmitt BM, Boron WF (2003) Role of glycosylation in the renal electrogenic Na^{+}-HCO_3^{-} cotransporter (NBCe1). Am J Physiol Renal Physiol 284:F1199–F1206

Choi I, Romero MF, Khandoudi N, Bril A, Boron WF (1999) Cloning and characterization of a human electrogenic Na+-HCO_3− cotransporter isoform (hhNBC). Am J Phys 276:C576–C584

Colmenares-Aguilar M, Mazzone A, Zhao W, Strege PR, Eisenman S, Silva J, Sha L, Shull GE, Romero MF, Farrugia G et al (2019) Immunoreactivity for the Na^{+}/HCO_3^{-} cotransporter (NBCe1, Slc4a4) in pacemaker interstitial cells of cajal is selectively reduced by knocking down Slc4a4 gene expression using an ETV1-cre driver in mouse. Neurogastroenterol Motil 31:13658

De Giusti VC, Orlowski A, Villa-Abrille MC, de Cingolani GEC, Casey JR, Alvarez BV, Aiello EA (2011) Antibodies against the cardiac sodium/bicarbonate co-transporter (NBCe1) as pharmacological tools. Br J Pharmacol 164:1976–1989

De Giusti VC, Ciancio MC, Orlowski A, Aiello EA (2013) Modulation of the cardiac sodium/ bicarbonate cotransporter by the renin angiotensin aldosterone system: pathophysiological consequences. Front Physiol 4:411

de Seigneux S, Malte H, Dimke H, Frøkiaer J, Nielsen S, Frische S (2007) Renal compensation to chronic hypoxic hypercapnia: downregulation of pendrin and adaptation of the proximal tubule. Am J Physiol Renal Physiol 292:F1256–F1266

Deda G, Ekim M, Güven A, Karagöl U, Tümer N (2001) Hypopotassemic paralysis: a rare presentation of proximal renal tubular acidosis. J Child Neurol 16:770–771

Delamere NA, Tamiya S (2009) Lens ion transport: from basic concepts to regulation of Na,K-ATPase activity. Exp Eye Res 88:140–143

Demirci FYK, Chang M-H, Mah TS, Romero MF, Gorin MB (2006) Proximal renal tubular acidosis and ocular pathology: a novel missense mutation in the gene (SLC4A4) for sodium bicarbonate cotransporter protein (NBCe1). Mol Vis 12:324–330

Dinour D, Chang M-H, Satoh J, Smith BL, Angle N, Knecht A, Serban I, Holtzman EJ, Romero MF (2004) A novel missense mutation in the sodium bicarbonate cotransporter (NBCe1/SLC4A4) causes proximal tubular acidosis and glaucoma through ion transport defects. J Biol Chem 279:52238–52246

Dorfman R, Li W, Sun L, Lin F, Wang Y, Sandford A, Paré PD, McKay K, Kayserova H, Piskackova T et al (2009) Modifier gene study of meconium ileus in cystic fibrosis: statistical considerations and gene mapping results. Hum Genet 126:763–778

Ducoudret O, Diakov A, Müller-Berger S, Romero MF, Frömter E (2001) The renal Na-HCO_3-cotransporter expressed in *Xenopus laevis* oocytes: inhibition by tenidap and benzamil and effect of temperature on transport rate and stoichiometry. Pflugers Arch 442:709–717

Espiritu DJD, Bernardo AA, Robey RB, Arruda JAL (2002) A central role for Pyk2-Src interaction in coupling diverse stimuli to increased epithelial NBC activity. Am J Physiol Renal Physiol 283:F663–F670

Espiritu DJD, Bernardo AA, Arruda JAL (2006) Role of NH_2 and COOH termini in targeting, stability, and activity of sodium bicarbonate cotransporter 1. Am J Physiol Renal Physiol 291: F588–F596

Fang L, Lee H-W, Chen C, Harris AN, Romero MF, Verlander JW, Weiner ID (2018) Expression of the B splice variant of NBCe1 (SLC4A4) in the mouse kidney. Am J Physiol Renal Physiol 315: F417–F428

Fantinelli JC, Orlowski A, Aiello EA, Mosca SM (2014) The electrogenic cardiac sodium bicarbonate co-transporter (NBCe1) contributes to the reperfusion injury. Cardiovasc Pathol 23:224–230

Feric M, Zhao B, Hoffert JD, Pisitkun T, Knepper MA (2011) Large-scale phosphoproteomic analysis of membrane proteins in renal proximal and distal tubule. Am J Physiol Cell Physiol 300:C755–C770

Gawenis LR, Bradford EM, Prasad V, Lorenz JN, Simpson JE, Clarke LL, Woo AL, Grisham C, Sanford LP, Doetschman T et al (2007) Colonic anion secretory defects and metabolic acidosis in mice lacking the NBC1 cotransporter. J Biol Chem 282:9042–9052

Gill HS, Boron WF (2006a) Preliminary X-ray diffraction analysis of the cytoplasmic N-terminal domain of the Na/HCO_3 cotransporter NBCe1-A. Acta Crystallogr Sect F Struct Biol Cryst Commun 62:534–537

Gill HS, Boron WF (2006b) Expression and purification of the cytoplasmic N-terminal domain of the Na/HCO_3 cotransporter NBCe1-A: structural insights from a generalized approach. Protein Expr Purif 49:228–234

Gill HS, Dutcher L, Boron WF, Patel S, Guay-Woodford LM (2013) X-ray diffraction studies on merohedrally twinned Δ1-62NtNBCe1-A crystals of the sodium/bicarbonate cotransporter. Acta Crystallogr Sect F Struct Biol Cryst Commun 69:796–799

Grichtchenko II, Romero MF, Boron WF (2000) Extracellular HCO_3^- dependence of Electrogenic Na^+/HCO_3^- cotransporters cloned from salamander and rat kidney. J Gen Physiol 115:533–546

Grichtchenko II, Choi I, Zhong X, Bray-Ward P, Russell JM, Boron WF (2001) Cloning, characterization, and chromosomal mapping of a human electroneutral Na^+-driven Cl-HCO_3 exchanger. J Biol Chem 276:8358–8363

Gross E, Hopfer U (1996) Activity and stoichiometry of Na^+:HCO_3^- cotransport in immortalized renal proximal tubule cells. J Membr Biol 152:245–252

Gross E, Abuladze N, Pushkin A, Kurtz I, Cotton CU (2001) The stoichiometry of the electrogenic sodium bicarbonate cotransporter pNBC1 in mouse pancreatic duct cells is 2 HCO_3^-:1 Na^+. J Physiol Lond 531:375–382

Horita S, Yamada H, Inatomi J, Moriyama N, Sekine T, Igarashi T, Endo Y, Dasouki M, Ekim M, Al-Gazali L et al (2005) Functional analysis of NBC1 mutants associated with proximal renal tubular acidosis and ocular abnormalities. J Am Soc Nephrol 16:2270–2278

Horita S, Simsek E, Simsek T, Yildirim N, Ishiura H, Nakamura M, Satoh N, Suzuki A, Tsukada H, Mizuno T et al (2018) SLC4A4 compound heterozygous mutations in exon-intron boundary regions presenting with severe proximal renal tubular acidosis and extrarenal symptoms coexisting with Turner's syndrome: a case report. BMC Med Genet 19:103

Huynh KW, Jiang J, Abuladze N, Tsirulnikov K, Kao L, Shao X, Newman D, Azimov R, Pushkin A, Zhou ZH et al (2018) CryoEM structure of the human SLC4A4 sodium-coupled acid-base transporter NBCe1. Nat Commun 9:900

Igarashi T, Inatomi J, Sekine T, Cha SH, Kanai Y, Kunimi M, Tsukamoto K, Satoh H, Shimadzu M, Tozawa F et al (1999) Mutations in *SLC4A4* cause permanent isolated proximal renal tubular acidosis with ocular abnormalities. Nat Genet 23:264–266

Igarashi T, Inatomi J, Sekine T, Seki G, Shimadzu M, Tozawa F, Takeshima Y, Takumi T, Takahashi T, Yoshikawa N et al (2001) Novel nonsense mutation in the Na^+/HCO_3^- cotransporter gene (SLC4A4) in a patient with permanent isolated proximal renal tubular acidosis and bilateral glaucoma. JASN 12:713–718

Inatomi J, Horita S, Braverman N, Sekine T, Yamada H, Suzuki Y, Kawahara K, Moriyama N, Kudo A, Kawakami H et al (2004) Mutational and functional analysis of SLC4A4 in a patient with proximal renal tubular acidosis. Pflugers Arch 448:438–444

Jacob P, Christiani S, Rossmann H, Lamprecht G, Vieillard-Baron D, Muller R, Gregor M, Seidler U (2000) Role of Na+/HCO3- cotransporter NBC1, Na+/H+ exchanger NHE1, and carbonic anhydrase in rabbit duodenal bicarbonate secretion. Gastroenterology 119:406–419

Jakab RL, Collaco AM, Ameen NA (2011) Physiological relevance of cell-specific distribution patterns of CFTR, NKCC1, NBCe1, and NHE3 along the crypt-villus axis in the intestine. Am J Physiol Gastrointest Liver Physiol 300:G82–G98. https://doi.org/10.1152/ajpgi.00245.2010

Jalali R, Guo J, Zandieh-Doulabi B, Bervoets TJM, Paine ML, Boron WF, Parker MD, Bijvelds MJC, Medina JF, DenBesten PK et al (2014) NBCe1 (SLC4A4) a potential pH regulator in enamel organ cells during enamel development in the mouse. Cell Tissue Res 358:433–442

Jentsch TJ, Schill BS, Schwartz P, Matthes H, Keller SK, Wiederholt M (1985) Kidney epithelial cells of monkey origin (BSC-1) express a sodium bicarbonate cotransport. Characterization by 22Na+ flux measurements. J Biol Chem 260:15554–15560

Kanaan A, Douglas RM, Alper SL, Boron WF, Haddad GG (2007) Effect of chronic elevated carbon dioxide on the expression of acid-base transporters in the neonatal and adult mouse. Am J Physiol Regul Integr Comp Physiol 293:R1294–R1302

Kao L, Sassani P, Azimov R, Pushkin A, Abuladze N, Peti-Peterdi J, Liu W, Newman D, Kurtz I (2008) Oligomeric structure and minimal functional unit of the electrogenic sodium bicarbonate cotransporter NBCe1-A. J Biol Chem 283:26782–26794

Kari JA, El Desoky SM, Singh AK, Gari MA, Kleta R, Bockenhauer D (2014) The case renal tubular acidosis and eye findings. Kidney Int 86:217–218

Khan AO, Basamh OS (2018) Pediatric primary calcific band keratopathy with or without glaucoma from biallelic SLC4A4 mutations. Ophthalmic Genet 39:425–427

Ko SBH, Shcheynikov N, Choi JY, Luo X, Ishibashi K, Thomas PJ, Kim JY, Kim KH, Lee MG, Naruse S et al (2002) A molecular mechanism for aberrant CFTR-dependent HCO_3^- transport in cystic fibrosis. EMBO J 21:5662–5672

Ko SBH, Zeng W, Dorwart MR, Luo X, Kim KH, Millen L, Goto H, Naruse S, Soyombo A, Thomas PJ et al (2004) Gating of CFTR by the STAS domain of SLC26 transporters. Nat Cell Biol 6:343–350

Kurtz I (2014) NBCe1 as a model carrier for understanding the structure–function properties of Na^+-coupled SLC4 transporters in health and disease. Pflugers Arch Eur J Physiol 466:1501–1516

Lacruz RS, Nanci A, White SN, Wen X, Wang H, Zalzal SF, Luong VQ, Schuetter VL, Conti PS, Kurtz I et al (2010) The sodium bicarbonate cotransporter (NBCe1) is essential for normal development of mouse dentition. J Biol Chem 285:24432–24438

Lane J, Wigham CG, Hodson SA (2000) A chloride-activated Na^+/HCO_3^--coupled transport activity in corneal endothelial membranes. Biophys J 78:2493–2498

Lee S-K, Grichtchenko II, Boron WF (2011a) Distinguishing HCO_3^- from CO_3^{2-} transport by NBCe1-A. FASEB J 25:656

Lee S-K, Boron WF, Parker MD (2011b) Relief of autoinhibition of the electrogenic Na-HCO3 cotransporter NBCe1-B: role of IRBIT vs. amino-terminal truncation. Am J Phys Cell Phys 302: C518–C526

Lee S-K, Boron WF, Parker MD (2013) Substrate specificity of the electrogenic sodium/bicarbonate cotransporter NBCe1-A (SLC4A4, variant A) from humans and rabbits. Am J Physiol Renal Physiol 304:F883–F899

Lee H-W, Osis G, Harris AN, Fang L, Romero MF, Handlogten ME, Verlander JW, Weiner ID (2018) NBCe1-A regulates proximal tubule ammonia metabolism under basal conditions and in response to metabolic acidosis. JASN 29:1182–1197

Li HC, Worrell RT, Matthews JB, Husseinzadeh H, Neumeier L, Petrovic S, Conforti L, Soleimani M (2004) Identification of a carboxyl-terminal motif essential for the targeting of Na^+-cotransporter NBC1 to the basolateral membrane. J Biol Chem 279:43190–43197

Li HC, Collier JH, Shawki A, Rudra JS, Li EY, Mackenzie B, Soleimani M (2009) Sequence- or position-specific mutations in the carboxyl-terminal FL motif of the kidney sodium bicarbonate cotransporter (NBC1) disrupt its basolateral targeting and α-helical structure. J Membr Biol 228:111

Liu X, Williams JB, Sumpter BR, Bevensee MO (2007) Inhibition of the Na/bicarbonate cotransporter NBCe1-A by diBAC oxonol dyes relative to niflumic acid and a stilbene. J Membr Biol 215:195–204

Liu Y, Xu J-Y, Wang D-K, Wang L, Chen L-M (2011) Cloning and identification of two novel NBCe1 splice variants from mouse reproductive tract tissues: a comparative study of NCBT genes. Genomics 98:112–119. https://doi.org/10.1016/j.ygeno.2011.04.010

Lo Y-F, Yang S-S, Seki G, Yamada H, Horita S, Yamazaki O, Fujita T, Usui T, Tsai J-D, Yu I-S et al (2011) Severe metabolic acidosis causes early lethality in NBC1 W516X knock-in mice as a model of human isolated proximal renal tubular acidosis. Kidney Int 79:730–741

Lu J, Boron WF (2007) Reversible and irreversible interactions of DIDS with the human electrogenic Na/HCO_3 cotransporter NBCe1-A: role of lysines in the KKMIK motif of TM5. Am J Physiol Cell Physiol 292:C1787–C1798

Ma SK, Kang JS, Bae EH, Choi C, Lee J, Kim SH, Choi KC, Kim SW (2008) Effects of volume depletion and $NaHCO_3$ loading on the expression of Na^+/H^+ exchanger isoform 3, Na^+: HCO_3^- cotransporter type 1 and nitric oxide synthase in rat kidney. Clin Exp Pharmacol Physiol 35:262–267

Marino CR, Jeanes V, Boron WF, Schmitt BM (1999) Expression and distribution of the Na^+-HCO_3^- cotransporter in human pancreas. Am J Phys 277:G487–G494

Maunsbach AB, Vorum H, Kwon T-H, Nielsen S, Simonsen B, Choi I, Schmitt BM, Boron WF, Aalkjaer, C. (2000) Immunoelectron microscopic localization of the electrogenic Na/HCO_3 cotransporter in rat and ambystoma kidney. JASN 11:2179–2189

May O, Yu H, Riederer B, Manns MP, Seidler U, Bachmann O (2014) Short-term regulation of murine colonic NBCe1-B (electrogenic Na+/HCO3(−) cotransporter) membrane expression and activity by protein kinase C. PLoS One 9:e92275

McAlear SD, Liu X, Williams JB, McNicholas-Bevensee CM, Bevensee MO (2006) Electrogenic Na/HCO_3 cotransporter (NBCe1) variants expressed in *Xenopus* oocytes: functional comparison and roles of the amino and carboxy termini. J Gen Physiol 127:639–658

Moss F, Lee S, Parker M, Boron W (2014) Distinguishing HCO3- from CO3= transport by the electrogenic Na/HCO3 cotransporter NBCe1 (SLC4A4). Faseb J 28:a1098.7

Myers EJ, Yuan L, Felmlee MA, Lin Y-Y, Jiang Y, Pei Y, Wang O, Li M, Xing X-P, Marshall A et al (2016) A novel mutant Na^+/HCO_3^- cotransporter NBCe1 in a case of compound-heterozygous inheritance of proximal renal tubular acidosis. J Physiol Lond 594:6267–6286

Park S, Hong JH, Ohana E, Muallem S (2012) The WNK/SPAK and IRBIT/PP1 pathways in epithelial fluid and electrolyte transport. Physiology (Bethesda) 27:291–299

Parker MD, Boron WF (2013) The divergence, actions, roles, and relatives of sodium-coupled bicarbonate transporters. Physiol Rev 93:803–959

Parker MD, Musa-Aziz R, Rojas JD, Choi I, Daly CM, Boron WF (2008) Characterization of human SLC4A10 as an electroneutral Na/HCO_3 cotransporter (NBCn2) with Cl^- self-exchange activity. J Biol Chem 283:12777–12788

Parker MD, Qin X, Williamson RC, Toye AM, Boron WF (2012) HCO_3^--independent conductance with a mutant Na^+/HCO_3^- cotransporter (SLC4A4) in a case of proximal renal tubular acidosis with hypokalaemic paralysis. J Physiol Lond 590:2009–2034

Patel N, Khan AO, Al-Saif M, Moghrabi WN, AlMaarik BM, Ibrahim N, Abdulwahab F, Hashem M, Alshidi T, Alobeid E et al (2017) A novel mechanism for variable phenotypic expressivity in Mendelian diseases uncovered by an AU-rich element (ARE)-creating mutation. Genome Biol 18:144

Perry C, Le H, Grichtchenko II (2007) ANG II and calmodulin/CaMKII regulate surface expression and functional activity of NBCe1 via separate means. Am J Physiol Renal Physiol 293:F68–F77

Praetorius J, Hager H, Nielsen S, Aalkjaer C, Friis UG, Ainsworth MA, Johansen T (2001) Molecular and functional evidence for electrogenic and electroneutral Na(+)-HCO(3)(-) cotransporters in murine duodenum. Am J Physiol Gastrointest Liver Physiol 280:G332–G343

Preisig PA, Alpern RJ (1988) Chronic metabolic acidosis causes an adaptation in the apical membrane Na/H antiporter and basolateral membrane $Na(HCO_3)_3$ symporter in the rat proximal convoluted tubule. J Clin Invest 82:1445–1453

Robey RB, Ruiz OS, Espiritu DJD, Ibañez VC, Kear FT, Noboa OA, Bernardo AA, Arruda JAL (2002) Angiotensin II stimulation of renal epithelial cell Na/HCO_3 cotransport activity: a central role for Src family kinase/classic MAPK pathway coupling. J Membr Biol 187:135–145

Romero MF, Hediger MA, Boulpaep EL, Boron WF (1997) Expression cloning and characterization of a renal electrogenic Na^+/HCO_3^- cotransporter. Nature 387:409–413

Romero MF, Fong P, Berger UV, Hediger MA, Boron WF (1998) Cloning and functional expression of rNBC, an electrogenic Na^+-HCO_3^- cotransporter from rat kidney. Am J Phys 274:F425–F432
Romero MF, Chang M-H, Mount DB (2010) Chapter 4. From cloning to structure, function, and regulation of chloride-dependent and independent bicarbonate transporters. In: Alvarez-Leefmans FJ, Delpire E (eds) Physiology and pathology of chloride transporters and channels in the nervous system. Academic Press, San Diego, pp 43–79
Romero MF, Chen A-P, Parker MD, Boron WF (2013) The SLC4 family of bicarbonate (HCO_3^-) transporters. Mol Asp Med 34:159–182
Romero MF, Holmes HL, Chowdhury UR, Hann CR, Chang M, Fautsch MP, Chen A-P (2014) Mutants in NBCe1A isoform have elevated intraocular pressure. Invest Ophthalmol Vis Sci 55:2415–2415
Ruiz OS, Qiu YY, Cardoso LR, Arruda JA (1997) Regulation of the renal Na-HCO_3 cotransporter: VII. Mechanism of the cholinergic stimulation. Kidney Int 51:1069–1077
Salerno EE, Patel SP, Marshall A, Marshall J, Alsufayan T, Mballo CSA, Quade BN, Parker MD (2019) Extrarenal signs of proximal renal tubular acidosis persist in nonacidemic Nbce1b/c-null mice. JASN 30:979–989
Sassani P, Pushkin A, Gross E, Gomer A, Abuladze N, Dukkipati R, Carpenito G, Kurtz I (2002) Functional characterization of NBC4: a new electrogenic sodium-bicarbonate cotransporter. Am J Physiol Cell Physiol 282:C408–C416
Satoh H, Moriyama N, Hara C, Yamada H, Horita S, Kunimi M, Tsukamoto K, Iso-O N, Inatomi J, Kawakami H et al (2003) Localization of Na^+-HCO_3^- cotransporter (NBC-1) variants in rat and human pancreas. Am J Physiol Cell Physiol 284:C729–C737
Schmitt BM, Biemesderfer D, Romero MF, Boulpaep EL, Boron WF (1999) Immunolocalization of the electrogenic Na^+-HCO_3^- cotransporter in mammalian and amphibian kidney. Am J Physiol Renal Physiol 276:F27–F38
Sciortino CM, Romero MF (1999) Cation and voltage dependence of rat kidney electrogenic Na^+-HCO_3^- cotransporter, rkNBC, expressed in oocytes. Am J Physiol-Renal Physiol 277:F611–F623
Seki G, Coppola S, Frömter E (1993) The Na^+-HCO_3^- cotransporter operates with a coupling ratio of 2 HCO_3^- to 1 Na^+ in isolated rabbit renal proximal tubule. Pflugers Arch 425:409–416
Seki G, Horita S, Suzuki M, Yamazaki O, Usui T, Nakamura M, Yamada H (2013) Molecular mechanisms of renal and extrarenal manifestations caused by inactivation of the electrogenic Na^+-HCO_3^- cotransporter NBCe1. Front Physiol 4:270
Sergeev M, Godin AG, Kao L, Abuladze N, Wiseman PW, Kurtz I (2012) Determination of membrane protein transporter oligomerization in native tissue using spatial fluorescence intensity fluctuation analysis. PLoS One 7:e36215
Shao XM, Kao L, Kurtz I (2014) A novel delta current method for transport stoichiometry estimation. BMC Biophys 7:14
Shiohara M, Igarashi T, Mori T, Komiyama A (2000) Genetic and long-term data on a patient with permanent isolated proximal renal tubular acidosis. Eur J Pediatr 159:892–894
Shirakabe K, Priori G, Yamada H, Ando H, Horita S, Fujita T, Fujimoto I, Mizutani A, Seki G, Mikoshiba K (2006) IRBIT, an inositol 1,4,5-trisphosphate receptor-binding protein, specifically binds to and activates pancreas-type Na^+/HCO_3^- cotransporter 1 (pNBC1). Proc Natl Acad Sci USA 103:9542–9547
Snead CM, Smith SM, Sadeghein N, Lacruz RS, Hu P, Kurtz I, Paine ML (2011) Identification of a pH-responsive DNA region upstream of the transcription start site of human NBCe1-B. Eur J Oral Sci 119(Suppl 1):136–141
Soleimani M, Aronson PS (1989) Ionic mechanism of Na+-HCO3- cotransport in rabbit renal basolateral membrane vesicles. J Biol Chem 264:18302–18308
Soyfoo MS, Bulur N, Virreira M, Louchami K, Lybaert P, Crutzen R, Perret J, Delporte C, Roussa E, Thevenod F et al (2009) Expression of the electrogenic Na^+-HCO_3^--cotransporters NBCe1-A and NBCe1-B in rat pancreatic islet cells. Endocrine 35:449–458

Suzuki M, Vaisbich MH, Yamada H, Horita S, Li Y, Sekine T, Moriyama N, Igarashi T, Endo Y, Cardoso TP et al (2008) Functional analysis of a novel missense NBC1 mutation and of other mutations causing proximal renal tubular acidosis. Pflugers Arch 455:583–593

Suzuki M, Paesschen WV, Stalmans I, Horita S, Yamada H, Bergmans BA, Legius E, Riant F, Jonghe PD, Li Y et al (2010) Defective membrane expression of the Na+-HCO3− cotransporter NBCe1 is associated with familial migraine. PNAS 107:15963–15968

Thornell IM, Bevensee MO (2015) Regulators of Slc4 bicarbonate transporter activity. Front Physiol 6:166

Thurtle-Schmidt BH, Stroud RM (2016) Structure of Bor1 supports an elevator transport mechanism for SLC4 anion exchangers. Proc Natl Acad Sci USA 113:10542–10546

Vachel L, Shcheynikov N, Yamazaki O, Fremder M, Ohana E, Son A, Shin DM, Yamazaki-Nakazawa A, Yang C-R, Knepper MA et al (2018) Modulation of Cl- signaling and ion transport by recruitment of kinases and phosphatases mediated by the regulatory protein IRBIT. Sci Signal 11. https://doi.org/10.1126/scisignal.aat5018

Virkki LV, Wilson DA, Vaughan-Jones RD, Boron WF (2002) Functional characterization of human NBC4 as an electrogenic Na^+-HCO_3^- cotransporter (NBCe2). Am J Physiol Cell Physiol 282:C1278–C1289

Wu J, McNicholas CM, Bevensee MO (2009) Phosphatidylinositol 4,5-bisphosphate (PIP2) stimulates the electrogenic Na/HCO3 cotransporter NBCe1-A expressed in Xenopus oocytes. Proc Natl Acad Sci USA 106:14150–14155

Yamaguchi S, Ishikawa T (2005) Electrophysiological characterization of native Na^+-HCO_3^- cotransporter current in bovine parotid acinar cells. J Physiol Lond 568:181–197

Yu H, Riederer B, Stieger N, Boron WF, Shull GE, Manns MP, Seidler UE, Bachmann O (2009) Secretagogue stimulation enhances NBCe1 (electrogenic Na^+/HCO_3^- cotransporter) surface expression in murine colonic crypts. Am J Physiol Gastrointest Liver Physiol 297:G1223–G1231

Yu Q, Liu X, Liu Y, Riederer B, Li T, Tian D-A, Tuo B, Shull G, Seidler U (2016) Defective small intestinal anion secretion, dipeptide absorption, and intestinal failure in suckling NBCe1-deficient mice. Pflugers Arch 468:1419–1432

Zhao W, Zhang L, Ermilov L, Mazzone A, Eisenman S, Colmenares-Aguilar M, Silva J, Shull GE, Romero MF, Sha L et al (2019) The role of Na^+/HCO_3^- co-transporter activity in electrical slow waves of the mouse intestine. Neurogastroenterol Motil 31:13658

Zhou Y, Boron WF (2008) Role of endogenously secreted angiotensin II in the CO_2-induced stimulation of HCO_3 reabsorption by renal proximal tubules. Am J Physiol Renal Physiol 294: F245–F252

Zhou Y, Bouyer P, Boron WF (2007) Role of the AT1A receptor in the CO_2-induced stimulation of HCO_3^- reabsorption by renal proximal tubules. Am J Physiol Renal Physiol 293:F110–F120

Zhu Q, Kao L, Azimov R, Abuladze N, Newman D, Pushkin A, Liu W, Chang C, Kurtz I (2010) Structural and functional characterization of the C-terminal transmembrane region of NBCe1-A. J Biol Chem 285:37178–37187

Zhu Q, Shao XM, Kao L, Azimov R, Weinstein AM, Newman D, Liu W, Kurtz I (2013) Missense mutation T485S alters NBCe1-A electrogenicity causing proximal renal tubular acidosis. Am J Physiol Cell Physiol 305:C392–C405

Zhu Q, Kao L, Azimov R, Abuladze N, Newman D, Kurtz I (2015) Interplay between disulfide bonding and N-glycosylation defines SLC4 Na^+-coupled transporter extracellular topography. J Biol Chem 290:5391–5404

Chapter 5
Na^+/H^+ Exchangers in Epithelia

Pawel R. Kiela, Hua Xu, and Fayez K. Ghishan

Abstract Cation/proton antiporters (CPAs), which include Na^+/H^+ exchangers, are evolutionarily ancient transporters present in most species from prokaryotes to higher eukaryotes. Many of them are expressed in epithelial cells or various organs, where they utilize the electrochemical gradient of one ion to transport another ion against its electrochemical gradient. In the intestinal and renal epithelia, NHEs are critical for vectorial transport of Na^+, HCO_3^- and water, and consequently for the systemic volume and acid–base homeostasis. They also contribute to nutrient absorption, cellular proliferation, migration, and apoptosis, and modulate extracellular milieu, e.g., to regulate the intestinal microbial microenvironment. Their dysregulation, or in some instances, mutations, contributes to the human disease pathogenesis and some of the well-characterized Na^+/H^+ exchangers have been considered as attractive targets for pharmacological inhibition. In this chapter, we provide an overview of the members of the CPA superfamily of cation/proton antiporters, with particular focus on their roles in epithelial cells, their expression patterns, intracellular localization, regulation, and function, primarily as determined by gene targeting studies.

Keywords NHE · Epithelial transport · Intracellular pH · Pathophysiology

5.1 Introduction

Most biological processes are regulated by pH homeostasis and ion concentrations. Therefore, it is not surprising that cation/proton antiporters (CPAs) are evolutionarily ancient transporters present in most if not all living species. They mediate the exchange of monovalent cations, primarily Na^+ and K^+, with one or two protons

P. R. Kiela · H. Xu · F. K. Ghishan (✉)
Department of Pediatrics, Steele Children's Research Center, University of Arizona, Tucson, AZ, USA
e-mail: pkiela@peds.arizona.edu; hxu@email.arizona.eu; fghishan@peds.arizona.edu

K. L. Hamilton, D. C. Devor (eds.), *Studies of Epithelial Transporters and Ion Channels*, Physiology in Health and Disease,
https://doi.org/10.1007/978-3-030-55454-5_5

across the plasma or organellar membranes. The importance of these antiporters is highlighted by the fact that many pathological conditions in humans, including hypertension, epilepsy, glaucoma, ischemia–reperfusion injury, gastric and renal diseases, diarrhea, and inflammatory bowel diseases, are related to a dysfunction of this transport mechanism. Membrane transport consistent with the Na^+/H^+ exchange mechanism was first reported by in the rat liver and cow heart mitochondria by Mitchell and Moyle (1967) and by Brierley et al. (1968) in the late 1960s. Similar descriptions in prokaryotic organisms soon followed (Harold and Papineau 1972; West and Mitchell 1974) and led to the cloning of the first Na^+/H^+ antiporter from *E. Coli*, later termed nhaA (Goldberg et al. 1987). Cloning of the first mammalian Na^+/H^+ exchanger by Sardet et al. was reported in the late 1980s (Sardet et al. 1988, 1989). Its molecular identification, initially described as a growth factor-activatable Na^+/H^+ antiporter and later termed NHE1, paved the way not only to the understanding of its function, structure, and regulation, but allowed for homology cloning approaches that led to an identification of a family of Na^+/H^+ antiporters expressed in both prokaryotes and eukaryotes. In the metazoan epithelial cells, the focus of this chapter, the roles of Na^+/H^+ exchange depend not only on the isoform in question, but also on its tissue distribution, and its cellular localization. In polarized epithelia, Na^+/H^+ exchangers can be fairly restricted e.g. NHE1 to the basolateral membrane, or more broadly expressed, e.g., NHE8. It is not uncommon that a specific isoform may show different cellular distribution in various tissues, or in different species. These aspects need to be considered when generalizing the functional importance of each exchanger, and when translating the findings from experimental models to humans.

5.2 Classification and Phylogeny of Na^+/H^+ Antiporters

The Saier Lab Group at the University of California, San Diego, maintains and curates the Transporter Classification DataBase, TCDB (http://www.tcdb.org), approved by The International Union of Biochemistry and Molecular Biology (IUBMB). This database classifies transport systems found in all living organisms on Earth and includes both functional and phylogenetic information (Saier et al. 2016). According to TCDB, all Na^+/H^+ exchangers expressed by the epithelial cells of metazoans are members of the cation/proton antiporter (CPA) superfamily, which can be subdivided into CPA1 and CPA2 families based primarily on ion selectivity and electrogenicity, with the common notion that CPA1s are electroneutral, while CPA2s are electrogenic exchangers, although this division has recently been debated (Masrati et al. 2018). Additional bacterial CPAs with monovalent cation: proton antiporter activity have been categorized into the Na^+ Transporting Mrp Superfamily.

Comprehensive in silico analysis of the evolutionary origins of eukaryotic Na^+/H^+ exchangers has been conducted by Brett et al. (2005) who showed that the eukaryotic CPA1 and CPA2 families are composed of five phylogenetically distinct

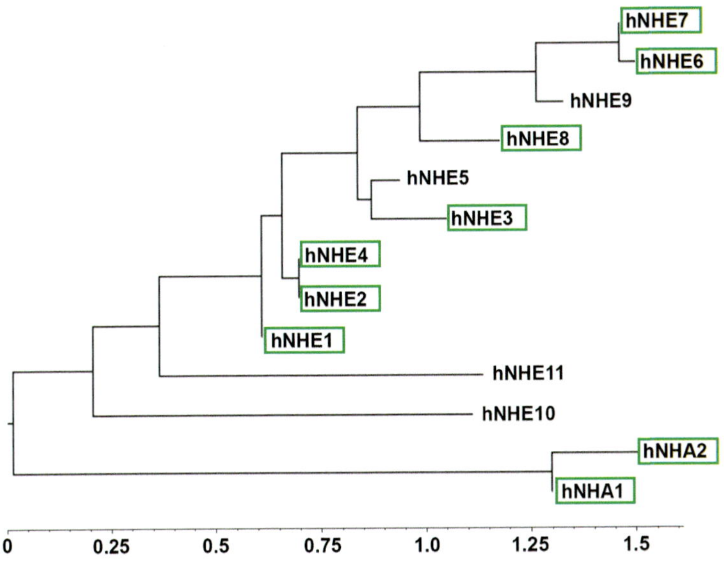

Fig. 5.1 Phylogenetic relationship of the human members of the CPA1 family of Na^+/H^+ exchangers. The following protein sequences were analyzed: NP_003038.2 (NHE1), NP_003039.2 (NHE2), NP_004165.2 (NHE3), NP_001011552.2 (NHE4), NP_004585.1 (NHE5), NP_001036002.1 (NHE6), NP_001244220.1 (NHE7), NP_056081.1 (NHE8), NP_775924.1 (NHE9), NP_898884.1 (NHE10), NP_848622.2 (NHE11), NP_631912.3 (NHA1), and NP_849155.2 (NHA2). The analysis was performed on the Phylogeny.fr platform and comprised the following steps: Sequences were aligned with MUSCLE (v3.8.31) configured for highest accuracy (MUSCLE with default settings); ambiguous regions (i.e., containing gaps and/or poorly aligned) were removed with Gblocks (v0.91b) using the default parameters; the phylogenetic tree was reconstructed using the maximum likelihood method implemented in the PhyML program (v3.1/3.0 aLRT); the data were imported in Newick format into FigTree 1.4.4 software for final presentation. Isoforms identified in epithelial cells are highlighted by the green boxes

clades that differ in subcellular location, drug sensitivity, cation selectivity, and sequence length. They also concluded that the eukaryotic NHE gene family originated as intracellular exchangers first seen in yeast, slime mold, and plant species and that they may have co-emerged with the plasma membrane Na^+/K^+-ATPase (Brett et al. 2005). An overview of the two superfamilies, CPA and the Na^+ Transporting MRP superfamily, with emphasis on the metazoan members, is presented in Table 5.1.

Among the CPA1 family, 11 isoforms of Na^+/H^+ exchangers. NHE1-NHE11, attracted the most attention among researchers and clinicians. They are encoded by their respective genes SLC9A1-SLC9A11 which share variable homology at the DNA as well as protein levels. SLC9A sub-family of mammalian Na^+/H^+ exchangers demonstrate considerable variation in their amino acid sequence, with amino acid identity ranging from under 13% (hNHA2 vs. hNHE6) to over 60% (hNHE6 vs. hNHE7). Nine of them have been detected in the epithelial cells in a range of tissues. Figure 5.1 depicts the phylogenetic relationships of the human NHEs and highlights the isoforms with demonstrated expression in epithelial cells.

Table 5.1 Classification of cation/proton antiporters according to the Transporter Classification DataBase, TCDB (http://www.tcdb.org)

	Family	Description	Proteins characterized in metazoa
CPA Superfamily	2.A.27 The Glutamate: Na^+ Symporter (ESS) Family	Na^+-dependent secondary carrier that transports L- and D-glutamate as well as the toxic analogues α-methyl glutamate and homocysteate. Widespread in bacteria but not in archaea or eukaryotes.	None
	2.A.36 The Monovalent Cation:Proton Antiporter-1 (CPA1) Family	Proteins derived from Gram-positive and Gram-negative bacteria, blue-green bacteria, archaea, yeast, plants and animals capable of catalyzing Na^+/H^+ exchange.	SLC9A1-SLC9A11 (NHE1-NHE11) of *Homo sapiens Rattus norvegicus, Mus musculus* SLC9B1 (NHA1) of *Homo sapiens* SLC9B2 (NHA2) of *Homo sapiens* mtsNHE of *Mus musculus* Cl^--dependent Nhe of *Rattus norvegicus* (putative) NheC of *Danio rerio* PBO-4 and Nhx-2 of *Caenorhabditis elegans* of *Caenorhabditis elegans* Nhe3 of *Aedes aegypti* NHE of *Penaeus vannamei*
	2.A.37 - The Monovalent Cation:Proton Antiporter-2 (CPA2) Family	Moderately large from bacteria, archaea and eukaryotes	TMCO3 of Homo sapiens
	2.A.70 - The Malonate: Na^+ Symporter (MSS) Family	Catalyze the electroneutral reversible uptake of H^+-malonate with one Na^+	None
	2.A.81 - The Aspartate: Alanine Exchanger (AAEx) Family	Bacterial Aspartate:Alanine Exchangers	None
	2.A.98 - The Putative Sulfate Exporter (PSE) Family	Cysteate- and taurine- - inducible sulfate efflux pumps found in diverse bacteria and archaea.	None
	3.B.1 - The Na^+-transporting Carboxylic	Catalyze decarboxylation of a substrate carboxylic acid and use the energy	None

(continued)

Table 5.1 (continued)

	Family	Description	Proteins characterized in metazoa
	Acid Decarboxylase (NaT-DC) Family	released to drive extrusion of one or two Na$^+$ ions from the cytoplasm of the cell	
Na$^+$ transporting MRP superfamily	2.A.63 The Monovalent Cation (K$^+$ or Na$^+$):Proton Antiporter-3 (CPA3) Family	Bacterial multicomponent K$^+$:H$^+$ and Na$^+$:H$^+$ antiporters	None
	3.D.1 - The H$^+$ or Na$^+$-translocating NADH Dehydrogenase (NDH) Family	NADH:ubiquinone oxido-reductases type I of bacteria and of eukaryotic mito-chondria and chloroplasts couple electron transfer to the electrogenic transport of H$^+$ or Na$^+$	None
	3.D.9 - The H$^+$-translocating $F_{420}H_2$ Dehydrogenase ($F_{420}H_2DH$) Family	F420H2:quinol oxidore-ductase from the methanogenic archaeon, Methanosarcina mazei Gö1, has been shown to act as a redox-driven H$^+$ pump	None

5.3 General Features of Epithelial Na$^+$/H$^+$ Exchangers

5.3.1 Membrane Topology and Functional Domains

Not all NHEs have been studied for their secondary structure and membrane topology. However, modeling predicts the hydrophobic and hydrophilic regions of these transporters to follow the same or very similar arrangement. Approximately 60% of the amphipathic N-terminus forms 10–12 membrane-spanning α-helices, which are relatively conserved among different isoforms. The hydrophilic C-terminus which forms a cytoplasmic tail is more isoform-specific and contains multiple phosphorylation sites and binding sites for scaffolding and regulatory proteins. Membrane topology of NHE1 and NHE3 has been scrutinized using cysteine substitution and accessibility (Wakabayashi et al. 2000a), C-terminal truncations (Cabado et al. 1996; Levine et al. 1995), glycosylation site analysis (Counillon et al. 1994; Tse et al. 1994), proteolytic cleavage (Shrode et al. 1998), and epitope immunolocalization (Biemesderfer et al. 1998; Khan 2001; Kurashima et al. 1998; Winkel et al. 1993). A typical topology of a Na$^+$/H$^+$ exchanger is depicted in Fig. 5.2 exemplified by NHE3. It is important to note that two studies on NHE1 and NHE3 did not conform with the general model of NHE membrane topology (Khan 2001; Biemesderfer et al. 1998) and suggested that at least some

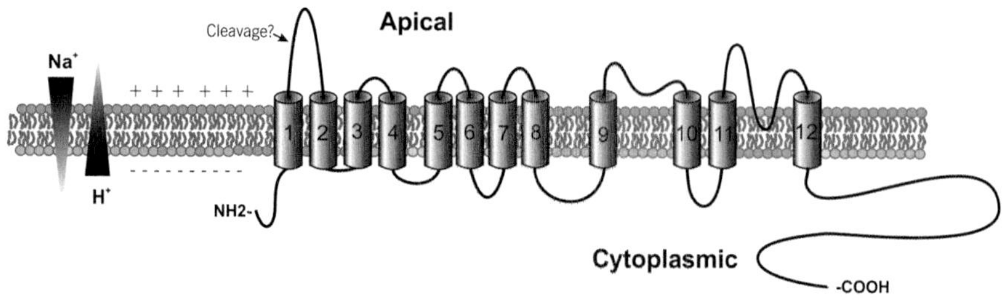

Fig. 5.2 Current general model of NHE membrane topology as exemplified by NHE3

epitopes located with the C-terminal tail may be extracellular. On the other hand, C-terminal tails of NHE1 or NHE3 do not have hydrophobic regions of sufficient length to form a membrane-spanning domain (15–20 amino acids). Moreover, the preponderance of evidence for multiple interactions of the C-terminus with cytoplasmic and cytoskeletal factors strongly argues against this concept (Fliegel 2019; Donowitz et al. 2009). Kemp et al. (2008) reviewed various models of NHE1 membrane topology.

If there is a cleavable signal peptide at the N-terminus, cleavage would likely occur before the experimentally confirmed glycosylation site in the first extracellular loop of NHE1 or NHE3 (Fig. 5.2). If this were the case, NHEs would have 11-, not 12-membrane-spanning domains. However, cysteine substitution studies demonstrated that the N-terminus of NHE1 is retained in the mature protein and remains in the cytosol (Wakabayashi et al. 2000a). Contrasting with this finding, the N-terminal signal peptide of the in vitro translated NHE3 is cleaved during microsomal processing (Zizak et al. 2000). These experimental findings suggest that caution should be taken when generalizing membrane topology predictions among the members of the NHE family.

The amphipathic region has been most extensively studied in the NHE1 isoform. Transmembrane domain 4 is crucial for NHE1 function. Within this region, the residues F^{161}, F^{162}, L^{163}, and G^{173} are critical for the affinity of NHE1 for Na^+ and/or its resistance to pharmacological inhibition (Counillon et al. 1993, 1997; Touret et al. 2001). P^{167} and P^{168} were also critical for NHE1 protein abundance, membrane targeting, and activity (Slepkov et al. 2004). Within the seventh transmembrane domain, Q^{262} and D^{267}, their charge and acidity were found indispensable for NHE1 activity (Murtazina et al. 2001). Transmembrane domain 9 also contains a sequence conferring sensitivity to inhibitors (Orlowski and Kandasamy 1996), with H^{349} being the most critical amino acid bestowing sensitivity to amiloride-related compounds (Wang et al. 1995). Y^{454} and R^{458} localized within the eleventh transmembrane domain are essential for proper trafficking of NHE1 to the cell surface (Wakabayashi et al. 2000b). Two glycines within this transmembrane domain (G^{455} and G^{456}) and arginine R^{440} in its neighboring intracellular loop 5 have also been implicated in pH sensing by NHE1 and are believed to constitute the putative pH_i sensor in NHE1.

The long cytosolic tail of NHE1 (amino acids 500–815) regulates the activity of the amphipathic domain and is itself a target of phosphorylation and interactions with scaffolding and regulatory proteins. Among protein kinases verified to phosphorylate NHE1 are Erk1/2, p90rsk, p160ROCK, p38, and an Nck-interacting kinase. Other regulatory proteins experimentally confirmed to physically interact with the NHE1 cytosolic domain are calmodulin, calcineurin homologous protein (CHP), tescalcin, and carbonic anhydrase II. These regulatory and physical interactions have been extensively reviewed by Putney et al. (2002), Slepkov and Fliegel (2002), Baumgartner et al. (2004), and Malo and Fliegel (2006). The cytoplasmic C-terminus of NHE3 has also been studied in great detail in terms of its regulatory binding partners, phosphorylation, and contribution to a dynamic regulation of NHE3 activity at the plasma membrane via trafficking between endosomal and plasmalemmal pools. These functions have been reviewed in greater detail by Donowitz et al. (2009) and Alexander and Grinstein (2009) and are discussed later in this chapter for each respective isoform.

5.3.2 *Energy Dependency*

All members of the CPA1 family of Na^+/H^+ exchangers are classified as electrochemical-potential-driven transporters. This means that cation fluxes via Na^+/H^+ exchange mechanisms are driven exclusively by the transmembrane substrate gradients and are only secondarily dependent on the cellular ATP as NHEs do not bind nor consume ATP directly. Nevertheless, despite a maintained transmembrane H^+ gradient, NHE1, NHE2, and NHE3 activities are significantly reduced upon cellular ATP depletion (Kapus et al. 1994; Bianchini et al. 1991). The mechanism responsible for this reduction appears to be isoform-specific. ATP depletion reduces NHE1 and NHE2 activity via impaired sensing of the intracellular pH (pHi), whereas NHE3 responds to energy depletion by both impaired pHi sensing and reduced maximal velocity of transport (V_{max}) (Levine et al. 1993). In the case of NHE1, the observed inhibition may be mediated by dephosphorylation and reduced association of phosphatidylinositol 4,5-bisphosphate (PIP2) with the plasma membrane. PIP2 was shown to associate with the C-terminus of NHE1 and support its optimal activity (Aharonovitz et al. 2000). The molecular mechanisms responsible for reduced NHE2 and NHE3 under ATP depletion has not been elucidated in detail. Cabado et al. (1996) performed a structure–function analysis of the NHE3 ATP dependence by examining progressive deletions of the C-terminal tail of NHE3 and found that even the most complete deletion of the C-terminal tail, which removed the majority of the consensus phosphorylation sites ($NHE3^{\Delta579}$), did not affect the sensitivity of this isoform to ATP depletion. This suggested that the effects of ATP depletion on NHE1 and NHE3 isoforms are likely mediated by a very different mechanism.

5.3.3 Sensitivity to Inhibitors

The development of Na^+/H^+ exchange inhibitors has been largely fueled by the attempts to reduce NHE1 activity in cardiac ischemia/reperfusion injury (Karmazyn et al. 2005), and subsequently as potential adjuncts to chemotherapy of cancers (Mihaila 2015; Masereel et al. 2003). Amiloride, a known natriuretic, became the first described NHE inhibitor (Benos 1982), albeit lack of specificity and its inhibitory effects on electrogenic Na^+ channels and the Na^+/Ca^{++} exchanger limit its research and clinical use. Among the tested NHE isoforms, NHE1 and NHE2 are the most sensitive to amiloride, whereas NHE3 and especially NHE4 are considered to be amiloride-resistant. The sensitivity of NHE8 and NHE9, two of the most recently cloned NHE isoforms, to currently known inhibitors has not been evaluated and only limited information on the sensitivity of NHE4 and NHE7 is available. Development of several pyrazine or phenyl derivatives of amiloride increased their potency toward NHEs, particularly NHE1, and more importantly increased their selectivity by eliminating the inhibitory potency toward the Na^+ channel and Na^+/Ca^{2+} exchangers. Of these molecules, DMA, EIPA, HOE-694, and HOE-642 are the most frequently used in experimental settings.

Several NHE inhibitors based on a bicyclic template have been introduced, such as zoniporide, SM-20550, BMS-284640, T-162559, or TY-12533. Other compounds not related to amiloride have also proven useful, especially S-3226, as the first NHE3-specific inhibitor (Schwark et al. 1998). Cimetidine, clonidine, and harmaline, although not frequently used, have also been reported to act as weak and nonspecific inhibitors of Na^+/H^+ exchange (Kulanthaivel et al. 1990). More recently, ligustrazine (2,3,5,6-tetramethylpyrazine) and its analogs, along with several other compounds (e.g., KR-32560, KR32570, or KR-33028) have been introduced as NHE1 inhibitors. It is important to point out that the reported IC_{50} values have been frequently derived from studies with forced expression NHEs in NHE-deficient fibroblasts, and significant differences in sensitivities of endogenous NHEs may not be uncommon (Kim et al. 2007).

Intestinal Na^+ absorption mediated by NHE3 has more recently become a target of commercial investigation, with the hope of developing a novel treatment for hypertension and/or constipation-predominant irritable bowel syndrome (IBS-C). Two oral nonabsorbable NHE3 inhibitors have been developed, SAR218034 (SAR) and tenapanor. Pharmacokinetic studies indicated that they did not cross the intestinal barrier in biologically active doses (Spencer et al. 2014; Linz et al. 2012). Tenapanor was well tolerated in a phase I clinical study (Johansson et al. 2016) and both inhibitors increased fecal and reduced urinary Na^+ concentrations in rodents and humans. Not surprisingly, both drugs caused an increase in luminal fluid resulting from increased Na^+, leading to loose stools (Spencer et al. 2014; Linz et al. 2012). In a spontaneously hypertensive rat model, SAR in conjunction with NaCl-laden drinking water markedly reduced systolic blood pressure (Linz et al. 2012). In a rat model of chronic kidney disease associated with hypertension, hypervolemia, cardiac hypertrophy, and arterial stiffening (salt-fed 5/6 nephrectomized rats), tenapanor reduced extracellular volume expansion, albuminuria, and blood pressure, in addition to promoting protective cardiorenal effects such as reducing left ventricular hypertrophy (Spencer et al. 2014). Both drugs show enhanced effects if administered in conjunction with an angiotensin-converting enzyme inhibitor, which was deemed important in cases in which hypertension could not be controlled by the administration of a single medication. However, at the time of this writing, the development of tenapanor and SAR, and the general concept of inhibition of intestinal Na^+/H^+ exchange as an antihypertensive strategy, appears to have been abandoned by Ardelyx (Fremont, CA) and Sanofi (Paris, France), respectively. Ardelyx continues the investigation into the use of tenapanor and its close analogs in patients with IBS-C and for the treatment of hyperphosphatemia in end-stage renal disease patients on dialysis (Zielinska et al. 2015; Labonte et al. 2015). In a phase 2, randomized, placebo-controlled efficacy and safety trial, 50 mg of tenapanor twice a day, the response rate for IBS-C symptoms such as pain, discomfort, bloating, cramping, and fullness, was significantly higher in the tenapanor than the placebo group, with diarrhea as the most frequent adverse effect (Chey et al. 2017). The concept behind the use of NHE3 inhibition in hyperphosphatemia is based on the described increased fecal P_i excretion and reduced urinary P_i excretion in NHE3 inhibitor-treated rats with chronic kidney disease with vascular calcification, in which tenapanor markedly reduced ectopic calcification and protected renal function (Labonte et al. 2015). The mechanism of this phenomenon is not clear. However, in hemodialysis patients, tenapanor indeed moderately but significantly lowered serum phosphate levels, albeit with diarrhea affecting as much as 68% of participants (Block et al. 2017). It is not yet evident whether targeting NHE3 would be more efficacious and have less adverse effects that the P_i binders currently used clinically.

5.4 CPA1/SLC9A: NHE1

NHE1, cloned in the J. Pouyssegur's lab at the University of Nice, France (Sardet et al. 1988, 1989) has become the prototypical eukaryotic Na^+/H^+ exchanger and a subject of over 1700 original publications, reviews (Fliegel 2019; Malo and Fliegel 2006; Meima et al. 2007; Orlowski and Grinstein 2004; Valles et al. 2015; Fliegel 2005), and monographs (Fliegel 1996). It is encoded by the gene SLC9A1 with cytogenetic location *1p36.11* in humans, and its allelic variant with G305R substitution has been linked with Lichtenstein–Knorr syndrome (autosomal recessive spinocerebellar ataxia) (Guissart et al. 2015). This variant leads to ca. 70% reduction in protein expression, hypoglycosylation, and mislocalization, with only about 2% residual activity compared to the wild-type protein (Guissart et al. 2015). Mammalian NHE1 is generally an 813–822 amino acid (~91 kDa) protein. It contains consensus sequences for both *N*- and *O*-linked glycosylation, and there is evidence that Asn-75 in the first extracellular loop of NHE1 is glycosylated, explaining the appearance of the mature 110 kDa form of NHE1 in Western blotting (Counillon et al. 1994).

5.4.1 Tissue Specificity and Subcellular Distribution

With its fundamental role to protect the cells from acidification, NHE1 is ubiquitously expressed in all cells and tissues albeit with some differences in mRNA and protein abundance (Fig. 5.3). This critical role of NHE1 isoform is highlighted by the very high homology among species. A great deal of effort has been placed on the cardiac and neuronal expression and function of NHE1. In the myocardium, NHE1 protein can be found at the plasma membrane and along the intercalated discs and transverse tubule system (Fliegel et al. 1991, 1993; Petrecca et al. 1999) and overactivation of NHE1 is known to contribute to heart disease, including acute ischemia–reperfusion damage and cardiac hypertrophy (Odunewu-Aderibigbe and Fliegel 2014; Wu and Kraut 2014; Wakabayashi et al. 2013). The roles of NHE1 in the neuronal function has also been relatively extensively studied, not only due to its genetic association with Lichtenstein–Knorr syndrome and the ataxia symptoms of NHE1-deficient mice (described later in this chapter) but also its contribution to enhanced neuronal excitability in epilepsy, and for similar reasons as in cardiac ischemia–reperfusion injury. In principle, excessive NHE activation during hypoxia–ischemia leads to intracellular Na^+ overload and an influx of Ca^{2+} through Na^+/Ca^{2+} exchanger which triggers the neurotoxic cascade. Hyperactivation of NHE1 is also thought to be an "overshot" in pHi normalization and an alkaline shift. This mechanism aimed at pHi recovery is deleterious to cell survival. In this chapter, focused on epithelial cell biology, we will not discuss neuronal NHE1

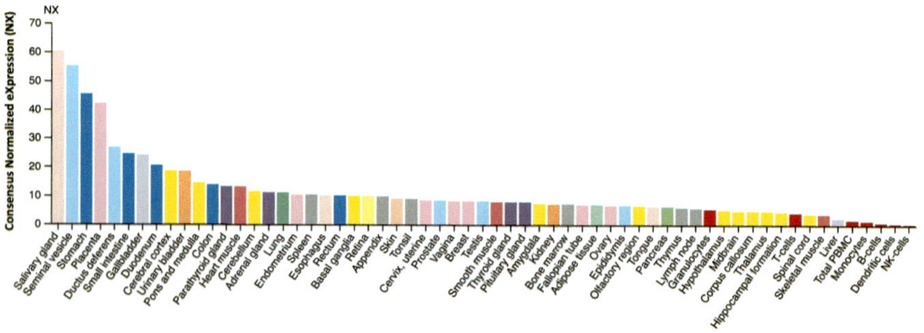

Fig. 5.3 NHE1 mRNA tissue specificity. Consensus Normalized eXpression (NX) levels for 55 tissue types and six blood cell types, created by combining the data from the three transcriptomics datasets (HPA, GTEx, and FANTOM5) using a normalization pipeline (The Human Protein Atlas). Color-coding is based on tissue groups, each consisting of tissues with functional features in common. Tissues sorted based on the mRNA expression level, from the highest (left) to the lowest (right). Image credit to The Human Protein Atlas (https://www.proteinatlas.org/ENSG00000090020-SLC9A1/tissue)

function in detail (with the exception of the epithelial cells of the choroid plexus) and the reader is referred to several review articles and the original citations therein (Uria-Avellanal and Robertson 2014; Zhao et al. 2016; Verma et al. 2015). NHE1 protein is predominantly localized to the plasma membrane. Dependent on the cell type, NHE1 tends to accumulate in distinct membrane domains. In polarized epithelial cells, NHE1 is expressed on the basolateral membrane (Bookstein et al. 1994a).

5.4.2 Physiological Roles

Although the primary function of the ubiquitously expressed NHE1 is to control cell volume and intracellular and pericellular pH, this isoform is also critically involved in complex biological processes like cell adhesion, cell migration, cell proliferation, and mechanosensation.

5.4.2.1 NHE1 in Epithelial Cell Adhesion and Migration

Polarized activity of ion channels and transporters has been proposed to contribute to cell migration in large part through local volume regulation. According to the model presented by Stock and Schwab (2006), NHE1, Cl^-/HCO_3^- exchanger AE2, aquaporin 1 (AQP1), and Na^+-HCO_3^- cotransporter NBC1 accumulate at the leading edge of migrating cells. Activation of NHE1 leads to local Na^+ and water uptake at the leading edge of the lamellipodium, and as swelling of the growing lamellipodium increases the tension of the plasma membrane, it activates Ca^{2+}-permeable mechanosensitive cation channels and a rise in intracellular Ca^{2+} concentration. Higher Ca^{2+} concentration leads to the activation of Ca^{2+}-sensitive K^+ channels IK1 at the trailing edge which leads to its shrinkage of the rear part of the cell. This happens in parallel with the activation of Ca^{2+}-sensitive contraction of the cortical actomyosin network at the trailing edge, and release of the focal adhesions. After volume loss and retraction of the trailing part, the Ca^{2+} influx stops, and the cytosolic Ca^{2+} returns to its equilibrium, which allows to cycle to repeat.

Stock et al. (2005) showed that in melanoma MV3 cells, extracellular acidification driven by NHE1 activation promoted cell migration. In these cells, cell adhesion, spreading, and migration depend to a large extent on the integrin $\alpha2\beta1$, processes sensitive to the extracellular pH. NHE1 and the integrins colocalize at the focal adhesion sites of the leading edges of lamellipodia (Grinstein et al. 1993; Plopper et al. 1995) where NHE1 could create a low pH "nanoenvironment" to influence the strength of cell adhesion. In MDA-MB-231 breast tumor cells, NHE1-mediated extracellular acidification also activated hyaluronidase 2 and cathepsin B, two pH-dependent enzymes capable of degradation of the extracellular matrix and promoting tumor cell invasion, and potentially metastasis (Bourguignon et al. 2004).

Independent of its ion transport function, NHE1 can function as a plasma membrane anchor for cortical actin through recruitment of the ERM (ezrin, radixin, moesin) family of actin-binding proteins (Denker et al. 2000; Denker and Barber 2002). Scaffolding by NHE1 aids in assembling signaling complexes that contribute to dynamic cytoskeletal reorganization and cell migration. ERM proteins are substrates for Rho-kinase 1 (ROCK1) and Nck-interacting kinase (NIK), both of which also activate NHE1 and it has been hypothesized that the colocalization and interaction of NHE1 with ERM proteins may facilitate a reciprocal modulation of cytoskeletal dynamics involved in cellular migration (Stock and Schwab 2006). Finally, NHE1 activity may also regulate the expression of genes involved in cell adhesion, ECM remodeling, and cell migration, although these have been analyzed in fibroblasts and not all may be related to the biology of normal or transformed epithelia (Putney and Barber 2004).

The above mechanisms of NHE1-stimulated cell migration were focused on subcellular differences in cell volume, pH, and cell-ECM interactions. However, collective cell migration as sheets also plays a critical role during embryonic development, in the restitution of normal epithelia, and even in cancer. A recent study by Jensen et al. (2019) with NHE1-overexpressing renal epithelial MDCK cells showed that NHE1 was concentrated not only in the lamellipodia and filopodia of leading cells in sheets of migrating cells but also in submarginal cell rows. Overexpression of NHE1 led to a more disorganized multicellular organization of MDCK grown as 3D cysts, but more importantly, without affecting cell proliferation, it increased the collective migration of cells in a wound-healing experiment and in response to epidermal growth factor (EGF). The authors speculated that NHE1 activity contributes to collective cell migration and epithelial morphogenesis. The results also suggested a role for the transporter in embryonic and early postnatal development, and potentially to epithelial wound healing. In an accompanying perspective, Sébastien Roger further speculated that during early carcinogenesis, expression and activity of NHE1 could disrupt epithelial tissues, drive tissue dysplasia, and promote epithelial-to-mesenchymal transition (Roger 2019). Although plausible scenarios, it is important to note that the null knockout of NHE1 in mice did not impair embryogenesis (Bell et al. 1999).

5.4.2.2 NHE1 in Epithelial Cell Cycle Regulation and Proliferation

Intracellular pH (pHi) is tightly regulated by the activity of several plasma membrane transporters and is kept at a near-neutral range (7.0–7.2) in most normal cells. Mitogen-induced cell proliferation is associated with an increase in pHi, and there is increasing evidence that hyperproliferation associated with cancer can be at least in part attributed to dysregulated pH homeostasis (both intracellular and pericellular pH). Pouysségur's group provided the first evidence of NHE1's involvement in cell proliferation by demonstrating reduced proliferative rate in NHE1-deficient fibroblasts at neutral and acidic pHi (Pouyssegur et al. 1984), later confirmed by Kapus et al. (1994). Following this description, work with pharmacological inhibition of

NHE1 demonstrated that NHE1 activity has a permissive rather than an obligatory role in promoting proliferation in response to mitogens in pancreatic exocrine cells (Delvaux et al. 1990), breast and adipose tumor cells (Horvat et al. 1992), or hepatic stellate cells (Benedetti et al. 2001). NHE1's ion-exchange activity, and not its function as a cytoskeletal anchor, was later found to be critical for promoting cellular proliferation in PS120 fibroblasts (Denker et al. 2000). The increased H$^+$ efflux and increased pHi as a result of NHE1 activation in mitogen-activated cells may work to reduce metabolic acid generated by glycolysis and lactic acid production. However, the relative contribution of changes in pHi versus cell volume in cell division has not been clearly established. Transit through the cell cycle is accompanied by increased cell volume. Achieving a critical size, with the contribution of NHE1, is believed to regulate entry into mitosis (Coelho and Leevers 2000). The mechanisms involved in the regulation of cell proliferation by pHi, including NHE1 activity, have been recently reviewed in more detail by Flinck et al. (2018).

Among epithelial cells, in HT-29 colonic carcinoma cells, extracellular zinc (Zn) acting through zinc-sensing receptor (ZnR) led to ERK1/2-mediated NHE1 activation (Azriel-Tamir et al. 2004). The results suggested that zinc may influence pH homeostasis in colonocytes and that the activation of ERK and NHE1 may explain in part the positive effects of Zn on cell proliferation. Jenkins et al. (2012) demonstrated that during morphogenesis, mammary branching induced by TGFα requires PI3K-dependent NHE1 activation and intracellular alkalization. Pharmacological inhibition of NHE1 disrupted TGFα-induced mammary branching morphogenesis in 3D in vitro culture and, surprisingly, induced extensive proliferation. In a follow-up article, the same group showed that inhibition of NHE1 in fully established branched mammary tissue resulted in a rapid (within 24 h) but reversible loss of branched architecture, albeit without changes in cell proliferation (Jenkins et al. 2014). These studies suggested that NHE1 functions to support not only mammary morphogenesis but also that it helps sustain mammary tissue architecture and a functional milk-producing gland. This phenomenon was not directly assessed in vivo, in part because NHE1$^{-/-}$ female mice surviving to be mated with NHE1$^{+/-}$ males die early postpartum (Bell et al. 1999).

There is a considerable amount of literature on the proliferation of cancer cells of epithelial and non-epithelial origin and the potential role for NHE1 inhibitors in cancer therapy (Tamtaji et al. 2019; Stock and Pedersen 2017; Aredia and Scovassi 2016). As an example of epithelial-derived cancer, gastric cancer cells (both from patients and established cell lines) express elevated levels of NHE1, and its knockdown or pharmacological inhibition suppresses gastric cancer cell proliferation by promoting G1/S and G2/M cell cycle phase transition (Xie et al. 2017). In addition, gastric migration, invasion, and markers of epithelial–mesenchymal transition (EMT) proteins were reduced by NHE1 inhibition, thus suggesting that NHE1 may be a useful target for adjuvant gastric cancer therapy.

5.4.2.3 NHE1 in Epithelial Cell Survival and Apoptosis

The effects of NHE1 activation on cell proliferation likely synergize with those promoting enhanced cell survival or inhibition of apoptosis (Lang et al. 2000, 2006). Since a decrease in pHi and decreased cell volume are both hallmarks of apoptosis, it is intuitive that NHE1 activity could provide the necessary counterbalance. In the early stages of apoptosis, there is a rapid and transient rise in cytosolic pH attributable to NHE1 activity induced by p38 MAP kinase, and dependent on the intact Ser^{726} and Ser^{729} residues of NHE1 (Grenier et al. 2008). This suggested that NHE1 activity indeed attempts to counteract cell death in the early stages of apoptosis. Such involvement has been described in several cell types, including cytotoxic T cells (Li and Eastman 1995), fibroblasts (Grenier et al. 2008), gastric cancer cells (Nagata et al. 2011), dendritic cells (Rotte et al. 2010), and it is believed to (at least in part) underly the neurological symptoms in $NHE1^{-/-}$ mice which show selective neuronal death in tissues with high metabolic activity, including the cerebellum and brainstem.

Apoptosis and the destruction of the tubular epithelial cells are hallmarks of chronic renal diseases. Contrary to lymphocytes, epithelial cells are relatively resistant to apoptosis under hypertonic conditions due to their capacity to expand their intracellular volume (Bortner and Cidlowski 1996). Nevertheless, in animal models of progressive renal disease, the proximal tubule segment is a site of abundant cell apoptosis with volume shrinkage, cytosolic acidification, and activation of pro-apoptotic caspases (Schelling et al. 1998; Schelling and Abu Jawdeh 2008). Collectively, the available data indicate that while NHE1 activation represents an initial defense against apoptotic stress, overwhelming or sustained proapoptotic stimuli may overcome the pro-survival roles and allow the cells to proceed toward apoptosis. Consistent with this model, in proximal tubule cell lines treated with staurosporine to induce apoptosis, there is a transient increase in NHE1 activity (within 30–60 min) followed by a quick decline (Wu et al. 2004), possibly related to NHE1 cleavage by caspase-3 (Wu et al. 2003). This "housekeeping" anti-apoptotic feature of NHE1 does not manifest under unstimulated conditions in vivo, as $NHE1^{-/-}$ mice or *Swe* mice with a premature stop codon and loss of NHE1 function do not develop spontaneous renal pathology. However, when *Swe* mice are treated with the podocyte toxin adriamycin, or with diabetes-inducing streptozotocin, they manifest with increased cortical tubular epithelial cell apoptosis as compared to their wild-type controls (Wu et al. 2003; Khan et al. 2006). Several excellent review articles discuss the function of NHE1 in the renal tubular epithelium in more detail (Valles et al. 2015; Schelling 2016; Parker et al. 2015; Schelling and Abu Jawdeh 2008).

In some instances, however, NHE1 overexpression or hyperactivity has been linked with the promotion of apoptosis. This has been reported in cardiomyocytes exposed to hypoxia and reoxygenation (Karki and Fliegel 2010), in LPS-treated endothelial cells (Cui et al. 2013; Zhao et al. 2012). The extensive research on the role of NHE1 in a hypertensive, hypertrophied, or diabetic myocardium, or during ischemia–reperfusion-induced myocardial damage suggests that in this tissue, increased NHE1 activity is maladaptive (Karki and Fliegel 2010; Odunewu-Aderibigbe and Fliegel 2014; Fliegel 2009). NHE1 activity may also be permissive in the intestinal epithelial barrier breach resulting from a burn injury. Yang et al. (2013) demonstrated a transient increase in epithelial NHE1 expression after burn injury, accompanied by increased mucosal permeability which was attenuated by NHE1 inhibitor, cariporide. Although apoptosis was not investigated in this study, it is plausible that it contributed to the findings. On the other hand, increased intestinal permeability resulting from ischemia–reperfusion was not reversed by NHE1 inhibition (Moeser et al. 2006).

NHE1 is the only plasma membrane Na^+/H^+ exchanger in the squamous esophageal epithelium (Shallat et al. 1995) where it offers protection from cell death and DNA damage induced by gastric acid in gastroesophageal reflux disease (GERD). Salivary epidermal growth factor (EGF) is cytoprotective for the acid-exposed esophageal cells and it utilizes the same PKC- and Ca^{2+}/calmodulin-dependent pathway as that involved in NHE1 activation (Fujiwara et al. 2006). In patients with low salivary EGF levels, GERD is associated with increased susceptibility to severe esophageal damage and increases the overall risk for the development of Barrett's esophagus (Marcinkiewicz et al. 1998; Gray et al. 1991). NHE1 expression is increased in GERD patients (Siddique and Khan 2003) and in Barrett's esophagus (Goldman et al. 2010), where it likely represents an adaptive mechanism to counteract the acute and chronic acid overload. However, bile acids present in the reflux chyme, via nitric oxide-mediated NHE1 inhibition, reduce the ability of the cells to control their pH_i, thus leading to increased DNA damage and either cell death or potentially to mutations and cancer progression (Goldman et al. 2010). In support of the latter, in Barrett's adenocarcinoma cell line, acid-induced NHE1 activity contributed to increased proliferation (Fitzgerald et al. 1998). It is thus plausible that while inhibition of NHE1 may be detrimental during the early stages of Barrett's esophagus via detrimental effects on cell survival and genomic integrity, pharmacological inhibition of NHE1 may be of therapeutic value in preventing progression from Barrett's to cancer, or in esophageal cancer therapy.

5.4.2.4 NHE1 in Epithelial Cell Mechanosensation

As we discussed above, NHE1 regulates cell shape, adhesion, proliferation, and migration, all of which are directly related to cell behavior in constrained space. NHE1 was also demonstrated to sense mechanical stress directly and act as a mechanosensor, presumably due to volume sensitivity dependent on the intact first extracellular loop (Kapus et al. 1994; Su et al. 2003). A report by Fuster et al. (2004) studied the role of NHE1 in the response to mechanical stress (by applying negative or positive pressure in the patch pipette) in carefully designed experiments that excluded the effects of osmolarity, changes in ATP, or of the actin cytoskeleton dynamics. NHE1 was reversibly activated by cell shrinkage and inhibited by cell swelling. Plasma membrane thinning with α-lysophosphatidylcholine and octyl-β-d-glucopyranoside reduced NHE1 activity, which was activated by cholesterol enrichment, which thickens membranes (Fuster et al. 2004). Although these studies were conducted in fibroblasts, more recent work conducted with renal epithelial cells supported the role of mechanical stress on NHE1 activity. The cell's resistance to an applied force depends on the Rho-dependent assembly of focal adhesions, where, as we describe in the section on cell adhesion and migration, NHE1 protein tends to concentrate (Ludwig et al. 2013). Bocanegra et al. (2014) studied the role of NHE1 on cell apoptosis induced by mechanical stretch in immortalized human proximal tubule HK-2 cells. Such regulation is of clinical relevance in the renal tubular epithelium, where retrograde pressure associated with unilateral ureteral obstruction leads to apoptosis and tubular atrophy, a hallmark of the progression of obstructive nephropathy. As we discussed above, NHE1 activation is thought to counteract apoptosis through the restoration of cell volume and cytosolic pH. Bocanegra et al. (2014) demonstrated that when mechanical stretch was applied to HK-2 cells, increased apoptosis was associated with diminished NHE1 expression and function, and with RhoA activation. NHE1 knockdown with siRNA mimicked the pro-apoptotic effects of mechanical stretch with decreased Bcl-2 protein expression and caspase 3 activation. Decreased NHE1 was associated with a higher ERK1/2 expression and decreased p38 activation. The authors ultimately concluded that mechanical stretch activates the RhoA signaling pathway, which leads to ERK1/2 inhibition and p38 activation, which downregulates NHE1 expression and activity and finally triggers caspase 3 activation and the induction of apoptosis (Bocanegra et al. 2014). The role of ERK and phosphorylation in modulating NHE1 activity is discussed below in the section on physiological regulation.

5.4.3 Physiological Regulation

It is important to note that the majority of the data related to NHE1 regulation are derived from non-epithelial cells, primarily from NHE-deficient transfected PS120 fibroblasts. As such, they may or may not apply universally to all cells, including epithelial cells, but non-transformed or neoplastic. The majority of modulators of NHE1 activity are thought to converge at the C-terminal cytosolic regulatory domain (Hendus-Altenburger et al. 2014). One regulatory modality involves the binding of regulatory proteins that control transport activity by altering the affinity of the transport domain for intracellular protons (Putney et al. 2002). Regulation of NHE1 is thought to occur through the interaction of multiple kinases and scaffolding proteins with the C-terminal cytoplasmic domain of NHE1 which leads to a conformational change of the amphipathic domain responsible for mediating Na^+/H^+ exchange.

5.4.3.1 pH Sensing

NHE1 is rapidly activated in response to various extracellular stimuli, including growth factors, Ca^{2+} releasing agonists, and by hyperosmotic stress. Activation of NHE1 by these stimuli results in a shift of the pH sensitivity curve such that NHE1 remains active in more alkaline pHi. Reciprocally, ATP depletion inhibits NHE1 by shifting its pHi-dependence to the acidic side as well as by reducing its V_{max}. Early studies with membrane vesicles suggested that NHE1 has a cytoplasmic pH-sensor site, distinct from the Na^+/H^+-exchange site (Aronson et al. 1982). This was later supported by more detailed mutational analyses (Wakabayashi et al. 2003). Several studies focused on the identification of the subdomains within NHE1 that affect the pHi-sensitivity in unstimulated conditions, regions that play essential roles in ATP depletion-induced inhibition of NHE1, and regions which are required for Ca^{2+}-induced activation. Ikeda et al. (1997) identified four subdomains in the C-terminal tail: domain I (amino acid (aa) 516–590/595), II (aa 596–635), III (aa 636–659), and IV (aa 660–815) as playing distinct roles. Subdomain I was critical for the sensing of pHi at baseline within a physiological pH range, whereas subdomain III, which overlapped with the calmodulin (CaM)-binding site (Wakabayashi et al. 1997), was autoinhibitory. Deletion of subdomain I also abolished the decrease of pHi-sensitivity induced by ATP depletion. The report also indicated a synergism between subdomains I and III in the full response to ATP depletion. More recent studies demonstrated that calcineurin B (CNB) homologous protein (CHP) may be a key factor interacting with the C-terminus of NHE1 (and NHE3) to regulate its pHi dependence (Pang et al. 2001; Pang et al. 2004). Ammar et al. (2006) performed a detailed analysis of this interaction, including solving the crystal structure of the complex, and suggested that CHP binding may stabilize the structure of the juxtamembrane domain of NHE1 by inducing a stable α-helix and maintaining NHE1 in the functional conformation and its responsiveness to changes in the pHi.

5.4.3.2 Phosphorylation

The distal 180 amino acids of the C-terminal tail of NHE1 include phosphorylation sites responsible for about 50% of the growth factor-induced activation of NHE1. The best-characterized protein kinases implicated in the regulation of NHE1 through phosphorylation of its cytosolic tail are β-Raf (Karki et al. 2011), p38 Mitogen-Activated Protein Kinase (MAPK) (Khaled et al. 2001), and protein kinase B/Akt (Snabaitis et al. 2008; Meima et al. 2009), although other kinases and phosphorylation sites have been identified as potentially important elements of NHE1 regulation via mass spectroscopy phosphoproteomic analyses. These have been summarized by Hendus-Altenburger et al. (2014).

The role of the ERK cascade chain involving the Raf kinases, MEK1/1, ERK1/2, and the ribosomal protein S6 kinase (RSKs) has been recently reviewed by Fliegel (2019) who highlighted several parallels between the importance of the ERK pathway and of the NHE1 functions in human health and pathology. These include their roles in cell growth and differentiation, carcinogenesis, cancer growth, and metastasis, and heart disease including cardiac hypertrophy and cardiac ischemia–reperfusion injury. Perhaps the most intriguing link between ERK signaling and NHE1 within the context of epithelia is provided by metastatic breast cancer. NHE1 activation is a known trigger for metastasis in triple-negative (negative for estrogen receptors, progesterone receptors, and excess HER2 protein) breast cancer cells (Amith and Fliegel 2013, 2017; Amith et al. 2015; Reshkin et al. 2013). Part of the pro-metastatic effects of NHE1 activation is believed to be related to protease activation by extracellular acidification, which facilitates the degradation and remodeling of the extracellular matrix (ECM), thus facilitating detachment and metastasis. Ribosomal protein S6 kinase alpha-1 (p90RSK) phosphorylates NHE1 at Ser^{703} in breast cancer cells, and substitution mutation of this residue changes the morphology, reduces the expression of mesenchymal marker vimentin, and reduces the invasiveness of MDA-MB-231 breast cancer cells (Amith et al. 2016). Consistent with this observation, pharmacological inhibition of p90RSK also reduced the metastatic potential of the invasive MDA-MB-231 cells. Further evidence for a causative relationship between ERK1/2 and p90RSK activities comes from the correlation of NHE1 expression and p90RSK activity in the primary metastatic basal triple-negative breast cancers (Amith et al. 2017). Moreover, constitutive activation of the ErbB2 tyrosine kinase receptor in MCF-7 breast cancer cells leads to increased ERK1/2 and p90RSK activity, increased phosphorylation of Ser^{703} of NHE1, and increased metastatic potential and cell motility (Lauritzen et al. 2012).

5.4.3.3 Endocytosis

NHE isoforms have been divided into those that cycle between recycling endosomes and plasma membrane, such as NHE3 and NHE5, and those that permanently reside on the plasma membrane, such as NHE1, NHE2, and NHE4 (Brett et al. 2005). However, one report utilizing fibroblast cells as a model showed that NHE1 undergoes ubiquitylation by the E3 ubiquitin ligase Nedd4-1 at the plasma membrane and is endocytosed in a β-arrestin-1-dependent fashion (Simonin and Fuster 2010). This mechanism has not been reproduced in other models, including epithelial cells, and it remains unclear if this mode of regulation applies to other systems.

5.4.3.4 Transcriptional Regulation

The regulation of NHE1 mRNA expression perhaps has been primarily studied in the myocardium (Slepkov and Fliegel 2002). Promoter regions of the human, mouse, rabbit, and pig SLC9A1 have been cloned and characterized to various extents (Blaurock et al. 1995; Dyck et al. 1995; Facanha et al. 2000; Miller et al. 1991). Its activity is largely dependent on AP2-like transcription factors (Dyck et al. 1995) as well as a poly (dA:dT) region of the promoter, which interacts with as yet unidentified nuclear protein (Yang et al. 1996). Regulation of NHE1 gene expression in epithelial cells has not been extensively studied, and the limited amount of data suggests that NHE1 is not regulated at the mRNA level under conditions when other Na^+/H^+ exchangers are, such as during metabolic acidosis (Lucioni et al. 2002), in microvillous inclusion disease (Michail et al. 1998), after small bowel resection (Musch et al. 2002), glucocorticoid administration (Cho et al. 1994), or during postnatal development (Collins et al. 1998b).

5.5 CPA1/SLC9A: NHE2

Rat, rabbit, and human sodium–hydrogen exchanger 2 (NHE2) cDNA was in the early 1990s (Collins et al. 1993; Wang et al. 1993; Tse et al. 1993; Ghishan et al. 1995; Malakooti et al. 1999). The SLC9A2 gene encoding NHE2 is located on chromosome 2 in the human, chromosome 9 in rats, and on chromosome 1 in mice (Pathak et al. 1996; Szpirer et al. 1994). The predicted molecular weight of the NHE2 protein is ~90 kDa, but in NHE2 cDNA-transfected PS120 fibroblasts, western blot detected a band at ~85 kDa. The NHE2 protein is modified by O-linked glycosylation in transfected PS120 cells (Tse et al. 1994). Functional characterization of NHE2 protein in stably transfected Chinese hamster ovary AP-1 cells showed that NHE2 has an apparent affinity constant for Na^+ (K_m Na^+) of ca. 50 mM, and for intracellular pH (Km pHi) at ca. pH 6.90. Li^+ and H^+, but not K^+, inhibits NHE2 activity.

5.5.1 *Tissue Specificity*

NHE2 mRNA expression is more restricted than that of NHE1. It is the highest in the gastrointestinal tract, but it is also expressed at varying levels in other tissues (Fig. 5.4). In the gut, NHE2 is expressed in the epithelial cells with the highest expression levels in the stomach and colon (Rossmann et al. 2001; Lee et al. 2000; Narins et al. 2004; Park et al. 1999). Among other epithelia, tissues, where NHE2 expression has been experimentally confirmed, are the kidney (Chambrey et al. 1998; Peti-Peterdi et al. 2000), acinar and ductal cells of the salivary gland (Park et al. 1999), epithelium of the excurrent system of the male reproductive tract (Leung et al. 2001), or the endometrium and placenta (Johansson et al. 2002; Wang et al. 2003).

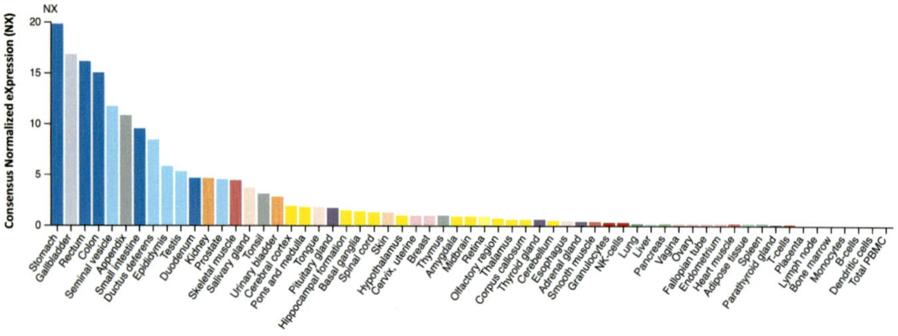

Fig. 5.4 NHE2 mRNA tissue specificity. Consensus Normalized eXpression (NX) levels for 55 tissue types and six blood cell types, created by combining the data from the three transcriptomics datasets (HPA, GTEx, and FANTOM5) using a normalization pipeline (The Human Protein Atlas). Color-coding is based on tissue groups, each consisting of tissues with functional features in common. Tissues sorted based on the mRNA expression level, from the highest (left) to the lowest (right). Image credit to The Human Protein Atlas (https://www.proteinatlas.org/ENSG00000115616-SLC9A2/tissue)

5.5.2 *Subcellular Distribution*

NHE2 protein is located at the apical membrane of enterocytes in the intestine with some species-dependent differences. In rabbits, NHE2 is detected at the apical membrane of the entire villus of the small intestine, the colonic surface cells, and the upper half of the crypt (Hoogerwerf et al. 1996). In mice, however, NHE2 is predominantly expressed in the intestinal crypts (Bachmann et al. 2004; Chu et al. 2002) where it is thought to contribute to the crypt pHi and volume homeostasis. In the gastric epithelium, NHE2 is thought to be expressed at the basolateral membrane of the gastric epithelial cells (Rossmann et al. 2001; Schultheis et al. 1998a), although one study detected the expression of NHE2 at the surface of the cells in the gastric corpus region (Xue et al. 2011). The potential consequences of this discrepancy are discussed in the section on the role of NHE2 in the gastric epithelium later in this chapter. In the epididymis, NHE2 is located at the apical side of the epithelium in the caput, corpus, and cauda regions (Chew et al. 2000). In the kidney, NHE2 is expressed at the apical membrane of the macula densa cells (Peti-Peterdi et al. 2000; Hanner et al. 2008), apical membrane of the epithelial cells in the medullary thick ascending limb (Sun et al. 1997). Chow et al. demonstrated that two proline-rich motifs in NHE2 protein (^{743}PPSVTPAP^{750} and ^{786}VPPKPPP^{792}), which can bind to SH3 domains, are critical for its apical membrane targeting. Mutants lacking the two domains preferentially sorted to the basolateral surface or accumulated intracellularly in transfected renal epithelial LLC-PK1 cells (Chow et al. 1999). Only one report suggested basolateral membrane distribution of NHE2 in the epithelium of the inner medullary collecting ducts but exclusively based on functional analysis and not immunohistochemical evidence (Sun et al. 1998).

5.5.3 *Physiological Regulation*

The maximal rate of NHE2-mediated Na^+/H^+ exchange (V_{max}) is stimulated by serum (in an ATP-dependent fashion), fibroblast growth factor (FGF), and protein kinase C activator phorbol myristate acetate (PMA) in PS120 fibroblasts (Levine et al. 1993). NHE2 is sensitive to ATP depletion which leads to a dramatic decrease in H^+ affinity as well as V_{max}, and the loss of the allosteric effect of H^+ (Levine et al. 1993). Thrombin increases NHE2 V_{max} without altering its Hill coefficient (Levine et al. 1993).

5.5.3.1 Developmental Regulation

Intestinal apical Na^+/H^+ activity is low during early postnatal development and gradually increases into adulthood in rats (Collins et al. 1997). The expression of NHE2 in the intestine gradually increases during postnatal development. It is the lowest at a young age and reaches its peak expression in adult life. Compared with 2-week-old rats, NHE2 protein is fourfold higher at weaning and sixfold higher in adults, while NHE2 mRNA increases threefold at weaning and fivefold in adults (Collins et al. 1998a). The mechanism of this regulation is likely mediated by changes in SLC9A2 gene transcription since nuclear run-on analyses showed an approximately twofold increase in transcription rates of adolescent rats compared to sucklings (Collins et al. 1998a). The expression pattern of NHE2 in the kidney is the opposite, where NHE2 mRNA and protein expression is the highest before weaning and declines into adulthood (Collins et al. 2000). However, these changes did not translate to functional differences.

5.5.3.2 Tissue-Specific Transcriptional Regulation

Human and rat NHE2 promoters have been cloned and characterized (Malakooti et al. 2001; Muller et al. 1998). The proximal promoter from both human and rat NHE2 genes lacks TATA and CAAT boxes and contains a GC-high region. Among the predicted putative transcriptional factor binding sites, only Sp1, AP-2, CACCC, NF-κB, and Oct-1 are conserved in the two species (Malakooti et al. 2001). The minimal promoter of the rat NHE2 gene was identified and its activity displayed tissue specificity. In transfected mouse renal epithelial cells (mIMCD-3), Sp1 transcriptional factor acted as an activator, while Sp3 and Sp4 functioned as inhibitors (Bai et al. 2001). In transfected intestinal epithelial cells (RIE and C2BBc1), both Sp1 and Sp3 transcription factors appeared to stimulate the minimal NHE2 promoter (Hua et al. 2007; Pearse et al. 2007).

5.5.3.3 Hormonal Regulation

EGF plays an important role during intestinal maturation. It is secreted predominantly by salivary glands, and EGF receptors are present along the intestinal tract (Chailler and Menard 1999). EGF not only enhances Na^+ absorption in the gastrointestinal tract (Ghishan et al. 1992; Martinez and Morse 2006; Opleta-Madsen et al. 1991) but also increases NHE activity in rat hepatocytes (Haimovici et al. 1994). In brush border membranes isolated from rat jejunal enterocytes, EGF treatment (200 ng/ml) modestly but significantly stimulated NHE2 activity by 1.8-fold (Ghishan et al. 1992), and in line with this observation, EGF administration in suckling rats increased jejunal but not renal NHE2 activity mRNA expression by about twofold (Xu et al. 2001). The increased NHE2 function was likely predominantly mediated by increased transcription since EGF also stimulated the activity of NHE2 gene promoter (Xu et al. 2001). Interestingly, EGF treatment did not increase the expression of NHE2 in the kidney or in the adult intestine, suggesting EGF-mediated upregulation of NHE2 expression may be tissue and/or age dependent (Xu et al. 2001; Falcone et al. 1999). While glucocorticoids do not affect NHE2 expression or activity on the studied epithelia, Zallocchi et al. (2003) showed that mineralocorticoids upregulate renal NHE2 expression and activity in the kidneys of adrenalectomized, aldosterone-treated rats.

5.5.3.4 Regulation by Osmolarity

The effects of hyperosmotic stress on NHE2 activity and expression have been studied in several cell systems. In NHE2 transfected PS120 fibroblasts, hyperosmolarity reduced NHE2 activity by reducing its maximal reaction velocity (V_{max}) without altering its Michaelis–Menten constant for intracellular H^+ (Nath et al. 1996). However, in the mouse inner medullary collecting duct cells (mIMCD-3), NHE2-transfected AP-1 fibroblasts, colonic crypt cells, and to some degree in the gastric in parietal and mucous cells, NHE2 was consistently stimulated by hyperosmotic conditions (Nagase et al. 1996; Kapus et al. 1994; Bachmann et al. 2004; Rossmann et al. 2001; Soleimani et al. 1994). Bai et al. (1999) mechanistically addressed the transcriptional component of increased NHE2 expression in response to hyperosmolarity. They identified a novel *cis*-acting element termed osmotic-responsive element (OsmoE), and a TonE-like sequence (the latter similar to that described for Na^+-Cl^--betaine transporter or Na^+-myoinositol cotransporter gene promoters), both of which were necessary for the activation of the rat NHE2 gene promoter by hyperosmolarity.

5.5.3.5 Posttranscriptional Regulation

The half-life of NHE2 protein is about 3 h, which is shorter than other NHE isoforms (e.g., 24 h for NHE1 and 14 h for NHE3) both in fibroblasts and Caco-2 colonic adenocarcinoma cells, and is regulated via lysosomal degradation (Cavet et al. 2001). Although NHE2 was shown to be *O*-glycosylated, inhibition of this modification had no effect on the initial rate of Na^+/H^+ exchange in PS120 cells transfected with NHE2 (Tse et al. 1994). NHE2 is also one of the residual plasma membrane proteins believed not to be regulated by endosomal recycling (Cavet et al. 2001).

5.5.4 Physiological Roles of NHE2: Lessons from Knockout Studies

Only one model of null NHE2 knockout has been described by Schultheis et al. (1998a) who targeted exon 2 of the mouse SLC9A2 gene. No models allowing for special or temporal control of NHE2 expression have been created to date.

5.5.4.1 Role of NHE2 in the Salivary Gland Epithelium

With the exception of one study (Park et al. 1999), NHE2 protein has been localized at the apical membrane of acinar and ductal cells in the parotid gland (He et al. 1997; Lee et al. 1998). Lack of NHE2 protein *in SLC9A2*$^{-/-}$ mice, did not alter pHi recovery in acid-loaded acinar cells (Evans et al. 1999), but in in vivo settings, it blunted pilocarpine-stimulated salivary secretion (Park et al. 2001). The authors speculated that NHE2, expressed on the apical membrane, is stimulated by the alkaline secretion in the acinar lumen and supports the basolateral NHE1 in the fine-tuning of pHi and the activity of the Ca^{2+}-stimulated CFTR chloride channel during muscarinic receptor-activated fluid secretion (Park et al. 2001). During the second stage of secretion, luminal NaCl is reabsorbed by the salivary gland ductal epithelial cells to generate a hypotonic saliva. The loss of neither NHE2 nor NHE3 (both ductal apical isoforms) affected the concentrations of Na^+, K^+, Cl^-, or osmolarity of saliva. The role of either exchanger cannot be fully dismissed; however, since the ductal epithelium of NHE2$^{-/-}$ and NHE3$^{-/-}$ mice upregulated the expression of ENac subunits α, β, and γ to compensate for the loss of the two apical NHE isoforms (Park et al. 2001).

5.5.4.2 Gastric Epithelium

Na^+/H^+ exchange has been implicated in the control of gastric mucosal pH_i homeostasis (Kaneko et al. 1992), cell volume regulation during secretory stimulation

(Sonnentag et al. 2000), and gastric epithelial restitution after mucosal injury (Joutsi et al. 1996; Yanaka et al. 2002). NHE2 is uniquely activated by alkaline extracellular pH (pHo) (Yu et al. 1993). Initially, without convincing immunolocalization studies of NHE2 in the stomach, it has been speculated that it is located at the basolateral membrane, where it can be activated by basolateral alkalinization during stimulated acid secretion by parietal cells to, along with NHE1, regulate pHi under high luminal acid load (Schultheis et al. 1998a). However, Xue et al. (2011) showed that NHE2 is located at the gastric surface epithelium. With the very steep inward proton gradient, NHE2 could either be dysfunctional (Yu et al. 1993) or potentially run in a reverse mode, resulting in cellular H^+ uptake and Na^+ extrusion. Although their data did not fully support such a conclusion (NHE2$^{-/-}$ mice had higher resting gastric surface pH), Xue et al. (2011) speculated that if the latter were the case, reverse NHE2 activity could lead to an increase in the surface pH to promote epithelial repair, at least in cells exposed to a less extreme acidic microenvironment and as long as the intracellular acid load could be handled by buffers or basolateral proton extrusion.

In the initial report on *SLC9A2* gene targeting, NHE2 deficient mice had normal growth and no obvious differences in their outward appearance or behavior (Schultheis et al. 1998a). However, NHE2$^{-/-}$ mice had a reduced number of parietal and chief cells in the oxyntic mucosa of the gastric corpus and decreased net acid secretion (Schultheis et al. 1998a). Xue et al. (2011) measured gastric epithelial restitution after laser-induced microscopic lesions and intravital microscopy. The report showed that NHE inhibition with luminal 5-(N-ethyl-N-isopropyl) amiloride or HOE694 slowed mucosal healing in both wild-type and NHE1$^{-/-}$ mice, but not in NHE2$^{-/-}$ or trefoil factor 2 (TFF2)-deficient mice. TFF3 did not improve the restitution when NHE2 was absent (NHE2$^{-/-}$) or pharmacologically inhibited, and the authors concluded that functional NHE2 (but not NHE1) is critical for gastric epithelial restitution and that trefoil factors may work to promote epithelial repair via induction of NHE2 (Xue et al. 2011; Matthis et al. 2019). The role of NHE2 in gastric cell migration (an integral element of mucosal wound healing) was also studied in vitro in the non-transformed rat gastric surface cell line RGM1. Surprisingly, while NHE1 promoted cell migration, lentiviral NHE2 overexpression increased the steady-state pHi and reduced the speed of restitution under low pH conditions, which was reversible by pharmacological NHE2 inhibition (Paehler Vor der Nolte et al. 2017). The reasons for these discrepancies are not entirely clear.

5.5.4.3 Intestinal and Colonic Epithelium

Donowitz et al. (1998) showed that in the chicken gut, NHE2 is the dominant NHE isoform mediating both basal and stimulated Na^+/H^+ exchange. Since NHE2 is also abundantly expressed in the mouse and human gut, it was surprising that NHE2 null knockout had no effect on the murine intestinal Na^+ absorption, suggesting a negligible role for this isoform in the intestinal epithelium (Schultheis et al. 1998a). An enhanced NHE3 expression in the colon of NHE2$^{-/-}$ mice was suggested to compensate for the loss of NHE2 activity (Bachmann et al. 2004).

However, colonic expression of NHE2 and NHE3 does not overlap; NHE2 is immunolocalized on apical membranes of wild-type crypts while NHE3 immunoreactivity is at the colonic surface, but not at the crypt base in NHE2$^{-/-}$ mice, thus making such compensation less likely. Alternative mechanisms to make up for the loss of NHE2 in the colonic crypts have been proposed by Guan et al. (2006). Additional loss of NHE2 in NHE2/3 double knockout mice had no additional effect on the impairment of systemic acid–base status or increase in aldosterone levels, and no apparent worsening of the diarrheal state (Ledoussal et al. 2001b). These results clearly demonstrated that intestinal NaCl, bicarbonate, and fluid absorption dependent on the apical Na^+/H^+ exchange is primarily due to NHE3 activity with little to no contribution from NHE2. Later, Xu et al. (2011) demonstrated that NHE8, with a more ubiquitous expression pattern in the gut epithelium, is upregulated in NHE2/3 double knockout female, but not male mice. It is plausible that the compensation for NHE2 knockout described by Guan et al. (2006) included the contribution of NHE8 in the colonic crypts.

Nowak et al. (2004) showed that pharmacological inhibition of NHE2 in Caco-2 cell monolayers stimulated an increase in transepithelial resistance. In the same study, NHE2 inhibition prevented the PMA-induced disruption of tight junctions and suggested a protective role of NHE2. In a model of intestinal ischemia injury in pigs, Moeser et al. (2006) demonstrated that inhibition of NHE2 (but not NHE1 or NHE3) stimulated the ileal restitution and barrier function recovery. Interestingly, in a mouse study from the same group, although histological assessment of recovery from intestinal ischemia showed no difference between WT controls and NHE2$^{-/-}$ mice, the latter had an increased small intestinal mucosal permeability during the post-ischemic recovery phase (Moeser et al. 2008). The localization of tight junction proteins ZO-1 and occludin was shifted to the detergent-soluble membrane fraction following the injury, and serine phosphorylation of occludin and claudin-1 was downregulated in the post-ischemic intestine of NHE2$^{-/-}$ mice compared with wild-type mice. The authors concluded that NHE2 plays different roles in murine and porcine mucosa in the context of barrier injury. It is also plausible that the interpretation of the study with pig ileum was confounded by the unknown sensitivities of pig NHE isoforms to the used inhibitors. The timing of the NHE2 inhibition/loss could provide an alternative explanation of this discrepancy. Acute in vitro inhibition of NHE2 in the injured ilea may be beneficial, while the chronic loss of NHE2 by gene targeting may induce changes in the epithelial cells making them more vulnerable to ischemic injury.

5.5.4.4 Pancreas

The expression of NHE2 in the acinar cells has not been verified, and its role is likely negligible since acini isolated from NHE2$^{-/-}$ mice recover from an intracellular acidification with kinetics similar to that seen in wild-type controls (Brown et al. 2003). In the pancreatic ductal epithelium, apical Na^+/H^+ exchange has been implicated in maintaining the relatively acidic pancreatic fluid collected under resting conditions (Zhao et al. 1994; Marteau et al. 1995). Both NHE2 and NHE3 are expressed on the apical membrane of the duct epithelial cells but NHE2$^{-/-}$ mice had unaltered luminal Na^+-dependent H^+ efflux in ducts (Lee et al. 2000).

5.5.4.5 Gallbladder Epithelium

Na^+/H^+ exchange has been best characterized in prairie dogs, believed to represent a good animal model of human gallstone formation. Primary culture of prairie dog gallbladder epithelial cells demonstrate H^+ gradient-dependent ^{22}Na uptake, mediated primarily by NHE2 (NHE1 6%, NHE2 66%, and NHE3 ~ 28% of total uptake), as determined by dimethylamiloride (DMA) and HOE-694 inhibition (Narins et al. 2004; Abedin et al. 2001). The significant role of Na^+/H^+ exchange in the epithelial Na^+ absorption in the gallbladder, as well as data suggesting increased gallbladder Na^+ and fluid absorption and NHE2/3 activity in the early stages of gallstone formation (Giurgiu et al. 1997; Conter et al. 1986; Narins et al. 2005) suggest a contribution of this transport modality in the pathogenesis of gallstone formation.

5.5.4.6 Renal Epithelium

Given the expression of NHE2 activity in the *macula densa* (MD) and renal distal convoluted tubules, several studies focused on the role of NHE2 in renal physiology and pathophysiology. MD is the part of the juxtaglomerular apparatus (JGA) responsible for sensing changes in salt concentration in the luminal fluid, generating and sending signals to the JGA that control renal blood flow and glomerular filtration rate and renin release. Tubular salt sensing by the MD has been shown to involve apical NaCl transport mechanisms including the furosemide-sensitive Na^+-K^+-$2Cl^-$ cotransporter NKCC2. Besides NKCC1, which is considered the primary NaCl entry mechanism, MD cells also express NHE2 on the apical side, and NHE4 on the basolateral side (Peti-Peterdi et al. 2000). It was though that NHE2 may contribute to salt sensing and the control of renin synthesis and release. Indeed, NHE2$^{-/-}$ mice showed a two- to fivefold increase in the number of renin-expressing cells in the afferent arteriole of JGA, increased renal cortical renin content and activity, increased plasma renin concentration, activation of ERK1/2, and elevated renal cortical cyclooxygenase-2 (COX-2) and microsomal prostaglandin E synthase (mPGES) expression (Hanner et al. 2008). A no-salt diet significantly increased renin content and activity in WT mice, but the response was blunted in NHE2$^{-/-}$ mice. The study concluded that NHE2 may participate in the MD feedback control of renin secretion and that NHE2-deficiency leads to MD cell shrinkage, activation of ERK1/2, COX-2, and the microsomal prostaglandin E synthase-1 (mPGES), well-characterized elements of the MD-PGE_2-renin release pathway (Hanner et al. 2008). Despite this finding, the NHE2$^{-/-}$ mice did not differ from their WT littermates with respect to blood pressure, plasma aldosterone levels, renal sodium excretion, or tubuloglomerular feedback (Schultheis et al. 1998a; Ledoussal et al. 2001a, b).

Angiotensin signaling also does not appear to alter NHE2 expression, localization, or function (Hatch and Freel 2008; Dixit et al. 2004).

A functional contribution of NHE2 to the acidification in the renal distal tubules was shown by Bailey et al. (2004). The authors postulated that NHE2 plays a major role in limiting urinary bicarbonate losses, particularly in states of high luminal bicarbonate load, mimicked by NHE3 deletion. In the collecting duct epithelium, aquaporin 2 (AQP2) promotes cellular proliferation at least in part, due to the downregulation of regulatory volume decrease (RVD) mechanisms when volume needs to be increased in order to proceed into the S phase (Di Giusto et al. 2012). NHE2 was shown to coordinate the cell volume regulation with AQP2 (Rivarola et al. 2017). In cortical collecting duct cell lines, NHE2 inhibition decreased cell proliferation and delayed cell cycle progression by slowing S and G_2/M phases, but only if AQP2 was expressed.

5.5.5 Role of NHE2 in Disease States

Although loss of NHE2 in mice displays abnormal parietal cell viability, no alterations in NHE2 expression are reported in human gastric disorders. TNFα and IFNγ inhibit NHE2 expression in the intestinal epithelium in vivo and in vitro (Amin et al. 2011; Rocha et al. 2001). These observations suggest that the alteration in NHE2 expression might be mediated by inflammation. Although based on the results, it was suggested that this could contribute to inflammation-associated diarrhea, in the context of negligible effects of NHE2 knockout on the intestinal luminal Na^+ handling, this seems unlikely. Contrary to $NHE3^{-/-}$ mice, NHE2-deficient mice are not hyper-susceptible to DSS (dextran sulfate sodium)-induced mucosal injury (Kiela et al. 2009). However, NHE2 knockout mice show delayed recovery from mesenteric ischemia with increased mucosal permeability and disrupted tight junctions (Moeser et al. 2008).

Enteropathogenic *E.coli* (EPEC) infection increased the activity of NHE2 several-fold in a type III secretion-dependent manner via activation of PKCα- and PKCε (Hecht et al. 2004; Hodges et al. 2006). The authors speculated increased NHE2 in the early stages of enteric infection may provide compensation to the increased luminal fluid resulting from inhibition of NHE3 activity, disruption of tight junctions, or inhibition of anion exchanger activity.

5.6 CPA1/SLC9A: NHE3

NHE3 was first cloned by Tse et al. and Orlowski et al. from rabbit and rat, respectively (Tse et al. 1992; Orlowski et al. 1992). Orlowski et al. also cloned a partial sequence of the human NHE3, which was later fully cloned and mapped to chromosome 5p15.3 (Brant et al. 1993, 1995). NHE3 is an 831–834 amino acid protein with a calculated molecular weight of ~93 kDa. Based on the presence of and conservation among species of the putative *N*-glycosylation sites, it was at first believed to be glycosylated, although subsequent work showed that while rabbit and pig renal NHE3 was indeed *N*-glycosylated (Soleimani et al. 1996; Bizal et al. 1996), rat or dog NHE3 was not (Bizal et al. 1996; Counillon et al. 1994). The secondary structure of NHE3 follows the general model for all members of the *Slc9a* family (Fig. 5.2). Its predominant expression in the gut and kidney epithelium as well as the results from studies with gene targeting pointed to a predominant physiological role of NHE3 in the intestinal and renal Na^+ (re)absorption.

5.6.1 Tissue Specificity

NHE3 is highly expressed in the gastrointestinal tract, kidney cortex, testes, ovaries, prostate, and thymus, with lower levels detected in the stomach, brain, and heart (Fig. 5.5) (Brant et al. 1995; Tse et al. 1992; Orlowski et al. 1992). NHE3 was also described in both the acinar and ductal cells of the salivary glands (Park et al. 1999), cholangiocytes and gallbladder epithelium where it has been implicated in bile formation and, similar to NHE2, possibly in the pathogenesis of gallstone formation (Mennone et al. 2001; Silviani et al. 1996; Colombani et al. 1996). In the rabbit and human gut, NHE3 expression is predominant in the ileum, (Dudeja et al. 1996; Hoogerwerf et al. 1996) and different reports focusing on specific segments of the gut showed some degree of species specificity in the rostrocaudal differences in NHE3 expression (Dudeja et al. 1996; Farkas et al. 2010; Hoogerwerf et al. 1996). NHE3 also displays a gradient in expression along the crypt–villous axis. In the small intestine, NHE3 is highly expressed in apical villous cells and not in the crypts

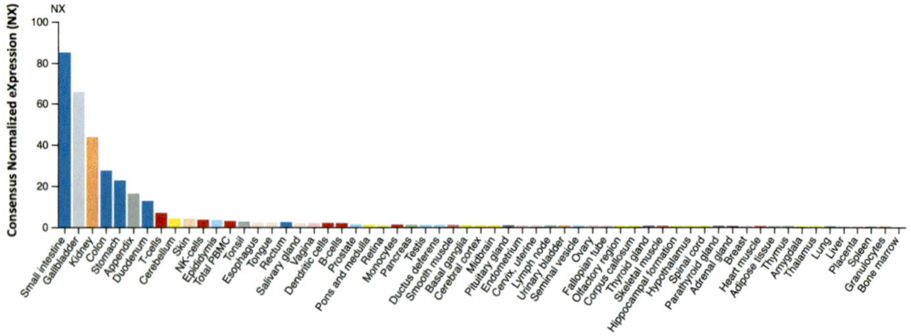

Fig. 5.5 NHE3 mRNA tissue specificity. Consensus Normalized eXpression (NX) levels for 55 tissue types and six blood cell types, created by combining the data from the three transcriptomics datasets (HPA, GTEx, and FANTOM5) using a normalization pipeline (The Human Protein Atlas). Color-coding is based on tissue groups, each consisting of tissues with functional features in common. Tissues sorted based on the mRNA expression level, from the highest (left) to the lowest (right). Image credit to The Human Protein Atlas (https://www.proteinatlas.org/ENSG00000066230-SLC9A3/tissue)

or goblet cells (Hoogerwerf et al. 1996). In the kidney, NHE3 is expressed by the epithelial cells of the proximal tubule (PT) and the thick ascending limb (Biemesderfer et al. 1997; Fenton et al. 2015).

5.6.2 *Subcellular Distribution*

In the epithelial cells of duodenum, jejunum, ileum, proximal colon, gall bladder, proximal tubule, and thick ascending limb as well as the proximal portion of the long descending thin limb of Henle, NHE3 is primarily present in the brush border/apical membrane (Zachos et al. 2005), both in the microvilli as well as in the intervillous clefts (Hoogerwerf et al. 1996). In intestinal and renal proximal tubule epithelial cells, a portion of the total NHE3 protein is localized to a diffused subapical pool and cycles between the plasma membrane and the endosomal compartment (Janecki et al. 1998, 1999; Biemesderfer et al. 1997). In the brush border membrane of the rabbit ileal enterocytes, NHE3 is equally distributed between the detergent-soluble and detergent-insoluble fractions (Li et al. 2001). The NHE3 pool within the detergent-insoluble fraction (cholesterol-enriched lipid microdomains, or lipid rafts) represents the predominant fraction regulated through endocytosis (Li et al. 2001).

5.6.3 *Physiological Regulation*

Epithelial NHE3 is highly regulated via transcriptional mechanisms, membrane targeting, and endosomal recycling, phosphorylation, and via its interaction with scaffolding and regulatory proteins within the C-terminal cytoplasmic domain. We will discuss some of the regulatory mechanisms as they pertain to the key functions of NHE3 in regulating intestinal and renal Na^+ (re)absorption.

5.6.3.1 Developmental Regulation

At the functional level, NHE3-mediated Na^+ uptake in rat jejunum is similar at 2 and 3 weeks of age and dramatically increased after weaning, at 6 weeks (Collins et al. 1997). Protein and mRNA levels did not fully correlate with activity; NHE3 expression was the lowest in 2-week-old suckling rats and increased in both 3- and 6-week-old rats, thus suggesting that postnatal changes in NHE3 expression and activity may be regulated at both transcriptional and posttranscriptional levels. Baum et al. (1995) showed that 1-week-old neonatal rabbits had approximately one-fourth of the NHE3 mRNA and protein abundance compared to adults and that NHE3 expression could be precociously stimulated by glucocorticoids. Later, our group showed a somewhat similar pattern in rats with NHE3 mRNA levels being lowest during the suckling period and highest at 3 weeks of age, corresponding with weaning and transit to solid food (Collins et al. 2000). However, this change was not reflected in protein expression or NHE3 activity.

5.6.3.2 Transcriptional Regulation

Two groups have cloned the rat NHE3 gene promoter (Kandasamy and Orlowski 1996; Cano 1996), and the discrepancies in the transcriptional start site were later resolved in our laboratory (Kiela et al. 2003). An atypical TATA box located −26/−31 bp upstream of the transcription start site mapped by Kandasamy and Orlowsky (1996) was not necessary and even detrimental for NHE3 promotor activity in intestinal epithelial cells, and a sequence at −20/+8-nt represented a functional transcriptional initiator. Three Sp transcription factor-binding sites within the −81 nt upstream region of NHE3 promoter are critical for the NHE3 promoter activity. Both Sp1 and Sp3 acted synergistically with GATA-5 bound to a GATA box within exon 1 to provide maximum promoter activity (Kiela et al. 2003). In renal epithelial MDCK cells, transcriptional activation of NHE3 gene promoter via cooperative binding of Stat3 and Sp1/Sp3 was necessary for the 3D dome formation, thus suggesting a role for NHE3 in the renal epithelial differentiation, and possibly morphogenesis (Su et al. 2009).

Human NHE3 gene promoter was cloned by Malakooti et al. (2002). In transiently transfected Cacco-2 cells, this group demonstrated a maximal promoter activity driven by the −95/+5 nt region, which contained several putative *cis*-elements for transcription factor IID (TF IID), CACCC, two Sp1 sites, and two AP-2 motifs, with the latter two transcription factors playing most critical roles in basal NHE3 promoter activity.

5.6.3.3 Hormonal Regulation of Epithelial NHE3

NHE3 is regulated by several humoral factors. The regulation of the small intestinal NHE3 by glucocorticoids has been shown to be not only segment-specific but also age-dependent (Kiela et al. 2000). Glucocorticoid responsiveness in the proximal small intestine was greatest in suckling animals and decreased with age, with no effect in adults. The distal small intestine/ileum was responsive only in adult rats and these varying responses correlated with the expression and ligand binding capacity of the glucocorticoid receptor in the small intestine. Studies by Cho et al. (1994) in rabbit small intestine and colon also showed that dexamethasone increased NHE3 mRNA expression in the ileum and proximal colon but not in the jejunum or distal colon. Conversely, adrenalectomy reduced NHE3 expression in the rat ileum and proximal colon but not in the jejunum. Two mechanisms for regulation of NHE3 expression and activity have been proposed. In the acute mode, glucocorticoid regulation of NHE3 involves posttranscriptional activation of the transporter by serum and glucocorticoid-induced protein kinase 1 (SGK1). The regulation may be biphasic, initially involving phosphorylation of the preexisting membrane NHE3 at Ser^{663} and subsequently by a mechanism whereby SGK1 and NHERF2 interact with the cytosolic tail of NHE3 to facilitate the translocation of the newly synthesized NHE3 to the cytoplasmic membrane (Yun 2003; Yun et al. 2002; Wang et al. 2005). SGK3 isoform also contributes to the effects of glucocorticoids on NHE3 activity (He et al. 2011). The second mechanism of NHE3 activation by glucocorticoids involves transcriptional regulation (Kandasamy and Orlowski 1996; Cano 1996).

NHE3 is also regulated by mineralocorticoids. Musch et al. (2008) showed that aldosterone stimulates membrane insertion of NHE3 followed by transcriptional activation. Both effects are dependent on the SGK1 pathway in Caco-2BBE cells. The stimulating effects of aldosterone were also demonstrated in the rat kidney (Eiam-Ong et al. 2017).

The first report on the regulation of NHE3 expression by the thyroid hormone showed increased expression of renal NHE3 in rats with hyperthyroidism (Azuma et al. 1996). Rat NHE3 gene promoter has putative thyroid receptor binding sites (Kandasamy and Orlowski 1996; Cano 1996) and NHE3 activity was induced by 3,5,3′-triiodothyronine (T3) via a transcriptional mechanism (Cano et al. 1999; Gupta et al. 2004).

Parathyroid hormone (PTH) acts on kidneys to alter urinary electrolyte and fluid excretion, with hyperparathyroidism leading to phosphaturia and, in some cases, bicarbonaturia. PTH inhibits NHE3 activity in a biphasic mode, acutely via activation of both protein kinase A (PKA) and protein kinase C (PKC) signaling pathways (Helmle-Kolb et al. 1990; Azarani et al. 1995), and chronically via decreased mRNA and protein expression (Girardi et al. 2000).

5.6.3.4 NHE3 Regulation by Short-Chain Fatty Acids

Short-chain fatty acids (SCFA), mainly butyrate, propionate, and acetate, are produced by microbial fermentation of dietary carbohydrates and they are important for gut health. Short-chain fatty acids are potent stimuli of intestinal sodium and water absorption (Binder and Mehta 1989; Choshniak and Mualem 1997; Krishnan et al. 1999; Ramakrishna et al. 1990; Ruppin et al. 1980) with butyrate being the most effective of all three SCFAs. Indeed, amylase resistant starch has been used as an additive to oral rehydration solutions and proved to be effective in reducing the diarrheal stool volume in cholera patients (Ramakrishna et al. 2000). This response is secondary to an increase in short-chain fatty acids driven by the fermentation of the amylase resistant starch in the colon. It has been suggested that short-chain fatty acid-mediated increase in sodium absorption is due to the coupling of two exchange mechanisms, Na^+/H^+ and $SCFA^-/Cl^-$ exchange. Our studies, as well as others, have shown that part of NHE regulation by SCFA is mediated by transcriptional induction of NHE3 gene expression (Musch et al. 2001; Kiela et al. 2001). The mechanism for induction of SCFA involves Ser/Thr kinase activity with a permissive role for protein kinase A (PKA), since the activation of the NHE3 gene promoter by butyrate was eliminated by PKA inhibitors or by overexpression of the dominant-negative mutant form of the regulatory subunit of PKA (Kiela et al. 2001). Further studies have determined that butyrate-induced phosphorylation of Sp1, acetylation of Sp3 and, a shift in their interactions with the proximal rat NHE3 promoter in favor of Sp3, a more potent inducer of NHE3 gene transcription (Kiela et al. 2007). A somewhat similar convergence of Sp1/Sp3 on the proximal NHE3 gene promoter was described for the human gene (Amin et al. 2007). In vivo, in rats fed a 5% pectin-supplemented diet for 2 days, NHE3 mRNA, protein, and activity were all increased in the colonic but not ileal epithelial cells and similar results were obtained with Caco-2/BBE cells treated with SCFAs in vitro (Musch et al. 2001). The effects of butyrate in the reversal of the negative effects of cholera toxin on NHE3 activity were described in the ileal epithelial cells, although it did not involve increases in NHE3 mRNA levels (Subramanya et al. 2007).

5.6.3.5 Serotonin

Serotonin, produced by the neuroendocrine cells in the gastrointestinal tract, plays a major role in regulating gastrointestinal motility, secretion, and absorption. Increased serotonin levels have been suggested to play a role in the pathophysiology of diarrhea secondary to carcinoid syndrome, in ulcerative colitis, and in irritable bowel syndrome. Serotonin reduced NHE3 activity via the 5-HT4 receptor in human intestinal epithelial cells, an effect associated with a PKC α-dependent decrease in the Sp1/Sp3 association with the proximal NHE3 gene promoter (Gill et al. 2005; Amin et al. 2009).

5.6.3.6 Metabolic Acidosis

Metabolic acidosis is a disturbance in the homeostasis of systemic pH. There are several types of metabolic acidosis: diabetic ketoacidosis associated with a buildup of ketone bodies; hyperchloremic acidosis associated with an excessive loss of sodium bicarbonate, e.g., in severe diarrhea; and lactic acidosis associated with alcohol consumption, cancer, extreme exercise, certain medications, kidney disease, or severe dehydration. The proximal tubules, the primary site of NHE3 expression in the kidney, also provide for ca. 70% of HCO_3^- reabsorption. $NaHCO_3$ reabsorption occurs in the proximal tubules with additional help from the thick ascending limb of the nephron. In chronic systemic acidosis, a compensatory HCO_3^- reabsorption is triggered via coupled apical Na^+/H^+ antiporter activity and basolateral electrogenic Na^+/HCO_3^- symporter (Preisig and Alpern 1988) and is associated with transcriptional activation of the NHE3 gene promoter and increased NHE3 mRNA and protein expression (Laghmani et al. 1997; Silva et al. 2012). This mechanism constitutes the conventional indirect pathway, in which NHE3 contributes protons to titrate luminal HCO_3^- and allows for CO_2 diffusion across the apical membrane. A novel, direct pathway for HCO_3^- reclamation was described by Guo et al. who showed a contribution of an electroneutral apical Na^+/HCO_3^- cotransporter NBCn2 (Guo et al. 2017). The role of Na^+/H^+ exchange in the renal regulation of systemic acid–base balance was reviewed by Bobulescu and Moe (2006).

Chronic metabolic acidosis also induces intestinal epithelial Na^+/H^+ exchange. Metabolic acidosis induced in rats by 5% ammonium chloride in drinking water increased ileal expression of NHE2 and NHE3 mRNA, protein, and activity (Lucioni et al. 2002). However, the precise mechanisms of these changes have not been described.

5.6.3.7 Intestinal Resection

The gastrointestinal tract adapts by increasing its transport capacity in the setting of intestinal resection as seen in short bowel syndrome. Indeed, in an animal model of intestinal resection, rat intestinal Na^+/H^+ exchange activity was shown to increase primarily in the segment distal from the resection (Sacks et al. 1993). This increase was associated with a threefold elevation of NHE3 mRNA and protein expression after a 50% proximal small bowel resection (Musch et al. 2002). Again, this increase was only seen in the ileal distal segment from the anastomosis, suggesting that dietary rather than humoral factors might be responsible.

5.6.3.8 Posttranscriptional Regulation of NHE3

Most of the studies on the acute regulation of NHE3 activity come from heterologous cell expression systems such as PS120 fibroblast, renal epithelial cells, and colon cancer cells. The key elements of NHE3 protein modification are located within the C-terminal regulatory domain (He and Yun 2010; Zachos et al. 2009; Donowitz et al. 2009; Alexander and Grinstein 2009). The C-terminus of NHE3 binds to several scaffolding proteins to form higher-order multi-protein complexes and NHE3, itself, can be considered a scaffold. Therefore, it does appear that NHE3 exists physiologically in large multi-protein complexes, which range from 400 kDa in the intracellular pool to 1000 kDa in the plasma membrane. A small putative, alpha-helical domain of NHE3, located between amino acid 586 and 605, was suggested to act as a “switch domain” to interact with at least seven other proteins to activate or inhibit NHE3. This dynamic complex assembly in association with cytoskeletal, endosomal recycling, and protein phosphorylation/dephosphorylation all act in concert to provide a mechanism for endosomal recycling and for the fine-tuning of NHE3 protein turnover and activity (Donowitz et al. 2009).

5.6.3.8.1 Role of Glycosylation in Regulation of NHE3 Activity

As we mentioned in the introduction, NHE3 is *N*-glycosylated in a species-dependent manner. Inhibition of *N*-glycosylation in tunicamycin-treated pig renal

LLC-PK cells significantly decreased NHE3 activity, as measured by pH-dependent ^{22}Na uptake and by Na-dependent pH_i recovery from an acid load (Soleimani et al. 1996). This decrease in NHE3 function in tunicamycin-treated cells was accompanied by an intracellular accumulation of seemingly unglycosylated forms of the protein and a conceivably compensatory threefold increase in NHE3 mRNA. Based on these studies, it has been postulated that glycosylation of porcine NHE3 plays a role in membrane trafficking and ultimately in NHE3 activity. On the other hand, deglycosylation of rabbit renal brush border protein did not impact acid-stimulated, amiloride-sensitive ^{22}Na influx into the vesicles (Bizal et al. 1996). Analogous in vivo experiments with NHE1 with mutated *N*-glycosylation sites, and with inhibition of *O*-glycosylation in NHE2, did not translate into detectable functional changes of the respective isoform (Tse et al. 1994; Counillon et al. 1994). Therefore, the physiological significance of NHE3 glycosylation is still unclear.

5.6.3.8.2 Regulation of NHE3 Activity by Endosomal Recycling

A state of dynamic equilibrium exists between the cell surface and the intracellular compartment. Internalization via clathrin-coated vesicles and exocytosis to the cytoplasmic membrane is driven by a phosphatidylinositol 3-kinase-dependent signaling pathway. Inhibition of PI3-kinase leads to decreased NHE3 activity correlating with the depletion of the plasma membrane pool of NHE3 protein (Kurashima et al. 1998), whereas activation of PI3-kinase or AKT in NHE3-expressing PS120 cells, where NHE3 is predominantly retained in the endosomal compartment at baseline, stimulates NHE3 and increases the percentage of NHE3 present on the plasma membrane (Lee-Kwon et al. 2001). EGF and FGF growth factors utilize the same PI3K-dependent pathway to stimulate NHE3 activity by increasing the surface protein pool (Donowitz et al. 2000; Janecki et al. 2000). A number of other factors increase the apical pool of NHE3, such as LPA (Lee-Kwon et al. 2003) or endothelin (Peng et al. 2001). On the other hand, decreased surface expression and activity of NHE3 have been associated with PKC activation (Janecki et al. 1998), and exposure to parathyroid hormone (Collazo et al. 2000) or dopamine (Hu et al. 2001). Casein kinase 2 (CK2), through its α-subunits, binds NHE3 at position 586–605 in the C-terminal tail and phosphorylates it at Ser^{719} (Sarker et al. 2008). Phosphorylation at Ser^{719} is responsible for ca. 70% of basal NHE3 activity. CK2 inhibition reduced NHE3 plasma membrane expression and activity, whereas CK2-mediated NHE3 phosphorylation increased its trafficking from the recycling pool to the plasma membrane, thus indicating that the majority of the CK2 effect occurs via regulation of NHE3 trafficking. These and other studies clearly suggest that NHE3 is dynamically redistributed between the subcellular compartment and the apical membrane. This mechanism of NHE3 regulation has been more extensively reviewed elsewhere (Alexander and Grinstein 2009; Donowitz et al. 2013).

5.6.3.8.3 Regulation of NHE3 Activity by Cyclic Nucleotides

cAMP regulates NHE3 activity via PKA-mediated phosphorylation mechanisms. NHE3 protein was phosphorylated by PKA in response to elevated intracellular cAMP at Ser^{552} and Ser^{605} (Zhao et al. 1999). This study demonstrated that phosphorylation of Ser^{552} participated in NHE3 response to cAMP, although this was not independently verified (Kurashima et al. 1997). NHE3 phosphorylation by PKA results in reduced V_{max} and decreased surface NHE3 protein distribution, which was presumably due to increased endocytosis and decreased exocytosis. In the setting of PKA activation, a multi-protein complex was recruited to the C-terminus of NHE3. This complex contained NHERF1, NHERF2, and a scaffolding protein, ezrin. When expressed in NHERF-deficient cells, both NHERF1 and NHERF2 restituted PKA-dependent NHE3 inhibition (Yun et al. 1997). NHERF1 and NHERF2 interact through their C-terminal 29 amino acids with the cytoskeleton-associated ezrin, which functions as a PKA-anchoring protein. Phosphorylation of NHE3 by PKA is therefore facilitated by bringing the catalytic subunit of PKA to the vicinity of the NHE3 cytoplasmic tail by a protein complex containing either of the two NHERF factors and ezrin. PDZ-binding protein, PDZK1, was also necessary for cAMP and calcium-mediated NHE3 regulation in mouse enterocytes (Cinar et al. 2007).

cGMP also inhibits NHE3 activity via the cGMP-dependent type II protein kinase, cGKII/PRKG2-dependent pathway. cGMP/cGKII-mediated rapid inhibition of rabbit NHE3. It was associated with decreased NHE3 plasma membrane abundance and required phosphorylation of Ser^{554}, Ser^{607}, and Ser^{663} (equivalent to mouse Ser^{552}, Ser^{605}, and Ser^{659}) (Chen et al. 2015). More recently, Avula et al. (2018) showed that NHE3 inhibition by cGMP (as well as elevated intracellular Ca^{2+}) in the mouse jejunum depended on the interaction with NHERF2/NHERF3 heterodimers.

5.6.3.8.4 NHE3 Regulation Via Association with the Cytoskeleton

An NHERF-mediated link with ezrin suggested the association of NHE3 with the cytoskeleton as a likely mechanism controlling NHE3 activity. Indeed, NHE3 was found to co-sediment with F-actin and pharmacological disruption of the cytoskeleton induced a profound inhibition of NHE3 activity (Kurashima et al. 1999). Inhibition of two kinases controlling cytoskeletal assembly, RhoA and ROK, also inhibited NHE3 activity in the Chinese hamster ovary (CHO) cells stably transfected with dominant-negative mutants of a respective kinase without altering NHE3 abundance in the cytoplasmic membrane (Szaszi et al. 2000b). Disruption of the actin cytoskeleton by hyperosmotic stress may be responsible for the decrease in NHE3 activity (Szaszi et al. 2000a). However, Hayashi et al. (2013) demonstrated unchanged NHE3 activity in epithelial cells lacking ezrin, which questioned the role of ezrin in maintaining NHE3 plasma membrane expression and activity. Therefore, ezrin/NHE3 interaction may not be an exclusive mechanism responsible for the retention of NHE3 at the apical plasma membrane or the interaction between NHE3 and the cytoskeleton.

5.6.4 *Physiological Roles of NHE3: Lessons from Knockout Studies*

The generation of null knockout of the SLC9A3 gene was reported by Schultheis et al. (1998b). To study the renal effects of NHE3 deficiency, Woo et al. (2003) created a partial rescue model by crossing NHE3-null knockout mice with transgenic mice expressing NHE3 in the intestinal epithelium under the control of the intestinal fatty acid-binding protein (IFABP) gene promoter. Later, a more refined tissue-specific knockout targeting the expression of NHE3 specifically in the renal epithelial cells of the segments 1 and 2 of the proximal tubules (SGL2-Cre x NHE3$^{fl/fl}$) was reported by Li et al. (2013) Finally, Fenton et al. (2015) described a model of "pan-tubular" NHE3 knockout by crossing Pax8-Cre mice with NHE3$^{fl/fl}$ mice to eliminate NHE3 expression in segments 1–3 of the proximal tubule and reduce NHE3 expression by 85% in the thick ascending limb.

5.6.4.1 Intestinal and Colonic Epithelium

Although the NHE3$^{-/-}$ mice experienced only mild diarrhea, an autopsy revealed that the small intestine, cecum, and proximal colon were enlarged, and the luminal content was more alkaline when compared to the WT mice. The pH of the colonic content of the NHE3 knockout mice was decreased relative to that in the cecum indicating that net hydrogen secretion occurred in the colon of the knockout mice, whereas net alkalization occurred in the colon of the WT mice (Schultheis et al. 1998b). Elevated epithelial Na^+ channel activity and colonic H^+/K^+-ATPase mRNA represented likely means of compensation, especially in the distal colon, and are likely the reason why these mice did not develop frank diarrhea and lethal dehydration. These initial findings clearly demonstrated that NHE3 was the major absorptive Na^+/H^+ exchanger in the gut.

Our group was the first to describe that conventionally housed NHE3$^{-/-}$ mice maintained on the original mixed genetic background (Black Swiss/129) developed bacterially mediated spontaneous distal colitis, which could be ameliorated by the administration of antibiotics (Laubitz et al. 2008). They also displayed remarkably high susceptibility to dextran sodium sulfate-induced mucosal injury (Kiela et al. 2009) and exacerbated colitis in IL10$^{-/-}$ mice (Larmonier et al. 2011). We have also described increased bacterial adhesion and bacterial translocation in the distal colon of NHE3$^{-/-}$ mice, findings later confirmed by our collaborators (Johansson et al. 2014). We hypothesized that the gut microbiota composition played an essential role in the pathogenesis of colitis in NHE3-deficient mice. Using 16S amplicon profiling, we demonstrated a significant decrease in alpha diversity of the luminal and mucosal microbiota of conventional NHE3$^{-/-}$ mice compared to wild-type mice, and significant differences in the taxonomic community composition. NHE3 deficiency was characterized by a reduction in pH-sensitive butyrate-producing Firmicutes families, *Lachnospiraceae*, and *Ruminococcaceae* (Clostridia clusters IV and XIV) with an expansion of inflammation-associated *Bacteroidaceae* (Larmonier et al. 2013). Re-derivation of NHE3$^{-/-}$ mice from a conventional to a barrier facility reduced the severity of colitis and decreased DSS susceptibility. Reintroduction of the conventional microflora to NHE3$^{-/-}$ mice from the barrier facility resulted in the restoration of the symptoms initially described in the conventional environment. NHE3 status, but not the degree of inflammation, was the most significant determinant of the gut microbiota in the model of adoptive T cell transfer colitis in Rag2/NHE3 double knockout mice, a strain with dramatically increased susceptibility to T cell-mediated colitis (Laubitz et al. 2016). Finally, fecal microbiome transplant partially, but effectively transmitted that susceptibility to germ-free Rag1$^{-/-}$ or IL10$^{-/-}$ mice (Harrison et al. 2018). NHE3 activity and/or expression are significantly reduced in Inflammatory Bowel Disease (IBD) patients (see later in the chapter). Thus, we believe that loss of NHE3 activity not only contributes to inflammation-associated diarrhea in IBD but that it also critically contributes to the severity of inflammation and disease outcome, at least in part via modulation of the gut microbial ecology.

5.6.4.2 Renal Tubular Epithelium

In the kidney, NHE3 is localized to apical membranes of the proximal tubule and to a lesser extent in the thick ascending limb of Henle, where it is responsible for reabsorbing large quantities of NaCl and HCO_3^-, along with water. NHE3 null knockout mice have severe absorptive defects both in the kidney and intestine, with symptoms of chronic volume depletion, low blood pressure, elevated expression of renin, and elevated serum aldosterone (Schultheis et al. 1998b). More targeted analyses with in situ microperfusion and micropuncture studies showed more definitively that reabsorption of HCO_3^- and water was reduced in the proximal tubule of NHE3$^{-/-}$ mice (Lorenz et al. 1999; Wang et al. 1999). There was a concern that extracellular fluid volume depletion itself, perhaps as the sum of the renal defect and the diarrheal state, may contribute to the observed diminished ability to reclaim NaCl in the kidneys of NHE3 null knockout mice. Similarly, it was unclear whether the observed reduction in glomerular filtration rate (GFR) in this mouse model might have been at least in part, due to systemic hypovolemia and hypotension. To address this problem, Woo et al. (2003) developed a transgenic mouse model in which NHE3 is expressed in the small intestine under the control of the IFABP gene promoter and crossed it with NHE3$^{-/-}$ mice (tgNHE3$^{-/-}$). In these mice, both dietary Na^+ restriction and Na^+ loading were better tolerated than in non-transgenic NHE3$^{-/-}$ mice. A high-salt diet led to a substantial reduction of aldosterone levels in tgNHE3$^{-/-}$ mice, indicating a partial correction of the decrease in the extracellular fluid volume. Importantly, tgNHE3$^{-/-}$ mice remained mildly hypotensive and had reduced GFR compared with tgNHE3^{WT} mice, which overexpressed NHE3 only in the intestine (Woo et al. 2003). To add fidelity to the system, two additional mouse models were created using Cre-lox recombination. In the first model, Cre is expressed under the control of the Sglt2 gene promoter and targets segments S1 and S2 of the proximal tubule, and deletion of NHE3 in the crossed NHE3$^{fl/fl}$ mice confirmed a reduced bicarbonate and H_2O reabsorption, mild metabolic acidosis, but no impairment in the secretion of ammonia ($NH3/NH_4^+$) into the lumen (Li et al. 2013). In the second model described by Fenton et al. (2015), Pax8-Cre mice targeted NHE3 for deletion in the entire tubule system and the collecting ducts. With this model, studies showed that renal NHE3 is not involved in natriuresis induced by caffeine (Fenton et al. 2015). However, at homeostasis, although their weight, plasma Na^+, K^+, aldosterone, osmolality or blood pH were unaffected, Pax8^{Cre}NHE3$^{fl/fl}$ mice had a higher fluid intake, lower GFR, and lower urine osmolality in conjunction with higher urinary pH and a trend for a higher Na^+ excretion (Fenton et al. 2017). As opposed to NHE3 null knockouts, in which dietary Na^+ restriction led to high mortality and dehydration, Pax8^{Cre}NHE3$^{fl/fl}$ mice tolerated NaCl restriction, and increased activity and/or expression of the Na^+-Cl^- cotransporter, NCC, or epithelial sodium channels (ENaC) has been speculated to be responsible for the related adaptive response (Fenton et al. 2017).

5.6.5 *Role of NHE3 in Disease States*

5.6.5.1 Congenital Sodium Diarrhea (CSD)

In 1994 (Holmberg and Perheentupa 1985) and 1995 (Booth et al. 1985), two independent groups described a rare autosomal diarrheal disorder presenting in early life with Na^+ diarrhea. It was presumed that this is related to a defective NHE3 based on a close phenotypical resemblance between this rare disease and symptoms displayed by NHE3-deficient mice. However, homozygosity mapping and multipoint linkage analysis studies in four candidate regions containing NHE1, NHE2, NHE3, and NHE5 genes showed that this disorder was not associated with these genes (Muller et al. 2000). However, Janecke et al. (2016) studied 18 patients with CSD from 16 families, and almost half of these patients had putative loss-of-function mutations (point, missense, and truncation) within the NHE3-encoding SLC9A3 gene. Four missense mutations (p.Arg382Gln, p.Ala311Val, p.Ala269Thr, and p.Ala127Thr) were tested in vitro and all but p.Ala127Thr (benign variant) resulted in decreased basal Na^+/H^+ exchange activity (Janecke et al. 2016). More recently, in a genome-wide SNP analysis of CSD patients, Heinz-Erian et al. (2009) identified a loss of function mutation in the SPINT2 gene, which encodes a Kunitz-type series protease inhibitor. These mutations are believed to be responsible for a third of CSD cases described as a syndromic form of CSD, but the relationship between SPINT2 and Na^+/H^+ exchange is not known. Activating mutations in the catalytic domain of the guanylate cyclase 2C gene (*GUCY2C*, which also serves as a heat-stable enterotoxin receptor) may account for 20% of cases of sporadic CSD (Fiskerstrand et al. 2012; Muller et al. 2016). These mutations are thought to be mechanistically linked to hyper-activation of the cystic fibrosis transfer conductance regulator (CFTR) and to inhibition of NHE3 function via increased intracellular cGMP and cGKII kinase-dependent mechanism (Fiskerstrand et al. 2012; Arshad and Visweswariah 2012; Chen et al. 2015; Gurney et al. 2017).

5.6.5.2 Infectious Diarrhea

Diarrheal disorders represent a significant problem, especially in the pediatric age group, with high morbidity and mortality worldwide. One mechanism of diarrhea induced by infectious agents is related to altered electrolyte flux with significant activation of chloride secretion via activation of cyclic AMP, cyclic GMP, and intracellular Ca^{2+}. The resulting activation of CFTR results in net Cl^- secretion in the gastrointestinal tract followed by Na^+ losses. The second mechanism is related to the inhibition of NHE3 activity, e.g., by enteropathogenic *Escherichia. Coli* (*E. coli*). Coupling of active Cl^- secretion and inhibition of Na^+ absorption results in dehydration, which carries significant morbidity and mortality in affected individuals. *Vibrio cholerae* infection is a prime example of secretory diarrhea, in which cholera toxin is capable of inhibition of NHE2 and NHE3 as well as active secretion of chloride through the CFTR. The resulting water losses in cholera can reach liters per day resulting in significant dehydration. Other examples of infectious agents that can cause diarrhea associated with reduced electroneutral Na^+ absorption include *Salmonella typhimurium*, *Shigella dysenteriae*, and *Campylobacter jejuni* (Khurana et al. 1991; Kaur et al. 1995; Kanwar et al. 1994).

Clostridium difficile (*C. difficile*) is a leading cause of nosocomial diarrhea and pseudomembranous colitis, which occurs in individuals being treated with antibiotics. *C. difficile* toxin B (TxB) inhibits Rho-family GTPases and alters the interaction between NHE3 and the microvillar actin cytoskeleton to facilitate its internalization and reduced function. TxB has also been shown to disrupt the stimulating effects of Rho guanine nucleotide exchanger factor 7, β-PIX, which normally, along with Shank-2, stabilizes NHE3 and activity at the apical membrane (Lee et al. 2010). Engevik et al. (2015) postulated that NHE3 inhibition by *C. difficile* is also related to an alteration of the intestinal environment and gut microbiota, which could facilitate the pathogen's colonization and expansion. They showed that, when compared with healthy controls, decreased expression of NHE3 in infected patients was correlated with increased stool Na^+ and pH. The changes in the microbiome included elevated relative abundance of *Bacteroidetes* and *Proteobacteria*, and decreased *Firmicutes* phyla.

There are a number of pathogenic *E. coli* strains, which are responsible for diarrheal outbreaks. These include enteropathogenic *E. coli* (EPEC), enterohemorrhagic *E. coli* (EHEC), enterotoxigenic *E. coli* (ETEC), enteroaggregative *E. coli* (EAEC), enteroinvasive *E. coli,* and diffusely adherent

E. coli (DAEC). A number of NHE-independent mechanisms contribute to the effects of EPEC are known (Lapointe et al. 2009). It is interesting, however, that in a cell culture model, significantly increased NHE1 and NHE2 apical activity with concurrent 50% inhibition of NHE3 (Hecht et al. 2004). Hodges et al. (2008) showed that the inhibitory effects of EPEC on NHE3 were dependent on EspF protein, a component of the *E. coli* type III secretion system. *E. coli* heat-stable enterotoxin (ST) is a major virulence factor of ETEC, which is the leading cause of childhood death from diarrhea and the leading cause of traveler's diarrhea, and has been known to inhibit NHE3 activity. Chen et al. (2019) demonstrated that in vivo and in vitro, ST rapidly increased intracellular cGMP, decreased the apical expression of NHE3, and inhibited its activity. The study implicated an ST-induced change in binding of the NHRF2/3 heterodimers with the C-terminal tail of NHE3 as partially responsible for the effects of ST on NHE3 activity. Another component of this inhibition, although not consistent in mice and Caco-2 cells, was elevated intracellular Ca^{2+}, which also depended on NHERF2/3.

In the setting of viral gastroenteritis, there is clear evidence to suggest that affected individuals develop diarrhea, but the diarrhea is not secretory in nature. Although less is known about the role of NHE3 in diarrhea during viral gastroenteritis, astroviral infections have been associated with Na^+ malabsorption, and Nighot et al. (2010) demonstrated that infection leads to decreased levels of NHE3 in the insoluble protein fraction in the enterocytes, presumably the apical membrane domains. In rotavirus-infected patients, expression of both NHE2 and NHE3 proteins was downregulated, and the remaining NHE3 was mislocalized (Baetz et al. 2016).

5.6.5.3 Inflammatory Bowel Diseases

Inflammatory Bowel Diseases (IBD) include Crohn's disease (CD) and ulcerative colitis (UC). Both conditions are characterized by a dysregulated immune response to normal microbial flora and disruption of the epithelial barrier, which may lead to bacterial translocation. Diarrhea is one of the most common symptoms of IBD patients and occurs in approximately 50% of acute flare-ups with Crohn's disease and in nearly all patients with ulcerative colitis (Seidler et al. 2006). The pathogenesis of diarrhea in IBD is complex and multi-factorial. Altered intestinal motility, abnormal epithelial iron transport, and defective NaCl absorption are some of the factors involved (Sandle 2005). The mechanisms involved in the inhibition of the epithelial Na^+/H^+ exchange in IBD are not well defined and the available data are inconsistent. IFNγ inhibits the expression and function of NHE3 in vivo and in vitro (Rocha et al. 2001). Inhibition of NHE3 expression and activity has also been described in several experimental models of colitis including IL-$2^{-/-}$ mice (Barmeyer et al. 2004) and in DSS- and TNBS (trinitrobenzene sulfonic acid)-induced colitis (Sullivan et al. 2009). In IL-$10^{-/-}$ mice, NHE3 activity in apical membranes of the isolated colonic crypts was decreased significantly without alteration of NHE3 expression and localization (Lenzen et al. 2012). In IBD patients,

Sullivan et al. (2009) described decreased NHE3 protein and mRNA expression in the sigmoid of a majority of the studied cases of active UC and/or CD, in ileal mucosal biopsies of active CD, as well as in ~50% of sigmoid biopsies from inactive UC or CD. However, Siddique et al. (2009) showed that CD and UC are associated with decreased NHE3 protein and activity but that NHE3 mRNA was reduced only in CD patients. It has been speculated that a decreased expression of two key NHE3-regulatory proteins, NHERF2 and PDZK1, may be responsible for decreased NHE3 activity (Lenzen et al. 2012). Additional work with UC patients showed a significant reduction of NHE3 activity despite the preserved protein and mRNA expression (Yeruva et al. 2010; Farkas et al. 2010). Decreased NHE3 expression and/or activity is likely not just related to inflammation-associated diarrhea, but also contributes to the gut microbial dysbiosis and the severity of mucosal inflammation, as we discussed above in the section on the roles of NHE3 in the intestinal and colonic epithelium. Importantly, among the small cohort of patients with congenital Na$^+$ diarrhea and mutations in the SLC9A3 or guanylate cyclase gene, some infants presented with symptoms of IBD, thus suggesting that the function of NHE3 may be one of the key elements defining the time of onset of IBD (Janecke et al. 2015, 2017).

5.6.5.4 Diabetic Diarrhea

Gastrointestinal complications are common in patients with diabetes mellitus. Indeed, diabetic diarrhea is seen in approximately 22% of patients with diabetes. The mechanisms responsible for the alterations in electrolyte imbalances are not well known. Recently, He et al. (2015) showed that NHE3, ezrin, NHERF1, and IRBIT/AHCYL1 (adenosylhomocysteinase like 1) form macrocomplexes, which are perturbed under diabetic conditions in streptozotocin-treated mice, and that insulin administration reconstituted these macrocomplexes and restored NHE3 expression and activity at the brush border membrane. Therefore, it does appear that defective NHE3 activity also plays a role in the pathogenesis of diabetic diarrhea.

5.7 CPA1/SLC9A: NHE4

NHE4 has been relatively under-studied. Initially cloned from the rat stomach cDNA library by Orlowski et al. (1992), rat NHE4 is a 717 amino acid (aa) protein with a calculated molecular weight of ~81.5 kDa. Human NHE4 is a little longer at 798 aa. Empirical analyses resulted in controversial findings regarding its size. When stably expressed in NHE-deficient PS120 fibroblasts as a fusion protein, polyclonal antibodies raised against a fusion protein containing 393–625 aa of the rat NHE4 protein recognized a band of approximately 100 kDa, suggesting posttranslational modifications, possibly glycosylation (Bookstein et al. 1994b). Monoclonal antibodies raised against a similar region (565–675 aa) of the rat NHE4 recognized a predominant band of approximately 65–70 kDa and two minor bands at 45–50 kDa, and

75 kDa (Pizzonia et al. 1998). Among the CPA1 family members, NHE4 protein shares the highest homology with NHE2 (Fig. 5.1). Although this was not experimentally verified, in silico hydropathy analyses indicate that the exchanger's membrane topology follows the membrane topology models established for the CPA1 family (Orlowski et al. 1992).

5.7.1 Tissue Specificity

NHE4 expression is relatively restricted. It is the most highly expressed in the gastric parietal and chief cells, and to a lesser extent in the mucous cells (Pizzonia et al. 1998; Rossmann et al. 2001). It is expressed at lower levels in the kidney (Bookstein et al. 1994b), where it is predominantly present in the epithelium of the thick ascending limb and distal convoluted tubule, and with weaker staining in the collecting ducts in the cortex to inner medulla and in the proximal tubules (Chambrey et al. 2001). Its expression and function have also been described in the cells of macula densa (Peti-Peterdi et al. 2000). In the pancreas, NHE4 expression was described in the zymogen granule membranes of the acinar cells (Roussa et al. 2001; Anderie et al. 1998), in the acinar and ductal cells of the salivary glands (Park et al. 1999; Oehlke et al. 2006). Expression of NHE4 in the small and large intestinal epithelium is controversial. In functional studies, Beltrán et al. (2008) described that NHE4 contributed to pHi regulation in T-84 human colonic carcinoma cells while Arena et al. (2012) described similar observations in the rat and human colonic crypts. These two reports did not attempt to demonstrate the physical presence of NHE4, and only a later study by Beltrán et al. (2015) showed NHE4 protein in T84 cells by Western blotting. The initial cloning study showed a detectable NHE4 transcript in the small and large intestine (Orlowski et al. 1992), but later studies with more specific cDNA probes, did not confirm the NHE4 expression in the rat jejunum or colon (Collins et al. 1998b; Bookstein et al. 1997), thus casting doubt on the significance of the findings with functional isoform-specific inhibition. NHE4 mRNA and protein expression were also described in the mouse endometrium (Wang et al. 2003) and in the rat endocervical epithelium (Ismail et al. 2015). Among non-epithelial cells, in situ hybridization showed high NHE4 mRNA expression in the neurons of the rat hippocampus (Bookstein et al. 1996).

5.7.2 Subcellular Distribution

In most of the reported cases, NHE4 protein was localized to the basolateral membrane. The two notable exceptions were the pancreatic acinar cells and endometrial cells, where NHE4 was also detected on the zymogen granule membrane, or both apical and basolateral membranes, respectively (Anderie et al. 1998; Wang et al. 2003).

5.7.3 Physiological Regulation

Very little is known about the regulation of NHE4 expression or activity in epithelial cells.

Renal expression of NHE4 mRNA is regulated during rat ontogeny with a peak at weaning age (3 weeks), and declines thereafter (Collins et al. 1998b). Metabolic acidosis led to NHE4 mRNA expression and transporter activity in the epithelial cells of the mouse medullary thick ascending limb of Henle's loop in the rat (Bourgeois et al. 2010). Interestingly, this regulation may be specific to renal epithelia. In the salivary gland, NHE4 was not regulated by either acute or chronic acid/base disturbance (systemic acidosis or alkalosis) (Oehlke et al. 2006). Although this study looked only at mRNA and protein expression and localization and not activity, this finding may be consistent with an earlier suggestion by Park et al. (1999) that in the salivary gland, NHE4 is not involved in pHi regulation in the acinar cells and may not even be active under most physiological conditions. Ismail et al. (2015) showed that phytoestrogen genistein could stimulate the expression of NHE4 mRNA and protein (along with NHE1 and NHE2) in the rat endocervical epithelium and suggested that genistein could help restore the cervicovaginal fluid pH that might help to prevent menopause-related cervicovaginal complications.

5.7.4 Physiological Relevance/Gene Targeting Studies

Due to its abundant expression in the parietal cells, NHE4 was assumed to participate in parietal cell physiology and gastric acid secretion. In mice with a targeted disruption of the *Slc9a4* gene, Gawenis et al. (2005) described hypochlorhydria independent of the animal age. In 9-week old mice, hypochlorhydria was accompanied by a reduced number of parietal cells due to apoptosis and necrosis and with a loss of mature chief cells, and an increased number of mucous and undifferentiated cells. This finding indicated that NHE4, normally coupled with the basolateral Cl^-/HCO_3^- exchanger AE2, plays an important role in regulating gastric acid secretion and the differentiation of the foveolar epithelium.

In the kidneys, ammonium ($NH4^+$) is the main component of urinary H^+ excretion. Secreted into the lumen of the proximal tubule cells as a product of primarily glutamine metabolism, ammonia reaches the medullary thick ascending limb of Henle's loop (TAL) where it creates an interstitial ammonia concentration gradient necessary for net $NH4^+$ secretion by the adjacent collecting duct. To cross the TAL epithelium, $NH4^+$ enters the cell via the apical NKCC2 cotransporter along with Cl^- and Na^+, the latter substituted for K^+. Since Na^+/H^+ exchangers can operate as $Na^+/$

$NH4^+$ exchangers in the kidneys (Blanchard et al. 1998), and NHE1 and NHE4 are co-expressed in the TAL epithelium, the latter became a candidate basolateral transporter for ammonia. Indeed, metabolic acidosis increased NHE4 mRNA expression in the murine TAL cells and enhanced renal NHE4 activity in rats (Bourgeois et al. 2010). The same study demonstrated that NHE4-deficient mice developed a compensated hyperchloremic metabolic acidosis, and showed an inappropriate urinary net acid excretion. During dietary-induced metabolic acidosis, $NHE4^{-/-}$ mice were unable to increase their urinary $NH4^+$ or net acid excretion compared to wild-type mice. The authors concluded that $NH4^+$ absorption by the TAL epithelium requires the presence of NHE4 and that its deficiency reduces the ability of TAL epithelial cells to create the cortico-papillary gradient of $NH3/NH4^+$ needed to excrete an acid load, thus contributing to systemic metabolic acidosis (Bourgeois et al. 2010).

5.7.5 *Role of NHE4 in Disease States*

Heat-stable STa enterotoxin from enterotoxigenic *E. coli* reduced NHE4 activity in human colonic adenocarcinoma T84 cell line and impaired the cell recovery from intracellular acidification (Beltran et al. 2015). Assuming NHE4 is indeed expressed in the human colon, the authors concluded that such inhibition, related to increased cAMP and increased PKA activity, may promote diarrhea during pathogenic *E. coli* infection.

5.8 CPA1/SLC9A: NHE5

Initial cloning reports of the human SLC9A5 gene encoding for NHE5 isoform showed an expression pattern limited to brain, testis, spleen, and skeletal muscle, and no expression in epithelial tissues (Klanke et al. 1995; Baird et al. 1999). This neuron-enriched Na^+/H^+ exchanger isoform is especially rich in the hippocampus, an important area for memory and plasticity, and it plays an important role in neurite growth and dendritic spine formation. In more recent studies, NHE5 deficiency in null-knockout mice led to improved learning and memory (Chen et al. 2017b). Due to its very limited expression pattern and lack of reported presence or function in the epithelia, this isoform will not be discussed in this chapter.

5.9 CPA1/SLC9A: NHE6

Nagase et al. (1996) cloned a putative cDNA KIAA0267 from the human myeloid cell line KG-1. Two years later, Numata et al. (1998) cloned a mitochondrial gene YDR456w with sequence homology to mammalian Na^+/H^+ exchangers, identified its high homology with a putative human gene (KIAA0267) and subsequently renamed this human homolog as NHE6 (Numata et al. 1998).

5.9.1 *Tissue Specificity*

NHE6 is ubiquitously expressed but its expression is the highest in mitochondria-rich tissues such as the brain, skeletal muscle, and heart (Numata et al. 1998).

5.9.2 *Subcellular Distribution*

Intracellular trafficking of NHE6 and its subcellular localization has been debated since the initial cloning of this isoform. In addition to the predominant expression in mitochondria-rich tissues, Numata et al. (1998) showed that NHE6 was targeted to the mitochondria when overexpressed in the epithelial HeLa cells (Numata et al. 1998). This was not confirmed in another heterologous system (transfected COS-7 cells) where NHE6 was targeted to the ER and secretory organelle membranes and not to mitochondria (Miyazaki et al. 2001). In support of these findings, Brett et al. (2002) showed that in transfected Chinese hamster ovary (CHO) fibroblasts and the renal epithelial opossum kidney (OK) cells, NHE6 transiently appeared on the plasma membrane, was not present in mitochondria but predominated in the endosomal recycling compartment, a location further confirmed by Nakamura et al. in COS-7 cells (Nakamura et al. 2005) and by Ohgaki et al. in HepG2 hepatocytes (Ohgaki et al. 2010). The trafficking of NHE6 between the plasma membrane and early endosomes appears to be regulated by association with RAC1 (receptor for activated C kinase 1) and the luminal pH of the recycling endosome was elevated in RACK1 knockdown HeLa cells (Ohgaki et al. 2008). In the mouse choroid plexus epithelial cells, NHE6 was immunolocalized to the luminal plasma membrane domain where it may contribute to the pH regulation of the cerebrospinal fluid (CSF) (Damkier et al. 2018).

5.9.3 Physiological Relevance/Gene Targeting Studies

Endosomal pH is important for regulating receptor–cargo interactions and for the trafficking of membrane proteins. Polarized hepatocytes rely on the transcellular transport of proteins and sphingolipids for the development of their apical (canalicular) cell surface domain. In HepG2, using dipeptidyl peptidase IV (DPPIV) as a model of basolateral to apical transcytosis, Ohgaki et al. (2010) showed that NHE6 knockdown altered endosomal pH and led to a reduction in the number of apical bile canalicular lumens, and a less efficient DPPIV retention at the canalicular membrane. The authors concluded that in hepatocytes, endosomal NHE6 regulates the development of the apical cell surface domain and highlighted the potential role of endosomal pH regulation in epithelial polarity (Ohgaki et al. 2010). In HeLa cells, NHE6 was colocalized with clathrin and NHE6 knockdown resulted in decreased early endocytic uptake of transferrin corresponding with decreased pH of transferrin-positive endosomes, while overexpression of NHE6 lead to opposite findings (Xinhan et al. 2011).

5.9.4 Role of NHE6 in Disease States

In colonic adenocarcinoma Caco-2 cells, rotavirus infection significantly reduced NHE6 mRNA and protein expression along with increased intracellular Ca^{2+}, and reduced calmodulin (CaM) and calmodulin kinase II (CaMKII) expression (Chen et al. 2017a). The authors suggested that this may contribute to rotavirus-induced diarrhea, although no additional studies have confirmed the role of NHE6 in transepithelial Na^+ transport in the gut. On the other hand, NHE6 may be involved in the regulation of chemoresistance in epithelial cancers. In the hypoxic tumor microenvironment, the interaction of NHE6 with RACK1 through a protein kinase C (PKC)-dependent mechanism leads to relocalization of NHE6 from the endosomal compartment to the plasma membrane, endosomal acidification, which combined with alkaline cytoplasmic pH leads to intravesicular drug trapping and chemoresistance (Lucien et al. 2017).

Human NHE6 cytogenetic location is at *Xq26.3* and mutations in SLC9A6 have been associated with Angelman syndrome-like X-linked mental retardation (XLMR) (Gilfillan et al. 2008), Christianson syndrome (Tzschach et al. 2011), and corticobasal degeneration with *tau* deposition (Garbern et al. 2010). Considering the role of NHE6 in regulating epithelial cell polarity, it was plausible that a similar neuronal polarity defect underlies these neurodegenerative disorders with mental retardation. For more information, the reader is referred to reviews covering this topic more extensively (Zhao et al. 2016; Sinajon et al. 2016; Kondapalli et al. 2014).

5.10 CPA1/SLC9A: NHE7

Numata and Orlowski (2001) cloned human NHE7 in 2001 and described it as a cation exchanger predominant in the trans-Golgi network (TGN), which regulates the organellar pH by transporting either Na^+ or K^+ in a pH-gradient dependent manner. They suggested its importance in the organellar pH regulation and in the posttranslational processing and sorting of newly synthesized protein cargo.

5.10.1 Tissue Specificity

NHE7 is ubiquitously expressed with the highest abundance in the putamen and occipital lobe of the brain, in skeletal muscle, and in a number of secretory tissues such as prostate, stomach, pancreas, pituitary gland, adrenal gland, thyroid gland, salivary gland, mammary gland, small intestine, and colon (Numata and Orlowski 2001).

5.10.2 Subcellular Distribution

NHE7 may be a unique member of the SLC9A gene family that dynamically shuttles between the trans-Golgi network (TGN), endosomes, and the plasma membrane. NHE7 protein was identified as restricted to a juxtanuclear compartment partially overlapping with α-mannosidase II-positive medial and trans-cisternae of the Golgi apparatus, in the TGN and in the mid-TGN (Numata and Orlowski 2001; Fukura et al. 2010; Nakamura et al. 2005). Lin et al. (2005) identified a physical interaction of SCAMP protein family members (SCAMP1, SCAMP2, and SCAMP5) with the cytoplasmic C-terminus of NHE7. The authors showed that mutagenesis of SCAMP2, which reduced its ability to bind NHE7, led to a redistribution of NHE7 away from TGN to a pool of scattered recycling vesicles (Lin et al. 2005). This appeared to be a specific effect as other markers of TGN (syntaxin 6) or Golgi cisternae (GM130) were not affected. A putative caveolin-binding motif $_{596}$WIFRLWYSF$_{604}$ was also located at the C-terminus of NHE7. Interestingly, while NHE7 could indeed bind caveolin, this association did not require this sequence, but rather another, unconventional motif in the C-terminus (Lin et al. 2007). The authors showed that in MCF7 breast adenocarcinoma cells, a small fraction of NHE7 was targeted to the caveolae/lipid rafts at the cell surface and subsequently internalized, a process blocked by inhibition of clathrin-dependent endocytosis. This finding demonstrated another, SCAMPs-independent recycling pathway of NHE7.

5.10.3 Physiological Regulation

The initial functional characterization of NHE7 defined it as a monovalent cation (Na^+ or K^+)/proton exchanger that localizes predominantly to the trans-Golgi network and this and later studies suggested that it contributes to the control of the luminal pH and cation composition in this cellular compartment (Numata and Orlowski 2001; Nakamura et al. 2005). Very little is known about the regulation of NHE7 expression or function beyond the trafficking and membrane distribution. Additional and related information was provided by Kagami et al. (2008) who showed physical interaction and colocalization with vimentin and actin in focal complexes using MDA-MB-231 breast cancer cells as a model. They also demonstrated NHE7 interaction at the lipid rafts with CD44, a cell surface glycoprotein receptor for hyaluronate (among other ligands), which was inducible with phorbol ester (PMA; nonspecific PKC activator) (Kagami et al. 2008). The same reports, in a limited in vivo xenograft experiment with athymic female nude mice, that NHE7-overexpressing cells resulted in larger and more invasive tumors, findings that collectively suggested that NHE7 enhances tumor progression (Onishi et al. 2012).

5.10.4 Physiological Relevance/Gene Targeting Studies

Although initially hypothesized to exchange cytosolic K^+ for H^+ and alkalinize vesicles, later, more precise analyses using transfected cells selected using the proton-killing technique and expressing NHE7 at the plasma membrane, showed that NHE7 transports Li^+ and Na^+, but not K^+, a process that was nonreversible in physiological conditions and was constitutively activated by cytosolic H^+ (Milosavljevic et al. 2014). These findings pointed out that NHE7 acts as a proton-loading transporter that acidifies intracellular vesicles rather than to allow for proton leak.

In addition to regulating TGN pH, or perhaps as its downstream consequence, NHE7 affects cancer cell behavior. NHE7 overexpression in breast cancer MDA-MB-231 cells enhanced cell overlay, cell–cell adhesion, invasion, and anchorage-independent tumor growth in vitro (Onishi et al. 2012). *Slc9a7* knockout mice have not been described in the literature to date.

5.10.5 Role of NHE7 in Disease States

Recently, a missense mutation (c.1543C > T:p.Leu515Phe) in the SLC9A7 gene has been associated with nonsyndromic X-linked intellectual disability with alteration of Golgi acidification and aberrant glycosylation (Khayat et al. 2019). In CHO cells transfected with the mutant cDNA, NHE7 was correctly targeted to TGN and post-

Golgi vesicles, though its N-linked glycosylation was impaired. Interestingly, a glycosylation defect was not limited to NHE7 but extended to other marker proteins, thus suggesting that NHE7 Leu515Phe mutation affects TGN/post-Golgi pH homeostasis and glycosylation of exported cargo, which may lead to pathophysiology and neurodevelopmental deficits (Khayat et al. 2019). No other pathologies associated with epithelial cell dysfunction have been reported as associated with altered NHE7 expression or function.

5.11 CPA1/SLC9A: NHE8

NHE8 cDNA was originally cloned from mouse kidney and rat intestine cDNA libraries (Xu et al. 2005; Goyal et al. 2003). This NHE isoform contains 576 amino acids and is highly conserved in mouse, rat, and human. Hydropathy analysis suggested that the membrane topology of NHE8 was similar to other NHEs, with 10–12 transmembrane domains and a short hydrophilic region at the C-terminus (Goyal et al. 2003). NHE8 protein was predicted to be ~64 kDa, size confirmed by western blotting in mouse, rat, and human intestine (Li et al. 2016; Wang et al. 2011; Xu et al. 2005). In COS-7 cells transfected with the rat NHE8 cDNA, the predominant band size of NHE8 was ~85 kDa, larger than predicted based on the open reading frame, and a minor band at ca. 60 kDa, similar to the murine NHE8 (Goyal et al. 2003). Inhibition of glycosylation with tunicamycin reduced the 85- and 60 kDa bands to one 52 kDa band, confirming *N*-glycosylation of the rat NHE8 (Goyal et al. 2003). Kinetic analysis revealed that NHE8 has an affinity for protons at ~pH 6.5 (Km pH_i) and an affinity for sodium at ~23 mM (Km Na^+). NHE8 had different sensitivities to known NHE inhibitors. It is insensitive to 1 μM HOE694, but its activity is significantly reduced at 10 μM. An NHE3-specific inhibitor, S3226, also inhibits NHE8 activity at a concentration of 80 μM (Xu et al. 2008). These observations may necessitate a more critical reevaluation of some of the earlier reports of specific isoform contribution to total Na^+/H^+ exchange in a given cell type/tissue published prior to NHE8 cloning.

5.11.1 *Tissue Specificity*

NHE8 expression appears to be semi-ubiquitous. In humans, NHE8 mRNA was detected in almost all tissues listed in the human 76-tissue array, and Northern blot detected highly abundant NHE8 transcript in the stomach, duodenum, and ascending colon (Xu et al. 2005). A similar pattern was also seen in mouse intestinal tissues where high NHE8 mRNA expression was detected in the jejunum, ileum, and colon (Xu et al. 2005, 2008), with especially high expression in the intestinal goblet cells (Xu et al. 2016) and the epithelial cells of the gastric pits (Xu et al. 2013). Mouse NHE8 mRNA was also highly expressed in the liver, skeletal muscle, kidney, and

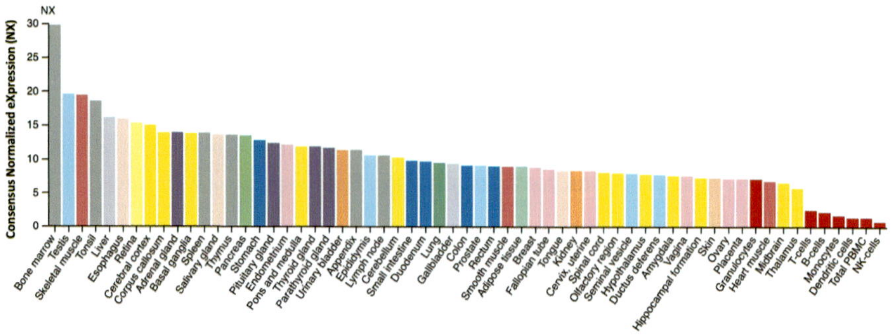

Fig. 5.6 NHE8 mRNA tissue specificity. Consensus Normalized eXpression (NX) levels for 55 tissue types and six blood cell types, created by combining the data from the three transcriptomics datasets (HPA, GTEx, and FANTOM5) using a normalization pipeline (The Human Protein Atlas). Color-coding is based on tissue groups, each consisting of tissues with functional features in common. Tissues sorted based on the mRNA expression level, from the highest (left) to the lowest (right). Image credit to The Human Protein Atlas (https://www.proteinatlas.org/ENSG00000197818-SLC9A8/tissue)

testes (Fig. 5.6) (Goyal et al. 2003). In situ hybridization detected the highest NHE8 mRNA expression in the proximal tubules in the outer stripe and surrounding the juxtamedullary glomeruli (Goyal et al. 2003). NHE8 protein was also detected and studied in the epithelial cells in the conjunctiva, the cornea, and the lacrimal gland both in human and mouse (Xu et al. 2015c), retinal pigment epithelium (Xia et al. 2015, 2018; Jadeja et al. 2015), and in the testicular Leydig cells (Xu et al. 2015a) and the developing acrosome of spermatids (Oberheide et al. 2017) in the seminiferous epithelium.

5.11.2 *Subcellular Distribution*

Due to the wide distribution of NHE8 in the system, the subcellular localization of NHE8 protein varies depending on tissue and cell types and may also be influenced in in vitro overexpression systems. In human NHE8 cDNA transfected COS-7 cells, NHE8 protein was expressed in the mid- to trans-Golgi compartments, but not at the plasma membrane (Nakamura et al. 2005). In NHE-deficient PS120 fibroblasts transfected with rat NHE8 cDNA, NHE8 protein was detected in both plasma membrane and in undefined intracellular compartments (Xu et al., unpublished data). In the mouse kidney, NHE8 protein was detected in the renal brush border membrane preparation (Goyal et al. 2003). In another report with transfected COS-7 cells, only the 85 kDa "mature" NHE8 protein was detected at the plasma membrane using surface biotinylation and polyclonal and monoclonal antibodies (Goyal et al. 2005). The same report showed that NHE8 is expressed on both the microvillar surface membrane and the megalin-positive intermicrovillar coated pits in the epithelial cells of proximal tubules (Goyal et al. 2005). In the rodent gastrointestinal tract, NHE8 protein is expressed on the apical membrane in the epithelial cells of fundic and pyloric glands in the stomach (Xu et al. 2013), in the epithelium of the small intestine and colon (Xu et al. 2005; Li et al. 2016). Similar localization has been reported in the human colon (Li et al. 2016; Xu et al. 2019).

In the testes, NHE8 protein predominates in the intracellular compartments overlapping with 58 K-positive Golgi immunostaining in Leydig cells in both humans and mice (Xu et al. 2015a) and is the predominantly Golgi-resident Na$^+$/H$^+$ exchanger in the developing acrosome of spermatids (Oberheide et al. 2017).

In the retinal pigment epithelium of the eye, NHE8 protein was detected in the ER, Golgi and intracellular vesicles (Jadeja et al. 2015). NHE8 was also detected on the plasma membrane of the epithelial cells in the conjunctiva, the cornea, and the lacrimal glands in humans and mice (Xu et al. 2015c).

The apparent diversity in the subcellular localization of NHE8 in various tissues and cell types may represent true variation in protein trafficking and the specialized function of this exchanger and the degree of polarity of the particular cell type. However, the possibility of methodological differences, varying antibodies and protocols used should also be considered as possible confounders.

5.11.3 Physiological Regulation

5.11.3.1 Developmental Regulation of Epithelial NHE8

NHE8 appears to be involved in the intestinal sodium absorption in early postnatal life and contributes to mucosal homeostasis in adults. In the immature intestine of sucking rodents, NHE8 expression is high while the expression of NHE2 and NHE3 is barely detectable (Xu et al. 2011). Postnatal growth is associated with a decreased NHE8 expression. NHE8 mRNA and protein display an age-dependent decrease in the small intestine in rats and mice (Xu et al. 2005), which is opposite to NHE2 and NHE3 whose expression and activity are increased after weaning (see sections on NHE2 and NHE3 earlier in this chapter). While NHE8 expression is downregulated in the adult small intestine, colonic NHE8 expression follows a different pattern. Mouse colonic NHE8 expression is low during the suckling period and it increases to reach a plateau after weaning (Xu et al. 2012). Follow-up studies suggested that the age-dependent regulation of NHE8 expression in the small intestine might be related to humoral factors affecting its expression, an aspect reviewed in the next section.

In the kidney, NHE8 is expressed at the apical membrane of the renal epithelial cells (Goyal et al. 2005). In the neonatal stage, renal NHE8 function is also required to adapt to metabolic acidosis (Pirojsakul et al. 2015). In rats, renal NHE8 mRNA expression remains unchanged throughout life but NHE8 protein expression on the brush border membranes decreases in adult rat kidney (Becker et al. 2007). Interestingly, total membrane-associated abundance of NHE8 appeared to increase during the same time, suggesting that it is the changes in the trafficking or recycling of NHE8 that determine the decreased NHE8 activity during development. In senile

(18-month-old) rats, on the other hand, both NHE8 protein and mRNA declined, compared with 3-month-old adults (Fiori et al. 2009).

5.11.3.2 Hormonal Regulation of Epithelial NHE8

Similar to other NHEs, the intestinal NHE8 is also subject to regulation by EGF. However, unlike NHE2, EGF treatment reduces NHE8 expression in the ileum in suckling rats at both protein and mRNA levels. A similar observation was also made in EGF-treated human intestinal epithelial Caco-2 cells. The effect of EGF inhibition on NHE8 expression is likely transcriptional and is associated with a reduced binding of Sp3 transcription factor to the basal promoter of the SLC9A8 gene (Xu et al. 2010a).

Glucocorticoids are important regulators of intestinal maturation and their endogenous levels surge around and after weaning (Henning 1978). A decrease of NHE8 expression in rat small intestine coincides with the timing of increased systemic glucocorticoid levels. Studies revealed that methylprednisolone administration in young rats precociously inhibited NHE8 expression, especially in the ileum. At the protein level, methylprednisolone treatment reduced NHE8 abundance in jejunum and ileum by 30% and 90%, respectively. At the mRNA level, methylprednisolone blocked NHE8 mRNA synthesis by 28% in the jejunum and 45% in the ileum. The effect of glucocorticoids on intestinal NHE8 expression occurs at the level of gene transcription and is associated with an enhanced binding of the inhibitory Pax5 transcription factor to the proximal NHE8 gene promoter (Xu et al. 2010b). In 7–10-day-old mice, dexamethasone also prematurely reduced the expression of renal NHE8, and observation confirmed in vitro with renal NRK cells, in which dexamethasone decreased NHE8 mRNA, total protein, apical protein abundance, and function of NHE8 (Joseph et al. 2013). Considering that renal NHE8 mRNA does not change from the suckling period into adulthood in rodents, the role of glucocorticoids in driving the growth-related decrease in NHE8 expression and activity is not entirely clear.

Somatostatin is an important neuropeptide produced by D-cells in the intestine, which also acts as a proabsorptive and antisecretory molecule. Somatostatin stimulates Na^+ absorption, an effect that may be at least in part mediated through its effects on NHE8 expression. Octreotide, a somatostatin analog, increased NHE8 protein expression at the epithelial brush border membrane by ~62% in the small intestine in vivo. In Caco-2 cells, somatostatin increased NHE8 activity by ~80% and NHE8 plasma membrane expression by ~86%. p38 mitogen-activated protein kinase (MAPK) activation via somatostatin receptor SSTR2 was involved in somatostatin-mediated NHE8 induction (Wang et al. 2011). In a follow-up study using a DSS colitis mouse model, octreotide treatment stimulated colonic NHE8 expression via activating both somatostatin receptor 2 (SSTR2) and somatostatin receptor 5 (SSTR5) and by suppressing ERK1/2 phosphorylation (Li et al. 2016). In an infectious colitis model, mice infected with *C. rodentium* had dramatically reduced NHE8 expression in the colonic epithelium, and octreotide treatment restored NHE8 expression in infected mice through activating somatostatin receptor 2 (SSTR2) and inhibiting p38 MAPK phosphorylation (Lei et al. 2018).

Thyroid hormone has been shown to regulate renal NHE8 expression. Administration of thyroid hormone to neonatal rats, before the physiological postnatal increase in serum thyroid hormone levels at 3 weeks of age, led to a precocious decrease in brush border membrane NHE8 expression and activity, but without concomitant changes in mRNA expression (Gattineni et al. 2008). The developmental decrease in apical NHE8 protein abundance was prevented in rats made hypothyroid at birth, thus suggesting that the thyroid hormone may be the key humoral factor responsible for postnatal changes in NHE8 expression in the kidney.

5.11.3.3 NHE8 Regulation by SCFAs

SCFAs have been known to stimulate sodium absorption in the intestine. Beside their role in regulating NHE3 expression and activity, SCFAs regulate NHE8 expression in the gut. Butyrate strongly induced NHE8 protein expression by threefold and NHE8 mRNA expression by 2.3-fold in Caco-2 cells. Butyrate stimulated NHE8 gene transcription in Caco-2 cells transiently transfected with the NHE8 gene promoter reporter constructs. Gel mobility shift assays demonstrated an enhanced Sp3 acetylation and binding to the human NHE8 basal gene promoter region upon butyrate treatment (Xu et al. 2009, 2015b).

5.11.4 Physiological Relevance/Gene Targeting Studies

NHE8 is not critical for embryonic development. Null NHE8 knockout mice display normal growth and are indistinguishable from their wild-type littermates (Xu et al. 2012; Jadeja et al. 2015). Although no gross phenotypical changes are observed in $NHE8^{-/-}$ mice, several changes have been reported at the tissue level under homeostatic conditions or under stress. Additional observations were made using mice carrying the *r15* mutation, identified from a screen of N-ethyl-N-nitrosourea (ENU)-induced mutant mice and described as leading to a retinal disorder (Xia et al. 2006). Genome-wide linkage analysis led to the identification of a missense mutation $M^{120}K$ in the NHE8-encoding SLC9A8 gene which was predicted to disrupt the NHE8 membrane domain structure, and to an effective loss of function (Xia et al. 2015).

5.11.4.1 NHE8 Contributes to Mucosal Protection in the Gut

NHE8 is expressed in the apical membrane in intestinal epithelial cells, suggesting a possible role in intestinal sodium absorption. Interestingly, $NHE8^{-/-}$ mice have normal serum Na^+ levels and have no evidence of diarrhea, indicating that NHE8's role in sodium absorption is not essential, or is compensated by an adaptive increase in NHE2 and NHE3 expression (Xu et al. 2012). In the murine stomach, NHE8 is located at the apical membrane of the glandular region. In the absence of NHE8, we observed a 13-fold increase in ulcer formation. NHE8$^{-/-}$ mice had reduced gastric mucosal surface pH and a decreased expression of the chloride anion exchanger, DRA (Xu et al. 2013). In the colon, loss of NHE8 resulted in decreased goblet cell numbers and mucin production along with altered mucosal surface pH and similarly decreased DRA expression (Xu et al. 2012). Reduced mucin production in NHE8$^{-/-}$ mice resulted in a disorganized mucin layer which enabled a closer interaction between epithelia and luminal bacteria (Wang et al. 2015; Liu et al. 2013). Consistent with these observations, siRNA-mediated NHE8 knockdown in Caco-2 cells increased mucosal adhesion of *Salmonella typhimurium*, with similar effects seen in vivo studies (Liu et al. 2013). Interestingly, NHE8 knockdown did not affect the adhesion of a probiotic strain *Lactobacillus plantarum JDM1*, suggesting at least some degree of selectivity that could not be explained by mucus production alone. The important role of NHE8 in mucosal protection was further confirmed in an AOM (azoxymethane)/DSS inflammation-associated colon cancer animal model, where NHE8$^{-/-}$ mice were highly susceptible, with 89% of NHE8$^{-/-}$ mice developing tumors as compared to only 9% of WT controls under the applied experimental conditions (Xu et al. 2019). In xenograft studies, NHE8-deficient HT29 cells generated larger tumors in NSG mice (Xu et al. 2019). We also confirmed that NHE8 is expressed in the intestinal stem cells, and loss of NHE8 expression leads to increased expression of Lgr5, β-catenin (CTNNB1), and cMyc. Therefore, loss of NHE8 expression could potentially enhance the Wnt/β-catenin pathway, which in turn may promote tumor cell growth in colorectal cancer (Xu et al. 2019).

Both clinical and experimental IBD is associated with reduced NHE8 expression. In endotoxemia or TNBS colitis, the expression of NHE8 protein and mRNA in the small intestine is reduced by 41% and 36%, respectively (Xu et al. 2009). In DSS colitis, the expression of NHE8 in the colon is decreased by 88% at the protein level and 52% at mRNA levels (Li et al. 2016). In *C. rodentium* infected mice, the expression of NHE8 is also reduced (Lei et al. 2018). In IBD patients, NHE8 protein was reduced by 55% in UC patients compared with control subjects (Li et al. 2016). Further studies confirmed that TNFα reduced NHE8 mRNA and protein expression in Caco-2 cells (Xu et al. 2009) and in HT29-MTX goblet-like cells (Xu et al. 2016). The mechanism in TNFα-mediated NHE8 repression was investigated in both cell lines, and the results implicate an inhibition of the Sp3 transcription factor binding to the NHE8 gene promoter (Xu et al. 2009, 2016).

5.11.4.2 Role of NHE8 in the Ocular Epithelium

NHE8 mRNA was detected in the eye and immunohistochemistry staining located NHE8 protein at the plasma membrane of the epithelial cells in the conjunctiva, the cornea, and the lacrimal gland both in humans and mice (Xu et al. 2015c). In retinal pigment epithelial cells, NHE8 was located in ER, Golgi, endosomes, and intracellular vesicles (Jadeja et al. 2015; Xia et al. 2015). NHE8$^{-/-}$ mice showed a reduction in tear formation along with alteration in several genes such as DRA, MMP9, TNFα, Sprr2h, and Tgm1 in the ocular surface epithelium (Xu et al. 2015c). These observations suggested a potential novel role for NHE8 in the pathogenesis of dry eye syndrome. Additionally, loss of NHE8 expression in mice resulted in abnormalities in retinal pigment epithelium (RPE), with a large amount of dark pigmented patches throughout the retina and possible retinal degeneration. Indeed, increased cell death was demonstrated in RPE cells in the absence of functional NHE8 (Jadeja et al. 2015). The same observation was made in mice with $M^{120}K$ point mutation that led to abnormalities in the retinal pigment epithelium and late-onset photoreceptor cell loss (Xia et al. 2015). Collectively, these studies suggested the role of NHE8 in regulating photoreceptor cell function in RPE as well as other roles in the lacrimal glands and the ocular surface epithelium.

5.11.4.3 Role of NHE8 in the Male Reproductive System

NHE8 was detected in the Golgi compartment in the interstitial Leydig cells (Xu et al. 2015a), as well as in the developing acrosome of spermatids (Oberheide et al. 2017). Male NHE8$^{-/-}$ mice are infertile (Oberheide et al. 2017; Xu et al. 2015a; Jadeja et al. 2015). Follow-up studies detected low testosterone levels in NHE8$^{-/-}$ male mice, which was likely the result of altered luteinizing hormone (LH) receptor expression in the Leydig cells. LH receptor expression was dramatically reduced in NHE8$^{-/-}$ testis and also in NHE8 siRNA-mediated knockdown in Leydig cells due to the disrupted LH receptor distribution (Xu et al. 2015a). In spermatids, NHE8 deficiency results in the failure of the formation of large acrosomal granules and the acrosomal caps. In spermatozoa, the loss of NHE8 leads to the complete loss of acrosomes (Oberheide et al. 2017).

5.11.5 Role of NHE8 in Disease States

Relatively little is known about the expression and activity of NHE8 in disease states, and its roles in the pathogenesis of human disorders beyond the findings implied by the knockout studies, as we described for the gut and ocular epithelium.

5.12 CPA1/SLC9B: NHA1 and NHA2

In addition to the better-known sodium/proton exchangers from the CPA1 family, the more recently discovered and less well understood CPA2 members (SLC9B), represented by two genes, Nha1 (SLC9B1) and Nha2 (SLC9B1) continue to gain recognition. Nha1 has been studied primarily in lower organisms. In *Drosophila*,

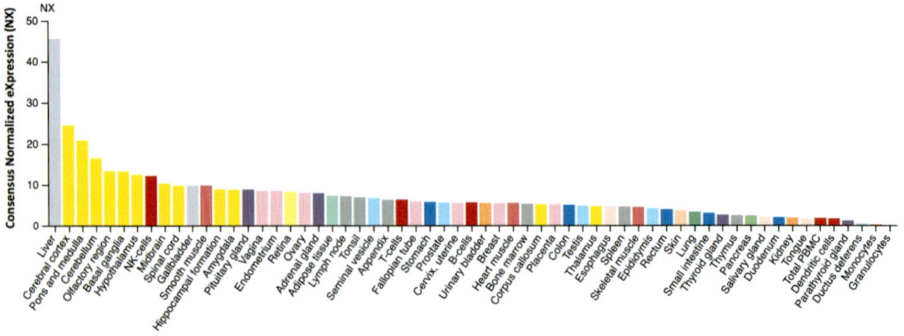

Fig. 5.7 NHA2 mRNA tissue specificity. Consensus Normalized eXpression (NX) levels for 55 tissue types and six blood cell types, created by combining the data from the three transcriptomics datasets (HPA, GTEx, and FANTOM5) using a normalization pipeline (The Human Protein Atlas). Color-coding is based on tissue groups, each consisting of tissues with functional features in common. Tissues sorted based on the mRNA expression level, from the highest (left) to the lowest (right). Image credit to The Human Protein Atlas (https://www.proteinatlas.org/ENSG00000164038-SLC9B2/tissue)

Nha1 and Nha2 are enriched and functionally significant in renal tubules; single RNAi knockdowns of either Nha1 or Nha2 reduce fruit fly survival and in combination were lethal (Chintapalli et al. 2015). However, the first cloning report of human NHA1 showed that it was a testis-specific exchanger with high homology to its paralogs in other mammals including mouse, rat, and the macaque (*Macaca fascicularis*) (Ye et al. 2006). NHA1 is localized on the principal piece of mouse sperm flagellum, and its knockout led to subfertility in male mice (Chen et al. 2016b). The same report showed a similar role for NHA2 in the regulation mouse fertility, and mice lacking both isoforms were completely infertile, with severely diminished sperm motility and attenuated soluble adenylyl cyclase (sAC)-cAMP signaling. Beyond these studies, the biologic function of NHA1 remains unknown and NHA1 has not been yet linked to human disease.

NHA2 is expressed more broadly and has been studied more extensively in animals, although the liver, the site of the highest expression of NHA2 in humans (Fig. 5.7), has not been the subject of investigation. Some of the initial descriptions of NHA2 focused on its high mitochondrial expression in differentiating osteoclasts, where it contributes to their differentiation and functions (Battaglino et al. 2008; Ha et al. 2008). NHA2-deficient mice with a gene-trap of intron 1 are born and develop normally and appear healthy without an obvious phenotype, and surprisingly, with normal bone density and structure (Hofstetter et al. 2010). Another group using a different, NHA2 gene-trapped mouse confirmed those findings (Charles et al. 2012). These findings suggested that NHA2 is dispensable for osteoclast differentiation and bone resorption in mice in vivo.

The subcellular localization of NHA2 has not been definitively established and the published data remains controversial. Fuster et al. (2008) characterized a monoclonal antibody against NHA2, which detected NHE2 in the kidney, where it co-localized with calbindin D28k in the distal convoluted tubule. Immunogold electron microscopy of rat distal convoluted tubule demonstrated NHA2 predominantly but not exclusively on the inner mitochondrial membrane, and NHE2 co-sedimented with mitochondrial membranes in differential centrifugation experiments. When expressed in NHE-deficient yeast, it protected the cells from high salinity (Na^+ or Li^+) in the presence of acidic extracellular pH, suggesting that at least in this organism, it works at the plasma membrane in the reverse mode (extruding Na^+ in exchange for extracellular protons) (Fuster et al. 2008; Xiang et al. 2007; Huang et al. 2010). When human NHA2 was expressed in polarized MDCK cells, it was localized to the apical membrane (Xiang et al. 2007). Based on these findings, NHA2 was postulated to be the Na^+/Li^+ countertransporter originally functionally described in the early 1980s in red blood cells and tentatively linked to the development of arterial hypertension and diabetes (Canessa et al. 1980). In osteoclasts, however, NHA2 was found to co-localize with the late endosomal and lysosomal marker LAMP1 and the V-ATPase α3 subunit, but not with mitochondrial markers or plasma membrane (Hofstetter et al. 2010). Those discrepancies in intracellular localization of NHA2 may represent true specialization of NHA2 in different cell lineages, although more research is needed to address that.

NHA2 was also found in the β cells of the endocrine pancreas where it resided in endosomes and synaptic-like microvesicles and regulated clathrin-dependent endocytosis (Deisl et al. 2013). NHA2 deficiency of pharmacological inhibition reduced sulfonylurea- and secretagogue-induced insulin secretion, although loss of NHA2 had no impact on the endosomal steady-state pH in β cells. It is possible that loss of NHA2 therefore may affect insulin secretion indirectly by interfering with clathrin-mediated endocytosis in β cells, although much of the underlying mechanisms remain unknown.

5.13 Other Epithelial Members of CPA1 Family

The nematode *Caenorhabditis elegans* (*C. elegans*) expresses nine isoforms of Na^+/H^+ exchangers, NHX-1 through -9. Four of the isoforms are believed to be expressed at the cell surface, and the rest are associated with the membranes of intracellular organelles of polarized epithelial cells: intestine, seam cells, hypodermal cells of the main body syncytium, and the excretory cell (Nehrke and Melvin 2002). NHX-1, initially though thought to be limited to the embryonic intestinal cells, is expressed intracellularly in the hypodermal and muscle cells over the entire length of the nematode, though the exact subcellular location is not yet known. Isoforms NHX-3, -5, -8, and -9 are also associated with intracellular membranes in *C. elegans*. NHX-2, -4, -6, and -7 are all expressed at the plasma membrane, with domain distribution dependent on the isoform and cell type. NHX-2 expression is restricted to the apical membrane of the intestine (Nehrke and Melvin 2002). This isoform has been postulated to be an ortholog of mammalian NHE3 and is functionally coupled with nutrient uptake through OPT-2, an oligopeptide transporter (Nehrke 2003). Interestingly, reduced *nhx-2* by RNAi led to a loss of fat stores in the nematode's intestine and a 40% increase in its longevity (Nehrke 2003). NHX-4 appears to be ubiquitous, with even distribution along the plasma membrane, although when expressed in polarized epithelial cells, NHX-4:GFP fusion protein is targeted primarily to the basolateral membrane, similar to NHE1 (Nehrke and Melvin 2002). NHX-6 is expressed along the length of the intestine, though its targeting to a specific membrane domain appears to be cell-specific: basolateral labeling in the most posterior and anterior ends of the intestine, and apical in the middle (Nehrke and Melvin 2002). NHX-7, encoded by the *nxh-7/pbo-4* gene, shares homology with human NHE1 (26% identity), and to a lesser extent with NHE2 and NHE3 (22% and 21%, respectively) (Beg et al. 2008). The carboxy tail of NHX-7/PBO-4 protein contains a predicted PIP2 binding domain, a calmodulin (CaM)-binding domain, and CamKII and PKC phosphorylation sites, similar to NHE1. PBO-4/NHX-7 is expressed at the basolateral site of the posterior intestine (Beg et al. 2008; Nehrke and Melvin 2002), and PBO-4-mediated acidification of the coelomic space is required to stimulate contraction of the posterior body muscles (Beg et al. 2008).

5.14 CPA2 Family: Transmembrane and Coiled-Coil Domain 3 (TMCO3)

The CPA2 family is a moderately large family of proton antiporters from bacteria, archaea, and eukaryotes (fungi and viridiplantae, or green algae). The only metazoan member of the family is the human transmembrane and coiled-coil domain 3 (TMCO3), a protein of 677 amino acids with 13 transmembrane domains (Chen et al. 2016a). TMCO3 is expressed in the human cornea, lens capsule, and choroid-retinal pigment epithelium, and heterozygous mutation c. − 129G > A in the 5′ UTR of *TMCO3* was associated with cornea guttata (dark spots on the corneal endothelium associated with Fuchs' dystrophy) and anterior polar cataracts (Chen et al. 2016a).

5.15 Conclusions

As we outlined in this chapter, epithelial (and non-epithelial) NHEs are involved in a wide array of biological processes from the organellar to systemic levels. While the roles of the major Na^+/H^+ exchangers such as NHE1 or NHE3 has attracted considerable attention, the physiological function of many other members of the CPA superfamily of antiporters remains ill-defined. Studies with knockout animals, both null and tissue-specific, have contributed significantly to our knowledge of their contribution to physiology and pathophysiology. As we continue to understand their roles in the pathophysiological processes better, it opens new avenues for the development of new and specific inhibitors or activations, to counteract the detrimental effects of hyperactivation or inhibition of these important transport proteins in human patients.

Acknowledgments The authors would like to thank Trudy Meckler for her editorial assistance.

References

Abedin MZ, Giurgiu DI, Abedin ZR, Peck EA, Su X, Smith PR (2001) Characterization of Na^+/H^+ exchanger isoform (NHE1, NH32 and NHE3) expression in prairie dog gallbladder. J Membr Biol 182:123–134

Aharonovitz O, Zaun HC, Balla T, York JD, Orlowski J, Grinstein S (2000) Intracellular pH regulation by Na^+/H^+ exchange requires phosphatidylinositol 4,5-bisphosphate. J Cell Biol 150:213–224

Alexander RT, Grinstein S (2009) Tethering, recycling and activation of the epithelial sodium-proton exchanger, NHE3. J Exp Biol 212:1630–1637. https://doi.org/10.1242/jeb.027375

Amin MR, Dudeja PK, Ramaswamy K, Malakooti J (2007) Involvement of Sp1 and Sp3 in differential regulation of human NHE3 promoter activity by sodium butyrate and IFN-gamma/TNF-alpha. Am J Physiol Gastrointest Liver Physiol 293:G374–G382. https://doi.org/10.1152/ajpgi.00128.2007

Amin MR, Ghannad L, Othman A, Gill RK, Dudeja PK, Ramaswamy K, Malakooti J (2009) Transcriptional regulation of the human Na^+/H^+ exchanger NHE3 by serotonin in intestinal epithelial cells. Biochem Biophys Res Commun 382:620–625. https://doi.org/10.1016/j.bbrc.2009.03.087

Amin MR, Orenuga T, Tyagi S, Dudeja PK, Ramaswamy K, Malakooti J (2011) Tumor necrosis factor-alpha represses the expression of NHE2 through NF-kappaB activation in intestinal epithelial cell model, C2BBe1. Inflamm Bowel Dis 17:720–731. https://doi.org/10.1002/ibd.21419

Amith SR, Fliegel L (2013) Regulation of the Na^+/H^+ exchanger (NHE1) in breast cancer metastasis. Cancer Res 73(4):1259–1264. https://doi.org/10.1158/0008-5472.CAN-12-4031

Amith SR, Fliegel L (2017) Na(+)/H(+) exchanger-mediated hydrogen ion extrusion as a carcinogenic signal in triple-negative breast cancer etiopathogenesis and prospects for its inhibition in therapeutics. Semin Cancer Biol 43:35–41. https://doi.org/10.1016/j.semcancer.2017.01.004

Amith SR, Wilkinson JM, Baksh S, Fliegel L (2015) The Na(+)/H(+) exchanger (NHE1) as a novel co-adjuvant target in paclitaxel therapy of triple-negative breast cancer cells. Oncotarget 6 (2):1262–1275. https://doi.org/10.18632/oncotarget.2860

Amith SR, Wilkinson JM, Fliegel L (2016) Na+/H+ exchanger NHE1 regulation modulates metastatic potential and epithelial-mesenchymal transition of triple-negative breast cancer cells. Oncotarget 7(16):21091–21113. https://doi.org/10.18632/oncotarget.8520

Amith SR, Vincent KM, Wilkinson JM, Postovit LM, Fliegel L (2017) Defining the Na+/H+ exchanger NHE1 interactome in triple-negative breast cancer cells. Cell Signal 29:69–77. https://doi.org/10.1016/j.cellsig.2016.10.005

Ammar YB, Takeda S, Hisamitsu T, Mori H, Wakabayashi S (2006) Crystal structure of CHP2 complexed with NHE1-cytosolic region and an implication for pH regulation. EMBO J 25 (11):2315–2325. https://doi.org/10.1038/sj.emboj.7601145

Anderie I, Blum R, Haase W, Grinstein S, Thevenod F (1998) Expression of NHE1 and NHE4 in rat pancreatic zymogen granule membranes. Biochem Biophys Res Commun 246(2):330–336. https://doi.org/10.1006/bbrc.1998.8611

Aredia F, Scovassi AI (2016) Manipulation of intracellular pH in cancer cells by NHE1 inhibitors. Protein Pept Lett 23(12):1123–1129

Arena EA, Longo WE, Roberts KE, Geibel P, Nateqi J, Brandstetter M, Geibel JP (2012) Functional role of NHE4 as a pH regulator in rat and human colonic crypts. Am J Physiol Cell Physiol 302(2):C412–C418. https://doi.org/10.1152/ajpcell.00163.2011

Aronson PS, Nee J, Suhm MA (1982) Modifier role of internal H+ in activating the Na^{+}-H^{+} exchanger in renal microvillus membrane vesicles. Nature 299(5879):161–163. https://doi.org/10.1038/299161a0

Arshad N, Visweswariah SS (2012) The multiple and enigmatic roles of guanylyl cyclase C in intestinal homeostasis. FEBS Lett 586(18):2835–2840. https://doi.org/10.1016/j.febslet.2012.07.028

Avula LR, Chen T, Kovbasnjuk O, Donowitz M (2018) Both NHERF3 and NHERF2 are necessary for multiple aspects of acute regulation of NHE3 by elevated Ca(2+), cGMP, and lysophosphatidic acid. Am J Physiol Gastrointest Liver Physiol 314(1):G81–G90. https://doi.org/10.1152/ajpgi.00140.2017

Azarani A, Goltzman D, Orlowski J (1995) Parathyroid hormone and parathyroid hormone-related peptide inhibit the apical Na^{+}/H^{+} exchanger NHE-3 isoform in renal cells (OK) via a dual signaling cascade involving protein kinase A and C. J Biol Chem 270(34):20004–20010

Azriel-Tamir H, Sharir H, Schwartz B, Hershfinkel M (2004) Extracellular zinc triggers ERK-dependent activation of Na^{+}/H^{+} exchange in colonocytes mediated by the zinc-sensing receptor. J Biol Chem 279(50):51804–51816. https://doi.org/10.1074/jbc.M406581200

Azuma KK, Balkovetz DF, Magyar CE, Lescale-Matys L, Zhang Y, Chambrey R, Warnock DG, McDonough AA (1996) Renal Na^{+}/H^{+} exchanger isoforms and their regulation by thyroid hormone. Am J Phys 270(2 Pt 1):C585–C592

Bachmann O, Riederer B, Rossmann H, Groos S, Schultheis PJ, Shull GE, Gregor M, Manns MP, Seidler U (2004) The Na+/H+ exchanger isoform 2 is the predominant NHE isoform in murine colonic crypts and its lack causes NHE3 upregulation. Am J Physiol Gastrointest Liver Physiol 287(1):G125–G133. https://doi.org/10.1152/ajpgi.00332.2003

Baetz NW, Gupta A, Kapoor A, Kovbasnjuk O, Turner J, Donowitz M, Estes M, Zachos NC (2016) Rotavirus infection in patients is associated with altered trafficking of apical membrane transport proteins. Gastroenterology 150(4):S113–S114

Bai L, Collins JF, Muller YL, Xu H, Kiela PR, Ghishan FK (1999) Characterization of cis-elements required for osmotic response of rat Na(+)/H(+) exchanger-2 (NHE-2) gene. Am J Phys 277 (4 Pt 2):R1112–R1119

Bai L, Collins JF, Xu H, Ghishan FK (2001) Transcriptional regulation of rat Na(+)/H(+) exchanger isoform-2 (NHE-2) gene by Sp1 transcription factor. Am J Physiol Cell Physiol 280(5):C1168–C1175

Bailey MA, Giebisch G, Abbiati T, Aronson PS, Gawenis LR, Shull GE, Wang T (2004) NHE2-mediated bicarbonate reabsorption in the distal tubule of NHE3 null mice. J Physiol 561 (Pt 3):765–775. https://doi.org/10.1113/jphysiol.2004.074716

Baird NR, Orlowski J, Szabo EZ, Zaun HC, Schultheis PJ, Menon AG, Shull GE (1999) Molecular cloning, genomic organization, and functional expression of Na^{+}/H^{+} exchanger isoform 5 (NHE5) from human brain. J Biol Chem 274(7):4377–4382

Barmeyer C, Harren M, Schmitz H, Heinzel-Pleines U, Mankertz J, Seidler U, Horak I, Wiedenmann B, Fromm M, Schulzke JD (2004) Mechanisms of diarrhea in the interleukin-2-deficient mouse model of colonic inflammation. Am J Physiol Gastrointest Liver Physiol 286 (2):G244–G252. https://doi.org/10.1152/ajpgi.00141.2003

Battaglino RA, Pham L, Morse LR, Vokes M, Sharma A, Odgren PR, Yang M, Sasaki H, Stashenko P (2008) NHA-oc/NHA2: a mitochondrial cation-proton antiporter selectively expressed in osteoclasts. Bone 42(1):180–192. https://doi.org/10.1016/j.bone.2007.09.046

Baum M, Biemesderfer D, Gentry D, Aronson PS (1995) Ontogeny of rabbit renal cortical NHE3 and NHE1: effect of glucocorticoids. Am J Phys 268(5 Pt 2):F815–F820

Baumgartner M, Patel H, Barber DL (2004) Na(+)/H(+) exchanger NHE1 as plasma membrane scaffold in the assembly of signaling complexes. Am J Physiol Cell Physiol 287(4):C844–C850. https://doi.org/10.1152/ajpcell.00094.2004

Becker AM, Zhang J, Goyal S, Dwarakanath V, Aronson PS, Moe OW, Baum M (2007) Ontogeny of NHE8 in the rat proximal tubule. Am J Physiol Renal Physiol 293(1):F255–F261. https://doi.org/10.1152/ajprenal.00400.2006

Beg AA, Ernstrom GG, Nix P, Davis MW, Jorgensen EM (2008) Protons act as a transmitter for muscle contraction in *C. elegans*. Cell 132(1):149–160. https://doi.org/10.1016/j.cell.2007.10.058

Bell SM, Schreiner CM, Schultheis PJ, Miller ML, Evans RL, Vorhees CV, Shull GE, Scott WJ (1999) Targeted disruption of the murine Nhe1 locus induces ataxia, growth retardation, and seizures. Am J Phys 276(4 Pt 1):C788–C795

Beltran AR, Ramirez MA, Carraro-Lacroix LR, Hiraki Y, Reboucas NA, Malnic G (2008) NHE1, NHE2, and NHE4 contribute to regulation of cell pH in T84 colon cancer cells. Pflugers Arch 455(5):799–810. https://doi.org/10.1007/s00424-007-0333-0

Beltran AR, Carraro-Lacroix LR, Bezerra CN, Cornejo M, Norambuena K, Toledo F, Araos J, Pardo F, Leiva A, Sanhueza C, Malnic G, Sobrevia L, Ramirez MA (2015) *Escherichia coli* heat-stable enterotoxin mediates Na^+/H^+ exchanger 4 inhibition involving cAMP in T84 human intestinal epithelial cells. PLoS One 10(12):e0146042. https://doi.org/10.1371/journal.pone.0146042

Benedetti A, Di Sario A, Casini A, Ridolfi F, Bendia E, Pigini P, Tonnini C, D'Ambrosio L, Feliciangeli G, Macarri G, Svegliati-Baroni G (2001) Inhibition of the NA(+)/H(+) exchanger reduces rat hepatic stellate cell activity and liver fibrosis: an in vitro and in vivo study. Gastroenterology 120(2):545–556. https://doi.org/10.1053/gast.2001.21203

Benos DJ (1982) Amiloride: a molecular probe of sodium transport in tissues and cells. Am J Phys 242(3):C131–C145

Bianchini L, Woodside M, Sardet C, Pouyssegur J, Takai A, Grinstein S (1991) Okadaic acid, a phosphatase inhibitor, induces activation and phosphorylation of the Na+/H+ antiport. J Biol Chem 266(23):15406–15413

Biemesderfer D, Rutherford PA, Nagy T, Pizzonia JH, Abu-Alfa AK, Aronson PS (1997) Monoclonal antibodies for high-resolution localization of NHE3 in adult and neonatal rat kidney. Am J Phys 273(2 Pt 2):F289–F299. https://doi.org/10.1152/ajprenal.1997.273.2.F289

Biemesderfer D, DeGray B, Aronson PS (1998) Membrane topology of NHE3. Epitopes within the carboxyl-terminal hydrophilic domain are exoplasmic. J Biol Chem 273(20):12391–12396

Binder HJ, Mehta P (1989) Short-chain fatty acids stimulate active sodium and chloride absorption in vitro in the rat distal colon. Gastroenterology 96(4):989–996

Bizal GL, Howard RL, Bookstein C, Rao MC, Chang EB, Soleimani M (1996) Glycosylation of the Na+/H+ exchanger isoform NHE-3 is species specific. J Lab Clin Med 128(3):304–312

Blanchard A, Eladari D, Leviel F, Tsimaratos M, Paillard M, Podevin RA (1998) NH4+ as a substrate for apical and basolateral Na(+)-H+ exchangers of thick ascending limbs of rat kidney: evidence from isolated membranes. J Physiol 506(Pt 3):689–698

Blaurock MC, Reboucas NA, Kusnezov JL, Igarashi P (1995) Phylogenetically conserved sequences in the promoter of the rabbit sodium-hydrogen exchanger isoform 1 gene (NHE1/SLC9A1). Biochim Biophys Acta 1262(2-3):159–163

Block GA, Rosenbaum DP, Leonsson-Zachrisson M, Astrand M, Johansson S, Knutsson M, Langkilde AM, Chertow GM (2017) Effect of Tenapanor on serum phosphate in patients receiving hemodialysis. J Am Soc Nephrol. https://doi.org/10.1681/ASN.2016080855

Bobulescu IA, Moe OW (2006) Na+/H+ exchangers in renal regulation of acid-base balance. Semin Nephrol 26(5):334–344. https://doi.org/10.1016/j.semnephrol.2006.07.001

Bocanegra V, Gil Lorenzo AF, Cacciamani V, Benardon ME, Costantino VV, Valles PG (2014) RhoA and MAPK signal transduction pathways regulate NHE1-dependent proximal tubule cell apoptosis after mechanical stretch. Am J Physiol Renal Physiol 307(7):F881–F889. https://doi.org/10.1152/ajprenal.00232.2014

Bookstein C, DePaoli AM, Xie Y, Niu P, Musch MW, Rao MC, Chang EB (1994a) Na+/H+ exchangers, NHE-1 and NHE-3, of rat intestine. Expression and localization. J Clin Invest 93(1):106–113

Bookstein C, Musch MW, DePaoli A, Xie Y, Villereal M, Rao MC, Chang EB (1994b) A unique sodium-hydrogen exchange isoform (NHE-4) of the inner medulla of the rat kidney is induced by hyperosmolarity. J Biol Chem 269(47):29704–29709

Bookstein C, Musch MW, DePaoli A, Xie Y, Rabenau K, Villereal M, Rao MC, Chang EB (1996) Characterization of the rat Na+/H+ exchanger isoform NHE4 and localization in rat hippocampus. Am J Phys 271(5 Pt 1):C1629–C1638. https://doi.org/10.1152/ajpcell.1996.271.5.C1629

Bookstein C, Xie Y, Rabenau K, Musch MW, McSwine RL, Rao MC, Chang EB (1997) Tissue distribution of Na+/H+ exchanger isoforms NHE2 and NHE4 in rat intestine and kidney. Am J Phys 273(5 Pt 1):C1496–C1505

Booth IW, Stange G, Murer H, Fenton TR, Milla PJ (1985) Defective jejunal brush-border Na+/H+ exchange: a cause of congenital secretory diarrhoea. Lancet 1(8437):1066–1069

Bortner CD, Cidlowski JA (1996) Absence of volume regulatory mechanisms contributes to the rapid activation of apoptosis in thymocytes. Am J Phys 271(3 Pt 1):C950–C961. https://doi.org/10.1152/ajpcell.1996.271.3.C950

Bourgeois S, Meer LV, Wootla B, Bloch-Faure M, Chambrey R, Shull GE, Gawenis LR, Houillier P (2010) NHE4 is critical for the renal handling of ammonia in rodents. J Clin Invest 120(6):1895–1904. https://doi.org/10.1172/JCI36581

Bourguignon LY, Singleton PA, Diedrich F, Stern R, Gilad E (2004) CD44 interaction with Na^+-H^+ exchanger (NHE1) creates acidic microenvironments leading to hyaluronidase-2 and cathepsin B activation and breast tumor cell invasion. J Biol Chem 279(26):26991–27007. https://doi.org/10.1074/jbc.M311838200

Brant SR, Bernstein M, Wasmuth JJ, Taylor EW, McPherson JD, Li X, Walker S, Pouyssegur J, Donowitz M, Tse CM (1993) Physical and genetic mapping of a human apical epithelial Na+/H + exchanger (NHE3) isoform to chromosome 5p15.3. Genomics 15(3):668–672

Brant SR, Yun CH, Donowitz M, Tse CM (1995) Cloning, tissue distribution, and functional analysis of the human Na^+/N^+ exchanger isoform, NHE3. Am J Phys 269(1 Pt 1):C198–C206

Brett CL, Wei Y, Donowitz M, Rao R (2002) Human Na(+)/H(+) exchanger isoform 6 is found in recycling endosomes of cells, not in mitochondria. Am J Physiol Cell Physiol 282(5):C1031–C1041. https://doi.org/10.1152/ajpcell.00420.2001

Brett CL, Donowitz M, Rao R (2005) Evolutionary origins of eukaryotic sodium/proton exchangers. Am J Physiol Cell Physiol 288(2):C223–C239. https://doi.org/10.1152/ajpcell.00360.2004

Brierley GP, Settlemire CT, Knight VA (1968) Ion transport by heart mitochondria. XI The spontaneous and induced permeability of heart mitochondria to cations. Arch Biochem Biophys 126(1):276–288. https://doi.org/10.1016/0003-9861(68)90584-5

Brown DA, Melvin JE, Yule DI (2003) Critical role for NHE1 in intracellular pH regulation in pancreatic acinar cells. Am J Physiol Gastrointest Liver Physiol 285(5):G804–G812

Cabado AG, Yu FH, Kapus A, Lukacs G, Grinstein S, Orlowski J (1996) Distinct structural domains confer cAMP sensitivity and ATP dependence to the Na+/H+ exchanger NHE3 isoform. J Biol Chem 271(7):3590–3599. https://doi.org/10.1074/jbc.271.7.3590

Canessa M, Adragna N, Solomon HS, Connolly TM, Tosteson DC (1980) Increased sodium-lithium countertransport in red cells of patients with essential hypertension. N Engl J Med 302(14):772–776. https://doi.org/10.1056/NEJM198004033021403

Cano A (1996) Characterization of the rat NHE3 promoter. Am J Phys 271(3 Pt 2):F629–F636. https://doi.org/10.1152/ajprenal.1996.271.3.F629

Cano A, Baum M, Moe OW (1999) Thyroid hormone stimulates the renal Na/H exchanger NHE3 by transcriptional activation. Am J Phys 276(1):C102–C108. https://doi.org/10.1152/ajpcell.1999.276.1.C102

Cavet ME, Akhter S, Murtazina R, Sanchez de Medina F, Tse CM, Donowitz M (2001) Half-lives of plasma membrane Na(+)/H(+) exchangers NHE1-3: plasma membrane NHE2 has a rapid rate of degradation. Am J Physiol Cell Physiol 281(6):C2039–C2048

Chailler P, Menard D (1999) Ontogeny of EGF receptors in the human gut. Front Biosci 4:D87–D101. https://doi.org/10.2741/chailler

Chambrey R, Warnock DG, Podevin RA, Bruneval P, Mandet C, Belair MF, Bariety J, Paillard M (1998) Immunolocalization of the Na+/H+ exchanger isoform NHE2 in rat kidney. Am J Phys 275(3):F379–F386. https://doi.org/10.1152/ajprenal.1998.275.3.F379

Chambrey R, St John PL, Eladari D, Quentin F, Warnock DG, Abrahamson DR, Podevin RA, Paillard M (2001) Localization and functional characterization of Na+/H+ exchanger isoform NHE4 in rat thick ascending limbs. Am J Physiol Renal Physiol 281(4):F707–F717. https://doi.org/10.1152/ajprenal.2001.281.4.F707

Charles JF, Coury F, Sulyanto R, Sitara D, Wu J, Brady N, Tsang K, Sigrist K, Tollefsen DM, He L, Storm D, Aliprantis AO (2012) The collection of NFATc1-dependent transcripts in the osteoclast includes numerous genes non-essential to physiologic bone resorption. Bone 51 (5):902–912. https://doi.org/10.1016/j.bone.2012.08.113

Chen T, Kocinsky HS, Cha B, Murtazina R, Yang J, Tse CM, Singh V, Cole R, Aronson PS, de Jonge H, Sarker R, Donowitz M (2015) Cyclic GMP kinase II (cGKII) inhibits NHE3 by altering its trafficking and phosphorylating NHE3 at three required sites: identification of a multifunctional phosphorylation site. J Biol Chem 290(4):1952–1965. https://doi.org/10.1074/jbc.M114.590174

Chen P, Hao X, Li W, Zhao X, Huang Y (2016a) Mutations in the TMCO3 gene are associated with cornea Guttata and anterior polar cataract. Sci Rep 6:31021. https://doi.org/10.1038/srep31021

Chen SR, Chen M, Deng SL, Hao XX, Wang XX, Liu YX (2016b) Sodium-hydrogen exchanger NHA1 and NHA2 control sperm motility and male fertility. Cell Death Dis 7:e2152. https://doi.org/10.1038/cddis.2016.65

Chen H, Song L, Li G, Chen W, Zhao S, Zhou R, Shi X, Peng Z, Zhao W (2017a) Human rotavirus strain Wa downregulates NHE1 and NHE6 expressions in rotavirus-infected Caco-2 cells. Virus Genes. https://doi.org/10.1007/s11262-017-1444-0

Chen X, Wang X, Tang L, Wang J, Shen C, Liu J, Lu S, Zhang H, Kuang Y, Fei J, Wang Z (2017b) Nhe5 deficiency enhances learning and memory via upregulating Bdnf/TrkB signaling in mice. Am J Med Genet B Neuropsychiatr Genet 174(8):828–838. https://doi.org/10.1002/ajmg.b.32600

Chen T, Lin R, Avula L, Sarker R, Yang J, Cha B, Tse CM, McNamara G, Seidler U, Waldman S, Snook A, Bijvelds MJC, de Jonge HR, Li X, Donowitz M (2019) NHERF3 is necessary for *E. coli* heat stable enterotoxin (ST)-induced inhibition of NHE3 - differences in signaling in mouse small intestine and Caco-2 cells. Am J Physiol Cell Physiol. https://doi.org/10.1152/ajpcell.00351.2018

Chew SB, Leung GP, Leung PY, Tse CM, Wong PY (2000) Polarized distribution of NHE1 and NHE2 in the rat epididymis. Biol Reprod 62(3):755–758. https://doi.org/10.1095/biolreprod62.3.755

Chey WD, Lembo AJ, Rosenbaum DP (2017) Tenapanor treatment of patients with constipation-predominant irritable bowel syndrome: a phase 2, randomized, placebo-controlled efficacy and safety trial. Am J Gastroenterol. https://doi.org/10.1038/ajg.2017.41

Chintapalli VR, Kato A, Henderson L, Hirata T, Woods DJ, Overend G, Davies SA, Romero MF, Dow JA (2015) Transport proteins NHA1 and NHA2 are essential for survival, but have distinct transport modalities. Proc Natl Acad Sci USA 112(37):11720–11725. https://doi.org/10.1073/pnas.1508031112

Cho JH, Musch MW, DePaoli AM, Bookstein CM, Xie Y, Burant CF, Rao MC, Chang EB (1994) Glucocorticoids regulate Na+/H+ exchange expression and activity in region- and tissue-specific manner. Am J Phys 267(3 Pt 1):C796–C803. https://doi.org/10.1152/ajpcell.1994.267.3.C796

Choshniak I, Mualem R (1997) SCFA and electrolyte absorption in the colon of three rodent species. Comp Biochem Physiol A Physiol 118(2):381–384

Chow CW, Woodside M, Demaurex N, Yu FH, Plant P, Rotin D, Grinstein S, Orlowski J (1999) Proline-rich motifs of the Na+/H+ exchanger 2 isoform. Binding of Src homology domain 3 and role in apical targeting in epithelia. J Biol Chem 274(15):10481–10488. https://doi.org/10.1074/jbc.274.15.10481

Chu J, Chu S, Montrose MH (2002) Apical Na+/H+ exchange near the base of mouse colonic crypts. Am J Physiol Cell Physiol 283(1):C358–C372. https://doi.org/10.1152/ajpcell.01380.2000

Cinar A, Chen M, Riederer B, Bachmann O, Wiemann M, Manns M, Kocher O, Seidler U (2007) NHE3 inhibition by cAMP and Ca2+ is abolished in PDZ-domain protein PDZK1-deficient murine enterocytes. J Physiol 581(Pt 3):1235–1246. https://doi.org/10.1113/jphysiol.2007.131722

Coelho CM, Leevers SJ (2000) Do growth and cell division rates determine cell size in multicellular organisms? J Cell Sci 113(Pt 17):2927–2934

Collazo R, Fan L, Hu MC, Zhao H, Wiederkehr MR, Moe OW (2000) Acute regulation of Na+/H+ exchanger NHE3 by parathyroid hormone via NHE3 phosphorylation and dynamin-dependent endocytosis. J Biol Chem 275(41):31601–31608

Collins JF, Honda T, Knobel S, Bulus NM, Conary J, DuBois R, Ghishan FK (1993) Molecular cloning, sequencing, tissue distribution, and functional expression of a Na+/H+ exchanger (NHE-2). Proc Natl Acad Sci USA 90(9):3938–3942. https://doi.org/10.1073/pnas.90.9.3938

Collins JF, Xu H, Kiela PR, Zeng J, Ghishan FK (1997) Functional and molecular characterization of NHE3 expression during ontogeny in rat jejunal epithelium. Am J Phys 273(6 Pt 1):C1937–C1946

Collins JF, Kiela PR, Xu H, Zeng J, Ghishan FK (1998a) Increased NHE2 expression in rat intestinal epithelium during ontogeny is transcriptionally mediated. Am J Phys 275(4): C1143–C1150. https://doi.org/10.1152/ajpcell.1998.275.4.C1143

Collins JF, Xu H, Kiela PR, Zeng J, Ghishan FK (1998b) Ontogeny of basolateral membrane sodium-hydrogen exchange (NHE) activity and mRNA expression of NHE-1 and NHE-4 in rat kidney and jejunum. Biochim Biophys Acta 1369(2):247–258

Collins JF, Kiela PR, Xu H, Ghishan FK (2000) Expression of rat, renal NHE2 and NHE3 during postnatal development. Biochim Biophys Acta 1464(1):7–17. https://doi.org/10.1016/s0005-2736(99)00241-2

Colombani V, Silviani V, Marteau C, Lerique B, Cartouzou G, Gerolami A (1996) Presence of the NHE3 isoform of the Na+/H+ exchanger in human gallbladder. Clin Sci (Lond) 91(2):209–212

Conter RL, Roslyn JJ, Porter-Fink V, DenBesten L (1986) Gallbladder absorption increases during early cholesterol gallstone formation. Am J Surg 151(1):184–191

Counillon L, Franchi A, Pouyssegur J (1993) A point mutation of the Na+/H+ exchanger gene (NHE1) and amplification of the mutated allele confer amiloride resistance upon chronic acidosis. Proc Natl Acad Sci USA 90(10):4508–4512

Counillon L, Pouyssegur J, Reithmeier RA (1994) The Na+/H+ exchanger NHE-1 possesses N- and O-linked glycosylation restricted to the first N-terminal extracellular domain. Biochemistry 33 (34):10463–10469

Counillon L, Noel J, Reithmeier RA, Pouyssegur J (1997) Random mutagenesis reveals a novel site involved in inhibitor interaction within the fourth transmembrane segment of the Na+/H+ exchanger-1. Biochemistry 36(10):2951–2959. https://doi.org/10.1021/bi9615405

Cui GM, Zhao YX, Zhang NN, Liu ZS, Sun WC, Peng QS (2013) Amiloride attenuates lipopolysaccharide-accelerated atherosclerosis via inhibition of NHE1-dependent endothelial cell apoptosis. Acta Pharmacol Sin 34(2):231–238. https://doi.org/10.1038/aps.2012.155

Damkier HH, Christensen HL, Christensen IB, Wu Q, Fenton RA, Praetorius J (2018) The murine choroid plexus epithelium expresses the 2Cl(−)/H(+) exchanger ClC-7 and Na(+)/H(+) exchanger NHE6 in the luminal membrane domain. Am J Physiol Cell Physiol 314(4):C439–C448. https://doi.org/10.1152/ajpcell.00145.2017

Deisl C, Simonin A, Anderegg M, Albano G, Kovacs G, Ackermann D, Moch H, Dolci W, Thorens B, Fuster DG (2013) Sodium/hydrogen exchanger NHA2 is critical for insulin secretion in beta-cells. Proc Natl Acad Sci USA 110(24):10004–10009. https://doi.org/10.1073/pnas.1220009110

Delvaux M, Bastie MJ, Chentoufi J, Cragoe EJ Jr, Vaysse N, Ribet A (1990) Amiloride and analogues inhibit Na(+)-H+ exchange and cell proliferation in AR42J pancreatic cell line. Am J Phys 259(5 Pt 1):G842–G849. https://doi.org/10.1152/ajpgi.1990.259.5.G842

Denker SP, Barber DL (2002) Cell migration requires both ion translocation and cytoskeletal anchoring by the Na-H exchanger NHE1. J Cell Biol 159(6):1087–1096. https://doi.org/10.1083/jcb.200208050

Denker SP, Huang DC, Orlowski J, Furthmayr H, Barber DL (2000) Direct binding of the Na–H exchanger NHE1 to ERM proteins regulates the cortical cytoskeleton and cell shape independently of H(+) translocation. Mol Cell 6(6):1425–1436

Di Giusto G, Flamenco P, Rivarola V, Fernandez J, Melamud L, Ford P, Capurro C (2012) Aquaporin 2-increased renal cell proliferation is associated with cell volume regulation. J Cell Biochem 113(12):3721–3729. https://doi.org/10.1002/jcb.24246

Dixit MP, Xu L, Xu H, Bai L, Collins JF, Ghishan FK (2004) Effect of angiotensin-II on renal Na+/H+ exchanger-NHE3 and NHE2. Biochim Biophys Acta 1664(1):38–44. https://doi.org/10.1016/j.bbamem.2004.03.011

Donowitz M, De La Horra C, Calonge ML, Wood IS, Dyer J, Gribble SM, De Medina FS, Tse CM, Shirazi-Beechey SP, Ilundain AA (1998) In birds, NHE2 is major brush-border Na+/H+ exchanger in colon and is increased by a low-NaCl diet. Am J Phys 274(6):R1659–R1669. https://doi.org/10.1152/ajpregu.1998.274.6.R1659

Donowitz M, Janecki A, Akhter S, Cavet ME, Sanchez F, Lamprecht G, Zizak M, Kwon WL, Khurana S, Yun CH, Tse CM (2000) Short-term regulation of NHE3 by EGF and protein kinase C but not protein kinase A involves vesicle trafficking in epithelial cells and fibroblasts. Ann N Y Acad Sci 915:30–42

Donowitz M, Mohan S, Zhu CX, Chen TE, Lin R, Cha B, Zachos NC, Murtazina R, Sarker R, Li X (2009) NHE3 regulatory complexes. J Exp Biol 212(Pt 11):1638–1646. https://doi.org/10.1242/jeb.028605

Donowitz M, Ming Tse C, Fuster D (2013) SLC9/NHE gene family, a plasma membrane and organellar family of Na(+)/H(+) exchangers. Mol Asp Med 34(2-3):236–251. https://doi.org/10.1016/j.mam.2012.05.001

Dudeja PK, Rao DD, Syed I, Joshi V, Dahdal RY, Gardner C, Risk MC, Schmidt L, Bavishi D, Kim KE, Harig JM, Goldstein JL, Layden TJ, Ramaswamy K (1996) Intestinal distribution of human Na+/H+ exchanger isoforms NHE-1, NHE-2, and NHE-3 mRNA. Am J Phys 271(3 Pt 1):G483–G493

Dyck JR, Silva NL, Fliegel L (1995) Activation of the Na+/H+ exchanger gene by the transcription factor AP-2. J Biol Chem 270(3):1375–1381

Eiam-Ong S, Chaipipat M, Manotham K, Eiam-Ong S (2017) Rapid action of aldosterone on protein levels of sodium-hydrogen exchangers and protein kinase C Beta isoforms in rat kidney. Int J Endocrinol 2017:2975853. https://doi.org/10.1155/2017/2975853

Engevik MA, Engevik KA, Yacyshyn MB, Wang J, Hassett DJ, Darien B, Yacyshyn BR, Worrell RT (2015) Human *Clostridium difficile* infection: inhibition of NHE3 and microbiota profile. Am J Physiol Gastrointest Liver Physiol 308(6):G497–G509. https://doi.org/10.1152/ajpgi.00090.2014

Evans RL, Bell SM, Schultheis PJ, Shull GE, Melvin JE (1999) Targeted disruption of the Nhe1 gene prevents muscarinic agonist-induced up-regulation of Na(+)/H(+) exchange in mouse parotid acinar cells. J Biol Chem 274(41):29025–29030

Facanha AL, dos Reis MC, Montero-Lomeli M (2000) Structural study of the porcine Na+/H+ exchanger NHE1 gene and its 5′-flanking region. Mol Cell Biochem 210(1–2):91–99

Falcone RA Jr, Shin CE, Stern LE, Wang Z, Erwin CR, Soleimani M, Warner BW (1999) Differential expression of ileal Na(+)/H(+) exchanger isoforms after enterectomy. J Surg Res 86(2):192–197. https://doi.org/10.1006/jsre.1999.5720

Farkas K, Yeruva S, Rakonczay Z Jr, Ludolph L, Molnar T, Nagy F, Szepes Z, Schnur A, Wittmann T, Hubricht J, Riederer B, Venglovecz V, Lazar G, Kiraly M, Zsembery A,

Varga G, Seidler U, Hegyi P (2010) New therapeutic targets in ulcerative colitis: the importance of ion transporters in the human colon. Inflamm Bowel Dis. https://doi.org/10.1002/ibd.21432
Fenton RA, Poulsen SB, de la Mora CS, Soleimani M, Busslinger M, Dominguez Rieg JA, Rieg T (2015) Caffeine-induced diuresis and natriuresis is independent of renal tubular NHE3. Am J Physiol Renal Physiol 308(12):F1409–F1420. https://doi.org/10.1152/ajprenal.00129.2015
Fenton RA, Poulsen SB, de la Mora CS, Soleimani M, Dominguez Rieg JA, Rieg T (2017) Renal tubular NHE3 is required in the maintenance of water and sodium chloride homeostasis. Kidney Int 92(2):397–414. https://doi.org/10.1016/j.kint.2017.02.001
Fiori M, Gras EG, Amorena C (2009) Decreased NHE8 isoform expression and defective acidification in proximal convoluted tubules of senile rats. Age (Dordr) 31(1):77–84. https://doi.org/10.1007/s11357-008-9083-y
Fiskerstrand T, Arshad N, Haukanes BI, Tronstad RR, Pham KD, Johansson S, Havik B, Tonder SL, Levy SE, Brackman D, Boman H, Biswas KH, Apold J, Hovdenak N, Visweswariah SS, Knappskog PM (2012) Familial diarrhea syndrome caused by an activating GUCY2C mutation. N Engl J Med 366(17):1586–1595. https://doi.org/10.1056/NEJMoa1110132
Fitzgerald RC, Omary MB, Triadafilopoulos G (1998) Altered sodium-hydrogen exchange activity is a mechanism for acid-induced hyperproliferation in Barrett's esophagus. Am J Phys 275(1 Pt 1):G47–G55
Fliegel L (1996) The Na^+/H^+ exchanger. R.G. Landes Company, Austin, TX
Fliegel L (2005) The Na+/H+ exchanger isoform 1. Int J Biochem Cell Biol 37(1):33–37. https://doi.org/10.1016/j.biocel.2004.02.006
Fliegel L (2009) Regulation of the Na(+)/H(+) exchanger in the healthy and diseased myocardium. Expert Opin Ther Targets 13(1):55–68. https://doi.org/10.1517/14728220802600707
Fliegel L (2019) Structural and functional changes in the Na(+)/H(+) exchanger isoform 1, induced by Erk1/2 phosphorylation. Int J Mol Sci 20(10). https://doi.org/10.3390/ijms20102378
Fliegel L, Sardet C, Pouyssegur J, Barr A (1991) Identification of the protein and cDNA of the cardiac Na+/H+ exchanger. FEBS Lett 279(1):25–29. https://doi.org/10.1016/0014-5793(91)80241-t
Fliegel L, Dyck JR, Wang H, Fong C, Haworth RS (1993) Cloning and analysis of the human myocardial Na+/H+ exchanger. Mol Cell Biochem 125(2):137–143. https://doi.org/10.1007/bf00936442
Flinck M, Kramer SH, Pedersen SF (2018) Roles of pH in control of cell proliferation. Acta Physiol (Oxf) 223(3):e13068. https://doi.org/10.1111/apha.13068
Fujiwara Y, Higuchi K, Takashima T, Hamaguchi M, Hayakawa T, Tominaga K, Watanabe T, Oshitani N, Shimada Y, Arakawa T (2006) Roles of epidermal growth factor and Na+/H+ exchanger-1 in esophageal epithelial defense against acid-induced injury. Am J Physiol Gastrointest Liver Physiol 290(4):G665–G673. https://doi.org/10.1152/ajpgi.00238.2005
Fukura N, Ohgaki R, Matsushita M, Nakamura N, Mitsui K, Kanazawa H (2010) A membrane-proximal region in the C-terminal tail of NHE7 is required for its distribution in the trans-Golgi network, distinct from NHE6 localization at endosomes. J Membr Biol 234(3):149–158. https://doi.org/10.1007/s00232-010-9242-9
Fuster D, Moe OW, Hilgemann DW (2004) Lipid- and mechanosensitivities of sodium/hydrogen exchangers analyzed by electrical methods. Proc Natl Acad Sci USA 101(28):10482–10487. https://doi.org/10.1073/pnas.0403930101
Fuster DG, Zhang J, Shi M, Bobulescu IA, Andersson S, Moe OW (2008) Characterization of the sodium/hydrogen exchanger NHA2. J Am Soc Nephrol 19(8):1547–1556. https://doi.org/10.1681/ASN.2007111245
Garbern JY, Neumann M, Trojanowski JQ, Lee VM, Feldman G, Norris JW, Friez MJ, Schwartz CE, Stevenson R, Sima AA (2010) A mutation affecting the sodium/proton exchanger, SLC9A6, causes mental retardation with tau deposition. Brain 133(Pt 5):1391–1402. https://doi.org/10.1093/brain/awq071

Gattineni J, Sas D, Dagan A, Dwarakanath V, Baum M (2008) Effect of thyroid hormone on the postnatal renal expression of NHE8. Am J Physiol Renal Physiol 294(1):F198–F204. https://doi.org/10.1152/ajprenal.00332.2007

Gawenis LR, Greeb JM, Prasad V, Grisham C, Sanford LP, Doetschman T, Andringa A, Miller ML, Shull GE (2005) Impaired gastric acid secretion in mice with a targeted disruption of the NHE4 Na+/H+ exchanger. J Biol Chem 280(13):12781–12789. https://doi.org/10.1074/jbc.M414118200

Ghishan FK, Kikuchi K, Riedel B (1992) Epidermal growth factor up-regulates intestinal Na+/H+ exchange activity. Proc Soc Exp Biol Med 201(3):289–295. https://doi.org/10.3181/00379727-201-43510

Ghishan FK, Knobel SM, Summar M (1995) Molecular cloning, sequencing, chromosomal localization, and tissue distribution of the human Na+/H+ exchanger (SLC9A2). Genomics 30 (1):25–30. https://doi.org/10.1006/geno.1995.0004

Gilfillan GD, Selmer KK, Roxrud I, Smith R, Kyllerman M, Eiklid K, Kroken M, Mattingsdal M, Egeland T, Stenmark H, Sjoholm H, Server A, Samuelsson L, Christianson A, Tarpey P, Whibley A, Stratton MR, Futreal PA, Teague J, Edkins S, Gecz J, Turner G, Raymond FL, Schwartz C, Stevenson RE, Undlien DE, Stromme P (2008) SLC9A6 mutations cause X-linked mental retardation, microcephaly, epilepsy, and ataxia, a phenotype mimicking Angelman syndrome. Am J Hum Genet 82(4):1003–1010. https://doi.org/10.1016/j.ajhg.2008.01.013

Gill RK, Saksena S, Tyagi S, Alrefai WA, Malakooti J, Sarwar Z, Turner JR, Ramaswamy K, Dudeja PK (2005) Serotonin inhibits Na+/H+ exchange activity via 5-HT4 receptors and activation of PKC alpha in human intestinal epithelial cells. Gastroenterology 128(4):962–974

Girardi AC, Titan SM, Malnic G, Reboucas NA (2000) Chronic effect of parathyroid hormone on NHE3 expression in rat renal proximal tubules. Kidney Int 58(4):1623–1631. https://doi.org/10.1046/j.1523-1755.2000.00323.x

Giurgiu DI, Saunders-Kirkwood KD, Roslyn JJ, Abedin MZ (1997) Sequential changes in biliary lipids and gallbladder ion transport during gallstone formation. Ann Surg 225(4):382–390

Goldberg EB, Arbel T, Chen J, Karpel R, Mackie GA, Schuldiner S, Padan E (1987) Characterization of a Na+/H+ antiporter gene of *Escherichia coli*. Proc Natl Acad Sci USA 84 (9):2615–2619. https://doi.org/10.1073/pnas.84.9.2615

Goldman A, Shahidullah M, Goldman D, Khailova L, Watts G, Delamere N, Dvorak K (2010) A novel mechanism of acid and bile acid-induced DNA damage involving Na+/H+ exchanger: implication for Barrett's oesophagus. Gut 59(12):1606–1616. https://doi.org/10.1136/gut.2010.213686

Goyal S, Vanden Heuvel G, Aronson PS (2003) Renal expression of novel Na+/H+ exchanger isoform NHE8. Am J Physiol Renal Physiol 284(3):F467–F473. https://doi.org/10.1152/ajprenal.00352.2002

Goyal S, Mentone S, Aronson PS (2005) Immunolocalization of NHE8 in rat kidney. Am J Physiol Renal Physiol 288(3):F530–F538. https://doi.org/10.1152/ajprenal.00229.2004

Gray MR, Donnelly RJ, Kingsnorth AN (1991) Role of salivary epidermal growth factor in the pathogenesis of Barrett's columnar lined oesophagus. Br J Surg 78(12):1461–1466

Grenier AL, Abu-ihweij K, Zhang G, Ruppert SM, Boohaker R, Slepkov ER, Pridemore K, Ren JJ, Fliegel L, Khaled AR (2008) Apoptosis-induced alkalinization by the Na+/H+ exchanger isoform 1 is mediated through phosphorylation of amino acids Ser726 and Ser729. Am J Physiol Cell Physiol 295(4):C883–C896. https://doi.org/10.1152/ajpcell.00574.2007

Grinstein S, Woodside M, Waddell TK, Downey GP, Orlowski J, Pouyssegur J, Wong DC, Foskett JK (1993) Focal localization of the NHE-1 isoform of the Na+/H+ antiport: assessment of effects on intracellular pH. EMBO J 12(13):5209–5218

Guan Y, Dong J, Tackett L, Meyer JW, Shull GE, Montrose MH (2006) NHE2 is the main apical NHE in mouse colonic crypts but an alternative Na⁺-dependent acid extrusion mechanism is upregulated in NHE2-null mice. Am J Physiol Gastrointest Liver Physiol 291(4):G689–G699. https://doi.org/10.1152/ajpgi.00342.2005

Guissart C, Li X, Leheup B, Drouot N, Montaut-Verient B, Raffo E, Jonveaux P, Roux AF, Claustres M, Fliegel L, Koenig M (2015) Mutation of SLC9A1, encoding the major Na(+)/H (+) exchanger, causes ataxia-deafness Lichtenstein-Knorr syndrome. Hum Mol Genet 24 (2):463–470. https://doi.org/10.1093/hmg/ddu461

Guo YM, Liu Y, Liu M, Wang JL, Xie ZD, Chen KJ, Wang DK, Occhipinti R, Boron WF, Chen LM (2017) Na(+)/HCO3(−) Cotransporter NBCn2 mediates HCO3(−) reclamation in the apical membrane of renal proximal tubules. J Am Soc Nephrol 28(8):2409–2419. https://doi.org/10.1681/ASN.2016080930

Gupta N, Dwarakanath V, Baum M (2004) Maturation of the Na+/H+ antiporter (NHE3) in the proximal tubule of the hypothyroid adrenalectomized rat. Am J Physiol Renal Physiol 287(3): F521–F527. https://doi.org/10.1152/ajprenal.00005.2004

Gurney MA, Laubitz D, Ghishan FK, Kiela PR (2017) Pathophysiology of intestinal Na^+/H^+ exchange. Cell Mol Gastroenterol Hepatol 3(1):27–40. https://doi.org/10.1016/j.jcmgh.2016.09.010

Ha BG, Hong JM, Park JY, Ha MH, Kim TH, Cho JY, Ryoo HM, Choi JY, Shin HI, Chun SY, Kim SY, Park EK (2008) Proteomic profile of osteoclast membrane proteins: identification of Na+/H + exchanger domain containing 2 and its role in osteoclast fusion. Proteomics 8(13):2625–2639. https://doi.org/10.1002/pmic.200701192

Haimovici J, Beck JS, Molla-Hosseini C, Vallerand D, Haddad P (1994) Different modulation of hepatocellular Na+/H+ exchange activity by insulin and EGF. Am J Phys 267(3 Pt 1):G364–G370. https://doi.org/10.1152/ajpgi.1994.267.3.G364

Hanner F, Chambrey R, Bourgeois S, Meer E, Mucsi I, Rosivall L, Shull GE, Lorenz JN, Eladari D, Peti-Peterdi J (2008) Increased renal renin content in mice lacking the Na+/H+ exchanger NHE2. Am J Physiol Renal Physiol 294(4):F937–F944. https://doi.org/10.1152/ajprenal.00591.2007

Harold FM, Papineau D (1972) Cation transport and electrogenesis by Streptococcus faecalis. II Proton and sodium extrusion. J Membr Biol 8(1):45–62

Harrison CA, Laubitz D, Ohland CL, Midura-Kiela MT, Patil K, Besselsen DG, Jamwal DR, Jobin C, Ghishan FK, Kiela PR (2018) Microbial dysbiosis associated with impaired intestinal Na(+)/H(+) exchange accelerates and exacerbates colitis in ex-germ free mice. Mucosal Immunol. https://doi.org/10.1038/s41385-018-0035-2

Hatch M, Freel RW (2008) Increased colonic sodium absorption in rats with chronic renal failure is partially mediated by AT1 receptor agonism. Am J Physiol Gastrointest Liver Physiol 295(2): G348–G356. https://doi.org/10.1152/ajpgi.00079.2008

Hayashi H, Tamura A, Krishnan D, Tsukita S, Suzuki Y, Kocinsky HS, Aronson PS, Orlowski J, Grinstein S, Alexander RT (2013) Ezrin is required for the functional regulation of the epithelial sodium proton exchanger, NHE3. PLoS One 8(2):e55623. https://doi.org/10.1371/journal.pone.0055623

He P, Yun CC (2010) Mechanisms of the regulation of the intestinal Na+/H+ exchanger NHE3. J Biomed Biotechnol 2010:238080. https://doi.org/10.1155/2010/238080

He X, Tse CM, Donowitz M, Alper SL, Gabriel SE, Baum BJ (1997) Polarized distribution of key membrane transport proteins in the rat submandibular gland. Pflugers Arch 433(3):260–268

He P, Lee SJ, Lin S, Seidler U, Lang F, Fejes-Toth G, Naray-Fejes-Toth A, Yun CC (2011) Serum- and glucocorticoid-induced kinase 3 in recycling endosomes mediates acute activation of Na+/H + exchanger NHE3 by glucocorticoids. Mol Biol Cell 22(20):3812–3825. https://doi.org/10.1091/mbc.E11-04-0328

He P, Zhao L, Zhu L, Weinman EJ, De Giorgio R, Koval M, Srinivasan S, Yun CC (2015) Restoration of Na+/H+ exchanger NHE3-containing macrocomplexes ameliorates diabetes-associated fluid loss. J Clin Invest 125(9):3519–3531. https://doi.org/10.1172/JCI79552

Hecht G, Hodges K, Gill RK, Kear F, Tyagi S, Malakooti J, Ramaswamy K, Dudeja PK (2004) Differential regulation of Na+/H+ exchange isoform activities by enteropathogenic *E. coli* in human intestinal epithelial cells. Am J Physiol Gastrointest Liver Physiol 287(2):G370–G378. https://doi.org/10.1152/ajpgi.00432.2003

Heinz-Erian P, Muller T, Krabichler B, Schranz M, Becker C, Ruschendorf F, Nurnberg P, Rossier B, Vujic M, Booth IW, Holmberg C, Wijmenga C, Grigelioniene G, Kneepkens CM, Rosipal S, Mistrik M, Kappler M, Michaud L, Doczy LC, Siu VM, Krantz M, Zoller H, Utermann G, Janecke AR (2009) Mutations in SPINT2 cause a syndromic form of congenital sodium diarrhea. Am J Hum Genet 84(2):188–196

Helmle-Kolb C, Montrose MH, Stange G, Murer H (1990) Regulation of Na+/H+ exchange in opossum kidney cells by parathyroid hormone, cyclic AMP and phorbol esters. Pflugers Arch 415(4):461–470. https://doi.org/10.1007/bf00373624

Hendus-Altenburger R, Kragelund BB, Pedersen SF (2014) Structural dynamics and regulation of the mammalian SLC9A family of Na(+)/H(+) exchangers. Curr Top Membr 73:69–148. https://doi.org/10.1016/B978-0-12-800223-0.00002-5

Henning SJ (1978) Plasma concentrations of total and free corticosterone during development in the rat. Am J Phys 235(5):E451–E456. https://doi.org/10.1152/ajpendo.1978.235.5.E451

Hodges K, Gill R, Ramaswamy K, Dudeja PK, Hecht G (2006) Rapid activation of Na+/H+ exchange by EPEC is PKC mediated. Am J Physiol Gastrointest Liver Physiol 291(5):G959–G968. https://doi.org/10.1152/ajpgi.00274.2005

Hodges K, Alto NM, Ramaswamy K, Dudeja PK, Hecht G (2008) The enteropathogenic *Escherichia coli* effector protein EspF decreases sodium hydrogen exchanger 3 activity. Cell Microbiol 10(8):1735–1745. https://doi.org/10.1111/j.1462-5822.2008.01163.x

Hofstetter W, Siegrist M, Simonin A, Bonny O, Fuster DG (2010) Sodium/hydrogen exchanger NHA2 in osteoclasts: subcellular localization and role in vitro and in vivo. Bone 47(2):331–340. https://doi.org/10.1016/j.bone.2010.04.605

Holmberg C, Perheentupa J (1985) Congenital Na+ diarrhea: a new type of secretory diarrhea. J Pediatr 106(1):56–61

Hoogerwerf WA, Tsao SC, Devuyst O, Levine SA, Yun CH, Yip JW, Cohen ME, Wilson PD, Lazenby AJ, Tse CM, Donowitz M (1996) NHE2 and NHE3 are human and rabbit intestinal brush-border proteins. Am J Phys 270(1 Pt 1):G29–G41

Horvat B, Taheri S, Salihagic A (1992) Tumour cell proliferation is abolished by inhibitors of Na+/H+ and HCO3-/Cl- exchange. Eur J Cancer 29A(1):132–137. https://doi.org/10.1016/0959-8049(93)90591-3

Hu MC, Fan L, Crowder LA, Karim-Jimenez Z, Murer H, Moe OW (2001) Dopamine acutely stimulates Na+/H+ exchanger (NHE3) endocytosis via clathrin-coated vesicles: dependence on protein kinase A-mediated NHE3 phosphorylation. J Biol Chem 276(29):26906–26915

Hua P, Xu H, Uno JK, Lipko MA, Dong J, Kiela PR, Ghishan FK (2007) Sp1 and Sp3 mediate NHE2 gene transcription in the intestinal epithelial cells. Am J Physiol Gastrointest Liver Physiol 293(1):G146–G153. https://doi.org/10.1152/ajpgi.00443.2006

Huang X, Morse LR, Xu Y, Zahradka J, Sychrova H, Stashenko P, Fan F, Battaglino RA (2010) Mutational analysis of NHAoc/NHA2 in *Saccharomyces cerevisiae*. Biochim Biophys Acta 1800(12):1241–1247. https://doi.org/10.1016/j.bbagen.2010.08.001

Ikeda T, Schmitt B, Pouyssegur J, Wakabayashi S, Shigekawa M (1997) Identification of cytoplasmic subdomains that control pH-sensing of the Na+/H+ exchanger (NHE1): pH-maintenance, ATP-sensitive, and flexible loop domains. J Biochem 121(2):295–303

Ismail N, Giribabu N, Muniandy S, Salleh N (2015) Enhanced expression of sodium hydrogen exchanger (NHE)-1, 2 and 4 in the cervix of ovariectomised rats by phytoestrogen genistein. Int J Med Sci 12(6):468–477. https://doi.org/10.7150/ijms.11210

Jadeja S, Barnard AR, McKie L, Cross SH, White JK, Sanger Mouse Genetics P, Robertson M, Budd PS, MacLaren RE, Jackson IJ (2015) Mouse slc9a8 mutants exhibit retinal defects due to retinal pigmented epithelium dysfunction. Invest Ophthalmol Vis Sci 56(5):3015–3026. https://doi.org/10.1167/iovs.14-15735

Janecke AR, Heinz-Erian P, Yin J, Petersen BS, Franke A, Lechner S, Fuchs I, Melancon S, Uhlig HH, Travis S, Marinier E, Perisic V, Ristic N, Gerner P, Booth IW, Wedenoja S, Baumgartner N, Vodopiutz J, Frechette-Duval MC, De Lafollie J, Persad R, Warner N, Tse CM, Sud K, Zachos NC, Sarker R, Zhu X, Muise AM, Zimmer KP, Witt H, Zoller H,

Donowitz M, Muller T (2015) Reduced sodium/proton exchanger NHE3 activity causes congenital sodium diarrhea. Hum Mol Genet 24(23):6614–6623. https://doi.org/10.1093/hmg/ddv367

Janecke AR, Heinz-Erian P, Muller T (2016) Congenital sodium diarrhea: a form of intractable diarrhea, with a link to inflammatory Bowel disease. J Pediatr Gastroenterol Nutr 63 (2):170–176. https://doi.org/10.1097/MPG.0000000000001139

Janecke AR, Heinz-Erian P, Muller T (2017) Reduced NHE3 activity results in congenital diarrhea and can predispose to inflammatory bowel disease. Am J Physiol Regul Integr Comp Physiol 312(3):R311. https://doi.org/10.1152/ajpregu.00545.2016

Janecki AJ, Montrose MH, Zimniak P, Zweibaum A, Tse CM, Khurana S, Donowitz M (1998) Subcellular redistribution is involved in acute regulation of the brush border Na+/H+ exchanger isoform 3 in human colon adenocarcinoma cell line Caco-2. Protein kinase C-mediated inhibition of the exchanger. J Biol Chem 273(15):8790–8798

Janecki AJ, Montrose MH, Tse CM, de Medina FS, Zweibaum A, Donowitz M (1999) Development of an endogenous epithelial Na(+)/H(+) exchanger (NHE3) in three clones of caco-2 cells. Am J Phys 277(2 Pt 1):G292–G305

Janecki AJ, Janecki M, Akhter S, Donowitz M (2000) Basic fibroblast growth factor stimulates surface expression and activity of Na(+)/H(+) exchanger NHE3 via mechanism involving phosphatidylinositol 3-kinase. J Biol Chem 275(11):8133–8142

Jenkins EC Jr, Debnath S, Gundry S, Gundry S, Uyar U, Fata JE (2012) Intracellular pH regulation by Na(+)/H(+) exchanger-1 (NHE1) is required for growth factor-induced mammary branching morphogenesis. Dev Biol 365(1):71–81. https://doi.org/10.1016/j.ydbio.2012.02.010

Jenkins EC Jr, Debnath S, Varriano S, Gundry S, Fata JE (2014) Na(+) /H(+) exchanger 1 (NHE1) function is necessary for maintaining mammary tissue architecture. Dev Dyn 243(2):229–242. https://doi.org/10.1002/dvdy.24032

Jensen HH, Pedersen GA, Morgen JJ, Parsons M, Pedersen SF, Nejsum LN (2019) The Na(+) /H(+) exchanger NHE1 localizes as clusters to cryptic lamellipodia and accelerates collective epithelial cell migration. J Physiol 597(3):849–867. https://doi.org/10.1113/JP277383

Johansson M, Glazier JD, Sibley CP, Jansson T, Powell TL (2002) Activity and protein expression of the Na+/H+ exchanger is reduced in syncytiotrophoblast microvillous plasma membranes isolated from preterm intrauterine growth restriction pregnancies. J Clin Endocrinol Metab 87 (12):5686–5694. https://doi.org/10.1210/jc.2002-020214

Johansson ME, Gustafsson JK, Holmen-Larsson J, Jabbar KS, Xia L, Xu H, Ghishan FK, Carvalho FA, Gewirtz AT, Sjovall H, Hansson GC (2014) Bacteria penetrate the normally impenetrable inner colon mucus layer in both murine colitis models and patients with ulcerative colitis. Gut 63 (2):281–291. https://doi.org/10.1136/gutjnl-2012-303207

Johansson S, Rosenbaum DP, Knutsson M, Leonsson-Zachrisson M (2016) A phase 1 study of the safety, tolerability, pharmacodynamics, and pharmacokinetics of tenapanor in healthy Japanese volunteers. Clin Exp Nephrol. https://doi.org/10.1007/s10157-016-1302-8

Joseph C, Gattineni J, Dwarakanath V, Baum M (2013) Glucocorticoids reduce renal NHE8 expression. Physiol Rep 1(2). https://doi.org/10.1002/phy2.31

Joutsi T, Paimela H, Bhowmik A, Kiviluoto T, Kivilaakso E (1996) Role of Na(+)-H(+)-antiport in restitution of isolated guinea pig gastric epithelium after superficial injury. Dig Dis Sci 41 (11):2187–2194

Kagami T, Chen S, Memar P, Choi M, Foster LJ, Numata M (2008) Identification and biochemical characterization of the SLC9A7 interactome. Mol Membr Biol 25(5):436–447. https://doi.org/10.1080/09687680802263046

Kandasamy RA, Orlowski J (1996) Genomic organization and glucocorticoid transcriptional activation of the rat Na+/H+ exchanger Nhe3 gene. J Biol Chem 271(18):10551–10559. https://doi.org/10.1074/jbc.271.18.10551

Kaneko K, Guth PH, Kaunitz JD (1992) Na+/H+ exchange regulates intracellular pH of rat gastric surface cells in vivo. Pflugers Arch 421(4):322–328

Kanwar RK, Ganguly NK, Kanwar JR, Kumar L, Walia BN (1994) Impairment of Na+,K(+)-ATPase activity following enterotoxigenic Campylobacter jejuni infection: changes in Na+, Cl- and 3-O-methyl-D-glucose transport in vitro, in rat ileum. FEMS Microbiol Lett 124 (3):381–385

Kapus A, Grinstein S, Wasan S, Kandasamy R, Orlowski J (1994) Functional characterization of three isoforms of the Na+/H+ exchanger stably expressed in Chinese hamster ovary cells. ATP dependence, osmotic sensitivity, and role in cell proliferation. J Biol Chem 269 (38):23544–23552

Karki P, Fliegel L (2010) Overexpression of the NHE1 isoform of the Na(+)/H (+) exchanger causes elevated apoptosis in isolated cardiomyocytes after hypoxia/reoxygenation challenge. Mol Cell Biochem 338(1-2):47–57. https://doi.org/10.1007/s11010-009-0337-5

Karki P, Li X, Schrama D, Fliegel L (2011) B-Raf associates with and activates the NHE1 isoform of the Na+/H+ exchanger. J Biol Chem 286(15):13096–13105. https://doi.org/10.1074/jbc.M110.165134

Karmazyn M, Sawyer M, Fliegel L (2005) The Na(+)/H(+) exchanger: a target for cardiac therapeutic intervention. Curr Drug Targets Cardiovasc Haematol Disord 5(4):323–335

Kaur T, Singh S, Gorowara S, Ganguly NK (1995) Role of enteric nervous system in Shigella dysenteriae type 1 toxin-induced fluid secretion in rabbit ileum. J Diarrhoeal Dis Res 13 (3):159–165

Kemp G, Young H, Fliegel L (2008) Structure and function of the human Na(+)/H(+) exchanger isoform 1. Channels (Austin) 2(5):329

Khaled AR, Moor AN, Li A, Kim K, Ferris DK, Muegge K, Fisher RJ, Fliegel L, Durum SK (2001) Trophic factor withdrawal: p38 mitogen-activated protein kinase activates NHE1, which induces intracellular alkalinization. Mol Cell Biol 21(22):7545–7557. https://doi.org/10.1128/MCB.21.22.7545-7557.2001

Khan I (2001) Topology of the C-terminus of sodium hydrogen exchanger isoform-1: presence of an extracellular epitope. Arch Biochem Biophys 391(1):25–29. https://doi.org/10.1006/abbi.2001.2387

Khan S, Wu KL, Sedor JR, Abu Jawdeh BG, Schelling JR (2006) The NHE1 Na+/H+ exchanger regulates cell survival by activating and targeting ezrin to specific plasma membrane domains. Cell Mol Biol (Noisy-le-Grand) 52(8):115–121

Khayat W, Hackett A, Shaw M, Ilie A, Dudding-Byth T, Kalscheuer VM, Christie L, Corbett MA, Juusola J, Friend KL, Kirmse BM, Gecz J, Field M, Orlowski J (2019) A recurrent missense variant in SLC9A7 causes nonsyndromic X-linked intellectual disability with alteration of Golgi acidification and aberrant glycosylation. Hum Mol Genet 28(4):598–614. https://doi.org/10.1093/hmg/ddy371

Khurana S, Ganguly NK, Khullar M, Panigrahi D, Walia BN (1991) Studies on the mechanism of Salmonella typhimurium enterotoxin-induced diarrhoea. Biochim Biophys Acta 1097 (3):171–176

Kiela PR, Guner YS, Xu H, Collins JF, Ghishan FK (2000) Age- and tissue-specific induction of NHE3 by glucocorticoids in the rat small intestine. Am J Physiol Cell Physiol 278(4):C629–C637

Kiela PR, Hines ER, Collins JF, Ghishan FK (2001) Regulation of the rat NHE3 gene promoter by sodium butyrate. Am J Physiol Gastrointest Liver Physiol 281(4):G947–G956

Kiela PR, LeSueur J, Collins JF, Ghishan FK (2003) Transcriptional regulation of the rat NHE3 gene. Functional interactions between GATA-5 and Sp family transcription factors. J Biol Chem 278(8):5659–5668. https://doi.org/10.1074/jbc.M209473200

Kiela PR, Kuscuoglu N, Midura AJ, Midura-Kiela MT, Larmonier CB, Lipko M, Ghishan FK (2007) Molecular mechanism of rat NHE3 gene promoter regulation by sodium butyrate. Am J Physiol Cell Physiol 293(1):C64–C74. https://doi.org/10.1152/ajpcell.00277.2006

Kiela PR, Laubitz D, Larmonier CB, Midura-Kiela MT, Lipko MA, Janikashvili N, Bai A, Thurston R, Ghishan FK (2009) Changes in mucosal homeostasis predispose NHE3 knockout mice to increased susceptibility to DSS-induced epithelial injury. Gastroenterology 137 (3):965–975. https://doi.org/10.1053/j.gastro.2009.05.043

Kim J, Jung YS, Han W, Kim MY, Namkung W, Lee BH, Yi KY, Yoo SE, Lee MG, Kim KH (2007) Pharmacodynamic characteristics and cardioprotective effects of new NHE1 inhibitors. Eur J Pharmacol 567(1–2):131–138

Klanke CA, Su YR, Callen DF, Wang Z, Meneton P, Baird N, Kandasamy RA, Orlowski J, Otterud BE, Leppert M et al (1995) Molecular cloning and physical and genetic mapping of a novel human Na+/H+ exchanger (NHE5/SLC9A5) to chromosome 16q22.1. Genomics 25 (3):615–622

Kondapalli KC, Prasad H, Rao R (2014) An inside job: how endosomal Na(+)/H(+) exchangers link to autism and neurological disease. Front Cell Neurosci 8:172. https://doi.org/10.3389/fncel.2014.00172

Krishnan S, Ramakrishna BS, Binder HJ (1999) Stimulation of sodium chloride absorption from secreting rat colon by short-chain fatty acids. Dig Dis Sci 44(9):1924–1930

Kulanthaivel P, Leibach FH, Mahesh VB, Cragoe EJ Jr, Ganapathy V (1990) The Na(+)-H+ exchanger of the placental brush-border membrane is pharmacologically distinct from that of the renal brush-border membrane. J Biol Chem 265(3):1249–1252

Kurashima K, Yu FH, Cabado AG, Szabo EZ, Grinstein S, Orlowski J (1997) Identification of sites required for down-regulation of Na+/H+ exchanger NHE3 activity by cAMP-dependent protein kinase. Phosphorylation-dependent and -independent mechanisms. J Biol Chem 272 (45):28672–28679. https://doi.org/10.1074/jbc.272.45.28672

Kurashima K, Szabo EZ, Lukacs G, Orlowski J, Grinstein S (1998) Endosomal recycling of the Na +/H+ exchanger NHE3 isoform is regulated by the phosphatidylinositol 3-kinase pathway. J Biol Chem 273(33):20828–20836. https://doi.org/10.1074/jbc.273.33.20828

Kurashima K, D'Souza S, Szaszi K, Ramjeesingh R, Orlowski J, Grinstein S (1999) The apical Na (+)/H(+) exchanger isoform NHE3 is regulated by the actin cytoskeleton. J Biol Chem 274 (42):29843–29849

Labonte ED, Carreras CW, Leadbetter MR, Kozuka K, Kohler J, Koo-McCoy S, He L, Dy E, Black D, Zhong Z, Langsetmo I, Spencer AG, Bell N, Deshpande D, Navre M, Lewis JG, Jacobs JW, Charmot D (2015) Gastrointestinal inhibition of sodium-hydrogen exchanger 3 reduces phosphorus absorption and protects against vascular calcification in CKD. J Am Soc Nephrol 26(5):1138–1149. https://doi.org/10.1681/ASN.2014030317

Laghmani K, Borensztein P, Ambuhl P, Froissart M, Bichara M, Moe OW, Alpern RJ, Paillard M (1997) Chronic metabolic acidosis enhances NHE-3 protein abundance and transport activity in the rat thick ascending limb by increasing NHE-3 mRNA. J Clin Invest 99(1):24–30. https://doi.org/10.1172/JCI119128

Lang F, Ritter M, Gamper N, Huber S, Fillon S, Tanneur V, Lepple-Wienhues A, Szabo I, Gulbins E (2000) Cell volume in the regulation of cell proliferation and apoptotic cell death. Cell Physiol Biochem 10(5-6):417–428. https://doi.org/10.1159/000016367

Lang F, Shumilina E, Ritter M, Gulbins E, Vereninov A, Huber SM (2006) Ion channels and cell volume in regulation of cell proliferation and apoptotic cell death. Contrib Nephrol 152:142–160. https://doi.org/10.1159/000096321

Lapointe TK, O'Connor PM, Buret AG (2009) The role of epithelial malfunction in the pathogenesis of enteropathogenic *E. coli*-induced diarrhea. Lab Investig 89(9):964–970. https://doi.org/10.1038/labinvest.2009.69

Larmonier CB, Laubitz D, Thurston RD, Bucknam AL, Hill FM, Midura-Kiela M, Ramalingam R, Kiela PR, Ghishan FK (2011) NHE3 modulates the severity of colitis in IL-10-deficient mice. Am J Physiol Gastrointest Liver Physiol 300(6):G998–G1009. https://doi.org/10.1152/ajpgi.00073.2011

Larmonier CB, Laubitz D, Hill FM, Shehab KW, Lipinski L, Midura-Kiela MT, McFadden RM, Ramalingam R, Hassan KA, Golebiewski M, Besselsen DG, Ghishan FK, Kiela PR (2013) Reduced colonic microbial diversity is associated with colitis in NHE3-deficient mice. Am J

Physiol Gastrointest Liver Physiol 305(10):G667–G677. https://doi.org/10.1152/ajpgi.00189.2013
Laubitz D, Larmonier CB, Bai A, Midura-Kiela MT, Lipko MA, Thurston RD, Kiela PR, Ghishan FK (2008) Colonic gene expression profile in NHE3-deficient mice: evidence for spontaneous distal colitis. Am J Physiol Gastrointest Liver Physiol 295(1):G63–G77. https://doi.org/10.1152/ajpgi.90207.2008
Laubitz D, Harrison CA, Midura-Kiela MT, Ramalingam R, Larmonier CB, Chase JH, Caporaso JG, Besselsen DG, Ghishan FK, Kiela PR (2016) Reduced epithelial Na+/H+ exchange drives gut microbial dysbiosis and promotes inflammatory response in T cell-mediated murine colitis. PLoS One 11(4):e0152044. https://doi.org/10.1371/journal.pone.0152044
Lauritzen G, Stock CM, Lemaire J, Lund SF, Jensen MF, Damsgaard B, Petersen KS, Wiwel M, Ronnov-Jessen L, Schwab A, Pedersen SF (2012) The Na+/H+ exchanger NHE1, but not the Na+, HCO3(−) cotransporter NBCn1, regulates motility of MCF7 breast cancer cells expressing constitutively active ErbB2. Cancer Lett 317(2):172–183. https://doi.org/10.1016/j.canlet.2011.11.023
Ledoussal C, Lorenz JN, Nieman ML, Soleimani M, Schultheis PJ, Shull GE (2001a) Renal salt wasting in mice lacking NHE3 Na+/H+ exchanger but not in mice lacking NHE2. Am J Physiol Renal Physiol 281(4):F718–F727
Ledoussal C, Woo AL, Miller ML, Shull GE (2001b) Loss of the NHE2 Na(+)/H(+) exchanger has no apparent effect on diarrheal state of NHE3-deficient mice. Am J Physiol Gastrointest Liver Physiol 281(6):G1385–G1396. https://doi.org/10.1152/ajpgi.2001.281.6.G1385
Lee MG, Schultheis PJ, Yan M, Shull GE, Bookstein C, Chang E, Tse M, Donowitz M, Park K, Muallem S (1998) Membrane-limited expression and regulation of Na+-H+ exchanger isoforms by P2 receptors in the rat submandibular gland duct. J Physiol 513(Pt 2):341–357
Lee MG, Ahn W, Choi JY, Luo X, Seo JT, Schultheis PJ, Shull GE, Kim KH, Muallem S (2000) Na(+)-dependent transporters mediate HCO(3)(−) salvage across the luminal membrane of the main pancreatic duct. J Clin Invest 105(11):1651–1658. https://doi.org/10.1172/JCI9207
Lee JS, Lee YM, Kim JY, Park HW, Grinstein S, Orlowski J, Kim E, Kim KH, Lee MG (2010) BetaPix up-regulates Na+/H+ exchanger 3 through a Shank2-mediated protein-protein interaction. J Biol Chem 285(11):8104–8113. https://doi.org/10.1074/jbc.M109.055079
Lee-Kwon W, Johns DC, Cha B, Cavet M, Park J, Tsichlis P, Donowitz M (2001) Constitutively active phosphatidylinositol 3-kinase and AKT are sufficient to stimulate the epithelial Na+/H+ exchanger 3. J Biol Chem 276(33):31296–31304. https://doi.org/10.1074/jbc.M103900200
Lee-Kwon W, Kawano K, Choi JW, Kim JH, Donowitz M (2003) Lysophosphatidic acid stimulates brush border Na+/H+ exchanger 3 (NHE3) activity by increasing its exocytosis by an NHE3 kinase A regulatory protein-dependent mechanism. J Biol Chem 278(19):16494–16501
Lei X, Cai L, Li X, Xu H, Geng C, Wang C (2018) Up-regulation of NHE8 by somatostatin ameliorates the diarrhea symptom in infectious colitis mice model. Korean J Physiol Pharmacol 22(3):269–275. https://doi.org/10.4196/kjpp.2018.22.3.269
Lenzen H, Lunnemann M, Bleich A, Manns MP, Seidler U, Jorns A (2012) Downregulation of the NHE3-binding PDZ-adaptor protein PDZK1 expression during cytokine-induced inflammation in interleukin-10-deficient mice. PLoS One 7(7):e40657. https://doi.org/10.1371/journal.pone.0040657
Leung GP, Tse CM, Chew SB, Wong PY (2001) Expression of multiple Na+/H+ exchanger isoforms in cultured epithelial cells from rat efferent duct and cauda epididymidis. Biol Reprod 64(2):482–490. https://doi.org/10.1095/biolreprod64.2.482
Levine SA, Montrose MH, Tse CM, Donowitz M (1993) Kinetics and regulation of three cloned mammalian Na+/H+ exchangers stably expressed in a fibroblast cell line. J Biol Chem 268(34):25527–25535
Levine SA, Nath SK, Yun CH, Yip JW, Montrose M, Donowitz M, Tse CM (1995) Separate C-terminal domains of the epithelial specific brush border Na+/H+ exchanger isoform NHE3 are involved in stimulation and inhibition by protein kinases/growth factors. J Biol Chem 270(23):13716–13725

Li J, Eastman A (1995) Apoptosis in an interleukin-2-dependent cytotoxic T lymphocyte cell line is associated with intracellular acidification. Role of the Na(+)/H(+)-antiport. J Biol Chem 270 (7):3203–3211. https://doi.org/10.1074/jbc.270.7.3203

Li X, Galli T, Leu S, Wade JB, Weinman EJ, Leung G, Cheong A, Louvard D, Donowitz M (2001) Na^+-H^+ exchanger 3 (NHE3) is present in lipid rafts in the rabbit ileal brush border: a role for rafts in trafficking and rapid stimulation of NHE3. J Physiol 537(Pt 2):537–552

Li HC, Du Z, Barone S, Rubera I, McDonough AA, Tauc M, Zahedi K, Wang T, Soleimani M (2013) Proximal tubule specific knockout of the Na(+)/H(+) exchanger NHE3: effects on bicarbonate absorption and ammonium excretion. J Mol Med (Berl) 91(8):951–963. https://doi.org/10.1007/s00109-013-1015-3

Li X, Cai L, Xu H, Geng C, Lu J, Tao L, Sun D, Ghishan FK, Wang C (2016) Somatostatin regulates NHE8 protein expression via the ERK1/2 MAPK pathway in DSS-induced colitis mice. Am J Physiol Gastrointest Liver Physiol 311(5):G954–G963. https://doi.org/10.1152/ajpgi.00239.2016

Lin PJ, Williams WP, Luu Y, Molday RS, Orlowski J, Numata M (2005) Secretory carrier membrane proteins interact and regulate trafficking of the organellar (Na^+,K^+)/H^+ exchanger NHE7. J Cell Sci 118(Pt 9):1885–1897. https://doi.org/10.1242/jcs.02315

Lin PJ, Williams WP, Kobiljski J, Numata M (2007) Caveolins bind to (Na^+, K^+)/H^+ exchanger NHE7 by a novel binding module. Cell Signal 19(5):978–988. https://doi.org/10.1016/j.cellsig.2006.11.006

Linz D, Wirth K, Linz W, Heuer HO, Frick W, Hofmeister A, Heinelt U, Arndt P, Schwahn U, Bohm M, Ruetten H (2012) Antihypertensive and laxative effects by pharmacological inhibition of sodium-proton-exchanger subtype 3-mediated sodium absorption in the gut. Hypertension 60 (6):1560–1567. https://doi.org/10.1161/HYPERTENSIONAHA.112.201590

Liu C, Xu H, Zhang B, Johansson ME, Li J, Hansson GC, Ghishan FK (2013) NHE8 plays an important role in mucosal protection via its effect on bacterial adhesion. Am J Physiol Cell Physiol 305(1):C121–C128. https://doi.org/10.1152/ajpcell.00101.2013

Lorenz JN, Schultheis PJ, Traynor T, Shull GE, Schnermann J (1999) Micropuncture analysis of single-nephron function in NHE3-deficient mice. Am J Phys 277(3 Pt 2):F447–F453

Lucien F, Pelletier PP, Lavoie RR, Lacroix JM, Roy S, Parent JL, Arsenault D, Harper K, Dubois CM (2017) Hypoxia-induced mobilization of NHE6 to the plasma membrane triggers endosome hyperacidification and chemoresistance. Nat Commun 8:15884. https://doi.org/10.1038/ncomms15884

Lucioni A, Womack C, Musch MW, Rocha FL, Bookstein C, Chang EB (2002) Metabolic acidosis in rats increases intestinal NHE2 and NHE3 expression and function. Am J Physiol Gastrointest Liver Physiol 283(1):G51–G56. https://doi.org/10.1152/ajpgi.00529.2001

Ludwig FT, Schwab A, Stock C (2013) The Na^+/H^+-exchanger (NHE1) generates pH nanodomains at focal adhesions. J Cell Physiol 228(6):1351–1358. https://doi.org/10.1002/jcp.24293

Malakooti J, Dahdal RY, Schmidt L, Layden TJ, Dudeja PK, Ramaswamy K (1999) Molecular cloning, tissue distribution, and functional expression of the human Na(+)/H(+) exchanger NHE2. Am J Phys 277(2 Pt 1):G383–G390

Malakooti J, Dahdal RY, Dudeja PK, Layden TJ, Ramaswamy K (2001) The human Na(+)/H(+) exchanger NHE2 gene: genomic organization and promoter characterization. Am J Physiol Gastrointest Liver Physiol 280(4):G763–G773

Malakooti J, Memark VC, Dudeja PK, Ramaswamy K (2002) Molecular cloning and functional analysis of the human Na(+)/H(+) exchanger NHE3 promoter. Am J Physiol Gastrointest Liver Physiol 282(3):G491–G500. https://doi.org/10.1152/ajpgi.00273.2001

Malo ME, Fliegel L (2006) Physiological role and regulation of the Na^+/H^+ exchanger. Can J Physiol Pharmacol 84(11):1081–1095. https://doi.org/10.1139/y06-065

Marcinkiewicz M, Grabowska SZ, Czyzewska E (1998) Role of epidermal growth factor (EGF) in oesophageal mucosal integrity. Curr Med Res Opin 14(3):145–153. https://doi.org/10.1185/03007999809113354

Marteau C, Silviani V, Ducroc R, Crotte C, Gerolami A (1995) Evidence for apical Na^+/H^+ exchanger in bovine main pancreatic duct. Dig Dis Sci 40(11):2336–2340
Martinez A, Morse MD (2006) Infrared diode laser spectroscopy of jet-cooled NiCO, Ni(CO)3 (13CO), and Ni(CO)3(C 18O). J Chem Phys 124(12):124316. https://doi.org/10.1063/1.2180249
Masereel B, Pochet L, Laeckmann D (2003) An overview of inhibitors of Na(+)/H(+) exchanger. Eur J Med Chem 38(6):547–554
Masrati G, Dwivedi M, Rimon A, Gluck-Margolin Y, Kessel A, Ashkenazy H, Mayrose I, Padan E, Ben-Tal N (2018) Broad phylogenetic analysis of cation/proton antiporters reveals transport determinants. Nat Commun 9(1):4205. https://doi.org/10.1038/s41467-018-06770-5
Matthis AL, Kaji I, Engevik KA, Akiba Y, Kaunitz JD, Montrose MH, Aihara E (2019) Deficient active transport activity in healing mucosa after mild gastric epithelial damage. Dig Dis Sci. https://doi.org/10.1007/s10620-019-05825-x
Meima ME, Mackley JR, Barber DL (2007) Beyond ion translocation: structural functions of the sodium-hydrogen exchanger isoform-1. Curr Opin Nephrol Hypertens 16(4):365–372. https://doi.org/10.1097/MNH.0b013e3281bd888d
Meima ME, Webb BA, Witkowska HE, Barber DL (2009) The sodium-hydrogen exchanger NHE1 is an Akt substrate necessary for actin filament reorganization by growth factors. J Biol Chem 284(39):26666–26675. https://doi.org/10.1074/jbc.M109.019448
Mennone A, Biemesderfer D, Negoianu D, Yang CL, Abbiati T, Schultheis PJ, Shull GE, Aronson PS, Boyer JL (2001) Role of sodium/hydrogen exchanger isoform NHE3 in fluid secretion and absorption in mouse and rat cholangiocytes. Am J Physiol Gastrointest Liver Physiol 280(2): G247–G254
Michail S, Collins JF, Xu H, Kaufman S, Vanderhoof J, Ghishan FK (1998) Abnormal expression of brush-border membrane transporters in the duodenal mucosa of two patients with microvillus inclusion disease. J Pediatr Gastroenterol Nutr 27(5):536–542
Mihaila RG (2015) A minireview on NHE1 inhibitors. A rediscovered hope in oncohematology. Biomed Pap Med Fac Univ Palacky Olomouc Czech Repub 159(4):519–526. https://doi.org/10.5507/bp.2015.060
Miller RT, Counillon L, Pages G, Lifton RP, Sardet C, Pouyssegur J (1991) Structure of the 5′-flanking regulatory region and gene for the human growth factor-activatable Na/H exchanger NHE-1. J Biol Chem 266(17):10813–10819
Milosavljevic N, Monet M, Lena I, Brau F, Lacas-Gervais S, Feliciangeli S, Counillon L, Poet M (2014) The intracellular Na(+)/H(+) exchanger NHE7 effects a Na(+)-coupled, but not K(+)-coupled proton-loading mechanism in endocytosis. Cell Rep 7(3):689–696. https://doi.org/10.1016/j.celrep.2014.03.054
Mitchell P, Moyle J (1967) Acid-base titration across the membrane system of rat-liver mitochondria. Catalysis by uncouplers. Biochem J 104(2):588–600. https://doi.org/10.1042/bj1040588
Miyazaki E, Sakaguchi M, Wakabayashi S, Shigekawa M, Mihara K (2001) NHE6 protein possesses a signal peptide destined for endoplasmic reticulum membrane and localizes in secretory organelles of the cell. J Biol Chem 276(52):49221–49227. https://doi.org/10.1074/jbc.M106267200
Moeser AJ, Nighot PK, Ryan KA, Wooten JG, Blikslager AT (2006) Prostaglandin-mediated inhibition of Na+/H+ exchanger isoform 2 stimulates recovery of barrier function in ischemia-injured intestine. Am J Physiol Gastrointest Liver Physiol 291(5):G885–G894. https://doi.org/10.1152/ajpgi.00380.2005
Moeser AJ, Nighot PK, Ryan KA, Simpson JE, Clarke LL, Blikslager AT (2008) Mice lacking the Na+/H+ exchanger 2 have impaired recovery of intestinal barrier function. Am J Physiol Gastrointest Liver Physiol 295(4):G791–G797. https://doi.org/10.1152/ajpgi.00538.2007
Muller YL, Collins JF, Bai L, Xu H, Ghishan FK (1998) Molecular cloning and characterization of the rat NHE-2 gene promoter. Biochim Biophys Acta 1442(2–3):314–319
Muller T, Wijmenga C, Phillips AD, Janecke A, Houwen RH, Fischer H, Ellemunter H, Fruhwirth M, Offner F, Hofer S, Muller W, Booth IW, Heinz-Erian P (2000) Congenital sodium

diarrhea is an autosomal recessive disorder of sodium/proton exchange but unrelated to known candidate genes. Gastroenterology 119(6):1506–1513

Muller T, Rasool I, Heinz-Erian P, Mildenberger E, Hulstrunk C, Muller A, Michaud L, Koot BG, Ballauff A, Vodopiutz J, Rosipal S, Petersen BS, Franke A, Fuchs I, Witt H, Zoller H, Janecke AR, Visweswariah SS (2016) Congenital secretory diarrhoea caused by activating germline mutations in GUCY2C. Gut 65(8):1306–1313. https://doi.org/10.1136/gutjnl-2015-309441

Murtazina R, Booth BJ, Bullis BL, Singh DN, Fliegel L (2001) Functional analysis of polar amino-acid residues in membrane associated regions of the NHE1 isoform of the mammalian Na+/H+ exchanger. Eur J Biochem 268(17):4674–4685

Musch MW, Bookstein C, Xie Y, Sellin JH, Chang EB (2001) SCFA increase intestinal Na absorption by induction of NHE3 in rat colon and human intestinal C2/bbe cells. Am J Physiol Gastrointest Liver Physiol 280(4):G687–G693

Musch MW, Bookstein C, Rocha F, Lucioni A, Ren H, Daniel J, Xie Y, McSwine RL, Rao MC, Alverdy J, Chang EB (2002) Region-specific adaptation of apical Na/H exchangers after extensive proximal small bowel resection. Am J Physiol Gastrointest Liver Physiol 283(4): G975–G985

Musch MW, Lucioni A, Chang EB (2008) Aldosterone regulation of intestinal Na absorption involves SGK-mediated changes in NHE3 and Na^+ pump activity. Am J Physiol Gastrointest Liver Physiol 295(5):G909–G919. https://doi.org/10.1152/ajpgi.90312.2008

Nagase T, Seki N, Ishikawa K, Ohira M, Kawarabayasi Y, Ohara O, Tanaka A, Kotani H, Miyajima N, Nomura N (1996) Prediction of the coding sequences of unidentified human genes. VI. The coding sequences of 80 new genes (KIAA0201-KIAA0280) deduced by analysis of cDNA clones from cell line KG-1 and brain. DNA Res 3(5):321–329

Nagata H, Che XF, Miyazawa K, Tomoda A, Konishi M, Ubukata H, Tabuchi T (2011) Rapid decrease of intracellular pH associated with inhibition of Na+/H+ exchanger precedes apoptotic events in the MNK45 and MNK74 gastric cancer cell lines treated with 2-aminophenoxazine-3-one. Oncol Rep 25(2):341–346. https://doi.org/10.3892/or.2010.1082

Nakamura N, Tanaka S, Teko Y, Mitsui K, Kanazawa H (2005) Four Na+/H+ exchanger isoforms are distributed to Golgi and post-Golgi compartments and are involved in organelle pH regulation. J Biol Chem 280(2):1561–1572. https://doi.org/10.1074/jbc.M410041200

Narins SC, Park EH, Ramakrishnan R, Garcia FU, Diven JN, Balin BJ, Hammond CJ, Sodam BR, Smith PR, Abedin MZ (2004) Functional characterization of Na(+)/H(+) exchangers in primary cultures of prairie dog gallbladder. J Membr Biol 197(2):123–134. https://doi.org/10.1007/s00232-003-0647-6

Narins SC, Ramakrishnan R, Park EH, Smith PR, Meyers WC, Abedin MZ (2005) Gallbladder Na+/H+ exchange activity is up-regulated prior to cholesterol crystal formation. Eur J Clin Investig 35(8):514–522. https://doi.org/10.1111/j.1365-2362.2005.01520.x

Nath SK, Hang CY, Levine SA, Yun CH, Montrose MH, Donowitz M, Tse CM (1996) Hyperosmolarity inhibits the Na+/H+ exchanger isoforms NHE2 and NHE3: an effect opposite to that on NHE1. Am J Phys 270(3 Pt 1):G431–G441

Nehrke K (2003) A reduction in intestinal cell pHi due to loss of the *Caenorhabditis elegans* Na+/H+ exchanger NHX-2 increases life span. J Biol Chem 278(45):44657–44666. https://doi.org/10.1074/jbc.M307351200

Nehrke K, Melvin JE (2002) The NHX family of Na^+-H^+ exchangers in *Caenorhabditis elegans*. J Biol Chem 277(32):29036–29044. https://doi.org/10.1074/jbc.M203200200

Nighot PK, Moeser A, Ali RA, Blikslager AT, Koci MD (2010) Astrovirus infection induces sodium malabsorption and redistributes sodium hydrogen exchanger expression. Virology 401(2):146–154. https://doi.org/10.1016/j.virol.2010.02.004

Nowak P, Blaheta R, Schuller A, Cinatl J, Wimmer-Greinecker G, Moritz A, Scholz M (2004) The Na+/H+ exchange inhibitor HOE642 prevents stress-induced epithelial barrier dysfunction. Int J Mol Med 14(2):175–178

Numata M, Orlowski J (2001) Molecular cloning and characterization of a novel $(Na^+,K^+)/H^+$ exchanger localized to the trans-Golgi network. J Biol Chem 276(20):17387–17394. https://doi.org/10.1074/jbc.M101319200

Numata M, Petrecca K, Lake N, Orlowski J (1998) Identification of a mitochondrial Na+/H+ exchanger. J Biol Chem 273(12):6951–6959

Oberheide K, Puchkov D, Jentsch TJ (2017) Loss of the Na(+)/H(+) exchanger NHE8 causes male infertility in mice by disrupting acrosome formation. J Biol Chem 292(26):10845–10854. https://doi.org/10.1074/jbc.M117.784108

Odunewu-Aderibigbe A, Fliegel L (2014) The Na(+)/H(+) exchanger and pH regulation in the heart. IUBMB Life 66(10):679–685. https://doi.org/10.1002/iub.1323

Oehlke O, Sprysch P, Rickmann M, Roussa E (2006) Na(+)/H(+) exchanger isoforms are differentially regulated in rat submandibular gland during acid/base disturbances in vivo. Cell Tissue Res 323(2):253–262. https://doi.org/10.1007/s00441-005-0055-6

Ohgaki R, Fukura N, Matsushita M, Mitsui K, Kanazawa H (2008) Cell surface levels of organellar Na+/H+ exchanger isoform 6 are regulated by interaction with RACK1. J Biol Chem 283 (7):4417–4429. https://doi.org/10.1074/jbc.M705146200

Ohgaki R, Matsushita M, Kanazawa H, Ogihara S, Hoekstra D, van Ijzendoorn SC (2010) The Na+/H+ exchanger NHE6 in the endosomal recycling system is involved in the development of apical bile canalicular surface domains in HepG2 cells. Mol Biol Cell 21(7):1293–1304. https://doi.org/10.1091/mbc.E09-09-0767

Onishi I, Lin PJ, Numata Y, Austin P, Cipollone J, Roberge M, Roskelley CD, Numata M (2012) Organellar (Na+, K+)/H+ exchanger NHE7 regulates cell adhesion, invasion and anchorage-independent growth of breast cancer MDA-MB-231 cells. Oncol Rep 27(2):311–317. https://doi.org/10.3892/or.2011.1542

Opleta-Madsen K, Hardin J, Gall DG (1991) Epidermal growth factor upregulates intestinal electrolyte and nutrient transport. Am J Phys 260(6 Pt 1):G807–G814. https://doi.org/10.1152/ajpgi.1991.260.6.G807

Orlowski J, Grinstein S (2004) Diversity of the mammalian sodium/proton exchanger SLC9 gene family. Pflugers Arch 447(5):549–565. https://doi.org/10.1007/s00424-003-1110-3

Orlowski J, Kandasamy RA (1996) Delineation of transmembrane domains of the Na+/H+ exchanger that confer sensitivity to pharmacological antagonists. J Biol Chem 271 (33):19922–19927

Orlowski J, Kandasamy RA, Shull GE (1992) Molecular cloning of putative members of the Na/H exchanger gene family. cDNA cloning, deduced amino acid sequence, and mRNA tissue expression of the rat Na/H exchanger NHE-1 and two structurally related proteins. J Biol Chem 267(13):9331–9339

Paehler Vor der Nolte A, Chodisetti G, Yuan Z, Busch F, Riederer B, Luo M, Yu Y, Menon MB, Schneider A, Stripecke R, Nikolovska K, Yeruva S, Seidler U (2017) Na^+/H^+ exchanger NHE1 and NHE2 have opposite effects on migration velocity in rat gastric surface cells. J Cell Physiol 232(7):1669–1680. https://doi.org/10.1002/jcp.25758

Pang T, Su X, Wakabayashi S, Shigekawa M (2001) Calcineurin homologous protein as an essential cofactor for Na^+/H^+ exchangers. J Biol Chem 276(20):17367–17372. https://doi.org/10.1074/jbc.M100296200

Pang T, Hisamitsu T, Mori H, Shigekawa M, Wakabayashi S (2004) Role of calcineurin B homologous protein in pH regulation by the Na^+/H^+ exchanger 1: tightly bound Ca^{2+} ions as important structural elements. Biochemistry 43(12):3628–3636. https://doi.org/10.1021/bi0360004

Park K, Olschowka JA, Richardson LA, Bookstein C, Chang EB, Melvin JE (1999) Expression of multiple Na+/H+ exchanger isoforms in rat parotid acinar and ductal cells. Am J Phys 276(2): G470–G478. https://doi.org/10.1152/ajpgi.1999.276.2.G470

Park K, Evans RL, Watson GE, Nehrke K, Richardson L, Bell SM, Schultheis PJ, Hand AR, Shull GE, Melvin JE (2001) Defective fluid secretion and NaCl absorption in the parotid glands of Na

+/H+ exchanger-deficient mice. J Biol Chem 276(29):27042–27050. https://doi.org/10.1074/jbc.M102901200
Parker MD, Myers EJ, Schelling JR (2015) Na^+-H^+ exchanger-1 (NHE1) regulation in kidney proximal tubule. Cell Mol Life Sci 72(11):2061–2074. https://doi.org/10.1007/s00018-015-1848-8
Pathak BG, Shull GE, Jenkins NA, Copeland NG (1996) Mouse chromosomal location of four Na/H exchanger isoform genes. Genomics 31(2):261–263. https://doi.org/10.1006/geno.1996.0045
Pearse I, Zhu YX, Murray EJ, Dudeja PK, Ramaswamy K, Malakooti J (2007) Sp1 and Sp3 control constitutive expression of the human NHE2 promoter by interactions with the proximal promoter and the transcription initiation site. Biochem J 407(1):101–111. https://doi.org/10.1042/BJ20070364
Peng Y, Amemiya M, Yang X, Fan L, Moe OW, Yin H, Preisig PA, Yanagisawa M, Alpern RJ (2001) ET(B) receptor activation causes exocytic insertion of NHE3 in OKP cells. Am J Physiol Renal Physiol 280(1):F34–F42
Peti-Peterdi J, Chambrey R, Bebok Z, Biemesderfer D, St John PL, Abrahamson DR, Warnock DG, Bell PD (2000) Macula densa Na(+)/H(+) exchange activities mediated by apical NHE2 and basolateral NHE4 isoforms. Am J Physiol Renal Physiol 278(3):F452–F463. https://doi.org/10.1152/ajprenal.2000.278.3.F452
Petrecca K, Atanasiu R, Grinstein S, Orlowski J, Shrier A (1999) Subcellular localization of the Na^+/H^+ exchanger NHE1 in rat myocardium. Am J Phys 276(2):H709–H717. https://doi.org/10.1152/ajpheart.1999.276.2.H709
Pirojsakul K, Gattineni J, Dwarakanath V, Baum M (2015) Renal NHE expression and activity in neonatal NHE3- and NHE8-null mice. Am J Physiol Renal Physiol 308(1):F31–F38. https://doi.org/10.1152/ajprenal.00492.2014
Pizzonia JH, Biemesderfer D, Abu-Alfa AK, Wu MS, Exner M, Isenring P, Igarashi P, Aronson PS (1998) Immunochemical characterization of Na^+/H^+ exchanger isoform NHE4. Am J Phys 275 (4 Pt 2):F510–F517
Plopper GE, McNamee HP, Dike LE, Bojanowski K, Ingber DE (1995) Convergence of integrin and growth factor receptor signaling pathways within the focal adhesion complex. Mol Biol Cell 6(10):1349–1365. https://doi.org/10.1091/mbc.6.10.1349
Pouyssegur J, Sardet C, Franchi A, L'Allemain G, Paris S (1984) A specific mutation abolishing Na^+/H^+ antiport activity in hamster fibroblasts precludes growth at neutral and acidic pH. Proc Natl Acad Sci USA 81(15):4833–4837. https://doi.org/10.1073/pnas.81.15.4833
Preisig PA, Alpern RJ (1988) Chronic metabolic acidosis causes an adaptation in the apical membrane Na/H antiporter and basolateral membrane Na(HCO3)3 symporter in the rat proximal convoluted tubule. J Clin Invest 82(4):1445–1453. https://doi.org/10.1172/JCI113750
Putney LK, Barber DL (2004) Expression profile of genes regulated by activity of the Na-H exchanger NHE1. BMC Genomics 5(1):46. https://doi.org/10.1186/1471-2164-5-46
Putney LK, Denker SP, Barber DL (2002) The changing face of the Na+/H+ exchanger, NHE1: structure, regulation, and cellular actions. Annu Rev Pharmacol Toxicol 42:527–552. https://doi.org/10.1146/annurev.pharmtox.42.092001.143801
Ramakrishna BS, Nance SH, Roberts-Thomson IC, Roediger WE (1990) The effects of enterotoxins and short-chain fatty acids on water and electrolyte fluxes in ileal and colonic loops in vivo in the rat. Digestion 45(2):93–101
Ramakrishna BS, Venkataraman S, Srinivasan P, Dash P, Young GP, Binder HJ (2000) Amylase-resistant starch plus oral rehydration solution for cholera. N Engl J Med 342(5):308–313. https://doi.org/10.1056/NEJM200002033420502
Reshkin SJ, Cardone RA, Harguindey S (2013) Na^+-H^+ exchanger, pH regulation and cancer. Recent Pat Anticancer Drug Discov 8(1):85–99
Rivarola V, Di Giusto G, Christensen MJ, Ford P, Capurro C (2017) AQP2-induced acceleration of renal cell proliferation involves the activation of a regulatory volume increase mechanism dependent on NHE2. J Cell Biochem 118(5):967–978. https://doi.org/10.1002/jcb.25602

Rocha F, Musch MW, Lishanskiy L, Bookstein C, Sugi K, Xie Y, Chang EB (2001) IFN-gamma downregulates expression of Na(+)/H(+) exchangers NHE2 and NHE3 in rat intestine and human Caco-2/bbe cells. Am J Physiol Cell Physiol 280(5):C1224–C1232

Roger S (2019) Na(+) /H(+) exchanger type-1: a shepherd for cellular transhumance. J Physiol 597 (3):661–662. https://doi.org/10.1113/JP277472

Rossmann H, Sonnentag T, Heinzmann A, Seidler B, Bachmann O, Vieillard-Baron D, Gregor M, Seidler U (2001) Differential expression and regulation of Na(+)/H(+) exchanger isoforms in rabbit parietal and mucous cells. Am J Physiol Gastrointest Liver Physiol 281(2):G447–G458. https://doi.org/10.1152/ajpgi.2001.281.2.G447

Rotte A, Pasham V, Eichenmuller M, Mahmud H, Xuan NT, Shumilina E, Gotz F, Lang F (2010) Effect of bacterial lipopolysaccharide on Na(+)/H(+) exchanger activity in dendritic cells. Cell Physiol Biochem 26(4–5):553–562. https://doi.org/10.1159/000322323

Roussa E, Alper SL, Thevenod F (2001) Immunolocalization of anion exchanger AE2, Na(+)/H(+) exchangers NHE1 and NHE4, and vacuolar type H(+)-ATPase in rat pancreas. J Histochem Cytochem 49(4):463–474. https://doi.org/10.1177/002215540104900406

Ruppin H, Bar-Meir S, Soergel KH, Wood CM, Schmitt MG Jr (1980) Absorption of short-chain fatty acids by the colon. Gastroenterology 78(6):1500–1507

Sacks AI, Acra SA, Dykes W, Polk DB, Barnard JA, Pietsch J, Ghishan FK (1993) Intestinal Na+/H + exchanger activity is up-regulated by bowel resection in the weanling rat. Pediatr Res 33 (3):215–220

Saier MH Jr, Reddy VS, Tsu BV, Ahmed MS, Li C, Moreno-Hagelsieb G (2016) The transporter classification database (TCDB): recent advances. Nucleic Acids Res 44(D1):D372–D379. https://doi.org/10.1093/nar/gkv1103

Sandle GI (2005) Pathogenesis of diarrhea in ulcerative colitis: new views on an old problem. J Clin Gastroenterol 39(4 Suppl 2):S49–S52

Sardet C, Franchi A, Pouyssegur J (1988) Molecular cloning of the growth-factor-activatable human Na+/H+ antiporter. Cold Spring Harb Symp Quant Biol 53(Pt 2):1011–1018

Sardet C, Franchi A, Pouyssegur J (1989) Molecular cloning, primary structure, and expression of the human growth factor-activatable Na^+/H^+ antiporter. Cell 56(2):271–280. https://doi.org/10.1016/0092-8674(89)90901-x

Sarker R, Gronborg M, Cha B, Mohan S, Chen Y, Pandey A, Litchfield D, Donowitz M, Li X (2008) Casein kinase 2 binds to the C terminus of Na^+/H^+ exchanger 3 (NHE3) and stimulates NHE3 basal activity by phosphorylating a separate site in NHE3. Mol Biol Cell 19 (9):3859–3870. https://doi.org/10.1091/mbc.E08-01-0019

Schelling JR (2016) Tubular atrophy in the pathogenesis of chronic kidney disease progression. Pediatr Nephrol 31(5):693–706. https://doi.org/10.1007/s00467-015-3169-4

Schelling JR, Abu Jawdeh BG (2008) Regulation of cell survival by Na+/H+ exchanger-1. Am J Physiol Renal Physiol 295(3):F625–F632. https://doi.org/10.1152/ajprenal.90212.2008

Schelling JR, Nkemere N, Kopp JB, Cleveland RP (1998) Fas-dependent fratricidal apoptosis is a mechanism of tubular epithelial cell deletion in chronic renal failure. Lab Investig 78 (7):813–824

Schultheis PJ, Clarke LL, Meneton P, Harline M, Boivin GP, Stemmermann G, Duffy JJ, Doetschman T, Miller ML, Shull GE (1998a) Targeted disruption of the murine Na+/H+ exchanger isoform 2 gene causes reduced viability of gastric parietal cells and loss of net acid secretion. J Clin Invest 101(6):1243–1253. https://doi.org/10.1172/JCI1249

Schultheis PJ, Clarke LL, Meneton P, Miller ML, Soleimani M, Gawenis LR, Riddle TM, Duffy JJ, Doetschman T, Wang T, Giebisch G, Aronson PS, Lorenz JN, Shull GE (1998b) Renal and intestinal absorptive defects in mice lacking the NHE3 Na+/H+ exchanger. Nat Genet 19 (3):282–285. https://doi.org/10.1038/969

Schwark JR, Jansen HW, Lang HJ, Krick W, Burckhardt G, Hropot M (1998) S3226, a novel inhibitor of Na+/H+ exchanger subtype 3 in various cell types. Pflugers Arch 436(5):797–800

Seidler U, Lenzen H, Cinar A, Tessema T, Bleich A, Riederer B (2006) Molecular mechanisms of disturbed electrolyte transport in intestinal inflammation. Ann N Y Acad Sci 1072:262–275. https://doi.org/10.1196/annals.1326.024

Shallat S, Schmidt L, Reaka A, Rao D, Chang EB, Rao MC, Ramaswamy K, Layden TJ (1995) NHE-1 isoform of the Na+/H+ antiport is expressed in the rat and rabbit esophagus. Gastroenterology 109(5):1421–1428

Shrode LD, Gan BS, D'Souza SJ, Orlowski J, Grinstein S (1998) Topological analysis of NHE1, the ubiquitous Na+/H+ exchanger using chymotryptic cleavage. Am J Phys 275(2 Pt 1):C431–C439

Siddique I, Khan I (2003) Regulation of Na/H exchanger-1 in gastroesophageal reflux disease: possible interaction of histamine receptor. Dig Dis Sci 48(9):1832–1838

Siddique I, Hasan F, Khan I (2009) Suppression of Na+/H+ exchanger isoform-3 in human inflammatory bowel disease: lack of reversal by 5′-aminosalicylate treatment. Scand J Gastroenterol 44(1):56–64. https://doi.org/10.1080/00365520802321253

Silva PH, Girardi AC, Neri EA, Reboucas NA (2012) Distinct mechanisms underlie adaptation of proximal tubule Na+/H+ exchanger isoform 3 in response to chronic metabolic and respiratory acidosis. Pflugers Arch 463(5):703–714. https://doi.org/10.1007/s00424-012-1092-0

Silviani V, Colombani V, Heyries L, Gerolami A, Cartouzou G, Marteau C (1996) Role of the NHE3 isoform of the Na+/H+ exchanger in sodium absorption by the rabbit gallbladder. Pflugers Arch 432(5):791–796

Simonin A, Fuster D (2010) Nedd4-1 and beta-arrestin-1 are key regulators of Na+/H+ exchanger 1 ubiquitylation, endocytosis, and function. J Biol Chem 285(49):38293–38303. https://doi.org/10.1074/jbc.M110.115089

Sinajon P, Verbaan D, So J (2016) The expanding phenotypic spectrum of female SLC9A6 mutation carriers: a case series and review of the literature. Hum Genet 135(8):841–850. https://doi.org/10.1007/s00439-016-1675-5

Slepkov E, Fliegel L (2002) Structure and function of the NHE1 isoform of the Na+/H+ exchanger. Biochem Cell Biol 80(5):499–508

Slepkov ER, Chow S, Lemieux MJ, Fliegel L (2004) Proline residues in transmembrane segment IV are critical for activity, expression and targeting of the Na+/H+ exchanger isoform 1. Biochem J 379(Pt 1):31–38

Snabaitis AK, Cuello F, Avkiran M (2008) Protein kinase B/Akt phosphorylates and inhibits the cardiac Na+/H+ exchanger NHE1. Circ Res 103(8):881–890. https://doi.org/10.1161/CIRCRESAHA.108.175877

Soleimani M, Singh G, Bizal GL, Gullans SR, McAteer JA (1994) Na+/H+ exchanger isoforms NHE-2 and NHE-1 in inner medullary collecting duct cells. Expression, functional localization, and differential regulation. J Biol Chem 269(45):27973–27978

Soleimani M, Singh G, Bookstein C, Rao MC, Chang EB, Dominguez JH (1996) Inhibition of glycosylation decreases Na+/H+ exchange activity, blocks NHE-3 transport to the membrane, and increases NHE-3 mRNA expression in LLC-PK1 cells. J Lab Clin Med 127(6):565–573

Sonnentag T, Siegel WK, Bachmann O, Rossmann H, Mack A, Wagner HJ, Gregor M, Seidler U (2000) Agonist-induced cytoplasmic volume changes in cultured rabbit parietal cells. Am J Physiol Gastrointest Liver Physiol 279(1):G40–G48

Spencer AG, Labonte ED, Rosenbaum DP, Plato CF, Carreras CW, Leadbetter MR, Kozuka K, Kohler J, Koo-McCoy S, He L, Bell N, Tabora J, Joly KM, Navre M, Jacobs JW, Charmot D (2014) Intestinal inhibition of the Na+/H+ exchanger 3 prevents cardiorenal damage in rats and inhibits Na+ uptake in humans. Sci Transl Med 6(227):227–236. https://doi.org/10.1126/scitranslmed.3007790

Stock C, Pedersen SF (2017) Roles of pH and the Na(+)/H(+) exchanger NHE1 in cancer: from cell biology and animal models to an emerging translational perspective? Semin Cancer Biol 43:5–16. https://doi.org/10.1016/j.semcancer.2016.12.001

Stock C, Schwab A (2006) Role of the Na/H exchanger NHE1 in cell migration. Acta Physiol (Oxf) 187(1–2):149–157. https://doi.org/10.1111/j.1748-1716.2006.01543.x

Stock C, Gassner B, Hauck CR, Arnold H, Mally S, Eble JA, Dieterich P, Schwab A (2005) Migration of human melanoma cells depends on extracellular pH and Na+/H+ exchange. J Physiol 567(Pt 1):225–238. https://doi.org/10.1113/jphysiol.2005.088344

Su X, Pang T, Wakabayashi S, Shigekawa M (2003) Evidence for involvement of the putative first extracellular loop in differential volume sensitivity of the Na+/H+ exchangers NHE1 and NHE2. Biochemistry 42(4):1086–1094. https://doi.org/10.1021/bi020427d

Su HW, Wang SW, Ghishan FK, Kiela PR, Tang MJ (2009) Cell confluency-induced Stat3 activation regulates NHE3 expression by recruiting Sp1 and Sp3 to the proximal NHE3 promoter region during epithelial dome formation. Am J Physiol Cell Physiol 296(1):C13–C24. https://doi.org/10.1152/ajpcell.00263.2008

Subramanya SB, Rajendran VM, Srinivasan P, Nanda Kumar NS, Ramakrishna BS, Binder HJ (2007) Differential regulation of cholera toxin-inhibited Na-H exchange isoforms by butyrate in rat ileum. Am J Physiol Gastrointest Liver Physiol 293(4):G857–G863. https://doi.org/10.1152/ajpgi.00462.2006

Sullivan S, Alex P, Dassopoulos T, Zachos NC, Iacobuzio-Donahue C, Donowitz M, Brant SR, Cuffari C, Harris ML, Datta LW, Conklin L, Chen Y, Li X (2009) Downregulation of sodium transporters and NHERF proteins in IBD patients and mouse colitis models: potential contributors to IBD-associated diarrhea. Inflamm Bowel Dis 15(2):261–274. https://doi.org/10.1002/ibd.20743

Sun AM, Liu Y, Dworkin LD, Tse CM, Donowitz M, Yip KP (1997) Na^+/H^+ exchanger isoform 2 (NHE2) is expressed in the apical membrane of the medullary thick ascending limb. J Membr Biol 160(1):85–90

Sun AM, Liu Y, Centracchio J, Dworkin LD (1998) Expression of Na+/H+ exchanger isoforms in inner segment of inner medullary collecting duct. J Membr Biol 164(3):293–300. https://doi.org/10.1007/s002329900414

Szaszi K, Grinstein S, Orlowski J, Kapus A (2000a) Regulation of the epithelial Na(+)/H(+) exchanger isoform by the cytoskeleton. Cell Physiol Biochem 10(5–6):265–272

Szaszi K, Kurashima K, Kapus A, Paulsen A, Kaibuchi K, Grinstein S, Orlowski J (2000b) RhoA and rho kinase regulate the epithelial Na+/H+ exchanger NHE3. Role of myosin light chain phosphorylation. J Biol Chem 275(37):28599–28606

Szpirer C, Szpirer J, Riviere M, Levan G, Orlowski J (1994) Chromosomal assignment of four genes encoding Na/H exchanger isoforms in human and rat. Mamm Genome 5(3):153–159. https://doi.org/10.1007/bf00352346

Tamtaji OR, Mirzaei H, Shamshirian A, Shamshirian D, Behnam M, Asemi Z (2019) New trends in glioma cancer therapy: targeting Na(+)/H (+) exchangers. J Cell Physiol. https://doi.org/10.1002/jcp.29014

Touret N, Poujeol P, Counillon L (2001) Second-site revertants of a low-sodium-affinity mutant of the Na+/H+ exchanger reveal the participation of TM4 into a highly constrained sodium-binding site. Biochemistry 40(16):5095–5101

Tse CM, Brant SR, Walker MS, Pouyssegur J, Donowitz M (1992) Cloning and sequencing of a rabbit cDNA encoding an intestinal and kidney-specific Na+/H+ exchanger isoform (NHE-3). J Biol Chem 267(13):9340–9346

Tse CM, Levine SA, Yun CH, Montrose MH, Little PJ, Pouyssegur J, Donowitz M (1993) Cloning and expression of a rabbit cDNA encoding a serum-activated ethylisopropylamiloride-resistant epithelial Na+/H+ exchanger isoform (NHE-2). J Biol Chem 268(16):11917–11924

Tse CM, Levine SA, Yun CH, Khurana S, Donowitz M (1994) Na^+/H^+ exchanger-2 is an O-linked but not an N-linked sialoglycoprotein. Biochemistry 33(44):12954–12961

Tzschach A, Ullmann R, Ahmed A, Martin T, Weber G, Decker-Schwering O, Pauly F, Shamdeen MG, Reith W, Oehl-Jaschkowitz B (2011) Christianson syndrome in a patient with an interstitial Xq26.3 deletion. Am J Med Genet A 155A(11):2771–2774. https://doi.org/10.1002/ajmg.a.34230

Uria-Avellanal C, Robertson NJ (2014) Na(+)/H(+) exchangers and intracellular pH in perinatal brain injury. Transl Stroke Res 5(1):79–98. https://doi.org/10.1007/s12975-013-0322-x

Valles PG, Bocanegra V, Gil Lorenzo A, Costantino VV (2015) Physiological functions and regulation of the Na^+/H^+ exchanger [NHE1] in renal tubule epithelial cells. Kidney Blood Press Res 40(5):452–466. https://doi.org/10.1159/000368521
Verma V, Bali A, Singh N, Jaggi AS (2015) Implications of sodium hydrogen exchangers in various brain diseases. J Basic Clin Physiol Pharmacol 26(5):417–426. https://doi.org/10.1515/jbcpp-2014-0117
Wakabayashi S, Ikeda T, Iwamoto T, Pouyssegur J, Shigekawa M (1997) Calmodulin-binding autoinhibitory domain controls "pH-sensing" in the Na+/H+ exchanger NHE1 through sequence-specific interaction. Biochemistry 36(42):12854–12861. https://doi.org/10.1021/bi9715472
Wakabayashi S, Pang T, Su X, Shigekawa M (2000a) A novel topology model of the human Na(+)/H(+) exchanger isoform 1. J Biol Chem 275(11):7942–7949
Wakabayashi S, Pang T, Su X, Shigekawa M (2000b) Second mutations rescue point mutant of the Na(+)/H(+) exchanger NHE1 showing defective surface expression. FEBS Lett 487(2):257–261
Wakabayashi S, Hisamitsu T, Pang T, Shigekawa M (2003) Kinetic dissection of two distinct proton binding sites in Na+/H+ exchangers by measurement of reverse mode reaction. J Biol Chem 278(44):43580–43585. https://doi.org/10.1074/jbc.M306690200
Wakabayashi S, Hisamitsu T, Nakamura TY (2013) Regulation of the cardiac Na(+)/H(+) exchanger in health and disease. J Mol Cell Cardiol 61:68–76. https://doi.org/10.1016/j.yjmcc.2013.02.007
Wang Z, Orlowski J, Shull GE (1993) Primary structure and functional expression of a novel gastrointestinal isoform of the rat Na/H exchanger. J Biol Chem 268(16):11925–11928
Wang D, Balkovetz DF, Warnock DG (1995) Mutational analysis of transmembrane histidines in the amiloride-sensitive Na+/H+ exchanger. Am J Phys 269(2 Pt 1):C392–C402
Wang T, Yang CL, Abbiati T, Schultheis PJ, Shull GE, Giebisch G, Aronson PS (1999) Mechanism of proximal tubule bicarbonate absorption in NHE3 null mice. Am J Phys 277(2 Pt 2):F298–F302
Wang XF, Yu MK, Lam SY, Leung KM, Jiang JL, Leung PS, Ko WH, Leung PY, Chew SB, Liu CQ, Tse CM, Chan HC (2003) Expression, immunolocalization, and functional activity of Na+/H+ exchanger isoforms in mouse endometrial epithelium. Biol Reprod 68(1):302–308
Wang D, Sun H, Lang F, Yun CC (2005) Activation of NHE3 by dexamethasone requires phosphorylation of NHE3 at Ser663 by SGK1. Am J Physiol Cell Physiol 289(4):C802–C810
Wang C, Xu H, Chen H, Li J, Zhang B, Tang C, Ghishan FK (2011) Somatostatin stimulates intestinal NHE8 expression via p38 MAPK pathway. Am J Physiol Cell Physiol 300(2):C375–C382. https://doi.org/10.1152/ajpcell.00421.2010
Wang A, Li J, Zhao Y, Johansson ME, Xu H, Ghishan FK (2015) Loss of NHE8 expression impairs intestinal mucosal integrity. Am J Physiol Gastrointest Liver Physiol 309(11):G855–G864. https://doi.org/10.1152/ajpgi.00278.2015
West IC, Mitchell P (1974) Proton/sodium ion antiport in *Escherichia coli*. Biochem J 144(1):87–90. https://doi.org/10.1042/bj1440087
Winkel GK, Sardet C, Pouyssegur J, Ives HE (1993) Role of cytoplasmic domain of the Na+/H+ exchanger in hormonal activation. J Biol Chem 268(5):3396–3400
Woo AL, Noonan WT, Schultheis PJ, Neumann JC, Manning PA, Lorenz JN, Shull GE (2003) Renal function in NHE3-deficient mice with transgenic rescue of small intestinal absorptive defect. Am J Physiol Renal Physiol 284(6):F1190–F1198. https://doi.org/10.1152/ajprenal.00418.2002
Wu D, Kraut JA (2014) Role of NHE1 in the cellular dysfunction of acute metabolic acidosis. Am J Nephrol 40(1):36–42. https://doi.org/10.1159/000364783
Wu KL, Khan S, Lakhe-Reddy S, Wang L, Jarad G, Miller RT, Konieczkowski M, Brown AM, Sedor JR, Schelling JR (2003) Renal tubular epithelial cell apoptosis is associated with caspase cleavage of the NHE1 Na+/H+ exchanger. Am J Physiol Renal Physiol 284(4):F829–F839. https://doi.org/10.1152/ajprenal.00314.2002

Wu KL, Khan S, Lakhe-Reddy S, Jarad G, Mukherjee A, Obejero-Paz CA, Konieczkowski M, Sedor JR, Schelling JR (2004) The NHE1 Na+/H+ exchanger recruits ezrin/radixin/moesin proteins to regulate Akt-dependent cell survival. J Biol Chem 279(25):26280–26286. https://doi.org/10.1074/jbc.M400814200

Xia CH, Liu H, Wang M, Cheung D, Park A, Yang Y, Du X, Chang B, Beutler B, Gong X (2006) Characterization of mouse mutants with abnormal RPE cells. Adv Exp Med Biol 572:95–100. https://doi.org/10.1007/0-387-32442-9_15

Xia CH, Liu H, Cheung D, Tang F, Chang B, Li M, Gong X (2015) NHE8 is essential for RPE cell polarity and photoreceptor survival. Sci Rep 5:9358. https://doi.org/10.1038/srep09358

Xia CH, Ferguson I, Li M, Kim A, Onishi A, Li L, Su B, Gong X (2018) Essential function of NHE8 in mouse retina demonstrated by AAV-mediated CRISPR/Cas9 knockdown. Exp Eye Res 176:29–39. https://doi.org/10.1016/j.exer.2018.06.026

Xiang M, Feng M, Muend S, Rao R (2007) A human Na+/H+ antiporter sharing evolutionary origins with bacterial NhaA may be a candidate gene for essential hypertension. Proc Natl Acad Sci USA 104(47):18677–18681. https://doi.org/10.1073/pnas.0707120104

Xie R, Wang H, Jin H, Wen G, Tuo B, Xu J (2017) NHE1 is upregulated in gastric cancer and regulates gastric cancer cell proliferation, migration and invasion. Oncol Rep 37(3):1451–1460. https://doi.org/10.3892/or.2017.5386

Xinhan L, Matsushita M, Numaza M, Taguchi A, Mitsui K, Kanazawa H (2011) Na+/H+ exchanger isoform 6 (NHE6/SLC9A6) is involved in clathrin-dependent endocytosis of transferrin. Am J Physiol Cell Physiol 301(6):C1431–C1444. https://doi.org/10.1152/ajpcell.00154.2011

Xu H, Collins JF, Bai L, Kiela PR, Lynch RM, Ghishan FK (2001) Epidermal growth factor regulation of rat NHE2 gene expression. Am J Physiol Cell Physiol 281(2):C504–C513

Xu H, Chen R, Ghishan FK (2005) Subcloning, localization, and expression of the rat intestinal sodium-hydrogen exchanger isoform 8. Am J Physiol Gastrointest Liver Physiol 289(1):G36–G41. https://doi.org/10.1152/ajpgi.00552.2004

Xu H, Chen H, Dong J, Lynch R, Ghishan FK (2008) Gastrointestinal distribution and kinetic characterization of the sodium-hydrogen exchanger isoform 8 (NHE8). Cell Physiol Biochem 21(1-3):109–116. https://doi.org/10.1159/000113752

Xu H, Chen H, Dong J, Li J, Chen R, Uno JK, Ghishan FK (2009) Tumor necrosis factor-α downregulates intestinal NHE8 expression by reducing basal promoter activity. Am J Physiol Cell Physiol 296(3):C489–C497. https://doi.org/10.1152/ajpcell.00482.2008

Xu H, Zhang B, Li J, Chen H, Tooley J, Ghishan FK (2010a) Epidermal growth factor inhibits intestinal NHE8 expression via reducing its basal transcription. Am J Physiol Cell Physiol 299(1):C51–C57. https://doi.org/10.1152/ajpcell.00081.2010

Xu H, Zhang B, Li J, Chen H, Wang C, Ghishan FK (2010b) Transcriptional inhibition of intestinal NHE8 expression by glucocorticoids involves Pax5. Am J Physiol Gastrointest Liver Physiol 299(4):G921–G927. https://doi.org/10.1152/ajpgi.00227.2010

Xu H, Li J, Chen R, Zhang B, Wang C, King N, Chen H, Ghishan FK (2011) NHE2X3 DKO mice exhibit gender-specific NHE8 compensation. Am J Physiol Gastrointest Liver Physiol 300(4):G647–G653. https://doi.org/10.1152/ajpgi.00546.2010

Xu H, Zhang B, Li J, Wang C, Chen H, Ghishan FK (2012) Impaired mucin synthesis and bicarbonate secretion in the colon of NHE8 knockout mice. Am J Physiol Gastrointest Liver Physiol 303(3):G335–G343. https://doi.org/10.1152/ajpgi.00146.2012

Xu H, Li J, Chen H, Wang C, Ghishan FK (2013) NHE8 plays important roles in gastric mucosal protection. Am J Physiol Gastrointest Liver Physiol 304(3):G257–G261. https://doi.org/10.1152/ajpgi.00433.2012

Xu H, Chen H, Li J, Zhao Y, Ghishan FK (2015a) Disruption of NHE8 expression impairs Leydig cell function in the testes. Am J Physiol Cell Physiol 308(4):C330–C338. https://doi.org/10.1152/ajpcell.00289.2014

Xu H, McCoy A, Li J, Zhao Y, Ghishan FK (2015b) Sodium butyrate stimulates NHE8 expression via its role on activating NHE8 basal promoter activity. Am J Physiol Gastrointest Liver Physiol 309(6):G500–G505. https://doi.org/10.1152/ajpgi.00194.2015

Xu H, Zhao Y, Li J, Wang M, Lian F, Gao M, Ghishan FK (2015c) Loss of NHE8 expression impairs ocular surface function in mice. Am J Physiol Cell Physiol 308(1):C79–C87. https://doi.org/10.1152/ajpcell.00296.2014
Xu H, Li Q, Zhao Y, Li J, Ghishan FK (2016) Intestinal NHE8 is highly expressed in goblet cells and its expression is subject to TNF-alpha regulation. Am J Physiol Gastrointest Liver Physiol 310(2):G64–G69. https://doi.org/10.1152/ajpgi.00367.2015
Xu H, Li J, Chen H, Ghishan FK (2019) NHE8 deficiency promotes colitis-associated Cancer in mice via expansion of Lgr5-expressing cells. Cell Mol Gastroenterol Hepatol 7(1):19–31. https://doi.org/10.1016/j.jcmgh.2018.08.005
Xue L, Aihara E, Wang TC, Montrose MH (2011) Trefoil factor 2 requires Na/H exchanger 2 activity to enhance mouse gastric epithelial repair. J Biol Chem 286(44):38375–38382. https://doi.org/10.1074/jbc.M111.268219
Yanaka A, Suzuki H, Shibahara T, Matsui H, Nakahara A, Tanaka N (2002) EGF promotes gastric mucosal restitution by activating Na(+)/H(+) exchange of epithelial cells. Am J Physiol Gastrointest Liver Physiol 282(5):G866–G876
Yang W, Wang H, Fliegel L (1996) Regulation of Na+/H+ exchanger gene expression. Role of a novel poly(dA.dT) element in regulation of the NHE1 promoter. J Biol Chem 271 (34):20444–20449
Yang X, Chen J, Bai H, Tao K, Zhou Q, Hou H, Hu D (2013) Inhibition of Na+/H+ exchanger 1 by cariporide reduces burn-induced intestinal barrier breakdown. Burns 39(8):1557–1564. https://doi.org/10.1016/j.burns.2013.04.007
Ye G, Chen C, Han D, Xiong X, Kong Y, Wan B, Yu L (2006) Cloning of a novel human NHEDC1 (Na+/H+ exchanger like domain containing 1) gene expressed specifically in testis. Mol Biol Rep 33(3):175–180. https://doi.org/10.1007/s11033-006-0010-y
Yeruva S, Farkas K, Hubricht J, Rode K, Riederer B, Bachmann O, Cinar A, Rakonczay Z, Molnar T, Nagy F, Wedemeyer J, Manns M, Raddatz D, Musch MW, Chang EB, Hegyi P, Seidler U (2010) Preserved Na(+)/H(+) exchanger isoform 3 expression and localization, but decreased NHE3 function indicate regulatory sodium transport defect in ulcerative colitis. Inflamm Bowel Dis 16(7):1149–1161. https://doi.org/10.1002/ibd.21183
Yu FH, Shull GE, Orlowski J (1993) Functional properties of the rat Na/H exchanger NHE-2 isoform expressed in Na/H exchanger-deficient Chinese hamster ovary cells. J Biol Chem 268 (34):25536–25541
Yun CC (2003) Concerted roles of SGK1 and the Na+/H+ exchanger regulatory factor 2 (NHERF2) in regulation of NHE3. Cell Physiol Biochem 13(1):29–40
Yun CH, Oh S, Zizak M, Steplock D, Tsao S, Tse CM, Weinman EJ, Donowitz M (1997) cAMP-mediated inhibition of the epithelial brush border Na+/H+ exchanger, NHE3, requires an associated regulatory protein. Proc Natl Acad Sci USA 94(7):3010–3015. https://doi.org/10.1073/pnas.94.7.3010
Yun CC, Chen Y, Lang F (2002) Glucocorticoid activation of Na(+)/H(+) exchanger isoform 3 revisited. The roles of SGK1 and NHERF2. J Biol Chem 277(10):7676–7683
Zachos NC, Tse M, Donowitz M (2005) Molecular physiology of intestinal Na+/H+ exchange. Annu Rev Physiol 67:411–443. https://doi.org/10.1146/annurev.physiol.67.031103.153004
Zachos NC, Kovbasnjuk O, Donowitz M (2009) Regulation of intestinal electroneutral sodium absorption and the brush border Na+/H+ exchanger by intracellular calcium. Ann N Y Acad Sci 1165:240–248. https://doi.org/10.1111/j.1749-6632.2009.04055.x
Zallocchi M, Igarreta P, Calvo JC, Reboucas NA, Damasco MC (2003) The mechanisms of brush border Na+/H+ exchanger activation by corticosteroids. Med Sci Monit 9(2):BR85–BR90
Zhao H, Star RA, Muallem S (1994) Membrane localization of H+ and HCO3- transporters in the rat pancreatic duct. J Gen Physiol 104(1):57–85
Zhao H, Wiederkehr MR, Fan L, Collazo RL, Crowder LA, Moe OW (1999) Acute inhibition of Na/H exchanger NHE-3 by cAMP. Role of protein kinase a and NHE-3 phosphoserines 552 and 605. J Biol Chem 274(7):3978–3987

Zhao Y, Cui G, Zhang N, Liu Z, Sun W, Peng Q (2012) Lipopolysaccharide induces endothelial cell apoptosis via activation of Na(+)/H(+) exchanger 1 and calpain-dependent degradation of Bcl-2. Biochem Biophys Res Commun 427(1):125–132. https://doi.org/10.1016/j.bbrc.2012.09.023
Zhao H, Carney KE, Falgoust L, Pan JW, Sun D, Zhang Z (2016) Emerging roles of Na(+)/H(+) exchangers in epilepsy and developmental brain disorders. Prog Neurobiol 138-140:19–35. https://doi.org/10.1016/j.pneurobio.2016.02.002
Zielinska M, Wasilewski A, Fichna J (2015) Tenapanor hydrochloride for the treatment of constipation-predominant irritable bowel syndrome. Expert Opin Invest Drugs 24 (8):1093–1099. https://doi.org/10.1517/13543784.2015.1054480
Zizak M, Cavet ME, Bayle D, Tse CM, Hallen S, Sachs G, Donowitz M (2000) Na(+)/H(+) exchanger NHE3 has 11 membrane spanning domains and a cleaved signal peptide: topology analysis using in vitro transcription/translation. Biochemistry 39(27):8102–8112

Chapter 6
Sugar Transport Across Epithelia

Donald D. F. Loo and Ernest M. Wright

Abstract The transport of D-glucose, D-galactose, and D-fructose across epithelial cells is mediated by SGLTs, GLUT5, and GLUT2 in the apical and/or basolateral membrane. The SGLTs (sodium–glucose cotransporters) are responsible for active glucose transport, while the GLUTs (facilitative glucose transporters) are responsible for passive glucose and fructose transport. The structure and function of each of these transport proteins are summarized, and then we highlight the similarities and differences between these two classes of membrane transporters. We next discuss their roles in sugar absorption in the intestine and glucose reabsorption from the glomerular filtrate in the kidney. Reference is made to genetic disorders of glucose transport in the intestine, Glucose–Galactose Malabsorption, and in the kidney, Familial Renal Glucosuria, and the Fanconi–Bickel syndrome, and the use of specific SGLT2 inhibitors to treat Type II Diabetes Mellitus.

Keywords GLUTs · SGLTs Structure · Kinetics · Inhibitors · Glucose · Galactose · Fructose · Uniport · Cotransport · Electrophysiology · *Xenopus laevis* oocytes · Small Intestine · Kidney · Brush border membrane · Basolateral membrane · PET imaging

6.1 Introduction

Glucose is the primary fuel for cells to grow and function. A question that has puzzled physiologists for over a century is how the sugar enters cells across the plasma membrane. Quite simply, how can glucose, a molecule with very low lipid solubility, cross the lipid plasma membrane at a sufficiently high rate to maintain cell metabolism? Studies on the kinetics and specificity of glucose transport into cells and tissues throughout the body revealed that there are two classes of transporters, passive and the active. The passive transporters were those where glucose transport

---

D. D. F. Loo · E. M. Wright (✉)
Department of Physiology, David Geffen School of Medicine at UCLA, Los Angeles, CA, USA
e-mail: ewright@mednet.ucla.edu

K. L. Hamilton, D. C. Devor (eds.), *Studies of Epithelial Transporters and Ion Channels*, Physiology in Health and Disease,
https://doi.org/10.1007/978-3-030-55454-5_6

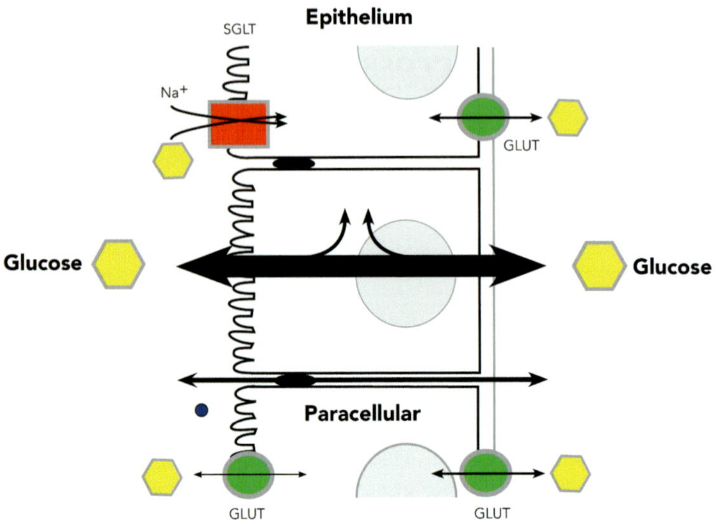

Fig. 6.1 Potential pathways for glucose transport across an epithelium. Glucose may be transported across an epithelium through the paracellular and cellular pathways. Glucose entering the cell from either surface may undergo the glycolytic metabolic pathway, but this can be experimentally minimized using non-metabolized glucose analogs such as 2-deoxy-D-glucose (2DOG), 3-O-methyl-D-glucose (3OMDG), and α-methyl-D-glucopyranoside (αMDG)

only occurred down the glucose concentration at a higher rate than simple diffusion, often called *facilitated diffusion*. This process was characterized as being specific for D-glucose and related natural pyranoses, saturable as a function of sugar concentration, reversible, and independent of sugar metabolism. The membrane proteins responsible are now known as *GLUTs* in the human gene family SLC2. The most well-known example is glucose transport into and out of human red blood cells by GLUT1. Active glucose transport shares many of the characteristics of facilitated transporters, but with the major difference in that transport may occur against a glucose concentration gradient, i.e., requires energy input. In 1960, Robert Crane proposed that the energy came from the Na^+ concentration gradient across the cell membrane, i.e., there is a direct coupling of sugar uptake and Na^+ transport down its concentration gradient, the Na^+/glucose cotransport hypothesis (see Wright et al. 2011). Since the Na^+ concentration gradient across cell membranes is generated and maintained by the Na^+/K^+-pump, this explains the link between "active" glucose transport and metabolism. Subsequently, the *Na^+/glucose cotransport hypothesis* was confirmed, and extended to include active transport systems in bacteria, plants, and animals, but in bacteria and plants, the driving cation is usually protons. The Na^+/glucose cotransporters are now known as *SGLTs* in the human gene family *SLC5*.

The central focus of this chapter is to explain glucose transport across epithelia by GLUTs, and SGLTs. Figure 6.1 shows the pathways for glucose transport across an epithelium. There are two potential routes, paracellular and cellular. In the paracellular route, the sugar bypasses the cell and diffuses through the "tight-junctions." In the cellular route, transport occurs across the apical and basolateral membranes in series. Glucose entering the cell may be metabolized, but in experimental studies, this complication is avoided by the use of non-metabolized glucose analogs. Glucose enters cells via either GLUTs or SGLTs. One example where only GLUTs are involved is in glucose transport across the blood–brain barrier endothelium where GLUT1 in the apical and basolateral membranes is responsible for the passive uptake of 120 grams of glucose into the brain each day. SGLTs are generally expressed only in the apical membrane, for example, in the small intestine and renal proximal tubule, where they are responsible for the absorption for 180 grams of glucose per day. This is a two-stage process where glucose is first accumulated in the epithelium by apical membrane SGLTs and then is probably transported out across the basolateral membrane by GLUTs.

In 2020, we have a fair understanding of the mechanisms of glucose transport by GLUTs and SGLTs, largely due to the cloning and expression of their genes in the 1980s, solving their atomic structure, and the development of sophisticated transport assays over the past two decades. In this chapter, we will first review the molecular biology, biochemistry, and structure of the GLUTs and SGLTs, and then discuss their physiology. The diffusion of glucose through the paracellular pathways is not considered, as there is little quantitative evidence for a physiological role. Consult Krug (2017) for recent information about ion and macromolecule permeation across junctions.

6.2 Gluts

6.2.1 Introduction

Early studies of glucose transport were largely limited to studies on the kinetics and selectivity of sugar uptakes into cells throughout the body (amply reviewed by Stein (1967)). Perhaps the best-characterized system was the human red cell, due to their availability, abundance, and ease of measuring glucose uptake, efflux, and exchange. In brief, the rate of transport for D-glucose is at least an order of magnitude higher than for other molecules of the same size and lipid solubility, e.g., L-glucose. The half-saturating concentration (K_m) for D-glucose is ~6 mM; there is a selectivity for monosaccharides, e.g., D-glucose (6 mM) > D-galactose (K_m 60 mM) > D-Xylose (K_m 110 mM). The red cell glucose transport system exhibits competition between hexoses, exchange, and counter flow. Various kinetic models were proposed, including the mobile carrier hypothesis by Widdas (1953). With the purification and reconstitution of the D-glucose transporter from human erythrocytes into proteoliposomes (Kasahara and Hinkle 1977), the involvement of a specific protein in sugar transport was firmly established. A simple 3-state alternating access model of facilitated diffusion is shown in Fig. 6.2. The sugar-binding site of the protein alternatively faces one side of the membrane or the other. In a transport cycle, the

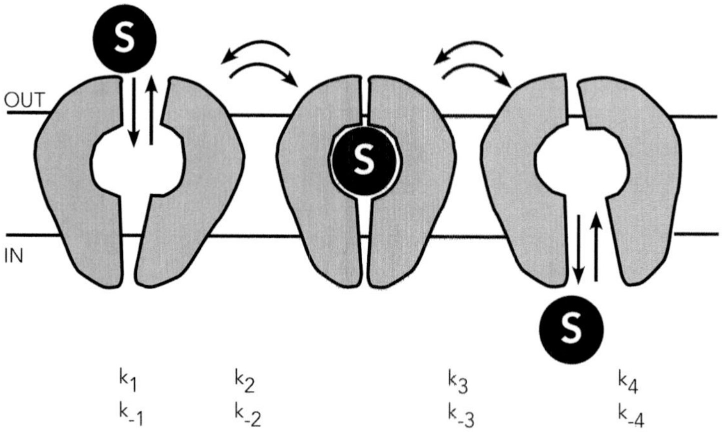

Fig. 6.2 A simple model of facilitated diffusion. The sugar (*S*) only crosses the plasma membrane through an integral membrane protein that can exist in two conformations, one open to the outside (left) and one open to the cytoplasm (right). Sugar is occluded from both sides (middle). Sugar binds to the open conformation with On and Off rates (k_1 and k_{-1}) that can be the same or different for the two conformations. The specificity of sugar binding depends on the architecture of the binding site. Once sugar binds, the binding site closes (k_2 and k_{-2}), before opening to the other face of the membrane (k_3 and k_{-3}) to permit the substrate to be released to the cytoplasm (k_4 and k_{-4}). Solving the structures of GLUTs has provided new insights into transport models (see below)

empty carrier opens to one side of the membrane where glucose can bind. The bound sugar is then occluded from one or both sides of the membrane. The sugar-binding site is next exposed to the opposite membrane surface. After sugar is released, the empty sugar-binding site returns to the original membrane surface.

The direction of net transport, in or out of a cell, depends on the concentration of sugar on each side of the membrane. The rate of transport depends on the difference in glucose concentration across the membrane, the K_D (dissociation constant) of sugar binding, and the turnover number of the transporter. In turn, the maximum rate of transport (V_{max}) depends on the number of transporters in the membrane. The actual rate depends on a minimum of eight different rate constants (see Fig. 6.2). In practice, it has been difficult to determine these rate constants. Instead, the kinetics of net facilitated sugar transport into or out of a cell are commonly fitted to phenomenological equations, such as the Michaelis–Menten equation: $V = V_{max}$ [S]/ $(1 + K_m/[S])$, where V is velocity, [S] is the sugar concentration, K_m is the Michaelis constant, and V_{max} is maximal velocity. K_m is the sugar concentration at 50% V_{max}. The red blood cell transporter, GLUT1, has been the subject of extensive investigations, where net transport and unidirectional influx and effluxes have all been measured as a function of both external and internal sugar concentrations. This has engendered vigorous discussions about the validity of the simple model (Fig. 6.2) and has resulted in even more complex models that have been difficult to evaluate (see, Lloyd et al. 2017; Naftalin 2018).

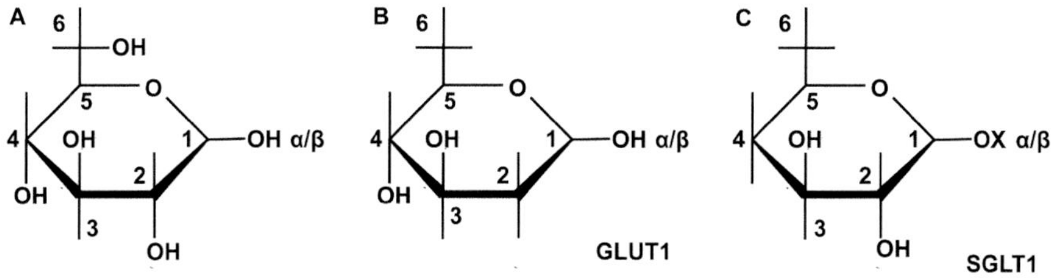

Fig. 6.3 The structure of D-glucose shown in a Haworth projection with the thicker line indicating atoms of the pyranose ring closer to the observer, Carbons number 1–6, and C number 1 is the anomeric carbon with the hydroxyl group in either the α or β position. The hydroxyl groups shown are in the equatorial plane of the pyranose ring. (**a**) D-glucose. (**b**) The minimum preferred substrate for GLUT1 (2-deoxy-D-glucose). (**c**) The preferred substrate for SGLT1, where X is either –H or –CH_3. That is, recognition by GLUT1 requires equatorial –OH groups at C1, C3, and C4, while SGLT1 requires –OH groups at C2 and C3

Glucose transport in tissues generally differ in their V_{max} and K_m values, and sugar selectivity. The explanation for this has largely been resolved by cloning the 14 transporters in the GLUT (SLC2) gene family. The preferred substrate for GLUTs 1, 2, and 3 is D-glucose (K_m 1–10 mM), whereas the preferred substrate for GLUT5 is fructose (K_m 6 mM). The sugar selectivity has mostly been studied by competition experiments due largely to their simplicity and the lack of specific assays for glucose analogs (see Barnett et al. 1973 for relative inhibitory constant K_i values for sugar). For GLUT1: hexoses are preferred over pentoses (K_i 1 mM for D-glucose and 10 mM for D-xylose); equatorial –OH groups are preferred at C#3 and C#4, (3-deoxy-D-glucose K_i 10 mM, and D-galactose, K_i 12 mM), and essential at C#1 (1-deoxy-glucose not recognized); –OH groups are not essential at C#2 and C#6, 2-deoxy-D-glucose (K_i 0.4 mM), and 6-deoxy-D-glucose (K_i 1 mM); and alkylation at C#1 and C#2 is not tolerated; α- and β- methyl- D-glucose; but is tolerated at C#2 and C#3 (2-O-methy-D-glucose, K_i 17 mM, and, 3-O-Methy-D-glucose, K_i 2 mM. The minimum glucose structure for high-affinity interaction with GLUT1 is shown in Fig. 6.3. Sugar selectivity is similar for GLUT2 and GLUT3 as revealed in the studies on the cloned transporters in heterologous expression systems and in proteoliposomes (Gould and Holman 1993; Deng et al. 2015). GLUT5 selectivity is different in that D-fructose is the preferred substrate (see Gould and Holman 1993; Nomura et al. 2015). Such studies on red blood cells, cloned transporters, and purified membrane vesicles have also revealed that external maltose (K_i 10 mM), phloretin (K_i 50 μM), and phlorizin (K_i 350 μM), and internal cytochalasin B (K_i 0.1 μM) inhibit GLUT1, and GLUT2, but not GLUT5 (Uldry and Thorens 2004; Deng et al. 2015; Kapoor et al. 2016).

6.2.2 Cloning

A major landmark was the cloning of the first human glucose transporter (GLUT1) by Mike Mueckler and colleagues (Mueckler et al. 1985). Sequencing revealed a protein of 492 amino acid residues and 12 predicted transmembrane helices. Since then, a total of 14 related members were identified by homology cloning (Thorens and Mueckler 2010). Figure 6.4 shows an alignment of the amino acid sequences and the predicted 12 transmembrane (TM1-TM12) and 5 cytosolic helices (ICH1-ICH5) for GLUTs 1, 2, and 5. There is high amino acid sequence homology between GLUT1, GLUT2, and GLUT5 (100, 55, and 41.7% identity), and the conserved residues are highlighted in black. The GLUTs belong to the huge major facilitator superfamily (MFS) with over 15,000 transporters found in bacteria, archaea, and

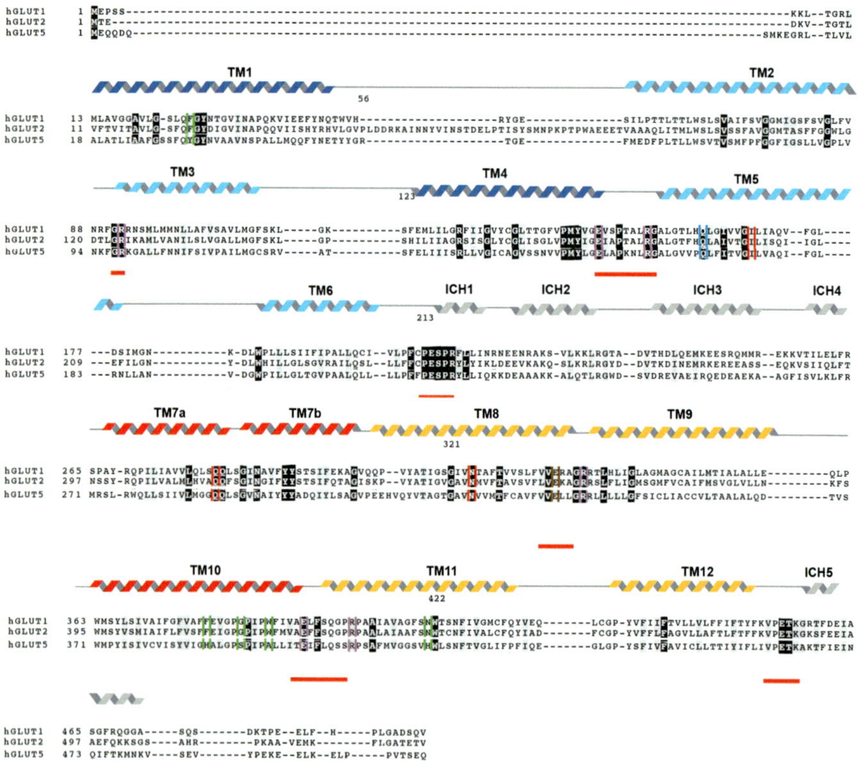

Fig. 6.4 Sequence alignment of GLUT1, GLUT2, and GLUT5. The proteins contain ~500 amino acids arranged in 12 transmembrane helices TM1–TM12 with the N- and C-termini facing the cytoplasm. Strictly conserved residues in the three proteins are highlighted in black, highly conserved residues are highlighted in gray, and those colored red have been thought to be involved in coordinating the sugar. There are four intracellular helical domains (ICH1–ICH4) located between TM6 and TM7, and C-terminal ICH5. Figure is extracted from Extended Data of Nomura et al. (2015)

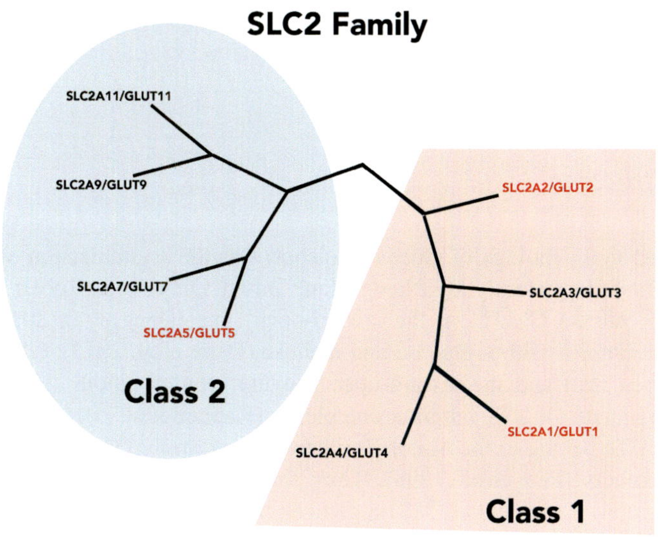

Fig. 6.5 THE SLC2 GENE Family. The phylogenetic tree of GLUT genes in *Homo Sapiens*. Eight members of the GLUT family belonging to two classes: Class 1 contains GLUT1/2/3/4; and Class 2 contains GLUT5/7/9/11. The remaining family members belong to Class 3, GLUT6/8/10/12/13 (not shown). Modified from Fig. 6.1 of Uldry and Thorens (2004)

eukarya, that can function as uniporters, cotransporters, and exchangers (Pao et al. 1998; Yan 2015, http://www.tcdb.org/tcdb)

The human SLC2 genes encode the GLUTs with 14 members are divided into three classes based on sequence similarity (Joost and Thorens 2001). GLUT1/2/3/4/14 are in *Class I*, GLUT5/7/9/11 in *Class II*, and GLUT6/8/10/12/13 in *Class III*. The GLUTs in epithelia belong to *Class I* and Class *II* (Fig. 6.5). Those in Class 1 are glucose transporters, while GLUT5 in Class 2 is a fructose transporter. Expression of the GLUT1 gene is ubiquitous in human tissues, but it plays especially important functions in erythrocytes and the blood–brain barrier. The GLUT2 gene is expressed in the liver and kidney, where it is important in glucose homeostasis. The major site of GLUT5 gene expression is in the small intestine, and the protein is restricted to the brush border membrane of enterocytes, where it is important in dietary fructose absorption (see Ferraris et al. 2018).

Following the cloning and sequencing of the class I and II members of the SLC2 gene family, there was a concerted effort to elucidate their secondary structure and search for sugar-binding sites. This was led in large part by Mike Mueckler, who first used glycosylation scanning mutagenesis to test the theoretical predicted 12 transmembrane helices (Hresko et al. 1994), and then carried out cysteine-scanning mutagenesis on each of the residues in each of the 12 transmembrane helices. The experimental readout he used was glucose uptake assays on mutants expressed in *Xenopus leavis* oocytes, and the inhibition produced by hydrophilic cysteine reagents (Mueckler and Makepeace 2009; Mueckler and Thorens 2013). This work culminated in a 2D model of the exofacial glucose-binding site and a low-resolution packing map of the 12 TMs.

6.2.3 Structure

A watershed in the biology of glucose transport was the crystallization of mammalian GLUTs independently by Nieng Yan, David Drew, and Robert Stroud in 2014–2016: human GLUT1 in the inward-open state (Deng et al. 2014); human GLUT3 in complex with D-glucose and maltose (Deng et al. 2015); GLUT5 in the outward-open (rat) and the inward-open (bovine) conformations (Nomura et al. 2015); and human GLUT1 inhibitor complexes (Kapoor et al. 2016).

The GLUTs belong to the Major Facilitator Superfamily (MFS) with more than 15,000 members (Pao et al. 1998). They share a common structural fold with 12 transmembrane segments (TMS) organized into two-folded domains, the amino (N) and carboxy (C). Within each domain, the six membrane segments are folded into a pair of "3 + 3" inverted repeats (Fig. 6.6). Between TM6 and TM7 is the

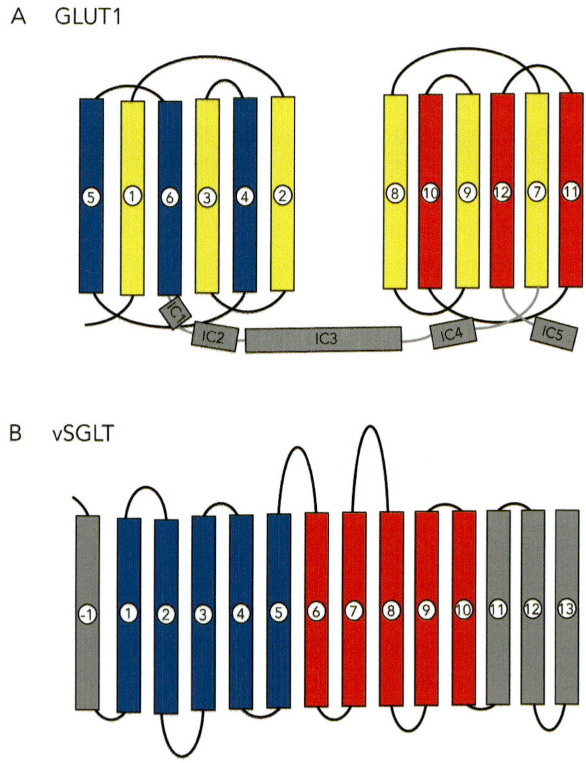

Fig. 6.6 Topology of GLUTs and SGLTs. (**a**). Structure of GLUTs. The 12 transmembrane helices are divided into N- and C-domains. TM1–TM6 form the N-, and TM7–TM12 the C-domain. TM4–6 (blue) is an inverted repeat of TM1–3 (yellow); and TM10–12 (red) is an inverted repeat of TM7–9 (yellow). The intracellular helical domains (CH1–4 and CH5) are colored black. (**b**) Structure of SGLTs. Transmembrane helices 1–5 and 6–10 are inverted repeats

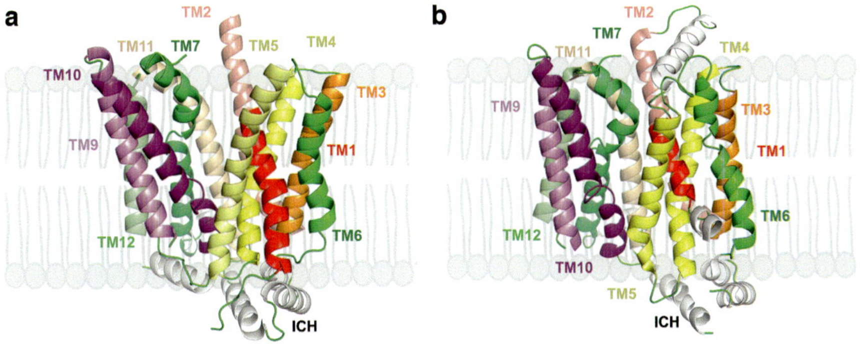

Fig. 6.7 Side views of the outward- and inward-open GLUT5. (**a**) Sideview of the out-open conformation of GLUT5, from PDB entry 4YBQ for rat GLUT5 (Nomura et al. 2015). The central cavity is open to the external compartment. TM7 and TM10 contain discontinuous regions. Residues in the discontinuous regions are involved in sugar coordination. In the outward-open conformation, the intracellular helices (ICH) form a barrier to entry into the internal compartment. (**b**) Sideview of the in-open conformation of GLUT5, from PDB 4YB9 for bovine GLUT5 (Nomura et al. 2015). Note the bending of the external end of TM7. TM8 has been removed for clarity in (**a** and **b**). In this and all subsequent figures structural models were prepared with PyMol (Delano 2002). The color scheme for transmembrane helices TM1–TM10 in all figures is based on the rainbow, ROYGBIV: Red TM 1–2, Orange TM 3, Yellow TM 4–5, Green TM 6–7, Blue TM 8, Violet TM 9, Purple TM 10, Wheat TM 11, and Pale green TM 12. In some figures, TM helices are removed for clarity

intracellular helical domain (ICH). The ICH contains 3–4 helices and another short cytoplasmic helix in the cytoplasmic C-terminus.

The transmembrane helices of TM7 and TM10 of the GLUTs contain discontinuous regions. In rat GLUT5, the discontinuous regions are between S290 and A299 in TM7, and G385–L396 in TM10. External to the discontinuous region in TM7 are three bulky conserved tyrosine residues (bovine Y296, Y297, Y298), and their movements between the outward and inward states result in blockage of the central vestibule to external sugar binding (see below).

In a transport cycle, the protein alternates between outward- and inward-facing states. The sugar-binding sites are either open or occluded, after substrate binding (Fig. 6.2). The GLUT states identified so far in crystal structures are outward-open, outward-occluded, inward-occluded, and inward-open (see Deng et al. 2014, 2015; Nomura et al. 2015). In common with other members of the MFS, e.g. LacY, the lactose transporter (Kaback and Guan 2019), it appears that there is an intermediate conformation where the substrate is occluded from the aqueous media on both sides of the membrane.

The most pronounced motions in the transition between outward- and inward-states, are: (1) a rocker switch rotation of the N- and C-domains, with the N-domain undergoing a 16° rigid-body rotation (Fig. 6.7a, b); (2) a local external gated-pore mechanism in the C-domain involving the extracellular end of TM7 and intracellular end of TM10. The external end of TM7 (external to the discontinuous region) consists of four helical turns, with three tyrosine residues (bovine Y296, Y297,

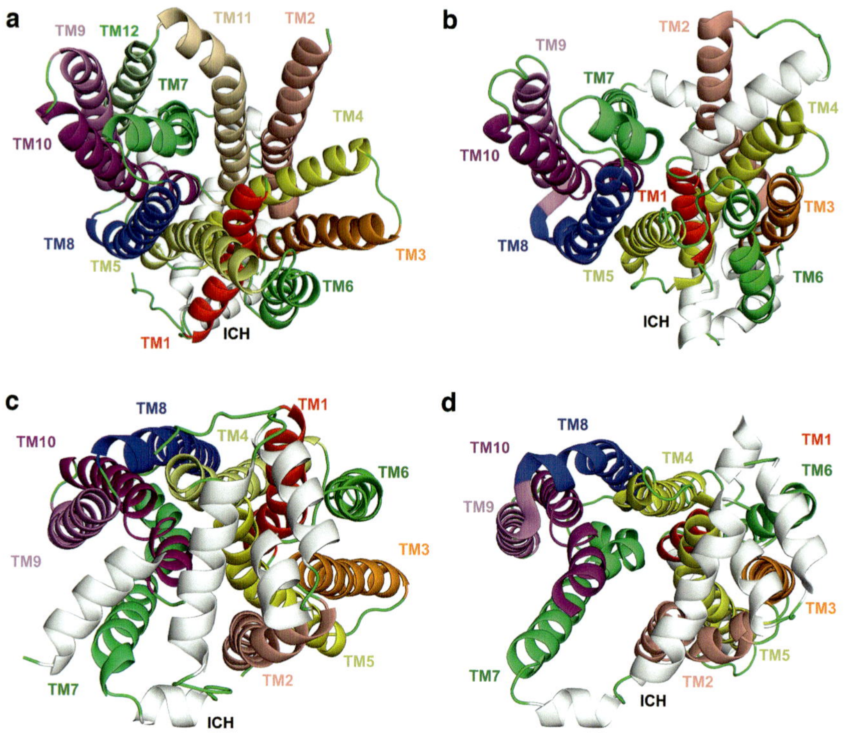

Fig. 6.8 External and internal views of the out-open and in-open conformations of GLUT5. (**a**) External view of the out-open conformation of GLUT5, from PDB 4YBQ for rat GLUT5 (Nomura et al. 2015). The central cavity, formed by the space between TMs 7, 10, 4, and 5, is accessible from the external compartment. (**b**) External view of the in-open conformation of GLUT5, from PDB 4YB9 for bovine GLUT5 (Nomura et al. 2015). External end of TM10 blocks entry to the central cavity. Note the changes in the external end of TM7 between (**a** and **b**). (**c**) Internal view of the out-open conformation of GLUT5, from PDB 4YBQ for rat GLUT5 (Nomura et al. 2015). ICHs blocks access to the vestibule from the internal compartment. (**d**) Internal view of the in-open conformation of GLUT5, from PDB 4YB9 for bovine GLUT5 (Nomura et al. 2015). Note the changes in the internal end of TM10 and ICH between (**c** and **d**). TM11 and TM12 are removed in (**b**, **c**, and **d**)

Y298). In going from outward-open to inward-open, the external end of TM7 bends and rotates clockwise (viewed from the external surface), thereby blocking external access to the central cavity (Fig. 6.7b). On the internal side, the intracellular end of TM10 shifts away from the substrate-binding site, to open the central cavity to the cytoplasm; and (3) the central intracellular helices (ICH) move and appear to act as an intracellular gate to the central cavity (Figs. 6.7 and 6.8).

Nieng Yan's group succeed in crystalizing glucose bound to GLUT3. Figure 6.9 shows the glucose-binding site of GLUT3 in the outward-occluded conformation. Mainly residues in the C-domain coordinate glucose: N286, Q281, and Q280 in TM7, N315 in TM8, and E378 and W386 in TM10. The exception is Q159 in TM5 of the N-domain. The coordinating residues are either polar (N286, Q280, Q281,

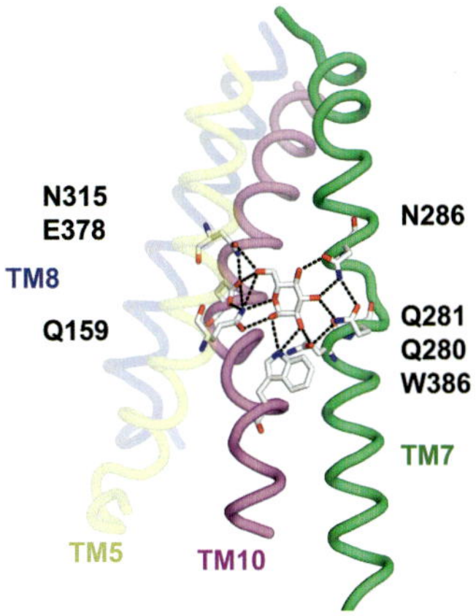

Fig. 6.9 D-glucose binding site of hGLUT3 in the outward-occluded conformation, from PDB file 4ZW9 (Deng et al. 2015). D-glucose is coordinated by Q159 in TM5, N286, Q281, and Q280 in TM7, N315 in TM8, and E378 and W386 in TM10, and lies perpendicular to the membrane plane

N315) or charged (E378, W386). The coordinating residues in TM7 and TM10 are located near the discontinuous regions of these helices. Although no other sugar-bound structures are yet available, these are conserved in GLUT1–5, with the exception of E378 and W378 that are both alanine in GLUT5, the fructose transporter (see Fig. 6.4). Interestingly, mutation of Q166 (Q159 in GLUT3) in GLUT5 to Q166E appears to convert GLUT5 into a glucose transporter with a K_D of 4 mM (Nomura et al. 2015).

Overall, the structural studies of GLUTs provide a powerful insight into the molecular mechanisms of sugar transport that were provided by simple models (see Fig. 6.2). They have shown that there are at least two major conformations of the structural homologs, outward- and inward-facing, and an occluded state where the sugar-binding site in the center of the protein exists in at least three states, outward- and inward-facing, occluded from the outside, inside and both sides. A major challenge yet to be overcome is the monitoring of the dynamics of sugar-binding and transport.

6.2.4 Inhibitors

A long-established tool in this field is the search for specific transport inhibitors to gain insights into transport mechanisms, and/or to develop drugs to modulate the rates of sugar transport into cells. For example, to block glucose transport into tumors, where the well-known Warburg effect that results in an increase in glucose uptake by GLUT1 to meet the metabolic demands for tumor growth. We highlight two inhibitors of the human red cell glucose transport, cytochalasin B and maltose. Cytochalasin B inhibits GLUT1–4, but not GLUT5, with IC_{50}s in the nano-molar range.

Stroud's group solved the crystal structure of cytochalasin B bound to human GLUT1 (Fig. 6.10) and produced homology models of inhibitor binding to other

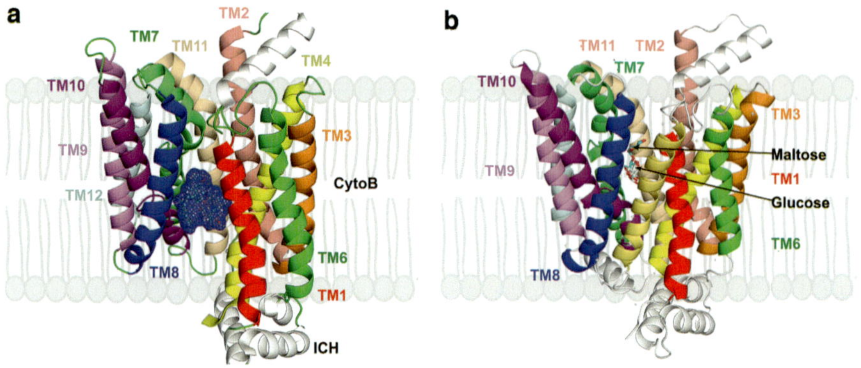

Fig. 6.10 (**a**) Cytochalasin B binding to GLUT1. The figure shows a side view of human GLUT1 with cytochalasin B (meshed surface) occupying the glucose binding site in the inward-open conformation. The central cavity is bounded by TM6, TM1, TM10, and TM11 (Kapoor et al. 2016). (**b**) Maltose binding to GLUT3. A side view of glucose and maltose bound to the outward-open conformation of GLUT3 (PDB 4ZWB and 4ZWC, Deng et al. 2015). In the outward-occluded conformation, both the α- and β-D-glucose are coordinated similarly with GLUT3

GLUTs (Kapoor et al. 2016). In each case, cytochalasin B was bound to the central cavity of the inward-open state of these transporters via interactions with Thr137, Gln282, Asn288, Gly384, Trp388, and Asn411. Furthermore, in high-throughput screens, they identified another class of inhibitors that bound to GLUT1 overlapping the cytochalasin B binding site, the same site with different coordinating residues. Homology modeling of GLUT2–4 in the inward-open conformation and calculation of inhibitor-binding energies, encourages the search for more specific inhibitors for members of the human GLUT transporters.

External maltose acts as a non-transported, competitive inhibitor of glucose uptake by GLUT1 with a K_i of 10 mM (see Carruthers and Helgerson 1991). Neither external sucrose nor lactose are efficient inhibitors. Yan's group (Deng et al. 2015) was successful in crystallizing a maltose–GLUT3 complex in two conformations, outward-open and outward-occluded. The outward-occluded structure was virtually identical to the outward-occluded structure of the glucose–GLUT3 complex. The second glucose of maltose completely overlaps glucose in the outward-open and outward-occluded structures. It, therefore, seems that that the first glucose in the disaccharide binds to the glucose binding site, but the additional sugar moiety in maltose prohibits transport through the GLUTs.

Although progress has been made in isolating natural specific GLUT5 inhibitors from plants, so far these are of low potency, 5–10 μM, and GLUT5 inhibitor complexes have not been crystalized (Ferraris et al. 2018).

6.2.5 *Summary*

Great strides have been made in our understanding of facilitated monosaccharide transport across biological membranes, with the first cloning of the 14 members of the human SLC2 gene family, and second solving the atomic structures of GLUT1, GLUT3, and GLUT5—with and without bound sugars and inhibitors. In terms of epithelial transport, we are still waiting on high-resolution structures of GLUT2, the most important player in renal and liver physiology. We next discuss the active glucose transporters, the Na^+/glucose cotransporters, and then we will consider the similarities and differences between the two classes of glucose transporters.

6.3 SGLTs

6.3.1 *Introduction*

Interest in glucose transport across epithelia originated largely in the nineteenth century with the realization that glucose in food is absorbed into the body by the small intestine, and that glucose is reabsorbed in the kidney after glomerular filtration. In human physiology, perhaps the most compelling evidence for a special epithelial glucose transport mechanism originates in Homer Smith's studies on the clearance of small molecules from the body (see Chasis et al. 1933). He found that glucose, unlike xylose, was completely reabsorbed from the glomerular filtrate, and this was blocked by the plant glycoside phlorizin. Similar studies on rodents demonstrated that glucose was completely absorbed from the small intestine. Understanding the mechanism of glucose transport in both the kidney and intestine was greatly facilitated over the next 30 years by the development of in vitro techniques and the use of non-metabolized glucose analogs such as 3-O-methyl-D-glucoside. This permitted measurement of sugar transport under well-controlled conditions and led to the concept of active sugar transport, i.e., absorption could occur against a sugar concentration gradient. The process was saturable, specific in that glucose was preferred over other monosaccharides, was blocked by metabolic inhibitors, and could occur against a glucose concentration gradient.

While many investigators contributed to the advances in the field, Robert Crane, in our opinion, deserves a lion's share of the credit for his pioneering work on the small intestine. For example, he demonstrated that the active step was at the brush border membrane, the minimum requirement of sugar transport was a hexose with an equatorial group at C#2 (see Fig. 6.3) and that neither phosphorylation nor any other metabolic conversion was involved in sugar transport (see Crane 1977). While others also reported that glucose transport was sodium-dependent, Crane was the first to postulate in 1960 that the driving force for "active" glucose transport was the sodium gradient across the brush border membrane, i.e., sodium–glucose cotransport, and showed that reversing the sodium gradient across the brush border membrane resulted in the reversal of active glucose transport (see Crane 1977). His work resulted in a paradigm shift in the transport field where it became accepted that the active transport of molecules and ions in both mammalian cells (and bacteria) could be explained by cotransport (see for example Schultz and Curran 1970).

The twentieth century landmarks in testing the sodium/glucose cotransport hypothesis were Ulrich Hopfer's experiments using intestinal brush border membrane vesicles to show that sodium gradients drive uphill glucose transport (Hopfer et al. 1973); and George Kimmich's experiment in chick enterocytes showing a strict stoichiometry between sodium and glucose transport (2/1) (Kimmich and Randles 1984). The outcome has been that the sodium/glucose cotransport hypothesis is now widely accepted.

As with the GLUTs (see above), the major advances in the field beginning in the mid-1980s have been the cloning of the SGLT family of genes (SLC5), their functional characterization in heterologous expression systems, and solving their atomic structure. Unexpected outcomes were that the SGLT gene family contains a diverse array of sodium cotransporters and that SGLT shares a structural fold with cotransporters and exchangers in different gene families, the LeuT or the 5-helix inverted repeat (5HIR) fold (see Wright et al. 2011; Abramson and Wright 2009). It follows that this diverse array of cotransporters and exchangers work by a common mechanism.

6.3.2 Cloning of the SGLTs

Following the cloning of GLUT1, our lab was stimulated to clone the rabbit intestinal sodium–glucose transporter. Along with Matthias Hediger, Tyson Ikeda, and Michael Coady, we were successful using a novel strategy, expression cloning, where an intestinal cDNA library was screened by functional transport assays in *Xenopus laevis* oocytes (Hediger et al. 1987). This was soon followed by homology cloning of the human intestinal transporter SGLT1 and the human renal isoform SGLT2 (Hediger et al. 1989; Wells et al. 1992). In total, there are 12 genes in the SLC5 gene family that code for glucose, myo-inositol, choline, multivitamin, short-chain fatty acid, and iodide cotransporters, and one glucose sensor (SGLT3) (Fig. 6.11). The function of these transporters was determined using transport assays on heterologous expression systems (see Wright et al. 2011). Here, we focus on SGLT1 and SGLT2.

Sequencing the cDNAs and modeling indicates that SGLT1 codes for a 73.5 kDa protein of 664 amino acids, a predicted molecular weight of 73.5 kDa with 14 transmembrane helices, TM-1 to TM13 (Fig. 6.12). As discussed below, the protein contains a core with an inverted repeat of five transmembrane helices (TM1–5 and TM6–10) (see Fig. 6.6), and the numbering of the helices has been revised to conform to the LeuT structural fold to facilitate easy comparisons between structural family members. Also highlighted are the residues involved in glucose binding and sodium binding at the Na_2 site (see below). SGLT2 codes for a protein of 672 amino

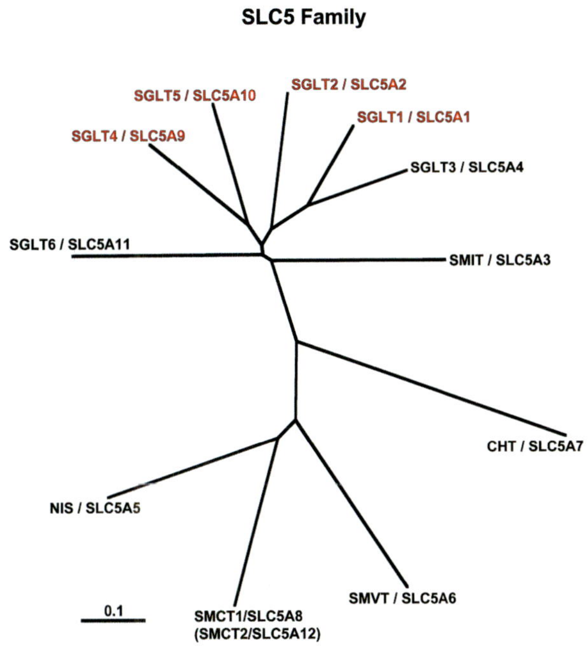

Fig. 6.11 SLC5 Family. An unrooted phylogenic tree of the 12 human members of the SLC5 gene family. SGLT6 is also known as SMIT2 (Na^+/myo-inositol cotransporter 2). SMIT, sodium myo-inositol; CHT, choline; SMVT, sodium multivitamin; SMCT, sodium monocarboxylic acid; NIS, sodium iodide cotransporters (from Wright et al. 2011)

acids and a predicted molecular weight of 72.9 kDa. There is a 59% amino acid identity between SGLT1 and SGLT, and the residues involved in glucose and sodium binding are conserved. We presume the secondary structures of SGLT2 and other members of the SLC5 family are similar to that of SGLT1.

6.3.3 *Kinetics of Sodium–Glucose Cotransport*

Rather than present a historical story, we describe our current model for SGLT1 and review the experimental data. What is unique about our experimental studies is that they were collected on homologous expression systems, *Xenopus laevis* oocytes and HEK293 cells, where transport of Na^+ and sugar was measured under strictly defined thermodynamic conditions, i.e., defined extracellular and intracellular Na^+ and sugar concentrations and membrane potentials. Another novel aspect is that Na^+/glucose cotransport transport may be recorded as Na^+ currents, and this enables a complete set of kinetics in a single cell (see Wright et al. 2011). First, we will cover the steady-state kinetics, and then show how the recording of pre-steady state kinetics opens the black box of transport kinetics.

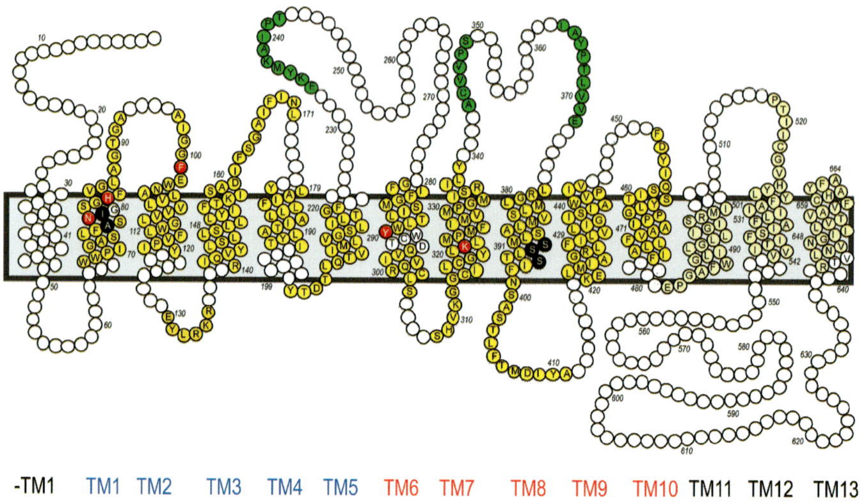

Fig. 6.12 Secondary structure of SGLT1. This model shows the sequence of the 664 residues arranged in 14 transmembrane helices with both the NH_2 and COOH termini facing the extracellular side of the plasma membrane. Highlighted are the locations of the helical domains based on the vSGLT crystal structure and residues involved in sugar binding and sodium binding at the Na_2 site (Faham et al. 2008). The numbering of the TMs has been revised to conform with the LeuT structural fold to allow easy comparisons between structural family members, i.e., TM1 through TM13 (see Wright et al. 2011)

In the model shown in Fig. 6.13, the monomeric SGLT1 protein in the plasma membrane can exist in a minimum of five different conformations (C1, C2, C3, C4, C6) where entry and exit to the glucose binding site in the middle of the protein is controlled by inner and outer gates. In the absence of ligands, the transporter exists in two closed-gate conformations C1 and C6. Upon addition of Na^+ to the outside solution, two Na^+ bind at the Na1 and Na_2 sites to open the outer gate to permit glucose binding. Closure of the outer gate means that glucose is now occluded from the out and inside solutions. Opening of the inner gate results in the simultaneous release of two Na^+ and one glucose to the inside solution (thereby eliminating state C5 in the earlier 6-state model). The closure of the inner gate and the transition of C6 to C1 returns the system to the initial starting point. The net result is that two Na^+ ions and one glucose molecule are transported from outside to inside for each turnover cycle. Since each of the reaction steps are considered reversible, the rate and direction of Na^+/glucose cotransport is simply determined by the Na^+ and glucose concentrations on each side of the membrane, and the electrical potential difference across the membrane. Another feature of this model is that the plant glycoside phlorizin act as a non-transported competitive inhibitor of glucose transport by binding to the outward-open conformation C2 (see below).

The fact that the transporter is electrogenic enables the experimenter to collect a complete set of kinetic data in a single cell such as an oocyte expressing SGLT1 (see Parent et al. 1992a, b; Wright et al. 2011). The validity of this method rests upon the

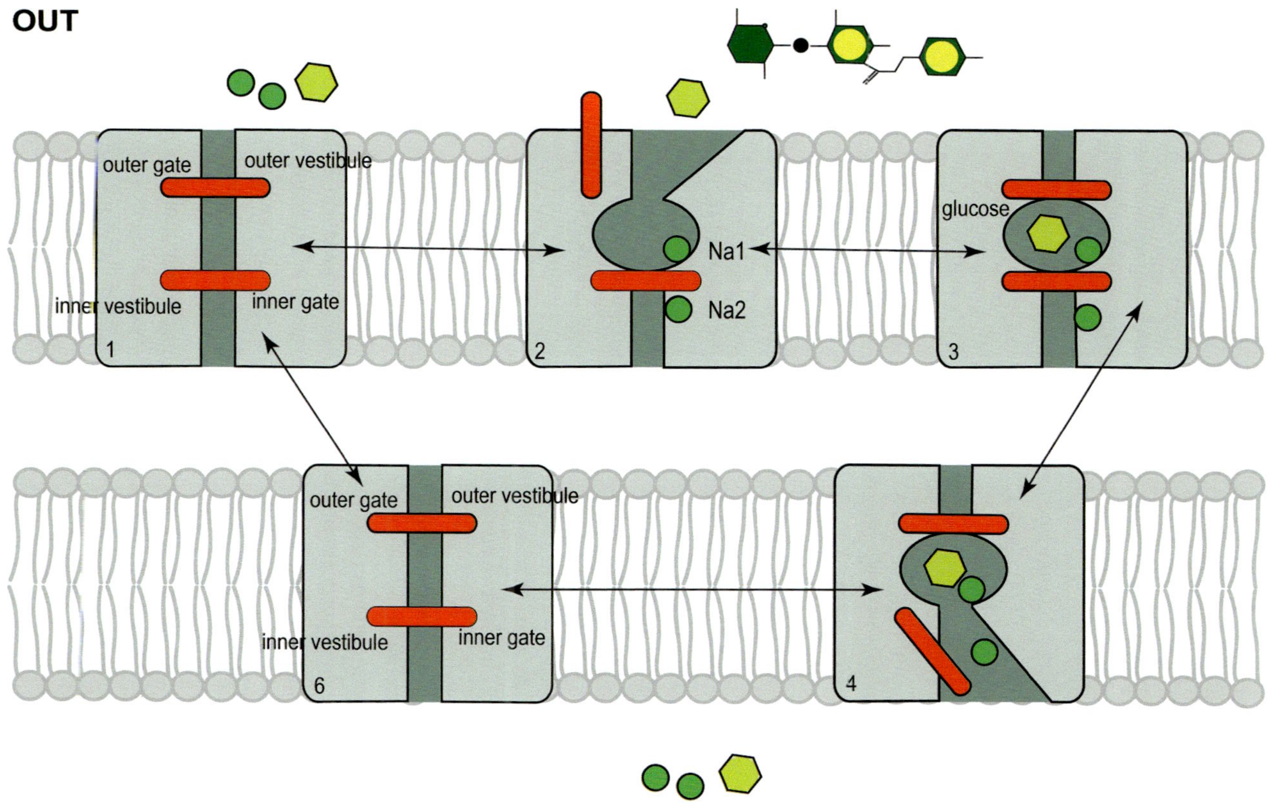

Fig. 6.13 Kinetic Model for SGLT1. Sodium (green circle) binds first to the extracellular side ("out"; state 1) to open the outer gate (state 2), permitting the sugar (glucose; yellow hexagon) to bind and be trapped in the bound site (state 3). The binding of both substrates induces a conformational change to an "inward facing" conformation, resulting in the opening of the inner gate (state 4) and the release of two Na^+ ions and one sugar into the cell interior. After the release of

facts that: Na^+-dependent glucose currents are only observed in cells expressing SGLT1; only sugars accepted by the SGLT1 transporter support these Na^+ currents; the glucose currents are strictly Na^+-dependent (only H^+, and to a lesser extend Li^+ can substitute for Na^+); and the glucose-dependent Na^+ currents are matched by glucose-dependent $^{22}Na^+$ influxes. The fact that the Na^+/glucose currents are quickly reversed on removing Na^+ and/or glucose means that Na^+/sugar currents may be measured as a function of Na^+ and sugar concentrations, at multiple voltages in a single experiment on a single cell to obtain the apparent affinities of Na^+ and sugar as a function of voltage.

An example of the data obtained on an experiment on single oocyte expressing human SGLT1 (hSGLT1) is shown in Fig. 6.14.

Such experiments show that:

1. Inward glucose transport is Na^+ dependent with a Hill coefficient for Na^+ of close to 2, and Na^+ transport is glucose-dependent with a glucose Hill coefficient of 1. Transport is inhibited by external phlorizin (K_i 0.3 μM).
2. At saturating $[Na^+]_o$, the $K_{0.5}$ for sugar (αMDG) decreases with voltage from 0 mV (10 mM) to −150 mV (0.2 mM), and at saturating $[\alpha MDG]_o$ the $K_{0.5}$ for Na^+ decreases from 45 m-equiv/l to 2 m-equivl/l at −150 mV. In both cases, the apparent affinities saturate between −100 and −150 mV.
3. At −150 mV the V_{max} for sugar (αMDG) is independent of $[Na^+]_o$ whereas V_{max} for Na^+ increased as a function of $[\alpha MDG]_o$. This leads to an ordered model for external Na^+ and sugar-binding to SGLT1, with Na^+ first, and has been extended to outward transport by *trans*-inhibition experiments where external glucose, but not Na^+, inhibited outward transport, i.e., glucose is first off in reverse transport (Adelman et al. 2016).
4. Experiments on right-side-out and inside-out membrane vesicles (Quick and Wright 2002), excised inside-out membrane patches from oocytes expressing SGLT1 (Eskandari et al. 2005), and whole-cell patch-clamp experiments on HEK293 cells expressing SGLT1 (Adelman et al. 2016) show that Na^+/glucose transport is reversible but asymmetric due to the stochastic binding and release of ligands on the internal surface.
5. Whole-cell patch-clamp studies by Chiara Ghezzi (Adelman et al. 2016) also have shown that with saturating Na^+ and glucose on each side of the membrane there is no net transport at zero voltage and that neither high internal Na^+ nor high internal glucose alone blocked inward transport. However, both high internal Na^+ and glucose together significantly reduced inward transport. These results are at odds with a transport model with mirror symmetry and point to the stochastic release of both ligands into the cytoplasm. This is supported by fitting the data to a 5-state kinetic model and by molecular dynamical simulations of Na^+ and sugar release from vSGLT (Adelman et al. 2016).

Fig. 6.13 (continued) both substrates, the inner gate closes to form the inward-facing ligand-free conformation (state 6). The cycle is completed by the change in conformation to the outward-facing ligand-free (state 1). The competitive inhibitor phlorizin binds to the external sodium-bound transporter. Figure adapted from Ghezzi et al. (2018)

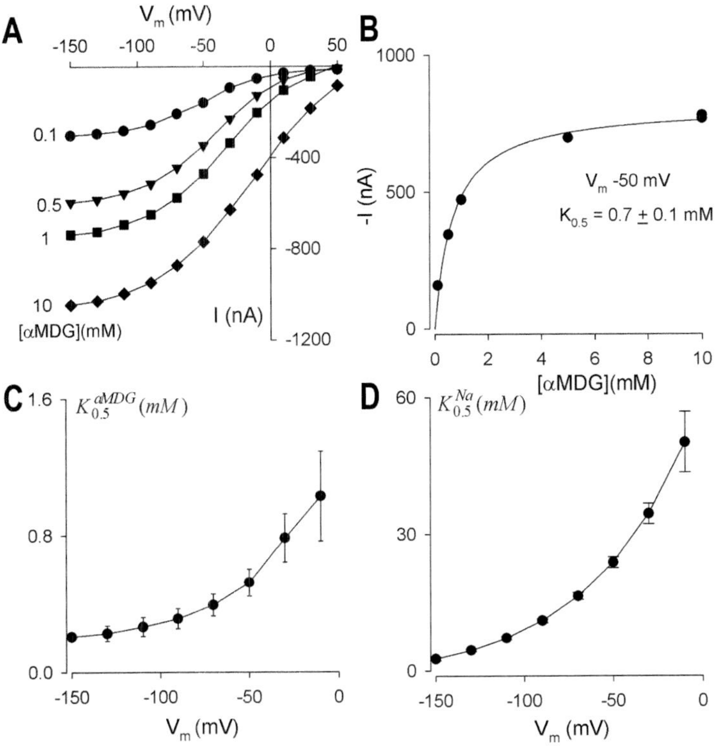

Fig. 6.14 Na^+/glucose currents recorded in a single oocyte expressing human SGLT1. αMDG currents were measured as a function of voltage, external αMDG, and Na^+ concentrations. No αMDG currents were recorded in the absence of external Na^+, but currents up to 1000 nA were observed in the presence of saturating Na^+ (100 mM) at voltages more negative than −100 mV. Under our experimental conditions, the internal sodium concentration was less than 10 mM and the internal sugar concentration was << 0.1 mM. Panel **a** shows the currents recorded at 0.1–10 mM αMDG in the presence of 100 mM Na^+. Note that the currents approached 0 nA at each sugar concentration at depolarizing voltages signaling no outward Na^+/glucose cotransport, and saturated beyond −100 mV. Panel **b** shows the currents recorded as a function of sugar concentration at a voltage of −50 mV. The current saturated with an apparent sugar affinity, $K_{0.5}$, of 0.7 mM. Panel **c** shows the apparent sugar affinity as a function of voltage. $K_{0.5}$ approached a limiting value of 0.2 mM at −150 mV. Panel **d** shows the apparent affinity for Na^+, $K_{0.5}$, as a function of voltage at an external sugar concentration of 10 mM. $K_{0.5}$ decreased with voltage to a limiting value of 0.8 mM at −150 mV. In this oocyte, the Hill coefficient for Na^+ was 1.5 and independent of voltage. Not shown are the data recorded for the maximum currents as a function of external Na^+ and sugar concentrations, where the I_{max} for sugar was independent of Na^+ from 10 to 100 mM Na^+, while the I_{max} for Na^+ was sugar concentration dependent. This indicates that Na^+ binds before sugar in inward Na^+/glucose cotransport. Taken from Wright et al. (2011)

An important takeaway from these studies is that it is essential to control the membrane potential in kinetic studies of electrogenic transporters.

A limiting factor in building quantitative kinetic models has been obtaining the rate constants required for a minimum of five (or six) states. The arbitrary approach has been to make assumptions about the order of the reaction sequence, the rate-limiting steps, and the voltage-dependent steps. Apart from the technical advances using heterologous expression systems, voltage-clamp, and whole-cell patch-clamp, a major break-through for SGLTs was the discovery of their capacitive currents. This enabled us to parse out the kinetics of the partial reactions in the absence of sugar, C2–C1–C6, the number of transporters in the cell membrane, and the K_d for external phlorizin binding.

6.3.3.1 SGLT1 Capacitive Currents

Electrophysiological studies of SGLT1 provided an entre into the sugar-free partial reactions, C2–C1–C6. Capacitive currents were first observed in SGLT1 voltage-jump experiments (see Parent et al. 1992a, b; Loo et al. 1993; Wright et al. 2011). These SGLT1 capacitive currents reflect the voltage-sensitivity of the conformational changes between C6 and C1, and between C1 and C2 (Fig. 6.13).

A summary of the evidence for SGLT1 capacitive currents include.

1. Rapid voltage-jumps on oocytes or HEK293 cells expressing SGLT1 exhibit slow transient currents, with half-time $t_{1/2}$ 2–10 ms, before reaching steady state (Fig. 6.16a, b). Integration of these transients gives the capacitive charge moved, Q (m-coulombs) for each voltage step, ON or OFF. Q is equal for the ON and OFF voltage steps, and Q as a function of voltage fits a Boltzmann relation with a maximal charge Q_{max} (the sum of Q at saturating hyper- and depolarizing voltages), mid-point voltage $V_{0.5}$ (the voltage at 0.5 Q_{max}), and apparent valence of the moveable charge z (obtained from the maximal slope of the Q vs voltage curve), see Fig. 6.16c; and Wright et al. 2011.
2. In any given cell expressing SGLT1 Q_{max} is proportional to the number of SGLT1 proteins in the plasma membrane (N), i.e. $N = Q_{max}\ x\ z$, where z is the apparent valence of the transporter. This is supported by varying the level of expression of SGLT1 in oocytes, recording Q_{max}, the area of the plasma membrane, and correlating the density of SGLT1 proteins in the plasma membrane by freeze-fracture electron microscopy (Zampighi et al. 1995). SGLT1 densities up to 10,000/(μm)^2 were observed.
3. Q_{max} is sodium-dependent (see theory section below), and in sodium, they are abolished by external phlorizin with a K_i 0.5 μM. This observation is interpreted as showing that phlorizin locks the transporter in a $C3Na_2Pz$ state.

6.3.3.2 Theory for hSGLT1 Capacitive Currents

We consider the determination of the kinetic rate constants, following closely the presentation by Loo et al. (2013). First, consider a three-state model for hSGLT1 where two external Na^+ ions bind simultaneously to the empty transporter in the absence of external sugar (Fig. 6.15). The empty transporter is assumed to be negatively charged and opens to the internal (C6) or external (C1) membrane

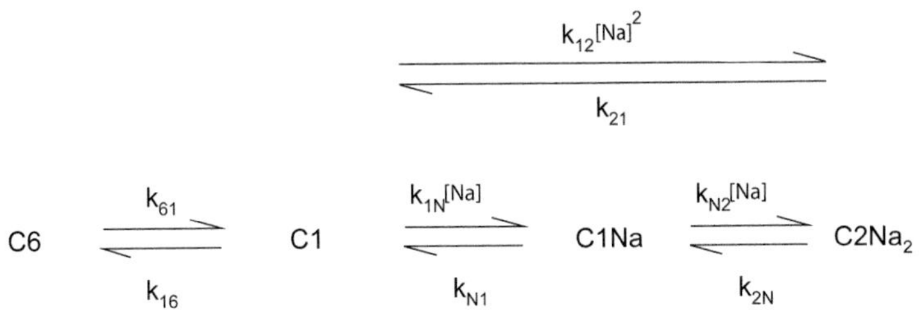

Fig. 6.15 Kinetic schemes for the transitions between states C6, C1, and C2 in the absence of sugar. In the lower scheme, we show the sequential binding of two external sodium ions to form C1Na and then $C2Na_2$, whereas, in the upper scheme, two sodium ions bind simultaneously to C1 to form $C2Na_2$ with a lumped sodium binding constant. We assume that each transition is voltage-dependent with symmetric energy barriers. The voltage-dependent rate constants are $k_{16} = k_{16}$ exp $(z_1VF/2RT)$, $k_{61} = k_{61} \exp(-z_1VF/2RT)$, $k_{12} = k_{12} ([Na^+]_o{}^n) \exp(-z_2VF/2RT)$, and $k_{21} = k_{21}$ exp $(z_2VF/2RT)$

surfaces depending on membrane voltage (V). Voltage also affects external Na^+ binding, and/or the conformational changes associated with it, thus the transporter is distributed between C6, C1, and $C2Na_n$ depending on voltage and $[Na^+]_o$.

We assume the rate constants k_{12}, k_{21}, k_{16}, k_{61} are dependent on voltage according to rate theory with symmetric energy barriers:

$$\begin{aligned} k_{16} &= \mathrm{k}_{16}\ \exp\left(z_1 VF/2RT\right) \\ k_{61} &= \mathrm{k}_{61}\ \exp\left(-z_1 VF/2RT\right) \\ k_{12} &= \mathrm{k}_{12}\ \left([Na^+]_o{}^n\right)\ \exp\left(-z_2 VF/2RT\right) \\ k_{21} &= \mathrm{k}_{21}\ \exp\left(z_2 VF/2RT\right) \end{aligned} \tag{6.1}$$

k_{16}, k_{61}, k_{12}, and k_{21} are voltage-independent rate constants. z_1 and z_2 represent the apparent valence of the moveable charge for the empty transporter (z_1) and Na^+ binding (z_2). F is the Faraday constant, R the gas constant, and T is the absolute temperature.

K_a (k_{16}/k_{61}), the equilibrium constant of the distribution of the empty transporter between states C6 and C1 (at 0 mV), and K_d ($(k_{21}/k_{12})^{1/n}$), the intrinsic Na^+ affinity at 0 mV is:

$$\begin{aligned} k_{16}/k_{61} &= K_a\ \exp\left(z_1 FV/RT\right) \\ k_{21}/k_{12} &= \left(k_{21}/k_{12}\right)\ \exp\left(z_2 FV/RT\right) = \left(K_d\right)^n\ \exp\left(z_2 FV/RT\right) \end{aligned} \tag{6.2}$$

In the presence of external Na^+, the distribution of hSGLT1 in the steady state is characterized by the midpoint voltage $V_{0.5}$, where the protein is equally distributed between $C2Na_n$ and C6 (see Fig. 6.16).

To obtain the $V_{0.5}$ vs. $[Na^+]_o$ relations, consider the charge moved (Q) when membrane voltage (V) is stepped from a very negative value (where all transporters are in $C2Na_n$) to a more positive value. Q is the sum of the charge transferred from $C2Na_n$ to C1 (each transition contributing a net charge z_2) and from $C2Na_n$ to C6

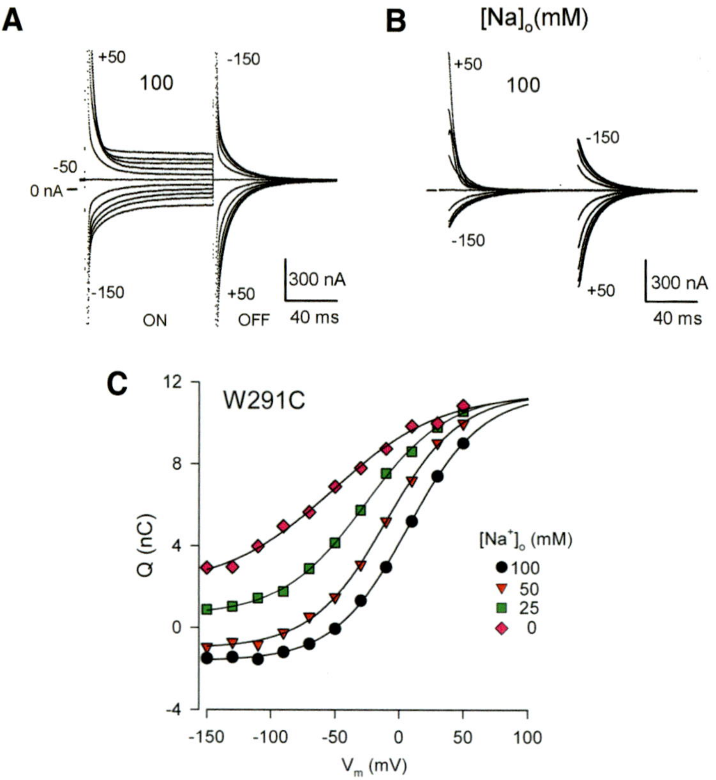

Fig. 6.16 hSGLT1 capacitive currents. (**a**) Total current records from an oocyte expressing WT hSGLT1 in 100 mM external NaCl buffer. Membrane potential was held at −50 mV (V_h) and stepped (ON) to various test voltages between +50 and − 150 mV for 100 ms before returning to V_h (OFF). (**b**) hSGLT1 capacitive currents at +50, −10, −90, and −150 mV. The transient currents were obtained from the total current by subtraction of the membrane bilayer capacitive and steady-state current. The charge Q was obtained from the integration of the SGLT1 transient currents. (**a** and **b**) are taken from Wright et al. (2011). (**c**) Characterization of charge movement for W291C-SGLT1.The data were collected on a single oocyte as the external sodium was varied from 0 to 100 mM. At each $[Na^+]_o$, the Q/V curves were fit to Boltzmann relations $((Q\text{-}Q_{hyp})/Q_{max} = 1/[1 + \exp(z(V_m - V_{0.5})F/RT])$ and are aligned at the depolarization limit (Loo et al. 2013). (**c**) taken from Loo et al. (2013)

(contributing a total charge $z = z_2 + z_1$). For a membrane with a transporter density N_T, the total charge is:

$$Q = N_T e(z_2[C1] + (z_1 + z_2)[C6]) \tag{6.3}$$

and maximal charge Q_{max} ($=N_T e(z_1 + z_2)$), where e is elementary charge). It may be shown (Loo et al. 2013) that $V_{0.5}$ as a function of $[Na^+]_o$ (Fig. 6.17) is described by:

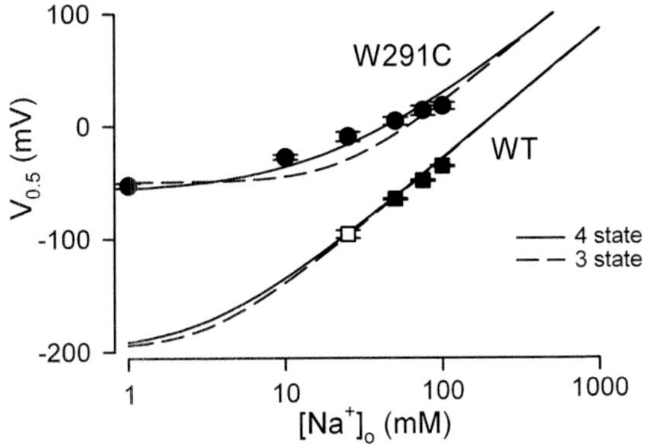

Fig. 6.17 Dependence of $V_{0.5}$ for Wild-Type (WT) hSGLT1 and Y291C on external sodium concentration (data from experiments shown in Fig. 6.16). The SE error bars are generally smaller than the size of the symbols. The curve are fit to the 3- and 4-state models of sodium binding. For WT the data is not sufficient to distinguish between the two models, but for Y291C there was a better fit to the 4-state model with K_{dA} and K_{dB} of 30 and 110 mM (Loo et al. 2013)

$$(z_1 + z_2)\left((1/K_a)e^{-z_1 uV_{0.5}} + \left([Na^+]_o{}^n/K_d{}^n K_a\right)e^{-(z_1+z_2)uV_{0.5}} - 1\right) - (2z_2/K_a)\,e^{-z_1 uV_{0.5}} = 0 \tag{6.4}$$

where $\mu = F/RT$. K_a and K_d can be estimated from Eq. 6.4. We consider the two limiting conditions:

1. In the absence of external Na^+, the empty transporter alternates between C6 and C1 depending on voltage. The voltage where C6 and C1 are equally distributed, $V_{0.5}$, is given by:

$$V_{0.5} = 2.303\ (RT/z_1F)\ \log\ (K_a) = 2.303\ (RT/z_1F)\ \log\ (k_{61}/k_{16}) \tag{6.5}$$

2. At high $[Na^+]_o$ (>> K_d), the transporter is distributed between C2Na$_n$ and C6 as a function of voltage. It can be shown from Eq. 6.4 that as $[Na^+]_o$ approaches saturation, $V_{0.5}$ is linearly related to log $[Na^+]_o$ with slope given by the ratio of the number (n) of Na^+ ions bound and the net charge (z) of the protein:

$$V_{0.5} = 2.303\ (n/z)\ (RT/F)\ \log\ [Na^+]_o \tag{6.6}$$

The theory can be extended to the sequential binding of 2 Na^+ ions (4-state model, Fig. 6.15), and the $V_{0.5}$ vs. $[Na^+]_o$ relation for sequential binding is given by:

$$(z_1 + z_2 + z_3)\left(1 - 0.5\left(1 + (1/K_a)e^{-z_1 uV_{0.5}} + \left([Na^+]_o/K_{dA}K_a\right)e^{-(z_1+z_2)uV_{0.5}} + \left([Na^+]_o{}^n/K_{dB}K_{dA}K_a\right)e^{-(z_1+z_2+z_3)uV_{0.5}}\right) + \left((z_2 + z_3)/K_a\right)e^{-z_1 uV_{0.5}} + \left([Na^+]_o z_3/K_{dA}K_a\right)e^{-(z_1+z_2)uV_{0.5}} = 0 \quad (6.7)$$

K_a is the equilibrium constant of the empty transporter, and K_{dA} ($= k_{N1}/k_{1N}$), and K_{dB} ($= k_{2N}/k_{N2}$) are the Na^+ affinities of the first and second sites (at 0 mV), respectively. z_1, z_2, and z_3 are the charge associated with the empty transporter, and the first and second Na^+ binding steps (see Fig. 6.15). K_{dA} and K_{dB} are related to the lumped K_d by $K_d = \sqrt{(K_{dA}K_{dB})}$.

The limiting behaviors of the $V_{0.5}$ vs. $[Na^+]_o$ relations (for the four-state model) are identical to that in the three-state model. In the absence of Na^+, $V_{0.5}$ depends on the apparent valence (z_1) of the empty transporter and the ratio of rate constants k_{16}/k_{61} (Eq. 6.4); and in saturating $[Na^+]_o$, $V_{0.5}$ is linearly related to log $[Na^+]_o$ with the slope given by the ratio (n/z) of the number of Na^+ ions bound (2) and the total charge z (Eq. 6.5). The difference lies in the approach to the limits. For the four-state model, as $[Na^+]_o$ approaches zero, the $V_{0.5}$ vs. $[Na^+]_o$ relations is governed by the affinity (K_{dA}) of the first Na^+ site. In contrast, as $[Na^+]_o$ nears saturation the approach to the maximal slope (115.6 mV/decade) is determined by the affinity (K_{dB}) of the second binding site.

K_a, k_{16}, and k_{61} were estimated from the $V_{0.5}$ and the time constants (τ_o) in the absence of Na^+ [$\tau_o = 1/(k_{16} + k_{61})$) and $K_a = k_{61}/k_{16}$]. The apparent valences z_1 and $z(= z_1 + z_2)$ were determined from the Q–V curves in the absence and presence of external Na^+. The Na^+ affinities (K_d, or K_{dA} and K_{dB}) were estimated by numerically fitting of the $V_{0.5}$ versus $[Na^+]_o$ data to (Eqs. 6.4 and 6.7) (see Fig. 6.17).

6.3.3.3 Presteady State Kinetics of W291C-SGLT1

By way of illustration, we focus on the kinetics of the W291C-SGLT1 mutant simply because the voltage-dependence of charge movement saturated between the practical limits of our voltage-clamp, −150 and +100 mV (Fig. 6.16c). At 100 mM external Na^+ the Q/V curve saturates with a $V_{0.5}$ of +12 mV and a z of 0.9. As the external Na^+ was reduced toward 0 mM, $V_{0.5}$ was −53 mV with a z of 0.6, and Q_{max} decreased from 10 to 3.5 nC. Using the time constants for the capacitive charge, 16 ms at 0 mV, give values of 21 and 43 s^{-1} for k_{16} and k_{61}. Although less reliable, we estimate the k_{16} and k_{61} values for wild-type SGLT1 to be 530 and 5 s^{-1}.

Plots of $V_{0.5}$ vs the external Na^+ concentration, Fig. 6.17, at saturating Na^+ concentrations show a limiting slope given by the ratio of the number of Na^+ ions bound and the valence of the protein. For both the mutant and wild-type transporter, the number of Na^+ ions bound is 2, which is consistent with other estimates of the Na^+/glucose coupling ratio of 2, e.g., from reversal potentials (Hummel et al. 2011). The $V_{0.5}$ versus Na^+ concentration curve for W291C-SGLT1 (Fig. 6.17) can be fitted to (Eqs. 6.4 and 6.5) to estimate the lumped sodium affinity K_d, and the affinities of the two Na^+-binding sites, K_{dA} and K_{dB}, 55, 30, and 110 mM. The data for wild-type SGLT1 cannot be resolved into the affinities for the two Na^+ sites, but for W291C the data can be resolved for the model with two sequential Na^+ binding sites, K_{dA} and K_{dB}, with values of 30 and 110 mM. A similar resolution of the two binding site affinities was obtained for other SGLT1 mutants (Y290C, Y290S, Y290F, Loo et al. 2013).

In summary, this analysis of capacitive SGLT1 currents shows that: two external Na^+ ions can bind to the protein in absence of sugar; provides estimates for the rate constants for voltage-dependent transitions between states 1 and 6; and the binding constants for the two external Na^+ ions.

6.3.3.4 Analysis of Models of SGLT1 Transport

Our approach was first described by Lucie Parent (Parent et al. 1992b) based on her steady-state and pre-steady state experiments on rabbit SGLT1 expressed in *Xenopus leavis* oocytes (Parent et al. 1992a). The model was chosen to minimize assumptions and the number of parameters based on the available experimental data. It was a six-state non-rapid equilibrium, ordered model with mirror symmetry, with a sodium to sugar stoichiometry of 2. We assumed that the valence of the unloaded transporters was −2, and the only voltage-dependent steps were the translocation of the empty carrier and the binding and dissociation of sodium. The presence of SGLT1 leak currents in the absence of sugar was taken as evidence for sodium uniport. Even with the assumption that two sodium ions bind to the transporter in one step, the six-state model equations contain 14 rate constants and an additional four voltage-dependent parameters. Global simulations of the steady-state kinetics for inward Na^+/glucose transport and the pre-steady state kinetics in the absence of sugar enabled us to obtain one set of parameters that quantitively accounted for the sodium and voltage-dependence of the glucose $K_{0.5}$ and V_{max}, the sugar and voltage-dependence of the sodium $K_{0.5}$ and V_{max}, the sigmoidal shape of the current/voltage curves, and the leak currents.

Further studies extended the model to human SGLT1 and confirmed or revised many of the assumptions. For example, the order of external Na^+ and sugar-binding, the stoichiometry of Na^+ to glucose transport, the density of SGLT1 proteins in the plasma membrane, and the reversibility of Na^+/glucose cotransport (see Wright et al. 2011) were confirmed. However, whole-cell patch-clamp experiment on HEK293 cells expressing hSGLT1, where inward and outwards currents were measured as a function of trans sodium and or glucose concentrations, have necessitated a rejection of the mirror-symmetry assumption (Adelman et al. 2016). These have shown that the internal release of ligands is stochastic and that the model can be reduced to a five-state model (see Fig. 6.13). The release step is electroneutral and rate-limiting at maximum transport rates (saturating external Na^+, sugar, and voltage) (Loo et al. 2006). The parameters that account for the five-state model are given in Table 6.1.

At this juncture, we have not extended modeling further as it would require extending the number of states and a subsequent increase in the number of

Table 6.1 Kinetic parameters for the five-state hSGLT1 transport model (Fig. 6.13)

$k_{12} = 50{,}000\ M^{-2}{\cdot}s^{-1}$	$k_{21} = 300\ s^{-1}$	$e_{12} = 0.3$	$e_{21} = 0.3$
$k_{16} = 600\ s^{-1}$	$k_{61} = 25\ s^{-1}$	$e_{16} = 0.7$	$e_{61} = 0.7$
$k_{23} = 45{,}000\ M^{-1}{\cdot}s^{-1}$	$k_{32} = 20\ s^{-1}$	$e_{23} = 0$	$e_{32} = 0$
$k_{34} = 50\ s^{-1}$	$k_{43} = 50\ s^{-1}$	$e_{34} = 0$	$e_{43} = 0$
$k_{46} = 10\ s^{-1}$	$k_{64} = 3{,}750{,}000\ M^{-3}{\cdot}s^{-1}$	$e_{46} = 0$	$e_{64} = 0$

Estimates for the rate constants for the transitions between the five states and the fraction of the electrical field for the voltage-dependent transitions between C1 and C6, and C1 and C2 (Adelman et al. 2016)

parameters. This includes consideration of sequential binding and dissociation of the two external sodium ions, and intermediate states between C1 and C6 (Loo et al. 2005, 2013). Such extensions increase the number of parameters and coefficients beyond our present ability to obtain the necessary additional experimental data to make the effort productive.

6.3.4 SGLT2

Although the gene for SGLT2 was cloned in 1992 (Wells et al. 1992), relatively little is known about the properties of this transporter despite the fact that it has been a very successful target for the treatment of Type 2 Diabetes Mellitus. The gene is almost exclusively expressed in the renal cortex where it is responsible for glucose reabsorption from the glomerular filtrate. The paucity of functional information arises from the fact that it is poorly expressed in heterologous expression systems such as *Xenopus laevis* oocytes or cultured cells studied at room temperature.

Overall, the properties of hSGLT2 are quite similar to SGLT1 at 37 °C, but with subtle differences such as a higher K_m for glucose, 5 versus 2 mM, a higher K_m for D-galactose, >100 versus 6 mM, and a higher affinity for phlorizin, K_i 11 vs 140 nM. Major differences are that the Na^+/glucose coupling ratio for SGLT2 is 1 versus 2 for SGLT1, and specific SGLT2 capacitive currents have not been resolved (Hummel et al. 2011, 2012). SGLT2 transport is reversible, and phlorizin and specific SGLT2 inhibitors only bind to the outward-facing conformation (Ghezzi et al. 2014; Bisignano et al. 2018). Similar selectivity and kinetics were reported for SGLT2 co-expressed with MAP17 in *Xenopus* oocytes at 22 °C (Coady et al. 2017). Estimates of SGLT2 turnover are similar to those for SGLT1 at 22 °C. MAP17 was essential for functional expression of SGLT2 in oocytes, but the explanation for this is not clear.

Given the high amino acid sequence identity between SGLT2 with SGLT1, 59%, and that the glucose binding site residues and the Na2 sodium site residues are conserved, at this time we assume their secondary and tertiary structures are similar, as is the 5-state kinetic model (Figs. 6.12, 6.13, 6.18, 6.19, 6.20).

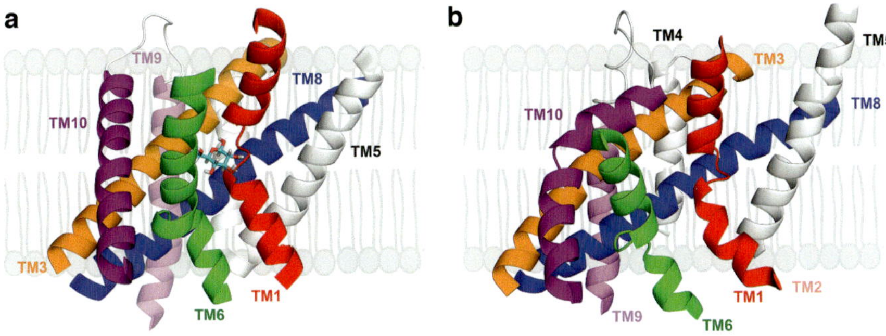

Fig. 6.18 Homology models of the outward- and inward-facing hSGLT1. (**a**) Sideview of hSGLT1 in the out-open conformation (based on SiaT PDB file 5NVA), and (**b**), the side view of hSGLT1 in the inward-facing apo conformation (based on vSGLT PDB file 2XQ2). TM2, TM7, TM11, and TM12–13 are removed for clarity. In transiting from the outward-facing to inward-facing, the external end of TM10 (purple) bends inward to block external access to the sugar vestibule, and this is accompanied by a bending of the internal half of TM1 (red) away from the vestibule, and a straightening of TM6 (green). D-glucose is coordinated by residues at or near the discontinuous regions of TM1 and TM6 in the outward-facing conformation (see Fig. 6.20)

6.3.5 *Structure*

Collaborations between the Wright and Abramson labs solved the crystal structures of a bacterial homolog of the mammalian SGLTs, vSGLT, in two states, galactose-bound and the apo state (Faham et al. 2008; Watanabe et al. 2010). The structures were modeled as a 14-transmembrane helical protein (TM-1 to TM13) with a 2D topology similar to that predicted from biochemical evidence (Fig. 6.12). A signature feature of this inward-open, galactose-bound occluded structure was that it contained a 5 + 5 transmembrane helical-repeat (TM 1–5 and TM 6–10) in an antiparallel orientation with a pseudo-symmetry axis in the membrane plane. There was no amino acid homology between the two repeats although the structures may be superimposed with an RMSD less than 4 Å. The first helix of each domain, TM1 and TM6, contains a discontinuous region in the middle of the membrane. The first two helices in each inverted-repeat formed a central 4-helix bundle, TM1, TM2, TM6, and TM7, and this core is surrounded by a scaffold bundle formed by TM3, TM4, TM8, and TM9. A sugar, galactose, was found in the discontinuous region of the core bundle interacting with residues on TM1, TM2, TM6, and TM10, and occluded from the external aqueous solutions by a pair of hydrophobic gates. Individual mutation of the residues interacting with galactose—Q428, Q69, E88, K294, N260, and Y263, to alanine dramatically reduced sodium-dependent sugar transport. A putative sodium binding site was formed close to the interactions between TM 1 and TM8, some 10 Å away from the sugar-binding site. Na^+ was thought to be coordinated with the carbonyl groups of A62, I65, and A361 and the hydroxyl group of S361 at a site now known as Na2. The external gate to the galactose binding site was predicted to be formed by hydrophobic residues M73,

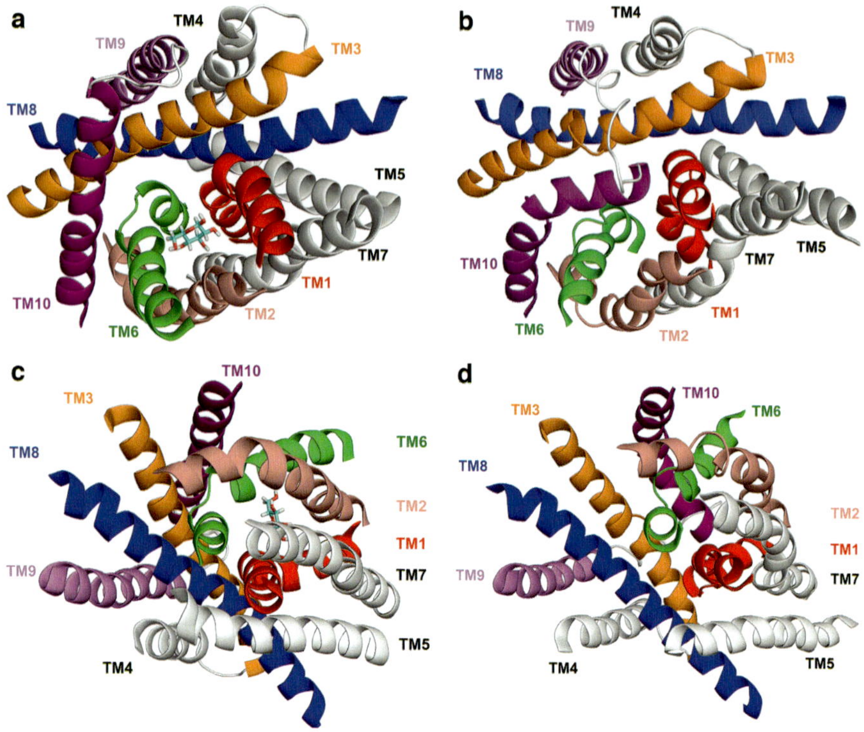

Fig. 6.19 External and internal views of hSGLT1 in the outward- and inward-facing conformations. (**a**) External view of hSGLT1 in the outward-open state (from SiaT PDB 5NVA). (**b**) External view of hSGLT1 in the inward-facing conformation (from vSGLT 2XQ2). The major difference between outward-open and inward-open states is the external end of TM10. (**c**) Internal view of the outward-open conformation of hSGLT1 (based on SiaT 5NVA). The external end of TM10 is straight. (**d**). Internal view of the inward-open conformation of hSGLT1 (based on vSGLT 2XQ2). Note the bending of the external end of TM10 and the absence of sugar. For clarity TM7, TM11–13 are removed

Y87, and F434, but this may include the external ends of TM1 and TM10, a protein mass formed by the between-helical ends of TMs 2, 3, 6, and 10 that interact with the extracellular loops between TM1 and TM2, TM7 and TM8, and TM9 and TM10. Sometimes this is referred to as a thick gate. The internal gate was predicted to be formed by Y263, flanked by Y262 and W264, on TM6. The inner gate opens to a large hydrophobic channel leading to the cytoplasm formed by the cytoplasmic halves of TM1, TM2, TM5, and TM7.

Initially, we were disappointed to learn that the structural fold of vSGLT was not unique, but the central core, 5 + 5 TM inverted repeat, was identical to that of bacterial sodium-dependent leucine cotransporter, LeuT, in the unrelated SLC6 gene family (Yamashita et al. 2005). Subsequently, we have learned that there is a superfamily of 5 + 5 TM inverted repeat transporters including those for amino acids, nucleobases, metal cations, and anions (Abramson and Wright 2009). The

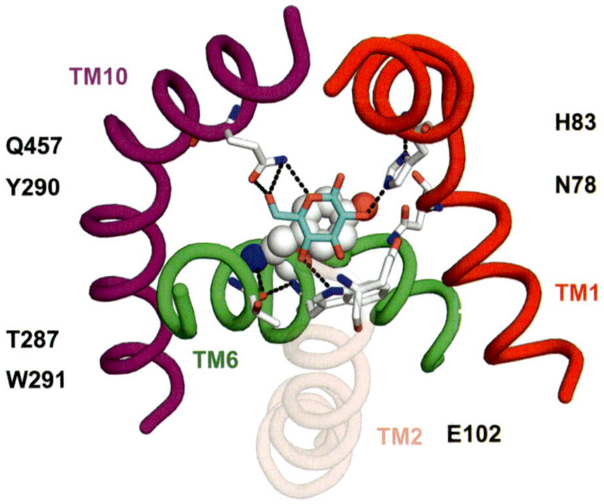

Fig. 6.20 The glucose-binding site of hSGLT1. D-glucose is coordinated with residues H83 and N78 on TM1, E102 in TM2, N287, Y290 and W291 on TM6, and Q457 on TM10. The pyranose ring is stacked on Y290 based on pi–pi interactions. Redrawn from the homology model based on PDB 5NVA

family of SGLT genes (also known as the SSS or SSF) is in a larger family of proteins in pro- and eukaryote organisms in the APC (amino acid/polyamine/organocation) class of membrane proteins. The 2.A.22 subfamily in the transporter classification database (TCDB, http://www.tcdb.org/tcdb) contains over 70 transporters and exchanger in seven different solute carrier (SLC) families (see also http://blanco.biomol.uci.edu/mpstruc/). There is no amino acid homology between the 5 + 5 inverted repeats in each protein, as well as no amino acid homology between the different members of the structural family. The substrate-binding site overlaps in the central cavity in each protein, and there is evidence that the sodium binding site, Na_2, is conserved (see Wright et al. 2011).

Collaboration with Michael Grabe on the molecular dynamics of vSGLT has provided novel insights into sodium sugar cotransport (Watanabe et al. 2010; Adelman et al. 2016). First, starting from the crystal structure it was found that galactose has dynamic interactions with water in the sugar-binding site, before escaping the inner gate and exiting with sodium to the cytoplasm down the cytoplasmic cavity between the inner ends of TM1, TM2, TM5, and TM7. However, repeated long microsecond runs showed a stochastic release of galactose and sodium from their binding sites into the cytoplasm. It was also noticed that is an occasional release of galactose to the external solution. This opening of the external gate, residues Y87, F424, and Q428, does not involve any change in the protein backbone, but rather is due to changes in their side-chain conformations. Calculations indicate that the open outer gate conformation only exists about 8% of the time to permit outward galactose escape.

As of 2020, there are no structures available for mammalian SGLT despite significant efforts by at least four groups. The bottleneck is the production of sufficient pure protein for crystallization trials and the fact that the protein is too small to be solved by cryo-electron-microscopic methods. However, in collaboration with Paola Bisignano and Michael Grabe, homology models have been made for

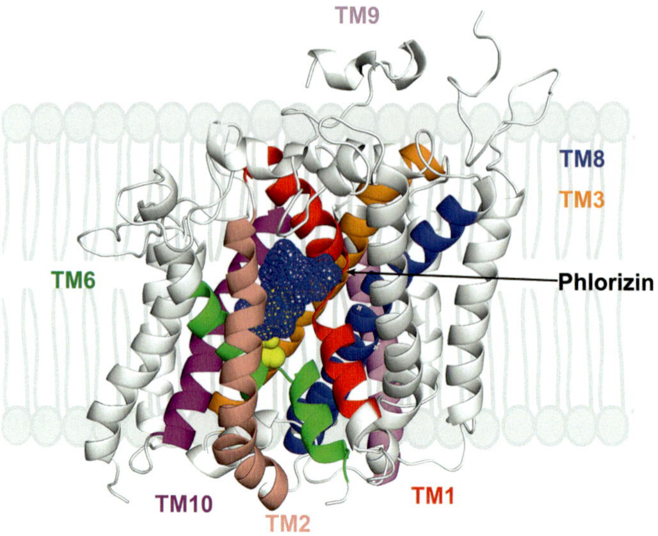

Fig. 6.21 Phlorizin binding. The space predicted to be occupied by phlorizin. Phlorizin is docked into the outward-open conformation of hSGLT1 based on SiaT 5NVA. The sugar moiety is in the same position as glucose and interacts with N78, H83, E102, Y290, W291, and K321 as in Fig. 6.20. The A-phenolic ring of phlorizin interacts with F101 (in TM2) and H83 (in TM1) via face-to-face aromatic pi–pi interactions (Bisignano et al. 2018)

hSGLT1 and hSGLT2 in two conformations, outward- and inward-facing open (Bisignano et al. 2018). The approach was to build the inward-facing model based on the inward-facing ligand-free structure of vSGLT, and the outward-facing conformation of the sodium sialic acid cotransporter SaiT. These proteins have 24–32% amino acid identity, and 46–60% similarity to SGLTs. The initial goal was to identify phlorizin and SGLT2 inhibitor binding sites and to test the predictions using select hSGLT1 mutants expressed in oocytes (Figs. 6.18, 6.19, 6.20 and 6.21).

Inspection of the model structures in Figs. 6.18 and 6.19 gives some rudimentary clues about the conformational changes of hSGLT1 that occur as the protein moves from the sodium-bound, outward-facing state C3 through C4–C6 (see Fig. 6.13). The obvious structural change is that the outer end of TM10 (purple) bends by 40° at P465/P466, and a straightening of TM6. The overall response is toward closing the outer gate as the transporter transitions to facing the inside. Information about the ligand release step first came come from comparing the inward-facing occluded structure of vSGLT (Faham et al. 2008) with the apo structure (Watanabe et al. 2010). The internal half of TM1 moves away from the center of the inner vestibule to open the pathway for Na^+ and sugar release into the cytoplasm. This also occurs in the models of hSGLT1 structure. What is now needed are high-resolution structures of the minimum 5-states of the transporter (Fig. 6.13), further molecular dynamical modeling, and biochemical and biophysical assays to follow the transitions between states in real time, e.g., Fig. 6.23.

The predicted glucose-hSGLT1 binding site recapitulates the sugar-binding site in the vSGLT crystal structure (Faham et al. 2008) and was verified by functional assays of hSGLT1 mutants expressed in *Xenopus laevis* oocytes (Sala-Rabanal et al. 2012). The putative Na2 sodium binding site residues are conserved in hSGLT1, A76, I79, S389, and S392, and these were also tested by mutational analysis (Sala-Rabanal et al. 2012; Loo et al. 2013). S389 was not involved, but S392C reduced the affinity of the Na2 site. We also measured Na^+ affinity in the absence of sugar for glucose binding site mutants (H83C, Y290C, W291C, and Q457C) using capacitive currents (see Figs. 6.15, 6.16, and 6.17), and surprisingly we recorded decreases in sodium affinity (Jiang et al. 2012; Loo et al. 2013). We suggested that these formed the Na1 site, but this was not confirmed by molecular studies (Bisignano et al. 2018).

Mutations of the triplet (W289, Y290, W291) conserved in the SLC5 family provided evidence on the importance of aromatic interactions in the transport cycle of SGLT1 (Jiang et al. 2012): (1) CH—pi stacking of Y290 with glucose; (2) T—pi stacking between Y290 and W291, and H-bonding between Y290 on TM6 and N78 on TM1; (3) H-bonding between W289 on TM6 and S149 or Y152 on TM3; and (4) the triad is important in determining the coupling of Na^+ and sugar transport. The coupling ratio was reduced from 2 to 1 with the Y290F and W291F mutants, while the binding of 2 Na^+ ions was still required for transport. However, this is not the whole story as the Na^+ to sugar coupling of SGLT2 was 1/1 despite the fact that the aromatic triplet is conserved.

The structural models identified the aqueous vestibules where glucose enters the binding site from the extracellular side and leaves for the cytoplasm side (see also, Faham et al. 2008; Watanabe et al. 2010). Two separate lines of evidence support these conclusions: the first comes from the analysis of phlorizin binding to the outward- and inward-facing models (Bisignano et al. 2018); and the second comes from biochemical and biophysical experiments on fluorophore labeling of the glucose binding site (Gorraitz et al. 2017).

6.3.5.1 Inhibitors

Phlorizin is a non-transported, competitive inhibitor of Na^+/glucose cotransport, that also blocks the voltage- and Na^+-dependent charge movements in the absence of sugar (see above, and Fig. 6.13). Phlorizin binds to the outward-open conformation of hSGLT1 with the glucose moiety in the sugar-binding site and the aglycone in an aqueous vestibule stretching from the sugar-binding site to the external surface (Fig. 6.21). The vestibule is bounded by the external ends of TM1, TM2, TM3, TM6, TM8, and TM10. Y290 has stacking interactions with glucose and phlorizin, and its -OH group H bonds to phlorizin. Mutating Y290 to Y290F decreases the affinity of phlorizin binding by eightfold, consistent with the loss of one- H-bond. In addition, F101 makes a pi–pi interaction with aromatic elements in phlorizin, and we found that the F101C mutation reduced the phlorizin affinity 170-fold with no effect on sugar transport. On the other hand, phlorizin did not bind to the inward-facing conformer of hSGLT1 (supplementary Fig. 6.1b, Bisignano et al. 2018), and this is

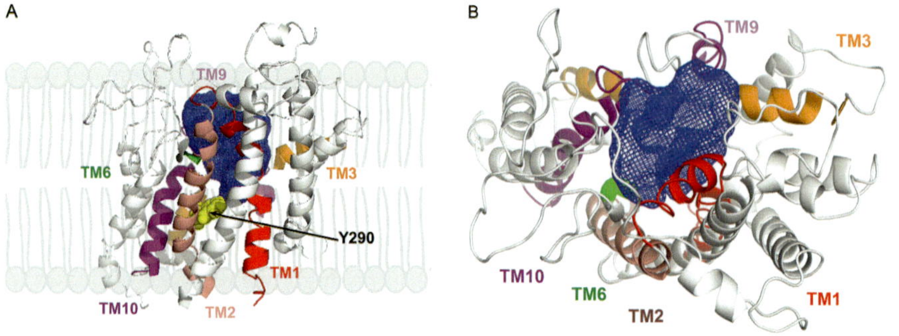

Fig. 6.22 External vestibule of hSGLT1. The space predicted to be occupied by MTS-TAMRA in the outward-facing conformation ($C2Na_2$) when covalently labeled at positions Y290C, T287C, H83C, and N78C in the glucose binding site. The fluorophore is bounded by the outer ends of TM1, TM2, TM3, TM6, TM8, TM9, and TM10 with a volume of 600 $Å^3$. Shown are the side (**a**) and top (**b**) views. From Gorraitz et al. (2017)

consistent with the lack of functional effects of cytoplasmic phlorizin (Ghezzi et al. 2014).

The second line of evidence for the location of the external aqueous vestibule comes from biochemical experiments to covalently tag the glucose site with fluorescent reagents (Gorraitz et al. 2017). Residues in the glucose binding site (Fig. 6.20) were each mutated in turn to cysteines, and the mutants were expressed in *Xenopus leavis* oocytes. Treating the oocytes, e.g., T287C-hSGLT1, with MTS-TAMRA ([2-((5(6)-tetramethylrhodamine) carboxylamino) ethyl methanethiosulfonate) in the presence of external sodium reversibly blocked cotransport, but not the voltage and Na^+-dependent conformational changes from C2 to C1 to C6. The labeled proteins were insensitive to the presence of external glucose, showing that access to the glucose binding site was restricted. Similar results were obtained with the N78C, H83C, and Y290C, and with TMR6M (tetramethylrhodamine-6-maleimide) and TRM5M (tetramethylrhodamine-5-maleimide), but there were no effects on the native hSGLT1 transporter. Molecular modeling of MTS-TAMRA and other fluorophores covalently bound to N78C, H83C, T287C, Y290C, and W291C showed that the fluorophores were in an aqueous vestibule surrounded by the outer ends of TM1, TM2, TM3, TM6, TM8, and TM10 (Fig. 6.22). This is the same location as phlorizin bound to the outward-facing conformation of hSGLT1 (see Fig. 6.21).

The dynamics of conformational changes between C2–C1–C6 covalently labeled at the glucose binding site of T287C-hSGLT1 with MTS-TAMRA have been recorded using voltage-clamp fluorometry (Gorraitz et al. 2017). As the transporter state was driven from $C2Na_2$ to C1 and then C6 by voltage, there was an increase in fluorescence intensity of TAMRA with a time course similar to the charge movements (Q) with fast (~1 ms) and slow (10–75 ms) time constants. The fluorescence changes were larger in external Na^+ than its absence (Fig. 6.23a vs 6.23d). We predict that the larger quench of fluorescence in Na^+ is due to an increase in solvent

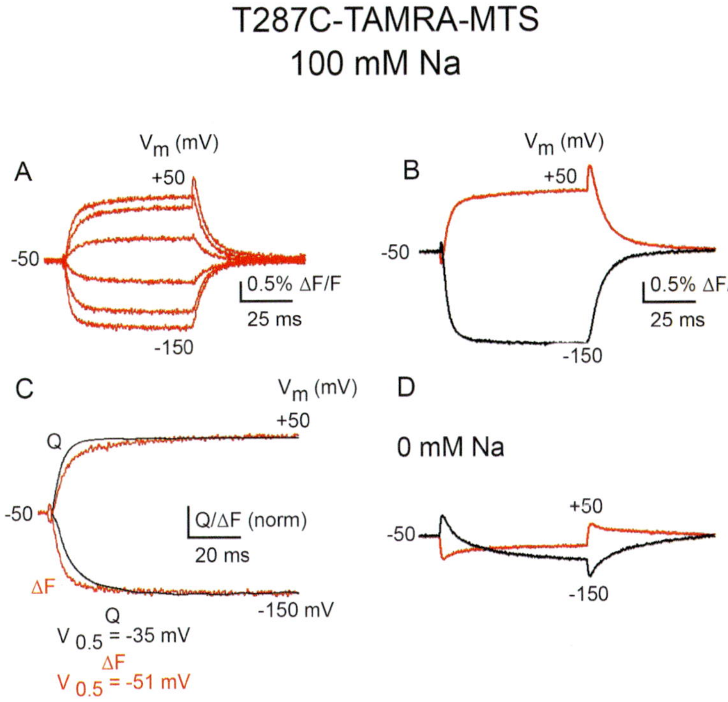

Fig. 6.23 Voltage clamp fluorometry of MTS-TAMRA labeled T287C-hSGLT1. The mutant was expressed in *Xenopus laevis* oocytes, T287C was labeled with MTS-TAMRA in external sodium, and the TAMRA fluorescence intensity changes (ΔF) and SGLT1 capacitive currents were recorded simultaneously when the membrane voltage was stepped from the holding potential, −50 mV, to hyperpolarizing and depolarizing potentials for 100 ms. (**a**) Fluorescence changes were recorded at voltage jumps between +50 and −150 mV. (**b**) For clarity only, the voltage jumps to +50 and −150 mV are shown. (**c**) Simultaneous charge (*Q*) and ΔF time courses are shown for the +50 and −150 mV pulses (normalized to agree at the end of the 100-ms pulse). (**d**) The fluorescence changes recorded in the absence of Na^+ (same experiment as **a**–**c**). Reproduced from Gorraitz et al. (2017)

access to the fluorophore and a more hydrophilic environment in the outward-open conformation ($C2Na_2$) with the opening of the outer gate, seen as a straightening of TM10. The voltage-induced changes in conformation are a reflection of the time scale for opening and closing the external gate, and transitions of the apo-transporter between the inward and outward states. We trust that other optical techniques will be employed to record the transitions between other steps in the transport cycle.

6.4 Intestinal and Renal Glucose Transport

6.4.1 Intestine

Having reviewed the properties of sugar transporters we now return to their specific relevance to glucose transport across epithelia (Fig. 6.1), specifically the small intestine and renal proximal tubule. In humans, approximately 180 grams of glucose is completely absorbed by the intestine each day, and a comparable amount is reabsorbed from the glomerular filtrate. In both cases, transport occurs uphill against the glucose concentration gradient. Up to 50 grams of fructose are absorbed by the intestine, but this is a passive process down its concentration gradient. Fructose reabsorption in the kidney is only likely to be a fraction of that for glucose, owing to low plasma fructose concentration (<<0.3 mM) and the small amount filtered.

The importance of the paracellular route is unclear given the selectivity of sugar absorption, and the lack of an "equivalent circuit analysis" to determine the relative amount of passive permeation through the cellular (#2 and #4) and paracellular (#8) pathways. Undoubtedly, some passive sugar permeation does occur through the junctions (see Krug 2017), but the physiological significance is unclear. A cellular route is required to explain the saturation kinetics, selectivity, competition, and effect of specific inhibitors such as phlorizin.

A central issue in quantitating sugar transport across the cellular pathway is the intracellular metabolism of glucose and fructose. In some species, more than 40% of glucose absorbed is converted to lactate. This has been addressed by the use of non-metabolized glucose analogs including: 3-O-methyl-D-Glucose (3OMG), a substrate for both GLUTs and SGLTs; 2-deoxy-D-glucose (2DOG), a substrate for GLUTs, but with the disadvantage that is metabolized to 2-deoxy-D-glucose-6-phosphate by intracellular hexokinase; and α-methyl-D-glucopyranose (αMDG), a substrate for SGLTs and a poor substrate for GLUT2 (see Sala-Rabanal et al. 2016). These substrates are commercially available as C14 or H3 labeled sugars, as is DOG as 2-F18–2-deoxy-D-glucose (2FDG). We have introduced an F18 derivative of αMDG, α-methy-4F18–4-deoxy-D-glucopyranoside (Me4FDG) (Wright et al. 2014). Advantages of F18-labeled sugars are that they decay by positron emission with a short half-life, 109 min. The F18-labeled sugars along with the use of Positron Emission Tomography (PET) permits us to noninvasively monitor their distribution in living animals and human subjects with a time scale of seconds and a spatial resolution of 6 mm.

The importance of SGLT1 in intestinal absorption in vivo has been documented by studies in mice and patients lacking the transporter, SGLT1$^{-/-}$ mice and patients with glucose–galactose malabsorption (GGM OMIM, 182380) (see Wright et al. 2011; Sala-Rabanal et al. 2018). In patients with GGM, mutations in the SGLT1 gene (SLC5A1) produce either truncated SGLT1 protein or mutant full-length protein that are not targeted correctly to the brush border membrane (see Wright et al. 2011). Such patients suffer glucose and galactose malabsorption with accompanying life-threatening diarrhea. Oral Glucose Tolerance Tests are abnormal in

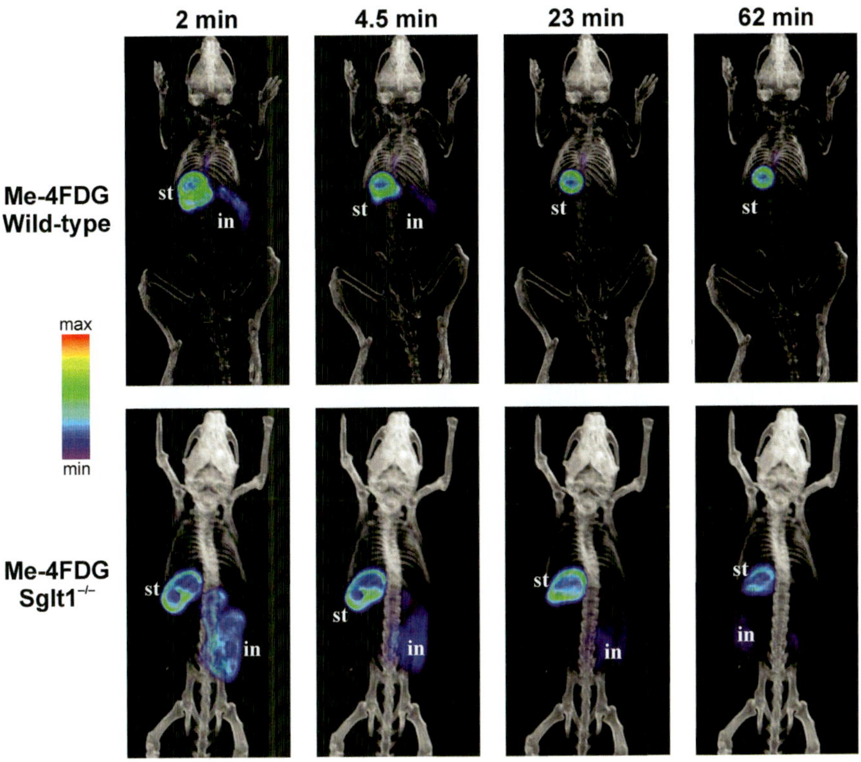

Fig. 6.24 Oral glucose tolerance in wild-type and SGLT$^{-/-}$ mice. The distribution of Me4FDG was followed for 60 min after gavage of 2 g glucose/g body weight containing 300 μCi of Me4FDG. A dose was slowly emptied for the stomach and rapidly absorbed in the upper small intestine. In wild-type mice (*upper panel*) the glucose entering the intestine was completely absorbed within 10 min, while in the SGLT$^{-/-}$ mouse (*lower panel*) it took 45 min to absorb the amount of glucose delivered into the small intestine. Taken from Sala-Rabanal et al. (2018)

both patients and SGLT1$^{-/-}$ mice. PET imaging of wild-type and SGLT1$^{-/-}$ mice with Me4FDG as a tracer showed glucose entering the small intestine was completely absorbed in less than 10 min in normal mice (Fig. 6.24a), while in SGLT1$^{-/-}$ mice absorption was much slower with a half-time of 45 min (Fig. 6.24b). This slow absorption in SGLT1$^{-/-}$ mice was not accompanied by an increase in blood glucose as the absorbed glucose was distributed to organs throughout the body. In the same study, Oral Glucose Tolerance tests with Me4FDG where phlorizin was included in the gavage solution also slowed the rate of glucose absorption in wild-type mice with no rise in blood glucose levels.

GLUT2 is unimportant for intestinal glucose absorption in both mice and human subjects. In GLUT2$^{-/-}$ mice oral glucose tolerance tests gave results that were identical to wild-type mice (Stümpel et al. 2001; Sala-Rabanal et al. 2018). Likewise, in patients with the Fanconi–Bickel Syndrome (OMIM 227810), who have severe

GLUT2 mutations, there is no deficit in intestinal glucose absorption although there was severe glucosuria (Santer et al. 2003). As there was no upregulation of other GLUT genes in the SGLT1$^{-/-}$ mice, it remains unclear how glucose exits from the intestinal epithelium across the basolateral membrane. The current dogma is that brush border SGLT1 and basolateral GLUT2 are responsible for glucose absorption in the intestine (Fig. 6.26), but as discussed above there is evidence in mice and man that GLUT2 is not essential. Nevertheless, the coupling between sodium and glucose transport leads to net salt and water absorption, as sodium entry with glucose across the brush border membrane is matched by sodium pumping out of the cell by the basolateral Na^+/K^+-pump. Water is coupled to sodium–glucose cotransport, either directly or indirectly by SGLT1, and we estimate that this accounts for 65% of the passive water flow and up to 80% of the glucose-stimulated water absorption (Loo et al. 2002; Zeuthen et al. 2016).

Fructose absorption across the intestinal epithelium is passive, and the current dogma is that GLUT5 in the brush border membrane (#6) and GLUT2 in the basolateral membrane (#7) are responsible (Fig. 6.25) (see Ferraris et al. 2018). Fructose absorption was eliminated in GLUT5$^{-/-}$ mice, but interestingly it was normal in GLUT2$^{-/-}$ mice (Stümpel et al. 2001). So, it appears that neither brush border nor basolateral GLUT2 plays an essential role in fructose absorption.

6.4.2 *Kidney*

In the human kidney all of the filtered load of glucose, 180 g/day, is completely reabsorbed and is blocked by phlorizin (Chasis et al. 1933). Glucose reabsorption has been located in the proximal tubule of nephrons. Normally SGLT2 expressed in the early proximal tubule is responsible for the bulk of glucose reabsorption, while SGLT1 in the late segments "mop" up the rest (Figs. 6.26 and 6.28). This is supported by observations in patients with familial renal glucosuria (OMIM 233100) with mutations in the SGLT2 genes (Santer and Calado 2010). Unfortunately, there are few studies that have addressed the transport defects in the mutant SGLT2s. In patients with truncated SGLT2 proteins the renal excretion of glucose often approaches the filtered load, but not in all cases. Patients and mice treated with specific SGLT2 inhibitors excrete glucose, but the maximum inhibition of glucose reabsorption is roughly only 50% (Rieg et al. 2014). The explanation was found in SGLT1$^{-/-}$ mice where the inhibitors completely blocked renal reabsorption.

In the early tubule of humans and mice, SGLT2 protein is expressed in the apical membrane, and GLUT2 in the basolateral membrane (Vrhovac et al. 2015, Ghezzi et al. 2017; Cramer et al. 1992). SGLT2 activity is blocked by external phlorizin and specific SGLT2 inhibitors (Ghezzi et al. 2014), indicating that the inhibitors are filtered by the glomerulus and then block glucose absorption from the tubule lumen (Ghezzi et al. 2017). Basolateral GLUT2 is required to complete glucose absorption in all segments of the proximal tubule (Figs. 6.26 and 6.28), as noted by the severe glucosuria in patients with Fanconi–Bickel syndrome.

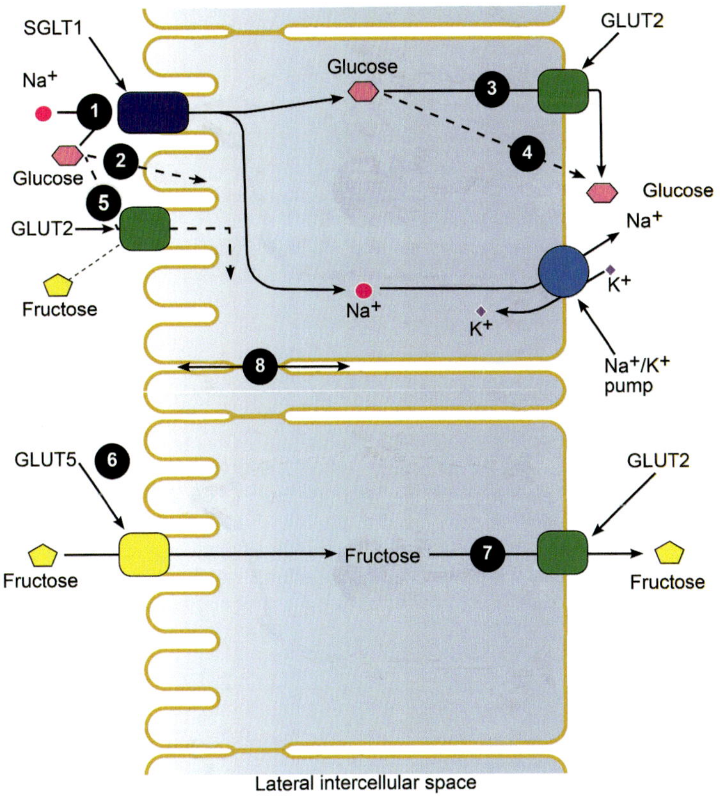

Fig. 6.25 Potential pathways for glucose and fructose absorption across intestinal and renal epithelia. There are two separate routes that sugars may take to cross the epithelium between the gut and the blood: paracellular where the sugar bypasses the cell across the tight junctions (#8), and cellular where the sugar goes through the cell passing in turn, the brush border and basolateral membranes. There are essentially three different possibilities for each membrane, passive diffusion (#2 and #4), SGLTs (#1), and GLUTs (#7 and #6)

The requirement for SGLTs and GLUT2 in renal glucose reabsorption has been confirmed in microPET studies of 2FDG and Me4FDG urinary excretion in wild-type, GLUT2$^{-/-}$, SGLT1$^{-/-}$, and SGLT2$^{-/-}$ mice (Fig. 6.27). The was no excretion of Me4FDG in normal mice but this SGLT-specific tracer was excreted in both SGLT1$^{-/-}$ and SGLT2$^{-/-}$ mice, as was the case on treating wild-type mice with the specific SGLT2 inhibitor dapagliflozin (Ghezzi et al. 2017). The maximum excretion of Me4FDG was observed in GLUT2$^{-/-}$ mice, implying that the normal absorption of Me4FDG requires the basolateral GLUT2 to exit from the epithelium across the basolateral membrane, even though Me4FDG is a poor substrate for this transporter (Sala-Rabanal et al. 2016). 2FDG is not a good substrate for either SGLT1 or SGLT2, and this accounts for the high rate of excretion of this tracer into the urinary bladder in normal mice.

The restricted expression of SGLT2 to the apical membrane of the early and mid-segments of the renal proximal tubule (Fig. 6.28), along with the high affinity of

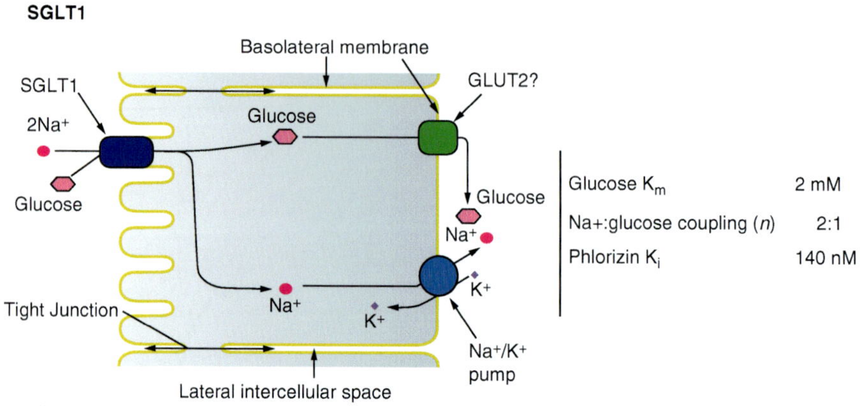

Fig. 6.26 Active glucose absorption across the intestine and the late proximal tubule. The first step is the accumulation of glucose in the epithelium by SGLT1 in the brush border membrane. This is driven by the sodium electrochemical potential gradient across the apical membrane, and cotransport results in two sodium ions entering the cell with glucose. The sodium gradient is maintained by the basolateral Na^+/K^+-pump that expels sodium into the blood. Studies on $GLUT2^{-/-}$ mice and patients with Fanconi–Bickel syndrome suggest that basolateral GLUT2 is not essential for intestinal glucose absorption, but it is for renal absorption

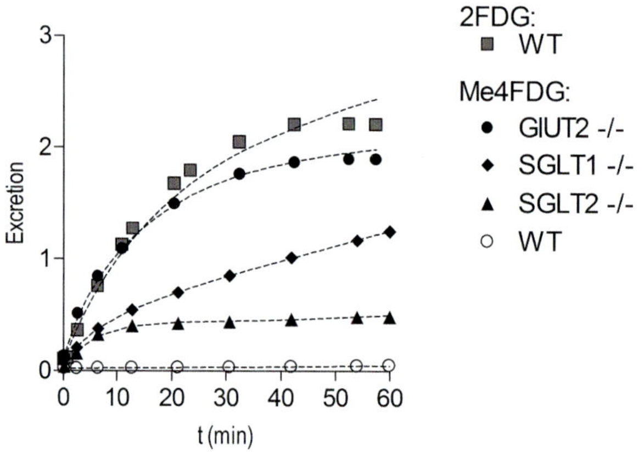

Fig. 6.27 Urinary excretion of the PET tracers 2FDG and Me4FDG in wild-type, $GLUT2^{-/-}$, $SGLT1^{-/-}$, and $SGLT2^{-/-}$ mice as determined by microPET. 300 μCi of F18 tracer was injected intravenously and the distribution within the whole mouse was followed for 60 min. The total rate of urinary bladder (MBq) was determined as a function of time. There was no significant metabolism of the tracer in blood or urine as measured by radio-thin-layer chromatography. Taken from Sala-Rabanal et al. 2016

SGLT2 for specific inhibitors such as dapagliflozin, K_i 6 nanomolar, account for their success in treating Type 2 Diabetes Mellitus (Gallo et al. 2015; Rieg and Vallon 2018). The oral drugs are absorbed, filtered by the glomerulus, and inhibit glucose

SGLT2

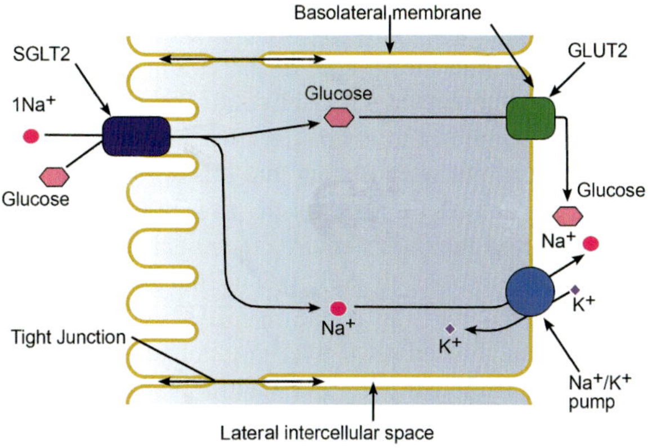

Glucose K_m (mM)	5 mM
Na+:glucose coupling (n)	1:1
Phlorizin K_i	11 nM

Fig. 6.28 Glucose reabsorption across the S1 and S2 segments of the renal proximal tubule. SGLT2 in the brush border membrane and GLUT2 in the basolateral membrane account for glucose reabsorption from the glomerular filtrate in this segment

reabsorption, thereby increasing glucose excretion and lowering blood glucose levels. This is a remarkable application of the fruits of basic physiology to improving the lives of millions of people.

6.5 Conclusions

6.5.1 *Similarities and Differences Between GLUTs and SGLTs*

Given the advances in the biology of GLUTs and SGLTs, it is appropriate to highlight the similarities and differences. First, the structural folds are different as GLUT2 and GLUT5 are members of the MFS structural family, while SGLTs belong to the APC structural family. There is some structural symmetry and functional similarity in that: (1) SGLTs have inverted repeats TM1-TM5 and TM6-TM10 while the GLUTs have M1-TM3 and TM4-TM6, and TM7-TM9 and TM10-TM12 (see Fig. 6.6); (2) the discontinuous helices TM1 and TM6 of SGLTs and TM7 and TM10 of GLUTs are involved in glucose coordination; and (3) the transition from outward-open to inward-open conformations is determined by local motions in the external and internal ends of transmembrane helices, in SGLTs the external TM10 and TM6 and internal TM1 and TM5, and in GLUTs external TM7 and internal TM10.

It is noteworthy that some members of the MFS are cotransporters, e.g., the H^+/lactose cotransporter (LacY), while the GLUTs are not. It is rather interesting that the transport kinetics of SGLT1 are very similar to those described for LacY (Kaback and Guan 2019). Both families operate by mechanisms where a single occluded sugar-binding site in the center of the protein, mainly coordinated by H-bonds, that alternately opens to each side of the plasma membrane. In the cotransporters, the opening and closing of the external and internal gates may be controlled by ions, Na^+ and/or H^+, and the membrane potential. As yet, it is not clear what determines the power stroke in uphill glucose transport, but we do know that this is related to the number of cation binding sites, e.g., 1 for SGLT2 and 2 for SGLT1. In both GLUTs and SGLTs, competitive non-transported inhibitors bind to the sugar-binding site and block transport: phlorizin from the outside in SGLTs and maltose from the outside in GLUTs, and Cytochalasin B from the inside in GLUTs 1–4 but not GLUT5. Phlorizin has been successfully used as a lead compound to develop SGLT2 drugs to treat Type 2 Diabetes Mellitus (Rieg and Vallon 2018), but so far there are no drugs to block glucose or fructose transport by specific GLUTs to treat cancer or obesity, e.g., GLUT1 or GLUT5. While GLUTs and SGLTs transport sugars, we still do not know what determines the selectivity of GLUT2 for glucose and GLUT5 for fructose, or the selectivity of SGLT2 for glucose over galactose, or the selectivity of SGLT2 over SGLT1 for diabetic drugs.

6.5.2 *Epithelial Sugar Transport*

It is self-evident that the transport of glucose and fructose across epithelia in the body is determined by the trafficking of GLUTs and SGLTs to one face of the cell or the other. In some cases, GLUT1 is directed to both plasma membranes of the endothelium that serves as the blood–brain barrier, while in the intestine and renal proximal tubule SGLTs are directed to the apical membrane and GLUT2 in the kidney to the basolateral membrane. In the former case, GLUT1 facilitates the uptake of some 100 grams of glucose into the brain, whereas in the latter the two transporters in series are responsible for the active reabsorption of glucose from the glomerular filtrate. In the case of fructose in the intestine, it is the expression of GLUT5 in the brush border, and possibly the basolateral membrane, that explains passive absorption. It is still an unsolved mystery what determines the polarized distribution of transporters in epithelial cells. A remaining question is why GLUT2 is not essential for the second stage of glucose and fructose absorption across the intestine. An ongoing challenge is to determine the quantitative importance of SGLTs and GLUTs in intestinal absorption and renal reabsorption of glucose, galactose, and fructose in human subjects in health and disease.

Acknowledgments This chapter is a very personal view of the current state of sugar transport where we have emphasized the advances from the Wright laboratory over the past 30 years. We apologize in advance for not fully recognizing the valuable contributions of other researchers around the world. We do acknowledge the many talented students, post-doctoral fellows, and collaborators who have made our advances possible, and these can be identified in our citations. We are particularly grateful for the funding from the NIH and other agencies that made our work possible. Finally, we acknowledge our debt to Wendy Ravenhill for drawing all the figures in this work.

References

Abramson J, Wright EM (2009) Structure and function of Na^+-symporters with inverted repeats. Curr Opin Struct Biol 19:425–432

Adelman JL, Ghezzi C, Bisignano P, Loo DD, Choe S, Abramson J, Rosenberg JM, Wright EM, Grabe M (2016) Stochastic steps in secondary active transport. Proc Natl Acad Sci U S A 113: E3960–E3966

Barnett JE, Holman GD, Munday KA (1973) Structural requirements for binding to the sugar-transport system of the human erythrocyte. Biochem J 131:211–221

Bisignano P, Ghezzi C, Jo H, Polizzi NF, Althoff T, Kalyanaraman C, Friemann R, Jacobson MP, Wright EM, Grabe M (2018) Inhibitor binding mode and allosteric regulation of Na^+-glucose symporters. Nat Commun 7:5245

Carruthers A, Helgerson AL (1991) Inhibition of sugar transport produced by ligands binding at opposite sides of the membrane. Evidence for simultaneous occupation of the carrier by maltose and cytochalasin B. iochemistry 16:3907–3915

Chasis H, Jolliffe N, Smith HW (1933) The action of phlorizin on the excretion of glucose, xylose, sucrose, creatinine and urea by man. J Clin Invest 12:1083–1090

Coady MJ, Wallendorff B, Lapointe JY (2017) Characterization of the transport activity of SGLT2/ MAP17, the renal low-affinity Na^+/glucose cotransporter. Am J Physiol Renal Physiol 313: F467–F474

Cramer SC, Pardridge WM, Hirayama BA, Wright EM (1992) Colocalization of GLUT2 glucose transporter, sodium/glucose cotransporter, and gamma-glutamyl transpeptidase in rat kidney with double-peroxidase immunocytochemistry. Diabetes 41:766–770

Crane RK (1977) The gradient hypothesis and other models of carrier-mediated active transport. Rev Physiol Biochem Pharmacol 78:99–159

DeLano WL (2002) The PyMOL molecular graphics system, http:/www.pymol.org

Deng D, Xu C, Sun P, Wu J, Yan C, Hu M, Yan N (2014) Crystal structure of the human glucose transporter GLUT1. Nature 510:121–125

Deng D, Sun P, Yan C, Ke M, Jiang X, Xiong L, Ren W, Hirata K, Yamamoto M, Fan S, Yan N (2015) Molecular basis of ligand recognition and transport by glucose transporters. Nature 526:391–396

Eskandari S, Wright EM, Loo DD (2005) Kinetics of the reverse mode of the Na^+/glucose cotransporter. J Membr Biol 204:23–32

Faham S, Watanabe A, Besserer GM, Cascio D, Specht A, Hirayama BA, Wright EM, Abramson J (2008) The crystal structure of a sodium galactose transporter reveals mechanistic insights into Na^+/sugar symport. Science 321:810–814

Ferraris RP, Choe JY, Patel CR (2018) Intestinal absorption of fructose. Annu Rev Nutrition 38:41–67

Gallo LA, Wright EM, Vallon V (2015) Probing SGLT2 is a therapeutic target for diabetes: basic physiology and consequences. Diab Vasc Dis Res 12:78–89

Ghezzi C, Hirayama BA, Gorraitz E, Loo DD, Liang Y, Wright EM (2014) SGLT2 inhibitors act from the extracellular surface of the cell membrane. Physiol Rep 2(6):e12058

Ghezzi C, Yu AS, Hirayama BA, Kepe V, Liu J, Scafoglio C, Powell DR, Huang SC, Satyamurthy N, Barrio JR, Wright EM (2017) Dapagliflozin binds specifically to sodium-glucose Cotransporter 2 in the proximal renal tubule. J Am Soc Nephrol 28:802–810

Ghezzi C, Loo DDF, Wright EM (2018) Physiology of renal glucose handling via SGLT1, SGLT2 and GLUT2. Diabetologia 61:2087–2097

Gorraitz E, Hirayama BA, Paz A, Wright EM, Loo DDF (2017) Active site voltage clamp fluorometry of the sodium glucose cotransporter hSGLT1. Proc Natl Acad Sci U S A 114: E9980–E9988

Gould GW, Holman GD (1993) The glucose transporter family: structure, function and tissue-specific expression. Biochem J 295:329–341

Hediger MA, Coady MJ, Ikeda TS, Wright EM (1987) Expression cloning and cDNA sequencing of the Na^+/glucose co-transporter. Nature 330:379–381

Hediger MA, Turk E, Wright EM (1989) Homology of the human intestinal Na^+/glucose and the *Escherichia coli* Na^+/proline cotransporter. Proc Natl Acad Sci U S A 86:5748–5752

Hopfer U, Nelson K, Perrotto J, Isselbacher KJ (1973) Glucose transport in isolated brush border membrane from rat small intestine. J Biol Chem 248:25–32

Hresko RC, Kruse M, Strube M, Mueckler M (1994) Topology of the Glut1 glucose transporter deduced from glycosylation scanning mutagenesis. J Biol Chem 269:20482–20488

Hummel CS, Lu C, Loo DD, Hirayama BA, Voss AA, Wright EM (2011) Glucose transport by human renal Na^+/D-glucose cotransporters SGLT1 and SGLT2 am. J Physiol cell Physiol 300: C14–C21

Hummel CS, Lu C, Liu J, Ghezzi C, Hirayama BA, Loo DD, Kepe V, Barrio JR, Wright EM (2012) Structural selectivity of human SGLT inhibitors. Am J Physiol Cell Physiol 302:C373–C382

Jiang X, Loo DD, Hirayama BA, Wright EM (2012) The importance of being aromatic: π interactions in sodium symporters. Biochemistry 51:9480–9487

Joost HG, Thorens B (2001) The extended GLUT-family of sugar/polyol transport facilitators: nomenclature, sequence characteristics, and potential function of its novel members. Mol Membr Biol 18:247–256

Kaback HR, Guan L (2019) It takes two to tango: the dance of the permease. J Gen Physiol 151:878–886

Kapoor K, Finer-Moore JS, Pedersen BP, Caboni L, Waight A, Hillig RC, Bringmann P, Heisler I, Muller T, Siebeneicher H, Stroud RM (2016) Mechanism of inhibition of human glucose transporter GLUT1 is conserved between cytochalasin B and phenylalanine amides. Proc Natl Acad Sci U S A 113:4711–4716

Kasahara M, Hinkle PC (1977) Reconstitution and purification of the D-glucose transporter from human erythrocytes. J Biol Chem 252:7384–7390

Kimmich GA, Randles J (1984) Sodium-sugar coupling stoichiometry in chick intestinal cells. Am J Phys 247:C74–C82

Krug SM (2017) Contribution of the tricellular tight junction to paracellular permeability in leaky and tight epithelia. Ann N Y Acad Sci 1397:219–230

Lloyd K, Ojelabi OA, De Zutter JK, Carruthers A (2017) Reconciling contradictory findings: glucose transporter 1 (GLUT1) functions as an oligomer of allosteric, alternating access transporters. J Biol Chem 292:21035–21046

Loo DD, Hazama A, Supplisson S, Turk E, Wright EM (1993) Relaxation kinetics of the Na^+/glucose cotransporter. Proc Natl Acad Sci U S A 90:5767–5771

Loo DD, Wright EM, Zeuthen T (2002) Water pumps. J Physiol 542:53–60

Loo DD, Hirayama BA, Cha A, Bezanilla F, Wright EM (2005) Perturbation analysis of the voltage-sensitive conformational changes of the Na^+/glucose cotransporter. J Gen Physiol 125:13–36

Loo DD, Hirayama BA, Karakossian MH, Meinild AK, Wright EM (2006) Conformational dynamics of hSGLT1 during Na^+/glucose cotransport. J Gen Physiol 128:701–720

Loo DD, Jiang X, Gorraitz E, Hirayama BA, Wright EM (2013) Functional identification and characterization of sodium binding sites in Na symporters. Proc Natl Acad Sci U S A 110: E4557–E4566

Mueckler M, Makepeace C (2009) Model of the exofacial substrate-binding site and helical folding of the human GLUT1 glucose transporter based on scanning mutagenesis. Biochemistry 48:5934–5942
Mueckler M, Thorens B (2013) The SLC2 (GLUT) family of membrane transporters. Mol Asp Med 34:121–138
Mueckler M, Caruso C, Baldwin SA, Panico M, Blench I, Morris HR, Allard WJ, Lienhard GE, Lodish HF (1985) Sequence and structure of a human glucose transporter. Science 229:941–945
Naftalin J (2018) A critique of the alternating access transporter model of uniport glucose transport. Biophys Rep 4:287–299
Nomura N, Verdon G, Kang HJ, Shimamura T, Nomura Y, Sonoda Y, Hussien SA, Qureshi AA, Coincon M, Sato Y, Abe H, Nakada-Nakada Y, Hino T, Arakawa T, Kusano-Arai O, Iwanari H, Murata T, Kobayashi T, Hamakubo T, Kasahara M, Iwata S, Drew D (2015) Structure and mechanism of the mammalian fructose transporter GLUT5. Nature 526:397–401
Pao SS, Paulsen IT, Saier MH Jr (1998) Major facilitator superfamily. Microbiol Mol Biol Rev 62:1–34
Parent L, Supplisson S, Loo DD, Wright EM (1992a) Electrogenic properties of the cloned Na^+/glucose cotransporter: I. Voltage-clamp studies. J Membr Biol 125:49–62
Parent L, Supplisson S, Loo DD, Wright EM (1992b) Electrogenic properties of the cloned Na^+/glucose cotransporter: II A transport model under nonrapid equilibrium conditions. J Membr Biol 125:63–79
Quick M, Wright EM (2002) Employing *Escherichia coli* to functionally express, purify, and characterize a human transporter. Proc Natl Acad Sci U S A 99:8597–8601
Rieg T, Vallon V (2018) Development of SGLT1 and SGLT2 inhibitors. Diabetologia 61:2079–2086
Rieg T, Masuda T, Gerasimova M, Mayoux E, Platt K, Powell DR, Thomson SC, Koepsell H, Vallon V (2014) Increase in SGLT1-mediated transport explains renal glucose reabsorption during genetic and pharmacological SGLT2 inhibition in euglycemia. Am J Physiol Renal Physiol 306:F188–F193
Sala-Rabanal M, Hirayama BA, Loo DD, Chaptal V, Abramson J, Wright EM (2012) Bridging the gap between structure and kinetics of human SGLT1. Am J Physiol Cell Physiol 302:C1293–C1305
Sala-Rabanal M, Hirayama BA, Ghezzi C, Liu J, Huang SC, Kepe V, Koepsell H, Yu A, Powell DR, Thorens B, Wright EM, Barrio JR (2016) Revisiting the physiological roles of SGLTs and GLUTs using positron emission tomography in mice. J Physiol 594:4425–4438
Sala-Rabanal M, Ghezzi C, Hirayama BA, Kepe V, Liu J, Barrio JR, Wright EM (2018) Intestinal absorption of glucose in mice as determined by positron emission tomography. J Physiol 596:2473–2489
Santer R, Calado J (2010) Familial renal glucosuria and SGLT2: from a mendelian trait to a therapeutic target. Clin J Am Soc Nephrol 5:133–141
Santer R, Hillebrand G, Steinmann B, Schaub J (2003) Intestinal glucose transport: evidence for a membrane traffic-based pathway in humans. Gastroenterology 124:34–39
Schultz SG, Curran PF (1970) Coupled transport of sodium and organic solutes. Physiol Rev 50:637–718
Stein WD (1967) The movement of molecules across cell membranes. Academic Press, New York
Stümpel F, Burcelin R, Jungermann K, Thorens B (2001) Normal kinetics of intestinal glucose absorption in the absence of GLUT2: evidence for a transport pathway requiring glucose phosphorylation and transfer into the endoplasmic reticulum. Proc Natl Acad Sci U S A 98:11330–11335
Thorens B, Mueckler M (2010) Glucose transporters in the 21st Century. Am J Physiol Endocrinol Metab 298:E141–E145
Uldry M, Thorens B (2004) The SLC2 family of facilitated hexose and polyol transporters. Pflugers Arch 447:480–489

Vrhovac I, Balen Eror D, Klessen D, Burger C, Breljak D, Kraus O, Radović N, Jadrijević S, Aleksic I, Walles T, Sauvant C, Sabolić I, Koepsell H (2015) Localizations of Na^+-D-glucose cotransporters SGLT1 and SGLT2 in human kidney and of SGLT1 in human small intestine, liver, lung, and heart. Pflugers Arch 467:1881–1898

Watanabe A, Choe S, Chaptal V, Rosenberg JM, Wright EM, Grabe M, Abramson J (2010) The mechanism of sodium and substrate release from the binding pocket of vSGLT. Nature 468:988–991

Wells RG, Pajor AM, Kanai Y, Turk E, Wright EM, Hediger MA (1992) Cloning of a human kidney cDNA with similarity to the sodium-glucose cotransporter. Am J Phys 263:F459–F465

Widdas WF (1953) Inability of diffusion to account for placental glucose transfer in the sheep and consideration of the kinetics of a possible carrier transfer. J Physiol 118:23–39

Wright EM, Loo DD, Hirayama BA (2011) Biology of human sodium glucose transporters. Physiol Rev 91:733–794

Wright EM, Barrio JR, Hirayama BA, Kepe V (2014) Tracers for monitoring the activity of Sodium/Glucose cotransporters in health and disease. United States Patent US 8,845,99 B2

Yamashita A, Singh SK, Kawate T, Jin Y, Gouaux E (2005) Crystal structure of a bacterial homologue of Na^+/Cl^--dependent neurotransmitter transporters. Nature 437:215–223

Yan N (2015) Structural biology of the major facilitator superfamily transporters. Annu Rev Biophys 44:257–283

Zampighi GA, Kreman M, Boorer KJ, Loo DD, Bezanilla F, Chandy G, Hall JE, Wright EM (1995) A method for determining the unitary functional capacity of cloned channels and transporters expressed in *Xenopus laevis* oocytes. J Membr Biol 148:65–78

Zeuthen T, Gorraitz E, Her K, Wright EM, Loo DD (2016) Structural and functional significance of water permeation through cotransporters. Proc Natl Acad Sci U S A 113:E6887–E6894

Chapter 7
Amino Acid Transporters of Epithelia

Simone M. Camargo, Nadège Poncet, and François Verrey

Abstract This chapter discusses amino acid transporters (carriers) of epithelial barriers and glands and their localization and function. We consider this array of solute carrier members from seven different SLC families with overlapping selectivity and different transport modes. Corresponding knockout mouse models are mentioned and we discuss then, in more detail, the transport in exocrine pancreas, small intestine, and kidney proximal tubule. In acinar cells of the pancreas that form a typical secretory epithelium, several antiporters with overlapping selectivity cooperate with Na^+ symporters, selective for nonessential amino acids, to actively accumulate all amino acids required for the synthesis of digestive enzymes and amino acid secretion. In kidney proximal tubule and small intestine epithelial cells, a mostly common set of transporters (re)absorbs amino acids from the lumen into the "*milieu intérieur*" (extracellular space). While luminal Na^+- and H^+-dependent symporters together with an antiporter actively import all amino acids across the apical membrane, basolateral uniporters selective for essential amino acids cooperate with broad selectivity antiporters to mediate the controlled efflux of amino acids. These latter transporters thereby maintain intracellular homeostasis including a high concentration of nonessential amino acids. This chapter further discusses the newest findings on the regulation of amino acid transporters and, in some cases, the effect of transporter ablation on organ function.

Keywords Amino acid transport · Solute carrier · Small intestine · Kidney proximal tubule · Pancreas acinar cells · Glandular epithelium

S. M. Camargo · N. Poncet · F. Verrey (✉)
Institute of Physiology, University of Zurich, Zurich, Switzerland
e-mail: simonemc@physiol.uzh.ch; nadege.poncet@uzh.ch; francois.verrey@uzh.ch

K. L. Hamilton, D. C. Devor (eds.), *Studies of Epithelial Transporters and Ion Channels*, Physiology in Health and Disease,
https://doi.org/10.1007/978-3-030-55454-5_7

7.1 Introduction

Epithelia form cellular barriers between the outside and the inside of the body, the so-called milieu intérieur. Additionally, epithelia form barriers inside of the body where they cover tissues or line cavities. They are formed by a single or multiple layers of asymmetric cells that generally lay with their basolateral membrane on a basement membrane. Thus, they form a border at the limit of an underlying tissue and fulfill a number of various functions, particularly protection, absorption, secretion, and sensing. Because they rest on a basement membrane and are not directly vascularized, epithelial cells strongly depend on fluid and solute exchange with the underlying tissue. This takes place across the basolateral cell membrane which, hence, needs to be equipped with appropriate transport proteins. The mere fact that epithelia face different luminal/external milieus and exert different functions explains their very distinct equipment with transport proteins at their luminal membranes and the even more complex combinations of transporters at their basolateral membranes. Indeed, the basolateral cell membrane not only represents an additional barrier that secreted or absorbed water and/or solutes need to cross, but in most cases, it is also the membrane through which epithelial cells need to import and export solutes and water to maintain their own intracellular milieu (housekeeping function).

Some cellular barriers are not formed by epithelial cells, but, for instance, by endothelial cells that separate blood from the extracellular space or mesothelial cells that form sheets separating organs from internal cavities. This review focuses, however, on amino acid transporters expressed in epithelia and their function. We first give an overview of amino acid transport in epithelial barriers and glands and briefly mention the blood-brain barrier endothelium. The review includes detailed lists of involved amino acid transporters (Table 7.1) and of published knockout mouse models (Table 7.2). Finally, we discuss in more detail the newest findings on amino acid transporter expression, function, and regulation in exocrine pancreas secretory cells and in small intestine and kidney proximal tubule transporting epithelia.

Table 7.1 Transporters expressed in epithelia

Luminal (apical) amino acid transporters						
Protein system ***gene***	Associated protein	Substrates (with highest affinity)	$K_{0.5}$ (of substrate with highest affinity)	Stoichiometry	Tissue distribution (major site of expression in human, RNA level)	Remarks
Symporters (cotransporters): Secondary active amino acid transport: Coupling with electrical and chemical gradient of cotransported ion. Some symporters additionally antiport (exchange) another ion						
EAAT3/ EAAC1 X^-_{AG} ***SLC1A1***		Anionic [b]AAs: [L- or D-Asp, L-Glu] (Kanai and Hediger 1992)	L-Asp: 6.5 µM (Kanai and Hediger 1992)	[b]1 1AAs$^-$,3Na^+,1H^+/1 K^+ (Kanai et al. 2013; Zerangue and Kavanaugh 1996)	Next to CNS: small intestine, kidney, epididymis, thyroid gland, lung, liver, prostate, etc. [a](HPrAt)	Antiport of K^+
ATB$^{0,+}$ $B^{0,+}$ ***SLC6A14***		All neutral and cationic AAs: [Ile, Leu, Met, Phe] (Sloan and Mager 1999)	Ile: 5 uM (Sloan and Mager 1999)	1AA,2Na^+,1Cl^- (Sloan and Mager 1999)	Respiratory tract > mammary gland > stomach, prostate, uterus, colon (Sloan and Mager 1999)	
B^0AT3 B^0 ***SLC6A18***	ACE2 (for mouse ortholog), ev. CLTRN (Vanslambrouck et al. 2010)	Most neutral AAs: [Ala, met, Ile, Val] (mouse ortholog)	Ile: 0.21 mM (for mouse ortholog) (Singer et al. 2009)	1AA,2Na^+,1Cl^- (presumably) (Singer et al. 2009)	Kidney (ev. small intestine) [a](HPrAt)	Frequent nonsense mutation producing truncated protein
B^0AT1 B^0 ***SLC6A19***	CLTRN (TMEM27) in kidney; ACE2 in intestine	All neutral AAs [Met, Leu, Ile, Val, Cys] (mouse) (Bohmer et al. 2005; Camargo et al. 2005)	Ile: 1 mM (Camargo et al. 2009)	1AA,1Na^+ (Bohmer et al. 2005; Camargo et al. 2005)	Small intestine, kidney > colon [a](HPrAt)	

(continued)

Table 7.1 (continued)

Protein system ***gene***	Associated protein	Substrates (with highest affinity)	$K_{0.5}$ (of substrate with highest affinity)	Stoichiometry	Tissue distribution (major site of expression in human, RNA level)	Remarks
Luminal (apical) amino acid transporters						
SIT1 IMINO ***SLC6A20***	In intestine: CLTRN (TMEM27); in intestine: presumably ACE2 (Vuille-dit-Bille et al. 2015)	Imino acid + betaine: [Pro, betaine] (Kowalczuk et al. 2005)	Pro: 0.13 mM (Kowalczuk et al. 2005)	1AA,2Na$^+$,1Cl$^-$ (Kowalczuk et al. 2005)	Intestine, gallbladder, lung, kidney [a](HPrAt)	
PAT2 IMINO ***SLC36A2***		Imino acid + small neutral AAs: [Pro, Gly, Ala] (Chen et al. 2003)	Gly: 0.5 mM (Chen et al. 2003)	1AA,1H$^+$ (Boll et al. 2002)	Kidney, skeletal muscle, testis [a](HPrAt)	
Antiporters (obligatory exchangers): Tertiary active amino acid transport (coupling with concentration gradient of exchanged amino acids)						
ASCT2/ EAAT5 ASC ***SLC1A5***		Nonessential neutral AAs: Ala, Ser, Cys, Thr, Gln, Asn (Broer and Fairweather 2018)	Ala, Ser, Cys, Thr: ~20 μM (Utsunomiya-Tate et al. 1996)	1AA,nNa$^+$/1AA, nNa$^+$ (Broer et al. 2000)	GI, kidney, prostate, adipose tissue [a](HPrAt)	
b$^{0,+}$AT b$^{0,+}$ ***SLC7A9***	rBAT *SLC3A1*	Cystine (CssC), [b]AAs$^+$, AAs, [Cys, Arg, Lys, Leu. Ala] (mouse) (Pfeiffer et al. 1999a)	L-Cystine: 41 μM, L-Arg: 72 μM, L-Leu 1.1 mM (mouse) (Pfeiffer et al. 1999a)	1AA$^{(+)}$ or CssC/1AA (Chillaron et al. 1996; Pfeiffer et al. 1999a)	Intestine, kidney (proximal tubule S1, S2) (HPrAt) (Nagamori et al. 2016)	
AGT1 x^-_{AGC} (≠classical) ***SLC7A13***	rBAT *SLC3A1*	Anionic AAs and cystine: Glu, Asp, Cystine (Nagamori et al. 2016)	Asp: 20 μM CssC: 68 μM (Matsuo et al. 2002; Nagamori et al. 2016)	1AA$^-$ or CssC/AA$^-$ (Nagamori et al. 2016)	Kidney [a](HPrAt) in mouse: Proximal tubule S3 (Nagamori et al. 2016)	

Basolateral amino acid transporters

Protein system ***gene***	Associated protein	Substrates (with highest affinity)	$K_{0.5}$ (of substrate with highest affinity)	Stoichiometry	Tissue distribution (major site of expression in human, RNA level)	Remarks
Symporters (cotransporters): Secondary active amino acid transport: Coupling with electrical and chemical gradient of cotransported ion. Some symporters additionally antiport (exchange) another ion						
EAAT1/ GLAST X^-_{AG} ***SLC1A3***		Anionic AAs: [L- or D-Asp, L-Glu] (Klockner et al. 1994)	L-Asp: 6 µM (Klockner et al. 1994)	1AA$^-$,3 Na^+,1 H^+/1 K^+ (Zerangue and Kavanaugh 1996)	Mainly brain, little in other tissue incl. Various epithelia: Salivary glands, transitional epithelium of the urinary tract, prostate, mammary gland [a](HPrAt)	Antiport of K^+
SNAT1 A ***SLC38A1***		Nonessential neutral AAs: [Glu, Ala, Asn, Cys, His, Ser] (Mackenzie and Erickson 2004)	Ala: 0.22 mM (Kowalczuk et al. 2005)	1AA,1Na^+ (Albers et al. 2001)	Ubiquitous, in particular endocrine glands, skeletal muscle, CNS, intestine [a](HPrAt)	
SNAT2 A ***SLC38A2***		Nonessential neutral AAs: [Asn, Gln, Ala, His, Ser, Pro] and Met (Chaudhry et al. 2002)	Ala: 0.19 mM (Yao et al. 2000)	1AA,1Na^+ (Broer 2014)	Ubiquitous, in particular endocrine glands, skin, liver, kidney, intestine [a](HPrAt)	Inducible
SNAT3 N ***SLC38A3***		Nonessential neutral AAs: [Gln, Asn, His] (Broer 2014)	Gln: 1.5 mM (Broer et al. 2002)	1AA,1Na^+/1H^+ (Broer 2014; Chaudhry et al. 2001)	Liver, pancreas, brain, skeletal and heart muscle, kidney, ovary, endocrine glands [a](HPrAt)	Antiport of H^+
SNAT4 A ***SLC38A4***		Nonessential neutral AAs and AA^+: [His, Arg, Ala, Asn, Lys, Gly] (Hatanaka et al. 2001)	Ala: 0.3 mM (Padmanabhan et al. 2012)	1AA,1Na^+ (Mackenzie and Erickson 2004)	Liver, perivenous hepatocytes (Hatanaka et al. 2001; Varoqui and Erickson 2002)	

(continued)

Table 7.1 (continued)

Basolateral amino acid transporters						
Protein system ***gene***	Associated protein	Substrates (with highest affinity)	$K_{0.5}$ (of substrate with highest affinity)	Stoichiometry	Tissue distribution (major site of expression in human, RNA level)	Remarks
Symporters (cotransporters): Secondary active amino acid transport: Coupling with electrical and chemical gradient of cotransported ion. Some symporters additionally antiport (exchange) another ion						
SNAT5 N ***SLC38A5***		Nonessential neutral AAs: [Gln, Asn, His, Ala, Ser] (Broer 2014)	Gln_ 3.28 mM (Master Thesis S. Krummenacher 2012)	1AA,1Na^+/1H^+ (Broer 2014)	Ubiquitous, in particular pancreas, endocrine, brain, immune cells, lung, skin, GI, kidney [a](HPrAt)	Antiport of H^+
Antiporters (obligatory exchangers): Tertiary active amino acid transport (coupling with concentration gradient of exchanged amino acids)						
ASCT1/ EAAT4 ASC ***SLC1A4***		Nonessential neutral AAs: [Ala, Ser, Cys, Thr, Pro] (Arriza et al. 1993; Pinilla-Tenas et al. 2003)	Cys: 29 µM Ala: 71 µM Ser: 88 µM (Arriza et al. 1993)	1AA,xNa^+/ 1AA,xNa^+ (Broer et al. 2000)	Brain, adrenal, pancreas [a](HPrAt)	
LAT1 L ***SLC7A5***	4F2hc *SLC3A2*	Essential neutral AAs: Phe, Leu, Ile, His, Trp, Tyr, Val, Met] > Gln (Kanai et al. 1998; Mastroberardino et al. 1998)	Leu: 32 µM His: 35 µM (Mastroberardino et al. 1998)	1AA/1AA (Meier et al. 2002)	Ubiquitous, in particular brain, testis, immune cells > GI [a](HPrAt)	
LAT2 L ***SLC7A8***	4F2hc *SLC3A2*	Broad spectr. neutral AAs: [Phe, His, Trp, Tyr, Ile, Val, Leu, Met, Gln] > Ala (mouse) (Rossier et al. 1999)	Phe: 12 µM Leu: 48 µM Ala: 167 µM Gln: 275 µM (mouse) (Rossier et al. 1999)	1AA/1AA (Meier et al. 2002)	Ubiquitous, in particular endocrine, kidney, intestine [a](HPrAt)	

Asc-1 asc *SLC7A10*	4F2hc *SLC3A2*	Small neutral AAs: [Gly, Ala, Ser, Thr, Cys] (also D-Ser) (Nakauchi et al. 2000)	Ala: 9 μM Ser: 23 μM (Nakauchi et al. 2000)	1AA/1AA (Verrey et al. 2004)	Placenta, kidney, brain, pancreas, sk. muscle, lung, liver, adipose t., intestine, pancreas, mammary gland (Nakauchi et al. 2000) [a](HPrAt)	
y^+LAT1 y^+L *SLC7A7*	4F2hc *SLC3A2*	Neutral AAs + Na^+ OR cationic AAs: Leu, Lys, Arg, Gln, Met, His (Pfeiffer et al. 1999b)	Presence of Na^+: Leu: 32 μM Arg: 341 μM (Pfeiffer et al. 1999b)	1AA,1Na^+/ 1AA^+ (Chillaron et al. 1996; Pfeiffer et al. 1999b)	Kidney, intestine, endocrine tissue, bone marrow and immune system, placenta, lungs [a](HPrAt)	
Uniporters: Facilitated diffusion pathway (electrochemical gradient of substrate represents driving force, in some cases trans-stimulation (e.g., CAT1))						
LAT4 L *SLC43A2*		Essential neutral AAs: [Ile, Leu, Val, Phe, Met] (Bodoy et al. 2005; Oparija et al. 2019)	Leu: 3.7 mM Phe: 4.7 mM (Bodoy et al. 2005), reported high affinity component not verified (Oparija et al. 2019)	1 AA	Placenta, kidney, small intestine, bone marrow and immune system, brain, lung [a](HPrAt), (Bodoy et al. 2005; Guetg et al. 2015)	
TAT1 T *SLC16A10*		Essential aromatic AAs: [Tyr, Trp, Phe, L-DOPA] (Kim et al. 2001; Ramadan et al. 2006)	Tyr: 2.6 mM Trp: 3.7 mM Phe: 7.0 mM (Kim et al. 2001) Phe: ~30 mM for influx and efflux (Ramadan et al. 2006)	1 AA	Ubiquitous, placenta, endocrine tissue, skin, intestine, kidney, liver, muscle [a](HPrAt), (Kim et al. 2001; Ramadan et al. 2006)	
CAT-1 y^+ *SLC7A1*		Cationic AAs: [Arg, Lys, Orn] (protonated His) (Closs et al. 2006)	AA^+: 0.1–0.4 mM (Verrey et al. 2004)	1 AA^+	Ubiquitous (cerebral cortex, esophagus, pancreas) [a](HPrAt), (Verrey et al. 2004)	

(continued)

Table 7.1 (continued)

Basolateral amino acid transporters						
Protein system ***gene***	Associated protein	Substrates (with highest affinity)	$K_{0.5}$ (of substrate with highest affinity)	Stoichiometry	Tissue distribution (major site of expression in human, RNA level)	Remarks
Symporters (cotransporters): Secondary active amino acid transport: Coupling with electrical and chemical gradient of cotransported ion. Some symporters additionally antiport (exchange) another ion						
CAT-2A y$^+$ ***SLC7A2***		Cationic AAs: [Arg, Lys, Orn] (protonated His) (Closs et al. 2006)	AA^+: 2–5 mM (Closs et al. 2006)	1 AA^+	Liver, endocrine glands > pancreas, skeletal muscle (Closs et al. 2006) [a](HPrAt)	Low affinity and not sensitive to trans-stimulation compared to CAT-1, CAT-2B and CAT-3

[a]The Human Protein Atlas. https://www.proteinatlas.org/

[b]AA, AA^-, AA^+: amino acid, anionic amino acid, cationic amino acid

Table 7.2 Transgenic animal models of amino acid and oligopeptide transporter and their accessory proteins

Protein (*Gene*)	KO model	Tissue analyzed	Phenotypes	References
EAAT3 (*slc1a1*)	Global	Brain and kidney	Dicarboxylic aminoaciduria (high urinary L-glutamate and aspartate excretion)	Peghini et al. (1997)
EAAT2 (*slc1a2*)	Global. β-cells conditional (EAAT2-RIP-Cre and EAAT2-IPF1-Cre). Liver conditional (EAAT2-Alb-Cre).	Brain β-cells Liver	Mice grow slowly, present spontaneous epileptic seizures and die early (6 weeks). β-cell knockout mouse body weight and glucose tolerance is not different from WT littermates. Liver-specific EAAT2 deletion does not affect body weight or liver morphology.	Tanaka et al. (1997) Zhou et al. (2014) Hu et al. (2018)
EAAT1 (*slc1a3*)	Global	Brain	Mice growth, sensory reflexes, and neurological functions are normal. They present signs of abnormal social behavior	Karlsson et al. (2009)
ASCT1 (*slc1a4*)	Global	Brain	Significant decrease in the volumes of the hippocampus and striatum. Mice display some degree of motor dysfunction along with impairments in learning and affective domain	Kaplan et al. (2018)
ASCT2 (*slc1a5*)	Global	Brain	Homozygotes are viable and fertile	Kaplan et al. (2018)
rBAT (*slc3a1*)	Global	Kidney	Dibasic aminoaciduria and high urinary cystine excretion causing kidney stone formation	Peters et al. (2003)
4F2hc (*slc3a2*)	Global. B lymphocytes conditional (Slc3a2-CD19-Cre) .	Embryo B and T cells Vascular smooth muscles Intestine	Total knockout is early embryonically lethal. Deletion from lymphocytes alters adaptive humoral immunity.	Tsumura et al. (2003) Cantor et al. (2009, 2011) Fogelstrand et al. (2009)

(continued)

Table 7.2 (continued)

Protein (*Gene*)	KO model	Tissue analyzed	Phenotypes	References
	T cells conditional (Slc3a2-dLck-Cre). Smooth muscles cells conditional (Slc3a2-SM22-α-Cre). Intestinal epithelial cells conditional (CD98-villin-Cre).		Vascular smooth muscle conditional knockout mouse presents decreased proliferation and vessel repair. Intestinal epithelial cells knockout attenuates susceptibility to colitis and colitis-associated cancer development.	Nguyen et al. (2011)
GLYT2 (*slc6a5*)	Global	Brain	Glycine uptake in brain stem and spinal cord membrane is reduced. Mice grow slowly, display a complex neuromotor phenotype and display premature death (2 weeks)	Gomeza et al. (2003b)
Taut (*slc6a6*)	Global	Kidney and retina	Mice have decreased transport of taurine in kidney, liver, heart, skeletal muscles, and eyes. They present normal renal function and severe retinal degeneration	Heller-Stilb et al. (2002), Huang et al. (2006)
CT1 (*slc6a8*)	Global	Brain and skeletal muscle	CT1 knockout (Crt $^{-/y}$) mice have a reduced body mass, but twofold increased body fat. They present increased energy expenditure, mitochondrial respiration in skeletal muscle fibers and hippocampal lysates. Additionally, the animals have learning and memory deficits	Perna et al. (2016), Skelton et al. (2011)
GLYT1 (*slc6a9*)	Global	Brain	Glycine uptake in forebrain and CNS caudal regions is massively reduced. Pups appear	Gomeza et al. (2003a)

(continued)

Table 7.2 (continued)

Protein (*Gene*)	KO model	Tissue analyzed	Phenotypes	References
			normal, but die on the day of birth	
$ATB^{0,+}$ (*Slc6a14*)	Global	Mammary gland, lungs, and colon	Mice are viable, fertile, and phenotypically normal with unchanged plasma amino acids concentration. Mammary gland morphology in knockout female mice is not different from WT. As expected, lungs and colon from knockouts do not express the transporter	Babu et al. (2015)
B^0AT2 (*slc6a15*)	Global	Brain	Low L-leucine accumulation in synaptosoms. Modest behavioral effects	Drgonova et al. (2007)
B^0AT3 (*slc6a18*)	Global	Kidney	Abnormal renal excretion of several neutral amino acids especially glycine	Quan et al. (2004), Singer et al. (2009)
Collectrin (*CLTRN*) (B^0AT1 kidney knockout)	CLTRN global	Kidney	Lacks renal B^0AT1 expression Massive neutral aminoaciduria without glucosuria or phosphaturia	Danilczyk et al. (2006), Malakauskas et al. (2007)
Ace2 (*ace2*) (B^0AT1 intestine knockout)	Ace2 global	Kidney and intestine	Lacks intestinal B^0AT1 expression. Impaired intestinal neutral amino acid transport	Camargo et al. (2009)
B^0AT1 (*slc6a19*)	Global	Kidney and intestine	Reduced body weight. Impaired neutral amino acid transport in Brush border vesicles	Broer et al. (2011)
CAT-1 (*slc7a1*)	Global	Newborn pups	Pups are smaller, have severe anemia, and die at the birthday	Perkins et al. (1997)
CAT-2 (*slc7a2*)	Global	Macrophages	Knockout mice are viable and fertile and have no gross abnormalities. The transport of arginine and NO production in macrophages is reduced	Nicholson et al. (2001)

(continued)

Table 7.2 (continued)

Protein (*Gene*)	KO model	Tissue analyzed	Phenotypes	References
LAT1 (*Slc7a5*)	Global. Muscle conditional (LAT1-MCK Cre). Global inducible knockout (LAT1-Rosa-Cre).	Embryo Skeletal muscle Small intestine, kidney and pancreas	Total knockout is early embryonically lethal. Muscle-specific Slc7a5 knockout present normal growth and normal muscle mass.	NP submitted Poncet et al. (2014) N. Poncet personal communication
y^{+}LAT1 (*slc7a7*)	Global	Embryo	Intrauterine growth retardation with very low survival rate. Mice presented bone developmental delay and spleen enlargement	Sperandeo et al. (2007)
LAT2 (*slc7a8*)	Global	Kidney and thyroid gland	High urinary neutral amino acids and normal thyroid hormone levels	Braun et al. (2011), Vilches et al. (2018)
$b^{0,+}$ (*slc7a9*)	Global	Kidneys	Dibasic aminoaciduria and high urinary cystine excretion causing kidney stone formation	Feliubadalo et al. (2003), Font-Llitjos et al. (2007)
Asc1 (*slc7a10*)	Global	Brain	Severe brain phenotype with tremors and seizures. Early postnatal death	Rutter et al. (2007), Xie et al. (2005)
xCT (*slc7a11*)	Global	Embryonic fibroblasts, plasma	High plasma cystine, low GSH	Sato et al. (2005)
PEPT1 (*slc15a1*)	Global	Intestine	Reduced intestinal absorption of glycyl-sarcosine	Hu et al. (2008), Nassl et al. (2011)
PEPT2 (*slc15a2*)	Global	Kidney and choroid plexus	Increase glycyl-sarcosine clearance	Ocheltree et al. (2005), Rubio-Aliaga et al. (2003), Shen et al. (2007)
TAT1 (*slc16a10*)	Global	Kidney, liver, muscles, small intestine	Increased urinary and plasma, muscle and kidney aromatic amino acid concentrations	Mariotta et al. (2012)
MCT12 (*slc16a12*)	Global—rats	Lenses	High urinary creatine excretion, no increase in cataract development	Abplanalp et al. (2013), Castorino et al. (2011)

(continued)

Table 7.2 (continued)

Protein (*Gene*)	KO model	Tissue analyzed	Phenotypes	References
SNAT2 (*slc38a2*)	Global	Lungs	SNAT2 knockout mice pups die with cyanotic dyspnea	Weidenfeld et al. (2017)
SNAT3 (*slc38a3*)	Global	Liver, brain, and kidney	Lethal postnatal at day 20. Mice show impaired growth, present ataxia, and impaired renal ammonia excretion	Chan et al. (2016), Ruderisch et al. (2011)
Nrf2-KO, (SNAT3 kidney knockdown)	nrf2 global KO	Kidney	Ablation of nrf2 reduces the expression of SNAT3 in kidney.	Lister et al. (2018)
SNAT5 (*slc38a5*)	Global	Pancreatic α-cells	Mice grow normally and have normal metabolic phenotype, but under stimulation they show reduced expansion of islets α cell mass	Kim et al. (2017)
LAT4 (*slc43a2*)	Global. Tissue-specific: small intestine (constitutive and inducible), and kidney tubule (inducible). Double KO: global inducible in TAT1 global KO background.	small intestine, kidney	Global: lethal postnatal before day 10. Small intestine: only mild absorption defect. Kidney tubule: broad amino aciduria, dramatically increased in double KO.	Guetg et al. (2015), Rajendran et al. (2020)

7.2 Epithelia Lining the Outside of the Body and Amino Acid Transport

The body is protected at its exterior by the skin, a stratified squamous epithelium whose surface is very small in comparison with the contact surface of our organism with the external world. This indeed includes, next to all skin glands, the large epithelial surfaces of tubular structures that are in contact with the outside of the body, namely the respiratory, intestinal, urinary (including the kidney), reproductive tracts, their accessory glands/organs, and the mammary glands.

As regards amino acid transport, cells of epithelia rely on the supply of essential amino acids coming from the milieu intérieur (extracellular space) of the body via their basolateral membrane where exchange of nonessential amino acids also needs to take place (Broer 2008; Broer and Fairweather 2018; Camargo et al. 2009; Makrides et al. 2014). Because of their various functions, epithelial cells have different needs and correspondingly their basolateral amino acid transport protein equipment is not uniform. As we will discuss below, epithelial cells of the proximal part of the small intestine and of the proximal kidney tubule are in a particular situation, since they are exposed intermittently (small intestine) or constantly (early segments of kidney proximal tubule) to a substantial amount of luminal amino acids and are equipped with a luminal amino acid uptake machinery.

An important general observation is that epithelial cells contain nonessential amino acids at a much higher concentration than present in the extracellular space. Therefore, nonessential amino acids cannot be imported into cells via uniporters against their concentration gradient, but are generally transported actively into cells via symporters using the energy provided by cotransported ions or eventually by antiporters that function as obligatory exchangers and thus do not mediate a net amino acid uptake (Fig. 7.1) (Rudnick et al. 2014; Verrey 2003). Some of these antiporters preferentially release intracellular nonessential amino acids to the outside in exchange for the uptake of essential amino acids (Meier et al. 2002; Nicklin et al. 2009; Verrey 2003). Many cells, in addition to symporters and antiporters, also express uniporters. These facilitated diffusion pathways are generally selective for essential amino acids and mediate their equilibration between extracellular space and intracellular compartment (Meier et al. 2002; Ramadan et al. 2006). This potentially directional-facilitated diffusion may mediate the recycling of essential amino acids, thereby allowing the net efflux (or influx) of nonessential amino acids via antiporters.

In this review, we will give a short overview of different epithelial barriers, before describing some in more details. In many cases, there is more information available on the expression of amino acid transporters in tumors/cancers-derived tissue than in the healthy original tissue. Nonetheless, this information will not be considered here, particularly because tumor transformation generally includes changes in amino acid transporter expression.

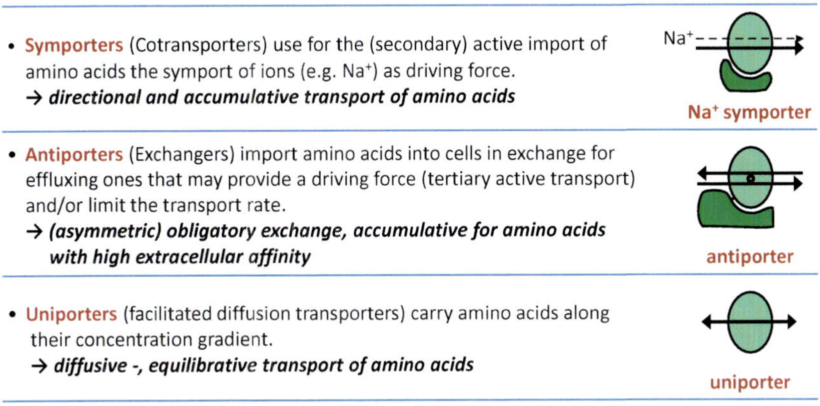

Fig. 7.1 The three major amino acid carrier types. The functional principle of these different transporter types and their alternative names are indicated. Additionally, to the depicted symporter, antiporter, and uniporter, some amino acid carriers display more complex transport schemas. For instance, some members of the SLC1 family (e.g., SLC1A1) function by amino acid-Na^+-H^+ symport coupled to K^+ antiport. Another more complex example is given by members of the SLC7 family (SLC7A6 and 7) that function as antiport accommodating in both directions either a cationic amino acid or symporting a neutral amino acid with a Na^+ ion

7.2.1 *Respiratory Tract*

Only little information is available on amino acid transport in respiratory tract epithelia. Of course, many amino acid transporter mRNAs have been detected in the lungs, but it is generally not possible to deduce whether they are expressed in respiratory epithelia. In any case, these epithelial cells need to have at their basolateral side the transporter equipment necessary for their housekeeping. One interesting publication by Sloan et al. showed that $ATB^{0,+}$ (SLC6A14), the sole active amino acid transporter that actively transports all neutral and cationic amino acid with mostly high affinity, is expressed on the apical membrane of epithelial cells throughout mouse airways from trachea to alveolar type I cell (Sloan et al. 2003). This localization of $ATB^{0,+}$ has been confirmed by using a corresponding knockout mouse model as negative control (Babu et al. 2015; Hatanaka et al. 2004). The authors proposed that, together with the peptide transporter PEPT2 (SLC15A2), this amino acid symporter clears the airway surface liquid from amino acids and small peptides, thereby playing a critical role in maintaining a low nutrient environment and consequently preventing bacterial growth.

7.2.2 *Gastrointestinal Tract*

The epithelia lining the digestive tube and the associated glands/organs fulfill very different functions such that a uniform description is not meaningful. To our knowledge, no luminal amino acid transporters have been described in the upper part of the digestive tract, including the stomach (Dave et al. 2004). Depending on their function, epithelial cells of the surface epithelium or of the glands may have different basolateral amino acid transporter setups. In particular, glandular cells specialized in the synthesis of secreted proteins, such as salivary glands that produce amylase or gastric chief cells that release pepsin and lipase, need a more developed amino acid uptake system. A special case is of course the liver, where hepatocytes exert not only a substantial secretory function via their canalicular membrane, but also have very important metabolic activities requiring much amino acid uptake and release at their basolateral side. Actually, the amino acid transporter set expressed by hepatocytes is not well-known. From experiments performed with TAT1 (SLC16A10) knockout mice, we learned that the aromatic amino acid uniporter TAT1 is expressed in the liver where it plays an important role by allowing hepatocytes to control the extracellular aromatic amino acid levels (Mariotta et al. 2012). Another amino acid transporter, SNAT3 (SLC38A3), was shown to be expressed in hepatocytes and play an important role for the metabolism of glutamine, urea, and ammonium (Balkrishna et al. 2014).

The small intestine is the site of oral nutrient absorption. For instance, ingested proteins are cleaved into small peptides and single amino acids in the intestinal lumen by the sequential action of gastric pepsin, pancreatic digestive enzyme, and then by brush border enzymes such as aminopeptidases, endopeptidases, carboxy-peptidases, dipeptidyl peptidase, and angiotensin converting enzymes ACE and ACE2. Di- and tripeptides are then taken up by the H^+-driven PEPT1 (SLC15A1) and subsequently cleaved into single amino acids within the enterocytes, whereas luminal single amino acids are absorbed into enterocytes via active amino acid transporters localized in the apical brush border membrane (Broer and Fairweather 2018; Nassl et al. 2011). Enterocytes not only transport amino acids, but are also an important site of amino acid metabolism. Amino acids metabolized/produced locally as well as amino acids taken up apically and not metabolized are then eventually released across the basolateral membrane into the extracellular space from where they are further transported to the liver via the portal circulation. The organization of the luminal and basolateral membrane amino acid transporters that together build the transepithelial amino acid transport machinery will be described in the next sections.

Intriguingly, transepithelial absorption of small solutes appears not to be limited to transcellular transport, but includes a substantial paracellular component as previously suggested by others and more recently by us regarding amino acids (Mariotta et al. 2012; Pappenheimer and Reiss 1987). Specifically, this paracellular leak pathway allows the absorption of a substantial proportion of small solutes when present at a sufficiently high concentration in the lumen and appears to be stimulated by ongoing active Na^+-dependent absorption. Interestingly, this pathway is more developed in animals with a shorter and lighter small intestine such as birds and bats, suggesting an evolutionary trade-off between efficacy of absorption and protection from potentially toxic ingested solutes (Garro et al. 2018; Karasov 2017; Pappenheimer and Reiss 1987). Active transcellular transport is, however, crucial for the absorption of the same solutes when present at lower concentrations.

In normal conditions, little amounts of protein and amino acid should reach the colon; nonetheless, the large intestine mucosa is to some extent able to absorb amino acids from the gut lumen. This may be useful in particular because the large intestine is the major site where microbiota further metabolizes the gut content and eventually releases, next to small chain fatty acids and vitamins, essential amino acids that may be absorbed. Analogous to airways, where amino acids are taken up with high affinity from the airway surface liquid, the broad range high affinity neutral and cationic amino acid transporter ATB$^{0,+}$ (SLC6A14) is expressed at the luminal surface of mucosal cells (Babu et al. 2015; Nakanishi et al. 2001; Ugawa et al. 2001). The peptide transporter PEPT1 (SLC15A1) has also been localized at the mucosal surface of colon, whereas the expression of other luminal amino acid transporters has only been postulated but not specifically shown (van der Wielen et al. 2017; Wuensch et al. 2013). Nonetheless, the major site of amino acid transport in the colon mucosa is likely the basolateral membrane of colonocytes where amino acid uptake and exchange ensure the housekeeping needs.

7.2.3 Urinary Tract

The urinary tract epithelia can be divided into two major categories, the kidney tubules that produce the urine from the glomerular filtrate and the urothelium, a transitional epithelium, which covers the inside of the ureters, bladder, and urethra, the structures that conduct, hold, and pass the urine. While little is known about the amino acid transporter setup of urothelium, amino acid transport along the kidney tubules has been studied extensively ((Makrides et al. 2014) and Refs therein). In humans, approximately one million glomeruli of each kidney produce together nearly 180 L of glomerular filtrate containing the same concentration of free amino acids as the plasma, and thus, about 50 g/24 h. Over 99% of these filtered amino acids are reabsorbed at the level of the proximal kidney tubule by a set of luminal and basolateral amino acid transporters that form a transcellular transport machinery resembling the one found in the small intestine mucosa. The common organization of these small intestine and kidney amino acid (re)absorption machineries and the specific elements characterizing the small intestine and the kidney proximal tubule transport machineries will be discussed in later sections.

The kidney tubules are composed of sequential segments that fulfill different functions. Nonetheless, all epithelial cells of the tubules face the renal extracellular space with their basolateral membrane. On this membrane, sets of amino acid transporters allow cells to maintain their intracellular amino acid concentration at an adequate level while fulfilling their specialized and housekeeping tasks. Depending on the cells' position and function along the tubule, these tasks may differ. As described in a later section, the proximal tubule, particularly in its convoluted early part, reabsorbs transcellularly the filtered amino acids, such that basolateral amino acid transporters have to mediate their cellular efflux. Luminal transporters have not been described in more distal tubule segments, and in a segment-specific transcriptome database, a search for typical luminal amino acid transporter genes only revealed the expression of EAAT3 (SLC1A1) in the thick ascending limb of the Henle loop (TAL) (Huling et al. 2012). The segmental and/or cellular specificity of basolateral amino acid transporters has not yet been studied systematically.

7.2.4 Reproductive Tract

Little is known about amino acid transport in epithelia lining the genital tract in human females and males. Most of the data available (gene expression) are from the whole organs and not specifically from the epithelia; except for the placenta in which amino acid transport has attracted much attention. Because of its availability after birth, mature human placenta can indeed readily be studied ex vivo and a substantial number of studies focusing on amino acid transport have been published. However, since the placenta is a complex structure that forms a barrier between two vascular beds, it does not really correspond to the usual definition of an epithelium. Thus, the readers are referred to thorough studies and reviews recently published elsewhere (Cleal et al. 2018; Vaughan et al. 2017).

7.2.5 Exocrine Gland Epithelial Cells

Epithelial exocrine secretory cells produce their secretions by merocrine (e.g., eccrine sweat glands, pancreatic acinar cells), apocrine (e.g., mammary gland), or holocrine (e.g., sebaceous gland) mode. In all cases, the basolateral amino acid transport machinery serves, next to its housekeeping role, to import the amino acids required for the secretion production. The numerous exocrine glands include skin-associated eccrine and apocrine sweat and sebaceous glands. They also include the mammary and lacrimal glands as well as glands associated with the female and male genital tracts such as Bartholin's glands and the prostate. There are also many glands within the GI tract, including the salivary glands, the exocrine part of the pancreas, and the liver that also has an important exocrine function. Additionally, there are many secreting cells lining the intestinal mucosa at its surface or within glandular structures. Such separated secreting cells are, for instance, the stomach parietal and chief cells, small intestinal paneth and goblet cells, etc.

From all these exocrine glandular epithelia, we will discuss briefly the mammary gland in the next paragraph. We will also describe more extensively the exocrine pancreas in a separate section below.

7.2.6 *Mammary Gland*

During lactation, the mammary gland epithelium produces milk proteins and thus needs to import large amounts of amino acids for their synthesis. Recent studies of mammary gland epithelial metabolism and transport mostly focus on animal species used for milk and/or meat production and less is known about human lactating mammary glands (Chen et al. 2018; Dai et al. 2018; Rezaei et al. 2016). It has, for instance, been shown that the lactating mammary gland of sows takes up much branched chain amino acids which are not only used for protein synthesis, but are to a large extent metabolized to nonessential amino acids, in particular glutamine, glutamate, and aspartate used for milk protein production and direct release in the milk (Li et al. 2009; Matsumoto et al. 2013; Smilowitz et al. 2013). The different studies investigating the expression of amino acid transporters in lactating mammary glands demonstrated the expression of a series of amino acid transporters that actively import amino acids into mammary gland epithelial cells. In particular, the nonessential neutral amino acid Na^+ symporters SNAT1 and SNAT2 (SLC38A1 and SLC38A2) were shown to be strongly expressed at the mRNA level. Also, the anionic amino acid transporter EAAT3 (SLC1A1) and the broad-spectrum high affinity symporter $ATB^{0,+}$ (SLC6A14) were shown to be substantially expressed in lactating sow mammary gland (Chen et al. 2018). Yet another highly expressed transporter that mediates the import of cationic amino acids driven by the membrane potential is the uniporter CAT1 (SLC7A1) (Chen et al. 2018; Xia et al. 2018). Additionally, some nonessential and essential neutral amino acids can be taken up by antiporters, shown to be strongly expressed in mammary epithelial cells and may, in exchange, release amino acids taken up by the active transporters mentioned before. These antiporters are, on the one hand, the two Na^+-dependent nonessential amino acid exchangers ASCT1 and ASCT2 (SLC1A4 and SLC1A5) and, on the other hand, the heterodimeric exchangers y^+LAT-2 (SLC7A6) as well as LAT2 and LAT1 (SLC7A8 and SLC7A5) (Dai et al. 2018; Lin et al. 2018; Xia et al. 2018; Chen et al. 2018). These latter antiporters preferentially import essential amino acids in exchange for intracellular cationic amino acids (y^+LAT2) or (nonessential) neutral amino acids (LAT1 and LAT2). The reader is directed to Chap. 6 of Volume 2 for additional discussion of amino acid transporters in mammary gland epithelium.

7.3 Examples of Epithelia Constituting Barriers Between Body Compartments

Epithelia also form barriers between different compartments. In the eye, the ciliary body epithelium separates the anterior chamber from the extracellular space of the body and secretes aqueous humor into the anterior chamber. The lens epithelium forms the border between the aqueous humor and the lens fibers, while the retinal pigmented epithelium separates the retina from the underlying choroid coat. The ciliary body epithelium has a particular structure: it is formed by a double layer of epithelial cells facing each other with their apical membranes where they are connected by gap junctions. Pigmented epithelial cells form the layer facing the stroma of the ciliary body and, on the aqueous humor side, there are the nonpigmented cells which are linked to each other by tight junctions and thus form the actual barrier between the extracellular space of the ciliary body and the aqueous humor. Regarding the transfer of amino acids across this double layer barrier, pigmented cells express the neutral amino acid transporter LAT2 at their basolateral side (Boiadjeva Knopfel et al. 2019) and it is likely that amino acids once in pigmented cells can diffuse to nonpigmented cells via their apical gap junctions. We have observed that these latter nonpigmented cells express the uniporter TAT1 at their basolateral, aqueous humor-facing side. However, both cell layers certainly express additionally a series of other transporters to support the transfer of all types of amino acids to and from the aqueous humor. Nonetheless, we have observed that LAT2 neutral amino acid exchanger plays a central role in maintaining aqueous humor and lens amino acid homeostasis. Additionally, it plays a major role at the level of the lens epithelium and we have observed that its defect is associated with cataract.

Yet another epithelial separation between body compartments is formed by the choroid plexus that produces the cerebrospinal fluid (CSF) at the level of brain ventricles. The basolateral surface of this cell monolayer faces the vascular side, while the luminal side faces the cerebrospinal fluid. Regarding amino acid transfer, this barrier has the important function of maintaining a large amino acid concentration difference between blood plasma and CSF where the level of all amino acids, but glutamine, is approximately tenfold lower than that in plasma (Dolgodilina et al. 2016). Although regarding this barrier the knowledge about amino acid transporter expression is not complete, it appears from RNA profiling data that the amino acid transporters SNAT3 (SLC38A3), ASC-1 (SLC7A10), SIT1 (SLC6A20), y^{+}LAT2 (SLC7A6), and EAAT3 (SLC1A1) are substantially expressed in the choroid plexus (Dahlin et al. 2009; Marques et al. 2011). While the latter anionic amino acid transporter EAAT3 (SLC1A1) has actually been localized at the CSF-facing side of the epithelium, the localization of the other amino acid transporters within the choroid plexus had not yet been reported (Akanuma et al. 2015). Our recently published experiments confirmed the mRNA expression of SNAT3, ASC-1, SIT1, y^{+}LAT2, and to a lesser extent, of EAAT3 in mouse choroid plexus and indicated that LAT2 (SLC7A8) and SNAT1 (SLC38A1) are also substantially expressed. Furthermore, immunofluorescence experiments showed that SNAT3 and LAT2 localize at the CSF-facing apical side of choroid plexus epithelial cells (Dolgodilina et al. 2020).

7.3.1 Barriers Formed by Endothelia

Endothelial barriers are briefly mentioned here, although they are generally not classified as epithelia, due to their mesoderm origin. However, the endothelia cellular layers also constitute a barrier between different compartments.

The blood-brain barrier is largely composed of endothelial cells and has the crucial function of separating the blood circulating in the brain capillaries from the brain interstitial fluid. This structure includes, next to the endothelial cells, astrocyte feet, pericytes, neurons, and the extracellular matrix. Importantly, the diffusion barrier between the brain capillary blood and the brain interstitial fluid is formed by the microvascular endothelial cells that are held together by tight junctions. The concentration difference of amino acids between blood plasma and brain interstitial fluid is even larger than the one between blood and cerebrospinal fluid. Indeed, many amino acids appear to be 30- to 100-fold less concentrated in brain interstitial fluid compared to blood plasma, with the exception of glutamine, glutamate, and serine, whose concentration is approximately tenfold lower in the brain interstitial space than in plasma. Amino acid transporters that are suggested to play an important role in this barrier are in particular LAT1 (SLC7A5) and SNAT3 (SLC38A3) (Dolgodilina et al. 2016; Lyck et al. 2009). Interestingly, to model the amino acid concentration fluctuations observed experimentally in the brain interstitial fluid in response to systemic administration of amino acids, we had to include the asymmetric nature of LAT1 (SLC7A5) with its low intracellular and high extracellular affinity for amino acids (Meier et al. 2018; Taslimifar et al. 2018).

After this brief overview of different epithelia, their amino acid transport requirements, and transport organization, we will now focus on a small number of special and important cases: first on pancreatic acinar cells, which are the secretory cells most actively synthesizing and secreting proteins, specifically digestive enzymes. This will be followed by sections on the two epithelial conduits performing the absorption and/or reabsorption of amino acids, namely the small intestine and the kidney proximal tubule.

7.4 Amino Acid Transporters in Exocrine Pancreas

7.4.1 Origin, Structure, and Function of the Exocrine Pancreas

The exocrine portion of the pancreas comprises 95% of the total organ mass and is mainly involved in the production and secretion of digestive enzymes. The majority of the exocrine pancreas consists of acinar cells organized as acini which represent its functional units. Acini are connected to a ductal network formed by centroacinar and ductal cells which conveys the pancreatic juice to the main biliary-pancreatic duct which then joins the duodenum (Steward and Ishiguro 2009). Acinar, centroacinar, and ductal cells, as well as islets cells, are derived from the same embryonic multipotent pancreatic progenitor cells. These pancreatic cells are derived from the dorsal and ventral portions of the endoderm. By day E10.5 in mouse embryo, they express the transcription factors, pancreatic and duodenal homeobox gene 1 (Pdx1), pancreatic transcription factor (Ptf1a), and sex-determining region Y-box 9 protein (Sox9). Subsequently, cells form a tubular structure with the tips giving rise to acinar cells, whereas the trunks are committed to ductal and endocrine cells (Cleveland et al. 2012; Stanger and Hebrok 2013). In the adult pancreas, Pdx1 is expressed only in islet β-cells, sox9 in ductal cells, and PTFa1 only in acinar cells. Secretory function of acinar cells is maintained by the expression of PTFa1 and transcription factor Mist1 (Jiang et al. 2016). Mature acinar cells are quiescent and do not have adult stem cells; so upon injury, acinar cells transiently dedifferentiate from secretory to an "embryonic pancreatic progenitor cell-like". These cells have a high proliferative capacity and express genes typically present only during embryonic pancreatic development (Jensen et al. 2005).

7.4.2 *Pancreatic Juice*

Acinar cells have the cellular machinery of a classical secretory epithelial cell. They are polarized with a predominant rough endoplasmic reticulum and Golgi apparatus and an apical membrane region packed with zymogen granules (ZG). Their main function is to synthesize digestive enzymes, store them in the granules, and secrete them upon stimulation (Koizumi et al. 1993; Konturek et al. 1995). Pancreatic acinar cells show one of the highest protein synthesis capacities of any secretory organ and have a high capacity to accumulate extracellular amino acids. The ductal cells secrete bicarbonate and water that, together with the digestive enzymes produced by acinar cells, form the pancreatic juice. Daily, acinar and ductal cells secrete approximately 5–20 g of digestive enzymes and up to 2 L of fluid. This high pH secretion released in the lumen of the duodenum neutralizes the gastric bolus and produces optimal conditions for proenzymes activation (Konturek et al. 1995; Singer and Niebergall-Roth 2009; Steward and Ishiguro 2009).

Meal ingestion stimulates the secretion of enzymes. When the stomach content reaches the small intestine, this triggers a series of neurohormonal and hormonal responses that control the pancreatic exocrine function. The presence of proteins, fat, acid chime, and mechanical distension induces the release of cholecystokinin (CCK) and secretin, both released on the blood side of specialized intestinal mucosal cells (enteroendocrine cells). CCK is produced by I cells and induces acinar cells to secrete digestive enzymes, whereas secretin is released by S cells and stimulates pancreatic duct cell water and bicarbonate secretion (Konturek et al. 1995; Singer and Niebergall-Roth 2009). Recently, fibroblast growth factor 21 (FGF21) was also shown to act as a secretagogue and to increase pancreatic protein and volume secretion. Acinar cells express FGF21 receptor and also highly express its ligand

FGF21, which they also secrete into pancreatic juice, suggesting that FGF21 may act as an autocrine secretagogue (Coate et al. 2017).

The pancreatic juice has a very complex composition with more than 200 proteins identified, but the digestive enzymes, secreted at the highest concentration, are the most representative proteins. The mRNA encoding secreted proteins constitute almost 70% of the total mRNA in pancreas (Danielsson et al. 2014). Digestive enzymes can be divided into groups according to their substrate specificity: the pro-proteolytic enzymes (trypsinogen, chymotrypsinogen, chymotrypsinogen-like (elastase), and procarboxypeptidase), enzymes involved in carbohydrate digestion (amylase and enolases), enzymes involved in lipid metabolism (lipases), and finally enzymes involved in ribonucleotide digestion (deoxyribo- and ribonucleases) (Doyle et al. 2012). The pancreatic juice also contains free amino acids (Araya et al. 2017; Fukushima et al. 2010). Their concentration is mostly lower than that in plasma with the exception of glutamate and aspartate which in contrast reach five- to sevenfold higher concentrations. In pancreatic acinar cells, as well as in lactating mammary gland epithelial cells, glutamate is synthesized from neutral amino acids (glutamine in acinar cells and leucine in mammary cells) and to a large extent accumulated in the cell and then secreted (Araya et al. 2017; Smilowitz et al. 2013). Glutamate secretion in the pancreatic juice is high even in unstimulated conditions, does not occur via zymogen granules, and is not stimulated by the hormones CCK and secretin. Glutamate and aspartate secretion may be mediated by membrane carriers or channels which have not yet been identified. Once secreted in the lumen of the duodenum, glutamate may be taken up by enterocytes and be used by these cells as their main source of energy. The high level of glutamate produced by the acinar cells from glutamine and released with the pancreatic juice into the intestinal lumen suggests an interorgan relationship between exocrine pancreas and small intestine for glutamine-glutamate utilization.

7.4.3 Amino Acid Import into Acinar Cells

The accumulation of glutamine and other amino acids was analyzed in vivo with radiolabeled species (^{14}C, ^{11}C, ^{18}F, and ^{123}I-amino acids). In humans and rodents, amino acid accumulation in pancreas was shown to be very rapid after intravenous administration, reaching for glutamine, tyrosine, methionine, and methylaminoisobutyric acid (MeAIB) highest concentrations compared to all other organs. Proline and the non-proteinogenic amino acid analog L-dopa were also shown to efficiently accumulate in pancreas (Bauwens et al. 2007; Borner et al. 2001; Dunphy et al. 2018; Hustinx et al. 2003; Kono et al. 2002; Mariotta et al. 2012; Qu et al. 2012; Ribeiro et al. 2005; Tolvanen et al. 2006).

The efficient import of amino acids in acinar cells is an active process displaying different affinities, ion dependence, and kinetic characteristics that suggest the presence of several transporters (Mailliard et al. 1995). Na^+-dependent and -independent mechanisms were described using in vitro and ex vivo pancreatic preparations (isolated acini, lobuli, pancreatic slices, and intact perfused pancreas). Another feature of the amino acids transported through the basolateral membrane observed in these experiments was the ability to be increased or trans-stimulated by other amino acids (Mann et al. 1986). After a broad screening, more than 20 amino acid transporters were detected at the mRNA level in murine pancreas (Rooman et al. 2013).

7.4.4 Neutral Amino Acids Transporters of Acinar Cells

From the transporters mRNAs detected in pancreas, five encode neutral amino acid carriers that have been shown to localize to the basolateral membrane of acinar cells: the Na^+-dependent nonessential amino acid transporters SNAT3 (SLC38A3) and SNAT5 (SLC38A5), the sodium-independent antiporters for large mostly essential neutral amino acids LAT1-4F2hc (SLC7A5-SLC3A2) and LAT2-4F2 (SLC7A8-SLC3A2) (Danielsson et al. 2014; Rooman et al. 2013), and the Na^+-dependent nonessential neutral amino acid exchanger ASCT1 (SLC1A4) (Fig. 7.2) (Hashimoto et al. 2004).

The sodium-dependent transporters SNAT3 (SLC38A3) and SNAT5 (SLC38A5) are localized to the basolateral membrane of acinar cells in the mouse and SNAT5 also to the same membrane in humans (Danielsson et al. 2014). SNAT5 (SLC38A5) and SNAT3 (SLC38A3) cotransport glutamine, histidine, and asparagine (SNAT5 also alanine and serine) with a sodium ion and counter-transport a proton (Mackenzie and Erickson 2004). SNAT5 mRNA has the highest expression in mouse and human pancreas (http://biogps.org); however, its ablation did not change mouse embryonic development, pancreas size, or plasma amino acids concentration (Kim et al. 2017). Since SNAT3 is also expressed in acinar cells and has a similar substrate selectivity as SNAT5, its expression might have compensated for the loss of SNAT5. SNAT3 is expressed in embryonic and adult liver as well as in kidney, where SNAT5 is not expressed. SNAT3 ablation causes mice to die early after weaning (Chan et al. 2016) such that its role in exocrine pancreas function could not be studied using this global knockout.

The sodium-independent antiporters for large neutral amino acids LAT1-4F2hc (SLC7A5-SLC3A2) and LAT2-4F2hc (SLC7A8-SLC3A2) were also localized to the basolateral membrane of acinar cells in mice, but no data is available for humans. LAT2 is expressed in epithelial cells as shown for kidney and small intestine (see below). LAT1 is crucial during embryonic development, and in adult life, it is highly expressed in brain (Fagerberg et al. 2014; Ohgaki et al. 2017). Both transporters have high affinity for aromatic amino acids (phenylalanine, histidine, tyrosine, tryptophan) and thyroid hormones (LAT2 also transports cysteine and LAT1 methionine) (Meier et al. 2002; Zevenbergen et al. 2015). As it is the case for SNAT3 and SNAT5, both transporters have very similar substrate selectivity, but it is not yet known if LAT1 and LAT2 transporters compensate each other in the pancreatic acinar cells in case of ablation of one of the transporters. The inducible global

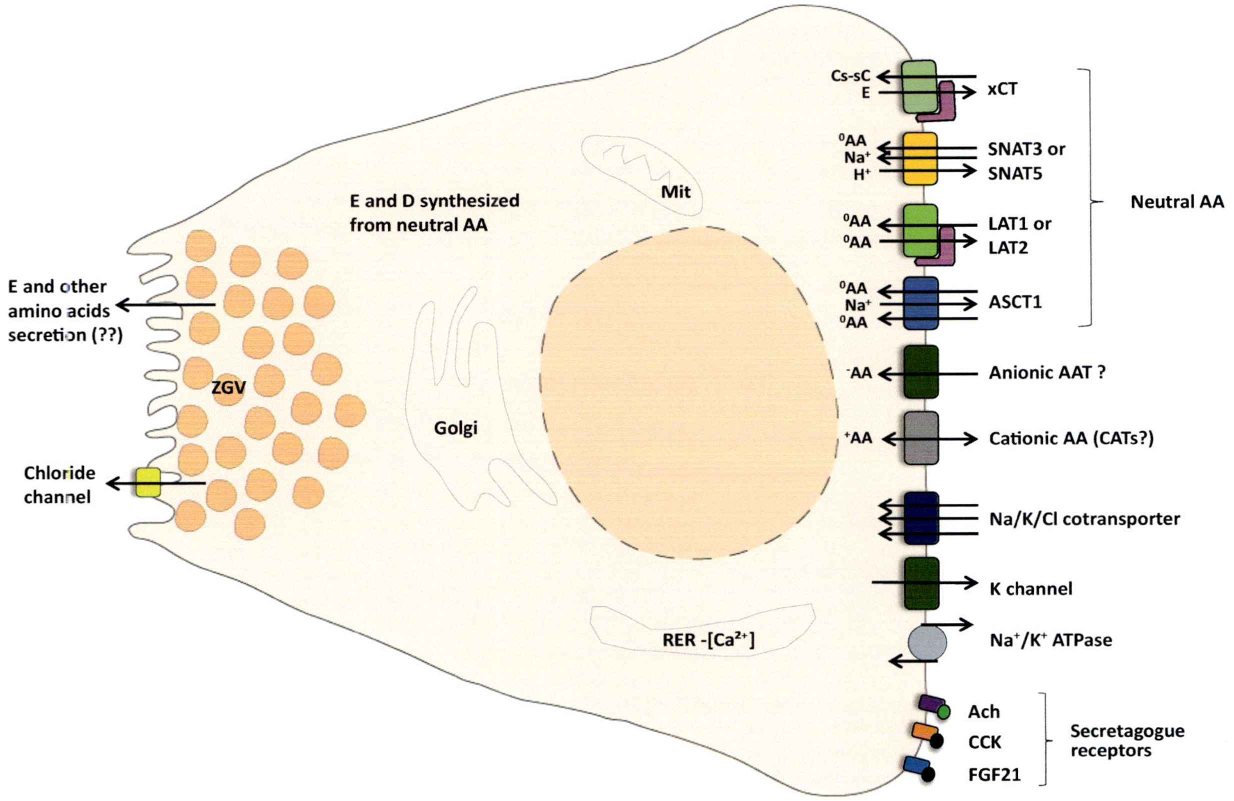

Fig. 7.2 Pancreatic acinar cells amino acid transporters. Acinar cells have a pyramidal shape and display a typical secretory machinery with well-developed rough endoplasmic reticulum (RER) and Golgi apparatus (Golgi) localized to the perinuclear region. In the apical region, digestive enzymes are stored in zymogen granules (ZG). Upon binding of the secretagogues Cholecystokinin (CCK), Acetylcholine (Ach), or fibroblast growth factor 21 (FGF21) to their receptors on the basolateral membrane (CCK/CCK1R, Ach/ mAChR1/3, FGF21/βKloto-FGFR1), an increase of the free cytoplasmatic calcium concentration

Fig. 7.2 (continued) ($[Ca^{2+}]_i$) originating from the RER (IP3-activated channels) induces the fusion of the ZG to the apical membrane and exocytosis of the secretory granules content. The ion flow, important for the transport and secretion, is maintained by the expression of pumps, channels, and transporters at the basolateral membrane (NaKATPase, Na/K/Chloride transporter (NKCC1-SLC12A2)), Potassium channel (MaxiK channels (Kcnma1) and mIK1 (Kcnn4)), and the apical membrane chloride channel (ANO1/tmem16a). The amino acid transporters are mainly expressed at the basolateral membrane and are responsible for the accumulation of substrates for the synthesis of digestive enzyme and proteins with housekeeping function; used as energy source (glutamine) and for the synthesis of anionic amino acids glutamate (E) and aspartate (D). Transporters for neutral amino acids localized at the basolateral membrane of the acinar cells are:—sodium-independent transporters for large neutral amino acids LAT1-4F2hc (SLC7A5-SLC3A2) and LAT2-4F2hc (SLC7A8-SLC3A2);—the exchanger for cystine and glutamate xCT-4F2hc (SLC7A11-SLC3A2) (in particular upon oxidative stress injury such as acute pancreatitis);—the Na^+-dependent and H^+-sensitive glutamine and small neutral amino acid transporter SNAT3 (SLC38A3) and SNAT5 (SLC38A5);—the Na^+-dependent exchanger for small neutral amino acids ASCT1 (SLC1A4);—presumably the cationic amino acid uniporters CAT1 (SLC7A1) and/or CAT2 (SLC7A2). Anionic amino acids are not efficiently imported into acinar cells and no candidate transporter was detected at the basolateral membrane. Glutamate and aspartate that are efficiently synthesized from neutral amino acids are accumulated in acinar cells and secreted into the pancreatic juice. The carrier(s) responsible for the apical secretion of glutamate and other amino acids is (are) not yet identified. Based on mRNA or proteomic analysis, other transporters should be expressed in acinar cells. They were discussed in the text, but not included in this figure. Keys: Cs-sC, cystine; ^{0}AA, neutral amino acids; $^+$AA, cationic amino acids; $^-$AA, anionic amino acids knockout mice for LAT1 (Rosa-Cre) have a normal pancreas morphology and did not show any compensatory upregulation of LAT2 in acinar cells (Nadège Poncet and Simone Camargo, personal communication), unlike observed in the case of the constitutive muscles-specific ablation of LAT1 (MCK-cre Slc7a5) where LAT2 mRNA was upregulated in muscle fibers (Poncet et al. 2014). The potential effect of LAT2 deficiency on LAT1 expression in the pancreas has not yet been studied.

The Na^+-dependent antiporter for neutral amino acids ASCT 1 (SLC1A4) was localized to the basolateral membrane of murine acinar cells (Hashimoto et al. 2004) and there is also evidence that it localizes to the basolateral membrane of human acinar cells (https://www.proteinatlas.org). It exchanges with high affinity neutral amino acids such as alanine, serine, cysteine, or threonine, but its role besides equilibrating the concentration of these different amino acids is not clear (Arriza et al. 1993). The other neutral amino acid exchanger of the SLC1 family, ASCT2 (SLC1A5), was shown to be expressed at the mRNA and protein levels in the mouse pancreas (Rooman et al. 2013; Zhou et al. 2014), but its subcellular localization is not known. It is expected that the recently developed ASCT2 knockout mouse model (Kaplan et al. 2018) will help solve the question of localization and function of this antiporter not only in pancreas, but also in kidney and intestine.

LAT1 and LAT2 have an important function for the accumulation of essential neutral amino acids in acinar cells. They display at the outside of the cell a high affinity for amino acids, in particular for essential ones, in contrast to a very low intracellular affinity. This situation favors the cytosolic accumulation of essential amino acids in exchange for the efflux of nonessential amino acids (Meier et al. 2002). Thus, for these exchange systems to function, (an)other transporter(s) with a directional transport mechanism need(s) to import nonessential amino acids into the cells which can then function as efflux substrates, analogous to what was described for the efflux of nonessential amino acids via LAT2 in small intestine and renal proximal tubule epithelial cells (see below). The Na^+-dependent cotransporters SNAT3 and SNAT5 are good candidates for this import function as they transport glutamine, histidine, serine, asparagine, and alanine with low affinity and high capacity. These nonessential amino acids transported into the cell by SNAT3 and SNAT5 can then be used as intracellular efflux substrates by LAT1 and LAT2 in exchange for the import of essential neutral amino acids not transported by the Na^+-dependent carriers such as branched chain amino acids, tryptophan, tyrosine, and methionine. The Na^+-dependent antiporter ASCT1, with its selectivity overlapping that of the active transporters SNAT3 and SNAT5, may broaden the range of nonessential amino acids that can be used as efflux substrates by LAT1 and LAT2.

In humans and mice, the amino acid analog MeAIB and proline are also efficiently accumulated in pancreas, although both compounds are not substrates for SNAT3, SNAT5, LAT1, LAT2, or ASCT1. MeAIB was shown to be efficiently transported by the Na^+-dependent transporter SNAT2 (SLC38A2). This sodium-amino acid symporter has a broader range of substrates compared to SNAT3 and SNAT5, as it additionally transports glycine, methionine, MeAIB, and the imino acid proline (Broer 2014; Mackenzie and Erickson 2004). Imino acids are also carried by the symporters PAT1 (SLC36A1) and SIT1 (SLC6A20), but evidences suggest they are not expressed in acinar cells. Specifically, PAT1 (SLC36A1), which belongs to the H^+-dependent SLC36 transporter family, was shown to be expressed in mouse pancreas at the mRNA level, but protein analysis suggests it is expressed rather in islets (Rooman et al. 2013; Zhou et al. 2014). The mRNA of the Na^+-dependent imino transporter SIT1 (SLC6A20), a member of the Na^+-dependent neutral amino acid and neurotransmitter transporter SLC6 family, is expressed at a very low level in pancreas (Rooman et al. 2013; Zhou et al. 2014). Whether SNAT2 (SLC38A2) localizes to acinar cells and is responsible for MeAIB and proline transport has not yet been investigated, although its mRNA and protein have been shown to be expressed in mouse pancreas (Rooman et al. 2013; Zhou et al. 2014). The recently developed SNAT2 (SLC38A2$^{-/-}$) knockout mouse (Weidenfeld et al. 2017) might be a useful tool to study these questions.

Accumulation of the oxidized form of the neutral amino acid cysteine also plays an important role in acinar cell homeostasis, in particular because of its role in glutathione (GSH) synthesis (Neuschwander-Tetri et al. 1997). The tripeptide GSH is synthesized in a two-step reaction from the amino acids, glutamate, glycine, and cysteine, and the intracellular concentration of cysteine generally represents the rate-limiting step for its synthesis. The transport of the oxidized form of cysteine, cystine, is accomplished by three antiporters from the SLC7 family, namely, AGT-rBAT (SLC7A13-SLC3A1), $b^{0,+}$-rBAT (SLC7A9-SLC3A1), and xCT-4F2hc (SLC7A11-SLC3A2) (Nagamori et al. 2016; Verrey et al. 2004). In contrast to AGT (SLC7A13) and $b^{0,+}$ (SLC7A9), which are (nearly) not expressed at mRNA level in pancreas, the cystine-glutamate exchanger xCT (SLC7A11) mRNA is detected at a substantial level and we observed that its protein is strongly expressed at the basolateral membrane of acinar cells upon stress-induced injury. Additionally, by using xCT knockout mice, we showed that xCT is the only pathway for cystine uptake in acinar cells (Simone Camargo, personal communication).

7.4.5 Cationic Amino Acids Transporters of Acinar Cells

Cationic amino acids are efficiently accumulated in isolated pancreas (Munoz et al. 1988; Sweiry et al. 1991). mRNAs and proteins of the cationic amino acid uniporters CAT1 (SLC7A1) and CAT2 (SLC7A2) and of the cationic amino acid/neutral amino acid + Na^+ antiporters y^+LAT1-4F2 (SLC7A7-SLC3A2) and y^+LAT2-4F2 (SLC7A6-SLC3A2) were found expressed in mouse pancreas (Rooman et al. 2013; Zhou et al. 2014). In kidney and small intestine, both CATs and y^+L-type transporters are localized to the basolateral membrane of epithelial cells, but their cellular and subcellular localization in pancreas have not yet been reported. It is, however, likely that the efficient cationic amino acid uptake mentioned above is mediated by CATs, since the driving force due to the membrane potential favors the influx of positively charged amino acids through these uniporters (Vanoaica et al. 2016). On the other hand, the efflux of cationic amino acids that can be trans-stimulated with neutral amino acids is probably mediated by a y^+L-type transporter

(Sweiry et al. 1991; Verrey et al. 2004). Interestingly, large doses of arginine, lysine, or ornithine injected intraperitoneally in mice and rats induce pancreatitis (Kui et al. 2015). The mechanism is not yet understood, but it has been suggested that the transport of the cationic amino acid into acinar cells triggers the injury (Cremades et al. 2016).

7.4.6 Anionic Amino Acid Transporters of Acinar Cells

Accumulation of anionic amino acids in mouse pancreas is not as efficient as that of neutral amino acids in ex vivo assays (Araya et al. 2017). Moreover, the screening for anionic transporters showed that the mRNA coding for excitatory amino acids transporter members of the SLC1 family had very low expression and they were not detected at the protein level (Araya et al. 2017; Rooman et al. 2013; Zhou et al. 2014). Other not yet characterized transporters may be expressed in acinar cells, but based on the functional and expression data, we postulate that acinar cells rely at least to a large extent on glutamate and aspartate synthesized from neutral amino acids. Acinar cells express all the enzymes necessary for anionic amino acids synthesis. Glutamate can be synthesized by the deamidation of glutamine by glutaminase (acinar cells express the liver isoform of glutaminase GLS2) and the transamination of α-ketoglutarate by the enzymes alanine (GPT1 and GPT2) and aspartate aminotransferases (GOT1 and GOT2). Additionally, Asparagine synthetase (ASNS) can synthesize glutamate and asparagine from glutamine and aspartate. Aspartate can be synthesized from glutamate by bidirectional enzyme GOT. By inhibiting pharmacologically GLS, GPT, and GOT, we could show that acinar cells preferentially synthesize glutamate from glutamine and alternatively from alanine when glutaminase was inhibited. Moreover, the presence of glutamine and alanine increased glutamate cellular accumulation and induced glutamate secretion (Araya et al. 2017). The synthesis has energetic and functional advantages for acinar cells since they use glutamine as their main source of energy, and the first step of this metabolism is the hydrolysis of glutamine to glutamate. Glutamate can be further metabolized to nonessential amino acids and aspartate, used in the synthesis of glutathione, metabolized to α-ketoglutarate to feed the TCA cycle, and finally secreted in the pancreatic juice.

To summarize, several neutral amino acid transporters have been shown to be expressed in pancreatic acinar cells, but it appears that several players are still missing. It is, however, interesting to observe that these cells express several transporters with overlapping substrate specificities and common mechanisms of transport (i.e., LAT1 and LAT2, SNAT3 and SNAT5), which could theoretically compensate for each other in case of loss of function. To our knowledge, there is no description of pancreatic diseases associated with variants in genes coding for amino acid transporters, but the research on amino acid transporters in pancreatic and other exocrine cells is still at an early stage. Further studies will be necessary to better understand the function and regulation of amino acid transporters in these cells in health and disease.

7.4.7 Regulation of Acinar Cell Amino Acid Transporters by Diet

The synthesis and secretory functions of acinar cells adapt to the diet. Indeed, fasting and protein-deficient diet reduce pancreatic volume, pancreatic protein synthesis, and enzyme secretion (Baumler et al. 2010; Crozier et al. 2009; Mann et al. 1986). The transcription of digestive enzymes is controlled by the pancreatic transcription factor PTFA1 (Hoang et al. 2016). The question whether PTFA1 complex is regulated by nutrient availability has not yet been addressed, but protein translation seems to be the main regulatory level for acinar cell adaptation to dietary manipulations (Graf et al. 2000; Hashimoto and Hara 2003). Fasting or protein-deficient diet reduced the translation of digestive enzymes without changing mRNA transcription (Graf et al. 2000; Sans et al. 2004). The refeeding after fasting normalized protein synthesis and organ growth. During refeeding, the translation initiation machinery was shown to be activated via the mTOR pathway (4EBP1 and S6K1 phosphorylation) (Guo et al. 2018a; Hashimoto and Hara 2003; Sans et al. 2004, 2006) and the inhibition of mTOR hindered cell growth and proliferation (Crozier et al. 2006, 2009). When leucine, a potent inducer of mTOR activity, was administered to fasted mice and rats, it efficiently induced the translation initiation machinery in acinar cells (Sans et al. 2006). In isolated acinar cells, leucine and phenylalanine were able to induce the translation initiation machinery and synthesis of amylase (Guo et al. 2018a, b). These data suggest that amino acid-induced mTOR activation seems to play an important role for the stimulation of protein synthesis in acinar cells.

Since protein translation in acinar cells decreases upon dietary protein shortage, it would be expected that amino acid accumulation also adapts to nutrient availability. Only a few studies analyzed the effect of dietary challenges on the accumulation of specific amino acids and on the expression of transporters in acinar cells and we will mention them below.

The effect of fasting on the accumulation of amino acids was shown to be time- and amino acid-dependent. For instance, an increase in the transport of phenylalanine was observed in pancreatic tissue of rats only after 72 h fasting. At that time point, the intra acinar cell concentration of aromatic and branched chain amino acids and of alanine was decreased, while that of other amino acids was similar to that in fed rats (Mann et al. 1986). The level of transporter expression after 72 h fasting has not been determined; however, after a shorter fasting time of 12 h in mice, we observed an increase of LAT1 (SLC7A5), xCT (SLC7A11), ASCT1 (SLC1A4), and ASCT2 (SLC1A5) mRNA levels (Rooman et al. 2013), but their protein expression and function were not assayed. Even though adaptation to fasting and refeeding is a very important physiological process for exocrine pancreas function, information on amino acids accumulation is still sparse.

When mice were challenged with protein-deficient diet for 8–12 days, the accumulation of phenylalanine, which is a substrate for the antiporters LAT1 and LAT2, and of glutamine, which is a substrate for the Na^+-dependent SNAT3 and SNAT5 carriers, was unchanged (Araya et al. 2017; Munoz et al. 1988). As regards the expression of these neutral amino acid transporters and of xCT (SLC7A11), which are localized at the basolateral membrane of acinar cells, only SNAT5 (SLC38A5) was significantly decreased at both protein and mRNA levels, when the animals were challenged with a protein-deprived diet (Araya et al. 2017). It is interesting to note that SNAT5 expression is under the control of PTFA1, like the digestive enzymes (Hoang et al. 2016; Masui et al. 2010), while LAT1, LAT2, SNAT3, and xCT expression can be regulated by the general control non-derepressible (GCN2-ATF4) pathway during amino acid starvation (Krokowski et al. 2013). Whether an activation of the GCN2-ATF4 pathway contributes to the steady state expression of these transporters was not analyzed in pancreatic acinar cells of these animals. As commented above, glutamine accumulation in acinar cells of animals challenged with protein-deficient diet was unchanged, despite the decrease in the expression of SNAT5 (SLC38A5). It might be that SNAT3, which is known to be expressed at the basolateral membrane of acinar cells, or SNAT2 and ASCT2 (SLC38A2 and SLC1A5), which have not yet been localized there, might have functionally compensated for SNAT5 loss. The maintenance of glutamine plays a central role for acinar cells, also because glutamine serves as their main energy source (Araya et al. 2017). The expression of transporters with overlapping substrate selectivity (SNAT3, SNAT5, and possibly SNAT2) may play a role in maintaining the function of these cells.

Enzyme secretion decreased under protein shortage (Graf et al. 2000; Kheroua and Belleville 1981; Kumar et al. 1975; Tandon et al. 1970), but refeeding intravenously and the transport of amino acids at the basolateral membrane of acinar cells did not reactivate enzyme secretion. In contrast, the delivery of free amino acids into the lumen of the duodenum indirectly induced the secretion of digestive enzymes via CCK release, in particular the aromatic amino acids phenylalanine and tryptophan which have the strongest effect on CCK and enzyme secretion (Meyer et al. 1976; Variyam et al. 1985). Additionally, secretagogues negatively affected amino acid transport on the basolateral membrane of acinar cells. Specifically, cholecystokinin and carbachol administered in physiological concentrations were shown to decrease the Na^+-dependent accumulation of the amino acid analogue aminoisobutyric acid (AIB), suggesting a decrease in SNAT2 activity (Bieger et al. 1977; Iwamoto and Williams 1980; Koizumi et al. 1993). It has, however, not yet been studied whether secretagogues influence the expression of amino acid transporters. In contrast, the secretion of free amino acids through the apical membrane of acinar cells into the pancreatic juice was shown to be not sensitive to the secretagogue CCK and to take place in the absence of stimulation. Additionally, the secretion of free amino acids into the pancreatic juice was not altered by the ingestion of protein-deficient diet (Araya et al. 2017) and could be stimulated by amino acid uptake as discussed before for the synthesis and secretion of glutamate.

Diet deficient in protein has been shown to decrease the synthesis and secretion of enzymes in acinar cells, but the expression of the transporters, the accumulation of amino acids, and their secretion were unchanged. This preservation of the cell transport machinery may allow the acinar cells to increase protein synthesis as soon as the nutrient shortage comes to an end without further adjustment. It may, however, be that an adaptation takes place upon more prolonged protein depletion.

7.4.8 Regulation of Amino Acid Transporters in Acinar Cells After Acute Injury

After acute pancreatic injury, the remaining acinar cells engage in a process involving phenotypical changes of tissue architecture, which is crucial for regeneration, known as metaplasia (Stanger and Hebrok 2013; Strobel et al. 2007). Recently, Willet and colleagues suggested that general stepwise metabolic phases take place during reprograming and regeneration (Willet et al. 2018). An initial downregulation of the mTOR pathway and of specialized functions would be followed by the activation of autophagy and dedifferentiation. During the next phase, proliferation and reactivation of the mTOR pathway would take place and finally the specialized functions restored. In these different phases, amino acid transporter expression may change to adapt to the changing demands of the cells.

Several models have been used to induce acute injury in exocrine pancreas and to study its regeneration. The induction of acute pancreatitis with supraphysiological doses of cerulein, an analogue of CCK, has been used in animal models to study the pathophysiology of acute pancreatitis since the 1970s (Lampel and Kern 1977; Lerch and Gorelick 2013). This treatment leads to an edematous and inflammatory injury with abnormal vacuolization followed by a considerable loss of acinar cells (Sakaguchi et al. 2006). In the cerulein mouse model, 1 week after the induction of pancreatitis, the morphology of the exocrine pancreas is nearly restored to the normal state.

In the regeneration process, a transient dedifferentiation of acinar cells towards a progenitor-like state can be identified by the upregulation of protein markers normally expressed in embryonic cells but excluded from adult acinar cells. These cells indeed express embryonic pancreatic progenitor markers such as Pdx1 and Sox9 and lose the expression of mature acinar cell markers like Ptf1a, MIST1, and digestive enzymes (Jensen et al. 2005; Pinho et al. 2011). Activation of the Hedgehog, Notch, and Wnt embryonic signaling pathways are important events for the regeneration of the exocrine pancreas in mice (Fendrich et al. 2008; Keefe et al. 2012; Molero et al. 2007; Siveke et al. 2008). These cells regain proliferative capacity and genes typically expressed in proliferating cells are strongly upregulated (Iovanna et al. 1992; Jensen et al. 2005; Pinho et al. 2011). Once the organ has recovered, expression of these genes returns to basal levels.

After the induction of acute pancreatitis, acinar cells decrease their protein synthesis rate and their accumulation of amino acids is variable (Fischer et al. 1995). For instance, the Na^+-dependent transport of the amino acid analogue α-aminobutiric acid (AIB), possibly mediated by SNAT2, was decreased, whereas the sodium-independent transport of leucine (substrate of LAT1 and LAT2) was increased (Adler et al. 1978). After injury, SNAT3, SNAT5, and LAT2 protein levels were decreased, while the expression of LAT1 was maintained at both mRNA and protein levels (Rooman et al. 2013). The stable expression of amino acid transporter LAT1 and the increase in leucine transport suggest that this antiporter may play an important role for the survival and dedifferentiation of the remaining acinar cells after the injury.

The expression and function of the antiporter xCT (SLC7A11) were increased in the early phase of injury and the transporter could be localized to the basolateral membrane of acinar cells. The transport of cystine and the expression of all enzymes involved in the de novo synthesis of the antioxidant GSH were increased after the induction of acute pancreatitis (Simone Camargo, personal communication).

Thus, in the earlier phase of injury, the transport and the use of amino acids by acinar cells appear to change, adjusting to a decrease in digestive enzyme synthesis and to an increased synthesis of the antioxidant GSH. In the next steps of regeneration, when proliferation, protein synthesis, and increased energy supply are needed, adaptation of the amino acid transport machinery may play a different role. We are currently analyzing amino acid transporter dynamics during regeneration and suggest that the characterization of their changes in expression may help understanding this process and its possible failure. Interestingly, although acute pancreatitis is reversible, recurrences and the presence of cells that fail to re-differentiate to acinar cells have been suggested to be involved in the development of chronic pancreatitis and Pancreatic Intraepithelial Neoplasia (PAnINs), a histologically well-defined precursor stage of invasive pancreatic ductal adenocarcinoma (Pinho et al. 2013).

7.5 Transepithelial Amino Acid Transport Machinery of Small Intestine and Kidney Proximal Tubule

Intestinal absorption of amino acids from ingested nutrients and renal reabsorption from primary urine filtrate into the extracellular space are evolutionarily very ancient. Absorption and reabsorption are performed by similar machineries mostly composed of the same transport proteins. In this section, we describe the transepithelial amino acid transport machinery and the involved transporters. Complementary information can be found in several other publications (Broer 2008; Broer and Fairweather 2018; Camargo et al. 2009; Makrides et al. 2014; Palacin et al. 1998; Vilches et al. 2018).

Amino acids are (re)absorbed nearly completely from the luminal content of these tubular structures via a two-step process. The first step is an active uptake through the apical membrane into the epithelial cells and the second step consists of their release through the basolateral membrane into the extracellular space, unless they are metabolized within the cell (Fig. 7.3). A major difference between the small intestine and the kidney proximal tubules (in humans ~1 million per kidney) is the irregular passage of nutrient boluses containing varying concentrations of amino acids in the

a

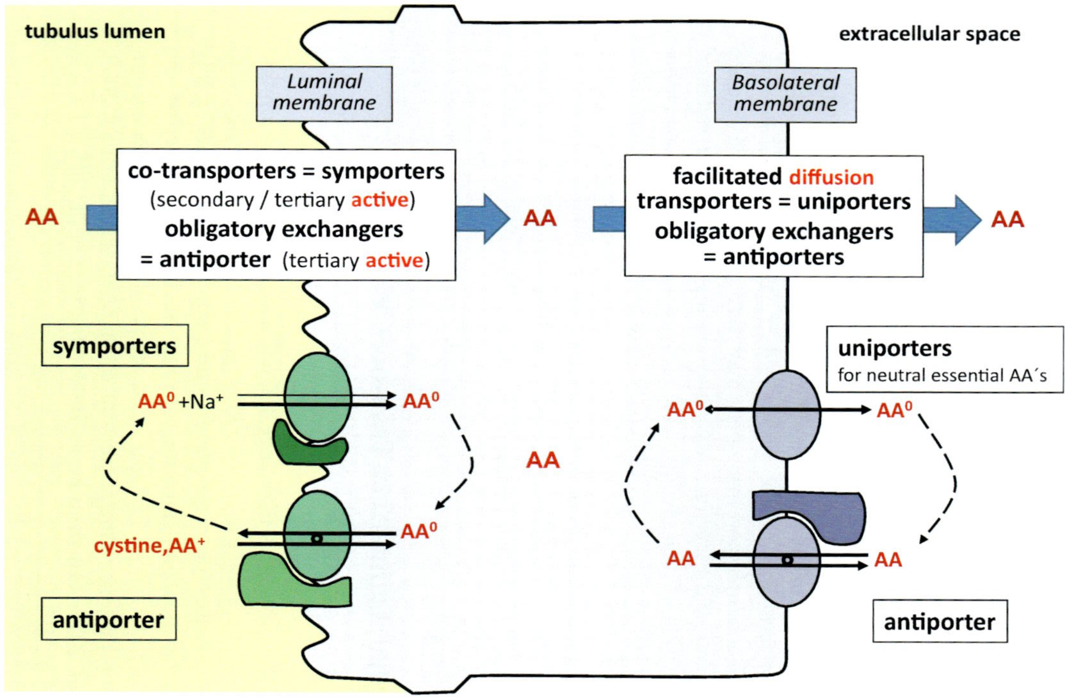

Fig. 7.3 Common mechanism of transcellular amino acid transport in small intestine and kidney proximal tubule epithelial cells. (**a**) Active luminal amino acid influx and mostly passive basolateral efflux. Luminal amino acids are actively transported possibly against their concentration gradient from the luminal compartment into the epithelial cells either by cotransport with Na^+ and/or H^+ (secondary active symport) or by exchange with other amino acids (tertiary active antiport). The transport out of the cell interior into the basolateral extracellular fluid is mediated for essential neutral amino acids to a large extent via facilitated diffusion pathways (uniport) and for nonessential neutral amino acids and cationic amino acids by exchangers (antiport). (**b**) Amino acid carriers involved in transepithelial amino acid transport. As indicated by a color code of the transporter names, most depicted transporters are expressed in both small intestinal enterocytes and kidney proximal tubule epithelial cells (red). Some transporters are expressed only in the kidney (blue) and generally display a higher affinity compared to the intestinal isoform and localize along the proximal kidney tubule distal to the lower affinity isoform. One transporter (PAT1-SLC36A1) and one associated protein (ACE2) are actively involved in transepithelial amino acid transport only in the small intestine and not in the kidney (lilac)

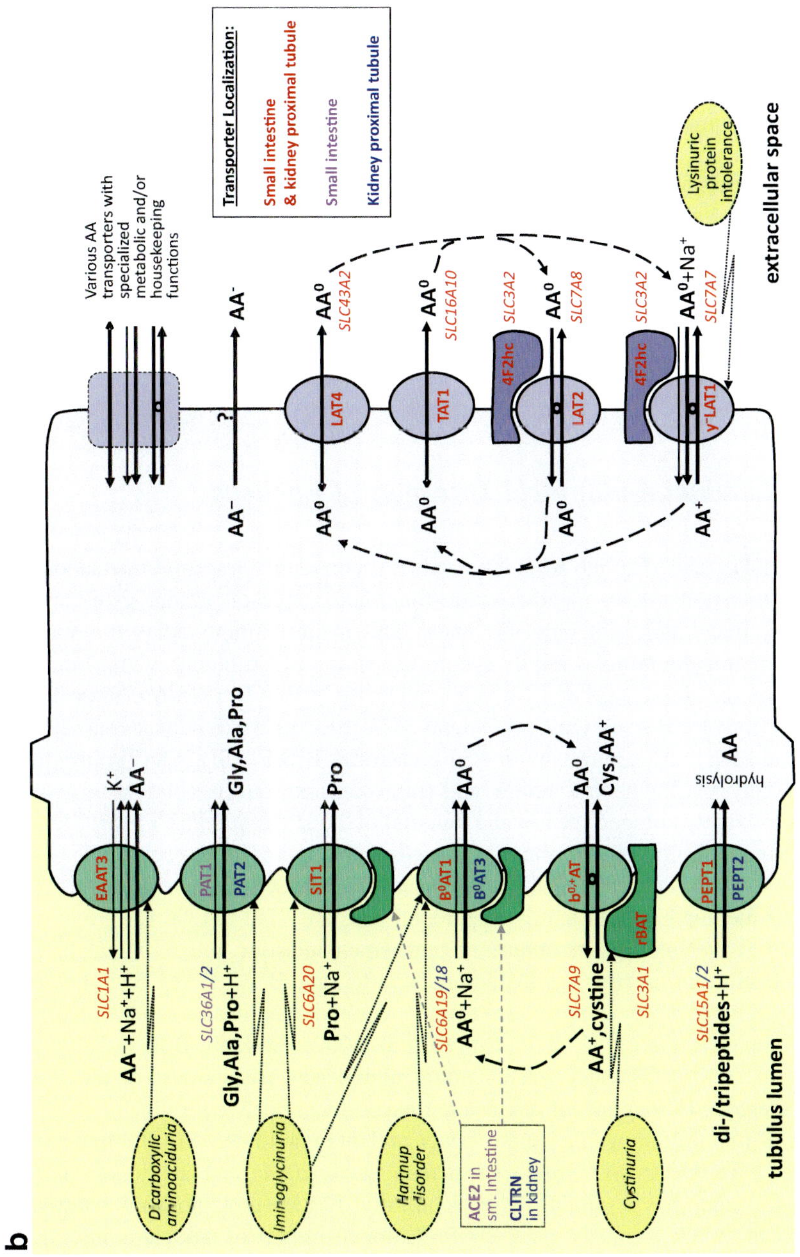

Fig. 7.3 (continued)

small intestine versus a relatively constant flow of primary urine filtrate containing a rather stable concentration of amino acids in the proximal tubule. Another major difference is the short length of the proximal tubules and thus the short contact time of the luminal fluid with the transporting epithelium. Thus, it is clear that the regulation of amino acid transport in proximal kidney tubule and small intestine fulfills different requirements. For instance, there is no need for major diurnal changes in amino acid transport capacity in the proximal tubule, unlike described below for the small intestine.

Yet another difference is the presence in the small intestine of a paracellular pathway that allows the absorption of a substantial proportion of small solutes when present at a sufficiently high concentration in the lumen. Interestingly, this pathway may be induced by ongoing Na^+-dependent absorption and is more developed in animals with a shorter and lighter small intestine such as birds and bats, suggesting an evolutionary trade-off between efficacy of absorption and protection from potentially toxic ingested solutes (Garro et al. 2018; Karasov 2017).

7.5.1 *Luminal Amino Acid Transport of Epithelia*

The luminal uptake of amino acids is an active, accumulative transport mediated by a series of transporters that together cover the whole range of proteinogenic amino acids and many non-proteinogenic ones. The neutral amino acids are mostly transported apically into the cell by symporters of the SLC6 family whose function is driven by the cotransport of 1 Na^+ ion in the case of B^0AT1 (SLC6A19) or 2 Na^+ and 1 Cl^- ions for B0AT3 (SLC6A18) and SIT1 (SLC6A20) (Rudnick et al. 2014). The broad selectivity neutral amino acid transporter B^0AT1 (SLC6A19) plays here an important role for neutral amino acid (re)absorption, but its ablation or loss of function is largely compensated by the function of the H^+-driven PEPT1 (SLC15A1) di- and tripeptides transporter and the paracellular pathway (Broer et al. 2011; Garro et al. 2018; Kleta et al. 2004; Rubio-Aliaga and Daniel 2008; Seow et al. 2004). The two other members of this family mentioned here are SIT1 (SLC6A20), which is expressed at the luminal membrane of both small intestine and proximal kidney epithelial cells and functions as IMINO transporter, and B^0AT3 (SLC6A18), which imports luminal, mostly small neutral amino acids and only localizes to the proximal kidney tubule (Romeo et al. 2006; Singer et al. 2009; Vuille-dit-Bille et al. 2015). Next to these three Na^+-dependent neutral amino acid transporters of the SLC6 family, there are H^+-dependent imino acid transporters selective for glycine, proline, and some non-proteinogenic amino acids and belong to the SLC36 family. In intestine, it is PAT1 (SLC36A1) (Anderson et al. 2004) and in kidneys PAT2 (SLC36A2) that has a slightly higher affinity for its proteinogenic substrates (Broer et al. 2008; Kennedy et al. 2005). Luminal import of anionic amino acids into small intestine and renal proximal tubule epithelial cells is mediated by EAAT3/ EAAC1 (SLC1A1) (Hu et al. 2018; Kanai and Hediger 1992). The last luminal amino acid transporter to be mentioned here is the antiporter $b^{0,+}AT1$ (SLC7A9). It belongs to the heterodimeric amino acid transporter family and is covalently associated with the glycoprotein rBAT (SLC3A1) (Pfeiffer et al. 1999a). $b^{0,+}$AT1 (SLC7A9) displays at the extracellular side a high affinity for L-cystine (cysteine dimer) as well as for cationic amino acids and transports some neutral amino acids like leucine, alanine, and tyrosine with a lower affinity. The affinity for intracellular exchange substrates has never been measured, but it has been inferred from electrophysiology and efflux experiments performed in *Xenopus* oocytes expressing exogenously injected rBAT and endogenous $b^{0,+}$AT1 that the preferential efflux exchange substrates are neutral amino acids that can recycle into the cell via B^0AT1 (Chillaron et al. 1996).

7.5.2 Basolateral Amino Acid Transport of Epithelia

Unlike the luminal uptake, the basolateral efflux of amino acids is not an active process and there are major differences in handling of essential and nonessential neutral, anionic (which are nonessential), and cationic amino acids (which are essential). A further complication is that, next to expressing a transepithelial amino acid transport machinery, epithelial cells of the small intestine and of the kidney proximal tubule also express at their basolateral membrane a set of transporters involved in housekeeping and/or metabolic functions.

Epithelial cells contain higher concentrations of nonessential amino acids compared with the extracellular space. Therefore, to maintain this concentration gradient, they do not express facilitated diffusion pathways permeable to nonessential amino acids, but amino acid antiporters (obligatory exchangers) that thus release nonessential (or essential) amino acids in the extracellular space only in exchange for the uptake of other amino acids. In the case of the heterodimeric amino acid antiporters, the favored extracellular substrates taken up are essential neutral amino acids. These essential amino acids can then be recycled back to the outside of the cell by parallel uniporters controlling essential amino acids efflux along their concentration gradient. Thus, it appears that the level of intracellular nonessential amino acids depends to a large extent on the intracellular amount of essential amino acids that are in equilibrium with their outside concentration. The transporters known to be involved in this concerted efflux mechanism are the two heterodimeric exchangers, LAT2 (SLC7A8) and y^+LAT1 (SLC7A7) and the two essential amino acid uniporters LAT4 (SLC43A2) and TAT1 (SLC16A10). The antiporter LAT2 favors the uptake of large essential neutral amino acids (Pineda et al. 1999; Rossier et al. 1999; Segawa et al. 1999). Given its high affinity for extracellular amino acids, LAT2 (SLC7A8) transport rate would be maximal if it was not limited by its intracellular substrates for which it displays a much lower affinity (Km mostly >>10 mM) (Meier et al. 2002). In this context, the efflux of (nonessential) amino acids with high intracellular concentrations appears to be favored. The basolateral heterodimeric antiporter y^+LAT1 (SLC7A7) has the particular capability of accommodating either a cationic amino acid or neutral amino acid together with Na^+. Extracellular concentrations and affinities favor the uptake of essential neutral amino acids (Pfeiffer et al. 1999b), and intracellular conditions seem to favor the efflux of cationic amino acids, although the actual intracellular affinities have not been measured/estimated (Bauch et al. 2003). Thus, the most prominent function of this exchanger is to efflux cations in exchange for essential neutral amino acids cotransported with Na^+. To summarize, when LAT2 (SLC7A8) and y^+LAT1 (SLC7A7) export, respectively, nonessential neutral amino acids and cationic amino acids, they also import essential neutral amino acids suggested to be effluxed by the two symmetrical uniporters, LAT4 (SLC43A2) and TAT1 (SLC16A10).

The question of anionic amino acids basolateral efflux is not solved yet. Intracellular concentrations of glutamate and aspartate are high compared to their extracellular concentrations and their negative charge increases the driving force for their efflux. In rat intestine, these amino acids are mostly metabolized within the enterocytes into other amino acids (Nakamura et al. 2013). However, it is likely that they may at least to some extent efflux from amino acid transporting epithelial cells via one or several transporters. It is unlikely that a basolateral efflux of anionic amino acids is mediated by 'reverse transport' across an excitatory amino acid transporter of the SLC1 family. Indeed, the driving force provided by the cotransport of 3 Na^+ and 1 H^+ and the antiport of 1 K^+ rather clearly favors anionic amino acid influx via these transporters (Zerangue and Kavanaugh 1996). Possible efflux pathways could instead be exchangers that may accommodate excitatory anionic amino acids. One possible candidate would be ASCT2 (SLC1A5), but its subcellular localization has been proposed to be rather luminal (Broer and Fairweather 2018; Oppedisano et al. 2007; Tetsuka et al. 2003). Other candidate pathways for basolateral anionic amino acid efflux would be, for instance, OATP4A1 (Lofthouse et al. 2018), or other anion exchangers or channels, provided they are expressed at the basolateral membrane of small intestine and/or kidney proximal tubule epithelial cells. The question of basolateral anionic amino acid efflux remains, however, open.

7.6 Small Intestine

7.6.1 Gastrointestinal Tract Origin and Formation

The gastrointestinal tract originates from the endoderm layer during gastrulation. In human, at the fourth week of gestation, it is composed of three regions that through various events of extension and rotation give rise to specific portions of the gastrointestinal tract (Lewis and Tam 2006). The foregut becomes mouth, esophagus, stomach, and the proximal part of the duodenum (plus spleen, liver gallbladder, and pancreas), the midgut gives rise to the remaining portion of the duodenum, jejunum ileum, caecum, ascending colon, and part of transverse colon, and the hindgut gives the remaining portion of the colon and the rectum. In mammalian small intestine, the mucosa is composed of a single layer of epithelial cells arranged in finger-like projections into the lumen. These villi increase the surface area for absorption and

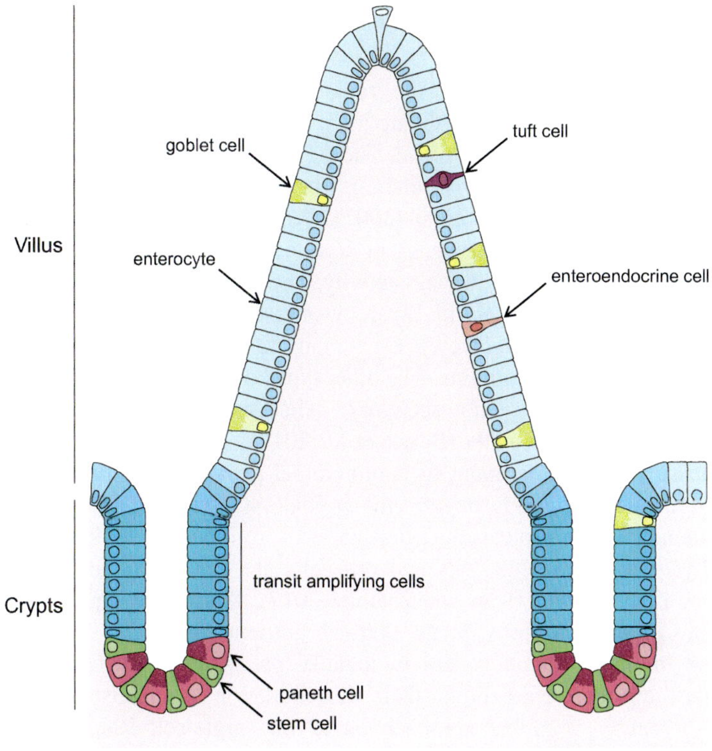

Fig. 7.4 Schematic representation of intestinal crypts and villus. Intestinal stem cells (ISCs) reside at the bottom of crypts where they intercalate with paneth cells. ISCs divide asymmetrically to renew the highly proliferative transit-amplifying cell population. These progenitors then commit to an absorptive or a secretory fate. The enterocytes constitute a specialized absorptive population; they express numerous amino acids, sugars, fatty acids, and micronutrients transporters. Enterocytes have a short lifespan and after about 5 days, they reach the top of the villus where they are eliminated by anoikis. This way, enterocytes are constantly renewed and allow an optimal nutrients absorption. Various cell types constitute the secretory lineage. Paneth cells migrate downwards to lay at the bottom of the crypts where they are thought to constitute a niche for the stem cells by paracrine secretion of wnt molecules. Paneth cells also secrete antimicrobial peptides in the intestinal lumen. The goblet cells are specialized secretory cells present along villi. They secrete mucins luminally to protect the intestinal wall from bacteria. The tuft cells are not as abundant in the small intestine; they protect against parasitic infection and express chemosensory receptors. There is a large variety of enteroendocrine cells spread along the small intestine. They release gastrointestinal hormones and peptides into the bloodstream and are thought to take part in luminal nutrient sensing

the cup-like structures between them, called crypts, contain stem cells and progenitors essential to maintain cell renewal along the villi (Fig. 7.4). Villus formation occurs during embryogenesis and the crypts appear in mouse pups only between postnatal days P4 and P10 by invagination (Sumigray et al. 2018).

7.6.2 Intestinal Cell Populations and Their Role

Stem cells residing at the bottom of the crypts give rise to a large variety of epithelial cells. The intestinal stem cells (ISC) are intercalated with the secretory paneth cells thought to provide the stem cell niche. The $Lgr5^+$ ISCs divide every 24 h to renew the transit-amplifying (TA) cell population, which proliferates and migrates towards the top of the villus (Barker et al. 2007). While migrating upwards, TA cells differentiate either into the absorptive or secretory lineage depending on the level of Notch and Wnt signaling pathways activity. The absorptive enterocyte population is the most abundant along the villi. Enterocytes harbor a vast number of transporters specific for amino acids (see below), sugars, fatty acids, and micronutrients to allow an optimal absorption of nutrients. The other types of specialized cells found in the intestinal epithelium are the secretory cells such as the paneth, goblet, enteroendocrine, and tuft cells (Basak et al. 2017). The two main secretory cell types release substantial amount of antimicrobial peptides into the lumen: paneth cells secrete defensins and lysozyme among others (Gassler 2017); and goblet cells secrete the mucus that protects the intestinal wall from bacteria (Birchenough et al. 2015). The enteroendocrine cells and tuft cells are considered to play a role in luminal nutrient sensing (Kaji and Kaunitz 2017). Altogether, these various cell populations interact to provide an optimal nutrient absorption. Nowadays, an increasing number of studies aim to identify precisely the different cell types composing intestinal epithelium, their differentiation mechanism, and specific markers. Recently, a global approach based on single-cell RNA-seq identified 15 clusters corresponding to distinct cell population signatures (Haber et al. 2017).

7.6.3 Intestinal Nutrient Absorption

The common elements of the small intestinal and renal proximal tubule amino acid (re)absorption machinery have been discussed above. In this section, we focus on small intestine-specific elements such as the axial localization and regulation of amino acid transporters expressed at the surface of enterocytes, the most numerous cells of the small intestine mucosa, responsible for the absorption of dietary nutrients. Indeed, when we eat, food is processed in our mouth and stomach before entering the small intestine where gastric and pancreatic enzymes cleave proteins into peptides and free amino acids, which can be taken up by enterocytes. In human, the small intestine is composed of duodenum (~25 cm), jejunum (~3 m), and ileum (~3 m) and the transit time through these portions is about 3–4 h. Jejunum and ileum have the largest absorbing surface area (each about 60 m^2) (Kararli 1995). In comparison, murine GI tract has a proportionally larger absorption surface in the jejunum (around 1 m^2) that constitutes the longest segment (>2/3 of the total small intestine length), but also because jejunum harbors long villi and a higher microvilli density compared to the other portions of the small intestine (Casteleyn et al. 2010).

Intestinal nutrient contact surface is so efficiently optimized that amino acid absorption appears to be nearly not saturable for some amino acids. For example, in pigs, glutamate absorptive capacity was estimated to be four-fold higher than the normal amount ingested daily (Janeczko et al. 2007). However, the absorption capacity varies between amino acids. For instance, we have previously observed that the proportion of gavaged labeled amino acids reaching the ileum with the bolus depends on the nature of the amino acid (Singer et al. 2012). And after an amino acid loading in piglets and human, the plasma concentration of different amino acids increases differentially, namely between 1.3- and 14-fold (Nuttall et al. 2006; Weber and Ehrlein 1998). These differences could to some extent be due to differential efficacies of their intestinal absorption. However, major differences are due to the preferential metabolism of some amino acids in the small intestine and their differential cellular uptake in the liver and the periphery after their release into the portal circulation.

We will now briefly review the amino acid transporters involved in transepithelial absorption, their axial distribution along the small intestine, and their regulation of expression and/or function when available.

7.6.4 Regulation of Small Intestine Amino Acid Transporters

Unlike in the kidney proximal tubule, the presence of amino acids in the small intestinal lumen is essentially limited to the periodic passage of the nutrient bolus that furthermore may contain variable amounts of amino acids. Thus, it appears economically meaningful that the small intestine does not maintain all the time a fully active amino acid transport machinery that consumes energy unnecessarily. It has been proposed and observed that there is a diurnal regulation of the intestinal amino acid transport machinery that is also entrained by food intake habits, and thus, adapts its functional capacity to the likelihood of amino acid-rich food intake (Hussain and Pan 2015; Jando et al. 2017; Oparija et al. 2019)(Oparija-Rogenmozere et al. 2020).

There are several regulatory mechanisms potentially responsible for the adaptation of the amino acid absorption machinery function to the availability or anticipation of luminal amino acids levels. We will mention some here before reviewing the known regulation of amino acids transporters. The first regulation level consists in the long-term adaptation of transport protein expression to the chronic dietary intake of amino acids/proteins, a situation we have observed for the luminal transporter B^0AT1 in rodents (Araya et al. 2018; Jando et al. 2017). Another type of regulation consists in the adaptation of the transport protein expression level to the diurnal rhythm of feeding. This possibility requires a good coordination with the actual feeding times and it costs the repeated synthesis and degradation of transport proteins. This type of regulation thus requires a rather short half-life of the regulated protein and mRNA. Such a situation has been described in the case of the intestinal glucose transporter SGLT1 and the peptide transporter PEPT1 whose mRNA and protein levels were shown to follow a similar rhythm that was entrained by the time of food intake (Pan et al. 2002, 2004). Considering this fast structural regulation of transporters, we may recall that, unlike in the renal proximal tubule, small intestinal epithelial cells have a short lifespan (~5 days) and thus long-term protein stability is not as energy-saving in this tissue as it is in kidney tubule. Another possibility to reduce the energy cost of transporters is to switch their activity status and/or subcellular localization according to the anticipated or real presence of food in the intestine. A well-documented example of acute transporter activity adaptation is the translocation of the glucose transporter SGLT1 to the luminal surface membrane upon substrate availability (Gorboulev et al. 2012). Such regulatory changes of transporter activity and/or subcellular localization can be signaled by phosphorylation changes. In the case of the basolateral amino acid transporter LAT4, we have for instance observed a diurnal phosphorylation switch which appears to favor transport at the time of food intake anticipation (Oparija et al. 2019) (Oparija-Rogenmozere et al. 2020).

7.6.5 Apical Amino Acid Transporters of Small Intestine

7.6.5.1 B^0AT1 (SLC6A19)

As mentioned above, B^0AT1 (SLC6A19) plays an important role by actively transporting into enterocytes all neutral amino acids in a Na^+-dependent manner, but with differential affinities (see review (Broer 2008)). In the small intestine, it requires the accessory protein ACE2 (Angiotensin converting enzyme 2) to localize to the brush border membrane (Camargo et al. 2009). B^0AT1 is expressed all along the small intestine with an increasing gradient towards the distal small intestine in mouse and rat (Jando et al. 2017; Romeo et al. 2006). As mentioned above, the expression of B^0AT1 is regulated in the long term. Rats fed a high protein diet for 7 days showed an increased B^0AT1 expression in the proximal small intestine and an increased isoleucine transport. On the contrary, mice fed for 8 days a protein-deficient diet showed a decreased B^0AT1 protein expression in jejunum, although it did not affect leucine transport (Araya et al. 2018; Jando et al. 2017). It may be that B^0AT1 expression is also regulated in the short term, although we were unable to verify it. Such a rapid regulation was suggested by experiments in which 60 min or 3 min luminal incubation with glutamine (10 mM) induced an increase of B^0AT1 mRNA in scraped mucosa and an increased amount of B^0AT1 protein at the brush border membranes, respectively (Ducroc et al. 2010). Taken together, these data suggest that the expression and activity of the luminal broad selectivity neutral amino acid transporter B^0AT1 are regulated by the dietary protein content in the long term and by the luminal presence of amino acids in the short term. However, it appears that the amplitude of these regulations is smaller than reported in the case of the sodium-glucose cotransporter SGLT1 mentioned above (Gorboulev et al. 2012).

7.6.5.2 SIT1 (SLC6A20)

SIT1 belongs to the same SLC family as B^0AT1 and functions as a Na^+- and Cl^--dependent IMINO transporter (Stevens and Wright 1985; Takanaga et al. 2005). In humans, SIT1 together with B^0AT1 and PAT1 mediate the transport of intestinal glycine and proline (Broer et al. 2008). SIT1 mRNA is expressed in all three parts of the small intestine and its protein was detected by immunostaining at the brush border membrane where it associates with ACE2 (Vuille-dit-Bille et al. 2015). In colon, SIT1 mRNA expression level is half that found in the small intestine. Interestingly, SIT1 protein is lacking in the small intestine of newborns (Meier et al. 2018). So far, we didn't find any report on its regulation.

7.6.5.3 $ATB^{0,+}$ (SLC6A14)

$ATB^{0,+}$is another member of the SLC6 family that transports with high affinity all neutral amino acids (especially large hydrophobic ones) and cationic amino acids in a Na^+- and Cl^--dependent manner (Sloan and Mager 1999). It does not appear to play a significant role in small intestine of most species, although it was clearly detected in piglet jejunum (Sun et al. 2015). Deposited microarray data suggest that its expression increases towards the distal part of the intestine (Broer and Fairweather 2018) and as shown by its substantial expression in mouse colon (Nakanishi et al. 2001; Ugawa et al. 2001). Very little is known about $ATB^{0,+}$ regulation, except that it is upregulated in pathologic contexts such as inflammation and cancer (Eriksson et al. 2008; Gupta et al. 2005).

7.6.5.4 PAT1 (SLC36A1)

PAT1 is a H^+-coupled and Na^+-independent low-affinity amino acid transporter that was first identified in neurons' lysosomal compartment and hence transports optimally at low pH (Sagne et al. 2001). PAT1 substrates are the imino acid proline, the small proteinogenic neutral amino acids alanine and glycine, the neurotransmitter GABA, the osmolyte taurine, and β-alanine (Boll et al. 2002). In the gastrointestinal tract, PAT1 is expressed in all three portions of the small intestine, with a peak in the jejunum, and exhibits a lower expression level in the colon. PAT1 has been visualized by immunostaining at the brush border membrane of rat ileum and human jejunum (Anderson et al. 2004). It cooperates with the apical Na^+/H^+ exchanger NHE3 to maintain the transmembrane H^+ electrochemical gradient. Therefore, PAT1 transport function is reduced when NHE3 activity is inhibited either by pharmacological means or via the activation of the cAMP/Protein Kinase A (PKA) pathway (Anderson and Thwaites 2005). High fat diet fed obese mice display an increase in alanine (twofold) and proline (threefold) concentration in the portal vein consistent with an upregulation of PAT1 expression (Do et al. 2014).

7.6.5.5 ASCT2 (SLC1A5)

The importance of the Na^+-dependent obligatory amino acid exchanger ASCT2 (SLC1A5), which transports alanine, serine, cysteine, threonine, and glutamine with high affinity, is not clear as regards the small intestine (Broer et al. 2000; Broer and Fairweather 2018; Scalise et al. 2018). It appears to be expressed along the gastrointestinal tract in the stomach, small intestine, and highest in colon and to localize at the apical membrane of enterocytes (Ducroc et al. 2010; Kirchhoff et al. 2006; Utsunomiya-Tate et al. 1996). Regarding the possible regulation of ASCT2, glutamine has been suggested to stimulate its translocation to the brush border membrane of rat enterocytes within 3 min and the satiety hormone leptin to downregulate it (Ducroc et al. 2010).

7.6.5.6 EAAT3 (EAAC1, SLC1A1)

EAAT3 is the main anionic amino acid (glutamate and aspartate) transporter in the small intestine and has been shown to transport also cysteine (Kanai et al. 1994; Zerangue and Kavanaugh 1996). It is suggested to be more highly expressed towards the distal portion of the small intestine (ileum) and appears to be specifically localized in the brush border membrane of enterocytes and crypt epithelial cells (Araya et al. 2018; Hu et al. 2018; Rome et al. 2002). Its transcript has been shown to be upregulated in the proximal parts of the small intestine of rats fed a protein-rich diet (Erickson et al. 1995). Similarly, its expression was increased at the membrane of rat jejunum after L-glutamate infusion (Mace et al. 2009). Interestingly, an increase in the protein expression of EAAT3 was also observed in mice fed a protein-deficient diet for 8 days without change in its transport capacity (Araya et al. 2018).

7.6.5.7 $b^{0,+}$AT (SLC7A9)

The catalytic subunit $b^{0,+}$AT (SLC7A9) is a Na^+-independent amino acid transporter that complexes with the heavy chain glycoprotein rBAT (SLC3A1) via a disulphide bond. This heteromeric amino acid transporter acts as an obligatory exchanger and has a high affinity for cationic amino acids and cystine (Broer and Fairweather 2018; Pfeiffer et al. 1999a; Pineda et al. 2004). Both subunits have a similar expression pattern showing increased mRNA expression towards the distal portion of the small intestine in mice (Dave et al. 2004). The heterodimer $b^{0,+}$AT-rBAT localizes at enterocytes apical membrane, more specifically at the brush border membrane where it absorbs luminal amino acids (Di Giacopo et al. 2013). It has been much studied in the renal context because mutations in either $b^{0,+}$AT or rBAT lead to cystinuria (see review by (Chillaron et al. 2010)). Until now, little has been done to study the regulation of $b^{0,+}$AT-rBAT expression in intestine, but interestingly, a dose-dependent increase of $b^{0,+}$AT (SLC7A9) mRNA expression was observed in jejunum and ileum of rats subjected daily to gastric infusion of L-theanine (Yan et al. 2017). Intestinal $b^{0,+}$AT-rBAT has been shown to be essential for the absorption of the Parkinson's disease medication L-dopa (Camargo et al. 2014).

7.6.6 Basolateral Amino Acid Transporters of Small Intestine

The complete set of basolateral amino acid transporters has not only the function of exporting amino acids imported luminally into the extracellular space (transepithelial amino acid transport machinery), but is also necessary for the import and/or export of amino acids needed for the cell metabolic activity and housekeeping functions. During the feeding phase, amino acids may be massively imported from the lumen into enterocytes and must be transferred efficiently to the circulating system. The apically transported amino acid glutamate is highly metabolized and used as the main source of energy. In contrast, when luminal amino acid levels are chronically low (animals fed protein-deficient diet), enterocytes take up glutamine across their basolateral membrane to supply their energy needs (Stoll et al. 1999; van der Schoor et al. 2001). As mentioned above, two main types of amino acid transporters cooperate at the basolateral membrane for the transepithelial transport of amino acids: antiporters (amino acids exchangers) and the uniporters (essential amino acids diffusion pathways). By recycling essential amino acids taken up by antiporters, the uniporters drive the basolateral efflux of other (nonessential) amino acids via these obligatory exchangers. Depending on the situation, enterocytes may also actively import amino acids from the extracellular space and thus express amino acid symporters from the SLC38 family.

7.6.7 Basolateral Antiporters of Small Intestine

7.6.7.1 LAT2-4F2hc (SLC7A8-SLC3A2)

LAT2-4F2hc is part of the epithelial amino acid transport machinery and is a system L transporter. It has a broad substrate selectivity and antiports (exchanges) all neutral amino acids except proline (Pineda et al. 1999; Rossier et al. 1999; Segawa et al. 1999). The catalytic light chain LAT2 forms a heteromeric complex with the 4F2 heavy chain (SLC3A2) which is stabilized by an extracellular disulfide bond (Pfeiffer et al. 1998). 4F2hc is required for the LAT2-4F2hc complex to translocate to the plasma membrane (Mastroberardino et al. 1998; Nakamura et al. 1999). LAT2-4F2hc is expressed along the small intestine with a higher expression level in jejunum and ileum (Dave et al. 2004), and it specifically localizes to the basolateral membrane (Rossier et al. 1999). LAT2-deficient mice only exhibit a mildly altered amino acid metabolism (Braun et al. 2011; Vilches et al. 2018).

7.6.7.2 y$^+$LAT1-4F2hc (SLC7A7-SLC3A2) and y$^+$LAT2-4F2hc (SLC7A6-SLC3A2)

In conditions of usual substrate concentrations, these antiporters export cationic amino acids against the import of neutral amino acid together with Na^+ (Pfeiffer et al. 1999b). The co-import of Na^+ not only compensates the charge of cationic amino acid efflux, but also provides some driving force for the import of neutral amino acids due to the Na^+ concentration gradient. In small intestine, y$^+$LAT1 plays an important role for the absorption of cationic amino acids. Importantly, the genetic defect of y$^+$LAT1 leads to the severe genetic disease lysinuric protein intolerance. Regarding regulation, it appears that y$^+$LAT1 is regulated in parallel with luminal b$^{0,+}$AT1 (SLC7A9) in jejunum of piglets. Indeed, lysine supplementation increased the mRNA expression of y$^+$LAT1 and b$^{0,+}$AT1 and decreased the mRNA expression of CAT1 (SLC7A1) (He et al. 2013).

7.6.8 *Basolateral Uniporters of Small Intestine*

7.6.8.1 LAT4 (SLC43A2)

LAT4 is a low-affinity symmetrical essential amino acid uniporter mediating the Na^+-independent efflux or import of branched chain amino acids, phenylalanine and methionine (Bodoy et al. 2005). It localizes at the basolateral membrane of villi within the whole length of the small intestine (Guetg et al. 2015). LAT4-deficient mice are born at the expected Mendelian inheritance ratio and show a very mild intrauterine growth retardation and a strongly reduced postnatal growth associated with signs of malnutrition, leading ultimately to death occurring before postnatal day P10 (Guetg et al. 2015). A recent study shows that this postnatal phenotype is not caused by the lack of LAT4 in intestine, where its absence leads only to a slight absorption defect accompanied by delayed gastrointestinal motility (Rajendran et al. 2020). Surprisingly, the expression of this intestinal LAT4 knockout in TAT1 defective mice induced no major effect (see below).

7.6.8.2 TAT1 (SLC16A10)

TAT1 is a low-affinity uniporter that can mediate a net Na^+-independent efflux or import of aromatic amino acids including L-DOPA. TAT1 is expressed at the basolateral membrane in all portions of the small intestine with a maximum expression level at the tip of villi (Kim et al. 2001; Ramadan et al. 2006). By using TAT1-deficient mice, we could verify its role in the basolateral transport of enterocytes. The efflux of phenylalanine and levodopa from the enterocytes was decreased in the knockout mice compared to their wild-type littermates (Camargo et al. 2014; Mariotta et al. 2012).

7.6.8.3 CAT-1 (SLC7A1)

This cationic amino acid uniporter does presumably not play a role for the small intestine transepithelial amino acid absorption machinery. However, it mediates the basolateral uptake of cationic amino acids into the epithelial cells. This process is driven by the membrane potential, as observed in HEK293T cells expressing an arginine FRET-based nanosensor (Vanoaica et al. 2016). CAT-1 thus contributes to maintaining cationic amino acid homeostasis in enterocytes in the absence of their absorption from the lumen. Supporting this interpretation, dietary lysine, shown in piglets' small intestine to increase the mRNA expression of the lysine efflux pathway y^+LAT1, was also shown to decrease CAT-1 mRNA expression (He et al. 2013).

7.6.9 Basolateral Symporters of Small Intestine

7.6.9.1 SNAT2 (SLC38A2)

As mentioned above, enterocytes eventually need to actively import amino acids from the extracellular space, depending on their metabolic situation and their possibility of importing amino acids from the lumen. The main transporter in charge of this basolateral uptake appears to be the symporter SNAT2 (SLC38A2) which, driven by Na^+ cotransport, imports small and hydrophilic neutral amino acids (glycine, proline, alanine, serine, cysteine, asparagine, glutamine, histidine, and methionine) (Yao et al. 2000). Indeed, glutamine uptake through the basolateral membrane has been shown to have system A characteristics, which led to the assumption that SNAT2 localizes there (Broer 2008, 2014; Taylor et al. 1989). In the case of piglet's small intestine, its presence has been documented (Li et al. 2015). Various SNAT2 regulation mechanisms have been highlighted in other cell types and SNAT2 has also been considered as having a nutrient sensor function. The *Slc38A2* gene is highly regulated, for instance, by amino acids availability, since it has an AARE (amino acid response element) allowing a GCN2/ATF4-dependent SNAT2 mRNA expression followed by a cap-independent SNAT2 translation when amino acids are scarce (Gaccioli et al. 2006; Palii et al. 2006). SNAT2 is also regulated posttranslationally by trafficking to the plasma membrane as observed in non-epithelial cell models (Hyde et al. 2007).

7.6.9.2 SNAT5 (SLC38A5)

SNAT5 is another member of the SLC38 family that besides cotransporting an amino acid with Na^+ (symport) also exchanges (antiport) a proton. This compensates the Na^+ charge movement and decreases the driving force for the uptake of the amino acid substrates, glutamine, asparagine, alanine, and serine. Thus, depending on pH and amino acid gradient, the transport may even be reverted (Baird et al. 2004). Interestingly, the level of SNAT5 (SLC38A5) transcript has been shown to be increased in epithelial cells of the duodenum of rats submitted to conditions of chronic acidosis (Wongdee et al. 2009).

7.6.10 Amino Acids Transporters in the Crypts of the Small Intestine

As described above, enterocytes are specialized cells expressing a complete amino acids transporter machinery necessary for nutrient transepithelial transport from the intestinal lumen to the extracellular space and circulatory system. These absorptive cells have a short lifespan and differentiate from the very active stem cell population residing in the crypts: the crypt base columnar (CBC) cells which do not perform transepithelial amino acid transport, but obtain amino acids by transport from the extracellular space. An interesting set of data obtained from pigeon embryos recapitulates precisely both amino acids transporter systems needed in the small intestine. On the one hand, transcripts of amino acids transporters involved in the transepithelial amino acids translocation from the lumen to the circulatory system increase during embryo development until the day of hatch. On the other hand, mRNAs encoding a set of basolateral amino acids transporter (EAAT2, LAT1, CAT1/2, SNAT1/2) are highest at E9 and then decrease until the day of hatch (Chen et al. 2015). These latter transporters are suggested to provide cells with amino acids from the extracellular space and may continue to play a role in stem cells or progenitors within the crypts.

Crypt base columnar cells divide asymmetrically to give rise to progenitor cells either moving upwards and ultimately differentiating into enterocytes or migrating towards the base of the crypts to become the secretory paneth cells. The $LGR5^{+}$ stem cells are intercalated between the Wnt-expressing paneth cells, which are suggested to play, in cooperation with the underlying mesenchymal cells, an important role as niche for the stem cells (Flanagan et al. 2018).

Interestingly, crypt-residing cells seem to be sensitive to nutrient availability. Indeed, caloric restriction increases stemness and promotes intestinal regeneration, while refeeding after starvation activates the mTORC1 pathway specifically in paneth cells (Yilmaz et al. 2012). Since some amino acids such as leucine, glutamine, and arginine are known to exert a permissive effect on the mTORC1 pathway (Jewell et al. 2013), it would be interesting to know which amino acids transporters are expressed in the basolateral membrane of CBC cells, thereby allowing them to respond to nutrient availability.

Unfortunately so far, little has been done to solve this question, but there is an increasing number of RNA-sequencing datasets obtained specifically from $LGR5^{+}$ stem cells, $CD24^{+}$ paneth cells, or via single-cell approaches (Haber et al. 2017; Munoz et al. 2012; Yu et al. 2018). A thorough analysis of these datasets highlighted the presence of some amino acid transporter transcripts sometimes confirmed by in situ hybridization or immunostaining.

One example is the glutamate/aspartate transporter EAAT2 (SLC1A2). It was detected in intestinal stem cells at the mRNA level and its protein was detected by immunostaining at the basolateral membrane of CBC cells (Iwanaga et al. 2005). This anionic amino acid transporter is, like other transporters of the SLC1 family, Na^+-, H^+-, and K^+-dependent. It has otherwise been shown to be highly expressed in the brain and moderately in glutamine-synthetase-positive hepatocytes (Hu et al. 2018).

The Na^+-dependent antiporter ASCT1 (SLC1A4), which has a high affinity for alanine, serine, and cysteine, was detected in paneth cells at the mRNA and protein levels (Haber et al. 2017; Hashimoto et al. 2004). This exchanger is also highly expressed in the brain, and mutations of the human *SLC1A4* gene were found associated with developmental delay and microcephaly, but nothing was reported about the small intestine (Damseh et al. 2015).

The essential amino acid antiporter LAT1 (SLC7A5) is widely expressed in various organs, with a high expression in the brain, but was not considered to be noticeably expressed in the small intestine until we detected it by qPCR and in situ hybridization enriched in the crypts. Its role in the CBC cells is currently under investigation, and it seems to be essential to maintain an adequate enterocyte differentiation and function (Poncet et al., manuscript in preparation).

These data on the amino acid transporters expression in small intestine crypt cells are still very scarce, but they suggest the presence of a variety of transporters at the membrane of $LGR5^+$ stem cells and secretory paneth cells. These two cell types have very different features, but they cooperate to generate and maintain the absorptive intestinal epithelium. The amino acids transporters they harbor at their plasma membrane appear to play an important role in this process. More investigations need to be carried out to understand it better.

7.7 Renal Reabsorption of Amino Acids Across the Proximal Tubule

The capability of the kidney tubule to efficiently recover valuable filtered solutes is observed very early during evolution and antedates animal life outside water. In mammals, amino acids reabsorption is mediated by transepithelial transport at the level of the proximal kidney tubule. Earlier micropuncture studies have shown that as much as 80% of this transport takes place at the level of the first half of the convoluted tubule (S1 segment), whereas less than 1% of the filtered amino acid load remains in the tubular fluid at the end of its last straight portion (S3 segment) (Silbernagl 1979). This section will focus on the proximal tubule, because it is the site of transepithelial amino acid reabsorption. Of course, all along the later tubule segments and in the collecting duct, there is an important basolateral amino acid transport activity which is presumably segment- and cell-type-specific, but of which not much is known. Only one of the basolateral amino acid transporters involved in proximal amino acid reabsorption has been detected at a substantial level in other tubule segments, the essential neutral amino acid uniporter LAT4 (SLC43A2). It is also strongly expressed in thick ascending limb and distal convoluted tubule where it probably functions to support the uptake rather than the efflux of amino acids.

Luminal amino acid transporters belonging to the amino acid reabsorption machinery of the proximal kidney tubule are not expressed to a significant level in other kidney tubule segments, with the exception of EAAT3 (SLC1A1) (Makrides et al. 2014; Shayakul et al. 1997). This anionic amino acid transporter is strongly expressed in the proximal tubule (in particular in S2 and S3) and has been detected at a lower level at luminal membranes of the thin descending limb of long-looped nephrons, the thick ascending limb, and the distal convoluted tubule. Its role in these segments is not yet clear, but it might be regulated in these segments by WNK (with-no-Lys kinase), SPAK (SPS1-related proline/alanine-rich kinase), and OSR1 (oxidative stress-responsive kinase 1) which are known stimulators of the Na^+-Cl^- (SLC12A3) and Na^+-K^+-$2Cl^-$ (SLC12A1) cotransporters (Borras et al. 2015; Nicholson and McGivan 1996).

The existence of a number of different luminal amino acid transporters has been suggested much before their molecular identification because of the discovery of various aminoacidurias, namely cystinuria, Hartnup disorder (neutral amino aciduria), iminoglycinuria, and dicarboxylic aminoaciduria as schematically indicated in Fig. 7.3. These disorders are not discussed here, but were extensively described in a number of reviews published in the past decade (Broer 2008; Camargo et al. 2013; Chillaron et al. 2010; Makrides et al. 2014). Interestingly, only one aminoaciduria is due to the defect of a basolateral transporter, namely lysinuric protein intolerance, which is caused by a defect of y^+LAT1 (SLC7A7) (Borsani et al. 1999; Torrents et al. 1999). No other aminoaciduria due to basolateral amino acid transporter defects has been described, and studies with knockout mouse models have revealed that this is likely due to the compensation by other transporters with partially overlapping amino acid selectivity (see below).

The renal proximal tubule role resembles much that of the small intestine; they both aim to recover maximally luminal amino acids and to prevent their loss. Accordingly, the transepithelial amino acid reabsorption machinery expressed in the proximal tubule is very similar to the absorption machinery of the small intestine and is mostly composed of the same transporters described in the sections on transepithelial amino acid transport and on small intestine (Fig. 7.3). In the present section, we will thus focus mainly on those characteristics that are specific to the proximal tubule.

7.7.1 Amino Acid Transporters of Proximal Kidney Tubule Not Expressed in Small Intestine

As mentioned before, the kidney proximal tubule shows a high expression of luminal amino acid transporters, in particular in its early segments (S1-S2), while its more distal segments (S2-S3) additionally express higher affinity transporters selective for di-/tripeptides (PEPT2 (SLC15A2)) and neutral amino acids (B^0AT3 (SLC6A18)) to prevent the loss of amino acids in the urine (Rubio-Aliaga and Daniel 2008; Singer et al. 2009).

7.7.1.1 PEPT2 (SLC15A2)

While the luminal low-affinity peptide transporter PEPT1 (SLC15A1) localizes mainly to S1, the early part of the proximal convoluted tubule, the higher affinity peptide transporter PEPT2 is expressed in S2 and S3 which are the later segments of the proximal tubule (Shen et al. 1999). The role of the low-affinity and high capacity transporter PEPT1 appears to be the uptake of the bulk of di- and tripeptides either filtered at the glomerulus or released from larger oligopeptides by surface-bound hydrolases along the early part of the proximal tubule. In contrast, the higher affinity and lower capacity PEPT2 may take up the residual di-/tripeptides from the primary urine in the later S2 and S3 segments of the proximal tubule (Daniel and Rubio-Aliaga 2003). Interestingly, the PEPT2 knockout mouse did not display any gross phenotype. A detailed analysis of this mouse model revealed, however, a defect in reabsorption of the dipeptide cys-gly originating from glutathione breakdown, and thus contributing to the resynthesis of glutathione in proximal tubule cells (Frey et al. 2007; Rubio-Aliaga et al. 2003).

7.7.1.2 B^0AT3 (SLC6A18)

An analogous situation is that of the luminal Na^+-dependent neutral amino acid transporters of the SLC6 family (Rudnick et al. 2014). The low-affinity broad selectivity neutral amino acid transporter B^0AT1 (SLC6A19) is expressed all along the proximal tubule with a gradient towards its early part. This symporter with the stoichiometry of 1 Na^+ per amino acid takes up nearly all neutral amino acids from the urinary filtrate, in particular at the level of the early proximal tubule segments (Camargo et al. 2005; Singer et al. 2009). Its genetic defect has been shown to cause Hartnup disorder, a condition exhibiting a high neutral aminoacid-uria (Kleta et al. 2004; Seow et al. 2004). A related transporter, B^0AT3 (SLC6A18), is also expressed at the luminal membrane of the proximal tubule, but with an opposed gradient of increasing levels toward the end of the proximal tubule. Its gene is localized in tandem with that of B^0AT1 (*SLC6A19*) on chromosome 5 in humans and the encoded neutral amino acid transporter B^0AT3 is ~50% identical with B^0AT1 and displays a slightly different selectivity spectrum with a much higher affinity for most neutral amino acids, in particular also for the small ones like alanine and glycine (Singer et al. 2009). Interestingly, its transport stoichiometry is also different from that of B^0AT1 as it cotransports with each amino acid 1 Cl^- and presumably 2 Na^+, providing a stronger driving force. A first study of a knockout mouse model had revealed a defect in glycine re-uptake associated with an increased blood pressure (Quan et al. 2004). A later study by our laboratory could only partially confirm the association with blood pressure, but demonstrated that the lack of B^0AT3 leads to a substantial broad-spectrum neutral amino aciduria involving all neutral amino acids but proline (Singer et al. 2009). The importance of this transporter for humans is questioned by the fact that nearly half of the Japanese population carries a variant that encodes a truncated transporter. This defect of B^0AT3 may, however, contribute to iminoaciduria (Broer et al. 2008). Additionally, the coincidence of a B^0AT3 defect with Hartnup disorder (B^0AT1 defect) would presumably worsen the aminoaciduria and possibly other symptoms.

7.7.1.3 PAT2 (SLC36A2)

Yet another difference of luminal transport protein between small intestine and kidney proximal tubule is the expression of the IMINO transporter PAT2 (SLC36A2) at the luminal brush border surface of the proximal tubule, versus the expression of PAT1 (SLC36A1) in the small intestine (Broer et al. 2008; Vanslambrouck et al. 2010). In this case, the transporter expressed in the kidney has a higher affinity for some of its substrates, compared to the one expressed in the small intestine. For instance, PAT2 transports proline with a $K_{0.5}$ of approximately 0.1 mM, whereas the typical substrate $K_{0.5}$ value of PAT1 for its substrates is around 2–10 mM. Both PAT1 and PAT2 cotransport H^+ and small neutral amino acids or imino acids with a 1:1 stoichiometry by Na^+-independent electrogenic transport (Anderson et al. 2004; Boll et al. 2002). Together with B^0AT1, B^0AT3, and SIT-1, PAT2 participates to renal proline and glycine reabsorption.

7.7.2 Lesson About Basolateral Transporter Cooperation from the Single and Double Transporter Knockout Mice

The role of different basolateral amino acid transporters in small intestine and kidney proximal tubule for the efflux of (re)absorbed amino acids has been addressed using mouse knockout models. The basolateral antiporter y^+LAT1 (SLC7A7) is a special case, since its defect causes lysinuric protein intolerance in humans. In mice, the ablation of this transporter, which in general mediates the efflux of cationic amino acids in exchange with neutral amino acids plus Na^+, resulted in the expected large cationic aminoaciduria known from lysinuric protein intolerance patients (Borsani et al. 1999; Torrents et al. 1999). However, it also led to a large neonatal lethality that prevented further studies of the epithelial transport defect and no study with tissue-specific knockout models has been reported so far (Sperandeo et al. 2007). The other basolateral antiporter, LAT2 (SLC7A8), exchanges neutral amino acids and displays at the outside a high affinity for the uptake of large neutral essential amino acid and at the inside a very low affinity for the efflux of neutral amino acids, of which presumably mostly nonessential ones are transported out (Meier et al. 2018; Rossier et al. 1999). Interestingly, the ablation of LAT2, the catalytic subunit of LAT2-4F2hc (SLC7A8-SLC3A2) heterodimer, led only to a mild aminoaciduria, notably increased under high protein diet (40%) and involving mostly neutral amino acids, but also some cationic amino acids and proline. Indeed, 10 out of 18 measured urinary amino acids displayed a statistically decreased tubular reabsorption under high protein diet (Braun et al. 2011; Vilches et al. 2018). It has been postulated that some redundancy and compensatory mechanisms at the level of the basolateral efflux of neutral amino acids underlie the quantitatively surprisingly small impact this knockout has on amino acid reabsorption.

Two uniporters, TAT1 (SLC16A10) and LAT4 (SLC43A2), each selective for a set of essential neutral amino acids, are known to be expressed at the basolateral membrane of small intestine and kidney proximal tubule (Guetg et al. 2015; Mariotta et al. 2012). They have been suggested to control the net basolateral efflux of cationic amino acids and nonessential neutral amino acids via the antiporters y^+LAT1 (SLC7A7) and LAT2 (SLC7A8) by recycling out of the cell the essential amino acids these antiporters take up in exchange (Makrides et al. 2014). The knockout of one of these transporters, LAT4 (SLC43A2, selective for branched chain amino acids, methionine and phenylalanine), is postnatally lethal (Guetg et al. 2015). The answer to the question whether the transport defect provoked by this ablation at the level of the intestinal and kidney epithelia is causative of this lethality appears to be no, in view of the observation that neither the lack of LAT4 in small intestine nor that in kidney tubule affects the postnatal development, but a combined lack of intestinal and kidney tubule LAT4 has not yet been tested (Rajendran et al. 2020). In contrast to LAT4, the knockout of TAT1 (SLC16A10) did not interfere with survival or reproduction of the mice (Mariotta et al. 2012). Interestingly, these mice displayed a high level of aromatic amino acids in plasma that is presumably due to a defect of their transport into hepatocytes that normally function as sink for these amino acids. TAT1 knockout mice also presented a small aminoaciduria that was exacerbated under high protein diet (40% protein). In this condition, tubular reabsorption of 13 out of 18 analyzed amino acids was decreased, including neutral amino acids, lysine and proline (Mariotta et al. 2012; Vilches et al. 2018). A very recent study demonstrated that the lack of kidney tubule LAT4 leads to a more substantial aminoaciduria than TAT1 knockout. This included amino acids that are substrates of the antiporter LAT2, demonstrating in vivo the functional cooperation of these two transporters (Rajendran et al. 2020). This study additionally demonstrated the major role that both basolateral antiporters LAT4 and TAT1 together play for tubular amino acid reabsorption, as a partial knockout of LAT4 in the background of TAT1 deficient mice led to a synergistic increase of the urinary amino acid loss.

The detailed analysis of the LAT2-TAT1 (SLC7A8-SLC16A10) double knockout mouse revealed how efficiently the basolateral transport machinery can compensate for the lack of specific transporters (Vilches et al. 2018). Indeed, on the one hand, double knockout mice displayed a substantial urinary loss of all neutral and cationic amino acids which was synergistically increased relative to the single knockout mice and thus supports the notion of a functional cooperation between TAT1 and LAT2 for the renal reabsorption of neutral amino acids. On the other hand, a surprisingly large percentage of tubular reabsorption of all amino acids remained in these double knockout animals, suggesting two possible compensatory mechanisms. One involves the participation of other transporters expressed in the basolateral membrane such as the uniporter LAT4 or SNAT3 (SLC38A3) that may, depending on the driving forces, efflux several neutral nonessential amino acids such as glutamine, histidine, alanine, and asparagine. The other possible compensation mechanism proposed is the reversion of y^+LAT1 (SLC7A7) transport activity that would in this case efflux neutral amino acids plus Na^+ against the influx of cationic amino acids. This mechanism may be supported by the observation that y^+LAT1 expression was increased in double knockout mice. The basolateral efflux of neutral amino acid via SNAT3 and y^+LAT1 becomes indeed possible if the intracellular neutral amino acid concentration is increased to a level that compensates the driving force of the inward Na^+ gradient. Another condition for the reversed function of y^+LAT1 to take place without dramatic cationic aminoaciduria is the existence of another basolateral efflux pathway for cationic amino acids. Such an efflux is likely possible through CAT uniporters that may mediate net cationic amino acids efflux when their intracellular concentration increases relative to the extracellular one to a level at which the driving force given by the chemical gradient exceeds that given by the membrane potential (Vanoaica et al. 2016). Taken together, these observations and the suggested compensatory mechanisms may explain why no neutral aminoaciduria has been detected in the case of a basolateral transport defect.

7.8 Conclusion

The molecular identification of most amino acid transporters and the better knowledge of their sites of expression have opened new research avenues toward a better understanding of their role at the cellular, organ, and organismal levels. It is of particular importance to understand the cooperation between transporters that together mediate the control of amino acid homeostasis in single cells as well as at the level of the organism and its extracellular space. These different homeostasis levels involve the function of various cellular barriers briefly discussed in this review, and of course, of the diverse amino acid transport machineries of single cells. The use of new approaches combining wet lab and computer-based methods will be required to study the complexity of amino acids transporter machineries, their interactions, and their underlying physiological regulatory networks (Panitchob et al. 2016; Taslimifar et al. 2017, 2018).

Acknowledgments The laboratory of FV, SCM, and NP is supported by the Swiss National Science Foundation grant #31_166430/1 to FV and the NCCR Kidney.CH.

References

Abplanalp J et al (2013) The cataract and glucosuria associated monocarboxylate transporter MCT12 is a new creatine transporter. Hum Mol Genet 22:3218–3226. https://doi.org/10.1093/hmg/ddt175

Adler G, Bieger W, Kern HF (1978) Amino acid transport in the rat exocrine pancreas. III. Effect of maximal and supramaximal hormonal stimulation in vivo. Cell Tissue Res 194:447–462

Akanuma S, Sakurai T, Tachikawa M, Kubo Y, Hosoya K (2015) Transporter-mediated L-glutamate elimination from cerebrospinal fluid: possible involvement of excitatory amino acid transporters expressed in ependymal cells and choroid plexus epithelial cells. Fluids Barriers CNS 12:11. https://doi.org/10.1186/s12987-015-0006-x

Albers A et al (2001) Na+ transport by the neural glutamine transporter ATA1. Pflugers Arch 443:92–101. https://doi.org/10.1007/s004240100663

Anderson CM, Thwaites DT (2005) Indirect regulation of the intestinal H+-coupled amino acid transporter hPAT1 (SLC36A1). J Cell Physiol 204:604–613. https://doi.org/10.1002/jcp.20337

Anderson CM et al (2004) H+/amino acid transporter 1 (PAT1) is the imino acid carrier: an intestinal nutrient/drug transporter in human and rat. Gastroenterology 127:1410–1422

Araya S et al (2017) Exocrine pancreas glutamate secretion help to sustain enterocyte nutritional needs under protein restriction. Am J Physiol Gastrointest Liver Physiol ajpgi001352017 https://doi.org/10.1152/ajpgi.00135.2017

Araya S et al (2018) Exocrine pancreas glutamate secretion help to sustain enterocyte nutritional needs under protein restriction. Am J Physiol Gastrointest Liver Physiol 314:G517–G536. https://doi.org/10.1152/ajpgi.00135.2017

Arriza JL, Kavanaugh MP, Fairman WA, Wu YN, Murdoch GH, North RA, Amara SG (1993) Cloning and expression of a human neutral amino acid transporter with structural similarity to the glutamate transporter gene family. J Biol Chem 268:15329–15332

Babu E, Bhutia YD, Ramachandran S, Gnanaprakasam JP, Prasad PD, Thangaraju M, Ganapathy V (2015) Deletion of the amino acid transporter Slc6a14 suppresses tumour growth in spontaneous mouse models of breast cancer. Biochem J 469:17–23. https://doi.org/10.1042/BJ20150437

Baird FE, Beattie KJ, Hyde AR, Ganapathy V, Rennie MJ, Taylor PM (2004) Bidirectional substrate fluxes through the system N (SNAT5) glutamine transporter may determine net glutamine flux in rat liver. J Physiol 559:367–381. https://doi.org/10.1113/jphysiol.2003.060293

Balkrishna S, Broer A, Welford SM, Hatzoglou M, Broer S (2014) Expression of glutamine transporter Slc38a3 (SNAT3) during acidosis is mediated by a different mechanism than tissue-specific expression. Cell Physiol Biochem 33:1591–1606. https://doi.org/10.1159/000358722

Barker N et al (2007) Identification of stem cells in small intestine and colon by marker gene Lgr5. Nature 449:1003. https://doi.org/10.1038/nature06196

Basak O, Beumer J, Wiebrands K, Seno H, van Oudenaarden A, Clevers H (2017) Induced quiescence of Lgr5+ stem cells in intestinal organoids enables differentiation of hormone-producing enteroendocrine cells. Cell Stem Cell 20:177–190 e174. https://doi.org/10.1016/j.stem.2016.11.001

Bauch C, Forster N, Loffing-Cueni D, Summa V, Verrey F (2003) Functional cooperation of epithelial heteromeric amino acid transporters expressed in Madin-Darby canine kidney cells. J Biol Chem 278:1316–1322. https://doi.org/10.1074/jbc.M210449200

Baumler MD, Koopmann MC, Thomas DD, Ney DM, Groblewski GE (2010) Intravenous or luminal amino acids are insufficient to maintain pancreatic growth and digestive enzyme expression in the absence of intact dietary protein. Am J Physiol Gastrointest Liver Physiol 299:G338–G347

Bauwens M, Keyaerts M, Lahoutte T, Kersemans K, Caveliers V, Bossuyt A, Mertens J (2007) Intra-individual comparison of the human biodistribution and dosimetry of the D and L isomers of 2-[123I]iodo-phenylalanine. Nucl Med Commun 28:823–828. https://doi.org/10.1097/MNM.0b013e3282e7d731

Bieger W, Peter S, Kern HF (1977) Amino acid transport in the rat exocrine pancreas. I Transport of neutral amino acids and their utilization in protein synthesis. Cell Tissue Res 180:31–44

Birchenough GM, Johansson ME, Gustafsson JK, Bergstrom JH, Hansson GC (2015) New developments in goblet cell mucus secretion and function. Mucosal Immunol 8:712–719. https://doi.org/10.1038/mi.2015.32

Bodoy S, Martin L, Zorzano A, Palacin M, Estevez R, Bertran J (2005) Identification of LAT4, a novel amino acid transporter with system L activity. J Biol Chem 280:12002–12011. https://doi.org/10.1074/jbc.M408638200

Bohmer C, Broer A, Munzinger M, Kowalczuk S, Rasko JE, Lang F, Broer S (2005) Characterization of mouse amino acid transporter B0AT1 (slc6a19). Biochem J 389:745–751. https://doi.org/10.1042/BJ20050083

Boiadjeva Knopfel E et al (2019) Dysfunctional LAT2 amino acid transporter is associated with cataract in mouse and humans. Front Physiol 10:688

Boll M, Foltz M, Rubio-Aliaga I, Kottra G, Daniel H (2002) Functional characterization of two novel mammalian electrogenic proton-dependent amino acid cotransporters. J Biol Chem 277:22966–22973. https://doi.org/10.1074/jbc.M200374200

Borner AR et al (2001) Whole-body kinetics and dosimetry of cis-4-[(18)F]fluoro-L-proline. Nucl Med Biol 28:287–292

Borras J, Salker MS, Elvira B, Warsi J, Fezai M, Hoseinzadeh Z, Lang F (2015) SPAK and OSR1 sensitivity of excitatory amino acid transporter EAAT3. Nephron 130:221–228. https://doi.org/10.1159/000433567

Borsani G et al (1999) SLC7A7, encoding a putative permease-related protein, is mutated in patients with lysinuric protein intolerance. Nat Genet 21:297–301. https://doi.org/10.1038/6815

Braun D et al (2011) Aminoaciduria, but normal thyroid hormone levels and signalling, in mice lacking the amino acid and thyroid hormone transporter Slc7a8. Biochem J 439:249–255. https://doi.org/10.1042/BJ20110759

Broer S (2008) Amino acid transport across mammalian intestinal and renal epithelia. Physiol Rev 88:249–286. https://doi.org/10.1152/physrev.00018.2006

Broer S (2014) The SLC38 family of sodium-amino acid co-transporters. Pflugers Arch 466:155–172. https://doi.org/10.1007/s00424-013-1393-y

Broer S, Fairweather SJ (2018) Amino acid transport across the mammalian intestine. Compr Physiol 9:343–373. https://doi.org/10.1002/cphy.c170041

Broer A, Wagner C, Lang F, Broer S (2000) Neutral amino acid transporter ASCT2 displays substrate-induced Na+ exchange and a substrate-gated anion conductance. Biochem J 346 (Pt 3):705–710

Broer A et al (2002) Regulation of the glutamine transporter SN1 by extracellular pH and intracellular sodium ions. J Physiol 539:3–14

Broer S et al (2008) Iminoglycinuria and hyperglycinuria are discrete human phenotypes resulting from complex mutations in proline and glycine transporters. J Clin Invest 118:3881–3892. https://doi.org/10.1172/JCI36625

Broer A et al (2011) Impaired nutrient signaling and body weight control in a Na+ neutral amino acid cotransporter (Slc6a19)-deficient mouse. J Biol Chem 286:26638–26651. https://doi.org/10.1074/jbc.M111.241323

Stevens BR, Wright EM (1985) Substrate specificity of the intestinal brush-border proline/sodium (IMINO) transporter. J Membr Biol 87(1):27–34

Camargo SM, Makrides V, Virkki LV, Forster IC, Verrey F (2005) Steady-state kinetic characterization of the mouse B(0)AT1 sodium-dependent neutral amino acid transporter. Pflugers Arch 451:338–348. https://doi.org/10.1007/s00424-005-1455-x

Camargo SM et al (2009) Tissue-specific amino acid transporter partners ACE2 and collectrin differentially interact with hartnup mutations. Gastroenterology 136:872–882. https://doi.org/10.1053/j.gastro.2008.10.055

Camargo SM, Makrides V, Kleta R, Verrey F (2013) Kidney transport of amino acids and oligopeptides, and aminoacidurias. In: Alpern RJ, Caplan M, Moe OW (eds) Seldin and Geibisch's the kidney. Elsevier, London, pp 2405–2423

Camargo SM et al (2014) The molecular mechanism of intestinal levodopa absorption and its possible implications for the treatment of Parkinson's disease. J Pharmacol Exp Ther 351:114–123. https://doi.org/10.1124/jpet.114.216317

Cantor J, Browne CD, Ruppert R, Feral CC, Fassler R, Rickert RC, Ginsberg MH (2009) CD98hc facilitates B cell proliferation and adaptive humoral immunity. Nat Immunol 10:412–419. https://doi.org/10.1038/ni.1712

Cantor J, Slepak M, Ege N, Chang JT, Ginsberg MH (2011) Loss of T cell CD98 H chain specifically ablates T cell clonal expansion and protects from autoimmunity. J Immunol 187:851–860. https://doi.org/10.4049/jimmunol.1100002

Casteleyn C, Rekecki A, Van der Aa A, Simoens P, Van den Broeck W (2010) Surface area assessment of the murine intestinal tract as a prerequisite for oral dose translation from mouse to man. Lab Anim 44:176–183. https://doi.org/10.1258/la.2009.009112

Castorino JJ et al (2011) Juvenile cataract-associated mutation of solute carrier SLC16A12 impairs trafficking of the protein to the plasma membrane. Invest Ophthalmol Vis Sci 52:6774–6784. https://doi.org/10.1167/iovs.10-6579

Chan K et al (2016) Loss of function mutation of the Slc38a3 glutamine transporter reveals its critical role for amino acid metabolism in the liver, brain, and kidney. Pflugers Arch 468:213–227. https://doi.org/10.1007/s00424-015-1742-0

Chaudhry FA et al (2001) Coupled and uncoupled proton movement by amino acid transport system. N EMBO J 20:7041–7051. https://doi.org/10.1093/emboj/20.24.7041

Chaudhry FA et al (2002) Glutamine uptake by neurons: interaction of protons with system a transporters. J Neurosci 22:62–72

Chen Z, Kennedy DJ, Wake KA, Zhuang L, Ganapathy V, Thwaites DT (2003) Structure, tissue expression pattern, and function of the amino acid transporter rat PAT2. Biochem Biophys Res Commun 304:747–754

Chen MX, Li XG, Yang JX, Gao CQ, Wang B, Wang XQ, Yan HC (2015) Growth of embryo and gene expression of nutrient transporters in the small intestine of the domestic pigeon (Columba livia). J Zhejiang Univ Sci B 16:511–523. https://doi.org/10.1631/jzus.B1400340

Chen F et al (2018) Regulation of amino acid transporters in the mammary gland from late pregnancy to peak lactation in the sow. J Anim Sci Biotechnol 9:35. https://doi.org/10.1186/s40104-018-0250-4

Chillaron J et al (1996) Obligatory amino acid exchange via systems bo,+-like and y+L-like. A tertiary active transport mechanism for renal reabsorption of cystine and dibasic amino acids. J Biol Chem 271:17761–17770

Chillaron J, Font-Llitjos M, Fort J, Zorzano A, Goldfarb DS, Nunes V, Palacin M (2010) Pathophysiology and treatment of cystinuria. Nat Rev Nephrol 6:424–434. https://doi.org/10.1038/nrneph.2010.69

Cleal JK, Lofthouse EM, Sengers BG, Lewis RM (2018) A systems perspective on placental amino acid transport. J Physiol 596:5511–5522. https://doi.org/10.1113/JP274883

Cleveland MH, Sawyer JM, Afelik S, Jensen J, Leach SD (2012) Exocrine ontogenies: on the development of pancreatic acinar, ductal and centroacinar cells. Semin Cell Dev Biol 23:711–719. https://doi.org/10.1016/j.semcdb.2012.06.008

Closs EI, Boissel JP, Habermeier A, Rotmann A (2006) Structure and function of cationic amino acid transporters (CATs). J Membr Biol 213:67–77. https://doi.org/10.1007/s00232-006-0875-7

Coate KC et al (2017) FGF21 is an exocrine pancreas secretagogue. Cell Metab 25:472–480. https://doi.org/10.1016/j.cmet.2016.12.004

Cremades A, Del Rio-Garcia J, Lambertos A, Lopez-Garcia C, Penafiel R (2016) Tissue-specific regulation of potassium homeostasis by high doses of cationic amino acids. Springerplus 5:616. https://doi.org/10.1186/s40064-016-2224-3

Crozier SJ, Sans MD, Guo L, D'Alecy LG, Williams JA (2006) Activation of the mTOR signalling pathway is required for pancreatic growth in protease-inhibitor-fed mice. J Physiol 573:775–786. https://doi.org/10.1113/jphysiol.2006.106914

Crozier SJ, D'Alecy LG, Ernst SA, Ginsburg LE, Williams JA (2009) Molecular mechanisms of pancreatic dysfunction induced by protein malnutrition. Gastroenterology 137:1093–1101, 1101 e1091–1093. https://doi.org/10.1053/j.gastro.2009.04.058

Dahlin A, Royall J, Hohmann JG, Wang J (2009) Expression profiling of the solute carrier gene family in the mouse brain. J Pharmacol Exp Ther 329:558–570. https://doi.org/10.1124/jpet.108.149831

Dai WT, Zou YX, White RR, Liu JX, Liu HY (2018) Transcriptomic profiles of the bovine mammary gland during lactation and the dry period. Funct Integr Genomics 18:125–140. https://doi.org/10.1007/s10142-017-0580-x

Damseh N et al (2015) Mutations in SLC1A4, encoding the brain serine transporter, are associated with developmental delay, microcephaly and hypomyelination. J Med Genet 52:541–547. https://doi.org/10.1136/jmedgenet-2015-103104

Daniel H, Rubio-Aliaga I (2003) An update on renal peptide transporters. Am J Physiol Renal Physiol 284:F885–F892. https://doi.org/10.1152/ajprenal.00123.2002

Danielsson A et al (2014) The human pancreas proteome defined by transcriptomics and antibody-based profiling. PLoS One 9:e115421. https://doi.org/10.1371/journal.pone.0115421

Danilczyk U et al (2006) Essential role for collectrin in renal amino acid transport. Nature 444:1088–1091

Dave MH, Schulz N, Zecevic M, Wagner CA, Verrey F (2004) Expression of heteromeric amino acid transporters along the murine intestine. J Physiol 558:597–610. https://doi.org/10.1113/jphysiol.2004.065037

Di Giacopo A et al (2013) Differential cystine and dibasic amino acid handling after loss of function of the amino acid transporter b0,+AT (Slc7a9) in mice. Am J Physiol Renal Physiol 305:F1645–F1655. https://doi.org/10.1152/ajprenal.00221.2013

Do TT et al (2014) Disturbed intestinal nitrogen homeostasis in a mouse model of high-fat diet-induced obesity and glucose intolerance. Am J Physiol Endocrinol Metab 306:E668–E680. https://doi.org/10.1152/ajpendo.00437.2013

Dolgodilina E, Camargo SM, Roth E, Herzog B, Nunes V, Palacin M, Verrey F (2020) Choroid plexus LAT2 and SNAT3 as partners in CSF amino acid homeostasis maintenance. Fluids Barriers CNS 17:17. https://doi.org/10.1186/s12987-020-0178-x

Dolgodilina E, Imobersteg S, Laczko E, Welt T, Verrey F, Makrides V (2016) Brain interstitial fluid glutamine homeostasis is controlled by blood-brain barrier SLC7A5/LAT1 amino acid transporter. J Cereb Blood Flow Metab 36:1929–1941. https://doi.org/10.1177/0271678X15609331

Doyle CJ et al (2012) The proteome of normal pancreatic juice. Pancreas 41:186–194. https://doi.org/10.1097/MPA.0b013e31822862f6

Drgonova J, Liu QR, Hall FS, Krieger RM, Uhl GR (2007) Deletion of v7-3 (SLC6A15) transporter allows assessment of its roles in synaptosomal proline uptake, leucine uptake and behaviors. Brain Res 1183:10–20. https://doi.org/10.1016/j.brainres.2007.09.001

Ducroc R, Sakar Y, Fanjul C, Barber A, Bado A, Lostao MP (2010) Luminal leptin inhibits L-glutamine transport in rat small intestine: involvement of ASCT2 and B0AT1. Am J Physiol Gastrointest Liver Physiol 299:G179–G185. https://doi.org/10.1152/ajpgi.00048.2010

Dunphy MPS et al (2018) In vivo PET assay of tumor glutamine flux and metabolism: in-human trial of (18)F-(2S,4R)-4-fluoroglutamine. Radiology 287:667–675. https://doi.org/10.1148/radiol.2017162610

Erickson RH, Gum JR Jr, Lindstrom MM, McKean D, Kim YS (1995) Regional expression and dietary regulation of rat small intestinal peptide and amino acid transporter mRNAs. Biochem Biophys Res Commun 216:249–257. https://doi.org/10.1006/bbrc.1995.2617
Eriksson A, Jennische E, Flach CF, Jorge A, Lange S (2008) Real-time PCR quantification analysis of five mucosal transcripts in patients with Crohn's disease. Eur J Gastroenterol Hepatol 20:290–296. https://doi.org/10.1097/MEG.0b013e3282f3557c
Fagerberg L et al (2014) Analysis of the human tissue-specific expression by genome-wide integration of transcriptomics and antibody-based proteomics. Mol Cell Proteomics 13:397–406. https://doi.org/10.1074/mcp.M113.035600
Feliubadalo L et al (2003) Slc7a9-deficient mice develop cystinuria non-I and cystine urolithiasis. Hum Mol Genet 12:2097–2108
Fendrich V et al (2008) Hedgehog signaling is required for effective regeneration of exocrine pancreas. Gastroenterology 135:621–631
Fischer H, Konturek JW, Szlachcic A, Konturek SJ, Domschke W (1995) Plasma amino acid consumption and pancreatic secretion during and after cerulein-induced pancreatitis in rats. Int J Pancreatol 18:127–134. https://doi.org/10.1007/BF02785886
Flanagan DJ, Austin CR, Vincan E, Phesse TJ (2018) Wnt signalling in gastrointestinal epithelial stem cells. Genes (Basel) 9. https://doi.org/10.3390/genes9040178
Fogelstrand P, Feral CC, Zargham R, Ginsberg MH (2009) Dependence of proliferative vascular smooth muscle cells on CD98hc (4F2hc, SLC3A2). J Exp Med 206:2397–2406. https://doi.org/10.1084/jem.20082845
Font-Llitjos M et al (2007) Slc7a9 knockout mouse is a good cystinuria model for antilithiasic pharmacological studies. Am J Physiol Renal Physiol 293:F732–F740
Frey IM et al (2007) Profiling at mRNA, protein, and metabolite levels reveals alterations in renal amino acid handling and glutathione metabolism in kidney tissue of Pept2−/− mice. Physiol Genomics 28:301–310. https://doi.org/10.1152/physiolgenomics.00193.2006
Fukushima D et al (2010) Glutamate exocrine dynamics augmented by plasma glutamine and the distribution of amino acid transporters of the rat pancreas. J Physiol Pharmacol 61:265–271
Gaccioli F et al (2006) Amino acid starvation induces the SNAT2 neutral amino acid transporter by a mechanism that involves eukaryotic initiation factor 2alpha phosphorylation and cap-independent translation. J Biol Chem 281:17929–17940. https://doi.org/10.1074/jbc.M600341200
Garro C, Brun A, Karasov WH, Caviedes-Vidal E (2018) Small intestinal epithelial permeability to water-soluble nutrients higher in passerine birds than in rodents. J Anim Physiol Anim Nutr (Berl) 102:1766–1773. https://doi.org/10.1111/jpn.12969
Gassler N (2017) Paneth cells in intestinal physiology and pathophysiology. World J Gastrointest Pathophysiol 8:150–160. https://doi.org/10.4291/wjgp.v8.i4.150
Gomeza J, Hulsmann S, Ohno K, Eulenburg V, Szoke K, Richter D, Betz H (2003a) Inactivation of the glycine transporter 1 gene discloses vital role of glial glycine uptake in glycinergic inhibition. Neuron 40:785–796
Gomeza J et al (2003b) Deletion of the mouse glycine transporter 2 results in a hyperekplexia phenotype and postnatal lethality. Neuron 40:797–806
Gorboulev V et al (2012) Na(+)-D-glucose cotransporter SGLT1 is pivotal for intestinal glucose absorption and glucose-dependent incretin secretion. Diabetes 61:187–196. https://doi.org/10.2337/db11-1029
Graf R, Valeri F, Gassmann R, Hailemariam S, Frick TW, Bimmler D (2000) Adaptive response of the rat pancreas to dietary substrates: parallel regulation of trypsinogen and pancreatic secretory trypsin inhibitor. Pancreas 21:181–190
Guetg A, Mariotta L, Bock L, Herzog B, Fingerhut R, Camargo SM, Verrey F (2015) Essential amino acid transporter Lat4 (Slc43a2) is required for mouse development. J Physiol 593:1273–1289. https://doi.org/10.1113/jphysiol.2014.283960

Guo L, Liang Z, Zheng C, Liu B, Yin Q, Cao Y, Yao J (2018a) Leucine affects alpha-amylase synthesis through PI3K/Akt-mTOR signaling pathways in pancreatic acinar cells of dairy calves. J Agric Food Chem 66:5149–5156. https://doi.org/10.1021/acs.jafc.8b01111
Guo L et al (2018b) Phenylalanine regulates initiation of digestive enzyme mRNA translation in pancreatic acinar cells and tissue segments in dairy calves. Biosci Rep 38. https://doi.org/10.1042/BSR20171189
Gupta N et al (2005) Upregulation of the amino acid transporter ATB0,+ (SLC6A14) in colorectal cancer and metastasis in humans. Biochim Biophys Acta 1741:215–223. https://doi.org/10.1016/j.bbadis.2005.04.002
Haber AL et al (2017) A single-cell survey of the small intestinal epithelium. Nature 551:333–339. https://doi.org/10.1038/nature24489
Hashimoto N, Hara H (2003) Dietary amino acids promote pancreatic protease synthesis at the translation stage in rats. J Nutr 133:3052–3057. https://doi.org/10.1093/jn/133.10.3052
Hashimoto Y, Sadamoto Y, Konno A, Kon Y, Iwanaga T (2004) Distribution of neutral amino acid transporter ASCT 1 in the non-neuronal tissues of mice. Jpn J Vet Res 52:113–124
Hatanaka T, Huang W, Ling R, Prasad PD, Sugawara M, Leibach FH, Ganapathy V (2001) Evidence for the transport of neutral as well as cationic amino acids by ATA3, a novel and liver-specific subtype of amino acid transport system A. Biochim Biophys Acta 1510:10–17
Hatanaka T et al (2004) Transport of amino acid-based prodrugs by the Na+- and Cl(-) -coupled amino acid transporter ATB0,+ and expression of the transporter in tissues amenable for drug delivery. J Pharmacol Exp Ther 308:1138–1147. https://doi.org/10.1124/jpet.103.057109
He L et al (2013) Effects of dietary L-lysine intake on the intestinal mucosa and expression of CAT genes in weaned piglets. Amino Acids 45:383–391. https://doi.org/10.1007/s00726-013-1514-0
Heller-Stilb B et al (2002) Disruption of the taurine transporter gene (taut) leads to retinal degeneration in mice. FASEB J 16:231–233. https://doi.org/10.1096/fj.01-0691fje
Hoang CQ et al (2016) Transcriptional maintenance of pancreatic acinar identity, differentiation, and homeostasis by PTF1A. Mol Cell Biol 36:3033–3047. https://doi.org/10.1128/MCB.00358-16
Hu Y, Smith DE, Ma K, Jappar D, Thomas W, Hillgren KM (2008) Targeted disruption of peptide transporter Pept1 gene in mice significantly reduces dipeptide absorption in intestine. Mol Pharm 5:1122–1130
Hu QX, Ottestad-Hansen S, Holmseth S, Hassel B, Danbolt NC, Zhou Y (2018) Expression of glutamate transporters in mouse liver, kidney, and intestine. J Histochem Cytochem 66:189–202. https://doi.org/10.1369/0022155417749828
Huang DY et al (2006) Impaired ability to increase water excretion in mice lacking the taurine transporter gene TAUT. Pflugers Arch 451:668–677. https://doi.org/10.1007/s00424-005-1499-y
Huling JC, Pisitkun T, Song JH, Yu MJ, Hoffert JD, Knepper MA (2012) Gene expression databases for kidney epithelial cells. Am J Physiol Ren Physiol 302:F401–F407. https://doi.org/10.1152/ajprenal.00457.2011
Hussain MM, Pan X (2015) Circadian regulation of macronutrient absorption. J Biol Rhyth 30:459–469. https://doi.org/10.1177/0748730415599081
Hustinx R et al (2003) Whole-body tumor imaging using PET and 2-18F-fluoro-L-tyrosine: preliminary evaluation and comparison with 18F-FDG. J Nucl Med 44:533–539
Hyde R, Cwiklinski EL, MacAulay K, Taylor PM, Hundal HS (2007) Distinct sensor pathways in the hierarchical control of SNAT2, a putative amino acid transceptor, by amino acid availability. J Biol Chem 282:19788–19798. https://doi.org/10.1074/jbc.M611520200
Iovanna JL, Lechene de la Porte P, Dagorn JC (1992) Expression of genes associated with dedifferentiation and cell proliferation during pancreatic regeneration following acute pancreatitis. Pancreas 7:712–718
Iwamoto Y, Williams JA (1980) Inhibition of pancreatic alpha-aminoisobutyric acid uptake by cholecystokinin and other secretagogues. Am J Physiol 238:G440–G444. https://doi.org/10.1152/ajpgi.1980.238.5.G440

Iwanaga T, Goto M, Watanabe M (2005) Cellular distribution of glutamate transporters in the gastrointestinal tract of mice: an immunohistochemical and in situ hybridization approach. Biomed Res 26:271–278

Jando J, Camargo SMR, Herzog B, Verrey F (2017) Expression and regulation of the neutral amino acid transporter B0AT1 in rat small intestine. PLoS One 12:e0184845. https://doi.org/10.1371/journal.pone.0184845

Janeczko MJ, Stoll B, Chang X, Guan X, Burrin DG (2007) Extensive gut metabolism limits the intestinal absorption of excessive supplemental dietary glutamate loads in infant pigs. J Nutr 137:2384–2390. https://doi.org/10.1093/jn/137.11.2384

Jensen JN, Cameron E, Garay MV, Starkey TW, Gianani R, Jensen J (2005) Recapitulation of elements of embryonic development in adult mouse pancreatic regeneration. Gastroenterology 128:728–741

Jewell JL, Russell RC, Guan K-L (2013) Amino acid signalling upstream of mTOR. Nat Rev Mol Cell Biol 14:133. https://doi.org/10.1038/nrm3522

Jiang M et al (2016) MIST1 and PTF1 collaborate in feed-forward regulatory loops that maintain the pancreatic acinar phenotype in adult mice. Mol Cell Biol. https://doi.org/10.1128/MCB.00370-16

Kaji I, Kaunitz JD (2017) Luminal chemosensing in the gastroduodenal mucosa. Curr Opin Gastroenterol 33:439–445. https://doi.org/10.1097/MOG.0000000000000396

Kanai Y, Hediger MA (1992) Primary structure and functional characterization of a high-affinity glutamate transporter. Nature 360:467–471. https://doi.org/10.1038/360467a0

Kanai Y, Stelzner M, Nussberger S, Khawaja S, Hebert SC, Smith CP, Hediger MA (1994) The neuronal and epithelial human high affinity glutamate transporter. Insights into structure and mechanism of transport. J Biol Chem 269:20599–20606

Kanai Y, Segawa H, Miyamoto K, Uchino H, Takeda E, Endou H (1998) Expression cloning and characterization of a transporter for large neutral amino acids activated by the heavy chain of 4F2 antigen (CD98). J Biol Chem 273:23629–23632

Kanai Y, Clemencon B, Simonin A, Leuenberger M, Lochner M, Weisstanner M, Hediger MA (2013) The SLC1 high-affinity glutamate and neutral amino acid transporter family. Mol Aspects Med 34:108–120. https://doi.org/10.1016/j.mam.2013.01.001

Kaplan E et al (2018) ASCT1 (Slc1a4) transporter is a physiologic regulator of brain d-serine and neurodevelopment. Proc Natl Acad Sci USA 115:9628–9633. https://doi.org/10.1073/pnas.1722677115

Kararli TT (1995) Comparison of the gastrointestinal anatomy, physiology, and biochemistry of humans and commonly used laboratory animals. Biopharm Drug Dispos 16:351–380

Karasov WH (2017) Integrative physiology of transcellular and paracellular intestinal absorption. J Exp Biol 220:2495–2501. https://doi.org/10.1242/jeb.144048

Karlsson RM, Tanaka K, Saksida LM, Bussey TJ, Heilig M, Holmes A (2009) Assessment of glutamate transporter GLAST (EAAT1)-deficient mice for phenotypes relevant to the negative and executive/cognitive symptoms of schizophrenia. Neuropsychopharmacology 34:1578–1589. https://doi.org/10.1038/npp.2008.215

Keefe MD, Wang H, De La OJ-P, Khan A, Firpo MA, Murtaugh LC (2012) Catenin is selectively required for the expansion and regeneration of mature pancreatic acinar cells in mice. Dis Mod Mech 5:503–514. https://doi.org/10.1242/dmm.007799

Kennedy DJ, Gatfield KM, Winpenny JP, Ganapathy V, Thwaites DT (2005) Substrate specificity and functional characterisation of the H+/amino acid transporter rat PAT2 (Slc36a2). Br J Pharmacol 144:28–41. https://doi.org/10.1038/sj.bjp.0706029

Kheroua O, Belleville J (1981) Behaviour of digestive enzymes in the pancreatic juice and pancreas of rats fed on a low-protein diet (3 p. 100 of cereal protein) then on a balanced diet (23.5 p. 100 of mixed protein). Reprod Nutr Dev 21:901–917

Kim DK, Kanai Y, Chairoungdua A, Matsuo H, Cha SH, Endou H (2001) Expression cloning of a Na+-independent aromatic amino acid transporter with structural similarity to H+/monocarboxylate transporters. J Biol Chem 276:17221–17228. https://doi.org/10.1074/jbc.M009462200

Kim J et al (2017) Amino acid transporter Slc38a5 controls glucagon receptor inhibition-induced pancreatic alpha cell hyperplasia in mice. Cell Metab 25:1348–1361 e1348. https://doi.org/10.1016/j.cmet.2017.05.006

Kirchhoff P et al (2006) An amino acid transporter involved in gastric acid secretion. Pflugers Arch 451:738–748. https://doi.org/10.1007/s00424-005-1507-2

Kleta R et al (2004) Mutations in SLC6A19, encoding B0AT1, cause Hartnup disorder. Nat Genet 36:999–1002. https://doi.org/10.1038/ng1405

Klockner U, Storck T, Conradt M, Stoffel W (1994) Functional properties and substrate specificity of the cloned L-glutamate/L-aspartate transporter GLAST-1 from rat brain expressed in Xenopus oocytes. J Neurosci 14:5759–5765

Koizumi T, Bamba T, Hosoda S (1993) Relationship between amino acid transport and protein synthesis in rat isolated pancreatic acini under stimulation with cholecystokinin. Gastroenterol Jpn 28:385–392

Kono T et al (2002) Clinical application of (11)C-methionine positron emission tomography for evaluation of pancreatic function. Pancreas 25:20–25

Konturek JW, Gabryelewicz A, Kulesza E, Konturek SJ, Domschke W (1995) Cholecystokinin (CCK) in the amino acid uptake and enzyme protein secretion by the pancreas in humans. Int J Pancreatol 17:55–61

Kowalczuk S, Broer A, Munzinger M, Tietze N, Klingel K, Broer S (2005) Molecular cloning of the mouse IMINO system: an Na+- and Cl—dependent proline transporter. Biochem J 386:417–422. https://doi.org/10.1042/BJ20050100

Krokowski D et al (2013) A self-defeating anabolic program leads to beta-cell apoptosis in endoplasmic reticulum stress-induced diabetes via regulation of amino acid flux. J Biol Chem 288:17202–17213. https://doi.org/10.1074/jbc.M113.466920

Kui B et al (2015) New insights into the methodology of L-arginine-induced acute pancreatitis. PLoS One 10:e0117588. https://doi.org/10.1371/journal.pone.0117588

Kumar R, Banks PA, George PK, Tandon BN (1975) Early recovery of exocrine pancreatic function in adult protein-calorie malnutrition. Gastroenterology 68:1593–1595

Lampel M, Kern HF (1977) Acute interstitial pancreatitis in the rat induced by excessive doses of a pancreatic secretagogue Virchows Archiv A. Pathol Anat Histol 373:97–117

Lerch MM, Gorelick FS (2013) Models of acute and chronic pancreatitis. Gastroenterology 144:1180–1193. https://doi.org/10.1053/j.gastro.2012.12.043

Lewis SL, Tam PP (2006) Definitive endoderm of the mouse embryo: formation, cell fates, and morphogenetic function. Dev Dyn 235:2315–2329. https://doi.org/10.1002/dvdy.20846

Li P, Knabe DA, Kim SW, Lynch CJ, Hutson SM, Wu G (2009) Lactating porcine mammary tissue catabolizes branched-chain amino acids for glutamine and aspartate synthesis. J Nutr 139:1502–1509. https://doi.org/10.3945/jn.109.105957

Li G et al (2015) Characterization and regulation of the amino acid transporter SNAT2 in the small intestine of piglets. PLoS One 10:e0128207. https://doi.org/10.1371/journal.pone.0128207

Lin Y, Duan X, Lv H, Yang Y, Liu Y, Gao X, Hou X (2018) The effects of L-type amino acid transporter 1 on milk protein synthesis in mammary glands of dairy cows. J Dairy Sci 101:1687–1696. https://doi.org/10.3168/jds.2017-13201

Lister A et al (2018) NRF2 regulates the glutamine transporter Slc38a3 (SNAT3) in kidney in response to metabolic acidosis. Sci Rep 8:5629. https://doi.org/10.1038/s41598-018-24000-2

Lofthouse EM, Cleal JK, O'Kelly IM, Sengers BG, Lewis RM (2018) Estrone sulphate uptake by the microvillous membrane of placental syncytiotrophoblast is coupled to glutamate efflux. Biochem Biophys Res Commun 506:237–242. https://doi.org/10.1016/j.bbrc.2018.10.074

Lyck R et al (2009) Culture-induced changes in blood-brain barrier transcriptome: implications for amino-acid transporters in vivo. J Cereb Blood Flow Metab 29:1491–1502. https://doi.org/10.1038/jcbfm.2009.72

Mace OJ et al (2009) An energy supply network of nutrient absorption coordinated by calcium and T1R taste receptors in rat small intestine. J Physiol 587:195–210. https://doi.org/10.1113/jphysiol.2008.159616

Mackenzie B, Erickson JD (2004) Sodium-coupled neutral amino acid (System N/A) transporters of the SLC38 gene family. Pflugers Arch 447:784–795
Mailliard ME, Stevens BR, Mann GE (1995) Amino acid transport by small intestinal, hepatic, and pancreatic epithelia. Gastroenterology 108:888–910
Makrides V, Camargo SM, Verrey F (2014) Transport of amino acids in the kidney. Compr Physiol 4:367–403. https://doi.org/10.1002/cphy.c130028
Malakauskas SM et al (2007) Aminoaciduria and altered renal expression of luminal amino acid transporters in mice lacking novel gene collectrin. Am J Physiol Renal Physiol 292:F533–F544
Mann GE, Munoz M, Peran S (1986) Fasting and refeeding modulate neutral amino acid transport activity in the basolateral membrane of the rat exocrine pancreatic epithelium: fasting-induced insulin insensitivity. Biochim Biophys Acta 862:119–126
Mariotta L et al (2012) T-type amino acid transporter TAT1 (Slc16a10) is essential for extracellular aromatic amino acid homeostasis control. J Physiol 590:6413–6424. https://doi.org/10.1113/jphysiol.2012.239574
Marques F et al (2011) Transcriptome signature of the adult mouse choroid plexus. Fluids Barriers CNS 8:10. https://doi.org/10.1186/2045-8118-8-10
Mastroberardino L, Spindler B, Pfeiffer R, Skelly PJ, Loffing J, Shoemaker CB, Verrey F (1998) Amino-acid transport by heterodimers of 4F2hc/CD98 and members of a permease family. Nature 395:288–291. https://doi.org/10.1038/26246
Masui T et al (2010) Replacement of Rbpj with Rbpjl in the PTF1 complex controls the final maturation of pancreatic acinar cells. Gastroenterology 139:270–280. https://doi.org/10.1053/j.gastro.2010.04.003
Matsumoto T et al (2013) Production of free glutamate in milk requires the leucine transporter LAT1. Am J Physiol Cell Physiol 305:C623–C631. https://doi.org/10.1152/ajpcell.00291.2012
Matsuo H et al (2002) Identification of a novel Na+-independent acidic amino acid transporter with structural similarity to the member of a heterodimeric amino acid transporter family associated with unknown heavy chains. J Biol Chem 277:21017–21026. https://doi.org/10.1074/jbc.M200019200
Meier C, Ristic Z, Klauser S, Verrey F (2002) Activation of system L heterodimeric amino acid exchangers by intracellular substrates. EMBO J 21:580–589
Meier C et al (2018) Intestinal IMINO transporter SIT1 is not expressed in human newborns. Am J Physiol Gastrointest Liver Physiol 315:G887–G895. https://doi.org/10.1152/ajpgi.00318.2017
Meyer JH, Kelly GA, Spingola LJ, Jones RS (1976) Canine gut receptors mediating pancreatic responses to luminal L-amino acids. Am J Phys 231:669–677. https://doi.org/10.1152/ajplegacy.1976.231.3.669
Molero X et al (2007) Pancreas transcription factor 1alpha expression is regulated in pancreatitis. Eur J Clin Investig 37:791–801. https://doi.org/10.1111/j.1365-2362.2007.01856.x
Munoz M, Emery PW, Peran S, Mann GE (1988) Dietary regulation of amino acid transport activity in the exocrine pancreatic epithelium. Biochim Biophys Acta 945:273–280
Munoz J et al (2012) The Lgr5 intestinal stem cell signature: robust expression of proposed quiescent '+4' cell markers. EMBO J 31:3079–3091. https://doi.org/10.1038/emboj.2012.166
Nagamori S et al (2016) Novel cystine transporter in renal proximal tubule identified as a missing partner of cystinuria-related plasma membrane protein rBAT/SLC3A1. Proc Natl Acad Sci USA 113:775–780. https://doi.org/10.1073/pnas.1519959113
Nakamura E et al (1999) 4F2 (CD98) heavy chain is associated covalently with an amino acid transporter and controls intracellular trafficking and membrane topology of 4F2 heterodimer. J Biol Chem 274:3009–3016
Nakamura H, Kawamata Y, Kuwahara T, Torii K, Sakai R (2013) Nitrogen in dietary glutamate is utilized exclusively for the synthesis of amino acids in the rat intestine. Am J Physiol Endocrinol Metab 304:E100–E108. https://doi.org/10.1152/ajpendo.00331.2012
Nakanishi T, Hatanaka T, Huang W, Prasad PD, Leibach FH, Ganapathy ME, Ganapathy V (2001) Na+- and Cl--coupled active transport of carnitine by the amino acid transporter ATB(0,+) from mouse colon expressed in HRPE cells and Xenopus oocytes. J Physiol 532:297–304

Nakauchi J et al (2000) Cloning and characterization of a human brain Na(+)-independent transporter for small neutral amino acids that transports D-serine with high affinity. Neurosci Lett 287:231–235
Nassl AM, Rubio-Aliaga I, Fenselau H, Marth MK, Kottra G, Daniel H (2011) Amino acid absorption and homeostasis in mice lacking the intestinal peptide transporter PEPT1. Am J Physiol Gastrointest Liver Physiol 301:G128–G137. https://doi.org/10.1152/ajpgi.00017.2011
Neuschwander-Tetri BA, Presti ME, Wells LD (1997) Glutathione synthesis in the exocrine pancreas. Pancreas 14:342–349
Nguyen HT et al (2011) CD98 expression modulates intestinal homeostasis, inflammation, and colitis-associated cancer in mice. J Clin Invest 121:1733–1747. https://doi.org/10.1172/JCI44631
Nicholson B, McGivan JD (1996) Induction of high affinity glutamate transport activity by amino acid deprivation in renal epithelial cells does not involve an increase in the amount of transporter protein. J Biol Chem 271:12159–12164
Nicholson B, Manner CK, Kleeman J, MacLeod CL (2001) Sustained nitric oxide production in macrophages requires the arginine transporter CAT2. J Biol Chem 276:15881–15885. https://doi.org/10.1074/jbc.M010030200
Nicklin P et al (2009) Bidirectional transport of amino acids regulates mTOR and autophagy. Cell 136:521–534. https://doi.org/10.1016/j.cell.2008.11.044
Nuttall FQ, Schweim KJ, Gannon MC (2006) Effect of orally administered phenylalanine with and without glucose on insulin, glucagon and glucose concentrations. Horm Metab Res 38:518–523. https://doi.org/10.1055/s-2006-949523
Ocheltree SM, Shen H, Hu Y, Keep RF, Smith DE (2005) Role and relevance of peptide transporter 2 (PEPT2) in the kidney and choroid plexus: in vivo studies with glycylsarcosine in wild-type and PEPT2 knockout mice. J Pharmacol Exp Ther 315:240–247
Ohgaki R et al (2017) Essential roles of L-type amino acid transporter 1 in syncytiotrophoblast development by presenting fusogenic 4F2hc Mol Cell Biol 37. https://doi.org/10.1128/MCB.00427-16
Oparija L, Rajendran A, Poncet N, Verrey F (2019) Anticipation of food intake induces phosphorylation switch to regulate basolateral amino acid transporter LAT4 (SLC43A2) function. J Physiol 597:521–542. https://doi.org/10.1113/JP276714
Oparija-Rogenmozere L, Rajendran A, Poncet N, Camargo SMR, Verrey F (2020) Phosphorylation of mouse intestinal basolateral amino acid uniporter LAT4 is controlled by foodentrained diurnal rhythm and dietary proteins. PLoS One 15:e0233863. https://doi.org/10.1371/journal.pone.0233863
Oppedisano F, Pochini L, Galluccio M, Indiveri C (2007) The glutamine/amino acid transporter (ASCT2) reconstituted in liposomes: transport mechanism, regulation by ATP and characterization of the glutamine/glutamate antiport. Biochim Biophys Acta 1768:291–298. https://doi.org/10.1016/j.bbamem.2006.09.002
Padmanabhan R, Gu S, Nicholson BJ, Jiang JX (2012) Residue cysteine 232 is important for substrate transport of neutral amino acid transporter, SNAT4 Int. J Biochem Mol Biol 3:374–383
Palacin M, Estevez R, Bertran J, Zorzano A (1998) Molecular biology of mammalian plasma membrane amino acid transporters. Physiol Rev 78:969–1054. https://doi.org/10.1152/physrev.1998.78.4.969
Palii SS, Thiaville MM, Pan YX, Zhong C, Kilberg MS (2006) Characterization of the amino acid response element within the human sodium-coupled neutral amino acid transporter 2 (SNAT2) system A transporter gene. Biochem J 395:517–527. https://doi.org/10.1042/BJ20051867
Pan X, Terada T, Irie M, Saito H, Inui K (2002) Diurnal rhythm of H+-peptide cotransporter in rat small intestine. Am J Physiol Gastrointest Liver Physiol 283:G57–G64. https://doi.org/10.1152/ajpgi.00545.2001
Pan X, Terada T, Okuda M, Inui K (2004) The diurnal rhythm of the intestinal transporters SGLT1 and PEPT1 is regulated by the feeding conditions in rats. J Nutr 134:2211–2215. https://doi.org/10.1093/jn/134.9.2211

Panitchob N et al (2016) Computational modelling of placental amino acid transfer as an integrated system. Biochim Biophys Acta 1858:1451–1461. https://doi.org/10.1016/j.bbamem.2016.03.028

Pappenheimer JR, Reiss KZ (1987) Contribution of solvent drag through intercellular junctions to absorption of nutrients by the small intestine of the rat. J Membr Biol 100:123–136

Peghini P, Janzen J, Stoffel W (1997) Glutamate transporter EAAC-1-deficient mice develop dicarboxylic aminoaciduria and behavioral abnormalities but no neurodegeneration. EMBO J 16:3822–3832. https://doi.org/10.1093/emboj/16.13.3822

Perkins CP et al (1997) Anemia and perinatal death result from loss of the murine ecotropic retrovirus receptor mCAT-1. Genes Dev 11:914–925

Perna MK et al (2016) Creatine transporter deficiency leads to increased whole body and cellular metabolism. Amino Acids 48:2057–2065. https://doi.org/10.1007/s00726-016-2291-3

Peters T et al (2003) A mouse model for cystinuria type I. Hum Mol Genet 12:2109–2120

Pfeiffer R, Spindler B, Loffing J, Skelly PJ, Shoemaker CB, Verrey F (1998) Functional heterodimeric amino acid transporters lacking cysteine residues involved in disulfide bond. FEBS Lett 439:157–162

Pfeiffer R et al (1999a) Luminal heterodimeric amino acid transporter defective in cystinuria. Mol Biol Cell 10:4135–4147. https://doi.org/10.1091/mbc.10.12.4135

Pfeiffer R, Rossier G, Spindler B, Meier C, Kuhn L, Verrey F (1999b) Amino acid transport of y+L-type by heterodimers of 4F2hc/CD98 and members of the glycoprotein-associated amino acid transporter family. EMBO J 18:49–57. https://doi.org/10.1093/emboj/18.1.49

Pineda M et al (1999) Identification of a membrane protein, LAT-2, that co-expresses with 4F2 heavy chain, an L-type amino acid transport activity with broad specificity for small and large zwitterionic amino acids. J Biol Chem 274:19738–19744

Pineda M et al (2004) Cystinuria-specific rBAT(R365W) mutation reveals two translocation pathways in the amino acid transporter rBAT-b0,+AT. Biochem J 377:665–674. https://doi.org/10.1042/BJ20030956

Pinho AV, Rooman I, Reichert M, Medts ND, Bouwens L, Rustgi AK, Real FX (2011) Adult pancreatic acinar cells dedifferentiate to an embryonic progenitor phenotype with concomitant activation of a senescence programme that is present in chronic pancreatitis. Gut 60:958–966. https://doi.org/10.1136/gut.2010.225920

Pinho AV, Chantrill L, Rooman I (2013) Chronic pancreatitis: a path to pancreatic cancer. Cancer Lett. https://doi.org/10.1016/j.canlet.2013.08.015

Pinilla-Tenas J, Barber A, Lostao MP (2003) Transport of proline and hydroxyproline by the neutral amino-acid exchanger ASCT1. J Membr Biol 195:27–32. https://doi.org/10.1007/s00232-003-2041-9

Poncet N et al (2014) The catalytic subunit of the system L1 amino acid transporter (slc7a5) facilitates nutrient signalling in mouse skeletal muscle. PLoS One 9:e89547. https://doi.org/10.1371/journal.pone.0089547

Qu W et al (2012) Preparation and characterization of L-[5-11C]-glutamine for metabolic imaging of tumors. J Nucl Med 53:98–105. https://doi.org/10.2967/jnumed.111.093831

Quan H et al (2004) Hypertension and impaired glycine handling in mice lacking the orphan transporter XT2. Mol Cell Biol 24:4166–4173

Rajendran A, Poncet N, Oparija-Rogenmozere L, Herzog B, Verrey F (2020) Tissue-specific deletion of mouse basolateral uniporter LAT4 (Slc43a2) reveals its crucial role in small intestine and kidney amino acid transport. J Physiol. https://doi.org/10.1113/JP28023

Ramadan T, Camargo SM, Summa V, Hunziker P, Chesnov S, Pos KM, Verrey F (2006) Basolateral aromatic amino acid transporter TAT1 (Slc16a10) functions as an efflux pathway. J Cell Physiol 206:771–779. https://doi.org/10.1002/jcp.20531

Rezaei R, Wu Z, Hou Y, Bazer FW, Wu G (2016) Amino acids and mammary gland development: nutritional implications for milk production and neonatal growth. J Anim Sci Biotechnol 7:20. https://doi.org/10.1186/s40104-016-0078-8

Ribeiro MJ et al (2005) Characterization of hyperinsulinism in infancy assessed with PET and 18F-fluoro-L-DOPA. J Nucl Med 46:560–566

Rome S et al (2002) The regionalization of PepT1, NBAT and EAAC1 transporters in the small intestine of rats are unchanged from birth to adulthood. J Nutr 132:1009–1011. https://doi.org/10.1093/jn/132.5.1009

Romeo E et al (2006) Luminal kidney and intestine SLC6 amino acid transporters of B0AT-cluster and their tissue distribution in *Mus musculus*. Am J Physiol Renal Physiol 290:F376–F383. https://doi.org/10.1152/ajprenal.00286.2005

Rooman I et al (2013) Amino acid transporters expression in acinar cells is changed during acute pancreatitis. Pancreatology 13:475–485. https://doi.org/10.1016/j.pan.2013.06.006

Rossier G, Meier C, Bauch C, Summa V, Sordat B, Verrey F, Kuhn LC (1999) LAT2, a new basolateral 4F2hc/CD98-associated amino acid transporter of kidney and intestine. J Biol Chem 274:34948–34954

Rubio-Aliaga I, Daniel H (2008) Peptide transporters and their roles in physiological processes and drug disposition. Xenobiotica 38:1022–1042. https://doi.org/10.1080/00498250701875254

Rubio-Aliaga I, Frey I, Boll M, Groneberg DA, Eichinger HM, Balling R, Daniel H (2003) Targeted disruption of the peptide transporter Pept2 gene in mice defines its physiological role in the kidney. Mol Cell Biol 23:3247–3252

Ruderisch N, Virgintino D, Makrides V, Verrey F (2011) Differential axial localization along the mouse brain vascular tree of luminal sodium-dependent glutamine transporters Snat1 and Snat3. J Cereb Blood Flow Metab 31:1637–1647

Rudnick G, Kramer R, Blakely RD, Murphy DL, Verrey F (2014) The SLC6 transporters: perspectives on structure, functions, regulation, and models for transporter dysfunction. Pflugers Arch 466:25–42. https://doi.org/10.1007/s00424-013-1410-1

Rutter AR, Fradley RL, Garrett EM, Chapman KL, Lawrence JM, Rosahl TW, Patel S (2007) Evidence from gene knockout studies implicates Asc-1 as the primary transporter mediating d-serine reuptake in the mouse CNS. Eur J Neurosci 25:1757–1766. https://doi.org/10.1111/j.1460-9568.2007.05446.x

Sagne C et al (2001) Identification and characterization of a lysosomal transporter for small neutral amino acids. Proc Natl Acad Sci USA 98:7206–7211. https://doi.org/10.1073/pnas.121183498

Sakaguchi Y, Inaba M, Kusafuka K, Okazaki K, Ikehara S (2006) Establishment of animal models for three types of pancreatitis and analyses of regeneration mechanisms. Pancreas 33:371–381. https://doi.org/10.1097/01.mpa.0000236734.39241.99

Sans MD, Lee SH, D'Alecy LG, Williams JA (2004) Feeding activates protein synthesis in mouse pancreas at the translational level without increase in mRNA. Am J Physiol Gastrointest Liver Physiol 287:G667–G675. https://doi.org/10.1152/ajpgi.00505.2003

Sans MD, Tashiro M, Vogel NL, Kimball SR, D'Alecy LG, Williams JA (2006) Leucine activates pancreatic translational machinery in rats and mice through mTOR independently of CCK and insulin. J Nutr 136:1792–1799. https://doi.org/10.1093/jn/136.7.1792

Sato H et al (2005) Redox imbalance in cystine/glutamate transporter-deficient mice. J Biol Chem 280:37423–37429. https://doi.org/10.1074/jbc.M506439200

Scalise M, Pochini L, Console L, Losso MA, Indiveri C (2018) The human SLC1A5 (ASCT2) amino acid transporter: from function to structure and role in cell biology. Front Cell Dev Biol 6:96. https://doi.org/10.3389/fcell.2018.00096

Segawa H, Fukasawa Y, Miyamoto K, Takeda E, Endou H, Kanai Y (1999) Identification and functional characterization of a Na+−independent neutral amino acid transporter with broad substrate selectivity. J Biol Chem 274:19745–19751

Seow HF, Broer S, Broer A, Bailey CG, Potter SJ, Cavanaugh JA, Rasko JE (2004) Hartnup disorder is caused by mutations in the gene encoding the neutral amino acid transporter SLC6A19. Nat Genet 36:1003–1007. https://doi.org/10.1038/ng1406

Shayakul C, Kanai Y, Lee WS, Brown D, Rothstein JD, Hediger MA (1997) Localization of the high-affinity glutamate transporter EAAC1 in rat kidney. Am J Phys 273:F1023–F1029

Shen H, Smith DE, Yang T, Huang YG, Schnermann JB, Brosius FC 3rd (1999) Localization of PEPT1 and PEPT2 proton-coupled oligopeptide transporter mRNA and protein in rat kidney. Am J Phys 276:F658–F665. https://doi.org/10.1152/ajprenal.1999.276.5.F658
Shen H, Ocheltree SM, Hu Y, Keep RF, Smith DE (2007) Impact of genetic knockout of PEPT2 on cefadroxil pharmacokinetics, renal tubular reabsorption, and brain penetration in mice. Drug Metab Dispos 35:1209–1216
Silbernagl S (1979) Renal transport of amino acids. Klin Wochenschr 57:1009–1019
Singer MV, Niebergall-Roth E (2009) Secretion from acinar cells of the exocrine pancreas: role of enteropancreatic reflexes and cholecystokinin. Cell Biol Int 33:1–9
Singer D et al (2009) Orphan transporter SLC6A18 is renal neutral amino acid transporter B0AT3. J Biol Chem 284:19953–19960. https://doi.org/10.1074/jbc.M109.011171
Singer D et al (2012) Defective intestinal amino acid absorption in Ace2 null mice. Am J Physiol Gastrointest Liver Physiol 303:G686–G695. https://doi.org/10.1152/ajpgi.00140.2012
Siveke JT, Lubeseder-Martellato C, Lee M, Mazur PK, Nakhai H, Radtke F, Schmid RM (2008) Notch signaling is required for exocrine regeneration after acute pancreatitis. Gastroenterology 134:544–555
Skelton MR, Schaefer TL, Graham DL, Degrauw TJ, Clark JF, Williams MT, Vorhees CV (2011) Creatine transporter (CrT; Slc6a8) knockout mice as a model of human CrT deficiency. PLoS One 6:e16187. https://doi.org/10.1371/journal.pone.0016187
Sloan JL, Mager S (1999) Cloning and functional expression of a human Na(+) and cl(−)-dependent neutral and cationic amino acid transporter B(0+). J Biol Chem 274:23740–23745
Sloan JL, Grubb BR, Mager S (2003) Expression of the amino acid transporter ATB 0+ in lung: possible role in luminal protein removal. Am J Physiol Lung Cell Mol Physiol 284:L39–L49. https://doi.org/10.1152/ajplung.00164.2002
Smilowitz JT, O'Sullivan A, Barile D, German JB, Lonnerdal B, Slupsky CM (2013) The human milk metabolome reveals diverse oligosaccharide profiles. J Nutr 143:1709–1718. https://doi.org/10.3945/jn.113.178772
Sperandeo MP et al (2007) Slc7a7 disruption causes fetal growth retardation by downregulating Igf1 in the mouse model of lysinuric protein intolerance. Am J Physiol Cell Physiol 293:C191–C198. https://doi.org/10.1152/ajpcell.00583.2006
Stanger BZ, Hebrok M (2013) Control of cell identity in pancreas development and regeneration. Gastroenterology 144:1170–1179. https://doi.org/10.1053/j.gastro.2013.01.074
Steward MC, Ishiguro H (2009) Molecular and cellular regulation of pancreatic duct cell function. Curr Opin Gastroenterol 25:447–453. https://doi.org/10.1097/MOG.0b013e32832e06ce
Stoll B, Burrin DG, Henry J, Yu H, Jahoor F, Reeds PJ (1999) Substrate oxidation by the portal drained viscera of fed piglets. Am J Phys 277:E168–E175. https://doi.org/10.1152/ajpendo.1999.277.1.E168
Strobel O et al (2007) In vivo lineage tracing defines the role of acinar-to-ductal transdifferentiation in inflammatory ductal metaplasia. Gastroenterology 133:1999–2009. https://doi.org/10.1053/j.gastro.2007.09.009
Sumigray KD, Terwilliger M, Lechler T (2018) Morphogenesis and compartmentalization of the intestinal. Crypt Dev Cell 45:183–197 e185. https://doi.org/10.1016/j.devcel.2018.03.024
Sun Y et al (2015) Dietary L-leucine supplementation enhances intestinal development in suckling piglets. Amino Acids 47:1517–1525. https://doi.org/10.1007/s00726-015-1985-2
Sweiry JH, Munoz M, Mann GE (1991) Cis-inhibition and trans-stimulation of cationic amino acid transport in the perfused rat pancreas. Am J Phys 261:C506–C514. https://doi.org/10.1152/ajpcell.1991.261.3.C506
Takanaga H, Mackenzie B, Suzuki Y, Hediger MA (2005) Identification of mammalian proline transporter SIT1 (SLC6A20) with characteristics of classical system imino. J Biol Chem 280:8974–8984. https://doi.org/10.1074/jbc.M413027200
Tanaka K et al (1997) Epilepsy and exacerbation of brain injury in mice lacking the glutamate transporter GLT-1. Science 276:1699–1702

Tandon BN, Banks PA, George PK, Sama SK, Ramachandran K, Gandhi PC (1970) Recovery of exocrine pancreatic function in adult protein-calorie malnutrition. Gastroenterology 58:358–362
Taslimifar M, Oparija L, Verrey F, Kurtcuoglu V, Olgac U, Makrides V (2017) Quantifying the relative contributions of different solute carriers to aggregate substrate transport. Sci Rep 7:40628. https://doi.org/10.1038/srep40628
Taslimifar M, Buoso S, Verrey F, Kurtcuoglu V (2018) Functional polarity of microvascular brain endothelial cells supported by neurovascular unit computational model of large neutral amino acid homeostasis. Front Physiol 9:171. https://doi.org/10.3389/fphys.2018.00171
Taylor PM, Egan CJ, Rennie MJ (1989) Transport of glutamine across blood-facing membranes of perfused rat jejunum. Am J Phys 256:E550–E558. https://doi.org/10.1152/ajpendo.1989.256.4.E550
Tetsuka K, Takanaga H, Ohtsuki S, Hosoya K, Terasaki T (2003) The l-isomer-selective transport of aspartic acid is mediated by ASCT2 at the blood-brain barrier. J Neurochem 87:891–901
Tolvanen T et al (2006) Human radiation dosimetry of [11C]MeAIB, a new tracer for imaging of system A amino acid transport. Eur J Nucl Med Mol Imaging 33:1178–1184. https://doi.org/10.1007/s00259-006-0096-4
Torrents D et al (1999) Identification of SLC7A7, encoding y+LAT-1, as the lysinuric protein intolerance gene. Nat Genet 21:293–296. https://doi.org/10.1038/6809
Tsumura H et al (2003) The targeted disruption of the CD98 gene results in embryonic lethality. Biochem Biophys Res Commun 308:847–851
Ugawa S, Sunouchi Y, Ueda T, Takahashi E, Saishin Y, Shimada S (2001) Characterization of a mouse colonic system B(0+) amino acid transporter related to amino acid absorption in colon. Am J Physiol Gastrointest Liver Physiol 281:G365–G370. https://doi.org/10.1152/ajpgi.2001.281.2.G365
Utsunomiya-Tate N, Endou H, Kanai Y (1996) Cloning and functional characterization of a system ASC-like Na+−dependent neutral amino acid transporter. J Biol Chem 271:14883–14890
van der Schoor SR, van Goudoever JB, Stoll B, Henry JF, Rosenberger JR, Burrin DG, Reeds PJ (2001) The pattern of intestinal substrate oxidation is altered by protein restriction in pigs. Gastroenterology 121:1167–1175
van der Wielen N, Moughan PJ, Mensink M (2017) Amino acid absorption in the large intestine of humans and porcine models. J Nutr 147:1493–1498. https://doi.org/10.3945/jn.117.248187
Vanoaica L, Behera A, Camargo SM, Forster IC, Verrey F (2016) Real-time functional characterization of cationic amino acid transporters using a new FRET sensor. Pflugers Arch 468:563–572. https://doi.org/10.1007/s00424-015-1754-9
Vanslambrouck JM, Broer A, Thavyogarajah T, Holst J, Bailey CG, Broer S, Rasko JE (2010) Renal imino acid and glycine transport system ontogeny and involvement in developmental iminoglycinuria. Biochem J 428:397–407. https://doi.org/10.1042/BJ20091667
Variyam EP, Fuller RK, Brown FM, Quallich LG (1985) Effect of parenteral amino acids on human pancreatic exocrine secretion. Dig Dis Sci 30:541–546
Varoqui H, Erickson JD (2002) Selective up-regulation of system a transporter mRNA in diabetic liver. Biochem Biophys Res Commun 290:903–908. https://doi.org/10.1006/bbrc.2001.6281
Vaughan OR, Rosario FJ, Powell TL, Jansson T (2017) Regulation of placental amino acid transport and fetal growth. Prog Mol Biol Transl Sci 145:217–251. https://doi.org/10.1016/bs.pmbts.2016.12.008
Verrey F (2003) System L: heteromeric exchangers of large, neutral amino acids involved in directional transport. Pflugers Arch 445:529–533. https://doi.org/10.1007/s00424-002-0973-z
Verrey F, Closs EI, Wagner CA, Palacin M, Endou H, Kanai Y (2004) CATs and HATs: the SLC7 family of amino acid transporters. Pflugers Arch 447:532–542. https://doi.org/10.1007/s00424-003-1086-z
Vilches C et al (2018) Cooperation of antiporter LAT2/CD98hc with uniporter TAT1 for renal reabsorption of neutral amino acids. J Am Soc Nephrol 29:1624–1635. https://doi.org/10.1681/ASN.2017111205

Vuille-dit-Bille RN et al (2015) Human intestine luminal ACE2 and amino acid transporter expression increased by ACE-inhibitors. Amino Acids 47:693–705. https://doi.org/10.1007/s00726-014-1889-6

Weber E, Ehrlein HJ (1998) Reserve capacities of the small intestine for absorption of energy. Am J Phys 275:R300–R307

Weidenfeld S, Chupin CJA, Rozowsky S, Kuebler WM (2017) Sodium-coupled neutral amino acid transporter SNAT2 counteracts edema formation and reduces autophagy and ER stress in acute lung injury. FASEB J 31:725.724. https://doi.org/10.1096/fasebj.31.1_supplement.725.4

Willet SG et al (2018) Regenerative proliferation of differentiated cells by mTORC1-dependent paligenosis. EMBO J 37. https://doi.org/10.15252/embj.201798311

Wongdee K, Teerapornpuntakit J, Riengrojpitak S, Krishnamra N, Charoenphandhu N (2009) Gene expression profile of duodenal epithelial cells in response to chronic metabolic acidosis. Mol Cell Biochem 321:173–188. https://doi.org/10.1007/s11010-008-9931-1

Wuensch T et al (2013) The peptide transporter PEPT1 is expressed in distal colon in rodents and humans and contributes to water absorption. Am J Physiol Gastrointest Liver Physiol 305:G66–G73. https://doi.org/10.1152/ajpgi.00491.2012

Xia W, Osorio JS, Yang Y, Liu D, Jiang MF (2018) Short communication: characterization of gene expression profiles related to yak milk protein synthesis during the lactation cycle. J Dairy Sci 101:11150–11158. https://doi.org/10.3168/jds.2018-14715

Xie X et al (2005) Lack of the alanine-serine-cysteine transporter 1 causes tremors, seizures, and early postnatal death in mice. Brain Res 1052:212–221. https://doi.org/10.1016/j.brainres.2005.06.039

Yan Q, Tong H, Tang S, Tan Z, Han X, Zhou C (2017) L-Theanine administration modulates the absorption of dietary nutrients and expression of transporters and receptors in the intestinal mucosa of rats. Biomed Res Int 2017:9747256. https://doi.org/10.1155/2017/9747256

Yao D, Mackenzie B, Ming H, Varoqui H, Zhu H, Hediger MA, Erickson JD (2000) A novel system A isoform mediating Na+/neutral amino acid cotransport. J Biol Chem 275:22790–22797. https://doi.org/10.1074/jbc.M002965200

Yilmaz ÖH et al (2012) mTORC1 in the Paneth cell niche couples intestinal stem-cell function to calorie intake. Nature 486:490. https://doi.org/10.1038/nature11163

Yu S et al (2018) Paneth cell multipotency induced by notch activation following injury cell. Stem Cell 23:46–59 e45. https://doi.org/10.1016/j.stem.2018.05.002

Zerangue N, Kavanaugh MP (1996) Flux coupling in a neuronal glutamate transporter. Nature 383:634–637. https://doi.org/10.1038/383634a0

Zevenbergen C et al (2015) Transport of iodothyronines by human L-type amino acid transporters. Endocrinology 156:4345–4355. https://doi.org/10.1210/en.2015-1140

Zhou Y et al (2014) Proteome analysis and conditional deletion of the EAAT2 glutamate transporter provide evidence against a role of EAAT2 in pancreatic insulin secretion in mice. J Biol Chem 289:1329–1344. https://doi.org/10.1074/jbc.M113.529065

Chapter 8
Structure-Dynamic and Regulatory Specificities of Epithelial Na^+/Ca^{2+} Exchangers

Daniel Khananshvili

Abstract A limited number of splice variants, derived from the three genes (NCX1–3) of the Na^+/Ca^{2+} exchanger family (NCX), are expressed in epithelial tissues while extruding a major amount of total Ca^{2+} (up to 70%) from the cell. The epithelial NCX isoform/splice variants are allosterically regulated by Ca^{2+}, Na^+, and H^+ ions; they exhibit different capacities for performing three major regulatory modes: (a) Ca^{2+}-dependent activation, (b) Na^+-dependent inactivation, and (c) Ca^{2+}-induced alleviation of Na^+-dependent inactivation. These allosteric modulations are mediated by two Ca^{2+}-binding regulatory domains (CBD1 and CBD2), where the splicing segment, exclusively allocated at CBD2, diversifies the regulatory features of both CBDs. During the last decade, vast progress has been made in understanding the molecular mechanisms underlying the ion-dependent regulation of tissue-specific NCX variants. The present review focuses on the structure-dynamic determinants governing the functional and regulatory features of NCX variants expressed in different types of epithelial tissues.

Keywords NCX · Na^+/Ca^{2+} exchanger · Isoform/splice variants · Ca^{2+} · Epithelia

8.1 Introduction

8.1.1 NCX as a Ubiquitous System for Ca^{2+} Extrusion

Ca^{2+} is a most versatile and ubiquitous secondary messenger, which maintains, regulates, and integrates nearly all cellular processes that take place in prokaryotic and eukaryotic cells (Carafoli 1987; Williams 1999; Berridge et al. 2003; Clapham 2007; Gifford et al. 2007; Krebs 2009; Carafoli and Krebs 2016). This requires a

D. Khananshvili (✉)
Department of Physiology and Pharmacology, Sackler School of Medicine, Tel-Aviv University, Tel-Aviv, Israel
e-mail: dhanan@post.tau.ac.il

K. L. Hamilton, D. C. Devor (eds.), *Studies of Epithelial Transporters and Ion Channels*, Physiology in Health and Disease,
https://doi.org/10.1007/978-3-030-55454-5_8

well-tuned dynamic orchestration of distinct Ca^{2+}-dependent events in functionally assorted cell types in time and space (Berridge et al. 2003; Clapham 2007). Toward fulfilling this requirement, the dynamic features of Ca^{2+}-binding proteins are structurally predefined to match Ca^{2+}-dependent events that last from microseconds to hours on occasions of distinct cell functions (Clapham 2007; Krebs 2009; Carafoli and Krebs 2016). The variability in functional dynamics is achieved by specific sets of toolbox proteins, which encode numerous gene isoforms and their splice variants to fulfill the dynamic features of Ca^{2+} signaling in a given cell type (Carafoli 1987; Clapham 2007; Gifford et al. 2007; Krebs 2009; Brini et al. 2014; Giladi et al. 2016b; Khananshvili 2016, 2017a).

Although Ca^{2+} binds to many thousands of proteins and regulates and integrates numerous biochemical and physiological activities, only a few Ca^{2+} transport systems control Ca^{2+} entry and exit (Carafoli 1987; Berridge et al. 2003; Clapham 2007). The plasma membrane (PM) Ca^{2+} channels mediate Ca^{2+} entry from the extracellular space into the cytosol in response to electric, ligand, or mechanical stimuli, whereas the PM Ca^{2+}-ATPase (PMCA) and Na^+/Ca^{2+} exchanger (NCX) proteins mediate the Ca^{2+} extrusion from the cell (Carafoli 1987, 1988; Blaustein and Lederer 1999; Brini et al. 2014; Khananshvili 2013, 2014; Lopreiato et al. 2014). Although the PMCA and NCX systems appear nearly in every mammalian cell, their contributions to Ca^{2+} extrusion are tissue-specific, and the functional and regulatory features of PMCA and NCX proteins are heavily modified by spawning a large number of gene isoform/splice variants (Philipson and Nicoll 2000; Lytton 2007; Lopreiato et al. 2014; Krebs 2009). Since Ca^{2+} extrusion is essential for proper handling of Ca^{2+} signaling and homeostasis, selective targeting of Ca^{2+} transporting proteins is a long-desired strategy for effective therapeutic treatment in many biomedical applications, albeit this intervention remains unrealized.

A thorough understanding of the structural determinants governing the functional and regulatory specificities of isoform/splice variants may provide new clues for predominant Ca^{2+} signaling pathways in distinct cell types. In addition, the relevant information may provide new opportunities for developing new pharmacological tools for tissue-specific targeting of distinct isoform/splice variants (Brini and Carafoli 2009; Brini et al. 2014; Giladi et al. 2016b; Khananshvili 2016, 2017a). This chapter aims to review the recent progress achieved in understanding the structure-dynamic determinants shaping the regulatory features of NCX isoform/splice variants. The major emphasis is on the comparison of NCX variants expressed in epithelial cells with counterpart variants expressed in the brain, cardiac, and skeletal tissues

8.1.2 Short History of NCX Discovery and Follow-Up Breakthroughs

In the late 1960s, three laboratories concurrently discovered NCX, while independently performing experiments with mammalian cardiac muscle (Reuter and Seitz 1968), invertebrate neurons (Baker and Blaustein 1968; Baker et al. 1969), and mammalian small intestinal epithelium (Martin and De Luca 1969). The follow-up studies with squid axon preparations provided very valuable information on NCX function and regulation, which is of general interest (DiPolo 1979; Blaustein and Lederer 1999; DiPolo and Beaugé 2006). Shortly after discovering the cell-membrane NCX, Carafoli and collaborators identified the Na^{+}/Ca^{2+} exchange activity in mitochondria, while suggesting that, in contrast, cell-membrane NCX, the mitochondrial NCX (later called NCXL) can transport either Na^{+} or Li^{+} in exchange with Ca^{2+} (Carafoli et al. 1974). Thirty-five years later, this unique feature for Li^{+}-transport was explored for disclosing the molecular identity of NCLX by Sekler and collaborators (Palty et al. 2010; Sekler 2015).

In the early 1980s, Reeves and coworkers applied biochemical approaches to assay the radioactive ion-fluxes in isolated preparations of cardiac sarcolemmal vesicles, which resulted in the discovery of an electrogenic ion-exchange with a stoichiometry of $3Na^{+}$:$1Ca^{2+}$ (Reeves and Hale 1984). In the early 1990s, the "Ping-Pong" mechanism was discovered for the transport cycle of Na^{+}/Ca^{2+} exchange, according to which the $3Na^{+}$ or Ca^{2+} ions alternatively bind to the NCX transport sites and Ca^{2+}- or Na^{+}-bound species are translocated in separate steps of the transport cycle across the membrane (Khananshvili 1990; Niggli and Lederer 1991; Hilgemann et al. 1991). At the interface of the late 1980s and early 1990s, Hilgemann and coworkers developed advanced techniques for the giant excised patch in conjunction with electrophysiological approaches for measuring NCX ion currents (Hilgemann 1990; Hilgemann et al. 1991, 1992a, b). This turned out to be a critical tool for understanding the structure-based regulatory mechanisms underlying eukaryotic NCX variants.

In 1988, Philipson and coworkers purified microgram quantities of NCX1 protein from isolated preparations of cardiac sarcolemma vesicles (Philipson et al. 1988), which subsequently enabled the cloning of the first NCX gene (NCX1) in 1990 (Nicoll et al. 1990). Shortly thereafter, two additional mammalian genes (isoforms), NCX2 (Li et al. 1994) and NCX3 (Nicoll et al. 1996b) were cloned with noticeable tissue-specific splice variants (Kofuji et al. 1994; Quednau et al. 1997, 2004; Linck et al. 1998, 2000; Lytton 2007). The follow-up experiments revealed that the cardiac, brain, and kidney isoform/splice variants have diverse regulatory responses to Ca^{2+} and Na^{+}, which might have a physiological relevance (Dyck et al. 1999; Dunn et al. 2002; Matsuoka 2004).

Early studies have shown that a large cytosolic loop-f (5 L6) of NCX (located between the transmembrane segments TM5 and TM6) contains a putative entity for regulatory Ca^{2+} binding (Levitsky et al. 1994; Matsuoka et al. 1993, 1995). By using NMR techniques, Hilge and collaborators identified two regulatory domains that bind Ca^{2+} (CBD1 and CBD2), exhibiting the characteristic immunoglobulin folding (Hilge et al. 2006). Shortly thereafter, the crystal structures of the isolated CBD1 (Nicoll et al. 2006), CBD2 (Besserer et al. 2007), and a two-domain tandem (CBD12) were resolved (Wu et al. 2011; Giladi et al. 2012a). The follow-up studies revealed that CBD1 comprises highly conserved Ca^{2+} binding sites (representing a primary allosteric sensor), whereas the splicing segment (exclusively located at CBD2) predefines the number of Ca^{2+} sites at CBD2 to modify the regulatory features of both CBDs (Hilge et al. 2009; Boyman et al. 2009; Giladi et al. 2012a, b, c, 2016b; Tal et al. 2016).

In conjunction with structure-based mutational studies, the specific roles of the CBD1 and CBD2 domains were resolved in tissue-specific NCX variants exhibiting characteristic differences in the Ca^{2+}-dependent activation and in the Ca^{2+}-induced alleviation of Na^{+}-dependent inactivation (Boyman et al. 2009; Ottolia et al. 2009, 2010; Giladi and Khananshvili 2013). The systematic applications of advanced biophysical approaches (including NMR, FRET, SAXS, and HDX-MS techniques) combined with structural studies allowed the identification of specific structure-dynamic determinants governing the regulatory specificities of tissue-specific NCX variants (for a review, see Giladi et al. 2016b; Khanashvili 2016, 2017a). Apparently Ca^{2+} binding to CBD1 results in interdomain tethering of CBDs, where a slow dissociation of "entrapped" (occluded) Ca^{2+} controls NCX activity and is secondarily modulated by a splicing segment (Giladi et al. 2012a, c; Khananshvili 2016, 2020). This kind of information provides new opportunities for selective pharmacological targeting of tissue-specific NCX variants, which is a long-desired intervention in many biomedical applications (Khananshvili 2013, 2014, 2016).

Until 2012, no structural information was available on the ion-transporting machinery for any NCX protein. Jiang and coworkers made a breakthrough discovery by revealing the X-ray structure of the archaeal NCX from *Methanococcus jannaschii* (NCX_Mj) (Liao et al. 2012, 2016). The crystal structure of NCX_Mj offers an ideal model for studying the ion-transport mechanisms for eukaryotic NCXs as well, since NCX_Mj lacks any regulatory domains while sharing with eukaryotic NCX variants highly conserved structure–functional elements controlling ion binding and transport (Marinelli et al. 2014; Giladi et al. 2016a; van Dijk et al. 2018). The future challenge is to resolve how the allosteric signal is diversified in distinct NCX variants and how the decoded signal is transmitted from the regulatory CBD domains to ion-transport sites over a distance of 80 Å, while retaining the regulatory specificity of a given isoform/splice variant.

8.2 Ca^{2+} Homeostasis in Epithelial Cells

8.2.1 Hallmark Features of Ca^{2+} Homeostasis in Epithelial Cells

In general, the epithelial cells form a single-cell layer (barrier) that separates the "internal" and "external" environments, where, according to a typical scenario (e.g., kidney and intestine), Ca^{2+} and other ions move from the apical (luminal) to the basolateral membrane toward their extrusion into the interstitial fluid, from which Ca^{2+} is absorbed into the blood circulation. The systemic efficiency of Ca^{2+} absorption into the blood circulation is regulated by the expression levels of the Ca^{2+} transporting proteins (including PMCA and NCX), controlled by specific hormones (e.g., PTH, $1{,}25(OH)_2D_3$, and calcitonin) in response to systemic signals (Brown 1991; Friedman and Gesek 1995; White et al. 1996; Dimke et al. 2011; Peng et al. 2018). These "long-lasting" modulatory modes of Ca^{2+} delivery to the blood circulation substantially differ among epithelial cells, since endocrine machinery differentially affects the expression levels of Ca^{2+} entry/exit proteins (Peng et al. 2003, 2018; Brini and Carafoli 2009; Centeno et al. 2011; Brini et al. 2014). Importantly, the extracellular $[Ca^{2+}]$ changes are rigorously "monitored" by the G-protein-coupled Ca^{2+}-sensing receptor (CaSR), which modulates hormonal secretions controlling the protein expression levels of Ca^{2+} entry/exit proteins and thus, "adjusts" Ca^{2+} absorption levels at the systemic level (van Cromphaut et al. 2001; Khundmiri et al. 2016; Peng et al. 2003, 2018).

The dynamic features of Ca^{2+} signaling in epithelial cells considerably differ from the dynamic patterns of Ca^{2+} signaling observed in excitable tissues (Friedman and Gesek 1995; White et al. 1996; Blaustein and Lederer 1999; Bers 2002, 2008). Moreover, the Ca^{2+} translocation rates across the epithelial cells may significantly vary at relatively small spikes of cytosolic $[Ca^{2+}]$ with short-lasting duration (Marhl et al. 2006; Rüdiger 2014; Verkhratsky and Parpura 2014; Brodskiy and Zartman 2018; Verkhratsky et al. 2018). Thus, the regulation of cell-specific Ca^{2+} entry/exit proteins must be suited to the dynamic integration (coupling) of Ca^{2+}-dependent signals involving cell division, migration, death, and differentiation among many others (Guillot and Lecuit 2013; Purvis and Lahav 2013; Rüdiger 2014; Plattner and Verkhratsky 2016; Brodskiy and Zartman 2018).

While equipped with specific sets of toolbox proteins, including Ca^{2+} entry channels (TRPV5, TRPV6, and $Ca_v1.3$), Ca^{2+} transporters (NCX1, NCX2, and NCX3), Ca^{2+} pumps (PMCA1 and PMCA4), and Ca^{2+} buffering proteins (calbindin-D_{9K} and calbindin-D_{28K}), each epithelial cell type characteristically handles the Ca^{2+} signaling dynamics. This certainly would require "arranging" cell-specific expression profiles of PMCA and NCX isoform/splice variants that must match well with Ca^{2+} signaling pathways in distinct cell types (Brini and Carafoli 2009; Guillot and Lecuit 2013; Purvis and Lahav 2013; Brodskiy and Zartman

2018). The challenge is to resolve how the isoform/splice variants of PMAC and NCX respond to Ca^{2+} waves (oscillations) in specific cell-types to shape epithelial functions (Brini and Carafoli 2009; Brini et al. 2014; Hong and Jeung 2013; Guillot and Lecuit 2013; van Loon et al. 2016; Wu et al. 2018; Brodskiy and Zartman 2018).

8.2.2 Ca^{2+} Entry, Buffering, and Exit in Epithelial Cells

Among several types of Ca^{2+} entry channels, involved in epithelial Ca^{2+} transport, TRPV5 and TRPV6 are the most suited for transcellular Ca^{2+} transport (Hoenderop et al. 1999, 2002; Peng et al. 2003, 2018; Dimke et al. 2011). Once Ca^{2+} enters into the epithelial cell (e.g., the kidney and intestine cells), the concentrations of free cytosolic Ca^{2+} are effectively buffered by Ca^{2+}-binding cytosolic proteins (Boros et al. 2009; Dimke et al. 2011; Peng et al. 2018). Under the steady-state conditions, the Ca^{2+} buffering proteins (e.g., calbindins) effectively translocate Ca^{2+} from the epical to the basolateral membrane, where the PMCA and NCX systems effectively extrude Ca^{2+} from the epithelial cells into the interstitial fluid. Thus, the Ca^{2+} buffering proteins, like calbindins, play an important role in transcellular Ca^{2+} transport, since they effectively decrease free Ca^{2+} concentrations and thereby keep the cytosolic $[Ca^{2+}]$ below the toxic levels (Hong and Jeung 2013; Dimke et al. 2011; van Loon et al. 2016).

TRPV5 and TRPV6 represent a major Ca^{2+}-entry system through the apical membrane of the epithelial cell (Hoenderop et al. 1999, 2002; Peng et al. 2003, 2018; Dimke et al. 2011; Stoerger and Flockerzi 2014). Notably, among the members of the TRP superfamily, TRPV5 and TRPV6 exhibit quite different properties, since they exhibit high selectivity for Ca^{2+} permeation, whereas the Ca^{2+} entry rates are tightly controlled by Ca^{2+}-dependent feedback inactivation involving many regulatory mechanisms. In general, TRPV5 and TRPV6 exhibit a robust response to 1,25-dihydroxivitamin D_3, which could be physiologically relevant for affecting the intestinal Ca^{2+} absorption, renal Ca^{2+} reabsorption, placental Ca^{2+} transfer to the fetus, Ca^{2+}-dependent secretory functions in glandular cells, and Ca^{2+} signaling in endothelial cells, among others (Dimke et al. 2011; Peng et al. 2018; Stoerger and Flockerzi 2014). TRPV6 is more broadly expressed in a variety of tissues such as the esophagus, stomach, small intestine, colon, kidney, placenta, pancreas, prostate, uterus, salivary gland, and sweat gland (Stoerger and Flockerzi 2014), whereas TRPV5 expression is confined to the distal convoluted tubule and to the connecting tubule of the kidney (Stoerger and Flockerzi 2014; Peng et al. 2018).

Two types of calbindins with different molecular weights, namely, calbindin-D_{9K} and calbindin-D_{28K} (having different numbers of Ca^{2+}-binding EF hand domains), are expressed in the mammalian intestine and kidney, respectively (Christakos et al. 1992; Hong and Jeung 2013). Besides their Ca^{2+} buffering role, calbindins modulate the Ca^{2+}-dependent regulation of TRPV5 and TRPV6 through their feedback interactions with TRPs (Lambers et al. 2006). For example, when apical Ca^{2+} entry exceeds the buffering capacity of calbindins, the binding of "extra" Ca^{2+} to calmodulin (CaM) results in the Ca^{2+}-dependent inhibition of TRPV5 or TRPV6 and provides feedback for Ca^{2+} handling in epithelial cells (Peng et al. 2003, 2018; Boros et al. 2009; Dimke et al. 2011). In contrast with TRPs, the interaction of Ca^{2+} with CaM activates Ca^{2+} transport through the PMCA pump and thereby accelerates Ca^{2+} extrusion from the cell (Brini et al. 2014; van Loon et al. 2016; Krebs 2009, 2017). Thus, the Ca^{2+} entry/exit balance is effectively controlled by CaM, while diversely affecting the Ca^{2+} transport rates through TRPs and PMCAs. Although CaM-dependent regulation of NCX has been reported (Chou et al. 2015) in a model system (HEK293T cells), the physiological relevance of NCX regulation by CaM has to be confirmed and further characterized by using physiologically more relevant cell lines. For further discussions of TRPV5 and TRPV6, interested readers are directed to Chap. 27 of this volume.

8.2.3 PMCA and NCX Control Ca^{2+} Extrusion in Mammalian Cells

The PMCA and NCX proteins represent a minor component in the total protein content of the plasma membrane (<0.1%) (Carafoli 1988; Philipson and Nicoll 2000; Khananshvili 2013, 2014). Despite this low protein abundance in the plasma membrane, Ca^{2+} extrusion in any mammalian cell (including the epithelial cells) is predominantly limited by two major systems: PMCA and NCX (Carafoli 1987, 1988; Berridge et al. 2003; Clapham 2007; Khananshvili 2013). In epithelial cells, the PMCA and NCX proteins are allocated at the basolateral membrane, while extruding Ca^{2+} from the cell into the interstitial fluid (Friedman and Gesek 1995; White et al. 1996; Dimke et al. 2011). However, in some specific epithelial cells (e.g., osteoblasts), the PCAM and NCX systems are asymmetrically distributed in the apical and basolateral membrane to support a certain physiological function. For example, in osteoblasts, PMCA does fringe the bone marrow compartment (Akisaka et al. 1988; Watson et al. 1989), whereas NCX faces the bone surface (Stains and Gay 1998; Jung et al. 2007). Thus, asymmetric distribution of NCX and PMCA suggests that NCX extrudes Ca^{2+} into the newly formed bone matrix, whereas PMCA is not involved (at least directly) in the mineralization process (Li et al. 2007; Sosnoski and Gay 2008; Albano et al. 2017).

Four gene isoforms encode PMCA proteins: PMCA1 and PMCA4 are ubiquitously distributed nearly in all mammalian cells, whereas PMCA2 and PMCA3 are mainly expressed in excitable tissues (Brini and Carafoli 2009; Krebs 2017). Among the three genes (encoding NCX proteins), NCX1 is the most ubiquitously distributed gene. The splice variants of NCX1 are expressed in a tissue-specific manner practically in every mammalian cell (Nicoll et al. 1990; Furman et al. 1993; Lee et al. 1994; Quednau et al. 1997, 2004; Zheng and Wang 2007; Linck et al. 1998, 2000;

Philipson and Nicoll 2000; Lytton 2007; van der Hagen et al. 2015). NCX2 is primarily expressed in the brain and spinal cord tissues (Li et al. 1994; Philipson and Nicoll 2000), although NCX2 was found in gastrointestinal, kidney, and glandular tissues (Quednau et al. 1997, 2004; Linck et al. 1998, 2000; Cantero-Recasens et al. 2019). Although NCX3 is chiefly expressed in the brain and smooth muscle (Quednau et al. 1997; Linck et al. 1998; Lindgren et al. 2005; Michel et al. 2014, 2017), NCX3 was also found in the intestine, bone, and glandular cells (Sosnoski and Gay 2008; Nishiyama et al. 2016, 2017; Albano et al. 2017).

8.2.4 *Partial Contributions of PMCA and NCX to Ca^{2+} Extrusion*

Early studies on the PMCA and NCX systems have suggested that these two systems might have distinct affinities and rates for Ca^{2+} transport (Carafoli 1987, 1988). The follow-up experiments, by using electrophysiological and biochemical approaches as well as different preparations of cell membranes, have shown that mammalian NCXs exhibit low affinity ($K_m = 10$–$25\ \mu M$) and high capacity ($k_{cat} = 1000$–$2500\text{-}s^{-1}$) for Ca^{2+} transport (Niggli and Lederer 1991; Hilgemann et al. 1991; Baazov et al. 1999; Khananshvili 2013, 2014). In contrast with NCX, the PMCA has a high affinity for Ca^{2+} transport ($K_m = 0.2$–$0.5\ \mu M$), whereas it exhibits a low capacity for Ca^{2+} transport rates ($k_{cat} = 5$–$25\ s^{-1}$) (Carafoli 1988; Rahamimoff et al. 1996; Brini and Carafoli 2009; Krebs 2009; Lopreiato et al. 2014). Notably, these fundamental differences in the "catalytic" capacities of PMCA and NCX are secondarily modified by specially designed regulatory domains, which are structurally different among isoform/splice variants and are genetically predefined. This structure-dependent diversification of regulatory properties plays a critical role in Ca^{2+}-dependent feedback (allosteric) "autoregulation" of PMCA and NCX systems (see below).

Owing to dramatic differences between the PMCA and NCX proteins regarding the Ca^{2+} affinity and transport rates, a "traditional" assumption was that the PMCA pump plays a housekeeping role in maintaining low cytosolic levels of Ca^{2+} (~0.1 μM) at "resting" conditions and thus, is responsible for maintaining the "basal" Ca^{2+} homeostasis. According to this scenario, NCX offers a rapid extrusion of Ca^{2+}, where "prompt" responses of NCX to large swings in the cytosolic $[Ca^{2+}]$ can effectively counteract Ca^{2+} oscillations, especially in excitable tissues (Carafoli 1987, 1988; Rahamimoff et al. 1996; Blaustein and Lederer 1999; Bers 2002, 2008). However, this viewpoint has been recently revised, since the molecular, functional, and regulatory properties of isoform/splice variants suggest that both Ca^{2+}-extruding systems (PMCA and NCX) can be involved in regulating basal and transient Ca^{2+} in response to distinct cellular signals (Lopreiato et al. 2014; Brini and Carafoli 2011; Brini et al. 2014).

It has been proposed that for many cell types (including the epithelial cells), PMCA is a major "housekeeping" system for Ca^{2+} extrusion from the cell, while predominantly contributing to total Ca^{2+} clearance. However, this "fundamental concept" requires a serious revision, since the cytosolic Ca^{2+} extrusion by PMCA might be minor (or even irrelevant) in light of accumulating findings suggesting that PMCA may play some other roles in cell function (Lopreiato et al. 2014; Brini and Carafoli 2011; Carafoli and Krebs 2016; Krebs 2017). For example, recent measurements revealed that at least 70% of basolateral Ca^{2+} extrusion is mediated by NCX, thereby supporting the notion that NCX is a major Ca^{2+} extrusion system at least in the kidney, pancreas, and bone cells (Sosnoski and Gay 2008; van der Hagen et al. 2015; Nishiyama et al. 2016; Albano et al. 2017; Herchuelz and Pachera 2018).

Cell-specific responses of PMCA and NCX to Ca^{2+} oscillations (waves) at functionally distinct microdomains of epithelial tissues can make major contributions to local and global Ca^{2+} signaling in epithelial and other cell types, although the underlying mechanism remains unresolved (Guillot and Lecuit 2013; Purvis and Lahav 2013; Giladi et al. 2016b; Brodskiy and Zartman 2018). This issue is especially interesting in light of recent findings suggesting that spatiotemporal tuning of Ca^{2+} dynamics plays a crucial role in the differentiation of epithelial cells during the development (for an updated review, see Brodskiy and Zartman 2018). This puts in a new perspective the importance of NCX regulation in response to Ca^{2+} swings in epithelial cells, since according to the previous view, the Ca^{2+}-dependent feedback mechanisms of NCX are physiologically "irrelevant" in non-excitable tissues, since the resting levels of cytosolic Ca^{2+} are rather steady. The challenge is to resolve how the cell-specific variants of NCX respond to Ca^{2+} spikes (waves) in a given cell-type and how the responses to global and local Ca^{2+} swings contribute to tissue-specific physiological events.

8.2.5 Autoregulation Controls the PMCA and NCX Activities

The Ca^{2+}-dependent regulation of Ca^{2+} toolbox proteins in a given cell type represents a general phenomenon called autoregulation, which covers a myriad of Ca^{2+}-binding proteins shaping the Ca^{2+} signaling pathways at the transcriptional and/or translational levels (Brini and Carafoli 2009). Autoregulation of the PMCA and NCX systems is of special interest from a physiological and mechanistic point of view, since Ca^{2+}-dependent feedback mechanisms are actively involved in fine-tuning Ca^{2+} extrusion either in excitable or non-excitable tissues. The reason for this is that any changes in the protein expression/regulation of PMCA and NCX activities might have immediate and/or long-term implications for the dynamic handling of Ca^{2+} homeostasis and Ca^{2+} signaling in distinct cell types (Carafoli 1987; Berridge et al. 2003; Clapham 2007). Thus, the cell-specific expression of isoform/splice variants can effectively diversify PMCA and NCX in response to cell-specific stimulus swaps during cell development, survival, hormone-induced activation, immune responses, and apoptosis, among many others. There some experimental indications that PMCA and NCX isoforms are characteristically expressed upon programmed cells that are remodeled together with other Ca^{2+} toolbox proteins (e.g., $InsP_3$-gated Ca^{2+} channel controlling Ca^{2+} release from ER) (Brini and Carafoli 2009). For example, in the cell survival model of granule cells, the expression levels of PMCA2, PMCA3, and NCX3 become slowly upregulated in time, whereas PMCA4 and NCX2 rapidly downregulate and then disappear at relatively steady levels of PMCA1 (Gomez-Villafuertes et al. 2005; Brini and Carafoli 2009; Krebs 2017).

The "short-term" autoregulation of PMCA- and NCX-mediated transport activities involves either the direct interactions of cytosolic Ca^{2+} with the regulatory domains of NCX (Hilge et al. 2006, 2009; Giladi et al. 2010, 2012a, b, c, 2016b; Tal et al. 2016; Khananshvili 2016, 2017a) or encompasses Ca^{2+}/calmodulin interactions with PMCA (Brini and Carafoli 2009; Carafoli and Krebs 2016; Krebs 2017). The Ca^{2+}/calmodulin interactions with PMCAs can dramatically elevate the Ca^{2+} affinity (up to sevenfold) and the transport rates (up to 25-fold) in an isoform/splice variant-dependent manner (Carafoli 1987, 1988; Brini and Carafoli 2009; Krebs 2017). There is no direct evidence for calmodulin-dependent activation of mammalian NCX in intact cells. Interestingly enough, the allosteric interactions of Ca^{2+} with the regulatory CBD1 and CBD2 domains of NCX elevate the transport rates up to 20-fold, while having rather minor (if any) effects on the apparent affinity (K_m) of Ca^{2+} transport in the cellular systems (Boyman et al. 2011; Ginsburg et al. 2013; Khananshvili 2016).

In contrast with the Ca^{2+} ions (causing the allosteric activation of NCX), two cytosolic ions, Na^+ and H^+, have a strong inhibitory effect on NCX-mediated transport activities (for review see DiPolo and Beaugé 2006; Giladi et al. 2016b; Khananshvili 2016, 2017a). The relevant regulatory modes might have physiological relevance for function and regulation of NCX under normal and altered physiological conditions (DiPolo 1979; Hilgemann et al. 1992a, b; Blaustein and Lederer 1999; Boyman et al. 2011). For example, the ischemia-induced acidosis in cardiomyocytes can effectively shut down the NCX-mediated arrhythmic currents raised at increased concentrations of cytosolic Ca^{2+} (Doering and Lederer 1994; Doering et al. 1996; Boyman et al. 2011).

Notably, while having opposite effects on the exchanger activity, the Ca^{2+}, Na^+, and H^+-dependent allosteric effects could be mechanistically related although the underlying molecular mechanism remain to be discovered (Hilgemann et al. 1992a, b; Rahamimoff et al. 1996; Matsuoka 2004; DiPolo and Beaugé 2006; Hryshko 2008; Lariccia and Amaroso 2018). Interestingly, the isoform/splice variants of NCX (expressed in a tissue-specific manner) exhibit characteristic profiles of ion-dependent allosteric regulation, which might have physiological relevance either in excitable or non-excitable cells (Hilgemann et al. 1992a, b; Matsuoka 2004).

8.3 NCX-Mediated Ca^{2+} Entry/Exit in Distinct Cell Types

8.3.1 *Electrogenic Stoichiometry of NCX-Mediated Ion-Exchange*

While exhibiting the tissue-specific expression profiles, the isoform/splice variants of NCX are actively involved in controlling the excitation–contraction coupling of the cardiac muscle, the long-term potentiation of the brain, blood pressure regulation, immunological responses, Ca^{2+} absorption by the intestine and kidney, bone formation, secretion of neurotransmitters and hormones, mitochondrial bioenergetics, and epithelial cell differentiation/development, among others. Thus, altered expression and/or regulation of NCXs contributes to numerous disorders and diseases, including heart failure, arrhythmia, cerebral ischemia, diabetes, hypertension, neurodegenerative diseases, muscle dystrophy, diarrhea, and colitis, among many other maladies (Blaustein and Lederer 1999; Annunziato et al. 2004, 2007; Khananshvili 2013, 2014).

The membrane-bound NCX transports $3Na^{+}$ ions in exchange with the $1Ca^{2+}$ ion, where the directionality of electrogenic net fluxes (Ca^{2+}entry or Ca^{2+} exit) depends on the intracellular and extracellular $[Na^{+}]$ and $[Ca^{2+}]$ as well as on the spatiotemporal features of membrane polarization/depolarization in a given cell type (Blaustein and Lederer 1999; Khananshvili 2013, 2014). Prokaryotic and eukaryotic NCX variants exhibit a comparable affinity for Ca^{2+} transport at cytosolic (K_m = 6–25 μM) and extracellular (K_m = 0.2–1.5 mM) sides (Hilgemann et al. 1992a, Khananshvili 1990, 1991; Matsuoka et al. 1995; Rahamimoff et al. 1996; Matsuoka 2004). Moreover, prokaryotic and eukaryotic NCXs share a common stoichiometry ($3Na^{+}$:$1Ca^{2+}$) for ion-exchange (Reeves and Hale 1984; Rahamimoff et al. 1996; Bers and Ginsburg 2007; Shlosman et al. 2018), thereby suggesting NCX variants might share a common structural basis for ion-transport machinery (Liao et al. 2012, 2016; Marinelli et al. 2014; Giladi et al. 2016a; van Dijk et al. 2018). Despite these similarities, the NCX-mediated turnover rates of the transport cycle vary from 0.5 s^{-1} to 2500 s^{-1} in prokaryotic and eukaryotic NCXs, which represent a primary requisite for matching the NCX-mediated rates of Ca^{2+} extrusion with cell-specific functional requirements (Niggli and Lederer 1991; Hilgemann et al. 1991; Baazov et al. 1999; Almagor et al. 2014). Thus, the allosteric regulation of variants is very instrumental for variable matching of NCX-mediated Ca^{2+} extrusion rates in distinct cell types (Blaustein and Lederer 1999; Giladi et al. 2016b; Khananshvili 2016, 2017a, b).

8.3.2 *Forward and Reverse Modes of Na^{+}/Ca^{2+} Exchange*

By definition, the electrogenic ion-exchange ($3Na^{+}$:1Ca) can occur either in the forward (Ca^{2+}-exit) or reverse (Ca^{2+}-entry) mode, depending on the $[Na^{+}]_o$, $[Na^{+}]_i$,

$[Ca^{2+}]_o$, $[Ca^{2+}]_i$, and membrane potential values (Reeves and Hale 1984; Rahamimoff et al. 1996; Blaustein and Lederer 1999; DiPolo and Beaugé 2006; Bers and Ginsburg 2007; Khananshvili 2014). Thus, the concentrations of transported ions at both sides of the membrane and the membrane potential determine not only the rates of NCX-mediated ion-exchange—also the directionality of net Ca^{2+} through NCX. At resting concentrations of $[Na^+]_o$, $[Na^+]_i$, $[Ca^{2+}]_o$, and $[Ca^{2+}]_i$, the electrochemical driving force for electrogenic exchange would be in the range of $E_{NCX} \approx -30$ mV. Thus, the directionality of net Ca^{2+} fluxes depends on the membrane potential (E_m) values with respect to E_{NCX} (Bers 2002, 2008; Blaustein and Lederer 1999; DiPolo and Beaugé 2006). Whenever E_m is more positive than E_{NCX} ($E_m > E_{NCX}$), the Ca^{2+} entry through NCX is favored, and when E_m is negative to E_{NCX} ($E_m < E_{NCX}$), the extrusion is preferred.

In excitable tissues (e.g., ventricle myocytes) the repetitive dynamic swings in the $[Na^+]_i$, $[Ca^{2+}]_i$, and membrane potential during the action potential swap the directionality of the net Ca^{2+} movements through NCX during the depolarization and repolarization alterations when E_m undergoes extensive alternations between -90 mV and +50 mV (Bers 2002, 2008; Bers and Ginsburg 2007). In addition, dynamic changes in $[Ca^{2+}]_i$ (up to 20-fold) and $[Na^+]_i$ (up to 2–3-fold) during the action potential determine the relative values of E_{NCX} and E_m and thus, define the directionality of Ca^{2+} movements through NCX (Blaustein and Lederer 1999; Bers 2002; Boyman et al. 2011).

8.3.3 *Functional Relevance of Ca^{2+} Flux Directionality Through NCX*

Historically, the Ca^{2+} extrusion (forward) mode of NCX has been considered as a major physiological mode either in excitable or non-excitable tissues, whereas the Ca^{2+}-entry through NCX was attributed to pathophysiological conditions (Blaustein and Lederer 1999; Khananshvili 2013, 2014). Even though the Ca^{2+} extrusion can be considered as a major "physiological" mode of NCX operation, the functional role of NCX-mediated Ca^{2+} entry should not be underestimated as well. The Ca^{2+} entry through NCX is particularly interesting regarding epithelial cells in which the resting membrane potential (E_m) approaches the reversal potential of NCX ($E_{NCX} \approx -30$ mV). Interestingly, in some epithelial cells, the basal E_m value could be more negative than that of E_{NCX}, whereas in other epithelial cells E_m could be characteristically more positive than E_{NCX} (Friedman and Gesek 1995; White et al. 1996). Moreover, even small changes in the basal values of E_m and/or E_{NCX} can reverse the directionality of net Ca^{2+} and/or the exchange rates. For example, at relatively stable values of extracellular $[Na^+]_o$, even twofold changes in cytosolic $[Na^+]_i$ significantly affect E_{NCX}, since any changes in $[Na]_i$ are powered in the third degree due to the stoichiometry of the $3Na^+$:$1Ca^{2+}$ exchange (Reeves and Hale 1984; Blaustein and Lederer 1999; Khananshvili 2014). Taking into account the measured values of E_m, one may note that NCX may operate either in the forward or reverse mode in epithelial cells such as rectal mucosa, trachea, alveolar, and colon epithelia, among others, since these cells have relatively positive values of E_m, which is expected to be very close to those of E_{NCX} (Zajac and Dolowy 2017). Thus, a given epithelial cell may support either the Ca^{2+} exit or Ca^{2+} entry mode through NCX, depending on the ionic conditions and membrane potential.

Notably, hyperpolarization or hypopolarization of basolateral membrane may alter the ion-exchange rates without altering the directionality of net Ca^{2+} fluxes (Friedman and Gesek 1995; White et al. 1996; Hoenderop et al. 2002). This may occur upon hormonal (e.g., PTH) or drug (e.g., thiazide)-induced hyperpolarization of the basolateral membrane, which in turn, can accelerate the Ca^{2+} extrusion rates through NCX without changing the directionality of net Ca^{2+} fluxes (White et al. 1996; Hoenderop et al. 2002). For example, thiazide drugs inhibit the apical Na^+-entry (by blocking the Na^+/Cl^- cotransporter at the distal tubule), which reduces the intracellular $[Na^+]_i$ levels and increases the electrochemical gradient (hyperpolarization) at the basolateral membrane with the subsequent acceleration of the Ca^{2+} extrusion rates through NCX (White et al. 1996; Hoenderop et al. 2002). Thus, thiazide drugs (in contrast with other diuretic drugs) increase the renal Ca^{2+}-reabsorption, while increasing the NCX-mediated excretion rates at the basolateral membrane and thus, decrease Ca^{2+} excretion into the urine.

8.4 Genetic Toolbox Shapes Regulatory Assets of NCX Variants

8.4.1 The NCX Gene Family Is a Branch of the Ca/CA Superfamily

The NCX gene family (including the prokaryotic and eukaryotic NCX orthologs) represents phylogenetically a very important division of proteins, which is a branch of a much larger family of transport proteins, called the superfamily of CaCA (Ca^{2+}/Cation) antiporters (Lytton 2007; On et al. 2008; Pittman and Hirschi 2016; Bode and O'Halloran 2018). While controlling the Ca^{2+} fluxes across the plasma membranes or intracellular compartments (e.g., mitochondria or nuclei) the CaCA proteins play a major role in handling the Ca^{2+} homeostasis from bacteria to humans. Strikingly, the Ca/CA proteins share structure–functional similarities in protein folding and ion-transport machinery (Liao et al. 2012; Nishizawa et al. 2013; Waight et al. 2013; Wu et al. 2013; Khananshvili, 2013, 2014; Taneja et al. 2016; van Dijk et al. 2018).

Three mammalian genes, *SLC8A1* (NCX1), *SLC8A2* (NCX2), and *SLC8A3* (NCX3) have been mapped to mouse chromosomes 17, 7, and 12, respectively, thereby revealing a dispersed gene distribution of NCX genes in mammalian species (Nicoll et al. 1990, 1996; Rahamimoff et al. 1996; Linck et al. 1998, 2000; Lee et al.

1994; Philipson and Nicoll 2000). Although an additional gene, *SLC8A4*, appears in teleost, amphibian, and reptilian species (Marshall et al. 2005; On et al. 2008), this gene has been secondarily and independently lost in mammals and birds during the evolution (Cai and Lytton 2004; Lytton 2007; Bode and O'Halloran 2018). Moreover, the *SLC8A4* gene was not detected in plants as well (Demaegd et al. 2014; Pittman and Hirschi 2016). At this end, it remains unclear why *SLC8A4* has been lost during the evolution, even though the experimental evidence suggests that the gene product of *SLC8A4* (NCX4) mediates Na^+/Ca^{2+} exchange activity, while a knockdown of NCX4 in zebrafish embryos results in altered left-right patterning (On et al. 2009). While ubiquitously expressed in different tissues, the highest levels of NCX4 expression occur in the brain and eyes. Interestingly enough, NCX4 has very special regulatory features, which may have physiological relevance. For example, NCX4 exhibits modest levels of Na^+-dependent inactivation and requires much higher levels of regulatory Ca^{2+} to activate outward exchange currents (On et al. 2009). NCX4 also exhibits extremely fast recovery from Na^+-dependent inactivation of outward currents (faster than any other isoform/splice variant of NCX tested until today). The physiological relevance of these exceptional regulatory properties of NCX4 and its evolutionary significance requires further resolution.

Similarly to the other members of the CaCA superfamily, the prokaryotic and eukaryotic NCX orthologs contain ten transmembrane helices (Liao et al. 2012; Waight et al. 2013; Wu et al. 2013; Ren and Philipson 2013; Nishizawa et al. 2013; Schnetkamp et al. 2014; Zhekova et al. 2016). In all Ca/CA proteins, a cytoplasmic loop (5 L6) between TM5 and TM6 connects two hubs, TM1–TM5 and TM6–TM10 (Fig. 8.1), where two hubs (clusters) are inversely oriented to each other, forming an inverted twofold "pseudo-symmetry" (Liao et al. 2012; Giladi et al. 2016a; van Dijk et al. 2018). Each hub in the CaCA proteins contains a highly conserved sequence motif (assigned as the α_1 and α_2 repeats). These proteins form the extracellular and cytosolic vestibules for ion passageway with an inversely oriented configuration, while referring to inverted twofold symmetry (Nicoll et al. 1990; 1994; Liao et al. 2012; Nishizawa et al. 2013; Waight et al. 2013; Wu et al. 2013; Taneja et al. 2016; Khananshvili 2020).

Despite the inverted twofold "symmetry" owned by prokaryotic and eukaryotic NCXs, the bidirectional movements of ions through NCX is an "asymmetric" process, exhibiting 10–500-fold differences between the cytosolic and extracellular values of Ca^{2+} K_m (Khananshvili et al. 1996; Almagor et al. 2014; Giladi et al. 2016a; van Dijk et al. 2018). This functional asymmetry seems to be essential for matching the physiologically "required" ion-transport rates and K_m values at opposite sides of the membrane (Almagor et al. 2014; Khananshvili 2013, 2014, 2016; Giladi et al. 2016a).

Even though all the Ca/CA proteins transport Ca^{2+} and share protein folding motifs, they exhibit a diverse selectivity to monovalent ions Na^+, K^+, Li^+, or H^+ (Carafoli et al. 1974; Palty et al. 2010; Emery et al. 2012; Schnetkamp et al. 2014; Sekler 2015; Refaeli et al. 2016; Bode and O'Halloran 2018). For example, the mitochondrial Na^+/Ca^{2+} exchanger (NCLX) can exchange Ca^{2+} with either Na^+ or Li^+, whereas the other NCXs exhibit high selectivity to Na^+ (Carafoli et al. 1974;

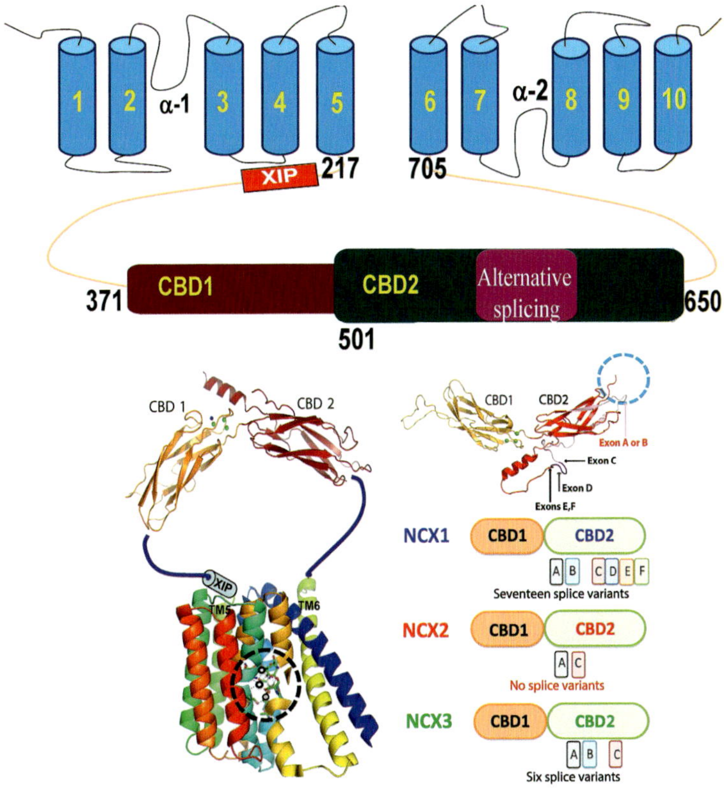

Fig. 8.1 Mammalian NCX topology and structural elements. Mammalian and eukaryotic NCX proteins comprise ten transmembrane segments (TM1–10). The cytosolic regulatory loop (5 L6) contains two Ca^{2+}-binding CBD domains, CBD1 and CBD2, which are connected with TM5 and TM6, respectively. The auto-inhibitory segment, XIP (20 amino acids), is connected with the C-terminal of the TM5. The crystal structure of prokaryotic NCX_Mj (3V5U) reveals the ion binding/transporting pocket, which can alternatively bind either $3Na^+$ or $1Ca^{2+}$ ion at the center of the protein (a doted black cycle). The CBD1 and CBD2 domains are connected through a very short linker (only five residues) to form a two-domain regulatory tandem (CBD12). The transmembrane helical structures were reproduced according to the crystal structure of archaeal NCX_Mj (3V5U), whereas the two-domain CBD12 structure was replicated according to the crystal structure of the brain splice variant CBD12-AD of NCX1 (3US9). In all eukaryotic NCX variants, the splice segment is exclusively allocated at the CBD2 domain. The mutually exclusive exons A and B encode strands E and F of CBD2 as well as the acidic segment of the Ca^{2+} binding site at the EF loop. The cassette exons (C, D, E, F) appear on the EF loop. The splice variants of NCX1 and NCX3 isoforms raise through a combination of mutually exclusive exon (either A or B) with cassette exons C, D, E, and F, whereas NCX2 does not undergo alternative splicing

Palty et al. 2010; Sekler 2015; Refaeli et al. 2016). The ion selectivity of Ca/CA proteins is a fascinating phenomenon, since these proteins might share general mechanisms for ion transport (Liao et al. 2012, 2016; Nishizawa et al. 2013; Waight et al. 2013; Wu et al. 2013; Giladi et al. 2016b; Khananshvili 2016; van Dijk et al. 2018).

8.4.2 Common and Distinct Features of NCX Topology Among NCX Variants

Protein products of three mammalian genes (isoforms), *SLC8A1* (NCX1), *SLC8A2* (NCX2), and *SLC8A3* (NCX3) contain 930–970 residues exhibiting ~70% sequence identity to one another (Nicoll et al. 1990, 1996; Li et al. 1994; Lee et al. 1994; Linck et al. 1998, 2000; Bode and O'Halloran 2018). Since mammalian NCXs contain the regulatory CBD1 and CBD2 domains, they are much bigger protein molecules than is the archaeal NCX_Mj, which contains "just" 301 residues (Liao et al. 2012). These differences in size between prokaryotic and eukaryotic NCXs account for evolutionary "supplementations" of regulatory domains (located on the 5 L6 loop) in eukaryotic NCX isoform/splice variants (Hilge et al. 2006, 2009; Cai and Lytton 2004; Lytton 2007; Emery et al. 2012; Demaegd et al. 2014; Pittman and Hirschi 2016). The 5 L6 loop of eukaryotic (but not of prokaryotic) NCXs contains two Ca^{2+}-binding regulatory domains, CBD1 and CBD2, which form a head-to-tail tandem of two domains (termed CBD12) through a very short interdomain linker containing only five residues (Levitsky et al. 1994; Hilge et al. 2006; Bode and O'Halloran 2018) (Fig. 8.1).

In mammalian NCX variants, the Ca^{2+} ion binding to CBD1 activates the ion-exchange activities. Moreover, the strength and duration of this allosteric regulation characteristically differ among distinct isoform/splices variants, expressed in a tissue-specific manner (Hilgemann et al. 1992a; Dyck et al. 1999; Dunn et al. 2002; Nicoll et al. 2006; Boyman et al. 2011). The Ca^{2+} binding to CBD2 alleviates Na^+-dependent inactivation caused by Na^+ interaction with the ion-transport domains but not with the regulatory CBD domains (Hilgemann et al. 1992a, b; Nicoll et al. 2006; Ottolia et al. 2009, 2010). Notably, in mammalian NCXs the splicing segment is exclusively associated with the CBD2 domain (Hilge et al. 2006, 2009), where the exon composition controls the affinity of Ca^{2+} binding at both CBDs as well as the number of Ca^{2+} binding sites at CBD2, which vary from zero to three (Giladi et al. 2010, 2012a, b, c; Tal et al. 2016). Thus, the isoform/splicing-controlled interactions of Ca^{2+} with CBDs serve as a very versatile tool for tuning the sensitivity and extent of NCX autoregulation. This exon-dependent tuning of regulatory properties in mammalian NCX variants is essential for matching the NCX responses to cell-specific swings in Ca^{2+} homeostasis.

Notably, mitochondrial NCLX does not contain any regulatory CBD domains (Palty et al. 2010; Palty and Sekler 2012), which underscores fundamental differences in the allosteric regulation of the plasma-membrane NCXs and mitochondrial NCLX (Palty and Sekler 2012; Boyman et al. 2013; Giladi et al. 2016b; Khananshvili 2016, 2017a). Interestingly, phosphorylation of Ser258 in NCLX or expression of phosphomimicking mutant (S258D) rescues NCLX activity from $\Delta\Psi_m$-driven allosteric inhibition (Kostic et al. 2018; Kostic and Sekler 2019). Interestingly, in pancreatic β-cells, glucose-driven $\Delta\Psi_m$ changes may govern mitochondrial Ca^{2+} signaling through NCLX regulation (Kostic et al. 2018). According to this scenario, a feedback control between metabolic activity and mitochondrial

Ca^{2+} signaling operates through the "safety valve" of NCLX phosphorylation, which rescues Ca^{2+} efflux in depolarized mitochondria (Kostic et al. 2018; Verkhratsky et al. 2018; Kostic and Sekler 2019).

Mammalian (NCX1–3) and *Drosophila* (CALX1) exchangers contain a positively charged auto-inhibitory XIP sequence (20 amino acids) with an α-helix structure, which is located at the N-terminus of the 5 L6 loop next to TM 5 (Li et al. 1991; Matsuoka et al. 1997) (Fig. 8.1). The site-directed mutagenesis, combined with patch-clamp and other biochemical approaches, strongly support the notion that the auto-inhibitory XIP region is somehow involved in the Na^+-dependent inactivation as well as in the PIP_2-induced alleviation of Na^+-dependent inactivation (Condrescu et al. 1995; Reeves and Condrescu 2003; Hryshko 2008). The structure–functional resolution of the underlying mechanisms involving the interactions of the XIP and PIP_2 domains with a putative Na^+-inhibitory site (probably located nearby the transport sites, but not at CBDs) remains to be highly challenging (see below).

8.4.3 *Structure-Related Functional Diversity of Tissues-Specific NCX Variants*

The NCX1, NCX2, and NCX3 genes (isoforms) and their splice variants are expressed in a tissue-specific manner to fulfill physiological demands in a given cell type (Nicoll et al. 1991; Philipson and Nicoll 2000; Lytton 2007; On et al. 2008, 2009; Bode and O'Halloran 2018). The challenge is to evaluate how these isoform/slice variants of NCX shape the Ca^{2+}-dependent events on the cellular, organ, and systemic levels in mammals (Linck et al. 1998, 2000; Blaustein and Lederer 1999; Khananshvili 2013, 2014; Giladi et al. 2016b). NCX1 is a ubiquitous isoform, the splice variants of which are expressed in a tissue-specific manner to support highly specialized physiological events (e.g., only one isoform/splice variant, NCX1-ACDEF is expressed in the cardiac muscle). Thus, specific splice variants of NCX1 are expressed in excitable and non-excitable tissues (Nicoll et al. 1990; Furman et al. 1993; Kofuji et al. 1993, 1994; Li et al. 1994; Linck et al. 1998, 2000; Philipson and Nicoll 2000; Van Eylen et al. 2001; Lytton 2007). The tissue-specific NCX variants are actively involved in cardiac contraction–relaxation, brain potentiation, kidney and intestinal Ca^{2+} absorption, bone formation, endothelial tonus, and pancreatic hormonal secretion in many other functions (for review see Blaustein and Lederer 1999; Bers 2002; Khananshvili 2013, 2014). Even though NCX2 is mainly expressed in the brain and spinal cord (Li et al. 1994; Quednau et al. 1997, 2004; Linck et al. 1998, 2000), significant amounts of NCX2 were also found in gastrointestinal, kidney, cochlear, and secretory cells (Nishiyama et al. 2016; Cantero-Recasens et al. 2019; Albano et al. 2017). NCX3 is chiefly expressed in the neuronal and smooth muscle tissues (Nicoll et al. 1996; Quednau et al. 1997, 2004; Linck et al. 1998, 2000; Sakaue et al. 2000; Michel et al. 2014, 2017). NCX3 can contribute to slow-twitch muscle contraction–relaxation and long-term potentiation in the hippocampus as well as can play an important role in neuronal excitotoxicity, brain stroke, and neuronal injuries (Gabellini et al. 2002; Minelli et al. 2007; Annunziato et al. 2007; Secondo et al. 2007). In gap with the previous concept (claiming the exclusive expression of NCX3 in neuronal tissues), several studies have identified significant levels of NCX3 in the adult and developmental epithelial cells, contributing to cell-specific physiological functions (e.g., Ca^{2+}-dependent bone formation) and cell cycling (Sosnoski and Gay 2008; Albano et al. 2017; Liu et al. 2018).

NCX1 and NCX3 undergo tissue-specific splicing, whereas no splice variants have been found for NCX2 (Kofuji et al. 1994; Li et al. 1994; Lee et al. 1994; Philipson and Nicoll 2000; Lytton 2007; On et al. 2008). The splice variants of NCX1 and NCX3 are generated by "mixing" six small exons (A, B, C, D, E, and F), exclusively located on CBD2 (Kofuji et al. 1993, 1994; Lee et al. 1994; Quednau et al. 1997, 2004; Linck et al. 1998, 2000; Philipson and Nicoll 2000; Fraysse et al. 2001; Gabellini et al. 2002; Lytton 2007) (Table 8.1). The mutually exclusive exon (A or B) alone or combined with "cassette" exons (C, D, E, and F) form a given isoform/splice variant expressed in a tissue-specific manner (Fig. 8.1). Notably, the cardiac and neuronal splice variants of NCX1 contain exon A, whereas the kidney, stomach, intestine, and skeletal muscle splice variants of NCX1 contain exon B (Table 8.1). At the posttranscriptional level, 17 splice variants of NCX1 arise by combining six exons (A–F), whereas only three exons (A, B, and C) generate a very few splice variants in NCX3 (Kofuji et al. 1994; Li et al. 1994; Lee et al. 1994; Philipson and Nicoll 2000; Fraysse et al. 2001; Gabellini et al. 2002; Lindgren et al. 2005; Khananshvili 2020).

The tissue-specific splicing of the NCX1 and NCX3 variants are of primary physiological relevance, since distinct exon combinations specifically shape the Ca^{2+}-sensing affinity and Ca^{2+} off rates at both CBDs (Besserer et al. 2007). The relevant regulatory mechanisms are controlled by exon combinations at CBD2, which structurally predefines the number of Ca^{2+} binding sites at CBD2 (Hilge et al. 2006, 2009; Boyman et al. 2009; Giladi et al. 2010, 2012c; Tal et al. 2016; Khananshvili 2017a, b). The underlying mechanisms determine the cell-specific regulatory features of NCX variants. For example, the Ca^{2+} interaction with CBD1 activates NCX1-mediated ion-exchange in the brain (AD), cardiac (ACDEF), and kidney (BD) variants, whereas the Ca^{2+}-induced relief of Na^+-dependent inactivation occurs in the cardiac and brain, but not in the kidney variants (Dyck et al. 1999; Hnatowich et al. 2012). Thus, the Na^+-dependent inactivation cannot be relieved by Ca^{2+} in the NCX1-BD (kidney or intestine) variant, since in NCX1 the B exon prevents Ca^{2+} from binding to CBD2 (Hilge et al. 2009; Giladi et al. 2010, 2012c).

In sharp contrast with NCX1, in NCX3 the B exon increases the stoichiometry of the Ca^{2+} binding sites at CBD2 of NCX3 (Hilge et al. 2009; Tal et al. 2016). Thus, the B exon-containing NCX3 variants have a capacity for performing the Ca^{2+}-induced relief of Na^+-dependent inactivation, although this regulatory mode can also be modified by some other cytosolic factors in epithelial (e.g., BHK) and other cell types (Hryshko 2008; Lariccia and Amoroso 2018). In contrast with NCX1 and

Table 8.1 Tissue-specific expression of NCX isoform/splice variants

Exon composition	NCX variants	Tissue-specific expression of NCX variants	References
NCX1			
ACDEF	NCX1.1	Heart, skeletal muscle	Nicoll et al. (1990), Quednau et al. (1997), Linck et al. (1998, 2000)
ACD	NCX1.6	Brain	Furman et al. (1993), Kofuji et al. (1994), Lee et al. (1994), Quednau et al. (1997)
ADF	NCX1.5	Brain, eye,	Furman et al. (1993), Quednau et al. (1997, 2004), Linck et al. (1998, 2000)
AD	NCX1.4	Brain, eye	Furman et al. (1993), Quednau et al. (1997, 2004)
BCD	NCX1.2	Kidney, skeletal muscle, pulmonary artery smooth muscle	Quednau et al. (1997), Zheng and Wang (2007), van der Hagen et al. (2015)
BDF	NCX1.7	Thymus, kidney, intestine, liver, adrenal gland, pancreas, endothelial cells, stomach, testes, skeletal muscle	Kofuji et al. (1993, 1994), Lee et al. (1994), Quednau et al. (1997, 2004), Zheng and Wang (2007), van der Hagen et al. (2015), Herchuelz and Pachera (2018)
BD	NCX1.3	Thymus, kidney, intestine, liver, adrenal gland, pancreas, endothelial cells, stomach, testes, skeletal muscle	Kofuji et al. (1993, 1994), Quednau et al. (1997, 2004), Linck et al. (1998, 2000), Zheng and Wang (2007), van der Hagen et al. (2015), Nishiyama et al. (2016), Herchuelz and Pachera (2018)
BDE	NCX1.9	Skeletal muscle, neuronal tissue	Quednau et al. (1997, 2004), Linck et al. (1998, 2000), Van Eylen et al. (2001)
NCX2			
AC		Skeletal muscle, brain, osteoclasts, secretory goblet cells, intestine	Li et al. (1994), Quednau et al. (1997), Linck et al. (2000), Sakaue et al. (2000), Fraysse et al. (2001), Nishiyama et al. (2016), Albano et al. (2017), Cantero-Recasens et al. (2019)
NCX3			
AC	NCX3.1	Skeletal muscle	Nicoll et al. (1996), Quednau et al. (1997, 2004), Fraysse et al. (2001), Michel et al. (2014, 2017)
B	NCX3.2	Brain, intestine, osteoblasts, osteoclasts	Quednau et al. (1997, 2004), Linck et al. (1998, 2000), Sakaue et al. (2000), Minelli et al. (2007), Li et al. (2007), Sosnoski and Gay (2008), Michel et al. (2014, 2017), Albano et al. (2017), Nishiyama et al. (2016, 2017)

(continued)

Table 8.1 (continued)

Exon composition	NCX variants	Tissue-specific expression of NCX variants	References
BC	NCX3.3	Brain, intestine, osteoblasts, osteoclasts	Quednau et al. (1997), Gabbellini et al. (2002), Li et al. (2007), Sosnoski and Gay (2008), Michel et al. (2014), Albano et al. (2017), Nishiyama et al. (2016, 2017)

NCX3, NCX2 does not show the Na^+-dependent inactivation mode and thus, the Ca^{2+}-induced relief of Na^+-dependent inactivation is physiologically irrelevant for NCX2 either in excitable or non-excitable tissues. The low-affinity binding of Ca^{2+} to CBD2 of NCX2 (Tal et al. 2016) can be explained by the notion that this low-affinity site is functionally invalid since the Ca^{2+}-induced alleviation of Na^+-dependent inactivation is irrelevant for NCX2 function (since NCX2 does not show the Na^+-dependent inactivation).

8.4.4 Epithelial NCX Variants

Early studies have shown that NCX1 and their splice variants are ubiquitously expressed in a tissue-specific manner, whereas the expression of NCX2 and NCX3 is limited to brain and skeletal muscle (Nicoll et al. 1990; Lee et al. 1994; Quednau et al. 1997, 2004). Only a few splice variants of NCX1 were found in the kidney, intestine, pancreas, osteoclasts, and endothelial and secretory cells (Szewczyk et al. 2007; Jia and Cui 2011; Andrikopoulos et al. 2011, 2017; Shubair et al. 2012; van der Hagen et al. 2015; Diaz de Barboza et al. 2015; Grover 2017; Cantero-Recasens et al. 2019; Lillo et al. 2018). Moreover, recent studies have shown that besides the neuronal tissues the NCX2 and NCX3 are also expressed in the intestine, osteoblast, and secretory tissues. The new findings provide an increasing body of evidence that cell-specific expression profiles of NCX isoform/splice variants may play differential roles in handling/shaping of epithelial Ca^{2+} homeostasis (Gabellini et al. 2002; Sosnoski and Gay 2008; Nishiyama et al. 2016; Albano et al. 2017; Grover 2017; Lillo et al. 2018; Cantero-Recasens et al. 2019). Important developments are expected in this direction in the near future.

Interestingly, the epithelial cells express specific patterns of NCX variants (Table 8.1), thereby supporting the notion that epithelial NCX variants, possessing specific regulatory features, are structurally predefined to fit cell-specific Ca^{2+} homeostasis and Ca^{2+} signaling in a given epithelial cell type to fulfill highly specialized functional activities. As can be seen from Table 8.1, the epithelial cells chiefly express NCX1-BD (NCX1.3), NCX1-BDF (NCX1.7), NCX2 (no splice variant), NCX3-B (NCX3.2), and NCX3-BC (NCX3.3). As discussed below in detail, the exon B-containing NCX1 and NCX3 splice variants as well as NCX2, expressed in epithelial tissues, exhibit specific regulatory features, regarding three major regulatory modes of mammalian NCX regulation: (a) Ca^{2+}-dependent activation, (b) Na^+-dependent inactivation, and (c) Ca^{2+}-induced alleviation of Na^+-dependent inactivation. During the last decade, huge progress has been made in better understanding the structure-dynamic determinants governing the ion-dependent regulation of invertebrate and mammalian NCX orthologs expressed in excitable or non-excitable tissues (for an updated review, see Giladi et al. 2016b, Khananshvili 2016, 2017a, 2020). Below, emphasis is on the NCX isoform/splice variants expressed in epithelial tissues, compared with the NCX variants expressed in the cardiac and brain tissues, with the goal of underscoring the underlying molecular mechanisms governing the regulatory specificities of NCX variants expressed in excitable and non-excitable tissues.

8.5 Ion-Dependent Regulation of Tissue-Specific NCX Variants

8.5.1 Allosteric Regulation of NCX Variants by Ca^{2+}, Na^+, and H^+

The mammalian NCX variants are chiefly regulated by cytosolic Ca^{2+}, Na^+, and H^+ ions (Hilgemann et al.1992a; Hryshko 2008; Khananshvili 2013, 2014); however, the underlying mechanisms and structure-dynamic determinants of ion interactions with NCX regulatory mechanisms are different but related (Giladi and Khananshvili 2013; Khananshvili 2016, 2017b). The ion-dependent regulation of epithelial NCX variants should be considered in the context of ion-transport mechanisms in epithelial cells, where the transcellular mechanisms of ion transport must be flexible enough to match the physiological demands at the cellular, organ, and systemic levels. In general terms, the transcellular translocation of Na^+ in the epithelial cells (e.g., in the kidney) largely controls the transport rates of other ions (including Ca^{2+}) as well as of metabolic solutes (e.g., glucose and amino acids). The fundamental physiological requirement is that the parallel translocation of transcellular Na^+ and Ca^{2+} fluxes across the epithelial cells must be delicately balanced; otherwise, some major physiological disturbances may occur. For example, the "overshoot" in the renal Na^+ reabsorption (e.g., due to increased filtered load of Na^+) may elevate transcellular Ca^{2+} absorption, where this tendency may become dangerous at a certain stage due to the development of systemic hypercalcemia. Thus, the $[Na^+]$-dependent inactivation of NCX (in conjunction with many other mechanisms) may play an important role in balancing the Ca^{2+} extrusion rates into the interstitial fluid of the kidney in order to lessen the tentative development of hypercalcemia. Therefore, it is not surprising that the epithelial cells express the exon B-containing NCX1 variants, which typically possess $[Na^+]$-dependent inactivation in the absence of Ca^{2+} or ATP/PIP_2-induced alleviation mechanisms (Asteggiano et al. 2001; Dyck et al. 1999; Dunn et al. 2002; Matsuoka 2004; Hryshko 2008; Barberian et al. 2009).

Thus, the Na^+-dependent inactivation of exon B-containing NCX1 variants may represent a very important feedback mechanism for matching the Na^+ and Ca^{2+} reabsorption rates in the kidney, intestine, and other epithelial tissues under conditions in which the rates of Ca^{2+} reabsorption rates must retain relatively steady with increased translocation of Na^+ across epithelial cells (van der Hagen et al. 2015).

In contrast with the NCX1 and NCX3 variants, NCX2 lacks Na^+-dependent inactivation (Hilgemann et al. 1992a, b; Matsuoka 2004; Lariccia and Amoroso 2018). This "Na^+-insensitive" mode of NCX2 operation is consistent with the low-affinity Ca^{2+} binding to CBD2 of NCX2 (K_d > 20 μM), which might be functionally invalid (Tal et al. 2016). Thus, the Ca^{2+} or ATP/PIP_2-induced alleviating mechanisms seem to be irrelevant for NCX2, either in excitable or non-excitable tissues. The present regulatory hallmarks of NCX2 are compatible with "Na^+-insensitive" entry or extrusion of Ca^{2+} via NCX2 toward matching distinct physiological requirements. For example, the reversal of the E_m and E_{NCX} relationships (e.g., from $E_m > E_{NCX}$ to $E_m < E_{NCX}$ or vice versa) can effectively reverse Ca^{2+} fluxes through NCX2 in a "Na^+-independent" manner. Notably, NCX2 appears in specific parts of the brain, while contributing to the injury-associated devastating effects of Na^+ and/or Ca^{2+} overload in neurons and glia (Annunziato et al. 2004, 2007; Secondo et al. 2007). Interestingly, NCX2 actively contributes to the motility of mouse ileum, while playing an important role in the development of diarrhea in the mouse model (Nishiyama et al. 2016, 2017).

Similar to NCX1, the NCX3 isoform also undergoes Na^+-dependent inactivation, thereby representing an important feedback mechanism for sensing the cytosolic [Na^+] changes (Matsuoka 2004; Michel et al. 2014, 2017; Lariccia and Amoroso 2018). However, a fundamental difference between the NCX1 and NCX3 variants is that the A and B exons have opposite effects on the Ca^{2+} binding capacity at CBD2 in NCX1 and NCX3, which, in turn, predefines the features of Ca^{2+}-induced alleviation of Na^+-dependent inactivation (Giladi et al. 2012b, 2017; Lee et al. 2016; Tal et al. 2016). More specifically, in NCX1, exon A predefines the binding of two Ca^{2+} ions to CBD2, whereas exon B prevents Ca^{2+} from binding to CBD2 (Hilge et al. 2009; Giladi et al. 2012b). In contrast to NCX1, exon A prevents Ca^{2+} from binding to CBD2 of NCX3, whereas exon B results in the binding of three Ca^{2+} ions to CBD2 of NCX3 (Tal et al. 2016; Giladi et al. 2017, Khananshvili 2016, 2017a). Thus, the exon B-containing NCX3 variants structurally control Ca^{2+} binding to CBD2 to ensure a mode of Ca^{2+}-induced alleviation of Na^+-dependent inactivation (Hilge et al. 2009; Giladi et al. 2016b).

8.5.2 Na^+-Dependent Inactivation and Its Alleviation in NCX Variants

In giant patches of cardiomyocytes or in oocytes with expressed NCX variants, a rise in the cytosolic [Na^+] quickly increases the NCX-mediated ion currents caused by the Na^+ interaction with the ion-transport sites. After reaching the peak values, the signal slowly decreases (for 20–60 seconds) before reaching much lower steady-state levels (Fig. 8.2). This regulatory mode, discovered by Hilgemann (termed the I_1-inactivation or Na^+-dependent inactivation), occurs in NCX1 and NCX3, but not in NCX2 (Hilgemann et al. 1992a, b; Nicoll et al. 2006; Besserer et al. 2007; Ottolia et al. 2009; Tal et al. 2016). It has been proposed that the "Na^+-inactivation site" is located on the "catenin-like domain" at the terminals of the cytosolic f-loop (5 L6) (Hilge et al. 2006, 2009), but this claim remains hypothetical in the absence of experimental evidence. Most probably, the Na^+ interaction with transport sites and/or nearby the ion-binding pocket results in a conformational "rearrangement" of the auto-inhibitory XIP domain (next to TM5), yielding a slow inactivation of NCX (Condrescu et al. 1995; Reeves and Condrescu 2003; Hryshko 2008; Hilgemann 2007). Notably, Na^+ does not interact with CBD domains nor competes

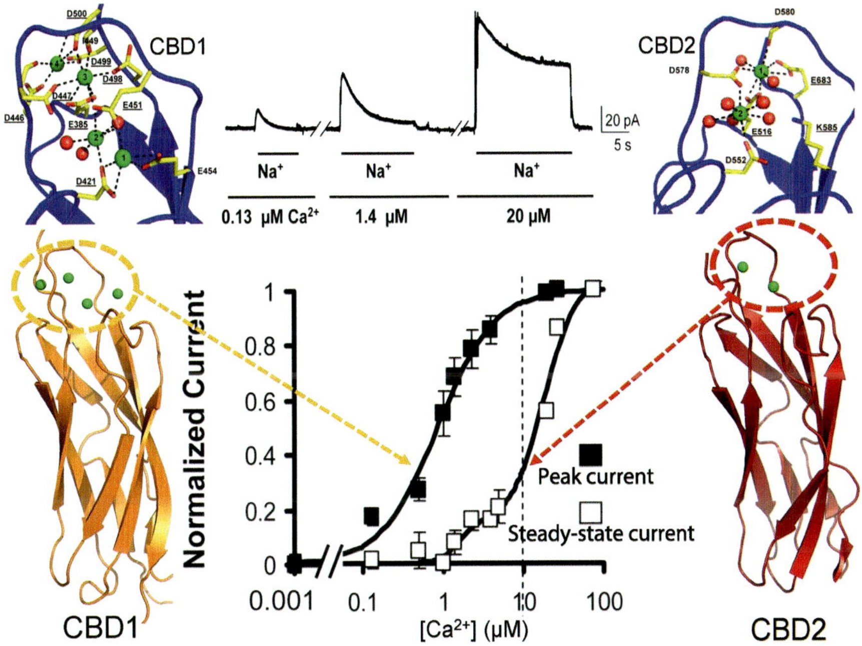

Fig. 8.2 Ca^{2+} and Na^+ dependent regulatory modes of mammalian NCX. For demonstration of Ca^{2+} and Na^+ dependent regulatory effects, typical patch-clam recordings are presented (for original data see Ottolia et al. 2009). In these experiments, cytosolic Na^+ is applied in the presence of varying concentrations of Ca^{2+} and NCX-mediated ion-currents are monitored in oocytes. Na^+ application initially activates NCX-mediated currents (peak current), followed by a slow decrease in the current (due to Na^+-dependent inactivation) until the ion current signal reaches a steady level (steady-state current). Higher $[Ca^{2+}]$ results in larger peak currents and in reduced Na^+-dependent inactivation (see the normalized data in respect with dotted line). Binding of Ca^{2+} ions to the Ca3-Ca4 sites of CBD1 (left panel) results in the activation of peak-current, whereas the binding of Ca^{2+} to the CaI site of CBD2 (right panel) results in enhancement of steady-state currents (for additional information see the text)

with Ca^{2+} ions for CBDs sites (Boyman et al. 2011). Moreover, the kinetics and amplitude of Na^+-dependent inactivation significantly differ in the cardiac (ACDEF), brain (AD), and kidney (BC) variants of NCX1 (Dyck et al. 1999; Hryshko 2008; Ottolia et al. 2009; Hnatowich et al. 2012), which could be physiologically relevant.

Two distinct mechanisms can alleviate the Na^+-dependent inactivation in NCX variants. The first mechanism refers to Ca^{2+} binding to CBD2 (Hilgemann et al. 1992a; Besserer et al. 2007; Ottolia et al. 2009; Hilge et al. 2009), and the second mechanism denotes PIP_2 interaction with a putative site nearby CBD2 (Hilgemann and Ball 1996; Reeves and Condrescu 2003; Barberián et al. 2009; Lariccia and Amoroso 2018). Interestingly, the tissue-specific variants of NCX1 and NCX3 exhibit distinct capacities for alleviating the Na^+-dependent inactivation. For example, CBD2 of the cardiac (ACDEF) and brain (AD) splice variants of NCX1 can bind two Ca^{2+} ions at CBD2, thus alleviating Na^+-dependent inactivation (Boyman et al. 2009; Hilge et al. 2009; Giladi et al. 2010, 2012a, b, c). In contrast, the lack of Ca^{2+} binding to CBD2 in the kidney (BD) splice variant of NCX1 disables the Ca^{2+}-induced alleviation of Na^+-dependent inactivation (Hilge et al. 2009; Giladi et al. 2012a, b, c).

The crucial fact is that the presence of exon A and B in either NCX1 or NCX3 structurally predefines not only the number of Ca^{2+} binding sites at CBD2 (from zero to three)—also the Ca^{2+} binding affinity (K_d) at CBD2. Notably, either in NCX1 or NCX3, CBD2 has at least a 10–500-fold lower affinity for Ca^{2+} binding, compared with the affinity of Ca3-Ca4 sites at CBD1 (Hilge et al. 2009; Boyman et al. 2009; Giladi et al. 2010, 2012a, c; Tal et al. 2016). Thus, the mutually exclusive presence of exon A or exon B differentially controls the capacity for Ca^{2+}-induced alleviation of Na^+-dependent inactivation while controlling Ca^2 binding to CBD2 (Hilge et al. 2009; Ottolia et al. 2009; Boyman et al. 2009; Giladi et al. 2010, 2012a, b, c, 2017). These differences in the Ca^{2+}-induced alleviation of Na^+-dependent inactivation among the NCX variants represent tissue-specific requirements for feedback "recovery" of NCX by Ca^{2+} upon Na^+-dependent inactivation (Dyck et al. 1999; Blaustein and Lederer 1999; Hnatowich et al. 2012; Khananshvili 2016, 2017a). Some additional regulatory mechanisms and cytosolic factors could be involved in regulating distinct NCX variants, expressed in a tissue-specific manner (Annunziato et al. 2004, 2007; Secondo et al. 2007; Hryshko 2008; Plain et al. 2017; Lariccia and Amoroso 2018), although the details of the underlying molecular mechanisms remain to be further resolved.

The ATP-induced "metabolic" alleviation of Na^+-dependent inactivation has been demonstrated for squid axon NCX and mammalian cardiac (NCX1-ACDEF) variants; however, there is no experimental evidence for ATP-induced relief of Na^+-dependent inactivation in the brain (NCX1-AD) and kidney (NCX1-BD) splice variants of NCX1 (Dyck et al. 1999; Hryshko 2008; Dibrov et al. 2006; Hnatowich et al. 2012). In the cardiac NCX1-ACDEF variant, the ATP-induced regulation takes place through the lipid kinase-dependent production of PIP_2 from ATP (Hilgemann 1990; Hilgemann and Ball 1996), although some alternative molecular mechanisms may exist for mediating the ATP-dependent activation in squid axon NCX (DiPolo and Beaugé 2006; Annunziato et al. 2004; Secondo et al. 2007; Hryshko 2008; Lariccia and Amoroso 2018).

The protein kinase-dependent phosphorylation of any NCX variant by ATP remains a highly controversial issue in the absence of "hard evidence" for NCX phosphorylation either in vitro or cellular system (Condrescu et al. 1995; Zhang and Hancox 2009; Wanichawan et al. 2011). Nevertheless, according to some recent reports, a direct phosphorylation of NCX and/or its regulatory protein could be an option for explaining complex behaviors of NCX regulation in intact cells (Michel et al. 2014, 2017; Lariccia and Amoroso 2018). Most importantly, distinct molecular mechanisms for kinase-dependent phosphorylation could be involved in regulating different isoform/splice variants of NCX. Even though the relevant possibilities require full attention, the application of complementary approaches (e.g., mass-spectrometry) is required for supporting any kinase-dependent regulatory mechanisms targeting the NCX variants. More dedicated experimentation is required in this respect, even though this fundamental issue awaits a resolution for nearly three decades.

Interestingly, PIP_2 can reverse the Na^+-dependent inactivation in the cardiac (NCX1-ACDEF), but not in the brain (NCX1-AD) splice variant of NCX1 (He et al. 2000; Hilgemann 2007), thereby suggesting that PIP_2-dependent regulation of NCX might be a tissue-specific event. An interesting possibility is that the PIP_2 and XIP (auto-inhibitory) domains interact with each other in a mechanism-dependent manner. More specifically, the Na^+ interaction with transport sites may "shuffle" the XIP domain to inhibit NCX, whereas Ca^{2+} or PIP_2 interaction with CBD2 may "unlock" the XIP-induced inhibitory effect by repositioning the inhibitory domain (He et al. 2000; Reeves and Condrescu 2003). Collectively, the Ca^{2+} or ATP/PIP_2-induced alleviation of Na^+-dependent inactivation mechanisms is irrelevant for NCX2 regulation, since these "recovering" mechanisms are not required owing to the lack of Na^+-dependent inactivation in NCX2 (Blaustein and Lederer 1999; He et al. 2000; Annunziato et al. 2004, 2007; Lariccia and Amoroso 2018).

8.5.3 *Ca^{2+}-Dependent Activation/Inactivation of NCX Variants*

The cytosolic Ca^{2+} interactions with the ion transport ($K_m = 5$–25 μM) and regulatory sites of CBD1 ($K_d = 0.2$–0.7 μM) and CBD2 ($K_d = 2$–100 μM) activate the mammalian NCX variants, although the underlying mechanisms of Ca^{2+}-dependent activation are different (Giladi et al. 2016b; Khananshvili 2016, 2017a, b). In patch-clamp experiments (with excised giant patches of cardiomyocytes or with oocytes expressing distinct NCX variants), the peak current increase refers to $[Ca^{2+}]_i$-dependent activation of NCX, whereas the ratio between steady-state and peak currents accounts for $[Ca^{2+}]_i$-induced relief of Na^+-dependent inactivation (Fig. 8.2) (Hilgemann et al. 1992a, b; Matsuoka et al. 1993; Matsuoka 2004). The Ca^{2+}

concentrations required for handling these two processes are characteristically different, thereby suggesting that regulatory Ca^{2+} interacts with two distinct sites at CBDs. A rise in the peak current occurs at much lower Ca^{2+} concentrations ($K_{0.5} \sim 0.2$ μM), representing the Ca^{2+} interactions with CBD1 (Fig. 8.2). Much higher Ca^{2+} concentrations ($K_{0.5} = 5$–10 μM) are required for elevating the steady-state to peak-current ratios representing the Ca^{2+} interactions with CBD2 (Hilgemann et al. 1992a; Nicoll et al. 2006; Chaptal et al. 2007; Besserer et al. 2007; Ottolia et al. 2009, 2010).

In intact mammalian cells (e.g., cardiomyocytes) the [Ca^{2+}]-dependent activation of NCX (due to Ca^{2+} binding to CBD1) is an extremely cooperative event, with a Hill coefficient of $n = 4$–8 (Maack et al. 2005; Boyman et al. 2011; Ginsburg et al. 2013). Notably, a much lower degree of cooperativity ($n \approx 2$) is observed for [Ca^{2+}]-dependent activation for the cardiac or brain splice variants of NCX1 expressed in oocytes (Nicoll et al. 2006; Besserer et al. 2007; Ottolia et al. 2009; John et al. 2011). The reasons for these discrepancies remain enigmatic, although it is possible that some regulatory mechanisms, operating in intact cells (e.g., squid axon or BHK cells), are lost in "non-physiological" preparations of excised patches (DiPolo and Beaugé 2006; Hurtado et al. 2006; Hilgemann 2007; Lariccia and Amaroso 2018). The [Ca^{2+}]-dependent allosteric activation of NCX, associated with occupation of a high-affinity Ca^{2+} sensor at CBD1, is a very powerful mode for NCX regulation, since the amplitude of the basal activity of NCX (displayed at resting levels of Ca^{2+}) can be magnified up to 20-fold within narrow-range changes in cytosolic [Ca^{2+}] (Boyman et al. 2011). Moreover, the occupation of a primary (high-affinity) allosteric sensor at CBD1 by Ca^{2+} is essential for retaining the ion-exchange activity (DiPolo 1979; Hilgemann et al. 1992a; Nicoll et al. 2006; Besserer et al. 2007; Ottolia et al. 2009, 2010).

The removal of cytosolic Ca^{2+} results in slow inactivation, thereby representing an I_2-inactivation (or Ca^{2+}-dependent inactivation) of NCX (Hilgemann et al. 1992a; Ottolia et al. 2009). Notably, the I_2-inactivation kinetics differs over 10–50-fold among NCX isoform/splice variants (Hilgemann et al. 1992a; Matsuoka et al. 1993, 1995; Dyck et al. 1999; Dunn et al. 2002; Hnatowich et al. 2012). Strikingly, the I_2-inactivation kinetics in full-size NCX variants and the Ca^{2+} dissociation rates from CBD1 in matching CBD12 variants are comparable in a given variant, thereby suggesting that the Ca^{2+} dissociation kinetics from the high-affinity sites of CBD1 represents I_2 inactivation (Giladi et al. 2010, 2012c, Tal et al. 2016). Collectively, these findings underscore the physiological relevance of I_2 inactivation kinetics representing biologically relevant hallmark features of mammalian NCX variants, expressed in a tissue-specific manner (Maxwell et al. 1999; Dyck et al. 1999; Dunn et al. 2002; Giladi et al. 2016b).

The NCX1-BD (NCX1.3) variant, expressed in the kidney, intestine, and other epithelial tissues (Quednau et al. 1997; Linck et al. 1998, 2000; van der Hagen et al. 2015), exhibits exceptional regulatory features, compared to the brain NCX1-AD and cardiac NCX1-ACDEF variants (Matsuoka et al. 1993, 1995; Dyck et al. 1999; Matsuoka 2004; Ottolia et al. 2009, 2010). More specifically, I_1 inactivation appears to be much more prominent for NCX1-BD than for the other variants, where this inactivation process resulted in steady-state currents that are less than 90%, in comparison with the current peak values (Dyck et al. 1999; Dunn et al. 2002; Hnatowich et al. 2012). This phenomenon represents the interaction of regulatory Ca^{2+} with CBD1 and cannot be explained by Ca^{2+} interaction, either with CBD2 (lacking the Ca^{2+} binding sites) or transport sites (Ottolia et al. 2009, 2010; Hilge et al. 2009; Giladi et al. 2010). Furthermore, regulatory Ca^{2+} is incapable of alleviating the Na^+-dependent inhibition in NCX1-BD, in sharp contrast with NCX1-ACDEF and NCX1-AD variants (Dyck et al. 1999; Ottolia et al. 2009; Hnatowich et al. 2012). This is because the exon B prevents the Ca^{2+} binding to CBD2 of NCX1, whereas the exon-A splice variants of NCX1 bind two Ca^{2+} ions at CBD2 (Boyman et al. 2009).

In light of the present considerations, one can conclude that the expression of the B exon in NCX1 adds a novel aspect for regulation, implying that Ca^{2+} interaction with a high-affinity primary sensor at CBD1 results in temporary activation of NCX1-BD (but lasts for only a few seconds). Moreover, even though the A- and B-containing splice variants of NCX1 exhibit an amplitude and kinetics comparable to Na^+-dependent inactivation, the epithelial NCX1-BD lacks the Ca^{2+}-induced alleviation of Na^+-dependent inactivation (Dyck et al. 1999; Dunn et al. 2002; Ottolia et al. 2009, 2010; van der Hagen et al. 2015). These observations are in good agreement with structural studies revealing that Ca^{2+} binding to CBD2 is prevented due to the structural contributions of exon B in NCX1 (Hilge et al. 2009; Giladi et al. 2012a, b). Thus, the $[Na^+]$ elevation inhibits the exon B-containing variants of NCX1 (expressed in the kidney, intestine, and other epithelial tissues, see Table 8.1) where the Ca^{2+}-dependent alleviation of Na^+-dependent inactivation is absent.

8.5.4 "Proton Block" and Related Regulatory Modules

Mammalian and squid axon NCX are exceedingly sensitive to cytosolic protons, where even a slight acidification of cytosolic pH from 7.2 to 6.9 results in nearly 90% inactivation of NCX (Baker and Blaustein 1968; DiPolo 1979; Doering and Lederer 1994; Doering et al. 1996; DiPolo and Beaugé 2006; Boyman et al. 2011). This phenomenon (known as a "proton block" mechanism) is of general interest since proton dependent regulation of NCX might play a critical role in handling Ca^{2+} homeostasis under acidosis and ischemia conditions either in excitable or non-excitable tissues. Even though protons may interact with transport and/or regulatory domains under physiologically relevant conditions, protons do not alter the ion-transport sites (Doering et al. 1996; DiPolo and Beaugé 2006; Boyman et al. 2011; John et al. 2018). Interestingly, in the absence of cytosolic Na^+ the acidic pH values have only a minor effect on NCX activity, whereas at basic pHs the cytosolic Na^+ does not induce inactivation (Doering et al. 1996; Blaustein and Lederer 1999; DiPolo and Beaugé 2006). Collectively, these findings suggest that proton and Na^+-dependent inhibitory mechanisms interrelate, even though without sharing a common site for ion binding (DiPolo and Beaugé 2006; Hryshko 2008; Boyman et al. 2011; John et al. 2018).

Patch-clamp experiments have identified two components in the proton-dependent inhibition of NCX, one of which is fast (and is not affected by the presence of other ions) and a second one, which is slow and requires cytoplasmic Na^+ (Doering and Lederer 1994; Doering et al. 1996). Notably, Ca^{2+} and H^+ (but not Na^+) can compete for common binding sites at CBDs, which is compatible with the Na^+-independent proton block of NCX (Boyman et al. 2011). Most probably, the protonation of Ca^{2+} binding sites at CBDs prevents the Ca^{2+}-dependent activation (through CBD1) as well as precludes a displacement of inhibitory Na^+ from the transport sites (due to proton competition with Ca^{2+} at CBD2). According to this scenario, protonation of CBDs can cause a dramatic shift in the $[Ca]_i$-dependent activation of NCX in intact cardiomyocytes, thereby decreasing the NCX activity even at high concentrations of cytosolic Ca^{2+} (Boyman et al. 2011). Thus, CBDs may function as a dual Ca^{2+}/pH sensor, whereas under acidosis/ischemia conditions the interaction of protons with CBDs can effectively reduce the affinity of Ca^{2+}-dependent activation toward the shutdown of NCX activity (Boyman et al. 2011; Giladi et al. 2015; Khananshvili 2016). This mechanism may prevent arrhythmogenic ion currents under ischemia/acidosis conditions, since elevated Ca^{2+} concentrations (caused by acidosis) can generate arrhythmogenic currents through the forward mode of NCX (Doering et al. 1996; Bers 2002; Boyman et al. 2011). The CBD-mediated proton block mechanisms seem to be less important (if relevant at all) for epithelial NCX1-BD, since exon B suppresses the Ca^{2+}-dependent regulatory effects by affecting the conformational dynamics of the CBD1 and CBD2 domains (Lee et al. 2016; Giladi et al. 2015, 2016b, 2017).

The combination of site-directed mutagenesis and electrophysiological approaches have identified pH-dependent allosteric inhibition of NCX, which could be distinct from the Na^+- and Ca^{2+}-dependent regulatory mechanisms, even though cytoplasmic Na^+ can affect the sensitivity of the cardiac NCX1-ACDEF to protons (John et al. 2018). Moreover, these studies have identified two histidine residues (His124 and His165), where His165 plays a predominant role in pH-dependent regulation of NCX. Interestingly, in cardiac NCX (NCX1-ACDEF), His165 is modeled to be allocated in the cytoplasmic loop 3 L4, connecting the TM3 and TM4 helices (John et al. 2018), whereas according to the crystal structure of archaeal NCX_Mj, the equivalent residue, Lys103, is located on TM4 near the cytosolic surface (Liao et al. 2012). Interestingly, H165A mutant renders NCX insensitive to either Na^+ or Ca^{2+} regulation (Ottolia et al. 2002; John et al. 2018), meaning that H165 could be a key element for transducing pH, Na^+, and Ca^{2+} regulatory signals. According to this proposal, His165 somehow mediates the propagation of allosteric messages from the inhibitory XIP region (involved in Na^+-dependent inactivation) and the Ca^{2+}-binding CBD domains to the transport sites. Consistent with this, alkaline pH abolishes both Na^+ and Ca^{2+} regulation (Hilgemann et al. 1992b; Doering and Lederer 1994; Doering et al. 1996), thereby imitating the H165A phenotype. Thus, the protonation of His165 may inhibit NCX directly via a mechanism that does not involve the structure-dynamic arrangements associated with the regulatory interactions of Na^+ or Ca^{2+} with protein (John et al. 2018). Notably, His165 is highly conserved among NCX1, NCX2, and NCX3 variants, thus suggesting that the relevant pH-dependent mechanisms could also be relevant for NCX isoform/splice variants expressed in epithelial tissues. Detailed structural information of mammalian NCX is required for elucidating the pH-dependent mechanisms of NCX.

8.6 Structural Basis of Regulatory Diversity in NCX Variants

8.6.1 High-Resolution Structures of CBD1 and CBD2 Domains

The high-resolution X-ray and NMR structures of the CBD1 and CBD2 domains depict a β-immunoglobulin (Ig)-like folding structure, where two antiparallel β-sheets (involving A-B-E and D-C-F-G strands) form a seven-strand β-sandwich motif (Hilge et al. 2006; Nicoll et al. 2006; Besserer et al. 2007; Wu et al. 2009a, b, 2011; Giladi et al. 2012a). The striking similarity between the folding structure of CBD1 and CBD2 stems from the fact that the overlay of the CBD1 and CBD2 crystal structures display nearly identical folding with RMSD = 1.3 Å. Moreover, all Ca^{2+} binding sites reside at C-terminal ends of distal loops (Nicoll et al. 2006; Besserer et al. 2007; Wu et al. 2011). Despite these structural similarities, CBD1 and CBD2 differ in a number of Ca^{2+} binding sites and in their coordination chemistry; however, these structural differences definitely have functional relevance (Hilge et al. 2009; Boyman et al. 2009; Wu et al. 2010). The Ig-like folding motif, exemplified in CBDs, is very similar to a very large group of regulatory proteins assigned as C_2-domains (Rizo and Sudhof 1998; Stahelin et al. 2005; Cho and Stahelin 2006; Haussinger et al. 2002). However, the canonical folding structure of C_2 proteins refers to an eight-strand β-sandwich motif instead of the seven-stranded one possessed by CBD domains (Burgoyne 2007; McCue et al. 2010). In CBD1 of NCX1, the Ca^{2+}-coordinating residues are allocated at the AB, CD, and EF loops (Levitsky et al. 1994; Nicoll et al. 2006; Wu et al. 2011). Similar to CBD1, the Ca^{2+}-coordinating cluster of CBD2 also contains the AB, CD, and EF loops. However, in addition to these three loops, the FG loop takes part in building the Ca^{2+}-binding cluster at CBD2 (Besserer et al. 2007; Chaptal et al. 2009; Wu et al. 2009a, b, 2011; Breukels and Vuister 2010). The additional structural contribution of the FG loop in Ca^{2+} coordination is crucial for determining the number of Ca^{2+} binding sites at CBD2, which in turn, is essential for diversifying the tissue-specific regulatory specificities of the NCX1 and NCX3 variants (Hilge et al. 2006, 2009; Besserer et al. 2007; Ottolia et al. 2009, 2010).

8.6.2 Structure–Functional Assignments of Ca^{2+}Binding Sites at CBDs

CBD1 of eukaryotic NCX orthologs contains four Ca^{2+} binding sites (Ca1-Ca4), whereas the number of Ca^{2+} binding sites at CBD2 varies from zero to three (CaI-CaIII) owing to an alternating splicing segment exclusively located on CBD2 (Nicoll et al. 2006; Hilge et al. 2009; Wu et al. 2011). The cluster of four Ca^{2+} binding sites in CBD1 of mammalian NCX1 is assembled in a parallelogram-like configuration, where the short distance (3.9–4.4 Å) between the Ca^{2+} binding sites allows the binding of four divalent cations within a very dense vicinity (Nicoll et al. 2006; Giladi et al. 2012a). This fascinating structural arrangement is due to polydentate coordination of Ca^{2+} ions by D500 and E451 residues (Figs. 8.2 and 8.4), which can ligate at once two and three Ca^{2+} ions, respectively. This closely located Ca^{2+} sites provide a structural basis for the cooperative binding of Ca^{2+} ions to CBD1, which is essential for an effective activation of NCX under physiologically relevant conditions when cytosolic [Ca^{2+}] undergoes relatively small swings (Boyman et al. 2011; Ginsburg et al. 2013).

Notably, the C3 and C4 sites of CBD1 have a high affinity ($K_d < 1$ μM) for Ca^{2+} binding, thereby representing a high-affinity "primary" allosteric sensor, which can mediate the Ca^{2+}-dependent activation of NCX1 at resting and exciting concentrations of cytosolic Ca^{2+} (Boyman et al. 2009; Giladi et al. 2010). In light of sequence similarities, the structural organization of CBD1 might be very similar (if not identical) in NCX2 and NCX3, although the CBD1 X-ray structures of these isoforms are currently unavailable. Although the Ca^{2+} affinity at the high-affinity sensor of CBD1 varies "only" 5–7-fold in the cardiac (ACDEF), brain (AD), and kidney (BD) splice variants of NCX1, there are up to 50–200-fold differences among variants in the Ca^{2+} dissociation rates (Giladi et al. 2010, 2012a, c, 2013). Accumulating evidence reveals that the slow dissociation kinetics of "occluded" Ca^{2+} (Boyman et. 2009; Giladi et al., 2010, 2012a, b, c) may represent I_2 inactivation kinetics (Dyck et al. 1999; Dunn et al. 2002; Ottolia et al. 2009, 2010), thus underscoring the physiological relevance of the underlying mechanisms (Giladi et al. 2016b; Tal et al. 2016; Khananshvili 2017a, b).

In the two-domain context (CBD12), only the high-affinity Ca3-Ca4 sites of CBD1 are involved in the interdomain tethering of CBDs and the slow dissociation of "occluded" Ca^{2+}, whereas the remaining two sites (Ca1 and Ca2) of CBD1 have at least 10–50 times lower affinity for Ca^{2+}, compared with the Ca3-Ca4 sites (Boyman et al. 2009; Giladi et al. 2010, 2013). Mutational studies in conjunction with patch-clamp measurements have shown that the Ca3-Ca4 sites are essential for Ca^{2+}-dependent activation of NCX, whereas the "low-affinity" Ca1 and Ca2 sites of CBD1 are not indispensable for mediating the Ca^{2+}-dependent activation (Nicoll et al. 2006, Ottolia et al. 2009; Giladi et al. 2013). Even though the low-affinity sites of CBD1 (Ca1 and Ca2) and of CBD2 (CaII) are not mandatory for regulatory activity, their occupation by Mg^{2+} can affect the affinity of Ca^{2+} sensing either at

CBD1 or CBD2 and thus, may play a physiological role as well (Boyman et al. 2009, Breukels et al. 2011; Giladi et al. 2013).

In contrast with CBD1, the Ca^{2+} binding sites of CBD are ~5.5 Å apart, where K585 (a homolog to E454 in CBD1) forms a salt-bridge with D552 and E648 in the absence of Ca^{2+} and thus, due to this reason, apo-CBD2 is structurally less flexible (Besserer et al. 2007; Chaptal et al. 2009; Giladi et al. 2012a). Thus, in the absence of bound Ca^{2+}, apo-CBD2 is structurally more stable than apo-CBD1. Notably, in CBD2, only a single residue (D578) bridges the CaI and CaII sites via a bidentate ligation (Figs. 8.3 and 8.4). The exon A-containing CBD2 variants of NCX1 (e.g., in the case of cardiac and brain splice variants) consist of two Ca^{2+}-binding sites, where the CaI ($K_d = 2–10$ μM) and CaII ($K_d > 20$ μM) sites exhibit moderate and low affinities for Ca^{2+} binding, respectively [Boyman et al. 2009; Giladi et al. 2010, 2012a). In sharp contrast with the exon-A effect, the exon B-containing CBD2 of NCX1 (e.g., the kidney splice variant) does not bind Ca^{2+} (Tal et al. 2016). Notably, CBD2 of NCX2 (which lacks the alternative splicing capacity) contains only one low-affinity site, which presumably has no regulatory relevance conditions (Tal et al. 2016). The exon A-containing CBD2 variants of NCX3 (e.g., the AC splice-variant expressed in skeletal muscle), does not bind Ca^{2+}, whereas the exon B-containing CBD2 of NCX3 (e.g., B or BC splice variants expressed in the brain) binds three Ca^{2+} ions with quite a low affinity (Hilge et al. 2009; Tal et al. 2016). Thus, the A and B exons play opposite roles in determining the number of Ca^{2+} sites of NCX1 and NCX3 at CBD2 (Khananshvili 2016, 2017a, b).

8.6.3 Exon-Related Modification of Structure-Dynamic Features

As discussed above, mammalian NCX variants exhibit two major modules of Ca^{2+}-dependent allosteric regulation: the first module refers to Ca^{2+}-dependent activation through CBD1 and the second module infers to Ca^{2+}-dependent alleviation of Na^+-dependent inactivation through CBD2 (Hilgemann et al. 1992a, b; Hilgemann 2007; Hryshko 2008; Khananshvili 2013, 2014). Recent structure-dynamic and functional studies have demonstrated that in NCX1, exon A (located at CBD2) moderately reduces the Ca^{2+} binding affinity at CBD1, while stabilizing the Ca^{2+}-binding sites at CBD2 (Boyman et al. 2009; Giladi et al. 2010, 2012a, b, c). In contrast with exon A, exon B increases the Ca^{2+} binding affinity of NCX1 at CBD1, while destabilizing the folding of Ca^{2+} binding sites at CBD2 (Hilge et al. 2009; Tal et al. 2016) and thereby prevents Ca^{2+} binding at CBD2 (Fig. 8.1).

Mutational studies, in conjunction with patch-clamp techniques, revealed that in mammalian NCX variants, the Ca^{2+} binding to the Ca3-Ca4 sites of CBD1 activates the NCX-mediated ion currents, whereas the Ca^{2+} binding to the CaI site of CBD2 alleviates the Na^+-dependent inactivation (Nicoll et al. 2006; Besserer et al. 2007; Ottolia et al. 2009; Chaptal et al. 2009). The kidney splice variant (NCX1-BD) does

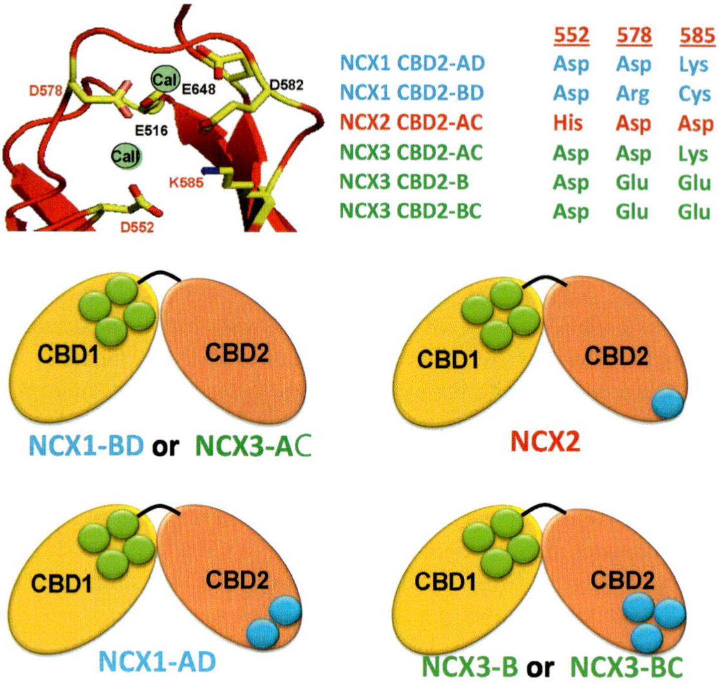

Fig. 8.3 Three critical positions within the splicing segment predefine the number of Ca^{2+} binding sites at CBD2. Exon A and B control the number of Ca^{2+} binding sites at CBD2 by editing the structure of E and F strands. The exon B-containing variants of NCX1 contain the arginine (instead of aspartate or glutamate) and cysteine (instead of lysine) at positions 578 and 585, respectively, which prevents Ca^{2+} binding to CBD2 by disabling the CaI site. The CBD2-AD variant of NCX1 retains its structural integrity even in the absence of Ca^{2+}, since K585 forms salt-bridges with two negatively charged residues involved at the CaI site of CBD2. In CBD2 of NCX2 (which does not undergo splicing), the replacement of D552 by histidine aborts a one Ca^{2+}-coordinating center, which aborts the Ca^{2+} binding capacity at the CaII site and reduces the Ca^{2+} affinity at the CaI site. In CBD2-B or CBD2-BC variant of NCX3, the replacement of 585 by glutamate generates three Ca^{2+} binding sites at CBD2, which is essential for Ca^{2+} induced alleviation of Na^+-dependent inhibition in NCX3

not bind Ca^{2+} at CBD2 (since the CaI site is structurally destabilized in CBD2); thus, it lacks Ca^{2+}-induced alleviation of Na^+-dependent inactivation (Ottolia et al. 2009; Chaptal et al. 2009; Boyman et al. 2009; Gilardi et al. 2010, 2012a, c; Tal et al. 2016) (Figs. 8.2 and 8.4).

Functional assignments of the low-affinity Ca^{2+} binding sites of CBD1 (Ca1 and Ca2) and CBD2 (CaII) are less clear for any isoform/splice variant. Nevertheless, accumulating evidence suggests that the Ca1 and Ca2 sites of CBD1 may serve as the Mg^{2+} rather than the Ca^{2+} sites, where a constitutive occupation of these sites by Mg^{2+} may modulate the affinity of the Ca3-Ca4 sites at CBD1 (Nicoll et al. 2006; Boyman et al. 2009, 2011; Giladi et al. 2013). In NCX1, the occupation of the Ca1-Ca2 sites of CBD1 by Mg^{2+} decreases the affinity at the nearby Ca3-Ca4 sites of

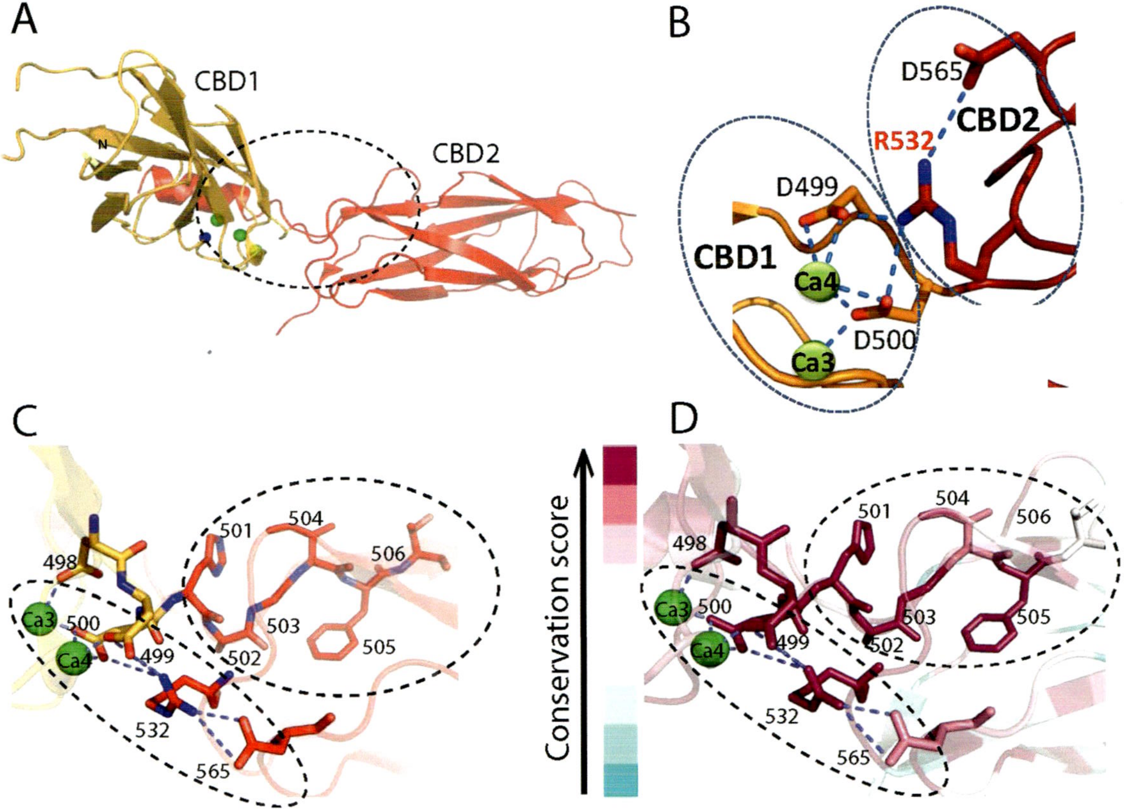

Fig. 8.4 Structural hallmarks at the interface of the CBD1 and CBD2 domains. (**a**) The X-ray-based structure of the CBD12-AD tandem (3US9). CBD1 and CBD2 domains are in orange and red, respectively. Green and blue spheres depict Ca^{2+} ions and water molecules, respectively. (**b**) Interdomain region contains a pivotal network of salt-bridges centered at R532 in CBD2. R532 takes a conformation that forms bifurcated hydrogen-bonded and non-hydrogen-bonded salt-

CBD1, whereas the occupation of the CaII site by Mg^{2+} increases the Ca^{2+} affinity at the CaI site of CBD2 (Boyman et al. 2009, Breukels et al. 2011; Giladi and Khananshvili 2013).

Owing to structural similarities among distinct NCX variants, the regulatory roles of the CBD1 Ca^{2+}-binding sites might be similar in the NCX1, NCX2, and NCX3 isoforms, although this statement requires further experimental confirmation. In any case, the presently available experimental data are consistent with the notion that the presence of exon A or B differentially governs the Ca^{2+}-binding stoichiometry/affinity at both CBD1 and CBD2. Most importantly, this seems to be essential for shaping and segregating distinct regulatory features in a given isoform/splice variant (Hilge et al. 2009; Ottolia et al. 2009; Giladi et al. 2010, 2012a, b; Tal et al. 2016; Khananshvili 2016). For instance, the Ca^{2+} binding to CBD1 of NCX1 activates the brain (NCX1-AD), cardiac (NCX1-ACDEF), or kidney (NCX1-BD) splice variants, although the Ca^{2+}-induced alleviation of Na^{+}-dependent inactivation (governed by CBD2) resides in the cardiac and brain variants. However, in the kidney variant (NCX1-BD) the Na^{+}-dependent inactivation cannot be alleviated by Ca^{2+} (Dyck et al. 1999; Ottolia et al. 2009; Hnatowich et al. 2012), since exon B prevents Ca^{2+} binding to CBD2 in the kidney variant (Hilge et al. 2009; Boyman et al. 2009; Giladi et al. 2010, 2012b).

These exon A- and B-dependent regulatory differences between NCX1 variants are aligned with the E and F strands of CBD2, contributing to Ca^{2+} binding at CBD2 (Hilge et al. 2006, 2009). The mutually exclusive exons (A or B) control the replacement of residues at three critical positions (e.g., Asp552, Asp578, and K585 in NCX1-AD), which determine the Ca^{2+} coordination mode (Fig. 8.3). For example, the kidney NCX1-BD variant contains Arg578 (instead of aspartate) and Cys585 (instead of lysine), which destabilizes CBD2 folding and precludes Ca^{2+} binding to CBD2 (Hilge et al. 2009; Giladi et al. 2010, 2012a, b).

The exon A-containing NCX1 variants retain their structural integrity even in the absence of Ca^{2+}, since K585 forms salt-bridges with two negatively charged residues located at the CaI site of CBD2 (Besserer et al. 2007; Chaptal et al. 2009). In CBD2 of NCX2 (which does not undergo splicing), the replacement of D552 by histidine aborts one of the two Ca^{2+}-coordinating atoms, which subsequently aborts the CaII site, while reducing the Ca^{2+} affinity of the CaI site (Hilge et al. 2009; Tal et al. 2016). In exon B-containing NCX3 variants, the replacement of K585 by glutamate

---

Fig. 8.4 (continued) bridges with D499, D500 in CBD1 and D565 in CBD2. D499 and D500 also participate in the coordination of the Ca3-Ca4 sites and thus, stabilize the two-domain interface. (**c**) The hydrophobic interfacial region contains residues from the Ca^{2+}-binding EF loop of CBD1, the linker and the FG loop of CBD2. F450 forms van der Waals interactions with H501, I628, A629, M631, and G632 and thereby represents a core residue for structure-dynamic coupling of CBDs. The dashed ellipses show the linker and Ca^{2+} dependent tethering of the CBD1 and CBD2 domains. (**d**) The interdomain residues are colored according to their conservation score (as calculated by Consurf software) by comparing multiple sequence alignments of CBD12 orthologues. These data underscore the evolutionary conservation of tightly packed structural elements involved in the building of the two-domain interface results in ligation of three Ca^{2+} ions at CBD2, since three negatively charges (Asp552, Glu578, and Glu585) predefine the Ca^{2+} binding at CBD2 (Hilge et al. 2009; Giladi et al. 2012b) (Fig. 8.3).

In sharp contrast to mammalian NCX variants, a *Drosophila* NCX ortholog has completely diverse regulatory responses to cytosolic Ca^{2+} (Hryshko et al. 1996; Omelchenko et al. 1998; Wu et al. 2009a, b, 2011; Chaptal et al. 2009; Ottolia et al. 2009). More specifically, in the *Drosophila* CALX1.1 splice variant, the Ca^{2+} binding to the Ca3-Ca4 sites of CBD1 inactivates the ion-transport activities, whereas in the second splice variant, CALX1.2, Ca^{2+} has no regulatory effect on transport activity (Hryshko et al. 1996; Omelchenko et al. 1998). Moreover, in both splice variants of CALX1, CBD2 does not bind Ca^{2+} and consequently, CALX lacks the Ca^{2+}-induced alleviation of Na^+-dependent inactivation (Wu et al. 2009a, b, 2010, 2011). Similar to mammalian NCX, CALX1 also undergoes alternative splicing only at CBD2, although the splicing segments of CALX1 are much shorter and differ by only five residues (Wu et al. 2009a, b, 2011). Structural studies revealed that these five residues in CALX1-CBD2 are located within an FG loop between the H1 α-helix and the β-strand, which is comparable with the cassette exons' (C, D, E, and F) positions in NCX (Wu et al. 2009a, b, 2011; Besserer et al. 2007; Giladi et al. 2012a; Khananshvili 2016, 2017a) (Figs. 8.3 and 8.4).

8.6.4 Synergistic Interactions Between the CBD1 and CBD2 Domains

Structure–functional and mutational studies have identified synergistic interactions between CBD1 and CBD2, which are capable of shaping not only the Ca^{2+}-sensing features at both CBDs but also the regulatory responses of NCX variants to cytosolic Ca^{2+} (Ottolia et al. 2009, 2010; Boyman et al. 2009; Giladi et al. 2010, 2013). Even though the structures of CBD1 in mammalian NCX and CALX (*Drosophila*) variants are nearly identical, the occupation of Ca3-Ca4 sites by Ca^{2+} activates (mammalian NCXs), inhibits (CALX1.1), or has no response (CALX1.2) (Hryshko et al. 1996; Dyck et al. 1998; Omelchenko et al. 1998; Nicoll et al. 2006; Ottolia et al. 2009; Wu et al. 2009a, b, 2010, 2011). Thus, the open question is: How can the occupation of the Ca3-Ca4 sites of CBD1 result in such diverse regulatory responses to Ca^{2+} in NCX and CALX and how could this be related to the synergistic interactions between the CBD1 and CBD2 domains. This question is especially intriguing in light of structural data indicating that the folding of CBD1, as well as the coordination chemistry of four Ca^{2+} binding sites at CBD1, is highly conserved among the CALX and NCX orthologs (Nicoll et al. 2006; Chaptal et al. 2007; Wu et al. 2009a, b, 2011; Giladi et al. 2012a).

A step toward better understanding the synergistic interactions between the CBD domains has been associated with applying stopped-flow techniques with the goal of monitoring the Ca^{2+} off-rates from the Ca3-Ca4 sites in the isolated preparations of

CBD1 and in the two-domain (CBD12) tandem (Boyman et al. 2009; Giladi et al. 2010, 2012a, b, c; Lee et al. 2016). Although the Ca^{2+} binding affinities (K_d) are comparable in the isolated preparations of CBD1 and in the two-domain tandem (CBD12), the rate constants (k_{off}) for Ca^{2+} dissociation from the Ca3-C4 sites are 30–50 times slower in CBD12 than in CBD1 (Boyman et al. 2009; Giladi et al. 2010, 2012a; Tal et al. 2016). These data clearly demonstrate that the linker-dependent interactions between the CBD1 and CBD2 domains generates the slow dissociation of occluded Ca^{2+} from the high-affinity allosteric sensor.

The slow dissociation of Ca^{2+} from the Ca3-Ca4 sites of CBD12 (considered as dissociation of "occluded" Ca^{2+}) has been detected for all tested preparations obtained from the NCX1, NCX2, NCX3, and CALX variants. Moreover, the dissociation kinetics of occluded Ca^{2+} characteristically varied from 0.05 s^{-1} to 12 s^{-1} among NCX and CALX variants (Boyman et al. 2009; Giladi et al. 2010, 2012a, b, c; Tal et al. 2016) (Fig. 8.5c). For example, the kinetics of occluded Ca^{2+} dissociation from the brain (CBD12-AD) and kidney (CBD12-BD) splice variants of NCX1 ($k_{off} = 0.5$ s^{-1}) is 5–10 times faster than the off-rates of occluded Ca^{2+} from the cardiac (CBD12-ACDEF) splice variant of NCX1 (Boyman et al. 2009; Giladi et al. 2010, 2012a, b, c). This might have a physiological relevance since the cardiac NCX (NCX1-ACDEF) must extrude the elevated levels of transient Ca^{2+} during much prolonged action potentials (that last 200–300 milliseconds) as compared with the neuronal (NCX1-AD, NCX2, NCX3-B, and NCX3-BC) and skeletal muscle (NCX3-AC) variants that operate at much shorter action potentials.

Similar to the NCX1 splice variants, the slow dissociation of occluded Ca^{2+} characteristically also varies among NCX2 and NCX3 variants (Tal et al. 2016; Giladi et al. 2017). For example, the NCX2-CBD12 and NCX3-CBD12 (B, BC or AC) variants share comparable affinities for Ca^{2+} at Ca3-C4 sites of CBD1, although the occluded Ca^{2+} dissociates ~tenfold slower in skeletal muscle (B or BC) than in the brain (AC) splice variants of NCX3 or in NCX2 (Tal et al. 2016; Giladi et al. 2017). Slower dissociation of occluded Ca^{2+} from the skeletal variant matches the physiological needs, since in myocytes NCX must clear up incomparably much more cytosolic Ca^{2+} than in neurons (Khananshvili 2013, 2014). Thus, accumulating evidence strongly supports the notion that the dissociation kinetics of occluded Ca^{2+} from CBD12 represents I_2 inactivation in full-length NCX variants, which is controlled by a splicing segment on the FG loop at CBD2 (Giladi et al. 2010, 2012a, b, c, 2017; Tal et al. 2016).

8.7 Conformational Dynamics of CBDs Are Characteristic Among NCX Variants

8.7.1 An Interdomain Linker Governs the Dynamic Coupling of CBDs

Mutational studies in conjunction with stopped-flow and patch-clamp approaches have demonstrated that a short interdomain linker (501-HAGIFT-506) between the

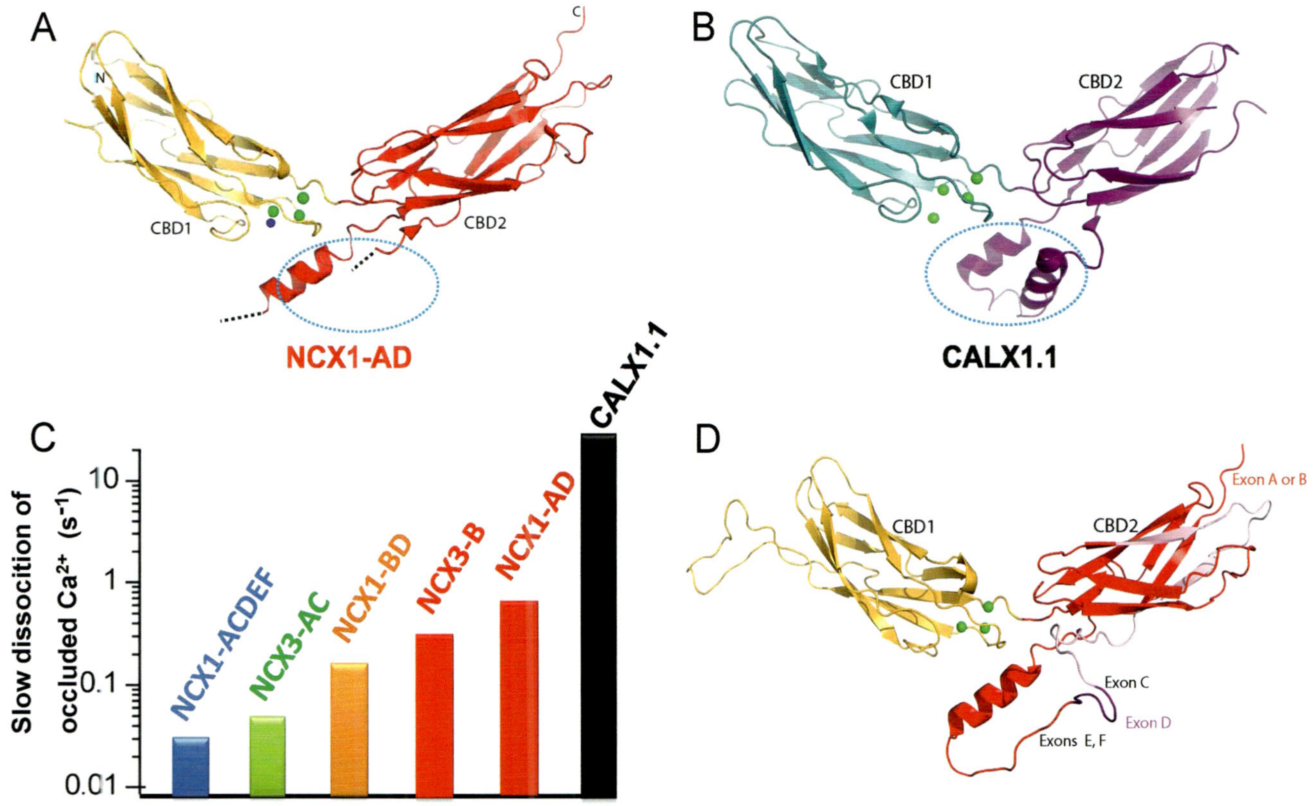

Fig. 8.5 Structural comparison of mammalian (NCX1-AD) and *Drosophila* (CALX1.1) CBD12 domains. The X-ray structures of NCX1-CBD12-AD (3US9) (**a**) and of CALX1.1-CBD12 (3RB5) (**b**) are presented to compare the α-helix folding at the two-domain interface. In contrast with NCX1-CBD2-AD, CALX1.1-CBD2 forms two-headed short helices (H1 and H2) nearly perpendicular to the plane of β-sheets. The fact that the α-helix regions of NCX1-AD and

CBD1 and CBD2 domains plays a critical role in setting up the synergistic interactions and functional coupling between the CBDs (Ottolia et al. 2009; Giladi et al. 2010, 2012a, b, c; Giladi and Khananshvili 2013). Notably, the interdomain CBD1-CBD2 linker is highly conserved among all known isoform/splice variants of NCX and CALX, since it is essential for providing a structural basis for Ca^{2+}-dependent tethering/rigidification of the CBD1 and CBD2 domains (Giladi et al. 2012a, c). Deletion or elongation of the CBD1-CBD2 linker accelerates (up to ~50-fold) the dissociation of occluded Ca^{2+}, while decreasing the Ca^{2+} binding affinity (up to ~tenfold) at the Ca3-Ca4 sites (Ottolia et al. 2009; Giladi et al. 2010, 2012a, b, c). Thus, the slow dissociation of occluded Ca^{2+} from the Ca3-Ca4 sites of CBD12 in NCX and CALX is controlled by the CBD1-CBD2 linker (Giladi et al. 2010, 2012a, b, c; Giladi and Khananshvili 2013; Wu et al. 2011).

Mutational studies in conjunction with stopped-flow and SAXS techniques demonstrated that G503 is the only residue in the linker, the mutation of which aborts the slow dissociation of occluded Ca^{2+} and CBD movements (Giladi et al. 2012c). Moreover, the crystal structures of NCX1-CBD12 (Giladi et al. 2012a) and CALX-CBD12 (Wu et al. 2011) indicate that the dihedral φ/ψ angles at position 503 are only allowed for the glycine residue, meaning that any other residue at this position in the linker would result in a steric clash of protein folding. Furthermore, patch-clamp experiments with full-size NCX have shown that mutations of either G503 in mammalian NCX1 or analogous G555 in CALX1.1 abort the Ca^{2+}-dependent regulation of NCX-mediated ion currents (Hryshko et al. 1996; Dyck et al. 1998; Omelchenko et al. 1998; Dunn et al. 2002). Thus, the interdomain linker encodes crucial structural information for Ca^{2+}-dependent regulation in NCX and CALX variants, which is essential for interdomain coupling of CBDs and for propagation of an allosteric message after decoding a regulatory signal upon Ca^{2+} binding at respective sites (Giladi et al. 2010, 2012b, c, 2013; Wu et al. 2009a, b, 2010, 2011). This statement is in an agreement with other observations obtained by

Fig. 8.5 (continued) CALX1.1 are very close to the interdomain linker and the Ca3-Ca4 sites of CBD1 is consistent with the notion that the α-helix region can differentially affect the dynamic features of Ca^{2+}-dependent tethering of CBDs. The relevant structural interactions might play a critical role in differential responses of mammalian NCXs and CALX1.1 to regulatory Ca^{2+} (namely, Ca^{2+} binding to the Ca3-Ca4 sites results in activation of NCX1-AD and in inhibition of CALX1.1). This interpretation is consistent with HDX-MS findings revealing that the α-helix region of NCX and CALX differentially affect the strength and span of backbone rigidification from the C-terminal of CBD1 toward the C-terminal of CBD2 upon Ca^{2+} binding to the Ca3-Ca4 sites (for additional information see the text). (**c**) Dissociation kinetics of occluded Ca^{2+} from CBD12 reveals up to 10–50-fold differences in the Ca^{2+} off-rates among different isoform/splice variants, thereby underscoring structure-related differences in the Ca^{2+} dependent inactivation kinetics. These data, obtained by stopped-flow techniques are in good correlation with I_2-inactivation kinetics, monitored by patch-clamp in oocytes expressing full-size NCX variants. (**d**) Structural assignment of six exons in the two-domain CBD12 tandem underscores a close vicinity of the splicing segment to the CBDs interface. The relevant structural arrangements "secondarily" modify Ca^{2+} binding/dissociation features at both CBDs

FRET (John et al. 2011), NMR (Salinas et al. 2011; Abiko et al. 2016; Yuan et al. 2018), SAXS (Giladi et al. 2012b, c, 2013), and HDX-MS (Lee et al. 2016; Giladi et al. 2015, 2016a, b, 2017).

8.7.2 Structural Bases for Positive, Negative, or no Response to Regulatory Ca^{2+}

The discovery of the two-domain tandem (CBD12) crystal structures (Wu et al. 2011; Giladi et al. 2012a) set up a platform for elucidating the structure-dynamic determinants for positive, negative, and no response to regulatory Ca^{2+} in NCX and CALX variants. The X-ray structures of the brain NCX1-CBD12-AD (Giladi et al. 2012a), CALX1.1-CBD12, and CALX1.2-CBD12 variants (Wu et al. 2009a, b, 2010, 2011) have enabled us to identify a relatively small contact area (~360 Å^2) between the CBDs with an interdomain angle of ~117°. These structural similarities between the NCX and CALX orthologs are intriguing in light of the diverse responses of NCX and CALX to regulatory Ca^{2+}. More specifically, Ca^{2+} binding to the Ca3-Ca4 sites of CBD1 activates NCX1-AD (and the other NCX variants), whereas in CALX the Ca^{2+} binding to the Ca3-Ca4 sites either inhibits (CALX1.1) or has no effect (CALX1.2) on CALX-mediated ion currents (Hryshko et al. 1996; Omelchenko et al. 1998; Ottolia et al. 2009; Wu et al. 2010). Notably, the isolated preparations of CBD1, CBD2, and CBD12 exhibit nearly identical coordination chemistry for Ca^{2+} ligation in the NCX and CALX variants with a few (but critical) exceptions. Namely, E385 coordinates only Ca3 in isolated CBD1, whereas this residue coordinates Ca2, Ca3, and Ca4 in the CBD12 of NCX1-AD, CALX1.1, or CALX1.2. Most importantly, D499 forms bidentate coordination with Ca4 in CBD12 (in contrast with monodentate coordination in isolated CBD1) and thereby "locks" Ca^{2+}-dependent tethering of CBDs (Giladi et al. 2012a, 2013).

According to the crystal structures of NCX1-CBD12 and CALX-CBD12, more than 20 residues are buried into the two-domain interface between the CBDs; however, three regions can be distinguished: the hydrophilic, hydrophobic, and loop/α-helix arrays (Wu et al. 2011; Giladi et al. 2012a). The hydrophilic region embraces a focal interdomain electrostatic network centered at R532 in CBD2, where R532 forms a bifurcated network of salt-bridges with D499 and D500 in CBD1 and D565 in CBD2 upon occupation of the Ca3-Ca4 sites by Ca^{2+}). Notably, this Ca^{2+}-mediated tethering structure (with D499 and D500 residues) contributes to the coordination of two Ca^{2+} at the Ca3-Ca4 sites, while concomitantly stabilizing the CBD interface (Giladi et al. 2012a). This interfacial region is highly conserved among NCX and CALX variants, thereby underscoring its functional relevance. Thus, this interdomain network of salt-bridges acts as the principal linchpin that holds the two CBDs together, which results in Ca^{2+} occlusion. This Ca^{2+}-governed interdomain tethering seems to be a general structural module for governing the interdomain coupling, although the rates of occluded Ca^{2+} dissociation are diversified by the splicing segment at CBD2 while showing a remarkable correlation with the I_2 inactivation kinetics in matching variants (Giladi et al. 2012a; Khananshvili 2016, 2017a, b).

The hydrophobic interfacial (interdomain) region contains residues from the Ca^{2+}-binding EF loop of CBD1, the linker, and the FG loop of CBD2, where F450 serves as a core residue, forming van der Waals interactions with H501, I628, A629, M631, and G632 (Giladi et al. 2012a, b). Most probably, this region limits linker flexibility via Ca^{2+}-dependent interaction of F450 with H501 (Giladi et al. 2012a, b, c, 2017). The hydrophobic interfacial region, formed between the CD and EF loops of CBD1 and the FG loop (consisting of different sets of exons) and the α-helix of CBD2 (next to the Ca3-Ca4 sites of CBD1), are engaged in important interactions, although the involved residues are quite inaccessible to the bulk phase. In NCX1-CBD12-AD, most of the FG loop of CBD2 is unstructured except for a short α-helix region (620–629) in the C-terminal portion of the FG loop and therefore, the side chains of canonical α-helix directly contribute to the CBD interface (Giladi et al. 2012a). In contrast, the FG loop of CALX1-CBD12 forms two-headed short helices (H1 and H2) nearly perpendicular to the plane of β-sheets (Wu et al. 2009a, b, 2010, 2011), which is strikingly different from the helix structure of NCX1-CBD12 (Giladi et al. 2012a) (Fig. 8.5). Because the α-helix regions of NCX and CALX are very close to the CBD1-CBD2 interdomain linker and the Ca3-Ca4 sites of CBD1, the diverse structural features of the α-helix region in NCX and CALX can differentially control the backbone rigidity of CBD2 (Salinas et al. 2011; Giladi et al. 2013, 2017). This may represent a structural basis for the differential responses of the NCX and CALX variants to regulatory Ca^{2+} (Fig. 8.5). In support of this, SAXS and HDX-MS data suggest that the propagation of Ca^{2+}-dependent rigidification within CBD2 may account for the regulatory differences between the NCX and CALX variants, while showing positive, negative, or no response to regulatory Ca^{2+} (Giladi et al. 2012a, b, c, 2013, 2017; Lee et al. 2016; Khananshvili 2016, 2017a, b).

Interestingly, the 585 position of CBD2 resides the valine residue (V636) in CALX, whereas, in mammalian NCXs, the 585 position inhabits lysine (NCX1-AD, NCX3-AC), cysteine (NCX1-BD), aspartate (NCX2-AC), and glutamate (NCX3-B, NCX3-BD) (Fig. 8.3). These "small" structural differences can predefine not only the number of Ca^{2+}-binding sites at CBD2 (as discussed above), but also the conformational stability of CBD2, which is relevant for alleviating the Na^+-dependent inactivation (Giladi et al. 2010, 2012a, b, c, 2015, 2016a, b; Lee et al. 2016; Tal et al. 2016). For example, the Ca^{2+} binding is practically absent (or is extremely reduced) in a number of CBD2 variants (NCX1-BD, NCX2-AC, NCX3-AC, CALX1.1, and CALX1.2), although the conformational profiles of backbone dynamics (as evaluated by HDX-MS) are strikingly different among NCX and NCLX variants (Lee et al. 2016; Giladi et al. 2015, 2016a, b, 2017; Tal et al. 2016). This may have physiological relevance. For example, in NCX2, the presence of histidine in position 552 (Table 8.1) may constitutively stabilize CBD2 conformation, which may own to a "constitutive resistance" in respect with Na^+-dependent inactivation (this may explain why the Na^+-dependent inhibition cannot be seen in patch-clamp experiments). In contrast with "Na^+-resistant" NCX2, the 585 substitution by cysteine (in the kidney NCX1-BD variant) or by valine (in the CALX variants) constitutively destabilize CBD2, thereby producing the "Na^+-sensitive" variants, which neither can resist to Na^+-dependent inactivation nor can alleviate the Na^+-dependent inactivation due to the absence of Ca^{2+} binding to CBD2.

8.7.3 *Ca^{2+}-Driven Tethering of CBDs Rigidifies the CBDs Movements*

Structure-dynamic data obtained by NMR (Johnson et al. 2008; Salinas et al. 2011; Abiko et al. 2016; Yuan et al. 2018), SAXS (Giladi et al. 2012a, c, 2013) and HDX-MS (Giladi et al. 2015, 2017) studies reveal a general model for Ca^{2+}-dependent activation of mammalian NCXs (Figs. 8.2 and 8.4). According to this model, the Ca^{2+}-dependent activation of NCX involves the Ca^{2+}-mediated tethering of CBDs (coupled with Ca^{2+} occlusion and the conformational constraint of CBD movements), whereas a slow dissociation of occluded Ca^{2+} leads to NCX inactivation (Fig. 8.5). This model is in a good agreement with stopped-flow, patch-clamp, and mutational studies, suggesting that slow dissociation of occluded Ca^{2+} (which varies over 50–200 times among NCX and CALX variants) represents dynamic coupling between CBDs (Nicoll et al. 2006; Besserer et al. 2007; Giladi et al. 2010, 2012b; Tal et al. 2016).

Structure-based mutational analyses of isolated CBD12 preparations (obtained from distinct NCX and CALX variants) provided further evidence for sharing structural elements that determine the slow dissociation of occluded Ca^{2+} with associated detachment of tethered CBDs. For example, mutation of R532 (in the BC loop of CBD2), which contributes to the interdomain network (by interacting with D499 and D500 at Ca3-Ca4 on CBD1) aborts the slow dissociation of occluded Ca^{2+} (Giladi et al. 2012a, b, c, 2013; Tal et al. 2016). Thus, R532 controls Ca^{2+} occlusion by stabilizing the interdomain tethering of CBDs (through the salt-bridge network), even though R532 is not directly involved in Ca^{2+} coordination (Fig. 8.4). Mutation of F540 also abolishes the slow dissociation of occluded Ca^{2+}, thereby revealing the functional role of the interdomain hydrophobic core in the regulatory coupling of CBDs (Salinas et al. 2011; Giladi et al. 2012a, 2015, 2017). According to these findings, the interdomain CBD1-CBD2 linker possesses the required flexibility (controlled by G503) to assist in Ca^{2+}-mediated tethering of CBDs, whereas the relay of CBDs (upon Ca^{2+} binding) stabilizes stochastic oscillations of the linker that restricts interdomain movements (Salinas et al. 2011; Giladi et al. 2012b, 2013, 2015; Abiko et al. 2016; Yuan et al. 2018). Most probably, the fast dissociation of the first Ca^{2+} ion from the Ca3 site enables occlusion of the second (remaining) Ca^{2+} ion at the Ca3-Ca4 sites, whereas the dissociation of the second (entrapped) Ca^{2+} ion is 20–50 times slower than the first Ca^{2+} ion (Giladi et al. 2010, 2012a, b, 2016b; Giladi and Khananshvili 2013). Thus, conserved interdomain linker and salt-bridge network determine the dynamic coupling between CBDs, while a nearby splice segment (located on the FG loop of CBD2) secondarily modifies the dynamic features (Giladi et al. 2016b; Khananshvili 2017a) (Fig. 8.4).

Crystal structures of isolated CALX1.1-CBD12 and CALX1.2-CBD12 revealed small differences in the interdomain angle (~8°) between the CBDs, which were suggested as the structure-governed basis for the differential regulatory responses to regulatory Ca^{2+} exhibited by full-size CALX1.1 and CALX1.2 (Wu et al. 2011). According to this proposal, the slight differences in the interdomain angle of CBD12 results in Ca^{2+}-dependent inhibition of CALX1.1 and no response to Ca^{2+} in CALX1.2. In disagreement with this proposal, however, the interdomain angle of Ca^{2+}-bound CBD12 is nearly identical for NCX1-CBD12-AD (117.4°) and CALX1.1-CBD12 (117.7°), meaning that the CBD alignment cannot account for the regulatory diversity in the NCX and CALX orthologs (Giladi et al. 2012a). These data reveal that the dynamic mechanisms of Ca^{2+}-dependent conformational changes are responsible for the positive, negative, and no response of NCX and CALX variants to regulatory Ca^{2+}.

8.7.4 *The "Population Shift" Mechanism Underlies the Dynamic Coupling of CBDs*

In search for the molecular mechanisms underlying the dynamic coupling of CBDs, advanced biophysical approaches were applied, including the NMR, SAXS, and HDX-MS techniques (for a review, see Khananshvili 2016, 2017a). Systematic investigations by using the SAXS and HDX-MS approaches have shown that the Ca^{2+}-bound conformational distributions of CALX1.1-CBD12 and CALX1.2-CBD12 are nearly identical to those of NCX1-CBD12, NCX2-CBD12, and NCX3-CBD12 (Giladi et al. 2012a, b, c, 2013, 2015, 2017; Lee et al. 2016). These experiments have shown that the occupation of the Ca3-Ca4 sites by Ca^{2+} ions shifts the distribution of conformational states toward more constraint and more populated conformational states. Moreover, Ca^{2+} rigidifies the dynamic movements of CBDs without making any significant changes in the interdomain distance or CBD alignment (Giladi et al. 2013). These findings support the "population shift mechanism," according to which Ca^{2+} shifts numerous conformational transitions over relatively small (incremental) energetic barriers in order to avoid large conformational changes. This mechanism may have physiological significance, since the induced fit (and an alternative mechanism to population shift) can take place under one of two scenarios: a high concentration of ligand or a high affinity of protein to ligand (Okazaki and Takada 2008; Boehr et al. 2009; Ma and Nussinov 2010; Ma et al. 2011). Neither of these conditions seems to be implemented for Ca^{2+}-dependent allosteric regulation of NCX.

Consistent with the population shift mechanism, NMR analysis of the NCX1 and CALX variants of CBD12 revealed that Ca^{2+} binding to the Ca3-Ca4 sites restricts the linkers' flexibility and CBD motions (Salinas et al. 2011; Abiko et al. 2016).

Thus, Ca^{2+} occlusion at the Ca3-Ca4 sites of the NCX and CALX variants rigidifies the backbone dynamics through the Ca^{2+}-dependent tethering of CBDs. The strength, location, and extent of Ca^{2+}-dependent rigidification differ in an exon-dependent manner, as has been shown by HDX-MS (Giladi et al. 2013, 2015, 2017; Lee et al. 2016). ^{15}N-relaxation NMR analyses of apo NCX1-CBD12-AD showed that two CBDs display nonlinear orientations, whereas Ca^{2+} binding restricts the linker's flexibility without altering the CBDs' alignment (Salinas et al. 2011). Similar ^{15}N-relaxation signals were observed for CALX1.1-CBD12 (Abiko et al. 2016), meaning that the "population shift" mechanism is common for Ca^{2+}-dependent conformational changes in the NCX and CALX variants. Thus, the dynamic coupling of CBDs involves Ca^{2+} occlusion (entrapping) at the two-domain interface, which is associated with Ca^{2+}-mediated tethering of CBDs through a network of interdomain salt-bridges, as outlined above.

In support of the population shift mechanism, the SAXS and HDX-MS analyses revealed that the Ca^{2+}-mediated tethering of CBDs rigidifies the CBD interface in all tested variants, although the strength, expansion, and remoteness of Ca^{2+}-dependent rigidification varies among the NCX1, NCX2, and NCX3 variants (Giladi et al. 2015, 2017; Lee et al. 2016). The SAXS analyses of isolated NCX1-CBD12 (AD, BD), NCX2-CBD12, and NCX3-CBD12 (B, BC) have shown that whereas the global structural parameters (e.g., the maximal intramolecular distance, the radius of gyration) are basically similar in the apo- and the Ca^{2+}-bound forms, Ca^{2+} induces redistribution in fractional conformational states (Giladi et al. 2015, 2017; Lee et al. 2016; Khananshvili 2016). Namely, the apo CBD12 forms undergo numerous conformational changess, whereas the Ca^{2+} binding results in narrower profiles of conformational distributions. More specifically, Ca^{2+} binding to the Ca3-Ca4 sites results in a population shift of conformational states, where more constrained conformational states become more populated at a dynamic equilibrium (Giladi et al. 2013; Giladi and Khananshvili 2013; Salinas et al. 2011).

In light of the present considerations, one can posit that the conformational stability of Ca^{2+}-mediated tethering of CBDs depends on the intrinsic folding energy of CBD2. This general mechanism, shared by NCX and CALX, is secondarily edited by the FG loop to shape intrinsic stability of CBD2 in an exon-dependent manner, which determines the dissociation kinetics of occluded Ca^{2+} from the Ca3-Ca4 sites (Giladi et al. 2012a, b, c, 2013, 2017; Lee et al. 2016; Khananshvili 2017a, b). This "editing" mechanism shapes the positive responses of NCX to regulatory Ca^{2+} in the cardiac, brain, and kidney splice variants (Matsuoka et al. 1995; Dyck et al. 1999; Maxwell et al. 1999; Nicoll et al. 2006; Chaptal et al. 2007; Ottolia et al. 2009, 2010).

8.7.5 *The Functional Relevance of CBDs Dynamic Coupling in NCXs*

Accumulating evidence brings to light a general mechanism for propagation of allosteric signals upon regulatory Ca^{2+} binding to NCX ortholog/isoform/splice variants, according to which the Ca^{2+}-dependent rigidification of backbone dynamics propagates from the Ca3-Ca4 sites of CBD1 toward CBD2 through the two-domain interface (Khananshvili 2016, 2017a, b) (Fig. 8.4). Mutational studies revealed that F450 (located within the hydrophobic core at the CBD interface) plays a role in preventing the signal propagating from the Ca3-Ca4 sites to the N-terminal of CBD1; therefore, the signal propagation is essentially conducted through CBD2 (Lee et al. 2016). Thus, the Ca^{2+} binding to the Ca3-Ca4 sites of CBD1 rigidifies CBD2 (but not CBD1), meaning that the Ca^{2+}-dependent allosteric signal propagates from the C-terminal of CBD1 toward the C-terminal of CBD2 (Giladi et al. 2015, 2017; Lee et al. 2016). In the framework of this general mechanism, the NCX1-CBD12-AD, CALX1.1-CBD12, and CALX1.2-CBD12 variants differ in the allocation, extent, and strength of Ca^{2+}-dependent rigidification in a given variant. Namely, in the Ca^{2+}-activated variant (NCX1-CBD12-AD) the backbone rigidification spans from the Ca3-Ca4 sites of CBD1, through the α-helix of CBD2 up to the C-terminal tip of CBD2 over a distance of 50 Å, whereas the Ca^{2+}-dependent rigidification stops at the CBD2 α-helix in the Ca^{2+}-inhibited variant (CALX1.1-CBD12) (Giladi et al. 2015, Lee et al. 2016). An intermediary picture emerges for CALX1.2-CBD12, which lacks a response to regulatory Ca^{2+}. In light of the present findings, one can conclude that regulatory differences (positive, negative, or no response to regulatory Ca^{2+}) result from the exon-controlled differential stabilization of specific local structures at the CBD interface and at the tip of CBD2 near the interdomain linker. Most probably, the Ca^{2+}-dependent rigidification of the CBD2 BC-loop and of the interdomain linker reduces translational movements of CBDs, whereas the rigidification of nearby segments of the β-strands at CBD2 restrict rotational movements within CBD2. More dedicated investigation is required for segregating the translational and rotational movements in the NCX and CALX variants.

In general, the splicing segments consist of intrinsically disordered regions that are unable to adopt a "discrete" and stable tertiary structure, although these disordered regions can accept a more stable conformation upon ligand binding toward "dynamic coupling" of biological functions (Babu et al. 2012; Buljan et al. 2013; Latysheva et al. 2015). Consistent with this concept, the Ca^{2+} binding to CBDs results in a population shift of numerous preexisting conformational states without undergoing a few large conformational changes (Giladi et al. 2013). Compiling the analyses of the NCX1-CBD12 and NCX3-CBD12 splice variants by using stopped-flow, equilibrium binding and HDX-MS approaches strongly supports the notion that the mutually exclusive exons (A and B) have diverse effects on the stability of the CBD1 and CBD2 domains in the NCX1 and NCX3 variants (Giladi et al. 2012a, 2013, 2015, 2017; Lee et al. 2016).

Notably, exon A and B play completely opposite roles in NCX1 and NCX3. For example, in NCX1 exon B increases the Ca^{2+} affinity at the Ca3-Ca4 sites of CBD1, while slowing down the dissociation of occluded Ca^{2+} from CBD1 and preventing the Ca^{2+} binding to CBD2 (Giladi et al. 2010, 2012a, b, c, 2017; Lee et al. 2016; Tal et al. 2016). In agreement with this, exon B effectively stabilizes the Ca^{2+} sites at CBD1, while destabilizing CBD2 and thus prevents Ca^{2+} binding to CBD2. Exon A of NCX1 effectively stabilizes two Ca^{2+} binding sites at CBD2, whereas it destabilizes the Ca^{2+} sites at CBD1. In contrast with NCX1, the roles of the A and B exons are completely diverse in NCX3 (Tal et al. 2016; Lee et al. 2016; Giladi et al. 2017). More specifically, in NCX3, exon A destabilizes the Ca^{2+} binding sites at CBD2 (with no Ca^{2+} binding to CBD2) and stabilizes the Ca^{2+} binding sites at CBD2. In contrast with this, exon B in NCX3 stabilizes CBD2 by generating three Ca^{2+} binding sites at CBD2 (since K585 is replaced by glutamate), whereas it destabilizes the Ca^{2+} binding at CBD1. Thus, the mutually exclusive exons (A and B) play opposite roles in the conformational stabilization of the Ca^{2+} binding sites in the CBD1 and CBD2 domains of NCX3 and NCX1 (Khananshvili 2016, 2017a, b).

In contrast with the mutually exclusive exons, the cassette exons (C, D, E, and F) do not affect the dynamic features of apo CBD1; instead, they rigidify the interdomain linker and CBD2 folding upon Ca^{2+} binding either in NCX1 or NCX3 (Lee et al. 2016; Giladi et al. 2015, 2017). For example, the FG loop/α-helix region of CBD2 is similarly stabilized in the kidney (NCX1-CBD12-BD) and cardiac (NCX1-CBD12-ACDEF) splice variants of NCX1, compared with the brain splice variant (NCX1-CBD12-AD). Thus, both mutually exclusive and cassette exons complementarily control the strength and extent of Ca^{2+}-dependent rigidification either at CBD1 or CBD2. These data support the notion that the genetically encoded actions of mutually exclusive and cassette exons complement each other toward the fine-tuning of Ca^{2+}-dependent activation (controlled by CBD1) and the Ca^{2+}-induced alleviation of Na^+-dependent inactivation (controlled by CBD2) either in NCX1 or NCX3 isoforms.

Notably, the gradual addition of cassette exons (C, D, E, and F) to exon A containing a splicing segment (e.g., in NCX1-CBD12-ACDEF) increasingly stabilizes and enhances the affinity of the Ca^{2+} binding sites at CBD1, which can potentially compensate for the destabilizing effect of exon A on the high-affinity Ca^{2+} sensor at CBD1. This phenomenon is fascinating as well as somewhat puzzling, since the existence of intrinsically disordered segments contradicts the high entropic cost that the system must incur upon ligand binding (Babu et al. 2012; Buljan et al. 2013; Latysheva et al. 2015). In principle, unfolded structures at the splicing segment can minimize the entropic cost upon ligand binding through an enthalpy compensation mechanism, although the underlying structure-dynamic events are not entirely clear (Buljan et al. 2013; Latysheva et al. 2015; Fuxreiter 2018; Sterne-Weiler et al. 2018). In general terms, the gradual lengthening of a splicing segment by adding the cassette exons to the mutually exclusive exon (e.g., in the cardiac variant, NCX1-CBD12-ACDEF) may compensate for the entropy loss by enthalpy changes, thereby enabling incremental changes in the affinity of the Ca^{2+}-dependent activation of mammalian NCX variants. How the enthalpy-driven entropy cost underlies the regulatory outcomes in the NCX and CALX variants remains to be discovered.

8.8 Conclusions

Three mammalian NCX genes (NCX1, NCX2, and NCX3) produce a limited number of splice variants in epithelial tissues (NCX1-BD, NCX1-BDF, NCX2, NCX3-B, NCX3-BC), while extruding up to 70% of total Ca^{2+}, transported across the basolateral membrane. The relevant NCX isoform/splice own to Ca^{2+}-dependent activation, whereas the NCX1 and NCX3 variants (but not NCX2) exhibit the Na^{+}-dependent inactivation. The NCX1 and NCX3 differ from each other in that that the NCX1-BD and NCX1-BDF variants lack the Ca^{2+}-induced alleviation of Na^{+}-dependent inactivation (since exon B prevents Ca^{2+} binding to CBD2 of NCX1) whereas the NCX3-B and NCX3-BC variants can perform the Ca^{2+}-induced alleviation of Na^{+}-dependent inactivation (since three Ca^{2+} ions bind to CBD2). These regulatory differences, exhibited by distinct isoform-splice variants, are structurally predefined by single-point replacement of amino acid residues (at three critical positions involved in the Ca^{+} coordination) within the exon B. The cassette exons (C, D, and F) secondarily modify the Ca^{2+} binding affinity and conformational dynamics at both regulatory domains (CBD1 and CBD2), where the slow dissociation of "occluded" Ca^{2+} from the two-domain interface controls the "inactivation" kinetics of a given variant. The underlying mechanisms of NCXs contributions to local and global Ca^{2+} signaling in the kidney, intestine, osteoblast, osteoclast, and secretory cells (among many other cell types) remain to be further resolved. The present achievements in understanding the structure-dynamic determinants governing the regulatory specificity of NCX variants provide new opportunities for a better understanding of cell-specific Ca^{2+} signaling and for structure-based pharmacological targeting of tissue-specific NCX variants. In the long-term run, this could be beneficial in many biomedical applications.

Acknowledgments This work was supported by the Israel Science Foundation Grant #1351/18 to DK. The financial support of the Shaltiel Foundation to DK is highly appreciated.

References

Abiko LA, Vitale PM, Favaro DC, Hauk P, Li DW, Yuan J, Bruschweiler-Li L, Salinas RK, Brüschweiler R (2016) Model for the allosteric regulation of the Na^{+}/Ca^{2+} exchanger (NCX). Proteins 84:580–590

Akisaka T, Yamamoto T, Gay CV (1988) Ultracytochemical investigation of calcium activated adenosine triphosphate (Ca^{2+}-ATPase) in chick tibia. J Bone Miner Res 3:19–25

Albano G, Dolder S, Siegrist M, Mercier-Zuber A, Auberson M, Stoudmann C, Hofstetter W, Bonny O, Fuster DG (2017) Pflugers Arch 469:225–233

Almagor L, Giladi M, van Dijk L, Buki T, Hiller R, Khananshvili D (2014) Functional asymmetry of bidirectional Ca^{2+}-movements in an archaeal sodium-calcium exchanger (NCX_Mj). Cell Calcium 56:276–284

Andrikopoulos P, Baba A, Matsuda T, Djamgoz MBA, Yaqoob MM, Eccles SA (2011) Ca^{2+} influx through reverse mode Na^{+}/Ca^{2+} exchange is critical for vascular endothelial growth factor-

mediated extracellular signal-regulated kinase (ERK) 1/2 activation and angiogenic functions of human endothelial cells. J Biol Chem 286:37919–37931
Andrikopoulos P, Eccles SA, Yaqoob MM (2017) Coupling between the TRPC3 ion channel and the NCX1 transporter contributed to VEGF-induced ERK1/2 activation and angiogenesis in human primary endothelial cells. Cell Signal 37:12–30
Annunziato L, Pignataro G, Di Renzo GF (2004) Pharmacology of brain Na+/Ca2+ exchanger: from molecular biology to therapeutic perspectives. Pharmacol Rev 56:633–654
Annunziato L et al (2007) ncx1, ncx2, and ncx3 gene product expression and function in neuronal anoxia and brain ischemia. Ann N Y Acad Sci 1099:413–426
Asteggiano C, Berberián G, Beaugé L (2001) Phosphatidylinositol-4,5-biphosphate bound to bovine cardiac Na/Ca exchanger displays a MgATP regulation similar to that of the exchange fluxes. Eur J Biochem 268:437–442
Baazov D, Wang X, Khananshvili D (1999) Time-resolved monitoring of electrogenic Na^+-Ca^{2+} exchange in the isolated cardiac sarcolemma vesicles by using a rapid-response fluorescent probe. Biochemistry 38:1435–1445
Babu MM, Kriwacki RW, Pappu RV (2012) Structural biology. Versatility from protein disorder. Science 337:1460–1464
Baker PF, Blaustein MP (1968) Sodium-dependent uptake of calcium by crab nerve. Biochim Biophys Acta 150:167–170
Baker PF, Blaustein MP, Hodgkin AL, Steinhardt RA (1969) The influence of calcium on sodium efflux in squid axons. J Physiol 200:431–458
Berberián G, Forcato D, Beaugé L (2009) Key role of a PtdIns-4,5P2 micro domain in ionic regulation of the mammalian heart Na^+/Ca^{2+} exchanger. Cell Calcium 45:546–553
Berridge MJ et al (2003) Calcium signalling: dynamics, homeostasis and remodelling. Nat Rev Mol Cell Biol 4:517–529
Bers DM (2002) Cardiac excitation-contraction coupling. Nature 415:198–205
Bers DM (2008) Calcium cycling and signaling in cardiac myocytes. Annu Rev Physiol 70:23–49
Bers DM, Ginsburg KS (2007) Na:Ca stoichiometry and cytosolic Ca-dependent activation of NCX in intact cardiomyocytes. Ann N Y Acad Sci 1099:326–338
Besserer GM, Ottolia M, Nicoll DA, Chaptal V, Cascio D, Philipson KD, Abramson J (2007) The second Ca^{2+}-binding domain of the Na^+ Ca^{2+} exchanger is essential for regulation: crystal structures and mutational analysis. Proc Natl Acad Sci U S A 104:18467–18472
Blaustein MP, Lederer WJ (1999) Sodium/calcium exchange: its physiological implications. Physiol Rev 79:763–854
Bode K, O'Halloran DM (2018) NCX-DB: a unified resource for integrative analysis of the sodium-calcium exchanger super-family. BMC Neurosci 19(1):19
Boehr DD, Nussinov R, Wright PE (2009) The role of dynamic conformational ensembles in biomolecular recognition. Nat Chem Biol 5:789–796
Boros S, Bindels RJ, Hoenderop JG (2009) Active Ca^{2+} reabsorption in the connecting tubule. Pflugers Arch 458:99–109
Boyman L, Mikhasenko H, Hiller R, Khananshvili D (2009) Kinetic and equilibrium properties of regulatory calcium sensors of NCX1 protein. J Biol Chem 284:6185–6193
Boyman L, Hagen BM, Giladi M, Hiller R, Lederer WJ, Khananshvili D (2011) Proton-sensing Ca^{2+} binding domains regulate the cardiac Na^+/Ca^{2+} exchanger. J Biol Chem 286:28811–28820
Boyman L, Williams GSB, Khananshvili D, Sekler I, Lederer WJ (2013) NCLX: the mitochondrial sodium- calcium exchanger. J Mol Cell Cardiol 59:205–213
Breukels V, Vuister GW (2010) Binding of calcium is sensed structurally and dynamically throughout the second calcium-binding domain of the sodium/calcium exchanger. Proteins 78:1813–1182
Breukels V, Konijnenberg A, Nabuurs SM, Touw WG, Vuister GW (2011) The second Ca^{2+}-binding domain of NCX1 binds Mg^{2+} with high affinity. Biochemistry 50:8804–8812
Brini M, Carafoli E (2009) Calcium pumps in health and disease. Physiol Rev 89:1341–1378

Brini M, Carafoli E (2011) The plasma membrane Ca^{2+} ATPase and the plasma membrane sodium-calcium exchanger cooperate in the regulation of cell calcium. Cold Spring Harb Perspect Biol 3 (2):a004168
Brini M, Calì T, Ottolini D, Carafoli E (2014) The plasma membrane calcium pump in health and disease. FEBS J 280:5385–5397
Brodskiy PA, Zartman JJ (2018) Calcium as a signal integrator in developing epithelial tissues. Phys Biol 15(051001):1–13
Brown EM (1991) Extracellular Ca^{2+} sensing, regulation of parathyroid cell function, and role of Ca^{2+} and other ions as extracellular (first) messengers. Physiol Rev 71:371–411
Buljan M, Chalancon G, Dunker AK, Bateman A, Balaji S, Fuxreiter M, Babu MM (2013) Alternative splicing of intrinsically disordered regions and rewiring of protein interactions. Curr Opin Struct Biol 23:443–450
Burgoyne RD (2007) Neuronal calcium sensor proteins: generating diversity in neuronal Ca^{2+} signalling. Nat Rev Neurosci 8:182–193
Cai X, Lytton J (2004) The cation/Ca^{2+} exchanger superfamily: phylogenetic analysis and structural implications. Mol Biol Evol 21:1692–1703
Cantero-Recasens G, Cristian M, Butnaru CM, Brouwers N, Mitrovic M, Valverde MA, Malhotra V (2019) Sodium channel TRPM4 and sodium/calcium exchangers (NCX) cooperate in the control of Ca^{2+}induced mucin secretion from goblet cells. J Biol Chem 294:816–826
Carafoli E (1987) Intracellular calcium homeostasis. Annu Rev Biochem 56:395–433
Carafoli E (1988) Membrane transport of calcium: an overview. Methods Enzymol 157:3–11
Carafoli E, Krebs J (2016) Why calcium? How calcium became the best communicator. J Biol Chem 291:20849–20857
Carafoli E, Tiozzo R, Lugli G, Crovetti F, Kratzing C (1974) The release of calcium from heart mitochondria by sodium. J Mol Cell Cardiol 6:361–371
Centeno V, Picotto G, Pérez A, Alisio A, de Talamoni NT (2011) Intestinal Na^{+}/Ca^{2+} exchanger protein and gene expression are regulated by $1,25(OH)_2D_3$ in vitamin D-deficient chicks. Arch Biochem Biophys 509:191–196
Chaptal V, Ottolia M, Mercado-Besserer G, Nicoll DA, Philipson KD et al (2009) Structure and functional analysis of a Ca^{2+} sensor mutant of the Na^{+}/Ca^{2+} exchanger. J Biol Chem 284:14688–14692
Cho W, Stahelin RV (2006) Membrane binding and subcellular localization of C2 domains. Biochem Biophys Acta 1761:838–849
Chou AC, Ju YT, Pan CY (2015) Calmodulin interacts with the sodium/calcium exchanger NCX1 to regulate activity. PLoS One 10(9):e013885
Christakos S, Gill R, Lee S, Li H (1992) Molecular aspects of the calbindins. J Nutr 122:678–682
Clapham DE (2007) Calcium signaling. Cell 131:1047–1058
Condrescu M, Gardner JP, Chernaya G, Aceto JF, Kroupis C, Reeves JP (1995) ATP-dependent regulation of sodium-calcium exchange in Chinese hamster ovary cells transfected with the bovine cardiac sodium-calcium exchanger. J Biol Chem 270:9137–9146
Demaegd D, Colinet AS, Deschamps A, Morsomme P (2014) Molecular evolution of a novel family of putative calcium transporters. PLoS One 9(6):e100851
Diaz de Barboza G, Guizzardi S, Tolosa de Talamoni N (2015) Molecular aspects of intestinal calcium absorption. World J Gastroenterol 21:7142–7154
Dimke H, Hoenderop JGJ, Bindels RJM (2011) Molecular basis of epithelial Ca^{2+} and Mg^{2+} transport: insights from the TRP channel family. J Physiol 589:1535–1542
DiPolo R (1979) Calcium influx in internally dialyzed squid giant axons. J Gen Physiol 73:91–113
DiPolo R, Beaugé L (2006) Sodium/calcium exchanger: influence of metabolic regulation on ion carrier interactions. Physiol Rev 86:155–203
Doering AE, Lederer WJ (1994) The action of Na^{+} as a cofactor in the inhibition by cytoplasmic protons of the cardiac Na^{+}-Ca^{2+} exchanger in the Guinea-pig. J Physiol 480:9–20
Doering AE, Eisner DA, Lederer WJ (1996) Cardiac Na-Ca exchange and pH. Ann N Y Acad Sci 779:182–198

Dunn J, Elias CL, Le HD, Omelchenko A, Hryshko LV, Lytton J (2002) The molecular determinants of ionic regulatory differences between brain and kidney Na^+/Ca^{2+} exchanger (NCX1) isoforms. J Biol Chem 277:33957–33962

Dyck C, Maxwell K, Buchko J, Trac M, Omelchenko A, Hnatowich M, Hryshko LV (1998) Structure-function analysis of CALX1.1, a Na^+-Ca^{2+} exchanger from Drosophila. Mutagenesis of ionic regulatory sites. J Biol Chem 273:12981–12987

Dyck C, Omelchenko A, Elias CL, Quednau BD, Philipson KD, Hnatowich M, Hryshko LV (1999) Ionic regulatory properties of brain and kidney splice variants of the NCX1 Na^+-Ca^{2+} exchanger. J Gen Physiol 114:701–711

Emery L, Whelan S, Hirschi KD, Pittman JK (2012) Protein phylogenetic analysis of Ca^{2+}/CA antiporters and insights into their evolution in plants. Front Plant Sci 3:1

Fraysse B, Rouaud T, Millour M, Fontaine-Pérus J, Gardahaut MF, Levitsky DO (2001) Expression of the Na^+/Ca^{2+} exchanger in skeletal muscle. Am J Physiol Cell Physiol 280:C146–C154

Friedman PA, Gesek FA (1995) Cellular calcium transport in renal epithelia: measurement, mechanisms, and regulation. Physiol Rev 75:429–471

Furman I, Cook O, Kasir J, Rahamimoff H (1993) Cloning of two isoforms of the rat brain Na^+-Ca^{2+} exchanger gene and their functional expression in HeLa cells. FEBS Lett 319:105–109

Fuxreiter M (2018) Fold or not to fold upon binding - does it really matter? Curr Opin Struct Biol 54:19–25

Gabellini N, Bortoluzzi S, Danieli GA, Carafoli E (2002) The human SLC8A3 gene and the tissue-specific $Na^+/Ca2^+$ exchanger 3 isoforms. Gene 298:1–7

Gifford JL, Walsh MP, Vogel HJ (2007) Structures and metal-ion-binding properties of the Ca^{2+}-binding helix-loop-helix EF-hand motifs. Biochem J 405:199–221

Giladi M, Boyman L, Mikhasenko H, Hiller R, Khananshvili D (2010) Essential role of the CBD1-CBD2 linker in slow dissociation of Ca^{2+} from the regulatory two-domain tandem of NCX1. J Biol Chem 285:28117–28125

Giladi M, Sasson Y, Fang X, Hiller R, Buki T, Wang YX, Hirsch JA, Khananshvili D (2012a) A common Ca^{2+}-driven interdomain module governs eukaryotic NCX regulation. PLoS One 7: e39985

Giladi M, Bohbot H, Buki T, Schulze DH, Hiller R, Khananshvili D (2012b) Dynamic features of allosteric Ca^{2+} sensor in tissue-specific NCX variants. Cell Calcium 51:478–485

Giladi M, Sasson Y, Fang X, Hiller R, Buki T, Wang YX, Hirsch JA, Khananshvili D (2012c) A common Ca^{2+}-driven interdomain module governs eukaryotic NCX regulation. PLoS One 7(6): e39985

Giladi M, Hiller R, Hirsch JA, Khananshvili D (2013) Population shift underlies Ca^{2+}-induced regulatory transitions in the sodium-calcium exchanger (NCX). J Biol Chem 288:23141–23149

Giladi M, Khananshvili D (2013) Molecular determinants of allosteric regulation in NCX proteins. Adv Exp Med Biol 961:35–48

Giladi M, Lee SY, Hiller R, Chung KY, Khananshvili D (2015) Structure-dynamic determinants governing a mode of regulatory response and propagation of allosteric signal in splice variants of Na^+/Ca^{2+} exchange (NCX) proteins. Biochem J 465:489–501

Giladi M, Almagor L, van Dijk L, Hiller R, Man P, Forest E, Khananshvili D (2016a) Asymmetric preorganization of inverted pair residues in the sodium–calcium exchanger. Sci Rep 6:20753

Giladi M, Tal I, Khananshvili D (2016b) Structural features of ion transport and allosteric regulation in sodium-calcium exchanger (NCX) proteins. Front Physiol 7:30

Giladi M, Lee SY, Ariely Y, Teldan Y, Granit R, Strulevich R, Haitin Y, Chung KY, Khananshvili D (2017) Structure based dynamic arrays in regulatory domains of sodium-calcium exchanger (NCX) isoforms. Sci Rep 7:993–1013

Ginsburg KS, Weber CR, Bers DM (2013) Cardiac Na^+–Ca^{2+} exchanger: dynamics of Ca^{2+}-dependent activation and deactivation in intact myocytes. J Physiol 591:2067–2086

Gomez-Villafuertes R, Torres B, Barrio J, Savignac M, Gabellini N, Rizzato F, Pintado B, Gutierrez-Adan A, Mellstrom B, Carafoli E, Naranjo JR (2005) Downstream regulatory element

antagonist modulator regulates Ca^{2+} homeostasis and viability in cerebellar neurons. J Neurosci 25:10822–10830
Grover AK (2017) Sodium–calcium exchanger in pig coronary artery. Adv Pharmacol 78:145–170
Guillot C, Lecuit T (2013) Mechanics of epithelial tissue homeostasis and morphogenesis. Science 340:1185–1189
Haussinger D, Ahrens T, Sass HJ, Pertz O, Engel J et al (2002) Calcium-dependent homoassociation of E-cadherin by NMR spectroscopy: changes in mobility, conformation and mapping of contact regions. J Mol Biol 324:823–839
He Z, Feng S, Tong Q, Hilgemann DW, Philipson KD (2000) Interaction of PIP_2 with the XIP region of the cardiac Na/Ca exchanger. Am J Physiol Cell Physiol 278:C661–C666
Herchuelz A, Pachera N (2018) The Na^+/Ca^{2+} exchanger and the plasma membrane Ca^{2+}-ATPase in β-cell function and diabetes. Neur Letters 663:72–78
Hilge M, Aelen J, Vuister GW (2006) Ca^{2+} regulation in the Na^+/Ca^{2+} exchanger involves two markedly different Ca^{2+} sensors. Mol Cell 22:15–25
Hilge M, Aelen J, Foarce A, Perrakis A, Vuister GW (2009) $Ca2^+$ regulation in the Na^+/Ca^{2+} exchanger features a dual electrostatic switch mechanism. Proc Natl Acad Sci U S A 106:14333–14338
Hilgemann DW (1990) Regulation and deregulation of cardiac Na^+-Ca^{2+} exchange in giant excised sarcolemmal membrane patches. Nature 344:242–245
Hilgemann DW (2007) Local PIP(2) signals: when, where, and how? Pflugers Arch 455:55–67
Hilgemann DW, Ball R (1996) Regulation of cardiac Na^+/Ca^{2+} exchange and MgATP potassium channels by PIP2. Science 273:956–959
Hilgemann DW, Nicoll DA, Philipson KD (1991) Charge movement during Na^+-translocation by native and cloned Na^+/Ca^{2+} exchanger. Nature 352:715–718
Hilgemann DW, Collins A, Matsuoka S (1992a) Steady-state and dynamic properties of cardiac Na/Ca. Secondary modulation by cytoplasmic calcium and ATP. J Gen Physiol 100:933–961
Hilgemann DW, Matsuoka S, Nagel GA, Collins A (1992b) Steady-state and dynamic properties of cardiac sodium-calcium exchange - sodium-dependent inactivation. J Gen Physiol 100:905–932
Hnatowich M, Le HD, DeMoissac D, Ranson K, Yurkov V, Gilchrist JS, Omelchenko A, Hryshko LV (2012) μ-Calpain-mediated deregulation of cardiac, brain, and kidney NCX1 splice variants. Cell Calcium 51:164–170
Hoenderop JG, van der Kemp AW, Hartog A, van de Graaf SF, van Os CH, Willems PH, Bindels RJ (1999) Molecular identification of the apical Ca^{2+} channel in 1,25-dihydroxyvitamin D_3-responsive epithelia. J Biol Chem 274:8375–8378
Hoenderop JGJ, Nilius B, Bindels RJM (2002) Molecular mechanism of active Ca^{2+} reabsorption in the distal nephron. Annu Rev Physiol 64:529–549
Hong EJ, Jeung EB (2013) Biological significance of calbindin-D9k within duodenal epithelium. Int J Mol Sci 4:23330–23340
Hryshko L (2008) What regulates Na^+/Ca^{2+} exchange? Focus on "sodium-dependent inactivation of sodium/calcium exchange in transfected Chinese hamster ovary cells". Am J Physiol Cell Physiol 295:C869–C871
Hryshko LV, Matsuoka S, Nicoll DA, Weiss JN, Schwarz EM, Benzer S, Philipson KD (1996) Anomalous regulation of the Drosophila Na^+-Ca^{2+} exchanger by Ca^{2+}. J Gen Physiol 108:67–74
Hurtado C, Prociuk M, Maddaford TG, Dibrov E, Mesaeli N, Hryshko LV, Pierce GN (2006) Cells expressing unique Na^+/Ca^{2+} exchange (NCX1) splice variants exhibit different susceptibilities to Ca^{2+} overload. Am J Physiol Heart Circ Physiol 290:H2155–H2162
Jia YH, Cui ZJ (2011) Tri-phasic modulation of ACh- and NA-maintained calcium plateau by high potassium in isolated mouse submandibular granular convoluted tubular cells. Arch Oral Biol 56:1347–1355
John SA, Ribalet B, Weiss JN, Philipson KD, Ottolia M (2011) Ca^{2+}-dependent structural rearrangements within Na^+-Ca^{2+} exchanger dimers. Proc Natl Acad Sci U S A 108:1699–1704

John SA, Kim B, Olcese R, Goldhaber JI, Ottolia M (2018) Molecular determinants of pH regulation in the cardiac Na^+-Ca^{2+} exchanger. J Gen Physiol 150:245–257
Johnson E, Bruschweiler-Li L, Showalter SA, Vuister GW, Zhang F, Brüschweiler R (2008) Structure and dynamics of Ca^{2+}-binding domain 1 of the Na^+/Ca^{2+} exchanger in the presence and in the absence of Ca^{2+}. J Mol Biol 377:945–955
Jung SY, Park YJ, Park YJ, Cha SH, Lee MZ, Suh CK (2007) Na^+-Ca^{2+} exchanger modulates Ca^{2+} content in intracellular Ca^{2+} stores in rat osteoblasts. Exp Mol Med 39:458–468
Khananshvili D (1990) Distinction between the two basic mechanisms of cation transport in the cardiac Na+-Ca2+ exchange system. Biochemistry 29:2437–2442
Khananshvili D (1991) Voltage-dependent modulation of ion binding and translocation in the cardiac Na+-Ca2+ exchange system. J Biol Chem 266:13764–13769
Khananshvili D (2013) The SLC8 gene family of sodium-calcium exchangers (NCX) - structure, function, and regulation in health and disease. Mol Asp Med 34:220–235
Khananshvili D (2014) Sodium-calcium exchangers (NCX): molecular hallmarks underlying the tissue-specific and systemic functions. Pflugers Arch 466:43–60
Khananshvili D (2016) Regulation of Ca^{2+}-ATPases, V-ATPases and F-ATPases. In: Chakraborti S, Dhalla NS (eds) Advances in biochemistry in health and disease, vol 14. Springer, London, pp 93–116
Khananshvili D (2017a) Structure-dynamic coupling through Ca^{2+}-binding regulatory domains of mammalian NCX isoform/splice variants. Adv Exp Med Biol 981:41–58
Khananshvili D (2017b) How a helix imposes palmitoylation of a membrane protein: what one can learn from NCX. J Biol Chem 292:10753–10754
Khananshvili D (2020) Basic and editing mechanisms underlying ion transport and regulation in NCX variants. Cell Calcium 85:102131
Khananshvili D, Weil-Maslansky E, Baazov D (1996) Kinetics and mechanism: modulation of ion transport in the cardiac sarcolemma sodium-calcium exchanger by protons, monovalent ions, and temperature. Ann NY Acad Sci USA 779:217–235
Khundmiri SJ, Murray RD, Lederer E (2016) PTH and vitamin D. Compr Physiol 6:561–601
Kofuji P, Lederer WJ, Schulze DH (1993) Na/Ca exchanger isoforms expressed in kidney. Am J Phys 265(4 Pt 2):F598–F603
Kofuji P, Lederer WJ, Schulze DH (1994) Mutually exclusive and cassette exons underlie alternatively spliced isoforms of the Na/Ca exchanger. J Biol Chem 269:5145–5149
Kostic M, Sekler I (2019) Functional properties and mode of regulation of the mitochondrial Na^+/Ca^{2+} exchanger, NCLX. Semin Cell Dev Biol S1084–9521(18):30273–30278
Kostic M, Katoshevski T, Sekler I (2018) Allosteric regulation of NCLX by mitochondrial membrane potential links the metabolic state and Ca^{2+} signaling in mitochondria. Cell Rep 25 (12):3465–3475.e4
Krebs J (2009) The influence of calcium signaling on the regulation of alternative splicing. Biochim Biophys Acta 1793:979–984
Krebs J (2017) The plasma membrane calcium pump (PMCA): Regulation of cytosolic Ca_{2+}, genetic diversities and its role in sub-plasma membrane microdomains. Adv Exp Med Biol 981:3–21
Lambers TT, Mahieu F, Oancea E, Hoofd L, de Lange F et al (2006) Calbindin-D_{28K} dynamically controls TRPV5-mediated Ca^{2+} transport. EMBO J 25:2978–2988
Lariccia V, Amoroso S (2018) Calcium- and ATP-dependent regulation of Na/Ca exchange function in BHK cells: Comparison of NCX1 and NCX3 exchangers. Cell Calcium 73:95–103
Latysheva NS, Flock T, Weatheritt RJ, Chavali S, Babu MM (2015) How do disordered regions achieve comparable functions to structured domains? Protein Sci 24:909–922
Lee SL, Yu AS, Lytton J (1994) Tissue-specific expression of Na^+-Ca^{2+} exchanger isoforms. J Biol Chem 269:14849–14852
Lee SY, Giladi M, Bohbot H, Hiller R, Chung KY, Khananshvili D (2016) Structure-dynamic basis of splicing dependent regulation in tissue-specific variants of the sodium-calcium exchanger (NCX1). FASEB J 30:1356–1366

Levitsky DO, Nicoll DA, Philipson KD (1994) Identification of the high affinity Ca^{2+}-binding domain of the cardiac Na^+-Ca^{2+} exchanger. J Biol Chem 269:22847–22852
Li Z, Nicoll DA, Collins A, Hilgemann DW et al (1991) Identification of a peptide inhibitor of the cardiac sarcolemmal Na^+-Ca^{2+} exchanger. J Biol Chem 266:1014–1020
Li Z, Matsuoka S, Hryshko LV, Nicoll DA, Bersohn MM, Burke EP, Lifton RP, Philipson KD (1994) Cloning of the NCX2 isoform of the plasma membrane Na^+-Ca^{2+} exchanger. J Biol Chem 269:17434–17439
Li JP, Kajiya H, Okamoto F, Nakao A, Iwamoto T, Okabe K (2007) Three Na^+/Ca^{2+} exchanger (NCX) variants are expressed in mouse osteoclasts and mediate calcium transport during bone resorption. Endocrinology 148:2116–2125
Liao J, Li H, Zeng W, Sauer DB, Belmares R, Jiang Y (2012) Structural insight into the ion-exchange mechanism of the sodium/calcium exchanger. Science 335:686–690
Liao J, Marinelli F, Lee C, Huang Y, Faraldo-Gómez JD, Jiang Y (2016) Mechanism of extracellular ion exchange and binding-site occlusion in a sodium/calcium exchanger. Nat Struct Mol Biol 23:590–599
Lillo MA, Gaete PS, Puebla M, Ardiles NM, Poblete I, Becerra A, Simon F, Figueroa XF (2018) Critical contribution of Na^+-Ca^{2+} exchanger to the Ca^{2+}-mediated vasodilation activated in endothelial cells of resistance arteries. FASEB J 32:2137–2147
Linck B, Qiu Z, He Z, Tong Q, Hilgemann DW, Philipson KD (1998) Functional comparison of the three isoforms of the Na^+/Ca^{2+} exchanger (NCX1, NCX2, NCX3). Am J Phys Cell Physiol 274: C415–C423
Linck B, Bokník P, Huke S, Kirchhefer U et al (2000) Functional properties of transgenic mouse hearts overexpressing both calsequestrin and the Na^+/Ca^{2+}exchanger. J Pharmacol Exp Ther 294:648–657
Lindgren RM, Zhao J, Heller S, Berglind H, Nistér M (2005) Molecular cloning and characterization of two novel truncated isoforms of human Na^+/Ca^{2+} exchanger 3, expressed in fetal brain. Gene 348:143–155
Liu T, Zhao J, Ibarra C, Garcia MU, Uhlén P, Nistér M (2018) Glycosylation controls sodium-calcium exchanger 3 sub-cellular localization during cell cycle. Eur J Cell Biol 97:190–203
Lopreiato R, Giacomello M, Carafoli E (2014) The plasma membrane calcium pump: new ways to look at an old enzyme. J Biol Chem 289:10261–10268
Lytton J (2007) Na^+/Ca^{2+} exchangers: three mammalian gene families control Ca^{2+} transport. Biochem J 406:365–382
Ma B, Nussinov R (2010) Enzyme dynamics point to stepwise conformational selection in catalysis. Curr Opin Chem Biol 14:652–659
Ma B, Tsai C-J, Haliloğlu T, Nussinov R (2011) Dynamic Allostery: linkers are not merely flexible. Structure 19:907–917
Maack C, Ganesan A, Sidor A, O'Rourke B (2005) Cardiac sodium-calcium exchanger is regulated by allosteric calcium and exchanger inhibitory peptide at distinct sites. Circ Res 96:91–99
Marhl M, Perc M, Schuster S (2006) A minimal model for decoding of time-limited Ca^{2+} oscillations. Biophys Chem 120:161–167
Marinelli F, Almagor L, Hiller R, Giladi M, Khananshvili D, Faraldo-Gomez JD (2014) Sodium recognition by the Na^+/Ca^{2+} exchanger in the outward-facing conformation. Proc Natl Acad Sci U S A 111:E5354–E5362
Marshall CR, Fox JA, Butland SL, Ouellette BF, Brinkman FS, Tibbits GF (2005) Phylogeny of Na^+-Ca^{2+} exchanger (NCX) genes from genomic data identifies new gene duplications and a new family member in fish species. Physiol Genomics 21:161–173
Martin DL, De Luca HF (1969) Influence of sodium on calcium transport by the rat small intestine. Am J Phys 216:1351–1359
Matsuoka S (2004) Forefront of Na^+/Ca^{2+} exchanger studies: regulation kinetics of Na^+/Ca^{2+} exchangers. J Pharmacol Sci 96:12–14

Matsuoka S, Nicoll DA, Reilly RF, Hilgemann DW, Philipson KD (1993) Initial localization of regulatory regions of the cardiac sarcolemmal Na^+-Ca^{2+} exchanger. Proc Natl Acad Sci U S A 90:3870–3874
Matsuoka S, Nicoll DA, Hryshko LV, Levitsky DO, Weiss JN, Philipson KD (1995) Regulation of the cardiac Na^+-Ca^{2+} exchanger by Ca^{2+}. Mutational analysis of the Ca^{2+} binding domain. J Gen Physiol 105:403–420
Matsuoka S, Nicoll DA, He Z, Philipson KD (1997) Regulation of cardiac Na^+-Ca^{2+} exchanger by the endogenous XIP region. J Gen Physiol 109:273–286
Maxwell K, Scott J, Omelchenko A, Lukas A, Lu L, Lu Y, Hnatowich M, Philipson KD, Hryshko LV (1999) Functional role of ionic regulation of Na^+/Ca^{2+} exchange assessed in transgenic mouse hearts. Am J Phys 277:H2212–H2221
McCue HV, Haynes LP, Burgoyne RD (2010) The diversity of calcium sensor proteins in the regulation of neuronal function. Cold Spring Harb Perspect Biol 2(8):a004085
Michel LY, Verkaart S, Koopman WJ, Willems PH, Hoenderop JG, Bindels RJ (2014) Function and regulation of the Na^+-Ca^{2+} exchanger NCX3 splice variants in brain and skeletal muscle. J Biol Chem 289:11293–11303
Michel LYM, Verkaart S, Latta F, Hoenderop JGJ, Bindels RJM (2017) Differential regulation of the Na $^+$-Ca^{2+} exchanger 3 (NCX3) by protein kinase PKC and PKA. Cell Calcium 65:52–62
Minelli A, Castaldo P, Gobbi P, Salucci S, Magi S, Amoroso S (2007) Cellular and subcellular localization of Na^+-Ca^{2+} exchanger protein isoforms, NCX1, NCX2, and NCX3 in cerebral cortex and hippocampus of adult rat. Cell Calcium 41:221–234
Nicoll DA, Longoni S, Philipson KD (1990) Molecular cloning and functional expression of the cardiac sarcolemmal Na^+-Ca^{2+} exchanger. Science 250:562–565
Nicoll DA, Quednau BD, Qui Z, Xia YR, Lusis AJ, Philipson KD (1996) Cloning of a third mammalian Na^+-Ca^{2+} exchanger, NCX3. J Biol Chem 271:24914–24921
Nicoll DA, Sawaya MR, Kwon S, Cascio D, Philipson KD, Abramson J (2006) The crystal structure of the primary Ca^{2+} sensor of the Na^+/Ca^{2+} exchanger reveals a novel Ca^{2+} binding motif. J Biol Chem 281:21577–21581
Niggli E, Lederer WJ (1991) Molecular operations of the sodium-calcium exchanger revealed by conformational currents. Nature 349:621–624
Nishiyama K, Azuma YT, Morioka A, Yoshida N, Teramoto M, Tanioka M, Kita S, Hayashi S, Nakajima H, Iwamoto T, Takeuchi T (2016) Roles of Na^+/Ca^{2+} exchanger isoforms NCX1 and NCX2 in motility in mouse ileum. Naunyn Schmiedeberg's Arch Pharmacol 389:1081–1090
Nishiyama K, Tanioka K, Azuma YT, Hayashi S, Fujimoto Y, Yoshida N, Kita S, Suzuki S, Nakajima H, Iwamoto T, Takeuchi T (2017) Na^+/Ca^{2+} exchanger contributes to stool transport in mice with experimental diarrhea. J Vet Med Sci 79:403–411
Nishizawa T, Kita S, Maturana AD, Furuya N, Hirata K, Kasuya G, Ogasawara S, Dohmae N, Iwamoto T, Ishitani R, Nureki O (2013) Structural basis for the counter-transport mechanism of a H+/Ca2+ exchanger. Science 341(6142):168–172
Okazaki K, Takada S (2008) Dynamic energy landscape view of coupled binding and protein conformational change: induced-fit versus population-shift mechanisms. Proc Natl Acad Sci U S A 105:11182–11187
Omelchenko A, Dyck C, Hnatowich M, Buchko J, Nicoll DA, Philipson KD, Hryshko LV (1998) Functional differences in ionic regulation between alternatively spliced isoforms of the Na^+-Ca^{2} $^+$ exchanger from *Drosophila melanogaster*. J Gen Physiol 111:691–702
On C, Marshall CR, Chen N, Moyes CD, Tibbits GF (2008) Gene structure evolution of the Na^+-Ca^{2+} exchanger (NCX) family. BMC Evol Biol 8:127–142
On C, Marshall CR, Perry SF, Le HD, Yurkov V, Omelchenko A, Hnatowich M, Hryshko LV, Tibbits GF (2009) Characterization of zebrafish (*Danio rerio*) NCX4: a novel NCX with distinct electrophysiological properties. Am J Physiol Cell Physiol 296:C173–C181
Ottolia M, Schumann S, Nicoll DA, Philipson KD (2002) Activation of the cardiac Na^+/Ca^{2+} exchanger by DEPC. Ann N Y Acad Sci 976:85–88

Ottolia M, Nicoll DA, Philipson KD (2009) Roles of two Ca^{2+}-binding domains in regulation of the cardiac Na^+-Ca^{2+} exchanger. J Biol Chem 284:32735–32741
Ottolia M, Nicoll DA, John S, Philipson KD (2010) Interactions between Ca^{2+} binding domains of the Na^+-Ca^{2+} exchanger and secondary regulation. Channels 4:1–4
Palty R, Sekler I (2012) The mitochondrial Na^+/Ca^{2+} exchanger. Cell Calcium 52:9–15
Palty R, Silverman WF, Hershfinkel M, Caporale T, Sensi SL, Parnis J, Nolte C, Fishman D, Shoshan-Barmatz V, Herrmann S, Khananshvili D, Sekler I (2010) NCLX is an essential component of mitochondrial Na^+/Ca^{2+} exchange. Proc Natl Acad Sci U S A 107:436–441
Peng J-B, Brown EM, Hediger MA (2003) Epithelial Ca^{2+} entry channels: transcellular Ca^{2+} transport and beyond. J Physiol 551:729–740
Peng JB, Suzuki Y, Gyimesi G, Hediger MA (2018) TRPV5 and TRPV6 calcium-selective channels. In: Kozak JA, Putney JW (eds) Calcium entry channels in non-excitable cells. CRC Press/Taylor & Francis, Boca Raton (FL), pp 241–274
Philipson KD, Nicoll DA (2000) Sodium-calcium exchange: a molecular perspective. Annu Rev Physiol 62:111–133
Philipson KD, Longoni S, Ward R (1988) Purification of the cardiac Na^+-Ca^{2+} exchange protein. Biochim Biophys Acta 945:298–306
Pittman JK, Hirschi KD (2016) Phylogenetic analysis and protein structure modelling identifies distinct Ca^{2+}/Cation antiporters and conservation of gene family structure within Arabidopsis and rice species. Rice (NY) 9(3):1–6
Plain F, Congreve SD, Yee RSZ, Kennedy J, Howie J, Kuo CW, Fraser NJ, Fuller W (2017) An amphipathic α-helix directs palmitoylation of the large intracellular loop of the sodium/calcium exchanger. J Biol Chem 292:10745–10752
Plattner H, Verkhratsky A (2016) Inseparable tandem: evolution chooses ATP and Ca^{2+} to control life, death and cellular signalling. Philos Trans R Soc Lond Ser B Biol Sci 371:1700
Purvis JE, Lahav G (2013) Encoding and decoding cellular information through signaling dynamics. Cell 152:945–956
Quednau BD, Nicoll DA, Philipson KD (1997) Tissue specificity and alternative splicing of the Na^+/Ca^{2+} exchanger isoforms NCX1, NCX2, and NCX3 in rat. Am J Physiol Cell Physiol 272: C1250–C1261
Quednau BD, Nicoll DA, Philipson KD (2004) The sodium/calcium exchanger family—SLC8. Pflugers Arch - Eur J Physiol 447:543–548
Rahamimoff H, Low W, Cook O, Furman I, Kasir J, Vatashski R (1996) The structural basis of Na^+-Ca^{2+} exchange activity. Ann N Y Acad Sci 779:29–36
Reeves JP, Condrescu M (2003) Allosteric activation of sodium-calcium exchange activity by calcium: persistence at low calcium concentrations. J Gen Physiol 122:621–639
Reeves JP, Hale CC (1984) The stoichiometry of the cardiac sodium-calcium exchange system. J Biol Chem 259:7733–7739
Refaeli B, Giladi M, Hiller R, Khananshvili D (2016) Structure-based engineering of lithium transporting capacity in the archaeal sodium-calcium exchanger (NCX_Mj). Biochemistry 55:1673–1676
Ren X, Philipson KD (2013) The topology of the cardiac Na^+/Ca^{2+} exchanger, NCX1. J Mol Cell Cardiol 57:68–71
Reuter H, Seitz N (1968) The dependence of calcium efflux from cardiac muscle on temperature and external ion composition. J Physiol 195:451–470
Rizo J, Sudhof TC (1998) C_2-domains, structure and function of a universal Ca^{2+}-binding domain. J Biol Chem 273:15879–15882
Rüdiger S (2014) Stochastic models of intracellular calcium signals. Phys Rep 534:39–87
Sakaue M, Nakamura H, Kaneko I, Kawasaki Y, Arakawa N, Hashimoto H, Koyama Y, Baba A, Matsuda T (2000) Na^+-Ca^{2+} exchanger isoforms in rat neuronal preparations: different changes in their expression during postnatal development. Brain Res 881:212–216

Salinas RK, Bruschweiler-Li L, Johnson E, Bruschweiler R (2011) Ca^{2+} binding alters the interdomain flexibility between the two cytoplasmic calcium-binding domains in the Na^{+}/Ca^{2+} exchanger. J Biol Chem 286:32123–32131

Schnetkamp PP, Jalloul AH, Liu G, Szerencsei RT (2014) The SLC24 family of K^{+}-dependent Na^{+}-Ca^{2+} exchangers: structure-function relationships. Curr Top Membr 73:263–287

Secondo A, Staiano IR, Scorziello A, Sirabella R, Boscia F, Adornetto A, Canzoniero LM, Di Renzo G, Annunziato L (2007) Ann N Y Acad Sci 1099:481–485

Sekler I (2015) Standing of giants shoulders the story of the mitochondrial Na^{+}-Ca^{2+} exchanger. Biochem Biophys Res Commun 460:50–52

Shlosman I, Marinelli F, Faraldo-Gómez JD, Mindell JA (2018) The prokaryotic Na^{+}/Ca^{2+} exchanger NCX_Mj transports Na^{+} and Ca^{2+} in a 3:1 stoichiometry. J Gen Physiol 150:51–65

Shubair M, Oriowo MA, Khan I (2012) Expression of alternatively spliced variants of Na–Ca-exchanger-1 in experimental colitis: role in reduced colonic contractility. Mol Cell Biochem 370:15–21

Sosnoski DM, Gay CV (2008) NCX3 is a major functional isoform of the sodium–calcium exchanger in osteoblasts. J Cell Biochem 103:1101–1110

Stahelin RV, Wang J, Blatner NR, Rafter JD, Murray D et al (2005) The origin of C1A-C2 interdomain interactions in protein kinase Cα. J Biol Chem 280:36452–36463

Stains JP, Gay CV (1998) Asymmetric distribution of functional sodium-calcium exchanger in primary osteoblasts. J Bone Miner Res 13(12):1862–1869

Sterne-Weiler T, Weatheritt RJ, Best AJ, Ha KCH, Blencowe BJ (2018) Efficient and accurate quantitative profiling of alternative splicing patterns of any complexity on a laptop. Mol Cell 72:187–200

Stoerger C, Flockerzi V (2014) The transient receptor potential cation channel subfamily V member 6 (TRPV6): genetics, biochemical properties, and functions of exceptional calcium channel proteins. Biochem Cell Biol 92:441–448

Szewczyk MM, Davis KA, Samson SE, Simpson F, Rangachari PK, Grover AK (2007) Ca^{2+}-pumps and Na^{+}-Ca^{2+}-exchangers in coronary artery endothelium versus smooth muscle. J Cell Mol Med 11:129–138

Tal I, Kozlovsky T, Brisker D, Giladi M, Khananshvili D (2016) Kinetic and equilibrium properties of regulatory Ca^{2+}-binding domains in sodium-calcium exchangers 2 and 3. Cell Calcium 59:181–188

Taneja M, Tyagi S, Sharma S, Upadhyay SK (2016) Ca^{2+}/Cation Antiporters (CaCA): identification, characterization and expression profiling in bread wheat (*Triticum aestivum* L). Front Plant Sci 7(1775):1–19

van Cromphaut SJ, Dewerchin M, Hoenderop JG, Stockmans I, Van HE, Kato S, Bindels RJ et al (2001) Duodenal calcium absorption in vitamin D receptor-knockout mice: functional and molecular aspects. Proc Natl Acad Sci U S A 98:13324–13329

van der Hagen EA, van Loon EP, Verkaart S, Latta F, Bindels RJ, Hoenderop JG (2015) The Na^{+}/Ca^{2+} Exchanger 1 (NCX1) Variant 3 as the Major Extrusion System in Renal Distal Tubular Transcellular Ca^{2+}-Transport. Nephron 131:145–152

van Dijk L, Giladi M, Refaeli B, Hiller R, Cheng MH, Bahar I, Khananshvili D (2018) Key residues controlling bidirectional ion movements in Na^{+}/Ca^{2+} exchanger. Cell Calcium 76:10–22

Van Eylen F, Bollen A, Herchuelz A (2001) J Endocrinol 168(3):517–526

Van Loon EPM, Little R, Prehar S, Bindels RJM, Cartwright EJ, Hoenderop JGJ (2016) Calcium extrusion pump PMCA4: a new player in renal calcium handling? PLoS One 11:e015348

Verkhratsky A, Parpura V (2014) Calcium signalling and calcium channels: evolution and general principles. Eur J Pharmacol 739:1–3

Verkhratsky A, Trebak M, Perocchi F, Khananshvili D, Sekler I (2018) Crosslink between calcium and sodium signalling. Exp Physiol 103:157–169

Waight AB, Pedersen BP, Schlessinger A, Bonomi M, Chau BH, Roe-Zurz Z, Risenmay AJ, Sali A, Stroud RM (2013) Structural basis for alternating access of a eukaryotic calcium/proton exchanger. Nature 499:107–110

Wanichawan P, Louch WE, Hortemo KH, Austbø B, Lunde PK, Scott JD, Sejersted OM, Carlson CR (2011) Full-length cardiac Na^+/Ca^{2+} exchanger 1 protein is not phosphorylated by protein kinase A. Am J Physiol Cell Physiol 300:C989–C997

Watson LP, Kang YH, Falk MC (1989) Cytochemical properties of osteoblast cell membrane domains. J Histochen Cytochem 37:1235–1246

White KE, Gesek FA, Friedman PA (1996) Structural and functional analysis of Na^+/Ca^{2+} exchange in distal convoluted tubule cells. Am J Physiol Renal Physiol 271:F560–F570

Williams RJP (1999) Calcium: the developing role of its chemistry in biological evolution. In: Carafoli E, Klee C (eds) Calcium as a cellular regulator. Oxford University Press, New York, NY, pp 3–27

Wu M, Wang M, Nix J, Hryshko LV, Zheng L (2009a) Crystal structure of CBD2 from the *Drosophila* Na^+/Ca^{2+} exchanger: diversity of Ca^{2+} regulation and its alternative splicing modification. J Mol Biol 387:104–112

Wu M, Wang M, Nix J, Hryshko LV, Zheng L (2009b) Crystal structure of CBD2 from the Drosophila Na^+/Ca^{2+} exchanger: diversity of Ca^{2+} regulation and its alternative splicing modification. J Mol Biol 387:104–112

Wu M, Le HD, Wang M, Yurkov V, Omelchenko A, Hnatowich M, Nix J, Hryshko LV, Zheng L (2010) Crystal structures of progressive Ca^{2+} binding states of the Ca^{2+} sensor Ca^{2+} binding domain 1 (CBD1) from the CALX Na^+/Ca^{2+} exchanger reveal incremental conformational transitions. J Biol Chem 285:2554–2561

Wu M, Tong S, Gonzalez J, Jayaraman V, Spudich JL, Zheng L (2011) Structural basis of the Ca^{2+} inhibitory mechanism of Drosophila Na^+/Ca^{2+} exchanger CALX and its modification by alternative splicing. Structure 19:1509–1517

Wu M, Tonga S, Walterspergerb S, Diederichsc K, Wang M, Zheng L (2013) Crystal structure of Ca^{2+}/H^+ antiporter protein YfkE reveals the mechanisms of Ca^{2+} efflux and its pH regulation. Proc Natl Acad. Sci USA 110:1367–11372

Wu X, Weng L, Zhang J, Liu X, Huang J (2018) The plasma membrane calcium ATPases in calcium signaling network. Curr Protein Pept Sci 19:813–822

Yuan J, Yuan C, Xie M, Yu L, Bruschweiler-Li L, Brüschweiler R (2018) The intracellular loop of the Na^+/Ca^{2+} exchanger contains an "Awareness ribbon"-shaped two-helix bundle domain. Biochemistry 57:5096–5104

Zajac M, Dolowy K (2017) Measurement of ion fluxes across epithelia. Prog Biophys Mol Biol 127:1–11

Zhang YH, Hancox JC (2009) Regulation of cardiac Na^+-Ca^{2+} exchanger activity by protein kinase phosphorylation–still a paradox. Cell Calcium 45:1–10

Zhekova HR, Zhao C, Schnetkamp PP, Noskov SY (2016) Characterization of the cation binding sites in the NCKX2 Na^+/Ca^{2+}-K^+ exchanger. Biochemistry 55:6445–6455

Zheng YM, Wang YX (2007) Sodium-calcium exchanger in pulmonary artery smooth muscle cells. Ann N Y Acad Sci 1099:427–435

Chapter 9
Urea Transporters in Health and Disease

Janet D. Klein and Jeff M. Sands

Abstract Urea transporters are a relatively recent discovery in the overall history of kidney physiology. Urea transport proteins were first proposed in the late 1980s when observations of urea permeability could not be explained by paracellular transport or lipid phase diffusion. The first urea transporter was cloned in 1992 and shortly thereafter a family of urea transporters was described. There are two major subgroups of the SLC14A family of urea transporters: *Slc14A2* (or UT-A) and *SLC14A1* (or UT-B). UT-A1 and UT-A3, which are expressed in the inner medullary collecting duct (IMCD) are crucial to the kidney's ability to concentrate urine. UT-A2, which is expressed in the thin descending limb is also involved since knock-out of any of these urea transporters results in a urine concentrating defect in mice. The regulation of urea transporter activity in the IMCD involves modification of UT-A1 and UT-A3 through phosphorylation, glycosylation, and ubiquitination, with subsequent effects on the accumulation of urea transporters in the apical plasma membrane. Long-term regulation of the IMCD urea transporters involves changes in protein abundance in response to changes in hydration status, low protein diets, or adrenal steroids. Vasopressin regulates UT-A1 and UT-A3 through PKA. Hypertonicity regulates UT-A1 through PKCα. The UT-B (*Slc14A1*) urea transporter was originally isolated from erythrocytes. In the kidney, UT-B is expressed primarily in the descending vasa recta. Urea transporters have been studied using animal models of disease including diabetes insipidus, diabetes mellitus, lithium administration, hypertension, and nephrotoxic drug responses. Several genetically engineered mouse models have been developed to study the different urea transporters. Urearetics, which are inhibitors of UT-A and/or UT-B, are being developed and tested as novel diuretic agents.

J. D. Klein · J. M. Sands (✉)
Renal Division, Department of Medicine and Department of Physiology, Emory University School of Medicine, Atlanta, GA, USA
e-mail: janet.klein@emory.edu; jeff.sands@emory.edu

K. L. Hamilton, D. C. Devor (eds.), *Studies of Epithelial Transporters and Ion Channels*, Physiology in Health and Disease,
https://doi.org/10.1007/978-3-030-55454-5_9

Keywords Urea · Urea transport · Vasopressin · Urine concentrating mechanism

9.1 Introduction

Urea ($O{=}C(-NH_2)_2$) is a highly polar small molecule that has a very low lipid solubility across artificial lipid bilayers (4×10^{-6} cm/s) that lack a urea transporter (Galluci et al. 1971). The permeability of urea across artificial lipid bilayers is quite low, but it is not zero. Thus, urea will diffuse across cell membranes and achieve equilibrium with sufficient time for equilibration. However, the transit time for tubular fluid through the kidney collecting duct, or for red blood cells through the medullary vasa recta, is too fast to allow urea concentrations to reach equilibrium solely by passive diffusion (reviewed in Klein et al. (2011)).

Facilitated transport of urea was initially proposed in rabbit and rat terminal inner medullary collecting ducts (IMCDs) in 1987 (Sands et al. 1987). Several physiologic studies functionally characterized vasopressin-regulated urea transport (reviewed in Klein et al. (2011)) and led to the expression cloning of the first urea transporter in 1993 (You et al. 1993). At present, two urea transporter genes have been cloned: the *Slc14A2* (UT-A) gene encodes 9 cDNA and 6 protein isoforms; and the *Slc14A1* (UT-B) gene encodes 2 isoforms (reviewed in Klein et al. (2011), Shayakul et al. (2013), Sands and Blount (2014)).

Since urea is freely permeable across cell membranes and is not osmotically active, why is a urea transporter necessary? The short answer: to move sufficient urea into the interstitium to promote the production of concentrated urine (Fig. 9.1).

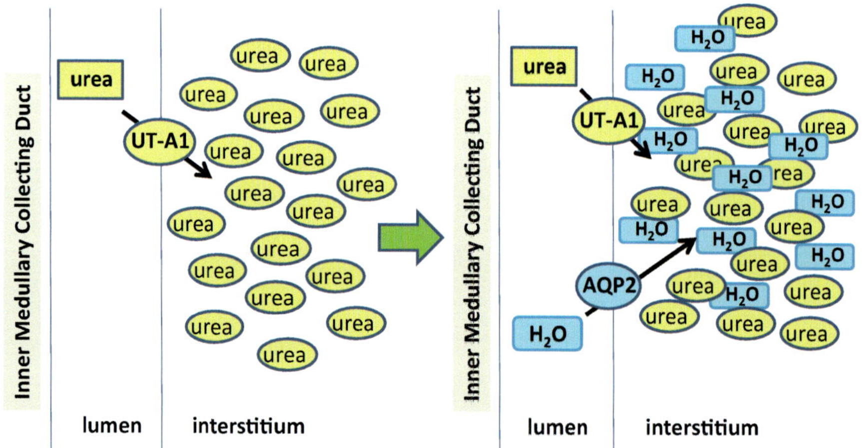

Fig. 9.1 Role of urea transporter in urine concentration mechanism. Urea is moved from the tubule lumen into the interstitium through urea transporters (UT-A1). The increased urea in the interstitium creates a more hypertonic milieu that promotes reabsorption of water through aquaporin-2 (AQP2). Removal of water from the lumen results in a more concentrated urine

Gamble and colleagues used clearance studies to demonstrate “an economy of water in renal function referable to urea,” i.e., more water is reabsorbed when urea is available than in the absence of urea (Gamble et al. 1934). Seven decades later, Fenton and colleagues confirmed Gamble’s findings using urea transporter knock-out mice, showing the importance of urea in the generation of a concentrated urine (Gamble et al. 1934; Fenton et al. 2006a). Maximal urine concentrating capacity is reduced in several mammalian species, including humans, during periods of protein deprivation, and concentrating ability is restored by urea infusion or protein refeeding (reviewed in Klein et al. (2011), Weiner et al. (2015)). Maximal urine concentrating ability is reduced in knock-out mice lacking: UT-A1/UT-A3 (Fenton et al. 2004, 2005; Jacob et al. 2008); UT-A2 (Uchida et al. 2005); UT-B (Yang et al. 2002; Yang and Verkman 2002; Klein et al. 2004); UT-A2 and UT-B (Lei et al. 2011); or all UTs (Jiang et al. 2017). The magnitude of the concentrating defect varies depending upon which urea transporter is knocked out. However, they all develop a urine concentrating defect, emphasizing the importance of urea and urea transporters to the production of concentrated urine in the inner medulla.

9.2 Regulation of Urea Transporters

9.2.1 Urea Transporters

The facilitated urea transporter gene families that have been cloned from mammals are listed in Table 9.1. UT-A1 is expressed in the apical plasma membrane of terminal IMCDs (reviewed in Klein et al. (2011)). UT-A3 is also expressed in terminal IMCDs, primarily in the basolateral plasma membrane but also in the apical plasma membrane in some studies. UT-A2 is expressed in thin descending limbs of the loop of Henle in both the apical and basolateral membranes (Wade et al. 2000; Lim et al. 2006). UT-A4 mRNA is detected in rat kidney medulla, although its protein has not been detected and UT-A4 mRNA has not been detected in mouse kidney. UT-A5 is expressed in testis (Fenton et al. 2000). UT-A6 is expressed in the human colon (Smith et al. 2004) and in Caco-2 cells (McGrane and Stewart 2016). Neither UT-A5 nor UT-A6 are expressed in the kidney. UT-B protein is expressed in erythrocytes and fenestrated endothelial cells in the descending vasa recta. UT-A1 exists as two glycoprotein forms of 117 and 97 kDa (Bradford et al. 2001). UT-A3 also exists as two glycoprotein forms of 65 and 45 kDa; the 65 kDa form is more stable and is enriched in lipid rafts (Su et al. 2012a).

Urea transporter evolution in vertebrates was analyzed in a recent review (LeMoine and Walsh 2015). Three homologs, UT-A, UT-C, and UT-D, are thought to have evolved from a single ancestral UT in piscine lineages, followed by a reduction to a single UT-A (LeMoine and Walsh 2015). UT-A1 reflects internal tandem duplication whereas a second duplication led to UT-B (LeMoine and Walsh 2015). This analysis proposes that a non-ornithine urea cycle-based production of urea is important for the generation of polyamines and neurotransmitters, and

Table 9.1 Facilitated urea transporter gene families in mammals

Gene name	Isoform name	RNA (kb)	Protein (kDa)	Vasopressin sensitive	Tissue location
Slc14a2	UT-A1	4.0	97, 117	Yes	IMCD
	UT-A1b	3.5			Medulla[a]
	UT-A2	2.9	55	No[b]	tDL, liver
	UT-A2b	2.5			Medulla[a], heart
	UT-A3	2.1	44, 67	Yes	IMCD
	UT-A3b	3.7			Medulla[a]
	UT-A4[c]	2.5	43	Yes	Medulla[a]
	UT-A5[d]	1.4			Testis
	UT-A6[e]	1.8			Colon
Slc14a1	UT-B1[f]	3.8	43		DVR, RBC[g]
	UT-B2	3.7	43–54	No	Bovine rumen

Isoform names are based upon the urea transporter nomenclature proposed in Sands et al. (1997); Vasopressin sensitive, urea flux is stimulated by vasopressin
IMCD inner medullary collecting duct, *tDL* thin descending limb
[a]*Medulla* (exact tubular location unknown)
[b]*No*, no in rat but yes in mouse
[c]Cloned from rat only
[d]Cloned from mouse only; *DVR* descending vasa recta, *RBC* red blood cells
[e]Cloned from human only
[f]Referred to as UT-B in text
[g]Also expressed in several other tissues and endothelial cells
Citations to original publications can be found in Klein et al. (2011)

provides an evolutionary explanation for urea transporter expression in extra-renal tissues (LeMoine and Walsh 2015).

9.3 Membrane Association and Transporter Activity

To transport urea, UT-A1 must be in the apical plasma membrane. Three components contribute to this aspect of regulation: regulation of trafficking and insertion into the membrane; regulation of retention and activity in the membrane; and regulation of removal from the membrane.

9.3.1 Trafficking and Insertion of UT-A1

9.3.1.1 Regulation by Vasopressin

In freshly isolated suspensions of rat IMCDs, the phosphorylation and apical plasma membrane accumulation of both UT-A1 and UT-A3 are increased by vasopressin

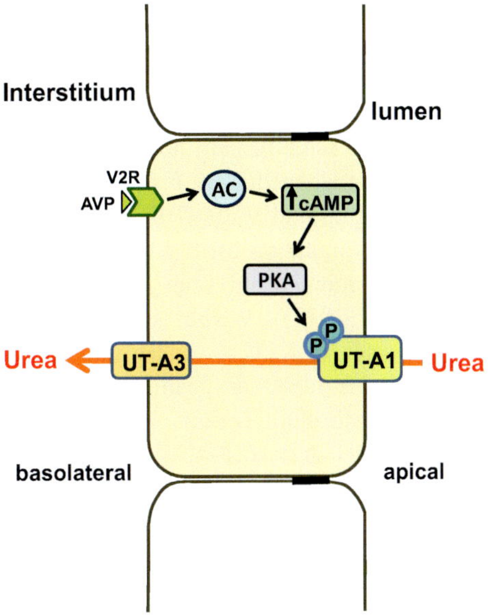

Fig. 9.2 Mechanism of urea transporter activation by vasopressin. Vasopressin (AVP) binds to the type 2 vasopressin receptor (V2R) and activates adenylyl cyclase (AC). AC increases cAMP, which activates cAMP-dependent protein kinase A (PKA) to phosphorylate UT-A1 on serines 486 and 499. Phosphorylation of UT-A1 facilitates insertion into the apical plasma membrane where it moves urea from the lumen into the cell. Urea exits through UT-A3 at the basolateral surface into the interstitium

(Zhang et al. 2002; Blount et al. 2007). Vasopressin phosphorylates UT-A1 at serines 486 and 499 (Blount et al. 2008a). Both phospho-S486-UT-A1 and phospho-S499-UT-A1 are primarily detected at the apical plasma membrane, although the two PKA phosphorylation events are independent from one another (Hoban et al. 2015). In contrast to UT-A1, the vasopressin-regulated phosphorylation site(s) in UT-A3 have not been determined, except that neither of the two conventional PKA consensus sites is involved (Smith et al. 2004). UT-A1 phosphorylation, apical plasma membrane accumulation, and urea transport are stimulated by vasopressin through two cAMP-dependent pathways (Fig. 9.2): PKA and Epac (exchange protein activated by cAMP) (Wang et al. 2009). Activation of the Epac pathway increases total UT-A1 phosphorylation but does not change the phosphorylation of either S486 or S499, suggesting that Epac is phosphorylating a different residue in UT-A1 than PKA (Hoban et al. 2015).

A phospho-proteomic study investigated the effect of a selective inhibitor of the type 2 vasopressin receptor (V2R), satavaptan, in a system-wide analysis (Hoffert et al. 2014). Satavaptan blocked V2R-mediated inhibition of proline-directed kinases and V2R-mediated activation of basophilic kinases (Hoffert et al. 2014). Thus, satavaptan and vasopressin affect many of the same signaling pathways, but in opposite directions (Hoffert et al. 2014).

In addition to the acute effects of vasopressin on UT-A1 and UT-A3 phosphorylation and plasma membrane accumulation, vasopressin exerts longer-term regulation through changes in urea transporter protein abundance; changes that do not appear to be regulated by transcription (reviewed in Klein et al. (2011), Klein (2014)). Brattleboro rats have central diabetes insipidus and are frequently used to investigate vasopressin effects since they have intact vasopressin receptors despite no endogenous vasopressin. Surprisingly, giving vasopressin to Brattleboro rats for 5 days decreases UT-A1 protein abundance in the inner medulla (Terris et al. 1998; Kim et al. 2004). However, UT-A1 protein abundance is significantly increased when vasopressin is administered for 12 days (Kim et al. 2004). The delayed increase in UT-A1 protein abundance is consistent with the time required for an increase in inner medullary urea content after vasopressin treatment of Brattleboro rats (Harrington and Valtin 1968). Analysis of the UT-A promoter I may provide a molecular explanation for the varied temporal response since it contains a tonicity enhancer element (TonE) but not a cAMP response element (CRE) (reviewed in Klein et al. (2011)). Vasopressin administration may initially increase Na-K-2Cl co-transporter (NKCC2) transcription in the thick ascending limb, which will increase inner medullary osmolality, and which in turn would increase UT-A1 transcription through TonE.

UT-A2, UT-A2b, UT-A3, and UT-A3b mRNA abundances are reduced in the inner medulla of water-loaded rats and rise in rats and mice that are water restricted or treated with vasopressin or dDAVP, a selective agonist of the V2R (reviewed in Klein et al. (2011)). Histochemistry also shows decreased UT-A2 and UT-A3 protein expression in 3 day water-loaded (vasopressin depleted) rats (Lim et al. 2006). Western blot shows decreased UT-A2 and UT-A3 protein abundances in rats that have been treated with furosemide. Conversely, UT-A2 and UT-A3 protein abundances are increased in the medulla of water-deprived or dDAVP-treated rats (reviewed in Klein et al. (2011)). UT-A2 regulation could involve transcriptional mechanisms stimulated by vasopressin since UT-A2 is under the control of UT-A promoter II, a unique internal promoter in the UT-A gene, which contains a CRE element (Nakayama et al. 2000). UT-A3 is under the control of UT-A promoter I, the same promoter as UT-A1, and which contains a tonicity-responsive enhancer (TonE) that could be mediated by tonicity-responsive transcription (Nakayama et al. 2000).

In inner medullary thin limb segments from Munich-Wistar rats, urea permeability was lower in the upper portion than in the lower portion of the thin descending limb or in the thin ascending limb (Nawata et al. 2014). Phloretin, a urea transport inhibitor (Chou and Knepper 1989) did not inhibit urea permeability in the upper portion of the thin descending limb, suggesting that urea transport in this segment is not mediated by UT-A2 (Nawata et al. 2014). A subsequent molecular cloning study identified two novel UT-A2 variants in the lower portion of the thin descending limb and in the thin ascending limb, UT-A2c and UT-A2d, as well as a variant of the sodium-glucose transporter 1, SGLT1a, that may mediate the high urea permeability (Nawata et al. 2015).

Loops of Henle are generally classified as either short- or long-looped, depending upon whether the loop extends into the inner medulla. An intermediate type of loop of Henle was recently identified (Kim et al. 2016). This intermediate loop extends into the inner medulla but turns within the first millimeter. UT-A2-positive and UT-A2-negative cells were intermingled in the type 1 epithelial cells of the intermediate loop. The UT-A2-positive thin descending limbs of the intermediate loops extended into the UT-A2-negative type 1 epithelial cells in the initial portion of the inner medulla, suggesting that intrarenal urea recycling may occur in the UT-A2-positive segment in the inner stripe of the outer medulla (Kim et al. 2016).

9.3.1.2 Regulation by Hyperosmolality

Urea transporters are also stimulated by hyperosmolality. Inner medullary interstitial osmolality is high during antidiuresis. In rat terminal IMCDs, hyperosmolality increases urea permeability, even in the absence of vasopressin, suggesting that it is an independent activator of urea transport (Sands and Schrader 1991; Gillin and Sands 1992; Kudo et al. 1992). In contrast to vasopressin, which stimulates urea permeability via increases in cAMP (Star et al. 1988), hyperosmolality stimulates urea permeability via activation of PKCα and intracellular calcium (Gillin et al. 1993; Kato et al. 2000; Wang et al. 2010, 2013). Similar to vasopressin, hyperosmolality increases the phosphorylation and plasma membrane accumulation of both UT-A1 and UT-A3 (Zhang et al. 2002; Klein et al. 2006a; Blount et al. 2007; Blessing et al. 2008).

UT-A1 is phosphorylated by PKC (Fig. 9.3), specifically by PKCα, at serine 494, which is not a PKA site (Wang et al. 2010, 2013; Klein et al. 2012; Blount et al. 2015). Two PKC activators, hyperosmolality and phorbol dibutyrate, increase UT-A1 phosphorylation at S494, while activators of either PKA or Epac do not (Blount et al. 2015). UT-A1 apical plasma membrane accumulation is increased by activation of both PKC and PKA, but not by PKC alone (Blount et al. 2015). These findings suggest that PKC may increase vasopressin-stimulated urea transport by phosphorylating UT-A1 at S494 and enhancing its retention in the apical plasma membrane (Blount et al. 2015).

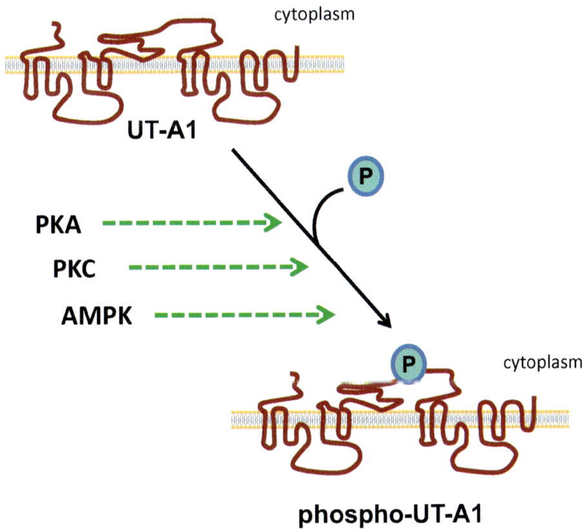

Fig. 9.3 Regulation of UT-A1 phosphorylation by multiple kinases. UT-A1 is phosphorylated at serines located in the large intracellular loop near the center of the protein. PKA phosphorylates at serines 486 and 499. PKC phosphorylates at serine 494 in response to hyperosmolar conditions. AMPK can stimulate phosphorylation at serine 486 and serine 499, but it is not known whether it has additional phosphorylation sites

There are two glycoprotein forms of UT-A1: 117 and 97 kDa (Bradford et al. 2001). UT-A1 accumulation in the apical plasma membrane is increased by overexpression or activation of PKCα through increases in UT-A1 sialylation (Li et al. 2015). PKCα knock-out mice have reduced levels of UT-A1 sialylation, an effect that is mediated by Src kinase (Li et al. 2015). PKC inhibition blocks the induction of UT-A1 sialylation by high glucose, suggesting that this pathway may be important in ameliorating the osmotic diuresis caused by glucosuria in patients with diabetes mellitus (Li et al. 2015). UT-A3 glycan sialylation is also enhanced by PKC and mediated by ST6GalI (Qian et al. 2016).

9.3.1.3 Regulation by Other Factors

Adenosine-activated monophosphate kinase (AMPK) is an energy-sensing serine/threonine kinase that is stimulated by hypoxia and osmotic stress (Pastor-Soler and Hallows 2012). AMPK activation increases urea and water transport through increases in UT-A1 and aquaporin-2 (AQP2) phosphorylation in IMCDs (Klein et al. 2016a). In two rodent models of congenital nephrogenic diabetes insipidus (NDI), rats treated with a V2R inhibitor, tolvaptan, or V2R knock-out mice, an AMPK activator, metformin, increased both UT-A1 and AQP2 protein abundances in the inner medulla, and increased urine osmolality (Efe et al. 2016). Thus, AMPK activation may represent a vasopressin-independent pathway to increase urine concentration and potentially treat congenital NDI (reviewed in Sands and Klein (2016)).

Ibuprofen and meloxicam are non-steroidal anti-inflammatory drugs (NSAIDs) that alter the phosphorylated forms of AQP2 (Ren et al. 2015). However, UT-A1 phosphorylation is unchanged by either medication in rat inner medulla (Ren et al. 2015).

Proteomic analysis of urea transporters and aquaporins has revealed a number of potential interacting proteins. This study compared proteins that were increased or decreased following vasopressin treatment of rat kidney. Proteins were collected by chemical cross-linking followed by immunoprecipitation and identified with LC-MS/MS analysis. The proteins that interact with urea transporters encompass numerous functional types from kinases and phosphatases to structural proteins to proteins involved in other posttranslational modifications such as ubiquitination or neddylation (Chou et al. 2018). Any of these modifying proteins could represent potential pathways for regulation.

9.3.2 Membrane Association and Activation

In lipid rafts, UT-A1 interacts with caveolin-1 ((Feng et al. 2009) and reviewed in Chen (2013)). UT-A1 endocytosis occurs by a dynamin-dependent process involving both caveolae and clathrin-coated pit pathways (Huang et al. 2010). Rab14 binds

UT-A1 and decreases UT-A1 urea transport activity by increasing clathrin-dependent UT-A1 endocytosis (Su et al. 2013b). UT-A1 co-localizes with cortical filamentous (F)-actin at the apical plasma membrane and in the subapical cytoplasm (Xu et al. 2012). Cortical F-actin regulates UT-A1 bioactivity by directly interacting with UT-A1, and controls UT-A1 cell surface expression by affecting both trafficking and endocytosis (Xu et al. 2012). Hepatocyte nuclear factor-1β (HNF-1β) is involved in UT-A1 vesicle trafficking to the apical plasma membrane as UT-A1 and collectrin are decreased in HNF-1β knock-out mice (Aboudehen et al. 2017). Vesicular trafficking of UT-A1 is linked to the SNARE machinery by Snapin (Mistry et al. 2007). Glycogen synthase kinase 3 (GSK3) affects UT-A1 intracellular redistribution and apical plasma membrane accumulation in response to hypertonicity (Li et al. 2016).

The 14-3-3 protein family consists of seven isoforms that regulate protein function by binding to phosphorylated serine or threonine residues. UT-A1 is bound by 14-3-3γ, and this binding is enhanced by PKA activation (Feng et al. 2015). 14-3-3γ increases UT-A1 ubiquitination and degradation by interacting with MDM2, an E3 ubiquitin ligase, and decreases urea transport (Feng et al. 2015). Thus, UT-A1 phosphorylation is increased by PKA activation, and UT-A1 degradation is increased by the subsequent binding of 14-3-3γ. If these two opposing effects occur sequentially, they could form a negative feedback mechanism to return the UT-A1 function to its basal state following stimulation by vasopressin (Feng et al. 2015). While the data showing these opposite effects of PKA are established, it remains to be determined whether there is a negative feedback loop or whether only one effect is dominant in vivo.

Collecting duct-specific deletion of an epithelial transcription factor, grainyhead-like 2 (GRHL2), in mice, results in nephrogenic diabetes insipidus (Hinze et al. 2018). However, there were no differences in either protein abundance or localization of UT-A1 or UT-A3 between control and Grhl2 collecting duct knock-out mice (Hinze et al. 2018).

9.3.3 Membrane Removal and Degradation

In addition to the regulation of UT-A1 by phosphorylation and glycosylation, UT-A1 is regulated by ubiquitination (reviewed in Chen (2014)). Both IMCD urea transporters, UT-A1 and UT-A3, are ubiquitinated (Stewart et al. 2008; Chen et al. 2008). The ubiquitin ligase MDM2 mediates UT-A1 ubiquitination and degradation (Chen et al. 2008). Activation of adenylyl cyclase with forskolin induces UT-A1 ubiquitination in Madin–Darby canine kidney (MDCK) cells that stably express UT-A1 (Su et al. 2012b). Forskolin induces mono-ubiquitination of UT-A1 and stimulates UT-A1 degradation through lysosomes (Su et al. 2013a). Thus, UT-A1 undergoes poly-ubiquitination under non-vasopressin stimulated conditions and mono-ubiquitination after vasopressin stimulation (Su et al. 2012b, 2013a).

9.4 Inner Medullary Architecture and Urea Transport

Inner medullary structure has important implications for the urinary concentrating mechanism ((Westrick et al. 2013) and reviewed in Pannabecker (2013)). In the inner medullary base, initial IMCDs form clusters that coalesce along the cortico-medullary axis (Kriz 1967; Kriz et al. 1972; Kriz and Bankir 1983; Lemley and Kriz 1987; Pannabecker and Dantzler 2004, 2007; Pannabecker et al. 2004, 2008b; Issaian et al. 2012; Layton et al. 2012a; Wei et al. 2015). Thin descending limbs, which are located at the periphery of these clusters, appear to form an asymmetric ring around each IMCD cluster, whereas thin ascending limbs are located more uniformly among the thin descending limbs and IMCDs (Pannabecker et al. 2008b; Yuan and Pannabecker 2010). In rats, approximately four ascending vasa recta surround each IMCD, with 1–2 thin ascending limbs lying between each ascending vasa recta and opposite to the IMCD (Gilbert and Pannabecker 2013). Thin limbs, both descending and ascending, enter and exit IMCD clusters to form an interstitial nodal space that runs axially through the inner medulla and that may carry urea, NaCl, and water (Gilbert and Pannabecker 2013) and may facilitate preferential mixing of solutes and water within these spaces (Layton et al. 2012b). In contrast to rodents, interstitial nodal spaces are relatively infrequent in humans (Wei et al. 2015).

Recently, hyperpolarized C-13 urea magnetic resonance imaging (MRI) techniques have been developed to facilitate in vivo assessment of urea gradients in the inner medulla (Von Morze et al. 2012; Reed et al. 2016). In rat, dynamic C-13 MRI following a bolus infusion of hyperpolarized C-13 urea showed significant signal differences between acute diuretic and antidiuretic states, with more rapid medullary enhancement during antidiuresis, consistent with known upregulation of UT-A1 (Von Morze et al. 2012). This technique was subsequently improved using C-13, N-15 urea T2 relaxometry, which can detect two steps in the urea handling process: glomerular filtration and IMCD urea transport mediated by UT-A1 and UT-A3 (Reed et al. 2016). This technique has also been used to study renal oxygen tension following acute kidney injury (Mariager et al. 2017) and the hemodynamic changes associated with glucagon infusion (Qi et al. 2019).

9.5 UT-B

The red blood cell urea transporter, UT-B, is also the Kidd blood group antigen (reviewed in Klein et al. (2011), Yang and Sands (2014), Ran et al. (2014), Yang (2014)). There are three antigens: Jka, Jkb, and Jk3. Patients who lack UT-B (JK-null) have a urine concentrating defect (Sands et al. 1992). Most of the patients lacking Kidd antigen are from Polynesia, New Zealand, or Finland (reviewed in Hamilton (2015), Lawicki et al. (2017)). However, novel mutations have recently been reported from a Caucasian woman of Polish-Czech descent (Ramsey et al. 2016),

China (Zhang et al. 2016), and Japan (Onodera et al. 2014). There is no association between Kidd phenotype and chronic kidney disease (Capriolli et al. 2017).

9.5.1 Functional Role of UT-B

Functionally, while all studies agree that UT-B transports urea, some studies also find that rat or mouse UT-B can function as a water channel when expressed in Xenopus oocytes (Yang and Verkman 1998; Yang et al. 2002). UT-B has a single channel water permeability similar to aquaporin 1 (Huang et al. 2017). However, a third study showed that rat UT-B specifically transports urea when a physiological expression level is achieved in oocytes, but higher levels of UT-B expression that are not physiological result in transport of both urea and water (Sidoux-Walter et al. 1999). A study of urea transport in red blood cells from *Amphiuma*, dog, and human showed that UT-B only transports urea and does not transport water (Brahm 2013). In contrast, a study of human UT-B concluded that it can transport water and should be considered a member of the water channel family (Azouzi et al. 2013). Another study also found that human UT-B can transport water and ammonia, in addition to urea, when expressed in *Xenopus* oocytes, and concludes that UT-B can function as a gas channel (Geyer et al. 2013). UT-B can act as an active water transporter in C6 glial cells (Ogami et al. 2006). UT-B does not play a role in hydroxyurea transport in patients with sickle cell anemia (Walker and Ofori-Acquah 2017).

9.5.2 Localization of UT-B

9.5.2.1 Kidney

In the kidney, UT-B is expressed in the fenestrated endothelial cells of the descending vasa recta in the outer medulla (Pallone et al. 1994). The inner medullary descending vasa recta also express UT-B and transport urea (Evans et al. 2015). UT-B protein expression is found in descending vasa recta in the human kidney, with expression decreasing with depth below the outer medulla (Wei et al. 2015). UT-B protein is also expressed in the papillary surface epithelium in the lower inner medulla, providing a paracellular pathway for urea transport across this epithelium (Wei et al. 2015). UT-B inhibition leads to the upregulation of the L-arginine-eNOS-nitric oxide pathway and plays an important role in vaso-relaxation and regulating blood pressure (Sun et al. 2016). The abundance of UT-B-positive descending vasa recta clusters is similar in mice with genetically attenuated, angiopoietin-Tie2 signaling, and control mice (Kenig-Kozlovsky et al. 2018).

The effect of vasopressin on UT-B expression varies in different studies. One study showed that long-term dDAVP infusion downregulates UT-B protein abundance (Trinh-Trang-Tan et al. 2002). In contrast, another study using histochemistry showed that 3 days of water loading increases UT-B protein expression without changing UT-B's subcellular distribution (Lim et al. 2006). The effect of vasopressin or dDAVP on UT-B mRNA varies with kidney region and duration of treatment (Promeneur et al. 1998). Vasopressin does not affect UT-B2 abundance (Tickle et al. 2009). UT-B and miR-200c are differentially regulated by dehydration in the outer versus inner medulla, with an inverse relationship between UT-B protein abundance and miR-200c expression (Wang et al. 2016).

Hyperpolarized C-13 urea MRI has been used to determine the kinetics of urea transport mediated by UT-B in human red blood cells on the sub-minute time scale (Pages et al. 2013). Cells expressing UT-B, or control cells, were implanted into mice to investigate the utility of a gene reporter system based upon UT-B and C-13 hyperpolarized urea (Patrick et al. 2015). The apparent diffusion coefficient of hyperpolarized urea was reduced by UT-B, suggesting that it has the potential to be used as a magnetic resonance-based gene reporter in vivo (Patrick et al. 2015). UT-B has been used as a gene reporter to measure water transport by proton MRI in transfected HEK cells (Schilling et al. 2017). Potentials of mean force for solute permeation were tested for use in measuring urea transport through UT-B. This technique was not found to be suitable for urea but was suitable for water transport through aquaporin 1 (Ariz-Extreme and Hub 2017).

9.5.2.2 Bladder

UT-B is expressed in human bladder, primarily the UT-B1 mRNA isoform, although some UT-B2 mRNA is also detected (Walpole et al. 2014). There is strong UT-B protein expression throughout all layers of the urothelium, except for the apical membrane of the outermost umbrella cells (Walpole et al. 2014). The American black bear, an animal that hibernates, has similar levels of UT-B expression in the urothelium as found in other mammals (Spector et al. 2015).

9.5.2.3 Gastrointestinal Tract/Rumen

UT-B is important for salvaging urea nitrogen from the bovine gastrointestinal tract, especially from the rumen (Coyle et al. 2016; Sacca et al. 2018). In Holstein calves, UT-B mediates the serosal-to-mucosal flux of urea across the isolated ruminal epithelium, in addition to aquaporins 3, 7, and 10 (Walpole et al. 2015). In addition to the rumen, UT-B protein is expressed in the bovine parotid salivary gland (Dix et al. 2013). UT-B has been posited as the major urea transporter in the human intestinal tract (Walpole et al. 2018).

Butyrate is thought to enhance urea recycling in the rumen (Agarwal et al. 2015). Butyrate infusion into the rumen does not affect nitrogen retention in growing sheep but may affect the redistribution of urea nitrogen fluxes (Agarwal et al. 2015). During the transition from milk-feeding to solid-feeding, UT-B and aquaporin-3 protein abundances, but not aquaporin-7, increase in the calves (Berends et al. 2014). Ruminal UT-B abundance increases with solid food intake, indicating that an increase in digested nitrogen and urea recycling contributes to the increase in nitrogen retention with increasing solid food intake in milk-fed calves (Berends et al. 2014). Both UT-B1 and UT-B2 mRNA transcripts are detected in sheep rumen, and urea transport is regulated by changes in ammonia concentrations and pH (Lu et al. 2014). An acidic pH and short-chain fatty acids increase UT-B mRNA and protein abundance in primary cultures of goat rumen epithelial cells (Lu et al. 2015). Feeding goats with a diet containing large amounts of nitrogen and non-fiber carbohydrates increases ruminal UT-B protein (Lu et al. 2015).

9.5.2.4 Choroid Plexus

A high salt diet (8% for 2 weeks) reduces UT-B mRNA and protein abundances in the choroid plexus of salt-sensitive Dahl S rats, but not in salt-resistant Dahl R rats (Guo et al. 2015). In contrast, the high salt diet did not alter UT-A mRNA abundance in either Dahl S or Dahl R rats (Guo et al. 2015).

9.6 Urea Transporter Structure

The crystal structure of two evolutionarily distant urea transporters has been solved (reviewed in Levin and Zhou (2014)). The first crystal structure for any urea transporter was the bacterium *Desulfovibrio vulgaris* (dv) urea transporter (dvUT), which is a homolog of mammalian kidney urea transporters (Levin et al. 2009). DvUT crystallizes as a homotrimer with each dvUT monomer displaying both helical and nonhelical loop regions (Levin et al. 2009). The second UT to be crystallized was the bovine UT-B (bUT-B) (Levin et al. 2012). UT-B is also a homotrimer and each monomer contains a urea conduction pore with a narrow selectivity filter (Levin et al. 2012). The trimeric structure of DvUT and bUT-B provides a nice model for the transport of urea since the monomers create a chute that will promote urea movement through the hydrophobic membrane (Levin et al. 2009). Binding of the urea analog dimethylthiourea, DMTU, at two sites within the crystal structure, suggest that the pores through which urea will pass are of sufficient size and organization that further structural shifts are not necessary for the passage of urea (reviewed in Knepper and Mindell (2009), Levin and Zhou (2014)).

9.7 Active Urea Transport

Active urea secretion into the pars recta has been proposed, based upon observations that the early distal tubular urea flow exceeds urea delivery by the proximal convoluted tubule and that the fractional excretion of urea may exceed values compatible with urea reabsorption in the proximal convoluted tubule (Layton and Bankir 2013). Mathematical simulation of active urea secretion suggests that it induces a urea-selective improvement in urine concentration by enhancing the efficiency of urea excretion without increasing urine flow rate (Layton and Bankir 2013). Although this active urea secretory transporter has not been cloned, active urea transporters have been cloned from unicellular organisms (reviewed in Bankir (2014)). Active urea transport mechanisms have been functionally characterized in the rat IMCD (Isozaki et al. 1993, 1994b; Sands et al. 1996; Kato and Sands 1998a, b, 1999), although these have not been cloned either (reviewed in Bankir (2014)).

9.8 Regulation of Urea Transporters in Health (Normal Physiology)

Urea transport can be influenced by several factors (Fig. 9.4). Angiotensin II increases UT-A1, while aging and adrenal steroids decrease UT-A1. The effects of these factors on urea transport led to the development of a new class of diuretics, named urearetics, which inhibit urea transport and are discussed in Sect. 9.9.

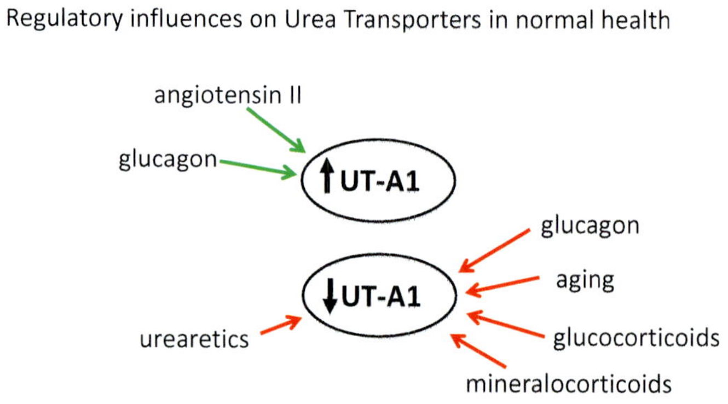

Fig. 9.4 Regulatory influences on urea transporters in normal health. Urea transport can be influenced by several factors. UT-A1 abundance can be decreased in response to glucocorticoids or mineralocorticoids. UT-A1 also decreases with aging. Glucagon has been reported to both decrease and increase UT-A1 abundance. Angiotensin II increases UT-A1 abundance. Urearetics are a new class of diuretics that inhibit the activity of UT-A1 but do not influence the protein abundance

9.8.1 Adrenal Steroids

Adrenalectomy produces a urine concentrating defect, although the mechanism by which this occurs is unclear (Schwartz and Kokko 1980; Jackson et al. 1983; Kamoi et al. 1993; Chen et al. 2005). In rats, dexamethasone, a glucocorticoid, increases the fractional excretion of urea (Knepper et al. 1975). Two studies examined the effect of dexamethasone on UT-A1 protein abundance and obtained opposite results. In one study, adrenalectomized rats were compared to sham-operated rats, each receiving a stress dose of dexamethasone or vehicle (Naruse et al. 1997). Adrenalectomy increased UT-A1 protein abundance in the inner medullary tip and urea permeability in terminal IMCDs when compared to sham-operated rats (Naruse et al. 1997). Administering dexamethasone to the adrenalectomized rats decreased UT-A1 protein abundance in the inner medullary tip and urea permeability in terminal IMCDs (Naruse et al. 1997). The second study employed a different model in which all rats were adrenalectomized and given aldosterone, either alone or in addition to dexamethasone (Li et al. 2008; Chen et al. 2005). Adrenalectomized, aldosterone replaced rats lacking glucocorticoids had lower UT-A1 and UT-A3 protein abundances than animals given dexamethasone (Chen et al. 2005; Li et al. 2008). The reason for the difference in UT-A1 responses between these studies is unclear, but may be due to the difference in animal models.

Administering stress dosages of dexamethasone to normal rats decreases UT-A1 and UT-A3 mRNA abundances in the inner medullary tip but does not change UT-A2 mRNA abundance (Peng et al. 2002). Dexamethasone also decreases the activity of UT-A promoter I (Peng et al. 2002), suggesting transcriptional regulation of UT-A1 and UT-A3 mRNA abundance. Consistent with transcriptional downregulation, dexamethasone treatment for 14 days resulted in an increase in urea excretion and a reduction in UT-A1 and UT-A3 abundances (Li et al. 2008).

Aldosterone, a mineralocorticoid, also decreases UT-A1 protein abundance in the inner medulla of adrenalectomized rats (Gertner et al. 2004). This decrease is blocked by spironolactone, a mineralocorticoid-receptor antagonist (Gertner et al. 2004). Spironolactone does not block the decrease in UT-A1 that results from dexamethasone, suggesting that each adrenal hormone works through its own receptor (Gertner et al. 2004). Both UT-A1 and UT-A3 protein abundances decrease when rats are fed a high salt diet to cause aldosterone-induced volume expansion (Wang et al. 2002).

The role of glucocorticoids and mineralocorticoids in the regulation of urine volume and salt intake was evaluated in men living in ultra-long-term controlled conditions. Increasing salt intake led to a decrease in the level of rhythmical mineralocorticoid release and an increase in rhythmical glucocorticoid release. Humans regulate water and osmolyte balance by rhythmical glucocorticoid and mineralocorticoid release, endogenous accrual of surplus body water, and precise surplus excretion. This water-conserving mechanism of dietary salt excretion relies on urea production by skeletal muscle and liver, and urea transporter-driven urea recycling by the kidney inner medulla. This natriuretic-ureotelic, water-conserving principle relies on metabolism-driven extracellular volume control and is regulated by concerted muscle, liver, and renal actions (Kitada et al. 2017; Rakova et al. 2017). Dexamethasone attenuates LPS-induced decreases in UT-A3 protein abundance in the hippocampus and brain cortical astrocytes (Du et al. 2014a). This change may be beneficial in future therapies for treating encephalopathy due to endotoxemia (Du et al. 2014a).

9.8.2 *Angiotensin II*

Urine osmolality increases following angiotensin II infusion into the renal artery (Faubert et al. 1987). Angiotensin-converting enzyme (ACE) or angiotensinogen knock-out mice, as well as pharmacological inhibition of the angiotensin II pathway with an ACE inhibitor, results in a reduction in urine concentrating ability (reviewed in Klein et al. (2011)). ACE knock-out mice have a normal-appearing medulla but still have a pronounced urine concentrating defect (Esther et al. 1997) and a reduction in UT-A1 protein abundance to 25% of the level seen in wild-type mice (Klein et al. 2002b). Administering angiotensin II to these mice for 2 weeks does not correct either the reduction in UT-A1 protein abundance or the urine concentrating defect (Klein et al. 2002b). The increase in urine osmolality in response to angiotensin II and the increase in UT-A1 phosphorylation by PKC in response to hypertonicity (Klein et al. 2012) suggest a synergy between vasopressin and angiotensin II in regulating the urea transporter. Surprisingly, there was no difference in the UT-A1 response to angiotensin II between PKCα knock-out mice and wild-type control mice (Thai et al. 2012).

9.8.3 *Glucagon*

Glucagon increases the fractional excretion of urea (Knepper et al. 1976; Ahloulay et al. 1992, 1995). However, glucagon's effect on IMCD urea transport has varied between studies. One study showed that glucagon decreases UT-A1 protein abundance and urea permeability in rat terminal IMCDs by stimulating a PKC signaling pathway (Yano et al. 2008). However, in other studies, glucagon did not alter basal or vasopressin-stimulated urea permeability, or cAMP production, in rat terminal IMCDs (Maeda et al. 1992; Isozaki et al. 1994a). Glucagon infusion improves the efficiency of urea excretion without stimulating any changes in blood flow or filtration, which supports the existence of a potential glucagon mediated active urea transport mechanism (Qi et al. 2019).

9.8.4 *Aging*

Urine concentrating ability is reduced during normal aging in humans and animals (reviewed in Sands (2012)). Aged rat kidneys show reduced levels of UT-A1, UT-A3, and UT-B protein abundances (Preisser et al. 2000; Combet et al. 2001; Trinh-Trang-Tan et al. 2003). Administering a supra-physiologic dose of dDAVP results in increased urine osmolality in aged rats. A supra-physiologic dose of dDAVP increases UT-A1, UT-A2, and UT-B protein abundances but not to the levels observed in younger rats (Combet et al. 2003). UT-A1, as well as NKCC2 and AQP2, are upregulated in aged rats treated with *N*-acetylcysteine, effects that are mediated by a reduction in oxidative stress and reduced renal inflammation (Shimizu et al. 2013).

9.9 Therapies Involving Urea Transport Inhibitors (Urearetics)

Inhibitors of UT-A and/or UT-B are attractive targets for the development of novel diuretics, especially, UT-A1, since it is located in the last portion of the IMCD and may have less risk for side-effects, such as hypokalemia, than conventional diuretics that act in more proximal portions of the nephron. Conventional diuretics target sodium transporters, so drugs inhibiting urea transporters (urearetics) could provide synergistic effects to those that inhibit sodium transport (reviewed in Sands (2013), Knepper and Miranda (2013), Denton et al. (2013), Verkman et al. (2014), Esteva-Font et al. (2015a), Klein and Sands (2016), Cheng et al. (2017)).

UT-A inhibitors were identified using a high-throughput assay in MDCK cells (Esteva-Font et al. 2013, 2014). UT-A1 selective and UT-A1/UT-B non-selective inhibitors were identified based upon a structure–activity analysis of over 400 analogs (Esteva-Font et al. 2013). The screen identified candidate compounds with a structure–activity profile that selectively inhibited UT-A by a noncompetitive mechanism with an IC50 of 1 μM (Esteva-Font et al. 2014). When an aryl-thiazole or γ-sultambenzosulfonamide was intravenously administered to rats, it resulted in a diuresis with more urea than salt excretion, even when the rats were given dDAVP (Esteva-Font et al. 2014). The diuresis, despite exogenous dDAVP, suggests that these UT-A inhibitors may be clinically useful in patients with high vasopressin levels and volume overload, such as patients with cirrhosis or congestive heart failure (Esteva-Font et al. 2014).

PU-14 is a thienoquinolin that causes diuresis by inhibition of both UT-A and UT-B (Li et al. 2013). The ability of PU-14 to inhibit UT-B may result in PU-14 being a novel therapy for hypertension (Sun et al. 2016). Structure–activity analysis was used to identify another thienoquinolin, PU-48, that is a more potent inhibitor of UT-A in the IMCD than PU-14 (Ren et al. 2014; Zhang et al. 2017). PU-48 causes diuresis in both wild-type and UT-B knock-out mice, indicating that its diuretic effect results from inhibition of UT-A (Ren et al. 2014). PU-48 inhibits urea permeability in perfused rat terminal IMCDs (Ren et al. 2014). The diuresis resulting from PU-48 administration does not change serum potassium, sodium, or chloride levels, supporting the hypothesis that a UT-A inhibitor acting in the IMCD will have fewer side-effects on electrolyte homeostasis than conventional diuretics (Ren et al. 2014). Discovery of the 1,2,4-triazoloquinoxaline family of compounds as inhibitors of both UT-A1 and UT-B was recently described (Lee et al. 2018). The IC50 for the most effective of these compounds is much lower for UT-A1 (150 nM) than for UT-B (2 μM) and exhibits increased molecular stability along with the increased sensitivity (Lee et al. 2018).

Four classes of compounds with UT-B inhibitory activity were identified using high-throughput virtual screening and used to predict molecules that may also inhibit human UT-B (Li et al. 2014b). By comparing UT-B from different species, a novel inhibitory mechanism for UT-B was postulated and the model suggests that phenylalanine 198 in mouse and rat UT-B might impede inhibitor–UT-B interactions (Li et al. 2014b). PU-1424 is a phenylphthalazine compound that inhibits both mouse and human UT-B with an IC50 in the submicromolar range (Ran et al. 2016). PU-1424 had no effect on urea transport in mouse red blood cells that lack UT-B (Ran et al. 2016). Two triazolothienopyrimidines inhibit UT-B (Liu et al. 2013). One compound with a 1,1-difluoroethyl-substitution, has improved microsomal stability (Liu et al. 2013). A small library based upon this compound was used to identify two novel compounds that inhibit UT-B with nanomolar range potency (Liu et al. 2013).

A cell-based high-throughput assay was used to identify 2,7-distributed fluorenones as urea transporter inhibitors (Lee et al. 2015). UT-A1 and UT-B were inhibited by the most potent compounds with an IC50 of 1 μM (Lee et al. 2015). The inhibitor binds to the urea transporter cytoplasmic domain at a site away from the putative urea binding site, based upon computational docking to a UT-A1 homology model (Lee et al. 2015).

Dimethylthiourea (DMTU) is a urea analog that inhibits UT-A1 and UT-B with an IC_{50} of 2–3 mM (Cil et al. 2015; Esteva-Font et al. 2015b). Chronic DMTU administration to rats results in a sustained and reversible reduction in urine osmolality, a threefold increase in urine volume, and a mild hypokalemia (Cil et al. 2015). Compared to rats treated with furosemide, a conventional diuretic, rats treated with DMTU have a larger diuresis with a reduction in urinary salt loss (Cil et al. 2015). DMTU treatment also prevents water retention and hyponatremia in a rat model of the syndrome of inappropriate antidiuretic hormone secretion (SIADH) (Cil et al. 2015). Thirty-six additional thiourea analogs were tested to identify more selective and potent urea transport inhibitors than DMTU (Esteva-Font et al. 2015b). The most potent compound, 3-nitrophenyl-thiourea, inhibited both UT-B and UT-A1 with an IC50 of 0.2 mM. Other analogs were identified that were relatively selective for UT-B or UT-A1 (Esteva-Font et al. 2015b).

The structural characteristics of *Desulfovibrio vulgaris* UT (dvUT), a bacterial urea transporter, were studied using a molecular dynamic simulation and computational approach, and three urea binding sites were identified (Wang et al. 2015).

Interestingly, these simulations found that dvUT is also permeable to water (Wang et al. 2015). DvUT was used as a model protein, along with the *Escherichia coli* L-arginine/agmatin anti-porter and lactose permease, to determine the detergent-binding capacity and phospholipid content of membrane proteins, with the ultimate goal of successfully crystallizing membrane proteins (Ilgu et al. 2014).

9.10 Urea Transporter Responses in Disease (Pathophysiology)

Diabetes mellitus, lithium administration, and hypertension are common human conditions that have been extensively studied in animal models. Changes in urea transporter abundance, phosphorylation, ubiquitination, and glycosylation have been identified (Fig. 9.5).

9.10.1 Diabetes Mellitus Type 1

Streptozotocin-treated rats are a model of type 1 diabetes mellitus. These rats develop an osmotic diuresis that is characteristic of type 1 diabetes mellitus, in addition to increases in urea excretion, corticosterone production, and plasma vasopressin levels (Brooks et al. 1989; Trinder et al. 1994; Mitch et al. 1999). UT-A1 protein abundance decreases at 3–5 days after inducing diabetes mellitus

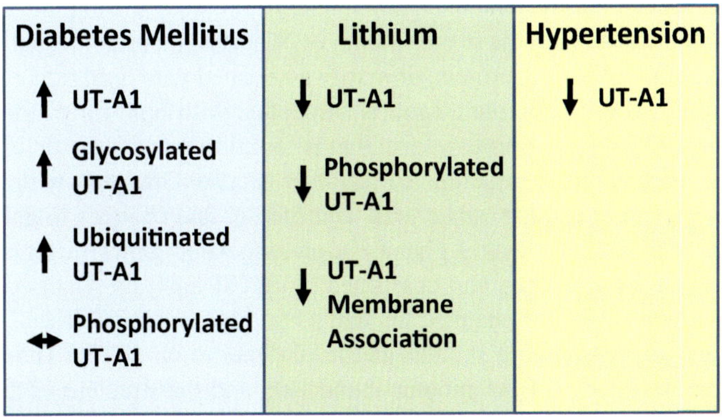

Fig. 9.5 Pathology that involves impaired urine concentrating ability. Three pathological conditions that show impaired urine concentrating ability are compared. Diabetes mellitus involves increases in UT-A1 abundance but not the degree of phosphorylation per protein. It includes increased UT-A1 glycosylation state and ubiquitination. Lithium treatment results in nephrogenic diabetes insipidus. The resulting polyuria involves decreases in UT-A1, phosphorylation of UT-A1, and UT-A1 membrane association. The hypertension-related urine concentrating defect involves decreased UT-A1 abundance

but not in rats that undergo adrenalectomy prior to streptozotocin (Klein et al. 1997; Kim et al. 2003). Administering dexamethasone to adrenalectomized rats with diabetes mellitus prevents the decrease in UT-A1 protein abundance, suggesting that glucocorticoids mediate the early decrease in UT-A1 protein (Klein et al. 1997).

Following the initial decrease in UT-A1 protein at 3–5 days after induction of diabetes mellitus, UT-A1 mRNA and protein increase over the next 10–21 days of diabetes mellitus, with increases in different parts of the inner medulla occurring at different times (Bardoux et al. 2001; Kim et al. 2003). In contrast, UT-A1 protein abundance does not increase in diabetic Brattleboro rats, indicating that vasopressin is necessary for this increase (Kim et al. 2004). Vasopressin repletion does not increase UT-A1 phosphorylation in Brattleboro rats, although it does in diabetic Brattleboro rats. Thus, vasopressin is necessary for the increase in UT-A1 protein abundance and phosphorylation that occurs at 10–21 days after diabetes mellitus is induced by streptozotocin (Kim et al. 2004).

The increase in UT-A1, as well as AQP2 and NKCC2, will reduce the loss of solute and water during the ongoing osmotic diuresis that occurs in uncontrolled diabetes mellitus. Dapagliflozin inhibits the type 2 sodium-glucose cotransporter, SGLT2. In streptozotocin-treated rats, dapagliflozin increased UT-A1 protein abundance at 7 and 14 days (Chen et al. 2016). Thus, dapagliflozin improves the compensatory change in UT-A1 protein abundance in diabetes mellitus (Chen et al. 2016).

The increase in UT-A1 protein abundance in the inner medulla of streptozotocin-treated rats is accompanied by a shift in the glycoprotein forms expressed in the inner medullary tip versus base (Kim et al. 2003; Klein et al. 2006a; Blount et al. 2008b). The base of the inner medulla, which expresses only the 97 kDa glycoprotein form of UT-A1 in normal rats, shows both the 97 and 117 kDa forms in the inner medulla of rats with diabetes mellitus (Kim et al. 2003; Klein et al. 2006a; Blount et al. 2008b). In addition, there is an increase in the amount of N-acetylglucosamine and sialic acid in the 117 kDa UT-A1 glycoprotein, primarily associated with lipid rafts (Chen et al. 2011). The 117 kDa glycoprotein form is associated with lipid rafts more than the mannose-rich 97 kDa glycoprotein form that is found in non-lipid rafts (Chen et al. 2011). The carbohydrate structure of UT-A1 is also changed, with increased amounts of glycan branching, sialic acid, and fucose, and changes in galectin proteins (Qian et al. 2015). Several glycosylation-associated genes differ in diabetic rats, as assessed by RNA-seq and confirmed by qPCR analysis (Qian et al. 2015). The types of genes that are changed include some glycosyltransferases, sialylation enzymes, and glycan-binding protein galectins (Qian et al. 2015). Thus, diabetes changes urea transport, UT-A1 protein abundance, and the structure of the glycans on UT-A1 (Qian et al. 2015).

Urea transport is increased in the initial IMCD in rats with streptozotocin-induced diabetes mellitus, as is the sensitivity to vasopressin (Pech et al. 2005). In normal rat inner medulla, only urea transport in the terminal IMCD is stimulated by vasopressin (Sands et al. 1987). In rats with diabetes mellitus, urea transport in both the initial and terminal IMCD is increased by vasopressin (Pech et al. 2005). This suggests that the expression of the 117 kDa glycoprotein form is important for vasopressin stimulation of urea transport (Pech et al. 2005).

In normal animals, vasopressin and angiotensin II synergistically stimulate urea permeability (Kato et al. 2000). In streptozotocin-treated rats in which angiotensin receptors are blocked with candesartan, there is an increase in both UT-A1 and UT-A3 proteins, apparently to reduce the loss of solute during uncontrolled diabetes mellitus (Blount et al. 2008b). Rats with diabetes mellitus treated with the nitric oxide synthase inhibitor L-NAME display a decrease in both UT-A1 and UT-A3 protein abundance (Cipriani et al. 2012). However, the diabetes mellitus-induced glycosylation of UT-A1 remained increased (Cipriani et al. 2012). Streptozotocin-treated mice that lack the prostaglandin E2 receptor EP3 have increased expression of UT-A1, as well as aquaporins 1 and 2, when compared to streptozotocin-treated wild-type mice (Hassouneh et al. 2016). Thus, EP3 contributes to the polyuria occurring in diabetes mellitus by decreasing the expression of UT-A1 and aquaporins (Hassouneh et al. 2016). UT-A2 protein is decreased in streptozotocin-treated rats (Bardoux et al. 2001). Since the abundance of UT-A2 mRNA is unaltered, the decrease in protein occurs posttranslationally and may reflect a general increase in protein degradation in the highly catabolic diabetic state.

9.10.2 Diabetes Mellitus Type 2

The obese Zucker rat model, which has a genetic mutation of the leptin receptor gene, is the most commonly used rat model of type 2 diabetes mellitus (Van Zwieten et al. 1996). UT-A1 protein abundance is dramatically decreased in obese Zucker rats compared to lean Zucker (nondiabetic) rats (Bickel et al. 2002). One of the differences between the streptozotocin-induced type 1, and the obese Zucker type 2 models, is that the Zucker rat model has pronounced hypertension (Van Zwieten et al. 1996), while the streptozotocin-induced diabetic rat has at most mild hypertension. UT-A1 protein is decreased in nondiabetic Sprague-Dawley rats that are treated with angiotensin II to induce hypertension (Klein et al. 2006b). Thus, the decrease in UT-A1 protein abundance in obese Zucker rats may be a response to hypertension rather than diabetes mellitus.

9.10.3 Lithium

Lithium is commonly used to treat patients suffering from manic-depressive (bipolar) disorder. An unfortunate consequence of long-term treatment with lithium is the development of nephrogenic diabetes insipidus (NDI) and an inability to concentrate urine (reviewed in Timmer and Sands (1999)). Lithium reduces inner medullary interstitial osmolality by reducing both interstitial urea and NaCl concentrations (Blount et al. 2010). Feeding lithium to rats results in a dramatic reduction in UT-A1 protein abundance in both inner medullary tip and base regions (Klein et al. 2002a), as well as a reduction in UT-A3 protein abundance in the inner medullary tip (Blount et al. 2010). This lithium-induced downregulation of urea transporters severely reduces transepithelial urea transport, resulting in disruption of the inner medullary osmotic gradient required for urine concentration. Lithium feeding also reduces the abundance of UT-B in the inner medullary base (Klein et al. 2002a) and decreases AQP2 abundance in the IMCD (Marples et al. 1995; Klein et al. 2002a).

Lithium enters collecting duct principal cells through ENaC, the epithelial sodium channel. Treating lithium-fed rats with amiloride, which inhibits ENaC, restores UT-A1 expression as well as urine concentrating ability (Bedford et al. 2008). Lithium increases the phosphodiesterase (PDE) concentration in kidney tissue (Dousa 1974). Administering the PDE5-inhibitor sildenafil to lithium-treated rats for 4 weeks results in normalization of UT-A1 and γ-ENaC levels, and polyuria is reduced (Sanches et al. 2012).

Discontinuing lithium therapy can reverse or eliminate some NDI symptoms in as little as 3 weeks. However, in many cases, NDI symptoms are not reversed by discontinuing lithium. UT-A1 and UT-A3 protein levels return to control values 14 days after stopping lithium in rats; however, even with both urea transporters returning to basal values of expression, urine urea levels remained significantly lower than prior to placing the rats on lithium (Blount et al. 2010). A proteomic analysis of kidney tissue from lithium-treated animals shows that several proteins involved in signaling cascades that impact urea transporter function and/or trafficking have altered expression after 14 days of lithium treatment (Nielsen et al. 2008). Immunohistochemical studies show that UT-A1 localization is unaffected by chronic lithium treatment, or after recovery from lithium treatment, suggesting that the trafficking machinery remains intact (Nielsen et al. 2008).

Lithium inhibits adenylyl cyclase activity, thereby reducing cAMP production. Lithium also increases cAMP-PDE activity, which further reduces the concentration of cAMP (Dousa 1974). PKA-mediated UT-A1 phosphorylation at serines 486 and 499 is necessary for UT-A1 movement to the apical membrane (Blount et al. 2008a). The lithium-mediated inhibition of cAMP in response to vasopressin reduces the phosphorylation of the urea transporters and thus prevents trafficking of UT-A1 to the membrane. Without membrane insertion of the urea transporters, the inner medullary osmotic gradient is reduced, leading to polyuria. Consistent with this mechanism, treating IMCD suspensions with a physiologic concentration of vasopressin stimulates urea transporter phosphorylation in normal rats, but does not increase UT-A1 phosphorylation in IMCD suspensions from lithium-fed rats (Klein et al. 2002a).

A global PKCα knock-out mouse has a urine concentrating defect (Yao et al. 2004; Klein et al. 2012). Surprisingly, PKCα knock-out mice are protected against lithium-induced NDI (Sim et al. 2014). In contrast to wild-type mice, lithium-treated PKCα knock-out mice have no polyuria after 5 days of lithium treatment and significantly less polyuria after 6 weeks, and UT-A1 protein abundance is unchanged at either time point (Sim et al. 2014). Thus, PKCα deletion prevents the development of severe NDI in mice (Sim et al. 2014).

9.10.4 *Hypertension*

Urea transport is increased in the terminal IMCD of salt-sensitive Dahl rats, and UT-A1 and UT-A3 protein abundances are increased (Fenton et al. 2003). The Dahl salt-sensitive rat also has an increased level of 11β-hydroxysteroid dehydrogenase type II, the enzyme that inactivates corticosterone (Fenton et al. 2003). This may be part of the mechanism for the change in urea transport since inactivating glucocorticoids lessens the repression of UT-A promoter I activity, thereby increasing transcription, and UT-A1 and UT-A3 mRNA and protein abundances (Peng et al. 2002).

In a pharmacological model of hypertension, a very high dose of angiotensin II was administered to rats. Blood pressure increased from 130 mmHg in control animals to 200 mmHg within 3 days of administering angiotensin II (Klein et al. 2006b). Inner medullary urea transporter protein is significantly decreased in the angiotensin II-treated rats (Klein et al. 2006b). Mineralocorticoids reduce UT-A1 levels, so angiotensin II-induced increases in aldosterone levels could be the mechanism for decreasing UT-A1 abundance. However, this seems unlikely to be the explanation since: (1) when spironolactone, which inhibits the mineralocorticoid receptor, is given in addition to angiotensin II, the decrease in UT-A1 is indistinguishable from the response to angiotensin II alone; (2) when blood pressure is increased using norepinephrine, instead of angiotensin II, UT-A1 abundance is still reduced. These studies suggest that high blood pressure decreases UT-A1 protein abundance (Klein et al. 2006b).

Male EP1 knock-out mice were bred with hypertensive TTRhRen mice to evaluate the role of blood pressure, since the EP1 knock-out mice are normotensive while the TTRhRen and EP1 knock-out/TTRhRen mice are hypertensive (Nasrallah et al. 2018). UT-A1 abundance was reduced in EP1 knock-out, TTRhRen, and EP1 knock-out/TTRhRen when compared to wild-type mice, suggesting that the reduction in UT-A1 was not related to blood pressure in these mice (Nasrallah et al. 2018).

9.11 Urea Transporter Responses in Renal Disease Models

Several animal models of less common human diseases that include a urine concentrating defect have been studied. Abnormalities in urea transporter abundance, phosphorylation, or membrane accumulation have been identified in different disease models (Fig. 9.6).

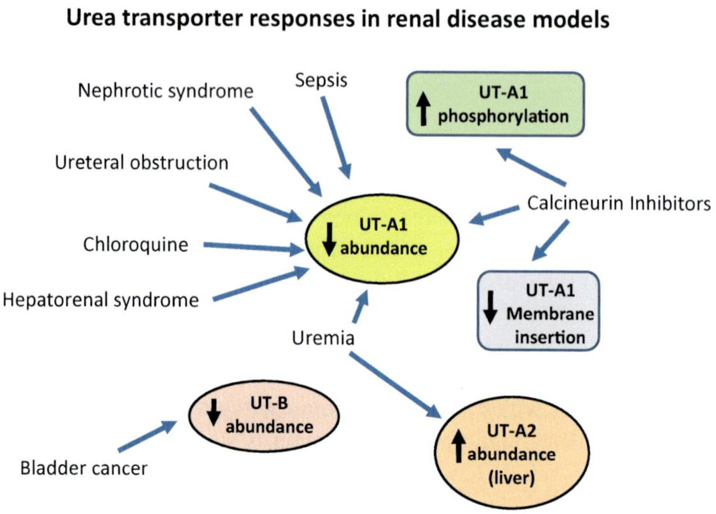

Fig. 9.6 Urea transporter responses in renal disease models. Most models of renal disease involve decreased UT-A1 abundance. Calcineurin inhibitors also cause increased UT-A1 phosphorylation but decreased UT-A1 membrane insertion. Uremia causes increased UT-A2 abundance with decreased UT-A1 abundance. Bladder cancer decreases UT-B abundance without changing UT-A1 abundance

9.11.1 Nephrotic Syndrome

Adriamycin, an anticancer chemotherapeutic agent, administration to rats for 3 weeks results in the development of proteinuria and decreased UT-A1 protein abundance in the inner medulla, compared to vehicle-treated rats (Fernández-Llama et al. 1998). UT-A1 protein abundance is reduced in non-pair-fed nephrotic animals, but not their pair-fed counterparts, suggesting that nutrition plays a role in the change in UT-A1 (Bou Matar et al. 2012). In adriamycin-induced nephrotic syndrome in rats: (1) NKCC2 is dramatically reduced; (2) the sodium-chloride co-transporter, NCC, and AQP2 are modestly reduced; and (3) active gamma-ENaC is increased. These changes in transporter abundances could be a response to an increase in sodium reabsorption in the cortical collecting duct along with disruption of the osmotic gradient along the medullary collecting duct (Bou Matar et al. 2012).

9.11.2 Calcineurin Inhibitors

Administering cyclosporine, an immunosuppressant medication used to prevent organ rejection following transplantation, frequently results in nephrotoxic side-effects. Urine concentrating ability is reduced in rats receiving cyclosporine for 4 weeks (Lim et al. 2004). Cyclosporine reduces UT-A2, UT-A3, and UT-B protein abundances but does not change UT-A1 protein abundance (Lim et al. 2004). Although cyclosporine does not change plasma vasopressin levels, vasopressin infusion does restore urine concentrating ability to cyclosporine treated animals (Lim et al. 2004). Cyclosporine treatment increases the number of TUNEL-positive (apoptotic) cells and the number of apoptotic cells correlates with UT-A3 protein expression (Lim et al. 2004).

Cyclosporine inhibits calcineurin, which is a protein phosphatase 2b. Tacrolimus (also known as FK506) also inhibits calcineurin and is a second immunosuppressant medication that is used to prevent organ rejection following transplantation. Tacrolimus results in the dephosphorylation of UT-A1 (Ilori et al. 2012). Although UT-A1 can be dephosphorylated by multiple phosphatases, the PKA-phosphorylated serine 486 residue, which is necessary for apical membrane trafficking, is dephosphorylated by tacrolimus (Ilori et al. 2012). Acute treatment with tacrolimus activates urea transport by promoting membrane accumulation of UT-A1 (Ilori et al. 2012). In contrast, chronic treatment with tacrolimus tends to lower the ability of phosphorylated UT-A1 to associate with the membrane and may contribute to the observed nephrotoxicity of the calcineurin inhibitors (Ilori et al. 2012).

9.11.3 Ureteral Obstruction

UT-A1, UT-A3, and UT-B protein abundances are reduced in the medulla of rats with either unilateral or bilateral ureteral obstruction (Li et al. 2004). Two weeks after the release of bilateral ureteral obstruction, all three urea transporter abundances remain reduced (Li et al. 2004).

9.11.4 Chloroquine

Chloroquine is a widely used anti-malaria drug that results in polyuria when used chronically. Chloroquine does not affect UT-A1 and UT-A3 abundances but does decrease aquaporin-2 and NKCC2 (von Bergen and Blount 2012). Chloroquine reduces cAMP production in the inner medulla and results in a marked reduction of UT-A1 and AQP2 in the apical plasma membrane in response to vasopressin in chloroquine-treated rats. This suggests that chloroquine-induced polyuria likely results from a reduction in cAMP production that alters UT-A1 trafficking (von Bergen and Blount 2012).

9.11.5 Sepsis

Sepsis often results in a urinary concentration defect. Inducing endotoxemia in rats results in reduced levels of UT-A1 and AQP2, as well as other transport proteins (Kuper et al. 2012). The decrease in these transport proteins correlates with a reduction in TonEBP/NFAT5 transcriptional activity and an increase in nitric oxide production (Kuper et al. 2012). Endotoxemia induced by LPS decreases UT-A3 protein abundance in the hippocampus and cortical astrocytes, and this decrease is partially restored by treatment with dexamethasone (Du et al. 2014a).

9.11.6 Hepatorenal Syndrome

Succinylated gelatin has been used to create a rat model of hepatorenal syndrome or abdominal compartment syndrome (Song et al. 2015). In the renal medulla of rats with hepatorenal syndrome, UT-A2 and UT-A3 mRNA abundances are reduced when compared to rats with cirrhosis (Song et al. 2015). In contrast, there was no difference in UT-A2, UT-A3, or UT-B mRNA abundances between rats with abdominal compartment syndrome and control rats (Song et al. 2015). UT-A1 protein abundance does not change in any of the groups of rats (Song et al. 2015).

9.11.7 Uremia

UT-A1 and UT-B are expressed in skin basal cells, skin sweat glands, and sweat ducts from normal and uremic patients (Xie et al. 2017). UT-A1 and UT-B mRNA abundances are increased in the skin and sweat glands of uremic patients (Xie et al. 2017). Control and uremic rats express UT-A2 and UT-B in their cutaneous structures (Keller et al. 2016). Uremic rats have twice the level of urea nitrogen in their sweat as control rats, with male uremic rats having a higher level of sweat urea nitrogen than female uremic rats (Keller et al. 2016). UT-A2 protein abundance is significantly increased in the liver of rats undergoing a 5/6th nephrectomy (Klein et al. 1999).

9.11.8 Bladder Cancer

While the physiological role for UT-B in the bladder has not been elucidated, a genome-wide association study found a link between a UT-B polymorphism and bladder cancer in a Northern India population (Singh et al. 2014). Several other studies also found an association between genetic variants in UT-B and bladder cancer risk (Koutros et al. 2013; Li et al. 2014a; Liu et al. 2016; Selinski et al. 2017; Hou et al. 2017a). A link between the UT-B1 gene and bladder cancer may be explained by the finding of UT-B protein in human bladder (reviewed in Atala (2015); Hou et al. (2017b)). In UT-B knock-out mice, there is DNA damage and apoptosis in bladder urothelial cells (Dong et al. 2013). High urinary urea concentrations, as can occur with a high protein diet, maybe carcinogenic in rat bladder (Liu et al. 2016). UT-B protein expression in urothelial carcinoma tumor cells is lower than in normal urothelial cells and is significantly reduced with increasing histological grade (Li et al. 2014a). Bladder cancer cells show a reduction in a longer form of UT-B and many tumor cells express a 24 nucleotide in-frame deletion in exon 4 (Hou et al. 2017a). Urine specific gravity may be useful as a molecular phenotype of bladder cancer associated with UT-B (Koutros et al. 2013).

9.12 Genetic Ablation of Urea Transporters

Several different urea transporter knock-out mice have been generated. They all have urine concentrating defects, although the magnitude of the defect varies with which transporter is missing (Fig. 9.7).

9.12.1 UT-B Knock-Out Mice

Humans lacking UT-B, the Kidd antigen, have a urine concentrating defect and are unable to increase urine osmolality above 800 mOsm/kg H_2O, even following overnight water deprivation and exogenous vasopressin administration (Sands et al. 1992). A UT-B knock-out mouse also has reduced urine concentrating ability, similar to humans, and can only concentrate their urine to 2400 mOsm/kg H_2O, compared to 3400 in a wild-type mouse (Yang et al. 2002). These findings support the hypothesis that urea transport in red blood cells is important for efficient countercurrent exchange (Macey 1984). Both red blood cells and descending vasa recta express UT-B protein and exhibit phloretin-inhibitable urea transport, suggesting that UT-B mediates urea transport in red blood cells and descending vasa recta (Macey 1984; Pallone et al. 1994; Evans et al. 2015; Wei et al. 2015).

UT-B knock-out mice demonstrate a functional impairment in early life, but as they age, they also demonstrate renal medullary atrophy that is not apparent at younger ages, suggesting that the severe polyuria and hydronephrosis contributes to the later structural impairment (Zhou et al. 2012). At 16 weeks of age, UT-B knock-out mice have an atrial-ventricular conduction block (Meng et al. 2009). A proteomics study identified 15 proteins involved in mitochondrial complexes I, III, IV, and V of the respiratory chain that are downregulated in UT-B knock-out mice (Du et al. 2014b). The downregulated proteins are likely to reduce electron transport chain activity, thereby leading to mitochondrial dysfunction, and may explain the

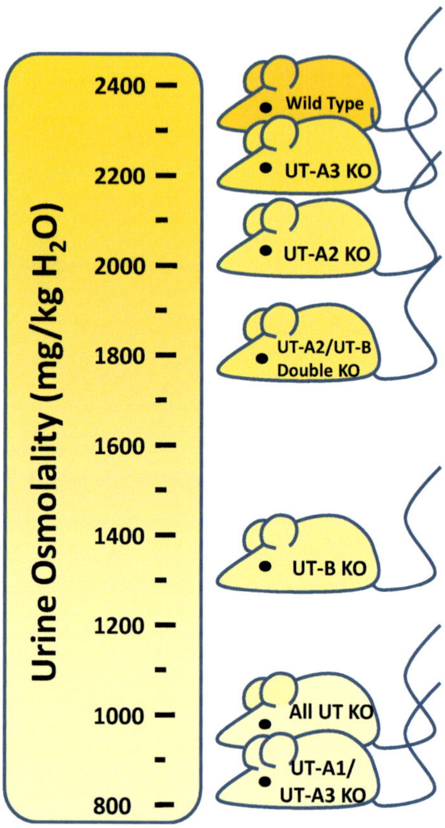

Fig. 9.7 Urine concentrating ability of urea transporter knock-out mice. Shown are urea transporter knock-out mice and their respective urine osmolalities (Jiang et al. 2017)

atrial-ventricular conduction block (Du et al. 2014b). UT-B knock-out improves the hemorheological properties of red blood cells, including increasing the red blood cell deformation index, small deformation index, loss of osmotic fragility, and high electrophoretic rate (Geng et al. 2017). UT-B knock-out mice exhibit depression-like behavior with the accumulation of urea and reduction in nitric oxide in the hippocampus (reviewed in Yang et al. (2014)).

UT-A2 protein abundance is increased while UT-A1 and UT-A3 abundances are unchanged in UT-B knock-out mice (Klein et al. 2004). This suggests the hypothesis that UT-A2 is upregulated to compensate for the loss of UT-B, since both UT-A2 and UT-B are involved in urea recycling, and maybe the explanation for the mild phenotype observed in UT-B knock-out mice and in humans lacking UT-B/Kidd antigen (Klein et al. 2004). However, this hypothesis was not supported by subsequent findings in a UT-B/UT-A2 double knock-out mouse (Lei et al. 2011).

9.12.2 UT-A2 and UT-B/UT-A2 Knock-Out Mice

UT-A2 knock-out mice have a urine concentrating defect that results from an impairment of urea recycling (Uchida et al. 2005). Since UT-A2 and UT-B are expressed in adjacent structures within the kidney, and because UT-B knock-out results in an increase in UT-A2 in the UT-B knock-out mice (Klein et al. 2004), a mouse lacking both UT-A2 and UT-B was generated to see if the relationship between urea handling by these two transporters could be elucidated (Lei et al. 2011). Surprisingly, UT-A2 knock-out appeared to partially correct the concentrating defect in mice lacking only UT-B (Lei et al. 2011). This study suggests that UT-A2 may function to move urea into stores in response to changes from diuresis to antidiuresis as part of an acute response, as opposed to playing a role in maintaining urea concentrations during the normal steady state (Lei et al. 2011).

9.12.3 UT-A1/UT-A3 and UT-A3 Knock-Out Mice

Mice with genetic knock-out of both of the IMCD urea transporters, UT-A1 and UT-A3, have a marked reduction in urine concentrating ability and inner medullary interstitial urea content, and also lack vasopressin-stimulated or phloretin-inhibitable urea transport in their IMCDs (Fenton et al. 2004, 2005) and reviewed in Fenton and Knepper (2007), Fenton (2008), Li et al. (2012), and Fenton and Yang (2014). Administering vasopressin to UT-A1/UT-A3 knock-out mice for 1 week does not increase urine osmolality despite AQP2 protein abundance increasing by the same percentage in UT-A1/UT-A3 knock-out and wild-type mice (Ilori et al. 2013). Thus, in the absence of UT-A1 and UT-A3, an increase in AQP2 protein abundance is not sufficient to increase urine concentrating ability (Ilori et al. 2013).

When UT-A1/UT-A3 knock-out mice are fed a low-protein diet, they can concentrate their urine almost as well as wild-type mice (Fenton et al. 2004), supporting the hypothesis that IMCD urea transporters contribute to urine concentrating ability by preventing urea-induced osmotic diuresis (Berliner et al. 1958). Urea within the collecting duct lumen is osmotically balanced by a high interstitial urea concentration (Berliner et al. 1958). If interstitial urea were unavailable to offset the osmotic effect of the luminal urea that is destined for excretion, then the interstitial NaCl concentration would need to be significantly higher (Berliner et al. 1958; Fenton et al. 2004).

Water restriction markedly reduced inner medullary tissue urea content in UT-A1/UT-A3 knock-out mice (Fenton et al. 2004). There is no measurable difference in medullary interstitial NaCl content between UT-A1/UT-A3 knock-out and wild-type mice (Fenton et al. 2004), a finding that has been interpreted as being inconsistent with the predictions of the passive mechanism hypothesis for urine concentration in the inner medulla (Fenton and Knepper 2007; Sands 2007). However, a mathematical modeling analysis of these same data concluded that the experimental results are precisely what one would predict for the passive mechanism hypothesis (Pannabecker et al. 2008a).

A mouse expressing only UT-A1 was created by transgenic restoration of UT-A1 into a UT-A1/UT-A3 knock-out mouse in order to determine the effect of expressing UT-A1 in the absence of UT-A3 (Klein et al. 2016b). The construct used to make this mouse included the 4.2 kb of the 5′-flanking region of the UT-A gene promoter α, which drives IMCD-specific expression (Fenton et al. 2006b). The mice expressing only UT-A1 (or lacking only UT-A3) had normal basal urea permeability in the IMCD, but unlike wild-type mice, vasopressin did not stimulate urea permeability (Klein et al. 2016b). Urine concentrating ability was restored in the UT-A1 only mice, even though they lacked UT-A3 (Klein et al. 2016b).

9.12.4 All UT Knock-Out Mice

A mouse lacking all urea transporters was generated by knocking out an 87 kb DNA fragment containing most of the UT-A and UT-B genes (Jiang et al. 2017). These mice have a marked urine concentrating defect. Blood pressure is reduced in the all UT knock-out mice. All UT knock-out also promoted the maturation of the male reproductive tract. Surprisingly, all UT knock-out mice did not manifest functional abnormalities in other, extrarenal tissues (Jiang et al. 2017).

9.13 Conclusions

Urea and urea transport proteins play critical roles in the urine concentrating mechanism in the inner medulla. Cloning the genes, developing polyclonal antibodies, and generating knock-out mice has elucidated the role of urea transporter in normal physiology and in pathophysiology. Vasopressin is the primary hormone that regulates urea transport. However non-vasopressin-mediated pathways that stimulate UT-A1 function may be important for the treatment of patients with nephrogenic diabetes insipidus. Additional research into the regulation of urea transporters may also be important for the development of urea transport inhibitors as urearetic agents that could be used as a novel class of diuretics. In particular, UT-A1 inhibitors may have fewer side-effects on serum electrolytes than conventional diuretics that inhibit sodium transport since they act upon the last segment of the nephron, the inner medullary collecting duct. Urea transport inhibitors may be especially useful in conditions associated with reduced circulating volume and hyponatremia due to non-osmotic release of vasopressin, such as cirrhosis or congestive heart failure.

Acknowledgment This review was supported by NIH grant R01-DK41707.

References

Aboudehen K, Noureddine L, Cobo-Stark P, Avdulov S, Farahani S, Gearhart MD, Bichet DG, Pontoglio M, Patel V, Igarashi P (2017) Hepatocyte nuclear factor-1beta regulates urinary concentration and response to hypertonicity. J Am Soc Nephrol 28:2887–2900. https://doi.org/10.1681/asn.2016101095

Agarwal U, Hu Q, Baldwin RL, Bequette BJ (2015) Role of rumen butyrate in regulation of nitrogen utilization and urea nitrogen kinetics in growing sheep. J Anim Sci 93:2382–2390. https://doi.org/10.2527/jas.2014-8738

Ahloulay M, Bouby N, Machet F, Kubrusly M, Coutaud C, Bankir L (1992) Effects of glucagon on glomerular filtration rate and urea and water excretion. Am J Physiol Renal Fluid Electrol Physiol 263:F24–F36

Ahloulay M, Déchaux M, Laborde K, Bankir L (1995) Influence of glucagon on GFR and on urea and electrolyte excretion: direct and indirect effects. Am J Physiol Renal Fluid Electrol Physiol 269:F225–F235

Ariz-Extreme I, Hub JS (2017) Potential of mean force calculations of solute permeation across UT-B and AQP1: a comparison between molecular dynamics and 3D-RISM. J Phys Chem B 121:1506–1519. https://doi.org/10.1021/acs.jpcb.6b11279

Atala A (2015) Re: expression and localization of a UT-B urea transporter in the human bladder. J Urol 194:592. https://doi.org/10.1016/j.juro.2015.05.016

Azouzi S, Gueroult M, Ripoche P, Genetet S, Colin Aronovicz Y, Le Van Kim C, Etchebest C, Mouro-Chanteloup I (2013) Energetic and molecular water permeation mechanisms of the human red blood cell urea transporter B. PLoS One 8:e82338. https://doi.org/10.1371/journal.pone.0082338

Bankir L (2014) Active urea transport in lower vertebrates and mammals. Subcell Biochem 73:193–226. https://doi.org/10.1007/978-94-017-9343-8_13

Bardoux P, Ahloulay M, Le Maout S, Bankir L, Trinh-Trang-Tan MM (2001) Aquaporin-2 and urea transporter-A1 are up-regulated in rats with type I diabetes mellitus. Diabetologia 44:637–645

Bedford JJ, Leader JP, Jing R, Walker LJ, Klein JD, Sands JM, Walker RJ (2008) Amiloride restores renal medullary osmolytes in lithium-induced nephrogenic diabetes insipidus. Am J Physiol Renal Physiol 294:F812–F820

Berends H, van den Borne JJ, Rojen BA, van Baal J, Gerrits WJ (2014) Urea recycling contributes to nitrogen retention in calves fed milk replacer and low-protein solid feed. J Nutr 144:1043–1049. https://doi.org/10.3945/jn.114.191353

Berliner RW, Levinsky NG, Davidson DG, Eden M (1958) Dilution and concentration of the urine and the action of antidiuretic hormone. Am J Med 24:730–744

Bickel CA, Knepper MA, Verbalis JG, Ecelbarger CA (2002) Dysregulation of renal salt and water transport proteins in diabetic Zucker rats. Kidney Int 61:2099–2110

Blessing NW, Blount MA, Sands JM, Martin CF, Klein JD (2008) Urea transporters UT-A1 and UT-A3 accumulate in the plasma membrane in response to increased hypertonicity. Am J Physiol Renal Physiol 295:F1336–F1341

Blount MA, Klein JD, Martin CF, Tchapyjnikov D, Sands JM (2007) Forskolin stimulates phosphorylation and membrane accumulation of UT-A3. Am J Physiol Renal Physiol 293:F1308–F1313

Blount MA, Mistry AC, Fröhlich O, Price SR, Chen G, Sands JM, Klein JD (2008a) Phosphorylation of UT-A1 urea transporter at serines 486 and 499 is important for vasopressin-regulated activity and membrane accumulation. Am J Physiol Renal Physiol 295:F295–F299

Blount MA, Sands JM, Kent KJ, Smith TD, Price SR, Klein JD (2008b) Candesartan augments compensatory changes in medullary transport proteins in the diabetic rat kidney. Am J Physiol Renal Physiol 285:F1448–F1452

Blount MA, Sim JH, Zhou R, Martin CF, Lu W, Sands JM, Klein JD (2010) The expression of transporters involved in urine concentration recover differently after ceasing lithium treatment. Am J Physiol Renal Physiol 298:F601–F608

Blount MA, Cipriani P, Redd SK, Ordas RJ, Black LN, Gumina DL, Hoban CA, Klein JD, Sands JM (2015) Activation of protein kinase Calpha increases phosphorylation of the UT-A1 urea transporter at serine 494 in the inner medullary collecting duct. Am J Physiol Cell Physiol 309: C608–C615. https://doi.org/10.1152/ajpcell.00171.2014

Bou Matar RN, Malik B, Wang XH, Martin CF, Eaton DC, Sands JM, Klein JD (2012) Protein abundance of urea transporters and aquaporin 2 change differently in nephrotic pair-fed vs. non-pair-fed rats. Am J Physiol Renal Physiol 302:F1545–F1553. https://doi.org/10.1152/ajprenal.00686.2011

Bradford AD, Terris J, Ecelbarger CA, Klein JD, Sands JM, Chou C-L, Knepper MA (2001) 97 and 117 kDa forms of the collecting duct urea transporter UT-A1 are due to different states of glycosylation. Am J Physiol Renal Physiol 281:F133–F143

Brahm J (2013) The permeability of red blood cells to chloride, urea and water. J Exp Biol 216:2238–2246. https://doi.org/10.1242/jeb.077941

Brooks DD, Nutting DF, Crofton JT, Share L (1989) Vasopressin in rats with genetic and streptozotocin-induced diabetes. Diabetes 38:54–57

Capriolli TV, Visentainer JEL, Sell AM (2017) Lack of association between Kidd blood group system and chronic kidney disease. Rev Bras Hematol Hemoter 39:301–305. https://doi.org/10.1016/j.bjhh.2017.05.007

Chen G (2013) New advances in urea transporter UT-A1 membrane trafficking. Int J Mol Sci 14:10674–10682. https://doi.org/10.3390/ijms140510674

Chen G (2014) Biochemical properties of urea transporters. Subcell Biochem 73:109–126. https://doi.org/10.1007/978-94-017-9343-8_7

Chen YC, Cadnapaphornchai MA, Summer SN, Falk S, Li C, Wang W, Schrier RW (2005) Molecular mechanisms of impaired urinary concentrating ability in glucocorticoid-deficient rats. J Am Soc Nephrol 16:2864–2871

Chen G, Huang H, Fröhlich O, Yang Y, Klein JD, Price SR, Sands JM (2008) MDM2 E3 ubiquitin ligase mediates UT-A1 urea transporter ubiquitination and degradation. Am J Physiol Renal Physiol 295:F1528–F1534

Chen G, Howe AG, Xu G, Frohlich O, Klein JD, Sands JM (2011) Mature N-linked glycans facilitate UT-A1 urea transporter lipid raft compartmentalization. FASEB J 25:4531–4539. https://doi.org/10.1096/fj.11-185991

Chen L, LaRocque LM, Efe O, Wang J, Sands JM, Klein JD (2016) Effect of dapagliflozin treatment on fluid and electrolytebalance in diabetic rats. Am J Med Sci 352:517–523. https://doi.org/10.1016/j.amjms.2016.08.015

Cheng CJ, Rodan AR, Huang CL (2017) Emerging targets of diuretic therapy. Clin Pharmacol Ther 102:420–435. https://doi.org/10.1002/cpt.754

Chou C-L, Knepper MA (1989) Inhibition of urea transport in inner medullary collecting duct by phloretin and urea analogues. Am J Physiol Renal Physiol 257:F359–F365

Chou CL, Hwang G, Hageman DJ, Han L, Agrawal P, Pisitkun T, Knepper MA (2018) Identification of UT-A1- and AQP2-interacting proteins in rat inner medullary collecting duct. Am J Physiol Cell Physiol 314:C99–C117. https://doi.org/10.1152/ajpcell.00082.2017

Cil O, Esteva-Font C, Tas ST, Su T, Lee S, Anderson MO, Ertunc M, Verkman AS (2015) Salt-sparing diuretic action of a water-soluble urea analog inhibitor of urea transporters UT-A and UT-B in rats. Kidney Int 88:311–320. https://doi.org/10.1038/ki.2015.138

Cipriani P, Kim SL, Klein JD, Sim JH, von Bergen TN, Blount MA (2012) The role of nitric oxide in the dysregulation of the urine concentration mechanism in diabetes mellitus. Front Physiol 3:176. https://doi.org/10.3389/fphys.2012.00176

Combet S, Teillet L, Geelen G, Pitrat B, Gobin R, Nielsen S, Trinh-Trang-Tan MM, Corman B, Verbavatz JM (2001) Food restriction prevents age-related polyuria by vasopressin-dependent recruitment of aquaporin-2. Am J Physiol Renal Physiol 281:F1123–F1131

Combet S, Geffroy N, Berthonaud V, Dick B, Teillet L, Verbavatz J-M, Corman B, Trinh-Trang-Tan M-M (2003) Correction of age-related polyuria by dDAVP: molecular analysis of aquaporins and urea transporters. Am J Physiol Renal Physiol 284:F199–F208

Coyle J, McDaid S, Walpole C, Stewart GS (2016) UT-B urea transporter localization in the bovine gastrointestinal tract. J Membr Biol 249:77–85. https://doi.org/10.1007/s00232-015-9850-5

Denton JS, Pao AC, Maduke M (2013) Novel diuretic targets. Am J Physiol Renal Physiol 305: F931–F942. https://doi.org/10.1152/ajprenal.00230.2013

Dix L, Ward DT, Stewart GS (2013) Short communication: urea transporter protein UT-B in the bovine parotid gland. J Dairy Sci 96(3):1685–1690. https://doi.org/10.3168/jds.2012-6230

Dong Z, Ran J, Zhou H, Chen J, Lei T, Wang W, Sun Y, Lin G, Bankir L, Yang B (2013) Urea transporter UT-B deletion induces DNA damage and apoptosis in mouse bladder urothelium. PLoS One 8:e76952. https://doi.org/10.1371/journal.pone.0076952

Dousa TP (1974) Interaction of lithium with vasopressin-sensitive cyclic AMP system of human renal medulla. Endocrinology 95:1359–1366

Du Y, Meng Y, Lv X, Guo L, Wang X, Su Z, Li L, Li N, Zhao S, Zhao L, Zhao X (2014a) Dexamethasone attenuates LPS-induced changes in expression of urea transporter and aquaporin proteins, ameliorating brain endotoxemia in mice. Int J Clin Exp Pathol 7:8443–8452

Du Y, Meng Y, Zhu J, Kang L, Jia X, Gui L, Zhang L, Ye M, Hu L, Zhao X, Gu J, Yang B, Zou H (2014b) Quantitative proteomic study of myocardial mitochondria in urea transporter B knock-out mice. Proteomics 14:2072–2083

Efe O, Klein JD, LaRocque LM, Ren H, Sands JM (2016) Metformin improves urine concentration in rodents with nephrogenic diabetes insipidus. JCI Insight 1:e88409

Esteva-Font C, Phuan P-W, Anderson MO, Verkman AS (2013) A small molecule screen identifies selective inhibitors of urea transporter UT-A. Chem Biol 20:1235–1244

Esteva-Font C, Cil O, Phuan PW, Su T, Lee S, Anderson MO, Verkman AS (2014) Diuresis and reduced urinary osmolality in rats produced by small-molecule UT-A-selective urea transport inhibitors. FASEB J 28:3878–3890. https://doi.org/10.1096/fj.14-253872

Esteva-Font C, Anderson MO, Verkman AS (2015a) Urea transporter proteins as targets for small-molecule diuretics. Nat Rev Nephrol 11:113–123. https://doi.org/10.1038/nrneph.2014.219

Esteva-Font C, Phuan PW, Lee S, Su T, Anderson MO, Verkman AS (2015b) Structure-activity analysis of thiourea analogs as inhibitors of UT-A and UT-B urea transporters. Biochim Biophys Acta 1848:1075–1080. https://doi.org/10.1016/j.bbamem.2015.01.004

Esther CR Jr, Marrero MB, Howard TE, Machaud A, Corvol P, Capecchi MR, Bernstein KE (1997) The critical role of tissue angiotensin-converting enzyme as revealed by gene targeting in mice. J Clin Invest 99:2375–2385

Evans KK, Nawata CM, Pannabecker TL (2015) Isolation and perfusion of rat inner medullary vasa recta. Am J Physiol Renal Physiol 309:F300–F304. https://doi.org/10.1152/ajprenal.00214.2015

Faubert PF, Chou SY, Porush JG, Byrd R (1987) Regulation of papillary plasma flow by angiotensin II. Kidney Int 32:472–478

Feng X, Huang H, Yang Y, Fröhlich O, Klein JD, Sands JM, Chen G (2009) Caveolin-1 directly interacts with UT-A1 urea transporter: the role of caveolae/lipid rafts in UT-A1 regulation at the cell membrane. Am J Physiol Renal Physiol 296:F1514–F1520

Feng X, Li Z, Du Y, Fu H, Klein JD, Cai H, Sands JM, Chen G (2015) Downregulation of urea transporter UT-A1 activity by 14-3-3 protein. Am J Physiol Renal Physiol 309:F71–F78. https://doi.org/10.1152/ajprenal.00546.2014

Fenton RA (2008) Urea transporters and renal function: lessons from knock-out mice. Curr Opin Nephrol Hypertens 17:513–518

Fenton RA, Knepper MA (2007) Urea and renal function in the 21st century: insights from knock-out mice. J Am Soc Nephrol 18:679–688

Fenton RA, Yang B (2014) Urea transporter knock-out mice and their renal phenotypes. Subcell Biochem 73:137–152. https://doi.org/10.1007/978-94-017-9343-8_9

Fenton RA, Howorth A, Cooper GJ, Meccariello R, Morris ID, Smith CP (2000) Molecular characterization of a novel UT-A urea transporter isoform (UT-A5) in testis. Am J Physiol Cell Physiol 279:C1425–C1C31
Fenton RA, Chou C-L, Ageloff S, Brandt W, Stokes JB III, Knepper M (2003) Increased collecting duct urea transporter expression in Dahl salt-sensitive rats. Am J Physiol Renal Physiol 285: F143–F151
Fenton RA, Chou C-L, Stewart GS, Smith CP, Knepper MA (2004) Urinary concentrating defect in mice with selective deletion of phloretin-sensitive urea transporters in the renal collecting duct. Proc Natl Acad Sci USA 101:7469–7474
Fenton RA, Flynn A, Shodeinde A, Smith CP, Schnermann J, Knepper MA (2005) Renal phenotype of UT-A urea transporter knock-out mice. J Am Soc Nephrol 16:1583–1592
Fenton RA, Chou CL, Sowersby H, Smith CP, Knepper MA (2006a) Gamble's "economy of water" revisited: studies in urea transporter knock-out mice. Am J Physiol Renal Physiol 291:F148–F154
Fenton RA, Shodeinde A, Knepper MA (2006b) UT-A urea transporter promoter, UT-Aalpha, targets principal cells of the renal inner medullary collecting duct. Am J Physiol Renal Physiol 290:F188–F195
Fernández-Llama P, Andrews P, Nielsen S, Ecelbarger CA, Knepper MA (1998) Impaired aquaporin and urea transporter expression in rats with adriamycin-induced nephrotic syndrome. Kidney Int 53:1244–1253
Galluci E, Micelli S, Lippe C (1971) Non-electrolyte permeability across thin lipid membranes. Arch Int Physiol Biochim 79:881–887
Gamble JL, McKhann CF, Butler AM, Tuthill E (1934) An economy of water in renal function referable to urea. Am J Phys 109:139–154
Geng X, Lei T, Zhou H, Yao W, Xin W, Yang B (2017) The knock-out of urea transporter-B improves the hemorheological properties of erythrocyte. Clin Hemorheol Microcirc 65:249–257. https://doi.org/10.3233/ch-16174
Gertner RA, Klein JD, Bailey JL, Kim D-U, Luo XH, Bagnasco SM, Sands JM (2004) Aldosterone decreases UT-A1 urea transporter expression via the mineralocorticoid receptor. J Am Soc Nephrol 15:558–565
Geyer RR, Musa-Aziz R, Enkavi G, Mahinthichaichan P, Tajkhorshid E, Boron WF (2013) Movement of NH3 through the human urea transporter B: a new gas channel. Am J Physiol Renal Physiol 304:F1447–F1457. https://doi.org/10.1152/ajprenal.00609.2012
Gilbert RL, Pannabecker TL (2013) Architecture of interstitial nodal spaces in the rodent renal inner medulla. Am J Physiol Renal Physiol 305:F745–F752. https://doi.org/10.1152/ajprenal.00239.2013
Gillin AG, Sands JM (1992) Characteristics of osmolarity-stimulated urea transport in rat IMCD. Am J Physiol Renal Physiol 262:F1061–F1067
Gillin AG, Star RA, Sands JM (1993) Osmolarity-stimulated urea transport in rat terminal IMCD: role of intracellular calcium. Am J Physiol Renal Physiol 265:F272–F277
Guo L, Meng J, Xuan C, Ge J, Sun W, O'Rourke ST, Sun C (2015) High salt-diet reduces SLC14A1 gene expression in the choroid plexus of Dahl salt sensitive rats. Biochem Biophys Res Commun 461:254–259. https://doi.org/10.1016/j.bbrc.2015.04.010
Hamilton JR (2015) Kidd blood group system: a review. Immunohematology 31:29–35
Harrington AR, Valtin H (1968) Impaired urinary concentration after vasopressin and its gradual correction in hypothalamic diabetes insipidus. J Clin Invest 47:502–510
Hassouneh R, Nasrallah R, Zimpelmann J, Gutsol A, Eckert D, Ghossein J, Burns KD, Hebert RL (2016) PGE2 receptor EP3 inhibits water reabsorption and contributes to polyuria and kidney injury in a streptozotocin-induced mouse model of diabetes. Diabetologia 59:1318–1328. https://doi.org/10.1007/s00125-016-3916-5
Hinze C, Ruffert J, Walentin K, Himmerkus N, Nikpey E, Tenstad O, Wiig H, Mutig K, Yurtdas ZY, Klein JD, Sands JM, Branchi F, Schumann M, Bachmann S, Bleich M, Schmidt-Ott KM

(2018) GRHL2 is required for collecting duct epithelial barrier function and renal osmoregulation. J Am Soc Nephrol 29:857–868. https://doi.org/10.1681/asn.2017030353
Hoban CA, Black LN, Ordas RJ, Gumina DL, Pulous FE, Sim JH, Sands JM, Blount MA (2015) Vasopressin regulation of multisite phosphorylation of UT-A1 in the inner medullary collecting duct. Am J Physiol Renal Physiol 308:F49–F55
Hoffert JD, Pisitkun T, Saeed F, Wilson JL, Knepper MA (2014) Global analysis of the effects of the V2 receptor antagonist satavaptan on protein phosphorylation in collecting duct. Am J Physiol Renal Physiol 306:410–421. https://doi.org/10.1152/ajprenal.00497.2013
Hou R, Alemozaffar M, Yang B, Sands JM, Kong X, Chen G (2017a) Identification of a novel UT-B urea transporter in human urothelial cancer. Front Physiol 8:245. https://doi.org/10.3389/fphys.2017.00245
Hou R, Kong X, Yang B, Xie Y, Chen G (2017b) SLC14A1: a novel target for human urothelial cancer. Clin Transl Oncol 19:1438–1446. https://doi.org/10.1007/s12094-017-1693-3
Huang H, Feng X, Zhuang J, Froehlich O, Klein JD, Cai H, Sands JM, Chen G (2010) Internalization of UT-A1 urea transporter is dynamin dependent and mediated by both caveolae and clathrin coated pit pathways. Am J Physiol Renal Physiol 299:F1389–F1395
Huang B, Wang H, Yang B (2017) Water transport mediated by other membrane proteins. Adv Exp Med Biol 969:251–261. https://doi.org/10.1007/978-94-024-1057-0_17
Ilgu H, Jeckelmann JM, Gachet MS, Boggavarapu R, Ucurum Z, Gertsch J, Fotiadis D (2014) Variation of the detergent-binding capacity and phospholipid content of membrane proteins when purified in different detergents. Biophys J 106:1660–1670. https://doi.org/10.1016/j.bpj.2014.02.024
Ilori TO, Wang Y, Blount MA, Martin CF, Sands JM, Klein JD (2012) Acute calcineurin inhibition with tacrolimus increases phosphorylated UT-A1. Am J Physiol Renal Physiol 302:F998–F1004. https://doi.org/10.1152/ajprenal.00358.2011
Ilori TO, Blount MA, Martin CF, Sands JM, Klein JD (2013) Urine concentration in the diabetic mouse requires both water and urea transporters. Am J Physiol Renal Physiol 304:F103–F111. https://doi.org/10.1152/ajprenal.00385.2012
Isozaki T, Verlander JW, Sands JM (1993) Low protein diet alters urea transport and cell structure in rat initial inner medullary collecting duct. J Clin Invest 92:2448–2457
Isozaki T, Gillin AG, Swanson CE, Sands JM (1994a) Protein restriction sequentially induces new urea transport processes in rat initial IMCDs. Am J Physiol Renal Physiol 266:F756–F761
Isozaki T, Lea JP, Tumlin JA, Sands JM (1994b) Sodium-dependent net urea transport in rat initial IMCDs. J Clin Invest 94:1513–1517
Issaian T, Urity VB, Dantzler WH, Pannabecker TL (2012) Architecture of vasa recta in the renal inner medulla of the desert rodent *Dipodomys merriami*: potential impact on the urine concentrating mechanism. Am J Physiol Regul Integr Comp Physiol 303:R748–R756. https://doi.org/10.1152/ajpregu.00300.2012
Jackson BA, Braun-Werness JL, Kusano E, Dousa TP (1983) Concentrating defect in the adrenalectomized rat. Abnormal vasopressin-sensitive cyclic adenosine monophosphate metabolism in the papillary collecting duct. J Clin Invest 72:997–1004
Jacob VA, Harbaugh CM, Dietz JR, Fenton RA, Kim SM, Castrop H, Schnermann J, Knepper MA, Chou CL, Anderson SA (2008) Magnetic resonance imaging of urea transporter knock-out mice shows renal pelvic abnormalities. Kidney Int 74:1202–1208
Jiang T, Li Y, Layton AT, Wang W, Sun Y, Li M, Zhou H, Yang B (2017) Generation and phenotypic analysis of mice lacking all urea transporters. Kidney Int 91:338–351. https://doi.org/10.1016/j.kint.2016.09.017
Kamoi K, Tamura T, Tanaka K, Ishikashi M, Yamagi T (1993) Hyponatremia and osmoregulation of thirst and vasopressin secretion in patients with adrenal insufficiency. J Clin Endocrin Metabol 77:1584–1588
Kato A, Sands JM (1998a) Active sodium-urea counter-transport is inducible in the basolateral membrane of rat renal initial inner medullary collecting ducts. J Clin Invest 102:1008–1015

Kato A, Sands JM (1998b) Evidence for sodium-dependent active urea secretion in the deepest subsegment of the rat inner medullary collecting duct. J Clin Invest 101:423–428
Kato A, Sands JM (1999) Urea transport processes are induced in rat IMCD subsegments when urine concentrating ability is reduced. Am J Phys 276:F62–F71
Kato A, Klein JD, Zhang C, Sands JM (2000) Angiotensin II increases vasopressin-stimulated facilitated urea permeability in rat terminal IMCDs. Am J Physiol Renal Physiol 279:F835–F840
Keller RW, Bailey JL, Wang Y, Klein JD, Sands JM (2016) Urea transporters and sweat response to uremia. Physiol Rep 4(11):e12825. https://doi.org/10.14814/phy2.12825
Kenig-Kozlovsky Y, Scott RP, Onay T, Carota IA, Thomson BR, Gil HJ, Ramirez V, Yamaguchi S, Tanna CE, Heinen S, Wu C, Stan RV, Klein JD, Sands JM, Oliver G, Quaggin SE (2018) Ascending vasa recta are angiopoietin/Tie2-dependent lymphatic-like vessels. J Am Soc Nephrol 29:1097–1107. https://doi.org/10.1681/asn.2017090962
Kim D-U, Sands JM, Klein JD (2003) Changes in renal medullary transport proteins during uncontrolled diabetes mellitus in rats. Am J Physiol Renal Physiol 285:F303–F309
Kim D-U, Sands JM, Klein JD (2004) Role of vasopressin in diabetes mellitus-induced changes in medullary transport proteins involved in urine concentration in Brattleboro rats. Am J Physiol Renal Physiol 286:F760–F766
Kim WY, Lee HW, Han KH, Nam SA, Choi A, Kim YK, Kim J (2016) Descending thin limb of the intermediate loop expresses both aquaporin 1 and urea transporter A2 in the mouse kidney. Histochem Cell Biol 146:1–12. https://doi.org/10.1007/s00418-016-1434-7
Kitada K, Daub S, Zhang Y, Klein JD, Nakano D, Pedchenko T, Lantier L, LaRocque LM, Marton A, Neubert P, Schroder A, Rakova N, Jantsch J, Dikalova AE, Dikalov SI, Harrison DG, Muller DN, Nishiyama A, Rauh M, Harris RC, Luft FC, Wassermann DH, Sands JM, Titze J (2017) High salt intake reprioritizes osmolyte and energy metabolism for body fluid conservation. J Clin Invest 127:1944–1959. https://doi.org/10.1172/jci88532
Klein JD (2014) Expression of urea transporters and their regulation. Subcell Biochem 73:79–107. https://doi.org/10.1007/978-94-017-9343-8_6
Klein JD, Sands JM (2016) Urea transport and clinical potential of urearetics. Curr Opin Nephrol Hypertens 25:444–451
Klein JD, Price SR, Bailey JL, Jacobs JD, Sands JM (1997) Glucocorticoids mediate a decrease in AVP-regulated urea transporter in diabetic rat inner medulla. Am J Physiol 273(6):F949–F953
Klein JD, Timmer RT, Rouillard P, Bailey JL, Sands JM (1999) UT-A urea transporter protein expressed in liver: upregulation by uremia. J Am Soc Nephrol 10:2076–2083
Klein JD, Gunn RB, Roberts BR, Sands JM (2002a) Down-regulation of urea transporters in the renal inner medulla of lithium-fed rats. Kidney Int 61:995–1002. https://doi.org/10.1046/j.1523-1755.2002.00210.x
Klein JD, Quach DL, Cole JM, Disher K, Mongiu AK, Wang X, Bernstein KE, Sands JM (2002b) Impaired urine concentration and the absence of tissue ACE: the involvement of medullary transport proteins. Am J Physiol Renal Physiol 283:F517–F524
Klein JD, Sands JM, Qian L, Wang X, Yang B (2004) Upregulation of urea transporter UT-A2 and water channels AQP2 and AQP3 in mice lacking urea transporter UT-B. J Am Soc Nephrol 15:1161–1167
Klein JD, Fröhlich O, Blount MA, Martin CF, Smith TD, Sands JM (2006a) Vasopressin increases plasma membrane accumulation of urea transporter UT-A1 in rat inner medullary collecting ducts. J Am Soc Nephrol 17:2680–2686
Klein JD, Murrell BP, Tucker S, Kim Y-H, Sands JM (2006b) Urea transporter UT-A1 and aquaporin-2 proteins decrease in response to angiotensin II or norepinephrine-induced acute hypertension. Am J Physiol Renal Physiol 291:F952–F959
Klein JD, Blount MA, Sands JM (2011) Urea transport in the kidney. Compr Physiol 1:699–729. https://doi.org/10.1002/cphy.c100030

Klein JD, Martin CF, Kent KJ, Sands JM (2012) Protein kinase C alpha mediates hypertonicity-stimulated increase in urea transporter phosphorylation in the inner medullary collecting duct. Am J Physiol Renal Physiol 302:F1098–F1103

Klein JD, Wang Y, Blount MA, Molina PA, LaRocque LM, Ruiz JA, Sands JM (2016a) Metformin, an AMPK activator, stimulates the phosphorylation of aquaporin 2 and urea transporter A1 in inner medullary collecting ducts. Am J Physiol Renal Physiol 310:F1008–F1012

Klein JD, Wang Y, Mistry A, LaRocque LM, Molina PA, Rogers RT, Blount MA, Sands JM (2016b) Transgenic restoration of urea transporter A1 confers maximal urinary concentration in the absence of urea transporter A3. J Am Soc Nephrol 27:1448–1455. https://doi.org/10.1681/asn.2014121267

Knepper MA, Mindell JA (2009) Molecular coin slots for urea. Nature 462:733–734

Knepper MA, Miranda CA (2013) Urea channel inhibitors: a new functional class of aquaretics. Kidney Int 83:991–993. https://doi.org/10.1038/ki.2013.94

Knepper MA, Danielson RA, Saidel GM, Johnston KH (1975) Effects of dietary protein restriction and glucocorticoid administration on urea excretion in rats. Kidney Int 8:303–315

Knepper MA, Gunter CV, Danielson RA (1976) Effects of glucagon on renal function in protein-deprived rats. Surg Forum 27:29–31

Koutros S, Baris D, Fischer A, Tang W, Garcia-Closas M, Karagas MR, Schwenn M, Johnson A, Figueroa J, Waddell R, Prokunina-Olsson L, Rothman N, Silverman DT (2013) Differential urinary specific gravity as a molecular phenotype of the bladder cancer genetic association in the urea transporter gene, SLC14A1. Int J Cancer 133:3008–3013. https://doi.org/10.1002/ijc.28325

Kriz W (1967) Der architektonische und funktionelle Aufbau der Rattenniere. Z Zellforsch 82:495–535

Kriz W, Bankir L (1983) Structural organization of the renal medullary counterflow system. Fed Proc 42:2379–2385

Kriz W, Schnermann J, Koepsell H (1972) The position of short and long loops of Henle in the rat kidney. Z Anat Entwickl Gesch 138:301–319

Kudo LH, César KR, Ping WC, Rocha AS (1992) Effect of peritubular hypertonicity on water and urea transport of inner medullary collecting duct. Am J Physiol Renal Fluid Electrol Physiol 262:F338–F347

Kuper C, Fraek ML, Muller HH, Beck FX, Neuhofer W (2012) Sepsis-induced urinary concentration defect is related to nitric oxide-dependent inactivation of TonEBP/NFAT5, which downregulates renal medullary solute transport proteins and aquaporin-2. Crit Care Med 40:1887–1895. https://doi.org/10.1097/CCM.0b013e31824e1186

Lawicki S, Covin RB, Powers AA (2017) The Kidd (JK) blood group system. Transfus Med Rev 31:165–172. https://doi.org/10.1016/j.tmrv.2016.10.003

Layton AT, Bankir L (2013) Impacts of active urea secretion into pars recta on urine concentration and urea excretion rate. Physiol Rep 1(3). https://doi.org/10.1002/phy2.34

Layton AT, Dantzler WH, Pannabecker TL (2012a) Urine concentrating mechanism: impact of vascular and tubular architecture and a proposed descending limb urea-Na^+ cotransporter. Am J Physiol Renal Physiol 302:F591–F605. https://doi.org/10.1152/ajprenal.00263.2011

Layton AT, Gilbert RL, Pannabecker TL (2012b) Isolated interstitial nodal spaces may facilitate preferential solute and fluid mixing in the rat renal inner medulla. Am J Physiol Renal Physiol 302:F830–F839. https://doi.org/10.1152/ajprenal.00539.2011

Lee S, Esteva-Font C, Phuan PW, Anderson MO, Verkman AS (2015) Discovery, synthesis and structure-activity analysis of symmetrical 2,7-disubstituted fluorenones as urea transporter inhibitors. Med Chem Commun 6:1278–1284. https://doi.org/10.1039/c5md00198f

Lee S, Cil O, Diez-Cecilia E, Anderson MO, Verkman AS (2018) Nanomolar-potency 1,2,4-Triazoloquinoxaline inhibitors of the kidney urea transporter UT-A1. J Med Chem 61:3209–3217. https://doi.org/10.1021/acs.jmedchem.8b00343

Lei T, Zhou L, Layton AT, Zhou H, Zhao X, Bankir L, Yang B (2011) Role of thin descending limb urea transport in renal urea handling and the urine concentrating mechanism. Am J Physiol Renal Physiol 301:F1251–F1259

Lemley KV, Kriz W (1987) Cycles and separations: the histotopography of the urinary concentrating process. Kidney Int 31:538–548

LeMoine CM, Walsh PJ (2015) Evolution of urea transporters in vertebrates: adaptation to urea's multiple roles and metabolic sources. J Exp Biol 218:1936–1945. https://doi.org/10.1242/jeb.114223

Levin EJ, Zhou M (2014) Structure of urea transporters. Subcell Biochem 73:65–78. https://doi.org/10.1007/978-94-017-9343-8_5

Levin EJ, Quick M, Zhou M (2009) Crystal structure of a bacterial homologue of the kidney urea transporter. Nature 462:757–762

Levin EJ, Cao Y, Enkavi G, Quick M, Pan Y, Tajkhorshid E, Zhou M (2012) Structure and permeation mechanism of a mammalian urea transporter. Proc Natl Acad Sci USA 109:11194–11199. https://doi.org/10.1073/pnas.1207362109

Li C, Klein JD, Wang W, Knepper MA, Nielsen S, Sands JM, Frokiaer J (2004) Altered expression of urea transporters in response to ureteral obstruction. Am J Physiol Renal Physiol 286:F1154–F1162. https://doi.org/10.1152/ajprenal.00453.2003

Li C, Wang W, Summer SN, Falk S, Schrier RW (2008) Downregulation of UT-A1/UT-A3 is associated with urinary concentrating defect in glucocorticoid-excess state. J Am Soc Nephrol 19:1975–1981

Li X, Chen G, Yang B (2012) Urea transporter physiology studied in knock-out mice. Front Physiol 3:217. https://doi.org/10.3389/fphys.2012.00217

Li F, Lei T, Zhu J, Wang W, Sun Y, Chen J, Dong Z, Zhou H, Yang B (2013) A novel small-molecule thienoquinolin urea transporter inhibitor acts as a potential diuretic. Kidney Int 83:1076–1086. https://doi.org/10.1038/ki.2013.62

Li C, Xue H, Lei Y, Zhu J, Yang B, Gai X (2014a) Clinical significance of the reduction of UT-B expression in urothelial carcinoma of the bladder. Pathol Res Pract 210(12):799–803. https://doi.org/10.1016/j.prp.2014.09.012

Li M, Tou WI, Zhou H, Li F, Ren H, Chen CY, Yang B (2014b) Developing hypothetical inhibition mechanism of novel urea transporter B inhibitor. Sci Rep 4:5775. https://doi.org/10.1038/srep05775

Li X, Yang B, Chen M, Klein JD, Sands JM, Chen G (2015) Activation of protein kinase C-alpha and Src kinase increases urea transporter A1 alpha-2, 6 sialylation. J Am Soc Nephrol 26:926–934. https://doi.org/10.1681/ASN.2014010026

Li YX, Huang Y, Liu S, Mao Y, Yuan CY, Yang X, Yao LJ (2016) Glycogen synthase kinase-3 modulates hyperosmotic-induced urea transporter A1 relocation in the inner medullary collecting duct cells. Nephron 133:71–79. https://doi.org/10.1159/000446158

Lim S-W, Li C, Sun B-K, Kim W-Y, Han K-H, Oh Y-W, Lee J-U, Kador PF, Knepper MA, Sands JM, Kim J, Yang C-W (2004) Long-term treatment with cyclosporine decreases aquaporins and urea transporters in rat kidney. Am J Physiol Renal Physiol 287:F139–F151

Lim S-W, Han K-H, Jung J-Y, Kim W-Y, Yang C-W, Sands JM, Knepper MA, Madsen KM, Kim J (2006) Ultrastructural localization of UT-A and UT-B in rat kidneys with different hydration status. Am J Physiol Regul Integr Comp Physiol 290:R479–R492

Liu Y, Esteva-Font C, Yao C, Phuan PW, Verkman AS, Anderson MO (2013) 1,1-Difluoroethyl-substituted triazolothienopyrimidines as inhibitors of a human urea transport protein (UT-B): new analogs and binding model. Bioorg Med Chem Lett 23:3338–3341. https://doi.org/10.1016/j.bmcl.2013.03.089

Liu M, Li M, Liu J, Wang H, Zhong D, Zhou H, Yang B (2016) Elevated urinary urea by high-protein diet could be one of the inducements of bladder disorders. J Transl Med 14:53. https://doi.org/10.1186/s12967-016-0809-9

Lu Z, Stumpff F, Deiner C, Rosendahl J, Braun H, Abdoun K, Aschenbach JR, Martens H (2014) Modulation of sheep ruminal urea transport by ammonia and pH. Am J Physiol Regul Integr Comp Physiol 307:R558–R570. https://doi.org/10.1152/ajpregu.00107.2014

Lu Z, Gui H, Yao L, Yan L, Martens H, Aschenbach JR, Shen Z (2015) Short-chain fatty acids and acidic pH upregulate UT-B, GPR41, and GPR4 in rumen epithelial cells of goats. Am J Physiol Regul Integr Comp Physiol 308:R283–R293. https://doi.org/10.1152/ajpregu.00323.2014

Macey RI (1984) Transport of water and urea in red blood cells. Am J Physiol Cell Physiol 246: C195–C203

Maeda Y, Terada Y, Nonoguchi H, Knepper MA (1992) Hormone and autocoid regulation of cAMP production in rat IMCD subsegments. Am J Physiol Renal Physiol 263:F319–F327

Mariager CO, Nielsen PM, Qi H, Ringgaard S, Laustsen C (2017) Hyperpolarized $^{13}C,^{15}N_2$-urea T_2 relaxation changes in acute kidney injury. Magn Reson Med 80:696–702. https://doi.org/10.1002/mrm.27050

Marples D, Christensen S, Christensen EI, Ottosen PD, Nielsen S (1995) Lithium-induced downregulation of Aquaporin-2 water channel expression in rat kidney medulla. J Clin Invest 95:1838–1845

McGrane A, Stewart G (2016) Hyperosmolality regulates UT-A6 urea transporter expression in the Caco-2 cell line. Physiol Rep 4(18):e12984. https://doi.org/10.14814/phy2.12984

Meng Y, Zhao C, Zhang X, Zhao H, Guo L, Lu B, Zhao X, Yang B (2009) Surface electrocardiogram and action potential in mice lacking urea transporter UT-B. Sci China C Life Sci 52:474–478. https://doi.org/10.1007/s11427-009-0047-y

Mistry AC, Mallick R, Fröhlich O, Klein JD, Rehm A, Chen G, Sands JM (2007) The UT-A1 urea transporter interacts with snapin, a snare-associated protein. J Biol Chem 282:30097–30106

Mitch WE, Bailey JL, Wang X, Jurkovitz C, Newby DN, Price SR (1999) Evaluation of signals activating ubiquitin-proteasome proteolysis in a model of muscle wasting. Am J Physiol Cell Physiol 276:C1132–C1138

Nakayama Y, Peng T, Sands JM, Bagnasco SM (2000) The TonE/TonEBP pathway mediates tonicity-responsive regulation of UT-A urea transporter expression. J Biol Chem 275:38275–38280

Naruse M, Klein JD, Ashkar ZM, Jacobs JD, Sands JM (1997) Glucocorticoids downregulate the rat vasopressin-regulated urea transporter in rat terminal inner medullary collecting ducts. J Am Soc Nephrol 8:517–523

Nasrallah R, Zimpelmann J, Eckert D, Ghossein J, Geddes S, Beique JC, Thibodeau JF, Kennedy CRJ, Burns KD, Hebert RL (2018) PGE2 EP1 receptor inhibits vasopressin-dependent water reabsorption and sodium transport in mouse collecting duct. Lab Investig 98:360–370. https://doi.org/10.1038/labinvest.2017.133

Nawata CM, Evans KK, Dantzler WH, Pannabecker TL (2014) Transepithelial water and urea permeabilities of isolated perfused Munich-Wistar rat inner medullary thin limbs of Henle's loop. Am J Physiol Renal Physiol 306:F123–F129. https://doi.org/10.1152/ajprenal.00491.2013

Nawata CM, Dantzler WH, Pannabecker TL (2015) Alternative channels for urea in the inner medulla of the rat kidney. Am J Physiol Renal Physiol 309:F916–F924. https://doi.org/10.1152/ajprenal.00392.2015

Nielsen J, Hoffert JD, Knepper MA, Agre P, Nielsen S, Fenton RA (2008) Proteomic analysis of lithium-induced nephrogenic diabetes insipidus: mechanisms for aquaporin 2 down-regulation and cellular proliferation. Proc Natl Acad Sci USA 105:3634–3639

Ogami A, Miyazaki H, Niisato N, Sugimoto T, Marunaka Y (2006) UT-B1 urea transporter plays a noble role as active water transporter in C6 glial cells. Biochem Biophys Res Commun 351:619–624

Onodera T, Sasaki K, Tsuneyama H, Isa K, Ogasawara K, Satake M, Tadokoro K, Uchikawa M (2014) JK null alleles identified from Japanese individuals with Jk(a-b-) phenotype. Vox Sang 106:382–384

Pages G, Puckeridge M, Liangfeng G, Tan YL, Jacob C, Garland M, Kuchel PW (2013) Transmembrane exchange of hyperpolarized ^{13}C-urea in human erythrocytes: subminute timescale kinetic analysis. Biophys J 105:1956–1966. https://doi.org/10.1016/j.bpj.2013.09.034
Pallone TL, Work J, Myers RL, Jamison RL (1994) Transport of sodium and urea in outer medullary descending vasa recta. J Clin Invest 93:212–222
Pannabecker TL (2013) Comparative physiology and architecture associated with the mammalian urine concentrating mechanism: role of inner medullary water and urea transport pathways in the rodent medulla. Am J Physioly Regul Integr Comp Physiol 304:R488–R503. https://doi.org/10.1152/ajpregu.00456.2012
Pannabecker TL, Dantzler WH (2004) Three-dimensional lateral and vertical relationships of inner medullary loops of Henle and collecting ducts. Am J Physiol Renal Physiol 287:F767–F774
Pannabecker TL, Dantzler WH (2007) Three-dimensional architecture of collecting ducts, loops of Henle, and blood vessels in the renal papilla. Am J Physiol Renal Physiol 293:F696–F704
Pannabecker TL, Abbott DE, Dantzler WH (2004) Three-dimensional functional reconstruction of inner medullary thin limbs of Henle's loop. Am J Physiol Renal Physiol 286:F38–F45
Pannabecker TL, Dantzler WH, Layton HE, Layton AT (2008a) Role of three-dimensional architecture in the urine concentrating mechanism of the rat renal inner medulla. Am J Physiol Renal Physiol 295:F1271–F1285
Pannabecker TL, Henderson CS, Dantzler WH (2008b) Quantitative analysis of functional reconstructions reveals lateral and axial zonation in the renal inner medulla. Am J Physiol Renal Physiol 294:F1306–F1314
Pastor-Soler NM, Hallows KR (2012) AMP-activated protein kinase regulation of kidney tubular transport. Curr Opin Nephrol Hypertens 21:523–533
Patrick PS, Kettunen MI, Tee SS, Rodrigues TB, Serrao E, Timm KN, McGuire S, Brindle KM (2015) Detection of transgene expression using hyperpolarized ^{13}C urea and diffusion-weighted magnetic resonance spectroscopy. Magn Reson Med 73:1401–1406. https://doi.org/10.1002/mrm.25254
Pech V, Klein JD, Kozlowski SD, Wall SM, Sands JM (2005) Vasopressin increases urea permeability in initial IMCDs from diabetic rats. Am J Physiol Renal Physiol 289:F531–F535
Peng T, Sands JM, Bagnasco SM (2002) Glucocorticoids inhibit transcription and expression of the rat UT-A urea transporter gene. Am J Physiol Renal Physiol 282:F853–F858
Preisser L, Teillet L, Aliotti S, Gobin R, Berthonaud V, Chevalier J, Corman B, Verbavatz JM (2000) Downregulation of aquaporin-2 and-3 in aging kidney is independent of V_2 vasopressin receptor. Am J Physiol Renal Physiol 279:F144–F152
Promeneur D, Bankir L, Hu MC, Trinh-Trang-Tan M-M (1998) Renal tubular and vascular urea transporters: influence of antidiuretic hormone on messenger RNA expression in Brattleboro rats. J Am Soc Nephrol 9:1359–1366
Qi H, Mariager CO, Nielsen PM, Schroeder M, Lindhardt J, Nørregaard R, Klein JD, Sands JM, Laustsen C (2019) Glucagon infusion alters the hyperpolarized ^{13}C-urea renal hemodynamic signature. NMR Biomed 32:e4028. https://doi.org/10.1002/nbm.4028
Qian X, Li X, Ilori TO, Klein JD, Hughey RP, Li CJ, Alli AA, Guo Z, Yu P, Song X, Chen G (2015) RNA-seq analysis of glycosylation related gene expression in STZ-induced diabetic rat kidney inner medulla. Front Physiol 6:274. https://doi.org/10.3389/fphys.2015.00274
Qian X, Sands JM, Song X, Chen G (2016) Modulation of kidney urea transporter UT-A3 activity by alpha2,6-sialylation. Pfluegers Arch 468:1161–1170
Rakova N, Kitada K, Lerchl K, Dahlmann A, Birukov A, Daub S, Kopp C, Pedchenko T, Zhang Y, Beck L, Johannes B, Marton A, Muller DN, Rauh M, Luft FC, Titze J (2017) Increased salt consumption induces body water conservation and decreases fluid intake. J Clin Invest 127:1932–1943. https://doi.org/10.1172/jci88530
Ramsey G, Sumugod RD, Lindholm PF, Zinni JG, Keller JA, Horn T, Keller MA (2016) A Caucasian JK*A/JK*B woman with Jk(a+b-) red blood cells, anti-Jkb, and a novel JK*B allele c.1038delG. Immunohematology 32:91–95

Ran J, Wang H, Hu T (2014) Clinical aspects of urea transporters. Subcellul Biochem 73:179–191. https://doi.org/10.1007/978-94-017-9343-8_12

Ran JH, Li M, Tou WI, Lei TL, Zhou H, Chen CY, Yang BX (2016) Phenylphthalazines as small-molecule inhibitors of urea transporter UT-B and their binding model. Acta Pharmacol Sin 37:973–983. https://doi.org/10.1038/aps.2016.4

Reed GD, von Morze C, Verkman AS, Koelsch BL, Chaumeil MM, Lustig M, Ronen SM, Bok RA, Sands JM, Larson PE, Wang ZJ, Larsen JH, Kurhanewicz J, Vigneron DB (2016) Imaging renal urea handling in rats at millimeter resolution using hyperpolarized magnetic resonance relaxometry. Tomography 2:125–135. https://doi.org/10.18383/j.tom.2016.00127

Ren H, Wang Y, Xing Y, Ran J, Liu M, Lei T, Zhou H, Li R, Sands JM, Yang B (2014) Thienoquinolins exert diuresis by strongly inhibiting UT-A urea transporters. Am J Physiol Renal Physiol 307:F1363–F1372. https://doi.org/10.1152/ajprenal.00421.2014

Ren H, Yang B, Molina PA, Sands JM, Klein JD (2015) NSAIDs alter phosphorylated forms of AQP2 in the inner medullary tip. PLoS One 10(10):e0141714

Sacca E, Corazzin M, Giannico F, Fabro C, Mason F, Spanghero M (2018) Effect of dietary nitrogen level and source on mRNA expression of urea transporters in the rumen epithelium of fattening bulls. Arch Anim Nutr 72:341–350. https://doi.org/10.1080/1745039x.2018.1507977

Sanches TR, Volpini RA, Massola Shimizu MH, Braganca AC, Oshiro-Monreal F, Seguro AC, Andrade L (2012) Sildenafil reduces polyuria in rats with lithium-induced NDI. Am J Physiol Renal Physiol 302:F216–F225. https://doi.org/10.1152/ajprenal.00439.2010

Sands JM (2007) Critical role of urea in the urine-concentrating mechanism. J Am Soc Nephrol 18:670–671

Sands JM (2012) Urine concentrating and diluting ability during aging. J Gerontol A Biol Med Sci 67:1352–1357. https://doi.org/10.1093/gerona/gls128

Sands JM (2013) Urea transporter inhibitors: en route to new diuretics. Chem Biol 20:1201–1202. https://doi.org/10.1016/j.chembiol.2013.10.003

Sands JM, Blount MA (2014) Genes and proteins of urea transporters. Subcell Biochem 73:45–63. https://doi.org/10.1007/978-94-017-9343-8_4

Sands JM, Klein JD (2016) Physiological insights into novel therapies for nephrogenic diabetes insipidus. Am J Physiol Renal Physiol 311:F1149–F1152. https://doi.org/10.1152/ajprenal.00418.2016

Sands JM, Schrader DC (1991) An independent effect of osmolality on urea transport in rat terminal IMCDs. J Clin Invest 88:137–142

Sands JM, Nonoguchi H, Knepper MA (1987) Vasopressin effects on urea and H_2O transport in inner medullary collecting duct subsegments. Am J Phys 253:F823–F832

Sands JM, Gargus JJ, Fröhlich O, Gunn RB, Kokko JP (1992) Urinary concentrating ability in patients with Jk(a-b-) blood type who lack carrier-mediated urea transport. J Am Soc Nephrol 2:1689–1696

Sands JM, Martial S, Isozaki T (1996) Active urea transport in the rat initial inner medullary collecting duct: functional characterization and initial expression cloning. Kidney Int 49:1611–1614

Sands JM, Timmer RT, Gunn RB (1997) Urea transporters in kidney and erythrocytes. Am J Phys 273:F321–F339

Schilling F, Ros S, Hu DE, D'Santos P, McGuire S, Mair R, Wright AJ, Mannion E, Franklin RJ, Neves AA, Brindle KM (2017) MRI measurements of reporter-mediated increases in transmembrane water exchange enable detection of a gene reporter. Nat Biotechnol 35:75–80. https://doi.org/10.1038/nbt.3714

Schwartz MJ, Kokko JP (1980) Urinary concentrating defect of adrenal insufficiency. Permissive role of adrenal steroids on the hydroosmotic response across the rabbit cortical collecting duct. J Clin Invest 66:234–242

Selinski S, Blaszkewicz M, Ickstadt K, Gerullis H, Otto T, Roth E, Volkert F, Ovsiannikov D, Moormann O, Banfi G, Nyirady P, Vermeulen SH, Garcia-Closas M, Figueroa JD, Johnson A, Karagas MR, Kogevinas M, Malats N, Schwenn M, Silverman DT, Koutros S, Rothman N,

Kiemeney LA, Hengstler JG, Golka K (2017) Identification and replication of the interplay of four genetic high-risk variants for urinary bladder cancer. Carcinogenesis 38:1167–1179. https://doi.org/10.1093/carcin/bgx102
Shayakul C, Clemencon B, Hediger MA (2013) The urea transporter family (SLC14): physiological, pathological and structural aspects. Mol Asp Med 34:313–322. https://doi.org/10.1016/j.mam.2012.12.003
Shimizu MHM, Volpini RA, de Bragança AC, Campos R, Canale D, Sanches TR, Andrade L, Seguro AC (2013) N-acetylcysteine attenuates renal alterations induced by senescence in the rat. Exp Gerontol 48:298–303. https://doi.org/10.1016/j.exger.2012.11.006
Sidoux-Walter F, Lucien N, Olivès B, Gobin R, Rousselet G, Kamsteeg EJ, Ripoche P, Deen PMT, Cartron JP, Bailly P (1999) At physiological expression levels the Kidd blood group/urea transporter protein is not a water channel. J Biol Chem 274:30228–30235
Sim JH, Himmel NJ, Redd SK, Pulous FE, Rogers RT, Black LN, Hong SM, von Bergen TN, Blount MA (2014) Absence of PKC-alpha attenuates lithium-induced nephrogenic diabetes insipidus. PLoS One 9(7):e101753. https://doi.org/10.1371/journal.pone.0101753
Singh V, Jaiswal PK, Mittal RD (2014) Replicative study of GWAS TP63C/T, TERTC/T, and SLC14A1C/T with susceptibility to bladder cancer in North Indians. Urol Oncol 32:1209–1214. https://doi.org/10.1016/j.urolonc.2014.05.013
Smith CP, Potter EA, Fenton RA, Stewart GS (2004) Characterization of a human colonic cDNA encoding a structurally novel urea transporter, UT-A6. Am J Physiol Cell Physiol 287:C1087–C1093
Song W, Qi X, Zhang W, Zhao Y, Cao Y, Wang F, Yang C (2015) Abnormal expression of urea transporter protein in a rat model of hepatorenal syndrome induced by succinylated gelatin. Med Sci Monit 21:2905–2911. https://doi.org/10.12659/msm.894232
Spector DA, Deng J, Coleman R, Wade JB (2015) The urothelium of a hibernator: the American black bear. Physiol Rep 3(6):12429. https://doi.org/10.14814/phy2.12429
Star RA, Nonoguchi H, Balaban R, Knepper MA (1988) Calcium and cyclic adenosine monophosphate as second messengers for vasopressin in the rat inner medullary collecting duct. J Clin Invest 81:1879–1888
Stewart GS, O'Brien JH, Smith CP (2008) Ubiquitination regulates the plasma membrane expression of renal UT-A urea transporters. Am J Physiol Cell Physiol 295:C121–C129
Su H, Carter CB, Frohlich O, Cummings RD, Chen G (2012a) Glycoforms of UT-A3 urea transporter with poly-N-acetyllactosamine glycosylation have enhanced transport activity. Am J Physiol Renal Physiol 303:F201–F208. https://doi.org/10.1152/ajprenal.00140.2012
Su H, Carter CB, Laur O, Sands JM, Chen G (2012b) Forskolin stimulation promotes urea transporter UT-A1 ubiquitination, endocytosis, and degradation in MDCK cells. Am J Physiol Renal Physiol 303:F1325–F1332. https://doi.org/10.1152/ajprenal.00248.2012
Su H, Chen M, Sands JM, Chen G (2013a) Activation of the cAMP/PKA pathway induces UT-A1 urea transporter monoubiquitination and targets it for lysosomal degradation. Am J Physiol Renal Physiol 305:F1775–F1782. https://doi.org/10.1152/ajprenal.00393.2013
Su H, Liu B, Frohlich O, Ma H, Sands JM, Chen G (2013b) Small GTPase Rab14 down-regulates UT-A1 urea transport activity through enhanced clathrin-dependent endocytosis. FASEB J 27:4100–4107
Sun Y, Lau CW, Jia Y, Li Y, Wang W, Ran J, Li F, Huang Y, Zhou H, Yang B (2016) Functional inhibition of urea transporter UT-B enhances endothelial-dependent vasodilatation and lowers blood pressure via L-arginine-endothelial nitric oxide synthase-nitric oxide pathway. Sci Rep 6:18697. https://doi.org/10.1038/srep18697
Terris J, Ecelbarger CA, Sands JM, Knepper MA (1998) Long-term regulation of collecting duct urea transporter proteins in rat. J Am Soc Nephrol 9:729–736
Thai TL, Blount MA, Klein JD, Sands JM (2012) Lack of protein kinase C-α leads to impaired urine concentrating ability and decreased aquaporin-2 in angiotensin II-induced hypertension. Am J Physiol Renal Physiol 303:F37–F44

Tickle P, Thistlethwaite A, Smith CP, Stewart GS (2009) Novel bUT-B2 urea transporter isoform is constitutively activated. Am J Physiol Regul Integr Comp Physiol 297:R323–R329

Timmer RT, Sands JM (1999) Lithium intoxication. J Am Soc Nephrol 10:666–674

Trinder D, Phillips PA, Stephenson JM, Risvanis J, Aminian A, Adam W, Cooper M, Johnston CI (1994) Vasopressin V_1 and V_2 receptors in diabetes mellitus. Am J Physiol Endocrinol Metab 266:E217–E223

Trinh-Trang-Tan M-M, Lasbennes F, Gane P, Roudier N, Ripoche P, Cartron J-P, Bailly P (2002) UT-B1 proteins in rat: tissue distribution and regulation by antidiuretic hormone in kidney. Am J Physiol Renal Physiol 283:F912–F922

Trinh-Trang-Tan MM, Geelen G, Teillet L, Corman B (2003) Urea transporter expression in aging kidney and brain during dehydration. Am J Physiol Regul Integr Comp Physiol 285:R1355–R1365

Uchida S, Sohara E, Rai T, Ikawa M, Okabe M, Sasaki S (2005) Impaired urea accumulation in the inner medulla of mice lacking the urea transporter UT-A2. Mol Cell Biol 25:7357–7363

Van Zwieten PA, Kam KL, Ijl AJ, Hendriks MGC, Beenen OHM, Pfaffendorf M (1996) Hypertensive diabetic ratsin pharmacological studies. Pharmacol Res 33:95–105

Verkman AS, Esteva-Font C, Cil O, Anderson MO, Li F, Li M, Lei T, Ren H, Yang B (2014) Small-molecule inhibitors of urea transporters. Subcell Biochem 73:165–177. https://doi.org/10.1007/978-94-017-9343-8_11

von Bergen TN, Blount MA (2012) Chronic use of chloroquine disrupts the urine concentration mechanism by lowering cAMP levels in the inner medulla. Am J Physiol Renal Physiol 303: F900–F905. https://doi.org/10.1152/ajprenal.00547.2011

Von Morze C, Bok RA, Sands JM, Kurhanewicz J, Vigneron DB (2012) Monitoring urea transport in rat kidney *in vivo* using hyperpolarized ^{13}C magnetic resonance imaging. Am J Physiol Renal Physiol 302:F1658–F1662

Wade JB, Lee AJ, Liu J, Ecelbarger CA, Mitchell C, Bradford AD, Terris J, Kim G-H, Knepper MA (2000) UT-A2: a 55 kDa urea transporter protein in thin descending limb of Henle's loop whose abundance is regulated by vasopressin. Am J Physiol Renal Physiol 278:F52–F62

Walker AL, Ofori-Acquah SF (2017) Sustained enhancement of OCTN1 transporter expression in association with hydroxyurea induced gamma-globin expression in erythroid progenitors. Exp Hematol 45:69–73.e62. https://doi.org/10.1016/j.exphem.2016.09.001

Walpole C, Farrell A, McGrane A, Stewart GS (2014) Expression and localization of a UT-B urea transporter in the human bladder. Am J Physiol Renal Physiol 307:F1088–F1094. https://doi.org/10.1152/ajprenal.00284.2014

Walpole ME, Schurmann BL, Gorka P, Penner GB, Loewen ME, Mutsvangwa T (2015) Serosal-to-mucosal urea flux across the isolated ruminal epithelium is mediated via urea transporter-B and aquaporins when Holstein calves are abruptly changed to a moderately fermentable diet. J Dairy Sci 98:1204–1213. https://doi.org/10.3168/jds.2014-8757

Walpole C, McGrane A, Al-Mousawi H, Winter D, Baird A, Stewart G (2018) Investigation of facilitative urea transporters in the human gastrointestinal tract. Physiol Rep 6(15):e13826. https://doi.org/10.14814/phy2.13826

Wang X-Y, Beutler K, Nielsen J, Nielsen S, Knepper MA, Masilamani S (2002) Decreased abundance of collecting duct urea transporters UT-A1 and UT-A3 with ECF volume expansion. Am J Physiol Renal Physiol 282:F577–F584

Wang Y, Klein JD, Blount MA, Martin CF, Kent KJ, Pech V, Wall SM, Sands JM (2009) Epac regulation of the UT-A1 urea transporter in rat IMCDs. J Am Soc Nephrol 20:2018–2024

Wang Y, Liedtke CM, Klein JD, Sands JM (2010) Protein kinase C regulates urea permeability in the rat inner medullary collecting duct. Am J Physiol Renal Physiol 299:F1401–F1406

Wang Y, Klein JD, Froehlich O, Sands JM (2013) Role of protein kinase C-alpha in hypertonicity-stimulated urea permeability in mouse inner medullary collecting ducts. Am J Physiol Renal Physiol 304:F233–F238. https://doi.org/10.1152/ajprenal.00484.2012

Wang Z, Yu T, Sang JP, Zou XW, Yan C, Zou X (2015) Computation and simulation of the structural characteristics of the kidney urea transporter and behaviors of urea transport. J Phys Chem B 119:5124–5131. https://doi.org/10.1021/jp511300u

Wang J, Wang XH, Wang H, Chen L, Klein JD, Sands JM (2016) Urea transporter B and microRNA-200c differ in kidney outer versus inner medulla following dehydration. Am J Med Sci 352:296–301. https://doi.org/10.1016/j.amjms.2016.06.003

Wei G, Rosen S, Dantzler WH, Pannabecker TL (2015) Architecture of the human renal inner medulla and functional implications. Am J Physiol Renal Physiol 309:F627–F637. https://doi.org/10.1152/ajprenal.00236.2015

Weiner ID, Mitch WE, Sands JM (2015) Urea and ammonia metabolism and the control of renal nitrogen excretion. Clin J Am Soc Nephrol 10:1444–1458. https://doi.org/10.2215/cjn.10311013

Westrick KY, Serack B, Dantzler WH, Pannabecker TL (2013) Axial compartmentation of descending and ascending thin limbs of Henle's loops. Am J Physiol Renal Physiol 304: F308–F316. https://doi.org/10.1152/ajprenal.00547.2012

Xie L, Jin L, Feng J, Lv J (2017) The expression of AQP5 and UTs in the sweat glands of uremic patients. Biomed Res Int 2017:8629783. https://doi.org/10.1155/2017/8629783

Xu G, Su H, Carter CB, Froehlich O, Chen G (2012) Depolymerization of cortical actin inhibits UT-A1 urea transporter endocytosis but promotes forskolin-stimulated membrane trafficking. Am J Physiol Cell Physiol 302:C1012–C1018

Yang B (2014) Transport characteristics of urea transporter-B. Subcell Biochem 73:127–135. https://doi.org/10.1007/978-94-017-9343-8_8

Yang B, Sands JM (2014) Urea transporters. Subcellular biochemistry, vol 78. Springer, New York

Yang BX, Verkman AS (1998) Urea transporter UT3 functions as an efficient water channel – direct evidence for a common water/urea pathway. J Biol Chem 273:9369–9372

Yang B, Verkman AS (2002) Analysis of double knock-out mice lacking aquaporin-1 and urea transporter UT-B. J Biol Chem 277:36782–36786

Yang B, Bankir L, Gillespie A, Epstein CJ, Verkman AS (2002) Urea-selective concentrating defect in transgenic mice lacking urea transporter UT-B. J Biol Chem 277:10633–10637

Yang B, Li X, Guo L, Meng Y, Dong Z, Zhao X (2014) Extrarenal phenotypes of the UT-B knock-out mouse. Subcell Biochem 73:153–164. https://doi.org/10.1007/978-94-017-9343-8_10

Yano Y, Rodrigues AC Jr, deBraganca AC, Andrade LC, Magaldi AJ (2008) PKC stimulated by glucagon decreases UT-A1 urea transporter expression in rat IMCD. Pflugers Arch 456:1229–1237

Yao LJ, Huang DY, Pfaff IL, Nie X, Leitges M, Vallon V (2004) Evidence for a role of protein kinase C-alpha in urine concentration. Am J Physiol Renal Physiol 287:F299–F304

You G, Smith CP, Kanai Y, Lee W-S, Stelzner M, Hediger MA (1993) Cloning and characterization of the vasopressin-regulated urea transporter. Nature 365:844–847

Yuan J, Pannabecker TL (2010) Architecture of inner medullary descending and ascending vasa recta: pathways for countercurrent exchange. Am J Physiol Renal Physiol 299:F265–F272

Zhang C, Sands JM, Klein JD (2002) Vasopressin rapidly increases phosphorylation of UT-A1 urea transporter in rat IMCDs through PKA. Am J Physiol Renal Physiol 282:F85–F90. https://doi.org/10.1152/ajprenal.00054.2001

Zhang A, Chi Q, Lin H, She Y (2016) Molecular genetic analysis of the Jk(a-b-) phenotype in Chinese: a novel silent recessive JK allele. Transfus Apheresis Sci 54:232–234. https://doi.org/10.1016/j.transci.2015.04.014

Zhang ZY, Wang X, Liu D, Zhang H, Zhang Q, Lu YY, Li P, Lou YQ, Yang BX, Lu C, Lou YX, Zhang GL (2017) Development and validation of an LC-MS/MS method for the determination of a novel thienoquinolin urea transporter inhibitor PU-48 in rat plasma and its application to a pharmacokinetic study. Biomed Chromat 32. https://doi.org/10.1002/bmc.4157

Zhou L, Meng Y, Lei T, Zhao D, Su J, Zhao X, Yang B (2012) UT-B-deficient mice develop renal dysfunction and structural damage. BMC Nephrol 13:6. https://doi.org/10.1186/1471-2369-13-6

Chapter 10
H,K-ATPases in Epithelia

Gilles Crambert

Abstract H,K-ATPases are plasma membrane ion pumps belonging to the P-type ATPase family. They are involved in multiple physiological functions, mainly through their activity of moving protons out of epithelial cells in exchange for K^+. There are two types of H,K-ATPases: type 1 is predominantly expressed in the stomach and corresponds to a 'pure' H,K-ATPase; type 2 has a broader tissue distribution and a more versatile ion selectivity. In this chapter, we will discuss in detail the functional and pharmacological characteristics of these ATPases and their physiological roles in the organism.

Keywords Ion pump · P-type ATPase · Acid production · Potassium balance

10.1 Introduction

10.1.1 The Family of P-Type ATPases

P-type ATPases belong to a family of molecular pumps that transport charged substrates across biological membranes according to a mechanism involving the formation of a transient covalent phosphorylated enzyme. There are 159 members of the P-type ATPase family that can be distinguished by their membrane topology, their substrates or some structural specificities defining five groups (type 1 to type 5) and several subgroups (Geering 2000; Palmgren and Nissen 2011). A similar catalytic cycle governs the transport activity of all P-type ATPases, and its description is known as the Albers–Post model (Post and Jolly 1957; see (Vedovato and Gadsby 2014)). It involves the passage from one conformation (E1, with a high-affinity binding site opened towards the cytoplasm) to another (E2, with a

G. Crambert (✉)
Laboratoire de Physiologie Rénale et Tubulopathies, Centre de Recherche des Cordeliers, INSERM, Sorbonne Université, Université de Paris, Paris, France
e-mail: gilles.crambert@crc.jussieu.fr

K. L. Hamilton, D. C. Devor (eds.), *Studies of Epithelial Transporters and Ion Channels*, Physiology in Health and Disease,
https://doi.org/10.1007/978-3-030-55454-5_10

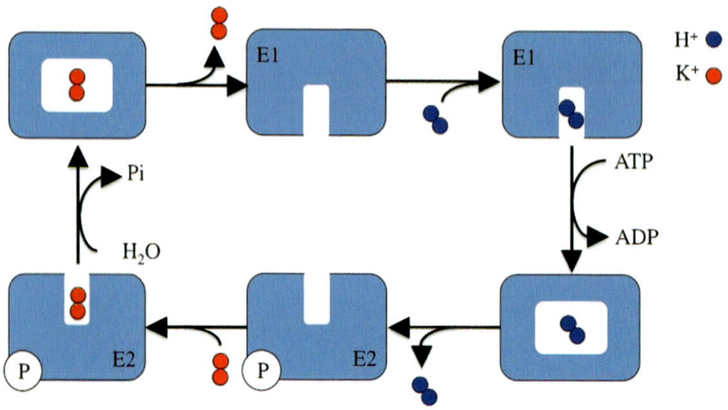

Fig. 10.1 Schematic representation of the Albers–Post catalytic cycle adapted for H,K-ATPase

high-affinity binding site opened towards the extracellular milieu) and the occlusion of the transported molecules into transmembrane domains (see Fig. 10.1). In the E1 conformation, the ion binding sites are exposed to the cytoplasm, and the pump binds ATP. The binding of the ion induces conformational changes, allowing hydrolysis of ATP and transfer of the released inorganic phosphate to an aspartate residue of the pump. These modifications lead to occlusion of the ion and to a rapid and strong conformational change that opens the binding site towards the extracellular side (E2 conformation) and allows the release of the 'cytoplasmic' ion outside the cell. This E2-P conformation is rather stable, and the catalytic cycle can be stopped at this stage in the absence of the 'extracellular' ion. When the extracellular ion binds to the E2-P conformation, it induces a conformational modification leading to release of the inorganic phosphate and occlusion of the ion. This occluded form rapidly returns to the E1 conformation with the release of the extracellular ion into the cytoplasm. It is generally admitted that ion pumps move ions against their concentration gradients and in doing so actually create these gradients. If we look closer at the Albers–Post model (Fig. 10.2), the movement of ions always follows the direction of the concentration gradient. Indeed, when 'cytoplasmic' ions bind to the empty binding site of the E1 conformation, they move from a high concentration compartment (cytoplasm) to a low concentration compartment (the empty binding site). The occlusion step is then fundamental because it artificially increases the concentration of the ion in the binding site compartment to an uncountable level, the ions being trapped in the protein structure with only a few molecules of water. Therefore, when the conformation changes again and exposes the binding site towards the extracellular compartment, the difference of concentration is still in favour of movement towards the exterior. This mechanism is obviously identical for the extracellular ion going to the cytoplasm. Therefore, this 'lock or floodgate system' allows the ions macroscopically to move against their concentration gradient, whereas they actually always move from the high to the low concentration compartment.

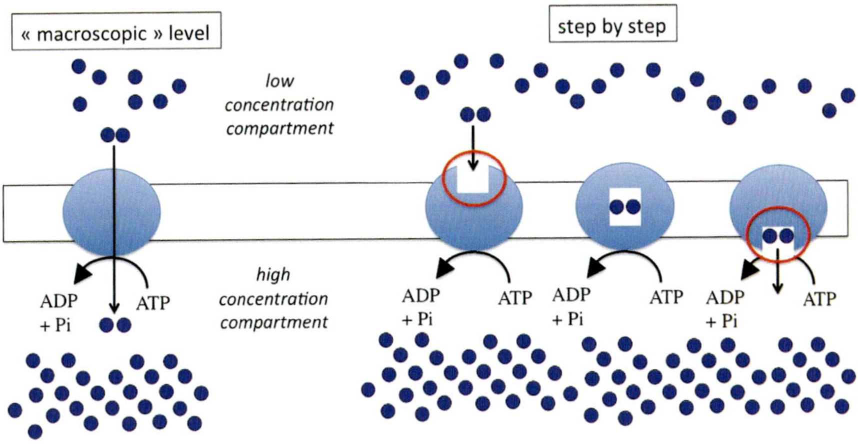

Fig. 10.2 Schematic representation of the course of ions from one compartment to another through H,K-ATPase, illustrating the concept of a lock permitting the ions to move according to their concentration gradient at every step of their transportation

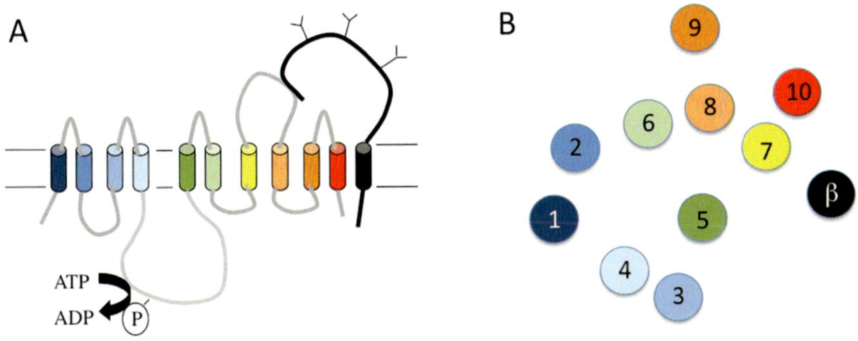

Fig. 10.3 (**a**) Schematic representation of the secondary structure of H,K-ATPases involving two membrane proteins: the catalytic α subunit (encoded by the gene *Atp4a* or *Atp12a*) and the chaperone-like β subunit (encoded by the gene *Atp1b1* or *Atp4b*). (**b**) Spatial organization of the transmembrane domains of H,K-ATPase in the cell membrane (seen from the cytoplasm). Adapted from Abe et al. (2018)

10.1.2 H,K-ATPases

H,K-ATPases (genes *Atp4a* and *Atp12a*), along with Na,K-ATPases (*Atp1a1*, *Atp1a2*, *Atp1a3* and *Atp1a4*), belong to the P2C-ATPase group (X,K-ATPases). All these X,K-ATPases exhibit a similar structure of ten transmembrane-spanning domains (Fig. 10.3) and require to be associated with another protein, the β subunit (*Atp1b1*, *Atp1b2*, *Atp1b3* and *Atp4b*), in order to be correctly folded and pass through the endoplasmic reticulum to the Golgi apparatus and possibly the plasma membrane (Geering et al. 1996). As shown by Abe et al. (2018), who provided the first crystal structure of HKA1, the β subunit is positioned at the periphery of the αβ complex, interacting with C-terminal transmembrane domains 7 and 10 of the α subunit (Abe et al. 2018). H,K-ATPases have a stoichiometry of $2H^+/2\ K^+/1ATP$ (Rabon et al. 1982) and are, therefore, electroneutral (Sachs et al. 1976; Burnay et al. 2001). However, because they catalyse the movement of ions in the membrane, some steps of the catalytic cycle are electrogenic and sensitive to the membrane potential. Thus, the H^+-transporting half-cycle of H,K-ATPase can be monitored using electrochromic fluorescent dye (van der Hijden et al. 1990).

10.1.3 Different H,K-ATPases for Different Physiological Functions

Two different H,K-ATPases associated with two different genes have been characterized: gastric H,K-ATPase or H,K-ATPase type 1 (HKA1) is encoded by the *Atp4a* gene, and its main tissue expression is in the stomach, where it is essential for gastric acid secretion, but its presence has also been documented in the kidney (see below); non-gastric H,K-ATPase (ngHKA), colonic H,K-ATPase (cHKA) or H,K-ATPase type 2 (HKA2) is encoded by the *Atp12a* gene. In the following sections, we will use the terminology HKA1 and HKA2 to distinguish between these two forms. Both HKA1 and HKA2 share around 65% of identity, but despite these similarities of sequence, they display strong differences in terms of structure (nature of the accompanying β subunit), ion specificity, pharmacology and physiological functions.

10.2 H,K-ATPase Type 1 (*Atp4a*)

10.2.1 Generality

Identification of the mechanism that catalyses acidification of gastric fluid started at the end of the nineteenth century with the establishment of gastric juice composition and anatomical characterization of the stomach. One of the first studies describing the mechanism of acidification identified ATPase in frog stomach (Kasbekar and Durbin 1965) that was not related to Na,K-ATPase (insensitive to ouabain). This observation was confirmed later in mammals using pig gastric mucosa (Forte et al. 1975). The relationship between pH and ATPase activity was observed using a microsomal fraction of canine stomach (Lee et al. 1974), which showed that the movement of H^+ ions was dependent and in opposition to the movement of K^+ ions. Soon after, Sachs et al. (1976) clearly established that this exchange of K^+ and H^+ ions, depending on ATP hydrolysis, was due to the same activity and used the term H,K-ATPase to describe it; they then confirmed the existence of H,K-ATPase in humans. They purified a 110-kDa protein that could be phosphorylated by $[\gamma^{32}P]$

ATP and dephosphorylated in the presence of K^+. Using an antibody raised against K^+-ATPase-enriched hog gastric vesicles, these authors localized H,K-ATPase in the parietal cells of human gastric mucosa (Saccomani et al. 1979). A few years later, cDNA encoding rat, hog and human gastric H,K-ATPases was cloned and characterized (Shull and Lingrel 1986; Maeda et al. 1988, 1990). More recently, the crystal structure of pig HKA1 bound to SCH28080 or vonoprazan has been resolved (Abe et al. 2018). Among the structural information, this study revealed that extensive polar interaction with a glutamate residue in the cation binding site lowers its pKa and permits the release of protons into the $pH < 2$ environment of the stomach.

As mentioned above, P2C-type ATPase requires an additional protein to associate with in order to be fully matured. In 1990, Forte's group observed that the same gastric vesicles displaying H,K-ATPase activity contained an abundant glycoprotein with molecular mass in the range 60–80 kDa (gp 60–80) (Okamoto et al. 1990). By analogy with Na,K-ATPase, they proposed that this protein could be the β subunit of H,K-ATPase. This was confirmed by showing the non-covalent association between gp 60–80 and the 110-kDa H,K-ATPase catalytic subunit. The cloning of gp 60–80 along with subsequent sequence analysis revealed the close proximity and high degree of similarity between gp 60–80 and the β subunit of Na,K-ATPase (Canfield et al. 1990), which was then named βHKA. The specificity of the α/β association was further tested in the *Xenopus* oocyte expression system. Thus, the co-expression of αNKA subunits and βHKA leads to a mature αβ complex with strongly altered kinetic properties (Jaisser et al. 1994; Geering et al. 2000). The reverse combination (αHKA1/βNKA) does not allow the αβ complex to be efficiently stabilized and therefore leads to its degradation (Geering et al. 2000).

In the first characterization of the proton pump present in hog stomach responsible for gastric acid secretion, Sachs et al. (1976) demonstrated that this function was borne by an electroneutral K-ATPase independent of Na^+. However, in non-physiological conditions under alkali pH (pH 8.5), HKA1 mediates K^+-dependent and SCH28080-sensitive Na^+ uptake (Polvani et al. 1989). Although not physiologically relevant, this observation demonstrates that HKA1 has the structural potential to transport Na^+ and H^+ in exchange for K^+.

10.2.2 Pharmacological Properties

All members of the X,K-ATPase family are sensitive to orthovanadate, but yet exhibit a specific pharmacological profile. Thus, HKA1 is not sensitive to NKA inhibitors such as cardiac glycosides or digitalis but are inhibited by substituted benzimidazole compounds such as omeprazole (Maton 1991). Omeprazole and its derivatives are non-charged molecules that can cross the plasma membrane and possess a sulphoxide group that needs to be activated by acidic pH to be reduced to a sulphonamide group, leading to the appearance of a reactive sulphur atom (Im et al. 1985). This sulphur atom can then established a disulphide bridge with cysteines present on the extracellular loops of HKA1, leading to an irreversible inhibition. The sequencing of an omeprazole-labelled tryptic digest fragment of hog HKA1 has suggested that Cys822 and Cys892 are reactive cysteines (Besancon et al. 1993). Directed mutagenesis of Cys822 confirmed its role in the omeprazole inhibitory effect (Lambrecht et al. 1998). Another cysteine, Cys322, has also been proposed by Morii et al. (1990). This mechanism of action makes omeprazole highly selective because it is only activated in hyperacidic conditions, as in the secretory canaliculus of the parietal cells. Another type of gastric antisecretory compound with a different mode of action was developed and characterized in the early 1980s (Long et al. 1983). Schering 28,080 (SCH28080) has an imidazopyridine structure and is a weak base that could allow its accumulation in the acidic space of the gastric gland (Wallmark et al. 1987). Although SCH28080 does not require an acidic pH to be activated, its action is dependent on pH, with a better value for the half-maximal inhibitory concentration (IC_{50}) at pH 6.5 than at pH 7.5. It has been proposed that SCH28080 competes with K^+ for the E2 state of the transporter (Wallmark et al. 1987; Keeling et al. 1988, 1989). Despite its potential as an anti-acid, SCH28080 was discontinued following a phase I test because of liver toxicity detected in preclinical studies (Long et al. 1983; Kaminski et al. 1987).

10.2.3 Physiological Roles

10.2.3.1 Acidification of the Gastric Fluid

HKA1-dependent gastric acidification (pH < 2) is a crucial step in the process of digestion because it activates pepsinogen into pepsin and thus triggers the degradation of ingested proteins and peptides. The gastric epithelium is composed of a variety of cells: enterochromaffin-like cells that produce histamine, goblet cells that produce mucus and parietal cells that abundantly express HKA1 (roughly 10% of parietal cell protein) (Fig. 10.4a). The parietal cells exhibit a complex internal system of organelles composed of tubulovesicles that contain HKA1. Interestingly, the presence of these tubulovesicles depends on the expression of α or β subunits of HKA1, indicating that this pump may also serve as a structural protein complex (Courtois-Coutry et al. 1997; Scarff et al. 1999). Upon activation, both the cholinergic system and endocrine secretion of gastrin stimulate the production of histamine by enterochromaffin-like cells. Paracrine secretion of histamine then activates H2 histamine receptors on neighbouring parietal cells. Following this receptor activation, there is a strong SNARE-dependent membrane remodelling (with a central role for syntaxin-3); (Ammar et al. 2002) and modification of the actin cytoskeleton. HKA1 residing in the tubulovesicular (tubulocisternal) network moves towards and fuses with the apically directed canaliculi (Forte et al. 1977) (Fig. 10.4b). Return to the rest situation seems to depend mainly on withdrawal of the secretagogue signal (Forte et al. 1977), inducing a resequestration of the apical membrane and the return of HKA1 back into its tubulovesicles. In this process, the β subunit of HKA1 (*Atp4b*) plays an important role, because it contains a tyrosine-based endocytosis signal.

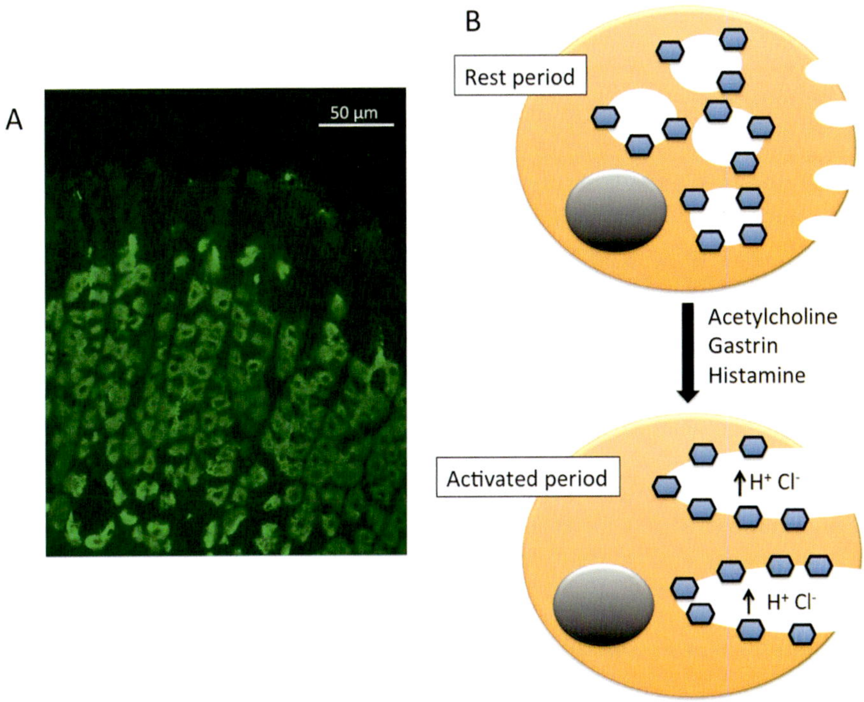

Fig. 10.4 (**a**) Localization of HKA1 in parietal cells of mouse stomach: personal data obtained following a protocol described by Crambert et al. (2005) and using monoclonal H,K-ATPase antibody provided by M. Caplan, Yale University, New Haven, USA. (**b**) Schematic representation of the fusion of the tubulovesicular network containing HKA1 with the apically directed canaliculi

Transgenic mice harbouring a mutation of this tyrosine residue into alanine display chronic gastric acidity and ulcers because the HKA1 internalization is impeded (Courtois-Coutry et al. 1997). In 2000, Spicer et al., who generated *Atp4a*-null mice, confirmed the importance of HKA1 in the generation of acidic gastric fluid and in the stomach structure. In addition to the acute regulation of HKA1 activity described above, gastric expression of the α subunit of HKA1 is under the control of oestrogen-related receptor γ (ERRγ) (Alaynick et al. 2010).

10.2.3.2 Renal Function of HKA1

The presence of HKA1 expression and activity has been localized in collecting ducts (Dherbecourt et al. 2006) and more precisely in the A- and B-type intercalated cells where it roughly contributes to 30% and 60% of the total proton extrusion, respectively (Lynch et al. 2008). This suggests that HKA1 may participate in the acid/base balance by contributing to the renal proton transport system. However, acid/base balance is not altered in HKA1-null mice (Spicer et al. 2000). HKA1 may also contribute to renal K^+ reabsorption, but K^+ is completely recycled at the apical membrane by Ba^{2+}-sensitive K^+ channels (Armitage and Wingo 1994). One possible physiological role for renal HKA1 is to modulate the activity of other transporters, which is what Charles Wingo's group recently described (Mironova et al. 2017) by showing that HKA1 modulates epithelial Na^+ channel activity. In the absence of HKA1, this channel activity remains highly activated in a high-Na^+ diet condition, leading to Na^+ retention and impaired dipsogenic response.

10.2.3.3 HKA1 and Embryonic Development

Left- and right-side-specific expression of genes in the early embryo determines the left–right asymmetry of organs in the adult body. In 2002, Michael Levin performed a pharmacological test on *Xenopus* and chick embryos and showed that omeprazole and SCH28080 induced the occurrence of heterotaxia (Levin et al. 2002). This result suggested the involvement of H,K-ATPase in this process, which was reinforced by the asymmetrical localization of HKA subunits at the four-cell stage of *Xenopus* embryo development (Levin et al. 2002), where they localized predominantly in the right ventral blastomere. Conversely, in chick embryos, HKA mRNA is expressed symmetrically, but the HKA activity is asymmetrical. Thus, in both species the asymmetrical function of HKA is required but not achieved through the same mechanism. At that time, the exact identity of H,K-ATPase was not completely certain, but the pharmacological profile was in good agreement with HKA1. The presence of HKA activity on the right side of the midline along with the expression of K channels Kir4.1 in both *Xenopus* and chick embryos results in asymmetrical hyperpolarization of the cell membrane. It was later determined that the asymmetrical hyperpolarization of cells in the right blastomere results in differential gene expression via serotonin- and Ca^{2+}-dependent processes (Levin et al. 2006; Raya and Izpisua Belmonte 2006). Despite some contradictions with the model elaborated by Michael Levin, the role of HKA1 in the left–right development processes of *Xenopus* embryos was further confirmed and linked with both canonical and noncanonical Wnt pathways (Walentek et al. 2012). It has to be pointed out, however, that *Atp4a*-null mice have not been reported to display any developmental defects (Spicer et al. 2000).

10.3 H,K-ATPase Type 2 (*Atp12a*)

10.3.1 Generality

Epithelial cells from the kidney, colon and bladder were shown to reabsorb K^+, but the nature of the system that promotes this process has remained unknown for many years. Renal physiologists first described an ouabain-sensitive K^+ pump located in the apical membrane of distal tubule cells from animals fed a low-K^+ diet (Malnic et al. 1964; Strieder et al. 1974; Hayashi and Katz 1987). In 1987 Doucet's group characterized this renal ouabain-sensitive K^+ transport as H,K-ATPase activity (Doucet and Marsy 1987; Okusa et al. 1992) but different from that observed in the stomach. In parallel, elucidation of the K^+ reabsorptive system present in rabbit colon was characterized, and this led to the identification of an apical ouabain-insensitive K-ATPase (Gustin and Goodman 1981, 1982; Wills and Biagi 1982). In guinea pig colon, apical K^+ reabsorption is coupled to H^+ secretion and mediated by an ouabain-sensitive ATPase (Suzuki and Kaneko 1989) that resembles the one present in the kidney after dietary K^+ restriction. These data and others strongly suggested the existence of a third type of X,K-ATPase in addition to NKA and HKA1. This hypothesis was finally confirmed when cDNA encoding for a novel form of H,K-ATPase was cloned from rat colon (Crowson and Shull 1992). This HKA2 is broadly expressed, at least at the mRNA level, in many different tissues. In addition to the colon, HKA2 has been found in the brain (the cortex, hypothalamus, choroid plexus), ovaries, uterus, vagina, placenta, epididymis, penis, lung, pancreas, skin (Crowson and Shull 1992; Pestov et al. 1998) and prostate (Pestov et al. 2002). In the kidney, HKA2 is expressed in connecting tubules and cortical and outer medullary collecting ducts (Marsy et al. 1996). Feeding rats a low-K^+ diet did not modify this localization profile but increased the level of expression. The presence of HKA2 mRNA and protein has been confirmed in the connecting tubules and cortical and outer medullary collecting ducts by many studies in different species (Fejes-Toth et al. 1999; Kraut et al. 2001; Verlander et al. 2001; Zhang et al. 2004). In human collecting ducts, HKA2 is more abundant in intercalated cells than in principal cells (Verlander et al. 2001). By measuring ouabain-sensitive proton fluxes on microdissected mouse collecting ducts, Lynch et al. (2010) demonstrated that HKA2 was present in both A- and B-type intercalated cells.

In contrast to NKA and HKA1, several β subunits may support the functional expression of αHKA2, such as βHKA (Grishin et al. 1996; Adams et al. 2001; Codina et al. 1996; Jaisser et al. 1992; Kone and Higham 1998; Modyanov et al. 1995) or βNKA (Modyanov et al. 1995; Codina et al. 1996; Asano et al. 1998; Cougnon et al. 1996, 1999). However, in the human cell line (HEK293), endogenous βNKA is not able to promote efficient maturation of HKA2, and the co-transfection of a βHKA subunit is necessary (Asano et al. 1998; Grishin et al. 1996; Reinhardt et al. 2000). Interestingly, the kinetic properties of HKA2 ($K_{1/2}$ for K^+ turnover) (Crambert et al. 2002) are not affected by the nature of the β subunit. In vivo, immunoprecipitation and immunodetection experiments proved that β_1NKA was associated with αHKA2 either in rat kidney (Codina et al. 1998; Kraut et al. 1998) or colon samples (Codina et al. 1998; Li et al. 2004). Moreover, clear labelling of β_1NKA is present at the apical side of colonocytes in WT mice, whereas this location disappears in HKA2-null animals (Scudieri et al. 2018). Similarly, in rat prostate gland, β_1NKA is present at the apical side of the cells associated with αHKA2 (Pestov et al. 2004). More recently, Galietta's group has shown the presence of HKA2 in the airway epithelium of normal and cystic fibrosis patients and demonstrates that β_1NKA supports this activity in this context (Scudieri et al. 2018). Finally, in βHKA-null mice, the colonic activity of HKA2 is preserved, whereas that of HKA1 is inhibited (Shao et al. 2010). Although αHKA2 may associate with β subunits of NKA or HKA, most of the in vivo data suggest that β_1NKA is the partner of αHKA2.

HKA2 is an electroneutral transporter (Burnay et al. 2001). This characteristic was shown to depend on the presence of lysine in the fifth transmembrane domain of both HKA1 and HKA2 (Burnay et al. 2003), and its replacement by alanine renders HKA2 electrogenic. However, in contrast to HKA1, HKA2 is a versatile pump that can transport either H^+ or Na^+ under physiological conditions. Thus, the transport of Rb^{86} (a radioactive surrogate for K^+) by HKA2 turns out to be more efficient in the presence of Na^+ (Grishin et al. 1996), and the calculated proton-to-Rb^{86} exchange ratio was not equal to 1, indicating that another ion should be transported since HKA2 is electroneutral. The ability of HKA2 to transport Na^+ was confirmed by measuring the rate of Na^+ efflux in transfected and non-transfected HEK293 cells (Grishin and Caplan 1998). This characteristic was also reported in the same year by Cougnon et al. (1998), who expressed rat αHKA2 and rabbit βHKA in *Xenopus* oocytes. A few years later, the intracellular apparent Na^+ affinity ($K_{1/2} = 9$ mM) for HKA2-mediated Rb^{86} uptake was determined (Crambert et al. 2002) and found to be very similar to that of Na,K-ATPase (Crambert et al. 2000). This value indicates that HKA2 may transport Na^+ in physiological conditions. In spite of these results obtained in heterologous expression systems, the activity responsible for apical reabsorption of K^+ in the colon was described to be Na-independent, SCH28080-sensitive K-ATPase (Gustin and Goodman 1981; Kaunitz and Sachs 1986; Belisario et al. 2010). However, in the kidney, this ability to transport Na^+ has a relevance that will be discussed in more detail in Sect. 10.3.3.2.

10.3.2 Pharmacological Properties of HKA2

The pharmacological profile of HKA2 is rather complicated and seems to depend on the species and on the tissues. Indeed, as mentioned above, already at the time of its first description, some investigators reported ouabain-sensitive and others ouabain-insensitive pumps. In *Xenopus* oocytes, rat HKA2-mediated $^{86}Rb^+$ flux is insensitive to SCH28080 (administered as a single dose of 500 μM) but is inhibited by ouabain, with *Ki* values that depend on the extracellular K^+ concentration (Cougnon et al. 1996; Codina et al. 1996). These results are in good agreement with the K-ATPase activity present in the rat distal colon that is, indeed, inhibited by ouabain but insensitive to SCH28080 (Sweiry and Binder 1990; Del Castillo et al. 1991). However, these findings differ from those obtained when rat HKA2 is expressed in Sf9 insect cells (Lee et al. 1995). Human αHKA2 when expressed in HEK cells (Grishin et al. 1996), insect Sf21 cells (Adams et al. 2001) and *Xenopus* oocytes (Crambert et al. 2002; Modyanov et al. 1995) is sensitive to ouabain (with *Ki* values in the tens of μM) and to SCH28080 (*Ki* around 100 μM). Mouse HKA2 exhibits a pharmacological profile that clearly depends on the tissue context. Thus, in cortical and outer medullary collecting ducts of mice under a low-K^+ diet, Dherbecourt et al.

(2006) showed that HKA2 was sensitive to both ouabain and SCH28080 by measuring K-ATPase activity. This result was confirmed by measurements of intracellular pH variations on microperfused cortical collecting ducts of mice under a normal diet (Lynch et al. 2010). However, the K-ATPase activity present in the colon of mice and attributable to HKA2 was insensitive to both SCH28080 and ouabain (Shao et al. 2010).

10.3.3 Physiological Roles of HKA2

10.3.3.1 HKA2 and K^+ Balance

We mentioned above that HKA2 expression was strongly stimulated by dietary K^+ restriction (Ahn et al. 1996; Marsy et al. 1996; Kraut et al. 1997). Moreover, Meneton et al. (1998) demonstrated without ambiguity the role of this transporter in response to hypokalaemia by investigating the phenotype of HKA2-null mice. However, the hormonal and cellular mechanisms involved in this stimulation have remained obscure for a long time. Some recent studies have started to answer this question. Wang and his colleagues have published a series of papers describing how K^+ depletion leads to the production of reactive oxygen species (ROS). These compounds activate different regulatory pathways leading to phosphorylation of Kir1.1 channels and inhibition of K^+ secretion (for a review, see Wang 2006). A possible link between this production of ROS and the stimulation of HKA2 is suggested by the finding that Nrf2, a ROS-induced transcription factor, is stimulated by K^+ depletion (Lee et al. 2012). Nrf2 is an antioxidant molecule that triggers protective pathways against oxidative stress. In their study, Lee et al. (2012) showed that overexpression of Nrf2 in HEK293 and CV-1 cells enhances the expression of HKA2. Pretreatment of these cells with a low-K^+ culture medium increased the endogenous expression of Nrf2 and HKA2. Transfection of a dominant negative Nrf2 abolished this low-K^+-mediated HKA2 expression. Expression of HKA2 could be directly controlled by the level of extracellular K^+ through the production of ROS.

Another process explaining the upregulation of HKA2 in response to K^+ depletion involves the production of progesterone by the adrenal glands (Elabida et al. 2011). This result, obtained with male mice, fits with the presence of different types of progesterone receptors along the nephron of male mice (Grimont et al. 2009), particularly in the collecting ducts. It was further demonstrated that progesterone stimulates the renal expression of HKA2 through a RU486-sensitive pathway, which strongly suggests the involvement of the nuclear progesterone receptor. Thus, the inhibition of progesterone action during K^+ restriction not only blunted the stimulation of HKA2 expression but also impeded the correct renal adaptation, leading to urinary loss of K^+. Recently, the presence of this regulatory system has been tested in men, either in healthy volunteers submitted to a mild K^+ restriction or to patients with Gitelman syndrome who display severe hypokalaemia. The results showed, at least in patients with Gitelman syndrome, that adrenal steroid production is strongly modified by severe K^+ depletion but the presence of CYP17 in human adrenal glands impedes the accumulation of progesterone (Blanchard et al. 2020).

In contrast to the condition of K^+ restriction, an increased K^+ intake should promote the inhibition of HKA2, which is precisely what El Moghrabi et al. (2010) observed; their results suggest that after an acute K^+ load, the tissue kallikrein level is increased and inhibits HKA2. In microperfused cortical collecting ducts of tissue from kallikrein null mice, transepithelial K^+ reabsorption is stimulated, which is correlated with the enhanced expression and activity of HKA2. The inability to reduce HKA2 activity after a K^+ load via the kallikrein pathway leads to a transient increase in the plasma K^+ level.

The link between HKA2 activity and the ability to retain K^+ was further illustrated in the context of gestation, a physiological situation known to require renal K^+ retention to cope with foetal development (Lindheimer et al. 1987). The renal expression and activity of HKA2 is strongly stimulated during gestation in mice and rats (Salhi et al. 2013; West et al. 2018), and the absence of HKA2 leads to a loss of K^+ during gestation and to gestational defects (decrease in fertility rate, increase in maternal mortality and decrease in the number of pups per litter) (Salhi et al. 2013).

10.3.3.2 HKA2 and Na^+ Balance

Few genomic data suggested that HKA2 might be involved in Na^+ homeostasis and the control of blood pressure. Thus, the renal expression of HKA2 was found to be lower in spontaneously hypertensive rats than in normotensive rats (Kinoshita et al. 2011), and human *Atp12a* gene polymorphism correlates with cardiovascular diseases (Knez et al. 2014; Evangelou et al. 2018). These suspicions were confirmed experimentally and showed that HKA2 function is involved in the mechanism of Na^+ homeostasis. Thus, its involvement in K^+ retention during dietary K^+ restriction (see above) also has an impact on the renal Na^+ transport system. Indeed, Walter et al. (Walter et al. 2016) showed that the absence of HKA2 in a low-K^+ diet condition is compensated by the development of hypovolaemia, which reduces the risk of hypokalaemia but leads to reduced blood pressure. HKA2 stimulation is therefore important in order to maintain both a normal plasma K^+ level and normal blood pressure in the K^+-restricted condition. As mentioned above, HKA2 has the characteristic of transporting Na^+ and protons in exchange for K^+. This property has remained without physiological relevance until it was shown to be involved in a novel pathway of Na^+ secretion through renal type A intercalated cells (Morla et al. 2016), where HKA2 serves as an apical Na,K-ATPase to promote Na^+ secretion in the lumen of the collecting duct. This pathway is under the control of the atrial natriuretic factor (Cheval et al. 2019), which promotes intracellular cGMP production. Interestingly, in this study Cheval et al. have shown that cGMP increases cell surface expression of HKA2 and modifies its transport properties by favouring the transport of Na^+ over protons. This indicates that the versatility of HKA2 is a controllable property that could be modulated to fit physiological demands.

10.3.3.3 HKA2 in Prostate

In rat and mouse prostate, HKA2 presents lobe-specific expression. It is absent in the ventral part but highly expressed in the lateral and dorsal lobes and in the coagulating gland, where it exhibits an apical localization (Pestov et al. 2002). In normal human prostate, HKA2 is expressed at the apical side of the epithelial cells (Streif et al. 2011). In benign prostate hyperplasia and prostate cancer, the HKA2 protein level is increased compared to healthy tissue but displays an altered cellular localization. By studying the prostate phenotype of HKA2-null mice, Pestov et al. (2006) established that HKA2 is responsible for acidification of the anterior prostate secretion: in the absence of HKA2, its pH is 0.8 units higher. This effect is not believed to impact the pH semen, but could be related to the ability of rodent semen to build a plug after copulation.

10.3.3.4 HKA2 in Pancreas

The pancreas produces and secretes a bicarbonate-rich fluid that conveys digestive hormones towards the intestine. The production of such fluid requires pancreatic duct cells to be equipped with H^+ and HCO_3^- transporters. Recently, Novak et al. (2011) identified both HKA1 and HKA2 in rat pancreatic ducts. An ammonium pulse in isolated pancreatic ducts, under conditions where most of the classical transporters (Na/H exchangers, Na/HCO_3^- co-transporters and anion exchanger 1) were inhibited, revealed SCH28080-sensitive intracellular pH recovery. In this study, HKA2 was localized at the apical and lateral sides of cells, which suggests that it may play a role in extruding protons generated by carbonic anhydrase activity, particularly in the serosa. HKA (1 or 2, or both) contributes significantly to fluid secretion in the pancreatic duct because HKA inhibitors drastically reduce the volume of secretin-stimulated fluid. Here, again, further experiments are necessary to clearly delineate the contribution of HKA1 and HKA2 to these processes.

10.3.3.5 HKA2 in Airway Epithelium

The presence of H,K-ATPase activity was first demonstrated in nasal airway epithelia from human patients with allergic rhinitis (Smith and Welsh 1993) and was proposed to participate in airway surface liquid acidification. In 2003, Coakley et al. showed that this activity was inhibited by ouabain and that human airway epithelial cells expressed the *Atp12a* gene (Coakley et al. 2003). Recently, HKA2 has emerged as an important pathogenic factor in cystic fibrosis and other chronic respiratory diseases (Shah et al. 2016; Min et al. 2017; Lennox et al. 2018). In cystic fibrosis patients, the expression of HKA2 at the apical side is strongly stimulated via a mechanism that may depend on the inflammation. Indeed, interleukins such as IL-4 and IL-13 strongly stimulate the expression of HKA2, particularly in non-ciliated mucus-producing cells (Lennox et al. 2018; Gorrieri et al. 2016). This upregulation leads to a decrease of airway surface liquid pH that favours bacterial growth through inhibition of antibiotic peptides (Simonin et al. 2019).

10.4 Conclusions

Through their functions in ion transport, H,K-ATPases participate in several fundamental physiological processes. HKA1 is well known to be responsible for gastric acidification and has become a very popular pharmacological target to reduce stomach acid. In addition, it now emerges, through the development of genomic strategies, that mutations of the *Atp4a* gene could be related to certain forms of gastric cancer (Calvete et al. 2015, 2017). It would be useful now to characterize the effects of these mutations on the function and expression of HKA1 and to extend these observations to other cancers or pathologies. Regarding HKA2, despite its broad expression in several epithelia, its involvement in pathologies such as cystic fibrosis and its very intriguing functional properties (Na^+ vs. H^+ transport), it remains poorly studied and still deserves its qualifier of the 'ugly duckling' of the P2C-ATPase family (Crambert 2014). However, recently HKA2 has emerged as a potential ion pump involved in different cancers (Jakab et al. 2014; Streif et al. 2011), and future studies should help to decipher its role in the context of cell proliferation.

Acknowledgements GC has been supported by a 'Fondation pour la Recherche Médicale' grant (DPC20171138949). I dedicate this chapter to my mentors Käthi Geering and Jean-Daniel Horisberger, passed away too early, who introduced me in science and shared with me their passion for ion transport and ion pumps during my PhD training and post-doctoral fellowship at the University of Lausanne, Switzerland.

References

Abe K, Irie K, Nakanishi H, Suzuki H, Fujiyoshi Y (2018) Crystal structures of the gastric proton pump. Nature 556:214–218

Adams G, Tillekeratne M, Yu C, Pestov NB, Modyanov NN (2001) Catalytic function of nongastric H,K-ATPase expressed in Sf-21 insect cells. Biochemistry 40:5765–5776

Ahn KY, Park KY, Kim KK, Kone BC (1996) Chronic hypokalemia enhances expression of the H^+-K^+-ATPase alpha 2-subunit gene in renal medulla. Am J Physiol Renal Physiol 271:F314–F321

Alaynick WA, Way JM, Wilson SA, Benson WG, Pei L, Downes M, Yu R, Jonker JW, Holt JA, Rajpal DK, Li H, Stuart J, McPherson R, Remlinger KS, Chang CY, McDonnell DP, Evans RM, Billin AN (2010) ERRgamma regulates cardiac, gastric, and renal potassium homeostasis. Mol Endocrinol 24:299–309

Ammar DA, Zhou R, Forte JG, Yao X (2002) Syntaxin 3 is required for cAMP-induced acid secretion: streptolysin O-permeabilized gastric gland model. Am J Physiol Gastrointest Liver Physiol 282:G23–G33

Armitage FE, Wingo CS (1994) Luminal acidification in K-replete OMCDi: contributions of H-K-ATPase and bafilomycin-A1-sensitive H-ATPase. Am J Phys 267:F450–F458
Asano S, Hoshina S, Nakaie Y, Watanabe T, Sato M, Suzuki Y, Takeguchi N (1998) Functional expression of putative H^+-K^+-ATPase from guinea pig distal colon. Am J Physiol Cell Physiol 275:C669–C674
Belisario DC, Rocafull MA, del Castillo JR (2010) Purification and characterization of the ouabain-sensitive H^+/K^+-ATPase from guinea-pig distal colon. Arch Biochem Biophys 496:21–23
Besancon M, Shin JM, Mercier F, Munson K, Miller M, Hersey S, Sachs G (1993) Membrane topology and omeprazole labeling of the gastric H^+,K^+-adenosinetriphosphatase. Biochemistry 32:2345–2355
Blanchard A, Brailly Tabard S, Lamaziere A, Bergerot D, Zhygalina V, Lorthioir A, Jacques A, Hourton D, Azizi M, Crambert G (2020) Adrenal adaptation in potassium-depleted men: role of progesterone? Nephrol Dial Transplant 35(11):1901–1908
Burnay M, Crambert G, Kharoubi-Hess S, Geering K, Horisberger JD (2001) *Bufo marinus* bladder H-K-ATPase carries out electroneutral ion transport. Am J Physiol Renal Physiol 281:F869–F874
Burnay M, Crambert G, Kharoubi-Hess S, Geering K, Horisberger JD (2003) Electrogenicity of Na, K- and H,K-ATPase activity and presence of a positively charged amino acid in the fifth transmembrane segment. J Biol Chem 278:19237–19244
Calvete O, Reyes J, Zuniga S, Paumard-Hernandez B, Fernandez V, Bujanda L, Rodriguez-Pinilla MS, Palacios J, Heine-Suner D, Banka S, Newman WG, Canamero M, Pritchard DM, Benitez J (2015) Exome sequencing identifies ATP4A gene as responsible of an atypical familial type I gastric neuroendocrine tumour. Hum Mol Genet 24:2914–2922
Calvete O, Herraiz M, Reyes J, Patino A, Benitez J (2017) A cumulative effect involving malfunction of the PTH1R and ATP4A genes explains a familial gastric neuroendocrine tumor with hypothyroidism and arthritis. Gastric Cancer 20:998–1003
Canfield VA, Okamoto CT, Chow D, Dorfman J, Gros P, Forte JG, Levenson R (1990) Cloning of the H,K-ATPase beta subunit. Tissue-specific expression, chromosomal assignment, and relationship to Na,K-ATPase beta subunits. J Biol Chem 265:19878–19884
Cheval L, Bakouh N, Walter C, Tembely D, Morla L, Escher G, Vogt B, Crambert G, Planelles G, Doucet A (2019) ANP-stimulated Na^+ secretion in the collecting duct prevents Na^+ retention in the renal adaptation to acid load. Am J Physiol Renal Physiol 317:F435–F443
Coakley RD, Grubb BR, Paradiso AM, Gatzy JT, Johnson LG, Kreda SM, O'Neal WK, Boucher RC (2003) Abnormal surface liquid pH regulation by cultured cystic fibrosis bronchial epithelium. Proc Natl Acad Sci USA 100:16083–16088
Codina J, Kone BC, Delmas-Mata JT, DuBose TD Jr (1996) Functional expression of the colonic H^+,K^+-ATPase alpha-subunit. Pharmacologic properties and assembly with X^+,K^+-ATPase beta-subunits. J Biol Chem 271:29759–29763
Codina J, Delmas-Mata JT, DuBose TD Jr (1998) The alpha-subunit of the colonic H^+,K^+-ATPase assembles with beta1-Na^+,K^+-ATPase in kidney and distal colon. J Biol Chem 273:7894–7899
Cougnon M, Planelles G, Crowson MS, Shull GE, Rossier BC, Jaisser F (1996) The rat distal colon P-ATPase alpha subunit encodes a ouabain-sensitive H^+,K^+-ATPase. J Biol Chem 271:7277–7280
Cougnon M, Bouyer P, Planelles G, Jaisser F (1998) Does the colonic H,K-ATPase also act as an Na,K-ATPase? Proc Natl Acad Sci U S A 95:6516–6520
Cougnon M, Bouyer P, Jaisser F, Edelman A, Planelles G (1999) Ammonium transport by the colonic H^+-K^+-ATPase expressed in xenopus oocytes. Am J Physiol Cell Physiol 277:C280–C287
Courtois-Coutry N, Roush D, Rajendran V, McCarthy JB, Geibel J, Kashgarian M, Caplan MJ (1997) A tyrosine-based signal targets H/K-ATPase to a regulated compartment and is required for the cessation of gastric acid secretion. Cell 90:501–510
Crambert G (2014) H-K-ATPase type 2: relevance for renal physiology and beyond. Am J Physiol Renal Physiol 306:F693–F700

Crambert G, Hasler U, Beggah AT, Yu C, Modyanov NN, Horisberger JD, Lelievre L, Geering K (2000) Transport and pharmacological properties of nine different human Na, K-ATPase isozymes. J Biol Chem 275:1976–1986

Crambert G, Horisberger JD, Modyanov NN, Geering K (2002) Human nongastric H+-K+-ATPase: transport properties of ATP1al1 assembled with different beta-subunits. Am J Physiol Cell Physiol 283:C305–C314

Crambert G, Li C, Claeys D, Geering K (2005) FXYD3 (Mat-8), a new regulator of Na,K-ATPase. Mol Biol Cell 16:2363–2371

Crowson MS, Shull GE (1992) Isolation and characterization of a cDNA encoding the putative distal colon H^+,K^+-ATPase. Similarity of deduced amino acid sequence to gastric H^+,K^+-ATPase and Na^+,K^+-ATPase and mRNA expression in distal colon, kidney, and uterus. J Biol Chem 267:13740–13748

Del Castillo JR, Rajendran VM, Binder HJ (1991) Apical membrane localization of ouabain-sensitive K^+-activated ATPase activities in rat distal colon. Am J Physiol Gastrointest Liver Physiol 261:G1005–G1011

Dherbecourt O, Cheval L, Bloch-Faure M, Meneton P, Doucet A (2006) Molecular identification of Sch28080-sensitive K-ATPase activities in the mouse kidney. Pflugers Arch 451:769–775

Doucet A, Marsy S (1987) Characterization of K-ATPase activity in distal nephron: stimulation by potassium depletion. Am J Physiol Renal Physiol 253:F418–F423

El Moghrabi S, Houillier P, Picard N, Sohet F, Wootla B, Bloch-Faure M, Leviel F, Cheval L, Frische S, Meneton P, Eladari D, Chambrey R (2010) Tissue kallikrein permits early renal adaptation to potassium load. Proc Natl Acad Sci USA 107:13526–13531

Elabida B, Edwards A, Salhi A, Azroyan A, Fodstad H, Meneton P, Doucet A, Bloch-Faure M, Crambert G (2011) Chronic potassium depletion increases adrenal progesterone production that is necessary for efficient renal retention of potassium. Kidney Int 80:256–262

Evangelou E, Warren HR, Mosen-Ansorena D, Mifsud B, Pazoki R, Gao H, Ntritsos G, Dimou N, Cabrera CP, Karaman I, Ng FL, Evangelou M, Witkowska K, Tzanis E, Hellwege JN, Giri A, Velez Edwards DR, Sun YV, Cho K, Gaziano JM, Wilson PWF, Tsao PS, Kovesdy CP, Esko T, Magi R, Milani L, Almgren P, Boutin T, Debette S, Ding J, Giulianini F, Holliday EG, Jackson AU, Li-Gao R, Lin WY, Luan J, Mangino M, Oldmeadow C, Prins BP, Qian Y, Sargurupremraj M, Shah N, Surendran P, Theriault S, Verweij N, Willems SM, Zhao JH, Amouyel P, Connell J, de Mutsert R, Doney ASF, Farrall M, Menni C, Morris AD, Noordam R, Pare G, Poulter NR, Shields DC, Stanton A, Thom S, Abecasis G, Amin N, Arking DE, Ayers KL, Barbieri CM, Batini C, Bis JC, Blake T, Bochud M, Boehnke M, Boerwinkle E, Boomsma DI, Bottinger EP, Braund PS, Brumat M, Campbell A, Campbell H, Chakravarti A, Chambers JC, Chauhan G, Ciullo M, Cocca M, Collins F, Cordell HJ, Davies G, de Borst MH, de Geus EJ, Deary IJ, Deelen J, Del Greco MF, Demirkale CY, Dorr M, Ehret GB, Elosua R, Enroth S, Erzurumluoglu AM, Ferreira T, Franberg M, Franco OH, Gandin I, Gasparini P, Giedraitis V, Gieger C, Girotto G, Goel A, Gow AJ, Gudnason V, Guo X, Gyllensten U, Hamsten A, Harris TB, Harris SE, Hartman CA, Havulinna AS, Hicks AA, Hofer E, Hofman A, Hottenga JJ, Huffman JE, Hwang SJ, Ingelsson E, James A, Jansen R, Jarvelin MR, Joehanes R, Johansson A, Johnson AD, Joshi PK, Jousilahti P, Jukema JW, Jula A, Kahonen M, Kathiresan S, Keavney BD, Khaw KT, Knekt P, Knight J, Kolcic I, Kooner JS, Koskinen S, Kristiansson K, Kutalik Z, Laan M, Larson M, Launer LJ, Lehne B, Lehtimaki T, Liewald DCM, Lin L, Lind L, Lindgren CM, Liu Y, Loos RJF, Lopez LM, Lu Y, Lyytikainen LP, Mahajan A, Mamasoula C, Marrugat J, Marten J, Milaneschi Y, Morgan A, Morris AP, Morrison AC, Munson PJ, Nalls MA, Nandakumar P, Nelson CP, Niiranen T, Nolte IM, Nutile T, Oldehinkel AJ, Oostra BA, O'Reilly PF, Org E, Padmanabhan S, Palmas W, Palotie A, Pattie A, Penninx B, Perola M, Peters A, Polasek O, Pramstaller PP, Nguyen QT, Raitakari OT, Ren M, Rettig R, Rice K, Ridker PM, Ried JS, Riese H, Ripatti S, Robino A, Rose LM, Rotter JI, Rudan I, Ruggiero D, Saba Y, Sala CF, Salomaa V, Samani NJ, Sarin AP, Schmidt R, Schmidt H, Shrine N, Siscovick D, Smith AV, Snieder H, Sober S, Sorice R, Starr JM, Stott DJ, Strachan DP, Strawbridge RJ, Sundstrom J, Swertz MA, Taylor KD, Teumer A,

Tobin MD, Tomaszewski M, Toniolo D, Traglia M, Trompet S, Tuomilehto J, Tzourio C, Uitterlinden AG, Vaez A, van der Most PJ, van Duijn CM, Vergnaud AC, Verwoert GC, Vitart V, Volker U, Vollenweider P, Vuckovic D, Watkins H, Wild SH, Willemsen G, Wilson JF, Wright AF, Yao J, Zemunik T, Zhang W, Attia JR, Butterworth AS, Chasman DI, Conen D, Cucca F, Danesh J, Hayward C, Howson JMM, Laakso M, Lakatta EG, Langenberg C, Melander O, Mook-Kanamori DO, Palmer CNA, Risch L, Scott RA, Scott RJ, Sever P, Spector TD, van der Harst P, Wareham NJ, Zeggini E, Levy D, Munroe PB, Newton-Cheh C, Brown MJ, Metspalu A, Hung AM, O'Donnell CJ, Edwards TL, Psaty BM, Tzoulaki I, Barnes MR, Wain LV, Elliott P, Caulfield MJ, Million Veteran P (2018) Genetic analysis of over 1 million people identifies 535 new loci associated with blood pressure traits. Nat Genet 50:1412–1425

Fejes-Toth G, Naray-Fejes-Toth A, Velazquez H (1999) Intrarenal distribution of the colonic H, K-ATPase mRNA in rabbit. Kidney Int 56:1029–1036

Forte JG, Ganser A, Beesley R, Forte TM (1975) Unique enzymes of purified microsomes from pig fundic mucosa. K^+-stimulated adenosine triphosphatase and K^+-stimulated pNPPase. Gastroenterology 69:175–189

Forte TM, Machen TE, Forte JG (1977) Ultrastructural changes in oxyntic cells associated with secretory function: a membrane-recycling hypothesis. Gastroenterology 73:941–955

Geering K (2000) Topogenic motifs in P-type ATPases. J Membr Biol 174:181–190

Geering K, Beggah A, Good P, Girardet S, Roy S, Schaer D, Jaunin P (1996) Oligomerization and maturation of Na,K-ATPase: functional interaction of the cytoplasmic NH2 terminus of the beta subunit with the alpha subunit. J Cell Biol 133:1193–1204

Geering K, Crambert G, Yu C, Korneenko TV, Pestov NB, Modyanov NN (2000) Intersubunit interactions in human X,K-ATPases: role of membrane domains M9 and M10 in the assembly process and association efficiency of human, nongastric H,K-ATPase alpha subunits (ATP1al1) with known beta subunits. Biochemistry 39:12688–12698

Gorrieri G, Scudieri P, Caci E, Schiavon M, Tomati V, Sirci F, Napolitano F, Carrella D, Gianotti A, Musante I, Favia M, Casavola V, Guerra L, Rea F, Ravazzolo R, Di Bernardo D, Galietta LJ (2016) Goblet cell hyperplasia requires high bicarbonate transport to support mucin release. Sci Rep 6:36016

Grimont A, Bloch-Faure M, El Abida B, Crambert G (2009) Mapping of sex hormone receptors and their modulators along the nephron of male and female mice. FEBS Lett 583:1644–1648

Grishin AV, Caplan MJ (1998) ATP1AL1, a member of the non-gastric H,K-ATPase family, functions as a sodium pump. J Biol Chem 273:27772–27778

Grishin AV, Bevensee MO, Modyanov NN, Rajendran V, Boron WF, Caplan MJ (1996) Functional expression of the cDNA encoded by the human ATP1AL1 gene. Am J Physiol Renal Physiol 271:F539–F551

Gustin MC, Goodman DB (1981) Isolation of brush-border membrane from the rabbit descending colon epithelium. Partial characterization of a unique K^+-activated ATPase. J Biol Chem 256:10651–10656

Gustin MC, Goodman DB (1982) Characterization of the phosphorylated intermediate of the K^+-ouabain-insensitive ATPase of the rabbit colon brush-border membrane. J Biol Chem 257:9629–9633

Hayashi M, Katz AI (1987) The kidney in potassium depletion. II. K^+ handling by the isolated perfused rat kidney. Am J Physiol Renal Physiol 252:F447–F452

Im WB, Sih JC, Blakeman DP, McGrath JP (1985) Omeprazole, a specific inhibitor of gastric (H^+-K^+)-ATPase, is a H^+-activated oxidizing agent of sulfhydryl groups. J Biol Chem 260:4591–4597

Jaisser F, Canessa CM, Horisberger JD, Rossier BC (1992) Primary sequence and functional expression of a novel ouabain-resistant Na,K-ATPase. The beta subunit modulates potassium activation of the Na,K-pump. J Biol Chem 267:16895–16903

Jaisser F, Jaunin P, Geering K, Rossier BC, Horisberger JD (1994) Modulation of the Na,K-pump function by beta subunit isoforms. J Gen Physiol 103:605–623

Jakab M, Hofer S, Ravasio A, Huber F, Schmidt S, Hitzl W, Geibel JP, Furst J, Ritter M (2014) The putative role of the non-gastric H^+/K^+-ATPase ATP12A (ATP1AL1) as anti-apoptotic ion transporter: effect of the H^+/K^+ ATPase inhibitor SCH28080 on butyrate-stimulated myelomonocytic HL-60 cells. Cell Physiol Biochem 34:1507–1526
Kaminski JJ, Hilbert JM, Pramanik BN, Solomon DM, Conn DJ, Rizvi RK, Elliott AJ, Guzik H, Lovey RG, Domalski MS et al (1987) Antiulcer agents. 2. Gastric antisecretory, cytoprotective, and metabolic properties of substituted imidazo[1,2-a]pyridines and analogues. J Med Chem 30:2031–2046
Kasbekar DK, Durbin RP (1965) An adenosine triphosphatase from frog gastric mucosa. Biochim Biophys Acta 105:472–482
Kaunitz JD, Sachs G (1986) Identification of a vanadate-sensitive potassium-dependent proton pump from rabbit colon. J Biol Chem 261:14005–14010
Keeling DJ, Laing SM, Senn-Bilfinger J (1988) SCH 28080 is a lumenally acting, K^+-site inhibitor of the gastric (H^+/K^+)-ATPase. Biochem Pharmacol 37:2231–2236
Keeling DJ, Fallowfield C, Lawrie KM, Saunders D, Richardson S, Ife RJ (1989) Photoaffinity labeling of the lumenal K+ site of the gastric $(H^+{+}K^+)$-ATPase. J Biol Chem 264:5552–5558
Kinoshita K, Ashenagar MS, Tabuchi M, Higashino H (2011) Whole rat DNA array survey for candidate genes related to hypertension in kidneys from three spontaneously hypertensive rat substrains at two stages of age and with hypotensive induction caused by hydralazine hydrochloride. Exp Ther Med 2:201–212
Knez J, Salvi E, Tikhonoff V, Stolarz-Skrzypek K, Ryabikov A, Thijs L, Braga D, Kloch-Badelek-M, Malyutina S, Casiglia E, Czarnecka D, Kawecka-Jaszcz K, Cusi D, Nawrot T, Staessen JA, Kuznetsova T (2014) Left ventricular diastolic function associated with common genetic variation in ATP12A in a general population. BMC Med Genet 15:121
Kone BC, Higham SC (1998) A novel N-terminal splice variant of the rat H^+-K^+-ATPase alpha2 subunit. Cloning, functional expression, and renal adaptive response to chronic hypokalemia. J Biol Chem 273:2543–2552
Kraut JA, Hiura J, Besancon M, Smolka A, Sachs G, Scott D (1997) Effect of hypokalemia on the abundance of HK alpha 1 and HK alpha 2 protein in the rat kidney. Am J Physiol Renal Physiol 272:F744–F750
Kraut JA, Hiura J, Shin JM, Smolka A, Sachs G, Scott D (1998) The Na^+-K^+-ATPase beta 1 subunit is associated with the HK alpha 2 protein in the rat kidney. Kidney Int 53:958–962
Kraut JA, Helander KG, Helander HF, Iroezi ND, Marcus EA, Sachs G (2001) Detection and localization of H^+-K^+-ATPase isoforms in human kidney. Am J Physiol Renal Physiol 281: F763–F768
Lambrecht N, Corbett Z, Bayle D, Karlish SJ, Sachs G (1998) Identification of the site of inhibition by omeprazole of a alpha-beta fusion protein of the H,K-ATPase using site-directed mutagenesis. J Biol Chem 273:13719–13728
Lee J, Simpson G, Scholes P (1974) An ATPase from dog gastric mucosa: changes of outer pH in suspensions of membrane vesicles accompanying ATP hydrolysis. Biochem Biophys Res Commun 60:825–832
Lee J, Rajendran VM, Mann AS, Kashgarian M, Binder HJ (1995) Functional expression and segmental localization of rat colonic K-adenosine triphosphatase. J Clin Invest 96:2002–2008
Lee CB, Lee YS, Lee JY, Lee SE, Ahn KY (2012) Nrf2 and Sp family synergistically enhance the expression of ion transporters in potassium-depleted conditions. J Nephrol 25:225–232
Lennox AT, Coburn SL, Leech JA, Heidrich EM, Kleyman TR, Wenzel SE, Pilewski JM, Corcoran TE, Myerburg MM (2018) ATP12A promotes mucus dysfunction during type 2 airway inflammation. Sci Rep 8:2109
Levin M, Thorlin T, Robinson KR, Nogi T, Mercola M (2002) Asymmetries in H^+/K^+-ATPase and cell membrane potentials comprise a very early step in left-right patterning. Cell 111:77–89
Levin M, Buznikov GA, Lauder JM (2006) Of minds and embryos: left-right asymmetry and the serotonergic controls of pre-neural morphogenesis. Dev Neurosci 28:171–185

Li J, Codina J, Petroske E, Werle MJ, Willingham MC, DuBose TD Jr (2004) The effect of beta-subunit assembly on function and localization of the colonic H^+,K^+-ATPase alpha-subunit. Kidney Int 66:1068–1075

Lindheimer MD, Richardson DA, Ehrlich EN, Katz AI (1987) Potassium homeostasis in pregnancy. J Reprod Med 32:517–522

Long JF, Chiu PJ, Derelanko MJ, Steinberg M (1983) Gastric antisecretory and cytoprotective activities of SCH 28080. J Pharmacol Exp Ther 226:114–120

Lynch IJ, Rudin A, Xia SL, Stow LR, Shull GE, Weiner ID, Cain BD, Wingo CS (2008) Impaired acid secretion in cortical collecting duct intercalated cells from H-K-ATPase-deficient mice: role of HKalpha isoforms. Am J Physiol Renal Physiol 294:F621–F627

Lynch IJ, Greenlee MM, Gumz ML, Rudin A, Xia SL, Wingo CS (2010) Heterogeneity of H-K-ATPase-mediated acid secretion along the mouse collecting duct. Am J Physiol Renal Physiol 298:F408–F415

Maeda M, Ishizaki J, Futai M (1988) cDNA cloning and sequence determination of pig gastric (H^++K^+)-ATPase. Biochem Biophys Res Commun 157:203–209

Maeda M, Oshiman K, Tamura S, Futai M (1990) Human gastric (H^++K^+)-ATPase gene. Similarity to (Na^++K^+)-ATPase genes in exon/intron organization but difference in control region. J Biol Chem 265:9027–9032

Malnic G, Klose RM, Giebisch G (1964) Micropuncture study of renal potassium excretion in the rat. Am J Phys 206:674–686

Marsy S, Elalouf JM, Doucet A (1996) Quantitative RT-PCR analysis of mRNAs encoding a colonic putative H,K-ATPase alpha subunit along the rat nephron: effect of K+ depletion. Pflugers Arch 432:494–500

Maton PN (1991) Omeprazole. N Engl J Med 324:965–975

Meneton P, Schultheis PJ, Greeb J, Nieman ML, Liu LH, Clarke LL, Duffy JJ, Doetschman T, Lorenz JN, Shull GE (1998) Increased sensitivity to K^+ deprivation in colonic H,K-ATPase-deficient mice. J Clin Invest 101:536–542

Min JY, Ocampo CJ, Stevens WW, Price CPE, Thompson CF, Homma T, Huang JH, Norton JE, Suh LA, Pothoven KL, Conley DB, Welch KC, Shintani-Smith S, Peters AT, Grammer LC 3rd, Harris KE, Hulse KE, Kato A, Modyanov NN, Kern RC, Schleimer RP, Tan BK (2017) Proton pump inhibitors decrease eotaxin-3/CCL26 expression in patients with chronic rhinosinusitis with nasal polyps: possible role of the nongastric H, K-ATPase. J Allergy Clin Immunol 139:130–141.e111

Mironova E, Lynch IJ, Berman JM, Gumz ML, Stockand JD, Wingo CS (2017) ENaC activity in the cortical collecting duct of HKalpha1 H^+,K^+-ATPase knockout mice is uncoupled from Na^+ intake. Am J Physiol Renal Physiol 312:F1073–F1080

Modyanov NN, Mathews PM, Grishin AV, Beguin P, Beggah AT, Rossier BC, Horisberger JD, Geering K (1995) Human ATP1AL1 gene encodes a ouabain-sensitive H-K-ATPase. Am J Physiol Cell Physiol 269:C992–C997

Morii M, Takata H, Takeguchi N (1990) Binding site of omeprazole in hog gastric H^+,K^+-ATPase. Biochem Biophys Res Commun 167:754–760

Morla L, Doucet A, Lamouroux C, Crambert G, Edwards A (2016) The renal cortical collecting duct: a secreting epithelium? J Physiol 5945991-6008

Novak I, Wang J, Henriksen KL, Haanes KA, Krabbe S, Nitschke R, Hede SE (2011) Pancreatic bicarbonate secretion involves two proton pumps. J Biol Chem 286:280–289

Okamoto CT, Karpilow JM, Smolka A, Forte JG (1990) Isolation and characterization of gastric microsomal glycoproteins. Evidence for a glycosylated beta-subunit of the H^+/K^+-ATPase. Biochim Biophys Acta 1037:360–372

Okusa MD, Unwin RJ, Velazquez H, Giebisch G, Wright FS (1992) Active potassium absorption by the renal distal tubule. Am J Physiol Renal Physiol 262:F488–F493

Palmgren MG, Nissen P (2011) P-type ATPases. Annu Rev Biophys 40:243–266

Pestov NB, Romanova LG, Korneenko TV, Egorov MV, Kostina MB, Sverdlov VE, Askari A, Shakhparonov MI, Modyanov NN (1998) Ouabain-sensitive H,K-ATPase: tissue-specific

expression of the mammalian genes encoding the catalytic alpha subunit. FEBS Lett 440:320–324
Pestov NB, Korneenko TV, Adams G, Tillekeratne M, Shakhparonov MI, Modyanov NN (2002) Nongastric H-K-ATPase in rodent prostate: lobe-specific expression and apical localization. Am J Physiol Cell Physiol 282:C907–C916
Pestov NB, Korneenko TV, Radkov R, Zhao H, Shakhparonov MI, Modyanov NN (2004) Identification of the beta-subunit for nongastric H-K-ATPase in rat anterior prostate. Am J Physiol Cell Physiol 286:C1229–C1237
Pestov NB, Korneenko TV, Shakhparonov MI, Shull GE, Modyanov NN (2006) Loss of acidification of anterior prostate fluids in Atp12a-null mutant mice indicates that nongastric H-K-ATPase functions as proton pump in vivo. Am J Physiol Cell Physiol 291:C366–C374
Polvani C, Sachs G, Blostein R (1989) Sodium ions as substitutes for protons in the gastric H, K-ATPase. J Biol Chem 264:17854–17859
Post RL, Jolly PC (1957) The linkage of sodium, potassium, and ammonium active transport across the human erythrocyte membrane. Biochim Biophys Acta 25:118–128
Rabon EC, McFall TL, Sachs G (1982) The gastric [H,K]ATPase:H^+/ATP stoichiometry. J Biol Chem 257:6296–6299
Raya A, Izpisua Belmonte JC (2006) Left-right asymmetry in the vertebrate embryo: from early information to higher-level integration. Nat Rev Genet 7:283–293
Reinhardt J, Grishin AV, Oberleithner H, Caplan MJ (2000) Differential localization of human nongastric H^+-K^+-ATPase ATP1AL1 in polarized renal epithelial cells. Am J Physiol Renal Physiol 279:F417–F425
Saccomani G, Chang HH, Mihas AA, Crago S, Sachs G (1979) An acid transporting enzyme in human gastric mucosa. J Clin Invest 64:627–635
Sachs G, Chang HH, Rabon E, Schackman R, Lewin M, Saccomani G (1976) A nonelectrogenic H^+ pump in plasma membranes of hog stomach. J Biol Chem 251:7690–7698
Salhi A, Lamouroux C, Pestov NB, Modyanov NN, Doucet A, Crambert G (2013) A link between fertility and K^+ homeostasis: role of the renal H,K-ATPase type 2. Pflugers Arch 465:1149–1158
Scarff KL, Judd LM, Toh BH, Gleeson PA, Van Driel IR (1999) Gastric H^+,K^+-adenosine triphosphatase beta subunit is required for normal function, development, and membrane structure of mouse parietal cells. Gastroenterology 117:605–618
Scudieri P, Musante I, Caci E, Venturini A, Morelli P, Walter C, Tosi D, Palleschi A, Martin-Vasallo P, Sermet-Gaudelus I, Planelles G, Crambert G, Galietta LJ (2018) Increased expression of ATP12A proton pump in cystic fibrosis airways. JCI Insight 3(20)
Shah VS, Meyerholz DK, Tang XX, Reznikov L, Abou Alaiwa M, Ernst SE, Karp PH, Wohlford-Lenane CL, Heilmann KP, Leidinger MR, Allen PD, Zabner J, McCray PB Jr, Ostedgaard LS, Stoltz DA, Randak CO, Welsh MJ (2016) Airway acidification initiates host defense abnormalities in cystic fibrosis mice. Science 351:503–507
Shao J, Gumz ML, Cain BD, Xia SL, Shull GE, van Driel IR, Wingo CS (2010) Pharmacological profiles of the murine gastric and colonic H,K-ATPases. Biochim Biophys Acta 1800:906–911
Shull GE, Lingrel JB (1986) Molecular cloning of the rat stomach (H^++K^+)-ATPase. J Biol Chem 261:16788–16791
Simonin J, Bille E, Crambert G, Noel S, Dreano E, Edwards A, Hatton A, Pranke I, Villeret B, Cottart CH, Vrel JP, Urbach V, Baatallah N, Hinzpeter A, Golec A, Touqui L, Nassif X, Galietta LJV, Planelles G, Sallenave JM, Edelman A, Sermet-Gaudelus I (2019) Airway surface liquid acidification initiates host defense abnormalities in cystic fibrosis. Sci Rep 9:6516
Smith JJ, Welsh MJ (1993) Fluid and electrolyte transport by cultured human airway epithelia. J Clin Invest 91:1590–1597
Spicer Z, Miller ML, Andringa A, Riddle TM, Duffy JJ, Doetschman T, Shull GE (2000) Stomachs of mice lacking the gastric H,K-ATPase alpha-subunit have achlorhydria, abnormal parietal cells, and ciliated metaplasia. J Biol Chem 275:21555–21565

Streif D, Iglseder E, Hauser-Kronberger C, Fink KG, Jakab M, Ritter M (2011) Expression of the non-gastric H^+/K^+ ATPase ATP12A in normal and pathological human prostate tissue. Cell Physiol Biochem 28:1287–1294

Strieder N, Khuri R, Wiederholt M, Giebisch G (1974) Studies on the renal action of ouabain in the rat. Effects in the non-diuretic state. Pflugers Arch 349:91–107

Suzuki Y, Kaneko K (1989) Ouabain-sensitive H^+-K^+ exchange mechanism in the apical membrane of guinea pig colon. Am J Physiol Gastrointest Liver Physiol 256:G979–G988

Sweiry JH, Binder HJ (1990) Active potassium absorption in rat distal colon. J Physiol 423:155–170

van der Hijden HT, Grell E, de Pont JJ, Bamberg E (1990) Demonstration of the electrogenicity of proton translocation during the phosphorylation step in gastric H^+K^+-ATPase. J Membr Biol 114:245–256

Vedovato N, Gadsby DC (2014) Route, mechanism, and implications of proton import during Na^+/K^+ exchange by native Na^+/K^+-ATPase pumps. J Gen Physiol 143:449–446

Verlander JW, Moudy RM, Campbell WG, Cain BD, Wingo CS (2001) Immunohistochemical localization of H-K-ATPase alpha(2c)-subunit in rabbit kidney. Am J Physiol Renal Physiol 281:F357–F365

Walentek P, Beyer T, Thumberger T, Schweickert A, Blum M (2012) ATP4a is required for Wnt-dependent Foxj1 expression and leftward flow in Xenopus left-right development. Cell Rep 1:516–527

Wallmark B, Briving C, Fryklund J, Munson K, Jackson R, Mendlein J, Rabon E, Sachs G (1987) Inhibition of gastric H^+,K^+-ATPase and acid secretion by SCH 28080, a substituted pyridyl (1,2a)imidazole. J Biol Chem 262:2077–2084

Walter C, Tanfous MB, Igoudjil K, Salhi A, Escher G, Crambert G (2016) H,K-ATPase type 2 contributes to salt-sensitive hypertension induced by K^+ restriction. Pflugers Arch 468:1673–1683

Wang WH (2006) Regulation of ROMK (Kir1.1) channels: new mechanisms and aspects. Am J Physiol Renal Physiol 290:F14–F19

West CA, Welling PA, West DA Jr, Coleman RA, Cheng KY, Chen C, DuBose TD Jr, Verlander JW, Baylis C, Gumz ML (2018) Renal and colonic potassium transporters in the pregnant rat. Am J Physiol Renal Physiol 314:F251–F259

Wills NK, Biagi B (1982) Active potassium transport by rabbit descending colon epithelium. J Membr Biol 64:195–203

Zhang W, Xia X, Zou L, Xu X, LeSage GD, Kone BC (2004) In vivo expression profile of a H^+-K^+-ATPase alpha2-subunit promoter-reporter transgene. Am J Physiol Renal Physiol 286:F1171–F1177

Chapter 11
Zinc Transporters Involved in Vectorial Zinc Transport in Intestinal Epithelial Cells

Yukina Nishito, Shuangyu Luo, and Taiho Kambe

Abstract Zinc is an essential trace element in humans. As zinc plays pivotal roles as a structural, catalytic, and signaling component of many proteins, its deficiency causes diverse pathophysiological symptoms. Although an adult human body contains only 2–3 g of zinc, humans have as many as 23 zinc transporters, highlighting that strict spatiotemporal regulation of both systemic and cellular zinc homeostases is undoubtedly crucial for good health. Recent progress understanding zinc metabolism has determined the importance of zinc transporters in physiology and pathology at a molecular level. Yet while specific zinc transporters are known to play distinct roles in either zinc uptake from the lumen or release into the portal bloodstream for delivery to the peripheral tissues, the molecular mechanisms underlying these processes remain only partially understood. This chapter describes the molecular mechanism underlying vectorial zinc transport in intestinal epithelial cells, with a focus on zinc transporters indispensable for zinc absorption, and compares and contrasts vectorial transepithelial zinc transport with vectorial transepithelial iron, copper, and manganese transport.

Keywords ZNT · ZIP · Intestinal epithelial cell · Solute carrier (SLC) family · Zinc · Iron · Copper · Manganese

11.1 Introduction

Zinc is an essential trace element for all life (Kambe et al. 2015; Hara et al. 2017). Zinc plays structural, catalytic, and signaling roles in proteins essential for processes including transcription and translation, enzymatic catalysis, intracellular signal transduction, lipid metabolism, growth factor release, and cell–cell communication.

Y. Nishito · S. Luo · T. Kambe (✉)
Division of Integrated Life Science, Graduate School of Biostudies, Kyoto University, Kyoto, Japan
e-mail: kambe.taiho.7z@kyoto-u.ac.jp

K. L. Hamilton, D. C. Devor (eds.), *Studies of Epithelial Transporters and Ion Channels*, Physiology in Health and Disease,
https://doi.org/10.1007/978-3-030-55454-5_11

Nevertheless, the adult human body contains 2–3 g of zinc, approximately 60% of which is present in skeletal muscle, 30% in bone, 5% in the skin and liver, and the remaining in other tissues (Kambe et al. 2015; Hara et al. 2017). Because of its essential roles, zinc deficiency symptoms include severe dermatitis, alopecia, diarrhea, mental disorders, and immune system dysfunction (Haase and Rink 2014). Why such diverse symptoms present themselves is unclear, although may partially be explained by our recent finding that zinc deficiency reduces extracellular ATP hydrolysis (Takeda et al. 2018).

Dietary zinc is absorbed along the intestinal tract, with highest uptake in the proximal small intestine and lowest toward the colon (Vancampen and Mitchell 1965). Central to systemic zinc homeostasis is absorption by several gastrointestinal zinc transporters found mainly in the small intestine. Intestinal zinc absorption follows three steps: uptake across the apical membrane (i.e., facing the intestinal lumen) into the enterocytes, transport to the basolateral side of the cell (i.e., facing the bloodstream), and release into the portal bloodstream for circulation to the peripheral tissues. Such vectorial (i.e., one-way) transport of zinc through enterocytes is fundamental to absorption of dietary zinc. In vertebrates, two zinc transporter family proteins play critical roles: Zn transporters (ZNT), members of solute carrier family 30 (SLC30A), export zinc from the cytoplasm out of the cell or into intracellular compartments, while Zrt-, Irt-related proteins (ZIP), members of the solute carrier family 39 (SLC39A), import zinc from intracellular compartments or outside the cell into the cytoplasm. Nine ZNT and 14 ZIP transporters have been identified in mammals (Fig. 11.1), all of which are highly conserved in their sequences and functions (Lichten and Cousins 2009; Kambe et al. 2015, 2016; Hara et al. 2017). Generally, ZNT transporters are thought to form homodimers to export zinc. Each protomer has six transmembrane domains (TMDs) (Fig. 11.2a) (Kambe et al. 2015; Hara et al. 2017). They belong to the Zn-cation diffusion facilitator (Zn-CDF) protein family (Kambe et al. 2017) and can be grouped into four subgroups, such as (i) ZNT1 and ZNT10; (ii) ZNT2, ZNT3, ZNT4, and ZNT8; (iii) ZNT5 and ZNT7, and (iv) ZNT6, based on their sequence similarity and subcellular localization (Fig. 11.1a) (Kambe et al. 2006, 2016). In contrast, ZIP transporters are thought to form homodimers, in which each protomer has eight TMDs, for importing zinc (Fig. 11.2b). They are divided into four subfamilies according to their phylogenetic relationships: ZIP-I (ZIP9), ZIP-II (ZIP1-ZIP3), gufA (ZIP11), and LIV-1 (ZIP4-ZIP8, ZIP10, ZIP12-ZIP14) (Fig. 11.1b). Members of the LIV-1 subfamily play diverse and important roles in various physiopathological responses (Kambe et al. 2015; Hara et al. 2017), which are characterized by the CPALLY (PAL) motif-containing domain (PCD) and the potential metalloprotease motif (HEXPHEXGD) in the TMD V (Taylor and Nicholson 2003; Zhang et al. 2017). Some ZIP transporters can transport iron and manganese in addition to zinc (Jenkitkasemwong et al. 2012), which suggests pivotal roles in cross talk among these trace elements and zinc.

Functional characterization of transporters in cell cultures, immunohistochemistry of the small intestine, and analysis of knockout mice have revealed that epithelial vectorial zinc transport requires ZIP transporters on the apical membrane and ZNT

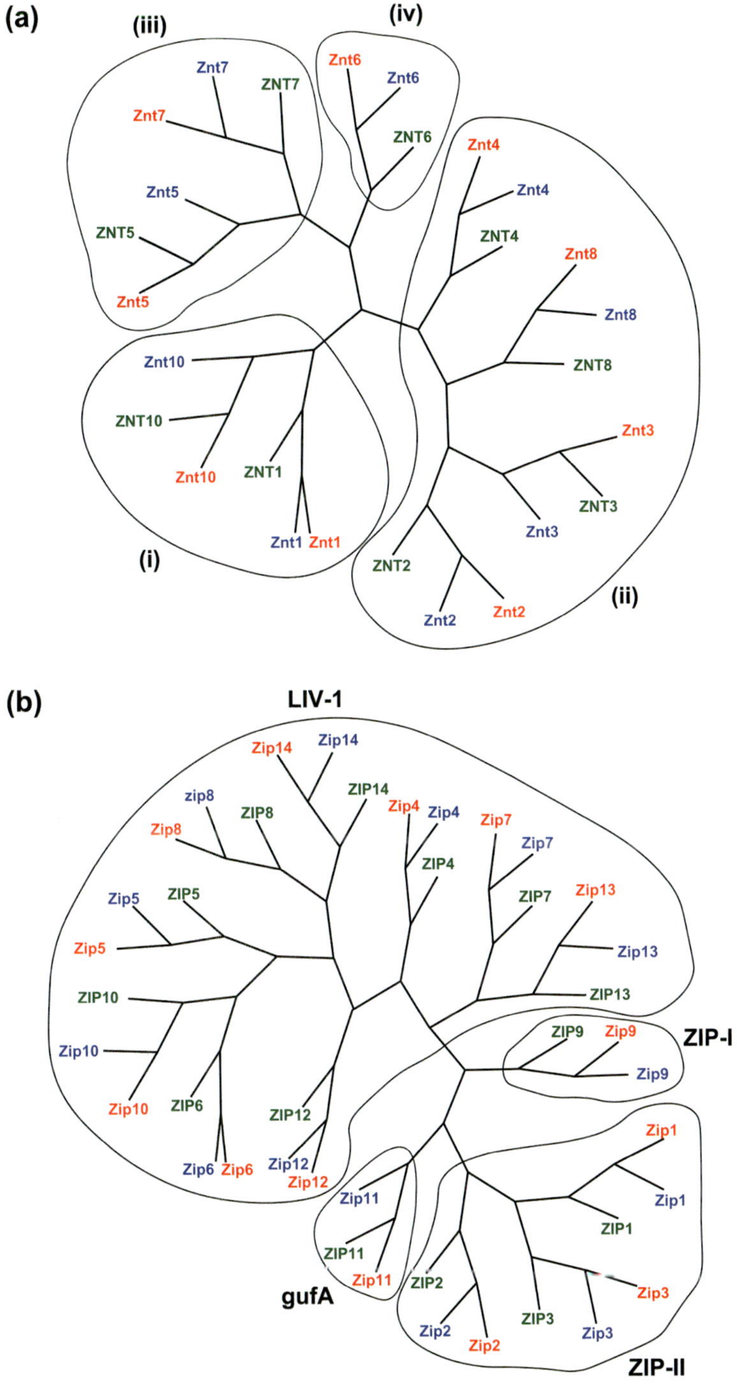

Fig. 11.1 Phylogenic relationships of ZNT and ZIP members among human, rat, and mice. ZNT members (**a**) and ZIP members (**b**) are shown. Green, human; blue, rat; and red, mouse. The neighbor-joining phylogenic tree was constructed using ClustalW protein alignment. Clusters are represented with different subgroups or subfamilies, which are designated according to the text

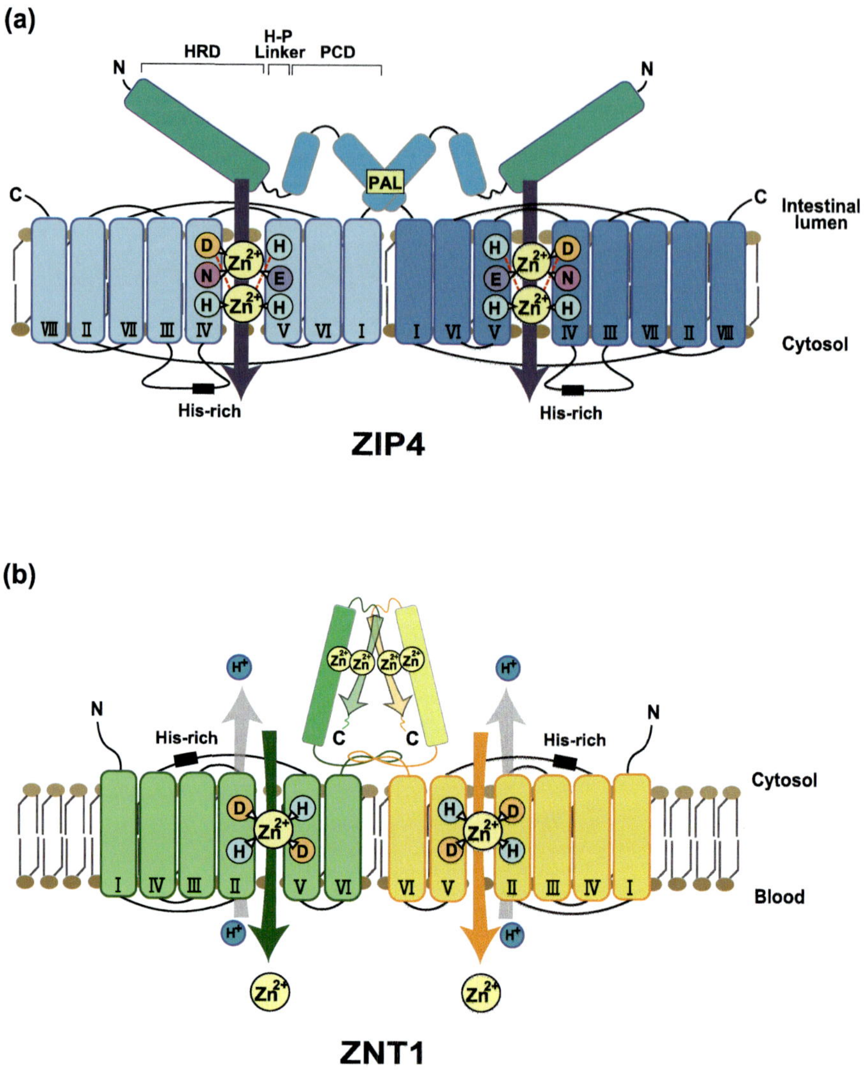

Fig. 11.2 Predicted membrane topology and structural features of ZIP4 and ZNT1. Similar topology is thought to be true for other ZNT and ZIP transporters. (**a**) The zinc transporter ZIP4, which localizes to the apical membrane of enterocytes during zinc deficiency, forms homodimers that import zinc. ZIP4 protomers likely have eight transmembrane domains (TMDs) consisting of a novel 3 + 2 + 3 TMD architecture, in which the first three TMDs (I–III) are symmetrically related to the last three TMDs (VI–VIII) by a pseudo-twofold axis (Zhang et al. 2017). The conserved amphipathic amino acid residues in TMD IV (histidine (H), asparagine (N), and aspartic acid (D)) and the potential metalloprotease motif (**H**EXP**H**EXGD) in TMD V (two histidines (H) and one glutamic acid (E) indicated in bold) form a binuclear metal center within the TMDs. ZIP4's long extracellular N-terminal portion is divided into two structural domains, the helix-rich domain (HRD) and the PAL motif-containing domain (PCD) that are connected by the H-P linker (Zhang et al. 2017). This extracellular portion is cleaved during zinc deficiency (Kambe and Andrews 2009). ZIP4 has a histidine-rich variable loop between TMDs III and IV that is involved in zinc-induced ubiquitination and degradation (Mao et al. 2007). (**b**) The zinc transporter ZNT1, which localizes to the basolateral membrane of enterocytes, forms homodimers that function as zinc/

transporters on the basolateral membrane. Essential for zinc absorption are apical ZIP4, which imports zinc from the intestinal lumen, and the relatively poorly characterized basolateral ZNT1, which exports zinc into the portal blood (Hashimoto and Kambe 2015). Unlike iron and copper, zinc is redox-inert and thus does not require a redox reaction prior to membrane transport by ZIP and ZNT family members (Kambe 2013; Nishito and Kambe 2018). Regulation of expression of both ZIP4 and ZNT1 is critically important for net zinc absorption.

This review summarizes current knowledge on the mechanism of intestinal zinc absorption with a focus on ZIP4 and ZNT1, discusses other zinc transporters that may be involved in the regulation of vectorial zinc transport in the enterocytes, and compares zinc transport to what is known about iron, copper, and manganese vectorial transport. Insights into the molecular basis of expression regulation, subcellular localization, and pathophysiological functions of ZIP and ZNT family members have been extensively reviewed recently in Kambe (2013) and Kambe et al. (2014a, b); interested readers are referred to these reviews.

Throughout this article, human gene and protein names, as well as common names, are fully capitalized, while only the first letter of murine gene and protein names is capitalized.

11.2 Zinc Import into Enterocytes Across the Apical Membrane

Several ZIP transporters are expressed in intestinal epithelial cells (Dufner-Beattie et al. 2003, 2004; Guthrie et al. 2015), of which ZIP4 is essential for uptake of dietary zinc from the intestinal lumen. ZIP4 is mainly expressed in intestinal epithelia of the duodenum and jejunum, the zinc absorption sites of vertebrates. Mutations in ZIP4 result in acrodermatitis enteropathica (AE), a rare genetic disorder of severe zinc deficiency due to defective zinc uptake (Barnes and Moynahan 1973; Kury et al. 2002; Wang et al. 2002), characterized by acral dermatitis, alopecia, diarrhea, and in some cases neuropsychological disturbances (Schmitt et al. 2009; Kambe et al. 2014a, b). These symptoms disappear upon administration of a daily oral zinc supplement (1–3 mg/kg/day of elemental zinc). Consistent with this, Zip4 is essential in mice: intestine-specific *Zip4*-knockout mice die unless fed a zinc supplement (Geiser et al. 2012, 2013a) (Table 11.1).

Understanding of ZIP4 has been facilitated by the Hu group's determination of the crystal structure of its extracellular N-terminal portion and the transmembrane

Fig. 11.2 (continued) proton antiporters to export zinc. ZNT1 likely has six TMDs divided into a compact four-helix bundle (TMDs I, II, IV, and V) and a two-helix pair (TMDs III and VI) (Coudray et al. 2013; Gupta et al. 2014; Lopez-Redondo et al. 2018). Biochemical studies show that its intramembranous zinc-binding site consists of two histidine (H) and two aspartic acid (D) residues

Table 11.1 Phenotype, or effects on the small intestine, of knocking out proteins in mice that are established or possibly involved in zinc absorption in intestinal epithelia

Protein name	Substrate	Phenotypes of intestine-specific knockout mouse
Zip4	Zinc	Rapid wasting and death, diminished function of the intestinal mucosa (Geiser et al. 2012)
Znt1	Zinc	*Embryonic lethal (Andrews et al. 2004)
Zip5	Zinc	~60% increase in pancreatic zinc after consumption of a zinc adequate diet, although no change in intestinal zinc. Increase in abundance of *Zip4* mRNA (Geiser et al. 2013a, b)
[a]Znt4	Zinc	*Systemic zinc deficiency after 8 months in *Znt4*-mutant (*lethal milk*) mice (Erway and Grider 1984)
[a]Znt7	Zinc	*Symptoms of dietary zinc deficiency, such as reduced food intake and poor body weight gain, although no sign of hair abnormality and dermatitis (Huang et al. 2007)
[a]Zip14	Zinc	*Greater intestinal zinc content, impaired barrier function in jejunum (Guthrie et al. 2015)
[a]Metallothionein	Zinc	*Increased intestinal and serum zinc concentrations after zinc administration (Davis et al. 1998)

[a]Only phenotypes related to the small intestine or intestinal absorption are shown; detailed phenotypes are described in Lichten and Cousins (2009), Kambe et al. (2015), Hara et al. (2017), and Aydemir and Cousins 2018)
*global knockout or mutant mouse

domain portion of a bacterial ZIP homolog (Zhang et al. 2016, 2017). As predicted computationally (Antala et al. 2015), the structure of the bacterial homolog suggests that ZIP4 has eight TMDs and form a homodimer. A binuclear metal center is found in each eight-TMD protomer, in which one zinc binding site seems crucial for zinc transport (Zhang et al. 2017) (Fig. 11.2a). The extracellular N-terminal half of the protein can also homodimerize alone (Zhang et al. 2016). Despite extensive understanding of structural information, the zinc transport mode of ZIP4 is still unknown.

Zinc tightly regulates ZIP4 expression, mainly posttranslationally, although zinc deficiency also contributes by stabilizing *Zip4* mRNA (Weaver et al. 2007). Apically localized ZIP4 escapes degradation during zinc deficiency and thus accumulates in the apical membrane (Dufner-Beattie et al. 2003; Weaver et al. 2007; Hashimoto et al. 2015; Hashimoto et al. 2016). If zinc is in excess, this surface-accumulated ZIP4 is rapidly endocytosed and degraded via the ubiquitin proteasome pathway (Kim et al. 2004; Wang et al. 2004a, b; Mao et al. 2007; Kambe et al. 2008; Hashimoto et al. 2015); ubiquitin-mediated degradation, but not endocytosis, requires a cytoplasmic histidine-rich cluster between TMD III and IV (Mao et al. 2007). Interestingly, under prolonged or severe zinc deficiency, the extracellular N-terminal domain of ZIP4 is processed by proteolytic cleavage, and the processed ZIP4 protein accumulates in the apical membrane (Kambe and Andrews 2009; Hashimoto et al. 2016). Introduction of mutations known to cause AE near the putative processing site impairs processing, suggesting that regulation of processing is important for zinc absorption (Kambe and Andrews 2009). The precise molecular mechanisms underlying ZIP4 modification associated with endocytosis, degradation, and processing require further investigation.

Ectopic expression of ZIP4 outside the small intestine is associated with cancers including pancreatic cancer, hepatic cancer, and glioma (Li et al. 2007; Weaver et al. 2010; Lin et al. 2013; Zhang et al. 2013; Yang et al. 2018). The role of ZIP4 in cancer cells is beginning to be understood, but whether its expression is regulated by zinc status in cancer cells remains to be studied.

11.3 Zinc Release from Enterocytes Across the Basolateral Membrane

ZNT1 is exclusively localized to the enterocyte basolateral membrane (McMahon and Cousins 1998; Yu et al. 2007; Hennigar and McClung 2016), in contrast to the majority of enterocyte ZNT transporters, which are localized to intracellular compartments (McMahon and Cousins 1998; Yu et al. 2007; Kelleher et al. 2011; Kambe et al. 2017). Zinc transported into the enterocytes by ZIP4 is believed to be exported across the basolateral membrane by ZNT1 into the portal blood. Zinc released by ZNT1 from epithelial cells into the bloodstream binds to albumin and α2-macroglobulin for delivery to the peripheral tissues. Gut-specific *Znt1*-silenced *Drosophila* accumulates zinc in the midgut, and zinc transport to the peripheral body parts is reduced; these defects are rescued by addition of human ZNT1 (Wang et al. 2009). As intestine-specific *Znt1*-knockout mice have yet to be generated (Table 11.1), this result is not confirmed in mammals.

Studies of bacterial homologs suggest that ZNT1 has six TMDs and forms homodimers (Fig. 11.2b) (Lu and Fu 2007; Lu et al. 2009; Coudray et al. 2013; Gupta et al. 2014); each monomer has a transmembrane metal-binding site, which consists of two histidine (H) and two aspartic acid (D) residues, thought to be specific for zinc (Nishito et al. 2016). ZNT1 and its homologs are Zn^{2+}/H^+ antiporters (Shusterman et al. 2014). In contrast to posttranscriptionally regulated ZIP4, *ZNT1* transcription is upregulated in zinc-sufficient conditions (Liuzzi et al. 2001) by binding of metal-response element-binding transcription factor-1 (MTF-1) to a metal-response element in *ZNT1*'s promoter region (Langmade et al. 2000; Hardyman et al. 2016). ZNT1 expression may be regulated to rapidly fine-tune zinc export across the basolateral membrane into the bloodstream in order to control the rate of zinc absorption. Interestingly, ZNT1 is degraded through a lysosome-mediated degradation pathway in response to hepcidin, a peptide hormone master regulator of iron homeostasis (Hennigar and McClung 2016; Nemeth et al. 2004). Further studies are needed to clarify how zinc and other stimuli regulate ZNT1 expression.

11.4 Transepithelial Transport of Zinc in Enterocytes

During iron and copper absorption, intestinal epithelial cells use specific chaperones for vectorial transepithelial metal ion transport from the apical to the basolateral membrane. Specifically, poly(C) binding protein 2 (PCBP2) traffics iron (Yanatori et al. 2014, 2016), while antioxidant-1 (ATOX1) traffics copper (Hamza et al. 2001) (see Sect. 11.6). Comparisons with iron and copper transport suggest that a chaperone is important for vectorial zinc transport. How, then, is vectorial zinc transport facilitated (Fig. 11.3)?

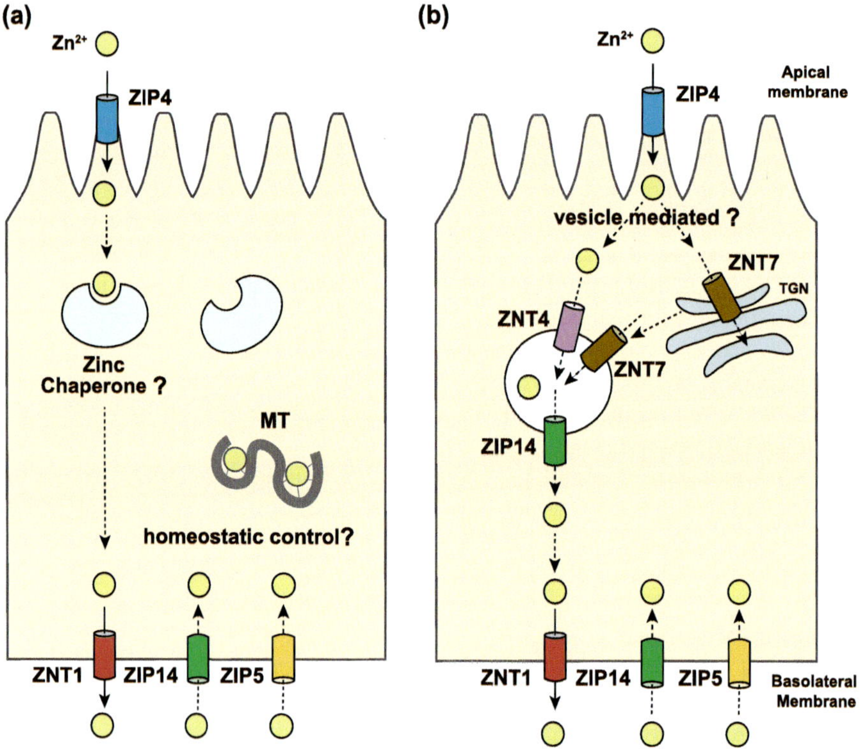

Fig. 11.3 Two possible mechanisms for vectorial zinc transport in intestinal epithelial cells. The zinc transporter ZIP4 imports dietary zinc across the enterocyte apical membrane for delivery to the basolateral membrane and release into the bloodstream by ZNT1 for delivery to the peripheral tissues. Transport to the basolateral membrane may be accomplished by one of two mechanisms. (**a**) After import by ZIP4, zinc may be bound by an unknown chaperone protein, possibly metallothionein (MT), for trafficking from the apical side to the basolateral side. (**b**) Alternatively, zinc may be transported into cytoplasmic vesicles to traffic it to the basolateral membrane; in this case, ZNT4 and ZNT7 may export zinc from the cytoplasm into vesicles, while ZIP14 may import zinc from the vesicles back into the cytoplasm. The *trans*-Golgi network (TGN)-located ZNT7 may contribute to this process. ZIP5 or ZIP14 expression levels in the basolateral membrane may modulate the rate of zinc absorption

Some studies have proposed the presence of zinc chaperones in the cells (Costello et al. 2004; Fujimoto et al. 2013). Metallothionein (MT) might be a potential candidate to act as zinc chaperone (Fig. 11.3a). Cytoplasmic, cysteine-rich MT binds and controls the concentration of cytoplasmic metals including zinc (Krezel and Maret 2008; Kimura and Kambe 2016). A study using *Mt*-KO mice and *Mt*-transgenic mice, however, showed that serum zinc concentration was inversely related to the level of Mt during elevated zinc intake (Davis et al. 1998), making the role of MT unclear (Table 11.1).

Vesicle-mediated transport is an alternative possible mechanism for apical-to-basolateral transepithelial zinc trafficking (Fig. 11.3b). In enterocytes, ZNT4, ZNT7, and ZIP14 are localized to cytoplasmic vesicles (McMahon and Cousins 1998; Yu et al. 2007). Indeed, *Znt4*-mutant mice (*lethal milk* mice) exhibit several typical features of zinc deficiency over 8 months of age (Erway and Grider 1984), and *Znt7*-knockout mice poorly absorb dietary zinc (Huang et al. 2007), implicating ZNT4 and ZNT7 in epithelial vectorial zinc transport (Table 11.1). Furthermore, zinc accumulates in the intestines of *Zip14*-knockout mice, likely due to loss of endosomal Zip14 (Guthrie et al. 2015), suggesting that the contribution of ZIP14 to transepithelial zinc transport regulation by mobilizing zinc from vesicles into the cytoplasm. Other as-yet-unidentified ZNT and ZIP transporters may also be involved in vesicle-mediated transepithelial zinc transport.

Further studies are evidently needed to better understand transepithelial vectorial transport of zinc in enterocytes.

11.5 Other Zinc Transporters Possibly Involved in Zinc Absorption

Although ZIP4 and ZNT1 are crucial for zinc absorption, zinc deficiency symptoms in AE patients are ameliorated by oral zinc supplementation (Lombeck et al. 1975), raising the possibility that other, yet-to-be-identified transporters may contribute to zinc absorption.

In enterocytes, ZIP5 and ZIP14 are highly expressed and localized to the basolateral membrane, functioning to import zinc into the cytoplasm from the bloodstream (Fig. 11.3a) (Dufner-Beattie et al. 2004; Wang et al. 2004a, b; Aydemir and Cousins 2018). ZIP5 is particularly interesting because its expression is reciprocal to ZIP4; while ZIP4 localizes apically when zinc is deficient, ZIP5 localizes basolaterally when zinc is in excess (Dufner-Beattie et al. 2004; Wang et al. 2004b). The mechanism underlying zinc-dependent ZIP5 expression differs to ZIP4; translational stall of *Zip5* mRNA is removed in response to zinc, mediated by conserved stem-loops and two miRNA seed sites in the 3′-untranslated region (Weaver and Andrews 2012). Intestine-specific *Zip5*-KO mice show an increase in *Zip4* mRNA, probably due to enterocyte zinc deficiency resulting from loss of Zip5-mediated zinc uptake from the blood (Table 11.1). Intestinal zinc, however, does not change, while pancreatic zinc is much higher than in control mice, likely due to the absence of enterocyte Zip5 causing an increased zinc burden to the body (Geiser et al. 2013b).

ZIP14 is involved in hypozincemia, liver regeneration (Aydemir and Cousins 2018), systemic growth (Hojyo et al. 2011), and cancer-induced cachexia (Wang et al. 2018). ZIP14 expression is drastically upregulated by proinflammatory conditions. *Zip14*-knockout mice show greater intestinal zinc accumulation. In addition to localization to enterocyte endosomes as described above, ZIP14 is also abundantly localized to the basolateral membrane (Guthrie et al. 2015) and thus may contribute to direct zinc uptake from the bloodstream into the cytoplasm. Moreover, ZIP14 in enterocytes is thought to be important for providing zinc for a regulatory role in maintenance of the intestinal barrier. More studies on zinc transporters and vectorial transport of zinc from the apical to the basolateral membrane are needed to better understand these processes. A schematic of the zinc absorption process is shown in Fig. 11.3.

11.6 Vectorial Transport of Other Trace Elements

While this article focuses on epithelial transport of zinc, we summarize iron, copper, and manganese transport here for comparison (see Figs. 11.4, 11.5, and 11.6). For further details, the reader is referred to recent publications (Lutsenko et al. 2007; Kim et al. 2008; Gulec and Collins 2014; Bogdan et al. 2016; Knutson 2017).

11.6.1 Iron Transport in the Enterocytes

Dietary non-heme ferrous iron (Fe^{2+}) is transported across the apical membrane into intestinal epithelial cells by divalent metal transporter 1 (DMT1); ferric iron (Fe^{3+}) must first be reduced to ferrous iron by ferrireductase duodenal cytochrome B (DCYTB) (Ganasen et al. 2018). DMT1 is an essential component in iron uptake, and conditional deletion of *Dmt1* in the intestine leads to severe hypochromic microcytic anemia (Table 11.2). An iron chaperone, PCBP2, shuttles imported iron either to the iron exporter ferroportin (FPN) in the basolateral membrane (Lane and Richardson 2014; Yanatori et al. 2014, 2016) or to the cytoplasmic iron storage protein, ferritin (Shi et al. 2008). PCBP2 is therefore crucial in vectorial transepithelial iron transport, which is mediated by binding to both DMT1 and FPN (Lane and Richardson 2014; Yanatori et al. 2014, 2016). Iron accumulation in cells together with severe iron deficiency anemia in conditional intestinal deletions of Fpn demonstrates that FPN is a major exporter of iron to the portal blood (Table 11.2). Ferrous iron exported by FPN is rapidly oxidized in the bloodstream to ferric iron by hephaestin (HEPH), a multicopper ferroxidase (Vulpe et al. 1999), and bound by transferrin for delivery to the peripheral tissues via circulation (Gulec and Collins 2014; Knutson 2017). Intestinal iron absorption is crucial for systemic iron

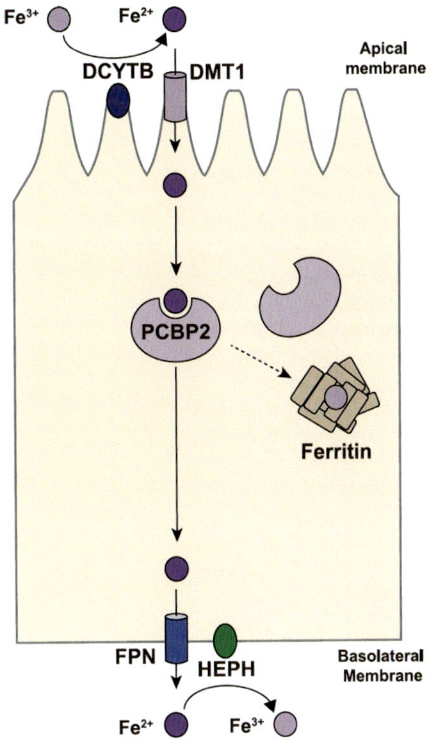

Fig. 11.4 Vectorial iron transport in intestinal epithelial cells. Divalent metal transporter 1 (DMT1) imports dietary non-heme ferrous iron (Fe^{2+}) into enterocytes across the apical membrane, after or concomitant with reduction from ferric iron (Fe^{3+}) by duodenal cytochrome B (DCYTB). After uptake, iron is bound to the iron chaperone poly(C) binding protein 2 (PCBP2) and trafficked to ferroportin (FPN) in the basolateral membrane for release into the bloodstream. The released iron is re-oxidizing to ferric iron by hephaestin (HEPH) and then bound by transferrin for delivery to the peripheral tissues. This illustration is simplified for comparison with zinc vectorial transport (Fig. 11.3)

homeostasis and is regulated by controlling iron release from cells, including intestinal epithelial cells, by FPN. FPN abundance is regulated by degradation by direct binding of hepcidin, the master hormone regulator of iron homeostasis. A schematic of the iron absorption process is shown in Fig. 11.4.

11.6.2 Copper Transport in the Enterocytes

Cuprous copper (Cu^{+}) is transported across the apical membrane into intestinal epithelial cells by copper transporter 1 (CTR1) (Nose et al. 2010); as with iron, cupric copper (Cu^{2+}) must first be reduced to cuprous copper, although the reducing enzyme is not known (Kim et al. 2008; Gulec and Collins 2014). CTR1's crucial role in copper absorption is demonstrated by severe copper deficiency in intestine-specific *Ctr1*-knockout mice (Table 11.2). Imported copper is trafficked to the copper-transporting ATPase ATP7A by metallochaperone ATOX1; disruption of Atox1 leads to significantly reduced copper export and copper deficiency symptoms (Hamza et al. 2001). ATP7A exports copper directly or indirectly across the basolateral membrane into the portal blood (Harrison et al. 1999): ATP7A is usually localized to the trans-Golgi network (Wang et al. 2011) as well as the basolateral

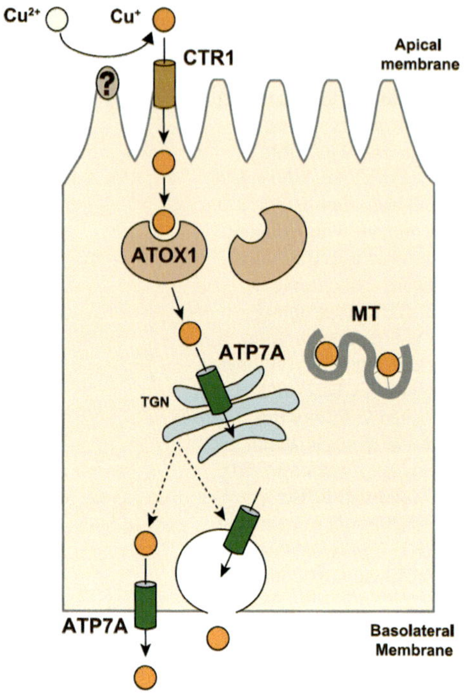

Fig. 11.5 Vectorial copper transport in intestinal epithelial cells. Copper transporter 1 (CTR1) imports dietary cuprous copper (Cu^{+}) into enterocytes across the apical membrane, after or concomitant with reduction from cupric copper (Cu^{2+}) by a metalloreductase. Imported copper is bound by antioxidant-1 (ATOX1), and transported to ATP7A, for either export across the basolateral membrane into the bloodstream or export into cytoplasmic vesicles for exocytosis into the bloodstream. Excess cytoplasmic copper may be bound by metallothionein (MT). This illustration is simplified for comparison with zinc vectorial transport (Fig. 11.3)

membrane (Kim et al. 2008), suggesting that ATP7A may either transport copper into vesicles for subsequent fusion with the basolateral membrane or directly export copper across the basolateral membrane. Mutation of *ATP7A* gene results in Menkes disease (Vulpe et al. 1993) characterized by systemic copper deficiency; intestine-specific *Atp7a*-knockout mice are similarly copper-deficient (Wang et al. 2012), highlighting the importance of ATP7A for copper export to the portal blood (Table 11.2). Excess copper in the cytoplasm is bound by MT, which may be involved in epithelial copper detoxification and transport (Kelly and Palmiter 1996). A schematic of the copper absorption process is shown in Fig. 11.5.

11.6.3 Manganese Transport in the Enterocytes

Manganese absorption by the intestinal epithelium is poorly understood (Jenkitkasemwong et al. 2012). Impaired manganese metabolism in *Dmt1*-mutant rats (*Belgrade* rats) suggests that DMT1 may be able to import manganese, in addition to iron, into enterocytes (Chua and Morgan 1997; Au et al. 2008), although a conflicting recent report using intestine-specific *Dmt1*-knockout mice indicated that Dmt1 is not critical for manganese absorption (Shawki et al. 2015). Export of manganese into the portal blood may be conducted by the iron exporter FPN (Madejczyk and Ballatori 2012), confirmed in a recent study using *Fpn*-mutant

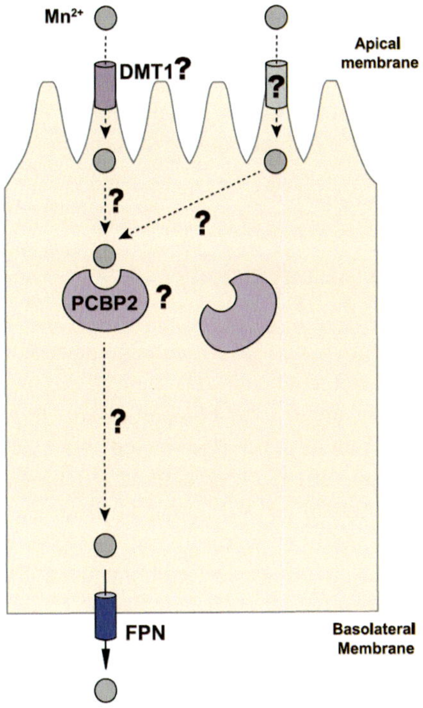

Fig. 11.6 Vectorial manganese transport in intestinal epithelial cells. Manganese transport is relatively poorly understood. Manganese may be imported across the apical membrane into enterocytes by divalent metal transporter 1 (DMT1), or an as-yet-unidentified protein. If DMT1 is the transporter, manganese would be in its divalent form. The iron transporter ferroportin (FPN) is likely involved in manganese export across the basolateral membrane into the bloodstream. Transepithelial transport of manganese has not yet been studied in detail, but might be regulated by poly (C) binding protein 2 (PCBP2) and its homologs, given the other parallels with iron metabolism

Table 11.2 Phenotypes, or effects on the small intestine, in knockout or mutant animals for transporters and chaperones involved in intestinal absorption of iron, copper, and manganese

Protein name	Substrate	Phenotypes of intestine-specific knockout mouse
Dmt1	Iron	Hypochromic microcytic anemia, splenomegaly, cardiomegaly (Shawki et al. 2015)
Fpn	Iron	Severe iron deficiency anemia (Donovan et al. 2005)
Pcbp2	Iron	*Embryonic lethal, due to combined cardiovascular and hematopoietic abnormalities (Ghanem et al. 2016)
Ctr1	Copper	Decreased copper accumulation in peripheral tissues, hepatic iron accumulation, cardiac hypertrophy, growth retardation (Nose et al. 2006)
Atp7a	Copper	Copper deficiency: growth impairment and neurological deterioration (Wang et al. 2012)
Atox1	Copper	*Growth failure: 45% of pups die before weaning (Hamza et al. 2001)
Metallothionein	Copper	*Exacerbation of the phenotypes (embryonic lethal) of *Atp7a*-mutant mice (Kelly and Palmiter 1996)
Dmt1	Manganese	*Impaired manganese metabolism in *Dmt1*-mutant (*Belgrade* mutant) rat (Chua and Morgan 1997)
Fpn	Manganese	*Reduced intestinal manganese absorption in *Fpn*-mutant (*flatiron* (ffe/+) mutant) mice (Seo and Wessling-Resnick 2015)

*global knockout or mutant animals

(*flatiron* (ffe/+) mutant) mice that implicated FPN in manganese absorption (Seo and Wessling-Resnick 2015). Considering analogy to vectorial transepithelial iron transport, the iron chaperone PCBP2 might play a role in transepithelial manganese transport. However, this possibility is unexplored. A schematic of what is currently known about manganese absorption is shown in Fig. 11.6.

11.7 Conclusions

Recent studies have gradually clarified the mechanism of zinc absorption by intestinal epithelial cells. The molecular basis underlying this process, however, remains poorly understood compared with that of iron and copper absorption and other better-understood zinc-related biophysiological areas (Ya et al. 2014; Hojyo and Fukada 2016; Ogawa et al. 2016; Kambe et al. 2017; Maywald et al. 2017; Hershfinkel 2018; Lopez and Skaar 2018). The understanding of molecular mechanism and regulation of zinc transport will benefit from determination of ZIP4 and ZNT1 structures. Determining what controls the subcellular localization of ZIP4, ZNT1, ZIP5, and ZIP14 will also be invaluable. Clarification of whether a humoral systemic regulator controls intestinal epithelial transport of zinc is another outstanding question; the recent finding that hepcidin may regulate ZNT1 expression is consistent with such systemic control (Hennigar and McClung 2016). Answering these questions will substantially contribute to understanding zinc transport in the intestinal epithelium and the physiology of zinc absorption, crucial to preventing zinc deficiency (Prasad et al. 2007; Tuerk and Fazel 2009; Penny 2013; Kumssa et al. 2015), contributing to good health from birth to old age (Haase et al. 2006; Yasuda and Tsutsui 2016; Golan et al. 2017; Wessels et al. 2017), and forming the basis for new research areas such as zinc-ome analysis, which interrelates systemic and cellular zinc metabolism, as demonstrated by our recent work on the impact of zinc deficiency on extracellular ATP metabolism (Takeda et al. 2018).

Acknowledgments This work was supported by JSPS KAKENHI Grant Numbers JP15H04501 (to T.K.) and 17J09455 (to Y.N.) and by the Ito Foundation (to T.K.).

References

Andrews GK, Wang H, Dey SK et al (2004) Mouse zinc transporter 1 gene provides an essential function during early embryonic development. Genesis 40:74–81

Antala S, Ovchinnikov S, Kamisetty H et al (2015) Computation and functional studies provide a model for the structure of the zinc transporter hZIP4. J Biol Chem 290:17796–17805

Au C, Benedetto A, Aschner M (2008) Manganese transport in eukaryotes: the role of DMT1. Neurotoxicology 29:569–576

Aydemir TB, Cousins RJ (2018) The multiple faces of the metal transporter ZIP14 (SLC39A14). J Nutr 148:174–184

Barnes PM, Moynahan EJ (1973) Zinc deficiency in acrodermatitis enteropathica: multiple dietary intolerance treated with synthetic diet. Proc R Soc Med 66:327–329
Bogdan AR, Miyazawa M, Hashimoto K et al (2016) Regulators of iron homeostasis: new players in metabolism, cell death, and disease. Trends Biochem Sci 41:274–286
Chua AC, Morgan EH (1997) Manganese metabolism is impaired in the Belgrade laboratory rat. J Comp Physiol B 167:361–369
Costello LC, Guan Z, Franklin RB et al (2004) Metallothionein can function as a chaperone for zinc uptake transport into prostate and liver mitochondria. J Inorg Biochem 98:664–666
Coudray N, Valvo S, Hu M et al (2013) Inward-facing conformation of the zinc transporter YiiP revealed by cryoelectron microscopy. Proc Natl Acad Sci USA 110:2140–2145
Davis SR, McMahon RJ, Cousins RJ (1998) Metallothionein knockout and transgenic mice exhibit altered intestinal processing of zinc with uniform zinc-dependent zinc transporter-1 expression. J Nutr 128:825–831
Donovan A, Lima CA, Pinkus JL et al (2005) The iron exporter ferroportin/Slc40a1 is essential for iron homeostasis. Cell Metab 1:191–200
Dufner-Beattie J, Wang F, Kuo YM et al (2003) The acrodermatitis enteropathica gene ZIP4 encodes a tissue-specific, zinc-regulated zinc transporter in mice. J Biol Chem 278:33474–33481
Dufner-Beattie J, Kuo YM, Gitschier J et al (2004) The adaptive response to dietary zinc in mice involves the differential cellular localization and zinc regulation of the zinc transporters ZIP4 and ZIP5. J Biol Chem 279:49082–49090
Erway LC, Grider A Jr (1984) Zinc metabolism in lethal-milk mice. Otolith, lactation, and aging effects. J Hered 75:480–484
Fujimoto S, Itsumura N, Tsuji T et al (2013) Cooperative functions of ZnT1, metallothionein and ZnT4 in the cytoplasm are required for full activation of TNAP in the early secretory pathway. PLoS One 8:e77445
Ganasen M, Togashi H, Takeda H et al (2018) Structural basis for promotion of duodenal iron absorption by enteric ferric reductase with ascorbate. Commun Biol 1:120
Geiser J, Venken KJ, De Lisle RC et al (2012) A mouse model of acrodermatitis enteropathica: loss of intestine zinc transporter ZIP4 (Slc39a4) disrupts the stem cell niche and intestine integrity. PLoS Genet 8:e1002766
Geiser J, De Lisle RC, Finkelstein D et al (2013a) Clioquinol synergistically augments rescue by zinc supplementation in a mouse model of acrodermatitis enteropathica. PLoS One 8(8):e72543
Geiser J, De Lisle RC, Andrews GK (2013b) The zinc transporter Zip5 (Slc39a5) regulates intestinal zinc excretion and protects the pancreas against zinc toxicity. PLoS One 8(11):e82149
Ghanem LR, Kromer A, Silverman IM et al (2016) The poly(C) binding protein Pcbp2 and its retrotransposed derivative pcbp1 are independently essential to mouse development. Mol Cell Biol 36:304–319
Golan Y, Kambe T, Assaraf YG (2017) The role of the zinc transporter SLC30A2/ZnT2 in transient neonatal zinc deficiency. Metallomics 9:1352–1366
Gulec S, Collins JF (2014) Molecular mediators governing iron-copper interactions. Annu Rev Nutr 34:95–116
Gupta S, Chai J, Cheng J et al (2014) Visualizing the kinetic power stroke that drives proton-coupled zinc (II) transport. Nature 512:101–104
Guthrie GJ, Aydemir TB, Troche C et al (2015) Influence of ZIP14 (slc39A14) on intestinal zinc processing and barrier function. Am J Physiol Gastrointest Liver Physiol 308:G171–G178
Haase H, Rink L (2014) Multiple impacts of zinc on immune function. Metallomics 6:1175–1180
Haase H, Mocchegiani E, Rink L (2006) Correlation between zinc status and immune function in the elderly. Biogerontology 7:421–428
Hamza I, Faisst A, Prohaska J et al (2001) The metallochaperone Atox1 plays a critical role in perinatal copper homeostasis. Proc Natl Acad Sci USA 98:6848–6852
Hara T, Takeda TA, Takagishi T et al (2017) Physiological roles of zinc transporters: molecular and genetic importance in zinc homeostasis. J Physiol Sci 67:283–301

Hardyman JE, Tyson J, Jackson KA et al (2016) Zinc sensing by metal-responsive transcription factor 1 (MTF1) controls metallothionein and ZnT1 expression to buffer the sensitivity of the transcriptome response to zinc. Metallomics 8:337–343

Harrison MD, Jones CE, Dameron CT (1999) Copper chaperones: function, structure and copper-binding properties. J Biol Inorg Chem 4:145–153

Hashimoto A, Kambe T (2015) Mg, Zn and Cu transport proteins: a brief overview from physiological and molecular perspectives. J Nutr Sci Vitaminol (Tokyo) 61:S116–S118

Hashimoto A, Ohkura K, Takahashi M et al (2015) Soybean extracts increase cell surface ZIP4 abundance and cellular zinc levels: a potential novel strategy to enhance zinc absorption by ZIP4 targeting. Biochem J 472:183–193

Hashimoto A, Nakagawa M, Tsujimura N et al (2016) Properties of Zip4 accumulation during zinc deficiency and its usefulness to evaluate zinc status: a study of the effects of zinc deficiency during lactation. Am J Physiol Regul Integr Comp Physiol 310:R459–R468

Hennigar SR, McClung JP (2016) Hepcidin attenuates zinc efflux in Caco-2 cells. J Nutr 146:1267–1273

Hershfinkel M (2018) The zinc sensing receptor, ZnR/GPR39, in health and disease. Int J Mol Sci 19:E439

Hojyo S, Fukada T (2016) Zinc transporters and signaling in physiology and pathogenesis. Arch Biochem Biophys 611:43–50

Hojyo S, Fukada T, Shimoda S et al (2011) The zinc transporter SLC39A14/ZIP14 controls G-protein coupled receptor-mediated signaling required for systemic growth. PLoS One 6: e18059

Huang L, Yu YY, Kirschke CP et al (2007) Znt7 (Slc30a7)-deficient mice display reduced body zinc status and body fat accumulation. J Biol Chem 282:37053–37063

Jenkitkasemwong S, Wang CY, Mackenzie B (2012) Physiologic implications of metal-ion transport by ZIP14 and ZIP8. Biometals 25:643–655

Kambe T (2013) Regulation of zinc transport. In: Culotta V, Scott RA (eds) Encyclopedia of inorganic and bioinorganic chemistry. Wiley, Chichester, pp 301–309

Kambe T, Andrews GK (2009) Novel proteolytic processing of the ectodomain of the zinc transporter ZIP4 (SLC39A4) during zinc deficiency is inhibited by acrodermatitis enteropathica mutations. Mol Cell Biol 29:129–139

Kambe T, Suzuki T, Nagao M et al (2006) Sequence similarity and functional relationship among eukaryotic ZIP and CDF transporters. Genom Proteom Bioinform 4:1–9

Kambe T, Geiser J, Lahner B et al (2008) Slc39a1 to 3 (subfamily II) Zip genes in mice have unique cell-specific functions during adaptation to zinc deficiency. Am J Physiol Regul Integr Comp Physiol 294:R1474–R1481

Kambe T, Hashimoto A, Fujimoto S (2014a) Current understanding of ZIP and ZnT zinc transporters in human health and diseases. Cell Mol Life Sci 71:3281–3295

Kambe T, Tsuji T, Fukue K (2014b) Zinc transport proteins and zinc signaling. In: Fukada T, Kambe T (eds) Zinc signals in cellular functions and disorders. Springer, Tokyo, pp 27–53

Kambe T, Tsuji T, Hashimoto A et al (2015) The physiological, biochemical, and molecular roles of zinc transporters in zinc homeostasis and metabolism. Physiol Rev 95:749–784

Kambe T, Nishito Y, Fukue K (2016) Zinc transporters in health and disease. In: Collins JF (ed) Molecular, genetic, and nutritional aspects of major and trace minerals. Elsevier, London, pp 283–291

Kambe T, Matsunaga M, Takeda TA (2017) Understanding the contribution of zinc transporters in the function of the early secretory pathway. Int J Mol Sci 18:E2179

Kelleher SL, McCormick NH, Velasquez V et al (2011) Zinc in specialized secretory tissues: roles in the pancreas, prostate, and mammary gland. Adv Nutr 2:101–111

Kelly EJ, Palmiter RD (1996) A murine model of Menkes disease reveals a physiological function of metallothionein. Nat Genet 13:219–222

Kim BE, Wang F, Dufner-Beattie J et al (2004) Zn^{2+}-stimulated endocytosis of the mZIP4 zinc transporter regulates its location at the plasma membrane. J Biol Chem 279:4523–4530

Kim BE, Nevitt T, Thiele DJ (2008) Mechanisms for copper acquisition, distribution and regulation. Nat Chem Biol 4:176–185
Kimura T, Kambe T (2016) The functions of metallothionein and ZIP and ZnT transporters: an overview and perspective. Int J Mol Sci 17:336
Knutson MD (2017) Iron transport proteins: gateways of cellular and systemic iron homeostasis. J Biol Chem 292:12735–12743
Krezel A, Maret W (2008) Thionein/metallothionein control Zn (II) availability and the activity of enzymes. J Biol Inorg Chem 13:401–409
Kumssa DB, Joy EJ, Ander EL et al (2015) Dietary calcium and zinc deficiency risks are decreasing but remain prevalent. Sci Rep 5:10974
Kury S, Dreno B, Bezieau S (2002) Identification of SLC39A4, a gene involved in acrodermatitis enteropathica. Nat Genet 31:239–240
Lane DJ, Richardson DR (2014) Chaperone turns gatekeeper: PCBP2 and DMT1 form an iron-transport pipeline. Biochem J 462:e1–e3
Langmade SJ, Ravindra R, Daniels PJ et al (2000) The transcription factor MTF-1 mediates metal regulation of the mouse ZnT1 gene. J Biol Chem 275:34803–34809
Li M, Zhang Y, Liu Z et al (2007) Aberrant expression of zinc transporter ZIP4 (SLC39A4) significantly contributes to human pancreatic cancer pathogenesis and progression. Proc Natl Acad Sci USA 104:18636–18641
Lichten LA, Cousins RJ (2009) Mammalian zinc transporters: nutritional and physiologic regulation. Annu Rev Nutr 29:153–176
Lin Y, Chen Y, Wang Y et al (2013) ZIP4 is a novel molecular marker for glioma. Neuro-Oncology 15:1008–1016
Liuzzi JP, Blanchard RK, Cousins RJ (2001) Differential regulation of zinc transporter 1, 2, and 4 mRNA expression by dietary zinc in rats. J Nutr 131:46–52
Lombeck T, Schnippering HG, Ritzl F et al (1975) Letter: absorption of zinc in acrodermatitis enteropathica. Lancet 1:855
Lopez CA, Skaar EP (2018) The impact of dietary transition metals on host-bacterial interactions. Cell Host Microbe 23:737–748
Lopez-Redondo ML, Coudray N, Zhang Z et al (2018) Structural basis for the alternating access mechanism of the cation diffusion facilitator YiiP. Proc Natl Acad Sci USA 115:3042–3047
Lu M, Fu D (2007) Structure of the zinc transporter YiiP. Science 317:1746–1748
Lu M, Chai J, Fu D (2009) Structural basis for autoregulation of the zinc transporter YiiP. Nat Struct Mol Biol 16:1063–1067
Lutsenko S, Barnes NL, Bartee MY et al (2007) Function and regulation of human copper-transporting ATPases. Physiol Rev 87:1011–1046
Madejczyk MS, Ballatori N (2012) The iron transporter ferroportin can also function as a manganese exporter. Biochim Biophys Acta 1818:651–657
Mao X, Kim BE, Wang F et al (2007) A histidine-rich cluster mediates the ubiquitination and degradation of the human zinc transporter, hZIP4, and protects against zinc cytotoxicity. J Biol Chem 282:6992–7000
Maywald M, Wessels I, Rink L (2017) Zinc signals and immunity. Int J Mol Sci 18:2222. https://doi.org/10.3390/ijms18102222
McMahon RJ, Cousins RJ (1998) Regulation of the zinc transporter ZnT-1 by dietary zinc. Proc Natl Acad Sci USA 95:4841–4846
Nemeth E, Tuttle MS, Powelson J et al (2004) Hepcidin regulates cellular iron efflux by binding to ferroportin and inducing its internalization. Science 306:2090–2093
Nishito Y, Kambe T (2018) Absorption mechanisms of iron, copper, and zinc: an overview. J Nutr Sci Vitaminol (Tokyo) 64:1–7
Nishito Y, Tsuji N, Fujishiro H et al (2016) Direct comparison of manganese detoxification/efflux proteins and molecular characterization of ZnT10 as a manganese transporter. J Biol Chem 291:14773–14787

Nose Y, Kim BE, Thiele DJ (2006) Ctr1 drives intestinal copper absorption and is essential for growth, iron metabolism, and neonatal cardiac function. Cell Metab 4:235–244

Nose Y, Wood LK, Kim BE et al (2010) Ctr1 is an apical copper transporter in mammalian intestinal epithelial cells in vivo that is controlled at the level of protein stability. J Biol Chem 285:32385–32392

Ogawa Y, Kawamura T, Shimada S (2016) Zinc and skin biology. Arch Biochem Biophys 611:113–119

Penny ME (2013) Zinc supplementation in public health. Ann Nutr Metab 62(Suppl 1):31–42

Prasad AS, Beck FW, Bao B et al (2007) Zinc supplementation decreases incidence of infections in the elderly: effect of zinc on generation of cytokines and oxidative stress. Am J Clin Nutr 85:837–844

Schmitt S, Kury S, Giraud M et al (2009) An update on mutations of the SLC39A4 gene in acrodermatitis enteropathica. Hum Mutat 30:926–933

Seo YA, Wessling-Resnick M (2015) Ferroportin deficiency impairs manganese metabolism in flatiron mice. FASEB J 29:2726–2733

Shawki A, Anthony SR, Nose Y et al (2015) Intestinal DMT1 is critical for iron absorption in the mouse but is not required for the absorption of copper or manganese. Am J Physiol Gastrointest Liver Physiol 309:G635–G647

Shi H, Bencze KZ, Stemmler TL et al (2008) A cytosolic iron chaperone that delivers iron to ferritin. Science 320:1207–1210

Shusterman E, Beharier O, Shiri L et al (2014) ZnT-1 extrudes zinc from mammalian cells functioning as a (Zn^{2+})/H^+ exchanger. Metallomics 6:1656–1663

Takeda TA, Miyazaki S, Kobayashi M et al (2018) Zinc deficiency causes delayed ATP clearance and adenosine generation in rats and cell culture models. Commun Biol 1:113

Taylor KM, Nicholson RI (2003) The LZT proteins; the LIV-1 subfamily of zinc transporters. Biochem Biophys Acta 1611:16–30

Tuerk MJ, Fazel N (2009) Zinc deficiency. Curr Opin Gastroenterol 25:136–143

Vancampen DR, Mitchell EA (1965) Absorption of Cu-64, Zn-65, Mo-99, and Fe-59 from ligated segments of the rat gastrointestinal tract. J Nutr 86:120–124

Vulpe C, Levinson B, Whitney S et al (1993) Isolation of a candidate gene for Menkes disease and evidence that it encodes a copper-transporting ATPase. Nat Genet 3:7–13

Vulpe CD, Kuo YM, Murphy TL et al (1999) Hephaestin, a ceruloplasmin homologue implicated in intestinal iron transport, is defective in the sla mouse. Nat Genet 21:195–199

Wang K, Zhou B, Kuo YM et al (2002) A novel member of a zinc transporter family is defective in acrodermatitis enteropathica. Am J Hum Genet 71:66–73

Wang F, Kim BE, Dufner-Beattie J et al (2004a) Acrodermatitis enteropathica mutations affect transport activity, localization and zinc-responsive trafficking of the mouse ZIP4 zinc transporter. Hum Mol Genet 13:563–571

Wang F, Kim BE, Petris MJ et al (2004b) The mammalian Zip5 protein is a zinc transporter that localizes to the basolateral surface of polarized cells. J Biol Chem 279:51433–51441

Wang X, Wu Y, Zhou B (2009) Dietary zinc absorption is mediated by ZnT1 in *Drosophila melanogaster*. FASEB J 23:2650–2661

Wang Y, Hodgkinson V, Zhu S et al (2011) Advances in the understanding of mammalian copper transporters. Adv Nutr 2:129–137

Wang Y, Zhu S, Hodgkinson V et al (2012) Maternofetal and neonatal copper requirements revealed by enterocyte-specific deletion of the Menkes disease protein. Am J Physiol Gastrointest Liver Physiol 303:G1236–G1244

Wang G, Biswas AK, Ma W et al (2018) Metastatic cancers promote cachexia through ZIP14 upregulation in skeletal muscle. Nat Med 24:770–781

Weaver BP, Andrews GK (2012) Regulation of zinc-responsive Slc39a5 (Zip5) translation is mediated by conserved elements in the 3′-untranslated region. Biometals 25:319–335

Weaver BP, Dufner-Beattie J, Kambe T et al (2007) Novel zinc-responsive post-transcriptional mechanisms reciprocally regulate expression of the mouse Slc39a4 and Slc39a5 zinc transporters (Zip4 and Zip5). Biol Chem 388:1301–1312

Weaver BP, Zhang Y, Hiscox S et al (2010) Zip4 (Slc39a4) expression is activated in hepatocellular carcinomas and functions to repress apoptosis, enhance cell cycle and increase migration. PLoS One 5:e13158

Wessels I, Maywald M, Rink L (2017) Zinc as a gatekeeper of immune function. Nutrients 9:1286

Ya R, Que EL, O'Halloran TV et al (2014) Zinc as a key meiotic cell-cycle regulator in the mammalian oocyte. In: Fukada T, Kambe T (eds) Zinc signals in cellular functions and disorders. Springer, Tokyo, pp 315–333

Yanatori I, Yasui Y, Tabuchi M et al (2014) Chaperone protein involved in transmembrane transport of iron. Biochem J 462:25–37

Yanatori I, Richardson DR, Imada K et al (2016) Iron export through the transporter ferroportin 1 is modulated by the iron chaperone PCBP2. J Biol Chem 291:17303–17318

Yang J, Zhang Z, Zhang Y et al (2018) ZIP4 promotes muscle wasting and cachexia in mice with orthotopic pancreatic tumors by stimulating RAB27B-regulated release of extracellular vesicles from cancer cells. Gastroenterology 156:722–734.e6. https://doi.org/10.1053/j.gastro.2018.10.026

Yasuda H, Tsutsui T (2016) Infants and elderlies are susceptible to zinc deficiency. Sci Rep 6:21850

Yu YY, Kirschke CP, Huang L (2007) Immunohistochemical analysis of ZnT1, 4, 5, 6, and 7 in the mouse gastrointestinal tract. J Histochem Cytochem 55:223–234

Zhang Y, Yang J, Cui X et al (2013) A novel epigenetic CREB-miR-373 axis mediates ZIP4-induced pancreatic cancer growth. EMBO Mol Med 5:1322–1334

Zhang T, Sui D, Hu J (2016) Structural insights of ZIP4 extracellular domain critical for optimal zinc transport. Nat Commun 7:11979

Zhang T, Liu J, Fellner M et al (2017) Crystal structures of a ZIP zinc transporter reveal a binuclear metal center in the transport pathway. Sci Adv 3:e1700344

Chapter 12
Properties, Structure, and Function of the Solute Carrier 26 Family of Anion Transporters

Boris M. Baranovski, Moran Fremder, and Ehud Ohana

Abstract The SLC26 family of anion transporters consists of ten members that display remarkable functional and substrate diversity. Mutations in several members of the family have been identified as causing a variety of human diseases and mouse phenotype when deleted. The family drew the attention and strong interest of epithelial biologists with the identification of the first elusive luminal Cl^-/HCO_3^- exchanger, which turned to be the third member of the family SLC26A3. Later, a fairly quick progress revealed that members of the family transport all halides, NO_3^-, SO_4^{2-}, oxalate, and formate among others. Members of the family can be grouped into three subgroups based on substrate selectivity and transport mode, the SO_4^{2-} transporters SLC26A1 and SLC26A2; the anion exchangers the $2Cl^-/1HCO_3^-$ SLC26A3, the $1Cl^-/1HCO_3^-$ SLC26A4, and the $1Cl^-/2HCO_3^-$ SLC26A6; and the Cl^- channels SLC26A7, SLC26A9, and SLC26A11. Another major leap in our understanding of SLC26 activity emerged from the analysis of the bacterial SLC26 homolog crystal structure that revealed the transmembrane architecture of the protein. This chapter discusses structural features, transport properties, and regulation of the transporters that are essential to understand their functions and roles in human diseases.

Keywords SLC26 transporters · Anions · Exchangers · Channels · Function · Regulation · Disease

---

B. M. Baranovski · M. Fremder · E. Ohana (✉)
Faculty of Health Sciences, Department of Clinical Biochemistry and Pharmacology, Ben Gurion University of the Negev, Beer Sheva, Israel
e-mail: ohanaeh@bgu.ac.il

K. L. Hamilton, D. C. Devor (eds.), *Studies of Epithelial Transporters and Ion Channels*, Physiology in Health and Disease,
https://doi.org/10.1007/978-3-030-55454-5_12

12.1 General Features of the SLC26 Transporters

This chapter discusses the properties and function of the anion transporting SLC26 family. Several SLC26 transporters are associated with human disease or mouse phenotype with implication to human disease. Members of the family transport diverse monovalent and polyvalent anions including Cl^-, HCO_3^-, I^-, SO_4^{2-}, oxalate, and formate. They display equally diverse modes of transport, with several functioning as electroneutral exchangers, electrogenic exchangers or bona fide ion channels, and even transporters that can function as both electrogenic exchangers and ion channels, depending on the substrate. As such, the SLC26 transporters are involved in various cellular functions in many cell types. In epithelia, their best documented and understood function is fluid and electrolyte secretion (Dorwart et al. 2008; Lee et al. 2012b; Park et al. 2012). In addition, several members of the SLC26 family play major roles in mediating and regulating transepithelial transport of metabolites such as oxalate, succinate, and citrate (Ohana et al. 2013b; Khamaysi et al. 2019).

Fluid and electrolyte secretion is the most crucial function of most epithelia and is driven by Cl^- absorption (or secretion). Another critical function of Cl^- transport is HCO_3^- secretion (or absorption). Together, the two functions control the volume, electrolyte composition, and pH of the secreted fluids in a variety of epithelia discussed in this book. A special example in which HCO_3^- secretion is of paramount importance is secretory glands, as discussed in detail by Blouquit-Laye and Chinet (2007) and Lee et al. (2012b). In general, vectorial Cl^- absorption and HCO_3^- secretion is mediated by HCO_3^- entry at the basolateral membrane, which is principally mediated by the ubiquitous Na^+-HCO_3^- cotransporter, NBCe1-B. Basolateral HCO_3^- influx by the Cl^-/HCO_3^- exchanger, AE2, coupled with the activity of the Na^+-H^+ exchanger, NHE1, mediates a small portion of transcellular HCO_3^- transport in secretory glands (Melvin et al. 2005; Steward et al. 2005; Zhao et al. 1994; Bachmann et al. 2006). However, HCO_3^- loading by AE2 appears to be equally important as NBCe1-B in other epithelia, like the airway serous glands (Huang et al. 2012; Shan et al. 2012). HCO_3^- then exits the luminal membrane by the combined action of the cystic fibrosis transmembrane conductance regulator (CFTR) and members of the SLC26 transporters, most commonly the $1Cl^-/2HCO_3^-$ exchanger SLC26A6 (Dorwart et al. 2008; Ko et al. 2004; Lee et al. 2012b). Accordingly, many SLC26 transporters are expressed in the luminal membrane and co-localize with CFTR.

The SLC26 transporter family is coded by 11 genes with SLC26A10 being a pseudogene in humans (Dorwart et al. 2008; Park et al. 2012). Further analysis of the SulP clad can be found in Shelden et al. (2010) and Compton et al. (2014). The SLC26 transporter family was established with the identification of SLC26A1 as a liver SO_4^{2-} transporter (Bissig et al. 1994). The second member to be identified was the SO_4^{2-} transporter SLC26A2 as the protein mutated in the human disease diastrophic dysplasia (DTDST; Hastbacka et al. 1994). The family grabbed the interest of epithelial biologists with the discovery that congenital Cl^- diarrhea is caused by mutations in SLC26A3 (Hoglund et al. 1996) and the subsequent demonstration that SLC26A3 is expressed in the luminal membrane of colonic epithelium and functions as a Cl^- and HCO_3^- transporter (Moseley et al. 1999). This was the first identification of the elusive luminal membrane transporter that mediates Cl^-/HCO_3^- exchange. The function of SLC26A3 as a coupled Cl^-/HCO_3^- exchanger was then described by expressing the protein in HEK cells (Melvin et al. 1999). SLC26A4 was then identified as the transporter mutated in Pendred syndrome (Everett et al. 1997) and was later shown to function as a $Cl^-/HCO_3^-/I^-$ exchanger (Shcheynikov et al. 2008). SLC26A5 (Prestin) was identified as the protein mediating electromotility of outer hair cells (Zheng et al. 2000). Other members of the family were identified by database searches (Dorwart et al. 2008).

12.1.1 Structure of the SLC26 Transporters

The elegant study of Geertsma et al. provided, for the first time, detailed structural information of the SLC26 family by solving and analyzing the crystal structure of the bacterial homolog, SLC26Dg, transmembrane domain (Geertsma et al. 2015) (Figs. 12.1 and 12.2).

The crystal structure revealed two inverted repeats of seven transmembrane segments in each monomer with high resemblance to the UraA transporter structure (Geertsma et al. 2015). In the SLC26Dg crystal structure, each monomer consists of two major domains—the core domain and the gate domain (Fig. 12.1b). Several residues on the core and gate domains are spatially oriented to form a substrate

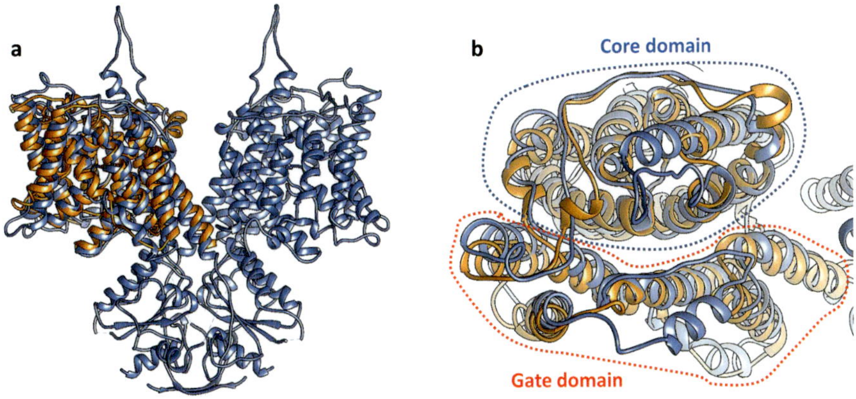

Fig. 12.1 The structures of SLC26A9 and SLC26Dg transmembrane domains are highly identical. The crystal structures of the murine SLC26A9 (PDB_ID: 6RTC; blue) and the bacterial SLC26Dg (PDB_ID: 5IOF; orange) were aligned using the UCSF Chimera 1.1 software. The alignment reveals a highly identical transmembrane domain fold, as shown in (**a**) (side view) and (**b**) (top view). The two major domains, namely, the core domain and gate domain, are indicated in (**b**) (dash)

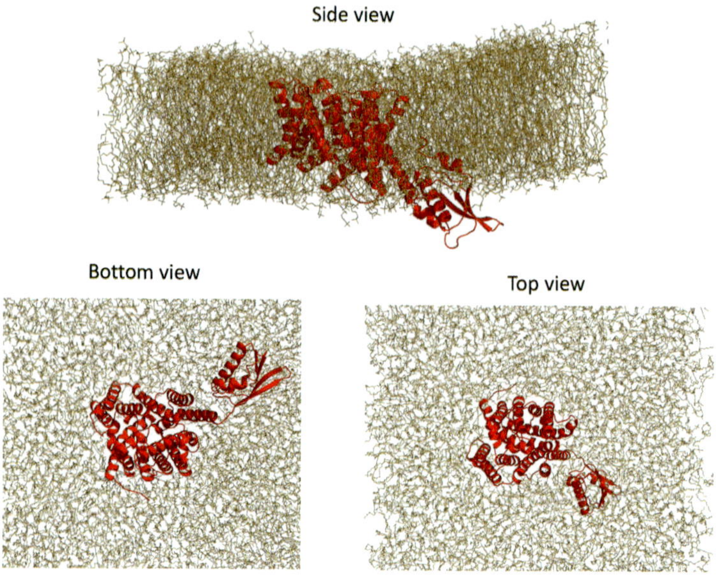

Fig. 12.2 The SLC26Dg orientation within the lipid bilayer. Molecular dynamics analysis of the SLC26Dg crystal structure (PDB_ID: 5IOF), reproduced from the MemProtMD website (Newport et al. 2019)

binding site in the center of each monomer. A putative SLC26 dimer is formed by interaction between the core domains. While the core domains are relatively stable, the transport function is achieved by movement of the gate domains, which alternately exposes the substrate binding site to the intracellular and extracellular environments. The structural similarity of SLC26 to UraA was also predicted by modeling and verified for SLC26A5 by cysteine accessibility scan (Gorbunov et al. 2014). Two conserved glutamate residues were suggested to affect ion permeation as part of a potential binding site (Gorbunov et al. 2014). Another conserved glutamate was suggested to determine the SLC26 mode of operation since mutations abolished SLC26A3 and SLC26A6 Cl^--coupled exchange function while retaining the uncoupled activity (Ohana et al. 2011). Yet, based on the structural analysis of SLC26Dg, the mechanism of the glutamate residue in controlling the SLC26 mode of operation is not clear. Gorbunov et al. suggested that this residue may be important for maintaining the stability of the core domain (Gorbunov et al. 2014). More recently, Chang et al. have determined the dimer structure of SLC26Dg (Chang et al. 2019). Their results indicated that monomeric SLC26Dg that lacks the STAS domain forms dimers when embedded into a membrane. The SLC26Dg dimer exhibited different membrane interface, compared to the SLC4 and SLC23 families, including different membrane dimer interface midpoint and small membrane interface. Based on comparison to other SLC26 proteins, it was suggested that the membrane dimer interface may be conserved throughout the SLC26 family. Notably, the authors showed that while the STAS domain contributes to dimer stabilization, it is not required for the dimerization of the protomers. Intriguingly, Walter et al. have recently reported a cryo-EM structure of the murine SLC26A9 (Walter et al. 2019) showing that the STAS domain directly mediates the interaction between SLC26A9 monomers (Fig. 12.1). The general role of STAS in the dimerization of all mammalian SLC26 transporters remains to be determined. However, it is conceivable that in different SLC26s, the STAS domain plays different functional roles. Indeed, SLC26A7 and SLC26A9 function as uncoupled Cl^- channels, while other members of the family are obligatory Cl^--dependent anion exchangers (Dorwart et al. 2008). Notably, the SLC26A9-STAS structure was not complete, since the intervening sequence and 44 residues of the C-terminus were deleted to achieve the structures (Walter et al. 2019).

Structural and domain analyses of bacterial SLC26 homologs established a cytoplasmic C-terminus STAS domain for the SLC26 transporter superfamily (Shelden et al. 2010; Compton et al. 2014; Ohana et al. 2011). The STAS domain was predicted based on similarity to the anti-sigma factor antagonist (Aravind and Koonin 2000), and the crystal structure of the SLC26A5 (Prestin) STAS domain was then solved (Pasqualetto et al. 2010). The SLC26Dg-based structure has a compact STAS domain, which is significantly different than the mammalian STAS, and its orientation suggests it is located within the inner leaflet of the membrane that is unlikely due to the hydrophilic nature SLC26Dg-STAS (Geertsma et al. 2015) (Fig. 12.2). The structure of the *Escherichia coli* (*E. coli*) YchM STAS domain in complex with acyl carrier protein (Babu et al. 2010) and of Rv1739c from *Mycobacterium tuberculosis* (Sharma et al. 2011) is also available. Therefore, unlike the relatively conserved sequence and structure of SLC26 proteins TMD, the SLC26 STAS domains appear to be highly variable and intrinsically disordered.

12.1.2 Regulation of the SLC26 Transporters

An important feature of the STAS domain in the mammalian transporter is the presence of an unstructured and divergent variable loop between the α1 helix and β3 strand (named intervening sequence, IVS) that is different in size among members of the family (Dorwart et al. 2008). The IVS appears to mediate interaction with and possible regulation of the SLC26 transporters by calmodulin (Keller et al. 2014). Of note is the variable length of the IVS in SLC26 transporters and that it is almost missing in SLC26A2 (Dorwart et al. 2008), raising the possibility of isoform-specific regulation of the SLC26 transporters by calmodulin.

A major role of the STAS domain is mediating the regulatory interaction of the SLC26 transporters with neighboring proteins within plasma membrane microdomains. The best established and understood form of regulation is the mutual regulation of SLC26 transporters and the Cl^- channel, CFTR, that is mediated by interaction of the STAS domain with the CFTR R domain. This was first demonstrated for interaction of SLC26A3 and SLC26A6 STAS domains with the R domain (Ko et al. 2004). Since its discovery, the interaction and in some cases the mutual

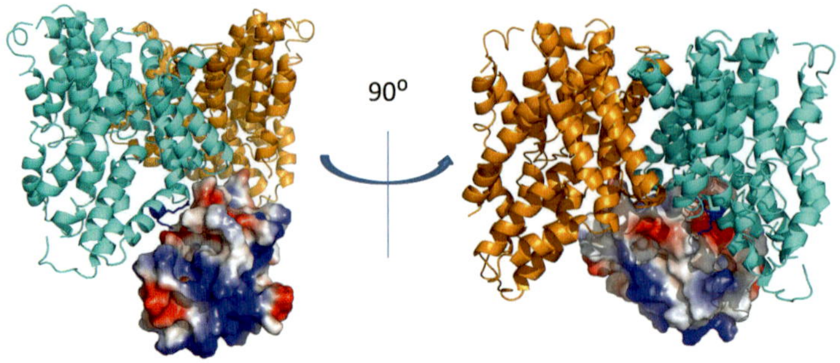

Fig. 12.3 Model of NaDC-1 and SLC26A6/STAS interaction sites. (**a** and **b**) Cartoon representation of the putative NaDC-1 (slc13a2) dimer (cyan and orange) and a surface representation of SLC26A6–STAS. The final image was generated using PyMOL software (Schrödinger, Germany)

regulation were shown for SLC26A4 (Shcheynikov et al. 2008; Garnett et al. 2011), SLC26A5 (Homma et al. 2010), SLC26A8 (Rode et al. 2012), and SLC26A9 (Bertrand et al. 2009), indicating a general mode of regulation of the SLC26 transporters by CFTR. The molecular details of interaction between the STAS and R domains have been examined biochemically and structurally. The R domain is intrinsically disordered and interacts with several partners (Bozoky et al. 2013a). The non-phosphorylated R domain avidly interacts with the nucleotide-binding domains (NBDs) of CFTR to inhibit channel activity, while phosphorylation of the R domain reduces its interaction with the NBDs and activates the channel (Baker et al. 2007; Bozoky et al. 2013b). A reciprocal situation is observed with the STAS domain, in which phosphorylation of the R domain increases its interaction with the STAS domain (Bozoky et al. 2013b; Ko et al. 2004) to activate the SLC26 transporters (Ko et al. 2004). Hence, the two transporters are mutually regulated by interaction of the R and STAS domains.

The SLC26 transporters are regulated by the IP_3 receptor binding protein released with IP_3 (IRBIT) (Ando et al. 2003, 2006) and the WNK/SPAK kinases (Park et al. 2013; Yang et al. 2009, 2011a). The WNK/SPAK pathway reduces surface expression of the transporters, while IRBIT recruits protein phosphatase 1 (PP1) to dephosphorylate the transporters, increase their surface expression, and activate them (Park et al. 2013; Yang et al. 2009, 2011a), probably by removal of autoinhibition (Hong et al. 2013). Moreover, IRBIT mediates the synergistic activation of the transporters by the cAMP and Ca^{2+} signaling pathways (Park et al. 2013; Ahuja et al. 2014). Recently, the molecular mechanism by which IRBIT recruits phosphatases and kinases, namely, PP1, SPAK, calcineurin, and calmodulin-kinase II, was reported (Vachel et al. 2018). This study revealed how IRBIT modulates the function of NBCe1-B that controls fluid and HCO_3^- secretion.

Other interactions of the STAS domain include interaction with the dicarboxylate transporter, NaDC-1 (Ohana et al. 2013a) (Fig. 12.3), and with the microtubule-associated protein MAP1S (Bai et al. 2010). The Slc26a6 STAS domain interacts with an NaDC-1 region that encompasses residues 45–118. More recently, the residues on both NaDC-1 and SLC26A6-STAS domain, which mediate protein binding, were identified. Specifically, a conserved positive residue (K107) on NaDC-1 was shown to play a role in the regulation of NaDC-1 and mediating interaction with a negative residue (E613) on SLC26A6-STAS (Khamaysi et al. 2019). Importantly, mutations which impair the charge of these residues abolish NaDC-1/SLC26A6 interaction and regulation. Based on the recently resolved crystal structure of a bacterial homolog of NaDC-1, vcINDY, the positively charged NaDC-1 residues are found on the H4c domain, which partially faces the intracellular domain and also interacts with the inner leaflet of the membrane bilayer. The STAS domain interacts with this region to stimulate SLC26A6 activity and at the same time strongly inhibit NaDC-1 activity (Fig. 12.3). This interaction operates in the kidney, intestine, and likely other tissues expressing the two proteins to regulate urinary or luminal citrate and succinate levels. While citrate protects against kidney stone formation (Ohana et al. 2013a), succinate regulates blood pressure. Numerous clinical studies suggest that elevated blood pressure increases the risk of developing kidney stones, yet the underlying mechanism was not known. Intriguingly, SLC26A6$^{-/-}$ mice form calcium oxalate kidney stones and are also hypertensive due to elevated transepithelial succinate/citrate absorption (Jiang et al. 2006; Khamaysi et al. 2019; Ohana et al. 2013a). This suggests that regulation of succinate/citrate homeostasis by SLC26 transporters may be the molecular mechanism underlying the association between kidney stones and hypertension.

Interaction of MAP1S with the STAS domain of SLC26A5 increases SLC26A5 surface expression and thus activity at the plasma membrane (Bai et al. 2010). Since MAP1S is ubiquitously expressed and interacts with the conserved STAS domain, it likely affects the activity of other SLC26 transporters. Interestingly, MAP1S appears to coordinate several components of the autophagic pathways to control autophagosome biogenesis and degradation (Xie et al. 2011). It should be of interest to determine if there is any relationship between the roles of MAP1S in the regulation of SLC26 transporters and autophagy and whether the two MAP1S roles are related.

The interactions of the STAS domain with other proteins described above are only the few known. It is highly plausible that the SLC26 transporter STAS domain interacts with other proteins to mediate additional modes of transporter regulation. In addition, most SLC26 transporters have C-terminus PDZ ligands that mediate their recruitment and assembly into complexes. The role of the PDZ ligands of several transporters in interaction with PDZ domains possessing scaffolds has been described, including their role in CFTR-SLC26A3 and CFTR-SLC26A6 interactions (Ko et al. 2004; Lamprecht and Seidler 2006; Lee et al. 2012a). Again, this increases the potential for interaction of the SLC26 transporters with other proteins that may involve domains other than the STAS domain.

12.1.3 *Cl^- Sensing by SLC26 Transporters*

Cl^- is the principal anion for all vertebrate cells. Several previous reports suggested that Cl^- is an essential signaling anion which regulates pivotal enzymes via binding to a specific regulatory site. Furthermore, chloride was shown to regulate several proteins, for example, the WNK1 kinase (Piala et al. 2014) and the NBC transporter (Shcheynikov et al. 2015). Moreover, SLC26A2 is allosterically regulated by extracellular chloride which lowers the affinity to $SO4^{2-}$ (Ohana et al. 2012). In addition, a Cl^- binding site was identified in the crystal structure of SLC26A5 which is regulated by intracellular Cl^- (Lolli et al. 2016). Therefore, SLC26 transporters appear to control the chloride-dependent transport by sensing Cl^- concentration in their environment. This has significant physiological implications since Cl^- concentrations may change considerably between cells and in response to external stimuli, as occurs, for example, in exocrine epithelia (Lee et al. 2012b).

12.2 Transport Properties of the SLC26 Transporters

Below, the SLC26 transporters are discussed in relation to their transported substrate and transport modes. Members of the first group are the SO_4^{2-} transporters, SLC26A1 and SLC26A2. The second group members are the exchangers, SLC26A3, SLC26A4, and SLC26A6. The third group members are the ion channels, SLC26A7, SLC26A9, and possibly SLC26A11 (Dorwart et al. 2008). The mammalian SLC26A5 functions as a motor protein that does not transport ions, although the invertebrate SLC26A5 functions as an exchanger, and their function is described elsewhere (Detro-Dassen et al. 2008; Schaechinger and Oliver 2007; Gorbunov et al. 2014). The transport mode and substrate specificity of SLC26A8 are not known at present, although mild Cl^-/SO_4^{2-}/oxalate transport has been reported (Lohi et al. 2002; Toure et al. 2001).

12.2.1 *The SO_4^{2-} Transporters*

SLC26A1 is expressed in the basolateral membrane and transports SO_4^{2-} (Karniski et al. 1998; Regeer et al. 2003) and oxalate (Xie et al. 2002), but not Cl^-, OH^-, or HCO_3^- (Karniski et al. 1998; Regeer et al. 2003) by an unknown transport mode. SO_4^{2-} transport by SLC26A1 is activated by extracellular halides and acidic pH (Xie et al. 2002), further indicating that SLC26A1 does not transport Cl^- and does not function as a SO_4^{2-}/Cl^- exchanger. The molecular feature of the interaction of SLC26A1 with halides has not been examined, but it may be mediated by a mechanism similar to that reported for activation of SLC26A2 by extracellular Cl^- (Ohana et al. 2012). The physiological role for SLC26A1 is not well understood, although some information became available from the SLC26A1$^{-/-}$ mice. That is, when fed with high oxalate, the SLC26A1$^{-/-}$ mice exhibit hyposulfatemia, hypersulfaturia, urolithiasis, and nephrocalcinosis (Dawson et al. 2010). These findings are consistent with a role of SLC26A1 in maintaining systemic SO_4^{2-} homeostasis and a role in oxalate clearance either by the intestine or the kidney.

SLC26A2 was first identified as the gene mutated in diastrophic dysplasia (Hastbacka et al. 1994). Subsequently, additional mutations associated with the disease were identified at different regions of the protein (Jackson et al. 2012). SLC26A2 is ubiquitous, but is expressed at high levels in cartilage, connective tissues, the adrenal cortex, and the apical membrane of several epithelia (Haila et al. 2001; Hastbacka et al. 1994; Chapman and Karniski 2010; Spyroglou et al. 2014). Expression in *Xenopus* oocytes showed that SLC26A2 functions as an electroneutral Cl^--regulated $SO_4^{2-}/Cl^-/OH^-$ exchanger (Ohana et al. 2012) and can also exchange Cl^- for oxalate (Ohana et al. 2012; Heneghan et al. 2010). The coupling of SO_4^{2-} transport to Cl^- or OH^- depends on the cellular Cl^- and pH gradients. For example, at acidic extracellular pH, SO_4^{2-} influx mostly coupled to OH^- efflux (or H^+ cotransport), while at high extracellular pH, the opposite holds. Interestingly, SLC26A2 activity is exquisitely regulated by extracellular Cl^-, which indicates that SLC26A2 has two distinct Cl^- binding sites, regulatory and catalytic. While the regulatory site remains elusive, at the catalytic site, Cl^- most likely interacts with a Phe residue found within a GFXXP motif (Ohana et al. 2012). When mutated to alanine, the affinity of SLC26A2 to Cl^- was attenuated, while SO_4^{2-} affinity was elevated further indicating that Cl^- competes with SO_4^{2-} for binding.

In bone, expression of SLC26A2 is not uniform but varies between growth plate zones (Park et al. 2014). Besides providing SO_4^{2-} for proteoglycan sulfation (Gualeni et al. 2013), SLC26A2 has multiple roles in chondrocyte biology, including cell proliferation, chondrocyte differentiation, and cell mass (Park et al. 2014). This is reflected in the multiple skeletal alterations observed in diastrophic dysplasia and in the mouse model of the disease (Forlino et al. 2005). The function of SLC26A2 may even be broader. That is, a recent study showed that SLC26A2 is expressed in the adrenal cortex and regulates aldosterone secretion (Spyroglou et al. 2014). A genome-wide association study found association between the aldosterone/renin ratio and a locus coding for SLC26A2. SLC26A2 suppresses aldosterone secretion as knockdown of SLC26A2 enhanced aldosterone secretion. These effects may not be direct, but rather through remodeling, steroidogenic, and Ca^{2+} signaling pathways in the knockdown cells (Spyroglou et al. 2014). Regulation of aldosterone levels and the potential effect on Ca^{2+} signaling can account for the multiple roles of SLC26A2 in the biology of chondrocytes and other cell types that express high levels of SLC26A2. This should be a very attractive topic to pursue in the future.

12.2.2 The Anion Exchangers

The SLC26 transporters that function as anion exchangers are SLC26A3, SLC26A4, and SLC26A6. The functions of these transporters were determined largely by expression in *Xenopus* oocytes, which is an ideal experimental system to characterize transport modes since it lacks many of the regulatory functions of mammalian cells and the large oocyte allows simultaneous monitoring of multiple ion concentrations and the membrane potential with the aid of ion-selective microelectrodes (Shcheynikov et al. 2006b). SLC26A3, SLC26A4, and SLC26A6 function as obligatory Cl^-/HCO_3^- exchangers, but with isoform-specific stoichiometry (Shcheynikov et al. 2006a, b, 2008). Examples of some of the transport modes mediated by each transporter are shown in Fig. 12.4.

Simultaneous measurement of Cl^-_{in} and pH_{in} in oocytes expressing SLC26A3 showed that SLC26A3 functions as a $2Cl^-/1HCO_3^-$ coupled exchanger (Fig. 12.4a) (Ko et al. 2002; Shcheynikov et al. 2006b; Ohana et al. 2009). However, whether SLC26A3 functions as a $2Cl^-/1HCO_3^-$ exchanger in vivo remains to be determined. Measurement of colonic SLC26A3-mediated Cl^- and HCO_3^- fluxes suggested that SLC26A3 may mediate electroneutral Cl^-/HCO_3^- exchange (Alper et al. 2011). Notably, the electrogenic properties of SLC26A3 are strongly supported by its function as a channel-like transporter when conducting NO_3^- and SCN^- (Ko et al. 2002; Shcheynikov et al. 2006b; Ohana et al. 2009). Figure 12.4b shows that SLC26A3 conducts several anions in the sequence and selectivity $SCN^- > NO_3^- > Cl^-$. A previous study identified a conserved glutamate in all SLC26 transporters that determines the properties of the gate by determining whether SLC26A3 or SLC26A6 (see below) functions as coupled or as uncoupled conductive transporter (Ohana et al. 2012).

SLC26A4 functions as an electroneutral $1Cl^-/1HCO_3^-$ exchanger (Fig. 12.4c, d) and Cl^-/I^- and Cl^-/HCO_3^- exchanger (Shcheynikov et al. 2008). SLC26A4 has particularly high affinity for I^-, and even in the presence of physiological intracellular and extracellular Cl^-, it transports I^- by mediating largely Cl^-/I^- exchange (Fig. 12.4d) and if HCO_3^- is present also I^-/HCO_3^- exchange (Shcheynikov et al. 2008; Pesce et al. 2012).

Slc26a6 functions as a $1Cl^-/2HCO_3^-$ exchanger, both in expression systems (Shcheynikov et al. 2006a, b, 2008; Xie et al. 2002; Ko et al. 2002) and in vivo (Song et al. 2012; Stewart et al. 2009; Kim et al. 2013). This is also illustrated in Fig. 12.4e. Measurement of the actual Cl^- and HCO_3^- fluxes resulted in an $1Cl^-/2HCO_3^-$ stoichiometry. Accordingly, removal of external Cl^- caused rapid and marked membrane hyperpolarization due to charge redistribution. All fluxes and the change in membrane potential were blocked by mutation of the conserved glutamate essential for the coupled transport (Ohana et al. 2009) and by 4,4-′-diisothiocyanatostilbene-2,2′-disulfonic acid (DIDS) (Shcheynikov et al. 2006b). The channel-like function was spared by the Glu/Ala mutation (Ohana et al. 2009). Notably, DIDS and its derivative H_2DIDS are nonselective inhibitors of anion transporters that affect Cl^- and HCO_3^- transporters including several members of

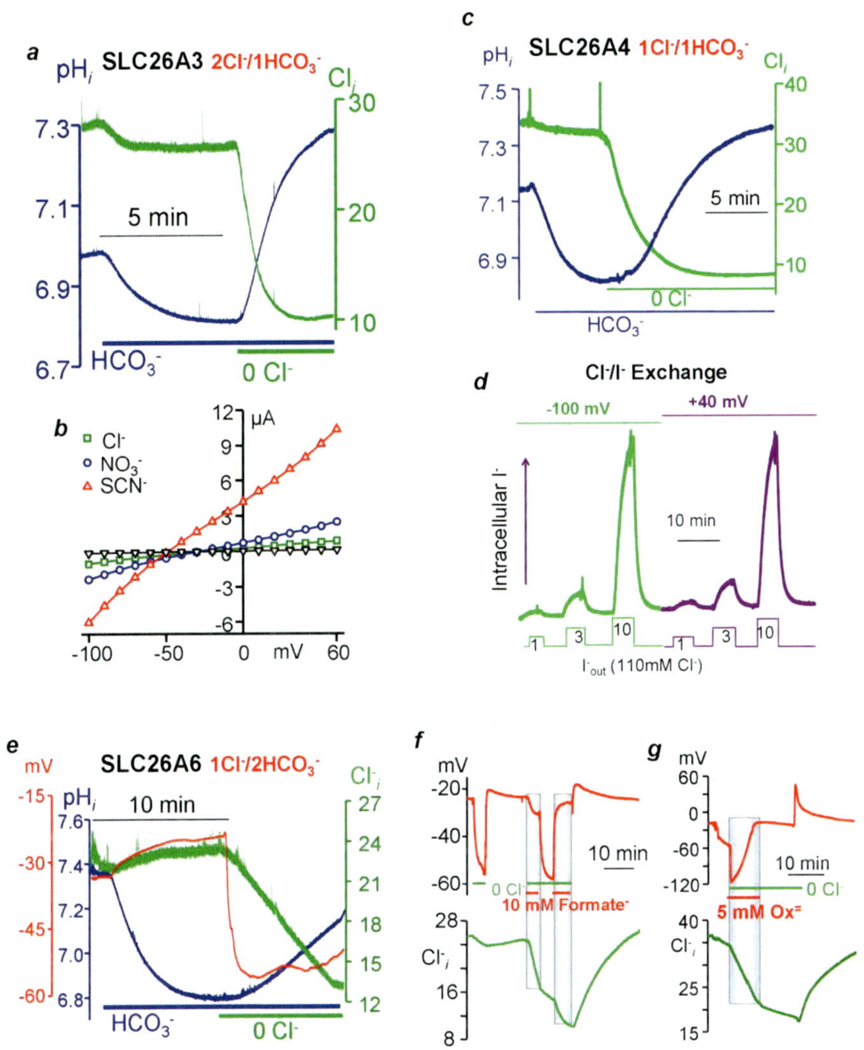

Fig. 12.4 Cl^-/HCO_3^- exchange and channel function of SLC26A3, SLC26A4, and SLC26A6. The results in (**a**), (**b**), (**e–g**) are reproduced from Ohana et al. (2011). (**a**) and (**b**) show the SLC26A3-mediated $2Cl^-/1HCO_3^-$ exchange (**a**) and the NO_3^- and SCN^- current (**b**). (**c**) and (**d**) show Cl^-/HCO_3^- exchange (**c**) and I^-/Cl^- exchange at two membrane potentials (**d**) by SLC26A4 (reproduced from Shcheynikov et al. (2008)). (**e**) shows the SLC26A6-mediated $1Cl^-/2HCO_3^-$ exchange and the associated change in membrane potential, and (**f**) and (**g**) show the SLC26A6-mediated formate (**f**) and oxalate (**g**) exchange with Cl^-

the SLC4 family (Choi 2012), of the SLC26 family (Alper and Sharma 2013), and of the ClC family (Matulef and Maduke 2005). In this respect, the activity of SLC26A3 is largely resistant to DIDS, which can be used as a tool to distinguish the contribution of the two exchangers when expressed in the same cells/tissues. Another study failed to report an electrogenic Cl^-/HCO_3^- exchange by SLC26A6 (Chernova et al. 2005).

Slc26a6 is the most versatile SLC26 transporter described so far. In addition to Cl^- and HCO_3^-, SLC26A6 also transports formate and oxalate in exchange for Cl^- (Fig. 12.4f, g, respectively; Ohana et al. 2009; Knauf et al. 2001; Jiang et al. 2002; Shcheynikov et al. 2006b). Interestingly, Cl^-/formate exchange is electroneutral and that may reflect fitting of only a single formate in the permeation pathway. Cl^-/oxalate exchange is electrogenic due to the oxalate charge, and, thus, SLC26A6 likely mediates an exchange of 1oxalate for 1Cl^-. Accordingly, SLC26A6 has a major role in intestinal and renal oxalate transport (Jiang et al. 2006) and in regulating dicarboxylate transport through the regulation of NaDC-1 (Ohana et al. 2013b). The combination of reduced SLC26A6-mediated intestinal oxalate excretion that increases urine oxalate load (Jiang et al. 2006) and inhibition of NaDC-1 by SLC26A6 that reduces urine citrate (Ohana et al. 2013b) results in Ca^{2+}-oxalate stone formation in the kidney.

Slc26a3 and SLC26A6 are present in the luminal membrane of several epithelia, although they do not always overlap (Haila et al. 2000; Hoglund et al. 1996, 2001; Lohi et al. 2002, 2003; Xia et al. 2013). A good example is the intestine, where SLC26A6 is more abundant in the proximal intestine, while SLC26A3 is expressed at high levels in the colon (Xia et al. 2013). Mutations in the *hSLC26A3* gene result in the disease congenital chloride diarrhea (Hoglund et al. 1996) with more than 60 known mutations and several founders (Wedenoja et al. 2011). Similarly, the SLC26A3$^{-/-}$ mouse exhibits Cl^--losing diarrhea and volume depletion (Schweinfest et al. 2006). As mentioned above, SLC26A3 interacts with and is activated by CFTR (Ko et al. 2004). In humans, impaired regulation of CFTR as a result of mutation in the SLC26A3-STAS domain leads to sperm motility defects and infertility (Wedenoja et al. 2017). Slc26a3 may also be regulated by IP_3 kinase through increased surface expression (Singla et al. 2010). In this respect, the surface expression of SLC26A3 (Park et al. 2010) and SLC26A6 (Park et al. 2013) is regulated by the WNK/SPAK pathway. It should be of interest to examine whether the regulation of the transporters surface expression by the two pathways is related or not.

An interesting alteration in SLC26A3 function is in the inflammatory disease ulcerative colitis, where mRNA (Yang et al. 1998) and protein (Farkas et al. 2011; Xiao et al. 2012) are reduced. This may be related to the reduced levels of SLC26A3 by inflammatory mediators (Malakooti et al. 2011). Another link of SLC26A3 to inflammatory diseases is the association of polymorphism in SLC26A3 with ulcerative colitis in Japanese patients (Asano et al. 2009) and with Crohn's disease in Korean patients (Yang et al. 2011b). A potential functional link between SLC26A3 and SLC26A6 function and inflammatory diseases is their central role in Cl^- absorption and HCO_3^- secretion in several epithelia, including the intestine (Jacob et al. 2002; Simpson et al. 2007; Xia et al. 2013), the pancreas (Wang et al. 2006; Song et al. 2012), and the salivary glands (Shcheynikov et al. 2008). Hence, previous work showed that intestinal HCO_3^- secretion is essential for mucosal integrity. Deletion of SLC26A3 eliminated mid-distal colonic HCO_3^- secretion and the associated fluid absorption, which led to a damaged, thin inner mucus layer and absence of stratified structures and consequently increased susceptibility to colitis (Xiao et al. 2014). Another option is that the tight regulation of the succinate transporter by SLC26A3 and SLC26A6 is impaired causing elevated succinate concentration (Khamaysi et al. 2019). Since succinate is emerging as a significant pro-inflammatory molecule (Tannahill et al. 2013), elevated succinate levels may lead to inflammatory processes.

SLC26A4 is expressed in the luminal membrane of several tissues and at high levels in the thyroid follicular cells, inner ear (Everett et al. 1997; Royaux et al. 2000), renal cortical collecting duct (Royaux et al. 2001; Soleimani et al. 2001), airway serous cells (Pedemonte et al. 2007), and the salivary gland ducts (Shcheynikov et al. 2008). The major role of SLC26A4 is transcellular I^- transport and Cl^-/HCO_3^- exchange. Mutations in SLC26A4 are associated with Pendred syndrome, which is characterized by impaired I^- organification in the thyroid and goiter (Campbell et al. 2001; Everett et al. 1997). This is likely due to impaired HCO_3^-/I^- exchange and limited I^- secretion into the follicular space.

Pendred syndrome is associated with hearing loss (Campbell et al. 2001; Coyle et al. 1996; Everett et al. 1997). This is likely due to impaired SLC26A4-mediated Cl^-/HCO_3^- exchange and altered endolymph pH. Indeed, SLC26A4 mediates HCO_3^- secretion by epithelial cells of the inner ear to alkalinize the pH of the endolymph. This function is absent in the SLC26A4$^{-/-}$ mouse with resultant reduced endocochlear potential due to reduced stria vascularis K^+ channel activity and accumulation of Na^+ (Wangemann et al. 2007; Li et al. 2013b). Interestingly, SLC26A4 is obligatory for development of normal hearing and balance only during early development. Hence, deletion of SLC26A4 at any time before P2 resulted in hearing and balance loss, but deletion of SLC26A4 at P2 or after did not result in deafness or loss of balance, which remained normal (Choi et al. 2011). Moreover, normal HCO_3^- secretion, endolymph pH, hearing, and balance can be rescued in the SLC26A4$^{-/-}$ mouse by expression of SLC26A4 exclusively in the endolymphatic sac (Li et al. 2013a).

Functions of SLC26A4 in other tissues include expression of SLC26A4 in the β intercalated cells of the cortical collecting duct where it has a significant role in renal acid-base homeostasis and the control of blood pressure (Wall 2005; Wall et al. 2004). Although basal blood pressure is normal in the SLC26A4$^{-/-}$ mice, their hypertensive response to treatment with angiotensin II (Verlander et al. 2011) or DOCA-salt is reduced (Verlander et al. 2003). In the absence of SLC26A4, the activity of ENaC is reduced (Pech et al. 2010), perhaps accounting for part of the reduced hypertension. Slc26a4 has a prominent role in the airway, in particular during inflammation such as in asthma and perhaps CF. Slc26a4 is induced by IL-13 stimulation, allergy, and rhinovirus infection, leading to increased thickness of airway surface liquid (Nakagami et al. 2008). Regulation of SLC26A4 expression by cytokines appears to be mediated by interferon-γ. Slc26a4 is upregulated in a mouse model of asthma and chronic lung disease to increase the levels of MUC5AC production and recruitment of neutrophils (Nakao et al. 2008). Whether the increased SLC26A4 is in response to the increased MUC5AC or is the cause remains to be determined using the SLC26A4$^{-/-}$ mice. If SLC26A4 is the cause of the MUC5AC increase, it can be a very promising target in asthma.

12.2.3 The Anion Channels

The two established SLC26 transporter ion channels are SLC26A7 and SLC26A9, with the possibility that SLC26A11 also functions as a channel. Examples of the currents mediated by these channels and their HCO_3^- permeability are shown in Fig. 12.5. Figure 12.5a shows measurements of SLC26A7 currents expressed in *Xenopus* oocytes and Fig. 12.5b the low HCO_3^- permeability of the channel. SLC26A7 functions as a Cl^- channel when expressed in *Xenopus* oocytes and in HEK cells and probably in parietal cells (Kim et al. 2005; Kosiek et al. 2007). In addition, SLC26A7 is expressed in Reissner's membrane epithelial cells where it functions as a Cl^- channel to participate in Cl^- transport by these cells (Kim et al. 2014). Slc26a7 shows low HCO_3^- permeability, and HCO_3^- does not appreciably affect channel function or compete with Cl^- (Fig. 12.5b). However, HCO_3^- increases the selectivity of SLC26A7 for Cl^- due to regulation of SLC26A7 by intracellular H^+ (Fig. 12.5b; Kim et al. 2005).

Slc26a7 was reported to be expressed in the basolateral membrane of parietal cells (Petrovic et al. 2003) where it may participate in Cl^- loading during acid secretion (Kosiek et al. 2007). On the other hand, SLC26A7 was localized to the subapical region in proximal tubule cells and on the basolateral surface of thick ascending limb cells in the kidney (Dudas et al. 2006), raising the possibility that SLC26A7 may have different roles in different cell types. The SLC26A7$^{-/-}$ mouse revealed several unexpected functions of SLC26A7. The mice developed distal renal tubular acidosis and gastric hypochlorhydria (Xu et al. 2009), suggesting a role in acid secretion by parietal cells and outer medullary collecting duct cells. The specific role of a basolateral Cl^- channel in acid secretion by the two cell types is not clear at present.

Following the initial demonstration that SLC26A9 expressed in *Xenopus* oocytes functions as a Cl^- channel that is regulated by the WNK kinases (Dorwart et al. 2007), it is now firmly established that SLC26A9 is a Cl^- channel with minimal or no OH^-/HCO_3^- permeability (Loriol et al. 2008; Bertrand et al. 2009; Chen et al. 2012). Figure 12.5c, d illustrates the Cl^- channel function of SLC26A9 and its low HCO_3^- permeability. However, further analysis reported that HCO_3^- slightly stimulates Cl^- channel activity by SLC26A9 (Loriol et al. 2008). Slc26a9 is expressed at high level in gastric gland surface epithelial cells (Xu et al. 2005) and in the parietal cells tubulovesicles (Xu et al. 2008). Based on loss of acid secretion in older SLC26A9$^{-/-}$ mice, it was suggested that SLC26A9 may be the elusive Cl^- channel mediating acid secretion. However, this needs to be further established as stimulated acid secretion in young SLC26A9$^{-/-}$ mice was normal and the achlorhydria developed with time as is damage to other cells in the gland, which may explain the reduction in the number of parietal cells and acid secretion with age (Xu et al. 2008).

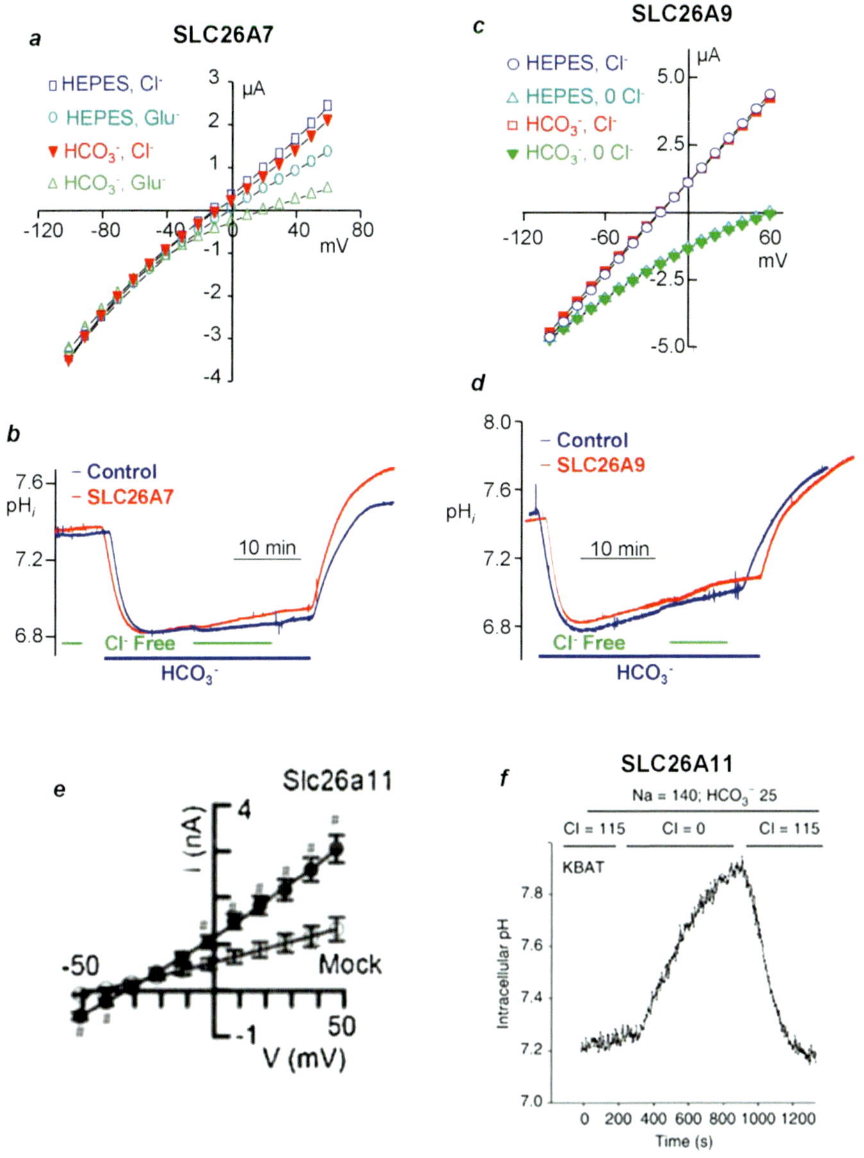

Fig. 12.5 Cl^- channel activity of SLC26A7 and SLC26A9 and SLC26A11. (**a**) shows the SLC26A7-mediated Cl^- current expressed in *Xenopus* oocytes, and (**b**) shows the minimal Cl^-/HCO_3^- exchange mediated by SLC26A7 (reproduced from Kim et al. (2005)). (**c**) shows the SLC26A9-mediated Cl^- current expressed in *Xenopus* oocytes, and (**d**) shows the minimal Cl^-/HCO_3^- exchange mediated by SLC26A9 (reproduced from Dorwart et al. (2007). (**e**) shows the SLC26A11-mediated Cl^- current expressed in *Xenopus* oocytes (reproduced from Rahmati et al. (2013)), and (**f**) shows the Cl^-/HCO_3^- exchange activity of SLC26A11 expressed in HEK cells (reproduced from Xu et al. (2011))

Another function of SLC26A9 is a potential role in the control of Cl^- secretion in the medullary collecting duct and consequently arterial blood pressure. The SLC26A9$^{-/-}$ mice are hypertensive, and the hypertension is exacerbated by high NaCl diet and water deprivation (Amlal et al. 2013).

Interest in SLC26A9 dramatically increased when a link between CF and SLC26A9 was established by a genome-wide association study with marked upregulation of SLC26A9 in cystic fibrosis (CF; Sun et al. 2012). SLC26A9 was found to be pleiotropic for meconium ileus and pancreatic damage (Li et al. 2014). SLC26A9 is also associated with CF-related diabetes (Blackman et al. 2013; Soave et al. 2014). Interestingly, two missense variants in *SLC26A9* were identified in heterozygote patients presenting with diffuse idiopathic bronchiectasis with one patient being Δ508F/+. The SLC26A9 mutations link to bronchiectasis eliminated Cl^- channel activity and, significantly, activation of CFTR by SLC26A9 (Bakouh et al. 2013). Interestingly, in a mouse model of CF, deletion of CFTR resulted in mild mortality. However, when SLC26A9 was deleted in these mice, mortality increased dramatically due to inhibition of intestinal secretion and likely the function of other organs affected in CF (Liu et al. 2014). Further evidence for a central role of SLC26A9 in the airway was obtained in SLC26A9$^{-/-}$ mice treated with IL-13 as an asthma model (Anagnostopoulou et al. 2012). In this study, IL-13-stimulated Cl^- secretion was absent in SLC26A9$^{-/-}$ mice that exhibited airway mucus obstruction. Moreover, the findings presented in this study indicated that an asthma-associated SLC26A9 polymorphism impaired protein expression. Strug et al. have shown that SLC26A9 modifies lung function only in individuals with CFTR mutations that affect channel gating and retain surface trafficking. Nevertheless, SLC26A9 failed to modify lung function in individuals homozygous for the F508del mutations which impairs CFTR function. Hence, airway function modification by SLC26A9 requires CFTR at the cell surface. Indeed, coexpression of SLC26A9 with the CFTR F508del mutation results in intracellular retention of SLC26A9. In the pancreas, SLC26A9 variants were reported to modify prenatal exocrine pancreatic damage in CF patients (Miller et al. 2015). These findings suggest that SLC26A9 is a potential candidate that may compensate for CFTR defects in multiple organs. The therapeutic potential of SLC26A9 has been extensively reviewed by Balazs et al. (Balazs and Mall 2018). Notably, in the future, the development compounds that potentiate SLC26A9 surface expression and function in a CFTR-independent manner may prove beneficial for CF treatment.

The function and physiological roles of SLC26A11 are not well understood. Only recently, some functional data became available. As shown in Fig. 12.5e, f, it was reported that when expressed in the plasma membrane, SLC26A11 can function as a Cl^- channel (Rahmati et al. 2013) and as HCO_3^- transporter (Xu et al. 2011). Yet, another study concluded that SLC26A11 has minimal HCO_3^- transport capacity (Stewart et al. 2011). The site(s) of SLC26A11 expression and its function are not certain. Immunolocalization in the kidney intercalated cells and in Purkinje neurons concluded plasma membrane localization (Rahmati et al. 2013; Xu et al. 2011), although the resolution of the images was not definitive to exclude mostly intracellular localization of the native SLC26A11. On the other hand, endogenous expression of SLC26A11 in the liver shows lysosomal localization similar to what is observed with transfected SLC26A11 (Della Valle et al. 2011; Schroder et al. 2010; Palmieri et al. 2011). Moreover, SLC26A11 expression is controlled by the master lysosomal biogenesis transcription factor TFEB (Sardiello et al. 2009; Palmieri et al. 2011). This raises the intriguing possibility that SLC26A11 is a lysosomal Cl^- channel that participates in the function of the lysosomes. It will be interesting to pursue this possibility.

12.3 Conclusion

This chapter illustrates the remarkable diversity of the SLC26 transporters and their multiple physiological roles. The solved structure of SLC26 homologs and multiple structural analyses of different SLC26 member STAS domains provide invaluable knowledge for future studies. Moreover, understanding regulation of the transporters is another fertile field of research. Finally, we are at the very beginning in appreciating the diverse physiological roles of the SLC26 transporters, let alone their tissue-specific function. The availability of mice knockout models and utilization of CRISPR technology for most of the SLC26 transporters should facilitate the latter.

Acknowledgments The authors wish to thank Dr. Shmuel Muallem for his critical assistance in writing this chapter. The authors' research is funded by the Israel Science Foundation grants No. 271/16 and 2164/16 to EO as well as the US-Israel Binational Science Foundation grant No. 2015003 to EO and SM.

References

Ahuja M, Jha A, Maleth J, Park S, Muallem S (2014) cAMP and Ca^{2+} signaling in secretory epithelia: crosstalk and synergism. Cell Calcium 55(6):385–393. https://doi.org/10.1016/j.ceca.2014.01.006

Alper SL, Sharma AK (2013) The SLC26 gene family of anion transporters and channels. Mol Asp Med 34(2–3):494–515. https://doi.org/10.1016/j.mam.2012.07.009

Alper SL, Stewart AK, Vandorpe DH, Clark JS, Horack RZ, Simpson JE, Walker NM, Clarke LL (2011) Native and recombinant Slc26a3 (downregulated in adenoma, Dra) do not exhibit properties of $2Cl^-/1HCO_3^-$ exchange. Am J Physiol Cell Physiol 300(2):C276–C286. https://doi.org/10.1152/ajpcell.00366.2010

Amlal H, Xu J, Barone S, Zahedi K, Soleimani M (2013) The chloride channel/transporter Slc26a9 regulates the systemic arterial pressure and renal chloride excretion. J Mol Med 91(5):561–572. https://doi.org/10.1007/s00109-012-0973-1

Anagnostopoulou P, Riederer B, Duerr J, Michel S, Binia A, Agrawal R, Liu X, Kalitzki K, Xiao F, Chen M, Schatterny J, Hartmann D, Thum T, Kabesch M, Soleimani M, Seidler U, Mall MA (2012) SLC26A9-mediated chloride secretion prevents mucus obstruction in airway inflammation. J Clin Invest 122(10):3629–3634. https://doi.org/10.1172/JCI60429

Ando H, Mizutani A, Matsu-ura T, Mikoshiba K (2003) IRBIT, a novel inositol 1,4,5-trisphosphate (IP3) receptor-binding protein, is released from the IP3 receptor upon IP3 binding to the receptor. J Biol Chem 278(12):10602–10612. https://doi.org/10.1074/jbc.M210119200

Ando H, Mizutani A, Kiefer H, Tsuzurugi D, Michikawa T, Mikoshiba K (2006) IRBIT suppresses IP3 receptor activity by competing with IP3 for the common binding site on the IP3 receptor. Mol Cell 22(6):795–806. https://doi.org/10.1016/j.molcel.2006.05.017

Aravind L, Koonin EV (2000) The STAS domain - a link between anion transporters and antisigma-factor antagonists. Curr Biol 10(2):R53–R55

Asano K, Matsushita T, Umeno J, Hosono N, Takahashi A, Kawaguchi T, Matsumoto T, Matsui T, Kakuta Y, Kinouchi Y, Shimosegawa T, Hosokawa M, Arimura Y, Shinomura Y, Kiyohara Y, Tsunoda T, Kamatani N, Iida M, Nakamura Y, Kubo M (2009) A genome-wide association study identifies three new susceptibility loci for ulcerative colitis in the Japanese population. Nat Genet 41(12):1325–1329. https://doi.org/10.1038/ng.482

Babu M, Greenblatt JF, Emili A, Strynadka NC, Reithmeier RA, Moraes TF (2010) Structure of a SLC26 anion transporter STAS domain in complex with acyl carrier protein: implications for *E. coli* YchM in fatty acid metabolism. Structure (London: 1993) 18(11):1450–1462. https://doi.org/10.1016/j.str.2010.08.015

Bachmann O, Reichelt D, Tuo B, Manns MP, Seidler U (2006) Carbachol increases Na^+-HCO_3^- cotransport activity in murine colonic crypts in a M3-, Ca^{2+}/calmodulin-, and PKC-dependent manner. Am J Physiol Gastrointest Liver Physiol 291(4):G650–G657. https://doi.org/10.1152/ajpgi.00376.2005

Bai JP, Surguchev A, Ogando Y, Song L, Bian S, Santos-Sacchi J, Navaratnam D (2010) Prestin surface expression and activity are augmented by interaction with MAP1S, a microtubule-associated protein. J Biol Chem 285(27):20834–20843. https://doi.org/10.1074/jbc.M110.117853

Baker JM, Hudson RP, Kanelis V, Choy WY, Thibodeau PH, Thomas PJ, Forman-Kay JD (2007) CFTR regulatory region interacts with NBD1 predominantly via multiple transient helices. Nat Struct Mol Biol 14(8):738–745. https://doi.org/10.1038/nsmb1278

Bakouh N, Bienvenu T, Thomas A, Ehrenfeld J, Liote H, Roussel D, Duquesnoy P, Farman N, Viel M, Cherif-Zahar B, Amselem S, Taam RA, Edelman A, Planelles G, Sermet-Gaudelus I (2013) Characterization of SLC26A9 in patients with CF-like lung disease. Hum Mutat 34(10):1404–1414. https://doi.org/10.1002/humu.22382

Balazs A, Mall MA (2018) Role of the SLC26A9 chloride channel as disease modifier and potential therapeutic target in cystic fibrosis. Front Pharmacol 9:1112. https://doi.org/10.3389/fphar.2018.01112

Bertrand CA, Zhang R, Pilewski JM, Frizzell RA (2009) SLC26A9 is a constitutively active, CFTR-regulated anion conductance in human bronchial epithelia. J Gen Physiol 133(4):421–438. https://doi.org/10.1085/jgp.200810097

Bissig M, Hagenbuch B, Stieger B, Koller T, Meier PJ (1994) Functional expression cloning of the canalicular sulfate transport system of rat hepatocytes. J Biol Chem 269(4):3017–3021

Blackman SM, Commander CW, Watson C, Arcara KM, Strug LJ, Stonebraker JR, Wright FA, Rommens JM, Sun L, Pace RG, Norris SA, Durie PR, Drumm ML, Knowles MR, Cutting GR (2013) Genetic modifiers of cystic fibrosis-related diabetes. Diabetes 62(10):3627–3635. https://doi.org/10.2337/db13-0510

Blouquit-Laye S, Chinet T (2007) Ion and liquid transport across the bronchiolar epithelium. Respir Physiol Neurobiol 159(3):278–282. https://doi.org/10.1016/j.resp.2007.03.007

Bozoky Z, Krzeminski M, Chong PA, Forman-Kay JD (2013a) Structural changes of CFTR R region upon phosphorylation: a plastic platform for intramolecular and intermolecular interactions. FEBS J 280(18):4407–4416. https://doi.org/10.1111/febs.12422

Bozoky Z, Krzeminski M, Muhandiram R, Birtley JR, Al-Zahrani A, Thomas PJ, Frizzell RA, Ford RC, Forman-Kay JD (2013b) Regulatory R region of the CFTR chloride channel is a dynamic integrator of phospho-dependent intra- and intermolecular interactions. Proc Natl Acad Sci USA 110(47):E4427–E4436. https://doi.org/10.1073/pnas.1315104110

Campbell C, Cucci RA, Prasad S, Green GE, Edeal JB, Galer CE, Karniski LP, Sheffield VC, Smith RJ (2001) Pendred syndrome, DFNB4, and PDS/SLC26A4 identification of eight novel mutations and possible genotype-phenotype correlations. Hum Mutat 17(5):403–411. https://doi.org/10.1002/humu.1116

Chang YN, Jaumann EA, Reichel K, Hartmann J, Oliver D, Hummer G, Joseph B, Geertsma ER (2019) Structural basis for functional interactions in dimers of SLC26 transporters. Nat Commun 10(1):2032. https://doi.org/10.1038/s41467-019-10001-w

Chapman JM, Karniski LP (2010) Protein localization of SLC26A2 (DTDST) in rat kidney. Histochem Cell Biol 133(5):541–547. https://doi.org/10.1007/s00418-010-0694-x

Chen AP, Chang MH, Romero MF (2012) Functional analysis of nonsynonymous single nucleotide polymorphisms in human SLC26A9. Hum Mutat 33(8):1275–1284. https://doi.org/10.1002/humu.22107

Chernova MN, Jiang L, Friedman DJ, Darman RB, Lohi H, Kere J, Vandorpe DH, Alper SL (2005) Functional comparison of mouse slc26a6 anion exchanger with human SLC26A6 polypeptide variants: differences in anion selectivity, regulation, and electrogenicity. J Biol Chem 280(9):8564–8580. https://doi.org/10.1074/jbc.M411703200

Choi I (2012) SLC4A transporters. Curr Top Membr 70:77–103. https://doi.org/10.1016/B978-0-12-394316-3.00003-X

Choi BY, Kim HM, Ito T, Lee KY, Li X, Monahan K, Wen Y, Wilson E, Kurima K, Saunders TL, Petralia RS, Wangemann P, Friedman TB, Griffith AJ (2011) Mouse model of enlarged vestibular aqueducts defines temporal requirement of Slc26a4 expression for hearing acquisition. J Clin Invest 121(11):4516–4525. https://doi.org/10.1172/JCI59353

Compton EL, Page K, Findlay HE, Haertlein M, Moulin M, Zachariae U, Norman DG, Gabel F, Javelle A (2014) Conserved structure and domain organisation amongst bacterial Slc26 transporters. Biochem J 463:297. https://doi.org/10.1042/BJ20130619

Coyle B, Coffey R, Armour JA, Gausden E, Hochberg Z, Grossman A, Britton K, Pembrey M, Reardon W, Trembath R (1996) Pendred syndrome (goitre and sensorineural hearing loss) maps to chromosome 7 in the region containing the nonsyndromic deafness gene DFNB4. Nat Genet 12(4):421–423. https://doi.org/10.1038/ng0496-421

Dawson PA, Russell CS, Lee S, McLeay SC, van Dongen JM, Cowley DM, Clarke LA, Markovich D (2010) Urolithiasis and hepatotoxicity are linked to the anion transporter Sat1 in mice. J Clin Invest 120(3):706–712. https://doi.org/10.1172/JCI31474

Della Valle MC, Sleat DE, Zheng H, Moore DF, Jadot M, Lobel P (2011) Classification of subcellular location by comparative proteomic analysis of native and density-shifted lysosomes. Mol Cell Proteomics 10(4):M110.006403. https://doi.org/10.1074/mcp.M110.006403

Detro-Dassen S, Schanzler M, Lauks H, Martin I, zu Berstenhorst SM, Nothmann D, Torres-Salazar D, Hidalgo P, Schmalzing G, Fahlke C (2008) Conserved dimeric subunit stoichiometry of SLC26 multifunctional anion exchangers. J Biol Chem 283(7):4177–4188. https://doi.org/10.1074/jbc.M704924200

Dorwart MR, Shcheynikov N, Wang Y, Stippec S, Muallem S (2007) SLC26A9 is a Cl(−) channel regulated by the WNK kinases. J Physiol 584(Pt 1):333–345. https://doi.org/10.1113/jphysiol.2007.135855

Dorwart MR, Shcheynikov N, Yang D, Muallem S (2008) The solute carrier 26 family of proteins in epithelial ion transport. Physiology (Bethesda) 23:104–114. https://doi.org/10.1152/physiol.00037.2007

Dudas PL, Mentone S, Greineder CF, Biemesderfer D, Aronson PS (2006) Immunolocalization of anion transporter Slc26a7 in mouse kidney. Am J Physiol Renal Physiol 290(4):F937–F945. https://doi.org/10.1152/ajprenal.00197.2004

Everett LA, Glaser B, Beck JC, Idol JR, Buchs A, Heyman M, Adawi F, Hazani E, Nassir E, Baxevanis AD, Sheffield VC, Green ED (1997) Pendred syndrome is caused by mutations in a putative sulphate transporter gene (PDS). Nat Genet 17(4):411–422. https://doi.org/10.1038/ng1297-411

Farkas K, Yeruva S, Rakonczay Z Jr, Ludolph L, Molnar T, Nagy F, Szepes Z, Schnur A, Wittmann T, Hubricht J, Riederer B, Venglovecz V, Lazar G, Kiraly M, Zsembery A, Varga G, Seidler U, Hegyi P (2011) New therapeutic targets in ulcerative colitis: the importance of ion transporters in the human colon. Inflamm Bowel Dis 17(4):884–898. https://doi.org/10.1002/ibd.21432

Forlino A, Piazza R, Tiveron C, Della Torre S, Tatangelo L, Bonafe L, Gualeni B, Romano A, Pecora F, Superti-Furga A, Cetta G, Rossi A (2005) A diastrophic dysplasia sulfate transporter (SLC26A2) mutant mouse: morphological and biochemical characterization of the resulting chondrodysplasia phenotype. Hum Mol Genet 14(6):859–871. https://doi.org/10.1093/hmg/ddi079

Garnett JP, Hickman E, Burrows R, Hegyi P, Tiszlavicz L, Cuthbert AW, Fong P, Gray MA (2011) Novel role for pendrin in orchestrating bicarbonate secretion in cystic fibrosis transmembrane conductance regulator (CFTR)-expressing airway serous cells. J Biol Chem 286 (47):41069–41082. https://doi.org/10.1074/jbc.M111.266734

Geertsma ER, Chang YN, Shaik FR, Neldner Y, Pardon E, Steyaert J, Dutzler R (2015) Structure of a prokaryotic fumarate transporter reveals the architecture of the SLC26 family. Nat Struct Mol Biol 22(10):803–808. https://doi.org/10.1038/nsmb.3091

Gorbunov D, Sturlese M, Nies F, Kluge M, Bellanda M, Battistutta R, Oliver D (2014) Molecular architecture and the structural basis for anion interaction in prestin and SLC26 transporters. Nat Commun 5:3622. https://doi.org/10.1038/ncomms4622

Gualeni B, de Vernejoul MC, Marty-Morieux C, De Leonardis F, Franchi M, Monti L, Forlino A, Houillier P, Rossi A, Geoffroy V (2013) Alteration of proteoglycan sulfation affects bone growth and remodeling. Bone 54(1):83–91. https://doi.org/10.1016/j.bone.2013.01.036

Haila S, Saarialho-Kere U, Karjalainen-Lindsberg ML, Lohi H, Airola K, Holmberg C, Hastbacka J, Kere J, Hoglund P (2000) The congenital chloride diarrhea gene is expressed in seminal vesicle, sweat gland, inflammatory colon epithelium, and in some dysplastic colon cells. Histochem Cell Biol 113(4):279–286

Haila S, Hastbacka J, Bohling T, Karjalainen-Lindsberg ML, Kere J, Saarialho-Kere U (2001) SLC26A2 (diastrophic dysplasia sulfate transporter) is expressed in developing and mature cartilage but also in other tissues and cell types. J Histochem Cytochem 49(8):973–982. https://doi.org/10.1177/002215540104900805

Hastbacka J, de la Chapelle A, Mahtani MM, Clines G, Reeve-Daly MP, Daly M, Hamilton BA, Kusumi K, Trivedi B, Weaver A et al (1994) The diastrophic dysplasia gene encodes a novel sulfate transporter: positional cloning by fine-structure linkage disequilibrium mapping. Cell 78 (6):1073–1087

Heneghan JF, Akhavein A, Salas MJ, Shmukler BE, Karniski LP, Vandorpe DH, Alper SL (2010) Regulated transport of sulfate and oxalate by SLC26A2/DTDST. Am J Physiol Cell Physiol 298 (6):C1363–C1375. https://doi.org/10.1152/ajpcell.00004.2010

Hoglund P, Haila S, Socha J, Tomaszewski L, Saarialho-Kere U, Karjalainen-Lindsberg ML, Airola K, Holmberg C, de la Chapelle A, Kere J (1996) Mutations of the down-regulated in adenoma (DRA) gene cause congenital chloride diarrhoea. Nat Genet 14(3):316–319. https://doi.org/10.1038/ng1196-316

Hoglund P, Sormaala M, Haila S, Socha J, Rajaram U, Scheurlen W, Sinaasappel M, de Jonge H, Holmberg C, Yoshikawa H, Kere J (2001) Identification of seven novel mutations including the first two genomic rearrangements in SLC26A3 mutated in congenital chloride diarrhea. Hum Mutat 18(3):233–242. https://doi.org/10.1002/humu.1179

Homma K, Miller KK, Anderson CT, Sengupta S, Du GG, Aguinaga S, Cheatham M, Dallos P, Zheng J (2010) Interaction between CFTR and prestin (SLC26A5). Biochim Biophys Acta 1798 (6):1029–1040. https://doi.org/10.1016/j.bbamem.2010.02.001

Hong JH, Yang D, Shcheynikov N, Ohana E, Shin DM, Muallem S (2013) Convergence of IRBIT, phosphatidylinositol (4,5) bisphosphate, and WNK/SPAK kinases in regulation of the Na^+-HCO_3^- cotransporters family. Proc Natl Acad Sci USA 110(10):4105–4110. https://doi.org/10.1073/pnas.1221410110

Huang J, Shan J, Kim D, Liao J, Evagelidis A, Alper SL, Hanrahan JW (2012) Basolateral chloride loading by the anion exchanger type 2: role in fluid secretion by the human airway epithelial cell line Calu-3. J Physiol 590(21):5299–5316. https://doi.org/10.1113/jphysiol.2012.236919

Jackson GC, Mittaz-Crettol L, Taylor JA, Mortier GR, Spranger J, Zabel B, Le Merrer M, Cormier-Daire V, Hall CM, Offiah A, Wright MJ, Savarirayan R, Nishimura G, Ramsden SC, Elles R, Bonafe L, Superti-Furga A, Unger S, Zankl A, Briggs MD (2012) Pseudoachondroplasia and multiple epiphyseal dysplasia: a 7-year comprehensive analysis of the known disease genes identify novel and recurrent mutations and provides an accurate assessment of their relative contribution. Hum Mutat 33(1):144–157. https://doi.org/10.1002/Humu.21611

Jacob P, Rossmann H, Lamprecht G, Kretz A, Neff C, Lin-Wu E, Gregor M, Groneberg DA, Kere J, Seidler U (2002) Down-regulated in adenoma mediates apical Cl^-/HCO_3^- exchange in rabbit, rat, and human duodenum. Gastroenterology 122(3):709–724. https://doi.org/10.1053/gast.2002.31875

Jiang Z, Grichtchenko II, Boron WF, Aronson PS (2002) Specificity of anion exchange mediated by mouse Slc26a6. J Biol Chem 277(37):33963–33967. https://doi.org/10.1074/jbc.M202660200

Jiang Z, Asplin JR, Evan AP, Rajendran VM, Velazquez H, Nottoli TP, Binder HJ, Aronson PS (2006) Calcium oxalate urolithiasis in mice lacking anion transporter Slc26a6. Nat Genet 38 (4):474–478. https://doi.org/10.1038/ng1762

Karniski LP, Lotscher M, Fucentese M, Hilfiker H, Biber J, Murer H (1998) Immunolocalization of sat-1 sulfate/oxalate/bicarbonate anion exchanger in the rat kidney. Am J Phys 275(1):F79–F87. https://doi.org/10.1152/ajprenal.1998.275.1.F79

Keller JP, Homma K, Duan C, Zheng J, Cheatham MA, Dallos P (2014) Functional regulation of the SLC26-family protein prestin by calcium/calmodulin. J Neurosci 34(4):1325–1332. https://doi.org/10.1523/jneurosci.4020-13.2014

Khamaysi A, Anbtawee-Jomaa S, Fremder M, Eini-Rider H, Shimshilashvili L, Aharon S, Aizenshtein E, Shlomi T, Noguchi A, Springer D, Moe OW, Shcheynikov N, Muallem S, Ohana E (2019) Systemic succinate homeostasis and local succinate signaling affect blood pressure and modify risks for calcium oxalate lithogenesis. J Am Soc Nephrol 30:381. https://doi.org/10.1681/ASN.2018030277

Kim KH, Shcheynikov N, Wang Y, Muallem S (2005) SLC26A7 is a Cl^- channel regulated by intracellular pH. J Biol Chem 280(8):6463–6470. https://doi.org/10.1074/jbc.M409162200

Kim HJ, Myers R, Sihn CR, Rafizadeh S, Zhang XD (2013) Slc26a6 functions as an electrogenic Cl^-/HCO_3^- exchanger in cardiac myocytes. Cardiovasc Res 100(3):383–391. https://doi.org/10.1093/cvr/cvt195

Kim KX, Sanneman JD, Kim HM, Harbidge DG, Xu J, Soleimani M, Wangemann P, Marcus DC (2014) Slc26a7 chloride channel activity and localization in mouse Reissner's membrane epithelium. PLoS One 9(5):e97191. https://doi.org/10.1371/journal.pone.0097191

Knauf F, Yang CL, Thomson RB, Mentone SA, Giebisch G, Aronson PS (2001) Identification of a chloride-formate exchanger expressed on the brush border membrane of renal proximal tubule cells. Proc Natl Acad Sci USA 98(16):9425–9430. https://doi.org/10.1073/pnas.141241098

Ko SB, Shcheynikov N, Choi JY, Luo X, Ishibashi K, Thomas PJ, Kim JY, Kim KH, Lee MG, Naruse S, Muallem S (2002) A molecular mechanism for aberrant CFTR-dependent HCO_3^- transport in cystic fibrosis. EMBO J 21(21):5662–5672

Ko SB, Zeng W, Dorwart MR, Luo X, Kim KH, Millen L, Goto H, Naruse S, Soyombo A, Thomas PJ, Muallem S (2004) Gating of CFTR by the STAS domain of SLC26 transporters. Nat Cell Biol 6(4):343–350. https://doi.org/10.1038/ncb1115

Kosiek O, Busque SM, Foller M, Shcheynikov N, Kirchhoff P, Bleich M, Muallem S, Geibel JP (2007) SLC26A7 can function as a chloride-loading mechanism in parietal cells. Pflugers Arch 454(6):989–998. https://doi.org/10.1007/s00424-007-0254-y

Lamprecht G, Seidler U (2006) The emerging role of PDZ adapter proteins for regulation of intestinal ion transport. Am J Physiol Gastrointest Liver Physiol 291(5):G766–G777. https://doi.org/10.1152/ajpgi.00135.2006

Lee JH, Nam JH, Park J, Kang DW, Kim JY, Lee MG, Yoon JS (2012a) Regulation of SLC26A3 activity by NHERF4 PDZ-mediated interaction. Cell Signal 24(9):1821–1830. https://doi.org/10.1016/j.cellsig.2012.05.010

Lee MG, Ohana E, Park HW, Yang D, Muallem S (2012b) Molecular mechanism of pancreatic and salivary gland fluid and HCO_3^- secretion. Physiol Rev 92(1):39–74. https://doi.org/10.1152/physrev.00011.2011

Li X, Sanneman JD, Harbidge DG, Zhou F, Ito T, Nelson R, Picard N, Chambrey R, Eladari D, Miesner T, Griffith AJ, Marcus DC, Wangemann P (2013a) SLC26A4 targeted to the endolymphatic sac rescues hearing and balance in Slc26a4 mutant mice. PLoS Genet 9(7):e1003641. https://doi.org/10.1371/journal.pgen.1003641

Li X, Zhou F, Marcus DC, Wangemann P (2013b) Endolymphatic Na^+ and K^+ concentrations during cochlear growth and enlargement in mice lacking Slc26a4/pendrin. PLoS One 8(5): e65977. https://doi.org/10.1371/journal.pone.0065977

Li W, Soave D, Miller MR, Keenan K, Lin F, Gong J, Chiang T, Stephenson AL, Durie P, Rommens J, Sun L, Strug LJ (2014) Unraveling the complex genetic model for cystic fibrosis: pleiotropic effects of modifier genes on early cystic fibrosis-related morbidities. Hum Genet 133 (2):151–161. https://doi.org/10.1007/s00439-013-1363-7

Liu X, Li T, Riederer B, Lenzen H, Ludolph L, Yeruva S, Tuo B, Soleimani M, Seidler U (2014) Loss of Slc26a9 anion transporter alters intestinal electrolyte and HCO transport and reduces survival in CFTR-deficient mice. Pflugers Arch 467:1261. https://doi.org/10.1007/s00424-014-1543-x

Lohi H, Kujala M, Makela S, Lehtonen E, Kestila M, Saarialho-Kere U, Markovich D, Kere J (2002) Functional characterization of three novel tissue-specific anion exchangers SLC26A7, -A8, and -A9. J Biol Chem 277(16):14246–14254. https://doi.org/10.1074/jbc.M111802200

Lohi H, Lamprecht G, Markovich D, Heil A, Kujala M, Seidler U, Kere J (2003) Isoforms of SLC26A6 mediate anion transport and have functional PDZ interaction domains. Am J Physiol Cell Physiol 284(3):C769–C779. https://doi.org/10.1152/ajpcell.00270.2002

Lolli G, Pasqualetto E, Costanzi E, Bonetto G, Battistutta R (2016) The STAS domain of mammalian SLC26A5 prestin harbours an anion-binding site. Biochem J 473(4):365–370. https://doi.org/10.1042/BJ20151089

Loriol C, Dulong S, Avella M, Gabillat N, Boulukos K, Borgese F, Ehrenfeld J (2008) Characterization of SLC26A9, facilitation of Cl(−) transport by bicarbonate. Cell Physiol Biochem 22 (1–4):15–30. https://doi.org/10.1159/000149780

Malakooti J, Saksena S, Gill RK, Dudeja PK (2011) Transcriptional regulation of the intestinal luminal Na^+ and Cl^- transporters. Biochem J 435(2):313–325. https://doi.org/10.1042/BJ20102062

Matulef K, Maduke M (2005) Side-dependent inhibition of a prokaryotic ClC by DIDS. Biophys J 89(3):1721–1730. https://doi.org/10.1529/biophysj.105.066522

Melvin JE, Park K, Richardson L, Schultheis PJ, Shull GE (1999) Mouse down-regulated in adenoma (DRA) is an intestinal Cl^-/HCO_3^- exchanger and is up-regulated in colon of mice lacking the NHE3 Na^+/H^+ exchanger. J Biol Chem 274(32):22855–22861. https://doi.org/10.1074/jbc.274.32.22855

Melvin JE, Yule D, Shuttleworth T, Begenisich T (2005) Regulation of fluid and electrolyte secretion in salivary gland acinar cells. Annu Rev Physiol 67:445–469. https://doi.org/10.1146/annurev.physiol.67.041703.084745

Miller MR, Soave D, Li W, Gong J, Pace RG, Boelle PY, Cutting GR, Drumm ML, Knowles MR, Sun L, Rommens JM, Accurso F, Durie PR, Corvol H, Levy H, Sontag MK, Strug LJ (2015) Variants in solute carrier SLC26A9 modify prenatal exocrine pancreatic damage in cystic fibrosis. J Pediatr 166(5):1152–1157. e1156. https://doi.org/10.1016/j.jpeds.2015.01.044

Moseley RH, Hoglund P, Wu GD, Silberg DG, Haila S, de la Chapelle A, Holmberg C, Kere J (1999) Downregulated in adenoma gene encodes a chloride transporter defective in congenital chloride diarrhea. Am J Phys 276(1):G185–G192. https://doi.org/10.1152/ajpgi.1999.276.1.G185

Nakagami Y, Favoreto S Jr, Zhen G, Park SW, Nguyenvu LT, Kuperman DA, Dolganov GM, Huang X, Boushey HA, Avila PC, Erle DJ (2008) The epithelial anion transporter pendrin is induced by allergy and rhinovirus infection, regulates airway surface liquid, and increases airway reactivity and inflammation in an asthma model. J Immunol 181(3):2203–2210

Nakao I, Kanaji S, Ohta S, Matsushita H, Arima K, Yuyama N, Yamaya M, Nakayama K, Kubo H, Watanabe M, Sagara H, Sugiyama K, Tanaka H, Toda S, Hayashi H, Inoue H, Hoshino T, Shiraki A, Inoue M, Suzuki K, Aizawa H, Okinami S, Nagai H, Hasegawa M, Fukuda T, Green ED, Izuhara K (2008) Identification of pendrin as a common mediator for mucus production in bronchial asthma and chronic obstructive pulmonary disease. J Immunol 180(9):6262–6269

Newport TD, Sansom MSP, Stansfeld PJ (2019) The MemProtMD database: a resource for membrane-embedded protein structures and their lipid interactions. Nucleic Acids Res 47 (D1):D390–D397. https://doi.org/10.1093/nar/gky1047

Ohana E, Yang D, Shcheynikov N, Muallem S (2009) Diverse transport modes by the solute carrier 26 family of anion transporters. J Physiol 587(Pt 10):2179–2185. https://doi.org/10.1113/jphysiol.2008.164863

Ohana E, Shcheynikov N, Yang D, So I, Muallem S (2011) Determinants of coupled transport and uncoupled current by the electrogenic SLC26 transporters. J Gen Physiol 137(2):239–251. https://doi.org/10.1085/jgp.201010531

Ohana E, Shcheynikov N, Park M, Muallem S (2012) Solute carrier family 26 member a2 (Slc26a2) protein functions as an electroneutral SOFormula/OH-/Cl- exchanger regulated by extracellular Cl. J Biol Chem 287(7):5122–5132. https://doi.org/10.1074/jbc.M111.297192

Ohana E, Shcheynikov N, Moe OW, Muallem S (2013a) SLC26A6 and NaDC-1 transporters interact to regulate oxalate and citrate homeostasis. J Am Soc Nephrol 24(10):1617–1626. https://doi.org/10.1681/ASN.2013010080

Ohana E, Shcheynikov N, Moe OW, Muallem S (2013b) SLC26A6 and NaDC-1 transporters interact to regulate oxalate and citrate homeostasis. J Am Soc Nephrol 24:1617. https://doi.org/10.1681/ASN.2013010080

Palmieri M, Impey S, Kang H, di Ronza A, Pelz C, Sardiello M, Ballabio A (2011) Characterization of the CLEAR network reveals an integrated control of cellular clearance pathways. Hum Mol Genet 20(19):3852–3866. https://doi.org/10.1093/hmg/ddr306

Park HW, Nam JH, Kim JY, Namkung W, Yoon JS, Lee JS, Kim KS, Venglovecz V, Gray MA, Kim KH, Lee MG (2010) Dynamic regulation of CFTR bicarbonate permeability by [Cl-]i and its role in pancreatic bicarbonate secretion. Gastroenterology 139(2):620–631. https://doi.org/10.1053/j.gastro.2010.04.004

Park S, Hong JH, Ohana E, Muallem S (2012) The WNK/SPAK and IRBIT/PP1 pathways in epithelial fluid and electrolyte transport. Physiology 27(5):291–299. https://doi.org/10.1152/physiol.00028.2012

Park S, Shcheynikov N, Hong JH, Zheng C, Suh SH, Kawaai K, Ando H, Mizutani A, Abe T, Kiyonari H, Seki G, Yule D, Mikoshiba K, Muallem S (2013) Irbit mediates synergy between Ca^{2+} and cAMP signaling pathways during epithelial transport in mice. Gastroenterology 145 (1):232–241. https://doi.org/10.1053/j.gastro.2013.03.047

Park M, Ohana E, Choi SY, Lee MS, Park JH, Muallem S (2014) Multiple roles of the $SO4^{2-}/Cl^{-}/OH^{-}$ exchanger protein Slc26a2 in chondrocyte functions. J Biol Chem 289(4):1993–2001. https://doi.org/10.1074/jbc.M113.503466

Pasqualetto E, Aiello R, Gesiot L, Bonetto G, Bellanda M, Battistutta R (2010) Structure of the cytosolic portion of the motor protein prestin and functional role of the STAS domain in SLC26/SulP anion transporters. J Mol Biol 400(3):448–462. https://doi.org/10.1016/j.jmb.2010.05.013

Pech V, Pham TD, Hong S, Weinstein AM, Spencer KB, Duke BJ, Walp E, Kim YH, Sutliff RL, Bao HF, Eaton DC, Wall SM (2010) Pendrin modulates ENaC function by changing luminal HCO_3^-. J Am Soc Nephrol 21(11):1928–1941. https://doi.org/10.1681/ASN.2009121257

Pedemonte N, Caci E, Sondo E, Caputo A, Rhoden K, Pfeffer U, Di Candia M, Bandettini R, Ravazzolo R, Zegarra-Moran O, Galietta LJ (2007) Thiocyanate transport in resting and IL-4-

stimulated human bronchial epithelial cells: role of pendrin and anion channels. J Immunol 178 (8):5144–5153

Pesce L, Bizhanova A, Caraballo JC, Westphal W, Butti ML, Comellas A, Kopp P (2012) TSH regulates pendrin membrane abundance and enhances iodide efflux in thyroid cells. Endocrinology 153(1):512–521. https://doi.org/10.1210/en.2011-1548

Petrovic S, Ju X, Barone S, Seidler U, Alper SL, Lohi H, Kere J, Soleimani M (2003) Identification of a basolateral Cl^-/HCO_3^- exchanger specific to gastric parietal cells. Am J Physiol Gastrointest Liver Physiol 284(6):G1093–G1103. https://doi.org/10.1152/ajpgi.00454.2002

Piala AT, Moon TM, Akella R, He H, Cobb MH, Goldsmith EJ (2014) Chloride sensing by WNK1 involves inhibition of autophosphorylation. Sci Signal 7(324):ra41. https://doi.org/10.1126/scisignal.2005050

Rahmati N, Kunzelmann K, Xu J, Barone S, Sirianant L, De Zeeuw CI, Soleimani M (2013) Slc26a11 is prominently expressed in the brain and functions as a chloride channel: expression in Purkinje cells and stimulation of V H^+-ATPase. Pflugers Arch 465(11):1583–1597. https://doi.org/10.1007/s00424-013-1300-6

Regeer RR, Lee A, Markovich D (2003) Characterization of the human sulfate anion transporter (hsat-1) protein and gene (SAT1; SLC26A1). DNA Cell Biol 22(2):107–117. https://doi.org/10.1089/104454903321515913

Rode B, Dirami T, Bakouh N, Rizk-Rabin M, Norez C, Lhuillier P, Lores P, Jollivet M, Melin P, Zvetkova I, Bienvenu T, Becq F, Planelles G, Edelman A, Gacon G, Toure A (2012) The testis anion transporter TAT1 (SLC26A8) physically and functionally interacts with the cystic fibrosis transmembrane conductance regulator channel: a potential role during sperm capacitation. Hum Mol Genet 21(6):1287–1298. https://doi.org/10.1093/hmg/ddr558

Royaux IE, Suzuki K, Mori A, Katoh R, Everett LA, Kohn LD, Green ED (2000) Pendrin, the protein encoded by the Pendred syndrome gene (PDS), is an apical porter of iodide in the thyroid and is regulated by thyroglobulin in FRTL-5 cells. Endocrinology 141(2):839–845. https://doi.org/10.1210/endo.141.2.7303

Royaux IE, Wall SM, Karniski LP, Everett LA, Suzuki K, Knepper MA, Green ED (2001) Pendrin, encoded by the Pendred syndrome gene, resides in the apical region of renal intercalated cells and mediates bicarbonate secretion. Proc Natl Acad Sci USA 98(7):4221–4226. https://doi.org/10.1073/pnas.071516798

Sardiello M, Palmieri M, di Ronza A, Medina DL, Valenza M, Gennarino VA, Di Malta C, Donaudy F, Embrione V, Polishchuk RS, Banfi S, Parenti G, Cattaneo E, Ballabio A (2009) A gene network regulating lysosomal biogenesis and function. Science 325(5939):473–477. https://doi.org/10.1126/science.1174447

Schaechinger TJ, Oliver D (2007) Nonmammalian orthologs of prestin (SLC26A5) are electrogenic divalent/chloride anion exchangers. Proc Natl Acad Sci 104(18):7693. https://doi.org/10.1073/pnas.0608583104

Schroder BA, Wrocklage C, Hasilik A, Saftig P (2010) The proteome of lysosomes. Proteomics 10 (22):4053–4076. https://doi.org/10.1002/pmic.201000196

Schweinfest CW, Spyropoulos DD, Henderson KW, Kim JH, Chapman JM, Barone S, Worrell RT, Wang Z, Soleimani M (2006) slc26a3 (dra)-deficient mice display chloride-losing diarrhea, enhanced colonic proliferation, and distinct up-regulation of ion transporters in the colon. J Biol Chem 281(49):37962–37971. https://doi.org/10.1074/jbc.M607527200

Shan J, Liao J, Huang J, Robert R, Palmer ML, Fahrenkrug SC, O'Grady SM, Hanrahan JW (2012) Bicarbonate-dependent chloride transport drives fluid secretion by the human airway epithelial cell line Calu-3. J Physiol 590(21):5273–5297. https://doi.org/10.1113/jphysiol.2012.236893

Sharma AK, Ye L, Baer CE, Shanmugasundaram K, Alber T, Alper SL, Rigby AC (2011) Solution structure of the guanine nucleotide-binding STAS domain of SLC26-related SulP protein Rv1739c from *Mycobacterium tuberculosis*. J Biol Chem 286(10):8534–8544. https://doi.org/10.1074/jbc.M110.165449

Shcheynikov N, Ko SB, Zeng W, Choi JY, Dorwart MR, Thomas PJ, Muallem S (2006a) Regulatory interaction between CFTR and the SLC26 transporters. Novartis Found Symp 273:177–186; discussion 186-192, 261-174

Shcheynikov N, Wang Y, Park M, Ko SB, Dorwart M, Naruse S, Thomas PJ, Muallem S (2006b) Coupling modes and stoichiometry of Cl^-/ HCO_3^- exchange by slc26a3 and slc26a6. J Gen Physiol 127(5):511–524. https://doi.org/10.1085/jgp.200509392

Shcheynikov N, Yang D, Wang Y, Zeng W, Karniski LP, So I, Wall SM, Muallem S (2008) The Slc26a4 transporter functions as an electroneutral Cl^-/I^-/ HCO_3^- exchanger: role of Slc26a4 and Slc26a6 in I^- and HCO_3^- secretion and in regulation of CFTR in the parotid duct. J Physiol 586(16):3813–3824. https://doi.org/10.1113/jphysiol.2008.154468

Shcheynikov N, Son A, Hong JH, Yamazaki O, Ohana E, Kurtz I, Shin DM, Muallem S (2015) Intracellular Cl- as a signaling ion that potently regulates Na^+/ HCO_3^- transporters. Proc Natl Acad Sci USA 112(3):E329–E337. https://doi.org/10.1073/pnas.1415673112

Shelden MC, Howitt SM, Price GD (2010) Membrane topology of the cyanobacterial bicarbonate transporter, BicA, a member of the SulP (SLC26A) family. Mol Membr Biol 27(1):12–22. https://doi.org/10.3109/09687680903400120

Simpson JE, Schweinfest CW, Shull GE, Gawenis LR, Walker NM, Boyle KT, Soleimani M, Clarke LL (2007) PAT-1 (Slc26a6) is the predominant apical membrane Cl^-/ HCO_3^- exchanger in the upper villous epithelium of the murine duodenum. Am J Physiol Gastrointest Liver Physiol 292(4):G1079–G1088. https://doi.org/10.1152/ajpgi.00354.2006

Singla A, Dwivedi A, Saksena S, Gill RK, Alrefai WA, Ramaswamy K, Dudeja PK (2010) Mechanisms of lysophosphatidic acid (LPA) mediated stimulation of intestinal apical Cl-/OH- exchange. Am J Physiol Gastrointest Liver Physiol 298(2):G182–G189. https://doi.org/10.1152/ajpgi.00345.2009

Soave D, Miller MR, Keenan K, Li W, Gong J, Ip W, Accurso F, Sun L, Rommens JM, Sontag M, Durie PR, Strug LJ (2014) Evidence for a causal relationship between early exocrine pancreatic disease and cystic fibrosis-related diabetes: a Mendelian randomization study. Diabetes 63 (6):2114–2119. https://doi.org/10.2337/db13-1464

Soleimani M, Greeley T, Petrovic S, Wang Z, Amlal H, Kopp P, Burnham CE (2001) Pendrin: an apical Cl^-/OH^-/HCO_3^- exchanger in the kidney cortex. Am J Physiol Renal Physiol 280(2): F356–F364. https://doi.org/10.1152/ajprenal.2001.280.2.F356

Song Y, Yamamoto A, Steward MC, Ko SB, Stewart AK, Soleimani M, Liu BC, Kondo T, Jin CX, Ishiguro H (2012) Deletion of Slc26a6 alters the stoichiometry of apical Cl-/HCO-3 exchange in mouse pancreatic duct. Am J Physiol Cell Physiol 303(8):C815–C824. https://doi.org/10.1152/ajpcell.00151.2012

Spyroglou A, Bozoglu T, Rawal R, De Leonardis F, Sterner C, Boulkroun S, Benecke AG, Monti L, Zennaro MC, Petersen AK, Doring A, Rossi A, Bidlingmaier M, Warth R, Gieger C, Reincke M, Beuschlein F (2014) Diastrophic dysplasia sulfate transporter (SLC26A2) is expressed in the adrenal cortex and regulates aldosterone secretion. Hypertension 63 (5):1102–1109. https://doi.org/10.1161/HYPERTENSIONAHA.113.02504

Steward MC, Ishiguro H, Case RM (2005) Mechanisms of bicarbonate secretion in the pancreatic duct. Annu Rev Physiol 67:377–409. https://doi.org/10.1146/annurev.physiol.67.031103.153247

Stewart AK, Yamamoto A, Nakakuki M, Kondo T, Alper SL, Ishiguro H (2009) Functional coupling of apical Cl^-/ HCO_3^- exchange with CFTR in stimulated HCO_3^- secretion by guinea pig interlobular pancreatic duct. Am J Physiol Gastrointest Liver Physiol 296(6):G1307–G1317. https://doi.org/10.1152/ajpgi.90697.2008

Stewart AK, Shmukler BE, Vandorpe DH, Reimold F, Heneghan JF, Nakakuki M, Akhavein A, Ko S, Ishiguro H, Alper SL (2011) SLC26 anion exchangers of guinea pig pancreatic duct: molecular cloning and functional characterization. Am J Physiol Cell Physiol 301(2):C289–C303. https://doi.org/10.1152/ajpcell.00089.2011

Sun L, Rommens JM, Corvol H, Li W, Li X, Chiang TA, Lin F, Dorfman R, Busson PF, Parekh RV, Zelenika D, Blackman SM, Corey M, Doshi VK, Henderson L, Naughton KM, O'Neal

WK, Pace RG, Stonebraker JR, Wood SD, Wright FA, Zielenski J, Clement A, Drumm ML, Boelle PY, Cutting GR, Knowles MR, Durie PR, Strug LJ (2012) Multiple apical plasma membrane constituents are associated with susceptibility to meconium ileus in individuals with cystic fibrosis. Nat Genet 44(5):562–569. https://doi.org/10.1038/ng.2221

Tannahill GM, Curtis AM, Adamik J, Palsson-McDermott EM, McGettrick AF, Goel G, Frezza C, Bernard NJ, Kelly B, Foley NH, Zheng L, Gardet A, Tong Z, Jany SS, Corr SC, Haneklaus M, Caffrey BE, Pierce K, Walmsley S, Beasley FC, Cummins E, Nizet V, Whyte M, Taylor CT, Lin H, Masters SL, Gottlieb E, Kelly VP, Clish C, Auron PE, Xavier RJ, O'Neill LA (2013) Succinate is an inflammatory signal that induces IL-1beta through HIF-1alpha. Nature 496 (7444):238–242. https://doi.org/10.1038/nature11986

Toure A, Morin L, Pineau C, Becq F, Dorseuil O, Gacon G (2001) Tat1, a novel sulfate transporter specifically expressed in human male germ cells and potentially linked to rhogtpase signaling. J Biol Chem 276(23):20309–20315. https://doi.org/10.1074/jbc.M011740200

Vachel L, Shcheynikov N, Yamazaki O, Fremder M, Ohana E, Son A, Shin DM, Yamazaki-Nakazawa A, Yang CR, Knepper MA, Muallem S (2018) Modulation of Cl(−) signaling and ion transport by recruitment of kinases and phosphatases mediated by the regulatory protein IRBIT. Sci Signal 11(554):eaat5018. https://doi.org/10.1126/scisignal.aat5018

Verlander JW, Hassell KA, Royaux IE, Glapion DM, Wang ME, Everett LA, Green ED, Wall SM (2003) Deoxycorticosterone upregulates PDS (Slc26a4) in mouse kidney: role of pendrin in mineralocorticoid-induced hypertension. Hypertension 42(3):356–362. https://doi.org/10.1161/01.HYP.0000088321.67254.B7

Verlander JW, Hong S, Pech V, Bailey JL, Agazatian D, Matthews SW, Coffman TM, Le T, Inagami T, Whitehill FM, Weiner ID, Farley DB, Kim YH, Wall SM (2011) Angiotensin II acts through the angiotensin 1a receptor to upregulate pendrin. Am J Physiol Renal Physiol 301(6): F1314–F1325. https://doi.org/10.1152/ajprenal.00114.2011

Wall S (2005) Recent advances in our understanding of intercalated cells. Curr Opin Nephrol Hypertens 14:480–484. https://doi.org/10.1097/01.mnh.0000168390.04520.06

Wall SM, Kim YH, Stanley L, Glapion DM, Everett LA, Green ED, Verlander JW (2004) NaCl restriction upregulates renal Slc26a4 through subcellular redistribution: role in Cl- conservation. Hypertension 44(6):982–987. https://doi.org/10.1161/01.Hyp.0000145863.96091.89

Walter JD, Sawicka M, Dutzler R (2019) Cryo-EM structures and functional characterization of murine Slc26a9 reveal mechanism of uncoupled chloride transport. Elife 8:e46986. https://doi.org/10.7554/eLife.46986

Wang Y, Soyombo AA, Shcheynikov N, Zeng W, Dorwart M, Marino CR, Thomas PJ, Muallem S (2006) Slc26a6 regulates CFTR activity in vivo to determine pancreatic duct HCO_3^- secretion: relevance to cystic fibrosis. EMBO J 25(21):5049–5057. https://doi.org/10.1038/sj.emboj.7601387

Wangemann P, Nakaya K, Wu T, Maganti RJ, Itza EM, Sanneman JD, Harbidge DG, Billings S, Marcus DC (2007) Loss of cochlear HCO_3^- secretion causes deafness via endolymphatic acidification and inhibition of Ca^{2+} reabsorption in a Pendred syndrome mouse model. Am J Physiol Renal Physiol 292(5):F1345–F1353. https://doi.org/10.1152/ajprenal.00487.2006

Wedenoja S, Pekansaari E, Hoglund P, Makela S, Holmberg C, Kere J (2011) Update on SLC26A3 mutations in congenital chloride diarrhea. Hum Mutat 32(7):715–722. https://doi.org/10.1002/humu.21498

Wedenoja S, Khamaysi A, Shimshilashvili L, Anbtawe-Jomaa S, Elomaa O, Toppari J, Hoglund P, Aittomaki K, Holmberg C, Hovatta O, Tapanainen JS, Ohana E, Kere J (2017) A missense mutation in SLC26A3 is associated with human male subfertility and impaired activation of CFTR. Sci Rep 7(1):14208. https://doi.org/10.1038/s41598-017-14606-3

Xia W, Yu Q, Riederer B, Singh AK, Engelhardt R, Yeruva S, Song P, Tian DA, Soleimani M, Seidler U (2013) The distinct roles of anion transporters Slc26a3 (DRA) and Slc26a6 (PAT-1) in fluid and electrolyte absorption in the murine small intestine. Pflugers Arch 466:1541. https://doi.org/10.1007/s00424-013-1381-2

Xiao F, Juric M, Li J, Riederer B, Yeruva S, Singh AK, Zheng L, Glage S, Kollias G, Dudeja P, Tian DA, Xu G, Zhu J, Bachmann O, Seidler U (2012) Loss of downregulated in adenoma (DRA) impairs mucosal HCO_3^- secretion in murine ileocolonic inflammation. Inflamm Bowel Dis 18(1):101–111. https://doi.org/10.1002/ibd.21744

Xiao F, Yu Q, Li J, Johansson ME, Singh AK, Xia W, Riederer B, Engelhardt R, Montrose M, Soleimani M, Tian DA, Xu G, Hansson GC, Seidler U (2014) Slc26a3 deficiency is associated with loss of colonic HCO_3^- secretion, absence of a firm mucus layer and barrier impairment in mice. Acta Physiol (Oxf) 211(1):161–175. https://doi.org/10.1111/apha.12220

Xie Q, Welch R, Mercado A, Romero MF, Mount DB (2002) Molecular characterization of the murine Slc26a6 anion exchanger: functional comparison with Slc26a1. Am J Physiol Renal Physiol 283(4):F826–F838. https://doi.org/10.1152/ajprenal.00079.2002

Xie R, Nguyen S, McKeehan K, Wang F, McKeehan WL, Liu L (2011) Microtubule-associated protein 1S (MAP1S) bridges autophagic components with microtubules and mitochondria to affect autophagosomal biogenesis and degradation. J Biol Chem 286(12):10367–10377. https://doi.org/10.1074/jbc.M110.206532

Xu J, Henriksnas J, Barone S, Witte D, Shull GE, Forte JG, Holm L, Soleimani M (2005) SLC26A9 is expressed in gastric surface epithelial cells, mediates Cl^-/ HCO_3^- exchange, and is inhibited by NH_4^+. Am J Physiol Cell Physiol 289(2):C493–C505. https://doi.org/10.1152/ajpcell.00030.2005

Xu J, Song P, Miller ML, Borgese F, Barone S, Riederer B, Wang Z, Alper SL, Forte JG, Shull GE, Ehrenfeld J, Seidler U, Soleimani M (2008) Deletion of the chloride transporter Slc26a9 causes loss of tubulovesicles in parietal cells and impairs acid secretion in the stomach. Proc Natl Acad Sci USA 105(46):17955–17960. https://doi.org/10.1073/pnas.0800616105

Xu J, Song P, Nakamura S, Miller M, Barone S, Alper SL, Riederer B, Bonhagen J, Arend LJ, Amlal H, Seidler U, Soleimani M (2009) Deletion of the chloride transporter slc26a7 causes distal renal tubular acidosis and impairs gastric acid secretion. J Biol Chem 284 (43):29470–29479. https://doi.org/10.1074/jbc.M109.044396

Xu J, Barone S, Li H, Holiday S, Zahedi K, Soleimani M (2011) Slc26a11, a chloride transporter, localizes with the vacuolar H^+-ATPase of A-intercalated cells of the kidney. Kidney Int 80 (9):926–937. https://doi.org/10.1038/ki.2011.196

Yang H, Jiang W, Furth EE, Wen X, Katz JP, Sellon RK, Silberg DG, Antalis TM, Schweinfest CW, Wu GD (1998) Intestinal inflammation reduces expression of DRA, a transporter responsible for congenital chloride diarrhea. Am J Phys 275(6 Pt 1):G1445–G1453

Yang D, Shcheynikov N, Zeng W, Ohana E, So I, Ando H, Mizutani A, Mikoshiba K, Muallem S (2009) IRBIT coordinates epithelial fluid and HCO_3^- secretion by stimulating the transporters pNBC1 and CFTR in the murine pancreatic duct. J Clin Invest 119(1):193–202. https://doi.org/10.1172/JCI36983

Yang D, Li Q, So I, Huang CL, Ando H, Mizutani A, Seki G, Mikoshiba K, Thomas PJ, Muallem S (2011a) IRBIT governs epithelial secretion in mice by antagonizing the WNK/SPAK kinase pathway. J Clin Invest 121(3):956–965. https://doi.org/10.1172/JCI43475

Yang SK, Jung Y, Kim H, Hong M, Ye BD, Song K (2011b) Association of FCGR2A, JAK2 or HNF4A variants with ulcerative colitis in Koreans. Dig Liver Dis 43(11):856–861. https://doi.org/10.1016/j.dld.2011.07.006

Zhao H, Star RA, Muallem S (1994) Membrane localization of H^+ and HCO_3^- transporters in the rat pancreatic duct. J Gen Physiol 104(1):57–85

Zheng J, Shen W, He DZ, Long KB, Madison LD, Dallos P (2000) Prestin is the motor protein of cochlear outer hair cells. Nature 405(6783):149–155. https://doi.org/10.1038/35012009

Chapter 13
ClC-2 Chloride Channels

John Cuppoletti, Danuta H. Malinowska, and Ryuji Ueno

Abstract This article focuses on activators and inhibitors of ClC-2. ClC-2 Cl^- channels from mouse, rabbit, rat, and human have been studied extensively in recombinant form by whole-cell patch-clamp and in single-channel studies. There are similarities in responses of these channels from different species to protons, ATP, fatty acids, and lubiprostone for the recombinant channels. However, there are differences in channel properties, notably in the activation of human and rabbit (but not rat or mouse) forms by PKA. Two PKA phosphorylation sites of human ClC-2 have been identified by mutagenesis. One is required for fatty acid and lubiprostone activation. Structural studies suggest that these sites interact with cystathionine-β-synthase (CBS) domains to regulate ClC-2. Single-channel studies have been carried out on recombinant ClC-2, cortical astrocytes, and A6 cells which have both ClC-2 and CFTR in the apical membrane. Methadone and GaTx2 differentially inhibit ClC-2, while DASU-02 differentially inhibits CFTR. CFTRinh172 and glibenclamide inhibit both. Knockdown of ion channels combined with pharmacology is more powerful than using pharmacological agents alone. Studies using human cells and tissues such as the intestine and lung have used various activators and inhibitors, and these studies are reviewed in detail. This review indicates that it is essential to verify the specificity of all agents experimentally to correctly interpret experimental results. Some studies provide insight into the possible roles of ClC-2 in physiological processes.

Keywords ClC-2 · ClC-2 phosphorylation · ClC-2 single channels · Lubiprostone · Intestinal chloride transport · Cystic fibrosis · CFTR · I_{sc} · Intestinal chloride · Airway chloride · Nasal potential difference

---

J. Cuppoletti (✉) · D. H. Malinowska
Department of Molecular and Cellular Physiology, University of Cincinnati, Cincinnati, OH, USA
e-mail: cuppolj@ucmail.uc.edu; Danuta.Malinowska@uc.edu

R. Ueno
VLP Therapeutics, Gaithersburg, MD, USA
e-mail: ktolliver@vlptherapeutics.com

K. L. Hamilton, D. C. Devor (eds.), *Studies of Epithelial Transporters and Ion Channels*, Physiology in Health and Disease,
https://doi.org/10.1007/978-3-030-55454-5_13

13.1 Introduction

This chapter will cover only ClC-2 Cl^- channels (except to provide a background of the ClC field based on the work of Miller and coworkers). ClC-2 is found in a variety of epithelial and non-epithelial tissues including the retina, testes, lung, stomach, intestine, and salivary ducts. This review of ClC-2 addresses the basic properties of recombinant ClC-2 from different species, ClC-2 functional studies in whole-cell current studies and in single-channel studies, ClC-2 in physiological processes, and activators and inhibitors of ClC-2 which have been shown to distinguish between ClC-2 and other channels such as CFTR. This review outlines some controversies in the ClC-2 literature, some of which appear to derive from differences in amino acid sequence of the proteins from different species, use of inhibitors or other chemical agents which have not been validated, and apparent differences in localization of ClC-2 in various tissues and species. Lubiprostone, a bicyclic fatty acid, is used to treat constipation by acting in the small intestine to cause fluid secretion. A number of studies using lubiprostone and forskolin to identify either ClC-2 or CFTR in Cl secretory function have appeared which are directed toward the repair of barrier properties and increasing fluid secretion in the intestine and in lung fluid. A number of articles have suggested that lubiprostone is an EP_4 receptor agonist and have suggested that the mechanism of action of lubiprostone is to activate CFTR instead of ClC-2. The lack of EP_4 receptor agonist activity and lack of effect of lubiprostone on human CFTR will be discussed. Thus, the present review will focus on the function and regulation of ClC-2 where it has been studied from various species and in epithelia from various tissues.

13.2 Review Articles

A comprehensive review of ClC-2 has recently appeared (Wang et al. 2017), and several reviews with a focus on ClC-2 Cl^- channelopathies are available (Bi et al. 2013; Poroca et al. 2017). The various classes of Cl^- channels have been reviewed (Duran et al. 2010). The biophysical properties of ClC Cl^- channels have been reviewed (Zifarelli and Pusch 2007; Fahlke 2004). The structures of ClC Cl^- channels have been reviewed (Dutzler 2004a, b, 2006, 2007; Chen 2005). Miller (2006) has reviewed ClC channel structure in light of findings that some of the ClC channels are proton-coupled Cl^- transporters. The molecular physiology, pathophysiology, and proposed channelopathies of ClC-type Cl^- channels have also been reviewed (Stauber et al. 2012; Jentsch et al. 2005). Unique ways to study various Cl^- channels against a background of numerous other Cl^- channels appear to be a promising approach in the future (Duan 2013). An excellent older review of the pharmacology of CFTR channel activity is recommended (Schultz et al. 1999). A very recent review by Linsdell (2014) has appeared which reviews the newest generation of CFTR inhibitors.

13.3 The Beginning: ClC-0

The *Torpedo* electroplax Cl^- channel, in impressive and pioneering studies, was purified and studied in bilayers as well as in patch-clamp of the protein reconstituted into liposomes. Voltage-dependent macroscopic currents were measured, ion selectivities were P_x/P_{Cl}^-: Cl^- (1.0), Br^- (0.58), and iodide and thiocyanate blocked the channel (Miller and White 1980). It was blocked by SITS and DIDS (White and Miller 1979). It was shown to be a dimeric channel of 10 and 20 pS with a closed state in 200 mM Cl^-, and the probability of forming an open channel was voltage dependent and increased with reduced pH (Hanke and Miller 1983). The channel was extensively characterized in a series of further studies (Miller 1982). The cloned *Torpedo* electroplax Cl^- channel protein was expressed in *Xenopus* oocytes and had essentially identical properties as that of the protein purified from electroplax (Bauer et al. 1991). It was later named ClC-0, and the properties of ClC-0 were a guide for studies of other ClC channels.

13.4 Properties of Rat ClC-2

13.4.1 Cloned Rat ClC-2

ClC-2 was initially cloned from rat (Thiemann et al. 1992). A construct expressed in *Xenopus* oocytes gave voltage-activated currents with ion selectivity of $Cl^- > Br^- > I^-$. ClC-2 was shown to be widely expressed in a variety of rat tissues including the skeletal muscle, heart, brain, lung, kidney pancreas, stomach, intestine, and liver by northern analysis. ClC-2 was also found in a large variety of cell lines from human sources. As was determined with ClC-0 (Hanke and Miller 1983), acidification altered gating and activated the channel (Jordt and Jentsch 1997).

13.4.2 ClC-2 in the Rat Airway

ClC-2 was found in the apical membranes of airway tissues of fetal rat lung (Murray et al. 1995), and it was downregulated after birth (Murray et al. 1996). It was postulated that ClC-2 might be involved in lung development. However, ClC-2$^{-/-}$ mice have normal lungs (Chu et al. 1999; Vij and Zeitlin 2006), suggesting a different role for rat lung ClC-2. The downregulation after birth in the rat may be mediated by the transcription factors Sp1 and Sp3 (Chu et al. 1999; Vij and Zeitlin 2006; Holmes et al. 2003). Growth factors can stimulate the expression of ClC-2 in rat lung (Blaisdell et al. 1999). Cl^- secretion, thought to involve ClC-2, can be stimulated by reduced pH in the fetal rat lung (Blaisdell et al. 2000). The fetal rat lung appears to be a good model for studies of ClC-2 in intact tissue. Transcriptional regulation by Sp1 and Sp3 and low pH activation might allow higher rates of Cl^- secretion needed to play a role as a rescue channel for cystic fibrosis (CF). The adult rat lung has low levels of ClC-2 (Murray et al. 1996) making it a poor model for functional studies. MacDonald et al. (2008), Schiffhauer et al. (2013), Joo et al. (2009), and Cuthbert (2011) have studied Cl^- transport in the adult airway. These and other model systems will be discussed in Sect. 13.9.

13.4.3 Cloned Human ClC-2

Human ClC-2 was first cloned from T84 cells by Cid et al. (1995). By northern analysis, it was found to be present in a large variety of human tissues including the pancreas, brain, heart, placenta, skeletal muscle, and kidney. Weak signals were also found in the lung and liver, in addition to being present in T84 cells (Cid et al. 1995). ClC-2 mRNA was also present in IB3-1 cells, an airway epithelial cell line from a CF patient. The investigators envisaged that this channel might be a potential rescue channel when CFTR is defective, using such techniques as gene replacement therapy or augmentation using ClC-2 as the alternative Cl^- secretory pathway. The authors noted that although human ClC-2 had 94% identity with the rat form, there were sequence differences between rat and human ClC-2 that might confer PKA activation and regulation by other kinases, including ten potential sites for phosphorylation by PKC in the human form. Human ClC-2 was characterized in IB3-1 cells and then overexpressed in those cells (Schwiebert et al. 1998). In IB3-1 cells, which contained a low level of ClC-2, the channel gave a linear I/V curve with $Cl^- > I^- >$ gluconate (Glu) anion selectivity. The channel was activated by low pH, with pH 3.79 the lowest pH tested. Time- and voltage-activated currents were minimal. However, upon overexpression of human ClC-2 in IB3-1 cells, ClC-2 gave inward rectification and time- and voltage-activated currents with pH activation and $Cl^- > I^- >$ Glu anion selectivity. Cadmium (Cd^{2+}) gave a concentration-dependent block of the currents in both IB3-1 cells and IB3-1 cells overexpressing human ClC-2. The authors suggested that pharmacological agents that might increase the acid sensitivity of the channel might lead to it being used as a rescue channel in CF.

The properties of the cloned human channel were similar to those observed in T84 cells in earlier studies by Fritsch and Edelman (1996). In this study, they showed that ClC-2 in T84 cells could be inhibited by treatment with the protein kinase C activator phorbol 12-myristate 13-acetate. Staurosporine and calphostin C increased activation of the channel. Neither forskolin nor myristoylated protein kinase A inhibitor fragment 5–24 amide (mPKI) was shown to have significant consistent effects in that study. They concluded that human ClC-2 could be modulated by phosphorylation, as later indicated from the consensus sequences identified by Cid et al. (1995). This group also showed that ClC-2 was affected by TNF-α (Bali et al. 2001) and membrane cholesterol (Hinzpeter et al. 2007). The distribution of ClC-2 in both human and rat tissues (Lipecka et al. 2002) was also studied. It was concluded that there were differences in expression of the channel between rat and human tissues, particularly in the intestine. Thus, findings in rat intestine might not extrapolate to human intestine. However, the human and rat airways appeared to have similar locations of ClC-2 expression.

13.5 Single-Channel Studies of ClC-2

Single-channel recordings of human ClC-2 were used in studies of channel gating (Stölting et al. 2013). The authors found one closed state and two open states in the presence of cytosolic ATP. The single-channel current in symmetrical 140 mM Cl^- was found to be approximately 0.23 pA at −100 mV. The same authors also carried out single-channel measurements of ClC-2 as part of a study of chimeras of human ClC-2 with ClC-1. ClC-2 showed similar single-channel current magnitudes and again showed double-barreled behavior (Stölting et al. 2014).

The cloned human channel was studied in lipid bilayers using 800 mM tetraethylammonium chloride (TEACl) to increase the small channel currents (Sherry et al. 1997). At −80 mV, the channel gave currents of approximately 1.5 pA, which would be correspondingly smaller if measured at the lower concentration of salt. PKA and acid activation of the human channel were demonstrated.

Single-channel recordings of rabbit ClC-2 were carried out in studies of a peptide toxin isolated from *Leiurus quinquestriatus hebraeus* venom, GaTx2 (gating modifier of anion channels 2), that inhibits ClC-2 channels in a voltage-dependent manner with an apparent K_D of ~20 pm (Thompson et al. 2009). Toxin inhibition is highly complex, being protocol dependent, showing poor inhibition of the open channel but affecting the slow gating of the channel. The single-channel currents from rabbit ClC-2 at −100 mV appear to be approximately 0.3 pA, slightly larger than the currents reported for human ClC-2.

Rabbit ClC-2 was cloned and studied at the single-channel level in lipid bilayers, using high concentrations of TEACl (800 mM) to increase channel currents (Malinowska et al. 1995). Channel records at −100 mV gave single-channel currents of approximately 2.2–3 pA at 800 mM TEACl. Given that the high concentration of Cl^- in this study was approximately 5.3 times the concentration used in the later study of rabbit ClC-2 by Thompson et al. (2009), these currents would be approximately similar in magnitude at the lower concentration of salt. PKA and low pH also activated rabbit ClC-2 (Malinowska et al. 1995).

Rat ClC-2 from astrocytes was studied at the single-channel level by Nobile et al. (2000), and conductance levels of 0.3 and 0.6 pA at −100 mV were observed, similar to the channel currents seen in human and rabbit ClC-2.

All of the above studies are somewhat limited in terms of detailed characterization of the single-channel characteristics of ClC-2 from any species, perhaps reflecting in part the difficulty of such measurements.

13.6 Single-Channel Studies of ClC-2 in the Presence of CFTR and Other Cl^- Channels

Single-channel studies provide a way to study ClC-2 in the presence of CFTR, since ClC-2 and CFTR have distinctly different biophysical properties at the single-channel level. A very complete study of the single-channel characteristics of A6 cell apical membrane ClC-2 and CFTR with an included comparison with human ClC-2 has appeared using lubiprostone to stimulate *Xenopus* A6 cell ClC-2 (Bao et al. 2008). The characteristics of A6 cell ClC-2 studied at the single-channel level were a unit conductance of 3–4 pS and a high degree of rectification, with voltage dependence of mean open time and open probability, both of which decreased with depolarization. The anion selectivity of A6 ClC-2 was $Cl^- > Br^- > NO_3^- > I^- > SCN^-$. ClC-2 was potently activated by lubiprostone ($EC_{50} = 69$ nM). CFTR was also present in A6 apical membranes but was activated by lubiprostone in a cAMP- independent manner with an EC_{50} about ten times higher than the EC_{50} for activation of ClC-2. During these studies, the authors determined that both diphenylcarboxylic acid and glibenclamide were blockers of both ClC-2 and CFTR. They found that glibenclamide was an open-channel blocker. When Cd^{2+} was added to the bath, it eliminated not only ClC-2 currents but also all other channel currents, suggesting a generally toxic effect of Cd^{2+}.

Human ClC-2 expressed in HEK293 cells as used previously (Tewari et al. 2000; Cuppoletti et al. 2001, 2004a, b, 2013) yielded single-channel currents activated by lubiprostone which were "indistinguishable from" those observed in *Xenopus* A6 cells (Bao et al. 2008).

Interestingly this was the first report and characterization of ClC-2 in the apical membrane of the widely studied *Xenopus* A6 cell line. Previous studies had studied apical membrane Cl^- transport by short-circuit current (I_{sc}) in Ussing chambers and I_{sc} fluctuation by noise analysis (Atia et al. 1999). The authors found evidence only for CFTR and a Ca^{2+}-activated Cl^- channel with no evidence of a Cl^- channel with the characteristics of ClC-2 in *Xenopus* A6 cells.

Single-channel studies of *Xenopus* A6 cell apical membranes (Price et al. 1996) showed that *Xenopus* CFTR when activated by forskolin had a slightly larger single-channel conductance than the human form and that the anion selectivity sequence varied from human. The authors did not mention finding ClC-2-like single channels suggesting that *Xenopus* ClC-2 is not activated by PKA and consistent with the lack of ClC-2 (or CFTR) channels in A6 cells observed by Bao et al. (2008) before the addition of lubiprostone. Supporting this view, CFTR and other Cl^- channels were not seen in A6 cells unless treated by forskolin in another single-channel study (Ling et al. 1997).

Thus, lubiprostone, used in the Bao et al. (2008) study, was the first to unmask A6 cell ClC-2. Previous studies showed that prostaglandins (which are metabolized to prostones by NAD^+PGDH) activated I_{sc} and Cl^- fluxes in *Xenopus* A6 cells (Keeler and Wong 1986). Studies using impedance measurements suggested that PGE_2 only activated Na^+ fluxes in Xenopus A6 cells (Paunescu and Helman 2001).

13.7 PKA Phosphorylation of ClC-2

Rabbit ClC-2 contains consensus phosphorylation sites for PKA and can be activated by PKA (Malinowska et al. 1995). Porcine choroid plexus contains ClC-2 activated by extracellular VIP and PKA catalytic subunit (Kajita et al. 2000). Cid et al. (1995) were the first to identify a PKA consensus sequence when the human ClC-2 was cloned. They noted that the site was not present in the rat sequence.

Recombinant human ClC-2 was studied in lipid bilayers and shown to be activated by the catalytic subunit of PKA plus ATP (Sherry et al. 1997). When expressed in HEK cells and studied by patch-clamp, human ClC-2 was activated by treatment of the cells with a combination of forskolin and IBMX. In another study, human ClC-2 was activated by arachidonic acid in a PKA-independent manner (Tewari et al. 2000). Forskolin-IBMX was also shown to activate human ClC-2 in an mPKI-sensitive manner, and both methadone and Cd^{2+} inhibited forskolin-IBMX-stimulated human ClC-2 (Cuppoletti et al. 2013).

A detailed study of two unique PKA consensus sites $RRAT_{655}$ and $RGET_{691}$ on human ClC-2 was undertaken by Cuppoletti et al. (2004a). In these studies, the effects of phosphorylation at one of the sites were studied in channels mutated at the other site at neutral and low extracellular pH. For each of the channels with mutated consensus phosphorylation sites, the effect of forskolin-IBMX at neutral and low pH was studied. In each case, activity of the mutant channels was confirmed by arachidonic acid activation, which is not dependent upon phosphorylation (Tewari et al. 2000). The effect of forskolin-IBMX plus phosphatase inhibitors was also tested to help ensure that the phosphodiesterase was inhibited and that phosphorylation of the sites was not differentially susceptible to phosphatase action. The double phosphorylation site mutant $RRAT(A)_{655}$ plus $RGET(A)_{691}$ lost PKA activation and also lost activation by low pH. Either $RRAT_{655}$ or $RGET_{691}$ was sufficient for activation, as long as phosphatase inhibitors were present. It was demonstrated that $RRAT_{655}(A)$ and $RGET_{691}(A)$ retained PKA activation at pH 7.4. $RGET_{691}$, but not $RRAT_{655}$, was required for PKA activation at pH 6. An $RGED_{691}$ mutant was constitutively active and could be further activated by PKA at low pH. These studies demonstrated that the consensus phosphorylation sites, $RRAT_{655}$ and $RGET_{691}$, were responsible for the observed PKA activation of human ClC-2 and showed that there was interplay between activation by phosphorylation and by low pH.

Rabbit ClC-2 also contains two consensus PKA phosphorylation sites RRQS (for PKA, CaMKII and PKC) and KRKS (for PKA and PKC) in the same general region of the channel shown to be responsible for PKA activation of the human channel. Activation of the native and recombinant rabbit ClC-2 channel was shown to be activated by ATP plus PKA catalytic subunit in planar lipid bilayers (Malinowska et al. 1995).

Alioth et al. (2007) showed that this region of the channel (where PKA phosphorylation sites are situated) contains organized structures, which appear to interact with the CBS domains, and they suggested that this mechanism is responsible for regulation of the channel. This same region is also phosphorylated in a cell cycle-dependent ClC anion channel (Denton et al. 2005). It has been recently suggested that phosphorylation in this region may be a general mechanism for regulation of ClC channels (Yamada et al. 2013; Miyazaki et al. 2012). Proteins involved in regulation of phosphorylation have been identified (Denton et al. 2005; Rutledge et al. 2002; Furukawa et al. 2002). GCK-3 is a kinase that has been shown to alter fast gating in *C. elegans* CLH-3b, an ortholog of ClC-2. GSK-3 binds to a specific site on CLH-3b and mutation of the binding site abolishes regulation by GCK-3 (Denton et al. 2005). Activation of CLH-3b during meiosis and oocyte swelling is also regulated by *C. elegans* PP1-type phosphatases CeGlC-7α and β (Rutledge et al. 2002). Furukawa et al. (2002) showed phosphorylation and functional regulation of rabbit ClC-2 by M cyclin-dependent protein kinase and that PP1α and PP1β physically interact with the COOH terminus of rabbit ClC-2.

Park et al. (2001) showed that rat ClC-2 could be phosphorylated by PKA, but without a change in the extent or rate of activation. These contrasting results are presumably because rat ClC-2 does not contain the same type of phosphorylation sites as those in human and rabbit in these regions (Cid et al. 1995; Sherry et al. 1997; Cuppoletti et al. 2004a). Mouse ClC-2 (which has the same sequence as the rat channel in this region) expressed in HEK293EBNA cells was not activated by forskolin-IBMX using the same conditions where human ClC-2 is activated (Cuppoletti et al., unpublished).

Porcine choroid plexus ClC-2 was activated by extracellular VIP as well as intracellular PKA catalytic subunit (Kajita et al. 2000), although the sites of PKA phosphorylation in porcine ClC-2 are not known.

13.8 Small-Molecule Activators of ClC-2

13.8.1 Protons

Protons were initially shown to activate ClC-0 (Hanke and Miller 1983). Protons also activate human ClC-2 currents (Sherry et al. 1997; Schwiebert et al. 1998; Tewari et al. 2000; Cuppoletti et al. 2001, 2004a; Stölting et al. 2014). Rabbit ClC-2 was activated by acid (Malinowska et al. 1995; Stroffekova et al. 1998). Chemical modification (amidation) resulted in activation and loss of further pH activation. Site-directed mutagenesis of the channel resulted in identification of residues 416–419, EELE, responsible for the pH effect (Stroffekova et al. 1998). EELE is located in the extracellular loop between transmembrane domains D8 and D9 and is conserved in both human and rabbit ClC-2, but not in rat ClC-2. Rat ClC-2 is also activated by low pH (Blaisdell et al. 2000). Jordt and Jentsch (1997) studied rat ClC-2 and showed that the cytoplasmic loop between D7 and D8 was an important region for acid activation. Protonation of a glutamate residue in the pore of guinea pig ClC-2 has been found to be important for gating (Niemeyer et al. 2003, 2009).

13.8.2 ATP

ATP has been shown to be important for endocytosis (Dhani et al. 2008). ATP depletion inhibits endocytosis of the channel (Dhani et al. 2008), but these effects appear to be through effects on proteins involved in intracellular trafficking.

Fritsch and Edelman (1996) showed that T84 cell ClC-2 could be measured with ATP in the pipette, but that omission of ATP or use of the non-hydrolyzable ATP analogue 5′-adenylimidodiphosphate accelerated activation kinetics and increased amplitude of the currents. Stölting et al. (2013) further investigated the ATP effect, showing that there was concentration-dependent slowing of the activation kinetics over the range of 1–5 mM ATP or ADP and the non-hydrolyzable 5′(β-γ-imido) triphosphate (but not AMP). ATP is known to bind to the CBS domains of the ClC channels (Meyer et al. 2007). Thus, ATP binding to the CBS domains slows gating of the channel (Stölting et al. 2013). A variety of natural mutants in the region of the CBS domain were also studied by Stölting et al. (2013), and these have altered gating expected for disruption of the ATP-/ADP-binding site. Studies on ATP/ADP and non-hydrolyzable analogues were carried out without fatty acid activators or phosphorylation to increase the activity of the human channel.

13.8.3 Activation by Fatty Acids and Lubiprostone

13.8.3.1 Lubiprostone Stimulation of Recombinant ClC-2

Lubiprostone, a bicyclic fatty acid, also activates human ClC-2 at low concentrations (EC_{50} approximately 20–30 nM; Cuppoletti et al. 2004b, 2013, 2014). These studies were carried out using recombinant human ClC-2 expressed in HEK293 cells (Cuppoletti et al. 2004b) and in a HEK293EBNA cells (Cuppoletti et al. 2013, 2014), which express a higher level of ClC-2 than HEK293 cells (Cuppoletti et al. 2013). ClC-2 currents in the latter cells give time- and voltage-dependent current recordings more readily than ClC-2 in HEK293 cells, and the currents were fourfold higher in control and lubiprostone-stimulated cells.

13.8.3.2 Lubiprostone Activation in Cell Lines Containing Both ClC-2 and CFTR

In T84 cell I_{sc} studies (Cuppoletti et al. 2004b), lubiprostone gave a similar EC_{50} as measured for recombinant human ClC-2 in HEK293 cells, but the basis for lubiprostone stimulation of I_{sc} of T84 cells was not determined in those studies. However, recent patch-clamp studies using various more selective inhibitors of ClC-2 and CFTR and knockdowns of ClC-2 and CFTR in T84 cells have allowed assignment of lubiprostone stimulation to ClC-2 (Cuppoletti et al. 2008b, 2014).

Since lubiprostone at concentrations up to 1 μM did not activate recombinant human CFTR in HEK293 cells (Cuppoletti et al. 2004b, 2013), it was concluded that T84 cell CFTR was not activated by lubiprostone. In later studies, Cuppoletti et al. (2008b) showed that in native T84 cells ($n = 6$), 20 nM lubiprostone significantly stimulated Cl^- currents ($P < 0.0025$) (control, -34.2 ± 3.4 pA/pF; with lubiprostone -78.2 ± 9.0 pA/pF). Cd^{2+} (500 μM) significantly ($P < 0.001$) inhibited the stimulated currents to -25.8 ± 3.5 pA/pF. It was further shown that knockdown of T84 ClC-2 with siRNA caused a loss of lubiprostone-stimulated Cl^- currents. Similar findings were demonstrated (Cuppoletti et al. 2014) where shRNA was used to knock down ClC-2 in T84 cells. When ClC-2 protein was reduced to very low levels, lubiprostone-stimulated currents were abolished compared to wild-type T84 cells. CFTR function was retained, stimulated by forskolin-IBMX, not affected by methadone, inhibited by DASU-02, and inhibited to very low levels by CFTRinh172. In contrast, when shRNA to CFTR was used to knock down CFTR in T84 cells, CFTR protein was reduced to very low levels, and lubiprostone-stimulated current was retained at levels comparable to those of the wild-type cells. The current was not inhibited by DASU-02 but reduced by methadone and further reduced by Cd^{2+}. Since human ClC-2 is also activated by PKA (Cuppoletti et al. 2004a, b, 2013, 2014), mPKI inhibited the response to forskolin-IBMX but did not affect lubiprostone stimulation of ClC-2 (Cuppoletti et al. 2013). Interestingly, treatment with mPKI reduced the control, basal (unstimulated) currents slightly without affecting lubiprostone stimulation suggesting that a part of the basal activity may be due to partially phosphorylated ClC-2. Similar effects were seen with treatment with H-89. H-89 results are unpublished, since H-89 has numerous nonspecific effects which cannot be controlled for without H89's inactive form, H-85, which is no longer available (see Sect. 13.12 on H-89). Thus, in T84 cells, CFTR does not contribute to lubiprostone-stimulated Cl^- currents, and ClC-2 activation is independent of the presence of CFTR, as demonstrated in HEK293 cells transfected separately with ClC-2 or CFTR (Cuppoletti et al. 2004b), in HEK293EBNA cells (Cuppoletti et al. 2013, 2014) transfected with ClC-2 alone, and in T84 cells lacking CFTR (Cuppoletti et al. 2014). Knockdown of ClC-2 in T84 cells causes loss of lubiprostone stimulation (Cuppoletti et al. 2014).

In studies of T84 cells, Bijvelds et al. (2009) found no inhibition by Cd^{2+}, which however has been shown to inhibit T84 lubiprostone-stimulated currents in patch-clamp studies (Cuppoletti et al. 2008b, 2013, 2014). It is not clear why Cd^{2+} responses were not seen in this study, especially since it is generally toxic to epithelial cells (Bao et al. 2008; Moeser et al. 2004, 2007). Bijvelds et al. (2009) also used CFTRinh172 assuming it is CFTR specific and found block of all Cl^- currents and ascribed all of the current activated by lubiprostone to CFTR. CFTRinh172 has now been shown to be equipotent in inhibiting both CFTR and ClC-2 (Cuppoletti et al. 2014). When lubiprostone and forskolin were used together, the experimental design did not take into account the fact that lubiprostone and forskolin-IBMX both stimulate human ClC-2 (Cuppoletti et al. 2004b, 2013, 2014). Yet they ascribed all of the current increases only to CFTR. Bijvelds et al. (2009) found cAMP was raised by lubiprostone, but to levels only about 30% of that of forskolin. Despite this, lubiprostone and forskolin gave similar activation of I_{sc}. This is not expected from the studies of Cartwright et al. (1985) showing that anion transport in T84 cells is proportional to cAMP levels over the ranges reported by Bijvelds et al. (2009). Neither Cuppoletti et al. (2008a) nor Bao et al. (2008) found increased cAMP levels with lubiprostone.

Lubiprostone similarly activated I_{sc} in Calu-3 cells, and knockdown of >95% of CFTR had no effect on lubiprostone stimulation with or without addition of the CFTR channel inhibitor GlyH-101 (*N*-(2-naphthalenyl)-[(3,5 dibromo-2,4-dihydroxy phenyl)methylene glycine hydrazide]) (MacVinish et al. 2007; Muanprasat et al. 2004). Thus, also in Calu-3 cells, lubiprostone does not stimulate CFTR, but rather it stimulated ClC-2. CFTR is not required for stimulation of ClC-2 by lubiprostone.

13.8.3.3 Site of Action of Lubiprostone in Stimulation of ClC-2

When one of two phosphorylation sites in human ClC-2 is mutated, $RGET_{691}$ > RGEA, the EC_{50} for lubiprostone shifted from 27.6 ± 4.5 nM ($n = 4$) to 445.8 ± 2.1 nM (Cuppoletti et al. 2004a). In the triple mutant $RGET_{691}$ > AGAA, complete loss of lubiprostone stimulation occurred (Cuppoletti et al. 2014, 2017), but forskolin-IBMX acting at $RRAT_{655}$ (Cuppoletti et al. 2004a) still activated human ClC-2 via PKA. Therefore, the $RGET_{691}$ site appears to be essential for lubiprostone stimulation. Greater detail is provided in Sect 13.8.3.6.

13.8.3.4 Lubiprostone Stimulation of Recombinant ClC-2 from Other Species

Lubiprostone also activated ClC-2 channels from other species, as determined by whole-cell patch-clamp of recombinant ClC-2 from various species expressed in HEK293EBNA cells (Cuppoletti et al. 2014). Rabbit ClC-2 gave an EC_{50} of 350 ± 37.9 nM ($n = 3$), and mouse ClC-2 gave an EC_{50} of 64.3 ± 3.7 nM ($n = 4$). By comparison, the EC_{50} for human ClC-2 expressed in these cells was 28.2 ± 2.2 nM ($n = 4$) (Cuppoletti et al. 2013). Norimatsu et al. (2012) were unable to activate rabbit ClC-2 with lubiprostone expressed in *Xenopus* oocytes. In contrast, rabbit ClC-2 expressed in HEK293 cells was stimulated by lubiprostone with an EC_{50} of approximately 350 nM (Cuppoletti et al. 2014). It is unclear why Norimatsu et al. (2012) were unable to observe activation of rabbit ClC-2 in the *Xenopus* oocyte expression system. It is possible that in this *Xenopus* system, ClC-2 was pre-activated by endogenous PKA before the addition of lubiprostone.

13.8.3.5 Lack of Involvement of PKA Phosphorylation in Lubiprostone Stimulation

Lubiprostone stimulation of human ClC-2 is independent of PKA, as determined from experiments where mPKI inhibited activation by forskolin-IBMX, but not activation by lubiprostone (Cuppoletti et al. 2004b, 2013). It has also been shown that a double phosphorylation knockout mutant of ClC-2 cannot be activated by forskolin-IBMX (Cuppoletti et al. 2004a) but is activated by 20 nM lubiprostone (Cuppoletti et al. 2008a). Wild-type ClC-2 control currents at −140 mV ($n = 6$) of -27.2 ± 3.5 pA/pF significantly ($P < 0.0005$) increased to -100.9 ± 6.5 pA/pF with lubiprostone, and 500 μM Cd^{2+} reduced the current to control levels. The mutant channel currents were also significantly ($P < 0.005$) increased by lubiprostone and inhibited by $CdCl_2$. Control currents were -27.1 ± 6.9 pA/pF which increased to -81.9 ± 5.1 pA/pF with lubiprostone and were reduced with $CdCl_2$ to -29.4 ± 3.7 pA/pF ($P < 0.005$) ($n = 6$). Thus, PKA phosphorylation is not involved in lubiprostone stimulation of human ClC-2.

It is clear from the above that human ClC-2 and not human CFTR was activated by lubiprostone and that PKA is not involved in lubiprostone stimulation. Bao et al. (2008) also carried out single-channel studies of the apical membrane of A6 epithelial cells and human ClC-2. The EC_{50} for A6 cell ClC-2 single channels was 69 ± 18 nM. However, in this study, there was also a non-cAMP-dependent activation of CFTR with an EC_{50} of 791 ± 273 nM; approximately tenfold higher than the EC_{50} for ClC-2. The mechanism whereby lubiprostone stimulates *Xenopus* A6 cell CFTR is not known.

13.8.3.6 Identification of Possible Fatty Acid Binding Site on Human ClC-2

Fatty acids including arachidonic acid activate human ClC-2 (Tewari et al. 2000; Cuppoletti et al. 2001, 2004a, 2013, 2014). Tewari et al. (2000) showed that recombinant human ClC-2 was activated by a variety of fatty acids including oleic acid (C:18 *cis*D9), elaidic acid (C:18 *trans*D9), arachidonic acid (AA; C:20 *cis*D5,8,11,14), and 5,8,11,14-eicosatetraynoic acid (ETYA; C:20 *trans*D5,8,11,14). Fatty acid activation of the human ClC-2 channel was not inhibited by mPKI or staurosporine and was therefore independent of PKA or PKC activation. Site-directed mutagenesis of one of the two PKA phosphorylation sites leads to ablation of fatty acid activation of ClC-2 (Cuppoletti et al. 2014, 2017) (see Sect. 13.8.3.3). More recently, Cuppoletti et al. (2017) investigated the role of a four-amino acid PKA activation site, $RGET_{691}$, in fatty acid activation of human ClC-2. Involvement of PKA, intracellular cAMP, EP_2, or EP_4 receptor agonist activity was also investigated. Wild-type (WT) and mutant (AGET, RGEA, and AGAA) hClC-2 expressed in HEK293EBNA cells were used. All the fatty acids examined (lubiprostone, cobiprostone, ETYA, oleic acid, and elaidic acid) caused significantly increasing

EC_{50}s (rightward shifts in concentration-dependent Cl^- current activation) in mutant compared to WT hClC-2 Cl^- channels. This occurred without altering time and voltage dependence, current-voltage rectification, or methadone inhibition of the channel. Fatty acids activated WT hClC-2 in the presence of the PKA inhibitor (mPKI) did not increase intracellular cAMP and were not agonists of EP_2 or EP_4 receptors. This study (Cuppoletti et al. 2017) suggests that $RGET_{691}$ of hClC-2 (possible binding site) plays a definitive, important functional role in fatty acid activation of hClC-2, providing the molecular basis for regulation of hClC-2 by fatty acids.

13.9 ClC-2 Function in Epithelia

Bösl et al. (2001) showed that ClC-$2^{-/-}$ mice showed only retinal and testes degeneration. Mice with a double CFTR and ClC-2 knockout (Zdebik et al. 2004) did not have increased mortality over $CFTR^{-/-}$ mice, and there were no anomalies in electrocardiograms or the morphology of the intestine, lung, or pancreas. ClC-$2^{-/-}$ mice crossed into a background of CFTR F508del/F508del lived longer than $CFTR^{F508del/F508del}$ mice. Increased Cl^- secretion was seen in the colon of ClC-$2^{-/-}$ mice. An important question then is whether ClC-2 has any additional functions, as might be expected with it being ubiquitously expressed.

Bösl et al. (2001) also reported that histamine-stimulated gastric acid secretion was normal in ClC-$2^{-/-}$ mice. This was somewhat unexpected, since ClC-$2^{-/-}$ mouse gastric mucosal histology was greatly disrupted. Parietal cell number, H^+/K^+ ATPase expression, and cytoplasmic tubulovesicles were all significantly reduced which would be expected to result in less HCl secretion in the ClC-$2^{-/-}$ mouse (Nighot et al. 2011, 2014, 2015). Indeed, Nighot et al. (2015) found that histamine stimulated $[H^+]$ of the gastric contents was significantly, 89%, lower and histamine/carbachol-stimulated gastric acid secretion was significantly, 84–95%, lower in the ClC-$2^{-/-}$ mouse compared to WT, while pepsinogen secretion was unaffected. ClC-$2^{-/-}$ mice also have significant loss of electroneutral absorption in the distal colon (Catalán et al. 2012). In addition, ClC-$2^{-/-}$ mice are more susceptible to damage of the intestine (Nighot et al. 2009). Thus, it appears that under certain conditions, ClC-2 may play a role in tissues other than the retina and testes. Additional examples of the importance of ClC-2 in epithelia will be discussed.

It has been shown by Gyömörey et al. (2000) that in the small intestine of mice, there exists a basal Cl^- current and a hyperpolarization-stimulated current which can be activated by hypertonic shock. These currents were sensitive to glibenclamide, an open-channel blocker of ClC-2 (Bao et al. 2008) as well as NPPB. These currents were present in both wild-type $CFTR^{+/+}$ and $CFTR^{-/-}$ mice. The channel had a distribution in the small intestine at the cellular level similar to that seen in Caco-2 cells (Mohammad-Panah et al. 2001), namely, predominantly at the tight junction complex between cells with diffuse labeling of the apical brush border membrane. Gyömörey et al. (2000) concluded that ClC-2 contributes to Cl^- secretion in both the wild-type and $CFTR^{-/-}$ mouse, as was concluded in studies of Caco-2 cells (Mohammad-Panah et al. 2001). Catalán et al. (2002, 2004) demonstrated that ClC-2 is expressed in the basolateral membrane of the mouse distal colon and indeed showed that the $ClC\text{-}2^{-/-}$ mouse has an absorptive defect in the distal colon (Catalán et al. 2012). It must be noted however that the basolateral location of ClC-2 in the colon does not mean that the work of others showing apical distribution in other parts of the intestine and lung are incorrect.

As described earlier (Sect. 13.8.3.2), Bijvelds et al. (2009) found no evidence for ClC-2 function in T84 cells based on the lack of effect of Cd^{2+}, the use of CFTRinh172 assumed to be CFTR specific and assuming that only CFTR is activated by forskolin. Using the same approaches, Bijvelds et al. (2009) also did not find evidence for ClC-2 in any experiments in healthy or CF rectum or ileum, $CFTR^{-/-}$ mice, or $CFTR^{F508del/F508del}$ mice. These findings are in conflict with those of Gyömörey et al. (2000), Catalán et al. (2012), Fei et al. (2009), and Moeser et al. (2004, 2007) who have readily demonstrated ClC-2 function in the intestine.

In another study in guinea pig intestine and colon, Fei et al. (2009) found that lubiprostone increased I_{sc} in the intestine, with an EC_{50} of 42.5 nM, and in the colon, with an EC_{50} of 31.7 nM. Lubiprostone responses were blocked by glibenclamide, which has been shown to be an open-channel blocker of ClC-2 (Bao et al. 2008) in addition to inhibiting CFTR. Antagonists of EP_1, EP_2, EP_3, or EP_4 receptors did not suppress I_{sc}, although they blocked PGE_2 responses. The finding by Fei et al. (2009) that an EP_4 antagonist, GW627368X, did not block I_{sc} responses conflict strongly with the results and conclusions of Bijvelds et al. (2009) where L161,982, also an EP_4 antagonist, inhibited all I_{sc} responses. This suggests that L161,982 at high concentrations will affect other prostaglandin receptors and may have unknown or nonspecific effects (see also Sect. 13.11). In contrast, the guinea pig appears to be well suited for studies of ClC-2 in intestinal function.

The possible involvement of ClC-2 in Cl^- transport function and barrier function has also been studied in the porcine intestine (Moeser et al. 2004, 2007). Ischemia, NSAIDS, and pathogens disrupt the intestinal epithelial barrier. PGE_2 has been shown to repair damaged porcine intestine and to increase Cl^- secretion as measured by I_{sc} in the small intestine (Moeser et al. 2004). Three Cl^- channels are present in the porcine intestine: Ca^{2+}-activated Cl^- channels, CFTR, and ClC-2. DNDS and DASU-02, inhibitors of Ca^{2+}-activated Cl^- channels (outwardly rectified), and CFTR, respectively, were used. Neither of these inhibitors prevented PGE_2-stimulated improvement in barrier function, as measured by transepithelial resistance (TER) or PGE_2-stimulated I_{sc}. However, Zn^{2+}, an inhibitor of ClC-2, inhibited both PGE_2-stimulated Cl^- transport and restitution of the barrier properties. Cd^{2+} was not used because histological examination showed that it caused damage to the tissue. ClC-2 channel protein was upregulated upon ischemic injury, while CFTR channel protein was downregulated. Occludin was found to immunoprecipitate with ClC-2 suggesting an association of ClC-2 with the tight junction. Immunogold electron microscopy showed ClC-2 localization at the tight junction during recovery. Immunofluorescence showed that ClC-2 was localized to the apical aspect of mucosal villi as well as in mucosal crypts, as indicated by others (Gyömörey et al.

2000; Mohammad-Panah et al. 2001; Moeser et al. 2004; Jin and Blikslager 2015). Thus, this porcine model of ischemic injury and recovery has demonstrated physiological roles for ClC-2 in barrier repair and associated Cl^- secretion. Careful selection of Cl^- channel inhibitors and activators was a strength of this study and will be instructive to the field.

In the same ischemia-injured porcine tissue, lubiprostone was also shown to give a concentration-dependent increase in TER and reduction of ^{3}H-mannitol flux. It also promoted recovery of ClC-2 to tight junctions. I_{sc} was stimulated by lubiprostone in a concentration-dependent manner, and the recovery of TER and I_{sc} increases were significantly correlated with lubiprostone treatment, whereas there was a poor correlation with PGE_2 treatment (Moeser et al. 2007). TER recovery and I_{sc} were blocked by Zn^{2+}, whereas DASU-02 was without effect, suggesting that ClC-2, and not CFTR, was responsible for the barrier repair and associated Cl^- secretion. The porcine intestine appears to be well suited for studies of the physiological functions of ClC-2.

In addition, it has been shown in a ClC-2$^{-/-}$ mouse model of ulcerative colitis that ClC-2 plays an important role in maintaining and improving recovery of loss of barrier function and maintaining tight junctions (Nighot et al. 2013). The model is used to mimic ulcerative colitis patient intestinal problems, and this result again suggests that mice might be useful for studies of ClC-2 function under some circumstances (Nighot et al. 2009). In the mouse intestine, ClC-2 appears to be essential for recovery of barrier function. In uninjured animals, ClC-2 knockout appears to improve intestinal barrier function (Nighot et al. 2009), but in the ischemic mouse intestine, ClC-2 appears to be essential for recovery of barrier function (Nighot and Blikslager 2010). Nighot and Blikslager (2012) showed that ClC-2 modulates tight junction function through its effects on caveolar trafficking of occludin. More recently, the findings of Nighot et al. (2017) using Caco 2 cells overexpressing human ClC-2 suggested that ClC-2 enhances tight junction barrier function via regulation of caveolin-1 and caveolae-mediated trafficking of occludin.

A comprehensive study by Joo et al. (2009) on the effects of lubiprostone on airway submucosal fluid secretion and surface epithelium I_{sc} in pigs, sheep, and man has appeared. According to the authors, "...a major reason for these studies was to examine ClC-2 channels as possible surrogates for CFTR using the putative ClC-2 activator lubiprostone in the airways..." and the authors cite the findings of Schwiebert et al. (1998) showing that ClC-2 currents can be obtained in human CF airway cells. Humans, pigs and sheep all responded to lubiprostone. The authors found that lubiprostone also stimulated anion currents at the apical surface epithelium with a K_d of 10.5 nM, with a single site of action, similar to that seen by others using recombinant human ClC-2 and T84 cells (Cuppoletti et al. 2004a, 2013). I_{sc} responses were sensitive to removal of Cl^- and were sensitive to 300 μM Cd^{2+}. I_{sc} responses to lubiprostone were reversible, and they were sensitive to bumetanide, to furosemide, and to the removal of Cl^-, thus likely emanating from Cl^- secretion. GlyH-101 caused a rightward shift in the concentration response curve, and the authors noted a personal communication from Dr. Verkman indicating that GlyH-101 may have some nonspecific actions, in addition to its action on CFTR.

GlyH-101 was previously shown by MacVinish et al. (2007) to not inhibit Cl^- secretion in Calu-3 cells when CFTR was ablated. Fluid secretion in airway glands of human, pigs, and sheep was also increased by lubiprostone. Removal of Cl^- and treatment with furosemide prevented lubiprostone stimulation of gland fluid secretion. In the presence of a maximal concentration of forskolin in pig, human, and sheep submucosal glands, lubiprostone caused a further increase in secretion, suggesting that forskolin and lubiprostone affect different targets. Gland fluid secretion was stimulated, but the increases were not significant in samples from two out of four CF patients and in two of three chronic obstructive pulmonary disease and idiopathic pulmonary fibrosis patient samples. No bronchoconstriction was noted with lubiprostone. Together, these results support the view that airway surface epithelium and glandular fluid secretion can be modulated by activators of ClC-2. These studies suggest that the tracheal surface epithelium and submucosal glands are a good model system for studies of activators of ClC-2 in the presence of normal or defective CFTR.

A contrasting conclusion was made by Cuthbert (2011). Lubiprostone was shown to stimulate sheep surface epithelium I_{sc} and submucosal gland secretion. Following the various protocols and approaches of Bijvelds et al. (2009), Cuthbert (2011) found that 10 μM L161,982 (an EP_4 antagonist), when added to the apical bathing solution, but not when added to the basolateral bathing solution of the surface epithelium, inhibited the lubiprostone response. 1 μM L161,982 shifted the concentration of lubiprostone required to achieve 50% activation by more than tenfold. However, L161,982 had no effect on lubiprostone-stimulated hClC-2 Cl^- currents but significantly decreased T84 cell barrier function as measured by TER and fluorescent dextran movement (Cuppoletti et al. 2017), indicating nonspecific effects of L161,982. Moreover, again following the approaches of Bijvelds et al. (2009), Cuthbert (2011) used 40 μM H-89, which blocked the lubiprostone response by only 50%. H-89 is discussed in Sect. 13.12, and experiments using H89 must be interpreted with great caution because of possible off-target or nonspecific effects which, as suggested above, may also apply to L161,982. Cuthbert (2011) further showed that the EP_4 agonist L-90288 inhibited the lubiprostone response, but not in a predictable manner. It is not known whether ovine ClC-2 is activated by PKA as is human ClC-2 (Cuppoletti et al. 2004a, 2014), so it is difficult to evaluate if L161,982 should prevent activation of ovine ClC-2 by lowering $[cAMP]_i$. It has been clearly shown that ClC-2 (non-CFTR Cl^- channel) is activated by lubiprostone in Calu-3 cells lacking CFTR (MacVinish et al. 2007). Others have shown that Calu-3 cells have EP_4 receptors which are coupled to anion transport (Joy and Cowley 2008). However, no test was done by Cuthbert (2011) to determine ClC-2 function in the airway. The interpretation of all of his experiments was that no functional ClC-2 was observed.

Acidic pH hyperpolarizes nasal potential difference (Uwaifo et al. 2006), suggesting the presence of ClC-2 in the nasal epithelia. A recent paper studied the effects of treatment of murine nasal epithelia with lubiprostone. Schiffhauer et al. (2013) suggest that CFTR and ClC-2 are activated by lubiprostone in mouse nasal epithelia. The authors treated mouse nasal epithelia with 20 μM lubiprostone. They used mice which conditionally (doxycycline regulated) overexpressed hClC-2 in the airway, CFTR$^{-/-}$ mice and ClC-2$^{-/-}$ mice. Human ClC-2 was found in the apical region of the doxycycline-treated mice and at lower levels in untreated mice and in the CFTR$^{-/-}$ mice. CFTR expression was similar in untreated, doxycycline-treated, and ClC-2$^{-/-}$ mice. Lubiprostone (20 μM) stimulated nasal Cl^- potential difference (PD) in doxycycline-untreated mice and stimulated nasal Cl^- PD to higher levels in the doxycycline-treated mice. Nasal Cl^- PD measurements were carried out in the presence of CFTRinh172, which may have decreased the response since it is now known that CFTRinh172 inhibits ClC-2 as well as CFTR (Cuppoletti et al. 2014). 200 nM GaTx2 inhibited the response in doxycycline-treated mice. Lubiprostone-stimulated nasal Cl^- PD in CFTR$^{-/-}$ mice and GaTx2 completely blocked the response. CFTRinh172 was not present in this experiment.

In addition, Schiffhauer et al. (2013) showed that lubiprostone also stimulated nasal Cl^- PD in ClC-2$^{-/-}$ mice. CFTRinh172 dose dependently inhibited nasal Cl^- PD in ClC-2$^{-/-}$ mice. The results of the study were interpreted to show that CFTR was activated by lubiprostone in the ClC-2$^{-/-}$ mice. This result was in conflict with findings in a previous study (MacDonald et al. 2008) where ClC-2$^{-/-}$ mice did not give a response to lubiprostone. This difference was not discussed and remains to be explained.

In the MacDonald et al. (2008) study, there was no effect of lubiprostone on ClC-2$^{-/-}$ mice, but wild-type and CFTR$^{-/-}$ mice showed activation consistent with stimulation of ClC-2. CFTRinh172 did not prevent the response, and it was concluded that CFTR was not required for the response to lubiprostone. The lack of effect of lubiprostone in ClC-2$^{-/-}$ mice suggests that CFTR is not activated by 20 μM lubiprostone in contrast to the effect of lubiprostone seen in the ClC-2$^{-/-}$ mouse in the Schiffhauer et al. (2013) study. It should be considered that at high concentrations (20 μM) of lubiprostone, there may be nonspecific effects. Further experiments using lower concentrations of lubiprostone might help clarify these conflicting results.

To reiterate, in the case of the doxycycline-treated mice overexpressing human ClC-2, 20 μM lubiprostone was used, which is 1000 times the EC_{50} of lubiprostone for human ClC-2 as determined in patch-clamp of human ClC-2 expressed in HEK293 cells (Cuppoletti et al. 2004a, 2013). Nevertheless, the results taken together suggest that ClC-2 is activated by lubiprostone, and the differences in the findings on CFTR in ClC2$^{-/-}$ mice between the two studies suggest that additional studies are needed to clarify whether indeed lubiprostone at these very high concentrations (20 μM) can have an effect on CFTR in addition to ClC-2. Also, future studies must take into account the equipotent inhibitory effect of CFTRinh172 on ClC-2, as well as CFTR (Cuppoletti et al. 2014).

Lubiprostone has been shown to cause a dose-dependent increase in Cl^--rich fluid in the small intestine of rats, without affecting the serum electrolyte levels (Ueno et al. 2004). Lubiprostone, given orally, is only known to reach the small intestine (Ueno R, personal communication).

13.10 Lubiprostone in Human Treatments

13.10.1 Chronic Idiopathic Constipation, Irritable Bowel Syndrome, and Opioid-Induced Constipation

Lubiprostone is approved for use in the treatment of chronic idiopathic constipation (Lembo et al. 2011; Barish et al. 2010; Johanson et al. 2008a). It is also used for treatment of irritable bowel syndrome (Chey et al. 2012; Carter and Scott 2009; Drossman et al. 2009; Johanson et al. 2008b) and opioid-induced constipation (Wong and Camilleri 2011; Cryer et al. 2014).

13.10.2 Effect of Lubiprostone in the Intestine of CF Patients

Pilot studies of the effect of lubiprostone on treatment of constipation in CF patients have been carried out (O'Brien et al. 2011). Lubiprostone improved overall symptoms of constipation in these CF patients.

13.11 Lubiprostone Is Not an EP_4 Receptor Agonist: Rather, It Is an EP_4 Receptor Antagonist

Bijvelds et al. (2009) and Cuthbert (2011) used L161,191 at high concentrations and found that it inhibited Cl^- transport as measured by I_{sc}. The K_i for blocking EP_4 receptors by L161,982 is 0.032 nM (Weinreb et al. 2006). Its specificity is 200 times greater for EP_4 than that for EP_{1-3} (K_i 1/4 6.4 μM) (Weinreb et al. 2006). Bijvelds et al. (2009) used concentrations of L161,982 as high as 20 μM, which is 625 times the IC_{50} for EP_4 receptor antagonism, and Cuthbert (2011) used L161,982 at concentrations as high as 10 μM. At these concentrations, L161,982 will also act upon EP_{1-3} receptors and could have other nonspecific effects. Thus, it is not possible to interpret the effects of L161,982 as being due to the binding of lubiprostone to EP_4 receptors alone. Other EP_4 receptor antagonists had no effect on lubiprostone stimulation of I_{sc} (Fei et al. 2009).

Lubiprostone is not an agonist of cloned human EP_4 receptors, but rather an antagonist of EP_4 receptors with an IC_{50} of 143 nM (Cuppoletti et al. 2008c, d). Also, lubiprostone activated ClC-2 whether expressed in HEK293 cells (Cuppoletti et al. 2004a, 2008a) or overexpressed in HEK293EBNA cells (Cuppoletti et al. 2013, 2014, 2017). These cells do not contain EP receptors, and for this reason, they are used for expression of EP receptors (Boie et al. 1997). Thus, lubiprostone binding to EP_4 receptors is not involved in lubiprostone stimulation of ClC-2 mediated Cl^- transport. Studies of the effects of 10–20 μM L161,982 on ClC-2 and CFTR in native cells and on recombinant ClC-2 and CFTR channel function and cell physiology in general would be required before it is possible to interpret L161,982 effects. In view of these reports, we examined the effects of L161,982 on the lubiprostone and forskolin/IBMX responses of recombinant ClC-2 and CFTR and knockdown T84 cells where only CFTR or ClC-2 is present. Lubiprostone stimulated ClC-2, but not CFTR in these systems, and forskolin/IBMX stimulated both ClC-2 and CFTR. In all cases there was no effect of L161,982 (unpublished data). Cuppoletti et al. (2017) reported that T84 cell barrier function was significantly decreased by L161,982, as measured by TER and fluorescent dextran movement while having no effect on lubiprostone-stimulated hClC-2 Cl^- currents. These nonspecific effects could explain many findings. The high concentrations used also mean that EP_{1-4} receptors would be occupied by L161,982. L161,982 effects clearly cannot be interpreted as antagonizing the binding of lubiprostone at only EP_4 receptors. In addition, other EP_4 antagonists do not inhibit lubiprostone-stimulated I_{sc} (Fei et al. 2009). To reiterate, lubiprostone at concentrations as high as 1 μM is not an agonist of cloned human EP_4 receptors, rather it is an antagonist (Cuppoletti et al. 2008a) (see also Sect. 13.9).

13.12 H-89 Effects

Bijvelds et al. (2009) studied the effect of treatment with 20 μM H-89 for 30 min on lubiprostone-stimulated I_{sc} in T84 cells and found partial (70%) inhibition, as we found in studies with mPKI (Cuppoletti et al. 2013) and H89 (see Sect. 13.8.3.2). Cuthbert (2011) studied the effect of 40 μM H-89 for 30 min on lubiprostone stimulation of ovine lung I_{sc} and also found partial (50%) inhibition. Both investigators assigned the effect to inhibition of CFTR. Strohmeier et al. (1995) and Hoque et al. (2010) both showed approximately 80–90% inhibition of forskolin-stimulated anion transport in T84 cells using 60 μM and 50 μM H-89, respectively. H-89 was much less effective in inhibition of anion transport in the Cuthbert (2011) studies. The inability of H-89 to more fully inhibit is consistent with lubiprostone stimulation of ClC-2 rather than CFTR.

We have examined the effect of a more specific PKA inhibitor, mPKI (Lochner and Moolman 2006), on lubiprostone-stimulated currents (Cuppoletti et al. 2004a, 2013). It does not inhibit lubiprostone-stimulated currents. mPKI readily inhibited forskolin-stimulated currents of HEK293EBNA cells containing only ClC-2 and also (not significantly) inhibited basal (non-stimulated) currents (Cuppoletti et al. 2013). Cuppoletti et al. (unpublished) found that H-89 behaves similarly and inhibits basal currents of recombinant ClC-2, but it did not inhibit lubiprostone-stimulated currents in HEK293EBNA cells containing only ClC-2. It is not possible to determine the specificity of H-89 without the use of H-85 (an inactive analogue) as explained below.

H-89 has been widely used to study ion channels. Park et al. (2007) showed that H-89 and its inactive analogue H-85 had similar effects, suggesting direct modulation of Ca^{2+}-activated K^+ currents (BK) with an EC_{50} of 0.5 μM H-89. Bijvelds et al.

(2009) used 40 times this concentration (20 μM), and Cuthbert (2011) used 80 times this concentration (40 μM). Similarly, Son et al. (2006) showed that H89 and H85 both inhibited an arterial Kv channel with a K_d of 1.2 μM. Sun Park et al. (2006) also found direct inhibition of K_{ATP} and K_{ir} channels with IC_{50}'s of 1.2 and 3.78 μM, respectively. Niisato et al. (1999) showed that epithelial Na^+ channels were instead activated by treatment with H89, H85, H7, and H8, whereas mPKI and KT-5720 decreased the currents. Marunaka and Niisato (2003) showed that H-89 stimulated Na^+ transport by transporting ENaC to the apical membrane. Thus, in the absence of testing whether there are effects of H-85 as well as with H-89 on ClC-2, it is not formally possible to interpret the effect of H-89. A review of the various effects of H-89 and the precautions that must be taken when interpreting the effects of this compound has appeared (Lochner and Moolman 2006).

13.13 Small-Molecule Effectors

13.13.1 Compounds That May Be Useful for Studies of ClC-2 in the Presence of CFTR and Other Cl^- Channels

GlyH-101 was discovered by high-throughput screening. It is water soluble up to 1 mM and is more active at positive voltages and has an IC_{50} of 5 μM. It inhibits CFTR-dependent nasal PDs in mice and was effective in blocking cholera toxin-induced intestinal fluid secretion (Muanprasat et al. 2004). It has also been used to differentiate between ClC-2 and CFTR in model systems. For example, GlyH-101 appears to have a more focused inhibitory effect on CFTR than ClC-2 based on the finding that GlyH-101 did not inhibit lubiprostone-stimulated currents thought to be ClC-2 currents before or after knockdown of CFTR (MacVinish et al. 2007). Similar findings were reported using GlyH-101 by Joo et al. (2009) in studies of airway surface I_{sc} and airway serous gland secretion. In that study, concentration-dependent stimulation by lubiprostone of I_{sc} persisted, but the concentration curves were shifted to the right, suggesting that GlyH-101 might also affect ClC-2 or another Cl^- channel and may not be CFTR specific. GlyH-101 may also affect mitochondrial function (Kelly et al. 2010). No studies have yet appeared to determine whether ClC-2 is affected by GlyH-101.

DASU-02 (N-(4-methylphenylsulfonyl)-N′-(4-trifluromethylphenyl)urea) is a CFTR inhibitor (Schultz et al. 1996, 1999). DASU-02 and several other related disulfonylurea analogues were tested on enterotoxin-induced porcine colonic secretion (O'Donnell et al. 2000). DASU-02 also inhibited *E. coli* EAST 1 toxin-induced I_{sc} in T84 cell cultures (Veilleux et al. 2008). DASU-02 was also used to study ClC-2 present with Ca^{2+}-activated Cl^- channels and CFTR (Moeser et al. 2004, 2007). Schultz et al. (1999) refer to unpublished data that showed that DASU-02 did not inhibit recombinant rabbit ClC-2 activated by forskolin-IBMX. We have recently shown that DASU-02 will inhibit recombinant human CFTR expressed in HEK293 cells. However, DASU-02 did not inhibit lubiprostone-stimulated Cl^- currents in HEK293EBNA cells expressing recombinant ClC-2 or in T84 cells where CFTR was knocked down with CFTR siRNA, but inhibited Cl^- currents in T84 cells where ClC-2 was knocked down with ClC-2 siRNA (Cuppoletti et al. 2014). Thus, DASU-02 was shown to be useful in studies which require separation of ClC-2 from CFTR in complex systems that might contain both channels.

Methadone has been shown to inhibit recombinant human ClC-2 without inhibition of recombinant human CFTR in HEK cell lines and Cl^- transport in T84 cell cultures (Cuppoletti et al. 2013). In that study, methadone inhibited lubiprostone-stimulated and PKA-activated recombinant human ClC-2 expressed in HEK293EBNA cells. Methadone also inhibited lubiprostone-stimulated I_{sc} in T84 cultures, as well as Cl^- currents measured by whole-cell patch-clamp of T84 cells in which CFTR was knocked down by siRNA, but not in T84 cells in which ClC-2 was knocked down by siRNA. Methadone was most effective when added before patch formation and before stimulation by drugs. To emphasize, it was much less effective when added after ClC-2 was activated by voltage or drugs. Methadone also prevented lubiprostone-induced protection against inflammatory cytokines (Cuppoletti et al. 2012). Methadone might also be useful in separating ClC-2 and CFTR effects in complex systems.

The *Leiurus quinquestriatus hebraeus* (scorpion) toxin GaTx2 has been shown to inhibit rabbit ClC-2 but not ClC-1, ClC-3, and ClC-4, CFTR, GABAc, Shaker B-IR, or Kv1.2 (Thompson et al. 2009). GaTx2 inhibits ClC-2 only from the extracytoplasmic face of the channel with an IC_{50} of 20 pM. The inhibition is partial, and it does not inhibit the activated channel (as is the case with methadone). Despite these limitations, the specificity of this drug for ClC-2 versus other specific channels is defined, and it has a very high affinity for ClC-2. GaTx2 has been shown to inhibit ClC-2 in nasal PD measurements in lubiprostone-stimulated $CFTR^{-/-}$ mice as well as in mice overexpressing human ClC-2 in the nasal epithelia (Schiffhauer et al. 2013).

13.13.2 Compounds That May Not Be Useful for Studies of ClC-2 in the Presence of CFTR and Other Cl^- Channels

Cd^{2+} has been shown to inhibit ClC-2 in a large number of studies, but it appears to have an effect on other channels, including CFTR in A6 cells, and is considered generally toxic to cells (Bao et al. 2008). Other inhibitors, for example, glibenclamide, an inhibitor of CFTR, also inhibit A6 cell ClC-2 (Bao et al. 2008). CFTRinh172 has now been shown to also be a potent inhibitor of human ClC-2 (Cuppoletti et al. 2014). Effects of a large number of other inhibitors of CFTR have been described, but the specificity of these inhibitors has not been examined and will not be discussed here. While methadone does not inhibit CFTR at concentrations up to 1 μM, it is known to inhibit numerous other channels (Cuppoletti et al. 2013). Only methadone and GaTx2 have been shown to be more effective on ClC-2 than CFTR and DASU-02 on CFTR and not ClC-2.

13.14 Summary

In summary, ClC-2 appears to have numerous physiological functions other than only involvement in the testes and retina as described initially in ClC-2$^{-/-}$ mice (Bösl et al. 2001). Other roles of ClC-2 are being discovered at a rapid pace. Inhibitors have now been identified and documented which might be useful for determination of ClC-2 function in the presence of CFTR. Knockouts and conditional knock-ins of ClC-2 in cells and tissues have been described. Combinations of pharmacological and genetic approaches will be useful to further delineate the functions of ClC-2 in different tissues and in different physiological contexts.

References

Alioth S, Meyer S, Dutzler R, Pervushin K (2007) The cytoplasmic domain of the chloride channel ClC-0: structural and dynamic characterization of flexible regions. J Mol Biol 369:1163–1169

Atia F, Zeiske W, van Driessche W (1999) Secretory apical Cl^- channels in A6 cells: possible control by cell Ca^{2+} and cAMP. Pflugers Arch 438:344–353

Bali M, Lipecka J, Edelman A, Fritsch J (2001) Regulation of ClC-2 chloride channels in T84 cells by TGF-alpha. Am J Physiol Cell Physiol 280:C1588–C1598

Bao HF, Liu L, Self J, Duke BJ, Ueno R, Eaton DC (2008) A synthetic prostone activates apical chloride channels in A6 epithelial cells. Am J Physiol Gastrointest Liver Physiol 295:G234–G251

Barish CF, Drossman D, Johanson JF, Ueno R (2010) Efficacy and safety of lubiprostone in patients with chronic constipation. Dig Dis Sci 55:1090–1097

Bauer CK, Steinmeyer K, Schwarz JR, Jentsch TJ (1991) Completely functional double-barreled chloride channel expressed from a single Torpedo cDNA. Proc Natl Acad Sci USA 88:11052–11056

Bi MM, Hong S, Zhou HY, Wang HW, Wang LN, Zheng YJ (2013) Chloride channelopathies of ClC-2. Int J Mol Sci 15:218–249

Bijvelds MJ, Bot AG, Escher JC, De Jonge HR (2009) Activation of intestinal Cl^- secretion by lubiprostone requires the cystic fibrosis transmembrane conductance regulator. Gastroenterology 137:976–985

Blaisdell CJ, Pellettieri JP, Loughlin CE, Chu S, Zeitlin PL (1999) Keratinocyte growth factor stimulates CLC-2 expression in primary fetal rat distal lung epithelial cells. Am J Respir Cell Mol Biol 20:842–847

Blaisdell CJ, Edmonds RD, Wang XT, Guggino S, Zeitlin PL (2000) pH-regulated chloride secretion in fetal lung epithelia. Am J Physiol Lung Cell Mol Physiol 278:L1248–L1255

Boie Y, Stocco R, Sawyer N, Slipetz DM, Ungrin MD, Neuschäfer-Rube F, Püchel GP, Metters KM, Abramovitz M (1997) Molecular cloning and characterization of the four rat prostaglandin E2 prostanoid receptor subtypes. Eur J Pharmacol 340:227–241

Bösl MR, Stein V, Hübner C, Zdebik AA, Jordt SE, Mukhopadhyay AK, Davidoff MS, Holstein AF, Jentsch TJ (2001) Male germ cells and photoreceptors, both dependent on close cell-cell interactions, degenerate upon ClC-2 Cl^- channel disruption. EMBO J 20:1289–1299
Carter NJ, Scott LJ (2009) Lubiprostone: in constipation-predominant irritable bowel syndrome. Drugs 69:1229–1237
Cartwright CA, McRoberts JA, Mandel KG, Dharmsathaphorn K (1985) Synergistic action of cyclic adenosine monophosphate- and calcium-mediated chloride secretion in a colonic epithelial cell line. J Clin Invest 76:1837–1842
Catalán M, Cornejo I, Figueroa CD, Niemeyer MI, Sepúlveda FV, Cid LP (2002) ClC-2 in guinea pig colon: mRNA, immunolabeling, and functional evidence for surface epithelium localization. Am J Physiol Gastrointest Liver Physiol 283:G1004–G1013
Catalán M, Niemeyer MI, Cid LP, Sepúlveda FV (2004) Basolateral ClC-2 chloride channels in surface colon epithelium: regulation by a direct effect of intracellular chloride. Gastroenterology 126:1104–1114
Catalán MA, Flores CA, González-Begne M, Zhang Y, Sepúlveda FV, Melvin JE (2012) Severe defects in absorptive ion transport in distal colons of mice that lack ClC-2 channels. Gastroenterology 142:346–354
Chen TY (2005) Structure and function of ClC channels. Annu Rev Physiol 67:809–839
Chey WD, Drossman DA, Johanson JF, Scott C, Panas RM, Ueno R (2012) Safety and patient outcomes with lubiprostone for up to 52 weeks in patients with irritable bowel syndrome with constipation. Aliment Pharmacol Ther 35:587–599
Chu S, Blaisdell CJ, Liu MZ, Zeitlin PL (1999) Perinatal regulation of the ClC-2 chloride channel in lung is mediated by Sp1 and Sp3. Am J Phys 276:L614–L624
Cid LP, Montrose-Rafizadeh C, Smith DI, Guggino WB, Cutting GR (1995) Cloning of a putative human voltage-gated chloride channel (CIC-2) cDNA widely expressed in human tissues. Hum Mol Genet 4:407–413
Cryer B, Katz S, Vallejo R, Popescu A, Ueno R (2014) A randomized study of lubiprostone for opioid-induced constipation in patients with chronic noncancer pain. Pain Med 15:1825. https://doi.org/10.1111/pme.12437
Cuppoletti J, Tewari KP, Sherry AM, Kupert EY, Malinowska DH (2001) ClC-2 Cl^- channels in human lung epithelia: activation by arachidonic acid, amidation, and acid-activated omeprazole. Am J Physiol Cell Physiol 281:C46–C54
Cuppoletti J, Tewari KP, Sherry AM, Ferrante CJ, Malinowska DH (2004a) Sites of protein kinase A activation of the human ClC-2 Cl^- channel. J Biol Chem 279:21849–21856
Cuppoletti J, Malinowska DH, Tewari KP, Li QJ, Sherry AM, Patchen ML, Ueno R (2004b) SPI-0211 activates T84 cell chloride transport and recombinant human ClC-2 chloride currents. Am J Physiol Cell Physiol 287:C1173–C1183
Cuppoletti J, Malinowska DH, Tewari K, Chakrabarti J, Mende K, Ueno R (2008a) Lubiprostone activation of Cl^- currents does not involve Ca^{2+}, cAMP, or PKA. Gastroenterology 134(Suppl 1):A581–A582
Cuppoletti J, Malinowska DH, Tewari KP, Chakrabarti J, Ueno R (2008b) ClC-2-targeted siRNA eliminates lubiprostone activation of Cl^- currents in T84 cells. Gastroenterology 134(Suppl 1): A-582
Cuppoletti J, Malinowska DH, Chakrabarti J, Ueno R (2008c) Effects of lubiprostone on human uterine smooth muscle cells. Prostaglandins Other Lipid Mediat 86:56–60
Cuppoletti J, Malinowska DH, Ueno R (2008d) Prostaglandin receptor activation properties of lubiprostone. Am J Gastroenterol 103(Suppl 1):S95–S118
Cuppoletti J, Blikslager AT, Chakrabarti J, Nighot PK, Malinowska DH (2012) Contrasting effects of linaclotide and lubiprostone on restitution of epithelial cell barrier properties and cellular homeostasis after exposure to cell stressors. BMC Pharmacol 12:3
Cuppoletti J, Chakrabarti J, Tewari K, Malinowska DH (2013) Methadone but not morphine inhibits lubiprostone stimulated Cl^- currents in T84 intestinal cells and recombinant human ClC-2, but not CFTR Cl^- currents. Cell Biochem Biophys 66:53–63

Cuppoletti J, Chakrabarti J, Tewari KP, Malinowska DH (2014) Differentiation between human ClC-2 and CFTR channels with pharmacological agents. Am J Physiol Cell Physiol 307:C479–C492

Cuppoletti J, Tewari KP, Chakrabarti J, Malinowska DH (2017) Identification of the fatty acid activation site on human ClC-2. Am J Physiol Cell Physiol 312:C707–C723

Cuthbert AW (2011) Lubiprostone targets prostanoid EP4 receptors in ovine airways. Br J Pharmacol 162:508–520

Denton J, Nehrke K, Yin X, Morrison R, Strange K (2005) GCK-3, a newly identified Ste20 kinase, binds to and regulates the activity of a cell cycle-dependent ClC anion channel. J Gen Physiol 125:113–125

Dhani SV, Kim Chiaw P, Huan LJ, Bear CE (2008) ATP depletion inhibits endocytosis of ClC-2. J Cell Physiol 214:273–280

Drossman DA, Chey WD, Johanson JF, Fass R, Scott C, Panas R, Ueno R (2009) Clinical trial: lubiprostone in patients with constipation-associated irritable bowel syndrome–results of two randomized, placebo-controlled studies. Aliment Pharmacol Ther 29:329–341

Duan DD (2013) Phenomics of cardiac chloride channels. Compr Physiol 3:667–692

Duran C, Thompson CH, Xiao Q, Hartzell HC (2010) Chloride channels: often enigmatic, rarely predictable. Annu Rev Physiol 72:95–121

Dutzler R (2004a) Structural basis for ion conduction and gating in ClC chloride channels. FEBS Lett 564:229–233

Dutzler R (2004b) The structural basis of ClC chloride channel function. Trends Neurosci 27:15–20

Dutzler R (2006) The ClC family of chloride channels and transporters. Curr Opin Struct Biol 16:439–446

Dutzler R (2007) A structural perspective on ClC channel and transporter function. FEBS Lett 581:2839–2844

Fahlke C (2004) Molecular physiology and pathophysiology of ClC-type chloride channels. Adv Mol Cell Physiol 32:189–217

Fei G, Wang YZ, Liu S, Hu HZ, Wang GD, Qu MH, Wang XY, Xia Y, Sun X, Bohn LM, Cooke HJ, Wood JD (2009) Stimulation of mucosal secretion by lubiprostone (SPI-0211) in guinea pig small intestine and colon. Am J Physiol Gastrointest Liver Physiol 296:G823–G832

Fritsch J, Edelman A (1996) Modulation of the hyperpolarization-activated Cl^- current in human intestinal T84 epithelial cells by phosphorylation. J Physiol 490:115–128

Furukawa T, Ogura T, Zheng YJ, Tsuchiya H, Nakaya H, Katayama Y, Inagaki N (2002) Phosphorylation and functional regulation of ClC-2 chloride channels expressed in *Xenopus* oocytes by M cyclin-dependent protein kinase. J Physiol 540:883–893

Gyömörey K, Yeger H, Ackerley C, Garami E, Bear CE (2000) Expression of the chloride channel ClC-2 in the murine small intestine epithelium. Am J Physiol Cell Physiol 279:C1787–C1794

Hanke W, Miller C (1983) Single chloride channels from Torpedo electroplax. Activation by protons. J Gen Physiol 82:25–45

Hinzpeter A, Fritsch J, Borot F, Trudel S, Vieu DL, Brouillard F, Baudouin-Legros M, Clain J, Edelman A, Ollero M (2007) Membrane cholesterol content modulates ClC-2 gating and sensitivity to oxidative stress. J Biol Chem 282:2423–2432

Holmes KW, Hales R, Chu S, Maxwell MJ, Mogayzel PJ Jr, Zeitlin PL (2003) Modulation of Sp1 and Sp3 in lung epithelial cells regulates ClC-2 chloride channel expression. Am J Respir Cell Mol Biol 29:499–505

Hoque KM, Woodward OM, van Rossum DB, Zachos NC, Chen L, Leung GP, Guggino WB, Guggino SE, Tse CM (2010) Epac1 mediates protein kinase A-independent mechanism of forskolin-activated intestinal chloride secretion. J Gen Physiol 135:43–58

Jentsch TJ, Poët M, Fuhrmann JC, Zdebik AA (2005) Physiological functions of CLC Cl^- channels gleaned from human genetic disease and mouse models. Annu Rev Physiol 67:779–807

Jin Y, Blikslager AT (2015) ClC-2 regulation of intestinal barrier function: translation of basic science to therapeutic target. Tiss Barr 3:e1105906-1-11

Johanson JF, Morton D, Geenen J, Ueno R (2008a) Multicenter, 4-week, double-blind, randomized, placebo-controlled trial of lubiprostone, a locally-acting type-2 chloride channel activator, in patients with chronic constipation. Am J Gastroenterol 103:170–177

Johanson JF, Drossman DA, Panas R, Wahle A, Ueno R (2008b) Clinical trial: phase 2 study of lubiprostone for irritable bowel syndrome with constipation. Aliment Pharmacol Ther 27:685–696

Joo NS, Wine JJ, Cuthbert AW (2009) Lubiprostone stimulates secretion from tracheal submucosal glands of sheep, pigs, and humans. Am J Physiol Lung Cell Mol Physiol 296:L811–L824

Jordt SE, Jentsch TJ (1997) Molecular dissection of gating in the ClC-2 chloride channel. EMBO J 16:1582–1592

Joy AP, Cowley EA (2008) 8-iso-PGE2 stimulates anion efflux from airway epithelial cells via the EP4 prostanoid receptor. Am J Respir Cell Mol Biol 38:143–152

Kajita H, Omori K, Matsuda H (2000) The chloride channel ClC-2 contributes to the inwardly rectifying Cl^- conductance in cultured porcine choroid plexus epithelial cells. J Physiol 523:313–324

Keeler R, Wong NL (1986) Evidence that prostaglandin E2 stimulates chloride secretion in cultured A6 renal epithelial cells. Am J Phys 250:F511–F515

Kelly M, Trudel S, Brouillard F, Bouillaud F, Colas J, Nguyen-Khoa T, Ollero M, Edelman A, Fritsch J (2010) Cystic fibrosis transmembrane regulator inhibitors CFTRinh-172 and GlyH-101 target mitochondrial functions, independently of chloride channel inhibition. J Pharmacol Exp Ther 333:60–69

Lembo AJ, Johanson JF, Parkman HP, Rao SS, Miner PB Jr, Ueno R (2011) Long-term safety and effectiveness of lubiprostone, a chloride channel (ClC-2) activator, in patients with chronic idiopathic constipation. Dig Dis Sci 56:2639–2645

Ling BN, Zuckerman JB, Lin C, Harte BJ, McNulty KA, Smith PR, Gomez LM, Worrell RT, Eaton DC, Kleyman TR (1997) Expression of the cystic fibrosis phenotype in a renal amphibian epithelial cell line. J Biol Chem 272:594–600

Linsdell P (2014) Cystic fibrosis transmembrane conductance regulator chloride channel blockers: pharmacological, biophysical and physiological relevance. World J Biol Chem 5:26–39

Lipecka J, Bali M, Thomas A, Fanen P, Edelman A, Fritsch J (2002) Distribution of ClC-2 chloride channel in rat and human epithelial tissues. Am J Physiol Cell Physiol 282:C805–C816

Lochner A, Moolman JA (2006) The many faces of H89: a review. Cardiovasc Drug Rev 24:261–274

MacDonald KD, McKenzie KR, Henderson MJ, Hawkins CE, Vij N, Zeitlin PL (2008) Lubiprostone activates non-CFTR-dependent respiratory epithelial chloride secretion in cystic fibrosis mice. Am J Physiol Lung Cell Mol Physiol 295:L933–L940

MacVinish LJ, Cope G, Ropenga A, Cuthbert AW (2007) Chloride transporting capability of Calu-3 epithelia following persistent knockdown of the cystic fibrosis transmembrane conductance regulator, CFTR. Br J Pharmacol 150:1055–1065

Malinowska DH, Kupert EY, Bahinski A, Sherry AM, Cuppoletti J (1995) Cloning, functional expression, and characterization of a PKA-activated gastric Cl^- channel. Am J Phys 268:C191–C200

Marunaka Y, Niisato N (2003) H89, an inhibitor of protein kinase A (PKA), stimulates Na^+ transport by translocating an epithelial Na^+ channel (ENaC) in fetal rat alveolar type II epithelium. Biochem Pharmacol 66:1083–1089

Meyer S, Savaresi S, Forster IC, Dutzler R (2007) Nucleotide recognition by the cytoplasmic domain of the human chloride transporter ClC-5. Nat Struct Mol Biol 14:60–67

Miller C (1982) Open-state substructure of single chloride channels from Torpedo electroplax. Philos Trans R Soc Lond Ser B Biol Sci 299:401–411

Miller C (2006) ClC chloride channels viewed through a transporter lens. Nature 440:484–489

Miller C, White MM (1980) A voltage-dependent chloride conductance channel from Torpedo electroplax membrane. Ann NY Acad Sci 341:534–551

Miyazaki H, Yamada T, Parton A, Morrison R, Kim S, Beth AH, Strange K (2012) CLC anion channel regulatory phosphorylation and conserved signal transduction domains. Biophys J 103:1706–1718

Moeser AJ, Haskell MM, Shifflett DE, Little D, Schultz BD, Blikslager AT (2004) ClC-2 chloride secretion mediates prostaglandin-induced recovery of barrier function in ischemia-injured porcine ileum. Gastroenterology 127:802–815

Moeser AJ, Nighot PK, Engelke KJ, Ueno R, Blikslager AT (2007) Recovery of mucosal barrier function in ischemic porcine ileum and colon is stimulated by a novel agonist of the ClC-2 chloride channel, lubiprostone. Am J Physiol Gastrointest Liver Physiol 292:G647–G656

Mohammad-Panah R, Gyomorey K, Rommens J, Choudhury M, Li C, Wang Y, Bear CE (2001) ClC-2 contributes to native chloride secretion by a human intestinal cell line, Caco-2. J Biol Chem 276:8306–8313

Muanprasat C, Sonawane ND, Salinas D, Taddei A, Galietta LJ, Verkman AS (2004) Discovery of glycine hydrazide pore-occluding CFTR inhibitors: mechanism, structure-activity analysis, and in vivo efficacy. J Gen Physiol 124:125–137

Murray CB, Morales MM, Flotte TR, McGrath-Morrow SA, Guggino WB, Zeitlin PL (1995) ClC-2: a developmentally dependent chloride channel expressed in the fetal lung and downregulated after birth. Am J Respir Cell Mol Biol 12:597–604

Murray CB, Chu S, Zeitlin PL (1996) Gestational and tissue-specific regulation of C1C-2 chloride channel expression. Am J Phys 271:L829–L837

Niemeyer MI, Cid LP, Züñniga L, Catalán M, Sepúlveda FV (2003) A conserved pore-lining glutamate as a voltage- and chloride-dependent gate in the ClC-2 chloride channel. J Physiol 553:873–879

Niemeyer MI, Cid LP, Yusef YR, Briones R, Sepúlveda FV (2009) Voltage-dependent and -independent titration of specific residues accounts for complex gating of a ClC chloride channel by extracellular protons. J Physiol 587:1387–1400

Nighot PK, Blikslager AT (2010) ClC-2 regulates mucosal barrier function associated with structural changes to the villus and epithelial tight junction. Am J Physiol Gastrointest Liver Physiol 299:G449–G456

Nighot PK, Blikslager AT (2012) Chloride channel ClC-2 modulates tight junction barrier function via intracellular trafficking of occludin. Am J Physiol Cell Physiol 302:C178–C187

Nighot PK, Moeser AJ, Ryan KA, Ghashghaei T, Blikslager AT (2009) ClC-2 is required for rapid restoration of epithelial tight junctions in ischemic-injured murine jejunum. Exp Cell Res 315:110–118

Nighot MP, Blikslager A, Nighot PK (2011) The chloride channel ClC-2 is involved in organization of murine gastric glands. Gastroenterology 140(Suppl 1):S93

Nighot P, Young K, Nighot M, Rawat M, Sung EJ, Maharshak N, Plevy SE, Ma T, Blikslager A (2013) Chloride channel ClC-2 is a key factor in the development of DSS-induced murine colitis. Inflamm Bowel Dis 13:2867–2877

Nighot MP, Nighot PK, Ma TY, Blikslager A (2014) Genetic absence of chloride channel ClC-2 results in disruption of organization and function of murine gastric glands. Gastroenterology 146:S505

Nighot MP, Nighot PK, Ma TY, Malinowska DH, Shull GE, Cuppoletti J, Blikslager AT (2015) Genetic ablation of the ClC-2 Cl^- channel disrupts mouse gastric parietal cell acid secretion. PLoS One 10(9):e0138174. https://doi.org/10.1371/journal.pone.0138174

Nighot PK, Leung L, Ma TY (2017) Chloride channel ClC-2 enhances intestinal epithelial tight junction barrier function via regulation of caveolin-1 and caveolar trafficking of occluding. Exp Cell Res 352:113–122

Niisato N, Ito Y, Marunaka Y (1999) Effects of PKA inhibitors, H-compounds, on epithelial Na^+ channels via PKA-independent mechanisms. Life Sci 65:109–114

Nobile M, Pusch M, Rapisarda C, Ferroni S (2000) Single-channel analysis of a ClC-2-like chloride conductance in cultured rat cortical astrocytes. FEBS Lett 479:10–14

Norimatsu Y, Moran AR, MacDonald KD (2012) Lubiprostone activates CFTR, but not ClC-2, via the prostaglandin receptor EP4. Biochem Biophys Res Commun 426:374–379

O'Brien CE, Anderson PJ, Stowe CD (2011) Lubiprostone for constipation in adults with cystic fibrosis: a pilot study. Ann Pharmacother 45:1061–1066

O'Donnell EK, Sedlacek RL, Singh AK, Schultz BD (2000) Inhibition of enterotoxin-induced porcine colonic secretion by diarylsulfonylureas in vitro. Am J Phys 279:G1104–G1112

Park K, Begenisich T, Melvin JE (2001) Protein kinase A activation phosphorylates the rat ClC-2 Cl^- channel but does not change activity. J Membr Biol 182:31–37

Park WS, Son YK, Kim N, Youm JB, Warda M, Ko JH, Ko EA, Kang SH, Kim E, Earm YE, Han J (2007) Direct modulation of Ca^{2+}-activated K^+ current by H-89 in rabbit coronary arterial smooth muscle cells. Vasc Pharmacol 46:105–113

Paunescu TG, Helman SI (2001) PGE2 activation of apical membrane Cl^- channels in A6 epithelia: impedance analysis. Biophys J 81:852–866

Poroca DR, Pelis RM, Chappe VM (2017) ClC channels and transporters: structure, physiological functions and implications in human chloride channelopathies. Front Pharm 8. https://doi.org/10.3389/fphar.2017.00151

Price MP, Ishihara H, Sheppard DN, Welsh MJ (1996) Function of *Xenopus* cystic fibrosis transmembrane conductance regulator (CFTR) Cl channels and use of human-*Xenopus* chimeras to investigate the pore properties of CFTR. J Biol Chem 271:25184–25191

Rutledge E, Denton J, Strange K (2002) Cell cycle- and swelling-induced activation of a *Caenorhabditis elegans* ClC channel is mediated by CeGLC-7alpha/beta phosphatases. J Cell Biol 158:435–444

Schiffhauer ES, Vij N, Kovbasnjuk O, Kang PW, Walker D, Lee S, Zeitlin PL (2013) Dual activation of CFTR and CLCN2 by lubiprostone in murine nasal epithelia. Am J Physiol Lung Cell Mol Physiol 304:L324–L331

Schultz BD, Singh AK, Frizzell RA, Bridges RJ (1996) Developing potent modulators of CFTR channel gating. Pediatr Pulmonol 13:258

Schultz BD, Singh AK, Devor DC, Bridges RJ (1999) Pharmacology of CFTR chloride channel activity. Physiol Rev 79:S109–S144

Schwiebert EM, Cid-Soto LP, Stafford D, Carter M, Blaisdell CJ, Zeitlin PL, Guggino WB, Cutting GR (1998) Analysis of ClC-2 channels as an alternative pathway for chloride conduction in cystic fibrosis airway cells. Proc Natl Acad Sci USA 95:3879–3884

Sherry AM, Stroffekova K, Knapp LM, Kupert EY, Cuppoletti J, Malinowska DH (1997) Characterization of the human pH- and PKA-activated ClC-2G(2α) Cl^- channel. Am J Phys 273: C384–C393

Son YK, Park WS, Kim SJ, Earm YE, Kim N, Youm JB, Warda M, Kim E, Han J (2006) Direct inhibition of a PKA inhibitor, H-89 on Kv channels in rabbit coronary arterial smooth muscle cells. Biochem Biophys Res Commun 341:931–937

Stauber T, Weinert S, Jentsch TJ (2012) Cell biology and physiology of CLC chloride channels and transporters. Compr Physiol 2:1701–1744

Stölting G, Teodorescu G, Begemann B, Schubert J, Nabbout R, Toliat MR, Sander T, Nürnberg P, Lerche H, Fahlke C (2013) Regulation of ClC-2 gating by intracellular ATP. Pflugers Arch 465:1423–1437

Stölting G, Fischer M, Fahlke C (2014) ClC-1 and ClC-2 form hetero-dimeric channels with novel protopore functions. Pflugers Arch 466(12):2191–2204

Stroffekova K, Kupert EY, Malinowska DH, Cuppoletti J (1998) Identification of the pH sensor and activation by chemical modification of the ClC-2G Cl^- channel. Am J Phys 275:C1113–C1123

Strohmeier GR, Reppert SM, Lencer WI, Madara JL (1995) The A2b adenosine receptor mediates cAMP responses to adenosine receptor agonists in human intestinal epithelia. J Biol Chem 270:2387–2394

Sun Park W, Kyoung Son Y, Kim N, Boum Youm J, Joo H, Warda M, Ko JH, Earm YE, Han J (2006) The protein kinase A inhibitor, H-89, directly inhibits KATP and Kir channels in rabbit coronary arterial smooth muscle cells. Biochem Biophys Res Commun 340:1104–1110

Tewari KP, Malinowska DH, Sherry AM, Cuppoletti J (2000) PKA and arachidonic acid activation of human recombinant ClC-2 chloride channels. Am J Physiol Cell Physiol 279:C40–C50

Thiemann A, Gründer S, Pusch M, Jentsch TJ (1992) A chloride channel widely expressed in epithelial and non-epithelial cells. Nature 356:57–60

Thompson CH, Olivetti PR, Fuller MD, Freeman CS, McMaster D, French RJ, Pohl J, Kubanek J, McCarty NA (2009) Isolation and characterization of a high affinity peptide inhibitor of ClC-2 chloride channels. J Biol Chem 284:26051–26062

Ueno R, Osama H, Habe T, Engelke K, Patchen M (2004) Oral SPI-0211 increases intestinal fluid secretion without altering serum electrolyte levels. Gastroenterology 126:A276

Uwaifo O, Bamford P, Zeitlin PL, Blaisdell CJ (2006) Acidic pH hyperpolarizes nasal potential difference. Pediatr Pulmonol 41:151–157

Veilleux S, Holt N, Schultz BD, Dubreuil JD (2008) *Escherichia coli* EAST1 toxin toxicity of variants 17-2 and O 42. Comp Immunol Microbiol Infect Dis 31:567–578

Vij N, Zeitlin PL (2006) Regulation of the ClC-2 lung epithelial chloride channel by glycosylation of SP1. Am J Respir Cell Mol Biol 34:754–759

Wang H, Xu M, Kong Q, Sun P, Yan F, Tiasn W, Wang X (2017) Research and progress on ClC-2 (review). Mol Med Rep 16:11–22

Weinreb M, Shamir D, Machwate M, Rodan GA, Harada S, Keila S (2006) Prostaglandin E2 (PGE2) increases the number of rat bone marrow osteogenic stromal cells (BMSC) via binding the EP4 receptor, activating sphingosine kinase and inhibiting caspase activity. Prostaglandins Leukot Essent Fatty Acids 75:81–90

White MM, Miller C (1979) A voltage-gated anion channel from the electric organ of *Torpedo californica*. J Biol Chem 254:10161–10166

Wong BS, Camilleri M (2011) Lubiprostone for the treatment of opioid-induced bowel dysfunction. Expert Opin Pharmacother 12:983–990

Yamada T, Bhate MP, Strange K (2013) Regulatory phosphorylation induces extracellular conformational changes in a CLC anion channel. Biophys J 104:1893–1904

Zdebik AA, Cuffe JE, Bertog M, Korbmacher C, Jentsch TJ (2004) Additional disruption of the ClC-2 Cl^- channel does not exacerbate the cystic fibrosis phenotype of cystic fibrosis transmembrane conductance regulator mouse models. J Biol Chem 279:22276–22283

Zifarelli G, Pusch M (2007) CLC chloride channels and transporters: a biophysical and physiological perspective. Rev Physiol Biochem Pharmacol 158:23–76

Chapter 14
The Role of the Endosomal Chloride/Proton Antiporter ClC-5 in Proximal Tubule Endocytosis and Kidney Physiology

Maddalena Comini and Giovanni Zifarelli

Abstract The chloride channel (CLC) protein family comprises ion channels and proton-coupled anion transporters with fundamental physiological roles in humans. Several properties of CLC proteins defy the rigid dichotomy between ion channels and transporters as these opposite thermodynamic mechanisms of transport are implemented in a very similar structural architecture. All the CLC transporters are expressed in intracellular organelles where they are somehow important for the ionic homeostasis of these compartments. However, their specific physiological role is still unclear. This chapter focuses on the biophysical properties and physiological role of the endosomal Cl^-/H^+ antiporter ClC-5 mutated in Dent's disease.

Keywords CLC proteins · ClC-5 · Endosomal physiology · Cl^- transport · Dent's disease · Proximal tubule

14.1 Introduction

14.1.1 The Physiological Relevance of Chloride Transport

Chloride is the most abundant anion in the human body, but for a long time its role in human physiology has been overlooked probably due to the initial development of the field of membrane transport around the topic of excitable cells, with a prominent role of sodium and potassium channels (Miller 2006). However, Cl^- ions are fundamental in many different physiological contexts. They ensure electroneutrality during Na^+ and K^+ transport across cellular membranes in different epithelia and in intracellular compartments. Owed to its high extracellular concentration, Cl^- serves, together with positively charged counterions, as an important osmolyte to drive water movement across cellular membranes and as a hyperpolarizing conductance in

M. Comini · G. Zifarelli (✉)
Department of Physiology, Anatomy and Genetics, University of Oxford, Oxford, UK
e-mail: maddalena.comini@dpag.ox.ac.uk; giovanni.zifarelli@dpag.ox.ac.uk

K. L. Hamilton, D. C. Devor (eds.), *Studies of Epithelial Transporters and Ion Channels*, Physiology in Health and Disease,
https://doi.org/10.1007/978-3-030-55454-5_14

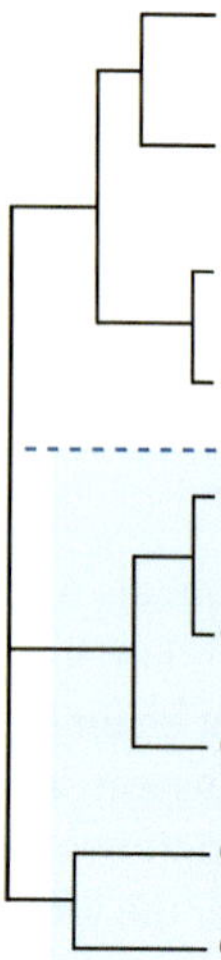

Name	β subunit	Tissue Expression	Function	Mouse model	Human disease
ClC-1		skeletal muscle	stabilization of membrane potential	myotonia congenita	recessive and dominant myotonia
ClC-2	GlialCAM	broad	transepithelial transport, ion homeostasis, regulation excitability	retinal degen. and testis leukodystrophy	aldosteronism, leukodystrophy
ClC-Ka	Barttin	kidney, inner ear	transepithelial transport	diabetes insipidus	?
ClC-Kb	Barttin	kidney, inner ear	transepithelial transport	renal salt loss	Bartter syndrome
ClC-3		broad (brain, kidney, liver)	acidification (late endosomes, SV, LDCV?) and ion homeostasis	neuronal and retinal degeneration	?
ClC-4		broad (brain, kidney, muscle)	ion homeostasis (endosomes)	no phenotype observed	intellectual disability, epilepsy
ClC-5		kidney and intestine	acidification (early/late endosomes) and ion homeostasis	impaired renal endocytosis	Dent's disease (proteinuria and kidney stones)
ClC-6		neuronal	ion homeostasis (late endosomes)	neuronal lysosomal storage disease	neuronal ceroid lipofuscinosis ?
ClC-7	Ostm1	broad	acidification (osteoclasts) and ion homeostasis (lysosomes)	osteopetrosis with neuronal and retinal degeneration	lysosomal storage disease, dom/reces osteopetrosis, neuronal and retinal degeneration

Fig. 14.1 Overview of the human CLC proteins. On the left, dendrogram showing that human CLCs can be grouped in three homology branches. The horizontal dashed line in blue separates ion channels and transporters as highlighted also by different background colours (white for ion channels and light blue for antiporters). The table indicates known *β*-subunits (in red), tissue distribution, known physiological functions, mouse and human pathologies observed upon disruption (mouse) or with mutations (humans) of the respective gene (adapted from Jentsch TJ and Pusch M, Physiol Rev. 98: 1493–1590, 2018)

several tissues. These functions are mediated by several unrelated Cl^- channel families which include the cystic fibrosis transmembrane conductance regulator (CFTR) (Riordan et al. 1989); pentameric ligand-gated anion channels, like GABA- and glycine-receptors (Grenningloh et al. 1987); the Ca^{2+}-activated Cl^- channels of the bestrophin (Sun et al. 2002; Qu et al. 2003) and TMEM16 (or Anoctamin) (Schroeder et al. 2008; Caputo et al. 2008; Yang et al. 2008) gene families; and finally the Leucine-rich repeat-containing protein 8 (LRRC8) proteins, which form volume-regulated anion channels (VRACs) (Voss et al. 2014). Underscoring their physiological relevance, most of these protein families are involved in genetic diseases. This chapter focuses on the role of the endosomal Cl^-/H^+ antiporter ClC-5 (gene name *CLCN5*) which belongs to the CLC protein family (Fig. 14.1) (Zifarelli and Pusch 2007; Stauber et al. 2012; Jentsch and Pusch 2018). We will first introduce the general structural architecture and mechanism of ion transport of CLC proteins and then illustrate in detail the biophysical properties and physiological roles of ClC-5 in the kidney and in other tissues.

14.1.2 The CLC Protein Family

ClC-0 was the first member of the CLC protein family to be identified at the molecular level by the group of Chris Miller, upon reconstitution of ray *Torpedo* electric organ membranes into lipid bilayers (White and Miller 1979). The distinct single channel behaviour of ClC-0 with two levels of conductance of identical amplitude suggested, more than 20 years before the first structure of a CLC protein was available, that the protein assembled as a dimer of identical subunits, each containing a voltage-dependent gating mechanism called protopore gate or fast gate (Miller 1982). In addition, there is a second gating mechanism which acts simultaneously on both subunits, defined common or slow gate because in ClC-0 its time course is order of magnitude slower than the fast gate (Miller and White 1984). These pioneering findings were followed by the cloning of the first human member of the CLC protein family in the lab of Thomas Jentsch, the skeletal muscle Cl^- channel ClC-1 (Jentsch et al. 1990) and, afterwards, of all the other human CLCs (Fig. 14.1). Personal accounts of Jentsch's and Miller's initial discoveries can be found in two recent reviews (Jentsch 2015; Miller 2015). More recently, it has been recognized that out of the nine human CLC proteins, ClC-1, ClC-2, and the kidney ClC-Ka and -Kb are ion channels, whereas ClC-3 to ClC-7 are coupled 2 Cl^-/1 H^+ antiporters. Several CLC proteins are mutated in human genetic diseases emphasizing their physiological relevance (Jentsch and Pusch 2018). The presence of channels and transporters, operating according to opposite thermodynamic basis in the same protein family, is one of the unique features of CLCs. Intriguingly, all the CLC ion channels are expressed in the plasma membrane, whereas all the antiporters are expressed in intracellular organelles or vesicles. Among the channels, ClC-1 is predominantly expressed in the skeletal muscle, and ClC-2 has a broad tissue distribution and associates with the accessory β-subunit glialCAM (glial cell adhesion molecule), which influences protein targeting and kinetic properties of ClC-2 (Jeworutzki et al. 2012; Hoegg-Beiler et al. 2014). ClC-Ka and ClC-Kb are tissue specific (kidney and ear) and require the β-subunit Barttin for proper functional expression (Estévez et al. 2001). Among the CLC antiporters, it has been reported that also ClC-7 requires the accessory β-subunit Ostm1 (osteoclastogenesis associated transmembrane protein 1) for proper functional expression (Lange et al. 2006; Leisle et al. 2011). However, very recent results seem to indicate that Ostm1 might be dispensable for the transport activity of ClC-7 (Pusch and Zifarelli, unpublished results).

14.2 Structures of CLC Channels and Transporters

There is now a substantial amount of structural information available on CLC proteins, either transporters (Dutzler et al. 2002, 2003; Feng et al. 2010) or ion channels (Park et al. 2017; Park and Mackinnon 2018) showing that the architecture

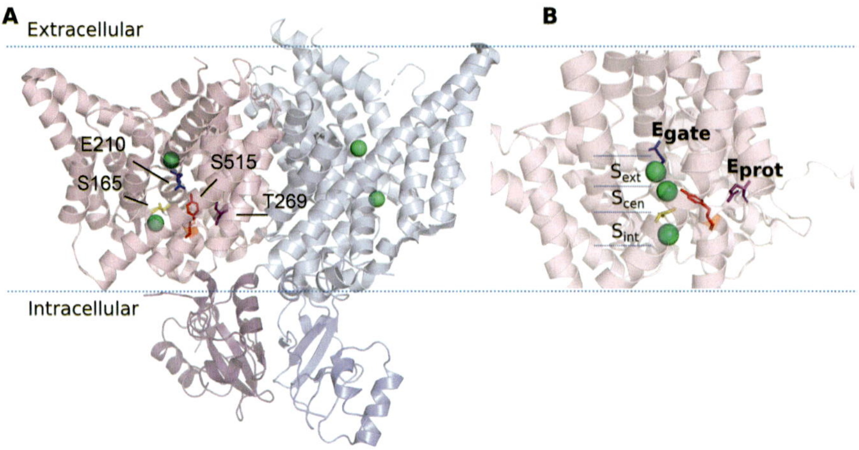

Fig. 14.2 Structural architecture of CLC proteins. (**a**) The crystal structure of the eukaryotic CmClC antiporter reveals the dimeric structure of CLC transporters viewed from the membrane plane (PDB: 3ORG). Each subunit of the dimer is depicted in different colours with the cytoplasmic C-terminal domains in darker shades. The structure indicates the Cl^- binding sites, S_{ext} and S_{int}, shown as green spheres and several residues important for transport function such as S165 (yellow), E210 (blue), T269 (purple) and Y515 (red) which in ClC-5 correspond to S168 (Ser_{cen}), E211 (E_{gate}), E268 (E_{prot}), and Y558 (Tyr_{cen}) of ClC-5, respectively. (**b**) The insert indicates the anion permeation pathway for one subunit of the mutant E148Q from the crystal structure of the bacterial EcClC-1 (PDB: 1OTU). The residues represented as sticks are S107 (yellow), Q148 (blue), E203 (purple) and Y445 (red) which in ClC-5 correspond to S168 (Ser_{cen}), E211 (E_{gate}), E268 (E_{prot}) and Y558 (Tyr_{cen}), respectively. The so-called gating glutamate, E211 (E_{gate}), and the proton glutamate, E268 (E_{prot}), of ClC-5 are indicated in bold. In the E148Q mutant, all three Cl^- binding sites (S_{ext}, S_{cen}, and S_{int}) are occupied by anions depicted as green spheres, and their position is highlighted by dashed lines

of transporters and channels is remarkably similar, in spite of the diverging functions. CLC proteins are dimers with an independent ion permeation pathway in each subunit. Three anion binding sites, S_{ext}, S_{cen} and S_{int}, mark the anion permeation pathway through each subunit. In the structure of WT EcClC-1 (Dutzler et al. 2002), S_{ext} is occupied by the side chain of the so called "gating glutamate", E_{gate} (Fig. 14.2). Protonation and displacement of the E_{gate} side chain is necessary for Cl^- and proton transport. In fact, neutralization of E_{gate} in CLC transporters results in uncoupled passive Cl^- flux (Leisle et al. 2011; Accardi and Miller 2004; Picollo and Pusch 2005; Scheel et al. 2005; Neagoe et al. 2010), and in channels it abolishes both fast and slow gating transitions (Jentsch and Pusch 2018). In the crystal structure of the EcClC-1 E148Q mutant (Dutzler et al. 2003), the Gln side chain points towards the extracellular space, and S_{ext} is occupied by a Cl^- ion, suggesting that this represents a conductive conformation of the permeation pathway. In the structure of CmClC (Feng et al. 2010), the side chain of E_{gate} occupies S_{cen}. S_{cen} is formed by the main chain nitrogen atoms of residues of the re-entrant loop alpha helices M and N (Ile 356 and Phe 357 in the bacterial EcClC-1) and by the side chains of two conserved residues, Ser_{cen} and Tyr_{cen}, forming a narrow constriction of the pore between the two chloride binding sites S_{cen} and S_{int} (Fig. 14.2). It has been proposed that these two residues form an intracellular gate, which regulates access to the central binding site from the intracellular solution in both CLC channels and transporters (Jayaram et al. 2008). Functional support for this notion came from a study in EcClC-1 in which mutations of Y445 with smaller amino acids produce a progressive uncoupling of Cl^-/H^+ transport, together with a Cl^- movement approaching diffusion-limited rates (Accardi et al. 2006; Lobet and Dutzler 2006; Jayaram et al. 2008). Further evidence, in favour of a role of Y445 as the intracellular gate has come from the finding that conformational changes outside the permeation pathway affect transport rate by a physical interaction with this residue (Basilio et al. 2014). Intriguingly, substitutions of the corresponding tyrosine residue in ClC-0 (Y512) do not affect the single-channel conductance (Accardi and Pusch 2003). The serine residue at S_{cen} is conserved in all mammalian CLCs and has a critical role in selectivity. Substitution with proline, the corresponding residue found in the plant NO_3^-/H^+ antiporter AtClC-a (De Angeli et al. 2006), turns ClC-5 into a NO_3^-/H^+ antiporter while preserving the 2:1 anion/H^+ stoichiometry (Zifarelli and Pusch 2009a). Consistent with this, NO_3^-/Cl^- relative current amplitude was increased by the same mutation in ClC-7 (Leisle et al. 2011), ClC-0 (Picollo et al. 2009; Bergsdorf et al. 2009) and ClC-6 (Neagoe et al. 2010). S_{ext} does not seem to be involved in transport stoichiometry, but at least in ClC-5 it can modulate anion selectivity as neutralization of a conserved lysine residue (K210), located close to the gating glutamate E211, inverted the WT selectivity of NO_3^- over Cl^- (De Stefano et al. 2011). The recent structures of two CLC channels, the human isoform of ClC-1 and the bovine isoform of the kidney ClC-Ka and -Kb channels, confirmed that the general structural architecture is the same in both channels and transporters but revealed an important difference for the central binding site of ClC-K (Park et al. 2017); here Ser_{cen} does not project into the permeation pathway like in the other structures, but rather sideways, leading to a removal of the constriction formed by Ser_{cen} and Tyr_{cen} in ClC-1 (Park and Mackinnon 2018) and in the transporters (Dutzler et al. 2002; Feng et al. 2010). The different conformation of Ser_{cen} in ClC-K might be a consequence of the presence of bulky hydrophobic residues in the upper part of S_{cen}, with a potential steric clash on the Ser_{cen} (Park et al. 2017; Lagostena et al. 2019).

14.2.1 Proton Transport

Extracellular protons can directly reach E_{gate} from the extracellular side. From the intracellular side, it is likely that protons reach E_{gate} when it occupies S_{cen} (Feng et al. 2010), but the precise H^+ pathway is still unclear. A relatively conserved glutamate residue, the proton glutamate, E_{prot}, located close to the intracellular solution, but not directly in the permeation pathway, is probably involved in shuttling protons from the intracellular side to S_{cen} (Accardi et al. 2005; Lim and Miller 2009), and it has been suggested that this process involves water wires in the intracellular cavity of the pore (Han et al. 2014). Importantly, a glutamate residue is not strictly necessary to serve this function, as in CmClC this residue is replaced by a threonine (Feng et al. 2010), and in the bacterial transporter homologue CkClC-2 (distantly related to EcClC-1) is an isoleucine (Phillips et al. 2012). In addition, in CLC-5 and other mammalian CLC transporters, mutations of E_{prot} preserve H^+ transport (Zdebik et al. 2008; Grieschat and Alekov 2012; Guzman et al. 2013). Regarding the proton glutamate, ClC-5 differs significantly from EcClC-1. Indeed, neutralizing E_{prot} (E203) in EcClC-1 eliminates H^+ transport and renders the transporter a passively conducting Cl^- conductance (Accardi et al. 2005). In contrast to this, in ClC-5, the E268A mutation completely abolishes steady-state transport of Cl^- and protons (Zdebik et al. 2008). The reason for this discrepancy is still unclear, but the behaviour of ClC-5 is in a more intuitive agreement with the idea of an obligatory coupling mechanism underlying the antiport activity of CLC transporters, which would posit that proton delivery via E_{prot} is a necessary step for the transport cycle. Interestingly, in ClC-5 the proton glutamate mutation E268A exhibits large "transient" capacitive currents that reflect partial reactions of the transport cycle (Smith and Lippiat 2010; Zifarelli et al. 2012).

14.2.2 CBS (Cystathionine Beta Synthase) Cytoplasmic Domains of ClC-5

ClC-5, like all eukaryotic CLCs, has an extended cytoplasmic C-terminal region that contains two CBS domains, CBS-1 and CBS-2, named after similar domains identified in cystathionine-synthetase (Bateman 1997). CBS domains are found in many different protein families as nucleotide binding modules. In the crystal structure of the isolated C-terminal fragment of ClC-5, ATP or AMP are bound in the cleft between the two CBS subdomains, engaging mostly the adenine base of the nucleotides, consistent with the similar binding affinity observed for ATP, ADP, and AMP (Meyer et al. 2007). In electrophysiological experiments, adenine nucleotides applied from the intracellular side in inside-out patch-clamp experiments on ClC-5 lead to a doubling of the current with an apparent affinity of $\sim$1 mM for ATP, AMP and ADP (Zifarelli and Pusch 2009b). Functional regulation by nucleotides was also shown for ClC-1 (Bennetts et al. 2007). The crystal structure of the eukaryotic CmClC showed that the CBS domains, and in particular CBS-2, come into close contact with the transmembrane region of the transporter, potentially affecting transport function (Feng et al. 2010; Duffield et al. 2003; Bennetts and Parker 2013). However, the similar effect of ATP, ADP and AMP suggests that nucleotide binding might not have any physiological relevance for ClC-5, even though it cannot be excluded that regulation by nucleotide binding might become relevant in vivo.

14.3 Gating

14.3.1 Gating Mechanism in the CLC Transporters

Two gating mechanisms operate in CLC channels: a fast gate, that can be mostly attributed to E_{gate} and controls the opening of each of the subunits independently (Miller 1982; Ludewig et al. 1996; Accardi and Pusch 2003); and a slow, or common gate, which acts simultaneously on both subunits (White and Miller 1979; Miller and White 1980; Pusch et al. 1997), although it has not been identified at the molecular level yet. Notably, in CLC channels, gating and permeation are not as clearly distinguishable as in other channel families, and Cl^- and other permeable and impermeable anions (Pusch et al. 1999; Rychkov et al. 1998), but also protons (Lísal and Maduke 2008), have an effect on gating. This has led to the suggestion that CLC channels are in fact broken Cl^-/H^+ antiporters (Lísal and Maduke 2008; Miller 2006). Intriguingly, a gating process is displayed also by some CLC transporters and consists in a mechanism that regulates the transitions between an actively transporting configuration and an inactive, non-transporting, state (Fig. 14.3). This concept is not typical in the description of transporters since the distinction between gating and permeation is in general very difficult in these proteins. In this regard, a

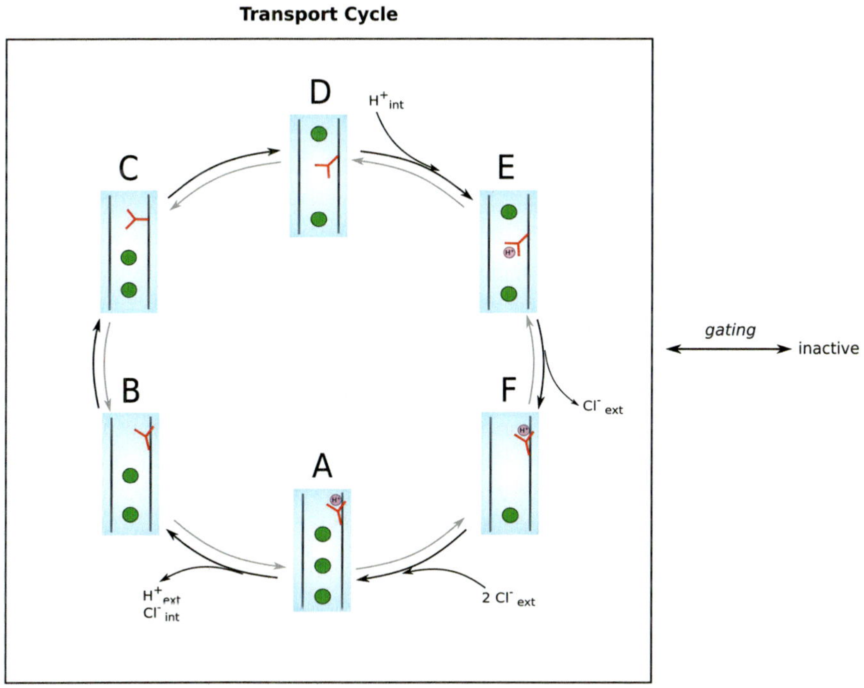

Fig. 14.3 Schematic representation of the ClC-5 transport cycle. Adapted from Pusch and Zifarelli (2014). See text for details

fundamental question is if the extreme outward rectification of ClC-5 (Steinmeyer et al. 1995), but also ClC-7 (Leisle et al. 2011), ClC-4 (Friedrich et al. 1999) and ClC-3 (Li et al. 2000), reflects a gating process or, rather, a rectifying transport cycle. Interestingly, the gating glutamate mutation E211A not only converts ClC-5 into a passive Cl^- conductance but also eliminates the rectification and any relaxation kinetics of the currents elicited by voltage steps, suggesting that at least the Cl^- pathway of the transporter is not intrinsically rectifying. The lack of "gating" kinetics in this uncoupled mutant parallels the behaviour of the equivalent E_{gate} mutation in the ClC-0 channel, which completely eliminates not only the fast gate but also the slow gate of the channel (Dutzler et al. 2003; Traverso et al. 2006). A first indication of the presence of a gating mechanism in ClC-5 came from the analysis of the spectral properties of the currents' noise, which showed a channel-like Lorentzian shape (Zdebik et al. 2008). This is consistent with a transporter that switches from active, transporting states to inactive states (Zdebik et al. 2008), in a manner similar to the opening of single channels, that is determined by the transition from closed to open state. The presence of a gating mechanism in ClC-5 was also suggested on the basis of the deactivation of the currents at positive potentials (Alekov and Fahlke 2009). However, the most direct evidence that a gating process in ClC-5 is responsible for its outward rectification was recently obtained studying the single point mutation D76H (De Stefano et al. 2013). D76 is one of the most conserved residues in ClC-5, and mutating it in ClC-0, ClC-1 or ClC-Kb leads to dramatic changes in gating properties (Fahlke et al. 1995; Ludewig et al. 1997; Picollo et al. 2004). The D76H mutant expressed in *Xenopus* oocytes significantly slowed the activation of outward currents and, particularly at acidic extracellular pH, displayed clear inwardly directed tail currents, after an activating voltage pulse. A detailed analysis of the tail currents revealed that they reflect a gating process that underlies, at least partially, the strong outward rectification of the transporter (De Stefano et al. 2013). Recently, a gating process was observed also for the lysosomal ClC-7/Ostm1 transporter (Leisle et al. 2011). Like ClC-5, also ClC-7 currents are strongly outwardly rectifying, but, in addition, they have an extremely slow activation kinetics at positive voltages and decay also quite slowly at negative voltages. It has been suggested that the gating process of ClC-7 corresponds to the common gate of CLC channels, i.e. involves conformational changes of both protopores (Ludwig et al. 2013). Physiologically, the gating process might be important because it restricts the activity of the transporter in a limited voltage range, but we still know very little about the membrane voltages across intracellular organelles, and in general whether the properties identified from the expression of ClC-5 and ClC-7 in the plasma membrane recapitulate the ones in native organelles.

14.4 Transport Mechanism

14.4.1 Stoichiometry

As already mentioned, CLC channels and antiporters share a very similar structural feature. In addition to the fact that the first members functionally investigated were Cl^- channels, this led to the initial assumption that all CLC proteins were channels. However, in 2004, the electrophysiological analysis of EcClC-1 (the first homologue to be crystallized) with a detailed study of the ionic dependence of the reversal potential of the currents showed that this protein was not a Cl^- channel, but instead a 2 Cl^-/1 H^+ antiporter (Accardi et al. 2004). It was later shown that also ClC-4 and ClC-5 (Picollo and Pusch 2005; Scheel et al. 2005), ClC-7 (Graves et al. 2008; Leisle et al. 2011), ClC-3 (Guzman et al. 2013) and ClC-6 (Neagoe et al. 2010) were in fact Cl^-/H^+ antiporters. The coupling ratio of Cl^- and H^+, i.e. the transport stoichiometry, has been much harder to dissect for ClC-5, because of the strong outward rectification of the currents which made measurements of reversal potential impossible. This problem was circumvented combining electrophysiological measurements with fluorescence-based measurements of proton flux, allowing the demonstration of a 2 Cl^-/1 H^+ stoichiometry also for ClC-5 (Zifarelli and Pusch 2009a).

14.4.2 Transport Cycle

The combination of structural and functional information on several CLC antiporters has allowed to propose at least a schematic model of their transport cycle applicable also to ClC-5 (Feng et al. 2010), and represented in Fig. 14.3. In state (A), the protonated side chain of E211 (in red) is displaced upwards from the permeation pathway, towards the extracellular space, and all the binding sites are occupied by Cl^- ions (green circles). E211 is then deprotonated and a Cl^- ion moves from S_{ext} to the extracellular space in state (B). The deprotonated E211 side chain then moves back to S_{ext} in state (C) and then to S_{cen} in state (D). In this conformation, the side chain of E211 can be protonated from the intracellular side in state (E) and moves to the extracellular space in state (F), while a Cl^- ion moves to the extracellular space. The empty binding sites can then be occupied by two Cl^- ions from the extracellular space. From one of these states, there is a gating transition to an "inactive" state, in which transport is not permitted. The "pre-steady state" currents, observed in the ClC-5 E268A mutant (Smith and Lippiat 2010; Zifarelli et al. 2012), correspond to transitions between states (C) and (E). A characteristic feature of this model is that Cl^-/H^+ coupling is obtained by a "kinetic" mechanism in which intrinsically uncoupled states must be short-lived. However, Accardi and colleagues proposed that coupling is a consequence of the thermodynamic stoichiometry of substrate binding (Picollo et al. 2012; Basilio et al. 2014).

14.5 Role of ClC-5 in Endosomal Physiology and Kidney Function

14.5.1 ClC-5 Localization in Renal Epithelia

ClC-5 is abundantly expressed in the kidney (mainly in the nephron), but it is also expressed in the intestine, thyroid and, in smaller amount, in liver and brain (Günther et al. 1998; Devuyst et al. 1999; Sakamoto et al. 1999; Steinmeyer et al. 1995; Van Den Hove et al. 2006). In the kidney, ClC-5 is expressed in all three different segments of the proximal tubule, in the collecting duct and, less abundantly, in the thick ascending limb of Henle's loop (Obermuller et al. 1998; Devuyst et al. 1999). In the proximal tubule, ClC-5 is mainly found in subapical endosomes of epithelial cells, below the brush border, with a small fraction also observed in the apical membrane (Günther et al. 1998; Devuyst et al. 1999). In subapical endosomes, ClC-5 co-localizes with the V-type ATPase (Fig. 14.4), and for a long time, it was thought to provide an electrical shunt for the proton pump, in order to allow proper acidification of endosomes (Günther et al. 1998; Sakamoto et al. 1999). Interestingly, in the collecting duct, ClC-5 co-localizes with V-ATPases in apical vesicles only in acid-secreting α-intercalated cells (Obermuller et al. 1998; Günther et al. 1998; Sakamoto et al. 1999), whereas it is mainly found in the cytoplasm in base-secreting β-intercalated cells (Günther et al. 1998; Sakamoto et al. 1999). In general, the physiological role of ClC-5 in these segments of the nephron is still unclear.

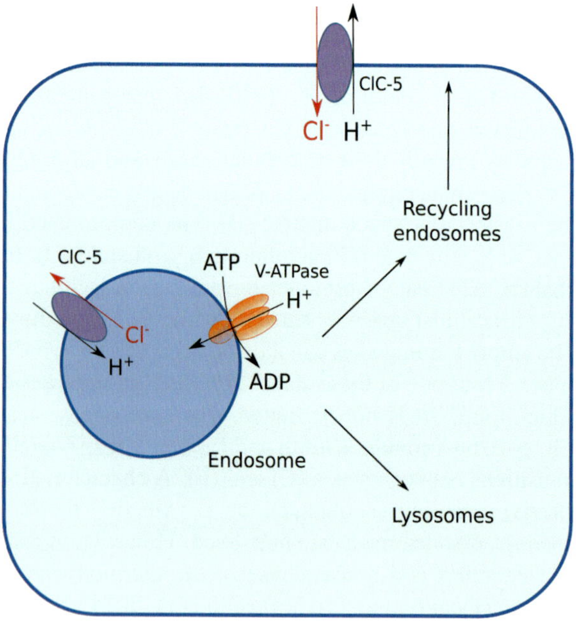

Fig. 14.4 Schematic representation of the putative role of ClC-5 in intracellular organelles' ionic homeostasis and trafficking. The V-type ATPase (in orange) and ClC-5 (in purple) are both expressed in early endosomes. After being endocytosed, ClC-5 can be transported to lysosomes, for degradation, and/or recycled back to the plasma membrane, where it will eventually take part in early endosomes formation. Adapted from Pusch and Zifarelli (2014)

14.5.2 Sorting and Degradation Processes of ClC-5

The mechanism regulating the subcellular targeting of ClC-5 remains still poorly understood. ClC-5 is predominantly localized in early endosomes of proximal tubule cells (Piwon et al. 2000; Suzuki et al. 2006) where it co-localizes with the GTPases Rab4 (Vandewalle et al. 2001) and Rab5a proteins (Devuyst et al. 1999; Günther et al. 1998; Sakamoto et al. 1999) (Luyckx et al. 1998). ClC-5 is also expressed in other intracellular organelles, such as recycling endosomes (Günther et al. 1998), partially overlapping with the closely related ClC-3 (Maritzen et al. 2008) and ClC-4 (Mohammad-Panah et al. 2009). Over the last decades, ClC-5 has been proposed to regulate the intracellular acidification along the endosomal pathway (Günther et al. 1998; Sakamoto et al. 1999) (Fig. 14.4). However, its physiological role in such organelles has not been fully elucidated so far. Concerning the degradation pathway of ClC-5, the linker region between the two intracellular CBS domains, contains a putative "PY motif", which is not conserved in the closely related ClC-3 and ClC-4 (Schwake et al. 2001). Such motif has been suggested as a specific target for ubiquitin-mediated degradation, at least in heterologous expression system (Schwake et al. 2001). This internalization process is probably activated upon interaction of the PY-motif with the E3 ubiquitin ligases (such as WWP2 and Nedd4-2) (Pirozzi et al. 1997). In line with such hypothesis, overexpression of a nonfunctional mutant of the WWP2 ligase or of ClC-5 with neutralizing mutations of the PY-motif in *Xenopus* oocytes leads to a doubling of the ClC-5-mediated currents (Schwake et al. 2001). In conclusion, the PY-motif appears to be important for both endocytotic uptake and degradation process (Schwake et al. 2001), as also suggested by in vitro ubiquitination experiments of ClC-5 (Hryciw et al. 2004). However, the biological role of the PY motif in protein sorting and degradation in vivo is still not entirely clear. Indeed, no effect was observed in the subcellular localization of ClC-5, as well as its endocytotic mechanism was apparently not altered in a *knock in* (KI) mouse model after disruption of the PY-motif (Rickheit et al. 2010).

14.5.3 ClC-5 and Dent's Disease

Dent's disease is an X-linked recessive chronic kidney malfunction, and a type of renal Fanconi disorder, defined for the first time in 1994 (Wrong et al. 1994; Fisher et al. 1994) and, at first, exclusively linked to mutations in the gene coding for ClC-5 (Steinmeyer et al. 1995; Lloyd et al. 1996). Dent's disease patients are affected by malfunctions of the proximal renal tubule, leading to progressive proteinuria, hyperphosphaturia, hypercalciuria and nephrolithiasis and consequent renal failure (Dent and Friedman 1964; Fisher et al. 1994; Lloyd et al. 1996; Pook et al. 1993). Female carriers show in general a milder phenotype possibly due to random X chromosome inactivation (Wrong et al. 1994; Reinhart et al. 1995). The most common Dent's disease symptoms concern defects in the reabsorption of proteins and salts in the proximal tubule, with consequent low molecular weight proteinuria (LMWP). Less consistently, patients present with excess of calcium in the urine (hypercalciuria) and calcium deposits in the kidney (nephrocalcinosis) (Cox et al. 1999). The symptoms typically appear in early childhood and progressively exacerbate potentially leading to end-stage renal disease in early to mid-adulthood. Secondary symptoms, such as the presence of rickets, have been additionally described in some patients (Igarashi et al. 1998). Interestingly, only 60% of patients with Dent's disease carry mutations in ClC-5. In 15% of patients, Dent's disease is instead correlated with mutations in the *OCRL1* gene (Hoopes et al. 2004). Intriguingly, in the remaining cases no mutations in either of the two genes have been described, suggesting the possible implication of other gene(s) (Hoopes et al. 2005; Wu et al. 2009). This aspect will be further discussed in Sect. 14.6.

14.5.4 ClC-5 Knockout Mice Reveal Dent's Disease Mechanism

The pathophysiological mechanism leading to Dent's disease has been extensively studied through the characterization of ClC-5 *knockout* (KO) animal models, independently generated by two research groups, and referred to as "Jentsch" and "Guggino" mice (Piwon et al. 2000; Wang et al. 2000).

14.5.4.1 Proteinuria

Low-molecular weight proteinuria seems to be caused by impaired endocytosis and uptake of proteins in the proximal renal tubule. A defect in megalin expression, as in the case of ClC-5 KO mice, might be caused by an impaired endosomal recycling (Maritzen et al. 2006; Van Den Hove et al. 2006). Indeed, in proximal tubule cells, the expression level of megalin was dramatically decreased after disruption of ClC-5 (Piwon et al. 2000). A similar effect was also observed for the megalin co-receptor cubulin in ClC-5 KO mice (Christensen et al. 2003). Megalin and cubulin are multiligand co-receptors localized on the apical brush-border membrane of proximal tubule epithelial cells (Christensen and Birn 2002). After ligand-binding, the complex is internalized, via a clathrin-mediated endocytotic process, in order to fuse with endosomal intermediates, such as early endosomes, to be finally degraded into lysosomes or recycled. Both receptors are involved in tubular uptake of albumin and other low molecular weight proteins. Cubilin, which was first identified as the receptor for the intrinsic factor and Vitamin B12 in the intestine, was subsequently shown to play the same role in the proximal tubule as well (Christensen and Birn 2002). Importantly, ClC-5 KO mice showed a defective trafficking in this nephron segment, with a parallel loss of megalin and cubulin at the apical brush-border membrane (Piwon et al. 2000; Christensen et al. 2003; Watanabe 2004), together with impaired lysosomal formation (Reed et al. 2010). After budding of clathrin-coated vesicles, such complexes need to be transported and dissociated along the endo-lysosomal pathway. This dissociation is facilitated by progressive endosomal acidification, which was initially proposed to be defective in ClC-5 KO mice. However, the mouse KI model (ClC-5^{unc}) harbouring a mutation of ClC-5 that turns it into a pure Cl^- conductance clearly indicates a different mechanism (Novarino et al. 2010). In fact, endosomal acidification was not impaired in ClC-5^{unc}, but still, endocytosis was dramatically reduced (Novarino et al. 2010). This suggests that the endosomal Cl^- concentration and the resulting coupled H^+ transport, rather than mere acidification, might be the most important factor regulating endocytosis.

14.5.4.2 Hypercalciuria

The two different mouse models of Dent's disease (Jentsch and Guggino) showed contrasting results in relation to hypercalciuria and nephrolithiasis. This is potentially explained by a complex regulation of calcium homeostasis. The impaired megalin-mediated endocytosis of parathyroid hormone (PTH) results in increased luminal concentration of PTH and thus enhanced uptake through the PTH receptors (PTH-R) leading to intensified 1α-hydroxylase activity. In the proximal renal tubule, 1α-hydroxylase converts 25(OH)$_2$-vitamin D_3 into its active form 1,25(OH)$_2$-vitamin D_3 and the increased production of active vitamin D_3 would lead to increased renal and intestinal calcium absorption (Scheinman 1998). However, the 25(OH)-vitamin D_3/BDP complex is known to be endocytosed via interaction with megalin (Nykjaer et al. 1999), a mechanism which is impaired in *Clcn5*$^{-/-}$ proximal tubule cells. In this scenario, a decreased uptake of vitamin D_3 precursor would balance the enhanced 1α-hydroxylase activity and might thus underlie the phenotype of the "Jentsch" mouse, which exhibited decreased 1,25(OH)$_2$-vitamin D_3 levels and no hypercalciuria (Piwon et al. 2000). A proposed model for hypercalciuria in Dent's disease patients is shown in Fig. 14.5.

14.5.4.3 Night Blindness

Another phenotype concerns the loss of retinol (vitamin A) and of its plasma binding protein (RBP) in the urine, due to altered endocytosis (Piwon et al. 2000). Not surprisingly, reabsorption of retinol-RBP complexes in proximal renal tubule cells is mediated by megalin (Christensen and Birn 2002). Physiologically, after uptake, RBP is transported to lysosomes for degradation, thus releasing vitamin A, which can be secreted into the bloodstream. A defective endocytotic process can thus lead to a decreased availability of vitamin A, finally resulting in night blindness, described in several cases of Dent's disease with early age onset (Sethi et al. 2009).

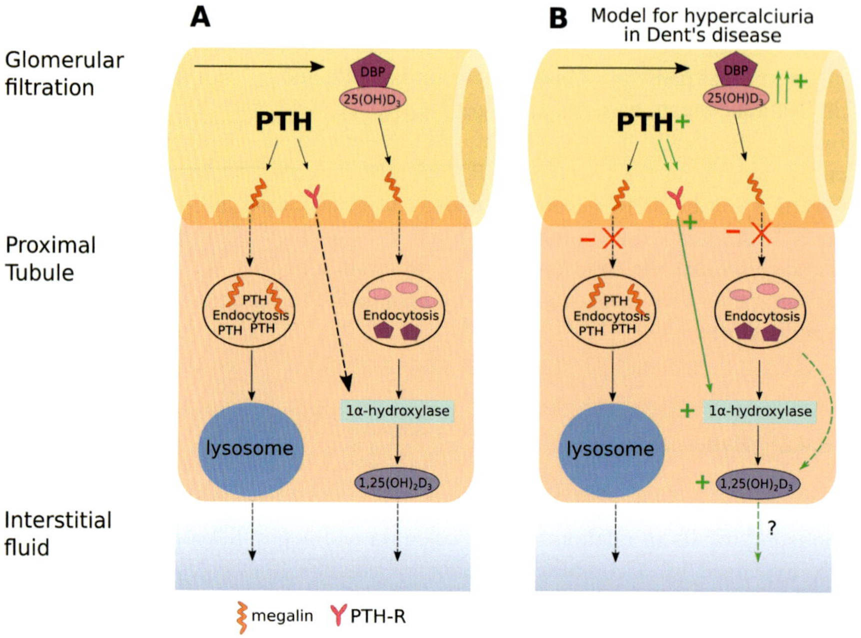

Fig. 14.5 Schematic model of hypercalciuria in Dent's disease. (**a**) Representation of physiological events characterizing proximal tubule re-absorption. After glomerular filtration, the parathyroid hormone (PTH) is endocytosed in a megalin-mediated process. Similarly, the inactive form of vitamin D (25(OH)D_3, pink spheres) is endocytosed, in complex with the vitamin D-binding protein (DBP, purple pentagons), via megalin binding. In normal conditions, the 25(OH)vitamin D_3 is then converted into its active form (1,25$(OH)_2D_3$) through the 1α-hydroxylase activity, in proximal renal tubule cells. Finally, the active vitamin D_3 can be transported in the interstitial fluid, thus regulating renal and intestinal calcium absorption. (**b**) Model of hypercalciuria in Dent's disease. Adapted from Maritzen et al. (2006). See text for details

14.5.4.4 Hyperphosphaturia

The increased level of PTH leads to enhanced expression of PTH receptors along the apical border of the proximal renal tubule. As result, the NaPi-2, sodium-phosphate cotransporter, is excessively endocytosed and degraded into lysosomes (Piwon et al. 2000). The decreased availability of NaPi-2 will then lead to a decreased absorption of phosphate, finally resulting in hyperphosphaturia (Piwon et al. 2000). A schematic model for the mechanism involved in hyperphosphaturia is represented in Fig. 14.6.

14.5.4.5 Altered Ion and Water Absorption

The increased level of the PTH hormone causes also altered regulation (and enhanced endocytosis) of NHE3, a Na^+/H^+ exchanger (Piwon et al. 2000). Therefore, a lower expression of NHE3 is associated with a lower absorption of sodium,

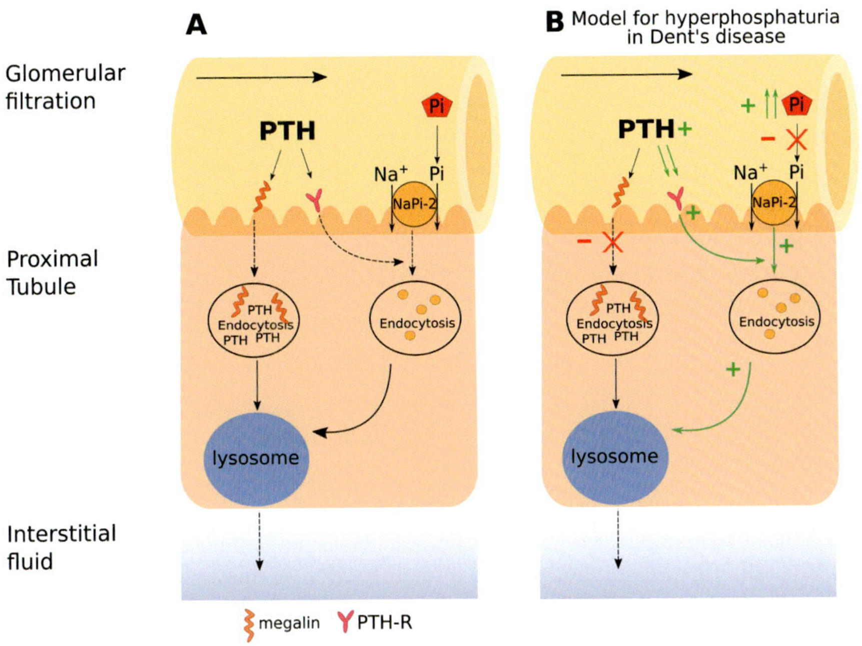

Fig. 14.6 Schematic model of hyperphosphaturia in Dent's disease. (**a**) In physiological conditions, after glomerular filtration, the PTH (parathyroid hormone) is endocytosed and recycled via megalin uptake in the proximal renal tubule cells and finally degraded into lysosomes. Its activity also regulates the uptake and degradation of the sodium-phosphate (NaPi-2) cotransporter. (**b**) In Dent's disease, megalin availability is strongly reduced due to its impaired recycling process mediated by ClC-5. Thus, PTH activity is dramatically enhanced, resulting in a higher internalization rate of NaPi-2 leading to reduced phosphate absorption, and finally higher phosphate level in the urine. Adapted from Maritzen et al. (2006)

bicarbonate and water in the proximal renal tubule, consistently with the increased urine volume (polyuria) and the altered ion concentration observed among the Dent's disease phenotype (Piwon et al. 2000). A plethora of several other symptoms, such as glycosuria, uricosuria, aminoaciduria, kaliuresis and impaired urinary acidification, have been described along the typical Dent's disease manifestations (Dent and Friedman 1964; Fisher et al. 1994; Lloyd et al. 1996; Pook et al. 1993). In such scenario, altered transport and protein trafficking might be among the intrinsic causes of such symptoms, which are still poorly characterized.

14.5.5 *Potential Binding Partner of ClC-5*

An intriguing hypothesis is that ClC-5 might participate in assembling endocytotic complexes, thus acting as scaffold protein. Indeed, several proteins have been already proposed as ClC-5 binding partners (Stauber et al. 2012), among which, the kinesin-like protein KIF3B kinase. Indeed, the transport of ClC-5-containing vesicles along the microtubules appears mediated by KIF3B, as shown through co-immunoprecipitation experiments and confirmed by laser scan confocal live cells imaging (Reed et al. 2010). In addition, HEK293 cells overexpressing KIF3B showed increased amplitude (~45%) of ClC-5 currents (Reed et al. 2010). Thus, absence of ClC-5 in renal epithelial cells might result in reduced ClC-5/KIF3B interaction, leading to impaired microtubular trafficking of megalin and cubulin endocytotic vesicles, causing defective protein re-absorption and Dent's disease renal manifestations (Reed et al. 2010). However, the lack of in vivo confirmation prevents from validating such hypothesis.

14.5.6 ClC-5 Mutations and Their Phenotypes

More than 100 mutations, including splice site, missense and nonsense mutations of ClC-5, have been implicated in Dent's disease (Mansour-Hendili et al. 2015; Pusch and Zifarelli 2014). Even though only a small fraction of missense mutations have been so far investigated in heterologous systems (Lloyd et al. 1996; Lourdel et al. 2012; Pusch and Zifarelli 2014) or patient-derived cells (Gorvin et al. 2013), a tentative genotype-phenotype correlation has been suggested (Gorvin et al. 2013). Some mutations, including S270R, G513E, R516W and I524K, are characterized by a strong reduction in current amplitude (Ludwig et al. 2005). This is caused by enhanced retention or degradation of ClC-5 in the endoplasmic reticulum (ER) as indicated by laser scan confocal microscopy, as shown in co-immunolabelling experiments with calnexin ER-specific antibody (Smith et al. 2009; Ludwig et al. 2005; Grand et al. 2009, Grand et al. 2011). The E527D causes defective activity of ClC-5, resulting in a strong impairment of endosomal acidification and abolished currents, with little effect on ClC-5 subcellular distribution, but this residue is relatively distant from the chloride and proton conduction pathways (Lloyd et al. 1997). Interestingly, the mutations G57V and R280P (both located close to the periphery of the subunit interface, facing the cytoplasm) show altered endosomal distribution (and reduced transport activity), but normal (for G57V) or enhanced (for R280P) endosomal acidification. In contrast to this, several mutations show a reduced (or even abolished) ion transport ability and impaired endosomal acidification, despite an unaltered cell surface expression. This is the case, for example, of the G212A mutant, which showed a normal plasma membrane distribution, comparable to that of WT ClC-5, but reduced transport current amplitudes (Grand et al. 2009) which is not too surprising considering that it is located close to the gating glutamate (E211) (Alekov 2015). Also the E267A mutant (Hoopes et al. 2004) leads to a much reduced current amplitude (Alekov 2015; Zdebik et al. 2008). In HEK cells both mutations altered the endosomal acidification observed upon expression of WT ClC-5 (Alekov 2015).

14.5.7 ClC-5 Mutations in Patient-Derived Cells Reveal Altered Endocytosis Without Effects on Endosomal Acidification

Two different groups found that for some mutations, altered endocytosis is not linked to defective endosomal acidification (Gorvin et al. 2013; Bignon et al. 2018). Gorvin et al. were able to generate conditionally immortalized proximal renal tubule cells (ciPTECs) derived from Dent's disease patients, carrying three ClC-5 mutations (30:insH, R637X and del132-241) that were conditionally immortalized (Gorvin et al. 2013). Albumin and transferrin endocytotic uptake was severely reduced in 30:insH and R637X ciPTECs mutants, while in del132-241, receptor-mediated endocytosis was completely abolished. Interestingly, live cell imaging of a GFP-tagged pH sensor associated with membrane vesicles (pHluorin-VAMP2) revealed unaltered endosomal acidification in 30:insH condition, while significantly more alkaline pH was measured in R637X and del132-241 ciPTECs.

14.6 Other Proteins Involved in Dent's Disease

As previously mentioned, 15% of Dent's disease patients have mutations in the OCRL1 rather than the ClC-5 gene (Hoopes et al. 2004). OCRL1 encodes for a protein with phosphoinositol 4,5-bisphosphate (PIP_2) 5-phosphatase activity, initially known to be mutated in the oculocerebrorenal Lowe syndrome. This is a multisystem disorder characterized by abnormalities affecting the eye, the nervous system and the kidney, where it leads to a Fanconi-type renal disorder (Lowe et al. 1952; Attree et al. 1992). In particular, some OCRL1 mutations have been described in Dent's disease of type 2, characterized by the same renal dysfunction occurring in Dent's disease of type 1, with no other additional symptoms (Hoopes et al. 2005). OCRL1 is a membrane-associated protein firstly found in the Golgi apparatus (Olivos-Glander et al. 1995; Dressman et al. 2000). More recently, OCRL1 has also been found in early endosomes where it is required for proper megalin trafficking (Vicinanza et al. 2011) but also in lysosomes (De Leo et al. 2016). Indeed, Lowe syndrome patients show reduced urine level of megalin and cubulin, consistent with a lower expression of these receptors on the apical membrane of the proximal renal tubule (Norden et al. 2002).

14.7 Summary

In conclusion, ClC-5 is an endosomal Cl^-/H^+ antiporter with a critical physiological role in proximal tubule endocytosis as highlighted by the fact that mutations in ClC-5 cause Dent's disease. The role of ClC-5 in the kidney has also been investigated in detail through KO and KI mice lines. The initial suggestion that ClC-5 provided only a Cl^- shunt conductance has been proved wrong, and it is now clear that it is the coupled flux of Cl^- and H^+ across the endosomal membrane that underlies its relevance. However, the specific role of this ClC-5-mediated coupled flux for endosomal physiology remains elusive.

References

Accardi A, Miller C (2004) Secondary active transport mediated by a prokaryotic homologue of ClC Cl^- channels. Nature 427:803–807

Accardi A, Pusch M (2003) Conformational changes in the pore of CLC-0. J Gen Physiol 122:277–293

Accardi A, Kolmakova-Partensky L, Williams C, Miller C (2004) Ionic currents mediated by a prokaryotic homologue of CLC Cl^- channels. J Gen Physiol 123:109–119

Accardi A, Walden M, Nguitragool W, Jayaram H, Williams C, Miller C (2005) Separate ion pathways in a Cl^-/H^+ exchanger. J Gen Physiol 126:563–570

Accardi A, Lobet S, Williams C, Miller C, Dutzler R (2006) Synergism between halide binding and proton transport in a CLC-type exchanger. J Mol Biol 362:691–699

Alekov AK (2015) Mutations associated with Dent's disease affect gating and voltage dependence of the human anion/proton exchanger ClC-5. Front Physiol 6:159

Alekov AK, Fahlke C (2009) Channel-like slippage modes in the human anion/proton exchanger ClC-4. J Gen Physiol 133:485–496

Attree O, Olivos IM, Okabe I, Bailey LC, Nelson DL, Lewis RA, Mcinnes RR, Nussbaum RL (1992) The Lowe's oculocerebrorenal syndrome gene encodes a protein highly homologous to inositol polyphosphate-5-phosphatase. Nature 358:239–242

Basilio D, Noack K, Picollo A, Accardi A (2014) Conformational changes required for H^+/Cl^- exchange mediated by a CLC transporter. Nat Struct Mol Biol 21:456–463

Bateman A (1997) The structure of a domain common to archaebacteria and the homocystinuria disease protein. Trends Biochem Sci 22:12–13

Bennetts B, Parker MW (2013) Molecular determinants of common gating of a ClC chloride channel. Nat Commun 4:1–11

Bennetts B, Parker MW, Cromer BA (2007) Inhibition of skeletal muscle CLC-1 chloride channels by low intracellular pH and ATP. J Biol Chem 282:32780–32791

Bergsdorf E-Y, Zdebik AA, Jentsch TJ (2009) Residues important for nitrate/proton coupling in plant and mammalian CLC transporters. J Biol Chem 284:11184–11193

Bignon Y, Alekov A, Frachon N, Lahuna O, Jean-Baptiste Doh-Egueli C, Deschenes G, Vargas-Poussou R, Lourdel S (2018) A novel CLCN5 pathogenic mutation supports Dent disease with normal endosomal acidification. Hum Mutat 39:1139–1149

Caputo A, Caci E, Ferrera L, Pedemonte N, Barsanti C, Sondo E et al (2008) TMEM16A, a membrane protein associated with calcium-dependent chloride channel activity. Science 322:590–594

Christensen EI, Birn H (2002) Megalin and cubilin: multifunctional endocytic receptors. Nat Rev Mol Cell Biol 3:256–266

Christensen EI, Devuyst O, Dom G, Nielsen R, Van Der Smissen P, Verroust P, Leruth M, Guggino WB, Courtoy PJ (2003) Loss of chloride channel ClC-5 impairs endocytosis by defective trafficking of megalin and cubilin in kidney proximal tubules. Proc Natl Acad Sci USA 100:8472–8477

Cox JP, Yamamoto K, Christie PT, Wooding C, Feest T, Flinter FA et al (1999) Renal chloride channel, CLCN5, mutations in Dent's disease. J Bone Miner Res 14:1536–1542

De Angeli A, Monachello D, Ephritikhine G, Frachisse JM, Thomine S, Gambale F, Barbier-Brygoo H (2006) The nitrate/proton antiporter AtCLCa mediates nitrate accumulation in plant vacuoles. Nature 442:939–942
De Leo MG, Staiano L, Vicinanza M, Luciani A, Carissimo A, Mutarelli M et al (2016) Autophagosome-lysosome fusion triggers a lysosomal response mediated by TLR9 and controlled by OCRL. Nat Cell Biol 18:839–850
De Stefano S, Pusch M, Zifarelli G (2011) Extracellular determinants of anion discrimination of the Cl^-/H^+ antiporter protein CLC-5. J Biol Chem 286:44134–44144
De Stefano S, Pusch M, Zifarelli G (2013) A single point mutation reveals gating of the human ClC-5 Cl^-/H^+ antiporter. J Physiol 591:5879–5893
Dent CE, Friedman M (1964) Hypercalcuric rickets associated with renal tubular damage. Arch Dis Child 39:240–249
Devuyst O, Christie PT, Courtoy PJ, Beauwens R, Thakker RV (1999) Intra-renal and subcellular distribution of the human chloride channel, CLC-5, reveals a pathophysiological basis for Dent's disease. Hum Mol Genet 8:247–257
Dressman MA, Olivos-Glander IM, Nussbaum RL, Suchy SF (2000) Ocrl1, a PtdIns(4,5)P(2) 5-phosphatase, is localized to the trans-Golgi network of fibroblasts and epithelial cells. J Histochem Cytochem 48:179–190
Duffield M, Rychkov G, Bretag A, Roberts M (2003) Involvement of helices at the dimer Interface in ClC-1 common gating. J Gen Physiol 121:149–161
Dutzler R, Campbell EB, Cadene M, Chait BT, Mackinnon R (2002) X-ray structure of a ClC chloride channel at 3.0 Å reveals the molecular basis of anion selectivity. Nature 415:287–294
Dutzler R, Campbell EB, Mackinnon R (2003) Gating the selectivity filter in ClC chloride channels. Science 300:108–112
Estévez R, Boettger T, Stein V, Birkenhäger R, Otto E, Hildebrandt F, Jentsch TJ (2001) Barttin is a Cl^- channel beta-subunit crucial for renal Cl^- reabsorption and inner ear K^+ secretion. Nature 414:558–561
Fahlke C, Rüdel R, Mitrovic N, Zhou M, George AL Jr (1995) An aspartic acid residue important for voltage-dependent gating of human muscle chloride channels. Neuron 15:463–472
Feng L, Campbell EB, Hsiung Y, Mackinnon R (2010) Structure of a eukaryotic CLC transporters defines an intermediate state in the transport cycle. Science 330:635–641
Fisher SE, Black GC, Lloyd SE, Hatchwell E, Wrong O, Thakker RV, Craig IW (1994) Isolation and partial characterization of a chloride channel gene which is expressed in kidney and is a candidate for Dent's disease (an X-linked hereditary nephrolithiasis). Hum Mol Genet 3:2053–2059
Friedrich T, Breiderhoff T, Jentsch TJ (1999) Mutational analysis demonstrates that ClC-4 and ClC-5 directly mediate plasma membrane currents. J Biol Chem 274:896–902
Gorvin CM, Wilmer MJ, Piret SE, Harding B, Van Den Heuvel LP, Wrong O et al (2013) Receptor-mediated endocytosis and endosomal acidification is impaired in proximal tubule epithelial cells of Dent disease patients. Proc Natl Acad Sci USA 110:7014–7019
Grand T, Mordasini D, L'hoste S, Pennaforte T, Genete M, Biyeyeme MJ et al (2009) Novel CLCN5 mutations in patients with Dent's disease result in altered ion currents or impaired exchanger processing. Kidney Int 76:999–1005
Grand T, L'hoste S, Mordasini D, Defontaine N, Keck M, Pennaforte T et al (2011) Heterogeneity in the processing of CLCN5 mutants related to Dent disease. Hum Mutat 32:476–483
Graves AR, Curran PK, Smith CL, Mindell JA (2008) The Cl^-/H^+ antiporter ClC-7 is the primary chloride permeation pathway in lysosomes. Nature 453:788–792
Grenningloh G, Gundelfinger E, Schmitt B, Betz H, Darlison MG, Barnard EA, Schofield PR, Seeburg PH (1987) Glycine vs GABA receptors. Nature 330:25–26
Grieschat M, Alekov AK (2012) Glutamate 268 regulates transport probability of the anion/proton exchanger ClC-5. J Biol Chem 287:8101–8109

Günther W, Lüchow A, Cluzeaud F, Vandewalle A, Jentsch TJ (1998) ClC-5, the chloride channel mutated in Dent's disease, colocalizes with the proton pump in endocytotically active kidney cells. Proc Natl Acad Sci USA 95:8075–8080

Guzman RE, Grieschat M, Fahlke C, Alekov AK (2013) ClC-3 is an intracellular chloride/proton exchanger with large voltage-dependent nonlinear capacitance. ACS Chem Neurosci 4:994–1003

Han W, Cheng RC, Maduke MC, Tajkhorshid E (2014) Water access points and hydration pathways in CLC H+/Cl- transporters. Proc Natl Acad Sci USA 111:1819–1824

Hoegg-Beiler MB, Sirisi S, Orozco IJ, Ferrer I, Hohensee S, Auberson M et al (2014) Disrupting MLC1 and GlialCAM and ClC-2 interactions in leukodystrophy entails glial chloride channel dysfunction. Nat Commun 5:3475

Hoopes RR Jr, Raja KM, Koich A, Hueber P, Reid R, Knohl SJ, Scheinman SJ (2004) Evidence for genetic heterogeneity in Dent's disease. Kidney Int 65:1615–1620

Hoopes RR Jr, Shrimpton AE, Knohl SJ, Hueber P, Hoppe B, Matyus J et al (2005) Dent disease with mutations in OCRL1. Am J Hum Genet 76:260–267

Hryciw DH, Ekberg J, Lee A, Lensink IL, Kumar S, Guggino WB, Cook DI, Pollock CA, Poronnik P (2004) Nedd4-2 functionally interacts with ClC-5: involvement in constitutive albumin endocytosis in proximal tubule cells. J Biol Chem 279:54996–55007

Igarashi T, Gunther W, Sekine T, Inatomi J, Shiraga H, Takahashi S et al (1998) Functional characterization of renal chloride channel, CLCN5, mutations associated with Dent's Japan disease. Kidney Int 54:1850–1856

Jayaram H, Accardi A, Wu F, Williams C, Miller C (2008) Ion permeation through a Cl^--selective channel designed from a CLC Cl^-/H^+ exchanger. Proc Natl Acad Sci USA 105:11194–11199

Jentsch TJ (2015) Discovery of CLC transport proteins: cloning, structure, function and pathophysiology. J Physiol 593:4091–4109

Jentsch TJ, Pusch M (2018) CLC Chloride channels and transporters: structure, function, physiology, and disease. Physiol Rev 98:1493–1590

Jentsch TJ, Steinmeyer K, Schwarz G (1990) Primary structure of *Torpedo marmorata* chloride channel isolated by expression cloning in Xenopus oocytes. Nature 348:510–514

Jeworutzki E, López-Hernández T, Capdevila-Nortes X, Sirisi S, Bengtsson L, Montolio M et al (2012) GlialCAM, a protein defective in a leukodystrophy, serves as a ClC-2 Cl^- channel auxiliary subunit. Neuron 73:951–961

Lagostena L, Zifarelli G, Picollo A (2019) New insights into the mechanism of NO_3^- selectivity in the human kidney chloride channel ClC-Ka and the CLC protein family. J Am Soc Nephr 30:293–302

Lange PF, Wartosch L, Jentsch TJ, Fuhrmann JC (2006) ClC-7 requires Ostm1 as a beta-subunit to support bone resorption and lysosomal function. Nature 440:220–223

Leisle L, Ludwig CF, Wagner FA, Jentsch TJ, Stauber T (2011) ClC-7 is a slowly voltage-gated $2Cl^-/1H^+$-exchanger and requires Ostm1 for transport activity. EMBO J 30:2140–2152

Li X, Shimada K, Showalter LA, Weinman SA (2000) Biophysical properties of ClC-3 differentiate it from swelling-activated chloride channels in Chinese hamster ovary-K1 cells. J Biol Chem 275:35994–35998

Lim HH, Miller C (2009) Intracellular proton-transfer mutants in a CLC Cl^-/H^+ exchanger. J Gen Physiol 133:131–138

Lísal J, Maduke M (2008) The ClC-0 chloride channel is a 'broken' Cl^-/H^+ antiporter. Nat Struct Mol Biol 15:805–810

Lloyd SE, Pearce SH, Fisher SE, Steinmeyer K, Schwappach B, Scheinman SJ et al (1996) A common molecular basis for three inherited kidney stone diseases. Nature 379:445–449

Lloyd SE, Günther W, Pearce SH, Thomson A, Bianchi ML, Bosio M et al (1997) Characterisation of renal chloride channel, CLCN5, mutations in hypercalciuric nephrolithiasis (kidney stones) disorders. Hum Mol Genet 6:1233–1239

Lobet S, Dutzler R (2006) Ion-binding properties of the ClC chloride selectivity filter. EMBO J 25:24–33

Lourdel S, Grand T, Burgos J, Gonzalez W, Sepulveda FV, Teulon J (2012) ClC-5 mutations associated with Dent's disease: a major role of the dimer interface. Pflugers Arch 463:247–256
Lowe CU, Terrey M, Mac LE (1952) Organic-aciduria, decreased renal ammonia production, hydrophthalmos, and mental retardation; a clinical entity. AMA Am J Dis Child 83:164–184
Ludewig U, Pusch M, Jentsch TJ (1996) Two physically distinct pores in the dimeric ClC-0 chloride channel. Nature 383:340–343
Ludewig U, Jentsch TJ, Pusch M (1997) Inward rectification in ClC-0 chloride channels caused by mutations in several protein regions. J Gen Physiol 110:165–171
Ludwig M, Doroszewicz J, Seyberth HW, Bokenkamp A, Balluch B, Nuutinen M, Utsch B, Waldegger S (2005) Functional evaluation of Dent's disease-causing mutations: implications for ClC-5 channel trafficking and internalization. Hum Genet 117:228–237
Ludwig CF, Ullrich F, Leisle L, Stauber T, Jentsch TJ (2013) Common gating of both CLC transporter subunits underlies voltage-dependent activation of the $2Cl^-/1H^+$ exchanger ClC-7/Ostm1. J Biol Chem 288:28611–28619
Luyckx VA, Goda FO, Mount DB, Nishio T, Hall A, Hebert SC, Hammond TG, Yu AS (1998) Intrarenal and subcellular localization of rat CLC5. Am J Phys 275:F761–FF69
Mansour-Hendili L, Blanchard A, Le Pottier N, Roncelin I, Lourdel S, Treard C et al (2015) Mutation update of the CLCN5 gene responsible for Dent disease 1. Hum Mutat 36:743–752
Maritzen T, Rickheit G, Schmitt A, Jentsch TJ (2006) Kidney-specific upregulation of vitamin D3 target genes in ClC-5 KO mice. Kidney Int 70:79–87
Maritzen T, Keating DJ, Neagoe I, Zdebik AA, Jentsch TJ (2008) Role of the vesicular chloride transporter ClC-3 in neuroendocrine tissue. J Neurosci 28:10587–10598
Meyer S, Savaresi S, Forster IC, Dutzler R (2007) Nucleotide recognition by the cytoplasmic domain of the human chloride transporter ClC-5. Nat Struct Mol Biol 14:60–67
Miller C (1982) Open-state substructure of single chloride channels from Torpedo electroplax. Philos Trans R Soc Lond Ser B Biol Sci 299:401–411
Miller C (2006) ClC chloride channels viewed through a transporter lens. Nature 440:484–489
Miller C (2015) In the beginning: a personal reminiscence on the origin and legacy of ClC-0, the 'Torpedo Cl^- channel'. J Physiol 593:4085–4090
Miller C, White MM (1980) A voltage-dependent chloride conductance channel from Torpedo electroplax membrane. Ann NY Acad Sci 341:534–551
Miller C, White MM (1984) Dimeric structure of single chloride channels from Torpedo electroplax. Proc Natl Acad Sci USA 81:2772–2775
Mohammad-Panah R, Wellhauser L, Steinberg BE, Wang Y, Huan LJ, Liu XD, Bear CE (2009) An essential role for ClC-4 in transferrin receptor function revealed in studies of fibroblasts derived from Clcn4-null mice. J Cell Sci 122:1229–1237
Neagoe I, Stauber T, Fidzinski P, Bergsdorf EY, Jentsch TJ (2010) The late endosomal ClC-6 mediates proton/chloride countertransport in heterologous plasma membrane expression. J Biol Chem 285:21689–21697
Norden AG, Lapsley M, Igarashi T, Kelleher CL, Lee PJ, Matsuyama T et al (2002) Urinary megalin deficiency implicates abnormal tubular endocytic function in Fanconi syndrome. J Am Soc Nephrol 13:125–133
Novarino G, Weinert S, Rickheit G, Jentsch TJ (2010) Endosomal chloride-proton exchange rather than chloride conductance is crucial for renal endocytosis. Science 328:1398–1401
Nykjaer A, Dragun D, Walther D, Vorum H, Jacobsen C, Herz J, Melsen F, Christensen EI, Willnow TE (1999) An endocytic pathway essential for renal uptake and activation of the steroid 25-(OH) vitamin D3. Cell 96:507–515
Obermuller N, Gretz N, Kriz W, Reilly RF, Witzgall R (1998) The swelling-activated chloride channel ClC-2, the chloride channel ClC-3, and ClC-5, a chloride channel mutated in kidney stone disease, are expressed in distinct subpopulations of renal epithelial cells. J Clin Invest 101:635–642
Olivos-Glander IM, Janne PA, Nussbaum RL (1995) The oculocerebrorenal syndrome gene product is a 105-kD protein localized to the Golgi complex. Am J Hum Genet 57:817–823

Park E, Mackinnon R (2018) Structure of the CLC-1 chloride channel from *Homo sapiens*. elife 7. https://doi.org/10.7554/eLife.36629
Park E, Campbell EB, Mackinnon R (2017) Structure of a CLC chloride ion channel by cryo-electron microscopy. Nature 541:500–505
Phillips S, Brammer AE, Rodriguez L, Lim HH, Stary-Weinzinger A, Matulef K (2012) Surprises from an unusual CLC homolog. Biophys J 103:L44–L46
Picollo A, Pusch M (2005) Chloride/proton antiporter activity of mammalian CLC proteins ClC-4 and ClC-5. Nature 436:420–423
Picollo A, Liantonio A, Didonna MP, Elia L, Camerino DC, Pusch M (2004) Molecular determinants of differential pore blocking of kidney CLC-K chloride channels. EMBO Rep 5:584–589
Picollo A, Malvezzi M, Houtman JC, Accardi A (2009) Basis of substrate binding and conservation of selectivity in the CLC family of channels and transporters. Nat Struct Mol Biol 16:1294–1301
Picollo A, Xu Y, Johner N, Berneche S, Accardi A (2012) Synergistic substrate binding determines the stoichiometry of transport of a prokaryotic H^+/Cl^- exchanger. Nat Struct Mol Biol 19:525–531, S1
Pirozzi G, Mcconnell SJ, Uveges AJ, Carter JM, Sparks AB, Kay BK, Fowlkes DM (1997) Identification of novel human WW domain-containing proteins by cloning of ligand targets. J Biol Chem 272:14611–14616
Piwon N, Günther W, Schwake M, Bösl MR, Jentsch TJ (2000) ClC-5 Cl^--channel disruption impairs endocytosis in a mouse model for Dent's disease. Nature 408:369–373
Pook MA, Wrong O, Wooding C, Norden AG, Feest TG, Thakker RV (1993) Dent's disease, a renal Fanconi syndrome with nephrocalcinosis and kidney stones, is associated with a microdeletion involving DXS255 and maps to Xp11.22. Hum Mol Genet 2:2129–2134
Pusch M, Zifarelli G (2014) ClC-5: physiological role and biophysical mechanisms. Cell Calcium 58:57–66
Pusch M, Ludewig U, Jentsch TJ (1997) Temperature dependence of fast and slow gating relaxations of ClC-0 chloride channels. J Gen Physiol 109:105–116
Pusch M, Jordt SE, Stein V, Jentsch TJ (1999) Chloride dependence of hyperpolarization-activated chloride channel gates. J Physiol 515:341–353
Qu Z, Wei RW, Mann W, Hartzell HC (2003) Two bestrophins cloned from *Xenopus laevis* oocytes express Ca^{2+}-activated Cl^- currents. J Biol Chem 278:49563–49572
Reed AA, Loh NY, Terryn S, Lippiat JD, Partridge C, Galvanovskis J et al (2010) CLC-5 and KIF3B interact to facilitate CLC-5 plasma membrane expression, endocytosis, and microtubular transport: relevance to pathophysiology of Dent's disease. Am J Physiol Renal Physiol 298: F365–F380
Reinhart SC, Norden AG, Lapsley M, Thakker RV, Pang J, Moses AM et al (1995) Characterization of carrier females and affected males with X-linked recessive nephrolithiasis. J Am Soc Nephrol 5:1451–1461
Rickheit G, Wartosch L, Schaffer S, Stobrawa SM, Novarino G, Weinert S, Jentsch TJ (2010) Role of ClC-5 in renal endocytosis is unique among ClC exchangers and does not require PY-motif-dependent ubiquitylation. J Biol Chem 285:17595–17603
Riordan JR, Rommens JM, Kerem B, Alon N, Rozmahel R, Grzelczak Z et al (1989) Identification of the cystic fibrosis gene: cloning and characterization of complementary DNA. Science 245:1066–1073
Rychkov GY, Pusch M, Roberts ML, Jentsch TJ, Bretag AH (1998) Permeation and block of the skeletal muscle chloride channel, ClC-1, by foreign anions. J Gen Physiol 111:653–665
Sakamoto H, Sado Y, Naito I, Kwon TH, Inoue S, Endo K et al (1999) Cellular and subcellular immunolocalization of ClC-5 channel in mouse kidney: colocalization with H^+-ATPase. Am J Phys 277:F957–F965
Scheel O, Zdebik AA, Lourdel S, Jentsch TJ (2005) Voltage-dependent electrogenic chloride/proton exchange by endosomal CLC proteins. Nature 436:424–427
Scheinman SJ (1998) X-linked hypercalciuric nephrolithiasis: clinical syndromes and chloride channel mutations. Kidney Int 53:3–17

Schroeder BC, Cheng T, Jan YN, Jan LY (2008) Expression cloning of TMEM16A as a calcium-activated chloride channel subunit. Cell 134:1019–1029

Schwake M, Friedrich T, Jentsch TJ (2001) An internalization signal in ClC-5, an endosomal Cl^- channel mutated in Dent's disease. J Biol Chem 276:12049–12054

Sethi SK, Ludwig M, Kabra M, Hari P, Bagga A (2009) Vitamin A responsive night blindness in Dent's disease. Pediatr Nephrol 24:1765–1770

Smith AJ, Lippiat JD (2010) Voltage-dependent charge movement associated with activation of the CLC-5 $2Cl^-/1H^+$ exchanger. FASEB J 24:3696–3705

Smith AJ, Reed AA, Loh NY, Thakker RV, Lippiat JD (2009) Characterization of Dent's disease mutations of CLC-5 reveals a correlation between functional and cell biological consequences and protein structure. Am J Physiol Renal Physiol 296:F390–F397

Stauber T, Weinert S, Jentsch TJ (2012) Cell biology and physiology of CLC chloride channels and transporters. Compr Physiol 2:1701–1744

Steinmeyer K, Schwappach B, Bens M, Vandewalle A, Jentsch TJ (1995) Cloning and functional expression of rat CLC-5, a chloride channel related to kidney disease. J Biol Chem 270:31172–31177

Sun H, Tsunenari T, Yau KW, Nathans J (2002) The vitelliform macular dystrophy protein defines a new family of chloride channels. Proc Natl Acad Sci USA 99:4008–4013

Suzuki T, Rai T, Hayama A, Sohara E, Suda S, Itoh T, Sasaki S, Uchida S (2006) Intracellular localization of ClC chloride channels and their ability to form hetero-oligomers. J Cell Physiol 206:792–798

Traverso S, Zifarelli G, Aiello R, Pusch M (2006) Proton sensing of CLC-0 mutant E166D. J Gen Physiol 127:51–66

Van Den Hove MF, Croizet-Berger K, Jouret F, Guggino SE, Guggino WB, Devuyst O, Courtoy PJ (2006) The loss of the chloride channel, ClC-5, delays apical iodide efflux and induces a euthyroid goiter in the mouse thyroid gland. Endocrinology 147:1287–1296

Vandewalle A, Cluzeaud F, Peng KC, Bens M, Lüchow A, Günther W, Jentsch TJ (2001) Tissue distribution and subcellular localization of the ClC-5 chloride channel in rat intestinal cells. Am J Physiol Cell Physiol 280:C373–C381

Vicinanza M, Di Campli A, Polishchuk E, Santoro M, Di Tullio G, Godi A et al (2011) OCRL controls trafficking through early endosomes via PtdIns4,5P(2)-dependent regulation of endosomal actin. EMBO J 30:4970–4985

Voss FK, Ullrich F, Münch J, Lazarow K, Lutter D, Mah N et al (2014) Identification of LRRC8 heteromers as an essential component of the volume-regulated anion channel VRAC. Science 344:634–638

Wang SS, Devuyst O, Courtoy PJ, Wang XT, Wang H, Wang Y, Thakker RV, Guggino S, Guggino WB (2000) Mice lacking renal chloride channel, CLC-5, are a model for Dent's disease, a nephrolithiasis disorder associated with defective receptor- mediated endocytosis. Hum Mol Genet 9:2937–2945

Watanabe T (2004) Megalin and proximal renal tubular dysfunction in Dent disease. Pediatr Nephrol 19:1305. author reply 1306

White MM, Miller C (1979) A voltage-gated anion channel from the electric organ of *Torpedo californica*. J Biol Chem 254:10161–10166

Wrong OM, Norden AG, Feest TG (1994) Dent's disease; a familial proximal renal tubular syndrome with low-molecular-weight proteinuria, hypercalciuria, nephrocalcinosis, metabolic bone disease, progressive renal failure and a marked male predominance. QJM 87:473–493

Wu F, Reed AA, Williams SE, Loh NY, Lippiat JD, Christie PT et al (2009) Mutational analysis of CLC-5, cofilin and CLC-4 in patients with Dent's disease. Nephron Physiol 112:53–62

Yang YD, Cho H, Koo JY, Tak MH, Cho Y, Shim WS et al (2008) TMEM16A confers receptor-activated calcium-dependent chloride conductance. Nature 455:1210–1215

Zdebik AA, Zifarelli G, Bergsdorf EY, Soliani P, Scheel O, Jentsch TJ, Pusch M (2008) Determinants of anion-proton coupling in mammalian endosomal CLC proteins. J Biol Chem 283:4219–4227

Zifarelli G, Pusch M (2007) CLC chloride channels and transporters: a biophysical and physiological perspective. Rev Physiol Biochem Pharmacol 158:23–76
Zifarelli G, Pusch M (2009a) Conversion of the 2 Cl^-/1 H^+ antiporter ClC-5 in a NO_3^-/H^+ antiporter by a single point mutation. EMBO J 28:175–182
Zifarelli G, Pusch M (2009b) Intracellular regulation of human ClC-5 by adenine nucleotides. EMBO Rep 10:1111–1116
Zifarelli G, De Stefano S, Zanardi I, Pusch M (2012) On the mechanism of gating charge movement of ClC-5, a human Cl^-/H^+ antiporter. Biophys J 102:2060–2069

Chapter 15
CFTR and Cystic Fibrosis: A Need for Personalized Medicine

Neil A. Bradbury

Abstract Cystic fibrosis (CF), a common lethal genetic disease, is caused by mutations in the cystic fibrosis transmembrane conductance regulator (*cftr*) gene, which codes for an epithelial anion channel. Although the identification of the gene in 1989 heralded hope for a therapy to treat the underlying protein defect, patients continued to be treated exclusively with drugs that address the symptoms of the disease (antibiotics for airway disease, pancreatic supplements to replace digestive enzymes, and anti-inflammatories to reduce airway inflammation) rather than treating the basic defect. Nevertheless, this approach has resulted in a marked improvement in the survival of patients with CF over the last few decades, such that the median predicted survival is now around 40 years of age. Since the discovery of the gene encoding CFTR and the identification of mutations underlying the defects in CFTR function, the hunt has been on to discover drugs that would correct the basic defect in the CFTR protein. Although almost 1900 CFTR mutations have been described in patients, they mostly fall into a few broad categories of disruption. Given the disparate ways in which CFTR mutations affect the CFTR protein however, it has become apparent that a single drug regimen will not be effective in treating all patients. Thus, a patient's genotype would have to be taken into account when deciding which drug would be appropriate to treat individual patients with CF. Individual or personalized medicine as a concept has been increasingly highlighted in recent years, yet with CF, personalized medicine is not a mere academic exercise but rather a necessity in order to effectively treat all the mutations that are found in CF patients. This chapter presents the current CFTR mutation classifications and shows how such classification is essential in the establishment of a mutation-specific targeted drug therapy for each individual with CF.

Keywords CFTR · Potentiator · Corrector · Amplifier · HBE · Genotype · Theratype · Mutations · High-throughput screening

---

N. A. Bradbury (✉)
Department of Physiology and Biophysics, Chicago Medical School, Rosalind Franklin University of Medicine and Science, North Chicago, IL, USA
e-mail: neil.bradbury@rosalindfranklin.edu

K. L. Hamilton, D. C. Devor (eds.), *Studies of Epithelial Transporters and Ion Channels*, Physiology in Health and Disease,
https://doi.org/10.1007/978-3-030-55454-5_15

15.1 Introduction

Cystic fibrosis (CF) is a fatal autosomal recessive disease affecting some 1:2500–4500 newborns among Caucasian populations (Davis et al. 1996). Although not completely absent from other ethnicities, the penetrance of CF occurs at a much lower rate within other racial groups. In African-American populations, for example, the penetrance of CF is around 1 in 17,000, whereas in Asian populations, such as Japan, the incidence is even lower at around 1 in 350,000 live births (Yamashiro et al. 1997). Although often thought of as a lung disease, CF in fact results in the dysfunction of multiple organs and tissues, causing recurrent sinopulmonary infections, pancreatic insufficiency (including both exocrine and endocrine pancreatic functions), bowel obstructions, liver disease, lipid abnormalities, and male infertility, as well as lowered bone calcification (Clague 2014; Elborn 2016; Hecker and Aris 2004; White et al. 2007). The overall pathology of CF is characterized by abnormal epithelial electrolyte transport of chloride, bicarbonate, and sodium, resulting in mucus accumulation in the airways, leading to excessive inflammation, airway remodeling, bronchiectasis, and ultimately death from respiratory failure and pulmonary hypertension (see Chap. 16 for more detailed information about the structural and mechanistic insights of CFTR) (Boat et al. 1989; Bradbury 2015; Elborn 2016; Nichols and Chmiel 2015). One of the cardinal symptoms of CF is the presence of high levels of salt in the sweat, a feature which has been associated with infant mortality for centuries. Medieval folklore avowed "*woe to the child, who when kissed on the forehead tastes salty; that child is bewitched and soon must die*" (Alonzo De Los Ruyzes De Fonteca 1606; Quinton 1999). CF was first comprehensively described in the 1930s by Dorothy Hansine Anderson (Anderson 1938), when she noted pancreatic lesions and congested airways in young children who had purportedly died from celiac disease. Subsequently, working with Dr. Paul di Sant'Agnese, the discovery was made that children whom Andersen had identified as having "cystic fibrosis of the pancreas" displayed markedly elevated sweat electrolyte levels, a finding that was to become a hallmark for the disease and the gold standard for diagnosis. Pioneering work by Dr. Paul Quinton in the late 1970s and early 1980s established that CF patients had elevated sweat electrolytes due to the absence of a chloride reabsorptive pathway in CF sweat ducts (Quinton 1983), although the identity of this pathway would remain unknown for the next few years. During the late 1980s, a major effort into identifying the genetic basis for CF culminated in the discovery of the cystic fibrosis transmembrane conductance regulator (*CFTR*) gene in 1989 (Kerem et al. 1989; Riordan et al. 1989) and the identification of a 3 base-pair deletion that was present in most patients with CF. The international effort marked the first time that a gene had been identified without any knowledge of the protein for which it coded.

15.2 Biology of CFTR

CF is a monogenic disorder affecting individuals predominantly of Caucasian descent. In Europe, around 1:3500 neonates are diagnosed annually with CF, with slightly more in the USA, where there is an incidence of around 1:2000–3000 live births (FOUNDATION 2015). Since CF is an autosomal recessive disease, a patient must inherit two mutant *CFTR* genes (alleles), one from each parent. The *CFTR* gene is located on the long arm of chromosome 7 at position q31.2 and comprises 27 exons, spanning over 190 kb. Following RNA splicing, the resultant *CFTR* mRNA is 6.5 kb, and all exons are required for normal CFTR function. The CFTR protein is composed of 1480 amino acids, which form a membrane-integrated anion channel localized predominantly to the apical membrane of epithelial cells (the sweat duct expresses CFTR in both the apical and basolateral membranes) (Cohn et al. 1991; Quinton 1983). CFTR is a member of a large adenosine triphosphate (ATP)-binding cassette (ABC) transporter family, which are generally responsible for the transport of small molecules in an ATP-dependent fashion. For example, P-glycoprotein (MDR1) confers resistance to chemotherapeutic drugs by "pumping" the drugs out of cells across the plasma membrane using energy from ATP hydrolysis (Choi and Yu 2014). Similarly, the ABC transport protein ALDP (*ABCD1*) is present in peroxisomal membranes and works to transport fatty acids into peroxisomes for degradation. Mutations in ALDP are associated with X-linked adrenoleukodystrophy, a condition brought to the attention of many through the Hollywood movie *Lorenzo's Oil*. As with all ABC transporters, CFTR is composed of two membrane-spanning domains (MSDs, each composed of six individual transmembrane segments), which form the anion pore, and two nucleotide-binding domains (NBDs). Unique to CFTR is the presence of a regulatory or R domain, linking the two halves (Fig. 15.1). It is now known that the NBDs do not act independently, but rather dimerize so that each NBD confers two half ATP-binding pockets. Recently, the structure of human CFTR has been determined by cryo-EM at a resolution of 3.9 Å (Liu et al. 2017) (Fig. 15.2). This certainly allows for in silico molecular docking of pharmacologic compounds into mutant CFTR, though translating such docking into therapeutic reagents is a mammoth task (de Ruyck et al. 2016).

Once the DNA encoding CFTR has been translated into mRNA, the mRNA leaves the nucleus and interacts with ribosomes that initiate the translation of CFTR within the endoplasmic reticulum (ER). During its synthesis, CFTR is co-translationally modified such that a core carbohydrate group is added to the fourth extracellular loop on each of two asparagine residues (designated Band B CFTR, as detected by immunoblot analysis). Within the ER, the complex co-translational folding process utilizes multiple accessory proteins, including membrane-associated calnexin, calreticulin, and other chaperone proteins (Bagdany et al. 2017; Harada et al. 2006; Kim and Skach 2012; Rosser et al. 2008) to cause CFTR to achieve its final 3D structure. During the folding process, appropriate interactions between the NBDs and cytoplasmic loops from the MSDs are required to help stabilize the final 3D conformation and yield an ER export competent protein.

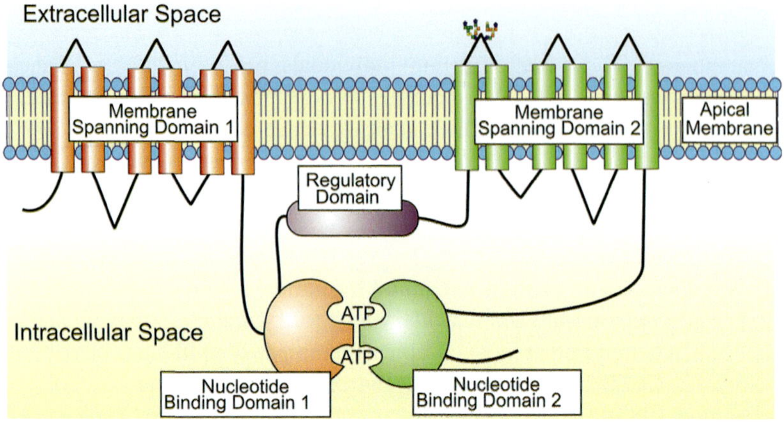

Fig. 15.1 Domain organization of CFTR in the plasma membrane. The cystic fibrosis transmembrane conductance regulator (CFTR) is composed of two repeated motifs, each containing a six-member helical membrane-spanning domain (MSD) and a nucleotide-binding domain (NBD). The regulatory (R) domain links the two halves and serves as a regulatory domain containing multiple phosphorylation consensus sites. Two N-linked carbohydrate moieties are added to the fourth extracellular loop. Activation of CFTR occurs upon phosphorylation of the R domain and binding of ATP to the NBDs

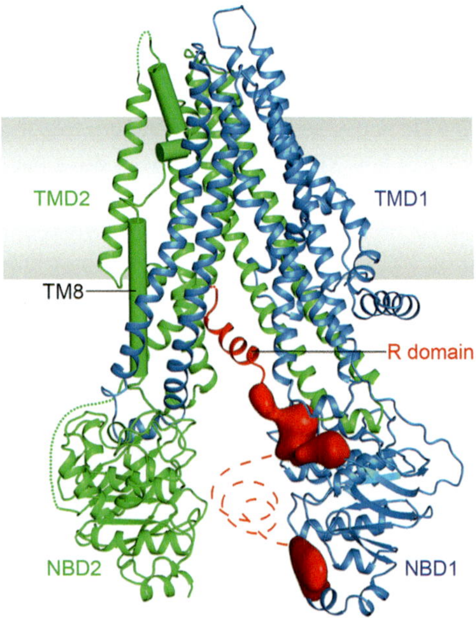

Fig. 15.2 Cryo-EM structure of human CFTR. Overall structure of human CFTR in the dephosphorylated ATP conformation. The EM densities shown in red correspond to unstructured regions within the R insertion of NBD1 and the R domain. Regions not resolved in the structure are shown as dashed lines for visualization purposes only. With permission from Liu et al. (2017)

This process is remarkably inefficient, with much of nascent CFTR protein failing to achieve the final conformation (though also see (Varga et al. 2004)). CFTR molecules that have achieved their final conformation bud off from the ER, and travel to the Golgi apparatus, where the conversion of the ER-appended mannose-enriched chain to a mature complex oligosaccharide occurs (Band C CFTR). CFTR again buds off from the Golgi and travels to the plasma membrane (Bidaud-Meynard et al. 2018; Yoo et al. 2002). In the plasma membrane, CFTR turns over through clathrin-mediated endocytosis at a rate of 10% per minute (Bradbury et al. 1994; Lukacs et al. 1997; Prince et al. 1999; Weixel and Bradbury 2000, 2001) followed by fairly efficient recycling back to the cell surface (Ameen et al. 2007; Picciano et al. 2003; Swiatecka-Urban et al. 2002). If CFTR fails to fold and stabilize appropriately in the ER (either through inefficient folding or because of mutations), misfolded CFTR is recognized and targeted for degradation by ER-associated degradation (ERAD), primarily involving the ubiquitin proteasome pathway (Brodsky and Skach 2011). Another later quality control process exists for CFTR at the plasma membrane, where poorly functioning or partially unfolded CFTR is ubiquitinated and targeted for proteosomal degradation (Sharma et al. 2004). Overall, CFTR appears to have a protein half-life of 12–24 h (Ward and Kopito 1994).

15.3 Clinical Manifestations of CFTR Mutations

Despite improvements in CF care over the past few decades, CF is still a debilitating disease, limiting what would otherwise be normal day-to-day activities. Indeed, many patients state that CF has a marked negative impact on work and career (Laborde-Casterot et al. 2012). Moreover, the life expectancy of a patient with CF in the USA is roughly half of that for the non-CF population (FOUNDATION 2015). At the physiological level, CFTR is involved in fluid and electrolyte transport across epithelial cells (Welsh and Ramsey 2001). The absence or dysfunction of CFTR results in altered ionic composition of epithelial secretions, causing thick dehydrated mucus in the airways and inspissated pancreatic enzymes in the exocrine pancreatic ducts (Welsh and Ramsey 2001). Shortly after the identification of CF as a unique disease, the association of CF with elevated sweat electrolytes was also documented. In normal individuals, the chloride (followed by sodium) present in the initial sweat produced by the secretory coil is reabsorbed by the duct cells in a CFTR-dependent manner, resulting in the appearance of a hypotonic fluid on the skin surface, evaporation of fluid from the cell surface helping to cool the body (Bradbury 2015). In CF individuals, this reabsorptive process is absent or significantly impaired leading to the appearance of elevated salt levels in the sweat. Note, sweat can be produced in the secretory coil in a CFTR-independent manner (usually with pilocarpine in the sweat test). The majority of patients diagnosed with CF have sweat chloride levels >60 mmol/L. Although a cutoff of >60 mmol/L is considered diagnostic of CF, many patients have significantly higher sweat chloride levels. A comparison of sweat chloride with predicted residual CFTR function based upon the

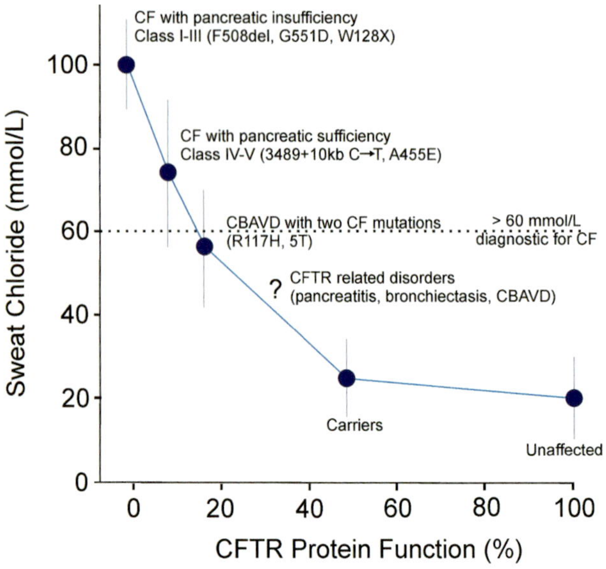

Fig. 15.3 Correlation of CFTR protein/function with sweat chloride levels. Patients with class I–III mutations have absent or minimal CFTR resulting in severe CF disease, with pancreatic insufficiency and highly elevated sweat electrolytes. Residual function mutations, class IV–V, are associated with milder disease, generally pancreatic sufficient and sweat electrolytes at (above or below) the diagnostic cutoff of 60 mmol/L. CF-related disorders with some CFTR dysfunction, for example, congenital bilateral absence of the vas deferens (CBAVD), present with intermediate or normal sweat electrolyte levels. Heterozygous individuals with one normal and one mutant *CFTR* allele and homozygous normal individuals are asymptomatic and have low sweat electrolyte levels. Adapted from Accurso et al. (2010)

specific mutation carried by a patient (see below) reveals a clear relation between genotype and phenotype (Fig. 15.3). Unaffected individuals are presumed to have 100 % CFTR function and show a mean sweat chloride of ~20 mmol/L. Carriers with one mutant allele show a small increase in sweat chloride (~26 mmol/L) compared to normal subjects. CF patients carrying one severe mutant allele and one mild allele show a sweat chloride of ~80 mmol/L, whereas CF patients with two severe mutant alleles have sweat chlorides >100 mmol/L (Rowe et al. 2007).

Exocrine pancreatic insufficiency is thought to occur in >90% of CF patients. Defective secretion of pancreatic enzymes into the small intestine results in fat malabsorption, leading to steatorrhea, reduced absorption of fat-soluble vitamins, and failure to thrive. Although not initially affected, the endocrine function of the pancreas can also be compromised leading to CF-related diabetes (CFRD), with around 25% of patients over the age of 25 displaying symptoms of CFRD (FOUNDATION 2015). At present, the mechanism for this endocrine disruption in CF is not fully known, but may stem from the presence of chronic low-level inflammation concomitant with exocrine pancreatic destruction. Nowadays, the most significant tissues, in terms of morbidity and mortality in CF, are the lungs and airways. CFTR

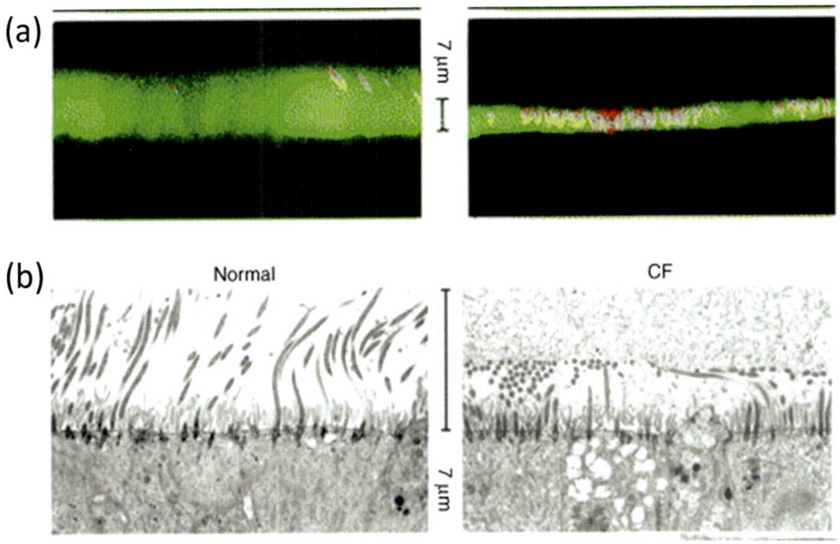

Fig. 15.4 Normal mucus clearance is abolished in the CF airway. (**a**) Confocal images of airway surface liquid (ASL) covering normal and CF airway surfaces. The aqueous component of the ASL is labeled with a green fluorescent dextran and the mucus layer with red fluorescent microspheres. The normal airway cells maintain an appropriate ASL height above the culture, whereas all available ASL has been absorbed from CF cultures, and the microsphere-labeled mucus layer has invaded the periciliary region. (**b**) Perfluorocarbon-osmium fixed cultures. In the normal cultures, a distinct periciliary layer is seen, with cilia fully extended. In CF cultures, the cilia are flattened, and the mucus later has invaded the periciliary space and adhered to the ciliary surfaces. Reproduced with permission from Matsui et al. (1998a)

is expressed in submucosal glands and in the apical surface of ciliated epithelial cells (Engelhardt et al. 1992; Kreda et al. 2005). Normal clearance and maintenance of sterility in the airways occurs through the interaction of cilia with the airway surface liquid (ASL), a periciliary layer of aqueous fluid and mucus that lines the respiratory tract (Boucher 2004). The correct depth of the ASL is achieved through the net effect of osmotic gradients generated by the efflux of chloride ions through CFTR, coupled with a small influx of sodium ions through the epithelial sodium channel (ENaC). Absence or dysfunction of CFTR in the airways reduces chloride efflux, which, coupled with increased sodium influx, leads to an osmotic imbalance such that the ASL becomes markedly dehydrated (i.e., reduced). This in turn leads to flattening of the cilia against the cell surface preventing their beating, which coupled with increased mucus viscosity impairs the normal clearance of the airways by the mucociliary escalator (Fig. 15.4). Over time this causes plugging of the airways with thick mucus and bacterial infections leading to inflammation and eventually bronchiectasis (Rowe et al. 2014, 2005; Webster and Tarran 2018).

15.4 Symptom-Based CF Therapies

Until recently, clinical care for CF patients focused exclusively on the delivery of therapies to relieve the symptoms associated with the disease. These are therapies that target the downstream consequences of CFTR dysfunction, but do not address CFTR itself. Nonetheless, such approaches have been highly successful. Therapies aimed at mucus obstruction, inflammation, and infection have certainly led to steady improvements in patient life expectancy (Fig. 15.5). Most people with CF (80–90% of patients) require pancreatic enzyme replacement therapy to prevent malnutrition, which was an important early improvement in CF therapy (Konstan et al. 2013; Levy et al. 1986). The current standard treatment for exocrine pancreatic insufficiency (EPI) is porcine pancreatic enzyme replacement pills, which consist of enzymes derived from pigs, and found in such preparations as Creon® (pancrelipase) marketed by AbbVie and Pancreaze® marketed by Janssen. An interesting recent Phase IIa clinical trial conducted by AzurRx BioPharma and Mayoly Spindler utilized an investigational treatment (MS1819-SD) which is a recombinant form of lipase cloned from the yeast *Yarrowia lipolytica* LIP2 gene. Although the study was aimed at patients with chronic pancreatitis rather than CF (both have markedly

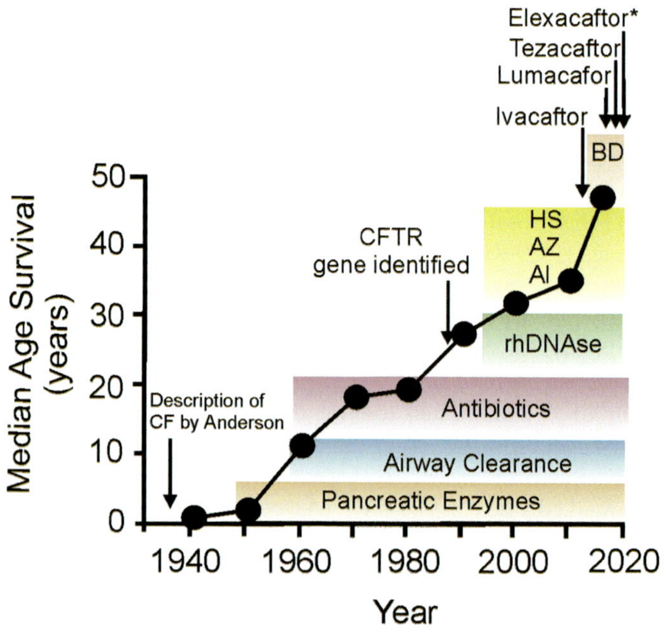

Fig. 15.5 Median survival in cystic fibrosis (CF). CF Survival over time, associated with milestones (*arrows*) and CF therapies (*boxes*). Steady increases in median survival have followed the introduction of new therapies and changes in care delivery. *AI* inhaled aztreonam, *AZ* azithromycin, *CFTR* cystic fibrosis transmembrane conductance regulator, *HS* hypertonic saline, *BD* drugs treating the basic defect, i.e., CFTR

reduced intestinal lipase activity), the outcomes are applicable to patients with CF. Indeed, a Phase2b clinical trial aimed at patients with CF is currently recruiting participants. MS1819-SD had a good safety profile and a significant increase in the coefficient of fat absorption. If FDA approved, MS1819-SD would become the first non-porcine product for EPI.

Early acquisition and chronic infection with *Pseudomonas aeruginosa* and *Staphylococcus aureus* have been targeted with antibiotics (Davis et al. 1996; Ratjen 2001). An important agent in this category is tobramycin marketed by Novartis which has proved remarkably effective in treating *Pseudomonas* infections. With Lupin Ltd. recently receiving FDA approval to market generic tobramycin, the cost of antibiotics for CF therapies should come down. Inhaling aztreonam has also proved useful in treating gram-negative bacterial infections like *Pseudomonas*. Interestingly, aztreonam is a beta-lactam antibiotic that was originally isolated from the bacterium *Chromobacterium violaceum* (Yaffe and Aranda 2010). Pulmozyme® (dornase alfa) (Pulmozyme 2005) marketed by Genentech is recombinant human DNase I, which has been shown to be effective in reducing the viscosity of purulent CF sputum (Shak et al. 1990; Thomson 1995). Pulmozyme reduces sputum viscosity by degrading DNA released from host immune cells and killed bacteria, but may also have a benefit on lung inflammation beyond DNA breakdown (Konstan and Ratjen 2012). In recent years, hypertonic saline has found a niche in improving patient quality of life (Elkins and Bye 2006), although compliance is difficult as hypertonic saline is fairly irritating, and patients are reluctant to add yet another component to their daily routines. Despite such therapies only treating symptoms, they nonetheless heralded an increase in life expectancy from around 5 years in the 1970s to about 35–47 years of age today (Fig. 15.5). Indeed, a newborn diagnosed with CF today can expect to live well into their fifth or sixth decade of life (Dodge et al. 2007). Although there is continued development of many exciting symptomatic therapies to treat patients with CF, this chapter will focus on strategies that impact the basic defect in CF, i.e., normalizing the function of mutant CFTR protein.

Despite the current focus on theratyping and personalized medicine (see below), one of the biggest concerns of patients with CF is the pulmonary damage done by chronic infections. Thus, there is still a need for novel and more potent antibiotics. In light of this, the North American Cystic Fibrosis Foundation announced in October 2018 that the foundation would invest $100 million over the next five years to develop reagents to fight infections in patients with CF (CFF 2018).

15.5 Classification of CFTR Mutations

Before the cloning of the CFTR gene in 1989 (Riordan et al. 1989), therapies, as discussed above, were based on treating symptoms and were indifferent to the nature of the genetic defect. Despite knowledge of the basic defect, symptom-based therapies have been remarkably beneficial in extending the life expectancy of

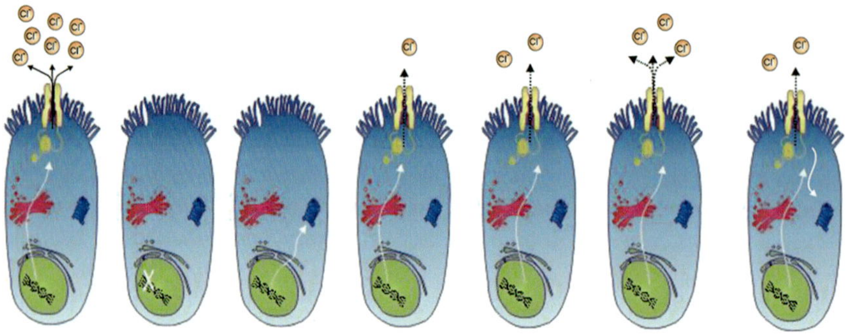

Normal	Class I	Class II	Class III	Class IV	Class V	Class VI
Defect	Defective Synthesis	Defective processing or maturation	Defective Regulation	Defective Conduction	Defective Synthesis	Defective Surface Stability
Therapy	Readthrough	Correctors (+ potentiators)	Potentiators	Potentiators	Potentiators Splice modulators	Trafficking modulators
Mutations (examples)	**G542X** **X1282X** R553X, E882X 621+1G>T	ΔF**508** N1303K ΔI507 R1066C	**G551D** G551S G178R G1244E	**R117H** R333W R347P R1070W	3272 6A>G A455E D565G 2849=1bk C>T	N287Y

Fig. 15.6 Classification of CFTR mutations according to protein production and function. Each of the five classes of mutations is described. An epithelial cell illustration depicts the localization of mutant CFTR protein relative to normal (wild-type) CFTR

patients with CF (Fig. 15.5). With the cloning of the gene, it was hoped that therapies aimed at treating the basic defect would soon emerge, though in fact it would take almost a quarter of a century for that dream to be realized. Although some 1900 sequence variations have thus far been described in *CFTR* (www.cftr2.org), they nonetheless fall into two broad categories: those affecting protein production and those affecting protein function. Given this division of mutational consequences, it is not surprising that different categories of drugs will likely be required to treat patients with CF, depending upon their unique genetic makeup. Moreover, despite the fact that this broad division in mutational consequences can accommodate most mutations, several mutations (including the common F508 del mutation) affect both protein production and function. As a refinement to the classification of CFTR mutations, Welsh and Smith (Welsh and Smith 1993) broke down mutations into four major subclasses, though currently six classes are recognized (Rowe et al. 2005; Zielenski et al. 1995) (Fig. 15.6). Of the 1900 CFTR mutations thus far identified, only ~20 are present at a frequency of >0.1% (Rogan et al. 2011). Indeed, excluding the common ΔF508 variant (which has an allelic frequency of around 90% (Genetic Analysis Consortium 1990)), only four CFTR mutations, G551D, W1282X, G542X, and N1301K, have a global prevalence of around 1–3% each (Clancy and Jain 2012;

Rogan et al. 2011). The class of CFTR mutation is not without impact, as there is differential disease severity associated with different classes of CFTR mutations (Accurso et al. 2010). Indeed, the ability (or inability) of a specific CFTR mutant to conduct chloride and/or bicarbonate ions has a strong correlation with overall clinical disease severity (Fig. 15.3). Although the most common gene mutation in Caucasians is the ΔF508 variant, the same mutation is only seen in 37% of affected African Americans diagnosed with CF; indeed, a number of mutations are specific to the African-American population, including the 3120+1 G→A mutation at a 12.2% prevalence, and A559T and 2307insA, both at 2% (McColley et al. 1991; Macek et al. 1997). This observation is recapitulated in other ethnic groups where certain mutations have a higher prevalence than is seen in other ethnicities.

15.5.1 *Class I Mutations Prevent the Production of Full-Length CFTR*

Class I mutations are typically characterized by a reduction in the quantity of CFTR produced and result in the complete or partial absence of CFTR protein. Such loss of CFTR protein typically arises from frameshift mutations (insertions or deletions), nonsense mutations (premature termination codons, PTC), or mRNA splicing defects (Rowntree and Harris 2003). For example, G542X, R553X, and W1282X are mutations generating PTC, resulting in premature termination of translation and the production of a truncated protein. Little if any functional CFTR protein is expressed with PTC mutations, and as such often results in severe disease (Derichs 2013; Lubamba et al. 2012). Mutations generating PTC can also lead to a marked loss in CFTR mRNA levels through nonsense-mediated decay (NMD) which degrades abnormal PTC containing mRNA species (Maquat 1995). The G542X mutation is the most prevalent mutation of its class worldwide, being expressed on at least one allele of up to 4 % of CF patients. Although the W1282X mutation is less common than G542X worldwide, it can reach a significantly higher frequency in specific populations, due to founder effects. Thus, 48% of all mutant CFTR alleles in Ashkenazi Jews in Israel carry the W1282X mutation (Shoshani et al. 1992). Similarly, 24% of CF patients in the French Reunion Island bear the Y122X mutation (Dugueperoux et al. 2003).

15.5.2 *Class II Mutations Alter the Intracellular Processing of CFTR*

Class II mutations are associated with defective protein processing, arising from misfolding and/or decreased stability of the 3D protein structure. Class II mutations result in a protein that is retained within the endoplasmic reticulum (ER), retro-translocated into the cytoplasm, and subject to proteosomal degradation (Cheng et al. 1990; Jensen et al. 1995; Ward et al. 1995). During initial translation, CFTR is co-translationally modified by core glycosylation (Band B CFTR). As CFTR progresses through the biosynthetic pathway, CFTR is fully glycosylated in the Golgi to a higher molecular weight B and C form. This shift in molecular weight can be assayed easily by western blot analysis and provides a simple indication of the mutation type and progress of CFTR through the cell. Depending upon the particular class II mutation, either a partial reduction (L206W) or a complete absence (F508del) of mature Band C CFTR is seen. Indeed, the most common mutation, occurring on ~70% of all mutant alleles and on ~90% of all CF patients, is the ΔF508 mutation. Initially thought to be purely a protein folding problem, subsequent crystallization of wild-type (wt) and F508del NBD1 revealed little difference in overall protein structure (Lewis et al. 2004). The F508del mutation arises from an in-frame deletion of three base pairs from two adjacent codons, resulting in the loss of one codon (for phenylalanine) and the codon for isoleucine at position 507 changed from the native ATC to ATT. The current model argues for an energetic and kinetic instability in ΔF508, causing a destabilized structure and an inability of NBD1 to interact appropriately with the fourth intracellular loop (ICL4) (Cui et al. 2007; He et al. 2010; Lubamba et al. 2012; Protasevich et al. 2010; Thibodeau et al. 2010; Thomas et al. 1991). This combination of defects has clear clinical implications, as correction of both NBD1 kinetics and NBD1/ICL4 surface stability will likely be required to restore ΔF508 processing (Mendoza et al. 2012; Rabeh et al. 2012).

15.5.3 Class III Mutations Alter CFTR Channel Regulation

Class III mutations are usually located within the ATP-binding domains of CFTR (NBD1, NBD2) and are referred to as *Gating* mutations. These are missense mutations that have little or no impact on protein processing (mutant proteins being inserted efficiently into the plasma membrane) but rather are resistant to normal activation and gating as a result of protein kinase phosphorylation (or can cause inappropriately rapid dephosphorylation and inactivation). The dominant mutation in this class is G551D, which results in a protein with an open probability (P_o) ~10-fold lower than that of wt CFTR (Bompadre et al. 2007; Curtis et al. 1991; Galietta et al. 2001; Illek et al. 1999; Lubamba et al. 2012). Class III mutations are associated with a severe clinical phenotype, characterized by pancreatic insufficiency and reduced pulmonary function (Cutting et al. 1990).

15.5.4 Class IV Mutations Alter CFTR Channel Conductance

Class IV mutations generally reside within those membrane-spanning domains proposed to be part of the pore-forming region. Class IV missense mutations produce a protein that is exported from the ER to the cell surface. The channel can be activated by kinase phosphorylation, but the channel has a reduced capacity to conduct anions through the pore. Since such mutations reduce the ability of the channel to conduct anions, rather than abolish conductance, and in other aspects the protein is "normal," it can be anticipated that class IV mutations are generally associated with milder forms of CF. In this class, the R117H mutation is the best characterized, yielding a protein that is properly processed and forms a cAMP-/PKA-dependent chloride channel with reduced conductance (Reddy and Quinton 2001; Sheppard et al. 1993). It should be noted, however, that the R117H protein also exhibits a lower open probability compared to wild-type CFTR, also engendering some class III characteristics to the channel (Reddy and Quinton 2001; Sheppard et al. 1993).

15.5.5 Class V Mutations Alter the Amount of Functional CFTR Protein

Class V mutations result in the generation of a fully functional CFTR channel; however, the level of protein expression is considerably reduced compared to wild type. Typically, these mutations occur in introns, close to splice sites, decreasing the efficiency of intron excision (Clain et al. 2005; Highsmith et al. 1997). Such splice mutations can result in complete or partial loss of an exon depending on how inefficient the pre-mRNA splicing becomes. If an exon is inefficiently incorporated into the mRNA, lower mRNA levels will be generated, yielding lower protein levels. The classic example of this kind of mutation is that leading to the skipping of exon 10. The incorporation or exclusion of this exon is correlated with a polymorphism in the poly-T tract upstream of the splice acceptor site. Thus, a poly-T tract of 7T or 9T results in up to 90% splicing efficiency (i.e., normal amounts of CFTR mRNA), whereas a poly-T tract of 5T reduces splicing efficiency to ~10–40% (i.e., only 10–40% of wt CFTR mRNA levels are generated) (Chu et al. 1992). Such mutations may not give rise to clinical CF, but may lead to other complications such as male infertility, as seen in congenital bilateral absence of the vas deferens (CBAVD) patients. A sixth class of mutation that decreases the stability of otherwise normal CFTR at the plasma membrane has been proposed (Haardt et al. 1999; Silvis et al. 2003). However, these mutations are now incorporated into class V mutations causing reductions in the amount of functional CFTR.

15.5.6 Class VI Mutations Alter Surface Retention

Class VI mutations have an increased turnover and retention within the plasma membrane (Haardt et al. 1999). A classic example is the N287Y mutation (Silvis et al. 2003), which generates an additional tyrosine-based internalization signal in the amino terminus, yielding a functional protein, albeit one with enhanced endocytic kinetics.

15.5.7 Multi-class Mutations

Although many mutations do in fact fall neatly within these five classes, some mutations cause multiple distinct defects. Indeed, this is the case for the common F508del mutation, where not only does the loss of phenylalanine 508 lead to inefficient processing and exit of ΔF508 from the ER, but even for mutant CFTR that does reach the cell surface, F508del displays channel defects characterized by a reduced open probability (P_o) and reduced stability of the protein at the cell surface (Cheng et al. 1990; Dalemans et al. 1991). The R117H mutation, although initially described as a class IV mutation with reduced channel conductance (Reddy and Quinton 2001; Sheppard et al. 1993), also exhibits a reduction in open probability, giving its characteristics associated with class III mutations. As discussed, CFTR conducts both chloride and bicarbonate ions and several mutations differentially affect the ability of CFTR to conduct chloride and bicarbonate. For example, the D648V mutation has little effect on chloride transport but reduces bicarbonate transport by 50–65% (Choi et al. 2001) which would put the mutation in the class IV conduction category. Interestingly, however, the D648V mutation also affects splicing of exon 13 by altering an exonic splicing enhancer (Aznarez et al. 2003), thus reducing mRNA levels which would also put the mutation also in class IV.

Although each affected allele generally harbors one mutation, some CF alleles possess two distinct mutations, referred to as complex alleles. As might be expected, the combined effects of the mutations lead to a more severe phenotype than would be expected from either mutation alone (Clain et al. 2001). For example, the effects of the R117H mutation can be modified by *cis* acting 5T/7T/9T polypyrimidine tract elements in exon 9. The R117H-7T genotype is associated with a milder phenotype even to the extent of a clinical presentation of male infertility, rather than cystic fibrosis. In contrast, the R117H-5T genotype causes elevated sweat electrolytes and clinical CF, which can be severe (Trujilliano et al. 2013). This mixing of class mutations led Lukacs and colleagues to create an expanded, morphed, classification of CFTR mutations (Veit et al. 2016) (Fig. 15.7), although the classic six model is still the dominant classification system employed.

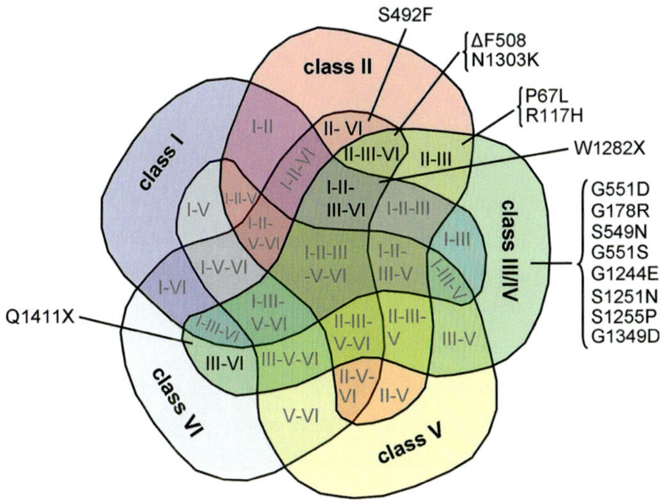

Fig. 15.7 Refined classification of CF mutations accounting for complex phenotypes of major CFTR cellular defect. The Venn diagram indicates all combinations of mutation classes with selected examples. Possible combinations without identified mutations are indicated in gray (Veit et al. 2016)

15.6 CFTR Mutants: What Needs to Be Fixed?

As can be appreciated, CFTR mutations and their clinical phenotype are very complex. Although all CF patients have mutations in their CFTR gene, some patients have very severe disease, whereas others have a milder disease presentation. Complicating this picture even further is the concept of modifier genes, which are genes that are different from CFTR, yet whose activity can modify the severity of the disease (Blackman et al. 2013; Cutting 2010). From the various mechanisms by which CFTR function can be compromised, it is clear that a single pharmacological approach is unlikely to be viable. A drug which is effective in activating class V mutations will have little efficacy on class I or II mutations. Therefore, minimally knowing a patient's genotype (i.e., the class of mutation expressed in that patient) will provide information on whether a particular drug treatment will be warranted in that patient. While some patients harbor the same mutation on each allele, this is not the case for all CF patients. Shown in Fig. 15.8 is an analysis of data taken from nearly 26,000 CF patients, where at least one mutant CF allele is known (Cystic Fibrosis Foundation 2012 Patient Registry). This highlights the complexity of the problem in treating all patients with CF. While F508del clearly predominates, it is also apparent that many patients have different mutations on each allele. Whether each allelic mutation will be amenable to the same pharmacotherapy or whether CFTR expressed from each allele will require a different drug again emphasizes the need to know the CFTR genetic makeup of each patient and tailoring of drug treatments to that individual patient.

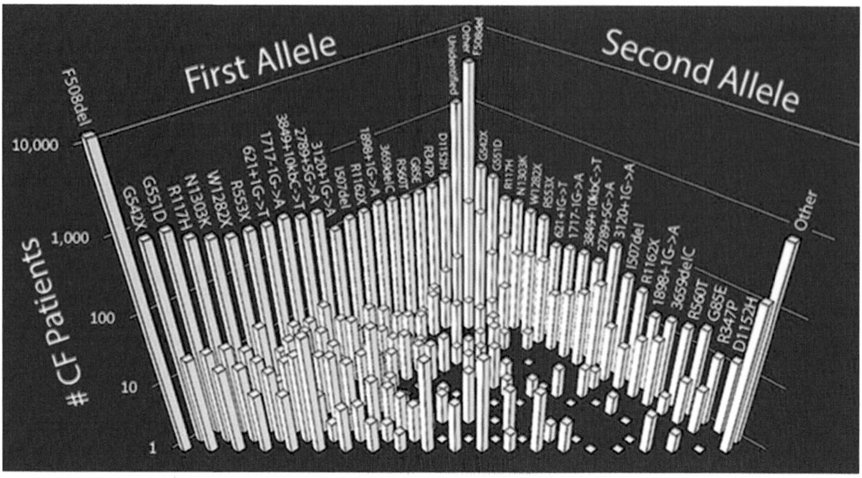

Fig. 15.8 Allelic variation among CF patients. Data are taken from 25,976 CF patients with at least one allele recorded in the US CF Foundation Patient Registry (2012). Note that the Y-axis is on a logarithmic scale

When considering how broken CFTR channels can be fixed, it is worthwhile noting what things are amenable to altering. Since CFTR is an ion channel, movement of ions through CFTR is governed by the formula $I = i \times N \times P_o$, where I is the overall current passing through all the CFTR molecules, i is the current through an individual CFTR molecule of each individual CFTR channel, N is the number of channels in the membrane, and P_o is the open probability of each channel. For example, nonsense mutations greatly impact N, whereas other mutations like G551D mostly impact P_o. Other mutations such as ΔF508 impact on both N and P_o.

15.7 Personalized Medicine, Bespoke Treatments, Precision Medicine, and Theratyping

Personalized medicine is a term used for treatment of patients based on individual clinical characterization and takes into account the variety of symptoms displayed by a patient. As such, personalized medicine has been performed on patients with CF, and indeed other diseases, for quite a long time. An archetypal example of such an approach in CF is the use of digestive enzymes, where the dose is adjusted to take into account an individual patient's characteristics with regard to amount and type of food ingested, body mass, growth rate, and type of enzyme used (Castellani and Assael 2016; Ozen and Duman 2016). A similar approach is also taken with respect to chronic pulmonary infections, and the varied lung flora and antibiotic resistance (Castellani and Assael 2016; Rutter et al. 2016). Thus, while personalized medicine does take into account an individual's condition, it could be argued that it does not fully take into account genetic variance for the same genetic disease.

Precision or bespoke therapy can be viewed as an extension of personalized medicine by understanding the genetic nuances of an individual's genetic makeup. In precision or bespoke therapy, molecular genetic information is included to help tailor a specific medicine to a specific genotype and also impinges on the interaction of the genetic defect with modifier genes, environmental factors, and lifestyle (Marson et al. 2015). Finally, theratyping can refer to determining which drugs will work on an individual patient by testing the drugs on patient tissue ex vivo, such as nasal or rectal tissues prior to prescribing to a patient (Clancy et al. 2018; Cutting 2015). Indeed, for patients with rare CFTR variants, especially variants that have not been fully explored, the cost of randomly trying corrector and potentiator drugs in the hope that "something might work" is prohibitive and may have to be used off-label. Pre-evaluating a patient's cells with drugs in a laboratory setting can provide a relatively quick and economically viable approach to determining the best drug regime for a particular patient. Both human nasal epithelial cells (hNE) and rectal biopsies are reasonably easy tissues to obtain for this kind of pre-screening approach (Clancy et al. 2018). Theratyping can also be used in a preclinical setting, by expressing CFTR variants in heterologous expression systems, such as Fischer rat thyroid (FRT) cells, and screening investigational new drugs (IND) against the variants prior to clinical trials.

Once a promising IND has been found, the compounds enter a series of clinical trials designed to evaluate both safety and efficacy. In Phase I, a small number of healthy volunteers receive the investigational drug under careful monitoring. Drug metabolism and safety profiles are assessed. In Phase II trials, patients are recruited, and safety during longer drug exposure, as well as efficacy in certain endpoints (changes in FEV_1, the forced expiratory volume in 1 second, number of pulmonary exacerbations, sweat chloride, etc.), is evaluated. Phase III trials are critical in determining whether an IND will reach FDA approval and market exposure. Longer safety and clinical efficacy determinations are paramount in this phase, and generally involve a large, often multicenter, patient population. Typically, Phase III trials are double-blind, randomized, placebo-controlled studies whose results cannot be evaluated until the end of the trial. Following productive Phase III trials, drugs are sent for approval by the US Food and Drug Administration (FDA). Often Phase IV or post-marketing surveillance is carried out to determine long-term efficacy and safety.

One of the biggest issues in moving drugs from the bench to the bedside is determining what clinical endpoints can (a) be evaluated and (b) be predictive (Clancy et al. 2018). Although nasal potential difference (NPD) measurements and sweat chloride determinations have been mainstays of CF clinical characteristics, they are not direct measurements of sole CFTR function and provide little specific information about CFTR chloride permeability. Moreover, the final analyses represent the concerted actions of complex transport processes, only one of which is CFTR. Even further from a direct assessment of CFTR is FEV_1, yet surprisingly this appears to be a rather good assessment of positive clinical outcome upon drug treatment.

15.8 Creating Drugs to Treat the Basic Defect in Cystic Fibrosis

15.8.1 Nucleic Acid Approaches

Strategies to treat the basic defect in CF can be classified into two main categories: targeting the genetic abnormality and pharmacological correction of the protein abnormality. Since CF is a monogenic autosomal recessive disease, the simple concept of a genetic strategy aimed at inserting a wt CFTR gene in mutant cells, independent of the particular CFTR mutation harbored by a patient, is appealing (Strug et al. 2018). Indeed, following the identification of the *CFTR* gene, there was great excitement among CF patients, parents, and caregivers with the hope that a gene therapy "cure" would be on the immediate horizon. CF gene therapy consists of delivering DNA or RNA nucleic acids encoding the CFTR protein to target cells. While CFTR cDNA needs to be delivered to the nucleus, CFTR mRNA only needs to be delivered to the cytoplasm, circumventing the need to overcome the nuclear barrier (Alton et al. 2016; Bridges and Bradbury 2018). Since the cloning of the CFTR gene, some 27 clinical trials involving 600 patients have been completed (Alton et al. 2016), yet despite the initial enthusiasm for this approach, the ability to deliver efficiently the gene to appropriate target tissues has remained elusive (Armstrong et al. 2014; Griesenbach and Alton 2012).

Since traditional gene therapy approaches to deliver CFTR cDNA, driven by a heterologous promoter, into airway cells have met with marginal success, and even when transduction occurs there is a steady loss of expression, the possibility to "repair" the endogenous mutant CFTR gene through so-called gene editing has recently garnered enthusiasm (Marangi and Pistritto 2018). This approach uses molecular scissors to cut the DNA helix at the point of a CFTR mutation and cause the correct wild-type sequence to be re-inserted (Cai et al. 2016; White et al. 2017). This repair approach has two major advantages over classic gene therapy. Firstly, the repaired gene remains under the control of endogenous in situ promoters, therefore ensuring appropriate expression and regulation within the cell. Secondly, depending on the delivery system employed, it obviates the inclusion of foreign nucleic acid material. However, in contrast to traditional gene therapy techniques in which all patients receive the same wt CFTR irrespective of their own particular genotype, CRISPR gene editing approaches will vary with each specific mutant genotype. Proof of concept for this approach was provided by studies in which gene editing successfully repaired the CFTR mutation in intestinal stem cell-derived organoids (Dey and Bradbury 2018) obtained from a patient with ΔF508 CFTR (Schwank et al. 2013). CFTR expression and function were restored using this approach, though the stability of the system was not evaluated. Although there are many benefits of CRISPR gene editing, there are challenges that need to be addressed. For example, off-target cutting remains a problem for Cas9 nickases. Mutant nickases have been generated to reduce "non-specific" DNA binding, and this has reduced the problem somewhat; however, our understanding of off-target effects remains poor.

A common class of errors in the CFTR gene involves disruption of splicing at one of the 27 exons present in the CFTR gene. Such mutations either reduce the efficiency of correct splicing or by generating a novel cryptic splice site within an intron (Spitali and Aartsma-Rus 2012). Antisense oligonucleotides (ASOs) utilize Watson–Crick base-pairing to bind to a specific RNA target, and sterically blocking access of splicing factors to the RNA sequence, and redirect splicing to an alternate site. Initial clinical trials using ASOs to treat spinal muscular atrophy (SMA) and Duchenne Muscular Dystrophy (DMD) showed modest effects (Aartsma-Rus et al. 2017; Bowerman et al. 2017), though they are now an accepted therapeutic intervention for type 1 SMA where ASOs enhance the inclusion of exon 7 of the Survival of Motor Neuron 2 (SMN2) gene. The utility of ASOs to treat CFTR splice mutations is more equivocal, and is still in the realm of experiment. The 3849 +10C>T mutation in CFTR creates a cryptic splice site and causes the inclusion of an extra 84bp exon, resulting in a premature stop codon in the final transcript (Friedman et al. 1999). In initial studies, although mRNA and Band C was detected in test cells, no functional data were presented (Friedman et al. 1999). More recently, the 2789+5G>A mutation has been targeted in a HEK293 expression system. Increased CFTR mRNA and protein expression was observed, and some iodide efflux data were provided (Igreja et al. 2016), though whether this can be formally attributed to CFTR remains to be determined.

The 3849+10kb C>T splice mutation in CFTR creates a de novo 5′ splice site resulting in the inclusion of a cryptic exon. This cryptic codon contains a stop codon, thereby causing the translation of a protein truncated at amino acid 1254. Single strand oligonucleotides (SSOs) have been designed to prevent such splicing (Michaels et al. 2018). Treatment of primary HBE cells with SSO increased appropriate splicing and stability of resultant mRNA. SSO corrected mRNA was translated into full-length wt CFTR that was measured as increased channel function following forskolin stimulation compared to DMSO-treated control cells. Application of the corrector (VX-770) in addition to SSOs slightly increased chloride secretion above that observed for SSO alone. These observations certainly give hope to those patients who harbor the roughly 12% of CFTR mutations that are splice mutations.

Although there has been a marked improvement in the generation of vectors and delivery systems for nucleic acids (Bridges and Bradbury 2018), strategies must also meet the challenge of the complexity of cellular sites in the lung that express CFTR (Jiang and Engelhardt 1998). The observation that CFTR is particularly abundant in submucosal glands of the airways also suggests that these cells, in addition to surface epithelia, need to be targeted. Moreover, recent studies identifying "ionocytes" as high CFTR expressing cells in the airways (Montoro et al. 2018; Plasschaert et al. 2018) have raised the question as to whether gene delivery and/or editing strategies need to focus on ionocytes or whether that cell type should be avoided. Of course, the concept of high expresser cells is not new, and they were first identified in the human intestine (Jakab et al. 2013).

15.8.2 Pharmacologic Approaches

With the reduction in enthusiasm for gene therapy (although this has been revived with CRISPR and ASOs), an alternative strategy based on pharmacological approaches was sought. In 1998, the Cystic Fibrosis Foundation (CFF) initiated a groundbreaking collaboration with pharmaceutical companies to support drug discovery efforts for correcting the basic defect in CF. So far, the CFF has invested more than $300 million to leverage early-phase CF-related drug discovery programs, an approach that culminated in the FDA approval of ivacaftor (Kalydeco, Vertex Pharmaceuticals) (Tables 15.1 and 15.2) as the first treatment for the basic defect in CF. The ideal drug to treat patients with CF would be one that normalizes multiple aspects of CFTR dysfunction, including defective protein folding, restricted ER exit, plasma membrane targeting, and surface stability; restores normal channel activation and gating; has no or minimal off-target effects; and can be taken as a single daily oral dose. Although remarkable advances in drug development for CF have been made in the past few years, such an ideal drug has not yet been realized. Nonetheless, drugs have been identified that restore some aspects of CFTR dysfunction. Before the generation of full-length cryo-EM structures of CFTR (Liu et al. 2017), crystal structures of NBD-1 (the site of the most common ΔF508 mutation) (Lewis et al. 2004, 2010) and homology models for CFTR (Mendoza and Thomas 2007; Serohijos et al. 2008) allowed for in silico-based virtual screening approaches to identify chemicals that would bind to mutant CFTR and restore folding and/or

Table 15.1 CFTR modulators

Name	Mode of action	Company	Approved
Ataluren	Read-through	PTC Therapeutics	
PTI-428	Expression amplifier	Proteostasis	
PTI-808	Potentiator	Proteostasis	
Ivacaftor (VX-770)	Potentiator	Vertex	X
VX-561 (CTP656)	Potentiator	Vertex/Concert	X
QBW251	Potentiator	Novartis	
GLPG1837	Potentiator	Galapagos[a]	
GLPG3067	Potentiator	Galapagos[a]/AbbVie	
GLPG2451	Potentiator	Galapagos[a]	
FDL169	Corrector	Flatley Labs	
GLPG2222/ABV2222	Corrector	Galapagos/AbbVie	
Lumacaftor (VX-809)	Corrector	Vertex	X
Tezacaftor (VX-661)	Corrector	Vertex	X
Olacaftor (VX-440)	Corrector	Vertex	
Bamocaftor (VX-659)	Corrector	Vertex	
Elexacaftor (VX-445)	Corrector	Vertex	[b]
GLPG2737	Corrector	Galapagos[a]	

[a]As of October 2018, AbbVie bought the rights to all Galapagos CFTR modulators
[b]Pending FDA approval as of August 2019

Table 15.2 Vertex reagents

Vertex number	Name	Tradename	Mode of action
VX-770	Ivacaftor	Kalydeco	Potentiator
VX-809	Lumacaftor		Corrector
VX-661	Tezacaftor		Corrector
VX-770 + VX-809	Ivacaftor + Lumacaftor	Orkambi	Potentiator/Corrector
VX-661 + VX-770	Tezacaftor + Ivacaftor	Symdeko	Potentiator/Corrector
VX-661 + Vx-770 +VX-445	Tezacaftor + Ivacaftor + Elexacaftor[a]		Potentiator/Corrector/ Corrector

[a]FDA approval pending as of August 2019

function. Using this strategy Kalid et al. (2010) identified three putative binding sites at the interfaces between the cytoplasmic domains that would be amenable to drug binding. Computer-based analysis, docking a virtual library of 496 candidates, has generated 15 potential compounds, although none has yet reached clinical trials.

High-throughput screening (HTS) for CF drugs initially relied on fluorescence methodologies, including the halide-sensitive yellow fluorescent protein (YFP), developed by Verkman (Jayaraman et al. 2000; Rowe and Verkman 2014), and a membrane potential-sensitive dye pair, developed by Molecular Devices Inc. for use on an FLIPR III plate reader (Gonzalez et al. 1999), and employed by Vertex Pharmaceuticals. Initial screening protocols relied on evaluating the effects of compounds not on all CFTR mutations, but rather on the most prevalent, i.e., ΔF508 and G551D, CFTR expressed in heterologous systems. Although useful, such cell models do not entirely recapitulate the target tissues in a CF patient, i.e., the airway epithelium. The development and implementation of primary airway epithelial cells as a model for drug screening have made an immense impact on the search for drugs to treat patients with CF. This advance has become particularly important, as several promising compounds which were effective in correcting CFTR in heterologous systems subsequently turned out to be ineffective when tested on airway cells, the major target tissue in CF patients. Airway epithelial cells, derived either from a CF patient at the time of lung transplant or from cadaveric wt tissues, are expanded in tissue culture prior to plating on permeable supports or grown at an air–liquid interface to recapitulate the native airway environment. Such cell monolayers develop an electrically tight epithelial layer containing ciliated epithelial cells. The advantage of this approach is that it facilitates the implementation of highly sensitive electrical measurements of CFTR function. Two dominant culture methods are at present utilized, one developed by the University of North Carolina (Matsui et al. 1998b), and one developed by Vertex Pharmaceuticals (Van Goor et al. 2009a). These two different culture methods do give rise to differences in the electrical properties of each monolayer, though which approach (if either) most faithfully recapitulates the native airway environment is not known. Nonetheless, it is clear that such models have proven to be invaluable in CF drug discovery and formed the basis of assays leading to the current FDA-approved drugs.

15.8.3 Suppressors of Premature Termination: Making More CFTR

For organisms employing the standard genetic code, including humans, translational termination occurs when one of the three STOP codons—UAA, UGA, or UAG—enters the ribosomal A site (Frischmeyer and Dietz 1999). Premature termination codons (PTC) arise when a single base-pair mutation generates an erroneous stop codon within the open reading frame of a gene. This can arise as a nonsense mutation, frameshift deletions or insertions, or from aberrant mRNA splicing that results in truncated reading frames. Approximately 10% of patients with CF exhibit a PTC mutation on at least one allele. Premature stop codons also have the effect of leading to rapid mRNA degradation through nonsense-mediated decay (NMD), as well as truncated non-functional proteins (Aslam et al. 2017; Du et al. 2008). The antimicrobial activities of aminoglycoside antibiotics arise from their ability to inhibit bacterial protein synthesis at high doses. At low doses, however, aminoglycosides bind inefficiently to eukaryotic ribosomes resulting in a translational read-through by the insertion of an amino acid at the stop codon (Fan-Minogue and Bedwell 2008; Martin et al. 1998). The potential for aminoglycosides in the treatment of PTC mutations (Class I mutations) in CFTR was first recognized in the late 1990s when studies by Bedwell (Bedwell et al. 1997) and Howard (Howard et al. 1996) using transient expression systems demonstrated that full-length CFTR could be detected following aminoglycoside treatment of cells expressing R1162X, G542X, and R553X CFTR constructs. Moreover, not only was full-length protein restored, but cAMP-activated chloride currents could also be observed. Studies using a *G542X-hCFTR* mouse (human G542X CFTR cDNA on a murine *Cftr*-null background) demonstrated that exposure to the aminoglycoside gentamycin increased the expression of full-length CFTR protein in the gut (the site of CF-related mortality in the mouse model) and was associated with the appearance of a cAMP-activated chloride conductance (Du et al. 2002). Proof-of-principle studies in patients with PTC in CFTR have been able to demonstrate increased CFTR channel activity, as measured by nasal PD (from basal potential difference 45 ± 8 to 34 ± 11 mV, $p < 0.005$) and by immunofluorescence staining of nasal cells using an antibody recognizing the carboxyl terminus of CFTR (Wilschanski et al. 2003). These studies applied topical (nasal) gentamicin for 2 weeks and resulted in a statistically significant improvement in CFTR function in treated groups compared to control. No effect was observed in patients homozygous for the F508del mutation. In studies with larger patient populations, however, a relative lack of efficacy of gentamicin has been reported (Rowe and Clancy 2009). In addition to minimal efficacy, chronic treatment with amino glycosides engenders unwarranted side effects. These include both acute and chronic renal nephrotoxicity (Cojocel et al. 1984a, b; Kang et al. 2000) and ototoxicity (Hiel et al. 1992; Tan et al. 2001; Zheng et al. 2001). One potential complication in addressing PTC mutations in CFTR is that different codons are present in different mutations. For example, the Y122X mutation bears the UAA (ochre) sequence, whereas the UGA (opal) sequence is present in the G542X and W1282X mutations. There is some suggestion that the Y122X ochre mutation is the most easily suppressed nonsense mutation (Sermet-Gaudelus et al. 2007); however, opal PTC mutations are 10-fold more frequent than ochre or amber (UAG) mutations (http://cftr2.org).

High-throughput screening (HTS) of compounds that promote read-through of PTC mutations in a luciferase-based reporter system led to the development of PTC124 (Ataluren, PTC Therapeutics, South Plainfield New Jersey). Ataluren, a 1,2,4-oxadiazole benzoic acid (Fig. 15.9) (Table 15.1), causes PTC read-through but appears to interact with mammalian ribosomes in a manner distinct from aminoglycosides (Welch et al. 2007); this may explain why Ataluren does not possess any antibacterial properties. Three Phase II clinical trials have reported mixed results: two studies reporting improvements in CFTR function (as assessed by nasal PD) and one study reporting no change in function (Clancy et al. 2006; Kerem et al. 2008; Sermet-Gaudelus et al. 2010). Randomized, double-blinded, placebo-controlled Phase III studies of Ataluren in patients $\geq$6 years of age failed to find any evidence of changes in FEV_1 or any secondary endpoint in treated subjects compared to control (Kerem et al. 2014). A further post hoc analysis of the data was performed excluding patients who were receiving chronic inhaled tobramycin which revealed a 5.7% improvement in FEV_1 and a 40% reduction in pulmonary exacerbations, compared to placebo. The authors proposed that since tobramycin and ataluren had the same target protein, tobramycin was interfering with ataluren's efficacy and preventing PTC read-through. Although there was a mild improvement in FEV_1, there was also a 15% increase in acute renal injury in the ataluren-treated group compared to control ($<$1%) (Kerem et al. 2014). A subsequent multicenter Phase II clinical trial in patients not receiving aminoglycoside therapy did not realize primary (FEV_1) or secondary endpoints (pulmonary exacerbations), and PTC Therapeutics elected not to pursue FDA approval of ataluren for PTC mutations in patients with CF (PTI 2017) (www.cff.org/research/DrugDevelopmentPipeline/#CFTRMODULATION).

A different HTS approach was taken by scientists at the Cystic Fibrosis Foundation Therapeutics (CFFT) labs, where an HRP fusion tag was inserted into the fourth extracellular loop of CFTR as a means of detecting surface expression (Liang et al. 2017). Approximately 85,000 compounds were screened against CFTR constructs bearing the Y122X and W1282X mutations. The most efficacious hit compounds were better than G418 in acting as read-through modulators. The top five W1282X hits increased surface CFTR to more than 20% of wild-type CFTR, compared to untreated controls. Whether such experiments performed using human codons in rat FRT cells will effectively translate into experiments in native human cells remains to be determined.

Potential problems and limitations to extending the use of compounds to treat PTC mutations include (1) specificity of read-through agents for mutant vs. authentic stop codons, (2) low levels of mRNA transcripts due to nonsense-mediated decay (NMD), (3) potential ototoxicity of aminoglycosides, and (4) generation of novel mutations through random amino acid insertion. The inappropriate read-through of authentic stop codons is likely more a theoretical than an actual problem. Structural

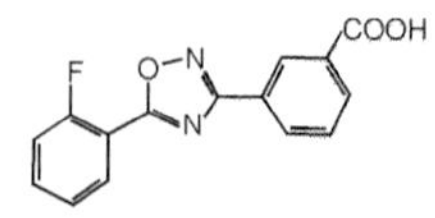

Ataluren (PTC124)

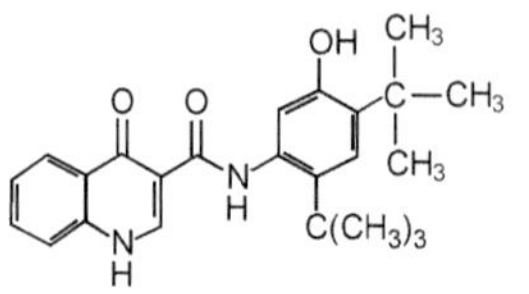

Ivacaftor (VX-770)

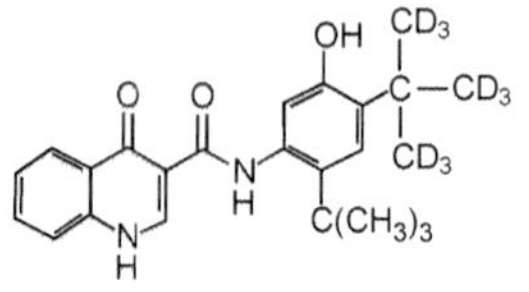

CTP-656

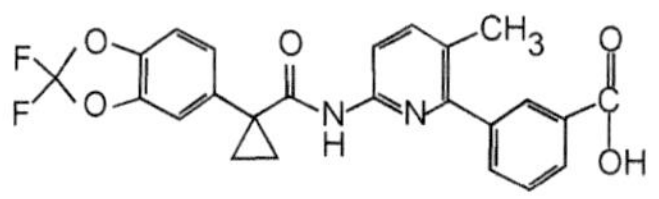

Lumacaftor (VX-809)

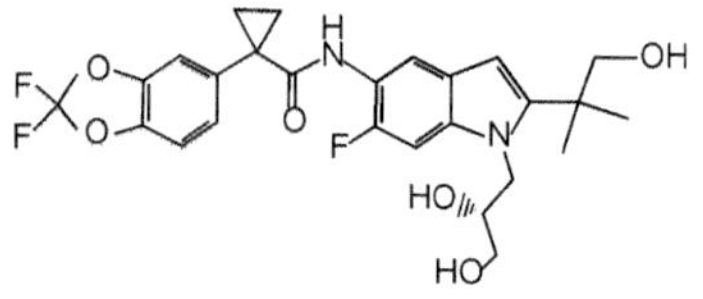

Tezacaftor (VX-661)

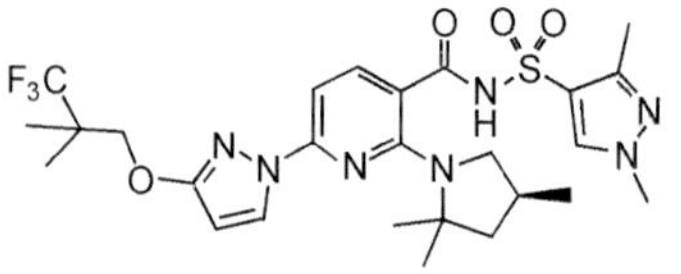

Elexacaftor (VX-445)

Fig. 15.9 CFTR modulators. Ataluren, a PTC read-through compound currently in Phase III trials for patients aged 6 years and above who have nonsense mutations. VX-770 (Ivacaftor, marketed as Kalydeco) is a CFTR potentiator used to treat patients with at least one copy of the G551D allele. VX-770 is FDA approved for patient use. Evaluation of Kalydeco for other class III mutations is underway. CTP-656 is a deuterated version of VX-770. A combination of the corrector VX-809 (lumacaftor) and XV-770 (ivacaftor) is marketed as orkambi and is FDA approved to treat patients with at least one ΔF508 mutation. Tezacaftor is a next-generation corrector, used in combination with lumacaftor as symdeko

differences between authentic stop codons and PTC mutations argue that read-through is limited to PTC mutations. For example, the 3^0 untranslated region (UTR) is a strong stimulus for translation termination at appropriate stop codons and is absent from PTC stop codons (Amrani et al. 2004). mRNA transcripts for PTC mutations tend to be expressed at low levels due to NMD (Maquat 2002); however, increased mRNA stabilization has been seen with administration of aminoglycosides in vitro (Bedwell et al. 1997). Ototoxicity has been linked to long-term use of aminoglycosides such as gentamycin, tobramycin, and kanamycin (Al-Malky et al.

2014). Read-through of PTC mutations obligatorily inserts an amino acid in the place of the mutant stop codon. The amino acid inserted appears to be somewhat random (with a preference for glutamine, tyrosine, cysteine, arginine, or tryptophan, at least in yeast) (Blanchet et al. 2014). Studies by Bedwell and colleagues (Xue et al. 2017) have determined which amino acids are inserted into the PTC site following G418-mediated read-through. Interestingly, the amino acid inserted depends upon the context of the PTC site. W1282X PTC (a UGA codon) incorporated leucine, cysteine, and tryptophan during W1282X suppression. In contrast, the G542X PTC (also a UGA codon) incorporated arginine, cysteine, and tryptophan during PTC suppression. All "full-length" CFTR proteins were functional, though none reached the level attributable to wild-type CFTR. Studies by Pranke and colleagues (Pranke et al. 2018) revealed that the actual PTC codon does make a difference in the efficacy of PTC suppression, with UGA being the optimal sequence. The North American Cystic Fibrosis Foundation is currently heavily invested in PTC therapies, as this is one of the larger categories of mutations, not really benefited much by correctors or potentiators (see below).

15.8.4 CFTR Potentiators: Opening a Sticky Gate

Potentiators are CFTR modulators that increase channel gating activity (increase P_o) leading to more time spent by the channel in the open, anion conducting, configuration (Kramer and Clancy 2016; Van Goor et al. 2009a). Class III and class IV mutations result in CFTR molecules that traffic through the biosynthetic pathway to the cell surface, but bear defects in gating and conductance, respectively (Rogan et al. 2011). The prototypical example of this class of mutation is G551D, a mutation present in around 5% of all mutant alleles. In contrast to the ΔF508 mutation, G551D CFTR molecules do reach the cell surface, and although they have a very low open probability, it is not zero (Bompadre et al. 2007; Curtis et al. 1991; Galietta et al. 2001; Illek et al. 1999). The observation that G551D was present at the cell surface, albeit with a significantly impaired open probability, raised the intriguing notion that maybe something could be done to encourage the channels to gate more effectively. Molecules that increase the activity of mutant CFTR have been known since shortly after the identification of the *CFTR* gene, when the phytoestrogens, genistein and isoflavone (Hwang et al. 1997), were identified as compounds capable of increasing the open probability of CFTR. The mechanisms of action of genistein are still unclear, but may include tyrosine kinase inhibition, inhibition of topoisomerase, or direct interaction with CFTR (Illek et al. 1995; Weinreich et al. 1997). Unfortunately, genistein also increases ENaC-mediated sodium absorption (Devor et al. 2000), which, along with the multiple known targets of genistein, makes it an unlikely clinical candidate for CF therapy.

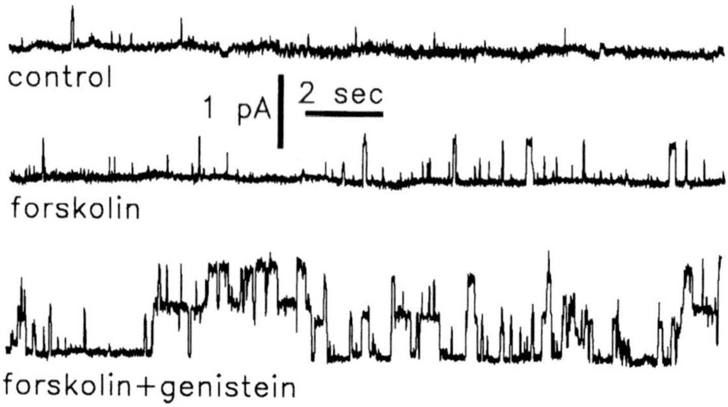

Fig. 15.10 G551D CFTR has a low P_o. Cell attached recording of G551D-CFTR (note the patch contained two channels). Reproduced with permission from Illek et al. (1999)

15.8.4.1 Ivacaftor (VX-770, Kalydeco™)

A remarkable collaboration between the CF Foundation and the pharmaceutical company, Vertex Pharmaceuticals (Cambridge, MA), came to a breakthrough in 2011 with the results of a positive clinical trial with ivacaftor (VX-770) (Fig. 15.9) (Table 15.1) in CF patients with the G551D mutation (impaired channel gating). Why did Vertex focus its initial attention of the fairly rare G551D mutation (~5% of patients) and not the common F508del variant (~90% of patients with at least one F508del allele)? The answer is fairly simple; it gave Vertex the best chance of success! In contrast to F508del, G551D CFTR actually does reach the plasma membrane. Indeed, early patch-clamp studies had shown that G551D CFTR could be activated by genistein (Illek et al. 1999) (Fig. 15.10). Ivacaftor (tradename Kalydeco) was identified following a high-throughput screen (HTS) utilizing a library of 228,000 compounds. Initial HTS assays were confirmed using short circuit current analysis (I_{sc}) and single channel patch-clamp analysis in FRT cells expressing G551D CFTR (Van Goor et al. 2009a). 10 μM ivacaftor increased the P_o of G551D CFTR by approximately sixfold ($p < 0.05$), findings subsequently validated in primary human bronchial epithelial cells (HBE) from G551D patients (Fig. 15.11a, b). Consistent with the need for mutant (G551D in this case) CFTR to be at the cell surface, ivacaftor only yielded small and variable improvement in chloride transport in HBE cells from F508del/F508del CFTR donor tissue. In addition, video microcopy showed that ciliary beat frequency in HBE cells increased to near wild-type levels, due to the increase in ASL height accompanying increased CFTR activity. Based on these preclinical data, one Phase II (Accurso et al. 2010) and three Phase III (Ramsey et al. 2011) clinical trials were initiated. In all studies, ivacaftor led to a marked and persistent improvement in lung function (as measured by forced expiratory volume, FEV_1) and other markers such as a reduction in sweat chloride and a movement of nasal PD out of the CF range. Excitingly in the Phase III

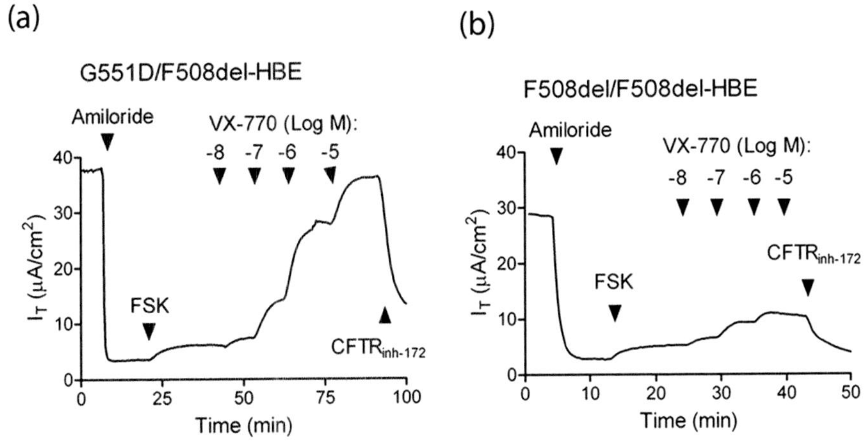

Fig. 15.11 VX-770 potentiation of CFTR-mediated chloride secretion in primary cultures of G551D/F808DEL HBE. The Ussing chamber technique was used to record transepithelial currents (I_T) resulting from CFTR-mediated chloride secretion. Currents were measured under a basolateral to apical chloride gradient, amiloride (30 μM) to block ENaC sodium currents, and forskolin (10 μM, FSK) to activate CFTR. The addition of $CFTR_{inh-172}$ (20 μM) to the apical surface was used to confirm that the I_T was CFTR mediated. (**a**) Current trace from G551D/ΔF508 HBE cells showing the efficacy of VX-770 on increasing CFTR currents. (**b**) Current trace from ΔF508/ΔF508 HBE showing minimal effect of VX-770 potentiator alone on chloride secretion. Reproduced with permission from Van Goor et al. (2009a, b)

trial, sweat chloride levels dropped by a mean of 42.3 mmol/L, many below an absolute value of 60 mmol/L into the normal range, showing partial restoration of sweat gland activity. Curiously, although the highest concentration of ivacaftor tested (250 mg) did cause a drop in sweat chloride levels, its effect was less than that of lower doses of ivacaftor (150 mg). Nasal potential difference studies also showed a change toward a normal level with a 5.5 mV change at the highest dose of ivacaftor and no effect of placebo. For lung function measurements, ivacaftor increased FEV_1 by greater than 10%. Given that patients were at least 18 years of age with long-standing lung disease, it is quite surprising that improved lung function should happen so rapidly, accompanied by a weight increase of >2.5 kg. On January 31, 2012, the US Food and Drug Administration (FDA) approved ivacaftor to treat patients with the G551D mutation (Van Goor et al. 2009a). Observational studies show that long-term treatment with ivacaftor slows the rate of lung function decline and also decreases the risk of pulmonary exacerbations (Guimbellot et al. 2018; Sawicki et al. 2015). The effectiveness of ivacaftor on the G551D gating mutation suggested that other gating mutants such as R117H may also benefit from the drug (Van Goor et al. 2012). A Phase III clinical trial using ivacaftor in children and adults with at least one R117H mutation was performed, but failed to achieve its primary endpoint (absolute change from baseline FEV_1 % predicted). However, a subset in patients at least 18 years of age ($n = 46$) did show an absolute mean improvement of FEV_1 of 5%. Significant improvements in secondary endpoints of sweat chloride and patient-reported respiratory symptoms convinced the FDA in December of 2014 to allow VX-770 for CF patients who are at least 6 years of age with the R117H mutation. Further in vitro data on other rare gating mutations (G178R, S549N, S549R, G551S, G1244E, S1251N, S1255P, and G1349D) with both conductance and gating defects (Char et al. 2014; Yu et al. 2012) suggested that ivacaftor would also be beneficial for CF patients bearing these mutations. In May 2017, the FDA expanded the approval of ivacaftor as a monotherapy for CF patients ≥2 years old possessing 1 of an additional 23 mutations. When Ivacaftor was first marketed in 2012, it came in the form of two pills a day and was priced at over $300,000 a year (Nocera 2015). Notably, even with the expansion of FDA approval of ivacaftor to now include other mutations resulting in about 10% of all patients with CF being covered, the price has not come down. In the absence of competition, the cost is unlikely to reduce any time soon.

The molecular mechanism underlying the effectiveness of ivacaftor in increasing channel open probability remains elusive, though initial work by Hwang and colleagues (Jih and Hwang 2013) argues that ivacaftor works by stabilizing a post-hydrolytic open state, leading to an uncoupling of ATP hydrolysis from the gating cycle and causing an increased residence time in the open state. Indeed, several studies have argued that ivacaftor binds to CFTR to prolong its open state in an ATP-independent manner (Eckford et al. 2012; Jih and Hwang 2013; Van Goor et al. 2009a). However, other patch-clamp studies have suggested that ivacaftor may work in both an ATP-independent and an ATP-dependent manner (Jih and Hwang 2013). Initial in silico molecular docking approaches suggested that potentiator molecules bound at the interface between NBD1 and NBD2, potentially stabilizing the interaction (Moran et al. 2005). However, ivacaftor is a very hydrophobic compound, and unlikely to bind to hydrophilic regions of CFTR, more probably interacting within the lipid bilayer regions of CFTR. Recently, comparisons of ivacaftor with the Galapagos/Abbvie potentiator GLPG1837 (see below) suggest that the binding site may be at a lipid–protein interface (Yeh et al. 2017), the authors arguing that since ivacaftor and GLP1837 bind competitively, all potentiators work through a common mechanism (Yeh et al. 2017). The Rowe laboratory has also attempted to elucidate potentiator mechanisms (Pyle et al. 2011), arguing that potentiation of cAMP-dependent CFTR activation is independent of R-domain phosphorylation. Along with data from Hwang and colleagues, who suggest that potentiators also work in the absence of ATP binding to the NBD dimers or ATP hydrolysis (Yeh et al. 2017), this suggests that G551D, at least, appears poised to gate in the absence of stimulation, just waiting for a potentiator to nudge it into activity. Although the G551D mutation is the second most prevalent CFTR mutation, it is only present in approximately 4–5% of all CF patients, and therefore, many patients will not benefit from ivacaftor alone. Initially, only approved for the treatment of patients with the G551D mutation, ivacaftor also appears to work on several other gating mutations, including G178R, S549R, G551S, G970R, G1244E, S1251N, S1255P, and G1349D (Yu et al. 2012); clinical trials for patients bearing these mutations are in progress. Interestingly, evidence also suggests that potentiators may also be of benefit in increasing the P_o of class II mutations, including F508del (Van Goor et al. 2009a), which have been brought to the cell surface. As a monotherapy in F508del patients, ivacaftor modestly reduced sweat chloride levels relative to baseline and placebo (<3 mM, $p < 0.05$), but did not improve lung function outcomes (Flume et al. 2012). While ivacaftor has provided a proof of concept that small molecules can interact with CFTR in a therapeutically beneficial manner, it is nonetheless only effective in a small number of patients. Indeed, since potentiators only work on CFTR protein that is present at the cell surface, it is clear that potentiators alone will be of no benefit to patients with either class I or class II mutations.

15.8.4.2 CTP-656

CTP-656, originally developed by Concert Pharmaceuticals Inc. (Lexington MA), is essentially ivacaftor in which nine hydrogen atoms from three methyl groups have been replaced with deuterium (Fig. 15.9) (Table 15.1). The rationale for deuterating drugs is to modulate and improve their pharmacokinetic profile, reducing their metabolism and excretion. Since the deuterium atom is similar to hydrogen, the biological activity of the drug is expected to remain the same, although the extended bioavailability should result in reduced doses or greater time between doses. A Phase I trial confirmed that CTP-656 had a reduced rate of clearance compared to ivacaftor, leading to a longer half-life (15 h) and greater plasma concentration with a threefold increase in the parent to metabolite ratio (Uttamsingh et al. 2016). Ivacaftor is taken twice a day, and with the extended half-life of CTP-656, it would be anticipated that the deuterated version would only need to be taken once a day, a major benefit for patients who are burdened with multiple daily medications. Concert Pharmaceuticals business model is to deuterate drugs, like ivacaftor, which have already received patent protection, and to enter into partnerships with the companies owning those patents. In 2017, Vertex bought the rights to CTP-656, now renamed VX-561.

15.8.4.3 GLPG1837

GLPG1837 is a potentiator developed by Galapagos. Preclinical data suggested that although GLPG1837 and ivacaftor had similar EC_{50}, GLPG1837 caused a greater increase in short circuit current across G551D HBE than ivacaftor (Van der Plas et al. 2018). The N1303K mutation is also a class II mutation, with a severe gating defect and reduced ATP-dependence. The open probability (P_o) of phosphorylated N1303K is ~0.03, compared to ~0.45 for wt CFTR, a P_o that can be increased by 17-fold in the presence of GLPG1837, compared to only an eightfold increase by ivacaftor (DeStefano et al. 2018). Three Phase I trials to determine oral bioavailability and safety in healthy volunteers have been completed, but the results have not been made public. Phase II trials (SAPHIRA 1) were established primarily to verify safety and tolerability and secondly to evaluate clinical endpoints (sweat chloride, pharmacokinetics, spirometry). Since patients in the study were already on ivacaftor, a washout period was required to generate a new baseline prior to GLPG1837 treatment. During the washout, FEV_1 decreased from 70–75% pre-washout to <70% post-washout and then approximately returned to the ivacaftor baseline FEV_1 on day 29 of GLPG1837 treatment (SAPHIRA1 2017). Reductions in sweat chloride were also observed at the highest doses. A small Phase II (SAPHIRAII) open label trial has also been reported in which six patients with the S1251N mutation (Class III) received a twice daily oral dose for 2 weeks followed by a higher dose for an additional 2 weeks. No safety issues were reported, but 4 of 5 subjects had a decrease in sweat chloride of ≥15 mmol/L, and one subject had a >50 mmol/L decrease on day 29 of GLPG1837 treatment (De Boeck et al. 2016). To date, no Phase II trials for GLPG1837 have been announced.

15.8.4.4 QBW251

QBW251 is a candidate potentiator developed by Novartis Pharmaceuticals. Preclinical in vitro studies have shown QBW251 to be an effective potentiator and that QBW251 in combination with lumacaftor (see Correctors, Sect. 15.8.5) was superior to ivacaftor in sustaining ΔF508 membrane expression and function. These data led to a Phase II study (NCT02190604) to evaluate the safety, tolerability, and pharmacodynamics of QBW251. Patients enrolled with class III and IV mutations showed good tolerance to the drug and had an increase in mean FEV_1 of 7.3% compared to placebo (Kazani et al. 2016). Additionally, sweat chloride decreased by an average of 16.9 mmol/L, although not into the normal range. Not surprisingly, QBW251 when tested on F508del patients showed no improvement in FEV_1, leading to an early termination of the clinical trial (Kazani et al. 2016).

15.8.4.5 Mechanism of Potentiator Action

Recently, comparisons of ivacaftor with the Galapagos/AbbVie potentiator GLPG1837 suggest that the binding site may be at a lipid–protein interface (Yeh et al. 2017), the authors arguing that since ivacaftor and GLP1837 bind competitively, all potentiators work through a common mechanism (Yeh et al. 2017). Chen and colleagues have reported cryo-electron microscopy structures of human CFTR in complex with GLPG1837 at 3.2 Å resolution, and with ivacaftor at 3.3 Å resolution (Liu et al. 2019). Although ivacaftor and GLPG1837 are structurally dissimilar, they bind at the same site within the transmembrane region of CFTR, interacting with similar residues S308, and Y304 through hydrogen bonds, and L233, F236, F305, A309, and F312 through hydrophobic interactions (Fig. 15.12). Ivacaftor binds to CFTR at a protein–lipid interface, docking into a cleft generated by transmembrane helices 4, 5, and 8, corresponding to a hinge region in TM8, a unique structural feature of CFTR not present in other ABC transporters (Zhang et al. 2018). The extracellular segment of TM8 rotates around this hinge region upon ATP binding, and it is likely that potentiators stabilize the open configuration of the hinge relative to the closed structure. Such actions would translate into an increased

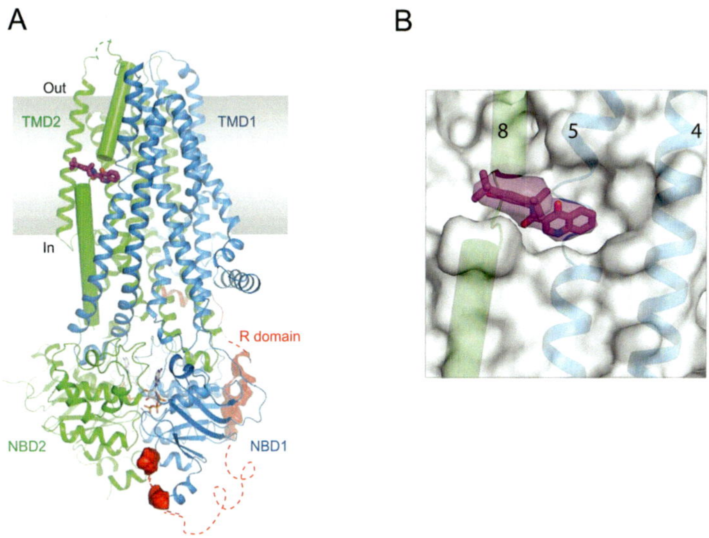

Fig. 15.12 Ivacaftor binds CFTR inside the membrane. (**a**) Overall structure of the phosphorylated, ATP-bound human CFTR in complex with Ivacaftor (shown in magenta), TMD1 and NBD1 are shown in blue. TMD2 and NBD2 are shown in green, and R domain in red. TM8 is highlighted in cylinder representation. (**b**) A magnified view of the Ivacaftor binding site. CFTR is shown as a transparent surface model with TM 4, 5, and 8 indicated. Ivacaftor is shown as a stick model together with the associated electron density. Reproduced with permission from Zhang et al. (2019)

opening rate in the presence of potentiators and a decreased closing rate. No gross structural changes in CFTR were induced by potentiator binding, though this is probably not surprising given that the potentiators increase open probability from 0.64 to 0.88, a drug-induced energy change in the closed equilibrium ($\Delta\Delta G$) of only 1 $k_B T$ (where k_B is the Boltzmann constant and T is temperature). The identification of a "*potentiator*" binding patch that does not require a precise chemical structure, but appears to be rather promiscuous as long as the right amino acids are bound through hydrogen bonds and hydrophobic interactions, suggests that in silico approaches to identifying CFTR modulators may finally prove beneficial.

15.8.5 CFTR Correctors: Solving Misguided Traffic

The holy grail of CF pharmacotherapy is the treatment of ΔF508 CFTR, which is present on around 90% of mutant alleles, and homozygous in around 70% of patients. The ΔF508 CFTR is an ER export incompetent mutation affecting *N*, the number of channels at the cell surface. In contrast to earlier assumptions, ΔF508 CFTR does not misfold per se, but rather adopts a relatively native conformation that is thermodynamically unstable (Protasevich et al. 2010; Wang et al. 2010), and disrupts the interface between NBD1 and the membrane-spanning domains (MSDs)

(Wang et al. 2018a). The cell biology of ΔF508 includes interruption of protein maturation and premature ER-associated degradation in the 26S proteosome. This results in little, if any, F508del CFTR ever reaching the plasma membrane (Cheng et al. 1990). Shortly after the cloning of the CFTR gene, and the identification of the F508del mutation, several strategies were identified that could increase the surface expression of the F508del protein. Restoration of ΔF508 could be achieved in cells kept at reduced temperature (Denning et al. 1992), or by exposure to a number of osmolytes (Howard et al. 2003; Sato et al. 1996). While none of these approaches are clinically practical, although one open parallel study in CF subjects using topical glycerol did fail to show any effect on nasal potential difference (Chadwick et al. 1998), they nonetheless show proof of concept that ΔF508 can be teased out to the cell surface given the right conditions. Butyric acid and its analogues were also shown to increase CFTR mRNA in heterologous F508del-expressing cells and to increase the expression of F508del at the cell surface. Interest in the potential use of the butyric acid analogue sodium 4-phenylbutyrate (4-PBA), which also increases F508del CFTR expression, arose because the compound was already approved for the clinical treatment of patients with urea cycle disorders (Iannitti and Palmieri 2011). Preclinical studies using primary human nasal epithelial cells expressing F508del/W1282X CFTR showed that forskolin-induced CFTR currents could be increased in the presence of 4-PBA (Rubenstein et al. 1997). In early pilot studies of 4-PBA in CF patients, a small change in nasal PD measurement, consistent with an increase in CFTR function, was observed, although no changes in sweat electrolytes were noted (Rubenstein and Zeitlin 1998). Although Phase I/II trials demonstrated a modest correction in F508del function, later studies revealed that the high doses of 4-PBA required to see these effects produced intolerable adverse effects (Rubenstein and Zeitlin 1998).

Curcumin, derived from the spice turmeric, is a SERCA pump inhibitor that reduces ER calcium concentrations. Treatment of cells expressing F508del CFTR with curcumin showed a modest increase in F508del expression, and F508del mice exposed to curcumin displayed a modest correction in nasal PD (Egan et al. 2004). However, subsequent evaluation of curcumin using human airway epithelial cells failed to replicate these findings (Grubb et al. 2006; Loo et al. 2004). Moreover, it appears that the doses of curcumin required to achieve FF508del correction in early cell models are not physiologically achievable in human subjects (Song et al. 2004). Another food-derived compound, resveratrol, found largely in the skins of red grapes, has also been proposed as a corrector compound to aid ΔF508 trafficking (Cho et al. 2018; Dhooghe et al. 2015; Hamdaoui et al. 2011), although the concentrations of resveratrol required far exceed bioavailability, and studies performed at clinically relevant doses failed to show any beneficial effect of resveratrol (Jai et al. 2015).

Despite studies showing the biochemical correction of F508del CFTR trafficking in response to various agents, it soon became clear that even if F508del got to the cell surface, it was still defective, displaying reduced gating and cell surface stability (Bebok et al. 2005; Heda et al. 2001; Okiyoneda et al. 2010; Van Goor et al. 2009a; Wang et al. 2005). These early investigations suggested that a corrector alone was unlikely to work as a monotherapy and that other compounds, such as potentiators, would likely be required. Nonetheless, the effectiveness of experimental manipulations to increase ΔF508 surface expression stimulated extensive high-throughput screening efforts to identify small molecule correctors that could mimic osmolytes and serve as "hits" and lead compounds in the development of pharmaceutical reagents to "correct" F508del trafficking (Hanrahan et al. 2013; Van Goor et al. 2009a; Verkman et al. 2006, 2015; Wang et al. 2018b). There is some evidence that correctors may bind directly to the nascent F508del polypeptide chain (He et al. 2013; Okiyoneda et al. 2013; Ren et al. 2013), although their precise binding sites are unknown. In addition, even if binding of correctors to CFTR is shown, it still requires experimental analysis to formally determine that this is the mechanism of action, and the correctors are not working on other aspects of the biosynthetic machinery (Balch et al. 2011).

15.8.5.1 FDL169

FDL169 is a small molecule being developed by Flatley Discovery Labs to correct F508del trafficking. In vitro studies using primary human bronchial epithelial cells (HBE) have shown that FDL169 can improve F508del trafficking and function, in the presence of ivacaftor, following forskolin stimulation (Zawistoski et al. 2016), although no increased open probability (P_o) of F508del was seen in the presence of FDL169. Of note, chronic treatment ≥24 h with FDL169, in the presence of ivacaftor, led to a mild 15% inhibition in F508del activity (Zawistoski et al. 2016). FDL169 has been reported to be effective on its own in increasing F508del trafficking to the cell surface by ~5-fold (similar to lumacaftor; see below) (Kwok et al. 2016). Thus far, three Phase I trials have been completed showing safety and tolerability of FDL169 in patients.

15.8.5.2 GLPG2222

GLPG2222 (also known as ABBV/GLPG2222) is a corrector compound developed from a collaboration between Galapagos and AbbVie and is a candidate for use in patients with at least one ΔF508 allele (Wang et al. 2018b). In vitro studies using GLPG2222 in CFBE41o-cells suggest that the compound is able to increase cell surface expression of HRP-tagged F508del CFTR. A Phase I randomized double-blind placebo-controlled study of a single dose of GLPG2222 did not reveal any adverse events and revealed favorable pharmacokinetics (Van de Steen et al. 2016). A recently completed Phase II trial (curiously called Albatross) was performed in subjects possessing one ΔF508 allele and one class III allele and already receiving ivacaftor. At the highest dose of GLPG2222 (along with ivacaftor) a 2.2% increase in FEV_1, along with a 6 mmol/L decrement in sweat chloride, was observed (CF Foundation 2017). Given the current efficacy of Vertex compounds, it is not likely that Galapagos compounds will see clinical use.

15.8.5.3 Lumacaftor

Lumacaftor, VX-809 (Fig. 15.9) (Table 15.1), is a corrector, and was developed by Vertex to increase the amount of ΔF508 trafficking to the cell surface (Van Goor et al. 2011). In vitro preclinical studies suggested that VX-809 could increase Band B to Band C conversion and lead to a marked increase in mature CFTR in homozygous ΔF508 primary HBE cells (Van Goor et al. 2011). In isolated HBE cells derived from ΔF508/ΔF508 lung tissue, lumacaftor improved transepithelial chloride secretion to approximately 15% of that observed from wt HBE cells, and the addition of ivacaftor further improved this signal (Van Goor et al. 2009b, 2011). In addition, increased expression of Band C CFTR can be observed in HBE cells exposed to VX-809 compared to DMSO control. In a Phase II randomized placebo-controlled clinical trial using lumacaftor as a monotherapy in patients homozygous for the ΔF508 mutation, median sweat chloride levels dropped from an average of 103.5 to 95.3 mmol/L in subjects receiving the highest lumacaftor dosage. However, there was no statistically significant improvement in nasal potential difference measurement, nor lung function (Clancy et al. 2012). Thus, current data suggest that there is little evidence to conclude that corrector monotherapy has clinically significant effects in CF patients homozygous for the F508del mutation (Southern et al. 2018).

15.8.5.4 Tezacaftor (VX-661)

Tezacaftor (VX-661) is a Vertex corrector proposed to increase the amount of ΔF508 CFTR at the cell surface (Fig. 15.9) (Table 15.1). Since the tezacaftor corrector is designed to be used in combination with a potentiator, there is little data available regarding the actions of tezacaftor alone. However, one Phase II randomized double-blinded placebo-controlled trial on patients homozygous for ΔF508 and >12 years old did have a tezacaftor monotherapy arm. Although the trial was primarily designed to test safety and tolerability, results from 28 days tezacaftor monotherapy showed a modest reduction in sweat chloride levels (−5.7 mmol/L, $p < 0.05$).

15.8.5.5 VX-440, VX-152, and VX-659

VX-440 and VX-152 are so-called next-generation correctors developed to provide increased ΔF508 CFTR surface expression. In preclinical trials on HBE derived from homozygous ΔF508 tissue, both VX-440 and 152 increased the amount of CFTR at the cell surface (Grootenhuis et al. 2016). The addition of lumacaftor further increased CFTR cell surface expression suggesting that VX-440 and 152 were working through a different or complementary mechanism to lumacaftor. Functional studies in Ussing chambers following 24-hour treatment with VX-440 or

VX-152 alone increased forskolin-stimulated chloride transport from a baseline of 6% of wt to 19% and 16% of wt, respectively. VX-659, also a next-generation corrector, is designed to be used in combination therapy, and no data on its monotherapy effects have been reported.

There is still no clear mechanism for the action of correctors on CFTR trafficking, nor formal evidence that correctors are even working on CFTR. The read-out for corrector studies is either the appearance of Band C or an increase in forskolin-stimulated chloride currents from human F508del airway tissues, both of which are fairly distal to the proposed effects on ER export of CFTR. Nor is it necessarily true that just because a corrector binds to CFTR the effect is due to such binding. It is conceivable that correctors could interact with other components of the ER export machinery to yield identical results. One could argue that since all correctors so far identified have been found in the absence of any mechanistic knowledge, it may be that mechanism is, to quote Dr. Robert Beall, former Chairman and CEO of the North American Cystic Fibrosis Foundation, "*nice to know, but not need to know*."

15.8.6 How Do Correctors Work?

Since correctors were identified using essentially a phenotypic screen, i.e., presence of CFTR at the cell surface, it is formally possible that correctors could not only bind to CFTR, but also to any of the components regulating the movement of CFTR along the biosynthetic pathway. For example, inhibition of RMA1 or CHIP ubiquitin ligase permits F508del CFTR to escape ER quality control and reach the plasma membrane (Grove et al. 2009). Alternatively, correctors may indeed bind directly to mutant CFTR, facilitating its passage to the cell surface. Cross-linking of chemically modified correctors has shown that VX-809 (Lumacaftor) does bind directly to F508del CFTR (Sinha et al. 2015), but the exact binding site remains unknown.

As with classification of CFTR mutations, a classification of correctors has been proposed based on the effect the correctors have on CFTR (Boinot et al. 2014; Hongyu et al. 2018). Class C1 correctors are those that stabilize the NBD1–ICL1 interface (He et al. 2013) and work early on in the biosynthetic processing of CFTR, as exemplified by Tezacaftor, ANNV/GLPG2222, ABV/GLPG2851, FDL-169, and PTI-801. Class C2 correctors (e.g., the bithiazole corrector 4a, VX-152, 440, 659, 445, ABBV/GLPG2737, ABBV/GLPG3221, and FD2052160), which likely target NBD2 and/or its interfaces (Okiyoneda et al. 2013), also likely act at a point later in the biosynthetic pathway than C1 correctors (Fig. 15.13).

15.8.7 Make It a Combo!

As noted, ΔF508 CFTR patients suffer from two problems: very little is ER export competent, and it also displays gating problems. Given these two separate issues, it is

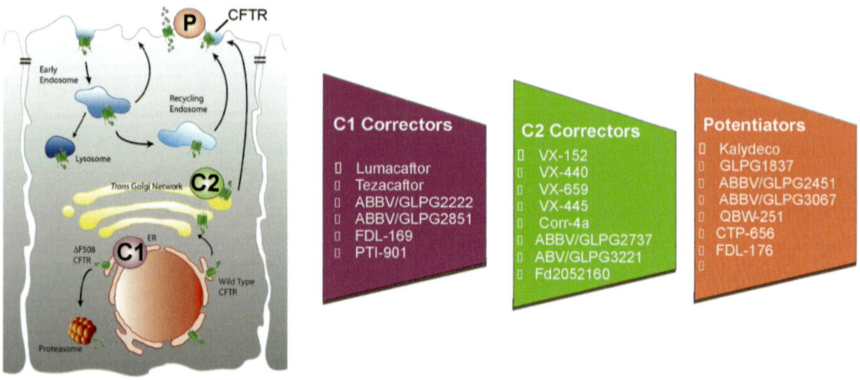

Fig. 15.13 CFTR modulator activity. CFTR modulators act as potentiators (P), C1-type correctors, and C2-type correctors. Adapted from Hongyu et al. (2018)

not surprising that two separate drugs would need to be employed to help CF patients with the F508del allele: a corrector to get F508del to the plasma membrane and a potentiator to keep it active.

15.8.7.1 ORKAMBI

Orkambi is a dual therapy composed of ivacaftor and lumacaftor together. Phase II clinical trials investigating the effect of the combination therapy showed a significant drop in sweat electrolytes (drop of 9.1 mmol/L) (Boyle et al. 2012) following twice daily ivacaftor (250 mg) and once daily lumacaftor (150 mg). For patients receiving only 150 mg ivacaftor, no drop in sweat electrolytes was observed. Curiously, an improvement in FEV_1 of 3.5% was observed in patients taking 150 mg ivacaftor, but not in those receiving 250 mg ivacaftor. Orkambi received FDA approval for use in adults in 2015 and in children in 2018. However, the clinical results with orkambi have been fairly weak giving modest to no improvement in FEV_1 (Boyle et al. 2014; Rowe et al. 2017b; Southern et al. 2018), although reduced sweat electrolytes were noted. Given the limited improvement in pulmonary function in the best responders, and a price tag of $258,000 per year (Pollack 2015), concerns over cost–benefit ratios have been raised. Although orkambi is approved and used in North America, Europe, and Australia, the British National Health Service (NHS) raised reservations concerning the high cost with marginal clinical benefit. Vertex CEO Dr. Jeff Leiden suggested that he would consider pulling the plug on Vertex's research and development sites in Oxford UK, unless the UK NHS agreed to pay what was being asked for orkambi (Adams and Owen 2018). Orkambi is still only prescribed in Britain on compassionate grounds and is not generally available.

Along with modest lung function improvements, orkambi is also associated with a transient increase in shortness of breath upon early exposure, with longer exposures leading to high blood pressure. These adverse effects were not seen with the symdeko combination (see below). Concern with the orkambi combination was also raised based on basic studies in cell model systems, which showed that ivacaftor interfered with the effect of lumacaftor on F508del channel function (Chin et al. 2018; Cholon et al. 2014; Veit et al. 2014), such that the ability of the corrector lumacaftor to increase cell surface expression of F508del in primary HBE cells, and HEK expression systems, was significantly compromised by the presence of the potentiator ivacaftor. Moreover, ivacaftor appeared to decrease the stability of wt CFTR. Thus, the efficacy of the orkambi combination may be limited due to antagonistic drug–drug interactions. Recent evidence suggests that the two drugs do not directly interfere with each other, but cause problems by increasing cytochrome-dependent metabolism of the drugs (Schneider 2018). Orkambi and its metabolites (and even ivacaftor alone) increase cytochrome P450 3A4 activity, leading to lowered plasma drug concentrations. Further concerns are raised that the Orkambi combination may not only impair F508del protein but may also cause problems for other transport proteins (Chin et al. 2018), where membrane stability of other transport proteins (SLC26A3, SLC6A14, and SLC26A) was also compromised. Such impairment seems to reside in the ivacaftor component, affecting lipid domains, since less hydrophobic ivacaftor analogues cause less of a problem. Certainly, it is important to acknowledge that ivacaftor as a monotherapy has a benefit to individuals with gating mutations. However, the above studies give some insights into mechanisms as to why the ivacaftor/lumacaftor combination orkambi has such weak effects when used as a combination therapy.

On a side note, it is interesting to note that orkambi can lead to false-positive signals for cannabinoid use during routine urinalysis of patients (Kissner et al. 2018). Thus, urine screening immunoassays in CF patients taking orkambi may require confirmatory testing.

15.8.7.2 Tezacaftor/Ivacaftor (Symdeko)

Tezacaftor, in combination with ivacaftor (symdeko), is a drug combination akin to orkambi, and designed to increase F508del trafficking and activity. An open label non-randomized study of tezacaftor (corrector) + ivacaftor (potentiator) + ciprofloxacin versus ivacaftor + ciprofloxacin was completed in 2014, though no data have yet been released. In Phase II trials with patients F508del/G551D, the combination of tezacaftor and ivacaftor showed a 4.6% increase in FEV1 compared to control, with no significant effect on pulmonary function when tezacaftor was used as a monotherapy (Pilewski et al. 2015; Vertex 2013). The EVOLVE study, completed in 2017, was a Phase II trial to evaluate tezacaftor taken every 24 h with ivacaftor taken every 12 h, versus placebo in homozygous F508del patients (Taylor-Cousar et al. 2017). Patients treated with the ivacaftor/tezacaftor combination had an increase in FEV_1 of 3.4% from baseline, which when combined with a decrease of 0.6% in the placebo group yielded a net increase in FEV_1 of 4%. There was also a reduction in the number of pulmonary exacerbations, with 122 events in the placebo group and only 78 in the ivacaftor/tezacaftor group. Importantly, chest tightness reported by patients taking orkambi was absent in patients taking the ivacaftor/tezacaftor combination (Grasemann 2017; Taylor-Cousar et al. 2017; Wainwright et al. 2015). In the EXPAND study, a large Phase III study with 234 patients with one F508del allele and one allele with partial function (NCT02392234) (Rowe et al. 2017a; Sala and Jain 2018), compared to placebo, the ivacaftor/tezacaftor group, and the ivacaftor monotherapy group, had a mean absolute increase in FEV_1 of 6.8% and 4.7%, respectively. Ivacaftor/tezacaftor compared to ivacaftor monotherapy showed a 2.1% increase in FEV_1. In 2018, the FDA approved the ivacaftor/tezacaftor combination, referred to as symdeko, for patients greater than 12 years of age who are homozygous for the F508del mutation or have one F508del mutation and another mutation responsive to symdeko. Importantly, the tezacaftor–ivacaftor combination does not induce cytochrome P450 activity, in the same way that the lumacaftor–ivacaftor combination induces activity. Although the cost of symdeko at $292,000 is 7% more expensive than orkambi, symdeko has a slightly better effectiveness and tolerability than orkambi. Indeed, the greater physician uptake of symdeko, tied to drug's broader label and greater efficacy over orkambi (4% increase in FEV_1 for symdeko vs. 2.8% increase in FEV_1 for orkambi), has led to a very modest increase in orkambi sales, and it is likely that orkambi will be further sidelined by the new triple combination currently under FDA investigation.

The modest effects of orkambi and symdeko relative to their pricing were criticized in a review of the Institute for Clinical and Economic Review (ICER) by the Midwest Comparative Effectiveness Public Advisory Council (CEPAC) (ICER 2018). The Council recommended that a 77% reduction in price for orkambi and symdeko would be more in line with its clinical value, where the commonly accepted threshold for cost-effectiveness of $100,00–$150,000 per quality-adjusted life year gained. Not surprisingly, Vertex disagreed with the CEPAC evaluation.

15.8.7.3 Triple Threat

Given the FDA approval of dual corrector/potentiators, there is now a move toward including a third compound into the mix, yielding a triple combination drug that would be taken by patients with CF, in the hope of covering ~90% of CFTR mutations. Vertex has taken the tack of using two correctors along with a potentiator in their triple compound. In contrast, Proteostasis Therapeutics has opted to employ three mechanistically distinct compounds: a corrector, a potentiator, and an amplifier (see Sect. 15.8.7). The next-generation CFTR corrector VX-659 in triple combination with tezacaftor and ivacaftor (VX-659–tezacaftor–ivacaftor) has recently undergone randomized, double-blind, placebo-controlled, multicenter Phase I clinical trials and displays an acceptable safety and side-effect profile. Treatment of patients with the F508del mutation on one allele and a minimal function (MF) mutation on the other allele, with VX-659–tezacaftor–ivacaftor, displayed an increase in FEV_1 of 13.3%. In patients homozygous for F508del and already receiving tezacaftor–ivacaftor, VX-659 led to a further increase in FEV_1 of 9.7% (Davies et al. 2018; Keating et al. 2018). A roughly 40 mmol/L decrement in sweat chloride was also observed for patients taking VX-659–tezacaftor–ivacaftor, irrespective of whether they were homozygous for the F508del mutation or possessed a ΔF508-MF genotype. As with all clinical trials, it is unclear whether the effects on lung function can be sustained for longer periods of treatment, nor what impact VX-659–tezacaftor–ivacaftor will have on pulmonary exacerbations, and other meaningful outcomes such as weight gain. Recently released Phase III studies on the Vertex triple combination of VX-659, Tezacaftor, and Ivacaftor were recently reported. There was an increase of 14% in FEV_1 after 4 weeks, in patients with one F508del allele and one low function allele. In patients with two F508del alleles, the mean absolute improvement in FEV_1 % predicted was smaller, at 10%. No adverse effects were reported in the press release, only the endpoints for the study.

In March 2019, Vertex became somewhat of a victim of its own success. The company has two triple combinations ready for FDA approval, but they both target the same population of patients. Indeed, the plans for the new triple combination are to expand Vertex's coverage of the CF patient population to everyone with at least one F508del allele. Two Phase III studies have now been completed in patients with one F508del mutation, both based upon symdeko (tezacaftor + ivacaftor). One study combined symdeko with VX-445, and one combined symdeko with VX-659. Patients receiving symdeko plus VX-445 saw a 13.8% increase in FEV_1 after four weeks compared to placebo control. Those treated with symdeko plus VX-659 showed a 14% increase in FEV_1 after four weeks compared to placebo control. In patients with two F508del mutations FEV_1 was improved by 10% in patients that added VX-659 to symdeko, compared to patients taking just symdeko. Similarly patients with two F508del alleles saw a 10% improvement in FEV_1 when taking VX-445 with symdeko, compared to those on symdeko alone. Certainly the four-week clinical trial results are sufficient for Vertex to take both triple combinations to the FDA, but Vertex decided to wait for the 24-week data for VX-445 and VX-650 before making a decision. In July 2019, Vertex sought FDA approval for marketing of VX-445, now elexacaftor, as part of a triple combination with tezacaftor and ivacaftor. The application included a request for priority review, which, if granted, would shorten the turnaround time for FDA decision from 12 months to 8 months. The submission was based on positive results of two global 24-week Phase III studies in 403 individuals with one F508del and one minimal function allele and a 4-week Phase III study with 107 patients homozygous for the F508del mutation. Improvement in lung function, as an endpoint assay, showed improvement in FEV1 of 14.3% in the 24-week trial, and by 10% in the 4-week trial. Secondary endpoints of reductions in sweat chloride were also noted.

15.8.8 *Proteostasis Therapeutics*

In October 2018, Proteostasis Therapeutics (PTI) announced the completion of a Phase I trial for a double-blind, placebo-controlled study for a proprietary dual combination therapy of PTI-801 and PTI-808 (PTI 2018). PTI-801 is a third-generation corrector, with PTI-808 a novel first-generation potentiator. In a completed study incorporating the PTI-801/808 combination, the drugs were found to be well tolerated, with no pulmonary exacerbations noted during the trial. The dual therapy achieved a statistically significant increase in FEV_1 (% predicted), with a maximum increase from baseline of +6.3% at day 7, and 5.9% at day 14. These results are indeed impressive since the current standard of care for F508del homozygous patients with orkambi (lumacaftor/ivacaftor duo) only leads to 2.41–2.65% increase in FEV_1. While the results from the initial PTI-801/808 combination are exciting, they were nonetheless obtained from a fairly small cohort of 21 patients. Therefore, confirmation with a larger population would be needed to confirm the results. In March 2019, Proteostasis announced the results of a larger double/triple Phase I clinical trial. In the double trial, PTI-801 and PTI-808 were combined, and in a triple trial, PTI-801 and PTI-808 were combined with the amplifier (see below) PTI-428. Both the double and the triple compound showed a 6–8% change in FEV_1 (Proteostasis 2019). The data from Proteostasis reporting a 6–8% improvement in FEV_1 was released a few days following Vertex's announcement of a 10–14% improvement in FEV_1 for their new triple combination trials before seeking FDA approval. Following the release of trial data from Proteostasis and Vertex, Proteostasis saw a 75% drop in its share price. Recruitment of patients for subsequent clinical trials would seem a steep climb for Proteostasis.

15.8.9 Galapagos/AbbVie

At the end of October 2018, AbbVie acquired all of Galapagos' CFTR modulators for $45 million with no royalty payments, following weak clinical trials that suggested Galapagos would have a hard time competing against Vertex. Although the results of the clinical trials were released in June of 2018, AbbVie did not obtain the compounds till October. The corrector GLPG 2327, when used as part of a triple combination, was expected to challenge Vertex's primacy, by achieving Kalydeco levels of efficacy, but it did not pan out. With the best combinations, Galapagos only achieved a 3% change in FEV_1, far below that required to threaten Vertex. For their part, AbbVie is playing up its expertise in medicinal chemistry to improve the modulators. Certainly, there is hope that the scaffolds acquired by AbbVie will prove amenable to modification and that competition will drive down the price of CFTR modulators.

15.8.10 Amplifiers

Correctors and potentiators only work on the CFTR molecules that have already been synthesized in the ER. What if there were a way to increase CFTR translation, giving more protein for correctors and potentiators to act upon? This is the concept of amplifiers, increasing the production of unfolded CFTR and providing more protein for other modulators to process. Amplifier compounds are predicted to improve the ability of emergent signal sequences to target ribosome to ER translocons, and slow down mRNA degradation. The big advantage of amplifiers is that they are theoretically independent of the CFTR mutation harbored in the mRNA (Miller et al. 2016). Currently, the only amplifier under clinical investigation is the Proteostasis compound PTI-428. Initial studies have shown a 2.5-fold increase in CFTR mRNA in healthy volunteers following 7 days of dosing with 150 mg PTI-428, similar to that observed in vitro in HBE cells expressing mutated (twofold) or normal CFTR (1.5-fold). This is exciting for patients whose mutations generate little CFTR mRNA or protein. The rare mutation c.3700 A>G (ΔI1234-R1239-CFTR) causes misprocessing and altered channel function and, importantly, very little mRNA or protein. Orkambi treatment alone of HBE bearing the c.3700 A>G mutation results in a modest response; however, the addition of the amplifier PTI-428 results in a significantly greater effect (Molinski et al. 2017). One area of concern is the unreported mechanism of action of amplifiers. Certainly, CFTR mRNA appears to be stabilized, but that was the only mRNA species reported. Is PTI-428 specific for CFTR mRNA or does it have a more global effect? No data on other genes have been reported. Given that mRNA stability and degradation is a key element in regulating cellular protein levels, one wonders what the potential long-term effects of globally stabilizing cellular mRNA species would be.

15.9 Alternative Approaches

An alternative strategy to combining a corrector drug with a potentiator drug is to combine both mechanisms of action into a single molecule. These compounds are called dual-acting CFTR modulators. Compared to multicomponent therapies, dual-acting modulators have several advantages. Pharmacokinetic/pharmacodynamic studies for single drugs are much simpler to conduct and more straightforward in terms of clinical trials (Morphy and Rankovic 2005). Joining a corrector compound (derived from CFTR corrector Corr-4a) with a potentiator moiety (derived from the phenylglycine potentiator PG01) using a hydrolysable ethylene glycol spacer, Mills and colleagues generated a hybrid dual-acting CFTR modulator (Mills et al. 2010). At least in FRT cells expressing ΔF508 CFTR, such an approach appears to have validity. However, because the two components are linked by a single cleavable linker, the compound does in fact break down into its constituent components and is, therefore, not a true dual-acting modulator. Screening over 100,000 compounds for a true dual-acting molecule, Verkman and colleagues identified six commercially available compounds that exhibited dual correcting and potentiating properties (Phuan et al. 2011). Although these compounds, especially CoPo-22, have an EC_{50} in the low micromolar range, their efficacy is still below that of other single-acting compounds. Whether a truly efficacious dual-acting compound will be identified remains to be determined.

15.9.1 PTI-801 + Orkambi

PTI have also completed a Phase I trial in which their corrector (PTI-801) was combined with Vertex's orkambi (Clinical Trial NCT03140527). In patients homozygous for the ΔF508 allele, PTI-801 was used in a dose escalation study, in concert with orkambi that patients were already taking (Proteostasis 2018). Following 14 days trial, sweat chloride dropped (82.8 mmol/L for orkambi + placebo to 55.8 mmol/L for orkambi + PTI-801, 400 mg). No significant change was observed in FEV_1, for patients taking PTI-801 in addition to orkambi. Interestingly, a small but significant increase in weight and BMI was noted for patients taking the triple combination, compared to orkambi alone.

15.10 How Much Is Enough?

A critical unanswered question in CF therapy is "how much correction do you need for long-term benefit?" On the surface, this seems a simple question, yet is surprisingly difficult to answer. The assumption is made that obligate heterozygotes (i.e., parents of children with CF) have half the amount of CFTR relative to non-carriers, yet this has not been extensively investigated. The other assumption made is that heterozygotes (even with half the amount of CFTR as non-CF carriers) are symptomless. This for the most part appears to be true, although recent studies suggest that there may be differences (Aeffner et al. 2013). Certainly heterozygotes and homozygous wt individuals appear to have similar CFTR function, as monitored by sweat electrolytes and nasal PD. Conversely, patients with two severe mutations, e.g., homozygous for F508del, have little or no detectable CFTR function or protein and have sweat electrolytes ~100 mM. Increasing CFTR protein/function, associated with milder mutations, is seen in pancreatic sufficient CF patients, with near-normal protein/function possibly in patients with idiopathic pancreatitis (Wilschanski et al. 2006). If sweat electrolyte levels are used as a biomarker, a value of ~70 mM would place a patient in a pancreatic sufficient-like state (known to be associated with a decade or so of increased life expectancy) (Davis et al. 2004) and ~60 mM essentially normal. However, sweat electrolytes, not normally a pathologic concern, appear to show little correlation with important clinical manifestations such as lung function (as measured by FEV_1). Even for those patients with residual CFTR function, lung disease including bronchiectasis with sinopulmonary infections and endocrine problems associated with CF-related diabetes can have devastating effects. Thus, it is clear that the present CFTR modulators will still necessitate treatment for residual disease. For CF researchers, the CF Foundation runs a program where many of the CFTR modulators are available for non-clinical investigations. This is a wonderful resource that allows for investigations of what can otherwise be extremely difficult molecules to obtain.

15.11 Future Perspectives

Until recently, a patient's CFTR genotype was more a research issue than a clinical issue, though the separation into severe and mild mutations would certainly impact prognosis. However, a patient's genotype had little impact upon treatment regimens. With the marketing of ivacaftor, all that changed. Now, it is clear that a physician must have a detailed knowledge of a patient's CFTR genotype in order to determine the most appropriate treatment; admittedly at present this comes down to whether a patient's mutations will benefit from potentiators such as ivacaftor or not. Although the FDA approval of ivacaftor was groundbreaking, it only affected a small subset of patients with G551D. Over the years, other gating mutations were added to the prescribing information, increasing the number of treatable patients. The FDA approval of CFTR correctors as lumacaftor and tezacaftor along with the ivacaftor potentiator further increased the number of treatable patients. FDA approval of triple therapies undergoing clinic trials will increase the percentage of patients eligible for treatment to around 90% of all CF patients, a remarkable feat achieved in a relatively few number of years. However, patients bearing nonsense mutations and those with splice mutations still await effective treatments for their condition. In a short period, we have therefore moved from a roughly one-size-fits-all approach to treating CF patients to a bespoke treatment based on a patients' DNA. The era of personalized medicine for CF has arrived, and it may not be too long before cystic fibrosis is no longer in the textbooks as a lethal genetic disease, but rather as a footnote on genetic traits.

Acknowledgments The author is grateful to Dr. Ashvani Singh, Principal Scientist AbbVie Inc, for helpful discussions and insights into medicinal chemistry. The author also acknowledges the help and support of the CF Research Group at the Chicago Medical School.

References

Aartsma-Rus A, Straub V, Hemmings R, Haas M, Schlosser-Weber G, Stoyanova-Beninska V, Mercuri E, Muntoni F, Sepodes B, Vroom E, Balabanov P (2017) Development of exon skipping therapies for Duchenne muscular dystrophy: a critical review and a perspective on the outstanding issues. Nucleic Acid Ther 27:251–259

Accurso FJ, Rowe SM, Clancy JP, Boyle MP, Dunitz JM, Durie PR, Sagel SD, Hornick DB, Konstan MW, Donaldson SH, Moss RB, Pilewski JM, Rubenstein RC, Uluer AZ, Aitken ML, Freedman SD, Rose LM, Mayer-Hamblett N, Dong Q, Zha J, Stone AJ, Olson ER, Ordonez CL, Campbell PW, Ashlock MA, Ramsey BW (2010) Effect of VX-770 in persons with cystic fibrosis and the G551D-CFTR mutation. N Engl J Med 363:1991–2003

Adams S, Owen G (2018) £11 million-a-year US pharmaceutical boss is accused of trying to 'blackmail' Theresa May into buying cystic fibrosis drug. https://www.dailymail.co.uk/news/article-5954789/11-million-year-pharmaceutical-boss-accused-trying-blackmail-Theresa-Mayhtml

Aeffner F, Abdulrahman B, Hickman-Davis JM, Janssen PM, Amer A, Bedwell DM, Sorscher EJ, Davis IC (2013) Heterozygosity for the F508del mutation in the cystic fibrosis transmembrane conductance regulator anion channel attenuates influenza severity. J Infect Dis 208:780–789

Al-Malky G, Suri R, Sirimanna T, Dawson SJ (2014) Normal hearing in a child with the m.1555A>G mutation despite repeated exposure to aminoglycosides. Has the penetrance of this pharmacogenetic interaction been overestimated? Int J Pediatr Otorhinolaryngol 78:969–973

Alonzo De Los Ruyzes De Fonteca J (1606) Diez previlegios para mgeres prenadas. Alcala de Henares

Alton EW, Boyd AC, Davies JC, Gill DR, Griesenbach U, Harrison PT, Henig N, Higgins T, Hyde SC, Innes JA, Korman MS (2016) Genetic medicines for CF: hype versus reality. Pediatr Pulmonol 51:S5–S17

Ameen N, Silvis M, Bradbury NA (2007) Endocytic trafficking of CFTR in health and disease. J Cyst Fibros 6:1–14

Amrani N, Ganesan R, Kervestin S, Mangus DA, Ghosh S, Jacobson A (2004) A faux 3′-UTR promotes aberrant termination and triggers nonsense-mediated mRNA decay. Nature 432:112–118

Anderson D (1938) Cystic fibrosis of the pancreas: clinical and pathological study. Am J Dis Child 56:344

Armstrong DK, Cunningham S, Davies JC, Alton EW (2014) Gene therapy in cystic fibrosis. Br Med J 99:465–468

Aslam A, Jahnke N, Remmington T, Southern KW (2017) Ataluren and similar compounds (specific therapies for premature termination codon class I mutations) for cystic fibrosis. Paediatr Respir Rev 24:32–34

Aznarez I, Chan EM, Zielenski J, Blencowe BJ, Tsui LC (2003) Characterization of disease-associated mutations affecting an exonic splicing enhancer and two cryptic splice sites in exon 13 of the cystic fibrosis transmembrane conductance regulator gene. Hum Mol Genet 12:2031–2040

Bagdany M, Veit G, Fukuda R, Avramescu RG, Okiyoneda T, Baaklini I, Singh J, Sovak G, Xu H, Apaja PM, Sattin S, Beitel LK, Roldan A, Colombo G, Balch W, Young JC, Lukacs GL (2017) Chaperones rescue the energetic landscape of mutant CFTR at single molecule and in cell. Nat Commun 8:398

Balch WE, Roth DM, Hutt DM (2011) Emergent properties of proteostasis in managing cystic fibrosis. Cold Spring Harb Perspect Biol 3:a004499

Bebok Z, Collawn JF, Wakefield J, Parker W, Li Y, Varga K, Sorscher EJ, Clancy JP (2005) Failure of cAMP agonists to activate rescued deltaF508 CFTR in CFBE41o- airway epithelial monolayers. J Physiol 569:601–615

Bedwell DM, Kaenjak A, Benos DJ, Bebok Z, Bubien JK, Hong J, Tousson A, Clancy JP, Sorscher EJ (1997) Suppression of a CFTR premature stop mutation in a bronchial epithelial cell line. Nat Med 3:1280–1284

Bidaud-Meynard A, Bossard F, Schnur A, Fukuda R, Veit G, Xu H, Lukacs GL (2018) Transcytosis maintains CFTR apical polarity in the face of constitutive and mutation-induced basolateral missorting. J Cell Sci 132:bioRxiv

Blackman SM, Commander CW, Watson C, Arcara KM, Strug LJ, Stonebraker JR, Wright FA, Rommens JM, Sun L, Pace RG, Norris SA, Durie PR, Drumm ML, Knowles MR, Cutting GR (2013) Genetic modifiers of cystic fibrosis-related diabetes. Diabetes 62:3627–3635

Blanchet S, Cornu D, Argentini M, Namy O (2014) New insights into the incorporation of natural suppressor tRNAs at stop codons in Saccharomyces cerevisiae. Nucleic Acids Res 42:10061–10072

Boat TF, Welsh MJ, Beadudet AL, Scriver CR (1989) Cystic fibrosis. In: Scriver CR, Beadudet AL, Valle D, Sly WS (eds) The metabolic basis of inherited disease, 8th edn. McGraw-Hill, New York, pp 2649–2680

Boinot C, Souchet MJ, Ferru-Clement R, Becq F (2014) Searching for combinations of small-molecule correctors to restore f508del-cystic fibrosis transmembrane conductance regulator function and processing. J Pharmacol Exp Ther 350: 624–634
Bompadre SG, Sohma Y, Li M, Hwang TC (2007) G551D and G1349D, two CF-associated mutations in the signature sequences of CFTR, exhibit distinct gating defects. J Gen Physiol 129:285–298
Boucher RC (2004) New concepts of the pathogenesis of cystic fibrosis lung disease. Eur Respir J 23:146–158
Bowerman M, Becker CG, Yanez-Munoz RJ, Ning K, Wood MJA, Gillingwater TH, Talbot K (2017) Therapeutic strategies for spinal muscular atrophy: SMN and beyond. Dis Model Mech 10:943–954
Boyle MP, Bell SC, Konstan MW (2012) The investigational CFTR corrector, VX-809 (Lumacaftor) co-administered with the oral potentiator ivacaftor improved CFTR and lung function in F508del homozygous patients: phase II study results. Pediatr Pulmonol 47:315
Boyle MP, Bell SC, Konstan MW, McColley SA, Rowe SM, Rietschel E, Huang X, Waltz D, Patel NR, Rodman D (2014) A CFTR corrector (lumacaftor) and a CFTR potentiator (ivacaftor) for treatment of patients with cystic fibrosis who have a phe508del CFTR mutation: a phase 2 randomised controlled trial. Lancet Respir Med 2:527–538
Bradbury NA (ed) (2015) Cystic fibrosis, vol 1. Encyclopedia of cell biology, 1st edn. Elsevier, London
Bradbury NA, Cohn JA, Venglarik CJ, Bridges RJ (1994) Biochemical and biophysical identification of cystic fibrosis transmembrane conductance regulator chloride channels as components of endocytic clathrin-coated vesicles. J Biol Chem 269:8296–8302
Bridges RJ, Bradbury NA (2018) Cystic fibrosis, cystic fibrosis transmembrane conductance regulator and Drugs: insights from cellular trafficking. In: Ulloa-Aguirre A, Tao Y-X (eds) Targeting trafficking in drugd evelopment, vol 245. Springer, New York
Brodsky JL, Skach WR (2011) Protein folding and quality control in the endoplasmic reticulum: Recent lessons from yeast and mammalian cell systems. Curr Opin Cell Biol 23:464–475
Cai L, Fisher AL, Huang H, Xie Z (2016) CRISPR-mediated genome editing and human diseases. Genes Dis 3:244–251
Castellani C, Assael BM (2016) Cystic fibrosis: a clinical view. Cell Mol Life Sci 74:129–140
CF Foundation (2018) NACFC2018—CFF announces $100 million 'Infection Research Initiative'. https://cysticfibrosisnewstoday.com/2018/10/19/nacfc2018-cff-announces-100-million-infection-research-initiative/
CF Foundation (2017) https://www.cff.org/Trials/finder/details/478/Phase-2a-study-of-Galapagos-GLPG2222-in-adults-with-CF
CF Foundation C (2015) Patient registry, annual report
Chadwick S, Browning JE, Stern M (1998) Nasal application of glycerol in DF508 cystic fibrosis patients. Thorax 53:A60
Char JE, Wolfe MH, Cho HJ, Park IH, Jeong JH, Frisbee E, Dunn C, Davies Z, Milla C, Moss RB, Thomas EA, Wine JJ (2014) A little CFTR goes a long way: CFTR-dependent sweat secretion from G551D and R117H-5T cystic fibrosis subjects taking ivacaftor. PLoS One 9:e88564
Cheng SH, Gregory RJ, Marshall J, Paul S, Souza DW, White GA, O'Riordan CR, Smith AE (1990) Defective intracellular transport and processing of CFTR is the molecular basis of most cystic fibrosis. Cell 63:827–834
Chin S, Hung M, Won A, Wu YS, Ahmadi S, Yang D, Elmallah S, Toutah K, Hamilton CM, Young RN, Viirre RD, Yip CM, Bear CE (2018) Lipophilicity of the cystic fibrosis drug, ivacaftor (VX-770), and its destabilizing effect on the major CF-causing mutation: F508del. Mol Pharmacol 94:917–925
Cho DY, Zhang S, Lazrak A, Grayson JW, Pena Garcia JA, Skinner DF, Lim DJ, Mackey C, Banks C, Matalon S, Woodworth BA (2018) Resveratrol and ivacaftor are additive G551D CFTR-channel potentiators: therapeutic implications for cystic fibrosis sinus disease. Int Forum Allergy Rhinol 9:100–105. https://doi.org/10.1002/alr.22202

Choi YH, Yu AM (2014) ABC transporters in multidrug resistance and pharmacokinetics, and strategies for drug development. Curr Pharm Des 20:793–807

Choi JY, Muallem D, Kiselyov K, Lee MG, Thomas PJ, Muallem S (2001) Aberrant CFTR-dependent HCO^{-}_{3} transport in mutations associated with cystic fibrosis. Nature 410:94–97

Cholon DM, Quinney NL, Fulcher ML, Esther CR Jr, Das J, Dokholyan NV, Randell SH, Boucher RC, Gentzsch M (2014) Potentiator ivacaftor abrogates pharmacological correction of DeltaF508 CFTR in cystic fibrosis. Sci Transl Med 6:246ra296

Chu CS, Trapnell BC, Curristin SM, Cutting G, Crystal RG (1992) Extensive posttranscriptional deletion of the coding sequences for part of nucleotide-binding fold 1 in respiratory epithelial mRNA transcripts of the cystic fibrosis transmembrane conductance regulator gene is not associated with the clinical manifestations of cystic fibrosis. J Clin Invest 90:785–790

Clague S (2014) Dorothy Hansine Andersen. Lancet Respir Med 2:184–185

Clain J, Fritsch J, Lehmann-Che J, Bali M, Arous M, Goossens M, Edelman A, Fanen P (2001) Two mild cystic fibrosis-associated mutations result in severe cystic fibrosis when combined in cis and reveal a residue important for cystic fibrosis transmembrane conductance regulator processing and function. J Biol Chem 276:9045–9049

Clain J, Lehmann-Che J, Dugueperoux I, Arous N, Girodon E, Legendre M, Goossens M, Edelman A, de Braekeleer M, Teulon J, Fanen P (2005) Misprocessing of the CFTR protein leads to mild cystic fibrosis phenotype. Hum Mutat 25:360–371

Clancy JP, Jain M (2012) Personalized medicine in cystic fibrosis: dawning of a new era. Am J Respir Crit Care Med 186:593–597

Clancy JP, Konstan MW, Rowe SM, Accurso FJ, Zeitlin PL, Hirawat S (2006) A phase II study of PTC124 in CF patients harboring premature stop mutations. North American Cystic Fibrosis Conference. Pediatr Pulmonol A269

Clancy JP, Rowe SM, Accurso FJ, Aitken ML, Amin RS, Ashlock MA, Ballmann M, Boyle MP, Bronsveld I, Campbell PW, De Boeck K, Donaldson SH, Dorkin HL, Dunitz JM, Durie PR, Jain M, Leonard A, McCoy KS, Moss RB, Pilewski JM, Rosenbluth DB, Rubenstein RC, Schechter MS, Botfield M, Ordonez CL, Spencer-Green GT, Vernillet L, Wisseh S, Yen K, Konstan MW (2012) Results of a phase IIa study of VX-809, an investigational CFTR corrector compound, in subjects with cystic fibrosis homozygous for the F508del-CFTR mutation. Thorax 67:12–18

Clancy JP, Cotton CU, Donaldson SH, Solomon GM, VanDevanter DR, Boyle MP, Gentzsch M, Nick JA, Illek B, Wallenburg JC, Sorscher EJ, Amaral MD, Beekman JM, Naren AP, Bridges RJ, Thomas PJ, Cutting G, Rowe S, Durmowicz AG, Mense M, Boeck KD, Skach W, Penland C, Joseloff E, Bihler H, Mahoney J, Borowitz D, Tuggle KL (2018) CFTR modulator theratyping: current status, gaps and future directions. J Cyst Fibros 18:22–34. https://doi.org/10.1016/j.jcf.2018.05.004

Cohn JA, Melhus O, Page LJ, Dittrich KL, Vigna SR (1991) CFTR: Development of high-affinity antibodies and localization in sweat gland. Biochem Biophys Res Commun 181:36–43

Cojocel C, Dociu N, Ceacmacudis E, Baumann K (1984a) Effects of aminoglycoside treatment on renal protein reabsorption and accumulation. Contrib Nephrol 42:196–201

Cojocel C, Docius N, Maita K, Smith JH, Hook JB (1984b) Renal ultrastructural and biochemical injuries induced by aminoglycosides. Environ Health Perspect 57:293–299

Cui L, Aleksandrov L, Chang XB, Hou YX, He L, Hegedus T, Gentzsch M, Aleksandrov A, Balch WE, Riordan JR (2007) Domain interdependence in the biosynthetic assembly of CFTR. J Mol Biol 365:981–994

Curtis A, Nelson R, Porteous M, Burn J, Bhattacharya SS (1991) Association of less common cystic fibrosis mutations with a mild phenotype. J Med Genet 28:34–37

Cutting GR (2010) Modifier genes in Mendelian disorders: the example of cystic fibrosis. Ann NY Acad Sci 1214:57–69

Cutting GR (2015) Cystic fibrosis genetics: from molecular understanding to clinical application. Nat Rev Genet 16:45–56

Cutting GR, Kasch LM, Rosenstein BJ, Zielenski J, Tsui LC, Antonarakis SE, Kazazian HH Jr (1990) A cluster of cystic fibrosis mutations in the first nucleotide-binding fold of the cystic fibrosis conductance regulator protein. Nature 346:366–369

Dalemans W, Barbry P, Champigny G, Jallat S, Dott K, Dreyer D, Crystal RG, Pavirani A, Lecocq JP, Lazdunski M (1991) Altered chloride ion channel kinetics associated with the delta F508 cystic fibrosis mutation. Nature 354:526–528

Davies JC, Moskowitz SM, Brown C, Horsley A, Mall MA, McKone EF, Plant BJ, Prais D, Ramsey BW, Taylor-Cousar JL, Tullis E, Uluer A, McKee CM, Robertson S, Shilling RA, Simard C, Van Goor F, Waltz D, Xuan F, Young T, Rowe SM (2018) VX-659-tezacaftor-ivacaftor in patients with cystic fibrosis and one or two Phe508del alleles. N Engl J Med 379:1599–1611

Davis PB, Drumm ML, Konstan MW (1996) Cystic fibrosis. Am J Respir Crit Care Med 1545:1229–1256

Davis PB, Schluchter MD, Konstan MW (2004) Relation of sweat chloride concentration to severity of lung disease in cystic fibrosis. Pediatr Pulmonol 38: 204–209

De Boeck C, Van Braeckel E, Van der Ent C (2016) GLPG1837 in subjects with cystic fibrosis (CF) and the S1251N mutation: results from a phase IIa study (SAPHIRA2). Pediatr Pulmonol S288

de Ruyck J, Brysbaert G, Blossey R, Lensink MF (2016) Molecular docking as a popular tool in drug design, an in silico travel. Adv Appl Bioinform Chem 9:1–11

Denning GM, Anderson MP, Amara JF, Marshall J, Smith AE, Welsh MJ (1992) Processing of mutant cystic fibrosis transmembrane conductance regulator is temperature-sensitive. Nature 358:761–764

Derichs N (2013) Targeting a genetic defect: cystic fibrosis transmembrane conductance regulator modulators in cystic fibrosis. Eur Respir Rev 22:58–65

DeStefano S, Gees M, Hwang TC (2018) Physiological and pharmacological characterization of the N1303K mutant CFTR. J Cyst Fibros 17:573–581

Devor DC, Bridges RJ, Pilewski JM (2000) Pharmacological modulation of ion transport across wild-type and DeltaF508 CFTR-expressing human bronchial epithelia. Am J Physiol Cell Physiol 279:C461–C479

Dey I, Bradbury NA (2018) Models for intestinal fluid transport. In: Levitan I, Delpire E, Rasgado-Flores H (eds) Cell volume regulation: current topics in membrane biology. Elsevier, London

Dhooghe B, Bouckaert C, Capron A, Wallemacq P, Leal T, Noel S (2015) Resveratrol increases F508del-CFTR dependent salivary secretion in cystic fibrosis mice. Biol Open 4:929–936. https://doi.org/10.1242/bio.010967

Dodge JA, Lewis PA, Stanton M, Wilsher J (2007) Cystic fibrosis mortality and survival in the UK: 1947-2003. Eur Respir J 29:522–526

Du M, Jones JR, Lanier J, Keeling KM, Lindsey JR, Tousson A, Bebok Z, Whitsett JA, Dey CR, Colledge WH, Evans MJ, Sorscher EJ, Bedwell DM (2002) Aminoglycoside suppression of a premature stop mutation in a Cftr-/- mouse carrying a human CFTR-G542X transgene. J Mol Med (Berl) 80:595–604

Du M, Liu X, Welch EM, Hirawat S, Peltz SW, Bedwell DM (2008) PTC124 is an orally bioavailable compound that promotes suppression of the human CFTR-G542X nonsense allele in a CF mouse model. Proc Natl Acad Sci USA 105:2064–2069

Dugueperoux I, Bellis G, Lesure JF, Renouil M, Flodrops H, De Braekeleer M (2003) Cystic fibrosis at the Reunion Island (France): spectrum of mutations and genotype-phenotype for the Y112X mutation. J Cyst Fibros 3:185–188

Eckford PD, Li C, Ramjeesingh M, Bear CE (2012) Cystic fibrosis transmembrane conductance regulator (CFTR) potentiator VX-770 (ivacaftor) opens the defective channel gate of mutant CFTR in a phosphorylation-dependent but ATP-independent manner. J Biol Chem 287:36639–36649

Egan ME, Pearson M, Weiner SA, Rajendran V, Rubin D, Glockner-Pagel J, Canny S, Du K, Lukacs GL, Caplan MJ (2004) Curcumin, a major constituent of turmeric, corrects cystic fibrosis defects. Science 304:600–602

Elborn JS (2016) Cystic fibrosis. Lancet 388:2519–2531

Elkins MR, Bye PT (2006) Inhaled hypertonic saline as a therapy for cystic fibrosis. Curr Opin Pulm Med 12:445–452

Engelhardt JF, Yankaskas JR, Ernst SA, Yang Y, Marino CR, Boucher RC, Cohn JA, Wilson JM (1992) Submucosal glands are the predominant site of CFTR expression in the human bronchus. Nat Genet 2:240–248

Fan-Minogue H, Bedwell DM (2008) Eukaryotic ribosomal RNA determinants of aminoglycoside resistance and their role in translational fidelity. RNA 14:148–157

Flume PA, Liou TG, Borowitz DS, Li H, Yen K, Ordonez CL, Geller DE (2012) Ivacaftor in subjects with cystic fibrosis who are homozygous for the F508del-CFTR mutation. Chest 142:718–724

Friedman KJ, Kole J, Cohn JA, Knowles MR, Silverman LM, Kole R (1999) Correction of aberrant splicing of the cystic fibrosis transmembrane conductance regulator (CFTR) gene by antisense oligonucleotides. J Biol Chem 274:36193–36199

Frischmeyer PA, Dietz HC (1999) Nonsense-mediated mRNA decay in health and disease. Hum Mol Genet 8:1893–1900

Galietta LJ, Springsteel MF, Eda M, Niedzinski EJ, By K, Haddadin MJ, Kurth MJ, Nantz MH, Verkman AS (2001) Novel CFTR chloride channel activators identified by screening of combinatorial libraries based on flavone and benzoquinolizinium lead compounds. J Biol Chem 276:19723–19728

Genetic Analysis Consortium C (1990) Worldwide survey of the DF508 mutation – report from the cystic fibrosis genetic analysis consortium. Am J Hum Genet 47:354–359

Gonzalez JE, Oades K, Leychkis Y, Harootunian A, Negulescu PA (1999) Cell-based assays and instrumentation for screening ion-channel targets. Drug Discov Today 4:431–439

Grasemann H (2017) CFTR Modulator therapy for cystic fibrosis. N Engl J Med 377:2085–2088

Griesenbach U, Alton EW (2012) Progress in gene and cell therapy for cystic fibrosis lung disease. Curr Pharm Des 18:642–662

Grootenhuis PD, Van Goor F, Hadida S, Burton B, Young T, Selkirk J, Chen W, Zhou J, Negulescu P (2016) Discovery and biological profile of next-generation CFTR correctors. Pediatr Pulmonol 51:Abs 188

Grove DE, Rosser MF, Ren HY, Naren AP, Cyr DM (2009) Mechanisms for rescue of correctable folding defects in CFTR Delta F508. Mol Biol Cell 20:4059–4069

Grubb BR, Gabriel SE, Mengos A, Gentzsch M, Randell SH, Van Heeckeren AM, Knowles MR, Drumm ML, Riordan JR, Boucher RC (2006) SERCA pump inhibitors do not correct biosynthetic arrest of deltaF508 CFTR in cystic fibrosis. Am J Respir Cell Mol Biol 34:355–363

Guimbellot J, Solomon GM, Baines A, Heltshe SL, VanDalfsen J, Joseloff E, Sagel SD, Rowe SM (2018) Effectiveness of ivacaftor in cystic fibrosis patients with non-G551D gating mutations. J Cyst Fibros S1569–1993

Haardt M, Benharouga M, Lechardeur D, Kartner N, Lukacs GL (1999) C-terminal truncations destabilize the cystic fibrosis transmembrane conductance regulator without impairing its biogenesis. A novel class of mutation. J Biol Chem 274:21873–21877

Hamdaoui N, Baudoin-Legros M, Kelly M, Aissat A, Moriceau S, Vieu DL, Colas J, Fritsch J, Edelman A, Planelles G (2011) Resveratrol rescues cAMP-dependent anionic transport in the cystic fibrosis pancreatic cell line. CFPAC1. Br J Pharmacol 163:876–886

Hanrahan JW, Sampson HM, Thomas DY (2013) Novel pharmacological strategies to treat cystic fibrosis. Trends Pharmacol Sci 34:119–125

Harada K, Okiyoneda T, Hashimoto Y, Ueno K, Nakamura K, Yamahira K, Suhgahara T, Shuto T, Wada I, Suico MA, Kaoi H (2006) Calreticulin negatively regulates the cell surface expression of cystic fibrosis transmembrane conductance regulator. J Biol Chem 281:12841–12848

He L, Aleksandrov LA, Cui L, Jensen TJ, Nesbitt KL, Riordan JR (2010) Restoration of domain folding and interdomain assembly by second-site supressors of the delta F508 mutation in CFTR. EMBO J 24:3103–3112
He L, Kota P, Aleksandrov AA, Cui L, Jensen T, Dokholyan NV, Riordan JR (2013) Correctors of DeltaF508 CFTR restore global conformational maturation without thermally stabilizing the mutant protein. FASEB J 27:536–545
Hecker TM, Aris RM (2004) Management of osteoporosis in adults with cystic fibrosis. Drugs 64:133–147
Heda GD, Tanwani M, Marino CR (2001) The Delta F508 mutation shortens the biochemical half-life of plasma membrane CFTR in polarized epithelial cells. Am J Physiol Cell Physiol 280: C166–C174
Hiel H, Bennani H, Erre JP, Aurousseau C, Aran JM (1992) Kinetics of gentamicin in cochlear hair cells after chronic treatment. Acta Otolaryngol 112:272–277
Highsmith WE Jr, Burch LH, Zhou Z, Olsen JC, Strong TV, Smith T, Friedman KJ, Silverman LM, Boucher RC, Collins FS, Knowles MR (1997) Identification of a splice site mutation (2789 +5 G > A) associated with small amounts of normal CFTR mRNA and mild cystic fibrosis. Hum Mutat 9:332–338
Hongyu L, Pesce E, Sheppard DN, Singh AK, Pedemonte N (2018) Therapeutic approaches to CFTR dysfunction: from discovery to drug development. J Cyst Fibros 17:S14–S21
Howard M, Frizzell RA, Bedwell DM (1996) Aminoglycoside antibiotics restore CFTR function by overcoming premature stop mutations. Nat Med 2:467–469
Howard M, Fischer H, Roux J, Santos BC, Gullans SR, Yancey PH, Welch WJ (2003) Mammalian osmolytes and S-nitrosoglutathione promote Delta F508 cystic fibrosis transmembrane conductance regulator (CFTR) protein maturation and function. J Biol Chem 278:35159–35167
Hwang TC, Wang F, Yang IC, Reenstra WW (1997) Genistein potentiates wild-type and delta F508-CFTR channel activity. Am J Physiol 273:C988–C998
Iannitti T, Palmieri B (2011) Clinical and experimental applications of sodium phenylbutyrate. Drugs R D 11:227–249
ICER (2018) Modulator treatments for cystic fibrosis: effectiveness and value. https://icer-review.org/wp-content/uploads/2017/10/CF_Final_Evidence_Report_06072018pdf
Igreja S, Clarke LA, Botelho HM, Marques L, Amaral MD (2016) Correction of a cystic fibrosis splicing mutation by antisense oligonucleotides. Hum Mutat 37:209–215
Illek B, Fischer H, Santos GF, Widdicombe JH, Machen TE, Reenstra WW (1995) cAMP-independent activation of CFTR Cl channels by the tyrosine kinase inhibitor genistein. Am J Physiol 268:C886–C893
Illek B, Zhang L, Lewis NC, Moss RB, Dong JY, Fischer H (1999) Defective function of the cystic fibrosis-causing missense mutation G551D is recovered by genistein. Am J Physiol 277:C833–C839
Jai Y, Shah K, Bridges RJ, Bradbury NA (2015) Evidence against resveratrol as a viable therapy for the rescue of defective DeltaF508 CFTR. Biochim Biophys Acta 1850:2377–2384
Jakab RL, Collaco AM, Ameen NA (2013) Characterization of CFTR high expresser cells in the intestine. Am J Physiol Gastrointest Liver Physiol 305:G453–G465
Jayaraman S, Haggie P, Wachter RM, Remington SJ, Verkman AS (2000) Mechanism and cellular applications of a green fluorescent protein-based halide sensor. J Biol Chem 275:6047–6050
Jensen TJ, Loo MA, Pind S, Williams DB, Goldberg AL, Riordan JR (1995) Multiple proteolytic systems, including the proteosome, contribute to CFTR processing. Cell 83:129–135
Jiang Q, Engelhardt JF (1998) Cellular heterogeneity of CFTR expression and function in the lung: implications for gene therapy of cystic fibrosis. Eur J Hum Genet 6:12–31
Jih KY, Hwang TC (2013) Vx-770 potentiates CFTR function by promoting decoupling between the gating cycle and ATP hydrolysis cycle. Proc Natl Acad Sci USA 110:4404–4409
Kalid O, Mense M, Fischman S, Shitrit A, Bihler H, Ben-Zeev E, Schutz N, Pedemonte N, Thomas PJ, Bridges RJ, Wetmore DR, Marantz Y, Senderowitz H (2010) Small molecule correctors of

F508del-CFTR discovered by structure-based virtual screening. J Comput Aided Mol Des 24:971–991

Kang HS, Kerstan D, Dai L, Ritchie G, Quamme GA (2000) Aminoglycosides inhibit hormone-stimulated Mg^{2+} uptake in mouse distal convoluted tubule cells. Can J Physiol Pharmacol 78:595–602

Kazani S, Alcantara J, Debonnett L, Doucet J, Jones IC, Kulmatycki K, Machineni S, Mostovy L, Nicholls I, Vegesna R, Verheijen J, Rowlands DJ (2016) Qbw251 is a safe and efficacious Cftr potentiator for patients with cystic fibrosis. Am J Respir Crit Care 193:A7789

Keating D, Marigowda G, Burr L, Daines C, Mall MA, McKone EF, Ramsey BW, Rowe SM, Sass LA, Tullis E, McKee CM, Moskowitz SM, Robertson S, Savage J, Simard C, Van Goor F, Waltz D, Xuan F, Young T, Taylor-Cousar JL (2018) VX-445-tezacaftor-ivacaftor in patients with cystic fibrosis and one or two Phe508del alleles. N Engl J Med 379(17):1599–1611. https://doi.org/10.1056/NEJMoa1807120

Kerem B, Rommens JM, Buchanan JA, Markiewicz D, Cox TK, Chakravarti A, Buchwald M, Tsui LC (1989) Identification of the cystic fibrosis gene: genetic analysis. Science 245:1073–1080

Kerem E, Hirawat S, Armoni S, Yaakov Y, Shoseyov D, Cohen M, Nissim-Rafinia M, Blau H, Rivlin J, Aviram M, Elfring GL, Northcutt VJ, Miller LL, Kerem B, Wilschanski M (2008) Effectiveness of PTC124 treatment of cystic fibrosis caused by nonsense mutations: a prospective phase II tria. Lancet 372:719–727

Kerem E, Konstan MW, De Boeck K, Accurso FJ, Sermet-Gaudelus I, Wilschanski M, Elborn JS, Melotti P, Bronsveld I, Fajac I, Malfroot A, Rosenbluth DB, Walker PA, McColley SA, Knoop C, Quattrucci S, Rietschel E, Zeitlin PL, Barth J, Elfring GL, Welch EM, Branstrom A, Spiegel RJ, Peltz SW, Ajayi T, Rowe SM (2014) Ataluren for the treatment of nonsense-mutation cystic fibrosis: a randomised, double-blind, placebo-controlled phase 3 trial. Lancet Respir Med 2:539–547

Kim SJ, Skach WR (2012) Mechanisms of CFTR folding at the endoplasmic reticulum. Front Pharmacol 3:201

Kissner D, LeFlore Y, Narayan SB, Marigowda G, Simard C, Le Camus C (2018) False-positive cannabinoid screens in adult cystic fibrosis patients treated with lumacaftor/ivacaftor. J Cyst Fibros. https://doi.org/10.1016/j.jcf.2018.08.006

Konstan MW, Ratjen F (2012) Effect of dornase alfa on inflammation and lung function: potential role in the early treatment of cystic fibrosis. J Cyst Fibros 11:78–83

Konstan MW, Accurso FJ, Nasr SZ, Ahrens RC, Graff GR (2013) Efficacy and safety of a unique enteric-coated bicarbonate-buffered pancreatic enzyme replacement therapy in children and adults with cystic fibrosis. Clin Investig (Lond) 3:723–729. https://doi.org/10.4155/cli.13.62

Kramer EL, Clancy JP (2016) CFTR modulator therapies in pediatric cystic fibrosis: focus on ivacaftor. Expert Opin Orphan Drugs 4:1033–1042

Kreda SM, Mall MMA, Rochelle L, Yankaskas JR, Riordan JR, Boucher RC (2005) Characterization of wild-type and deltaF508 cystic fibrosis transmembrane conductance regulator in human respiratory epithelia. Mol Biol Cell 16:2154–2167

Kwok I, Miranda L, Bhatt P, Bailey V, Dasgupta V, Chin A, An J, Bresilla W, Krouse ME, Cole B, Fitzpatrick R (2016) Effect of novel F508del-CFTR modulator combination FDL169 and FDL176 on expression and functional activity. Pediatr Pulmonol 53:Abs48

Laborde-Casterot H, Donnay C, Chapron J, Burgel PR, Kanaan R, Honore I, Dusser D, Choudat D, Hubert D (2012) Employment and work disability in adults with cystic fibrosis. J Cyst Fibros 11:137–143

Levy L, Durie P, Pencharz P, Corey M (1986) Prognostic factors associated with patient survival during nutritional rehabilitatino in malnourished children and adolescents with cystic fibrosis. J Pediatr Gastroenterol 5:97–102

Lewis HA, Buchanan SG, Burley SK, Conners K, Dickey M, Dorwart M, Fowler R, Gao X, Guggino WB, Hendrickson WA, Hunt JF, Kearins MC, Lorimer D, Maloney PC, Post KW, Rajashankar KR, Rutter ME, Sauder JM, Shriver S, Thibodeau PH, Thomas P, Zhang M,

Zhao X, Emtage S (2004) Structure of nucleotide-binding domain 1 of the cystic fibrosis transmembrane conductance regulator. EMBO J 23:282–293
Lewis HA, Wang C, Zhao X, Hamuro Y, Conners K, Kearins MC, Lu F, Sauder JM, Molnar KS, Coales SJ, Maloney PC, Guggino WB, Wetmore DR, Weber PC, Hunt JF (2010) Structure and dynamics of NBD1 from CFTR characterized using crystallography and hydrogen/deuterium exchange mass spectrometry. J Mol Biol 396:406–430
Liang F, Shang H, Jordan NJ, Wong E, Mercadante D, Saltz J, Mahiou J, Bihler H, Mense M (2017) High-throughput screening for readthrough modulators of CFTR PTC mutations. SLAS Technol 22:315–324
Liu F, Zhang Z, Csanady L, Gadsby DC, Chen J (2017) Molecular structure of the human CFTR ion. Channel Cell 169:85–95
Liu F, Zhang Z, Levit A, Levring J, Touhara KK, Shoichet BK, Chen J (2019) Structural identification of a hotspot on CFTR for potentiation. Science 364:1184–1188
Loo TW, Bartlett MC, Clarke DM (2004) Thapsigargin or curcumin does not promote maturation of processing mutants of the ABC transporters, CFTR, and P-glycoprotein. Biochem Biophys Res Commun 325:580–585
Lubamba B, Dhooghe B, Noel S, Leal T (2012) Cystic fibrosis: insight into CFTR pathophysiology and pharmacotherapy. Clin Biochem 45:1132–1144
Lukacs GL, Segal G, Kartner N, Grinstein S, Zhang F (1997) Constitutive internalization of cystic fibrosis transmembrane conductance regulator occurs via clathrin-dependent endocytosis and is regulated by protein phosphorylation. Biochem J 328(Pt 2):353–361
Macek M Jr, Mackova A, Hamosh A, Hilman BC, Selden RF, Lucotte G, Friedman KJ, Knowles MR, Rosenstein BJ, Cutting GR (1997) Identification of common cystic fibrosis mutations in African-Americans with cystic fibrosis increases the detection rate to 75%. Am J Hum Genet 60:1122–1127
Maquat LE (1995) When cells stop making sense: effects of nonsense codons on RNA metabolism in vertebrate cells. RNA 1:453–465
Maquat LE (2002) Molecular biology. Skiing toward nonstop mRNA decay. Science 295:2221–2222
Marangi M, Pistritto G (2018) Innovative therapeutic strategies for cystic fibrosis: moving forward to CRISPR technique. Front Pharmacol 9:396
Marson FAL, Bertuzzo CS, Ribeiro JD (2015) Personalized drug therapy in cystic fibrosis: from fiction to reality. Curr Drug Targets 16:1007–2017
Martin BL, Luo S, Kintanar A, Chen M, Graves DJ (1998) Effect of citrulline for arginine replacement on the structure and turnover of phosphopeptide substrates of protein phosphatase-1. Arch Biochem Biophys 359:179–191
Matsui H, Grubb BR, Tarran R, Randell SH, Gatzy JT, Davis CW, Boucher RC (1998a) Evidence for periciliary liquid layer depletion, not abnormal ion composition, in the pathogenesis of cystic fibrosis airways disease. Cell 95:1005–1015
Matsui H, Randell SH, Peretti SW, Davis CW, Boucher RC (1998b) Coordinated clearance of periciliary liquid and mucus from airway surfaces. J Clin Invest 102:1125–1131
McColley SA, Rosenstein BJ, Cutting GR (1991) Differences in expression of cystic fibrosis in blacks and whites. Am J Dis Child 145:94–97
Mendoza JL, Thomas PJ (2007) Building an understanding of cystic fibrosis on the foundation of ABC transporter structures. J Bioenerg Biomembr 39:499–505
Mendoza JL, Schmidt A, Li Q, Nuvaga E, Barrett T, Bridges RJ, Feranchak AP, Brautigam CA, Thomas PJ (2012) Requirements for efficient correction of DeltaF508 CFTR revealed by analyses of evolved sequences. Cell 148:164–174
Michaels W, Bridges RJ, Hastings M (2018) Splice switching antisense oligonucleotides for the treatment of cystic fibrosis. Pediatr Pulmonol 53:Abs 196
Miller JP, Drew L, Green O, Villella A, McEwan B, Patel N, Qiu D, Bhalla A, Bastos C, Parks D, Giuliano K, Longo K, Ivarsson M, Munoz B, Lee P-S, Mehmet H, Haeberlein M (2016) CFTR

amplifiers: a new class of CFTR modulator that complements the substrate limitations of other CF therapeutic modalities. Am J Respir Crit Care Med 193:absA557

Mills AD, Yoo C, Butler JD, Yang B, Verkman AS, Kurth MJ (2010) Design and synthesis of a hybrid potentiator-corrector agonist of the cystic fibrosis mutant protein DeltaF508-CFTR. Bioorg Med Chem Lett 20:87–91

Molinski SV, Ahmadi S, Ip W, Ouyang H, Villella A, Miller JP, Lee PS, Kulleperuma K, Du K, Di Paola M, Eckford PD, Laselva O, Huan LJ, Wellhauser L, Li E, Ray PN, Pomes R, Moraes TJ, Gonska T, Ratjen F, Bear CE (2017) Orkambi(R) and amplifier co-therapy improves function from a rare CFTR mutation in gene-edited cells and patient tissue. EMBO Mol Med 9:1224–1243

Montoro DT, Haber AL, Biton M, Vinarsky V, Lin B, Birket SE, Yuan F, Chen S, Leung HM, Villoria J, Rogel N, Burgin G, Tsankov AM, Waghray A, Slyper M, Waldman J, Nguyen L, Dionne D, Rozenblatt-Rosen O, Tata PR, Mou H, Shivaraju M, Bihler H, Mense M, Tearney GJ, Rowe SM, Engelhardt JF, Regev A, Rajagopal J (2018) A revised airway epithelial hierarchy includes CFTR-expressing ionocytes. Nature 560:319–324

Moran O, Galietta LJ, Zegarra-Moran O (2005) Binding site of activators of the cystic fibrosis transmembrane conductance regulator in the nucleotide binding domains. Cell Mol Life Sci 62:446–460

Morphy R, Rankovic Z (2005) Designed multiple ligands. An emerging drug discovery paradigm. J Med Chem 48:6523–6543

Nichols DP, Chmiel JF (2015) Inflammation and its genesis in cystic fibrosis. Pediatr Pulmonol 50 (Suppl 40):S39–S56

Nocera J (2015) The $300,000 drug. New York Times, July 18. https://www.nytimes.com/2014/07/19/opinion/joe-nocera-cystic-fibrosis-drug-price.html

Okiyoneda T, Barriere H, Bagdany M, Rabeh WM, Du K, Hohfeld J, Young JC, Lukacs GL (2010) Peripheral protein quality control removes unfolded CFTR from the plasma membrane. Science 329:805–810

Okiyoneda T, Veit G, Dekkers JF, Bagdany M, Soya N, Xu H, Roldan A, Verkman AS, Kurth M, Simon A, Hegedus T, Beekman JM, Lukacs GL (2013) Mechanism-based corrector combination restores DeltaF508-CFTR folding and function. Nat Chem Biol 9:444–454

Ozen AY, Duman DG (2016) Pancreatic involvement in cystic fibrosis. Minerva Med 107:427–436

Phuan PW, Yang B, Knapp JM, Wood AB, Lukacs GL, Kurth MJ, Verkman AS (2011) Cyanoquinolines with independent corrector and potentiator activities restore DeltaPhe508-cystic fibrosis transmembrane conductance regulator chloride channel function in cystic fibrosis. Mol Pharmacol 80:683–693

Picciano JA, Ameen N, Grant BD, Bradbury NA (2003) Rme-1 regulates the recycling of the cystic fibrosis transmembrane conductance regulator. Am J Physiol Cell Physiol 285:C1009–C1018

Pilewski JM, Cooke J, Lekstrom-Himes J, Donaldson S (2015) VX-661 in combination with ivacaftor in patients with cystic fibrosis and the F508del-CFTR mutation. J Cystic Fibrosis 14:S1

Plasschaert LW, Zilionis R, Choo-Wing R, Savova V, Knehr J, Roma G, Klein AM, Jaffe AB (2018) A single-cell atlas of the airway epithelium reveals the CFTR-rich pulmonary ionocyte. Nature 560:377–381

Pollack A (2015) Orkambi, a new cystic fibrosis drug, wins F.D.A. approval. New York Times, July 2. http://www.nytimes.com/2015/07/03/business/orkambi-a-new-cystic-fibrosis-drug-wins-fda-approval.html. Accessed 20 July 2015

Pranke I, Bidou L, Martin N, Blanchet S, Hatton A, Karri S, Cornu D, Costes B, Chevalier B, Tondelier D, Girodon E, Coupet M, Edelman A, Fanen P, Namy O, Sermet-Gaudelus I, Hinzpeter A (2018) Factors influencing readthrough therapy for frequent cystic fibrosis premature termination codons. ERJ Open Res 4. https://doi.org/10.1183/23120541.00080-2017

Prince LS, Peter K, Hatton SR, Zaliauskiene L, Cotlin LF, Clancy JP, Marchase RB, Collawn JF (1999) Efficient endocytosis of the cystic fibrosis transmembrane conductance regulator requires a tyrosine-based signal. J Biol Chem 274:3602–3609

Protasevich I, Yang Z, Wang C, Atwell S, Zhao X, Emtage S, Wetmore DR, Hunt JF, Brouillette CG (2010) Thermal unfolding studies show the disease causing F508del mutation in CFTR thermodynamically destabilizes nucleotide-binding domain. Protein Sci 19:1917–1931

Proteostasis (2018) Proteostasis announces positive data from ongoing phase 1 study of PTI-801 in cystic fibrosis patients on background Orkambi® therapy. http://ir.proteostasis.com/news-releases/news-release-details/proteostasis-announces-positive-data-ongoing-phase-1-study-pti

Proteostatsis (2019) New data set release. http://ir.proteostasis.com/news-releases/news-release-details/proteostasis-therapeutics-announces-broad-new-dataset

PTI (2017) PTC Therapeutics announces results from pivotal phase 3 clinical trial of ataluren in patients living with nonsense mutation cystic fibrosis. http://ir.ptcbio.com/releasedetailcfm?releaseid=1015471

PTI (2018) Proteostasis therapeutics announces positive preliminary results from proprietary doublet in cystic fibrosis patients. http://ir.proteostasis.com/news-releases/news-release-details/proteostasis-therapeutics-announces-positive-preliminary-results

Pulmozyme (2005) Pulmozyme (dornase alfa) Inhalation Solution. Genentech, South San Francisco

Pyle LC, Ehrhardt A, Mitchell LH, Fan L, Ren A, Naren AP, Li Y, Clancy JP, Bolger GB, Sorscher EJ, Rowe SM (2011) Regulatory domain phosphorylation to distinguish the mechanistic basis underlying acute CFTR modulators. Am J Physiol Lung Cell Mol Physiol 301:L587–L597

Quinton PM (1983) Chloride impermeability in cystic fibrosis. Nature 301:421–422

Quinton PM (1999) Physiological basis of cystic fibrosis: a historical perspective. Physiol Rev 79: S3–S22

Rabeh WM, Bossard F, Xu H, Okiyoneda T, Bagdany M, Mulvihill CM, Du K, di Bernardo S, Liu Y, Konermann L, Roldan A, Lukacs GL (2012) Correction of both NBD1 energetics and domain interface is required to restore ΔF508 CFTR folding and function. Cell 148:150–163

Ramsey BW, Davies J, McElvaney NG, Tullis E, Bell SC, Drevinek P, Griese M, McKone EF, Wainwright CE, Konstan MW, Moss R, Ratjen F, Sermet-Gaudelus I, Rowe SM, Dong Q, Rodriguez S, Yen K, Ordonez C, Elborn JS (2011) A CFTR potentiator in patients with cystic fibrosis and the G551D mutation. N Engl J Med 365:1663–1672

Ratjen F (2001) Changes in strategies for optimal antibacterial therapy in cystic fibrosis. Int J Antimicrob Agents 17:93–96

Reddy MM, Quinton PM (2001) Selective activation of cystic fibrosis transmembrane conductance regulator Cl^- and HCO_3^- conductances. JOP 2:212–218

Ren HY, Grove DE, De La Rosa O, Houck SA, Sopha P, Van Goor F, Hoffman BJ, Cyr DM (2013) VX-809 corrects folding defects in cystic fibrosis transmembrane conductance regulator protein through action on membrane-spanning domain 1. Mol Biol Cell 24:3016–3024

Riordan JR, Rommens JM, Kerem B, Alon N, Rozmahel R, Grzelczak Z, Zielenski J, Lok S, Plavsic N, Chou JL et al (1989) Identification of the cystic fibrosis gene: cloning and characterization of complementary DNA. Science 245:1066–1073

Rogan MP, Stoltz DA, Hornick DB (2011) Cystic fibrosis transmembrane conductance regulator intracellular processing, trafficking, and opportunities for mutation-specific treatment. Chest 139:1480–1490

Rosser MF, Grove DE, Chen L, Cyr DM (2008) Assembly and misassembly of cystic fibrosis transmembrane conductance regulator: folding defects caused by deletion of F508 occur before and after the calnexin-dependent association of membrane spanning domain (MSD) 1 and MSD2. Mol Biol Cell 1911:4570–4579

Rowe SM, Clancy JP (2009) Pharmaceuticals targeting nonsense mutations in genetic diseases: progress in development. BioDrugs 23:165–174

Rowe SM, Verkman AS (2014) Cystic fibrosis transmembrane regulator correctors and potentiators. In: Riordan JR, Boucher RC, Quinton PM (eds) Cold Spring Harbor perspectives in medicine. Cold Spring Harbor Laboratory Press, Cold Spring Harbor

Rowe SM, Miller S, Sorscher E (2005) Cystic fibrosis. N Engl J Med 352:1992–2001

Rowe SM, Accurso FJ, Clancy JP (2007) Detection of cystic fibrosis transmembrane conductance regulator activity in early phase clinical trials. Proc Am Thorac Soc 4:387–398

Rowe SM, Heltshe SL, Gonska T, Donaldson SH, Borowitz D, Gelfond D, Sagel SD, Khan U, Mayer-Hamblett N, Van Dalfsen JM, Joseloff E, Ramsey BW (2014) Clinical mechanism of the cystic fibrosis transmembrane conductance regulator potentiator ivacaftor in G551D-mediated cystic fibrosis. Am J Respir Crit Care Med 190:175–184

Rowe SM, Daines C, Ringshausen FC, Kerem E, Wilson J, Tullis E, Nair N, Simard C, Han L, Ingenito EP, McKee C, Lekstrom-Himes J, Davies JC (2017a) Tezacaftor-Ivacaftor in residual-function heterozygotes with cystic fibrosis. N Engl J Med 377:2024–2035

Rowe SM, McColley SA, Rietschel E, Li X, Bell SC, Konstan MW, Marigowda G, Waltz D, Boyle MP (2017b) Lumacaftor/Ivacaftor treatment of patients with cystic fibrosis heterozygous for F508del-CFTR. Ann Am Thorac Soc 14:213–219

Rowntree RK, Harris A (2003) The phenotypic consequences of CFTR mutations. Am Hum Genet 67:471–485

Rubenstein RC, Zeitlin PL (1998) A pilot clinical trial of oral sodium 4-phenylbutyrate (Buphenyl) in deltaF508-homozygous cystic fibrosis patients: partial restoration of nasal epithelial CFTR function. Am J Respir Crit Care Med 157:484–490

Rubenstein RC, Egan ME, Zeitlin PL (1997) In vitro pharmacologic restoration of CFTR-mediated chloride transport with sodium 4-phenylbutyrate in cystic fibrosis epithelial cells containing delta F508-CFTR. J Clin Invest 100:2457–2465

Rutter WC, Burgess DR, Burgess DS (2016) Increasing incidence of multidrug resistance among cystic fibrosis respiratpry bacterial isolates. Microb Drug Resist 23:51–55

Sala MA, Jain M (2018) Tezacaftor for the treatment of cystic fibrosis. Expert Rev Respir Med 12:725–732. https://doi.org/10.1080/17476348.2018.1507741

SAPHIRA1 (2017) GLPG1837 in subjects with cystic fibrosis (CF) and the G551D mutation: results from a Phase II study (Saphira 1). http://www.glpg.com/docs/view/593e66d0a49a4-en

Sato S, Ward CL, Krouse ME, Wine JJ, Kopito RR (1996) Glycerol reverses the misfolding phenotype of the most common cystic fibrosis mutation. J Biol Chem 271:635–638

Sawicki GS, McKone EF, Pasta DJ, Millar SJ, Wagener JS, Johnson CA, Konstan MW (2015) Sustained Benefit from ivacaftor demonstrated by combining clinical trial and cystic fibrosis patient registry data. Am J Respir Crit Care Med 192:836–842

Schneider EK (2018) Cytochrome P450 3A4 Induction: Lumacaftor versus ivacaftor potentially resulting in significantly reduced plasma concentration of ivacaftor. Drug Metab Lett 12:71–74

Schwank G, Koo BK, Sasselli V, Dekkers JF, Heo I, Demircan T, Sasaki N, Boymans S, Cuppen E, van der Ent CK, Nieuwenhuis EE, Beekman JM, Clevers H (2013) Functional repair of CFTR by CRISPR/Cas9 in intestinal stem cell organoids of cystic fibrosis patients. Cell Stem Cell 13:653–658

Sermet-Gaudelus I, Renouil M, Fajac A, Bidou L, Parbaille B, Pierrot S, Davy N, Bismuth E, Reinert P, Lenoir G, Lesure JF, Rousset JP, Edelman A (2007) In vitro prediction of stop-codon suppression by intravenous gentamicin in patients with cystic fibrosis: a pilot study. BMC Med 5:5

Sermet-Gaudelus I, Boeck KD, Casimir GJ, Vermeulen F, Leal T, Mogenet A, Roussel D, Fritsch J, Hanssens L, Hirawat S, Miller NL, Constantine S, Reha A, Ajayi T, Elfring GL, Miller LL (2010) Ataluren (PTC124) induces cystic fibrosis transmembrane conductance regulator protein expression and activity in children with nonsense mutation cystic fibrosis. Am J Respir Crit Care Med 182:1262–1272

Serohijos AW, Hegedus T, Aleksandrov AA, He L, Cui L, Dokholyan NV, Riordan JR (2008) Phenylalanine-508 mediates a cytoplasmic-membrane domain contact in the CFTR 3D structure crucial to assembly and channel function. Proc Natl Acad Sci USA 105:3256–3261

Shak S, Capon DJ, Hellmiss R, Marsters SA, Baker CL (1990) Recombinant human DNase I reduces the viscosity of cystic fibrosis sputum. Proc Natl Acad Sci USA 87:9188–9192

Sharma M, Pampinella F, Nemes C, Benharouga M, So J, Du K, Bache KG, Papsin B, Zerangue N, Stenmark H, Lukacs GL (2004) Misfolding diverts CFTR from recycling to degradation: quality control at early endosomes. J Cell Biol 164:923–933

Sheppard DN, Rich DP, Ostedgaard LS, Gregory RJ, Smith AE, Welsh MJ (1993) Mutations in CFTR associated with mild-disease-form Cl^- channels with altered pore properties. Nature 362:160–164
Shoshani T, Augarten A, Gazit E, Bashan N, Yahav Y, Rivlin Y, Tal A, Seret H, Yaar L, Kerem E, Bat-Sheva K (1992) Association of a nonsense mutation (W1282X), the most common mutation in the Ashkenazi Jewish cystic fibrosis patients in Israel, with presentation of severe disease. Am J Hum Genet 50:222–228
Silvis MR, Picciano JA, Bertrand C, Weixel K, Bridges RJ, Bradbury NA (2003) A mutation in the cystic fibrosis transmembrane conductance regulator generates a novel internalization sequence and enhances endocytic rates. J Biol Chem 278:11554–11560
Sinha C, Zhang W, Moon CS, Actis M, Yarlagadda S, Arora K, Woodroofe K, Clancy JP, Lin S, Ziady AG, Frizzell R, Fuji N, Naren AP (2015) Capturing the direct binding of CFTR correctors to CFTR by using Click chemistry. Chembiochem 16:2017–2022
Song Y, Sonawane ND, Salinas D, Qian L, Pedemonte N, Galietta LJ, Verkman AS (2004) Evidence against the rescue of defective DeltaF508-CFTR cellular processing by curcumin in cell culture and mouse models. J Biol Chem 279:40629–40633
Southern KW, Patel S, Sinha IP, Nevitt SJ (2018) Correctors (specific therapies for class II CFTR mutations) for cystic fibrosis. Cochrane Database Syst Rev 8:CD010966
Spitali P, Aartsma-Rus A (2012) Splice modulating therapies for human disease. Cell 148:1085–1088
Strug LJ, Stephenson AL, Panjwani N, Harris A (2018) Recent advances in developing therapeutics for cystic fibrosis. Hum Mol Genet 27:R173–R186
Swiatecka-Urban A, Duhaime M, Coutermarsh B, Karlson KH, Collawn J, Milewski M, Cutting GR, Guggino WB, Langford G, Stanton BA (2002) PDZ domain interaction controls the endocytic recycling of the cystic fibrosis transmembrane conductance regulator. J Biol Chem 277:40099–40105
Tan CT, Lee SY, Yao CJ, Liu SH, Lin-Shiau SY (2001) Effects of gentamicin and pH on [Ca2+]i in apical and basal outer hair cells from guinea pigs. Hear Res 154:81–87
Taylor-Cousar JL, Munck A, McKone EF, van der Ent CK, Moeller A, Simard C, Wang LT, Ingenito EP, McKee C, Lu Y, Lekstrom-Himes J, Elborn JS (2017) Tezacaftor-Ivacaftor in patients with cystic fibrosis homozygous for Phe508del. N Engl J Med 377:2013–2023
Thibodeau PH, Richardson JM 3rd, Wang W, Millen L, Watson J, Mendoza JL, Du K, Fischman S, Senderowitz H, Lukacs GL, Kirk K, Thomas PJ (2010) The cystic fibrosis-causing mutation deltaF508 affects multiple steps in cystic fibrosis transmembrane conductance regulator biogenesis. J Biol Chem 285:35825–35835
Thomas PJ, Shenbagamurthi P, Ysern X, Pedersen PL (1991) Cystic fibrosis transmembrane conductance regulator: nucleotide binding to a synthetic peptide. Science 251:555–557
Thomson AH (1995) Human recombinant DNase in cystic fibrosis. J R Soc Med 88(Suppl 25):24–29
Trujilliano D, Ramos MD, Gonzalez J, Tornador C, Sotillo F, Escaramis G, Ossowski SLA, Casais T, Estavill X (2013) Next generation diagnostics of cystic fibrosis and CFTR-related disorders by targeted multiplex high-coverage resequencing of CFTR. J Med Genet 50:455–462
Uttamsingh V, Pilja L, Grotbeck B, Brummel CL, Uddin N, Harbeson SL, Braman V, Cassella J (2016) CTP-656 tablet confirmed superiority of pharmacokinetic profile relative to Kalydeco in Phase 1 clinical trials. J Cystic Fibrosis 10:S22
Van de Steen O, Namour F, Kanters D, Gheyle L, Geller DE, de Kock H, Vanhoutte FP (2016) Safety, tolerability and pharmacokinetics of a novel CFTR corrector molecule GLPG2222 in healthy volunteers. Pediatr Pulmonol 252
Van der Plas SE, Kelgtermans H, De Munck T, Martina SLX, Dropsit S, Quinton E, De Blieck A, Joannesse C, Tomaskovic L, Jans M, Christophe T, van der Aar E, Borgonovi M, Nelles L, Gees M, Stouten P, Van Der Schueren J, Mammoliti O, Conrath K, Andrews M (2018) Discovery of N-(3-Carbamoyl-5,5,7,7-tetramethyl-5,7-dihydro-4H-thieno[2,3-c]pyran-2-yl)-lH-pyr azole-5-carboxamide (GLPG1837), a novel potentiator which can open Class III Mutant

Cystic Fibrosis Transmembrane Conductance Regulator (CFTR) channels to a high extent. J Med Chem 61:1425–1435
Van Goor F, Hadida S, Grootenhuis PD, Burton B, Cao D, Neuberger T, Turnbull A, Singh A, Joubran J, Hazlewood A, Zhou J, McCartney J, Arumugam V, Decker C, Yang J, Young C, Olson ER, Wine JJ, Frizzell RA, Ashlock M, Negulescu P (2009a) Rescue of CF airway epithelial cell function in vitro by a CFTR potentiator, VX-770. Proc Natl Acad Sci USA 106:18825–18830
Van Goor F, Hadida S, Grootenhuis PD, Burton B, Stack JH, Cao D, Neuberger T, Singh AK, Olson ER, Wine JJ, Frizzell R, Ashlock M, Negulescu P (2009b) VX-809, a CFTR corrector, increases the cell surface density of functional F508del-CFTR in pre-clinical models of cystic fibrosis. Pediatr Pulmonol 44:154–155
Van Goor F, Hadida S, Grootenhuis PD, Burton B, Stack JH, Straley KS, Decker CJ, Miller M, McCartney J, Olson ER, Wine JJ, Frizzell RA, Ashlock M, Negulescu PA (2011) Correction of the F508del-CFTR protein processing defect in vitro by the investigational drug VX-809. Proc Natl Acad Sci USA 108:18843–18848
Van Goor F, Yu H, Burton B, Huang C, Hoffman B (2012) Ivacaftor potentiates multiplemutant cystic fibrosis transconductance regulator (CFTR) forms. Pediatr Pulmonol 47:233
Varga K, Jurkuvenaite A, Wakefield J, Hong JS, Guimbellot JS, Venglarik CJ, Niraj A, Mazur M, Sorscher EJ, Collawn JF, Bebok Z (2004) Efficient intracellular processing of the endogenous cystic fibrosis transmembrane conductance regulator in epithelial cell lines. J Biol Chem 279:22578–22584
Veit G, Avramescu RG, Perdomo D, Phuan PW, Bagdany M, Apaja PM, Borot F, Szollosi D, Wu YS, Finkbeiner WE, Hegedus T, Verkman AS, Lukacs GL (2014) Some gating potentiators, including VX-770, diminish DeltaF508-CFTR functional expression. Sci Transl Med 6:246ra297
Veit G, Avramescu RG, Chiang AN, Houchk SA, Cai Z, Peters KW, Hong JS, Pollard H, Guggino WB, Balch WE, Skach WR, Cutting G, Frizzell R, Sheppard D, Cyr D, Sorscher E, Brodsky JL, Lukacs GL (2016) From CFTR biology toward combinatorial pharmacotherapy: expanded classification of cystic fibrosis mutations. Mol Biol Cell 27:424–433
Verkman AS, Lukacs GL, Galietta LJ (2006) CFTR chloride channel drug discovery—inhibitors as antidiarrheals and activators for therapy of cystic fibrosis. Curr Pharm Des 12:2235–2247
Verkman AS, Edelman A, Amaral M, Mall MA, Beekman JM, Meiners T, Galietta LJ, Bear CE (2015) Finding new drugs to enhance anion secretion in cystic fibrosis: Toward suitable systems for better drug screening. Report on the pre-conference meeting to the 12th ECFS Basic Science Conference, Albufeira, 25-28 March 2015. J Cyst Fibros 14:700–705
Vertex (2013) Treatment with VX-661 and ivacaftor in a phase 2 study resulted in statistically significant improvements in lung function in people with cystic fibrosis who have two copies of the F508del mutation. https://investors.vrtx.com/news-releases/news-release-details/treatment-vx-661-and-ivacaftor-phase-2-study-resulted?ReleaseID=757597
Wainwright CE, Elborn JS, Ramsey BW, Marigowda G, Huang X, Cipolli M, Colombo C, Davies JC, De Boeck K, Flume PA, Konstan MW, McColley SA, McCoy K, McKone EF, Munck A, Ratjen F, Rowe SM, Waltz D, Boyle MP (2015) Lumacaftor-Ivacaftor in patients with cystic fibrosis homozygous for Phe508del CFTR. N Engl J Med 373:220–231
Wang W, Li G, Clancy JP, Kirk KL (2005) Activating cystic fibrosis transmembrane conductance regulator channels with pore blocker analogs. J Biol Chem 280:23622–23630
Wang C, Protasevich I, Yang Z, Seehausen D, Skalak T, Zhao X, Atwell S, Spencer Emtage J, Wetmore DR, Brouillette CG, Hunt JF (2010) Integrated biophysical studies implicate partial unfolding of NBD1 of CFTR in the molecular pathogenesis of F508del cystic fibrosis. Protein Sci 19:1932–1947
Wang C, Aleksandrov AA, Yang Z, Forouhar F, Proctor EA, Kota P, An J, Kaplan A, Khazanov N, Boel G, Stockwell BR, Senderowitz H, Dokholyan NV, Riordan JR, Brouillette CG, Hunt JF (2018a) Ligand binding to a remote site thermodynamically corrects the F508del mutation in the

human cystic fibrosis transmembrane conductance regulator. J Biol Chem 293 (46):17685–17704

Wang X, Liu B, Searle X, Yeung C, Bogdan A, Greszler S, Singh A, Fan Y, Swensen AM, Vortherms T, Balut C, Jia Y, Desino K, Gao W, Yong H, Tse C, Kym P (2018b) Discovery of 4-[(2R,4R)-4-({[1-(2,2-Difluoro-1,3-benzodioxol-5-yl)cyclopropyl]carbonyl}amino)-7-(difluoromethoxy)-3,4-dihydro-2H-chromen-2-yl]benzoic Acid (ABBV/GLPG-2222), a potent cystic fibrosis transmembrane conductance regulator (CFTR) corrector for the treatment of cystic fibrosis. J Med Chem 61:1436–1449

Ward CL, Kopito RR (1994) Intracellular turnover of cystic fibrosis transmembrane conductance regulator. Inefficient processing and rapid degradation of wild-type and mutant proteins. J Biol Chem 26941:25710–25718

Ward CL, Omura S, Kopito RR (1995) Degradation of CFTR by the ubiquitin-proteasome pathway. Cell 83:121–127

Webster MJ, Tarran R (2018) Slippery when wet: airway surface liquid homeostasis and mucus hydration. In: Levitan I, Delpire E, Rasgado-Flores H (eds) Cell volume regulation, vol 81. Academic, London, pp 293–336

Weinreich F, Wood PG, Riordan JR, Nagel G (1997) Direct action of genistein on CFTR. Pflugers Arch 434:484–491

Weixel KM, Bradbury NA (2000) The carboxyl terminus of the cystic fibrosis transmembrane conductance regulator binds to AP-2 clathrin adaptors. J Biol Chem 275:3655–3660

Weixel KM, Bradbury NA (2001) Mu 2 binding directs the cystic fibrosis transmembrane conductance regulator to the clathrin-mediated endocytic pathway. J Biol Chem 276:46251–46259

Welch EM, Barton ER, Zhuo J, Tomizawa Y, Friesen WJ, Trifillis P, Paushkin S, Patel M, Trotta CR, Hwang S, Wilde RG, Karp G, Takasugi J, Chen G, Jones S, Ren H, Moon YC, Corson D, Turpoff AA, Campbell JA, Conn MM, Khan A, Almstead NG, Hedrick J, Mollin A, Risher N, Weetall M, Yeh S, Branstrom AA, Colacino JM, Babiak J, Ju WD, Hirawat S, Northcutt VJ, Miller LL, Spatrick P, He F, Kawana M, Feng H, Jacobson A, Peltz SW, Sweeney HL (2007) PTC124 targets genetic disorders caused by nonsense mutations. Nature 447:87–91

Welsh M, Ramsey B (2001) Cystic Fibrosis. In: Scriver C, Beaudet A, Valle D (eds) The metabolic and molecular basis of inherited disease, vol 3. McGraw-Hill, New York, pp 5121–5188

Welsh MJ, Smith AE (1993) Molecular mechanisms of CFTR chloride channel dysfunction in cystic fibrosis. Cell 73:1251–1254

White NM, Jiang D, Burgess JD, Bederman IR, Previs SF, Kelley TJ (2007) Altered cholesterol homeostasis in cultured and in vivo models of cystic fibrosis. Am J Physiol Lung Cell Mol Physiol 292:L476–L486

White MK, Kaminski R, Young WB, Roehm PC, Khalili K (2017) CRISPR editing technology in biological and biomedical investigation. J Cell Biochem 118:3586–3594

Wilschanski M, Yahav Y, Yaacov Y, Blau H, Bentur L, Rivlin J, Aviram M, Bdolah-Abram T, Bebok Z, Shushi L, Kerem B, Kerem E (2003) Gentamicin-induced correction of CFTR function in patients with cystic fibrosis and CFTR stop mutations. N Engl J Med 349:1433–1441

Wilschanski M, Dupuis A, Ellis L, Jarvi K, Zielenski J, Tullis E, Martin S, Corey M, Tsui LC, Durie P (2006) Mutations in the cystic fibrosis transmembrane regulator gene and in vivo transepithelial potentials. Am J Respir Crit Care Med 174:787–794

Xue X, Mutyam V, Thakerar A, Mobley J, Bridges RJ, Rowe SM, Keeling KM, Bedwell DM (2017) Identification of the amino acids inserted during suppression of CFTR nonsense mutations and determination of their functional consequences. Hum Mol Genet 26:3116–3129

Yaffe SJ, Aranda JV (2010) Neonatal and pediatric pharmacology: therapeutic principles in practice. Lippincott Williams and Wilkins, Philadelphia, PA

Yamashiro Y, Shimizu T, Oguchi S, Shioya T, Nagata S, Ohtsuka Y (1997) The estimated incidence of cystic fibrosis in Japan. J Pediatr Gastroenterol Nutr 24:544–547

Yeh HI, Sohma Y, Conrath K, Hwang TC (2017) A common mechanism for CFTR potentiators. J Gen Physiol 149:1105–1118

Yoo JS, Moyer BD, Bannykh S, Yoo HM, Riordan JR, Balch WE (2002) Non-conventional trafficking of the cystic fibrosis transmembrane conductance regulator through the early secretory pathway. J Biol Chem 277:11401–11409

Yu H, Burton B, Huang CJ, Worley J, Cao D, Johnson JP Jr, Urrutia A, Joubran J, Seepersaud S, Sussky K, Hoffman BJ, Van Goor F (2012) Ivacaftor potentiation of multiple CFTR channels with gating mutations. J Cyst Fibros 11:237–245

Zawistoski M, Sui J, Ordonez C, Mai V, Liu ET, Li T, Kwok I, Kolodziej J, Kanawade A, Fitzpatrick R, Deshpande A, Dasgupta A, Cole B, Chin JL, Bresilla C, Bailey V, An W, Krouse ME (2016) Properties of a novel F508del-CFTR corrector FDL169. J Cyst Fibros 15:559–560

Zhang Z, Liu F, Chen J (2018) Molecular structure of the ATP-bound, phosphorylated human CFTR. 115: 12757–12762

Zheng Y, Schachern PA, Sone M, Papapella MM (2001) Aminoglycoside ototoxicity. Otol Neurotol 22:266–268

Zielenski J, Patrizio P, Corey M, Handelin B, Markiewicz D, Asch R, Tsui LC (1995) CFTR gene variant for patients with congenital absence of vas deferens. Am J Hum Genet 57:958–960

Chapter 16
Molecular Physiology and Pharmacology of the Cystic Fibrosis Transmembrane Conductance Regulator

Majid K. Al Salmani, Elvira Sondo, Corina Balut, David N. Sheppard, Ashvani K. Singh, and Nicoletta Pedemonte

Abstract Defective epithelial ion transport is the hallmark of the common life-limiting genetic disease cystic fibrosis (CF). CF is caused by dysfunction of the cystic fibrosis transmembrane conductance regulator (CFTR), an ATP-binding cassette transporter, which functions as an epithelial anion channel regulated by phosphorylation and cycles of ATP binding and hydrolysis. Here, we review selectively the structure and function of CFTR, including how the membrane-spanning domains form an anion-selective pore and the roles of the regulatory domain and nucleotide-binding domains in channel gating. We discuss the development of combination therapy for CF mutations, including CFTR correctors, which rescue misfolding to transport mutant CFTR to the plasma membrane and CFTR potentiators, which increase CFTR activity by enhancing channel gating. Finally, we highlight studies of CFTR processing and membrane trafficking, which identify proteostasis regulators that have the potential to lead to innovative new therapies for CF. Thus, understanding of CFTR has driven new insights into epithelial ion transport and transformed the treatment of CF.

Keywords CFTR Cl^- channel · Cystic fibrosis · Channel regulation · CFTR potentiators · CFTR correctors · Proteostasis regulators

---

M. K. Al Salmani
Natural and Medical Sciences Research Centre, University of Nizwa, Nizwa, Oman
e-mail: majid.al.salmani@unizwa.edu.om

E. Sondo · N. Pedemonte
U.O.C. Genetica Medica, Istituto Giannina Gaslini, Genova, Italy
e-mail: nicoletta.pedemonte@unige.it

C. Balut · A. K. Singh
AbbVie, Inc., North Chicago, IL, USA
e-mail: corina.balut@abbvie.com; ashvani.singh@abbvie.com

D. N. Sheppard (✉)
School of Physiology, Pharmacology and Neuroscience, University of Bristol, Bristol, UK
e-mail: D.N.Sheppard@bristol.ac.uk

K. L. Hamilton, D. C. Devor (eds.), *Studies of Epithelial Transporters and Ion Channels*, Physiology in Health and Disease,
https://doi.org/10.1007/978-3-030-55454-5_16

16.1 Introduction

The genetic disease cystic fibrosis (CF) is caused by loss-of-function mutations in the cystic fibrosis transmembrane conductance regulator (CFTR) (Riordan et al. 1989; Ratjen et al. 2015; Elborn 2016). The *CFTR* gene is positioned on chromosome 7q31.2 and contains 27 exons that span 250 kb (Tsui and Dorfman 2013). It encodes a 6.5 kb mRNA that provides instructions for the synthesis of the CFTR protein composed of 1480 amino acids (Tsui and Dorfman 2013). Assembled from two membrane-spanning domains (MSDs), two nucleotide-binding domains (NBDs), and a regulatory domain (RD), the CFTR protein is a member of the ATP-binding cassette (ABC) transporter superfamily (Riordan et al. 1989; Holland et al. 2003). ABC transporters are found in all organisms, where they perform diverse roles. Most ABC transporters utilize the energy of ATP binding and hydrolysis to actively pump diverse substrates across cellular membranes either as importers or as exporters (Holland et al. 2003). A few ABC transporters [e.g., the sulfonylurea receptor; SUR1 (ABCC8; pancreas) and SUR2 (ABCC9; heart and muscle)] regulate associated ion channels (Inagaki et al. 1995, 1996). However, only CFTR (ABCC7) forms an ion channel (Anderson et al. 1991b; Bear et al. 1992).

Here, we review selectively the molecular physiology and pharmacology of CFTR. To understand its physiological role and dysfunction in CF, we explore the function of the individual domains from which CFTR is assembled. To learn about the development of CFTR modulator combination therapy for CF, we consider the mechanism of action of CFTR correctors and potentiators. To investigate the therapeutic potential of proteostasis regulators, we discuss the CFTR-interacting proteins that orchestrate its intracellular synthesis, transport to, and stability at the plasma membrane. Optimal therapy for CF might be achieved with innovative CFTR modulator combination therapy alone or together with proteostasis regulators.

16.2 Roles of CFTR in Epithelial Ion Transport and Host Defense

CFTR is predominantly expressed in the apical membrane of epithelial tissues lining ducts and tubes throughout the body, where it functions as a tightly regulated anion-selective channel and regulates the quantity and composition of epithelial secretions (Frizzell and Hanrahan 2012; Saint-Criq and Gray 2017). It plays a crucial role in epithelial physiology, which influences development, growth, innate immunity, and fecundity. Examination of the physiological roles of different epithelial tissues reveals tissue-specific differences in CFTR function, which have consequences for the development of disease.

16.2.1 The Sweat Gland

A distinguishing feature of the CF sweat gland is the absence of pathological changes initiated by CFTR dysfunction (Kartner et al. 1992). By preventing the reabsorption of NaCl, loss of CFTR function causes salty sweat (i.e., $[Cl^-] > 60$ mmol/L), the hallmark used clinically to diagnose the disease (Knowles and Durie 2002; De Boeck et al. 2006; Bombieri et al. 2011).

CFTR is predominantly expressed in the duct of the sweat gland, a water-impermeable epithelium that absorbs NaCl to generate hypotonic sweat (Welsh et al. 2001; Quinton 2007). For two reasons, CFTR expression in the sweat duct is distinctive. First, CFTR is localized to both the apical and basolateral membranes of sweat duct epithelial cells. In other epithelia, CFTR is found exclusively in the apical membrane (Welsh et al. 2001; Quinton 2007). Second, in the sweat duct CFTR absorbs Cl^- in contrast to its secretory role in other epithelial tissues (Welsh et al. 2001; Quinton 2007).

16.2.2 The Pancreas, Intestine, Hepatobiliary System, and Reproductive Tissues

Ducts and tubes in these organs transport a variety of protein-rich cargoes. The efficient movement of these cargoes is achieved by the lubrication of epithelial surfaces by CFTR-driven salt and water secretion (Kopelman et al. 1985; Welsh et al. 2001; Frizzell and Hanrahan 2012). CFTR dysfunction disrupts transepithelial ion transport leading to the stasis of protein cargoes. The plugging of ducts that results initiates a pathogenic cascade characterized by abundant mucus producing cells, striking mucus accumulation, mild-to-moderate inflammation, and tissue remodeling (Meyerholz et al. 2010). In its most extreme form, this type of tissue damage causes the destruction and atrophy of the vas deferens in utero leading to congenital bilateral absence of the vas deferens (Welsh et al. 2001). Thus, the rate of movement of protein-rich cargoes along ducts and tubes is intimately related to tissue damage in CF (Kopelman et al. 1985).

Some epithelia, most notably the pancreas, secrete copious amounts of HCO_3^--rich fluids (Lee et al. 2012). Although the CFTR pore is permeable to HCO_3^- (Poulsen et al. 1994), most epithelial HCO_3^- secretion is achieved by the coordinated activity of CFTR and anion (Cl^-/HCO_3^-) exchangers located in the apical membrane of duct-lining epithelial cells (Frizzell and Hanrahan 2012). These anion exchangers are members of the SLC26 family of anion transporters (Group 2) (Frizzell and Hanrahan 2012). Interestingly, CFTR and Group 2 SLC26 transporters (e.g., SLC26A6) assemble with scaffolding proteins to form a macromolecular complex, explaining how activation of the cAMP signaling pathway stimulates epithelial HCO_3^- secretion (Ko et al. 2004).

For three reasons, loss of CFTR-dependent HCO_3^- secretion is intimately linked to CF pathophysiology. First, neutralization of acid. The reduced HCO_3^- concentration of epithelial secretions decreases pH. Acidic epithelial secretions damage duct-lining epithelial cells initiating tissue damage (Uc et al. 2012). Second, mucus expansion. HCO_3^- is crucial for mucus to unravel (Quinton 2010). In mucus producing cells, mucins are stored in granules in a highly condensed form with their negative charges shielded by cations. The unraveling of mucus to form molecular strings requires the neutralization of bound cations by HCO_3^- (Quinton 2010). In CF, the absence of HCO_3^- prevents mucin expansion, resulting in the development of viscous mucus (Quinton 2010). Third, bacterial killing. Chronic bacterial infection is the hallmark of CF lung disease (Ratjen et al. 2015; Stoltz et al. 2015). Loss of CFTR-dependent HCO_3^- secretion inhibits bacterial killing by antimicrobial factors (Pezzulo et al. 2012).

16.2.3 The Respiratory Airways

Lung disease in CF has two components: first, defective epithelial ion transport, which prevents normal mucociliary clearance, leading to the obstruction of the respiratory airways by viscous mucus (Quinton 2010; Ratjen et al. 2015), and second, there is a localized failure of the host defense system in the respiratory airways, leading to persistent bacterial infections (Ratjen et al. 2015; Stoltz et al. 2015). Figure 16.1 illustrates the roles of CFTR in epithelial ion transport and host defense.

Loss of CFTR-mediated anion transport impacts mucociliary clearance in several ways. First, failure of CFTR-mediated Cl^- secretion reduces the quantity of airway surface liquid (ASL), preventing mucociliary clearance and hindering the delivery of antimicrobial factors to ASL (Ratjen et al. 2015; Stoltz et al. 2015; Widdicombe and Wine 2015). Loss of ASL is aggravated by Na^+ and H_2O absorption driven by the epithelial Na^+ channel (ENaC) (Boucher 2007). Second, loss of CFTR-mediated Cl^- secretion alters osmotic forces causing the concentration of mucins, the collapse of cilia, and the failure of mucociliary clearance (Boucher 2007; Ratjen et al. 2015). Third, the failure of CFTR-dependent HCO_3^- secretion leads to abnormal mucus (see Sect. 16.2.2) (Quinton 2010). Fourth, the detachment of mucins from submucosal glands and Goblet cells is abnormal (Hansson 2019).

Loss of CFTR-dependent HCO_3^- secretion in CF leads to acidification of ASL (Pezzulo et al. 2012). This change in ASL pH leads to the inactivation of antimicrobial peptides, preventing the rapid killing of inhaled bacteria (Pezzulo et al. 2012). Host defense is further compromised by loss of CFTR-dependent SCN^- transport, which prevents the production of bactericidal $OSCN^-$ by the dual oxidase/lactoperoxidase defense system (Fischer 2009). Thus, CFTR dysfunction has multiple impacts on the respiratory airways, which predispose to bacterial infection. Bacterial infection reduces lung function by initiating a vicious cycle of

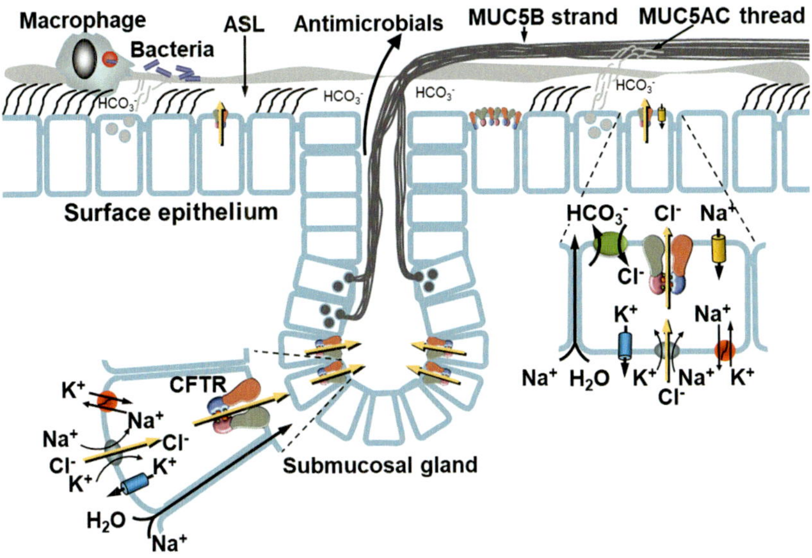

Fig. 16.1 Role of CFTR in transepithelial ion transport and host defense in the respiratory airways. The schematic representation shows a thin layer of airway surface liquid (ASL) covering the surface epithelium and a submucosal gland. In the surface epithelium, CFTR is abundantly expressed in rare pulmonary ionocytes and at lower levels in the abundant basal and club cells (Montoro et al. 2018; Plasschaert et al. 2018). The activity of CFTR, the Cl^-/HCO_3^- exchanger SLC26A4, and the epithelial Na^+ channel (ENaC) in the apical membrane controls the quantity and composition of ASL. In surface epithelial cells, CFTR-dependent HCO_3^- secretion is mediated both directly by CFTR (Stoltz et al. 2015) and indirectly by the Cl^-/HCO_3^- exchanger SLC26A4 (Kim et al. 2019). Submucosal glands deliver antimicrobial factors and mucus strands composed of MUC5B to the airway surface (Stoltz et al. 2015; Ostedgaard et al. 2017; Hansson 2019). By contrast, Goblet cells produce mucus threads composed of MUC5AC (Ostedgaard et al. 2017; Hansson 2019). The inserts show cellular models of transepithelial fluid and electrolyte transport in submucosal glands and surface epithelial cells. Modified with permission from Wang et al. (2014) and Ostedgaard et al. (2017)

inflammation and remodeling that destroys lung tissue leading to end-stage lung disease (Mall and Hartl 2014; Ratjen et al. 2015).

16.2.4 The Kidney

CFTR is abundantly expressed in the kidney, and Cl^- channels resembling CFTR are found along the nephron (Husted et al. 1995; Morales et al. 1996). However, CF patients do not demonstrate major changes in renal function, leading to speculation that other renal Cl^- channels compensate for loss of CFTR function in the kidney (Stanton 1997).

CFTR plays several roles in the kidney. First, it mediates transepithelial anion movements. In the proximal tubule it reabsorbs Cl^-, in the inner medullary collecting duct, it secretes Cl^-, and in principal cells of the cortical collecting duct, it either absorbs or secretes Cl^- (Stanton 1997). However, the most striking example of CFTR-mediated Cl^- transport in the kidney is provided by autosomal dominant polycystic kidney disease (ADPKD), the most common single gene disorder to affect the kidney (Bergmann et al. 2018). In ADPKD, nonphysiological CFTR activity contributes to the insidious growth of fluid-filled epithelial cysts, which destroy kidney tissue leading to renal failure (Davidow et al. 1996; Hanaoka et al. 1996). Thus, one strategy to prevent cyst enlargement in ADPKD is to inhibit CFTR. Consistent with this idea, Yang et al. (2008) demonstrated that CFTR inhibitors reduce cyst formation and decrease kidney weight in mouse models of ADPKD. Similarly, some individuals with ADPKD and CF have reduced kidney volumes with fewer smaller cysts (Xu et al. 2006), while F508del, the most common CF mutation (see Sect. 16.6), disrupts renal cyst formation and growth by inhibiting fluid accumulation without compromising epithelial integrity (Li et al. 2012).

Second, CFTR plays a role in receptor-mediated endocytosis in the proximal tubule and its absence causes low-molecular-weight proteinuria (Jouret et al. 2007). Jouret et al. (2007) demonstrated that CFTR stabilizes expression of the multiligand receptor cubilin in the apical membrane of the proximal tubule. Because CFTR also colocalizes with the Cl^-/H^+ exchanger ClC-5 and the vacuolar H^+-ATPase in endosomes (Jouret et al. 2007), it might also contribute to the control of endosomal pH. Thus, CFTR dysfunction has subtle, but important consequences for kidney function.

Beyond the scope of this chapter is the expression of CFTR in diverse non-epithelial cells, including cells of the cardiovascular and musculoskeletal systems (Nagel et al. 1992; Sprague et al. 1998; Robert et al. 2005; Bonvin et al. 2008; Divangahi et al. 2009; Le Henaff et al. 2014). Loss of CFTR function in non-epithelial cells contributes to the wide spectrum of disease afflicting CF patients.

16.3 Molecular Architecture of CFTR

Figure 16.2 illustrates the single-channel behavior of wild-type human CFTR in an excised inside-out membrane patch. The biophysical properties and regulation of this channel readily distinguish it from other anion channels. They include (1) regulation by cAMP-dependent phosphorylation and intracellular nucleotides (Tabcharani et al. 1991; Anderson et al. 1991a); (2) linear current–voltage (i–V) relationship (Fig. 16.2); (3) small single-channel conductance (6–10 pS) (Fig. 16.2); (4) selectivity for anions over cations (Anderson et al. 1991b); (5) anion permeability sequence $Br^- \geq Cl^- > I^-$ (Anderson et al. 1991b); (6) time- and voltage-independent gating (Fig. 16.2); and (7) pharmacological profile (e.g., potentiation by ivacaftor and inhibition by $CFTR_{inh}$-172) (Ma et al. 2002; Van Goor et al. 2009). Some of these characteristics vary across species (Bose et al. 2015). They are determined by the function of the MSDs, NBDs, and R domain from which CFTR is assembled.

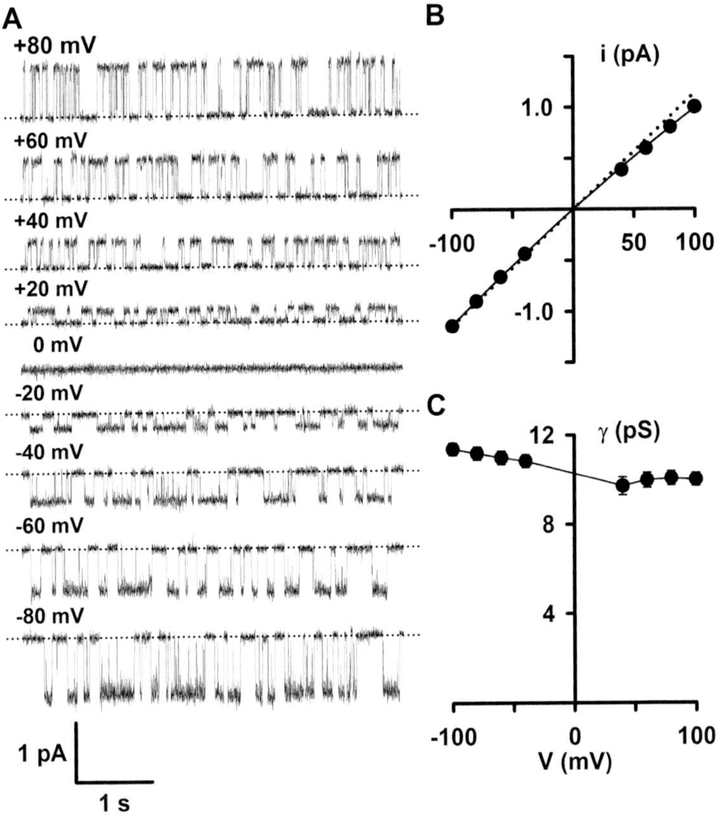

Fig. 16.2 The single-channel behavior of wild-type human CFTR. (**a**) Representative recordings of a single CFTR Cl^- channel in an inside-out membrane patch excised from a C127 recorded at the indicated voltages. The membrane patch was bathed in symmetrical 147 mM *N*-methyl-D-glucamine (NMDG)Cl with ATP (1 mM) and PKA (75 nM) continuously present in the intracellular solution; temperature was 37 °C. Dotted lines indicate the closed channel state with downward and upward deflections corresponding to channel openings at negative and positive voltages, respectively. (**b**) Single-channel i–V relationship of wild-type human CFTR. Data are means ± SEM ($n = 10$). (**c**) Relationship between chord conductance and voltage for the data shown in (**b**). Chord conductance was calculated by dividing unitary current by the difference between the applied voltage and the reversal potential. (**b**) and (**c**) are reproduced from Cai et al. (2003) by permission of the Rockefeller University Press

CFTR is composed of two MSD–NBD motifs linked by a unique RD (Riordan et al. 1989). Initial insights into the relationship between CFTR structure and function were provided by functional studies of site-directed mutations (for a review, see Sheppard and Welsh 1999). These studies demonstrated that the MSDs form a pore shaped like an asymmetric hourglass with a deep intracellular vestibule and a shallow extracellular vestibule separated by a constriction. They also showed that anion flow through this pore was gated by cycles of ATP binding and hydrolysis at the NBDs, while phosphorylation of the RD was a prerequisite for channel opening.

Early structural studies of CFTR focused on individual domains. They revealed that the RD is unstructured and the NBDs form a head-to-tail dimer with two ATP-binding sites located at the dimer interface (Ostedgaard et al. 2000; Lewis et al. 2004). Analysis of initial low-resolution structures of CFTR protein (Rosenberg et al. 2004) revealed that the quaternary structure of CFTR is monomeric and provided a tantalizing snapshot of the organization of the pore consistent with the interpretation of functional data (Sheppard and Welsh 1999). Following the elucidation of the atomic structure of the ABC transporter Sav1866, the multidrug transporter of *Staphylococcus aureus* (Dawson and Locher 2006), 3D models of the CFTR's MSD-NBD motifs were developed (Serohijos et al. 2008; Mornon et al. 2008). These 3D models provided several important insights into CFTR structure and function. First, the organization of the transmembrane segments in the MSDs to form two units, each comprising four segments from one MSD and two from the other (Serohijos et al. 2008; Mornon et al. 2008) (Fig. 16.3). Second, the arrangement of the intracellular loops (ICLs) of each MSD to contact both NBDs (e.g., ICL1 (MSD1) with NBD1 and ICL2 (MSD1) with NBD2). This organization of the ICLs suggests that communication between the NBDs and MSDs is both vertical and orthogonal (e.g., NBD1-MSD1 and NBD1-MSD2) (Serohijos et al. 2008; Mornon et al. 2008) (Fig. 16.3). Third, the coupling helix, the short α-helix, at the extremity of ICL4 interacts with the surface of NBD1 in the region of F508 (Serohijos et al. 2008; Mornon et al. 2008) (Fig. 16.3). This observation argues that F508del affects a critical residue in the CFTR gating pathway (see Sect. 16.4).

After decades of effort by many groups (for a review, see Hunt et al. 2013), Zhang and Chen (2016) solved the first atomic resolution structure of CFTR by cryo-electron microscopy—dephosphorylated zebrafish CFTR in the absence of ATP. Further structures rapidly followed including human CFTR in dephosphorylated, phosphorylated, and drug-bound configurations (Liu et al. 2017, 2019; Zhang et al. 2018b) (Fig. 16.3). Comparison of these structures reveals high degrees of structural similarity despite only 54% amino acid identity between human and zebrafish CFTR (Zhang and Chen 2016; Liu et al. 2017; Zhang et al. 2017, 2018b). By contrast, zebrafish CFTR exhibits notable differences in function from human CFTR, which are not accounted for by codon optimization (Zhang et al. 2018a). Thus, subtle structural differences between CFTR homologues have the potential to alter markedly CFTR function (Bose et al. 2015).

The atomic resolution structures of CFTR have provided important new insights into the relationship between CFTR structure and function, which embolden mechanistic studies of CF mutations and small molecule CFTR modulators. They include the following: first, the N-terminus of CFTR, which forms a previously unrecognized "lasso"-shaped motif that wraps around transmembrane segments 2 (M2), M6, M10, and M11 just below the level of the plasma membrane where it might regulate channel gating (Liu et al. 2017; Zhang et al. 2018b) (Fig. 16.3). Second, the RD "wedges" between the NBDs in the quiescent state, but disengages from them upon phosphorylation to promote dimerization of the NBDs and hence channel opening (Fay et al. 2018; Zhang et al. 2018b) (Fig. 16.3). Third, a locally unstructured region in M8, not found in other ABC transporters, located close to

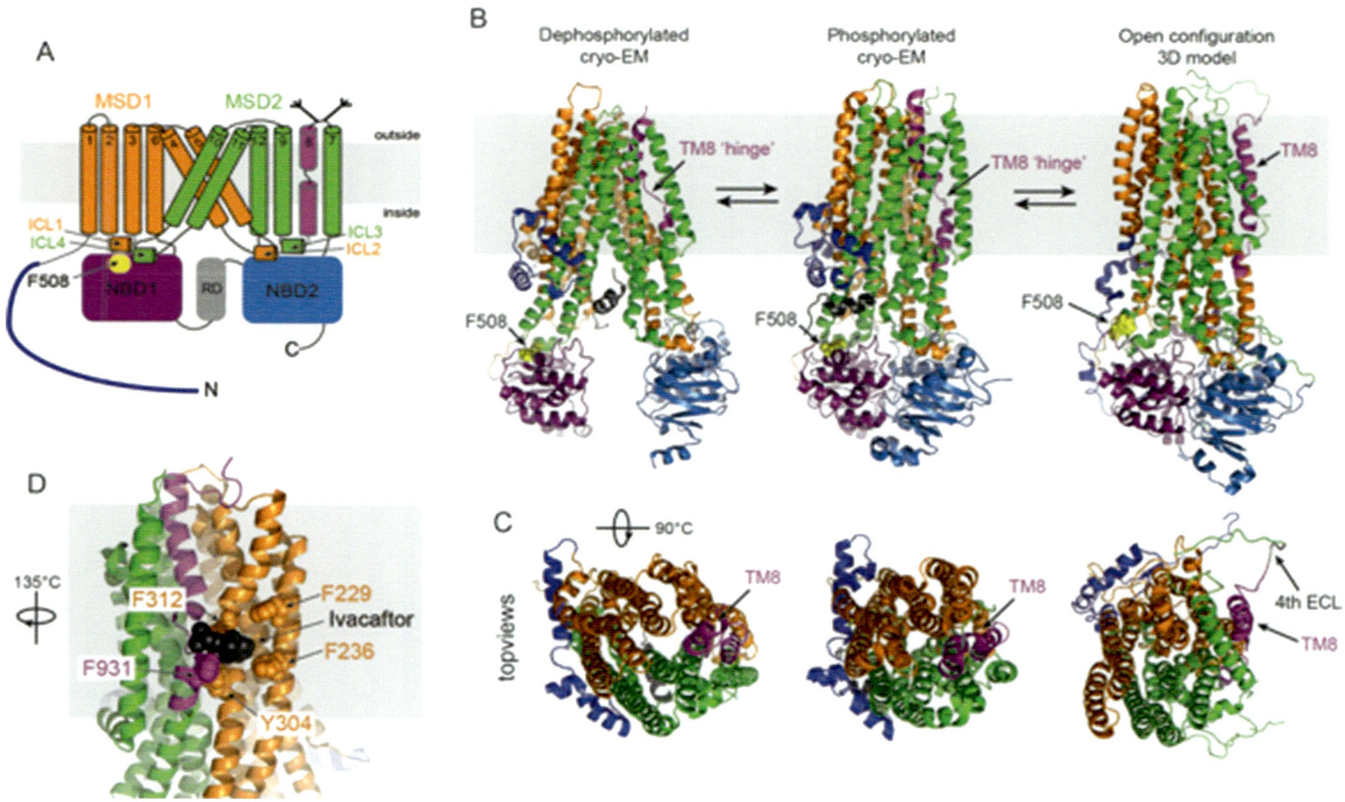

Fig. 16.3 Cryo-EM structures and 3D models of the CFTR Cl^- channel. (**a**) Schematic representation of the domain organization of CFTR. *MSD* membrane-spanning domain, *NBD* nucleotide-binding domain, *RD* regulatory domain. The position of the plasma membrane (gray) and the F508 residue are indicated; numbers identify individual transmembrane segments (M1–M12). Two *N*-glycans are present on the fourth extracellular loop (ECL) between M7 and M8. Short horizontal cylinders represent the coupling helices of the intracellular loops (ICLs). (**b** and **c**) Orthogonal (top) and extracellular (bottom) views of PyMOL

transmembrane segments that gate the channel pore (Liu et al. 2017; Zhang et al. 2018b) (Fig. 16.3). Fourth, a binding site for CFTR potentiators, small molecules that enhance channel gating (see Sect. 16.7.1). Of note, this drug-binding site is located at the interface between the MSDs and the plasma membrane, close to the unstructured region of M8 (Liu et al. 2019; Yeh et al. 2019) (Fig. 16.3). For further information about CFTR structure and function, we refer the reader to recent excellent comprehensive reviews by Hwang et al. (2018) and Csanády et al. (2019). Below, we review briefly the structural contribution of each CFTR domain to the formation of a functional ion channel.

16.3.1 The Membrane-Spanning Domains

In most ABC transporters, the membrane-spanning domains (MSDs) translocate diverse substrates across cell membranes by forming a pathway with two gates that alternatively accesses either side of the membrane (Holland et al. 2003). By contrast, in CFTR the MSDs form an anion-selective pore with a single gate (Hwang et al. 2018). The evolution of CFTR from an active transporter into an ion channel is a consequence of structural changes in its MSDs. The degraded ABC transporter model of CFTR function (Chen and Hwang 2008) argues that CFTR became an ion channel by atrophy of one of the two gates required to control substrate shuttling across cell membranes by the alternating access model. In support of this model, the pore constriction located toward the extracellular side of the plasma membrane is the location of a channel gate (Gao and Hwang 2015; Zhang and Chen 2016; Liu et al. 2017; Zhang et al. 2018b). Interestingly, the location of the intracellular gate in other ABC transporters is the site of a lateral tunnel through which anions access the CFTR pore from the cytoplasm (Zhang et al. 2017) (Fig. 16.4). Thus, CFTR became an ion channel by loss of its intracellular gate.

The architecture of the CFTR pore is asymmetric in two respects. First, the two MSDs do not contribute transmembrane segments equally to the lining of the channel pore. The dominant role of M6 in determining the pore properties of CFTR has been long known (Sheppard and Welsh 1999; McCarty 2000; Linsdell 2006). Building on this result, functional studies using the substituted cysteine accessibility method (SCAM) demonstrated that residues along the lengths of M6 and M12 line the pore, while sections of M1, M3, M4, M5, M9, M10, and M11

Fig. 16.3 (continued) representations of the cryo-EM structures of human CFTR (left, dephosphorylated [PDB id: 5UAK]; center, phosphorylated [PDB id: 6MSM]) and 3D model of the open-channel configuration (Hoffmann et al. 2018) with domains color-coded as for the schematic in (**a**). (**d**) Magnified view of the ivacaftor-bound cryo-EM structure [PDB id: 6O2P] rotated 135° compared to the structures in (**b**) (top) to illustrate amino acid residues that interact with ivacaftor (Liu et al. 2019; Yeh et al. 2019). Figure reproduced with permission of B Kleizen and the *Journal of Cystic Fibrosis*

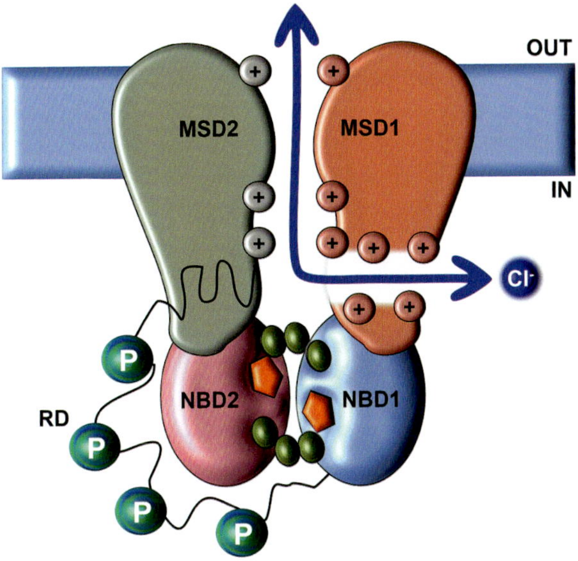

Fig. 16.4 Architecture of the CFTR pore. Schematic representation of the CFTR pore based on structural and functional data. On the intracellular side of the membrane, anions enter the pore through a lateral tunnel within the ICLs beneath MSD1 connected to the intracellular vestibule, whereas they exit the pore on the extracellular side directly from the extracellular vestibule. The pore is lined by positively charged arginine and lysine residues, which stream anions through the channel. ATP is represented as three dark green balls linked to a pentagon

contribute some pore-lining residues (for a review, see Hwang et al. 2018). These data argue that MSD1 is dominant in specifying the pore properties of CFTR. Second, the CFTR pore resembles an asymmetric hourglass with a deep intracellular vestibule and a shallow extracellular vestibule separated by a constriction near the extracellular end of the pore (for a review, see Linsdell 2017). On the outside of the channel, anions readily enter the extracellular vestibule and are funneled toward the pore constriction (Zhang and Chen 2016; Linsdell 2017; Liu et al. 2017; Zhang et al. 2018b) (Fig. 16.4). By contrast, the intracellular vestibule is not directly accessible from the cytoplasm (Zhang and Chen 2016; Linsdell 2017; Liu et al. 2017; Zhang et al. 2018b). Instead, anions access the intracellular vestibule through a lateral tunnel within the ICLs beneath M4 and M6, running parallel to the plane of the lipid bilayer (Li et al. 2018b; Zhang et al. 2018b; Hoffmann et al. 2018) (Fig. 16.4). At the tunnel's mouth and within the intra- and extracellular vestibules positively charged arginine and lysine residues maximize anion flow toward the pore constriction by electrostatic repulsion (Zhang and Chen 2016; Linsdell 2017; Liu et al. 2017; Li et al. 2018b; Zhang et al. 2018b; Hwang et al. 2018) (Fig. 16.4).

The pore constriction, the narrowest part of the channel, determines the selectivity of CFTR. Studies using anions of different sizes as molecular calipers determined that the narrowest part of the CFTR pore is ~5.3 Å based on the size of permeant and impermeant anions (Linsdell and Hanrahan 1998). SCAM data argue that the narrowest part of the pore occurs between L102–I106 (M1) and T338–S341 (M6) (Hwang et al. 2018). Consistent with this idea, mutation of F337 and T338 altered dramatically the anion permeability of CFTR, whereas mutation of other residues had less marked effects (Linsdell et al. 1997, 2000; Gong et al. 2002). Studies of polyatomic anions of known dimensions demonstrated that the anion permeability sequence of CFTR follows a lyotropic sequence (Smith et al. 1999; Linsdell et al. 2000). This suggests that anion permeability is determined by the hydration energy of anions with large, weakly hydrated anions being most permeant. Similarly, anion binding exhibits a lyotropic sequence with large anions binding tighter to the CFTR pore, explaining why these anions avidly block Cl^- permeation (Gong et al. 2002).

A striking feature of the CFTR pore is that anion selectivity is "dynamic," with the pore switching between conformations permeable to Cl^- and large anions (e.g., glutathione and HCO_3^-) in the presence of different factors [e.g., glutamate (Reddy and Quinton 2003) and non-hydrolyzable ATP analogues (Kogan et al. 2003)] or following phosphorylation by the $[Cl^-]_i$-sensitive with-no-lysine kinase 1 (WNK1)-oxidative stress-response kinase 1 (OSR1)/sterile20/SPS1-related proline/alanine-rich kinase (SPAK) pathway (Park et al. 2010). Among the $[Cl^-]_i$-sensitive kinase cascade, WNK1 alone switches CFTR to a HCO_3^--selective channel when $[Cl^-]_i$ falls below 10 mM (Kim et al. 2020). WNK1 physically interacts with the N-terminus of CFTR, leading, by an unknown mechanism, to dilation of the CFTR pore, which reduces the energy barrier for HCO_3^- permeation (Jun et al. 2016; Kim et al. 2020). This switch to a HCO_3^--selective CFTR pore might sustain the high rates of HCO_3^- secretion by pancreatic duct cells necessary to achieve HCO_3^--rich pancreatic juice (Kim et al. 2020).

Using a channel-permeant thiol probe of similar size to Cl^- to modify pore-lining residues, Gao and Hwang (2015) identified residues in M6 that were modified in the presence but not absence of ATP. The authors interpreted their data to suggest that residues between I344 and T338 in M6 are part of an ATP-controlled gate. Building on these data, Zhang and Chen (2016) interpreted the structure of zebrafish CFTR to suggest that residues toward the extracellular end of M1, M6, M8, and M12 form the gate with the discontinuous region of M6 acting as a flexible hinge to open and close the pore. Upon opening, the channel pore is stabilized by salt bridges formed between residues in different transmembrane segments [e.g., R347 (M6)–D924 (M8) (Cotten and Welsh 1999) and R352 (M6)–D993 (M9) (Cui et al. 2008)]. For a discussion of the sequence of conformation changes which gate the pore, see Sect. 16.4.

16.3.2 The Regulatory Domain

The two MSD-NBD motifs of many eukaryotic ABC transporters are joined by disordered linker domains (Ford et al. 2020). In CFTR, this domain forms a unique structure, the regulatory domain (RD), which plays a pivotal role by controlling channel function (Rich et al. 1991; Cheng et al. 1991; Winter and Welsh 1997). The RD has several distinguishing features. First, multiple consensus phosphorylation sequences (Riordan et al. 1989). Second, many charged amino acids (Riordan et al. 1989). Third, no stable secondary and tertiary structures (Ostedgaard et al. 2000; Baker et al. 2007; Liu et al. 2017; Zhang et al. 2018b). With these characteristics, the RD is an intrinsically disordered hub for intra- and intermolecular protein interactions (Bozoky et al. 2013a). Through multisite dynamic interactions with different domains of CFTR, the RD generates graded responses to phosphorylation, which control CFTR activity (Ostedgaard et al. 2000; Bozoky et al. 2013b). Moreover, by engaging diverse binding partners, such as the sulfate transporter and anti-sigma antagonist (STAS) domain of SLC26 transporters and calmodulin (Ko et al. 2004; Bozoky et al. 2013b, 2017), the RD modulates epithelial ion transport.

The phosphorylation status of the RD and hence CFTR activity are tightly controlled by the balance of protein kinase and phosphatase activity within cells (Gadsby and Nairn 1999). Consistent with the dominant role of cAMP agonists in stimulating CFTR-mediated transepithelial ion transport (Anderson et al. 1992), protein kinase A (PKA) is the most important kinase responsible for RD phosphorylation and channel activation (Berger et al. 1993). However, other kinases including protein kinase C (PKC), the type II cGMP-dependent protein kinase, and tyrosine kinases phosphorylate CFTR and stimulate channel activity (Tabcharani et al. 1991; French et al. 1995; Billet et al. 2015). For example, it is now recognized that CFTR is a major target for Ca^{2+} agonists in epithelial tissues, which mediate their action, in part, through the tyrosine kinases proto-oncogene tyrosine protein kinase (p60c-Src) and proline-rich tyrosine kinase 2 (Pyk2) [for a review, see Billet and Hanrahan (2013)]. Of note, CFTR phosphorylation by PKC accelerates channel activation by PKA and augments the magnitude of CFTR activity achieved (Tabcharani et al. 1991), suggesting hierarchical phosphorylation of the RD by different protein kinases. Dephosphorylation of the RD by protein phosphatases is cell-type specific (Gadsby and Nairn 1999). In airway and intestinal epithelia, CFTR is dephosphorylated by protein phosphatase 2C (PP2C), whereas in the sweat duct, PP1 and PP2A mediate CFTR deactivation (Reddy and Quinton 1996; Travis et al. 1997).

Amino acid sequence analysis identified numerous consensus phosphorylation sites for PKA on the RD (Riordan et al. 1989). Herculean efforts demonstrated that the elimination of 15 PKA phosphorylation sites (located at S422, S660, S670, S686, T690, S700, S712, S737, S753, S768, T787, T788, S790, S795, and S813) was required to render CFTR Cl^- channels unresponsive to PKA-dependent phosphorylation (Hegedus et al. 2009). Among these 15 PKA phosphorylation sites, for two reasons, those at S660, S700, S737, S795, and S813 were identified as the critical sites for CFTR regulation by PKA. First, cAMP agonists phosphorylated these sites in vivo (Cheng et al. 1991). Second, the largest decrement in CFTR activity occurred with the simultaneous mutation of S660, S737, S795, and S813 (Chang et al. 1993; Rich et al. 1993). Intriguingly, site-directed mutation of the consensus phosphorylation sites at S737 and S768 increased CFTR Cl^- currents, suggesting that phosphorylation of these sites inhibits CFTR activity (Wilkinson et al. 1997; Baldursson et al. 2000). However, using the PKA-unresponsive CFTR construct, Hegedus et al. (2009) demonstrated that phosphorylation of S737 and S768 stimulated channel activity. Thus, the data argue that PKA stimulation of CFTR is redundant with no individual phosphoserine residue being essential. They also suggest that there are context-dependent functional interactions between different phosphoserines to precisely regulate CFTR activity.

The RD was originally proposed to regulate CFTR by keeping the channel closed at rest (Rich et al. 1991). Consistent with this idea, structural studies revealed that the unphosphorylated RD is located between the two MSD–NBD motifs, contacting NBD1, MSD2, NBD2, and the C-terminus (Chong et al. 2013; Zhang and Chen 2016; Liu et al. 2017). Channel inhibition is relieved by phosphorylation, addition of negative charge to the RD, or RD deletion (Rich et al. 1991, 1993; Chang et al. 1993). Upon phosphorylation, the RD disengages from between the NBDs and binds the lasso motif, allowing ATP to initiate NBD dimerization (Mense et al. 2006; Fay et al. 2018; Zhang et al. 2018b) (Fig. 16.3). Of note, the activity of phosphorylated CFTR greatly exceeded constructs, which did not require phosphorylation to open in the presence of ATP (Chang et al. 1993; Rich et al. 1993). These data suggest that the phosphorylated RD might stimulate CFTR activity. Consistent with this idea, phosphorylation augments ATP binding and hydrolysis at the NBDs to accelerate the rate of channel opening (Li et al. 1996; Winter and Welsh 1997; Liu et al. 2017). Thus, the RD does not function as an "on-off" switch, but as a dynamic hub mediating multisite interactions with the MSDs and NBDs to control ATP-dependent channel gating (Ostedgaard et al. 2000; Bozoky et al. 2013b).

16.3.3 The Nucleotide-Binding Domains

The nucleotide-binding domains (NBDs) of ABC transporters are globular cytoplasmic proteins composed of three subdomains: the F1-type ATP-binding core, the ABC α, and the ABC β, which are distinguished by conserved amino acid sequences (Holland et al. 2003; Lewis et al. 2004). In most ABC transporters, the NBDs form a common engine that utilizes the energy of ATP binding and hydrolysis to pump diverse substrates through a translocation pathway formed by the MSDs (Holland et al. 2003). By contrast, in CFTR ATP binding and hydrolysis at the NBDs gates a transmembrane pore formed by the MSDs, acting as a timing mechanism to control anion flux (Hwang and Sheppard 2009; Hunt et al. 2013).

The NBDs function as a head-to-tail dimer with the F1-type ATP-binding core subdomain of one NBD interacting with the ABC α subdomain of the opposing NBD (Lewis et al. 2004; Liu et al. 2017; Zhang et al. 2018b) (Fig. 16.3). The interaction of these subdomains forms two ATP-binding sites at the NBD1:NBD2 dimer interface. ATP-binding site 1 is formed by the Walker A and B motifs of NBD1 and the ABC signature or LSGGQ motif of NBD2, while ATP-binding site 2 is formed by the Walker A and B motifs of NBD2 and the LSGGQ motif of NBD1 (Lewis et al. 2004; Liu et al. 2017; Zhang et al. 2018b). Functional studies demonstrate that the association of NBD1 and NBD2 is required for optimal ATPase activity and channel gating (Vergani et al. 2003, 2005; Kidd et al. 2004). To transduce conformational changes in the NBDs to gating of the channel pore, the NBDs contain surface hydrophobic grooves where the coupling helices of the intracellular loops dock precisely (Serohijos et al. 2008; Mornon et al. 2008; Liu et al. 2017; Zhang et al. 2018b). Thus, formation of the NBD1:NBD2 dimer acts as a mechanism to convert the energy of ATP binding to conformational changes in the MSDs that open the channel gate and allow passive Cl^- flow through the CFTR pore down a favorable electrochemical gradient (Vergani et al. 2005).

Many studies have revealed asymmetries in the structure and function of CFTR's NBDs (for a review, see Sheppard and Welsh 1999), which reflect differences in the function of the two ATP-binding sites. Biochemical studies demonstrated that NBD1 (i.e., site 1) stably binds nucleotides, whereas NBD2 (i.e., site 2) rapidly hydrolyzes them (Aleksandrov et al. 2002; Basso et al. 2003). Amino acid sequence analysis identified two critical changes in site 1 which render it catalytically inactive. First, the Walker B motif of NBD1 lacks a catalytic base at position S573 (Lewis et al. 2004). Elimination of negative charge at E1371, the equivalent residue in site 2, disables channel closure (Vergani et al. 2005). Second, the ABC signature motif in site 2 has a notably different amino acid sequence, LSHGH (Lewis et al. 2004). By contrast, site 2 is a canonical ATP-binding site and is catalytically active with a turnover rate of ~1 s^{-1} (Li et al. 1996; Liu et al. 2017).

By analyzing the distribution of CFTR open-channel durations, Csanády et al. (2010) concluded that there is strict coupling between ATP binding and hydrolysis at the NBDs and gating of the channel pore. However, Jih et al. (2012) interpreted the gating behavior of a pore-lining CFTR mutation to propose that more than one ATP molecule might be hydrolyzed during an open-channel burst. Based on this observation, the authors developed the energetic coupling model of CFTR channel gating (for a review, see Jih and Hwang 2012). This gating model predicts that conformational changes in the NBDs are not strictly coupled to gating of the CFTR pore. It has been used with success to explain the mechanism of action of the CFTR potentiator ivacaftor (Jih and Hwang 2013) (see Sect. 16.7.1).

The structural and enzymatic differences between the two ATP-binding sites have implications for the gating behavior of CFTR. These differences will be discussed further below (Sect. 16.4). Overall, the NBDs act as an engine that utilizes the energy derived from ATP binding and hydrolysis to power dynamic structural rearrangements in the MSDs which gate the channel pore.

16.3.4 The Amino and Carboxyl Termini

Biochemical and functional studies of CFTR constructs bearing site-directed mutations and deletions in the N-terminus of CFTR identified a cluster of acidic residues (D47, E51, E54, and D58) that control channel regulation by PKA-dependent phosphorylation (Naren et al. 1999) and the stability of the open-channel configuration, which is regulated by the NBDs (Fu et al. 2001). Although Naren et al. (1999) demonstrated that the N-terminus and RD physically interact, the structural basis of this interaction remained obscure until atomic resolution structures of CFTR were solved (Zhang and Chen 2016; Liu et al. 2017; Zhang et al. 2018b). Zhang and Chen (2016) demonstrated that the N-terminus of CFTR contains a lasso motif, so-called because it forms a circular noose resembling a cowboy lasso. This motif wraps around M2, M6, M10, and M11 just below the level of the plasma membrane (Liu et al. 2017; Zhang et al. 2018b) (Fig. 16.3). In phosphorylated ATP-bound CFTR, the lasso motif forms molecular interactions with the RD, but not when the RD is dephosphorylated (Liu et al. 2017; Zhang et al. 2018b). Thus, the RD mediates interactions between the N-terminus and NBDs to regulate channel function.

Like the RD (see Sect. 16.3.2), the N-terminus is a site of intermolecular protein interactions. Naren et al. (1998) demonstrated that syntaxin 1A, a regulator of membrane trafficking, inhibits CFTR Cl^- currents by interacting with the N-terminus, including sequences now recognized to form the lasso motif (Zhang and Chen 2016). Other syntaxins, including syntaxins 6 and 16, interact with the N-terminus of CFTR to regulate its journey to the plasma membrane and modulate its surface expression through recycling (Gee et al. 2010; Tang et al. 2011). Of note, the extreme N-terminus of CFTR interacts with filamin proteins of the cytoskeleton, which bind actin to regulate the plasma membrane stability of CFTR (Thelin et al. 2007) (Fig. 16.5).

At the extreme C-terminus of CFTR, there is a PDZ-binding domain that mediates the interaction of CFTR with PDZ-domain proteins, such as the Na^+/H^+ exchanger regulatory factor isoform-1 (NHERF1) (Guggino and Stanton 2006) (Fig. 16.5). Short et al. (1998) demonstrated that NHERF1 is localized at the apical membrane of human airway epithelia and interacts with CFTR. Based on these and other data, Short et al. (1998) proposed a model of the CFTR macromolecular signaling complex. In this model, NHERF1 binds the C-terminus of CFTR through a PDZ-binding domain at its N-terminus (termed PDZ1) and interacts with ezrin via an ezrin/radixin/moesin (ERM)-binding domain at its C-terminus (Fig. 16.5). Because ezrin is an actin-binding protein, protein–protein interactions facilitated by NHERF1 tether CFTR to the actin cytoskeleton at the apical membrane of polarized epithelial cells (Short et al. 1998) (Fig. 16.5). The interaction of CFTR with NHERF1 has several important consequences for CFTR regulation. First, the RD of CFTR is positioned close to PKA because ezrin acts as an anchoring protein for PKA (Short et al. 1998) (Fig. 16.5). Second, it localizes CFTR to sub-apical pools of cAMP, which are confined by the cytoskeleton (Monterisi et al. 2012) (Fig. 16.5). Third, NHERF1 also binds the β_2-adrenergic receptor through a second PDZ-binding domain (termed PDZ2) (Naren et al. 2003) (Fig. 16.5). Thus, activation of the β_2-adrenergic receptor causes G_s to stimulate the adenylate cyclase leading to a local rise in cAMP and hence the phosphorylation of CFTR by PKA (Naren et al. 2003).

The C-terminus also plays a pivotal role in coupling CFTR activity to cell metabolism, vital for the homeostasis of epithelial tissues. The metabolic-sensing

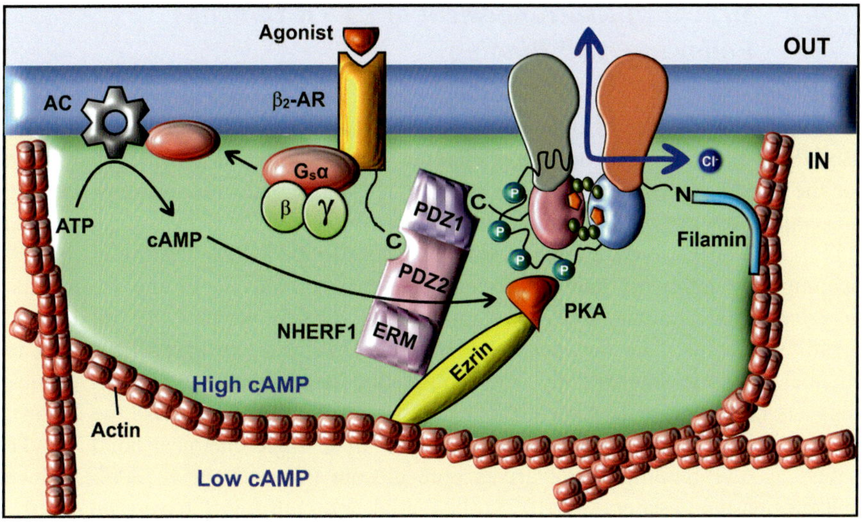

Fig. 16.5 Regulation of CFTR activity by the β_2-adrenergic receptor in airway epithelia. The schematic shows the regulation of CFTR by a macromolecular signaling complex and the cAMP signaling pathway. *AC* adenylate cyclase, *β_2-AR* b2-adrenergic receptor, *ERM* ezrin/radixin/moesin, *G_s* stimulatory GTP-binding protein, *NHERF1* Na^+/H^+ exchanger regulatory factor isoform-1, *PDZ* postsynaptic density protein (PSD95), *Drosophila* disc large tumor suppressor (DlgA), and zona occludens 1 (zo-1) protein; PKA, protein kinase A. Modified with permission from Guggino and Stanton 2006 and Monterisi et al. (2013)

kinase AMP-activated protein kinase (AMKP) conserves cellular energy by responding to changes in the [AMP]/[ATP] ratio during cell stress to increase catabolism and decrease anabolism (Herzig and Shaw 2018). Using a yeast two-hybrid screen to identify proteins that interact with the C-terminus of CFTR, Hallows et al. (2000) identified the α_1 (catalytic) subunit of AMPK as a CFTR-interacting protein, colocalized at the apical membrane of epithelia. Phosphorylation of CFTR by AMPK inhibits CFTR activation by PKA (King et al. 2009).

Thus, the N- and C-termini mediate intra- and intermolecular protein interactions, critical for regulating CFTR expression and function. They anchor CFTR to the cytoskeleton, assemble it into a macromolecular signaling complex, which tightly controls channel activity, and balance transmembrane anion flow with cell metabolism.

16.4 The Gating Pathway of CFTR

The CFTR gating pathway is the sequence of conformational changes initiated by ATP binding at the NBDs, which lead to opening of the CFTR pore and anion flow through the channel. In this section, we review the structural rearrangements that underlie CFTR gating and the accompanying steps that form the CFTR gating cycle.

16.4.1 Structural Rearrangement of CFTR Domains Following ATP Binding

Prior to the emergence of high-resolution CFTR structures, Csanády et al. (2006) used the temperature dependence of channel gating to propose an energetic profile for the gating cycle of CFTR. The authors' analysis argued that channel opening proceeds as a wave of conformational change, which is initiated by formation of an ATP-bound NBD1:NBD2 dimer and then sweeps to the MSDs, leading to channel opening. By applying rate-equilibrium free-energy analysis to the hydrolysis-deficient CFTR mutant D1370N, Sorum et al. (2015) verified this model and identified the interface between the NBDs and MSDs as an energetic barrier to channel opening. Interestingly, the temperature dependence of channel closure suggests that large-scale protein movements are not required to close the channel (Csanády et al. 2006). Instead, the hydrolysis of a single chemical bond in ATP bound at ATP-binding site 2 drives pore closure (Csanády et al. 2006), which precedes partial (Tsai et al. 2010; Szollosi et al. 2011) or full (Chaves and Gadsby 2015) disassembly of the NBD1:NBD2 dimer interface.

Comparison of the dephosphorylated ATP-free and phosphorylated ATP-bound structures of human and zebrafish CFTR illuminates into how ATP binding by the NBDs opens the channel pore (Zhang and Chen 2016; Liu et al. 2017; Zhang et al. 2017, 2018b). In the dephosphorylated ATP-free structures, the two MSD-NBD motifs form an inverted V shape with the MSDs locked together near the extracellular side of the membrane and the NBDs held apart by the RD (Zhang and Chen 2016; Liu et al. 2017) (Fig. 16.6). Following its phosphorylation, the RD is displaced and ATP drives NBD dimerization, leading to conformational changes in the MSDs, which result in the two MSD-NBD motifs aligning vertically to each other (Vergani et al. 2005; Mense et al. 2006; Zhang et al. 2017, 2018b) (Fig. 16.6). These movements of MSD–NBD motifs represent two-part rigid body dynamics: NBD1 moves with the lasso motif and M1-M3, M6, and M10-M11, whereas NBD2 moves with M4-M5, M7-M9, and M12 (Zhang et al. 2017). Building on these data, molecular dynamics simulations using 3D models suggest that M8 swings out to open the CFTR pore with its motion coordinated with the movement of M6, M7, and M12 (Hoffmann et al. 2018; Kleizen et al. 2020) (Fig. 16.3).

The reorientation of the MSDs during the transition from dephosphorylated ATP-free to phosphorylated ATP-bound structures packs transmembrane segments together to shield the CFTR pore from the lipid bilayer (Zhang et al. 2017, 2018b). Consistent with functional data (Bai et al. 2011), these movements cause the intracellular vestibule to narrow (Zhang et al. 2017). However, in both phosphorylated ATP-bound structures, the CFTR pore is too narrow in the region of the pore constriction for Cl^- permeation (Zhang et al. 2017, 2018b). Pore occlusion in these CFTR structures might represent a transient state where the NBDs are dimerized, but the channel pore is occluded. Potentially, this transient state might be equivalent to the brief channel closures that occur during busts of channel opening (e.g., Cai et al.

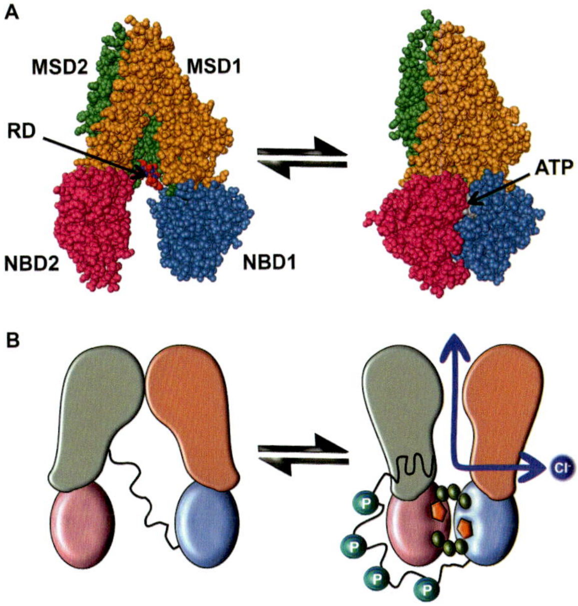

Fig. 16.6 Reorganization of CFTR domains upon phosphorylation and ATP binding. (**a**) The structure of human CFTR in the absence (left) and presence (right) of RD phosphorylation and ATP binding at the NBDs (PDB ids: 5UAK and 6MSM). (**b**) Schematic representation of the two CFTR structures presented in A showing the organization of the different domains, the ATP-binding sites, and the gating movements driven by ATP binding. For further information, see the text and Fig. 16.4

2003). Thus, a different structure might represent the open-channel configuration observed in electrophysiological studies (Kleizen et al. 2020).

16.4.2 The CFTR Gating Cycle

CFTR is a ligand-gated ion channel with the architecture of an ABC transporter (Hwang et al. 2018). Cycles of ATP binding and hydrolysis at its NBDs gate a transmembrane pore to control anion flow (Vergani et al. 2003, 2005). Thus, unique to CFTR, its ligand is consumed to render the gating cycle irreversible. Figure 16.7 shows a simplified representation of the CFTR gating cycle based on biochemical, functional, and structural data. The cycle has several characteristics, including (1) ATP binding at two sites is required for channel opening (Anderson and Welsh 1992; Lewis et al. 2004); (2) the CFTR pore can close even when both ATP-binding sites are occupied (Sorum et al. 2015; Zhang et al. 2017); (3) only one ATP-binding site hydrolyzes ATP; the other tightly retains ATP for multiple gating cycles (Aleksandrov et al. 2002; Basso et al. 2003; Vergani et al. 2005); (4) ATP hydrolysis initiates channel closure (Gunderson and Kopito 1994; Carson et al. 1995; Vergani et al. 2003); and (5) there is a short-lived post-hydrolytic open state (Gunderson and Kopito 1995; Jih et al. 2012). For comprehensive descriptions of the CFTR gating cycle, see Jih and Hwang (2012), Hwang et al. (2018), and Csanády et al. (2019).

Many studies support the basic organization of the CFTR gating cycle described in Fig. 16.7. ATP-binding site 1 binds ATP with higher affinity than ATP-binding

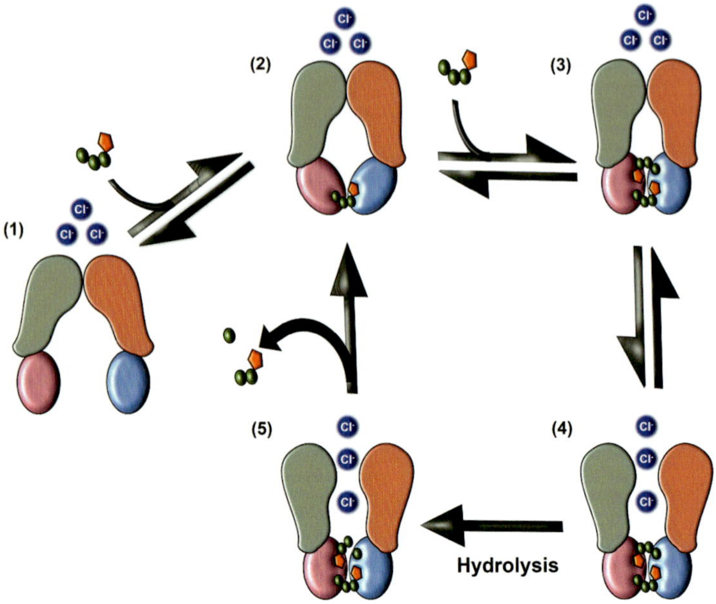

Fig. 16.7 Simplified model of the CFTR gating cycle showing major structural changes. The gating cycle contains three reversible and two irreversible steps: (1) Phosphorylated CFTR is closed in the absence of ATP. ATP binds to ATP-binding site 1 to form a partial NBD dimer. (2) ATP then binds to ATP-binding site 2 to form the full NBD1:NBD2 dimer stabilized by two molecules of ATP. (3) The NBD1:NBD2 dimer signals to the channel pore by a conformational wave that sweeps from the NBDs through the ICLs to the gate located at the pore constriction. Gate opening allows Cl^- flow through CFTR. (4) Hydrolysis of ATP at ATP-binding site 2 leads irreversibly to a transient post-hydrolytic state where the products of ATP hydrolysis remain bound to ATP-binding site 2 and the pore open. (5) Dissociation of the products of ATP hydrolysis from ATP-binding site 2 leads irreversibly to the partial NBD dimer conformation where ATP occupies ATP-binding site 1 and the pore is closed (step 2). From step 2, a new gating cycle is initiated by ATP binding at site 2. For further information, see the text and Fig. 16.4

site 2 (Aleksandrov et al. 2002; Basso et al. 2003). Moreover, mutations that prevent ATP binding and/or NBD dimerization disrupt channel gating. For example, the CF mutations G551D and G1349D affect equivalent key residues in the LSGGQ motifs of ATP-binding sites 1 and 2, respectively, which form critical contacts with the γ-phosphate of ATP (Hwang and Sheppard 2009). These mutations disrupt severely channel gating by decreasing greatly the frequency of channel openings (Li et al. 1996; Cai et al. 2006; Bompadre et al. 2007). Of note, G551D all but eliminates the ATP dependence of channel gating, converting ATP-binding site 2 into a site where ATP inhibits channel gating, while G1349D retains only limited ATP sensitivity (Bompadre et al. 2007; Lin et al. 2014). Because mutation of equivalent residues in the related ABC transporter MRP1 had little effect on ATP binding (Ren et al. 2004), Bompadre et al. (2007) interpreted their data to suggest that G551D and G1349D disrupt CFTR function by hindering NBD dimerization.

The importance of ATP hydrolysis for the CFTR gating cycle is evident from many studies. CFTR is not efficiently gated by non-hydrolyzable ATP analogues (e.g., AMP–PNP and ATPγS) applied in the absence of ATP (Anderson et al. 1991a; Gunderson and Kopito 1994; Aleksandrov et al. 2000; Vergani et al. 2003). Similarly, mutations that disrupt ATP hydrolysis (e.g., K1250A and E1371S) alter dramatically the gating behavior of CFTR by prolonging the duration of channel openings (Carson et al. 1995; Vergani et al. 2003). The prolonged openings of hydrolysis-deficient mutants were interpreted to suggest that ATP hydrolysis might act as a mechanism to renew the gating cycle by emptying the ATP-binding sites (Vergani et al. 2003). But, in addition, ATP hydrolysis transforms a reversible gating cycle into an irreversible one. In this way, enzymatic activity controls ATP-driven channel activity.

The first evidence for a post-hydrolytic open state was the appearance of two discrete current levels termed O_1 and O_2 within individual bursts of channel openings (Gunderson and Kopito 1995). Several lines of evidence suggested that O_1 represents a pre-hydrolytic open state and O_2 a post-hydrolytic open state: (1) the transition to O_2 was essentially irreversible; (2) non-hydrolyzable ATP analogues occasionally locked channel openings in O_1; and (3) mutations that inhibited ATP hydrolysis greatly prolonged the duration of the O_1 state (Gunderson and Kopito 1995). Gunderson and Kopito (1995) interpreted these results to suggest that the transition to the O_2 state required ATP hydrolysis. Ishihara and Welsh (1997) then demonstrated that the observed difference in current amplitude of the O_1 and O_2 states represented conformational changes in the channel pore, which altered binding of the biological buffer 3-(*N*-morpholino)propanesulfonic acid (MOPS) within the intracellular vestibule. Thus, the conformation of the channel pore was modified by ATP hydrolysis at the NBDs. Interestingly, similar hydrolysis-dependent gating transitions were captured using the CFTR mutation R352C, which affects a pore-lining residue in M6 (Jih et al. 2012). Through kinetic analyses of this CFTR mutation, Jih et al. (2012) developed the energetic coupling model of CFTR channel gating. This model permits the consumption of more than one molecule of ATP per gating cycle in contrast to the model developed by Csanády et al. (2010).

The fact that CFTR consumes its ligand during the gating cycle makes it unique among ion channels. Thus, based on studies that support coupling between CFTR enzymatic activity and the gating cycle, CFTR gating involves at least one irreversible step and would operate away from thermodynamic equilibrium (Gunderson and Kopito 1995; Zeltwanger et al. 1999; Csanády et al. 2010). However, some studies suggest that CFTR gating operates through an allosteric mechanism, which is thermodynamically reversible (e.g., Aleksandrov and Riordan 1998; Aleksandrov et al. 2000, 2009). To reconcile these different models of channel gating, Kirk and Wang (2011) proposed a unified gating scheme which integrates the allosterism of ligand-gated ion channels with the enzymatic activity of CFTR, an ion channel evolved from an active transporter. In this unified gating scheme, both reversible and irreversible gating cycles govern CFTR function.

In summary, the gating cycle of CFTR is driven by the free energy of ATP binding and hydrolysis. ATP binding is a rate-limiting step for channel opening, while ATP hydrolysis and the release of hydrolytic products are rate-limiting steps for channel closure. The two ATP-binding sites cooperate to form an NBD dimer, which is essential for CFTR opening. ATP hydrolysis induces structural changes that are sensed within the CFTR pore, leading to gate closure.

16.5 CFTR Biogenesis and Plasma Membrane Expression

The biogenesis of CFTR begins with the production of CFTR mRNA in the nucleus and its delivery to the ribosome for translation. The ribosome takes about 9 min to complete the synthesis of the CFTR peptide during which time co-translational folding inserts transmembrane segments of the MSDs into the lipid bilayer of the endoplasmic reticulum (ER) and initiates assembly of the NBDs and RD (Ward and Kopito 1994; Kim and Skach 2012; Farinha and Canato 2017). Final assembly and packing of the different domains into the tertiary structure of CFTR takes 30–90 min and is completed prior to exit from the ER (Ward and Kopito 1994; Kim and Skach 2012; Farinha and Canato 2017). The remainder of CFTR's journey to the plasma membrane takes 2–3 h and includes its maturation in the Golgi apparatus (Ward and Kopito 1994; Kim and Skach 2012; Farinha and Canato 2017). Once delivered to the plasma membrane, CFTR remains active for ~48 h as judged by its functional half-life (Lukacs et al. 1993). It is then endocytosed and either recycled to the plasma membrane or degraded by the lysosome (Okiyoneda et al. 2010; Farinha and Canato 2017). The plasma membrane expression of CFTR is therefore determined by membrane delivery, stability, and retrieval, which are tightly controlled by a network of CFTR-interacting proteins (Balch et al. 2008; Sondo et al. 2017) (see Sect. 16.8).

During maturation in the Golgi apparatus, complex sugars are added to two *N*-glycosylation consensus sequences located at N894 and N900 in ECL4 (Gregory et al. 1991; Ward and Kopito 1994). Glycosylation does not alter the intracellular transport nor channel function of CFTR (Gregory et al. 1991; Chang et al. 2008). However, it is a valuable tool to evaluate the processing and plasma membrane expression of CFTR. Figure 16.8 shows an immunoblot of CFTR heterologously expressed in the CF bronchial epithelial cell line CFBE41o$^-$. Band B (~140 kDa) represents the immature, core-glycosylated form of the protein with endoglycosidase H-sensitive simple sugars, which is found in the ER (Cheng et al. 1990) (Fig. 16.8). By contrast, band C (~160 kDa) represents the mature, fully glycosylated form of the protein decorated with *N*-glycanase-sensitive complex sugars, which are added in the Golgi apparatus (Cheng et al. 1990) (Fig. 16.8). As exemplified by F508del, many CF mutations disrupt CFTR processing, leading to the absence of band C (Cheng et al. 1990; Veit et al. 2016) (Fig. 16.8). Monitoring the glycosylation status of CFTR was pivotal in the identification of small molecules that rescue the processing and plasma membrane expression of mutant CFTR (Van Goor et al. 2011) (Fig. 16.8) (see Sect. 16.7.2).

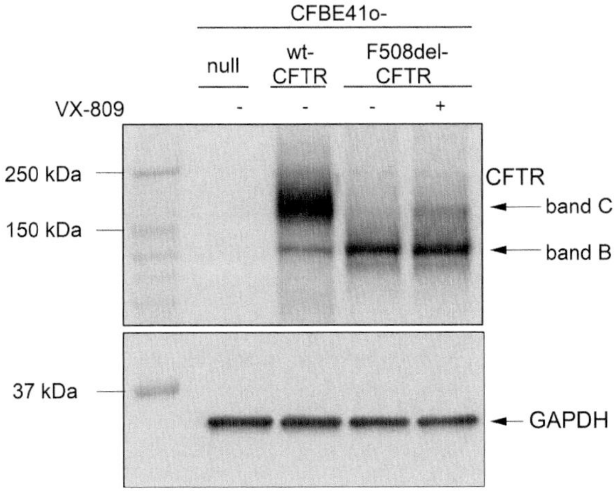

Fig. 16.8 Lumacaftor improves the processing of CFTR protein. Immunoblots of CFTR protein from CFBE41o$^-$ cells stably expressing wild-type and F508del-CFTR are shown. F508del-CFTR-expressing CFBE41o$^-$ cells were treated with lumacaftor (VX-809; 3 μM) for 24 h at 37 °C before cells were lysed and immunoblots performed. Lysates from untransfected (null) and wild-type CFTR-expressing CFBE41o$^-$ cells were used as controls. CFTR was detected with the mouse anti-CFTR monoclonal antibody (596), which recognizes a region of NBD2 (1204–1211; Cui et al. 2007). Arrows indicate the positions of the band B (immature) and band C (mature) forms of CFTR. Glyceraldehyde 3-phosphate dehydrogenase (GAPDH) was used as a loading control

16.6 CFTR Mutations

CFTR is a highly polymorphic human gene with more than 2000 mutations (http://www.genet.sickkids.on.ca/app). Most mutations are extremely rare and not all cause disease (Sosnay et al. 2013; Cutting 2015). The mutation p.Phe508del (more commonly referred to as F508del or ΔF508) is by far the most common cause of CF with ~70% of CF patients worldwide homozygous for the mutation and ~90% heterozygous (Bobadilla et al. 2002; Bell et al. 2015). Some other CF mutations occur at high frequencies in certain populations as a result of founder effects, but none approach the global dominance of the F508del mutation (Bobadilla et al. 2002; Bell et al. 2015). To explain its high frequency, it has been speculated that one copy of F508del confers a survival advantage, possibly against diarrheal disease (Gabriel et al. 1994; Pier et al. 1998) or lactose intolerance (Modiano et al. 2007).

Comparatively few of the 2000 mutations in the *CFTR* gene have been unequivocally linked to CF (Sosnay et al. 2013). The clinical impact of the remaining mutations in the *CFTR* gene is unclear. These mutations might cause CF (classical or non-classical/atypical), a CFTR-related disorder, or have no known clinical consequence (Knowles and Durie 2002; De Boeck et al. 2006; Bombieri et al. 2011). However, the relationship between CF mutations and the clinical presentation of CF is complex and demonstrates tissue-specific variation in the contribution of CF

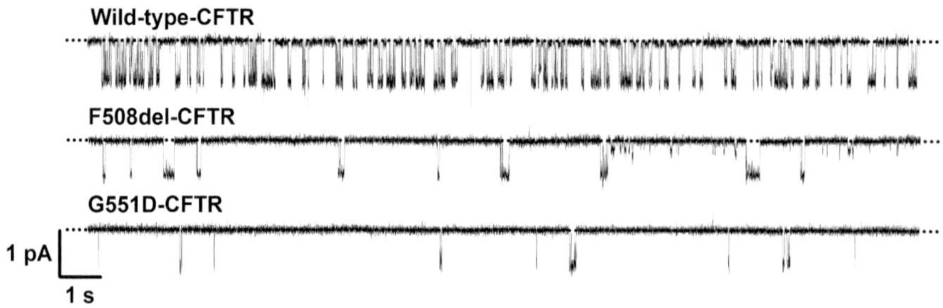

Fig. 16.9 The CF mutations F508del and G551D disrupt channel gating. Representative recordings of wild-type, low-temperature-rescued F508del- and G551D-CFTR Cl^- channels in membrane patches excised from baby hamster kidney (wild-type and F508del) and Fischer rat thyroid (G551D) cells stably expressing CFTR constructs. Membrane patches were voltage-clamped at −50 mV, a large Cl^- concentration gradient ($[Cl^-]_i$, 147 mM; $[Cl^-]_e$, 10 mM) was imposed across the membrane, and ATP (1 mM) and PKA (75 nM) were continuously present in the intracellular solution; temperature was 37 °C. Dotted lines indicate the closed channel state and downward deflections correspond to channel openings. For wild-type and F508del-CFTR, the membrane patches contained one active channel, but for G551D-CFTR, the number of active channels is unknown. Reproduced from Meng et al. (2017) by permission of the American Society from Biochemistry and Molecular Biology

mutations, modifier genes, and the environment (Cutting 2015). For information about the clinical consequences of specific CF mutations, see the Clinical and Functional Translation of CFTR (CFTR2) website (https://cftr2.org/).

Biochemical and function studies demonstrate that CF mutations affect the expression and stability of CFTR protein at the plasma membrane and its function as an epithelial anion channel (for a review, see Wang et al. 2014 and Chap. 15). Six general mechanisms of CFTR dysfunction have been identified: defective protein production (class I), defective protein processing (class II), defective channel regulation (class III); defective channel conduction (class IV), reduced protein synthesis (class V), and reduced protein stability (class VI) (Welsh and Smith 1993; Zielenski and Tsui 1995; Haardt et al. 1999). To identify CF mutations unresponsive to mutation-specific therapies (see Sect. 16.7), De Boeck and Amaral (2016) introduced class VII (no mRNA).

Some CF mutations are associated with a single mechanism of CFTR dysfunction [e.g., the class III mutation p.Gly551Asp (G551D), which severely disrupts CFTR channel gating (Li et al. 1996; Cai et al. 2006; Bompadre et al. 2007)] (Fig. 16.9). However, many CF mutations, quite likely the majority, cause CFTR dysfunction by multiple mechanisms (Wang et al. 2014). For example, F508del is a temperature-sensitive folding defect, which disrupts CFTR processing and intracellular transport to the plasma membrane (Cheng et al. 1990; Denning et al. 1992) (Fig. 16.8). But, in addition, the mutation destabilizes any CFTR protein that reaches the plasma membrane and interferes with channel gating, reducing greatly the frequency of channel opening (Dalemans et al. 1991; Lukacs et al. 1993) (Figs. 16.9 and 16.10). To accommodate the complex effects of CF mutations on CFTR, Veit et al. (2016) proposed a revised combinatorial classification consisting of 31 possible classes of

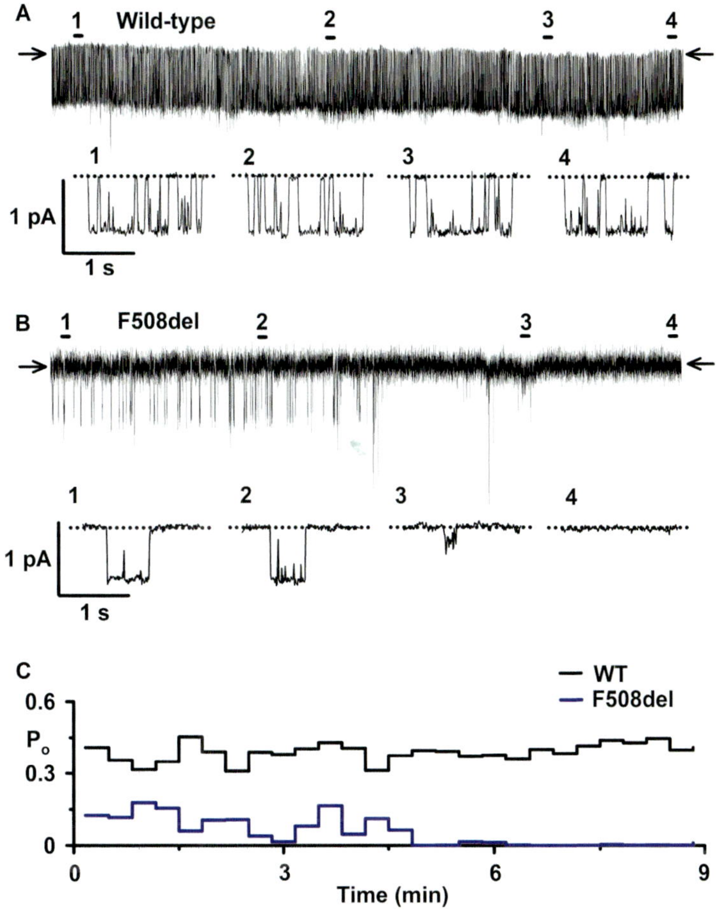

Fig. 16.10 F508del-CFTR is unstable in cell-free membrane patches at 37 °C. (**a** and **b**) Representative 9-min single-channel recordings of wild-type and low-temperature-rescued F508del-CFTR in membrane patches excised from baby hamster kidney cells stably expressing CFTR constructs following full channel activation. ATP (1 mM) and PKA (75 nM) were continuously present in the intracellular solution; temperature was 37 °C. Four 2-s single-channel recordings labeled 1–4 indicated by bars are displayed on an expanded timescale below the 9-min recordings to illustrate the stability of wild-type CFTR and the instability of F508del-CFTR. Arrows and dotted lines indicate the closed channel state and downward deflections correspond to channel openings. (**c**) The open probability (P_o) time courses for the recordings shown in (**a**) and (**b**). P_o values were calculated in 20-s intervals. Reproduced from Meng et al. (2017) by permission of the American Society from Biochemistry and Molecular Biology

mutations, including the original six classes, but with classes III and IV combined into a single class. In this revised classification, F508del is a class II–III–VI mutation and G551D is a class III/IV mutation (Veit et al. 2016). Importantly, this revised classification provides a framework to classify CF mutations according to the molecular-based therapies required to restore CFTR function (termed theratypes) (Cutting 2015).

16.7 Mutation-Specific Therapies for CF

Up until 2012, therapies for CF were directed exclusively against disease symptoms (for a review, see Ratjen et al. 2015; Elborn 2016). However, since 2012, drug therapies targeting mutant CFTR have become available to most people with CF (Ramsey et al. 2011; Middleton et al. 2019; Heijerman et al. 2019). Here, we selectively review two types of small molecules: first, CFTR potentiators, which increase open probability (P_o) by enhancing channel gating following CFTR phosphorylation by PKA (Verkman and Galietta 2009; Jih et al. 2017). Second, CFTR correctors, which overcome the processing defect of F508del-CFTR to deliver the mutant protein to the plasma membrane and hence increase channel number (Verkman and Galietta 2009; Mijnders et al. 2017). Combination therapy with correctors and potentiators is now transforming the clinical management of CF (Hanrahan et al. 2017; Bell et al. 2019). For reviews of mutation-independent strategies to develop therapeutics applicable to all CF mutations, we refer the reader to Alton et al. (2016) and Li et al. (2017).

16.7.1 CFTR Potentiators

The first clinically approved drug therapy to target CF mutations was the CFTR potentiator ivacaftor (Kalydeco®; VX-770; Vertex Pharmaceuticals, Boston, MA, USA) (Van Goor et al. 2009; Ramsey et al. 2011) (Fig. 16.11). In individuals with the gating mutation G551D, the clinical impact of ivacaftor was dramatic and led to improvements not seen in previous clinical trials of new therapies targeting the symptoms of CF. They included a 10% increase in lung function (as measured by forced expiratory volume in 1 s (FEV_1)) after just 2 weeks in adult patients many with irreversible lung damage; the frequency of hospitalizations for treatment of lung infections, a measure of disease stability, improved by 55%; individuals treated with the drug gained weight (>2.5 kg over 48 weeks) likely as a result of normalization of intestinal pH by CFTR-dependent HCO_3^- secretion; and sweat chloride levels decreased by almost 50 mmol/L to fall below the diagnostic level for CF (Ramsey et al. 2011; Rowe et al. 2014). However, the most surprising outcome of ivacaftor treatment was witnessed with 1–2 year-old children with CF gating mutations. In these individuals, ivacaftor protected exocrine pancreatic function, suggesting that early treatment with the drug might delay or minimize irreversible damage to the pancreas (Rosenfeld et al. 2018). Consistent with this idea, using ferrets homozygous for the G551D mutation, Sun et al. (2019) demonstrated that in utero and sustained postnatal ivacaftor administration protected against organ damage. Thus, early pharmacological intervention might prevent disease development in some cases of CF.

At the time of ivacaftor's clinical approval, its mechanism of action was unknown. Studies of other CFTR potentiators, most notably genistein, suggested

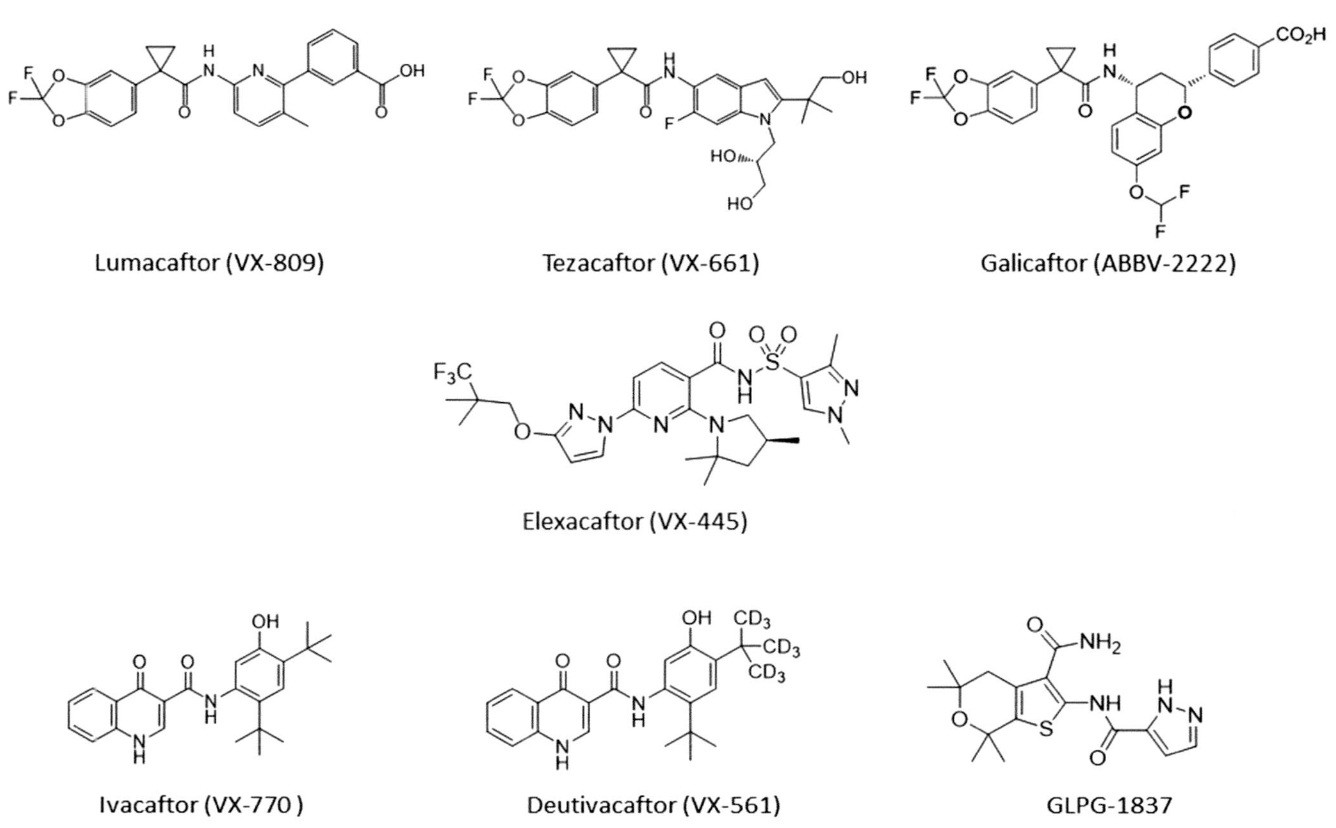

Fig. 16.11 Chemical structures of clinically approved CFTR modulators. Orkambi® combines lumacaftor and ivacaftor, Symdeko®/Symkevi® combines tezacaftor and ivacaftor, and Trikafta™ combines tezacaftor, elexacaftor, and ivacaftor. Also shown are the chemical structures of the CFTR potentiators deutivacaftor (deuterated ivacaftor; VX-561) and GLPG1837 and the CFTR corrector galicaftor (ABBV-2222)

that they might enhance CFTR activity by interacting with the NBD1:NBD2 dimer to modify ATP-dependent channel gating (Ai et al. 2004; Moran et al. 2005; Cai et al. 2006). However, several lines of evidence soon argued that this was not the case for ivacaftor. First, ivacaftor potentiated ATP-independent channel gating and was without effect on the ATPase activity of CFTR (Eckford et al. 2012; Jih and Hwang 2013). These results and its potentiation of a CFTR construct lacking NBD2 (Yeh et al. 2015) argued persuasively that ivacaftor does not interact with the ATP-binding sites of CFTR. Second, ivacaftor potentiation of ΔR-CFTR (Jih and Hwang 2013) indicated that the drug does not bind to the RD. Third, the accumulation of ivacaftor in the inner leaflet of the lipid bilayer and the reduced or lack of sensitivity of some CF mutations in the MSDs to ivacaftor (Van Goor et al. 2014; Baroni et al. 2014; van Willigen et al. 2019) suggested that the drug's binding site was located in the MSDs. Consistent with this idea, the ivacaftor binding site is located at the MSD–lipid interface and involves residues from M4, M5, and the unstructured region of M8 (Liu et al. 2019; Yeh et al. 2019) (Fig. 16.3d). Such a site of action explains the drug's effects on the kinetics of channel gating and its rescue of CF mutations throughout the structure of CFTR (Jih and Hwang 2013; Van Goor et al. 2014). Interestingly, GLPG1837 (Galapagos NV, Mechelen, Belgium), a small molecule with a distinct chemical structure (Fig. 16.11), binds to the same site as ivacaftor to potentiate CFTR channel gating by a common mechanism (Yeh et al. 2017, 2019; Liu et al. 2019).

The success of ivacaftor has stimulated the clinical development of other CFTR potentiators, including ABBV3067 (AbbVie, Inc., North Chicago, IL, USA), FDL-176 (Flatley Discovery Lab, Charlestown, MA, USA), and PTI-808 (Proteostasis Therapeutics, Inc., Boston, MA, USA) (for a review, see Kym et al. 2018). With the exception of VX-561 (deuterated ivacaftor; Vertex Pharmaceuticals) (Kym et al. 2018) (Fig. 16.11), the binding sites and mechanism of action of these potentiators are either unknown or undisclosed. Small molecules with distinct binding sites to ivacaftor might be used together with ivacaftor to enhance further CFTR channel gating. Initial studies of combinations of potentiators were unsuccessful (PG-01 and SF-01; Caputo et al. 2009) or used a channel blocker (ivacaftor and NPPB; Lin et al. 2016). More recent efforts have identified two classes of CFTR potentiators (class I: ivacaftor and GLPG1837; class II: apigenin and ASP-11), which enhance greatly the activity of CF mutations when used together as "co-potentiators" (Phuan et al. 2018, 2019; Veit et al. 2019). Such agents might be used with difficult to treat rare CF mutations, but they would also likely benefit most CF mutations by boosting further channel activity.

16.7.2 CFTR Correctors

Two types of small molecules have been identified that overcome protein misfolding to traffic mutant CFTR to the plasma membrane: CFTR correctors and proteostasis regulators (Balch et al. 2008; Verkman and Galietta 2009) (Fig. 16.12). Here, we

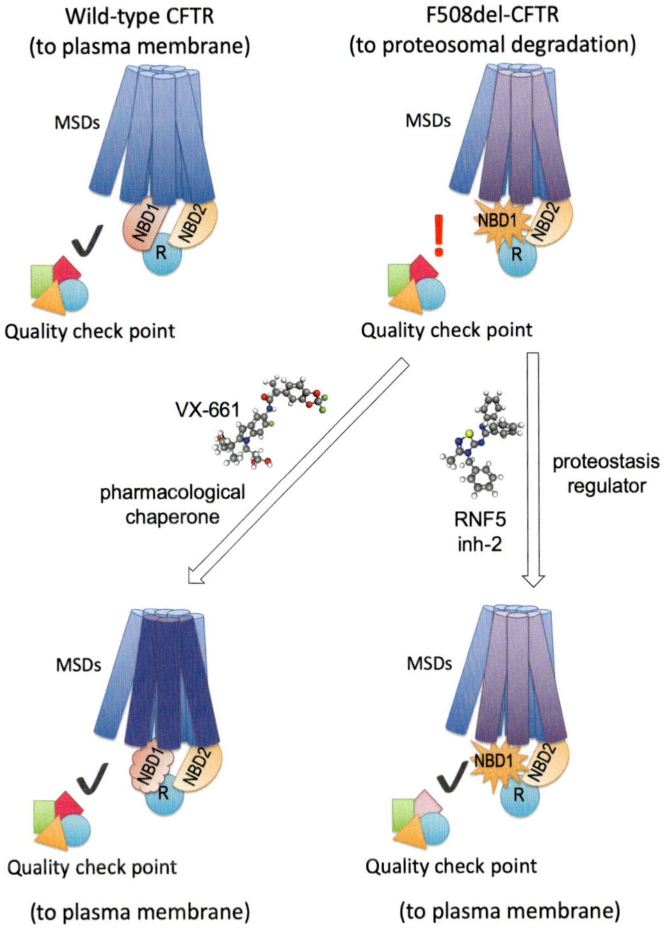

Fig. 16.12 Rescue of F508del-CFTR by pharmacological chaperones and proteostasis regulators. Pharmacological chaperones (e.g., tezacaftor [VX-661]) interact directly with the mutant protein to improve folding and prevent detection by quality control checkpoints. As a result, the mutant protein is processed and trafficked to the plasma membrane. Proteostasis regulators (e.g., ring finger protein 5 inhibitor-2 [RNF5 inh-2]) modify the cellular environment to achieve more favorable conditions for processing of the mutant protein. *MSD* membrane-spanning domain, *NBD* nucleotide-binding domain, *RD* regulatory domain

discuss selectively CFTR correctors. For a discussion of proteostasis regulators, see Sect. 16.8. CFTR correctors are pharmacological chaperones that likely interact directly with CFTR by acting as substrate mimics or active site inhibitors. By improving CFTR folding, CFTR correctors allow mutant CFTR to escape cellular quality control checkpoints and traffic to the plasma membrane (Lukacs and Verkman 2012).

The first CFTR corrector to be clinically approved was lumacaftor (VX-809; Vertex Pharmaceuticals) (Van Goor et al. 2011) (Fig. 16.11). Treatment of cells heterologously expressing F508del-CFTR with lumacaftor converted immature core-glycosylated CFTR protein (band B) to mature, fully glycosylated CFTR protein (band C) (Fig. 16.8). These data suggested that lumacaftor corrected the folding and intracellular transport of some F508del-CFTR protein. When compared to other CFTR correctors (e.g., corr-4a (Pedemonte et al. 2005)), lumacaftor demonstrated improved selectivity and greater efficacy, restoring to CF bronchial epithelia (genotype: F508del/F508del) transepithelial ion transport function equivalent to 14% that of non-CF bronchial epithelia (Van Goor et al. 2011). However, in clinical trials lumacaftor, by itself, failed to improve lung function (Clancy et al. 2012).

Mendoza et al. (2012) and Rabeh et al. (2012) independently demonstrated why lumacaftor, by itself, lacked clinical efficacy. By showing that no CFTR corrector had efficacy >15% wild-type CFTR function, Mendoza et al. (2012) identified an apparent efficacy ceiling for F508del-CFTR rescue by CFTR correctors. To elucidate the mechanistic basis of this efficacy ceiling, the authors identified amino acid positions statistically coupled to position 508 in the amino acid sequences of ABC transporters and evaluated the consequences of mutations at these positions on NBD1 folding and CFTR maturation. Using revertant mutations that corrected F508del-induced defects in NBD1 folding or the interaction of NBD1 with ICL4 (e.g., Serohijos et al. 2008; Thibodeau et al. 2010), Mendoza et al. (2012) demonstrated that only combinations of revertant mutations, which rescue both defects, restored robust expression and function to F508del-CFTR. Identical conclusions were obtained by Rabeh et al. (2012) through analyses of the thermodynamics of NBD1 folding and the stability of the NBD1:ICL4 domain interface. Building on these studies, Okiyoneda et al. (2013) identified three classes of CFTR correctors: class I (e.g., lumacaftor) target the MSD-NBD interface, class II (e.g., corr-4a) target NBD2, and class III (e.g., glycerol) target NBD1. Using high-throughput screening, Veit et al. (2018) identified small molecules from the three different classes that act synergistically to restore 50–100% wild-type CFTR function to F508del and rare mutations located in different domains of CFTR through structural allostery.

16.7.3 Combination Therapy with CFTR Correctors and Potentiators

For several reasons, combination therapy with CFTR correctors and potentiators has emerged as the treatment of choice for most CF mutations. First, the multiple effects of CF variants on CFTR expression and function (Veit et al. 2016). Second, the failure of either lumacaftor or ivacaftor to achieve clinical benefit when used by themselves (Clancy et al. 2012; Flume et al. 2012). Third, the inability to identify dual-acting small molecules (CFTR corrector-potentiators), which rescue both the plasma membrane expression of mutant CFTR and enhance its activity (Kalid et al. 2010; Pedemonte et al. 2011; Phuan et al. 2011; Liu et al. 2018). When tested in CF patients homozygous for F508del, combination therapy with lumacaftor and ivacaftor (Orkambi®; Vertex Pharmaceuticals) increased FEV_1 by ~3%, but improved disease stability by 30–40% (Wainwright et al. 2015), leading in 2015 to regulatory approval for the use of Orkambi® in this group of CF patients. However, prohibitive drug–drug interactions and adverse respiratory events in some CF patients treated with Orkambi® stimulated the development of new CFTR potentiators. Tezacaftor (VX-661) (Fig. 16.11) is a close analogue of lumacaftor with improved metabolism and pharmacokinetics, but reduced likelihood of drug–drug interactions (Kym et al. 2018). Combination therapy with tezacaftor and ivacaftor (Symdeko®/Symkevi®; Vertex Pharmaceuticals) not only was of greater benefit to CF patients homozygous for F508del than Orkambi®, but also improved lung function in CF patients with F508del and a CF mutation associated with residual function (Rowe et al. 2017; Taylor-Cousar et al. 2017).

Trikafta™ (elexacaftor (VX-445)—tezacaftor–ivacaftor; Vertex Pharmaceuticals), the latest combination therapy to receive regulatory approval, has the potential to benefit ~90% of CF patients, including those with F508del and CF mutations associated with minimal function (Middleton et al. 2019; Heijerman et al. 2019) (for the chemical structure of elexacaftor, see Fig. 16.11). Trikafta™ improved lung function and sweat chloride concentration in CF patients homozygous for F508del by similar magnitudes to the actions of ivacaftor on CF patients with G551D (Heijerman et al. 2019). Encouragingly, Trikafta™ achieved the greatest improvement in lung function (14% increase in FEV_1) in CF patients with F508del and a minimal function mutation (Middleton et al. 2019). In this group of patients, it also reduced the sweat chloride concentration below the diagnostic threshold, improved disease stability, and led to weight gain (Middleton et al. 2019). Taken together, the data demonstrate that Trikafta™ is a highly effective therapy that will likely benefit most people with CF.

The mechanistic basis for the success of Trikafta™ is the use of two CFTR correctors: C1-type (tezacaftor) and C2-type (elexacaftor), with complementary mechanisms of action to deliver mutant CFTR protein to the plasma membrane in combination with a potentiator (ivacaftor) to enhance channel activity (Keating et al. 2018). C1-type correctors target early folding events, whereas C2-type correctors have novel mechanisms of action (Li et al. 2018a). Consistent with this idea, Laselva et al. (2019) demonstrated that lumacaftor and AC1 (X281602; AbbVie Inc., North Chicago, IL, USA) target MSD1, whereas AC2-1 (X281632; AbbVie Inc.) interacts with MSD2 and AC2-2 (X300549; AbbVie Inc.) targets NBD2. In combination with a nonsense-mediated decay inhibitor and a CFTR potentiator, AC1 and AC2-2 restored greatest CFTR function to the CF nonsense mutation c.3846G > A (W1282X) (Laselva et al. 2019).

Studies of CFTR protein processing and single-channel behavior provide molecular explanations for the action of CFTR modulators. CFTR-mediated transepithelial Cl^- current is determined by the product of the number of CFTR Cl^- channels in the apical membrane (N), the current flowing through an open channel (i), and its open probability (P_o): $I = N \times i \times P_o$. At therapeutic concentrations (e.g., Van Goor et al. 2011; Jih and Hwang 2013), clinically approved CFTR correctors and potentiators

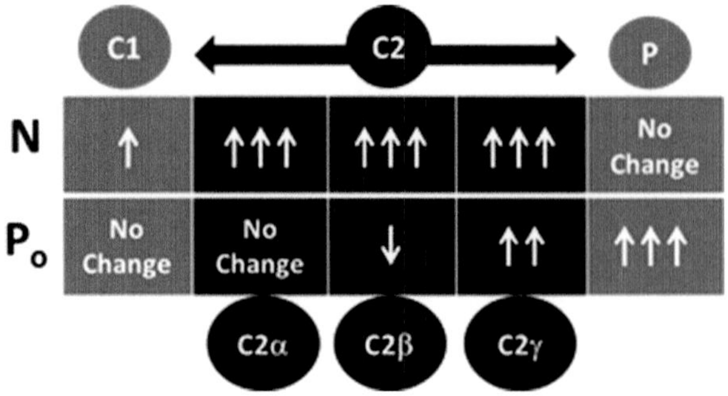

Fig. 16.13 Proposed mode of action of C1- and C2-type CFTR correctors. The schematic shows how CFTR correctors and potentiators alter the number of CFTR Cl^- channels at the plasma membrane (N) and their open probability (P_o) to modulate CFTR-mediated transepithelial ion transport measured with the Ussing chamber and TECC assays. *C* corrector, *P* potentiator

are without effect on the single-channel current amplitude (i) of CFTR. Thus, compound efficacy in in vitro assays, such as the Ussing chamber technique and the transepithelial Cl^- conductance (TECC) assay, is determined by N and P_o (Fig. 16.13). The amount of rescued CFTR protein (N) is determined qualitatively by Western blotting using either human bronchial epithelial cells that endogenously express CFTR or cell lines expressing CFTR heterologously, whereas P_o is determined by single-channel recording using excised inside-out membrane patches from cells expressing CFTR heterologously (for further discussion, see Sheppard et al. 1995).

The C1-type correctors identified to date increase N without altering P_o (e.g., lumacaftor (Van Goor et al. 2011, but see Kopeikin et al. 2014; Meng et al. 2017); ABBV-2222 (Singh et al. 2020) (Figs. 16.11 and 16.13)). Figure 16.14a demonstrates that incubation of CFBE cells stably expressing F508del-CFTR (clone DG3) with the C1-type CFTR corrector ABBV-2222 increased the production of mature CFTR protein without improving channel activity. However, AbbVie's proprietary CFTR potentiator restored wild-type CFTR levels of channel activity (as measured by P_o) to F508del-CFTR rescued by the C1-type corrector ABBV-2222 (Fig. 16.14b).

At least three different forms of C2-type correctors additive to C1-type correctors can be conceptualized based on their different effects on P_o (Singh et al. 2018) (Fig. 16.13). First, C2α-type CFTR correctors exhibit additivity for F508del-CFTR rescue with C1-type CFTR correctors. The combination of these two CFTR correctors only affects the number of CFTR Cl^- channels delivered to the plasma membrane (N), not channel activity (P_o) (Fig. 16.13). Like the action of C1-type CFTR correctors (Fig. 16.14), the activity of F508del-CFTR rescued by the combination of C1 and C2α-type CFTR correctors is enhanced robustly by CFTR potentiators (Fig. 16.15). Figure 16.15a demonstrates that treatment of CFBE cells stably expressing F508del-CFTR (clone DG3) with the C1-type CFTR corrector ABBV-2222 and a C2α-type CFTR corrector (AbbVie proprietary CFTR corrector) rescued

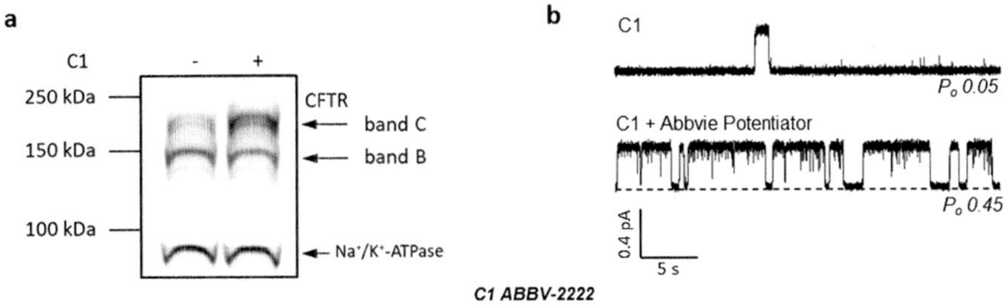

Fig. 16.14 C1-type CFTR correctors improve CFTR protein processing without altering channel activity. (**a**) Immunoblot of CFTR protein from CFBE cells stably expressing F508del-CFTR (clone DG3). Prior to cell lysis, the cells were treated with the C1-type CFTR corrector ABBV-2222 (0.5 μM) or its vehicle control DMSO (0.1% $v \cdot v^{-1}$) for 24 h at 37 °C. CFTR was detected with the mouse anti-CFTR monoclonal antibody (596) (Cui et al. 2007). Arrows indicate the positions of the band B (immature) and band C (mature) forms of CFTR. The α subunit of the Na^+, K^+-ATPase was used as a loading control. (**b**) Single-channel recordings of F508del-CFTR from membrane patches excised from CHO cells treated with the C1-type corrector ABBV-2222 using similar conditions to A. Following channel activation by PKA-dependent phosphorylation, the recordings were acquired in the absence and presence of AbbVie's proprietary CFTR potentiator (3 μM) using the experimental conditions described in Yeh et al. (2019). The dashed line indicates where channels are closed and upward deflections of the traces correspond to channel openings. The open probability (P_o) of the recordings is indicated

noticeably greater amounts of mature CFTR protein than ABBV-2222 alone. Although there was no difference in the amount of forskolin-stimulated CFTR-mediated Cl^- current between primary cultures of human bronchial epithelia (hBE) (genotype: F508del/F508del) treated with C1-type and C2α-type CFTR correctors compared to C1-type alone, AbbVie's proprietary CFTR potentiator generated markedly greater CFTR-mediated transepithelial Cl^- current when the same cells were treated with both CFTR correctors and these Cl^- currents exhibited stability unlike those treated with just the C1-type CFTR corrector ABBV-2222 (Fig. 16.15b).

Second, C2β-type CFTR correctors show additivity with C1-type CFTR correctors for rescue of F508del-CFTR protein expression, but inhibit CFTR activity (Figs. 16.13 and 16.16). Several lines of evidence support this idea. First, the amount of forskolin-stimulated F508del-CFTR Cl^- current achieved with the combination of the C1-type CFTR corrector ABBV-2222 and a C2β-type CFTR corrector was less than that of ABBV-2222 and higher concentrations of AbbVie's proprietary CFTR potentiator were required to potentiate F508del-CFTR activity (Fig. 16.16b). Second, removal of a C2β-type CFTR corrector enhanced the magnitude of F508del-CFTR Cl^- currents recorded using excised inside-out membrane patches (Fig. 16.16c). Third, acute addition of GLPG2737, a CFTR corrector from the same chemical class as the C2β-type, reduced the P_o of F508del-CFTR Cl^- channels in excised inside-out membrane patches by decreasing the frequency of channel openings (de Wilde et al. 2019). Taken together, the data demonstrate that the combination of C1- and C2β-type CFTR correctors greatly enhances the number

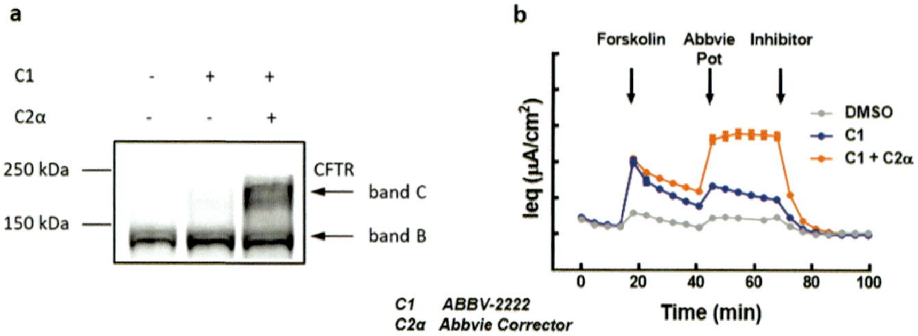

Fig. 16.15 Correction with C1- and C2α-type CFTR correctors together increases greatly the amount of rescued F508del-CFTR available for potentiation. (**a**) Immunoblot of CFTR protein from CFBE cells stably expressing F508del-CFTR (clone DG3). Prior to cell lysis, cells were treated with either the vehicle control DMSO (0.1% $v \cdot v^{-1}$), the C1-type CFTR corrector ABBV-2222 (0.5 μM), or the combination of ABBV-2222 and a C2α-type CFTR corrector (AbbVie proprietary CFTR corrector) (1 μM) for 24 h at 37 °C. CFTR was detected with the mouse anti-CFTR monoclonal antibody (596) (Cui et al. 2007). Arrows indicate the positions of the band B (immature) and band C (mature) forms of CFTR. (**b**) Time course of equivalent current (I_{eq}) recorded from primary hBE (genotype: F508del/F508del) treated with C1- and C2α-type CFTR correctors using similar conditions to A. The vertical arrows indicate the sequential and cumulative addition of forskolin (Fsk; 10 μM), AbbVie's proprietary CFTR potentiator (*P*; 0.5 μM), and the CFTR inhibitor $CFTR_{inh}$-172 (Inhib; 20 μM) to the apical solution bathing epithelia. I_{eq} was recorded using the TECC assay using the experimental conditions described in Singh et al. (2020). For proprietary reasons, values of I_{eq} are not disclosed

of F508del-CFTR Cl^- channels delivered to the plasma membrane, but inhibits their activity.

Third, C2γ-type CFTR correctors enhance both the expression of F508del-CFTR protein at the plasma membrane and its channel activity when used with C1-type CFTR correctors (Fig. 16.13). Figure 16.17a demonstrates that treatment of CFBE cells stably expressing F508del-CFTR (clone DG3) with the C1-type CFTR corrector ABBV-2222 and a C2γ-type CFTR corrector (AbbVie proprietary CFTR corrector) generated markedly greater amounts of mature CFTR protein than ABBV-2222 alone. Moreover, the amount of forskolin-stimulated CFTR-mediated Cl^- current was noticeably greater than that in primary cultures of hBE (genotype: F508del/F508del) treated with ABBV-2222 alone and AbbVie's proprietary CFTR potentiator achieved little further enhancement of current magnitude (Fig. 16.17b). These effects of the C2γ-type CFTR corrector (AbbVie proprietary CFTR corrector) are reminiscent of CFTR corrector-potentiators (Kalid et al. 2010; Pedemonte et al. 2011; Phuan et al. 2011; Liu et al. 2018) and suggest that it has dual activity.

In summary, Trikafta, the combination of two CFTR correctors with complementary mechanisms of action and a CFTR potentiator, is a highly effective therapy that promises to benefit most people with CF (Keating et al. 2018; Middleton et al. 2019; Heijerman et al. 2019). However, evidence of disease progression in CFTR modulator-treated patients (for a review, see Kleizen et al. 2020) suggests that additional CFTR modulators are required. Mechanistic studies of CFTR correctors

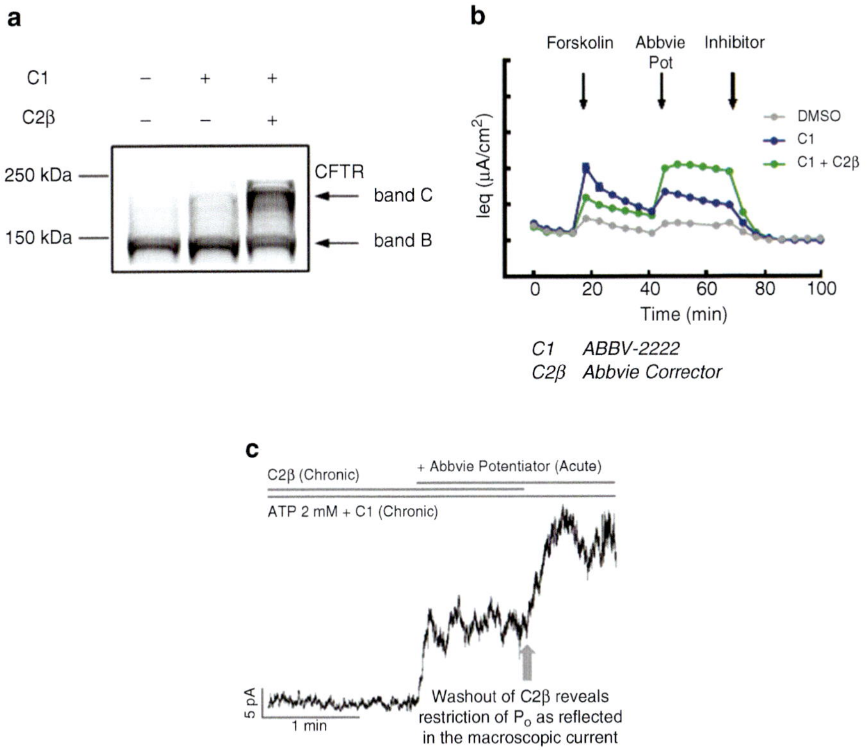

Fig. 16.16 C2β-type CFTR correctors rescue F508del-CFTR protein expression but inhibit CFTR activity. (**a**) Immunoblot of CFTR protein from CFBE cells stably expressing F508del-CFTR (clone DG3). Prior to cell lysis, cells were treated with the vehicle control DMSO (0.1% v·v^{-1}), the C1-type CFTR corrector ABBV-2222 (0.5 μM), or the combination of ABBV-2222 and a C2β-type CFTR corrector (AbbVie proprietary CFTR corrector) (3 μM) for 24 h at 37 °C. CFTR was detected with the mouse anti-CFTR monoclonal antibody (596) (Cui et al. 2007). Arrows indicate the positions of the band B (immature) and band C (mature) forms of CFTR. (**b**) Time course of equivalent current (I_{eq}) recorded from primary hBE (genotype: F508del/F508del) treated with C1- and C2β-type CFTR correctors using similar conditions to A. The vertical arrows indicate the sequential and cumulative addition of forskolin (Fsk; 10 μM), AbbVie's proprietary CFTR potentiator (*P*; 1.5 μM), and the CFTR inhibitor CFTR$_{inh}$-172 (Inhib; 20 μM) to the apical solution bathing epithelia; for other details, see Fig. 16.15. (**c**) F508del-CFTR Cl^- current recording from an inside-out membrane patch excised from a CHO cell expressing F508del-CFTR treated with the C1-type CFTR corrector ABBV-2222 (0.5 μM) and a C2β-type CFTR corrector (AbbVie proprietary CFTR corrector) (1 μM) for 24 h at 37 °C. Following channel activation by PKA-dependent phosphorylation, the recording was acquired using the experimental conditions described in Yeh et al. 2019). The horizontal bars indicate the presence of ATP (2 mM) and AbbVie's proprietary CFTR potentiator (3 μM) in the intracellular solution. The vertical red arrow indicates when the C2β-type CFTR corrector (AbbVie proprietary CFTR corrector) was washed from the recording chamber

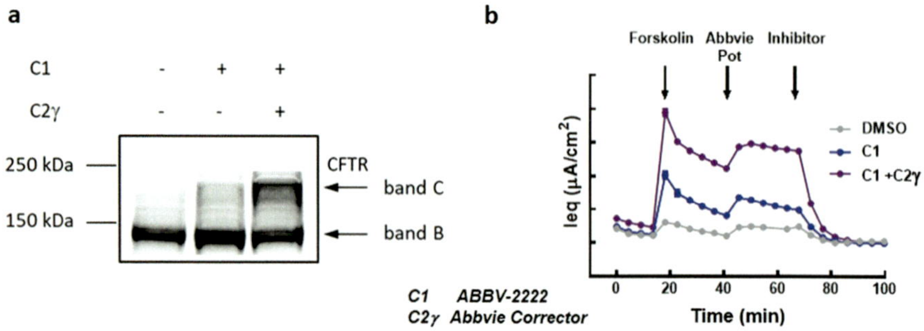

Fig. 16.17 Correction with C1- and C2γ-type CFTR correctors together enhance greatly F508del-CFTR channel activity. (**a**) Immunoblot of CFTR protein from CFBE cells stably expressing F508del-CFTR (clone DG3). Prior to cell lysis, cells were treated with either the vehicle control DMSO (0.1% $v{\cdot}v^{-1}$), the C1-type CFTR corrector ABBV-2222 (0.5 μM), or the combination of ABBV-2222 and a C2γ-type CFTR corrector (AbbVie proprietary CFTR corrector) (3 μM) for 24 h at 37 °C. CFTR was detected with the mouse anti-CFTR monoclonal antibody (596) (Cui et al. 2007). Arrows indicate the positions of the band B (immature) and band C (mature) forms of CFTR. (**b**) Time course of equivalent current (I_{eq}) recorded from primary hBE (genotype: F508del/F508del) treated with C1- and C2γ-type CFTR correctors using similar conditions to A. The vertical arrows indicate the sequential and cumulative addition of forskolin (Fsk; 10 μM), AbbVie's proprietary CFTR potentiator (*P*; 0.5 μM), and the CFTR inhibitor $CFTR_{inh}$-172 (Inhib; 20 μM) to the apical solution bathing epithelia; for other details, see Fig. 16.15

are the foundation for their development. These small molecules might be used alone or together with proteostasis modulators to achieve maximal therapeutic benefit.

16.8 Rescuing the Plasma Membrane Expression of CF Mutants with Proteostasis Regulators

Proteostasis regulators are molecules that modulate cellular homeostasis to regulate specific cellular functions (Balch et al. 2008). By targeting CFTR-interacting proteins, proteostasis regulators have beneficial effects on CFTR biosynthesis, its trafficking to, and its expression at the plasma membrane (Balch et al. 2008; Sondo et al. 2017) (Fig. 16.12). Proteostasis regulators might modulate either the expression or activity of CFTR-interacting proteins, including components of the ER and plasma membrane quality control systems (Balch et al. 2008; Sondo et al. 2017). Alternatively, they might act in a more general and indirect way, by regulating proteins (and pathways) that are not directly involved in CFTR biogenesis, but whose modulation results in increased CFTR processing (Balch et al. 2008; Sondo et al. 2017). Several CFTR-interacting proteins have already been identified, which represent potential CF drug targets (for a review, see Sondo et al. 2017). However, better understanding of the molecular events underlying CFTR processing, trafficking, expression, and degradation might reveal new targets, leading to the development of more effective CF therapeutics.

16.8.1 Constituents of the ER Quality Control Machinery as Possible Drug Targets for F508del-CFTR Rescue

Defective folding of F508del-CFTR is recognized by the ER quality control (ERQC) system, leading to its retention in the ER and targeting for degradation by the ubiquitin proteasome system (UPS) (Farinha and Amaral 2005; Riordan 2008) (Fig. 16.18). The ERQC pathway known as ER-associated degradation (ERAD) disposes of newly synthesized proteins, which fail to achieve their correct conformation. ERAD includes the recognition and targeting of misfolded substrates to the retrotranslocation machinery, transport into the cytosol, and degradation by the UPS (for a review, see Needham et al. 2019). Various proteins, located at four or more different checkpoints, are implicated in ERAD, including chaperones, components of the retro-translocon, proteins that extract ubiquitinated substrates from the ER membrane, and cytosolic- and membrane-associated E3 ubiquitin ligases (Farinha et al. 2002; Farinha and Amaral 2005; Roxo-Rosa et al. 2006b; Wang et al. 2008; El Khouri et al. 2013; Matsumura et al. 2013; Farinha and Canato 2017; Hou et al. 2018; Huang et al. 2019). At the first two checkpoints, CFTR interacts with molecular chaperones that promote/monitor CFTR folding, whereas the last two checkpoints recognize trafficking/exit signals to allow CFTR exit from the ER by incorporation into vesicles (Farinha et al. 2002; Farinha and Amaral 2005; Roxo-Rosa et al. 2006b; Matsumura et al. 2011; Saxena et al. 2012; Farinha and Canato 2017; Cui et al. 2019; Doonan et al. 2019; Huang et al. 2019).

At the first checkpoint, the status of CFTR folding is assessed as it emerges from the ribosome by Hsp70 machinery, which includes several co-chaperones (Youker et al. 2004; Matsumura et al. 2011; Kim and Skach 2012; Saxena et al. 2012; Matsumura et al. 2013) (Fig. 16.18). The Hsp70-interacting co-chaperone determines the fate of CFTR with recruitment of Hsp40 promoting CFTR folding, but enlistment of CHIP initiating CFTR degradation (Yang et al. 1993; Meacham et al. 2001; Farinha et al. 2002; Youker et al. 2004; Matsumura et al. 2013). At the second checkpoint, CFTR folding is interrogated through interaction of CFTR glycans with calnexin, leading to glycoprotein-ER-associated degradation (GERAD) of mutant CFTR (Farinha and Amaral 2005; Rosser et al. 2008) (Fig. 16.18). The third checkpoint detects retention motifs, the arginine-framed tripeptides (AFTs), which are exposed in misfolded proteins (Roxo-Rosa et al. 2006b; Cheng and Guggino 2013) (Fig. 16.18). Finally, the fourth checkpoint recognizes a diacidic code, a specific exit code that mediates CFTR incorporation into COPII vesicles. Export of CFTR is highly sensitive to mutation of this strongly conserved exit code (Chang et al. 1999; Wang et al. 2004; Roy et al. 2010; Farinha et al. 2013) (Fig. 16.18).

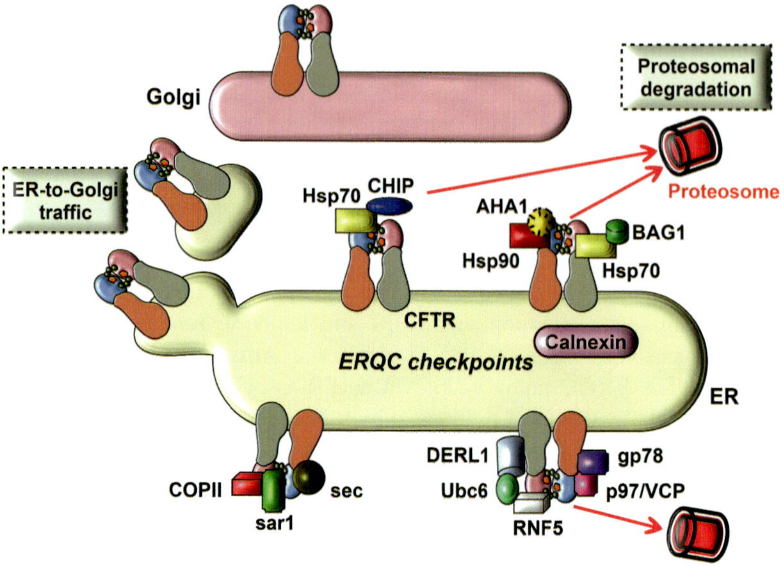

Fig. 16.18 Proteins of ER quality control checkpoints as putative drug targets. The schematic shows some of the proteins that constitute ER quality control checkpoints, which recognize misfolded mutant CFTR during protein synthesis. Selected proteostasis regulators that represent targets for CF drug development are indicated. *AHA1* activator of Hsp90 ATPase activity 1, *BAG1* BAG family molecular chaperone regulator 1, *CHIP* Hsp70-interacting protein, *COPII* Coat protein complex II, *DERL1* Derlin-1, *gp78* glycoprotein 78, *Hsp70* 70 kDa heat shock proteins, *Hsp90* 90 kDa heat shock proteins, *p97/VCP* 97 kDa valosin containing protein, *RNF5* ring finger protein 5, *sar1* small guanosine triphosphate (GTP)-binding protein secretion-associated Ras-related protein 1, *sec* protein transport proteins belonging to Sec family, constituting the inner COPII coat subcomplex, *Ubc6* ubiquitin-conjugating enzyme E2 6

During the early stages of their assembly, wild-type and F508del-CFTR appear to assume similar conformations (Zhang et al. 1998), but soon after F508del-CFTR folding stops, leading to its retention in the ER, retrotranslocation to the cytosol, and degradation by the proteasome (for a review, see Nakatsukasa and Brodsky 2008; Needham et al. 2019). Degradation of mutant CFTR involves a complex web of interactions between CFTR and cytosolic and ER lumen chaperones, membrane-localized ubiquitin enzymes, and cytosolic ubiquitin degradation systems (Younger et al. 2006; Wang et al. 2006). A major role is played by an ER membrane-associated ubiquitin ligase complex, containing the E3 ubiquitin ligase RNF5 (also called RMA1), the E2 Ubc6e, and Derlin-1, which cooperates with Hsp70 machinery to triage wild-type and F508del-CFTR (Younger et al. 2006; Hou et al. 2018) (Fig. 16.18). Other cofactors, including gp78, BAP31, and p97, are then responsible for delivering the ubiquitinated CFTR to the proteasome (Goldstein et al. 2007; Morito et al. 2008; Wang et al. 2008) (Fig. 16.18). Several of these proteins constitute drug targets to rescue mutant CFTR.

Derlin-1 and its degradation complex partner, p97, interact physically with wild-type CFTR to reduce its maturation (Sun et al. 2006). RNA interference knockdown experiments demonstrated that when Derlin-1 levels were reduced, wild-type and F508del-CFTR expression increased 10- to 15-fold (Younger et al. 2006; Sun et al. 2006). However, reduction of Derlin-1 expression did not promote the maturation of F508del-CFTR (Sun et al. 2006), a result reminiscent of the effects of inhibiting either the ubiquitination or proteasome-mediated degradation of F508del-CFTR (Jensen et al. 1995; Ward et al. 1995). An explanation for the lack of F508del-CFTR maturation when Derlin-1 is inhibited is the presence of additional downstream checkpoints after the Derlin-1 checkpoint (Sun and Brodsky 2019). Experiments using CFTR fragments suggest that MSD1 is degraded by Derlin-1, whereas MSD1-NBD1 has some resistance to Derlin-1-mediated degradation (Sun et al. 2006), consistent with the stabilizing effect of CFTR domain–domain interactions (Okiyoneda and Lukacs 2012).

The ER-associated E3 ligase RNF5 plays a critical role in selecting F508del-CFTR for premature degradation (Younger et al. 2006), working in concert with other proteins to recognize misfolded F508del-CFTR and initiate its degradation (Fig. 16.18). For example, in association with Derlin-1, RNF5 begins to ubiquitinate CFTR (Younger et al. 2006). Recognition of CFTR folding defects by the RNF5 E3 complex occurs during CFTR translation (Younger et al. 2006). Polyubiquitination of CFTR then occurs through recruitment of the E4-like enzyme gp78 that sequentially cooperates with RNF5, and this interaction is possibly mediated by Derlin (s) (Morito et al. 2008).

Using immortalized bronchial epithelial cells and mouse models of CF, Tomati et al. (2015) demonstrated that RNF5 is a drug target for CF therapy. Like the effects of Derlin-1 knockdown (Sun et al. 2006), RNF5 silencing increased the amount of immature, but not mature F508del-CFTR protein in immortalized bronchial epithelial cells (Tomati et al. 2015). Surprisingly, functional studies using these cells revealed that RNF5 suppression enhanced strongly F508del-CFTR activity (Tomati et al. 2015). To explain these data, the authors examined which form of CFTR was expressed at the plasma membrane upon RNF5 silencing. Previous work had identified an unconventional secretory pathway dependent on Golgi reassembly stacking proteins (GRASPs), which trafficked immature, core-glycosylated CFTR to the plasma membrane (Gee et al. 2011). Interestingly, Tomati et al. (2015) observed additive/synergistic effects of combining the CFTR corrector lumacaftor with RNF5 knockdown, suggesting that each manipulation affects distinct pools of mutant CFTR. Blocking F508del-CFTR degradation by silencing RNF5 trafficked immature F508del-CFTR protein to the plasma membrane by the GRASPs-dependent pathway. By contrast, lumacaftor promoted F508del-CFTR maturation and its transport to the plasma membrane by the conventional secretory pathway (Tomati et al. 2015). To evaluate the therapeutic relevance of RNF5 suppression in vivo, Tomati et al. (2015) generated CF mice with the F508del mutation, which lacked RNF5. Suppression of RNF5 in vivo ameliorated the CF phenotype in F508del-CFTR mice, including weight loss, gastrointestinal pathology, and defective epithelial ion transport (Wilke et al. 2011; Tomati et al. 2015). Taken together, the data argue strongly that drug-like small molecule inhibitors of RNF5 have therapeutic potential for the treatment of CF.

Using a computational approach, Sondo et al. (2018) identified the RNF5 inhibitor inh-2, which rescues F508del-CFTR in different cellular models, including primary bronchial epithelial cells, by decreasing its ubiquitination and stabilizing the mature form of the protein. To demonstrate convincingly that inh-2 targets RNF5, Sondo et al. (2018) investigated the effects of inh-2 on ATG4B and Paxillin, two cellular targets of RNF5 (Didier et al. 2003; Kuang et al. 2012). In both cases, inh-2 modulated protein function by reducing RNF5-mediated ubiquitination (Sondo et al. 2018) (Fig. 16.19). Although pharmacological inhibition of RNF5 has a significant impact on different cellular pathways (Fig. 16.19), mice lacking RNF5 expression were viable (Delaunay et al. 2008). This suggests that inhibition of RNF5 in vivo might be exploited therapeutically. In addition, RNF5 knockout mice are resistant to group A Streptococcus infection, in part, due to enhanced autophagy-mediated clearance of invading bacteria by macrophages (Kuang et al. 2012). This represents a positive secondary effect of RNF5 inhibition for CF patients with chronic lung infections.

To identify CFTR-interacting proteins in the late ERQC checkpoints, which determine ER exit, transcriptomic- and proteomic-based approaches have been utilized to analyze global gene and protein expression patterns and map the CFTR interactome (Roxo-Rosa et al. 2006a; Wang et al. 2006; Gomes-Alves et al. 2010; Clarke et al. 2013; Rauniyar et al. 2014; Pankow et al. 2015; Reilly et al. 2017). For example, Canato et al. (2018) profiled the proteomic interactions of F508del-CFTR and F508del-4RK-CFTR cells to investigate the cellular machinery required to decode the AFTs, which are exposed by F508del-CFTR misfolding. The authors found that the F508del-CFTR interactome is enriched in proteins involved in RNA processing, but was deficient in those responsible for epithelial integrity. Among the proteins with increased affinity for F508del-CFTR, Canato et al. (2018) identified the kinesin-14 family member KIFC1, a kinesin motor protein involved in vesicle transport. Interestingly, downregulation or inhibition of KIFC1 stabilized the immature form of F508del-CFTR by decreasing its degradation rate (Canato et al. 2018). Thus, proteomic-based studies can identify innovative targets to exploit for the development of new therapeutics for CF.

16.8.2 Heat Shock Proteins and Co-chaperones: Helping CFTR to Fold (or to Degrade?)

During co- and posttranslational processing, CFTR binds to several cytosolic molecular chaperones (Hsp70, Hsp40, and Hsp90), which facilitate its folding and degrade misfolded proteins (Wang et al. 2006; Matsumura et al. 2011; Saxena et al. 2012; Young 2014; Huang et al. 2019) (Fig. 16.18). For example, Wang et al. (2006) demonstrated that Hsp90 co-chaperones, such as Aha1, which stimulates Hsp90 ATPase activity (Lotz et al. 2003), modulate the Hsp90-dependent stability of CFTR protein folding in the ER. The authors proposed that failure of F508del-CFTR to

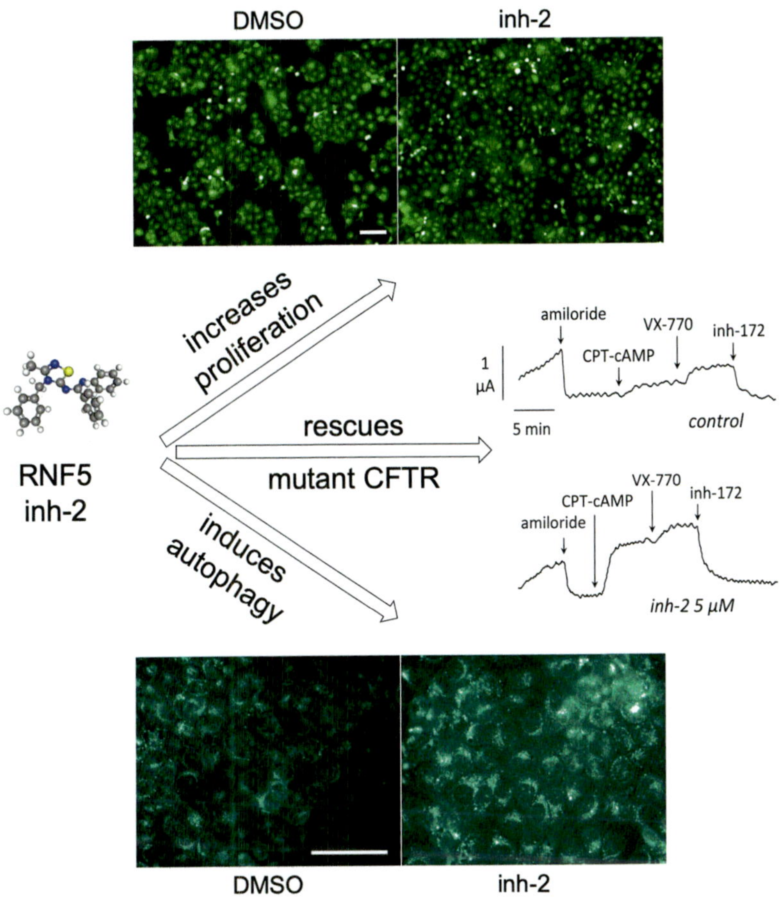

Fig. 16.19 Biological activities of RNF5 inhibitor-2. By reducing RNF5-mediated ubiquitination of target proteins, RNF5 inhibitor-2 (RNF5 inh-2) modulates several cellular processes. RNF5 inh-2 rescues F508del-CFTR activity in human primary bronchial epithelia by inhibiting CFTR ubiquitination and preventing its degradation. Representative recordings show F508del-CFTR-mediated transepithelial Cl^- currents in control and RNF5 inh-2-treated primary cultures of human bronchial epithelia. Arrows indicate the sequential and cumulative addition of amiloride (10 μM), CPT-cAMP (100 μM), ivacaftor (VX-770; 1 μM), and $CFTR_{inh}$-172 (inh-172; 5 μM) to the apical solution bathing the epithelia. In addition, RNF5 inh-2 induces autophagy and increases cell proliferation, by reducing ubiquitination of ATG4B and paxillin, respectively. Upper panel: representative images showing the difference in cell proliferation observed in YFP-expressing CFBE41o$^-$ cells treated with vehicle alone (DMSO) or with RNF5 inh-2 (5 μM) for 72 h at 37 °C. Lower panel: representative images showing the appearance of autophagic vacuoles (visualized using monodansylcadaverine probe) in CFBE41o$^-$ cells treated with vehicle alone (DMSO) or with RNF5 inh-2 (5 μM) for 6 h at 37 °C; upper and lower panel scale bars, 100 μm

achieve an energetically favorable fold in response to the steady-state dynamics of the chaperone folding environment (the "chaperome") disrupts interactions within the secretory pathway, leading to disease development. To elucidate how Aha1 stimulates the ATPase activity of Hsp90, Wang et al. (2006) and Koulov et al. (2010) used innovative mass spectrometry approaches to demonstrate that the N- and C-terminal domains of Aha1 cooperatively bind across the dimer interface of Hsp90 to modulate its ATP hydrolysis cycle and substrate activity in vivo. The data suggest that Aha1 activity integrates chaperone function with substrate folding energetics, thereby protecting transient folding intermediates in vivo, which might contribute to protein misfolding disorders, such as CF, when destabilized (Koulov et al. 2010).

The Hsp70 chaperone system was demonstrated to regulate CFTR folding (Loo et al. 1998), but lack of specific inhibitors precluded identification of its precise role. Soon afterward, Hsc70 and its co-chaperone DNAJA1 were found to associate with wild-type and F508del-CFTR (Meacham et al. 1999), with greater binding to F508del-CFTR suggesting that the chaperones were trying to refold the mutant protein. Consistent with this idea, siRNA knockdown of DNAJA1 increased misfolding and reduced the amount of both wild-type and F508del-CFTR (Grove et al. 2011). By contrast, overexpression of Hsp70 and the co-chaperone DNAJB1 induced modest improvements in trafficking and stabilization of F508del-CFTR (Choo-Kang and Zeitlin 2001; Farinha et al. 2002).

CFTR is degraded at both the ER and cell periphery by the E3 ubiquitin ligase CHIP complexed with Hsc/Hsp70 (Meacham et al. 2001). At the ER level, CHIP acts in concert with the ER transmembrane E3 ligases gp78 and RNF5/RMA1 to detect misfolded proteins (Younger et al. 2006; Morito et al. 2008) (Fig. 16.18). Hsc/Hsp70-CHIP likely recognize misfolding of cytosolic protein domains, while gp78 and RNF5/RMA1 are more sensitive to protein misfolding in the membrane (Younger et al. 2006; Morito et al. 2008). A parallel role for CHIP was found at the plasma membrane. When F508del-CFTR protein is transported to the plasma membrane by low-temperature incubation, mutant CFTR is more rapidly internalized by endocytosis than wild-type CFTR, leading to its degradation in the lysosomes. Using a functional genomics approach based on siRNA screening, Okiyoneda et al. (2010) identified CHIP as the E3 ligase responsible for initiating internalization by polyubiquitinating F508del-CFTR. The role of CHIP at the plasma membrane is notable because it had been thought to only act on secretory-pathway proteins within the ER (Matsumura et al. 2013). It also demonstrates the important role of Hsc/Hsp70-CHIP complexes in determining the fate of mutant CFTR.

Co-chaperone DNAJB12 is an ER transmembrane protein that interacts with Hsc70 (Yamamoto et al. 2010). DNAJB12 overexpression promotes ERAD of wild-type and F508del-CFTR, whereas DNAJB12 knockdown protects F508del-CFTR from degradation, without affecting trafficking (Yamamoto et al. 2010; Grove et al. 2011). In addition, DNAJB12 enhances the interaction of F508del-CFTR with RNF5/RMA1 and gp78, but not with CHIP (Grove et al. 2011). Because co-chaperone-induced ERAD might prevent the rescue of F508del-CFTR trafficking, both pro-folding and anti-degradation strategies will likely be needed for therapeutic benefit.

Nucleotide exchange factors (NEFs), which promote the release of ADP from Hsp70, the re-binding of ATP, and the dissociation of Hsp70 from its substrate, also regulate CFTR degradation. For example, BAG1 impairs Hsc/Hsp70-assisted CFTR folding and CHIP-mediated degradation through its NEF mechanism, which promotes the dissociation of Hsc/Hsp70 from substrates, rather than interactions with CHIP (Matsumura et al. 2011) (Fig. 16.18). Other NEFs, including Hspbp1 and Hsp110, associate with CFTR and influence its degradation. Hspbp1 binds CHIP directly and when overexpressed protects wild-type and mutant CFTR from degradation, while Hspbp1 knockdown impairs CFTR trafficking (Alberti et al. 2004). By contrast, Hsp110 knockdown reduces co-translational degradation of both wild-type and F508del-CFTR, while its overexpression promotes degradation, but at the same time increases F508del-CFTR trafficking (Saxena et al. 2012). Interestingly, Hsp110 can associate with F508del-CFTR after trafficking to the plasma membrane to prolong the lifetime of the mature protein (Saxena et al. 2012), an effect that is possibly due to its potential substrate binding activity (Kampinga and Craig 2010; Young 2014).

The identification of Hsp70 inhibitors has raised interest in it as a drug target. Park et al. (2009) demonstrated that the glycosphingolipid sulfogalactosyl ceramide (SGC) and its water-soluble mimic adamantylSGC (adaSGC) prevented the degradation of F508del-CFTR by inhibiting the ATPase activity of Hsc70, leading to augmented maturation of F508del-CFTR. In a screen for drugs that promoted apoptosis, Cho et al. (2011) identified an imidazole-based compound, apoptozole, whose mechanism of action involves interaction with the Hsc/Hsp70 complex and Hsc70 inhibition. The authors demonstrated that apoptozole increased F508del-CFTR trafficking to the plasma membrane and enhanced channel function, but had limited effect on internalization of the mutant protein from the plasma membrane. Interestingly, modulation of chaperone and co-chaperone recruitment and binding to F508del-CFTR also appears to be a "secondary effect" of combination therapy with CFTR correctors. When used together, C4 (corr-4a) (Pedemonte et al. 2005) and C18 (VRT-534; an analog of lumacaftor) (Eckford et al. 2014) not only rescued more F508del-CFTR protein, but also modified the proteostatic network (Lopes-Pacheco et al. 2017). Together, C4 and C18 decreased F508del-CFTR binding to the ERAD proteins Hsp27 and Hsp40, allowing the stabilized CFTR protein to reach the plasma membrane (Lopes-Pacheco et al. 2017). Similarly, the combination of the Hsp70 inhibitor MKT077 and the CFTR corrector lumacaftor enhanced the plasma membrane expression and function of F508del-CFTR (Kim Chiaw et al. 2019). Thus, pharmacological modulation of the Hsp70 complex in combination with small molecule CFTR correctors improves the folding of mutant CFTR.

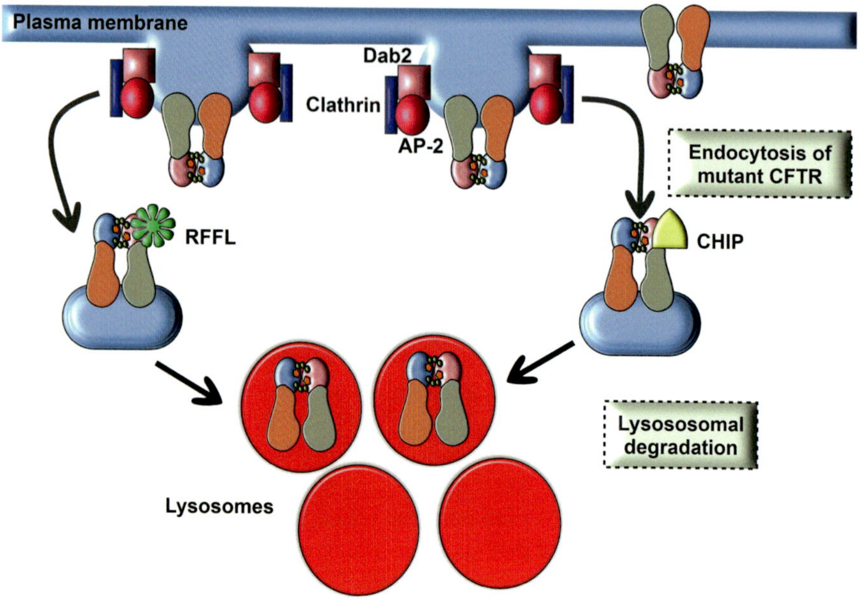

Fig. 16.20 Proteins of peripheral quality control checkpoints as putative drug targets. The schematic shows some proteins that constitute peripheral quality control checkpoints, which mediate the endocytosis of CFTR and its degradation in lysosomes. Selected proteostasis regulators that represent targets for CF drug development are indicated. *AP-2* adaptor protein 2, *CHIP* Hsp70-interacting protein, *Dab2* disabled-2, *RFFL* ring finger and FYVE-like domain containing E3-ubiquitin protein ligase

16.8.3 Constituents of the Peripheral Quality Control Machinery as Possible Drug Targets for F508del-CFTR Rescue

The amount of CFTR at the plasma membrane is regulated by endocytosis and recycling (Guggino and Stanton 2006) (Fig. 16.20). Endocytosis of CFTR is mediated by clathrin and requires two closely related but distinct processes: assembly of the clathrin coat and recruitment of cargo proteins (Guggino and Stanton 2006) (Fig. 16.20). Several endocytic adaptors have been identified, but their roles are incompletely defined and there is evidence for tissue- and cell-specific differences (Traub 2003). The assembly polypeptide-2 complex (AP-2) is the prototypical endocytic adaptor responsible for optimal clathrin coat formation, and it is composed of $\alpha 2$, $\beta 2$, $\mu 2$, and $\sigma 2$ subunits (Traub 2003). Disabled-2 (Dab2) is a multidomain clathrin-associated sorting protein, which like AP-2, facilitates endocytosis by organizing clathrin assembly and by recruiting cargo and other adaptor proteins (Motley et al. 2003; Fu et al. 2012). Both Dab2 and AP-2 play roles in CFTR endocytosis in cells that endogenously express the channel (Fig. 16.20). In human airway epithelial cells, knockdown of $\mu 2$ adaptin by more than 90% resulted in only a twofold reduction in CFTR endocytosis (Fu et al. 2012), suggesting that this process is AP-2 independent (Cihil et al. 2012). By contrast, knockdown of Dab2 markedly decreased CFTR endocytosis (Cihil et al. 2012). However, in intestinal epithelial cells, Dab2 does not play a direct role in CFTR endocytosis, but appears to promote CFTR endocytosis by working in concert with AP-2 (Madden and Swiatecka-Urban 2012). Furthermore, the rapid turnover of rescued F508del-CFTR at the plasma membrane involves internalization by both AP-2 and Dab2 and Dab2 targeting of ubiquitinated rescued F508del-CFTR to the lysosome (Fu et al. 2015). Like the effects of CFTR correctors on the proteostatic network (Lopes-Pacheco et al. 2017), lumacaftor increased the plasma membrane stability of rescued F508del-CFTR by inhibiting CFTR ubiquitination (Fu et al. 2015) (Fig. 16.20).

Based on current knowledge, targeting Dab2 might represent a therapeutic strategy to control the limited plasma membrane stability of rescued F508del-CFTR. Previous data have shown that a peptide, which binds the Dab2 DH peptide-binding pocket, regulates the abundance of CFTR at the plasma membrane (Cihil et al. 2012). Peptides or drug-like small molecules, which inhibit the Dab2 DH peptide-binding pocket, do not interfere with endocytosis because they do not target AP-2-mediated interactions. Thus, these peptides and small molecules might represent new pharmacological tools to correct the post-maturation trafficking defects of F508del-CFTR.

Ubiquitination of rescued F508del-CFTR at the plasma membrane occurs through the recruitment of different ligases with distinct roles (Sharma et al. 2004; Okiyoneda et al. 2010; Fu et al. 2015) (Fig. 16.20). For example, MARCH2, C-Cbl, and Nedd4–2 ubiquitinate CFTR regardless of its folding status (Caohuy et al. 2009; Ye et al. 2010; Cheng and Guggino 2013), whereas CHIP-UbcH5c selectively recognizes and ubiquitinates conformationally defective CFTR through Hsc70/Hsp90 and co-chaperone complexes (Okiyoneda et al. 2010). However, ablation of the E3 ligases CHIP and c-Cbl only partially inhibited rapid elimination of rescued F508del-CFTR from the plasma membrane (Okiyoneda et al. 2010; Ye et al. 2010; Fu et al. 2015), suggesting the presence of additional ligases responsible for the detection of rescued, but still partially misfolded, F508del-CFTR. By screening a library of siRNA molecules targeting more than 600 different ligases, Okiyoneda et al. (2018) identified the E3 ligase RFFL as one of the primary E3 enzymes responsible for chaperone-independent ubiquitination and peripheral QC of misfolded CFTR, possibly by recognizing the unfolded regions of CFTR, including NBD1 at the plasma membrane and in endosomes (Okiyoneda et al. 2018) (Fig. 16.20). Interestingly, RFFL inhibition led to significant improvement in the rescue of F508del-CFTR by lumacaftor (Okiyoneda et al. 2018). Thus, pharmacological modulation of peripheral quality control might constitute a strategy to prolong the residence time at the plasma membrane of misfolded CF mutants rescued by CFTR correctors.

To summarize, proteostasis regulators that target either the chaperome to improve CFTR folding or key molecular components of cellular quality control mechanisms to prevent CFTR degradation have therapeutic potential individually, in combination or together with CFTR modulators. However, since proteostasis regulators are not specific CFTR modulators, there is the potential for adverse off-target effects. Thus, the application of proteostasis regulators as CF therapeutics requires a better understanding of their biology and careful evaluation of the balance between their beneficial and harmful actions.

16.9 Conclusion

CFTR plays a pivotal role in epithelial physiology as its dysfunction in disease testifies. The identification and cloning of CFTR over 30 years ago initiated a concerted effort to understand how its expression and function are impacted by CF mutations. These studies have led to great advances in our understanding of the physiology of epithelial ion transport. Importantly, they provided an assured foundation for drug discovery and development efforts that have transformed the treatment of CF: ivacaftor for CF patients with gating mutations and Trikafta for CF patients with F508del and a minimal function mutation. Building on this success, future studies aim to develop even more effective therapies for CF through deep mechanistic insights. For further information about CFTR and clinical aspects of CF, interested readers are directed to Chap. 15.

Acknowledgments We thank our laboratory colleagues and collaborators for stimulating discussions and assistance and B Kleizen for the generous gift of Fig. 16.3. We are very grateful to T-C Hwang for patch-clamp data (Figs. 16.14 and 16.16) and RJ Bridges for guidance and support. Work in the authors' laboratories discussed in this manuscript was supported by a postgraduate scholarship from the Government of Oman and grant numbers GRSP/OMC/15/004 and BFP/RGP/HSS/18/030 from the Research Council of Oman (MKAS), Cystic Fibrosis Foundation Therapeutics, Cystic Fibrosis Trust and Medical Research Council (DNS), AbbVie's CF program (CB and AKS) and Fondazione per la Ricerca sulla Fibrosi Cistica, Italian Ministry of Health (Cinque per mille and Ricerca Corrente—Linea1), and Cystic Fibrosis Foundation (NP).

Conflict of Interest CB and AKS are employees of AbbVie, Inc. DNS is the recipient of a Vertex Innovation Award from Vertex Pharmaceuticals (Europe) Ltd. All other authors have no conflicts of interest to declare.

Author Contributions All authors drafted the chapter or revised it critically for important intellectual content. All authors approved the final version of the chapter.

References

Ai T, Bompadre SG, Wang X, Hu S, Li M, Hwang TC (2004) Capsaicin potentiates wild-type and mutant cystic fibrosis transmembrane conductance regulator chloride-channel currents. Mol Pharmacol 65:1415–1426. https://doi.org/10.1124/mol.65.6.1415

Alberti S, Böhse K, Arndt V, Schmitz A, Höhfeld J (2004) The cochaperone HspBP1 inhibits the CHIP ubiquitin ligase and stimulates the maturation of the cystic fibrosis transmembrane conductance regulator. Mol Biol Cell 15:4003–4010. https://doi.org/10.1091/mbc.e04-04-0293
Aleksandrov AA, Riordan JR (1998) Regulation of CFTR ion channel gating by MgATP. FEBS Lett 431:97–101. https://doi.org/10.1016/S0014-5793(98)00713-3
Aleksandrov AA, Chang X, Aleksandrov L, Riordan JR (2000) The non-hydrolytic pathway of cystic fibrosis transmembrane conductance regulator ion channel gating. J Physiol 528:259–265. https://doi.org/10.1111/j.1469-7793.2000.00259.x
Aleksandrov L, Aleksandrov AA, Chang XB, Riordan JR (2002) The first nucleotide binding domain of cystic fibrosis transmembrane conductance regulator is a site of stable nucleotide interaction, whereas the second is a site of rapid turnover. J Biol Chem 277:15419–15425. https://doi.org/10.1074/jbc.M111713200
Aleksandrov AA, Cui L, Riordan JR (2009) Relationship between nucleotide binding and ion channel gating in cystic fibrosis transmembrane conductance regulator. J Physiol 587:2875–2886. https://doi.org/10.1113/jphysiol.2009.170258
Alton EWFW, Boyd AC, Davies JC, Gill DR, Griesenbach U, Harrison PT, Henig N, Higgins T, Hyde SC, Innes JA, Korman MSD (2016) Genetic medicines for CF: hype versus reality. Pediatr Pulmonol 51:S5–S17. https://doi.org/10.1002/ppul.23543
Anderson MP, Welsh MJ (1992) Regulation by ATP and ADP of CFTR chloride channels that contain mutant nucleotide-binding domains. Science 257:1701–1704. https://doi.org/10.1126/science.1382316
Anderson MP, Berger HA, Rich DP, Gregory RJ, Smith AE, Welsh MJ (1991a) Nucleoside triphosphates are required to open the CFTR chloride channel. Cell 67:775–784. https://doi.org/10.1016/0092-8674(91)90072-7
Anderson MP, Gregory RJ, Thompson S, Souza DW, Paul S, Mulligan RC, Smith AE, Welsh MJ (1991b) Demonstration that CFTR is a chloride channel by alteration of its anion selectivity. Science 253:202–205. https://doi.org/10.1126/science.1712984
Anderson MP, Sheppard DN, Berger HA, Welsh MJ (1992) Chloride channels in the apical membrane of normal and cystic fibrosis airway and intestinal epithelia. Am J Physiol Lung Cell Mol Physiol 263:L1–L14. https://doi.org/10.1152/ajplung.1992.263.1.l1
Bai Y, Li M, Hwang TC (2011) Structural basis for the channel function of a degraded ABC transporter, CFTR (ABCC7). J Gen Physiol 138:495–507. https://doi.org/10.1085/jgp.201110705
Baker JMR, Hudson RP, Kanelis V, Choy WY, Thibodeau PH, Thomas PJ, Forman-Kay JD (2007) CFTR regulatory region interacts with NBD1 predominantly via multiple transient helices. Nat Struct Mol Biol 14:738–745. https://doi.org/10.1038/nsmb1278
Balch WE, Morimoto RI, Dillin A, Kelly JW (2008) Adapting proteostasis for disease intervention. Science 319:916–919. https://doi.org/10.1126/science.1141448
Baldursson O, Berger HA, Welsh MJ (2000) Contribution of R domain phosphoserines to the function of CFTR studied in Fischer rat thyroid epithelia. Am J Physiol Lung Cell Mol Physiol 279:L835–L841. https://doi.org/10.1152/ajplung.2000.279.5.l835
Baroni D, Zegarra-Moran O, Svensson A, Moran O (2014) Direct interaction of a CFTR potentiator and a CFTR corrector with phospholipid bilayers. Eur Biophys J 43:341–346. https://doi.org/10.1007/s00249-014-0956-y
Basso C, Vergani P, Nairn AC, Gadsby DC (2003) Prolonged nonhydrolytic interaction of nucleotide with CFTR's NH_2-terminal nucleotide binding domain and its role in channel gating. J Gen Physiol 122:333–348. https://doi.org/10.1085/jgp.200308798
Bear CE, Li CH, Kartner N, Bridges RJ, Jensen TJ, Ramjeesingh M, Riordan JR (1992) Purification and functional reconstitution of the cystic fibrosis transmembrane conductance regulator (CFTR). Cell 68:809–818. https://doi.org/10.1016/0092-8674(92)90155-6
Bell SC, De Boeck K, Amaral MD (2015) New pharmacological approaches for cystic fibrosis: promises, progress, pitfalls. Pharmacol Ther 145:19–34. https://doi.org/10.1016/j.pharmthera.2014.06.005

Bell SC, Mall MA, Gutierrez H, Macek M, Madge S, Davies JC, Burgel PR, Tullis E, Castaños C, Castellani C, Byrnes CA, Cathcart F, Chotirmall SH, Cosgriff R, Eichler I, Fajac I, Goss CH, Drevinek P, Farrell PM, Gravelle AM, Havermans T, Mayer-Hamblett N, Kashirskaya N, Kerem E, Mathew JL, McKone EF, Naehrlich L, Nasr SZ, Oates GR, O'Neill C, Pypops U, Raraigh KS, Rowe SM, Southern KW, Sivam S, Stephenson AL, Zampoli M, Ratjen F (2019) The future of cystic fibrosis care: a global perspective. Lancet Respir Med 8:65–124. https://doi.org/10.1016/s2213-2600(19)30337-6

Berger HA, Travis SM, Welsh MJ (1993) Regulation of the cystic fibrosis transmembrane conductance regulator Cl^- channel by specific protein kinases and protein phosphatases. J Biol Chem 268:2037–2047

Bergmann C, Guay-Woodford LM, Harris PC, Horie S, Peters DJM, Torres VE (2018) Polycystic kidney disease. Nat Rev Dis Primers 4:50. https://doi.org/10.1038/s41572-018-0047-y

Billet A, Hanrahan JW (2013) The secret life of CFTR as a calcium-activated chloride channel. J Physiol 591:5273–5278. https://doi.org/10.1113/jphysiol.2013.261909

Billet A, Jia Y, Jensen T, Riordan JR, Hanrahan JW (2015) Regulation of the cystic fibrosis transmembrane conductance regulator anion channel by tyrosine phosphorylation. FASEB J 29:3945–3953. https://doi.org/10.1096/fj.15-273151

Bobadilla JL, Macek M Jr, Fine JP, Farrell PM (2002) Cystic fibrosis: a worldwide analysis of CFTR mutations—correlation with incidence data and application to screening. Hum Mutat 19:575–606. https://doi.org/10.1002/humu.10041

Bombieri C, Claustres M, De Boeck K, Derichs N, Dodge J, Girodon E, Sermet I, Schwarz M, Tzetis M, Wilschanski M, Bareil C, Bilton D, Castellani C, Cuppens H, Cutting GR, Drevínek P, Farrell P, Elborn JS, Jarvi K, Kerem B, Kerem E, Knowles M, Macek M Jr, Munck A, Radojkovic D, Seia M, Sheppard DN, Southern KW, Stuhrmann M, Tullis E, Zielenski J, Pignatti PF, Ferec C (2011) Recommendations for the classification of diseases as CFTR-related disorders. J Cyst Fibros 10:S86–S102. https://doi.org/10.1016/S1569-1993(11)60014-3

Bompadre SG, Sohma Y, Li M, Hwang TC (2007) G551D and G1349D, two CF-associated mutations in the signature sequences of CFTR, exhibit distinct gating defects. J Gen Physiol 129:285–298. https://doi.org/10.1085/jgp.200609667

Bonvin E, Le Rouzic P, Bernaudin JF, Cottart CH, Vandebrouck C, Crié A, Leal T, Clement A, Bonora M (2008) Congenital tracheal malformation in cystic fibrosis transmembrane conductance regulator-deficient mice. J Physiol 586:3231–3243. https://doi.org/10.1113/jphysiol.2008.150763

Bose SJ, Scott-Ward TS, Cai Z, Sheppard DN (2015) Exploiting species differences to understand the CFTR Cl^- channel. Biochem Soc Trans 43:975–982. https://doi.org/10.1042/BST20150129

Boucher RC (2007) Airway surface dehydration in cystic fibrosis: pathogenesis and therapy. Annu Rev Med 58:157–170. https://doi.org/10.1146/annurev.med.58.071905.105316

Bozoky Z, Krzeminski M, Chong PA, Forman-Kay JD (2013a) Structural changes of CFTR R region upon phosphorylation: a plastic platform for intramolecular and intermolecular interactions. FEBS J 280:4407–4416. https://doi.org/10.1111/febs.12422

Bozoky Z, Krzeminski M, Muhandiram R, Birtley JR, Al-Zahrani A, Thomas PJ, Frizzell RA, Ford RC, Forman-Kay JD (2013b) Regulatory R region of the CFTR chloride channel is a dynamic integrator of phospho-dependent intra- and intermolecular interactions. Proc Natl Acad Sci USA 110:E4427–E4436. https://doi.org/10.1073/pnas.1315104110

Bozoky Z, Ahmadi S, Milman T, Kim TH, Du K, Di Paola M, Pasyk S, Pekhletski R, Keller JP, Bear CE, Forman-Kay JD (2017) Synergy of cAMP and calcium signaling pathways in CFTR regulation. Proc Natl Acad Sci USA 114:E2086–E2095. https://doi.org/10.1073/pnas.1613546114

Cai Z, Scott-Ward TS, Sheppard DN (2003) Voltage-dependent gating of the cystic fibrosis transmembrane conductance regulator Cl^- channel. J Gen Physiol 122:605–620. https://doi.org/10.1085/jgp.200308921

Cai Z, Taddei A, Sheppard DN (2006) Differential sensitivity of the cystic fibrosis (CF)-associated mutants G551D and G1349D to potentiators of the cystic fibrosis transmembrane conductance regulator (CFTR) Cl^- channel. J Biol Chem 281:1970–1977. https://doi.org/10.1074/jbc.M510576200

Canato S, Santos JD, Carvalho AS, Aloria K, Amaral MD, Matthiesen R, Falcao AO, Farinha CM (2018) Proteomic interaction profiling reveals KIFC1 as a factor involved in early targeting of F508del-CFTR to degradation. Cell Mol Life Sci 75:4495–4509. https://doi.org/10.1007/s00018-018-2896-7

Caohuy H, Jozwik C, Pollard HB (2009) Rescue of ΔF508-CFTR by the SGK1/Nedd4-2 signaling pathway. J Biol Chem 284:25241–25253. https://doi.org/10.1074/jbc.M109.035345

Caputo A, Hinzpeter A, Caci E, Pedemonte N, Arous N, Di Duca M, Zegarra-Moran O, Fanen P, Galietta LJ (2009) Mutation-specific potency and efficacy of cystic fibrosis transmembrane conductance regulator chloride channel potentiators. J Pharmacol Exp Ther 330:783–791. https://doi.org/10.1124/jpet.109.154146

Carson MR, Travis SM, Welsh MJ (1995) The two nucleotide-binding domains of cystic fibrosis transmembrane conductance regulator (CFTR) have distinct functions in controlling channel activity. J Biol Chem 270:1711–1717. https://doi.org/10.1074/jbc.270.4.1711

Chang XB, Tabcharani JA, Hou YX, Jensen TJ, Kartner N, Alon N, Hanrahan JW, Riordan JR (1993) Protein kinase A (PKA) still activates CFTR chloride channel after mutagenesis of all 10 PKA consensus phosphorylation sites. J Biol Chem 268:11304–11311

Chang XB, Cui L, Hou YX, Jensen TJ, Aleksandrov AA, Mengos A, Riordan JR (1999) Removal of multiple arginine-framed trafficking signals overcomes misprocessing of ΔF508 CFTR present in most patients with cystic fibrosis. Mol Cell 4:137–142. https://doi.org/10.1016/S1097-2765(00)80196-3

Chang XB, Mengos A, Hou YX, Cui L, Jensen TJ, Aleksandrov A, Riordan JR, Gentzsch M (2008) Role of N-linked oligosaccharides in the biosynthetic processing of the cystic fibrosis membrane conductance regulator. J Cell Sci 121:2814–2823. https://doi.org/10.1242/jcs.028951

Chaves LA, Gadsby DC (2015) Cysteine accessibility probes timing and extent of NBD separation along the dimer interface in gating CFTR channels. J Gen Physiol 145:261–283. https://doi.org/10.1085/jgp.201411347

Chen TY, Hwang TC (2008) CLC-0 and CFTR: chloride channels evolved from transporters. Physiol Rev 88:351–387. https://doi.org/10.1152/physrev.00058.2006

Cheng J, Guggino W (2013) Ubiquitination and degradation of CFTR by the E3 ubiquitin ligase MARCH2 through its association with adaptor proteins CAL and STX6. PLoS One 8:e68001. https://doi.org/10.1371/journal.pone.0068001

Cheng SH, Gregory RJ, Marshall J, Paul S, Souza DW, White GA, O'Riordan CR, Smith AE (1990) Defective intracellular transport and processing of CFTR is the molecular basis of most cystic fibrosis. Cell 63:827–834. https://doi.org/10.1016/0092-8674(90)90148-8

Cheng SH, Rich DP, Marshall J, Gregory RJ, Welsh MJ, Smith AE (1991) Phosphorylation of the R domain by cAMP-dependent protein kinase regulates the CFTR chloride channel. Cell 66:1027–1036. https://doi.org/10.1016/0092-8674(91)90446-6

Cho HJ, Gee HY, Baek KH, Ko SK, Park JM, Lee H, Kim ND, Lee MG, Shin I (2011) A small molecule that binds to an ATPase domain of Hsc70 promotes membrane trafficking of mutant cystic fibrosis transmembrane conductance regulator. J Am Chem Soc 133:20267–20276. https://doi.org/10.1021/ja206762p

Chong PA, Kota P, Dokholyan NV, Forman-Kay JD (2013) Dynamics intrinsic to cystic fibrosis transmembrane conductance regulator function and stability. Cold Spring Harb Perspect Med 3: a009522. https://doi.org/10.1101/cshperspect.a009522

Choo-Kang LR, Zeitlin PL (2001) Induction of HSP70 promotes ΔF508 CFTR trafficking. Am J Physiol Lung Cell Mol Physiol 281:L58–L68. https://doi.org/10.1152/ajplung.2001.281.1.L58

Cihil KM, Ellinger P, Fellows A, Stolz DB, Madden DR, Swiatecka-Urban A (2012) Disabled-2 protein facilitates assembly polypeptide-2-independent recruitment of cystic fibrosis transmembrane conductance regulator to endocytic vesicles in polarized human airway epithelial cells. J Biol Chem 287:15087–15099. https://doi.org/10.1074/jbc.M112.341875

Clancy JP, Rowe SM, Accurso FJ, Aitken ML, Amin RS, Ashlock MA, Ballmann M, Boyle MP, Bronsveld I, Campbell PW, De Boeck K, Donaldson SH, Dorkin HL, Dunitz JM, Durie PR, Jain M, Leonard A, McCoy KS, Moss RB, Pilewski JM, Rosenbluth DB, Rubenstein RC, Schechter MS, Botfield M, Ordoñez CL, Spencer-Green GT, Vernillet L, Wisseh S, Yen K, Konstan MW (2012) Results of a phase IIa study of VX-809, an investigational CFTR corrector compound, in subjects with cystic fibrosis homozygous for the F508del-CFTR mutation. Thorax 67:12–18. https://doi.org/10.1136/thoraxjnl-2011-200393

Clarke LA, Sousa L, Barreto C, Amaral MD (2013) Changes in transcriptome of native nasal epithelium expressing F508del-CFTR and intersecting data from comparable studies. Respir Res 14:38. https://doi.org/10.1186/1465-9921-14-38

Cotten JF, Welsh MJ (1999) Cystic fibrosis-associated mutations at arginine 347 alter the pore architecture of CFTR. Evidence for disruption of a salt bridge. J Biol Chem 274:5429–5435. https://doi.org/10.1074/jbc.274.9.5429

Csanády L, Nairn AC, Gadsby DC (2006) Thermodynamics of CFTR channel gating: a spreading conformational change initiates an irreversible gating cycle. J Gen Physiol 128:523–533. https://doi.org/10.1085/jgp.200609558

Csanády L, Vergani P, Gadsby DC (2010) Strict coupling between CFTR's catalytic cycle and gating of its Cl^- ion pore revealed by distributions of open channel burst durations. Proc Natl Acad Sci USA 107:1241–1246. https://doi.org/10.1073/pnas.0911061107

Csanády L, Vergani P, Gadsby DC (2019) Structure, gating, and regulation of the CFTR anion channel. Physiol Rev 99:707–738. https://doi.org/10.1152/physrev.00007.2018

Cui L, Aleksandrov L, Chang XB, Hou YX, He L, Hegedus T, Gentzsch M, Aleksandrov A, Balch WE, Riordan JR (2007) Domain interdependence in the biosynthetic assembly of CFTR. J Mol Biol 365:981–994. https://doi.org/10.1016/j.jmb.2006.10.086

Cui G, Zhang ZR, O'Brien ARW, Song B, McCarty NA (2008) Mutations at arginine 352 alter the pore architecture of CFTR. J Membr Biol 222:91–106. https://doi.org/10.1007/s00232-008-9105-9

Cui Y, Parashar S, Zahoor M, Needham PG, Mari M, Zhu M, Chen S, Ho HC, Reggiori F, Farhan H, Brodsky JL, Ferro-Novick S (2019) A COPII subunit acts with an autophagy receptor to target endoplasmic reticulum for degradation. Science 365:53–60. https://doi.org/10.1126/science.aau9263

Cutting GR (2015) Cystic fibrosis genetics: from molecular understanding to clinical application. Nat Rev Genet 16:45–56. https://doi.org/10.1038/nrg3849

Dalemans W, Barbry P, Champigny G, Jallat S, Dott K, Dreyer D, Crystal RG, Pavirani A, Lecocq JP, Lazdunski M (1991) Altered chloride ion channel kinetics associated with the ΔF508 cystic fibrosis mutation. Nature 354:526–528. https://doi.org/10.1038/354526a0

Davidow CJ, Maser RL, Rome LA, Calvet JP, Grantham JJ (1996) The cystic fibrosis transmembrane conductance regulator mediates transepithelial fluid secretion by human autosomal dominant polycystic kidney disease epithelium in vitro. Kidney Int 50:208–218. https://doi.org/10.1038/ki.1996.304

Dawson RJP, Locher KP (2006) Structure of a bacterial multidrug ABC transporter. Nature 443:180–185. https://doi.org/10.1038/nature05155

De Boeck K, Amaral MD (2016) Progress in therapies for cystic fibrosis. Lancet Respir Med 4:662–674. https://doi.org/10.1016/S2213-2600(16)00023-0

De Boeck K, Wilschanski M, Castellani C, Taylor C, Cuppens H, Dodge J, Sinaasappel M, Diagnostic Working Group (2006) Cystic fibrosis: terminology and diagnostic algorithms. Thorax 61:627–635. https://doi.org/10.1136/thx.2005.043539

de Wilde G, Gees M, Musch S, Verdonck K, Jans M, Wesse AS, Singh AK, Hwang TC, Christophe T, Pizzonero M, Van der Plas S, Desroy N, Cowart M, Stouten P, Nelles L, Conrath K (2019) Identification of GLPG/ABBV-2737, a novel class of corrector, which exerts functional synergy with other CFTR modulators. Front Pharmacol 10:514. https://doi.org/10.3389/fphar.2019.00514

Delaunay A, Bromberg KD, Hayashi Y, Mirabella M, Burch D, Kirkwood B, Serra C, Malicdan MC, Mizisin AP, Morosetti R, Broccolini A, Guo LT, Jones SN, Lira SA, Puri PL, Shelton GD, Ronai Z (2008) The ER-bound RING finger protein 5 (RNF5/RMA1) causes degenerative myopathy in transgenic mice and is deregulated in inclusion body myositis. PLoS One 3:e1609. https://doi.org/10.1371/journal.pone.0001609

Denning GM, Anderson MP, Amara JF, Marshall J, Smith AE, Welsh MJ (1992) Processing of mutant cystic fibrosis transmembrane conductance regulator is temperature-sensitive. Nature 358:761–764. https://doi.org/10.1038/358761a0

Didier C, Broday L, Bhoumik A, Israeli S, Takahashi S, Nakayama K, Thomas SM, Turner CE, Henderson S, Sabe H, Ronai Z (2003) RNF5, a RING finger protein that regulates cell motility by targeting paxillin ubiquitination and altered localization. Mol Cell Biol 23:5331–5345. https://doi.org/10.1128/mcb.23.15.5331-5345.2003

Divangahi M, Balghi H, Danialou G, Comtois AS, Demoule A, Ernest S, Haston C, Robert R, Hanrahan JW, Radzioch D, Petrof BJ (2009) Lack of CFTR in skeletal muscle predisposes to muscle wasting and diaphragm muscle pump failure in cystic fibrosis mice. PLoS Genet 5: e1000586. https://doi.org/10.1371/journal.pgen.1000586

Doonan LM, Guerriero CJ, Preston GM, Buck TM, Khazanov N, Fisher EA, Senderowitz H, Brodsky JL (2019) Hsp104 facilitates the endoplasmic-reticulum–associated degradation of disease-associated and aggregation-prone substrates. Protein Sci 28:1290–1306. https://doi.org/10.1002/pro.3636

Eckford PDW, Li C, Ramjeesingh M, Bear CE (2012) Cystic fibrosis transmembrane conductance regulator (CFTR) potentiator VX-770 (ivacaftor) opens the defective channel gate of mutant CFTR in a phosphorylation-dependent but ATP-independent manner. J Biol Chem 287:36639–36649. https://doi.org/10.1074/jbc.M112.393637

Eckford PDW, Ramjeesingh M, Molinski S, Pasyk S, Dekkers JF, Li C, Ahmadi S, Ip W, Chung TE, Du K, Yeger H, Beekman J, Gonska T, Bear CE (2014) VX-809 and related corrector compounds exhibit secondary activity stabilizing active F508del-CFTR after its partial rescue to the cell surface. Chem Biol 21:666–678. https://doi.org/10.1016/j.chembiol.2014.02.021

El Khouri E, Le Pavec G, Toledano MB, Delaunay-Moisan A (2013) RNF185 is a novel E3 ligase of endoplasmic reticulum-associated degradation (ERAD) that targets cystic fibrosis transmembrane conductance regulator (CFTR). J Biol Chem 288:31177–31191. https://doi.org/10.1074/jbc.M113.470500

Elborn JS (2016) Cystic fibrosis. Lancet 388:2519–2531. https://doi.org/10.1016/S0140-6736(16)00576-6

Farinha CM, Amaral MD (2005) Most F508del-CFTR is targeted to degradation at an early folding checkpoint and independently of calnexin. Mol Cell Biol 25:5242–5252. https://doi.org/10.1128/mcb.25.12.5242-5252.2005

Farinha CM, Canato S (2017) From the endoplasmic reticulum to the plasma membrane: mechanisms of CFTR folding and trafficking. Cell Mol Life Sci 74:39–55. https://doi.org/10.1007/s00018-016-2387-7

Farinha CM, Nogueira P, Mendes F, Penque D, Amaral MD (2002) The human DnaJ homologue (Hdj)-1/heat-shock protein (Hsp) 40 co-chaperone is required for the in vivo stabilization of the cystic fibrosis transmembrane conductance regulator by Hsp70. Biochem J 366:797–806. https://doi.org/10.1042/BJ20011717

Farinha CM, King-Underwood J, Sousa M, Correia AR, Henriques BJ, Roxo-Rosa M, Da Paula AC, Williams J, Hirst S, Gomes CM, Amaral MD (2013) Revertants, low temperature, and correctors reveal the mechanism of F508del-CFTR rescue by VX-809 and suggest multiple agents for full correction. Chem Biol 20:943–955. https://doi.org/10.1016/j.chembiol.2013.06.004

Fay JF, Aleksandrov LA, Jensen TJ, Cui LL, Kousouros JN, He L, Aleksandrov AA, Gingerich DS, Riordan JR, Chen JZ (2018) Cryo-EM visualization of an active high open probability CFTR anion channel. Biochemistry 57:6234–6246. https://doi.org/10.1021/acs.biochem.8b00763
Fischer H (2009) Mechanisms and function of DUOX in epithelia of the lung. Antioxid Redox Signal 11:2453–2465. https://doi.org/10.1089/ars.2009.2558
Flume PA, Liou TG, Borowitz DS, Li H, Yen K, Ordoñez CL, Geller DE, VX 08-770-104 Study Group (2012) Ivacaftor in subjects with cystic fibrosis who are homozygous for the F508del-CFTR mutation. Chest 142:718–724. https://doi.org/10.1378/chest.11-2672
Ford RC, Marshall-Sabey D, Schuetz J (2020) Linker domains: why ABC transporters 'live in fragments no longer'. Trends Biochem Sci 45:137–148. https://doi.org/10.1016/j.tibs.2019.11.004
French PJ, Bijman J, Edixhoven M, Vaandrager AB, Scholte BJ, Lohmann SM, Nairn AC, de Jonge HR (1995) Isotype-specific activation of cystic fibrosis transmembrane conductance regulator-chloride channels by cGMP-dependent protein kinase II. J Biol Chem 270:26626–26631. https://doi.org/10.1074/jbc.270.44.26626
Frizzell RA, Hanrahan JW (2012) Physiology of epithelial chloride and fluid secretion. Cold Spring Harb Perspect Med 2:a009563. https://doi.org/10.1101/cshperspect.a009563
Fu J, Ji HL, Naren AP, Kirk KL (2001) A cluster of negative charges at the amino terminal tail of CFTR regulates ATP-dependent channel gating. J Physiol 536:459–470. https://doi.org/10.1111/j.1469-7793.2001.0459c.xd
Fu L, Rab A, Tang LP, Rowe SM, Bebok Z, Collawn JF (2012) Dab2 is a key regulator of endocytosis and post-endocytic trafficking of the cystic fibrosis transmembrane conductance regulator. Biochem J 441:633–643. https://doi.org/10.1042/BJ20111566
Fu L, Rab A, Lp T, Bebok Z, Rowe SM, Bartoszewski R, Collawn JF (2015) ΔF508 CFTR surface stability is regulated by DAB2 and CHIP-mediated ubiquitination in post-endocytic compartments. PLoS One 10:e0123131. https://doi.org/10.1371/journal.pone.0123131
Gabriel SE, Brigman KN, Koller BH, Boucher RC, Stutts MJ (1994) Cystic fibrosis heterozygote resistance to cholera toxin in the cystic fibrosis mouse model. Science 266:107–109. https://doi.org/10.1126/science.7524148
Gadsby DC, Nairn AC (1999) Control of CFTR channel gating by phosphorylation and nucleotide hydrolysis. Physiol Rev 79:S77–S107. https://doi.org/10.1152/physrev.1999.79.1.S77
Gao X, Hwang TC (2015) Localizing a gate in CFTR. Proc Natl Acad Sci USA 112:2461–2466. https://doi.org/10.1073/pnas.1420676112
Gee HY, Tang BL, Kim KH, Lee MG (2010) Syntaxin 16 binds to cystic fibrosis transmembrane conductance regulator and regulates its membrane trafficking in epithelial cells. J Biol Chem 285:35519–35527. https://doi.org/10.1074/jbc.M110.162438
Gee HY, Noh SH, Tang BL, Kim KH, Lee MG (2011) Rescue of ΔF508-CFTR trafficking via a GRASP-dependent unconventional secretion pathway. Cell 146:746–760. https://doi.org/10.1016/j.cell.2011.07.021
Goldstein RF, Niraj A, Sanderson TP, Wilson LS, Rab A, Kim H, Bebok Z, Collawn JF (2007) VCP/p97 AAA-ATPase does not interact with the endogenous wild-type cystic fibrosis transmembrane conductance regulator. Am J Respir Cell Mol Biol 36:706–714. https://doi.org/10.1165/rcmb.2006-0365OC
Gomes-Alves P, Couto F, Pesquita C, Coelho AV, Penque D (2010) Rescue of F508del-CFTR by RXR motif inactivation triggers proteome modulation associated with the unfolded protein response. Biochim Biophys Acta 1804:856–865. https://doi.org/10.1016/j.bbapap.2009.12.013
Gong X, Burbridge SM, Cowley EA, Linsdell P (2002) Molecular determinants of $Au(CN)_2^-$ binding and permeability within the cystic fibrosis transmembrane conductance regulator Cl^- channel pore. J Physiol 540:39–47. https://doi.org/10.1113/jphysiol.2001.013235
Gregory RJ, Rich DP, Cheng SH, Souza DW, Paul S, Manavalan P, Anderson MP, Welsh MJ, Smith AE (1991) Maturation and function of cystic fibrosis transmembrane conductance regulator variants bearing mutations in putative nucleotide-binding domains 1 and 2. Mol Cell Biol 11:3886–3893. https://doi.org/10.1128/mcb.11.8.3886

Grove DE, Fan CY, Ren HY, Cyr DM (2011) The endoplasmic reticulum-associated Hsp40 DNAJB12 and Hsc70 cooperate to facilitate RMA1 E3-dependent degradation of nascent CFTRΔF508. Mol Biol Cell 22:301–314. https://doi.org/10.1091/mbc.E10-09-0760

Guggino WB, Stanton BA (2006) New insights into cystic fibrosis: molecular switches that regulate CFTR. Nat Rev Mol Cell Biol 7:426–436. https://doi.org/10.1038/nrm1949

Gunderson KL, Kopito RR (1994) Effects of pyrophosphate and nucleotide analogs suggest a role for ATP hydrolysis in cystic fibrosis transmembrane regulator channel gating. J Biol Chem 269:19349–19353

Gunderson KL, Kopito RR (1995) Conformational states of CFTR associated with channel gating: the role ATP binding and hydrolysis. Cell 82:231–239. https://doi.org/10.1016/0092-8674(95)90310-0

Haardt M, Benharouga M, Lechardeur D, Kartner N, Lukacs GL (1999) C-terminal truncations destabilize the cystic fibrosis transmembrane conductance regulator without impairing its biogenesis. A novel class of mutation. J Biol Chem 274:21873–21877. https://doi.org/10.1074/jbc.274.31.21873

Hallows KR, Raghuram V, Kemp BE, Witters LA, Foskett JK (2000) Inhibition of cystic fibrosis transmembrane conductance regulator by novel interaction with the metabolic sensor AMP-activated protein kinase. J Clin Invest 105:1711–1721. https://doi.org/10.1172/JCI9622

Hanaoka K, Devuyst O, Schwiebert EM, Wilson PD, Guggino WB (1996) A role for CFTR in human autosomal dominant polycystic kidney disease. Am J Physiol Cell Physiol 270:C389–C399. https://doi.org/10.1152/ajpcell.1996.270.1.C389

Hanrahan JW, Matthes E, Carlile G, Thomas DY (2017) Corrector combination therapies for F508del-CFTR. Curr Opin Pharmacol 34:105–111. https://doi.org/10.1016/j.coph.2017.09.016

Hansson GC (2019) Mucus and mucins in diseases of the intestinal and respiratory tracts. J Intern Med 285:479–490. https://doi.org/10.1111/joim.12910

Hegedus T, Aleksandrov A, Mengos A, Cui L, Jensen TJ, Riordan JR (2009) Role of individual R domain phosphorylation sites in CFTR regulation by protein kinase A. Biochim Biophys Acta 1788:1341–1349. https://doi.org/10.1016/j.bbamem.2009.03.015

Heijerman HGM, McKone EF, Downey DG, Van Braeckel E, Rowe SM, Tullis E, Mall MA, Welter JJ, Ramsey BW, McKee CM, Marigowda G, Moskowitz SM, Waltz D, Sosnay PR, Simard C, Ahluwalia N, Xuan F, Zhang Y, Taylor-Cousar JL, McCoy KS, VX17-445-103 Trial Group (2019) Efficacy and safety of the elexacaftor plus tezacaftor plus ivacaftor combination regimen in people with cystic fibrosis homozygous for the F508del mutation: a double-blind, randomised, phase 3 trial. Lancet 394:1940–1948. https://doi.org/10.1016/S0140-6736(19)32597-8

Herzig S, Shaw RJ (2018) AMPK: Guardian of metabolism and mitochondrial homeostasis. Nat Rev Mol Cell Biol 19:121–135. https://doi.org/10.1038/nrm.2017.95

Hoffmann B, Elbahnsi A, Lehn P, Décout JL, Pietrucci F, Mornon JP, Callebaut I (2018) Combining theoretical and experimental data to decipher CFTR 3D structures and functions. Cell Mol Life Sci 75:3829–3855. https://doi.org/10.1007/s00018-018-2835-7

Holland IB, Cole SPC, Kuchler K, Higgins CF (2003) ABC proteins: from bacteria to man. Academic, London

Hou X, Wei H, Rajagopalan C, Jiang H, Wu Q, Zaman K, Xie Y, Sun F (2018) Dissection of the role of VIMP in endoplasmic reticulum-associated degradation of CFTRΔF508. Sci Rep 8:4764. https://doi.org/10.1038/s41598-018-23284-8

Huang Y, Arora K, Mun KS, Yang F, Moon C, Yarlagadda S, Jegga A, Weaver T, Naren AP (2019) Targeting DNAJB9, a novel ER luminal co-chaperone, to rescue ΔF508-CFTR. Sci Rep 9:9808. https://doi.org/10.1038/s41598-019-46161-4

Hunt JF, Wang C, Ford RC (2013) Cystic fibrosis transmembrane conductance regulator (ABCC7) structure. Cold Spring Harb Perspect Med 3:a009514. https://doi.org/10.1101/cshperspect.a009514

Husted RF, Volk KA, Sigmund RD, Stokes JB (1995) Anion secretion by the inner medullary collecting duct. Evidence for involvement of the cystic fibrosis transmembrane conductance regulator. J Clin Invest 95:644–650. https://doi.org/10.1172/jci117709

Hwang TC, Sheppard DN (2009) Gating of the CFTR Cl^- channel by ATP-driven nucleotide-binding domain dimerisation. J Physiol 587:2151–2161. https://doi.org/10.1113/jphysiol.2009.171595

Hwang TC, Yeh JT, Zhang J, Yu YC, Yeh HI, Destefano S (2018) Structural mechanisms of CFTR function and dysfunction. J Gen Physiol 150:539–570. https://doi.org/10.1085/jgp.201711946

Inagaki N, Gonoi T, Clement JP IV, Namba N, Inazawa J, Gonzalez G, Aguilar-Bryan L, Seino S, Bryan J (1995) Reconstitution of I_{KATP}: an inward rectifier subunit plus the sulfonylurea receptor. Science 270:1166–1170. https://doi.org/10.1126/science.270.5239.1166

Inagaki N, Gonoi T, Clement JP IV, Wang CZ, Aguilar-Bryan L, Bryan J, Seino S (1996) A family of sulfonylurea receptors determines the pharmacological properties of ATP-sensitive K^+ channels. Neuron 16:1011–1017. https://doi.org/10.1016/S0896-6273(00)80124-5

Ishihara H, Welsh MJ (1997) Block by MOPS reveals a conformation change in the CFTR pore produced by ATP hydrolysis. Am J Physiol Cell Physiol 273:C1278–C1289. https://doi.org/10.1152/ajpcell.1997.273.4.C1278

Jensen TJ, Loo MA, Pind S, Williams DB, Goldberg AL, Riordan JR (1995) Multiple proteolytic systems, including the proteasome, contribute to CFTR processing. Cell 83:129–135. https://doi.org/10.1016/0092-8674(95)90241-4

Jih KY, Hwang TC (2012) Nonequilibrium gating of CFTR on an equilibrium theme. Physiology (Bethesda) 27:351–361. https://doi.org/10.1152/physiol.00026.2012

Jih KY, Hwang TC (2013) Vx-770 potentiates CFTR function by promoting decoupling between the gating cycle and ATP hydrolysis cycle. Proc Natl Acad Sci USA 110:4404–4409. https://doi.org/10.1073/pnas.1215982110

Jih KY, Sohma Y, Hwang TC (2012) Nonintegral stoichiometry in CFTR gating revealed by a pore-lining mutation. J Gen Physiol 140:347–359. https://doi.org/10.1085/jgp.201210834

Jih KY, Lin WY, Sohma Y, Hwang TC (2017) CFTR potentiators: from bench to bedside. Curr Opin Pharmacol 34:98–104. https://doi.org/10.1016/j.coph.2017.09.015

Jouret F, Bernard A, Hermans C, Dom G, Terryn S, Leal T, Lebecque P, Cassiman JJ, Scholte BJ, de Jonge HR, Courtoy PJ, Devuyst O (2007) Cystic fibrosis is associated with a defect in apical receptor-mediated endocytosis in mouse and human kidney. J Am Soc Nephrol 18:707–718. https://doi.org/10.1681/ASN.2006030269

Jun I, Cheng MH, Sim E, Jung J, Suh BL, Kim Y, Son H, Park K, Kim CH, Yoon JH, Whitcomb DC, Bahar I, Lee MG (2016) Pore dilatation increases the bicarbonate permeability of CFTR, ANO1 and glycine receptor anion channels. J Physiol 594:2929–2955. https://doi.org/10.1113/JP271311

Kalid O, Mense M, Fischman S, Shitrit A, Bihler H, Ben-Zeev E, Schutz N, Pedemonte N, Thomas PJ, Bridges RJ, Wetmore DR, Marantz Y, Senderowitz H (2010) Small molecule correctors of F508del-CFTR discovered by structure-based virtual screening. J Comput Aided Mol Des 24:971–991. https://doi.org/10.1007/s10822-010-9390-0

Kampinga HH, Craig EA (2010) The HSP70 chaperone machinery: J proteins as drivers of functional specificity. Nat Rev Mol Cell Biol 11:579–592. https://doi.org/10.1038/nrm2941

Kartner N, Augustinas O, Jensen TJ, Naismith AL, Riordan JR (1992) Mislocalization of ΔF508 CFTR in cystic fibrosis sweat gland. Nat Genet 1:321–327. https://doi.org/10.1038/ng0892-321

Keating D, Marigowda G, Burr L, Daines C, Mall MA, McKone EF, Ramsey BW, Rowe SM, Sass LA, Tullis E, CM MK, Moskowitz SM, Robertson S, Savage J, Simard C, Van Goor F, Waltz D, Xuan F, Young T, Taylor-Cousar JL, VX16-445-001 Study Group (2018) VX-445–tezacaftor–ivacaftor in patients with cystic fibrosis and one or two Phe508del alleles. N Engl J Med 379:1612–1620. https://doi.org/10.1056/NEJMoa1807120

Kidd JF, Ramjeesingh M, Stratford F, Huan LJ, Bear CE (2004) A heteromeric complex of the two nucleotide binding domains of cystic fibrosis transmembrane conductance regulator (CFTR) mediates ATPase activity. J Biol Chem 279:41664–41669. https://doi.org/10.1074/jbc.M407666200

Kim Chiaw P, Hantouche C, Wong MJH, Matthes E, Robert R, Hanrahan JW, Shrier A, Young JC (2019) Hsp70 and DNAJA2 limit CFTR levels through degradation. PLoS One 14:e0220984. https://doi.org/10.1371/journal.pone.0220984

Kim SJ, Skach WR (2012) Mechanisms of CFTR folding at the endoplasmic reticulum. Front Pharmacol 3:201. https://doi.org/10.3389/fphar.2012.00201

Kim D, Huang J, Billet A, Abu-Arish A, Goepp J, Matthes E, Tewfik MA, Frenkiel S, Hanrahan JW (2019) Pendrin mediates bicarbonate secretion and enhances cystic fibrosis transmembrane conductance regulator function in airway surface epithelia. Am J Respir Cell Mol Biol 60:705–716. https://doi.org/10.1165/rcmb.2018-0158OC

Kim Y, Jun I, Shin DH, Yoon JG, Piao H, Jung J, Park HW, Cheng MH, Bahar I, Whitcomb DC, Lee MG (2020) Regulation of CFTR bicarbonate channel activity by WNK1: implications for pancreatitis and CFTR-related disorders. Cell Mol Gastroenterol Hepatol 9:79–103. https://doi.org/10.1016/j.jcmgh.2019.09.003

King JD Jr, Fitch AC, Lee JK, McCane JE, Mak DOD, Foskett JK, Hallows KR (2009) AMP-activated protein kinase phosphorylation of the R domain inhibits PKA stimulation of CFTR. Am J Physiol Cell Physiol 297:C94–C101. https://doi.org/10.1152/ajpcell.00677.2008

Kirk KL, Wang W (2011) A unified view of cystic fibrosis transmembrane conductance regulator (CFTR) gating: combining the allosterism of a ligand-gated channel with the enzymatic activity of an ATP-binding cassette (ABC) transporter. J Biol Chem 286:12813–12819. https://doi.org/10.1074/jbc.R111.219634

Kleizen B, Hunt JF, Callebaut I, Hwang TC, Sermet-Gaudelus I, Hafkemeyer S, Sheppard DN (2020) CFTR: new insights into structure and function and implications for modulation by small molecules. J Cyst Fibros 19:S19–S24. https://doi.org/10.1016/j.jcf.2019.10.021

Knowles MR, Durie PR (2002) What is cystic fibrosis? N Engl J Med 347:439–442. https://doi.org/10.1056/NEJMe020070

Ko SBH, Zeng W, Dorwart MR, Luo X, Kim KH, Millen L, Goto H, Naruse S, Soyombo A, Thomas PJ, Muallem S (2004) Gating of CFTR by the STAS domain of SLC26 transporters. Nat Cell Biol 6:343–350. https://doi.org/10.1038/ncb1115

Kogan I, Ramjeesingh M, Li C, Kidd JF, Wang Y, Leslie EM, Cole SPC, Bear CE (2003) CFTR directly mediates nucleotide-regulated glutathione flux. EMBO J 22:1981–1989. https://doi.org/10.1093/emboj/cdg194

Kopeikin Z, Yuksek Z, Yang HY, Bompadre SG (2014) Combined effects of VX-770 and VX-809 on several functional abnormalities of F508del-CFTR channels. J Cyst Fibros 13:508–514. https://doi.org/10.1016/j.jcf.2014.04.003

Kopelman H, Durie P, Gaskin K, Weizman Z, Forstner G (1985) Pancreatic fluid secretion and protein hyperconcentration in cystic fibrosis. N Engl J Med 312:329–334. https://doi.org/10.1056/NEJM198502073120601

Koulov AV, LaPointe P, Lu B, Razvi A, Coppinger J, Dong MQ, Matteson J, Laister R, Arrowsmith C, Yates JR III, Balch WE (2010) Biological and structural basis for Aha1 regulation of Hsp90 ATPase activity in maintaining proteostasis in the human disease cystic fibrosis. Mol Biol Cell 21:871–884. https://doi.org/10.1091/mbc.E09-12-1017

Kuang E, Okumura CYM, Sheffy-Levin S, Varsano T, Shu VC, Qi J, Niesman IR, Yang HJ, López-Otín C, Yang WY, Reed JC, Broday L, Nizet V, Ronai ZA (2012) Regulation of ATG4B stability by RNF5 limits basal levels of autophagy and influences susceptibility to bacterial infection. PLoS Genet 8:e1003007. https://doi.org/10.1371/journal.pgen.1003007

Kym PR, Wang X, Pizzonero M, Van der Plas SE (2018) Recent progress in the discovery and development of small-molecule modulators of CFTR. Prog Med Chem 57:235–276. https://doi.org/10.1016/bs.pmch.2018.01.001

Laselva O, Eckford PD, Bartlett C, Ouyang H, Gunawardena TN, Gonska T, Moraes TJ, Bear CE (2019) Functional rescue of c.3846G>A (W1282X) in patient-derived nasal cultures achieved

by inhibition of nonsense mediated decay and protein modulators with complementary mechanisms of action. J Cyst Fibros. https://doi.org/10.1016/j.jcf.2019.12.001
Le Henaff C, Haÿ E, Velard F, Marty C, Tabary O, Marie PJ, Jacquot JP (2014) Enhanced F508del-CFTR channel activity ameliorates bone pathology in murine cystic fibrosis. Am J Pathol 184:1132–1141. https://doi.org/10.1016/j.ajpath.2013.12.027
Lee MG, Ohana E, Park HW, Yang D, Muallem S (2012) Molecular mechanism of pancreatic and salivary gland fluid and HCO_3^- secretion. Physiol Rev 92:39–74. https://doi.org/10.1152/physrev.00011.2011
Lewis HA, Buchanan SG, Burley SK, Conners K, Dickey M, Dorwart M, Fowler R, Gao X, Guggino WB, Hendrickson WA, Hunt JF, Kearins MC, Lorimer D, Maloney PC, Post KW, Rajashankar KR, Rutter ME, Sauder JM, Shriver S, Thibodeau PH, Thomas PJ, Zhang M, Zhao X, Emtage S (2004) Structure of nucleotide-binding domain 1 of the cystic fibrosis transmembrane conductance regulator. EMBO J 23:282–293. https://doi.org/10.1038/sj.emboj.7600040
Li C, Ramjeesingh M, Wang W, Garami E, Hewryk M, Lee D, Rommens JM, Galley K, Bear CE (1996) ATPase activity of the cystic fibrosis transmembrane conductance regulator. J Biol Chem 271:28463–28468. https://doi.org/10.1074/jbc.271.45.28463
Li H, Yang W, Mendes F, Amaral MD, Sheppard DN (2012) Impact of the cystic fibrosis mutation F508del-CFTR on renal cyst formation and growth. Am J Physiol Renal Physiol 303:F1176–F1186. https://doi.org/10.1152/ajprenal.00130.2012
Li H, Salomon JJ, Sheppard DN, Mall MA, Galietta LJ (2017) Bypassing CFTR dysfunction in cystic fibrosis with alternative pathways for anion transport. Curr Opin Pharmacol 34:91–97. https://doi.org/10.1016/j.coph.2017.10.002
Li H, Pesce E, Sheppard DN, Singh AK, Pedemonte N (2018a) Therapeutic approaches to CFTR dysfunction: from discovery to drug development. J Cyst Fibros 17:S14–S21. https://doi.org/10.1016/j.jcf.2017.08.013
Li MS, Cowley EA, El Hiani Y, Linsdell P (2018b) Functional organization of cytoplasmic portals controlling access to the cystic fibrosis transmembrane conductance regulator (CFTR) chloride channel pore. J Biol Chem 293:5649–5658. https://doi.org/10.1074/jbc.RA117.001373
Lin WY, Jih KY, Hwang TC (2014) A single amino acid substitution in CFTR converts ATP to an inhibitory ligand. J Gen Physiol 144:311–320. https://doi.org/10.1085/jgp.201411247
Lin WY, Sohma Y, Hwang TC (2016) Synergistic potentiation of cystic fibrosis transmembrane conductance regulator gating by two chemically distinct potentiators, ivacaftor (VX-770) and 5-nitro-2-(3-Phenylpropylamino) benzoate. Mol Pharmacol 90:275–285. https://doi.org/10.1124/mol.116.104570
Linsdell P (2006) Mechanism of chloride permeation in the cystic fibrosis transmembrane conductance regulator chloride channel. Exp Physiol 91:123–129. https://doi.org/10.1113/expphysiol.2005.031757
Linsdell P (2017) Architecture and functional properties of the CFTR channel pore. Cell Mol Life Sci 74:67–83. https://doi.org/10.1007/s00018-016-2389-5
Linsdell P, Hanrahan JW (1998) Adenosine triphosphate-dependent asymmetry of anion permeation in the cystic fibrosis transmembrane conductance regulator chloride channel. J Gen Physiol 111:601–614. https://doi.org/10.1085/jgp.111.4.601
Linsdell P, Tabcharani JA, Rommens JM, Hou YX, Chang XB, Tsui LC, Riordan JR, Hanrahan JW (1997) Permeability of wild-type and mutant cystic fibrosis transmembrane conductance regulator chloride channels to polyatomic anions. J Gen Physiol 110:355–364. https://doi.org/10.1085/jgp.110.4.355
Linsdell P, Evagelidis A, Hanrahan JW (2000) Molecular determinants of anion selectivity in the cystic fibrosis transmembrane conductance regulator chloride channel pore. Biophys J 78:2973–2982. https://doi.org/10.1016/S0006-3495(00)76836-6
Liu F, Zhang Z, Csanády L, Gadsby DC, Chen J (2017) Molecular structure of the human CFTR ion channel. Cell 169:85–95.e8. https://doi.org/10.1016/j.cell.2017.02.024

Liu J, Bihler H, Farinha CM, Awatade NT, Romão AM, Mercadante D, Cheng Y, Musisi I, Jantarajit W, Wang Y, Cai Z, Amaral MD, Mense M, Sheppard DN (2018) Partial rescue of F508del-cystic fibrosis transmembrane conductance regulator channel gating with modest improvement of protein processing, but not stability, by a dual-acting small molecule. Br J Pharmacol 175:1017–1038. https://doi.org/10.1111/bph.14141

Liu F, Zhang Z, Levit A, Levring J, Touhara KK, Shoichet BK, Chen J (2019) Structural identification of a hotspot on CFTR for potentiation. Science 364:1184–1188. https://doi.org/10.1126/science.aaw7611

Loo MA, Jensen TJ, Cui L, Hou Y, Chang XB, Riordan JR (1998) Perturbation of Hsp90 interaction with nascent CFTR prevents its maturation and accelerates its degradation by the proteasome. EMBO J 17:6879–6887. https://doi.org/10.1093/emboj/17.23.6879

Lopes-Pacheco M, Boinot C, Sabirzhanova I, Rapino D, Cebotaru L (2017) Combination of correctors rescues CFTR transmembrane-domain mutants by mitigating their interactions with proteostasis. Cell Physiol Biochem 41:2194–2210. https://doi.org/10.1159/000475578

Lotz GP, Lin H, Harst A, Obermann WMJ (2003) Aha1 binds to the middle domain of Hsp90, contributes to client protein activation, and stimulates the ATPase activity of the molecular chaperone. J Biol Chem 278:17228–17235. https://doi.org/10.1074/jbc.M212761200

Lukacs GL, Verkman AS (2012) CFTR: folding, misfolding and correcting the ΔF508 conformational defect. Trends Mol Med 18:81–91. https://doi.org/10.1016/j.molmed.2011.10.003

Lukacs GL, Chang XB, Bear C, Kartner N, Mohamed A, Riordan JR, Grinstein S (1993) The ΔF508 mutation decreases the stability of cystic fibrosis transmembrane conductance regulator in the plasma membrane. Determination of functional half-lives on transfected cells. J Biol Chem 268:21592–21598

Ma T, Thiagarajah JR, Yang H, Sonawane ND, Folli C, Galietta LJV, Verkman AS (2002) Thiazolidinone CFTR inhibitor identified by high-throughput screening blocks cholera toxin–induced intestinal fluid secretion. J Clin Invest 110:1651–1658. https://doi.org/10.1172/JCI16112

Madden DR, Swiatecka-Urban A (2012) Tissue-specific control of CFTR endocytosis by Dab2. Commun Integr Biol 5:473–476. https://doi.org/10.4161/cib.21375

Mall MA, Hartl D (2014) CFTR: cystic fibrosis and beyond. Eur Respir J 44:1042–1054. https://doi.org/10.1183/09031936.00228013

Matsumura Y, David LL, Skach WR (2011) Role of Hsc70 binding cycle in CFTR folding and endoplasmic reticulum-associated degradation. Mol Biol Cell 22:2797–2809. https://doi.org/10.1091/mbc.E11-02-0137

Matsumura Y, Sakai J, Skach WR (2013) Endoplasmic reticulum protein quality control is determined by cooperative interactions between Hsp/c70 protein and the CHIP E3 ligase. J Biol Chem 288:31069–31079. https://doi.org/10.1074/jbc.M113.479345

McCarty NA (2000) Permeation through the CFTR chloride channel. J Exp Biol 203:1947–1962

Meacham GC, Lu Z, King S, Sorscher E, Tousson A, Cyr DM (1999) The Hdj-2/Hsc70 chaperone pair facilitates early steps in CFTR biogenesis. EMBO J 18:1492–1505. https://doi.org/10.1093/emboj/18.6.1492

Meacham GC, Patterson C, Zhang W, Younger JM, Cyr DM (2001) The Hsc70 co-chaperone CHIP targets immature CFTR for proteasomal degradation. Nat Cell Biol 3:100–105. https://doi.org/10.1038/35050509

Mendoza JL, Schmidt A, Li Q, Nuvaga E, Barrett T, Bridges RJ, Feranchak AP, Brautigam CA, Thomas PJ (2012) Requirements for efficient correction of ΔF508 CFTR revealed by analyses of evolved sequences. Cell 148:164–174. https://doi.org/10.1016/j.cell.2011.11.023

Meng X, Wang Y, Wang X, Wrennall JA, Rimington TL, Li H, Cai Z, Ford RC, Sheppard DN (2017) Two small molecules restore stability to a subpopulation of the cystic fibrosis transmembrane conductance regulator with the predominant disease-causing mutation. J Biol Chem 292:3706–3719. https://doi.org/10.1074/jbc.M116.751537

Mense M, Vergani P, White DM, Altberg G, Nairn AC, Gadsby DC (2006) In vivo phosphorylation of CFTR promotes formation of a nucleotide-binding domain heterodimer. EMBO J 25:4728–4739. https://doi.org/10.1038/sj.emboj.7601373

Meyerholz DK, Stoltz DA, Pezzulo AA, Welsh MJ (2010) Pathology of gastrointestinal organs in a porcine model of cystic fibrosis. Am J Pathol 176:1377–1389. https://doi.org/10.2353/ajpath.2010.090849

Middleton PG, Mall MA, Dřevínek P, Lands LC, McKone EF, Polineni D, Ramsey BW, Taylor-Cousar JL, Tullis E, Vermeulen F, Marigowda G, McKee CM, Moskowitz SM, Nair N, Savage J, Simard C, Tian S, Waltz D, Xuan F, Rowe SM, Jain R, VX17-445-102 Study Group (2019) Elexacaftor-tezacaftor-ivacaftor for cystic fibrosis with a single Phe508del allele. N Engl J Med 381:1809–1819. https://doi.org/10.1056/NEJMoa1908639

Mijnders M, Kleizen B, Braakman I (2017) Correcting CFTR folding defects by small-molecule correctors to cure cystic fibrosis. Curr Opin Pharmacol 34:83–90. https://doi.org/10.1016/j.coph.2017.09.014

Modiano G, Ciminelli BM, Pignatti PF (2007) Cystic fibrosis and lactase persistence: a possible correlation. Eur J Hum Genet 15:255–259. https://doi.org/10.1038/sj.ejhg.5201749

Monterisi S, Favia M, Guerra L, Cardone RA, Marzulli D, Reshkin SJ, Casavola V, Zaccolo M (2012) CFTR regulation in human airway epithelial cells requires integrity of the actin cytoskeleton and compartmentalized cAMP and PKA activity. J Cell Sci 125:1106–1117. https://doi.org/10.1242/jcs.089086

Monterisi S, Casavola V, Zaccolo M (2013) Local modulation of cystic fibrosis conductance regulator: cytoskeleton and compartmentalized cAMP signalling. Br J Pharmacol 169:1–9. https://doi.org/10.1111/bph.12017

Montoro DT, Haber AL, Biton M, Vinarsky V, Lin B, Birket SE, Yuan F, Chen S, Leung HM, Villoria J, Rogel N, Burgin G, Tsankov AM, Waghray A, Slyper M, Waldman J, Nguyen L, Dionne D, Rozenblatt-Rosen O, Tata PR, Mou H, Shivaraju M, Bihler H, Mense M, Tearney GJ, Rowe SM, Engelhardt JF, Regev A, Rajagopal J (2018) A revised airway epithelial hierarchy includes CFTR-expressing ionocytes. Nature 560:319–324. https://doi.org/10.1038/s41586-018-0393-7

Morales MM, Carroll TP, Morita T, Schwiebert EM, Devuyst O, Wilson PD, Lopes AG, Stanton BA, Dietz HC, Cutting GR, Guggino WB (1996) Both the wild type and a functional isoform of CFTR are expressed in kidney. Am J Physiol Renal Physiol 270:F1038–F1048. https://doi.org/10.1152/ajprenal.1996.270.6.f1038

Moran O, Galietta LJV, Zegarra-Moran O (2005) Binding site of activators of the cystic fibrosis transmembrane conductance regulator in the nucleotide binding domains. Cell Mol Life Sci 62:446–460. https://doi.org/10.1007/s00018-004-4422-3

Morito D, Hirao K, Oda Y, Hosokawa N, Tokunaga F, Cyr DM, Tanaka K, Iwai K, Nagata K (2008) Gp78 cooperates with RMA1 in endoplasmic reticulum-associated degradation of CFTRΔF508. Mol Biol Cell 19:1328–1336. https://doi.org/10.1091/mbc.E07-06-0601

Mornon JP, Lehn P, Callebaut I (2008) Atomic model of human cystic fibrosis transmembrane conductance regulator: membrane-spanning domains and coupling interfaces. Cell Mol Life Sci 65:2594–2612. https://doi.org/10.1007/s00018-008-8249-1

Motley A, Bright NA, Seaman MNJ, Robinson MS (2003) Clathrin-mediated endocytosis in AP-2-depleted cells. J Cell Biol 162:909–918. https://doi.org/10.1083/jcb.200305145

Nagel G, Hwang TC, Nastiuk KL, Nairn AC, Gadsby DC (1992) The protein kinase A-regulated cardiac Cl^- channel resembles the cystic fibrosis transmembrane conductance regulator. Nature 360:81–84. https://doi.org/10.1038/360081a0

Nakatsukasa K, Brodsky JL (2008) The recognition and retrotranslocation of misfolded proteins from the endoplasmic reticulum. Traffic 9:861–870. https://doi.org/10.1111/j.1600-0854.2008.00729.x

Naren AP, Quick MW, Collawn JF, Nelson DJ, Kirk KL (1998) Syntaxin 1A inhibits CFTR chloride channels by means of domain-specific protein-protein interactions. Proc Natl Acad Sci USA 95:10972–10977. https://doi.org/10.1073/pnas.95.18.10972

Naren AP, Cormet-Boyaka E, Fu J, Villain M, Blalock JE, Quick MW, Kirk KL (1999) CFTR chloride channel regulation by an interdomain interaction. Science 286:544–548. https://doi.org/10.1126/science.286.5439.544

Naren AP, Cobb B, Li C, Roy K, Nelson D, Heda GD, Liao J, Kirk KL, Sorscher EJ, Hanrahan J, Clancy JP (2003) A macromolecular complex of β_2 adrenergic receptor, CFTR, and ezrin/radixin/moesin-binding phosphoprotein 50 is regulated by PKA. Proc Natl Acad Sci USA 100:342–346. https://doi.org/10.1073/pnas.0135434100

Needham PG, Guerriero CJ, Brodsky JL (2019) Chaperoning endoplasmic reticulum – associated degradation (ERAD) and protein conformational diseases. Cold Spring Harb Perspect Biol 11: a033928. https://doi.org/10.1101/cshperspect.a033928

Okiyoneda T, Lukacs GL (2012) Fixing cystic fibrosis by correcting CFTR domain assembly. J Cell Biol 199:199–204. https://doi.org/10.1083/jcb.201208083

Okiyoneda T, Barrière H, Bagdány M, Rabeh WM, Du K, Höhfeld J, Young JC, Lukacs GL (2010) Peripheral protein quality control removes unfolded CFTR from the plasma membrane. Science 329:805–810. https://doi.org/10.1126/science.1191542

Okiyoneda T, Veit G, Dekkers JF, Bagdany M, Soya N, Xu H, Roldan A, Verkman AS, Kurth M, Simon A, Hegedus T, Beekman JM, Lukacs GL (2013) Mechanism-based corrector combination restores ΔF508-CFTR folding and function. Nat Chem Biol 9:444–454. https://doi.org/10.1038/nchembio.1253

Okiyoneda T, Veit G, Sakai R, Aki M, Fujihara T, Higashi M, Susuki-Miyata S, Miyata M, Fukuda N, Yoshida A, Xu H, Apaja PM, Lukacs GL (2018) Chaperone-independent peripheral quality control of CFTR by RFFL E3 ligase. Dev Cell 44:694–708.e7. https://doi.org/10.1016/j.devcel.2018.02.001

Ostedgaard LS, Baldursson O, Vermeer DW, Welsh MJ, Robertson AD (2000) A functional R domain from cystic fibrosis transmembrane conductance regulator is predominantly unstructured in solution. Proc Natl Acad Sci USA 97:5657–5662. https://doi.org/10.1073/pnas.100588797

Ostedgaard LS, Moninger TO, McMenimen JD, Sawin NM, Parker CP, Thornell IM, Powers LS, Gansemer ND, Bouzek DC, Cook DP, Meyerholz DK, Abou Alaiwa MH, Stoltz DA, Welsh MJ (2017) Gel-forming mucins form distinct morphologic structures in airways. Proc Natl Acad Sci USA 114:6842–6847. https://doi.org/10.1073/pnas.1703228114

Pankow S, Bamberger C, Calzolari D, Martínez-Bartolomé S, Lavallée-Adam M, Balch WE, Yates JR III (2015) ΔF508 CFTR interactome remodelling promotes rescue of cystic fibrosis. Nature 528:510–516. https://doi.org/10.1038/nature15729

Park HJ, Mylvaganum M, McPherson A, Fewell SW, Brodsky JL, Lingwood CA (2009) A soluble sulfogalactosyl ceramide mimic promotes ΔF508 CFTR escape from endoplasmic reticulum associated degradation. Chem Biol 16:461–470. https://doi.org/10.1016/j.chembiol.2009.02.014

Park HW, Nam JH, Kim JY, Namkung W, Yoon JS, Lee JS, Kim KS, Venglovecz V, Gray MA, Kim KH, Lee MG (2010) Dynamic regulation of CFTR bicarbonate permeability by $[Cl^-]_i$ and its role in pancreatic bicarbonate secretion. Gastroenterology 139:620–631. https://doi.org/10.1053/j.gastro.2010.04.004

Pedemonte N, Lukacs GL, Du K, Caci E, Zegarra-Moran O, Galietta LJV, Verkman AS (2005) Small-molecule correctors of defective ΔF508-CFTR cellular processing identified by high-throughput screening. J Clin Invest 115:2564–2571. https://doi.org/10.1172/JCI24898

Pedemonte N, Tomati V, Sondo E, Caci E, Millo E, Armirotti A, Damonte G, Zegarra-Moran O, Galietta LJV (2011) Dual activity of aminoarylthiazoles on the trafficking and gating defects of the cystic fibrosis transmembrane conductance regulator chloride channel caused by cystic fibrosis mutations. J Biol Chem 286:15215–15226. https://doi.org/10.1074/jbc.M110.184267

Pezzulo AA, Tang XX, Hoegger MJ, Abou Alaiwa MH, Ramachandran S, Moninger TO, Karp PH, Wohlford-Lenane CL, Haagsman HP, van Eijk M, Bánfi B, Horswill AR, Stoltz DA, McCray PB Jr, Welsh MJ, Zabner J (2012) Reduced airway surface pH impairs bacterial killing in the porcine cystic fibrosis lung. Nature 487:109–113. https://doi.org/10.1038/nature11130

Phuan PW, Yang B, Knapp JM, Wood AB, Lukacs GL, Kurth MJ, Verkman AS (2011) Cyanoquinolines with independent corrector and potentiator activities restore ΔPhe508-cystic fibrosis transmembrane conductance regulator chloride channel function in cystic fibrosis. Mol Pharmacol 80:683–693. https://doi.org/10.1124/mol.111.073056

Phuan PW, Son JH, Tan JA, Li C, Musante I, Zlock L, Nielson DW, Finkbeiner WE, Kurth MJ, Galietta LJ, Haggie PM, Verkman AS (2018) Combination potentiator ('co-potentiator') therapy for CF caused by CFTR mutants, including N1303K, that are poorly responsive to single potentiators. J Cyst Fibros 17:595–606. https://doi.org/10.1016/j.jcf.2018.05.010

Phuan PW, Tan JA, Rivera AA, Zlock L, Nielson DW, Finkbeiner WE, Haggie PM, Verkman AS (2019) Nanomolar-potency 'co-potentiator' therapy for cystic fibrosis caused by a defined subset of minimal function CFTR mutants. Sci Rep 9:17640. https://doi.org/10.1038/s41598-019-54158-2

Pier GB, Grout M, Zaidi T, Meluleni G, Mueschenborn SS, Banting G, Ratcliff R, Evans MJ, Colledge WH (1998) Salmonella typhi uses CFTR to enter intestinal epithelial cells. Nature 393:79–82. https://doi.org/10.1038/30006

Plasschaert LW, Žilionis R, Choo-Wing R, Savova V, Knehr J, Roma G, Klein AM, Jaffe AB (2018) A single-cell atlas of the airway epithelium reveals the CFTR-rich pulmonary ionocyte. Nature 560:377–381. https://doi.org/10.1038/s41586-018-0394-6

Poulsen JH, Fischer H, Illek B, Machen TE (1994) Bicarbonate conductance and pH regulatory capability of cystic fibrosis transmembrane conductance regulator. Proc Natl Acad Sci USA 91:5340–5344. https://doi.org/10.1073/pnas.91.12.5340

Quinton PM (2007) Cystic fibrosis: lessons from the sweat gland. Physiology (Bethesda) 22:212–225. https://doi.org/10.1152/physiol.00041.2006

Quinton PM (2010) Role of epithelial HCO_3^- transport in mucin secretion: lessons from cystic fibrosis. Am J Physiol Cell Physiol 299:C1222–C1233. https://doi.org/10.1152/ajpcell.00362.2010

Rabeh WM, Bossard F, Xu H, Okiyoneda T, Bagdany M, Mulvihill CM, Du K, di Bernardo S, Liu Y, Konermann L, Roldan A, Lukacs GL (2012) Correction of both NBD1 energetics and domain interface is required to restore ΔF508 CFTR folding and function. Cell 148:150–163. https://doi.org/10.1016/j.cell.2011.11.024

Ramsey BW, Davies J, McElvaney NG, Tullis E, Bell SC, Dřevínek P, Griese M, McKone EF, Wainwright CE, Konstan MW, Moss R, Ratjen F, Sermet-Gaudelus I, Rowe SM, Dong Q, Rodriguez S, Yen K, Ordoñez C, Elborn JS; VX08-770-102 Study Group (2011) A CFTR potentiator in patients with cystic fibrosis and the G551D mutation. N Engl J Med 365:1663–1672. https://doi.org/10.1056/NEJMoa1105185

Ratjen F, Bell SC, Rowe SM, Goss CH, Quittner AL, Bush A (2015) Cystic fibrosis. Nat Rev Dis Primers 1:15010. https://doi.org/10.1038/nrdp.2015.10

Rauniyar N, Gupta V, Balch WE, Yates JR III (2014) Quantitative proteomic profiling reveals differentially regulated proteins in cystic fibrosis cells. J Proteome Res 13:4668–4675. https://doi.org/10.1021/pr500370g

Reddy MM, Quinton PM (1996) Deactivation of CFTR-Cl conductance by endogenous phosphatases in the native sweat duct. Am J Physiol Cell Physiol 270:C474–C480. https://doi.org/10.1152/ajpcell.1996.270.2.c474

Reddy MM, Quinton PM (2003) Control of dynamic CFTR selectivity by glutamate and ATP in epithelial cells. Nature 423:756–760. https://doi.org/10.1038/nature01694

Reilly R, Mroz MS, Dempsey E, Wynne K, Keely SJ, McKone EF, Hiebel C, Behl C, Coppinger JA (2017) Targeting the PI3K/Akt/mTOR signalling pathway in cystic fibrosis. Sci Rep 7:7642. https://doi.org/10.1038/s41598-017-06588-z

Ren XQ, Furukawa T, Haraguchi M, Sumizawa T, Aoki S, Kobayashi M, Akiyama SI (2004) Function of the ABC signature sequences in the human multidrug resistance protein 1. Mol Pharmacol 65:1536–1542. https://doi.org/10.1124/mol.65.6.1536

Rich DP, Gregory RJ, Anderson MP, Manavalan P, Smith AE, Welsh MJ (1991) Effect of deleting the R domain on CFTR-generated chloride channels. Science 253:205–207. https://doi.org/10.1126/science.1712985
Rich DP, Berger HA, Cheng SH, Travis SM, Saxena M, Smith AE, Welsh MJ (1993) Regulation of the cystic fibrosis transmembrane conductance regulator Cl^- channel by negative charge in the R domain. J Biol Chem 268:20259–20267
Riordan JR (2008) CFTR function and prospects for therapy. Annu Rev Biochem 77:701–726. https://doi.org/10.1146/annurev.biochem.75.103004.142532
Riordan JR, Rommens JM, Kerem B, Alon N, Rozmahel R, Grzelczak Z, Zielenski J, Lok S, Plavsic N, Chou JL, Drumm ML, Iannuzzi MC, Collins FS, Tsui L-C (1989) Identification of the cystic fibrosis gene: cloning and characterization of complementary DNA. Science 245:1066–1073
Robert R, Norez C, Becq F (2005) Disruption of CFTR chloride channel alters mechanical properties and cAMP-dependent Cl^- transport of mouse aortic smooth muscle cells. J Physiol 568:483–495. https://doi.org/10.1113/jphysiol.2005.085019
Rosenberg MF, Kamis AB, Aleksandrov LA, Ford RC, Riordan JR (2004) Purification and crystallisation of the cystic fibrosis transmembrane conductance regulator (CFTR). J Biol Chem 279:39051–39057. https://doi.org/10.1074/jbc.M407434200
Rosenfeld M, Wainwright CE, Higgins M, Wang LT, McKee C, Campbell D, Tian S, Schneider J, Cunningham S, Davies JC, ARRIVAL Study Group (2018) Ivacaftor treatment of cystic fibrosis in children aged 12 to <24 months and with a CFTR gating mutation (ARRIVAL): a phase 3 single-arm study. Lancet Respir Med 6:545–553. https://doi.org/10.1016/S2213-2600(18)30202-9
Rosser MFN, Grove DE, Chen L, Cyr DM (2008) Assembly and misassembly of cystic fibrosis transmembrane conductance regulator: folding defects caused by deletion of F508 occur before and after the calnexin-dependent association of membrane spanning domain (MSD) 1 and MSD2. Mol Biol Cell 19:4570–4579. https://doi.org/10.1091/mbc.E08-04-0357
Rowe SM, Heltshe SL, Gonska T, Donaldson SH, Borowitz D, Gelfond D, Sagel SD, Khan U, Mayer-Hamblett N, Van Dalfsen JM, Joseloff E, Ramsey BW, GOAL Investigators of the Cystic Fibrosis Foundation Therapeutics Development Network (2014) Clinical mechanism of the cystic fibrosis transmembrane conductance regulator potentiator ivacaftor in G551D-mediated cystic fibrosis. Am J Respir Crit Care Med 190:175–184. https://doi.org/10.1164/rccm.201404-0703OC
Rowe SM, Daines C, Ringshausen FC, Kerem E, Wilson J, Tullis E, Nair N, Simard C, Han L, Ingenito EP, McKee C, Lekstrom-Himes J, Davies JC (2017) Tezacaftor–ivacaftor in residual-function heterozygotes with cystic fibrosis. N Engl J Med 377:2024–2035. https://doi.org/10.1056/NEJMoa1709847
Roxo-Rosa M, da Costa G, Luider TM, Scholte BJ, Coelho AV, Amaral MD, Penque D (2006a) Proteomic analysis of nasal cells from cystic fibrosis patients and non-cystic fibrosis control individuals: search for novel biomarkers of cystic fibrosis lung disease. Proteomics 6:2314–2325. https://doi.org/10.1002/pmic.200500273
Roxo-Rosa M, Xu Z, Schmidt A, Neto M, Cai Z, Soares CM, Sheppard DN, Amaral MD (2006b) Revertant mutants G550E and 4RK rescue cystic fibrosis mutants in the first nucleotide-binding domain of CFTR by different mechanisms. Proc Natl Acad Sci USA 103:17891–17896. https://doi.org/10.1073/pnas.0608312103
Roy G, Chalfin EM, Saxena A, Wang X (2010) Interplay between ER exit code and domain conformation in CFTR misprocessing and rescue. Mol Biol Cell 21:597–609. https://doi.org/10.1091/mbc.E09-05-0427
Saint-Criq V, Gray MA (2017) Role of CFTR in epithelial physiology. Cell Mol Life Sci 74:93–115. https://doi.org/10.1007/s00018-016-2391-y
Saxena A, Banasavadi-Siddegowda YK, Fan Y, Bhattacharya S, Roy G, Giovannucci DR, Frizzell RA, Wang X (2012) Human heat shock protein 105/110 kDa (Hsp105/110) regulates biogenesis

and quality control of misfolded cystic fibrosis transmembrane conductance regulator at multiple levels. J Biol Chem 287:19158–19170. https://doi.org/10.1074/jbc.M111.297580
Serohijos AWR, Hegedus T, Aleksandrov AA, He L, Cui L, Dokholyan NV, Riordan JR (2008) Phenylalanine-508 mediates a cytoplasmic-membrane domain contact in the CFTR 3D structure crucial to assembly and channel function. Proc Natl Acad Sci USA 105:3256–3261. https://doi.org/10.1073/pnas.0800254105
Sharma M, Pampinella F, Nemes C, Benharouga M, So J, Du K, Bache KG, Papsin B, Zerangue N, Stenmark H, Lukacs GL (2004) Misfolding diverts CFTR from recycling to degradation: quality control at early endosomes. J Cell Biol 164:923–933. https://doi.org/10.1083/jcb.200312018
Sheppard DN, Welsh MJ (1999) Structure and function of the CFTR chloride channel. Physiol Rev 79:S23–S45. https://doi.org/10.1152/physrev.1999.79.1.S23
Sheppard DN, Ostedgaard LS, Winter MC, Welsh MJ (1995) Mechanism of dysfunction of two nucleotide binding domain mutations in cystic fibrosis transmembrane conductance regulator that are associated with pancreatic sufficiency. EMBO J 14:876–883
Short DB, Trotter KW, Reczek D, Kreda SM, Bretscher A, Boucher RC, Stutts MJ, Milgram SL (1998) An apical PDZ protein anchors the cystic fibrosis transmembrane conductance regulator to the cytoskeleton. J Biol Chem 273:19797–19801. https://doi.org/10.1074/jbc.273.31.19797
Singh AK, Hwang TC, Balut CM, Alani S, Fan Y, Swensen AM, Vortherms TA, Neelands TR, Wang X, Conrath K, Tse C (2018) Deciphering possible mechanisms of action of CFTR correctors – "An AbbVie and Galapagos experience". Pediatr Pulmonol 53:S255–S256. https://doi.org/10.1002/ppul.24152
Singh AK, Fan Y, Balut C, Alani S, Manelli AM, Swensen AM, Jia Y, Neelands TR, Vortherms TA, Liu B, Searle XB, Wang X, Gao W, Hwang TC, Ren HY, Cyr D, Kym PR, Conrath K, Tse C (2020) Biological characterization of F508delCFTR protein processing by the CFTR corrector ABBV-2222/GLPG2222. J Pharmacol Exp Ther 372:107–118. https://doi.org/10.1124/jpet.119.261800
Smith SS, Steinle ED, Meyerhoff ME, Dawson DC (1999) Cystic fibrosis transmembrane conductance regulator: physical basis for lyotropic anion selectivity patterns. J Gen Physiol 114:799–817. https://doi.org/10.1085/jgp.114.6.799
Sondo E, Pesce E, Tomati V, Marini M, Pedemonte N (2017) RNF5, DAB2 and friends: novel drug targets for cystic fibrosis. Curr Pharm Des 23:176–186. https://doi.org/10.2174/1381612822666161006161033
Sondo E, Falchi F, Caci E, Ferrera L, Giacomini E, Pesce E, Tomati V, Mandrup Bertozzi S, Goldoni L, Armirotti A, Ravazzolo R, Cavalli A, Pedemonte N (2018) Pharmacological inhibition of the ubiquitin ligase RNF5 rescues F508del-CFTR in cystic fibrosis airway epithelia. Cell Chem Biol 25:891–905.e8. https://doi.org/10.1016/j.chembiol.2018.04.010
Sorum B, Czégé D, Csanády L (2015) Timing of CFTR pore opening and structure of its transition state. Cell 163:724–733. https://doi.org/10.1016/j.cell.2015.09.052
Sosnay PR, Siklosi KR, Van Goor F, Kaniecki K, Yu H, Sharma N, Ramalho AS, Amaral MD, Dorfman R, Zielenski J, Masica DL, Karchin R, Millen L, Thomas PJ, Patrinos GP, Corey M, Lewis MH, Rommens JM, Castellani C, Penland CM, Cutting GR (2013) Defining the disease liability of variants in the cystic fibrosis transmembrane conductance regulator gene. Nat Genet 45:1160–1167. https://doi.org/10.1038/ng.2745
Sprague RS, Ellsworth ML, Stephenson AH, Kleinhenz ME, Lonigro AJ (1998) Deformation-induced ATP release from red blood cells requires CFTR activity. Am J Physiol Heart Circ Physiol 275:H1726–H1732. https://doi.org/10.1152/ajpheart.1998.275.5.h1726
Stanton BA (1997) Cystic fibrosis transmembrane conductance regulator (CFTR) and renal function. Wien Klin Wochenschr 109:457–464
Stoltz DA, Meyerholz DK, Welsh MJ (2015) Origins of cystic fibrosis lung disease. N Engl J Med 372:351–362. https://doi.org/10.1056/NEJMra1300109
Sun Z, Brodsky JL (2019) Protein quality control in the secretory pathway. J Cell Biol 218:3171–3187. https://doi.org/10.1083/jcb.201906047

Sun F, Zhang R, Gong X, Geng X, Drain PF, Frizzell RA (2006) Derlin-1 promotes the efficient degradation of the cystic fibrosis transmembrane conductance regulator (CFTR) and CFTR folding mutants. J Biol Chem 281:36856–36863. https://doi.org/10.1074/jbc.M607085200

Sun X, Yi Y, Yan Z, Rosen BH, Liang B, Winter MC, Evans TIA, Rotti PG, Yang Y, Gray JS, Park SY, Zhou W, Zhang Y, Moll SR, Woody L, Tran DM, Jiang L, Vonk AM, Beekman JM, Negulescu P, Van Goor F, Fiorino DF, Gibson-Corley KN, Engelhardt JF (2019) In utero and postnatal VX-770 administration rescues multiorgan disease in a ferret model of cystic fibrosis. Sci Transl Med 11:eaau7531. https://doi.org/10.1126/scitranslmed.aau7531

Szollosi A, Muallem DR, Csanády L, Vergani P (2011) Mutant cycles at CFTR's non-canonical ATP-binding site support little interface separation during gating. J Gen Physiol 137:549–562. https://doi.org/10.1085/jgp.201110608

Tabcharani JA, Chang XB, Riordan JR, Hanrahan JW (1991) Phosphorylation-regulated Cl^- channel in CHO cells stably expressing the cystic fibrosis gene. Nature 352:628–631. https://doi.org/10.1038/352628a0

Tang BL, Gee HY, Lee MG (2011) The cystic fibrosis transmembrane conductance regulator's expanding SNARE interactome. Traffic 12:364–371. https://doi.org/10.1111/j.1600-0854.2011.01161.x

Taylor-Cousar JL, Munck A, McKone EF, van der Ent CK, Moeller A, Simard C, Wang LT, Ingenito EP, McKee C, Lu Y, Lekstrom-Himes J, Elborn JS (2017) Tezacaftor–ivacaftor in patients with cystic fibrosis homozygous for Phe508del. N Engl J Med 377:2013–2023. https://doi.org/10.1056/NEJMoa1709846

Thelin WR, Chen Y, Gentzsch M, Kreda SM, Sallee JL, Scarlett CO, Borchers CH, Jacobson K, Stutts MJ, Milgram SL (2007) Direct interaction with filamins modulates the stability and plasma membrane expression of CFTR. J Clin Invest 117:364–374. https://doi.org/10.1172/JCI30376

Thibodeau PH, Richardson JM III, Wang W, Millen L, Watson J, Mendoza JL, Du K, Fischman S, Senderowitz H, Lukacs GL, Kirk K, Thomas PJ (2010) The cystic fibrosis-causing mutation ΔF508 affects multiple steps in cystic fibrosis transmembrane conductance regulator biogenesis. J Biol Chem 285:35825–35835. https://doi.org/10.1074/jbc.M110.131623

Tomati V, Sondo E, Armirotti A, Caci E, Pesce E, Marini M, Gianotti A, Jeon YJ, Cilli M, Pistorio A, Mastracci L, Ravazzolo R, Scholte B, Ronai Z, Galietta LJV, Pedemonte N (2015) Genetic inhibition of the ubiquitin ligase Rnf5 attenuates phenotypes associated to F508del cystic fibrosis mutation. Sci Rep 5:12138. https://doi.org/10.1038/srep12138

Traub LM (2003) Sorting it out: AP-2 and alternate clathrin adaptors in endocytic cargo selection. J Cell Biol 163:203–208. https://doi.org/10.1083/jcb.200309175

Travis SM, Berger HA, Welsh MJ (1997) Protein phosphatase 2C dephosphorylates and inactivates cystic fibrosis transmembrane conductance regulator. Proc Natl Acad Sci USA 94:11055–11060. https://doi.org/10.1073/pnas.94.20.11055

Tsai MF, Li M, Hwang TC (2010) Stable ATP binding mediated by a partial NBD dimer of the CFTR chloride channel. J Gen Physiol 135:399–414. https://doi.org/10.1085/jgp.201010399

Tsui LC, Dorfman R (2013) The cystic fibrosis gene: a molecular genetic perspective. Cold Spring Harb Perspect Med 3:a009472. https://doi.org/10.1101/cshperspect.a009472

Uc A, Giriyappa R, Meyerholz DK, Griffin M, Ostedgaard LS, Tang XX, Abu-El-Haija M, Stoltz DA, Ludwig P, Pezzulo A, Abu-El-Haija M, Taft P, Welsh MJ (2012) Pancreatic and biliary secretion are both altered in cystic fibrosis pigs. Am J Physiol Gastrointest Liver Physiol 303:G961–G968. https://doi.org/10.1152/ajpgi.00030.2012

Van Goor F, Hadida S, Grootenhuis PDJ, Burton B, Cao D, Neuberger T, Turnbull A, Singh A, Joubran J, Hazlewood A, Zhou J, McCartney J, Arumugam V, Decker C, Yang J, Young C, Olson ER, Wine JJ, Frizzell RA, Ashlock M, Negulescu P (2009) Rescue of CF airway epithelial cell function in vitro by a CFTR potentiator, VX-770. Proc Natl Acad Sci USA 106:18825–18830. https://doi.org/10.1073/pnas.0904709106

Van Goor F, Hadida S, Grootenhuis PDJ, Burton B, Stack JH, Straley KS, Decker CJ, Miller M, McCartney J, Olson ER, Wine JJ, Frizzell RA, Ashlock M, Negulescu PA (2011) Correction of

the F508del-CFTR protein processing defect in vitro by the investigational drug VX-809. Proc Natl Acad Sci USA 108:18843–18848. https://doi.org/10.1073/pnas.1105787108
Van Goor F, Yu H, Burton B, Hoffman BJ (2014) Effect of ivacaftor on CFTR forms with missense mutations associated with defects in protein processing or function. J Cyst Fibros 13:29–36. https://doi.org/10.1016/j.jcf.2013.06.008
van Willigen M, Vonk AM, Yeoh HY, Kruisselbrink E, Kleizen B, van der Ent CK, Egmond MR, de Jonge HR, Braakman I, Beekman JM, van der Sluijs P (2019) Folding-function relationship of the most common cystic fibrosis-causing CFTR conductance mutants. Life Sci Alliance 2: e201800172. https://doi.org/10.26508/lsa.201800172
Veit G, Avramescu RG, Chiang AN, Houck SA, Cai Z, Peters KW, Hong JS, Pollard HB, Guggino WB, Balch WE, Skach WR, Cutting GR, Frizzell RA, Sheppard DN, Cyr DM, Sorscher EJ, Brodsky JL, Lukacs GL (2016) From CFTR biology toward combinatorial pharmacotherapy: expanded classification of cystic fibrosis mutations. Mol Biol Cell 27:424–433. https://doi.org/10.1091/mbc.E14-04-0935
Veit G, Xu H, Dreano E, Avramescu RG, Bagdany M, Beitel LK, Roldan A, Hancock MA, Lay C, Li W, Morin K, Gao S, Mak PA, Ainscow E, Orth AP, McNamara P, Edelman A, Frenkiel S, Matouk E, Sermet-Gaudelus I, Barnes WG, Lukacs GL (2018) Structure-guided combination therapy to potently improve the function of mutant CFTRs. Nat Med 24:1732–1742. https://doi.org/10.1038/s41591-018-0200-x
Veit G, Da Fonte DF, Avramescu RG, Premchandar A, Bagdany M, Xu H, Bensinger D, Stubba D, Schmidt B, Matouk E, Lukacs GL (2019) Mutation-specific dual potentiators maximize rescue of CFTR gating mutants. J Cyst Fibros. https://doi.org/10.1016/j.jcf.2019.10.011
Vergani P, Nairn AC, Gadsby DC (2003) On the mechanism of MgATP-dependent gating of CFTR Cl^- channels. J Gen Physiol 121:17–36. https://doi.org/10.1085/jgp.20028673
Vergani P, Lockless SW, Nairn AC, Gadsby DC (2005) CFTR channel opening by ATP-driven tight dimerization of its nucleotide-binding domains. Nature 433:876–880. https://doi.org/10.1038/nature03313
Verkman AS, Galietta LJV (2009) Chloride channels as drug targets. Nat Rev Drug Discov 8:153–171. https://doi.org/10.1038/nrd2780
Wainwright CE, Elborn JS, Ramsey BW, Marigowda G, Huang X, Cipolli M, Colombo C, Davies JC, De Boeck K, Flume PA, Konstan MW, McColley SA, McCoy K, McKone EF, Munck A, Ratjen F, Rowe SM, Waltz D, Boyle MP, TRAFFIC Study Group; TRANSPORT Study Group (2015) Lumacaftor–ivacaftor in patients with cystic fibrosis homozygous for Phe508del CFTR. N Engl J Med 373:220–231. https://doi.org/10.1056/NEJMoa1409547
Wang X, Matteson J, An Y, Moyer B, Yoo JS, Bannykh S, Wilson IA, Riordan JR, Balch WE (2004) COPII-dependent export of cystic fibrosis transmembrane conductance regulator from the ER uses di-acidic exit code. J Cell Biol 167:65–74. https://doi.org/10.1083/jcb.200401035
Wang X, Venable J, LaPointe P, Hutt DM, Koulov AV, Coppinger J, Gurkan C, Kellner W, Matteson J, Plutner H, Riordan JR, Kelly JW, Yates JR III, Balch WE (2006) Hsp90 cochaperone Aha1 downregulation rescues misfolding of CFTR in cystic fibrosis. Cell 127:803–815. https://doi.org/10.1016/j.cell.2006.09.043
Wang B, Heath-Engel H, Zhang D, Nguyen N, Thomas DY, Hanrahan JW, Shore GC (2008) BAP31 interacts with Sec61 translocons and promotes retrotranslocation of CFTRΔF508 via the derlin-1 complex. Cell 133:1080–1092. https://doi.org/10.1016/j.cell.2008.04.042
Wang Y, Wrennall JA, Cai Z, Li H, Sheppard DN (2014) Understanding how cystic fibrosis mutations disrupt CFTR function: from single molecules to animal models. Int J Biochem Cell Biol 52:47–57. https://doi.org/10.1016/j.biocel.2014.04.001
Ward CL, Kopito RR (1994) Intracellular turnover of cystic fibrosis transmembrane conductance regulator. Inefficient processing and rapid degradation of wild-type and mutant proteins. J Biol Chem 269:25710–25718
Ward CL, Omura S, Kopito RR (1995) Degradation of CFTR by the ubiquitin-proteasome pathway. Cell 83:121–127. https://doi.org/10.1016/0092-8674(95)90240-6
Welsh MJ, Smith AE (1993) Molecular mechanisms of CFTR chloride channel dysfunction in cystic fibrosis. Cell 73:1251–1254. https://doi.org/10.1016/0092-8674(93)90353-r

Welsh MJ, Ramsey BW, Accurso F, Cutting GR (2001) Cystic fibrosis. In: Scriver CR, Beaudet AL, Sly WS, Valle D (eds) The metabolic and molecular basis of inherited disease. McGraw-Hill, New York, pp 5121–5188

Widdicombe JH, Wine JJ (2015) Airway gland structure and function. Physiol Rev 95:1241–1319. https://doi.org/10.1152/physrev.00039.2014

Wilke M, Buijs-Offerman RM, Aarbiou J, Colledge WH, Sheppard DN, Touqui L, Bot A, Jorna H, de Jonge HR, Scholte BJ (2011) Mouse models of cystic fibrosis: phenotypic analysis and research applications. J Cyst Fibros 10:S152–S171. https://doi.org/10.1016/S1569-1993(11)60020-9

Wilkinson DJ, Strong TV, Mansoura MK, Wood DL, Smith SS, Collins FS, Dawson DC (1997) CFTR activation: additive effects of stimulatory and inhibitory phosphorylation sites in the R domain. Am J Physiol Lung Cell Mol Physiol 273:L127–L133. https://doi.org/10.1152/ajplung.1997.273.1.l127

Winter MC, Welsh MJ (1997) Stimulation of CFTR activity by its phosphorylated R domain. Nature 389:294–296. https://doi.org/10.1038/38514

Xu N, Glockner JF, Rossetti S, Babovich-Vuksanovic D, Harris PC, Torres VE (2006) Autosomal dominant polycystic kidney disease coexisting with cystic fibrosis. J Nephrol 19:529–534

Yamamoto YH, Kimura T, Momohara S, Takeuchi M, Tani T, Kimata Y, Kadokura H, Kohno K (2010) A novel ER J-protein DNAJB12 accelerates ER-associated degradation of membrane proteins including CFTR. Cell Struct Funct 35:107–116. https://doi.org/10.1247/csf.10023

Yang Y, Janich S, Cohn JA, Wilson JM (1993) The common variant of cystic fibrosis transmembrane conductance regulator is recognized by hsp70 and degraded in a pre-Golgi nonlysosomal compartment. Proc Natl Acad Sci USA 90:9480–9484. https://doi.org/10.1073/pnas.90.20.9480

Yang B, Sonawane ND, Zhao D, Somlo S, Verkman AS (2008) Small-molecule CFTR inhibitors slow cyst growth in polycystic kidney disease. J Am Soc Nephrol 19:1300–1310. https://doi.org/10.1681/ASN.2007070828

Ye S, Cihil K, Stolz DB, Pilewski JM, Stanton BA, Swiatecka-Urban A (2010) C-Cbl facilitates endocytosis and lysosomal degradation of cystic fibrosis transmembrane conductance regulator in human airway epithelial cells. J Biol Chem 285:27008–27018. https://doi.org/10.1074/jbc.M110.139881

Yeh HI, Yeh JT, Hwang TC (2015) Modulation of CFTR gating by permeant ions. J Gen Physiol 145:47–60. https://doi.org/10.1085/jgp.201411272

Yeh HI, Sohma Y, Conrath K, Hwang TC (2017) A common mechanism for CFTR potentiators. J Gen Physiol 149:1105–1118. https://doi.org/10.1085/jgp.201711886

Yeh HI, Qiu L, Sohma Y, Conrath K, Zou X, Hwang TC (2019) Identifying the molecular target sites for CFTR potentiators GLPG1837 and VX-770. J Gen Physiol 151:912–928. https://doi.org/10.1085/jgp.201912360

Youker RT, Walsh P, Beilharz T, Lithgow T, Brodsky JL (2004) Distinct roles for the Hsp40 and Hsp90 molecular chaperones during cystic fibrosis transmembrane conductance regulator degradation in yeast. Mol Biol Cell 15:4787–4797. https://doi.org/10.1091/mbc.E04-07-0584

Young JC (2014) The role of the cytosolic HSP70 chaperone system in diseases caused by misfolding and aberrant trafficking of ion channels. Dis Model Mech 7:319–329. https://doi.org/10.1242/dmm.014001

Younger JM, Chen L, Ren HY, Rosser MFN, Turnbull EL, Fan CY, Patterson C, Cyr DM (2006) Sequential quality-control checkpoints triage misfolded cystic fibrosis transmembrane conductance regulator. Cell 126:571–582. https://doi.org/10.1016/j.cell.2006.06.041

Zeltwanger S, Wang F, Wang GT, Gillis KD, Hwang TC (1999) Gating of cystic fibrosis transmembrane conductance regulator chloride channels by adenosine triphosphate hydrolysis. Quantitative analysis of a cyclic gating scheme. J Gen Physiol 113:541–554. https://doi.org/10.1085/jgp.113.4.541

Zhang Z, Chen J (2016) Atomic structure of the cystic fibrosis transmembrane conductance regulator. Cell 167:1586–1597.e9. https://doi.org/10.1016/j.cell.2016.11.014

Zhang F, Kartner N, Lukacs GL (1998) Limited proteolysis as a probe for arrested conformational maturation of ΔF508 CFTR. Nat Struct Biol 5:180–183. https://doi.org/10.1038/nsb0398-180
Zhang Z, Liu F, Chen J (2017) Conformational changes of CFTR upon phosphorylation and ATP binding. Cell 170:483–491.e8. https://doi.org/10.1016/j.cell.2017.06.041
Zhang J, Yu YC, Yeh JT, Hwang TC (2018a) Functional characterization reveals that zebrafish CFTR prefers to occupy closed channel conformations. PLoS One 13:e0209862. https://doi.org/10.1371/journal.pone.0209862
Zhang Z, Liu F, Chen J (2018b) Molecular structure of the ATP-bound, phosphorylated human CFTR. Proc Natl Acad Sci USA 115:12757–12762. https://doi.org/10.1073/pnas.1815287115
Zielenski J, Tsui LC (1995) Cystic fibrosis: genotypic and phenotypic variations. Annu Rev Genet 29:777–807. https://doi.org/10.1146/annurev.ge.29.120195.004021

Chapter 17
TMEM16 Proteins (Anoctamins) in Epithelia

Paolo Scudieri and Luis J. V. Galietta

Abstract The members of Transmembrane 16 (TMEM16) family, also known as anoctamins, are membrane proteins regulated by cytosolic Ca^{2+}. TMEM16A (ANO1) and TMEM16B (ANO2) are Ca^{2+}-activated Cl^- channels, whereas other anoctamins, such as TMEM16F (ANO6), are associated with Ca^{2+}-activated scrambling of membrane glycerophospholipids and sphingolipids. TMEM16F also shows ion channel activity although at very high Ca^{2+} concentrations. In airway, salivary, biliary, pancreatic, and intestinal epithelial cells, TMEM16A is involved in Cl^- and bicarbonate secretion across the apical membrane. This function is directly mediated by TMEM16A as an apical channel. However, TMEM16A may also have an active role in the organization and regulation of compartments involved in Ca^{2+} signaling. In airway epithelial cells, TMEM16A expression is strongly stimulated by Th2 cytokines, IL-4 and IL-13. This upregulation occurs in mucus-producing goblet cells thus indicating a particular role of TMEM16A in mucus release and expansion.

Keywords Chloride channel · Chloride secretion · Airway epithelium · Intestinal epithelium · CFTR · Cystic fibrosis · Mucociliary clearance · Lipid scramblase · Intracellular calcium

P. Scudieri
Telethon Institute of Genetics and Medicine, Pozzuoli, NA, Italy
e-mail: p.scudieri@tigem.it

L. J. V. Galietta (✉)
Telethon Institute of Genetics and Medicine and DISMET, University of Naples "Federico II", Pozzuoli, NA, Italy
e-mail: l.galietta@tigem.it

K. L. Hamilton, D. C. Devor (eds.), *Studies of Epithelial Transporters and Ion Channels*, Physiology in Health and Disease,
https://doi.org/10.1007/978-3-030-55454-5_17

17.1 General Characteristics of TMEM16 Proteins

In the last years, TMEM16 proteins, also known as anoctamins, have appeared as an important class of macromolecules residing in cell membranes and involved in a variety of physiological functions. In 2008, TMEM16A (ANO1) and TMEM16B (ANO2) were found to act as Ca^{2+}-activated Cl^- channels, CaCCs (Caputo et al. 2008; Schroeder et al. 2008; Yang et al. 2008). Therefore, it was presumed that the other eight members of the family were also involved in ion transport. However, a few years later another study revealed that TMEM16F (ANO6) expression is associated with scramblase activity, i.e., translocation of phosphatidylserine from one leaflet to the other of plasma membrane (Suzuki et al. 2010). Later, it was reported that other members of the family (TMEM16C, TMEM16D, TMEM16G, and TMEM16J) are also involved in glycerophospholipid and sphingolipid scrambling (Suzuki et al. 2013). Furthermore, it has been shown that a fungal TMEM16 protein has both ion channel and lipid scramblase function (Malvezzi et al. 2013), although the ion transport activity appears more as a byproduct of lipid scrambling. Hence, the TMEM16 family may include proteins with diversified and multiple functions. In a recent phylogenetic analysis, it appears that TMEM16 homologs are present in many eukaryotic genomes, but the full set of ten anoctamin genes is only present in vertebrates (Pelz et al. 2018). Furthermore, functional analysis of a Dictyostelium discoideum TMEM16 protein revealed lipid scrambling but not Ca^{2+}-regulated Cl^- transport (Pelz et al. 2018). It is probable that lipid scrambling was the original function of these proteins and that later, during evolution, specialized ion transport ability has been acquired by some anoctamins.

The precise role and characteristics of each member of the family are not clear yet. TMEM16 proteins seem to share a common structural organization (Fig. 17.1a) with multiple transmembrane segments and the amino- and carboxy-termini protruding into the cytosol (Pedemonte and Galietta 2014). Most, if not all, TMEM16 proteins are also regulated by cytosolic Ca^{2+} concentrations. TMEM16A is activated when Ca^{2+} reaches concentrations in the submicromolar range (Caputo et al. 2008; Schroeder et al. 2008; Yang et al. 2008) (Fig. 17.1b). TMEM16B instead requires values above 1 μM (Pifferi et al. 2009). TMEM16F is even less sensitive, with activation of membrane currents occurring at 10 μM or higher (Grubb et al. 2013; Shimizu et al. 2013; Yang et al. 2012; Scudieri et al. 2015). Despite the different Ca^{2+} sensitivity, the mechanism of activation may involve the binding of Ca^{2+} to five highly conserved acidic amino acid residues localized in transmembrane domains TM6, TM7, and TM8. Mutagenesis of these residues in TMEM16A and TMEM16F (Yang et al. 2012; Yu et al. 2012; Tien et al. 2014), as well as in the fungal TMEM16 (Malvezzi et al. 2013), markedly decreases the apparent affinity for Ca^{2+}. The presence of the acidic residues in all TMEM16 family members suggests that these proteins are all regulated by Ca^{2+}.

TMEM16 proteins have different patterns of expression among organs and tissues. TMEM16B, C, and D have a predominant expression in neuronal cells. In this respect, TMEM16B appears to code for the Ca^{2+}-dependent Cl^- channel of

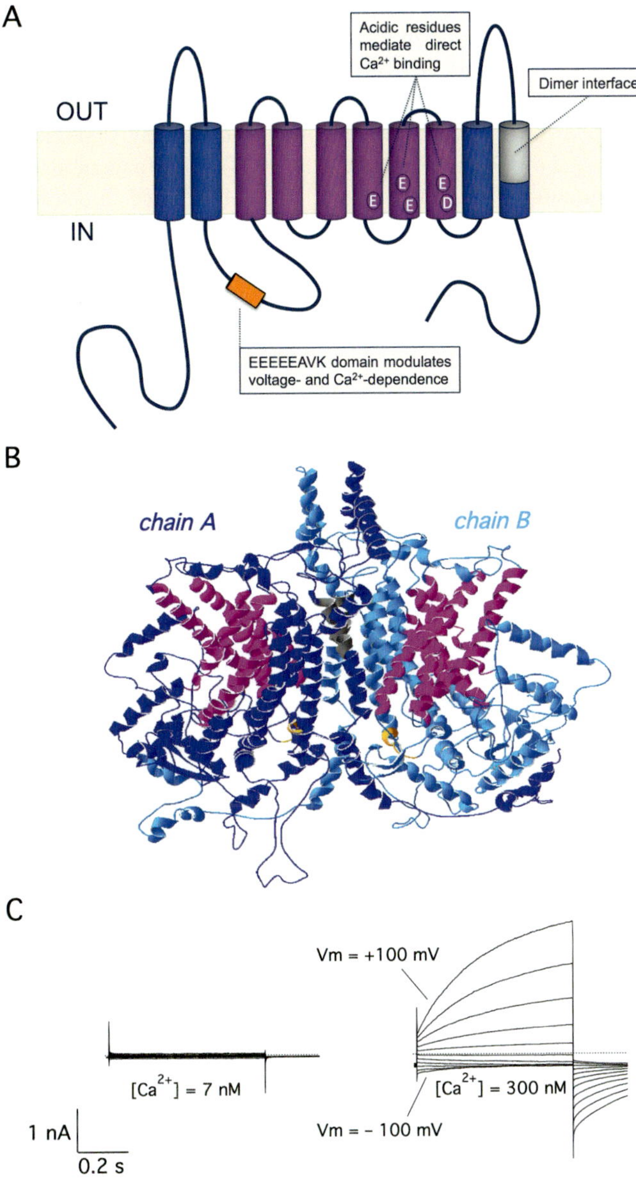

Fig. 17.1 TMEM16A/ANO1 as a Ca^{2+}-activated Cl^- channel. Simplified model (**a**) and 3D structure (**b**) of TMEM16A. Two TMEM16A proteins assemble to form the native channel (**b**). The protein has ten transmembrane segments (TM) with both amino- and carboxy-termini exposed to the intracellular environment. The transmembrane region contains both the channel pore and the Ca^{2+} binding site. The former is constituted by transmembrane segments, from third to eighth (labeled in violet in the cartoons). The binding of Ca^{2+} is coordinated by at least 5 acidic residues localized in TM6 (Glu654), TM7 (Glu702 and Glu705), and TM8 (Glu734 and Asp738). A Glu-Glu-Glu-Glu-Glu-Ala-Val-Lys (EEEEEAVK) sequence (positions 444–451, labeled in orange) in the first intracellular loop is also important for channel opening (Xiao et al. 2011). The outer part of TM10 (labeled in grey) is involved in the formation of TMEM16A homodimers

olfactory neurons and photoreceptors (Stephan et al. 2009; Stöhr et al. 2009). Instead, TMEM16C is involved in the regulation of a Na^+-activated K^+ channel in neurons (Huang et al. 2013). TMEM16E has a selective expression in skeletal muscle and bone. Indeed, mutations in TMEM16E cause two different human genetic diseases, a form of limb girdle muscular dystrophy (Bolduc et al. 2010) and gnatodiaphyseal dysplasia (Tsutsumi et al. 2004). Other anoctamins have a more ubiquitous distribution. This is the case of TMEM16F, and to a lower extent, of TMEM16K (Schreiber et al. 2010). According to various studies and gene expression databases, TMEM16F is the anoctamin with the highest expression in all tissues (Schreiber et al. 2010).

Although present also in specific types of neurons and in some smooth muscle cells, TMEM16A is the anoctamin with a particular localization in epithelial cells (Huang et al. 2009; Yang et al. 2008). This suggests a direct role in transepithelial anion secretion. Elevation of cytosolic Ca^{2+} concentrations by physiological agonists is expected to open TMEM16A channels in the apical membrane. This event creates a conductive pathway for Cl^- and bicarbonate that leave the cell creating an electrochemical driving force for parallel Na^+ and water transport. In agreement with this model, TMEM16A protein has been detected in the apical membrane of many epithelial cells (Scudieri et al. 2012; Huang et al. 2009; Huang et al. 2012; Romanenko et al. 2010). However, as discussed in the following sections, TMEM16A has also been found in the basolateral membrane (He et al. 2011; Schreiber et al. 2015). This localization implies a different and still to be clarified function.

Among all TMEM16 proteins, TMEM16A is the most studied (Pedemonte and Galietta 2014). TMEM16A, but probably all members of the TMEM16 family, exists as homodimers with each subunit consisting of an extracellular domain, a transmembrane unit, and cytosolic N- and C-terminal domains (Fig. 17.1a and b) (Sheridan et al. 2011; Fallah et al. 2011; Brunner et al. 2014; Scudieri et al. 2016; Paulino et al. 2017a, b; Dang et al. 2017). According to the molecular structures determined by cryo-EM and X-ray for mTMEM16A and other TMEM16 orthologs, the dimer interface is formed mainly by interactions between the outer part of TM10 of adjacent subunits (Brunner et al. 2014; Paulino et al. 2017a, b; Dang et al. 2017) and not by the N-terminal domains as previously proposed on the basis of biochemical experiments (Tien et al. 2013). On the extracellular side, the loops connecting transmembrane segment 1–2 and 9–10 form a domain stabilized by disulfide bridges involving six cysteines that are strictly necessary for channel activity (Yu et al. 2012; Paulino et al. 2017b). The transmembrane unit is the most conserved region among

Fig. 17.1 (continued) (Brunner et al. 2014; Paulino et al. 2017a; Dang et al. 2017). (**c**) The figure shows superimposed membrane currents elicited in cells expressing TMEM16A by application of membrane voltages in the range +100 to −100 mV. Currents were recorded with two intracellular Ca^{2+} concentrations, 7 nM and 300 nM. With low Ca^{2+}, no currents are activated at any membrane voltage. At high Ca^{2+}, large currents are detected, with increasing activation as the membrane potential is made more positive the different members of the TMEM16 family. It is composed of ten membrane-spanning helices (TM1–10) harboring the channel pore and the Ca^{2+} binding site (Brunner et al. 2014; Paulino et al. 2017a, b; Dang et al. 2017). Importantly, as previously suggested by functional studies (Jeng et al. 2016; Lim et al. 2016), each subunit contains a separate and independent channel pore (Paulino et al. 2017a; Dang et al. 2017). The pore is surrounded by transmembrane helices TM3-TM8 forming an hourglass-shaped cavity, with highly hydrophilic large intracellular and small extracellular vestibules separated by a narrow amphiphilic neck region (Paulino et al. 2017b; Dang et al. 2017). Mutagenesis studies identified a large number of possible pore-lining residues having a role in the anion selectivity of TMEM16A channel, reflecting the presence of an extended selectivity filter (Paulino et al. 2017a; Dang et al. 2017). Close to the upper part of the intracellular vestibule there is the Ca^{2+} binding site (Fig. 17.1a), formed by five acidic residues localized in transmembrane helices TM6 (Glu654), TM7 (Glu702 and Glu705), and TM8 (Glu734 and Asp738) (Yu et al. 2012; Tien et al. 2014; Brunner et al. 2014; Paulino et al. 2017a; Paulino et al. 2017b, Dang et al. 2017). Purification of mTMEM16A protein in the presence and absence of Ca^{2+}, combined with molecular dynamics simulation and functional studies, started to gain insight into the gating mechanism of the channel, and identified the TM6 as a major gating component of TMEM16A channel (Paulino et al. 2017b; Peters et al. 2018). Indeed, TM6, contributing residues to both the pore and the Ca^{2+} binding site, undergoes pronounced conformational changes upon Ca^{2+} binding: through a hinge movement around a glycine residue (G644 in the *ac* isoform of mTMEM16A) the inner half part of TM6 moves from TM4 to TM7, enlarging the neck region of the pore and thus favoring the entry of permeating ions (Paulino et al. 2017b; Peters et al. 2018). Interestingly, mutation of residues located in the hinge region of TM6 produced a channel with constitutive activity in the absence of Ca^{2+} (Paulino et al. 2017b; Peters et al. 2018). A similar behavior was observed also by perturbing the first part of the TMEM16A C-terminus, close to the last transmembrane segment, where it forms a complex with the region corresponding to the first intracellular loop connecting TM2 and TM3 (Scudieri et al. 2016). The latter is another region of TMEM16A protein with important functional properties: it is located close to the cytosolic opening of the pore and contains a segment of eight amino acids (Glu-Glu-Glu-Glu-Glu-Ala-Val-Lys) that is involved in the control of both Ca^{2+}- and voltage sensitivity (Ferrera et al. 2009; Xiao et al. 2011).

17.2 Regulation of TMEM16A Channel Activity by Ca^{2+} Signaling

As discussed in the previous section, TMEM16A activity is directly induced by binding of Ca^{2+} to specific sites of the protein. Membrane potential depolarization also activates the channel (Fig. 17.1c), possibly by cooperating with the conformational changes induced by Ca^{2+}. Calmodulin (CaM) was postulated to be involved in the regulation of TMEM16A by cytosolic Ca^{2+} but the results obtained so far are conflicting. In a series of papers, it was proposed that CaM controls activation, deactivation, and/or ion permeability of TMEM16A and TMEM16B channels (Tian et al. 2011; Jung et al. 2013; Vocke et al. 2013; Yang et al. 2014). However, the involvement of CaM was questioned in other studies (Yu et al. 2014a, b; Terashima et al. 2013). In particular, purification of TMEM16A and incorporation in liposomes was enough to obtain Ca^{2+}-dependent Cl^- transport in the absence of CaM (Terashima et al. 2013). It has been reported that alternatively spliced TMEM16A segments have a role in the binding to CaM (Vocke et al. 2013; Yang et al. 2014). Therefore, a difference in the expression of TMEM16A variants among studies could be responsible, at least in part, for the discrepant results.

In theory, all signals leading to cytosolic Ca^{2+} increase could lead to TMEM16A activation. However, this does not seem to be the case as there is evidence of compartmentalized Ca^{2+} signals that specifically activate TMEM16A. In some cell types, TMEM16A appears to be activated by release of Ca^{2+} from the endoplasmic reticulum (Fig. 17.2a) and not by Ca^{2+} influx through store-operated Ca^{2+} entry (SOCE) channels or through voltage-dependent Ca^{2+} channels (Jin et al. 2013; Cabrita et al. 2017). This type of specificity is due to the formation of microdomains where TMEM16A and inositol-triphosphate receptors are close to each other and probably physically interact (Jin et al. 2013). These microdomains also include plasma membrane receptors and all other proteins (e.g., phospholipase C) that are involved in the release of Ca^{2+} from intracellular stores. In these cell types, SOCE does not directly activate TMEM16A because ORAI channels are relatively distant. However, a dependence on SOCE may indirectly occur through a "tunneling" mechanism (Fig. 17.2b). This is based on Ca^{2+} that enters the cells through Orai channels, is taken up by the SERCA pump, and moves in the endoplasmic reticulum lumen to reach distant sites where it is released by the inositol triphosphate receptor to activate TMEM16A (Courjaret and Machaca 2014; Petersen et al. 2017). SOCE-dependent activation of TMEM16A appears particularly important in secretory cells of sweat glands (Concepcion et al. 2016), although it is not clear if this occurs through the tunneling mechanism or through a more direct connection between ORAI channels and TMEM16A.

Another Ca^{2+}-dependent mechanism controlling TMEM16A activity involves channels of the TRP superfamily (Fig. 17.2c). For example, TRPV4-dependent Ca^{2+} influx activates TMEM16A in choroid plexus epithelial cells (Takayama et al. 2014) and in lacrimal and salivary glands (Derouiche et al. 2018). TRPC1 is also involved in controlling TMEM16A in salivary glands. TRPC1-mediated Ca^{2+} influx and TMEM16A activation could occur as a result of Ca^{2+} store depletion (Sun et al. 2015). Coupling of TRPV6 and TMEM16A occurs in epithelial cells of rat epididymis (Gao et al. 2016). In thyroid cells, TRPC2 is instead the channel involved in activation of TMEM16A (Viitanen et al. 2013).

In conclusion, the relative sensitivity of TMEM16A to different Ca^{2+} signaling mechanisms is probably controlled by the specific proteome of each cell type. In this respect, it has to be emphasized that TMEM16A protein does not have a passive role

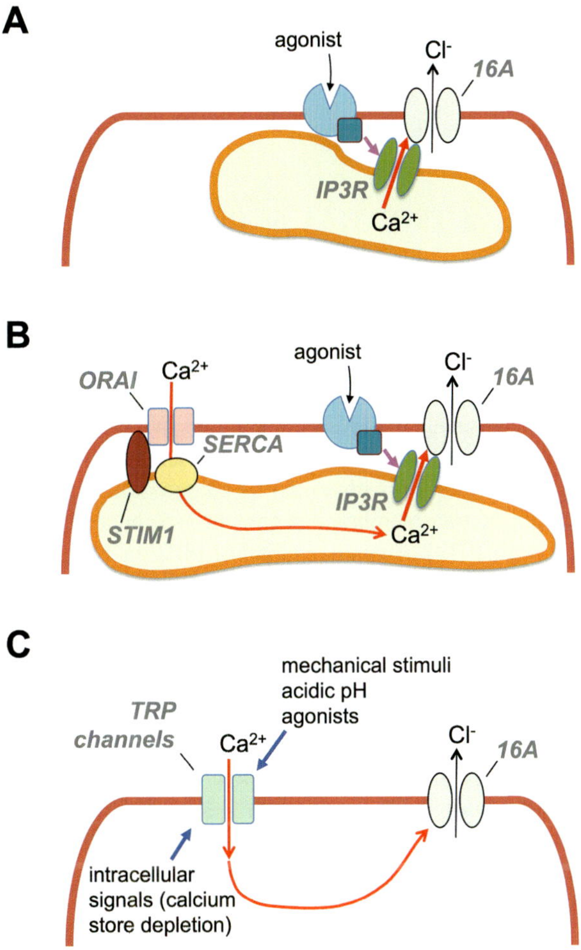

Fig. 17.2 Ca^{2+} signaling mechanisms involved in TMEM16A channel regulation. (**a**) TMEM16A may be activated by release of Ca^{2+} from intracellular stores with minimal contribution of Ca^{2+} entry from extracellular space. This mechanism implies a compartmentalization of the various proteins taking part to the signaling cascade, particularly TMEM16A and the inositol triphosphate receptor (IP3R), which have been found to physically interact. (**b**) In some cell systems, Ca^{2+} release from the intracellular store is connected via a "tunneling" mechanism to store-operated Ca^{2+} entry (SOCE). In fact, the depletion of the Ca^{2+} store caused by receptor-activated Ca^{2+} release is communicated to ORAI channel via STIM1 sensor. The opening or ORAI allows entry of Ca^{2+} that is taken up by the SERCA pump and diffuses to the site where IP3R is localized. In this way, indirect activation of TMEM16A by extracellular Ca^{2+} occurs through diffusion in the endoplasmic reticulum lumen. (**c**) TMEM16A may be also directly activated by extracellular Ca^{2+} entry through TRP channels, namely TRPV1, TRPV4, TRPV6, TRPC1, and or TRPC1. Depending on the channel type, entry of Ca^{2+} can be constitutive, activated by extracellular stimuli, or by intracellular regulatory processes

in this process but may actively participate in the formation of Ca^{2+} microdomains by working as a tether between plasma membrane and endoplasmic reticulum (Jin et al. 2013; Cabrita et al. 2017; Lérias et al. 2018). This issue is further discussed in Sect. 17.5.

17.3 Regulation of TMEM16A by PIP2

Interestingly, various studies have reported that phosphatidylinositol (4,5)-biphosphate (PIP2) has a modulatory role on TMEM16A activity. In particular, it was found that PIP2 cooperates with Ca^{2+} by keeping the TMEM16A channel in an active state (Le et al. 2019). During long-lasting activation by high cytosolic Ca^{2+} concentrations, a decrease in PIP2 levels close to TMEM16A protein causes channel pore collapse and therefore desensitization. This effect is due to a direct interaction of negatively charged PIP2 head with positively charged amino acid residues present at the cytosolic side of transmembrane segments 3–5, hence separate from Ca^{2+}-binding sites (Le et al. 2019). A positive effect of PIP2 on TMEM16A activity was also found in two additional studies (De Jesús-Pérez et al. 2018; Ta et al. 2017). In contrast, other authors found that PIP2 has an inhibitory effect on TMEM16A (Pritchard et al. 2014). The reasons for this discrepancy are not clear. However, it appears that further investigation of PIP2 role is quite important to elucidate the events that are triggered by receptor-mediated activation of TMEM16A in a physiological context. Indeed, most receptors leading to TMEM16A activation by intracellular Ca^{2+} mobilization involve activation of phospholipase C and consequently PIP2 hydrolysis. Therefore, if PIP2 is required to keep TMEM16A active, its decay during receptor stimulation would represent a negative feedback involving channel desensitization.

17.4 TMEM16A in Airway Surface Epithelial Cells

One of the three studies that first reported TMEM16A as a Ca^{2+}-activated Cl^- channel was specifically focused on airway epithelial cells (Caputo et al. 2008). Actually, the initial observation that led to TMEM16A discovery was the strong increase in Ca^{2+}-dependent Cl^- secretion caused by treatment of airway epithelial cells in vitro with Th2 cytokines, namely IL-4 and IL-13 (Danahay et al. 2002; Galietta et al. 2002). This observation suggested a possible mechanism based on increased expression of the gene coding for the Ca^{2+}-activated Cl^- channel. Starting from this hypothesis, TMEM16A was identified with a strategy based on global gene expression profiling of cells treated with IL-4, to identify genes coding for membrane proteins of unknown function, followed by silencing of the candidate genes with siRNA and evaluation of Cl^- channel function (Caputo et al. 2008). Silencing of TMEM16A in various epithelial cell types, including human bronchial epithelial cells and pancreatic CFPAC-1 cells, consistently inhibited Ca^{2+}-dependent anion transport. Other studies subsequently confirmed that TMEM16A is expressed in airway epithelial cells and involved in electrolyte/fluid secretion. In particular, with the generation of a conventional knockout mouse, it was possible to demonstrate that TMEM16a loss-of-function causes a reduced anion secretion and impaired mucociliary clearance in the airways (Ousingsawat et al. 2009; Rock et al. 2009). Interestingly, the knockout mouse showed also a malformation of the trachea at birth, i.e., tracheomalacia (Rock et al. 2008). It is not clear if this defect results from defective TMEM16A function in the epithelium or in the smooth muscle/connective tissue.

IL-4 and IL-13 are cytokines particularly involved in the Th2 immune response that underlies asthma. In the airway epithelium, Th2 cytokines promote mucous metaplasia, i.e., a marked increase in the number of mucus-secreting goblet cells and a parallel decrease in ciliated cells (Curran and Cohn 2010; Park et al. 2007) (Fig. 17.3a). The induction of TMEM16A by Th2 cytokines suggested a relationship with mucus production. Indeed, two studies revealed that TMEM16A is particularly expressed in secretory (club) and goblet cells with minimal localization in ciliated cells (Huang et al. 2012; Scudieri et al. 2012). This finding was obtained in cultured cells treated with Th2 cytokines, in mice with experimentally induced Th2 airway inflammation, and in asthmatic patients. More recently, IL-4 was reported to increase ATP- and histamine-dependent Cl^- secretion in cultured nasal epithelial cells (Kang et al. 2017) and IL-13 was found to induce TMEM16A overexpression with a mechanism involving NF-kB (Lin et al. 2015). Furthermore, TMEM16A protein expression was enhanced in the nasal epithelium of patients with allergic rhinitis, thus further confirming the association of TMEM16A with Th2 inflammation (Kang et al. 2017).

TMEM16A upregulation is not only associated to asthma and Th2 cytokines. It was found that TMEM16A expression along with mucus metaplasia can be induced in vitro by treatment with bacterial components, particularly pyocyanin, and detected in vivo in the airways of cystic fibrosis (CF) patients which are chronically colonized by bacteria at an advanced state of the disease (Caci et al. 2015). These findings suggest that infection by itself can induce TMEM16A upregulation. In general, it can be concluded that chronic obstructive pulmonary diseases involving mucus hypersecretion, such as asthma, CF, and COPD, are generally associated with TMEM16A upregulation in epithelial cells. This association could indicate that TMEM16A function is closely linked to mucus production and/or release. In this respect, it has been shown that secretion of anions, particularly of bicarbonate, is required to support mucin release and expansion from surface goblet cells and submucosal glands (Garcia et al. 2009; Gustafsson et al. 2012; Hoegger et al. 2014), and that TMEM16A is permeable to bicarbonate (Jung et al. 2013). Therefore, upregulation of TMEM16A could be a mechanism to adjust anion secretion levels to increased production of mucus. Actually, treatment of bronchial epithelial cells with IL-4 causes a general upregulation of several other anion channels and transporters including SLC26A4 (pendrin) and the SLC12A2 (NKCC1) cotransporter (Gorrieri et al. 2016; Scudieri et al. 2018).

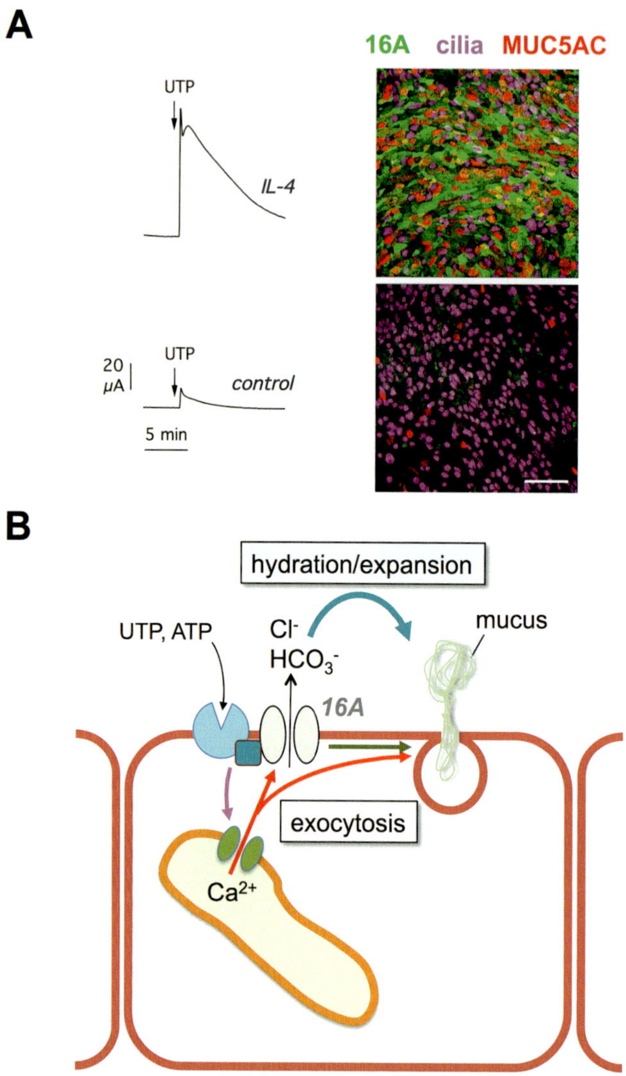

Fig. 17.3 Role of TMEM16A in airway epithelial cells. (**a**) TMEM16A expression and function is upregulated by Th2 cytokines, such as IL-4, or other stimuli. In airway epithelial cells, IL-4 causes a marked upregulation of Cl^- secretion as demonstrated by the enhanced response to purinergic agonists in short-circuit current recordings (left). This effect is due to enhanced expression of TMEM16A protein as indicated by immunofluorescence (right; scale bar 50 μm). IL-4 also induces mucous metaplasia, i.e. increase in the number of goblet cells positive for the MUC5AC mucin. (**b**) Upregulation of TMEM16A in mucous metaplasia may be important to boost the ability of the epithelium to secrete Cl^- and bicarbonate. These anions could be particularly required to hydrate and allow expansion of mucus. However, TMEM16A may be also involved in mucus exocytosis. Therefore, depending on the main function of TMEM16A, therapeutic approaches to prevent the clogging of the airways in chronic obstructive pulmonary diseases could require TMEM16A activation (to favor mucociliary clearance) or inhibition (to block mucus release)

It is important to note that airway epithelia express the cystic fibrosis transmembrane conductance regulator (CFTR) protein as another type of Cl^- channel (Riordan 2008). CFTR, which is regulated by cAMP-dependent phosphorylation, is absolutely important to maintain the human airway surface clean and microbial-free. In patients with CF, loss of CFTR function in the airways causes impaired mucociliary clearance, accumulation of dense mucus, and loss of bactericidal activity (Hoegger et al. 2014; Matsui et al. 1998; Pezzulo et al. 2012). Because of their similar activity as anion channels, it may be argued that CFTR and TMEM16A have a complementary or overlapping function in the respiratory system. Actually, as reported above, generation of conventional TMEM16A knockout mice (i.e., with ablation of TMEM16A expression in the whole body) resulted in impaired mucociliary clearance (Ousingsawat et al. 2009; Rock et al. 2009) thus mimicking to some extent the CF phenotype. However, general TMEM16A knockout mice are characterized by very compromised vitality, possibly due to tracheal malformation or block of gastrointestinal peristalsis. Therefore, experiments could only be done in the first days of life and it was not possible to establish if altered mucociliary clearance was directly caused by TMEM16A loss in the airway epithelium or by a general illness of the animal. Therefore, more recently, conditional TMEM16A knockout mice have been generated by using the FoxJ1 promoter to drive the expression of the Cre recombinase. This approach is not ideal since FoxJ1 is selective for ciliated cells whereas TMEM16A is prevalently expressed in secretory/goblet cells. Despite this limitation, the results obtained with these animals have been quite surprising and intriguing. First, ablation of TMEM16A led to the unexpected decrease of both Ca^{2+}- and cAMP-activated Cl^- currents (Benedetto et al. 2017). Since cAMP-dependent Cl^- secretion in airway epithelial cells is mediated by CFTR, these authors postulated that TMEM16A is important for CFTR function. Actually, they found that conditional TMEM16A knockout resulted in mislocalization of CFTR protein to intracellular compartments. A second type of unexpected finding was that TMEM16A knockout mice showed an abnormal accumulation of mucus in airway epithelial cells as evidenced by PAS staining (Benedetto et al. 2019a). Apparently, the loss of TMEM16A led to an impairment of basal secretion causing retention of mucus in intracellular compartments. Furthermore, mucus discharge by cholinergic agonists was unaffected whereas that by purinergic agonists was impaired (Benedetto et al. 2019a). These authors concluded that TMEM16A is in some way involved in the regulation of basal and stimulated mucus exocytosis and that pharmacological inhibition of TMEM16A, thus mimicking the effect of genetic knockout, could help to prevent mucus discharge and airway obstruction in CF and other chronic obstructive pulmonary diseases. They also argued that pharmacological activation of TMEM16A, to induce a compensatory Cl^- secretion in CF, could actually be pejorative because it could lead to more mucus release and further airway clogging (Kunzelmann et al. 2019). To prove this concept, Cabrita and colleagues (Cabrita et al. 2019) tested niclosamide, a recently discovered TMEM16A inhibitor (Miner et al. 2019), on mucus metaplasia induced in mice by Th2 inflammation. Intriguingly, treatment with niclosamide but not TMEM16A knockout prevented the increase in mucus production elicited by inflammation (Cabrita et al. 2019). To explain these results, it can be hypothesized that TMEM16A has an important role in the induction of mucus metaplasia. This mechanism can be blocked with niclosamide because it acts on all cells but not with the TMEM16A-defective animal model used by the authors because the gene knockout is presumed to only occur in ciliated cells. However, it cannot be excluded that niclosamide actually works with a mechanism not involving TMEM16A. In general, these intriguing findings point out the need of generating additional animal models in which TMEM16A is selectively ablated in non-ciliated mucus-producing cells. These models will be important to elucidate the role of TMEM16A in the airway epithelium and to define which therapeutic strategy needs to be adopted to treat CF, asthma, and other respiratory diseases. On one hand, stimulation of TMEM16A function (i.e., Cl^-/bicarbonate secretion) would be theoretically useful to hydrate airway surface, favor mucus expansion, and support mucociliary clearance. On the other hand, inhibition of TMEM16A participation to Ca^{2+} signaling and exocytosis could prevent accumulation of mucus in the airway lumen (Fig. 17.3b).

17.5 TMEM16A in the Intestine

There are conflicting results about the expression, localization, and role of TMEM16A in the intestinal epithelium. Indeed, two separate studies found evidence for localization of TMEM16A protein in the basolateral and not in the apical membrane (He et al. 2011; Schreiber et al. 2015). However, these two studies arrived to different conclusions about the contribution of TMEM16A to Cl^- secretion. In the first one, the use of a TMEM16A inhibitor, $CaCC_{inh}$-A01, resulted in increased Cl^- secretion (He et al. 2011). The authors concluded that opening of basolateral TMEM16A channels favors K^+ secretion and limits the rate of Cl^- secretion. This latter effect would occur because Cl^- entering the cells by the action of NKCC1 cotransporter would be redirected to interstitial fluids thus reducing the driving force for the Cl^- exit at the apical side (Fig. 17.4a). In the second study, TMEM16A contribution to intestinal ion transport was investigated by generating a conditional knockout mouse, with selective ablation of TMEM16A gene in intestinal epithelial cells using the villin promoter (Schreiber et al. 2015). Ca^{2+}-dependent Cl^- secretion elicited by the cholinergic agonist carbachol (CCH) was significantly diminished in the ileum and large intestine of TMEM16A$^{-/-}$ animals. Importantly, this effect was due to a reduction of Ca^{2+} mobilization by CCH (Schreiber et al. 2015). Therefore, based on these results, TMEM16A would be important to support Cl^- secretion not as an apical membrane-conductive pathway but as a facilitator of the Ca^{2+} increase that is needed to activate basolateral K^+ channels (Fig. 17.4b). The link between TMEM16A and Ca^{2+} increase is not clear, but Schreiber et al. (2015) postulated a role in the release of Ca^{2+} from the endoplasmic reticulum (ER). A possibility is that TMEM16A tethers the ER to the plasma membrane thus facilitating the release of Ca^{2+} from intracellular stores and the activation of K^+ channels (Schreiber et al. 2015). In this respect, Ist2 protein, a TMEM16A yeast homologue, is involved in the

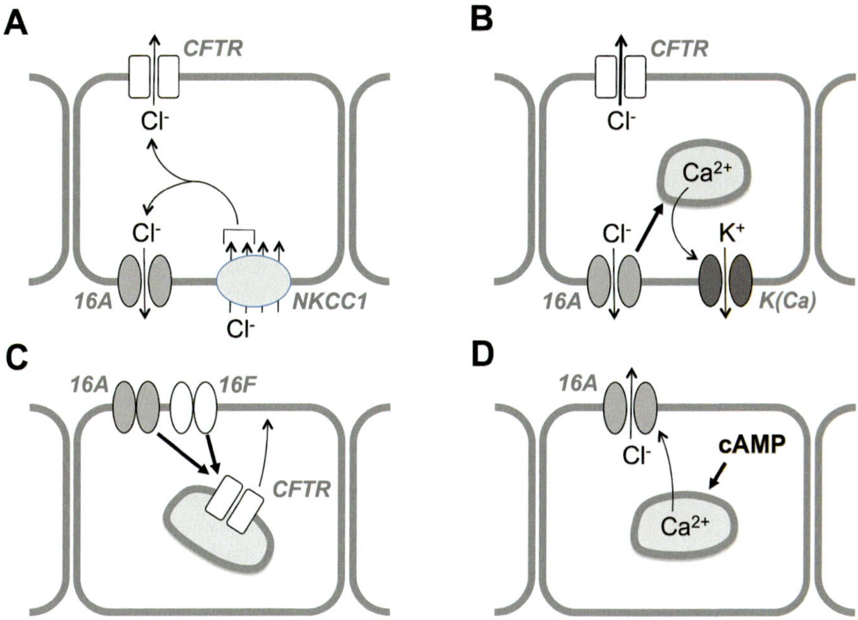

Fig. 17.4 Possible models for TMEM16A function in intestinal epithelial cells. (**a**) TMEM16A was found to be localized in the basolateral membrane and to control the ability of the epithelium to secrete Cl^-. Indeed, Cl^- entering the cells through the NKCC1 cotransporter would be recycled back to interstitial fluids preventing Cl^- exit through apical (CFTR) channels. (**b**) An alternative model assumes that TMEM16A is in the basolateral membrane but with a more prevalent role as a "facilitator" of Ca^{2+} signals controlling Ca^{2+}-activated K^+ channels. Activation of these channels causes membrane potential hyperpolarization and hence increased driving force for Cl^- secretion through apical membrane. (**c**) As a protein involved in controlling exocytic trafficking, TMEM16A (and TMEM16F) would be important to control the extent of CFTR protein expression in the apical membrane and therefore, indirectly, Cl^- secretion. (**d**) Finally, TMEM16A could directly mediate Cl^- exit at the apical side. To explain a cAMP-dependent Cl^- secretion mediated by TMEM16A and not by CFTR, it can be considered that intracellular cAMP elevation may lead to release of Ca^{2+} from intracellular stores

formation of contact sites between ER and plasma membrane (Wolf et al. 2012). Furthermore, other studies indeed indicate that TMEM16A, besides being a Ca^{2+}-regulated channel can also orchestrate the Ca^{2+} signaling process by acting as a scaffolding protein (Jin et al. 2013; Cabrita et al. 2017; Lérias et al. 2018). It should be taken into consideration that the mechanism proposed by Schreiber et al. (2015) would imply that the positive effect of TMEM16A on Cl^- secretion as a facilitator of Ca^{2+} signal is prevalent over the negative effect resulting from Cl^- exiting through the basolateral membrane.

More recently, Benedetto and colleagues (Benedetto et al. 2017) found evidence of another mechanism through which TMEM16A may influence intestinal Cl^- secretion. Indeed, by using mice with conditional knockout of TMEM16A in intestinal epithelial cells, they found mislocalization of CFTR protein in intracellular compartments and, consequently, a significantly reduced cAMP-dependent Cl^- secretion. Their conclusion was that TMEM16A controls correct trafficking of CFTR to the plasma membrane (Fig. 17.4c), possibly through an effect on Ca^{2+} signaling. Interestingly, in a later study, TMEM16F (ANO6) was also found to be important for CFTR trafficking and function (Benedetto et al. 2019b). Indeed, TMEM16F knockout mice showed reduced cAMP-dependent Cl^- secretion, particularly in jejunum where TMEM16A is much less expressed. An effect of TMEM16F on CFTR trafficking was also observed in lymphocytes and Calu-3 cells. It was postulated that both TMEM16A and TMEM16F affect CFTR expression in the plasma membrane by influencing exocytosis (Fig. 17.4c).

In another recent study, Lee and colleagues (Lee et al. 2019) also found that TMEM16A ablation in intestinal epithelial cells does not only abolish Ca^{2+}- but also cAMP-dependent Cl^- secretion. TMEM16A knockout also reduced the fluid secretion induced by cholera toxin, an effect that is classically attributed to CFTR activation. However, they did not find evidence of altered CFTR protein expression. Furthermore, they found that MONNA, a TMEM16A inhibitor, inhibits cAMP-activated currents. To explain their results, Lee and colleagues postulated that TMEM16A in the apical membrane is a main route for both Ca^{2+}- and cAMP-activated Cl^- secretion in intestinal epithelial cells (Fig. 17.4d). In this respect, they considered that intracellular cAMP elevation may lead to Ca^{2+} increase and hence to TMEM16A activation. In agreement with an important role of TMEM16A in intestinal epithelium physiology, Lee and colleagues observed that knockout mice had mucosal thickening and submucosal edema (Lee et al. 2019).

A relationship between TMEM16A and intestinal Cl^- secretion was also found by Rottgen et al. (2018). These authors found that the decrease in Ca^{2+}-activated Cl^- secretion elicited in a mouse model of chronic colitis is associated with a downregulation of TMEM16A but not CFTR protein expression.

In contrast to the results described above, other studies have found no evidence of TMEM16A involvement in intestinal Cl^- secretion. Using conditional TMEM16A knockout animals, Vega et al. (2019) did not observe a decrease in CCH- and ATP-activated Cl^- secretion. Furthermore, there were no apparent intestinal mucus abnormalities with respect to control animals. Also, the vitality of TMEM16A-defective animals was not affected. In contrast, CFTR knockout had a marked inhibitory effect on Cl^- secretion and altered mucus properties. In another study, knockout of TMEM16A abolished CCH- but not cAMP-induced Cl^- secretion (Ousingsawat et al. 2009).

It is difficult to reconcile the different results obtained for TMEM16A in intestinal epithelial cells. They may derive from different experimental approaches, pharmacological tools, and animal models. In this respect, it is important to note that TMEM16A expression, function, and localization may be affected by the age of the animal. For example, the pro-secretory effect of NSP4 rotavirus toxin, which acts by increasing intracellular Ca^{2+}, is negligible at 1–3 days, high at 10–14 days, and diminished at 30–40 days (Morris et al. 1999; Ousingsawat et al. 2011). Subcellular localization of TMEM16A may also be affected by age, with more prevalent basolateral localization in older animals (Kunzelmann et al. 2011). Furthermore, there are issues regarding the use of TMEM16A inhibitors. The specificity of these compounds needs to be demonstrated. For example, the $CaCC_{inh}$-A01 (De La Fuente et al. 2008), which is widely used in many studies, was found to also inhibit CFTR (Gianotti et al. 2016).

In conclusion, further studies are needed to clarify the role of TMEM16A in intestinal physiology and pathology.

17.6 Expression and Function of TMEM16A in Exocrine Glands

TMEM16A appears to directly contribute to Cl^- secretion in many types of exocrine glands by providing a route for Cl^- and bicarbonate exit in the apical membrane.

In addition to airway surface epithelial cells, discussed in Sect. 17.4, submucosal glands of the airways also show expression of TMEM16A protein (Fischer et al. 2010; Lee and Foskett 2010). In particular, TMEM16A was found to be more abundant in mucous than in serous cells, a finding consistent with a relationship between TMEM16A-dependent anion secretion and mucus secretion (Fischer et al. 2010).

In salivary glands, TMEM16A is typically characterized by strong expression. In vivo silencing of TMEM16A in mice by injection of siRNA resulted in decreased salivary fluid secretion (Yang et al. 2008). Also, salivary gland cells from Tmem16a knockout mice were devoid of Ca^{2+}-activated Cl^- currents and of carbachol-induced Cl^- efflux (Romanenko et al. 2010). As discussed in Sect. 17.2, one of the mechanisms controlling TMEM16A activity in salivary glands is Ca^{2+} influx through TRP channels (Derouiche et al. 2018; Sun et al. 2015).

TMEM16A protein was also found in human, rat, and mouse biliary epithelial cells where it probably contributes to the volume and composition of bile (Dutta et al. 2011, 2013). TMEM16A channel activity underlies the Ca^{2+}-dependent Cl^- and fluid secretion elicited by purinergic agonists with a mechanism involving regulation by protein kinase C alpha (Dutta et al. 2016). These regulatory mechanisms, particularly ATP release by mechanical stimuli, are probably involved in activation of TMEM16A in biliary ducts by shear stress (Dutta et al. 2013). Another important stimulus for TMEM16A-dependent biliary secretion is represented by bile acids (Li et al. 2018). Ursodeoxycholic and taurodeoxycholic acid promoted TMEM16A activation with a mechanism involving ATP release and autocrine stimulation of purinergic receptors. This process would couple bile acid secretion by hepatocytes with induction of fluid and bicarbonate secretion by cholangiocytes. Interestingly, treatment of biliary epithelial cells with IL-4 increased TMEM16A expression and function (Dutta et al. 2011). This result indicates that upregulation by Th2 cytokines is not restricted to the airway epithelium but is a more general phenomenon.

TMEM16A is presumably expressed also in exocrine pancreas. This assumption is based on the finding that TMEM16A is highly expressed in several cell lines derived from the pancreatic ducts such as CFPAC-1, AsPC-1, BxPC-3, and Capan-1 (Caputo et al. 2008; Wang et al. 2013; Sauter et al. 2015). In the duct, TMEM16A could represent a route for the Ca^{2+}-dependent regulation of bicarbonate secretion. However, the pancreatic duct cell lines used in the different studies are derived from cancers of the exocrine pancreas and therefore it is not clear if part of the high expression is due to the tumoral transformation. Possible evidence that TMEM16A contributes to exocrine pancreatic function comes from a study in which the TMEM16A inhibitor $T16_{inh}$-A01 blocked stimulated bicarbonate secretion by isolated pancreatic acini (Han et al. 2016).

Finally, as discussed in Sect. 17.2, TMEM16A is involved in sweat secretion with a mechanism involving Ca^{2+} influx by SOCE (Concepcion et al. 2016). Indeed, individuals with mutations affecting *ORAI1* and *STIM1* genes, main components of SOCE, suffer from anhidrosis and hyperthermia at high ambient temperatures due to impaired TMEM16A activation and sweat secretion (Concepcion et al. 2016).

17.7 TMEM16A in Kidney

TMEM16A was detected in human and mouse kidney, with a predominantly apical membrane localization in proximal tubular epithelial cells and additional expression in podocytes and the collecting duct (Faria et al. 2014; Svenningsen et al. 2014; Schenk et al. 2018). TMEM16A was also found expressed in the renal cyst-lining epithelium of individuals affected by autosomal dominant polycystic kidney disease (ADPKD) and in in vitro cyst models (Buchholz et al. 2014). Interestingly, the in vitro models revealed a dynamic cellular localization of TMEM16A controlled by cAMP. Indeed, non-stimulated cysts derived from principal like-MDCK cells (PLCs) showed a diffuse cytosolic and plasma membrane TMEM16A expression, whereas stimulation with forskolin led to luminal membrane localization of TMEM16A protein.

Polycystic kidney disease is characterized by multiple renal cysts that gradually enlarge leading to compression of the adjacent normal tissue and decline of the renal function. Enlargement of kidney cysts is due to both proliferation of the cyst-lining cells and fluid secretion into the cyst lumen. Besides a cAMP-dependent Cl^- secretion through CFTR, that represents the major driving force, it was thought that an additional ion channel might also contribute to electrolyte secretion that leads to cyst enlargement. Buchholz and colleagues showed indeed that MDCK cells display Ca^{2+}-dependent Cl^- secretion in response to purinergic stimulation and that knockdown of TMEM16A protein expression or pharmacological inhibition of TMEM16A function abolish this process (Buchholz et al. 2014). Importantly, knockdown or inhibition of TMEM16A also resulted in reduced cyst growth in PLCs and in the metanephric kidney cyst model. Furthermore, knockdown of TMEM16A also decreased the number of cyst proliferating cells, a finding consistent with the involvement of TMEM16A in promoting cell proliferation (Qu et al. 2014).

Important information about TMEM16A function in kidney was obtained through the generation of kidney-specific knockout mice (Faria et al. 2014; Schenk et al. 2018). Nephron-specific ablation of TMEM16A reduced glomeruli number and morphology (Schenk et al. 2018). Furthermore, selective ablation of TMEM16A in tubular epithelial cells, but not in podocytes, affects renal function (Faria et al. 2014). Tubular knockout of TMEM16A seems associated with a defect in proximal tubular protein reabsorption that leads to accumulation of large intracellular vesicles in proximal tubular epithelial cells, enhanced urine protein excretion with macro-albuminuria, and reduced urine electrolyte concentrations (Faria et al. 2014). Proximal tubular epithelial cells from wild-type animals exhibit Cl^- currents activated by purinergic stimulation and inhibited by the CaCC inhibitor, tannic acid. Cl^- currents in these cells, but not in cells derived from the Tmem16a$^{-/-}$ animal, were also activated by extracellular acidification, thus suggesting a role of TMEM16A during proton secretion (Faria et al. 2014).

17.8 Regulation of TMEM16A Expression

The mechanisms controlling tissue-specific TMEM16A expression are still not known. As discussed in other sections, TMEM16A expression is markedly induced by Th2 cytokines (Caputo et al. 2008; Dutta et al. 2011; Gorrieri et al. 2016; Scudieri et al. 2012) or by bacterial components (Caci et al. 2015). TMEM16A function is also enhanced in intestinal epithelial cells by epidermal growth factor (Mroz and Keely 2012). This phenomenon is due to upregulation of TMEM16A mRNA and protein levels resulting in enhanced intestinal Cl^- secretion. Evidence was provided in favor of TMEM16A as an apical Cl^- channel (Mroz and Keely 2012). Whether upregulation by Th2 cytokines and epidermal growth factor shares common signaling mechanisms needs to be clarified.

Interestingly, TMEM16A is also regulated by testosterone in prostate epithelial cells (Cha et al. 2015). In this respect, three androgen-responsive elements were found in the promoter region of the *TMEM16A* gene. Of note, inhibition of TMEM16A with a pharmacological agent or by gene silencing inhibited prostate enlargement thus indicating TMEM16A as target in benign prostatic hyperplasia.

There is also evidence of TMEM16A expression regulation by microRNAs. A recent study found that miR-9 binds to the 3′ untranslated region of the TMEM16A mRNA causing repression of protein expression (Sonneville et al. 2017). These authors found enhanced levels of miR-9 in CF epithelial cells and proposed that the resulting inhibition of TMEM16A could contribute to CF lung disease.

17.9 Other Functions of TMEM16A

TMEM16A expression has been linked to cancer in several studies since it seems to affect cell proliferation and/or migration (Britschgi et al. 2013; Qu et al. 2014; Sauter et al. 2015). A comprehensive review on this topic has been recently published (Wang et al. 2017). The role of TMEM16A in cancer may be related to its activity as a Cl^- channel but also, and maybe more probably, to its role as a scaffolding protein. Transport of Cl^- mediated by TMEM16A and other anion channel and transporters could underlie changes in cell volume required for cell migration. However, changes in intracellular Cl^- concentration due to TMEM16A gating could also have a signaling role. For example, both CFTR and TMEM16A seem to affect secretion of the pro-inflammatory and chemoattractant cytokine IL-8, possibly by their effect on intracellular Cl^- levels (Veit et al. 2012). Furthermore, control of cytosolic Cl^- by TMEM16A appear to regulate PIP2 organization, endocytic trafficking, and epithelial morphogenesis (He et al. 2017).

17.10 TMEM16F and Other Anoctamins in Epithelial Cells

As discussed in Sect. 17.1, TMEM16A and TMEM16B are the members of the anoctamin family with a more direct involvement in ion transport as Cl^- channels. TMEM16F expression is clearly associated with Ca^{2+}-activated scrambling of phosphatidylserine in the plasma membrane (Suzuki et al. 2010). It has been shown that TMEM16F also has ion channel activity. However, the characteristics of the ion currents generated by TMEM16F are not clear. In one study, TMEM16F was reported to form a cation-selective channel (Yang et al. 2012). In other studies, TMEM16F was instead found to carry anions although with low selectivity (Grubb et al. 2013; Shimizu et al. 2013; Scudieri et al. 2015). The reason for these discrepancies is not clear. Another intriguing characteristic is the activation at nonphysiological high Ca^{2+} concentrations and the rectification of the current–voltage relationship. Regarding this latter aspect, TMEM16F currents are mostly seen at very positive membrane potentials that are probably not experienced by epithelial cells. Therefore, it is probable that TMEM16F and other anoctamins with scramblase activity contribute to epithelial cells' function not by directly mediating ion transport but through other mechanisms (Scudieri et al. 2015).

Because of its high expression in all cell types, TMEM16F is expected to have a physiological function also in epithelial cells (Schreiber et al. 2010). However, patients affected by Scott syndrome do not show any abnormality besides a defect in blood coagulation (Weiss 1994). This problem is due to reduced externalization of phosphatidylserine on the surface of platelets. The coagulation defect was also confirmed in a TMEM16F knockout mouse (Yang et al. 2012). Besides a prolonged bleeding time, this animal did not show other defects thus indicating that the absence of TMEM16F in epithelial cells is without clear consequences. In a second type of knockout mouse, TMEM16F loss resulted in a significant defect in osteoblast function with severe skeletal abnormalities (Ehlen et al. 2013) but without a clear dysfunction in epithelial cells. As a scramblase, TMEM16F may be involved in a variety of cell functions such as control of endocytic trafficking. As discussed in Sect. 17.5, TMEM16F was reported to control CFTR trafficking, particularly in the jejunum and to be able to compensate TMEM16A loss in other tissues (Benedetto et al. 2019b). To explain the absence of an epithelial phenotype in TMEM16F$^{-/-}$ animals, it is possible to hypothesize that other anoctamins compensate for the absence of TMEM16F.

17.11 Perspectives

The discovery of anoctamins has provided the molecular basis for Ca^{2+}-activated Cl^- transport and phospholipid scrambling. However, future studies, also based on animal models and human genetic diseases, may reveal novel and unsuspected functions for this family of membrane proteins. In particular, there is the intriguing possibility that several anoctamins work as tethers between different cell membranes. Besides TMEM16A and TMEM16F, already discussed in other sections, TMEM16H (ANO8) is also involved in the formation of contact sites between plasma membrane and endoplasmic reticulum that are important for Ca^{2+} signaling (Jha et al. 2019).

A particular important issue is the importance of TMEM16A and other anoctamins as possible therapeutic targets in a variety of human diseases. Regarding TMEM16A, it will be particularly important to finally establish if activation or inhibition is the appropriate approach to treat chronic obstructive respiratory diseases such as cystic fibrosis and asthma (Kunzelmann et al. 2019).

Acknowledgements Our work is supported by Telethon Foundation, Cystic Fibrosis Foundation, and Fondazione Italiana Fibrosi Cistica.

References

Benedetto R, Ousingsawat J, Wanitchakool P, Zhang Y, Holtzman MJ, Amaral M, Rock JR, Schreiber R, Kunzelmann K (2017) Epithelial chloride transport by CFTR requires TMEM16A. Sci Rep 7:12397

Benedetto R, Cabrita I, Schreiber R, Kunzelmann K (2019a) TMEM16A is indispensable for basal mucus secretion in airways and intestine. FASEB J 33:4502–4512

Benedetto R, Ousingsawat J, Cabrita I, Pinto M, Lérias JR, Wanitchakool P, Schreiber R, Kunzelmann K (2019b) Plasma membrane-localized TMEM16 proteins are indispensable for expression of CFTR. J Mol Med (Berl) 97:711–722

Bolduc V, Marlow G, Boycott KM, Saleki K, Inoue H, Kroon J, Itakura M, Robitaille Y, Parent L, Baas F, Mizuta K, Kamata N, Richard I, Linssen WH, Mahjneh I, de Visser M, Bashir R, Brais B (2010) Recessive mutations in the putative calcium-activated chloride channel Anoctamin

5 cause proximal LGMD2L and distal MMD3 muscular dystrophies. Am J Hum Genet 86:213–221

Britschgi A, Bill A, Brinkhaus H, Rothwell C, Clay I, Duss S, Rebhan M, Raman P, Guy CT, Wetzel K, George E, Popa MO, Lilley S, Choudhury H, Gosling M, Wang L, Fitzgerald S, Borawski J, Baffoe J, Labow M, Gaither LA, Bentires-Alj M (2013) Calcium-activated chloride channel ANO1 promotes breast cancer progression by activating EGFR and CAMK signaling. Proc Natl Acad Sci U S A 110:E1026–E1034

Brunner JD, Lim NK, Schenck S, Duerst A, Dutzler R (2014) X-ray structure of a calcium-activated TMEM16 lipid scramblase. Nature 516(7530):207–212

Buchholz B, Faria D, Schley G, Schreiber R, Eckardt KU, Kunzelmann K (2014) Anoctamin 1 induces calcium-activated chloride secretion and proliferation of renal cyst–forming epithelial cells. Kidney Int 85:1058–1067

Cabrita I, Benedetto R, Fonseca A, Wanitchakool P, Sirianant L, Skryabin BV, Schenk LK, Pavenstädt H, Schreiber R, Kunzelmann K (2017) Differential effects of anoctamins on intracellular calcium signals. FASEB J 31:2123–2134

Cabrita I, Benedetto R, Schreiber R, Kunzelmann K (2019) Niclosamide repurposed for the treatment of inflammatory airway disease. JCI Insight 4(15):128414. https://doi.org/10.1172/jci.insight.128414

Caci E, Scudieri P, Di Carlo E, Morelli P, Bruno S, De Fino I, Bragonzi A, Gianotti A, Sondo E, Ferrera L, Palleschi A, Santambrogio L, Ravazzolo R, Galietta LJ (2015) Upregulation of TMEM16A protein in bronchial epithelial cells by bacterial pyocyanin. PLoS One 10:e0131775

Caputo A, Caci E, Ferrera L, Pedemonte N, Barsanti C, Sondo E, Pfeffer U, Ravazzolo R, Zegarra-Moran O, Galietta LJ (2008) TMEM16A, a membrane protein associated with calcium-dependent chloride channel activity. Science 322:590–594

Cha JY, Wee J, Jung J, Jang Y, Lee B, Hong GS, Chang BC, Choi YL, Shin YK, Min HY, Lee HY, Na TY, Lee MO, Oh U (2015) Anoctamin 1 (TMEM16A) is essential for testosterone-induced prostate hyperplasia. Proc Natl Acad Sci U S A 112:9722–9727

Concepcion AR, Vaeth M, Wagner LE 2nd, Eckstein M, Hecht L, Yang J, Crottes D, Seidl M, Shin HP, Weidinger C, Cameron S, Turvey SE, Issekutz T, Meyts I, Lacruz RS, Cuk M, Yule DI, Feske S (2016) Store-operated Ca^{2+} entry regulates Ca^{2+}-activated chloride channels and eccrine sweat gland function. J Clin Invest 126:4303–4318

Courjaret R, Machaca K (2014) Mid-range Ca^{2+} signalling mediated by functional coupling between store-operated Ca^{2+} entry and IP3-dependent Ca^{2+} release. Nat Commun 5:3916

Curran DR, Cohn L (2010) Advances in mucous cell metaplasia: a plug for mucus as a therapeutic focus in chronic airway disease. Am J Respir Cell Mol Biol 42:268–275

Danahay H, Atherton H, Jones G, Bridges RJ, Poll CT (2002) Interleukin-13 induces a hypersecretory ion transport phenotype in human bronchial epithelial cells. Am J Physiol Lung Cell Mol Physiol 282:L226–L236

Dang S, Feng S, Tien J, Peters CJ, Bulkley D, Lolicato M, Zhao J, Zuberbühler K, Ye W, Qi L, Chen T, Craik CS, Jan YN, Minor DL Jr, Cheng Y, Jan LY (2017) Cryo-EM structures of the TMEM16A calcium-activated chloride channel. Nature 552(7685):426–429

De Jesús-Pérez JJ, Cruz-Rangel S, Espino-Saldaña ÁE, Martínez-Torres A, Qu Z, Hartzell HC, Corral-Fernandez NE, Pérez-Cornejo P, Arreola J (2018) Phosphatidylinositol 4,5-bisphosphate, cholesterol, and fatty acids modulate the calcium-activated chloride channel TMEM16A (ANO1). Biochim Biophys Acta Mol Cell Biol Lipids 1863:299–312

De La Fuente R, Namkung W, Mills A, Verkman AS (2008) Small-molecule screen identifies inhibitors of a human intestinal calcium-activated chloride channel. Mol Pharmacol 73:758–768

Derouiche S, Takayama Y, Murakami M, Tominaga M (2018) TRPV4 heats up ANO1-dependent exocrine gland fluid secretion. FASEB J 32:1841–1854

Dutta AK, Khimji AK, Kresge C, Bugde A, Dougherty M, Esser V, Ueno Y, Glaser SS, Alpini G, Rockey DC, Feranchak AP (2011) Identification and functional characterization of TMEM16A, a Ca^{2+}-activated Cl^- channel activated by extracellular nucleotides, in biliary epithelium. J Biol Chem 286:766–776

Dutta AK, Woo K, Khimji AK, Kresge C, Feranchak AP (2013) Mechanosensitive Cl^- secretion in biliary epithelium mediated through TMEM16A. Am J Physiol Gastrointest Liver Physiol 304: G87–G98

Dutta AK, Khimji AK, Liu S, Karamysheva Z, Fujita A, Kresge C, Rockey DC, Feranchak AP (2016) PKCα regulates TMEM16A-mediated Cl^- secretion in human biliary cells. Am J Physiol Gastrointest Liver Physiol 310:G34–G42

Ehlen HW, Chinenkova M, Moser M, Munter HM, Krause Y, Gross S, Brachvogel B, Wuelling M, Kornak U, Vortkamp A (2013) Inactivation of anoctamin-6/Tmem16f, a regulator of phosphatidylserine scrambling in osteoblasts, leads to decreased mineral deposition in skeletal tissues. J Bone Miner Res 28:246–259

Fallah G, Römer T, Detro-Dassen S, Braam U, Markwardt F, Schmalzing G (2011) TMEM16A(a)/ anoctamin-1 shares a homodimeric architecture with CLC chloride channels. Mol Cell Proteomics 10:1–10

Faria D, Rock JR, Romao AM, Schweda F, Bandulik S, Witzgall R, Schlatter E, Heitzmann D, Pavenstädt H, Herrmann E, Kunzelmann K, Schreiber R (2014) The calcium-activated chloride channel Anoctamin 1 contributes to the regulation of renal function. Kidney Int 85:1369–1381

Ferrera L, Caputo A, Ubby I, Bussani E, Zegarra-Moran O, Ravazzolo R, Pagani F, Galietta LJ (2009) Regulation of TMEM16A chloride channel properties by alternative splicing. J Biol Chem 284:33360–33368

Fischer H, Illek B, Sachs L, Finkbeiner WE, Widdicombe JH (2010) CFTR and calcium-activated chloride channels in primary cultures of human airway gland cells of serous or mucous phenotype. Am J Physiol Lung Cell Mol Physiol 299:L585–L594

Galietta LJ, Pagesy P, Folli C, Caci E, Romio L, Costes B, Nicolis E, Cabrini G, Goossens M, Ravazzolo R, Zegarra-Moran O (2002) IL-4 is a potent modulator of ion transport in the human bronchial epithelium in vitro. J Immunol 168:839–845

Gao DY, Zhang BL, Leung MC, Au SC, Wong PY, Shum WW (2016) Coupling of TRPV6 and TMEM16A in epithelial principal cells of the rat epididymis. J Gen Physiol 148:161–182

Garcia MA, Yang N, Quinton PM (2009) Normal mouse intestinal mucus release requires cystic fibrosis transmembrane regulator-dependent bicarbonate secretion. J Clin Invest 119:2613–2622

Gianotti A, Ferrera L, Philp AR, Caci E, Zegarra-Moran O, Galietta LJ, Flores CA (2016) Pharmacological analysis of epithelial chloride secretion mechanisms in adult murine airways. Eur J Pharmacol 781:100–108

Gorrieri G, Scudieri P, Caci E, Schiavon M, Tomati V, Sirci F, Napolitano F, Carrella D, Gianotti A, Musante I, Favia M, Casavola V, Guerra L, Rea F, Ravazzolo R, Di Bernardo D, Galietta LJ (2016) Goblet cell hyperplasia requires high bicarbonate transport to support mucin release. Sci Rep 6:36016

Grubb S, Poulsen KA, Juul CA, Kyed T, Klausen TK, Larsen EH, Hoffmann EK (2013) TMEM16F (Anoctamin 6), an anion channel of delayed Ca^{2+} activation. J Gen Physiol 141:585–560

Gustafsson JK, Ermund A, Ambort D, Johansson ME, Nilsson HE, Thorell K, Hebert H, Sjövall H, Hansson GC (2012) Bicarbonate and functional CFTR channel are required for proper mucin secretion and link cystic fibrosis with its mucus phenotype. J Exp Med 209:1263–1272

Han Y, Shewan AM, Thorn P (2016) HCO_3^- transport through anoctamin/transmembrane protein ANO1/TMEM16A in pancreatic acinar cells regulates luminal pH. J Biol Chem 291:20345–20352

He Q, Halm ST, Zhang J, Halm DR (2011) Activation of the basolateral membrane Cl^- conductance essential for electrogenic K^+ secretion suppresses electrogenic Cl^- secretion. Exp Physiol 96.305–316

He M, Ye W, Wang WJ, Sison ES, Jan YN, Jan LY (2017) Cytoplasmic Cl^- couples membrane remodeling to epithelial morphogenesis. Proc Natl Acad Sci U S A 114:E11161–E11169

Hoegger MJ, Fischer AJ, McMenimen JD, Ostedgaard LS, Tucker AJ, Awadalla MA, Moninger TO, Michalski AS, Hoffman EA, Zabner J, Stoltz DA, Welsh MJ (2014) Cystic fibrosis.

Impaired mucus detachment disrupts mucociliary transport in a piglet model of cystic fibrosis. Science 345:818–822
Huang F, Rock JR, Harfe BD, Cheng T, Huang X, Jan YN, Jan LY (2009) Studies on expression and function of the TMEM16A calcium-activated chloride channel. Proc Natl Acad Sci U S A 106:21413–21418
Huang F, Zhang H, Wu M, Yang H, Kudo M, Peters CJ, Woodruff PG, Solberg OD, Donne ML, Huang X, Sheppard D, Fahy JV, Wolters PJ, Hogan BL, Finkbeiner WE, Li M, Jan YN, Jan LY, Rock JR (2012) Calcium-activated chloride channel TMEM16A modulates mucin secretion and airway smooth muscle contraction. Proc Natl Acad Sci U S A 109:16354–16359
Huang F, Wang X, Ostertag EM, Nuwal T, Huang B, Jan YN, Basbaum AI, Jan LY (2013) TMEM16C facilitates Na^+-activated K^+ currents in rat sensory neurons and regulates pain processing. Nat Neurosci 16:1284–1290
Jeng G, Aggarwal M, Yu WP, Chen TY (2016) Independent activation of distinct pores in dimeric TMEM16A channels. J Gen Physiol 148(5):393–404
Jha A, Chung WY, Vachel L, Maleth J, Lake S, Zhang G, Ahuja M, Muallem S (2019) Anoctamin 8 tethers endoplasmic reticulum and plasma membrane for assembly of Ca^{2+} signaling complexes at the ER/PM compartment. EMBO J 38(12):e101452. https://doi.org/10.15252/embj.2018101452
Jin X, Shah S, Liu Y, Zhang H, Lees M, Fu Z, Lippiat JD, Beech DJ, Sivaprasadarao A, Baldwin SA, Zhang H, Gamper N (2013) Activation of the Cl^- channel ANO1 by localized calcium signals in nociceptive sensory neurons requires coupling with the IP3 receptor. Sci Signal 6: ra73. https://doi.org/10.1126/scisignal.2004184
Jung J, Nam JH, Park HW, Oh U, Yoon JH, Lee MG (2013) Dynamic modulation of ANO1/TMEM16A HCO_3^- permeability by Ca^{2+}/calmodulin. Proc Natl Acad Sci U S A 110:360–365
Kang JW, Lee YH, Kang MJ, Lee HJ, Oh R, Min HJ, Namkung W, Choi JY, Lee SN, Kim CH, Yoon JH, Cho HJ (2017) Synergistic mucus secretion by histamine and IL-4 through TMEM16A in airway epithelium. Am J Physiol Lung Cell Mol Physiol 313:L466–L476
Kunzelmann K, Tian Y, Martins JR, Faria D, Kongsuphol P, Ousingsawat J, Thevenod F, Roussa E, Rock J, Schreiber R (2011) Anoctamins. Pflugers Arch 462:195–208
Kunzelmann K, Ousingsawat J, Cabrita I, Doušová T, Bähr A, Janda M, Schreiber R, Benedetto R (2019) TMEM16A in cystic fibrosis: activating or inhibiting? Front Pharmacol 10:3. https://doi.org/10.3389/fphar.2019.00003
Le SC, Jia Z, Chen J, Yang H (2019) Molecular basis of PIP2-dependent regulation of the Ca^{2+}-activated chloride channel TMEM16A. Nat Commun 10:3769
Lee RJ, Foskett JK (2010) Mechanisms of Ca^{2+}-stimulated fluid secretion by porcine bronchial submucosal gland serous acinar cells. Am J Physiol Lung Cell Mol Physiol 298:L210–L231
Lee B, Hong GS, Lee SH, Kim H, Kim A, Hwang EM, Kim J, Lee MG, Yang JY, Kweon MN, Tse CM, Mark D, Oh U (2019) Anoctamin 1/TMEM16A controls intestinal Cl^- secretion induced by carbachol and cholera toxin. Exp Mol Med 51:91. https://doi.org/10.1038/s12276-019-0287-2
Lérias JR, Pinto MC, Botelho HM, Awatade NT, Quaresma MC, Silva IAL, Wanitchakool P, Schreiber R, Pepperkok R, Kunzelmann K, Amaral MD (2018) A novel microscopy-based assay identifies extended synaptotagmin-1 (ESYT1) as a positive regulator of anoctamin 1 traffic. Biochim Biophys Acta, Mol Cell Res 1865:421–431
Li Q, Dutta A, Kresge C, Bugde A, Feranchak AP (2018) Bile acids stimulate cholangiocyte fluid secretion by activation of transmembrane member 16A Cl^- channels. Hepatology 68:187–199
Lim NK, Lam AK, Dutzler R (2016) Independent activation of ion conduction pores in the double-barreled calcium-activated chloride channel TMEM16A. J Gen Physiol 148(5):375–392
Lin J, Jiang Y, Li L, Liu Y, Tang H, Jiang D (2015) TMEM16A mediates the hypersecretion of mucus induced by interleukin-13. Exp Cell Res 334:260–269
Malvezzi M, Chalat M, Janjusevic R, Picollo A, Terashima H, Menon AK, Accardi A (2013) Ca^{2+}-dependent phospholipid scrambling by a reconstituted TMEM16 ion channel. Nat Commun 4:2367

Matsui H, Randell SH, Peretti SW, Davis CW, Boucher RC (1998) Coordinated clearance of periciliary liquid and mucus from airway surfaces. J Clin Invest 102:1125–1131

Miner K, Labitzke K, Liu B, Wang P, Henckels K, Gaida K, Elliott R, Chen JJ, Liu L, Leith A, Trueblood E, Hensley K, Xia XZ, Homann O, Bennett B, Fiorino M, Whoriskey J, Yu G, Escobar S, Wong M, Born TL, Budelsky A, Comeau M, Smith D, Phillips J, Johnston JA, McGivern JG, Weikl K, Powers D, Kunzelmann K, Mohn D, Hochheimer A, Sullivan JK (2019) Drug repurposing: the anthelmintics niclosamide and nitazoxanide are potent TMEM16A antagonists that fully bronchodilate airways. Front Pharmacol 10:51. https://doi.org/10.3389/fphar.2019.00051

Morris AP, Scott JK, Ball JM, Zeng CQ, O'Neal WK, Estes MK (1999) NSP4 elicits age-dependent diarrhea and Ca^{2+} mediated I^- influx into intestinal crypts of CF mice. Am J Phys 277:G431–G444

Mroz MS, Keely SJ (2012) Epidermal growth factor chronically upregulates Ca^{2+}-dependent Cl^- conductance and TMEM16A expression in intestinal epithelial cells. J Physiol 590:1907–1920

Ousingsawat J, Martins JR, Schreiber R, Rock JR, Harfe BD, Kunzelmann K (2009) Loss of TMEM16A causes a defect in epithelial Ca^{2+}-dependent chloride transport. J Biol Chem 284:28698–28703

Ousingsawat J, Mirza M, Tian Y, Roussa E, Schreiber R, Cook DI, Kunzelmann K (2011) Rotavirus toxin NSP4 induces diarrhea by activation of TMEM16A and inhibition of Na+ absorption. Pflugers Arch 461:579–589

Park KS, Korfhagen TR, Bruno MD, Kitzmiller JA, Wan H, Wert SE, Khurana Hershey GK, Chen G, Whitsett JA (2007) SPDEF regulates goblet cell hyperplasia in the airway epithelium. J Clin Invest 117:978–988

Paulino C, Neldner Y, Lam AK, Kalienkova V, Brunner JD, Schenck S, Dutzler R (2017a) Structural basis for anion conduction in the calcium-activated chloride channel TMEM16A. elife 6:e26232

Paulino C, Kalienkova V, Lam AKM, Neldner Y, Dutzler R (2017b) Activation mechanism of the calcium-activated chloride channel TMEM16A revealed by cryo-EM. Nature 552 (7685):421–425

Pedemonte N, Galietta LJ (2014) Structure and function of TMEM16 proteins (anoctamins). Physiol Rev 94:419–459

Pelz T, Drose DR, Fleck D, Henkel B, Ackels T, Spehr M, Neuhaus EM (2018) An ancestral TMEM16 homolog from Dictyostelium discoideum forms a scramblase. PLoS One 13: e0191219

Peters CJ, Gilchrist JM, Tien J, Bethel NP, Qi L, Chen T, Wang L, Jan YN, Grabe M, Jan LY (2018) The sixth transmembrane segment is a major gating component of the TMEM16A calcium-activated chloride channel. Neuron 97(5):1063–1077

Petersen OH, Courjaret R, Machaca K (2017) Ca^{2+} tunnelling through the ER lumen as a mechanism for delivering Ca^{2+} entering via store-operated Ca^{2+} channels to specific target sites. J Physiol 595:2999–3014

Pezzulo AA, Tang XX, Hoegger MJ, Alaiwa MH, Ramachandran S, Moninger TO, Karp PH, Wohlford-Lenane CL, Haagsman HP, van Eijk M, Bánfi B, Horswill AR, Stoltz DA, McCray PB Jr, Welsh MJ, Zabner J (2012) Reduced airway surface pH impairs bacterial killing in the porcine cystic fibrosis lung. Nature 487:109–113

Pifferi S, Dibattista M, Menini A (2009) TMEM16B induces chloride currents activated by calcium in mammalian cells. Pflugers Arch 458:1023–1038

Pritchard HA, Leblanc N, Albert AP, Greenwood IA (2014) Inhibitory role of phosphatidylinositol 4,5-bisphosphate on TMEM16A-encoded calcium-activated chloride channels in rat pulmonary artery. Br J Pharmacol 171:4311–4321

Qu Z, Yao W, Yao R, Liu X, Yu K, Hartzell C (2014) The Ca^{2+}-activated Cl^- channel, ANO1 (TMEM16A), is a double-edged sword in cell proliferation and tumorigenesis. Cancer Med 3:453–461

Riordan JR (2008) CFTR function and prospects for therapy. Annu Rev Biochem 77:701–726

Rock JR, Futtner CR, Harfe BD (2008) The transmembrane protein TMEM16A is required for normal development of the murine trachea. Dev Biol 321:141–149

Rock JR, O'Neal WK, Gabriel SE, Randell SH, Harfe BD, Boucher RC, Grubb BR (2009) Transmembrane protein 16A (TMEM16A) is a Ca^{2+}-regulated Cl^{-} secretory channel in mouse airways. J Biol Chem 284:14875–14880

Romanenko VG, Catalán MA, Brown DA, Putzier I, Hartzell HC, Marmorstein AD, Gonzalez-Begne M, Rock JR, Harfe BD, Melvin JE (2010) Tmem16A encodes the Ca^{2+}-activated Cl^{-} channel in mouse submandibular salivary gland acinar cells. J Biol Chem 285:12990–13001

Rottgen TS, Nickerson AJ, Minor EA, Stewart AB, Harold AD, Rajendran VM (2018) Dextran sulfate sodium-induced chronic colitis attenuates Ca^{2+}-activated Cl^{-} secretion in murine colon by downregulating TMEM16A. Am J Physiol Cell Physiol 315:C10–C20

Sauter DR, Novak I, Pedersen SF, Larsen EH, Hoffmann EK (2015) ANO1 (TMEM16A) in pancreatic ductal adenocarcinoma (PDAC). Pflugers Arch 467(7):1495–1508

Schenk LK, Buchholz B, Henke SF, Michgehl U, Daniel C, Amann K, Kunzelmann K, Pavenstädt H (2018) Nephron-specific knockout of TMEM16A leads to reduced number of glomeruli and albuminuria. Am J Physiol Renal Physiol 315(6):F1777–F1786

Schreiber R, Uliyakina I, Kongsuphol P, Warth R, Mirza M, Martins JR, Kunzelmann K (2010) Expression and function of epithelial anoctamins. J Biol Chem 285:7838–7845

Schreiber R, Faria D, Skryabin BV, Wanitchakool P, Rock JR, Kunzelmann K (2015) Anoctamins support calcium-dependent chloride secretion by facilitating calcium signaling in adult mouse intestine. Pflugers Arch 467(6):1203–1213

Schroeder BC, Cheng T, Jan YN, Jan LY (2008) Expression cloning of TMEM16A as a calcium-activated chloride channel subunit. Cell 134:1019–1029

Scudieri P, Caci E, Bruno S, Ferrera L, Schiavon M, Sondo E, Tomati V, Gianotti A, Zegarra-Moran O, Pedemonte N, Rea F, Ravazzolo R, Galietta LJ (2012) Association of TMEM16A chloride channel overexpression with airway goblet cell metaplasia. J Physiol 590:6141–6155

Scudieri P, Caci E, Venturini A, Sondo E, Pianigiani G, Marchetti C, Ravazzolo R, Pagani F, Galietta LJ (2015) Ion channel and lipid scramblase activity associated with expression of TMEM16F/ANO6 isoforms. J Physiol 593(17):3829–3848

Scudieri P, Musante I, Gianotti A, Moran O, Galietta LJ (2016) Intermolecular interactions in the TMEM16A dimer controlling channel activity. Sci Rep 6:38788

Scudieri P, Musante I, Caci E, Venturini A, Morelli P, Walter C, Tosi D, Palleschi A, Martin-Vasallo P, Sermet-Gaudelus I, Planelles G, Crambert G, Galietta LJ (2018) Increased expression of ATP12A proton pump in cystic fibrosis airways. JCI Insight 3(20). https://doi.org/10.1172/jci.insight.123616

Sheridan JT, Worthington EN, Yu K, Gabriel SE, Hartzell HC, Tarran R (2011) Characterization of the oligomeric structure of the Ca^{2+}-activated Cl^{-} channel Ano1/TMEM16A. J Biol Chem 286:1381–1388

Shimizu T, Iehara T, Sato K, Fujii T, Sakai H, Okada Y (2013) TMEM16F is a component of a Ca^{2+}-activated Cl^{-} channel but not a volume-sensitive outwardly rectifying Cl^{-} channel. Am J Physiol Cell Physiol 304:C748–C759

Sonneville F, Ruffin M, Coraux C, Rousselet N, Le Rouzic P, Blouquit-Laye S, Corvol H, Tabary O (2017) MicroRNA-9 downregulates the ANO1 chloride channel and contributes to cystic fibrosis lung pathology. Nat Commun 8:710

Stephan AB, Shum EY, Hirsh S, Cygnar KD, Reisert J, Zhao H (2009) ANO2 is the cilial calcium-activated chloride channel that may mediate olfactory amplification. Proc Natl Acad Sci U S A 106:11776–11781

Stöhr H, Heisig JB, Benz PM, Schöberl S, Milenkovic VM, Strauss O, Aartsen WM, Wijnholds J, Weber BH, Schulz HL (2009) TMEM16B, a novel protein with calcium-dependent chloride channel activity, associates with a presynaptic protein complex in photoreceptor terminals. J Neurosci 29:6809–6818

Sun Y, Birnbaumer L, Singh BB (2015) TRPC1 regulates calcium-activated chloride channels in salivary gland cells. J Cell Physiol 230:2848–2856

Suzuki J, Umeda M, Sims PJ, Nagata S (2010) Calcium-dependent phospholipid scrambling by TMEM16F. Nature 468:834–838

Suzuki J, Fujii T, Imao T, Ishihara K, Kuba H, Nagata S (2013) Calcium-dependent phospholipid scramblase activity of TMEM16 protein family members. J Biol Chem 288:13305–13316

Svenningsen P, Nielsen MR, Marcussen N, Walter S, Jensen BL (2014) TMEM16A is a Ca^{2+}-activated Cl^- channel expressed in the renal collecting duct. Acta Physiol 212:166–174

Ta CM, Acheson KE, Rorsman NJG, Jongkind RC, Tammaro P (2017) Contrasting effects of phosphatidylinositol 4,5-bisphosphate on cloned TMEM16A and TMEM16B channels. Br J Pharmacol 174:2984–2999

Takayama Y, Shibasaki K, Suzuki Y, Yamanaka A, Tominaga M (2014) Modulation of water efflux through functional interaction between TRPV4 and TMEM16A/anoctamin 1. FASEB J 28:2238–2248

Terashima H, Picollo A, Accardi A (2013) Purified TMEM16A is sufficient to form Ca^{2+}-activated Cl^- channels. Proc Natl Acad Sci U S A 110:19354–19359

Tian Y, Kongsuphol P, Hug M, Ousingsawat J, Witzgall R, Schreiber R, Kunzelmann K (2011) Calmodulin-dependent activation of the epithelial calcium-dependent chloride channel TMEM16A. FASEB J 25:1–11

Tien J, Lee HY, Minor DL Jr, Jan YN, Jan LY (2013) Identification of a dimerization domain in the TMEM16A calcium-activated chloride channel (CaCC). Proc Natl Acad Sci U S A 110:6352–6357

Tien J, Peters CJ, Wong XM, Cheng T, Jan YN, Jan LY, Yang H (2014) A comprehensive search for calcium binding sites critical for TMEM16A calcium-activated chloride channel activity. elife 3:e02772

Tsutsumi S, Kamata N, Vokes TJ, Maruoka Y, Nakakuki K, Enomoto S, Omura K, Amagasa T, Nagayama M, Saito-Ohara F, Inazawa J, Moritani M, Yamaoka T, Inoue H, Itakura M (2004) The novel gene encoding a putative transmembrane protein is mutated in gnathodiaphyseal dysplasia (GDD). Am J Hum Genet 74:1255–1261

Vega G, Guequén A, Johansson MEV, Arike L, Martínez-Abad B, Scudieri P, Pedemonte N, Millar-Büchner P, Philp AR, Galietta LJ, Hansson GC, Flores CA (2019) Normal calcium-activated anion secretion in a mouse selectively lacking TMEM16A in intestinal epithelium. Front Physiol 10:694

Veit G, Bossard F, Goepp J, Verkman AS, Galietta LJ, Hanrahan JW, Lukacs GL (2012) Proinflammatory cytokine secretion is suppressed by TMEM16A or CFTR channel activity in human cystic fibrosis bronchial epithelia. Mol Biol Cell 23:4188–4202

Viitanen TM, Sukumaran P, Löf C, Törnquist K (2013) Functional coupling of TRPC2 cation channels and the calcium-activated anion channels in rat thyroid cells: implications for iodide homeostasis. J Cell Physiol 228:814–823

Vocke K, Dauner K, Hahn A, Ulbrich A, Broecker J, Keller S, Frings S, Möhrlen F (2013) Calmodulin-dependent activation and inactivation of anoctamin calcium-gated chloride channels. J Gen Physiol 142:381–404

Wang J, Haanes KA, Novak I (2013) Purinergic regulation of CFTR and Ca^{2+}-activated Cl^- channels and K^+ channels in human pancreatic duct epithelium. Am J Physiol Cell Physiol 304:C673–C684

Wang H, Zou L, Ma K, Yu J, Wu H, Wei M, Xiao Q (2017) Cell-specific mechanisms of TMEM16A Ca^{2+}-activated chloride channel in cancer. Mol Cancer 16:152

Weiss HJ (1994) Scott syndrome: a disorder of platelet coagulant activity. Semin Hematol 31:312–319

Wolf W, Kilic A, Schrul B, Lorenz H, Schwappach B, Seedorf M (2012) Yeast Ist2 recruits the endoplasmic reticulum to the plasma membrane and creates a ribosome-free membrane microcompartment. PLoS One 7:e39703

Xiao Q, Yu K, Perez-Cornejo P, Cui Y, Arreola J, Hartzell HC (2011) Voltage- and calcium-dependent gating of TMEM16A/Ano1 chloride channels are physically coupled by the first intracellular loop. Proc Natl Acad Sci U S A 108:8891–8896

Yang YD, Cho H, Koo JY, Tak MH, Cho Y, Shim WS, Park SP, Lee J, Lee B, Kim BM, Raouf R, Shin YK, Oh U (2008) TMEM16A confers receptor-activated calcium-dependent chloride conductance. Nature 455:1210–1215

Yang H, Kim A, David T, Palmer D, Jin T, Tien J, Huang F, Cheng T, Coughlin SR, Jan YN, Jan LY (2012) TMEM16F forms a Ca^{2+}-activated cation channel required for lipid scrambling in platelets during blood coagulation. Cell 151:111–122

Yang T, Hendrickson WA, Colecraft HM (2014) Preassociated apocalmodulin mediates Ca^{2+}-dependent sensitization of activation and inactivation of TMEM16A/16B Ca^{2+}-gated Cl^- channels. Proc Natl Acad Sci U S A 111:18213–18218

Yu K, Duran C, Qu Z, Cui YY, Hartzell HC (2012) Explaining calcium-dependent gating of anoctamin-1 chloride channels requires a revised topology. Circ Res 110:990–999

Yu K, Zhu J, Qu Z, Cui YY, Hartzell HC (2014a) Activation of the Ano1 (TMEM16A) chloride channel by calcium is not mediated by calmodulin. J Gen Physiol 143:253–267

Yu Y, Kuan AS, Chen TY (2014b) Calcium-calmodulin does not alter the anion permeability of the mouse TMEM16A calcium-activated chloride channel. J Gen Physiol 144:115–124

Chapter 18
Epithelial Sodium Channels (ENaC)

Chang Song, He-Ping Ma, and Douglas C. Eaton

Abstract Amiloride-sensitive epithelial sodium (Na^+) channels (ENaC) play a crucial role in Na^+ transport and fluid reabsorption in the kidney, lung, colon, vasculature, and immune system. ENaC is made up of three homologous subunits, α, β, and γ, in a trimeric structure. The magnitude of ENaC-mediated Na^+ transport in epithelial cells depends on the average open probability of the channels and the number of channels on the apical surface of epithelial cells. The open probability of ENaC depends upon the association of ENaC with inositol lipid phosphates and association with a collection of chaperone proteins including calmodulin and MARCKS (myristoylated alanine-rich C kinase substrate). MARCKS, when associated with the membrane, sequesters PIP_2 and increases open probability and channel stability. The number of channels in the apical membrane, in turn, depends upon a balance between the rate of ENaC insertion and the rate of removal from the apical membrane. To this end, ENaC is associated with an additional collection of proteins, the ENaC regulatory complex (ERC) that contains proteins including SGK1, GILZ1, ERK, and Nedd4-2 that regulate channel stability in the membrane. The ERC under basal conditions contains several molecules including ERK and Nedd4-2 that inhibit ENaC activity. The C-terminal domain of all three subunits is intracellular and contains a proline-rich motif (PPxY) to which one isoform of the ubiquitin ligase, Nedd4-2 (neural-precursor-cell-expressed-developmentally-down-regulated-protein), binds to these domains in ENaC, thereby promoting ubiquitin conjugation of the channel subunits followed by internalization and degradation or recycling. In the presence of aldosterone, several proteins including SGK1, GILZ1, and MARCKS are recruited to the complex and stimulate ENaC activity. In addition, aldosterone promotes insertion of channels from intracellular pools and transcription/translation of new ENaC. Other agents that alter ENaC activity work by

C. Song · H.-P. Ma
Department of Physiology, Emory University School of Medicine, Atlanta, GA, USA
e-mail: chang.song@emory.edu; heping.ma@emory.edu

D. C. Eaton (✉)
Department of Medicine, Emory University School of Medicine, Atlanta, GA, USA
e-mail: deaton@emory.edu

K. L. Hamilton, D. C. Devor (eds.), *Studies of Epithelial Transporters and Ion Channels*, Physiology in Health and Disease,
https://doi.org/10.1007/978-3-030-55454-5_18

modulating PIP_2 presentation to the channel or by altering the rate of retrieval, recycling, insertion, or transcription/translation of ENaC.

Keywords ENaC · Epithelial sodium channels · aldosterone · ENaC structure · ENaC regulation · ENaC regulatory complex · ENaC trafficking · ENaC pathology · MARCKS · Hypertension · Kidney · Lung

18.1 Introduction

Epithelial tissues, in general, are designed to form a barrier between two body compartments and to facilitate transport from one compartment to the other, thereby regulating the composition of the compartments. Many epithelial tissues are responsible for movement of salt and water to maintain total body salt and water balance. This regulation of salt and water balance involves primary active transport of sodium (Na^+) and secondary active transport of chloride (Cl^-) followed by, in eukaryotes, osmotically driven water movement. That is, water movement depends upon prior transport of osmotic equivalents, usually salt. This implies that the proximate event in controlling blood osmolarity, blood pressure, and blood volume is sodium transport. The major hormonally controlled mechanism for sodium transport are epithelial sodium channels (ENaC). Therefore, regulation of water and salt balance requires the ability to regulate both salt reabsorption and salt secretion (Garty and Benos 1988; Garty and Palmer 1997; Rossier 1997; Rossier et al. 2002). Besides control of blood pressure, ENaC-mediated sodium transport in the lungs is responsible for normal fluid clearance from the alveolar space (Hummler et al. 1997; Matalon and O'Brodovich 1999) and consequently for normal gas exchange in the lungs. In vascular endothelium, ENaC is responsible for maintaining vascular tone and contributing to control of blood pressure.

Not surprisingly then, abnormalities of ENaC function or ENaC regulation have been linked to disorders of total body Na^+ homeostasis, blood volume, blood pressure, and lung fluid balance (Boucher 2004; Garty and Palmer 1997; Hummler et al. 1996; Mall et al. 2004). ENaCs are also expressed in other tissues: distal colon, sweat ducts, salivary ducts, inner ear, lingual epithelium, keratinocytes, lymphocytes, vascular smooth muscle, endothelium, and in the eye (epithelia within retina, lens, and pigmented ciliary body and iris) (Brouard et al. 1999; Bubien et al. 2001; Couloigner et al. 2001; Drummond et al. 2004; Grunder et al. 2001; Kretz et al. 1999; Mirshahi et al. 1999). The role of ENaC within some of these tissues is, at present, unclear. We will describe the properties of ENaC and its regulation.

18.1.1 Mechanism of Salt Transport Across Epithelial Tissues

This concept of epithelial salt and water transport is generally associated with both salt concentration gradients and electrical gradients across epithelial tissue. The ability of one epithelial tissue, the living frog skin, to generate an electrical potential has been known since the time of Emil du Bois Reymond in 1857 (Du Bois-Reymond et al. 1857). The origin of the potential was unknown at the time but was described as a "vital force" since it was not present in dead frog skin. There were many additional studies of the transepithelial potential (Dean and Gatty 1937; Francis and Pumphrey 1933; Orbeli 1910), but the origin of the potential was unclear until the work of Ussing and Zerhan (1951), later codified by Koefoed-Johnsen and Ussing (1958). They unambiguously showed that the frog skin bioelectric potential was due to active sodium transport mediated by a two-step process in which "...sodium movement through the outer [apical] border is passive, but highly selective..." and sodium movement across the basolateral membrane is via active transport of sodium.

Koefoed-Johnsen and Ussing's cellular model of epithelial sodium absorption has been repeatedly confirmed in all sodium-transporting epithelial tissues. These tissues include amphibian and mammalian urinary bladder, distal colon, various segments of the mammalian distal nephron, respiratory epithelia, sweat ducts, and numerous cultured cell lines derived from various tight absorptive epithelial tissues.

18.1.2 The Sodium-Selective Entry Pathway Is Blocked by the Diuretic, Amiloride, and Is an Ion Channel

In Ussing's work, the nature of the entry step remained unclear, but in 1967, a potassium-sparing diuretic, amiloride, was approved. Amiloride, after apical application, could reversibly reduce the frog skin potential close to zero (Bentley 1968; Dorge and Nagel 1970; Nagel and Dorge 1970). Many of these results strongly suggested that the sodium entry pathway was an ion-conducting channel protein. Initial microelectrode recordings from several epithelia lent weight to the idea that sodium moved down its electrochemical gradient through an ion channel (Fromter and Gebler 1977; Lewis et al. 1976; Wills et al. 1979). This idea was further strengthened by fluctuation analysis by Lindemann (Lindemann and Van Driessche 1977) which strongly suggested sodium-specific membrane channels blocked by amiloride in frog skin. The experimental evidence of amiloride-inhibitable epithelial Na^+ channels was finally confirmed by patch-clamp studies by Hamilton and Eaton (1985; Hamilton et al. 1986) and Palmer and Frindt (1986). Over 35 years after Koefoed-Johnsen and Ussing published their model, molecular studies by Canessa et al. (1993, 1994b) identified the genes for the epithelial Na^+ channel, now simply known as ENaC.

18.2 Molecular Properties of ENaC

ENaC consists of three homologous subunits, α, β, and γ, presumably arranged in a trimeric stoichiometry (Kashlan et al. 2011; Stockand et al. 2008) (Fig. 18.1). cDNAs for epithelial Na channels were originally described by two groups of investigators who both used expression cloning methods to isolate α-ENaC cDNA from rat distal colon (Canessa et al. 1993; Lingueglia et al. 1993). Canessa (Canessa et al. 1994b) subsequently identified by complementation the cDNAs for β- and γ-rENaC (*r*at *E*pithelial *Na C*hannel). Co-injection into *Xenopus* oocytes of cRNA prepared from all three cDNAs induces expression of amiloride-sensitive Na channels (Ahn et al. 1999; Canessa et al. 1994b; George et al. 1989; Hinton and Eaton 1989; Rossier et al. 1994) with characteristics virtually identical to those observed in renal cells in culture (Eaton and Hamilton 1988) or renal cortical collecting tubule principal cells (Palmer and Frindt 1988). The three subunits show limited sequence homology to each other although, across species, the same subunit has significant amino acid sequence homology (between 70% and 90%), even though there are some regions which show only limited homology. Other subunits, the so-called δ and ε subunits, have been described which, when expressed in heterologous expression systems, can by themselves or when associated with β and γ (and possibly α) subunits produce cationic channels (Babini et al. 2003; Biasio et al. 2004; Chang and McDonald 2007; Ji et al. 2004, 2006; Palmer and Frindt 1988). The ε subunit has only been reported in *Xenopus* renal cells once (Babini et al. 2003) and bears no significant homology to any proteins other than those in *Xenopus* (where the homology is less than 30%). It has not been cloned from any other genus than *Xenopus*. Its properties are poorly described; however recently, based upon homology considerations, this subunit has been reclassified as *Xenopus* δ (Althaus et al.

- Na^+ channels are composed of three structurally related subunits
  - α, β and γ
- Trimeric complex
  - 1α / 1β / 1γ
- Members of the ENaC/degenerin family
- Found in epithelia tissues:
  - Aldosterone-sensitive distal nephron
  - Airway and alveoli
  - Distal colon
- But expressed in numerous other sites

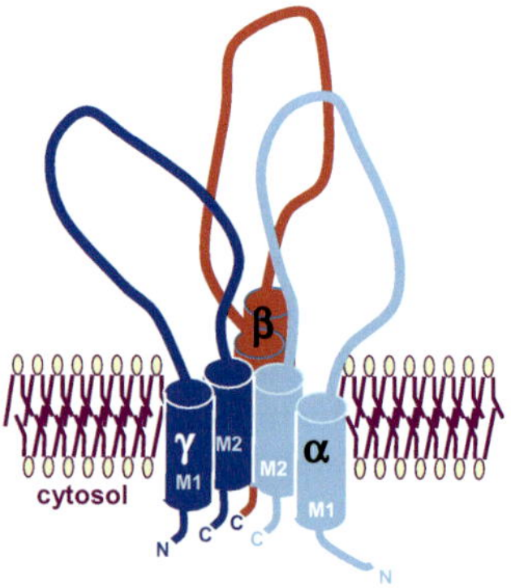

Fig. 18.1 ENaC structure. ENaC is composed of three structurally related subunits: α, β, and γ in a trimeric complex 1α/1β/1γ. ENaC is a member of the ENaC/degenerin family and is found in epithelia tissues including: aldosterone-sensitive distal nephron; airway and alveoli; distal colon; but also expressed in numerous other sites

2007). The δ subunit originally appeared to be primate specific (human and chimp) although there are apparently predicted homologs in mammals and a few non-mammals only a few of which have been successfully cloned. Its channel-forming properties are similar to the α subunit when expressed in heterologous expression systems (Chang et al. 2011; Giraldez et al. 2012; Haerteis et al. 2009; Ji et al. 2012; Schwagerus et al. 2015). The highest expression levels of δ-ENaC mRNA are in brain, testis, ovary, and pancreas, suggesting that the primary function of δ-ENaC may not involve sodium transport in epithelial tissue. This surmise is re-enforced by the observation that δ-ENaC cannot compensate for the loss of α-ENaC in the lung or kidney in α-ENaC knockout mice which die shortly after birth from a failure to clear fluid from their lungs (Hummler et al. 1996; Schwagerus et al. 2015). However, it has recently been functionally described in human lung cells, but its role is unclear (Schwagerus et al. 2015).

Although other combinations may exist (Fyfe and Canessa 1998; McNicholas and Canessa 1997), channels composed of the three subunits, α, β, and γ, have the following characteristic features: the channel is highly selective for sodium over potassium (more than 100:1), has a small single channel conductance of 4–6 pS; is inhibited by the diuretic, amiloride, at sub-micromolar concentrations; and is typically found in the apical membranes of sodium-transporting epithelial cells. Each subunit has only two predicted membrane-spanning domains, an unusual topology for an ion channel. Topologic analyses suggest that the amino and carboxy terminal regions of the ENaCs are cytoplasmic, and that each subunit has a large extracellular loop present between the membrane-spanning regions (Canessa 1996; Fyfe and Canessa 1998; Schild and Kellenberger 2001). The α-, β-, and γ-ENaC have cysteine-rich regions within putative extracellular loops which are postulated to form intersubunit disulfide bonds which are important in oligomerization of Na^+ channels (Collier et al. 2014; Sheng et al. 2005). The three ENaC subunits also possess proline-rich regions within their intracellular C-terminal domains that are important for interaction with accessory proteins that regulate the stability of functional channels in the apical membrane of renal cells (Canessa et al. 1994a; Collier et al. 2014; Debonneville et al. 2001; Rotin et al. 2000; Staub et al. 1997a, 2000).

18.3 Structure of ENaC

ENaC subunits are members of the ENaC/Degenerin gene family. Based on the homotrimeric structure of a closely related gene family member, acid sensing ion channel 1 (ASIC1) (Jasti et al. 2007), several investigators suggested that functional ENaC containing α, β, and γ subunits would also form a heterotrimer (Kashlan and Kleyman 2011; Stewart et al. 2011; Stockand et al. 2008). Recently, the structure deduced from cryo-electron microscopy showed that ENaC was indeed a heterotrimer (Noreng et al. 2018).

Each subunit has two membrane-spanning helices (TM1 and TM2) resulting in cytoplasmic amino and carboxyl termini (Canessa et al. 1994a; Renard et al. 1994).

The cytoplasmic domains have several sites of post-translational modification, protein–protein, and protein–lipid interactions that affect channel gating or trafficking. The structure and function of ENaC have been interpreted based on similarities to the crystal structure of the homologous acid-sensing ion channel 1 (ASIC1) and the cryo-EM structure of ENaC, itself (Noreng et al. 2018). A large extracellular region lies between the transmembrane-spanning domains. Based on homology to ASIC1, this region is organized in distinct domains with well-conserved β-sheet domains at the center of the fold and poorly conserved α-helical domains surrounding the protein core (Fig. 18.2) (Kashlan et al. 2011; Kashlan and Kleyman 2011). This extracellular domain contains sites for ligand-dependent gating, proteolytic activation, and mechanosensitivity. The proposed structure of the extracellular and pore regions allows some understanding of channel assembly, processing, and transduction of external environment. Unfortunately, current models do not provide information about the structure of the cytosolic domains so that the mechanisms by which post-translational events modify channel structure remains unclear.

The extracellular structure of ENaC and ASIC1 looks like an outstretched hand containing a ball with so-called finger, thumb, palm, knuckle, and β-ball domains (Jasti et al. 2007; Noreng et al. 2018). There appears to be structural homology between ENaC subunits and ASIC1 in most of these extracellular domains (Kashlan et al. 2011; Kashlan and Kleyman 2011). ENaC subunits have a predicted secondary structure that matches the observed ASIC1a structures in most regions of the extracellular region (for a thorough discussion, see Kashlan et al. 2011; Kashlan and Kleyman 2011) (Fig. 18.2). Two of these domains, the palm and β-ball, are β strands and are in close proximity to the lipid bilayer. More peripheral domains, the thumb, finger, and knuckle, are primarily α helices (Jasti et al. 2007; Stockand et al. 2008; Collier and Snyder 2009, 2011; Chen et al. 2011, 2013; Shi et al. 2011, 2012a, b; Stewart et al. 2011; Kashlan et al. 2011, 2015; Noreng et al. 2018).

Despite different suggestions for ENaC subunit stoichiometry, the similarity between the structure of ENaC and ASIC1a strongly implied that a heterotrimeric subunit stoichiometry was most likely. Subsequent resolution of ENaC channels with atomic force microscopy appears to confirm the heterotrimeric structure (Stewart et al. 2011). If ENaC is a heterotrimer, then the question arises of the radial organization. Is ENaC an α-β-γ trimer or an α-γ-β trimer? Work examining Cl^--dependent inhibition of ENaC (Collier and Snyder 2009) shows that ENaC is likely an α-γ-β trimer (clockwise, viewed from above). Based on Cu^{2+} binding at subunit interfaces, Chen et al. (2011) suggested a similar arrangement.

The TM1 and TM2 helices that form the ENaC pore are homologous to their ASIC1 counterparts; the sequence identity for the TM2 domain is considerably higher than TM1 (Stockand et al. 2008). This apparent similarity and structure of the pore regions is supported by the accessibility of TM2 cysteine residues to hydrophilic cysteine-modifying reagents (Kellenberger et al. 2001, 1999a, b; Li et al. 2003; Schild et al. 1997; Sheng et al. 2001, 2005).

One characteristic of the ENaC pore is its remarkably high selectivity for Li^+ and Na^+ over K^+. The exact selectivity for Na^+ over K^+ is difficult to determine, but it is at least 100:1 and may be substantially higher. While ENaC does transport some K^+,

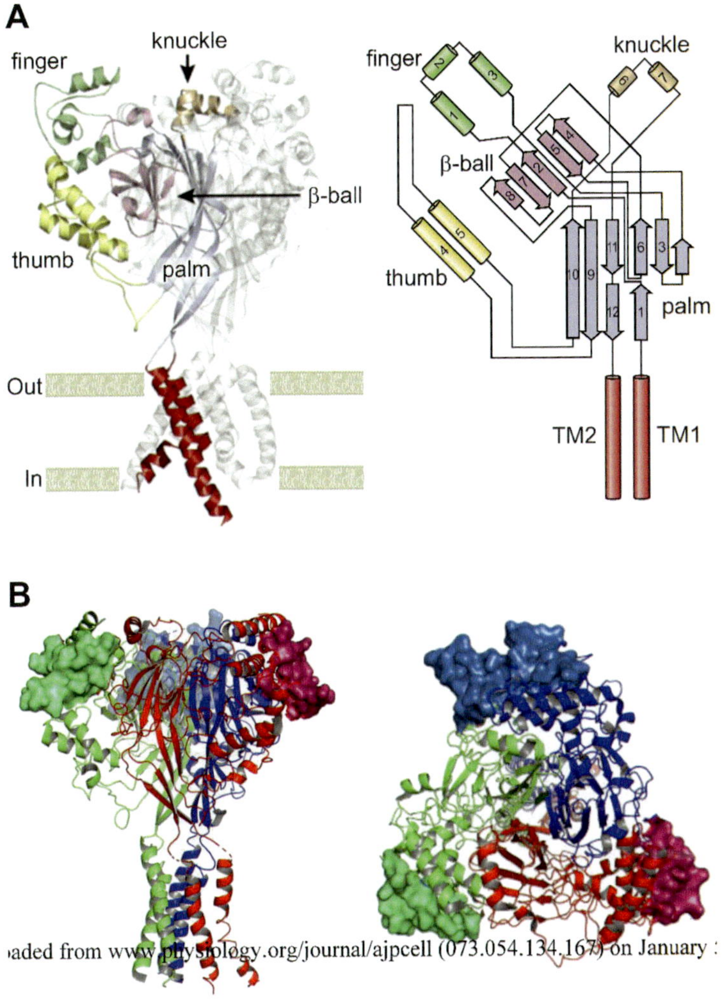

Fig. 18.2 ASIC1 and ENaC structures. Top left, ribbon illustration of the ASIC1 homotrimer, highlighting the extracellular and transmembrane regions of one subunit. Discrete domains within the extracellular region are labeled, including the proximal palm and b-ball formed by b strands, and the peripheral a helical thumb, finger, and knuckle domains. Transmembrane helices are indicated in red. Top right, organization of an ASIC1 subunit. Contiguous peripheral domains with α helices (cylinders) arise from non-contiguous proximal domains with b strands (arrows). Bottom, ribbon illustration of a human ENaC heterotrimer. Residues in the gating relief of inhibition by proteolysis (GRIP) domains that are unique to ENaC, and contain the a and g subunit imbedded inhibitory tracts are shown as surface rendering. Red (α subunit), blue (β subunit), and green (γ subunit) are shown. Side (left) and top (right) views are shown. ASIC1 and ENaC structures are based on pdb codes 3HGC and 6BQN (1, 2). Top figures were modified from (4). Editorial note: Springer or APS art staff can access this figure from a recent AJP review (3). 1. Jasti J, Furukawa H, Gonzales EB, and Gouaux E. Structure of acid-sensing ion channel 1 at 1.9 A resolution and low pH. *Nature* 449:316–323, 2007. 2. Kashlan OB, and Kleyman TR. ENaC structure and function in the wake of a resolved structure of a family member. *Am J Physiol Renal Physiol* 301:F684–F696, 2011. 3. Kleyman TR, and Eaton DC. Regulating ENaC's gate. *Am J Physiol Cell Physiol* 2019.

ENaC is absolutely impermeable to larger ions (Palmer 1982). ENaC's high selectivity is unique among the members of the ENaC/degenerin family of proteins (Favre et al. 1996; Kellenberger et al. 1999b; Sheng et al. 2005). Examination of point mutations in the pore region determined that a Gly/Ser-X-Ser tract in TM2 of each ENaC subunit was responsible for determining ENaC selectivity (Kellenberger et al. 2001, 1999b; Li et al. 2003; Sheng et al. 2000; Snyder et al. 1999).

The proposed model of the ENaC extracellular domain also provides some insight into the site for amiloride binding. In particular, mutations within the Gly/Ser-X-Ser tract affected amiloride affinity (Kellenberger et al. 2001, 1999a, b; Li et al. 2003; Sheng et al. 2000; Waldmann et al. 1995). Sheng et al. and Kashlan et al. suggest that the pyrazine ring of amiloride binds at αSer583/βGly525/γGly537 and the guanidinium group is positioned at the outer mouth of the Gly/Ser-X-Ser selectivity tract (Kashlan and Kleyman 2011; Kashlan et al. 2006; Sheng et al. 2005).

18.3.1 Alternative Structures for ENaC Family Members

ENaC in the kidney and colon are heterotrimers as described above. The original motivation for concluding that ENaC was likely to be a heterotrimer come from the crystal structure of another member of the ENaC/Degenerin family of proteins: the ASIC1a protein that forms a homotrimer (Jasti et al. 2007). In fact, when ENaC structure was resolved (Noreng et al. 2018), the structures were quite similar. This raised the question of whether subunits from other members of the family might form alternative channels. It was already known that a so-called delta subunit could substitute for alpha in the trimer to produce a channel that had significant differences from the more traditional α-β-γ channel (Chang et al. 2011; Giraldez et al. 2012; Ji et al. 2012; Kevin et al. 2013; Rauh et al. 2017; Wesch et al. 2012). The answer to the question of alternative structures came from an unexpected observation. In lung alveoli, salt and water reabsorption keep the alveoli from flooding in the face of fluid movement out of the pulmonary circulation (Eaton et al. 2009). Conventional wisdom suggested the reabsorption was via epithelial Na^+ channels (ENaC), but if all Na^+ reabsorption was via ENaC, then amiloride should block reabsorption. But amiloride blocks only about half of the reabsorption (O'Brodovich et al. 2008). Single channel measurements from alveolar epithelial cells show why only half is blocked: ENaC channels are present, but another channel is present at the same frequency that has a higher conductance, is nonselective for Na^+ over K^+, and different kinetics of opening than ENaC (Jain et al. 2001; Johnson et al. 2006; Lazrak et al. 2000; Matalon and O'Brodovich 1999; Yue et al. 1993). These high conductance channels contain α-ENaC since knocking down α-ENaC reduces the

---

Fig. 18.2 (continued) 4. Noreng S, Bharadwaj A, Posert R, Yoshioka C, and Baconguis I. Structure of the human epithelial sodium channel by cryo-electron microscopy. *Elife* 7: 2018

number of high conductance channels (knocking down β- or γ-ENaC has no effect) (Jain et al. 1999). No high conductance channels are detectable in single channel patches on lung slices from ASIC1a knockout mice (Trac et al. 2017). Thus, it appears that high conductance channels are composed of, at a minimum, one α-ENaC subunit and one or more Acid-Sensing Ion Channel 1a (ASIC1a) subunits. Recent work with atomic force microscopy supports this idea (Jeggle et al. 2015).

The significance of the two types of channels is that the regulation of the high conductance channels is dramatically different than low conductance channels (Downs et al. 2012; Eaton et al. 2009; Johnson et al. 2006; Matalon et al. 1996; Yue et al. 1993, 1994). Therefore, treatments to reduce alveolar flooding based on the known properties of ENaC may be counterproductive because high conductance channels are regulated so differently.

18.4 Cellular Regulation of ENaC

Sodium transport across Na^+ reabsorbing, tight epithelia such as the distal nephron and colon is the major factor determining total body Na^+ levels and sodium transport in lung epithelium is the major determinant of lung fluid balance. Sodium reabsorption in these epithelia is a two-step process. First, sodium enters renal cells from the tubular lumen through ENaC in the apical membrane before it is actively transported out of the cell by the basolateral Na,K-ATPase. Normally, there is excess capacity of the basolateral Na,K-ATPase so that the activity of apical ENaC determines the overall rate of sodium transport. Therefore, sodium transport is described by the flux of sodium through ENaC (measured as sodium current) across the apical membrane given as:

$$I_{\mathrm{Na}} = i\mathrm{NP_o} \tag{18.1}$$

Where I_{Na} is total sodium flux, i is the flux through a single open ENaC, N is the number of ENaC in the surface membrane, and $\mathrm{P_o}$ is the probability that a channel is open. In theory, total sodium transport could be regulated by altering any of the factors in the equation above; but, in practice, sodium transport is almost always regulated by altering the probability that a channel is open or by regulating the number of functional channels in the apical membrane.

The total number of ENaC subunits within the cytosol of epithelial cells is relatively large compared with the number in the apical membrane (that can functionally transport sodium into the cell). Therefore, in polarized epithelia, it appears that assembly and trafficking of ENaC subunits out of the endoplasmic reticulum is an inefficient process. Estimates of the relative amounts of ENaC in the cell interior vs the plasma membrane of epithelial cells vary enormously probably at least partially because of the type of the cells being studied. In transfected systems, only 1% or 2% of ENaC is outside the ER; in native cells, it appears to be much higher (10–30%) (Cheng et al. 1998; Hughey et al. 2002, 2004c; Prince and Welsh

1998, 1999). The large variability in the native cells appears to depend upon the hormonal state of the cell; e.g., EGFR ligands can dramatically reduce ENaC outside the ER and also total cellular ENaC (Liu et al. 2009; Tong and Stockand 2005). The exact point in the trafficking pathway at which ENaC subunits are assembled into a functional channel is unclear. Nonetheless, one of the mechanisms by which ENaC functional activity can be regulated is by altering the rate of delivery of assembled ENaC to the surface membrane. This mechanism has been suggested as the mechanism by which the hormones, vasopressin and, possibly, aldosterone alter sodium transport (Eaton et al. 2001).

18.4.1 Small Molecule Agents that Modify ENaC Activity

18.4.1.1 Amiloride

As mentioned above (Sect. 18.1.2), amiloride has played a central role in our understanding of ENaC. Until the advent of single channel measurements, ENaC current was defined as that component of transepithelial or whole-cell current that could be blocked by amiloride. In the 1980s, Palmer found that amiloride and other large cations blocked apical Na^+ channels in a voltage-dependent manner, suggesting that amiloride entered the channel pore and coulombically blocked sodium movement (Palmer 1984). The cloning of ENaC subunits allowed a further characterization of the amiloride-binding site. Schild et al. (1997) described mutations in all three subunits that reduced the efficacy of amiloride block and the mutation sites were only accessible from the extracellular space. When Cys residues were introduced at these sites, they were susceptible to chemical modification by methanethiosulfonate compounds from the extracellular space (Carattino et al. 2005a; Kashlan et al. 2005; Li et al. 2003; Sheng et al. 2001, 2005; Snyder et al. 1999) implying that amiloride entered a relatively wide channel mouth before blocking the channel at its narrowest point.

18.4.1.2 Ion Activity

18.4.1.2.1 Sodium Ion Activity

ENaC activity is regulated by changes in both extracellular and intracellular sodium. This was first suggested by MacRobbie and Ussing (MacRobbie and Ussing 1961) to explain why, when the basolateral Na^+, K^+-ATPase was inhibited, that frog skin epithelial cells did not swell as expected. They suggested that there was a mechanism by which intracellular sodium could limit the apical entry of sodium and thereby prevent sodium loading of cells. Subsequently, it was recognized that there were two categories of effect. A rapid change in sodium permeability or channel activity when extracellular sodium is changed has been called sodium self-inhibition (Fuchs et al.

1977). A slower change (over hours) in channel activity caused by changes in intracellular sodium concentration has been called feedback inhibition.

Both self-inhibition and feedback inhibition make strong evolutionary and physiological sense. ENaC plays a key role in the regulation of extracellular fluid volume and blood pressure. Therefore, from a regulatory standpoint, sodium transport through ENaC should respond to signals related to blood pressure and ECF volume. If net sodium transport changed dramatically in response to changes in lumenal sodium or intracellular sodium, the system would be unstable and subject to oscillatory changes unrelated to blood pressure or ECF volume. Self-inhibition and feedback inhibition provide a short-term and a long-term damping effect on overall transport that tend to keep transport relatively constant and to match apical entry with basolateral exit, thereby preventing large changes in intracellular sodium activity and epithelial cell volume (Palmer et al. 1998; Turnheim 1991).

18.4.1.2.2 Sodium Self-Inhibition

Self-inhibition is the ability of extracellular sodium (or lithium) to alter the activity of ENaC independent of changes in intracellular sodium activity (Lindemann 1984). Self-inhibition will reduce the permeability of the apical membrane when lumenal sodium concentration increases. In frog skin, when sodium concentration on the pond side of the frog skin is rapidly increased, short-circuit current rapidly increases and then decreases in a matter of seconds (Fuchs et al. 1977). The phenomenon was suggested to be due to the presence of an allosteric receptor for sodium separate from the pore that could reduce channel activity. When sodium interacts with these extracellular sites, it drives allosteric changes to the channel gate to reduce channel open probability (Kashlan et al. 2015; Kleyman et al. 2018). Such a response would allow ENaC in the distal nephron to adjust rates of sodium reabsorption consistent with changes in urinary Na^+ concentration (Kleyman et al. 2018).

The Na^+ self-inhibition response is tightly linked to ENaC processing by proteases. Treatment of ENaC with trypsin eliminated self-inhibition suggesting that self-inhibition is reduced in proteolytically cleaved channels. Noncleaved channels have low activity that appears to result from enhanced channel inhibition by external sodium (Sheng et al. 2006).

The mechanism for self-inhibition has been illuminated by considerations of the proposed structure of the channel (see Sect. 18.3). The structural model of the extracellular region of ENaC (Fig. 18.3) has clefts with acidic residues. On the basis of this model, Sheng et al. (Sheng et al. 2004a) identified histidine residues in these acidic clefts that modulate sodium self-inhibition.

The extracellular regions of the α and γ subunits have specific amino acid residues which when mutated alter the Na^+ self-inhibition and channel activity (Sheng et al. 2004b, 2006, 2007; Shi et al. 2012a, 2011, 2013; Maarouf et al. 2009; Winarski et al. 2010; Kashlan et al. 2011). These mutations likely directly participate in sodium binding, or produce allosteric transitions induced by sodium binding. Sites in the extracellular loop between the β6 and β7 strands and in the vicinity of an α subunit

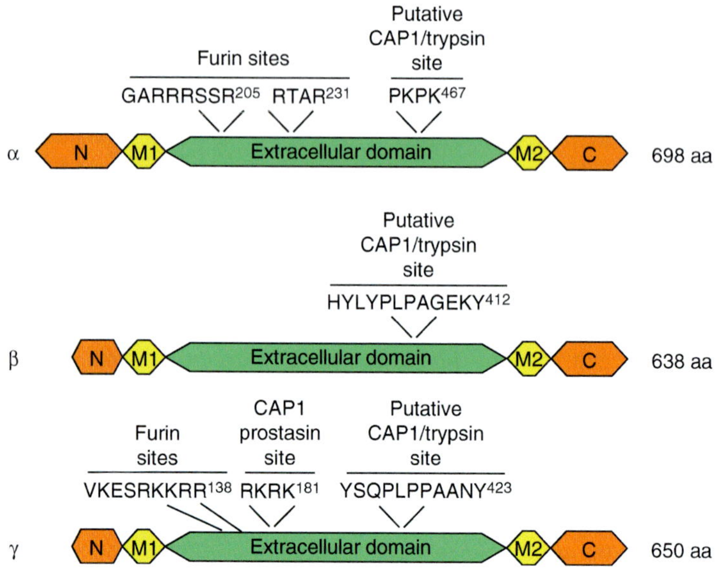

Fig. 18.3 Schematic representation of the rat ENaC subunits and their identified and putative sites for furin, CAP1, trypsin, and prostasin cleavage. M1, M2: transmembrane domains; N, C: intracellular amino- and carboxy-termini, respectively. From Ahmed Chraibi and Stéphane Renauld (2013). New Insights into the Epithelial Sodium Channel Using Directed Mutagenesis, Genetic Manipulation of DNA and Protein - Examples from Current Research, Dr. David Figurski (Ed.), ISBN: 978-953-51-0994-5, InTech, DOI: 10.5772/50309. Available from: http://www.intechopen.com/books/genetic-manipulation-of-dna-and-protein-examples-from-current-research/new-insights-into-the-epithelial-sodium-channel-using-directed-mutagenesis. Open access

acidic cleft alter cation selectivity of the Na^+ self-inhibition response implying that residues in this region are a binding site for Na^+ (Kashlan et al. 2015; Kleyman et al. 2018).

There have now been many single nucleotide polymorphisms that alter sodium self-inhibition. It remains to be seen if any of these loss of self-inhibition mutations are also associated with human hypertension.

18.4.1.2.3 Sodium Feedback Inhibition

Several investigators initially found little evidence for a direct effect of intracellular sodium on channel activity (Eaton 1981; Palmer 1985). However, in mouse mandibular duct cells, a G protein-dependent decrease in Na^+ permeability was observed when intracellular sodium was increased during whole cell patch-clamp recording (Komwatana et al. 1996). This is consistent with prior work in excised patches from renal cells in culture, in which it had been shown that G protein activation could inhibit ENaC activity (Ohara et al. 1993). The same investigators, again using mouse mandibular duct cells, showed that increases in intracellular sodium could activate the ubiquitin ligase, Nedd4-2, and Nedd4-2 in turn could reduce ENaC activity by promoting channel internalization (see below) (Dinudom et al. 2001).

However, the original situation described by MacRobbie and Ussing was more definitively examined by Silver et al. (1993) in cell-attached patches on tubules that were split open to expose the luminal surface and ouabain was added to the bath to increase cell Na^+. Channel activity decreased dramatically, the mechanism of inhibition did not appear to be a direct effect of intracellular sodium, but rather appeared to involve increased intracellular calcium since calcium measured with Fura-2 increased and removal of calcium eliminated the effects of ouabain and increased sodium. Increasing intracellular calcium by apical addition of the calcium ionophore, ionomycin, produces a similar inhibition of ENaC (Palmer and Frindt 1987; Thai et al. 2014).

18.4.1.2.4 Chloride Ion Activity

Collier and Snyder (2011; Collier et al. 2014) have shown that Cl^- inhibits ENaC currents with an apparent K_i of 30 mM, suggesting that urinary Cl^- has a physiological role in modulating ENaC activity in the collecting duct. External Cl^- inhibits ENaC by reducing P_o with an apparent time course similar to the other allosteric inhibitors of ENaC. This inhibitory effect has a degree of anion selectivity, i.e., Cl^-, I^-, and Br^- inhibition of ENaC is approximately the same, and F^- and larger sulfates and phosphates have little effect.

The ASIC1 crystal structure revealed a bound Cl^- at each of the three subunit interfaces, with residues from the thumb and palm domains of different subunits defining each binding site (Kashlan and Kleyman 2011). Collier and Snyder (2011; Collier et al. 2014) examined whether homologous sites within ENaC are involved in Cl^- binding (Fig. 18.3). Mutations of key residues at these sites had differing effects on Cl^- inhibition as well as Na^+ self-inhibition. Mutants at the αγ interface had little effect on Cl^- inhibition, mutants at the βγ interface reduced Cl^- inhibition, and mutants at the αβ interface reduced both Cl^- inhibition and Na^+ self-inhibition. These mutations suggested a clockwise α-γ-β subunit arrangement when viewed from the extracellular side.

18.4.1.2.5 Hydrogen Ion

Protons regulate electrogenic sodium absorption in a variety of epithelia, including the cortical collecting duct, frog skin, and urinary bladder. When various combinations of α-, β-, and γ-subunits of ENaC are expressed in *Xenopus*, ENaC activity was reduced by intracellular acidification and increased by intracellular alkalization. Changes in extracellular pH had no effect on ENaC (Chalfant et al. 1999). A more direct examination of the effects of pH used patches excised from the apical membrane of rat principal cells and varied the cytosolic pH. ENaC open probability increased with alkalization of the cytosolic side of the patch increasing approximately eightfold between pH 6.4 and 7.4 (Palmer and Frindt 1987).

The effects of pH may be physiologically relevant. Although there is little direct effect of extracellular pH in oocytes or in excised membrane patches, because there are other pathways for the movement of hydrogen ion, intracellular pH often tracks extracellular pH particularly in the presence of CO_2 or bicarbonate (Clark and Eaton 1983; Ifshin et al. 1983). In a specific example, pendrin in collecting duct intercalated cells secretes bicarbonate into the lumen which increases ENaC activity in adjacent principal cells, presumably by alkalization of the principal cell interiors (Pech et al. 2010).

18.4.1.2.6 Calcium Ion Activity

As already mentioned above, there is ample evidence that increasing intracellular calcium reduces ENaC activity (Thai et al. 2014; Palmer and Frindt 1987; Ishikawa et al. 1998; Wang and Chan 2000; Poulsen et al. 2005; Taruno et al. 2008; Alli et al. 2012a). Most of these studies probably involved an indirect effect of calcium on ENaC. The question remains of whether there is a direct effect. In excised, inside-out patches, varying the free Ca^{2+} concentration of the solution bathing the cytoplasmic side of patches had no measurable effect on channel activity even at nonphysiologically high calcium activities (Palmer and Frindt 1987).

18.4.1.2.7 Heavy Metal Ions

For many years, heavy metals have been known to alter sodium transport. In fact, prior to the availability of amiloride, mercuric chloride was used as an inhibitor of apical sodium entry in frog skin (Linderholm 1952; Lodin et al. 1963). But the effects of heavy metal poisoning on the kidney are profound and often involve failure to reabsorb sodium (Barbier et al. 2005). Yu et al. (2007b) examined the effects of Zn^{2+}, Ni^{2+}, Cu^{2+}, Pb^{2+}, Cd^{2+}, and Hg^{2+} on the activity of single ENaC in a renal epithelial cell line. None of the cations changed the single-channel conductance. However, ENaC activity was decreased by Cd^{2+} and Hg^{2+} and increased by Cu^{2+}, Zn^{2+}, and Ni^{2+} but was not changed by Pb^{2+}. Cu^{2+} increased channel density and stimulated open probability by reducing sodium self-inhibition as others have also shown (Chen et al. 2011; Sheng et al. 2002). Ni^{2+} binds histidine residues and the effect of the histidine-modifying reagent, diethyl pyrocarbonate, mimicked that of Ni^{2+}. This result is contrary to the result of (Sheng et al. 2002) who found that Ni^{2+} inhibited by binding to specific extracellular histidines. On the other hand, several of the heavy metals that inhibit ENaC (especially Hg^{2+}) strongly bind cysteines and cysteines are known to be important for subunit assembly (Sheng et al. 2002, 2007).

18.4.2 Post-Translational Modifications of ENaC

Post-translational modifications are a common mechanism to regulate proteins, especially membrane proteins. ENaC is no exception: it is modified by at least six separate modifications.

18.4.2.1 Acetylation

Butler et al. (2015) showed that α-, β-, and γ-ENaC are substrates for lysine acetylation. Trichostatin A (TSA), a histone deacetylase inhibitor, enhanced ENaC acetylation and increased ENaC abundance in the total cell lysate and at the cell surface and increased transepithelial current in tissue culture and native renal epithelial cells. ENaC abundance and activity was reduced by overexpression of histone deacetylase 7 (HDAC7) and was increased by silencing HDAC7. There was also a reciprocal relationship between acetylation and ubiquitination; TSA reduced ENaC ubiquitination whereas HDAC7 increased ubiquitination. Trichostatin A blocks aldosterone-SGK-1 in cortical collecting duct cells and thereby blocks Nedd4-mediated ubiquitination (Mansley et al. 2019). By reducing ENaC ubiquitination, TSA decreased the rate of ENaC degradation. Thus, by antagonizing ENaC ubiquitination, acetylation decreases ENaC internalization and increases sodium absorption. This stabilizes ENaC, and hence, increases its abundance at the cell surface.

18.4.2.2 Methylation

Methylation is a modification that in some sense is analogous to phosphorylation except that the rate of demethylation reactions are much slower making methylation more persistent than phosphorylation; there are several enzymes that promote addition of methyl groups to isoprenylcysteine, aspartate, leucine, or lysine (Lanouette et al. 2014; Mato et al. 2002; Morgunkova and Barlev 2006; Wei et al. 2014; Zakrzewicz et al. 2012). Direct methylation increases the activity of ENaC reconstituted into lipid bilayers (Ismailov et al. 1994). In a tissue model of renal principal cells, inhibiting all protein methylation with 3-Deazaadenosine dramatically reduced ENaC single channel activity (Becchetti et al. 2000). 3-Deazaadenosine also blocked the increase in channel open probability associated with addition of aldosterone. A methyl donor, S-adenosyl-l-methionine, activated ENaC in excised, inside-out patches. The combination of the methyl donor with intracellular GTP stabilized ENaC channel activity in excised patches for relatively long periods of time.

Isoprenylcysteine-O-carboxyl methyltransferase (pcMTase) is responsible for methylation of terminal cysteines modified by isoprenyl groups and the enzyme can be detected immunologically in renal principal cells. Aldosterone increased pcMTase activity without increasing pcMTase protein levels. These results suggest that a critical transmethylation is necessary for aldosterone-mediated increases in Na^+ transport (Stockand et al. 1999b, 2000b). Aldosterone, via pcMTase, does induce methylation of the small G protein, K-Ras, and membrane-associated K-Ras does activate ENaC. However, methylation also has a more direct role in ENaC activation. Rokaw et al. (1998) showed that, in cultured mouse cortical collecting duct epithelial cells, the target for aldosterone-induced methylation was β-ENaC and that β-ENaC is a substrate for methylation by an enzyme other than pcMTase contained in the mouse CCD cells. Subsequently, Edinger et al. (2006) identified a protein, the product of gene ENSMUSG00000021773, that was expressed in CCD cells and was responsible for β-ENaC methylation. This protein product is catechol O-methyltransferase domain-containing protein 1 (Accession: Q8BIG7) (Carninci et al. 2005).

18.4.2.3 Ubiquitination

Ubiquitin is a highly conserved 76-amino-acid polypeptide with a role in a variety of cellular functions. Ubiquitin plays a major role in protein degradation because it serves as a tag for internalization and degradation of membrane proteins (Hershko and Ciechanover 1998; Hicke 1997, 2001). Ubiquitin is covalently linked to substrate proteins via the ε-amino group of lysine residues. Ubiquitination of proteins generally requires three different enzymatic activities mediated by either two or three enzymes. First, a ubiquitin-activating enzyme (E1) activates ubiquitin by forming a thioester bond in an ATP-dependent reaction. Ubiquitin is then transferred to an ubiquitin-conjugating enzyme (E2), followed by the addition of ubiquitin to target proteins by an ubiquitin ligase (E3). Single ubiquitin molecules can be added to different lysine residues on the target protein so there will be one ubiquitin per lysine on multiple lysines; this modification is known as monoubiquitination or multiubiquitination. Alternatively, additional ubiquitins can be added to ubiquitins already on lysines resulting in a chain of multiple ubiquitins, a process known as polyubiquitination (Hershko and Ciechanover 1998).

Ubiquitination of membrane proteins promotes internalization; but once internalized, the degradative pathway that proteins follow appears to depend upon whether the proteins are monoubiquitinated or polyubiquitinated. Polyubiquitination directs proteins to the proteasome for degradation (Coux et al. 1996; Hershko and Ciechanover 1998; Mitch and Goldberg 1996); monoubiquitination is a signal for endocytosis and lysosomal degradation (Rotin et al. 2000; Bonifacino and Weissman 1998; Pickart 2000a, b; Pickart and VanDemark 2000; Mimnaugh et al. 1999; Thrower et al. 2000; van Kerkhof and Strous 2001; van Kerkhof et al. 2001; Yu and Kopito 1999). Proteins degraded by the proteasome tend to be associated with lipid rafts and internalized via caveolae (Minshall et al. 2003; Felley-Bosco et al. 2002; Waterman and Yarden 2001). Once internalized, the targeted protein is unfolded, ubiquitin is removed from the proteins by deubiquitinating enzymes, and the proteins are cleaved to small peptides by proteasomal proteolytic activity (Hershko and Ciechanover 1998; Hicke 2001; Weissman 2001).

In contrast, monoubiquitinated membrane proteins are internalized and degraded by proteases in lysosomes (Hicke 1997, 2001; Jeffers et al. 1997). Ubiquitination is not a prerequisite for degradation of all proteins by the lysosome; however, several receptors and ENaC must undergo ubiquitination before degradation (Jeffers et al. 1997; Hicke 2001; van Kerkhof et al. 2001). In the process of internalization via clathrin-mediated endocytosis, ubiquitinated proteins pass through several endosomal compartments to lysosomes. The first compartment is an early endosome (also referred to as recycling endosome). Proteins from this compartment can be de-ubiquitinated and recycled back to the plasma membrane. If early endosomes are not recycled back to the plasma membrane, then they become late endosomes by acquiring Rab7-GTP protein which finally fuse with lysosomes that contain proteolytic enzymes which degrade internalized plasma membrane proteins (Van IJzendoorn et al. 2000; Van IJzendoorn and Mostov 2000).

18.4.2.4 A Specific Ubiquitin Ligase Isoform, Nedd4-2, Ubiquitinates ENaC

Ubiquitin conjugation is required for ENaC internalization and subsequent degradation. Ubiquitin-conjugation of ENaC subunits requires a specific ubiquitin ligase, Nedd4 (*n*eural-precursor-cell-*e*xpressed-*d*evelopmentally-*d*own-regulated protein). Nedd4 contains an E6-AP carboxyl terminus (Hect) domain that is homologous to other ubiquitin ligases, three WW domains in the rat and mouse protein or four WW domains in human and *Xenopus* protein, and a calcium/lipid-binding domain (CaLB/C2) (Staub and Rotin 1996). In humans, Nedd4 has nine known isoforms (for reviews, see (Rotin and Kumar 2009; Ingham et al. 2004; Yang and Kumar 2010)). Nedd4-2 appears to be the isoform responsible for physiologically relevant ubiquitination of ENaC in all mammalian epithelial cells (Snyder et al. 2004b), but other E3 ligases can modify ENaC. WWP2 is expressed in human adult and embryonic epithelial tissues and transepithelial current is inhibited when coexpressed with ENaC subunits in Fischer rat thyroid epithelial cells (McDonald et al. 2002). Nonetheless, there is direct evidence that Nedd4-2 is the ligase responsible for ENaC ubiquitination and that Nedd4-2 is associated with membrane ENaC molecules (Zhou et al. 2007; Snyder 2005; Malik et al. 2005b). Nedd4-2 co-localizes with ENaC in renal cortical and outer medulla collecting duct principal cells and in airway epithelia (Staub et al. 1997b). Nedd4-2 apparently requires direct association with ENaC to allow ubiquitin conjugation as suggested by the direct binding of Nedd4 to all three ENaC subunits (Staub et al. 1996; Dinudom et al. 1998; Debonneville et al. 2001). Interaction of the Nedd4-2 WW domain with the PY motif of ENaC β subunits involves contact between tyrosine 618 and leucine 621 and WW domain of Nedd4-2 (Kanelis et al. 2001). Nedd4-2 WW domain binding to ENaC β and γ subunits is practically irreversible because of a very slow dissociation rate when determined by surface plasmon resonance (Asher et al. 2001).

Co-expression of ENaC and Nedd4-2 reduces whole cell current and the amount of ENaC subunit protein at the cell surface. ENaC mutants lacking the PPxY motif are unaffected by co-expression of Nedd4-2 (Abriel et al. 1999; Goulet et al. 1998). In renal cells in culture, Nedd4-2 co-immunoprecipitates with α- and β-ENaC and is associated with apical membrane proteins (Malik et al. 2005b).

One reason for alternative degradative pathways, lysosomal vs. proteasomal, could be different levels of ENaC protein expression: in native cells, the surface expression of ENaC channels ranges from about 40 channels per cell in mouse cortical collecting duct cells (M1) to as low as 4 per cell in the *Xenopus* renal cell line (A6) (with alveolar type 2 cells somewhere in between) whereas in transfected cells, surface expression levels may exceed several thousand per cell. The large number in the membrane overwhelms the normal internalization and degradative machinery of the proteasome and promotes clathrin-mediated endocytosis and degradation via a lysosomal pathway at a rate that is independent of proteasome activity. In addition, the large excess of ENaC protein may limit the extent of ubiquitination, thus promoting monoubiquitination (and lysosomal degradation) rather than polyubiquitination (and proteosomal degradation).

Although it has yet to be demonstrated for α-, β-, or γ-ENaC, ubiquitination of the δ subunit can be promoted by interactions of the COMMD1 (formerly MURR1) protein with Nedd4-2 (Chang and McDonald 2007; Biasio et al. 2004). COMMD1 plays numerous other roles in cellular regulation so the significance of its regulation of the δ subunit remains to be determined (as does the role of the δ subunit itself in epithelial function).

18.4.2.5 De-ubiquitination of ENaC

All of the mechanisms mentioned above involve changes in the affinity of Nedd4-2 for ENaC with the consequent change in the rate of ubiquitination and the rate of internalization. If ENaC subunits remain ubiquitinated, they will be degraded by the proteasome or the lysosome, but ubiquitins can be removed from ENaC. USP8 removes ubiquitin from ENaC before ENaC leaves the membrane (Zhou et al. 2013). After internalization, in early endosomes, the deubiquitinating enzyme, UCH-L3, appears to remove approximately half of the ubiquitin from internalized subunits (Butterworth et al. 2007). At some later point in the endosomal progression, another de-ubiquitinating enzyme, USP2-45, can be induced by steroids and promote removal of ubiquitin to allow functional ENaC to return to the surface membrane (Fakitsas et al. 2007).

18.4.2.6 Phosphorylation

Phosphorylation of ENaC was first described by (Shimkets et al. 1998) as a possible mechanism for the regulation of ENaC. Phosphorylation is central to regulating ENaC ubiquitination and degradation (Bhalla and Hallows 2008; Snyder 2009;

Soundararajan et al. 2010a). Phosphorylation of Nedd4-2 controls ENaC ubiquitination (Debonneville et al. 2001).

G protein receptor kinase 2 (GRK2) phosphorylates Nedd4-2 to reduce interaction with ENaC and prevent ubiquitination and degradation (Sanchez-Perez et al. 2007). Nedd4-2 function is also inhibited by cAMP-dependent-protein-kinase-A (PKA) phosphorylation (Snyder et al. 2004a). Transmitter agents which increase cAMP (vasopressin in the kidney, β-adrenergic agents in the lung) stimulate ENaC activity. The stimulation has been widely attributed to PKA-induced ENaC insertion, but clearly increases of cAMP might also control ENaC function by regulating Nedd4-2 function. This implies that there is again a concerted response with one action of the transmitter agents promoting addition of channels to the membrane and the other action preventing their removal.

Phosphorylation of ENaC itself plays a role in altering Nedd4 ubiquitination. MAPK phosphorylation of ENaC promotes Nedd4-2 binding. Adenosine-monophosphate-activated kinase (AMPK) promotes Nedd4-2 retrieval of ENaC and AMPK knockout mice have excess ENaC activity (Almaca et al. 2009). Whether AMPK has its effect by phosphorylating ENaC or Nedd4 remains to be determined. In contrast, casein kinase 2 phosphorylates ENaC and prevents interaction with Nedd4 (Bachhuber et al. 2008).

Protein kinase C (PKC), which may be stimulated through extracellular purinergic receptor activation, inhibits both ENaC P_o and surface expression (Stockand et al. 2000a; Booth and Stockand 2003). The change in surface expression is due to signaling through the mitogen-activated protein kinase/ERK pathway. The change in open probability is due to PKC phosphorylation of MARCKS (myristoylated alanine-rich C-kinase substrate). The phosphorylation inactivates MARCKS and reduces MARCKS ability to present PIP_2 to ENaC. Reduced PIP_2 reduces ENaC open probability (Alli et al. 2012a) (and see below).

ERK2 and casein kinase 2 (CK2) directly phosphorylate the COOH-termini of β- and γ-ENaC and promote Nedd4-2-mediated ENaC internalization (Shi et al. 2002a, b; Berman et al. 2018). A functional polymorphism in the COOH-terminal tail of human α-ENaC also prompted investigation of its regulation by kinase-mediated changes in ENaC trafficking (Samaha et al. 2004). The COOH-terminal tail of α-ENaC is regulated by CK1 or PKC-δ to increase insertion of channels in the plasma membrane (Yan et al. 2007, 2006). SGK1 also directly phosphorylates and stimulates α-ENaC (Diakov and Korbmacher 2004; Thomas et al. 2011).

Some kinases can indirectly regulate ENaC trafficking. IκB kinase-β interacts with ENaC to increase cell surface expression and Na^+ current in CCD cells (Edinger et al. 2009; Lebowitz et al. 2004). WNK1 (with No lysine [K] kinase 1) was demonstrated to alter SGK1 activity and hence stimulate ENaC (Xu et al. 2005). In contrast, WNK4 inhibits ENaC and the inhibition is not mediated through an action of SGK1 or Nedd4-2 (Yu et al. 2013). Akt (similar to SGK1) also disrupts ENaC–Nedd4-2 binding and hence stimulates ENaC-mediated Na^+ current in response to insulin (Lee et al. 2007a; Holzman et al. 2007). Protein kinase A was shown previously to inhibit Nedd4-2 (Snyder et al. 2004a) (similar to SGK1) but has also now been suggested to modulate tyrosine kinase activity to increase apical expression of ENaC (Woollhead and Baines 2006). Protein kinase D also rapidly promotes insertion of ENaC into the plasma membrane (McEneaney et al. 2007). The metabolic sensor AMP-activated kinase (AMPK), which is sensitive to a host of external stimuli and intracellular energy stores, indirectly inhibits ENaC surface expression by promoting an interaction between Nedd4-2 and ENaC (Carattino et al. 2005a; Bhalla et al. 2006a).

18.4.2.7 Proteolysis

Originally, it was shown that the serine protease inhibitor, aprotinin, a nonselective serine protease inhibitor, could reduce Na^+ transport across toad urinary bladder (Orce et al. 1980). Subsequently, the serine protease, trypsin, was shown to activate ENaC in cultured renal cells (Vallet et al. 1997). Expression cloning identified channel-activating peptidase 1, also called prostasin, as a factor that increases ENaC currents when co-expressed in oocytes. There followed an identification of a series of serine and metaloproteases that can activate ENaC. Initially, it was not clear whether ENaC itself was the target of proteases, but (Masilamani et al. 1999) showed direct biochemical evidence for cleavage by showing the apparent molecular weight of the γ subunit shifts from approximately 85 to approximately 70 kD in rats treated with aldosterone or a low-salt diet. Patch-clamp studies showed that proteolytic cleavage substantially increased ENaC single-channel open probability from "near-silent" to frequently open (Caldwell et al. 2004). It was also reported that furin, a member of the pro-protein convertase family thought to reside in the trans-Golgi network, was important in cleaving the α subunit at two sites and the γ subunit at one site in the extracellular loops (Hughey et al. 2003, 2004a, b) (Fig. 18.3). Furin-mediated cleavage of α-ENaC activates ENaC by releasing an inhibitory peptide from the extracellular loop (Carattino et al. 2006). Furin-dependent proteolysis of α-ENaC activates the channel by relieving Na^+ self-inhibition, thereby increasing Na^+ transport at higher extracellular Na^+ concentrations (Carattino et al. 2006; Sheng et al. 2006; Hughey et al. 2007). Recently, it was also shown that prostasin, likely present in the extracellular apical compartment, also cleaves γ-ENaC at a site in the extracellular loop near furin cleavage sites, which as with the dual furin cleavage sites in α-ENaC, releases an inhibitory peptide that further activates ENaC (Carattino et al. 2006; Bruns et al. 2007). However, there are many other proteases that cleave the subunit on the C-terminal side of the furin site, releasing an imbedded inhibitory tract of >40 residues and converting channels to a high open probability state (Bruns et al. 2007). These proteases include prostasin, transmembrane serine protease 4, matriptase, cathepsin B and S, neutrophil and pancreatic elastase, kallikrein, urokinase, plasmin, and specific bacterial proteases (Vallet et al. 1997; Bruns et al. 2007; Caldwell et al. 2005; Adebamiro et al. 2007; Harris et al. 2007; Vuagniaux et al. 2002; Garcia-Caballero et al. 2008; Passero et al. 2008, 2012; Alli et al. 2012b; Tan et al. 2014; Patel et al. 2012; Picard et al. 2008; Svenningsen et al. 2009; Ji et al. 2015).

Proteolysis of the channel per se is not strictly necessary for activation, because deletion of these inhibitory tracts in α and γ subunits is sufficient to enhance channel activity without any cleavage. Also, addition of the inhibitory peptides to wild-type ENaC inhibits the channel (Carattino et al. 2006, 2008).

Apparently, γ subunit cleavage has a dominant role in channel activation (Carattino et al. 2006). Proteolytic processing and channel activation appear to be a stepwise process. Non-cleaved channels have a low open probability. Channels that have been cleaved by furin in which the α subunit has lost its inhibitory tract and the γ subunit has been cleaved once have an intermediate open probability. After α and γ channels have both lost their inhibitory tracts, ENaC channels containing the cleaved subunits have the capacity to range in open probability from very low to very high depending upon secondary regulatory factors as discussed elsewhere in this chapter.

Based on the newly described molecular structure (Noreng et al. 2018), the α and γ subunit inhibitory tracts are at the interface of the thumb and finger domains of the subunits, regions that are not present in other members of the ENaC/degenerin family (Kashlan et al. 2010, 2012; Balchak et al. 2018).

It is relatively easy to show in heterologous expression systems that proteolysis can activate ENaC. It is more difficult to show that this is true *in vivo*. Administering aprotinin to mice produced natriuresis and is consistent with protease-dependent activation of ENaC *in vivo* (Bohnert et al. 2018). Selective knockout of the serine protease, prostasin, in alveolae reduced fluid clearance consistent with a loss of ENaC-mediated reabsorption of salt and water (Planes et al. 2010) and a colonic knockout led to a reduction in colonic potential difference (Malsure et al. 2014).

If cleavage activates channels *in vivo*, then the biochemical appearance of cleaved subunits should correspond to channel activation; but Frindt, Palmer, and coworkers found that in rats on a low salt diet cleaved subunits appeared shortly after starting the diet, the increase in channel activity was delayed by days (Frindt et al. 2018). This result may not be surprising if proteolysis primarily alters the number of functional channels in the membrane, but trafficking to the membrane and the open probability of channel once in the membrane is independent of proteolysis.

Both processed (complex glycosylated with cleaved subunits; active) and nonprocessed (noncleaved; inactive) forms of ENaC subunits exist in cells and tissues expressing endogenous ENaC (Hughey et al. 2004a; Ergonul et al. 2006; Masilamani et al. 1999). Inactive, near-silent channels at the cell surface are probably uncleaved and may serve as a channel pool that can be activated by extracellular proteases. Evidence suggests the extent of ENaC proteolysis is dependent on channel residency time in the plasma membrane, because channels with mutations that block ubiquitin-dependent endocytosis remain on the surface longer, have an increased proportion of mature (cleaved) subunits, and have a correspondingly higher activity (Carattino et al. 2008).

There is at least one additional level of regulation since there are endogenous protease inhibitors, such as protease nexin 1 (Wakida et al. 2006). Indeed, studies suggest the proteolytic state of ENaC subunits reflect a balance between the expression of activating proteases and protease inhibitors (Myerburg et al. 2006). ENaC regulation by proteases and protease inhibitors is likely to be physiologically significant in several organs, including lung and kidney. For reviews, see (Soundararajan et al. 2010a; Hughey et al. 2007; Kleyman et al. 2006; Kleyman and Eaton 2019; Ray and Kleyman 2015).

18.4.2.8 ENaC's Role in Nephrotic Syndrome

Proteinuria in nephrotic syndrome is associated with sodium retention and edema. Studies from mice, rats, and humans have shown that the sodium retention depends on urinary serine proteases and that it can be reduced by ENaC blockers (Hinrichs et al. 2019). ENaC subunit proteolysis has been described during extracellular volume depletion, decreased effective arterial volume such as heart failure and with aldosterone administration in normal volume states. Furin cleaves the α subunit twice, but only cleaves the γ subunit once. In the kidney, the second γ subunit cleavage is mediated by urinary proteases particularly in nephrotic syndrome (Bohnert et al. 2018; Buhl et al. 2012; Ji et al. 2015). In nephrotic syndrome, the protease precursor plasminogen is filtered through damaged glomeruli after which tubular urokinase (or possibly other urinary proteases) converts plasminogen to the active protease plasmin. Plasmin cleaves the γ subunit and activates the channel (Svenningsen et al. 2009; Passero et al. 2008; Ray and Kleyman 2015); thereby, contributing to urinary sodium retention in nephrotic syndrome. Plasmin may also influence ENaC activity by interacting with other proteases, such as prostasin.

Whether ENaC activation in nephrotic syndrome is the primary mechanism for sodium retention is still not completely clear; however, plasminogen and plasmin can be detected in the urine of nephrotic patients, and the amounts in urine correlate with amounts of urinary albumin (Buhl et al. 2012; Svenningsen et al. 2015; Unruh et al. 2017; Hinrichs et al. 2018; Ray 2018).

One would expect that, if ENaC activation is the primary cause of sodium retention in nephrotic syndrome, then a generic ENaC inhibitor like amiloride or triamterene would reduce reabsorption and lead to natriuresis and reductions in weight and blood pressure in humans with nephrotic syndrome. Short-term (2 days) administration of amiloride to diabetics with nephropathy and controls led to a natriuresis in both groups (Andersen et al. 2015). When the effects of hydrochlorothiazide and amiloride were compared in nine subjects with type 2 diabetes and proteinuria, there were no differences between the two diuretics (Unruh et al. 2017), but the study was stopped after two individuals developed severe hyperkalemia and acute kidney injury while on amiloride. Other reports have described variable results with the major negative effects being severe hyperkalemia and acute kidney injury, but with several studies showing positive reductions in sodium and fluid balance in response to amiloride or triamterene (Hinrichs et al. 2018; Yamaguchi et al. 2018). These studies suggest that, in some instances nephrotic and other proteinuric disease patients with associated sodium retention and hypertension, an ENaC inhibitor (amiloride or triamterene) may alleviate sodium retention and low-renin hypertension.

While plasmin is a primary suspect in the development of nephrotic syndrome, other proteases likely play a role. For example, cathepsin B cleaves ENaC (Alli et al. 2012b) and in the kidney, cathepsin B increases ENaC activity leading to hypertension early in nephrotic syndrome. Inhibition of cathepsin B prevented nephrotic hypertension (Larionov et al. 2019).

Other factors might also be involved. High salt diet is often time a precursor to kidney disease, but apparently the effects of high salt can be exacerbated by abnormalities in the ENaC processing machinery. Knockout of Nedd4-2 coupled with a high salt diet in mice rapidly leads to progressive kidney injury (Henshall et al. 2017) and nephrotic syndrome (Manning et al. 2018).

18.4.3 Transmitter and Humoral Agents That Modify ENaC Activity

ENaC is regulated by a large variety of agents; these include transmitters interacting with G protein-coupled receptors (e.g., purinergic, adrenergic, and dopaminergic agents), circulating hormones (e.g., glucocorticoids, angiotensin, and eicosanoids), chemokines (e.g., TNF-α, TGF-β, interleukins), and reactive oxygen and nitrogen species (e.g., superoxide and nitric oxide).

18.4.3.1 Regulation of ENaC by Adrenergic Agents

Lung epithelial cells are a well-known target for adrenergic transmitters, but the kidney contains a more diverse set of adrenergic receptors. There are few reports of alpha-adrenergic regulation of ENaC. However, recently Mansley et al. (2015) described an increase in ENaC-mediated transepithelial current in response to norepinephrine applied to confluent mCCDcl1 murine cortical collecting duct cells. This effect was concentration dependent and mediated via basolateral α2-adrenoceptors.

Sympathetic innervation of the kidney tends to be mostly on the proximal nephron and juxtaglomerular apparatus (Badve et al. 2011) with, at most, modest effects on ENaC in the distal nephron. On the other hand, β-adrenergic stimulation of ENaC in the lungs is one of the primary mechanisms controlling ENaC-mediated lung fluid clearance (Fronius 2013; Davis and Matalon 2007). Activation of β_2 receptors on alveolar cells stimulates adenylyl cyclase that, in turn, increases intracellular cyclic AMP levels and increases the density of ENaC in the apical membrane. The effects of increased cAMP are totally blocked by the β-antagonist, propranolol, and by the protein kinase A (PKA) blocker, H89 (Chen et al. 2002; Eaton et al. 2004; Downs et al. 2012).

18.4.3.2 Regulation of ENaC by Vasopressin

Excessive sodium reabsorption by the kidney has long been known to participate in the pathogenesis of some forms of hypertension. In the kidney, final control of NaCl reabsorption takes place in the distal nephron through ENaC. Although aldosterone is usually thought to be the main hormone regulating ENaC activity, several studies in animal models and in humans highlight the important effect of vasopressin on ENaC regulation and sodium transport (Kortenoeven et al. 2015). The selective V2 receptor agonist, dDAVP, not only increases urine osmolality and reduces urine flow rate but also reduces sodium excretion in rats and humans. Chronic V2 receptor stimulation increases blood pressure in rats, and blood pressure and urine concentration are significantly correlated in healthy humans. Thus, excessive vasopressin-dependent ENaC stimulation could be a risk factor for sodium retention and high blood pressure (Bankir et al. 2005, 2010; Nicco et al. 2001).

Marunaka and Eaton (1991) examined single amiloride-blockable Na^+ channels in membrane patches from cultured distal nephron cells before and after treatment with arginine vasopressin. Vasopressin increases ENaC activity by increasing the number of ENaC channels with only a modest increase in the open probability of individual channels. Cells pretreated with cholera toxin or dibutyryl-cAMP to increase intracellular cAMP appeared similar to cells treated with vasopressin; that is, the number of Na^+ channels per patch increased with a small effect on the open probability of individual Na^+ channels. These observations suggest that vasopressin works by increasing cAMP and promoting insertion of clusters of new sodium channels. Subsequent work by Snyder (Snyder et al. 2004a) on cells transfected with labeled channels confirmed this idea. He examined the appearance of modified channels at the surface membrane and found that the number increased in response to cAMP. Also, inhibition of vesicle trafficking by incubating epithelia at 15 °C prevented the cAMP-mediated stimulation of ENaC. Morris et al. (Morris and Schafer 2002) used more quantitative methods to examine the effects of vasopressin. They concluded that the increased density of ENaC subunits in the apical membrane can account completely for the transepithelial current increase produced by cAMP. However, their work made several assumptions and a direct measure of ENaC open probability using patch-clamp methods in split open mouse collecting duct showed that there was a rapid increase in ENaC open probability immediately after addition of vasopressin followed by an increase in channel density (Bugaj et al. 2009). In addition, stimulated production of cAMP with forskolin and phosphodiesterase inhibitors also enhanced ENaC open probability. The stimulatory effect of cAMP was dependent on the presence of intact PY motifs in the C termini of the channel (Yang et al. 2006). The change in open probability was not due to direct phosphorylation of ENaC by protein kinase A, but rather involved ERK1/2 phosphorylation of two sites, T613 in β-ENaC and T623 in γ-ENaC. Phosphorylation of these sites enhances the interaction of ENaC and Nedd4 (Shi et al. 2002a). Thus, an increase in intracellular cAMP appears to promote dephosphorylation of the two ERK sites, reducing channel retrieval and increasing open probability by reducing ENaC/Nedd4 interaction.

Vasopressin stimulates ENaC activity in adrenalectomized mice (Mironova et al. 2012). This implies that the ability of vasopressin to stimulate ENaC is independent of aldosterone. Vasopressin modulates sodium reabsorption in the collecting duct through adenylyl cyclase-stimulated cyclic AMP. The specific isoform involved in vasopressin-stimulated sodium transport appears to be adenylyl cyclase type VI since mice deficient in this isoform in the principal cells of the collecting duct are no longer responsive to vasopressin (Roos et al. 2013).

The observations on vasopressin-stimulated sodium reabsorption have important physiological consequences. Ordinarily, vasopressin is viewed as only affecting water reabsorption through aquaporins, but concomitant vasopressin-induced increases in salt reabsorption will enhance the driving force for water reabsorption. Thus, while vasopressin decreases free water excretion, it does this, in part, by promoting ENaC-mediated sodium reabsorption (Stockand 2010; Snyder et al. 2004a).

18.4.3.3 Regulation of ENaC by Purinergic Agonists

Epithelial tissues expressing ENaC often also contain an intrinsic (autocrine/paracrine) mechanism of ENaC regulation. In the best example of this intrinsic regulation, circulating aldosterone can be excessively high, but ENaC activity in the kidney remains stagnant after an initial increase (Knox et al. 1980). This phenomenon is referred to as aldosterone escape. While many factors are thought to mediate this response, one is an intrinsic regulatory mechanism that occurs with an increase in renal perfusion pressure. The source of this intrinsic ENaC inhibition in the presence of high aldosterone remained mysterious until (Stockand et al. 2010) found that mice lacking $P2Y_2$ purinergic receptors that bind locally produced ATP have impaired aldosterone escape. While other mechanisms such as atrial natriuretic peptide, arginine vasopressin, and even other channels such as the sodium chloride cotransporter (NCC) have been suggested to mediate aldosterone escape (Kelly and Nelson 1987; Stockand et al. 2010; Wang et al. 2001b), it is clear that locally produced ATP participates in this response. While the best studied example is that of renal epithelia, ATP is produced intrinsically in many tissues that express ENaC and has varying actions on the channel as listed below. The distal nephron contains a robust inhibitory purinergic signaling system that regulates ENaC. ATP is released as a paracrine agent that activates metabotropic $P2Y_2$ purinergic receptors to inhibit ENaC to match urinary sodium excretion to dietary sodium intake and, thereby, control blood pressure within a normal range despite large changes in dietary sodium. Loss of purinergic inhibition of ENaC increases blood pressure by causing inappropriate sodium reabsorption while stimulation of the $P2Y_2$ receptor promotes natriuresis and a decrease in blood pressure (Mironova et al. 2015).

18.4.3.3.1 Mechanisms of Apical ATP Release in Epithelia

In the kidney, apical ATP release is well studied. ATP is released from the apical surface of CCD cells in response to an increase in tubular flow or renal perfusion pressure (Hovater et al. 2008). While ENaC is expressed on the apical surface of principal cells, connexin 30 hemichannels are expressed on the apical surface of intercalated cells (McCulloch et al. 2005). Placing an ATP sensor cell next to a principal or intercalated cell from an isolated mouse cortical collecting duct (CCD) showed that ATP is produced by the intercalated and not the principal cells (Sipos et al. 2009). Furthermore, when this sensor was placed close to an intercalated cell on a CCD isolated from a connexin 30 knockout mouse, no ATP release was sensed. CCD cells may also release ATP packaged into vesicles that are stimulated to release their contents upon increases in intracellular calcium or hypotonicity (Bjaelde et al. 2013).

In pulmonary epithelia, the generally accepted primary mechanism of ATP release from the apical surface of the cell involves Pannexin 1 (Panx1), a channel expressed on the apical pole of airway epithelia. In human airway epithelia, inhibiting Panx1 through knockdown or pharmacology impaired ATP secretion both under basal and hypotonic (ATP stimulating) conditions (Ransford et al. 2009). Panx1-induced ATP release is thought to be a response to mechanical and hypotonic stress (Ohbuchi et al. 2014). Panx1-mediated ATP only accounts for 60% of ATP release both under basal and stimulated conditions. The remainder of ATP release is likely mediated by vesicular stores of ATP maintained in goblet cells that are released by a Ca^{2+}-dependent mechanism (Lazarowski et al. 2003). This process is thought to mediate increased ATP release under pathological conditions such as inflammation (Okada et al. 2013). Many other mediators of airway ATP release have been suggested over the years including CFTR, VDAC1, and connexins, but all of these have been proven not to directly facilitate ATP production.

18.4.3.3.2 Mechanisms of Basal ATP Release in Epithelia

The mechanisms underlying serosal ATP release in ENaC-expressing epithelia are not as thoroughly characterized as apical pathways. In the renal CCD, ATP release is stimulated by a decrease in serum osmolality (Schwiebert and Zsembery 2003) and in the airway, ATP release occurs downstream of mechanotransduction pathways (Homolya et al. 2000). The molecular mechanisms underlying this ATP release and the channels that the ATP travels through on the basolateral side of these epithelia are yet to be identified.

18.4.3.3.3 Purinergic Receptor Families

Purinergic (nucleotide binding) receptor classes include P1 receptors which bind adenosine and P2 receptors which bind ATP (Burnstock 2006). The P2 receptors are divided into several families which include P2X, Y, Z, U, and T. P2X and Y regulate ENaC as detailed below. P2Y receptors are metabotropic, signaling through G protein-coupled receptors of the Gq class, activating phospholipase C (PLC) and generating IP3 and DAG. P2X receptors are ionotropic, meaning they are ligand-gated (ATP being the ligand) ion channels capable of transporting a variety of cations including Na^+ and Ca^{2+}. P2X receptors can function as monomers or in homo- or heterodimeric complexes.

18.4.3.3.4 Effects of Apical ATP on ENaC in the Kidney

In the distal nephron where ENaC is expressed, P2Y and P2X receptors are expressed on the apical and basolateral surface. On the apical surface, $P2Y_2$, $P2Y_4$, $P2X_4$, and $P2X_6$ play some role in ENaC regulation. The vast majority of the work detailing the effects of purinergic signaling on renal ENaC function has focused on $P2Y_2$ receptor-mediated ENaC inhibition. Early evidence for an effect of ATP on renal ENaC came from *Xenopus* distal nephron cells (A6 line) (Ma et al. 2002). A non-hydrolyzable ATP analog, ATPγS, applied to the apical surface of these cells, decreased the probability of opening of ENaC. Furthermore, ENaC activity only increased in response to stretch when the P2 receptor inhibitor suramin was added to the patch pipet, indicating that tonic inhibition by P2 receptors prevent a stretch response in this cell line. The effects of ATP on ENaC in this model are dependent on phospholipase C (PLC) activity. $P2Y_2$ receptors activate PLC which cleaves PIP_2 to produce IP3 and DAG. These molecules can work together to activate protein kinase C (PKC) and decrease available PIP_2 in the process. Since PIP_2 typically functions to hold ENaC into the membrane and Ca^{2+} can inhibit ENaC by a variety of mechanisms, it was important to determine whether decreasing PIP_2 or increasing intracellular Ca^{2+} and activating PKC is inhibiting ENaC. Pochynyuk et al. (2008c) showed that inhibiting PLC attenuated ENaC's response to apical ATP whereas inhibiting PKC had no effect, suggesting that the decrease in PIP_2 mediated the ATP response but activating PKC had no effect. Contrary to the generally accepted view that $P2Y_2$ inhibits renal ENaC, however, mice lacking $P2Y_2$ receptors have decreased ENaC expression (but increased P_o of the channels that are still present in the membrane) and salt-resistant hypertension (but ENaC did not appropriately downregulate in response to high salt diet) (Pochynyuk et al. 2008a, 2010). This response is explained by suppressed renin angiotensin aldosterone system activation. These mice had an enhanced response to increasing plasma aldosterone, however, suggesting that they could not "escape" from aldosterone and that $P2Y_2$ receptors participate in aldosterone escape. Stockand et al. (2010) showed that $P2Y_2$ KO mice did in fact have impaired ability to escape from high aldosterone levels. The compensatory changes in RAAS in $P2Y_2$ knockout mice make it difficult to assess purinergic effects on renal ENaC function in this model. To address the role of luminal ATP in countering aldosterone-induced ENaC stimulation, transgenic mice that globally overexpress the ectonucleotidase NTPDase1 were created. These mice have decreased extracellular ATP throughout all tissues. When aldosterone is clamped in these animals, they show impaired natriuresis on a high salt diet which correlated with an inability to downregulate ENaC suggesting that ATP-induced ENaC inhibition allows the body to maintain sodium balance during periods of high salt intake (Zhang et al. 2015).

In addition to $P2Y_2$, $P2Y_4$, $P2X_4$, and $P2X_6$ are all expressed on the apical membrane of distal nephron principal cells and have been reported to have inhibitory effects on ENaC activity. Interestingly, however, in the presence of low extracellular Na^+, P2X channels on the apical membrane may stimulate ENaC via a PI3 kinase-dependent mechanism (Wildman et al. 2009).

18.4.3.3.5 Effects of Basolateral ATP on ENaC in the Kidney

While luminal renal ATP production is thought to help the kidney escape from the effects of aldosterone, serosal ATP has been postulated to mediate aldosterone's effects. In 1963, Edelman et al. (1963) showed that the effects of aldosterone on ENaC were dependent on ATP production. It was not until 50 years later that this phenomenon was revisited. Gorelick et al. (2005) found that aldosterone stimulates ATP release on the basolateral side of *Xenopus* cortical collecting duct A6 cells and that this basolateral ATP causes cell contraction, cytoskeletal reorganization, and ENaC activation. The same group used a complicated pharmacological approach to determine that this response was likely mediated by basolateral $P2X_4$ receptor-mediated increases in PI3 kinase (Zhang et al. 2007b). However, cell attached patch-clamp techniques and a dominant negative $P2X_4$ receptor construct showed that basolaterally expressed $P2X_4$ channels enhance ENaC activity (Thai et al. 2014). This result contrasts with the report that $P2X_4$ knockout mice have hypertension on a low sodium diet, but the hypertension is not associated with excess sodium reabsorption (Craigie et al. 2018). The authors speculate that global knockout is affecting vascular tone.

$P2Y_2$ receptors are also expressed on the basolateral surface of ENaC-expressing principal cells and have an inhibitory effect on amiloride-sensitive current in M1 mouse collecting duct principal cells (Cuffe et al. 2000). This effect is independent of increases in intracellular Ca^{2+}, but is interestingly dependent, in part, on changes in intracellular pH (Thomas et al. 2001).

18.4.3.3.6 Effects of ATP on ENaC in Airway

While the effect of serosal ATP on ENaC in the airway has not been studied, apical ATP can be stimulatory or inhibitory. Cellular products of *Pseudomonas* inhibit ENaC in type II alveolar cells by stimulating ATP release and activation of P2Y receptors (Kunzelmann et al. 2005). Lipopolysaccharide was identified as the trigger for ATP release and, unlike in kidney where $P2Y_2$ receptors inhibit ENaC by PLC but not Ca^{2+} or PKC, both a decrease in PIP_2 availability and an increase in PKC activation mediate the effects of apical ATP on ENaC in type II alveolar cells

(Boncoeur et al. 2010). In type I alveolar cells, on the other hand, apical P2Y stimulation increases short-circuit current and this effect is mediated by ENaC activation rather than inhibition (Yang et al. 2009). In intact rabbit airway epithelium, ENaC is inhibited by ATP via a P2Y receptor-mediated pathway that involves increases in intracellular Ca^{2+} (Poulsen et al. 2005). Interestingly, this same study found no effect of increasing basolateral ATP on short-circuit current suggesting that interstitial ATP may not effect ENaC activity in airway epithelium.

18.4.3.3.7 ENaC Regulation by Purinergic Receptors in Other Tissues

The general consensus in most tissues where ENaC is expressed is that apical stimulation of purinergic receptors inhibits ENaC. Experiments looking at the effects of basolateral ATP on ENaC have not been done. In the distal colon, ENaC is inhibited by apical addition of ATP (Matos et al. 2007). This effect is mediated by $P2Y_2$ receptors and mice lacking these receptors show no change in colonic sodium uptake in response to ATP. Apical $P2Y_4$ receptors inhibit ENaC in Reissner's membrane of the chochlea by a mechanism that is PLC-dependent (Kim et al. 2010). ENaC in mouse endometrial epithelium is also inhibited by apical ATP via a P2Y receptor-mediated mechanism but in this tissue, the inhibition is Ca^{2+}-dependent (Wang and Chan 2000). In primary cultures of mammary epithelia, basolateral P2Y receptors stimulate ENaC by a mechanism that is dependent on intracellular Ca^{2+} increases (Lee et al. 2007b).

18.4.3.4 Regulation of ENaC by Dopamine

Dopamine increases lung liquid clearance under basal conditions and in situations where edema accompanies lung injury. Alveolar epithelial cells appear to contain both type 1 and 2 dopamine receptors (D1 and D2). Stimulation of D1 or D2 receptors on the basolateral surface of T2 cells activates Na,K-ATPase, but some of the increase in lung fluid clearance is mediated by D1 activation of apical ENaC. Apical D1 receptors stimulate the production of cyclic AMP but do not activate protein kinase A. Rather cAMP, through a complicated signaling pathway that involves activation of the small G protein, rap-1, by cAMP activation of a GDP exchange protein, EPAC, finally increases the activity of individual ENaC proteins (Helms et al. 2006a, b). The effect of dopamine on both ENaC and (Na,K)-ATPase suggests a coordinated response of both transporters to promote maximal increases in transport and alveolar fluid clearance. The action of dopamine also demonstrates that signaling within alveolar cells is compartmentalized. Basolateral production of cAMP by β-adrenergic receptors produces a protein kinase A-dependent increase in the number of ENaC in the apical membrane with little or no change in the activity of individual channels. On the other hand, the apical production of cAMP by D1 receptors produces a protein kinase A-independent increase in the activity of individual channels with little or no change in the number of channels.

18.4.3.5 Regulation of ENaC by Cholinergic Agonists

There are few reports about the action of cholinergic agonists on ENaC in epithelial tissues. Acetylcholine receptors (AChRs) are present in the lung in the same cells as ENaC (Racke et al. 2006) and airway surface liquid contains acetylcholine (Yang et al. 2009). Takemura et al. (2013) using cell-attached patch-clamp recordings found that, in alveolar type 2 cells, the muscarinic agonists, carbachol and oxotremorine, activated ENaC in a dose-dependent manner but that nicotine did not. Carbachol-induced activation of ENaC was blocked by atropine. Western blotting, PCR, and immunohistochemistry showed that muscarinic M2 and M3 receptors but not nicotinic receptors were present in alveolar type 2 cells (Takemura et al. 2013) and airway epithelial cells (Yang et al. 2009). The small G protein, RhoA, and its activated form, GTP-RhoA, increased in response to carbachol and the increase was reduced by pretreatment with atropine. A Rho-associated protein kinase inhibitor, Y-27632, blocked carbachol-induced ENaC activity. Interestingly, while carbachol activates ENaC in alveolar cells, it does not appear to activate ENaC in tracheal epithelial cells but does activate other ion channels (Chen et al. 2006).

18.4.3.6 Regulation of ENaC by Angiotensin II

Angiotensin II contributes to the regulation of total body salt and water balance; therefore, it would seem evolutionarily sensible for it to also regulate ENaC. In fact, in biochemical studies, angiotensin II treatment increased the amount of ENaC subunit proteins and angiotensin II receptor antagonists like candesartan decreased ENaC protein. Immunocytochemistry confirmed the increase in β- and γ-ENaC protein abundance and demonstrated candesartan-induced ENaC internalization in collecting duct cells (Beutler et al. 2003). When intracellular Na^+ concentration was used as a measure of Na^+ transport across the apical membrane, angiotensin II added to the lumen of isolated cortical collecting ducts increased intracellular Na^+ and apical sodium permeability (Peti-Peterdi et al. 2002). Angiotensin II added to the basolateral surface also increased intracellular Na^+; however, luminal addition was much more effective. An amiloride analog, benzamil, almost completely inhibited the elevations in intracellular Na^+. These results suggest that angiotensin II directly stimulates ENaC activity in the CCD.

A more direct examination of the effects of angiotensin II on ENaC combined patch-clamp electrophysiology and immunohistochemistry in freshly isolated split-opened distal nephrons of mice to determine the mechanism and molecular signaling pathway of Angiotensin II regulation of ENaC (Mamenko et al. 2012). Angiotensin II acutely increases ENaC open probability and prolonged exposure induces translocation of α-ENaC to the apical membrane (Beutler et al. 2003). The actions of angiotensin II are independent of the effects of aldosterone on ENaC. Although angiotensin II produces large increases in intracellular calcium in smooth muscle cells, it does not in the renal epithelial cells. It does, however, produce an increase in reactive oxygen species sufficient to explain the increase in ENaC open probability (see below). Inhibition of NADPH oxidase with apocynin abolished the angiotensin II-mediated increases in ENaC open probability.

The results show that agonist binding to the angiotensin II receptor regulates ENaC abundance and functional activity, implying a physiological role for angiotensin II to increase sodium reabsorption in the face of reduced blood pressure or blood volume.

18.4.3.7 Regulation of ENaC by Hydrogen Sulfide

Hydrogen sulfide decreases β-adrenergic agonist-stimulated lung liquid clearance by inhibiting ENaC-mediated transepithelial sodium absorption (Agne et al. 2015). Application of an H_2S donor reduced both baseline and adrenergic-stimulated amiloride-sensitive alveolar fluid clearance. It also blocked amiloride-sensitive transepithelial current in lung cells in tissue culture. H_2S prevents the stimulation of ENaC by cAMP/PKA and, thereby, inhibits the increased absorption of β-adrenergic agonists on lung liquid clearance. Pathologically, H_2S can produce pulmonary edema by inhibiting ENaC (Jiang et al. 2015).

18.4.3.8 Regulation of ENaC by Sex Hormones

Female sex predisposes individuals to poorer outcomes from respiratory disorders like cystic fibrosis and influenza-associated pneumonia (Carey et al. 2007; Zeitlin 2008; Stephenson et al. 2011; Olesen et al. 2010; Serfling et al. 1967; Klein et al. 2010; Organization 2008). Both pathologies have abnormal alveolar fluid clearance and reduced ENaC activity. Recently, two sex hormones, estradiol and prolactin, were shown to directly regulate expression and activity of alveolar ENaC (Greenlee et al. 2013). Overnight exposure to estradiol increased ENaC channel activity (NP_o) through an increase in channel open probability (P_o) and an increased number of patches with observable channel activity. Apical plasma membrane abundance of the ENaC α-subunit more than doubled in response to estradiol as determined by cell surface biotinylation. α-ENaC in the apical membrane was approximately threefold greater in lungs from female rats in proestrus, when serum estradiol is greatest, compared with diestrus, when it is lowest. These effects were mediated by the G protein-coupled estrogen receptor (Gper). In contrast to estradiol, there was no effect of progesterone (Greenlee et al. 2013).

Overnight exposure to the peptide, prolactin, also increased ENaC channel NP_o through an increase in channel open probability and an increased number of patches with observable channel activity. Inhibition of protein kinase A with H-89 abolished the effect of prolactin on ENaC and prolactin also increased cyclic AMP (cAMP) in a renal principal cell line, consistent with signaling through the cAMP-dependent PKA pathway. These observations show that alveolar ENaC is at least partially responsible for the sex differences observed in the pathogenesis of several pulmonary diseases (Greenlee et al. 2015).

18.4.3.9 Regulation of ENaC by Reactive Oxygen Species (ROS)

All cells produce reactive oxygen species (ROS) as part of their normal mitochondrial metabolism. Reactive Oxygen Species are radical or molecular species whose physical–chemical properties are well described. They are produced from molecular oxygen; the successive 4 steps of reaction involve the addition of 1 electron (or alternatively for the second and fourth reactions, 2 H^+). All of the reactions are equilibrium reactions even though equilibrium tends to lie far to the right (Eq. 18.2).

$$\begin{array}{cccccc} O_2 \rightleftarrows & O\cdot_2^- \rightleftarrows & H_2O_2 \rightleftarrows & HO\cdot \ + & OH^- \rightleftarrows & 2H_2O \\ \text{Free} & \text{Super-} & \text{Hydrogen} & \text{Hydroxyl} & \text{Hydroxyl} & \text{Water} \\ \text{oxygen} & \text{Oxide} & \text{Peroxide} & \text{Radical} & \text{Ion} & \end{array} \quad (18.2)$$

The three primary species, i.e., the superoxide anion (O2•–), hydrogen peroxide (H_2O_2), and the hydroxyl radical (HO•), are called reactive oxygen species because they are oxygen-containing compounds with reactive properties. In the past, ROS have generally been considered damaging to cells (particularly oxidative nucleotide and protein damage) and exposure of cells to high concentrations of ROS under pathological circumstances for extended periods of time does result in inflammation and fibrosis and finally programmed cell death. However, this damage is usually associated with high levels of ROS production that exceed the cellular or organismic ability to maintain ROS at low levels compatible with normal cellular function. Interestingly, low concentrations of some ROS, notably hydrogen peroxide and superoxide, rather than producing adverse events, appear to act as signaling molecules that detect the oxidative state of cells and play a significant role in regulating ENaC activity (Downs and Helms 2013; Gonzalez-Vicente et al. 2019).

ENaC is generally considered responsible for alveolar liquid clearance, because genetic deletion of its α subunit in neonatal mice prevents them from clearing fluid from their lungs, leading to respiratory distress and death shortly after birth (Hummler et al. 1996). The observation that single nucleotide polymorphisms rs4149570 and rs7956915 of α-ENaC are associated with neonatal respiratory distress syndrome and lung fluid absorption disorders indicates that α-ENaC also has a significant role in neonatal fluid clearance in man (Li et al. 2015). As the organ ordinarily exposed to the highest levels of oxygen with concomitant production of ROS, ENaC in the lung is susceptible to varying levels of ROS. Some of the cells containing ENaC, like alveolar type 1 cells, may have especially high levels of reactive oxygen species. In a physiological context, the cells appear to cope with the high level of ROS by producing reducing agents like glutathione. When glutathione is oxidized, it can no longer neutralize ROS with a concomitant increase in ENaC activity (Downs et al. 2015). For example, besides the endogenously derived ROS, lung cells, especially alveolar type 1 cells, are also exposed to relatively high levels of free oxygen in inspired air that can lead to production of additional ROS such that physiological concentrations of H_2O_2 in the alveolar space in man are in the 1–-10 μmolar range (Corradi et al. 2008). Physiological concentrations of endogenous, DUOX1/2-mediated, and exogenous H_2O_2 increase ENaC activity (Helms et al. 2008; Ma 2011; Fischer 2009) by reducing ENaC ubiquitination and subsequent degradation. H_2O_2 stimulation of ENaC appears to involve activation of PI3-kinases that produce the anionic phospholipids, phosphatidylinositol-4,5-bisphosphate (PIP_2), and phosphatidylinositol-3,4,5-trisphosphate (PIP_3) which increase ENaC open probability and alveolar liquid clearance (Kooijman et al. 2011; Pochynyuk et al. 2006; Yue et al. 2002). However, bacterial pneumonia can produce near-millimolar concentrations of H_2O_2 that activate ERK which suppress α-ENaC transcription (Wang et al. 2000; Xu and Chu 2007). This can lead to alveolar flooding and lethal hypoxemia.

The importance of ROS-dependent signaling is underscored by remembering that shortly after birth, the newborn lung must change from a fluid-secreting to fluid-absorbing organ suitable for independent breathing. Perinatal exposure of lung cells to oxygen-induced ROS could be one signal for initiating the active transport of Na^+ ions necessary to promote fluid clearance from the newborn lung. Cell culture models of the alveolar epithelium do show that increases in oxygen tension enhance ENaC activity. ENaC's oxygen sensitivity could be a direct effect of oxygen; however, changes in oxygen tension also produce changes in ROS, such as superoxide anion (O_2^-) and there does appear to be ROS regulation of ENaC since addition of a cell-permeable, O_2^- scavenger [2,2,6,6-tetramethylpiperidine 1-oxyl (TEMPOL)] significantly decreases ENaC activity (Yu et al. 2007a; Rafii et al. 1998). In fetal distal lung epithelial (FDLE) cells, Na^+ transport increases 6 h after an increase in oxygen tension (up to 100 mmHg) (Baines et al. 2001). Besides the direct effect on Na^+ transport, ROS can influence transport properties of the alveolar epithelium in the long term by altering ENaC expression. Long-term exposure (48 h) to oxygen or ROS increases ENaC-subunit promoter activity (Baines et al. 2001) and following extended exposure to high O_2 tension, total ENaC protein increases (Baines et al. 2001; Rafii et al. 1998; Thome et al. 2003). The promoter regions of ENaC subunits have NF-κB and AP-1 response elements (Bremner et al. 2002; Rafii et al. 1998) and both transcription factors respond to the redox state of the cell. Transcription factor regulation of ENaC gene expression in response to ROS could account for ROS-induced increases in ENaC protein.

There are several sources of ROS within cells, some endogenous, some exogenous. NADPH oxidases are one of these sources. There are seven different isoforms of NADPH oxidases (Nox1–5, Duox1, and Duox2). Nox4 is the predominant form in the kidney, although Nox2 is also expressed. Nox4 has been implicated in the basal production of ROS in the kidney and in pathologic conditions such as diabetic nephropathy and CKD, upregulation of Nox4 may be important in renal oxidative stress and kidney injury (Sedeek et al. 2013).

Recent patch-clamp and biochemical studies show that ROS stimulates ENaC activity in both native collecting ducts and tissue culture models of principal cells. ROS is apparently the final step in signaling mechanisms that modify ENaC activity. Components of the renin-angiotensin-aldosterone system including aldosterone (Trac et al. 2013; Yu et al. 2007a), angiotensin II (Sun et al. 2012), and the prorenin receptor (Lu et al. 2016) stimulate ENaC, at least partially, via ROS in tubules and tissue culture models of renal cells. Activating protein kinase C with phorbol ester or superoxide donors mimicked the effect of angiotensin II while PKC inhibition or superoxide scavengers blocked ang II stimulation of ENaC (Sun et al. 2012). In contrast to the effects of superoxide, ENaC stimulation by the prorenin receptor appeared to be mediated by H_2O_2 (Lu et al. 2016). In addition, ENaC activity is acutely increased by EGF, insulin and IGF-1 and is associated with a comparable increase in ROS production (Ilatovskaya et al. 2013). Pretreatment with the generic NADPH oxidase inhibitor, apocyanin, reduced both H_2O_2 production and the increase in ENaC activity. Knocking down a key NADPH oxidase subunit (Rac1) also eliminated EGF activation of ENaC (Ilatovskaya et al. 2013).

Aldosterone is normally considered to produce its ENaC stimulatory effects by inducing gene transcription, but one of the immediate, early events is an increase in superoxide and an increase in ENaC open probability (Yu et al. 2007a). These events appear to be mediated by activation of NOX2 and NOX4 (Trac et al. 2013) which in turn may be activated by the JAK/STAT pathway (Krylatov et al. 2018).

These observations imply that aldosterone, angiotensin II, EGF, insulin, and IGF-1 signaling pathways converge at a common ENaC stimulatory pathway mediated by ROS production.

Care should be taken in attributing specific stimulatory effects to specific species of ROS (especially H_2O_2 and superoxide anion). Both the SOD-mediated conversion of superoxide to H_2O_2 and the catalase-mediated conversion of H_2O_2 to water are equilibrium reactions even though equilibrium tends to lie far to the right (favoring final production of water). Nonetheless, reductions in the concentrations of any specific species will drive both the reverse reaction and the forward reaction. For example, in Eq. 18.2, a significant increase in the formation of H_2O_2 (through NOX4 activation) will lead to a large increase in hydroxyl radical (and then rapidly water). But it will also lead to a smaller, but significant increase in superoxide. This may explain why inhibition of a $H_2O_2^-$ producing NADPH oxidase (e.g., NOX4) can lead to effects that appear to be mediated by superoxide. Care should also be taken in examining the effects of ROS scavengers. Not only can they shift the equilibrium of the ROS degradation reactions, but they are rarely completely specific for a single ROS species.

Pathways for the production of ROS appears to be cell type dependent: different cells have different signaling molecules that lead to different ROS responses. There appear to be several canonical pathways which are activated to a greater or lesser extent in different cell types and lead to ENaC activation. The most ubiquitous is activation of PI-3-kinase. PI-3-Kinase can be directly activated by ROS (Brennan-Minnella et al. 2013; Barthel and Klotz 2005; Martindale and Holbrook 2002; Sugden and Clerk 2006) and its metabolic product 3,4,5-phosphatidylinositol *tris*-phosphate (PIP_3) activates ENaC. The phosphatase, PTEN, is also activated by ROS and converts PIP_3 to 4,5-phosphatidylinositol *bis*-phosphate (PIP_2) (Seo et al. 2005) which also activates ENaC (Blazer-Yost et al. 2003; Pochynyuk et al. 2006; Yue et al. 2002). The question remains: where does the activating ROS come from? Many investigators have suggested that protein kinase C activates NADPH oxidase to produce superoxide or H_2O_2 (Sun et al. 2012; White et al. 2009; Koike et al. 2007; Hong et al. 2010; Asano et al. 2012; Xia et al. 2008; Fontayne et al. 2002). Others have suggested that ROS activates PKC (Chen et al. 2010; Gong et al. 2017; Steinberg 2015; Bouwman et al. 2004; Novalija et al. 2003; Shibukawa et al. 2003; Talior et al. 2003). Probably both views are correct. PKC consists of at least 10 different isotypes grouped into typical (α, βI, βII, and γ), novel (δ, ε, η, and θ), and atypical (ζ and λ) forms (Ohno and Nishizuka 2002) which are often not well distinguished by the pharmacological methods used in many experiments: either inhibitors are generic and so do not distinguish between isotypes (e.g., GF109203X) or they are purportedly selective but do not distinguish well between the isotypes (e.g., staurosporine) or they are selective for specific isotypes, but block other unrelated signal pathways (e.g., rottlerin). This is important for two reasons: first, different isotypes interact with ROS-generating enzymes in profoundly different ways, and second, the distribution of the isotype is cell type specific. In general, typical and some atypical isotypes can activate NOX isoforms by phosphorylating critical NOX components (Fontayne et al. 2002; Xia et al. 2008). Novel isotypes, particularly δ, are activated by ROS.

In renal principal cells, activation of PKCα decreases ENaC activity greatly (Booth and Stockand 2003; Stockand et al. 2000a) by promoting ENaC internalization and degradation and, as might be expected, knockout of PKCα causes an *increase* in ENaC activity (Bao et al. 2014). PKCα is the only typical or novel PKC isotype in principal cells (Kim et al. 2006) and knocking it out does not appear to cause any significant change in cellular ROS.

This lack of ROS effect in principal cells contrasts markedly with alveolar epithelial cells where PKCα knockout *decreases* ENaC activity (Xu and Chu 2007; Eaton et al. 2014). Alveolar cells contain PKC-α, -β, -δ, -η, and -ζ. Ordinarily, PKCα regulates superoxide levels in alveolar cells by stimulating activity and expression of superoxide dismutase (SOD). In the absence of PKCα, SOD activity decreases and ROS increases. ROS strongly activates PKCδ which initiates a signaling cascade that inhibits ENaC (See Fig. 18.4).

In addition to the NADPH oxidases, mitochondria are a major source of cellular ROS as a natural result of respiration and ATP production. Ordinarily, mitochondrial ROS accumulation is limited by the action of cellular enzymes, superoxide dismutase, catalase, and aldehyde dehydrogenase. However, mitochondrial defects can lead to elevated ROS concentrations which stimulate signaling mechanism to correct the excess ROS and in the worst case to pathological damage due to high levels of ROS. Voltage-dependent anion channels (VDACs) are thought to facilitate superoxide transport out of mitochondria (Han et al. 2003). There are 3 isoforms of VDAC, VDAC1-3. VDAC3 knockout mice have morphological abnormalities in their mitochondria. They also have higher systolic blood pressures than wild-type

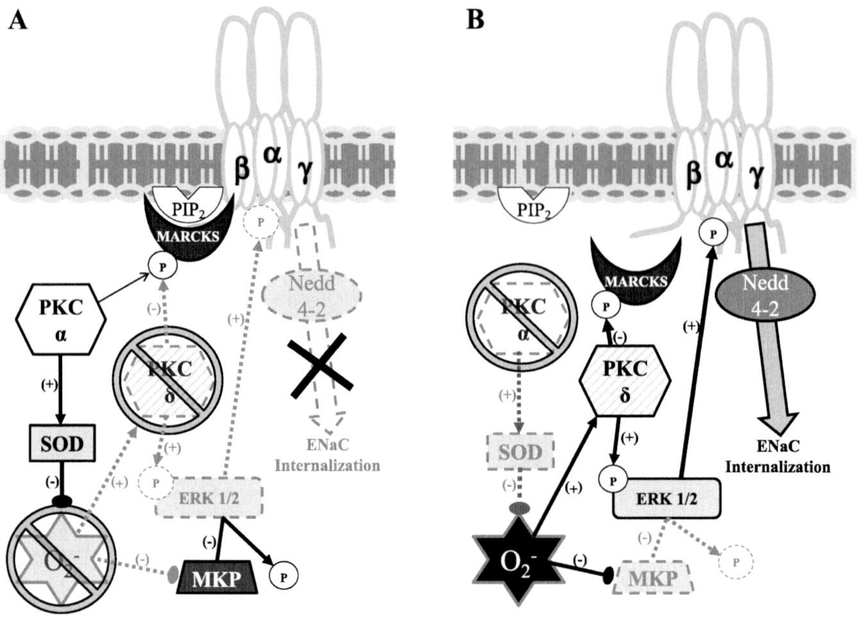

Fig. 18.4 Schematic diagram of alveolar PKC signaling in wild-type and PKC-α knockout mice. Panel **a** shows the wild-type lung. PKC-α is active, stimulating SOD, which breaks down ROS. In the absence of ROS, PKC-δ is inactive, and does not phosphorylate MARCKS protein, which allows MARCKS to sequester PIP_2 near ENaC in the membrane increasing ENaC open probability. In addition, ERK 1/2 is not phosphorylated by PKC-δ, so that ENaC density at the membrane is increased due to decreased internalization. In addition, without high levels of superoxide in the lung MKP is active, which dephosphorylates ERK 1/2 leading to reduced apical expression of ENaC. Panel **b** shows that in the knockout lung, in the absence of PKC-α, SOD is less active leading to elevated ROS. ROS activates PKC-δ, and inhibits MKP. The combination leads to increased ERK1/2 phosphorylation via PKC-δ, and decreased dephosphorylation via MKP, which together activate ERK. ERK in turn phosphorylates ENaC, which promotes interaction with Nedd4-2 causing the ubiquitination and subsequent internalization of ENaC. ROS activation of PKC-δ promotes phosphorylation of MARCKS protein; which, when phosphorylated, leaves the membrane and does not sequester PIP_2 in proximity to ENaC. Without this interaction with PIP_2, ENaC open probability is decreased. Adapted from Eaton AF, Yue Q, Eaton DC, and Bao HF. ENaC activity and expression is decreased in the lungs of protein kinase C-alpha knockout mice. *Am J Physiol Lung Cell Mol Physiol* 307: L374–385, 2014

mice when fed a high-salt diet, and benzamil, an ENaC inhibitor, normalizes their blood pressure. High-salt diet also increased ENaC open probability in isolated cortical collecting ducts from VDAC3 knockout mice in contrast to a reduction in open probability in wild-type mice. VDAC3 knockout mice on a normal diet have higher resting levels of ROS, and high salt increased ROS (measured by dihydroethidium fluorescence) by the same amount in cortical collecting ducts from wild-type and VDAC3 knockout mice, thus leaving the knockout animals with overall higher ROS. Systemic administration of tempol and mito-tempol (a mitochondrial-specific superoxide scavenger) prevented the high-salt-induced increase in systolic blood pressure in VDAC3 knockout mice with no effect in wild-type mice. Mito-tempol fed to the mice only reduced ENaC P_o in principal cells of VDAC3 knockout but not wild-type mice. These results imply that knocking out VDAC3 increases overall ROS production and oxidative stress, leading to ENaC activation causing salt-sensitive hypertension (Zou et al. 2018).

Exogenous agents that modify ENaC activity can do so through ROS production. Acute exposure of a renal cell tissue culture model to ethanol activates ENaC by significantly increasing both ENaC open probability and the active ENaC density in the membrane. The effects of ethanol were mimicked by acetaldehyde, the first metabolic product of ethanol. The effects of ethanol on ENaC were abolished by a superoxide scavenger, 4-hydroxy-2,2,6,6-tetramethylpiperidinyloxy (TEMPOL). Consistent with an effect of ethanol-induced reactive oxygen species (ROS) on ENaC, ethanol and acetaldehyde elevated intracellular ROS (Bao et al. 2012). The same effects are observed in intact alveolar cells. ROS production increased in rat lung slices and *in vivo* mouse lung in response to acute EtOH exposure. ROS production was Rac1 dependent (Downs and Helms 2013).

18.4.3.10 Regulation of ENaC by Nitric Oxide

Nitric oxide (NO) is a highly diffusible, short-lived free radical that is synthesized in cells from L-arginine in a reaction catalyzed by NO synthase (NOS). Nitric oxide is involved in the regulation of sodium balance. Nitric oxide and, to a lesser extent, peroxynitrite ($ONOO^-$ formed from the reaction of O_2^- and NO) are produced by many different cell types. There is a large literature on the role of NO in chronic kidney disease and pulmonary vascular disease (Baylis 2008, 2012; Yamagishi and Matsui 2011; Yousefipour et al. 2010). In terms of ENaC-mediated lung fluid reabsorption, nitric oxide is known for its deleterious effects: elevated levels of NO metabolites have been detected in pulmonary edema fluid (Zhu et al. 2001), and several studies report nitric oxide inhibition of ENaC activity (Guo et al. 1998; Helms et al. 2005b; Nielsen et al. 2000; Ruckes-Nilges et al. 2000; Yu et al. 2007a). Direct nitric oxide inhibition of ENaC is certainly possible, given that there are several tyrosine and cysteine residues on the transmembrane domains and extracellular loops of ENaC subunits, respectively, that could be nitrated or nitrosylated directly by NO; but there is currently no strong evidence for NO-mediated modification of ENaC. On the other hand, the more traditional NO signaling pathway involving soluble guanylate cyclase (sGC) activation and production of cyclic guanosine 3′,5′-monophosphate (cGMP) does inhibit ENaC function (Jain et al. 1998). This pathway also inhibits ENaC in a tissue culture model of renal principal cells (Guo et al. 2013). All of these studies examined the effects of relatively high concentrations of NO in which NO reduced ENaC open probability. On the other hand, if NO is completely removed by treatment with an iNOS inhibitor from airway (by IV infusion) or lung cells in culture for several hours, ENaC channel density is reduced (Hardiman et al. 2004). This implies that NO derived from iNOS under basal conditions is necessary for amiloride-sensitive Na^+ transport across lung epithelial cells and modulates the amount of α- and γ-ENaC via post-transcriptional, cGMP-independent mechanisms or gene regulation.

18.4.3.11 Interaction of NO and ROS Signaling in ENaC Regulation

Since nitric oxide is an inhibitor of ENaC activity, it might be necessary for epithelial cells to decrease the production or activity of NO under edematous conditions. In fact, the bioavailability of nitric oxide can be limited directly by ROS. Increased O_2^- prevents the immediate NO inhibition of ENaC that is normally observed in Na-transporting epithelia (Yu et al. 2007a). This effect is presumably due to the reaction of O_2^- with NO to form peroxynitrite. This may be an important mechanism regulating lung fluid balance, since AT1 cells appear to generate very high levels of O_2^- compared to AT2 cells (Yu et al. 2007a; Helms et al. 2008). This explains why exogenous NO given to reduce pulmonary vascular resistance and increase lung perfusion does not cause edema: NO presumably reduces ENaC activity in AT2 cells, but ENaC activity is unaffected in AT1 cells since NO is scavenged by the high levels of superoxide.

18.4.3.12 Regulation of ENaC by Endothelin

The endothelins are a family of three structurally similar 21-amino acid peptides. Endothelin-1 and -2 activate two G-protein-coupled receptors, ETA and ETB, with equal affinity, while endothelin-3 has a lower affinity for ETA. Originally, endothelin was described as an endothelium-derived, smooth muscle constrictive paracrine (Hickey et al. 1985) and subsequently identified as ET-1 (Yanagisawa et al. 1988). Endothelins are most commonly associated with regulation of vascular tone, but endothelin receptors are present in many other tissues. Endothelin is produced by all renal tubular cells at a level higher than any other cell type in the body. The production by of ET-1 by the inner medullary collecting duct is much higher than other parts of the nephron (Speed et al. 2015). Human and rat renal tubular cells, vascular smooth muscle cells, glomerular mesangial cells, and many other cell types express high levels of both ETA and ETB receptors (Bouallegue et al. 2007; Kohan et al. 2011; Orth et al. 2000; Sorokin and Staruschenko 2015).

The specific effects of endothelin-1 on ENaC were first described in a tissue culture model of principal cells. Gallego and Ling (Gallego and Ling 1996) using patch-clamp methods showed that sub-nanomolar concentrations of ET-1 inhibited ENaC, ETB agonists also inhibited ENaC. ETB, but not ETA, antagonists blocked the inhibition. In fact, at higher doses of ET1 in the presence of an ETB antagonist, ET1 increased channel activity. Patch-clamp studies of ENaC activity in isolated split-opened collecting ducts showed that ET-1 decreases ENaC P_o via ETB receptors by about threefold within 5 minutes. The signaling pathways involve Src tyrosine kinase and MAPK1/2 signaling (Bugaj et al. 2008). Electrophysiological studies in collecting ducts from collecting-duct-specific ET-1, ETA, ETB, and

ETA/ETB knockout mice showed that ENaC regulation by ET-1 via ETB receptors produces a lowering of blood pressure and urinary sodium loss in mammalian collecting ducts (Bugaj et al. 2012a; Sorokin and Staruschenko 2015). In addition, NO appears to control ET-1-mediated inhibition of ENaC (Kohan 2013) since ET-1 increases NOS expression and NO production in the collecting ducts (Hyndman et al. 2015; Nakano and Pollock 2012; Pollock and Pollock 2011; Schneider et al. 2008). NO inhibits ENaC in cultured cells (Yu et al. 2007a). Also, high flow rates in the nephron after high salt intake induce ET-1 production by the collecting ducts and promote nitric oxide-dependent urinary sodium loss through ENaC inhibition (Lyon-Roberts et al. 2011). Finally, activation of purinergic receptors inhibits ENaC (Bugaj et al. 2012b; Kunzelmann et al. 2005; Pochynyuk et al. 2008a, c, 2010; Vallon and Rieg 2011; Vallon et al. 2012), but there also appears to be a link between purinergic receptor activation and endothelin production within the renal collecting duct to promote additional urinary sodium loss (Gohar et al. 2017).

Recent evidence also indicates that obesity and sex hormones regulate the renal ET-1 system differently in men and women, with estrogen suppressing renal ET-1 production and testosterone upregulating ET-1 production (Gohar and Pollock 2018). Obesity and diabetes stiffen the vasculature, with females more adversely affected than males. Since an increase in arterial stiffness is associated with cardiovascular disease, the increased predisposition of women with obesity and diabetes to arterial stiffening likely accounts for their heightened risk for cardiovascular pathology (Padilla et al. 2019).

18.4.3.13 ENaC and Diabetes

Diabetes produces significant changes in renal sodium transport. The high blood glucose and high blood insulin associated with diabetes lead to changes in the expression and activity of the different sodium transporters located along the nephron. In particular, ENaC activity in the distal nephron and collecting duct is increased. Studies in humans and animals have shown that diabetes alters ENaC regulation through multiple pathways, resulting in upregulation of both ENaC activity and abundance which may reflect or contribute to the pathophysiology of diabetic nephropathy and hypertension. In diabetic animals, the expression of full-length and cleaved forms of α-ENaC, β-ENaC, and the cleaved form of γ-ENaC are significantly increased compared to nondiabetic littermates (Eriguchi et al. 2018; Song et al. 2003). There is a significant positive correlation between full-length γ-ENaC and blood glucose level, while the cleavage of γ-ENaC is weakly related to blood glucose level in diabetic rats (Song et al. 2003). Rosiglitazone, a widely used treatment for type 2 diabetes, can decrease the expression of cortical β and γ ENaC in the obese Zucker rat (Riazi et al. 2006). Full-length and cleaved fragments of γ-ENaC can be detected in the urinary exosomes of patients with diabetes, while the abundance of distally cleaved γ-ENaC by the extracellular serine proteases plasmin or prostasin is only observed in patients with diabetic nephropathy (Andersen et al. 2015). The cleaved fragment of γ-ENaC is sufficient to stimulate

ENaC activity under conditions of sodium retention and urinary protein excretion in diabetic nephropathy (Svenningsen et al. 2012).

Diabetic high blood glucose and insulin resistance are increasingly common population pathologies. Insulin and insulin-like growth factor-1 (IGF-1) can acutely stimulate ENaC activity as shown in both patch-clamp studies and transepithelial current measurements on kidney epithelial cell lines (Blazer-Yost et al. 1998; Markadieu et al. 2004; Marunaka et al. 1992). Studies have shown that insulin stimulates ENaC trafficking to the lateral membrane and increases the density of the pool of apical plasma membrane ENaC (Blazer-Yost et al. 2003, 2004). Within 1 min of insulin stimulation, ENaC migrates from an intracellular vesicular pool to the apical and lateral membranes (Blazer-Yost et al. 2003).

In *in vivo* studies, acute intraperitoneal administration of insulin to mice increases the abundance of all three ENaC subunits in the apical membrane of principal cells, indicating trafficking of ENaC subunits in the cortical collecting duct (Tiwari et al. 2007). Moreover, Pavlov et al. (2013) found that ENaC activity decreases after deletion of insulin receptors from the collecting duct in mice.

One proposed mechanism for the involvement of ENaC in diabetes is that serum and glucocorticoid-regulated kinase-1 (SGK-1) mediates the effects of insulin on ENaC activity (Gonzalez-Rodriguez et al. 2007; Kamynina and Staub 2002; Pearce 2001; Schwab et al. 2008). Studies have shown that SGK-1 is a powerful stimulator of ENaC by interacting with ENaC and increasing the activity of ENaC in *Xenopus* oocytes (Lang et al. 2000), A6 cells (Alvarez de la Rosa and Canessa 2003; Alvarez de la Rosa et al. 2004; Thomas et al. 2011), and cortical collecting duct cells (Naray-Fejes-Toth et al. 2004). SGK-1 is activated by insulin via phosphatidylinositol 3-kinase (PI3K), 3-phosphoinositide-dependent kinase PDK,1 and mTOR (Lang et al. 2014; Wang et al. 2001a). In human cortical collecting duct cells *in vitro*, the expression of both SGK-1 and ENaC is upregulated under the condition of high levels of extracellular glucose, and is correlated to the increase in sodium uptake at the single-cell level (Hills et al. 2006). In addition, another study confirmed that in mouse cortical collecting duct cells, insulin increases ENaC activity through a significant increase in SGK-1 activity and cellular PI-3-K activity rather than an increase in SGK1 expression (Mansley et al. 2016). In an analysis of single nucleotide polymorphisms in three independent European populations, SGK-1 has been shown to be related to insulin secretion and was associated with differences in diabetes risk (Friedrich et al. 2008). A novel SGK1 inhibitor (EMD638683) has been considered as a possible treatment for hypertension in type 2 diabetes and metabolic syndrome (Ackermann et al. 2011).

On the other hand, oxidative stress, which is a result of overproduction of oxidative free radicals and ROS is involved in the high blood glucose, high blood insulin and insulin resistance in diabetes, and diabetic nephropathy (Newsholme et al. 2016). ROS induced by high blood glucose contributes to the processes of sodium retention and volume expansion in diabetes as a result of NADPH oxidase and mitochondrial dysfunction (Jha et al. 2016). The accumulation of ROS is thought to play a major role in the albuminuria and progression of diabetic kidney injury because of podocyte damage (Fakhruddin et al. 2017). As pointed out above, there has been a long-standing association between ROS and ENaC activity: ROS strongly increases ENaC sodium transport activity (Goodson et al. 2012; Ilatovskaya et al. 2013).

One hallmark of diabetes is formation of reactive advanced glycation end-products (RAGE). Advanced glycation end-products (AGEs) are modified proteins or lipids and heterogeneous compounds as a result of prolonged hyperglycemia. Addition of RAGE to the basolateral surface of A6 cells increased ENaC NP_o, an effect apparently due to H_2O_2. RAGE also increased total ROS, an effect prevented by the reducing agent NaHS. Inhibition of catalase caused a similar increase in ROS that was prevented by NaHS. The effects of RAGE application were unaffected by treatment with the superoxide scavenger (Wang et al. 2015). This suggests that H_2O_2 mediates at least part of RAGE's ability to stimulate ENaC, since tempol can scavenge H_2O_2, but it does so poorly.

H_2O_2 (Ma 2011; Zhang et al. 2013a) and RAGE (Wang et al. 2015) raised phosphatidylinositol trisphosphate (PIP_3) levels and ENaC activity. An inhibitor of PI-3-kinase blocked the ability of RAGE and H_2O_2 to increase ENaC activity. When cultured tubular epithelial cells were treated with AGEs, AGEs enhanced ENaC mRNA and protein expression and increased ENaC activity and sodium uptake. AGEs-related ENaC regulation is via the SGK-1 pathway, the inhibition of catalase, and stimulating intracellular ROS production (Chang et al. 2007; Wang et al. 2015).

Since the combined impact of hypertension and diabetes can increase the risk of the development of kidney and cardiac disease, understanding the role of ENaC in producing diabetic pathology may lead to potential treatment targets for diabetic kidney disease.

18.4.4 Regulation of ENaC by Pathogens, Cytokines, and Chemokines

The lung is on the front line of the bodies' defenses against airborne pathogens and many cytokines are mobilized in the lung not only by pathogens, but also by lung cells, themselves. Unfortunately, many of these cytokines when present at high concentrations can be cytotoxic. In particular, ENaC-mediated lung fluid clearance can be negatively impacted both by pathogens and cytokines.

Direct interaction with pathogens, such as influenza, measles, and respiratory syncytial virus, acutely reduces ENaC activity (Chen et al. 2004b, 2009a; Le Goffic et al. 2002; Trac et al. 2017). Chronic deleterious effects of pathogens are often associated with production of high levels of ROS by the pathogens (Londino et al. 2017). For example, *Streptococcus pneumoniae* can endogenously generate millimolar levels of H_2O_2 (McLeod and Gordon 1922; Pericone et al. 2003). Additional ROS results from host inflammatory responses. Generation of ROS is a conserved strategy of host phagocytic cells, primarily neutrophils, monocytes, and macrophages, to clear bacteria at the infection site. Bacteria can be engulfed and enclosed in phagosomes, into which superoxide is released by activated NADPH oxidase 2 (Nox-2) (Winterbourn et al. 2016; Winterbourn and Kettle 2013; Hu et al. 2019). Additionally, inflammatory activation of the airway epithelium can lead to local nitric oxide (NO) production further reducing ENaC activity and fluid transport (Belshe 1999; Chen et al. 2004b; Doyle et al. 1994; Greenberg 2002; Helms et al. 2005b; Matalon et al. 2003). ROS activates ENaC at relatively low concentrations, but at the much higher concentrations associated with pathogen-induced cytokine production, ENaC is strongly inhibited (Helms et al. 2008; Snyder 2012; Yu et al. 2007a).

18.4.4.1 Regulation of ENaC by TNF-α

TNF-α is associated with pulmonary edema and ARDS (Hamacher et al. 2002; Horgan et al. 1993). Monocytes and macrophages produce significant TNF-α, but it is also produced by alveolar epithelial cells following lipopolysaccharide stimulation (McRitchie et al. 2000). In freshly isolated AT2 cells, TNF-α decreased α- and γ-ENaC mRNA and protein levels and reduced amiloride-sensitive transepithelial current (Bao et al. 2007; Barmeyer et al. 2004; Bergann et al. 2009; Dagenais et al. 2004, 2006; DiPetrillo et al. 2004; Yamagata et al. 2009).

Interestingly, TNF-α not only contains a receptor binding domain, but also a lectin-like domain which is spatially distinct from the receptor binding site (Czikora et al. 2014, 2017; Lucas et al. 2016; Yang et al. 2010). TNF-α produces an opposite response when there is binding of the lectin-like domain to certain oligosaccharides at high concentrations of TNF-α. This process increases Na^+ uptake in AT2 cells and may account for the observed differences in TNF-α responses (Braun et al. 2005; Rezaiguia et al. 1997; Tillie-Leblond et al. 2002).

18.4.4.2 Regulation of ENaC by the Interleukins

TNFα binding to its type 1 receptor activates NF-κB which increases cytokine (IL-1, IL-4, IL-6, IL-8, IL-13) and chemokine production. Several interleukins increase in the early stages of ALI; however, the most well-studied is IL-1β. IL-1β levels are higher in pulmonary lavage fluids compared to serum suggesting that there is a local, pulmonary source for IL-1β (Meduri et al. 1995; Pugin et al. 1999). IL-1β has been shown to directly affect ENaC expression. Incubation with IL-1β reduced ENaC mRNA protein expression and reduced apical ENaC protein and amiloride-sensitive transepithelial current and Na^+ flux (Roux et al. 2005). An IL-1-induced increase in ENaC activity appears to be mediated via NF-κB and ERK1/2 and p38 MAPK signaling pathways (Mustafa et al. 2019).

Other studies have tried to reverse IL-1β effects. *In vitro* modeling suggests that reduction of IL-1β, via suppressor of cytokine signaling (SOCS-1), can rescue the IL-1β-mediated suppression of ENaC channels. In the kidney, IL-1β strongly inhibited ENaC activity in the medullary collecting duct (Husted et al. 1998).

ENaC regulation by IL-4 and IL-13. Studies in airway epithelial cells from human bronchi suggest that these cytokines may also alter ion transport. IL-4 significantly reduced ENaC subunits γ and β without a change in α-ENaC levels. IL-4 and IL-13 treatment both reduced amiloride-sensitive short-circuit current (Galietta et al. 2002).

18.4.4.3 Regulation of ENaC Via TGF-β1

Transforming growth factor (TGF-β1) is a pathogenic cytokine, which has been implicated in the early phase of ALI and as a predictor of ALI progression to ARDS (Hamacher et al. 2002; Wagener et al. 2015). Increased TGF-β1 levels in ARDS patients are associated with poor prognosis and is associated with a reduced ability of glucocorticoids to reduce pulmonary edema (Fahy et al. 2003; Wakefield et al. 1995). TGF-β1 was shown to regulate ENaC by Frank and colleagues who showed that TGF-β1 reduced amiloride-sensitive Na^+ transport in lung epithelial cells (Frank et al. 2003). *In vivo* studies showed that TGF-β1 reduces vectorial Na^+ and water transport (Frank et al. 2003; Pittet et al. 2001). TGF-β was also found to have an integral role in ENaC trafficking (Peters et al. 2014). In the renal collecting duct, TGF-β also strongly inhibits steroid action by blocking the ability of aldosterone to increase ENaC activity (Husted et al. 2000).

18.4.4.4 Regulation of ENaC by Interferon-γ

During inflammation, INF-γ is secreted by multiple immune cells, but mostly by T-lymphocytes. INF-γ increases NO production via inducible nitric oxide synthase (iNOS) so INF-γ might be expected to reduce ENaC activity, and in fact, studies using human bronchial epithelial cells showed that INF-γ treatment significantly reduced transepithelial Na^+ transport in normal human BECs (Galietta et al. 2000).

18.4.4.5 Cytokines in the Kidney

In the kidney, ENaC are expressed in renal dendritic cells where they appear to respond to inflammation and blood pressure and may have a role in linking a high-sodium diet, inflammation, and hypertension (Barbaro et al. 2017) since rodents fed a high-salt diet accumulate sodium in the renal interstitium even though serum sodium is unchanged (Machnik et al. 2009; Titze and Machnik 2010; Wiig et al. 2013, 2018). ENaC in dendritic cells respond to increases in extracellular sodium resulting in calcium influx and activation of protein kinase C and NADPH oxidase with consequent superoxide. This leads to T-cell activation, release of inflammatory cytokines, and an angiotensin II-dependent increase in blood pressure (BP) (Barbaro et al. 2017; Van Beusecum et al. 2019).

18.5 Regulation of ENaC Density in the Apical Membrane

As mentioned in Sect. 18.4, sodium reabsorption through ENaC can be altered in two ways: by changing the open probability of channels already in the membrane or by increasing the number of functional channels in the membrane. These two mechanisms are not mutually exclusive and some combination of the two does occur in response to several agents which alter ENaC activity (e.g., aldosterone, vasopressin, sex hormones). Nonetheless, because altering membrane channel density is a major means to alter sodium transport, it is important to understand how ENaC is inserted into the surface membrane and the mechanisms for ENaC trafficking. This section will deal with trafficking of ENaC to the membrane, ENaC recycling, and to a lesser extent, ENaC retrieval (since it was dealt with in Sect. 18.4 above where we discussed the role of ubiquitination in retrieval of ENaC from the surface membrane). There have been several reviews on ENaC trafficking (Butterworth 2010; Myerburg et al. 2010; Butterworth et al. 2009; Snyder 2005).

18.5.1 ENaC Trafficking to the Apical Membrane

As a starting point, Fig. 18.5 (Butterworth et al. 2009) is a schematic diagram depicting common, well-characterized trafficking routes found in many epithelial cells. In many ways ENaC trafficking recapitulates these generic pathways (for an additional review, see (Buck and Brodsky 2018)).

Experiments in heterologous expression systems show that individual ENaC subunits can traffic from the ER to the cell surface, but this process is inefficient (Weisz et al. 2000; Prince and Welsh 1998; Valentijn et al. 1998; Firsov et al. 1996). Efficient trafficking of ENaC out of the ER requires assembly of the heterotrimeric complex. Any excess subunits not part of a heterotrimer are targeted for degradation in the proteasome (Valentijn et al. 1998; Staub et al. 1997a). The mechanism by which subunits are restricted to the ER in the absence of assembly may be that all three subunits are required for proper protein folding and assembly (Duncan and Mata 2011; Buck and Brodsky 2018). Alternatively, ER retention signals might prevent release of individual subunits from the ER until a complex is formed between all three subunits, similar to trafficking of K^+(ATP) channel complexes (Zerangue et al. 1999).

The rate of insertion of ENaC in the membrane has been controversial with lifetimes varying from minutes to hours. The explanation for this may lie in the results of de la Rosa and colleagues who measured tagged ENaC subunit entry into the membrane and fluorescent recovery after photobleaching to conclude that there were two separate pathways to the membrane: a fast process that involved fast recycling of endocytosed ENaC from a subapical compartment with parallel incorporation of newly synthesized channels at a much slower rate via the Golgi (Gonzalez-Montelongo et al. 2016). Palmer, Frindt, and their coworkers have,

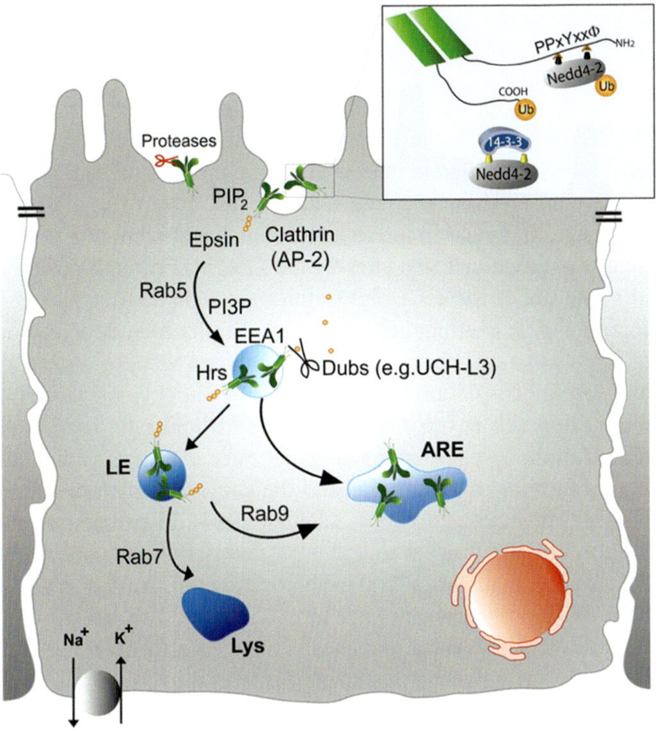

Fig. 18.5 Schematic diagram outlining steps in epithelial sodium channel (ENaC) endocytosis and progress through early endosomes as it is recycled. ENaC internalization is initiated by the binding of Nedd4-2 to the intracellular N terminus (NH2) of each subunit and the addition of ubiquitin (Ub) moieties to the COOH tail (COOH) of the ENaC subunits (enlarged inset). Binding of 14-3-3 proteins to phosphorylated Nedd4-2 (yellow symbols on Nedd4-2) prevents Nedd4-2 interaction with the PY motifs on ENaC. Following ubiquitination, the channel is internalized by clathrin-dependent endocytosis through an interaction with epsin and passes through an early endosomal compartment (EEA1 and Hrs positive). If ENaC is deubiquitinated by deubiquitinating enzymes (Dubs), it will be transferred to the apical recycling endosome (ARE) for recycling to the apical membrane. If it remains ubiquitinated, it is likely trafficked via the late endosomes (LE) to the lysosomes (Lys) for degradation. From studies on other transporters, it is likely that Rab proteins may be involved in rescuing ENaC from the LE back to the ARE or facilitating its movement to Lys for degradation. From Michael B. Butterworth et al. Am J Physiol Renal Physiol 2009;296:F10–F24

using electrophysiological and biochemical methods, *in vivo* and on isolated tubules, examined ENaC trafficking and like de la Rosa have found different rates of insertion (Frindt et al. 2016, 2018). These results are also consistent with the lifetimes measured by (Yu et al. 2008) for trimeric, functional channels delivered from the Golgi.

As with many other proteins, after heterotrimeric assembly, ENaC is core glycosylated in the endoplasmic reticulum with more complex modifications and additions to the glycans beyond the ER in the cis- and trans-Golgi (Ergonul et al.

2006; Rotin et al. 2001; Prince and Welsh 1998). There has been a suggestion that there is an alternative population of ENaC that appears to bypass Golgi processing and channels with "immature" n-linked glycans have been observed in cultured renal cells (Hughey et al. 2004c; Yu et al. 2008). The alternative interpretation of these results is that, like the ribophorin proteins (Rosenfeld et al. 1984), not all the ER glycosylation sites are modified in the Golgi. The α-mannosidases that are the initial step in Golgi glycan modification and addition are extremely sensitive to steric restrictions (Stanley 2011) and in the absence of mannose trimming no other glycan processing can take place at these sites in the Golgi. If enough are modified, the proteins will move out of the cis-Golgi to the trans-Golgi and subsequently to the apical membrane, but an examination of their sensitivity to endoglycosidase H will suggest there are "immature" channels in the membrane because some sites will still have high mannose glycosylation and be cleaved by endoglycosidase H.

There is also some controversy about the role of lipid rafts in the delivery and stability of ENaC at the apical membrane. Lipid rafts are domains that contain high concentrations of inositol phospholipids and the enzymes which produce the lipids (Tong et al. 2008; Matsuura et al. 2007; Li et al. 2004). There seems little question that some fraction of ENaC traffics to and is present in the apical membrane in rafts (Balut et al. 2006; Hill et al. 2002, 2007; Shlyonsky et al. 2003; Reifenberger et al. 2014; Alli et al. 2012a) that contain high levels of inositol phospholipids and other charged lipids (Bao et al. 2007; Ma and Eaton 2005; Ma et al. 2002, 2004). The work of Hill et al. (2002, 2007) has implied a role for lipid rafts in trafficking ENaC.

Removal of cholesterol from the surface membranes of kidney epithelial cells to disrupt lipid rafts resulted in the absence of any active channels in the membrane and a failure of cells to constitutively deliver ENaC to the apical surface. When exogenous cholesterol was added to the apical membrane, channel activity increased (Wang et al. 2008b). However, the regulated trafficking of ENaC in response to cAMP remained unaffected (Hill et al. 2007). Other studies found little alteration in ENaC regulation when lipid rafts were disrupted from the apical surface, but substantial increase or decrease when cholesterol was added or removed from the basolateral surface (West and Blazer-Yost 2005). The authors of these studies suggested that their results implied that functional ENaC was probably not in lipid rafts; but an alternative interpretation is that modification of cholesterol content from the apical surface is difficult. Cholesterol in apical rafts is only present in the inner leaflet of the membrane and, therefore is accessible only with difficulty from the apical surface of the cells. On the other hand, there is a transporter, ABCA1, on the basolateral surface that equilibrates cholesterol between the outer and inner leaflets (Albrecht and Viturro 2007). Since the inner leaflets of both apical and basolateral membranes are contiguous, cholesterol will diffuse laterally from the basolateral to apical membranes, thus, making it appear that only basolateral cholesterol is important. Also, if ENaC from the recycling pool is associated with lipid rafts in the recycling vesicles, then, when they are inserted in response to cAMP, ENaC will still be functional (although one might expect the lifetimes of the channels to be less). Rafts are dynamic, continuously forming and dissipating. Therefore, it should not be surprising to see non-raft ENaC even if it originally reached the surface in a raft.

Artificial disruption of lipid rafts reduces ENaC activity, but the prevalence of lipid rafts also depends upon hormonal factors (e.g., tumor necrosis factor α {Bαo, 2007 #941}, activators of protein kinase C (Alli et al. 2012a), or PTEN tumor suppressor (Seo et al. 2005). Inositol phospholipids stabilize functional ENaC and are necessary for channel trafficking and channel gating as well (Pochynyuk et al. 2005, 2007b; Staruschenko et al. 2007; Ma et al. 2007). Overexpression of caveolin-1, a signature protein of caveolae, which promotes formation of caveolae and internalization reduces the activity and membrane surface expression of ENaC (Lee et al. 2009). One interesting possibility consistent with these results is that ENaC is more stable in rafts. Only when it leaves the raft environment can it be ubiquitinated, retrieved, and degraded. The fact that the ENaC Regulatory Complex (see below) is associated with rafts, lends some credence to this idea.

A potential mechanism linking lipid rafts to apical membrane traffic may involve the phosphatidyl-4-phosphate adaptor protein 2 (FAPP2) (Vieira et al. 2005, 2006; Godi et al. 2004; D'Angelo et al. 2012). FAPP2 is involved in glycolipid production and targeting in the trans-Golgi that could allow the formation of lipid rafts. FAPP2 links 4-PIP and activated Arf1 in the trans-Golgi (D'Angelo et al. 2007) and may facilitate the clustering of apical membrane proteins in lipid rafts (Bhalla et al. 2005; Godi et al. 2004). Disruption of FAPP2 function leads to a failure of cells to deliver membrane proteins to the apical membrane.

As with many other proteins, ENaC-containing vesicles likely use the cytoskeleton and molecular motors to move from the trans-Golgi to the apical membrane (Weisz and Rodriguez-Boulan 2009). When either actin or tubulin is disrupted using pharmacological agents, there is a generalized failure to deliver ENaC to the apical membrane (Butterworth et al. 2005). However, ENaC density in the apical membrane does not change rapidly, but rather the transport activity is reduced (Reifenberger et al. 2014). It therefore appears likely that actin may play an important role in ENaC trafficking and may stabilize ENaC in lipid rafts (Zuckerman et al. 1999; Smith et al. 1991; Mazzochi et al. 2006).

A more specific schematic of ENaC trafficking is shown in Fig. 18.6 (Butterworth et al. 2009). This figure shows a role of SNARE proteins in the final vesicle fusion events that would deliver ENaC to the apical membrane. Disruption of SNAREs or SNARE-binding proteins that facilitate SNARE complex formation prevent the fusion of ENaC vesicles with the apical membrane. SNARE proteins also likely play a role in delivery of ENaC to the apical membrane. The delivery of transmembrane proteins to the apical surface of epithelial cells following synthesis involves sorting and packaging in the Golgi and forward traffic from the trans-Golgi network to the apical membrane (Stoops and Caplan 2014; Weisz and Rodriguez-Boulan 2009; Folsch et al. 2009). In addition, a direct association between syntaxin and the cytoplasmic domains of ENaC has been reported (Saxena et al. 1999, 2006, 2007; Berdiev et al. 2004; Condliffe et al. 2003, 2004; Peters et al. 2001; Qi et al. 1999). Several accessory proteins are essential to ENaC trafficking. These include small GTPases, Rab and Rho proteins (Butterworth et al. 2012; Butterworth 2010), and motor proteins to move membrane vesicles to the desired compartment or apical surface (Weisz and Rodriguez-Boulan 2009). The role of Rab proteins is

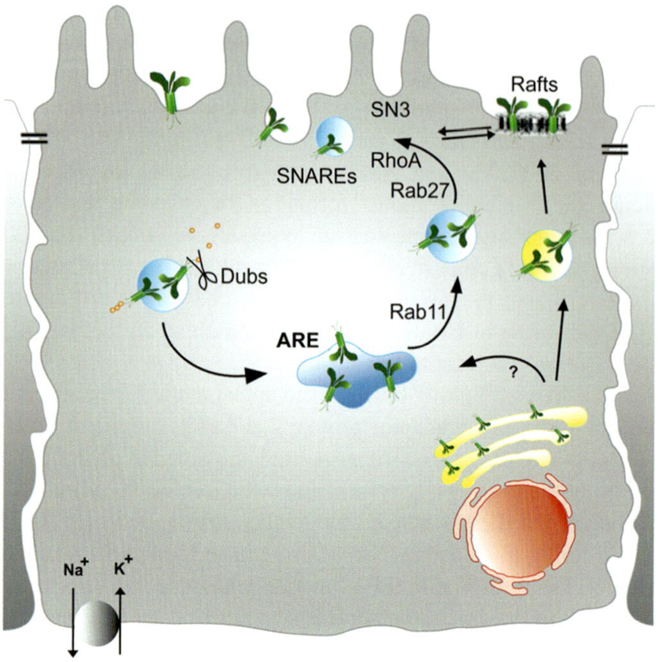

Fig. 18.6 Schematic diagram outlining the steps in ENaC recycling and delivery to the apical membrane. Following synthesis and assembly (yellow Golgi apparatus), channels are transported to the apical surface (yellow vesicle) and can be localized in lipid raft membrane domains (Rafts). It is also possible that ENaC enters the regulated recycling pathway directly (? and arrow), but no support for this idea has yet emerged. Those channels retrieved from the apical surface (see Fig. 18.1) and delivered to the ARE are trafficked back to the apical membrane through the involvement of Rab proteins (most likely Rab11, 27). Delivery of ENaC to the apical surface is regulated by RhoA and sorting nexin 3 (SN3) and vesicle fusion is mediated by SNARE proteins (SNAREs). Details of the mechanisms of ENaC delivery are provided in the text. From Michael B. Butterworth et al. Am J Physiol Renal Physiol 2009;296:F10–F24

complicated by the interplay between Rab-GEFS and Rab-GAPs. A specific example of this is the Rab-GAP, AS160. In renal epithelia, (Liang et al. 2010) showed that aldosterone increased AS160 protein expression threefold. In the absence of aldosterone, AS160 overexpression increased total ENaC expression but did not increase apical membrane ENaC or amiloride-sensitive Na^+ current, but when AS160 was overexpressed, aldosterone was more effective at increasing apical ENaC and transepithelial ENaC current. Aldosterone promoted AS160 binding to the steroid-induced 14-3-3 isoforms, β and ε. These findings suggest that AS160 stabilizes ENaC in an intracellular compartment under basal conditions, and that aldosterone-dependent AS160 phosphorylation permits ENaC forward trafficking to the apical membrane to augment Na^+ absorption.

ENaC activity is increased by anionic phospholipid phosphate phosphatidylinositol 4,5-bisphosphate (PIP_2). PIP_2 hydrolysis by phospholipase C

(PLC) reduces amounts of PIP_2 in the membrane of renal cells. Myelin and lymphocyte (Mal) protein is a lipid raft-associated protein that regulates delivery of proteins to lipid rafts. Apparently, Mal protein negatively regulates renal ENaC activity by stabilizing PLC protein expression at the luminal plasma membrane, presumably in lipid rafts. Mal colocalizes with PLC-beta3 in lipid rafts and positively regulates its protein expression, thereby reducing PIP_2 and decreasing ENaC activity (Tuna et al. 2019).

When the vesicles do arrive at the apical membrane, SNARE protein family members promote docking and fusion of vesicles (Szule and Coorssen 2003; Duman and Forte 2003). ENaC trafficking may be unique in that it appears to involve a direct interaction between cargo (ENaC) and syntaxins (Saxena et al. 2007; Condliffe et al. 2003). Besides altering ENaC trafficking, syntaxin 1A inhibits ENaC by altering its gating (Berdiev et al. 2004; Qi et al. 1999).

18.5.2 ENaC Leaves the Apical Membrane

As we discussed previously, the first step in ENaC internalization is ubiquitination by Nedd4-2 (Malik et al. 2001, 2005a, 2006; Eaton et al. 2010). The next step in ENaC (and other membrane proteins) endocytosis is the association of ENaC with specialized structures on the membrane surface, chlathrin-coated pits, or caveolae: the two types of structures tend to promote endocytosis from different regions of the surface membrane. Chlathrin-coated pits form from areas of the membrane that include normal regions of the membrane; caveolae tend to form from lipid rafts. In general, membrane proteins in lipid rafts that are internalized via caveolae are likely to be degraded by the proteasome while membrane proteins in non-raft areas of the membrane tend to be internalized via chlathrin-coated pits and degraded in lysosomes. The mechanism for ENaC retrieval remains ambiguous; however, in oocytes at least, ENaC appears to be internalized by a dynamin-2-dependent process into either chlathrin-coated pits or caveolae (Shimkets et al. 1997). At this point, the addition of ubiquitin targets ENaC for internalization. However, ENaC can be de-ubiquitinated by the de-ubiquitinating enzyme, USP2-45 (ubiquitin-specific protease 2) (Fakitsas et al. 2007; Ruffieux-Daidie et al. 2008; Verrey et al. 2008). USP2-45 associates with Nedd4-2 to reverse its effects. If ENaC is de-ubiquitinated at this point, it remains in the membrane (Oberfeld et al. 2011).

If not de-ubiquitinated, ENaC is internalized into an EEA1-positive early endosomal compartment. ENaC has several alternative fates. In the endosome, ubiquitinated ENaC is detected and sorted to lysosomes by Hrs (hepatocyte growth factor-regulated tyrosine kinase substrate) (Zhou et al. 2010). Together with STAM (signal-transducing adaptor molecule), Hrs forms a complex called ESCRT-0- (Raiborg and Stenmark 2009; Piper and Luzio 2007). Disruption of this complex reduces ENaC degradation and increases its recycling to the cell surface (Zhou et al. 2010). If it remains ubiquitinated, it will be recognized by the ESCRT machinery and proceeds to late endosomes and then into lysosome or proteasomes for degradation. However, several de-ubiquitinating enzymes can remove ubiquitin in the early endosome. The de-ubiquitinating enzyme, USP8 (ubiquitin-specific peptidase 8), binds to the ESCRT-0 complex through an interaction with STAM (Kato et al. 2000). USP8 regulates ENaC trafficking selectively at the endocytic sorting step and prevents it from progressing to degradative compartments.

The evidence for a recycling pool is relatively clear-cut. Many investigators have shown that cyclic-AMP stimulation promotes ENaC trafficking to the apical surface to increase channel number and transcellular ENaC current (Robins et al. 2013; Yang et al. 2006; Butterworth et al. 2005; Morris and Schafer 2002; Marunaka and Eaton 1991). Edinger et al. (2012) subsequently showed that ENaC plays a role in establishing and maintaining the pool of vesicles that respond to cAMP stimulation and allow ENaC exocytosis. The ENaC channels responsive to cAMP are derived in part from a recycling pool of channels (Fig. 18.6), although it is likely that a pool of newly synthesized channels also contributes (Butterworth et al. 2001).

18.6 An Extended ENaC Regulatory Complex

Membrane proteins have an inherent problem: if they are free to move in the plane of the membrane, interactions with regulatory molecules will be hit-or-miss in a true Markovian statistical sense. The evolutionary solution to this problem is conceptually simple: put the effector molecule and all the relevant signaling molecules together into a stable complex tethered together through direct interaction of chaperone proteins or localized to sub-domains of the membrane (for reviews, see (Kapus and Janmey 2013; Cho and Stahelin 2005; Calder and Yaqoob 2007)). ENaC is part of a surprisingly large complex of signaling molecules that are intrinsic to its normal function and regulation. In a previous section, we have described the interaction between ENaC and Nedd4-2, the initial interacting ENaC partner. The initial description of a large complex of proteins associated with ENaC was based on biochemical isolation and purification (Benos et al. 1987). Under nonreducing conditions, these authors described a protein band of approximately 730 kDa. After reduction, they resolved the complex into five major polypeptide bands with apparent average molecular weights of 315, 149, 95, 71, and 55 kDa and suggested that the 150 kDa band contained the amiloride-binding domain; i.e., that it was ENaC. They did not formally report on the identity of the other bands. Subsequently, the concept of a large ENaC regulatory complex (ERC) in which the components are identifiable is due to Pearce and his colleagues (Pearce et al. 2015; Soundararajan et al. 2009, 2010a, 2012a, b; Bhalla et al. 2006b) who described many ENaC interacting proteins including SGK1 (Pearce 2001; Chen et al. 1999). Not surprisingly, some of the same regulatory molecules we have already discussed are part of or can be recruited to the ERC. Both the small G protein Ras and its downstream cascade of Raf, MEK, and ERK are part of the complex (Soundararajan et al. 2010a). ENaC is inhibited by the Raf-MEK-ERK1/2 MAPK pathway (Grossmann et al. 2004; Falin et al. 2005; Nicod et al. 2002), but another aldosterone-induced protein, glucocorticoid-induced leucine zipper protein 1 (GILZ1) binds to and inhibits the initial member of the cascade, Raf-1, leading to the relief of the ERK1/2 inhibition of ENaC. GILZ1 associates with other members of the ERC, in particular with SGK1 and Nedd4-2 (Soundararajan et al. 2009) (Fig. 18.7). The ERC is dynamic in the sense that, in the absence of aldosterone with little SGK1 or GILZ1 in cells, the complex likely consists of only Nedd4-2 and Raf-1, along with other components of the Raf-MEK-ERK signaling pathway that are tightly associated with Raf-1. In the presence of aldosterone, GILZ1 associates with SGK1 and this association prevents SGK1 from entering endoplasmic-reticulum-associated protein degradation pathway where misfolded proteins of the endoplasmic reticulum (i.e., ones not associated with a chaperone like SGK1 and GILZ1) are ubiquitinated and degraded. Together, GILZ1 and SGK1 move into the complex. Once there, they cooperatively inhibit Nedd4-2 and Raf-1 and stimulate ENaC surface expression and activity (Soundararajan et al. 2009).

However, there are other components of the ERC. One protein, connector enhancer of kinase suppressor of Ras isoform 3 (CNK3) acts as a scaffold on which the other members of the ERC assemble and are organized (Soundararajan et al. 2012c). CNK family members function as protein scaffolds, regulating the activity and the subcellular localization of RAS-activated RAF. Mammals have 3 CNK proteins (CNK1, CNK2, and CNK3). All of the CNK members contain a PSD-95/DLG-1/ZO-1 (PDZ) domain, and, with the exception of CNK3, a PH domain (http://www.ncbi.nlm.nih.gov/Structure/cdd/cddsrv.cgi?uid=cd01260). Soundararajan et al. (2012a, b, c) show that aldosterone induces the formation of a 1.0–1.2-MDa plasma membrane complex which contains CNK3. CNK3 physically interacts with ENaC, Nedd4-2, and SGK1 and stimulates ENaC function in a PDZ domain-dependent manner. It is interesting that CNK3 is the one member of the family that does not contain a pH domain since many of the other components of the ERC associate with the membrane by interacting with membrane inositol lipid phosphate (Fig. 18.8).

Even when associated with GILZ1, SGK1 will be inactive in the absence of phosphorylation at at least two sites. 3-Phosphoinositide-dependent protein kinase (PDK) phosphorylates at one site (Gleason et al. 2015). They show that mTOR (mammalian target for rapamycin) as part of a complex (mTORC2) regulates phosphorylation and activation SGK1 by promoting a second phosphorylation by GILZ1. Using patch-clamp studies on cortical tubule apical membranes, they showed that mTOR inhibition markedly reduces ENaC activity, but not the activity of other principal cell channels (Lang and Pearce 2016).

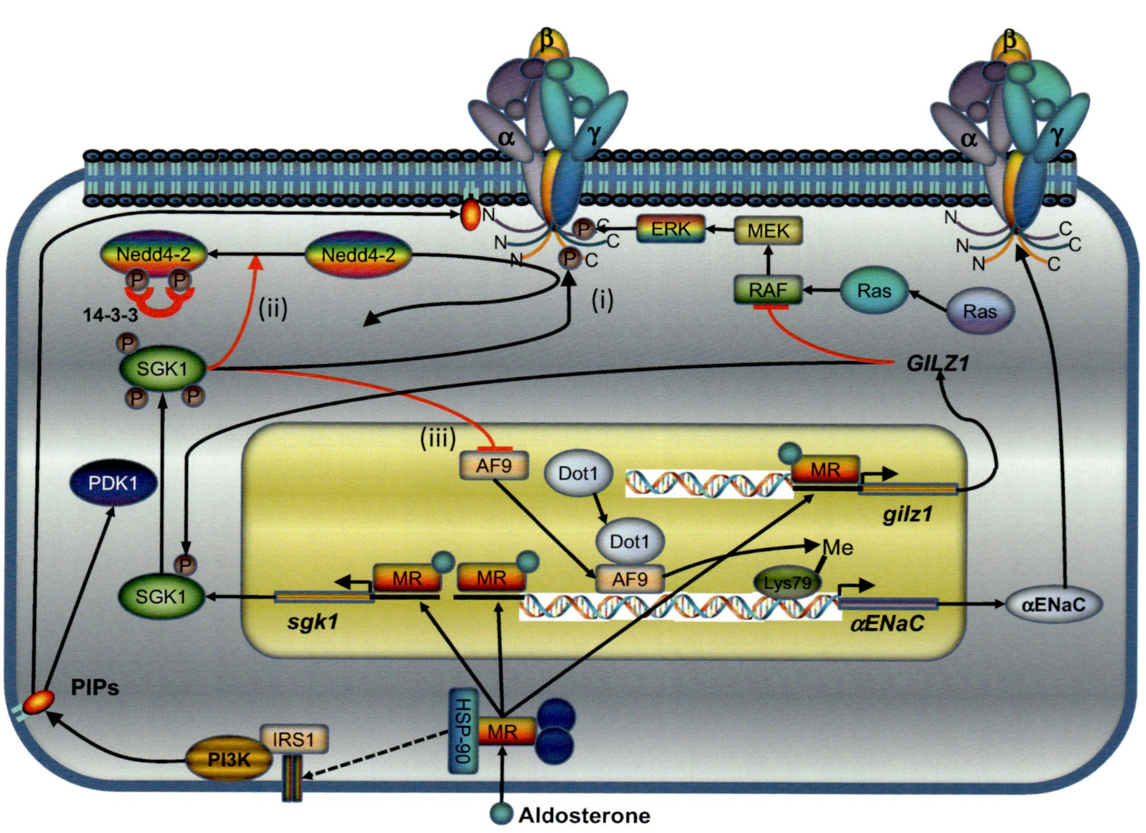

Fig. 18.7 SGK1 acts at multiple levels to regulate ENaC-mediated Na^+ transport in renal collecting duct cells. At low aldosterone (Aldo) levels, ENaCα gene transcription is low because of tonic inhibition by a complex of Dot1a and AF9, which methylates Lys79 on histone H3. At the same time, Nedd4-2 is active and maintains plasma membrane ENaCs at low levels through ubiquitinylation, which results in channel internalization and degradation. With an elevation in

aldosterone levels, activated MR translocates to the nucleus, binds to MREs in the SGK1 5′ regulatory region, and rapidly stimulates SGK1 transcription (solid arrow). ENaCα is not initially responsive to aldosterone-activated MR because the Dot1a-AF9 complex maintains surrounding chromatin in a hypermethylated, stable state, which blocks entry of transcription factors, including MR itself, to their DNA-binding sites (dashed arrow). SGK1 catalytic activity is stimulated by PI3K-dependent generation of 3-phosphorylated phosphoinositides (PtdIns), which activate phosphoinositide-dependent protein kinase 1 (PDK1) and ultimately SGK1 itself. SGK1 activates ENaC through phosphorylation of several key substrates: the channel itself (i), resulting in increased open probability; Nedd4-2 (ii), resulting in inhibition of Nedd4-2 through recruitment of inhibitory 14-3-3 chaperone proteins and subsequent decreased channel internalization and degradation; and AF9 (iii), resulting in inhibition of the Dot1a-AF9 complex and diminished H3 hypermethylation in the vicinity of the ENaCα promoter. This latter effect is predicted to enhance MR access to chromatin, binding to MREs, and stimulation of transcription. IRS1, insulin receptor substrate 1; Meth, methylation; α, β, γ, ENaCα, -β, and -γ subunits, respectively. Modified from Bhalla V et al. Am J Physiol Renal Physiol 2006;291:F714–F721

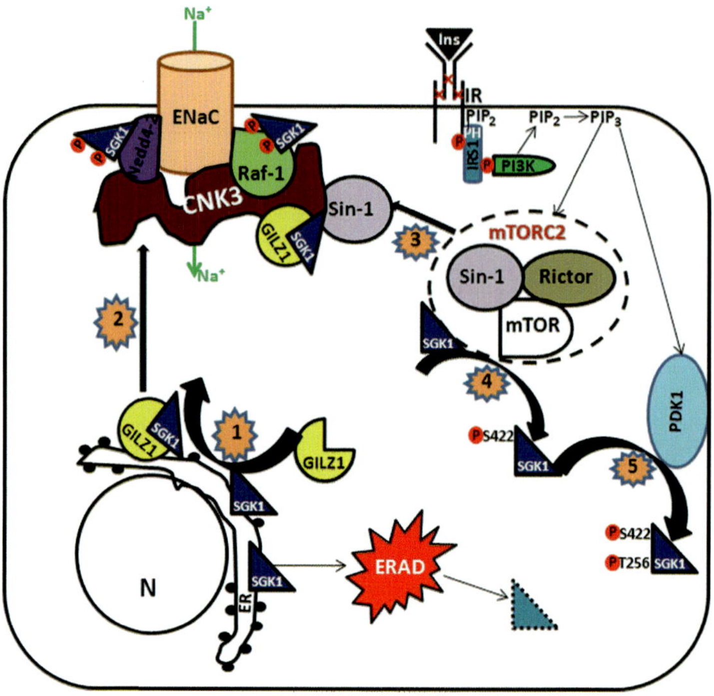

Fig. 18.8 Hypothetical model for SGK1 activation and regulation of ENaC by the ENaC-regulatory complex (or ERC). ENaC regulation occurs within a unique multi-protein complex called the ERC. The various components of the complex are regulated by diverse signaling networks, including steroid receptor-, PI3K-, mTOR-, and Raf-MEK-ERK-dependent pathways. Aldosterone coordinately induces the expression of SGK1, GILZ1, as well as the scaffold protein CNK3, among many others. Highlights 1 through 5 depict sequential steps in the selective recruitment and activation of SGK1 (not meant to depict spatial/subcellular localization). GILZ1 protects SGK1 from rapid endoplasmic reticulum-associated degradation and redirects it away from the endoplasmic reticulum and into the ERC that possibly includes CNK3. CNK3 provides a platform for organizing various components of the ENaC-regulatory machinery by virtue of its ability to interact with multiple proteins including mSin1. mTOR couples with Rictor and mSin1 to form the functional TORC2 complex, which phosphorylates and activates SGK1 at S422. Subsequent phosphorylation of SGK1 at T256 by PDK1 is required for its full activation. It is possible that activation of SGK1 occurs at the ERC, and that mTORC2 is recruited to the ERC via mSin1. It is also possible that SGK1 is brought into the ERC after it has been activated by mTORC2 and PDK1 elsewhere. ENaC, epithelial Na^+ channel; ERAD, endoplasmic reticulum-associated degradation; mTOR, mammalian target of rapamycin; PDK1, phosphoinositide-dependent kinase-1; SGK1, serum- and-glucocorticoid-induced-kinase-1. From (with permission) Soundararajan R1, Lu M, Pearce D. Organization of the ENaC-regulatory machinery. Crit Rev. Biochem Mol Biol. 2012 Jul–Aug;47(4):349–59. 10.3109/10409238.2012.678285

18.6.1 The ERC and Regulation of Na Channels by Inositol Lipids

Besides the elements described by Pearce and his coworkers, there may well be other elements of the ERC associated with membrane lipids. Phosphatidylinositol phosphates (PIPs) are known to regulate epithelial sodium channels (ENaC) (Pochynyuk et al. 2008b; Yue et al. 2002; Blazer-Yost et al. 1999). Phosphatidylinositol 4,5-bisphosphate (PIP_2) and phosphatidylinositol 3,4,5-trisphosphate (PIP_3) are trace, but ubiquitous, components of eukaryotic membrane phospholipids. The traditional view of phosphoinositide metabolism suggests that phosphatidylinositol is sequentially phosphorylated by phosphatidylinositol 4-kinase and phosphatidylinositol 5-kinase to produce 4,5-PIP_2. Once formed, PIP_2 can be further phosphorylated to produce phosphatidylinositol 3,4,5-PIP_3 by phosphatidylinositol 3-kinase or PIP_2 can be hydrolyzed by phospholipase C to produce 1,4,5-inositol trisphosphate (IP3) and diacylglycerol (DAG). Until recently, it was assumed that the sole purpose of PIP_2 production was to act as a precursor for IP3 and DAG which activate protein kinase C. It now seems clear that PIP_2 is a signaling molecule in and of itself that can act directly to activate other effector and signaling proteins and localize them at their sites of action. In particular, PIP_2 modulates the activity of ENaC apparently through interactions with hydrophobic and positively charged residues in the cytosolic tails of the β and γ subunits. Lipid binding assays and coimmunoprecipitation showed that the amino-terminal domain of the β- and γ-subunits of *Xenopus* ENaC can directly bind to phosphatidylinositol 4,5-bisphosphate (PIP_2), phosphatidylinositol 3,4,5-trisphosphate (PIP_3), and phosphatidic acid (PA) (Alli et al. 2012a). Other experiments defined the regions in the ENaC subunits that interacted with PIP_2 and PIP_3 (Zhang et al. 2010; Pochynyuk et al. 2005, 2006, 2007a, b, 2008b; Ma et al. 2002, 2007; Staruschenko et al. 2007; Helms et al. 2005a; Ma and Eaton 2005; Tong et al. 2004; Yue et al. 2002).

18.6.2 The Role of MARCKS Protein

Besides binding to ENaC subunits, additional direct binding assays showed that various PIPs can bind strongly to another native protein, myristoylated alanine-rich C-kinase substrate (MARCKS) or its closely related isoform MARCKS-like protein 1 (MLP-1), but weakly or not at all to a mutant form of MARCKS (Alli et al. 2012a). MARCKS and MLP-1 consist of a myristoylated amino-terminal domain and a basic effector domain containing multiple hydrophobic and basic residues, three sites for PKC phosphorylation, filamentous (F)-actin binding, and calcium/calmodulin binding (Yamauchi et al. 2003; Hartwig et al. 1992). All of our remarks concerning MARCKS apply equally well to MLP-1. MARCKS reversibly associates with the plasma membrane through hydrophobic and electrostatic interactions of its myristoylation and basic effector domain, respectively (Gambhir et al. 2004;

Wang et al. 2004; McLaughlin et al. 2002; McLaughlin and Aderem 1995). MARCKS cross-links actin, but binding is regulated by PKC phosphorylation and calmodulin binding (McLaughlin et al. 2005; Arbuzova et al. 1998; Kim et al. 1994). PKC phosphorylates serine residues within the basic effector domain of MARCKS, leading to the translocation of MARCKS from the membrane to the cytoplasm. Phosphorylation prevents the electrostatic interaction of the effector region MARCKS with plasma membrane PIP_2 after which MARCKS enters the cytosol (George and Blackshear 1992; Taniguchi and Manenti 1993; Kim et al. 1994). In epithelial cells, MARCKS also cross-links actin (Yarmola et al. 2001) and acts as a source of PIP_2 for other membrane proteins (including ENaC, SGK1, and PDK) (Fig. 18.9).

Confocal microscopy demonstrated colocalization of MARCKS and PIP_2 (Alli et al. 2012a). Confocal microscopy also showed that in renal principal cells in culture that MARCKS redistributes from the apical membrane to the cytoplasm after PMA-induced MARCKS phosphorylation or ionomycin-induced calmodulin activation. When MARCKS is de-phosphorylated (i.e., at the membrane), it can present PIP_2 to ENaC and activate ENaC. In PKC-α knockout mice with no constitutive phosphorylation of MARCKS, ENaC is persistently activated (Bao et al. 2014). Thus, another component of the ERC is MARCKS and calmodulin which move from the complex to the cytosol if MARCKS is phosphorylated (Fig. 18.10). Besides phosphorylation, MARCKS, like ENaC, can be controlled by proteolysis. Calpain-2 cleaves MARCKS toward the C-terminus to allow it to maintain its presence at the apical membrane, and thereby, increase its activity (Montgomery et al. 2017).

18.6.3 ENaC Is Activated by Cysteine-Palmitoylation

Cysteine-palmitoylation is a reversible linkage of palmitic acid to cytoplasmic cysteine residues on proteins. Post-translational modification by palmitoylation is another mechanism by which lipid molecules activate ENaC (Mueller et al. 2010; Mukherjee et al. 2014). The β and γ subunits of ENaC are palmitoylated at cysteine residues. Pharmacologically or genetically preventing palmitoylation reduced channel open probability, but channel density in the membrane was unaffected (Mueller et al. 2010; Mukherjee et al. 2014). γ subunit palmitoylation was more important than β (Mukherjee et al. 2014). The mechanism by which palmitoylation activates ENaC is unclear, but has been proposed to act through changes in intracellular calcium, ROS, and PI-3-kinase activity (Wang et al. 2018). There are several enzymes that transfer palmitate to protein; only 5 of 23 appear to palmitoylate and activate ENaC (Mukherjee et al. 2017).

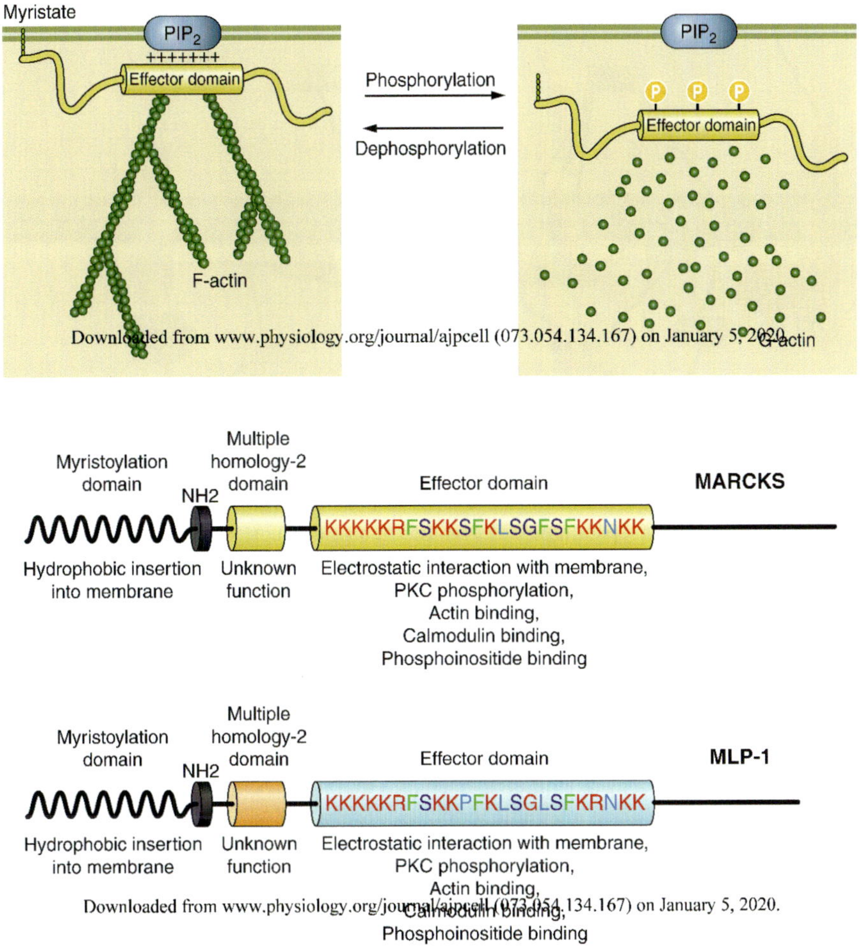

Fig. 18.9 MARCKS function. Above: Dephospho-MARCKS binds to the plasma membrane via a myristoyl group and electrostatic interaction between the effector basic residues and membrane PIP_2. It sequesters three PIP_2 per MARCKS and cross-links actin. Phospho-MARCKS detaches from the plasma membrane to enter the cytosol and disrupts actin filaments and leaves PIP_2 subject to hydrolysis. Below: MARCKS and MLP-1 structure showing the effector domain sequence with phosphorylation sites in red. It mainly consists of basic amino acids (K and R) and has serine residues which can be phosphorylated by protein kinase C and by Rho kinase (ROCK). Adapted from Kleyman TR, and Eaton DC. Regulating ENaC's gate. *Am J Physiol Cell Physiol* 2019

18.6.4 Role of the Cytoskeleton

MARCKS or MLP-1 acts as an organizing center for the actin cytoskeleton and CNK3 has PSD domain that could interact with the MARCKS-organized cytoskeletal proteins. Numerous reports have linked cytoskeleton-associated proteins with the regulation of ENaC activity (Awayda and Subramanyam 1998; Zuckerman et al.

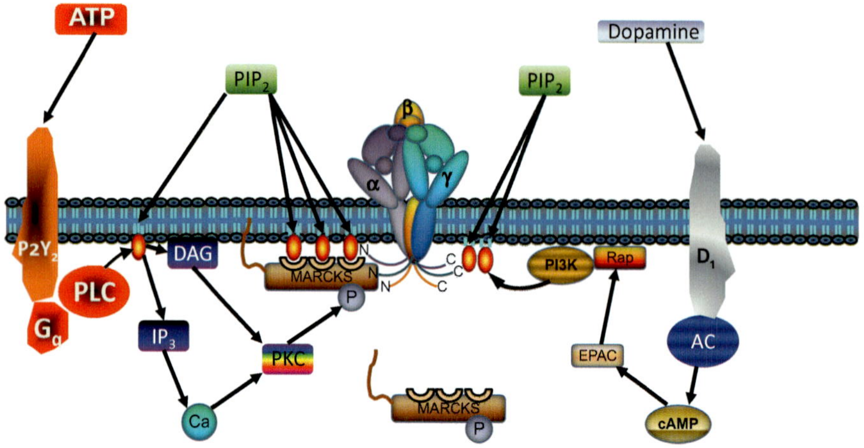

Fig. 18.10 Role of G protein-linked receptors and MARCKS in regulating ENaC activity. In the absence of purinergic ligands, MARCKS sequesters PIP_2 near ENaC to increase ENaC open probability. Two receptors, purinergic p2Y2 on the left and dopaminergic D1 on the right couple to phospholipase C or adenylyl cyclase. In the absence of purinergic ligands, MARCKS sequesters PIP_2 near ENaC to increase ENaC open probability. In the presence of P2Y2 ligand, the receptor activates protein kinase c that phosphorylates MARCKS. MARCKS leaves the membrane and ENaC is no longer stimulated by PIP_2. In the presence of a dopaminergic ligand, cAMP is produced that activates EPAC and finally PI-3-K that produces inositol lipid phosphates that stimulate ENaC

1999; Karpushev et al. 2010a, b; Ilatovskaya et al. 2011; Wang et al. 2013). Recently, Butterworth and coworkers have described a role for the aldosterone-induced cytoskeletal protein, ankyrin G as a regulator of ENaC delivery and density at the apical membrane (Klemens et al. 2017). Also, Reifenberger et al. (Reifenberger et al. 2014) showed that cytochalasin E or latrunculin treatment for 60 min can disrupt the integrity of the actin cytoskeleton in cultured renal principal cells. In single-channel patch-clamp experiments, there were few active channels in treated vs untreated cells. However, the number of ENaC subunits on the surface of the cells measured by *in situ* biotinylation was unchanged by disruption of the cytoskeleton. Analysis of those patches in which ENaC single channels could be observed after cytochalasin treatment had significantly decreased channel open probability. The subcellular expression of fodrin changed significantly, and several protein elements of the cytoskeleton decreased at least twofold after 60 min of cytochalasin E treatment. Cytochalasin E treatment also disrupted the association between ENaC and MARCKS. These results suggest that disruption of the actin cytoskeleton can reduce ENaC activity by destabilizing the ENaC-MARCKS complex, reducing the availability of PIP_2 to activate ENaC, and possibly destabilizing the entire ERC.

18.7 ENaC Mechanosensitivity

ENaC mechanosensitivity is not surprising given that the ENaC/deg. family originally evolved in the first metazoans as mechanosensitive channels such as Mecs and Degs (Studer et al. 2011). Whether ENaC is activated by membrane stretch caused by hydrostatic pressure or cell swelling remains controversial (Fronius and Clauss 2008); however, fluid flow under physiologically relevant conditions increases ENaC open probability in perfused rabbit collecting ducts and in ENaC-expressing *Xenopus* oocytes (Carattino et al. 2004, 2005b; Satlin et al. 2001; Althaus et al. 2007). As pointed out above, the cytoskeleton regulates ENaC, as the COOH-terminus of α-ENaC interacts directly with the actin cytoskeleton (Mazzochi et al. 2006) providing a mechanism to transduce flow shear to intracellular events.

18.8 ENaC in Endothelial and Vascular Smooth Muscle Cells

As befits its name and going back many years to the original work of Ussing and his colleagues (Koefoed-Johnsen and Ussing 1953; Koefoed-Johnsen et al. 1952; Ussing 1953; Ussing and Zerahn 1951), ENaC was traditionally thought to be exclusively in specific, tight sodium-transporting epithelia in the kidneys, colon, and sweat glands where it was considered to be the main determinant of sodium homeostasis. However, spironolactone inhibition of intracellular aldosterone receptors and, more directly, ENaC blockers suggested that there were amiloride-sensitive and aldosterone-inducible channels in endothelium and vascular smooth muscle (Golestaneh et al. 2001; Kusche-Vihrog et al. 2008, 2010; Wang et al. 2009). It is now possible to directly demonstrate by biochemical, molecular biological, and electrophysiological methods ENaC subunits and ENaC channels (Drummond 2012; Ge et al. 2012; Chung et al. 2013; Jeggle et al. 2013; Kusche-Vihrog et al. 2014a, b; Drummond and Stec 2015). ENaC appears to have two major functions in endothelium. First, α-ENaC activity increases the mechanical stiffness of the endothelial cell cortex, a layer 50–200 nm beneath the plasma membrane (Fels et al. 2012; Jeggle et al. 2013; Sowers et al. 2019; Tarjus et al. 2017); increased stiffness reduces production of endothelium-derived vasodilator nitric oxide (NO) (Fels et al. 2010a, 2012) which directly increases the tone of the closely associated vascular smooth muscle cells (Ge et al. 2012; Korte et al. 2014; Kusche-Vihrog et al. 2014b). The increase in tone can increase peripheral vascular resistance with a concomitant elevation in arterial blood pressure. Second, ENaC is an active target for circulating aldosterone and adrenergic transmitters which links total body sodium balance to endothelial function and vascular resistance and blood pressure.

Aldosterone acts on cells of the vascular system via genomic and non-genomic pathways (Fels et al. 2010b). Endothelium expresses mineralocorticoid receptors (Caprio et al. 2008; Wildling et al. 2009). Aldosterone binding to its surface receptor can produce a rapid response necessary for short-term vascular resistance response (Golestaneh et al. 2001; Jia et al. 2018; Korte et al. 2014; Kusche-Vihrog et al. 2010) and also aldosterone binding to the classical MR can produce a slow response necessary for long-term maintenance of mean blood pressure and total body sodium balance (Chen et al. 2004a; Golestaneh et al. 2001; Jia et al. 2018; Kusche-Vihrog et al. 2010). Electrophysiological measurements in vascular endothelial cells show that amiloride and benzamil inhibit ENaC (Wang et al. 2009; Korte et al. 2012; Perez et al. 2009). However, in endothelial cells, amiloride blocks sodium uptake better than benzamil which is different than the relative potencies in epithelial tissues (Perez et al. 2009). This could suggest the presence of nonclassical ENaC channels in the endothelium such as the ASIC-α-ENaC channels (described above). Endothelial ENaC channels also differ in their response to extracellular sodium: in vascular endothelium, in contrast to the kidney, the expression and membrane abundance of ENaC were increased by high extracellular sodium concentrations (>145 mM) (Korte et al. 2012; Oberleithner et al. 2007) instead of being reduced as a result of the well-described feedback inhibition shown in epithelial tissues.

Other observations imply a role, at least, for α-ENaC. When ENaC is either acutely inhibited by amiloride or the α subunit has been genetically deleted, arteries are less responsive to the vasodilator acetylcholine. Dahl salt-sensitive rats given a high-salt diet increased mesenteric artery ENaC expression and activity, and were less responsive to acetylcholine. The responsiveness was restored with acute amiloride treatment (Sowers et al. 2019; Sternak et al. 2018). The observations suggest that ENaC regulates acetylcholine responsiveness, and that ENaC abnormalities may mediate vascular dysfunction seen in the Dahl salt-sensitive rats.

18.9 The Molecular Basis for ENaC Regulation by Aldosterone

Aldosterone is the primary hormonal mechanism for regulating ENaC. In an evolutionary sense, ENaC and aldosterone are likely to have coevolved. Aldosterone dates from approximately 400 million years ago (Mya) at approximately the time of divergence of teleosts and tetrapods (Fournier et al. 2012; Colombo et al. 2006). Aldosterone synthase, encoded by a member of the P450 family, CYP11B2, arose by duplication of the CYP11B gene in the sarcopterygian or lobe-finned fish/tetrapod line after its divergence from the actinopterygian or ray-finned fish line 420 Mya. This was before the beginning of the colonization of land by tetrapods in the late Devonian period, around 370 Mya. The fact that aldosterone was already present in Dipnoi, which occupy an evolutionary transition between water- and air-breathing but are fully aquatic, suggests that the role of this steroid was to potentiate the corticoid response to hypoxia, rather than to prevent dehydration out of the water (Colombo et al. 2006). But it would also be consistent with the requirements of sarcopterygii and early tetrapods to regulate salt balance separately from water balance in an estuarine or partially aquatic environment. Members of the ENaC/degenerin family are much older dating from the very first metazoans (Studer et al. 2011); however, the advent of recognizable ENaC function probably also dates from the late teleosts (Wilson et al. 2000) and sequences for all three subunits are detectable in dipnoid osteicthys. On the other hand, a well-developed ability to link blood pressure and blood volume to sodium reabsorption was only finally developed in mammals (Studer et al. 2011). In states of volume depletion, decreased renal perfusion results in release of renin from the juxtaglomerular apparatus, activating the renin-angiotensin-aldosterone pathway. Most of the actions of aldosterone require gene transcription (Garty 1986) fournier, although some actions only require receptor binding (Holzman et al. 2007) and a few are non-genomic (Losel et al. 2002; Funder 2001; Falkenstein and Wehling 2000; Christ et al. 1995; Baker and Katsu 2019). Binding of aldosterone to the mineralocorticoid receptor activates transcription of a variety of genes, which increases ENaC current in two phases: an initial phase which increases transport four- to six-fold in the first 0 to 6 h and a late phase which requires 12–48 h and increases transport another three- to four-fold (Garty 1986). The mechanisms for the early and late phases appear to be different. Aldosterone, like other steroid hormones, enters target cells (in the case of the kidney, a major target are principal cells of the collecting duct) and binds to cytosolic, mineralocorticoid receptor complexes. In mammalian principal cells, mineralocorticoid receptors are protected from activation by the high levels of circulating glucocorticoids, by the enzyme, 11-β-hydroxysteroid dehydrogenase that alters glucocorticoids to a form that is unable to bind aldosterone receptors (Monder 1991; Naray-Fejes-Toth et al. 1991). After some rearrangement, the aldosterone-bound receptor acts as a DNA-binding protein which targets mineralocorticoid response elements (MREs) on genetic DNA (Horisberger and Rossier 1992; Funder 1997). Binding to the response elements alters gene expression (Horisberger and Rossier 1992). Increases in Na^+ transport can be measured within the first half hour of exposure to aldosterone, and this increase is dependent on gene transcription and translation with the gene products generically referred to as aldosterone-induced proteins (AIPs). Aldosterone and increased intracellular sodium also promote increased sodium pump expression in the long term.

Many attempts using a variety of differential screening methods have been made to identify the aldosterone-induced gene products (Spindler et al. 1997; Mastroberardino et al. 1998; Verrey 1998; Chen et al. 1999; Spindler and Verrey 1999; Pearce et al. 2003). These attempts have met with varying levels of success, but the relationship between the genes identified and the mechanism by which aldosterone increases Na^+ transport was often unclear. Originally, because of the observation that protein synthesis was required for aldosterone to increase sodium transport, it was hypothesized that aldosterone induced ENaC synthesis and insertion. While the evidence from electrophysiological measurements remains somewhat controversial, all of the available biochemical evidence suggest that the number of sodium channel proteins in the apical membrane does not change in the early period after aldosterone treatment (at least in the first two to four hours when the increase in sodium transport is most dramatic) (Kleyman et al. 1989, 1992; Tousson et al. 1989; Helman et al. 1996). More evidence also supports the view that the early effects of aldosterone are indirect and that the proteins which are synthesized are modulatory proteins which convert poorly transporting ENaC in the apical membrane into ones which readily transport sodium (Chen et al. 1999; Verrey et al. 2000; Loffing et al. 2001).

The mechanism for the long-term effect of aldosterone to increase sodium transport is also not completely clear. There is evidence in a variety of tissues that the total cellular amount of ENaC subunits may increase; however, whether this is due to an increase in translation is controversial since changes in message levels are quite variable in different tissues. Therefore, the long-term changes could be due to an increase in translation, an increase in trafficking of subunits to the membrane, post-translational modifications of channels in the membrane, or a reduction in degradation of membrane channels.

On cursory examination, the magnitude of the changes induced by aldosterone is surprising; the expression of so many genes and activity of so many proteins is somewhat overwhelming. However, the overall effects of aldosterone are exactly what one would expect for a graded physiological response. The actions can be summarized as four successive molecular events. (1) Activate existing channels by increasing their open probability; (2) increase channel density by reducing the rate at which channels are removed from the membrane; (3) increase the rate at which channels are inserted into the membrane from cellular pools; and (4) increase transcription and translation to add to the total amount of ENaC in the membrane and cell.

18.9.1 Activate Existing Channels by Increasing Their Open Probability

Likely the most important early event is activation of PI-3-kinase to produce more inositol lipid phosphates (Blazer-Yost et al. 1999; Pochynyuk et al. 2006, 2007b, 2008b; Staruschenko et al. 2007; Tong et al. 2004; Wang et al. 2008a). This will increase ENaC open probability (Frindt and Palmer 2015; Yue et al. 2002; Kemendy et al. 1992). Another important early event is expression and activation of K-Ras (Mastroberardino et al. 1998; Verrey et al. 2000). K-Ras is an early gene, but K-Ras requires post-translational modification to activate ENaC. Aldosterone-induced methylation may directly activate ENaC (Rokaw et al. 1998; Edinger et al. 2006), but it also promotes methylation of K-Ras, an essential step in its association with the membrane and activation (Al-Baldawi et al. 2000; Becchetti et al. 2000; Stockand et al. 1999b, 2000b). The isoprenyl-cysteine-O-carboxy-methyl-transferase responsible for K-Ras2A methylation is not an aldosterone-induced protein, but the intrinsic activity of the enzyme does increase in response to aldosterone (by protein kinase C phosphorylation). In addition, the activity of the other enzyme intimately involved in protein methylation, SAH hydrolase, is also increased by aldosterone (Stockand et al. 1999a, 2001). Once active, K-Ras couples PI-3-kinase to ENaC (Staruschenko et al. 2004, 2005) and increases open probability. Thus, at least part of the aldosterone-induced short-term increase in ENaC activity is due to an increase in anionic lipids that directly activate existing ENaC channels.

18.9.2 Keep Existing (and Any New) ENaC in the Membrane

ENaC internalization and degradation involves Nedd4-2 ubiquitination (Staub et al. 1997a; Staub and Rotin 1997). Aldosterone controls Nedd4-2 activity by promoting expression of SGK1 (Chen et al. 1999; Verrey et al. 2000; Bhargava et al. 2001; Debonneville et al. 2001; Loffing et al. 2001; Pearce 2001; Snyder et al. 2002, 2004a; Zhou and Snyder 2005). SGK1 is a kinase that phosphorylates Nedd4-2 at two locations after which Nedd4-2 is inactivated by interaction with the scaffold protein, 14-3-3σ (Bhalla et al. 2005). SGK1 is normally inactive after translation and is activated by phosphorylation at three or four sites. The first phosphorylation is by the aldosterone-induced chaperone protein, GILZ1. GILZ1 also protects SGK1 from ERAD proteasomal degradation (Soundararajan et al. 2010b, 2012a). The second phosphorylation is by PDK1 (3-phosphoinositide dependent kinase) (Faletti et al. 2002). PDK1 is activated by PI-3-kinase production of phosphatidyl inositol phosphates (produced in step 1) (Hemmings and Restuccia 2012). Subsequently, mTORC2 (mTOR complex 2) phosphorylates the hydrophobic motif of SGK1 (Lu et al. 2010; Gleason et al. 2015; Garcia-Martinez and Alessi 2008). mTORC2 activation may be part of an aldosterone-induced long-term proliferative response (Alessi et al. 2009). Finally, for full activation, SGK1 may be additionally phosphorylated by another kinase, WNK1 (Chen et al. 2009b).

Nedd4-2 interaction with ENaC is normally promoted by ERK phosphorylation of ENaC (Soundararajan et al. 2009, 2012a). It is unclear whether SGK1-mediated inactivation of Nedd4-2 is complete, but, besides altering SGK1 activity, GILZ1 also inhibits ERK activation (Soundararajan et al. 2009, 2012a). There may be additional kinases that also affect Nedd4-2 activity, but their relationship to aldosterone activation of ENaC is unclear (Hallows et al. 2010).

18.9.3 Recruit More Active Channels to the Membrane

There are two processes by which aldosterone recruits active channels to the membrane: recruitment from cytosolic recycling pools and promoting proteolytic cleavage to convert inactive to active channels.

Volume depletion and aldosterone administration are associated with an increase in the proteolytic processing of ENaC subunits in rodent kidney (Kleyman et al. 2006; Masilamani et al. 1999; Ergonul et al. 2006). An increase in the residency time of channels at the plasma membrane in response to aldosterone may contribute to the increase in ENaC subunit proteolysis, since non-cleaved channels at the plasma membrane will have more time to be processed by proteases (Knight et al. 2006). Altering the protease/protease inhibitor balance may also contribute to the increase in ENaC subunit proteolysis observed with volume depletion and aldosterone administration (Narikiyo et al. 2002).

Aldosterone does promote incorporation of ENaC from subapical pools of channels (Loffing et al. 2000; O'Neill et al. 2008). The mechanisms by which this occurs is often unclear; however, a few cases are clearer. AS160 phosphorylation permits ENaC forward trafficking to the apical membrane to augment Na^+ absorption (Liang et al. 2010). Also, there is more trafficking from a subapical pool of channels (Edinger et al. 2012).

18.9.4 Transcribe and Translate New ENaC Subunits

Aldosterone controls the transcription of ENaC subunits (Mick et al. 2001; Thomas et al. 1996, 1999, 2002; Sayegh et al. 1999; Auerbach et al. 2000) and this transcription leads to increased numbers of functional channels in the apical membrane. Only recently have some of the specifics of the transcriptional control been described. Zhang et al. (2013b) have shown that aldosterone activates α-ENaC transcription by reducing expression of Dot1a (disruption of telomeric silencing 1) and Af9 (ALL1-fused gene from chromosome 9 protein) and by impairing Dot1a-Af9 interaction. Members of the Dot1 family including mouse Dot1l specifically methylate histone H3 K79 (van Leeuwen et al. 2002; Zhang et al. 2004). Dot1a is highly expressed in kidney (Zhang et al. 2004). Under basal conditions, Dot1a and Af9 form a repression complex, which binds to specific sites of the α-ENaC promoter, leading to targeted histone H3 K79 hypermethylation and repression of α-ENaC. Aldosterone attenuates the effects of the Dot1a-Af9 complex by reducing expression of Dot1a and Af9, and Sgk1 impairs Dot1a-Af9 interaction by phosphorylating Af9 (Zhang et al. 2007a), leading to histone H3 K79 hypomethylation of the α-ENaC promoter and release of α-ENaC repression. In addition, the mineralocorticoid receptor competes with Dot1a to bind Af9.

18.9.5 Regulation of ENaC by MicroRNAs

The mineralocorticoid hormone aldosterone is part of the renin-angiotensin-aldosterone system, the parts of which work in concert to regulate total body sodium balance and blood pressure. Aldosterone is the final component in the system and stimulates the production of several proteins that regulate ENaC activity (as described above). Aldosterone can either work though a G protein-coupled membrane receptor (GPER) membrane receptor (Ding et al. 2015; Feldman et al. 2014; Gros et al. 2011a, b, 2013; Mihailidou et al. 2019) or through the traditional cytosolic mineralocorticoid receptor (Fuller et al. 2017, 2019). The membrane receptor can produce very rapid signaling events through cytosolic signaling pathways (Mihailidou et al. 2019); the cytosolic receptor can produce rapid events (Holzman et al. 2007), but is usually considered to be a transcriptional agent. Aldosterone binding promotes movement of the receptor to the nucleus where it can promote transcription and translation of proteins involved in regulation of sodium balance including ENaC. This is a relatively traditional role for aldosterone, but more recent evidence suggests that aldosterone can also induce the transcription of noncoding RNAs (for reviews, see (Butterworth and Alvarez de la Rosa 2019; Butterworth 2015, 2018)). Small noncoding RNA, so-called microRNAs or miRNAs are regulated by aldosterone stimulation. As a separate mechanism for controlling ENaC transcription/translation, Edinger et al. (2014) originally described regulation of microRNA expression by aldosterone in both cultured mCCD and isolated primary distal nephron principal cells. Reducing the expression of these miRs increased epithelial sodium channel (ENaC)-mediated sodium transport in mCCD cells. MicroRNAs can control protein expression at all steps in the renin-angiotensin-aldosterone-signaling (RAAS) system (Liu et al. 2017; Jacobs et al. 2016; Qi et al. 2015; Edinger et al. 2014). Therefore, miRNAs can fine-tune sodium transport protein levels in cells of the aldosterone-sensitive distal nephron and act as a signal integrator.

Other ENaC containing tissues also appear to have miRNAs regulating sodium balance. ENaC in both adult and developing lung is the target for miRNA regulation of sodium transport and fluid balance (Ding et al. 2017; Qin et al. 2016; Tamarapu Parthasarathy et al. 2012).

18.9.6 Summary

Thus, aldosterone, the hormone primarily responsible for altering epithelial Na transport, apparently accomplishes the increase in many ways. Initially, as a rapid response, aldosterone promotes the production of PIP_2 (and PIP_3) which rapidly increases the activity of ENaCs that are already in the surface membrane. Later, aldosterone apparently increases the number of functional channels in the surface membrane, at least in part, by reducing the degradation rate of ENaC and allowing an increase in the surface membrane pool of functional ENaC. Finally, aldosterone promotes an increase in ENaC transcription and translation. And all of these processes are fine-tuned by the production of miRNA.

Acknowledgments This work was supported by DHHS R01 DK110409 to DCE.

References

Abriel H, Loffing J, Rebhun JF, Pratt JH, Schild L, Horisberger JD, Rotin D, Staub O (1999) Defective regulation of the epithelial Na+ channel by Nedd4 in Liddle's syndrome. J Clin Invest 103(5):667–673. https://doi.org/10.1172/JCI5713

Ackermann TF, Boini KM, Beier N, Scholz W, Fuchss T, Lang F (2011) EMD638683, a novel SGK inhibitor with antihypertensive potency. Cell Physiol Biochem 28(1):137–146. https://doi.org/10.1159/000331722

Adebamiro A, Cheng Y, Rao US, Danahay H, Bridges RJ (2007) A segment of gamma ENaC mediates elastase activation of Na+ transport. J Gen Physiol 130(6):611–629

Agne AM, Baldin JP, Benjamin AR, Orogo-Wenn MC, Wichmann L, Olson KR, Walters DV, Althaus M (2015) Hydrogen sulfide decreases beta-adrenergic agonist-stimulated lung liquid clearance by inhibiting ENaC-mediated transepithelial sodium absorption. Am J Physiol Regul Integr Comp Physiol 308(7):R636–R649. https://doi.org/10.1152/ajpregu.00489.2014

Ahn YJ, Brooker DR, Kosari F, Harte BJ, Li J, Mackler SA, Kleyman TR (1999) Cloning and functional expression of the mouse epithelial sodium channel. Am J Phys 277(1 Pt 2):F121–F129

Al-Baldawi NF, Stockand JD, Al Khalili OK, Yue G, Eaton DC (2000) Aldosterone induces *Ras* methylation in A6 epithelia. Am J Physiol Cell Physiol 279(2):C429–C439

Albrecht C, Viturro E (2007) The ABCA subfamily—gene and protein structures, functions and associated hereditary diseases. Pflugers Arch 453(5):581–589

Alessi DR, Pearce LR, García-Martínez JM (2009) New insights into mTOR signaling: mTORC2 and beyond. Sci Signal 2(67):e27–e27

Alli AA, Bao HF, Alli AA, Aldrugh Y, Song JZ, Ma HP, Yu L, Al-Khalili O, Eaton DC (2012a) Phosphatidylinositol phosphate-dependent regulation of Xenopus ENaC by MARCKS protein. Am J Physiol Renal Physiol 303(6):F800–F811. https://doi.org/10.1152/ajprenal.00703.2011

Alli AA, Song JZ, Al-Khalili O, Bao HF, Ma HP, Alli AA, Eaton DC (2012b) Cathepsin B is secreted apically from Xenopus 2F3 cells and cleaves the epithelial sodium channel (ENaC) to increase its activity. J Biol Chem 287(36):30073–30083. https://doi.org/10.1074/jbc.M111.338574

Almaca J, Kongsuphol P, Hieke B, Ousingsawat J, Viollet B, Schreiber R, Amaral MD, Kunzelmann K (2009) AMPK controls epithelial Na(+) channels through Nedd4-2 and causes an epithelial phenotype when mutated. Pflugers Arch 458(4):713–721. https://doi.org/10.1007/s00424-009-0660-4

Althaus M, Bogdan R, Clauss WG, Fronius M (2007) Mechano-sensitivity of epithelial sodium channels (ENaCs): laminar shear stress increases ion channel open probability. FASEB J 21(10):2389–2399

Alvarez de la Rosa D, Canessa CM (2003) Role of SGK in hormonal regulation of epithelial sodium channel in A6 cells. Am J Physiol Cell Physiol 284(2):C404–C414. https://doi.org/10.1152/ajpcell.00398.2002

Alvarez de la Rosa D, Paunescu TG, Els WJ, Helman SI, Canessa CM (2004) Mechanisms of regulation of epithelial sodium channel by SGK1 in A6 cells. J Gen Physiol 124(4):395–407. https://doi.org/10.1085/jgp.200409120

Andersen H, Friis UG, Hansen PB, Svenningsen P, Henriksen JE, Jensen BL (2015) Diabetic nephropathy is associated with increased urine excretion of proteases plasmin, prostasin and urokinase and activation of amiloride-sensitive current in collecting duct cells. Nephrol Dial Transplant 30(5):781–789. https://doi.org/10.1093/ndt/gfu402

Arbuzova A, Murray D, McLaughlin SG (1998) MARCKS, membranes, and calmodulin: kinetics of their interaction. Biochim Biophys Acta 1376(3):369–379. doi:S0304-4157(98)00011-2

Asano H, Horinouchi T, Mai Y, Sawada O, Fujii S, Nishiya T, Minami M, Katayama T, Iwanaga T, Terada K, Miwa S (2012) Nicotine- and tar-free cigarette smoke induces cell damage through reactive oxygen species newly generated by PKC-dependent activation of NADPH oxidase. J Pharmacol Sci 118(2):275–287. https://doi.org/10.1254/jphs.11166fp

Asher C, Chigaev A, Garty H (2001) Characterization of interactions between Nedd4 and b and gENaC using surface plasmon resonance. Biochem Biophys Res Commun 286(5):1228–1231

Auerbach SD, Loftus RW, Itani OA, Thomas CP (2000) Human amiloride-sensitive epithelial Na+ channel gamma subunit promoter: functional analysis and identification of a polypurine-polypyrimidine tract with the potential for triplex DNA formation. Biochem J 347 (Pt 1):105–114

Awayda MS, Subramanyam M (1998) Regulation of the epithelial Na^+ channel by membrane tension. J Gen Physiol 112(2):97–111

Babini E, Geisler HS, Siba M, Grunder S (2003) A new subunit of the epithelial Na+ channel identifies regions involved in Na+ self-inhibition. J Biol Chem 278(31):28418–28426. https://doi.org/10.1074/jbc.M301315200

Bachhuber T, Almaca J, Aldehni F, Mehta A, Amaral MD, Schreiber R, Kunzelmann K (2008) Regulation of the epithelial Na+ channel by the protein kinase CK2. J Biol Chem 283 (19):13225–13232. https://doi.org/10.1074/jbc.M704532200

Badve SV, Roberts MA, Hawley CM, Cass A, Garg AX, Krum H, Tonkin A, Perkovic V (2011) Effects of beta-adrenergic antagonists in patients with chronic kidney disease a systematic review and meta-analysis. J Am Coll Cardiol 58(11):1152–1161

Baines DL, Ramminger SJ, Collett A, Haddad JJ, Best OG, Land SC, Olver RE, Wilson SM (2001) Oxygen-evoked Na+ transport in rat fetal distal lung epithelial cells. J Physiol 532 (Pt 1):105-113

Baker ME, Katsu Y (2019) Evolution of the mineralocorticoid receptor. Vitam Horm 109:17–36. https://doi.org/10.1016/bs.vh.2018.10.009

Balchak DM, Thompson RN, Kashlan OB (2018) The epithelial Na(+) channel gamma subunit autoinhibitory tract suppresses channel activity by binding the gamma subunit's finger-thumb domain interface. J Biol Chem 293(42):16217–16225. https://doi.org/10.1074/jbc.RA118.004362

Balut C, Steels P, Radu M, Ameloot M, Driessche WV, Jans D (2006) Membrane cholesterol extraction decreases Na+ transport in A6 renal epithelia. Am J Physiol Cell Physiol 290 (1):87–94

Bankir L, Fernandes S, Bardoux P, Bouby N, Bichet DG (2005) Vasopressin-V2 receptor stimulation reduces sodium excretion in healthy humans. J Am Soc Nephrol 16(7):1920–1928. https://doi.org/10.1681/ASN.2004121079

Bankir L, Bichet DG, Bouby N (2010) Vasopressin V2 receptors, ENaC, and sodium reabsorption: a risk factor for hypertension? Am J Physiol Renal Physiol 299(5):F917–F928. https://doi.org/10.1152/ajprenal.00413.2010

Bao HF, Zhang ZR, Liang YY, Ma JJ, Eaton DC, Ma HP (2007) Ceramide mediates inhibition of the renal epithelial sodium channel by tumor necrosis factor-alpha through protein kinase C. Am J Physiol Renal Physiol 293(4):F1178–F1186. https://doi.org/10.1152/ajprenal.00153.2007

Bao HF, Song JZ, Duke BJ, Ma HP, Denson DD, Eaton DC (2012) Ethanol stimulates epithelial sodium channels by elevating reactive oxygen species. Am J Physiol Cell Physiol 303(11): C1129–C1138. https://doi.org/10.1152/ajpcell.00139.2012

Bao HF, Thai TL, Yue Q, Ma HP, Eaton AF, Cai H, Klein JD, Sands JM, Eaton DC (2014) ENaC activity is increased in isolated, split-open cortical collecting ducts from protein kinase Calpha knockout mice. Am J Physiol Renal Physiol 306(3):F309–F320. https://doi.org/10.1152/ajprenal.00519.2013

Barbaro NR, Foss JD, Kryshtal DO, Tsyba N, Kumaresan S, Xiao L, Mernaugh RL, Itani HA, Loperena R, Chen W, Dikalov S, Titze JM, Knollmann BC, Harrison DG, Kirabo A (2017) Dendritic cell amiloride-sensitive channels mediate sodium-induced inflammation and hypertension. Cell Rep 21(4):1009–1020. https://doi.org/10.1016/j.celrep.2017.10.002

Barbier O, Jacquillet G, Tauc M, Cougnon M, Poujeol P (2005) Effect of heavy metals on, and handling by, the kidney. Nephron Physiol 99(4):105–110

Barmeyer C, Amasheh S, Tavalali S, Mankertz J, Zeitz M, Fromm M, Schulzke JD (2004) IL-1beta and TNFalpha regulate sodium absorption in rat distal colon. Biochem Biophys Res Commun 317(2):500–507. https://doi.org/10.1016/j.bbrc.2004.03.072
Barthel A, Klotz LO (2005) Phosphoinositide 3-kinase signaling in the cellular response to oxidative stress. Biol Chem 386(3):207–216. https://doi.org/10.1515/bc.2005.026
Baylis C (2008) Nitric oxide deficiency in chronic kidney disease. Am J Physiol Renal Physiol 294 (1):F1–F9. https://doi.org/10.1152/ajprenal.00424.2007
Baylis C (2012) Nitric oxide synthase derangements and hypertension in kidney disease. Curr Opin Nephrol Hypertens 21:1):1–1):6. https://doi.org/10.1097/MNH.0b013e32834d54ca
Becchetti A, Kemendy AE, Stockand JD, Sariban-Sohraby S, Eaton DC (2000) Methylation increases the open probability of the epithelial sodium channel in A6 epithelia. J Biol Chem 275(22):16550–16559. https://doi.org/10.1074/jbc.M000954200
Belshe RB (1999) Influenza prevention and treatment: current practices and new horizons. Ann Intern Med 131(8):621–624
Benos DJ, Saccomani G, Sariban-Sohraby S (1987) The epithelial sodium channel. Subunit number and location of the amiloride binding site. J Biol Chem 262(22):10613–10618
Bentley PJ (1968) Amiloride: a potent inhibitor of sodium transport across the toad bladder. J Physiol 195(2):317–330
Berdiev BK, Jovov B, Tucker WC, Naren AP, Fuller CM, Chapman ER, Benos DJ (2004) ENaC subunit-subunit interactions and inhibition by syntaxin 1A. Am J Physiol Renal Physiol 286(6): F1100–F1106
Bergann T, Zeissig S, Fromm A, Richter JF, Fromm M, Schulzke JD (2009) Glucocorticoids and tumor necrosis factor-alpha synergize to induce absorption by the epithelial sodium channel in the colon. Gastroenterology 136(3):933–942. https://doi.org/10.1053/j.gastro.2008.12.008
Berman JM, Mironova E, Stockand JD (2018) Physiological regulation of the epithelial Na(+) channel by casein kinase II. Am J Physiol Renal Physiol 314(3):F367–f372. https://doi.org/10.1152/ajprenal.00469.2017
Beutler KT, Masilamani S, Turban S, Nielsen J, Brooks HL, Ageloff S, Fenton RA, Packer RK, Knepper MA (2003) Long-term regulation of ENaC expression in kidney by angiotensin II. Hypertension 41(5):1143–1150. https://doi.org/10.1161/01.HYP.0000066129.12106.E2
Bhalla V, Hallows KR (2008) Mechanisms of ENaC regulation and clinical implications. J Am Soc Nephrol 19(10):1845–1854. https://doi.org/10.1681/ASN.2008020225
Bhalla V, Daidie D, Li H, Pao AC, LaGrange LP, Wang J, Vandewalle A, Stockand JD, Staub O, Pearce D (2005) Serum- and glucocorticoid-regulated kinase 1 regulates ubiquitin ligase neural precursor cell-expressed, developmentally down-regulated protein 4-2 by inducing interaction with 14-3-3. Mol Endocrinol 19(12):3073–3084
Bhalla V, Oyster NM, Fitch AC, Wijngaarden MA, Neumann D, Schlattner U, Pearce D, Hallows KR (2006a) AMP-activated kinase inhibits the epithelial Na+ channel through functional regulation of the ubiquitin ligase Nedd4-2. J Biol Chem 281(36):26159–26169. https://doi.org/10.1074/jbc.M606045200
Bhalla V, Soundararajan R, Pao AC, Li H, Pearce D (2006b) Disinhibitory pathways for control of sodium transport: regulation of ENaC by SGK1 and GILZ. Am J Physiol Renal Physiol 291(4): F714–F721. https://doi.org/10.1152/ajprenal.00061.2006
Bhargava A, Fullerton MJ, Myles K, Purdy TM, Funder JW, Pearce D, Cole TJ (2001) The serum- and glucocorticoid-induced kinase is a physiological mediator of aldosterone action. Endocrinology 142(4):1587–1594
Biasio W, Chang T, McIntosh CJ, McDonald FJ (2004) Identification of Murr1 as a regulator of the human delta epithelial sodium channel. J Biol Chem 279(7):5429–5434. https://doi.org/10.1074/jbc.M311155200
Bjaelde RG, Arnadottir SS, Overgaard MT, Leipziger J, Praetorius HA (2013) Renal epithelial cells can release ATP by vesicular fusion. Paper presented at the Frontiers in physiology. Front Physiol

Blazer-Yost BL, Liu X, Helman SI (1998) Hormonal regulation of ENaCs: insulin and aldosterone. Am J Phys 274(5):C1373–C1379. https://doi.org/10.1152/ajpcell.1998.274.5.C1373
Blazer-Yost BL, Punescu TG, Helman SI, Lee KD, Vlahos CJ (1999) Phosphoinositide 3-kinase is required for aldosterone-regulated sodium reabsorption. Am J Phys 277(3 Pt 1):C531–C536
Blazer-Yost BL, Esterman MA, Vlahos CJ (2003) Insulin-stimulated trafficking of ENaC in renal cells requires PI 3-kinase activity. Am J Physiol Cell Physiol 284(6):C1645–C1653. https://doi.org/10.1152/ajpcell.00372.2002
Blazer-Yost BL, Vahle JC, Byars JM, Bacallao RL (2004) Real-time three-dimensional imaging of lipid signal transduction: apical membrane insertion of epithelial Na(+) channels. Am J Physiol Cell Physiol 287(6):C1569–C1576. https://doi.org/10.1152/ajpcell.00226.2004
Bohnert BN, Menacher M, Janessa A, Worn M, Schork A, Daiminger S, Kalbacher H, Haring HU, Daniel C, Amann K, Sure F, Bertog M, Haerteis S, Korbmacher C, Artunc F (2018) Aprotinin prevents proteolytic epithelial sodium channel (ENaC) activation and volume retention in nephrotic syndrome. Kidney Int 93(1):159–172. https://doi.org/10.1016/j.kint.2017.07.023
Boncoeur E, Tardif V, Tessier MC, Morneau F, Lavoie J, Gendreau-Berthiaume E, Grygorczyk R, Dagenais A, Berthiaume Y (2010) Modulation of epithelial sodium channel activity by lipopolysaccharide in alveolar type II cells: involvement of purinergic signaling. Am J Physiol Lung Cell Mol Physiol 298(3):L417–L426. https://doi.org/10.1152/ajplung.00170.2009
Bonifacino JS, Weissman AM (1998) Ubiquitin and the control of protein fate in the secretory and endocytic pathways. Annu Rev Cell Dev Biol 14:19–57
Booth RE, Stockand JD (2003) Targeted degradation of ENaC in response to PKC activation of the ERK1/2 cascade. Am J Physiol Renal Physiol 284(5):F938–F947
Bouallegue A, Daou GB, Srivastava AK (2007) Endothelin-1-induced signaling pathways in vascular smooth muscle cells. Curr Vasc Pharmacol 5(1):45–52. https://doi.org/10.2174/157016107779317161
Boucher RC (2004) New concepts of the pathogenesis of cystic fibrosis lung disease. Eur Respir J 23(1):146–158. https://doi.org/10.1183/09031936.03.00057003
Bouwman RA, Musters RJ, van Beek-Harmsen BJ, de Lange JJ, Boer C (2004) Reactive oxygen species precede protein kinase C-delta activation independent of adenosine triphosphate-sensitive mitochondrial channel opening in sevoflurane-induced cardioprotection. Anesthesiology 100(3):506–514. https://doi.org/10.1097/00000542-200403000-00008
Braun C, Hamacher J, Morel DR, Wendel A, Lucas R (2005) Dichotomal role of TNF in experimental pulmonary edema reabsorption. J Immunol 175(5):3402–3408
Bremner HR, Freywald T, O'Brodovich HM, Otulakowski G (2002) Promoter analysis of the gene encoding the beta-subunit of the rat amiloride-sensitive epithelial sodium channel. Am J Physiol Lung Cell Mol Physiol 282(1):L124–L134
Brennan-Minnella AM, Shen Y, El-Benna J, Swanson RA (2013) Phosphoinositide 3-kinase couples NMDA receptors to superoxide release in excitotoxic neuronal death. Cell Death Dis 4:e580. https://doi.org/10.1038/cddis.2013.111
Brouard M, Casado M, Djelidi S, Barrandon Y, Farman N (1999) Epithelial sodium channel in human epidermal keratinocytes: expression of its subunits and relation to sodium transport and differentiation. J Cell Sci 112(Pt 19):3343–3352
Bruns JB, Carattino MD, Sheng S, Maarouf AB, Weisz OA, Pilewski JM, Hughey RP, Kleyman TR (2007) Epithelial Na+ channels are fully activated by furin- and prostasin-dependent release of an inhibitory peptide from the gamma-subunit. J Biol Chem 282(9):6153–6160. https://doi.org/10.1074/jbc.M610636200
Bubien JK, Watson B, Khan MA, Langloh AL, Fuller CM, Berdiev B, Tousson A, Benos DJ (2001) Expression and regulation of normal and polymorphic epithelial sodium channel by human lymphocytes. J Biol Chem 276(11):8557–8566. https://doi.org/10.1074/jbc.M008886200
Buck TM, Brodsky JL (2018) Epithelial sodium channel biogenesis and quality control in the early secretory pathway. Curr Opin Nephrol Hypertens 27(5):364–372. https://doi.org/10.1097/mnh.0000000000000438

Bugaj V, Pochynyuk O, Mironova E, Vandewalle A, Medina JL, Stockand JD (2008) Regulation of the epithelial Na+ channel by endothelin-1 in rat collecting duct. Am J Physiol Renal Physiol 295(4):F1063–F1070. https://doi.org/10.1152/ajprenal.90321.2008

Bugaj V, Pochynyuk O, Stockand JD (2009) Activation of the epithelial Na+ channel in the collecting duct by vasopressin contributes to water reabsorption. Am J Physiol Renal Physiol 297(5):F1411–F1418. https://doi.org/10.1152/ajprenal.00371.2009

Bugaj V, Mironova E, Kohan DE, Stockand JD (2012a) Collecting duct-specific endothelin B receptor knockout increases ENaC activity. Am J Physiol Cell Physiol 302(1):C188–C194. https://doi.org/10.1152/ajpcell.00301.2011

Bugaj V, Sansom SC, Wen D, Hatcher LI, Stockand JD, Mironova E (2012b) Flow-sensitive K+-coupled ATP secretion modulates activity of the epithelial Na+ channel in the distal nephron. J Biol Chem 287(46):38552–38558. https://doi.org/10.1074/jbc.M112.408476

Buhl KB, Friis UG, Svenningsen P, Gulaveerasingam A, Ovesen P, Frederiksen-Moller B, Jespersen B, Bistrup C, Jensen BL (2012) Urinary plasmin activates collecting duct ENaC current in preeclampsia. Hypertension 60(5):1346–1351. https://doi.org/10.1161/HYPERTENSIONAHA.112.198879

Burnstock G (2006) Purinergic signalling—an overview. Paper presented at the Novartis Foundation symposium. Novartis Found Symp

Butler PL, Staruschenko A, Snyder PM (2015) Acetylation stimulates the epithelial sodium channel by reducing its ubiquitination and degradation. J Biol Chem 290(20):12497–12503. https://doi.org/10.1074/jbc.M114.635540

Butterworth MB (2010) Regulation of the epithelial sodium channel (ENaC) by membrane trafficking. Biochim Biophys Acta 1802(12):1166–1177

Butterworth MB (2015) MicroRNAs and the regulation of aldosterone signaling in the kidney. Am J Physiol Cell Physiol 308(7):C521–C527. https://doi.org/10.1152/ajpcell.00026.2015

Butterworth MB (2018) Role of microRNAs in aldosterone signaling. Curr Opin Nephrol Hypertens 27(5):390–394. https://doi.org/10.1097/mnh.0000000000000440

Butterworth MB, Alvarez de la Rosa D (2019) Regulation of aldosterone signaling by microRNAs. Vitam Horm 109:69–103. https://doi.org/10.1016/bs.vh.2018.09.002

Butterworth MB, Helman SI, Els WJ (2001) cAMP-sensitive endocytic trafficking in A6 epithelia. Am J Physiol Cell Physiol 280(4):C752–C762

Butterworth MB, Edinger RS, Johnson JP, Frizzell RA (2005) Acute ENaC stimulation by cAMP in a kidney cell line is mediated by exocytic insertion from a recycling channel pool. J Gen Physiol 125(1):81–101

Butterworth MB, Edinger RS, Ovaa H, Burg D, Johnson JP, Frizzell RA (2007) The deubiquitinating enzyme UCH-L3 regulates the apical membrane recycling of the epithelial sodium channel. J Biol Chem 282(52):37885–37893

Butterworth MB, Edinger RS, Frizzell RA, Johnson JP (2009) Regulation of the epithelial sodium channel by membrane trafficking. Am J Physiol Renal Physiol 296(1):F10–F24

Butterworth MB, Edinger RS, Silvis MR, Gallo LI, Liang X, Apodaca G, Frizzell RA, Johnson JP (2012) Rab11b regulates the trafficking and recycling of the epithelial sodium channel (ENaC). Am J Physiol Renal Physiol 302(5):F581–F590

Calder PC, Yaqoob P (2007) Lipid rafts—composition, characterization, and controversies. J Nutr 137(3):545–547

Caldwell RA, Boucher RC, Stutts MJ (2004) Serine protease activation of near-silent epithelial Na+ channels. Am J Physiol Cell Physiol 286(1):C190–C194. https://doi.org/10.1152/ajpcell.00342.2003

Caldwell RA, Boucher RC, Stutts MJ (2005) Neutrophil elastase activates near-silent epithelial Na+ channels and increases airway epithelial Na+ transport. Am J Physiol Lung Cell Mol Physiol 288(5):L813–L819

Canessa CM (1996) What is new about the structure of the epithelial Na^+ channel? News Physiol Sci 11(5):195–201

Canessa CM, Horisberger JD, Rossier BC (1993) Epithelial sodium channel related to proteins involved in neurodegeneration. Nature 361:467–470

Canessa CM, Merillat AM, Rossier BC (1994a) Membrane topology of the epithelial sodium channel in intact cells. Am J Phys 267(6 Pt 1):C1682–C1690

Canessa CM, Schild L, Buell G, Thorens B, Gautschi I, Horisberger JD, Rossier BC (1994b) Amiloride-sensitive epithelial Na+ channel is made of three homologous subunits. Nature 367 (6462):463–467

Caprio M, Newfell BG, la Sala A, Baur W, Fabbri A, Rosano G, Mendelsohn ME, Jaffe IZ (2008) Functional mineralocorticoid receptors in human vascular endothelial cells regulate intercellular adhesion molecule-1 expression and promote leukocyte adhesion. Circ Res 102 (11):1359–1367. https://doi.org/10.1161/CIRCRESAHA.108.174235

Carattino MD, Sheng S, Kleyman TR (2004) Epithelial Na+ channels are activated by laminar shear stress. J Biol Chem 279(6):4120–4126. https://doi.org/10.1074/jbc.M311783200

Carattino MD, Edinger RS, Grieser HJ, Wise R, Neumann D, Schlattner U, Johnson JP, Kleyman TR, Hallows KR (2005a) Epithelial sodium channel inhibition by AMP-activated protein kinase in oocytes and polarized renal epithelial cells. J Biol Chem 280(18):17608–17616

Carattino MD, Sheng S, Kleyman TR (2005b) Mutations in the pore region modify epithelial sodium channel gating by shear stress. J Biol Chem 280(6):4393–4401. https://doi.org/10.1074/jbc.M413123200

Carattino MD, Sheng S, Bruns JB, Pilewski JM, Hughey RP, Kleyman TR (2006) The epithelial Na + channel is inhibited by a peptide derived from proteolytic processing of its alpha subunit. J Biol Chem 281(27):18901–18907

Carattino MD, Passero CJ, Steren CA, Maarouf AB, Pilewski JM, Myerburg MM, Hughey RP, Kleyman TR (2008) Defining an inhibitory domain in the alpha-subunit of the epithelial sodium channel. Am J Physiol Renal Physiol 294(1):F47–F52. https://doi.org/10.1152/ajprenal.00399.2007

Carey MA, Card JW, Voltz JW, Germolec DR, Korach KS, Zeldin DC (2007) The impact of sex and sex hormones on lung physiology and disease: lessons from animal studies. Paper presented at the American journal of physiology. Lung cellular and molecular physiology. Am J Physiol Lung Cell Mol Physiol

Carninci P, Kasukawa T, Katayama S, Gough J, Frith MC, Maeda N, Oyama R, Ravasi T, Lenhard B, Wells C, Kodzius R, Shimokawa K, Bajic VB, Brenner SE, Batalov S, Forrest AR, Zavolan M, Davis MJ, Wilming LG, Aidinis V, Allen JE, Ambesi-Impiombato A, Apweiler R, Aturaliya RN, Bailey TL, Bansal M, Baxter L, Beisel KW, Bersano T, Bono H, Chalk AM, Chiu KP, Choudhary V, Christoffels A, Clutterbuck DR, Crowe ML, Dalla E, Dalrymple BP, de BB, Della GG, di BD, Down T, Engstrom P, Fagiolini M, Faulkner G, Fletcher CF, Fukushima T, Furuno M, Futaki S, Gariboldi M, Georgii-Hemming P, Gingeras TR, Gojobori T, Green RE, Gustincich S, Harbers M, Hayashi Y, Hensch TK, Hirokawa N, Hill D, Huminiecki L, Iacono M, Ikeo K, Iwama A, Ishikawa T, Jakt M, Kanapin A, Katoh M, Kawasawa Y, Kelso J, Kitamura H, Kitano H, Kollias G, Krishnan SP, Kruger A, Kummerfeld SK, Kurochkin IV, Lareau LF, Lazarevic D, Lipovich L, Liu J, Liuni S, McWilliam S, Madan BM, Madera M, Marchionni L, Matsuda H, Matsuzawa S, Miki H, Mignone F, Miyake S, Morris K, Mottagui-Tabar S, Mulder N, Nakano N, Nakauchi H, Ng P, Nilsson R, Nishiguchi S, Nishikawa S, Nori F, Ohara O, Okazaki Y, Orlando V, Pang KC, Pavan WJ, Pavesi G, Pesole G, Petrovsky N, Piazza S, Reed J, Reid JF, Ring BZ, Ringwald M, Rost B, Ruan Y, Salzberg SL, Sandelin A, Schneider C, Schonbach C, Sekiguchi K, Semple CA, Seno S, Sessa L, Sheng Y, Shibata Y, Shimada H, Shimada K, Silva D, Sinclair B, Sperling S, Stupka E, Sugiura K, Sultana R, Takenaka Y, Taki K, Tammoja K, Tan SL, Tang S, Taylor MS, Tegner J, Teichmann SA, Ueda HR, van NE, Verardo R, Wei CL, YAGI K, Yamanishi H, Zabarovsky E, Zhu S, Zimmer A, Hide W, Bult C, Grimmond SM, Teasdale RD, Liu ET, Brusic V, Quackenbush J, Wahlestedt C, Mattick JS, Hume DA, Kai C, Sasaki D, Tomaru Y, Fukuda S, Kanamori-Katayama M, Suzuki M, Aoki J, Arakawa T, Iida J, Imamura K, Itoh M, Kato T, Kawaji H, Kawagashira N, Kawashima T, Kojima M, Kondo S, Konno H, Nakano K, Ninomiya N,

Nishio T, Okada M, Plessy C, Shibata K, Shiraki T, Suzuki S, Tagami M, Waki K, Watahiki A, Okamura-Oho Y, Suzuki H, Kawai J, Hayashizaki Y (2005) The transcriptional landscape of the mammalian genome. Science 309(5740):1559–1563

Chalfant ML, Denton JS, Berdiev BK, Ismailov II, Benos DJ, Stanton BA (1999) Intracellular H+ regulates the alpha-subunit of ENaC, the epithelial Na+ channel. Am J Phys 276(2 Pt 1):C477–C486

Chang T, McDonald FJ (2007) COMMD1 downregulates d epithelial sodium channel (dENaC) through trafficking and ubiquitination. FASEB J 21:Ib465

Chang CT, Wu MS, Tian YC, Chen KH, Yu CC, Liao CH, Hung CC, Yang CW (2007) Enhancement of epithelial sodium channel expression in renal cortical collecting ducts cells by advanced glycation end products. Nephrol Dial Transplant 22(3):722–731. https://doi.org/10.1093/ndt/gfl668

Chang T, Ke Y, Ly K, McDonald FJ (2011) COMMD1 regulates the delta epithelial sodium channel (deltaENaC) through trafficking and ubiquitination. Biochem Biophys Res Commun 411(3):506–511. https://doi.org/10.1016/j.bbrc.2011.06.149

Chen SY, Bhargava A, Mastroberardino L, Meijer OC, Wang J, Buse P, Firestone GL, Verrey F, Pearce D (1999) Epithelial sodium channel regulated by aldosterone-induced protein sgk. Proc Natl Acad Sci USA 96(5):2514–2519

Chen XJ, Eaton DC, Jain L (2002) Beta-adrenergic regulation of amiloride-sensitive lung sodium channels. Am J Physiol Lung Cell Mol Physiol 282(4):609–620

Chen W, Valamanesh F, Mirshahi T, Soria J, Tang R, Agarwal MK, Mirshahi M (2004a) Aldosterone signaling modifies capillary formation by human bone marrow endothelial cells. Vasc Pharmacol 40(6):269–277. https://doi.org/10.1016/j.vph.2003.08.003

Chen XJ, Seth S, Yue G, Kamat P, Compans RW, Guidot D, Brown LA, Eaton DC, Jain L (2004b) Influenza virus inhibits ENaC and lung fluid clearance. Am J Physiol Lung Cell Mol Physiol 287(2):L366–L373. https://doi.org/10.1152/ajplung.00011.2004

Chen J, Chen Z, Chintagari NR, Bhaskaran M, Jin N, Narasaraju T, Liu L (2006) Alveolar type I cells protect rat lung epithelium from oxidative injury. J Physiol 572(Pt 3):625–638

Chen L, Song W, Davis IC, Shrestha K, Schwiebert E, Sullender WM, Matalon S (2009a) Inhibition of Na+ transport in lung epithelial cells by respiratory syncytial virus infection. Am J Respir Cell Mol Biol 40(5):588–600. https://doi.org/10.1165/rcmb.2008-0034OC

Chen W, Chen Y, Xu BE, Juang YC, Stippec S, Zhao Y, Cobb MH (2009b) Regulation of a third conserved phosphorylation site in SGK1. J Biol Chem 284(6):3453–3460. https://doi.org/10.1074/jbc.M807502200

Chen CL, Chan PC, Wang SH, Pan YR, Chen HC (2010) Elevated expression of protein kinase C delta induces cell scattering upon serum deprivation. J Cell Sci 123(Pt 17):2901–2913. https://doi.org/10.1242/jcs.069765

Chen J, Myerburg MM, Passero CJ, Winarski KL, Sheng S (2011) External Cu2+ inhibits human epithelial Na+ channels by binding at a subunit interface of extracellular domains. J Biol Chem 286(31):27436–27446

Chen J, Kleyman TR, Sheng S (2013) Gain-of-function variant of the human epithelial sodium channel. Am J Physiol Renal Physiol 304(2):F207–F213. https://doi.org/10.1152/ajprenal.00563.2012

Cheng C, Prince LS, Snyder PM, Welsh MJ (1998) Assembly of the epithelial Na^+ channel evaluated using sucrose gradient sedimentation analysis. J Biol Chem 273(35):22693–22700

Cho W, Stahelin RV (2005) Membrane-protein interactions in cell signaling and membrane trafficking. Annu Rev Biophys Biomol Struct 34:119–151

Christ M, Meyer C, Sippel K, Wehling M (1995) Rapid aldosterone signaling in vascular smooth muscle cells: involvement of phospholipase C, diacylglycerol and protein kinase C alpha. Biochem Biophys Res Commun 213(1):123–129. https://doi.org/10.1006/bbrc.1995.2106

Chung WS, Weissman JL, Farley J, Drummond HA (2013) betaENaC is required for whole cell mechanically gated currents in renal vascular smooth muscle cells. Am J Physiol Renal Physiol 304(12):F1428–F1437. https://doi.org/10.1152/ajprenal.00444.2012

Clark MA, Eaton DC (1983) Effect of CO2 on neurons of the house cricket, Acheta domestica. J Neurobiol 14(3):237–250

Collier DM, Snyder PM (2009) Extracellular chloride regulates the epithelial sodium channel. J Biol Chem 284(43):29320–29325

Collier DM, Snyder PM (2011) Identification of epithelial Na+ channel (ENaC) intersubunit Cl- inhibitory residues suggests a trimeric alpha gamma beta channel architecture. J Biol Chem 286 (8):6027–6032

Collier DM, Tomkovicz VR, Peterson ZJ, Benson CJ, Snyder PM (2014) Intersubunit conformational changes mediate epithelial sodium channel gating. J Gen Physiol 144(4):337–348

Colombo L, Dalla Valle L, Fiore C, Armanini D, Belvedere P (2006) Aldosterone and the conquest of land. J Endocrinol Investig 29(4):373–379. https://doi.org/10.1007/bf03344112

Condliffe SB, Carattino MD, Frizzell RA, Zhang H (2003) Syntaxin 1A regulates ENaC via domain-specific interactions. J Biol Chem 278(15):12796–12804

Condliffe SB, Zhang H, Frizzell RA (2004) Syntaxin 1A regulates ENaC channel activity. J Biol Chem 279(11):10085–10092

Corradi M, Pignatti P, Brunetti G, Goldoni M, Caglieri A, Nava S, Moscato G, Balbi B (2008) Comparison between exhaled and bronchoalveolar lavage levels of hydrogen peroxide in patients with diffuse interstitial lung diseases. Acta Biomed 79(Suppl 1):73–78

Couloigner V, Fay M, Djelidi S, Farman N, Escoubet B, Runembert I, Sterkers O, Friedlander G, Ferrary E (2001) Location and function of the epithelial Na channel in the cochlea. Am J Physiol Renal Physiol 280(2):F214–F222. https://doi.org/10.1152/ajprenal.2001.280.2.F214

Coux O, Tanaka K, Goldberg AL (1996) Structure and functions of the 20S and 26S proteasomes. Annu Rev Biochem 65:801–847

Craigie E, Menzies RI, Larsen CK, Jacquillet G, Carrel M, Wildman SS, Loffing J, Leipziger J, Shirley DG, Bailey MA, Unwin RJ (2018) The renal and blood pressure response to low sodium diet in P2X4 receptor knockout mice. Physiol Rep 6(20):e13899. https://doi.org/10.14814/phy2.13899

Cuffe JE, Bielfeld-Ackermann A, Thomas J, Leipziger J, Korbmacher C (2000) ATP stimulates Cl- secretion and reduces amiloride-sensitive Na+ absorption in M-1 mouse cortical collecting duct cells. Paper presented at the J. Physiol. (London) J Physiol

Czikora I, Alli A, Bao HF, Kaftan D, Sridhar S, Apell HJ, Gorshkov B, White R, Zimmermann A, Wendel A, Pauly-Evers M, Hamacher J, Garcia-Gabay I, Fischer B, Verin A, Bagi Z, Pittet JF, Shabbir W, Lemmens-Gruber R, Chakraborty T, Lazrak A, Matthay MA, Eaton DC, Lucas R (2014) A novel tumor necrosis factor-mediated mechanism of direct epithelial sodium channel activation. Am J Respir Crit Care Med 190(5):522–532. https://doi.org/10.1164/rccm.201405-0833OC

Czikora I, Alli AA, Sridhar S, Matthay MA, Pillich H, Hudel M, Berisha B, Gorshkov B, Romero MJ, Gonzales J, Wu G, Huo Y, Su Y, Verin AD, Fulton D, Chakraborty T, Eaton DC, Lucas R (2017) Epithelial sodium channel-alpha mediates the protective effect of the TNF-derived TIP peptide in pneumolysin-induced endothelial barrier dysfunction. Front Immunol 8:842. https://doi.org/10.3389/fimmu.2017.00842

Dagenais A, Frechette R, Yamagata Y, Yamagata T, Carmel JF, Clermont ME, Brochiero E, Masse C, Berthiaume Y (2004) Downregulation of ENaC activity and expression by TNF-alpha in alveolar epithelial cells. Am J Physiol Lung Cell Mol Physiol 286(2):L301–L311. https://doi.org/10.1152/ajplung.00326.2002

Dagenais A, Frechette R, Clermont ME, Masse C, Prive A, Brochiero E, Berthiaume Y (2006) Dexamethasone inhibits the action of TNF on ENaC expression and activity. Am J Physiol Lung Cell Mol Physiol 291(6):L1220–L1231. https://doi.org/10.1152/ajplung.00511.2005

D'Angelo G, Polishchuk E, Di TG, Santoro M, Di CA, Godi A, West G, Bielawski J, Chuang CC, van der Spoel AC, Platt FM, Hannun YA, Polishchuk R, Mattjus P, De Matteis MA (2007) Glycosphingolipid synthesis requires FAPP2 transfer of glucosylceramide. Nature 449 (7158):62–67

D'Angelo G, Rega LR, De Matteis MA (2012) Connecting vesicular transport with lipid synthesis: FAPP2(). Biochim Biophys Acta 1821(8):1089–1095

Davis IC, Matalon S (2007) Epithelial sodium channels in the adult lung—important modulators of pulmonary health and disease. Adv Exp Med Biol 618:127–140

Dean RB, Gatty O (1937) The bioelectrical properties of frog skin. Trans Faraday Soc 33 (0):1040–1046. https://doi.org/10.1039/TF9373301040

Debonneville C, Flores SY, Kamynina E, Plant PJ, Tauxe C, Thomas MA, Munster C, Chraibi A, Pratt JH, Horisberger JD, Pearce D, Loffing J, Staub O (2001) Phosphorylation of Nedd4-2 by Sgk1 regulates epithelial Na(+) channel cell surface expression. EMBO J 20(24):7052–7059. https://doi.org/10.1093/emboj/20.24.7052

Diakov A, Korbmacher C (2004) A novel pathway of epithelial sodium channel activation involves a serum- and glucocorticoid-inducible kinase consensus motif in the C terminus of the channel's alpha-subunit. J Biol Chem 279(37):38134–38142. https://doi.org/10.1074/jbc.M403260200

Ding Q, Hussain Y, Chorazyczewski J, Gros R, Feldman RD (2015) GPER-independent effects of estrogen in rat aortic vascular endothelial cells. Mol Cell Endocrinol 399:60–68. https://doi.org/10.1016/j.mce.2014.07.023

Ding Y, Zhao R, Zhao X, Matthay MA, Nie HG, Ji HL (2017) ENaCs as both effectors and regulators of MiRNAs in lung epithelial development and regeneration. Cell Physiol Biochem 44(3):1120–1132. https://doi.org/10.1159/000485417

Dinudom A, Harvey KF, Komwatana P, Young JA, Kumar S, Cook DI (1998) Nedd4 mediates control of an epithelial Na+ channel in salivary duct cells by cytosolic Na+. Proc Natl Acad Sci USA 95(12):7169–7173

Dinudom A, Harvey KF, Komwatana P, Jolliffe CN, Young JA, Kumar S, Cook DI (2001) Roles of the C termini of a-, b-, and g-subunits of epithelial Na^+ channels (ENaC) in regulating ENaC and mediating its inhibition by cytosolic Na^+. J Biol Chem 276(17):13744–13749

DiPetrillo K, Coutermarsh B, Soucy N, Hwa J, Gesek F (2004) Tumor necrosis factor induces sodium retention in diabetic rats through sequential effects on distal tubule cells. Kidney Int 65 (5):1676–1683. https://doi.org/10.1111/j.1523-1755.2004.00606.x

Dorge A, Nagel W (1970) Effect of amiloride on sodium transport in frog skin. II. Sodium transport pool and unidirectional fluxes. Pflugers Arch 321(2):91–101

Downs CA, Helms MN (2013) Regulation of ion transport by oxidants. Am J Physiol Lung Cell Mol Physiol 305(9):L595–L603. https://doi.org/10.1152/ajplung.00212.2013

Downs CA, Kriener LH, Yu L, Eaton DC, Jain L, Helms MN (2012) beta-Adrenergic agonists differentially regulate highly selective and nonselective epithelial sodium channels to promote alveolar fluid clearance in vivo. Am J Physiol Lung Cell Mol Physiol 302(11):L1167–L1178. https://doi.org/10.1152/ajplung.00038.2012

Downs CA, Kreiner L, Zhao XM, Trac P, Johnson NM, Hansen JM, Brown LA, Helms MN (2015) Oxidized glutathione (GSSG) inhibits epithelial sodium channel activity in primary alveolar epithelial cells. Am J Physiol Lung Cell Mol Physiol 308(9):L943–L952. https://doi.org/10.1152/ajplung.00213.2014

Doyle WJ, Skoner DP, Hayden F, Buchman CA, Seroky JT, Fireman P (1994) Nasal and otologic effects of experimental influenza A virus infection. Ann Otol Rhinol Laryngol 103(1):59–69

Drummond HA (2012) betaENaC is a molecular component of a VSMC mechanotransducer that contributes to renal blood flow regulation, protection from renal injury, and hypertension. Front Physiol 3:341. https://doi.org/10.3389/fphys.2012.00341

Drummond HA, Stec DE (2015) betaENaC acts as a mechanosensor in renal vascular smooth muscle cells that contributes to renal myogenic blood flow regulation, protection from renal injury and hypertension. J Nephrol Res 1(1):1–9. https://doi.org/10.17554/j.issn.2410-0579.2015.01.12

Drummond HA, Gebremedhin D, Harder DR (2004) Degenerin/epithelial Na+ channel proteins: components of a vascular mechanosensor. Hypertension 44(5):643–648. https://doi.org/10.1161/01.HYP.0000144465.56360.ad

Du Bois-Reymond E, Churchill J, Jones B (eds) (1857) On animal electricity being an abstract of the discoveries of Emil du Bois-Reymond, member of the Academy of Sciences of Berlin, etc. etc. John Churchill, Princes Street, Soho, London

Duman JG, Forte JG (2003) What is the role of SNARE proteins in membrane fusion? Am J Physiol Cell Physiol 285(2):C237–C249

Duncan CD, Mata J (2011) Widespread cotranslational formation of protein complexes. PLoS Genet 7(12):e1002398. https://doi.org/10.1371/journal.pgen.1002398

Eaton DC (1981) Intracellular sodium ion activity and sodium transport in rabbit urinary bladder. J Physiol 316:527–544

Eaton DC, Hamilton KL (1988) The amiloride-blockable sodium channel of epithelial tissue. Ion Channels 1:251–282

Eaton DC, Malik B, Saxena NC, Al Khalili OK, Yue G (2001) Mechanisms of Aldosterone's action on epithelial Na^+ transport. J Membr Biol 184(3):313–319

Eaton DC, Chen XJ, Ramosevac S, Matalon S, Jain L (2004) Regulation of Na+ channels in lung alveolar type II epithelial cells. Proc Am Thorac Soc 1(1):10–16

Eaton DC, Helms MN, Koval M, Bao HF, Jain L (2009) The contribution of epithelial sodium channels to alveolar function in health and disease. Annu Rev Physiol 71:403–423. https://doi.org/10.1146/annurev.physiol.010908.163250

Eaton DC, Malik B, Bao HF, Yu L, Jain L (2010) Regulation of epithelial sodium channel trafficking by ubiquitination. Proc Am Thorac Soc 7(1):54–64. https://doi.org/10.1513/pats.200909-096JS

Eaton AF, Yue Q, Eaton DC, Bao HF (2014) ENaC activity and expression is decreased in the lungs of protein kinase C-alpha knockout mice. Am J Physiol Lung Cell Mol Physiol 307(5):L374–L385. https://doi.org/10.1152/ajplung.00040.2014

Edelman IS, Bogoroch R, Porter GA (1963) On the mechanism of action of aldosterone on sodium transport: the role of protein synthesis. Paper presented at the Proc Natl Acad Sci USA

Edinger RS, Yospin J, Perry C, Kleyman TR, Johnson JP (2006) Regulation of epithelial Na+ channels (ENaC) by methylation: a novel methyltransferase stimulates ENaC activity. J Biol Chem 281(14):9110–9117

Edinger RS, Lebowitz J, Li H, Alzamora R, Wang H, Johnson JP, Hallows KR (2009) Functional regulation of the epithelial Na+ channel by I{kappa}B kinase-{beta} occurs via phosphorylation of the ubiquitin ligase Nedd4-2. J Biol Chem 284(1):150–157. https://doi.org/10.1074/jbc.M807358200

Edinger RS, Bertrand CA, Rondandino C, Apodaca GA, Johnson JP, Butterworth MB (2012) The epithelial sodium channel (ENaC) establishes a trafficking vesicle pool responsible for its regulation. PLoS One 7(9):e46593

Edinger RS, Coronnello C, Bodnar AJ, Labarca M, Bhalla V, LaFramboise WA, Benos PV, Ho J, Johnson JP, Butterworth MB (2014) Aldosterone regulates microRNAs in the cortical collecting duct to alter sodium transport. J Am Soc Nephrol 25(11):2445–2457. https://doi.org/10.1681/asn.2013090931

Ergonul Z, Frindt G, Palmer LG (2006) Regulation of maturation and processing of ENaC subunits in the rat kidney. Am J Physiol Renal Physiol 291(3):F683–F693. https://doi.org/10.1152/ajprenal.00422.2005

Eriguchi M, Bernstein EA, Veiras LC, Khan Z, Cao DY, Fuchs S, McDonough AA, Toblli JE, Gonzalez-Villalobos RA, Bernstein KE, Giani JF (2018) The absence of the ACE N-domain decreases renal inflammation and facilitates sodium excretion during diabetic kidney disease. J Am Soc Nephrol 29(10):2546–2561. https://doi.org/10.1681/asn.2018030323

Fahy RJ, Lichtenberger F, McKeegan CB, Nuovo GJ, Marsh CB, Wewers MD (2003) The acute respiratory distress syndrome: a role for transforming growth factor-beta 1. Am J Respir Cell Mol Biol 28(4):499–503. https://doi.org/10.1165/rcmb.2002-0092OC

Fakhruddin S, Alanazi W, Jackson KE (2017) Diabetes-induced reactive oxygen species: mechanism of their generation and role in renal injury. J Diabetes Res 2017:8379327. https://doi.org/10.1155/2017/8379327

Fakitsas P, Adam G, Daidie D, van Bemmelen MX, Fouladkou F, Patrignani A, Wagner U, Warth R, Camargo SM, Staub O, Verrey F (2007) Early aldosterone-induced gene product regulates the epithelial sodium channel by deubiquitylation. J Am Soc Nephrol 18 (4):1084–1092

Faletti CJ, Perrotti N, Taylor SI, Blazer-Yost BL (2002) sgk: an essential convergence point for peptide and steroid hormone regulation of ENaC-mediated Na+ transport. Am J Physiol Cell Physiol 282(3):C494–C500

Falin R, Veizis IE, Cotton CU (2005) A role for ERK1/2 in EGF- and ATP-dependent regulation of amiloride-sensitive sodium absorption. Am J Physiol Cell Physiol 288(5):C1003–C1011. https://doi.org/10.1152/ajpcell.00213.2004

Falkenstein E, Wehling M (2000) Nongenomically initiated steroid actions. Eur J Clin Investig 30 (Suppl 3):51–54. https://doi.org/10.1046/j.1365-2362.2000.0300s3051.x

Favre I, Moczydlowski E, Schild L (1996) On the structural basis for ionic selectivity among Na+, K+, and Ca2+ in the voltage-gated sodium channel. Biophys J 71(6):3110–3125. https://doi.org/10.1016/S0006-3495(96)79505-X

Feldman RD, Gros R, Ding Q, Hussain Y, Ban MR, McIntyre AD, Hegele RA (2014) A common hypofunctional genetic variant of GPER is associated with increased blood pressure in women. Br J Clin Pharmacol 78(6):1441–1452. https://doi.org/10.1111/bcp.12471

Felley-Bosco E, Bender F, Quest AF (2002) Caveolin-1-mediated post-transcriptional regulation of inducible nitric oxide synthase in human colon carcinoma cells. Biol Res 35(2):169–176

Fels J, Callies C, Kusche-Vihrog K, Oberleithner H (2010a) Nitric oxide release follows endothelial nanomechanics and not vice versa. Pflugers Arch 460(5):915–923. https://doi.org/10.1007/s00424-010-0871-8

Fels J, Oberleithner H, Kusche-Vihrog K (2010b) Menage a trois: aldosterone, sodium and nitric oxide in vascular endothelium. Biochim Biophys Acta 1802(12):1193–1202. https://doi.org/10.1016/j.bbadis.2010.03.006

Fels J, Jeggle P, Kusche-Vihrog K, Oberleithner H (2012) Cortical actin nanodynamics determines nitric oxide release in vascular endothelium. PLoS One 7(7):e41520. https://doi.org/10.1371/journal.pone.0041520

Firsov D, Schild L, Gautschi I, Merillat AM, Schneeberger E, Rossier BC (1996) Cell surface expression of the epithelial Na channel and a mutant causing Liddle syndrome: a quantitative approach. Proc Natl Acad Sci USA 93(26):15370–15375

Fischer H (2009) Mechanisms and function of DUOX in epithelia of the lung. Antioxid Redox Signal 11(10):2453–2465. https://doi.org/10.1089/ars.2009.2558

Folsch H, Mattila PE, Weisz OA (2009) Taking the scenic route: biosynthetic traffic to the plasma membrane in polarized epithelial cells. Traffic 10(8):972–981

Fontayne A, Dang PMC, Gougerot-Pocidalo MA, Benna J (2002) Phosphorylation of p47(phox) sites by pkc alpha, beta II, delta and zeta effect on binding to p22(phox) and on NADPH oxidase activation. Free Radic Biol Med 33:S26–S26

Fournier D, Luft FC, Bader M, Ganten D, Andrade-Navarro MA (2012) Emergence and evolution of the renin-angiotensin-aldosterone system. J Mol Med (Berl) 90(5):495–508. https://doi.org/10.1007/s00109-012-0894-z

Francis WL, Pumphrey RJ (1933) The electrical properties of prog skin. Part I Introductory 10 (4):379–385

Frank J, Roux J, Kawakatsu H, Su G, Dagenais A, Berthiaume Y, Howard M, Canessa CM, Fang X, Sheppard D, Matthay MA, Pittet JF (2003) Transforming growth factor-beta1 decreases expression of the epithelial sodium channel alphaENaC and alveolar epithelial vectorial sodium and fluid transport via an ERK1/2-dependent mechanism. J Biol Chem 278(45):43939–43950. https://doi.org/10.1074/jbc.M304882200

Friedrich B, Weyrich P, Stancakova A, Wang J, Kuusisto J, Laakso M, Sesti G, Succurro E, Smith U, Hansen T, Pedersen O, Machicao F, Schafer S, Lang F, Risler T, Ullrich S, Stefan N, Fritsche A, Haring HU (2008) Variance of the SGK1 gene is associated with insulin secretion in

different European populations: results from the TUEF, EUGENE2, and METSIM studies. PLoS One 3(11):e3506. https://doi.org/10.1371/journal.pone.0003506
Frindt G, Palmer LG (2015) Acute effects of aldosterone on the epithelial Na channel in rat kidney. Am J Physiol Renal Physiol 308(6):F572–F578. https://doi.org/10.1152/ajprenal.00585.2014
Frindt G, Gravotta D, Palmer LG (2016) Regulation of ENaC trafficking in rat kidney. J Gen Physiol 147(3):217–227. https://doi.org/10.1085/jgp.201511533
Frindt G, Yang L, Bamberg K, Palmer LG (2018) Na restriction activates epithelial Na channels in rat kidney through two mechanisms and decreases distal Na(+) delivery. J Physiol 596 (16):3585–3602. https://doi.org/10.1113/jp275988
Fromter E, Gebler B (1977) Electrical properties of amphibian urinary bladder epithelia. III. The cell membrane resistances and the effect of amiloride. Pflugers Arch 371(1–2):99–108
Fronius M (2013) Treatment of pulmonary edema by ENaC activators/stimulators. Curr Mol Pharmacol 6(1):13–27
Fronius M, Clauss W (2008) Mechano-sensitivity of ENaC: may the (shear) force be with you. Pflugers Arch – Eur J Physiol 455(5):775–785
Fuchs W, Larsen EH, Lindemann B (1977) Current-voltage curve of sodium channels and concentration dependence of sodium permeability in frog skin. J Physiol 267(1):137–166
Fuller PJ, Yang J, Young MJ (2017) 30 years of the mineralocorticoid receptor: coregulators as mediators of mineralocorticoid receptor signalling diversity. J Endocrinol 234(1):T23–T34. https://doi.org/10.1530/JOE-17-0060
Fuller PJ, Yang J, Young MJ (2019) Mechanisms of mineralocorticoid receptor signaling. Vitam Horm 109:37–68. https://doi.org/10.1016/bs.vh.2018.09.004
Funder JW (1997) Glucocorticoid and mineralocorticoid receptors: biology and clinical relevance. Annu Rev Med 48:231–240
Funder JW (2001) Non-genomic actions of aldosterone: role in hypertension. Curr Opin Nephrol Hypertens 10(2):227–230. https://doi.org/10.1097/00041552-200103000-00011
Fyfe GK, Canessa CM (1998) Subunit composition determines the single channel kinetics of the epithelial sodium channel. J Gen Physiol 112:423–432
Galietta LJ, Folli C, Marchetti C, Romano L, Carpani D, Conese M, Zegarra-Moran O (2000) Modification of transepithelial ion transport in human cultured bronchial epithelial cells by interferon-gamma. Am J Physiol Lung Cell Mol Physiol 278(6):L1186–L1194
Galietta LJ, Pagesy P, Folli C, Caci E, Romio L, Costes B, Nicolis E, Cabrini G, Goossens M, Ravazzolo R, Zegarra-Moran O (2002) IL-4 is a potent modulator of ion transport in the human bronchial epithelium in vitro. J Immunol 168(2):839–845
Gallego MS, Ling BN (1996) Regulation of amiloride-sensitive Na+ channels by endothelin-1 in distal nephron cells. Am J Phys 271(2 Pt 2):F451–F460. https://doi.org/10.1152/ajprenal.1996.271.2.F451
Gambhir A, Hangyas-Mihalyne G, Zaitseva I, Cafiso DS, Wang J, Murray D, Pentyala SN, Smith SO, McLaughlin SG (2004) Electrostatic sequestration of PIP2 on phospholipid membranes by basic/aromatic regions of proteins. Biophys J 86(4):2188–2207. https://doi.org/10.1016/S0006-3495(04)74278-2
Garcia-Caballero A, Dang Y, He H, Stutts MJ (2008) ENaC proteolytic regulation by channel-activating protease 2. J Gen Physiol 132(5):521–535. https://doi.org/10.1085/jgp.200810030
Garcia-Martinez JM, Alessi DR (2008) mTOR complex 2 (mTORC2) controls hydrophobic motif phosphorylation and activation of serum- and glucocorticoid-induced protein kinase 1 (SGK1). Biochem J 416(3):375–385
Garty H (1986) Mechanisms of aldosterone action in tight epithelia. J Membr Biol 90(3):193–205. https://doi.org/10.1007/bf01870126
Garty H, Benos DJ (1988) Characteristics and regulatory mechanisms of the amiloride-blockable Na+ channel. Physiol Rev 68(2):309–373. https://doi.org/10.1152/physrev.1988.68.2.309
Garty H, Palmer LG (1997) Epithelial sodium channels: function, structure, and regulation. Physiol Rev 77(2):359–396. https://doi.org/10.1152/physrev.1997.77.2.359

Ge Y, Gannon K, Gousset M, Liu R, Murphey B, Drummond HA (2012) Impaired myogenic constriction of the renal afferent arteriole in a mouse model of reduced betaENaC expression. Am J Physiol Renal Physiol 302(11):F1486–F1493. https://doi.org/10.1152/ajprenal.00638.2011

George DJ, Blackshear PJ (1992) Membrane association of the myristoylated alanine-rich C kinase substrate (MARCKS) protein appears to involve myristate-dependent binding in the absence of a myristoyl protein receptor. J Biol Chem 267(34):24879–24885

George AL Jr, Staub O, Geering K, Rossier BC, Kleyman TR, Kraehenbuhl JP (1989) Functional expression of the amiloride-sensitive sodium channel in Xenopus oocytes. Proc Natl Acad Sci USA 86(18):7295–7298

Giraldez T, Rojas P, Jou J, Flores C, Alvarez de la Rosa D (2012) The epithelial sodium channel delta-subunit: new notes for an old song. Am J Physiol Renal Physiol 303(3):F328–F338. https://doi.org/10.1152/ajprenal.00116.2012

Gleason CE, Frindt G, Cheng CJ, Ng M, Kidwai A, Rashmi P, Lang F, Baum M, Palmer LG, Pearce D (2015) mTORC2 regulates renal tubule sodium uptake by promoting ENaC activity. J Clin Invest 125(1):117–128. https://doi.org/10.1172/jci73935

Godi A, Campli AD, Konstantakopoulos A, Tullio GD, Alessi DR, Kular GS, Daniele T, Marra P, Lucocq JM, Matteis MAD (2004) FAPPs control Golgi-to-cell-surface membrane traffic by binding to ARF and PtdIns(4)P. Nat Cell Biol 6(5):393–404

Gohar EY, Pollock DM (2018) Sex-specific contributions of endothelin to hypertension. Curr Hypertens Rep 20(7):58. https://doi.org/10.1007/s11906-018-0856-0

Gohar EY, Kasztan M, Pollock DM (2017) Interplay between renal endothelin and purinergic signaling systems. Am J Physiol Renal Physiol 313(3):F666–f668. https://doi.org/10.1152/ajprenal.00639.2016

Golestaneh N, Klein C, Valamanesh F, Suarez G, Agarwal MK, Mirshahi M (2001) Mineralocorticoid receptor-mediated signaling regulates the ion gated sodium channel in vascular endothelial cells and requires an intact cytoskeleton. Biochem Biophys Res Commun 280(5):1300–1306. https://doi.org/10.1006/bbrc.2001.4275

Gong J, Park M, Steinberg SF (2017) Cleavage alters the molecular determinants of protein kinase C-delta catalytic activity. Mol Cell Biol 37(20). https://doi.org/10.1128/mcb.00324-17

Gonzalez-Montelongo R, Barros F, Alvarez de la Rosa D, Giraldez T (2016) Plasma membrane insertion of epithelial sodium channels occurs with dual kinetics. Pflugers Arch 468(5):859–870. https://doi.org/10.1007/s00424-016-1799-4

Gonzalez-Rodriguez E, Gaeggeler HP, Rossier BC (2007) IGF-1 vs insulin: respective roles in modulating sodium transport via the PI-3 kinase/Sgk1 pathway in a cortical collecting duct cell line. Kidney Int 71(2):116–125. https://doi.org/10.1038/sj.ki.5002018

Gonzalez-Vicente A, Hong NJ, Garvin JL (2019) Effects of reactive oxygen species on renal tubular transport. Am J Physiol Renal Physiol 317:F444–F455. https://doi.org/10.1152/ajprenal.00604.2018

Goodson P, Kumar A, Jain L, Kundu K, Murthy N, Koval M, Helms MN (2012) Nadph oxidase regulates alveolar epithelial sodium channel activity and lung fluid balance in vivo via O(−)(2) signaling. Am J Physiol Lung Cell Mol Physiol 302(4):L410–L419. https://doi.org/10.1152/ajplung.00260.2011

Gorelik J, Zhang Y, Sanchez D, Shevchuk A, Frolenkov G, Lab M, Klenerman D, Edwards C, Korchev Y (2005) Aldosterone acts via an ATP autocrine/paracrine system: the Edelman ATP hypothesis revisited. Paper presented at the Proc Natl Acad Sci USA

Goulet CC, Volk KA, Adams CM, Prince LS, Stokes JB, Snyder PM (1998) Inhibition of the epithelial Na^+ channel by interaction of Nedd4 with a PY motif deleted in Liddle's syndrome. J Biol Chem 273(45):30012–30017

Greenberg SB (2002) Respiratory viral infections in adults. Curr Opin Pulm Med 8(3):201–208

Greenlee MM, Mitzelfelt JD, Yu L, Yue Q, Duke BJ, Harrell CS, Neigh GN, Eaton DC (2013) Estradiol activates epithelial sodium channels in rat alveolar cells through the G protein-coupled estrogen receptor. Am J Physiol Lung Cell Mol Physiol 305(11):L878–L889. https://doi.org/10.1152/ajplung.00008.2013

Greenlee MM, Mitzelfelt JD, Duke BJ, Al-Khalili OK, Bao HF, Eaton DC (2015) Prolactin stimulates sodium and chloride ion channels in A6 renal epithelial cells. Am J Physiol Renal Physiol 308(7):F697–F705. https://doi.org/10.1152/ajprenal.00270.2014

Gros R, Ding Q, Davis M, Shaikh R, Liu B, Chorazyczewski J, Pickering JG, Feldman RD (2011a) Delineating the receptor mechanisms underlying the rapid vascular contractile effects of aldosterone and estradiol. Can J Physiol Pharmacol 89(9):655–663. https://doi.org/10.1139/Y11-062

Gros R, Ding Q, Sklar LA, Prossnitz EE, Arterburn JB, Chorazyczewski J, Feldman RD (2011b) GPR30 expression is required for the mineralocorticoid receptor-independent rapid vascular effects of aldosterone. Hypertension 57(3):442–451. https://doi.org/10.1161/HYPERTENSIONAHA.110.161653

Gros R, Ding Q, Liu B, Chorazyczewski J, Feldman RD (2013) Aldosterone mediates its rapid effects in vascular endothelial cells through GPER activation. Am J Physiol Cell Physiol 304(6):C532–C540. https://doi.org/10.1152/ajpcell.00203.2012

Grossmann C, Freudinger R, Mildenberger S, Krug AW, Gekle M (2004) Evidence for epidermal growth factor receptor as negative-feedback control in aldosterone-induced Na+ reabsorption. Am J Physiol Renal Physiol 286(6):F1226–F1231. https://doi.org/10.1152/ajprenal.00378.2003

Grunder S, Muller A, Ruppersberg JP (2001) Developmental and cellular expression pattern of epithelial sodium channel alpha, beta and gamma subunits in the inner ear of the rat. Eur J Neurosci 13(4):641–648. https://doi.org/10.1046/j.1460-9568.2001.01426.x

Guo Y, Duvall MD, Crow JP, Matalon S (1998) Nitric oxide inhibits Na^+ absorption across cultured alveolar type II monolayers. Am J Phys 274(3 Pt 1):L369–L377

Guo LJ, Alli AA, Eaton DC, Bao HF (2013) ENaC is regulated by natriuretic peptide receptor-dependent cGMP signaling. Am J Physiol Renal Physiol 304(7):F930–F937. https://doi.org/10.1152/ajprenal.00638.2012

Haerteis S, Krueger B, Korbmacher C, Rauh R (2009) The delta-subunit of the epithelial sodium channel (ENaC) enhances channel activity and alters proteolytic ENaC activation. J Biol Chem 284(42):29024–29040

Hallows KR, Bhalla V, Oyster NM, Wijngaarden MA, Lee JK, Li H, Chandran S, Xia X, Huang Z, Chalkley RJ, Burlingame AL, Pearce D (2010) Phosphopeptide screen uncovers novel phosphorylation sites of Nedd4-2 that potentiate its inhibition of the epithelial Na+ channel. J Biol Chem 285(28):21671–21678. https://doi.org/10.1074/jbc.M109.084731

Hamacher J, Lucas R, Lijnen HR, Buschke S, Dunant Y, Wendel A, Grau GE, Suter PM, Ricou B (2002) Tumor necrosis factor-alpha and angiostatin are mediators of endothelial cytotoxicity in bronchoalveolar lavages of patients with acute respiratory distress syndrome. Am J Respir Crit Care Med 166(5):651–656. https://doi.org/10.1164/rccm.2109004

Hamilton KL, Eaton DC (1985) Single-channel recordings from amiloride-sensitive epithelial sodium channel. Am J Phys 249(3 Pt 1):200–207

Hamilton KL, Good DW, Eaton DC (1986) Single sodium channel events in apical membrane of rat cortical duct (CCD). Kidney Int 29:397

Han D, Antunes F, Canali R, Rettori D, Cadenas E (2003) Voltage-dependent anion channels control the release of the superoxide anion from mitochondria to cytosol. J Biol Chem 278(8):5557–5563. https://doi.org/10.1074/jbc.M210269200

Hardiman KM, McNicholas-Bevensee CM, Fortenberry J, Myles CT, Malik B, Eaton DC, Matalon S (2004) Regulation of amiloride-sensitive Na(+) transport by basal nitric oxide. Am J Respir Cell Mol Biol 30(5):720–728. https://doi.org/10.1165/rcmb.2003-0325OC

Harris M, Firsov D, Vuagniaux G, Stutts MJ, Rossier BC (2007) A novel neutrophil elastase inhibitor prevents elastase activation and surface cleavage of the epithelial sodium channel expressed in *Xenopus laevis* oocytes. J Biol Chem 282(1):58–64

Hartwig JH, Thelen M, Resen A, Janmey PA, Nairn AC, Aderem A (1992) MARCKS is an actin filament crosslinking protein regulated by protein kinase C and calcium-calmodulin. Nature 356 (6370):618–622

Helman SI, Liu X, Blazer-Yost BL (1996) Early response of A6 epithelia to aldosterone is mediated by vesicle trafficking of apical Na^+ channels and not open probability. FASEB J 10:452–452

Helms MN, Liu L, Liang YY, Al-Khalili O, Vandewalle A, Saxena S, Eaton DC, Ma HP (2005a) Phosphatidylinositol 3,4,5-trisphosphate mediates aldosterone stimulation of epithelial sodium channel (ENaC) and interacts with gamma-ENaC. J Biol Chem 280(49):40885–40891

Helms MN, Yu L, Malik B, Kleinhenz DJ, Hart CM, Eaton DC (2005b) Role of SGK1 in nitric oxide inhibition of ENaC in Na+−transporting epithelia. Am J Physiol Cell Physiol 289(3): C717–C726. https://doi.org/10.1152/ajpcell.00006.2005

Helms MN, Chen XJ, Ramosevac S, Eaton DC, Jain L (2006a) Dopamine regulation of amiloride-sensitive sodium channels in lung cells. Am J Physiol Lung Cell Mol Physiol 290(4):L710–L722. https://doi.org/10.1152/ajplung.00486.2004

Helms MN, Self J, Bao HF, Job LC, Jain L, Eaton DC (2006b) Dopamine activates amiloride-sensitive sodium channels in alveolar type I cells in lung slice preparations. Am J Physiol Lung Cell Mol Physiol 291(4):L610–L618. https://doi.org/10.1152/ajplung.00426.2005

Helms MN, Jain L, Self JL, Eaton DC (2008) Redox regulation of epithelial sodium channels examined in alveolar type 1 and 2 cells patch-clamped in lung slice tissue. J Biol Chem 283 (33):22875–22883. https://doi.org/10.1074/jbc.M801363200

Hemmings BA, Restuccia DF (2012) PI3K-PKB/Akt pathway. Cold Spring Harb Perspect Biol 4 (9):a011189

Henshall TL, Manning JA, Alfassy OS, Goel P, Boase NA, Kawabe H, Kumar S (2017) Deletion of Nedd4-2 results in progressive kidney disease in mice. Cell Death Differ 24(12):2150–2160. https://doi.org/10.1038/cdd.2017.137

Hershko A, Ciechanover A (1998) The ubiquitin system. Annu Rev Biochem 67:425–479

Hicke L (1997) Ubiquitin-dependent internalization and down-regulation of plasma membrane proteins. FASEB J 11(14):1215–1226

Hicke L (2001) Protein regulation by monoubiquitin. Nat Rev Mol Cell Biol 2(3):195–201

Hickey KA, Rubanyi G, Paul RJ, Highsmith RF (1985) Characterization of a coronary vasoconstrictor produced by cultured endothelial cells. Am J Phys 248(5 Pt 1):C550–C556. https://doi.org/10.1152/ajpcell.1985.248.5.C550

Hill WG, An B, Johnson JP (2002) Endogenously expressed epithelial sodium channel is present in lipid rafts in A6 cells. J Biol Chem 277(37):33541–33544

Hill WG, Butterworth MB, Wang H, Edinger RS, Lebowitz J, Peters KW, Frizzell RA, Johnson JP (2007) The epithelial sodium channel (ENaC) traffics to apical membrane in lipid rafts in mouse cortical collecting duct cells. J Biol Chem 282(52):37402–37411. https://doi.org/10.1074/jbc.M704084200

Hills CE, Bland R, Bennett J, Ronco PM, Squires PE (2006) High glucose up-regulates ENaC and SGK1 expression in HCD-cells. Cell Physiol Biochem 18(6):337–346. https://doi.org/10.1159/000097611

Hinrichs GR, Michelsen JS, Zachar R, Friis UG, Svenningsen P, Birn H, Bistrup C, Jensen BL (2018) Albuminuria in kidney transplant recipients is associated with increased urinary serine proteases and activation of the epithelial sodium channel. Am J Physiol Renal Physiol 315(1): F151–f160. https://doi.org/10.1152/ajprenal.00545.2017

Hinrichs GR, Jensen BL, Svenningsen P (2019) Mechanisms of sodium retention in nephrotic syndrome. Curr Opin Nephrol Hypertens. https://doi.org/10.1097/mnh.0000000000000578

Hinton CF, Eaton DC (1989) Expression of amiloride-blockable Na channels in *Xenopus* oocytes. Am J Phys 257:C825–C829

Holzman JL, Liu L, Duke BJ, Kemendy AE, Eaton DC (2007) Transactivation of the IGF-1R by aldosterone. Am J Physiol Renal Physiol 292(4):F1219–F1228. https://doi.org/10.1152/ajprenal.00214.2006

Homolya L, Steinberg TH, Boucher RC (2000) Cell to cell communication in response to mechanical stress via bilateral release of ATP and UTP in polarized epithelia. J Cell Biol 150 (6):1349–1360

Hong NJ, Silva GB, Garvin JL (2010) PKC-alpha mediates flow-stimulated superoxide production in thick ascending limbs. Am J Physiol Renal Physiol 298(4):F885–F891. https://doi.org/10.1152/ajprenal.00543.2009

Horgan MJ, Palace GP, Everitt JE, Malik AB (1993) TNF-alpha release in endotoxemia contributes to neutrophil-dependent pulmonary edema. Am J Phys 264(4 Pt 2):H1161–H1165

Horisberger JD, Rossier BC (1992) Aldosterone regulation of gene transcription leading to control of ion transport. Hypertension 19(3):221–227

Hovater MB, Olteanu D, Hanson EL, Cheng NL, Siroky B, Fintha A, Komlosi P, Liu W, Satlin LM, Bell PD, Yoder BK, Schwiebert EM (2008) Loss of apical monocilia on collecting duct principal cells impairs ATP secretion across the apical cell surface and ATP-dependent and flow-induced calcium signals. Paper presented at the Purinergic signalling Purinergic Signal

Hu L, Zachariae ED, Larsen UG, Vilhardt F, Petersen SV (2019) The dynamic uptake and release of SOD3 from intracellular stores in macrophages modulates the inflammatory response. Redox Biol 26:101268. https://doi.org/10.1016/j.redox.2019.101268

Hughey RP, Kinlough CL, Bruns JB, Mueller GM, Poland PA, Kleyman TR (2002) Expression of functional epithelial sodium channels (ENaC) is dependent on a lectin-specific quality control pathway in the endoplasmic reticulum. Mol Biol Cell 13:235A–236A

Hughey RP, Mueller GM, Bruns JB, Kinlough CL, Poland PA, Harkleroad KL, Carattino MD, Kleyman TR (2003) Maturation of the epithelial Na+ channel involves proteolytic processing of the alpha- and gamma-subunits. J Biol Chem 278(39):37073–37082

Hughey RP, Bruns JB, Kinlough CL, Harkleroad KL, Poland PA, Carattino MD, Mueller GM, Kleyman TR (2004a) Furin-dependent processing of the epithelial sodium channel (ENaC). FASEB J 18(5):A1040–A1040

Hughey RP, Bruns JB, Kinlough CL, Harkleroad KL, Tong Q, Carattino MD, Johnson JP, Stockand JD, Kleyman TR (2004b) Epithelial sodium channels are activated by furin-dependent proteolysis. J Biol Chem 279(18):18111–18114

Hughey RP, Bruns JB, Kinlough CL, Kleyman TR (2004c) Distinct pools of epithelial sodium channels are expressed at the plasma membrane. J Biol Chem 279(47):48491–48494

Hughey RP, Carattino MD, Kleyman TR (2007) Role of proteolysis in the activation of epithelial sodium channels. Curr Opin Nephrol Hypertens 16(5):444–450. https://doi.org/10.1097/MNH.0b013e32821f6072

Hummler E, Barker PM, Gatzy J, Beermann F, Verdumo C, Schmidt A, Boucher RC, Rossier BC (1996) Early death due to defective neonatal lung liquid clearance in alpha- ENaC-deficient mice. Nat Genet 12(3):325–328

Hummler E, Barker P, Talbot C, Wang Q, Verdumo C, Grubb B, Gatzy J, Burnier M, Horisberger JD, Beermann F, Boucher R, Rossier BC (1997) A mouse model for the renal salt-wasting syndrome pseudohypoaldosteronism. Proc Natl Acad Sci USA 94(21):11710–11715. https://doi.org/10.1073/pnas.94.21.11710

Husted RF, Zhang C, Stokes JB (1998) Concerted actions of IL-1beta inhibit Na+ absorption and stimulate anion secretion by IMCD cells. Am J Phys 275(6):F946–F954. https://doi.org/10.1152/ajprenal.1998.275.6.F946

Husted RF, Sigmund RD, Stokes JB (2000) Mechanisms of inactivation of the action of aldosterone on collecting duct by TGF-beta. Am J Physiol Renal Physiol 278(3):F425–F433. https://doi.org/10.1152/ajprenal.2000.278.3.F425

Hyndman KA, Bugaj V, Mironova E, Stockand JD, Pollock JS (2015) NOS1-dependent negative feedback regulation of the epithelial sodium channel in the collecting duct. Am J Physiol Renal Physiol 308(3):F244–F251. https://doi.org/10.1152/ajprenal.00596.2013

Ifshin MS, Johnson KE, Eaton DC (1983) Acid pH and weak acids induce Na--Cl cotransport in the rabbit urinary bladder. J Membr Biol 76:151–164

Ilatovskaya DV, Pavlov TS, Levchenko V, Negulyaev YA, Staruschenko A (2011) Cortical actin binding protein cortactin mediates ENaC activity via Arp2/3 complex. FASEB J 25 (8):2688–2699. https://doi.org/10.1096/fj.10-167262
Ilatovskaya DV, Pavlov TS, Levchenko V, Staruschenko A (2013) ROS production as a common mechanism of ENaC regulation by EGF, insulin, and IGF-1. Am J Physiol Cell Physiol 304(1): C102–C111. https://doi.org/10.1152/ajpcell.00231.2012
Ingham RJ, Gish G, Pawson T (2004) The Nedd4 family of E3 ubiquitin ligases: functional diversity within a common modular architecture. Oncogene 23(11):1972–1984
Ishikawa T, Marunaka Y, Rotin D (1998) Electrophysiological characterization of the rat epithelial Na+ channel (rENaC) expressed in MDCK cells. Effects of Na+ and Ca2+. J Gen Physiol 111:825–846
Ismailov II, McDuffie JH, Sariban-Sohraby S, Johnson JP, Benos DJ (1994) Carboxyl methylation activates purified renal amiloride-sensitive Na^+ channels in planar lipid bilayers. J Biol Chem 269(35):22193–22197
Jacobs ME, Kathpalia PP, Chen Y, Thomas SV, Noonan EJ, Pao AC (2016) SGK1 regulation by miR-466g in cortical collecting duct cells. Am J Physiol Renal Physiol 310(11):F1251–F1257. https://doi.org/10.1152/ajprenal.00024.2016
Jain L, Chen XJ, Brown LA, Eaton DC (1998) Nitric oxide inhibits lung sodium transport through a cGMP-mediated inhibition of epithelial cation channels. Am J Phys 274(4 Pt 1):L475–L484
Jain L, Chen XJ, Malik B, Al-Khalili O, Eaton DC (1999) Antisense oligonucleotides against the alpha-subunit of ENaC decrease lung epithelial cation-channel activity. Am J Phys 276(6): L1046–L1051. https://doi.org/10.1152/ajplung.1999.276.6.L1046
Jain L, Chen XJ, Ramosevac S, Brown LA, Eaton DC (2001) Expression of highly selective sodium channels in alveolar type II cells is determined by culture conditions. Am J Physiol Lung Cell Mol Physiol 280(4):L646–L658. https://doi.org/10.1152/ajplung.2001.280.4.L646
Jasti J, Furukawa H, Gonzales EB, Gouaux E (2007) Structure of acid-sensing ion channel 1 at 1.9 A resolution and low pH. Nature 449(7160):316–323
Jeffers M, Taylor GA, Weidner KM, Omura S, VandeWoude GF (1997) Degradation of the Met tyrosine kinase receptor by the ubiquitin-proteasome pathway. Mol Cell Biol 17(2):799–808
Jeggle P, Callies C, Tarjus A, Fassot C, Fels J, Oberleithner H, Jaisser F, Kusche-Vihrog K (2013) Epithelial sodium channel stiffens the vascular endothelium in vitro and in Liddle mice. Hypertension 61(5):1053–1059. https://doi.org/10.1161/HYPERTENSIONAHA.111.199455
Jeggle P, Smith ES, Stewart AP, Haerteis S, Korbmacher C, Edwardson JM (2015) Atomic force microscopy imaging reveals the formation of ASIC/ENaC cross-clade ion channels. Biochem Biophys Res Commun 464(1):38–44. https://doi.org/10.1016/j.bbrc.2015.05.091
Jha JC, Banal C, Chow BS, Cooper ME, Jandeleit-Dahm K (2016) Diabetes and kidney disease: role of oxidative stress. Antioxid Redox Signal 25(12):657–684. https://doi.org/10.1089/ars.2016.6664
Ji HL, Bishop LR, Anderson SJ, Fuller CM, Benos DJ (2004) The role of Pre-H2 domains of alpha- and delta-epithelial Na+ channels in ion permeation, conductance, and amiloride sensitivity. J Biol Chem 279(9):8428–8440. https://doi.org/10.1074/jbc.M312012200
Ji HL, Su XF, Kedar S, Li J, Barbry P, Smith PR, Matalon S, Benos DJ (2006) delta-subunit confers novel biophysical features to alpha beta gamma-human ENaC via a physical interaction. J Biol Chem
Ji HL, Zhao RZ, Chen ZX, Shetty S, Idell S, Matalon S (2012) delta ENaC: a novel divergent amiloride-inhibitable sodium channel. Am J Physiol Lung Cell Mol Physiol 303(12):L1013–L1026. https://doi.org/10.1152/ajplung.00206.2012
Ji HL, Zhao R, Komissarov AA, Chang Y, Liu Y, Matthay MA (2015) Proteolytic regulation of epithelial sodium channels by urokinase plasminogen activator: cutting edge and cleavage sites. J Biol Chem 290(9):5241–5255. https://doi.org/10.1074/jbc.M114.623496
Jia G, Habibi J, Aroor AR, Hill MA, Yang Y, Whaley-Connell A, Jaisser F, Sowers JR (2018) Epithelial sodium channel in aldosterone-induced endothelium stiffness and aortic dysfunction. Hypertension 72(3):731–738. https://doi.org/10.1161/HYPERTENSIONAHA.118.11339

Jiang L, Wang Y, Su C, Sun H, Zhang H, Zhu B, Zhang H, Xiao H, Wang J, Zhang J (2015) Epithelial sodium channel is involved in H2S-induced acute pulmonary edema. Inhal Toxicol 27 (12):613–620. https://doi.org/10.3109/08958378.2015.1048909
Johnson MD, Bao HF, Helms MN, Chen XJ, Tigue Z, Jain L, Dobbs LG, Eaton DC (2006) Functional ion channels in pulmonary alveolar type I cells support a role for type I cells in lung ion transport. Proc Natl Acad Sci USA 103(13):4964–4969. https://doi.org/10.1073/pnas.0600855103
Kamynina E, Staub O (2002) Concerted action of ENaC, Nedd4-2, and Sgk1 in transepithelial Na (+) transport. Am J Physiol Renal Physiol 283(3):F377–F387. https://doi.org/10.1152/ajprenal.00143.2002
Kanelis V, Rotin D, Forman-Kay JD (2001) Solution structure of a Nedd4 WW domain-ENaC peptide complex. !Lost Data 8(5):407–412
Kapus A, Janmey P (2013) Plasma membrane – cortical cytoskeleton interactions: a cell biology approach with biophysical considerations. Compr Physiol 3(3):1231–1281
Karpushev AV, Ilatovskaya DV, Pavlov TS, Negulyaev YA, Staruschenko A (2010a) Intact cytoskeleton is required for small G protein dependent activation of the epithelial Na+ channel. PLoS One 5(1):e8827. https://doi.org/10.1371/journal.pone.0008827
Karpushev AV, Ilatovskaya DV, Staruschenko A (2010b) The actin cytoskeleton and small G protein RhoA are not involved in flow-dependent activation of ENaC. BMC Res Notes 3:210. https://doi.org/10.1186/1756-0500-3-210
Kashlan OB, Kleyman TR (2011) ENaC structure and function in the wake of a resolved structure of a family member. Am J Physiol Renal Physiol 301(4):F684–F696
Kashlan OB, Sheng S, Kleyman TR (2005) On the interaction between amiloride and its putative alpha-subunit epithelial Na+ channel binding site. J Biol Chem 280(28):26206–26215. https://doi.org/10.1074/jbc.M503500200
Kashlan OB, Maarouf AB, Kussius C, Denshaw RM, Blumenthal KM, Kleyman TR (2006) Distinct structural elements in the first membrane-spanning segment of the epithelial sodium channel. J Biol Chem 281(41):30455–30462. https://doi.org/10.1074/jbc.M604615200
Kashlan OB, Boyd CR, Argyropoulos C, Okumura S, Hughey RP, Grabe M, Kleyman TR (2010) Allosteric inhibition of the epithelial Na+ channel through peptide binding at peripheral finger and thumb domains. J Biol Chem 285(45):35216–35223. https://doi.org/10.1074/jbc.M110.167064
Kashlan OB, Adelman JL, Okumura S, Blobner BM, Zuzek Z, Hughey RP, Kleyman TR, Grabe M (2011) Constraint-based, homology model of the extracellular domain of the epithelial Na+ channel alpha subunit reveals a mechanism of channel activation by proteases. J Biol Chem 286 (1):649–660
Kashlan OB, Blobner BM, Zuzek Z, Carattino MD, Kleyman TR (2012) Inhibitory tract traps the epithelial Na+ channel in a low activity conformation. J Biol Chem 287(24):20720–20726. https://doi.org/10.1074/jbc.M112.358218
Kashlan OB, Blobner BM, Zuzek Z, Tolino M, Kleyman TR (2015) Na+ inhibits the epithelial Na+ channel by binding to a site in an extracellular acidic cleft. J Biol Chem 290(1):568–576. https://doi.org/10.1074/jbc.M114.606152
Kato M, Miyazawa K, Kitamura N (2000) A deubiquitinating enzyme UBPY interacts with the Src homology 3 domain of Hrs-binding protein via a novel binding motif PX(V/I)(D/N)RXXKP. J Biol Chem 275(48):37481–37487
Kellenberger S, Gautschi I, Schild L (1999a) A single point mutation in the pore region of the epithelial Na+ channel changes ion selectivity by modifying molecular sieving. Proc Natl Acad Sci USA 96(7):4170–4175
Kellenberger S, Hoffmann-Pochon N, Gautschi I, Schneeberger E, Schild L (1999b) On the molecular basis of ion permeation in the epithelial Na+ channel. J Gen Physiol 114(1):13–30
Kellenberger S, Auberson M, Gautschi I, Schneeberger E, Schild L (2001) Permeability properties of ENaC selectivity filter mutants. J Gen Physiol 118(6):679–692

Kelly TM, Nelson DH (1987) Sodium excretion and atrial natriuretic peptide levels during mineralocorticoid administration. A mechanism for the escape from hyperaldosteronism. Paper presented at the Endocr Res

Kemendy AE, Kleyman TR, Eaton DC (1992) Aldosterone alters the open probability of amiloride-blockable sodium channels in A6 epithelia. Am J Phys 263(4 Pt 1):C825–C837. https://doi.org/10.1152/ajpcell.1992.263.4.C825

Kevin LY, McIntosh CJ, Biasio W, Liu Y, Ke Y, Olson DR, Miller JH, Page R, Snyder PM, McDonald FJ (2013) Regulation of the delta and alpha epithelial sodium channel (ENaC) by ubiquitination and Nedd8. J Cell Physiol 228(11):2190–2201. https://doi.org/10.1002/jcp.24390

Kim J, Shishido T, Jiang X, Aderem A, McLaughlin S (1994) Phosphorylation, high ionic strength, and calmodulin reverse the binding of MARCKS to phospholipid vesicles. J Biol Chem 269 (45):28214–28219

Kim WY, Jung JH, Park EY, Yang CW, Kim H, Nielsen S, Madsen KM, Kim J (2006) Expression of protein kinase C isoenzymes alpha, betaI, and delta in subtypes of intercalated cells of mouse kidney. Am J Physiol Renal Physiol 291(5):F1052–F1060. https://doi.org/10.1152/ajprenal.00016.2006

Kim CH, Kim HY, Lee HS, Chang SO, Oh SH, Lee JH (2010) P2Y4-mediated regulation of Na+ absorption in the Reissner's membrane of the cochlea. Paper presented at the The Journal of Neuroscience: the official journal of the Society for Neuroscience

Klein SL, Passaretti C, Anker M, Olukoya P, Pekosz A (2010) The impact of sex, gender and pregnancy on 2009 H1N1 disease. Paper presented at the Biology of Sex Differences

Klemens CA, Edinger RS, Kightlinger L, Liu X, Butterworth MB (2017) Ankyrin G expression regulates apical delivery of the epithelial sodium channel (ENaC). J Biol Chem 292(1):375–385. https://doi.org/10.1074/jbc.M116.753616

Kleyman TR, Eaton DC (2019) Regulating ENaC's gate. Am J Physiol Cell Physiol. https://doi.org/10.1152/ajpcell.00418.2019

Kleyman TR, Cragoe EJ Jr, Kraehenbuhl JP (1989) The cellular pool of Na+ channels in the amphibian cell line A6 is not altered by mineralocorticoids. Analysis using a new photoactive amiloride analog in combination with anti-amiloride antibodies. J Biol Chem 264 (20):11995–12000

Kleyman TR, Coupaye-Gerard B, Ernst SA (1992) Aldosterone does not alter apical cell-surface expression of epithelial Na+ channels in the amphibian cell line A6. J Biol Chem 267 (14):9622–9628

Kleyman TR, Myerburg MM, Hughey RP (2006) Regulation of ENaCs by proteases: an increasingly complex story. Kidney Int 70(8):1391–1392

Kleyman TR, Kashlan OB, Hughey RP (2018) Epithelial Na(+) channel regulation by extracellular and intracellular factors. Annu Rev Physiol 80:263–281. https://doi.org/10.1146/annurev-physiol-021317-121143

Knight KK, Olson DR, Zhou R, Snyder PM (2006) Liddle's syndrome mutations increase Na^+ transport through dual effects on epithelial Na^+ channel surface expression and proteolytic cleavage. !Lost Data 103(8):2805–2808

Knox FG, Burnett JC, Jr., Kohan DE, Spielman WS, Strand JC (1980) Escape from the sodium-retaining effects of mineralocorticoids. Paper presented at the Kidney Int

Koefoed-Johnsen V, Ussing HH (1953) The contributions of diffusion and flow to the passage of D2O through living membranes; effect of neurohypophyseal hormone on isolated anuran skin. Acta Physiol Scand 28(1):60–76. https://doi.org/10.1111/j.1748-1716.1953.tb00959.x

Koefoed-Johnsen V, Ussing HH (1958) The nature of the frog skin potential. Acta Physiol Scand 42 (3-4):298–308. https://doi.org/10.1111/j.1748-1716.1958.tb01563.x

Koefoed-Johnsen V, Ussing HH, Zerahn K (1952) The origin of the short-circuit current in the adrenaline stimulated frog skin. Acta Physiol Scand 27(1):38–48. https://doi.org/10.1111/j.1748-1716.1953.tb00922.x

Kohan DE (2013) Role of collecting duct endothelin in control of renal function and blood pressure. Am J Physiol Regul Integr Comp Physiol 305(7):R659–R668. https://doi.org/10.1152/ajpregu.00345.2013

Kohan DE, Inscho EW, Wesson D, Pollock DM (2011) Physiology of endothelin and the kidney. Compr Physiol 1(2):883–919. https://doi.org/10.1002/cphy.c100039

Koike N, Takamura T, Kaneko S (2007) Induction of reactive oxygen species from isolated rat glomeruli by protein kinase C activation and TNF-alpha stimulation, and effects of a phosphodiesterase inhibitor. Life Sci 80(18):1721–1728. https://doi.org/10.1016/j.lfs.2007.02.001

Komwatana P, Dinudom A, Young JA, Cook DI (1996) Cytosolic Na^+ controls an epithelial Na^+ channel via the G_o guanine nucleotide-binding regulatory protein. Proc Natl Acad Sci USA 93 (15):8107–8111

Kooijman EE, Kuzenko SR, Gong D, Best MD, Folkesson HG (2011) Phosphatidylinositol 4,5-bisphosphate stimulates alveolar epithelial fluid clearance in male and female adult rats. Paper presented at the American journal of physiology. Lung cellular and molecular physiology. Am J Physiol Lung Cell Mol Physiol

Korte S, Wiesinger A, Straeter AS, Peters W, Oberleithner H, Kusche-Vihrog K (2012) Firewall function of the endothelial glycocalyx in the regulation of sodium homeostasis. Pflugers Arch 463(2):269–278. https://doi.org/10.1007/s00424-011-1038-y

Korte S, Strater AS, Druppel V, Oberleithner H, Jeggle P, Grossmann C, Fobker M, Nofer JR, Brand E, Kusche-Vihrog K (2014) Feedforward activation of endothelial ENaC by high sodium. FASEB J 28(9):4015–4025. https://doi.org/10.1096/fj.14-250282

Kortenoeven ML, Pedersen NB, Rosenbaek LL, Fenton RA (2015) Vasopressin regulation of sodium transport in the distal nephron and collecting duct. Am J Physiol Renal Physiol 309 (4):F280–F299. https://doi.org/10.1152/ajprenal.00093.2015

Kretz O, Barbry P, Bock R, Lindemann B (1999) Differential expression of RNA and protein of the three pore-forming subunits of the amiloride-sensitive epithelial sodium channel in taste buds of the rat. J Histochem Cytochem 47(1):51–64. https://doi.org/10.1177/002215549904700106

Krylatov AV, Maslov LN, Voronkov NS, Boshchenko AA, Popov SV, Gomez L, Wang H, Jaggi AS, Downey JM (2018) Reactive oxygen species as intracellular signaling molecules in the cardiovascular system. Curr Cardiol Rev 14(4):290–300. https://doi.org/10.2174/1573403x14666180702152436

Kunzelmann K, Bachhuber T, Regeer R, Markovich D, Sun J, Schreiber R (2005) Purinergic inhibition of the epithelial Na+ transport via hydrolysis of PIP2. FASEB J 19(1):142–143. https://doi.org/10.1096/fj.04-2314fje

Kusche-Vihrog K, Sobczak K, Bangel N, Wilhelmi M, Nechyporuk-Zloy V, Schwab A, Schillers H, Oberleithner H (2008) Aldosterone and amiloride alter ENaC abundance in vascular endothelium. Pflugers Arch 455(5):849–857. https://doi.org/10.1007/s00424-007-0341-0

Kusche-Vihrog K, Callies C, Fels J, Oberleithner H (2010) The epithelial sodium channel (ENaC): mediator of the aldosterone response in the vascular endothelium? Steroids 75(8-9):544–549. https://doi.org/10.1016/j.steroids.2009.09.003

Kusche-Vihrog K, Jeggle P, Oberleithner H (2014a) The role of ENaC in vascular endothelium. Pflugers Arch – Eur J Physiol 466(5):851–859. https://doi.org/10.1007/s00424-013-1356-3

Kusche-Vihrog K, Tarjus A, Fels J, Jaisser F (2014b) The epithelial Na+ channel: a new player in the vasculature. Curr Opin Nephrol Hypertens 23(2):143–148. https://doi.org/10.1097/01.mnh.0000441054.88962.2c

Lang F, Pearce D (2016) Regulation of the epithelial Na+ channel by the mTORC2/SGK1 pathway. Nephrol Dial Transplant 31(2):200–205. https://doi.org/10.1093/ndt/gfv270

Lang F, Klingel K, Wagner CA, Stegen C, Warntges S, Friedrich B, Lanzendorfer M, Melzig J, Moschen I, Steuer S, Waldegger S, Sauter M, Paulmichl M, Gerke V, Risler T, Gamba G, Capasso G, Kandolf R, Hebert SC, Massry SG, Broer S (2000) Deranged transcriptional regulation of cell-volume-sensitive kinase hSGK in diabetic nephropathy. Proc Natl Acad Sci USA 97(14):8157–8162. https://doi.org/10.1073/pnas.97.14.8157

Lang F, Stournaras C, Alesutan I (2014) Regulation of transport across cell membranes by the serum- and glucocorticoid-inducible kinase SGK1. Mol Membr Biol 31(1):29–36. https://doi.org/10.3109/09687688.2013.874598

Lanouette S, Mongeon V, Figeys D, Couture JF (2014) The functional diversity of protein lysine methylation. Mol Syst Biol 10:724. https://doi.org/10.1002/msb.134974

Larionov A, Dahlke E, Kunke M, Zanon Rodriguez L, Schiessl IM, Magnin JL, Kern U, Alli AA, Mollet G, Schilling O, Castrop H, Theilig F (2019) Cathepsin B increases ENaC activity leading to hypertension early in nephrotic syndrome. J Cell Mol Med 23(10):6543–6553. https://doi.org/10.1111/jcmm.14387

Lazarowski ER, Boucher RC, Harden TK (2003) Mechanisms of release of nucleotides and integration of their action as P2X- and P2Y-receptor activating molecules. Mol Pharmacol 64 (4):785–795

Lazrak A, Samanta A, Venetsanou K, Barbry P, Matalon S (2000) Modification of biophysical properties of lung epithelial Na(+) channels by dexamethasone. Am J Physiol Cell Physiol 279 (3):C762–C770. https://doi.org/10.1152/ajpcell.2000.279.3.C762

Le Goffic R, Mouchel T, Aubry F, Patard J-J, Ruffault A, Jégou B, Samson M (2002) Production of the chemokines monocyte chemotactic protein-1, regulated on activation normal T cell expressed and secreted protein, growth-related oncogene, and interferon-γ-inducible protein-10 is induced by the sendai virus in human and rat testicular cells. Endocrinology 143 (4):1434–1440. https://doi.org/10.1210/endo.143.4.8735

Lebowitz J, Edinger RS, An B, Perry CJ, Onate S, Kleyman TR, Johnson JP (2004) Ikappab kinase-beta (ikkbeta) modulation of epithelial sodium channel activity J Biol Chem 279 (40):41985–41990. https://doi.org/10.1074/jbc.M403923200

Lee IH, Dinudom A, Sanchez-Perez A, Kumar S, Cook DI (2007a) Akt mediates the effect of insulin on epithelial sodium channels by inhibiting Nedd4-2. J Biol Chem 282 (41):29866–29873. https://doi.org/10.1074/jbc.M701923200

Lee SY, Palmer ML, Maniak PJ, Jang SH, Ryu PD, O'Grady SM (2007b) P2Y receptor regulation of sodium transport in human mammary epithelial cells. Paper presented at the American Journal of Physiology

Lee IH, Campbell CR, Song SH, Day ML, Kumar S, Cook DI, Dinudom A (2009) The activity of the epithelial sodium channels is regulated by caveolin-1 via a Nedd4-2-dependent mechanism. J Biol Chem 284(19):12663–12669. https://doi.org/10.1074/jbc.M809737200

Lewis SA, Eaton DC, Diamond JM (1976) The mechanism of Na^+ transport by rabbit urinary bladder. J Membr Biol 28:41–70

Li J, Sheng S, Perry CJ, Kleyman TR (2003) Asymmetric organization of the pore region of the epithelial sodium channel. J Biol Chem 278(16):13867–13874

Li X, Leu S, Cheong A, Zhang H, Baibakov B, Shih C, Birnbaum MJ, Donowitz M (2004) Akt2, phosphatidylinositol 3-kinase, and PTEN are in lipid rafts of intestinal cells: role in absorption and differentiation. Gastroenterology 126(1):122–135

Li W, Long C, Renjun L, Zhangxue H, Yin H, Wanwei L, Juan M, Yuan S (2015) Association of SCNN1A Single Nucleotide Polymorphisms with neonatal respiratory distress syndrome. Sci Rep 5:17317. https://doi.org/10.1038/srep17317

Liang X, Butterworth MB, Peters KW, Frizzell RA (2010) AS160 modulates aldosterone-stimulated epithelial sodium channel forward trafficking. Mol Biol Cell 21(12):2024–2033

Lindemann B (1984) Fluctuation analysis of sodium channels in epithelia. Annu Rev Physiol 46:497–515. https://doi.org/10.1146/annurev.ph.46.030184.002433

Lindemann B, Van Driessche W (1977) Sodium-specific membrane channels of frog skin are pores: current fluctuations reveal high turnover. Science 195:292–294

Linderholm H (1952) Active transport of ions through frog skin with special reference to the action of certain diuretics; a study of the relation between electrical properties, the flux of labelled ions, and respiration. Acta Physiol Scand Suppl 27(97):1–144

Linguegia E, Voilley N, Waldmann R, Lazdunski M, Barbry P (1993) Expression cloning of an epithelial amiloride-sensitive Na+ channel. A new channel type with homologies to *Caenorhabditis elegans* degenerins. FEBS Lett 318(1):95–99

Liu L, Duke BJ, Malik B, Yue Q, Eaton DC (2009) Biphasic regulation of ENaC by TGF-{alpha} and EGF in renal epithelial cells. Am J Physiol Renal Physiol 296(6):F1417–F1427. https://doi.org/10.1152/ajprenal.90337.2008

Liu X, Edinger RS, Klemens CA, Phua YL, Bodnar AJ, LaFramboise WA, Ho J, Butterworth MB (2017) A microRNA cluster miR-23-24-27 is upregulated by aldosterone in the distal kidney nephron where it alters sodium transport. J Cell Physiol 232(6):1306–1317. https://doi.org/10.1002/jcp.25599

Lodin Z, Janacek K, Mueller JF (1963) The localization and effects of mercury bound by the surviving frog skin. J Cell Physiol 62:215–230

Loffing J, Pietri L, Aregger F, Bloch-Faure M, Ziegler U, Meneton P, Rossier BC, Kaissling B (2000) Differential subcellular localization of ENaC subunits in mouse kidney in response to high- and low-Na diets. Am J Physiol Ren Physiol 279(2):F252–F258

Loffing J, Zecevic M, Feraille E, Kaissling B, Asher C, Rossier BC, Firestone GL, Pearce D, Verrey F (2001) Aldosterone induces rapid apical translocation of ENaC in early portion of renal collecting system: possible role of SGK. Am J Physiol Renal Physiol 280(4):F675–F682

Londino JD, Lazrak A, Collawn JF, Bebok Z, Harrod KS, Matalon S (2017) Influenza virus infection alters ion channel function of airway and alveolar cells: mechanisms and physiological sequelae. Am J Physiol Lung Cell Mol Physiol 313(5):L845–l858. https://doi.org/10.1152/ajplung.00244.2017

Losel RM, Feuring M, Falkenstein E, Wehling M (2002) Nongenomic effects of aldosterone: cellular aspects and clinical implications. Steroids 67(6):493–498. https://doi.org/10.1016/s0039-128x(01)00176-3

Lu M, Wang J, Jones KT, Ives HE, Feldman ME, Yao LJ, Shokat KM, Ashrafi K, Pearce D (2010) mTOR complex-2 activates ENaC by phosphorylating SGK1. J Am Soc Nephrol 21 (5):811–818

Lu X, Wang F, Liu M, Yang KT, Nau A, Kohan DE, Reese V, Richardson RS, Yang T (2016) Activation of ENaC in collecting duct cells by prorenin and its receptor PRR: involvement of Nox4-derived hydrogen peroxide. Am J Physiol Renal Physiol 310(11):F1243–F1250. https://doi.org/10.1152/ajprenal.00492.2015

Lucas R, Yue Q, Alli A, Duke BJ, Al-Khalili O, Thai TL, Hamacher J, Sridhar S, Lebedyeva I, Su H, Tzotzos S, Fischer B, Gameiro AF, Loose M, Chakraborty T, Shabbir W, Aufy M, Lemmens-Gruber R, Eaton DC, Czikora I (2016) The lectin-like domain of TNF increases ENaC open probability through a novel site at the interface between the second transmembrane and C-terminal domains of the alpha-subunit. J Biol Chem 291(45):23440–23451. https://doi.org/10.1074/jbc.M116.718163

Lyon-Roberts B, Strait KA, van Peursem E, Kittikulsuth W, Pollock JS, Pollock DM, Kohan DE (2011) Flow regulation of collecting duct endothelin-1 production. Am J Physiol Renal Physiol 300(3):F650–F656. https://doi.org/10.1152/ajprenal.00530.2010

Ma HP (2011) Hydrogen peroxide stimulates the epithelial sodium channel through a phosphatidylinositide 3-kinase-dependent pathway. J Biol Chem 286(37):32444–32453. https://doi.org/10.1074/jbc.M111.254102

Ma HP, Eaton DC (2005) Acute regulation of epithelial sodium channel by anionic phospholipids. J Am Soc Nephrol 16(11):3182–3187

Ma HP, Saxena S, Warnock DG (2002) Anionic phospholipids regulate native and expressed ENaC. J Biol Chem 277:7641–7644

Ma HP, Al-Khalili O, Ramosevac S, Saxena S, Liang YY, Warnock DG, Eaton DC (2004) Steroids and exogenous gamma-ENaC subunit modulate cation channels formed by alpha-ENaC in human B lymphocytes. J Biol Chem 279(32):33206–33212

Ma HP, Chou CF, Wei SP, Eaton DC (2007) Regulation of the epithelial sodium channel by phosphatidylinositides: experiments, implications, and speculations. Pflugers Arch 455 (1):169–180. https://doi.org/10.1007/s00424-007-0294-3

Maarouf AB, Sheng N, Chen J, Winarski KL, Okumura S, Carattino MD, Boyd CR, Kleyman TR, Sheng S (2009) Novel determinants of epithelial sodium channel gating within extracellular thumb domains. J Biol Chem 284(12):7756–7765. https://doi.org/10.1074/jbc.M807060200

Machnik A, Neuhofer W, Jantsch J, Dahlmann A, Tammela T, Machura K, Park JK, Beck FX, Muller DN, Derer W, Goss J, Ziomber A, Dietsch P, Wagner H, van Rooijen N, Kurtz A, Hilgers KF, Alitalo K, Eckardt KU, Luft FC, Kerjaschki D, Titze J (2009) Macrophages regulate salt-dependent volume and blood pressure by a vascular endothelial growth factor-C-dependent buffering mechanism. Nat Med 15(5):545–552. https://doi.org/10.1038/nm.1960

MacRobbie EAC, Ussing HH (1961) Osmotic behavior of the epithelial cells of frog skin. Acta Physiol Scand 53:348–365

Malik B, Schlanger L, Al-Khalili O, Bao HF, Yue G, Price SR, Mitch WE, Eaton DC (2001) ENaC degradation in A6 cells by the ubiquitin-proteosome proteolytic pathway. Paper presented at the The Journal of biological chemistry

Malik B, Yue Q, Yue G, Chen XJ, Price SR, Mitch WE, Eaton DC (2005a) Role of Nedd4-2 and polyubiquitination in epithelial sodium channel degradation in untransfected renal A6 cells expressing endogenous ENaC subunits. Paper presented at the American journal of physiology. Renal Physiology

Malik B, Yue Q, Yue GC, Ma HP, Bao HF, Eaton DC (2005b) Growth conditions determine ENaC channel type. FASEB J 19(5):A1179–A1179

Malik B, Price SR, Mitch WE, Yue Q, Eaton DC (2006) Regulation of epithelial sodium channels by the ubiquitin-proteasome proteolytic pathway. Am J Physiol Renal Physiol 290(6):F1285–F1294

Mall M, Grubb BR, Harkema JR, O'Neal WK, Boucher RC (2004) Increased airway epithelial Na+ absorption produces cystic fibrosis-like lung disease in mice. Nat Med 10(5):487–493. https://doi.org/10.1038/nm1028

Malsure S, Wang Q, Charles RP, Sergi C, Perrier R, Christensen BM, Maillard M, Rossier BC, Hummler E (2014) Colon-specific deletion of epithelial sodium channel causes sodium loss and aldosterone resistance. J Am Soc Nephrol 25(7):1453–1464. https://doi.org/10.1681/ASN.2013090936

Mamenko M, Zaika O, Ilatovskaya DV, Staruschenko A, Pochynyuk O (2012) Angiotensin II increases activity of the epithelial Na+ channel (ENaC) in distal nephron additively to aldosterone. J Biol Chem 287(1):660–671. https://doi.org/10.1074/jbc.M111.298919

Manning JA, Henshall TL, Kumar S (2018) NEDD4-2-dependent control of Na(+) homeostasis and renal disease. Cell Cycle 17(1):1–2. https://doi.org/10.1080/15384101.2017.1386514

Mansley MK, Neuhuber W, Korbmacher C, Bertog M (2015) Norepinephrine stimulates the epithelial Na+ channel in cortical collecting duct cells via alpha2-adrenoceptors. Am J Physiol Renal Physiol 308(5):F450–F458. https://doi.org/10.1152/ajprenal.00548.2014

Mansley MK, Watt GB, Francis SL, Walker DJ, Land SC, Bailey MA, Wilson SM (2016) Dexamethasone and insulin activate serum and glucocorticoid-inducible kinase 1 (SGK1) via different molecular mechanisms in cortical collecting duct cells. Physiol Rep 4(10). https://doi.org/10.14814/phy2.12792

Mansley MK, Roe AJ, Francis SL, Gill JH, Bailey MA, Wilson SM (2019) Trichostatin A blocks aldosterone-induced Na(+) transport and control of serum- and glucocorticoid-inducible kinase 1 in cortical collecting duct cells. Br J Pharmacol. https://doi.org/10.1111/bph.14837

Markadieu N, Blero D, Boom A, Erneux C, Beauwens R (2004) Phosphatidylinositol 3,4,5-trisphosphate: an early mediator of insulin-stimulated sodium transport in A6 cells. Am J Physiol Renal Physiol 287(2):F319–F328. https://doi.org/10.1152/ajprenal.00314.2003

Martindale JL, Holbrook NJ (2002) Cellular response to oxidative stress: signaling for suicide and survival. J Cell Physiol 192(1):1–15. https://doi.org/10.1002/jcp.10119

Marunaka Y, Eaton DC (1991) Effects of vasopressin and cAMP on single amiloride-blockable Na channels. Am J Phys 260(5 Pt 1):C1071–C1084

Marunaka Y, Hagiwara N, Tohda H (1992) Insulin activates single amiloride-blockable Na channels in a distal nephron cell line (A6). Am J Phys 263(3 Pt 2):F392–F400. https://doi.org/10.1152/ajprenal.1992.263.3.F392

Masilamani S, Kim GH, Mitchell C, Wade JB, Knepper MA (1999) Aldosterone-mediated regulation of ENaC alpha, beta, and gamma subunit proteins in rat kidney. J Clin Invest 104(7):R19–R23. https://doi.org/10.1172/jci7840

Mastroberardino L, Spindler B, Forster I, Loffing J, Assandri R, May A, Verrey F (1998) Ras pathway activates epithelial Na+ channel and decreases its surface expression in Xenopus oocytes. Mol Biol Cell 9(12):3417–3427

Matalon S, O'Brodovich H (1999) Sodium channels in alveolar epithelial cells: molecular characterization, biophysical properties, and physiological significance. Annu Rev Physiol 61:627–661. https://doi.org/10.1146/annurev.physiol.61.1.627

Matalon S, Benos DJ, Jackson RM (1996) Biophysical and molecular properties of amiloride-inhibitable Na+ channels in alveolar epithelial cells. Am J Phys 271(1 Pt 1):L1–L22. https://doi.org/10.1152/ajplung.1996.271.1.L1

Matalon S, Hardiman KM, Jain L, Eaton DC, Kotlikoff M, Eu JP, Sun J, Meissner G, Stamler JS (2003) Regulation of ion channel structure and function by reactive oxygen-nitrogen species. Am J Physiol Lung Cell Mol Physiol 285(6):L1184–L1189. https://doi.org/10.1152/ajplung.00281.2003

Mato JM, Corrales FJ, Lu SC, Avila MA (2002) S-Adenosylmethionine: a control switch that regulates liver function. FASEB J 16(1):15–26

Matos JE, Sorensen MV, Geyti CS, Robaye B, Boeynaems JM, Leipziger J (2007) Distal colonic Na(+) absorption inhibited by luminal P2Y(2) receptors. Paper presented at the Pflugers Archiv: European journal of physiology. Pflugers Arch

Matsuura D, Taguchi K, Yagisawa H, Maekawa S (2007) Lipid components in the detergent-resistant membrane microdomain (DRM) obtained from the synaptic plasma membrane of rat brain. Neurosci Lett 423(2):158–161

Mazzochi C, Benos DJ, Smith PR (2006) Interaction of epithelial ion channels with the actin-based cytoskeleton. Am J Physiol Renal Physiol 291(6):F1113–F1122

McCulloch F, Chambrey R, Eladari D, Peti-Peterdi J (2005) Localization of connexin 30 in the luminal membrane of cells in the distal nephron. Paper presented at the American journal of physiology. Renal physiology. Am J Physiol Renal Physiol

McDonald FJ, Western AH, McNeil JD, Thomas BC, Olson DR, Snyder PM (2002) Ubiquitin-protein ligase WWP2 binds to and downregulates the epithelial Na^+ channel. Am J Physiol (Renal, Fluid Electrolyte Physiol) 283(3):F431–F436

McEneaney V, Harvey BJ, Thomas W (2007) Aldosterone rapidly activates protein kinase D via a mineralocorticoid receptor/EGFR trans-activation pathway in the M1 kidney CCD cell line. J Steroid Biochem Mol Biol 107(3-5):180–190. https://doi.org/10.1016/j.jsbmb.2007.03.043

McLaughlin SG, Aderem A (1995) The myristoyl-electrostatic switch: a modulator of reversible protein-membrane interactions. Trends Biochem Sci 20(7):272–276

McLaughlin SG, Wang J, Gambhir A, Murray D (2002) PIP(2) and proteins: interactions, organization, and information flow. Annu Rev Biophys Biomol Struct 31:151–175. https://doi.org/10.1146/annurev.biophys.31.082901.134259

McLaughlin S, Hangyas-Mihalyne G, Zaitseva I, Golebiewska U (2005) Reversible – through calmodulin – electrostatic interactions between basic residues on proteins and acidic lipids in the plasma membrane. Biochem Soc Symp 72:189–198

McLeod JW, Gordon J (1922) Production of hydrogen peroxide by bacteria. Biochem J 16(4):499–506. https://doi.org/10.1042/bj0160499

McNicholas CM, Canessa CM (1997) Diversity of channels generated by different combinations of epithelial sodium channel subunits. J Gen Physiol 109(6):681–692

McRitchie DI, Isowa N, Edelson JD, Xavier AM, Cai L, Man HY, Wang YT, Keshavjee SH, Slutsky AS, Liu M (2000) Production of tumour necrosis factor alpha by primary cultured rat alveolar epithelial cells. Cytokine 12(6):644–654. https://doi.org/10.1006/cyto.1999.0656

Meduri GU, Kohler G, Headley S, Tolley E, Stentz F, Postlethwaite A (1995) Inflammatory cytokines in the BAL of patients with ARDS. Persistent elevation over time predicts poor outcome. Chest 108(5):1303–1314

Mick VE, Itani OA, Loftus RW, Husted RF, Schmidt TJ, Thomas CP (2001) The alpha-subunit of the epithelial sodium channel is an aldosterone-induced transcript in mammalian collecting ducts, and this transcriptional response is mediated via distinct cis-elements in the 5′-flanking region of the gene. Mol Endocrinol 15(4):575–588

Mihailidou AS, Tzakos AG, Ashton AW (2019) Non-genomic effects of aldosterone. Vitam Horm 109:133–149. https://doi.org/10.1016/bs.vh.2018.12.001

Mimnaugh EG, Bonvini P, Neckers L (1999) The measurement of ubiquitin and ubiquitinated proteins. Electrophoresis 20(2):418–428

Minshall RD, Sessa WC, Stan RV, Anderson RG, Malik AB (2003) Caveolin regulation of endothelial function. Am J Physiol Lung Cell Mol Physiol 285(6):L1179–L1183

Mironova E, Bugaj V, Roos KP, Kohan DE, Stockand JD (2012) Aldosterone-independent regulation of the epithelial Na+ channel (ENaC) by vasopressin in adrenalectomized mice. Proc Natl Acad Sci USA 109(25):10095–10100

Mironova E, Boiko N, Bugaj V, Kucher V, Stockand JD (2015) Regulation of Na+ excretion and arterial blood pressure by purinergic signalling intrinsic to the distal nephron: consequences and mechanisms. Acta Physiol (Oxf) 213(1):213–221. https://doi.org/10.1111/apha.12372

Mirshahi M, Nicolas C, Mirshahi S, Golestaneh N, d'Hermies F, Agarwal MK (1999) Immunochemical analysis of the sodium channel in rodent and human eye. Exp Eye Res 69(1):21–32. https://doi.org/10.1006/exer.1999.0675

Mitch WE, Goldberg AL (1996) Mechanisms of muscle wasting. The role of the ubiquitin-proteasome pathway. N Engl J Med 335(25):1897–1905

Monder C (1991) Heterogeneity of 11 beta-hydroxysteroid dehydrogenase in rat tissues. J Steroid Biochem Mol Biol 40:533–536

Montgomery DS, Yu L, Ghazi ZM, Thai TL, Al-Khalili O, Ma HP, Eaton DC, Alli AA (2017) ENaC activity is regulated by calpain-2 proteolysis of MARCKS proteins. Am J Physiol Cell Physiol 313(1):C42–c53. https://doi.org/10.1152/ajpcell.00244.2016

Morgunkova A, Barlev NA (2006) Lysine methylation goes global. Cell Cycle 5(12):1308–1312

Morris RG, Schafer JA (2002) cAMP increases density of ENaC subunits in the apical membrane of MDCK cells in direct proportion to amiloride-sensitive Na(+) transport. PG – J Gen Physiol 120 (1)

Mueller GM, Maarouf AB, Kinlough CL, Sheng N, Kashlan OB, Okumura S, Luthy S, Kleyman TR, Hughey RP (2010) Cys palmitoylation of the beta subunit modulates gating of the epithelial sodium channel. J Biol Chem 285(40):30453–30462. https://doi.org/10.1074/jbc.M110.151845

Mukherjee A, Mueller GM, Kinlough CL, Sheng N, Wang Z, Mustafa SA, Kashlan OB, Kleyman TR, Hughey RP (2014) Cysteine palmitoylation of the gamma subunit has a dominant role in modulating activity of the epithelial sodium channel. J Biol Chem 289(20):14351–14359. https://doi.org/10.1074/jbc.M113.526020

Mukherjee A, Wang Z, Kinlough CL, Poland PA, Marciszyn AL, Montalbetti N, Carattino MD, Butterworth MB, Kleyman TR, Hughey RP (2017) Specific palmitoyltransferases associate with and activate the epithelial sodium channel. J Biol Chem 292(10):4152–4163. https://doi.org/10.1074/jbc.M117.776146

Mustafa SB, Hernandez TF, Johnson-Pais TL, Kumar PA, Petershack JA, Henson BM, Seidner SR (2019) IL-1 promotes alpha-epithelial Sodium Channel (alpha-ENaC) expression in murine lung epithelial cells: involvement of NF-kappaB. J Cell Commun Signal. https://doi.org/10.1007/s12079-019-00533-7

Myerburg MM, McKenna EE, McKenna EE, Peters KW, Frizzell RA, Kleyman TR, Pilewski JM (2006) Airway surface liquid volume regulates ENaC by altering the serine protease-protease inhibitor balance: a mechanism for sodium hyperabsorption in cystic fibrosis. J Biol Chem 281 (38):27942–27949. https://doi.org/10.1074/jbc.M606449200

Myerburg MM, Harvey PR, Heidrich EM, Pilewski JM, Butterworth MB (2010) Acute regulation of the epithelial sodium channel in airway epithelia by proteases and trafficking. Am J Respir Cell Mol Biol 43(6):712–719

Nagel W, Dorge A (1970) Effect of Amiloride on sodium transport of frog skin. I. Action on intracellular sodium content. Pflugers Arch 317(1):84–92

Nakano D, Pollock D (2012) New concepts in endothelin control of sodium balance. Clin Exp Pharmacol Physiol 39(1):104–110. https://doi.org/10.1111/j.1440-1681.2011.05517.x

Naray-Fejes-Toth A, Watlington CO, Fejes-Toth G (1991) 11b-hydroxysteroid dehydrogenase activity in the renal target cells of aldosterone. Endocrinology 129:17–21

Naray-Fejes-Toth A, Helms MN, Stokes JB, Fejes-Toth G (2004) Regulation of sodium transport in mammalian collecting duct cells by aldosterone-induced kinase, SGK1: structure/function studies. Mol Cell Endocrinol 217(1-2):197–202. https://doi.org/10.1016/j.mce.2003.10.043

Narikiyo T, Kitamura K, Adachi M, Miyoshi T, Iwashita K, Shiraishi N, Nonoguchi H, Chen LM, Chai KX, Chao J, Tomita K (2002) Regulation of prostasin by aldosterone in the kidney. J Clin Invest 109(3):401–408. https://doi.org/10.1172/JCI13229

Newsholme P, Cruzat VF, Keane KN, Carlessi R, de Bittencourt PI Jr (2016) Molecular mechanisms of ROS production and oxidative stress in diabetes. Biochem J 473(24):4527–4550. https://doi.org/10.1042/bcj20160503c

Nicco C, Wittner M, DiStefano A, Jounier S, Bankir L, Bouby N (2001) Chronic exposure to vasopressin upregulates ENaC and sodium transport in the rat renal collecting duct and lung. Hypertension 38(5):1143–1149

Nicod M, Michlig S, Flahaut M, Salinas M, Fowler JN, Horisberger JD, Rossier BC, Firsov D (2002) A novel vasopressin-induced transcript promotes MAP kinase activation and ENaC downregulation. EMBO J 21(19):5109–5117

Nielsen VG, Baird MS, Chen L, Matalon S (2000) DETANONOate, a nitric oxide donor, decreases amiloride-sensitive alveolar fluid clearance in rabbits. Am J Respir Crit Care Med 161(4 Pt 1):1154–1160

Noreng S, Bharadwaj A, Posert R, Yoshioka C, Baconguis I (2018) Structure of the human epithelial sodium channel by cryo-electron microscopy. elife 7. https://doi.org/10.7554/eLife.39340

Novalija E, Kevin LG, Camara AK, Bosnjak ZJ, Kampine JP, Stowe DF (2003) Reactive oxygen species precede the epsilon isoform of protein kinase C in the anesthetic preconditioning signaling cascade. Anesthesiology 99(2):421–428. https://doi.org/10.1097/00000542-200308000-00024

Oberfeld B, Ruffieux-Daidie D, Vitagliano JJ, Pos KM, Verrey F, Staub O (2011) Ubiquitin-specific protease 2-45 (Usp2-45) binds to epithelial Na+ channel (ENaC)-ubiquitylating enzyme Nedd4-2. Am J Physiol Renal Physiol 301(1):F189–F196

Oberleithner H, Riethmuller C, Schillers H, MacGregor GA, de Wardener HE, Hausberg M (2007) Plasma sodium stiffens vascular endothelium and reduces nitric oxide release. Proc Natl Acad Sci USA 104(41):16281–16286. https://doi.org/10.1073/pnas.0707791104

O'Brodovich H, Yang P, Gandhi S, Otulakowski G (2008) Amiloride-insensitive Na+ and fluid absorption in the mammalian distal lung. Am J Physiol Lung Cell Mol Physiol 294(3):L401–L408. https://doi.org/10.1152/ajplung.00431.2007

Ohara A, Matsunaga H, Eaton DC (1993) G protein activation inhibits amiloride-blockable highly selective sodium channels in A6 cells. Am J Phys 264:C352–C360

Ohbuchi T, Takenaga F, Hohchi N, Wakasugi T, Ueta Y, Suzuki H (2014) Possible contribution of pannexin-1 to ATP release in human upper airway epithelia. Paper presented at the Physiological reports

Ohno S, Nishizuka Y (2002) Protein kinase C isotypes and their specific functions: prologue. J Biochem 132(4):509–511. https://doi.org/10.1093/oxfordjournals.jbchem.a003249

Okada SF, Ribeiro CM, Sesma JI, Seminario-Vidal L, Abdullah LH, van Heusden C, Lazarowski ER, Boucher RC (2013) Inflammation promotes airway epithelial ATP release via calcium-dependent vesicular pathways. Paper presented at the American Journal of Respiratory Cell and Molecular Biology

Olesen HV, Pressler T, Hjelte L, Mared L, Lindblad A, Knudsen PK, Laerum BN, Johannesson M (2010) Gender differences in the Scandinavian cystic fibrosis population. Paper presented at the Pediatric pulmonology

O'Neill H, Lebeck J, Collins PB, Kwon TH, Frokiaer J, Nielsen S (2008) Aldosterone-mediated apical targeting of ENaC subunits is blunted in rats with streptozotocin-induced diabetes mellitus. Nephrol Dial Transplant 23(5):1546–1555. https://doi.org/10.1093/ndt/gfm814

Orbeli AL (1910) The electrical properties of frog skin. Zeitschrift fuer Biologie 59:329–339

Orce GG, Castillo GA, Margolius HS (1980) Inhibition of short-circuit current in toad urinary bladder by inhibitors of glandular kallikrein. Am J Phys 239(5):F459–F465. https://doi.org/10.1152/ajprenal.1980.239.5.F459

Organization WH (2008) Update: WHO-confirmed human cases of avian influenza A (H5N1) infection, November 2003–May 2008. Paper presented at the Releve epidemiologique hebdomadaire/Section d'hygiene du Secretariat de la Societe des Nations = Weekly epidemiological record/Health Section of the Secretariat of the League of Nations. Wkly Epidemiol Rec

Orth SR, Amann K, Gehlen F, Unger L, Wagner J, Raschack M, Ritz E (2000) Adult human mesangial cells (HMCs) express endothelin-B-receptors which mediate endothelin-1-induced cell growth. J Cardiovasc Pharmacol 36(5 Suppl 1):S232–S237. https://doi.org/10.1097/00005344-200036051-00069

Padilla J, Woodford ML, Lastra-Gonzalez G, Martinez-Diaz V, Fujie S, Yang Y, Lising AMC, Ramirez-Perez FI, Aroor AR, Morales-Quinones M, Ghiarone T, Whaley-Connell A, Martinez-Lemus LA, Hill MA, Manrique-Acevedo C (2019) Sexual dimorphism in obesity-associated endothelial ENaC activity and stiffening in mice. Endocrinology 160(12):2918–2928. https://doi.org/10.1210/en.2019-00483

Palmer LG (1982) Ion selectivity of the apical membrane Na channel in the toad urinary bladder. J Membr Biol 67:91–98

Palmer LG (1984) Voltage-dependent block by amiloride and other monovalent cations of apical Na channels in the toad urinary bladder. J Membr Biol 80:153–165

Palmer LG (1985) Modulation of apical Na permeability of the toad urinary bladder by intracellular Na, Ca, and H. J Membr Biol 83:57–69

Palmer LG, Frindt G (1986) Amiloride-sensitive Na channels from the apical membrane of the rat cortical collecting tubule. Proc Natl Acad Sci USA 83(8):2767–2770

Palmer LG, Frindt G (1987) Effects of cell Ca and pH on Na channels from rat cortical collecting tubule. Am J Phys 253(2 Pt 2):F333–F339

Palmer LG, Frindt G (1988) Conductance and gating of epithelial Na channels from rat cortical collecting tubule. Effects of luminal Na and Li. J Gen Physiol 92(1):121–138

Palmer LG, Sackin H, Frindt G (1998) Regulation of Na^+ channels by luminal Na^+ in rat cortical collecting tubule. J Physiol 509(Pt 1):151–162

Passero CJ, Mueller GM, Rondon-Berrios H, Tofovic SP, Hughey RP, Kleyman TR (2008) Plasmin activates epithelial Na+ channels by cleaving the gamma subunit. J Biol Chem 283(52):36586–36591. https://doi.org/10.1074/jbc.M805676200

Passero CJ, Mueller GM, Myerburg MM, Carattino MD, Hughey RP, Kleyman TR (2012) TMPRSS4-dependent activation of the epithelial sodium channel requires cleavage of the gamma-subunit distal to the furin cleavage site. Am J Physiol Renal Physiol 302(1):F1–F8. https://doi.org/10.1152/ajprenal.00330.2011

Patel AB, Chao J, Palmer LG (2012) Tissue kallikrein activation of the epithelial Na channel. Am J Physiol Renal Physiol 303(4):F540–F550. https://doi.org/10.1152/ajprenal.00133.2012

Pavlov TS, Ilatovskaya DV, Levchenko V, Li L, Ecelbarger CM, Staruschenko A (2013) Regulation of ENaC in mice lacking renal insulin receptors in the collecting duct. FASEB J 27(7):2723–2732. https://doi.org/10.1096/fj.12-223792

Pearce D (2001) The role of SGK1 in hormone-regulated sodium transport. Trends Endocrinol Metab 12(8):341–347. https://doi.org/10.1016/s1043-2760(01)00439-8
Pearce D, Bhargava A, Cole TJ (2003) Aldosterone: Its receptor, target genes, and actions, vol 66. Vitamins and hormones-advances in research and applications
Pearce D, Soundararajan R, Trimpert C, Kashlan OB, Deen PM, Kohan DE (2015) Collecting duct principal cell transport processes and their regulation. Clin J Am Soc Nephrol 10(1):135–146. https://doi.org/10.2215/cjn.05760513
Pech V, Pham TD, Hong S, Weinstein AM, Spencer KB, Duke BJ, Walp E, Kim YH, Sutliff RL, Bao HF, Eaton DC, Wall SM (2010) Pendrin modulates ENaC function by changing luminal HCO3−. J Am Soc Nephrol 21(11):1928–1941. https://doi.org/10.1681/ASN.2009121257
Perez FR, Venegas F, Gonzalez M, Andres S, Vallejos C, Riquelme G, Sierralta J, Michea L (2009) Endothelial epithelial sodium channel inhibition activates endothelial nitric oxide synthase via phosphoinositide 3-kinase/Akt in small-diameter mesenteric arteries. Hypertension 53 (6):1000–1007. https://doi.org/10.1161/HYPERTENSIONAHA.108.128520
Pericone CD, Park S, Imlay JA, Weiser JN (2003) Factors contributing to hydrogen peroxide resistance in *Streptococcus pneumoniae* include pyruvate oxidase (SpxB) and avoidance of the toxic effects of the fenton reaction. J Bacteriol 185(23):6815–6825. https://doi.org/10.1128/jb.185.23.6815-6825.2003
Peters KW, Qi J, Johnson JP, Watkins SC, Frizzell RA (2001) Role of snare proteins in CFTR and ENaC trafficking. Pflugers Arch 443(Suppl 1):S65–S69
Peters DM, Vadasz I, Wujak L, Wygrecka M, Olschewski A, Becker C, Herold S, Papp R, Mayer K, Rummel S, Brandes RP, Gunther A, Waldegger S, Eickelberg O, Seeger W, Morty RE (2014) TGF-beta directs trafficking of the epithelial sodium channel ENaC which has implications for ion and fluid transport in acute lung injury. Proc Natl Acad Sci USA 111(3):E374–E383. https://doi.org/10.1073/pnas.1306798111
Peti-Peterdi J, Warnock DG, Bell PD (2002) Angiotensin II directly stimulates ENaC activity in the cortical collecting duct via AT(1) receptors. J Am Soc Nephrol 13(5):1131–1135
Picard N, Eladari D, El Moghrabi S, Planes C, Bourgeois S, Houillier P, Wang Q, Burnier M, Deschenes G, Knepper MA, Meneton P, Chambrey R (2008) Defective ENaC processing and function in tissue kallikrein-deficient mice. J Biol Chem 283(8):4602–4611. https://doi.org/10.1074/jbc.M705664200
Pickart CM (2000a) Ubiquitin biology: an old dog learns an old trick. Nat Cell Biol 2(8):E139–E141
Pickart CM (2000b) Ubiquitin in chains. Trends Biochem Sci 25(11):544–548
Pickart CM, VanDemark AP (2000) Opening doors into the proteasome. Nat Struct Biol 7 (11):999–1001
Piper RC, Luzio JP (2007) Ubiquitin-dependent sorting of integral membrane proteins for degradation in lysosomes. Curr Opin Cell Biol 19(4):459–465
Pittet JF, Griffiths MJ, Geiser T, Kaminski N, Dalton SL, Huang X, Brown LA, Gotwals PJ, Koteliansky VE, Matthay MA, Sheppard D (2001) TGF-beta is a critical mediator of acute lung injury. J Clin Invest 107(12):1537–1544. https://doi.org/10.1172/JCI11963
Planes C, Randrianarison NH, Charles RP, Frateschi S, Cluzeaud F, Vuagniaux G, Soler P, Clerici C, Rossier BC, Hummler E (2010) ENaC-mediated alveolar fluid clearance and lung fluid balance depend on the channel-activating protease 1. EMBO Mol Med 2(1):26–37. https://doi.org/10.1002/emmm.200900050
Pochynyuk O, Staruschenko A, Tong Q, Medina J, Stockand JD (2005) Identification of a functional phosphatidylinositol 3,4,5-trisphosphate binding site in the epithelial Na+ channel. J Biol Chem 280(45):37565–37571. https://doi.org/10.1074/jbc.M509071200
Pochynyuk O, Tong Q, Staruschenko A, Ma HP, Stockand JD (2006) Regulation of the epithelial Na^+ channel (ENaC) by phosphatidylinositides. Am J Physiol Renal Physiol 290(5):F949–F957
Pochynyuk O, Tong Q, Medina J, Vandewalle A, Staruschenko A, Bugaj V, Stockand JD (2007a) Molecular determinants of PI(4,5)P2 and PI(3,4,5)P3 regulation of the epithelial Na+ channel. J Gen Physiol 130(4):399–413. https://doi.org/10.1085/jgp.200709800

Pochynyuk O, Tong Q, Staruschenko A, Stockand JD (2007b) Binding and direct activation of the epithelial Na+ channel (ENaC) by phosphatidylinositides. J Physiol 580(Pt 2):365–372. https://doi.org/10.1113/jphysiol.2006.127449
Pochynyuk O, Bugaj V, Rieg T, Insel PA, Mironova E, Vallon V, Stockand JD (2008a) Paracrine regulation of the epithelial Na+ channel in the mammalian collecting duct by purinergic P2Y2 receptor tone. J Biol Chem 283(52):36599–36607. https://doi.org/10.1074/jbc.M807129200
Pochynyuk O, Bugaj V, Stockand JD (2008b) Physiologic regulation of the epithelial sodium channel by phosphatidylinositides. Curr Opin Nephrol Hypertens 17(5):533–540. https://doi.org/10.1097/MNH.0b013e328308fff3
Pochynyuk O, Bugaj V, Vandewalle A, Stockand JD (2008c) Purinergic control of apical plasma membrane PI(4,5)P2 levels sets ENaC activity in principal cells. Am J Physiol Renal Physiol 294(1):F38–F46. https://doi.org/10.1152/ajprenal.00403.2007
Pochynyuk O, Rieg T, Bugaj V, Schroth J, Fridman A, Boss GR, Insel PA, Stockand JD, Vallon V (2010) Dietary Na+ inhibits the open probability of the epithelial sodium channel in the kidney by enhancing apical P2Y2-receptor tone. FASEB J 24(6):2056–2065. https://doi.org/10.1096/fj.09-151506
Pollock JS, Pollock DM (2011) Endothelin, nitric oxide, and reactive oxygen species in diabetic kidney disease. Contrib Nephrol 172:149–159. https://doi.org/10.1159/000329054
Poulsen AN, Klausen TL, Pedersen PS, Willumsen NJ, Frederiksen O (2005) Regulation of ion transport via apical purinergic receptors in intact rabbit airway epithelium. Paper presented at the Pflugers Archiv: European Journal of Physiology
Prince LS, Welsh MJ (1998) Cell surface expression and biosynthesis of epithelial Na+ channels. Biochem J 336(Pt 3):705–710
Prince LS, Welsh MJ (1999) Effect of subunit composition and Liddle's syndrome mutations on biosynthesis of ENaC. Am J Phys 276(6 Pt 1):C1346–C1351
Pugin J, Verghese G, Widmer MC, Matthay MA (1999) The alveolar space is the site of intense inflammatory and profibrotic reactions in the early phase of acute respiratory distress syndrome. Crit Care Med 27(2):304–312
Qi J, Peters KW, Liu C, Wang JM, Edinger RS, Johnson JP, Watkins SC, Frizzell RA (1999) Regulation of the amiloride-sensitive epithelial sodium channel by syntaxin 1A. J Biol Chem 274(43):30345–30348
Qi W, Li H, Cai XH, Gu JQ, Meng J, Xie HQ, Zhang JL, Chen J, Jin XG, Tang Q, Hao Y, Gao Y, Wen AQ, Xue XY, Gao Smith F, Jin SW (2015) Lipoxin A4 activates alveolar epithelial sodium channel gamma via the microRNA-21/PTEN/AKT pathway in lipopolysaccharide-induced inflammatory lung injury. Lab Investig 95(11):1258–1268. https://doi.org/10.1038/labinvest.2015.109
Qin K, Zhong X, Wang D (2016) MicroRNA-7-5p regulates human alveolar epithelial sodium channels by targeting the mTORC2/SGK-1 signaling pathway. Exp Lung Res 42(5):237–244. https://doi.org/10.1080/01902148.2016.1197347
Racke K, Juergens UR, Matthiesen S (2006) Control by cholinergic mechanisms. Eur J Pharmacol 533(1–3):57–68. https://doi.org/10.1016/j.ejphar.2005.12.050
Rafii B, Tanswell AK, Otulakowski G, Pitkanen O, Belcastro-Taylor R, O'Brodovich H (1998) O2-induced ENaC expression is associated with NF-kappaB activation and blocked by superoxide scavenger. Am J Phys 275(4 Pt 1):L764–L770
Raiborg C, Stenmark H (2009) The ESCRT machinery in endosomal sorting of ubiquitylated membrane proteins. Nature 458(7237):445–452
Ransford GA, Fregien N, Qiu F, Dahl G, Conner GE, Salathe M (2009) Pannexin 1 contributes to ATP release in airway epithelia. Paper presented at the American Journal of Respiratory Cell and Molecular Biology
Rauh R, Hoerner C, Korbmacher C (2017) deltabetagamma-ENaC is inhibited by CFTR but stimulated by cAMP in *Xenopus laevis* oocytes. Am J Physiol Lung Cell Mol Physiol 312(2):L277–l287. https://doi.org/10.1152/ajplung.00375.2016

Ray EC (2018) ENaC blockade in proteinuria-associated extracellular fluid volume overload – effective but risky. Physiol Rep 6(17):e13835. https://doi.org/10.14814/phy2.13835

Ray EC, Kleyman TR (2015) Cutting it out: ENaC processing in the human nephron. J Am Soc Nephrol 26(1):1–3. https://doi.org/10.1681/asn.2014060618

Reifenberger MS, Yu L, Bao HF, Duke BJ, Liu BC, Ma HP, Alli AA, Eaton DC, Alli AA (2014) Cytochalasin E alters the cytoskeleton and decreases ENaC activity in Xenopus 2F3 cells. Am J Physiol Renal Physiol 307(1):F86–F95. https://doi.org/10.1152/ajprenal.00251.2013

Renard S, Lingueglia E, Voilley N, Lazdunski M, Barbry P (1994) Biochemical analysis of the membrane topology of the amiloride-sensitive Na+ channel. J Biol Chem 269 (17):12981–12986

Rezaiguia S, Garat C, Delclaux C, Meignan M, Fleury J, Legrand P, Matthay MA, Jayr C (1997) Acute bacterial pneumonia in rats increases alveolar epithelial fluid clearance by a tumor necrosis factor-alpha-dependent mechanism. J Clin Invest 99(2):325–335. https://doi.org/10.1172/JCI119161

Riazi S, Khan O, Tiwari S, Hu X, Ecelbarger CA (2006) Rosiglitazone regulates ENaC and Na-K-2Cl cotransporter (NKCC2) abundance in the obese Zucker rat. Am J Nephrol 26(3):245–257. https://doi.org/10.1159/000093783

Robins GG, Maclennan KA, Boot-Handford RP, Sandle GI (2013) Rapid stimulation of human renal ENaC by cAMP in *Xenopus laevis* oocytes. J Physiol Biochem 69(3):419–427. https://doi.org/10.1007/s13105-012-0223-6

Rokaw MD, Wang JM, Edinger RS, Weisz OA, Hui D, Middleton P, Shlyonsky V, Berdiev BK, Ismailov II, Eaton DC, Benos DJ, Johnson JP (1998) Carboxylmethylation of the b subunit of xENaC regulates channel activity. J Biol Chem 273:28746–28784

Roos KP, Bugaj V, Mironova E, Stockand JD, Ramkumar N, Rees S, Kohan DE (2013) Adenylyl cyclase VI mediates vasopressin-stimulated ENaC activity. J Am Soc Nephrol 24(2):218–227. https://doi.org/10.1681/ASN.2012050449

Rosenfeld MG, Marcantonio EE, Hakimi J, Ort VM, Atkinson PH, Sabatini D, Kreibich G (1984) Biosynthesis and processing of ribophorins in the endoplasmic reticulum. J Cell Biol 99 (3):1076–1082

Rossier BC (1997) Lose salt and gain a friend! A tribute to Gerhard Giebisch. Wien Klin Wochenschr 109(12–13):504–506

Rossier BC, Canessa CM, Schild L, Horisberger JD (1994) Epithelial sodium channels. Curr Opin Nephrol Hypertens 3(5):487–496

Rossier BC, Pradervand S, Schild L, Hummler E (2002) Epithelial sodium channel and the control of sodium balance: interaction between genetic and environmental factors. Annu Rev Physiol 64:877–897. https://doi.org/10.1146/annurev.physiol.64.082101.143243

Rotin D, Kumar S (2009) Physiological functions of the HECT family of ubiquitin ligases. !Lost Data 10(6):398–409. https://doi.org/10.1038/nrm2690

Rotin D, Staub O, Haguenauer-Tsapis R (2000) Ubiquitination and endocytosis of plasma membrane proteins: role of Nedd4/Rsp5p family of ubiquitin-protein ligases. J Membr Biol 176 (1):1–17. https://doi.org/10.1007/s00232001079

Rotin D, Kanelis V, Schild L (2001) Trafficking and cell surface stability of ENaC. Am J Physiol (Renal, Fluid Electrolyte Physiol) 281(3):F391–F399

Roux J, Kawakatsu H, Gartland B, Pespeni M, Sheppard D, Matthay MA, Canessa CM, Pittet JF (2005) Interleukin-1beta decreases expression of the epithelial sodium channel alpha-subunit in alveolar epithelial cells via a p38 MAPK-dependent signaling pathway. J Biol Chem 280 (19):18579–18589. https://doi.org/10.1074/jbc.M410561200

Ruckes-Nilges C, Lindemann H, Klimek T, Glanz H, Weber WM (2000) Nitric oxide has no beneficial effects on ion transport defects in cystic fibrosis human nasal epithelium. Pflugers Arch 441(1):133–137

Ruffieux-Daidie D, Poirot O, Boulkroun S, Verrey F, Kellenberger S, Staub O (2008) Deubiquitylation regulates activation and proteolytic cleavage of ENaC. J Am Soc Nephrol 19(11):2170–2180

Samaha FF, Rubenstein RC, Yan W, Ramkumar M, Levy DI, Ahn YJ, Sheng S, Kleyman TR (2004) Functional polymorphism in the carboxyl terminus of the alpha-subunit of the human epithelial sodium channel. J Biol Chem 279(23):23900–23907. https://doi.org/10.1074/jbc.M401941200
Sanchez-Perez A, Kumar S, Cook DI (2007) GRK2 interacts with and phosphorylates Nedd4 and Nedd4-2. Biochem Biophys Res Commun 359(3):611–615. https://doi.org/10.1016/j.bbrc.2007.05.134
Satlin LM, Sheng S, Woda CB, Kleyman TR (2001) Epithelial Na(+) channels are regulated by flow. Am J Physiol Renal Physiol 280(6):F1010–F1018
Saxena S, Quick MW, Tousson A, Oh Y, Warnock DG (1999) Interaction of syntaxins with the amiloride-sensitive epithelial sodium channel. J Biol Chem 274(30):20812–20817
Saxena SK, George CM, Pinskiy V, McConnell B (2006) Epithelial sodium channel is regulated by SNAP-23/syntaxin 1A interplay. Biochem Biophys Res Commun 343(4):1279–1285
Saxena SK, Singh M, Kaur S, George C (2007) Distinct domain-dependent effect of syntaxin1A on amiloride-sensitive sodium channel (ENaC) currents in HT-29 colonic epithelial cells. Int J Biol Sci 3(1):47–56
Sayegh R, Auerbach SD, Li X, Loftus RW, Husted RF, Stokes JB, Thomas CP (1999) Glucocorticoid induction of epithelial sodium channel expression in lung and renal epithelia occurs via trans-activation of a hormone response element in the 5′-flanking region of the human epithelial sodium channel alpha subunit gene. J Biol Chem 274(18):12431–12437. https://doi.org/10.1074/jbc.274.18.12431
Schild L, Kellenberger S (2001) Structure function relationships of ENaC and its role in sodium handling. Adv Exp Med Biol 502:305–314
Schild L, Schneeberger E, Gautschi I, Firsov D (1997) Identification of amino acid residues in the alpha, beta, and gamma subunits of the epithelial sodium channel (ENaC) involved in amiloride block and ion permeation. J Gen Physiol 109(1):15–26
Schneider MP, Ge Y, Pollock DM, Pollock JS, Kohan DE (2008) Collecting duct-derived endothelin regulates arterial pressure and Na excretion via nitric oxide. Hypertension 51 (6):1605–1610. https://doi.org/10.1161/hypertensionaha.107.108126
Schwab M, Lupescu A, Mota M, Mota E, Frey A, Simon P, Mertens PR, Floege J, Luft F, Asante-Poku S, Schaeffeler E, Lang F (2008) Association of SGK1 gene polymorphisms with type 2 diabetes. Cell Physiol Biochem 21(1-3):151–160. https://doi.org/10.1159/000113757
Schwagerus E, Sladek S, Buckley ST, Armas-Capote N, Alvarez de la Rosa D, Harvey BJ, Fischer H, Illek B, Huwer H, Schneider-Daum N, Lehr CM, Ehrhardt C (2015) Expression and function of the epithelial sodium channel delta-subunit in human respiratory epithelial cells in vitro. Pflugers Arch 467(11):2257–2273. https://doi.org/10.1007/s00424-015-1693-5
Schwiebert EM, Zsembery A (2003) Extracellular ATP as a signaling molecule for epithelial cells. Biochim Biophys Acta 1615(1–2):7–32
Sedeek M, Nasrallah R, Touyz RM, Hebert RL (2013) NADPH oxidases, reactive oxygen species, and the kidney: friend and foe. J Am Soc Nephrol 24(10):1512–1518. https://doi.org/10.1681/asn.2012111112
Seo JH, Ahn Y, Lee SR, Yeol Yeo C, Chung Hur K (2005) The major target of the endogenously generated reactive oxygen species in response to insulin stimulation is phosphatase and tensin homolog and not phosphoinositide-3 kinase (PI-3 kinase) in the PI-3 kinase/Akt pathway. Mol Biol Cell 16(1):348–357. https://doi.org/10.1091/mbc.e04-05-0369
Serfling RE, Sherman IL, Houseworth WJ (1967) Excess pneumonia-influenza mortality by age and sex in three major influenza A2 epidemics, United States, 1957–58, 1960 and 1963. Paper presented at the American Journal of Epidemiology
Sheng S, Li J, Mcnulty KA, Avery D, Kleyman TR (2000) Characterization of the selectivity filter of the epithelial sodium channel. J Biol Chem 275(12):8572–8581
Sheng S, Li J, Mcnulty KA, Kieber-Emmons T, Kleyman TR (2001) Epithelial sodium channel pore region. structure and role in gating. J Biol Chem 276(2):1326–1334

Sheng S, Perry CJ, Kleyman TR (2002) External nickel inhibits epithelial sodium channel by binding to histidine residues within the extracellular domains of alpha and gamma subunits and reducing channel open probability. J Biol Chem 277(51):50098–50111. https://doi.org/10.1074/jbc.M209975200

Sheng S, Bruns JB, Kleyman TR (2004a) Extracellular histidine residues crucial for Na^+ self-inhibition of epithelial Na^+ channels. J Biol Chem 279(11):9743–9749. https://doi.org/10.1074/jbc.M311952200

Sheng S, Perry CJ, Kleyman TR (2004b) Extracellular Zn2+ activates epithelial Na+ channels by eliminating Na+ self-inhibition. J Biol Chem 279(30):31687–31696. https://doi.org/10.1074/jbc.M405224200

Sheng S, Perry CJ, Kashlan OB, Kleyman TR (2005) Side chain orientation of residues lining the selectivity filter of epithelial Na+ channels. J Biol Chem 280(9):8513–8522. https://doi.org/10.1074/jbc.M413880200

Sheng S, Carattino MD, Bruns JB, Hughey RP, Kleyman TR (2006) Furin cleavage activates the epithelial Na+ channel by relieving Na+ self-inhibition. Am J Physiol Renal Physiol 290(6):F1488–F1496

Sheng S, Maarouf AB, Bruns JB, Hughey RP, Kleyman TR (2007) Functional role of extracellular loop cysteine residues of the epithelial Na+ channel in Na+ self-inhibition. J Biol Chem 282(28):20180–20190. https://doi.org/10.1074/jbc.M611761200

Shi H, Asher C, Chigaev A, Yung Y, Reuveny E, Seger R, Garty H (2002a) Interactions of beta and gamma ENaC with Nedd4 can be facilitated by an ERK-mediated phosphorylation. J Biol Chem 277(16):13539–13547

Shi H, Asher C, Yung Y, Kligman L, Reuveny E, Seger R, Garty H (2002b) Casein kinase 2 specifically binds to and phosphorylates the carboxy termini of ENaC subunits. Eur J Biochem 269(18):4551–4558

Shi S, Ghosh DD, Okumura S, Carattino MD, Kashlan OB, Sheng S, Kleyman TR (2011) Base of the thumb domain modulates epithelial sodium channel gating. J Biol Chem 286(17):14753–14761

Shi S, Blobner BM, Kashlan OB, Kleyman TR (2012a) Extracellular finger domain modulates the response of the epithelial sodium channel to shear stress. J Biol Chem 287(19):15439–15444

Shi S, Carattino MD, Kleyman TR (2012b) Role of the wrist domain in the response of the epithelial sodium channel to external stimuli. J Biol Chem 287(53):44027–44035. https://doi.org/10.1074/jbc.M112.421743

Shi S, Carattino MD, Hughey RP, Kleyman TR (2013) ENaC regulation by proteases and shear stress. Curr Mol Pharmacol 6(1):28–34

Shibukawa Y, Takahashi M, Laffont I, Honke K, Taniguchi N (2003) Down-regulation of hydrogen peroxide-induced PKC delta activation in N-acetylglucosaminyltransferase III-transfected HeLaS3 cells. J Biol Chem 278(5):3197–3203. https://doi.org/10.1074/jbc.M207870200

Shimkets RA, Lifton RP, Canessa CM (1997) The activity of the epithelial sodium channel is regulated by clathrin-mediated endocytosis. J Biol Chem 272(41):25537–25541

Shimkets RA, Lifton R, Canessa CM (1998) In vivo phosphorylation of the epithelial sodium channel. Proc Natl Acad Sci USA 95:3301–3305

Shlyonsky VG, Mies F, Sariban-Sohraby S (2003) Epithelial sodium channel activity in detergent-resistant membrane microdomains. Am J Physiol Renal Physiol 284(1):F182–F188. https://doi.org/10.1152/ajprenal.00216.2002

Silver RB, Frindt G, Windhager EE, Palmer LG (1993) Feedback regulation of Na channels in rat CCT. I. Effects of inhibition of Na pump. Am J Phys 264:F557–F564

Sipos A, Vargas SL, Toma I, Hanner F, Willecke K, Peti-Peterdi J (2009) Connexin 30 deficiency impairs renal tubular ATP release and pressure natriuresis. Paper presented at the Journal of the American Society of Nephrology: JASN

Smith PR, Saccomani G, Joe EH, Angelides KJ, Benos DJ (1991) Amiloride-sensitive sodium channel is linked to the cytoskeleton in renal epithelial cells. Proc Natl Acad Sci USA 88(16):6971–6975

Snyder PM (2005) Minireview: regulation of epithelial Na+ channel trafficking. Endocrinology 146 (12):5079–5085

Snyder PM (2009) Down-regulating destruction: phosphorylation regulates the E3 ubiquitin ligase Nedd4-2. Sci Signal 2(79):e41. https://doi.org/10.1126/scisignal.279pe41

Snyder PM (2012) Intoxicated Na(+) channels. Focus on "ethanol stimulates epithelial sodium channels by elevating reactive oxygen species". Am J Physiol Cell Physiol 303(11):C1125–C1126. https://doi.org/10.1152/ajpcell.00301.2012

Snyder PM, Olson DR, Bucher DB (1999) A pore segment in DEG/ENaC Na(+) channels. J Biol Chem 274(40):28484–28490

Snyder PM, Olson DR, Thomas BC (2002) Serum and glucocorticoid-regulated kinase modulates Nedd4-2-mediated inhibition of the epithelial Na+ channel. J Biol Chem 277(1):5–8. https://doi.org/10.1074/jbc.C100623200

Snyder PM, Olson DR, Kabra R, Zhou R, Steines JC (2004a) cAMP and serum and glucocorticoid-inducible kinase (SGK) regulate the epithelial Na(+) channel through convergent phosphorylation of Nedd4-2. J Biol Chem 279(44):45753–45758. https://doi.org/10.1074/jbc.M407858200

Snyder PM, Steines JC, Olson DR (2004b) Relative contribution of Nedd4 and Nedd4-2 to ENaC regulation in epithelia determined by RNA interference. J Biol Chem 279(6):5042–5046

Song J, Knepper MA, Verbalis JG, Ecelbarger CA (2003) Increased renal ENaC subunit and sodium transporter abundances in streptozotocin-induced type 1 diabetes. Am J Physiol Renal Physiol 285(6):F1125–F1137. https://doi.org/10.1152/ajprenal.00143.2003

Sorokin A, Staruschenko A (2015) Inhibition of ENaC by endothelin-1. Vitam Horm 98:155–187. https://doi.org/10.1016/bs.vh.2015.01.001

Soundararajan R, Melters D, Shih IC, Wang J, Pearce D (2009) Epithelial sodium channel regulated by differential composition of a signaling complex. Proc Natl Acad Sci USA 106 (19):7804–7809. https://doi.org/10.1073/pnas.0809892106

Soundararajan R, Pearce D, Hughey RP, Kleyman TR (2010a) Role of epithelial sodium channels and their regulators in hypertension. J Biol Chem 285(40):30363–30369. https://doi.org/10.1074/jbc.R110.155341

Soundararajan R, Wang J, Melters D, Pearce D (2010b) Glucocorticoid-induced Leucine zipper 1 stimulates the epithelial sodium channel by regulating serum- and glucocorticoid-induced kinase 1 stability and subcellular localization. J Biol Chem 285(51):39905–39913

Soundararajan R, Lu M, Pearce D (2012a) Organization of the ENaC-regulatory machinery. Crit Rev Biochem Mol Biol 47(4):349–359

Soundararajan R, Pearce D, Ziera T (2012b) The role of the ENaC-regulatory complex in aldosterone-mediated sodium transport. Mol Cell Endocrinol 350(2):242–247

Soundararajan R, Ziera T, Koo E, Ling K, Wang J, Borden SA, Pearce D (2012c) Scaffold protein connector enhancer of kinase suppressor of Ras isoform 3 (CNK3) coordinates assembly of a multiprotein epithelial sodium channel (ENaC)-regulatory complex. J Biol Chem 287 (39):33014–33025

Sowers JR, Habibi J, Aroor AR, Yang Y, Lastra G, Hill MA, Whaley-Connell A, Jaisser F, Jia G (2019) Epithelial sodium channels in endothelial cells mediate diet-induced endothelium stiffness and impaired vascular relaxation in obese female mice. Metabolism 99:57–66. https://doi.org/10.1016/j.metabol.2019.153946

Speed JS, Fox BM, Johnston JG, Pollock DM (2015) Endothelin and renal ion and water transport. Semin Nephrol 35(2):137–144. https://doi.org/10.1016/j.semnephrol.2015.02.003

Spindler B, Verrey F (1999) Aldosterone action: induction of p21(ras) and fra-2 and transcription-independent decrease in myc, jun, and fos. Am J Phys 276(5 Pt 1):C1154–C1161

Spindler B, Mastroberardino L, Custer M, Verrey F (1997) Characterization of early aldosterone-induced RNAs identified in A6 kidney epithelia. Pflugers Arch 434(3):323–331

Stanley P (2011) Golgi glycosylation. Cold Spring Harb Perspect Biol 3(4)

Staruschenko A, Patel P, Tong Q, Medina JL, Stockand JD (2004) Ras activates the epithelial Na(+) channel through phosphoinositide 3-OH kinase signaling. J Biol Chem 279(36):37771–37778. https://doi.org/10.1074/jbc.M402176200

Staruschenko A, Pochynyuk OM, Tong Q, Stockand JD (2005) Ras couples phosphoinositide 3-OH kinase to the epithelial Na^+ channel. Biochim Biophys Acta 1669(2):108–115

Staruschenko A, Pochynyuk O, Vandewalle A, Bugaj V, Stockand JD (2007) Acute regulation of the epithelial Na+ channel by phosphatidylinositide 3-OH kinase signaling in native collecting duct principal cells. J Am Soc Nephrol 18(6):1652–1661. https://doi.org/10.1681/ASN.2007010020

Staub O, Rotin D (1996) WW domains. Structure 4(5):495–499

Staub O, Rotin D (1997) Regulation of ion transport by protein-protein interaction domains. Curr Opin Nephrol Hypertens 6(5):447–454

Staub O, Dho S, Henry P, Correa J, Ishikawa T, McGlade J, Rotin D (1996) WW domains of Nedd4 bind to the proline-rich PY motifs in the epithelial Na+ channel deleted in Liddle's syndrome. EMBO J 15(10):2371–2380

Staub O, Gautschi I, Ishikawa T, Breitschopf K, Ciechanover A, Schild L, Rotin D (1997a) Regulation of stability and function of the epithelial Na^+ channel (ENaC) by ubiquitination. EMBO J 16(21):6325–6336

Staub O, Yeger H, Plant PJ, Kim H, Ernst SA, Rotin D (1997b) Immunolocalization of the ubiquitin-protein ligase Nedd4 in tissues expressing the epithelial Na+ channel (ENaC). Am J Phys 272(6 Pt 1):C1871–C1880

Staub O, Abriel H, Plant P, Ishikawa T, Kanelis V, Saleki R, Horisberger JD, Schild L, Rotin D (2000) Regulation of the epithelial Na^+ channel by Nedd4 and ubiquitination. Kidney Int 57 (3):809–815

Steinberg SF (2015) Mechanisms for redox-regulation of protein kinase C. Front Pharmacol 6:128. https://doi.org/10.3389/fphar.2015.00128

Stephenson A, Hux J, Tullis E, Austin PC, Corey M, Ray J (2011) Higher risk of hospitalization among females with cystic fibrosis. Paper presented at the Journal of cystic fibrosis: official journal of the European Cystic Fibrosis Society

Sternak M, Bar A, Adamski MG, Mohaissen T, Marczyk B, Kieronska A, Stojak M, Kus K, Tarjus A, Jaisser F, Chlopicki S (2018) The deletion of endothelial sodium channel alpha (alphaENaC) impairs endothelium-dependent vasodilation and endothelial barrier integrity in endotoxemia in vivo. Front Pharmacol 9:178. https://doi.org/10.3389/fphar.2018.00178

Stewart AP, Haerteis S, Diakov A, Korbmacher C, Edwardson JM (2011) Atomic force microscopy reveals the architecture of the epithelial sodium channel (ENaC). J Biol Chem 286 (37):31944–31952

Stockand JD (2010) Vasopressin regulation of renal sodium excretion. Kidney Int 78(9):849–856. https://doi.org/10.1038/ki.2010.276

Stockand JD, Al-Baldawi NF, Al-Khalili OK, Worrell RT, Eaton DC (1999a) S-adenosyl-L-homocysteine hydrolase regulates aldosterone-induced Na+ transport. J Biol Chem 274 (6):3842–3850

Stockand JD, Edinger RS, Al-Baldawi NF, Sariban-Sohraby S, Al-Khalili OK, Eaton DC, Johnson JP (1999b) Isoprenylcysteine-O-carboxyl methyltransferase regulates aldosterone-sensitive Na^+ reabsorption. J Biol Chem 274(38):26912–26916

Stockand JD, Bao HF, Schenck J, Malik B, Middleton P, Schlanger LE, Eaton DC (2000a) Differential effects of protein kinase C on the levels of epithelial Na+ channel subunit proteins. J Biol Chem 275(33):25760–25765. https://doi.org/10.1074/jbc.M003615200

Stockand JD, Edinger RS, Eaton DC, Johnson JP (2000b) Toward understanding the role of methylation in aldosterone-sensitive $Na(^+)$ transport. News Physiol Sci 15:161–165

Stockand JD, Zeltwanger S, Bao HF, Becchetti A, Worrell RT, Eaton DC (2001) S-adenosyl-L-homocysteine hydrolase is necessary for aldosterone-induced activity of epithelial $Na(^+)$ channels. Am J Physiol Cell Physiol 281(3):C773–C785

Stockand JD, Staruschenko A, Pochynyuk O, Booth RE, Silverthorn DU (2008) Insight toward epithelial Na^+ channel mechanism revealed by the acid-sensing ion channel 1 structure. IUBMB Life 60(9):620–628. https://doi.org/10.1002/iub.89

Stockand JD, Mironova E, Bugaj V, Rieg T, Insel PA, Vallon V, Peti-Peterdi J, Pochynyuk O (2010) Purinergic inhibition of ENaC produces aldosterone escape. Paper presented at the Journal of the American Society of Nephrology: JASN

Stoops EH, Caplan MJ (2014) Trafficking to the apical and basolateral membranes in polarized epithelial cells. J Am Soc Nephrol 25(7):1375–1386

Studer RA, Person E, Robinson-Rechavi M, Rossier BC (2011) Evolution of the epithelial sodium channel and the sodium pump as limiting factors of aldosterone action on sodium transport. Physiol Genomics 43(13):844–854. https://doi.org/10.1152/physiolgenomics.00002.2011

Sugden PH, Clerk A (2006) Oxidative stress and growth-regulating intracellular signaling pathways in cardiac myocytes. Antioxid Redox Signal 8(11–12):2111–2124. https://doi.org/10.1089/ars.2006.8.2111

Sun P, Yue P, Wang WH (2012) Angiotensin II stimulates epithelial sodium channels in the cortical collecting duct of the rat kidney. Am J Physiol Renal Physiol 302(6):F679–F687. https://doi.org/10.1152/ajprenal.00368.2011

Svenningsen P, Bistrup C, Friis UG, Bertog M, Haerteis S, Krueger B, Stubbe J, Jensen ON, Thiesson HC, Uhrenholt TR, Jespersen B, Jensen BL, Korbmacher C, Skott O (2009) Plasmin in nephrotic urine activates the epithelial sodium channel. J Am Soc Nephrol 20(2):299–310. https://doi.org/10.1681/ASN.2008040364

Svenningsen P, Skott O, Jensen BL (2012) Proteinuric diseases with sodium retention: is plasmin the link? Clin Exp Pharmacol Physiol 39(1):117–124. https://doi.org/10.1111/j.1440-1681.2011.05524.x

Svenningsen P, Andersen H, Nielsen LH, Jensen BL (2015) Urinary serine proteases and activation of ENaC in kidney – implications for physiological renal salt handling and hypertensive disorders with albuminuria. Pflugers Arch 467(3):531–542. https://doi.org/10.1007/s00424-014-1661-5

Szule JA, Coorssen JR (2003) Revisiting the role of SNAREs in exocytosis and membrane fusion. Biochim Biophy Acta (BBA) Mol Cell Res 1641(2–3):121–135

Takemura Y, Helms MN, Eaton AF, Self J, Ramosevac S, Jain L, Bao HF, Eaton DC (2013) Cholinergic regulation of epithelial sodium channels in rat alveolar type 2 epithelial cells. Am J Physiol Lung Cell Mol Physiol 304(6):L428–L437. https://doi.org/10.1152/ajplung.00129.2012

Talior I, Yarkoni M, Bashan N, Eldar-Finkelman H (2003) Increased glucose uptake promotes oxidative stress and PKC-delta activation in adipocytes of obese, insulin-resistant mice. Am J Physiol Endocrinol Metab 285(2):E295–E302. https://doi.org/10.1152/ajpendo.00044.2003

Tamarapu Parthasarathy P, Galam L, Huynh B, Yunus A, Abuelenen T, Castillo A, Kollongod Ramanathan G, Cox R Jr, Kolliputi N (2012) MicroRNA 16 modulates epithelial sodium channel in human alveolar epithelial cells. Biochem Biophys Res Commun 426(2):203–208. https://doi.org/10.1016/j.bbrc.2012.08.063

Tan CD, Hobbs C, Sameni M, Sloane BF, Stutts MJ, Tarran R (2014) Cathepsin B contributes to Na + hyperabsorption in cystic fibrosis airway epithelial cultures. J Physiol 592(Pt 23):5251–5268. https://doi.org/10.1113/jphysiol.2013.267286

Taniguchi H, Manenti S (1993) Interaction of myristoylated alanine-rich protein kinase C substrate (MARCKS) with membrane phospholipids. J Biol Chem 268(14):9960–9963

Tarjus A, Maase M, Jeggle P, Martinez-Martinez E, Fassot C, Loufrani L, Henrion D, Hansen PBL, Kusche-Vihrog K, Jaisser F (2017) The endothelial alphaENaC contributes to vascular endothelial function in vivo. PLoS One 12(9):e0185319. https://doi.org/10.1371/journal.pone.0185319

Taruno A, Niisato N, Marunaka Y (2008) Intracellular calcium plays a role as the second messenger of hypotonic stress in gene regulation of SGK1 and ENaC in renal epithelial A6 cells. Am J Physiol Renal Physiol 294(1):F177–F186. https://doi.org/10.1152/ajprenal.00250.2007

Thai TL, Yu L, Eaton DC, Duke BJ, Al-Khalili O, Lam HY, Ma H, Bao HF (2014) Basolateral P2X (4)channels stimulate ENaC activity in Xenopus cortical collecting duct A6 cells. Am J Physiol Renal Physiol 307(7):F806–F813. https://doi.org/10.1152/ajprenal.00350.2013

Thomas CP, Doggett NA, Fisher R, Stokes JB (1996) Genomic organization and the 5′ flanking region of the gamma subunit of the human amiloride-sensitive epithelial sodium channel. J Biol Chem 271(42):26062–26066. https://doi.org/10.1074/jbc.271.42.26062
Thomas CP, Auerbach SD, Zhang C, Stokes JB (1999) The structure of the rat amiloride-sensitive epithelial sodium channel gamma subunit gene and functional analysis of its promoter. Gene 228(1–2):111–122. https://doi.org/10.1016/s0378-1119(99)00016-5
Thomas J, Deetjen P, Ko WH, Jacobi C, Leipziger J (2001) P2Y(2) receptor-mediated inhibition of amiloride-sensitive short circuit current in M-1 mouse cortical collecting duct cells. Paper presented at the The Journal of Membrane Biology
Thomas CP, Zhou J, Liu KZ, Mick VE, MacLaughlin E, Knowles M (2002) Systemic pseudohypoaldosteronism from deletion of the promoter region of the human Beta epithelial na(+) channel subunit. Am J Respir Cell Mol Biol 27(3):314–319
Thomas SV, Kathpalia PP, Rajagopal M, Charlton C, Zhang J, Eaton DC, Helms MN, Pao AC (2011) Epithelial sodium channel regulation by cell surface-associated serum- and glucocorticoid-regulated kinase 1. J Biol Chem 286(37):32074–32085. https://doi.org/10.1074/jbc.M111.278283
Thome UH, Davis IC, Nguyen SV, Shelton BJ, Matalon S (2003) Modulation of sodium transport in fetal alveolar epithelial cells by oxygen and corticosterone. Am J Physiol Lung Cell Mol Physiol 284(2):L376–L385
Thrower JS, Hoffman L, Rechsteiner M, Pickart CM (2000) Recognition of the polyubiquitin proteolytic signal. EMBO J 19(1):94–102
Tillie-Leblond I, Guery BP, Janin A, Leberre R, Just N, Pittet JF, Tonnel AB, Gosset P (2002) Chronic bronchial allergic inflammation increases alveolar liquid clearance by TNF-alpha-dependent mechanism. Am J Physiol Lung Cell Mol Physiol 283(6):L1303–L1309. https://doi.org/10.1152/ajplung.00147.2002
Titze J, Machnik A (2010) Sodium sensing in the interstitium and relationship to hypertension. Curr Opin Nephrol Hypertens 19(4):385–392. https://doi.org/10.1097/MNH.0b013e32833aeb3b
Tiwari S, Nordquist L, Halagappa VK, Ecelbarger CA (2007) Trafficking of ENaC subunits in response to acute insulin in mouse kidney. Am J Physiol Renal Physiol 293(1):F178–F185. https://doi.org/10.1152/ajprenal.00447.2006
Tong Q, Stockand JD (2005) Receptor tyrosine kinases mediate epithelial Na(+) channel inhibition by epidermal growth factor. Am J Physiol Renal Physiol 288(1):F150–F161. https://doi.org/10.1152/ajprenal.00261.2004
Tong Q, Gamper N, Medina JL, Shapiro MS, Stockand JD (2004) Direct activation of the epithelial Na(+) channel by phosphatidylinositol 3,4,5-trisphosphate and phosphatidylinositol 3,4-bisphosphate produced by phosphoinositide 3-OH kinase. J Biol Chem 279(21):22654–22663. https://doi.org/10.1074/jbc.M401004200
Tong J, Nguyen L, Vidal A, Simon SA, Skene JH, McIntosh TJ (2008) Role of GAP-43 in sequestering phosphatidylinositol 4,5-bisphosphate to Raft bilayers. Biophys J 94(1):125–133
Tousson A, Alley CD, Sorscher EJ, Brinkley BR, Benos DJ (1989) Immunochemical localization of amiloride-sensitive sodium channels in sodium-transporting epithelia. J Cell Sci 93 (Pt 2):349–362
Trac D, Liu B, Pao AC, Thomas SV, Park M, Downs CA, Ma HP, Helms MN (2013) Fulvene-5 inhibition of Nadph oxidases attenuates activation of epithelial sodium channels in A6 distal nephron cells. Am J Physiol Renal Physiol 305(7):F995–F1005. https://doi.org/10.1152/ajprenal.00098.2013
Trac PT, Thai TL, Linck V, Zou L, Greenlee M, Yue Q, Al-Khalili O, Alli AA, Eaton AF, Eaton DC (2017) Alveolar nonselective channels are ASIC1a/alpha-ENaC channels and contribute to AFC. Am J Physiol Lung Cell Mol Physiol 312(6):L797–L811. https://doi.org/10.1152/ajplung.00379.2016
Tuna KM, Liu BC, Yue Q, Ghazi ZM, Ma HP, Eaton DC, Alli AA (2019) Mal protein stabilizes luminal membrane PLC-beta3 and negatively regulates ENaC in mouse cortical collecting duct

cells. Am J Physiol Renal Physiol 317(4):F986–F995. https://doi.org/10.1152/ajprenal.00446.2018
Turnheim K (1991) Intrinsic regulation of apical sodium entry in epithelia. Physiol Rev 71:429–445
Unruh ML, Pankratz VS, Demko JE, Ray EC, Hughey RP, Kleyman TR (2017) Trial of amiloride in type 2 diabetes with proteinuria. Kidney Int Rep 2(5):893–904. https://doi.org/10.1016/j.ekir.2017.05.008
Ussing HH (1953) Transport through biological membranes. Annu Rev Physiol 15:1–20. https://doi.org/10.1146/annurev.ph.15.030153.000245
Ussing HH, Zerahn K (1951) Active transport of sodium as the source of electric current in the short-circuited isolated frog skin. Acta Physiol Scand 23(2–3):110–127. https://doi.org/10.1111/j.1748-1716.1951.tb00800.x
Valentijn JA, Fyfe GK, Canessa CM (1998) Biosynthesis and processing of epithelial sodium channels in Xenopus oocytes. J Biol Chem 273(46):30344–30351
Vallet V, Chraibi A, Gaeggeler HP, Horisberger JD, Rossier BC (1997) An epithelial serine protease activates the amiloride-sensitive sodium channel. Nature 389(6651):607–610. https://doi.org/10.1038/39329
Vallon V, Rieg T (2011) Regulation of renal NaCl and water transport by the ATP/UTP/P2Y2 receptor system. Am J Physiol Renal Physiol 301(3):F463–F475. https://doi.org/10.1152/ajprenal.00236.2011
Vallon V, Stockand J, Rieg T (2012) P2Y receptors and kidney function. Wiley Interdiscip Rev Membr Transp Signal 1(6):731–742. https://doi.org/10.1002/wmts.61
Van Beusecum JP, Barbaro NR, McDowell Z, Aden LA, Xiao L, Pandey AK, Itani HA, Himmel LE, Harrison DG, Kirabo A (2019) High salt activates CD11c(+) Antigen-presenting cells via SGK (serum glucocorticoid kinase) 1 to promote renal inflammation and salt-sensitive hypertension. Hypertension 74(3):555–563. https://doi.org/10.1161/hypertensionaha.119.12761
Van IJzendoorn SC, Mostov KE (2000) Connecting apical endocytosis to the intracellular traffic infrastructure in polarized hepatocytes. Gastroenterology 119(6):1791–1794
Van IJzendoorn SC, Maier O, Van Der Wouden JM, Hoekstra D (2000) The subapical compartment and its role in intracellular trafficking and cell polarity. J Cell Physiol 184(2):151–160
van Kerkhof P, Strous GJ (2001) The ubiquitin-proteasome pathway regulates lysosomal degradation of the growth hormone receptor and its ligand. Biochem Soc Trans 29(Pt 4):488–493
van Kerkhof P, Alves dos Santos CM, Sachse M, Klumperman J, Bu G, Strous GJ (2001) Proteasome inhibitors block a late step in lysosomal transport of selected membrane but not soluble proteins. Mol Biol Cell 12(8):2556–2566
van Leeuwen F, Gafken PR, Gottschling DE (2002) Dot1p modulates silencing in yeast by methylation of the nucleosome core. Cell 109(6):745–756
Verrey F (1998) Early aldosterone effects. Exp Nephrol 6(4):294–301
Verrey F, Pearce D, Pfeiffer R, Spindler B, Mastroberardino L, Summa V, Zecevic M (2000) Pleiotropic action of aldosterone in epithelia mediated by transcription and post-transcription mechanisms. Kidney Int 57(4):1277–1282
Verrey F, Fakitsas P, Adam G, Staub O (2008) Early transcriptional control of ENaC (de)-ubiquitylation by aldosterone. Kidney Int 73(6):691–696
Vieira OV, Verkade P, Manninen A, Simons K (2005) FAPP2 is involved in the transport of apical cargo in polarized MDCK cells. J Cell Biol 170(4):521–526
Vieira OV, Gaus K, Verkade P, Fullekrug J, Vaz WLC, Simons K (2006) FAPP2, cilium formation, and compartmentalization of the apical membrane in polarized Madin–Darby canine kidney (MDCK) cells. Proc Natl Acad Sci 103(49):18556–18561
Vuagniaux G, Vallet V, Jaeger NF, Hummler E, Rossier BC (2002) Synergistic activation of ENaC by three membrane-bound channel-activating serine proteases (mCAP1, mCAP2, and mCAP3) and serum- and glucocorticoid-regulated kinase (Sgk1) in Xenopus Oocytes. J Gen Physiol 120 (2):191–201
Wagener BM, Roux J, Carles M, Pittet JF (2015) Synergistic inhibition of beta2-adrenergic receptor-mediated alveolar epithelial fluid transport by interleukin-8 and transforming growth

factor-beta. Anesthesiology 122(5):1084–1092. https://doi.org/10.1097/ALN.0000000000000595

Wakefield LM, Letterio JJ, Chen T, Danielpour D, Allison RS, Pai LH, Denicoff AM, Noone MH, Cowan KH, O'Shaughnessy JA et al (1995) Transforming growth factor-beta1 circulates in normal human plasma and is unchanged in advanced metastatic breast cancer. Clin Cancer Res 1 (1):129–136

Wakida N, Kitamura K, Tuyen DG, Maekawa A, Miyoshi T, Adachi M, Shiraishi N, Ko T, Ha V, Nonoguchi H, Tomita K (2006) Inhibition of prostasin-induced ENaC activities by PN-1 and regulation of PN-1 expression by TGF-beta1 and aldosterone. Kidney Int 70(8):1432–1438. https://doi.org/10.1038/sj.ki.5001787

Waldmann R, Champigny G, Lazdunski M (1995) Functional degenerin-containing chimeras identify residues essential for amiloride-sensitive Na+ channel function. J Biol Chem 270 (20):11735–11737

Wang XF, Chan HC (2000) Adenosine triphosphate induces inhibition of Na(+) absorption in mouse endometrial epithelium: a Ca(2+)-dependent mechanism. Paper presented at the Biol Repro Biol Reprod

Wang HC, Zentner MD, Deng HT, Kim KJ, Wu R, Yang PC, Ann DK (2000) Oxidative stress disrupts glucocorticoid hormone-dependent transcription of the amiloride-sensitive epithelial sodium channel alpha-subunit in lung epithelial cells through ERK-dependent and thioredoxin-sensitive pathways. J Biol Chem 275(12):8600–8609

Wang J, Barbry P, Maiyar AC, Rozansky DJ, Bhargava A, Leong M, Firestone GL, Pearce D (2001a) SGK integrates insulin and mineralocorticoid regulation of epithelial sodium transport. Am J Physiol Renal Physiol 280(2):F303–F313. https://doi.org/10.1152/ajprenal.2001.280.2.F303

Wang XY, Masilamani S, Nielsen J, Kwon TH, Brooks HL, Nielsen S, Knepper MA (2001b) The renal thiazide-sensitive Na-Cl cotransporter as mediator of the aldosterone-escape phenomenon. Paper presented at the The Journal of Clinical Investigation

Wang J, Gambhir A, McLaughlin SG, Murray D (2004) A computational model for the electrostatic sequestration of PI(4,5)P2 by membrane-adsorbed basic peptides. Biophys J 86(4):1969–1986. https://doi.org/10.1016/S0006-3495(04)74260-5

Wang J, Knight ZA, Fiedler D, Williams O, Shokat KM, Pearce D (2008a) Activity of the p110-alpha subunit of phosphatidylinositol-3-kinase is required for activation of epithelial sodium transport. Am J Physiol Renal Physiol 295(3):F843–F850

Wang J, Zhang ZR, Chou CF, Liang YY, Gu Y, Ma HP (2008b) Cyclosporine stimulates the renal epithelial sodium channel by elevating cholesterol. Am J Physiol Renal Physiol. https://doi.org/10.1152/ajprenal.90647.2008

Wang S, Meng F, Mohan S, Champaneri B, Gu Y (2009) Functional ENaC channels expressed in endothelial cells: a new candidate for mediating shear force. Microcirculation 16(3):276–287. https://doi.org/10.1080/10739680802653150

Wang Q, Dai XQ, Li Q, Tuli J, Liang G, Li SS, Chen XZ (2013) Filamin interacts with epithelial sodium channel and inhibits its channel function. J Biol Chem 288(1):264–273. https://doi.org/10.1074/jbc.M112.396408

Wang Q, Song B, Jiang S, Liang C, Chen X, Shi J, Li X, Sun Y, Wu M, Zhao D, Zhang ZR, Ma HP (2015) Hydrogen sulfide prevents advanced glycation end-products induced activation of the epithelial sodium channel. Oxidative Med Cell Longev 2015:976848. https://doi.org/10.1155/2015/976848

Wang QS, Liang C, Niu N, Yang X, Chen X, Song BL, Yu CJ, Wu MM, Zhang ZR, Ma HP (2018) Palmitate stimulates the epithelial sodium channel by elevating intracellular calcium, reactive oxygen species, and phosphoinositide 3-kinase activity. Oxidative Med Cell Longev 2018:7560610. https://doi.org/10.1155/2018/7560610

Waterman H, Yarden Y (2001) Molecular mechanisms underlying endocytosis and sorting of ErbB receptor tyrosine kinases. FEBS Lett 490(3):142–152

Wei H, Mundade R, Lange KC, Lu T (2014) Protein arginine methylation of non-histone proteins and its role in diseases. Cell Cycle 13(1):32–41

Weissman AM (2001) Themes and variations on ubiquitylation. Nat Rev Mol Cell Biol 2 (3):169–178

Weisz OA, Rodriguez-Boulan E (2009) Apical trafficking in epithelial cells: signals, clusters and motors. J Cell Sci 122(Pt 23):4253–4266

Weisz OA, Wang JM, Edinger RS, Johnson JP (2000) Non-coordinate regulation of endogenous epithelial sodium channel (ENaC) subunit expression at the apical membrane of A6 cells in response to various transporting conditions. J Biol Chem 275(51):39886–39893. https://doi.org/10.1074/jbc.M003822200

Wesch D, Althaus M, Miranda P, Cruz-Muros I, Fronius M, Gonzalez-Hernandez T, Clauss WG, Alvarez de la Rosa D, Giraldez T (2012) Differential N termini in epithelial Na+ channel delta-subunit isoforms modulate channel trafficking to the membrane. Am J Physiol Cell Physiol 302 (6):C868–C879. https://doi.org/10.1152/ajpcell.00255.2011

West TA, Blazer-Yost BL (2005) Modulation of basal and peptide hormone-stimulated Na^+ transport by membrane cholesterol content in the A6 epithelial cell line. Cell Physiol Biochem 16(4–6):263–270

White CN, Figtree GA, Liu CC, Garcia A, Hamilton EJ, Chia KK, Rasmussen HH (2009) Angiotensin II inhibits the Na+-K+ pump via PKC-dependent activation of NADPH oxidase. Am J Physiol Cell Physiol 296(4):C693–C700. https://doi.org/10.1152/ajpcell.00648.2008

Wiig H, Schroder A, Neuhofer W, Jantsch J, Kopp C, Karlsen TV, Boschmann M, Goss J, Bry M, Rakova N, Dahlmann A, Brenner S, Tenstad O, Nurmi H, Mervaala E, Wagner H, Beck FX, Muller DN, Kerjaschki D, Luft FC, Harrison DG, Alitalo K, Titze J (2013) Immune cells control skin lymphatic electrolyte homeostasis and blood pressure. J Clin Invest 123(7):2803–2815. https://doi.org/10.1172/JCI60113

Wiig H, Luft FC, Titze JM (2018) The interstitium conducts extrarenal storage of sodium and represents a third compartment essential for extracellular volume and blood pressure homeostasis. Acta Physiol (Oxf) 222 (3). https://doi.org/10.1111/apha.13006

Wildling L, Hinterdorfer P, Kusche-Vihrog K, Treffner Y, Oberleithner H (2009) Aldosterone receptor sites on plasma membrane of human vascular endothelium detected by a mechanical nanosensor. Pflugers Arch 458(2):223–230. https://doi.org/10.1007/s00424-008-0615-1

Wildman SS, Kang ES, King BF (2009) ENaC, renal sodium excretion and extracellular ATP. Purinergic Signal 5(4):481–489. https://doi.org/10.1007/s11302-009-9150-6

Wills NK, Lewis SA, Eaton DC (1979) Active and passive properties of rabbit descending colon: a microelectrode and nystatin study. J Membr Biol 45:81–108

Wilson JM, Laurent P, Tufts BL, Benos DJ, Donowitz M, Vogl AW, Randall DJ (2000) NaCl uptake by the branchial epithelium in freshwater teleost fish: an immunological approach to ion-transport protein localization. J Exp Biol 203(Pt 15):2279–2296

Winarski KL, Sheng N, Chen J, Kleyman TR, Sheng S (2010) Extracellular allosteric regulatory subdomain within the gamma subunit of the epithelial Na+ channel. J Biol Chem 285 (34):26088–26096. https://doi.org/10.1074/jbc.M110.149963

Winterbourn CC, Kettle AJ (2013) Redox reactions and microbial killing in the neutrophil phagosome. Antioxid Redox Signal 18(6):642–660. https://doi.org/10.1089/ars.2012.4827

Winterbourn CC, Kettle AJ, Hampton MB (2016) Reactive oxygen species and neutrophil function. Annu Rev Biochem 85:765–792. https://doi.org/10.1146/annurev-biochem-060815-014442

Woollhead AM, Baines DL (2006) Forskolin-induced cell shrinkage and apical translocation of functional enhanced green fluorescent protein-human alphaENaC in H441 lung epithelial cell monolayers. J Biol Chem 281(8):5158–5168. https://doi.org/10.1074/jbc.M509947200

Xia L, Wang H, Munk S, Kwan J, Goldberg HJ, Fantus IG, Whiteside CI (2008) High glucose activates PKC-zeta and NADPH oxidase through autocrine TGF-beta1 signaling in mesangial cells. Am J Physiol Renal Physiol 295(6):F1705–F1714. https://doi.org/10.1152/ajprenal.00043.2008

Xu H, Chu S (2007) ENaC alpha-subunit variants are expressed in lung epithelial cells and are suppressed by oxidative stress. Am J Physiol Lung Cell Mol Physiol 293(6):L1454–L1462. https://doi.org/10.1152/ajplung.00248.2007

Xu B, Stippec S, Chu PY, Lazrak A, Li XJ, Lee BH, English JM, Ortega B, Huang CL, Cobb MH (2005) WNK1 activates SGK1 to regulate the epithelial sodium channel. Proc Natl Acad Sci USA 102(29):10315–10320

Yamagata T, Yamagata Y, Nishimoto T, Hirano T, Nakanishi M, Minakata Y, Ichinose M, Dagenais A, Berthiaume Y (2009) The regulation of amiloride-sensitive epithelial sodium channels by tumor necrosis factor-alpha in injured lungs and alveolar type II cells. Respir Physiol Neurobiol 166(1):16–23. https://doi.org/10.1016/j.resp.2008.12.008

Yamagishi S, Matsui T (2011) Nitric oxide, a janus-faced therapeutic target for diabetic microangiopathy-Friend or foe? Pharmacol Res 64(3):187–194. https://doi.org/10.1016/j.phrs.2011.05.009

Yamaguchi E, Yoshikawa K, Nakaya I, Kato K, Miyasato Y, Nakagawa T, Kakizoe Y, Mukoyama M, Soma J (2018) Liddle's-like syndrome associated with nephrotic syndrome secondary to membranous nephropathy: the first case report. BMC Nephrol 19(1):122. https://doi.org/10.1186/s12882-018-0916-3

Yamauchi E, Nakatsu T, Matsubara M, Kato H, Taniguchi H (2003) Crystal structure of a MARCKS peptide containing the calmodulin-binding domain in complex with Ca2+–calmodulin. Nat Struct Mol Biol 10(3):226–231

Yan W, Suaud L, Kleyman TR, Rubenstein RC (2006) Differential modulation of a polymorphism in the COOH terminus of the alpha-subunit of the human epithelial sodium channel by protein kinase Cdelta. Am J Physiol Renal Physiol 290(2):F279–F288. https://doi.org/10.1152/ajprenal.00277.2005

Yan W, Spruce L, Rosenblatt MM, Kleyman TR, Rubenstein RC (2007) Intracellular trafficking of a polymorphism in the COOH terminus of the alpha-subunit of the human epithelial sodium channel is modulated by casein kinase 1. Am J Physiol Renal Physiol 293(3):F868–F876. https://doi.org/10.1152/ajprenal.00194.2007

Yanagisawa M, Kurihara H, Kimura S, Goto K, Masaki T (1988) A novel peptide vasoconstrictor, endothelin, is produced by vascular endothelium and modulates smooth muscle Ca2+ channels. J Hypertens Suppl 6(4):S188–S191. https://doi.org/10.1097/00004872-198812040-00056

Yang B, Kumar S (2010) Nedd4 and Nedd4-2: closely related ubiquitin-protein ligases with distinct physiological functions. Cell Death Differ 17(1):68–77

Yang LM, Rinke R, Korbmacher C (2006) Stimulation of the epithelial sodium channel (ENaC) by cAMP involves putative ERK phosphorylation sites in the C termini of the channel's beta- and gamma-subunit. J Biol Chem 281(15):9859–9868

Yang C, Su L, Wang Y, Liu L (2009) UTP regulation of ion transport in alveolar epithelial cells involves distinct mechanisms. Paper presented at the American journal of physiology. Lung cellular and molecular physiology

Yang G, Hamacher J, Gorshkov B, White R, Sridhar S, Verin A, Chakraborty T, Lucas R (2010) The dual role of TNF in pulmonary edema. J Cardiovasc Dis Res 1(1):29–36. https://doi.org/10.4103/0975-3583.59983

Yarmola EG, Edison AS, Lenox RH, Bubb MR (2001) Actin filament cross-linking by MARCKS: characterization of two actin-binding sites within the phosphorylation site domain. J Biol Chem 276(25):22351–22358

Yousefipour Z, Oyekan A, Newaz M (2010) Interaction of oxidative stress, nitric oxide and peroxisome proliferator activated receptor gamma in acute renal failure. Pharmacol Ther 125(3):436–445. https://doi.org/10.1016/j.pharmthera.2009.12.004

Yu H, Kopito RR (1999) The role of multiubiquitination in dislocation and degradation of the alpha subunit of the T cell antigen receptor. J Biol Chem 274(52):36852–36858

Yu L, Bao HF, Self JL, Eaton DC, Helms MN (2007a) Aldosterone-induced increases in superoxide production counters nitric oxide inhibition of epithelial Na channel activity in A6 distal nephron

cells. Am J Physiol Renal Physiol 293(5):F1666–F1677. https://doi.org/10.1152/ajprenal.00444.2006
Yu L, Eaton DC, Helms MN (2007b) Effect of divalent heavy metals on epithelial Na+ channels in A6 cells. Am J Physiol Renal Physiol 293(1):F236–F244. https://doi.org/10.1152/ajprenal.00002.2007
Yu L, Helms MN, Yue Q, Eaton DC (2008) Single-channel analysis of functional epithelial sodium channel (ENaC) stability at the apical membrane of A6 distal kidney cells. Am J Physiol Renal Physiol 295(5):F1519–F1527. https://doi.org/10.1152/ajprenal.00605.2007
Yu L, Cai H, Yue Q, Alli AA, Wang D, Al-Khalili O, Bao HF, Eaton DC (2013) WNK4 inhibition of ENaC is independent of Nedd4-2-mediated ENaC ubiquitination. Am J Physiol Renal Physiol 305(1):F31–F41. https://doi.org/10.1152/ajprenal.00652.2012
Yue G, Hu P, Oh Y, Jilling T, Shoemaker RL, Benos DJ, Cragoe EJ Jr, Matalon S (1993) Culture-induced alterations in alveolar type II cell Na+ conductance. Am J Phys 265(3 Pt 1):C630–C640. https://doi.org/10.1152/ajpcell.1993.265.3.C630
Yue G, Shoemaker RL, Matalon S (1994) Regulation of low-amiloride-affinity sodium channels in alveolar type II cells. Am J Phys 267(1 Pt 1):L94–L100. https://doi.org/10.1152/ajplung.1994.267.1.L94
Yue G, Malik B, Yue G, Eaton DC (2002) Phosphatidylinositol 4,5-bisphosphate (PIP2) stimulates epithelial sodium channel activity in A6 cells. J Biol Chem 277(14):11965–11969
Zakrzewicz D, Zakrzewicz A, Preissner KT, Markart P, Wygrecka M (2012) Protein Arginine Methyltransferases (PRMTs): promising targets for the treatment of pulmonary disorders. Int J Mol Sci 13(10):12383–12400
Zeitlin PL (2008) Cystic fibrosis and estrogens: a perfect storm. Paper presented at the The Journal of Clinical Investigation
Zerangue N, Schwappach B, Jan YN, Jan LY (1999) A new ER trafficking signal regulates the subunit stoichiometry of plasma membrane K(ATP) channels. Neuron 22(3):537–548
Zhang W, Hayashizaki Y, Kone BC (2004) Structure and regulation of the mDot1 gene, a mouse histone H3 methyltransferase. Biochem J 377(Pt 3):641–651
Zhang W, Xia X, Reisenauer MR, Rieg T, Lang F, Kuhl D, Vallon V, Kone BC (2007a) Aldosterone-induced Sgk1 relieves Dot1a-Af9-mediated transcriptional repression of epithelial Na+ channel. J Clin Invest 117(3):773–783
Zhang Y, Sanchez D, Gorelik J, Klenerman D, Lab M, Edwards C, Korchev Y (2007b) Basolateral P2X4-like receptors regulate the extracellular ATP-stimulated epithelial Na+ channel activity in renal epithelia. Paper presented at the American journal of physiology
Zhang ZR, Chou CF, Wang J, Liang YY, Ma HP (2010) Anionic phospholipids differentially regulate the epithelial sodium channel (ENaC) by interacting with alpha, beta, and gamma ENaC subunits. Pflugers Arch 459(3):377–387. https://doi.org/10.1007/s00424-009-0733-4
Zhang J, Chen S, Liu H, Zhang B, Zhao Y, Ma K, Zhao D, Wang Q, Ma H, Zhang Z (2013a) Hydrogen sulfide prevents hydrogen peroxide-induced activation of epithelial sodium channel through a PTEN/PI(3,4,5)P3 dependent pathway. PLoS One 8(5):e64304. https://doi.org/10.1371/journal.pone.0064304
Zhang X, Zhou Q, Chen L, Berger S, Wu H, Xiao Z, Pearce D, Zhou X, Zhang W (2013b) Mineralocorticoid receptor antagonizes Dot1a-Af9 complex to increase alphaENaC transcription. Am J Physiol Renal Physiol 305(10):F1436–F1444
Zhang Y, Robson SC, Morris KL, Heiney KM, Dwyer KM, Kishore BK, Ecelbarger CM (2015) Impaired natriuretic response to high-NaCl diet plus aldosterone infusion in mice over-expressing human CD39, an ectonucleotidase (NTPDase1). Paper presented at the American Journal of Physiology. Renal Physiology
Zhou R, Snyder PM (2005) Nedd4-2 phosphorylation induces serum and glucocorticoid-regulated kinase (SGK) ubiquitination and degradation. J Biol Chem 280(6):4518–4523. https://doi.org/10.1074/jbc.M411053200
Zhou R, Patel SV, Snyder PM (2007) Nedd4-2 catalyzes ubiquitination and degradation of cell surface ENaC. J Biol Chem 282(28):20207–20212. https://doi.org/10.1074/jbc.M611329200

Zhou R, Kabra R, Olson DR, Piper RC, Snyder PM (2010) Hrs controls sorting of the epithelial Na+ channel between endosomal degradation and recycling pathways. J Biol Chem 285 (40):30523–30530. https://doi.org/10.1074/jbc.M110.150755

Zhou R, Tomkovicz VR, Butler PL, Ochoa LA, Peterson ZJ, Snyder PM (2013) Ubiquitin-specific peptidase 8 (USP8) regulates endosomal trafficking of the epithelial Na+ channel. J Biol Chem 288(8):5389–5397

Zhu S, Ware LB, Geiser T, Matthay MA, Matalon S (2001) Increased levels of nitrate and surfactant protein a nitration in the pulmonary edema fluid of patients with acute lung injury. Am J Respir Crit Care Med 163(1):166–172

Zou L, Linck V, Zhai YJ, Galarza-Paez L, Li L, Yue Q, Al-Khalili O, Bao HF, Ma HP, Thai TL, Jiao J, Eaton DC (2018) Knockout of mitochondrial voltage-dependent anion channel type 3 increases reactive oxygen species (ROS) levels and alters renal sodium transport. J Biol Chem 293(5):1666–1675. https://doi.org/10.1074/jbc.M117.798645

Zuckerman JB, Chen X, Jacobs JD, Hu B, Kleyman TR, Smith PR (1999) Association of the epithelial sodium channel with Apx and alpha-spectrin in A6 renal epithelial cells. J Biol Chem 274(33):23286–23295

Chapter 19
ROMK and Bartter Syndrome Type 2

Paul G. Welling

Abstract The *R*enal *O*uter *M*edullary *P*(*K*)otassium Channel, ROMK (aka Kir1.1, encoded by the KCNJ1 gene), is the founding member of the inwardly rectifying potassium channel family. It is primarily expressed in the kidney where it transports potassium into the pro-urine, important for maintaining salt reabsorption in the thick ascending limb, and potassium-homeostasis. Loss-of-function mutations in ROMK cause Bartter syndrome (type 2), a familial salt-losing nephropathy. This review provides a bedside-to-bench understanding of the disease, tracking the clinical phenotype to defects in tubule transport to the mutations in the channel.

Keywords ROMK · KCNJ1 · Potassium · Hypokalemia · Hyperkalemia · Blood pressure · Salt · Hypertension · Hypotension · Thick ascending limb of Henle's loop · Kidney · Channel

19.1 Introduction

Bartter syndrome (Bartter et al. 1962) is a rare kidney disorder caused by impaired salt reabsorption in the thick ascending limb of Henle's loop (Fig. 19.1). Typically, the disease is manifested in urinary salt wasting, hypokalemic alkalosis, hypotension, hyperreninism, hyperaldosteronism, hypercalciuria, occasional hypomagnesemia, and a constellation of secondary sequelae, similar to those observed with loop-diuretic toxicity. The antenatal or infantile form (ABS) presents with maternal polyhydramnios and severe, life-threatening intravascular volume contraction, hyperprostaglandinuria, and failure to thrive in the postnatal period (Seyberth and Schlingmann 2011). The Gitelman variant of Bartter syndrome, caused by loss-of-function mutations in the thiazide-sensitive sodium-chloride cotransporter,

P. G. Welling (✉)
Departments of Medicine and Physiology, Nephrology Division, Johns Hopkins Medical School, Baltimore, MD, USA
e-mail: pwellin1@jhmi.edu

K. L. Hamilton, D. C. Devor (eds.), *Studies of Epithelial Transporters and Ion Channels*, Physiology in Health and Disease,
https://doi.org/10.1007/978-3-030-55454-5_19

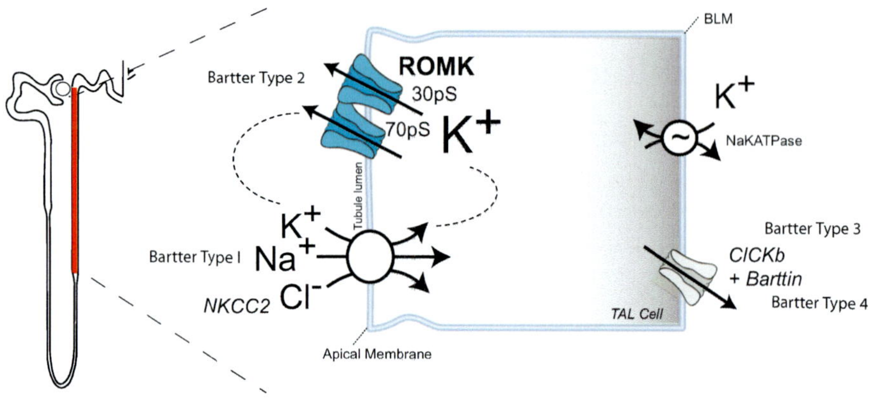

Fig. 19.1 The thick ascending limb of Henle's loop reabsorbs ~20% of the filtered load of sodium-chloride through a transcellular transport pathway. The major elements of the sodium-chloride reabsorptive machine are shown. Loss-of-function mutations in each of them cause Bartter syndrome

SLC12A3 (OMIM 600968), is characterized by milder salt wasting and hypotension, hypocalciuria, and more profound hypomagnesemia than classic Bartter syndrome.

ABS is usually inherited through autosomal recessive transmission, but has heterogenous genetic underpinnings. Remarkably, loss-of-function mutations in any one of the four major components of the NaCl transport machinery in the thick ascending limb (TAL) of the loop of Henle can produce ABS (Fig. 19.1). These include the $Na^+/K^+/2Cl^-$ cotransporter, NKCC2 (SCL12a1) (Simon et al. 1996), the Renal Outer Medullary Potassium Channel, ROMK (Kir1.1 of the KCJ1 gene) (Simon et al. 1997a; Karolyi 1997), the basolateral Cl^- channel, CLCNKB (Simon et al. 1997b; Konrad et al. 2000), and the accessory subunit of CLCNKB, BSND (Birkenhager et al. 2001). Gain-of-function mutations in the Calcium-sensing G-protein coupled receptor, CASR, also lead to a Bartter phenotype in addition to hypocalcemia and hypoparathyroidism (Watanabe et al. 2002; Vargas-Poussou et al. 2002). Still, the genetic bases for as many as 20% of patients with ABS remain unknown. Identification of these may reveal new transport mechanisms in the TAL. For example, Melanoma Antigen-Family D2 MAGE-D2, which was found to control the expression of NKCC2 in the developing fetus, was discovered because mutations in MAGE-D2 cause an X-linked form of ABS, characterized by severe, early onset polyhydramnios, premature birth, and high fetal mortality. A new classification scheme of Bartter Subtypes has been suggested (Besouw et al. 2019) and is further refined here to include CASR and Bartter syndrome 6 (Table 19.1), which the On line Inheritance in Man (OMIM) only classifies as autosomal dominant hypocalcemia.

This review focuses on Type II Bartter Syndrome (BS II) and the ROMK. Loss of function mutations in ROMK cause ABS, although late-onset BS II with mild clinical manifestations has been described (Gollasch et al. 2017).

Table 19.1 Bartter syndrome subtypes

Type	Gene	Chromosome	OMIM	Unique features
BARS1	SLC12a1	15q21.1	#601678	
BARS 2	KCNJ1	11q24.3	#241200	Transient Hyperkalemia followed by mild hypokalemia
BARS 3	CLCNKB	1p36.13	#607364	Hypocalciuria
BARS4a	BSND	1p32.3	#602522	Sensorineural deafness
BARS4b	CLCNKA +B	1p36.13	#613090	Sensorineural deafness, mixed Gitelman syndrome
BARS 5	MAGE-D2	Xp11	#300971	Symptoms Resolve after Birth, except hypercalciuria
BARS6	CASR	3q21.1	#601198	Autosomal dominant hypocalcemia1/ hypoparathyroidism

19.2 The ROMK Channel

ROMK (aka Kir 1.1 or KCNJ1, (Ho et al. 1993)) is the founding member of the inward-rectifier "Kir" type potassium channel family (Welling and Ho 2009), which is encoded by KCN "J" type genes. Each ROMK subunit has two transmembrane spanning regions and a large cytoplasmic domain, typical of Kir channels. The transmembrane domains of four ROMK subunits assemble to create a canonical potassium conduction pathway, which includes the potassium selectivity filter and a regulated gate structure. Cytoplasmic N- and C-termini of each subunit fold to create a large regulatory structure with a centrally located pore (see Fig. 20.2). Missense mutations that are associated with Bartter syndrome are located in each of these major domains (Fig. 19.2).

Like other inward-rectifiers, ROMK exhibits a nonlinear current-voltage relationship, distinguished by a larger inward current than outward current. Nevertheless, ROMK exhibits the weakest inward-rectification of all members of the Kir-family. In fact, ROMK channels carry significant outward potassium flux under physiological conditions. Together with their high open-probability (Po = 0.9), ROMK channels support robust potassium efflux at physiological membrane potentials. These properties make ROMK channels ideally suited for their physiological role in secreting potassium into the pro-urine.

ROMK channels have three major functions in the kidney. In the TAL, ROMK channels are expressed on the apical membrane and mediate the K^+ efflux into the tubule lumen to help sustain the turnover of NKCC2 for NaCl transport (see Sect. 19.3). They also contribute to the transepithelial current flow and potential difference in the TAL, important for paracellular Na^+ and divalent cation reabsorption. In the late distal tubule, connecting tubule, and collecting duct, ROMK channels function in the final step of urinary potassium excretion, a highly regulated process that ensures that potassium is precisely excreted into the urine to match variations in dietary potassium intake for K^+ homeostasis (Hebert et al. 2005; Welling 2013).

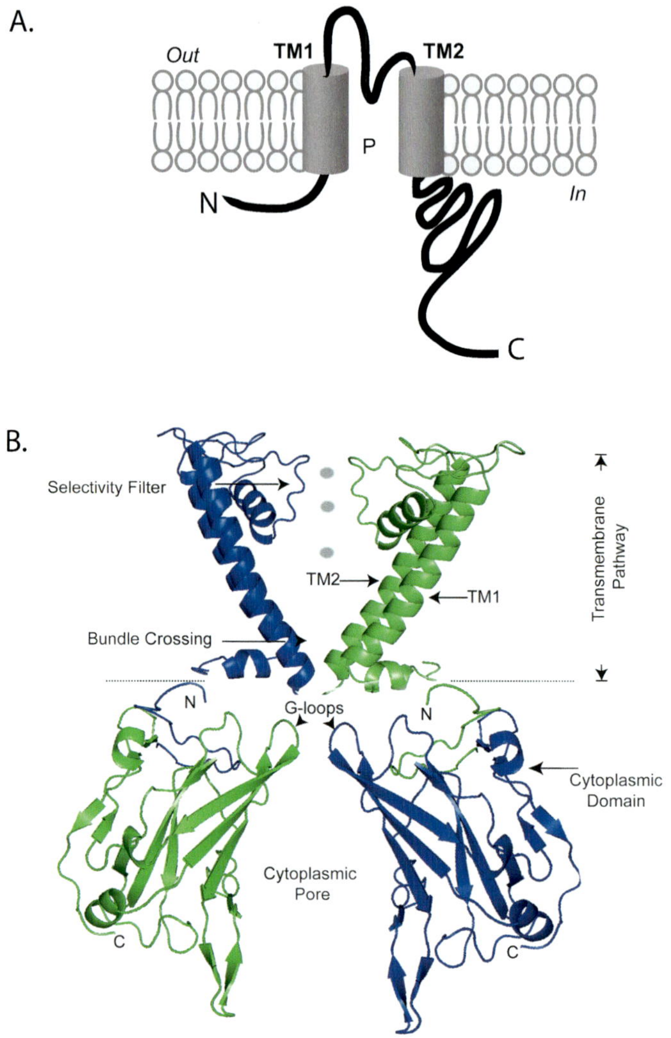

Fig. 19.2 ROMK Structure. (**a**) Cartoon of a ROMK channel subunit, with a two-transmembrane body-plan, typical of inwardly rectifying potassium channels. (**b**) Atomic resolution model of ROMK based on the structure of Kir2.2 (pdb 3JYC) and Kir3.1 (pdb 2QKS). Two subunits are shown. Front and back subunits have been removed to see the potassium permeation pathway

Patch clamp studies of the apical membrane in split-open TAL cells and connecting tubule (CNT) and cortical collecting (CCD) principal cells revealed that the predominate apical K_{ir} channels in the TAL and CCD have biophysical and pharmacological properties closely matching recombinant ROMK (~single channel conductance 35pS, high open-probability) (Palmer and Frindt 1999; Wang et al. 1990; Bleich et al. 1990). Ablation of the ROMK gene (Lorenz et al. 2002) eliminates the 35-pS channel in the mouse TAL and CCD (Lu et al. 2002), providing strong evidence that ROMK encodes the renal secretory channel. Another potassium channel on the TAL apical membrane has been described that exhibits ROMK-like gating characteristics, but has a higher single channel conductance (70pS) (Wang 1994). This channel is also lost in the ROMK null mouse, strongly suggesting that ROMK is a critical subunit of the higher conductance channel (Lu et al. 2002). The identity of the conductance modifier is presently unknown.

There are at least three different ROMK isoforms, each differing in the extreme N-terminus (Welling and Ho 2009). ROMK1 has a 19 amino acid sequence that is longer than ROMK2, and ROMK3 has a 16 amino acid sequence that is longer than ROMK2. The function of these isoform-specific regions remains unknown. ROMK1 is predominately expressed in late distal convoluted tubule, connecting tubule and collecting duct, and genetic ablation of the ROMK1 form does not produce a Bartter's phenotype, but rather urinary potassium retention (Dong et al. 2016). ROMK2 and ROMK3 are expressed in the TAL.

19.3 ROMK Function in the TAL

Elegant physiologic studies in the isolated perfused tubule provide a context to understand why ROMK channels are required for salt reabsorption in the TAL (Greger and Schlatter 1981), and how their dysfunction leads to Bartter Syndrome. Normally, the thick ascending limb reabsorbs ~20% of the filtered load of sodium-chloride through the multistep process shown in Fig. 19.1. In this scheme, NaCl, along with potassium, is transported from the tubule lumen into the cell by the apically localized $Na^+/K^+/2Cl^-$ cotransporter, NKCC2 (product of the SLC12A1 gene); sodium is then actively transported across the basolateral membrane through the Na/K ATPase, while Cl- diffuses down a favorable electrochemical gradient into the interstitium through the basolaterally located CLCNKB channels. Apical potassium channels, formed by ROMK (Lorenz et al. 2002; Lu et al. 2002), allow the obligatory uptake of potassium through NKCC2, to recycle across the apical membrane. In fact, it has been believed that the ROMK-mediated recycling process is necessary to supply a sufficient level of potassium to drive the turnover of NKCC2 (Greger and Schlatter 1981; Hebert et al. 1984; Hebert and Andreoli 1984; Hebert 2003). Consistent with this, in the isolated perfused tubule preparation, sodium reabsorption is critically dependent on the secretory movement of potassium (Greger and Schlatter 1981).

The discovery that loss-of-function mutations in ROMK cause Bartter syndrome, together with subsequent studies in whole ROMK gene knockout mice (Lu et al. 2002; Lorenz et al. 2002; Bailey et al. 2006), which recapitulate the disease, provided solid evidence of the transport mechanism. Nevertheless, mathematical modeling of salt transport in the thick ascending limb in vivo reveals a wrinkle in the transport scheme (Weinstein 2010). Remarkably, when physiological levels of NH_4^+ are included in the model, it fails to capture the inhibition of salt transport in type II Bartter syndrome. In this case, the model reveals there is little impact of luminal membrane K^+ permeability on overall Na^+ reabsorption because NH_4^+ can take the place of K^+ to maintain NKCC2 turnover. By contrast, sodium reabsorption in the isolated perfused TAL preparation requires potassium secretion (Greger and Schlatter 1981) because NH_4^+ is not typically included.

Viewed in this light, it seems likely that ROMK does more to sustain sodium transport than provide potassium to the tubule lumen. Two mechanisms should be considered. The ROMK channel contributes to the generation of a lumen-positive transepithelial potential, which drives the passive reabsorptive movement of cations through the paracellular pathway. In the absence of the ROMK, the passive reabsorptive process, which accounts for about a third of the total reabsorptive flux of sodium, will be impaired. Secondly, the expression of NKCC2 is compromised in ROMK knockout mice (Wagner et al. 2008), suggesting ROMK and NKCC2 might be more directly linked together. The basis for ROMK-NKCC2 coupling is not known, but it may relate to the mutual dependence of ROMK and NKCC2 for a glycosylphosphatidylinositol-anchored uromodulin (aka Tamm Horsfall protein) to be efficiently expressed on the apical membrane (Mutig et al. 2011; Renigunta et al. 2011).

19.4 Transient Hyperkalemia, A Unique Phenotype of Bartter Type II

Characteristically, profound hypokalemia occurs in Bartter syndrome as a consequence of urinary potassium wasting. This happens because factors that normally stimulate potassium secretion in the distal nephron [increased delivery of sodium, increased tubular fluid flow rate, and elevation of aldosterone (Welling 2013)] are pathologically activated when sodium reabsorption in the TAL is impaired. Additionally, reduced potassium absorption in the TAL can account for significant potassium losses in Bartter syndrome (Bailey et al. 2006).

In remarkable contrast, newborns with Bartter type II exhibit profound hyperkalemia (Brochard et al. 2009; Finer et al. 2003; Peters et al. 2003) opposite of the classic Bartter syndrome phenotype of urinary potassium wasting and hypokalemia. This most likely occurs because loss of ROMK function in the distal nephron leaves the neonatal kidney without a means to efficiently excrete potassium. The hyperkalemia is transient as plasma potassium returns to normal levels within a few weeks after birth, and then persistent mild hypokalemia can follow throughout life. Free-flow micropuncture and stationary microperfusion studies in ROMK knockout mice revealed that the renal potassium excretion is restored in Type II Bartter by compensatory overexpression of BK channels (Bailey et al. 2006), similar to the compensatory upregulation of ROMK channels that preserves potassium secretion in BK channel knockout animals (Rieg et al. 2007). BK channels are especially suited to stimulate potassium excretion in the high-flow, high-sodium delivery setting of Bartter syndrome (Bailey et al. 2006) because they are activated by flow and membrane depolarization (Liu et al. 2007; Pluznick et al. 2003). In the mouse model of type II Bartter syndrome, urinary potassium wasting is further exaggerated by reduced potassium reabsorption in the TAL (Bailey et al. 2006).

19.5 Carriers of Bartter Mutations Are Protected from Hypertension

Recent genetic and physiological studies indicate that rare mutations in ROMK also contribute to blood pressure variation in the general population (Ji et al. 2008; Fang et al. 2010). In a large-scale exome sequencing effort in the Framingham heart study cohort, Ji et al. determined that carriers of known or inferred loss-of-function mutations in ROMK (as well as NKCC2, and Gitelman-associated mutations in the thiazide-sensitive sodium-chloride cotransporter, see Chap. 3, Volume 3) exhibited a significantly lower blood pressure throughout their life than the rest of the cohort. The trait conferred protection from the development of hypertension (Ji et al. 2008). Subsequent physiological studies revealed that these rare mutations do, in fact, cause a loss of ROMK (Fang et al. 2010), NKCC2, or NCC function (Acuna et al. 2011; Monette et al. 2011), verifying and reinforcing the role of functional variants in kidney salt handling genes in blood pressure. In the case of ROMK, one defective allele appears sufficient to reduce the sodium reabsorption in the thick ascending limb enough to shift the blood pressure "set-point" for sodium excretion and confer a protective trait. These observations have inspired the development of ROMK inhibitors as a new class of antikaliuretic diuretics (Garcia et al. 2014; Tang et al. 2012; Bhave et al. 2011).

The discovery that rare independent mutations in renal salt handling genes, such as ROMK, contribute to blood pressure variation in the general population has important implications for the genetic bases of hypertension (Subramanya and Welling 2011). Because the ROMK, NKCC2, and NCC mutants represent only a small fraction of variant genes that alter blood pressure and many more remain to be discovered, Ji et al. reasonably argued that the combined effects of many rare independent mutations will likely account for a substantial fraction of blood pressure variation in the general population (Ji et al. 2008). With numerous studies indicating that the high heritability of the blood pressure trait cannot be easily explained by common polymorphisms, even when combined into a polygenic risk, the view of Ji et al. makes sense. It will be interesting to learn how many of these genes impinge on the regulation of ROMK.

19.6 Bartter Mutations in ROMK, Overview

Over 35 different Bartter disease mutations have been identified in ROMK. Several introduce nonsense codons or frameshifts that produce truncated proteins and obvious loss-of-function. Over half of the missense mutations reduce or even eliminate surface expression of ROMK (Peters et al. 2003). At least a subset of these appears to cause misfolding and, thus, becomes substrates for the endoplasmic reticulum (ER)-associated degradation (ERAD) pathway, suggesting therapeutic strategies that focus on correcting deficiencies in ROMK folding (O'Donnell et al. 2017a, b). Other BSII mutations alter ROMK channel opening by disrupting the potassium permeation pathway or by upsetting the way the channel is regulated by phosphorylation, PIP2, or intracellular pH. The underlying mechanisms are discussed below.

19.7 Bartter Mutations in the Potassium Permeation Structure

Like all known potassium channels, the transmembrane linkers of each ROMK subunit project into the membrane to form the narrowest part of the open pore and create the potassium selectivity filter around the conserved sequence, "T[V/I]GYG" (Heginbotham et al. 1994). The sequence adopts a strand conformation where backbone carbonyl oxygens point into the pore, effectively mimicking the inner hydration shell of potassium but not of sodium (MacKinnon 2004). In this way, the selectivity filter creates energetically favorable sites for potassium to selectively diffuse from the aqueous environment into the pore. A Bartter mutation, $I_{142}T$, in the second residue of the selectivity sequence blocks channel function, presumably by collapsing the potassium permeation pathway (Peters et al. 2003). Other mutations in the vicinity do the same (Fig. 19.3a). In some cases, mutations in this region change ion permeation. In the related Kir 3.4 channel (KCNJ5), for example, mutations that are in or near the conserved sequence cause the channel to lose potassium selectivity and gain sodium permeation (Choi et al. 2011). These gain-of-function mutations underlie the pathogenesis of aldosterone-producing adrenal adenoma formation in some patients.

19.8 Bartter Mutations That Alter Regulated Gates

PKA phosphorylation (Xu et al. 1996; McNicholas et al. 1994) and PIP2 binding (Huang et al. 1998; Huang 2007) are required to maintain ROMK in a high open-probability state, while intracellular acidification (Zhou et al. 1994) and ATP binding close the channel (McNicholas et al. 1996; Ruknudin et al. 1998). According to

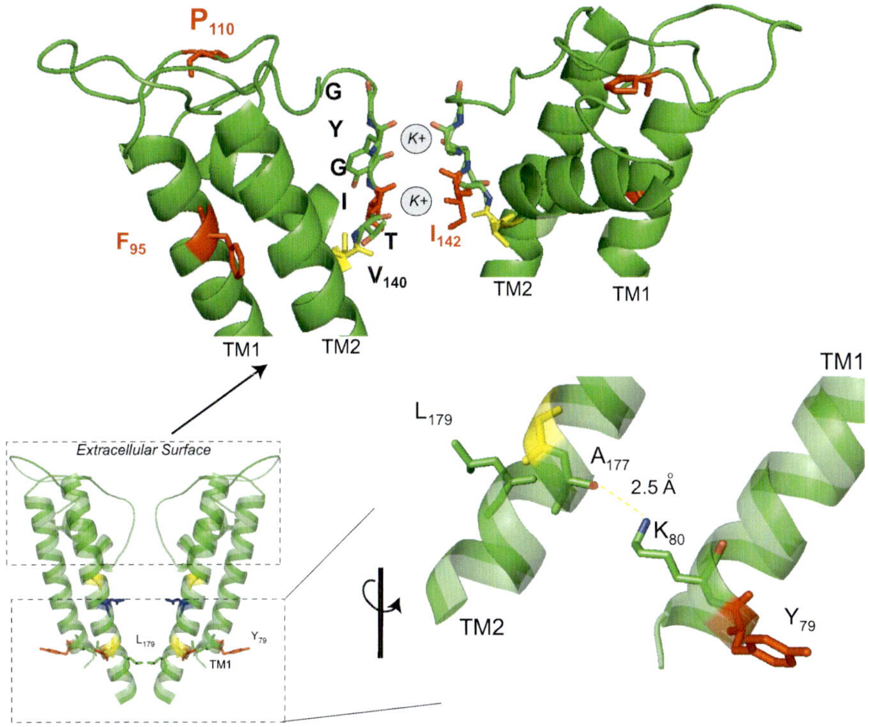

Fig. 19.3 Bartter mutations affect the transmembrane potassium permeation pathway at the selectivity filter (*top*) and the regulated gate (*bottom*). Residues that are mutated in Bartter syndrome are shown in *red*

current understanding, the effects of phosphorylation and ligand binding within the cytoplasmic domain are transmitted to gating structures, located in the transmembrane helices, and the apex of the cytoplasmic domain. Bartter mutations affect these gates (Fig. 19.3).

Regulated channel opening of the transmembrane gate is governed by movement of the end of second transmembrane domain (TM2), which pulls the putative gate residue, L_{179}, (Sackin et al. 2005; Nanazashvili et al. 2007) away from the pore (Fig. 19.3). This is influenced by interactions between the two-transmembrane helices, involving A_{177} on TM2 and K_{80} on TM1 (Rapedius et al. 2006, 2007). A Bartter mutation that replaces A_{177} with threonine (Peters et al. 2003) disrupts the interaction and causes a profound alkaline shift in the pH-sensitivity of ROMK (Rapedius et al. 2006), effectively turning off channel function at physiological pH_i. Another Bartter mutation, involving a nearby lipid-facing residue, Y_{79}, at the base of TM1, disrupts regulated channel opening in a similar way (Peters et al. 2003). Interestingly, mutations in the comparable residues of the Kir6.2 K_{ATP} channel alter ATP-dependent gating and produce neonatal diabetes (Hattersley and Ashcroft 2005). It would seem that residues at the base of TM1 and TM2 strongly influence the energetics of regulated K^+ channel opening, making these structures especially vulnerable to disease-causing mutations. Furthermore, inhibitory effects of Bartter mutations in other regions of the channel, especially those that influence pH and PIP2-dependent gating, can be rescued by a second-site suppressor mutation, $K_{80}M$, in the transmembrane gate (Fig. 20.3) (Flagg et al. 2002; Schulte et al. 1999), highlighting closed-TM gate stabilization as a broad mechanism for loss of ROMK channel function in Bartter disease.

An adjacent gating structure involves a narrow opening at the apex of the cytoplasmic pore, formed by the coalescence of four identical loop structures from each subunit (Fig. 20.2), called the G-loop. The structure is believed to create a flexible diffusion barrier between the cytoplasmic and transmembrane pore. G-loop residues, which face into the central axis of the cytoplasmic pore, have been implicated in channel gating (Pegan et al. 2005). One of these residues is affected by a Bartter mutation ($A_{306}T$), which disrupts potassium conduction (Peters et al. 2003). Disease-causing mutations in other K_{ir} channels also affect the G-loop, underscoring the general importance of this structure in the inward-rectifiers (Ma et al. 2007; Pegan et al. 2006).

19.9 Bartter Mutations That Disrupt Channel Modulation

The large cytoplasmic domain structure has an enormous solvent-exposed surface area, important for docking modulators (PIP2, pH) and serving as a substrate for PKA phosphorylation. A number of Bartter mutations affect residues in this domain, and these mutations alter the way the modulators control channel gating.

19.9.1 PIP2

The X-ray structure of the Kir 2.2 channel in complex with a derivative of PIP2 revealed that PIP2 binds to residues at the top of the cytoplasmic domain, and upon binding, an adjacent linker contracts to move the entire cytoplasmic domain toward the membrane and pull the inner helix gate open (Hansen et al. 2011). Systematic mutagenesis studies in ROMK and other K_{ir} channels (Lopes et al. 2002; Zeng et al. 2002) support a common mechanism. Indeed, these studies revealed that PIP2-dependent gating is determined by a group of shared residues near or within the PIP2 binding surface and gating transducer that was identified in Kir 2.2. Many of the Bartter mutations ($C_{49}Y$, $I_{51}T$, $P_{166}S$, $R_{169}H$, $A_{214}V$, and $L_{220}F$) are found in the vicinity of this region and disrupt PIP2-dependent gating (Lopes et al. 2002; Fang et al. 2010). Similar observations with disease-causing mutations in Kir2.1 (Andersen-Tawil Disease, (Plaster et al. 2001)) have been interpreted to indicate that a decrease in channel-PIP2 interactions underlies many Kir channelopathies (Lopes et al. 2002).

The effects of several Bartter type II mutations [(P_{166}S, R_{169}H, L_{220}F (Fang et al. 2010; Srivastava et al. 2013)] are not observed until PIP_2 hydrolysis is stimulated (Fig. 20.4). Wild-type ROMK channels are relatively resistant to physiological changes in PIP_2 levels because of their high apparent binding affinity for PIP_2 (Lopes et al. 2002; Huang 2007). By contrast, the P166S, R169H mutations destabilize PIP2 binding, and this causes channels bearing these Bartter mutations to rapidly close in response to activation of GPCR-Gq-PLC signaling, which stimulates PIP2 hydrolysis (Fang et al. 2010). Astonishingly, P_{166} and R_{169} reside in the PIP2 binding site, flanking the two residues that interact with the phosphates on the phosphatidylinositol moiety of PIP2 (Fig. 20.4). Although L_{220}F also causes a loss of function that is conditional on PIP2 hydrolysis, this Bartter mutation is located near the G-loop and presumably destabilizes the PIP2-dependent gating mechanism rather than the PIP2 binding site as have been shown in other inwardly rectifying potassium channels (Meng et al. 2016) (Fig. 19.4).

19.9.2 PKA Phosphorylation Sites

ROMK is maintained in a high open-probability state by phosphorylation of two key PKA sites in the cytoplasmic domain (Xu et al. 1996; McNicholas et al. 1994). Two Bartter disease-causing mutations replace these phosphorylatable residues [S_{219}R, (Simon et al. 1997a) and S_{313}C (Starremans et al. 2002)], and thus, alter the regulated gating process.

19.9.3 pH-Dependent Gating

ROMK channels close in response to cytoplasmic acidification with an apparent pK_a near neutral pH (Welling and Ho 2009). Many Bartter disease-associated mutations in the cytoplasmic domain shift the pK_a in the alkaline direction and cause channel closure at physiological pH (Flagg et al. 1999, 2002; Schulte et al. 1998, 1999; Peters et al. 2003). Several of the affected residues were once believed to form the pH sensor (Schulte et al. 1999), but the atomic structures of related channels indicate that this is not true (Rapedius et al. 2006). The precise location of the pH sensor remains elusive, but likely involves titratable residues in the cytoplasmic domain which stabilize the TM gate in an open configuration. Recent mutagenesis studies reveal that the pH-sensing is coupled to gating by a large network of residues that link the cytoplasmic domain to the TM gate (Bollepalli et al. 2014). Several Bartter mutants (e.g., Asp74, Ala177, and Ala306) likely increase pH-sensitivity by destabilizing contacts in the network.

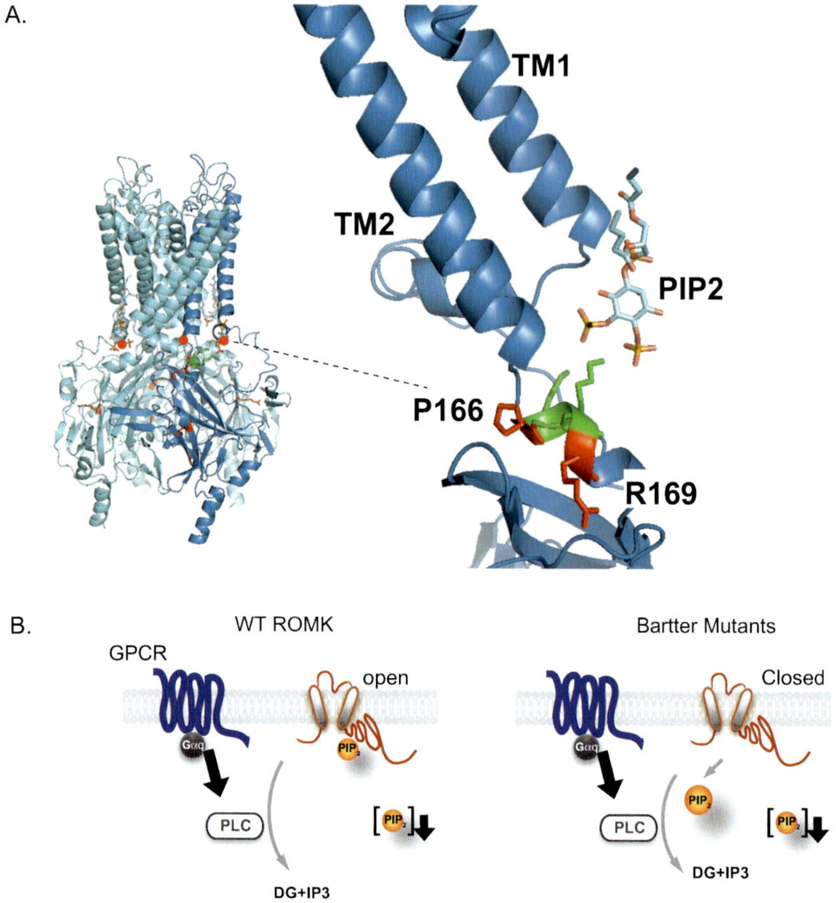

Fig. 19.4 Bartter mutations near the PIP2 binding site can be conditionally manifested. (**a**) Atomic resolution model of ROMK with PIP2 based on structure of Kir2.2 in complex with PIP2 (pdb 3SPG), showing positions of P166 and R169, flanking the basic residues that bind PIP2. Bartter mutations at these positions destabilize PIP2 binding. (**b**) Channels bearing P166S, R169H mutations are open until activation of GPCR-Gq-PLC signaling simulates PIP2 hydrolysis, while wild-type (WT) ROMK stay open because they have a tight binding affinity for PIP2

19.10 Summary

The discovery that loss-of-function mutations in ROMK cause Bartter disease type 2, together with studies in ROMK knockout mice, provided airtight evidence that ROMK is necessary for sodium-chloride reabsorption in the thick ascending limb of Henle's loop and to control potassium balance. It has been suggested that ROMK mediates a potassium recycling process in the TAL that is necessary to sustain the turnover of NKCC2, but further studies will be required to more precisely establish the physiological mechanism underlying the requirement for ROMK in TAL salt reabsorption. Nevertheless, efforts to understand the underlying molecular mechanisms for loss of channel function have revealed and reinforced new ideas about the structural basis for channel function and regulation.

References

Acuna R, Martinez-de-la-Maza L, Ponce-Coria J, Vazquez N, Ortal-Vite P, Pacheco-Alvarez D, Bobadilla NA, Gamba G (2011) Rare mutations in SLC12A1 and SLC12A3 protect against hypertension by reducing the activity of renal salt cotransporters. J Hypertens 29:475–483

Bailey MA, Cantone A, Yan Q, MacGregor GG, Leng Q, Amorim JB, Wang T, Hebert SC, Giebisch G, Malnic G (2006) Maxi-K channels contribute to urinary potassium excretion in the ROMK-deficient mouse model of Type II Bartter's syndrome and in adaptation to a high-K diet. Kidney Int 70:51–59

Bartter FC, Pronove P, Gill JR Jr, MacCardle RC (1962) Hyperplasia of the juxtaglomerular complex with hyperaldosteronism and hypokalemic alkalosis: a new syndrome. Am J Med 33:811–828

Besouw MTP, Kleta R, Bockenhauer D (2019) Bartter and Gitelman syndromes: questions of class. Pediatr Nephrol. https://doi.org/10.1007/s00467-019-04371-y

Bhave G, Chauder BA, Liu W, Dawson ES, Kadakia R, Nguyen TT, Lewis LM, Meiler J, Weaver CD, Satlin LM, Lindsley CW, Denton JS (2011) Development of a selective small-molecule inhibitor of Kir1.1, the renal outer medullary potassium channel. Mol Pharmacol 79:42–50

Birkenhager R, Otto E, Schurmann MJ, Vollmer M, Ruf EM, Maier-Lutz I, Beekmann F, Fekete A, Omran H, Feldmann D, Milford DV, Jeck N, Konrad M, Landau D, Knoers NV, Antignac C, Sudbrak R, Kispert A, Hildebrandt F (2001) Mutation of BSND causes Bartter syndrome with sensorineural deafness and kidney failure. Nat Genet 29:310–314

Bleich M, Schlatter E, Greger R (1990) The luminal K+ channel of the thick ascending limb of Henle's loop. Pflugers Arch 415:449–460

Bollepalli MK, Fowler PW, Rapedius M, Shang L, Sansom MS, Tucker SJ, Baukrowitz T (2014) State-dependent network connectivity determines gating in a K^+ channel. Structure 22:1037–1046

Brochard K, Boyer O, Blanchard A, Loirat C, Niaudet P, Macher MA, Deschenes G, Bensman A, Decramer S, Cochat P, Morin D, Broux F, Caillez M, Guyot C, Novo R, Jeunemaitre X, Vargas-Poussou R (2009) Phenotype-genotype correlation in antenatal and neonatal variants of Bartter syndrome. Nephrol Dial Transplant 24:1455–1464

Choi M, Scholl UI, Yue P, Bjorklund P, Zhao B, Nelson-Williams C, Ji W, Cho Y, Patel A, Men CJ, Lolis E, Wisgerhof MV, Geller DS, Mane S, Hellman P, Westin G, Akerstrom G, Wang W, Carling T, Lifton RP (2011) K^+ channel mutations in adrenal aldosterone-producing adenomas and hereditary hypertension. Science 331:768–772

Dong K, Yan Q, Lu M, Wan L, Hu H, Guo J, Boulpaep E, Wang W, Giebisch G, Hebert SC, Wang T (2016) Romk1 knockout mice do not produce bartter phenotype but exhibit impaired K excretion. J Biol Chem 291:5259–5269

Fang L, Li D, Welling PA (2010) Hypertension resistance polymorphisms in ROMK (Kir1.1) alter channel function by different mechanisms. Am J Physiol Renal Physiol 299:F1359–F1364

Finer G, Shalev H, Birk OS, Galron D, Jeck N, Sinai-Treiman L, Landau D (2003) Transient neonatal hyperkalemia in the antenatal (ROMK defective) Bartter syndrome. J Pediatr 142:318–323

Flagg TP, Tate M, Merot J, Welling PA (1999) A mutation linked with Bartter's syndrome locks Kir 1.1a (ROMK1) channels in a closed state. J Gen Physiol 114:685–700

Flagg TP, Yoo D, Sciortino CM, Tate M, Romero MF, Welling PA (2002) Molecular mechanism of a COOH-terminal gating determinant in the ROMK channel revealed by a Bartter's disease mutation. J Physiol 544:351–362

Garcia ML, Priest BT, Alonso-Galicia M, Zhou X, Felix JP, Brochu RM, Bailey T, Thomas-Fowlkes B, Liu J, Swensen A, Pai LY, Xiao J, Hernandez M, Hoagland K, Owens K, Tang H, de Jesus RK, Roy S, Kaczorowski GJ, Pasternak A (2014) Pharmacologic inhibition of the renal outer medullary potassium channel causes diuresis and natriuresis in the absence of kaliuresis. J Pharmacol Exp Ther 348:153–164

Gollasch B, Anistan YM, Canaan-Kuhl S, Gollasch M (2017) Late-onset Bartter syndrome type II. Clin Kidney J 10:594–599

Greger R, Schlatter E (1981) Presence of luminal K^+, a prerequisite for active NaCl transport in the cortical thick ascending limb of Henle's loop of rabbit kidney. Pflugers Arch 392:92–94

Hansen SB, Tao X, MacKinnon R (2011) Structural basis of PIP2 activation of the classical inward rectifier K^+ channel Kir2.2. Nature 477:495–498

Hattersley AT, Ashcroft FM (2005) Activating mutations in Kir6.2 and neonatal diabetes: new clinical syndromes, new scientific insights, and new therapy. Diabetes 54:2503–2513

Hebert SC (2003) Bartter syndrome. Curr Opin Nephrol Hypertens 12:527–532

Hebert SC, Andreoli TE (1984) Effects of antidiuretic hormone on cellular conductive pathways in mouse medullary thick ascending limbs of Henle: II. Determinants of the ADH-mediated increases in transepithelial voltage and in net Cl-absorption. J Membr Biol 80:221–233

Hebert SC, Friedman PA, Andreoli TE (1984) Effects of antidiuretic hormone on cellular conductive pathways in mouse medullary thick ascending limbs of Henle: I. ADH increases transcellular conductance pathways. J Membr Biol 80:201–219

Hebert SC, Desir G, Giebisch G, Wang W (2005) Molecular diversity and regulation of renal potassium channels. Physiol Rev 85:319–371

Heginbotham L, Lu Z, Abramson T, MacKinnon R (1994) Mutations in the K+ channel signature sequence. Biophys J 66:1061–1067

Ho K, Nichols CG, Lederer WJ, Lytton J, Vassilev PM, Kanazirska MV, Hebert SC (1993) Cloning and expression of an inwardly rectifying ATP-regulated potassium channel. Nature 362:31–38

Huang CL (2007) Complex roles of PIP2 in the regulation of ion channels and transporters. Am J Physiol Renal Physiol 293:F1761–F1765

Huang CL, Feng S, Hilgemann DW (1998) Direct activation of inward rectifier potassium channels by PIP2 and its stabilization by Gbetagamma. Nature 391:803–886

Ji W, Foo JN, O'Roak BJ, Zhao H, Larson MG, Simon DB, Newton-Cheh C, State MW, Levy D, Lifton RP (2008) Rare independent mutations in renal salt handling genes contribute to blood pressure variation. Nat Genet 40:592–599

Karolyi L, I. C. S. G. f. B.-l. S. (1997) Mutations in the gene encoding the inwardly-rectifying renal potassium channel, ROMK, cause the antenatal variant of Bartter syndrome: evidence for genetic heterogeneity. International Collaborative Study Group for Bartter-like Syndromes. Hum Mol Genet 6:17–26

Konrad M, Vollmer M, Lemmink HH, van den Heuvel LP, Jeck N, Vargas-Poussou R, Lakings A, Ruf R, Deschenes G, Antignac C, Guay-Woodford L, Knoers NV, Seyberth HW, Feldmann D, Hildebrandt F (2000) Mutations in the chloride channel gene CLCNKB as a cause of classic Bartter syndrome. J Am Soc Nephrol 11:1449–1459

Liu W, Morimoto T, Woda C, Kleyman TR, Satlin LM (2007) $Ca2^+$ dependence of flow-stimulated K secretion in the mammalian cortical collecting duct. Am J Physiol Renal Physiol 293:F227–F235

Lopes CM, Zhang H, Rohacs T, Jin T, Yang J, Logothetis DE (2002) Alterations in conserved Kir channel-PIP2 interactions underlie channelopathies. Neuron 34:933–944

Lorenz JN, Baird NR, Judd LM, Noonan WT, Andringa A, Doetschman T, Manning PA, Liu LH, Miller ML, Shull GE (2002) Impaired renal NaCl absorption in mice lacking the ROMK potassium channel, a model for type II Bartter's syndrome. J Biol Chem 277:37871–37880

Lu M, Wang T, Yan Q, Yang X, Dong K, Knepper MA, Wang W, Giebisch G, Shull GE, Hebert SC (2002) Absence of small conductance K+ channel (SK) activity in apical membranes of thick ascending limb and cortical collecting duct in ROMK (Bartter's) knockout mice. J Biol Chem 277:37881–37887

Ma D, Tang XD, Rogers TB, Welling PA (2007) An Andersen-Tawil syndrome mutation in Kir2.1 (V302M) alters the G-loop cytoplasmic K^+ conduction pathway. J Biol Chem 282:5781–5789

MacKinnon R (2004) Nobel Lecture. Potassium channels and the atomic basis of selective ion conduction. Biosci Rep 24:75–100

McNicholas CM, Wang W, Ho K, Hebert SC, Giebisch G (1994) Regulation of ROMK1 K^+ channel activity involves phosphorylation processes. Proc Natl Acad Sci USA 91:8077–8081

McNicholas CM, Guggino WB, Schwiebert EM, Hebert SC, Giebisch G, Egan ME (1996) Sensitivity of a renal K^+ channel (ROMK2) to the inhibitory sulfonylurea compound glibenclamide is enhanced by coexpression with the ATP-binding cassette transporter cystic fibrosis transmembrane regulator. Proc Natl Acad Sci USA 93:8083–8088

Meng XY, Liu S, Cui M, Zhou R, Logothetis DE (2016) The molecular mechanism of opening the Helix Bundle Crossing (HBC) gate of a Kir channel. Sci Rep 6:29399

Monette MY, Rinehart J, Lifton RP, Forbush B (2011) Rare mutations in the human Na-K-Cl cotransporter (NKCC2) associated with lower blood pressure exhibit impaired processing and transport function. Am J Physiol Renal Physiol 300:840–847

Mutig K, Kahl T, Saritas T, Godes M, Persson P, Bates J, Raffi H, Rampoldi L, Uchida S, Hille C, Dosche C, Kumar S, Castaneda-Bueno M, Gamba G, Bachmann S (2011) Activation of the bumetanide-sensitive Na^+,K^+,$2Cl^-$ cotransporter (NKCC2) is facilitated by Tamm-Horsfall protein in a chloride-sensitive manner. J Biol Chem 286:30200–30210

Nanazashvili M, Li H, Palmer LG, Walters DE, Sackin H (2007) Moving the pH gate of the Kir1.1 inward rectifier channel. Channels (Austin) 1:21–28

O'Donnell BM, Mackie TD, Brodsky JL (2017a) Linking chanelopathies with endoplasmic reticulum associated degradation. Channels (Austin) 11:499–501

O'Donnell BM, Mackie TD, Subramanya AR, Brodsky JL (2017b) Endoplasmic reticulum-associated degradation of the renal potassium channel, ROMK, leads to type II Bartter syndrome. J Biol Chem 292:12813–12827

Palmer LG, Frindt G (1999) Regulation of apical K channels in rat cortical collecting tubule during changes in dietary K intake. Am J Physiol 277:F805–FF12

Pegan S, Arrabit C, Zhou W, Kwiatkowski W, Collins A, Slesinger PA, Choe S (2005) Cytoplasmic domain structures of Kir2.1 and Kir3.1 show sites for modulating gating and rectification. Nat Neurosci 8:279–287

Pegan S, Arrabit C, Slesinger PA, Choe S (2006) Andersen's syndrome mutation effects on the structure and assembly of the cytoplasmic domains of Kir2.1. Biochemistry 45:8599–8606

Peters M, Ermert S, Jeck N, Derst C, Pechmann U, Weber S, Schlingmann KP, Seyberth HW, Waldegger S, Konrad M (2003) Classification and rescue of ROMK mutations underlying hyperprostaglandin E syndrome/antenatal Bartter syndrome. Kidney Int 64:923–932

Plaster NM, Tawil R, Tristani-Firouzi M, Canun S, Bendahhou S, Tsunoda A, Donaldson MR, Iannaccone ST, Brunt E, Barohn R, Clark J, Deymeer F, George AL Jr, Fish FA, Hahn A, Nitu A, Ozdemir C, Serdaroglu P, Subramony SH, Wolfe G, Fu YH, Ptacek LJ (2001) Mutations in Kir2.1 cause the developmental and episodic electrical phenotypes of Andersen's syndrome. Cell 105:511–519

Pluznick JL, Wei P, Carmines PK, Sansom SC (2003) Renal fluid and electrolyte handling in BKCa-beta1 −/− mice. Am J Physiol Renal Physiol 284:F1274–F1279

Rapedius M, Haider S, Browne KF, Shang L, Sansom MS, Baukrowitz T, Tucker SJ (2006) Structural and functional analysis of the putative pH sensor in the Kir1.1 (ROMK) potassium channel. EMBO Rep 7:611–616

Rapedius M, Fowler PW, Shang L, Sansom MS, Tucker SJ, Baukrowitz T (2007) H bonding at the helix-bundle crossing controls gating in Kir potassium channels. Neuron 55:602–614

Renigunta A, Renigunta V, Saritas T, Decher N, Mutig K, Waldegger S (2011) Tamm-Horsfall glycoprotein interacts with renal outer medullary potassium channel ROMK2 and regulates its function. J Biol Chem 286:2224–2235
Rieg T, Vallon V, Sausbier M, Sausbier U, Kaissling B, Ruth P, Osswald H (2007) The role of the BK channel in potassium homeostasis and flow-induced renal potassium excretion. Kidney Int 72:566–573
Ruknudin A, Schulze DH, Sullivan SK, Lederer WJ, Welling PA (1998) Novel subunit composition in a renal Katp Channel (Kir1.1a/CFTR). J Biol Chem 273:14165–14171
Sackin H, Nanazashvili M, Palmer LG, Krambis M, Walters DE (2005) Structural locus of the pH gate in the Kir1.1 inward rectifier channel. Biophys J 88:2597–2606
Schulte U, Hahn H, Wiesinger H, Ruppersberg JP, Fakler B (1998) pH-dependent gating of ROMK (Kir1.1) channels involves conformational changes in both N and C termini. J Biol Chem 273:34575–34579
Schulte U, Hahn H, Konrad M, Jeck N, Derst C, Wild K, Weidemann S, Ruppersberg JP, Fakler B, Ludwig J (1999) pH gating of ROMK (K(ir)1.1) channels: control by an Arg-Lys-Arg triad disrupted in antenatal Bartter syndrome. Proc Natl Acad Sci USA 96:15298–15303
Seyberth HW, Schlingmann KP (2011) Bartter- and Gitelman-like syndromes: salt-losing tubulopathies with loop or DCT defects. Pediatr Nephrol 26:1789–1802
Simon DB, Karet FE, Hamdan JM, DiPietro A, Sanjad SA, Lifton RP (1996) Bartter's syndrome, hypokalaemic alkalosis with hypercalciuria, is caused by mutations in the Na-K-2Cl cotransporter NKCC2. Nat Genet 13:183–188
Simon D, Karet F, Rodriguez-Soriano J, Hamdan J, DiPietro A, Sanjad S, Lifton R (1997a) Genetic heterogeneity of Bartter's syndrome revealed by mutations in the K+ channel, ROMK. Nat Genet 14:152–156
Simon DB, Bindra RS, Mansfield TA, Nelson-Williams C, Mendonca E, Stone R, Schurman S, Nayir A, Alpay H, Bakkaloglu A, Rodriguez-Soriano J, Morales JM, Sanjad SA, Taylor CM, Pilz D, Brem A, Trachtman H, Griswold W, Richard GA, John E, Lifton RP (1997b) Mutations in the chloride channel gene, CLCNKB, cause Bartter's syndrome type III. Nat Genet 17:171–118
Srivastava S, Li D, Edwards N, Hynes AM, Wood K, Al-Hamed M, Wroe AC, Reaich D, Moochhala SH, Welling PA, Sayer JA (2013) Identification of compound heterozygous KCNJ1 mutations (encoding ROMK) in a kindred with Bartter's syndrome and a functional analysis of their pathogenicity. Physiol Rep 1:e00160
Starremans PG, van der Kemp AW, Knoers NV, van den Heuvel LP, Bindels RJ (2002) Functional implications of mutations in the human renal outer medullary potassium channel (ROMK2) identified in Bartter syndrome. Pflugers Archiv 443:466–472
Subramanya AR, Welling PA (2011) Toward an understanding of hypertension resistance. Am J Physiol Renal Physiol 300:F838–F839
Tang H, Walsh SP, Yan Y, de Jesus RK, Shahripour A, Teumelsan N, Zhu Y, Ha S, Owens KA, Thomas-Fowlkes BS, Felix JP, Liu J, Kohler M, Priest BT, Bailey T, Brochu R, Alonso-Galicia M, Kaczorowski GJ, Roy S, Yang L, Mills SG, Garcia ML, Pasternak A (2012) Discovery of selective small molecule ROMK inhibitors as potential new mechanism diuretics. ACS Med Chem Lett 3:367–372
Vargas-Poussou R, Huang C, Hulin P, Houillier P, Jeunemaitre X, Paillard M, Planelles G, Dechaux M, Miller RT, Antignac C (2002) Functional characterization of a calcium-sensing receptor mutation in severe autosomal dominant hypocalcemia with a Bartter-like syndrome. J Am Soc Nephrol 13:259–266
Wagner CA, Loffing-Cueni D, Yan Q, Schulz N, Fakitsas P, Carrel M, Wang T, Verrey F, Geibel JP, Giebisch G, Hebert SC, Loffing J (2008) Mouse model of type II Bartter's syndrome. II. Altered expression of renal sodium- and water-transporting proteins. Am J Physiol Renal Physiol 294:F1373–F1380
Wang WH (1994) Two types of K^+ channel in thick ascending limb of rat kidney. Am J Physiol 267:F599–F605

Wang WH, Schwab A, Giebisch G (1990) Regulation of small-conductance K^+ channel in apical membrane of rat cortical collecting tubule. Am J Phys 259:F494–F502

Watanabe S, Fukumoto S, Chang H, Takeuchi Y, Hasegawa Y, Okazaki R, Chikatsu N, Fujita T (2002) Association between activating mutations of calcium-sensing receptor and Bartter's syndrome. Lancet 360:692–694

Weinstein AM (2010) A mathematical model of rat ascending Henle limb. III. Tubular function. Am J Physiol Renal Physiol 298:F543–F556

Welling PA (2013) Regulation of renal potassium secretion: molecular mechanisms. Semin Nephrol 33:215–228

Welling PA, Ho K (2009) A comprehensive guide to the ROMK potassium channel: form and function in health and disease. Am J Physiol Renal Physiol 297:F849–F863

Xu ZC, Yang Y, Hebert SC (1996) Phosphorylation of the ATP-sensitive, inwardly rectifying K+ channel, ROMK, by cyclic AMP-dependent protein kinase. J Biol Chem 271:9313–9319

Zeng WZ, Liou HH, Krishna UM, Falck JR, Huang CL (2002) Structural determinants and specificities for ROMK1-phosphoinositide interaction. Am J Physiol Renal Physiol 282: F826–34

Zhou H, Tate SS, Palmer LG (1994) Primary structure and functional properties of an epithelial K channel. Am J Physiol 266:C809–C824

Chapter 20
Inwardly Rectifying K^+ Channel 4.1 Regulates Renal K^+ Excretion in the Aldosterone-Sensitive Distal Nephron

Wen-Hui Wang and Dao-Hong Lin

Abstract The inwardly rectifying potassium channel 4.1 (Kir4.1, KCNJ10) in the aldosterone-sensitive distal nephron (ASDN) plays a key role in the regulation of Na^+ and K^+ transport by sensing the dietary K^+ and Na^+ intake. The K^+ channel is also the target for angiotensin II, bradykinin, and norepinephrine which are known to regulate the renal Na^+ and K^+ transport in the ASDN. Since Kir4.1 plays a critical role in the regulation of renal K^+ excretion and in maintaining K^+ homeostasis, the main focus of this chapter is to review the role of Kir4.1 in the regulation of renal K^+ excretion. Finally, because Kir4.1/Kir5.1 (encoded by KCNJ16) heterotetramer is a more physiologically relevant form of the K^+ channel in the ASDN than Kir4.1 homotetramer, the role of the Kir4.1/Kir5.1 heterotetramer in the regulation of membrane transport is discussed in the chapter together with Kir4.1 homotetramer.

Keywords KCNJ10 · KCNJ16 · Na-Cl cotransport · K^+ excretion · SeSAME/EAST syndrome

20.1 Introduction

The inwardly rectifying K^+ channel 4.1 (Kir4.1) is encoded by *KCNJ10* which was first cloned from a brain cDNA library (Bond et al. 1994; Bredt et al. 1995; Takumi et al. 1995) and late from the kidney (Shuck et al. 1997). The amino acid sequence of Kir4.1 has 53% identity with Kir1.1, 43% with Kir2.1, 43% with Kir3.1, and 30% with Kir5.1. Kir4.1 is highly expressed in the brain, inner ear, retinal tissue, stomach, and kidney (Kofuji et al. 2000; Neusch et al. 2001; Fujita et al. 2002; Lourdel et al. 2002; Rozengurt et al. 2003). Loss-of-function-mutations of Kir4.1 cause EAST/SeSAME syndrome in humans (seizures, sensorineural deafness, ataxia, mental retardation, and electrolyte imbalance) (Bockenhauer et al. 2009; Scholl et al.

W.-H. Wang (✉) · D.-H. Lin
Department of Pharmacology, New York Medical College, Valhalla, NY, USA
e-mail: Wenhui_wang@nymc.edu; Daohong_lin@nymc.edu

K. L. Hamilton, D. C. Devor (eds.), *Studies of Epithelial Transporters and Ion Channels*, Physiology in Health and Disease,
https://doi.org/10.1007/978-3-030-55454-5_20

2009; Reichold et al. 2010; Bandulik et al. 2011). The renal phenotypes caused by Kir4.1 mutations include a mild Na^+ wasting, hypomagnesemia, metabolic alkalosis, and hypokalemia, suggesting the role of Kir4.1 in the regulation of renal membrane transport (Scholl et al. 2009). Thus, the main aim of this chapter is to address the role of Kir4.1 in the regulation of membrane transport in the kidney but not to discuss the function and structure relationship of Kir4.1, although it is an important topic. Moreover, considering that Kir4.1 plays a critical role in the regulation of renal K^+ excretion and in maintaining K^+ homeostasis, the role of Kir4.1 in the regulation of renal K^+ excretion is a main focus of this chapter, although Kir4.1 is also involved in the regulation of Na^+ and Mg^{2+} transport (Ferre et al. 2011; Houillier 2014; Wu et al. 2019). Finally, because Kir4.1/Kir5.1 (encoded by *KCNJ16*) heterotetramer is a more physiologically relevant form of the K^+ channel than Kir4.1 homotetramer in the kidney (Lourdel et al. 2002; Lachheb et al. 2008; Zaika et al. 2013; Zhang et al. 2014), the role of the Kir4.1/Kir5.1 heterotetramer in the regulation of membrane transport is discussed in the chapter together with Kir4.1 homotetramer.

20.2 Kir4.1/Kir5.1 Forms the Basolateral K^+ Channel in the ASDN

Histochemical staining shows that both Kir4.1 and Kir5.1 are exclusively expressed in the renal cortex (Tucker et al. 2000; Derst et al. 2001; Cuevas et al. 2017). Furthermore, morphological examinations show that Kir4.1 is expressed only in the basolateral membrane of the late part of cortical thick ascending limb (cTAL), the distal convoluted tubule (DCT), connecting tubule (CNT), and early portion of cortical collecting duct (CCD) (Lourdel et al. 2002; Zhang et al. 2014, 2015; Su et al. 2016). Figure 20.1 is a scheme illustrating the expression pattern of Kir4.1 and Kir5.1 together with Na-Cl cotransporter (NCC), epithelial Na^+ channel (ENaC), type II Na-K-Cl cotransporter (NKCC2), and Kir1.1 (ROMK) from the cTAL to the CCD. It is apparent that Kir4.1 and Kir5.1 are co-expressed in the basolateral membrane of the aldosterone-sensitive distal nephron (ASDN). The basolateral location of Kir4.1 in the kidney may require a scaffolding protein, *m*embrane-*a*ssociated *g*uanylate kinase with *i*nverted domain structure 1 (MAGI-1a-long) (Tanemoto et al. 2008). The notion that MAGI-1a may play a role in sorting Kir4.1 to the basolateral membrane has been supported by immunostaining images showing that MAGI and Kir4.1 are colocalized in the basolateral membrane of renal tubules. The role of MAGI in sorting Kir4.1 has also been indicated *in vitro* experiments showing that Kir4.1 is expressed in the basolateral membrane of MDCK cells cotransfected with MAGI-1a-long. It has been shown that MAGI-1a-long interacts with the PSD-95/Dlg/ZO-1(PDZ)-binding motif of Kir4.1 at the C-terminus for the basolateral membrane location and that the disruption of C-terminal PDZ binding motif affects the basolateral location of the K^+ channel (Tanemoto et al. 2004, 2005). Moreover, phosphorylation of serine residue 377 in

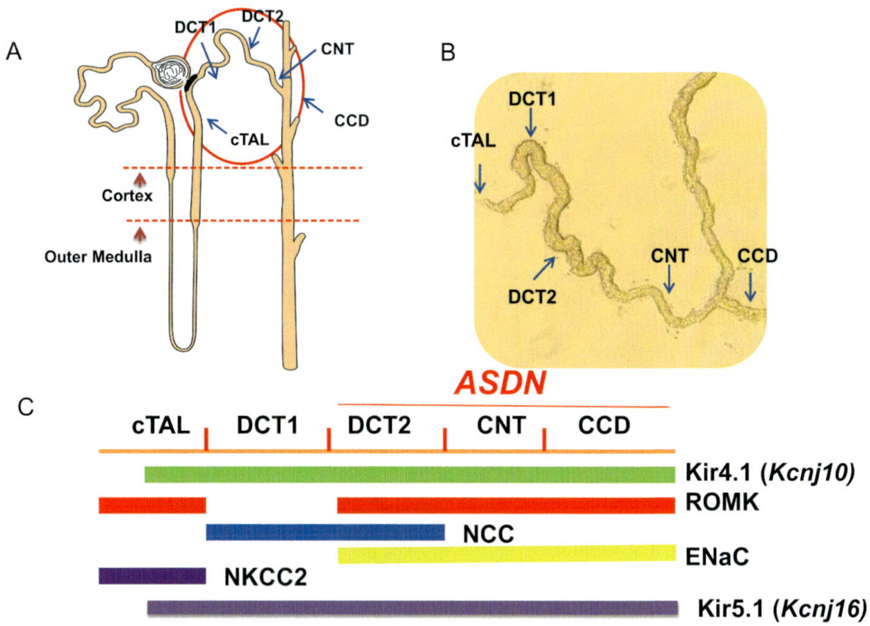

Fig. 20.1 The expression of Kir4.1 is detected in the basolateral membrane from the cortical thick ascending limb (cTAL), distal convoluted tubule (DCT), connecting tubule (CNT), and cortical collecting duct (CCD). (**a**) A scheme of nephron indicates the position of the cTAL, DCT, CNT, and CCD, and a red oval indicates the nephron segment where Kir4.1 is expressed. (**b**) An isolated tubule shows late cTAL, DCT, CNT to the CCD where Kir4.1 is expressed in the basolateral membrane. (**c**) A scheme illustrating the expression of Kir4.1, ROMK, NCC, ENaC, NKCC2 and Kir5.1 from cTAL to the CCD

the PDZ-binding motif of Kir4.1 decreased the association of MAGI-1a-long with Kir4.1 and inhibited its basolateral sorting in MDCK cells (Tanemoto et al. 2005).

Immunostaining examination reveals that Kir5.1 is also expressed in the basolateral membrane of the proximal tubule, TAL, DCT, CNT, and CCD (Tucker et al. 2000; Lachheb et al. 2008). However, the basolateral expression of Kir5.1 in the DCT, CNT, or CCD requires the presence of Kir4.1. The immunostaining images in Kir4.1 knockout mice showed that in the absence of Kir4.1, the expression of Kir5.1 was located in the perinuclear region but not on the plasma membrane (Fig. 20.2), suggesting that Kir5.1 trafficking to the plasma membrane was blocked in the absence of Kir4.1 (Zhang et al. 2014). Moreover, the previous study has reported that the expression of Kir5.1 alone failed to generate the functional K^+ channels in the plasma membrane in vitro (Pessia et al. 1996). The notion that Kir5.1 alone cannot form a functional K^+ channel is also supported by patch-clamp experiments showing that the K^+ channel activity was completely absent in the basolateral membrane of the DCT of global or kidney-specific Kir4.1 knockout mice (Zhang et al. 2014). Figure 20.3a is a whole-cell recording showing that the inwardly rectifying K^+ currents are recorded in the DCT of WT mice but not in Ks-Kir4.1 KO

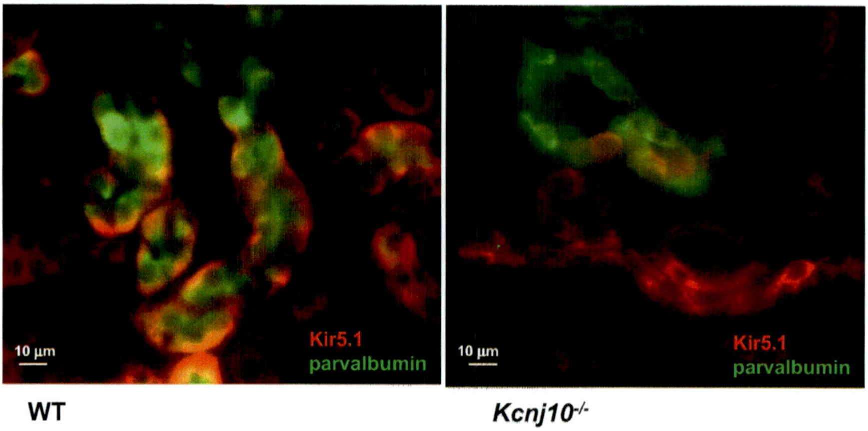

Fig. 20.2 Kir4.1 interacts with Kir5.1 to form a heterotetramer in the basolateral membrane from the cTAL, the DCT to the CCD. However, Kir5.1 is not able to be expressed in the basolateral membrane in the absence of Kir4.1. Immunofluorescence images show the expression of Kir5.1 (red) in the renal cortex of WT and *Kcnj10*$^{-/-}$ mice. The parvalbumin (green) was used as a mark of the DCT. Note: the membrane staining of Kir5.1 is absent in *Kcnj10*$^{-/-}$ mice

mice. Thus, Kir4.1 is solely responsible for providing the K^+ conductance of the Kir4.1/Kir5.1 heterotetramer in the DCT (Zhang et al. 2014).

While Kir4.1 alone can form homotetramer (a 20–25 pS K^+ channel) *in vivo* and *in vitro* (Pessia et al. 1996; Paulais et al. 2011; Palygin et al. 2017), Kir.4.1 prefers interacting with Kir5.1 to form a functional 40–45 pS heterotetrameric K^+ channel in the basolateral membrane of the native tissue under physiological conditions (Lachheb et al. 2008; Zhang et al. 2013, 2014). Figure 20.3b is a channel recording showing 20 pS K^+ channel activity in the basolateral membrane of the DCT of Kir5.1 knockout mouse and 40 pS K^+ channel activity in the basolateral membrane of the DCT of a WT mouse. Figure 20.3b is a current (I) and voltage (V) relationship of 20 pS and 40 pS K^+ channel, respectively. For the interaction of Kir4.1 with Kir5.1, glutamate residue 177 in Kir4.1 C-terminus located after second transmembrane segment has been demonstrated to be required for forming a Kir4.1/Kir5.1 heterotetramer (Konstas et al. 2003). The Kir4.1 homotetramer shows modest inward rectification, whereas Kir4.1/Kir5.1 heterotetramer exhibits stronger rectification than Kir4.1. The modest inward rectification of Kir4.1 homotetramer may be due to partial permeation of spermine of the K^+ channel, thereby diminishing spermine-induced blockade of outward K^+ currents (Kucheryavykh et al. 2007).

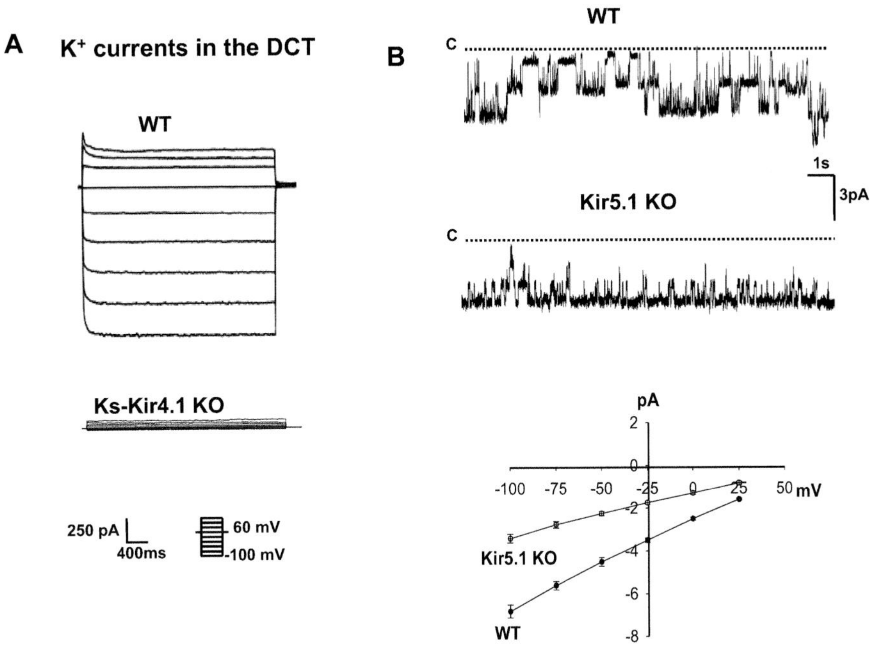

Fig. 20.3 Kir4.1 is the only component of Kir4.1/Kir5.1 heterotetramer for providing K^+ permeability (conductance). In the presence of Kir5.1, Kir4.1 interacts with Kir5.1 to form a 40 pS K^+ channel while Kir4.1 forms a 20 pS homeotetramer K^+ channel in the absence of Kir5.1. (**a**) A whole-cell recording showing the Ba^{2+}-sensitive K^+ currents in the DCT1 of WT and Ks-Kir4.1 KO mice, showing that the deletion of Kir4.1 abolishes the K^+ conductance in the DCT, suggesting that Kir4.1 is the only component of Kir4.1/Kir5.1 heterotetramer for providing K^+ permeability. (**b**) A single channel recording in a cell-attached patch showing the basolateral 40 pS K^+ channel activity (Kir4.1/Kir5.1 heterotetramer) in the DCT of WT mice and the basolateral 20 pS K^+ channel activity (Kir4.1 homotetramer) in the DCT of Kir5.1 knockout mice. The holding potential was zero mV and the channel closed level is indicated by a dotted line and "C." The I/V curve of the basolateral K^+ channel of the DCT is shown in the bottom of the panel

20.3 Kir5.1 Is a Regulatory Subunit for Kir4.1/Kir5.1 Heterotetramer

While the role of Kir4.1 in the heterotetramer is to provide K^+ permeability, a large body of evidence has indicated that Kir5.1 may be served as a regulatory subunit for Kir4.1/Kir5.1 heterotetramer (Tucker et al. 2000; Tanemoto et al. 2000; Pessia et al. 2001; Casamassima et al. 2003; Lachheb et al. 2008; Wang et al. 2018b). Kir5.1 has been shown to play a role in the regulation of pH sensitivity of the Kir4.1/Kir5.1 heterotetramer because Kir4.1/Kir5.1 heterotetramer is more pH sensitive than Kir4.1 homotetramer (Yang et al. 2000; Tucker et al. 2000; Tanemoto et al. 2000; Pessia et al. 2001; Casamassima et al. 2003; Lachheb et al. 2008). It has been demonstrated that reduced cell pH from 7.0 to 6.5 completely blocked the Kir4.1/

Kir5.1 activity, but it only partially inhibited Kir4.1 homotetramer (Tucker et al. 2000). The notion that Kir5.1 plays a role in the regulation of pH sensitivity of the native basolateral K^+ channels in the kidney is also confirmed by the finding that pH sensitivity of the 20–24 pS K^+ channel (Kir4.1 homotetramer) in the DCT of Kir5.1 knockout mice was reduced (Paulais et al. 2011). Two studies have demonstrated that arginine residue (R230), histidine residue 24 (H24), H38, H49 H72 H255, H259, H293, H308, and H311 in Kir5.1 and that lysine residue (K67) in the N-terminus and H190 in c-terminus of Kir4.1 are involved in the regulation of pH sensitivity of Kir4.1/Kir5.1 heterotetramer (Pessia et al. 2001; Casamassima et al. 2003). The mechanism by which Kir5.1/Kir4.1 heterotetramer is more pH sensitive than Kir4.1 homotetramer may be induced by inter-subunit interactions produced by heteropolymerisation of Kir5.1 with Kir4.1 (Pessia et al. 2001; Casamassima et al. 2003).

In addition, Kir5.1 may play a role in the regulation of the expression of the basolateral K^+ channels in the kidney. Although the expression of Kir5.1 has been reported to increase the expression of Kir4.1 *in vitro* (Pessia et al. 1996), the deletion of Kir5.1 actually increased the expression of Kir4.1 in vivo rather than decreased Kir4.1 expression (Paulais et al. 2011; Wang et al. 2018b), suggesting that Kir5.1 may exert an inhibitory effect on Kir4.1 expression in the native renal tubules. In this regard, it has been reported that the biophysical properties of the homotetramer of Kir4.1A167V, a mutant which causes EAST syndrome, are different from the heterotetramer of Kir5.1/Kir4.1A167V (Parrock et al. 2013). It has been shown that K^+ currents in oocytes injected with Kir4.1A167V were significantly larger than in oocytes coinjected with Kir5.1 mRNA, suggesting the role of Kir5.1 as a regulatory subunit. Recently, Wang et al. (2018b) have demonstrated that Kir5.1 is associated with E3 ubiquitin ligase, Nedd4-2, to its C-terminus, thereby regulating the ubiquitination of Kir4.1 (Wang et al. 2018b). Furthermore, the deletion of Kir5.1 or Nedd4-2 significantly increased Kir4.1 expression and K^+ currents in the DCT, suggesting that Kir5.1 is involved in regulating the ubiquitination of Kir4.1 in the DCT. These results suggest that Kir5.1 is a regulatory subunit for Kir4.1/Kir5.1 heterotetramer and plays a role in determining Kir4.1 activity in the kidney.

20.4 Regulation of Kir4.1 and Kir5.1 in the Kidney

It has been reported that hepatocyte nuclear factor 1 homeobox B (HNF1β), a transcription factor, plays a role in the regulation of Kir5.1 expression. A specific and conserved HNF1β site has been identified in the promoter region of Kcnj16 gene and coexpression of HNF1β increases *KCNJ16* transcription activity detected with luciferase-promoter assay (Kompatscher et al. 2017). In contrast, no effect on *KCNJ16* transcription has been detected with the expression of Hnf1βp.Lys156Glu, a mutant identified in a patient with autosomal dominant tubulointerstitial kidney disease characterized by renal cysts and severe electrolyte disorder including hypomagnesemia and hypokalemia. Moreover, knockdown of Hnf1β in mouse DCT cells has been reported to decrease the expression of Kir5.1 and Kir4.1 by 38% and 37%, respectively (Kompatscher et al. 2017). Also, the deletion of HNF1β in the mouse kidney has been shown to reduce *Kcnj16* and *Kcnj10* transcripts by 78% and 83%, respectively, suggesting that HNF1β is a transcriptional activator for *Kcnj16*. However, it is not clear whether a decrease in *Kcnj10* expression in kidney-specific HNF1β knockout mice is due to the inhibition of Kir5.1 expression since the deletion of Kir5.1 in the mice actually stimulates Kir4.1 expression in the kidney (Paulais et al. 2011; Palygin et al. 2017). In addition, because Kir5.1 is also expressed in the PCT and forms an ATP-sensitive K^+ channel (possible a Kir4.2/Kir5.1 heterotetramer) in the basolateral membrane (Tsuchiya et al. 1992; Tucker et al. 2000), a compromised membrane transport in proximal tubules may also contribute to the phenotypes observed in patients with HNF1β mutations.

Like other inwardly rectifying K^+ channels, PtdIns(4,5)P_2 has been shown to regulate Kir4.1/Kir5.1 activity and PtdIns(4,5)P_2-mediated modulation of the heterotetramer affects the K^+ channel sensitivity to pH- and Mg^{2+}-induced inhibition (Yang et al. 2000; Du et al. 2004). Dopamine has been shown to inhibit Kir4.1/Kir5.1 activity in the CCD by a protein kinase C-dependent mechanism (Zaika et al. 2013). In addition, it has been reported that Kir4.1/Kir5.1 in the CCD is stimulated by insulin/IGF1 (Zaika et al. 2015). Calcium-sensing receptor (CaSR) has been shown to be coimmunoprecipitated with Kir4.1 and Kir5.1 in HEK293 cells. Moreover, the expression of CaSR decreased the surface expression of Kir4.1, suggesting the role of CaSR in modulating the activity of the basolateral 40 pS K^+ channels (Huang et al. 2007; Cha et al. 2011). Considering that gain-of-function-mutations of CaSR have been shown to cause Bartter-like syndrome characterized by metabolic alkalosis and hypokalemia (Vargas-Poussou et al. 2002), it is conceivable that the phenotypes related to CaSR mutations are at least in part induced by the abnormal regulation of basolateral Kir4.1/Kir5.1 channel in the DCT.

Kir4.1 is a substrate of src-family protein tyrosine kinase (SFK) and the coexpression of c-Src phosphorylates Kir4.1 at Tyr^9 (Zhang et al. 2013). The SFK-induced tyrosine phosphorylation of Kir4.1 plays a role in the stimulation of the basolateral 40 pS K^+ channel in the DCT because the inhibition of SFK suppresses the 40 pS K^+ channel activity in the DCT. Caveolin-1 plays an important role in the regulation of Kir4.1/Kir5.1 expression in the plasma membrane of DCT (Wang et al. 2015). Wang et al. have observed that the deletion of caveolin-1 significantly inhibited the basolateral K^+ channel (Kir4.1/Kir5.1 heterotetramer) activity in the DCT and that Kir4.1 expression was downregulated in caveolin-1 knockout mice. In addition, the cell membrane potential of DCT was depolarized in cavolin-1 knockout mice. Although the mechanism by which the deletion of caveolin-1 decreases the Kir4.1 activity in the DCT is not completely clear, one possible mechanism is that cavolin-1 is required for the stimulatory effect of c-Src tyrosine kinase on Kir4.1. This notion is supported by the observation that c-Src stimulates Kir4.1 channel activity only in the presence of cavolin-1. Also, immunostaining demonstrates that cavolin-1 is highly expressed in the basolateral membrane of the DCT and collocalized with Kir4.1. Furthermore, SFK may be involved in mediating the effect of $CD8^+$-T cells on Kir4.1 in cultured mouse DCT cells (Liu et al. 2017). It has been shown that co-culture of $CD8^+$-T cells with mouse DCT cells increases the expression of Kir4.1 of the DCT cells. Moreover, the effect of $CD8^+$-T cells on Kir4.1 can be mimicked by applying reactive oxygen species (ROS), but it is inhibited by SFK siRNA. This finding suggests that the interaction of the DCT cells with $CD8^+$-T cells may activate ROS-SFK signaling in the DCT cells thereby stimulating Kir4.1.

Dietary K^+ intake has been shown to regulate Kir4.1/Kir5.1 (Tomilin et al. 2018; Wang et al. 2018a). However, the effect of increased dietary K^+ intake on Kir4.1/Kir5.1 is different between the DCT and CCD. Increased dietary K^+ intake inhibits Kir4.1/Kir5.1 in the DCT, whereas it stimulates the basolateral K^+ channels in the CCD. The different responses of Kir4.1/Kir5.1 between DCT and CCD to high K^+ intake have a physiological significance since high K^+-induced inhibition of Kir4.1/Kir5.1 in the DCT is essential for the inhibition of NCC whereas high K^+-induced stimulation of Kir4.1/Kir5.1 in the CCD should increase the driving force of K^+ excretion by increasing the negativity of the membrane potential in the CCD. High aldosterone level induced by high K^+ diet may be responsible for the stimulation of Kir4.1/Kir5.1 in the CCD, but it may have no effect on the K^+ channel in the DCT. This possibility is supported by our unpublished observations that the effect of dietary K^+ intake on the basolateral Kir4.1/Kir5.1 in the DCT is intact in the absence of mineralocorticoid receptor. A relevant sample for the discriminated effect of aldosterone on ion channels is ENaC activity which is aldosterone-dependent in the CCD but not in the DCT (Nesterov et al. 2012).

20.5 Kir4.1/Kir5.1 Determines the Membrane Potential of the DCT

Because Kir4.1/Kir5.1 heterotetramer is highly expressed in the basolateral membrane from cTAL to the CCD, it is conceivable that the heterotetramer should determine the negativity of the membrane potential of the corresponding renal tubule cells. Indeed, previous experiments performed in the native cTAL, the DCT and the CCD have confirmed that Kir4.1/Kir5.1 participates in generating the membrane potential (Zhang et al. 2014, 2015; Su et al. 2016). However, the contribution of Kir4.1 to the basolateral K^+ conductance is varied from cTAL to CCD. For instance, Kir4.1 may contribute only partially to the basolateral K^+ conductance in the cTAL and in the CCD. This notion is strongly suggested by the finding that the deletion of Kir4.1 did not completely eliminate the basolateral K^+ conductance and only modestly depolarized the basolateral membrane in the cTAL and CCD (Zhang et al. 2015; Su et al. 2016). The possibility that K^+ channels other than Kir4.1 contribute to the basolateral K^+ conductance in the cTAL and in the CCD has been confirmed by single-channel patch-clamp experiments in which several types of K^+ channels other than Kir4.1 have been detected (Lu and Wang 1996; Paulais et al. 2006; Zhang et al. 2015; Su et al. 2016). In contrast, Kir4.1 is the only type of K^+

channel responsible for the basolateral K^+ conductance in the entire DCT since the deletion of Kir4.1 completely eliminates the basolateral K^+ conductance and strongly depolarizes the membrane (Zhang et al. 2014; Cuevas et al. 2017). Thus, Kir4.1/Kir5.1 heterotetramer determines the basolateral K^+ conductance in the DCT and plays a key role in generating negative membrane potential in the DCT (Zhang et al. 2014).

The DCT is an initial nephron segment of the ASDN which plays a key role in mediating the hormone-regulated Na^+ absorption and in the regulation of renal K^+ excretion. Because Kir4.1 expression is overlapped with NCC, ENaC and ROMK along the ASDN (Fig. 20.1), the activity of Kir4.1 should affect the membrane transport in the ASDN. Indeed, recent studies have convincingly indicated that Kir4.1 presence is essential for the regulation of renal K^+ excretion (Terker et al. 2015; Cuevas et al. 2017; Wu et al. 2018; Wang et al. 2018a). For understanding the role of Kir4.1 in the regulation of K^+ excretion and in maintaining K^+ homeostasis with a perspective view, a brief overview is introduced in the following section to discuss the role between renal and extra-renal interaction in keeping normal plasma K^+ levels and to describe renal K^+ transport along the different nephron segments.

20.6 Regulation of K^+ Homeostasis by Kidney and Extrarenal Factors

Hyperkalemia or hypokalemia is known to cause cardiac arrhythmias and to interfere with normal neuron function and muscle contraction. On the other hand, keeping a high intracellular K^+ concentration is also critical for cell growth and differentiation. It is well established that K^+ homeostasis is achieved by the coordination between kidney-mediated K^+ secretion and extrarenal K^+ handling (Youn and McDonough 2009). Under physiological conditions, the total K^+ content in the extracellular fluid is approximately 70 meqM while the intracellular K^+ content is approximately 3500–4000 megM (Fig. 20.4). However, it is well known that K^+ ions are constantly shifted between the intracellular and extracellular fluids during increasing dietary K^+ intake or during K^+ restriction. Several factors which are known to enhance K^+ shifting from the extracellular to intracellular fluid are β2-adrenergics, insulin, and aldosterone which stimulate K^+ uptake by activating Na-K-ATPase. The role of insulin in mediating K^+ shifting after meals is well studied. An increased glucose uptake and an elevation of dietary K^+ intake stimulate insulin secretion which activates Na-K-ATPase in skeletal muscle and liver cells, thereby transferring K^+ from extracellular to intracellular fluid (Bia and DeFronzo 1981). This shift from the extracellular space to the intracellular environment minimizes the rise of plasma K^+ after each meal. However, K^+ up-taken by skeletal muscle and liver cells during the meal is eventually released into the extracellular fluid again and these K^+ ions have to be eliminated from extracellular fluid in order to maintain the plasma K^+ in a normal range. Although the distal colon is able to secrete K^+ when renal K^+ secretion

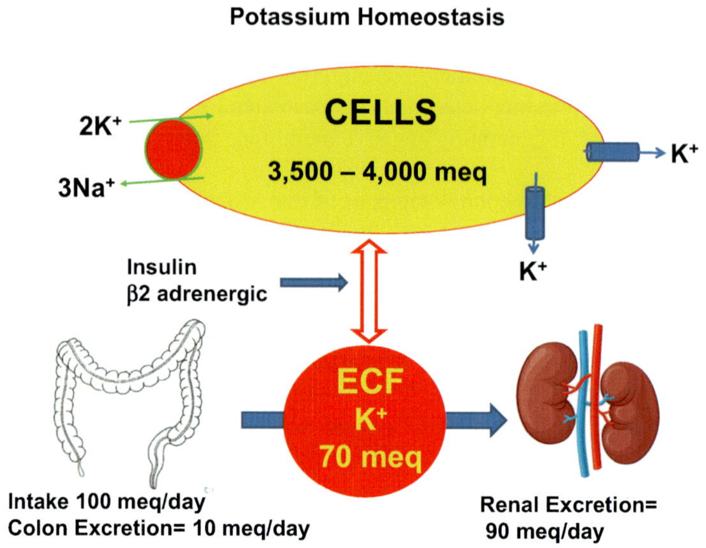

Fig. 20.4 K^+ homeostasis is maintained by kidney and extrarenal factors, and the kidney plays a key role in regulating K^+ excretion in response to dietary K^+ intake. A scheme illustrates the role of kidney and extrarenal organs in the regulation of K^+ homeostasis. K^+ ions are constantly shifted between the intracellular and extracellular fluids during increasing dietary K^+ intake or during K^+ restriction. Two factors which are known to enhance K^+ shifting from the extracellular to intracellular fluid are β2-adrenergics and insulin which stimulate K^+ uptake by activating Na-K-ATPase. This shift from the extracellular space to the intracellular environment minimizes the rise of plasma K^+ after each meal. However, K^+ up-taken by extrarenal organs is eventually released into the extracellular fluid again and these K^+ ions have to be eliminated from extracellular fluid in order to maintain the plasma K^+ in a normal range. The distal colon is able to secrete 10% of dietary K^+ intake, while the kidney plays an essential role in excreting 90% of dietary K^+ intake for maintaining K^+ homeostasis

is compromised or when the dietary K^+ intake is increased (Fisher et al. 1976; Bastl et al. 1977), the kidney plays an essential role in excreting 90% of dietary K^+ intake for maintaining K^+ homeostasis.

The synchronized action among the kidney, colon, and skeletal muscle also plays an important role in maintaining plasma K^+ within a relative normal range during K^+ restriction, at least, for a limited time frame. K^+ restriction not only stimulates K^+ absorption in the renal distal nephron segment (Giebisch 1998) but also favors K^+ shifting from skeletal muscles to extracellular fluids by decreasing the surface numbers of Na-K-ATPase (Azuma et al. 1991; Thompson and McDonough 1996). K^+ depletion also attenuates the response of skeletal muscle to insulin and inhibits insulin-induced K^+ shifting from extracellular to intracellular fluid (Choi et al. 2001). K^+ depletion has also been demonstrated to stimulate the expression of H-K-ATPase in the colon and to increase intestinal K^+ absorption (Codina et al. 1997). Thus, the interaction between renal and extra-renal regulation is essential for maintaining plasma K^+ in a relative normal range. However, the kidney is ultimately responsible for maintaining K^+ homeostasis and for excretion of 90% dietary K^+ content through glomerular filtration, reabsorption, and secretion along the nephron segment.

20.7 Renal K^+ Transport Along the Nephron Segment

K^+ transport in the kidney includes the K^+ absorption and the K^+ excretion: K^+ absorption mainly takes place in the proximal tubule and TAL under physiological conditions, while K^+ excretion occurs in ASDN including the late DCT (DCT2), CNT, and CCD.

20.7.1 Proximal Convoluted Tubule (PCT)

Approximately 60–70% of filtered K^+ is absorbed in the PCT, and this amount is similar to the proportion of Na^+ and water absorption (Giebisch 1998). K^+ reabsorption in the PCT is largely passive and proceeds mainly through paracellular pathways (Kaufman and Hamburger 1985; Kibble et al. 1995; Weinstein and Windhager 2001). However, the observation that the inhibition of K^+ channel activity with Ba^{2+} diminished the transepithelial K^+ absorption suggests that a small fractional K^+ may also be reabsorbed via transcellular pathways (Kibble et al. 1995). The main driving force for the passive paracellular K^+ absorption is solvent drag and the lumen-positive transepithelial potential generated by transepithelial Cl^- reabsorption in the late proximal tubule (Fromter 1974; Shirley et al. 1998). The PCT is responsible for the reabsorption of 70% of filtered NaCl and 100% of filtered glucose and amino acids. The absorption of these solutes is followed by water movement across the paracellular pathway. Because several Na^+-coupled transport processes are electrogenic in the PCT, an alteration in the membrane potential may affect the Na^+-coupled membrane transport and the fluid absorption. Consequently, changes in Na^+-dependent membrane transport may indirectly alter paracellular K^+ absorption because a significant fraction of filtered K^+ is reabsorbed by the solute-dependent water movement across a high K^+ permeable paracellular pathway. This solvent drag may be responsible for some paracellular K^+ reabsorption (Bomsztyk and Wright 1986; Weinstein 1986; Wareing et al. 1995). Also, water absorption in the proximal tubule creates a favorable K^+ concentration gradient from the lumen to the peritubular fluid, thereby facilitating paracellular K^+ reabsorption. In addition, a passive K^+ diffusion occurs by a favorable electrochemical gradient along the proximal tubule. This notion is further indicated by the finding that K^+ absorption in the proximal tubule is very sensitive to alterations in transepithelial voltage and K^+ concentrations in the lumen (Bomsztyk and Wright 1986).

The mechanism of K^+ absorption via transcellular pathways is not completely understood. Although K^+ could be reabsorbed from lumen into the cell by

H-K-ATPase in the apical membrane (Younes-Ibrahim et al. 1995), this possibility is not supported by the finding that no significant K^+ reabsorption occurs after inhibition of passive K^+ transport in the PCT (Kaufman and Hamburger 1985). Several studies have demonstrated that the electrochemical gradient for K^+ across the apical membrane could favor K^+ secretion because it is above K^+ equilibrium potential (Biagi et al. 1981a, b). However, it is unlikely that a significant net K^+ secretion occurs in the proximal tubule because of low open probability of apical K^+ channels in the PCT, including Ca^{2+}-dependent large-conductance K^+ channel and voltage-gated K^+ channels (Kcna10 and Kcnq1) (Zweifach et al. 1992; Vallon et al. 2001; Yao et al. 2002).

20.7.2 Thick Ascending Limb (TAL)

Approximately 25% of filtered K^+ is reabsorbed in the TAL. Moreover, K^+ depletion enhances while a high K^+ intake decreases the K^+ reabsorption in the TAL (Unwin et al. 1994). Recently, it has been reported that TAL may be able to excrete K^+ in the animals fed with high K^+ and Na^+-deficient diet (Wang et al. 2017). For the absorption, K^+ enters the cell via NKCC2 across the apical membrane and by the basolateral membrane by Na-K-ATPase (Greger et al. 1983; Greger and Schlatter 1983; Hebert and Andreoli 1984; Gamba et al. 1994), while K^+ leaves the cells across the basolateral membrane by KCl, $KHCO_3$ cotransporters, and K^+ channels (Borensztein et al. 1991; Leviel et al. 1992; Paulais et al. 2006). Apical K^+ channels are responsible for K^+ recycling which is essential for maintaining the function of the NKCC2 (Hebert and Andreoli 1984; Greger 1985). K^+ recycling not only provides an adequate supply of K^+ for NKCC2 but it also participates in generating the lumen-positive transepithelial potential which is an important driving force for passive paracellular K^+, Na^+ and divalent cation absorption (Greger 1985; Giebisch 1998). Inhibition of K^+ recycling impairs Na^+ reabsorption in the TAL (Greger 1985; Giebisch 1998; Huang et al. 2000) and causes type II Bartter's syndromes characterized by salt-wasting phenotype (Simon et al. 1996a). Electrophysiological studies have identified three types of K^+ channels in the apical membrane of the TAL of the rat kidney (Taniguchi and Guggino 1989; Bleich et al. 1990; Wang et al. 1990; Wang 1994): a Ca^{2+}-dependent large-conductance, a ROMK-like 30 pS, and a 70 pS. Because both 30 pS and 70 pS K^+ channels are absent in the TAL of $Kcnj1^{-/-}$ mice, this suggests that ROMK is a key component of the both 30 pS and 70 pS K^+ channel (Lu et al. 2002).

20.7.3 Distal Convoluted Tubule (DCT)

Early studies have reported that the DCT is able to secrete K^+, although the capacitance of K^+ secretion in the DCT is less than that in the CCD (Schnermann

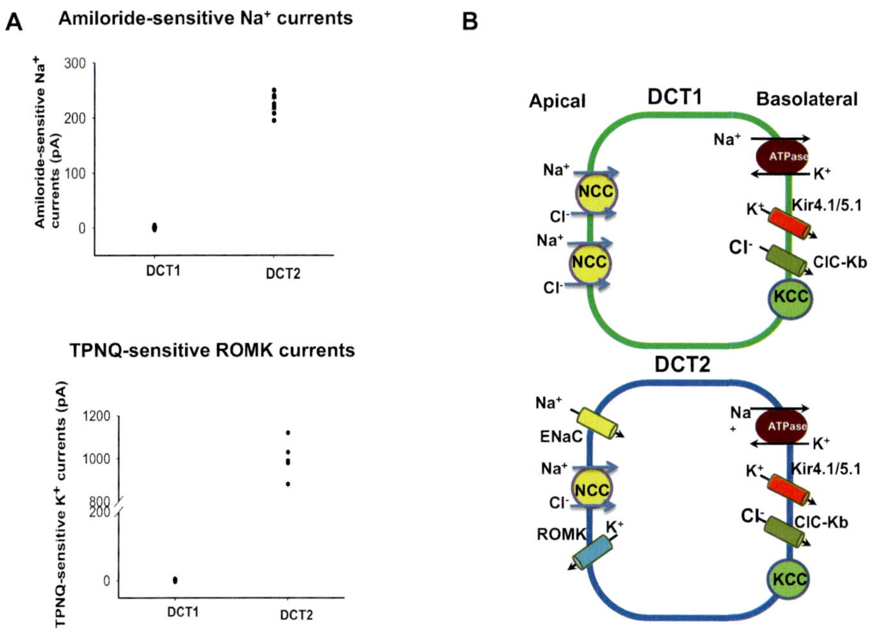

Fig. 20.5 Epithelial Na^+ channel (ENaC) and Kir1.1 (ROMK) activity are present in the apical membrane of later DCT (DCT2) but not in the early portion of the DCT (DCT1). (**a**) A scatter graph showing amiloride-sensitive Na^+ (ENaC) currents (top panel) and TPNQ-sensitive K^+ (ROMK) currents (bottom panel) in the DCT1 and DCT2. (**b**) A cell model illustrating the Na^+ transport mechanism in the DCT1 and DCT2. Note that NCC is expressed in the apical membrane of both DCT1 and DCT2

et al. 1987). It is now well established that K^+ secretion takes place only in the late portion of the DCT (DCT2) but not in early portion of the DCT (DCT1). This notion is supported by the finding that the expression and the activity of ENaC and ROMK have been detected in the DCT2 (Wade et al. 2011; Duan et al. 2019). Figure 20.5a is a graph showing that amiloride-sensitive Na^+ currents (ENaC) and TPNQ-sensitive K^+ currents (ROMK) are detected in the DCT2 but not in the DCT1. Figure 20.5b is a scheme illustrating the cell models for the DCT1 and DCT2, respectively. While NCC is expressed in the apical membrane throughout the DCT (Obermuller et al. 1995; Gamba 1999), ROMK and ENaC have been shown to be highly expressed in the apical membrane of the DCT2 (Schmitt et al. 1999; Wade et al. 2011). For secretion, K^+ enters the cell across the basolateral membrane by Na-K-ATPase and diffuses across the apical membrane via ROMK channel. A favorable electrochemical gradient generated by Na^+ absorption through ENaC provides the driving force for K^+ secretion in the DCT2.

20.7.4 Connecting Tubule (CNT) and Cortical Collecting Duct (CCD)

The CNT/CCD is responsible for the final regulation of renal K^+ secretion under normal conditions (Giebisch 1998). However, the CCD can also reabsorb K^+ when K^+ intake is restricted (Giebisch 1998). Two morphological and functional distinct cells, principal (PC) and intercalated cells (IC) are present in the CCD. The classical view is that PCs are responsible for K^+ secretion whereas ICs reabsorb K^+ (O'Neil 1981; O'Neil and Hayhurst 1985; Giebisch 1998). However, this view has been challenged by the possibility that ICs may also be able to secrete K^+ during hyperkalemia. It has been reported that Ca^{2+}-dependent big-conductance K^+ channels (BK) in ICs are responsible for the flow-stimulated K^+ secretion in the CCD since the deletion of BK channels in ICs impairs the flow-stimulated K^+ excretion (Woda et al. 2001; Najjar et al. 2005; Morimoto et al. 2006). K^+ secretion in PCs takes place by a two-step process: K^+ enters the cell via the basolateral Na-K-ATPase and is secreted into the lumen through apical K^+ channels along a favorable electrochemical gradient. This K^+ electrochemical gradient augments when Na^+ transport is enhanced by aldosterone or increased luminal Na^+ delivery (Schafer et al. 1990). K^+ can also enter the cell across the basolateral membrane via K^+ channels in deoxycorticosterone acetate (DOCA)-treated animals because mineralocorticoids stimulate the activity of Na-K-ATPase and hyperpolarize the basolateral membrane exceeding the K^+ equilibrium potential (Sansom and O'Neil 1986). K^+ secretion in the CCD depends on ENaC activity, however, a Na^+ channel independent K^+ secretion capability has been reported in the collecting duct from high K^+ adapted animals (Frindt and Palmer 2009). In this regard, it is possible that BK-dependent K^+ excretion may be an ENaC-independent K^+ excretion mechanism in the CCD. It has been shown that high flow rate in the CCD increases the intracellular Ca^{2+} level which stimulates the BK channel in IC thereby enhancing K^+ excretion in the CCD (Woda et al. 2002; Liu et al. 2003). Thus, it is apparent that both Na^+ and fluid volume delivery to the ASDN are two important factors controlling the K^+ excretion in the ASDN. In this regard, the DCT segment plays a key role in determining Na^+ and fluid volume delivery to the late portion of ASDN.

20.8 NCC Regulates Renal K^+ Excretion

Recent development in the field strongly indicates that NCC plays a key role in the regulation of K^+ homeostasis and renal K^+ excretion by controlling the amount of Na^+ and fluid volume delivery to the late portion of the ASDN (Simon et al. 1996b; Lalioti et al. 2006; Louis-Dit-Picard et al. 2012; Boyden et al. 2012). The importance of NCC in maintaining K^+ homeostasis is demonstrated in patients with Gitelman's syndrome and pseudohypoaldosteronism type 2 (PHAII) syndrome or familial hyperkalemic hypertension. Gitelman's Syndrome, which is the most common inherited tubular disease, is resulted from loss-of-function mutations of the *SLC12A3* gene encoding the thiazide-sensitive NCC and is characterized by hypokalemia and metabolic alkalosis (Simon et al. 1996b). In contrast, PHAII syndrome is caused by high intrinsic NCC activity and is characterized by hypertension, metabolic acidosis, and hyperkalemia (Pathare et al. 2013). The view that high NCC activity in patients with PHAII syndrome is the cause inducing the electrolyte disturbance is strongly indicated by the fact that the syndrome is completely corrected by thiazide diuretics. The upregulation of NCC activity in patients with PHAII syndrome is the result of mutations in with-no-lysine kinase 1 (WNK1), WNK4, klech-like 3 (KLHL3) and cullin 3 (CUL3) (Wilson et al. 2001; Lalioti et al. 2006; Boyden et al. 2012; Louis-Dit-Picard et al. 2012; Shibata et al. 2013; McCormick et al. 2014). Because high NCC activity augments Na^+ absorption in the DCT, the Na^+ delivery to the distal nephron including CNT and CCD is reduced thereby decreasing the driving force for the K^+ secretion. On the other hand, low NCC activity in patients with Gitelman's syndrome decreases Na^+ absorption in the DCT and increases the Na^+ delivery to the distal nephron, thereby increasing K^+ excretion in the ASDN.

The notion that abnormal NCC activity impairs renal K^+ excretion is also supported in the mouse model (TgWNK4^{PhAII} knockin mice), the mice harboring PHAII WNK4 mutant by genomic manipulation (Lalioti et al. 2006). These mice have developed typical PHAII phenotypes characterized by hypertension, hyperkalemia, and hypercalciuria due to high NCC expression and increased Na^+ transport in the DCT. In addition, it has been observed that ENaC and ROMK activity in the DCT2 and CNT segments are also significantly decreased in TgWNK4^{PHAII} knockin mice in comparison to the WT animal (Zhang et al. 2016). These findings have strongly suggested that the thiazide-sensitive NCC plays an essential role in the modulation of renal K^+ excretion. Thus, factors which affect NCC activity should have a significant effect on the renal K^+ excretion and K^+ homeostasis. In this regard, a large body of evidence indicates that the basolateral K^+ channel, Kir4.1/Kir5.1 heterotetramer, plays a key role in controlling NCC activity in the DCT.

20.9 Role of Kir4.1/Kir5.1 in the Regulation of NCC

The role of Kir4.1 in the regulation of NCC has been demonstrated in both human and in mouse models by the finding that the loss-of-function-of-mutations of Kir4.1 in the kidney inhibit NCC function (Scholl et al. 2009; Bockenhauer et al. 2009; Reichold et al. 2010; Zhang et al. 2014; Cuevas et al. 2017; Malik et al. 2018; Wang et al. 2018a). The mechanisms by which the mutations impair the function of Kir4.1 function are different (Sala-Rabanal et al. 2010; Williams et al. 2010; Parrock et al. 2013; Tanemoto et al. 2014; Mendez-Gonzalez et al. 2016). For instance, T65R, T164I and R297C mutations increase the pH sensitivity of the K^+ channel such that the K^+ channel opens only at more alkalized pH than at physiological relevant ranges (Sala-Rabanal et al. 2010). On the other hand, A167V mutation has been shown to decrease membrane expression of Kir4.1 (Tanemoto et al. 2014). However, regardless of molecular mechanisms by which the mutations impair the K^+ channel function, the net effect caused by Kir4.1 mutations is the inhibition of thiazide-sensitive NCC. The renal phenotypes in patients with EAST syndrome include mild salt waste, hypokalemic metabolic alkalosis, hypomagnesemia, and elevated levels of renin and aldosterone (Bockenhauer et al. 2009; Scholl et al. 2009; Reichold et al. 2010). Thus, the phenotype of loss-of-function mutations of Kir4.1 resembles Gitelman's syndrome (Simon et al. 1996b). The inhibition of NCC increases the delivery of Na^+ and fluid volume to the ASDN and stimulates K^+ excretion from DCT2 to the CCD, thereby causing urinary K^+ wasting and hypokalemia. Thus, the loss-of-function-mutations of Kir4.1 generate the same phenotype as loss-of-function-mutations of NCC, indicating an intrinsic relationship between Kir4.1 and NCC.

The notion that the basolateral K^+ channel activity in the DCT controls NCC activity is also suggested in several animal models including *Kcnj10*$^{-/-}$ mice, kidney-specific Kir4.1 conditional knockout (Ks-Kir4.1 KO) mice, and cavelolin-1 knockout mice. While *Kcnj10*$^{-/-}$ mice were viable during the birth, their growth stopped within one week after birth. Thus, the mice could not survive more than two weeks after the birth. The study performed in postnatal 7–10 days *Kcnj10*$^{-/-}$ mice has demonstrated that the deletion of Kir4.1 inhibits the expression and activity of NCC (Zhang et al. 2014). Moreover, the deletion of Kir4.1 has almost completely abolished the basolateral K^+ conductance in the DCT and depolarized the DCT membrane. The notion that Kir4.1 is solely responsible for the basolateral K^+ conductance in the DCT has also been confirmed in the studies performed in Ks-Kir4.1 KO mice (Cuevas et al. 2017; Wang et al. 2018a).

The major phenotypes of Ks-Kir4.1 KO mice were mild urinary Na^+ and Mg^{2+} wasting, severe urinary K^+ wasting, hypokalemia, and metabolic alkalosis (Cuevas et al. 2017). Further analysis has demonstrated that the deletion of Kir4.1 severely impaired the expression of NCC in the DCT and abolished thiazide-induced natriuretic effect, suggesting that NCC function was compromised in Ks-Kir4.1 KO mice. In contrast, the deletion of Kir4.1 increased the expression of ENaC-α and ENaC-γ in the kidney, suggesting a compensatory action following the down-regulation of NCC in the DCT. Similar phenotypes have also been found in caveolin-1 knockout mice in which the Kir4.1 activity was inhibited, thereby cavolin-1 knockout mice having a functional inhibition of Kir4.1 (Wang et al. 2015). Consequently, the renal phenotype of caveolin-1 knockout mice is similar to Kir4.1 knockout mice including hypokalemia. Thus, the mouse models have recapitulated the renal phenotype of EAST/SeSAME syndrome in human, indicating the role of Kir4.1 in the regulation of the NCC and in the regulation of renal K^+ excretion and K^+ homeostasis.

20.10 The Mechanism of Kir4.1 Regulating NCC

The mechanism by which the deletion of Kir4.1 in the kidney inhibits NCC is most likely mediated by Cl^--sensitive WNK pathway (Cuevas et al. 2017). It is possible that the inhibition of the basolateral K^+ conductance decreases Cl^- exit across the basolateral membrane, thereby suppressing WNK activity in the DCT. It is well established that high intracellular Cl^- concentrations (Cl^-_{in}) inhibit the activity of WNK by regulating auto-inhibition and autophosphorylation (Piala et al. 2014). Because WNK activity is an upstream protein kinase for the activation of ste20-proline-alanine-rich protein kinase (SPAK) and oxidation-sensitive response kinase (OSR), two kinases which stimulate NCC activity (McCormick et al. 2011; Castaeda-Bueno et al. 2012; Thastrup et al. 2012; Grimm et al. 2017), high Cl^-_{in} levels should inhibit SPAK/OSR, thereby decreasing NCC phosphorylation. A decrease in NCC phosphorylation enhances the internalization and ubiquitination of NCC (Rosenbaek et al. 2014). The notion that Kir4.1 activity in the DCT determines the expression and activity of thiazide-sensitive NCC activity by regulating Cl^-_{in} is supported by two observations: (1) Terker et al. (2015) have demonstrated that the changes in the intracellular Cl^- concentration induced by membrane voltage are responsible for the inhibition of NCC activity in DCT cells expressing loss-function-of-Kir4.1 mutants; (2) Experiments performed in both Ks-Kir4.1 KO mice and *Kcnj10*$^{-/-}$ mice have demonstrated that the Cl^- conductance in the DCT was inhibited (Zhang et al. 2014; Cuevas et al. 2017). Moreover, Kir4.1 may also play a role in the regulation of NKCC2 in the TAL since the deletion of Kir4.1 also decreased NKCC2 activity (Terker et al. 2018). However, because the K^+ channels other than Kir4.1 are also expressed in the basolateral membrane of the cTAL (Zhang et al. 2015), the inhibition of NKCC2 in KS-Kir4.1 KO mice was less severe than NCC.

Figure 20.6 is a cell model illustrating the mechanism by which the loss-of-function-mutations of Kir4.1 inhibit NCC (Su and Wang 2016). The inhibition of Kir4.1 decreases the basolateral K^+ conductance in the DCT and depolarizes the membrane potential. Because Cl^- movement across the basolateral membrane is an electrogenic process, the depolarization in the DCT inhibits Cl^- exit and increases Cl^-_{in}, thereby inhibiting WNK activity. A decrease in WNK activity is expected to suppress SPAK/OSR activity and NCC phosphorylation, thereby inhibiting NCC activity. On the other hand, stimulation of Kir4.1/Kir5.1 should hyperpolarize the membrane of the DCT and increases NCC activity. Thus, through Cl^-_{in} the basolateral Kir4.1/Kir5.1 activity controls the apical NCC activity.

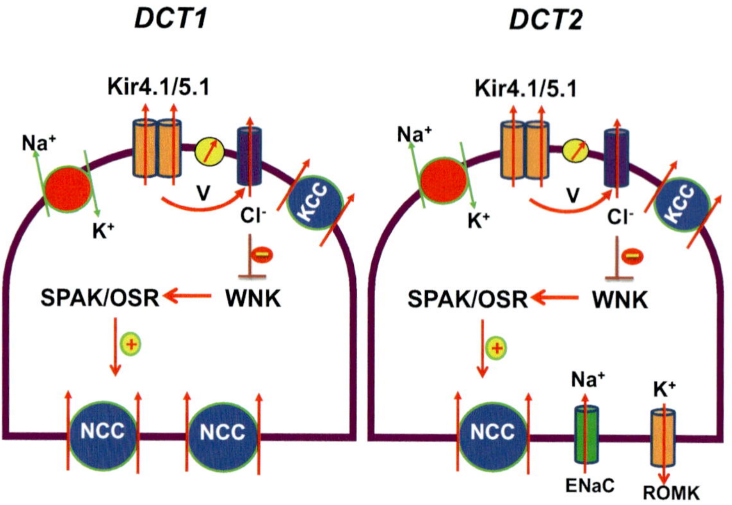

Fig. 20.6 A cell scheme illustrating the mechanism by which Kir4.1/Kir5.1 activity in the basolateral membrane is associated with NCC activity in the apical membrane. The activation of Kir4.1/Kir5.1 in the basolateral membrane of the DCT hyperpolarizes the membrane thereby stimulating Cl^- exit through ClC-Kb. Since WNK is a Cl^--sensitive kinase, the hyperpolarization is expected to decrease the intracellular Cl^- concentration, thereby activating WNK which should further stimulate SPAK. Because SPAK-induced phosphorylation of NCC activates the cotransporter, Kir4.1/Kir5.1 activity regulates the NCC activity. Abbreviations: WNK (with no lysine kinase), SPAK (ste20 proline-alanine rich protein kinase), OSR (oxidation-sensitive response kinase) and V (cell voltage)

20.11 Kir4.1/Kir5.1 Is Essential for Dietary K^+ Intake-Induced Regulation of NCC

A large body of evidence has demonstrated that dietary K^+ intake regulates the NCC activity such that decreased dietary K^+ intake stimulates whereas increased dietary K^+ intake inhibits NCC activity (van der Lubbe et al. 2013; Sorensen et al. 2013; Rengarajan et al. 2014; Castaneda-Bueno et al. 2014; Terker et al. 2015, 2016; Yang et al. 2018; Wang et al. 2018a). The dietary K^+ intake-induced regulation of NCC activity is essential for either preserving K^+ during hypokalemia or facilitating renal K^+ excretion during hyperkalemia (Ellison et al. 2015). Using phosphorylated NCC as an index of NCC activation, Terker et al. (2016) have elegantly demonstrated that steep relationship between plasma K^+ and NCC activity falls in the physiologically relevant ranges from 3.5 to 4.5 mM plasma K^+, suggesting the importance of NCC in maintaining K^+ homeostasis. High K^+-induced inhibition of NCC should stimulate K^+ secretion by augmenting the Na^+ and volume delivery to the CNT and CCD. Because high K^+ intake also stimulates aldosterone secretion and activates apical ENaC, the combined actions of increased Na^+ delivery and upregulated ENaC should enhance renal K^+ excretion. Conversely, low dietary K^+ intake increases

NCC activity and decreases Na^+ and volume delivery to the ASDN, thereby inhibiting K^+ excretion. Therefore, dietary K^+-intake-induced regulation of NCC activity plays a key role in maintaining K^+ homeostasis and renal K^+ excretion.

The finding that Kir4.1/Kir5.1 activity in the DCT is associated with NCC function has a physiological significance. It has been suggested that the cell membrane potential of the DCT may be responsible for the effect of dietary K^+ intake on NCC (Terker et al. 2015). Indeed, performing electrophysiological studies in the mouse DCT, Wang et al. (2018a) have demonstrated that high K^+ intake depolarizes the membrane of the DCT, whereas low K^+ intake hyperpolarizes membrane of the DCT (Fig. 20.7a). The effect of dietary K^+ intake on the membrane potential of the DCT may be induced by either changing the basolateral K^+ conductance (permeability) or altering extracellular (plasma) K^+ concentrations. However, plasma K^+ concentrations are normally varied in a very narrow range even during increased or decreased dietary K^+ intake. Thus, it is unlikely that high K^+ intake-induced depolarization of the DCT is mainly due to the increase in plasma K^+ concentrations or that low K^+-intake-induced hyperpolarization is due to decreased plasma K^+ concentrations. Thus, it is conceivable the change in basolateral K^+ permeability is mainly responsible for dietary K^+ intake-induced alteration of DCT membrane potential. The notion that changes in K^+ permeability in the DCT are responsible for the hyperpolarization during decreased dietary K^+ intake or depolarization during increased dietary K^+ intake is indicated by patch-clamp experiments: Fig. 20.7b is a channel recording and Fig. 20.7c is a whole-cell recording showing that high K^+ intake significantly decreases while low K^+ intake increases basolateral 40 pS K^+ channel activity in the DCT and whole-cell K^+ conductance. The changes in the basolateral K^+ permeability and DCT membrane potential induced by dietary K^+ intake completely depend on the Kir4.1 since the deletion of Kir4.1 not only abolished basolateral K^+ conductance but also the effect of dietary K^+ intake on the membrane potential (Wang et al. 2018a). Also, the observation that deletion of Kir4.1 in the DCT not only inhibits NCC but also abolishes the effect of dietary K^+ intake on NCC expression and activity strongly suggests that the effect of dietary K^+ intake on NCC is associated with basolateral Kir4.1/Kir5.1 activity in the DCT. From the inspection of Fig. 20.8, it is apparent that decreased dietary K^+ intake increases while increased dietary K^+ intake decreased the expression of pNCC and tNCC. However, the deletion of Kir4.1 abolishes the effect of dietary K^+ intake on the expression of pNCC and tNCC. Thus, the basolateral Kir4.1/5.1 plays a key role in mediating the effect of dietary K^+ intake on NCC.

20.12 Signaling in the Regulation of Kir4.1/Kir5.1 and NCC

It has been reported that angiotensin type II receptor (AT2R), bradykinin type II receptor (BK2R), and β-adrenergic receptor are involved in the regulation of the basolateral Kir4.1/Kir5.1 in the DCT.

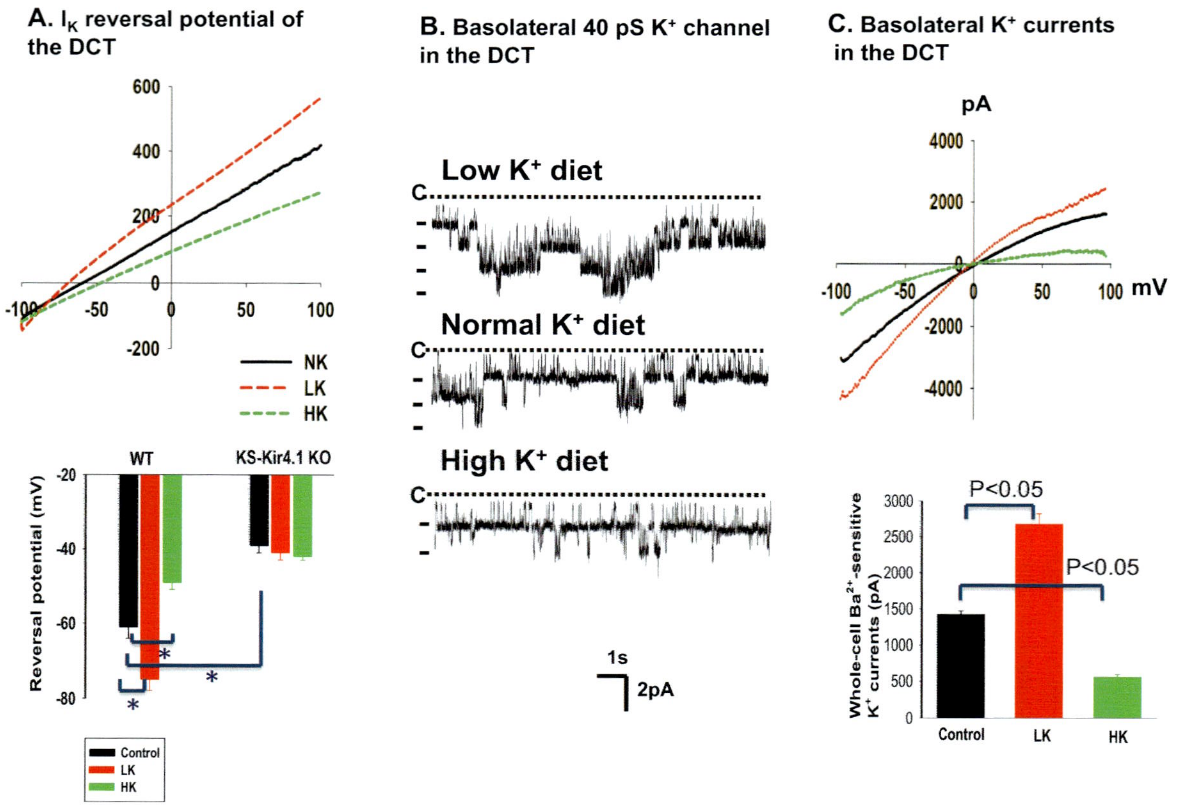

Fig. 20.7 The dietary K^+ intake regulates the Kir4.1/Kir5.1 heterotetramer activity in the DCT such as LK intake stimulates and HK inhibits basolateral K^+ channels. (**a**) A perforated whole-cell recording showing K current (I_K) reversal potential in the DCT cells of WT mice on an NK (black), an LK (Red), or a HK diet (green) for 4 days. A bar graph summarizes the results of six similar experiments performed in WT and KS-Kir4.1 KO mice, respectively. The significance is indicated by "*". For the measurement of I_K reversal potential, the bath solution contains 140 mM NaCl and 5 mM KCl while the pipette solution has 140 mM

KCl. (**b**) A single channel recording shows the basolateral 40 pS K^+ channel activity in the DCT of control mice (considered as WT) on a low K^+ (LK), normal K^+ (NK), or a high K^+ (HK) diet for 4 days, respectively. For single channel recording, the DCT was bathed in a solution containing (mM) 140 NaCl, 5 KCl 2 $MgCl_2$, 1.8 mM $CaCl_2$, and 10 HEPES (pH = 7.4) and the pipette solution contains 140 KCl, 2 $MgCl_2$, 1 EGTA, and 5 HEPES. (**c**) A set of recording shows Ba^{2+}-sensitive K^+ currents measured with RAMP protocol from −100 to 100 mV in the DCT of WT mice on an NK (black), LK (red) or HK diet (green). A bar graph summarizes the values measured at −60 mV with whole-cell recording. For the whole-cell recording, a symmetric 140 mM KCl solution has been used for the bath and the pipette

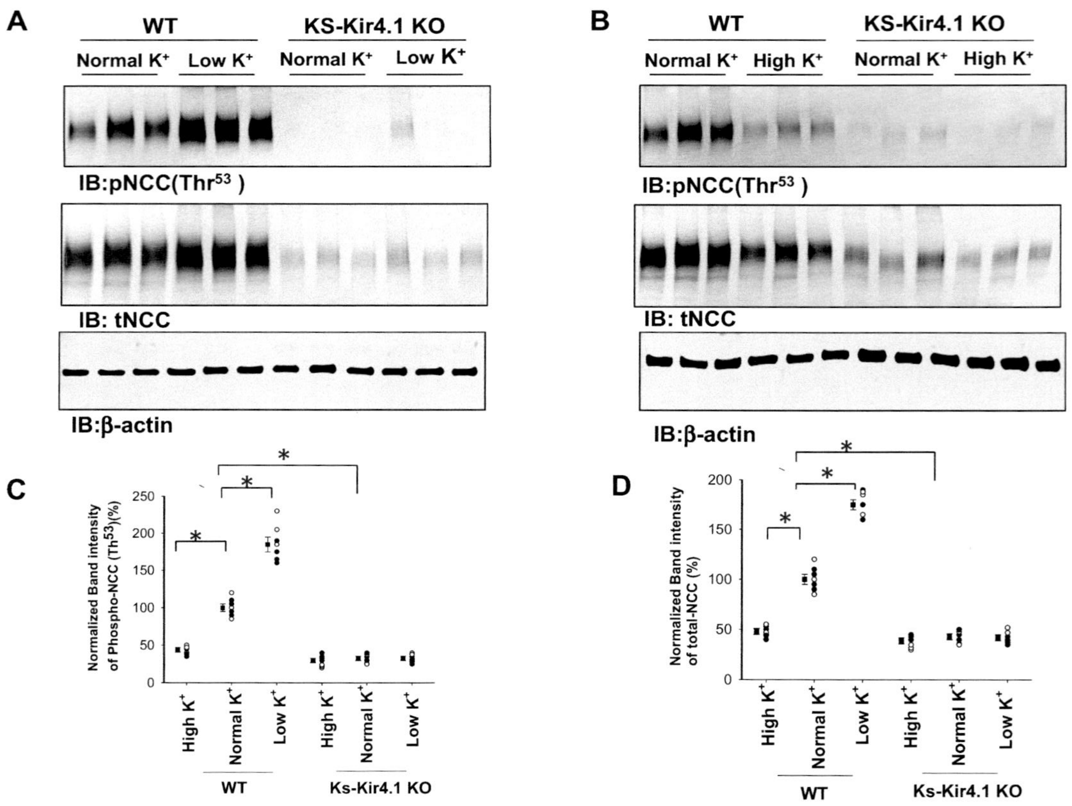

Fig. 20.8 Dietary K^+ intake regulates the expression of phosphor-NCC (pNCC) and total NCC (tNCC) in WT mice such that high K^+ (HK) intake decreases while low K^+ (LK) intake increases the expression of pNCC and tNCC. However, the regulatory effect of dietary K^+ intake on NCC is abolished in the absence

of Kir4.1 in the DCT. Immunoblot shows the expression of pNCC and tNCC in WT and Ks-Kir4.1 KO mice on an NK or an LK diet for 4 days (**a**) and on an NK or a HK diet for 4 days (**b**). A scatter graph summarizes the normalized band intensity of above experiments for pNCC (**c**) and tNCC (**d**), respectively. Mean value and SEM are shown in the left side of each plot. "*" indicates the significant difference

20.12.1 Role of AT2R and BK2R in the Regulation of Kir4.1 and NCC

Two studies have suggested that AT2R and BK2R are involved in mediating the effect of high dietary K^+ intake on the basolateral K^+ channels in the DCT. BK2R is highly expressed in the apical membrane and basolateral membrane of the DCT. Moreover, increased dietary K^+ intake has been shown to augment the expression of BK2R and bradykinin levels (Jin et al. 1999; Suzuki et al. 2000), suggesting the possible role of BK2R in the stimulation of K^+ excretion in the ASDN. This notion is supported by the finding that bradykinin infusion inhibits the expression of phosphor-NCC and total NCC and diminishes the thiazide-induced natriuresis (Zhang et al. 2018), suggesting the inhibition of NCC function. Since the infusion of bradykinin has no effect on NCC activity in Ks-Kir4.1 KO mice, the results indicate Kir4.1 is required for the effect of bradykinin on NCC. Indeed, bradykinin inhibits Kir4.1/Kir5.1 in the DCT and bradykinin infusion depolarizes the DCT membrane. Moreover, the effect of bradykinin on Kir4.1/Kir5.1 is completely blocked by the specific BK2R antagonist, suggesting that BK2R is involved in the inhibition of the basolateral Kir4.1/Kir5.1 in the DCT. Thus, it is most likely that BK2R is involved in mediating the inhibitory effect of high dietary K^+ intake on Kir4.1/Kir5.1 in the DCT.

In addition, AT2R may be also involved in mediating the effect of increased dietary K^+ intake on Kir4.1 and NCC. Increased dietary K^+ intake has been shown to stimulate the expression of AT2R (Wei et al. 2014). We have observed that the AT2R-inhibition significantly increases the basolateral K^+ conductance in the DCT, hyperpolarizes the DCT membrane, and augments NCC activity (Wu et al. 2018). Consequently, the animals treated with AT2R-antagonist have hyperkalemia. Because the stimulatory effect of AT2R-inhibition on NCC is absent in KS-Kir4.1 KO mice, it suggests that the inhibition of AT2R-induced activation of Kir4.1/Kir5.1 is required for the upregulation of NCC by AT2R-antagonist. The notion that AT2R may be involved in mediating the effect of high K^+ intake on Kir4.1/Kir5.1 in the DCT is further indicated by the finding that application of the AT2R agonist decreases the basolateral K^+ channels in the DCT. Thus, these studies suggest that both AT2R and BK2R may be involved in the regulation of renal K^+ excretion by targeting the basolateral Kir4.1/Kir5.1 in the DCT. Figure 20.9 is a cartoon illustrating the role of AT2R and BK2R in mediating the effect of increased dietary K^+ intake on renal K excretion (Su et al. 2019). The stimulation of AT2R and BK2R depolarizes the cell membrane of the DCT and inhibits NCC thereby increasing the delivery of Na^+ and fluid volume to the late ASDN and stimulating K^+ excretion.

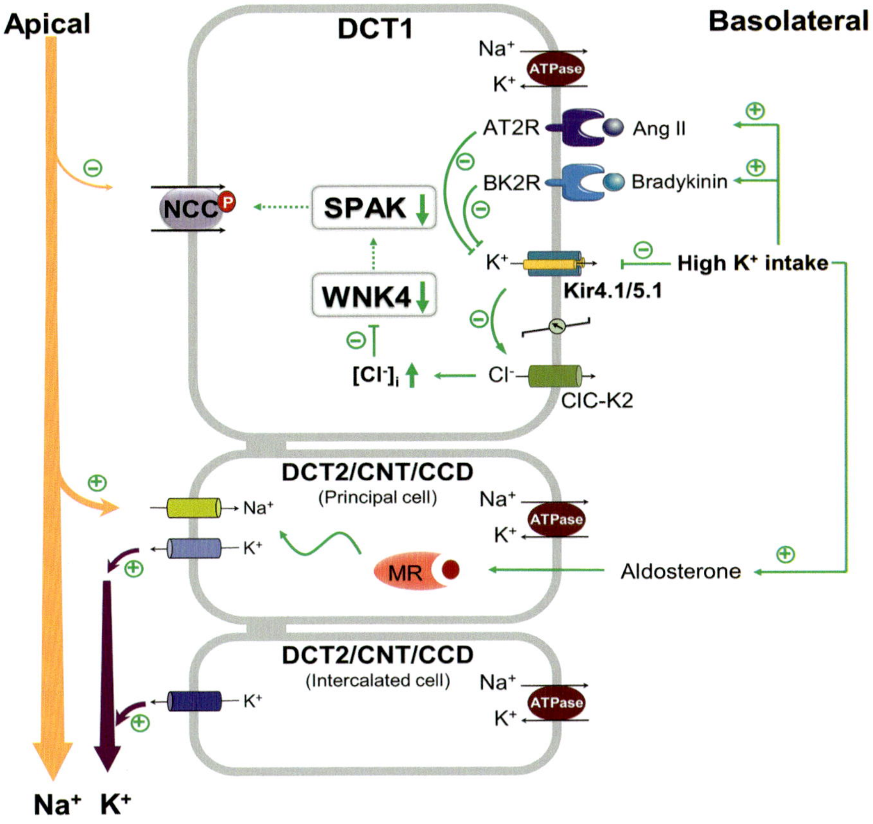

Fig. 20.9 Schematic model illustrating the role of Kir4.1/5.1 in mediating the effect dietary K^+ intake on NCC activity. High K^+ intake (green arrow or symbol) decreases Kir4.1/5.1 channel activity, which subsequently inhibits WNK-SPAK pathway and NCC activity in the DCT cells. The sodium reabsorption and potassium excretion rate increases along the downstream segments. AT2R, angiotensin II type 2 receptor; BK2R, bradykinin type 2 receptor; MR, mineralocorticoid receptor. The dotted line indicates diminished effects

20.12.2 *Role of Kir4.1 in Mediating the Adrenergic Receptor-Induced Stimulation of NCC*

The sympathetic-stimulation is one important factor causing salt-sensitive hypertension and that the hypertension induced by sympathetic-stimulation may be partially due to high NCC activity (Mu et al. 2011; Foss et al. 2013; Terker et al. 2014). Terker et al. (2014) have demonstrating that the mice acutely treated with norepinephrine (NE) or isoproterenol (ISO) increased NCC phosphorylation. Since the activation of NCC induced by sympathetic-stimulation was also observed in angiotensin II type 1 receptor (AT1R) knockout mice, this indicated that AT1R was not involved in mediating sympathetic stimulation of NCC. Furthermore, the sympathetic-stimulation has been reported to activate WNK4 through epigenetic modulation (Mu et al. 2011).

We have demonstrated that the stimulation of Kir4.1/Kir5.1 activity in the DCT is also essential for the effect of sympathetic-stimulation on NCC (Duan et al. 2019). First, NE-induced stimulation of NCC activity is absent in the DCT of Ks-Kir4.1 KO mice. Second, NE-induced stimulation of NCC is closely correlated with NE-induced hyperpolarization of DCT membrane. Thus, NE-induced stimulation of NCC activity may be, at least in part, the result of NE-induced activation of basolateral Kir4.1/Kir5.1 and hyperpolarization. The effect of sympathetic-stimulation on the basolateral Kir4.1/Kir5.1 is mediated by stimulating β-adrenoceptor. First, the specific β-adrenergic receptor agonists but not α-adrenergic receptor agonists mimic the effect of NE on the K^+ channel in the DCT. Second, the stimulatory effect of NE on the K^+ channel is blocked by propranolol. Finally, NE-induced stimulation of the basolateral 40 pS K^+ channel is mimicked by membrane-permeable cAMP analogue and abolished by inhibiting protein kinase A, suggesting the role of β-adrenoceptor in the stimulation of the basolateral Kir4.1/Kir5.1 channel in the DCT. Thus, the stimulation of Kir4.1/Kir5.1 in the DCT by sympathetic activity is a key target for the activation of NCC by NE.

20.13 Kir4.1/Kir5.1 Is Essential for the Effect of Dietary Na^+ Intake on NCC

Previous studies have demonstrated that dietary Na^+ intake also regulates NCC activity in the DCT such that low sodium (LS) intake stimulates NCC whereas high sodium (HS) intake inhibits NCC activity (Chiga et al. 2008). Recent study has suggested that the basolateral Kir4.1/Kir5.1 in the DCT is also responsible for the effect of dietary Na^+ intake on NCC (Wu et al. 2019). It has been showed that LS intake stimulates the basolateral Kir4.1/Kir5.1 heterotetramer in the DCT, increases basolateral K^+ conductance, and hyperpolarizes the membrane (Wu et al. 2019). Conversely, HS intake inhibits the K^+ channel, decreases the basolateral K^+ currents, and depolarizes the membrane (Wu et al. 2019). Thus, LS-induced high NCC activity is associated with upregulation of Kir4.1 activity in the DCT while HS-induced inhibition of NCC activity is associated with diminished Kir4.1 activity. The role of Kir4.1/Kir5.1 in mediating the effect of dietary Na^+ intake on NCC is strongly suggested by the finding that the deletion of Kir4.1 not only decreases NCC but also abrogates LS-induced stimulation of NCC, suggesting the basolateral Kir4.1/Kir5.1 activity in the DCT is essential for the effect of dietary Na^+ intake on NCC (Wu et al. 2019).

20.14 Role of Kir4.1/Kir5.1 in Aldosterone Paradox

The classic view of K$^+$ excretion is focused on ENaC and secretory K$^+$ channels including ROMK and Ca^{2+}-activated BK in the ASDN where Na$^+$ absorption through ENaC provides the lumen-negative driving force for K$^+$ secretion (Giebisch 1998). Aldosterone is considered to be the hormone regulating renal K$^+$ excretion by enhancing Na$^+$ absorption through ENaC. High K$^+$ intake is known to stimulate aldosterone synthesis and increases its plasma level. However, high K$^+$ intake causes natriuresis and increases renal K$^+$ excretion simultaneously despite high aldosterone level. On the other hand, LS also stimulates aldosterone levels, but it stimulates Na$^+$ absorption without increasing K$^+$ excretion. The finding high K$^+$ intake inhibits NCC by suppressing Kir4.1/Kir5.1 in the DCT, whereas LS stimulates NCC by activating Kir4.1/Kir5.1 explains the puzzle of "aldosterone paradox." For instance, LS intake-induced stimulation of the basolateral Kir4.1/Kir5.1 in the DCT should increase NCC activity and enhance Na$^+$ absorption in the DCT, thereby decreasing the delivery of Na$^+$ and fluid volume into the late ASDN. Thus, although ENaC activity in the CCD is upregulated, Na$^+$ absorption through ENaC in the CCD is minimal, thereby effectively suppressing K$^+$ excretion in CNT and CCD. On the other hand, high K$^+$ intake-induced inhibition of the basolateral Kir4.1/Kir5.1 in the DCT suppresses NCC and increases Na$^+$ and volume delivery to the late ASDN. Because aldosterone stimulates ENaC in the ASDN, NCC inhibition promotes ENaC-dependent and flow-stimulated K$^+$ secretion in the ASDN without increasing net Na$^+$ absorption. Thus, Kir4.1/Kir5.1 in the DCT works in concert with aldosterone either to stimulate K$^+$ secretion during high dietary K$^+$ intake or to inhibit K$^+$ excretion during Na$^+$ restriction.

20.15 Conclusion Remarks

Kir4.1 is expressed in the basolateral membrane from cTAL to early CCD, and the Kir4.1/Kir5.1 heterotetramer is the only type of the basolateral K$^+$ channel in the DCT. Recent progress in studying the regulation of renal Na$^+$ and K$^+$ transport indicates that Kir4.1/Kir5.1 heterotetramer plays a key role in the regulation of NCC activity. Kir4.1/Kir5.1 heterotetramer is a key player to sense the dietary K$^+$ and Na$^+$ intake and is the target of the hormones which are known to regulate the renal Na$^+$ and K$^+$ transport. Kir4.1/Kir5.1 serves as a gateway to transduce the effects of angiotensin II, bradykinin, and sympathetic stimulation on NCC through a Cl$^-$-sensitive WNK pathway.

Acknowledgments This work is supported by NIH grant DK 54983 (WHW) and DK 115366 (LDH). The author thanks Ms. Gail Anderson for editing the manuscript.

Disclosure

None.

References

Azuma KK, Hensley CB, Putnam DS, McDonough AA (1991) Hypokalemia decreases Na^+-K^+-ATPase alpha2- but not alpha1-isoform abundance in heart, muscle, and brain. Am J Physiol Cell Physiol 260:C958–C964

Bandulik S, Schmidt K, Bockenhauer D, Zdebik AA, Humberg E, Kleta R, Warth R, Reichold M (2011) The salt-wasting phenotype of EAST syndrome, a disease with multifaceted symptoms linked to the KCNJ10 K channel. Pflugers Arch 461:423–435

Bastl C, Hayslett JP, Binder HJ (1977) Increased large intestinal secretion of potassium in renal insufficiency. Kidney Int 12:9–16

Bia MJ, DeFronzo RA (1981) Extrarenal potassium homeostasis. Am J Phys 240:F257–F268

Biagi B, Kubota T, Sohtell M, Giebisch G (1981a) Intracellular potentials in rabbit proximal tubules perfused in vitro. Am J Phys 240:F200–F210

Biagi B, Sohtell M, Giebisch G (1981b) Intracellular potassium activity in the rabbit proximal straight tubule. Am J Phys 241:F677–F686

Bleich M, Schlatter E, Greger R (1990) The luminal K^+ channel of the thick ascending limb of Henle's loop. Pflugers Arch 415:449–460

Bockenhauer D, Feather S, Stanescu HC, Bandulik S, Zdebik AA, Reichold M, Tobin J, Lieberer E, Sterner C, Landoure G, Arora R, Sirimanna T, Thompson D, Cross JH, van't Hoff W, Al Masri O, Tullus K, Yeung S, Anikster Y, Klootwijk E, Hubank M, Dillon MJ, Heitzmann D, Arcos-Burgos M, Knepper MA, Dobbie A, Gahl WA, Warth R, Sheridan E, Kleta R (2009) Epilepsy, ataxia, sensorineural deafness, tubulopathy, and KCNJ10 mutations. N Engl J Med 360:1960–1970

Bomsztyk K, Wright FS (1986) Dependence of ion fluxes on fluid transport by rat proximal tubule. Am J Phys 250:F680–F689

Bond CT, Pessia M, Xia XM, Lagrutta A, Kavanaugh MP, Adelman JP (1994) Cloning and expression of a family of inward rectifier potassium channels. Receptors Channels 2:183–191

Borensztein P, Leviel F, Froissart M, Houillier P, Poggioli J, Marty E, Bichara M, Paillard M (1991) Mechanisms of H^+/HCO^{3-} transport in the medullary thick ascending limb of rat kidney. Kidney Int Suppl 33:S43–S46

Boyden LM, Choi M, Choate KA, Nelson-Williams CJ, Farhi A, Toka HR, Tikhonova IR, Bjornson R, Mane SM, Colussi G, Lebel M, Gordon RD, Semmekrot BA, Poujol A, Valimaki MJ, De Ferrari ME, Sanjad SA, Gutkin M, Karet FE, Tucci JR, Stockigt JR, Keppler-Noreuil KM, Porter CC, Anand SK, Whiteford ML, Davis ID, Dewar SB, Bettinelli A, Fadrowski JJ, Belsha CW, Hunley TE, Nelson RD, Trachtman H, Cole TRP, Pinsk M, Bockenhauer D, Shenoy M, Vaidyanathan P, Foreman JW, Rasoulpour M, Thameem F, Al-Shahrouri HZ, Radhakrishnan J, Gharavi AG, Goilav B, Lifton RP (2012) Mutations in kelch-like 3 and cullin 3 cause hypertension and electrolyte abnormalities. Nature 482:98–102

Bredt DS, Wang TL, Cohen NA, Guggino WB, Snyder SH (1995) Cloning and expression of two brain-specific inwardly rectifying potassium channels. Proc Natl Acad Sci USA 92:6753–6757

Casamassima M, D'Adamo MC, Pessia M, Tucker SJ (2003) Identification of a heteromeric interaction that influences the rectification, gating, and pH sensitivity of Kir4.1/Kir5.1 potassium channels. J Biol Chem 278:43533–43540

Castaeda-Bueno M, Cervantes-Prez LG, Vzquez N, Uribe N, Kantesaria S, Morla L, Bobadilla NA, Doucet A, Alessi DR, Gamba G (2012) Activation of the renal Na:Cl cotransporter by angiotensin II is a WNK4-dependent process. Proc Natl Acad Sci USA 109:7929–7934

Castaneda-Bueno M, Cervantes-Perez LG, Rojas-Vega L, Arroyo-Garza I, Vazquez N, Moreno E, Gamba G (2014) Modulation of NCC activity by low and high K^+ intake: insights into the signaling pathways involved. Am J Physiol Renal Physiol 306:F1507–F1519

Cha SK, Huang C, Ding Y, Qi X, Huang CL, Miller RT (2011) Calcium-sensing receptor decreases cell surface expression of the inwardly rectifying K^+ channel Kir4.1. J Biol Chem 286:1828–1835

Chiga M, Rai T, Yang SS, Ohta A, Takizawa T, Sasaki S, Uchida S (2008) Dietary salt regulates the phosphorylation of OSR1//SPAK kinases and the sodium chloride cotransporter through aldosterone. Kidney Int 74:1403–1409
Choi CSC, Thompson CB, Leong PKK, McDonough AA, Youn JH (2001) Short-term K^+ deprivation provokes insulin resistance of cellular K^+ uptake revealed with the K+ clamp. Am J Physiol Renal Physiol 280:F95–F102
Codina J, Pressley TA, DuBose TD Jr (1997) Effect of chronic hypokalemia on H-K-ATPase in rat colon. Am J Phys 272:F22–F30
Cuevas CA, Su XT, Wang MX, Terker AS, Lin DH, McCormick JA, Yang C-L, Ellison DH, Wang WH (2017) Potassium sensing by renal distal tubules requires Kir4.1. J Am Soc Nephrol 28:1814–1825
Derst C, Karschin C, Wischmeyer E, Hirsch JR, Preisig-Muller R, Rajan S, Engel H, Grzeschik KH, Daut J, Karschin A (2001) Genetic and functional linkage of Kir5.1 and Kir2.1 channel subunits. FEBS Lett 491:305–311
Du X, Zhang H, Lopes C, Mirshahi T, Rohacs T, Logothetis DE (2004) Characteristic interactions with phosphatidylinositol 4,5-bisphosphate determine regulation of Kir channels by diverse modulators. J Biol Chem 279:37271–37281
Duan X, Gu L, Xiao Y, Gao Z, Wu P, Zhang Y, Meng X, Wang J, Zhang D, Lin D, Wang W, Gu R (2019) Norepinephrine-induced stimulation of Kir4.1/Kir5.1 is required for the activation of NaCl transporter in distal convoluted tubule. Hypertension 73:112–120
Ellison DH, Terker AS, Gamba G (2015) Potassium and its discontents: new insight, new treatments. J Am Soc Nephrol 27:981–989
Ferre S, Hoenderop JJ, Bindels RJ (2011) Role of the distal convoluted tubule in renal Mg^{2+} handling: molecular lessons from inherited hypomagnesemia. Magnes Res 24:s101–s108
Fisher KA, Binder HJ, Hayslett JP (1976) Potassium secretion by colonic mucosal cells after potassium adaptation. Am J Phys 231:987–994
Foss JD, Fink GD, Osborn JW (2013) Reversal of genetic salt-sensitive hypertension by targeted sympathetic ablation. Hypertension 61:806–811
Frindt G, Palmer LG (2009) K^+ secretion in the rat kidney: Na^+ channel-dependent and -independent mechanisms. Am J Physiol Renal Physiol 297:F389–F396
Fromter E (1974) Free-flow potential profile along rat kidney proximal tubule. Pflugers Arch 351:69–83
Fujita A, Horio Y, Higashi K, Mouri T, Hata F, Takeguchi N, Kurachi Y (2002) Specific localization of an inwardly rectifying K^+ channel, Kir4.1, at the apical membrane of rat gastric parietal cells; its possible involvement in K^+ recycling for the H^+-K^+-pump. J Physiol 540:85–92
Gamba G (1999) Molecular biology of distal nephron sodium transport mechanisms. Kidney Int 56:1606–1622
Gamba G, Miyanoshita A, Lombardi M, Lytton J, Lee WS, Hediger MA, Hebert SC (1994) Molecular cloning, primary structure, and characterization of two members of the mammalian electroneutral sodium-(potassium)-chloride cotransporter family expressed in kidney. J Biol Chem 269:17713–17722
Giebisch G (1998) Renal potassium transport:mechanisms and regulation. Am J Phys 274:F817–F833
Greger R (1985) Ion transport mechanisms in thick ascending limb of Henle's loop of mammalian nephron. Physiol Rev 65:760–797
Greger R, Schlatter E (1983) Properties of the basolateral membrane of the cortical thick ascending limb of Henle's loopof rabbit kidney. Pflugers Arch 396:325–334
Greger R, Schlatter E, Lang F (1983) Evidence for electroneutral sodium chloride cotransport in the cortical thick ascending limb of Henle's loop of rabbit kidney. Pflugers Arch 396:308–314
Grimm PR, Coleman R, Delpire E, Welling PA (2017) Constitutively active SPAK causes hyperkalemia by activating NCC and remodeling distal tubules. J Am Soc Nephrol 28:2555–2563

Hebert SC, Andreoli TE (1984) Control of NaCl transport in the thick ascending limb. Am J Phys 246:F745–F756
Houillier P (2014) Mechanisms and regulation of renal magnesium transport. Annu Rev Physiol 76:411–430
Huang DY, Osswald H, Vallon V (2000) Sodium reabsorption in thic ascending limb of Henle's loop: effect of potassium channel blockade in vivo. Br J Pharmacol 130:1255–1266
Huang C, Sindic A, Hill CE, Hujer KM, Chan KW, Sassen M, Wu Z, Kurachi Y, Nielsen S, Romero MF, Miller RT (2007) Interaction of the Ca^{2+}-sensing receptor with the inwardly rectifying potassium channels Kir4.1 and Kir4.2 results in inhibition of channel function. Am J Physiol Renal Physiol 292:F1073–F1081
Jin L, Chao L, Chao J (1999) Potassium supplement upregulates the expression of renal kallikrein and bradykinin B2 receptor in SHR. Am J Physiol Renal Physiol 276:F476–F484
Kaufman JS, Hamburger RJ (1985) Passive potassium transport in the proximal convoluted tubule. Am J Phys 248:F228–F232
Kibble JD, Wareing M, Wilson RW, Green R (1995) Effect of barium on potassium diffusion across the proximal convoluted tubule of the anesthetized rat. Am J Phys 268:F778–F783
Kofuji P, Ceelen P, Zahs KR, Surbeck LW, Lester HA, Newman EA (2000) Genetic inactivation of an inwardly rectifying potassium channel (Kir4.1 Subunit) in mice: phenotypic impact in retina. J Neurosci 20:5733–5740
Kompatscher A, de Baaij JHF, Aboudehen K, Hoefnagels APWM, Igarashi P, Bindels RJM, Veenstra GJC, Hoenderop JGJ (2017) Loss of transcriptional activation of the potassium channel Kir5.1 by HNF1+¦ drives autosomal dominant tubulointerstitial kidney disease. Kidney Int 92:1145–1156
Konstas AA, Korbmacher C, Tucker SJ (2003) Identification of domains which control the heteromeric assembly of Kir5.1/Kir4.0 potassium channels. Am J Physiol Cell Physiol 284: C910–C917
Kucheryavykh YV, Pearson WL, Kurata HT, Eaton MJ, Skatchkov SN, Nichols CG (2007) Polyamine permeation and rectification of Kir4.1 channels. Channels 1:172–178
Lachheb S, Cluzeaud F, Bens M, Genete M, Hibino H, Lourdel S, Kurachi Y, Vandewalle A, Teulon J, Paulais M (2008) Kir4.1/Kir5.1 channel forms the major K^+ channel in the basolateral membrane of mouse renal collecting duct principal cells. Am J Physiol Renal Physiol 294: F1398–F1407
Lalioti MD, Zhang J, Volkman HM, Kahle KT, Hoffmann KE, Toka HR, Nelson-Williams C, Ellison DH, Flavell R, Booth CJ, Lu Y, Geller DS, Lifton RP (2006) Wnk4 controls blood pressure and potassium homeostasis via regulation of mass and activity of the distal convoluted tubule. Nat Genet 38:1124–1132
Leviel F, Borensztein P, Houillier P, Paillard M, Bichara M (1992) Electroneutral K^+/HCO_3^- cotransport in cells of medullary thick ascending limb of rat kidney. J Clin Invest 90:869–878
Liu W, Xu S, Woda C, Kim P, Weinbaum S, Satlin LM (2003) Effect of flow and stretch of the [Ca^{2+}] in response of principal and intercalated cells in cortical collecting duct. Am J Physiol Renal Physiol 285:F998–F1012
Liu Y, Rafferty TM, Rhee SW, Webber JS, Song L, Ko B, Hoover RS, He B, Mu S (2017) CD8+ T cells stimulate Na-Cl co-transporter NCC in distal convoluted tubules leading to salt-sensitive hypertension. Nat Commun 8:14037
Louis-Dit-Picard H, Barc J, Trujillano D, Miserey-Lenkei S, Bouatia-Naji N, Pylypenko O, Beaurain G, Bonnefond A, Sand O, Simian C, Vidal-Petiot E, Soukaseum C, Mandet C, Broux F, Chabre O, Delahousse M, Esnault V, Fiquet B, Houillier P, Bagnis CI, Koenig J, Konrad M, Landais P, Mourani C, Niaudet P, Probst V, Thauvin C, Unwin RJ, Soroka SD, Ehret G, Ossowski S, Caulfield M, International Consortium for Blood Pressure (ICBP), Bruneval P, Estivill X, Froguel P, Hadchouel J, Schott JJ, Jeunemaitre X (2012) KLHL3 mutations cause familial hyperkalemic hypertension by impairing ion transport in the distal nephron. Nat Genet 44:456–460

Lourdel S, Paulais M, Cluzeaud F, Bens M, Tanemoto M, Kurachi Y, Vandewalle A, Teulon J (2002) An inward rectifier K^+ channel at the basolateral membrane of the mouse distal convoluted tubule: similarities with Kir4-Kir5.1 heteromeric channels. J Physiol 538:391–404

Lu M, Wang WH (1996) Nitric oxide regulates the low-conductance K^+ channel in basolateral membrane of cortical collecting duct. Am J Phys 270:C1336–C1342

Lu M, Wang T, Yan Q, Yang X, Dong K, Knepper MA, Wang W, Giebisch G, Shull GE, Hebert SC (2002) Absence of small-conductance K^+ channel (SK) activity in apical membranes of thick ascending limb and cortical collecting duct in ROMK (Bartter's) knockout mice. J Biol Chem 277:37881–37887

Malik S, Lambert E, Zhang J, Wang T, Clark HL, Cypress M, Goldman BI, Porter GA, Pena S, Nino W, Gray DA (2018) Potassium conservation is impaired in mice with reduced renal expression of Kir4.1. Am J Physiol Renal Physiol 315:F1271–F1282

McCormick JA, Mutig K, Nelson JH, Saritas T, Hoorn EJ, Yang C-L, Rogers S, Curry J, Delpire E, Bachmann S, Ellison DH (2011) A SPAK isoform switch modulates renal salt transport and blood pressure. Cell Metab 14:352–364

McCormick JA, Yang CL, Zhang C, Davidge B, Blankenstein KI, Terker AS, Yarbrough B, Meermeier NP, Park HJ, McCully B, West M, Borschewski A, Himmerkus N, Bleich M, Bachmann S, Mutig K, Argaiz ER, Gamba G, Singer JD, Ellison DH (2014) Hyperkalemic hypertension-associated cullin 3 promotes WNK signaling by degrading KLHL3. J Clin Invest 124:4723–4736

Mendez-Gonzalez MP, Kucheryavykh YV, Zayas-Santiago A, Velez-Carrasco W, Maldonado-Martinez G, Cubano LA, Nichols CG, Skatchkov SN, Eaton MJ (2016) Novel KCNJ10 gene variations compromise function of inwardly rectifying potassium channel 4.1. J Biol Chem 291:7716–7726

Morimoto T, Liu W, Woda C, Carattino MD, Wei Y, Hughey RP, Apodaca G, Satlin LM, Kleyman TR (2006) Mechanism underlying flow stimulation of sodium absorption in the mammalian collecting duct. Am J Physiol Renal Physiol 291:F663–F669

Mu S, Shimosawa T, Ogura S, Wang H, Uetake Y, Kawakami-Mori F, Marumo T, Yatomi Y, Geller DS, Tanaka H, Fujita T (2011) Epigenetic modulation of the renal β-adrenergic-WNK4 pathway in salt-sensitive hypertension. Nat Med 17:573–580

Najjar F, Zhou H, Morimoto T, Bruns JB, Li HS, Liu W, Kleyman TR, Satlin LM (2005) Dietary K^+ regulates apical membrane expression of maxi-K channels in rabbit cortical collecting duct. Am J Physiol Renal Physiol 289:F922–F932

Nesterov V, Dahlmann A, Krueger BK, Bertog M, Loffing J, Korbmacher C (2012) Aldosterone-dependent and -independent regulation of the epithelial sodium channel (ENaC) in mouse distal nephron. Am J Physiol Renal Physiol 303:F1289–F1299

Neusch C, Rozengurt N, Jacobs RE, Lester HA, Kofuji P (2001) Kir4.1 Potassium channel subunit Is crucial for oligodendrocyte development and in vivo myelination. J Neurosci 21:5429–5438

O'Neil RG (1981) Potassium secretion by the cortical collecting tubule. Fed Proc 40:2403–2407

O'Neil RG, Hayhurst AR (1985) Functional differentiation of cell types of cortical collecting duct. Am J Phys 248:449–453

Obermuller N, Bernstein P, Velazquez H, Reilly R, Moser D, Ellison DH, Bachmann S (1995) Expression of the thiazide-sensitive Na-Cl cotransporter in rat and human kidney. Am J Phys 269:F900–F910

Palygin O, Levchenko V, Ilatovskaya DV, Pavlov TS, Pochynyuk OM, Jacob HJ, Geurts AM, Hodges MR, Staruschenko A (2017) Essential role of Kir5.1 channels in renal salt handling and blood pressure control. JCI Insight 2:pil 92331

Parrock S, Hussain S, Issler N, Differ AM, Lench N, Guarino S, Oosterveld MJS, Keijzer-Veen M, Brilstra E, van Wieringen H, Konijnenberg AY, Amin-Rasip S, Dumitriu S, Klootwijk E, Knoers N, Bockenhauer D, Kleta R, Zdebik AA (2013) KCNJ10 mutations display differential sensitivity to heteromerisation with KCNJ16. Nephron Physiol 123:7–14

Pathare G, Hoenderop JGJ, Bindels RJM, San-Cristobal P (2013) A molecular update on pseudohypoaldosteronism type II. Am J Physiol Renal Physiol 305:F1513–F1520

Paulais M, Lachheb S, Teulon J (2006) A Na^+ and Cl-activated K^+ Channel in the thick ascending limb of mouse kidney. J Gen Physiol 127:205–215

Paulais M, Bloch-Faure M, Picard N, Jacques T, Ramakrishnan SK, Keck M, Sohet F, Eladari D, Houillier P, Sp L, Teulon J, Tucker SJ (2011) Renal phenotype in mice lacking the Kir5.1 (Kcnj16) K+ channel subunit contrasts with that observed in SeSAME/EAST syndrome. Proc Natl Acad Sci USA 108:10361–10366

Pessia M, Tucker SJ, Lee K, Bond CT, Adelman JP (1996) Subunit positional effects revealed by novel heteromeric inwardly rectifying K^+ channels. EMBO J 15:2980–2987

Pessia M, Imbrici P, D'Adamo MC, Salvatore L, Tucker SJ (2001) Differential pH sensitivity of Kir4.1 and Kir4.2 potassium channels and their modulation by heteropolymerisation with Kir5.1. J Physiol 532:359–367

Piala AT, Moon TM, Akella R, He HX, Cobb MH, Goldsmith EJ (2014) Chloride sensing by WNK1 involves inhibition of autophosphosphorylation. Sci Signal 7:ra41

Reichold M, Zdebik AA, Lieberer E, Rapedius M, Schmidt K, Bandulik S, Sterner C, Tegtmeier I, Penton D, Baukrowitz T, Hulton SA, Witzgall R, Ben Zeev B, Howie AJ, Kleta R, Bockenhauer D, Warth R (2010) KCNJ10 gene mutations causing EAST syndrome (epilepsy, ataxia, sensorineural deafness, and tubulopathy) disrupt channel function. Proc Natl Acad Sci USA 107:14490–14495

Rengarajan S, Lee DH, Oh YT, Delpire E, Youn JH, McDonough AA (2014) Increasing plasma K+ by intravenous potassium infusion reduces NCC phosphorylation and drives kaliuresis and natriuresis. Am J Physiol Renal Physiol 306:F1059–F1068

Rosenbaek LL, Kortenoeven MLA, Aroankins TS, Fenton RA (2014) Phosphorylation decreases ubiquitylation of the thiazide-sensitive cotransporter NCC and subsequent clathrin-mediated endocytosis. J Biol Chem 289:13347–13361

Rozengurt N, Lopez I, Chiu CS, Kofuji P, Lester HA, Neusch C (2003) Time course of inner ear degeneration and deafness in mice lacking the Kir4.1 potassium channel subunit. Hear Res 177:71–80

Sala-Rabanal M, Kucheryavykh LY, Skatchkov SN, Eaton MJ, Nichols CG (2010) Molecular mechanisms of EAST/SeSAME syndrome mutations in Kir4.1 (KCNJ10). J Biol Chem 285:36040–36048

Sansom SC, O'Neil RG (1986) Effects of mineralocorticoids on transport properties of the cortical collecting duct basolateral membrane. Am J Phys 251:743–757

Schafer JA, Troutman SL, Schlatter E (1990) Vasopressin and mineralocorticoid increase apical membrane driving force for K+ secretion in rat CCD. Am J Phys 258:F199–F210

Schmitt R, Ellison DH, Farman N, Rossier BC, Reilly RF, Reeves WB, Oberbaumer I, Tapp R, Bachmann S (1999) Developmental expression of sodium entry pathways in rat nephron. Am J Phys 276:F367–F381

Schnermann J, Steipe B, Briggs JP (1987) In situ studies of distal convoluted tubule in rat. II. K secretion. Am J Phys 252:970–976

Scholl UI, Choi M, Liu T, Ramaekers VT, Hausler MG, Grimmer J, Tobe SW, Farhi A, Nelson-Williams C, Lifton RP (2009) Seizures, sensorineural deafness, ataxia, mental retardation, and electrolyte imbalance (SeSAME syndrome) caused by mutations in KCNJ10. Proc Natl Acad Sci USA 106:5842–5847

Shibata S, Zhang J, Puthumana J, Stone KL, Lifton RP (2013) Kelch-like 3 and Cullin 3 regulate electrolyte homeostasis via ubiquitination and degradation of WNK4. Proc Natl Acad Sci USA 110:7838–7843

Shirley DG, Walter SJ, Folkerd EJ, Unwin RJ, Bailey MA (1998) Transepithelial electrochemical gradients in the proximal convoluted tubule during potassium depletion the rat. J Physiol 513:551–557

Shuck ME, Piser TM, Block JH, Slightom JL, Lee KS, Bienkowski MJ (1997) Cloning and characterization of two K^+ inward rectifier (K_{ir}) 1.1 potassium channel homologs from human kidney ($K_{ir}1.2$ and $K_{ir}1.3$). J Biol Chem 272:586–593

Simon DB, Karet FE, Rodriguez J, Hamdan JH, DiPietro A, Trachtman H, Sanjad SA, Lifton RP (1996a) Genetic heterogeneity of Bartter's syndrome revealed by mutations in the K channel, ROMK. Nat Genet 14:152–156

Simon DB, Nelson-Williams C, Bia MJ, Ellison D, Karet FE, Molina AM, Vaara I, Iwata F, Cushner HM, Koolen M, Gainza F, Gitelman HJ, Lifton RP (1996b) Gitelman's variant of Bartter's syndrome inherited hypokalaemic alkalosis, is caused by mutations in the thiazide-sensitive Na-Cl cotranspoter. Nat Genet 12:24–30

Sorensen MV, Grossmann S, Roesinger M, Gresko N, Todkar AP, Barmettler G, Ziegler U, Odermatt A, Loffing-Cueni D, Loffing J (2013) Rapid dephosphorylation of the renal sodium chloride cotransporter in response to oral potassium intake in mice. Kidney Int 83:811–824

Su XT, Wang WH (2016) The expression, regulation, and function of Kir4.1 (Kcnj10) in the ammalian kidney. Am J Physiol Renal Physiol 311:F12–F15

Su XT, Zhang C, Wang L, Gu R, Lin DH, Wang WH (2016) Disruption of KCNJ10 (Kir4.1) stimulates the expression of ENaC in the collecting duct. Am J Physiol Renal Physiol 310: F985–F993

Su XT, Ellison DH, Wang WH (2019) Kir4.1/Kir5.1 in the DCT plays a role in the regulation of renal K^+ excretion. Am J Physiol Renal Physiol 316:F582–F586

Suzuki T, Katori M, Fujita T, Kumagai Y, Majima M (2000) Involvement of the renal kallikrein-kinin system in K(+)-induced diuresis and natriuresis in anesthetized rats. Eur J Pharmacol 399:223–227

Takumi T, Ishii T, Horio Y, Morishige K, Takahashi N, Yamada M, Yamashita T, Kiyama H, Sohmiya K, Nakanishi S, Kurachi Y (1995) A novel ATP-dependent inward rectifier potassium channel expressed predominantly in glial cells. J Biol Chem 270:16339–16346

Tanemoto M, Kittaka N, Inanobe A, Kurachi Y (2000) *In vivo* formation of a proton-sensitive K^+ channel by heteromeric subunit assembly of Kir5.1 with Kir4.1. J Physiol 525:587–592

Tanemoto M, Abe T, Onogawa T, Ito S (2004) PDZ binding motif-dependent localization of K+ channel on the basolateral side in distal tubules. Am J Physiol Renal Physiol 287:F1148–F1153

Tanemoto M, Abe T, Ito S (2005) PDZ-binding and Di-hydrophobic motifs regulate distribution of Kir4.1 channels in renal cells. J Am Soc Nephrol 16:2608–2614

Tanemoto M, Toyohara T, Abe T, Ito S (2008) MAGI-1a functions as a scaffolding protein for the distal renal tubular basolateral K+ channels. J Biol Chem 283:12241–12247

Tanemoto M, Abe T, Uchida S, Kawahara K (2014) Mislocalization of K^+ channels causes the renal salt wasting in EAST/SeSAME syndrome. FEBS Lett 588:899–905

Taniguchi J, Guggino WB (1989) Membrane stretch: a physiological stimulator of Ca^{2+}-activated K^+ channels in thick ascending limb. Am J Phys 257(Renal 26):F347–F352

Terker AS, Yang CL, McCormick JA, Meermeier NP, Rogers SL, Grossmann S, Trompf K, Delpire E, Loffing J, Ellison DH (2014) Sympathetic stimulation of thiazide-sensitive sodium chloride cotransport in the generation of salt-sensitive hypertension. Hypertension 64:178–184

Terker A-S, Zhang C, McCormick J-A, Lazelle R-A, Zhang C, Meermeier N-P, Siler D-A, Park H-J, Fu Y, Cohen D-M, Weinstein A-M, Wang WH, Yang CL, Ellison D-H (2015) Potassium modulates electrolyte balance and blood pressure through effects on distal cell voltage and chloride. Cell Metab 21:39–50

Terker AS, Zhang C, Erspamer KJ, Gamba G, Yang CL, Ellison DH (2016) Unique chloride-sensing properties of WNK4 permit the distal nephron to modulate potassium homeostasis. Kidney Int 89:127–134

Terker AS, Ferdaus MZ, Cornelius R, Erspamer KJ, Su XT, Miller L, McCormick JA, Wang W-H, Gamba G, Yang C-L, Ellison DH (2018) With no lysine kinase 4 modulates sodium potassium 2 chloride cotransporter activity in vivo. Am J Physiol Renal Physiol 315:F781–F790

Thastrup J, Rafiqi F, Vitari A, Pozo-Guisado E, Deak M, Mehellou Y, Alessi D (2012) SPAK/OSR1 regulate NKCC1 and WNK activity: analysis of WNK isoform interactions and activation by T-loop trans-autophosphorylation. Biochem J 441:325–337

Thompson CB, McDonough AA (1996) Skeletal muscle Na,K-ATPase alpha and beta subunit protein levels respond to hypokalemic challenge with isoform and muscle type specificity. J Biol Chem 271:32653–32658

Tomilin VN, Zaika O, Subramanya AR, Pochynyuk O (2018) Dietary K^+ and Cl^- independently regulate basolateral conductance in principal and intercalated cells of the collecting duct. Pflugers Archiv Eur J Physiol 470:339–353

Tsuchiya K, Wang W, Giebisch G, Welling PA (1992) ATP is a coupling modulator of parallel Na, K-ATPase-K-channel activity in the renal proximal tubule. Proc Natl Acad Sci USA 89:6418–6422

Tucker SJ, Imbrici P, Salvatore L, D'Adamo MC, Pessia M (2000) pH dependence of the inwardly rectifying potassium channel, Kir5.1, and localization in renal tubular epithelia. J Biol Chem 275:16404–16407

Unwin R, Capasso G, Giebisch G (1994) Potassium and sodium transport along he loop of Henle: effects of altered dietary potassium intake. Kidney Int 46:1092–1099

Vallon V, Grahammer F, Richter K, Bleich M, Lang F, Barhanin J, Voelker H, Warth R (2001) Role of KCNE1-dependent K+ fluxes in mouse proximal tubule. J Am Soc Nephrol 12:2003–2011

van der Lubbe N, Moes AD, Rosenbaek LL, Schoep S, Meima ME, Danser AHJ, Fenton RA, Zietse R, Hoorn EJ (2013) K^+-induced natriuresis is preserved during Na^+ depletion and accompanied by inhibition of the Na^+-Cl^- cotransporter. Am J Physiol Renal Physiol 305: F1177–F1188

Vargas-Poussou R, Huang C, Hulin P, Houillier P, Jeunemaitre X, Paillard M, Planelles G, Dechaux M, Miller RT, Antignac C (2002) Functional characterization of a calcium-sensing receptor mutation in severe autosomal dominant hypocalcemia with a Bartter-like syndrome. J Am Soc Nephrol 13:2259–2266

Wade JB, Fang L, Coleman RA, Liu J, Grimm PR, Wang T, Welling PA (2011) Differential regulation of ROMK (Kir1.1) in distal nephron segments by dietary potassium. Am J Physiol Renal Physiol 300:F1385–F1393

Wang WH (1994) Two types of K^+ channel in thick ascending limb of rat kidney. Am J Phys 267: F599–F605

Wang WH, White S, Geibel J, Giebisch G (1990) A potassium channel in the apical membrane of the rabbit thick ascending limb of Henle's loop. Am J Phys 258:F244–F253

Wang L, Zhang C, Su X, Lin DH, Wang W (2015) Caveolin-1 deficiency inhibits the basolateral K^+ channels in the distal convoluted tubule and impairs renal K^+ and Mg^{2+} transport. J Am Soc Nephrol 26:2678–2690

Wang B, Wen D, Li H, Wang-France J, Sansom SC (2017) Net K secretion in the thick ascending limb of mice on a low-Na, high-K diet. Kidney Int 92:864–875

Wang MX, Cuevas-Gallardo C, Su XT, Wu P, Gao Z-X, Lin DH, McCormick JA, Yang CL, Wang WH, Ellison DH (2018a) Potassium (K^+) intake modulates NCC activity via the K+ channel. Kir4.1. Kidney Int 93:893–902

Wang MX, Su XT, Wu P, Gao ZX, Wang WH, Staub O, Lin DH (2018b) Kir5.1 regulates Nedd4-2-mediated ubiquitination of Kir4.1 in distal nephron. Am J Physiol Renal Physiol 315:F986–F996

Wareing M, Wilson RW, Kibble JD, Green R (1995) Estimated potassium reflection coefficient in perfused proximal convoluted tubules of the anaesthetized rat in vivo. J Physiol 488:153–161

Wei Y, Liao Y, Zavilowitz B, Ren J, Liu W, Chan P, Rohatgi R, Estilo G, Jackson EK, Wang WH, Satlin LM (2014) Angiotensin II type 2 receptor regulates ROMK-like K^+ channel activity in the renal cortical collecting duct during high dietary K+ adaptation. Am J Physiol Renal Physiol 307:F833–F843

Weinstein AM (1986) A mathematical model of the rat proximal tubule. Am J Physiol 250:F860–F873

Weinstein AM, Windhager EE (2001) The paracellular shunt of proximal tubule. J Membr Biol 184:241–245

Williams DM, Lopes CMB, Rosenhouse-Dantsker A, Connelly HL, Matavel A, Uchi J, McBeath E, Gray DA (2010) Molecular basis of decreased Kir4.1 function in SeSAME/EAST syndrome. J Am Soc Nephrol 21:2117–2129

Wilson FH, Disse-Nicodeme S, Choate KA, Ishikawa K, Nelson-Williams C, Desitter I, Gunel M, Milford DV, Lipkin GW, Achard JM, Feely MP, Dussol B, Berland Y, Unwin RJ, Mayan H, Simon DB, Farfel Z, Jeunemaitre X, Lifton RP (2001) Human hypertension caused by mutations in WNK kinases. Science 293:1107–1112

Woda CB, Bragin A, Kleyman TR, Satlin LM (2001) Flow-dependent K^+ secretion in the cortical collecting duct is mediated by a maxi-K channel. Am J Physiol Renal Physiol 280:F786–F793

Woda CB, Leite M Jr, Rohatgi R, Satlin LM (2002) Effects of luminal flow and nucleotides on $[Ca^{2+}]_i$ in rabbit cortical collecting duct. Am J Physiol Renal Physiol 283:F437–F446

Wu P, Gao Z-X, Duan X, Su X-T, Wang MX, Lin D-H, Gu RM, Wang WH (2018) AT2R-mediated regulation of Na-Cl cotransporter (NCC) and renal K excretion depends on the K channel, Kir4.1. Hypertension 71:622–630

Wu P, Gao ZX, Su XT, Wang MX, Wang WH, Lin DH (2019) Kir4.1/Kir5.1 activity is essential for dietary sodium intak-induced modulation of Na-Cl cotransporter. J Am Soc Nephrol 30:216–227

Yang Z, Xu H, Cui N, Qu Z, Chanchevalap S, Shen W, Jiang C (2000) Biophysical and molecular mechanisms underlying the modulation of heteromeric Kir4.1-Kir5.1 channels by CO_2 and pH. J Gen Physiol 116:33–45

Yang L, Xu S, Guo X, Uchida S, Weinstein AM, Wang T, Palmer LG (2018) Regulation of renal Na transporters in response to dietary K. Am J Physiol Renal Physiol 315:F1032–F1041

Yao X, Tian S, Chan HY, Biemesderfer D, Desir GV (2002) Expression of KCNA10, a voltage-gated K channel, in glomerular endothelium and at the apical membrane of the renal proximal tubule. J Am Soc Nephrol 13:2839

Youn JH, McDonough AA (2009) Recent advances in understanding integrative control of potassium homeostasis. Annu Rev Physiol 71:381–401

Younes-Ibrahim M, Barlet-Bas C, Buffin-Meyer B, Cheval L, Rajerison R, Doucet A (1995) Ouabain-sensitive and -insensitive K-ATPases in rat nephron: effect of K depletion. Am J Phys 268:F1141–F1147

Zaika OL, Mamenko M, Palygin O, Boukelmoune N, Staruschenko A, Pochynyuk O (2013) Direct inhibition of basolateral Kir4.1/5.1 and Kir4.1 channels in the cortical collecting duct by dopamine. Am J Physiol Renal Physiol 305:F1277–F1287

Zaika O, Palygin O, Tomilin V, Mamenko M, Staruschenko A, Pochynyuk O (2015) Insulin and IGF-1 activate Kir4.1/5.1 channels in cortical collecting duct principal cells to control basolateral membrane voltage. Am J Physiol Renal Physiol 310:F311–F321

Zhang C, Wang L, Thomas S, Wang K, Lin DH, Rinehart J, Wang WH (2013) Src-family protein tyrosine kinase regulates the basolateral K channel in the distal convoluted tubule (DCT) by phosphorylation of KCNJ10. J Biol Chem 288:26135–26146

Zhang C, Wang L, Zhang J, Su X-T, Lin DH, Scholl UI, Giebisch G, Lifton RP, Wang WH (2014) KCNJ10 determines the expression of the apical Na-Cl cotransporter (NCC) in the early distal convoluted tubule (DCT1). Proc Natl Acad Sci USA 111:11864–11869

Zhang C, Wang L, Su XT, Lin DH, Wang WH (2015) KCNJ10 (Kir4.1) is expressed in the basolateral membrane of the cortical thick ascending limb. Am J Physiol Renal Physiol 308: F1288–F1296

Zhang C, Wang L, Su XT, Zhang J, Lin DH, Wang WH (2016) ENaC and ROMK activity are inhibited in the DCT2/CNT of $TgWnk4^{PHAII}$ mice. Am J Physiol Renal Physiol 312:F682–F688

Zhang D, Gao Z, Vio CP, Xiao Y, Wu P, Zhang H, Guo X, Meng X, Gu L, Wang J, Duan X, Lin D, Wang W, Gu R (2018) Bradykinin stimulates renal Na^+ and K^+ excretion by inhibiting the K^+ Channel (Kir4.1) in the distal convoluted tubule. Hypertension 72:361–369

Zweifach A, Desir GV, Aronson PS, Giebisch G (1992) Inhibition of Ca-activated K channels from renal microvillus membrane vesicles by amiloride analogs. J Membr Biol 128:115–122

Chapter 21
Small-Molecule Pharmacology of Epithelial Inward Rectifier Potassium Channels

Sujay V. Kharade and Jerod S. Denton

Abstract Inward rectifier potassium (Kir) channels play fundamental roles in epithelial fluid and electrolyte transport in several organ systems and represent novel drug targets for human diseases. The loss-of-function mutations in genes encoding Kir1.1 (ROMK) and Kir4.1 give rise to renal tubule phenotypes that resemble those in patients taking diuretics. This discovery has greatly motivated drug discovery efforts aimed at developing novel classes of diuretics targeting these channels. In many cases, however, even a rudimentary understanding of the functions of Kir channels in epithelia is lacking due, in part, to the lack of precise pharmacological tools for specifically modulating the activity of Kir channels in complex biological cells and tissues. This barrier is gradually disappearing with the development of Kir channel-specific modulators that can be used to explore the physiology of Kir channels. The goals of this chapter are threefold: (1) Provide an overview of the expression patterns and known functions of Kir channels in epithelial cells, (2) provide a summary of the small-molecule pharmacology of Kir channels expressed in epithelial cells to help guide the design of pharmacological experiments for dissecting their functions, and (3) discuss in detail the ongoing efforts to develop small-molecule antagonists of Kir1.1 and Kir4.1 for use as diuretics to serve as a model for Kir channel-directed drug discovery.

Keywords ROMK · Kir4.1 · Kir7.1 · Drug discovery · Thallium flux · Medicinal chemistry

S. V. Kharade
Department of Anesthesiology, Vanderbilt University Medical Center, Nashville, TN, USA
e-mail: sujay.v.kharade@vumc.org

J. S. Denton (✉)
Department of Pharmacology, Institute of Chemical Biology, Vanderbilt University, Nashville, TN, USA
e-mail: jerod.s.denton@vumc.org

K. L. Hamilton, D. C. Devor (eds.), *Studies of Epithelial Transporters and Ion Channels*, Physiology in Health and Disease,
https://doi.org/10.1007/978-3-030-55454-5_21

21.1 Overview of Inward Rectifier Potassium Channel Structure and Function

The Kir channel family comprises sixteen genes (denoted *KCNJx*) in humans that encode seven distinct sub-families of proteinaceous subunits (denoted Kir1-7), many of which are expressed in epithelial cells of various organs (Table 21.1). The simplest functional unit of Kir channels is a tetramer, which is formed by the assembly of four membrane-spanning subunits around an aqueous pore (Fig. 21.1). In most cases, a clear understanding of the functions, regulatory pathways, and therapeutic potential of Kir channels is unknown due to the lack of selective pharmacological tools that can discriminate between individual family members. This challenge is exacerbated by the fact that some Kir channels can be heteromeric assemblies of different subunits (e.g., Ki4.1/Kir5.1—see below), which increases the functional, regulatory, and pharmacological diversity of the channel family. Thus, an important and challenging goal in the field that will help elucidate the integrative physiology and druggability of Kir channels will be development of

Table 21.1 Kir channel distribution in mammalian epithelial cells

Kir Channel	Organ (Cell type)	References
Kir1.1 (ROMK)	Kidney (TAL, DCT, CCD)	Boim et al. (1995), Hebert et al. (2005)
Kir2.1	Mammary gland (Mammary secretory)	Kamikawa and Ishikawa (2014)
Kir2.3	Kidney (CCD)	Millar et al. (2006), Welling (1997)
Kir4.1	Kidney (DCT, CCD)	Lachheb et al. (2008), Lourdel et al. (2002), Paulais et al. (2011), Scholl et al. (2009), Zaika et al. (2013)
	Cochlea (stria vascularis)	Hibino et al. (1997), Rozengurt et al. (2003)
	Stomach (Parietal cells)	Fujita et al. (2002)
Kir5.1	Kidney (DCT)	Lachheb et al. (2008), Lourdel et al. (2002), Paulais et al. (2011), Zaika et al. (2013)
Kir6.1/ SUR2B	Lung (alveolar type II)	Leroy et al. (2004), Monaghan et al. (1997), Sakuma et al. (1998)
Kir7.1	Brain (choroid Plexus)	Doring et al. (1998), Krapivinsky et al. (1998)
	Retina (RPE)	Nakamura et al. (1999), Shimura et al. (2001), Yang et al. (2008), Yasuda et al. (2003)
	Kidney (PT, TAL, CD, DCT)	Derst et al. (2001), Nakamura et al. (1999), Ookata et al. (2000)
	Intestine	Nakamura et al. (1999), Partiseti et al. (1998)
	Stomach	Nakamura et al. (1999)
	Thyroid (Follicular cells)	Nakamura et al. (1999)

Abbreviations: *PT* proximal tubule, *TAL* thick ascending limb, *DCT* distal convoluted tubule, *DCT* cortical collecting duct

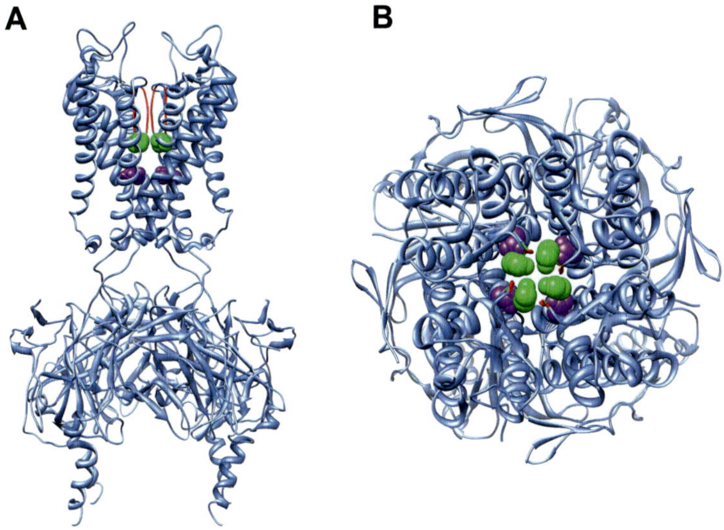

Fig. 21.1 Kir channel structure and small-molecule binding sites. (**a**) Side and (**b**) top-down view of a ribbon diagram of chicken Kir2.2 crystal structure (Tao et al. 2009). The selectivity filter has been colored red. Amino acids corresponding to those implicated in small-molecule binding in ROMK (N171-purple) and Kir4.1 (E158—purple; T128—green) are shown as space-filling models for clarity. The image was generated from protein databank file PDB 3JYC using Chimera software (Pettersen et al. 2004). See text for details

precise pharmacological tools that can discriminate between myriad Kir channel subtypes expressed in epithelia and other tissues.

Under almost all physiological conditions, K^+ ions follow an outwardly directed electrochemical gradient created by the Na^+-K^+-ATPase. As will be discussed below, the movement of K^+ ions through inward rectifier K^+ channels helps energize substrate- or electrophysiologically coupled transport processes in epithelia that contribute to tissue, organ, and systemic fluid and electrolyte balance (Fig. 21.2). The term "inward rectification" refers to the electrophysiological behavior of Kir channels under voltage clamp conditions. By measuring current at voltages more positive and more negative than the Nernst electrochemical potential for K^+ (E_K), one observes larger *inward* current amplitude (i.e., at voltages more negative than E_K) than *outward* current amplitude (i.e., at voltages more positive than E_K). This is in contrast to the behavior of voltage-gated K^+ (K_v) channels, which typically carry only outward current following channel opening at depolarized potentials. The mechanism underlying inward rectification in Kir channels has been studied extensively (Fakler et al. 1995; Ficker et al. 1994; Hibino et al. 2010; Kurata et al. 2007; Lopatin et al. 1995; Lu and MacKinnon 1994; Nichols et al. 1994; Nishida and MacKinnon 2002; Pegan et al. 2005; Wible et al. 1994) and is fundamentally different from that giving rise to outward rectification in K_v channels (Hille 1992). In Kir channels, at voltages more positive than E_K, intracellular polyvalent cations such as Mg^{2+} and polyamines (e.g., spermine) enter the pore in a steeply voltage-

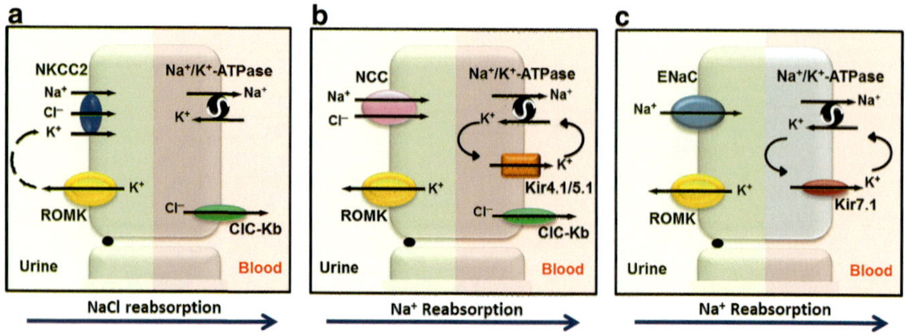

Fig. 21.2 Role of renal Kir channels in renal solute transport. Schematic of different nephron segments where Kir channels are thought to play an important role in transepithelial Na^+ reabsorption and K^+ secretion. (**a**) Thick Ascending Limb: Loop diuretic target Na^+-K^+-$2Cl^-$ co-transporter (NKCC2) is the major pathway for the reabsorption of NaCl across the apical membrane. However, since luminal K^+ concentration is relatively low compared to Na^+ and Cl^- concentrations, K^+ must be recycled in the opposite direction through the renal outer medullary potassium (ROMK) channel to maintain the robust activity of NKCC2 and efficient NaCl reabsorption. On the basolateral membrane, Na^+ is taken back in to the blood by the Na^+-K^+-ATPase pump, while Cl^- makes its way back through ClC-Kb. (**b**) Distal Convoluted Tubule: The thiazide sensitive Na^+-Cl^- co-transporter (NCC) mediates NaCl entry across the apical membrane whereas K^+ secretion is mediated by ROMK. On the basolateral membrane, K^+ recycling via Kir4.1/5.1 maintains optimal activity of the Na^+-K^+-ATPase since the extracellular K^+ concentration is low and helps Na^+ reabsorption. Kir7.1 (not shown) is thought to be also involved in the basolateral K^+ recycling. ClC-Kb mediates Cl^- exit across the basolateral membrane. (**c**) Cortical collecting duct: The amiloride sensitive epithelial sodium channel (ENaC) is responsible for apical Na^+ entry; however, to negate the resulting electro-positivity and to maintain its activity, ROMK mediates the K^+ secretion. On the basolateral side, the Na^+ is taken back by Na^+-K^+-ATPase and Kir7.1 is thought to be responsible for the necessary K^+ recycling in this part of the nephron

dependent manner and physically occlude the ion conduction pathway. Negatively charged residues that mediate these electrostatic interactions with blocking cations have been identified through mutagenesis and crystallography experiments. The identity and location of these so-called "rectification residues" will be revisited later in the chapter, since some of these appear to contribute to selective small-molecule binding sites in Kir channels.

21.2 The Renal Outer Medullary Potassium Channel (ROMK, Kir1.1, *KCNJ1*)

21.2.1 Overview of ROMK Expression and Functions in Renal Tubule Epithelia

Kir1.1, which is commonly referred to as the *R*enal *O*uter *M*edullary K^+ (ROMK) channel, is the founding member of the Kir channel family. The complementary

DNA (cDNA) of the gene encoding ROMK (i.e., *KCNJ1*) was cloned by two independent groups more than 20 years ago. Dr. Steven Hebert's group cloned splice variant 1 (i.e., ROMK1 or Kir1.1a) of the channel from the inner stripe of the outer medulla of rat kidneys and showed that it forms a barium (Ba^{2+})-inhibitable K^+ channel when heterologously expressed in *Xenopus* oocytes (Ho et al. 1993). Dr. Larry Palmer's group subsequently reported cloning of the ROMK2 splice variant, which lacks the first 19 N-terminal amino acids found in ROMK1 (Zhou et al. 1994). A ROMK3 splice variant was later reported by Hebert and colleagues and contains a unique stretch of 26 N-terminal amino acids (Boim et al. 1995). All three splice variants express robust, weakly rectifying K^+ currents that are functionally indistinguishable. We will therefore not differentiate between the splice variants and simply refer to them collectively as ROMK for the remainder of the chapter.

ROMK is abundantly expressed in the thick ascending limb (TAL) of Henle's loop and collecting duct (CD) of the nephron, but is also expressed at lower levels in the distal convoluted tubule (DCT) and connecting tubule (CNT) (Boim et al. 1995). In the TAL, ROMK forms an ~30 picoSiemens (pS) channel in the luminal membrane that underlies to the so-called secretory K^+ (SK) channel, first identified with patch-clamp electrophysiology (Wang et al. 1990). The activity of ROMK is essential for NaCl reabsorption in the TAL, as well as distal nephron water reabsorption required to generate concentrated urine (Fig. 21.2). The first step in NaCl reabsorption in the TAL is catalyzed by the apical membrane Na^+-K^+-2 Cl^- co-transporter, NKCC2, which is the molecular target for the loop diuretics furosemide and bumetanide. Because the concentration of K^+ in the luminal fluid is lower than that of Na^+ and Cl^-, K^+ must be continually recycled across the luminal membrane by ROMK to maintain the activity of NKCC2 (Fig. 21.2). Inhibition of ROMK with the non-specific blocker Ba^{2+} reduces NaCl reabsorption in the TAL (Greger and Schlatter 1981), essentially by starving NKCC2 of substrate K^+ ions needed for its activity. This observation provided some of the first evidence that pharmacological inhibition of ROMK in the TAL may represent a target for a novel class of loop diuretic by indirectly inhibiting NKCC2.

In the aldosterone-sensitive distal nephron, ROMK comprises the major constitutive secretory pathway for K^+ (Giebisch 1995; Wang et al. 1997; Welling and Ho 2009). K^+ secretion in principal cells of the CD occurs by a two-step process (Fig. 21.2). First, K^+ is transported from blood across the basolateral membrane by the Na^+-K^+-ATPase. K^+ ions subsequently move down their electrochemical gradient and cross the apical membrane via ROMK channels. K^+ secretion through ROMK makes the apical membrane potential more negative, thereby increasing the driving force for Na^+ reabsorption by the *E*pithelial *Na*$^+$ *C*hannel (ENaC). Consistent with this model, inhibition of ROMK with Ba^{2+} depolarizes the apical membrane potential to limit the ENaC-mediated Na^+ reabsorption (Koeppen et al. 1983). This observation suggests that inhibition of ROMK activity in the distal nephron may have two consequences: Firstly, a reduction in Na^+ reabsorption through ENaC may contribute to the natriuresis caused by inhibition of NaCl reabsorption in the TAL. Secondly, and importantly, ROMK inhibition may limit the loss of K^+ in the urine by blocking one of the two known pathways for K^+

secretion. The second pathway is mediated by flow-stimulated, Ca^{2+}-activated K^+ channels (Bailey et al. 2006; Liu et al. 2007). The relative contribution of these two pathways during chronic pharmacological ROMK inhibition has yet to be determined experimentally.

21.2.2 Genetic Validation of ROMK as a Diuretic Target

The first genetic validation of ROMK as a diuretic target came from the discovery by Dr. Richard Lifton and colleagues of autosomal recessive loss-of-function (LOF) mutations in *KCNJ1* in patients with type II Bartter syndrome (Simon et al. 1996). The clinical manifestations of Type II Bartter's syndrome include symptoms that would be expected from loss of ROMK activity in the nephron: namely, hypernatremia, hypokalemic alkalosis, polyuria, and low blood pressure (Bartter et al. 1962). In 2008, Lifton and colleagues found that heterozygous carriers of *KCNJ1* mutations in the Framingham Heart Cohort have lower blood pressure, are protected from development of hypertension with age, and, importantly, exhibit no overt clinical symptoms of type II Bartter syndrome (Ji et al. 2008). Taken together, these data suggest that partial ROMK inhibition may lead to favorable clinical outcomes, but that complete inhibition may lead to pathological salt and water wasting. However, given that a majority of *KCNJ1*$^{-/-}$ mice exhibit hydronephrosis and renal insufficiency (Lu et al. 2002), it is conceivable that a genetic loss of ROMK function may result in abnormal kidney development and function that would not occur in genetically normal individuals taking a medication that inhibits ROMK activity. With the recent development of drug-like inhibitors of ROMK (see below), the answer to this question may soon be revealed.

21.2.3 ROMK Drug Discovery

A discussion of the small-molecule pharmacology for ROMK begins with the efforts of Clark and colleagues, who first demonstrated that administration of glyburide—a sulfonylurea K_{ATP} channel inhibitor—induced a dose-dependent stimulation of renal Na^+ excretion without affecting K^+ excretion in rats (Clark et al. 1993). Wang and colleagues later showed that this was associated with a reduction in Na^+ and K^+ transport in the TALH and CD via inhibition of ROMK (Wang et al. 1995a; Wang et al. 1995b). However, whether the high doses of glyburide used in these experiments altered renal tubule transport through inhibition of other transport mechanisms is unclear. Assessing whether specific inhibition of ROMK function will elicit natriuresis and diuresis without inducing kaliuresis would require development of more potent and selective small-molecule inhibitors.

In 2006, our group at Vanderbilt began developing a high-throughput screening (HTS) platform to enable the discovery of novel inhibitors of ROMK. Because single-cell, voltage-clamp electrophysiology is too slow and laborious to enable robust drug discovery, we set out to develop an assay that can be performed in 384-well plates to dramatically improve throughput. The assay we chose takes advantage of the permeability of the ROMK channel pore to the K^+ congener thallium (Tl^+) and utilizes an intracellular dye whose fluorescence emission increases dose-dependently with Tl^+ concentration. There are Tl^+ sensitive dyes that are commercially available, including BTC, FluoZin-2, FluxOR (all from Life Technologies), and Thallos (WaveFront Biosciences, Franklin, TN). We have developed robust assays using FluoZin-2 (Lewis et al. 2009; Raphemot et al. 2011, 2013a, b) and Thallos (Kharade et al. 2018), but have recently begun using Thallos for all of our Kir channel assays because of its lower cost. The Tl^+ flux assay is performed using T-REx-HEK-293 (Life Technologies) cells that express ROMK from a tetracycline-inducible promoter. The use of an inducible system serves at least two important purposes. First, inducible ROMK expression limits cell line degeneration and loss of assay performance that can be caused by the constitutive overexpression of some Kir channels. Secondly, an inducible expression provides a convenient way to test compounds for activity toward endogenous Tl^+ flux pathways, such as the Na^+-K^+-ATPase and voltage-gated K^+ channels expressed in HEK-293 cells. The assay is technically straightforward and performed essentially as follows. After loading cells with the acetoxymethylester form of Thallos, the cells are pre-incubated with small molecules (one compound per well) for 20 min and the plate is moved to the stage of a kinetic imaging plate reader capable of measuring fluorescence from each well of the plate simultaneously (e.g., Fig. 21.3a). The assay is initiated by adding a Tl^+ stimulus buffer to every well of the plate, while monitoring fluorescence emission. As shown in Fig. 21.3b, extracellular Tl^+ addition stimulates robust Tl^+ flux in cells induced overnight with tetracycline, but not in uninduced cells not expressing ROMK or in induced cells that have been pre-incubated with the peptide ROMK inhibitor tertiapin-Q (TPNQ) (Fig. 21.3b). In a HTS campaign, ROMK modulators are identified by monitoring the slope of fluorescence change over time: inhibitors decrease the slope by inhibiting ROMK-dependent Tl^+ flux (e.g., Fig. 21.3a, well N17), whereas activators increase Tl^+ flux. A step-by-step description of how to develop and validate a Kir channel Tl^+ flux assay for HTS has been described elsewhere (Raphemot et al. 2013a).

Using the Tl^+ flux assay, we performed a screen of approximately 225,000 compounds from the NIH Molecular Libraries Small-Molecule Repository (MLSMR) for pharmacological modulators of ROMK (Lewis et al. 2009). The results of the screen have been deposited in a searchable format in the National Center for Biotechnology Information PubChem Bioassay database (https://pubchem.ncbi.nlm.nih.gov). Compounds were screened at a nominal concentration of 10 μM and selected as "hits" if they were associated with a change in a well's fluorescence that differed by greater than 3 standard deviations from the mean fluorescence of control wells. The hits were picked for confirmation testing in tetracycline-induced and uninduced cells, and those exhibiting activity in uninduced cells were excluded from further analysis because they were likely to also affect endogenous Tl^+ flux pathways. ROMK-dependent modulators were re-ordered as

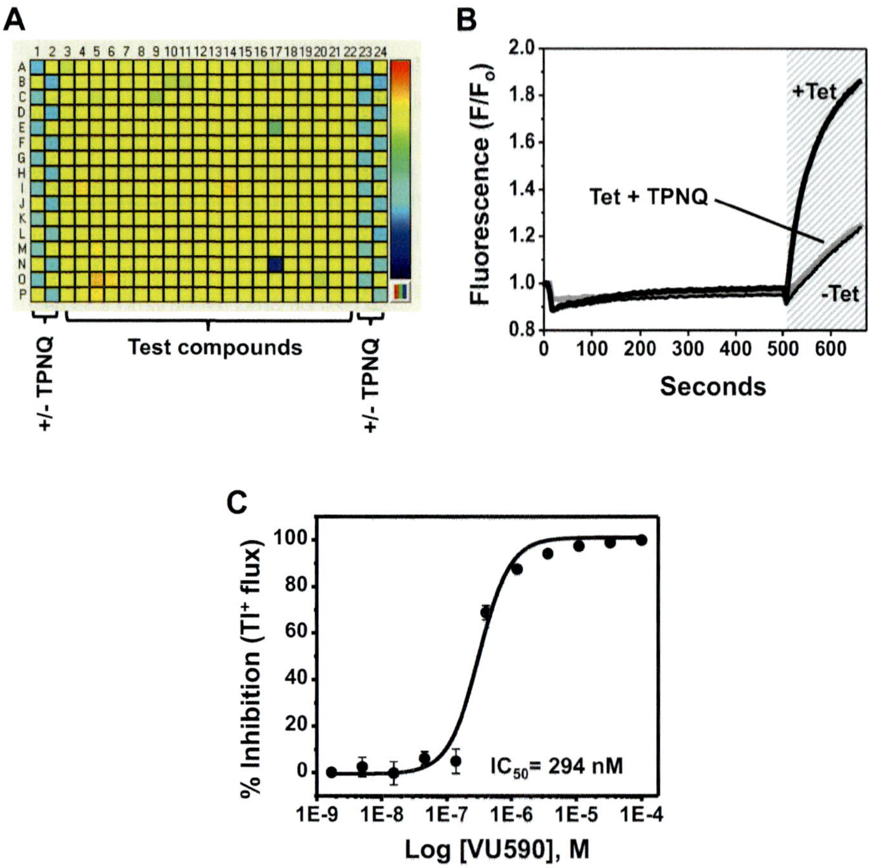

Fig. 21.3 Tl^+ flux assay of ROMK function for HTS and inhibition by VU590. (**a**) Representative fluorescence "heat map" recorded from a 384-well plate containing 20,000 T-REx-HEK293-ROMK cells/well induced overnight with tetracycline. Fluorescence intensity has been pseudocolored using the color scale shown to the right of the plate, with cooler colors (blues) indicating lower fluorescence and hotter colors (yellow and red) indicating high fluorescence and enhanced Tl^+ flux. Every other well of columns 1–2 and 23–24 have been treated with the ROMK inhibitor TPNQ to provide a measure of unblocked and blocked ROMK-mediated Tl^+ flux for quality control purposes. Each well in columns 3–22 have been treated with 10 μM test compound (one per well). Note that column N17 received an inhibitory small molecule, leading to lower rate of Tl^+ flux and fluorescence. This compound would be identified as a "hit" and picked for subsequent analysis. (**b**) Timecourse of Tl^+-induced (stippled box) changes in fluorescence in T-REx-HEK293-ROMK cells that were cultured overnight without tetracycline (−Tet) or with tetracycline (+Tet), or with tetracycline and pre-treated with TPNQ (Tet + TPNQ). Cells that were left uninduced or induced and treated with TPNQ exhibit low background Tl^+ flux, whereas those that were induced to express ROMK and left unblocked exhibit robust Tl^+ flux. (**c**) Dose-dependent inhibition of ROMK-mediated Tl^+ flux by the small-molecule inhibitor VU590. **b** and **c** are reproduced with permission from (Lewis et al. 2009)

dry powders, dissolved freshly in DMSO, and used for testing in concentration-response experiments against ROMK and other Kir channels.

The first publicly disclosed small-molecule inhibitor of ROMK was 7,13-*Bis* [(4-nitrophenyl)methyl]-1,4,10-trioxa-7,13-diazacyclopentadecane, which we termed VU590 (Lewis et al. 2009), consists of two nitro-phenyl groups flanking a nitrogen- and oxygen-rich crown ether moiety. It was almost rejected as a "false-positive" because crown ethers are well known to chelate alkali metals such as K^+, raising the possibility that the activity observed in the Tl^+ flux assay was due to chelation of Tl^+ and not frank inhibition of ROMK. However, patch-clamp experiments showed that VU590 is indeed a bona fide inhibitor of ROMK that dose-dependently inhibits the channel with a 50% inhibition concentration (IC_{50}) of approximately 300 nM (Fig. 21.3c) (Lewis et al. 2009). Moreover, VU590 was moderately selective. At a concentration of 10 μM, VU590 had no effect on the closely related Kir channels Kir2.1 or Kir4.1, providing at least a 30-fold selectivity window for ROMK over these channels. Additional experiments revealed that VU590 was not completely selective for ROMK, as it inhibited Kir7.1 with an IC_{50} of ~8 μM. This presented a problem for the development of a small-molecule probe that could be used to investigate the effects of pharmacological inhibition of ROMK on renal tubule electrolyte transport, because both ROMK and Kir7.1 are expressed in K^+-secreting principal cells of the collecting duct (see below for discussion of Kir7.1). Thus, any effects of VU590 on renal tubule function could be attributed to inhibition of ROMK, Kir7.1, or both Kir channels.

The pharmacological activity of a small-molecule can be refined by making changes to its chemical structure and hence ability to engage its primary target potently and specifically. This is often done by synthesizing a library of structurally distinct analogs and testing them for activity against the primary target (e.g., ROMK), as well as off-targets that are to be avoided (e.g., Kir7.1). Establishing so-called *S*tructure-*A*ctivity *R*elationships (SAR) can reveal which fragments of a compound form are important for ligand-receptor interactions and help guide subsequent steps in an iterative cycle of lead optimization. In the case of VU590, however, we found that virtually any change to its chemical structure led to a dramatic loss of potency toward ROMK. For example, migrating both nitro groups from the 4-position to the 3- or 2-position led to a complete loss of ROMK activity. Replacing the nitro groups with 4-chlorine, 4-thiomethyl, or 4-methyl afforded only weak ROMK inhibitors with IC_{50} values ranging between 10 and 30 μM. Despite our ability to improve the potency of VU590, this limited round of chemistry revealed that the presence and position of the nitro groups are essential for the potency of VU590 toward ROMK.

Having developed an understanding of the SAR between VU590 and ROMK, we turned our attention to another bis-nitro compound identified in the primary screen: 1,3-bis(5-nitro-1*H*-benzo[*d*]imidazol-2-yl)propane (BNBI) (Fig. 21.4). Although BNBI is relatively weak inhibitor of ROMK (IC_{50} ~ 8 μM), it exhibits no activity toward Kir7.1 at concentrations up to 100 μM. In an effort to improve the potency of BNBI while maintaining selectivity over Kir7.1, we focused on the linker region, which differs from that of VU590 by the notable absence of an ether oxygen

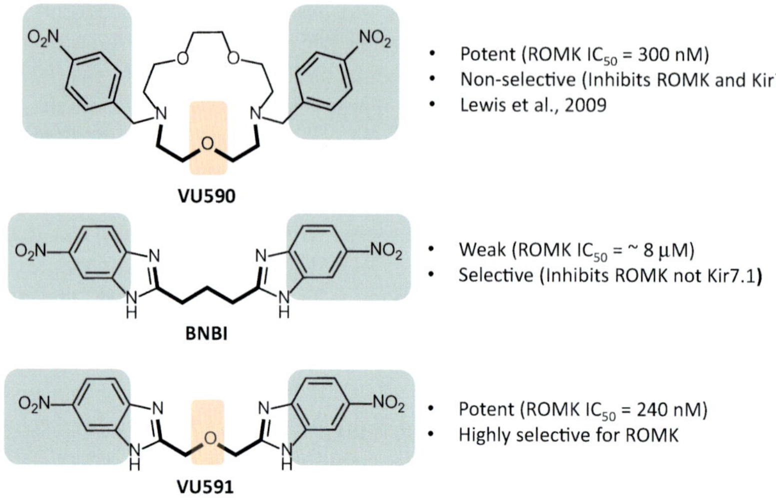

Fig. 21.4 Chemical structures and pharmacological properties of VU590, BNBI, and VU591. Nitro groups and ether oxygen, which are critical for ROMK inhibitory activity, are highlighted with gray and red boxes, respectively. See text for details

(Table 21.2). Remarkably, the addition of a single ether oxygen improved the potency of BNBI by ~30-fold (IC_{50} = 240 nM). This BNBI analog was termed VU591.

The selectivity of VU591 was characterized in a series of patch-clamp experiments on HEK-293 cells transfected with plasmids encoding mostly human inward rectifier K^+ channels, including Kir2.1, Kir2.3, Kir3.1/3.2, Kir4.1, Kir6.2/SUR1, and Kir7.1, as well as the voltage-gated K^+ channels Kv1.3 and Kv11.1, and the voltage- and calcium-activated, large-conductance "BK" K^+ channel Slo1/β1. Kir2.3, Kir4.1, Kv1.3, and Slo1/β1 were important off-targets to consider because these channels contribute to K^+ transport in the nephron. The selectivity of VU591 was also characterized in a series of radioligand-displacement assays against 68 neuronal and cardiac ion channels, transporters, and receptors. With the exception of Kir6.2/SUR1, hERG (for *h*uman *E*AG-*R*elated *G*ene), Dopamine $D_{4.2}$ receptor, $GABA_A$ receptor, dopamine transporter, and norepinephrine transporter, all of which were only weakly sensitive to VU591, the ancillary pharmacology of VU591 was clean (Bhave et al. 2011).

A number of studies have suggested that ROMK exists as part of a multi-protein complex in renal tubule epithelial cells (McNicholas et al. 1996; Ruknudin et al. 1998; Tanemoto et al. 2000; Yoo et al. 2004), raising the possibility that protein-protein interactions could obscure or alter the binding site for inhibitory small-molecules such as VU591. Because all of the discovery and characterization work on VU591 had been performed on heterologously expressed ROMK channels in HEK293 fibroblasts, we tested if VU591 inhibits native ROMK channels expressed in the collecting duct. Single rat collecting duct tubules were isolated and perfused

Table 21.2 Inhibitors of epithelial Kir channels

Compound	Structure	Kir Target	IC_{50}/EC_{50} (Max)	Selectivity	References
3	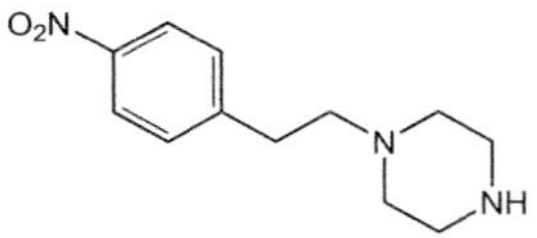	Kir1.1	NA	NA	Tang et al. (2012)
5	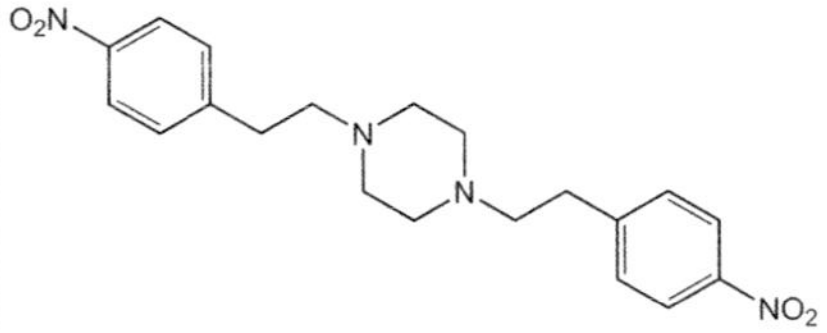	Kir1.1	52 nM	>> Kir 2.1, 2.3 Kir7.1 hERG blocker	Tang et al. (2012)
13	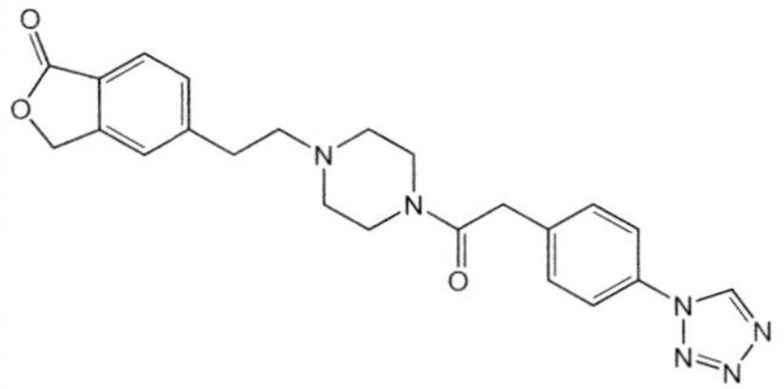	Kir1.1	60 nM	>> hERG	Tang et al. (2012)
18	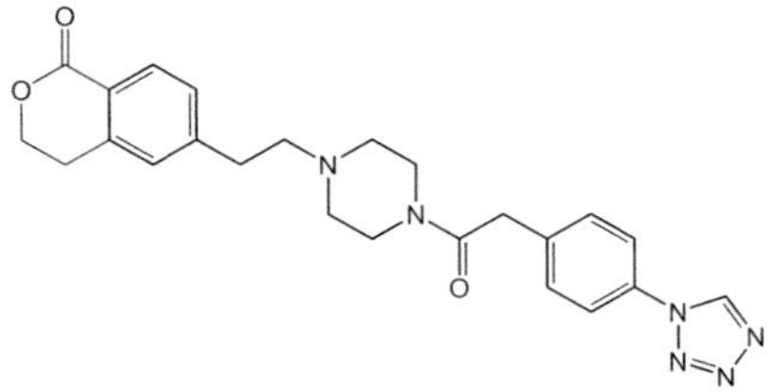	Kir1.1	60 nM	>> hERG	Tang et al. (2012)

(continued)

Table 21.2 (continued)

Compound	Structure	Kir Target	IC_{50}/EC_{50} (Max)	Selectivity	References
30	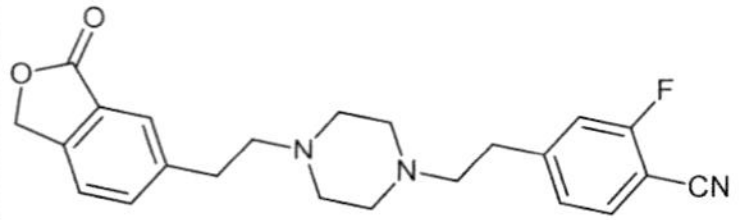	Kir1.1	30-49[a] nM	> Kir 2.1, 2.3 >> Kir4.1, Kir7.1 hERG	Tang et al. (2012)
VU590	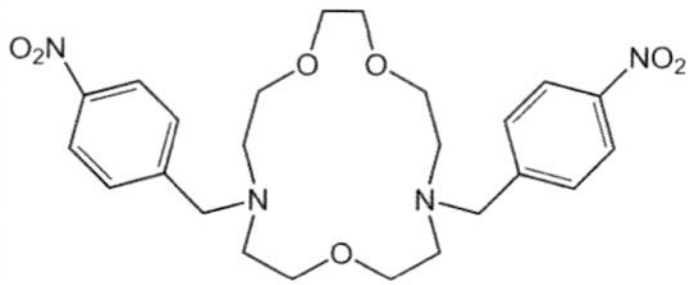	Kir1.1	0.29 μM[b] (100%)	> Kir7.1 (IC_{50} ~ 8 μM)	Lewis et al. (2009)
VU591	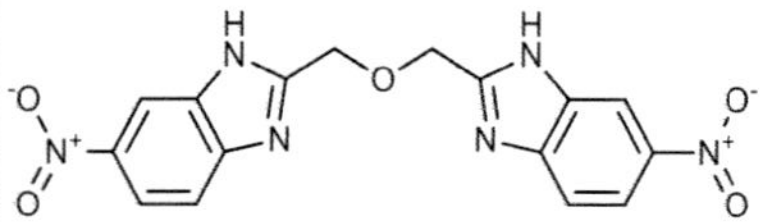	Kir1.1	0.30 μM[b] (100%)	> Kir2.1, Kir2.3, Kir4.1	Bhave et al. (2011)
Pregnenolone sulfate	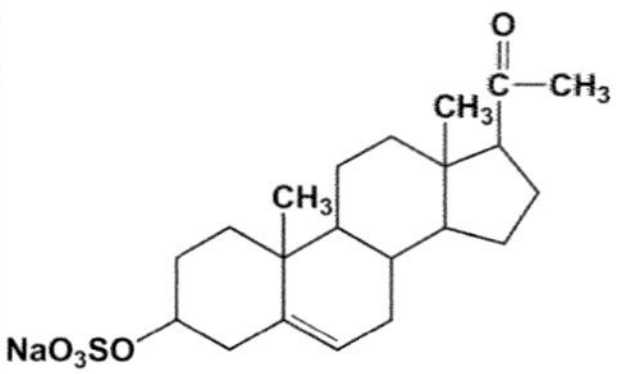	Kir2.3	~15 μM (60%)	>Kir2.1/2.3 Kir2.2/2.3	Kobayashi et al. (2009)
VU717	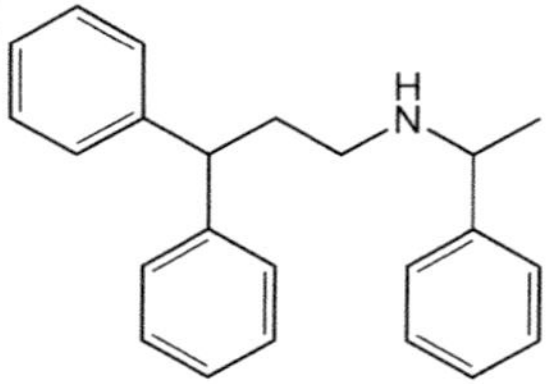	Kir4.1	~6.0 μM	NA	Raphemot et al. (2013a)

VU0134992	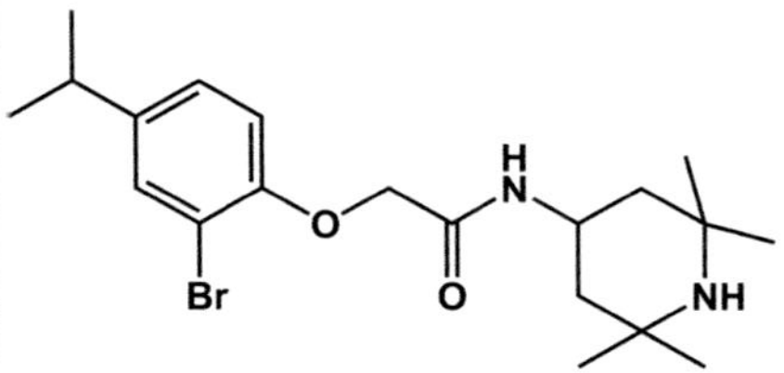	Kir4.1	1 μM	>Kir4.1/5.1, Kir1.1., Kir2.2/2.2	Kharade et al. (2018)
Nortriptyline	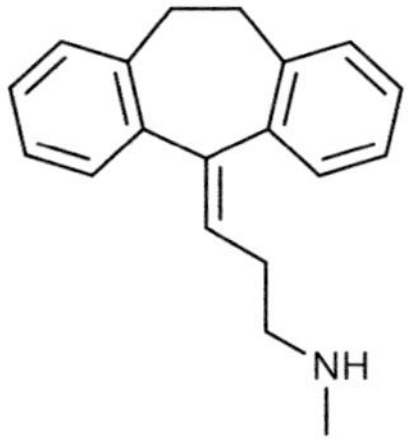	Kir4.1	38 μM	> Kir 1.1, 2.1	Su et al. (2007)
ML418	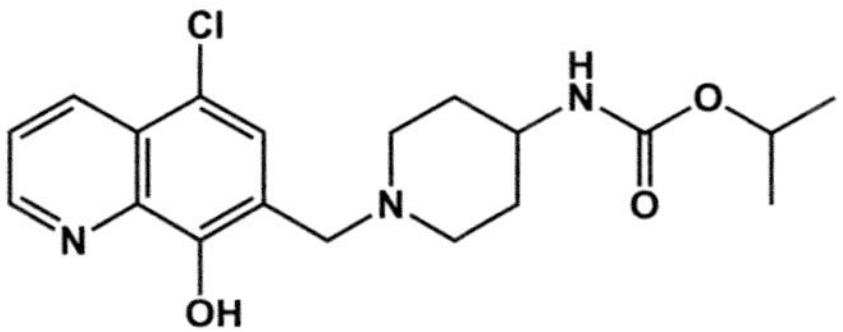	Kir57.1	310 nM	>Kir1.1,2.1,2.2,2.3,3.1/3.2,4.1	Swale et al. (2016)
Fluoxetine	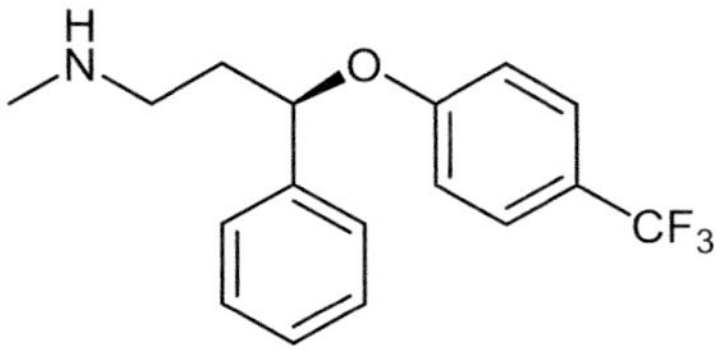	Kir3	7–13[a] μM (58–74%)	> Kir 1.1, Kir 2.1 Known SSRI	Kobayashi et al. (2004), Ohno et al. (2007)
		Kir4.1	15.2 μM (100%)		
		Kir4.1	16–38 μM; voltage dependent (100%)		

(continued)

Table 21.2 (continued)

Compound	Structure	Kir Target	IC_{50}/EC_{50} (Max)	Selectivity	References
VU573	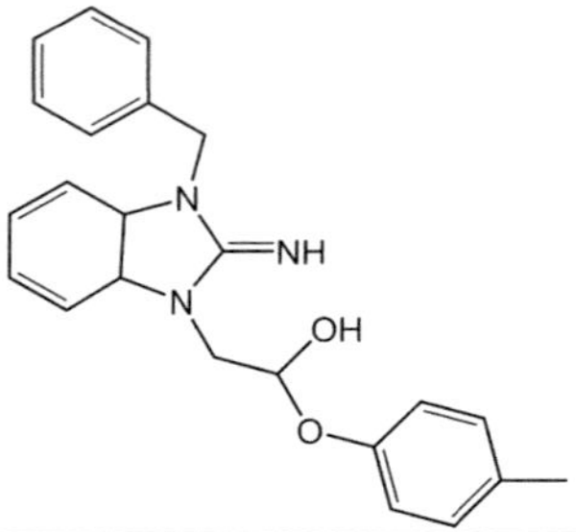	Kir2.3	0.8 μM (70%)	>Kir1.1, Kir2.1	Raphemot et al. (2011)
		Kir3.x	1.3 μM (80%)		
		Kir7.1	0.9 μM (70%)		

[a]Indicates results from experiments on Kir channels heterologously expressed in *Xenopus* oocytes, which have been reported to yield significantly higher IC_{50} values when compared to mammalian cells (Matsuda et al. 2006)
[b]Indicates results obtained from Tl^+ flux experiments

with a buffer containing DMSO (solvent control) or 10 μM VU591 for 10 min, after which the concentration of Na^+, K^+, and Cl^- in the perfusate was analyzed so that the rate of ion flux could be calculated. Over a 10-min collection period, VU591 inhibited net K^+ secretion by approximately 50%, which is consistent with inhibition of ROMK function in this nephron segment.

The metabolic stability of VU591 was assessed *in vitro* by measuring its rate of degradation in the presence of human or rat liver microsomes. These studies revealed that VU591 was more stable than the clinically used drugs propranolol and nifedipine, suggesting it may escape first-pass metabolism in the liver after administration in animal models. VU591 exhibits high serum protein binding (98% and 99% in rat and human, respectively), which could limit its ability to engage and inhibit ROMK *in vivo* (Bhave et al. 2011). Indeed, oral administration to volume-loaded rats had no effect on renal excretion (Kharade et al. 2016).

Merck Research Laboratories screened approximately 1.5 million compounds from their internal small-molecule library for inhibitors of ROMK. The group employed a Fluorescence Resonance Energy Transfer (FRET) assay that reports changes in membrane potential using the dye pair DiSBAC2(3) and CC2-DMPE, which are commercially available from Life Technologies. They selected one compound, termed compound ***3***, for development due to its moderate activity toward ROMK ($IC_{50} \sim 5$ μM) and lack of activity toward Kir2.1 and Kir2.3 at doses up to 100 μM in radiolabeled rubidium uptake or Tl^+ flux assays. Interestingly, purification of ***3*** by HPLC led to a complete loss of ROMK inhibitory activity. A detailed analysis of the sample by mass spectrometry revealed a minor sample impurity with a molecular weight that was larger than expected for the screening hit, raising the possibility that the active compound was actually a symmetrical, bis-nitro derivative of ***3***. The suspected compound was synthesized and shown to inhibit ROMK with an $IC_{50} = 52$ nM and exhibits greater than 1900-fold selectivity over Kir2.1 and Kir2.3. A liability of the new compound, termed ***5***, was that it inhibits the cardiac K^+ channel Kv11.1, more commonly known as hERG with single-nanomolar affinity ($IC_{50} = 5$ nM). Because inhibition of hERG can delay repolarization of ventricular action potentials leading to cardiac arrhythmias and sudden death, Merck set out to develop an analog of ***5*** with reduced potency toward hERG.

Similar to what was observed for VU590 (Lewis et al. 2009), the initial lead optimization campaign on ***5*** revealed that the nitro groups extending from the 4′ position of the phenyl ring are critically important for potency toward ROMK. Removing one or both nitro groups or replacing them with chloride, methoxy, trifluoromethyl, or methylsulfone led to a loss of inhibitory activity. Interestingly, shortening the distance between the nitro groups by one carbon resulted in a complete loss of activity, indicating that a precise geometrical configuration between the nitro groups and binding site in ROMK is required for high-affinity block.

After extensive medicinal chemistry efforts, the Merck team was ultimately successful in replacing the nitro groups and developing a compound, termed ***30*** (Table 21.1), that preferentially inhibits ROMK ($IC_{50} = 55$ nM) over hERG ($IC_{50} = 170$ nM) and the inward rectifiers Kir2.1, Kir2.3, Kir4.1, and Kir7.1 ($IC_{50} > 100$ μM). Compound ***30*** also exhibits pharmacokinetic properties that should make it useful for *in vivo* testing, although no such studies have been published to date. In a subsequent paper (Tang et al. 2013) the Merck group reported a series of ***5*** derivatives (termed ***13*** and ***18***; Table 21.2) exhibiting even better selectivity for ROMK (IC_{50} ~ 60 nM) over hERG (IC_{50} = 5–20 μM). Importantly, oral administration of either compound to volume-loaded Sprague-Dawley rats led to a significant natriuresis and diuresis that matched or exceeded that observed with the control diuretic hydrochlorothiazide. This was the first demonstration that acute ROMK inhibition with a pharmacological antagonist could induce urinary salt and water excretion.

A fundamental question, however, is, "will ROMK inhibition will lead to an increased K^+ excretion during diuresis?" Indeed, as noted earlier, a genetic loss of ROMK in Bartter syndrome is associated with hypokalemia due to enhanced urinary K^+ excretion, a phenotype that is also observed in ROMK knockout mice. The residual pathway for K^+ secretion in the absence of ROMK appears to be mediated by Ca^{2+}-activated BK channels, which are stimulated by increased delivery of fluid from the TAL to the distal nephron. To begin to address this issue, the Merck group treated normotensive rats and dogs with compound ***13*** and measured urinary volume and electrolyte concentrations over a 4-hour period (Garcia et al. 2014). Once again, the effects of compound ***13*** on renal function were comparable to those observed with hydrochlorothiazide. An important distinction, however, is that, unlike hydrochlorothiazide, ***13*** did not stimulate K^+ excretion over the study period. Taken together, these data strongly support the idea that ROMK represents a molecular target for a novel class of diuretic.

Because ROMK inhibitors have been predicted to behave as "dual-site" diuretics that will inhibit sodium transport in both the TAL and CCD, in contrast to "single-site" conventional diuretics (i.e., loop, thiazide, potassium-sparing), our group carried out a series of *in vivo* pharmacology experiments utilizing *13* to begin assessing where along the nephron this compounds acts. Volume-loaded rats were dosed with *13* either alone or in combination with diuretics targeting the TAL (bumetanide), DCT (hydrochlorothiazide; HCTZ), or the CCD (amiloride or benzamil), and urine volume, sodium, and potassium concentration were measured over a 6-hour period. We reasoned that if *13* inhibits sodium reabsorption in the TAL, for example, then co-administration of bumetanide will inhibit the natriuretic response to *13* when administered alone. However, if *13* also inhibits sodium reabsorption in the CCD, then bumetanide and *13* will have a synergistic effect on urine production. As predicted from the lack of ROMK expression in the DCT of rats fed a normal-potassium diet, *13* and HCTZ had a synergistic effect on urine volume, sodium, and potassium, indicating that HCTZ and *13* act on distinct nephron segments. However, co-administration of *13* and bumetanide did not increase urine production beyond that observed when the compounds were administered alone. We interpret these data to mean that both compounds inhibit sodium in the same nephron segment, and that *13* has minimal, if any, effect on sodium reabsorption in the CCD. Consistent with the latter, we found that in the presence of amiloride or benzamil, *13* induced robust diuresis and natriuresis above that observed with potassium-sparing diuretics. These results—coupled with our observations that (1) *13* effectively blocks the potassium wasting induced by bumetanide, and (2) has minimal effect on potassium-wasting induced by HCTZ (a measure of low-induced potassium secretion in the CCD)—suggests that the major site of action of *13* is the TAL.

Looking ahead, it will be important to examine the effects of chronic ROMK inhibition on fluid and electrolyte (particularly K^+) balance and interactions with commonly used diuretics. These studies will reveal how potential remodeling of renal tubule ion channels and transporters influences the long-term efficacy of ROMK inhibitors, and helps inform the best combination of diuretics to prescribe clinically to meet target blood pressure in hypertensive patients.

21.2.4 ROMK Inhibitor Molecular Mechanisms of Action

Unlike many drug targets, Kir channels are relatively small (~350 amino acids per subunit) and architecturally simple proteins that lack modulatable gating mechanisms found in many voltage-gated K^+ channels that are common targets of inhibitory toxins and small-molecules. From a screen of 225,000 compounds at Vanderbilt, VU590 was the only sub-micromolar inhibitor identified. Merck reported that only a "few" tractable hits were identified in their 1.5 million compound screen. Taken together, these low hit rates support the idea that Kir channels are difficult targets for developing potent small-molecule inhibitors. Furthermore, Kir channels share 40–60% amino acid identity with each other, raising the question of whether a selective small-molecule inhibitor of ROMK could be developed. Our development of VU591 and Merck's development of their class of inhibitor show that it is indeed possible to develop sharp pharmacological tools for Kir channels, but it requires screening a large number of compounds—and luck.

The first clues to the location of a potential binding site of VU590 arose from a series of patch-clamp experiments in which we observed the onset of block was associated with increased current rectification of ROMK (Fig. 21.5a). As noted earlier, rectification is caused by block of outward current by intracellular Mg^{2+} and polyamines. The increased rectification in the presence of VU590 suggested that VU590 was being swept into the pore at positive voltages and inducing rectification. As a more direct test of pore block, we determined if inhibition of ROMK exhibits voltage dependence. Under a physiological K^+ gradient, where E_K is approximately -80 mV, ROMK was completely inhibited by VU590 at every voltage except -180 mV, where a small, time-dependent inward current was observed (Fig. 21.5b). A much larger current could be evoked at -120 mV after increasing the extracellular K^+ concentration and hence electrochemical driving force for K^+ entry (Fig. 21.5c). These observations are consistent with a classical "knock-off" effect, whereby VU590 occupying a binding site in the ion conduction pathway can be displaced by inwardly directed K^+ ions. Similar results were observed for VU591, suggesting both Vanderbilt inhibitors are pore blockers (Bhave et al. 2011; Lewis et al. 2009).

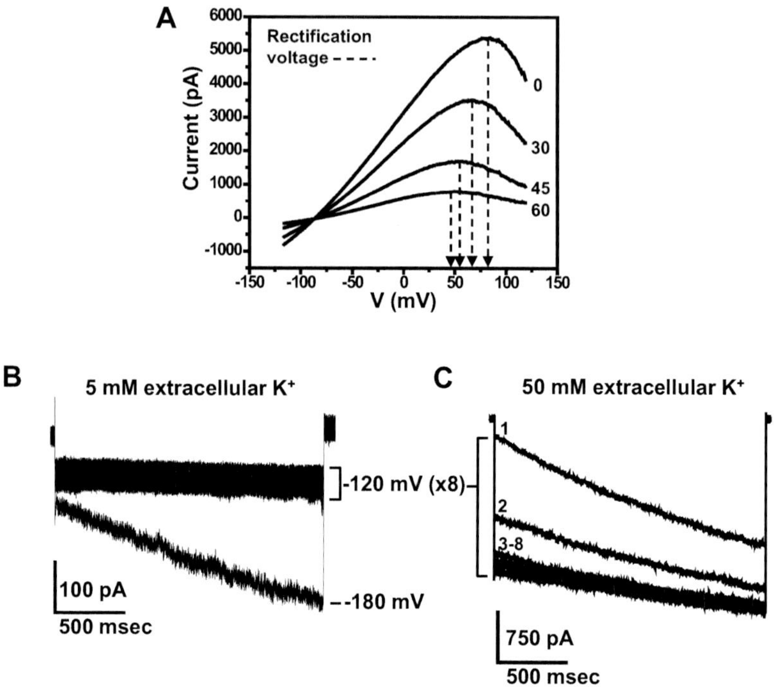

Fig. 21.5 Knockoff of VU590 from the ROMK channel pore. (**a**) Representative ROMK current traces evoked from voltage ramps before and during the onset of block by 10 μM VU590. The time (in sec) after VU590 application is shown next to each current trace. The arrows are drawn at the point of current inflection to indicate the rectification voltage. (**b**) ROMK current traces recorded from a cell incubated with 5 mM K^+ and 10 μM VU590 after achieving maximal block. Eight repetitive pulses to −120 mV had virtually no effect on current amplitude. However, a step to −180 mV led to time-dependent unblock of the channel. (**c**) Elevation of extracellular K^+ to 50 mM progressively displaces VU590 during repetitive steps (labeled 1–8) to −120 mV, indicating that the VU590 binding site is likely to be in the ion-conduction pathway. Reproduced with permission from (Lewis et al. 2009)

We recently confirmed this hypothesis and localized the binding site for both VU590 and VU591 to the pore of ROMK (Kharade et al. 2017; Swale et al. 2015). In an effort to determine where VU591 binds, we created ten different comparative homology models of ROMK based on published crystal structures of different Kir channels that were available at the time. Our goal was to simulate the conformational flexibility of ROMK channels by creating structurally distinct homology models for in silico docking computations. We docked 300 conformers of VU591 into the ten ROMK models, for a total of approximately 200,000 simulations. Cluster analysis was used to identify the lowest energy binding sites in the ROMK model. These simulations identified two likely binding sites in the membrane-spanning pore; an upper site and a lower site. Scanning mutagenesis ruled out residues in the lower site but identified N171 and V168 in the upper site as being critical for VU591-

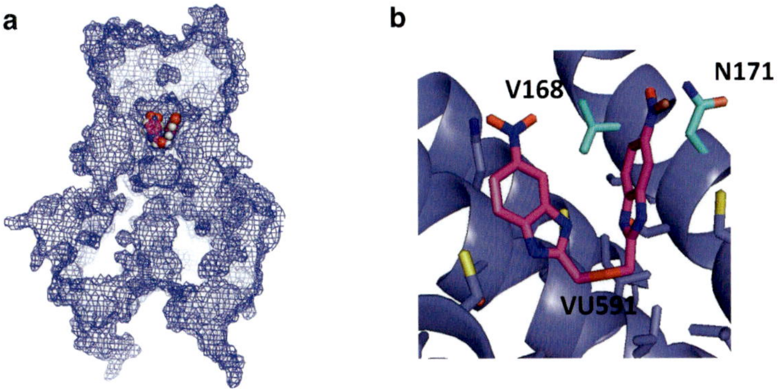

Fig. 21.6 VU591 docked into ROMK model. (**a**) VU591 docked into the "upper site" of a ROMK homology model. (**b**) Magnified view showing horseshoe configuration of VU591 possibly enabled by the linker ether oxygen and close apposition of nitro groups to N171 and V168, which are essential for VU591 potency. Reproduced with permission from (Swale et al. 2015)

dependent block. Close inspection of the docking model revealed that the nitro groups of VU591 appear to mediate interactions with N171 and V168. Furthermore, the ether oxygen in the linker of VU591 may allow the compound to assume a horseshoe shape needed for high-affinity interactions with the channel pore (Swale et al. 2015) (Fig. 21.6).

Given the strong voltage dependence of ROMK block by VU590, we employed scanning mutagenesis to evaluate pore-lining residues for interactions with VU590 (Kharade et al. 2017). As we observed for VU591, mutation of N171 to aspartate (N171D) virtually abolished block of the channel by VU590, whereas mutation of V168 had no apparent effect. An alignment of ROMK and Kir7.1 reveals that the residue in Kir7.1 that corresponds to N171 in ROMK is a glutamate residue (E149), leading us to hypothesize that mutation of E149 to an uncharged residue would improve potency of VU590 toward Kir7.1. To our surprise, however, mutation of E149 to glutamine (E158Q) led to a loss, not an increase, of VU590 potency. In addition, we found that mutation of threonine 153 to cysteine, valine, or serine that reduce constrained polarity at this site improved potency of VU590 and VU714 (discussed below) but not that of ML418—a higher-affinity analog of VU714 suggesting that the polar side chain of T153 creates a barrier to low-affinity ligands that interact with E149 and A150. Reverse mutations in ROMK suggested that this mechanism is conserved in other Kir channels. This study revealed a new role of pore polarity in Kir channel inhibitor pharmacology (Kharade et al. 2017).

21.3 Kir4.1 (*KCNJ10*) and Kir4.1/5.1 (*KCNJ16*)

21.3.1 Are Kir4.1-Containg Channels Basolateral Membrane Diuretic Targets?

Loop (i.e., furosemide, bumetanide) and thiazide (i.e., hydrochlorothiazide) diuretics work by inhibiting their respective molecular targets in the luminal membrane of the TAL and DCT, respectively. In order to reach their target sites, the diuretics must first be secreted into the tubule fluid by proximal tubule cells following their transepithelial transport via organic anion transporters and multidrug resistance proteins. Competitive interactions between loop and thiazide diuretics with certain antibiotics, non-steroidal anti-inflammatory drugs, anti-viral drugs, as well as organic acids in the setting of renal failure, can limit the secretion of these diuretics into the tubule fluid and hence their efficacy and predictability at lowering blood volume and pressure. Diuretics acting on basolateral targets that can be reached directly from the blood would presumably avoid this limitation and provide clinicians with more options for managing blood pressure in the setting of other co-morbidities. A growing body of genetic and physiological data suggests that Kir4.1 homomeric and/or Kir4.1/5.1 heteromeric channels may represent such a target.

Soon after the cDNA encoding Kir4.1 was cloned (Bond et al. 1994; Takumi et al. 1995), Ito and colleagues (Ito et al. 1996) demonstrated that the Kir4.1 protein was immunolocalized to the basolateral membrane of what was probably the DCT, connecting tubule and principal cells of the CD. This was the first time that any molecularly defined K^+ channel was shown to exhibit polarized expression in the basolateral membrane of renal tubule epithelial cells. Tucker and colleagues (Tucker et al. 2000) subsequently showed that Kir5.1, which is electrically silent when expressed alone, is also localized to the distal nephron. This raised the intriguing notion, which was subsequently borne out by a series of elegant physiological and genetic studies, that Kir4.1/Kir5.1 heteromeric channels play critical roles in the regulation of electrolyte transport in the nephron. These studies also provided genetic validation for Kir4.1/5.1 channels as putative diuretic targets.

21.3.2 Genetic Validation of Kir4.1 as a Diuretic Target

The first piece of genetic evidence that Kir4.1 is essential for urine concentration by the kidney came from the discovery by two independent groups that inactivating mutations in the Kir4.1-encoding gene, *KCNJ10*, give rise to severe salt and water wasting in patients with EAST or SeSAME syndrome (Bockenhauer et al. 2009; Scholl et al. 2009). The disease also presents with severe neurological symptoms, including seizures, ataxia, and intellectual disability, which reflects the loss of homomeric Kir4.1 expression in astrocytes and glia of the central nervous system.

The renal phenotype associated with EAST/SeSAME syndrome is thought to reflect loss of Kir4.1 activity primarily in the DCT; however, recent evidence suggests that the loss of channel activity in the CCD may also contribute to the pathophysiology of the disease. In the DCT, NaCl is reabsorbed from the tubule fluid across the luminal plasma membrane via the electroneutral NaCl co-transporter, NCC. The steep Na^+ concentration gradient across the luminal membrane is maintained by the activity of the basolateral Na^+-K^+-ATPase, which pumps 3 Na^+ ions out in exchange for 2 K^+ ions in, during each catalytic cycle. Chloride ions entering the cell across the luminal membrane move down their electrochemical gradient and exit the basolateral membrane via ClC-Kb (Fig. 21.2). Homomeric Kir4.1 and/or heteromeric Kir4.1/5.1 channels facilitate net NaCl reabsorption in the DCT in two important ways. First, they recycle K^+ ions across the basolateral membrane to help sustain robust activity of the Na^+-K^+-ATPase. Second, they hyperpolarize the basolateral membrane potential and thus increase the electrochemical driving force for Cl^- exit. Thus, a loss of Kir4.1 expression in SeSAME/EAST syndrome would reduce NaCl reabsorption in the DCT, leading to a loss of Na^+ and water in the urine. In principle, pharmacological inhibition of Kir4.1-containing channels in the DCT should recapitulate the renal consequences of thiazide diuretics, which directly inhibit NCC.

A potential caveat of extrapolating from genetic studies to justify a molecular target for drug discovery is that the phenotypes may be associated with mechanisms that are triggered to compensate for the loss of the gene product during development. It is therefore beneficial to have independent verification using, for example, acute manipulation of target function in genetically wild type animals, to further validate the target for drug discovery. Zaika et al. (2013) showed that homomeric Kir4.1 and heteromeric Kir4.1/5.1 channels are functionally coupled to a D2 dopaminergic signaling pathway in principal cells of the CCD. Cell-attached patch-clamp recordings from the basolateral membrane revealed that acute dopamine application led to a rapid inhibition of Kir4.1 homomers and Kir4.1/5.1 heteromers and depolarization of the basolateral membrane. This in turn would reduce the electrochemical driving force for Na^+ reabsorption through ENaC in the luminal membrane and potentially explain the natriuretic actions of dopamine in the CCD. Taken together, these data support the notion that Kir4.1-containing channels in the basolateral membrane of the DCT and CCD represent unexploited drug targets for a novel class of diuretic.

21.3.3 Kir4.1 Pharmacology

There are currently no nanomolar affinity Kir4.1 or Kir4.1/5.1 small-molecule inhibitors available for exploring the therapeutic potential of these channels for the treatment of hypertension. However, homomeric Kir4.1 is blocked by the selective serotonin reuptake inhibitor (SSRI) fluoxetine (IC_{50} = 15 μM; (Ohno et al. 2007) and tricyclic antidepressants (TCA) nortriptyline (IC_{50} = 28 μM; (Su et al. 2007) (Table 21.2). These compounds appear to be somewhat selective for Kir4.1 within the Kir channel family since they exhibit only weak inhibitory activity toward Kir1.1

and Kir2.1 (Ohno et al. 2007; Su et al. 2007). Similar to what was observed for VU590- and VU591-dependent block of ROMK (Bhave et al. 2011; Lewis et al. 2009), fluoxetine and nortriptyline exhibit voltage-dependent knock-off from the Kir4.1 channel pore. Mutation of T128 and E158 weakens block of Kir4.1 by fluoxetine and nortriptyline, suggesting these residues participate in the pore blocking mechanism (Furutani et al. 2009). Fluoxetine sensitivity can be conferred to ROMK by mutating N171, which is equivalent to E158 in Kir4.1, to an E or a D (N171D/E), providing further support for E158 playing a role in the fluoxetine mechanism of action. Computational modeling simulations suggest that both drugs can interact with E158 and T128 on diagonally apposed subunits of the tetramer (Furutani et al. 2009), raising the possibility that these drugs may be specific for homomeric Kir4.1 vs. heteromeric Kir4.1/5.1 channels since Kir5.1 has an asparagine residue (N161) at the equivalent position of Kir4.1-E158.

In an effort to develop more potent and selective inhibitors of Kir4.1, we initially screened a library of 3655 compounds using a Tl^+ flux assay similar to the one developed for ROMK (discussed above). Sixteen confirmed Kir4.1 inhibitors were identified in the screen, the most potent of which was 3,3-Diphenyl-*N*-(1-phenylethyl)propan-1-amine, which we termed VU717 (Raphemot et al. 2013a). An automated patch-clamp electrophysiology assay was developed around the IonFlux HT bench top platform (Molecular Devices), which enables 32 patch-clamp recordings to be made simultaneously, to improve throughput and facilitate compound characterization. In both Tl^+ flux and electrophysiology assays, VU717 inhibited Kir4.1 with an IC_{50} of ~6 μM, which represents a modest improvement in potency over fluoxetine and nortriptyline. The continued development of improved inhibitors targeting Kir4.1 and Kir4.1/5.1 channels will provide critically needed tools for exploring the therapeutic potential of these channels in hypertension and other disorders.

We subsequently carried out a larger-scale HTS campaign and reported discovery and characterization of VU0134992, the first *in vivo*-active inhibitor of renal Kir4.1 channels (Kharade et al. 2018). The HTS campaign for Kir4.1 involved screening of 76,575 structurally diverse compounds using a Tl^+ flux assay similar to the one developed for ROMK (discussed above). The primary screen yielded 640 hit compounds which were retested for confirmation and counter screening against uninduced and tetracycline-induced T-Rex-HEK-293-Kir4.1 cells. Subsequently, 16 authentic inhibitors of Kir4.1 dependent Tl^+ flux were identified (0.02% hit rate). VU0134992 was selected for further testing due to its superior potency, selectivity, and chemical tractability. In gold-standard patch-clamp electrophysiology experiments, VU0134992 markedly inhibited Kir4.1 with an IC_{50} of 0.97 μM. Because Kir4.1 forms heteromeric channels with Kir5.1 in the kidney (Lachheb et al. 2008; Tucker et al. 2000; Zhang et al. 2014, 2015), we evaluated the activity of VU0134992 toward Kir4.1-5.1 concatemeric channels. VU0134992 inhibits Kir4.1-5.1 concatemeric channels with an IC_{50} of 9.05 μM at −120 mV, providing approximately nine-fold selectivity toward Kir4.1 over Kir4.1-5.1 channels (Kharade et al. 2018).

Similar to VU590- and VU591-dependent block of ROMK (Bhave et al. 2011; Lewis et al. 2009), VU0134992 exhibits voltage-dependent knock-off from the Kir4.1 channel pore. Molecular modeling, site-directed mutagenesis paired with patch-clamp electrophysiology identified pore-lining E158 and I159 as critical residues for block of the channel (Kharade et al. 2018). Interestingly, E158 is also involved in the Kir4.1 block by fluoxetine and nortriptyline and is equivalent to N171 in ROMK providing further support for E158 playing a role in the VU0134992 mechanism of action. On the other hand, presence of asparagine residue (N161) in Kir5.1 at the equivalent position of Kir4.1-E158 could also explain the preference of VU0134992 toward Kir4.1 over Kir4.1-5.1 channels.

As noted above, an emerging body of physiological and genetic data from rodents and humans has implicated heteromeric Kir4.1-5.1 channels as molecular targets for a novel class of diuretics that could potentially be used to circumvent loop diuretic resistance. The discovery of VU0134992 allowed for the first time to pharmacologically test whether renal Kir4.1-containing channels represent "druggable" targets. As postulated, orally administered VU0134992 induced diuresis, natriuresis, and kaliuresis in volume-loaded rats (Fig. 21.7) (Kharade et al. 2018). Thus, VU0134992 represents the first subtype-preferring Kir4.1 inhibitor that is active *in vivo*. Efforts to understand the molecular basis of this selectivity and to identify VU0134992 analogs with improved potency and selectivity toward Kir4.1 and Kir4.1/5.1 channels are underway.

21.4 Kir7.1 (*KCNJ13*)

21.4.1 Overview of Kir7.1 Expression and Function

The gene encoding Kir7.1 (i.e., *KCNJ13*) was originally cloned from human fetal brain by the laboratory of Dr. David Clapham in 1998 (Krapivinsky et al. 1998) and subsequently identified in epithelial cells of the eye, intestine, stomach, thyroid, and kidney (Table 21.1), where it is thought to play important roles in regulation of epithelial salt and water homeostasis (Kusaka et al. 2001; Nakamura et al. 1999).

The most comprehensive understanding of Kir7.1 physiology has come from studies of retinal pigment epithelial (RPE) cells of the eye, which are polarized epithelial cells that carry out several important supportive functions required for vision (Strauss 2005). Kir7.1 activity is critical for electrolyte and water transport across the RPE. Water produced by metabolically active neurons and photoreceptors must be continuously transported from the subretinal space, across RPE cells, and into the blood. Water transport is driven primarily by a transepithelial Cl^- gradient created by apical membrane Na^+-K^+-$2Cl^-$ co-transporter, which is energized by an inwardly directed Na^+ gradient established by the Na^+-K^+-ATPase. Unlike most polarized epithelial cells, the Na^+-K^+-ATPase of RPE cells is localized to the apical membrane and helps maintain robust transport activity of the Na^+-K^+-$2Cl^-$ co-transporter by pumping Na^+ ions into the subretinal space. Kir7.1 is also

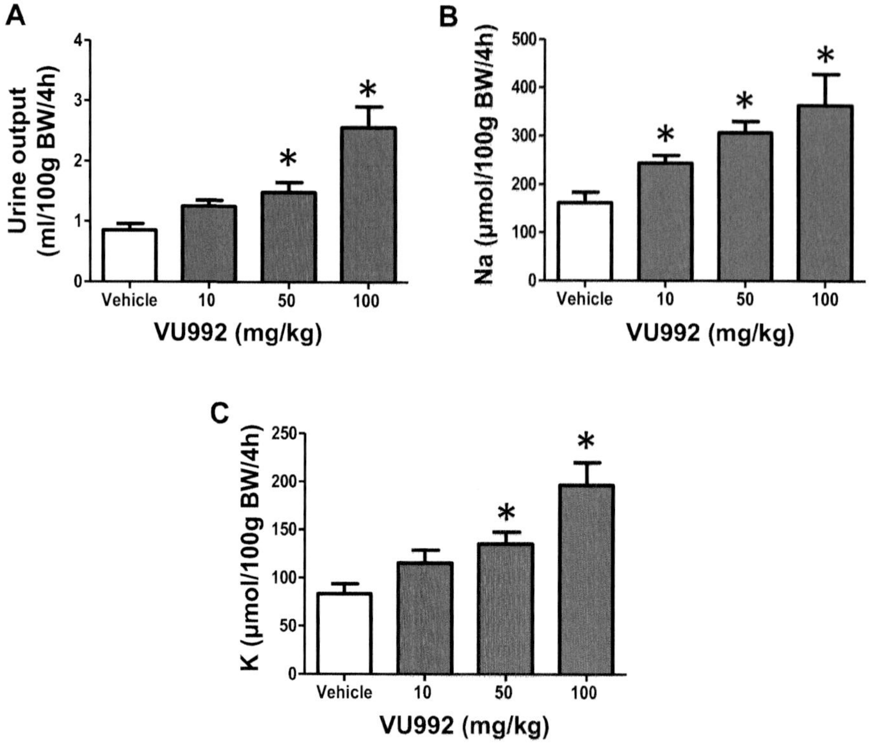

Fig. 21.7 Kir4.1 inhibitor VU992 induces diuresis in rats. Oral administration of VU992 to volume-loaded rats induces the dose-dependent increase in **a**. urine output, **b**. urinary sodium, and **c**. urinary potassium. These are the first pharmacological data to show proof-of-concept that renal Kir4.1-containing channels are viable diuretic targets. Reproduced with permission from (Kharade et al. 2018)

expressed in the apical membrane and recycles substrate K^+ across the apical membrane to help sustain the activity of the Na^+-K^+-$2Cl^-$ co-transporter. Consistent with playing an important role in RPE transport functions, a LOF mutation (R162W) in *KCNJ13* has been associated with the rare disorder Snowflake Vitreoretinal Degeneration (Hejtmancik et al. 2008; Pattnaik et al. 2013).

In the kidney, Kir7.1 exhibits an interesting expression pattern and regulated expression, suggesting the channel could play an important role in renal function. Kir7.1 is expressed on the basolateral membrane of multiple nephron segments, including DCT, CCD, and to a lesser extent, TAL (Ookata et al. 2000). The functions of Kir7.1 in the nephron are purely speculative at this point, but may be analogous to and perhaps overlapping with that of Kir4.1/5.1. Thus, in the DCT, Kir7.1 may help energize the Na^+-K^+-ATPase and in turn maintain a favorable driving force for Na^+ reabsorption by the NaCl co-transporter. In the CCD, Kir7.1 may contribute to a favorable electrochemical gradient for K^+ excretion and Na^+ reabsorption (Fig. 21.2). Indeed the observation that renal expression of Kir7.1 is increased with dietary K^+ is consistent with a role of this channel in regulating K^+ excretion (Ookata et al. 2000). Due to its expression in the DCT and CCD, it is intriguing to speculate that pharmacological inhibitors of the channel may recapitulate some of clinical symptoms of thiazide and K^+-sparing diuretics by inhibiting Kir7.1 in these segments, respectively. However, progress toward developing the pharmacology of Kir7.1 thus far has been limited.

21.4.2 Kir7.1 Pharmacology

At present, the small-molecule pharmacology of Kir7.1 is rudimentary and limited to three inhibitors developed by our group (Table 21.2). As noted earlier, the ROMK antagonist VU590 exhibits moderately potent activity toward Kir7.1 (IC_{50} ~ 8 μM; (Lewis et al. 2009). VU590 was recently used to help uncover a role of Kir7.1 in regulation of uterine muscle tone during pregnancy and implicates Kir7.1 as a novel target for dystocia and postpartum hemorrhage (McCloskey et al. 2014). Another inhibitor, termed VU573, was discovered as a weak inhibitor of ROMK in the 225,000 compound HTS and shown in secondary assays to be a preferential inhibitor of Kir2.3 (IC_{50} ~ 1 μM), Kir3.1/3.2 (IC_{50} ~ 1.3 μM), and Kir7.1 (IC_{50} ~ 1.7 μM; (Raphemot et al. 2011). A lead optimization campaign was initiated to improve the potency and selectivity of VU573 for any of the three channels. A library of 26 VU573 analogs was synthesized and tested in the hope of identifying compounds with improved pharmacology. To improve the throughput of compound testing, we attempted to develop a Tl^+ flux assay using standard methods. However, despite several attempts using several different polyclonal and monoclonal cell lines and experimental conditions, we were unable to convincingly demonstrate tetracycline-inducible (i.e., Kir7.1 channel-dependent) Tl^+ flux. Unlike other members of the Kir channel family, Kir7.1 contains a methionine (M) residue at position 125 (M125) near the extracellular pore mouth (Fig. 21.8a1–a2). This amino acid confers unique functional properties to the channel, including reduced sensitivity to extracellular K^+ and Ba^{2+} and lower unitary conductance (Doring et al. 1998; Krapivinsky et al. 1998). We considered the possibility that mutation of M125 to an arginine (M125R), which is found at the equivalent position in every other family member (Fig. 21.8b), would increase Tl^+ flux through the channel. Indeed, as shown in Fig. 21.6c1–c2, Kir7.1 (M125R) exhibited robust tetracycline-inducible Tl^+ flux that could be readily measured in 384-well format (Raphemot et al. 2011). The assay reports dose-dependent inhibition of the channel (Fig. 21.9c), enabling HTS and rapid assessment of compound potency to support lead optimization. Furthermore, the M125R mutation has no effect on the pharmacological sensitivity of Kir7.1 compared to the wild type channel (Fig. 21.9d).

The third Kir7.1 inhibitor developed by our group, which currently represents the state-of-the-art for this channel, is termed ML418 (Swale et al. 2016). The parent compound (VU714) that led to the subsequent development with medicinal chemistry of ML418 was discovered in a small-scale screen of 5230 compounds from the

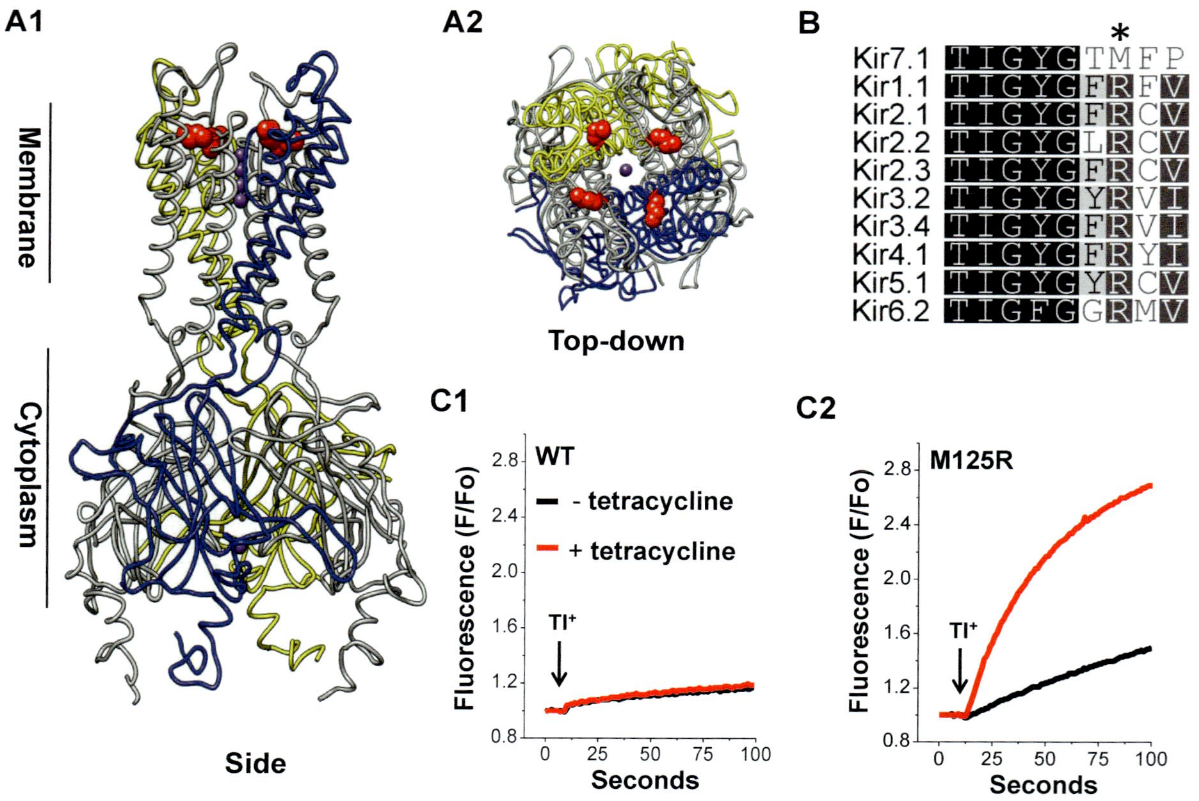

Fig. 21.8 Development of Kir7.1 (M125R) Tl^+ flux assay. (**a**1) Side and (**a**2) top-down views of a Kir7.1 model with M125 highlighted in red. (**b**) Alignment showing M125 is unique to Kir7.1. (**c**1–**c**2) M125R but not wild-type (WT) Kir7.1 mediates robust Tl^+ flux in tetracycline-induced cells. See text for details

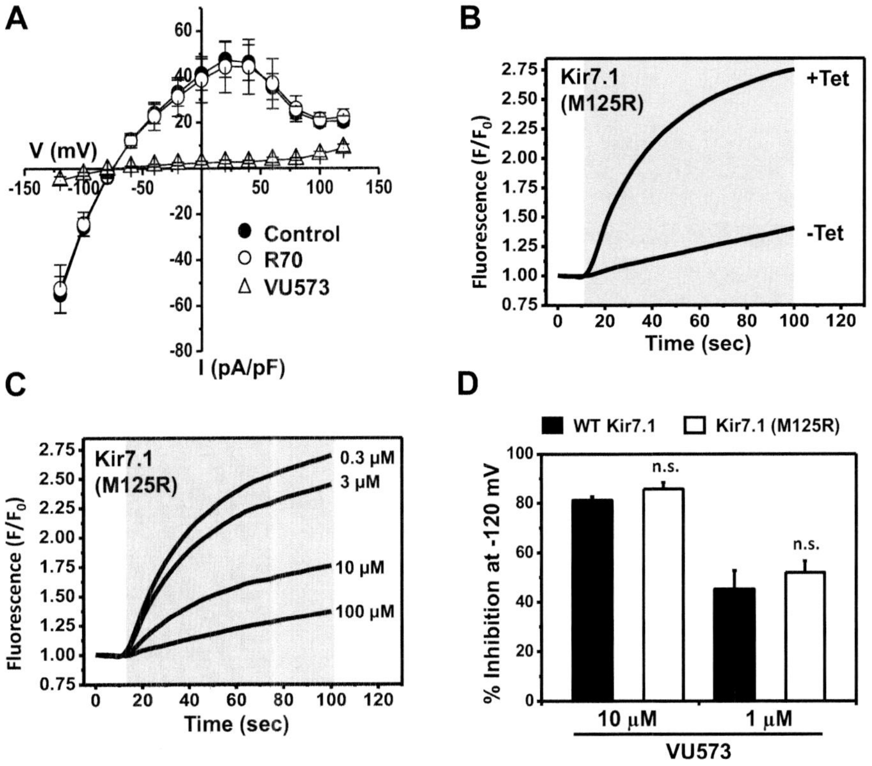

Fig. 21.9 Development of the VU573 series of Kir channel antagonists. (**a**) Kir7.1 current traces recorded in the absence or presence of 10 μM R70 or VU573 suggests that R70 is an inactive analog. (**b**) Tl^+ flux-dependent FluoZin-2 fluorescence recorded from polyclonal T-REx-HEK-293-Kir7.1 (M125R) cells cultured overnight in absence (−Tet) or presence (+Tet) of tetracycline. The fluorescence emission was recorded before and after the addition of extracellular Tl^+ (shaded box). (**c**) Pre-treatment of induced cells expressing Kir7.1 with VU573 show dose dependent reduction in the FluoZin-2 fluorescence. (**d**) VU573 causes equal pharmacological inhibition of the Kir7.1-wild type and the M125R mutant. Modified from (Raphemot et al. 2011) with permission

VICB library. The Tl^+-permeable Kir7.1-M125R mutant was used as a surrogate for the wild type channel (discussed above). VU714 inhibits Kir7.1 with an IC_{50} of 1.5 μM and is only moderately selective for Kir7.1 over Kir4.1 (IC_{50} = 13 μM), Kir1.1 (IC_{50} = 16 μM) and Kir6.1/SUR1 (IC_{50} = 30 μM). Site-directed mutagenesis analysis of the membrane-spanning pore of Kir7.1 identified E149 and A150 as being critical for high-affinity block of the channel. Medicinal chemistry efforts to improve on the potency and selectivity of the VU714 scaffold were successful and led to the development of ML418, which inhibits Kir7.1 with an IC_{50} of 310 nM, is at least 23-fold selective for Kir7.1 over Kir1.1, Kir2.1, Kir2.2, Kir2.3 and Kir4.1, and exhibits clean ancillary pharmacology against 58 GPCRs, ion channels, and transporters, including the cardiac liability potassium channel, hERG. The potency and selectivity of ML418, in addition to its pharmacokinetic profile, should make ML418 useful for probing the physiology of Kir7.1 in the periphery and brain.

21.5 Kir2.3 (*KCNJ4*)

21.5.1 A Basolateral Channel with no Known Functions

Dr. Paul Welling cloned and heterologously expressed a cDNA encoding the inward rectifier Kir2.3 (IRK3) from the mouse M1 CCD cell line and showed that it forms a K^+ current with biophysical and regulatory properties similar to that of the native small-conductance K^+ channel expressed in the basolateral membrane of the CCD (Welling 1997). When expressed in polarized Madin-Darby Canine Kidney epithelial cells, Kir2.3 is directed to the basolateral membrane via a C-terminal PDZ-interacting domain (Le Maout et al. 2001), further suggesting that Kir2.3 contributes to the basolateral K^+ conductance in the CCD. Millar et al. (2006) used Ba^{2+} as a Kir channel probe to establish a pharmacological "fingerprint" of the macroscopic K^+ currents expressed in principal cells of the mouse CCD. The authors concluded that the voltage-dependent and kinetic properties of Ba^{2+} block of the current are consistent with the corresponding properties of Kir2.3 (Millar et al. 2006). However, there are at least 4 distinct Kir channels (Kir1.1, Kir2.3, Kir4.1, and Kir7.1) expressed in principal cells of the CCD (Table 21.1). Studies aimed at determining if Kir2.3 is indeed functionally expressed in the CCD and exploring its physiological functions in the nephron will benefit from the judicious use of sharper pharmacological tools capable of discriminating Kir2.3 from other Kir channels.

21.5.2 Kir2.3 Pharmacology

As is the case with most members of the Kir channel family, the small-molecule pharmacology of Kir2.3 is still rudimentary. However, one could envision experiments in which VU591 and the VU573 series described earlier in the chapter could be used to identify Kir2.3-dependent currents in patch-clamp experiments, as follows. As noted above, VU573 inhibits the known CCD Kir channels with the rank-order potency: Kir2.3 = Kir7.1 ($IC_{50} \sim 1$ μM) > ROMK ($IC_{50} \sim 15$ μM) > > Kir4.1 ($IC_{50} > 30$ μM). In the presence of the ROMK inhibitor VU591, which should preserve currents carried by Kir2.3, Kir4.1, and Kir7.1, application of VU573 will preferentially inhibit Kir2.3 and Kir7.1. The relative contributions of Kir2.3 (and Kir7.1) to the VU573-inhibitable component could be separated using R70 or R5C, the VU573 analog that inhibits Kir2.3, but has no apparent activity toward Kir7.1 (discussed above; (Raphemot et al. 2011). In principle, a response to R5C in the presence of VU591 should provide pharmacological evidence of functional Kir2.3 channels in the CCD.

Another tool compound that could be used to identify functional Kir2.3 channels in the distal nephron is pregnenolone sulfate (PREGS) (Table 21.2), a neurosteroid that also acutely potentiates currents through Kir2.3 at micromolar concentrations (Kobayashi et al. 2009). Although PREGS has no apparent effect on ROMK, its activity toward Kir4.1, Kir4.1/5.1, and Kir7.1 has not yet been reported and should be evaluated before utilizing in the nephron to study Kir2.3.

21.6 Conclusions and Future Perspectives

Since ROMK was cloned more than 20 years ago, the rate of discoveries implicating Kir channels in fundamentally important physiological processes in epithelial and other tissues has far outpaced the development of sharp pharmacological tools for exploring the biology and potential druggability of inward rectifiers. With the recent development of small-molecule modulators of Kir channels expressed in various epithelial tissues, the pharmacological toolkit for dissecting the relative contributions of Kir channels to epithelial cell physiology is expanding. However, while these success stories demonstrate proof-of-concept that potent and highly selective small-molecule modulators can be developed using conventional drug discovery approaches, there is still a lot of work to be done. Notably for studies of epithelial Kir channel physiology, the development of safe, efficacious, nanomolar-affinity inhibitors of homomeric Kir4.1, heteromeric Kir4.1/5.1, and Kir7.1 will provide critically needed tools for examining the therapeutic potential of these channels for the treatment of hypertension and edema. Furthermore, and importantly, the development of potent and selective inhibitors has created unprecedented opportunities for finally assessing whether ROMK and Kir4.1 represent viable therapeutic targets for disorders of excessive extracellular fluid balance.

References

Bailey MA, Cantone A, Yan Q, MacGregor GG, Leng Q, Amorim JB, Wang T, Hebert SC, Giebisch G, Malnic G (2006) Maxi-K channels contribute to urinary potassium excretion in the ROMK-deficient mouse model of Type II Bartter's syndrome and in adaptation to a high-K diet. Kidney Int 70:51–59

Bartter FC, Pronove P, Gill JR Jr, Maccardle RC (1962) Hyperplasia of the juxtaglomerular complex with hyperaldosteronism and hypokalemic alkalosis. A new syndrome. Am J Med 33:811–828

Bhave G, Chauder BA, Liu W, Dawson ES, Kadakia R, Nguyen TT, Lewis LM, Meiler J, Weaver CD, Satlin LM, Lindsley CW, Denton JS (2011) Development of a selective small-molecule inhibitor of Kir1.1, the renal outer medullary potassium channel. Mol Pharmacol 79:42–50

Bockenhauer D, Feather S, Stanescu HC, Bandulik S, Zdebik AA, Reichold M, Tobin J, Lieberer E, Sterner C, Landoure G, Arora R, Sirimanna T, Thompson D, Cross JH, van't Hoff W, Al

Masri O, Tullus K, Yeung S, Anikster Y, Klootwijk E, Hubank M, Dillon MJ, Heitzmann D, Arcos-Burgos M, Knepper MA, Dobbie A, Gahl WA, Warth R, Sheridan E, Kleta R (2009) Epilepsy, ataxia, sensorineural deafness, tubulopathy, and KCNJ10 mutations. N Engl J Med 360:1960–1970
Boim MA, Ho K, Shuck ME, Bienkowski MJ, Block JH, Slightom JL, Yang Y, Brenner BM, Hebert SC (1995) ROMK inwardly rectifying ATP-sensitive K^+ channel. II. Cloning and distribution of alternative forms. Am J Phys 268:F1132–F1140
Bond CT, Pessia M, Xia XM, Lagrutta A, Kavanaugh MP, Adelman JP (1994) Cloning and expression of a family of inward rectifier potassium channels. Receptors Channels 2:183–191
Clark MA, Humphrey SJ, Smith MP, Ludens JH (1993) Unique natriuretic properties of the ATP-sensitive K(+)-channel blocker glyburide in conscious rats. J Pharmacol Exp Ther 265:933–937
Derst C, Hirsch JR, Preisig-Muller R, Wischmeyer E, Karschin A, Doring F, Thomzig A, Veh RW, Schlatter E, Kummer W, Daut J (2001) Cellular localization of the potassium channel Kir7.1 in guinea pig and human kidney. Kidney Int 59:2197–2205
Doring F, Derst C, Wischmeyer E, Karschin C, Schneggenburger R, Daut J, Karschin A (1998) The epithelial inward rectifier channel Kir7.1 displays unusual K^+ permeation properties. J Neurosci 18:8625–8636
Fakler B, Brandle U, Glowatzki E, Weidemann S, Zenner HP, Ruppersberg JP (1995) Strong voltage-dependent inward rectification of inward rectifier K^+ channels is caused by intracellular spermine. Cell 80:149–154
Ficker E, Taglialatela M, Wible BA, Henley CM, Brown AM (1994) Spermine and spermidine as gating molecules for inward rectifier K+ channels. Science 266:1068–1072
Fujita A, Horio Y, Higashi K, Mouri T, Hata F, Takeguchi N, Kurachi Y (2002) Specific localization of an inwardly rectifying K(+) channel, Kir4.1, at the apical membrane of rat gastric parietal cells; its possible involvement in K(+) recycling for the H(+)-K(+)-pump. J Physiol 540:85–92
Furutani K, Ohno Y, Inanobe A, Hibino H, Kurachi Y (2009) Mutational and in silico analyses for antidepressant block of astroglial inward-rectifier Kir4.1 channel. Mol Pharmacol 75:1287–1295
Garcia ML, Priest BT, Alonso-Galicia M, Zhou X, Felix JP, Brochu RM, Bailey T, Thomas-Fowlkes B, Liu J, Swensen A, Pai LY, Xiao J, Hernandez M, Hoagland K, Owens K, Tang H, de Jesus RK, Roy S, Kaczorowski GJ, Pasternak A (2014) Pharmacologic inhibition of the renal outer medullary potassium channel causes diuresis and natriuresis in the absence of kaliuresis. J Pharmacol Exp Ther 348:153–164
Giebisch G (1995) Renal potassium channels: an overview. Kidney Int 48:1004–1009
Greger R, Schlatter E (1981) Presence of luminal K^+, a prerequisite for active NaCl transport in the cortical thick ascending limb of Henle's loop of rabbit kidney. Pflugers Arch 392:92–94
Hebert SC, Desir G, Giebisch G, Wang W (2005) Molecular diversity and regulation of renal potassium channels. Physiol Rev 85:319–371
Hejtmancik JF, Jiao X, Li A, Sergeev YV, Ding X, Sharma AK, Chan CC, Medina I, Edwards AO (2008) Mutations in KCNJ13 cause autosomal-dominant snowflake vitreoretinal degeneration. Am J Hum Genet 82:174–180
Hibino H, Horio Y, Inanobe A, Doi K, Ito M, Yamada M, Gotow T, Uchiyama Y, Kawamura M, Kubo T, Kurachi Y (1997) An ATP-dependent inwardly rectifying potassium channel, KAB-2 (Kir4. 1), in cochlear stria vascularis of inner ear: its specific subcellular localization and correlation with the formation of endocochlear potential. J Neurosci 17:4711–4721
Hibino H, Inanobe A, Furutani K, Murakami S, Findlay I, Kurachi Y (2010) Inwardly rectifying potassium channels: their structure, function, and physiological roles. Physiol Rev 90:291–366
Hille B (1992) Ionic channels of excitable membranes. Sunderland Associates, Sunderland, MA
Ho K, Nichols CG, Lederer WJ, Lytton J, Vassilev PM, Kanazirska MV, Hebert SC (1993) Cloning and expression of an inwardly rectifying ATP-regulated potassium channel. Nature 362:31–38

Ito M, Inanobe A, Horio Y, Hibino H, Isomoto S, Ito H, Mori K, Tonosaki A, Tomoike H, Kurachi Y (1996) Immunolocalization of an inwardly rectifying K^+ channel, K(AB)-2 (Kir4.1), in the basolateral membrane of renal distal tubular epithelia. FEBS Lett 388:11–15

Ji W, Foo JN, O'Roak BJ, Zhao H, Larson MG, Simon DB, Newton-Cheh C, State MW, Levy D, Lifton RP (2008) Rare independent mutations in renal salt handling genes contribute to blood pressure variation. Nat Genet 40:592–599

Kamikawa A, Ishikawa T (2014) Functional expression of a Kir2.1-like inwardly rectifying potassium channel in mouse mammary secretory cells. Am J Physiol Cell Physiol 306:C230–C240

Kharade SV, Flores D, Byers FW, Morrison RD, Daniels S, Lindsley CW, Satlin LM, Denton JS (2016) ROMK inhibitor actions in the kidney probed with diuretics. Am J Physiol – Ren Physiol 310(8):F732–F737. PMCID: PMC4835929

Kharade S, Sheehan J, Figueroa E, Meiler J, Denton J (2017) Pore polarity and charge determine differential block of Kir1.1 and Kir7.1 potassium channels by the small-molecule inhibitor VU590. Mol Pharmacol 92:338–346

Kharade SV, Kurata H, Bender AM, Blobaum AL, Figueroa EE, Duran A, Kramer M, Days E, Vinson P, Flores D, Satlin LM, Meiler J, Weaver CD, Lindsley CW, Hopkins CR, Denton JS (2018) Discovery, characterization, and effects on renal fluid and electrolyte excretion of the Kir4.1 potassium channel pore blocker, VU0134992. Mol Pharmacol 94:926–937

Kobayashi T, Washiyama K, Ikeda K (2004) Modulators of G protein-activated inwardly rectifying K^+ channels: potentially therapeutic agents for addictive drug users. Ann NY Acad Sci 1025:590–594

Kobayashi T, Washiyama K, Ikeda K (2009) Pregnenolone sulfate potentiates the inwardly rectifying K channel Kir2.3. PLoS One 4:e6311

Koeppen BM, Biagi BA, Giebisch GH (1983) Intracellular microelectrode characterization of the rabbit cortical collecting duct. Am J Phys 244:F35–F47

Krapivinsky G, Medina I, Eng L, Krapivinsky L, Yang Y, Clapham DE (1998) A novel inward rectifier K+ channel with unique pore properties. Neuron 20:995–1005

Kurata HT, Cheng WW, Arrabit C, Slesinger PA, Nichols CG (2007) The role of the cytoplasmic pore in inward rectification of Kir2.1 channels. J Gen Physiol 130:145–155

Kusaka S, Inanobe A, Fujita A, Makino Y, Tanemoto M, Matsushita K, Tano Y, Kurachi Y (2001) Functional Kir7.1 channels localized at the root of apical processes in rat retinal pigment epithelium. J Physiol 531:27–36

Lachheb S, Cluzeaud F, Bens M, Genete M, Hibino H, Lourdel S, Kurachi Y, Vandewalle A, Teulon J, Paulais M (2008) Kir4.1/Kir5.1 channel forms the major K^+ channel in the basolateral membrane of mouse renal collecting duct principal cells. Am J Physiol Renal Physiol 294: F1398–F1407

Le Maout S, Welling PA, Brejon M, Olsen O, Merot J (2001) Basolateral membrane expression of a K^+ channel, Kir 2.3, is directed by a cytoplasmic COOH-terminal domain. Proc Natl Acad Sci USA 98:10475–10480

Leroy C, Dagenais A, Berthiaume Y, Brochiero E (2004) Molecular identity and function in transepithelial transport of K(ATP) channels in alveolar epithelial cells. Am J Physiol Lung Cell Mol Physiol 286:L1027–L1037

Lewis LM, Bhave G, Chauder BA, Banerjee S, Lornsen KA, Redha R, Fallen K, Lindsley CW, Weaver CD, Denton JS (2009) High-throughput screening reveals a small-molecule inhibitor of the renal outer medullary potassium channel and Kir7.1. Mol Pharmacol 76:1094–1103

Liu W, Morimoto T, Woda C, Kleyman TR, Satlin LM (2007) Ca^{2+} dependence of flow-stimulated K secretion in the mammalian cortical collecting duct. Am J Physiol Renal Physiol 293:F227–F235

Lopatin AN, Makhina EN, Nichols CG (1995) The mechanism of inward rectification of potassium channels: "Long-pore plugging" by cytoplasmic polyamines. J Gen Physiol 106:923–955

Lourdel S, Paulais M, Cluzeaud F, Bens M, Tanemoto M, Kurachi Y, Vandewalle A, Teulon J (2002) An inward rectifier K(+) channel at the basolateral membrane of the mouse distal convoluted tubule: similarities with Kir4-Kir5.1 heteromeric channels. J Physiol 538:391–404
Lu Z, MacKinnon R (1994) Electrostatic tuning of Mg2+ affinity in an inward-rectifier K^+ channel. Nature 371:243–246
Lu M, Wang T, Yan Q, Yang X, Dong K, Knepper MA, Wang W, Giebisch G, Shull GE, Hebert SC (2002) Absence of small conductance K+ channel (SK) activity in apical membranes of thick ascending limb and cortical collecting duct in ROMK (Bartter's) knockout mice. J Biol Chem 277:37881–37887
Matsuda T, Ito M, Ishimaru S, Tsuruoka N, Saito T, Iida-Tanaka N, Hashimoto N, Yamashita T, Tsuruzoe N, Tanaka H, Shigenobu K (2006) Blockade by NIP-142, an antiarrhythmic agent, of carbachol-induced atrial action potential shortening and GIRK1/4 channel. J Pharmacol Sci 101:303–310
McCloskey C, Rada C, Bailey E, McCavera S, Van den Berg H, Atia J, Rand DA, Shmygol A, Chan Y-W, Quenby S, Brosens JJ, Vatish M, Zhang J, Denton JS, Taggart MJ, Kettleborough C, Tickle D, Jerman J, Wright P, Dale T, Kanumilli S, Trezise DJ, Thornton S, Brown P, Catalano R, Lin N, England SK, Blanks AM (2014) The inwardly rectifying K^+ channel Kir7.1 controls uterine excitability throughout pregnancy. EMBO Mol Med 6:1161–1174
McNicholas CM, Guggino WB, Schwiebert EM, Hebert SC, Giebisch G, Egan ME (1996) Sensitivity of a renal K+ channel (ROMK2) to the inhibitory sulfonylurea compound glibenclamide is enhanced by coexpression with the ATP-binding cassette transporter cystic fibrosis transmembrane regulator. Proc Natl Acad Sci USA 93:8083–8088
Millar ID, Taylor HC, Cooper GJ, Kibble JD, Robson L (2006) A Kir2.3-like K+ conductance in mouse cortical collecting duct principal cells. J Membr Biol 211:173–184
Monaghan AS, Baines DL, Kemp PJ, Olver RE (1997) Inwardly rectifying K+ currents of alveolar type II cells isolated from fetal guinea-pig lung: regulation by G protein- and Mg2+-dependent pathways. Pflugers Arch 433:294–303
Nakamura N, Suzuki Y, Sakuta H, Ookata K, Kawahara K, Hirose S (1999) Inwardly rectifying K^+ channel Kir7.1 is highly expressed in thyroid follicular cells, intestinal epithelial cells and choroid plexus epithelial cells: implication for a functional coupling with Na+,K+-ATPase. Biochem J 342(Pt 2):329–336
Nichols CG, Ho K, Hebert S (1994) Mg^{2+}-dependent inward rectification of ROMK1 potassium channels expressed in Xenopus oocytes. J Physiol 476:399–409
Nishida M, MacKinnon R (2002) Structural basis of inward rectification: cytoplasmic pore of the G protein-gated inward rectifier GIRK1 at 1.8 A resolution. Cell 111:957–965
Ohno Y, Hibino H, Lossin C, Inanobe A, Kurachi Y (2007) Inhibition of astroglial Kir4.1 channels by selective serotonin reuptake inhibitors. Brain Res 1178:44–51
Ookata K, Tojo A, Suzuki Y, Nakamura N, Kimura K, Wilcox CS, Hirose S (2000) Localization of inward rectifier potassium channel Kir7.1 in the basolateral membrane of distal nephron and collecting duct. J Am Soc Nephrol 11:1987–1994
Partiseti M, Collura V, Agnel M, Culouscou JM, Graham D (1998) Cloning and characterization of a novel human inwardly rectifying potassium channel predominantly expressed in small intestine. FEBS Lett 434:171–176
Pattnaik BR, Tokarz S, Asuma MP, Schroeder T, Sharma A, Mitchell JC, Edwards AO, Pillers DA (2013) Snowflake vitreoretinal degeneration (SVD) mutation R162W provides new insights into Kir7.1 ion channel structure and function. PLoS One 8:e71744
Paulais M, Bloch-Faure M, Picard N, Jacques T, Ramakrishnan SK, Keck M, Sohet F, Eladari D, Houillier P, Lourdel S, Teulon J, Tucker SJ (2011) Renal phenotype in mice lacking the Kir5.1 (Kcnj16) K^+ channel subunit contrasts with that observed in SeSAME/EAST syndrome. Proc Natl Acad Sci USA 108:10361–10366

Pegan S, Arrabit C, Zhou W, Kwiatkowski W, Collins A, Slesinger PA, Choe S (2005) Cytoplasmic domain structures of Kir2.1 and Kir3.1 show sites for modulating gating and rectification. Nat Neurosci 8:279–287

Pettersen EF, Goddard TD, Huang CC, Couch GS, Greenblatt DM, Meng EC, Ferrin TE (2004) UCSF Chimera – a visualization system for exploratory research and analysis. J Comput Chem 25:1605–1612

Raphemot R, Lonergan DF, Nguyen TT, Utley T, Lewis LM, Kadakia R, Weaver CD, Gogliotti R, Hopkins C, Lindsley CW, Denton JS (2011) Discovery, characterization, and structure-activity relationships of an inhibitor of inward rectifier potassium (Kir) channels with preference for kir2.3, kir3.x, and kir7.1. Front Pharmacol 2:75

Raphemot R, Kadakia RJ, Olsen ML, Banerjee S, Days E, Smith SS, Weaver CD, Denton JS (2013a) Development and validation of fluorescence-based and automated patch clamp-based functional assays for the inward rectifier potassium channel kir4.1. Assay Drug Dev Technol 11:532–543

Raphemot R, Weaver CD, Denton JS (2013b) High-throughput screening for small-molecule modulators of inward rectifier potassium channels. J Vis Exp 71:4209. https://doi.org/10.3791/4209

Rozengurt N, Lopez I, Chiu CS, Kofuji P, Lester HA, Neusch C (2003) Time course of inner ear degeneration and deafness in mice lacking the Kir4.1 potassium channel subunit. Hear Res 177:71–80

Ruknudin A, Schulze DH, Sullivan SK, Lederer WJ, Welling PA (1998) Novel subunit composition of a renal epithelial KATP channel. J Biol Chem 273:14165–14171

Sakuma T, Takahashi K, Ohya N, Nakada T, Matthay MA (1998) Effects of ATP-sensitive potassium channel opener on potassium transport and alveolar fluid clearance in the resected human lung. Pharmacol Toxicol 83:16–22

Scholl UI, Choi M, Liu T, Ramaekers VT, Hausler MG, Grimmer J, Tobe SW, Farhi A, Nelson-Williams C, Lifton RP (2009) Seizures, sensorineural deafness, ataxia, mental retardation, and electrolyte imbalance (SeSAME syndrome) caused by mutations in KCNJ10. Proc Natl Acad Sci USA 106:5842–5847

Shimura M, Yuan Y, Chang JT, Zhang S, Campochiaro PA, Zack DJ, Hughes BA (2001) Expression and permeation properties of the K(+) channel Kir7.1 in the retinal pigment epithelium. J Physiol 531:329–346

Simon DB, Karet FE, Rodriguez-Soriano J, Hamdan JH, DiPietro A, Trachtman H, Sanjad SA, Lifton RP (1996) Genetic heterogeneity of Bartter's syndrome revealed by mutations in the K+ channel, ROMK. Nat Genet 14:152–156

Strauss O (2005) The retinal pigment epithelium in visual function. Physiol Rev 85:845–881

Su S, Ohno Y, Lossin C, Hibino H, Inanobe A, Kurachi Y (2007) Inhibition of astroglial inwardly rectifying Kir4.1 channels by a tricyclic antidepressant, nortriptyline. J Pharmacol Exp Ther 320:573–580

Swale DR, Sheehan JH, Banerjee S, Husni AS, Nguyen TT, Meiler J, Denton JS (2015) Computational and functional analyses of a small-molecule binding site in ROMK. Biophys J 108:1094–1103

Swale DR, Kurata H, Kharade SV, Sheehan J, Raphemot R, Voigtritter KR, Figueroa EE, Meiler J, Blobaum AL, Lindsley CW, Hopkins CR, Denton JS (2016) ML418: The first selective, sub-micromolar pore blocker of Kir7.1 potassium channels. ACS Chem Neurosci 7:1013–1023

Takumi T, Ishii T, Horio Y, Morishige K, Takahashi N, Yamada M, Yamashita T, Kiyama H, Sohmiya K, Nakanishi S et al (1995) A novel ATP-dependent inward rectifier potassium channel expressed predominantly in glial cells. J Biol Chem 270:16339–16346

Tanemoto M, Vanoye CG, Dong K, Welch R, Abe T, Hebert SC, Xu JZ (2000) Rat homolog of sulfonylurea receptor 2B determines glibenclamide sensitivity of ROMK2 in *Xenopus laevis* oocyte. Am J Physiol Renal Physiol 278:F659–F666

Tang H, Walsh SP, Yan Y, de Jesus RK, Shahripour A, Teumelsan N, Zhu Y, Ha S, Owens KA, Thomas-Fowlkes BS, Felix JP, Liu J, Kohler M, Priest BT, Bailey T, Brochu R, Alonso-

Galicia M, Kaczorowski GJ, Roy S, Yang L, Mills SG, Garcia ML, Pasternak A (2012) Discovery of selective small molecule ROMK inhibitors as potential new mechanism diuretics. ACS Med Chem Lett 3:367–372

Tang H, de Jesus RK, Walsh SP, Zhu Y, Yan Y, Priest BT, Swensen AM, Alonso-Galicia M, Felix JP, Brochu RM, Bailey T, Thomas-Fowlkes B, Zhou X, Pai LY, Hampton C, Hernandez M, Owens K, Roy S, Kaczorowski GJ, Yang L, Garcia ML, Pasternak A (2013) Discovery of a novel sub-class of ROMK channel inhibitors typified by 5-(2-(4-(2-(4-(1H-Tetrazol-1-yl)phenyl)acetyl)piperazin-1-yl)ethyl)isobenzofuran- 1(3H)-one. Bioorg Med Chem Lett 23:5829–5832

Tao X, Avalos JL, Chen J, MacKinnon R (2009) Crystal structure of the eukaryotic strong inward-rectifier K+ channel Kir2.2 at 3.1 A resolution. Science 326:1668–1674

Tucker SJ, Imbrici P, Salvatore L, D'Adamo MC, Pessia M (2000) pH dependence of the inwardly rectifying potassium channel, Kir5.1, and localization in renal tubular epithelia. J Biol Chem 275:16404–16407

Wang WH, White S, Geibel J, Giebisch G (1990) A potassium channel in the apical membrane of rabbit thick ascending limb of Henle's loop. Am J Phys 258:F244–F253

Wang T, Wang WH, Klein-Robbenhaar G, Giebisch G (1995a) Effects of a novel KATP channel blocker on renal tubule function and K channel activity. J Pharmacol Exp Ther 273:1382–1389

Wang T, Wang WH, Klein-Robbenhaar G, Giebisch G (1995b) Effects of glyburide on renal tubule transport and potassium-channel activity. Ren Physiol Biochem 18:169–182

Wang W, Hebert SC, Giebisch G (1997) Renal K^+ channels: structure and function. Annu Rev Physiol 59:413–436

Welling PA (1997) Primary structure and functional expression of a cortical collecting duct Kir channel. Am J Phys 273:F825–F836

Welling PA, Ho K (2009) A comprehensive guide to the ROMK potassium channel: form and function in health and disease. Am J Physiol Renal Physiol 297:F849–F863

Wible BA, Taglialatela M, Ficker E, Brown AM (1994) Gating of inwardly rectifying K+ channels localized to a single negatively charged residue. Nature 371:246–249

Yang D, Zhang X, Hughes BA (2008) Expression of inwardly rectifying potassium channel subunits in native human retinal pigment epithelium. Exp Eye Res 87:176–183

Yasuda K, Shimura M, Nakazawa T, Sato H, Tomita H, Sugano E, Tamai M (2003) Expression and functional properties of unique inward rectifier K^+ channel Kir7.1 in the porcine iris and retinal pigment epithelium. Curr Eye Res 27:279–287

Yoo D, Flagg TP, Olsen O, Raghuram V, Foskett JK, Welling PA (2004) Assembly and trafficking of a multiprotein ROMK (Kir 1.1) channel complex by PDZ interactions. J Biol Chem 279:6863–6873

Zaika OL, Mamenko M, Palygin O, Boukelmoune N, Staruschenko A, Pochynyuk O (2013) Direct inhibition of basolateral Kir4.1/5.1 and Kir4.1 channels in the cortical collecting duct by dopamine. Am J Physiol Ren Physiol 305(9):F1277–F1287. PMCID: PMC3840222

Zhang C, Wang L, Zhang J, Su XT, Lin DH, Scholl UI, Giebisch G, Lifton RP, Wang WH (2014) KCNJ10 determines the expression of the apical Na-Cl cotransporter (NCC) in the early distal convoluted tubule (DCT1). Proc Natl Acad Sci USA 111:11864–11869

Zhang C, Wang L, Su XT, Lin DH, Wang WH (2015) KCNJ10 (Kir4.1) is expressed in the basolateral membrane of the cortical thick ascending limb. Am J Physiol Renal Physiol 308: F1288–F1296

Zhou H, Tate SS, Palmer LG (1994) Primary structure and functional properties of an epithelial K channel. Am J Phys 266:C809–C824

Chapter 22
KCa3.1 in Epithelia

Daniel C. Devor, Patrick H. Thibodeau, and Kirk L. Hamilton

Abstract Calcium-mediated agonists have long been known to stimulate transepithelial ion and fluid transport across a wide array of epithelial tissues. A key component of this response are the Ca^{2+}-activated K^+ channels, whose activation results in a hyperpolarization of the basolateral and apical membranes, thereby maintaining the electrochemical driving force for ion transport. In 1997, the intermediate conductance, Ca^{2+}-activated K^+ channel, KCa3.1 was cloned and subsequently shown to be localized to both the basolateral and apical membranes of secretory epithelia where it is activated by Ca^{2+}-mediated agonists. Herein, we review the data confirming the critical role that KCa3.1 plays in transepithelial ion transport as well as the regulation, gating, trafficking of this channel, and this channel's role in cell proliferation. Very recently, the cryo-EM structure of human KCa3.1 was solved in both the closed and activated states; providing novel insight into the Ca^{2+}-dependent gating of this family of channels. Finally, KCa3.1 has been recently linked to two separate diseases. That is, the rare anemia, hereditary xerocytosis, was recently shown to be caused by mutations in KCa3.1 that result in a shift in the Ca^{2+}-dependent gating of the channel. In addition, KCa3.1 was very recently shown to be a modifier gene for cystic fibrosis (CF). Thus, we summarize the evidence for the role of KCa3.1 in both hereditary xerocytosis and epithelial diseases.

D. C. Devor (✉)
Department of Cell Biology, University of Pittsburgh, School of Medicine, Pittsburgh, PA, USA
e-mail: dd2@pitt.edu

P. H. Thibodeau
Department of Microbiology and Molecular Genetics, University of Pittsburgh, School of Medicine, Pittsburgh, PA, USA
e-mail: thibodea@pitt.edu

K. L. Hamilton
Department of Physiology, School of Biomedical Sciences, University of Otago, Dunedin, New Zealand
e-mail: kirk.hamilton@otago.ac.nz

K. L. Hamilton, D. C. Devor (eds.), *Studies of Epithelial Transporters and Ion Channels*, Physiology in Health and Disease,
https://doi.org/10.1007/978-3-030-55454-5_22

Keywords KCa3.1 · IK1 · SK4 · KCNN4 · Epithelia

22.1 Introduction

22.1.1 Early Evidence for Ca^{2+}-mediated Regulation of Transepithelial Transport

Evidence that parasympathetic stimulation is important in intestinal salt and water secretion was first obtained by several investigators in the 1880s when they demonstrated injection of the muscarinic agonist pilocarpine, in large doses, could induce colonic secretion (Florey et al. 1941). While this early work suggested an important role for parasympathetic innervation of the intestine, very little additional work was done in this area until the 1940s. In a classic paper, Wright et al. (1940) demonstrated direct vagal stimulation resulted in the flow of intestinal juices that continued only as long as the stimulation was maintained. It was further shown this vagally-induced secretion was blocked by the classic muscarinic antagonist, atropine (Wright et al. 1940). This potentially fruitful field of investigation again lays dormant until the 1960s when Tidball (1961) demonstrated, in the anesthetized dog, administration of the cholinergic agonist bethanecol resulted in a change in the direction of intestinal water and Cl^- movement from absorption to secretion. It was further demonstrated Cl^- was being secreted against both electrical and chemical gradients, demonstrating secretion must be an active process (Tidball 1961). Similar increases in the transepithelial potential difference (PD_{te}) and Cl^- transport were reported for both *in vivo* (Hardcastle and Eggenton 1973; Hubel 1976, 1977) and *in vitro* (Hardcastle and Eggenton 1973) rat intestinal preparations. Isaacs et al. (1976) confirmed the above results in a stripped ileal mucosa preparation from human intestine and further demonstrated, using isotopic flux measurements, the increase in PD_{te} was due to an increase in the unidirectional flux of Cl^- from serosa to mucosa. Although the above results clearly demonstrated parasympathetic stimulation was involved in secretion, all of these results were from small intestinal preparations. It was not until 1977 that acetylcholine (Ach) was shown to cause fluid and electrolyte secretion from the colon, both *in vivo* and *in vitro* (Browning et al. 1977).

Initial evidence that sympathetic stimulation of ion and water secretion was due to direct innervation of epithelia came from staining for acetylcholinesterase, thereby localizing the nerve endings at the mucosa (Jacobowitz 1965; Browning et al. 1977; Isaacs et al. 1976). The functional release of neurotransmitter in the proximity of colonic epithelial cells was demonstrated by showing the uptake of radiolabeled choline, and the subsequent synthesis and release of labeled Ach (Wu et al. 1982). This localization of functional nerve endings at the mucosa suggested the physiological response associated with parasympathetic stimulation was due to the actions of Ach directly on the epithelial cells. This was confirmed by Rimele et al. (1981) when they demonstrated the existence of muscarinic receptors on intestinal epithelial

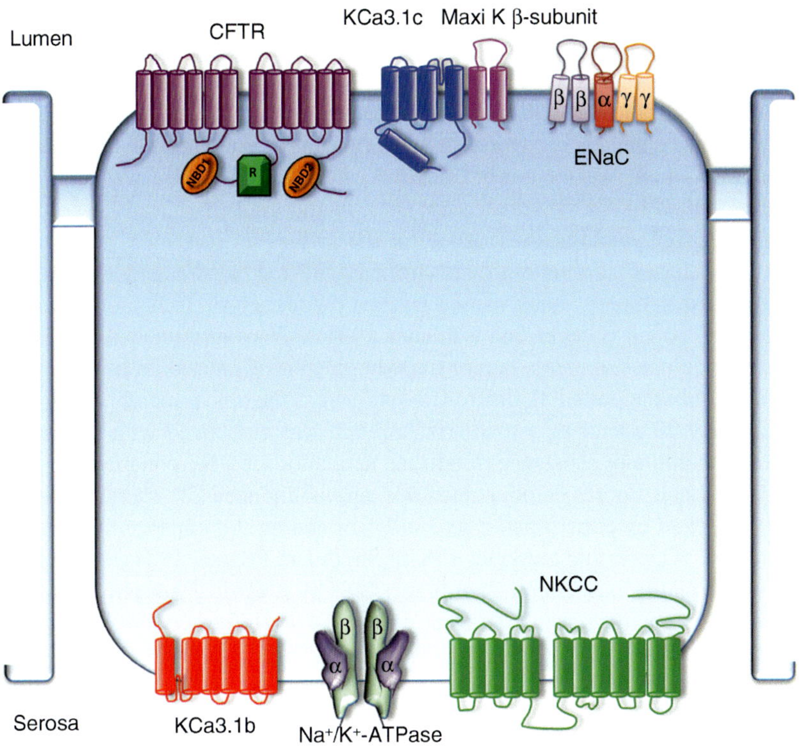

Fig. 22.1 Generalized epithelial cell model that includes the transport proteins for both Cl^- secretion and Na^+ absorption of epithelial cells. For Cl^- secretion, the uptake of Cl^- across the basolateral membrane is via the Na^+-K^+-$2Cl^-$ cotransporter (NKCC). Once stimulated, Cl^- moves passively across the apical membrane through Cl^- channels. The recycling of Na^+ and K^+ across the basolateral membrane (BM) is performed by Na^+/K^+-ATPase and K^+ channels, respectively. For Na^+ absorption, Na^+ is transported down its electrochemical gradient via ENaC and Na^+ exits the BM by the Na^+/K^+-ATPase. Activation of KCa3.1 in the apical and basolateral membranes results in hyperpolarization that maintains the electrochemical driving force for both Cl^- secretion and Na^+ absorption. CFTR = cystic fibrosis transmembrane conductance regulator; R = Regulatory Domain; NBD = Nuclear Binding Domain; ENaC = Epithelial Na^+ channel; KCa3.1c = apical KCa3.1 channel; KCa3.1b = basolateral KCa3.1; Maxi-K β-subunit. See the text for details

cells. Zimmerman and Binder (1982) further demonstrated the ability of muscarinic agonists to increase the short-circuit current (I_{sc}) was directly related to agonist binding to specific muscarinic receptors on colonic epithelial cells.

While these early studies confirmed parasympathetic stimulation of the intestine resulted in the direct stimulation of Cl^- and fluid secretion from the epithelia, a great many additional studies were required to elucidate the mechanisms by which this fluid secretion occurred, as detailed in earlier chapters in this volume. Drawing upon a wealth of electrophysiological data, a general Cl^- secretory model has been elucidated (Welsh et al. 1982; Reuss et al. 1983; Greger et al. 1984; Suzuki and Petersen 1985) as shown in Fig. 22.1 This model includes (a) the uptake of Cl^- from the interstitial fluid by an electroneutral Na^+-K^+-$2Cl^-$ cotransporter, (b) the recycling of Na^+ and K^+ across the basolateral (serosal) membrane by the Na^+/K^+-ATPase, and K^+ channel, respectively, and (c) the movement of Cl^- across the apical (mucosal) membrane through a Cl^- channel by passive diffusion. Each of these transporters and channels is covered in detail in additional chapters in this volume. In the present chapter, we focus our attention on the basolateral, Ca^{2+}-activated K^+ channel activated by Ca^{2+}-mediated agonists, including Ach.

Early electrophysiological evidence for the role of an increased basolateral K^+ conductance during stimulation came from cAMP-mediated agonists, rather than Ca^{2+}-mediated agonists, using canine trachea (Welsh et al. 1982) and later spiny dogfish rectal gland (Greger and Schlatter 1984a). In canine trachea, intracellular microelectrode measurements demonstrated epinephrine caused an initial depolarization of membrane potential, due to Cl^- exit across the apical membrane, followed approximately 20 s later by a repolarization that was associated with a decrease in basolateral membrane resistance due to the activation of a K^+ conductance (Welsh et al. 1982). Later studies demonstrated this agonist-induced Cl^- secretory response could be blocked by either raising serosal K^+ or adding the nonspecific K^+ channel blocker, Ba^{2+} (Greger and Schlatter 1984b; Smith and Frizzell 1984), both of which will depolarize the basolateral membrane and, by electrical coupling, the apical membrane, thereby decreasing the electrochemical driving force for Cl^- exit across the apical membrane.

22.1.2 *Basolateral Membrane Ca^{2+}-activated K^+ Channels*

The first direct demonstration of a basolateral Ca^{2+}-activated K^+ conductance in epithelial cells was made by Maruyama et al. (1983a, b) in salivary acinar cells. The relatively easy access of the basolateral membrane of acinar cells to patch-clamp electrodes allowed others to quickly follow up on these observations by identifying additional Ca^{2+}-activated K^+ channels. Petersen and colleagues (Maruyama et al. 1983a, b; Maruyama and Petersen 1984) identified a 200 pS K^+ channel activated by Ach and cholecystokinin in pig pancreatic acinar cells and later identified a 50 pS channel in the basolateral membrane. Early on, Ca^{2+}-activated K^+ channels were also identified in lacrimal (Trautmann and Marty 1984) and parotid (Foskett et al. 1989) acinar cells. In contrast to acinar cells, basolateral membrane Ca^{2+}-activated K^+ channels were only identified later in intestinal cells. Chang and Dawson (1988) identified a Ca^{2+}-activated K^+ conductance in permeabilized turtle colon, whereas Sepulveda and Mason (1985) characterized 36 pS and 90 pS Ca^{2+}-activated K^+ conductances in rabbit enterocytes, while Morris et al. (1986) identified a 250 pS K^+ channel in rat enterocytes. However, these initial observations were all made on absorptive cells of the intestine rather than the secretory crypts. Subsequently, Loo and Kaunitz (1989) patch-clamped the basolateral membrane of crypt cells from rabbit distal colon and identified a 130 pS K^+ channel that was activated by both Ca^{2+} and cAMP. Furthermore, Burckhardt and Gogelein (1992) identified a 12 pS K^+

channel in rat distal colonic crypts. In 1993, we identified a Ca^{2+}-activated K^+ channel in the T84 colonic cell line that was activated by Ach and taurodeoxycholate (Devor and Frizzell 1993). We had previously shown bile acids increase intracellular Ca^{2+} via an IP_3-mediated mechanism, such that this activation of KCa3.1 was identical to the Ach-mediated activation of the channel. Furthermore, this channel was inwardly rectified, with chord conductances of 10 pS at +100 mV and 35 pS at −100 mV in symmetric K^+. This rectification was independent of voltage and the channel was blocked by extracellular charybdotoxin (CTX) in a voltage-dependent manner (Devor and Frizzell 1993). This colonic Ca^{2+}-activated K^+ channel appeared to be identical to that described by Welsh and colleagues in tracheal epithelia (Welsh and McCann 1985; McCann et al. 1990) in which they characterized a channel with a slope conductance of 19 pS at 0 mV that was activated by intracellular Ca^{2+} and blocked by extracellular CTX. To the best of our knowledge, these studies represented the first identification of the Ca^{2+}-activated K^+ channel in epithelia that would eventually be realized to be the KCNN4 gene product, KCa3.1 (syn, IK1, SK4, and KCa4).

22.2 Cloning of KCa3.1

As is apparent from the above, there are multiple forms of Ca^{2+}-activated K^+ channels. These have historically been grouped into (a) small conductance, SK channels with single-channel conductances of ~5–20 pS; many of which are sensitive to block by the bee venom, apamin. (b) The intermediate conductance, IK channels having single-channel conductances of ~20–80 pS and which are blocked by clotrimazole and charybdotoxin. (c) Finally, the large, or Maxi-K channels (also called BK), having single-channel conductances of ~100–250 pS that are blocked by charybdotoxin, iberiotoxin, and paxilline. The Maxi-K channels (KCa1.1) have been discussed in Chap. 23 (this volume) and will not be discussed herein. In 1996, Kohler et al. described the cloning of the SK family of channels, including SK1, SK2, and SK3, now referred to as KCa2.1, KCa2.2, and KCa2.3, respectively. These channels were predicted to have a classic 6 transmembrane domain architecture with both the N- and C-termini being cytosolic and the pore region being between the fifth and sixth transmembrane domains (S5 and S6), similar to the voltage-gated K^+ channels (K_v), although these channels reside on distinct evolutionary branches. Kohler et al. (1996) demonstrated these SK channels were Ca^{2+}-dependent, being half-maximally activated at ~600–700 nM intracellular Ca^{2+} with a very steep dependence on Ca^{2+}, having Hill coefficients of between 4 and 5. These authors further demonstrated the SK channels were voltage-independent and that SK2 and SK3 were apamin sensitive, whereas SK1 was insensitive to apamin block. In total, these characteristics were completely consistent with these channels being responsible for the wide array of physiological events attributed to SK channels previously. As KCa2.x (SK) channels have rarely been reported to be expressed in epithelia as of the writing of this chapter these channels will not be discussed in any detail, but

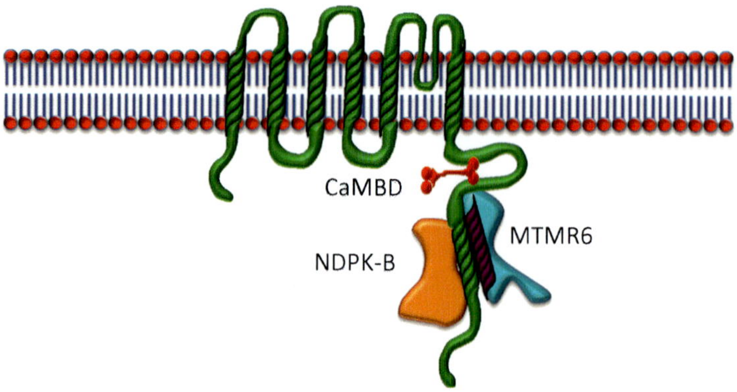

Fig. 22.2 Schematic of KCa3.1 consisting of six transmembrane domains (TMD) with both the N- and C-termini within the intracellular compartment. The pore loop region of the channel lies between TMDs S5 and S6. Additionally, regulatory subunits which have been demonstrated to have a role in the modulation of KCa3.1 include the calmodulin binding domain (CaMBD), myotubularin 6 (MTMR6, a lipid (PI(3)P) phosphatase), and nucleoside diphosphate kinase B (NDPK-B, a histidine kinase). See the text for further details

rather we will contrast some of their attributes with KCa3.1 where appropriate. Interested readers are referred to Adelman et al. (2012) for further details about KCa2.x channels.

Based on the SK channel clones, Joiner et al. (1997) and Ishii et al. (1997) screened the expressed sequence tag database and identified a fourth family member. Joiner et al. (1997) referred to this clone as SK4, based on its homology to the rest of the SK family members, whereas Ishii et al. (1997) dubbed this channel IK1 based on this clone having both conductance and blocker characteristics historically associated with IK channels. Logsdon et al. (1997) also cloned this channel from T lymphocytes and referred to it as KCa4. Based on the standard nomenclature derived from the International Union of Pharmacology (IUPHAR; Gutman et al. 2003), this channel is now referred to as KCa3.1. Full-length KCa3.1 is 427 amino acids in length with an architecture identical to that for the SK and K_v channels (Fig. 22.2). Indeed, KCa3.1 is ~40% identical to the KCa2.x channels, with the highest degree of homology being in the transmembrane domains, pore region, and the proximal C-terminus (Joiner et al. 1997; Ishii et al. 1997). In contrast, the cytosolic N-terminus and distal C-terminus exhibit virtually no homology within the gene family and numerous unique regulatory and trafficking motifs have been identified in this region of KCa3.1, as detailed below. KCa3.1 was shown to activate in response to increasing intracellular Ca^{2+}, with a half-maximal concentration reported anywhere between 100 and 600 nM and a Hill coefficient of ~2 (Joiner et al. 1997; Ishii et al. 1997; Gerlach et al. 2000; Bailey et al. 2010). Note these rather disparate $K_{0.5}$ values for Ca^{2+} are likely associated with differences in post-translational modifications (see below). At the single channel level, KCa3.1 was shown to exhibit slight inward rectification in symmetric K^+, having chord conductances of ~10 pS at +100 mV and ~35 pS at −100 mV (Fig. 22.3), similar to what

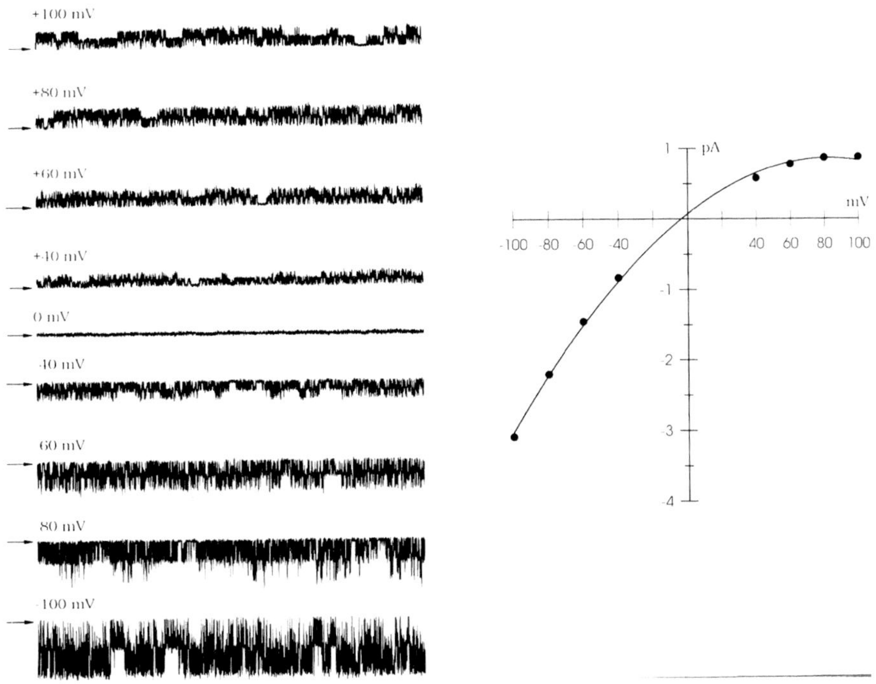

Fig. 22.3 Single-channel recording and current-voltage relationship of KCa3.1. Left panel: Single-channel recordings of KCa3.1 from an inside-out patch recordings made at the indicated voltages (referenced to the inside surface of the membrane patch) in symmetric 150 mM K^+-gluconate solutions. The bath Ca^{2+} concentration was 400 nM free Ca^{2+}. Arrows indicate the closed state of the channel. Right panel: Average current–voltage relationship of KCa3.1 recorded in symmetric 150 mM K^+-gluconate ($n = 4$). From Devor et al. (1999); used with permission from Rockefeller Press

had previously been reported for IK channels in epithelia (McCann et al. 1990; Devor and Frizzell 1993). Furthermore, in these initial reports, it was demonstrated KCa3.1 was blocked by clotrimazole and charybdotoxin (Fig. 22.4), known blockers of IK channels in epithelia (McCann et al. 1990; Devor and Frizzell 1993; Rufo et al. 1996; Devor et al. 1997). Importantly, transcripts for KCa3.1 were found in numerous epithelial containing tissues, including placenta, lung, trachea, salivary gland, kidney, pancreas, colon, bladder, stomach, and prostate (Joiner et al. 1997; Ishii et al. 1997; Logsdon et al. 1997; Jensen et al. 1998). Warth et al. (1999) and Gerlach et al. (2000) subsequently identified KCa3.1 by RT-PCR and Northern blot, respectively, in the T84 cell line, a human colonic crypt cell model. Gerlach et al. (2000) also identified KCa3.1 in Calu 3 cells, which are a model for serous cells from human airway. These results clearly implicated KCa3.1 as being the previously characterized IK1 channel in airway and colonic epithelia, although additional studies would be required to confirm this speculation.

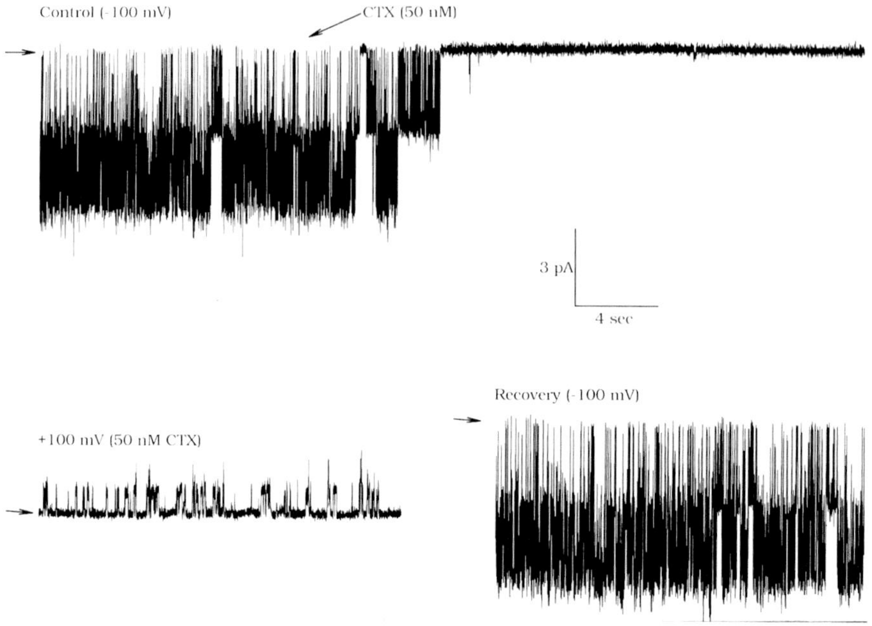

Fig. 22.4 Inhibition of KCa3.1 by charybdotoxin (CTX). Upper panel: Addition of CTX (50 nM) to the extracellular surface of KCa3.1 in an excised outside-out patch held at −100 mV resulted in complete inhibition of the channel within 5 s. Lower Left panel: The inhibitory effect of CTX was reversed at +100 mV due to knock-off of the CTX by K^+ flowing from the inside of the channel to the outside. Lower Right panel: The inhibition of KCa3.1 was totally reversed upon washout of the CTX (−100 mV). Modified from Devor et al. (1999)

22.3 Role of Basolateral KCa3.1 in Transepithelial Ion Transport

As is clear from the above, Ca^{2+}-dependent K^+ channels are required in the basolateral membrane to maintain the electrochemical driving force for transepithelial Cl^- secretion during Ca^{2+}-mediated agonist secretion. Below, we will detail the experimental results that have led to our understanding of the role of KCa3.1 in the basolateral membrane of colonic, airway, salivary acinar, and pancreatic duct epithelia.

22.3.1 KCa3.1 in the Basolateral Membrane of Intestinal Epithelium

An important breakthrough in the study of Ca^{2+}-mediated transepithelial Cl^- secretion came from the work of Dharmsathaphorn and colleagues when they characterized the T84 cell line as a colonic crypt model system, which formed high resistance monolayers capable of vectorial electrolyte transport (Dharmsathaphorn et al. 1984, 1985, 1989; Weymer et al. 1985; Dharmsathaphorn and Pandol 1986; Mandell et al. 1986; Madara et al. 1987; Wasserman et al. 1988). Using $^{86}Rb^{+}$ efflux measurements (note that Rb^{+} is often used as a surrogate for K^{+} in these studies), Dharmsathaphorn and Pandol (1986) demonstrated the cAMP-mediated agonists, vasoactive intestinal peptide (VIP), and prostaglandin E_1, as well as the Ca^{2+}-mediated agonist, carbachol increased K^{+} efflux across the basolateral membrane, while having no effect on K^{+} efflux across the apical membrane, in T84 cells. These authors further demonstrated Ba^{2+} blocked the VIP-stimulated basolateral K^{+} efflux pathway, whereas the carbachol-dependent pathway was insensitive to Ba^{2+} block. Finally, these two K^{+} efflux pathways were additive in nature, indicative of them being unique conductances. Duffey and colleagues (Devor et al. 1990, 1991; Devor and Duffey 1992) characterized this carbachol-induced K^{+} conductance using the whole-cell patch-clamp technique; demonstrating it was activated by the increase in intracellular Ca^{2+} induced by carbachol and insensitive to block by Ba^{2+} as predicted based on the $^{86}Rb^{+}$ efflux experiments of Dharmsathaphorn and Pandol (1986). Further studies by Devor and Frizzell (1993) and Devor et al. (1993), using on-cell and excised patch-clamp techniques, defined the basolateral membrane, carbachol-activated K^{+} channel in T84 cells as being Ca^{2+}-dependent, inwardly rectifying and blocked by extracellular charybdotoxin.

Further evidence for the role of this channel in transepithelial Cl^{-} secretion came when Devor et al. (1996a, b) characterized the first known pharmacological opener of this channel, 1-ethyl-2-benzimidazolinone (1-EBIO). 1-EBIO was subsequently shown to activate KCa3.1 by numerous labs (Devor et al. 1996a, b; Pedersen et al. 1999; Syme et al. 2000; Jensen et al. 2001), as detailed in Chap. 24 (this volume), thus the pharmacology of these modulators will not be discussed herein. Critically, 1-EBIO induced a sustained transepithelial Cl^{-} secretory response in T84 cells that was blocked by the basolateral addition of charybdotoxin, with an inhibitory constant of 3.6 nM, indicative of IK (KCa3.1) being the basolateral K^{+} channel activated. In contrast, the chromanol, 293B, an inhibitor of the cAMP-dependent K^{+} channel (now known to be KCNQ1) had no effect on this 1-EBIO-induced current. These results, as illustrated in Fig. 22.5, clearly demonstrated, using selective K^{+} channel blockers, that the Ca^{2+} and cAMP-mediated basolateral K^{+} conductances were distinct, as initially proposed based upon their Ba^{2+} sensitivity. Similar results were observed in rat colonic epithelium, consistent with this IK channel playing a key role in native *ex vivo* tissue (Devor et al. 1996a, b). Another critical piece of evidence for these IK channels being the basolateral K^{+} channel activated by Ca^{2+}-dependent agonists came when Rufo et al. (1996) demonstrated clotrimazole inhibited transepithelial Cl^{-} secretion across T84 cells via an inhibition of $^{86}Rb^{+}$ efflux across the basolateral membrane with no effect on other transporters critical to this Cl^{-} secretory process. These authors further demonstrated clotrimazole inhibited the basolateral K^{+} conductance in T84 cells using nystatin-permeabilized monolayers (Rufo et al. 1997). Finally, Devor et al. (1997) demonstrated a direct effect of clotrimazole on IK channels in excised patch-clamp experiments from T84

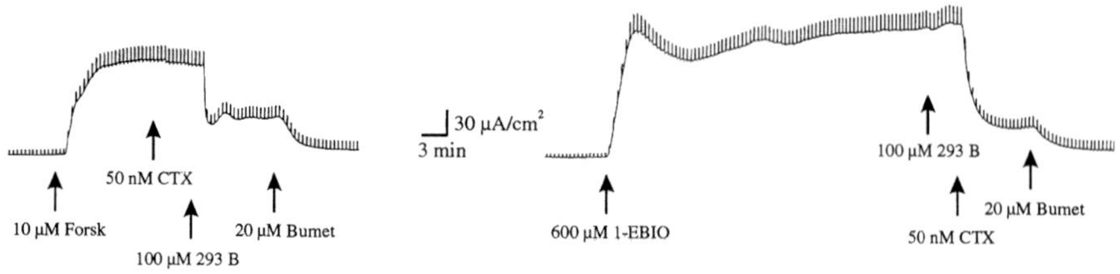

Fig. 22.5 The effects of activation and inhibition of endogenous KCa3.1 in T84 cell monolayers as measured with the short-circuit current (I_{sc}) technique using Ussing chambers. Left panel: Forskolin (10 μM, a cAMP-dependent agonist) stimulated a sustained I_{sc} that was not inhibited by CTX (50 nM), however, the stimulated I_{sc} was partially blocked by 293B (100 μM, serosal addition, an inhibitor of cAMP-dependent K^+ channels). Bumetanide (20 μM, serosal addition, inhibitor of the Na^+-K^+-$2Cl^-$ cotransporter) blocked the remainder of the stimulated I_{sc}. Right panel: 1-EBIO (600 μM, a KCa3.1 agonist) activated a prolonged I_{sc} that was not inhibited by 293B, but was partially inhibited by CTX and further blocked by bumetanide. Taken together, these results demonstrate that cAMP- and Ca^{2+}-mediated agonists activate basolateral K^+ channels with unique pharmacologies. See the text for additional details. From Devor et al. (1996a); used with permission from the American Physiological Society

cells and further showed clotrimazole inhibited the 1-EBIO-induced Cl^- secretory response. Subsequently, Hamilton et al. (1999) demonstrated 1-EBIO stimulated transepithelial Cl^- secretion across mouse jejunum and further showed 1-EBIO directly activated IK channels in the basolateral membrane of isolated jejunal crypts using the on-cell patch-clamp technique. Similarly, Warth et al. (1999) demonstrated in rat and rabbit colonic mucosa that 1-EBIO stimulated clotrimazole-sensitive Cl^- secretion as well as activated IK channels in excised patches from the basolateral membrane of rat colonic crypts. These authors also demonstrated antisense probes to KCa3.1 attenuated the carbachol-induced Cl^- secretory current across T84 cells, confirming the critical role of this channel in Ca^{2+}-mediated Cl^- secretion in these cells (Warth et al. 1999). In an additional series of studies, Cuthbert and colleagues (Cuthbert et al. 1999; Cuthbert 2001; MacVinish et al. 2001) demonstrated 1-EBIO stimulated Cl^- secretion across mouse colon and this was dependent upon a charybdotoxin-sensitive K^+ conductance in the basolateral membrane. As noted above, subsequent to the cloning of KCa3.1 and the realization, this was the IK channel previously reported, Warth et al. (1999) and Gerlach et al. (2000) confirmed by RT-PCR and Northern blot, respectively, that this was the channel expressed in T84 cells. Finally, Flores et al. (2007) carried out studies on KCa3.1 knockout mice and demonstrated Ca^{2+}-mediated Cl^- secretion was completely eliminated in both distal colon and small intestinal epithelium, which resulted in a marked reduction in the water content in the stools. In total, these studies unequivocally demonstrate that KCa3.1 is the basolateral membrane K^+ channel activated by Ca^{2+}-mediated agonists as a means of maintaining transepithelial Cl^- secretion across intestinal epithelia.

22.3.2 *KCa3.1 in the Basolateral Membrane of Airway Epithelium*

Similar to colonic epithelia, early microelectrode data demonstrated cAMP-mediated agonists activate a basolateral K^+ conductance necessary for the maintenance of transepithelial Cl^- secretion (Welsh et al. 1982) and that Ba^{2+} applied to the basolateral side blocked this conductance, resulting in a diminution of Cl^- secretion (Greger and Schlatter 1984b; Smith and Frizzell 1984). Welsh and McCann (1985) first characterized the Ca^{2+}-activated K^+ channel in primary cultures of canine tracheal epithelial cells; demonstrating it was activated by epinephrine, which increases Ca^{2+} in tracheal epithelia. These authors further showed increasing Ca^{2+} induced $^{86}Rb^+$ efflux from subconfluent canine tracheal epithelial cells that was blocked by charybdotoxin. Additional patch-clamp studies showed the K^+ channel was blocked by charybdotoxin (McCann et al. 1990). Finally, these authors (McCann and Welsh 1990) confirmed these studies on primary cultures of canine tracheal epithelial cells grown on semipermeable supports, thereby demonstrating this IK channel (called KCLIC by these authors) was the basolateral membrane Ca^{2+}-activated K^+ channel in airway epithelia.

With the identification of 1-EBIO as a direct activator of IK channels, Devor et al. (1996a, b) demonstrated this compound stimulated transepithelial Cl^- secretion across murine tracheal epithelium. Subsequent studies in both the serous cell model cell line, Calu-3 (Devor et al. 1999), as well as primary cultures of human bronchial epithelia (Devor et al. 2000), demonstrated 1-EBIO, as well as the higher affinity analogue, DCEBIO (5,6-dichloro-1-ethyl-1,3-dihydro-2H-benzimidazol-2-one) (Singh et al. 2001) stimulated transepithelial Cl^- secretion that was blocked by basolateral addition of either charybdotoxin or clotrimazole. Additional patch-clamp (Devor et al. 1999) and Northern blot analysis (Gerlach et al. 2000) confirmed expression of KCa3.1 in Calu-3 cells. Subsequent studies by Cowley and Linsdell (2002) and Mall et al. (2003) confirmed expression of KCa3.1 in Calu-3 cells and native human nasal epithelia from normal and cystic fibrosis (CF) patient samples by RT-PCR, respectively. Similar studies on the 16HBE14o- (Bernard et al. 1993) and H441 (Wilson et al. 2006) human airway cell lines confirmed expression of KCa3.1 by RT-PCR and that this channel was activated by 1-EBIO resulting in sustained transepithelial Cl^- secretion that was blocked by clotrimazole. More recently, Arthur et al. (2015) confirmed expression of KCa3.1 in primary cultures of human bronchial epithelia (HBE) and further showed expression was increased in asthmatic compared to healthy HBEs. In total, these studies clearly demonstrate expression of KCa3.1 in the basolateral membrane of airway epithelia and that activation of this channel is required for Ca^{2+}-mediated Cl^- secretion.

Interestingly, Smith and Welsh (1992) demonstrated increasing cAMP stimulated HCO_3^- secretion across normal airway epithelia but not in CF airway epithelia, while Ashton et al. (1991) suggested pancreatic duct epithelial cells could be differentially stimulated to secrete either Cl^- or HCO_3^-. In this regard, it has been shown the cystic fibrosis transmembrane conductance regulator (CFTR) conducts both Cl^- and HCO_3^- (Gray et al. 1990; Linsdell et al. 1997; Ishiguro et al. 2009). Furthermore, Devor et al. (1999) found activation of KCa3.1 acted as an "ionic switch"; dictating whether the human airway serous cell line, Calu-3 secreted HCO_3^- or Cl^-. That is, stimulation of Calu-3 cells with forskolin, to increase cAMP, resulted in a HCO_3^- secretory response as assessed by both the lack of bumetanide sensitivity and isotopic flux experiments. In contrast, stimulating Calu-3 cells with the KCa3.1 opener, 1-EBIO, resulted in Cl^- secretion. Finally, stimulation with both forskolin and 1-EBIO resulted in a primarily Cl^- secretory response with a decrease in HCO_3^- secretion relative to forskolin alone. Thus, these results suggest activation of basolateral KCa3.1 and the concomitant basolateral hyperpolarization dictates the anion transport properties of the epithelium.

Finally, Devor et al. (2000) showed activation of basolateral KCa3.1 also increased the amiloride-sensitive transepithelial Na^+ absorption across primary cultures of HBE expressing F508del CFTR. These results indicate the KCa3.1-induced hyperpolarization was sufficient to drive Na^+ entry though the epithelial Na^+ channel, ENaC in the apical membrane of HBEs. Gao et al. (2001) confirmed 1-EBIO increased transepithelial Na^+ transport in monolayers of the F508del CFTR-expressing human airway cell line, CFT1. These authors further showed 1-EBIO did not activate apical ENaC in basolateral membrane amphotericin B-permeabilized CFT1 monolayers or in *Xenopus* oocytes expressing ENaC, confirming activation of basolateral KCa3.1 results in an increased transepithelial Na^+ absorption. In contrast to these studies, Devor et al. (1996a) demonstrated 1-EBIO did not influence Na^+ transport in either rat colonic or murine tracheal epithelia. These apparently disparate results are likely the result of the unique physiologies of human airway versus mouse airway and rat colon such that the rate-limiting conductance for Na^+ absorption is different in these cell types.

22.3.3 KCa3.1 in the Basolateral Membrane of Salivary Acinar and Pancreatic Duct Epithelium

Salivary acinar cells secrete electrolytes and fluid by mechanisms similar to those described above and recently reviewed by Catalan et al. (2014). Importantly, both Maxi-K (KCa1.1) and KCa3.1 channels have been molecularly described in human and rodent salivary gland acinar cells (Nehrke et al. 2003; Begenisich et al. 2004). Surprisingly, genetic knockout of either KCa1.1 or KCa3.1 alone in mice had no effect on salivary gland fluid secretion, indicating expression of either of these channels is sufficient to maintain fluid secretion (Begenisich et al. 2004; Romanenko et al. 2006). This result was explained by the observation that knockout of either of these channels did not affect the ability of the acinar cells to hyperpolarize toward the K^+ reversal potential (E_K). In contrast, mice lacking both KCa1.1 and KCa3.1 channels exhibited a severely reduced fluid secretion rate in response to Ca^{2+}-mediated agonists and this was paralleled by an inhibition of the membrane hyperpolarization toward E_K (Romanenko et al. 2006). These results clearly demonstrate a role for KCa3.1 in salivary acinar cell physiology, but also demonstrate these cells can bypass deficits in KCa3.1 function by recruiting KCa1.1 channels.

Finally, Nguyen and Moody (1998) characterized an 1-EBIO-activated and charybdotoxin- and clotrimazole-inhibited IK channel in primary cultures of dog pancreatic duct epithelial cells localized to the basolateral membrane based on both $^{86}Rb^+$ efflux studies as well as blocker sidedness. Similar to the above, this channel has now been positively identified as KCa3.1 in rodent and human pancreatic duct cell lines (Capan-1, PANC-1, and CFPAC-1) as well as in ducts isolated from rodent pancreas (Hede et al. 2005; Hayashi et al. 2012; Wang et al. 2013). In these cells, DCEBIO stimulated, and clotrimazole inhibited, anion secretion, although it was not determined whether this was HCO_3^- or Cl^- secretion. As Ashton et al. (1991) have suggested pancreatic duct epithelial cells can be differentially stimulated to secrete either Cl^- or HCO_3^-, it will be important to determine whether activation of KCa3.1 regulates this process as was described in human airway (Devor et al. 2000). The study of the regulation and function of KCa3.1 in the pancreatic ducts is still in its early stages and the next few years are likely to shed additional light on this important subject.

22.4 Role of KCa3.1 in the Apical Membrane

In addition to K^+ channels playing a pivotal role in the basolateral membrane to modulate transepithelial Cl^- secretion, it is also well known that epithelia of the airway, pancreas, kidneys, parotid gland, and intestine secrete K^+ across the apical membrane, as detailed in Chap. 10 (Volume 1). While the K^+ channels that have been historically shown to be involved in this K^+ secretory process include the Maxi-K (KCa1.1, KCNMA1; Chap. 23, this volume) and ROMK (Kir1.1, KCNJ1; Chap. 19, this volume) channels, evidence has accumulated to implicate KCa3.1 in this process as well. Indeed, Bernard et al. (1993) demonstrated the KCa3.1 blocker, clotrimazole (Rufo et al. 1996; Devor et al. 1997; Wulff and Castle 2010), inhibited $^{86}Rb^+$ efflux across both the basolateral and apical membranes of the human bronchial cell line, 16HBE14o-. Furthermore, 1-EBIO stimulated $^{86}Rb^+$ efflux across both basolateral and apical membranes in a clotrimazole-dependent manner, implicating a role for KCa3.1 in the apical membrane of airway epithelia (Bernard et al. 1993). Similarly, Joiner et al. (2003) demonstrated the apical addition of clotrimazole completely inhibited the carbachol-induced $^{86}Rb^+$ secretion across rat proximal colon. These authors also isolated apical versus basolateral membrane and demonstrated that KCa3.1 was expressed in both membrane fractions by immunoblot. Furthermore, it was demonstrated that rats deprived of K^+ in their diet exhibited a loss of KCa3.1 mRNA that correlated with $^{86}Rb^+$ secretion being abolished (Joiner et al. 2003). At the same time, Neylon and colleagues (Furness et al. 2003) demonstrated, by immunofluorescence localization, that KCa3.1 was expressed in both apical and basolateral membranes of rat intestine. Finally,

Almassy et al. (2012) demonstrated expression of KCa3.1 in the apical membrane of parotid acinar cells. In these experiments, caged Ca^{2+} was released in close proximity to the apical membrane while simultaneously recording whole-cell K^+ currents. These authors demonstrated release of Ca^{2+} exclusively adjacent to the apical membrane resulted in the activation of K^+ currents that were partially blocked by both paxilline and TRAM-34, inhibitors of Maxi-K and KCa3.1 channels, respectively. Furthermore, in parotid acinar cells from Maxi-K null mice, apical TRAM-34-sensitive KCa3.1 currents were still observed following release of Ca^{2+} at the apical membrane, providing compelling evidence for KCa3.1 in the apical membrane of parotic acinar cells. Taken together, these results implicate KCa3.1 in K^+ secretion across the apical membrane of various epithelial tissues.

While there are clear data supporting a role for KCa3.1 in both the basolateral and apical membranes of secretory epithelia, these data beg the question of how this channel is targeted to both apical and basolateral membranes? The answer to this question began to come into focus in 2010 when Rajendran and colleagues (Barmeyer et al. 2010) identified a splice variant of KCa3.1, KCNN4c from rat colon lacking exon 2, which encodes 29 amino acids, including 5 amino acids in the S1–S2 extracellular linker as well as all of S2 and two amino acids in the proposed intracellular S2–S3 linker. As expected, deletion of S2 resulted in a channel that failed to traffic the plasma membrane when heterologously expressed. However, when KCNN4c was co-expressed with the Maxi-K β1-subunit, the channel was expressed at the membrane in *Xenopus* oocytes (see Fig. 22.2, for an example). KCNN4c was confirmed to form a functional channel by measuring $^{86}Rb^+$ efflux from occytes activated by the KCa3.1 channel opener, DCEBIO (Singh et al. 2001) and inhibited by the KCa3.1 blocker, TRAM-34 (Wulff et al. 2000), although the sensitivity of TRAM-34 block was significantly reduced in the KCNN4c channel, being 0.6 μM for full-length KCa3.1 and 7.8 μM for KCNN4c plus β1-subunit (Barmeyer et al. 2010). The recent cryo-EM structure of KCa3.1 (Lee and MacKinnon 2018) shows that calmodulin (CaM) interacts with both S1 and S2, which extend well into the cytosolic domain. As KCNN4c lacks S2, this is likely to influence this domain-domain association resulting in altered pharmacology (see below for a detailed discussion of the structure of KCa3.1). Further studies demonstrated that the KCNN4c/β1 channels had single-channel properties that are similar to full-length KCa3.1, i.e., they display inward rectification and have a chord conductance of 31 pS in symmetric K^+ (Basalingappa et al. 2011). Interestingly, these authors also demonstrated that KCNN4c did not co-assemble and, therefore, inhibit expression of the full-length KCa3.1 channel (Barmeyer et al. 2010). Importantly, immunoblots of apical versus basolateral membranes demonstrated an ~37 kDa band in the apical membrane lysates, while an ~40 kDa band was detected in the basolateral membrane lysates, suggesting this lower molecular mass product corresponds to KCNN4c being expressed in the apical membrane of rat colonic epithelia (Barmeyer et al. 2010).

Further studies by Singh et al. (2012) demonstrated a Na^+-free diet, which increases circulating aldosterone levels, increases KCNN4c mRNA and protein expression in rat distal colon. These authors also demonstrated, in the presence of aldosterone, a TRAM-34-dependent K^+ secretory pathway was revealed, consistent with an increase in apical membrane expression of KCNN4c under these conditions. More recent studies have identified mineralocorticoid response elements in KCNN4 introns that respond to aldosterone (O'Hara et al. 2014). In an exciting series of studies, Hoque and colleagues demonstrated that binding of cAMP to exchange protein directly activated by cAMP (Epac) activates a PKA-independent, Ca^{2+}-dependent Cl^- secretory process in T84 cells that involves stimulation of Rap2-phospholipase Cε (Hoque et al. 2010). More recently, these authors demonstrated Epac1 regulates the apical surface expression of KCNN4c in T84 cells (Sheikh et al. 2013). That is, knockdown of Epac1 both significantly decreased the TRAM-34-dependent apical potassium conductance (G_K) and also resulted in the redistribution of KCNN4c into subapical vesicles. Furthermore, inhibitors of both RhoA (GGT1298) and Rho-associated kinase (ROCK; H1152) reduced apical G_K and ROCK inhibition also caused KCNN4c to redistribute out of the apical membrane (Sheikh et al. 2013). To our knowledge, these studies represent the first demonstration of a dynamic redistribution of KCNN4c into and out of the apical membrane of polarized epithelia.

Very recently, Preston et al. (2018) postulated KCNN4c was expressed in the apical membrane of the porcine choroid plexus epithelial cell line, PCP-R, where it plays a role in K^+ secretion. That is, activation of the Ca^{2+} channel, TRPV4 stimulated K^+ secretion across the apical membrane and RT-PCR confirmed expression of KCa2.2 and KCa3.1. The increased K^+ conductance was not blocked by SK inhibitors, whereas large concentrations of TRAM34 (25 μM) were required to inhibit this response, consistent with KCNN4c. However, KCNN4c was not confirmed by sequencing and thus, this awaits further verification.

22.5 Gating of KCa3.1

The gating of an ion channel can be simply thought of as the process by which the pore transitions between nonconducting (closed) and conducting (open) states. As outlined above, KCa3.1 gating is dictated by changes in the intracellular Ca^{2+} concentration. The mechanism by which Ca^{2+} binding is translated to KCa3.1 gating has received considerable attention, as detailed below.

22.5.1 Structure of KCa3.1

In a landmark paper for the KCa3.1 field, Lee and MacKinnon (2018) solved the cryo-EM structure in the closed and activated states to a resolution of 3.4 Å and 3.5 Å, respectively (Fig. 22.6a, b). These structures confirmed the 4-fold symmetry of KCa3.1 in which the S1–S4 transmembrane helices interact with the pore (S5–S6) from the same subunit. While this arrangement is the same as that observed for BK

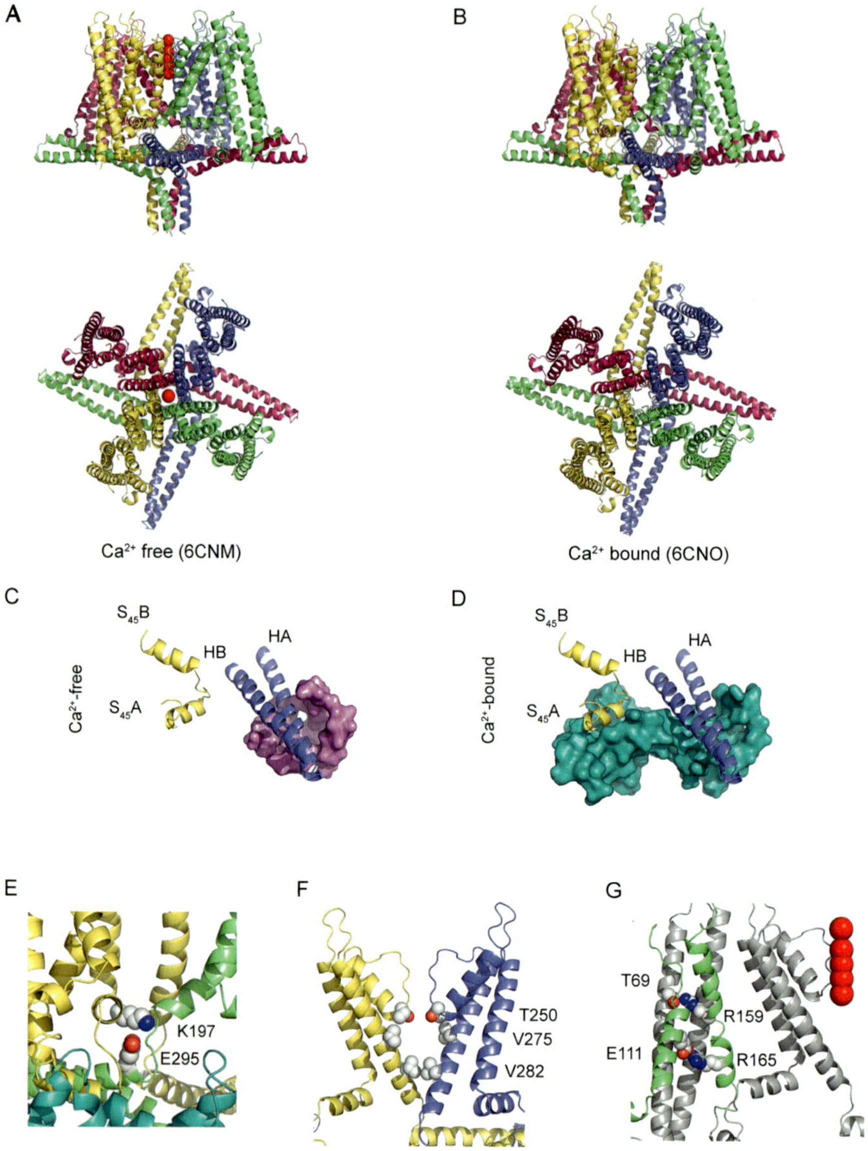

Fig. 22.6 Structures of the KCa3.1 and CaM complex. Structures of the KCa3.1-CaM complex were solved by cryo-EM in the presence and absence of Ca^{2+}. (**a**) Two views of the Ca^{2+}-free KCa3.1 channel are shown, rotated by 90°, with the individual monomers of the channel tetramer colored differently. Top, a view of the channel from within the membrane; bottom, a view of the channel from the extracellular space. The visualized K^+ ions in the pore are shown as red spheres. (**b**) The Ca^{2+}-bound channel is shown as in **a**. (**c**) The interactions of CaM with the HA and HB loops are shown in the Ca^{2+}-free state. A single lobe of CaM is visualized in the EM maps in the absence of Ca^{2+}. CaM is shown as a magenta surface representation with the KCa3.1 helices shown as cartoons and labelled. (**d**) The interactions of CaM with the KCa3.1 channel is shown in the Ca^{2+}-bound state. Two lobes of the CaM are visualized under these conditions, interacting with the HA and HB helices, as well as $S_{45}A$. CaM is shown as a teal surface representation with the KCa3.1 helices colored and labelled as in **c**. (**e**) The positions of K197 and E295 are shown, making a salt

channels (Tao et al. 2017; Hite et al. 2017), it is different from that seen for the K_v channels, in which S1–S4 from one subunit associates with the pore from a neighboring subunit (Long et al. 2005; Sun and MacKinnon 2017). Early on, in an attempt to clarify the Ca^{2+}-dependent gating mechanism of the KCa3.1/KCa2.x channels, Adelman and colleagues (Xia et al. 1998; Keen et al. 1999) demonstrated CaM bound constitutively to the proximal C-terminus of the KCa2.x channels at the CaMBD just distal to S6. These authors further demonstrated that one CaM was bound to each of the four subunits of the tetrameric channel and that it was specifically the binding of Ca^{2+} to the N-lobe E-F hands of CaM that induced channel gating via a conformational change in CaM that was transmitted to the KCa2.x channel α-subunit. In contrast, the C-lobe E-F hands of CaM did not bind Ca^{2+} when it was associated with KCa2.x channels. Fanger et al. (1999) and Li et al. (2009), similarly demonstrated that KCa3.1 channels were constitutively bound to CaM that resulted in their Ca^{2+}-dependent gating. The recent cryo-EM structure confirms these previous studies (Lee and MacKinnon 2018) (Fig. 22.6c, d). Indeed, Lee and MacKinnon (2018) further showed the C-terminus of KCa3.1 is composed of three α-helical domains, termed HA, HB, and HC (Fig. 22.6c, d). HA and HB run nearly parallel to the plane of the membrane and form the CaMBD, as described (Xia et al. 1998; Keen et al. 1999).

In contrast to HA and HB, HC forms a coiled-coil domain that runs perpendicular to the plane of the membrane at the center of the channel (Lee and MacKinnon 2018) (Fig. 22.6a, b). We previously demonstrated a leucine zipper in the C-terminus of KCa3.1 was capable of self-assembly and this is required for the correct trafficking of the channel to the plasma membrane (Syme et al. 2003), consistent with these structural results. We also demonstrated small deletions of the distal C-terminus of KCa3.1 (last 26 amino acids) result in a channel, which fails to correctly fold and traffic to the plasma membrane (Syme et al. 2003). Unfortunately, the last 41 amino acids of KCa3.1 are invisible in the cryo-EM structure and thus what they may interact with remains unknown. It is interesting to note that mutation of this conserved leucine zipper in KCa2.3 had no effect on channel expression or localization, suggesting either the structure of this region of the channel or its interacting partners may be unique amongst the gene family members (Syme et al. 2003). More recently, the structure of amino acid residues Met376-Leu415 of KCa3.1 was solved, including the leucine zipper, which formed a four-helix bundle (Ji et al. 2018).

It is also critical to point out that while K_v and BK channels have a short S4–S5 linker, KCa3.1 has an extended S4–S5 linker composed of two α-helices, termed $S_{45}A$ and $S_{45}B$ (Lee and MacKinnon 2018). $S_{45}B$ is wedged between HA and S6,

Fig. 22.6 (continued) bridge critical for channel assembly and function. The K197 and E295 residues are located on adjacent channel subunits, which are colored as in **a** and **b**. (**f**) The positions of critical residues found in the K^+-channel pore are shown with sidechains represented as spheres (see text for details). Two subunits of the channel have been removed for clarity. (**g**) Salt bridges between the basic residues in S4 are shown. Arg159 and Arg165 interact with T69 (S2) and E111 (S3), respectively. The sidechains are represented as spheres thereby providing structural contacts between the CaMBD and the pore. Upon Ca^{2+} binding to the CaM N-lobes, CaM associates with HA and HC from an adjacent subunit together with $S_{45}A$, thereby forming an extensive interaction domain (Fig. 22.6c, d). It is this association between CaM and $S_{45}A$ that forms the critical interacting domain for gating. That is, binding of the CaM N-lobe to $S_{45}A$ pulls this helix downward, which in turn causes $S_{45}B$ to move outward away from the pore, thereby expanding the S6 helical bundle resulting in pore opening (Fig. 22.6c, d). Crucially, this is significantly different from the mechanism proposed by Schumacher et al. (2001, 2004), based on the crystal structure of CaM bound to the CaMBD in both the apo and calcified states. Based on these studies, it was proposed the binding of Ca^{2+} to the N-lobes of CaM in KCa2.2 causes the formation of a dimeric complex between two adjacent CaMBD; initiating a >90° rotation of the S6 helices leading to the opening of the channel pore. Thus, the channel was proposed to become a 2-fold symmetric dimer of dimers to create this rotary force. However, the recent cryo-EM structure, which shows a more complete view of the channel, demonstrates the channel maintains 4-fold symmetry during the Ca^{2+}-dependent gating process. We previously performed a partial alanine-scan of the $S_{45}B$ region of KCa3.1 and identified Lys197 as a critical amino acid required for the proper folding and plasma membrane targeting of the channel (Jones et al. 2005). The recent structure confirms Lys197 in $S_{45}B$ forms hydrogen bonds with Glu295 in HA, thereby "gluing" these structural elements together (Fig. 22.6e); Lee and MacKinnon 2018). Additional key aspects of the recently described structure will be used to inform our discussion of gating and trafficking below.

22.5.2 A Gating Model for KCa3.1

The first attempt to define the open-closed gating kinetics of KCa3.1 was carried out by Grygorczyk and Schwarz (1985) using the single channel patch-clamp technique on human red blood cells. These authors identified a single open state, as assessed by fitting an open-time (τ_o) distribution plot to a single exponential, as well as two closed states. However, the authors speculated an additional long closed time (τ_c) that was >60 ms in duration that could not be resolved during their recordings to fully account for their entire data set. τ_o was shown to increase with increasing Ca^{2+}, whereas the two τ_c did not significantly depend on Ca^{2+}, although the mean closed time was decreased by increasing Ca^{2+}, indicative of the long, unresolved τ_c being reduced in the duration. In 1992, Leinders et al. (1992) again evaluated the gating kinetics of KCa3.1 in human erythrocytes. These authors found that both the τ_o and τ_c distributions were fit to two exponentials, indicating two open, as well as two closed states. However, these authors similarly proposed a long-lived third τ_c to account for all of their data. While Leinders et al. (1992) found no dependence of τ_o on Ca^{2+}, distinct from Grygorczyk and Schwarz (1985), they proposed that the long-lived third τ_c was dependent on Ca^{2+}, similar to Grygorczyk and Schwarz (1985). Dunn (1998) conducted a gating study on KCa3.1 in human erythrocytes and identified two τ_o and three τ_c, indicative of two open and three closed gating steps. Indeed, this third τ_c averaged 155 ms in agreement with previous reports. However, Dunn did not evaluate the effect of Ca^{2+} on these gating kinetics. Syme et al. (2000) studied KCa3.1 heterologously expressed in HEK cells and also identified two τ_o and three τ_c. However, based on an analysis of the relative amplitudes of each closed component, the data they could resolve would have accounted for a P_o of 0.3, when the single channel patches use to obtain the data had an actual P_o of 0.18. Thus, similar to the authors above, Syme et al. (2000) concluded that a longer, fourth τ_c of ~200 ms, that could not be resolved, must be present. Again, these authors did not evaluate the effect of changing Ca^{2+} on these gating parameters.

A significant advancement in our understanding of the gating of the KCNN gene family members came in 1998 when Hirschberg et al. (1998) studied the single channel gating kinetics of KCa2.2 heterologously expressed in *Xenopus* oocytes. An analysis of the single channel behavior revealed two τ_o and three τ_c. Furthermore, it was demonstrated that neither of the τ_o nor the shorter τ_c were Ca^{2+}-dependent. Rather, it was the long τ_c that was shown to decrease with increasing concentrations of Ca^{2+}, similar to what was described by Grygorczyk and Schwarz (1985) for KCa3.1. When Hirschberg et al. (1998) attempted to mathematically model the KCa2.2 channel gating kinetics they found that a model with two τ_o and three τ_c would only yield a reasonable fit to their experimental data if they assumed that the derived rate constants had a nonlinear dependence on Ca^{2+} concentration. To obtain the desired linear dependence on the Ca^{2+} concentration these authors required a fourth τ_c, that was not experimentally resolved, similar to what had been described for KCa3.1. The six-state open-closed gating model for KCa2.2 is shown Fig. 22.7a with the forward transitions between closed states being Ca^{2+} dependent.

Bailey et al. (2010) undertook a study to model the Ca^{2+}-dependence of KCa3.1 gating behavior based upon the model of Hirschberg et al. (1998) described above. In contrast to previous studies on single channel patches, Bailey et al. (2010) utilized macropatches containing hundreds or thousands of channels heterologously expressed in HEK cells and carried out rapid step changes in Ca^{2+} (~5 ms) from 0.4 to 10 μM, after which the macroscopic currents were fit to a series of exponentials and gating kinetics modeled. Consistent with other reports, the activation kinetics of KCa3.1 were found to be Ca^{2+} dependent, whereas the deactivation kinetics were Ca^{2+} independent. Using the six-state model of Hirschberg et al. (1998), Bailey et al. (2010) obtained an accurate description of their entire data set except for the activation rate at saturating Ca^{2+} (Fig. 22.7c). Based on this, these authors assumed that the Ca^{2+}-dependent steps in KCa3.1 gating depended nonlinearly on Ca^{2+} concentration such that the Ca^{2+}-binding rate constants could saturate at high Ca^{2+} concentrations. With this assumption, these authors were able to fit their entire data set (Fig. 22.7b). However, a further sensitivity analysis to evaluate the importance of each parameter in the model revealed that the data could be fit with a simpler 4-state kinetic model that excluded the final C–C and C–O transitions shown in the box in Fig. 22.7a such that the data were fit with a single τ_o and three τ_c with the Ca^{2+}-dependent rate constants saturating at high Ca^{2+}. Finally, these authors reported unpublished observations suggesting that the gating behavior

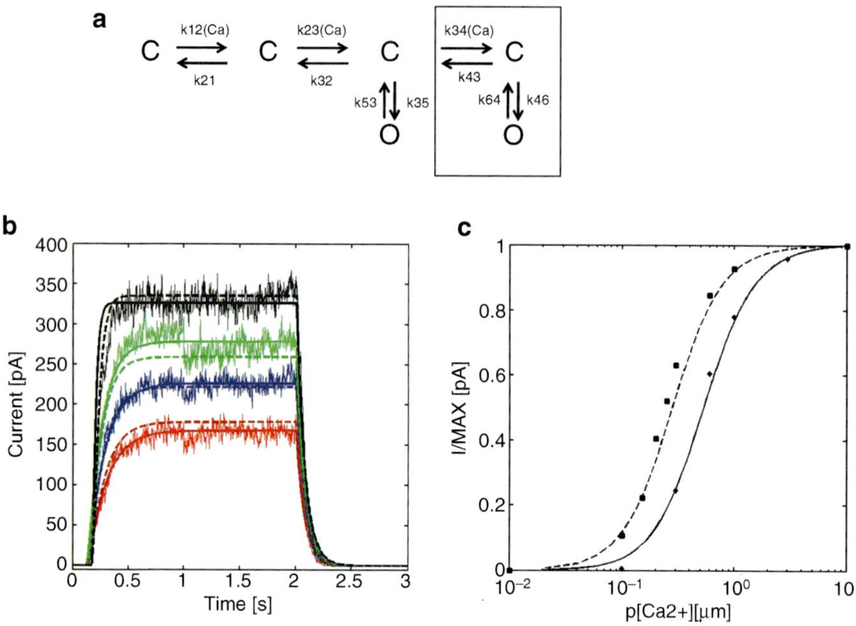

Fig. 22.7 Kinetic model for KCa3.1 gating. (**a**) The gating scheme to describe the activation and deactivation kinetics of KCa2.3 required a six-state kinetic model with four closed states and two open states, with the forward transitions between closed states dependent upon Ca^{2+} and the other states Ca^{2+}-independent (Hirschberg et al. 1998). The box indicates kinetic states that are not required to fit the gating kinetics of KCa3.1 as determined by sensitivity analysis (not shown), such that the gating of KCa3.1 was described by three closed and one open state. (**b**) Representative records for activation and deactivation with two variations of the model shown in the Upper panel. The colors refer to varying concentrations of Ca^{2+}; red = 0.5, blue = 0.7, green = 1.0 and black = 10 μM Ca^{2+}. The dashed line indicates the fitting assuming Ca^{2+}-dependent rate constants have a nonlinear dependence on Ca^{2+} concentration, $k = A \cdot [Ca]/(B + [Ca])$. The solid line represents the fit assuming Ca^{2+}-dependent rate constants have a linear dependence on Ca^{2+} concentration, $k = A \cdot [Ca]$. (**c**) The model can be used to predict the shift in the apparent Ca^{2+} affinity of PCMBS. Plot of normalized <I> current against the corresponding Ca^{2+}_i for KCa3.1 (diamonds) and KCa3.1 + PCMBS (squares). These data demonstrate that a four-state kinetic model accurately explains the Ca^{2+}-dependent gating of KCa3.1. See the text for additional details. From Bailey et al. 2010; used with permission from Rockefeller Press

of KCa2.2 could also be described by a simpler 4-state gating scheme with the Ca^{2+}-dependent transitions being nonlinear. The nonlinear dependence of the opening transitions on Ca^{2+} observed by Bailey et al. (2010) compared to the linear dependence proposed by Hirschberg et al. (1998) may be based on the fact that Bailey et al. (2010) utilized 10 μM Ca^{2+} as a saturating level and this may begin to produce channel block (Ledoux et al. 2008). Whether this becomes important in physiological or pathophysiological situations is unclear.

22.5.3 KCa3.1 Pore Architecture and Allostery

The work of Simoes et al. (2002) was the first to examine the pore architecture of KCa3.1. These authors used the substituted cysteine accessibility method (SCAM) that combines site-directed mutagenesis and chemical modification with thiol specific reagents to analyze the access of introduced cysteines (Falke et al. 1988; Karlin and Akabas 1998; Roberts et al. 1986; Akabas et al. 1992, 1994; Kuner et al. 1996). Thus, by combining SCAM with the patch-clamp technique, gating-sensitive amino acids can be identified by alterations in channel gating (Holmgren et al. 1996). In this regard, Simoes et al. (2002) identified two regions (Val275-Val282 and Ala283-Ala286) along S6 of KCa3.1 that display functional differences to MTSET ([2-(trimethylammonium) ethyl] methanethiosulfonate bromide) binding. That is, in the open state, the addition of MTSET caused inhibition at positions Val275, Thr278, and Val282, implying that these residues line the lumen of the pore. In both channel open and closed configurations the Ala283-Ala286 region is accessible to MTSET, suggesting this region is not embedded in the membrane and participates in the conformational changes in CaM/CaMBD to open the KCa3.1 channel's pore.

Additionally, Klein et al. (2007) studied the KCa3.1 channel pore structure in the closed configuration and found that the modification rates of MTSEA (2-aminoethyl methanethiosulfonate hydrobromide), with a diameter of 4.6 Å, for a residue located in the central cavity (Val275Cys) compared to a residue located at the C-terminal end of S6 (Ala286Cys) were found to differ by less than 7-fold, whereas experiments performed with MTSET, with a diameter of 5.8 Å, resulted in a modification rate 10^3–10^4 faster for the cysteine located in the C-terminal end compared to the cysteine in the central cavity. Modification rates of Val275Cys using Et-Hg^+ (4.1 Å diameter) and Ag^+ (2.55 Å diameter), which are smaller than MTSET, were found to be closed/open state independent, while modification rates for MTSET were 10^3 times faster for the open compared to the closed state. Based on these results, the authors proposed the closed structure of KCa3.1 can be represented by a narrow passage centered at Val282, connecting the channel inner cavity to the cytosolic medium, instead of the inverted teepee-like structure described for KcsA channel (Doyle et al. 1998). The recent structure of KCa3.1 (Lee and MacKinnon 2018) confirms that Val282 from each of the four S6 helices forms a constricted gate with a radius of less than 1 Å, which will not allow the passage of K^+ (Fig. 22.6f).

The substitution of Val188 (Val282 in KCa3.1) by less hydrophobic residues in the GIRK2 channel (Yi et al. 2001) or Pro475 (Ala283 in KCa3.1) by a more hydrophilic residue in the *Shaker* channel (Sukhareva et al. 2003), resulted in GIRK2 channels that were constitutively active or in *Shaker* channels that were unstable in the closed configuration. Using a Glycine (Gly) scan analysis, Garneau et al. (2009) observed that the substitutions Ala279Gly and Val282Gly in S6 of KCa3.1 resulted in the channel being constitutively active in zero Ca^{2+}. In contrast, when residues between Cys276 and Ala286, as well as Ala279 to Val282, were substituted with Gly constitutive activation of KCa3.1 was not observed. These results demonstrate that hydrophobic interactions involving Val282 are key determinants of KCa3.1 gating. The authors further demonstrated that Ag^+ enjoyed free access to the channel cavity in both an ion-conducting mutant, Val275Cys/Val282Gly, as well as in the closed channel mutant, Val275Cys, further arguing against the activation gate of KCa3.1 being localized to the distal end of S6 (Garneau et al. 2009).

Although KCa3.1 has four (4) cysteines in S6 (Cys267, 269, 276, and 277), Simoes et al. (2002) demonstrated that MTSET only transiently inhibited KCa3.1 current, indicative of these cysteines being inaccessible for modification, i.e., these cysteines face away from the pore lumen. Subsequently, Bailey et al. (2010) evaluated the effect of another cysteine modifying agent, parachloromercuribenze sulfonate (PCMBS) on KCa3.1 channel activity. In contrast to the MTS reagent, PCMBS activated KCa3.1 by inducing both an increase in current at saturating Ca^{2+} concentrations, as well as a leftward-shift in the Ca^{2+} activation curve to higher affinity. These results led the authors to conclude that PCMBS activated KCa3.1 via both Ca^{2+}-independent and -dependent pathways as reflected in the changes in both $P_{o(max)}$ and apparent Ca^{2+} affinity, respectively. Mutational analysis demonstrated that Cys276 was required for the PCMBS-mediated activation of KCa3.1. These authors further carried out a partial tryptophan scan of S6 and showed that both Leu281Trp and Val282Trp substitutions resulted in dramatic shifts in apparent Ca^{2+} affinity, similar to what was observed with PCMBS (Bailey et al. 2010). In contrast to Val282, which points into the water-filled inner vestibule, Leu281 is directed away from the central cavity. The recent cryo-EM structure predicts Leu281 in S6 makes intra-subunit contacts with Leu208 in S5 and inter-subunit contacts with Cys276 in S6 (Lee and MacKinnon 2018). As the Cys276Trp mutation did not affect Ca^{2+} affinity (Bailey et al. 2010) this suggests intra-subunit S6/S5 interactions are capable of modulating KCa3.1 gating, consistent with $S_{45}B$ being directly coupled to this domain, although this is yet to be evaluated. In total, the results from Bailey et al. (2010) and Sauvé and colleagues (Simoes et al. 2002; Klein et al. 2007; Garneau et al. 2009) demonstrate that both luminal and nonluminal amino acids along S6 participate in the activation mechanism of KCa3.1 and that the C-terminal region of S6 must be allosterically coupled to the activation gate. It should be noted that mutations at V282 have recently been directly linked to the rare hereditary anemia, hereditary xerocytosis (HX) (Andolfo et al. 2015; Glogowska et al. 2015). Indeed, the two different mutations described (Val282Met and Val282Glu) have unique disease-causing mechanisms, with Val282Met being active in 0 Ca^{2+} and Val282Glu displaying a shift in Ca^{2+} dependence (Rapetti-Mauss et al. 2016; Rivera et al. 2017), as suggested by the above studies (also, see below).

An interesting characteristic of KCa3.1 gating is that the $P_{o(max)}$, at saturating Ca^{2+} concentrations, is ~0.1–0.2 compared with a $P_{o(max)}$ of ~0.8 for the KCa2.2 channel (Hirschberg et al. 1998). The recent cryo-EM structure may provide insight into this conundrum. That is, in the Ca^{2+} bound (activated I) state the channel gate, formed by Val282, is still only ~1.6 Å in radius, which will not allow the flow of K^+ (Lee and MacKinnon 2018). Using 3-D classification, these authors identified a second activated state (activated II), at a resolution of 4.7 Å, with a pore radius of ~3.5 Å at Val282—sufficient to allow the passage of K^+. This state was achieved by a further movement of $S_{45}A$, and thus $S_{45}B$, away from the pore by an additional 2 Å. As this activated II state was observed less frequently, this suggests this conducting, activated II state is entered into at a low frequency, consistent with the low P_o observed for KCa3.1 in saturating Ca^{2+}.

Morales et al. (2013) undertook a series of mutagenesis studies designed to identify amino acids in the CaMBD that are critical to determining KCa3.1 $P_{o\ (max)}$. These authors demonstrated that hydrophobic effects at Ser367 contribute to $P_{o(max)}$, whereas electrostatic interactions involving Arg362 and Glu363 regulate channel activation rate (τ_{on}). That is, the Ser367Cys mutation resulted in an increase in $P_{o(max)}$ from 0.22 to 0.62 coupled with an increase in channel open time and a decrease in channel closed time. Additional substitutions with nonpolar amino acids of greater surface area (Thr, Leu, Trp) also increased $P_{o(max)}$, whereas substitution with an amino acid of similar surface area (Ala) had no effect. It was also demonstrated that neutralizing the charged amino acids at positions Arg362 and Glu363 resulted in dramatic shifts in τ_{on}, indicative of amino acids not involved in KCa3.1-CaM binding are important in channel activation.

22.5.4 Role of KCa3.1 in Gating of Maxi-K (KCa1.1)

Interestingly, the gating of KCa3.1 has also been shown to directly regulate the gating of maxi-K (KCa1.1) channels. That is, Thompson and Begenisich (2006) showed activation of KCa3.1 channels inhibits maxi-K channels in acinar cells, where the two channels are endogenously expressed, and in heterologous expression systems. Similar reports were described in human parotid (Nakamoto et al. 2007) and in mouse submandibular glands (Romanenko et al. 2007). The inhibition of maxi-K channels occurred whether the KCa3.1 channels were activated by direct perfusion of the cell with elevated Ca^{2+}, by muscarinic receptor stimulation, or by the exogenous chemical activator DCEBIO at low Ca^{2+} concentration. In addition, the inhibition of maxi-K currents was blocked by TRAM-34, an inhibitor of KCa3.1 (Thompson and Begenisich 2006; Romanenko et al. 2009). Quantitative analysis led these authors to postulate that each maxi-K channel may be surrounded by four KCa3.1 channels and it would be inhibited if any one of these KCa3.1 channels opens (Thompson and Begenisich 2006). Furthermore, since KCa3.1-induced inhibition of maxi-K channels takes place in excised membrane patches, the authors proposed a close interaction between the two channels without any intermediaries such that block of the ion flow through the pore in the maxi-K channel may occur by inserting the cytoplasmic N-terminal domains of KCa3.1 (Thompson and Begenisich 2006), as was described for two maxi-K channel β subunits (β2 and β3) (Wallner et al. 1999; Xia et al. 1999; Zhang et al. 2006; Li et al. 2007). In this regard, Thompson and Begenisich (2009) observed that the N-terminus of KCa3.1 shares general similarity with the N-terminal regions of these two maxi-K β subunits, the N-terminal region peptide of KCa3.1 blocked maxi-K channels and that activation of KCa3.1 competed with the N-terminus peptide for block of maxi-K channels.

In total, these data resulted in these authors proposing a model where KCa3.1 channels are located sufficiently close to the maxi-K channels such that activation of KCa3.1 results in a conformational change in their N-termini, which inserts through a cytoplasmic side portal in the BK channel, thereby inhibiting permeation through the maxi-K pore.

22.6 Regulation of KCa3.1

As detailed above, the main regulator of KCa3.1 gating is Ca^{2+} binding to calmodulin constitutively associated with the CaMBD. However, shortly after the mechanism by which Ca^{2+}-dependent gating was beginning to be unraveled, it was shown that phosphorylation modified this Ca^{2+}-dependent gating process. That is, Gerlach et al. (2000) first demonstrated a role for kinases in the regulation of KCa3.1 by showing that hydrolyzable ATP analogues, but not other nucleoside triphosphates, activated the channel via an increase in $P_{o(max)}$ without changing the apparent Ca^{2+} sensitivity. Addition of alkaline phosphatase eliminated this ATP-dependent activation of KCa3.1, indicative of a phosphorylation event. Indeed, this ATP-dependent activation was shown to be mediated through PKA when KCa3.1 was heterologously expressed in *Xenopus* oocytes or when endogenously expressed in T84 cells, but was independent of PKA when heterologously expressed in HEK cells, suggesting additional kinases may also regulate KCa3.1. In contrast to KCa3.1, KCa2.3 was not activated by ATP under similar conditions. Thus, using a series of KCa3.1/KCa2.3 chimeras, these authors further demonstrated the kinase-dependent activation of KCa3.1 could be localized to a 14 amino acid domain within the C-terminus (Arg^{355}–Met^{368}), distal to the CaMBD (Gerlach et al. 2001). Additional studies by Jones et al. (2007) indicated the ATP-dependent activation of KCa3.1 also depended upon an N-terminal RKR motif. That is, mutation of $^{15}RKR^{17}$ to alanines resulted in a channel that was no longer sensitive to ATP and alkaline phosphatase no longer reduced channel activity. Finally, the open probability (P_o) of these mutated channels was reduced 4-fold compared to WT KCa3.1, suggesting the ATP/kinase-dependent regulation of KCa3.1 observed by these authors is dependent upon a close N/C-terminal association. In this regard, it has similarly been demonstrated that KCa2.2 associates with both protein kinase CK2 as well as protein phosphatase 2A (PP2A), in a manner that brings the N- and C-termini into close proximity, such that they phosphorylate and dephosphorylate the associated CaM, respectively, resulting in a shift in the apparent Ca^{2+}-sensitivity of the channel (Bildl et al. 2004; Allen et al. 2007).

Additional studies have shown PKC regulates KCa3.1 function (Wulf and Schwab 2002), although this appears to be an indirect effect as mutation of the four conserved PKC phosphorylation sites has no effect on activation (Gerlach et al. 2000). Also, AMP-activated protein kinase has been shown to directly interact with the distal C-terminus of KCa3.1 between Asp^{380} and Ala^{400}, where activation of AMP-activated protein kinase results in an inhibition of KCa3.1 function and subsequent inhibition of transepithelial Cl^- secretion (Klein et al. 2009). In additional studies, Skolnik and colleagues (Srivastava et al. 2005) utilized a yeast two-hybrid approach to identify the myotubularin, MTMR6, a lipid (PI(3)P) phosphatase, as an interacting protein with a *C. elegans* KCa2.x (KCNL) family member. This interaction occurred via the channel coiled-coil domain, a region Syme et al. (2003) had previously shown to be required for proper channel assembly and trafficking (see below). Srivastava et al. (2006) further demonstrated this MTMR6-dependent regulation required the same 14 amino acid region of KCa3.1 previously shown to be required for ATP/kinase-dependent activation (Gerlach et al. 2001) (Fig. 22.2). In an exciting next step, Skolnik and colleagues demonstrated nucleoside diphosphate kinase B (NDPK-B) (Fig. 22.2), a histidine kinase, directly binds and activates KCa3.1 by phosphorylating His^{358} and this is reversed by protein histidine phosphatase (Srivastava et al. 2008). More recently, these authors demonstrated histidine phosphorylation activates KCa3.1 by antagonizing the copper-mediated inhibition of the channel (Srivastava et al. 2016). In addition, the recent crystal structure of the leucine-zipper region of KCa3.1 identified a copper ion binding to each of the four His389 positions within the leucine zipper, suggesting the possibility that copper ions may also bind to the inhibitory His358 position (Ji et al. 2018). Furthermore, NDPK-B knockout mice exhibit a reduced KCa3.1 activity (Di et al. 2010), while T cells deficient in intracellular copper demonstrate increased histidine phosphorylation of KCa3.1 resulting in increased channel activity (Srivastava et al. 2016), indicative of this pathway being functional *in vivo*, as well. Finally, it has also been demonstrated that the nitric oxide (NO) donor, sodium nitroprusside activated KCa3.1 channels in interstitial cells of Cajal (Zhu et al. 2007). It was subsequently shown that NO activates KCa3.1 via the PKG pathway in human dermal fibroblasts (Bae et al. 2014). Ferreira et al. (2015) subsequently demonstrated elevating cGMP increased KCa3.1 currents in intact microglia via an increase in reactive oxygen species (ROS) production and this was prevented by CaMKII inhibitors. As this could not be replicated in excised patches, it suggests this regulation may be indirect. In total, these results suggest that a kinase/phosphatase protein regulatory complex exists for the careful regulation of KCa3.1.

Devor and Frizzell (1998) demonstrated KCa3.1 channels in the T84 epithelial cell line were inhibited by numerous fatty acids, including arachidonic acid with an apparent K_i of 425 nM. Indeed, in intact epithelial monolayers, inhibition of cytosolic phospholipase A2 resulted in a significant increase in Ca^{2+}-mediated agonist transepithelial Cl^- secretion, suggesting that agonist-mediated generation of arachidonic acid may blunt the response to these agonists. Subsequently, Hamilton et al. (2003) identified two amino acids in the pore of KCa3.1 responsible for the arachidonic-mediated inhibition of KCa3.1 and further demonstrated introduction of these amino acids into KCa2.2 resulted in this channel being sensitive to arachidonic acid block. Interestingly, these two amino acids, Thr250 and Val275 had previously been shown to be responsible for the clotrimazole-dependent inhibition of KCa3.1 (Wulff et al. 2001). As shown by the recent cryo-EM structure (Lee and MacKinnon 2018), these amino acids form a pocket in the water-filled inner vestibule of the pore where clotrimazole and arachidonic acid can bind

(Fig. 22.6f). Clotrimazole and arachidonic acid have significantly different structures with clotrimazole being round with a diameter of 9 Å, while arachidonic acid is generally a U-shaped molecule with a height of 12.8 Å and a width of 11 Å. This result suggests this site may have significant flexibility or that clotrimazole and arachidonic acid are accessing different conformational states—perhaps activated I and activated II. The recent observation that Senicapoc fails to block the Val282Glu HX-causing mutation (Rapetti-Mauss et al. 2016) is consistent with this region of the channel being flexible enough to accommodate different chemical structures.

22.7 Trafficking of KCa3.1

The physiological response of an ion channel is dependent upon both its gating, i.e., the likelihood the channel is the open or closed state (P_o) as well as the number of channels (N) at the cell surface. The N of channels is in turn determined by the relative rates of trafficking along the anterograde and retrograde pathways. Studies over the past 15 years on the trafficking of KCa3.1 channels have revealed key molecular components in the assembly, delivery to the plasma membrane, and subsequent internalization and degradation of these channels. A schematic representation of the known intracellular trafficking pathways for KCa3.1 is presented in Fig. 22.8.

22.7.1 Anterograde Trafficking of KCa3.1

Prior to being delivered to the plasma membrane via the biosynthetic route, KCa3.1 must be correctly folded and assembled into tetramers in the endoplasmic reticulum (ER; Ellgaard et al. 1999; Papazian 1999; Deutsch 2002). Since the cloning of the KCa2.x/KCa3.1 channels, numerous investigators have identified motifs or regions within the N- and C-termini of KCa3.1 required for proper subunit folding and association. In an initial series of studies, Kaczmarek and colleagues (Joiner et al. 2001) and Adelman and colleagues (Lee et al. 2003) demonstrated that, apart from its role in channel gating, constitutive association of CaM with the proximal C-terminal CaMBD of KCa3.1 and KCa2.x channels regulates the surface expression of these channels by regulating their multimerization. In addition to the CaMBD, the distal C-termini of KCa3.1 and KCa2.x channels contain a leucine zipper motif that was first identified by Khanna et al. (1999) as being required for the expression of a hyperpolarizing K^+ conductance. The recent cryo-EM structure (Lee and MacKinnon 2018) shows this leucine zipper is part of HC and forms a coiled-coil in the center of the channel that runs perpendicular to the membrane plane (Fig. 22.6a, b). Devor and colleagues (Syme et al. 2003) demonstrated that although this C-terminal leucine zipper is conserved in KCa3.1 and KCa2.3 only KCa3.1 expression was affected by mutations in the leucine zipper. These results suggest a

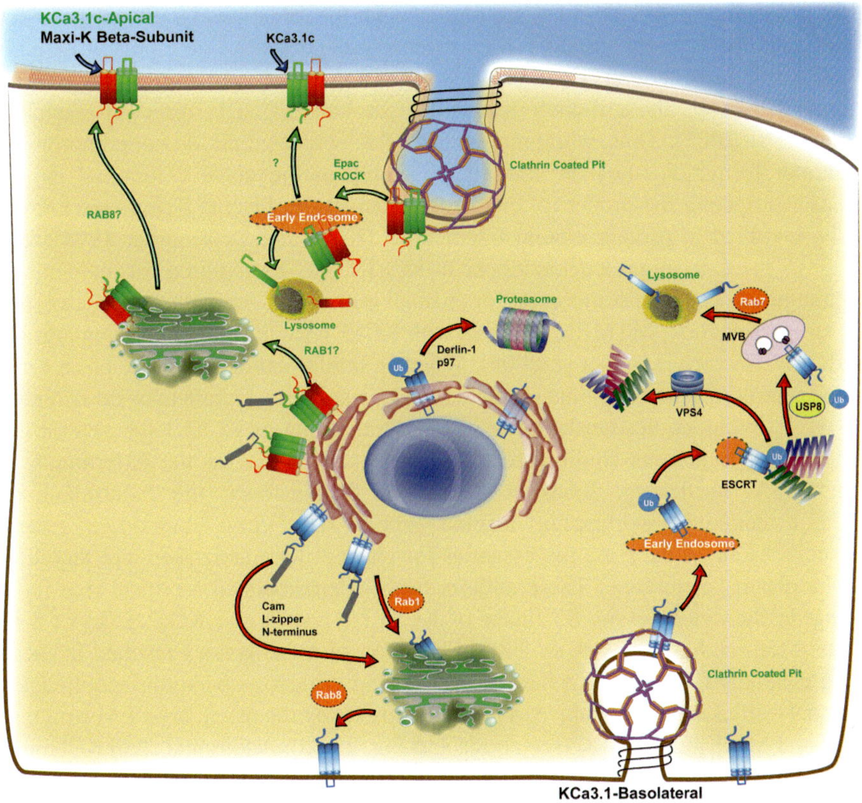

Fig. 22.8 Schematic of anterograde and retrograde trafficking pathways for KCa3.1 channels (KCa3.1-Basolateral and KCa3.1c-Apical). The important motifs and modulatory proteins in the channel assembly are shown. The molecular players of the anterograde trafficking of KCa channels are still emerging. KCa3.1-basolateral (KCa3.1b, blue channel) channels are trafficked to the basolateral membrane via Rab1- and Rab8-dependent pathways with additional trafficking motifs within N-terminus leucine-zipper (L-zipper). The forward trafficking of KCa3.1c-Apical (green channel) and Maxi K β-subunit (red channel) to the apical membrane is not known. Hence, the involvement of Rab1 and Rab8 in apical membrane targeting is not known. The retrograde trafficking of both KCa3.1-basolateral and KCa3.1c-Apical is via clathrin coated pit mechanisms. Endocytosed KCa3.1b is quickly targeted for lysosomal degradation through an endosomal sorting complexes required for transport (ESCRT)- and Rab7-dependent pathways. Ubiquitylation (Ub) facilitates the targeting of KCa3.1b to the lysosomes, and USP8 (ubiquitin specific protease 8) modulates the rate of KCa3.1b degradation by deubiquitylating KCa3.1b prior to delivery to the lysosome. In contrast, KCa3.1c-Apical uses an Epac- and ROCK-dependent pathway on route to early endosome. Whether KCa3.1c-Apical is recycled is not known. And, the mechanism(s) for lysosomal degradation is also currently not known. Epac = Exchange Protein directly activated by cAMP; ROCK = Rho-associated kinase; MSV = multivesicular body; calmodulin = Cam; VPS4 = Vacuolar protein sorting-associated protein 4. See the text for details

clear divergence in the structural requirements for the folding and trafficking of KCa family members. These authors further demonstrated that the C-terminus of KCa3.1 was capable of co-assembly and that this was abrogated by mutations in the leucine zipper. However, these mutations did not preclude KCa3.1 from self-assembly (Syme et al. 2003). Thus, while mutations of the leucine zipper do not compromise multimer formation, they do alter the assembly of the distal C-terminal tail of KCa3.1, affecting the proper folding of the channel. As leucine zippers are known to be involved in protein-protein interactions (Kobe and Deisenhofer 1994), this suggests the C-terminal leucine zipper in KCa3.1 may be required for interactions with regulatory proteins important for channel trafficking. As mentioned above, the PI(3)P phosphatase, MTMR6, interacts with KCa3.1 via the coiled-coil domains on both proteins (Srivastava et al. 2005). Whether similar interactions take place early in the anterograde pathway that regulate KCa3.1 trafficking remain to be determined.

While initial studies on the assembly and trafficking of KCa3.1 focused on the C-terminus, numerous studies have also identified a role for the N-terminus in anterograde trafficking. Jones et al. (2004) demonstrated the N-terminus of KCa3.1 contains overlapping leucine zipper (Leu^{18}/Leu^{25}/Leu^{32}/Leu^{39}) and dileucine (Leu^{18}/Leu^{19}) motifs essential for channel tetramerization and targeting to the plasma membrane. These authors further demonstrated a critical role for a lysine in the cytosolic S4–S5 linker of KCa3.1 (Lys^{197}) and KCa2.3 (Lys^{453}) in protein trafficking (Jones et al. 2005). Mutation of these lysines resulted in these KCa channels failing to exit the ER, suggesting critical associations required for channel trafficking are disrupted; a result confirmed by the recent cryo-EM structure (Lee and MacKinnon 2018) (see above and Fig. 22.6d). Additional studies demonstrated misfolded KCa3.1 and KCa2.3 channels are polyubiquitylated prior to degradation (Gao et al. 2008). These authors further demonstrated a role for the ER integral membrane protein, Derlin-1 and the cytosolic chaperone AAA-ATPase p97 in the ER translocation, protein dislocation and targeting of misfolded and polyubiquitinated KCa3.1 and KCa2.3 channels for proteasomal degradation (Fig. 22.8).

The anterograde trafficking of KCa channels requires they be properly assembled into tetramers. It is generally believed the 6 TMD KCNN gene family members (KCa2.x, KCa3.1) are evolutionarily older than the voltage-gated K^+ channels (K_V) (Anderson and Greenberg 2001) and from the crystal structure data of K_V channels it is clear the additional 4 TM domains (S1–S4), that were appended to the central pore (S5-Pore-S6) to form the 6-TMD channels, form a unique structural domain (Lee et al. 2005). A fundamental difference between the K_V and KCa3.1/KCa2.x channels is that the KCa channels exhibit no voltage dependence to their gating. The voltage dependence of K_V channels is conferred by a series of 4–7 positively charged amino acids in S4 as well as critical negatively charged amino acids in S2 and S3 (Aggarwal and MacKinnon 1996; Seoh et al. 1996). In addition to conferring voltage dependence, Papazian and colleagues (Papazian et al. 1995; Tiwari-Woodruff et al. 1997, 2000; Myers et al. 2004) demonstrated these charged amino acids are essential for channel biogenesis. That is, key salt bridges are required to maintain the fundamental fold of this S1–S4 domain. Indeed, it has been shown these electrostatic interactions are required at the initial folding process as the nascent polypeptide chain is being formed prior to entering the ER (Sato et al. 2002, 2003). Although KCa3.1 and KCa2.x channels are voltage-independent, these channels conserve two critical arginines in S4 as well as a glutamic acid in S3. Devor and colleagues (Gao et al. 2008) demonstrated mutation of these charged amino acids in S3 or S4 resulted in the rapid degradation of KCa3.1, KCa2.3, and KCa2.2 suggesting that, similar to K_V channels, key electrostatic interactions required for folding of the S1–S4 domain are abrogated, resulting in channel degradation. The observation that S4 arginines, as well as an S3 glutamic acid, share a conserved function between K_V channels and the ligand-gated KCa channels led the authors to speculate that the role of these charged amino acids in establishing the proper folding of the S1–S4 domain evolved prior to the establishment of voltage dependence and that the basic structure of the S1–S4 domain is conserved between the K_V and KCa channels. Indeed, the recent cryo-EM structure (Lee and MacKinnon 2018) confirms that Glu111 in S3 forms a salt-bridge with Arg165 in S4 (Fig. 22.6g), consistent with the observations of Gao et al. (2008).

As noted, the splice variant KCNN4c is missing 29 amino acids, including the entire second transmembrane domain (S2) and is only expressed at the cell surface in the presence of the Maxi-K β1-subunit (Barmeyer et al. 2010) (Fig. 22.8). Thus, a key remaining question is what is the structural topology of KCNN4c? That is, does the S1 segment of KCNN4c still act as a TMD or is it perhaps cytosolic in the final folded KCNN4c structure? In this way, the two TMDs of the β1-subunit may substitute for S1–S2 in KCNN4c. In the case of the KAT1, Shaker-type K^+ channel, S1 and S2 exhibited typical topogenic properties in which S1 is inserted into the ER membrane followed by the insertion of S2 (Sato et al. 2002). In contrast, in $K_V1.3$, the S2 segment appears to function as the signal sequence to establish topology such that S1 remains cytosolic and translocation into the Sec61 translocon and the ER membrane is initiated by S2 (Tu et al. 2000). Similarly, in the CFTR Cl^- channel, the second TMD initiates translocation rather than TMD1 as a means of establishing proper channel topology (Lu et al. 1998; Sadlish and Skach 2004). It has also been demonstrated in the KAT1 channel that S3 and S4 are post-translationally inserted into the ER membrane following release of the positively charged S4 from the ribosome (Sato et al. 2002). Indeed, this was shown to be dependent upon the known S3-S4 salt-bridge interactions in this channel (Sato et al. 2002). As noted, we have shown these S3–S4 interactions exist in KCa3.1, although this channel is voltage-independent (Gao et al. 2008). Thus, interactions between S3–S4 may similarly be required for the insertion of this domain into the membrane in KCa3.1. Finally, it has been shown that the S5-P-S6 domain can be integrated independent of the S1–S4 TMDs of KAT1 and $K_V1.3$, consistent with these regions of the channels being unique structural domains (Tu et al. 2000; Sato et al. 2002). Based on these previous studies, it is interesting to speculate that S1 of KCNN4c remains in the cytosol as it is typically inserted into the translocon and ER membrane together with S2. The subsequent translation of S3 and S4 results in salt-bridge pairings such that these TMDs are inserted into the membrane, thereby establishing the proper topology of the channel, with the S5-P-S6 region being inserted as in the full-length channel. Future studies will undoubtedly answer these interesting questions concerning KCNN4c biogenesis in addition to defining how and where the β1-subunit associates with KCNN4c to divert the channel from the degradative pathway along the biosynthetic route.

22.7.2 Retrograde Trafficking of KCa3.1

Compared to the extensive studies on the KCa channels anterograde trafficking, the mechanisms of endocytosis and downstream sorting of these channels have only recently begun to be unraveled. To begin to elucidate these mechanisms, Devor and colleagues (Gao et al. 2010) developed a strategy for monitoring the endocytic routes of KCa3.1 based on the work of Ting and colleagues (Howarth et al. 2005). For these studies, the biotin ligase acceptor peptide (BLAP) sequence was inserted into the second extracellular loop of KCa3.1, followed by a fast and specific biotinylation of the channels at the cell surface using recombinant biotin ligase (BirA) (Gao et al. 2010; Balut et al. 2010a). Addition of fluorophore-conjugated streptavidin allows the fate of the endocytosed channels to be addressed using a combination of biochemical and imaging techniques. Using these approaches, Balut et al. (2010a) demonstrated KCa3.1 is endocytosed from the plasma membrane within 60–90 min. This endocytic step of KCa3.1 has been proposed to be clathrin-dependent by Schwab et al. (2012) in migrating MDCK cells. In addition, a dileucine motif present in the C-terminus of KCa3.1 (Leu^{344}/Leu^{345}) appears to be important for channel internalization (Schwab et al. 2012), consistent with adaptor protein binding and clathrin-dependent endocytosis. Subsequent to endocytosis, KCa3.1 is targeted for lysosomal degradation via a Rab7- and MVB (multivesicular body)/ESCRT (endosomal sorting complex required for transport)-dependent pathway (Balut et al. 2010a) (Fig. 22.8). By combining BLAP-tagged KCa3.1 with TUBEs (tandem ubiquitin binding entities) and deubiquitylase Chip methodologies, Balut et al. (2011) further demonstrated polyubiquitylation mediates the targeting of internalized KCa3.1 to the lysosomes and also that USP8 (ubiquitin-specific protease) regulates the rate of KCa3.1 degradation by deubiquitylating KCa3.1 prior to lysosomal delivery (Fig. 22.8). Importantly, it has been shown that KCa3.1 endocytosis can be reduced by blocking the ubiquitin activating enzyme E1 (Balut et al. 2010b, 2011), suggesting ubiquitylation of either the channel itself or associated adaptors functions as a signal for KCa3.1 endocytosis. In contrast to our data on the KCNN family member KCa2.3, which rapidly recycles back to the plasma membrane following endocytosis (Gao et al. 2010), we did not observe recycling of KCa3.1 following endocytosis suggesting this channel is targeted for degradation following endocytosis. The plausibility of a ubiquitin-dependent pathway is supported by data showing that in CD4 T cells, the tripartite motif containing protein 27 (TRIM27)-dependent ubiquitylation of the class II PI3K-C2β kinase acts as a negative regulator of KCa3.1 activation (Cai et al. 2011). Thus, TRIM27 functions as an E3 ligase and mediates lysine 48 polyubiquitylation of PI3K-C2β kinase, leading to an inhibition of the PI3K enzyme activity. This leads to decreased levels of PI(3)P which, as discussed above, results in a decreased histidine phosphorylation and activation of KCa3.1 by NDPK-B (Cai et al. 2011).

22.7.3 *Plasma Membrane Targeting of KCa3.1 in Polarized Cells*

In contrast to the studies above, the trafficking of KCa3.1 in polarized cells has only just begun to be investigated. In T lymphocytes, it has been shown that KCa3.1 is evenly distributed on the plasma membrane but rapidly translocates to the immunological synapse upon antigen stimulation as part of the signaling complex that facilitates T cell activation (Nicolaou et al. 2007). It has also been shown that a KCa3.1 isoform lacking the N-terminus is retained within the cytosol and acts as a dominant negative on the full-length channel in human lymphoid tissues (Ohya et al. 2011). This is consistent with the previously published role of the N-terminus in KCa3.1 trafficking (Jones et al. 2004). Additionally, it was demonstrated KCa3.1 localizes with F-actin to the immunological synapse, emphasizing the role of cytoskeleton in providing directionality for channel movement (Nicolaou et al. 2007). An interesting observation is signaling molecules such as PKA and PKC, which are known regulators of KCa3.1 function, as discussed above, accumulate in the immunological synapse on T cell activation, as well (Bi et al. 2001; Skalhegg et al. 1994; Zhou et al. 2004), and this close proximity could provide a regulatory mechanism for KCa3.1 function at the synapse.

To date, very few studies that have explored the trafficking of KCa3.1 in polarized epithelia. While it has been shown that in migrating MDCK cells KCa3.1 displays a polarized distribution by accumulating at the leading edge of the lamellipodium (Schwab et al. 2006), these cells do not demonstrate classic apical-to-basolateral polarization. We recently utilized polarized MDCK, Caco-2, and FRT cells, grown on semipermeable Transwell® supports, to investigate the anterograde and retrograde trafficking of full-length KCa3.1 in polarized epithelia (Bertuccio et al. 2014). Given that Rabs 1, 2, and 6 have been shown to be involved in the trafficking of proteins from the ER to Golgi, while Rabs 8 and 10 are known to be involved in the trafficking of proteins from the Golgi to the plasma membrane (Martinez and Goud 1998; Babbey et al. 2006; Zhang et al. 2009; Dong et al. 2010), we determined whether any of these Rabs played a crucial role in the anterograde trafficking of KCa3.1 in polarized epithelia. In this regard, we demonstrated Rab1 and Rab8 are required for the ER/Golgi exit of KCa3.1 and subsequent trafficking to the basolateral membrane (Fig. 22.8), whereas Rabs 2, 6, and 10 did not play a role in this trafficking step (Bertuccio et al. 2014). In contrast, trafficking of KCa2.3 was unaffected by siRNA-mediated knockdown or overexpression of dominant negatives of these Rabs, thereby confirming the specificity of these Rabs within the gene family. It has also been shown the sorting of numerous proteins to the basolateral domain of polarized epithelia is the result of an association with the μ1 subunit of the adaptor protein-1 complex (AP-1) (Ohno et al. 1998; Muth and Caplan 2003; Cancino et al. 2007). The AP-1 complex is the only clathrin-associated adaptor complex implicated in basolateral sorting in polarized epithelia (Futter et al. 1998). Importantly, while μ1B is expressed in MDCK cells, it is not expressed in LLC-PK_1 cells and the lack of the μ1B subunit in LLC-PK_1 cells has been shown to result in the mistargeting of some basolateral proteins to the apical surface while stable expression of μ1B restores proper basolateral targeting (Fölsch et al. 1999). However, we demonstrated correct basolateral targeting of KCa3.1 in LLC-PK1 cells, indicating μ1B is not required for basolateral sorting of KCa3.1 (Bertuccio et al. 2014). It has also been shown proteins destined for the basolateral membrane can traffic either: (1) from the Golgi to the basolateral membrane directly or (2) from the Golgi through recycling endosomes to the basolateral membrane (Fuller et al. 1985; Orzech et al. 2000). To determine whether KCa3.1 traffics via recycling endosomes we carried out a series of studies to block trafficking through the recycling endosomes. We demonstrated that KCa3.1 did not traffic to the basolateral membrane via transferrin- and RME-1-positive recycling endosomes, suggesting this channel traffics directly to the basolateral membrane from the Golgi (Bertuccio et al. 2014). More recently, we showed correct targeting to the basolateral membrane required an intact actin cytoskeleton and microtubule network with the Myosin-Vc motor playing an instrumental role in moving KCa3.1-containing vesicles to the basolateral membrane (Farquhar et al. 2017). Finally, we demonstrated, similar to our studies in nonpolarized cells, KCa3.1 is endocytosed from the basolateral membrane of polarized MDCK, Caco-2 and FRT cells in a ubiquitin-dependent manner and that the channels are then targeted to the lysosome and proteasome for degradation (Bertuccio et al. 2014; Lee et al. 2017).

Whether the correct apical targeting of KCNN4c similarly requires Rabs 1 and 8 and whether this splice variant of KCa3.1 is directly targeted to the apical membrane or via the apical recycling endosomal compartment has yet to be addressed (Fig. 22.8). As detailed above, apical KCNN4c expression is regulated by knockdown of Epac1 with KCNN4c being redistributed to subapical vesicles. Similarly, inhibition of RhoA and ROCK induced a redistribution of KCNN4c out of the apical membrane (Sheikh et al. 2013). Whether KCNN4c is capable of recycling back to the apical membrane from this vesicular pool, whether this involves ubiquitylation and de-ubiquitylation and whether this channel is targeted for lysosomal or proteasomal degradation remain open questions yet to be addressed.

22.8 Role of KCa3.1 in the Cell Cycle, Cell Proliferation, and Cancer Biology

KCa3.1 plays crucial roles in a number of cellular functions including the cell cycle and cell proliferation of epithelial cells. It has been known, for some time, that intracellular Ca^{2+}, $[Ca^{2+}]_i$, levels alter cell proliferation (Hazelton et al. 1979). Welsh and McCann (1985) and McCann and Welsh (1990), provided the first evidence, in airway epithelia, of the Ca^{2+}-dependent K^{+} channel that is now known to be KCa3.1. Additionally, Nilius and Wohlrab (1992), using human melanoma cells, provided the first evidence linking changes in the electrochemical gradient for Ca^{2+} entry into the cell due to membrane hyperpolarization and K^{+} channels. Huang and Rane (1994) provided one of the earliest reports demonstrating cell proliferation of NIH 3T3 cells was attenuated by charybdotoxin, an inhibitor of Ca^{2+}-activated K^{+} channels, including KCa3.1. Eventually, Wonderlin and colleagues (Strobl et al. 1995; Wonderlin et al. 1995; Wonderlin and Strobl 1996) suggested Ca^{2+}-activated K^{+} channels might play a significant role in regulation of the cell through the cell cycle. Therefore, evidence has been mounting for some time that Ca^{2+}-activated K^{+} channels, such as KCa3.1, plays a role in the cell cycle and cell proliferation of epithelial cells.

During cell proliferation, a cell progresses through various phases of the cell cycle (G_0, G_1, S, G_2, and M). Interestingly, the $Ca^{2+}{}_i$ of epithelial cells increase during the early and late G_1 phases of the cell cycle (Ouadid-Ahidouch et al. 2004). Indeed, Ouadid-Ahidouch et al. (2004) reported that mRNA levels of KCa3.1 are altered during different phases of the cell cycle, reaching a maximum in cells synchronized at the end of G_1. Also, pharmacological block of KCa3.1 inhibits cell growth suggesting KCa3.1 plays a critical role in cell proliferation (Parihar et al. 2003; Ouadid-Ahidouch et al. 2004) Therefore, what is the link between the K^{+} channel, Ca^{2+}, cell cycle, and cell progression? In a nonexcitable cell, such as an epithelial cell, mitogens activate Ca^{2+} entry into the cell via a process that begins with the production of inositol 1,4,5-triphosphate (IP_3) that subsequently activates the IP_3 receptor (IP_3R), a multimeric Ca^{2+} channel of the endoplasmic reticulum (ER) membrane. This results in a quick Ca^{2+} spike as Ca^{2+} exits the ER (Schreiber 2005). The Ca^{2+} spike is limited by the amount of Ca^{2+} stored in the ER. However, the initial Ca^{2+} spike stimulates Ca^{2+}-activated K^{+} channels (e.g., KCa3.1) that allow K^{+} exit from the cell causing the cell voltage to become hyperpolarized. The hyperpolarization results in an increased driving force for Ca^{2+} entry into the cell via a membrane store-operated capacitive Ca^{2+} entry (SOC) step (Schreiber 2005) via ORAI1 (Hammadi et al. 2012) and/or ORAI3 (Faouzi et al. 2011) channels and through TRP Ca^{2+}-channels, TRPC1 (El Hiani et al. 2009) and TRPV6. This entry of external Ca^{2+} maintains the elevated $[Ca^{2+}]_i$, which continues to activate K^{+} channels. The coupled action of the cellular hyperpolarization via K^{+} channels and increased $[Ca^{2+}]_i$ has been demonstrated to enhance passing of cells through G_0/G_1 and into the S phase, resulting in cell proliferation (Schreiber 2005; Wonderlin and Strobl 1996; Wonderlin et al. 1995). The current thought is that KCa3.1 is activated during the late G_1 phase and G_2/M phases of the cell cycle (Lallet-Daher et al. 2009; Lai et al. 2011).

The basis of our appreciation of the possible role of KCa3.1 in cell proliferation in epithelial cells has arisen from cell differentiation and blocker pharmacology studies of lymphocytes. Cahalan and coworkers (Decoursey et al. 1984) first provided evidence for the relationship between K^{+} channels and the proliferative processes of T lymphocytes. Originally, the characterization of KCa3.1 in cell activation/

proliferation was demonstrated in erythroid cells, T-lymphoblasts and T-lymphocytes. Alper and colleagues (Vandorpe et al. 1998) reported, in mouse embryonic stem cells (ES), that mRNA for mIK1 (mouse KCa3.1) was maintained at low levels in uninduced ES cells, however, exhibited sustained high mRNA levels during erythroid differentiation of ES cells. Further, they demonstrated that increasing concentrations of clotrimazole (even as low as 10 nM) inhibited cell proliferation of human peripheral blood stem cells, thus, these studies provide strong evidence that KCa3.1 participates in cell differentiation and proliferation (Vandorpe et al. 1998).

The information available about the role of KCa3.1 and the cell cycle and cell proliferation in epithelial cells is limited, though moving forward. Strobl and colleagues (Wang et al. 1998), using quinidine (a generic K^+ channel blocker), provided early evidence that K^+ channels might participate in the G phase of the cell cycle of MCF-7 breast cancer epithelial cells. Afterward, Ouadid-Ahidouch et al. (2004) demonstrated that the current density of KCa3.1 increased in MCF-7 cells (a breast cancer cell line) that had been synchronized at the end of the G_1 or S phase when compared to those cells in the early G_1 phase of the cell cycle. Additionally, they reported that the increase in current density was in congruence with increased mRNA levels of KCa3.1 and the enhanced negative membrane potential observed. To examine whether KCa3.1 influenced cell progression by altering the $[Ca^{2+}]_i$, Ouadid-Ahidouch et al. (2004) examined whether there was a correlation between effects of $[Ca^{2+}]_i$ on K^+ channel activity. They reported that $[Ca^{2+}]_i$ was lower in the early G_1 phase (49 $\pm$ 6 nM) when the activity of KCa3.1 was low and the membrane potential was depolarized. Additionally, the $[Ca^{2+}]_i$ level was higher (240 $\pm$ 23 nM) in cells arrested at the end of G_1 and $[Ca^{2+}]_i$ reached high levels as cells progressing into the S phase (323 $\pm$ 20 nM) when the activity of KCa3.1 was high and the membrane was hyperpolarized. Furthermore, Ouadid-Ahidouch et al. (2004) demonstrated that clotrimazole caused a depolarization of membrane potential of cells arrested at the end of G_1 and during the S phase, providing evidence of active KCa3.1, while clotrimazole had no effect on the membrane potential of cells arrested in early G_1 phase. Finally, the treatment of MCF-7 cells with clotrimazole and econazole (another inhibitor of KCa3.1) caused a dose-dependent reduction in cell proliferation, and these inhibitors resulted in an increase of cells in the G_1 phase and a decrease in the number of cells moving into the S phase (Ouadid-Ahidouch et al. 2004). In agreement, Parihar et al. (2003), using LNCaP and PC-3 (prostate cancer epithelial cells) reported that both 1-EBIO and riluzole (activators of KCa3.1) increased cell proliferation of both types of prostate cells and resulted in decreased cell proliferation when administered in the presence of charybdotoxin and clotrimazole. Additionally, Lai et al. (2011) reported that inhibition of KCa3.1 accumulates cells in the G_2/M phase in LoVo human colon cancer cells and enhances the phosphorylation of Cdc2. Finally, Tao et al. (2008) demonstrated the KCa3.1 channels participate in volume regulation in mouse mesenchymal stem cells and the channels modulate the regulation of the cell cycle (G_0/G_1 transition or the G_1/S boundary) in a cyclin E- and cyclin D1-dependent manner. These studies highlight the importance of KCa3.1' in the cell cycle of certain epithelial cell types.

There have been advances in our understanding of the "link" between KCa3.1 and the Ca^{2+} entry step into cells in the cell proliferation process. Lallet-Daher et al. (2009) examined aspects of the control of Ca^{2+} entry in human prostate cancer cell proliferation using LNCaP and PC-3 prostate cells that has furthered our understanding of the role of KCa3.1 in cell proliferation. They examined the mechanism by which the inhibition of KCa3.1 prevented cell proliferation by examining the changes in expression of $p21^{Cip1}$ (cyclin-dependent kinase inhibitor 1) that is known to participate in the G1 phase (Ghiani et al. 1999). Indeed, when KCa3.1 was inhibited by TRAM-34, the expression (mRNA) of $p21^{Cip1}$ of both LNCaP and PC-3 cells was increased by 2- to 4-fold, and they also reported a similar fold-increase in $p21^{Cip1}$ when they knocked down KCa3.1 with siRNA. Furthermore, Lallet-Daher et al. (2009) provided a role of the Ca^{2+} channel, TRPV6 in the signaling pathway, when they demonstrated an association of TRPV6 and KCa3.1 by co-immunoprecipitation. Furthermore, they also reported a novel role of TRPV6 in maintaining the hyperpolarization-activated Ca^{2+} entry by using an siRNA approach to knockdown TRPV6 and Ca^{2+} entry was not stimulated by activation of KCa3.1. These data certainly suggested a "close" association of KCa3.1 and TRPV6 in the regulation of cell proliferation. Finally, Ouadid-Ahdouch and colleagues (Dhennin-Duthille et al. 2011) and others (Bolanz et al. 2008) have now identified TRPV6 in MCF-7 cells, while Dhennin-Duthille et al. (2011) reported highly expressed levels of TRPV6 in human breast ductal adenocarcinoma cells compared to adjacent nontumoral tissue. Therefore, the link between KCa3.1, TRPV6 and cell proliferation is gaining strength.

The role of KCa3.1 in cell proliferation and hence cancer biology has been further highlighted by several recent exciting studies. In this regard, Liu et al. (2017) showed KCa3.1 mRNA and protein were overexpressed in cervical cancer tissues and either TRAM34 inhibition or siRNA-mediated KCa3.1 knockdown suppressed cervical cancer cell proliferation. Similarly, Zhang et al. (2016) demonstrated KCa3.1 expression was increased in various breast cancer subtypes compared to nontumor breast tissue and that TRAM34- or siRNA-mediated knockdown of KCa3.1 inhibited cell proliferation and migration while promoting apoptosis. Similar results were obtained in a murine breast cancer cell model (Steudel et al. 2017). Ouadid-Ahidouch and colleagues (Faouzi et al. 2016) further demonstrated RNAi-mediated knockdown of KCa3.1 decreased cell proliferation due to cell cycle arrest in the G_1 phase of breast cancer cells while showing a correlation between KCa3.1 mRNA expression and poor patient prognosis through data mining of public domain data sets. Similar studies have shown a role for KCa3.1 in the progression of hepatocellular carcinoma using HepG2 cells (Liu et al. 2015), others have reported increased cell proliferation and migration of pancreatic cells in pancreatic ductal adenocarcinoma (Bonito et al. 2016), and human ovarian cancer cell migration (Robles-Martinez et al. 2017). Additionally, Song and coworkers (Du et al. 2019) reported that KCa3.1 stimulated the cell cycle progression, migration, and invasion of hepatocellular carcinoma cells by the activation of S-phase protein kinase 2, which promoted the degradation of p21 and p27 that targeted reelin (an extracellular glycoprotein) to stimulate epithelial-mesenchymal transition.

Finally, a role for miRNAs was recently demonstrated in the regulation of KCa3.1 expression in cancer (Chen et al. 2016). That is, miR-497-5p was shown to bind to the KCa3.1 mRNA 3′ untranslated region and this miRNA was downregulated in human angiosarcomas compared to capillary hemangiomas. Transfection of an miR-497-5p mimic inhibited cell proliferation, cell cycle progression, and invasion, while also inhibiting tumor growth in an *in vivo* angiosarcoma xenograft model via downregulation of KCa3.1 expression. These results suggest KCa3.1 and its regulation will continue to be a critical target in the cell proliferation and cancer fields over the coming years. Further information about KCa3.1 and its potential role in cancer can be found in the following review articles by Panyi et al. (2014), D'Alessandro et al. (2018) and Mohr et al. (2019).

There is an exciting caveat to the KCa3.1 and the cell proliferation story as Bruce and coworkers (Millership et al. 2011) reported that both a nonfunctional pore mutant KCa3.1 ($KCa3.1^{GYG/AAA}$) and a nontrafficking mutant KCa3.1 ($KCa3.1^{L18A/L25A}$; Jones et al. 2004) resulted in enhanced cell proliferation using a heterologous expression system (human embryonic kidney cells, HEK). These results indicated that KCa3.1 induced cell proliferation, but did not require membrane localized KCa3.1. Interestingly, others have reported that functionally expressed KCa3.1 in human gliomas did not alter cell proliferation (Abdullaev et al. 2010). This represents a shift in how we envision KCa3.1's involvement of cell proliferation; indicating further studies are required to further explore exactly how KCa3.1 regulates this aspect of cell biology.

For additional information regarding roles of K^+ channels in the cell cycle and cell proliferation, readers are directed to reviews by Ouadid-Ahidouch and Ahidouch (2013), Girault and Brochiero (2014), Urrego et al. (2014), and Weiger and Hermann (2014).

22.9 Role of KCa3.1 in Disease

22.9.1 Role of KCa3.1 in Hemolytic Anemia

While this review focuses on the role of KCa3.1 in epithelia, it is critical to highlight the recent developments that have taken place in defining disease-causing mutations of this channel in the hemolytic anemias. That is, what is now known as KCa3.1 was originally described by Gardos (1958) in red blood cells (RBC) as a Ca^{2+}-dependent K^+ efflux pathway. This "Gardos" channel is now known to play a key role in ion homeostasis and volume regulation in RBC (Cala 1983; Canessa 1991). Activation of KCa3.1, via an increase in intracellular Ca^{2+}, leads to K^+ and water efflux and hence RBC dehydration. This activation of KCa3.1 occurs in response to local membrane deformations in RBC via an increase in intracellular Ca^{2+}, likely as the cells transit the microcirculation (Dyrda et al. 2010). The increased content of Ca^{2+} in RBC in hereditary anemias, such as thalassemia and sickle cell disease (SCD), may be caused by this stretch-activated increase in cellular Ca^{2+} (Dyrda et al. 2010).

Importantly, the polymerization of hemoglobin S (HbS) in sickle RBC is dramatically influenced by hydration status such that KCa3.1 has become a target for SCD therapy. In both animal models and patients with SCD, inhibition of KCa3.1 blocked K^+ loss and prevented HbS polymerization (Brugnara et al. 1993; De Franceschi et al. 1994). Clinical trials with Senicapoc, a high-affinity inhibitor of KCa3.1, demonstrated reduced RBC dehydration and hemolysis, but were stopped early due to a lack of efficacy in terms of sickling crisis events (Ataga et al. 2008, 2011).

More recently, hereditary xerocytosis (HX), also called hereditary dehydrated stomatocytosis, an autosomal dominantly inherited disease was shown to be caused by mutations in the KCNN4 gene, which encodes KCa3.1 (Rapetti-Mauss et al. 2015; Andolfo et al. 2015; Glogowska et al. 2015). Two different apparent hot-spots for mutations were identified, including in the CaMBD (Arg352His) (Rapetti-Mauss et al. 2015; Andolfo et al. 2015; Fermo et al. 2017) and in the pore-lining S6 transmembrane domain (Val282Met, Val282Glu) (Andolfo et al. 2015; Glogowska et al. 2015). Analysis of the Arg352His mutation revealed that it caused a shift in the apparent Ca^{2+} affinity of KCa3.1 from 950 nM to 210 nM (Rapetti-Mauss et al. 2015). This would result in KCa3.1 being significantly more active at lower Ca^{2+} concentrations, thereby promoting K^+ and water loss and hence dehydration—leading to disease. Analysis of the Val282 mutations, which forms the inner gate as discussed above (Fig. 22.6c), suggests Val282Met and Val282Glu have different mechanisms of action. That is, Val282Met is active in resting Ca^{2+} concentrations, having an ~8-fold increase in channel activity during on-cell recording (Rivera et al. 2017) and its activity is not increased between 0.25 μM and 1 μM Ca^{2+} during whole-cell recording (Rapetti-Mauss et al. 2016). This is similar to what Garneau et al. (2009) described for mutations at Val282 (see above). In contrast, Val282Glu remains Ca^{2+} sensitive, as assessed by whole-cell recordings (Rapetti-Mauss et al. 2016), indicative of a difference in the underlying mechanisms of action for these two disease-causing mutations. Indeed, we previously demonstrated that Val282Trp mutation results in a significant leftward shift in KCa3.1 apparent Ca^{2+} affinity (Bailey et al. 2010); demonstrating different side-chain substitutions at this position have unique mechanisms of action (see above). Importantly, Rapetti-Mauss et al. (2016) demonstrated that the KCa3.1 inhibitor, Senicapoc blocked Val282Met with similar affinity to WT channels, whereas the Val282Glu mutation was essentially insensitive to Senicapoc. This raises the specter that disease-causing mutations in KCa3.1 will need to be individually assessed for blocker/opener pharmacology to determine their therapeutic potential in the future. Very recently, Alper and colleagues (Rivera et al. 2019) suggested KCa3.1 activation in HX, and hence chronic red blood cell dehydration and intracellular cation imbalance, may lead to decreased channel activity of KCa3.1.

22.9.2 Role of KCa3.1 in Epithelial Diseases

While mutations in both the C-terminus and S6 of KCa3.1 have now been directly linked to HX, these mutations have not been directly linked to diseases of epithelia, as of this publication. However, given the critical role of KCa3.1 in maintaining transepithelial ion and fluid transport, it is not surprising that this channel has been suggested to play a role in the etiology of various diseases (Table 22.1). In the near future, it will be critical to determine whether the mutations described in HX can be linked to perturbations in transepithelial ion and fluid transport.

As detailed above, and in other chapters in this volume, the human colon absorbs Na^+, Cl^-, and water and simultaneously secretes K^+ and HCO_3^- (Sandle 1998). In the distal colon, the electrogenic absorption of Na^+ generates a lumen-negative electrical potential difference (PD_{te}) and it is known that the downregulation of Na^+ absorption promotes the pathogenesis of diarrhea in patients with ulcerative colitis (UC) (Sandle et al. 1986; Hawker et al. 1980; Amasheh et al. 2004; Greig et al. 2004). As indicated above, basolateral KCa3.1 plays a critical role in maintaining transepithelial ion transport by maintaining this transepithelial PD_{te}. Thus, Sandle et al. (1990) have shown the changes in transepithelial PD_{te} associated with the inflamed human colon are at least partially the result of a depolarized basolateral membrane. Indeed, a variant of the KCa3.1 gene, KCNN4, has been reported to be associated with ileal Crohn's disease (Simms et al. 2010). In this regard, Al-Hazza et al. (2012) demonstrated that the reduced Na^+ absorption in UC was associated with an ~75% decrease in basolateral KCa3.1 channel expression and activity in patients with UC, whereas crypts from patients with Crohn's, collagenous and infective colitis showed identical levels of KCa3.1 channel protein to those of control patients.

Table 22.1 KCa3.1 and its role in epithelial diseases

Organ	Disease	Function	References
Kidney	ADPKD	TRAM-34 blocks Cl^- secretion and ameliorates cyst formation and enlargement *in vitro*	Albaqumi et al. (2008)
Kidney	Diabetic nephropathy	TRAM-34 inhibited fibroblast proliferation	Huang et al. (2013)
Pancreas	Pancreas ductal adenocarcinoma	TRAM-34 blocks the proliferation of these pancreatic cells	Jager et al. (2004)
Colon	Ulcerative colitis	Decreased basolateral expression and activity of KCa3.1 channel	Al-Hazza et al. (2012)
Distal colon and rectum	Inflammatory bowel diseases	Blockade of KCa3.1 in T-cells decreased inflammatory cell infiltration and tissue destruction	Al-Hazza et al. (2012), Di et al. (2010)
Respiratory tract	Cystic fibrosis and chronic obstructive pulmonary disease (COPD)	Activation of basolateral KCa3.1 with 1-EBIO or DCEBIO stimulates transepithelial Cl^- secretion	Devor et al. (1999, 2000), Singh et al. (2001), Roth et al. (2011)

Bridges and colleagues (Devor et al. 1999, 2000; Singh et al. 2001) demonstrated that pharmacological activation of basolateral KCa3.1 with 1-EBIO or DCEBIO in airway epithelia from both wild-type CFTR-expressing and CF airway stimulates transepithelial Cl^- secretion. These results led the authors to propose pharmacological modulation of basolateral KCa3.1 as a potential therapeutic strategy in airway diseases, including CF and chronic obstructive pulmonary disease (COPD). Indeed, Roth et al. (2011) demonstrated that activation of basolateral KCa3.1 with 1-EBIO potentiated the Cl^- secretory response to cAMP in rectal biopsy tissues from CF patients compared to age-matched controls, thereby supporting this proposal. More recently, Arthur et al. (2015) showed KCa3.1 expression was increased in asthmatic compared to healthy airway epithelium *in situ*. These authors further demonstrated KCa3.1 currents were larger in asthmatic compared to healthy human bronchial epithelium cultured *in vitro*, further suggesting KCa3.1 may be a novel therapeutic target for airway disease.

It has long been recognized that mutations in CFTR only partially explain the clinical heterogeneity of CF disease. Indeed, meconium ileus is a severe intestinal obstruction that occurs in only 15–20% of CF patients at birth. As there is a familial recurrence rate of 29%, this suggests other genetic factors are involved in the expression of the meconium ileus phenotype (Kerem et al. 1989). That is, there are likely modifier genes that can alter the course of CF disease independent of CF genotype. Tsui and colleagues (Zielenski et al. 1999; Rozmahel et al. 1996) mapped this potential modifier gene to chromosome 19q13 in humans and the proximal region of mouse chromosome 7; a syntenic region to that in humans, and found this correlated with intestinal but not pulmonary phenotype. Indeed, the survival of $CFTR^{-/-}$ mice was dependent on strain, with increased survival correlating to increased Ca^{2+}-dependent Cl^- secretion (Rozmahel et al. 1996)—a process in which KCa3.1 plays a critical role. These authors went on to demonstrate meconium ileus was associated with polymorphic markers in the KCa3.1 gene, KCNN4 (Zielenski et al. 2004). These results strongly implicated KCa3.1 as a modifier gene for meconium ileus in CF. Very recently, Philp et al. (2018) directly explored this possibility by creating double $CFTR^{-/-}$/$KCa3.1^{-/-}$ knockout mice, as they reasoned knockout of KCa3.1 would exacerbate the $CFTR^{-/-}$ phenotype if KCa3.1 were a modifier gene for CF. Surprisingly, addition of the $KCa3.1^{-/-}$ genotype nearly abolished the lethality of the $CFTR^{-/-}$ genotype in mice and this was not due to an improvement in intestinal secretory function. Rather, $KCa3.1^{-/-}$ mice displayed a corrected circulating TNF-α compared to the elevated levels associated with the $CFTR^{-/-}$ phenotype while also reducing the mast cell increase observed in $CFTR^{-/-}$ mice. Additional studies demonstrated the KCNN4 modifier gene phenotype was STAT6-dependent. These results suggest KCa3.1 may still be a target for CF therapeutics, albeit this effect is not associated with the channels' ion transport properties. As noted above, the ion transport properties of KCa3.1 were similarly not important in the channels' regulation of cell proliferation.

As noted above, Ca^{2+}-activated K^+ channels are known to mediate the secretion of fluid and electrolytes in epithelial cells of pancreatic ducts (Nguyen and Moody 1998) and this channel has been identified as KCa3.1 (Jung et al. 2006; Hayashi et al.

2012). Whether KCa3.1 channels are essential in the dysfunction of pancreatic epithelial transport in cystic fibrosis remains unknown, however. In additional studies, it has been demonstrated that the mRNA levels of KCa3.1 are upregulated in patients with pancreas ductal adenocarcinoma and inhibitors of KCa3.1 block the proliferation of these pancreatic cells (Jager et al. 2004).

In the kidneys, autosomal-dominant polycystic kidney disease (ADPKD) is characterized by the progressive development and enlargement of multiple bilateral fluid-filled cysts. Evidence indicates that cyst formation in ADPKD is produced by tubular cell proliferation, anomalies in the extracellular matrix, and a net transepithelial Cl^- secretion toward the cyst lumen (Ye and Grantham 1993; Li et al. 2004; Wallace et al. 1996; Mangos et al. 2010). Thus, a clear objective in ADPKD treatment is to slow or stop the growth of the cysts (Grantham et al. 2006a, b). Based on the known role of KCa3.1 in maintaining transepithelial Cl^- secretion, it was demonstrated that inhibition of KCa3.1 by TRAM-34 blocks Cl^- secretion and ameliorates cyst formation and enlargement *in vitro* (Albaqumi et al. 2008). A second kidney disease in which epithelial KCa3.1 has been implicated is diabetic nephropathy, which is a chronic complication that affects ~30% of patients with diabetes mellitus (Reddy et al. 2008). In this case, Huang et al. (2013) demonstrated that KCa3.1 protein expression was upregulated in proximal tubular cells from patients with diabetic nephropathy (Huang et al. 2013). These authors similarly observed an increase in KCa3.1 expression in mouse models of diabetes and found that administration of TRAM-34 significantly reversed the increased KCa3.1 expression in the proximal tubule. More recently, these authors further demonstrated that subcutaneous TRAM34 reduced albuminuria, decreased inflammatory markers, and reversed extracellular matrix deposition in kidneys from $eNOS^{-/-}$ mice administered streptozotocin to induce diabetes (Huang et al. 2018). Thus, inhibition of KCa3.1 may provide a novel approach in the treatment of diabetic nephropathy.

While KCa3.1 has also been shown to play a critical role in cells of the immune system and the targeting of this channel in T-cells has shown promise for the treatment of various diseases associated with lung, intestine and kidney, these studies have not been discussed in this chapter due to our emphasis on epithelial function. However, given the recent demonstration that KCa3.1 acts as a modifier gene for CF by altering TNF-α and mast cell numbers associated with the immune response (Philp et al. 2018), we would direct the reader to several excellent reviews on this subject (Wulff and Zhorov 2008; Wulff and Castle 2010; Huang et al. 2014).

22.10 Conclusions

Based on the above, it is clear that basolateral membrane-localized KCa3.1 plays a critical role in maintaining the electrochemical driving force for transepithelial ion and fluid transport across a wide array of epithelial tissues. More recent evidence also points to the crucial role of this channel in maintaining a hyperpolarized apical membrane potential, although additional studies are required to fully understand its regulation, trafficking, and function in this membrane. The study of any ion channel is dramatically augmented by having both activators and inhibitors such that its physiology can be explored and it is clear that the identification of both activators (1-EBIO, DCEBIO) and inhibitors (clotrimazole) prior to the cloning of KCa3.1 undoubtedly propelled our rapid understanding of the role of these channels in epithelial physiology. The identification of both higher specificity and affinity activators and inhibitors, as reviewed in Chap. 24 (this volume), has further fostered this advance. The elucidation of the structure of KCa3.1 will undoubtedly serve as a critical turning point in our understanding of the structure-function of this class of channels. In particular, the mechanism by which Ca^{2+} binding to CaM is translated through the S4–S5 linker to induce gating, as well as direct novel studies to identify the binding site for pharmacological openers of this channel that may ultimately lead to novel therapeutics are expected to be advanced by the recent structure. Furthermore, the recent demonstrations that mutations in KCa3.1 are genetically linked to hereditary xerocytosis and that KCa3.1 acts as a modifier gene for CF will surely spur a significant increase in studies designed to assess the therapeutic potential of this channel in a wide array of diseases over the next decade.

Acknowledgments The authors would like to thank David Hess for the trafficking model (Fig. 22.8). DCD has been supported by National Institutes of Health grants (HL092157), and KLH has been supported by an UORG grant and an OSMS Strategic Research award from the University of Otago.

References

Abdullaev IF, Rudkouskaya A, Mongin AA, Kuo Y-H (2010) Calcium-activated potassium channels BK and IK1 are functionally expressed in human gliomas but do not regulate cell proliferation. PLoS One 5:e12304. https://doi.org/10.1371/journal.pone.0012304

Adelman JP, Maylie J, Sah P (2012) Small-conductance Ca^{2+}-activated K^+ channels: form and function. Annu Rev Physiol 75:245–269

Aggarwal SK, MacKinnon R (1996) Contribution of the S4 segment to gating charge in the Shaker K^+ channel. Neuron 16:1169–1177

Akabas MH, Stauffer DA, Xu M, Karlin A (1992) Acetylcholine receptor channel structure probed in cysteine-substitution mutants. Science 258:307–310

Akabas MH, Kaufmann C, Archdeacon P, Karlin A (1994) Identification of acetylcholine receptor channel-lining residues in the entire M2 segment of the alpha subunit. Neuron 13:919–927

Albaqumi M, Srivastava S, Li Z, Zhdnova O, Wulff H, Itani O, Wallace DP, Skolnik EY (2008) KCa3.1 potassium channels are critical for cAMP-dependent chloride secretion and cyst growth in autosomal-dominant polycystic kidney disease. Kidney Int 74:740–749

Al-Hazza A, Linley JE, Aziz Q, Maclennan KA, Hunter M, Sandle GI (2012) Potential role of reduced basolateral potassium (IKCa3.1) channel expression in the pathogenesis of diarrhoea in ulcerative colitis. J Pathol 226:463–470

Allen D, Fakler B, Maylie J, Adelman JP (2007) Organization and regulation of small conductance Ca^{2+}-activated K^+ channel multiprotein complexes. J Neurosci 27:2369–2376

Almassy J, Won JH, Begenisich TB, Yule DI (2012) Apical Ca^{2+}-activated potassium channels in mouse parotid acinar cells. J Gen Physiol 139:121–133

Amasheh S, Barmeyer C, Koch CS, Tavalali S, Mankertz J, Epple HJ, Gehring MM, Florian P, Kroesen AJ, Zeitz M, Fromm M, Schulzke JD (2004) Cytokine-dependent transcriptional down regulation of epithelial sodium channel in ulcerative colitis. Gastroenterology 126:1711–1720
Anderson PA, Greenberg RM (2001) Phylogeny of ion channels: clues to structure and function. Comp Biochem Physiol B Biochem Mol Biol 129:17–28
Andolfo I, Russo R, Manna F, Shmukler BF, Gambale A, Vitiello G, De Rosa G, Brugnara C, Alper SL, Snyder LM, Iolascon A (2015) Novel Gardos channel mutations linked to dehydrated hereditary stomatocytosis (xerocytosis). Am J Hematol 90:921–926
Arthur GK, Duffy SM, Roach KM, Hirst RA, Shikotra A, Gaillard EA, Bradding P (2015) $K_{Ca}3.1$ K^+ channel expression and function in human bronchial epithelial cells. PLoS One 10(12): e0145259. https://doi.org/10.1371/journal.pone.0145259
Ashton N, Argent BE, Green R (1991) Characteristics of fluid secretion from isolated rat pancreatic ducts stimulated with secretin and bombesin. J Physiol Lond 435:533–546
Ataga KI, Smith WR, De Castro LM, Swerdlow P, Saunthararajah Y, Castro O, Vichinsky E, Kutlar A, Orringer EP, Rigdon GC, Stocker JW (2008) Efficacy and safety of the Gardos channel blocker, senicapoc (ICA-17043), in patients with sickle cell anemia. Blood 11:3991–3997
Ataga KI, Reid M, Ballas SK, Yasin Z, Bigelow C, St. James L, Smith WR, Galacteros F, Kutlar A, Hull JH, Stocker JW (2011) Improvements in haemolysis and indicators of erythrocyte survival do not correlate with acute vaso-occlusive crises in patients with sickle cell disease: a phase III randomized, pacebo-controlled, doouoble-blinded study of the Gardos channel blocker senicapoc (ICA-17043). Br J Haemotol 153:92–104
Babbey CM, Ahktar N, Wang E, Chen CC, Grant BD, Dunn KW (2006) Rab10 regulates membrane transport through early endosomes of polarized Madin-Darby canine kidney cells. Mol Biol Cell 17:3156–3175
Bae H, Lee HJ, Kim K, Kim JH, Kim T, Ko JH, Bang H, Lim I (2014) The stimulating effects of nitric oxide on intermediate conductance Ca^{2+}-activated K^+ channels in human dermal fibroblasts through PKG pathways but not PKA pathways. Chin J Physiol 57:137–151
Bailey MA, Grabe M, Devor DC (2010) Characterization of the PCMBS-dependent modification of KCa3.1 channel gating. J Gen Physiol 136:367–387
Balut CM, Gao Y, Murray SA, Thibodeau PH, Devor DC (2010a) ESCRT-dependent targeting of plasma membrane localized KCa3.1 to the lysosomes. Am J Physiol Cell Physiol 299:C1015–C1027
Balut CM, Gao Y, Luke C, Devor DC (2010b) Immunofluorescence-based assay to identify modulators of the number of plasma membrane KCa3.1 channels. Future Med Chem 2:707–713
Balut CM, Loch CM, Devor DC (2011) Role of ubiquitylation and USP8-dependent deubiquitylation in the endocytosis and lysosomal targeting of plasma membrane KCa3.1. FASEB J 25:3938–3948
Barmeyer C, Rahner C, Yang Y, Sigworth FJ, Binder HJ, Rajendran VM (2010) Cloning and identification of tissue-specific expression of KCNN4 splice variants in rat colon. Am J Physiol Cell Physiol 299:C251–C263
Basalingappa KM, Rajendran VM, Wonderlin WF (2011) Characteristics of Kcnn4 channels in the apical membranes of an intestinal cell line. Am J Physiol Gastrointest Liver Physiol 301:G905–G911
Begenisich T, Nakamoto T, Ovitt CE, Nehrke K, Brugnara C, Alper SL, Melvin JE (2004) Physiological roles of the intermediate conductance, Ca^{2+}-activated potassium channel KCNN4. J Biol Chem 46:47681–47687
Bernard K, Bogliolo O, Soriani J, Ehrenfeld J (1993) Modulation of chloride secretion by basolateral SK4-like channels in a human bronchial cell line. J Membr Biol 196:15–31
Bertuccio CA, Lee S-L, Wu G, Butterworth MB, Hamilton KL, Devor DC (2014) Anterograde trafficking of KCa3.1 in polarized epithelia is Rab1- and Rab8-dependent and recycling endosome-independent. PLoS One 9:e92013. https://doi.org/10.1371/journal.pone.0092013

Bi K, Tanaka Y, Coudronniere N, Sugie K, Hong S, van Stipdonk MJ, Altman A (2001) Antigen-induced translocation of PKC-theta to membrane rafts is required for T cell activation. Nat Immunol 2:556–563

Bildl W, Strassmaier T, Thurm H, Andersen J, Eble S, Oliver D, Knipper M, Mann M, Schulte U, Adelman JP, Fakler B (2004) Protein kinase CK2 is coassembled with small conductance Ca^{2+}-activated K^+ channels and regulates channel gating. Neuron 43:847–858

Bolanz KA, Hediger MA, Landowski CP (2008) The role of TRPV6 in breast carcinogenesis. Mol Cancer Ther 7:271–279

Bonito B, Sauter DRP, Schwab A, Djamgoz MBA, Novak I (2016) KCa3.1 (IK) modulates pancreatic cancer cell migrations, invasion and proliferation: anomalous effects on TRAM-34. Pflugers Arch 468:1865–1875

Browning JG, Hardcastle J, Hardcastle PT, Sanford PA (1977) The role of acetylcholine in the regulation of ion transport by rat colon mucosa. J Physiol 272:737–754

Brugnara C, de Franceschi L, Alper SL (1993) Inhibition of Ca^{2+}-dependent K^+ transport and cell dehydration in sickle erythrocytes by clotrimazole and other imidazole derivatives. J Clin Invest 92:520–526

Burckhardt B-C, Gogelein H (1992) Small and maxi K^+ channels in the basolateral membrane of isolated crypts from rat distal colon: single-channel and slow whole-cell recordings. Pflügers Arch 420:54–60

Cai X, Srivastava S, Sun Y, Li Z, Wu H, Zuvela-Jelaska L, Li J, Salamon RS, Backer JM, Skolnik EY (2011) Tripartite motif containing protein 27 negatively regulates CD4 T cells by ubiquitinating and inhibiting the class II PI3K-C2beta. Proc Natl Acad Sci USA 108:20072–20077

Cala PM (1983) Cell volume regulation by *Amphiuma* red blood cells. J Gen Physiol 82:761–784

Cancino J, Torrealba C, Soza A, Yuseff MI, Gravotta D, Henklein P, Rodriguez-Boulan E, Gonzalez A (2007) Antibody to AP1B adaptor blocks biosynthetic and recycling routes of basolateral proteins at recycling endosomes. Mol Biol Cell 18:4872–4884

Canessa M (1991) Red cell volume-related ion transport systems in hemaglobinopathies. Hematol/Oncol Clin N Am 5:495–516

Catalan MA, Pena-Munzenmayer G, Melvin JE (2014) Ca^{2+}-dependent K^+ channels in exocrine salivary glands. Cell Calcium 55:362–368

Chang D, Dawson DC (1988) Digitonin-permeabilized colonic cell layers: Demonstration of calcium-activated basolateral K^+ and Cl^- conductances. J Gen Physiol 92:281–306

Chen Y, Kuang D, Zhao X, Chen D, Wang X, Yang Q, Wan J, Zhu Y, Wang Y, Zhang S, Wang Y, Tang Q, Masuzawa M, Wang G, Duan Y (2016) miR-497-5p inhibits cell proliferation and invasion by targeting KCa3.1 in angiosarcoma. Oncotarget 7(36):58148–58160

Cowley EA, Linsdell P (2002) Characterization of basolateral K^+ channels underlying anion secretion in the human airway cell line Calu-3. J Physiol 538:747–757

Cuthbert AW (2001) Assessment of CFTR chloride channel openers in intact normal and cystic fibrosis murine epithelia. Br J Pharmacol 132:659–668

Cuthbert AW, Hickman ME, Thorn P, MacVinish LJ (1999) Activation of Ca^{2+}- and cAMP-sensitive K^+ channels in murine colonic epithelia by 1-ethyl-2-benzimidazolone. Am J Physiol Cell Physiol 277:C111–C120

D'Alessandro G, Limatola C, Catalano M (2018) Functional roles of the Ca^{2+}-activated K^+ channel, KCa3.1, in brain tumours. Curr Neuropharmacol 16:636–643

De Franceschi L, Saadane N, Trudel M, Alper SL, Brugnara C, Beuzard Y (1994) Treatment with oral clotrimazole blocks Ca^{2+}-activated K^+ transport and reverses erythrocyte dehydration in transgenic SAD mice. A model for therapy of sickle cell disease. J Clin Invest 93:1670–1676

DeCoursey TE, Chandy KG, Gupta S, Cahalan MD (1984) Voltage-gated K^+ channels in human T lymphocytes: a role in mitogenesis? Nature 307:465–468

Deutsch C (2002) Potassium channel ontogeny. Annu Rev Physiol 64:19–46

Devor DC, Duffey ME (1992) Carbachol induces K^+, Cl^- and nonselective cation conductances in T84 cells: a perforated patch-clamp study. Am J Physiol Cell Physiol 263:C780–C787

Devor DC, Frizzell RA (1993) Calcium-mediated agonists activate an inwardly rectified K^+ channel in colonic secretory cells. Am J Physiol Cell Physiol 265:C1271–C1280

Devor DC, Frizzell RA (1998) Modulation of K^+ channels by arachidonic acid in T84 cells. I. Inhibition of the Ca^{2+}-dependent K^+ channel. Am J Physiol Cell Physiol 274:C138–C148

Devor DC, Simasko SM, Duffey ME (1990) Carbachol induces oscillations of membrane potassium conductance in a colonic cell line, T84. Am J Physiol Cell Physiol 258:C318–C326

Devor DC, Ahmed Z, Duffey ME (1991) Cholinergic stiumlation produces oscillations of cytosolic Ca^{2+} in a secretory epithelial cell line, T84. Am J Physiol Cell Physiol 260:C598–C608

Devor DC, Sekar MC, Frizzell RA, Duffey ME (1993) Taurodeoxycholate activates K^+ and Cl^- conductances via an IP_3-mediated release of Ca^{2+} from intracellular stores. J Clin Invest 92:2173–2181

Devor DC, Singh AK, Frizzell RA, Bridges RJ (1996a) Modulation of Cl^- secretion by benzimidazolones. I. Direct activation of a Ca^{2+}-dependent K^+ channel. Am J Physiol Lung Cell Mol Physiol 271:L775–L784

Devor DC, Singh AK, Bridges RJ, Frizzell RA (1996b) Modulation of Cl^- secretion by benzimidazolones. II. Coordinate regulation of apical G_{Cl} and basolateral G_K. Am J Physiol Lung Cell Mol Physiol 271:L785–L795

Devor DC, Singh AK, Gerlach AC, Frizzell RA, Bridges RJ (1997) Inhibition of intestinal Cl^- secretion by clotrimazole: direct effect on basolateral membrane K^+ channels. Am J Physiol Cell Physiol 273:C531–C540

Devor DC, Singh AK, Lambert LC, DeLuca A, Frizzell RA, Bridges RJ (1999) Bicarbonate and chloride secretion in Calu-3 human airway epithelial cells. J Gen Physiol 113:743–760

Devor DC, Bridges RJ, Pilewski JM (2000) Pharmacological modulation of ion transport across wild-type and ΔF508 CFTR-expressing human bronchial epithelia. Am J Physiol Cell Physiol 279:C461–C479

Dharmsathaphorn K, Pandol SJ (1986) Mechanism of chloride secretion induced by carbachol in a colonic epithelial cell line. J Clin Invest 77:348–354

Dharmsathaphorn K, McRoberts JA, Mandel KG, Tisdale LD, Masui H (1984) A human colonic tumor cell line that maintains vectorial electrolyte transport. Am J Physiol Gastrointest Liver Physiol 246:G204–G208

Dharmsathaphorn K, Mandel KG, Masui H, McRoberts JA (1985) Vasoactive intestinal polypeptide-induced chloride secretion by a colonic epithelial cell line: direct participation of a basolaterally localized Na^+, K^+, Cl^- cotransport system. J Clin Invest 75:462–471

Dharmsathaphorn K, Cohn J, Beuerlein G (1989) Multiple calcium-mediated effector mechanisms regulate chloride secretory responses in T84 cells. Am J Physiol Cell Physiol 256:C1224–C1230

Dhennin-Duthille I, Gautier M, Faouzi M, Guilbert A, Brevet M, Vaudry D, Ahidouch A, Sevestre H, Ouadid-Ahidouch H (2011) High expression of transient receptor potential channels in human breast cancer epithelial cells and tissues: correlation with pathological parameters. Cell Physiol Biochem 28:813–822

Di L, Srivastava S, Zhdanova O, Sun Y, Li Z, Skolnik EY (2010) Nucleoside diphosphate kinase B knock-out mice have impaired activation of the K^+ channel KCa3.1, resulting in defective T cell activation. J Biol Chem 285:38765–38771

Dong C, Yang L, Zhang X, Gu H, Lam ML, Claycomb WC, Xia H, Wu G (2010) Rab8 interacts with distinct motifs in α_{2B}- and β_2-adrenergic receptors and differentially modulates their transport. J Biol Chem 285:20369–20380

Doyle DA, Cabral JM, Pfuetzner RA, Kuo A, Gulbis JM, Cohen SL, Chait BT, MacKinnon R (1998) The structure of the potassium channel: molecular basis of K^+ conduction and selectivity. Science 280:69–77

Du Y, Song W, Chen J, Chen H, Xuan Z, Zhao L, Chen J, Jin C, Zhou M, Tuo B, Zhao Y, Zheng S, Song P (2019) The potassium channel KCa3.1 promotes cell proliferation by activating SKP2 and metastasis through the EMT pathway in hepatocellular carcinoma. Int J Cancer 145:503–516

Dunn PM (1998) The action of blocking agents applied to the inner face of Ca^{2+}-activated K^+ channels from human erythrocytes. J Membr Biol 165:133–143

Dyrda A, Cytlak U, Ciuraszkiewicz A, Lipinska A, Cueff A, Bouyer G, Egee S, Bennekou P, Lew VL, Thomas SLY (2010) Local membrane deformations activate Ca^{2+}-dependent K^+ and anionic currents in intact human red blood cells. PLoS One 5:e9447. https://doi.org/10.1371/journal.pone.0009447

El Hiani Y, Ahidouch A, Lehen'kyi V, Hague F, Gouilleux F, Mentarverri KS, Lassoued K, Brûlé O-A (2009) Extracellular signal-regulated kinases 1 and 2 and TRPC1 channels are required for calcium-sensing receptor-stimulated MCF-7 breast cancer cell proliferation. Cell Physiol Biochem 23:335–346

Ellgaard L, Molinari M, Helenius A (1999) Setting the standards: quality control in the secretory pathway. Science 286:1882–1888

Falke JJ, Dernburg AF, Sternberg DA, Zalkin N, Milligan DL, Koshland DE Jr (1988) Structure of a bacterial sensory receptor. A site-directed sulfhydryl study. J Biol Chem 263:14850–14858

Fanger CM, Ghanshani S, Logsdon NJ, Rauer H, Kalman K, Zhou J, Beckingham K, Chandy KG, Cahalan MD, Aiyar J (1999) Calmodulin mediates calcium-dependent activation of the intermediate conductance KCa channel, IKCa1. J Biol Chem 274:5746–5754

Faouzi M, Hague F, Potier M, Ahidouch A, Sevestre H, Ouadid-Ahidouch H (2011) Down-regulation of Orai3 arrests cell-cycle progression and induces apoptosis in breast cancer cells but not in normal breast epithelial cells. J Cell Physiol 226:542–551

Faouzi M, Hague F, Geerts D, Ay A-S, Potier-Cartereau M, Ahidouch A, Ouadid-Ahidouch H (2016) Functional cooperation between KCa3.1 and TRPC1 channels in human breast cancer: Role in cell proliferation and patient prognosis. Oncotarget 7:36419–36435

Farquhar RE, Rodrigues E, Hamilton KL (2017) The role of the cytoskeleton and Myosin-Vc in the targeting of KCa3.1 to the basolateral membrane of polarized epithelial cells. Front Physiol 7 (article 639). https://doi.org/10.3389/fphys.2016.00639

Fermo E, Bogdanova A, Petkova-Kirova P, Zaninoni A, Marcello PA, Makhro A, Hanggi P, Hertz L, Danielczok J, Vercellati C, Mirra N, Zanella A, Cortelezzi A, Barcellini W, Kaestner L, Bianchi P (2017) 'Gardos channelopathy': a variant of hereditary stomatocytosis with complex molecular regulation. Sci Rep 7:1744. https://doi.org/10.1038/s41598-017-11591-w

Ferreira R, Wong R, Schlichter LC (2015) KCa3.1/IK1 channel regulation by cGMP-dependent protein kinase (PKG) via reactive oxygen species and CaMKII in microglia: an immune modulating feedback system? Front Immunol 6 (article 153). https://doi.org/10.3389/fimmu.2015.00153

Flores CA, Melvin JE, Figueroa CD, Sepulveda FV (2007) Abolition of Ca^{2+}-mediated intestinal anion secretion and increased stool dehydration in mice lacking the intermediate conductance Ca^{2+}-dependent K^+ channel KCNN4. J Physiol (London) 583:705–717

Florey HW, Wright RD, Jennings MA (1941) The secretions of the intestine. Physiol Rev 21:36–69

Fölsch H, Ohno H, Bonifacino JS, Mellman I (1999) A novel clathrin adaptor complex mediates basolateral targeting in polarized epithelial cells. Cell 99:189–198

Foskett JK, Gunter-Smith PJ, Melvin JE, Turner RJ (1989) Physiological localization of an agonist-sensitive pool of Ca^{2+} in parotid acinar cells. Proc Natl Acad Sci USA 86:161–117

Fuller SD, Bravo R, Simons K (1985) An enzymatic assay reveals that proteins destined for the apical or basolateral domains of an epithelial cell line share the same late Golgi compartments. EMBO J 4:297–307

Furness JB, Robbins HL, Selmer IS, Hunne B, Chen MX, Hicks GA, Moore S, Neylon CB (2003) Expression of intermediate conductance potassium channel immunoreactivity in neurons and epithelial cells of rat gastrointestinal tract. Cell Tissue Res 314:179–189

Futter CE, Gibson A, Allchin EH, Maxwell S, Ruddock LJ, Odorizzi G, Domingo D, Trowbridge IS, Hopkins CR (1998) In polarized MDCK cells basolateral vesicles arise from clathrin-gamma-adaptin-coated domains on endosomal tubules. J Cell Biol 141:611–623

Gao L, Yankaskas JR, Fuller CM, Sorscher EJ, Matalon S, Forman HJ, Venglarik CJ (2001) Chlorzoxazone or 1-EBIO increases Na^+ absorption across cystic fibrosis airway epithelial cells. Am J Physiol Lung Cell Mol Physiol 281:L1123–L1129

Gao Y, Chotoo CK, Balut CM, Sun F, Bailey MA, Devor DC (2008) Role of S3 and S4 transmembrane domain charged amino acids in channel biogenesis and gating of KCa2.3 and KCa3.1. J Biol Chem 283:9049–9059

Gao Y, Balut CM, Bailey MA, Patino-Lopez G, Shaw S, Devor DC (2010) Recycling of the Ca^{2+}-activated K^+ channel, KCa2.3, is dependent upon RME-1, Rab35/EPI64C, and an N-terminal domain. J Biol Chem 285:17938–17953

Gardos G (1958) The function of calcium in the potassium permeability of human erythrocytes. Biochim Biophys Acta 30:653–654

Garneau L, Klein H, Banderali U, Longpré-Lauzon A, Parent L, Sauvé R (2009) Hydrophobic interactions as key determinants to the KCa3.1 channel closed configuration. An analysis of KCa3.1 mutants constitutively active in zero Ca^{2+}. J Biol Chem 284:389–403

Gerlach AC, Gangopadhyay NN, Devor DC (2000) Kinase-dependent regulation of the intermediate conductance, calcium-dependent potassium channel, hIK1. J Biol Chem 275:585–598

Gerlach AC, Syme CA, Giltinan L, Adelman JP, Devor DC (2001) ATP-dependent activation of the intermediate conductance, Ca^{2+}-activated K^+ channel, hIK1, is conferred by a C-terminal domain. J Biol Chem 276:10963–10970

Ghiani CA, Eisen A, Yuan X, DePinho RA, McBain CJ, Gallo V (1999) Neurotransmitter receptor activation triggers $p27^{Kip1}$ and $p21^{CIP1}$ accumulation and G1 cell cycle arrest in oligodendrocyte progenitors. Development 126:1077–1090

Girault A, Brochiero E (2014) Evidence of K^+ channel function in epithelial cell migration, proliferation, and repair. Am J Physiol Cell Physiol 306:C307–C319

Glogowska E, Lezon-Geyda K, Maksimova Y, Schulz VP, Gallagher PG (2015) Mutations in the Gardos channel (KCNN4) are associated with hereditary xerocytosis. Blood 126:1281–1284

Grantham JJ, Chapman AB, Torres VE (2006a) Volume progression in autosomal dominant polycystic kidney disease: the major factor determining clinical outcomes. Clin J Am Soc Nephrol 1:148–157

Grantham JJ, Torres VE, Chapman AB, Guay-Woodford LM, Bae KT, King BF Jr, Wetzel LH, Baumgarten DA, Kenney PJ, Harris PC, Klahr S, Bennet WM, Hirscvhman GN, Meyers CM, Zhang X, Zhu F, Miller JP, Investigators CRISP (2006b) Volume progression in polycystic kidney disease. N Engl J Med 354:2122–2130

Gray MA, Pollard CE, Harris A, Coleman L, Greenwell JR, Argent BE (1990) Anion selectivity and block of the small conductance chloride channel on pancreatic duct cells. Am J Physiol Cell Physiol 259:C752–C761

Greger R, Schlatter E (1984a) Mechanism of NaCl secretion in the rectal gland of spiny dogfish (*Squalus acanthias*) I. Experiments in isolated *in vitro* perfused rectal gland tubules. Pflügers Arch 402:63–75

Greger R, Schlatter E (1984b) Mechanism of NaCl secretion in the rectal gland of spiny dogfish (*Squalus acanthias*) II. Effects of inhibitors. Pflügers Arch 402:364–375

Greger R, Schlatter E, Wang F, Forrest JN Jr (1984) Mechanism of NaCl secretion in rectal gland tubules of spiny dogfish (*Squalus acanthias*) III. Effects of stimulation of secretion by cyclic AMP. Pflügers Arch 402:376–384

Greig ER, Boot-Handford RP, Mani V, Sandle GI (2004) Decreased expression of apical Na^+ channels and basolateral Na^+,K^+-ATPase in ulcerative colitis. J Pathol 204:84–92

Grygorczyk R, Schwarz W (1985) Ca^{2+}-activated K^+ permeability in human erythrocytes: modulation of single-channel events. Eur Biophys J 12:57–65

Gutman GA, Chandy KG, Adelman JP, Aiyar J, Bayliss DA, Clapham DE, Covarriubias M, Desir GV, Furuichi K, Ganetzky B, Garcia ML, Grissmer S, Jan LY, Karschin A, Kim D,

Kuperschmidt S, Kurachi Y, Lazdunski M, Lesage F, Lester HA, McKinnon D, Nichols CG, O'Kelly I, Robbins J, Robertson GA, Rudy B, Sanguinetti M, Seino S, Stuehmer W, Tamkun MM, Vandenberg CA, Wei A, Wulff H, Wymore RS, International Union of Pharmacology (2003) International Union of Pharmacology. XLI. Compendium of voltage-gated ion channels: potassium channels. Pharmacol Rev 55:583–586

Hamilton KL, Meads L, Butt AG (1999) 1-EBIO stimulates Cl^- secretion by activating a basolateral K^+ channel in the mouse jejunum. Pflügers Arch 439:158–166

Hamilton KL, Syme CA, Devor DC (2003) Molecular localization of the inhibitory arachidonic acid binding site to the pore of hIK1. J Biol Chem 278:16690–16697

Hammadi M, Chopin V, Matifat F, Dhennin-Duthille I, Chasseraud M, Sevestre H, Ouadid-Ahidouch H (2012) Human ether à-gogo K^+ channel 1 (hEag1) regulates MDA-MB-231 breast cancer cell migration through Orai1-dependent calcium entry. J Cell Physiol 227:3837–3384

Hardcastle PT, Eggenton J (1973) The effect of acetylcholine on the electrical activity of intestinal epithelial cells. Biochim Biophys Acta 298:95–100

Hawker PC, McKay JS, Turnberg LA (1980) Electrolyte transport across colonic mucosa from patients with inflammatory bowel disease. Gastroenterology 79:508–511

Hayashi M, Wang J, Hede SE, Novak I (2012) An intermediate-conductance Ca^{2+}-activated K^+ channel is important for secretion in pancreatic duct cells. Am J Physiol Cell Physiol 303:C151–C159

Hazelton B, Mitchell B, Tupper J (1979) Calcium, magnesium, and growth control in the WI-38 human fibroblast cell. J Cell Biol 83:487–498

Hede SE, Amstrup J, Klaerke DA, Novak I (2005) $P2Y_2$ and $P2Y_4$ receptors regulate pancreatic Ca^{2+}-activated K^+ channel differently. Pflügers Arch 450:429–436

Hirschberg B, Maylie J, Adelman JP, Marrion NV (1998) Gating of recombinant small-conductance Ca-activated K^+ channels by calcium. J Gen Physiol 111:565–581

Hite RK, Tao X, MacKinnon R (2017) Structural basis for gating the high-conductance Ca^{2+}-activated K^+ channel. Nature 541:52–57

Holmgren M, Liu Y, Xu Y, Yellen G (1996) On the use of thiol-modifying agents to determine channel topology. Neuropharmacology 35:797–804

Hoque KM, Woodward OM, van Rossum DB, Zachos NC, Chen L, Leung GP, Guggino WB, Guggino SE, Tse CM (2010) Epac1 mediates protein kinase A-independent mechanism of forskolin-activated intestinal chloride secretion. J Gen Physiol 135:43–58

Howarth M, Takao K, Hayashi Y, Ting AY (2005) Targeting quantum dots to surface proteins in living cells with biotin ligase. Proc Natl Acad Sci USA 102:7583–7588

Huang Y, Rane SG (1994) Potassium channel induction by the Ras/Raf signal transduction cascade. J Biol Chem 269:31183–31189

Huang C, Shen S, Ma Q, Chen J, Gill A, Pollock CA, Chen X-M (2013) Blockade of KCa3.1 ameliorates renal fibrosis through the TGF-β1/Smad pathway in diabetic mice. Diabetes 62:2923–2934

Huang C, Pollock CA, Chen XM (2014) Role of potassium channel KCa3.1 in diabetic nephropathy. Clin Sci (Lond) 127:423–433

Huang C, Zhang L, Shi Y, Yi H, Zhao Y, Chen J, Pollock CA, Chen X-M (2018) The KCa3.1 blocker TRAM34 reverses renal damage in a mouse model of established diabetic nephropathy. PLoS One 13(2):e0192800. https://doi.org/10.1371/journal.pone.0192800

Hubel KA (1976) Intestinal ion transport: effect of norepinephrine, pilocarpine, and atropine. Am J Phys 231:252–257

Hubel KA (1977) Effects of bethanacol on intestinal ion transport in the rat. Proc Soc Exp Biol Med 154:41–44

Isaacs PET, Corbett CL, Riley AK, Hawker PC, Turnberg LA (1976) *In vitro* behavior of human intestinal mucosa. J Clin Invest 58:535–542

Ishiguro H, Steward MC, Naruse S, Ko SBH, Goto H, Case RM, Kondo T, Yamamoto A (2009) CFTR functions as a bicarbonate channel in pancreatic duct cells. J Gen Physiol 133:315–326

Ishii TM, Silvia C, Hirschberg B, Bond CT, Adelman JP (1997) A human intermediate conductance calcium-activate potassium channel. Proc Natl Acad Sci USA 94:11651–11656

Jacobowitz D (1965) Histochemical studies of the autonomic innervation of the gut. J Pharmacol Exp Ther 149:358–364

Jager H, Dreker T, Buck A, Giehl K, Gress T, Grissmer S (2004) Blockage of intermediate-conductance Ca^{2+}activated K^+ channels inhibit human pancreatic cancer cell growth in vitro. Mol Pharmacol 65:630–638

Jensen BS, Strobaek D, Christophersen P, Jorgensen TD, Hansen C, Silahtaroglu A, Olesen SP, Ahring PK (1998) Characterization of the cloned human intermediate-conductance Ca^{2+}-activated K^+ channel. Am J Physiol Cell Physiol 275:C848–C856

Jensen BS, Strobaek D, Olesen SP, Christophersen P (2001) The Ca^{2+}-activated K^+ channel of intermediate conductance: a molecular target for novel treatments? Curr Drug Targets 2:401–422

Ji T, Corbalan-Garcia S, Hubbard SR (2018) Crystal structure of the C-terminal four-helix bundle of the potassium channel KCa3.1. PLoS One 13(6):e0199942. https://doi.org/10.1371/journal.pone.0199942

Joiner WJ, Wang L-Y, Tang MD, Kaczmarek LK (1997) hSK4, a member of a novel subfamily of calcium-activated potassium channels. Proc Natl Acad Sci USA 94:11013–11018

Joiner WJ, Khanna R, Schlichter LC, Kaczmarek LK (2001) Calmodulin regulates assembly and trafficking of SK4/IK1 Ca^{2+}-activated K^+ channels. J Biol Chem 276:37980–37985

Joiner WJ, Basavappa S, Vidyasagar S, Nehrke K, Krishnan S, Binder HJ, Boulpaep EL, Rajendran VM (2003) Active K^+ secretion through multiple K_{Ca}-type channels and regulation by IK_{Ca} channels in rat proximal colon. Am J Physiol Liver Physiol 285:G185–G196

Jones HM, Hamilton KL, Papworth GD, Syme CA, Watkins SC, Bradbury NA, Devor DC (2004) Role of the NH_2 terminus in the assembly and trafficking of the intermediate conductance Ca^{2+}-activated K^+ channel hIK1. J Biol Chem 279:15531–15540

Jones HM, Hamilton KL, Devor DC (2005) Role of an S4-S5 linker lysine in the trafficking of the Ca^{2+}-activated K^+ channels IK1 and SK3. J Biol Chem 280:37257–37265

Jones HM, Bailey MA, Baty CJ, MacGregor GG, Syme CA, Hamilton KL, Devor DC (2007) An NH_2-terminal multi-basic RKR motif is required for the ATP-dependent regulation of hIK1. Channels (Austin) 2:80–91

Jung SR, Kim K, Hille B, Nguyen TD, Koh DS (2006) Pattern of Ca^{2+} increase determines the type of secretory mechanism activated in dog pancreatic duct epithelial cells. J Physiol 576:163–178

Karlin A, Akabas MH (1998) Substituted-cysteine accessibility method. Methods Enzymol 293:123–145

Keen JE, Khawaled R, Farrens DL, Neelands T, Rivard A, Bond CT, Janowsky A, Fakler B, Adelman JP, Maylie J (1999) Domains responsible for constitutive and Ca^{2+}-dependent interactions between calmodulin and small conductance Ca^{2+}-activated potassium channels. J Neurosci 19:8830–8838

Kerem E, Corey M, Kerem B, Durie P, Tsui LC, Levison H (1989) Clinical and genetic comparisons of patients with cystic fibrosis, with or without meconium ileus. J Pediatr 114:767–773

Khanna R, Chang MC, Joiner WJ, Kaczmarek LK, Schlichter LC (1999) hSK4/hIK1, a calmodulin-binding KCa channel in human T lymphocytes. Roles in proliferation and volume regulation. J Biol Chem 274:14838–14849

Klein H, Garneau L, Banderali U, Simoes M, Parent L, Sauvé R (2007) Structural determinants of the closed KCa3.1 channel pore in relation to channel gating: results from a substituted cysteine accessibility analysis. J Gen Physiol 129:299–315

Klein H, Garneau L, Trinh NT, Prive A, Dionne F, Goupil E, Thuringer D, Parent L, Brochiero E, Sauve R (2009) Inhibition of the KCa3.1 channels by AMP-activated protein kinase in human airway epithelial cells. Am J Physiol Cell Physiol 296:C285–C295

Kobe B, Deisenhofer J (1994) The leucine-rich repeat: a versatile binding motif. Trends Biochem Sci 19:415–421

Kohler R, Hirschberg B, Bond CT, Kinzie JM, Marrion NV, Maylie J, Adelman JP (1996) Small-conductance, calcium-activated potassium channels from mammlian brian. Science 273:1709–1714

Kuner T, Wollmuth LP, Karlin A, Seeburg PH, Sakmann B (1996) Structure of the NMDA receptor channel M2 segment inferred from the accessibility of substituted cysteines. Neuron 17:343–352

Lai W, Chen S, Wu H, Guan Y, Liu L, Zeng Y, Zhao H, Jiang J, Chu Z (2011) PRL-3 promotes the proliferation of LoVo cells via the upregulation of KCNN4 channels. Oncol Rep 26:909–917

Lallet-Daher H, Roudbaraki M, Bavencoffe A, Mariot P, Gackière F, Bidaux G, Urbain R, Gosset P, Delcourt P, Fleurisse L, Slomianny C, Dewailly E, Mauroy B, Bonnal JL, Skryma R, Prevarskaya N (2009) Intermediate-conductance Ca^{2+}-activated K^+ channels (IKCa1) regulate human prostate cancer cell proliferation through a close control of calcium entry. Oncogene 28:1792–1806

Ledoux J, Bonex AD, Nelson MT (2008) Ca^{2+}-activated K^+ channels in murine endothelial cells: block by intracellular calcium and magnesium. J Gen Physiol 131:125–135

Lee C-H, MacKinnon R (2018) Activation mechanism of a human SK-calmodulin channel complex elucidated by cryo-EM structures. Science 360:508–513

Lee WS, Ngo-Anh TJ, Bruening-Wright A, Maylie J, Adelman JP (2003) Small conductance Ca^{2+}-activated K^+ channels and calmodulin: cell surface expression and gating. J Biol Chem 278:25940–25946

Lee S-Y, Lee A, Chen J, MacKinnon R (2005) Structure of the KvAP voltage-dependent K^+ channel and its dependence on the lipid domain. Proc Natl Acad Sci USA 102:15441–15446

Lee BS-L, Devor DC, Hamilton KL (2017) Modulation of retrograde trafficking of KCa3.1 in a polarized epithelium. Front Physiol 8(489). https://doi.org/10.3389/fphys.2017.00489

Leinders T, van Kleef RG, Vijverberg HP (1992) Single Ca^{2+}-activated K^+ channels in human erythrocytes: Ca^{2+} dependence of opening frequency but not of open lifetimes. Biochim Biophys Acta 1112:67–74

Li H, Findlay IA, Sheppard DN (2004) The relationship between cell proliferation, Cl^- secretion, and renal cyst growth: a study using CFTR inhibitors. Kidney Int 66:1926–1938

Li H, Yao J, Tong X, Guo Z, Wu Y, Sun L, Pan N, Wu H, Xu T, Ding J (2007) Interaction sites between the *Slo1* pore and the NH_2 terminus of the β_2 subunit, probed with a three-residue sensor. J Biol Chem 282:17720–17728

Li W, Halling DB, Hall AW, Aldrich RW (2009) EF hands at the N-lobe of calmodulin are required for both SK channel gating and stable SK-calmodulin interaction. J Gen Physiol 134:281–293

Linsdell P, Tabcharani JA, Hanrahan JW (1997) Multi-ion mechanism for ion permeation and block in the cystic fibrosis transmembrane conductance regulator chloride channel. J Gen Physiol 110:365–377

Liu Y, Zhao L, Ma W, Cao X, Chen H, Feng D, Liang J, Yin K, Jiang X (2015) The blockage of KCa3.1 channel inhibited proliferation, migration and promoted apoptosis of human hepatocellular carcinoma cells. J Cancer 6:643–651

Liu L, Zhan P, Nie D, Fan L, Lin H, Gao L, Mao X (2017) Intermediate-conductance calcium-activated potassium channel (IKCa1) is upregulated and promotes cell proliferation in cervical cancer. Med Sci Monit Basic Res 23:45–57

Logsdon NJ, Kang J, Togo JA, Christian EP, Aiyar J (1997) A novel gene, hKCa4, encodes the calcium-activated potassium channel in human T lymphocytes. J Biol Chem 272:32723–32726

Long SB, Campbell EB, MacKinnon R (2005) Crystal structure of a mammalian voltage-dependent Shaker family K^+ channel. Science 309:897–903

Loo DDF, Kaunitz JD (1989) Ca^{2+} and cAMP activate K^+ channels in the basolateral membrane of crypt cells isolated from rabbit distal colon. J Membr Biol 110:19–28

Lu Y, Xiong X, Helm A, Kimani K, Bragin A, Skach WR (1998) Co- and posttranslational translocation mechanisms direct cystic fibrosis transmembrane conductance regulator N terminus transmembrane assembly. J Biol Chem 273:568–576

MacVinish LJ, Keogh J, Cuthbert AW (2001) EBIO, an agent causing maintained epithelial chloride secretion by co-ordinate actions at both apical and basolateral membranes. Pflügers Arch 443:S127–S131
Madara JL, Stafford J, Dharmsathaphorn K, Carlson S (1987) Structural analysis of a human intestinal epithelial cell line. Gastroenterology 92:1133–1145
Mall M, Gonska T, Thomas J, Schreiber R, Seyedwitz HH, Kuehr J, Brandis M, Kunzelmann K (2003) Modulation of Ca^{2+}-activated Cl^- secretion by basolateral K^+ channels in human normal and cystic fibrosis airway epithelia. Pediatr Res 53:608–618
Mandell KG, Dharmsathaphorn K, McRoberts JA (1986) Characterization of a cAMP-activated Cl^- transport pathway in the apical membrane of a human colonic epithelial cell line. J Biol Chem 261:704–712
Mangos S, Lam PY, Zhao A, Liu Y, Mudumana S, Vasilyev A, Liu A, Drummond IA (2010) The ADPKD genes pkd1a/b and pkd2 regulate extracellular matrix formation. Dis Model Mech 3:354–365
Martinez O, Goud B (1998) Rab proteins. Biochim Biophys Acta 1404:101–112
Maruyama Y, Petersen OH (1984) Control of K^+ conductance by cholecystokinin and Ca^{2+} in single pancreatic acinar cells studied by the patch-clamp technique. J Membr Biol 79:293–300
Maruyama Y, Gallacher DV, Petersen OH (1983a) Voltage and Ca^{2+}-activated K^+ channel in basolateral acinar cell membranes of mammalian salivary glands. Nature 302:827–829
Maruyama Y, Petersen OH, Flanagan P, Pearson GT (1983b) Quantification of Ca^{2+}-activated K^+ channels under hormonal control in pig pancreas acinar cells. Nature 305:228–232
McCann JD, Welsh MJ (1990) Basolateral K^+ channels in airway epithelia II. Role in Cl^- secretion and evidence for two types of K^+ channel. Am J Physiol Lung Cell Mol Physiol 258:L343–L348
McCann JD, Matsuda J, Garcia M, Kaczorowski G, Welsh MJ (1990) Basolateral K^+ channels in airway epithelia I. Regulation by Ca^{2+} and block by charybdotoxin. Am J Physiol Lung Cell Mol Physiol 258:L334–L342
Millership J, Devor DC, Hamilton KL, Balut CM, Bruce JIE, Fearson IM (2011) Calcium activated K^+ channels increase cell proliferation independent of K^+ conductance. Am J Physiol Cell Physiol 300:C792–C802
Mohr CJ, Steudel FA, Gross D, Ruth P, Lo W-L, Hoppe R, Schroth W, Brauch H, Huber SM, Lukowski R (2019) Cancer-associated intermediate conductance Ca^{2+}-activated K^+ channel $K_{Ca}3.1$. Cancers (Basel) 11:109. https://doi.org/10.3390/cancers11010109
Morales P, Garneau L, Klein H, Lavoie MF, Parent L, Sauvé R (2013) Contribution of the KCa3.1 channel-calmodulin interactions to the regulation of the KCa3.1 gating process. J Gen Physiol 142:37–60
Morris AP, Gallacher DV, Lee JAC (1986) A large conductance, voltage-and calcium-activated K^+ channel in the basolateral membrane of rat enterocytes. FEBS Lett 206:87–92
Muth TR, Caplan MJ (2003) Transport protein trafficking in polarized cells. Annu Rev Cell Dev Biol 19:333–366
Myers MP, Khanna R, Lee EJ, Papazian DM (2004) Voltage sensor mutations differentially target misfolded K^+ channel subunits to proteasomal and non-proteasomal disposal pathways. FEBS Lett 568:110–116
Nakamoto T, Srivastava A, Romanenko VG, Ovitt CE, Perez-Cornejo P, Arreola J, Begenisich T, Melvin JE (2007) Functional and molecular characterization of the fluid secretion mechanism in human parotid acinar cells. Am J Physiol Regul Integr Comp Physiol 292:R2380–R2390
Nehrke K, Quinn CC, Begenisich T (2003) Molecular identification of Ca^{2+}-activated K^+ channels in parotid acinar cells. Am J Physiol Cell Physiol 284:C535–C546
Nguyen TD, Moody MW (1998) Calcium-activated potassium conductances on cultured nontransformed dog pancreatic duct epithelial cells. Pancreas 17:348–358
Nicolaou SA, Neumeier L, Peng Y, Devor DC, Conforti L (2007) The Ca^{2+}-activated K^+ channel KCa3.1 compartmentalizes in the immunological synapse of human T lymphocytes. Am J Physiol Cell Physiol 292:C1431–C1439

Nilius B, Wohlrab W (1992) Potassium channels and regulation of proliferation of human melanoma cells. J Physiol 445:537–548
O'Hara B, de la Rosa DA, Rajendra VM (2014) Multiple mineralocorticoid response elements localized in different introns regulate intermediate conductance K^+ (KCNN4) channel expression in the rat distal colon. PLoS One 9:e98695. https://doi.org/10.1371/journal.pone.0098695
Ohno H, Aguilar RC, Yeh D, Taura D, Saito T (1998) The medium subunits of adaptor complexes recognize distinct but overlapping sets of tyrosine-based sorting signals. J Biol Chem 273:25915–25921
Ohya S, Niwa S, Yanagi A, Fukuyo Y, Yamamura H, Imaizumi Y (2011) Involvement of dominant-negative spliced variants of the intermediate conductance Ca^{2+}-activated K^+ channel, $K_{Ca}3.1$, in immune function of lymphoid cells. J Biol Chem 286:16940–16952
Orzech E, Cohen S, Weiss A, Aroeti B (2000) Interactions between the exocytic and endocytic pathways in polarized Madin–Darby canine kidney cells. J Biol Chem 275:15207–15219
Ouadid-Ahidouch H, Ahidouch A (2013) K^+ channels and cell cycle progression in tumor cells. Front Physiol 4:220. https://doi.org/10.3389/fphys.2013.00220
Ouadid-Ahidouch H, Roudbaraki M, Delcourt P, Ahidouch A, Joury N, Prevarskaya N (2004) Functional and molecular identification of intermediate-conductance Ca^{2+}-activated K^+ channels in breast cancer cells: association with cell cycle progression. Am J Physiol Cell Physiol 287:C125–C134
Panyi G, Beeton C, Felipe A (2014) Ion channels and anti-cancer immunity. Phil Trans R Soc B 369:20130106. https://doi.org/10.1098/rstb.2013.0106
Papazian DM (1999) Potassium channels: some assembly required. Neuron 23:7–10
Papazian DM, Shao XM, Seoh S-A, Mock AF, Huang Y, Wainstock DH (1995) Electrostatic interactions of S4 voltage sensor in Shaker K^+ channel. Neuron 14:1293–1301
Parihar AS, Coghlan MJ, Gopalakrishnan M, Shieh C-C (2003) Effects of intermediate-conductance Ca^{2+}-activated K^+ channel modulators on human prostate cancer cell proliferation. Eur J Pharmacol 417:157–164
Pedersen KA, Schroder RL, Skaaning-Jensen B, Strobeak D, Olesen SP, Christophersen P (1999) Activation of the human intermediate-conductance Ca^{2+}-activated K^+ channel by 1-ethyl-2-benzimidazolinone is strongly Ca^{2+}-dependent. Biochim Biophys Acta 1420:231–240
Philp AR, Riquelme TT, Millar-Büchner P, González R, Sepúlevda FV, Cid LP, Flores CA (2018) *Kcnn4* is a modifier gene of intestinal cystic fibrosis preventing lethality in the *Cftr*-F508del mouse. Sci Rep 8:9320. https://doi.org/10.1038/s41598-018-27465-3
Preston D, Simpson S, Halm D, Hochstetler A, Schwerk C, Schroten H, Blazer-Yost BL (2018) Activation of TRPV4 stimulates ion flux in a porcine choroid plexus cell line. Am J Physiol Cell Physiol 315:C357–C366
Rapetti-Mauss R, Lacoste C, Picard V, Guitton C, Lombard E, Loosveld M, Nivaggioni V, Dasliva N, Salgado D, Desvignes JP, Beroud C, Viout P, Bernard M, Soriani O, Vinti H, Lacroze V, Feneant-Thibault M, Thuret I, Guizouarn H, Badens C (2015) A mutation in the Gardos channel is associated with hereditary xerocytosis. Blood 126:1273–1280
Rapetti-Mauss R, Soriani O, Vinti H, Badens C, Guizouarn H (2016) Senicapoc: a potent candidate for the treatment of a subset of hereditary xerocytosis caused by mutations in the Gardos channel. Haematologica 101:e431–e435
Reddy GR, Kotlyarevska K, Ransom RF, Menon RK (2008) The podocyte and diabetes mellitus: is the podocyte the key to the origins of diabetic nephropathy? Curr Opin Nephrol Hypertens 17:32–36
Reuss L, Reinach P, Weinman SA, Grady TP (1983) Intracellular ion activities and Cl^- transport mechanisms in bullfrog corneal epithelium. Am J Physiol Cell Physiol 244:C336–C347
Rimele TJ, O'Dorisio MS, Gaginella TS (1981) Evidence for muscarinic receptors on rat colonic epithelial cells: binding of [^{3}H]quinuclidinyl benzilate. J Pharmacol Exp Ther 218:426–434
Rivera A, Vandorpe DH, Shmukler BE, Gallagher DR, Fikry CC, Kuypers FA, Brugnara C, Snyder LM, Alper SL (2017) Erythrocytes from hereditary xerocytosis patients heterozygous for

KCNN4 V282M exhibit increased spontaneous Gardos channel-like activity inhibited by senicapoc. Am J Hematol 92:E108–E110. https://doi.org/10.1002/ajh.24716
Rivera A, Vandorpe DH, Shmukler BE, Andolfi I, Lolascon A, Archer NM, Shabani E, Auerbach M, Hamerschlak N, Morton J, Wohlgemuth JG, Brugnara C, Snyder LM, Alper SL (2019) Erythrocyte ion content and dehydration modulate maximal Gardos channel activity in KCNN4 V282M/+ hereditary xerocytosis (HXD) red cells. Am J Physiol Cell Physiol. https://doi.org/10.1152/ajpcell.00074.2019
Roberts DD, Lewis SD, Ballou DP, Olson ST, Shafer JA (1986) Reactivity of small thiolate anions and cysteine-25 in papain toward methyl methanethiosulfonate. Biochemistry 25:5595–5601
Robles-Martinez L, Garay E, Martel-Gallegos MG, Cisneros-Mejorado A, Perez-Montiel D, Lara A, Arellano RO (2017) $K_{Ca}3.1$ activation via $P2y_2$ purinergic receptors promotes human ovarian cancer cell (Skov-3) migration. Scientific Reports 7:4340. https://doi.org/10.1038/s41598-017-04292-6
Romanenko V, Nakamoto T, Srivastava A, Melvin JE, Begenisich T (2006) Molecular identification and molecular roles of parotid acinar maxi-K channels. J Physiol 281:27964–27972
Romanenko VG, Nakamoto T, Srivastava A, Begenisich T, Melvin JE (2007) Regulation of membrane potential and fluid secretion by Ca^{2+}-activated K^+ channels in mouse submandibular glands. J Physiol 581:801–817
Romanenko VG, Roser KS, Melvin JE, Begenisich T (2009) The role of cholesterol and the cytoskeleton in the interaction between IK1 and maxi-K channels. Am J Physiol Cell Physiol 296:C878–C888
Roth EK, Hirtz S, Duerr J, Wenning D, Eichler I, Seydwitz HH, Amaral MD, Mall MA (2011) The K^+ channel opener 1-EBIO potentiates residual function of mutant CFTR in rectal biopsies from cystic fibrosis patients. PLoS One 6:e24445. https://doi.org/10.1371/journal.pone.0024445
Rozmahel R, Wilschanski M, Matin A, Plyte S, Oliver M, Auerbach W, Moore A, Forstner J, Durie P, Nadeau J, Bear C, Tsui LC (1996) Modulation of disease severity in cystic fibrosis transmembrane conducctance regulator deficient mice by a secondary genetic factor. Nat Genet 12:280–287
Rufo PA, Jiang L, Moe SJ, Brugnara C, Alper SL, Lencer WI (1996) The antifungal antibiotic, clotrimazole, inhibits Cl^- secretion by polarized monolayers of human colonic epithelial cells. J Clin Invest 98:2066–2075
Rufo PA, Merlin D, Riegler M, Ferguson-Maltzman MH, Dickinson BL, Brugnara C, Alper SL, Lencer WI (1997) The antifungal antibiotic, clotrimazole, inhibits chloride secretion by human intestinal T84 cells via blockade of distinct basolateral K^+ conductances. Demonstration of efficacy in intact rabbit colon and in an in vivo mouse model of cholera. J Clin Invest 100:3111–3120
Sadlish H, Skach WR (2004) Biogenesis of CFTR and other polytopic membrane proteins: new roles for the ribosome-translocon complex. J Membr Biol 202:115–126
Sandle GI (1998) Salt and water absorption in the human colon: a modern appraisal. Gut 43:294–299
Sandle GI, Wills NK, Alles W, Binder HJ (1986) Electrophysiology of the human colon: evidence of segmental heterogeneity. Gut 27:999–1005
Sandle GI, Higgs N, Crowe P, Marsh MM, Venkatesan S, Peters TJ (1990) Cellular basis for defective electrolyte transport in inflamed human colon. Gastroenterology 99:97–105
Sato Y, Sakaguchi M, Goshima S, Nakamura T, Uozumi N (2002) Integration of Shaker-type K^+ channel, KAT1, into the endoplasmic reticulum membrane: synergistic insertion of voltage-sensing segments, S3-S4, and independent insertion of pore-forming segments, S5-P-S6. Proc Natl Acad Sci USA 99:60–65
Sato Y, Sakaguchi M, Goshima S, Nakamura T, Uozumi N (2003) Molecular dissection of the contribution of negatively and positively charged residues in S2, S3, and S4 to the final membrane topology of the voltage sensor in the K^+ channel, KAT1. J Biol Chem 278:13227–13234

Schreiber R (2005) Ca^{2+} signaling, intracellular pH and cell volume in cell proliferation. J Membr Biol 205:129–137
Schumacher MA, Rivard AF, Bachinger HP, Adelman JP (2001) Structure of the gating domain of a Ca^{2+}-activated K^+ channel complexed with Ca^{2+}/calmodulin. Nature 410:1120–1124
Schumacher MA, Crum M, Miller MC (2004) Crystal structures of apocalmodulin and an apocalmodulin/SK potassium channel gating domain complex. Structure 12:849–860
Schwab A, Wulf A, Schulz C, Kessler W, Nechyporuk-Zloy V, Romer M, Reinhardt J, Weinhold D, Dieterich P, Stock C, Hebert SC (2006) Subcellular distribution of calcium-sensitive potassium channels (IK1) in migrating cells. J Cell Physiol 206:86–94
Schwab A, Nechyporuk-Zloy V, Gassner B, Schulz C, Kessler W, Mally S, Romer M, Stock C (2012) Dynamic redistribution of calcium sensitive potassium channels ($hK_{Ca}3.1$) in migrating cells. J Cell Physiol 227:686–696
Seoh S-A, Sigg D, Papazian DM, Bezanilla F (1996) Voltage-sensing residues in the S2 and S4 segments of the Shaker K^+ channel. Neuron 16:1159–1167
Sepulveda FV, Mason WT (1985) Single channel recordings obtained from basolateral membranes of isolated rabbit enterocytes. FEBS Lett 191:87–91
Sheikh IA, Koley H, Chakrabarti MJ, Hoque KM (2013) The EZpac1 signaling pathway regulates Cl^- secretion via modulation of apical KCNN4c channels in diarrhea. J Biol Chem 288:20404–20415
Simms LA, Doecke JD, Roberts RL, Fowler EV, Zhao ZZ, McGuckin MA, Huang N, Hayward NK, Webb PM, Whiteman DC, Cavanaugh JA, McCallum R, Florin TH, Barclay ML, Gearry RB, Merriman TR, Montgomery GW, Radford-Smith GL (2010) KCNN4 gene variant is associated with ileal Crohn's Disease in the Australian and New Zealand population. Am J Gastroenterol 105:2209–2217
Simoes M, Garneau L, Klein H, Banderali U, Hobeila F, Roux B, Parent L, Sauvé R (2002) Cysteine mutagenesis and computer modeling of the S6 region of an intermediate conductance IKCa channel. J Gen Physiol 120:99–116
Singh S, Syme CA, Singh AK, Devor DC, Bridges RJ (2001) Benzimidazolone activators of chloride secretion: potential therapeutics for cystic fibrosis and chronic obstructive pulmonary disease. J Pharmacol Exp Ther 296:600–611
Singh SK, O'Hara B, Talukder JR, Rajendran VM (2012) Aldosterone induces active K^+ secretion by enhancing mucosal expression of Kcnn4c and Kcnma1 channels in rat distal colon. Am J Physiol Cell Physiol 302:C1353–C1360
Skalhegg BS, Tasken K, Hansson V, Huitfeldt HS, Jahnsen T, Lea T (1994) Location of cAMP-dependent protein kinase type I with the TCR-CD3 complex. Science 263:84–87
Smith PL, Frizzell RA (1984) Chloride secretion by canine tracheal epithelium IV. Basolateral membrane K permeability parallels secretion rate. J Membr Biol 77:187–199
Smith JJ, Welsh MJ (1992) cAMP stimulates bicarbonate secretion across normal, but not cystic fibrosis airway epithelia. J Clin Invest 89:1148–1153
Srivastava S, Li Z, Lin L, Liu G, Ko K, Coetzee WA, Skolnik EY (2005) The phosphatidylinositol 3-phosphate phosphatase myotubularin- related protein 6 (MTMR6) is a negative regulator of the Ca^{2+}-activated K^+ channel KCa3.1. Mol Cell Biol 25:3630–3638
Srivastava S, Choudhury P, Li Z, Liu G, Nadkarni V, Ko K, Coetzee WA, Skolnik EY (2006) Phosphatidylinositol 3-phosphate indirectly activates KCa3.1 via 14 amino acids in the carboxy terminus of KCa3.1. Mol Biol Cell 17:146–154
Srivastava S, Zhdanova O, Di L, Li Z, Albaqumi M, Wulff H, Skolnik EY (2008) Protein histidine phosphatase 1 negatively regulates CD4 T cells by inhibiting the K^+ channel KCa3.1. Proc Natl Acad Sci USA 105:14442–14446
Srivastava S, Panda S, Li Z, Fuhs SR, Hunter T, Thieler DJ, Hubbard SR, Skolnik EY (2016) Histidine phosphorylation relieves copper inhibition in the mammalian potassium channel KCa3.1. eLife 5. https://doi.org/10.7554/elife.16093e16093
Steudel FA, Mohr CJ, Stegen B, Nguyen HY, Barnert A, Steinle M, Beer-Hammer S, Koch P, Lo W-Y, Schroth W, Hoppe R, Brauch H, Ruth P, Huber SM, Lukowski R (2017) SK4 channels

modulate Ca^{2+} signalling and cell cylce progression in murine breast cancer. Mol Oncol 11:1172–1188
Strobl JS, Wonderlin WF, Flynn DC (1995) Mitogenic signal transduction in human breast cancer cells. Gen Pharmacol 26:1643–1649
Sukhareva M, Hackos DH, Swartz KJ (2003) Constitutive activation of the Shaker Kv channel. J Gen Physiol 122:541–556
Sun J, MacKinnon R (2017) Cryo-EM structure of a KCNQ1/CaM complex reveals insights into congenital long QT syndrome. Cell 169:1042–1050
Suzuki K, Petersen OH (1985) The effects of Na^+ and Cl^- removal and of loop diuretics on acetylcholine-evoked membrane potential changes in mouse lacrimal acinar cells. Q J Exp Physiol 70:437–445
Syme CA, Gerlach AC, Singh AK, Devor DC (2000) Pharmacological activation of cloned intermediate- and small-conductance Ca^{2+}-activated K^+ channels. Am J Physiol Cell Physiol 278:C570–C581
Syme CA, Hamilton KL, Jones HM, Gerlach AC, Giltinan L, Papworth GD, Watkins SC, Bradbury NA, Devor DC (2003) Trafficking of the Ca^{2+}-activated K^+ channel, hIK1, is dependent upon a C-terminal leucine zipper. J Biol Chem 278:8476–8486
Tao R, Lau C-P, Tse H-F, Li R-G (2008) Regulation of cell proliferation by intermediate Ca^{2+}-activated potassium and volume-sensitive chloride channels in mouse mesenchymal stem cells. Am J Physiol Cell Physiol 295:C1409–C1416
Tao X, Hite RK, MacKinnon R (2017) Cryo-EM structure of the open high-conductance Ca^{2+}-activated K^+ channel. Nature 541:46–51
Thompson J, Begenisich T (2006) Membrane-delimited inhibition of maxi-K channel activity by the intermediate conductance Ca^{2+}-activated K channel. J Gen Physiol 127:159–169
Thompson J, Begenisich T (2009) Mechanistic details of BK channel inhibition by the intermediate conductance, Ca^{2+}-activated K channel. Channels 3:194–204
Tidball CS (1961) Active chloride transport during intestinal secretion. Am J Phys 200:309–312
Tiwari-Woodruff SK, Schulteis CT, Mock AF, Papazian DM (1997) Electrostatic interactions between transmembrane segments mediate folding of Shaker K^+ channel subunits. Biophys J 72:1489–1500
Tiwari-Woodruff SK, Lin MA, Schulteis CT, Papazian DM (2000) Voltage-dependent structural interactions in the Shaker K^+ channel. J Gen Physiol 115:123–138
Trautmann A, Marty A (1984) Activation of Ca-dependent K channels by carbamylcholine in rat lacrimal glands. Proc Natl Acad Sci USA 81:611–615
Tu L, Wang J, Helm A, Skach WR, Deutsch C (2000) Transmembrane biogenesis of Kv1.3. Biochemistry 39:824–836
Urrego D, Tomczak AP, Zahed F, Stühmer W, Pardo LA (2014) Potassium channels in cell cycle and cell proliferation. Philos Trans R Soc B 369(1638):20130094. https://doi.org/10.1098/rstb.2013.0094
Vandorpe DH, Shmukler BE, Jiang L, Lim B, Maylie AJA, de Franceschi L, Cappellini MD, Brugnara C, Alper SL (1998) cDNA cloning and functional characterization of the mouse Ca^{2+}-gated K^+ channel, mIK1. Role in regulatory volume decrease and erythroid differentiation. J Biol Chem 273:21542–21553
Wallace DP, Grantham JJ, Sullivan LP (1996) Chloride and fluid secretion by cultured human polycystic kidney cells. Kidney Int 50:1327–1336
Wallner M, Meera P, Toro L (1999) Molecular basis of fast inactivation in voltage and Ca^{2+}-activated K^+ channels: a transmembrane β-subunit homolog. Proc Natl Acad Sci USA 96:4137–4142
Wang S, Melkoumian Z, Woodfork KA, Cather C, Davidson AG, Wonderlin WF, Strobl SJ (1998) Evidence for an early G1 ionic event necessary for cell cycle progression and survival in the MCF-7 human breast carcinoma cell line. J Cell Physiol 176:456–464

Wang J, Kristian A, Novak I (2013) Purinergic regulation of CFTR and Ca^{2+}-activated Cl^- channels and K^+ channels in human pancreatic duct epithelium. Am J Physiol Cell Physiol 304:C673–C684

Warth R, Hamm K, Bleich M, Kunzelmann K, von Hahn T, Schreiber R, Ullrich E, Mengel M, Trautmann N, Kindle P, Schwab A, Greger R (1999) Molecular and functional characterization of the small Ca^{2+}-regulated K^+ channel (rSK4) of colonic crypts. Pflügers Arch 438:437–444

Wasserman SI, Barrett KI, Huott PA, Beuerlein G, Kagnoff MF, Dharmsathaphorn K (1988) Immune-related intestinal Cl^- secretion. I. Effect of histamine on the T84 cell line. Am J Physiol Cell Physiol 254:C53–C62

Weiger TM, Hermann A (2014) Cell proliferation, potassium channels, polyamines and their interactions: a mini review. Amino Acids 46:681–688

Welsh MJ, McCann JD (1985) Intracellular calcium regulates basolateral potassium channels in a chloride-secreting epithelium. Proc Natl Acad Sci USA 82:8823–8826

Welsh MJ, Smith PL, Frizzell RA (1982) Chloride secretion by canine tracheal epithelium: II. The cellular electrical potential profile. J Membr Biol 70:227–238

Weymer A, Huott P, Liu W, McRoberts JA, Dharmsathaphorn K (1985) Chloride secretory mechanism induced by prostaglandin E_1 in a colonic epithelial cell line. J Clin Invest 76:1828–1836

Wilson SM, Brown SG, McTavish N, McNeill RP, Husband EM, Inglis SK, Olver RE, Clunes MT (2006) Expression of intermediate-conductance, Ca^{2+}-activated K^+ channel (KCNN4) in H441 human distal airway epithelial cells. Am J Physiol Lung Cell Mol Physiol 291:L957–L965

Wonderlin WF, Strobl JS (1996) Potassium channels, proliferation and G1 progression. J Membr Biol 154:91–107

Wonderlin WF, Woodfork KA, Strobl JS (1995) Changes in membrane potential during the progression of MCF-7 human mammary tumor cells through the cell cycle. J Cell Physiol 165:177–185

Wright RD, Jennings MA, Florey HW, Lium R (1940) The influence of nerves and drugs on secretion by the small intestine and an investigation of the enzymes in intestinal juice. Q J Exp Physiol 30:73–120

Wu ZC, Kisslinger SD, Gaginella TS (1982) Functional evidence for the presence of cholinergic nerve endings in the colonic mucosa of the rat. J Pharmacol Exp Ther 221:664–669

Wulf A, Schwab A (2002) Regulation of a calcium-sensitive K^+ channel (cIK1) by protein kinase C. J Membr Biol 187:71–79

Wulff H, Castle NA (2010) Therapeutic potential of KCa3.1 blockers: recent advances and promising trends. Expert Rev Clin Pharmacol 3:385–396

Wulff H, Zhorov BS (2008) K^+ channel modulators for the treatment of neurological disorders and autoimmune diseases. Chem Rev 108:1744–1773

Wulff H, Miller MJ, Hansel W, Grissmer S, Cahalan MD, Chandy KG (2000) Design of a potent and selective inhibitor of the intermediate-conductance Ca^{2+}-activated K^+ channel: a potential immunosuppressant. Proc Natl Acad Sci USA 97:8151–8156

Wulff H, Gutman GA, Cahalan MD, Chandy KG (2001) Delineation of the clotrimazole/TRAM-34 binding site on the intermediate conductance calcium-activated potassium channel, IKCa1. J Biol Chem 276:32040–32045

Xia XM, Fakler B, Rivard A, Wayman G, Johnson-Pais T, Keen JE, Ishii T, Hirschberg B, Bond CT, Lutsenko S, Maylie J, Adelman JP (1998) Mechanism of calcium gating in small-conductance calcium-activated potassium channels. Nature 395:503–507

Xia XM, Ding JP, Lingle CJ (1999) Molecular basis for the inactivation of Ca^{2+}- and voltage-dependent BK channels in adrenal chromaffin cells and rat insulinoma tumor cells. J Neurosci 19:5255–5264

Ye M, Grantham JJ (1993) The secretion of fluid by renal cysts from patients with autosomal dominant polycystic kidney disease. N Engl J Med 329:310–311

Yi BA, Lin YF, Jan YN, Jan LY (2001) Yeast screen for constitutively active mutant G protein-activated potassium channels. Neuron 29:657–667

Zhang Z, Zhou Y, Ding JP, Xia XM, Lingle CJ (2006) A limited access compartment between the pore domain and cytosolic domain of the BK channel. J Neurosci 26:11833–11843
Zhang X, Wang G, Dupre DJ, Feng Y, Robitaille M, Lazartigues E, Feng YH, Hebert TE, Wu G (2009) Rab1 GTPase and dimerization in the cell surface expression of angiotensin II type 2 receptor. J Pharmacol Exp Ther 330:109–117
Zhang P, Yang X, Yin Q, Yi J, Shen W, Zhao L, Zhu Z, Liu J (2016) Inhibition of SK4 potassium channels suppresses cell proliferation, migration and the epithelial-mesenchymal transition in triple-negative breast cancer cells. PLoS One 11(4):e0154471. https://doi.org/10.1371/journal.pone.0154471
Zhou W, Vergara L, Konig R (2004) T cell receptor induced intracellular redistribution of type I protein kinase A. Immunology 113:453–459
Zhu Y, Ye J, Huizinga JD (2007) Clotrimazole-sensitive K^+ currents regulate pacemaker activity in interstitial cells of Cajal. Am J Physiol Gastrointest Liver Physiol 292:G1715–G1725
Zielenski J, Corey M, Rozmahel R, Markiewicz D, Aznarez I, Casals T, Larriba S, Mercier B, Cutting GR, Krebsova A, Macek M Jr, Langfelder-Schwind E, Marshall BC, DeCelie-Germana-J, Claustres M, Palacio A, Bal J, Nowakowska A, Ferec C, Estivill X, Durie P, Tsui LC (1999) Detection of a cystic fibrosis modifier locus for meconium ileus on human chromosome 19q13. Nat Genet 22:128–129
Zielenski J, Markiewicz D, Yuan X, Patel M, Sun L, Aznarez I, Tsui L (2004) P0915 meconium ileus in cystic fibrosis is associated with polymorphic markers in the calcium-activatted potassium channel (KCNN4) gene. J Pediatr Gastroenterol Nutr 39:S405–S406
Zimmerman TW, Binder HJ (1982) Muscarinic receptors on rat isolated colonic epithelial cells: a correlation between inhibition of [^{3}H]quinuclidinyl benzilate binding and alteration in ion transport. Gastroenterology 83:1244–1251

Chapter 23
BK Channels in Epithelia

Ryan J. Cornelius, Jun Wang-France, and Steven C. Sansom

Abstract Large, Ca-activated K^+ channels (BK) are comprised of the pore-forming α subunit (BK-α; encoded by *slo1* gene), and can be influenced by one of four β subunits (β1–4) and often a γ subunit of the extracellular leucine-rich-repeat-only (Elron) cluster. The γ subunit, LRRC26, yields the most substantial left shift (-140 mV) in voltage-activation and endows BK-α the capacity to drive fluid secretion at resting cell [Ca^{2+}]. Epithelial BK channels have been studied in the distal nephron of kidneys and in a variety of extrarenal, fluid-secreting epithelia, such as the gastrointestinal tract, exocrine glands, and pulmonary cells. In the distal nephron, BK-α/β1 resides in connecting tubule principal cells (CNT), where they are activated by high fluid flow and BK-α/β4 is localized in intercalated cells (IC) where they mediate K^+ secretion in mice on a high K^+, alkaline diet. In goblet cells of the colon, BK mediate K^+ secretion to generate fluid volume as part of a neurogenic-stimulated response to infection or bacterial imbalance. BK activity in the colon is also enhanced by aldosterone and can play a role to maintain K^+ balance in end-stage renal disease. In exocrine glands, BK secrete K^+ in conjunction with HCO_3^- to generate an alkaline fluid that neutralizes the acidic contents of the mouth or stomach. In pulmonary epithelia, BK mediate K^+ secretion as a counter cation to Cl^- to drive secretion of water for proper mucociliary activity. BK also reside in the basolateral membranes of some epithelial cells, where they recycle cellular K^+ that enters via Na-K-ATPase.

Keywords Maxi K^+ · Distal nephron · Colon · Exocrine glands · K^+ secretion

R. J. Cornelius
Division of Nephrology and Hypertension, Department of Medicine, Oregon Health and Science University, Portland, OR, USA
e-mail: cornelir@ohsu.edu

J. Wang-France · S. C. Sansom (✉)
Department of Cellular and Integrative Physiology, Nebraska Medical Center, Omaha, NE, USA
e-mail: j.wang@unmc.edu; ssansom@unmc.edu

K. L. Hamilton, D. C. Devor (eds.), *Studies of Epithelial Transporters and Ion Channels*, Physiology in Health and Disease,
https://doi.org/10.1007/978-3-030-55454-5_23

23.1 Introduction

Large, Ca^{2+}- and voltage-activated K^+ channels (BK) are expressed in epithelial cells lining the gastrointestinal track, exocrine glands, pulmonary cell, and renal distal tubules. The pore-forming channels (BK-α) have a single channel conductance ranging from 120 to 250 pS and are encoded by a single gene (*slo1; KCNMA1*) (Adelman et al. 1992; Atkinson et al. 1991). Two splice variants have been studied prevalently: stress-regulated exon (STREX) with 58 inserted amino acids and ZERO, with no inserted amino acids. The BK-α is associated with at least four different ancillary (β) subunits (BK-β1–4) (Contreras et al. 2012; Gonzalez-Perez and Lingle 2019). A more recently discovered family of γ subunits of the extracellular leucine-rich-repeat-only (Elron) cluster includes LRRC26, LRRC38, LRRC52, LRRC55, LRTM1, and LRTM2 as important modulators of BK-α (Yan and Aldrich 2010) (Yan and Aldrich 2012).

The BK-β1 subunit is predominantly expressed with BK-α in smooth muscle cells (Knaus et al. 1994) and the BK-β4 is predominantly expressed with BK-α in brain (Behrens et al. 2000). However, any of the four β subunits can be associated with BK-α in the different types of epithelial cells. Although the BK-α is able to conduct K^+ current in the absence of a β subunit in expression systems, it is unclear whether the BK-α functions solo in epithelial cells in vivo. The variable roles of the β subunits include: bestowing increased sensitivity of the BK-α to Ca^{2+} and voltage, promoting the membrane insertion of BK-α, and modifying the sensitivity to toxins. The γ subunit can also modify BK-α activity. Most notably, interaction with LRRC26 creates a −140 mV shift in the half-activation potential of BK, resulting in activation at resting voltage conditions without an increase in intracellular $[Ca^{2+}]$ (Yan and Aldrich 2010; Zhang and Yan 2014). LRRC26 activity is found in a variety of secretory epithelial cells, such as the acinar cells of lacrimal and salivary glands and goblet cells of intestine and colon (Yang et al. 2017).

BK can reside in either the apical or basolateral membranes of epithelial cells. In the apical membrane, BK mediate K^+ secretion (Holtzclaw et al. 2011). In the basolateral membrane, BK participate in the governance of the cell membrane potential. Although quiescent at resting membrane potentials, BK are activated by a combination of increased intracellular $[Ca^{2+}]$ and depolarizing cell membrane potentials (Cox et al. 1997). In addition, the ZERO variant is activated by protein kinase A phosphorylation (Tian et al. 2001a; Tian et al. 2003). When apical BK are activated, K^+ is driven from the cell to the lumen due to its electrochemical gradient. The basolateral Na^+-K^+-ATPase is the primary driving force that increases K^+ above electrochemical equilibrium. This also sets up gradients for Na^+, Cl^-, and K^+ to enter the cell via the Na^+-K^+-$2Cl^-$ cotransporter (NKCC1) located in the basolateral membrane of a Cl^- secreting cell. After elevating Cl^- above its electrochemical equilibrium potential, Cl^- exerts a depolarizing force across the apical membrane for K^+ to exit the cell.

This chapter will report studies of BK as a K^+ secretory channel in epithelial cells that secrete a K^+-rich fluid. Although other K^+ channels have been shown to have secretory roles, most, if not all, K^+-secreting epithelial cells utilize BK to mediate K^+ secretion. There are many differences and similarities in the mechanisms for regulating K^+ secretion in the renal and extrarenal epithelia. In some of the extrarenal cells, BK respond to sympathetic stimulation with increases in cAMP (Van Jr et al. 2005). However, such adrenergic stimulation of K^+ secretion in renal cells has not been established. On the other hand, BK channels in the colon respond to aldosterone, much like BK in the distal tubule (Sorensen and Leipziger 2009).

23.2 BK-Mediated K^+ Secretion in Renal Epithelia

Electrophysiological experiments utilizing microelectrodes revealed that the rabbit cortical collecting duct (CCD) possess a large apical membrane K^+ conductance, which is enhanced in magnitude when the animals are given a synthetic mineralocorticoid (Sansom and O'Neil 1985). Subsequently, BK were the first ion-selective channels identified in mammalian renal epithelial cells using the patch clamp technique applied to split-open isolated rabbit CCD (Hunter et al. 1984). After the gene (*slo1*) was cloned, transcript for BK was found in the distal nephron segments (Morita et al. 1997) and BK-α protein was identified in the connecting tubule (CNT) and CCD using immunohistochemistry (IHC) (Holtzclaw et al. 2010a).

The function of BK to secrete K^+ became uncertain when the apical BK of rat CCD served to extrude K^+ as a volume regulator during exposure to hypoosmotic solutions (Stoner and Morley 1995). However, renal BK-mediated K^+ secretion was subsequently demonstrated by raising the $[K^+]$ of the solution bathing the isolated Ambystoma (amphibian) collecting duct (Stoner and Viggiano 1999). Several years later, the isolated rabbit cortical collecting duct displayed BK-mediated K^+ secretion when perfused at high fluid flows of 6 nl/min, but not at low fluid flows of 0.5 nl/min (Satlin et al. 2001). The mechanism for flow-stimulated BK was partly due to a shear stress-induced increase in intracellular $[Ca^{2+}]$ (Woda et al. 2002). More recent studies showed that fluid flow-induced bending of the primary cilium of principal cells activates BK-α/β1 (Carrisoza-Gaytan et al. 2017).

The availability of mice with genetically ablated BK-α (BK-α-KO) and BK-β subunits enabled in vivo studies of BK-mediated K^+ secretion and revealed the roles of the subunits. In the first in vivo study, a mouse with a genetic ablation of the BK-β1 subunit (β1-KO) exhibited a diminished capacity to excrete K^+ when fluid flow was increased by volume expansion (Pluznick et al. 2003). The discovery that the BK-β1 subunit was localized to the apical membrane of the principal cells of the mouse CNT suggested a secretory role for the BK-α/β1 under high flow conditions (Pluznick et al. 2005). BK-α-KO also exhibited defective K^+ secretion when flow was increased by pharmacological blockade of V2 vasopressin receptors (Rieg et al. 2007). BK-α-KO excreted K^+ at a rate similar to WT but the plasma [aldosterone] was very high, presumably to compensate with a higher expression of Na^+-K^+-ATPase and epithelial Na^+ channels (ENaC). In the absence of BK, the renal outer medullary K^+ channel (ROMK; Kir1.1), also identified in the CNT/CCD (Frindt and

Palmer 2004) (Mennitt et al. 1997), is upregulated and secretes K^+ in high dietary K^+ conditions (Sansom and Welling 2007; Wade et al. 2011).

A micropuncture study of ROMK-KO also indicated that high distal fluid flow stimulates BK-mediated K^+ secretion (Bailey et al. 2006). A primary role of ROMK is to mediate apical K^+ recycling in the thick ascending limb. Therefore, ROMK-KO display Bartter's syndrome with increased tubular fluid flow due to a dysfunctional concentrating mechanism. ROMK-KO exhibited increased iberiotoxin-sensitive K^+ secretion in the late distal tubule of mice. This was additional in vivo evidence that high distal flow stimulates BK-mediated K^+ secretion. It was notable that the micropuncture study also showed that high K^+ consumption induced BK-mediated K^+ secretion (Bailey et al. 2006). The high distal flow may be linked to the high K^+ diet, which in itself causes a fourfold increase in urinary output in mice (Cornelius et al. 2015). Transient receptor potential villinoid 4 (TRPV4) may be the flow sensor that delivers Ca^+ to BK in the CCD. TRPV4-KO mice exhibit reduced BK channel activity in the split-open CCD and a reduction in K^+ excretion when given a high K^+ diet (Mamenko et al. 2017).

Consuming a high K^+ diet results in increased mRNA and protein expression of BK-α and the β subunits (Holtzclaw et al. 2011; Najjar et al. 2005). β1-KO exhibited normal K^+ homeostasis when placed on a regular diet, but when placed on a high K^+ diet, the plasma [K^+] and [aldosterone] increased and mean arterial blood pressure was elevated by 20 mmHg (Grimm et al. 2009b). The failure to excrete substantial K^+ resulted in elevated plasma [K^+] and a compensating increase in ENaC-mediated Na^+ reabsorption that was accompanied by Cl^- reabsorption instead of K^+ secretion. Accumulating Na^+ and Cl^- reabsorption led to volume expansion and hypertension.

BK-β1 was localized in the apical membrane of CNT cells (the equivalent of principal cells of the CCD) and β1-KO clearly exhibited a K^+ secretory defect when given a high K^+ diet; however, both patch clamping of rat split-open tubules (Pacha et al. 1991) and IHC (Fig. 23.1) localized the preponderance of BK-α to the intercalated cells (IC) of CNT and CCD. IC are the acid/base transporting cells of the distal nephron. Unlike principal cells, which have the BK-β1 subunit, the BK-β4 subunit is expressed with BK-α in IC (Grimm et al. 2007). The localization of BK-α/β4 to IC was interesting because the PC were considered the Na^+ and K^+ transporting cells with an abundance of basolateral Na^+-K^+-ATPase that increased with a high K^+ diet. Intercalated cells contain an abundance of H^+-ATPase but nearly undetectable Na^+-K^+-ATPase. The PC and IC are not physically coupled; there are no reports of gap junctions in the CCD. However, the two cell types are linked by the high luminal electronegative field (transepithelial potential; Vte), which influences electrical transport of both PC and IC cells.

Mice with a knockout of the β4 subunit (β4-KO) exhibited attenuated K^+ secretion when placed on a high K^+ diet. However, the defective K^+ excretion was demonstrated only when the high K^+ diet was associated with an alkaline load that increased the urine pH to >8.0 (Cornelius et al. 2012). Luminal HCO_3^- is more resistant than Cl^- to reabsorb through tight junctions, thereby enhancing the electronegative lumen potential that drives K^+ secretion. In addition, with carbonic anhydrase IV localized to the apical membrane of alpha-IC (Schwartz et al. 2000),

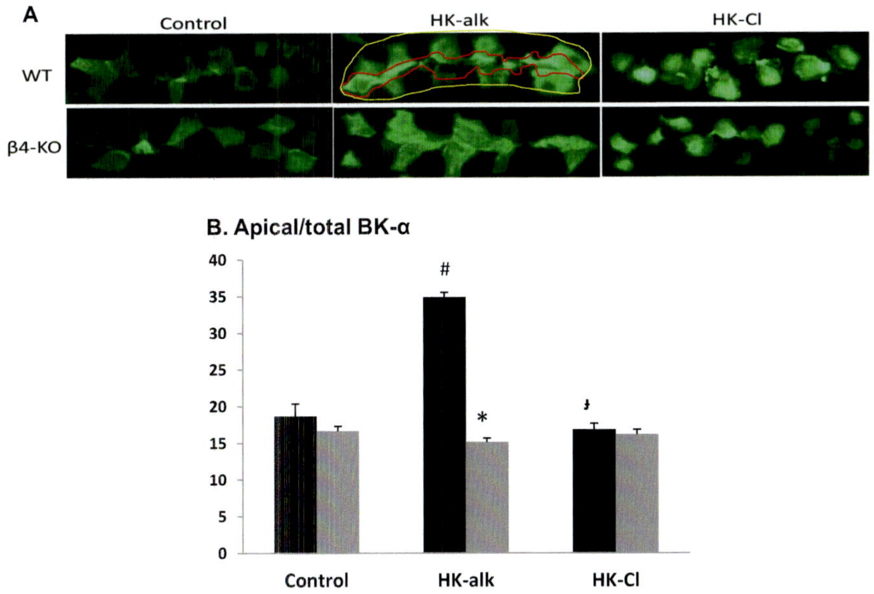

Fig. 23.1 Immunohistochemical analysis of CCD for BK-α in wild type (WT) and β4KO on control, high K^+ alkaline (HK-alk) and high K^+ chloride (HK-Cl) diet. (**a**) Representative stainings reveal BK-α in the CCD of WT and β4-KO on control, HK-alk, and HK-Cl diets. The outlines in upper middle panel show how fluorescence intensity was measured. The areas occupied by the red and yellow lines represent apical and total BK-α staining, respectively (magnification: ×40). (**b**) Summary bar plots of quantitated fluorescence for proportion of apical/total BK-α expression levels. Values are mean ± SEM. $*P < 0.01$ vs WT; $^{\#}P < 0.01$ vs control; $^{J}P < 0.01$ vs HK-alk; $n \geq 9$ in each group. (Adapted from Wen et al.) (Wen et al. 2013)

the gradient for secreted H^+ via V-ATPase will be immediately reduced, thereby creating an alkaline intracellular environment and reversing the gradients for Na^+-H^+ exchange (Chambrey et al. 2001; Orlowski and Grinstein 2004) in the basolateral membrane. In essence, the V-ATPase creates the driving force by creating the extrusion of Na^+ via the Na^+-H^+ exchanger 4 (NHE4), yielding a chemical Na^+ driving force for basolateral K^+ entry via NKCC1 (Liu et al. 2011), the secretory isoform of the Na^+-K^+-$2Cl^-$ cotransporter (Fig. 23.2).

Additional studies revealed the separate roles for high K^+ consumption and alkaline loading on BK channel expression in IC. The high K^+ load, regardless of accompanying acid or base, increased the expression level of BK-α in IC; however, the BK-β4 subunit was required for the BK-α to traffic to the apical membrane. BK-β4 expression is increased by alkalinity and prevents the lysosomal breakdown of BK-α, thereby increasing its time in the apical membrane (Wen et al. 2013; Wen et al. 2014). These results explained the attenuated K^+ excretion of β4-KO on the alkaline high K^+ diet.

The role of aldosterone in BK-mediated K^+ secretion was best demonstrated when mice were given a low Na^+, high K^+ diet (LNaHK). In these mice, the

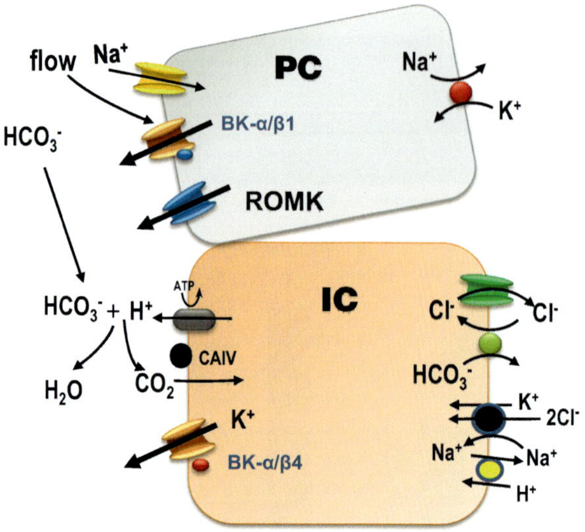

Fig. 23.2 Previously defined transporters of the aldosterone-sensitive distal nephron explain how flow and an alkaline urine load, generated from more proximal segments, influence BK-mediated K^+ secretion to accomplish K^+ for Na^+ exchange >2:3 as dictated by the Na-K-ATPase. Flow activates BK-α/β1 by localized increases in intracellular Ca^{2+}, mediated by bending of the primary cilium. BK-α/β4-mediated secretion from IC is supported by apical carbonic anhydrase IV (CAIV), which immediately removes the electrical force of H^+ secreted from V-ATPase. The H^+ gradient created by V-ATPase causes a reversal of the basolateral Na-H exchange to create a Na^+ chemical driving force for K^+ to enter the cell via the NKCC1

aldosterone levels increased to more than 3000 pg/ml, which is ten-fold the plasma [aldosterone] of mice on a regular diet. When LNaHK mice were adrenalectomized, P[K^+] was greater than 9 mM and mice survived only 1 day. When LNaHK mice were adrenalectomized and placed on a low dose of aldosterone to achieve plasma [aldosterone] of approximately 500 pg/ml, the BK expression was near basal levels and the rate of urinary K^+ excretion only increased to 25% of sham values (Cornelius et al. 2015).

In the mouse medullary collecting duct cells, BK-α/β4 are activated by high distal fluid flows and mediate a reduction of cellular K^+ content and volume (Holtzclaw et al. 2010b). The shear stress from high fluid flow also reduces the size of Madin-Darby canine kidney (MDCK)-C11 cells (Holtzclaw et al. 2010b), which are a subclone that predominantly express an IC phenotype (Gekle et al. 1994). The K^+ exiting the cell may have been a counter cation for ATP release from connexins in the apical membrane. ATP activates BK via purinergic receptors, thereby completing a positive, feed-forward loop (Bugaj et al. 2012). In support of this notion, β4-KO, unlike WT, do not exhibit cell volume reduction during high urinary flows generated from mice on high K^+ diets (Holtzclaw et al. 2010a). Moreover, when incorporating BK-β4 siRNA, MDCK-C11 cells fail to exhibit a shear-stress-induced loss of volume (Holtzclaw et al. 2010b).

The BK-β1 was detected in the apical membrane of the CNT; however, BK-α was only faintly detected in the same cells by IHC, leading to the speculation that BK-α in PC was a different splice variant from that of IC. Unlike MDCK-C11, which contain BK-α/β4, the PC-like MDCK-C7 cells contain BK-β1 but not BK-α (Holtzclaw et al. 2010b). The split-open patch technique revealed that native PC contained BK, which were quiescent unless the MAPK pathway was inhibited (Li et al. 2006). Moreover, PGE2 inhibit BK of PC by inhibiting MAPK (Jin et al. 2007).

In isolated perfused rabbit CCDs, BK was constitutively inhibited by protein kinase A (PKA) but activated at high flow via the protein kinase C (PKC) pathway (Liu et al. 2009). Inhibition by PKA is consistent with the notion that the BK of CCDs are of the STREX splice variant. Activation by PKC is consistent with the muscarinic activation of BK, which also occurs in salivary glands (see below). Because the intracellular [Ca^{2+}] would need to approach 10 μM to activate BK at negative cell potentials, the additional activation by PKC can explain how BK opens to secrete K^+ during high distal fluid flows. RT-PCR indicates low expression of LRRC26 mRNA in the kidney, however, LRRC52 expression is moderately high (Yang et al. 2011). LRRC52 shifts the activation potential of *Slo3* of the BK family in testis by −50 mV. Studies have not yet determined the influence of LRRC52 on BK in the distal nephron of the kidney.

WNK (*w*ith *n*o *l*ysine) kinases are important regulators of Na^+ and K^+ balance in the distal nephron (Welling et al. 2010; Wade et al. 2006). Mutations in WNK1 and WNK4 cause the disease familial hyperkalemic hypertension. At least three studies have shown that WNK4 inhibits BK (Wang et al. 2013; Yue et al. 2013; Zhuang et al. 2011) with activation of the lysosomal degradation of BK as the proposed mechanism (Zhuang et al. 2011).

23.3 BK-Mediated K^+ Secretion in the Colon

The colon is primarily responsible for intestinal water absorption, a process achieved by the surface cells, which contain epithelial Na^+ channels (ENaC) that mediate the Na^+-absorption that osmotically drives transepithelial water transport (Sorensen et al. 2008). However, water secretion is accomplished by the crypt cells and is mainly elicited as a pathophysiological response to infections or an imbalance of intestinal bacteria. The fluid-secreting intestinal crypt cells are comprised of two types, the columnar cells and the goblet cells. The columnar cells secrete water driven by cystic fibrosis transmembrane conductance regulator (CFTR)-mediated Cl^- secretion. The goblet cells secrete a mucous solution driven by BK-mediated K^+ secretion (Fig. 23.3) (Linley et al. 2014).

Organisms eliminate intestinal infection by a rapid clearing response that is performed by enteric nervous discharge. The cholinergic innervation activates primarily CFTR-mediated Cl^- secretion but also K^+ secretion. These ions create the electrogenic and osmotic forces to drive water secretion (Rechkemmer et al.

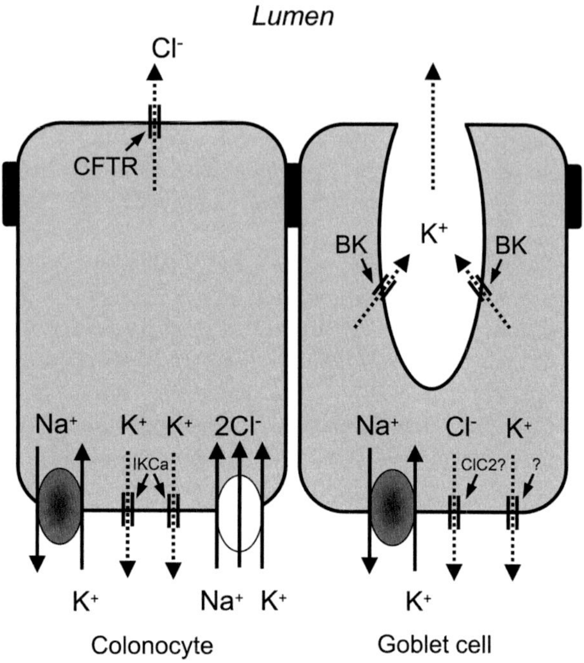

Fig. 23.3 Mechanism of BK-mediated K$^+$ secretion in the colon. The BK-mediated K$^+$ secretion in the apical membrane of goblet cells is linked to the CFTR-mediated Cl$^-$ secretion in the apical membrane of colonocytes. (Adapted from Linley, J. et al.) (Linley et al. 2014)

1996). Colonic pseudo-obstruction produces a predominant response to a sympathetic β-adrenergic increase in cAMP that initiates K$^+$ secretion and produces intestinal [K$^+$] of near 150 mM (Van Dinter et al. 2005). Persistent intestinal diarrhea often results in severe hypokalemia (Field 2003) that can be corrected by intestinal resection (Yusuf et al. 1999).

BK channels are the primary mediators of intestinal K$^+$ secretion. A model of ulcerative colitis presents with enhanced colonic K$^+$ secretion (Kanthesh et al. 2013) and BK expression (Kanthesh et al. 2013; Sandle et al. 2007). Specific BK inhibitors, such as iberiotoxin and paxilline, show that BK are the exclusive secretors of K$^+$ in the distal colon (Matos et al. 2007). Although BK have been reported in the apical membranes of surface cells (Hay-Schmidt et al. 2003), other studies find BK only in the crypt cells (Sausbier et al. 2006; Sorensen et al. 2010b). BK activity is enhanced in crypt cells under various conditions such as end-stage renal disease (Mathialahan et al. 2005), secretory diarrhea (Simon et al. 2008), and ulcerative colitis (Sandle et al. 2007). The role for BK-mediated K$^+$ secretion in mice was demonstrated in vivo with BK-α-KO, which exhibited a substantially reduced K$^+$ secretory response to β-adrenergic stimulation (Sorensen et al. 2010b).

Under normal physiological conditions, the gastrointestinal tract is responsible for approximately 10% of K$^+$ excretion and can have a role to maintain K$^+$ homeostasis. Intestinal resection does not result in hyperkalemia because the kidneys will elevate their capacity to secrete K$^+$. Despite having a minimal role to maintain K$^+$ homeostasis, the colonic BK are still regulated by aldosterone, which increases

BK-α expression (Sorensen et al. 2010a) and BK-mediated K^+ secretion in the distal colon (Sorensen et al. 2008). Moreover, patients with end-stage renal disease exhibit enhanced gastrointestinal K^+ secretion (Sandle et al. 2007) and their plasma $[K^+]$ can be reduced by excreting a K^+-rich intestinal fluid (Sandle et al. 1986).

Colonic enterocytes contain the STREX and ZERO splice variants (Kanthesh et al. 2013). The ZERO variant is phosphorylated and activated by PKA (Tian et al. 2001b), which is consistent with its role in the K^+ secretory response to sympathetic stimulation. RT-PCR has identified the BK-β1 and BK-β4 as the predominating β subunits in mouse (Flores et al. 2007) and guinea pig distal colon (Zhang et al. 2012). These are the same subunits associated with BK-α in the mouse distal nephron. However, all four β subunits (Sandle et al. 2007) and the gamma subunit, LRR26 (Yang et al. 2017), have been described in human sigmoid colon. The roles of the β subunits, with respect to modulating colonic BK-α, have not been determined. As in other K^+-rich, fluid-secreting epithelial cells, LRRC26 would shift the activation curve to enhance BK activity at normal cell potentials and intracellular $[Ca^{2+}]$.

23.4 BK-Mediated K^+ Secretion in Exocrine Glands

The salivary and pancreatic exocrine glands contain similar transport processes (Lee et al. 2012). The salivary glands secrete a solution rich in HCO_3^- to initiate the digestion process and prevent acid-induced dental erosion. The salivary glands include the sublingual, submandibular, and parotid and comprise two major cell types, acinar and ductal cells. The acinar cells secrete a watery fluid rich in Na^+ and Cl^-. The ductal cells reabsorb the Na^+ and Cl^- and secrete K^+ and HCO_3^- to yield final concentrations of 2 mM Na^+, 135 mM K^+, 55 mM HCO_3^-, and 78 mM Cl^- (Lee et al. 2012). Studies with BK-α-KO revealed that apical BK channels mediate K^+ secretion in striated and excretory ductal cells of submandibular glands (Nakamoto et al. 2008). The parasympathetic nerves, via muscarinic M3 receptors, and the sympathetic nerves, via β-adrenergic generation of cAMP, have roles in both the acinar and ductal secretory mechanisms.

The pancreas is an exocrine gland, which also contains acinar and ductal cells. The ductal cells secrete HCO_3^- in order to neutralize the gastric acid in the small intestine. Immunostaining revealed BK-α in the apical membrane of the ductal cells [see Fig. 23.4; (Venglovecz et al. 2011)]. HCO_3^- secretion is attenuated when BK are blocked by apically applied iberiotoxin, indicating a link between HCO_3^- secretion and BK-mediated K^+ secretion (Venglovecz et al. 2011). In ductal cells, the bile acid chenodeoxycholate (CDC) activates BK from the apical side in low concentrations (100 μM) via IP3-mediated release of $[Ca^{2+}]$ from intracellular stores (Venglovecz et al. 2011). These investigators speculated that CDC-induced BK-mediated fluid secretion in the pancreas was only evoked during a "hyper-secretion" condition, activated to preemptively combat the consequences of disease resulting from compromised bile secretion.

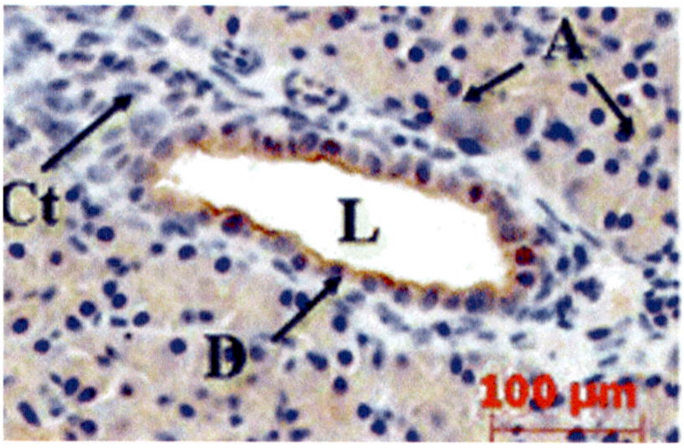

Fig. 23.4 Immunohistochemical analysis with polyclonal antibody raised against amino acids 1097–1196 of the BK-α. The immunostaining revealed apical expression of BK channels in the intralobular/interlobular pancreatic ducts. Staining is undetectable in basolateral membranes of ductal cells or acinar cells. (Adapted from Venglovecz et al.) (Venglovecz et al. 2011)

23.5 Pulmonary Epithelia

Much like the colon and exocrine glands, the dynamics of pulmonary mucociliary clearance depends on ionic secretion that drives transepithelial water transport. Pulmonary BK channels were first identified in cultured pulmonary cells (A549) using patch clamp techniques (Ridge et al. 1997). Although water secretion, driven by CFTR-mediated Cl^- secretion, has been well-established, more recent studies have shown that BK channels mediate K^+ secretion to promote the pulmonary water for proper mucociliary beating (Manzanares et al. 2015). In normal human bronchial epithelial cells (NHBE), ATP activates BK-mediated K^+ secretion (ManzanareS et al. 2011), with K^+ serving as a counter cation for Cl^- secretion. Inhibition of BK reduces the volume of the liquid surface (ManzanareS et al. 2011). Pulmonary BK channels are most likely segregated to the ciliary cells, rather than basal or goblet cells (ManzanareS et al. 2011; Kis et al. 2016). Quantitative PCR reveals high expression of the β2 and β4 subunits and low expression of β1 and β3 (Kis et al. 2016).

Cigarette smoke is a toxin that may reduce pulmonary volume through p38-mediated reduction in BK channel activity (Sailland et al. 2017). The cytokine, IFN-γ, shown to decrease pulmonary fluid secretion, increases mRNA of BK-β2 and decreases BK-β4. The γ subunit, LRRC26, which normally associates with BK-α in NHBE, is dissociated due to IFN-γ treatment (Manzanares et al. 2014). TGF-β, which is elevated in the lungs of many patients with cystic fibrosis, is associated with pulmonary dehydration due to its inhibition of BK (Manzanares et al. 2015). Overexpression of LRRC26 in cultured pulmonary cells with the Δ508 cystic fibrosis mutation rescues TGF-β-induced dehydration (Manzanares et al. 2015). Figure 23.5, from Kis et al. (Kis et al. 2016), shows a hypothetical model depicting the association between the α, β, and γ subunits of BK in pulmonary epithelia.

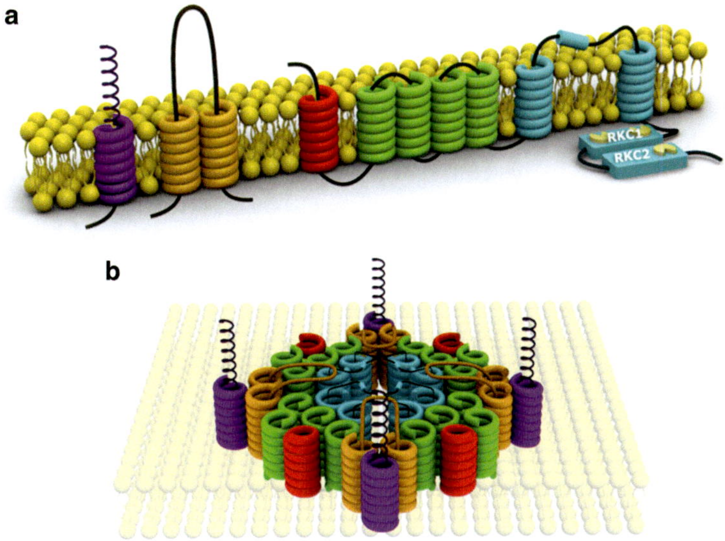

Fig. 23.5 (**a**) Hypothetical structure of the pulmonary epithelial BK channel with associated β and γ subunits, as published by Kis et al. (2016) for BK in pulmonary cells. The α subunit forms the voltage sensor domain (green), the gate domain (blue), the cytosolic, Ca^{2+}-binding site (yellow), and the S0 segment (red). The β subunits are in orange and the leucine-rich repeat-containing 26 protein (LRRC26) γ subunits are depicted in purple. (Adapted from Kis et al.) (Kis et al. 2016)

23.6 BK in the Basolateral Membranes of Epithelia

BK is also localized in basolateral membranes of some epithelia, where it influences either Na^+ absorption or Cl^- secretion. BK mediate recycling of the K^+ entering the cell via Na^+-K^+-ATPase across the basolateral membrane, thereby creating a negative cellular potential that drives either lumen to cell Na^+ uptake or cell to lumen Cl^- extrusion. In the mouse, recycling K^+ via basolateral BK augments Cl^- secretion in the proximal colon and Na^+ absorption in the surface cells of the distal colon (Puntheeranurak et al. 2007). However, in nasal epithelia, outward basolateral BK currents support both Na^+ absorption and Cl^- secretion in the same cell (Clarke et al. 1992). Basolateral BK-mediated K^+ recycling and apical CFTR-mediated Cl^- secretion are initiated by muscarinic ligands, such as acetylcholine and bradykinin (Clarke et al. 1992). Activation of phospholipase C, with the generation of IP3, induces release of stores of Ca^{2+}, which activate BK.

The BK-β1 subunit was also described with IHC in the basolateral membrane of renal connecting tubule cells when mice were subjected to a low Na^+ diet (Grimm et al. 2009a). Compared with wild-type mice, the genetically ablated BK-β1 became extremely dehydrated and Na^+ and volume depleted. These results indicated that the BK-α/β1 trafficked to the basolateral membrane of CNT cells to mediate K^+ recycling, hyperpolarize the basolateral membrane, and support ENaC-mediated Na^+ reabsorption while preventing K^+ secretion. The basolateral localization of

BK-β1 with a low Na^+ diet contrasted with the localization of BK-β1 in the apical membrane of principal cells in mice on a high K^+ diet.

At the luminal sides of epithelial cells, depolarizing potentials and high fluid flow are cooperative activators of apical BK. BK in the basolateral membranes, however, are presented with hyperpolarizing membrane potentials. A study of BK channels in pancreatic acinar cells, in which BK reside in the basolateral membrane, addressed the issue of BK quiescence at resting membrane potentials (Romanenko et al. 2010). When observed in heterologous expression systems or neuronal cells at a cell $[Ca^{2+}]$ of 500 nM, BK-α require a membrane potential of more than 75 mV to initiate activation to an open probability >0.1; however, with the same 500 nM intracellular $[Ca^{2+}]$, native BK in parotid cells are active at cell potentials of −25 mV. Thus, in native epithelial cells, BK may be activated by components not found in neuronal cells or heterologous expression systems. The association of BK-α with subunits BK-β1 or BK-β4 in parotid cells (Nehrke et al. 2003) can shift the potential gating in high intracellular Ca^{2+} by 60 mV, a value still less than the 140 mV left shift in activation of parotid BK by LRRC26. It remains to be determined if LRRC26 is associated with BK-α and BK-β1 in the basolateral membranes of epithelial cells.

The details on the role of basolateral BK channels in Na^+ and Cl^- transport, beyond the original observational studies, have not been forthcoming. Basolateral channels are difficult to study by patch clamp techniques due to the basement membrane barrier. In addition, studies with BK knockout mice have indicated considerable redundancy of basolateral K^+ channels, with other families of upregulated K^+ channels in the absence of BK to control membrane potential. Most studies indicate that the intermediate conductance K^+ channels (IK, KCa2.3), which are activated by Ca^{2+} and not membrane potential, are the predominant channels that mediate recycling of K^+ across the basolateral membrane (Linley et al. 2014). Nevertheless, substantial evidence has supported a role for basolateral BK to facilitate Na^+ reabsorption and Cl^- secretion in some secretory epithelial cells.

23.7 Summary

BK channels are ubiquitously expressed in cells throughout the body and have a variety of roles in epithelial cells. In the colon, exocrine glands, and lungs, BK-mediated K^+ secretion is electrogenic and drives the transport of other ions and fluid. Secretion of K^+ generates a high volume in reaction to infection in the colon, and associates with HCO_3^- secretion to generate an alkaline environment in exocrine glands. In pulmonary epithelia, BK-mediated K^+ secretion is an important component of hydration for mucociliary cleaning. Unlike the distal nephron of the kidneys, BK-mediated K^+ secretion that generates fluid is independent of ENaC-mediated Na^+ absorption. In the colon, Na^+ absorbs via ENaC in surface cells and reduces volume while K^+ is secreted via BK channels in goblet cells to enhance volume and eliminate toxins.

Aldosterone enhances the expressions of colonic ENaC, BK, and Na-K-ATPase (Zemanova and Pacha 1998), indicating that the colon has a role to maintain both Na^+ and K^+ homeostasis. Colonic BK-mediated K^+ secretion is cholinergically and adrenergically stimulated, in different cells than ENaC-mediated Na^+ absorption. However, as illustrated for the kidney in Fig. 23.2, K^+ is secreted from the same (principal) cell as ENaC-mediated Na^+ reabsorption with high flow, but also from an adjacent intercalated cell in animals consuming a high K^+, alkaline diet. A high ratio of K^+ secreted per Na^+ reabsorbed can be accomplished using the driving forces generated by Na-K-ATPase of the PC and V-ATPase of the IC.

The γ subunit, LRRC26, shifts the half-activation potential of BK-α by −140 mV and is highly expressed in all epithelial cells that generate a high volume. Whether a γ subunit of the LLRC family similarly associates with BK in the distal nephron, where K^+ homeostasis is primarily regulated, remains to be determined.

References

Adelman JP, Shen KZ, Kavanaugh MP, Warren RA, Wu YN, Lagrutta A, Bond CT, North RA (1992) Calcium-activated potassium channels expressed from cloned complementary DNAs. Neuro 9:209–216

Atkinson NS, Robertson GA, Ganetzky B (1991) A component of calcium-activated potassium channels encoded by the *Drosophila slo* locus. Science 253:551–555

Bailey M, Cantone A, Yan Q, Macgregor GG, Leng Q, Amorim JB, Wang T, Hebert SC, Giebisch G, Malnic G (2006) Maxi-K channels contribute to urinary potassium excretion in the ROMK-deficient mouse model of type II Bartter's syndrome and in adaptation to a high-K diet. Kidney Int 70:51–59

Behrens R, Nolting A, Reimann F, Schwarz M, Waldschütz R, Pongs O (2000) HKCNMB3 and hKCNMB4, cloning and characterization of two members of the large-conductance calcium-activated potassium channel β subunit family. FEBS Lett 474:99–106

Bugaj V, Sansom SC, Wen D, Hatcher LI, Stockand JD, Mironova E (2012) Flow-sensitive K^+-coupled ATP secretion modulates activity of the epithelial Na^+ channel in the distal nephron. J Biol Chem 287:38552–38558

Carrisoza-Gaytan R, Wang L, Schreck C, Kleyman TR, Wang WH, Satlin LM (2017) The mechanosensitive BKalpha/beta1 channel localizes to cilia of principal cells in rabbit cortical collecting duct (CCD). Am J Physiol Renal Physio 312:F143–F156

Chambrey R, St John PL, Eladari D, Quentin F, Warnock DG, Abrahamson DR, Podevin RA, Paillard M (2001) Localization and functional characterization of Na^+/H^+ exchanger isoform NHE4 in rat thick ascending limbs. Am J Physiol Renal Physiol 281:F707–F717

Clarke LL, Paradiso AM, Mason SJ, Boucher RC (1992) Effects of bradykinin on Na^+ and Cl^- transport in human nasal epithelium. Am J Physiol Cell Physiol 262:C644–C655

Contreras GF, Neely A, Alvarez O, Gonzalez C, Latorre R (2012) Modulation of BK channel voltage gating by different auxiliary beta subunits. Proc Natl Acad Sci U S A 109:18991–18996

Cornelius RJ, Wen D, Hatcher LI, Sansom SC (2012) Bicarbonate promotes BK-alpha/beta4-mediated K excretion in the renal distal nephron. Am J Physiol Renal Physiol 303:F1563–F1571

Cornelius RJ, Wen D, Li H, Yuan Y, Wang-France J, Warner PC, Sansom SC (2015) Low Na, high K diet and the role of aldosterone in BK-mediated K excretion. PLoS One 10:e0115515

Cox DH, Cui J, Aldrich RW (1997) Allosteric gating of a large conductance Ca-activated K^+ channel. J Gen Physiol 110:257–281

Field M (2003) Intestinal ion transport and the pathophysiology of diarrhea. J Clin Invest 111:931–943

Flores CA, Melvin JE, Figueroa CD, Sepulveda FV (2007) Abolition of Ca^{2+}-mediated intestinal anion secretion and increased stool dehydration in mice lacking the intermediate conductance Ca^{2+}-dependent K^+ channel Kcnn4. J Physiol 583:705–717

Frindt G, Palmer LG (2004) Apical potassium channels in the rat connecting tubule. Am J Physiol Renal Physiol 287:F1030–F1037

Gekle M, Wunsch S, Oberleithner H, Silbernagl S (1994) Characterization of two MDCK-cell subtypes as a model system to study principal cell and intercalated cell properties. Pflügers Arch 428:157–162

Gonzalez-Perez V, Lingle CJ (2019) Regulation of BK channels by beta and gamma subunits. Annu Rev Physiol 81:113–137

Grimm PR, Foutz RM, Brenner R, Sansom SC (2007) Identification and localization of BK-beta subunits in the distal nephron of the mouse kidney. Am J Physiol Renal Physiol 293:F350–F359

Grimm PR, Irsik DL, Liu L, Holtzclaw JD, Sansom SC (2009a) Role of BKbeta1 in Na^+ reabsorption by cortical collecting ducts of Na+−deprived mice. Am J Physiol Renal Physiol 297:F420–F428

Grimm PR, Irsik DL, Settles DC, Holtzclaw JD, Sansom SC (2009b) Hypertension of Kcnmb1$^{-/-}$ is linked to deficient K secretion and aldosteronism. Proc Natl Acad Sci U S A 106:11800–11805

Hay-Schmidt A, Grunnet M, Abrahamse SL, Knaus HG, Klaerke DA (2003) Localization of Ca^{2+}−activated big-conductance K^+ channels in rabbit distal colon. Pflügers Arch 446:61–68

Holtzclaw JD, Grimm PR, Sansom SC (2010a) Intercalated cell BK-alpha/beta4 channels modulate sodium and potassium handling during potassium adaptation. J Am Soc Nephrol 21:634–645

Holtzclaw JD, Liu L, Grimm PR, Sansom SC (2010b) Shear stress-induced volume decrease in C11-MDCK cells by BK-alpha/beta4. Am J Physiol Renal Physiol 29:F507–F516

Holtzclaw JD, Grimm PR, Sansom SC (2011) Role of BK channels in hypertension and potassium secretion. Curr Opin Nephrol Hypertens 20:512–517

Hunter M, Lopes AG, Boulpaep EL, Cohen B, Giebisch G (1984) Single channel recordings of calcium-activated potassium channels in the apical membrane of rabbit cortical collecting tubules. Proc Natl Acad Sci U S A 81:4237–4239

Jin Y, Wang Z, Zhang Y, Yang B, Wang WH (2007) PGE2 inhibits apical K channels in the CCD through activation of the MAPK pathway. Am J Physiol Renal Physiol 293:F1299–F1307

Kanthesh BM, Sandle GI, Rajendran VM (2013) Enhanced K^+ secretion in dextran sulfate-induced colitis reflects upregulation of large conductance apical K^+ channels (BK; Kcnma1). Am J Physiol Cell Physiol 305:C972–C980

Kis A, Krick S, Baumlin N, Salathe M (2016) Airway hydration, apical K^+ secretion, and the large-conductance, Ca^{2+}-activated and voltage-dependent potassium (BK) channel. Ann Am Thorac Soc 13(Suppl 2):S163–S168

Knaus H-G, Garcia-Calvo M, Kaczorowski GJ, Garcia ML (1994) Subunit composition of the high conductance calcium-activated potassium channel from smooth muscle, a representative of the *mSlo* and *slowpoke* family of potassium channels. J Biol Chem 269:3921–3924

Lee MG, Ohana E, Park HW, Yang D, Muallem S (2012) Molecular mechanism of pancreatic and salivary gland fluid and HCO_3 secretion. Physiol Rev 92:39–74

Li D, Wang Z, Sun P, Jin Y, Lin DH, Hebert SC, Giebisch G, Wang WH (2006) Inhibition of MAPK stimulates the Ca^{2+}-dependent big-conductance K channels in cortical collecting duct. Proc Natl Acad Sci U S A 103:19569–19574

Linley J, Loganathan A, Kopanati S, Sandle GI, Hunter M (2014) Evidence that two distinct crypt cell types secrete chloride and potassium in human colon. Gut 63:472–479

Liu W, Wei Y, Sun P, Wang WH, Kleyman TR, Satlin LM (2009) Mechanoregulation of BK channel activity in the mammalian cortical collecting duct: role of protein kinases a and C. Am J Physiol Renal Physiol 297:F904–F915

Liu W, Schreck C, Coleman RA, Wade JB, Hernandez Y, Zavilowitz B, Warth R, Kleyman TR, Satlin LM (2011) Role of NKCC in BK channel-mediated net K^+ secretion in the CCD. Am J Physiol Renal Physiol 301:F1088–F1097

Mamenko MV, Boukelmoune N, Tomilin VN, Zaika OL, Jensen VB, O'Neil RG, Pochynyuk OM (2017) The renal TRPV4 channel is essential for adaptation to increased dietary potassium. Kidney Int 91:1398–1409

ManzanareS D, Gonzalez C, Ivonnet P, Chen RS, Valencia-Gattas M, Conner GE, Larsson HP, Salathe M (2011) Functional apical large conductance, Ca^{2+}-activated, and voltage-dependent K^+ channels are required for maintenance of airway surface liquid volume. J Biol Chem 286:19830–19839

Manzanares D, Srinivasan M, Salathe ST, Ivonnet P, Baumlin N, Dennis JS, Conner GE, Salathe M (2014) IFN-gamma reduction of large conductance, Ca^{2+} activated, voltage-dependent K^+ (BK) channel activity in airway epithelial cells leads to mucociliary dysfunction. Am J Physiol Lung Cell Mol Physiol 306:L453–L462

Manzanares D, Krick S, Baumlin N, Dennis JS, Tyrrell J, Tarran R, Salathe M (2015) Airway surface dehydration by transforming growth factor beta (TGF-beta) in cystic fibrosis is due to decreased function of a voltage-dependent potassium channel and can be rescued by the drug pirfenidone. J Biol Chem 290:25710–25716

Mathialahan T, Maclennan KA, Sandle LN, Verbeke C, Sandle GI (2005) Enhanced large intestinal potassium permeability in end-stage renal disease. J Pathol 206:46–51

Matos JE, Sausbier M, Beranek G, Sausbier U, Ruth P, Leipziger J (2007) Role of cholinergic-activated KCa1.1 (BK), KCa3.1 (SK4) and KV7.1 (KCNQ1) channels in mouse colonic Cl^- secretion. Acta Physiol (Oxf) 189:251–258

Mennitt PA, Wade JB, Ecelbarger CA, Palmer LG, Frindt G (1997) Localization of ROMK channels in the rat kidney. J Am Soc Nephrol 8:1823–1830

Morita T, Hanaoka K, Morales MM, Montrose-Rafizadeh C, Guggino WB (1997) Cloning and characterization of maxi K^+ channel alpha-subunit in rabbit kidney. Am J Physiol Renal Physiol 273:F615–F624

Najjar F, Zhou H, Morimoto T, Bruns JB, Li HS, Liu W, Kleyman TR, Satlin LM (2005) Dietary K^+ regulates apical membrane expression of maxi-K channels in rabbit cortical collecting duct. Am J Physiol Renal Physiol 289:F922–F932

Nakamoto T, Romanenko VG, Takahashi A, Begenisich T, Melvin JE (2008) Apical maxi-K (KCa1.1) channels mediate K^+ secretion by the mouse submandibular exocrine gland. Am J Physiol Cell Physiol 294:C810–C819

Nehrke K, Quinn CC, Begenisich T (2003) Molecular identification of Ca^{2+}-activated K^+ channels in parotid acinar cells. Am J Physiol Cell Physiol 284:C535–C546

Orlowski J, Grinstein S (2004) Diversity of the mammalian sodium/proton exchanger SLC9 gene family. Pflügers Arch 447:549–565

Pacha J, Frindt G, Sackin H, Palmer LG (1991) Apical maxi K channels in intercalated cells of CCT. Am J Physiol Renal Physiol 261:F696–F705

Pluznick JL, Wei P, Carmines PK, Sansom SC (2003) Renal fluid and electrolyte handling in BKCa-beta1$^{-/-}$ mice. Am J Physiol Renal Physiol 284:F1274–F1279

Pluznick JL, Wei P, Grimm PR, Sansom SC (2005) BK-{beta}1 subunit: immunolocalization in the mammalian connecting tubule and its role in the kaliuretic response to volume expansion. Am J Physiol Renal Physiol 288:F846–F854

Puntheeranurak S, Schreiber R, Spitzner M, Ousingsawat J, Krishnamra N, Kunzelmann K (2007) Control of ion transport in mouse proximal and distal colon by prolactin. Cell Physiol Biochem 19:77–88

Rechkemmer G, Frizzell RA, Halm DR (1996) Active potassium transport across Guinea-pig distal colon: action of secretagogues. J Physiol 493:485–502

Ridge FP, Duszyk M, French AS (1997) A large conductance, Ca^{2+}-activated K^+ channel in a human lung epithelial cell line (A549). Biochim Biophys Acta 1327:249–258

Rieg T, Vallon V, Sausbier M, Sausbier U, Kaissling B, Ruth P, Osswald H (2007) The role of the BK channel in potassium homeostasis and flow-induced renal potassium excretion. Kidney Int 72:566–573

Romanenko VG, Thompson J, Begenisich T (2010) Ca^{2+}-activated K channels in parotid acinar cells: the functional basis for the hyperpolarized activation of BK channels. Channels (Austin) 4:278–288

Sailland J, Grosche A, Baumlin N, Dennis JS, Schmid A, Krick S, Salathe M (2017) Role of Smad3 and p38 Signalling in cigarette smoke-induced CFTR and BK dysfunction in primary Hhuman bronchial airway epithelial cells. Sci Rep 7:10506

Sandle GI, Gaiger E, Tapster S, Goodship TH (1986) Enhanced rectal potassium secretion in chronic renal insufficiency: evidence for large intestinal potassium adaptation in man. Clin Sci (Lond) 71:393–401

Sandle GI, Perry MD, Mathialahan T, Linley JE, Robinson P, Hunter M, Maclennan KA (2007) Altered cryptal expression of luminal potassium (BK) channels in ulcerative colitis. J Patho 212:66–73

Sansom SC, O'Neil RG (1985) Mineralocorticoid regulation of apical cell membrane Na^+ and K^+transport of the cortical collecting duct. Am J Phys 248:858–868

Sansom SC, Welling PA (2007) Two channels for one job. Kidney Int 72:529–530

Satlin LM, Sheng S, Woda CB, Kleyman TR (2001) Epithelial Na^+ channels are regulated by flow. Am J Physiol Renal Physiol 280:F1010–F1018

Sausbier M, Matos JE, Sausbier U, Beranek G, Arntz C, Neuhuber W, Ruth P, Leipziger J (2006) Distal colonic K^+ secretion occurs via BK channels. J Am Soc Nephrol 17:1275–1282

Schwartz GJ, Kittelberger AM, Barnhart DA, Vijayakumar S (2000) Carbonic anhydrase IV is expressed in H(+)-secreting cells of rabbit kidney. Am J Physiol Renal Physiol 278:F894–F904

Simon M, Duong JP, Mallet V, Jian R, Maclennan KA, Sandle GI, Marteau P (2008) Over-expression of colonic K^+ channels associated with severe potassium secretory diarrhoea after haemorrhagic shock. Nephrol Dial Transplant 23:3350–3352

Sorensen MV, Leipziger J (2009) The essential role of luminal BK channels in distal colonic K^+ secretion. J Med Invest 56(Suppl):301

Sorensen MV, Matos JE, Sausbier M, Sausbier U, Ruth P, Praetorius HA, Leipziger J (2008) Aldosterone increases KCa1.1 (BK) channel-mediated colonic K^+ secretion. J Physiol 586:4251–4264

Sorensen MV, Matos JE, Praetorius HA, Leipziger J (2010a) Colonic potassium handling. Pflügers Arch 459:645–656

Sorensen MV, Sausbier M, Ruth P, Seidler U, Riederer B, Praetorius HA, Leipziger J (2010b) Adrenaline-induced colonic K^+ secretion is mediated by KCa1.1 (BK) channels. J Physiol 588:1763–1777

Stoner LC, Morley GE (1995) Effect of basolateral or apical hyposmolarity on apical maxi K channels of everted rat collecting tubule. Am J Physiol Renal Physiol 268:F569–F580

Stoner LC, Viggiano SC (1999) Elevation of basolateral K^+ induces K^+ secretion by apical maxi K^+ channels in Ambystoma collecting tubule. Am J Phys 276:R616–R621

Tian L, Duncan RR, Hammond MS, Coghill LS, Wen H, Rusinova R, Clark AG, Levitan IB, Shipston MJ (2001a) Alternative splicing switches potassium channel sensitivity to protein phosphorylation. J Biol Chem 276:7717–7720

Tian L, Hammon MS, Florance H, Antoni FA, SHIPSTON MJ (2001b) Alternative splicing determines sensitivity of murine calcium-activated potassium channels to glucocorticoids. J Physiol 537:57–68

Tian L, Coghill LS, Macdonald SH, Armstrong DL, Shipston MJ (2003) Leucine zipper domain targets cAMP-dependent protein kinase to mammalian BK channels. J Biol Chem 278:8669–8677

Van Dinter TGJ, Fuerst FC, Richardson CT, Ana CA, Polter DE, Fordtran JS, Binder HJ (2005) Stimulated active potassium secretion in a patient with colonic pseudo-obstruction: a new mechanism of secretory diarrhea. Gastroenterology 129:1268–1273

Van D Jr, Fuerst FC, Richardson CT, Ana CA, Polter DE, Fordtran JS, Binder HJ (2005) Stimulated active potassium secretion in a patient with colonic pseudo-obstruction: a new mechanism of secretory diarrhea. Gastroenterology 129:1268–1273

Venglovecz V, Hegyi P, Rakonczay Z Jr, Tiszlavicz L, NardI A, Grunnet M, Gray MA (2011) Pathophysiological relevance of apical large-conductance Ca^{2+}-activated potassium channels in pancreatic duct epithelial cells. Gut 60:361–369

Wade JB, Fang L, Liu J, Li D, Yang CL, Subramanya AR, Maouyo D, Mason A, Ellison DH, Welling PA (2006) WNK1 kinase isoform switch regulates renal potassium excretion. Proc Natl Acad Sci U S A 103:8558–8563

Wade JB, Fang L, Coleman RA, Liu J, Grimm PR, Wang T, Welling PA (2011) Differential regulation of ROMK (Kir1.1) in distal nephron segments by dietary potassium. Am J Physiol Renal Physiol 300:F1385–F1393

Wang Z, Subramanya AR, Satlin LM, Pastor-Soler NM, Carattino MD, Kleyman TR (2013) Regulation of large-conductance Ca^{2+}-activated K^{+} channels by WNK4 kinase. Am J Physiol Cell Physiol 305:C846–C853

Welling PA, Chang YP, Delpire E, Wade JB (2010) Multigene kinase network, kidney transport, and salt in essential hypertension. Kidney Int 77:1063–1069

Wen D, Cornelius RJ, Yuan Y, Sansom SC (2013) Regulation of BK-alpha expression in the distal nephron by aldosterone and urine pH. Am J Physiol Renal Physiol 305:F463–F476

Wen D, Cornelius RJ, Sansom SC (2014) Interacting influence of diuretics and diet on BK channel-regulated K homeostasis. Curr Opin Pharmacol 15C:28–32

Woda CB, Leite M Jr, Rohatgi R, Satlin LM (2002) Effects of luminal flow and nucleotides on [Ca2 +](i) in rabbit cortical collecting duct. Am J Physiol Renal Physiol 283:F437–F446

Yan J, Aldrich RW (2010) LRRC26 auxiliary protein allows BK channel activation at resting voltage without calcium. Nature 466:513–516

Yan J, Aldrich RW (2012) BK potassium channel modulation by leucine-rich repeat-containing proteins. Proc Natl Acad Sci U S A 109:7917–7922

Yang C, Zeng XH, Zhou Y, Xia XM, Lingle CJ (2011) LRRC52 (leucine-rich-repeat-containing protein 52), a testis-specific auxiliary subunit of the alkalization-activated Slo3 channel. Proc Natl Acad Sci U S A 108:19419–19424

Yang C, Gonzalez-Perez V, Mukaibo T, Melvin JE, Xia XM, Lingle CJ (2017) Knockout of the LRRC26 subunit reveals a primary role of LRRC26-containing BK channels in secretory epithelial cells. Proc Natl Acad Sci U S A 114:E3739–E3747

Yue P, Zhang C, Lin DH, Sun P, Wang WH (2013) WNK4 inhibits Ca^{2+}-activated big-conductance potassium channels (BK) via mitogen-activated protein kinase-dependent pathway. Biochim Biophys Acta 1833:2101–2110

Yusuf TE, Soemijarsih M, Arpaia A, Goldberg SL, Sottile VM (1999) Chronic microscopic enterocolitis with severe hypokalemia responding to subtotal colectomy. J Clin Gastroenterol 29:284–288

Zemanova Z, Pacha J (1998) Localization of Na,K-ATPase activity in developing rat distal colon: role of corticosteroids. Mech Ageing Dev 101:129–143

Zhang J, Yan J (2014) Regulation of BK channels by auxiliary gamma subunits. Front Physiol 5:401

Zhang J, Halm ST, Halm DR (2012) Role of the BK channel (KCa1.1) during activation of electrogenic K^{+} secretion in Guinea pig distal colon. Am J Physiol Gastrointest Liver Physiol 303:G1322–G1334

Zhuang J, Zhang X, Wang D, Li J, Zhou B, Shz GD, Denson DD, Eaton DC, Cai H (2011) WNK4 kinase inhibits maxi K channel activity by a kinase-dependent mechanism. Am J Physiol Renal Physiol 301:F410–F419

Chapter 24
Recent Developments in the Pharmacology of Epithelial Ca^{2+}-Activated K^+ Channels

Antonio Nardi, Søren-Peter Olesen, and Palle Christophersen

Abstract Calcium (Ca^{2+})-activated Potassium (K^+) channels (K_{Ca}) in epithelia serve important functions in fluid and salt secretion and may be attractive targets for drug development for epithelial disorders, such as cystic fibrosis, diarrhoea, COPD, polycystic kidney disease and glaucoma. Two very different types of K_{Ca} channels are generally found in epithelia: The big conductance, Ca^{2+}-activated K^+ channel (BK, $K_{Ca}1.1$), and the intermediate conductance, Ca^{2+}-activated K^+ channel (IK, $K_{Ca}3.1$). These channels are differentially expressed in various cells and tissues and serve different physiological and potentially also pathophysiological functions in epithelia. The present chapter aims at giving a brief review of the physiology and function of $K_{Ca}1.1$ and $K_{Ca}3.1$ channels, a description of the classical pharmacology of these channels, an in-depth overview of the status of drug discovery and development programmes in the pharmaceutical industry as well as academia, based on both literature and patent references, and it ends with a discussion of recent advances in the understanding of the molecular pharmacology of these channels. The review is comprehensive, and the focus is throughout ion channel pharmacology with exemplary cases from the epithelial as well as non-epithelial fields.

Keywords MaxiK · $K_{Ca}1.1$ · IK channel · $K_{Ca}3.1$ · Large conductance Ca^{2+}-activated K^+ channel · Intermediate conductance Ca^{2+}-activated K^+ channel · Iberiotoxin

A. Nardi
Italfarmaco, Milan, Italy

S.-P. Olesen
University of Copenhagen, Panum Institute, Copenhagen, Denmark
e-mail: spto@sund.ku.dk

P. Christophersen (✉)
Saniona A/S, Glostrup, Denmark
e-mail: palle.christophersen@saniona.com

K. L. Hamilton, D. C. Devor (eds.), *Studies of Epithelial Transporters and Ion Channels*, Physiology in Health and Disease,
https://doi.org/10.1007/978-3-030-55454-5_24

24.1 Introduction

24.1.1 K_{Ca} Channel Expression and Function in Epithelia

The presence of Ca^{2+}-activated K^+ channels has been demonstrated in most epithelia, either by physiological or expression studies. SK ($K_{Ca}2.1$), IK ($K_{Ca}3.1$) and BK ($K_{Ca}1.1$) channels have been reported, but the most comprehensive expression across different types of epithelia is the $K_{Ca}3.1$ and $K_{Ca}1.1$ channels with physiological roles varying from one epithelium to another. $K_{Ca}3.1$ channels are broadly expressed in both secretory epithelia like the colonic crypts and exocrine glands as well as in non-secretory epithelia like the skin and the bladder epithelium (Thompson-Vest et al. 2006). In secretory epithelia, $K_{Ca}3.1$ channels are often expressed on the basolateral membrane, where their activation hyperpolarizes the cell and thereby—in cooperation with other K^+ channels—helps maintain a large driving force for Cl^- secretion across the luminal membrane via chloride channels, such as the cAMP-activated CFTR and the Ca^{2+}-activated TMEM16A (Schroeder et al. 2008; Caputo et al. 2008). Pharmacological activation of $K_{Ca}3.1$ channels by compounds such as 1-EBIO (for full chemical names please see the figure legends associated with a detailed description of each pharmacological agent below) and chlorzoxazone was initially suggested as a way to increase Cl^- and water secretion across epithelia (Devor et al. 1996) and even as a possible therapeutic principle for restoration of epithelial transport in patients suffering from cystic fibrosis (Mall et al. 2003; Roth et al. 2011). Rationalized by a similar line of arguments, $K_{Ca}3.1$ channel inhibition was shown by the use of $K_{Ca}3.1$ channel knockout mice and pharmacological inhibitors to be a realistic opportunity for treatment of diarrhoea (Rufo et al. 1997). Physiologically, activation of epithelial $K_{Ca}3.1$ channels is coupled to various secretagogues, such as ATP acting via P2Y receptors (Dutta et al. 2009; Hayashi et al. 2011; Bardou et al. 2009) or acetylcholine acting via muscarinic receptors (Matos et al. 2007). In non-secretory epithelia, the role of $K_{Ca}3.1$ channels is less clear but they may participate in epithelial cell volume regulation and thereby protect, for example bladder and kidney epithelial cells against osmotic stress, or they may be involved in the controlled proliferation, repair and differentiation of stratified epithelia (Denda et al. 2007; Manaves et al. 2004; Koegel and Alzheimer 2001; Koegel et al. 2003).

$K_{Ca}1.1$ channels are also found in secretory epithelia and are typically expressed luminally where they contribute to K^+ secretion in parallel with Cl^-. In the airways, epithelial $K_{Ca}1.1$ channels can be activated by secreted ATP via P2Y receptors on the luminal membrane, which probably helps preserve a well-hydrated surface liquid and a low viscosity, which is important for normal cilia beat rate and function (Manzanares et al. 2011). $K_{Ca}1.1$ channels also play an important role for global K^+ homeostasis via K^+ excretion into the intestinal lumen of the distal colon. That is, upon feeding a K^+-rich diet to mice, a $K_{Ca}1.1$ splice variant sensitive to cAMP regulation was specifically up-regulated by aldosterone and gave rise to an increased adrenaline-dependent K^+ secretory response (Sorensen et al. 2008, 2010; Larsen et al. 2017). $K_{Ca}1.1$ channels are also functionally expressed in the ciliary nonpigmented epithelium of the eye, where they may be important for the regulation of aqueous humour secretion (Ryan et al. 1998; Cullinane et al. 2002).

24.1.2 Basic Properties of $K_{Ca}1.1$ and $K_{Ca}3.1$ Channels

BK channels ($K_{Ca}1.1$) are named according to their large (Big) unitary conductance of approximately 250 pS as measured electrophysiologically in symmetric K^+ solutions using patch-clamp or lipid bilayer techniques. Despite their large single-channel conductance, $K_{Ca}1.1$ channels are highly K^+ selective, with the reversal potential following closely the Nernst potential for potassium ion when Na^+ and K^+ concentration gradients are varied. In contrast to many other potassium channels, the single-channel IV relation in symmetric K^+ is linear over a broad voltage span ($\pm$ 100 mV) and only exhibits weak outward rectification ("constant field-like rectification") with varied Na^+/K^+ ratios in the extracellular phase. $K_{Ca}1.1$ channel gating is basically controlled by two factors: The intracellular Ca^{2+} concentration and the membrane voltage (V_m). At "resting" levels of Ca^{2+} and V_m, the $K_{Ca}1.1$ channels are effectively closed, whereas the combination of increased intracellular Ca^{2+} and depolarized membrane potential activates the channel. A systematic analysis reveals that the open state probability dependency of V_m follows classic Boltzmann equation curves as most K_v channels, which are shifted gradually in the hyperpolarizing direction by increasing intracellular Ca^{2+} concentrations (Barrett et al. 1982; Latorre et al. 1982).

$K_{Ca}1.1$ channels are encoded by the *KCNMA1* gene, and the human protein is 1182 amino acids long (Dworetzky et al. 1994). This is assembled as a homotetramer in the membrane and constitutes the functional $K_{Ca}1.1$ channel with the central K^+ selective pore created at the interface of the subunits. In analogy to other K^+ channels, the re-entrant P-loop between TM5 and TM6 contains the GYG sequence that forms the main selectivity filter of the pore. The extremely long C-termini are clearly exposed to the intracellular side of the membrane and contain a pair of homology RCK domains regulating the K^+ conductance. These latter domains create a "gating ring" structure (Wu et al. 2010) upon which Ca^{2+} binding occurs directly at Ca^{2+} binding "bowls" (Schreiber and Salkoff 1997), i.e. molecular regions that are rich in negatively charged amino acids. Selective binding of Ca^{2+} shortens the C-terminal-RCK1 linker, which seems to be central for opening of the channel gate (Javaherian et al. 2011), which itself is positioned in the pore region (Chen et al. 2014). In contrast to the C-terminus, it was debated whether the N-terminus is extracellular or intracellular. It is now clear that the $K_{Ca}1.1$ channel is unique among potassium channels in being a 7 TM channel with an extra transmembrane helix called TM0 preceding the equivalent TM1–4 in Kv channels. In analogy with K_v channels, the voltage-sensing part of the BK channel is constituted by TM0-TM4, with the highly charged TM4 likely playing a central role in voltage sensing, but with additionally important characteristics imposed by the participation of TM0 (Pantazis et al. 2010a, 2010b; Niu et al. 2013).

Understanding the nature of the cooperation of the voltage- and Ca^{2+}-sensing parts of the protein represents a major challenge for the molecular biophysics of $K_{Ca}1.1$ channels and is also a major challenge for rationalizing the pharmacology of the gating process. The TM0 of the $K_{Ca}1.1$ channel alpha-subunit plays a unique role in the co-assembly and interactions with the β-subunits (β1–4; encoded by *KCNMB*1–4) (Wallner et al. 1996), and the channel also interacts with leucine-rich repeat-containing proteins (LRRC26, -52,-55,-38), which are now considered as $K_{Ca}1.1$ γ-subunits 1–4 (Zhang and Yan 2014; Yan and Aldrich 2010). Both β- and γ-subunits can influence both the gating and pharmacology of the channel (see later). Adding to $K_{Ca}1.1$ heterogeneity is the fact that several splice variants exist as well (McCobb et al. 1995).

IK channels ($K_{Ca}3.1$) are so named because their single-channel conductance is Intermediate (approximately 30 pS at 0 mV and symmetric K^+ concentrations) between the high conductance of $K_{Ca}3.1$ channels and the small conductance of SK channels ($K_{Ca}2.x$). As for the $K_{Ca}1.1$ channel, the $K_{Ca}3.1$ channel pore is highly selective for K^+ over Na^+, but in contrast to $K_{Ca}1.1$ channels, the single-channel IV curve reveals a prominent inward rectification in symmetric K^+. Also, in contrast to $K_{Ca}1.1$ channels, the gating of $K_{Ca}3.1$ channels is not influenced by V_m, rather it is solely dependent on the intracellular Ca^{2+} concentration: At "resting" Ca^{2+} levels the channel is closed, but it is sensitive to much smaller increases (nM) in intracellular Ca^{2+} concentration than the $K_{Ca}1.1$ channel (μM, at normal ranges of membrane potentials) and may even be more sensitive to low Ca^{2+} fluctuations than $K_{Ca}2.x$ channels (Ishii et al. 1997; Joiner et al. 1997).

The $K_{Ca}3.1$ channel belongs to the same gene family as the $K_{Ca}2.x$ channels (the *KCNN* family, encoded by *KCNN4*) with which it shares many functional and physiological traits. The $K_{Ca}3.1$ channel alpha subunit (427 aa) (Ishii et al. 1997; Joiner et al. 1997) is a 6 TM protein only remotely related to the $K_{Ca}1.1$ channel except for sharing the overall K^+ channel topology, at least around the pore region. The $K_{Ca}3.1$ alpha subunit does not have any binding site for Ca^{2+} but relies on constitutively bound calmodulin (Cam) at the C-terminal, which acts as the Ca^{2+} sensor. Upon binding of Ca^{2+} to calmodulin, a conformational change is transferred via the TM4-TM5 linker to the pore region, where the physical gate is positioned in $K_{Ca}3.1$ and $K_{Ca}2.x$ channels (Garneau et al. 2014; Bruening-Wright et al. 2007). The understanding of this transformation is pivotal to the understanding of gating pharmacology (selectivity, site- and mode-of-actions) of both $K_{Ca}3.1$ and $K_{Ca}2.x$ channels (see later section). For additional information about KCa3.1, the readers are directed to Chap. 22 of this volume.

24.1.3 Basic Pharmacology of $K_{Ca}1.1$ and $K_{Ca}3.1$ Channels

The pivotal description of $K_{Ca}1.1$ channels in 1981–1982 by the groups of Karl Magleby and Ramon Latorre initiated decades of extensive characterizations at the single-channel level, including detailed descriptions of inhibition by inorganic ions, like Cs^+ and Ba^{2+} as well as small organic ions like TEA^+. The main conclusions from this purely biophysical (non-structural) approach to $K_{Ca}1.1$ channel pharmacology were that the positively charged inhibitors mimicked the K-ion and were able to pass some way into the permeation pore, where they were immobilized at a selectivity filter and thereby prevented the normal flow of K-ions through the channel. In other words, these inhibitors are all blockers of the open channel. Whereas easily comprehended for the monovalent alkali metal ions, Ba^{2+} might intuitively have been thought of as a competitive Ca^{2+} analogue preventing channel opening at the Ca^{2+} binding site. This was, however, not the case! (Miller 1987; Neyton and Miller 1988). Being "substrate analogues" for the K-ion, it is not very surprising that the blocking mechanism of these ions gave important insights to specific binding sites positioned deep in the pore in the electrical field generated by the membrane potential. Remarkably good quantitative descriptions of different blocking phenomena, such as voltage-dependent block (increased binding affinity with increased potential), "fast" and "slow" blockers (*apparent* reduction in current vs. *apparent* change of gating; high vs. low blocker k_{off}) as well as sidedness of action (different blocker affinity from inside vs. outside of channel) were achieved (Kehl 1996). In addition to the blocker ions being biophysical "probes," the studies clearly pointed towards the existence of unique pharmacological sites in the open BK channel pore region, since both TEA^+ added from the outside and Ba^{2+} added from the inside exhibited much more potent block of BK channels than of most other K^+ channels including other Ca^{2+}-activated K^+ channels (Brown et al. 1988; Lang and Ritchie 1990; Rae et al. 1990).

The first described high-affinity peptide blocker of the $K_{Ca}1.1$ channel was charybdotoxin, a constituent from the venom of the scorpion *Leiurus quinquestriatus hebraeus* (Miller et al. 1985). Charybdotoxin (ChTx) was shown to bind to the outer pore mouth, thus causing blocked periods lasting several seconds in the single-channel record. Closer analysis revealed a simple bimolecular reaction mechanism with k_{on} proportional to blocker concentration and a concentration-independent k_{off} (Anderson et al. 1988). Despite its high affinity, charybdotoxin proved not to be a specific blocker of BK channels, since several K_v channels and also the $K_{Ca}3.1$ channel were blocked with very high affinity. Chemically modified charybdotoxin was nonetheless very important in the early days for defining $K_{Ca}1.1$ channel pharmacology, since a binding assay—developed by scientists at Merck—based on the ^{125}I-labelled peptide, picked up $K_{Ca}1.1$ modulators with radically different chemical structures and modulatory properties, including certain secondary metabolites of "ryegrass staggers," fungi which lie behind outbreaks of frequent tremors and convulsions to livestock cattle in New Zealand and Australia. These tremorgenic indole alkaloids, which are steroid-like structures, interacted in complex ways with ^{125}I-ChTx binding to smooth muscle membranes and became the first potent class of non-peptide $K_{Ca}1.1$ channel inhibitors identified (Knaus et al. 1994). Interestingly, the functional mode-of-actions of these inhibitors were more complex than the simple blocking mechanism shown by the peptide inhibitors, since they clearly influenced gating parameters (Strobaek et al. 1996; Sanchez and McManus 1996) and are thus better categorized as negative gating modulators than pore blockers. Another class of molecules also displacing ^{125}I-ChTx was the dehydrosoyasaponin-1 (DHS-1), a secondary metabolite of the medical herb *Desmodium adscendens*. DHS-1 was, in contrast to the indole alkaloids, identified in functional assays as a very potent activator of $K_{Ca}1.1$ channels (McManus et al. 1993) acting exclusively from the inside of the membrane and only in the presence of the beta1 subunit. Shortly thereafter, the first synthetic molecule (NS1619, see later) with $K_{Ca}1.1$ activating properties was described (Olesen et al. 1994). In contrast to DHS-1, NS1619 easily crosses biological membranes, works independently of whether it is added from the outside or inside of the membrane and does not require the beta subunit. Despite its much lower potency, NS1619 became the most widely used tool compound for investigation of positive modulation of $K_{Ca}1.1$ channels for many years, and also inspired several lead optimization programmes in the pharmaceutical industry. These original positive modulators did not make it into the clinical development phases and have also largely been replaced nowadays by more adequate research compounds as will be discussed in more details in the section to follow.

Whereas the $K_{Ca}1.1$ channel was only recognized after the development of single-channel recording techniques, the description of Ca^{2+}-activated K^+ fluxes and hyperpolarizations mediated by the erythrocyte $K_{Ca}3.1$ channel (Hoffman et al. 2003; Lew and Ferreira 1976; Heinz and Passow 1980; Vestergaard-Bogind et al. 1985) were studied for decades before its characterization at the single-channel level. Therefore, it is not particularly surprising that various low-affinity compounds were identified more-or-less by serendipity during the years. One important example is cetiedil, a drug with vasodilative properties, which was shown to counteract sickling of erythrocytes from sickle cell patients and also to inhibit the Ca^{2+}-activated K^+ fluxes (Berkowitz and Orringer 1981). Despite the fact that cetiedil entered clinical trials for sickle cell anaemia that was later discontinued, it was the identification of two other compound classes as $K_{Ca}3.1$ channel inhibitors that became the guides for further optimization towards clinical useful compounds: the triarylmethane (TRAM) class of antimycotics (Alvarez et al. 1992) and the dihydropyridine class of antihypertensives (Ellory et al. 1994) (see later). Potent class representatives are clotrimazole and nitrendipine. Despite its limited usefulness in vivo, clotrimazole was shown to reduce Cl^- secretion from colon epithelia cells by inhibiting the basolateral K^+ conductance and to inhibit cholera toxin-induced diarrhoea in mice (Rufo et al. 1997).

In contrast to the inhibitor field, the multiple erythrocyte K^+ transport studies did not reveal clear leads for the development of activators of $K_{Ca}3.1$ channels, with propranolol as a possible (and still unexplored) exception (Schwarz et al. 1989). The pivotal observation in that area rather came from the study of basic epithelial transport. Daniel Devor investigated the action of NS1619-like molecules (NS1619 per se does *not* work; Balut et al. 2012), i.e. benzimidazolones on Cl^- secretion across colon and tracheal cell layers, and found compelling evidence for a direct activating effect of the compound 1-EBIO on the basolateral Ca^{2+}-activated K^+ conductance and thus on overall transepithelial Cl^- secretion (Devor et al. 1996). The biophysical properties of the epithelial 1-EBIO-sensitive channel were indistinguishable from the cloned $K_{Ca}3.1$ channel (Ishii et al. 1997) which was soon after also shown to be sensitive to 1-EBIO (Jensen et al. 1998). Interestingly, several clinically used drugs with an overall structure like 1-EBIO additionally activated the $K_{Ca}3.1$ channel and stimulated epithelial Cl^- secretion (Singh et al. 2000; Syme et al. 2000). These compounds also activate $K_{Ca}2.x$ channels, albeit with approximately ten-fold lower potency.

24.2 $K_{Ca}1.1$ Channel Modulator Chemistry

24.2.1 $K_{Ca}1.1$ Channel Activators

Consistent with the wealth of publications highlighted in the previous sections suggesting that the pharmacological modulation of $K_{Ca}1.1$ channels is a clear opportunity to develop new therapeutics, several research efforts have been directed towards the identification of $K_{Ca}1.1$ channel activators and inhibitors (Nardi and Olesen 2008). As a result, a variety of pharmacological agents have been identified, although very little success has been made in bringing these forward to clinical settings, neither in epithelia related disorders nor in other diseases. With regard to $K_{Ca}1.1$ activators, the attrition rate has been exceptionally high and it culminated in only a few development compounds, i.e. andolast, NS8, BMS204352, BMS223131 (Fig. 24.1) and TA1702 (chemical structure undisclosed) (Nardi and Olesen 2008).

NS8 (Fig. 24.1) (Parihar et al. 2003; La Fuente et al. 2014) was a pyrrole development compound designed by Nippon Shinyaku. Notwithstanding its low

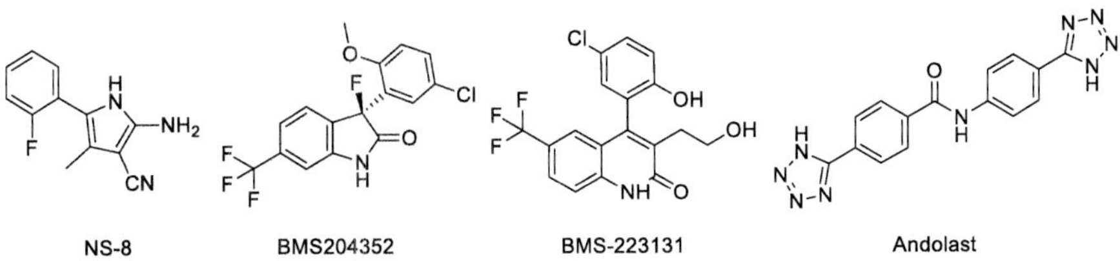

Fig. 24.1 Chemical structures of NS8, BMS204352, BMS-223131—$K_{Ca}1.1$ channel openers progressed to clinical development and discontinued. Andolast is a topical and supposedly a non-selective BK channel opener. Previously in phase III for respiratory diseases. **NS-8**: 2-amino-5-(2-fluorophenyl)-4-methyl-1H-pyrrole-3-carbonitrile. **BMS204352**: (S)-3-(5-chloro-2-methoxyphenyl)-3-fluoro-6-(trifluoromethyl)indolin-2-one. **BMS-223131**: 4-(5-chloro-2-hydroxyphenyl)-3-(2-hydroxyethyl)-6-(trifluoromethyl)quinolin-2(1H)-one. **Andolast**: N-(4-(1H-tetrazol-5-yl)phenyl)-4-(1H-tetrazol-5-yl)benzamide

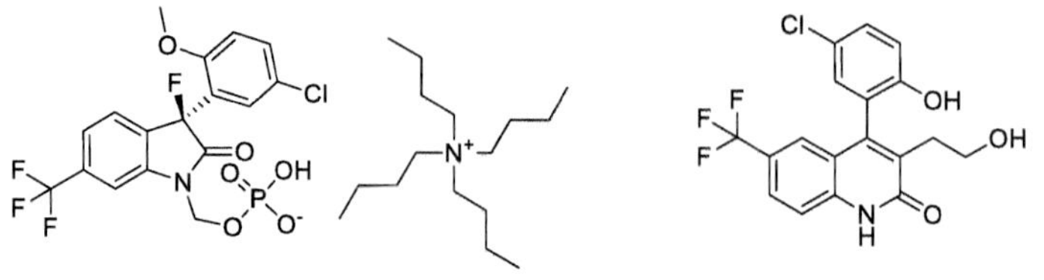

Fig. 24.2 BMS-554216 and BMS223131, chemical entities structurally-related to BMS204352 (Fig. 24.1). **BMS-554216**: tetrabutylammonium (S)-(3-(5-chloro-2-methoxyphenyl)-3-fluoro-2-oxo-6-(trifluoromethyl)indolin-1-yl)methyl hydrogenphosphate. **BMS-223131**: 4-(5-chloro-2-hydroxyphenyl)-3-(2-hydroxyethyl)-6-(trifluoromethyl)quinolin-2(1H)-one

potency at the channel and functional effects in the micromolar range, the in vivo pharmacological profile in preclinical settings has prompted the European licensee Apogepha Arzneimittel to develop it first up to phase II for the potential treatment of overactive bladder and to later discontinue it due to a lack of efficacy at the expected therapeutic dosage in a proof-of-concept clinical trial (Akada et al. 2002; Nakamura et al. 1996, 1999; Malysz et al. 2004; Tanaka et al. 2003). Further structural optimization of NS8 either via expansion of the 5-membered ring into a 6-membered ring (Harada et al. 2003) or via its fusion with other rings (Turner et al. 2003, 2004) did afford more potent derivative compounds but they did not emerge up to the clinical stage.

BMS204352 (Fig. 24.1) (MaxiPost™) is a relatively non-selective BK channel activator whose efficacy as a neuroprotective agent in in vivo pharmacological models of stroke was demonstrated in a series of studies (Hewawasam et al. 1997; Gribkoff et al. 2001; Mackay 2001; Krishna et al. 2002a, b, c; Jensen 2002; Zhang et al. 2005). Congruous with this experimental evidence, and the rationale that a $K_{Ca}1.1$ channel activator might prevent the ischemic cell depolarization and excitotoxicity, Bristol-Myers Squibb progressed the compound in phase III for the treatment of acute ischemic stroke. The drug failed to show superiority over placebo, a result that was disappointingly also confirmed in a Magnetic Resonance Imaging sub-study comparing the effect of 1 mg/Kg of MaxiPost with placebo in more than 100 patients (Warach et al. 2002). The poor aqueous solubility of BMS204352 has considerably limited the administrable dose by intravenous infusion, and this has been claimed as one of the potential explanations for lack of efficacy. Pro-drugs of BMS204352 bearing diverse solubilizing moieties on the amidic nitrogen (Gillman et al. 2003; Gillman and Bocchino 2004; Starrett et al. 2005), including BMS554216 (Fig. 24.2), were subsequently explored. The latter compound has been reported as being more water-soluble, able to induce higher brain concentration in preclinical species, and to be efficacious in the same species in reducing oedema after traumatic brain injury (Starrett 2004). This compound may have entered the clinical phase at some point, yet its further development has not been reported recently, and it is presumed to be halted. Additional efforts around the same chemical space of BMS204352 capitalized more structurally-related $K_{Ca}1.1$ openers that were evaluated pre-clinically and clinically by the same company as new therapeutics for urinary incontinence, erectile dysfunction and bowel disorders. Among several others, BMS223131 was synthesized and profiled (Hewawasam et al. 2003, 2004; Boy et al. 2004, 2005a, b; Hewawasam et al. 2005; Boy 2005), and notwithstanding the modest ability to increase the $K_{Ca}1.1$ current at 20 μM by about four-fold, it showed iberiotoxin-sensitive efficacy in an in vivo rat model of erectile function (Hewawasam et al. 2004) and reduced stress-induced colonic motility and visceral nociception (Sivarao et al. 2005). The same compound was later acquired in an in-licensing deal by Xention (XEN-D0401) and, albeit initially reported in the company clinical pipeline for the treatment of overactive bladder, no further developments have been announced. One development issue that was clearly identified and reported in the public domain was the inhibitory action on the cytochrome P450 (CYP) 2C9 isoform (IC_{50} of 1.7 μM), so that a medicinal chemistry effort aimed at reducing the CYP inhibitory properties while preserving the BK channel activity has been attempted (Vrudhula et al. 2005). Additional complications to this series of molecules may also be their lack of selectivity between different K-channels, as demonstrated for BMS204352 with respect to Kv7 channels (Schroder et al. 2001) and later confirmed in a number of papers.

TA1702 (chemical structure not disclosed) is a $K_{Ca}1.1$-activator that GlaxoSmithKline (GSK) acquired from Tanabe Seiyaku and was in clinical development (Phase I, 2005) for the treatment of overactive bladder (Nardi and Olesen 2008). A couple of years later, with no public disclosures on the scientific grounds for this decision, the compound was again withdrawn both from the product development pipeline of Mitsubishi Tanabe Pharma Corporation and GSK (Nardi and Olesen 2008).

Andolast® (or CR2039) (Fig. 24.1), as dry powder inhaled formulation, was developed (phase III) by Rottapharm, a multinational pharmaceutical group based in Italy, for the potential treatment of bronchial asthma, allergic rhinitis and chronic obstructive pulmonary disease (COPD) (Nardi and Olesen 2008). Andolast was reported to be more efficacious than placebo in a clinical study of mild to moderate asthma (Malerba et al. 2015). It is believed that the $K_{Ca}1.1$-opening activity is an important mechanism of action for its efficacy as shown in preclinical models (Rovati et al. 2012) in defiance of diverse pharmacological properties and action at different cellular levels (Makovec et al. 1990, 1992; Malerba et al. 2009). Andolast is currently not mentioned at the company's homepage and is probably not developed further.

Contradictory to the poor clinical success of selective $K_{Ca}1.1$ channel openers, a wealth of very diverse late clinical or marketed compounds have been reported to possess $K_{Ca}1.1$ channel activation as an add-on or ancillary mechanism, including Puerarin or Kakonein (Phase II, treatment of alcohol abuse; Sun et al. 2007), Cilostazol (Launched, intermittent claudication, Pletaal®; Wu et al. 2004), Zonisamide (Launched, Excegran®, epilepsy; Huang et al. 2007), Riluzole (Launched, Amyotrophic lateral sclerosis, Rilutek®; Wang et al. 2008), Unoprostone (Launched, Glaucome, Rescula®; Cuppoletti et al. 2012), Acetazolamide (Launched, epilepsy etc., Diamox®; Tricarico et al. 2004), as well as a number of cox-2 inhibitors, such as Rofecoxib (Launched, arthritis etc., Celebrex®; Aihara et al.

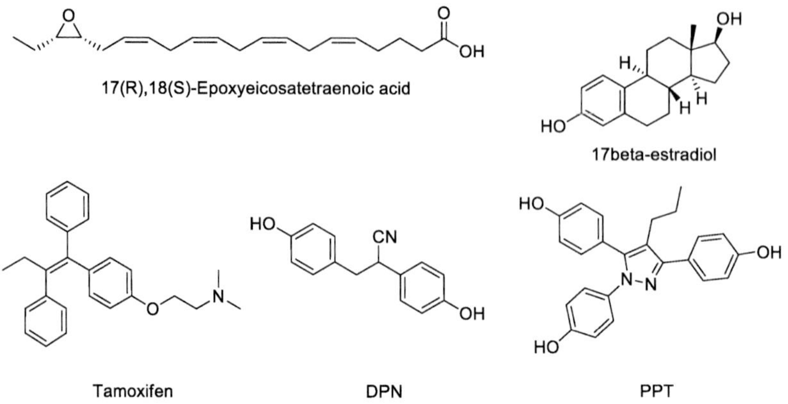

Fig. 24.3 Endogenous $K_{Ca}1.1$-channel modulators and structural analogues: 17(R),18(S)-Epoxyeicosatetraenoic acid, 17β-oestradiol, tamoxifen, DPN, PPT. **17(R),18(S)-Epoxyeicosatetraenoic acid**: (5Z,8Z,11Z,14Z)-16-((2R,3S)-3-ethyloxiran-2-yl)hexadeca-5,8,11,14-tetraenoic acid. **17beta-oestradiol**: (8R,9S,13S,14S,17S)-13-methyl-7,8,9,11,12,13,14,15,16,17-decahydro-6H-cyclopenta [a]phenanthrene-3,17-diol. **Tamoxifen**: (Z)-2-(4-(1,2-diphenylbut-1-en-1-yl)phenoxy)-N,N-dimethylethanamine. **DPN**: 2,3-bis(4-hydroxyphenyl)propanenitrile. **PPT**: 4,4′,4″-(4-propyl-1H-pyrazole-1,3,5-triyl)triphenol

2004), and Vinpocetine (Launched, cognitive disorders etc., Cavinton®; Wu et al. 2001). Moreover, activation of $K_{Ca}1.1$ channels can be produced by a multiplicity of endogenous molecules, including polyunsaturated arachidonic acid, metabolites of lipoxygenase (Duerson et al. 1996) and cytochrome P450 epoxygenase (Morin et al. 2007; Hercule et al. 2007; Lu et al. 2001; Benoit et al. 2001; Fukao et al. 2001), which have also been employed as starting points for further chemical derivatization in the synthesis of new epoxyeicosatrienoic acid derivatives (Yang et al. 2007; Falck et al. 2003, 2011; Lauterbach et al. 2002; Schunck et al. 2010; Lucas et al. 2010; Hercule et al. 2007) (Fig. 24.3). The finding that 17β-oestradiol (Fig. 24.3), an endogenous sex hormone, is a direct and β-subunit-dependent activator of $K_{Ca}1.1$ channels (De Wet et al. 2006; Valverde et al. 1999) prompted additional studies leading to the discovery of the $K_{Ca}1.1$ opening properties of a number of estrogenic compounds without steroid structure, including tamoxifen (Coiret et al. 2007; Dick et al. 2001, 2002; Dick and Sanders 2001) and close analogues (Sha et al. 2005), DPN and PPT (Fig. 24.3; Wang et al. 2007) as well as steroid structures with or without hormonal activity, such as lithocholate (Bukiya et al. 2007) and synthetic analogues (Bukiya et al. 2013), testosterone and corticosterone (King et al. 2006; Unemoto et al. 2007). Nitric oxide (NO) is one more example of an endogenous mediator exerting $K_{Ca}1.1$ activating properties directly and indirectly (Carrier et al. 1997; Gragasin et al. 2004; Archer et al. 1994; Haburcak et al. 1997; Bolotina et al. 1994; Lee et al. 2006), and it theoretically opens the opportunity to endow any drug with a donor of nitric oxide to induce "physiological" $K_{Ca}1.1$ channel activation (Vaali et al. 1998).

Natural products have also been demonstrated to be a rich source of $K_{Ca}1.1$ channel openers, which have been the object of comprehensive reviews (Nardi et al. 2003; Nardi and Olesen 2008). Polyphenolic compounds, such as Puerarin (Sun et al. 2007), Mallotoxin (Zakharov et al. 2005), trans-Resveratrol (Wu 2003; Sheu et al. 2008), and several other flavonoids or terpene compounds, such as Pimaric acid (PiMA; Imaizumi and Ohwada 2002; Imaizumi et al. 2002) and dehydroabietic acid (Ohwada et al. 2003), are clearly privileged scaffolds whose promiscuous pharmacological profile has led to several synthetic efforts to improve upon both potency and selectivity (Ohwada et al. 2003; Li et al. 2003c; Rittenhouse et al. 1997; Tashima et al. 2006; Owada et al. 2007; Cui et al. 2008) and also the well-profiled Cym04 (Fig. 24.4; Cui et al. 2010; Gessner et al. 2012). Very recently, a new series of flavonoids isolated from the medical herb *Sophora flavescens* have been characterized as $K_{Ca}1.1$ activators, with Kuraridinol (Fig. 24.4) being the most potent (Lee et al. 2018). The activating effect of mallotoxin was shown to be essentially absent in the presence of $K_{Ca}1.1$ γ-subunits (Guan et al. 2017), and it remains to be seen if this counts for other flavonoids as well.

Whereas pharmacological selectivity and potency have certainly been the limit for preclinical studies of endogenous substances, natural compounds and the original prototype phenolic $K_{Ca}1.1$ activators (NS004, NS1619, NS1608; Nardi and Olesen 2008), a more potent and more selective pharmacological tool NS11021 (Bentzen et al. 2007), equipped also with its corresponding negative control NS13558 (Bentzen et al. 2010), has recently been disclosed, and is available commercially by several vendors (Tocris, Sigma; Fig. 24.5). NS11021, originally discovered at NeuroSearch, currently appears to be the pharmacological BK channel activating tool of choice (EC_{50}: 400 nM) given the widespread use among different research groups (Wulf-Johansson et al. 2010; Wojtovich et al. 2013; Venglovecz et al. 2011; Liu et al. 2014; Layne et al. 2010; Kun et al. 2009; Kiraly et al. 2013; Borchert et al. 2013; Bentzen et al. 2009; Bednarczyk et al. 2013; Aon et al. 2010; Zaidman et al. 2017; Lui et al. 2019). A common feature for these latter molecules, as well as for most of the known $K_{Ca}1.1$ activators, is the presence of an acidic functionality, especially carboxylic acids. This plethora includes, for instance CTBIC (Fig. 24.5; Gormemis et al. 2005), a compound able to reduce bladder voiding, relax bladder smooth muscle and inhibit intestinal motility (Dela Pena et al. 2009a, b) and whose carboxylic acid group has been suggested to be critical for its binding mode (Lee et al. 2012).

Carboxylic acids are, however, regarded with caution in current drug development as they may act as acylators either via glucuronidation or acyl CoA formation, which may constitute bioactivation pathways that lead to toxicity (Sallustio et al. 2000; Li et al. 2003a, b). Tetrazole moieties, present for instance in NS11021, are common bioisostere for carboxylate in view of their planar structure and ability to ionize at physiological pH (Pinter et al. 2011; Biot et al. 2004; Song et al. 2013). Whereas tetrazoles may offer several advantages over the carboxylic acids, such as the ability to circumvent some of the classical phase II biotransformation of carboxylic acids or the higher lipophilicity leading to a higher permeability, they may nonetheless suffer from similar undesirable properties, such as high protein binding

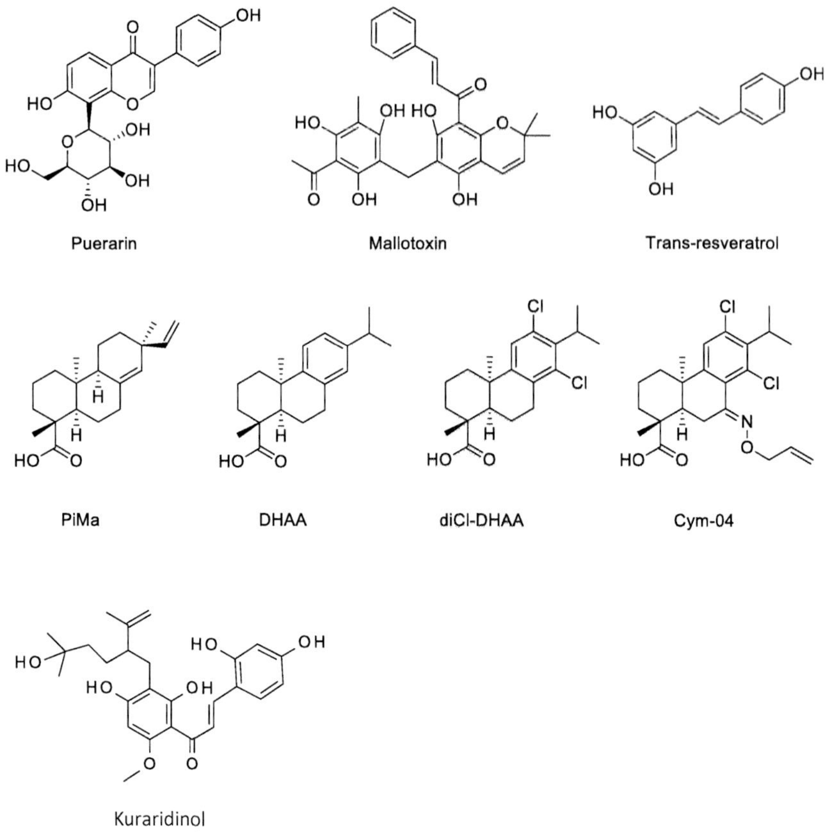

Fig. 24.4 Natural $K_{Ca}1.1$ channel activators Puerarin, Mallotoxin, Trans-resveratrol, PiMa, DHAA and the synthetic derivatives diCl-DHAA and Cym-04. **Puerarin**: 7-hydroxy-3-(4-hydroxyphenyl)-8-((2S,3R,4R,5S,6R)-3,4,5-trihydroxy-6-(hydroxymethyl)tetrahydro-2H-pyran-2-yl)-4H-chromen-4-one. **Mallotoxin**: (E)-1-(6-(3-acetyl-2,4,6-trihydroxy-5-methylbenzyl)-5,7-dihydroxy-2,2-dimethyl-2H-chromen-8-yl)-3-phenylprop-2-en-1-one. **Trans-resveratrol:** (E)-5-(4-hydroxystyryl) benzene-1,3-diol. **PiMa**: (1R,4aS,4bS,7S,10aR)-1,4a,7-trimethyl-7-vinyl-1,2,3,4,4a,4b,5,6,7,9,10,10a-dodecahydrophenanthrene-1-carboxylic acid. **DHAA**: (1R,4aR,10aR)-7-isopropyl-1,4a-dimethyl-1,2,3,4,4a,9,10,10a-octahydrophenanthrene-1-carboxylic acid. **diCl-DHAA**: (1R,4aR,10aR)-6,8-dichloro-7-isopropyl-1,4a-dimethyl-1,2,3,4,4a,9,10,10a-octahydrophenanthrene-1-carboxylic acid. **Cym-04**: (1R,4aR,10aR,E)-9-((allyloxy)imino)-6,8-dichloro-7-isopropyl-1,4a-dimethyl-1,2,3,4,4a,9,10,10a-octahydrophenanthrene-1-carboxylic acid. **Kuraridinol**: (E)-1-[2,4-dihydroxy-3-(5-hydroxy-5-methyl-2-prop-1-en-2-ylhexyl)-6-methoxyphenyl]-3-(2,4-dihydroxyphenyl)prop-2-en-1-one

and limited extravascular tissue distribution (Maillard et al. 2000; Treiber et al. 2003; Davi et al. 2000; Laeis et al. 2001). With this in mind, NeuroSearch recently launched a high-throughput screening campaign employing a new FLIPR-based Tl^{+}-flux assay as a primary screen and an automated QPatch assay as a secondary screen in order to identify and profile small-molecule $K_{Ca}1.1$ activators without any acidic functionalities present in the molecule (Nausch et al. 2014). These efforts

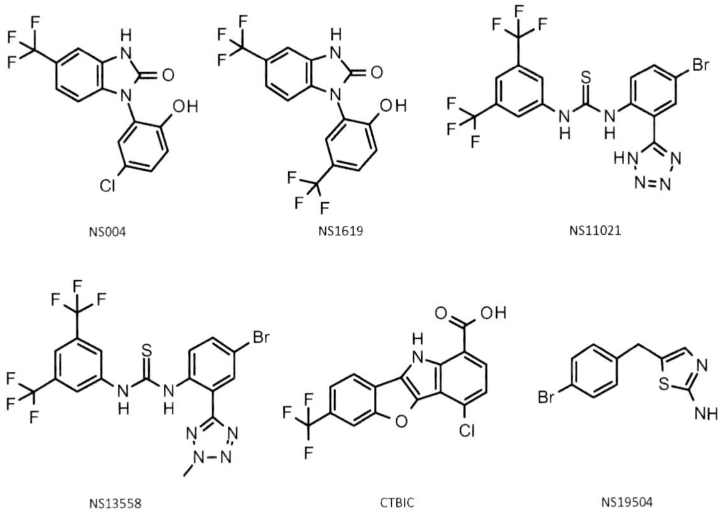

Fig. 24.5 $K_{Ca}1.1$ activator tool compounds: the former and outclassed NS004 and NS1619, the current NS11021 and the newly-identified NS19504. **NS004**: 1-(5-chloro-2-hydroxyphenyl)-5-(trifluoromethyl)-1H-benzo[d]imidazol-2(3H)-one. **NS1619**: 1-(2-hydroxy-5-(trifluoromethyl)-phenyl)-5-(trifluoromethyl)-1H-benzo[d]imidazol-2(3H)-one. **NS11021**: 1-(3,5-bis (trifluoromethyl)phenyl)-3-(4-bromo-2-(1H-tetrazol-5-yl)phenyl)thiourea. **NS13558**: 1-(3,5-bis (trifluoromethyl)phenyl)-3-(4-bromo-2-(2-methyl-2H-tetrazol-5-yl)phenyl)thiourea. **CTBIC**: 4-chloro-7-(trifluoromethyl)-10H-benzofuro[3,2-b]indole-1-carboxylic acid. **NS19504**: 5-(4-bromobenzyl)thiazol-2-amine

materialized in a low molecular weight and structurally-different commercial molecule, namely NS19504 (Fig. 24.5), which activated the $K_{Ca}1.1$ channel with an EC_{50} value of 11.0 μM in the FLIPR assay, produced activation from a concentration of 0.3 μM, and left-shifted the voltage activation curve by 60 mV at 10 μM, in manual and automated (QPatch) patch-clamp assays. Furthermore, NS19504 showed an iberiotoxin-sensitive reduction of spontaneous phasic contractions of guinea pig urinary bladder strips, providing further support for the role of $K_{Ca}1.1$ channels in urinary bladder function. This completely distinct chemotype may now offer a new pharmacological tool with very diverse chemical-physical properties and a new starting point for a lead optimization campaign.

Finally, in the last few years, and since the most comprehensive review on $K_{Ca}1.1$ channel modulators (Nardi and Olesen 2008), relatively few small-molecule openers have been identified. Among these are a set of novel anthraquinone analogues (named the GoSlo-SR family; Roy et al. 2012), structurally similar to compounds previously reported as purinoceptor antagonists (Weyler et al. 2008; Baqi et al. 2009). This set of molecules was tested using the inside-out patch-clamp technique in rabbit bladder smooth muscle cells while holding patches at −60 mV. The best compound emerging ($EC_{50} = 2.3$ μM) was 5r or GoSlo-SR-5-44, able to shift the voltage-activation curve by more than −100 mV in a concentration-dependent manner and in the presence of a low concentration of calcium ions (100 nM).

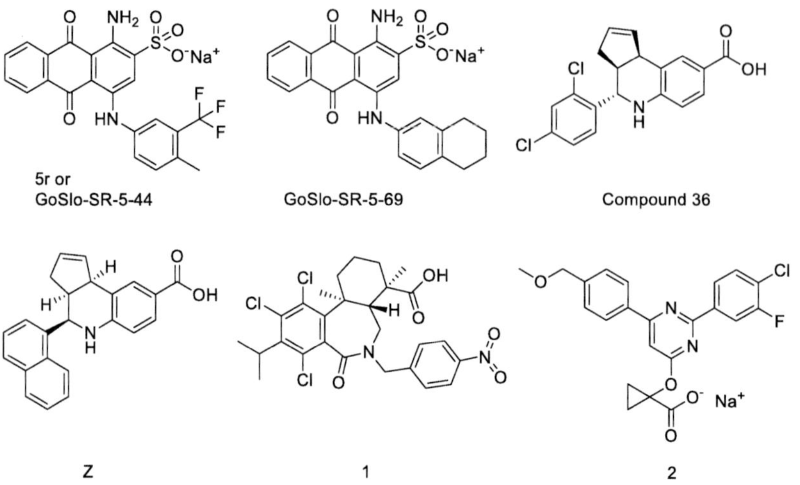

Fig. 24.6 Recently-disclosed $K_{Ca}1.1$ activators. **5r or GoSlo-SR-5-44**: sodium 1-amino-4-((4-methyl-3-(trifluoromethyl)phenyl)amino)-9,10-dioxo-9,10-dihydroanthracene-2-sulfonate. **GoSlo-SR-5-69**: sodium 1-amino-9,10-dioxo-4-((5,6,7,8-tetrahydronaphthalen-2-yl)amino)-9,10-dihydroanthracene-2-sulfonate. **Compound 36**: (3aS,4S,9bR)-4-(2,4-dichlorophenyl)-3a,4,5,9b-tetrahydro-3H-cyclopenta[c]quinoline-8-carboxylic acid. **Z**: (3aS,4R,9bR)-4-(naphthalen-1-yl)-3a,4,5,9b-tetrahydro-3H-cyclopenta[c]quinoline-8-carboxylic acid. **1**: (4R,4aR,11bS)-8,10,11-trichloro-9-isopropyl-4,11b-dimethyl-6-(4-nitrobenzyl)-7-oxo-2,3,4,4a,5,6,7,11b-octahydro-1H-dibenzo[c,e]azepine-4-carboxylic acid. **2**: sodium 1-((2-(4-chloro-3-fluorophenyl)-6-(4-(methoxymethyl)phenyl)pyrimidin-4-yl)oxy)cyclopropanecarboxylate

GoSlo-SR-5-69 is from the same chemical class of GoSlo-SR-5-44, both appearing in the same patent applications (Hollywood et al. 2012a, b), but fully disclosed only later (Roy et al. 2014) as an optimized analogue with an EC_{50} of 251 nM. Other more recent compounds still characterized by a carboxylic acid moiety were divulged by Amgen and Merck. The Merck tetrahydroquinoline 36 was able to produce an ~seven-fold increase of the maximum current observed compared to control at 3 μM compound concentration, whereas the Amgen tetrahydroquinoline Z displayed an EC_{50} of 109 nM. These compounds were able to inhibit spontaneous neuronal firing in an electrophysiological model of migraine (Gore et al. 2010) and to induce smooth muscle relaxation (Ponte et al. 2012). Finally, more carboxylic acid derivatives were identified by Japanese groups from the university of Tokyo (no biological data; Ohwada et al. 2009) and from Mitsubishi Tanabe Pharma (Compound 1 and Compound 2, Fig. 24.6). The $K_{Ca}1.1$ activator ($EC_{50} = 0.37$ μM) described by this latter group did not display the cyclooxygenase-inhibitory activity ($IC_{50} > 30$ μM) common for this chemical template and, administered at 10 mg/kg i.v. in rats, it significantly (> 50%) reduced the frequency of rhythmic bladder contractions induced by iberiotoxin (Tsuzuki and Hirai 2009).

24.2.2 $K_{Ca}1.1$ Channel Inhibitors

$K_{Ca}1.1$ channel inhibition has been shown to be a secondary effect of several marketed drugs in the same manner as is the case for $K_{Ca}1.1$ channel activators. The most noteworthy of these inhibitors are probably verapamil (Harper et al. 2001), ketamine (Hayashi et al. 2011; Klockgether-Radke et al. 2005; Denson et al. 1994; Denson and Eaton 1994), simvastatin (Seto et al. 2007), clotrimazole (Wu et al. 1999) and ketoconazole (Wu et al. 1999). Ketoconazole may be particularly interesting, since this molecule (in contrast to clotrimazole) does not potently inhibit $K_{Ca}3.1$ channels (Jensen et al. 1999). Furthermore, synthetic analogues and partial structures of ketoconazole have interestingly afforded compounds with either activating or inhibiting $K_{Ca}1.1$ channel activity (Power et al. 2006), confirming the idea that known drugs may serve well as useful template structures.

The most valuable and used inhibiting tool compounds for in vitro and ex vivo pharmacological studies are, however, $K_{Ca}1.1$ inhibitors from natural sources. Of these, the most frequently used is certainly iberiotoxin (Kunze et al. 1994; Kozlowski et al. 1996; Haghdoost-Yazdi et al. 2008), a 37-amino acid peptide isolated from the venom of the scorpion *Buthus tamulus* (Galvez et al. 1990), with affinity (K_i) at the channel down to the single-digit nanomolar range. Blockade of $K_{Ca}1.1$ channels by iberiotoxin is promoted by a very selective binding to the external mouth of the channel that physically blocks the conduction pathway (Candia et al. 1992). A remarkable selectivity feature of iberiotoxin is that it is inactive on $K_{Ca}1.1$ channels associated with the neuronal beta4 subunit (Wang et al. 2006). Several other peptidic scorpion toxins have been reported to similarly block the $K_{Ca}1.1$ channel, although with different degrees of selectivity and mechanisms of action, including charybdotoxin (Tang et al. 2010), slotoxin (Garcia-Valdes et al. 2001), BmP09 (Yao et al. 2005) and martentoxin (Shi et al. 2008; Ji et al. 2003; Wang et al. 2013; Tao et al. 2012).

Indole-diterpenes of natural origin are among the most selective and potent non-peptidic $K_{Ca}1.1$ channel inhibitors. This group includes paxilline ($IC_{50:}$ 1.9 nM) (Sanchez and McManus 1996; Strobaek et al. 1996), verruculogen, paspalicine, paspalitrem C and penitrem A (DeFarias et al. 1996; Gribkoff et al. 1996; Knaus et al. 1994). Structurally similar tremorgenic indole alkaloids have been claimed to be useful as a topical treatment of glaucoma and intraocular hypertension via inhibition of potassium channel conductance in the ciliary epithelium (Adorante et al. 1996). Their ability to reduce intraocular pressure in proof of concept studies in rabbit comparably to the β-blocker timolol catalysed the interest for scientists at Merck to establish a fluorescence resonance energy transfer (FRET)/voltage ion probe reader (VIPR) assay to identify small-molecule inhibitors of $K_{Ca}1.1$ channels with potency in the low, double-digit nanomolar range. This assay has been presumably used to identify and/or profile a large array of compounds and similar diterpene frameworks lacking tremorgenic liability while retaining high affinity for $K_{Ca}1.1$ channels. Such compounds have been subsequently patented for the same therapeutic benefit in an impressive number of patents (Goetz et al. 2003; Goetz and

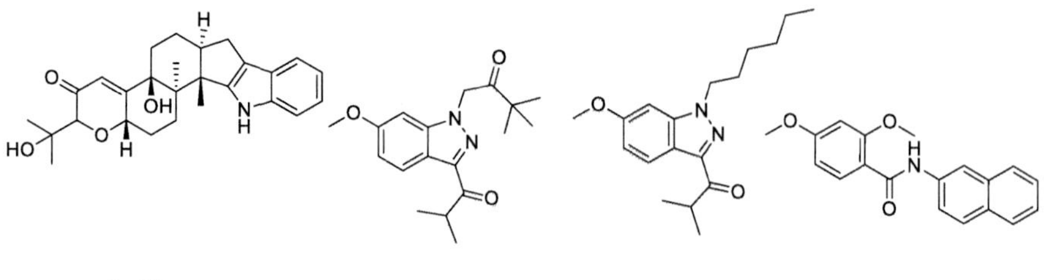

Fig. 24.7 Naturally-occurring (Paxilline) and synthetic small-molecule (compound 3, HMIMP, A272651) $K_{Ca}1.1$-channel inhibitors. **Paxilline**: (4bS,6aS,12bS,12cR,14aS)-4b-hydroxy-2-(2-hydroxypropan-2-yl)-12b,12c-dimethyl-5,6,6a,7,12,12b,12c,13,14,14a-decahydro-2H-chromeno[5′,6′:6,7]indeno[1,2-b]indol-3(4bH)-one. **3**: 1-(3-isobutyryl-6-methoxy-1H-indazol-1-yl)-3,3-dimethylbutan-2-one. **HMIMP**: 1-(1-hexyl-6-methoxy-1H-indazol-3-yl)-2-methylpropan-1-one. **A272651**: 2,4-dimethoxy-N-(naphthalen-2-yl)benzamide

Kaczorowski 2004; Garcia et al. 2001; Doherty et al. 2007, 2008a, b; Gao and Shen 2006; Doherty and Shen 2006; Chen et al. 2004a, 2005a, b; Boyd et al. 2005; Doherty et al. 2004a, b). One particular compound from these studies (Compound 3, Fig. 24.7) has later been the specific object of optimization for ophthalmic formulation (Miller and Yu 2008), leading to the supposition that this compound might have been the frontrunner emerging from this drug discovery programme. It is not currently known whether this compound was further advanced to clinical settings prior to the supposed discontinuation of the entire research project.

Furthermore, other synthetic small molecules have been identified. The small-molecule A272651 (Fig. 24.7; Shieh et al. 2007), disclosed by Abbott, was to our knowledge the first small molecule $K_{Ca}1.1$ channel inhibitor to be identified and fully described in a scientific article. However, with an IC_{50} of about 5 μM, a potency >1000-fold lower than that of iberiotoxin and paxilline, its value as a pharmacological tool compound appears questionable, a conclusion that is reinforced by the absence of any additional studies or publications with this very compound in pubmed/medline. Novel indole, indazole, imidazole, benzofuran, benzothiophene, benzimidazole, oxoquinoline, 1,2-dihydroquinolin-2-one, 1,2-dihydroquinoxalin-2-one and 1,2-dihydronaphthyridin-2-one derivatives as $K_{Ca}1.1$ inhibitors had been earlier reported in the patent literature by Merck, but poorly described from a pharmacological standpoint (Doherty et al. 2007; Chen et al. 2003, 2004b, 2005a, b; Doherty and Shen 2005, 2006; Gao and Shen 2006). One such small molecule, HMIMP (Fig. 24.7), has been interestingly profiled in a study carried out later by GlaxoSmithKline and identified as a highly potent ($IC_{50} = 2$ nM) and beta-subunit independent $K_{Ca}1.1$ channel inhibitor (Zeng et al. 2008). While the potency is comparable to iberiotoxin, the mechanism of action is surely very different, with the small molecule acting allosterically and able to inhibit the $K_{Ca}1.1$ channel from either side of the membrane (Zeng et al. 2008).

24.3 $K_{Ca}3.1$ Channel Modulator Chemistry

24.3.1 $K_{Ca}3.1$ Channel Activators

Whereas small-molecule $K_{Ca}1.1$ channel inhibitors and, as described in the next section, $K_{Ca}3.1$ channel inhibitors usually display a high degree of selectivity towards the other subtypes of calcium-activated potassium channels, the first generation $K_{Ca}3.1$ small-molecule activators usually possess only limited selectivity as they are typically embedded with $K_{Ca}2.x$ activating properties. One such molecule is the benzimidazolone 1-EBIO (Devor et al. 1996; Singh et al. 2000), with an EC_{50} of 30 μM (Wulff et al. 2007) at $K_{Ca}3.1$ and an EC_{50} of 589 μM on $K_{Ca}2.2$ in a whole-cell patch-clamp assay (Hougaard et al. 2009; Fig. 24.8). This compound was originally suggested to have potential therapeutic benefit for disease characterized by mucus-congested airways such as cystic fibrosis and chronic obstructive pulmonary disease (Singh et al. 2001) as well as for disease characterized by altered neuronal excitability (Garduno et al. 2005; Anderson et al. 2006; Kobayashi et al. 2008). Improvement of potency and selectivity of 1-EBIO has been the object of several medicinal chemistry campaigns. DC-EBIO (Singh et al. 2000; Fig. 24.8) is one analogue identified in a structure-activity relationship study, namely the 5,6-dichloro-1-ethyl analogue, displaying an improved (EC_{50}: 750 nM) potency profile over other differently-decorated benzimidazolones. Furthermore, installation of an oxime moiety in the corresponding indolin-2-one nucleus and retention of a dichloro pattern (6,7) afforded NS309 (Fig. 24.8; Strobaek et al. 2004), which demonstrated to be some 30-fold more potent than DC-EBIO in a back-to-back comparison. NS309, although featuring poor metabolic stability in in vivo settings, due to its high potency and excellent penetration into various tissues, has served as a

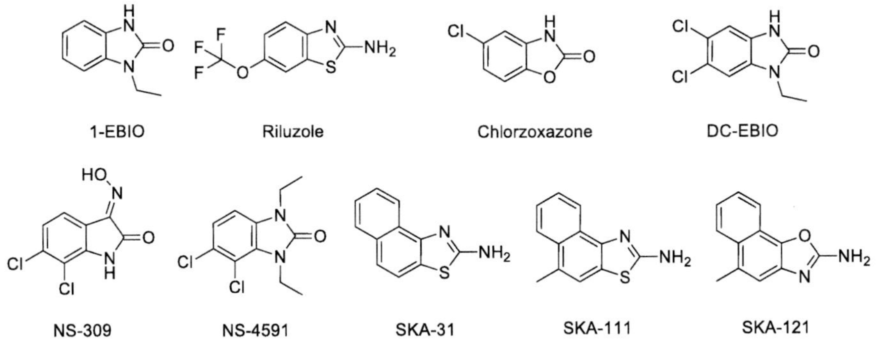

Fig. 24.8 $K_{Ca}3.1$ activators with very different degrees of subtype selectivity. **1-EBIO**: 1-ethyl-1H-benzo[d]imidazol-2(3H)-one. **Riluzole**: 6-(trifluoromethoxy)benzo[d]thiazol-2-amine. **Chlorzoxazone**: 5-chlorobenzo[d]oxazol-2(3H)-one. **DC-EBIO**: 5,6-dichloro-1-ethyl-1H-benzo[d]imidazol-2(3H)-one. **NS309**: (E)-6,7-dichloro-3-(hydroxyimino)indolin-2-one. **NS4591**: 4,5-dichloro-1,3-diethyl-1H-benzo[d]imidazol-2(3H)-one. **SKA-31**: naphtho[1,2-d]thiazol-2-amine. **SKA-111**: 5-methylnaphtho[1,2-d]thiazol-2-amine. **SKA-121**: 5-methylnaphtho[2,1-d]oxazol-2-amine

valuable tool compound for demonstrating participation of $K_{Ca}3.1/K_{Ca}2.x$ channels in many in vitro and ex vivo experimental set-ups. Recently, it has been used to advocate the involvement of $K_{Ca}3.1/K_{Ca}2.x$ channels in the regulation of detrusor smooth muscle contractions (Parajuli et al. 2013), for demonstration of the pivotal role of $K_{Ca}3.1/K_{Ca}2.x$ channels in epithelial and endothelial-dependent smooth muscle relaxation (Kroigaard et al. 2012; Brondum et al. 2010), and for establishing $K_{Ca}2.x$ channel activation for stabilization of cerebellar Purkinje cell activity as neuro-protective and anti-ataxic principle in the cerebellum (Kasumu et al. 2012; Tara et al. 2018). Based on these observations, a new quest for novel positive modulators of $K_{Ca}3.1/K_{Ca}2.x$ channels with improved bioavailability led to the identification of NS4591 (Fig. 24.8). In rats, the bioavailability of NS4591 after oral dosing was estimated to be 97% (Hougaard et al. 2009). In whole-cell patch-clamp experiments, NS4591 doubled $K_{Ca}3.1$-mediated currents at a concentration of 45 nM, whereas 530 nM was required for doubling of $K_{Ca}2.3$-mediated currents under the same experimental conditions. Furthermore, in ex vivo settings, NS4591 inhibited the firing of action potentials in primary bladder afferent DRG neurons and reduced carbachol-induced contractions of bladder rings. Consistent with this in vitro and ex vivo profile, NS4591 alleviated in vivo bladder overactivity induced by capsaicin in awake rats as well as by acetic acid in anaesthetized cats, further supporting the hypothesis that $K_{Ca}3.1/K_{Ca}2.x$ channels play an important role in bladder function and suggest that positive modulators of $K_{Ca}3.1/K_{Ca}2.x$ channels may be of benefit for the therapeutic treatment of overactive bladder.

Analogue small-molecules, such as methylxanthines (theophylline, IBMX and caffeine; Schroder et al. 2000), the marketed glutamate-release inhibitor riluzole (Grunnet et al. 2001; Cao et al. 2001) and the marketed chlorzoxazone (Syme et al. 2000; Fig. 24.8), characterized by relatively-simple bicyclic chemical frameworks, have also exhibited similar and poor level of potency and specificity for the $K_{Ca}3.1$ channel activation. The diversity of therapeutic indications addressed by riluzole underlies the wide range of mechanisms of action embedded in this molecule: first launched in 1996, it is approved as a treatment for amyotrophic lateral sclerosis and has been in clinical development for additional indications as diverse as bipolar disorders (phase II), multiple sclerosis (phase II), malignant melanoma (phase I), ataxia (Phase II), etc.

Improvement of potency and selectivity of riluzole favouring $K_{Ca}3.1$ channel activation over Na_v channel inhibition have been attempted and achieved by the compound SKA-31 (Fig. 24.8; Sankaranarayanan et al. 2009). After screening 55 riluzole analogues, scientists at the University of California identified this latter molecule as a more potent (ten-fold, EC_{50}, 250 nM) and more selective $K_{Ca}3.1$ channel opener, including ten-fold selectivity over $K_{Ca}2.x$ and other ion channels. This in vitro profile, combined with an acceptable pharmacokinetic profile, has made it possible to use it as a tool compound in several studies and to probe the therapeutic potential of $K_{Ca}3.1$ channel opening. Via endothelial $K_{Ca}3.1$ and $K_{Ca}2.3$ activation, SKA-31 strengthens the EDH response (Khaddaj-Mallat et al. 2018) and causes profound vasodilation in animal models of cardiovascular diseases, including

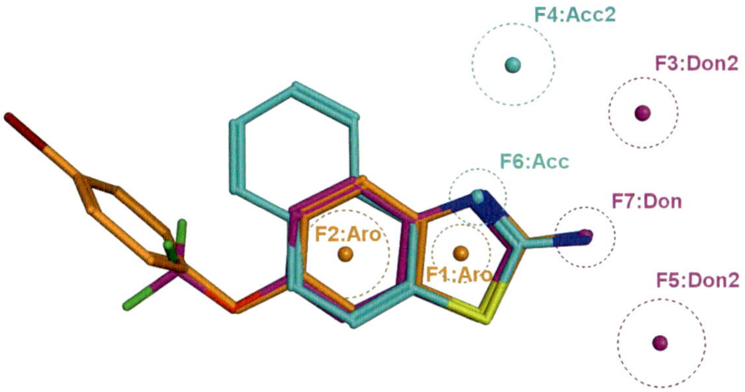

Fig. 24.9 Standard modelling techniques have been applied to generate, overlay and visualize the pharmacophore features of the compounds NS19504 (orange), riluzole (magenta) and SKA-31 (cyan) as implemented in MOE2013.08 (Molecular Operating Environment 2014, CCG, Montreal, Canada). Figure generated by Dr. Achim Kless (Grünenthal). **NS19504**: 5-(4-bromobenzyl)thiazol-2-amine. **Riluzole**: 6-(trifluoromethoxy)benzo[d]thiazol-2-amine. **SKA-31**: naphtho[1,2-d]thiazol-2-amine

primary hypertension (Kloza et al. 2019; Mishra et al. 2016) and preservation of coronary flow in diabetic myocardium (Mishra et al. 2014).

Nonetheless, the need for a yet more potent and selective $K_{Ca}3.1$ (over $K_{Ca}2.x$) remains in order to investigate the potential of the $K_{Ca}3.1$ channel as target for novel endothelial antihypertensives and to further dissect the functional effects of $K_{Ca}3.1$ channel activation from those of $K_{Ca}2.x$ channel activation in the above as well as in other studies (Radtke et al. 2013). Analogues of SKA-31, namely SKA-111 and SKA-121 (Fig. 24.8), have been very recently identified as better in vitro/ex vivo tool $K_{Ca}3.1$ activators (EC_{50}: 100 nM for both; Coleman et al. 2014; Oliván-Viguera et al. 2016). Surprisingly, SKA-111 showed a 120-fold selectivity for $K_{Ca}3.1$ over $K_{Ca}2.x$, despite the difference of only a methyl substituent when compared to SKA-31. SKA-121 carries instead an additional structural difference of the core, with the thiazole being replaced by an oxazole ring. In common for all (SKA-31, SKA-111, SKA-121) the retention of a primary amino group at the terminal and right-hand site as an evidently-important pharmacophoric feature. Interestingly, such a primary amino functionality is present in both $K_{Ca}3.1$ activators (riluzole, SKA-31, SKA-111, SKA-121) and $K_{Ca}1.1$ activators (NS19504), suggesting the intriguing possibility of this group serving as an anchor point across different calcium-activated potassium channels and the possibility of fine-tuning the specificity via molecular decoration and the molecular diversification of the left-hand site of the molecules. This is clearly represented in Fig. 24.9, where NS19504 (orange), riluzole (magenta) and SKA-31 (cyan) have been overlaid and the potential common pharmacophoric features (aromatics, accepton/hydrogen bond donors) have been highlighted.

24.3.2 $K_{Ca}3.1$ Channel Inhibitors

A number of peptide toxins, encompassing maurotoxin, charybdotoxin and OSK1, isolated from venoms of animals such as scorpions have demonstrated nanomolar potency as $K_{Ca}3.1$ inhibitors (Wei et al. 2005; Mouhat et al. 2005; Jaravine et al. 1997). These toxins, however, not additionally featured with a specificity that is ideally required for chemical tools. Interestingly, arachidonic acid, an endogenous, small molecule signalling entity produced enzymatically from membrane phospholipids, directly inhibits the $K_{Ca}3.1$ channel from colonic T84 cells (Devor and Frizzell 1998). Despite being functionally selective over $K_{Ca}2$ channels (Hamilton et al. 2003), arachidonic acid is not an ideal drug template either, which have allowed synthetic small molecules to take centre stage in most of $K_{Ca}3.1$ pharmacological studies and drug discovery programmes.

The potent antifungal imidazole clotrimazole (IC_{50}: 70–250 nM, Wu et al. 1999; Brugnara et al. 1996; McNaughton-Smith et al. 2008) originally developed at Bayer, and especially TRAM-34 (Wulff et al. 2001; Fig. 24.10), designed at the University of California Irvine, (IC_{50}: 8.5 nM, Strobaek et al. 2013; Wulff et al. 2000) are the prototype small-molecule agents that currently serve as gold reference standards for the pharmacological inhibition of $K_{Ca}3.1$ channels. Despite TRAM-34 having much less inhibitory activity at CYP3A4 enzymes than clotrimazole (Wulff et al. 2000), a broader test involving several CYP isoforms has revealed some activity (both stimulatory and inhibitory) in the μM range (Agarwal et al. 2013).

On the opposite end of the spectrum of small-molecule $K_{Ca}3.1$ inhibitor agents is the Icagen compound ICA-17043 (Senicapoc, Fig. 24.10; IC_{50}: 11 nM, rubidium efflux assay; Stocker et al. 2003; Ataga and Stocker 2009; Ataga et al. 2006, 2008), a drug candidate that has been advanced to clinical Phase III as an orally available drug for the chronic treatment of sickle cell disease (Brugnara and de Franceschi 2006), but failed due to lack of effect on the number of sickle cell crises, which was the primary endpoint (Ataga et al. 2011). The ability of Senicapoc to reverse antigen-induced changes in late-phase airway resistance and airway hyper-reactivity in a sheep model of asthma (Van Der Velden et al. 2013) triggered Icagen to perform two

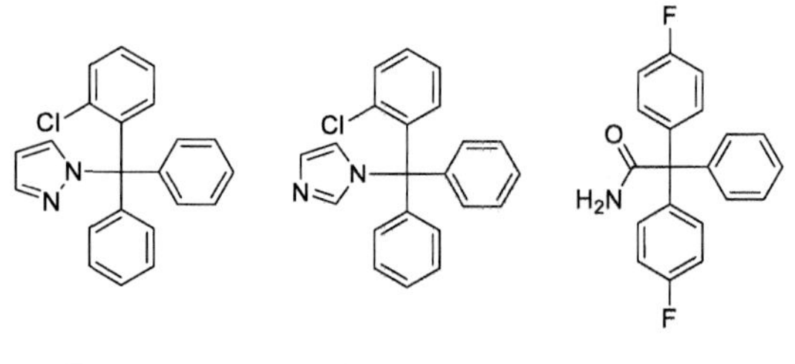

TRAM-34 Clotrimazole ICA-17043, Senicapoc

Fig. 24.10 Traditional $K_{Ca}3.1$ channel inhibitors. **TRAM-34**: 1-((2-chlorophenyl) diphenylmethyl)-1H-pyrazole. **Clotrimazole**: 1-((2-chlorophenyl)diphenylmethyl)-1H-imidazole. **ICA-17043, Senicapoc**: 2,2-bis(4-fluorophenyl)-2-phenylacetamide

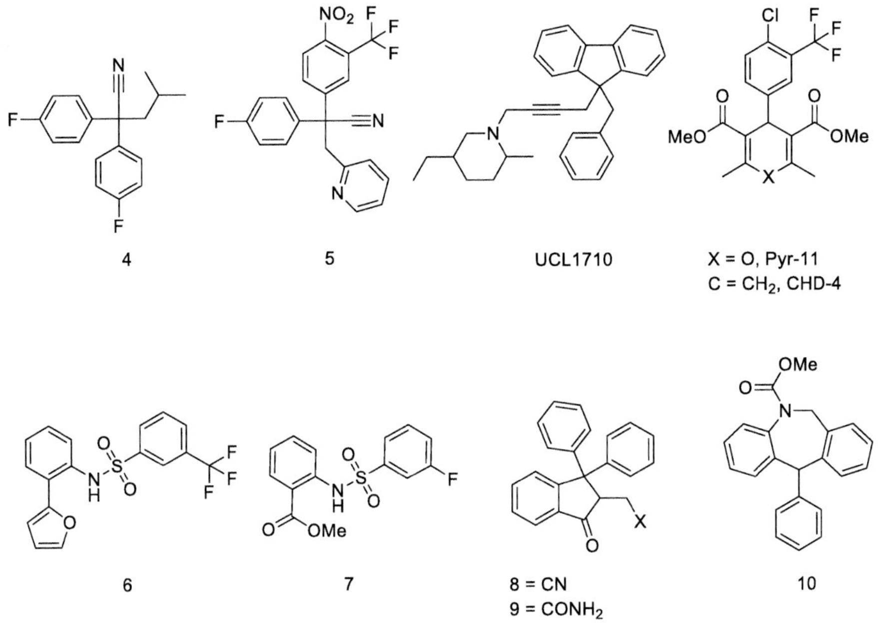

Fig. 24.11 $K_{Ca}3.1$ channel inhibitors characterized by the traditional TRAM motif (top) as well as structurally-diversified molecular frameworks. **4**: 2,2-bis(4-fluorophenyl)-4-methylpentanenitrile. **5**: 2-(4-fluorophenyl)-2-(4-nitro-3-(trifluoromethyl)phenyl)-3-(pyridin-2-yl)propanenitrile. **UCL1710**: 1-(4-(9-benzyl-9H-fluoren-9-yl)but-2-yn-1-yl)-5-ethyl-2-methylpiperidine. **Pyr-11**: dimethyl 4-(4-chloro-3-(trifluoromethyl)phenyl)-2,6-dimethyl-4H-pyran-3,5-dicarboxylate. **CHD-4**: dimethyl 4′-chloro-3,5-dimethyl-3′-(trifluoromethyl)-1,4-dihydro-[1,1′-biphenyl]-2,6-dicarboxylate. **6**: N-(2-(furan-2-yl)phenyl)-3-(trifluoromethyl)benzenesulfonamide. **7**: methyl 2-(3-fluorophenylsulfonamido)benzoate. **8**: 2-(3-oxo-1,1-diphenyl-2,3-dihydro-1H-inden-2-yl)acetonitrile. **9**: 2-(3-oxo-1,1-diphenyl-2,3-dihydro-1H-inden-2-yl)acetamide. **10**: methyl 11-phenyl-6,11-dihydro-5H-dibenzo[b,e]azepine-5-carboxylate

small Ph 2a studies in asthma, showing effect in allergen-induced asthma, but not asthma elicited by exercise, before the company was acquired by Pfizer. Senicapoc was then deposited as PF-05416266 in NIH's "National Center for Advancing Translational Research" library, making it available for investigator-initiated clinical trials. Senicapoc is currently considered by Springworks Therapeutics for the rare anaemia disease hereditary xerocytosis, which can be caused by gain of function mutations in *KCNN4* (Andolfo et al. 2015) and by University of California for Alzheimer's disease, which is rationalized by high expression of $K_{Ca}3.1$ in microglia (Jin et al. 2019).

Senicapoc still belongs to the TRAM class of compounds in which the heteroaromatic rings of clotrimazole and TRAM-34 have been replaced by a carboxamide ($CONH_2$) function, thus emphasizing the limited chemical diversity of available $K_{Ca}3.1$ inhibitors. In fact, a patent application from NeuroSearch illustrates a large series of di- and triarylmethane derivatives (an exemplary compound 4 is reported in Fig. 24.11; Demnitz et al. 2003), with the most preferred compounds inhibiting $K_{Ca}3.1$ channels in electrophysiology experiments in the "low micromolar and nanomolar range." In one more follow-up patent application, NeuroSearch again presented a structurally-related diarylmethane derivative (in Fig. 24.11 the exemplary Compound 5), with a K_d value for inhibition of $hK_{Ca}3.1$ channels of 65 nM (Demnitz et al. 2005). These and other analogues have been the subject of other reviews, and the reader is referred to these for further detailed background (Wulff and Castle 2010; Wulff et al. 2007).

Molecular diversification of these pharmacological tools would allow the scientific community to appraise whether the chemical space represented by these triarylmethane compounds is the reason for failure in delivering successful drugs as well as to potentially elude a number typical pharmaceutical and druggability issues for this class of compounds, e.g. limited freedom to operate, low solubility and high plasma protein binding.

Medicinal chemists had actually witnessed in the past some efforts in this direction. From experimental work based on the pioneer $K_{Ca}3.1$ channel inhibitor cetiedil (Berkowitz and Orringer 1981), scientists at University College London identified compounds such as UCL1710 (Fig. 24.11; Roxburgh et al. 2001; Narenjkar et al. 2004) with inhibitory potency of 300 nM. A close inspection of the chemical structures described so far and depicted in Figs. 24.10 and 24.11 does, however, highlight a common motif preserved, i.e. two or three, frequently electron deficient, aromatic rings compactly arranged around a "tight core" of one or two carbon atoms, with a fourth substituent being slightly more "versatile," as in UCL1710.

Furthermore, the observation that the $Ca_v1.x$ (L-type) Ca^{2+} channel inhibitor nifedipine inhibited the $K_{Ca}3.1$ channel in the low micromolar range (Kaji 1990) also inspired efforts culminating in the 4H-pyran (Pyr-11; Urbahns et al. 2003) and the 1,4-cyclohexadiene (CHD-4; Nardi et al. 2008; Fig. 24.11), able to inhibit $K_{Ca}3.1$ with IC_{50} values of 8 and 1.5 nM, respectively. Additionally, Icagen has disclosed a series of novel diarylsulfonamide $K_{Ca}3.1$ inhibitors for use in glaucoma, sickle cell disease, abnormal cell proliferation and inflammation (Atkinson et al. 2004). A common feature of these compounds is the left-hand aniline-derived aryl group usually substituted by one or more groups, of which one is invariably either a methyl ester or a presumably isosteric five-membered heterocycle ortho to the sulfonyl nitrogen. On the right-hand sulfonyl aryl ring, one or more electronegative halogen or trifluoromethyl substituents imply the necessity for electron deficiency. Although no discrete potencies are quoted in the patent, compounds in Fig. 24.11 are two examples (compounds 6 and 7) of more potent derivatives as determined in $^{86}Rb^+$-efflux assays from red blood cells ($IC_{50} < 500$ nM). Finally, also the Brugnara group demonstrated the switching from the "classical" triarylmethane chemotype to more restricted diphenyl indanones (Bellott et al. 1999b; Alper et al. 1999b); Compounds 8 and 9, Fig. 24.11) or 11-phenyl-dibenzazepines (Alper et al. 1999a; Bellott et al. 1999a; exemplary Compound 10, Fig. 24.11), yield new classes of $K_{Ca}3.1$ inhibitors, which can be considered conformationally restricted triarylmethanes.

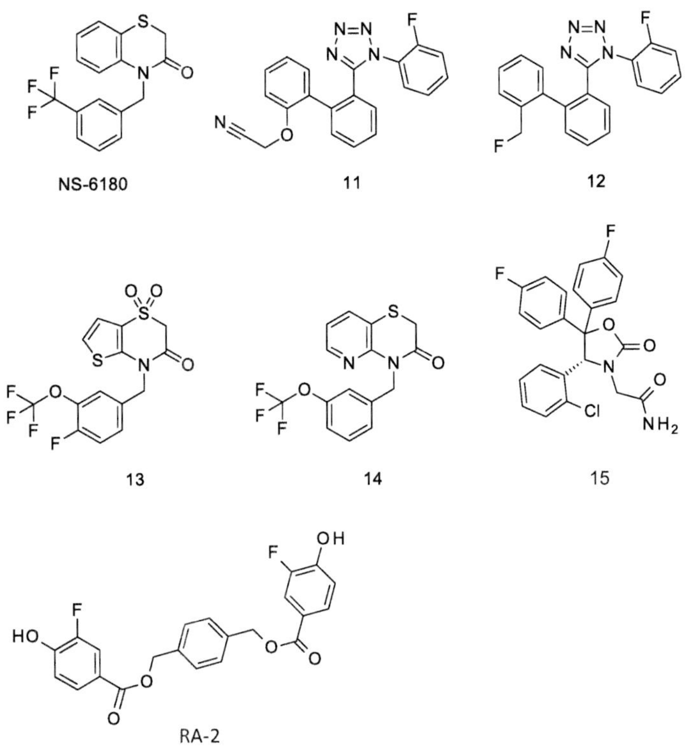

Fig. 24.12 Innovative molecular frameworks, such as the thiazinone compounds identified at NeuroSearch (Saniona A/S) and Boehringer Ingelheim Pharma as $K_{Ca}3.1$ channel inhibitors (NS6180 and 13,14, respectively). 11 and 12 show chemical structures of tetrazole $K_{Ca}3.1$ inhibitors, built on a novel molecular backbone with pharmacophoric features resembling the traditional TRAM motif. 15 shows a 3,4 disubstituted oxazolidinone from Roche, a novel variation of the TRAM motif. **NS-6180**: 4-(3-(trifluoromethyl)benzyl)-2H-benzo[b][1,4]thiazin-3(4H)-one. **11**: 2-((2′-(1-(2-fluorophenyl)-1H-tetrazol-5-yl)-[1,1′-biphenyl]-2-yl)oxy)acetonitril. **12**: 5-(2-′-(fluoromethyl)-[1,1′-biphenyl]-2-yl)-1-(2-fluorophenyl)-1H-tetrazole. **13**: 4-(4-fluoro-3-(trifluoromethoxy)benzyl)-2H-thieno[3,2-b][1,4]thiazin-3(4H)-one 1,1-dioxide. **14**: 4-(3-(trifluoromethoxy)benzyl)-2H-pyrido[3,2-b][1,4]thiazin-3(4H)-one. **15**: 2-[(S)-4-(2-Chloro-phenyl)-5,5-bis-(4-fluoro-phenyl)-2-oxo-oxazolidin-3-yl]-acetamide. **RA-2**: 1,3-phenylenebis (methylene) bis(3-fluoro-4-hydroxybenzoate)

More recently, NeuroSearch (nowadays Saniona A/S) has significantly addressed the need of identification of new chemical scaffolds by describing a 384-well high-throughput Tl^{+} influx assay that was capable of screening rapidly more than 200,000 compounds (Jorgensen et al. 2013) and to single out several new chemical series of inhibitors as well as activators of $K_{Ca}3.1$ channels, including the inhibitor NS6180 (Fig. 24.12), making its first appearance in the public domain in February 2012, at the 56th Annual Biophysical Society Meeting in San Diego and being fully disclosed in a publication shortly thereafter (Strobaek et al. 2013). This compound showed anti-inflammatory actions in a model of inflammatory bowel disease, possibly by a local action in the gut. NS6180 is a commercial benzothiazinone inhibitor that is equipotent to TRAM-34 (IC_{50}: 10–20 nM) and that has no structural resemblance to TRAM analogues or previous $K_{Ca}3.1$ inhibitors. One more chemical series by Saniona A/S is represented by the tetrazole derivatives (Demnitz and Jorgensen 2014) exemplified by Compounds 11 and 12 in Fig. 24.12 with IC_{50} of 16 nM and 17 nM, respectively. From these, it may be concluded that the tetrazole ring may serve a double role, both as bioisosteric replacement of the carboxamide moiety of TRAM-34 and as a platform to correctly position the three classic aromatic rings in the binding space.

Contextually with its commercial origin, neither the benzothiazinone compound NS6180 nor its molecular framework currently is not yet included in any published patent application. Similar fused thiazin-3-ones have recently been claimed by Boehringer Ingelheim Pharma GmbH & Co (Priority date Jun 21, 2012) as $K_{Ca}3.1$ inhibitors, also identified in a FLIPR assay (Burke et al. 2013). Two of the exemplified compounds (Compounds 13, 14, Fig. 24.12) had in such an assay an IC_{50} value of 39 and 52 nM, respectively, and they have been reported to be potentially useful for several inflammatory and autoimmune diseases, including psoriasis, asthma and inflammatory bowel disease. It remains to be seen whether these minor structural changes are able to exhibit advantages over NS6180, such as its extremely low bioavailability upon oral dosing in preclinical settings (Strobaek et al. 2013). Roche has published a patent on 3,4 disubstituted oxazolidinones (Green and Wang 2014) as inhibitors of $K_{Ca}3.1$ with EC_{50} values as low as 15 nM (see Fig. 24.12). This chemical series can be seen as an interesting variation over the TRAM motive but further investigations are needed to see if this represents an improvement of the basic druggability properties. Finally, academia has recently contributed to the field by the fluoro-di-benzoate RA-2 (Fig. 24.12, which surprisingly inhibits both $K_{Ca}3.1$ and $K_{Ca}2.x$ with essentially the same potency (17 nM), in strong contrast to all other $K_{Ca}3.1$ inhibitors, which are highly selective. In accordance with its broad selectivity, RA-2 potently inhibits the EDHF response (Oliván-Viguera et al. 2015).

24.4 Interaction Sites for $K_{Ca}1.1$/$K_{Ca}3.1$ Channel Modulators

$K_{Ca}1.1$ channel activators interact with several different sites on $K_{Ca}1.1$ channels as seen from their various dependencies on beta-subunit co-expression and splice variants (Saito et al. 1997). A recent analysis revealed that Cym04 (Fig. 24.4) and NS1619 (Fig. 24.5), which both act on the $K_{Ca}1.1$ alpha subunit per se, are inactive on a splice variant comprising the linker between the S6 and the RKC1 domain in the C-terminal (Gessner et al. 2012). Since mutational analysis revealed that Cym04 specifically lost its activating effect on $K_{Ca}1.1$ channels upon deletion of two amino acids in the distal linker (Phe336 and Ser335 or Ser337), whereas amino acid substitutions were ineffective, it was suggested that the effect of Cym04 was to shorten the S6-RKC1 linker, thereby facilitating force transduction from the C-terminal "gating ring" to the channel gate, much in line with the suggested mechanism for gating by Ca^{2+}. However, the actual binding site for Cym04 was not finally revealed by these studies. It is noteworthy that mallotoxin (Fig. 24.4), another $K_{Ca}1.1$ channel activator, acted independently of the S6-RKC linker in the same study, alluding to the existence of other sites for positive modulators. One example of this is the benzofuroindole CTBIC (Fig. 24.5) site positioned at the extracellular side of the channel between two adjacent alpha subunits (Lee et al. 2012). It is very intriguing that pharmacological gating modulation can occur at such remotely placed sites which may reflect that the physiological gating process involves many parts of the channel protein (Chen et al. 2014). Concerning $K_{Ca}1.1$ inhibitors, the potency of HMIMP has been shown to decrease by a factor of 10 by the Thr(352)Ser mutation (Gordon et al. 2010), whereas no information is yet available for the site of actions of the inhibitors of the tremorgenic indole alkaloid class.

The binding site for the triarylmethane class (clotrimazole, a.o.) of $K_{Ca}3.1$ channel inhibitors was originally delineated by a combined chimeric and site-directed mutagenesis approach, taking advantage of the more than 1000-fold affinity difference between the very closely related $K_{Ca}3.1$ and $K_{Ca}2.x$ channels (Wulff et al. 2001). The 3D structure of the entire $K_{Ca}3.1$ protein was recently solved by cryoEM (Lee and MacKinnon 2018), which has now allowed a much more accurate modelling of binding sites and interaction modes (for a recent review, see Brown et al. 2019). Replacement of just two amino acids, Thr(250) and Val(275), in the $K_{Ca}3.1$ channel with the corresponding residues in $K_{Ca}2.3$ completely abolished activity of TRAM-34 and clotrimazole. Conversely, these two amino acids also yielded full TRAM-34 sensitivity when incorporated in $K_{Ca}2.x$ channels. Thr(250) and Val(275) are positioned in extracellular end of S6 and in the pore helix, respectively, just below the K^{+} selectivity filter and are therefore only accessible from the inside of the channel as shown by membrane impermeable TRAMs. Modelling reveals that TRAM-34 and senicapoc both interact hydrophobically with Val(275) from all four subunits, whereas differences exist in their interaction with Thr(250), since TRAM-34 forms hydrogen bonds to just one subunit via its pyrazole moiety, whereas senicapoc interacts via the amide moiety with two subunits. Benzothiazinones, a novel structural class of high potency $K_{Ca}3.1$ inhibitors exemplified by the compound NS6180, were also shown to be obligatory dependent on these two amino acids (Strobaek et al. 2013). However, in contrast to the TRAMs, NS6180 only interacts with residues from two subunits, but probably still works by inhibiting K^{+} conduction. (It is interesting that the "TRAM site" on the $K_{Ca}3.1$ channel was later shown to correspond exactly to the negative gating modulator site on $K_{Ca}2.x$ channels defined by the 5-aminobenzimidazole compounds, such as NS8593 (Jenkins et al. 2011)). In contrast, the dihydropyridine inhibitor nifedipine acts at another site, which is located in the pore fenestration region via interaction with the residues Thr(212) in S5 and Val(272) in S6, a position in which the bound molecule will probably not directly interfere with ion permeation. Most likely dihydropyridines stabilize the channel in a non-conducting conformation (Nguyen et al. 2017). Surprisingly, cyclohexadiene and phenylpyran inhibitors, which are derived by rational drug design from dihydropyridines (Urbahns et al. 2003), behave as classical $K_{Ca}3.1$ inhibitors at the TRAM site. The complexity of small molecule interactions with the pore region of $K_{Ca}3.1$ is probably most clearly demonstrated by arachidonic acid, which clearly depends on the same amino acids as the TRAM molecules, despite very low resemblance to these molecules (Hamilton et al. 2003).

Use of 1-EBIO in a mutational approach founded the notion that the activator interaction site was in the C-terminal of $K_{Ca}3.1$ and $K_{Ca}2.x$ channels (Pedarzani et al. 2001). Successful co-crystallization of several activators with $K_{Ca}2.x$ C-terminal peptides and Cam (Zhang et al. 2012, 2013) indicated that $K_{Ca}2.x/K_{Ca}3.1$ activator binding-site was positioned between Cam and the CamBD. With the full 3D structure in hand (Lee and MacKinnon 2018) and a highly plausible scheme for the gating process, this theory is now widely rejected in favour of a binding site placed between the Cam N lobe and the TM4-TM5 linker, in a helical region called $S_{45}A$ with which the Ca^{2+}-loaded Cam binds and opens the channel. Mutational analysis and modelling have revealed the SKA-111 interacts with Ser(181) and Leu (185) in the TM4-TM5 linker and with the Met(51), Glu(54), and Met(71) on Cam, the same amino acids as found in the C-term/Cam crystal mentioned above (Brown et al. 2019). It remains to be formally proven that these conclusions hold for other chemical classes of activators as well (details will most certainly be different), and that the same concept holds for $K_{Ca}2.3$ selective activators (Hougaard et al. 2007; Kasumu et al. 2012). For some $K_{Ca}2.1$ preferring modulators, which have a common interaction site for positive and negative modulators in the transmembrane regions (TM5; Hougaard et al. 2009, 2012), this will not be the case, and therefore, other positive modulator sites probably also exist on $K_{Ca}3.1$ channels, although the hitherto identified $K_{Ca}2.1$ channel modulators working at this site only exhibit low affinity for the $K_{Ca}3.1$ channel.

24.5 Conclusions and Perspectives

The preceding paragraphs have outlined the established classical pharmacology of $K_{Ca}3.1$ and $K_{Ca}1.1$ channels and described the various attempts done in both academia and the pharmaceutical industry to expand the chemical scope within this field, both with the aim of target validation and drug development. Target validating type of experiments where tool compounds have been used in combination with knockout and transgenic animal studies point towards a yet unexploited potential for use of both activators and inhibitors of K_{Ca} channels for therapeutic uses in serious epithelial diseases, such as $K_{Ca}3.1$ channel inhibitors for treating cAMP-dependent secretion and cyst growth in autosomal-dominant polycystic kidney disease (Albaqumi et al. 2008), $K_{Ca}1.1$ channel inhibitors for glaucoma and $K_{Ca}3.1$ activators for cystic fibrosis. In spite of this, it is noteworthy that a relatively low number of realized industrial drug discovery/development programmes have been specifically targeted against correcting pathological conditions via interaction with the epithelial K_{Ca} channels. An outstanding exception is probably the $K_{Ca}1.1$ channel inhibitor programme from Merck against glaucoma. On the other hand, several of these industrial programmes have been directed towards conditions such as asthma, chronic obstructive pulmonary disease or inflammatory bowel disease, where the active participation of the epithelium in the complex disease process cannot be excluded, although the primary therapeutic actions may well be via K_{Ca} channels expressed in other cells, for example direct smooth muscle relaxation in overactive bladder (i.e. $K_{Ca}1.1$ channel activators) or direct inhibition of immune cells mediating the inflammatory responses in asthma and IBD (i.e. $K_{Ca}3.1$ channel inhibitors). Thus, for a number of both novel and classic $K_{Ca}1.1$ and $K_{Ca}3.1$ channel modulators, we are still awaiting profiling of the direct effects on the epithelial channels as well as the corresponding change in epithelial physiology in the relevant disease models.

As the review describes, the pharmaceutical efforts within the K_{Ca} channel field have resulted in a wealth of structures with direct, or in some cases indirect, actions on the channels. Understanding the nature of how this wealth of molecular frameworks ultimately interacts with the channel proteins or with their associated subunits to produce unique modulatory effects is only beginning to be uncovered. Similarly, knowledge of the long-term effects on K_{Ca} channel expression and trafficking is only sporadically known, although epithelial pharmacology has delivered a clear example of down-regulation of $K_{Ca}3.1$ channels in response to activators (Manaves et al. 2004; Koegel et al. 2003). The limited body of knowledge of the structural determinants underlying ligand-receptor interaction within the K_{Ca} channels as opposed to other families of ion channel targets or potassium channel types has surely hampered the original drug design of truly innovative K_{Ca} modulators by means of structural biology, virtual screening and/or fragment-based approaches. This outstanding evidence, together with lack, until recently, of established high-throughput screening assays formats and the omnipresence of large chemical libraries often optimized for classical targets such as G protein-coupled receptor or kinases may have represented a key deterrent for several pharmas and big pharmas to enter the field and, by consequence, may have limited the full exploitation of the field itself. This becomes, for instance, very evident when considering the overall limited chemical diversity that has emerged so far, especially in the field of $K_{Ca}1.1$ and $K_{Ca}3.1$ channel activators, where several drug discovery programmes appear too often guided by a patent busting approach and with the resulting molecules displaying very similar structural features and similar druggability hurdles. As more information on structural biology is emerging (Yuan et al. 2010; Lee and MacKinnon 2018), high-throughput screening assay formats for both acute and lasting modulatory effects are being published (Jorgensen et al. 2013; Balut et al. 2010), and molecular pharmacology approaches based on mutations are made available (Strobaek et al. 2013; Nausch et al. 2014), we would expect the scenario to significantly change in the short-term future. It is with this in mind that we look forward to a larger exploration of the chemical space of $K_{Ca}3.1$ and $K_{Ca}1.1$ channel modulators and the disclosure of more potent, specific and druggable pharmacological agents to allow new exciting insight into the therapeutic potential of epithelial K_{Ca} channels in the patient population.

References

Adorante JS, Woldemussie E, Ruiz G (1996) Method for reducing intraocular pressure in the mammalian eye by administration of potassium channel blockers. WO 1996033719

Agarwal JJ, Zhu Y, Zhang QY, Mongin AA, Hough LB (2013) TRAM-34, a putatively selective blocker of intermediate-conductance, calcium-activated potassium channels, inhibits cytochrome P450 activity. PLoS One 8(5):e63028

Aihara H, Hosaka T, Kashiwagi T, Kohnomi S, Kono RTCC (2004) Calcium-activated K channel activator EP 1400243

Akada K, Izumi S, Yasuura H (2002) Sustained release oral preparations WO 2001070221

Albaqumi M, Srivastava S, Li Z, Zhdnova O, Wulff H, Itani O, Wallace DP, Skolnik EY (2008) KCa3.1 potassium channels are critical for cAMP-dependent chloride secretion and cyst growth in autosomal-dominant polycystic kidney disease. Kidney Int 74(6):740–749

Alper S, Bellot EM, Brugnara C, Clifford JJ, Froimowitz M, Gao YD, Haidar RM, Halperin J, Kelleher EW, Kher FM (1999a) Use of substituted 11-phenyl-dibenzazepine compounds for the treatment or prevention of sickle cell disease, inflammatory diseases characterized by abnormal cell proliferation, diarrhea and scour. WO 1999026628

Alper S, Bellot EM, Brugnara C, Clifford JJ, Froimowitz M, Gao YD, Haidar RM, Halperin J, Kelleher EW, Kher FM (1999b) Use of substituted diphenyl indanone, indane and indole compounds for the treatment or prevention of sickle cell disease, inflammatory diseases characterized by abnormal cell proliferation, diarrhea and scours WO1999026624

Alvarez J, Montero M, Garcia-Sancho J (1992) High affinity inhibition of Ca^{2+}-dependent K^+ channels by cytochrome P-450 inhibitors. J Biol Chem 267(17):11789–11793

Anderson CS, MacKinnon R, Smith C, Miller C (1988) Charybdotoxin block of single Ca^{2+}-activated K^+ channels. Effects of channel gating, voltage, and ionic strength. J Gen Physiol 91 (3):317–333

Anderson NJ, Slough S, Watson WP (2006) In vivo characterisation of the small-conductance KCa (SK) channel activator 1-ethyl-2-benzimidazolinone (1-EBIO) as a potential anticonvulsant. Eur J Pharmacol 546(1–3):48–53

Andolfo I, Russo R, Manna F, Shmukler BE, Gambale A, Vitiello G, De Rosa G, Brugnara C, Alper SL, Snyder LM, Iolascon A (2015) Novel Gardos channel mutations linked to dehydrated hereditary stomatocytosis (xerocytosis). Am J Hematol 90(10):921–926

Aon MA, Cortassa S, Wei AC, Grunnet M, O'Rourke B (2010) Energetic performance is improved by specific activation of K^+ fluxes through K(Ca) channels in heart mitochondria. Biochim Biophys Acta 1797(1):71–80

Archer SL, Huang JM, Hampl V, Nelson DP, Shultz PJ, Weir EK (1994) Nitric oxide and cGMP cause vasorelaxation by activation of a charybdotoxin-sensitive K channel by cGMP-dependent protein kinase. Proc Natl Acad Sci U S A 91(16):7583–7587

Ataga KI, Stocker J (2009) Senicapoc (ICA-17043): a potential therapy for the prevention and treatment of hemolysis-associated complications in sickle cell anemia. Expert Opin Investig Drugs 18(2):231–239

Ataga KI, Orringer EP, Styles L, Vichinsky EP, Swerdlow P, Davis GA, Desimone PA, Stocker JW (2006) Dose-escalation study of ICA-17043 in patients with sickle cell disease. Pharmacotherapy 26(11):1557–1564

Ataga KI, Smith WR, De Castro LM, Swerdlow P, Saunthararajah Y, Castro O, Vichinsky E, Kutlar A, Orringer EP, Rigdon GC, Stocker JW, Investigators ICA (2008) Efficacy and safety of

the Gardos channel blocker, senicapoc (ICA-17043), in patients with sickle cell anemia. Blood 111(8):3991–3997
Ataga KI, Reid M, Ballas SK, Yasin Z, Bigelow C, James LS, Smith WR, Galacteros F, Kutlar A, Hull JH, Stocker JW, Investigators ICAS (2011) Improvements in haemolysis and indicators of erythrocyte survival do not correlate with acute vaso-occlusive crises in patients with sickle cell disease: a phase III randomized, placebo-controlled, double-blind study of the Gardos channel blocker senicapoc (ICA-17043). Br J Haematol 153(1):92–104
Atkinson RN, Mcnaughton-Smith GA, Reed AD (2004) Sulfonamides as potassium channel blockers. WO 2004016221
Balut CM, Gao Y, Luke C, Devor DC (2010) Immunofluorescence-based assay to identify modulators of the number of plasma membrane KCa3.1 channels. Future Med Chem 2 (5):707–713
Balut CM, Hamilton KL, Devor DC (2012) Trafficking of intermediate (KCa3.1) and small (KCa2. x) conductance, Ca^{2+}-activated K^{+} channels: a novel target for medicinal chemistry efforts? Chem Med Chem 7(10):1741–1755
Baqi Y, Weyler S, Iqbal J, Zimmermann H, Muller CE (2009) Structure-activity relationships of anthraquinone derivatives derived from bromaminic acid as inhibitors of ectonucleoside triphosphate diphosphohydrolases (E-NTPDases). Purinergic Signalling 5(1):91–106
Bardou O, Trinh NT, Brochiero E (2009) Molecular diversity and function of K^{+} channels in airway and alveolar epithelial cells. Am J Physiol Lung Cell Mol Physiol 296(2):L145–L155
Barrett JN, Magleby KL, Pallotta BS (1982) Properties of single calcium-activated potassium channels in cultured rat muscle. J Physiol 331:211–230
Bednarczyk P, Koziel A, Jarmuszkiewicz W, Szewczyk A (2013) Large-conductance Ca^{2+}-activated potassium channel in mitochondria of endothelial EA.hy926 cells. Am J Phys Heart Circ Phys 304(11):H1415–H1427
Bellott EM, Brugnara C, Clifford JJ, Fluckiger R, Gao YD, Haidar RM, Halperin J, Kelleher EW, Lombardy RJ, Moussa AM (1999a) Substituted 11-phenyl-dibenzazepine compounds useful for the treatment or prevention of diseases characterized by abnormal cell proliferation. WO 1999026929
Bellott EM, Brugnara C, Clifford JJ, Fluckiger R, Gao YD, Haidar RM, Halperin J, Kelleher EW, Lombardy RJ, Moussa AM (1999b) Substituted diphenyl indanone, indane and indole compounds and analogues thereof useful for the treatment or prevention of diseases characterized by abnormal cell proliferation. WO 1999026611
Benoit C, Renaudon B, Salvail D, Rousseau E (2001) EETs relax airway smooth muscle via an EpDHF effect: BK(Ca) channel activation and hyperpolarization. Am J Physiol Lung Cell Mol Physiol 280(5):L965–L973
Bentzen BH, Nardi A, Calloe K, Madsen LS, Olesen SP, Grunnet M (2007) The small molecule NS11021 is a potent and specific activator of Ca^{2+}-activated big-conductance K^{+} channels. Mol Pharmacol 72(4):1033–1044
Bentzen BH, Osadchii O, Jespersen T, Hansen RS, Olesen SP, Grunnet M (2009) Activation of big conductance Ca^{2+}-activated K^{+} channels (BK) protects the heart against ischemia-reperfusion injury. Pflugers Archiv: Eur J Physiol 457(5):979–988
Bentzen BH, Andersen RW, Olesen SP, Grunnet M, Nardi A (2010) Synthesis and characterisation of NS13558: a new important tool for addressing KCa1.1 channel function ex vivo. Naunyn Schmiedeberg's Arch Pharmacol 381(3):271–283
Berkowitz LR, Orringer EP (1981) Effect of cetiedil, an in vitro antisickling agent, on erythrocyte membrane cation permeability. J Clin Invest 68(5):1215–1220
Biot C, Bauer H, Schirmer RH, Davioud-Charvet E (2004) 5-substituted tetrazoles as bioisosteres of carboxylic acids. Bioisosterism and mechanistic studies on glutathione reductase inhibitors as antimalarials. J Med Chem 47(24):5972–5983
Bolotina VM, Najibi S, Palacino JJ, Pagano PJ, Cohen RA (1994) Nitric oxide directly activates calcium-dependent potassium channels in vascular smooth muscle. Nature 368(6474):850–853

Borchert GH, Hlavackova M, Kolar F (2013) Pharmacological activation of mitochondrial BK (Ca) channels protects isolated cardiomyocytes against simulated reperfusion-induced injury. Exp Biol Med 238(2):233–241

Boy K (2005) Such as N-[4-[[4-(5-chloro-2-hydroxyphenyl)-1,2-dihydro-2-oxo-6-(trifluoromethyl)-3-quinolinyl]thio]phenyl]-4-morpholineacetamide; potassium channel modulators. US 20050124611

Boy KM, Guernon JM, Sit SY, Xie K, Hewawasam P, Boissard CG, Dworetzky SI, Natale J, Gribkoff VK, Lodge N, Starrett JE Jr (2004) 3-Thio-quinolinone maxi-K openers for the treatment of erectile dysfunction. Bioorg Med Chem Lett 14(20):5089–5093

Boy K, Hewawasam P, Sit SY (2005a) 3-thia-4-arylquinolin-2-one derivatives. US 20050080105

Boy K, Hewawasam P, Sit SY, Xie K (2005b) Administering a therapeutically effective amount of 3-thia-4-(2-oxy/hydroxyphenyl)quinolin-2-one derivative to a patient to treat erectile dysfunction or irritable bowel syndrome. US 20050176763

Boyd EA, Fisher MH, Garcia ML, Kaczorowski GJ, Meinke PT, Parsons WH, Price S, Stibbard J (2005) Ophthalmic compositions for treating ocular hypertension. WO 2004087051

Brondum E, Kold-Petersen H, Simonsen U, Aalkjaer C (2010) NS309 restores EDHF-type relaxation in mesenteric small arteries from type 2 diabetic ZDF rats. Br J Pharmacol 159(1):154–165

Brown PD, Loo DD, Wright EM (1988) Ca^{2+}-activated K^{+} channels in the apical membrane of Necturus choroid plexus. J Membr Biol 105(3):207–219

Brown BM, Shim H, Christophersen P, Wulff H (2019) Pharmacology of small- and intermediate-conductance calcium-activated potassium channels. Annu Rev Pharmacol Toxicol. https://doi.org/10.1146/annurev-pharmtox-010919-023420

Bruening-Wright A, Lee WS, Adelman JP, Maylie J (2007) Evidence for a deep pore activation gate in small conductance Ca^{2+}-activated K^{+} channels. J Gen Physiol 130(6):601–610

Brugnara C, de Franceschi L (2006) Clinical trials of new therapeutic pharmacology for sickle cell disease. Sante 16(4):263–268

Brugnara C, Gee B, Armsby CC, Kurth S, Sakamoto M, Rifai N, Alper SL, Platt OS (1996) Therapy with oral clotrimazole induces inhibition of the Gardos channel and reduction of erythrocyte dehydration in patients with sickle cell disease. J Clin Invest 97(5):1227–1234

Bukiya AN, Liu J, Toro L, Dopico AM (2007) Beta1 (KCNMB1) subunits mediate lithocholate activation of large-conductance Ca^{2+}-activated K^{+} channels and dilation in small, resistance-size arteries. Mol Pharmacol 72(2):359–369

Bukiya AN, McMillan JE, Fedinec AL, Patil SA, Miller DD, Leffler CW, Parrill AL, Dopico AM (2013) Cerebrovascular dilation via selective targeting of the cholane steroid-recognition site in the BK channel beta1-subunit by a novel nonsteroidal agent. Mol Pharmacol 83(5):1030–1044

Burke MJ, Mckibben B, Tschantz MA (2013) Fused thiazin-3-ones as kca3.1 inhibitors. WO 2013191984

Candia S, Garcia ML, Latorre R (1992) Mode of action of iberiotoxin, a potent blocker of the large conductance Ca^{2+}-activated K^{+} channel. Biophys J 63(2):583–590

Cao Y, Dreixler JC, Roizen JD, Roberts MT, Houamed KM (2001) Modulation of recombinant small-conductance Ca^{2+}-activated K^{+} channels by the muscle relaxant chlorzoxazone and structurally related compounds. J Pharmacol Exp Ther 296(3):683–689

Caputo A, Caci E, Ferrera L, Pedemonte N, Barsanti C, Sondo E, Pfeffer U, Ravazzolo R, Zegarra-Moran O, Galietta LJ (2008) TMEM16A, a membrane protein associated with calcium-dependent chloride channel activity. Science 322(5901):590–594

Carrier GO, Fuchs LC, Winecoff AP, Giulumian AD, White RE (1997) Nitrovasodilators relax mesenteric microvessels by cGMP-induced stimulation of Ca-activated K channels. Am J Phys 273(1 Pt 2):H76–H84

Chen MH, Doherty JB, Liu L, Meinke PT, Natarajan R, Parsons WH, Shen DM, Shu M, Stelmach JE, Wisnoski D (2003) Ophthalmic compositions for treating ocular hypertension. WO 2003105847

Chen MH, Doherty JB, Liu L, Natarajan SR, Shen DM, Tynebor RM (2004a) Ophthalmic compositions for treating ocular hypertension. WO 2004043354

Chen MH, Doherty JB, Liu L, Natarajan SR, Tynebor RM (2004b) Ophthalmic compositions for treating ocular hypertension. WO 2004043933

Chen MH, Doherty JB, Liu L, Natarajan S, Tynebor RM (2005a) Ophthalmic compositions for treating ocular hypertension. WO 2005026128

Chen MH, Doherty JB, Liu L, Natarajan SR, Shen DM, Shu M (2005b) Ophthalmic compositions for treating ocular hypertension. WO 2005002520

Chen X, Yan J, Aldrich RW (2014) BK channel opening involves side-chain reorientation of multiple deep-pore residues. Proc Natl Acad Sci U S A 111(1):E79–E88

Coiret G, Borowiec AS, Mariot P, Ouadid-Ahidouch H, Matifat F (2007) The antiestrogen tamoxifen activates BK channels and stimulates proliferation of MCF-7 breast cancer cells. Mol Pharmacol 71(3):843–851

Coleman N, Brown BM, Olivan-Viguera A, Singh V, Olmstead MM, Valero MS, Kohler R, Wulff H (2014) New positive KCa Channel gating modulators with selectivity for KCa3.1. Mol Pharmacol 86(3):342–357

Cui YM, Yasutomi E, Otani Y, Yoshinaga T, Ido K, Sawada K, Ohwada T (2008) Design, synthesis and characterization of podocarpate derivatives as openers of BK channels. Bioorg Med Chem Lett 18(19):5197–5200

Cui YM, Yasutomi E, Otani Y, Ido K, Yoshinaga T, Sawada K, Ohwada T (2010) Design, synthesis, and characterization of BK channel openers based on oximation of abietane diterpene derivatives. Bioorg Med Chem 18(24):8642–8659

Cullinane AB, Leung PS, Ortego J, Coca-Prados M, Harvey BJ (2002) Renin-angiotensin system expression and secretory function in cultured human ciliary body non-pigmented epithelium. Br J Ophthalmol 86(6):676–683

Cuppoletti J, Malinowska DH, Tewari KP, Chakrabarti J, Ueno R (2012) Unoprostone isopropyl and metabolite M1 activate BK channels and prevent ET-1-induced $[Ca^{2+}]i$ increases in human trabecular meshwork and smooth muscle. Invest Ophthalmol Vis Sci 53(9):5178–5189

Davi H, Tronquet C, Miscoria G, Perrier L, DuPont P, Caix J, Simiand J, Berger Y (2000) Disposition of irbesartan, an angiotensin II AT1-receptor antagonist, in mice, rats, rabbits, and macaques. Drug Metab Dispos 28(1):79–88

De Wet H, Allen M, Holmes C, Stobbart M, Lippiat JD, Callaghan R (2006) Modulation of the BK channel by estrogens: examination at single channel level. Mol Membr Biol 23(5):420–429

DeFarias FP, Carvalho MF, Lee SH, Kaczorowski GJ, Suarez-Kurtz G (1996) Effects of the K^+ channel blockers paspalitrem-C and paxilline on mammalian smooth muscle. Eur J Pharmacol 314(1–2):123–128

Dela Pena IC, Yoon SY, Kim SM, Lee GS, Park CS, Kim YC, Cheong JH (2009a) Inhibition of intestinal motility by the putative BK(Ca) channel opener LDD175. Arch Pharm Res 32 (3):413–420

Dela Pena IC, Yoon SY, Kim SM, Lee GS, Ryu JH, Park CS, Kim YC, Cheong JH (2009b) Bladder-relaxant properties of the novel benzofuroindole analogue LDD175. Pharmacology 83 (6):367–378

Demnitz J, Jorgensen S (2014) Novel tetrazole derivatives and their use as potassium channel modulators. WO 2014001363

Demnitz J, Gouliaev AH, Sloek FA, Teuber L (2003) Potassium channel modulators. WO 2003059873

Demnitz J, Madsen LS, Olesen SP, Stroebaek D (2005) Diarylmethyl derivatives as potassium channel modulators. WO 2005003094

Denda M, Tsutsumi M, Inoue K, Crumrine D, Feingold KR, Elias PM (2007) Potassium channel openers accelerate epidermal barrier recovery. Br J Dermatol 157(5):888–893

Denson DD, Eaton DC (1994) Ketamine inhibition of large conductance Ca^{2+}-activated K^+ channels is modulated by intracellular Ca^{2+}. Am J Phys 267(5 Pt 1):C1452–C1458

Denson DD, Duchatelle P, Eaton DC (1994) The effect of racemic ketamine on the large conductance Ca^{2+}-activated potassium (BK) channels in GH3 cells. Brain Res 638(1–2):61–68

Devor DC, Frizzell RA (1998) Modulation of K^+ channels by arachidonic acid in T84 cells. I. Inhibition of the Ca^{2+}-dependent K^+ channel. Am J Phys 274(1):C138–C148

Devor DC, Singh AK, Frizzell RA, Bridges RJ (1996) Modulation of Cl^- secretion by benzimidazolones. I. Direct activation of a Ca^{2+}-dependent K^+ channel. Am J Phys 271(5 Pt 1):L775–L784

Dick GM, Sanders KM (2001) (Xeno)estrogen sensitivity of smooth muscle BK channels conferred by the regulatory beta1 subunit: a study of beta1 knockout mice. J Biol Chem 276 (48):44835–44840

Dick GM, Rossow CF, Smirnov S, Horowitz B, Sanders KM (2001) Tamoxifen activates smooth muscle BK channels through the regulatory beta 1 subunit. J Biol Chem 276(37):34594–34599

Dick GM, Hunter AC, Sanders KM (2002) Ethylbromide tamoxifen, a membrane-impermeant antiestrogen, activates smooth muscle calcium-activated large-conductance potassium channels from the extracellular side. Mol Pharmacol 61(5):1105–1113

Doherty JB, Shen DM (2005) Ophthalmic compositions for treating ocular hypertension. WO 2005020917

Doherty JB, Shen DM (2006) Ophthalmic compositions for treating ocular hypertension. WO 2006020003

Doherty J, Chen MH, Liu L, Natarajan S, Tynebor R (2004a) Ophthalmic compositions for treating ocular hypertension. US 20040097575

Doherty JB, Chen MH, Liu L, Natarajan SR, Tynebor RM (2004b) Ophthalmic compositions for treating ocular hypertension. WO 2004043932

Doherty JB, Dong-Ming S, Min S, Fengqi Z (2007) Ophthalmic compositions for treating ocular hypertension. WO 2007108968

Doherty JB, Dong-Ming S, Min S (2008a) Ophthalmic compositions for treating ocular hypertension. WO 2008030390

Doherty JB, Natarajan SR, Shen DM, Zhang F (2008b) Ophthalmic compositions for treating ocular hypertension. WO 2007146136

Duerson K, White RE, Jiang F, Schonbrunn A, Armstrong DL (1996) Somatostatin stimulates BKCa channels in rat pituitary tumor cells through lipoxygenase metabolites of arachidonic acid. Neuropharmacology 35(7):949–961

Dutta AK, Khimji AK, Sathe M, Kresge C, Parameswara V, Esser V, Rockey DC, Feranchak AP (2009) Identification and functional characterization of the intermediate-conductance Ca^{2+}-activated K^+ channel (IK-1) in biliary epithelium. Am J Physiol Gastrointest Liver Physiol 297(5):G1009–G1018

Dworetzky SI, Trojnacki JT, Gribkoff VK (1994) Cloning and expression of a human large-conductance calcium-activated potassium channel. Brain Res Mol Brain Res 27(1):189–193

Ellory JC, Culliford SJ, Smith PA, Wolowyk MW, Knaus EE (1994) Specific inhibition of Ca-activated K channels in red cells by selected dihydropyridine derivatives. Br J Pharmacol 111(3):903–905

Falck JR, Reddy LM, Reddy YK, Bondlela M, Krishna UM, Ji Y, Sun J, Liao JK (2003) 11,12-epoxyeicosatrienoic acid (11,12-EET): structural determinants for inhibition of TNF-alpha-induced VCAM-1 expression. Bioorg Med Chem Lett 13(22):4011–4014

Falck JR, Wallukat G, Puli N, Goli M, Arnold C, Konkel A, Rothe M, Fischer R, Muller DN, Schunck WH (2011) 17(R),18(S)-epoxyeicosatetraenoic acid, a potent eicosapentaenoic acid (EPA) derived regulator of cardiomyocyte contraction: structure-activity relationships and stable analogues. J Med Chem 54(12):4109–4118

Fukao M, Mason HS, Kenyon JL, Horowitz B, Keef KD (2001) Regulation of BK(Ca) channels expressed in human embryonic kidney 293 cells by epoxyeicosatrienoic acid. Mol Pharmacol 59 (1):16–23

Galvez A, Gimenez-Gallego G, Reuben JP, Roy-Contancin L, Feigenbaum P, Kaczorowski GJ, Garcia ML (1990) Purification and characterization of a unique, potent, peptidyl probe for the high conductance calcium-activated potassium channel from venom of the scorpion Buthus tamulus. J Biol Chem 265(19):11083–11090

Gao YD, Shen DM (2006) Ophthalmic compositions for treating ocular hypertension. WO 2006044232

Garcia M, Kaczorowski G, McManus O (2001) Method for treating ocular hypertension. US 20010047025

Garcia-Valdes J, Zamudio FZ, Toro L, Possani LD (2001) Slotoxin, alphaKTx1.11, a new scorpion peptide blocker of MaxiK channels that differentiates between alpha and alpha+beta (beta1 or beta4) complexes. FEBS Lett 505(3):369–373

Garduno J, Galvan E, Fernandez de Sevilla D, Buno W (2005) 1-Ethyl-2-benzimidazolinone (EBIO) suppresses epileptiform activity in in vitro hippocampus. Neuropharmacology 49 (3):376–388

Garneau L, Klein H, Lavoie MF, Brochiero E, Parent L, Sauve R (2014) Aromatic-aromatic interactions between residues in KCa3.1 pore helix and S5 transmembrane segment control the channel gating process. J Gen Physiol 143(2):289–307

Gessner G, Cui YM, Otani Y, Ohwada T, Soom M, Hoshi T, Heinemann SH (2012) Molecular mechanism of pharmacological activation of BK channels. Proc Natl Acad Sci U S A 109 (9):3552–3557

Gillman K, Bocchino D (2004) Useful in therapy of disorders which are responsive to the opening of potassium channels such as ischemia, stroke, convulsions, asthma, epilepsy, irritable bowel syndrome, migraine, traumatic brain injury, spinal cord injury, sexual dysfunction. US 20040152646

Gillman KW, Hewawasam P, Schmitz WD, Lopez OD, Starrett JE, Provencal DP (2003) Phosphate prodrugs of fluorooxindoles. WO 2003080047

Goetz MA, Kaczorowski GJ (2004) Compositions and methods for treating glaucoma and ocular hypertension. WO 2003077845

Goetz MA, Kaczorowski GJ, Monaghan RL, Strohl WR, Tkacz JS (2003) Novel maxi-k channel blockers, methods of use and process for making the same. WO 2003105868

Gordon E, Semus SF, Lozinskaya IM, Lin Z, Xu X (2010) Characterizing the role of Thr352 in the inhibition of the large conductance Ca^{2+}-activated K^{+} channels by 1-[1-hexyl-6-(methyloxy)-1H-indazol-3-yl]-2-methyl-1-propanone. J Pharmacol Exp Ther 334(2):402–409

Gore VK, Ma VV, Yin R, Ligutti J, Immke D, Doherty EM, Norman MH (2010) Structure-activity relationship (SAR) investigations of tetrahydroquinolines as BKCa agonists. Bioorg Med Chem Lett 20(12):3573–3578

Gormemis AE, Ha TS, Im I, Jung KY, Lee JY, Park CS, Kim YC (2005) Benzofuroindole analogues as potent BK(Ca) channel openers. Chembiochem: Eur J Chem Biol 6 (10):1745–1748

Gragasin FS, Michelakis ED, Hogan A, Moudgil R, Hashimoto K, Wu X, Bonnet S, Haromy A, Archer SL (2004) The neurovascular mechanism of clitoral erection: nitric oxide and cGMP-stimulated activation of BKCa channels. FASEB J 18(12):1382–1391

Green L, Wang H (2014) 3,4-disubstituted oxazolidinone derivatives and their use as inhibitors of the calcium activated potassium channel. WO2014067861

Gribkoff VK, Lum-Ragan JT, Boissard CG, Post-Munson DJ, Meanwell NA, Starrett JE Jr, Kozlowski ES, Romine JL, Trojnacki JT, McKay MC, Zhong J, Dworetzky SI (1996) Effects of channel modulators on cloned large-conductance calcium-activated potassium channels. Mol Pharmacol 50(1):206–217

Gribkoff VK, Starrett JE Jr, Dworetzky SI, Hewawasam P, Boissard CG, Cook DA, Frantz SW, Heman K, Hibbard JR, Huston K, Johnson G, Krishnan BS, Kinney GG, Lombardo LA, Meanwell NA, Molinoff PB, Myers RA, Moon SL, Ortiz A, Pajor L, Pieschl RL, Post-Munson DJ, Signor LJ, Srinivas N, Taber MT, Thalody G, Trojnacki JT, Wiener H, Yeleswaram K, Yeola SW (2001) Targeting acute ischemic stroke with a calcium-sensitive opener of maxi-K potassium channels. Nat Med 7(4):471–477

Grunnet M, Jespersen T, Angelo K, Frokjaer-Jensen C, Klaerke DA, Olesen SP, Jensen BS (2001) Pharmacological modulation of SK3 channels. Neuropharmacology 40(7):879–887

Guan X, Li Q, Yan J (2017) Relationship between auxiliary gamma subunits and mallotoxin on BK channel modulation. Sci Rep 7:42240

Haburcak M, Wei L, Viana F, Prenen J, Droogmans G, Nilius B (1997) Calcium-activated potassium channels in cultured human endothelial cells are not directly modulated by nitric oxide. Cell Calcium 21(4):291–300

Haghdoost-Yazdi H, Janahmadi M, Behzadi G (2008) Iberiotoxin-sensitive large conductance Ca^{2+} –dependent K^+ (BK) channels regulate the spike configuration in the burst firing of cerebellar Purkinje neurons. Brain Res 1212:1–8

Hamilton KL, Syme CA, Devor DC (2003) Molecular localization of the inhibitory arachidonic acid binding site to the pore of hIK1. J Biol Chem 278(19):16690–16697

Harada H, Watanuki S, Takuwa T, Kawaguchi K, Okazaki T, Hirano Y, Saitoh C (2003) Medicine comprising dicyanopyridine derivative. US 20030232860

Harper AA, Catacuzzeno L, Trequattrini C, Petris A, Franciolini F (2001) Verapamil block of large-conductance Ca-activated K channels in rat aortic myocytes. J Membr Biol 179(2):103–111

Hayashi Y, Kawaji K, Sun L, Zhang X, Koyano K, Yokoyama T, Kohsaka S, Inoue K, Nakanishi H (2011) Microglial Ca^{2+}-activated K^+ channels are possible molecular targets for the analgesic effects of S-ketamine on neuropathic pain. J Neurosci 31(48):17370–17382

Heinz A, Passow H (1980) Role of external potassium in the calcium-induced potassium efflux from human red blood cell ghosts. J Membr Biol 57(2):119–131

Hercule HC, Salanova B, Essin K, Honeck H, Falck JR, Sausbier M, Ruth P, Schunck WH, Luft FC, Gollasch M (2007) The vasodilator 17,18-epoxyeicosatetraenoic acid targets the pore-forming BK alpha channel subunit in rodents. Exp Physiol 92(6):1067–1076

Hewawasam P, Meanwell NA, Gribkoff VK (1997) Useful in reducing neuronal damage during ischemic stroke. US 5602169

Hewawasam P, Fan W, Ding M, Flint K, Cook D, Goggins GD, Myers RA, Gribkoff VK, Boissard CG, Dworetzky SI, Starrett JE Jr, Lodge NJ (2003) 4-Aryl-3-(hydroxyalkyl)quinolin-2-ones: novel maxi-K channel opening relaxants of corporal smooth muscle targeted for erectile dysfunction. J Med Chem 46(14):2819–2822

Hewawasam P, Fan W, Cook DA, Newberry KS, Boissard CG, Gribkoff VK, Starrett J, Lodge NJ (2004) 4-Aryl-3-(mercapto)quinolin-2-ones: novel maxi-K channel opening relaxants of corporal smooth muscle. Bioorg Med Chem Lett 14(17):4479–4482

Hewawasam P, Sit SY, Starrett J (2005) Potassium channel modulators. US 20050080082

Hoffman JF, Joiner W, Nehrke K, Potapova O, Foye K, Wickrema A (2003) The hSK4 (KCNN4) isoform is the Ca^{2+}-activated K^+ channel (Gardos channel) in human red blood cells. Proc Natl Acad Sci U S A 100(12):7366–7371

Hollywood MA, Mchale N, Roy S, Sergeant GP, Thornbury K (2012a) Anthraquinone compounds and their uses. WO 2012035122

Hollywood MA, Thornbury K, Sergeant GP, Mchale N, Roy S (2012b) Anthraquinone compounds and their uses. EP 2439200

Hougaard C, Eriksen BL, Jørgensen S, Johansen TH, Dyhring T, Madsen LS, Strøbaek D, Christophersen P (2007) Selective positive modulation of the SK3 and SK2 subtypes of small conductance Ca^{2+}-activated K^+ channels. Br J Pharmacol 151(5):655–665

Hougaard C, Fraser MO, Chien C, Bookout A, Katofiasc M, Jensen BS, Rode F, Bitsch-Norhave J, Teuber L, Thor KB, Strobaek D, Burgard EC, Ronn LC (2009) A positive modulator of K Ca 2 and K Ca 3 channels, 4,5-dichloro-1,3-diethyl-1,3-dihydro-benzoimidazol-2-one (NS4591), inhibits bladder afferent firing in vitro and bladder overactivity in vivo. J Pharmacol Exp Ther 328(1):28–39

Hougaard C, Hammami S, Eriksen BL, Sorensen US, Jensen ML, Strobaek D, Christophersen P (2012) Evidence for a common pharmacological interaction site on K(Ca)2 channels providing both selective activation and selective inhibition of the human K(Ca)2.1 subtype. Mol Pharmacol 81(2):210–219

Huang CW, Huang CC, Wu SN (2007) Activation by zonisamide, a newer antiepileptic drug, of large-conductance calcium-activated potassium channel in differentiated hippocampal neuron-derived H19-7 cells. J Pharmacol Exp Ther 321(1):98–106
Imaizumi Y, Ohwada T (2002) Potassium channel opener. WO 2002087559
Imaizumi Y, Sakamoto K, Yamada A, Hotta A, Ohya S, Muraki K, Uchiyama M, Ohwada T (2002) Molecular basis of pimarane compounds as novel activators of large-conductance Ca^{2+}-activated K^{+} channel alpha-subunit. Mol Pharmacol 62(4):836–846
Ishii TM, Silvia C, Hirschberg B, Bond CT, Adelman JP, Maylie J (1997) A human intermediate conductance calcium-activated potassium channel. Proc Natl Acad Sci U S A 94 (21):11651–11656
Jaravine VA, Nolde DE, Reibarkh MJ, Korolkova YV, Kozlov SA, Pluzhnikov KA, Grishin EV, Arseniev AS (1997) Three-dimensional structure of toxin OSK1 from Orthochirus scrobiculosus scorpion venom. Biochemistry 36(6):1223–1232
Javaherian AD, Yusifov T, Pantazis A, Franklin S, Gandhi CS, Olcese R (2011) Metal-driven operation of the human large-conductance voltage- and Ca^{2+}-dependent potassium channel (BK) gating ring apparatus. J Biol Chem 286(23):20701–20709
Jenkins DP, Strobaek D, Hougaard C, Jensen ML, Hummel R, Sorensen US, Christophersen P, Wulff H (2011) Negative gating modulation by (R)-N-(benzimidazol-2-yl)-1,2,3,4-tetrahydro-1-naphthylamine (NS8593) depends on residues in the inner pore vestibule: pharmacological evidence of deep-pore gating of K(Ca)2 channels. Mol Pharmacol 79(6):899–909
Jensen BS (2002) BMS-204352: a potassium channel opener developed for the treatment of stroke. CNS Drug Rev 8(4):353–360
Jensen BS, Strobaek D, Christophersen P, Jorgensen TD, Hansen C, Silahtaroglu A, Olesen SP, Ahring PK (1998) Characterization of the cloned human intermediate-conductance Ca^{2+}-activated K^{+} channel. Am J Phys 275(3 Pt 1):C848–C856
Jensen BS, Odum N, Jorgensen NK, Christophersen P, Olesen SP (1999) Inhibition of T cell proliferation by selective block of Ca^{2+}-activated K^{+} channels. Proc Natl Acad Sci U S A 96 (19):10917–10921
Ji YH, Wang WX, Ye JG, He LL, Li YJ, Yan YP, Zhou Z (2003) Martentoxin, a novel K^{+}-channel-blocking peptide: purification, cDNA and genomic cloning, and electrophysiological and pharmacological characterization. J Neurochem 84(2):325–335
Jin LW, Lucente JD, Nguyen HM, Singh V, Singh L, Chavez M, Bushong T, Wulff H, Maezawa I (2019 Mar 18) Repurposing the KCa3.1 inhibitor senicapoc for Alzheimer's disease. Ann Clin Transl Neurol 6(4):723–738
Joiner WJ, Wang LY, Tang MD, Kaczmarek LK (1997) hSK4, a member of a novel subfamily of calcium-activated potassium channels. Proc Natl Acad Sci U S A 94(20):11013–11018
Jorgensen S, Dyhring T, Brown DT, Strobaek D, Christophersen P, Demnitz J (2013) A high-throughput screening campaign for detection of Ca^{2+}-activated K^{+} channel activators and inhibitors using a fluorometric imaging plate reader-based Tl^{+}-influx assay. Assay Drug Dev Technol 11(3):163–172
Kaji DM (1990) Nifedipine inhibits calcium-activated K transport in human erythrocytes. Am J Phys 259(2 Pt 1):C332–C339
Kasumu AW, Hougaard C, Rode F, Jacobsen TA, Sabatier JM, Eriksen BL, Strobaek D, Liang X, Egorova P, Vorontsova D, Christophersen P, Ronn LC, Bezprozvanny I (2012) Selective positive modulator of calcium-activated potassium channels exerts beneficial effects in a mouse model of spinocerebellar ataxia type 2. Chem Biol 19(10):1340–1353
Kehl SJ (1996) Block of BK (maxi K) channels of rat pituitary melanotrophs by Na^{+} and other alkali metal ions. Pflugers Archiv: Eur J Physiol 432(4):623–629
Khaddaj-Mallat R, Mathew John C, Braun AP (2018) SKA-31, an activator of endothelial Ca^{2+}-activated K^{+} channels evokes robust vasodilation in rat mesenteric arteries. Eur J Pharmacol 831:60–67

King JT, Lovell PV, Rishniw M, Kotlikoff MI, Zeeman ML, McCobb DP (2006) Beta2 and beta4 subunits of BK channels confer differential sensitivity to acute modulation by steroid hormones. J Neurophysiol 95(5):2878–2888

Kiraly I, Pataricza J, Bajory Z, Simonsen U, Varro A, Papp JG, Pajor L, Kun A (2013) Involvement of large-conductance Ca^{2+} –activated K^+ channels in both nitric oxide and endothelium-derived hyperpolarization-type relaxation in human penile small arteries. Basic Clin Pharmacol Toxicol 113(1):19–24

Klockgether-Radke AP, Huneck S, Meyberg S, Neumann P, Hellige G (2005) Ketamine enantiomers differentially relax isolated coronary artery rings. Eur J Anaesthesiol 22(3):215–221

Kloza M, Baranowska-Kuczko M, Toczek M, Kusaczuk M, Sadowska O, Kasacka I, Kozłowska H (2019) Modulation of cardiovascular function in primary hypertension in rat by SKA-31, an activator of KCa2.X and KCa3.1 channels. Int J Mol Sci 20(17)

Knaus HG, McManus OB, Lee SH, Schmalhofer WA, Garcia-Calvo M, Helms LM, Sanchez M, Giangiacomo K, Reuben JP, Smith AB 3rd et al (1994) Tremorgenic indole alkaloids potently inhibit smooth muscle high-conductance calcium-activated potassium channels. Biochemistry 33(19):5819–5828

Kobayashi K, Nishizawa Y, Sawada K, Ogura H, Miyabe M (2008) K^+-channel openers suppress epileptiform activities induced by 4-aminopyridine in cultured rat hippocampal neurons. J Pharmacol Sci 108(4):517–528

Koegel H, Alzheimer C (2001) Expression and biological significance of Ca^{2+}-activated ion channels in human keratinocytes. FASEB J 15(1):145–154

Koegel H, Kaesler S, Burgstahler R, Werner S, Alzheimer C (2003) Unexpected down-regulation of the hIK1 Ca^{2+}-activated K^+ channel by its opener 1-ethyl-2-benzimidazolinone in HaCaT keratinocytes. Inverse effects on cell growth and proliferation. J Biol Chem 278(5):3323–3330

Kozlowski ES, Johnson G, Dischino DD, Dworetzky SI, Boissard CG, Gribkoff VK (1996) Synthesis and biological evaluation of an iodinated iberiotoxin analogue, [mono-iodo-Tyr5, Phe36]-iberiotoxin. Int J Pept Protein Res 48(2):194–199

Krishna R, Palme H, Zeng J, Srinivas N (2002a) Effect of dose and input rate on the brain penetration of BMS-204352 following intravenous administration to rats. Biopharm Drug Dispos 23(6):227–231

Krishna R, Shah VR, Mantha S, Vachharajani NN, Srinivas N (2002b) Pharmacokinetics and dose proportionality of BMS-204352 after intravenous administration to dogs. Biopharm Drug Dispos 23(2):83–86

Krishna R, Shah VR, Srinivas N (2002c) Pharmacokinetics and dose proportionality of BMS-204352 after intraarterial administration to rats. Biopharm Drug Dispos 23(6):233–237

Kroigaard C, Dalsgaard T, Nielsen G, Laursen BE, Pilegaard H, Kohler R, Simonsen U (2012) Activation of endothelial and epithelial K(Ca) 2.3 calcium-activated potassium channels by NS309 relaxes human small pulmonary arteries and bronchioles. Br J Pharmacol 167(1):37–47

Kun A, Matchkov VV, Stankevicius E, Nardi A, Hughes AD, Kirkeby HJ, Demnitz J, Simonsen U (2009) NS11021, a novel opener of large-conductance Ca^{2+}-activated K^+ channels, enhances erectile responses in rats. Br J Pharmacol 158(6):1465–1476

Kunze WA, Bornstein JC, Furness JB, Hendriks R, Stephenson DS (1994) Charybdotoxin and iberiotoxin but not apamin abolish the slow after-hyperpolarization in myenteric plexus neurons. Pflugers Archiv: Eur J Physiol 428(3–4):300–306

La Fuente JM, Fernandez A, Cuevas P, Gonzalez-Corrochano R, Chen M, Angulo J (2014) Stimulation of large-conductance calcium-activated potassium channels inhibits neurogenic contraction of human bladder from patients with urinary symptoms and reverses acetic acid-induced bladder hyperactivity in rats. Eur J Pharmacol 735:68–76

Laeis P, Puchler K, Kirch W (2001) The pharmacokinetic and metabolic profile of olmesartan medoxomil limits the risk of clinically relevant drug interaction. J Hypertens Suppl 19(1):S21–S32

Lang DG, Ritchie AK (1990) Tetraethylammonium blockade of apamin-sensitive and insensitive Ca^{2+}-activated K^+ channels in a pituitary cell line. J Physiol 425:117–132

Larsen CK, Praetorius HA, Leipziger J, Sorensen MV (2017) Intact colonic KCa1.1 channel activity in KCNMB2 knockout mice. Physiol Rep 5(5):pii:e13179
Latorre R, Vergara C, Hidalgo C (1982) Reconstitution in planar lipid bilayers of a Ca^{2+}-dependent K^{+} channel from transverse tubule membranes isolated from rabbit skeletal muscle. Proc Natl Acad Sci U S A 79(3):805–809
Lauterbach B, Barbosa-Sicard E, Wang MH, Honeck H, Kargel E, Theuer J, Schwartzman ML, Haller H, Luft FC, Gollasch M, Schunck WH (2002) Cytochrome P450-dependent eicosapentaenoic acid metabolites are novel BK channel activators. Hypertension 39(2 Pt 2):609–613
Layne JJ, Nausch B, Olesen SP, Nelson MT (2010) BK channel activation by NS11021 decreases excitability and contractility of urinary bladder smooth muscle. Am J Physiol Regul Integr Comp Physiol 298(2):R378–R384
Lee CH, MacKinnon R (2018) Activation mechanism of a human SK-calmodulin channel complex elucidated by cryo-EM structures. Science 360(6388):508–513
Lee JE, Kwak J, Suh CK, Shin JH (2006) Dual effects of nitric oxide on the large conductance calcium-activated potassium channels of rat brain. J Biochem Mol Biol 39(1):91–96
Lee BC, Lim HH, Kim S, Youn HS, Lee Y, Kim YC, Eom SH, Lee KW, Park CS (2012) Localization of a site of action for benzofuroindole-induced potentiation of BKCa channels. Mol Pharmacol 82(2):143–155
Lee S, Choi JS, Park CS (2018) Direct activation of the Large-conductance calcium-activated Potassium Channel by flavonoids isolated from *Sophora flavescens*. Biol Pharm Bull 41 (8):1295–1298
Lew VL, Ferreira HG (1976) Variable Ca sensitivity of a K-selective channel in intact red-cell membranes. Nature 263(5575):336–338
Li C, Grillo MP, Benet LZ (2003a) In vitro studies on the chemical reactivity of 2,4-dichlorophenoxyacetyl-S-acyl-CoA thioester. Toxicol Appl Pharmacol 187(2):101–109
Li C, Olurinde MO, Hodges LM, Grillo MP, Benet LZ (2003b) Covalent binding of 2-phenylpropionyl-S-acyl-CoA thioester to tissue proteins in vitro. Drug Metab Dispos 31 (6):727–730
Li Y, Johnson G, Romine JL, Meanwell NA, Martin SW, Dworetzky SI, Boissard CG, Gribkoff VK, Starrett JE Jr (2003c) Novel openers of Ca^{2+}-dependent large-conductance potassium channels: symmetrical pharmacophore and electrophysiological evaluation of bisphenols. Bioorg Med Chem Lett 13(8):1437–1439
Liu H, Xin T, He W, Li F, Su ZQ (2014) Myelinated ah-type trigeminal ganglion neurons in female rats: neuroexcitability, chemosensitivity to histamine, and potential clinical impact. Neurosci Lett 567:74–79
Liu NN, Xie H, Xiang-Wei WS, Gao K, Wang TS, Jiang YW (2019) The absence of NIPA2 enhances neural excitability through BK (big potassium) channels. CNS Neurosci Ther 25 (8):865–875
Lu T, Katakam PV, VanRollins M, Weintraub NL, Spector AA, Lee HC (2001) Dihydroxyeicosatrienoic acids are potent activators of Ca^{2+}-activated K^{+} channels in isolated rat coronary arterial myocytes. J Physiol 534(Pt 3):651–667
Lucas D, Goulitquer S, Marienhagen J, Fer M, Dreano Y, Schwaneberg U, Amet Y, Corcos L (2010) Stereoselective epoxidation of the last double bond of polyunsaturated fatty acids by human cytochromes P450. J Lipid Res 51(5):1125–1133
Mackay KB (2001) BMS-204352 (Bristol Myers Squibb). Curr Opin Investig Drugs 2(6):820–823
Maillard MP, Rossat J, Brunner HR, Burnier M (2000) Tasosartan, enoltasosartan, and angiotensin II receptor blockade: the confounding role of protein binding. J Pharmacol Exp Ther 295 (2):649–654
Makovec F, Peris W, Rovati AL (1990) Derivatives of n-phenylbenzamide with anti-ulcer and anti-allergy activity and a method for their preparation. WO 1990009989

Makovec F, Peris W, Revel L, Giovanetti R, Redaelli D, Rovati LC (1992) Antiallergic and cytoprotective activity of new N-phenylbenzamido acid derivatives. J Med Chem 35 (20):3633–3640
Malerba M, Mennuni L, Piepoli T, Caselli G, Makovec F, Rovati LC, D'Amato M, Ferrari F (2009) Andolast acts at different cellular levels to inhibit immunoglobulin E synthesis. Int J Immunopathol Pharmacol 22(1):85–94
Malerba M, Amato MD, Radaeli A, Giacovelli G, Rovati L, Arshad SH, Holgate ST (2015) Efficacy of andolast in mild to moderate asthma: a randomized, controlled, double-blind multicenter study (the andast trial). Curr Pharm Des. 21(26):3835–3843
Mall M, Gonska T, Thomas J, Schreiber R, Seydewitz HH, Kuehr J, Brandis M, Kunzelmann K (2003) Modulation of Ca^{2+}-activated Cl^- secretion by basolateral K^+ channels in human normal and cystic fibrosis airway epithelia. Pediatr Res 53(4):608–618
Malysz J, Buckner SA, Daza AV, Milicic I, Perez-Medrano A, Gopalakrishnan M (2004) Functional characterization of large conductance calcium-activated K^+ channel openers in bladder and vascular smooth muscle. Naunyn Schmiedeberg's Arch Pharmacol 369(5):481–489
Manaves V, Qin W, Bauer AL, Rossie S, Kobayashi M, Rane SG (2004) Calcium and vitamin D increase mRNA levels for the growth control hIK1 channel in human epidermal keratinocytes but functional channels are not observed. BMC Dermatol 4:7
Manzanares D, Gonzalez C, Ivonnet P, Chen RS, Valencia-Gattas M, Conner GE, Larsson HP, Salathe M (2011) Functional apical large conductance, Ca^{2+}-activated, and voltage-dependent K^+ channels are required for maintenance of airway surface liquid volume. J Biol Chem 286 (22):19830–19839
Matos JE, Sausbier M, Beranek G, Sausbier U, Ruth P, Leipziger J (2007) Role of cholinergic-activated KCa1.1 (BK), KCa3.1 (SK4) and KV7.1 (KCNQ1) channels in mouse colonic Cl^- secretion. Acta Physiol 189(3):251–258
McCobb DP, Fowler NL, Featherstone T, Lingle CJ, Saito M, Krause JE, Salkoff L (1995) A human calcium-activated potassium channel gene expressed in vascular smooth muscle. Am J Phys 269(3 Pt 2):H767–H777
McManus OB, Harris GH, Giangiacomo KM, Feigenbaum P, Reuben JP, Addy ME, Burka JF, Kaczorowski GJ, Garcia ML (1993) An activator of calcium-dependent potassium channels isolated from a medicinal herb. Biochemistry 32(24):6128–6133
McNaughton-Smith GA, Burns JF, Stocker JW, Rigdon GC, Creech C, Arrington S, Shelton T, de Franceschi L (2008) Novel inhibitors of the Gardos channel for the treatment of sickle cell disease. J Med Chem 51(4):976–982
Miller C (1987) Trapping single ions inside single ion channels. Biophys J 52(1):123–126
Miller O, Yu R (2008) Topical ophthalmic formulations. WO 2002087559
Miller C, Moczydlowski E, Latorre R, Phillips M (1985) Charybdotoxin, a protein inhibitor of single Ca^{2+}-activated K^+ channels from mammalian skeletal muscle. Nature 313 (6000):316–318
Mishra RC, Wulff H, Cole WC, Braun AP (2014) A pharmacologic activator of endothelial KCa channels enhances coronary flow in the hearts of type 2 diabetic rats. J Mol Cell Cardiol 72:364–373
Mishra RC, Mitchell JR, Gibbons-Kroeker C, Wulff H, Belenkie I, Tyberg JV, Braun AP (2016) A pharmacologic activator of endothelial KCa channels increases systemic conductance and reduces arterial pressure in an anesthetized pig model. Vasc Pharmacol 79:24–31
Morin C, Sirois M, Echave V, Gomes MM, Rousseau E (2007) Functional effects of 20-HETE on human bronchi: hyperpolarization and relaxation due to BKCa channel activation. Am J Physiol Lung Cell Mol Physiol 293(4):L1037–L1044
Mouhat S, Visan V, Ananthakrishnan S, Wulff H, Andreotti N, Grissmer S, Darbon H, De Waard M, Sabatier JM (2005) K^+ channel types targeted by synthetic OSK1, a toxin from Orthochirus scrobiculosus scorpion venom. Biochem J 385(Pt 1):95–104
Nakamura A, Tanaka M, Tsuda M (1996) Pyrrole derivatives and medicinal composition. WO 1996040634

Nakamura A, Tanaka M, Tsuda M (1999) Potassium channel activators. WO 1999036068
Nardi A, Olesen SP (2008) BK channel modulators: a comprehensive overview. Curr Med Chem 15 (11):1126–1146
Nardi A, Calderone V, Chericoni S, Morelli I (2003) Natural modulators of large-conductance calcium-activated potassium channels. Planta Med 69(10):885–892
Nardi A, Demnitz, D., Garcia, M.L., Polosa, R. (2008) Potassium channels as drug targets for therapeutic intervention in respiratory diseases. Expert Opin Ther Pat 18 (12):1361–1384
Narenjkar J, el Assem SK, Ganellin CR (2004) Inhibition of the antigen-induced activation of RBL-2H3 cells by cetiedil and some of its analogues. Eur J Pharmacol 483(2–3):107–116
Nausch B, Rode F, Jorgensen S, Nardi A, Korsgaard MP, Hougaard C, Brown WD, Bonev AD, Dyhring T, Strobaek D, Olesen SP, Christophersen P, Grunnet M, Nelson MT, Ronn LC (2014) NS19504: a novel BK channel activator with relaxing effect on bladder smooth muscle spontaneous phasic contractions. J Pharmacol Exp Ther
Neyton J, Miller C (1988) Discrete Ba^{2+} block as a probe of ion occupancy and pore structure in the high-conductance Ca^{2+} –activated K^+ channel. J Gen Physiol 92(5):569–586
Nguyen HM, Singh V, Pressly B, Jenkins DP, Wulff H, Yarov-Yarovoy V (2017) Structural insights into the atomistic mechanisms of action of small molecule inhibitors targeting the KCa3.1 channel pore. Mol Pharmacol 91(4):392–402
Niu X, Liu G, Wu RS, Chudasama N, Zakharov SI, Karlin A, Marx SO (2013) Orientations and proximities of the extracellular ends of transmembrane helices S0 and S4 in open and closed BK potassium channels. PLoS One 8(3):e58335
Ohwada T, Nonomura T, Maki K, Sakamoto K, Ohya S, Muraki K, Imaizumi Y (2003) Dehydroabietic acid derivatives as a novel scaffold for large-conductance calcium-activated K^+ channel openers. Bioorg Med Chem Lett 13(22):3971–3974
Ohwada T, Sakamaki Y, Kabasawa Y, Yasutomi E (2009) Potassium channel opener. WO 2009110468
Olesen SP, Munch E, Moldt P, Drejer J (1994) Selective activation of Ca^{2+}-dependent K^+ channels by novel benzimidazolone. Eur J Pharmacol 251(1):53–59
Oliván-Viguera A, Valero MS, Coleman N, Brown BM, Laría C, Murillo MD, Gálvez JA, Díaz-de-Villegas MD, Wulff H, Badorrey R, Köhler R (2015) A novel pan-negative-gating modulator of KCa2/3 channels, fluoro-di-benzoate, RA-2, inhibits endothelium-derived hyperpolarization-type relaxation in coronary artery and produces bradycardia in vivo. Mol Pharmacol 87 (2):338–348
Oliván-Viguera A, Valero MS, Pinilla E, Amor S, García-Villalón ÁL, Coleman N, Laría C, Calvín-Tienza V, García-Otín ÁL, Fernández-Fernández JM, Murillo MD, Gálvez JA, Díaz-de-Villegas MD, Badorrey R, Simonsen U, Rivera L, Wulff H, Köhler R (2016) Vascular reactivity profile of novel KCa 3.1-selective positive-gating modulators in the coronary vascular bed. Basic Clin Pharmacol Toxicol 119(2):184–192
Owada T, Tajima T, Chokai Y, Akaha S (2007) Potassium channel opening agent. JP 2007186480
Pantazis A, Gudzenko V, Savalli N, Sigg D, Olcese R (2010a) Operation of the voltage sensor of a human voltage- and Ca^{2+}-activated K^+ channel. Proc Natl Acad Sci U S A 107(9):4459–4464
Pantazis A, Kohanteb AP, Olcese R (2010b) Relative motion of transmembrane segments S0 and S4 during voltage sensor activation in the human BK(Ca) channel. J Gen Physiol 136 (6):645–657
Parajuli SP, Hristov KL, Soder RP, Kellett WF, Petkov GV (2013) NS309 decreases rat detrusor smooth muscle membrane potential and phasic contractions by activating SK3 channels. Br J Pharmacol 168(7):1611–1625
Parihar AS, Groebe DR, Scott VE, Feng J, Zhang XF, Warrior U, Gopalakrishnan M, Shieh CC (2003) Functional analysis of large conductance Ca^{2+}-activated K^+ channels: ion flux studies by atomic absorption spectrometry. Assay Drug Dev Technol 1(5):647–654
Pedarzani P, Mosbacher J, Rivard A, Cingolani LA, Oliver D, Stocker M, Adelman JP, Fakler B (2001) Control of electrical activity in central neurons by modulating the gating of small conductance Ca^{2+}-activated K^+ channels. J Biol Chem 276(13):9762–9769

Pinter T, Jana S, Courtemanche RJ, Hof F (2011) Recognition properties of carboxylic acid bioisosteres: anion binding by tetrazoles, aryl sulfonamides, and acyl sulfonamides on a calix [4]arene scaffold. J Org Chem 76(10):3733–3741

Ponte CG, McManus OB, Schmalhofer WA, Shen DM, Dai G, Stevenson A, Sur S, Shah T, Kiss L, Shu M, Doherty JB, Nargund R, Kaczorowski GJ, Suarez-Kurtz G, Garcia ML (2012) Selective, direct activation of high-conductance, calcium-activated potassium channels causes smooth muscle relaxation. Mol Pharmacol 81(4):567–577

Power EC, Ganellin CR, Benton DC (2006) Partial structures of ketoconazole as modulators of the large conductance calcium-activated potassium channel (BK(Ca)). Bioorg Med Chem Lett 16 (4):887–890

Radtke J, Schmidt K, Wulff H, Kohler R, de Wit C (2013) Activation of KCa3.1 by SKA-31 induces arteriolar dilatation and lowers blood pressure in normo- and hypertensive connexin40-deficient mice. Br J Pharmacol 170(2):293–303

Rae JL, Dewey J, Rae JS, Cooper K (1990) A maxi calcium-activated potassium channel from chick lens epithelium. Curr Eye Res 9(9):847–861

Rittenhouse AR, Parker C, Brugnara C, Morgan KG, Alper SL (1997) Inhibition of maxi-K currents in ferret portal vein smooth muscle cells by the antifungal clotrimazole. Am J Phys 273(1 Pt 1): C45–C56

Roth EK, Hirtz S, Duerr J, Wenning D, Eichler I, Seydewitz HH, Amaral MD, Mall MA (2011) The K^+ channel opener 1-EBIO potentiates residual function of mutant CFTR in rectal biopsies from cystic fibrosis patients. PLoS One 6(8):e24445

Rovati LC, Giordani A, Caselli G (2012) N-phenylbenzamide derivatives as drugs for the treatment of copd. CA 2514373

Roxburgh CJ, Ganellin CR, Athmani S, Bisi A, Quaglia W, Benton DC, Shiner MA, Malik-Hall M, Haylett DG, Jenkinson DH (2001) Synthesis and structure-activity relationships of cetiedil analogues as blockers of the Ca^{2+}-activated K^+ permeability of erythrocytes. J Med Chem 44 (20):3244–3253

Roy S, Morayo Akande A, Large RJ, Webb TI, Camarasu C, Sergeant GP, McHale NG, Thornbury KD, Hollywood MA (2012) Structure-activity relationships of a novel group of large-conductance Ca^{2+}-activated K^+ (BK) channel modulators: the GoSlo-SR family. ChemMedChem 7(10):1763–1769

Roy S, Large RJ, Akande AM, Kshatri A, Webb TI, Domene C, Sergeant GP, McHale NG, Thornbury KD, Hollywood MA (2014) Development of GoSlo-SR-5-69, a potent activator of large conductance Ca^{2+}-activated K^+ (BK) channels. Eur J Med Chem 75:426–437

Rufo PA, Merlin D, Riegler M, Ferguson-Maltzman MH, Dickinson BL, Brugnara C, Alper SL, Lencer WI (1997) The antifungal antibiotic, clotrimazole, inhibits chloride secretion by human intestinal T84 cells via blockade of distinct basolateral K^+ conductances. Demonstration of efficacy in intact rabbit colon and in an in vivo mouse model of cholera. J Clin Invest 100 (12):3111–3120

Ryan JS, Tao QP, Kelly ME (1998) Adrenergic regulation of calcium-activated potassium current in cultured rabbit pigmented ciliary epithelial cells. J Physiol 511(Pt 1):145–157

Saito M, Nelson C, Salkoff L, Lingle CJ (1997) A cysteine-rich domain defined by a novel exon in aslo variant in rat adrenal chromaffin cells and PC12 cells. J Biol Chem 272(18):11710–11717

Sallustio BC, Sabordo L, Evans AM, Nation RL (2000) Hepatic disposition of electrophilic acyl glucuronide conjugates. Curr Drug Metab 1(2):163–180

Sanchez M, McManus OB (1996) Paxilline inhibition of the alpha-subunit of the high-conductance calcium-activated potassium channel. Neuropharmacology 35(7):963–968

Sankaranarayanan A, Raman G, Busch C, Schultz T, Zimin PI, Hoyer J, Kohler R, Wulff H (2009) Naphtho[1,2-d]thiazol-2-ylamine (SKA-31), a new activator of KCa2 and KCa3.1 potassium channels, potentiates the endothelium-derived hyperpolarizing factor response and lowers blood pressure. Mol Pharmacol 75(2):281–295

Schreiber M, Salkoff L (1997) A novel calcium-sensing domain in the BK channel. Biophys J 73 (3):1355–1363

Schroder RL, Jensen BS, Strobaek D, Olesen SP, Christophersen P (2000) Activation of the human, intermediate-conductance, Ca^{2+}-activated K^+ channel by methylxanthines. Pflugers Archiv: Eur J Physiol 440(6):809–818

Schroder RL, Jespersen T, Christophersen P, Strobaek D, Jensen BS, Olesen SP (2001) KCNQ4 channel activation by BMS-204352 and retigabine. Neuropharmacology 40(7):888–898

Schroeder BC, Cheng T, Jan YN, Jan LY (2008) Expression cloning of TMEM16A as a calcium-activated chloride channel subunit. Cell 134(6):1019–1029

Schunck WH, Wallukat G, Fischer R, Schmidt C, MÜLLER DN, Puli N, Falck JR (2010) Novel eicosanoid derivatives. WO 2010081683

Schwarz W, Keim H, Fehlau R, Fuhrmann GF (1989) Modulation of the Ca^{2+}- or Pb^{2+}-activated K^+-selective channels in human red cells. I. Effects of propranolol. Biochimica et Biophysica Acta 978(1):32–36

Seto SW, Au AL, Lam TY, Chim SS, Lee SM, Wan S, Tjiu DC, Shigemura N, Yim AP, Chan SW, Tsui SK, Leung GP, Kwan YW (2007) Modulation by simvastatin of iberiotoxin-sensitive, Ca^{2+}-activated K^+ channels of porcine coronary artery smooth muscle cells. Br J Pharmacol 151 (7):987–997

Sha Y, Tashima T, Mochizuki Y, Toriumi Y, Adachi-Akahane S, Nonomura T, Cheng M, Ohwada T (2005) Compounds structurally related to tamoxifen as openers of large-conductance calcium-activated K^+ channel. Chem Pharm Bull 53(10):1372–1373

Sheu SJ, Bee YS, Chen CH (2008) Resveratrol and large-conductance calcium-activated potassium channels in the protection of human retinal pigment epithelial cells. Journal of Ocular Pharmacology and Therapeutics: the Official Journal of the Association for Ocular Pharmacology and Therapeutics 24(6):551–555

Shi J, He HQ, Zhao R, Duan YH, Chen J, Chen Y, Yang J, Zhang JW, Shu XQ, Zheng P, Ji YH (2008) Inhibition of martentoxin on neuronal BK channel subtype (alpha+beta4): implications for a novel interaction model. Biophys J 94(9):3706–3713

Shieh CC, Turner SC, Zhang XF, Milicic I, Parihar A, Jinkerson T, Wilkins J, Buckner SA, Gopalakrishnan M (2007) A-272651, a nonpeptidic blocker of large-conductance Ca^{2+}-activated K^+ channels, modulates bladder smooth muscle contractility and neuronal action potentials. Br J Pharmacol 151(6):798–806

Singh AK, Devor DC, Gerlach AC, Gondor M, Pilewski JM, Bridges RJ (2000) Stimulation of Cl^- secretion by chlorzoxazone. J Pharmacol Exp Ther 292(2):778–787

Singh S, Syme CA, Singh AK, Devor DC, Bridges RJ (2001) Benzimidazolone activators of chloride secretion: potential therapeutics for cystic fibrosis and chronic obstructive pulmonary disease. J Pharmacol Exp Ther 296(2):600–611

Sivarao DV, Newberry K, Langdon S, Lee AV, Hewawasam P, Plym MJ, Signor L, Myers R, Lodge NJ (2005) Effect of 4-(5-chloro-2-hydroxyphenyl)-3-(2-hydroxyethyl)-6-(trifluoromethyl)-quinolin-2(1H)-one (BMS-223131), a novel opener of large conductance Ca^{2+}-activated K^+ (maxi-K) channels on normal and stress-aggravated colonic motility and visceral nociception. J Pharmacol Exp Ther 313(2):840–847

Song WH, Liu MM, Zhong DW, Zhu YL, Bosscher M, Zhou L, Ye DY, Yuan ZH (2013) Tetrazole and triazole as bioisosteres of carboxylic acid: discovery of diketo tetrazoles and diketo triazoles as anti-HCV agents. Bioorg Med Chem Lett 23(16):4528–4531

Sorensen MV, Matos JE, Sausbier M, Sausbier U, Ruth P, Praetorius HA, Leipziger J (2008) Aldosterone increases KCa1.1 (BK) channel-mediated colonic K^+ secretion. J Physiol 586 (Pt 17):4251–4264

Sorensen MV, Sausbier M, Ruth P, Seidler U, Riederer B, Praetorius HA, Leipziger J (2010) Adrenaline-induced colonic K^+ secretion is mediated by KCa1.1 (BK) channels. J Physiol 588 (Pt 10):1763–1777

Starrett JEJ, Gillmsan KW, Bocchino DM, et al. (2004) Maxi-K openers as neuroprotective agents. In: 227th ACS Natl Meet, March 28–April 1, Anaheim

Starrett J, Lopez O, Hewawasam P, Ding M (2005) N-substituted prodrugs of fluorooxindoles US 20050203089

Stocker JW, De Franceschi L, McNaughton-Smith GA, Corrocher R, Beuzard Y, Brugnara C (2003) ICA-17043, a novel Gardos channel blocker, prevents sickled red blood cell dehydration in vitro and in vivo in SAD mice. Blood 101(6):2412–2418

Strobaek D, Christophersen P, Holm NR, Moldt P, Ahring PK, Johansen TE, Olesen SP (1996) Modulation of the Ca^{2+}-dependent K^+ channel, hslo, by the substituted diphenylurea NS 1608, paxilline and internal Ca^{2+}. Neuropharmacology 35(7):903–914

Strobaek D, Teuber L, Jorgensen TD, Ahring PK, Kjaer K, Hansen RS, Olesen SP, Christophersen P, Skaaning-Jensen B (2004) Activation of human IK and SK Ca^{2+} –activated K^+ channels by NS309 (6,7-dichloro-1H-indole-2,3-dione 3-oxime). Biochim Biophys Acta 1665(1–2):1–5

Strobaek D, Brown DT, Jenkins DP, Chen YJ, Coleman N, Ando Y, Chiu P, Jorgensen S, Demnitz J, Wulff H, Christophersen P (2013) NS6180, a new K(Ca) 3.1 channel inhibitor prevents T-cell activation and inflammation in a rat model of inflammatory bowel disease. Br J Pharmacol 168(2):432–444

Sun XH, Ding JP, Li H, Pan N, Gan L, Yang XL, Xu HB (2007) Activation of large-conductance calcium-activated potassium channels by puerarin: the underlying mechanism of puerarin-mediated vasodilation. J Pharmacol Exp Ther 323(1):391–397

Syme CA, Gerlach AC, Singh AK, Devor DC (2000) Pharmacological activation of cloned intermediate- and small-conductance Ca^{2+}-activated K^+ channels. Am J Physiol Cell Physiol 278(3):C570–C581

Tanaka M, Sasaki Y, Kimura Y, Fukui T, Hamada K, Ukai Y (2003) A novel pyrrole derivative, NS-8, suppresses the rat micturition reflex by inhibiting afferent pelvic nerve activity. BJU Int 92(9):1031–1036

Tang QY, Zhang Z, Xia XM, Lingle CJ (2010) Block of mouse Slo1 and Slo3 K^+ channels by CTX, IbTX, TEA, 4-AP and quinidine. Channels 4(1):22–41

Tao J, Shi J, Liu ZR, Ji YH (2012) Martentoxin: a unique ligand of BK channels. Sheng li xue bao: [Acta physiologica Sinica] 64(4):355–364

Tara E, Vitenzon A, Hess E, Khodakhah K (2018) Aberrant cerebellar Purkinje cell activity as the cause of motor attacks in a mouse model of episodic ataxia type 2. Dis Model Mech 11(9)

Tashima T, Toriumi Y, Mochizuki Y, Nonomura T, Nagaoka S, Furukawa K, Tsuru H, Adachi-Akahane S, Ohwada T (2006) Design, synthesis, and BK channel-opening activity of hexahydrodibenzazepinone derivatives. Bioorg Med Chem 14(23):8014–8031

Thompson-Vest N, Shimizu Y, Hunne B, Furness JB (2006) The distribution of intermediate-conductance, calcium-activated, potassium (IK) channels in epithelial cells. J Anat 208 (2):219–229

Treiber A, Van Giersbergen PL, Dingemanse J (2003) In vivo and in vitro disposition profile of tezosentan, an intravenous dual endothelin receptor antagonist, in humans. Xenobiotica 33 (4):399–414

Tricarico D, Barbieri M, Mele A, Carbonara G, Camerino DC (2004) Carbonic anhydrase inhibitors are specific openers of skeletal muscle BK channel of K^+-deficient rats. FASEB J 18 (6):760–761

Tsuzuki Y, Hirai M (2009) Dérivés de pyrimidine, pyridine et triazine en tant qu'ouvreurs de canaux maxi-k. WO 2009125870

Turner SC, Carroll WA, White TK, Gopalakrishnan M, Coghlan MJ, Shieh CC, Zhang XF, Parihar AS, Buckner SA, Milicic I, Sullivan JP (2003) The discovery of a new class of large-conductance Ca^{2+}-activated K^+ channel opener targeted for overactive bladder: synthesis and structure-activity relationships of 2-amino-4-azaindoles. Bioorg Med Chem Lett 13 (12):2003–2007

Turner S, White T, Carroll W (2004) Pyrrolopyridine potassium channel openers. US 20040192697

Unemoto T, Matsushita M, Tamura K, Tanaka Y, Koike K, Kogo H (2007) Role of BK channels in testosterone-induced relaxation of the aorta in spontaneously hypertensive rats. Biol Pharm Bull 30(8):1477–1480

Urbahns K, Horvath E, Stasch JP, Mauler F (2003) 4-phenyl-4H-pyrans as IK(Ca) channel blockers. Bioorg Med Chem Lett 13(16):2637–2639

Vaali K, Li L, Paakkari I, Vapaatalo H (1998) Relaxing effects of NO donors on Guinea pig trachea in vitro are mediated by calcium-sensitive potassium channels. J Pharmacol Exp Ther 286 (1):110–114

Valverde MA, Rojas P, Amigo J, Cosmelli D, Orio P, Bahamonde MI, Mann GE, Vergara C, Latorre R (1999) Acute activation of maxi-K channels (hSlo) by estradiol binding to the beta subunit. Science 285(5435):1929–1931

Van Der Velden J, Sum G, Barker D, Koumoundouros E, Barcham G, Wulff H, Castle N, Bradding P, Snibson K (2013) K(Ca)3.1 channel-blockade attenuates airway pathophysiology in a sheep model of chronic asthma. PLoS One 8(6):e66886

Venglovecz V, Hegyi P, Rakonczay Z Jr, Tiszlavicz L, Nardi A, Grunnet M, Gray MA (2011) Pathophysiological relevance of apical large-conductance Ca^{2+}-activated potassium channels in pancreatic duct epithelial cells. Gut 60(3):361–369

Vestergaard-Bogind B, Stampe P, Christophersen P (1985) Single-file diffusion through the Ca^{2+}-activated K^+ channel of human red cells. J Membr Biol 88(1):67–75

Vrudhula VM, Dasgupta B, Boissard CG, Gribkoff VK, Santone KS, Dalterio RA, Lodge NJ, Starrett JE Jr (2005) Analogs of a potent maxi-K potassium channel opener with an improved inhibitory profile toward cytochrome P450 isozymes. Bioorg Med Chem Lett 15 (19):4286–4290

Wallner M, Meera P, Toro L (1996) Determinant for beta-subunit regulation in high-conductance voltage-activated and Ca^{2+}-sensitive K^+ channels: an additional transmembrane region at the N terminus. Proc Natl Acad Sci U S A 93(25):14922–14927

Wang B, Rothberg BS, Brenner R (2006) Mechanism of beta4 subunit modulation of BK channels. J Gen Physiol 127(4):449–465

Wang YJ, Lin MW, Wu SN, Sung RJ (2007) The activation by estrogen receptor agonists of the BK (Ca)-channel in human cardiac fibroblasts. Biochem Pharmacol 73(9):1347–1357

Wang YJ, Lin MW, Lin AA, Wu SN (2008) Riluzole-induced block of voltage-gated Na^+ current and activation of BKCa channels in cultured differentiated human skeletal muscle cells. Life Sci 82(1–2):11–20

Wang J, Qian W, Zhu Q, Chen J, Huan F, Gao R, Xiao H (2013) Martentoxin, a large-conductance Ca^{2+}-activated K^+ channel inhibitor, attenuated TNF-alpha-induced nitric oxide release by human umbilical vein endothelial cells. J Biomed Res 27(5):386–393

Warach S, Hacke W, Hsu C, Luby M, Sullivan M, Noonan T, Lin CY, Fernandes L, Brunell R, Bozik M (2002) Effect of MaxiPost on ischemic lesions in patients with acute stroke: The POST-010 MRI substudy. 33(1):383 (vol 1. By: STROKE, Meeting Abstract: P117 Published: JAN 2002, p 383)

Wei AD, Gutman GA, Aldrich R, Chandy KG, Grissmer S, Wulff H (2005) International Union of Pharmacology. LII Nomenclature and molecular relationships of calcium-activated potassium channels. Pharmacol Rev 57(4):463–472

Weyler S, Baqi Y, Hillmann P, Kaulich M, Hunder AM, Muller IA, Muller CE (2008) Combinatorial synthesis of anilinoanthraquinone derivatives and evaluation as non-nucleotide-derived P2Y2 receptor antagonists. Bioorg Med Chem Lett 18(1):223–227

Wojtovich AP, Nadtochiy SM, Urciuoli WR, Smith CO, Grunnet M, Nehrke K, Brookes PS (2013) A non-cardiomyocyte autonomous mechanism of cardioprotection involving the SLO1 BK channel. PeerJ 1:e48. https://doi.org/10.7717/peerj.48

Wu SN (2003) Large-conductance Ca^{2+}- activated K^+ channels:physiological role and pharmacology. Curr Med Chem 10(8):649–661

Wu SN, Li HF, Jan CR, Shen AY (1999) Inhibition of Ca^{2+}-activated K^+ current by clotrimazole in rat anterior pituitary GH3 cells. Neuropharmacology 38(7):979–989

Wu SN, Li HF, Chiang HT (2001) Vinpocetine-induced stimulation of calcium-activated potassium currents in rat pituitary GH3 cells. Biochem Pharmacol 61(7):877–892

Wu SN, Liu SI, Huang MH (2004) Cilostazol, an inhibitor of type 3 phosphodiesterase, stimulates large-conductance, calcium-activated potassium channels in pituitary GH3 cells and pheochromocytoma PC12 cells. Endocrinology 145(3):1175–1184

Wu Y, Yang Y, Ye S, Jiang Y (2010) Structure of the gating ring from the human large-conductance Ca^{2+}-gated K^{+} channel. Nature 466(7304):393–397

Wulff H, Castle NA (2010) Therapeutic potential of KCa3.1 blockers: recent advances and promising trends. Expert Rev Clin Pharmacol 3(3):385–396

Wulff H, Miller MJ, Hansel W, Grissmer S, Cahalan MD, Chandy KG (2000) Design of a potent and selective inhibitor of the intermediate-conductance Ca^{2+}-activated K^{+} channel, IKCa1: a potential immunosuppressant. Proc Natl Acad Sci U S A 97(14):8151–8156

Wulff H, Gutman GA, Cahalan MD, Chandy KG (2001) Delineation of the clotrimazole/TRAM-34 binding site on the intermediate conductance calcium-activated potassium channel, IKCa1. J Biol Chem 276(34):32040–32045

Wulff H, Kolski-Andreaco A, Sankaranarayanan A, Sabatier JM, Shakkottai V (2007) Modulators of small- and intermediate-conductance calcium-activated potassium channels and their therapeutic indications. Curr Med Chem 14(13):1437–1457

Wulf-Johansson H, Amrutkar DV, Hay-Schmidt A, Poulsen AN, Klaerke DA, Olesen J, Jansen-Olesen I (2010) Localization of large conductance calcium-activated potassium channels and their effect on calcitonin gene-related peptide release in the rat trigemino-neuronal pathway. Neuroscience 167(4):1091–1102

Yan J, Aldrich RW (2010) LRRC26 auxiliary protein allows BK channel activation at resting voltage without calcium. Nature 466(7305):513–516

Yang W, Holmes BB, Gopal VR, Kishore RV, Sangras B, Yi XY, Falck JR, Campbell WB (2007) Characterization of 14,15-epoxyeicosatrienoyl-sulfonamides as 14,15-epoxyeicosatrienoic acid agonists: use for studies of metabolism and ligand binding. J Pharmacol Exp Ther 321 (3):1023–1031

Yao J, Chen X, Li H, Zhou Y, Yao L, Wu G, Chen X, Zhang N, Zhou Z, Xu T, Wu H, Ding J (2005) BmP09, a "long chain" scorpion peptide blocker of BK channels. J Biol Chem 280 (15):14819–14828

Yuan P, Leonetti MD, Pico AR, Hsiung Y, MacKinnon R (2010) Structure of the human BK channel Ca^{2+}-activation apparatus at 3.0 a resolution. Science 329(5988):182–186

Zaidman NA, Panoskaltsis-Mortari A, O'Grady SM (2017) Large-conductance Ca^{2+} –activated K^{+} channel activation by apical P2Y receptor agonists requires hydrocortisone in differentiated airway epithelium. J Physiol 595(14):4631–4645

Zakharov SI, Morrow JP, Liu G, Yang L, Marx SO (2005) Activation of the BK (SLO1) potassium channel by mallotoxin. J Biol Chem 280(35):30882–30887

Zeng H, Gordon E, Lin Z, Lozinskaya IM, Willette RN, Xu X (2008) 1-[1-Hexyl-6-(methyloxy)-1H-indazol-3-yl]-2-methyl-1-propanone, a potent and highly selective small molecule blocker of the large-conductance voltage-gated and calcium-dependent K^{+} channel. J Pharmacol Exp Ther 327(1):168–177

Zhang J, Yan J (2014, Oct 15) Regulation of BK channels by auxiliary γ subunits. Front Physiol 5:401

Zhang D, Krishna R, Wang L, Zeng J, Mitroka J, Dai R, Narasimhan N, Reeves RA, Srinivas NR, Klunk LJ (2005) Metabolism, pharmacokinetics, and protein covalent binding of radiolabeled MaxiPost (BMS-204352) in humans. Drug Metab Dispos 33(1):83–93

Zhang M, Pascal JM, Schumann M, Armen RS, Zhang JF (2012) Identification of the functional binding pocket for compounds targeting small-conductance Ca^{2+}-activated potassium channels. Nat Commun 3:1021

Zhang M, Pascal JM, Zhang JF (2013) Unstructured to structured transition of an intrinsically disordered protein peptide in coupling Ca^{2+}-sensing and SK channel activation. Proc Natl Acad Sci U S A 110(12):4828–4833

Chapter 25
KCNE Regulation of KCNQ Channels

Geoffrey W. Abbott

Abstract Voltage-gated potassium (Kv) channels are perhaps best known for their roles in the action potentials of excitable cells, but they also fulfill a diverse array of vital roles in polarized epithelial cells. Much of the research and current knowledge in this area centers on the KCNQ1 potassium channel and the KCNE family of regulatory proteins. By exploiting the functional versatility endowed by forming complexes with single transmembrane-spanning KCNE subunits, KCNQ1 is able to serve a plethora of different roles in polarized epithelia and other tissues. This chapter is focused on the epithelial functions of KCNE proteins, with a major emphasis on regulation of KCNQ1 as this dominates our present knowledge in this field. In addition to outlining what is known about epithelial KCNE physiology, the chapter also covers KCNE pathobiology, as determined from human disease associations and *Kcne* knockout mouse studies.

Keywords Choroid plexus · Diabetes · Gastric acid · Hypothyroidism · KCNQ1 · MinK-related peptides

25.1 Introduction

Ion channels provide aqueous conduits across cell membranes that permit transmembrane movement of aqueous ions across the hydrophobic interior of the lipid bilayer. Rather than active or facilitated transport, channels instead utilize the electrochemical gradient of the ion involved, yet can still achieve ion movement through the channel pore at rates close to the diffusion limit. Crucially, some ion channels can also achieve impressive ion selectivity ratios, some estimated at 100:1, even for K^+ over the smaller but similarly charged Na^+ (Hille et al. 1999). Ion

G. W. Abbott (✉)
Bioelectricity Laboratory, Irvine Hall 249, Department of Physiology and Biophysics, School of Medicine, University of California, Irvine, CA, USA
e-mail: abbottg@uci.edu

K. L. Hamilton, D. C. Devor (eds.), *Studies of Epithelial Transporters and Ion Channels*, Physiology in Health and Disease,
https://doi.org/10.1007/978-3-030-55454-5_25

channels come in various guises, and can be classified by their primary sequence (there are many different ion channel families organized by sequence similarity), and also by their ion selectivity (e.g., K^+ versus Na^+) and how they are activated, or gated (e.g., voltage or ligand). Voltage-gated ion channels are perhaps best known for their roles in excitable cell processes including electrical signaling in the brain, pumping of the heart, and skeletal muscle contraction. However, many voltage-gated ion channels also play essential roles in non-excitable tissues, including in polarized epithelial cells.

25.1.1 Voltage-Dependent Ion Channel Functional Architecture

Voltage-gated ion channels all share the common motif of pseudo-fourfold symmetry, achieved differently according to the channel family. Voltage-gated sodium (Nav) and calcium (Cav) channels comprise one contiguous pore-forming (α) subunit with four repeating units encoded by a single gene; the sequences of each repeating unit are actually slightly different from one another (especially in the loops between membrane-spanning segments) (Noda et al. 1986; Numa and Noda 1986). Voltage-gated potassium (Kv) channels possess 4 repeating units, each of which is a separate gene product; a tetramer of Kv channel α subunits forms co-translationally. For all the above, each unit comprises six transmembrane segments, with segments 1–4 (S1–S4) forming one voltage-sensing domain (VSD), and S5 and S6 contributing from each unit to form one-quarter of an interlocking pore structure (Fig. 25.1a, b) (Papazian et al. 1987; Tempel et al. 1987; MacKinnon 1991; Doyle et al. 1998; Long et al. 2005a).

The VSD is the defining feature of voltage-gated ion channels, and each incorporates a transmembrane helix (S4) containing periodic basic residues (Catterall 1995). The basic residues sense membrane potential and the VSD, therefore, shifts position relative to the membrane electric field in response to membrane potential changes (Fig. 25.1a–c). This conformational shift is communicated to the ion channel pore module via an intracellular linker region, termed the S4–5 linker in voltage-gated potassium (Kv channels), and probably also direct interactions between elements of the pore module and S4 (Bezanilla and Perozo 2003). Other parts of the VSD (especially S1 and S2) bear acidic residues that shield or stabilize S4 as it moves to its open state position following membrane depolarization (Cuello et al. 2004; DeCaen et al. 2009). Voltage-dependent ion channel gating is therefore electromechanically coupled to the plasma membrane potential (Fig. 25.1c) (Long et al. 2005b).

Once the voltage-gated ion channel pore opens, ion selectivity is paramount. In K^+ channels, selectivity over similar ions including Na^+ occurs via backbone carbonyl oxygen atoms of glycine residues within the canonical GYGD K^+ selectivity filter motif. This forms a pseudo-hydration shell that perfectly accommodates

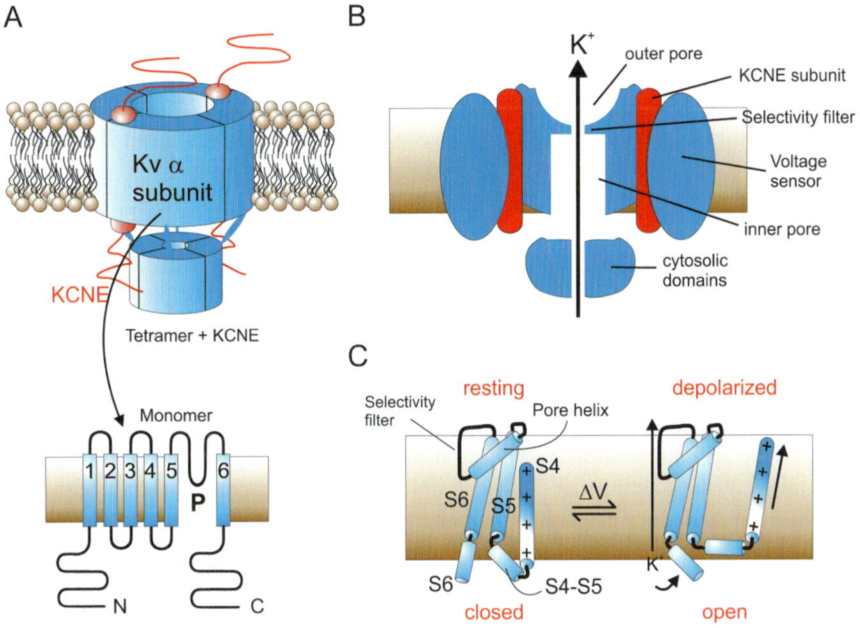

Fig. 25.1 Kv-KCNE channel functional architecture. (**a**) *Upper*, cartoon of a Kv α-KCNE complex; *lower*, topology diagram of a Kv α subunit monomer. (**b**) Cartoon section through a Kv-KCNE channel complex showing main features. (**c**) Cartoon of voltage gating depicting the S4–S6 portions of a Kv α subunit monomer and an approximation of the conformational shifts that may occur upon membrane depolarization. Adapted from Abbott (2014)

K^+ but that is slightly too big to efficiently coordinate Na^+. Rapid diffusion of K^+ ions through the K^+ channel pore involves binding and unbinding of ions to the pore elements, which evolved to accommodate K^+ at the expense of the other physiologically abundant ions (Doyle et al. 1998; Ranganathan et al. 1996; Zhou and MacKinnon 2003).

All voltage-gated ion channel α subunit assemblies also form complexes with one or more non-pore-forming, ancillary subunits (McCrossan and Abbott 2004; Panaghie and Abbott 2006; Pongs and Schwarz 2010). The focus of this chapter is on a class of regulatory subunits, the KCNE proteins, which endow voltage-gated potassium (Kv) channels with functional properties required to fulfil a variety of roles in diverse epithelia.

25.1.2 *Background to KCNE Proteins*

The first *KCNE* gene was discovered in 1988 when Takumi and colleagues performed functional expression studies on fractionated rat kidney mRNA using oocytes of the African clawed toad, *Xenopus laevis*. A fraction containing shorter-length mRNAs generated a very slow-activating, voltage-dependent K^+ current, leading to the cloning of a single transmembrane domain protein termed MinK (for minimal K^+ channel) which was assumed to form a noncanonical, small but K^+-selective Kv channel (Takumi et al. 1991; Takumi et al. 1988). MinK protein is now more typically referred to by its gene name, KCNE1, italicized if referring specifically to the gene rather than the protein, as per convention.

The current formed by *KCNE1* cRNA injection resembles the cardiac slow-activating K^+ current, I_{Ks}, important in human and guinea pig ventricular repolarization (Busch and Lang 1993), but the single-transmembrane-domain architecture was unprecedented among K^+ channels. In 1996, the human KCNQ1 (originally termed KvLQT1) Kv channel α subunit was cloned and discovered to underlie the LQT1 form of the cardiac arrhythmia, Long QT syndrome (LQTS) (Wang et al. 1996). At that point, it was discovered that KCNE1 does not form channels by itself (Blumenthal and Kaczmarek 1994; Wang and Goldstein 1995). Rather, it upregulates an endogenous *Xenopus* oocyte KCNQ1 channel current (Barhanin et al. 1996; Sanguinetti et al. 1996). KCNQ1–KCNE1 channels were discovered to be the molecular correlate of human and guinea pig cardiac I_{Ks}, and mutations in human *KCNE1* were discovered to cause the LQT5 form of inherited LQTS (Splawski et al. 1997a, b). LQT5 gene variants generally cause loss-of-function of KCNQ1–KCNE1 channels, and in some cases also impair the function of two other Kv channels with which KCNE1 can form complexes, hERG and Kv2.1 (Bianchi et al. 1999; McCrossan et al. 2009; McDonald et al. 1997).

Later, by BLAST searching online databases with stretches of KCNE1 sequence that were known to be functionally important, several expressed sequence tags (ESTs) were discovered that bore resemblance to KCNE1; we termed their protein products the MinK-related peptides (MiRPs). As for MinK, the MiRPs are now typically referred to by their gene names (KCNE2-5) and this nomenclature will be utilized herein to avoid confusion. We cloned three *KCNE* genes (*KCNE2, 3,* and *4*) and screened for their α subunit partners using co-expression in *Xenopus* oocytes coupled with two-electrode voltage clamp electrophysiology (Abbott et al. 1999). We discovered that the *KCNE2* gene product (MiRP1, now termed KCNE2) alters the functional attributes of co-expressed hERG (KCNH2), a Kv α subunit required for human ventricular myocyte repolarization (Abbott et al. 1999). An additional protein, KCNE1-like or KCNE1-L (now more commonly referred to as KCNE5) (Piccini et al. 1999) was discovered around the same time as our first report on the cloning of *KCNE2*, *KCNE3,* and *KCNE4* genes (Abbott et al. 1999). *KCNE* genes share comparatively little homology or identity with one another, but each possesses a single TM segment and a consensus PKC phosphorylation site in the intracellular, membrane-proximal region (Fig. 25.2). *KCNE2* and *KCNE1* may have arisen from a gene duplication event, because human *KCNE2* is located on chromosome 21q22.1, about 79 kb from and in the opposite orientation to *KCNE1*. The *KCNE1* and *KCNE2* open reading frames share 34% identity (Abbott et al. 1999).

KCNE proteins regulate all aspects of Kv channel biology, beginning with channel biogenesis through to functional properties including channel gating, selectivity, and conductivity. KCNEs also impact channel internalization from the cell

```
hKCNE3 MHRVAALGPPSVRGRRVQRANGRPGALLRGRASTESSPTSIPVAMETTNGTETWYESLHA
HKCNE5 -----------------------MNCSES--QRLRTLLSRLLLELHHRGNASG-------
hKCNE4 ----------------MHFLTIYPNCSSGVVRAQSRTEQKNPLGLDDLGIQNLGQTVSLA
HKCNE2 -----------------------MSTLSNFTQTLEDVFRRIFITYMDNWRQ---------
HKCNE1 -------------------------MILSNTTAVTPFLTKLWQETVQQGG----------

hKCNE3 VLKALNATLHSNLLCRPGPGLGPDNQTEERRASLPGRD-DNSYM---YILFVMFLFAVTV
HKCNE5 --------LGAGPRPSMGMGVVPDPFVGREVTSAKGDD---AYL---YILLIMIFYACLA
hKCNE4 PAVEAASMLKMEPLNSTHPGTAASSSPLESRAAGGGSGNGNEYF---YILVVMSFYGIFL
HKCNE2 ------------------------NTTAEQEALQAKVDAENFYYVILYLMVMIGMFSFII
HKCNE1 ------------------------NMSGLARRSPRSSDGKLEAL---YVLMVLGFFGFFT
                                              .         *::.:: ::.

hKCNE3 GSLILGYTRSRKV-DKRSDPYHVYIKNR------------VSMI----------------
HKCNE5 GGLILAYTRSRKLVEAKDEPSQACAEHE--WAPGGALTADAEAAAGSQA-----------
hKCNE4 IGIMLGYMKSKRR-EKKSSLLLLYKDEERLWGEAMKPLPVVSGLRSVQVPLMLNMLQESV
HKCNE2 VAILVSTVKSKRR-EHSNDPYHQYIVED--WQEKYK--SQILNLEESKA-----------
HKCNE1 LGIMLSYIRSKKL-EHSNDPFNVYIESD-AWQEKDKAYVQARVLESYRSCY---------
        .:::.  .*.. : ..

hKCNE3 -------------------------------------------------------------
HKCNE5 -----------------------EGRRQLASEGLPALAQGAERV-----------------
hKCNE4 APALSCTLCSMEGDSVSSESSSPDVHLTIQEEGADDELEETSETPLNESSEGSSENIHQNS
HKCNE2 -----------------------TIHENIGAAGFKMSP-----------------------
HKCNE1 -----------------------VVENHLAIEOPNTHLPETKPSP----------------
```

Fig. 25.2 Human KCNE gene family sequence alignment. Sequence alignment (using EMBL EBI MUSCLE, with display order indicative of closest sequence identities) of the human KCNE family. *Underlined*, newly discovered exon 1-encoded portions of KCNE3 and KCNE4; *yellow background*, predicted single transmembrane segment; *cyan background*, consensus PKC phosphorylation sites; gray background: calmodulin-binding motif of KCNE4

membrane (McCrossan and Abbott 2004; Kanda et al. 2011a, b, c). KCNEs form heteromeric complexes with Kv α subunits, with a stoichiometry of 4α:2KCNE subunits (Chen et al. 2003; Plant et al. 2014) elucidated for KCNQ1–KCNE1, with some groups also reporting flexible stoichiometry (Yu et al. 2013) (Fig. 25.1a). Electrophysiological studies of heterologously expressed channels in vitro together with biochemical and molecular biological approaches such as site-directed mutagenesis have helped to elucidate much of what we know about KCNE physiology. However, much of what we know about the native functions of KCNE subunits and the pathophysiological outcomes of their dysfunction has been learned from human and mouse genetics studies.

Each KCNE isoform can regulate more than one Kv α subunit isoform (McCrossan and Abbott 2004), and many of the Kv α subunit isoforms can each be regulated by more than one KCNE isoform; this may even occur simultaneously. All five human KCNEs can regulate KCNQ1, with a diverse array of functional outcomes, many of which are important in epithelial biology (Abbott 2014). It has also been suggested that KCNQ1–KCNE1–KCNE2 (tripartite) complexes form in the heart (Jiang et al. 2009).

25.1.3 KCNQ1: The Primary KCNE Partner in Epithelial Biology

There are five known human KCNQ genes within the larger 40-member Kv α subunit gene family (Abbott 2014). The KCNQ1 Kv α subunit is highly studied because of its association with human disease, varied roles in physiology, and impressive functional repertoire. KCNQ2–5 are also under intense investigation, primarily for their roles in the brain (mainly KCNQ2, 3, and 5) (Biervert et al. 1998; Klinger et al. 2011; Singh et al. 1998; Tzingounis et al. 2010; Wang et al. 1998), vasculature (KCNQ4 and 5) (Yeung et al. 2007) and auditory system (KCNQ4) (Kubisch et al. 1999). KCNQ1 is unique among the KCNQ gene products and perhaps all known voltage-gated ion channel pore-forming (α) subunits as it can be converted to a constitutively active channel with minimal voltage dependence across the physiological voltage range. This conversion arises from co-assembly with either KCNE2 or KCNE3 (Barhanin et al. 1996; Sanguinetti et al. 1996; Schroeder et al. 2000; Tinel et al. 2000).

While KCNE subunits can interact with many different Kv α subunits, KCNQ1 exhibits the widest known range of functional outcomes from this co-assembly (McCrossan and Abbott 2004). This (among other features) allows KCNQ1 to perform an especially diverse range of functional roles, spanning both excitable and non-excitable cells. The focus of this chapter is the epithelial biology of KCNQ1–KCNE and other Kv-KCNE channel complexes.

25.2 KCNQ1–KCNE1 Channels in Epithelial Biology

25.2.1 KCNQ1–KCNE1 in Auditory Epithelium

KCNQ1 loss-of-function gene variants cause LQTS in human beings because the delay in ventricular myocyte repolarization caused by reduced KCNQ1 activity (and therefore diminished I_{Ks} current) delays ventricular repolarization. This in turn can cause the *torsades de pointe* electrocardiogram phenomenon and ultimately the often lethal ventricular fibrillation (Jackman et al. 1984, 1988). Adult mice are not a useful model to study the role of KCNQ1 function in the heart because adult mouse ventricular myocytes do not express much Kcnq1, Kcne1, or I_{Ks} current. In contrast, adult *Kcnq1* knockout (*Kcnq1*$^{-/-}$) mice have been a very useful tool for investigating the roles of KCNQ1 in polarized epithelial cells, and the associated pathobiology. This includes the auditory dysfunction associated with Jervell and Lange-Nielsen Syndrome (JLNS). In humans, JLNS is most often an autosomal recessive disorder, caused by *KCNQ1* or *KCNE1* loss-of-function mutations being present in both alleles. JLNS presents as severe LQTS coupled with sensorineural deafness, the latter of which has been studied intensively using the *Kcnq1*$^{-/-}$ mouse (Lee et al. 2000; Rivas and Francis 2005). It is important to mention that the more common, autosomal dominant, form of KCNQ1-linked LQTS (sometimes referred to as Romano-Ward Syndrome) is not usually accompanied by hearing loss (Neyroud et al. 1997; Wang et al. 1996).

JLNS arises from KCNQ1 and/or KCNE1 mutations because KCNQ1–KCNE1 complexes are important for function of the auditory epithelium (Volume 2, - Chapter 8). To understand the physiology of this role, it is first important to understand how KCNE1 modulates KCNQ1 function. The most obvious result of KCNQ1–KCNE1 co-assembly is that KCNE1 slows KCNQ1 activation 5–10-fold (Sesti and Goldstein 1998; Yang et al. 1997) (Fig. 25.3a). Various mechanistic models exist for this slowed activation. Some suggest that KCNE1 simply dramatically slows S4 movement (Nakajo and Kubo 2007) (Ruscic et al. 2013), while others contend KCNE1 directly slows KCNQ1 pore opening (Rocheleau et al. 2006). Another theory is that KCNE1 imposes the necessity for movement of multiple voltage sensors before KCNQ1 channels open, whereas homomeric KCNQ1 can pass current with only a single voltage sensor out of the four per channel tetramer being activated (Osteen et al. 2010, 2012). This model suggested that KCNE1 both slowed voltage sensor movement and imposed the need for multiple voltage sensors to move, for channel opening (Osteen et al. 2012). To achieve these effects, it is thought that KCNEs lie in a groove close to both the pore and the VSD (Fig. 25.3b, c).

KCNE1 also positive-shifts the voltage dependence of KCNQ1 activation, probably by interacting with S4 and changing the way that S4 responds to voltage (Strutz-Seebohm et al. 2011; Wu et al. 2010a, b). This decreases the relative current at a given voltage, but to counteract this KCNE1 also increases KCNQ1 unitary conductance (Sesti and Goldstein 1998). Moreover, KCNE1 eliminates KCNQ1 slow inactivation (Tristani-Firouzi and Sanguinetti 1998).

KCNQ1–KCNE1 permits K^+ recycling in the endolymph of the inner ear (Bleich and Warth 2000; Casimiro et al. 2001), explaining why disruption of either *KCNQ1* (Lee et al. 2000; Neyroud et al. 1997; Wollnik et al. 1997) or *KCNE1* (Splawski et al. 1997a, b; Schulze-Bahr et al. 1997) in man or mouse can cause deafness. The capacity of KCNQ1–KCNE1 to sustain current for long periods without inactivating, because KCNE1 prevents KCNQ1 inactivation, may be important for KCNQ1–KCNE1 function in auditory epithelium. While KCNQ1–KCNE1 activation is strongly voltage-dependent, the channel opens at membrane potentials positive to –40 mV and because of the lack of inactivation it remains open at these voltages. The membrane potential across the apical membrane of inner ear dark cells and strial marginal cells is between 0 and +10 mV (Offner et al. 1987). KCNQ1–KCNE1 is therefore constitutively active in this environment, and generates the predominant (and perhaps sole) K^+ secretion pathway in this location (Wangemann 2002). *Kcnq1* deletion in mice revealed that loss of the KCNQ1–KCNE1 current also causes morphological abnormalities because of reduction in the volume of endolymph (Casimiro et al. 2001).

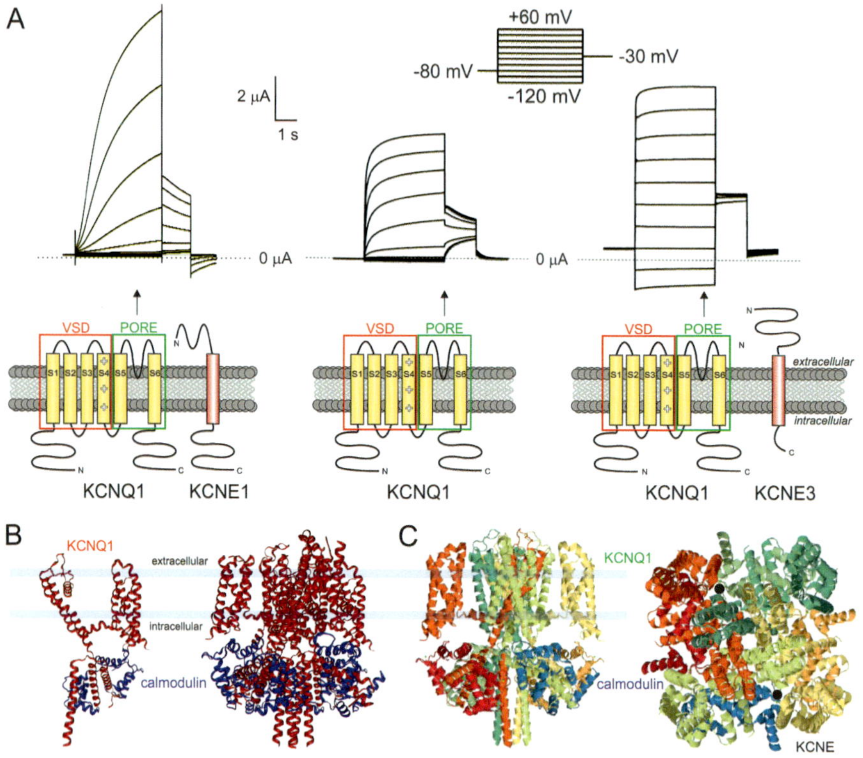

Fig. 25.3 KCNQ1-KCNE channel structure and function. (**a**) Two-electrode voltage-clamp recordings (Abbott 2016a) from *Xenopus* oocytes expressing KCNQ1 alone (center), or with KCNE1 (left) or KCNE3 (right). *Inset*, voltage protocol inset. Subunit topologies are shown beneath corresponding traces. VSD, voltage-sensing domain. (**b**) High-resolution structures obtained by cryo-EM of *Xenopus* KCNQ1 (Sun and MacKinnon 2017) (red) bound to calmodulin (blue); *left*, one subunit; *right*, tetramer. View from inside membrane perspective. (**c**) A tetramer of KCNQ1 (Sun and MacKinnon 2017) is viewed from inside the membrane (*left*) or outside the cell (*right*) perspectives. Individual subunits are colored orange, green, light green, and gold (KCNQ1) versus red and blue (calmodulin). Putative position of KCNEs (Kroncke et al. 2016) (black circles) shown in right-hand image. For all figures, structures were obtained from the Rutgers/UC San Diego/SDSC (RCSB) protein data bank and viewed using Jsmol (Javascript) (this panel) or NGL Viewer (Rose and Hildebrand 2015; Rose et al. 2017) (all other structure images). Adapted from Abbott (2017)

25.2.2 The Possible Roles of KCNQ1 and KCNE1 in the Kidneys

KCNQ1 and KCNE1 are both expressed in mammalian kidney, but their potential role there is still debated. Mouse kidneys express Kcnq1 and Kcne1, in the proximal tubule luminal membrane. Kcne1 knockout in mice reduces late proximal and early distal tubular fluid K^+ concentration, suggesting a possible role for Kcne1 in mediating K^+ flux to the lumen in this region. Electrophysiological analyses of isolated perfused proximal tubules supported this hypothesis, suggesting a role for the Kcne1-linked K^+ current in maintaining constant membrane polarization to counteract electrogenic sodium-coupled transport mechanisms. Consistent with this, *Kcne1*$^{-/-}$ mice exhibit abnormally high urinary excretion of fluid, Na^+, Cl^-, and glucose (Vallon et al. 2001). Female *Kcne1*$^{-/-}$ mice also exhibit reduced NaCl consumption when offered 150 mM NaCl solution (Puchalski et al. 2001), potentially indicating a perturbation in renal salt handling leading to an adverse reaction to high salt; it could also arise from disruption of a role for Kcne1 in the salivary glands. Kcne1 was also reported to modulate swelling-activated K^+ and Cl^- currents in primary cultured proximal convoluted tubule epithelial cells, suggesting a role in cell volume regulation in this tissue, although the current functional properties and pharmacology suggested against it being generated by KCNQ1–KCNE1 channels (Barriere et al. 2003; Millar et al. 2004). Counterintuitively, the KCNQ1-preferential antagonist, chromanol 293B, had similar functional effects in the proximal tubule to those of *Kcne1* deletion, but this may reflect nonspecific action of chromanol 293B at the dosage used (100 μM).

The cell-specific expression, biological functions, and α subunit partners of KCNE1 in the kidneys remain a matter of debate, and it may be that KCNE1 serves a role there but with a different α subunit to KCNQ1 (Barriere et al. 2003; Millar et al. 2004; Neal et al. 2011; Vallon et al. 2001, 2005). Indeed, *Kcnq1*$^{-/-}$ mice, when supplied with standard food and water, show normal Na^+ and glucose urinary excretion, in contrast to *Kcne1* knockouts (Vallon et al. 2005). Kcnq1 is thought to only be important in mouse kidney when under the stress of increased substrate load, when apically expressed Kcnq1 maintains a suitable driving force to facilitate Na^+ reabsorption in the proximal tubule. In the mouse, therefore, Kcne1 and Kcnq1 are probably each performing a physiological role in the proximal tubules of the kidney, very likely in different channel complexes, and their function appears to only become important in conditions of physiological stress.

25.3 KCNQ1–KCNE2: Constitutively Active at Hyperpolarized Potentials Despite Its Voltage Sensor

The effects of KCNE2 and KCNE3 on KCNQ1 differ sharply from those of KCNE1 on KCNQ1, and this is also reflected in their contrasting physiological functions. Macroscopic currents recorded from KCNQ1–KCNE2 channels show a constitutively active yet low conductance channel (Fig. 25.4a) (Tinel et al. 2000), properties that suggest it will be incredibly challenging to reliably record recognizable unitary KCNQ1–KCNE2 currents. Hence, unitary current recordings or noise variance analysis of KCNQ1–KCNE2 currents are thus far lacking in the literature. The

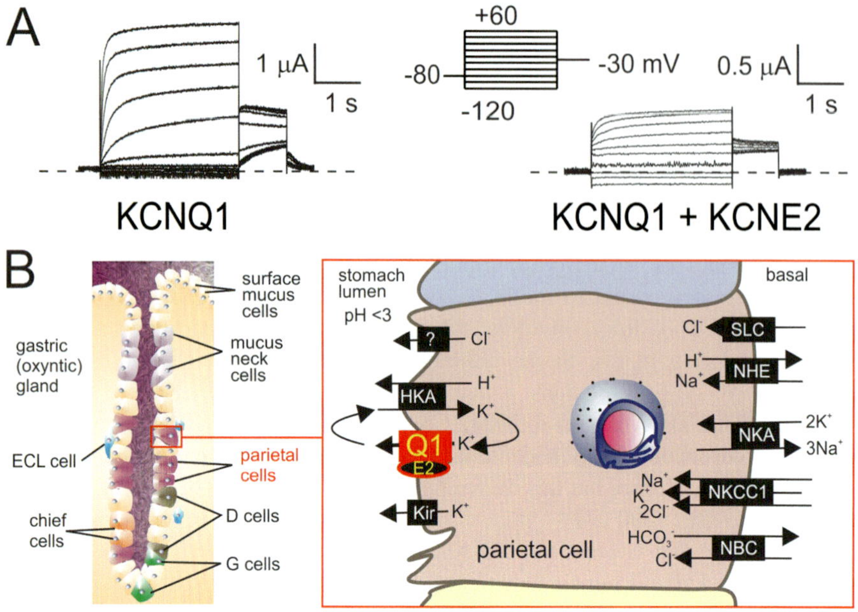

Fig. 25.4 KCNQ1 and KCNE2 form constitutively active gastric K^+ channels. (**a**) Effects of KCNE2 on KCNQ1 in *Xenopus* oocyte expression studies measured by two-electrode voltage clamp with 4 mM KCl bath solution. Voltage protocol is shown at top. KCNE2 greatly slows KCNQ1 deactivation and/or left-shifts its voltage dependence of activation, and decreases current density. (**b**) The role of KCNQ1–KCNE2 in parietal cells. D cell, somatostatin-producing cell; E2, KCNE2; ECL, enterochromaffin-like; G cell, Guard cell; HKA, H^+/K^+-ATPase; Kir, inward rectifier K^+ channels; NBC, Na^+/HCO_3^- co-transporter; NHE, sodium/hydrogen exchanger; NKA, Na^+/K^+-ATPase; NKCC1, $Na^+/K^+/2Cl^-$ co-transporter 1; Q1, KCNQ1; SLC, solute carrier transporter. Adapted from Abbott (2015)

ability to constitutively activate across the physiological voltage range is thought to be essential for the functions KCNQ1–KCNE2 serves in polarized epithelial cells.

The ability of KCNE2 and KCNE3 to convert KCNQ1 to a constitutively active "leak" channel is unique among the S4 superfamily of voltage-gated ion channels: no other voltage-gated ion channels are known to be completely convertible. The leucine-rich repeat (LRR)-containing protein 26, LRRC26, negative-shifts activation of voltage- and Ca^{2+}-activated BK channel gating by –140 mV, but these heteromers are still closed at –80 mV (Yan and Aldrich 2010, 2012). By the same token, KCNE2 and KCNE3 each interact with several other Kv channels but do not lock them open, although KCNE3 left-shifts the voltage dependence of Kv3.4 activation by –45 mV, a significant hyperpolarizing shift that permits function of this channel at subthreshold voltages in skeletal muscle (Abbott et al. 2001, 2006; Abbott and Goldstein 2002; McCrossan and Abbott 2004).

25.3.1 KCNQ1–KCNE2 in the Gastric Epithelium

In gastric parietal cells, located within the oxyntic glands of the stomach epithelium, apical KCNQ1–KCNE2 channels provide a K^+ recycling conduit to replenish luminal K^+ that entered the parietal cells through the apical gastric H^+K^+ATPase that acidifies the stomach lumen (Fig. 25.4b) (Grahammer et al. 2001a; Lee et al. 2000; Roepke et al. 2006). Discovery of the gastric role of the KCNQ1 α subunit was a surprise made possible by mouse genetics. Lee and colleagues found that *Kcnq1*$^{-/-}$ mice exhibited gastric hyperplasia, and that this stemmed from an inability to secrete gastric acid (Lee et al. 2000). One of the aspects that made this so surprising was that KCNQ1 forms voltage-gated channels—how could it function efficiently in a non-excitable cell such as the parietal cell, which secretes gastric acid through its apically located H^+/K^+-ATPase? Several subsequent studies implicated regulation of KCNQ1 by KCNE2, which had recently been found to impart striking changes in gating attributes upon KCNQ1 (Tinel et al. 2000; Dedek and Waldegger 2001; Grahammer et al. 2001a; Ohya et al. 2002; Heitzmann et al. 2004).

Correspondingly, germline deletion of *Kcne2* in mice likewise causes prominent gastric dysfunction. *Kcne2*$^{-/-}$ mice are achlorhydric, with a resting gastric pH of 6.5 and no measurable lowering of stomach pH when stimulated with histamine. Using pH-sensitive dyes to dynamically monitor pH in individual parietal cells ex vivo, we demonstrated that proton-loaded parietal cells from *Kcne2*$^{-/-}$ mice cannot restore their pH, which remains essentially constant upon histamine or carbachol stimulation; in contrast, wild-type parietal cells restore their pH at a rate of almost 0.1 pH/minute after cessation of proton loading (Roepke et al. 2006). Parietal cell morphology, and in particular the secretory canaliculus, is severely disrupted by *Kcne2* deletion. The mechanism for loss of gastric acid secretion in *Kcne2*$^{-/-}$ mice is not secondary to disrupted cellular morphology, however, but vice versa; in *Kcne2*$^{+/-}$ (heterozygous) mice, parietal cell morphology appears normal and yet there is still an ~50% disruption of gastric acid secretion. (Roepke et al. 2006).

To perform their role in gastric glands, KCNQ1–KCNE2 channels must stay open at negative voltages (probably largely between –40 and –20 mV in parietal cells) and continue to function while withstanding the highly acidic extracellular environment of the gastric pits (pH 2–3). This constitutive activation is most likely achieved, at least in part, by KCNE2 favoring the activated conformation of the voltage sensor, a phenomenon that has been studied in much greater detail for KCNQ1–KCNE3 channels because their larger currents facilitate electrophysiological and mutagenesis-based structure–function studies (see below). In contrast, the KCNQ1 current augmentation that occurs in response to low extracellular pH is uniquely endowed by KCNE2. Homomeric KCNQ1 channels are inhibited by extracellular protons, manifested as positive-shifting of the voltage dependence of activation and slowed gating. KCNQ1–KCNE1 channels also exhibit inhibition but are less sensitive to pH, whereas KCNQ1–KCNE2 current, crucially, is potentiated by low extracellular pH. We know that this characteristic, endowed by the KCNE2 extracellular N-terminus and neighboring region of the transmembrane domain, is not a general feature of constitutively active KCNQ1 channels because KCNQ1–KCNE3 channels are largely voltage-independent but insensitive to pH (see below) (Heitzmann et al. 2004, 2007). One might imagine that the pH dependence of KCNQ1–KCNE2 channels evolved because it gave a selective advantage in terms of greater ability to maintain proton secretion from the gastric pits for longer and thus more fully digest complex, protein-rich foods; this has not, however, been formally demonstrated.

Extracellular potassium has also been found to inhibit KCNQ1 activity, beyond levels predicted by the Goldman-Hodgkin-Katz flux equation (Larsen et al. 2011). This inhibition occurs within a physiologically important range (IC_{50} of 6 mM), which in human serum would constitute mild hyperkalemia. The mechanism for this inhibition was discovered to be an increased proportion of KCNQ1 channels in the inactivated state. Channels formed by KCNQ1 and either KCNE1 or KCNE3 do not inactivate; accordingly, these channel complexes are insensitive to the inhibitory effects of extracellular K^+. Thus, extracellular K^+ may play a physiological role in regulation of some channels formed with KCNQ1, by promoting and/or stabilizing the inactivated state (Larsen et al. 2011). This could be potentially useful in limiting K^+ efflux (beyond the effects of reduced electrochemical gradient) in scenarios in which the extracellular space is K^+ replete, such as in the deep pits of oxyntic glands in the gastric lumen, into which K^+ is recycled from parietal cells by the KCNQ1–KCNE2 complex. Accordingly, like homomeric KCNQ1 but unlike KCNQ1–KCNE1 and KCNQ1–KCNE3 channels, KCNQ1–KCNE2 channels are reportedly inhibited by extracellular K^+ (Larsen et al. 2011).

The most striking gross pathological change in the *Kcne2*$^{-/-}$ mouse line is that the stomachs of the mice are massively hyperplastic (Roepke et al. 2006). By 12 months, *Kcne2* deletion has resulted in an eightfold increase in stomach mass. The stomach wall thickens several-fold and develops cysts, a condition termed gastritis cystica profunda (Roepke et al. 2010) that is also observed in people, most commonly following gastric surgery, and that predisposes to gastric cancer. *KCNE2* expression also is disrupted and reduced in human gastric carcinoma and adenocarcinoma (Roepke et al. 2010). Following these discoveries in *Kcne2*$^{-/-}$ mice, areas of cysts in gastritis cystica profunda in a human patient with gastric adenocarcinoma were discovered to exhibit reduced KCNE2 expression (Kuwahara et al. 2013).

Impaired gastric acid secretion permits bacterial overgrowth and subsequent inflammation, which can develop into metaplasia and gastric cancer; accordingly, we observed neoplasia in *Kcne2*$^{-/-}$ mouse gastric tissue (Roepke et al. 2010). This may not arise solely from loss of gastric acid secretion: it was previously shown that KCNE2 downregulation also increases human gastric cancer cell proliferation in vitro, irrespective of pH. In both that in vitro study, and in our in vivo study of *Kcne2* deletion, increased nuclear cyclin D1 localization occurred (Yanglin et al. 2007; Roepke et al. 2010). The findings suggest KCNE2 can directly influence the cell cycle; whether this involves regulation of a channel remains unresolved.

Although the loss of KCNE2 expression has been observed in human gastric cancers and around gastric cysts, association between human hypochlorhydria and KCNE2 polymorphisms has not yet been reported. However, human KCNQ1 mutations associate with impaired gastric acid secretion, iron deficiency anemia stemming from hypochlorhydria (disrupting gastric iron absorption) and gastric cancer. Longer electrocardiogram QT intervals were found in one family to correlate with elevated serum gastrin, a biomarker of diminished gastric acid secretion (Tranebjaerg et al. 1993; Rice et al. 2011; Winbo et al. 2013). *Kcne2*$^{-/-}$ mice also exhibit iron-deficiency anemia (Hu et al. 2014b); it will be of interest to determine whether human *KCNE2* polymorphisms, as observed for KCNQ1, associate with anemia.

25.3.2 KCNQ1–KCNE2 in the Thyroid Epithelium

Mouse pups lacking *Kcne2* exhibit prenatal mortality, retarded growth, alopecia, and cardiac hypertrophy (Roepke et al. 2009). These traits are more severe in pups from *Kcne2*$^{-/-}$ dams; even wild-type pups from *Kcne2*$^{-/-}$ dams exhibit this phenotype. These phenomena arise because KCNQ1–KCNE2 is expressed in the basolateral membrane of thyroid epithelial cells and is required for optimal production of thyroid hormone by the thyroid gland (Fig. 25.5a, b) (Roepke et al. 2009).

At the basolateral side of thyroid cells, the sodium-coupled iodide symporter (NIS) sequesters iodide in the thyroid gland, utilizing the downhill sodium gradient to transport iodide ions (I^-) into thyroid cells (Riedel et al. 2001). KCNQ1–KCNE2 is required for efficient function of NIS in this fashion, and thus for efficient I^- uptake. NIS appears to function adequately without KCNQ1–KCNE2 in young virgin adult mice, which appear normal and euthyroid. In contrast, *Kcne2*$^{-/-}$ mice that are gestating or lactating are not able to produce sufficient thyroid hormone,

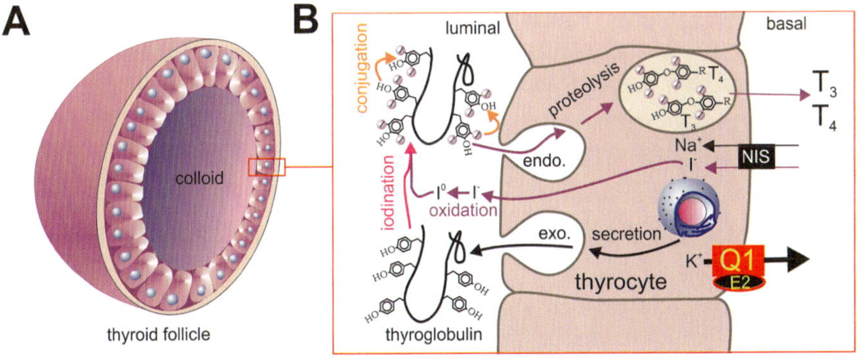

Fig. 25.5 KCNQ1–KCNE2 in thyroid cells. (**a**) Representation of a thyroid follicle, skirted by thyroid epithelial cells. (**b**) Thyroid hormone (T_3 and T_4) biosynthesis requires I^- to pass across the thyrocyte from the blood into the colloid, where it is oxidized, and organified by incorporation into thyroglobulin (iodination and conjugation). Channels formed by a complex of KCNQ1 and KCNE2, on the thyrocyte basolateral membrane, facilitate the efficient function of the basolateral sodium iodide symporter (NIS). Adapted from Abbott (2012)

probably because of increased thyroid hormone requirements. We observed the impaired ability of the thyroid gland in these mice to take up sufficient iodide, using positron emission tomography to track ^{124}I movement (Purtell et al. 2012; Roepke et al. 2009). We found that pups of *Kcne2*$^{-/-}$ dams were also severely affected, exhibiting cardiac hypertrophy, growth retardation, and alopecia, largely because the dams' hypothyroidism also impaired milk ejection. This was rescuable using oxytocin injection of the dams in the short term, or in the long term with thyroid hormone supplementation during gestation and lactation. *Kcne2*$^{-/-}$ pups surrogated to *Kcne2*$^{+/+}$ dams after birth (i.e., separated from their mothers and instead allowed to suckle from wild-type dams) also develop normally because they have sufficient access to milk. Conversely, *Kcne2*$^{+/+}$ pups surrogated to *Kcne2*$^{-/-}$ dams develop features characteristic of hypothyroidism (Roepke et al. 2009).

We do not yet fully understand the necessity for KCNQ1–KCNE2 channels for optimal NIS function, but several mechanistic insights have been elucidated. First, KCNQ1–KCNE2 is not needed for I^- organification (the process of incorporation of I^- into thyroglobulin), but is required for I^- uptake. Second, inhibition of KCNQ1–KCNE2 in the FRTL-5 rat thyroid cell line, by the relatively specific KCNQ1-specific blocker (–)-[3R,4S]-chromanol 293B, diminishes I^- sequestration by thyroid cells via Na^+-dependent I^- uptake by NIS, but does not alter Na^+-dependent nicotinate uptake by the sodium-coupled monocarboxylate transporter (SMCT). This suggests specificity of the effects of KCNQ1–KCNE2 on NIS and not simply an effect arising from maintenance of the whole-cell transmembrane Na^+ gradient for all Na^+-dependent transporters. Third, KCNQ1 requires co-assembly with KCNE2 for its role in the thyroid, a conclusion based on the findings that *Kcne2*$^{-/-}$ mice have similar thyroid hormone insufficiency to *Kcnq1*$^{-/-}$ mice (Frohlich et al. 2011), and that a current with the functional properties of KCNQ1–KCNE2 (sensitivity to the KCNQ antagonist, XE991, combined with a linear current–voltage relationship) is evident in FRTL-5 cells. Fourth, thyroid-stimulating hormone (TSH) or its major downstream effector, cyclic adenosine monophosphate (cAMP), each increase expression of both KCNQ1 and KCNE2 proteins in FRTL-5 cells (Purtell et al. 2012; Roepke et al. 2009).

25.3.3 KCNQ1–KCNE2 in the Choroid Plexus Epithelium

KCNQ1–KCNE2 channels are expressed in the apical membrane of the choroid plexus epithelium (Volume 2, Chapter 10), which is the blood–CSF barrier and primary site of cerebrospinal fluid (CSF) production and secretion (Fig. 25.6a, b) (Roepke et al. 2011a). *Kcne2*$^{-/-}$ mice exhibit increased outward KCNQ1 current, hyperpolarized membrane potential, a switch in KCNQ1 localization (from apical to basolateral) and a relatively minor shift in CSF chloride concentration (14% increase) compared to *Kcne2*$^{+/+}$ (wild type) mice. The increased KCNQ1 current can be explained by loss of KCNE2 from the complex, as KCNE2 reduces KCNQ1

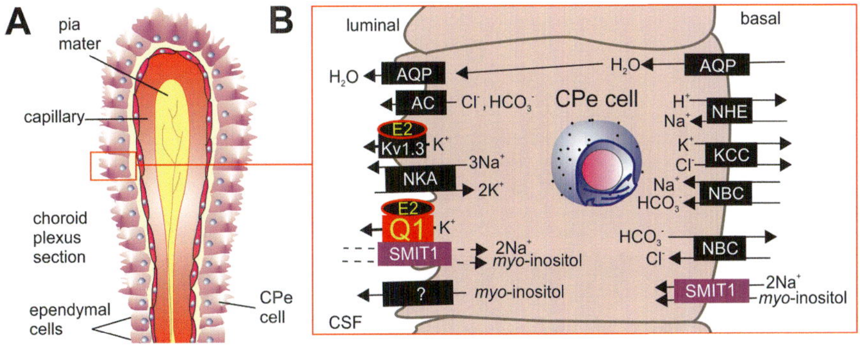

Fig. 25.6 KCNQ1–KCNE2 in choroid plexus epithelial cells. (**a**) Representation of the choroid plexus epithelium (CPe). (**b**) KCNE2 regulates KCNQ1–SMIT1 complexes and KCNA3 channels in the CPe apical membrane. Adapted from Abbott (2012)

macroscopic outward currents (Roepke et al. 2011a). *Kcne2* deletion also causes switching of KCNQ1 to the basolateral side of the choroid plexus epithelium, a phenomenon also observed in parietal cells (Roepke et al. 2011a) and which is discussed below in more detail, in the trafficking section. In addition to modulation of KCNQ1, electrophysiological analyses comparing the effects of *Kcne2* deletion on choroid plexus epithelial cells revealed alteration of currents sensitive to margatoxin, but not dendrotoxin, suggestive of regulation of KCNA3 (Kv1.3), but not KCNA1 (Kv1.1), by KCNE2 in the choroid plexus (Roepke et al. 2011a). As for KCNQ1 complexes, Kv1.3–KCNE2 complexes were confirmed by immunofluorescence and co-immunoprecipitation studies (Abbott et al. 2014).

Kcne2$^{-/-}$ mice are prone to handling-induced seizures, and also have an increased susceptibility to pentylenetetrazole-induced seizures, showing shorter latency to first seizure, increased seizure severity and increased mortality compared to their wild-type littermates (Abbott et al. 2014). The small changes we observed in CSF composition with *Kcne2* deletion seemed insufficient to generate increased seizure predisposition, and CSF potassium concentration and pH, two criteria that can dictate seizure susceptibility, were unaltered in *Kcne2*$^{-/-}$ mice (Roepke et al. 2011a). We therefore used mass spectrometry-based metabolomics to looked for possible changes in CSF metabolites, and discovered that CSF levels of the cyclic polyol *myo*-inositol are reduced in *Kcne2*$^{-/-}$ mice (Abbott et al. 2014).

CSF *myo*-inositol levels are highly regulated, and *myo*-inositol is actively concentrated in the CSF from the blood, primarily across the choroid plexus epithelium. It had previously been suggested that this transport occurs via basolateral uptake from the blood through sodium-dependent *myo*-inositol transporters, probably SMIT1, encoded by *SLC5A3*, although SMIT1 protein had not been visualized in the choroid plexus epithelium (Guo et al. 1997; Spector and Lorenzo 1975). We discovered that SMIT1 (Fig. 25.7a) is expressed at both the basolateral and apical membranes of the choroid plexus epithelium, an observation verified by the use of *Slc5a3*$^{-/-}$ mouse choroid plexus epithelium and kidney tissue as a negative control,

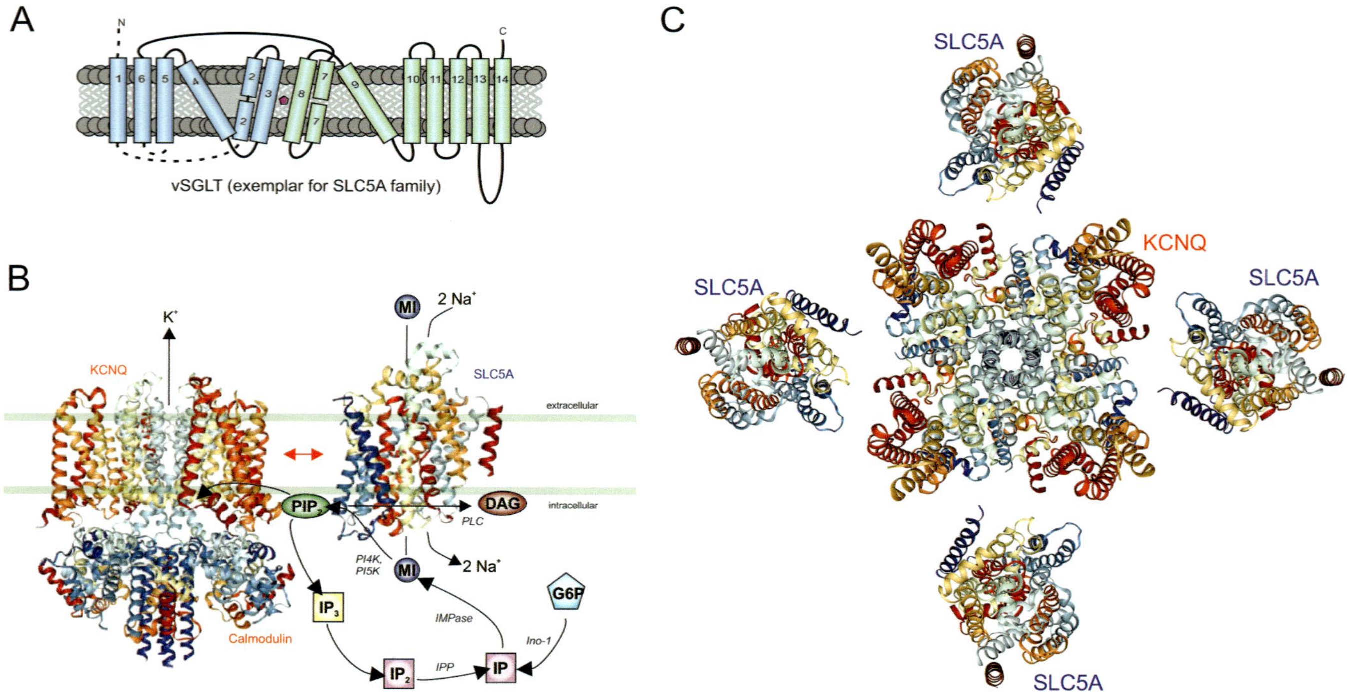

Fig. 25.7 SMIT1–KCNQ1 complexes. (**a**) Transmembrane topology of vSGLT, a representative of the SLC5A transporter family (Faham et al. 2008). Substrate (magenta) shown in binding site. (**b**) *Xenopus* KCNQ1 structure with calmodulin bound (Sun and MacKinnon 2017) shown alongside vSGLT (Faham et al. 2008) in place of SMIT1. Green lines approximate membrane boundaries. Also shown is the *myo*-inositol (MI) pathway, indicating how SMIT-transported *myo*-inositol can generate PIP_2 which then regulates KCNQ gating. Abbreviations: PI4K, phosphatidylinositol 4-kinase; PI5K, phosphatidylinositol 5-kinase; PLC, phospholipase C; DAG, diacylglycerol; IP_3, inositol(1,4,5)-triphosphate, IP_2, inositol(1,4)-bisphosphate; IP, inositol monophosphate; IPP, inositol-1,4 bisphosphate 1-phosphatase; IMPase, inositol-1(or 4)-monophosphatase; G6P, glucose 6-phosphate; Ino-1, inositol synthase. (**c**) Hypothetical 4:4 stoichiometry (KCNQ1:SLC5A) viewed from outside the cell. Adapted from Abbott (2017)

with wild-type kidney tissue as a positive control. We also discovered that SMIT1, KCNE2, and KCNQ1 form complexes together (Figs. 25.6b and 25.7b, c) (Abbott et al. 2014). *Myo*-inositol is one of the principal osmolytes in the body, and is also a precursor for signaling molecules including phosphatidylinositol phosphates. These include PIP_2, which regulates the function of many ion channels, including KCNQ1, suggesting dual modes of reciprocal regulation within KCNQ1–SMIT1–KCNE2 complexes (Fig. 25.7b) (Abbott et al. 2014).

KCNQ1–KCNE2 channels actually inhibit SMIT1 transport activity, reducing *myo*-inositol uptake, when studied in *Xenopus* oocytes (Abbott et al. 2014). Studies are ongoing, but at this point we conclude that SMIT1 is sensitive to the conformation of KCNQ1–KCNE2, and something about the conformation adopted by KCNQ1 with co-assembled KCNE2, inhibits SMIT1 *myo*-inositol efflux activity. KCNQ1-specific inhibitors also inhibit co-assembled SMIT1 activity, suggesting that K^+ efflux through KCNQ1 helps SMIT1 bring in more substrate, but only when the KCNQ1 conformation is amenable. The dual basolateral and apical localization of SMIT1 in the choroid plexus epithelium, the apical location of SMIT1–KCNQ1–KCNE2 complexes, and the observation that KCNE2 inhibits transport activity in these complexes, suggests that apical channel–transporter complexes may regulate CSF *myo*-inositol to ensure it does not get too high, or at times when variations in composition might be needed for neuroprotection or other reasons. Removal of KCNE2 from these complexes could therefore result in overactivity, and thus too much *myo*-inositol being removed from the CSF, hence the reduction in CSF *myo*-inositol in the *Kcne2*$^{-/-}$ mice. More recently, we discovered that SMIT1 physically couples specifically to the pore module of a related channel KCNQ2, with which SMIT1 co-assembles in nodes of Ranvier and neuronal axon initial segments in rodent brain (Neverisky and Abbott 2017). In addition, we found that SMIT1 alters the pore conformation, gating kinetics, and pharmacology of KCNQ1, KCNQ1–KCNE1, KCNQ2, KCNQ3, and KCNQ2/3 channels (Manville et al. 2017).

25.3.4 KCNE2 in the Pancreas

KCNQ1 and KCNE2 are each expressed in pancreatic β-cells, but the possible roles of each there may be complex, especially for KCNQ1. We recently found that *Kcne2* deletion causes Type 2 diabetes mellitus (T2DM), with *Kcne2*$^{-/-}$ mice exhibiting glucose intolerance as early as 5 weeks of age. *Kcne2* deletion also caused pancreatic transcriptome changes consistent with T2DM, including endoplasmic reticulum stress, inflammation, and hyperproliferation (Lee et al. 2017). We found that *Kcne2* deletion diminished pancreatic β-cell insulin secretion in vitro up to eightfold and also inhibited β-cell peak outward K^+ current at positive membrane potentials, in addition to negative-shifting its voltage dependence and slowing inactivation. KCNE2 is therefore required for normal β-cell electrical activity and insulin secretion. KCNE2 may regulate more than one K^+ channel type in pancreatic β cells, with the T2DM-linked KCNQ1 α subunit being one candidate (Lee et al. 2017).

Prior findings from overexpression studies utilizing the MIN6 mouse β-cell line suggested that KCNQ1 limits insulin secretion from pancreatic β-cells (Yamagata et al. 2011), although it is possible that KCNQ1 overexpression does not reflect normal physiologic mechanisms of pancreatic function. The WTC-dfk rat, which harbors a KCNQ1 intragenic deletion, shows pancreatic metaplasia (Kuwamura et al. 2008), although, again, this does not necessarily imply direct disruption of a primary role in the pancreas. More convincingly, KCNQ1 antagonists HMR1556 and chromanol 293B inhibit whole-cell outward K^+ currents, alter action potential duration and frequency, and increase insulin secretion from the rat β-cell-derived INS-1 cell line (Ullrich et al. 2005). In addition, single nucleotide polymorphisms (SNPs) within KCNQ1 are strongly linked to T2DM (Lee et al. 2008; Liu et al. 2009; Unoki et al. 2008; Yasuda et al. 2008; Zhou et al. 2010), gestational diabetes (Zhou et al. 2009), insulin secretion (Mussig et al. 2009) and impaired fasting glucose (Qi et al. 2009) and efficacy of diabetic medications (Yu et al. 2011) in multiple human populations, highly consistent with a link between altered KCNQ1 function and human pancreatic dysfunction.

Human pancreatic β-cells carrying human T2DM susceptibility-linked *KCNQ1* gene variants were also found to exhibit reduced depolarization-evoked insulin exocytosis and altered granule docking, strengthening the link (Rosengren et al. 2012). Interestingly, results from *Kcnq1* knockout studies indicated that *Kcnq1*$^{-/-}$ mice exhibit lower fed and fasted plasma glucose and insulin levels, and Kcnq1 appears to act against insulin-stimulated cellular K^+ uptake in the mouse pancreas. In this way, *Kcnq1* deletion influences the downstream effects of insulin on glucose metabolism, but in a manner that does not necessarily involve direct disruption of a primary role in the pancreas. Instead, Kcnq1 may tune the cellular glucose uptake of tissues including the liver, skeletal muscle, lungs, and kidneys (Boini et al. 2009). This could mean that KCNE2 modulates other channels in the exclusion of or in addition to KCNQ1 in the pancreas. In mice and humans, KCNQ1 is expressed in so many tissues that the effects on glucose control and handling of KCNQ1 disruption are anticipated to be multifactorial in both mechanism and outcome, perhaps involving both the pancreas and other organs. It is also that *KCNQ1* locus SNPs influence T2DM via a nearby gene, the imprinted expression of which is influenced by *KCNQ1* or *KCNQ1OT1,* cyclin-dependent kinase inhibitor (*CDKN1C*) being one candidate (Travers et al. 2013).

25.4 KCNQ1–KCNE3: A Highly Studied, Constitutively Active Channel

As for KCNE2, KCNE3 also converts KCNQ1 to a constitutively open channel, and unlike KCNE2, KCNE3 permits KCNQ1 to maintain a macroscopic outward current density at levels equivalent to or higher than those generated by homomeric KCNQ1 (Schroeder et al. 2000). KCNE3 appears to favor constitutive activation of KCNQ1 by locking open its voltage sensor rather than by directly locking open the pore independent of the voltage sensor (Nakajo and Kubo 2007; Panaghie and Abbott 2007). We identified critical KCNE3 residues at the membrane-proximal region of the extracellular portion of KCNE3 (D54 and D55) that interact with KCNQ1-S4 as part of this process (Choi and Abbott 2010). This work followed from studies in the McDonald lab showing that residues in the KCNE3 transmembrane domain are also essential for KCNQ1–KCNE3 constitutive activation (Melman et al. 2001, 2002).

Interestingly, KCNQ1–KCNE3 channel currents typically exhibit a voltage-dependent fraction, and that fraction is proportionately much greater when heterologously expressed in mammalian cell lines including human embryonic kidney (HEK) cells, versus when expressed in *Xenopus laevis* oocytes (Abbott and Goldstein 2002; Melman et al. 2001, 2002; Panaghie et al. 2006). It is possible that this arises from regulatory differences or even different subunit compositions between the two types of expression system, but this has not been experimentally confirmed. However, *Xenopus* oocytes are known to express endogenous KCNE proteins that can influence the results of electrophysiological studies of Kv channels (Anantharam et al. 2003; Gordon et al. 2006). Recently, the author discovered an additional human KCNE3 coding exon that appears to only be present in primates; this adds 44 residues to bring the full-length protein (KCNE3L) to 147 amino acids. KCNE3L still constitutively activates KCNQ1 but the currents are smaller compared to KCNQ1-KCNE3S channels (Abbott 2016b).

25.4.1 KCNQ1–KCNE3 in the Intestinal Epithelium

KCNQ1 co-localizes with KCNE3 in the basolateral membrane of colonic crypts (Preston et al. 2010; Schroeder et al. 2000). Studies utilizing a *Kcne3*$^{-/-}$ mouse line (Preston et al. 2010) confirmed the prior report (Schroeder et al. 2000) that KCNQ1–KCNE3 is required for regulation of cAMP-stimulated chloride secretion. All available evidence for the importance of this process is from the mouse, and specifically the *Kcne3*$^{-/-}$ mouse line, and there are currently no reported human disease associations linking KCNE3 (or KCNQ1) mutations to altered intestinal Cl^- secretion (Preston et al. 2010; Schroeder et al. 2000). Basolateral recycling of K^+ through KCNQ1–KCNE3 channels, together with the increased driving force generated by K^+ efflux through this heteromer, promote electrogenic Cl^- secretion across the intestinal epithelium (Bleich and Warth 2000; Boucherot et al. 2001; Dedek and Waldegger 2001). *Kcne3* deletion in mice reduces intestinal epithelial Cl^- secretion without altering KCNQ1 localization or abundance, suggesting that the modification of KCNQ1 gating (conversion from voltage-dependent activation to constitutive activation) by KCNE3 is a primary factor in the necessity of KCNE3 in colonic KCNQ1 channel complexes (Preston et al. 2010). By locking KCNQ1 open, KCNE3, therefore, diversifies the functional capabilities of this multifunctional Kv channel α subunit, permitting it to function in a non-excitable cell type.

KCNE3 also adds a further level of complexity to the colonic current in that KCNE3 is regulated by estrogen (O'Mahony et al. 2009), which downregulates KCNE3 and thus increases the relative proportion of homomeric KCNQ1, which does not function so efficiently in non-excitable cells because it requires cellular depolarization for efficient activation. Thus, during the estrous cycle, the high levels of estrogen attained in proestrus cause water retention in female mammals partly by removing the capacity of KCNQ1 to operate, and therefore facilitate cAMP-stimulated Cl^- secretion, at negative voltages. Colonic crypts from the distal colon of male rats show higher KCNE3 expression, and >twofold higher association of KCNQ1 and KCNE3, than those from female rats. Strikingly, estrogen causes dissociation of KCNE3 from KCNQ1 within minutes (Rapetti-Mauss et al. 2013; Alzamora et al. 2011). Mutation to alanine of the intracellular, membrane-proximal serine PKC phosphorylation site, KCNE3-S82, blunts the effects of 17β-estradiol on KCNQ1–KCNE3 in vitro and also caused rapid rundown (Alzamora et al. 2011). KCNE3-S82 phosphorylation is also crucial for KCNE3 modulation of KCNC4 (Kv3.4) activation gating (Abbott et al. 2006), suggesting that the importance of this KCNE3 site and its role in control of gating through phosphorylation are conserved between different channel complexes.

25.4.2 KCNQ1–KCNE3 in the Mammary Epithelium

KCNQ1–KCNE3 channels may also play a role in the mammary epithelium (vanTol et al. 2007). Pharmacological inhibition of KCNQ1 channels in the MCF-7 human mammary epithelium cell line eliminates the ability to respond to hypoosmotic stress initiated by switching from an isosmotic to a hypoosmotic extracellular solution. This ability is likely to be highly important in mammary epithelia, as milk is rich both in K^+ and in impermeable solutes such as lactose. KCNQ1 appears to be localized to the mammary epithelial cell apical membrane, and a KCNQ1-like current in MCF-7 cells is activated by hypoosmotic extracellular solutions. These findings were recapitulated in baby hamster kidney cells with KCNQ1 heterologously expressed alone or with KCNE3, which is also detected in MCF-7 cells (vanTol et al. 2007).

It is also possible that KCNQ1 interacts with other KCNE subunits in the mammary epithelium; for instance, KCNE1 and KCNE2 are both reportedly expressed there (vanTol et al. 2007). *Kcne2*$^{-/-}$ mice have a milk ejection defect that was able to be rescued acutely by oxytocin injection, but this defect was found to be attributable to the hypothyroidism exhibited by lactating *Kcne2*$^{-/-}$ dams, and was preventable using thyroid hormone treatment (Roepke et al. 2009). It is reasonable to assume that KCNQ1 is required for mammary epithelial cell volume regulation, probably via K^+ secretion. In mammary epithelial cells, KCNQ1 may co-assemble with one or more of the KCNE subunits, but KCNQ1 does not appear to require this co-assembly for cell volume regulation, based on data from heterologous expression studies (Hammami et al. 2009; vanTol et al. 2007).

25.4.3 KCNQ1–KCNE3 in the Airway Epithelium

Kcne3 deletion in mice impairs transepithelial electrogenic Cl^- secretion in murine trachea, most likely because of conversion of constitutively active, cAMP-sensitive KCNQ1–KCNE3 back to homomeric, voltage-dependent KCNQ1 (Preston et al. 2010). Several studies have implicated KCNQ1 in functional roles associated with the basolateral membrane of the airway epithelium. Accumulating evidence, including results from characterization of *Kcne3*$^{-/-}$ mice, suggests that the KCNQ1–KCNE3 complex contributes to the activity of a basolateral airway epithelial cell K^+ current essential for normal cAMP-stimulated Cl^- secretion (Grahammer et al. 2001b; Kim et al. 2007; Preston et al. 2010; Schreiber et al. 2002). In the airways, this Cl^- secretion is important for normal mucous production and secretion, as evidenced by the overproduction defect in mucus secretion caused by CFTR channel mutations in cystic fibrosis (Kreda et al. 2012).

In human airway epithelium, the role of KCNQs may be different from that in mice. Cultured and freshly isolated human bronchial epithelial cells, and also the human airway epithelial cell line Calu-3, express KCNQ1, 3, and 5 (transcript and protein); *KCNE1-3* and to a lesser extent *KCNE4* transcripts were also detected in both cultured and freshly isolated human bronchial epithelial cells. KCNQ1 was localized to both the apical and basolateral membranes; KCNQ3 and 5 were only detected in the apical membrane of human bronchial epithelial cells. In addition, functional studies utilizing KCNQ1-preferring blocker chromanol 293 and KCNQ family-wide blocker XE991 and the less potent derivative linopirdine, suggested that KCNQ3 and/or 5, and to a lesser extent KCNQ1, regulate airway epithelium short-circuit current (I_{sc}), cAMP- and Ca^{2+}-stimulated anion secretion, and airway surface liquid [K^+] (Moser et al. 2008; Namkung et al. 2009). KCNQ3 and KCNQ5, but not KCNQ1, were also found to regulate apical Na^+ flux in the H441 human lung adenocarcinoma cell line, a model of absorptive airway epithelium (Greenwood et al. 2009a). KCNE1 was previously suggested to regulate KCNQ1 in mouse airway epithelia, but a subsequent paper contained three lines of evidence supporting, instead, a role for KCNE3 (Grahammer et al. 2001b). First, *Kcne1*-null mice had no measurable defect in tracheal epithelial Cl^- secretion. Second, KCNQ1 but not KCNE1 was detected in the airway epithelium basolateral membrane. Third, KCNE3 and KCNQ1 messages but not KCNE1 messages were detected in tracheal epithelial cells by RT-PCR. Basolateral KCNQ1–KCNE3 was suggested in this study to not only regulate Cl^- secretion, but also support Na^+ reabsorption, in mouse airway epithelium (Grahammer et al. 2001b).

Significantly, a human genomic DNA sequence variant 84-kb upstream of *KCNE2* has been found to associate with lung dysfunction (Soler Artigas et al. 2011), and we recently detected *Kcne2* transcript expression in whole lung preparations from wild-type mice, using tissue from *Kcne2*$^{-/-}$ mice as a negative control (Hu et al. 2014a). In addition, other groups detected KCNE2 in the Calu-3 human airway cell line (Cowley and Linsdell 2002). The potential role for KCNQ1–KCNE2 in lung epithelia is unknown but under investigation.

25.5 Epithelial Roles for KCNE4

The KCNE4 subunit is expressed in human and mouse heart, brain, skeletal muscle, uterus, testis, kidney, liver, lymphocytes, and placenta (Abbott and Goldstein 1998; Grunnet et al. 2002; Manderfield and George 2008). KCNE4 possesses the longest intracellular domain among the mammalian KCNE family (Abbott and Goldstein 1998), suggestive of a role for this domain in mediating interactions with intracellular regulatory proteins or in sensing other aspects of the intracellular milieu. Mouse and human variants of KCNE4 are reported to strongly (>90%) inhibit KCNQ1 activity in heterologous co-expression experiments in *Xenopus laevis* oocytes, as assayed by two-electrode voltage clamp (Grunnet et al. 2002). Human KCNE4 was recently found to be longer than previously thought, with the discovery of an additional 5′ coding exon encoding 51 amino acids and not present in the mouse gene, bringing the full-length human protein to 221 residues (Abbott 2016b).

25.5.1 KCNE4 in the Kidney

KCNE4 inhibits KCNQ1 function but the physiological relevance of this is not known; *Kcne4* deletion impairs ventricular myocyte repolarization in mice because of loss of Kv1.5 and Kv4.2 currents (Crump et al. 2016). In epithelial cells, however, KCNE4 regulates the Ca^{2+}-activated K^+ channel, BK (Levy et al. 2008). KCNE4 co-localizes with BK channel α subunits in the apical membranes of intercalated cells in the medulla and the renal cortex of the rat kidney. Human KCNE4 downregulates BK current threefold without altering unitary conductance, but rather by disfavoring channel opening and also accelerating BK channel protein degradation (Levy et al. 2008).

BK channels are known to regulate flow-dependent K^+ secretion in rabbit kidney, where BK expression is downregulated by a low K^+ diet (Woda et al. 2001). KCNE4 may contribute to this or other regulatory processes controlling renal BK channel activity, although this has not yet been reported (Levy et al. 2008). Neither has the possibility of 5α-dihydrotestosterone (5α-DHT) regulation of renal KCNE4 yet been pursued, although it is known that 5α-DHT upregulates *Kcne4* expression in mouse heart (Crump et al. 2016). Interestingly, 5β-DHT (and to a lesser extent, testosterone and 5α-DHT) was found to directly inhibit BK channels in excised inside-out membrane patches from anterior pituitary rat tumor (GH3) cells, while testosterone, but neither form of DHT, was found to augment the activity of channels formed by the short isoform of the BK channel α subunit expressed in HEK-293 cells (Suarez et al. 2015). Androgens might therefore have complex effects on BK–KCNE4 channel complexes, depending on the tissue. It is still not known whether DHT regulation of KCNE4 is direct, genomic, or via an intermediary protein such as a KCNE4 regulating transcription factor, expression of which could be regulated by DHT (Crump et al. 2016). KCNQ1 is also expressed in the kidney, and therefore

KCNE4 could have more than one renal function (Vallon et al. 2005; Zheng et al. 2007). Renal effects of *Kcne4* deletion in mice have yet to be reported.

25.5.2 *KCNE4 in the Uterus*

KCNE4 is highly expressed in the uterus and may possibly regulate the various KCNQ isoforms expressed there, although functional studies have not been reported (Grunnet et al. 2002). While the activity of mERG1 channels was previously found to diminish toward the end of gestation in mice, mERG1 transcript and protein expression were unaltered, suggesting the possibility of regulation of mERG1 activity by another factor (Greenwood et al. 2009b). Increases in expression of both KCNE2 and KCNE4 were found to coincide with the reduced mERG1 activity (Greenwood et al. 2009b). The KCNE2 increase is more likely to be responsible for altered mERG1 activity, given its known ability to reduce hERG current density (Abbott et al. 1999), whereas KCNE4 was not found to alter hERG current in heterologous expression studies in *Xenopus* oocytes (Grunnet et al. 2002). However, it is possible that KCNE4 might regulate hERG in the mammalian uterus, while failing to exert functional effects in *Xenopus* oocytes. Elucidation of the mechanisms underlying reduced ERG activity late in pregnancy is an important goal because it could potentially contribute to the onset of labor (Greenwood et al. 2009b).

Human KCNE4L transcript is most readily detectable in the uterus, followed by the ureter, spleen, placenta, lymph node, adrenal, and heart (enriched in atria) out of 48 tissues tested by real-time qPCR. Yet, aside from the heart, roles in these tissues have yet to be determined (Abbott 2016b). Likewise, aside from in the heart, the role of KCNE5 in native physiology has been little studied. Briefly, *KCNE5* is expressed in the human heart and inhibits KCNQ1 in heterologous expression studies by shifting its voltage dependence of activation >140 mV more positive (Angelo et al. 2002; Bendahhou et al. 2005). Deletion of *Kcne5*, which is an X-linked gene, in mice causes increased ventricular Kv currents (specifically Kv2.1, Kv1.5, and Kv4 channel activity) and ventricular arrhythmia, consistent with its association with human Brugada syndrome (David et al. 2019).

25.6 Trafficking of KCNQ1–KCNE Channel Complexes

In most cell types, and particularly in polarized epithelial cells, the precise targeting of channel subunits is all-important. In this final section, the impact of KCNE subunits versus other protein classes in regulating KCNQ targeting in various cell types is reviewed.

25.6.1 RAB- and Clathrin-Dependent Trafficking of KCNQ1

A RAB5-dependent pathway is one mechanism mediating KCNQ1 endocytosis from the membrane into early endosomes; conversely, a pathway involving the small GTPase RAB11 mediates KCNQ1 recycling from endosomes back into the plasma membrane, at least in cardiac myocytes (Seebohm et al. 2007). The RAB11-dependent pathway is modulated by the serum glucocorticoid kinase SGK1. The SGKs are activated by stress hormones such as the glucocorticoid cortisol through the hypothalamic–pituitary–adrenal axis. SGK1-3 each activate KCNQ1–KCNE1 channels when expressed in *Xenopus* oocytes, independent of endogenous Na^+/K^+ATPase activity (Embark et al. 2003). An important underlying mechanism for SGK1 augmentation of KCNQ1–KCNE1 current is the enhancement of channel exocytosis by the small GTPase RAB11. SGK1 activates PIKfyve, a $PI(_5)$kinase, generating $PI_{(3,5)}P_2$, which in turn enhances RAB11-dependent KCNQ1–KCNE1 surface expression (Seebohm et al. 2007). In contrast, KCNQ1–KCNE3 channels appear to be insensitive to SGK1-3 (Strutz-Seebohm et al. 2009). In resting parietal cells, most of the KCNQ1 protein pool is localized in a RAB11-positive compartment, as opposed to the gastric H^+/K^+ATPase, which is found in tubulovesicular structures. When the parietal cell is stimulated, both proteins move to the apically located secretory canaliculi, sufficiently proximal to one another for robust Förster resonance energy transfer (FRET) (Nguyen et al. 2013). RAB11-dependent exocytosis of KCNQ1 in response to certain stimuli is therefore a mechanism common to cardiac myocytes and at least one polarized epithelial cell type.

Trafficking studies from the Harvey lab built on their prior discovery that estrogen uncoupled KCNE3 from KCNQ1, and showed that estrogen stimulated clathrin-mediated endocytosis of KCNQ1 (in the HT29cl.19A colonic cell line) via interaction with the AP-2 adaptor protein. Rather than being degraded after internalization, KCNQ1 was recycled from early endosomes back to the membrane via a RAB4- and RAB11-dependent pathway. This process was PKCδ and AMP-dependent kinase-dependent, and was activated by phosphorylation in an estrogen-dependent signaling cascade. Nedd4.2 was also found to promote KCNQ1 internalization in HT29cl.19A colonic cells, again in response to estrogen (Rapetti-Mauss et al. 2013). AMP-dependent kinase was also found to increase KCNQ1 internalization and subsequent degradation in lysosomes, in MDCK cells and in *Xenopus* oocytes, by activating Nedd4-2; in the former, this process occurred during the early stages of cell polarization (Andersen et al. 2011, 2012). Mechanisms of KCNQ1 internalization and recycling are summarized in Fig. 25.8.

In addition to the RAB-dependent processes, we uncovered a KCNE1-dependent internalization mechanism for KCNQ1 in vitro, and confirmed its occurrence in isolated guinea pig ventricular myocytes (Xu et al. 2009). We found that KCNE1 coordinates clathrin-mediated endocytosis of KCNQ1–KCNE1 channels, and demonstrated that this process is stimulated by PKC phosphorylation of KCNE1–S102 (Kanda et al. 2011c). We also detected this process occurring in neonatal mouse myocytes, but have yet to investigate its potential role in polarized epithelial cells.

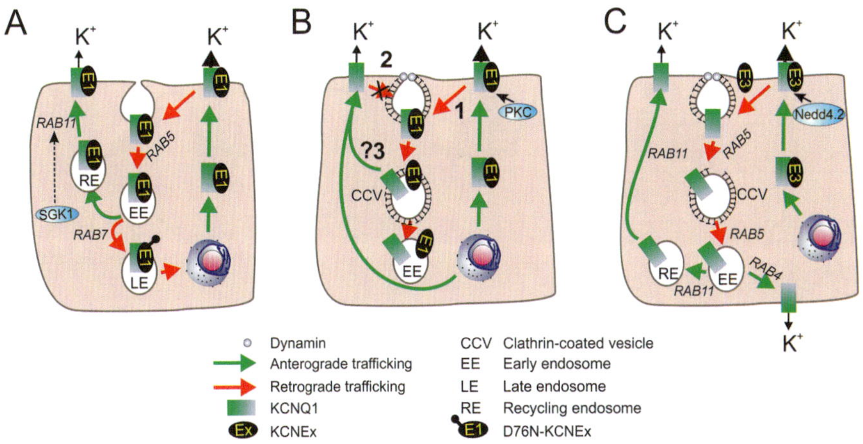

Fig. 25.8 KCNE-dependent and independent mechanisms of KCNQ1 internalization and recycling. (**a**) KCNQ1–KCNE1 complexes can be internalized and recycled by Rab-dependent pathways, with SGK1 stimulating RAB11-dependent recycling in response to stress. I_{Ks} channels formed with loss-of-function LQTS mutant D76N-KCNE1 are degraded via a RAB7-dependent pathway (Seebohm et al. 2007, 2008). (**b**) PKC stimulates KCNE1-dependent clathrin-mediated endocytosis of KCNQ1–KCNE1 (1); in contrast, homomeric KCNQ1 is not internalized by this pathway (2). This process could favor accumulation of non-KCNE1 channels containing KCNQ1 (3), perhaps in complexes with other KCNE subunits (Kanda et al. 2011c; Xu et al. 2009). (**c**) In the colon, Nedd4.2 (in response to estrogen) stimulates KCNQ1 internalization, leaving KCNE3 at the cell surface. KCNQ1 can be recycled by rapid (Rab4-dependent) or slow (Rab11-dependent) pathways (Rapetti-Mauss et al. 2013). Adapted from Abbott (2014)

25.6.2 KCNE-Dependent Polarized Trafficking of Epithelial KCNQ1

KCNQ1–KCNE2 channels are apically expressed in parietal cells, enabling them to return to the stomach lumen potassium ions that have entered through the apical H^+/K^+ATPase during gastric acidification (Dedek and Waldegger 2001; Grahammer et al. 2001a; Roepke et al. 2006). KCNQ1–KCNE2 is also apically expressed in the choroid plexus epithelium (Roepke et al. 2011a). Yet, in thyroid epithelial cells, KCNQ1–KCNE2 complexes are basolateral, where they are important for efficient iodide uptake by NIS (Roepke et al. 2009). KCNQ1–KCNE3 complexes are also basolaterally located, in colonic crypts and airway epithelia, where they regulate cAMP-stimulated chloride secretion (Preston et al. 2010); KCNQ1 has been detected at both the apical and basolateral membranes in primary human bronchial epithelia, and at the apical membrane in Calu-3 cells, although the KCNE partner in these cell types is not known (Moser et al. 2008; Namkung et al. 2009). The findings for KCNQ1–KCNE2 and KCNQ1–KCNE3 complexes might suggest that cellular environment dictates KCNQ1 location regardless of KCNE co-expression, but this is not the case. Initial studies did indeed appear to show a lack of KCNE influence on KCNQ1 localization in MDCK cells (Jespersen et al. 2004). However, further cell

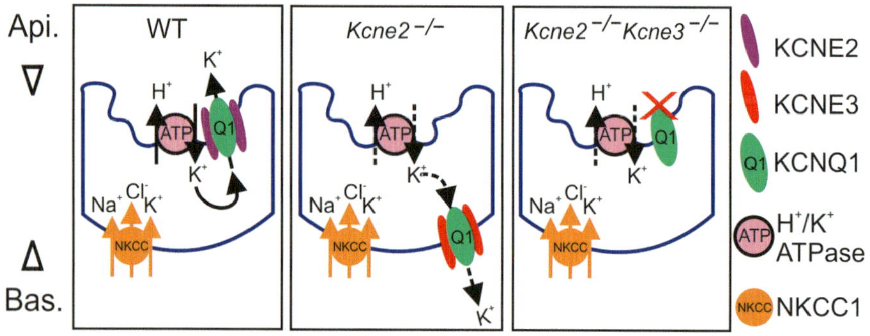

Fig. 25.9 Model of the effects of *Kcne2/Kcne3* knockout in gastric parietal cells. Model showing the putative short-circuit K^+ current in parietal cells of *Kcne2*$^{-/-}$ mice (*middle panel*) compared to *Kcne2*$^{+/+}$ mice (*left-hand panel*). This arises from the switch in KCNQ1 trafficking from apical (*Kcne2*$^{+/+}$, left-hand panel) to basolateral because of KCNE3 upregulation (*middle panel*). In *Kcne2*$^{-/-}$ *Kcne3*$^{-/-}$ mice, KCNQ1 is apical but nonfunctional because it lacks KCNE2 and KCNE3, which confer constitutive activation and limit inactivation (*right-hand panel*). Adapted from Roepke et al. (2011b)

studies revealed that KCNQ1 is by default enriched at the basolateral membrane in this kidney cell line, but that KCNE1 can apically reroute KCNQ1 if certain KCNQ1 basolateral sorting signals are mutated (KCNQ1-L38A/L40A and KCNQ1-Y51A) (David et al. 2013). The mutants had previously been characterized for their effect on homomeric KCNQ1 channels: KCNQ1-Y51A subunits lose their polarity and are expressed at both apical and basolateral membranes; KCNQ1-L38A/L40A is redirected to endosomal compartments instead of reaching the cell surface (Jespersen et al. 2004). In sum, the findings indicated the potential influence of KCNE subunits in KCNQ1 polarized trafficking (for example, in cells lacking the proteins necessary to decode the KCNQ1 basolateral targeting information) and also the potential for flexible KCNQ1 targeting (David et al. 2013).

Following the first MDCK-based investigation of KCNE–KCNQ1 trafficking (Jespersen et al. 2004), we studied the influence of Kcne2 and Kcne3 on Kcnq1 localization in parietal cells using single- and double-knockout mice (Roepke et al. 2011b). In *Kcne2*$^{-/-}$ mice, Kcnq1 targeted abnormally, to the basolateral membrane of parietal cells; in addition, Kcne3 transcript and protein expression was increased. In double-knockout *Kcne2*$^{-/-}$*Kcne3*$^{-/-}$ mice, apical expression of Kcnq1 was restored, but the gastric hypertrophy was worse than that of single-knockout *Kcne2*$^{-/-}$ mice, as was the extent of the hypochlorhydria we observed compared to heterozygous *Kcne2*$^{+/-}$ mice (Roepke et al. 2011b). This led to the following conclusions: KCNE2 is not required for KCNQ1 apical localization in normal parietal cells, but in the absence of KCNE2, abnormally upregulated KCNE3 can hijack KCNQ1 to the basolateral side (Fig. 25.9). In addition, homomeric KCNQ1 cannot function sufficiently well to support gastric acid secretion even if at the apical membrane, probably because of two reasons. First, KCNE2 is required for KCNQ1 to be constitutively active and to not inactivate; and KCNE2 ensures that KCNQ1 is activated, not inhibited, by extracellular protons. Second, it was better for gastric function for parietal cells to express basolaterally located KCNQ1-KCNE3 channels than apically located homomeric KCNQ1 channels (Roepke et al. 2011b). This may be because basolateral KCNQ1–KCNE3 channels permit exit of some K^+ (albeit basolaterally) (Fig. 25.9) and thus lessen K^+ accumulation in the parietal cells, which is detrimental to H^+/K^+ATPase function; homomeric KCNQ1 channels were likely to be completely nonfunctional for the reasons outlined above, despite their favorable location (Roepke et al. 2011b). This recapitulates the situation further down the digestive tract; in the normal intestine, basolateral KCNQ1-KCNE3 channels regulate cAMP-stimulated chloride secretion (Schroeder et al. 2000). Could this remodeling be related to other forms of gastric metaplasia also characterized by the adoption of intestinal characteristics?

Similar to effects in parietal cells, *Kcne2* deletion also results in relocation of KCNQ1 from the apical to the basolateral membrane in the choroid plexus epithelium. Similar results were also observed for KCNA3 (Kv1.3), another Kv α subunit partner of KCNE2 in the choroid plexus epithelium (Roepke et al. 2011a). The various localizations and roles of KCNQ1–KCNE2 channels in polarized epithelia are summarized in Figs. 25.4, 25.5, and 25.6.

Possible mechanisms underlying some aspects of KCNE-dependent KCNQ1 trafficking, involving differing membrane compositions, have also been investigated. KCNQ1 and KCNE3, but not other KCNE subunits, were found to each be enriched in caveolin-rich membrane fractions when expressed alone, suggesting targeting to sphingolipid-cholesterol-enriched (lipid raft) microdomains. In addition, KCNE2 and KCNE5 each efficiently reached the cell surface whereas KCNQ1 and the other KCNEs were found to be less efficiently trafficked to the surface. KCNQ1–KCNE1, and KCNQ1–KCNE2 complexes were found to be enriched in lipid rafts, while other KCNQ1–KCNE complexes were not. The results suggest that KCNQ1 and each of the five KCNE isoforms differentially influence one another's membrane subdomain location (at least in HEK cells) (Roura-Ferrer et al. 2010).

25.7 Conclusions

KCNE proteins are vital to epithelial biology, with current knowledge indicating that their predominant role in these systems is to modify the function of KCNQ1. Often, the role of KCNQ1 appears to be to support or facilitate the activity of one or more ion or solute transporters or pumps. Examples of epithelial processes assisted by KCNQ1–KCNE complexes are gastric acid secretion, cAMP-stimulated Cl^- secretion, thyroid hormone production, *myo*-inositol transport, and perhaps insulin secretion. Diseases associated with KCNE disruption can be complex because of the multiple tissues in which each *KCNE* gene is expressed, and the wide number of ion channels that they can co-assemble with and modulate, in polarized epithelia and also in excitable cells and endothelial cells. Further progress in this area will be aided by tissue-specific *Kcne* gene knockouts and pharmacological agents or modified toxins that can distinguish between different KCNE–Kv channel complexes. Indeed, recent progress has been made in this regard. KCNE1 (but not the other KCNEs) sensitizes KCNQ1 to inhibitory adamantane derivatives that show no effect on homomeric KCNQ1 (Wrobel et al. 2016); KCNE1 also potentiates the KCNQ1-opening effects of the plant compound, mallotoxin (De Silva et al. 2018; Matschke et al. 2016). Conversely, KCNE1 completely desensitizes KCNQ1 to effects of the chemical opener, zinc pyrithione (Gao et al. 2008).

Acknowledgment The author is grateful for financial support from the US National Institutes of Health awards GM115189, GM130377, and DK41544.

References

Abbott GW (2012) KCNE2 and the K^+ channel: the tail wagging the dog. Channels (Austin) 6(1)

Abbott GW (2014) Biology of the KCNQ1 potassium channel. New J Sci 2014 (Article ID 237431):26. https://doi.org/10.1155/2014/237431

Abbott GW (2015) The KCNE2 K(+) channel regulatory subunit: ubiquitous influence, complex pathobiology. Gene 569(2):162–172. https://doi.org/10.1016/j.gene.2015.06.061

Abbott GW (2016a) KCNE1 and KCNE3: the yin and yang of voltage-gated K(+) channel regulation. Gene 576(1 Pt 1):1–13. https://doi.org/10.1016/j.gene.2015.09.059

Abbott GW (2016b) Novel exon 1 protein-coding regions N-terminally extend human KCNE3 and KCNE4. FASEB J 30(8):2959–2969. https://doi.org/10.1096/fj.201600467R

Abbott GW (2017) Chansporter complexes in cell signaling. FEBS Lett 591(17):2556–2576. https://doi.org/10.1002/1873-3468.12755

Abbott GW, Goldstein SA (1998) A superfamily of small potassium channel subunits: form and function of the MinK-related peptides (MiRPs). Q Rev Biophys 31(4):357–398

Abbott GW, Goldstein SA (2002) Disease-associated mutations in KCNE potassium channel subunits (MiRPs) reveal promiscuous disruption of multiple currents and conservation of mechanism. FASEB J 16(3):390–400. https://doi.org/10.1096/fj.01-0520hyp

Abbott GW, Sesti F, Splawski I, Buck ME, Lehmann MH, Timothy KW, Keating MT, Goldstein SA (1999) MiRP1 forms IKr potassium channels with HERG and is associated with cardiac arrhythmia. Cell 97(2):175–187

Abbott GW, Butler MH, Bendahhou S, Dalakas MC, Ptacek LJ, Goldstein SA (2001) MiRP2 forms potassium channels in skeletal muscle with Kv3.4 and is associated with periodic paralysis. Cell 104(2):217–231

Abbott GW, Butler MH, Goldstein SA (2006) Phosphorylation and protonation of neighboring MiRP2 sites: function and pathophysiology of MiRP2-Kv3.4 potassium channels in periodic paralysis. FASEB J 20(2):293–301. https://doi.org/10.1096/fj.05-5070com

Abbott GW, Tai KK, Neverisky DL, Hansler A, Hu Z, Roepke TK, Lerner DJ, Chen Q, Liu L, Zupan B, Toth M, Haynes R, Huang X, Demirbas D, Buccafusca R, Gross SS, Kanda VA, Berry GT (2014) KCNQ1, KCNE2, and Na+-coupled solute transporters form reciprocally regulating complexes that affect neuronal excitability. Sci Signal 7 (315):ra22. https://doi.org/10.1126/scisignal.2005025

Alzamora R, O'Mahony F, Bustos V, Rapetti-Mauss R, Urbach V, Cid LP, Sepulveda FV, Harvey BJ (2011) Sexual dimorphism and oestrogen regulation of KCNE3 expression modulates the functional properties of KCNQ1 K(+) channels. J Physiol 589(Pt 21):5091–5107. https://doi.org/10.1113/jphysiol.2011.215772

Anantharam A, Lewis A, Panaghie G, Gordon E, McCrossan ZA, Lerner DJ, Abbott GW (2003) RNA interference reveals that endogenous Xenopus MinK-related peptides govern mammalian

K+ channel function in oocyte expression studies. J Biol Chem 278(14):11739–11745. https://doi.org/10.1074/jbc.M212751200

Andersen MN, Olesen SP, Rasmussen HB (2011) Kv7.1 surface expression is regulated by epithelial cell polarization. Am J Physiol Cell Physiol 300(4):C814–C824. https://doi.org/10.1152/ajpcell.00390.2010

Andersen MN, Krzystanek K, Jespersen T, Olesen SP, Rasmussen HB (2012) AMP-activated protein kinase downregulates Kv7.1 cell surface expression. Traffic 13(1):143–156. https://doi.org/10.1111/j.1600-0854.2011.01295.x

Angelo K, Jespersen T, Grunnet M, Nielsen MS, Klaerke DA, Olesen SP (2002) KCNE5 induces time- and voltage-dependent modulation of the KCNQ1 current. Biophys J 83(4):1997–2006. https://doi.org/10.1016/S0006-3495(02)73961-1

Barhanin J, Lesage F, Guillemare E, Fink M, Lazdunski M, Romey G (1996) K(V)LQT1 and lsK (minK) proteins associate to form the I(Ks) cardiac potassium current. Nature 384(6604):78–80. https://doi.org/10.1038/384078a0

Barriere H, Rubera I, Belfodil R, Tauc M, Tonnerieux N, Poujeol C, Barhanin J, Poujeol P (2003) Swelling-activated chloride and potassium conductance in primary cultures of mouse proximal tubules. Implication of KCNE1 protein. J Membr Biol 193(3):153–170. https://doi.org/10.1007/s00232-003-2014-z

Bendahhou S, Marionneau C, Haurogne K, Larroque MM, Derand R, Szuts V, Escande D, Demolombe S, Barhanin J (2005) In vitro molecular interactions and distribution of KCNE family with KCNQ1 in the human heart. Cardiovasc Res 67(3):529–538. https://doi.org/10.1016/j.cardiores.2005.02.014

Bezanilla F, Perozo E (2003) The voltage sensor and the gate in ion channels. Adv Protein Chem 63:211–241

Bianchi L, Shen Z, Dennis AT, Priori SG, Napolitano C, Ronchetti E, Bryskin R, Schwartz PJ, Brown AM (1999) Cellular dysfunction of LQT5-minK mutants: abnormalities of IKs, IKr and trafficking in long QT syndrome. Hum Mol Genet 8(8):1499–1507

Biervert C, Schroeder BC, Kubisch C, Berkovic SF, Propping P, Jentsch TJ, Steinlein OK (1998) A potassium channel mutation in neonatal human epilepsy. Science 279(5349):403–406

Bleich M, Warth R (2000) The very small-conductance K+ channel KvLQT1 and epithelial function. Pflugers Archiv 440(2):202–206

Blumenthal EM, Kaczmarek LK (1994) The minK potassium channel exists in functional and nonfunctional forms when expressed in the plasma membrane of Xenopus oocytes. J Neurosci 14(5 Pt 2):3097–3105

Boini KM, Graf D, Hennige AM, Koka S, Kempe DS, Wang K, Ackermann TF, Foller M, Vallon V, Pfeifer K, Schleicher E, Ullrich S, Haring HU, Haussinger D, Lang F (2009) Enhanced insulin sensitivity of gene-targeted mice lacking functional KCNQ1. Am J Physiol Regulat Integr Comp Physiol 296(6):R1695–R1701. https://doi.org/10.1152/ajpregu.90839.2008

Boucherot A, Schreiber R, Kunzelmann K (2001) Regulation and properties of KCNQ1 (K(V)LQT1) and impact of the cystic fibrosis transmembrane conductance regulator. J Membr Biol 182(1):39–47

Busch AE, Lang F (1993) Effects of [Ca2+]i and temperature on minK channels expressed in Xenopus oocytes. FEBS Lett 334(2):221–224

Casimiro MC, Knollmann BC, Ebert SN, Vary JC Jr, Greene AE, Franz MR, Grinberg A, Huang SP, Pfeifer K (2001) Targeted disruption of the Kcnq1 gene produces a mouse model of Jervell and Lange-Nielsen Syndrome. Proc Natl Acad Sci U S A 98(5):2526–2531. https://doi.org/10.1073/pnas.041398998

Catterall WA (1995) Structure and function of voltage-gated ion channels. Annu Rev Biochem 64:493–531. https://doi.org/10.1146/annurev.bi.64.070195.002425

Chen H, Kim LA, Rajan S, Xu S, Goldstein SA (2003) Charybdotoxin binding in the I(Ks) pore demonstrates two MinK subunits in each channel complex. Neuron 40(1):15–23

Choi E, Abbott GW (2010) A shared mechanism for lipid- and beta-subunit-coordinated stabilization of the activated K+ channel voltage sensor. FASEB J 24(5):1518–1524. https://doi.org/10.1096/fj.09-145219

Cowley EA, Linsdell P (2002) Characterization of basolateral K+ channels underlying anion secretion in the human airway cell line Calu-3. J Physiol 538(Pt 3):747–757

Crump SM, Hu Z, Kant R, Levy DI, Goldstein SA, Abbott GW (2016) Kcne4 deletion sex- and age-specifically impairs cardiac repolarization in mice. FASEB J 30(1):360–369. https://doi.org/10.1096/fj.15-278754

Cuello LG, Cortes DM, Perozo E (2004) Molecular architecture of the KvAP voltage-dependent K+ channel in a lipid bilayer. Science 306(5695):491–495. https://doi.org/10.1126/science.1101373

David JP, Andersen MN, Olesen SP, Rasmussen HB, Schmitt N (2013) Trafficking of the IKs -complex in MDCK cells: site of subunit assembly and determinants of polarized localization. Traffic 14(4):399–411. https://doi.org/10.1111/tra.12042

David JP, Lisewski U, Crump SM, Jepps TA, Bocksteins E, Wilck N, Lossie J, Roepke TK, Schmitt N, Abbott GW (2019) Deletion in mice of X-linked, Brugada syndrome- and atrial fibrillation-associated Kcne5 augments ventricular KV currents and predisposes to ventricular arrhythmia. FASEB J 33(2):2537–2552. https://doi.org/10.1096/fj.201800502R

DeCaen PG, Yarov-Yarovoy V, Sharp EM, Scheuer T, Catterall WA (2009) Sequential formation of ion pairs during activation of a sodium channel voltage sensor. Proc Natl Acad Sci USA 106 (52):22498–22503. https://doi.org/10.1073/pnas.0912307106

Dedek K, Waldegger S (2001) Colocalization of KCNQ1/KCNE channel subunits in the mouse gastrointestinal tract. Pflugers Arch 442(6):896–902

De Silva AM, Manville RW, Abbott GW (2018) Deconstruction of an African folk medicine uncovers a novel molecular strategy for therapeutic potassium channel activation. Sci Adv 4 (11):eaav0824. https://doi.org/10.1126/sciadv.aav0824

Doyle DA, Morais Cabral J, Pfuetzner RA, Kuo A, Gulbis JM, Cohen SL, Chait BT, MacKinnon R (1998) The structure of the potassium channel: molecular basis of K+ conduction and selectivity. Science 280(5360):69–77

Embark HM, Bohmer C, Vallon V, Luft F, Lang F (2003) Regulation of KCNE1-dependent K(+) current by the serum and glucocorticoid-inducible kinase (SGK) isoforms. Pflugers Arch 445 (5):601–606. https://doi.org/10.1007/s00424-002-0982-y

Faham S, Watanabe A, Besserer GM, Cascio D, Specht A, Hirayama BA, Wright EM, Abramson J (2008) The crystal structure of a sodium galactose transporter reveals mechanistic insights into Na+/sugar symport. Science 321(5890):810–814. https://doi.org/10.1126/science.1160406

Frohlich H, Boini KM, Seebohm G, Strutz-Seebohm N, Ureche ON, Foller M, Eichenmuller M, Shumilina E, Pathare G, Singh AK, Seidler U, Pfeifer KE, Lang F (2011) Hypothyroidism of gene-targeted mice lacking Kcnq1. Pflugers Arch 461(1):45–52. https://doi.org/10.1007/s00424-010-0890-5

Gao Z, Xiong Q, Sun H, Li M (2008) Desensitization of chemical activation by auxiliary subunits: convergence of molecular determinants critical for augmenting KCNQ1 potassium channels. J Biol Chem 283(33):22649–22658. https://doi.org/10.1074/jbc.M802426200

Gordon E, Roepke TK, Abbott GW (2006) Endogenous KCNE subunits govern Kv2.1 K+ channel activation kinetics in Xenopus oocyte studies. Biophys J 90(4):1223–1231. https://doi.org/10.1529/biophysj.105.072504

Grahammer F, Herling AW, Lang HJ, Schmitt-Graff A, Wittekindt OH, Nitschke R, Bleich M, Barhanin J, Warth R (2001a) The cardiac K+ channel KCNQ1 is essential for gastric acid secretion. Gastroenterology 120(6):1363–1371

Grahammer F, Warth R, Barhanin J, Bleich M, Hug MJ (2001b) The small conductance K+ channel, KCNQ1: expression, function, and subunit composition in murine trachea. J Biol Chem 276(45):42268–42275. https://doi.org/10.1074/jbc.M105014200

Greenwood IA, Yeung SY, Hettiarachi S, Andersson M, Baines DL (2009a) KCNQ-encoded channels regulate Na+ transport across H441 lung epithelial cells. Pflugers Arch 457 (4):785–794. https://doi.org/10.1007/s00424-008-0557-7

Greenwood IA, Yeung SY, Tribe RM, Ohya S (2009b) Loss of functional K+ channels encoded by ether-a-go-go-related genes in mouse myometrium prior to labour onset. J Physiol 587 (Pt 10):2313–2326. https://doi.org/10.1113/jphysiol.2009.171272

Grunnet M, Jespersen T, Rasmussen HB, Ljungstrom T, Jorgensen NK, Olesen SP, Klaerke DA (2002) KCNE4 is an inhibitory subunit to the KCNQ1 channel. J Physiol 542(Pt 1):119–130

Guo W, Shimada S, Tajiri H, Yamauchi A, Yamashita T, Okada S, Tohyama M (1997) Developmental regulation of Na+ / myo-inositol cotransporter gene expression. Brain Res Mol Brain Res 51(1–2):91–96

Hammami S, Willumsen NJ, Olsen HL, Morera FJ, Latorre R, Klaerke DA (2009) Cell volume and membrane stretch independently control K+ channel activity. J Physiol 587(Pt 10):2225–2231. https://doi.org/10.1113/jphysiol.2008.163550

Heitzmann D, Grahammer F, von Hahn T, Schmitt-Graff A, Romeo E, Nitschke R, Gerlach U, Lang HJ, Verrey F, Barhanin J, Warth R (2004) Heteromeric KCNE2/KCNQ1 potassium channels in the luminal membrane of gastric parietal cells. J Physiol 561(Pt 2):547–557. https://doi.org/10.1113/jphysiol.2004.075168

Heitzmann D, Koren V, Wagner M, Sterner C, Reichold M, Tegtmeier I, Volk T, Warth R (2007) KCNE beta subunits determine pH sensitivity of KCNQ1 potassium channels. Cellular Physiol Biochem 19(1–4):21–32. https://doi.org/10.1159/000099189

Hille B, Armstrong CM, MacKinnon R (1999) Ion channels: from idea to reality. Nat Med 5 (10):1105–1109. https://doi.org/10.1038/13415

Hu Z, Kant R, Anand M, King EC, Krogh-Madsen T, Christini DJ, Abbott GW (2014a) Kcne2 deletion creates a multisystem syndrome predisposing to sudden cardiac death. Circ Cardiovasc Genet 7(1):33–42. https://doi.org/10.1161/CIRCGENETICS.113.000315

Hu Z, Kant R, Anand M, King EC, Krogh-Madsen T, Christini DJ, Abbott GW (2014b) Kcne2 deletion creates a multisystem syndrome predisposing to sudden cardiac death. Circ Cardiovasc Genet. https://doi.org/10.1161/CIRCGENETICS.113.000315

Jackman WM, Clark M, Friday KJ, Aliot EM, Anderson J, Lazzara R (1984) Ventricular tachyarrhythmias in the long QT syndromes. Med Clin North Am 68(5):1079–1109

Jackman WM, Friday KJ, Anderson JL, Aliot EM, Clark M, Lazzara R (1988) The long QT syndromes: a critical review, new clinical observations and a unifying hypothesis. Prog Cardiovasc Dis 31(2):115–172

Jespersen T, Rasmussen HB, Grunnet M, Jensen HS, Angelo K, Dupuis DS, Vogel LK, Jorgensen NK, Klaerke DA, Olesen SP (2004) Basolateral localisation of KCNQ1 potassium channels in MDCK cells: molecular identification of an N-terminal targeting motif. J Cell Sci 117 (Pt 19):4517–4526. https://doi.org/10.1242/jcs.01318

Jiang M, Xu X, Wang Y, Toyoda F, Liu XS, Zhang M, Robinson RB, Tseng GN (2009) Dynamic partnership between KCNQ1 and KCNE1 and influence on cardiac IKs current amplitude by KCNE2. J Biol Chem 284(24):16452–16462. https://doi.org/10.1074/jbc.M808262200

Kanda VA, Lewis A, Xu X, Abbott GW (2011a) KCNE1 and KCNE2 inhibit forward trafficking of homomeric N-type voltage-gated potassium channels. Biophys J 101(6):1354–1363

Kanda VA, Lewis A, Xu X, Abbott GW (2011b) KCNE1 and KCNE2 provide a checkpoint governing voltage-gated potassium channel alpha-subunit composition. Biophys J 101 (6):1364–1375. https://doi.org/10.1016/j.bpj.2011.08.014

Kanda VA, Purtell K, Abbott GW (2011c) Protein kinase C downregulates I(Ks) by stimulating KCNQ1-KCNE1 potassium channel endocytosis. Heart Rhythm 8(10):1641–1647. https://doi.org/10.1016/j.hrthm.2011.04.034

Kim JK, Yoo HY, Kim SJ, Hwang YS, Han J, Kim JA, Kim CS, Cho HS (2007) Effects of sevoflurane on the cAMP-induced short-circuit current in mouse tracheal epithelium and recombinant Cl- (CFTR) and K+ (KCNQ1) channels. Br J Anaesth 99(2):245–251. https://doi.org/10.1093/bja/aem123

Klinger F, Gould G, Boehm S, Shapiro MS (2011) Distribution of M-channel subunits KCNQ2 and KCNQ3 in rat hippocampus. NeuroImage 58(3):761–769. https://doi.org/10.1016/j.neuroimage.2011.07.003

Kreda SM, Davis CW, Rose MC (2012) CFTR, mucins, and mucus obstruction in cystic fibrosis. Cold Spring Harb Perspect Med 2(9):a009589. https://doi.org/10.1101/cshperspect.a009589

Kroncke BM, Van Horn WD, Smith J, Kang C, Welch RC, Song Y, Nannemann DP, Taylor KC, Sisco NJ, George AL Jr, Meiler J, Vanoye CG, Sanders CR (2016) Structural basis for KCNE3 modulation of potassium recycling in epithelia. Sci Adv 2(9):e1501228. https://doi.org/10.1126/sciadv.1501228

Kubisch C, Schroeder BC, Friedrich T, Lutjohann B, El-Amraoui A, Marlin S, Petit C, Jentsch TJ (1999) KCNQ4, a novel potassium channel expressed in sensory outer hair cells, is mutated in dominant deafness. Cell 96(3):437–446

Kuwahara N, Kitazawa R, Fujiishi K, Nagai Y, Haraguchi R, Kitazawa S (2013) Gastric adenocarcinoma arising in gastritis cystica profunda presenting with selective loss of KCNE2 expression. World J Gastroenterol 19(8):1314–1317. https://doi.org/10.3748/wjg.v19.i8.1314

Kuwamura M, Okajima R, Yamate J, Kotani T, Kuramoto T, Serikawa T (2008) Pancreatic metaplasia in the gastro-achlorhydria in WTC-dfk rat, a potassium channel Kcnq1 mutant. Vet Pathol 45(4):586–591. https://doi.org/10.1354/vp.45-4-586

Larsen AP, Steffensen AB, Grunnet M, Olesen SP (2011) Extracellular potassium inhibits Kv7.1 potassium channels by stabilizing an inactivated state. Biophys J 101(4):818–827. https://doi.org/10.1016/j.bpj.2011.06.034

Lee MP, Ravenel JD, Hu RJ, Lustig LR, Tomaselli G, Berger RD, Brandenburg SA, Litzi TJ, Bunton TE, Limb C, Francis H, Gorelikow M, Gu H, Washington K, Argani P, Goldenring JR, Coffey RJ, Feinberg AP (2000) Targeted disruption of the Kvlqt1 gene causes deafness and gastric hyperplasia in mice. J Clin Invest 106(12):1447–1455. https://doi.org/10.1172/JCI10897

Lee YH, Kang ES, Kim SH, Han SJ, Kim CH, Kim HJ, Ahn CW, Cha BS, Nam M, Nam CM, Lee HC (2008) Association between polymorphisms in SLC30A8, HHEX, CDKN2A/B, IGF2BP2, FTO, WFS1, CDKAL1, KCNQ1 and type 2 diabetes in the Korean population. J Hum Genet 53(11-12):991–998. https://doi.org/10.1007/s10038-008-0341-8

Lee SM, Baik J, Nguyen D, Nguyen V, Liu S, Hu Z, Abbott GW (2017) Kcne2 deletion impairs insulin secretion and causes type 2 diabetes mellitus. FASEB J 31(6):2674–2685. https://doi.org/10.1096/fj.201601347

Levy DI, Wanderling S, Biemesderfer D, Goldstein SA (2008) MiRP3 acts as an accessory subunit with the BK potassium channel. Am J Physiol Renal Physiol 295(2):F380–F387. https://doi.org/10.1152/ajprenal.00598.2007

Liu Y, Zhou DZ, Zhang D, Chen Z, Zhao T, Zhang Z, Ning M, Hu X, Yang YF, Zhang ZF, Yu L, He L, Xu H (2009) Variants in KCNQ1 are associated with susceptibility to type 2 diabetes in the population of mainland China. Diabetologia 52(7):1315–1321. https://doi.org/10.1007/s00125-009-1375-y

Long SB, Campbell EB, Mackinnon R (2005a) Crystal structure of a mammalian voltage-dependent Shaker family K+ channel. Science 309(5736):897–903. https://doi.org/10.1126/science.1116269

Long SB, Campbell EB, Mackinnon R (2005b) Voltage sensor of Kv1.2: structural basis of electromechanical coupling. Science 309(5736):903–908. https://doi.org/10.1126/science.1116270

MacKinnon R (1991) Determination of the subunit stoichiometry of a voltage-activated potassium channel. Nature 350(6315):232–235. https://doi.org/10.1038/350232a0

Manderfield LJ, George AL Jr (2008) KCNE4 can co-associate with the I(Ks) (KCNQ1-KCNE1) channel complex. FEBS J 275(6):1336–1349. https://doi.org/10.1111/j.1742-4658.2008.06294.x

Manville RW, Neverisky DL, Abbott GW (2017) SMIT1 modifies KCNQ channel function and pharmacology by physical interaction with the pore. Biophys J 113(3):613–626. https://doi.org/10.1016/j.bpj.2017.06.055

Matschke V, Piccini I, Schubert J, Wrobel E, Lang F, Matschke J, Amedonu E, Meuth SG, Strunker T, Strutz-Seebohm N, Greber B, Scherkenbeck J, Seebohm G (2016) The natural plant product rottlerin activates Kv7.1/KCNE1 channels. Cell Physiol Biochem 40 (6):1549–1558. https://doi.org/10.1159/000453205

McCrossan ZA, Abbott GW (2004) The MinK-related peptides. Neuropharmacology 47 (6):787–821. https://doi.org/10.1016/j.neuropharm.2004.06.018

McCrossan ZA, Roepke TK, Lewis A, Panaghie G, Abbott GW (2009) Regulation of the Kv2.1 potassium channel by MinK and MiRP1. J Membr Biol 228(1):1–14. https://doi.org/10.1007/s00232-009-9154-8

McDonald TV, Yu Z, Ming Z, Palma E, Meyers MB, Wang KW, Goldstein SA, Fishman GI (1997) A minK-HERG complex regulates the cardiac potassium current I(Kr). Nature 388 (6639):289–292. https://doi.org/10.1038/40882

Melman YF, Domenech A, de la Luna S, McDonald TV (2001) Structural determinants of KvLQT1 control by the KCNE family of proteins. J Biol Chem 276(9):6439–6444. https://doi.org/10.1074/jbc.M010713200

Melman YF, Krumerman A, McDonald TV (2002) A single transmembrane site in the KCNE-encoded proteins controls the specificity of KvLQT1 channel gating. J Biol Chem 277 (28):25187–25194. https://doi.org/10.1074/jbc.M200564200

Millar ID, Hartley JA, Haigh C, Grace AA, White SJ, Kibble JD, Robson L (2004) Volume regulation is defective in renal proximal tubule cells isolated from KCNE1 knockout mice. Exp Physiol 89(2):173–180. https://doi.org/10.1113/expphysiol.2003.026674

Moser SL, Harron SA, Crack J, Fawcett JP, Cowley EA (2008) Multiple KCNQ potassium channel subtypes mediate basal anion secretion from the human airway epithelial cell line Calu-3. J Membr Biol 221(3):153–163. https://doi.org/10.1007/s00232-008-9093-9

Mussig K, Staiger H, Machicao F, Kirchhoff K, Guthoff M, Schafer SA, Kantartzis K, Silbernagel G, Stefan N, Holst JJ, Gallwitz B, Haring HU, Fritsche A (2009) Association of type 2 diabetes candidate polymorphisms in KCNQ1 with incretin and insulin secretion. Diabetes 58(7):1715–1720. https://doi.org/10.2337/db08-1589

Nakajo K, Kubo Y (2007) KCNE1 and KCNE3 stabilize and/or slow voltage sensing S4 segment of KCNQ1 channel. J Gen Physiol 130(3):269–281. https://doi.org/10.1085/jgp.200709805

Namkung W, Song Y, Mills AD, Padmawar P, Finkbeiner WE, Verkman AS (2009) In situ measurement of airway surface liquid [K+] using a ratioable K+-sensitive fluorescent dye. J Biol Chem 284(23):15916–15926. https://doi.org/10.1074/jbc.M808021200

Neal AM, Taylor HC, Millar ID, Kibble JD, White SJ, Robson L (2011) Renal defects in KCNE1 knockout mice are mimicked by chromanol 293B in vivo: identification of a KCNE1-regulated K+ conductance in the proximal tubule. J Physiol 589(Pt 14):3595–3609. https://doi.org/10.1113/jphysiol.2011.209155

Neverisky DL, Abbott GW (2017) KCNQ-SMIT complex formation facilitates ion channel-solute transporter cross talk. FASEB J 31(7):2828–2838. https://doi.org/10.1096/fj.201601334R

Neyroud N, Tesson F, Denjoy I, Leibovici M, Donger C, Barhanin J, Faure S, Gary F, Coumel P, Petit C, Schwartz K, Guicheney P (1997) A novel mutation in the potassium channel gene KVLQT1 causes the Jervell and Lange-Nielsen cardioauditory syndrome. Nat Genet 15 (2):186–189. https://doi.org/10.1038/ng0297-186

Nguyen N, Kozer-Gorevich N, Gliddon BL, Smolka AJ, Clayton AH, Gleeson PA, van Driel IR (2013) Independent trafficking of the KCNQ1 K+ channel and H+-K+-ATPase in gastric parietal cells from mice. Am J Physiol Gastrointest Liver Physiol 304(2):G157–G166. https://doi.org/10.1152/ajpgi.00346.2012

Noda M, Ikeda T, Suzuki H, Takeshima H, Takahashi T, Kuno M, Numa S (1986) Expression of functional sodium channels from cloned cDNA. Nature 322(6082):826–828. https://doi.org/10.1038/322826a0

Numa S, Noda M (1986) Molecular structure of sodium channels. Ann N Y Acad Sci 479:338–355

Offner FF, Dallos P, Cheatham MA (1987) Positive endocochlear potential: mechanism of production by marginal cells of stria vascularis. Hear Res 29(2–3):117–124

Ohya S, Asakura K, Muraki K, Watanabe M, Imaizumi Y (2002) Molecular and functional characterization of ERG, KCNQ, and KCNE subtypes in rat stomach smooth muscle. Am J Physiol Gastrointest Liver Physiol 282(2):G277–G287. https://doi.org/10.1152/ajpgi.00200.2001

O'Mahony F, Thomas W, Harvey BJ (2009) Novel female sex-dependent actions of oestrogen in the intestine. J Physiol 587(Pt 21):5039–5044. https://doi.org/10.1113/jphysiol.2009.177972

Osteen JD, Gonzalez C, Sampson KJ, Iyer V, Rebolledo S, Larsson HP, Kass RS (2010) KCNE1 alters the voltage sensor movements necessary to open the KCNQ1 channel gate. Proc Natl Acad Sci U S A 107(52):22710–22715. https://doi.org/10.1073/pnas.1016300108

Osteen JD, Barro-Soria R, Robey S, Sampson KJ, Kass RS, Larsson HP (2012) Allosteric gating mechanism underlies the flexible gating of KCNQ1 potassium channels. Proc Natl Acad Sci U S A 109(18):7103–7108. https://doi.org/10.1073/pnas.1201582109

Panaghie G, Abbott GW (2006) The impact of ancillary subunits on small-molecule interactions with voltage-gated potassium channels. Curr Pharm Design 12(18):2285–2302

Panaghie G, Abbott GW (2007) The role of S4 charges in voltage-dependent and voltage-independent KCNQ1 potassium channel complexes. J Gen Physiol 129(2):121–133. https://doi.org/10.1085/jgp.200609612

Panaghie G, Tai KK, Abbott GW (2006) Interaction of KCNE subunits with the KCNQ1 K+ channel pore. J Physiol 570(Pt 3):455–467. https://doi.org/10.1113/jphysiol.2005.100644

Papazian DM, Schwarz TL, Tempel BL, Jan YN, Jan LY (1987) Cloning of genomic and complementary DNA from Shaker, a putative potassium channel gene from Drosophila. Science 237(4816):749–753

Piccini M, Vitelli F, Seri M, Galietta LJ, Moran O, Bulfone A, Banfi S, Pober B, Renieri A (1999) KCNE1-like gene is deleted in AMME contiguous gene syndrome: identification and characterization of the human and mouse homologs. Genomics 60(3):251–257. https://doi.org/10.1006/geno.1999.5904

Plant LD, Xiong D, Dai H, Goldstein SA (2014) Individual IKs channels at the surface of mammalian cells contain two KCNE1 accessory subunits. Proc Natl Acad Sci U S A. https://doi.org/10.1073/pnas.1323548111

Pongs O, Schwarz JR (2010) Ancillary subunits associated with voltage-dependent K+ channels. Physiol Rev 90(2):755–796. https://doi.org/10.1152/physrev.00020.2009

Preston P, Wartosch L, Gunzel D, Fromm M, Kongsuphol P, Ousingsawat J, Kunzelmann K, Barhanin J, Warth R, Jentsch TJ (2010) Disruption of the K+ channel beta-subunit KCNE3 reveals an important role in intestinal and tracheal Cl- transport. J Biol Chem 285(10):7165–7175. https://doi.org/10.1074/jbc.M109.047829

Puchalski RB, Kelly E, Bachmanov AA, Brazier SP, Kuang J, Arrighi I, Barhanin J, Tordoff MG (2001) NaCl consumption is attenuated in female KCNE1 null mutant mice. Physiol Behav 74(3):267–276

Purtell K, Paroder-Belenitsky M, Reyna-Neyra A, Nicola JP, Koba W, Fine E, Carrasco N, Abbott GW (2012) The KCNQ1-KCNE2 K+ channel is required for adequate thyroid I- uptake. FASEB J. https://doi.org/10.1096/fj.12-206110

Qi Q, Li H, Loos RJ, Liu C, Wu Y, Hu FB, Wu H, Lu L, Yu Z, Lin X (2009) Common variants in KCNQ1 are associated with type 2 diabetes and impaired fasting glucose in a Chinese Han population. Hum Mol Genet 18(18):3508–3515. https://doi.org/10.1093/hmg/ddp294

Ranganathan R, Lewis JH, MacKinnon R (1996) Spatial localization of the K+ channel selectivity filter by mutant cycle-based structure analysis. Neuron 16(1):131–139

Rapetti-Mauss R, O'Mahony F, Sepulveda FV, Urbach V, Harvey BJ (2013) Oestrogen promotes KCNQ1 potassium channel endocytosis and postendocytic trafficking in colonic epithelium. J Physiol 591(Pt 11):2813–2831. https://doi.org/10.1113/jphysiol.2013.251678

Rice KS, Dickson G, Lane M, Crawford J, Chung SK, Rees MI, Shelling AN, Love DR, Skinner JR (2011) Elevated serum gastrin levels in Jervell and Lange-Nielsen syndrome: a marker of severe KCNQ1 dysfunction? Heart Rhythm 8(4):551–554. https://doi.org/10.1016/j.hrthm.2010.11.039

Riedel C, Dohan O, De la Vieja A, Ginter CS, Carrasco N (2001) Journey of the iodide transporter NIS: from its molecular identification to its clinical role in cancer. Trends Biochem Sci 26 (8):490–496

Rivas A, Francis HW (2005) Inner ear abnormalities in a Kcnq1 (Kvlqt1) knockout mouse: a model of Jervell and Lange-Nielsen syndrome. Otol Neurotol 26(3):415–424

Rocheleau JM, Gage SD, Kobertz WR (2006) Secondary structure of a KCNE cytoplasmic domain. J Gen Physiol 128(6):721–729. https://doi.org/10.1085/jgp.200609657

Roepke TK, Anantharam A, Kirchhoff P, Busque SM, Young JB, Geibel JP, Lerner DJ, Abbott GW (2006) The KCNE2 potassium channel ancillary subunit is essential for gastric acid secretion. J Biol Chem 281(33):23740–23747. https://doi.org/10.1074/jbc.M604155200

Roepke TK, King EC, Reyna-Neyra A, Paroder M, Purtell K, Koba W, Fine E, Lerner DJ, Carrasco N, Abbott GW (2009) Kcne2 deletion uncovers its crucial role in thyroid hormone biosynthesis. Nat Med 15(10):1186–1194. https://doi.org/10.1038/nm.2029

Roepke TK, Purtell K, King EC, La Perle KM, Lerner DJ, Abbott GW (2010) Targeted deletion of Kcne2 causes gastritis cystica profunda and gastric neoplasia. PLoS One 5(7):e11451. https://doi.org/10.1371/journal.pone.0011451

Roepke TK, Kanda VA, Purtell K, King EC, Lerner DJ, Abbott GW (2011a) KCNE2 forms potassium channels with KCNA3 and KCNQ1 in the choroid plexus epithelium. FASEB J 25 (12):4264–4273

Roepke TK, King EC, Purtell K, Kanda VA, Lerner DJ, Abbott GW (2011b) Genetic dissection reveals unexpected influence of beta subunits on KCNQ1 K+ channel polarized trafficking in vivo. FASEB J 25(2):727–736. https://doi.org/10.1096/fj.10-173682

Rose AS, Hildebrand PW (2015) NGL Viewer: a web application for molecular visualization. Nucleic Acids Res 43(W1):W576–W579. https://doi.org/10.1093/nar/gkv402

Rose PW, Prlic A, Altunkaya A, Bi C, Bradley AR, Christie CH, Costanzo LD, Duarte JM, Dutta S, Feng Z, Green RK, Goodsell DS, Hudson B, Kalro T, Lowe R, Peisach E, Randle C, Rose AS, Shao C, Tao YP, Valasatava Y, Voigt M, Westbrook JD, Woo J, Yang H, Young JY, Zardecki C, Berman HM, Burley SK (2017) The RCSB protein data bank: integrative view of protein, gene and 3D structural information. Nucleic Acids Res 45(D1):D271–D281. https://doi.org/10.1093/nar/gkw1000

Rosengren AH, Braun M, Mahdi T, Andersson SA, Travers ME, Shigeto M, Zhang E, Almgren P, Ladenvall C, Axelsson AS, Edlund A, Pedersen MG, Jonsson A, Ramracheya R, Tang Y, Walker JN, Barrett A, Johnson PR, Lyssenko V, McCarthy MI, Groop L, Salehi A, Gloyn AL, Renstrom E, Rorsman P, Eliasson L (2012) Reduced insulin exocytosis in human pancreatic beta-cells with gene variants linked to type 2 diabetes. Diabetes 61(7):1726–1733. https://doi.org/10.2337/db11-1516

Roura-Ferrer M, Sole L, Oliveras A, Dahan R, Bielanska J, Villarroel A, Comes N, Felipe A (2010) Impact of KCNE subunits on KCNQ1 (Kv7.1) channel membrane surface targeting. J Cell Physiol 225(3):692–700. https://doi.org/10.1002/jcp.22265

Ruscic KJ, Miceli F, Villalba-Galea CA, Dai H, Mishina Y, Bezanilla F, Goldstein SA (2013) IKs channels open slowly because KCNE1 accessory subunits slow the movement of S4 voltage sensors in KCNQ1 pore-forming subunits. Proc Natl Acad Sci U S A 110(7):E559–E566. https://doi.org/10.1073/pnas.1222616110

Sanguinetti MC, Curran ME, Zou A, Shen J, Spector PS, Atkinson DL, Keating MT (1996) Coassembly of K(V)LQT1 and minK (IsK) proteins to form cardiac I(Ks) potassium channel. Nature 384(6604):80–83. https://doi.org/10.1038/384080a0

Schreiber R, Murle B, Sun J, Kunzelmann K (2002) Electrolyte transport in the mouse trachea: no evidence for a contribution of luminal K(+) conductance. J Membr Biol 189(2):143–151. https://doi.org/10.1007/s00232-002-1009-5

Schroeder BC, Waldegger S, Fehr S, Bleich M, Warth R, Greger R, Jentsch TJ (2000) A constitutively open potassium channel formed by KCNQ1 and KCNE3. Nature 403 (6766):196–199. https://doi.org/10.1038/35003200

Schulze-Bahr E, Wang Q, Wedekind H, Haverkamp W, Chen Q, Sun Y, Rubie C, Hordt M, Towbin JA, Borggrefe M, Assmann G, Qu X, Somberg JC, Breithardt G, Oberti C, Funke H (1997) KCNE1 mutations cause jervell and Lange-Nielsen syndrome. Nat Genet 17(3):267–268. https://doi.org/10.1038/ng1197-267

Seebohm G, Strutz-Seebohm N, Birkin R, Dell G, Bucci C, Spinosa MR, Baltaev R, Mack AF, Korniychuk G, Choudhury A, Marks D, Pagano RE, Attali B, Pfeufer A, Kass RS, Sanguinetti MC, Tavare JM, Lang F (2007) Regulation of endocytic recycling of KCNQ1/KCNE1 potassium channels. Circ Res 100(5):686–692. https://doi.org/10.1161/01.RES.0000260250.83824.8f

Seebohm G, Strutz-Seebohm N, Ureche ON, Henrion U, Baltaev R, Mack AF, Korniychuk G, Steinke K, Tapken D, Pfeufer A, Kaab S, Bucci C, Attali B, Merot J, Tavare JM, Hoppe UC, Sanguinetti MC, Lang F (2008) Long QT syndrome-associated mutations in KCNQ1 and KCNE1 subunits disrupt normal endosomal recycling of IKs channels. Circ Res 103 (12):1451–1457. https://doi.org/10.1161/CIRCRESAHA.108.177360

Sesti F, Goldstein SA (1998) Single-channel characteristics of wild-type IKs channels and channels formed with two minK mutants that cause long QT syndrome. J Gen Physiol 112(6):651–663

Singh NA, Charlier C, Stauffer D, DuPont BR, Leach RJ, Melis R, Ronen GM, Bjerre I, Quattlebaum T, Murphy JV, McHarg ML, Gagnon D, Rosales TO, Peiffer A, Anderson VE, Leppert M (1998) A novel potassium channel gene, KCNQ2, is mutated in an inherited epilepsy of newborns. Nat Genet 18(1):25–29. https://doi.org/10.1038/ng0198-25

Soler Artigas M, Loth DW, Wain LV, Gharib SA, Obeidat M, Tang W, Zhai G, Zhao JH, Smith AV, Huffman JE, Albrecht E, Jackson CM, Evans DM, Cadby G, Fornage M, Manichaikul A, Lopez LM, Johnson T, Aldrich MC, Aspelund T, Barroso I, Campbell H, Cassano PA, Couper DJ, Eiriksdottir G, Franceschini N, Garcia M, Gieger C, Gislason GK, Grkovic I, Hammond CJ, Hancock DB, Harris TB, Ramasamy A, Heckbert SR, Heliovaara M, Homuth G, Hysi PG, James AL, Jankovic S, Joubert BR, Karrasch S, Klopp N, Koch B, Kritchevsky SB, Launer LJ, Liu Y, Loehr LR, Lohman K, Loos RJ, Lumley T, Al Balushi KA, Ang WQ, Barr RG, Beilby J, Blakey JD, Boban M, Boraska V, Brisman J, Britton JR, Brusselle GG, Cooper C, Curjuric I, Dahgam S, Deary IJ, Ebrahim S, Eijgelsheim M, Francks C, Gaysina D, Granell R, Gu X, Hankinson JL, Hardy R, Harris SE, Henderson J, Henry A, Hingorani AD, Hofman A, Holt PG, Hui J, Hunter ML, Imboden M, Jameson KA, Kerr SM, Kolcic I, Kronenberg F, Liu JZ, Marchini J, McKeever T, Morris AD, Olin AC, Porteous DJ, Postma DS, Rich SS, Ring SM, Rivadeneira F, Rochat T, Sayer AA, Sayers I, Sly PD, Smith GD, Sood A, Starr JM, Uitterlinden AG, Vonk JM, Wannamethee SG, Whincup PH, Wijmenga C, Williams OD, Wong A, Mangino M, Marciante KD, McArdle WL, Meibohm B, Morrison AC, North KE, Omenaas E, Palmer LJ, Pietilainen KH, Pin I, Pola Sbreve Ek O, Pouta A, Psaty BM, Hartikainen AL, Rantanen T, Ripatti S, Rotter JI, Rudan I, Rudnicka AR, Schulz H, Shin SY, Spector TD, Surakka I, Vitart V, Volzke H, Wareham NJ, Warrington NM, Wichmann HE, Wild SH, Wilk JB, Wjst M, Wright AF, Zgaga L, Zemunik T, Pennell CE, Nyberg F, Kuh D, Holloway JW, Boezen HM, Lawlor DA, Morris RW, Probst-Hensch N, Kaprio J, Wilson JF, Hayward C, Kahonen M, Heinrich J, Musk AW, Jarvis DL, Glaser S, Jarvelin MR, Ch Stricker BH, Elliott P, O'Connor GT, Strachan DP, London SJ, Hall IP, Gudnason V, Tobin MD (2011) Genome-wide association and large-scale follow up identifies 16 new loci influencing lung function. Nat Genet 43(11):1082–1090. https://doi.org/10.1038/ng.941

Spector R, Lorenzo AV (1975) Myo-inositol transport in the central nervous system. Am J Physiol 228(5):1510–1518

Splawski I, Timothy KW, Vincent GM, Atkinson DL, Keating MT (1997a) Molecular basis of the long-QT syndrome associated with deafness. N Engl J Med 336(22):1562–1567. https://doi.org/10.1056/NEJM199705293362204

Splawski I, Tristani-Firouzi M, Lehmann MH, Sanguinetti MC, Keating MT (1997b) Mutations in the hminK gene cause long QT syndrome and suppress IKs function. Nat Genet 17(3):338–340. https://doi.org/10.1038/ng1197-338

Strutz-Seebohm N, Henrion U, Steinke K, Tapken D, Lang F, Seebohm G (2009) Serum- and glucocorticoid-inducible kinases (SGK) regulate KCNQ1/KCNE potassium channels. Channels (Austin) 3(2):88–90

Strutz-Seebohm N, Pusch M, Wolf S, Stoll R, Tapken D, Gerwert K, Attali B, Seebohm G (2011) Structural basis of slow activation gating in the cardiac I Ks channel complex. Cell Physiol Biochem 27(5):443–452. https://doi.org/10.1159/000329965

Suarez L, Bilal U, Bordallo J, Cantabrana B, Sanchez M (2015) Androgens block outward potassium currents and decrease spontaneous action potentials in GH3 cells. Naunyn-Schmiedebergs Arch Pharmacol 388(1):67–78. https://doi.org/10.1007/s00210-014-1057-2

Sun J, MacKinnon R (2017) Cryo-EM structure of a KCNQ1/CaM complex reveals insights into congenital long QT syndrome. Cell 169(6):1042–1050.e1049. https://doi.org/10.1016/j.cell.2017.05.019

Takumi T, Ohkubo H, Nakanishi S (1988) Cloning of a membrane protein that induces a slow voltage-gated potassium current. Science 242(4881):1042–1045

Takumi T, Moriyoshi K, Aramori I, Ishii T, Oiki S, Okada Y, Ohkubo H, Nakanishi S (1991) Alteration of channel activities and gating by mutations of slow ISK potassium channel. J Biol Chem 266(33):22192–22198

Tempel BL, Papazian DM, Schwarz TL, Jan YN, Jan LY (1987) Sequence of a probable potassium channel component encoded at Shaker locus of Drosophila. Science 237(4816):770–775

Tinel N, Diochot S, Borsotto M, Lazdunski M, Barhanin J (2000) KCNE2 confers background current characteristics to the cardiac KCNQ1 potassium channel. EMBO J 19(23):6326–6330. https://doi.org/10.1093/emboj/19.23.6326

Tranebjaerg L, Samson RA, Green GE (1993) Jervell and Lange-Nielsen syndrome. In: Pagon RA, Adam MP, Ardinger HH et al (eds) GeneReviews(R). Seattle, WA

Travers ME, Mackay DJ, Dekker Nitert M, Morris AP, Lindgren CM, Berry A, Johnson PR, Hanley N, Groop LC, McCarthy MI, Gloyn AL (2013) Insights into the molecular mechanism for type 2 diabetes susceptibility at the KCNQ1 locus from temporal changes in imprinting status in human islets. Diabetes 62(3):987–992. https://doi.org/10.2337/db12-0819

Tristani-Firouzi M, Sanguinetti MC (1998) Voltage-dependent inactivation of the human K+ channel KvLQT1 is eliminated by association with minimal K+ channel (minK) subunits. J Physiol 510(Pt 1):37–45

Tzingounis AV, Heidenreich M, Kharkovets T, Spitzmaul G, Jensen HS, Nicoll RA, Jentsch TJ (2010) The KCNQ5 potassium channel mediates a component of the after hyperpolarization current in mouse hippocampus. Proc Natl Acad Sci USA 107(22):10232–10237. https://doi.org/10.1073/pnas.1004644107

Ullrich S, Su J, Ranta F, Wittekindt OH, Ris F, Rosler M, Gerlach U, Heitzmann D, Warth R, Lang F (2005) Effects of I(Ks) channel inhibitors in insulin-secreting INS-1 cells. Pflugers Arch 451(3):428–436. https://doi.org/10.1007/s00424-005-1479-2

Unoki H, Takahashi A, Kawaguchi T, Hara K, Horikoshi M, Andersen G, Ng DP, Holmkvist J, Borch-Johnsen K, Jorgensen T, Sandbaek A, Lauritzen T, Hansen T, Nurbaya S, Tsunoda T, Kubo M, Babazono T, Hirose H, Hayashi M, Iwamoto Y, Kashiwagi A, Kaku K, Kawamori R, Tai ES, Pedersen O, Kamatani N, Kadowaki T, Kikkawa R, Nakamura Y, Maeda S (2008) SNPs in KCNQ1 are associated with susceptibility to type 2 diabetes in East Asian and European populations. Nat Genet 40(9):1098–1102. https://doi.org/10.1038/ng.208

Vallon V, Grahammer F, Richter K, Bleich M, Lang F, Barhanin J, Volkl H, Warth R (2001) Role of KCNE1-dependent K+ fluxes in mouse proximal tubule. J Am Soc Nephrol 12(10):2003–2011

Vallon V, Grahammer F, Volkl H, Sandu CD, Richter K, Rexhepaj R, Gerlach U, Rong Q, Pfeifer K, Lang F (2005) KCNQ1-dependent transport in renal and gastrointestinal epithelia. Proc Natl Acad Sci U S A 102(49):17864–17869. https://doi.org/10.1073/pnas.0505860102

vanTol BL, Missan S, Crack J, Moser S, Baldridge WH, Linsdell P, Cowley EA (2007) Contribution of KCNQ1 to the regulatory volume decrease in the human mammary epithelial cell line

MCF-7. Am J Physiol Cell Physiol 293(3):C1010–C1019. https://doi.org/10.1152/ajpcell.00071.2007

Wang KW, Goldstein SA (1995) Subunit composition of minK potassium channels. Neuron 14 (6):1303–1309

Wang Q, Curran ME, Splawski I, Burn TC, Millholland JM, VanRaay TJ, Shen J, Timothy KW, Vincent GM, de Jager T, Schwartz PJ, Toubin JA, Moss AJ, Atkinson DL, Landes GM, Connors TD, Keating MT (1996) Positional cloning of a novel potassium channel gene: KVLQT1 mutations cause cardiac arrhythmias. Nat Genet 12(1):17–23. https://doi.org/10.1038/ng0196-17

Wang HS, Pan Z, Shi W, Brown BS, Wymore RS, Cohen IS, Dixon JE, McKinnon D (1998) KCNQ2 and KCNQ3 potassium channel subunits: molecular correlates of the M-channel. Science 282(5395):1890–1893

Wangemann P (2002) K+ cycling and the endocochlear potential. Hear Res 165(1–2):1–9

Winbo A, Sandstrom O, Palmqvist R, Rydberg A (2013) Iron-deficiency anaemia, gastric hyperplasia, and elevated gastrin levels due to potassium channel dysfunction in the Jervell and Lange-Nielsen syndrome. Cardiol Young 23(3):325–334. https://doi.org/10.1017/S1047951112001060

Woda CB, Bragin A, Kleyman TR, Satlin LM (2001) Flow-dependent K+ secretion in the cortical collecting duct is mediated by a maxi-K channel. Am J Physiol Renal Physiol 280(5):F786–F793

Wollnik B, Schroeder BC, Kubisch C, Esperer HD, Wieacker P, Jentsch TJ (1997) Pathophysiological mechanisms of dominant and recessive KVLQT1 K+ channel mutations found in inherited cardiac arrhythmias. Hum Mol Genet 6(11):1943–1949

Wrobel E, Rothenberg I, Krisp C, Hundt F, Fraenzel B, Eckey K, Linders JT, Gallacher DJ, Towart R, Pott L, Pusch M, Yang T, Roden DM, Kurata HT, Schulze-Bahr E, Strutz-Seebohm-N, Wolters D, Seebohm G (2016) KCNE1 induces fenestration in the Kv7.1/KCNE1 channel complex that allows for highly specific pharmacological targeting. Nat Commun 7:12795. https://doi.org/10.1038/ncomms12795

Wu D, Delaloye K, Zaydman MA, Nekouzadeh A, Rudy Y, Cui J (2010a) State-dependent electrostatic interactions of S4 arginines with E1 in S2 during Kv7.1 activation. J Gen Physiol 135(6):595–606. https://doi.org/10.1085/jgp.201010408

Wu D, Pan H, Delaloye K, Cui J (2010b) KCNE1 remodels the voltage sensor of Kv7.1 to modulate channel function. Biophys J 99(11):3599–3608. https://doi.org/10.1016/j.bpj.2010.10.018

Xu X, Kanda VA, Choi E, Panaghie G, Roepke TK, Gaeta SA, Christini DJ, Lerner DJ, Abbott GW (2009) MinK-dependent internalization of the IKs potassium channel. Cardiovasc Res 82 (3):430–438. https://doi.org/10.1093/cvr/cvp047

Yamagata K, Senokuchi T, Lu M, Takemoto M, Fazlul Karim M, Go C, Sato Y, Hatta M, Yoshizawa T, Araki E, Miyazaki J, Song WJ (2011) Voltage-gated K+ channel KCNQ1 regulates insulin secretion in MIN6 beta-cell line. Biochem Biophys Res Commun 407 (3):620–625. https://doi.org/10.1016/j.bbrc.2011.03.083

Yan J, Aldrich RW (2010) LRRC26 auxiliary protein allows BK channel activation at resting voltage without calcium. Nature 466(7305):513–516. https://doi.org/10.1038/nature09162

Yan J, Aldrich RW (2012) BK potassium channel modulation by leucine-rich repeat-containing proteins. Proc Natl Acad Sci U S A 109(20):7917–7922. https://doi.org/10.1073/pnas.1205435109

Yang WP, Levesque PC, Little WA, Conder ML, Shalaby FY, Blanar MA (1997) KvLQT1, a voltage-gated potassium channel responsible for human cardiac arrhythmias. Proc Natl Acad Sci U S A 94(8):4017–4021

Yanglin P, Lina Z, Zhiguo L, Na L, Haifeng J, Guoyun Z, Jie L, Jun W, Tao L, Li S, Taidong Q, Jianhong W, Daiming F (2007) KCNE2, a down-regulated gene identified by in silico analysis, suppressed proliferation of gastric cancer cells. Cancer Lett 246(1–2):129–138. https://doi.org/10.1016/j.canlet.2006.02.010

Yasuda K, Miyake K, Horikawa Y, Hara K, Osawa H, Furuta H, Hirota Y, Mori H, Jonsson A, Sato Y, Yamagata K, Hinokio Y, Wang HY, Tanahashi T, Nakamura N, Oka Y, Iwasaki N, Iwamoto Y, Yamada Y, Seino Y, Maegawa H, Kashiwagi A, Takeda J, Maeda E, Shin HD, Cho YM, Park KS, Lee HK, Ng MC, Ma RC, So WY, Chan JC, Lyssenko V, Tuomi T, Nilsson P, Groop L, Kamatani N, Sekine A, Nakamura Y, Yamamoto K, Yoshida T, Tokunaga K, Itakura M, Makino H, Nanjo K, Kadowaki T, Kasuga M (2008) Variants in KCNQ1 are associated with susceptibility to type 2 diabetes mellitus. Nat Genet 40(9):1092–1097. https://doi.org/10.1038/ng.207

Yeung SY, Pucovsky V, Moffatt JD, Saldanha L, Schwake M, Ohya S, Greenwood IA (2007) Molecular expression and pharmacological identification of a role for K(v)7 channels in murine vascular reactivity. Br J Pharmacol 151(6):758–770. https://doi.org/10.1038/sj.bjp.0707284

Yu W, Hu C, Zhang R, Wang C, Qin W, Lu J, Jiang F, Tang S, Bao Y, Xiang K, Jia W (2011) Effects of KCNQ1 polymorphisms on the therapeutic efficacy of oral antidiabetic drugs in Chinese patients with type 2 diabetes. Clin Pharmacol Therap 89(3):437–442. https://doi.org/10.1038/clpt.2010.351

Yu H, Lin Z, Mattmann ME, Zou B, Terrenoire C, Zhang H, Wu M, McManus OB, Kass RS, Lindsley CW, Hopkins CR, Li M (2013) Dynamic subunit stoichiometry confers a progressive continuum of pharmacological sensitivity by KCNQ potassium channels. Proc Natl Acad Sci U S A 110(21):8732–8737. https://doi.org/10.1073/pnas.1300684110

Zheng W, Verlander JW, Lynch IJ, Cash M, Shao J, Stow LR, Cain BD, Weiner ID, Wall SM, Wingo CS (2007) Cellular distribution of the potassium channel KCNQ1 in normal mouse kidney. Am J Physiol Renal Physiol 292(1):F456–F466. https://doi.org/10.1152/ajprenal.00087.2006

Zhou Y, MacKinnon R (2003) The occupancy of ions in the K+ selectivity filter: charge balance and coupling of ion binding to a protein conformational change underlie high conduction rates. J Mol Biol 333(5):965–975

Zhou Q, Zhang K, Li W, Liu JT, Hong J, Qin SW, Ping F, Sun ML, Nie M (2009) Association of KCNQ1 gene polymorphism with gestational diabetes mellitus in a Chinese population. Diabetologia 52(11):2466–2468. https://doi.org/10.1007/s00125-009-1500-y

Zhou JB, Yang JK, Zhao L, Xin Z (2010) Variants in KCNQ1, AP3S1, MAN2A1, and ALDH7A1 and the risk of type 2 diabetes in the Chinese Northern Han population: a case-control study and meta-analysis. Med Sci Monit 16(6):BR179–BR183

Chapter 26
Orai Channels

Trevor J. Shuttleworth

Abstract The highly calcium-selective ion channels formed by the Orai proteins represent a principal route for the agonist-induced entry of extracellular calcium in non-excitable cells, a process that is necessary for the generation of the calcium signals involved in the initiation and regulation of a multitude of diverse cellular responses. Consequently, their expression and activities play a major role in the essential functions of a wide range of diverse epithelial tissues. In marked contrast to the voltage-gated calcium channels of excitable cells, the molecular components of these channels (the Orai proteins) and their activation and regulation were only identified a little over 13 years ago. Because of this, there is still much to learn about the details of their unique biophysical properties, modes of activation, and functional roles.

Keywords Calcium release-activated calcium (CRAC) channel · Arachidonate-regulated calcium (ARC) channel · Stromal interacting molecule (STIM) · Calcium signals · Calcium oscillations

26.1 Introduction

The origins of our understanding of agonist-activated calcium entry pathways in non-excitable cells actually began some 20 years before the discovery of the Orai proteins, with the characterization of the so-called "capacitative" entry of calcium by Putney (Putney 1986; Takemura and Putney 1989). In this, it was proposed that the agonist-induced entry of calcium was specifically initiated as a result of a depletion of endoplasmic reticulum calcium stores—normally as a result of the generation of inositol 1,4,5-trisphosphate (InsP3) subsequent to appropriate receptor activation.

T. J. Shuttleworth (✉)
Department of Pharmacology and Physiology, University of Rochester Medical Center, New York, USA
e-mail: trevor_shuttleworth@urmc.rochester.edu

K. L. Hamilton, D. C. Devor (eds.), *Studies of Epithelial Transporters and Ion Channels*, Physiology in Health and Disease,
https://doi.org/10.1007/978-3-030-55454-5_26

Initially, it was suggested that the entering calcium passed directly into the empty stores situated close to the entry sites in the plasma membrane. However, it was subsequently shown that the calcium actually entered directly into the cytosol (Kwan and Putney 1990; Putney 1990), and the mechanism became known as "capacitative, or store-operated, calcium entry." Notably, most of these early studies were performed on epithelial cells—specifically those in the acini of various exocrine glands (e.g., parotid and lacrimal glands). Again, it took a further 6 years before the basic biophysical features of the conductance responsible, named I_{CRAC} (calcium release-activated calcium current), were first described (Hoth and Penner 1992, 1993; Zweifach and Lewis 1993). Over the subsequent years, considerable information was obtained on the properties and behavior of these channels, but the molecular basis of their composition and activation remained unknown. Only in 2005 was the 685-residue, single transmembrane domain protein known as stromal-interacting molecule 1 (STIM1, also originally known as GOK) identified as the essential sensor of the depletion of calcium levels in the endoplasmic reticulum (ER) and regulator of the CRAC channel activity (Roos et al. 2005; Liou et al. 2005; Zhang et al. 2005). This was followed a year later by the discovery of the novel protein Orai1 (together with its close relatives Orai2 and Orai3) as the actual channel-forming molecule (Feske et al. 2006; Vig et al. 2006b; Zhang et al. 2006; Prakriya et al. 2006; Yeromin et al. 2006).

Consequently, the "Orai channel" field is still relatively young, and there is much yet to be learnt about this important family of calcium channels that clearly play key roles in a diverse range of critical cellular activities. In particular, it is now clear that the Orai proteins can form additional endogenous channels that are distinct from the store-operated CRAC channels, and which act in entirely unique ways to modulate agonist-induced calcium signals in cells. Therefore, this chapter will review the studies describing the properties, structure, and regulation of the well-known store-operated Orai channels (CRAC channels), as well as those of the less extensively studied *store-independent* Orai channels such as the arachidonic acid-regulated (ARC) channels. In addition, the evidence indicating the specific roles of these channels in both the normal physiological functions, as well as those in pathological conditions, of epithelial tissue will be discussed.

26.2 Store-Operated CRAC Channels

26.2.1 Biophysical Properties

The biophysical properties of the CRAC channels have been comprehensively studied over the past 25 years or so, and will only be briefly summarized here. More detailed information can be found in numerous excellent reviews (Parekh and Putney 2005; Prakriya 2009; McNally and Prakriya 2012; Derler et al. 2012; Prakriya and Lewis 2015; Amcheslavsky et al. 2015; Gudlur and Hogan 2017). As already noted, the initial biophysical characterization of the store-operated CRAC

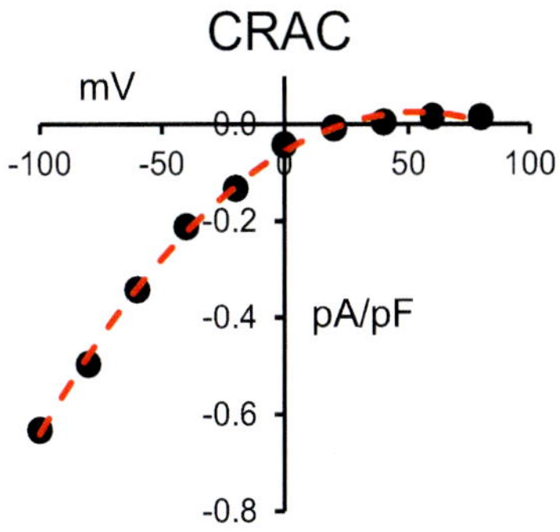

Fig. 26.1 Representative current–voltage relationship of endogenous CRAC channels. Whole-cell patch clamp current values were recorded in HEK293 cells over the voltage range of −100 mV to +80 mV, with an external medium containing 10 mM calcium (see Thompson and Shuttleworth 2015 for details)

channels occurred several years before their molecular identity was eventually revealed. As such, these early studies established the essential biophysical nature of these channels as highly calcium-selective, low-conductance channels whose gating is not membrane voltage dependent. Thus, in whole-cell patch clamp studies these channels display marked inward rectification, with inward currents of around 0.5–1.2 pA/pF at −80 mV, depending on cell type, but with no significant outward currents at membrane potentials up to +60 mV (Fig. 26.1). Brief pulses to negative potentials reveal a significant, highly localized, calcium-dependent fast inactivation (Hoth and Penner 1993; Zweifach and Lewis 1995). Unlike voltage-gated calcium channels, permeability to other divalent cations particularly barium and strontium is significantly lower than that for calcium. This high selectivity for calcium displayed by these channels is further demonstrated by the fact that replacing all extracellular Na^+ and K^+ with $NMDG^+$ results in only a very minor reduction in the inward current, indicating an approximately 1000-fold selectivity for calcium ions over sodium ions. In addition, the currents are effectively blocked by low concentrations of lanthanum (Hoth and Penner 1993), nickel (Zweifach and Lewis 1993), or gadolinium (Broad et al. 1999). Similar to voltage-gated calcium channels, complete removal of external divalent cations results in the appearance of inward monovalent currents that are some 8–10-fold larger than the Ca^{2+} currents in the same cells. These monovalent currents decay over the subsequent 30 s or so, and display an approximate 8-fold selectivity for Na^+ over Cs^+ (Prakriya and Lewis 2002). With regard to the single channel calcium conductance of these CRAC channels, it is clear that it is exceptionally small, and can only be estimated by noise analysis of the larger monovalent currents measured in the absence of external divalent ions. Such approaches result in an estimated unitary CRAC channel Ca^{2+} conductance of ~20 fS (Zweifach and Lewis 1993; Prakriya and Lewis 2002), a value that is some 500-fold smaller than that reported for voltage-gated calcium channels.

26.2.2 Molecular Identity of the CRAC Channel

Prior to the discovery of the Orai proteins in 2006, various molecular entities had been proposed as forming the low-conductance, highly calcium-selective channels responsible for agonist-induced calcium entry in non-excitable cells. Principal among these were members of the TRPC family, an idea largely based on the belief that the sustained portion of the PLC-dependent responses to light in *Drosophila* eyes was induced as a result of calcium store depletion, and that these were impaired in animals bearing the *trp* mutations (Hardie and Minke 1992). Subsequent studies, however, revealed that the activation of the *Drosophila* TRP channels was not dependent on calcium store depletion (Hardie et al. 2001). More importantly, it is clear that the biophysical properties of the channels formed from the TRP proteins are entirely distinct from those of the true highly calcium-selective store-operated CRAC channels—a fact that was confirmed by the subsequent discovery of the Orai family of proteins. The key studies in this long awaited development involved combinations of a rigorous gene mapping of a family displaying immunodeficiency that was attributed to the absence of functional CRAC channel activity (Feske et al. 2006), along with whole genome screening of *Drosophila* S2 cells (Feske et al. 2006; Vig et al. 2006b; Zhang et al. 2006). Although these studies revealed that the Orai protein family includes two additional members (Orai2 and Orai3), and all three members are known to exist in mammals, it is clear that the endogenous CRAC channel is formed exclusively from Orai1 subunits. Importantly, although studies involving the overexpression of Orai2 and Orai3 have indicated that these can also form store-operated conductances, they display biophysical and pharmacological features that are generally significantly distinct from those seen with endogenous CRAC channels (Vig et al. 2006a; Lis et al. 2007)—see Sect. 26.4 for more details.

Orai1 is an approximately 300-residue protein comprised of four transmembrane domains with both N-terminal and C-terminal sequences located in the cytosolic domain (Fig. 26.2a). Various approaches have been utilized to examine how the

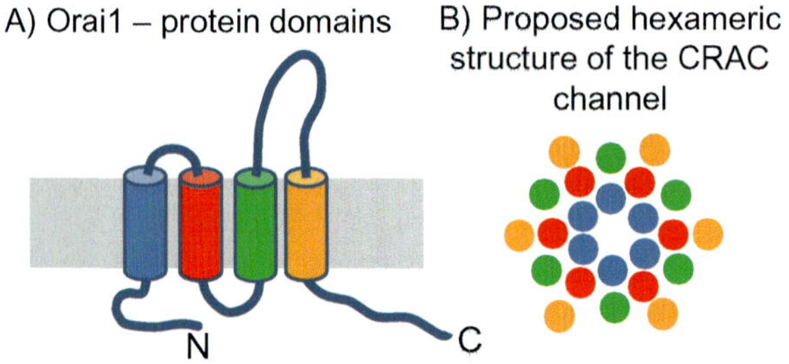

Fig. 26.2 Cartoon illustrating the basic structural features of the Orai1 molecule. (**a**) Illustrated are the four transmembrane domains, and intracellular N and C termini. Note the extended second extracellular loop. (**b**) A diagram indicating the proposed hexameric CRAC channel construct with the transmembrane domains arranged as proposed, following the equivalent coloring of the relevant transmembrane domains of the individual subunits shown in (**a**)

Orai1 subunits assemble to form the functional endogenous CRAC channels. For example, we used an approach involving the expression of concatenated assemblies of different numbers of Orai1 subunits, either with or without the co-expression of a dominant-negative Orai1(Glu106Gln) mutant monomer (Mignen et al. 2008b). The anticipation was that, if any of the expressed concatemers required the incorporation of additional subunits to form a functional CRAC channel, the presence of the co-expressed dominant-negative mutant would result in significantly reduced store-operated currents. Consequently, although expression of each of the concatemers alone resulted in conductances showing all the features of endogenous CRAC channels, co-expression of the dominant-negative Orai1 monomer significantly reduced each of these currents except those seen with the concatenated Orai1 tetramer (Mignen et al. 2008b). This tetrameric stoichiometry of the functional channel was further supported by additional independent studies from several groups involving a variety of diverse approaches, including single-molecule photobleaching (Ji et al. 2008; Penna et al. 2008; Demuro et al. 2011), high-resolution electron microscopy (Maruyama et al. 2009), and single-molecule brightness analysis (Madl et al. 2010). However, in marked contrast to these various findings, a report examining the X-ray structure of the purified *Drosophila* Orai channel demonstrated a hexameric assembly (Hou et al. 2012), with the four transmembrane domains of each individual Orai protein distributed as indicated in Fig. 26.2b. In this report, functional analysis of the hexameric channel was limited to the demonstration that (a) a constitutively active mutant version incorporated into liposomes allowed the passage of Na^+ ions but not $NMDG^+$ in the absence of external divalent cations, (b) this was absent in the Lys163Trp mutant (equivalent to the Arg91Trp mutant in mammalian Orai1), and (c) could be blocked by relatively high concentrations of Gd^{3+} (300 nM). However, a subsequent examination of a concatenated Orai1 hexameric channel demonstrated that this construct did not result in a store-operated conductance with properties that match endogenous CRAC channels (Thompson and Shuttleworth 2013a). Specifically, such a channel failed to duplicate the high Ca^{2+} selectivity of the CRAC channels, displaying marked permeability to Na^+ (and, to a somewhat lesser extent, Cs^+) even in the presence of high external Ca^{2+} (10 mM). Of course, it should always be remembered that the concatenated construct approach is not without its own problems, perhaps most critically as they relate to the nature of the sequences inserted in the construct to link the subunits. However, even given such potential concerns, it would still seem problematic to explain why, if the CRAC channel truly is a hexamer, the store-operated Ca^{2+} currents seen on expression of a concatenated Orai1 tetramer are completely unaffected by the co-expression of a dominant-negative Orai1(Glu106Gln) construct (Mignen et al. 2008b). Moreover, the expressed Orai1 tetramer duplicates essentially all the key biophysical properties of the endogenous CRAC channel, while a concatenated Orai1 hexamer does not (Thompson and Shuttleworth 2013a).

In an attempt to reconcile these obviously contradictory findings, a study from the Gill group (Cai et al. 2016) reported that inhibition of CRAC channel currents by incorporation of a mutant E106A subunit in a tetrameric construct was only seen when the mutation was located in either of the first two N-terminal sites in the construct, whereas the same mutation incorporated in either of the two C-terminal sites failed to significantly affect channel activity. In contrast, incorporation of the Glu106Ala mutant in any position in the hexameric construct negatively impacted ionomycin-mediated calcium signals to varying degrees, ranging from 80% inhibition, to 50%. Moreover, a subsequent detailed study of Orai channel structure and function from the Lewis group (Yen, et al. 2016) also supported a hexameric structure, and suggested that the previously proposed tetrameric structure was a result of the application of inappropriately short subunit linkers (i.e., 6 residues versus 24 residues). Clearly, it would seem that definitive evaluation of the true conformation of the native CRAC channel must await further detailed examination of the overall crystal structure of the endogenous CRAC channel, and perhaps involving the application of additional structural methodologies such as cryo-EM, etc.

26.2.3 STIM1 and CRAC Channel Activation

Initially, a variety of diverse "agents" and/or mechanisms were proposed as providing the essential link between the depletion of the ER calcium stores and the activation of the CRAC channels in the plasma membrane. These included the involvement of a "second" or "third" messenger (e.g., InsP4), an undefined diffusible calcium influx factor ("CIF"), or via a direct interaction with the InsP3 receptors in the ER. However, it is now clear that the "stromal-interacting molecule" STIM1 located in the ER membrane is the key molecule linking ER calcium store depletion with the activation of the CRAC channels.

Structurally, STIM1 consists of an N-terminal sequence of some 210 residues located within the ER, a single transmembrane domain spanning the ER membrane, and an extensive cytosolic domain (~450 residues) that includes a series of three coiled-coil sequences (CC1-3) (Fig. 26.3). Critical to the function of STIM1 as the

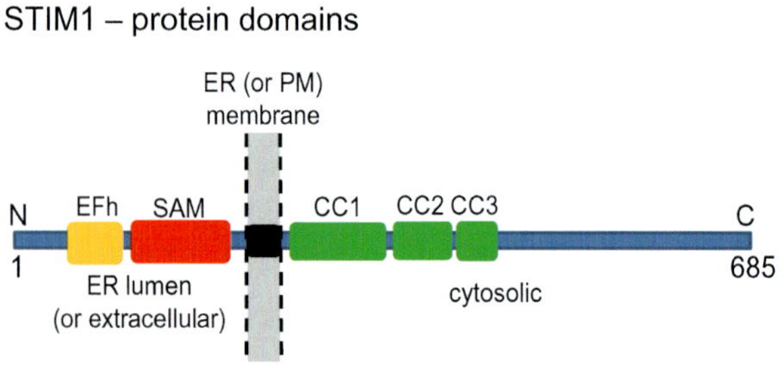

Fig. 26.3 Diagram illustrating the overall structural sequence of the STIM1 protein. The calcium-binding EF-hand domain, and SAM (sterile α-motif) domains are the key features of the STIM1 molecule residing in the ER lumen. The extensive cytosolic region includes the three coiled-coil domains (CC1–CC3) of which the CC2 and CC3 domains are critical features involved in the actual activation of Orai

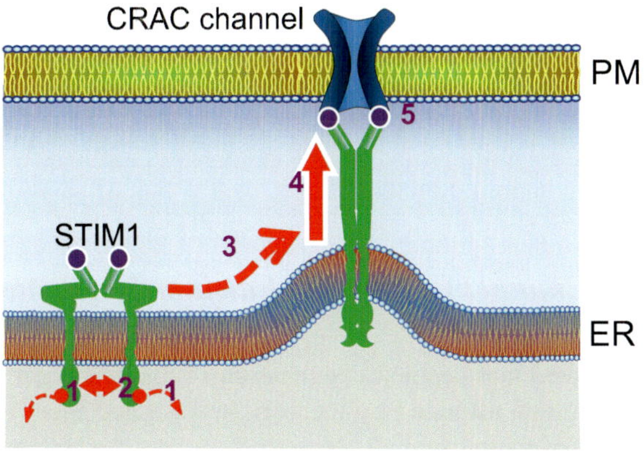

Fig. 26.4 Diagram illustrating the mechanism underlying the activation of the CRAC channels following depletion of endoplasmic reticulum (ER) calcium stores. (#1) A fall in calcium ion concentration within the ER store results in the loss of calcium ions (red) from the EF-hand domains of STIM1 (green) located in the ER membrane. This induces the oligomerization of the STIM1 molecules and transformation into an extended conformation (#2), followed by the translocation of the STIM1 oligomers within the ER membranes (#3) to regions that lie close to the plasma membrane (PM) (#4). Here, the STIM1 molecules interact with and trap the CRAC channels in the plasma membrane (#5), resulting in their activation

activator of store-operated CRAC channels is a calcium-binding EF-hand motif located in the N-terminal sequence within the ER. Under resting conditions, calcium levels within the ER are sufficient to maintain this EF-hand domain in the calcium-bound condition, and STIM1 exists as a monomer with its cytosolic CC1-3 regions in a compact, folded conformation. Lowering of ER calcium levels—typically as a result of agonist-induced activation of InsP3 receptors in the ER membrane (Zhang et al. 2005; Liou et al. 2005)—induces the loss of calcium from the EF-hand domains in the luminal N-terminal portion of STIM1 resulting in their dimerization. If the ER calcium levels remain depleted, then the STIM1 dimers oligomerize (Stathopulos et al. 2006; Liou et al. 2007; Covington et al. 2010), followed by their translocation, within the ER membrane domain, to regions located close to the plasma membrane (ER-PM "junctions") (Wu et al. 2006) (Fig. 26.4). Here, the cytosolic regions of the STIM1 oligomers (specifically the CC1 domains) transition into an extended conformation that enables the CC2 and, more specifically, the CC3 domains to interact with, and trap, the Orai1 molecules that form the CRAC channel in the plasma membrane—a process that can be visualized by the formation of discrete puncta representing the co-localization of fluorescently tagged STIM1 and Orai1 molecules at the cell surface (Luik et al. 2006; Xu et al. 2006). Finally, the actual induction of CRAC channel activity involves the extended cytosolic portion of the STIM1 molecule, incorporating the CC2 and CC3 domains, that is generally known as the "CRAC activation domain" (CAD, residues 342–448) (Park et al. 2009), or "STIM-Orai activating region" (SOAR, residues 344–442) (Yuan et al. 2009). Here, the CC2 and CC3 domains of the individual STIM1 molecules form an antiparallel V-shaped structure (Fig. 26.4) which unfolds to allow interaction between the apical region of the STIM1 CC2 domains and the intracellular C-terminal region of Orai1, resulting in the activation of the CRAC channels (Kawasaki et al. 2009; Park et al. 2009; Yuan et al. 2009; Zhou et al. 2013; Hirve et al. 2018).

26.2.4 CRAC Channel Gating and Calcium Permeation

Here it should be noted that detailed examination of the actual gating of the Orai channel is significantly constrained by the very low conductance (typically in the femtoSiemen range) of these channels. Nevertheless, and despite the obvious controversy already noted above, the crystal structure of the CRAC channel certainly provided potentially useful information regarding specific molecular features of the actual channel pore that are likely important for the selectivity and permeation of Ca^{2+} ions. Most particular of these is the positioning of the glutamate residue (Glu106) at the apical end of the first transmembrane domain of Orai1, as it is the binding of calcium ions at this residue that forms the critical selectivity filter for calcium entry. The remainder of the channel pore is formed by the extended cytosolic region of the first transmembrane domain (residues Leu74-Arg91), named the ETON (extended transmembrane Orai1 N-terminal) region, that forms both an essential binding surface and a gate for Orai1 activation by STIM1. Thus, current models of the channel activation suggest that permeation is achieved via a flexing movement of the Orai sequence (specifically the TM1 helixes) internally to residue Arg91, resulting in a displacement of this sequence such that it no longer blocks the pore (Derler et al. 2013, 2018).

As to how STIM1 induces the gating of the CRAC channel, it is clear that the initial interactions between STIM1 and the channel involve the extended cytosolic region of transmembrane four (TM4) of the CRAC channel (Park et al. 2009; Muik et al. 2008), while actual channel gating involves interactions between the previously noted STIM1 CC2 region (aa344–390) and the Orai C-termini (aa272–292). Additional studies have proposed a subsequent interaction between STIM1 and the Orai N-termini that is necessary to result in the full activation of the channel (Park et al. 2009; McNally et al. 2013; Palty and Isacoff 2016), although this remains a controversial issue (Zhou et al. 2016). More detailed analyses on channel gating can be found in the paper by Gudlur (Gudlur and Hogan 2017).

Finally, it should be noted that the above focusses on the activation of Orai1 by STIM1. However, STIM1 is just one member of the STIM family, and various cells also express STIM2—a related molecule that has been shown to display significant functional differences compared to STIM1. Principal among these are the relative sensitivity to changes in ER calcium concentrations. For example, it has been shown that, when co-expressed with Orai1, activation of cellular calcium influx by STIM2 occurs at reduced ER calcium levels that are significantly smaller than those seen with STIM1 activation (Brandman et al. 2007), and that STIM2 displays a reduced efficacy of coupling to Orai1 (Wang et al. 2014). Together, such features have led to the suggestion that the principal role of STIM2 is in the regulation of basal ER and cytosolic calcium levels (Brandman et al. 2007; Oh-Hora et al. 2008).

26.3 Store-Independent ARC Channels

Based on the information described so far, it is abundantly clear that store-operated calcium entry mediated by the CRAC channels can provide both the calcium required for replenishing depleted endoplasmic reticulum calcium stores, and a calcium signal for regulating various calcium-dependent intracellular biochemical events. Despite this, the idea that CRAC channel activity might not be able to explain *all* modes of agonist-induced cytosolic calcium entry in non-excitable cells began with a consideration of the underlying mechanism inherent in the store-operated model of calcium entry. In particular, the essential dependence on an InsP3-induced depletion of the ER calcium store to a level that was sufficient to subsequently induce the activation of the calcium entry pathway represented a strict sequence of events that might impose some temporal constraints on the activation of calcium entry. Clearly, inherent in this system is the fact that the InsP3-induced release of calcium from the ER calcium stores must precede the onset of calcium entry. Initial studies designed to examine this involved the application of NaF and $AlCl_3$ to induce a gradual generation of the G-protein activating fluoroaluminate ion (AlF_4^-), thereby slowing the rate of InsP3 generation. Curiously, these experiments revealed a clear increased entry of calcium that significantly preceded any detectable release of calcium from intracellular stores (Shuttleworth (1990)). Consistent with this, subsequent studies showed that the agonist-induced rate of calcium entry was independent of the cyclical emptying and refilling of the agonist-sensitive Ca^{2+} pool during $[Ca^{2+}]_i$ oscillations, and that the relevant Ca^{2+} entry pathway remained activated for extended periods following the inhibition of oscillations under conditions in which agonist-sensitive stores could be shown to be full (Shuttleworth and Thompson 1996a, b). Moreover, later studies from the Lewis group (Luik et al. 2008), examining the relationship between ER calcium levels and the consequent CRAC channel currents, indicated that channel activity was only detected when ER calcium stores had been reduced by some 40–50%. Finally, Malli et al. (2008) showed that, while the initial oligomerization of STIM1 in the ER membrane was maximal within ~100 s, the subsequent clustering of STIM1 at the plasma membrane necessary for activation of the Orai channels took an additional 5 min to reach completion.

Taken together, these findings were obviously difficult to reconcile with the release of stored Ca^{2+} being the *necessary prerequisite* for the induction of Ca^{2+} entry, as discussed in more detail elsewhere (Shuttleworth 1999; Shuttleworth and Mignen 2003). Of course, one possibility is that the Ca^{2+} entry process is activated by a minor subset of the cellular ER store whose depletion cannot be readily detected, as was indeed suggested at the time (Ribeiro and Putney 1996; Parekh et al. 1997; Huang and Putney 1998). Nevertheless, the above findings certainly raised the possibility of an agonist-activated Ca^{2+} entry pathway that was distinct from the store-operated, CRAC channel-mediated pathway. The search for the identity of such a store-*independent* Ca^{2+} entry channel, and its mode of activation, eventually lead to the discovery of a unique arachidonic acid-regulated channel, named the "ARC channel" (Shuttleworth 1996; Shuttleworth and Thompson 1998; Shuttleworth and Thompson 1999; Mignen and Shuttleworth 2000). Interestingly, like the CRAC channel, the ARC channel subsequently proved to be a new, but distinct, member of the Orai channel family.

26.3.1 Biophysical Properties

The basic biophysical properties of the store-*independent* Orai channels known as the ARC channels are very similar to those noted above for the CRAC channels. They show the same markedly inwardly rectifying I/V curve with negligible outward currents at voltages up to +60 mV, and their whole-cell current magnitudes are generally similar to that of the CRAC channels when measured in the same cells (Fig. 26.5). In addition, like the CRAC channels, calcium currents through the ARC channels are profoundly inhibited by low micromolar concentrations (<5 μM) of gadolinium (Mignen et al. 2003).

However, despite the close similarity between the store-operated CRAC channels and the store-independent ARC channels, some clear differences do exist. The most obvious of these is the fact that agonist-induced activation of the ARC channels is entirely independent of any depletion of intracellular calcium stores, and is actually mediated by increased intracellular levels of arachidonic acid (Mignen and Shuttleworth 2000), an effect that can be duplicated by addition of low levels (i.e., ~5–8 μM) of exogenous arachidonic acid (Shuttleworth 1996). Biophysical examination of ARC channel currents indicates that they do not display the clear calcium-dependent fast inactivation seen in CRAC channels (Mignen and Shuttleworth 2000), and recordings of the monovalent currents in ARC channels, as measured

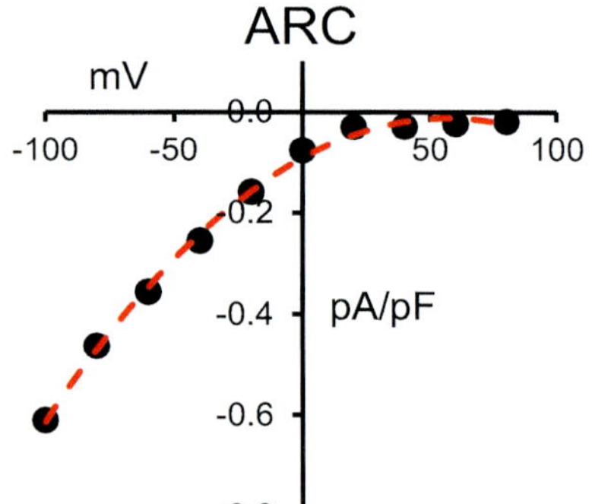

Fig. 26.5 Representative current–voltage relationship of endogenous ARC channels. Whole-cell patch clamp current values were recorded in HEK293 cells over the voltage range of −100 mV to +80 mV, in an external solution containing 10 mM CaCl (see Shuttleworth 2009 for details)

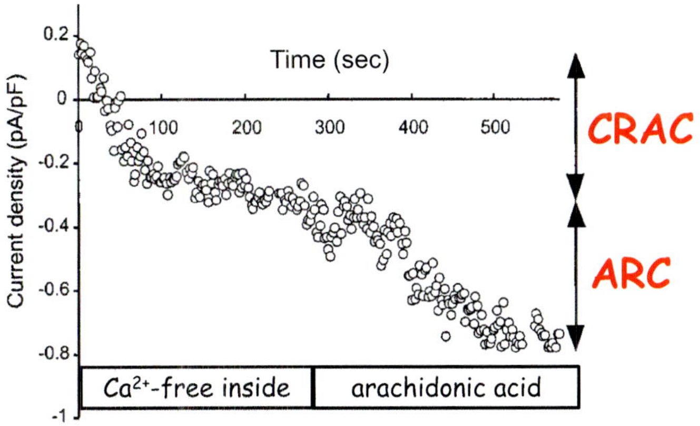

Fig. 26.6 Representative recording of maximal Orai channel currents induced by the depletion of intracellular calcium stores to activate CRAC channels, followed by the additional current induced by application of exogenous arachidonic acid. The data represent whole-cell currents measured at −80 mV in m3-HEK cells using a calcium-free pipette solution to deplete intracellular calcium stores and maximally activate CRAC channel currents. This was followed by exogenous arachidonic acid (8 μM), added as indicated (redrawn from Mignen and Shuttleworth 2000)

in the absence of external divalent cations, show a modest permeability to Cs^+ ions that is typically not seen in CRAC channel currents under the same conditions (Mignen et al. 2003). Although the above differences between the CRAC channel and ARC channel currents indicate that they may result from the existence of two distinct channel entities, the fact that the differences are rather subtle could potentially reflect the activity of the same channel under different modes of regulation. Critically, however, it was shown that, providing cytosolic calcium levels are maintained at approximately resting levels (i.e., ~100 nM), the two currents are strictly additive within the same cell (Fig. 26.6), thereby demonstrating that these two biophysically very similar conductances are indeed entirely separate entities (Mignen and Shuttleworth 2000)—a fact subsequently confirmed by their distinct molecular compositions (see Sect. 26.3.3).

26.3.2 STIM1 and ARC Channel Activation

As already noted, the molecular characterization of CRAC channel-mediated calcium entry began with the identification of STIM1 as the essential sensor for the depletion of ER calcium stores. Given this, the clear indication was that this molecule would likely play no part in the activation of the strictly *store-independent* ARC channels. Surprisingly, however, it was shown that STIM1 was equally essential for the activation of the ARC channels (Mignen et al. 2007). However, subsequent studies revealed that such activation depended not on the pool of STIM1 in the ER, but rather on the minor pool (typically ~15–20%) of STIM1 located in the

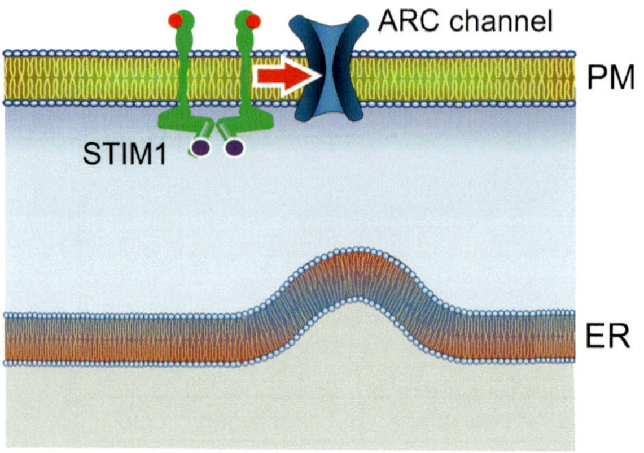

Fig. 26.7 Diagram illustrating the mechanism underlying the activation of the ARC channels by plasma membrane-located STIM1. Here, the EF hands of plasma membrane located STIM1 (red) are located in the extracellular medium and, physiologically, would never normally lose their bound calcium. Moreover, the calcium-bound PM-located STIM1 must interact with the ARC channels located *in the same membrane*, to induce the arachidonic acid-induced entry of extracellular calcium (compare with CRAC channel activation, see Fig. 26.4)

plasma membrane (Mignen et al. 2007) (Fig. 26.7). Interestingly, this rather unexpected finding was reconciled by the realization that STIM1 was, in fact, originally identified as a protein constitutively residing in the plasma membrane (Manji et al. 2000). Nevertheless, this did raise the question as to how STIM1 in the plasma membrane operated to induce the activation of the ARC channels *in the same membrane*. Most immediately obvious were the implications of the location in the extracellular environment of the N-terminal portion of STIM1 with its Ca^{2+}-binding EF hand. Here, given its measured affinity for Ca^{2+} of ~0.2–0.3 mM (Stathopulos et al. 2006; Luik et al. 2008), the EF hand of STIM1 located in the plasma membrane would never normally lose its bound Ca^{2+}, an event that is the essential initiating step in the activation of the CRAC channels on store depletion. Consequently, in marked contrast to the CRAC channels, the N-terminal portion of STIM1 would be unlikely to have any significant role in the activation of the ARC channels. This was confirmed by experiments involving the expression of a construct comprising just the cytosolic region of STIM1 anchored to the inner face of plasma membrane via an N-terminally attached sequence based on the SH4 domain of Lck kinase (Lck-STIM1-C construct). Here, addition of an appropriate agonist (e.g., carbachol) that would normally be able to activate both ARC and CRAC channels now resulted in the exclusive activation of the ARC channels (Thompson and Shuttleworth 2013b). Moreover, and again in marked contrast to the situation with CRAC channels where STIM1 in the ER transiently interacts with the channels in the plasma membrane, the pool of STIM1 located in the plasma membrane is apparently constitutively associated with the ARC channels even under resting conditions (Thompson and Shuttleworth 2013b).

Further studies revealed clear differences in the detailed interactions of specific regions of STIM1 in the activation of the ARC channels versus those in the CRAC channels. For example, selective mutations in a so-called "inhibitory sequence" in the CC1 domain of STIM1 (residues 310–337)—specifically Glu318Ala-Glu320Ala and Glu22Ala—result in the constitutive activation of CRAC channels in the absence of any depletion of ER calcium stores (Korzeniowski et al. 2010; Yang et al. 2012). However, these same mutations not only fail to induce any constitutive activation of the ARC channels, but also prevent the ability of arachidonic acid to activate the channels (Thompson and Shuttleworth 2013b). Together these, and several other studies, demonstrated that, although both CRAC and ARC channels share the requirement for STIM1 for their activation, they rely on distinct pools of the total cellular STIM1, and the underlying mechanism of activation is entirely different.

26.3.3 Molecular Basis of ARC Channel Structure

Given the close biophysical similarity of currents through ARC channels and CRAC channels, together with their common dependence on STIM1 for activation—albeit via an entirely different mechanism—the finding that the ARC channels are also formed from Orai proteins was not surprising. Thus, co-expression of combinations of the three Orai subtypes, followed by examination for the appearance of AA-activated currents with properties consistent with endogenous ARC channels, revealed a requirement for both Orai1 and Orai3, but apparently not Orai2, in forming functional ARC channels (Mignen et al. 2008a). Based on the original understanding prior to 2012 that CRAC was a tetramer, we used the concatenated approach to show that all possible Orai1/Orai3 combinations in a tetrameric assembly gave predominantly store-operated currents, rather than AA-activated currents. However, expanding this same approach to eventually examine Orai1/Orai3 pentamers revealed that expression of a pentamer uniquely comprised of three Orai1 subunits and two Orai3 subunits gave large AA-activated currents that duplicated all the essential biophysical and pharmacological properties of the endogenous ARC channels (Mignen et al. 2009). Finally, experiments involving the incorporation of individual Orai1/3 "chimeric" subunits into the concatenated pentamer showed that AA-activated ARC channel activity specifically depended on the presence of a minimum of two copies of the N-terminal Orai3 region within the pentameric assembly (Thompson et al. 2010).

26.3.4 Molecular Basis for the Selective Activation of ARC Channels

As noted, the molecular basis of the structure of both ARC and CRAC channels is very similar (i.e., both are formed by Orai proteins), as are the key components of

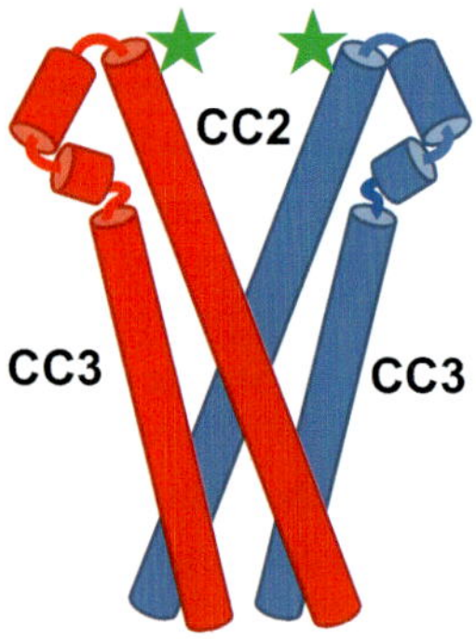

Fig. 26.8 Cartoon of the CC2–CC3 dimer region of the STIM1 molecule that is essential in the STIM1-induced activation of both CRAC and ARC channels. Highlighted by the green "stars" are the location of the critical Thr389 residue (T389), the phosphorylation status of which determines whether CRAC channel (un-phosphorylated Thr389) or ARC channel (phosphorylated Thr389) activity is activated. (Modified from Yang et al. 2012)

their activation (i.e., they are both activated by STIM1). Yet, as we have seen, the manner in which these two channels operate is entirely distinct, and their activities are strictly additive. Clearly, the activation of these two closely related Orai channels—and the role of STIM1 in such activation—must be fundamentally distinct. Critical in this distinction is the fact that the plasma membrane-located pool of STIM1 that is essential for ARC channel activation exists with its EF-hand domain located in the extracellular environment where it would never normally lose its bound calcium (Fig. 26.7). In contrast, the ability of ER-located STIM1 to interact with the plasma membrane-located CRAC channels requires the loss of calcium from this EF hand, resulting in a profound extension of the cytosolic domains of STIM1. Importantly, any similar extension of the cytosolic regions of *plasma membrane-located* STIM1 would take the key regions responsible for activation away from the ARC channels that are also located in the plasma membrane.

Key to the resolution of this contrast were previous studies that had shown that ARC channel activity was specifically dependent on a PKA-dependent phosphorylation, and was inhibited by a calcineurin-dependent dephosphorylation (Mignen et al. 2005). Subsequent studies revealed that this key phosphorylation was mediated by the activity of an A-kinase anchoring protein (AKAP79), and specifically targeted a single threonine residue (Thr389) located at the apex of each of the paired CC2 regions of the STIM1 dimer (Thompson and Shuttleworth 2015) (Fig. 26.8). Consistent with this, mutation of the Thr389 residue to a glutamate (Thr389Glu) to mimic phosphorylation resulted in normal arachidonic acid-mediated ARC channel activity, but the complete loss of store-operated CRAC channel activity. Correspondingly, incorporation of a Thr389Ala mutation to prevent phosphorylation resulted in the exclusive activity of the CRAC channels (Thompson and Shuttleworth 2015) (Fig. 26.9). Further studies designed to examine the specific basis for this phosphorylation-mediated "switch" in the distinct activities of these two channels utilized the incorporation of selective single-residue mutations along

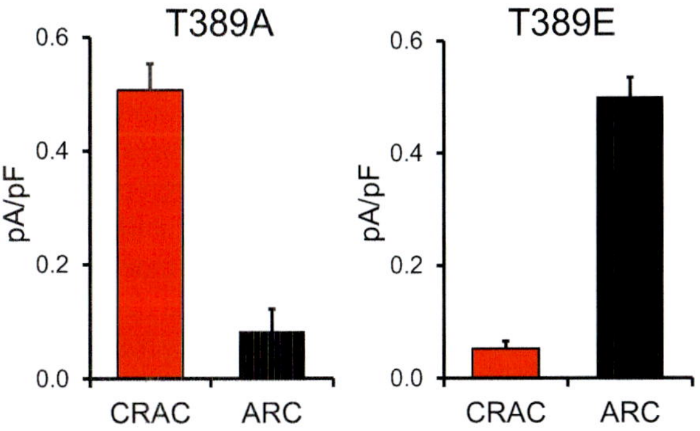

Fig. 26.9 Data showing the profound effect of phosphorylation status of the STIM1 Thr389 (T389) residue on CRAC versus ARC channel activities. STIM1 bearing the Thr389Ala (T389A) mutation that prevents phosphorylation of this residue allows normal agonist-activated CRAC channel activity, but negligible ARC channel activity. In contrast, inclusion of the phosphomimetic Thr389Glu (T389E) mutation in STIM1 results in full arachidonic acid-induced ARC channel activity, but negligible CRAC channel activity

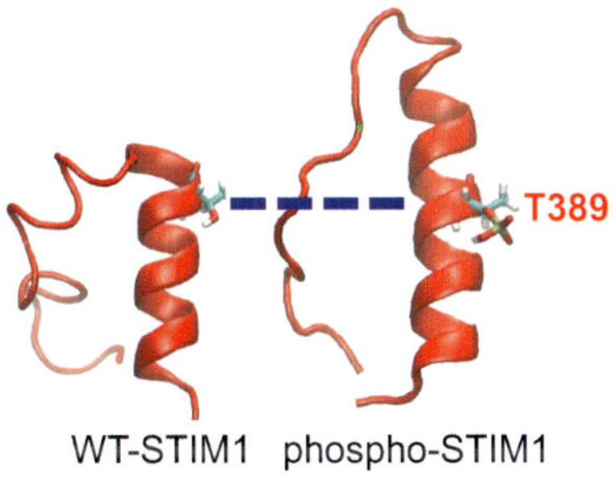

Fig. 26.10 The structural effect of phosphorylation of the Thr389 (T389) residue in STIM1. Molecular dynamics simulation and analyses reveal that phosphorylation of the Thr389 (T389) residue in STIM1 results in a modest, but significant extension of the CC2 portion of STIM1. It is specifically this extension of apical sequence of the CC2 domain that determines the selective activation of ARC channel activity over that of the CRAC channels. (Modified from Thompson et al. 2018)

with detailed biophysical and molecular dynamic analyses (Thompson et al. 2018). These studies revealed that the phosphorylation of the Thr389 residue induced an approximately four-residue extension of the apical coiled-coil region of the CC2 domain of STIM1 (Fig. 26.10). Critically, it was revealed that it was this "physical/structural" change induced by phosphorylation of the Thr389 residue—rather than the phosphorylation per se—that actually determined the specific activity of the ARC channels over the CRAC channels (Thompson et al. 2018).

26.4 Other Orai Channels

As noted above, several studies have shown that the expression of different Orai proteins in cells can result in the appearance of conductances with a variety of unique features. Principal among these are the conductances seen on the overexpression of Orai3, where the resulting channels display the unique ability to be activated either in a store-dependent manner by STIM1 or in a store-independent manner by the CRAC channel inhibitor 2-aminoethoxydiphenyl borate (2-APB) (Lis et al. 2007). Curiously, these distinct modes of activation result in profound differences in the selectivity of the Orai3 channel, with STIM1 activation producing a Ca^{2+}-selective conductance, while activation by 2-APB results in a nonselective conductance (DeHaven et al. 2008; Lis et al. 2007; Schindl et al. 2008; Zhang et al. 2008; Peinelt et al. 2008). However, in the context of the discussion here, it is important to note that the unique features of these "Orai3 channels" have yet to be seen in any *endogenous* conductances. Consequently, their relevance in any physiological context is currently unclear.

More relevant are the few examples where Orai3 has been shown to be a component of endogenous conductances that appear to display features distinct from the ARC channels described above. Perhaps most notable of these are the store-independent conductances named "LRC channels" whose activation uniquely depends, not on arachidonic acid itself, but rather on a metabolite of arachidonic acid namely leukotriene C_4 (Gonzalez-Cobos et al. 2013; Zhang et al. 2013). The basic biophysical features of these channels are apparently generally indistinguishable from those of the ARC channels, and they share the inability to be inhibited by the rather promiscuous drug 2-APB (Zhang et al. 2013). However, the reported whole-cell current magnitudes through these LRC channels are extremely small (Zhang et al. 2014), equivalent to only ~2–3% of the corresponding currents through ARC channels recorded under similar conditions (Mignen et al. 2003).

26.5 Physiological Roles of Orai Channels

The original concept of "store-operated calcium entry" (SOCE)— or, as it was originally described, "capacitative Ca^{2+} entry"—formulated by Putney, was actually based on studies of epithelial tissue, namely the secretory cells of the lachrymal and salivary glands (Parod and Putney 1978; Putney 1986; Takemura et al. 1989; Takemura and Putney 1989; Putney et al. 1998). However, the subsequent identification of the key molecular "players" in this process, namely STIM1 and the Orai proteins, essentially resulted from studies of immune cells, most notably in relation to the condition of severe combined immune deficiency (SCID) (Feske et al. 2006; Shaw and Feske 2012). However, it is clear that the principal role of Orai channel activation—in both CRAC and ARC channels—is in the generation of appropriate signals for the regulation of a wide variety of critical cellular activities. These include the generation of agonist-induced calcium oscillations that in themselves control multiple key functions including cell differentiation, gene expression (Dolmetsch et al. 1997), smooth muscle contraction (Trebak 2012), allergic responses (Di Capite et al. 2011), cardiac hypertrophy (Hulot et al. 2011), and platelet activation and thrombus formation (Braun et al. 2009).

Nevertheless, the fundamental dependence of ER Ca^{2+} store depletion for the activation of the CRAC channels would suggest that a primary role of these channels is likely to be related to the maintenance of the Ca^{2+} content of these stores. Such a role is clearly important for cellular viability, as a sustained depletion of ER calcium levels will result in the disruption of normal protein folding and covalent modification, lipid biosynthesis, etc., resulting in the induction of the "ER stress response" and, ultimately, apoptosis. Indeed, the demonstration that effective CRAC channel activation requires a rather profound and sustained depletion of ER calcium levels (Parekh et al. 1997) only lends support for such a role, and would certainly explain the widespread expression of these channels in diverse cell types.

Although less comprehensively examined, evidence indicates that the store-*independent* ARC channels are also present in a wide range of different cell types, including various exocrine secretory cells such as salivary and pancreatic acinar cells (Mignen et al. 2005) and avian nasal gland cells (Shuttleworth 1996), as well as the insulin-secreting pancreatic β-cells (Yeung-Yam-Wah et al. 2010), endothelial cells (Fiorio Pla and Munaron 2001), and astrocytes (Yang et al. 2005). Moreover, these channels are also present in a variety of different cell lines, including HEK cells (Mignen and Shuttleworth 2000), HeLa, COS and RBL cells (Mignen et al. 2003), SH-SY5Y cells (Yoo et al. 2000), and K562 cells (Li et al. 2008). Again, such widespread cellular expression clearly suggests that ARC channel function plays important roles in a multitude of essential cellular activities.

Critically, in any discussion of the physiological roles of the Orai channels it is important to remember that the activation of the CRAC channels versus the ARC channels depends on two entirely discrete and spatially distinct pools of STIM1. Consequently, this enables these two biophysically similar and molecularly related conductances to operate in entirely separate ways to independently induce distinct physiological responses. However, in general terms, it is clear that a principal role of Orai channels, both the CRAC and ARC channels, is in the agonist-induced generation of the cytosolic calcium signals that are essential for the initiation and regulation of a wide range of critical processes, including gene transcription, the activation of certain adenylyl cyclases, kinases and phosphatases, as well as the regulation of the cell cycle, cell proliferation, and apoptosis. With regard to the actual nature of these agonist-activated Ca^{2+} signals, studies show that maximal activation of the relevant receptors typically results in a sustained elevation of cytosolic Ca^{2+} concentration. However, because such sustained elevations in cytosolic Ca^{2+} above a certain value can typically induce cell death, such "amplitude-based" signals provide only a limited ability to generate signals that are able to sufficiently discriminate between the various appropriate Ca^{2+}-dependent effector systems in cells. With the development of systems enabling examination of agonist-induced responses in single cells, it became clear that physiologically relevant signals generally took the form of a series of Ca^{2+} oscillations (Berridge 1990). This, together with the uncovering of various downstream systems capable of responding to such *frequency-based* signals, revealed that such responses are able to generate distinct "digital" signals over an almost infinite range (Thomas et al. 1996).

As for the specific role of Orai channels in such signals, it was originally thought that the CRAC channels, with their dependence on Ca^{2+} store depletion for activation, were required to both replenish Ca^{2+} lost from the cell and recharge the ER stores between each oscillation. However, several lines of evidence were difficult to reconcile with such an effect. For example, as already noted above, studies indicated that measurable receptor-activated calcium entry can frequently be seen to begin within as little as 3–5 s of agonist addition, and that the receptor-activated rate of calcium entry during an oscillatory calcium signal is often independent of the cyclical emptying and refilling of the intracellular agonist-sensitive calcium pool (Shuttleworth 1996). Moreover, subsequent studies indicated that Ca^{2+} entry also played a key role in actually "driving" the intrinsic oscillatory mechanism itself. For example, oscillatory Ca^{2+} signals typically cease almost immediately on removing extracellular Ca^{2+}, yet intracellular Ca^{2+} stores are not obviously depleted under these conditions, and that the oscillations return without any appreciable delay on the restoration of external Ca^{2+} (Martin and Shuttleworth 1994; Morgan and Jacob 1996). Such findings clearly raised questions as to how such activity could be achieved by the store-operated CRAC channels (Shuttleworth 1999), although this has been challenged (Bird and Putney 2005; Bird et al. 2009). Subsequently, more direct evidence for an essential role of the store-independent ARC channels in providing a major pathway for agonist-induced Ca^{2+} entry during oscillatory signals was revealed in a range of different cell types (Shuttleworth 1996; Mignen et al. 2001; Mignen et al. 2005).

Nevertheless, the question remained as to how the ARC channel-mediated Ca^{2+} entry could act to drive and/or modulate such oscillatory responses. Subsequent studies revealed that one way in which this is achieved involves the Ca^{2+} entering via ARC channels inducing a specific, and highly local, activation of a membrane-located phospholipase C (PLC-δ3) (Thompson and Shuttleworth 2011). This ARC channel-induced activation of PLC-δ3 results in the generation of additional InsP3, thereby increasing the frequency of the Ca^{2+} oscillations and, presumably, the modulation of the intensity and/or nature of downstream effects (Fig. 26.11). Critically, neither the similar activation of Ca^{2+} entry via the CRAC channels nor global increases in cytosolic calcium levels over the physiological range were able to duplicate this activity. In this context it is perhaps relevant to note that in evolutionary terms, calcium-induced activation of PLCδ, resulting in increased InsP3 levels, actually preceded the more widely recognized PLCβ-mediated pathway (Allen et al. 1997; Rebecchi and Pentyala 2000).

Of more direct relevance in the context of this volume with its focus on epithelial tissues, several subsequent studies have revealed the critical role of Orai channel activity in a wide range of epithelial tissues, including the regulation of secretory activities in the exocrine cells of the salivary glands (Melvin et al. 1991; Ambudkar 2014), the pancreas (Petersen and Tepikin 2008), sweat glands (Concepcion et al.

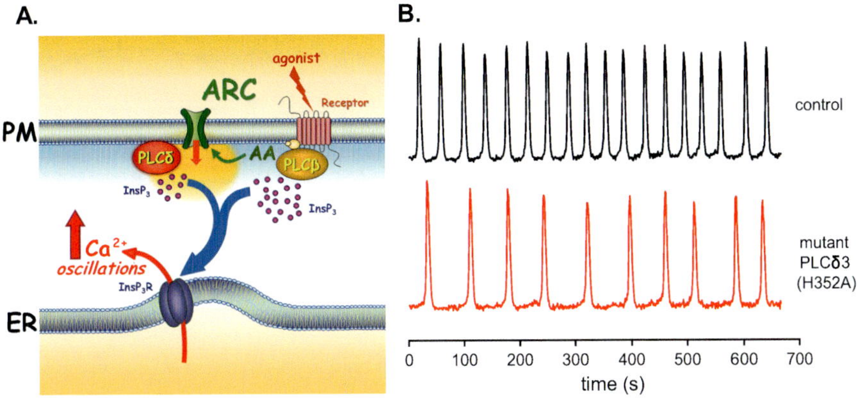

Fig. 26.11 (**a**) Diagram illustrating how the local stimulation of PLCδ by calcium entering exclusively via the ARC channels might act to modulate the frequency of agonist-induced calcium oscillations. The ARC channel-mediated, calcium-dependent activation of PLCδ results in the generation of $InsP_3$ that adds to that produced by the direct agonist-induced activation of PLCβ, thereby effectively raising total $InsP_3$ levels and increasing the oscillatory behavior of the $InsP_3$ receptors on the ER. (**b**) Figure showing the effect of expression of a catalytically dead mutant PLCδ3 on agonist-induced ARC channel-mediated calcium oscillations in HEK cells. See Thompson and Shuttleworth (2011) for further details

2016; Concepcion and Feske 2017), and mammary glands (McAndrew et al. 2011). Moreover, additional evidence of the important role of Orai channels in epithelial cell function is supported by the extensive reports of the involvement of these channels in various epithelial cancers—most notably breast, pancreatic, prostate, cervical, and colorectal cancers (Yang et al. 2009; Flourakis et al. 2010; Chen et al. 2011; Faouzi et al. 2011; McAndrew et al. 2011; Vashisht et al. 2015). Finally, there are also examples where an impaired Orai channel function can impact epithelial function in an indirect way. For example, knockdown of STIM1 (and STIM2) and the consequent loss of Orai channel function have been shown to impact salivary gland exocrine function via the induction of the autoimmune condition known as Sjögren's syndrome, resulting in destruction of the secretory tissue (Cheng et al. 2012).

An important point of note is that the highly polarized nature of epithelial tissues provides the opportunity for entirely discrete signaling process at the opposite ends of the cells, many of which involve Orai channels. An example of this is the generation of distinct apical versus basal Ca^{2+} signals by activation of store-operated Ca^{2+} channels (presumably CRAC channels) by purinergic receptors in human nasal epithelial monolayers (Paradiso et al. 1995). In a similar way, the expression of distinct Ca^{2+} entry channel types in the opposite membrane domains of such polarized tissues can result in distinct responses (Ko et al. 1996), and apical agonist receptors can activate a store-operated Ca^{2+} entry that is exclusively located in the basolateral membrane (Bahra et al. 2004).

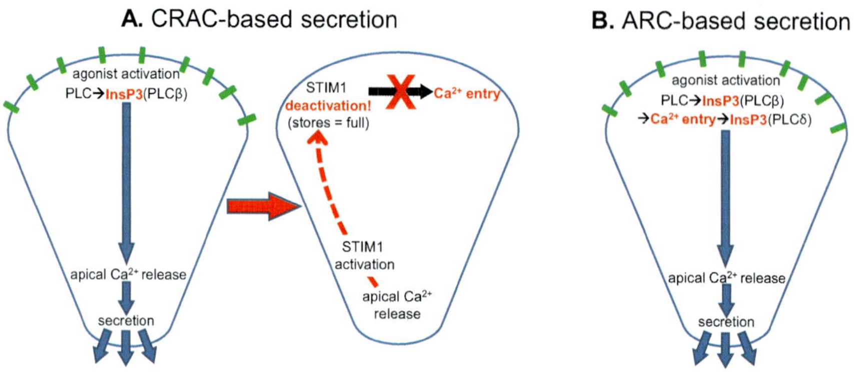

Fig. 26.12 Diagrams comparing the potential regulation of exocrine secretory activity in a pancreatic acinar cell by CRAC channel-mediated and ARC channel-mediated pathways. (**a**) The CRAC channel pathway depends on depletion of ER calcium stores, which typically occurs *only* in the apical part of the cell. STIM1 activated in this region of the ER, would then have to translocate to the more basal regions in order to activate CRAC channels. However, ER calcium stores would be unlikely to be depleted in this region, potentially inducing the *deactivation* of the translocated STIM1. (**b**) In contrast, the *direct* agonist activation of the ARC channels would generate $InsP_3$ which would diffuse to the apical ER stores, inducing the release of calcium to drive the secretory process

As to the distinct respective roles of store-operated and store-independent Ca^{2+} entry, activated via the CRAC and ARC channels, respectively, this is perhaps most clearly revealed in the acinar cells of the pancreas. Here, the role of cytosolic Ca^{2+} signals in both physiological (Petersen and Tepikin 2008) and pathological (Kruger et al. 2000; Petersen and Sutton 2006; Criddle et al. 2007; Petersen et al. 2009) responses is well established, with both types of response being dependent on the entry of Ca^{2+} in the exocrine pancreatic secretory cells (Yule and Gallacher 1988; Raraty et al. 2000). Consistent with the discussion above, the physiologically relevant secretory responses in these cells are driven by oscillatory increases in cytosolic Ca^{2+} signals. However, such oscillatory signals typically originate in, and are often confined to, the extreme apical region of the cell (Kasai et al. 1993; Thorn et al. 1993). In contrast, the principal site for any agonist-induced Ca^{2+} entry would seem to be in the more basal regions of the cells, where the relevant agonist receptors are located. For a mechanism dependent on store depletion to activate such entry channels would require that activated STIM1, located in the ER membrane of the depleted apical portion of the store, could somehow translocate through the ER in regions where such depletion has not occurred, while still retaining its ability to interact with and activate the store-dependent CRAC channels located in the plasma membrane of the basal regions of the cell (Fig. 26.12a). Moreover, the entering Ca^{2+} would then have to traverse the cell, without significant buffering, to interact with InsP3 receptors in the apical ER to impact the frequency of Ca^{2+} oscillations in this region. In contrast, if the relevant source of Ca^{2+} entry was via the store-independent ARC channels, their agonist-induced activation could generate additional InsP3—for example, via the PLCδ-mediated mechanism outlined above—which could then readily diffuse to the apical region of the cell without significant degradation—to act on the InsP3 receptors and modulate the resulting oscillatory Ca^{2+} signals responsible for controlling secretion (Fig. 26.12b). Consistent with such a role is the demonstration that, indeed, it is the store-independent ARC channels that provide the predominant source of receptor-induced Ca^{2+} entry at the low agonist concentrations that result in oscillatory Ca^{2+} signals in these cells (Mignen et al. 2005). Moreover, in contrast to this specific role of the ARC channels in regulating the physiologically relevant oscillatory Ca^{2+} signals associated with secretion, the sustained Ca^{2+} signals associated with activation of the CRAC channels following the depletion of ER Ca^{2+} stores, have been shown to result in the premature activation of cellular trypsinogen, and subsequent cellular disruption and necrosis, that is associated with the induction of pancreatitis (Gerasimenko et al. 2013)—an example of where CRAC channel activity is uniquely associated with pathological rather than physiological responses.

An additional example of an important, although perhaps unexpected, role of Orai channels in epithelial function comes from studies investigating the basis for the reported gender differences in the disease cystic fibrosis. Here, females display a markedly poorer general prognosis with a 70-to-30% difference in the appearance of early onset COPD (chronic obstructive pulmonary disease), an increased risk of asthma, and a significantly decreased average life span (Rosenfeld et al. 1997; Carey et al. 2007). Examination of the possible basis for such gender differences in cystic fibrosis revealed that estrogen, acting at the ESR1 receptor, impairs normal STIM1 function in the bronchial epithelial cells as a result of a dephosphorylation of a serine residue (Ser575) in the cytosolic portion of the molecule (Coakley et al. 2008). This dephosphorylation inhibits the normal Ca^{2+} store depletion-induced oligomerization of STIM1 and its ability to interact with Orai1, leading to a reduced CRAC channel activity (Sheridan et al. 2013). As a result, the normal activation of transepithelial secretion of chloride ions through Ca^{2+}-activated chloride channels is impaired, leading to a reduced fluid secretion by the bronchial epithelium, increasing the susceptibility for lung infections, etc. (Coakley et al. 2008).

Finally, it should be noted that attempts to apply pharmacological approaches to delineate the physiological roles of Orai channels have met with only limited success, in that most of the agents currently available and most frequently used to inhibit these channels—namely SKF-96365, 2-aminoethoxydiphenyl borate (2-APB), and pyrazole derivatives (YM-58483, BTP2, Pyr2)—suffer from poor specificity, displaying a variety of effects on other channels particularly members of the various TRP families including TRPC3/5/6/7, TRPV1/2/3, and TRPM4/6/7. More important is a critical, but almost universally overlooked, problem arising in any attempt to develop appropriate drugs that target these channels. Thus, the current emphasis on CRAC channels, together with our increasing knowledge of the detailed molecular nature of the Orai1 molecules that form these channels and the STIM1 that regulates their activity, would seem to suggest an opportunity to directly target these molecules. Indeed, several major pharmaceutical companies are actively involved in just such a task. However, it is important to note that such approaches potentially overlook a fundamental and critical problem—namely that both Orai1 and STIM1 are also essential components of the co-existing, *but functionally entirely distinct*, ARC channels. Consequently, any attempts to target these proteins would *not* be selective for CRAC channel activity—or, indeed, ARC channel activity—alone. True selectivity for either of these Orai channels will require a far more subtle and incisive approach, most likely based on a detailed understanding of the distinct modes of activation of these co-existing channels.

26.6 Conclusions

It should be clear from the above that the highly calcium-selective, inwardly rectifying channels encoded by the Orai family of proteins represent a vitally important component of cellular calcium signaling in a diverse range of cell types. While the critical role of the CRAC channels in a variety of immunological and allergic reactions has essentially driven the majority of studies to date, it is becoming increasingly clear that these, as well as other endogenous Orai channels, are widely expressed in a variety of different cell types including those in epithelial tissues. Consequently, although the study of their roles in such tissues is somewhat complicated by their characteristically small whole-cell currents, the demonstrated agonist-induced calcium signals in these cell types indicates that they are likely to be of critical importance. However, in this context, it is important to raise a note of caution. Thus, in almost all cells studied to date, the currents generated by the Orai proteins are uniquely small (i.e., whole-cell currents of ~0.5–1.0 pA/pF at −80 mV, and estimated single channel currents in the 10–20 fS range) (Zweifach and Lewis 1993; Prakriya and Lewis 2002). Because of this, examination of their contributions to various key functions is often problematic. A common response to this is to overexpress the relevant proteins (STIM1 and/or Orai) to facilitate the recording of their activities. However, given the increasing number of reports indicating a plethora of different proteins that act to modulate and/or influence the activity of these channels, it would seem that such "selective overexpression" of just STIM and/or Orai might result in significant misinterpretation of their natural functions in cells. In any event, it is clear that there is still much to be learnt about their activities in such tissues under both physiological and pathophysiological conditions.

Acknowledgements The studies reported here from the author's laboratory were supported by NIH grant GM040557 to TJS. The excellent technical assistance of Ms. Jill Thompson is also acknowledged.

References

Allen V, Swigart P, Cheung R, Cockcroft S, Katan M (1997) Regulation of inositol lipid-specific phospholipase Cδ by changes in Ca^{2+} ion concentrations. Biochem J 327:545–552

Ambudkar IS (2014) Ca^{2+} signaling and regulation of fluid secretion in salivary gland acinar cells. Cell Calcium 55:297–305
Amcheslavsky A, Wood ML, Yeromin AV, Parker I, Freites JA, Tobias DJ, Cahalan MD (2015) Molecular biophysics of Orai store-operated Ca^{2+} channels. Biophys J 108:237–246
Bahra P, Mesher J, Li S, Poll CT, Danahay H (2004) P2Y2-receptor-mediated activation of a contralateral, lanthanide-sensitive calcium entry pathway in the human airway epithelium. Br J Pharmacol 143:91–98
Berridge MJ (1990) Calcium oscillations. J Biol Chem 265:9583–9586
Bird GS, Putney JW Jr (2005) Capacitative calcium entry supports calcium oscillations in human embryonic kidney cells. J Physiol 562:697–706
Bird GS, Hwang SY, Smyth JT, Fukushima M, Boyles RR, Puney JW Jr (2009) STIM1 is a calcium sensor specialized for digital signaling. Curr Biol 19:1724–1729
Brandman O, Liou J, Park WS, Meyer T (2007) STIM2 is a feedback regulator that stabilizes basal cytosolic and endoplasmic reticulum Ca^{2+} levels. Cell 131:1327–1339
Braun A, Varga-Szabo D, Kleinschnitz C, Pleines I, Bender M, Austinat M, Bösl M, Stoll G, Nieswandt B (2009) Orai1 (CRACM1) is the platelet SOC channel and essential for pathological thrombus formation. Blood 113:2056–2063
Broad LM, Cannon TR, Taylor CW (1999) A non-capacitative pathway activated by arachidonic acid is the major Ca^{2+} entry mechanism in rat A7r5 smooth muscle cells stimulated with low concentrations of vasopressin. J Physiol 517:121–134
Cai X, Zhou Y, Nwokonko RM, Loktionova NA, Wang X, Xin P, Trebak M, Wang Y, Gill DL (2016) The Orai1 store-operated calcium channel functions as a hexamer. J Biol Chem 291:25764–25775
Carey MA, Card JW, Voltz JW, Voltz JW, Arbes SJ Jr, Germolec DR, Korach KS, Zeldin DC (2007) It's all about sex: gender, lung development and lung disease. Trends Endocrinol Metab 18:308–313
Chen YF, Chiu WT, Chen YT, Lin PY, Huang HJ, Chou CY, Chang HC, Tang MJ, Shen MR (2011) Calcium store sensor stromal-interaction molecule 1-dependent signaling plays an important role in cervical cancer growth, migration, and angiogenesis. Proc Natl Acad Sci U S A 108:15225–15230
Cheng KT, Alevizos I, Liu X, Swaim WD, Yin H, Feske S, Oh-Hora M, Ambudkar IS (2012) STIM1 and STIM2 protein deficiency in T lymphocytes underlies development of the exocrine gland autoimmune disease, Sjogren's syndrome. Proc Natl Acad Sci U S A 109:14544–14549
Coakley RD, Sun H, Clunes LA, Rasmussen JE, Stackhouse JR, Okada SF, Fricks I, Young SL, Tarran R (2008) 17beta-estradiol inhibits Ca^{2+}-dependent homeostasis of airway surface liquid volume in human cystic fibrosis airway epithelia. J Clin Invest 118:4025–4035
Concepcion AR, Feske S (2017) Regulation of epithelial ion transport in exocrine glands by store-operated Ca^{2+} entry. Cell Calcium 63:53–59
Concepcion AR, Vaeth M, Wagner LE, Eckstein M, Hecht L, Yang J, Crottes D, Seidl M, Shin HP, Weidinger C, Cameron S, Turvey SE, Issekutz T, Lacruz RS MI, Cuk M, Yule DI, Feske S (2016) Store-operated Ca^{2+} entry regulates Ca^{2+}-activated chloride channels and eccrine sweat gland function. J Clin Invest 126:4303–4318
Covington ED, Wu MM, Lewis RS (2010) Essential role for the CRAC activation domain in store-dependent oligomerization of STIM1. Mol Biol Cell 21:1897–1907
Criddle DN, McLaughlin E, Murphy JA, Petersen OH, Sutton R (2007) The pancreas misled: signals to pancreatitis. Pancreatology 7:436–446
DeHaven WI, Smyth JT, Boyles RR, Bird GS, Putney JW Jr (2008) Complex actions of 2-aminoethyldiphenyl borate on store-operated calcium entry. J Biol Chem 283:19265–19273
Demuro A, Penna A, Safrina O, Yeromin AV, Amcheslavsky A, Cahalan MD, Parker I (2011) Subunit stoichiometry of human Orai1 and Orai3 channels in closed and open states. Proc Natl Acad Sci U S A 108:17832–17837
Derler I, Schindl R, Fritsch R, Fritsch R, Romanin C (2012) Gating and permeation of Orai channels. Front Biosci 17:1304–1322

Derler I, Plenk P, Fahrner M, Muik M, Jardin I, Schindl R, Gruber HJ, Groschner K, Romanin C (2013) The extended transmembrane Orai1 N-terminal (ETON) region combines binding interface and gate for Orai1 activation by STIM1. J Biol Chem 288:29025–29034

Derler I, Butorac C, Krizova A, Stadlbauer M, Muik M, Fahrner M, Frischauf I, Romanin C (2018) Authentic CRAC channel activity requires STIM1 and the conserved portion of the Orai N terminus. J Biol Chem 293:1259–1270

Di Capite JL, Bates GJ, Parekh AB (2011) Mast cell CRAC channel as a novel therapeutic target in allergy. Curr Opin Allergy Clin Immunol 11:33–38

Dolmetsch RE, Lewis RS, Goodnow CC, Healy JI (1997) Differential activation of transcription factors induced by Ca^{2+} response amplitude and duration. Nature 386:855–858

Faouzi M, Hague F, Potier M, Ahidouch A, Sevestre H, Ouadid-Ahidouch H (2011) Down-regulation of Orai3 arrests cell-cycle progression and induces apoptosis in breast cancer cells but not in normal breast epithelial cells. J Cell Physiol 226:542–551

Feske S, Gwack Y, Prakriya M, Srikanth S, Puppel SH, Tanasa B, Hogan PG, Lewis RS, Daly M, Rao A (2006) A mutation in Orai1 causes immune deficiency by abrogating CRAC channel function. Nature 441:179–185

Fiorio Pla A, Munaron L (2001) Calcium influx, arachidonic acid, and control of endothelial cell proliferation. Cell Calcium 30:235–244

Flourakis M, Lehen'kyi V, Beck B, Raphaël M, Vandenberghe M, Abeele FV, Roudbaraki M, Lepage G, Mauroy B, Romanin C, Shuba Y, Skryma R, Prevarskaya N (2010) Orai1 contributes to the establishment of an apoptosis-resistant phenotype in prostate cancer cells. Cell Death Dis 1:e75

Gerasimenko JV, Gryshchenko O, Ferdek PE, Stapleton E, Hébert TO, Bychkova S, Peng S, Begg M, Gerasimenko OV, Petersen OH (2013) Ca^{2+} release-activated Ca^{2+} channel blockade as a potential tool in antipancreatitis therapy. Proc Natl Acad Sci U S A 110:13186–13191

Gonzalez-Cobos JC, Zhang X, Zhang W, Ruhle B, Motiani RK, Schindl R, Muik M, Spinelli AM, Bisaillon JM, Shinde AV, Fahrner M, Singer HA, Matrougui K, Barroso M, Romanin C, Trebak M (2013) Store-independent Orai1/3 channels activated by intracrine leukotriene C4: role in neointimal hyperplasia. Circ Res 112:1013–1025

Gudlur A, Hogan PG (2017) The STIM-Orai pathway: Orai, the pore-forming subunit of the CRAC channel. Adv Exp Med Biol 993:39–57

Hardie RC, Minke B (1992) The trp gene is essential for a light-activated Ca^{2+} channel in Drosophila photoreceptors. Neuron 8:643–651

Hardie RC, Raghu P, Moore S, Juusola M, Baines RA, Sweeney ST (2001) Calcium influx via TRP channels is required to maintain PIP2 levels in Drosophila photoreceptors. Neuron 30:149–159

Hirve N, Rajanikanth V, Hogan PG, Gudlur A (2018) Coiled-coil formation conveys a STIM1 signal from ER lumen to cytoplasm. Cell Rep 22:72–83

Hoth M, Penner R (1992) Depletion of intracellular calcium stores activates a calcium current in mast cells. Nature 355:353–356

Hoth M, Penner R (1993) Calcium release-activated calcium current in rat mast cells. J Physiol 465:359–386

Hou X, Pedi L, Diver MM, Long SB (2012) Crystal structure of the calcium release-activated calcium channel Orai. Science 338:1308–1313

Huang Y, Putney JW Jr (1998) Relationship between intracellular calcium store depletion and calcium release-activated calcium current in a mast cell line (RBL-1). J Biol Chem 273:19554–19559

Hulot JS, Fauconnier J, Ramanujam D, Chaanine A, Aubart F, Sassi Y, Merkle S, Cazorla O, Ouillé A, Dupuis M, Hadri L, Jeong D, Mühlstedt S, Schmitt J, Braun A, Bénard L, Saliba Y, Laggerbauer B, Nieswandt B, Lacampagne A, Hajjar RJ, Lompré AM, Engelhardt S (2011) Critical role for stromal interaction molecule 1 in cardiac hypertrophy. Circulation 124:796–805

Ji W, Xu P, Li Z, Lu J, Liu L, Zhan Y, Chen Y, Hille B, Xu T, Chen L (2008) Functional stoichiometry of the unitary calcium-release-activated calcium channel. Proc Natl Acad Sci U S A 105:13668–13673

Kasai H, Li YX, Miyashita Y (1993) Subcellular distribution of Ca^{2+} release channels underlying Ca^{2+} waves and oscillations in exocrine pancreas. Cell 74:669–677

Kawasaki T, Lange I, Feske S (2009) A minimal regulatory domain in the C terminus of STIM1 binds to and activates ORAI1 CRAC channels. Biochem Biophys Res Commun 385:49–54

Ko WH, Chan HC, Wong PY (1996) Anion secretion induced by capacitative Ca^{2+} entry through apical and basolateral membranes of cultured equine sweat gland epithelium. J Physiol 497:19–29

Korzeniowski MK, Manjarres IM, Varnai P, Balla T (2010) Activation of STIM1-Orai1 involves an intramolecular switching mechanism. Sci Signal 3:ra82

Kruger B, Albrecht E, Lerch MM (2000) The role of intracellular calcium signaling in premature protease activation and the onset of pancreatitis. Am J Pathol 157:43–50

Kwan CY, Putney JW Jr (1990) Uptake and intracellular sequestration of divalent cations in resting and methacholine-stimulated mouse lacrimal acinar cells. Dissociation by Sr^{2+} and Ba^{2+} of agonist-stimulated divalent cation entry from the refilling of the agonist- sensitive intracellular pool. J Biol Chem 265:678–684

Li YS, Wu P, Zhou XY, Chen JG, Cai L, Wang F, Xu LM, Zhang XL, Chen Y, Liu SJ, Huang YP, Ye DY (2008) Formyl-peptide receptor like 1: a potent mediator of the Ca^{2+} release-activated Ca^{2+} current ICRAC. Arch Biochem Biophys 478:110–118

Liou J, Kim ML, Heo WD, Jones JT, Myers JW, Ferrell JE Jr, Meyer T (2005) STIM is a Ca^{2+} sensor essential for Ca^{2+}-store-depletion-triggered Ca^{2+} influx. Curr Biol 15:1235–1241

Liou J, Fivaz M, Inoue T, Meyer T (2007) Live-cell imaging reveals sequential oligomerization and local plasma membrane targeting of stromal interaction molecule 1 after Ca^{2+} store depletion. Proc Natl Acad Sci U S A 104:9301–9306

Lis A, Peinelt C, Beck A, Parvez S, Monteilh-Zoller M, Fleig A, Penner R (2007) CRACM1, CRACM2, and CRACM3 are store-operated Ca^{2+} channels with distinct functional properties. Curr Biol 17:794–800

Luik RM, Wu MM, Buchanan J, Lewis RM (2006) The elementary unit of store-operated Ca^{2+} entry: local activation of CRAC channels by STIM1 at ER-plasma membrane junctions. J Cell Biol 174:815–825

Luik RM, Wang B, Prakriya M, Wu MM, Lewis RS (2008) Oligomerization of STIM1 couples ER calcium depletion to CRAC channel activation. Nature 454:538–542

Madl J, Weghuber J, Fritsch R, Derler I, Fahrner M, Frischauf I, Lackner B, Romanin C, Schütz GJ (2010) Resting state Orai1 diffuses as homotetramer in the plasma membrane of live mammalian cells. J Biol Chem 285:41135–41142

Malli R, Naghdi S, Romanin C, Graier WF (2008) Cytosolic Ca^{2+} prevents the subplasmalemmal clustering of STIM1: an intrinsic mechanism to avoid Ca^{2+} overload. J Cell Sci 121:3133–3139

Manji SS, Parker NJ, Williams RT, van Stekelenburg L, Pearson RB, Dziadek M, Smith PJ (2000) STIM1: a novel phosphoprotein located at the cell surface. Biochim Biophys Acta 1481:147–155

Martin SC, Shuttleworth TJ (1994) Ca^{2+} influx drives agonist-activated $[Ca^{2+}]_i$ oscillations in an exocrine cell. FEBS Lett 352:32–36

Maruyama Y, Ogura T, Mio K, Kato K, Kaneko T, Kiyonaka S, Mori Y, Sato C (2009) Tetrameric Orai1 is a teardrop-shaped molecule with a long, tapered cytoplasmic domain. J Biol Chem 284:13676–13685

McAndrew D, Grice DM, Peters AA, Davis FM, Stewart T, Rice M, Smart CE, Brown MA, Kenny PA, Roberts-Thompson SJ, Monteith GR (2011) ORAI1-mediated calcium influx in lactation and in breast cancer. Mol Cancer Ther 10:448–460

McNally BA, Prakriya M (2012) Permeation, selectivity and gating in store-operated CRAC channels. J Physiol 590:4179–4191

McNally BA, Somasundaram A, Jairaman A, Jairaman A, Yamashita M, Prakriya M (2013) The C- and N-terminal STIM1 binding sites on Orai1 are required for both trapping and gating CRAC channels. J Physiol 591:2833–2850

Melvin JL, Koek L, Zhang GH (1991) A capacitative Ca^{2+} influx is required for sustained fluid secretion in sublingual mucous acini. Am J Phys 261:G1043–G1050
Mignen O, Shuttleworth TJ (2000) I_{ARC}, a novel arachidonate-regulated, noncapacitative Ca^{2+} entry channel. J Biol Chem 275:9114–9119
Mignen O, Thompson JL, Shuttleworth TJ (2001) Reciprocal regulation of capacitative and arachidonate-regulated noncapacitative Ca^{2+} entry pathways. J Biol Chem 276:35676–35683
Mignen O, Thompson JL, Shuttleworth TJ (2003) Ca^{2+} selectivity and fatty acid specificity of the noncapacitative, arachidonate-regulated Ca^{2+} (ARC) channels. J Biol Chem 278:10174–10181
Mignen O, Thompson JL, Yule DI, Shuttleworth TJ (2005) Agonist activation of arachidonate-regulated Ca^{2+}-selective (ARC) channels in murine parotid and pancreatic acinar cells. J Physiol 564:791–801
Mignen O, Thompson JL, Shuttleworth TJ (2007) STIM1 regulates Ca^{2+} entry via arachidonate-regulated Ca^{2+}-selective (ARC) channels without store depletion or translocation to the plasma membrane. J Physiol 579:703–715
Mignen O, Thompson JL, Shuttleworth TJ (2008a) Both Orai1 and Orai3 are essential components of the arachidonate-regulated Ca^{2+}-selective (ARC) channels. J Physiol 586:185–195
Mignen O, Thompson JL, Shuttleworth TJ (2008b) Orai1 subunit stoichiometry of the mammalian CRAC channel pore. J Physiol 586:419–425
Mignen O, Thompson JL, Shuttleworth TJ (2009) The molecular architecture of the arachidonate-regulated Ca^{2+}-selective ARC channel is a pentameric assembly of Orai1 and Orai3 subunits. J Physiol 587:4181–4197
Morgan AJ, Jacob R (1996) Ca^{2+} influx does more than provide releasable Ca^{2+} to maintain repetitive spiking in human umbilical vein endothelial cells. Biochem J 320:505–517
Muik M, Frischauf I, Derler I, Fahrner M, Bergsmann J, Eder P, Schindl R, Hesch C, Polzinger B, Fritsch R, Kahr H, Madl J, Gruber H, Groschner K, Romanin C (2008) Dynamic coupling of the putative coiled-coil domain of ORAI1 with STIM1 mediates ORAI1 channel activation. J Biol Chem 283:8014–8022
Oh-Hora M, Yamashita M, Hogan PG, Sharma S, Lamperti E, Chung W, Prakriya M, Feske S, Rao A (2008) Dual functions for the endoplasmic reticulum calcium sensors STIM1 and STIM2 in T cell activation and tolerance. Nat Immunol 9:432–443
Palty R, Isacoff EY (2016) Cooperative binding of stromal interaction molecule 1 (STIM1) to the N and C termini of calcium release-activated calcium modulator 1 (Orai1). J Biol Chem 29:334–341
Paradiso AM, Mason SJ, Lazarowski ER, Boucher RC (1995) Membrane-restricted regulation of Ca^{2+} release and influx in polarized epithelia. Nature 377:643–646
Parekh AB, Putney JW Jr (2005) Store-operated calcium channels. Physiol Rev 85:757–810
Parekh AB, Fleig A, Penner R (1997) The store-operated calcium current I_{CRAC}: nonlinear activation by $InsP_3$ and dissociation from calcium release. Cell 89:973–980
Park CY, Hoover PJ, Mullins FM, Bachhawat P, Covington ED, Raunser S, Walz T, Garcia KC, Dolmetsch RE, Lewis RS (2009) STIM1 clusters and activates CRAC channels via direct binding of a cytosolic domain to Orai1. Cell 136:876–890
Parod RJ, Putney JW Jr (1978) The role of calcium in the receptor mediated control of potassium permeability in the rat lacrimal gland. J Physiol 281:371–381
Peinelt C, Lis A, Beck A, Fleig A, Penner R (2008) 2-Aminoethoxydiphenyl borate directly facilitates and indirectly inhibits STIM1-dependent gating of CRAC channels. J Physiol 586:3061–3073
Penna A, Demuro A, Yeromin AV, Zhang SL, Safrina O, Parker I, Cahalan MD (2008) The CRAC channel consists of a tetramer formed by Stim-induced dimerization of Orai dimers. Nature 456:116–120
Petersen OH, Sutton R (2006) Ca^{2+} signalling and pancreatitis: effects of alcohol, bile and coffee. Trends Pharmacol Sci 27:113–120
Petersen OH, Tepikin AV (2008) Polarized calcium signaling in exocrine gland cells. Annu Rev Physiol 70:273–299

Petersen OH, Tepikin AV, Gerasimenko JV, Gerasimenko OV, Sutton R, Criddle DN (2009) Fatty acids, alcohol and fatty acid ethyl esters: toxic Ca^{2+} signal generation and pancreatitis. Cell Calcium 45:634–642
Prakriya M (2009) The molecular physiology of CRAC channels. Immunol Rev 231:88–98
Prakriya M, Lewis RS (2002) Separation and characterization of currents through store-operated CRAC channels and Mg^{2+}-inhibited cation (MIC) channels. J Gen Physiol 119:487–507
Prakriya M, Lewis RS (2015) Store-operated calcium currents. Physiol Rev 95:1383–1436
Prakriya M, Feske S, Gwack Y, Srikanth S, Rao A, Hogan PG (2006) Orai1 is an essential pore subunit of the CRAC channel. Nature 443:230–233
Putney JW Jr (1986) A model for receptor-regulated calcium entry. Cell Calcium 7:1–12
Putney JW Jr (1990) Capacitative calcium entry revisited. Cell Calcium 11:611–624
Putney JW Jr, Huang Y, Bird GS (1998) Calcium signalling in lacrimal acinar cells. Adv Exp Med Biol 438:123–128
Raraty M, Ward J, Erdemli G, Vaillant C, Neoptolemos JP, Sutton R, Petersen OH (2000) Calcium-dependent enzyme activation and vacuole formation in the apical granular region of pancreatic acinar cells. Proc Natl Acad Sci U S A 97:13126–13131
Rebecchi MJ, Pentyala SN (2000) Structure, function, and control of phosphoinositide-specific phospholipase C. Physiol Rev 80:1291–1335
Ribeiro CM, Putney JW (1996) Differential effects of protein kinase C activation on calcium storage and capacitative calcium entry in NIH 3T3 cells. J Biol Chem 271:21522–21528
Roos J, DiGregorio PJ, Yeromin AV, Ohlsen K, Lioudyno M, Zhang S, Safrina O, Kozak JA, Wagner SL, Cahalan MD, Veliçelebi G, Stauderman KA (2005) STIM1, an essential and conserved component of store-operated Ca^{2+} channel function. J Cell Biol 169:435–445
Rosenfeld M, Davis R, FitzSimmons S, Pepe M, Ramsey B (1997) Gender gap in cystic fibrosis mortality. Am J Epidem 145:794–803
Schindl R, Bergsmann J, Frischauf I, Derler I, Fahrner M, Muik M, Fritsch R, Groschner K, Romanin C (2008) 2-aminoethoxydiphenyl borate alters selectivity of Orai3 channels by increasing their pore size. J Biol Chem 283:20261–20267
Shaw PJ, Feske S (2012) Physiological and pathophysiological functions of SOCE in the immune system. Front Biosci 4:2253–2268
Sheridan JT, Gilmore RC, Watson MJ, Archer CB, Tarran R (2013) 17β-estradiol inhibits phosphorylation of stromal interaction molecule 1 (STIM1) protein: implication for store-operated calcium entry and chronic lung diseases. J Biol Chem 288:33509–33518
Shuttleworth TJ (1990) Fluoroaluminate activation of different components of the calcium signal in an exocrine cell. Biochem J 269:417–422
Shuttleworth TJ (1996) Arachidonic acid activates the noncapacitative entry of Ca^{2+} during $[Ca^{2+}]_i$ oscillations. J Biol Chem 271:21720–21725
Shuttleworth TJ (1999) What drives calcium entry during $[Ca^{2+}]_i$ oscillations?--challenging the capacitative model. Cell Calcium 25:237–246
Shuttleworth TJ (2009) Arachidonic acid, ARC channels, and Orai proteins. Cell Calcium 45:602–610
Shuttleworth TJ, Mignen O (2003) Calcium entry and the control of calcium oscillations. Biochem Soc Trans 31:916–919
Shuttleworth TJ, Thompson JL (1996a) Ca^{2+} entry modulates oscillation frequency by triggering Ca^{2+} release. Biochem J 313:815–819
Shuttleworth TJ, Thompson JL (1996b) Evidence for a non-capacitative Ca^{2+} entry during $[Ca^{2+}]$ oscillations. Biochem J 316:819–824
Shuttleworth TJ, Thompson JL (1998) Muscarinic receptor activation of arachidonate-mediated Ca^{2+} entry in HEK293 cells is independent of phospholipase C. J Biol Chem 273:32636–32643
Shuttleworth TJ, Thompson JL (1999) Discriminating between capacitative and arachidonate-activated Ca^{2+} entry pathways in HEK293 cells. J Biol Chem 274:31174–31178

Stathopulos PB, Li GY, Plevin MJ, Ames JB, Ikura M (2006) Stored Ca^{2+} depletion-induced oligomerization of stromal interaction molecule 1 (STIM1) via the EF-SAM region: an initiation mechanism for capacitive Ca^{2+} entry. J Biol Chem 281:35855–35862
Takemura H, Putney JW Jr (1989) Capacitative calcium entry in parotid acinar cells. Biochem J 258:409–412
Takemura H, Hughes AR, Thastrup O, Putney JW (1989) Activation of calcium entry by the tumor promoter, thapsigargin, in parotid acinar cells. Evidence that an intracellular calcium pool, and not an inositol phosphate, regulates calcium fluxes at the plasma membrane. J Biol Chem 264:12266–12271
Thomas AP, Bird GS, Hajnoczky G, Robb-Gaspers LD, Putney JW Jr (1996) Spatial and temporal aspects of cellular calcium signaling. FASEB J 10:1505–1517
Thompson JL, Shuttleworth TJ (2011) Orai channel-dependent activation of phospholipase C-δ: a novel mechanism for the effects of calcium entry on calcium oscillations. J Physiol 589:5057–5069
Thompson JL, Shuttleworth TJ (2013a) How many Orai's does it take to make a CRAC channel? Sci Rep 3:1961
Thompson JL, Shuttleworth TJ (2013b) Molecular basis of activation of the arachidonate-regulated Ca^{2+} (ARC) channel, a store-independent Orai channel, by plasma membrane STIM1. J Physiol 591:3507–3523
Thompson JL, Shuttleworth TJ (2015) Anchoring protein AKAP79-mediated PKA phosphorylation of STIM1 determines selective activation of the ARC channel, a store-independent Orai channel. J Physiol 593:559–572
Thompson JL, Mignen O, Shuttleworth TJ (2010) The N-terminal domain of Orai3 determines selectivity for activation of the store-independent ARC channel by arachidonic acid. Channels 4:398–410
Thompson JL, Zhao Y, Stathopulos PB, Grossfield A, Shuttleworth TJ (2018) Phosphorylation-mediated structural changes within the SOAR domain of stromal interaction molecule 1 enable specific activation of distinct Orai channels. J Biol Chem 293:3145–3155
Thorn P, Lawrie AM, Smith PM, Gallacher DV, Petersen OH (1993) Local and global cytosolic Ca^{2+} oscillations in exocrine cells evoked by agonists and inositol trisphosphate. Cell 74:661–668
Trebak M (2012) STIM/Orai signalling complexes in vascular smooth muscle. J Physiol 590:4201–4208
Vashisht A, Trebak M, Motiani RK (2015) STIM and Orai proteins as novel targets for cancer therapy: a review in the theme: cell and molecular processes in cancer metastasis. Am J Physiol Cell Physiol 309:C457–C469
Vig M, Beck A, Billingsley JM, Lis A, Parvez S, Peinelt C, Koomoa DL, Soboloff J, Gill DL, Fleig A, Kinet JP, Penner R (2006a) CRACM1 multimers form the ion-selective pore of the CRAC channel. Curr Biol 16:2073–2079
Vig M, Peinelt C, Beck A, Koomoa DL, Rabah D, Koblan-Huberson M, Kraft S, Turner H, Fleig A, Penner R, Kinet JP (2006b) CRACM1 is a plasma membrane protein essential for store-operated Ca^{2+} entry. Science 312:1220–1223
Wang X, Wang Y, Zhou Y, Hendron E, Mancarella S, Andrake MD, Rothberg BS, Soboloff J, Gill DL (2014) Distinct Orai-coupling domains in STIM1 and STIM2 define the Orai-activating site. Nat Commun 5:3183
Wu MM, Buchanan J, Luik RM, Lewis RS (2006) Ca^{2+} store depletion causes STIM1 to accumulate in ER regions closely associated with the plasma membrane. J Cell Biol 174:803–813
Xu P, Lu J, Li Z, Yu X, Chen L, Xu T (2006) Aggregation of STIM1 underneath the plasma membrane induces clustering of Orai1. Biochem Biophys Res Commun 350:969–976
Yang KT, Chen WP, Chang WL, Su MJ, Tsai KL (2005) Arachidonic acid inhibits capacitative Ca^{2+} entry and activates non-capacitative Ca^{2+} entry in cultured astrocytes. Biochem Biophys Res Commun 331:603–613
Yang S, Zhang JJ, Huang XY (2009) Orai1 and STIM1 are critical for breast tumor cell migration and metastasis. Cancer Cell 15:124–134

Yang X, Jin H, Cai X, Li S, Shen Y (2012) Structural and mechanistic insights into the activation of stromal interaction molecule 1 (STIM1). Proc Natl Acad Sci U S A 109:5657–5662

Yen M, Lokteva LA, Lewis RS (2016) Functional analysis of Orai1 concatemers supports a hexameric stoichiometry for the CRAC channel. Biophys J 111:1897–1907

Yeromin AV, Zhang SL, Jiang W, Yu Y, Safrina O, Cahalan MD (2006) Molecular identification of the CRAC channel by altered ion selectivity in a mutant of Orai. Nature 443:226–229

Yeung-Yam-Wah V, Lee AK, Tse FW, Tse A (2010) Arachidonic acid stimulates extracellular Ca^{2+} entry in rat pancreatic β cells via activation of the noncapacitative arachidonate-regulated Ca^{2+} (ARC) channels. Cell Calcium 47:77–83

Yoo AS, Cheng I, Chung S, Grenfell TZ, Lee H, Pack-Chung E, Handler M, Shen J, Xia W, Tesco G, Saunders AJ, Ding K, Frosch MP, Tanzi RE, Kim TW (2000) Presenilin-mediated modulation of capacitative calcium entry. Neuron 27:561–572

Yuan JP, Zeng W, Dorwart MR, Choi YJ, Worley PF, Muallem S (2009) SOAR and the polybasic STIM1 domains gate and regulate Orai channels. Nat Cell Biol 11:337–343

Yule DI, Gallacher DV (1988) Oscillations of cytosolic calcium in single pancreatic acinar cells stimulated by acetylcholine. FEBS Lett 239:358–362

Zhang SL, Yu Y, Roos J, Kozak JA, Deerinck TJ, Ellisman MH, Stauderman KA, Cahalan MD (2005) STIM1 is a Ca^{2+} sensor that activates CRAC channels and migrates from the Ca^{2+} store to the plasma membrane. Nature 437:902–905

Zhang SL, Yeromin AV, Zhang XHYY, Safrina O, Penna A, Roos J, Stauderman KA, Cahalan MD (2006) Genome-wide RNAi screen of Ca^{2+} influx identifies genes that regulate Ca^{2+} release-activated Ca^{2+} channel activity. Proc Natl Acad Sci U S A 103:9357–9362

Zhang SL, Kozak JA, Jiang W, Yeromin AV, Chen J, Yu Y, Penna A, Shen W, Chi V, Cahalan MD (2008) Store-dependent and -independent modes regulating Ca^{2+} release-activated Ca^{2+} channel activity of human Orai1 and Orai3. J Biol Chem 283:17662–17671

Zhang X, Gonzalez-Cobos JC, Schindl R, Muik M, Ruhle B, Motiani RK, Bisaillon JM, Zhang W, Fahrner M, Barroso M, Matrougui K, Romanin C, Trebak M (2013) Mechanisms of STIM1 activation of store-independent leukotriene C4-regulated Ca^{2+} channels. Mol Cell Biol 33:3715–3723

Zhang X, Zhang W, Gonzalez-Cobos JC, Jardin I, Romanin C, Matrougui K, Trebak M (2014) Complex role of STIM1 in the activation of store-independent Orai1/3 channels. J Gen Physiol 143:345–359

Zhou Y, Srinivasan P, Razavi S, Seymour S, Meraner P, Gudlur A, Stathopulos PB, Ikura M, Rao A, Hogan PG (2013) Initial activation of STIM1, the regulator of store-operated calcium entry. Nat Struct Mol Biol 20:973–981

Zhou Y, Cai X, Loktionova NA, Wang X, Nwokonko RM, Wang X, Wang Y, Rothberg BS, Trebak M, Gill DL (2016) The STIM1-binding site nexus remotely controls Orai1 channel gating. Nat Commun 7:13725

Zweifach A, Lewis RS (1993) Mitogen-regulated Ca^{2+} current of T lymphocytes is activated by depletion of intracellular Ca^{2+} stores. Proc Natl Acad Sci U S A 90:6295–6299

Zweifach A, Lewis RS (1995) Rapid inactivation of depletion-activated calcium current (I_{CRAC}) due to local calcium feedback. J Gen Physiol 105:209–226

Chapter 27
TRP Channels in Renal Epithelia

Viktor N. Tomilin, Oleg Zaika, and Oleh Pochynyuk

Abstract Invariability of the internal milieu of the body is essential for normal physiology. Kidneys are imperative for systemic homeostasis and blood pressure control by tuning water-solute transport in the renal tubular epithelium in response to a vast array of dietary, endocrine, and mechanical inputs. Over the last two decades, functional expression of different transient receptor potential (TRP) channels with highly specialized mechanisms of activation to a broad range of external stimuli has been identified in renal epithelia. Here, TRP channels have been demonstrated to significantly contribute to renal physiology and pathology. Recent advances in creating transgenic animals lacking TRP channels and in elucidating high-resolution near-atomic TRP channel structures enabled to integrate TRP-mediated cellular processes with changes in kidney function and urinary production. It is anticipated that modulation of TRP channels activity will be instrumental in treatment of various renal and cardiovascular pathologies.

Keywords Electrolyte Transport · Mechanosensitivity · Channelopathies · Calcium · Magnesium · Renal Injury

27.1 Introduction

Transient receptor potential (TRP) channels inherited their name after a casual but undoubtedly seminal observation in 1969 when a spontaneous recessive mutation in chromosome 3 of *Drosophila melanogaster* caused a defective transient electroretinogram response to a continuous stimulation with light (Cosens and Manning 1969). The mutant was thereafter known as "transient receptor potential." The founding member of TRP family was first cloned from the rhabdomeric membranes

V. N. Tomilin · O. Zaika · O. Pochynyuk (✉)
Department of Integrative Biology and Pharmacology, University of Texas Health Science Center, Houston, USA
e-mail: Oleh.M.Pochynyuk@uth.tmc.edu

K. L. Hamilton, D. C. Devor (eds.), *Studies of Epithelial Transporters and Ion Channels*, Physiology in Health and Disease,
https://doi.org/10.1007/978-3-030-55454-5_27

of the photoreceptor cells as a 1275-amino acid protein with eight putative transmembrane segments (two of them are not truly transmembrane forming so-called P-loop) and was further characterized as a Ca^{2+}-permeable channel using whole-cell patch clamp in *Drosophila* photoreceptors (Montell and Rubin 1989; Hardie and Minke 1992). Since then, multiple TRP and TRP-related genes have been identified in all organisms from *fungi* and lower invertebrates to mammals (Venkatachalam and Montell 2007; Nilius and Owsianik 2011).

TRP are cation-selective channels allowing passage of both monovalent (Na^+, K^+) and divalent (Ca^{2+}, Mg^{2+}) physiologically relevant ions (Venkatachalam and Montell 2007). However, TRP-dependent Ca^{2+} influx mediates the majority of TRP functions at the cellular and systemic levels (Nilius and Owsianik 2011). TRP channels demonstrate a remarkable ability to be activated by a whole spectrum of physical (shear stress, osmotic swelling/shrinkage, cold and hot temperatures) and chemical (biological agents, signaling molecules, and so on) stimuli (Venkatachalam and Montell 2007). Thus, TRP channels are thought to orchestrate cellular responses/sensitivity to environmental changes by translating them into the form of elevated intracellular Ca^{2+} concentration ($[Ca^{2+}]_i$). TRP channels are ubiquitously expressed in both excitable and nonexcitable cells, where they contribute to a large number of physiological processes, including nociception, inflammation, control of vascular tone, bone metabolism, ion transport, to name a few (Venkatachalam and Montell 2007). Their dysfunction underlies a variety of pathologies in clinic, including autosomal dominant polycystic kidney disease (ADPKD) (Mochizuki et al. 1996), focal segmental glomerular sclerosis (FSGS) (Reiser et al. 2005; Winn et al. 2005), hypomagnesemia with secondary hypocalcemia (HSH) (Schlingmann et al. 2002; Walder et al. 2002; Lainez et al. 2014), and Charcot-Marie-Tooth disease type 2C (Nilius and Voets 2013).

Based on amino acid sequence homology, TRP channels can be further divided into TRPC (canonical, originally classical), TRPV (vanilloid), TRPM (melastatin), TRPP (polycystin), TRPA (ankyrin), TRPML (mucolipin), and TRPN (no mechanoreceptor potential C, only found in invertebrates and zebrafish) subfamilies (Fig. 27.1) (Venkatachalam and Montell 2007; Nilius and Owsianik 2011). The current "golden age" of cryo-electron microscopy (cryo-EM) allowed to obtain near-atomic high-resolution 3-D structures of different TRP channels, providing invaluable insights into their molecular organization, ion permeability/selectivity, and gating mechanisms (Liao et al. 2013, 2014; Wilkes et al. 2017; Zhang et al. 2017; Deng et al. 2018; Duan et al. 2018; Fan et al. 2018; Tang et al. 2018). All TRP proteins contain six transmembrane domains (S1–S6) with intracellular amino (N)- and carboxyl (C)-termini of variable length and domain composition (Fig. 27.2a) (Venkatachalam and Montell 2007; Nilius and Owsianik 2011). The functional channel is a symmetric tetramer with S1-S4 domains (distantly related to the analogous domains in voltage-gated ion channels) surrounding the permeation pathway formed by S5 and S6 domains of each subunit (Liao et al. 2013). In addition, heteromers consisting of different TRP subunits were commonly reported in the literature (Schaefer 2005; Stewart et al. 2010; Fischer and Edwardson 2014), thus bringing another layer of complexity to the versatile nature of TRP channel

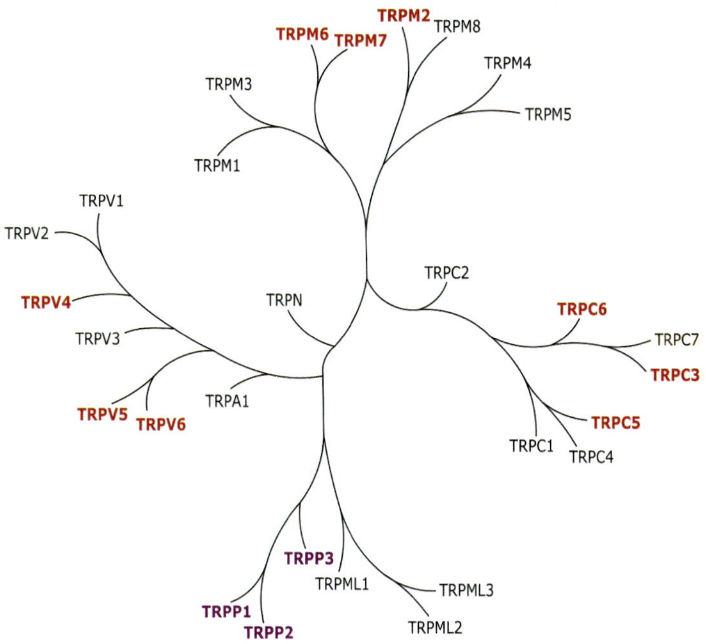

Fig. 27.1 Schematic presentation of TRP channel superfamily phylogenic tree. TRP channels expressed in the renal epithelia are highlighted in red

superfamily. The multiple available structures of TRP channels suggest the presence of a lower gate-blocking ion passage and an upper gate forming a selectivity filter with a wide central vestibule in the middle (Fig. 27.2b) (Liao et al. 2013; Deng et al. 2018; Fan et al. 2018). During a transition between open and closed states, both gates undergo conformational changes suggesting a dual gating mechanism (Cao et al. 2013; Liao et al. 2013). Furthermore, amphipathic TRP helix (the conserved among TRP channels a 23–25-amino-acid-long region located just after S6, also known as TRP box) is parallel to the plasma membrane inner leaflet. It is thought to connect S1-S4 and pore-forming domains likely serving to orchestrate channel gating in response to mechanical and chemical stimuli (Liao et al. 2013; Deng et al. 2018). The N-terminus contains a variable number (from 3–4 in TRPCs, 6 in TRPVs, and up to 15 in TRPA) of ankyrin repeat domains (ARD), which are important for proper positioning of the pre-S1 linker domain and for intersubunit interactions (Fig. 27.2a). Furthermore, ARD are likely involved in interactions with cytoskeleton and intracellular signaling molecules, such as adenosine triphosphate (ATP) and calmodulin to regulate channel gating (Liedtke et al. 2000; Lishko et al. 2007). TRP channels are also commonly associated with membrane phospholipids, most notably with phosphatidylinositol 4,5 bisphosphate (PI(4,5)P_2 or PIP_2). It is reported that PI (4,5)P_2 binding can elicit both stimulatory and inhibitory actions on different TRP channels (reviewed in (Rohacs 2014)). In addition, TRP channels undergo covalent post-translational modifications (including N-linked glycosylation, phosphorylation,

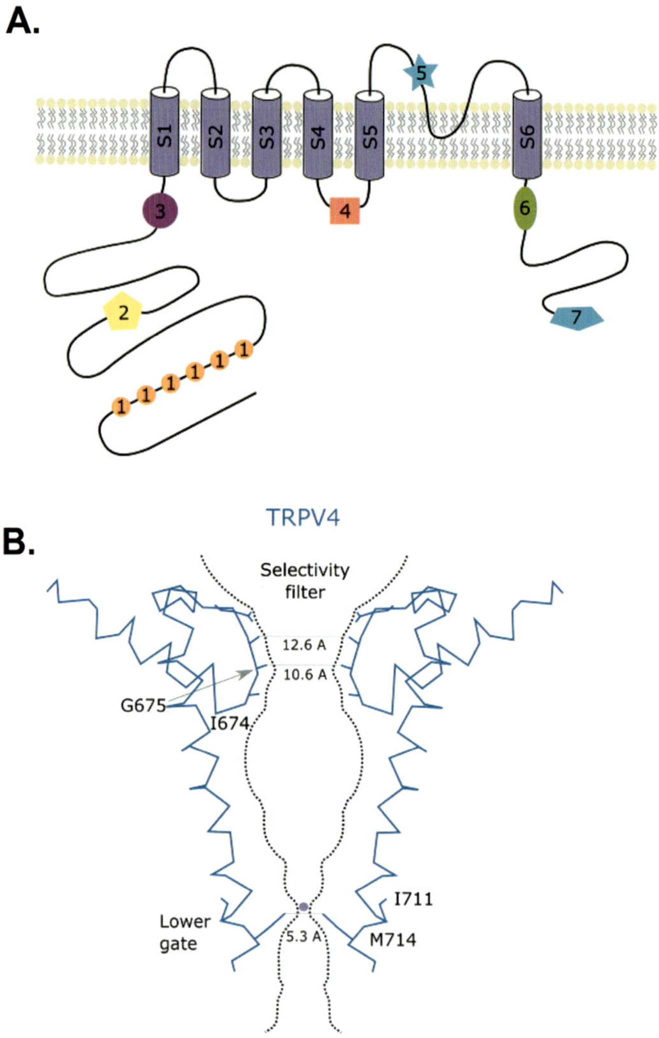

Fig. 27.2 TRP channel structure. (**a**) Schematic representation of major functional domains of a TRP channel subunit. (**S1–S6**) transmembrane domains 1–6, respectively. (**1**) ankyrin repeat domains, (**2**) linker domain, (**3**) pre-S1 helix, (**4**) S4–S5 linker, (**5**) pore helix, (**6**) TRP domain, 7—C-terminal domain. (**b**) The pore structure of TRPV4 channel outlining the upper selectivity filter and the lower permeation gate. Adapted from (Deng et al. 2018)

covalent modification of cysteine residues) causing profound effects on their gating, subcellular localization, and trafficking to the plasma membrane (reviewed in (Voolstra and Huber 2014)).

Kidneys are highly dynamic paired organs essential to maintain invariability of the internal milieu of the body. Even subtle changes in renal function due to genetic defects or polymorphisms commonly exert profound pathophysiological

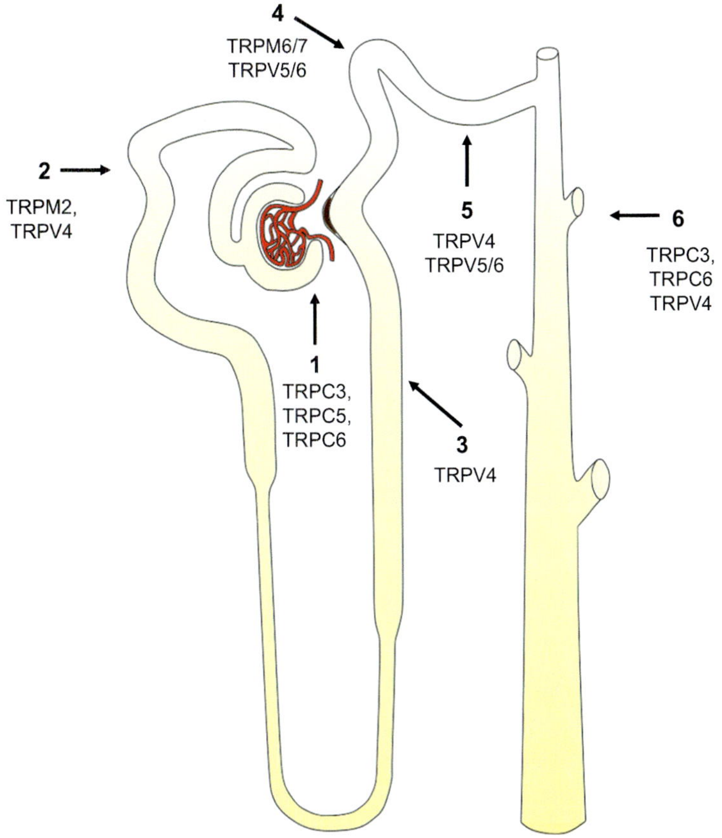

Fig. 27.3 TRP channel expression in the renal epithelium. Schematic representation of an individual renal nephron showing different segments and names of TRP channels reported to be expressed in the respective segment. (**1**) glomerulus, (**2**) proximal tubule, (**3**) thick ascending limb, (**4**) distal convoluted tubule, (**5**) connecting tubule, (**6**) collecting duct

ramifications resulting in systemic imbalance of electrolytes and blood pressure distortions (Lifton et al. 2001). The typical urine volume (1–1.5 L/day in humans) is only less than 1% of the filtered plasma volume by the glomeruli, whereas the remaining 99% of the filtrate is returned to circulation via the process of reabsorption by the renal tubule (see Fig. 27.3 for renal tubule structure). The proximal tubule (PT) segments and the loop of Henle are responsible for constitutive reabsorption of nearly 90% of the filtered load. In contrast, the distal segments, including the distal convoluted tubule (DCT), the connecting tubule (CNT), and the collecting duct (CD), contribute collectively up to 10% of the total reabsorption and are the sites of finetuning of water and electrolyte balance in response to dietary and endocrine inputs (Staruschenko 2012; Curthoys and Moe 2014; Mount 2014; Subramanya and Ellison 2014). Overall, kidneys represent a nexus of various environmental and systemic endocrine factors, which align nicely with the versatility of TRP channel activation by extracellular stimuli. Indeed, multiple TRP channels are abundantly expressed in both renal vasculature and renal tubule epithelium (Fig. 27.3). The expression patterns vary and could be very specific for some TRPs (TRPM2, TRPM6, and TRPV5), whereas other channels (TRPV4, TRPC3, and TRPC6) have a broad distribution along the renal tubule likely reflecting their distinct roles in the kidney. We next attempt to summarize in alphabetical order the available experimental and clinical evidence about TRP channels with the most defined functions in the renal tubular epithelium emphasizing their physiological relevance and hereditary Mendelian disorders due to channel mutations (so-called TRPpathies). TRPP subfamily of channels (polycystins) is the focus of Chap. 29 of this volume, and thus will be only briefly mentioned here to avoid redundancy.

27.2 Contribution of TRPC3 to Arginine Vasopressin (AVP) Signaling in the Collecting Duct

TRPC3 (originally HTRP3) is a nonselective cation-permeable ion channel displaying double rectification with estimated single-channel conductance of 66 pS in overexpression systems (Zhu et al. 1996; Hurst et al. 1998). It is a member of the “canonical” subfamily of TRP channels, which has the highest degree of homology with the first cloned TRP in *Drosophila melanogaster* (Montell and Rubin 1989). Similar to other TRPCs, TRPC3 is primarily activated in response to stimulation of various G-protein-coupled receptors (GPCR) linked to phospholipase C (PLC) cascade (Kiselyov and Patterson 2009). It has been well demonstrated that PLC-dependent generation of diacylglycerol (DAG) increases TRPC3 activity in native cells and in overexpression systems (Hofmann et al. 1999; Trebak et al. 2003a, b). However, it is unknown whether DAG directly binds to the channel or activates the channel via an intermediate pathway. Inositol triphosphate (IP_3)-dependent depletion of the endoplasmic reticulum stores can also activate TRPC3, thus triggering the store-operated Ca^{2+} entry (SOCE) (Vazquez et al. 2003). Recently determined cryo-EM structure of the full-length human TRPC3 revealed unique features of the channel providing mechanistic insights into its activation and gating (Fan et al. 2018). The unusual perpendicular position of TRP-helix, which is disengaged from the pore domain, points to a specific mode of channel gating. S3 segment is nearly twice as long as the S3 in any other DAG-insensitive TRPC channels (such as TRPC1, TRPC4, and TRPC5). It forms an extracellular pocket-like domain potentially serving as a sensor of external stimuli. In addition, two putative lipid-binding sites were identified: in the large elbow-like pre-S1 domain and in the lateral fenestration of the pore domain, wedged between the P-loop and S6 of adjacent subunit (Fan et al. 2018). It is plausible to speculate that DAG-dependent activation of the channel involves its direct binding to these sites. Future studies are needed to test this hypothesis.

TRPC3 is abundantly expressed in brain, mostly in cerebellar Purkinje and in olfactory mitral cells; cardiomyocytes, and smooth muscle cells. The channel participates in a variety of $[Ca^{2+}]_i$-signaling mechanisms in response to GPCR activation in these tissues (Zhu et al. 1996; Dietrich et al. 2005; Lichtenegger and Groschner 2014; Kitajima et al. 2016). In the kidney, TRPC3 expression is apparent in the renal vasculature (Facemire et al. 2004), glomerulus, and collecting duct system (Goel et al. 2006). However, our understanding of TRPC3 physiology in the kidney is rather fragmental. Activation of the calcium sensing receptor (CaSR) by high extracellular Ca^{2+} augmented proliferation of glomerular mesangial cells via a mechanism involving stimulation of TRPC3-dependent Ca^{2+} influx (Meng et al. 2014). Unilateral ureteral obstruction increased TRPC3 expression in fibroblasts of obstructed kidneys and this was associated with an increased Ca^{2+} entry, augmented phosphorylation of the extracellular signal-regulated kinase 1/2 (ERK1/2), and fibroblast proliferation contributing to renal fibrosis. Inhibition and genetic deletion of TRPC3 ameliorated tubule-interstitial damage and renal fibrosis (Saliba et al. 2015) potentially implicating its role in ischemia-reperfusion kidney injury and acute kidney failure. Furthermore, TRPC3 expression in monocytes was significantly higher in patients with chronic kidney disease and inversely correlated with estimated glomerular filtration rate (GFR) and serum Ca^{2+} concentration (Liu et al. 2011).

Antibody-based detection of TRPC3 in kidney sections revealed that the channel is present in both apical plasma membrane and intracellular compartments of the collecting duct aquaporin 2 (AQP2)-positive principal cells (Goel et al. 2007). Intravenous injection of arginine vasopressin (AVP) caused redistribution of both TRPC3- and AQP2-reporting fluorescent signals to the apical membrane (Goel et al. 2007). The signaling mechanism involved stimulation of vasopressin V2 receptors (V2R) and activation of the adenylyl cyclase (AC)/cAMP/protein kinase A (PKA)-signaling cascade (Fig. 27.4) (Goel et al. 2010). It is currently unknown whether TRPC3 participates in AVP-dependent water reabsorption in the collecting duct by controlling $[Ca^{2+}]_i$ levels. It was shown that Ca^{2+}-permeable Orai1, but not TRPC3 channel, mediated SOCE in split-opened murine collecting ducts to regulate AVP-dependent AQP2 trafficking and expression (Mamenko et al. 2016). AVP caused translocation of TRPC3 to the plasma membrane of LLC-PK1 cells in the absence of AQP2 expression (Goel et al. 2007), potentially indicating a mechanism other than the control of AQP2-dependent water transport. Interestingly, stimulation of purinergic GPCR with ATP promoted apical-to-basolateral TRPC3-dependent Ca^{2+} flux in cultured IMCD-3 cells (Goel and Schilling 2010). While usually neglected, collecting duct performs notable Ca^{2+} reabsorption (Bengele et al. 1980; Magaldi et al. 1989). It is possible that AVP stimulates TRPC3-mediated Ca^{2+} reabsorption to decrease a risk of nephrocalcinosis and urolithiasis by reducing luminal Ca^{2+} levels and preventing Ca^{2+} precipitation in the luminal fluid during antidiuretic conditions. Consistently, Williams-Beuren syndrome (WBS), a neurodevelopmental disorder commonly associated with hypercalcemia of unknown origin, has been recently linked to upregulation of TRPC3 expression due to deletion of the gene encoding the transcription factor TFII-I (Letavernier et al. 2012). TFII-I

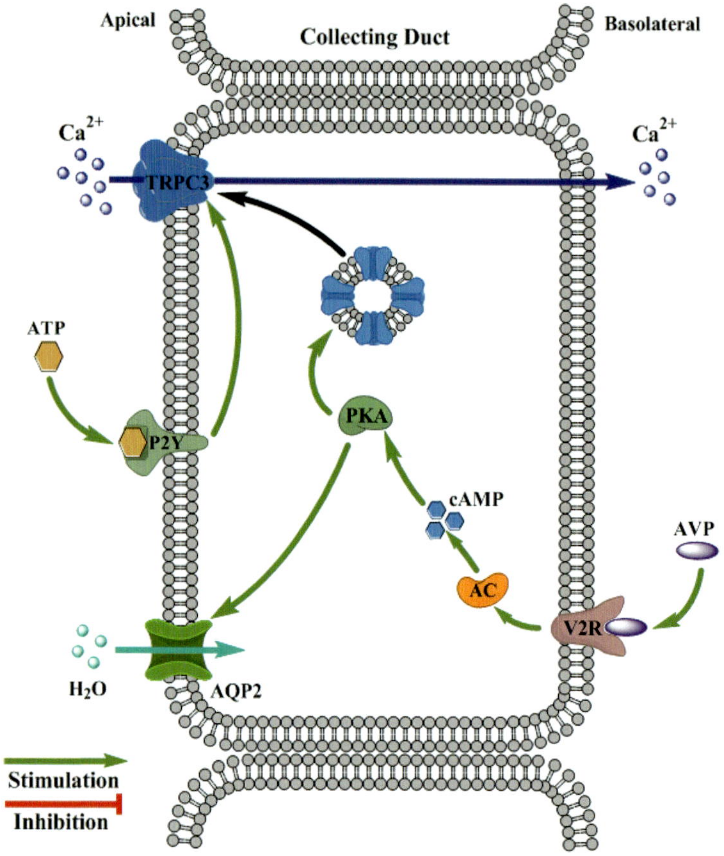

Fig. 27.4 Proposed TRPC3 function and regulation in collecting duct cells. *ATP* adenosine triphosphate, *P2Y* purinergic receptor type Y, *AQP2* aquaporin water channel type 2, *AVP* arginine vasopressin, *V2R* vasopressin receptor type 2, *AC* adenylyl cyclase, *cAMP* cyclic adenosine monophosphate, *PKA* protein kinase A

suppresses cell-surface accumulation of TRPC3. It was further shown that TRPC3 expression was high in collecting duct cells of human kidney biopsies, and it was markedly upregulated in intestinal cells of WBS patient, potentially implicating TRPC3 in regulation of systemic Ca^{2+} homeostasis (Letavernier et al. 2012). Future studies are needed to thoroughly investigate the contribution of TRPC3 in renal Ca^{2+} and water handling.

27.3 Role of TRPC6/TRPC5 in Glomerular Filtration and Podocyte Function

TRPC6 is a tetrameric nonselective cation-permeable channel of approximately 37 pS conductance sharing many features with its close relatives, TRPC3 and TRPC7 (Hofmann et al. 1999; Harteneck and Gollasch 2011). It is well known that TRPC6 can be activated in response to stimulation of GPCR in a DAG-dependent manner (Hofmann et al. 1999; Shi et al. 2004; Estacion et al. 2006). Recently resolved human TRPC6 channel structure demonstrated the classical four subunit TRP channel architecture (Tang et al. 2018). The permeability pore (the outer selectivity filter and inner permeability gate) was virtually identical to that of TRPC3. It is also worth noting that the S3 helix of TRPC6 is longer by about three turns at the extracellular space side creating so-called protrusion (Tang et al. 2018). This is closely reminiscent of the respective pocket-like domain in TRPC3 (Fan et al. 2018), likely indicating distinct structural characteristics of DAG-sensitive TRPC channels. Interestingly, TRPC6 can form functional heteromers with TRPC3 and TRPC7, but does not readily do so with DAG-insensitive TRPC4 and TRPC5 or members of other TRP channel subfamilies (Hofmann et al. 1999, 2002).

TRPC6 is widely expressed across the body, including central nervous system, smooth muscle, immune cells, etc., where it participates in a variety of Ca^{2+}-dependent cellular processes (Bouron et al. 2016). In the kidney, TRPC6 expression is restricted to glomerulus and the collecting duct system (see Fig. 27.3) (Goel et al. 2006). While nothing is known about the relevance of TRPC6 in the latter segment, abundant experimental and clinical evidence suggest the critical role of the channel in maintaining integrity of the glomerular filtration barrier, as will be discussed later. The barrier is formed by the fenestrated endothelium of the glomerular capillaries, negatively charged glomerular basement membrane (GBM, also known as basal lamina), and the slit diaphragms formed by foot process of the podocytes, highly specialized epithelial cells enveloping the capillaries. Loss of podocytes or podocyte effacement during different pathologies (for example, diabetic nephropathy) compromises the filtration barrier leading to proteinuria and gradual decline in kidney function (Kim et al. 2001; Neal et al. 2005; Winn et al. 2005; Tanner et al. 2009; Mundel and Reiser 2010; Peti-Peterdi and Sipos 2010; Pollak et al. 2014). Changes in hydrostatic pressure in glomerular capillaries elicit mechanical stress on podocytes leading to elevated $[Ca^{2+}]_i$ (Endlich et al. 2001; Friedrich et al. 2006). Real-time in vivo multiphoton imaging directly demonstrated the correlation between changes in $[Ca^{2+}]_i$ in podocytes and contraction of the respective afferent arteriole (Peti-Peterdi and Sipos 2010). It appears that the activity of Ca^{2+}-permeable TRPC6 channel significantly contributes to the elevated $[Ca^{2+}]_i$ in podocytes in response to mechanical and chemical stimuli (Reiser et al. 2005). During normal physiology, TRPC6-mediated Ca^{2+} influx was proposed to regulate activity of the large conductance Ca^{2+}-activated K^+ (BK) channels in podocytes to control membrane voltage (Fig. 27.5) (Kim et al. 2013). The pathophysiological relevance of TRPC6 Ca^{2+} signaling has been revealed, when TRPC6 gain-of-function mutations

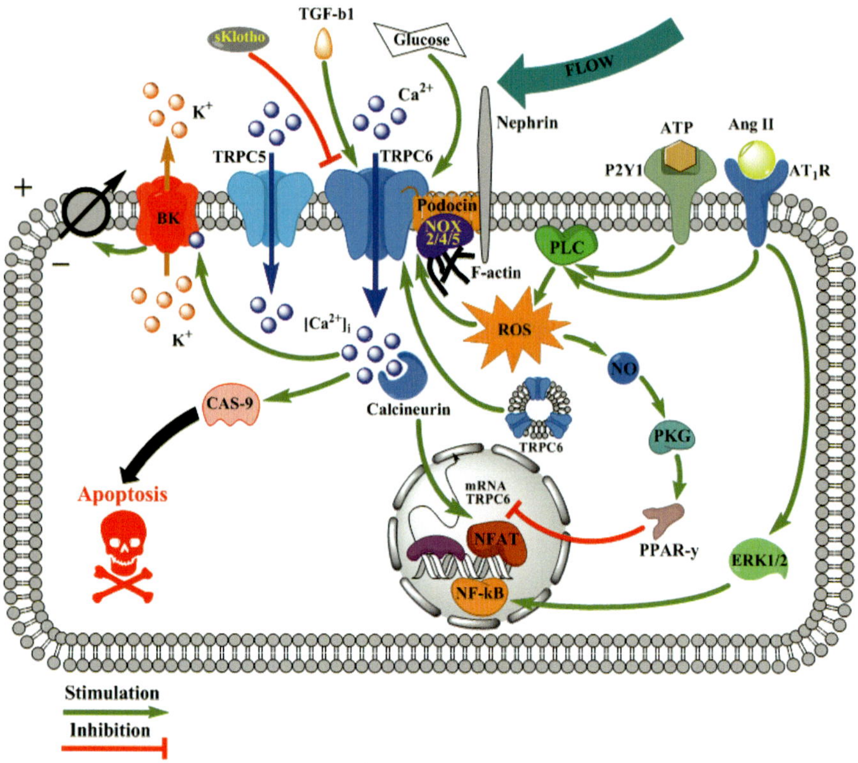

Fig. 27.5 Physiological role and molecular mechanisms of TRPC6 channel regulation in glomerular podocytes. *Ang II* Angiotensin II, *AT_1R* Angiotensin II receptor type 1, *ATP* adenosine triphosphate, *P2Y1* purinergic receptor type Y1, *TGF-β1* transforming growth factor beta 1, *sKlotho* soluble Klotho, *ERK1/2* extracellular signal-regulated kinases, *PLC* phospholipase C, *NOX2/4/5* NADPH oxidase isoforms 2, 4, and 5 respectively, *ROS* reactive oxygen species, *BK* Ca^{2+}-activated big conductance K^+ channel, *CAS-9* caspase type 9, *NO* nitric oxide, *PKG* protein kinase G, *PPAR-γ* peroxisome proliferator-activated receptor gamma, *NFAT* nuclear factor of activated T-cells, *NF-kB* nuclear factor kappa-light-chain-enhancer of activated B cells. Membrane voltage is shown as a gray circle with arrow

were linked to the focal and segmental glomerulosclerosis (FSGS) (Reiser et al. 2005; Winn et al. 2005). FSGS is characterized by hypertension, proteinuria, hypoalbuminemia, and decreased renal function progressing to the end-stage renal disease in approximately 40% of patients. So far, over 20 different FSGS-causing gain-of-function mutations in TRPC6 has been identified (Heeringa et al. 2009; Santin et al. 2009; Zhu et al. 2009; Gigante et al. 2011; Buscher et al. 2012; Mir et al. 2012; Obeidova et al. 2012; Hofstra et al. 2013; Kuang et al. 2013; Zhang et al. 2013a). The functional analysis demonstrated both increased plasma membrane channel expression and augmented TRPC6 gating (i.e., open probability) (Reiser et al. 2005; Winn et al. 2005; Heeringa et al. 2009). These mutations cause a slow onset of FSGS (16–61 years). It was further shown that the magnitude of TRPC6 potentiation correlated with the severity of disease progression (Dryer and Reiser 2010). While the exact molecular mechanism is not fully elucidated, it is likely that sustained hyperactivation of TRPC6 channels results in $[Ca^{2+}]_i$ overload, leading to the decreased population of podocytes due to their limited capacity for regeneration (Ilatovskaya et al. 2015). It is anticipated that development of TRPC6 selective inhibitors, such as larixyl carbamate, might be of clinical relevance to treat FSGS (Urban et al. 2017). Consistently, podocyte-specific overexpression of TRPC6 (wild-type and FSGS-causing mutants) induced glomerular lesions and led to measurable proteinuria, although the phenotype was mild and had a late onset (Krall et al. 2010). Paradoxically, a recent study found that five FSGS-causing TRPC6 mutations (N125S, L395A, G757D, L780P, and R895L) led to decreased channel activity upon overexpression in HEK293 cells, whereas channel expression and trafficking to the plasma membrane were unaffected (Riehle et al. 2016). Moreover, these mutations were associated with a remarkably early onset of FSGS (1–7 years), as opposed to the late onset reported for the gain-of-function mutations (Riehle et al. 2016). The nature of such dualism of TRPC6 activity on FSGS occurrence is not known and requires further investigation. One possibility is that TRPC6-mediated Ca^{2+} influx might be necessary for podocyte development and maturation, which would be consistent with the early onset of the disease due to TRPC6 loss-of-function. On the contrary, TRPC6-deficient rodent models (which recapitulate the state of TRPC6 loss-of-function) exhibit protection against glomerular damage and do not demonstrate noticeable signs of glomerular damage/sclerosis at the baseline (Ilatovskaya et al. 2018; Kim et al. 2018; Spires et al. 2018).

In addition to TRPC6, a role for another TRPC member, TRPC5 has been proposed in regulation of $[Ca^{2+}]_i$ dynamics in podocytes (Dryer and Reiser 2010; Schaldecker et al. 2013). TRPC5- and TRPC6-mediated Ca^{2+} influx exerted opposite effects on cytoskeleton rearrangements in podocytes (Greka and Mundel 2011). It is further proposed that TRPC6 has a homeostatic function, whereas activation of TRPC5 leads to pathological maladaptation resulting in podocyte damage. Selective inhibition (with AC1903) or genetic deletion of TRPC5 protected from lipopolysaccharide (LPS)-induced albuminuria and angiotensin II (Ang II)-dependent proteinuric kidney disease in mice (Schaldecker et al. 2013; Zhou et al. 2017). On the contrary, transgenic mice overexpressing either wild-type or inactive TRPC5 did not develop proteinuria at the baseline and upon pharmacological TRPC5 stimulation with Englerin A (Wang et al. 2018). Increased TRPC5 expression also had no measurable effect on LPS-induced proteinuria (Wang et al. 2018). The nature of these discrepancies is currently unknown.

Functional analysis revealed low basal TRPC6 activity in podocytes, which can be markedly stimulated by post-translational modifications (such as glycosylation) and by mechanical stress of the plasma membrane (Dietrich et al. 2003; Hisatsune et al. 2004; Spassova et al. 2006). It was shown that interaction with podocin, a protein expressed specifically in podocyte foot processes/slit diaphragm (Roselli et al. 2002), is essential for mechanosensitive TRPC6-dependent Ca^{2+} influx (Huber et al. 2006, 2007). Importantly, mutations of podocin were also linked to FSGS development (Woroniecki and Kopp 2007). Knockdown of podocin markedly increased mechanosensitive TRPC6 activation in cultured podocytes, basically recapitulating channel hyperactivity during FSGS (Anderson et al. 2013). Several other signaling molecules, including nephrin (Huber et al. 2001; Reiser et al. 2005), NADPH oxidase isoform 2 (NOX2) (Kim et al. 2013), phospholipase C (PLC)-γ1 (Kanda et al. 2011), multiple PKC isoforms (Ambrus et al. 2015), F-actin (Endlich et al. 2001; Jiang et al. 2011), synaptopodin (Yu et al. 2016) can also modulate TRPC6 activity during mechanical or endocrine stimulation (Fig. 27.5). Of interest, TRPC6 N143S mutation abolished channel activation by hypoosmotic stretch but did not affect TRPC6 regulation by DAG-dependent pathways (Wilson and Dryer 2014), suggesting a complex polymodal nature of channel gating.

Inappropriately elevated TRPC6 activity has been shown to play detrimental roles in glomeruli during different pathophysiological states (Ilatovskaya and Staruschenko 2015). Ang II is the principal hormone of the renin-angiotensin system, which is known to elicit numerous actions in multiple tissues to regulate systemic water-electrolyte balance and control blood pressure (Zaman et al. 2002; Kobori et al. 2007; Seva et al. 2013). Chronically elevated Ang II levels are commonly associated with hypertension and end-organ damage (Zaman et al. 2002; Kobori et al. 2007; Seva et al. 2013). Abundant clinical and basic evidence demonstrate glomerulosclerosis and proteinuria in the presence of chronically elevated Ang II (reviewed in (Ma and Fogo 2001; Kagami 2012; Navar 2014)). Ang II increased TRPC6 expression in podocytes via stimulation of the ERK1/2 pathway and subsequent translocation of the nuclear factor kappa-light-chain-enhancer of activated B cells (NF-κB) to the nucleus (Zhang et al. 2009) (Fig. 27.5). Activation of TRPC6 and NF-κB pathways has been also shown to mediate podocyte apoptosis in adriamycin-induced renal damage (Zhang et al. 2016). In addition, TRPC6-dependent Ca^{2+} influx led to the activation of the Ca^{2+}-dependent protein phosphatase, calcineurin, and its substrate nuclear factor of activated T cells (NFAT). NFAT, in turn, increased TRPC6 mRNA levels to drive TRPC6 expression causing sustained elevations of $[Ca^{2+}]_i$ levels and podocyte damage (Zhang et al. 2009). Ang II can also acutely stimulate single-channel TRPC6 activity leading to increased $[Ca^{2+}]_i$ in freshly isolated podocytes (Zhang et al. 2009; Nijenhuis et al. 2011; Anderson et al. 2014; Ilatovskaya et al. 2014). The signaling pathway involved activation of angiotensin type 1 receptors (AT_1R), stimulation of PLC followed by production of reactive oxygen species (ROS) (Anderson et al. 2014; Ilatovskaya et al. 2018). AT_1R blocker, losartan, decreased TRPC6 activity and expression in podocytes and ameliorated glomerulosclerosis and proteinuria (Chi et al. 2015).

Diabetic nephropathy is a progressive renal disease characterized by podocyte loss, glomerulosclerosis, proteinuria, and decline in kidney function affecting 20–40% of patients suffering from diabetes mellitus (Jefferson et al. 2008; Diez-Sampedro et al. 2011). Streptozotocin (STZ)-mediated induction of type I diabetes mellitus increased TRPC6 expression and activity in podocytes of Munich-Wistar and Dahl salt-sensitive rats (Sonneveld et al. 2014; Ilatovskaya et al. 2015; Spires et al. 2018). TRPC6 deletion attenuated inappropriately high basal $[Ca^{2+}]_i$ levels and podocyte damage but did not have a significant effect on proteinuria (Spires et al. 2018). It was further proposed that increased Ang II levels contribute to the augmented TRPC6 activity during diabetes (Ilatovskaya et al. 2015). Elevated circulating glucose levels can per se increase TRPC6 mRNA levels and TRPC6-dependent Ca^{2+} influx in podocytes (Sonneveld et al. 2014). This mechanism seems to be selective for TRPC6 because expression of TRPC1 and TRPC5 was unaffected (Sonneveld et al. 2014). Treatment of podocytes with high glucose also upregulated renin mRNA expression leading to increased angiotensin production (Durvasula and Shankland 2008). This suggests that high glucose and Ang II converge to stimulate TRPC6 activity during diabetes (Durvasula and Shankland 2008; Sonneveld et al. 2014; Thilo et al. 2014). Excessive generation of ROS, particularly H_2O_2, is viewed as a critical downstream signal of both Ang II and high glucose in TRPC6 overactivation in podocytes during diabetic nephropathy. Consistently, ROS scavengers, such as Tempol, decreased proteinuria (Luan et al. 2012; Jha et al. 2017). Deletion of NADPH oxidase 4 (NOX4) significantly attenuated Ang II-dependent H_2O_2 production and $[Ca^{2+}]_i$ responses in rat podocytes (Ilatovskaya et al. 2018). NOX5 isoform (which is absent in rat/mouse genome) was reported to be critical for glomerular ROS production in humans during diabetic nephropathy (Jha et al. 2017). In addition, application of Ang II and H_2O_2 caused increased nitric oxide (NO) production in podocytes. This pathway was impaired during diabetes, which can also contribute to glomerular damage (Palygin et al. 2018). Stimulation of NO-dependent signaling, specifically cGMP-activated protein kinase G (PKG) and downstream peroxisome proliferator-activated receptor γ (PPAR-γ) with sildenafil, decreased TRPC6 expression and activity and ameliorated podocyte injury and proteinuria (Sonneveld et al. 2017). It is interesting that elevated circulating glucose levels decreased TRPC6 expression via unknown mechanism in contractile mesangial cells, which regulate blood flow across glomerular capillaries (Graham et al. 2007). This decreased TRPC6-dependent $[Ca^{2+}]_i$ signaling in response to Ang II. TRPC6 deficiency and impaired contractility of the mesangial cells might contribute to the hyperfiltration during early stages of diabetes mellitus. Consistently, TRPC6 knockout mice have significantly higher GFR, which was at least partially attributed to the lack of TRPC6-dependent $[Ca^{2+}]_i$ signaling and reduced contraction tone of mesangial cells (Li et al. 2017a).

Several other pathways have been demonstrated to regulate TRPC6 activity in podocytes (Fig. 27.5). Mechanical stress is known to cause ATP release from renal epithelial cells (Vallon 2008). ATP stimulated TRPC6 activity likely via P2Y1 receptors (Ilatovskaya et al. 2013), generation of ROS, which can lead to Ca^{2+}-dependent changes in gene expression, effacement of podocyte foot processes, and glomerular injury (Roshanravan and Dryer 2014). Transforming growth factor β1 (TGF-β1) is known to contribute significantly to the pathogenesis of glomerulosclerosis (Lee and Song 2010). TGF-β1 treatment promoted apoptosis of podocytes by upregulating desmin and caspase-9 levels, which was alleviated by TRPC6 knockdown (Huang et al. 2017). Application of soluble form of glycoprotein Klotho (see also TRPV5/6 section) decreased TRPC6-dependent Ca^{2+} influx in cultured podocytes by inhibiting phosphoinositide 3 kinase (PI3-K)-dependent exocytosis of the channel (Kim et al. 2017). In vivo administration of soluble Klotho reduced albuminuria in a mouse model overexpressing TRPC6, suggesting its important role in maintaining filtration barrier (Kim et al. 2017). Renoprotective effects of Klotho were, at least partially, attributed to inhibition of TRPC6 (Wu et al. 2017).

27.4 TRPM2 in the Proximal Tubule Exacerbates Ischemia Reperfusion Injury

TRPM2 (formerly known as LTRPC2 and TRPC7) is a nonselective Ca^{2+}-permeable cation channel widely expressed in multiple tissues, including brain, cardiovascular system, and kidneys (Perraud et al. 2001; Gao et al. 2014). TRPM2 activation requires binding of cytosolic ADP-ribose (ADPR) to the NUDT9H domain in the carboxyl terminus (Perraud et al. 2001). In addition, Ca^{2+} and $PI(4,5)P_2$ are also essential for channel activity (Du et al. 2009a; Toth and Csanady 2012). TRPM2 activity can be also stimulated by ROS, such as micromolar concentrations of H_2O_2 (Toth and Csanady 2010). Simply put, TRPM2 is proposed to serve as a ROS sensor in multiple tissues. The mechanism of TRPM2 activation by ROS is not fully elucidated but likely occurs in an ADPR-dependent manner (Perraud et al. 2005; Toth and Csanady 2010). Cryo-EM structure of *Nematostella vectensis* TRPM2 (62% homology with human TRPM2) directly demonstrated classical TRP channel tetrameric organization (Zhang et al. 2018). At the same time, unique structural features included a larger and more negatively charged pore consistently with high TRPM2 conductance (approximately 75 pS at physiological pH (Du et al. 2009b)), the intracellular Ca^{2+}-binding sites were connected to the pore and cytosol, explaining the strong activation of TRPM2 by Ca^{2+} on both sides. However, structure of the terminal ADPR-binding NUDT9H domain was not resolved.

At present, little is known about functional roles of TRPM2 in renal epithelium. Immunohistochemical evidence suggested almost exclusive expression of the channel in renal proximal tubules (see Fig. 27.3) and the specificity of a TRPM2-reporting signal was verified in $TRPM2^{-/-}$ mice (Gao et al. 2014). However, TRPM2 staining was diffuse in the cytosol without apparent plasma membrane localization. It is unclear whether the channel is functional in proximal tubule cells, since no evidence of TRPM2 activity is available. However, TRPM2 deficiency was shown to be protective against renal ischemia reperfusion injury (Gao et al. 2014). This manifested as improved kidney function/lower serum creatinine levels, reduced histologic damage, suppression of proapoptotic pathways, and reduced inflammation. It was proposed that activation of TRPM2 by initial ROS or other ischemia-accompanying factors led to activation of the small G-protein, Rac1, to further induce oxidative stress. It is not known whether TRPM2 increases $[Ca^{2+}]_i$ levels in proximal tubule cells to activate Rac1, since no TRPM2 translocation to the plasma membrane was detected (Gao et al. 2014). While TRPM2 is known to be also expressed in lysosomes, it is questionable whether its activity can significantly alter cytosolic $[Ca^{2+}]_i$ levels. Paradoxically, deficiency of another

Ca^{2+}-permeable TRP channel, TRPV4, in the proximal tubule cells exacerbated renal tubular damage of AQP1-positive proximal tubule cells (Mannaa et al. 2018), suggesting that TRPV4-dependent Ca^{2+} influx is beneficial for cell survival during ischemia. It is puzzling that TRPM2-dependent Ca^{2+} influx is detrimental, whereas TRPV4-dependent Ca^{2+} influx is beneficial for proximal tubule cells during ischemia-reperfusion injury. Experimental evidence also suggested a role for TRPM2 during ischemia injury in extrarenal tissues, such as brain and heart. However, both protective and detrimental effects of TRPM2 deletion on ischemia-reperfusion injury were reported (Alim et al. 2013; Hiroi et al. 2013; Miller et al. 2013). It is clear that future more direct studies are necessary to solve the physiological role of TRPM2 in the kidney.

27.5 TRPM6/TRPM7 as Mediators of Mg^{2+} Transport in the Distal Convoluted Tubule

TRPM6 is a voltage-insensitive, constitutively active cation-permeable channel (Chubanov and Gudermann 2014). TRPM6, along with its close relative TRPM7 (52% homology with TRPM6), are the only TRP channels with a higher selectivity to Mg^{2+} over Ca^{2+} as a conducting divalent cation (permeability of Mg^{2+} to Ca^{2+} is 5:1) (Voets et al. 2004). Another unique feature of both channels is the presence of a serine/threonine alpha-kinase domain fused to C-terminus. Thus, TRPM6 and TRPM7 are often referred to as chanzymes (a chimera of channel and enzyme) (Riazanova et al. 2001; Montell 2003). While the exact function of this kinase domain remains debatable, the proposed roles include signaling to the nucleus (Krapivinsky et al. 2017), proper translocation to the plasma membrane (Matsushita et al. 2005), and inhibition of the channels by intracellular Mg^{2+} (Voets et al. 2004). The latter mechanism likely involves a direct regulation of the kinase domain of the channel by Mg-ATP (Thebault et al. 2008).

Proper control of Mg^{2+} homeostasis is essential for health (de Baaij et al. 2015). Mg^{2+} plays an important role in a variety of physiological processes in the brain, heart, skeletal muscles, etc. Moreover, intracellular Mg^{2+} is critical for enzymatic reactions, synthesis, and proper folding of proteins and polynucleotides (de Baaij et al. 2015). The total amount of Mg^{2+} in the body is determined by the balance between intestinal Mg^{2+} uptake and urinary Mg^{2+} excretion. Hypermagnesemia (serum Mg^{2+} levels above 1.1 mM/l) is present in patients suffering from nausea, vomiting, lethargy, headaches, etc. Increased magnesium could also lead to cardiac defects and brachycardia in extreme cases. Hypomagnesemia (serum Mg^{2+} levels below 0.4 mM/l) may cause cardiac arrhythmias, tetany, and seizures. Since Mg^{2+} absorption in the intestine is poorly controlled depending primarily on dietary intake (Schweigel and Martens 2000; Lameris et al. 2015), it is thought that the kidneys play the central role in governing systemic Mg^{2+} balance (Ferre et al. 2011).

Proximal tubule and thick ascending limb reabsorb approximately 90% of the filtered Mg^{2+} through paracellular diffusion. This process is constitutive and driven by favorable transepithelial voltage generated by active Na^{+} transport at these sites. The remaining approximately 10% is reabsorbed in a controlled transcellular manner by the distal convoluted tubule. Distal convoluted tubule can be further divided into aldosterone-insensitive DCT1 (early DCT) and aldosterone-responsive DCT2 (late DCT). Mg^{2+} reabsorption is greater in DCT1 than in DCT2 mostly due to a more favorable electrochemical gradient for the ion in the former. Since no appreciable amount of Mg^{2+} can be reabsorbed in the downstream tubular segments, transport in the distal convoluted tubule determines urinary Mg^{2+} levels. In the kidney, Mg^{2+}-permeable TRPM6 channel expresses exclusively on the apical side of the distal convoluted tubule cells strongly suggesting its specific role in Mg^{2+} reabsorption (Fig. 27.6) (Schlingmann et al. 2002; Voets et al. 2004; Schlingmann et al. 2007). Indeed, it is now well accepted that TRPM6 serves as the chief pathway for Mg^{2+} entry across the apical membrane in DCT1 and DCT2 cells (Woudenberg-Vrenken et al. 2009). TRPM6 is also expressed in intestinal epithelium, where it is similarly involved in Mg^{2+} uptake (Voets et al. 2004). Loss-of-function mutations in TRPM6 underlie hereditary hypomagnesemia with secondary hypocalcemia (HSH) in humans (Schlingmann et al. 2002; Walder et al. 2002; Lainez et al. 2014), thereby capitalizing the essential role of the channel in systemic Mg^{2+} homeostasis. Since that, different HSH mutations of TRPM6 have been reported (Astor et al. 2015; Coulter et al. 2015; Patel et al. 2016). HSH is characterized by reduced intestinal Mg^{2+} absorption, increased urinary Mg^{2+} levels, and hypoparathyroidism causing secondary hypocalcemia. Systemic Mg^{2+} deficiency during HSH also leads to convulsions, seizures, and tetany. Dietary Mg^{2+} supplementation, while did not completely restore normal plasma Mg^{2+} levels, partially mitigated the symptoms of HSH (Schlingmann et al. 2007; Quamme 2008). It is interesting that TRPM7 is also expressed in the distal convoluted tubule and intestine epithelium pointing to its similar role in Mg^{2+} transport (Schlingmann et al. 2007). It was further proposed that both channels exist primarily as heteromers (Schmitz et al. 2005; Li et al. 2006). TRPM6/TRPM7 channels have constitutively high activity in the presence of physiological levels of Mg^{2+} and ATP (Ferioli et al. 2017). However, no HSH-causing mutations in TRPM7 were identified so far. Nevertheless, two TRPM7 point mutation variants (p.C721G and p.R902C) have been recently linked to macrothrombocytopenia with accompanying atrial fibrillation (Stritt et al. 2016). The impaired proplatelet formation was associated with cytoskeletal alterations, which can be rescued by Mg^{2+} supplementation.

TRPM6 and TRPM7 knockout mice are embryonically lethal suggesting indispensable roles of the channels during embryogenesis (Jin et al. 2008; Walder et al. 2009). It was clarified that the loss of TRPM6 activity in placental SynT-I and yolk sac endoderm cells was responsible for embryonic mortality (Chubanov et al. 2016). $TRPM6^{+/-}$ animals develop normally and display mild hypomagnesemia in the presence of normal urinary Mg^{2+} excretion (Woudenberg-Vrenken et al. 2011). Conditional TRPM6 inactivation in adult mice caused impaired growth, defective energy balance, shorter life span, and profound Mg^{2+} deficit, which can be corrected

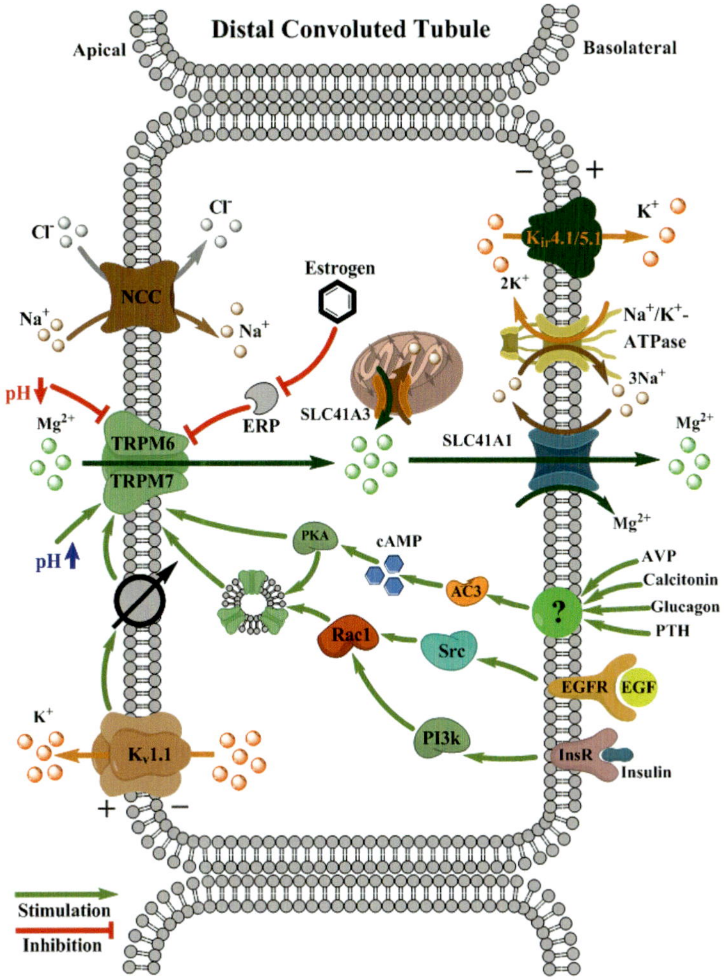

Fig. 27.6 Proposed role of TRPM6 channel in transcellular Mg^{2+} reabsorption in distal convoluted tubule cells. Description of TRPM6 and TRPM7 represents functional heteromerization of both channels. *NCC* thiazide-sensitive sodium chloride cotransporter, SLC12A3, *K_v1.1* potassium voltage-gated channel subfamily A member 1, *ERP* estrogen responsive protein, *AC3* adenylyl cyclase type 3, *cAMP* cyclic adenosine monophosphate, *PKA* protein kinase A, *Src* proto-oncogene encoding tyrosine kinase, *PI3K* Phosphatidylinositol-4,5-bisphosphate 3-kinase, *Rac1* small G protein Ras-related C3 botulinum toxin substrate 1, *K_{ir}4.1/5.1* inward rectifying heteromeric K^+ channel consisting of 4.1 and 5.1 subunits, *SLC41A1* solute carrier Mg^{2+}/Na^+ antiporter family 41 member 1, *SLC41A3* solute carrier Mg^{2+}/Na^+ antiporter family 41 member 3 (expressed in mitochondria), *AVP* arginine vasopressin, *PTH* parathyroid hormone, *InsR* insulin receptor, *EGF* epidermal growth factor, *EGFR* epidermal growth factor receptor. Membrane voltage is shown as a gray circle with arrow

by dietary Mg^{2+} supplementation (Chubanov et al. 2016). Somewhat surprising, this was primarily associated with decreased Mg^{2+} uptake by the intestine, whereas TRPM6 deletion in the kidney had only a minor effect on systemic Mg^{2+} balance. It is clear that future studies have to resolve uncertainties related to TRPM6 contribution in renal versus extrarenal tissues.

TRPM6-mediated Mg^{2+} reabsorption strongly depends on transepithelial voltage in the distal convoluted tubule. Unlike other ions, only a minimal difference exists between luminal, intracellular, and interstitial Mg^{2+} concentrations. It was shown that electrogenic K^+ efflux mediated by the apically localized voltage-gated $K_V1.1$ channel creates lumen-positive voltage in the distal convoluted tubule to drive TRPM6-dependent Mg^{2+} entry (Fig. 27.6). Loss-of-function mutations of $K_V1.1$ resulted in autosomal dominant hypomagnesemia in humans (Glaudemans et al. 2009; van der Wijst et al. 2010). The mechanism of basolateral Mg^{2+} exit remains unclear. Strong negative basolateral voltage (in the range of −60 to −70 mV), generated by the activity of Na^+/K^+ ATPase and inward rectifying $K_{ir}4.1/5.1$ (KCNJ10/16) potassium channel (Lourdel et al. 2002; Cuevas et al. 2017), precludes direct Mg^{2+} exit. It was proposed that basolateral Mg^{2+} extrusion is mediated by Na^+/Mg^{2+} antiporters and is driven by Na^+ gradient created by Na^+/K^+ ATPase (Guther et al. 1984). Consistently, loss-of-function mutations in the dominant Na^+-transporting protein, thiazide-sensitive Na-Cl cotransporter (NCC, SLC12A3) cause Gitelman's syndrome in human associated with hypomagnesemia (Simon et al. 1996; Ellison 2000). Moreover, mutations in KCNJ10, essential for generation of the basolateral membrane voltage, cause EAST/SESAME syndrome, leading to renal Mg^{2+} wasting and hypomagnesemia (Bockenhauer et al. 2009; Scholl et al. 2009), at least partially due to disruption of the basolateral Mg^{2+} exit in the distal convoluted tubule. SLC41A1 (Na^+/Mg^{2+} exchanger 4 (Kolisek et al. 2012)) was shown to be predominantly expressed in the distal convoluted tubule of human kidney sections (Hurd et al. 2013). While it is possible that SLC41A1 mediates basolateral Mg^{2+} exit, its precise contribution is not established. Deletion of SLC41A1 caused a cystic kidney disease phenotypically similar to nephronophthisis (Hurd et al. 2013), thus making it difficult to assess renal magnesium handling due to numerous structural abnormalities. Another member of SLC41 family, SLC41A3 (Na^+/Mg^{2+} antiporter), is also enriched in the distal convoluted tubule and its expression is regulated by dietary Mg^{2+} intake (de Baaij et al. 2016). SLC41A3$^{-/-}$ mice develop hypomagnesemia in the presence of normal urinary Mg^{2+} levels indicative of renal Mg^{2+} leak (de Baaij et al. 2016). Renal TRPM6 levels were unchanged, whereas intestinal Mg^{2+} transporters, including TRPM6, were upregulated in SLC41A3$^{-/-}$ mice. Interestingly, it was further demonstrated that SLC41A3 Na^+/Mg^{2+} antiporter mediates Mg^{2+} exit from mitochondria (Mastrototaro et al. 2016). SLC41A3 did not generate whole-cell Mg^{2+} current upon overexpression in HEK293 cells (de Baaij et al. 2016), suggesting its predominant intracellular localization. Future studies have to investigate how disruption of SLC41A3-dependent Mg^{2+} handling by mitochondria is translated into disturbed systemic Mg^{2+} balance.

Several hormones, including estrogen, insulin, and EGF, augment TRPM6 activity and expression in the distal convoluted tubule (Groenestege et al. 2006; Thebault et al. 2009; Nair et al. 2012) (Fig. 27.6). Estrogen increased TRPM6-dependent Mg^{2+} influx by removing repressive estrogen responsive protein, which blocks TRPM6 but not TRPM7 activity via interaction with its kinase domain (Cao et al. 2009). Epidermal growth factor (EGF) increased TRPM6 activity and surface expression by targeting its basolateral receptor, EGFR, and activation of the Src-Rac1-signaling pathway (Thebault et al. 2009). Interestingly, hypermagnesuria is common during treatment with cetuximab, an EGFR inhibitor, used in anticancer chemotherapy (Schrag et al. 2005). Similarly, another anticancer drug cisplatin downregulated the EGF and TRPM6 mRNA levels and augmented urinary fractional excretion of Mg^{2+} (Ledeganck et al. 2013). Insulin increased apical TRPM6 levels via a signaling cascade involving Rac1 and phosphoinositide 3-kinase (Nair et al. 2012). Interestingly, a genome-wide meta-analysis of Mg^{2+} homeostasis from 9099 individuals found two loci located in the vicinity of gene encoding TRPM6. The results revealed a significant correlation between urinary Mg^{2+} and fasting insulin levels, potentially linking systemic Mg^{2+} deficiency to insulin resistance and obesity (Corre et al. 2018). Recent study also found a nonredundant role of adenylyl cyclase 3 (AC3)-cAMP-PKA-signaling pathway in controlling TRPM6 abundance in the distal convoluted tubule (Blanchard et al. 2016). Specifically, increased cAMP was able to stimulate activity of TRPM6 mutant (S1583A), which was insensitive to actions of insulin. Renal tubule selective $AC3^{-/-}$ mice fed a regular diet did not show differences in serum Mg^{2+} values but had increased urinary Mg^{2+} levels. Switching to low Mg^{2+} diet resulted in higher urinary Mg^{2+} levels in the first 2 days and significantly lower serum Mg^{2+} at day 7 (Blanchard et al. 2016). It is quite possible that AC3-dependent pathway might be utilized in response to hormones reported to regulate Mg^{2+} reabsorption, such as PTH, glucagon, calcitonin, and AVP (Dai et al. 2001). Finally, TRPM6 activity is modulated by intracellular and extracellular pH. Systemic acidosis reduces TRPM6 levels favoring urinary Mg^{2+} wasting (Lennon and Piering 1970; Blumberg et al. 1998; Nijenhuis et al. 2006). In contrast, alkalosis causes increased tubular Mg^{2+} reabsorption (Wong et al. 1986; Nijenhuis et al. 2006).

27.6 TRPV4 and Mechanosensitivity in the Renal Tubule

TRPV4 is one of the most intensively studied mechanosensitive channels (Nilius et al. 2004; Everaerts et al. 2010a). Numerous loss-of-function and gain-of-function mutations of the channel have been linked to various clinical syndromes, including Charcot-Marie-Tooth disease type 2C, skeletal dysplasia, and osteoarthropathies (reviewed by (Nilius and Voets 2013)). TRPV4 exhibits so-called polymodal gating and can be activated by a broad spectrum of stimuli, including flow-induced shear stress, osmotic cell swelling, warm temperatures above 25 °C, polysaturated fatty acids, and 4α-phorbol ester derivatives (O'Neil and Heller 2005; Everaerts et al.

2010a; Guilak et al. 2010). Robust activation of TRPV4 by mechanical forces was demonstrated in cells natively expressing the channel (Gao et al. 2003; Wu et al. 2007) and upon TRPV4 overexpression (Wu et al. 2007; Tomilin et al. 2018). It has to be noted that TRPV4 is not considered as a "true" mechanosensitive channel (Christensen and Corey 2007; Nilius and Honore 2012). TRPV4 stimulation by mechanical stress does not fulfill the criterion of rapid activation kinetics (Nilius and Honore 2012). Increased TRPV4 activity by plasma membrane stretching or by shear stress was observed with a few seconds' delay (Nilius et al. 2004; Wu et al. 2007; Everaerts et al. 2010b; Berrout et al. 2012; Mamenko et al. 2013). In contrast, direct mechanosensitive channels (such as Piezo, TREK, and TRAAK) responded to such stimuli in the millisecond range (Nilius and Honore 2012). The recently elucidated crystal structure of *Xenopus tropicalis* TRPV4 (sharing 78% sequence identity with human TRPV4) provided important insights into its mechanism of gating and mechanoactivation (Deng et al. 2018). While the overall TRPV4 architecture is similar to that of other TRPV channels (TRPV1 (Liao et al. 2013), TRPV2 (Huynh et al. 2016), and TRPV6 (McGoldrick et al. 2018; Singh et al. 2018), see also Fig. 27.2), the packing interface between S1 and S4 transmembrane domains and pore domain is unique among TRP channels and distantly related voltage-gated ion channels. Specifically, organization of S1–S4 region engages both S3 and S4 transmembrane domains to form extensive interactions with the pore at S5–S6. This indicates that perturbations of S3 due to mechanical forces or chemical agents will directly propagate to S6 causing opening of the lower gate (Deng et al. 2018).

Functional TRPV4 channel exhibits moderate selectivity for Ca^{2+} having P_{Ca}/P_{Na} in the range of 6–10 (Voets et al. 2002). A single-channel conductance of TRPV4 is 90–100 pS for outward currents and 50–60 pS for inward currents in overexpression systems (Watanabe et al. 2002, 2003a, b). Activation of TRPV4 by mechanical and chemical stimuli resulted in substantial Ca^{2+} influx across the plasma membrane with subsequent activation of intracellular Ca^{2+}-dependent pathways. In addition to the aforementioned unique organization of transmembrane domains allowing mechanical control of the lower gate, functional studies also demonstrated a critical role of N-terminus in opening of the channel. Deletion of the ARDs drastically impaired activation of TRPV4 by mechanical forces likely due to disruption of its interaction with cytoskeleton (Liedtke et al. 2000). In addition, proline-rich domain (PRD), located just prior to ARD, interacts with PACSIN3 protein to preclude TRPV4 activation by hypotonic cell swelling (Cuajungco et al. 2006; D'Hoedt et al. 2008). PRD is not present in other TRPV channels (D'Hoedt et al. 2008) and, therefore, might also contribute to the unique TRPV4 properties. Interestingly, different stimuli could induce discrete mechanisms of TRPV4 activation. Blockade of phospholipase A2 (PLA_2) disrupted activation of TRPV4 by hypotonicity but not by heat and 4α 12,13-phorbol didecanoate (4α-PDD) (Vriens et al. 2004). Conversely, point mutation Y555A in S3 domain did not affect TRPV4 activation by osmolarity but abolished stimulatory effects of temperature and 4α-PDD (Vriens et al. 2004).

TRPV4 was cloned as a mammalian homolog of the *Caenorhabditis elegans* osmosensory protein OSM-9 and demonstrated activation by hypotonicity-induced cell swelling (Liedtke et al. 2000; Wissenbach et al. 2000; Delany et al. 2001). TRPV4 is widely expressed in the body, including lung, spleen, skin, sweat glands; as well as in endothelia, heart, skeletal muscles; sensory neurons, such as dorsal root ganglia (DRG), circumventricular organs, organum vasculosum laminae terminalis (OVLT); osteoblasts and chondrocytes (Liedtke et al. 2000; Liedtke and Friedman 2003; Muramatsu et al. 2007; Venkatachalam and Montell 2007; Masuyama et al. 2008). However, its expression is the most abundant in the kidney (Liedtke et al. 2000). Considering its osmosensitive nature and prominent renal expression, TRPV4 was thought to govern water handling by the kidney. TRPV4 knockout mice were able to maintain normal serum osmolarity in unstressed conditions but demonstrated impaired responses to both hyper- and hypotonic stimuli (Liedtke and Friedman 2003; Mizuno et al. 2003). However, this was not attributed to the defect in the renal water handling (Liedtke and Friedman 2003). TRPV4$^{-/-}$ mice and their wild-type (WT) littermates had comparable urine volumes and osmolality at the baseline. Water deprivation similarly decreased urine volumes and increased urine osmolality in both groups. Finally, WT and TRPV4$^{-/-}$ mice had indistinguishable AQP2 water channel total and plasma membrane levels (Janas et al. 2016). It was concluded that TRPV4 did not play a major role in renal water handling or was efficiently compensated upon deletion. The observed abnormalities in systemic water homeostasis in TRPV4$^{-/-}$ mice were likely related to the extrarenal factors, such as a disturbed release of vasopressin, sensation of thirst, and central osmoregulation (Liedtke and Friedman 2003; Janas et al. 2016). However, it was also suggested that TRPV1 rather than TRPV4 is critical for central osmosensitivity (Ciura et al. 2011). Alternative hypothesis implicated that TRPV4-expressing thoracic DRG neurons innervating hepatic blood vessels were able to sense physiologically relevant small changes in plasma osmolality and this mechanism was abrogated upon TRPV4 deletion (Lechner et al. 2011). It is clear that additional studies are need to accurately dissect the contribution of TRPV4 in systemic water balance.

TRPV4 is expressed in several segments of the renal nephron (Fig. 27.3). TRPV4 mRNA was found in the convoluted and straight segments of the proximal tubule and the highest levels were observed in the AQP2-positive connecting tubule and collecting ducts (Janas et al. 2016). In contrast, TRPV4 was not detectable in the glomeruli, the loop of Henle, and the early distal convoluted tubule (Janas et al. 2016), whereas earlier study reported TRPV4 mRNA from the ascending limb to the collecting duct, but not in the proximal tubule (Tian et al. 2004). Using a combination of $[Ca^{2+}]_i$ imaging, TRPV4 antibodies with verification in TRPV4 knockouts, and TRPV4 *LacZ* reporter mice, functional channel expression was found in both principal and intercalated cells of the connecting tubule and the collecting duct, as well as in the proximal tubule (Berrout et al. 2012; Janas et al. 2016). In summary, TRPV4 is highly expressed in the connecting tubule, the collecting duct and, to a lesser degree, in the proximal tubule (Fig. 27.3). More direct evidence is necessary to verify channel expression in other tubular segments, such as thick ascending limb. While out of the scope of this chapter, TRPV4 might also be expressed in the renal vasculature with the proposed role in the vasodilation and regulation of myogenic response (Chen et al. 2015; Soni et al. 2017). However, the physiological relevance of TRPV4 in the renal vasculature is not clear, since TRPV4$^{-/-}$ mice have comparable with WT plasma creatinine levels and urinary output (Janas et al. 2016) suggestive of unchanged GFR.

Recent studies greatly improved our understanding of the TRPV4 functional role in the renal tubule. While TRPV4 does not seem to participate in the regulation of osmotically driven water reabsorption (Liedtke and Friedman 2003; Janas et al. 2016), channel activity is indispensable for sensing changes in tubular flow particularly in the distal tubular segments. Increased tubular flow exerts shear stress, transmural pressure, and stretch of the epithelial cells. Numerous studies reported elevated $[Ca^{2+}]_i$ in native and cultured renal epithelial cells in response to increased flow over the apical plasma membrane (Woda et al. 2002; Liu et al. 2003, 2005, 2009; Satlin et al. 2006; Weinbaum et al. 2010; Mamenko et al. 2011; Berrout et al. 2012). Using ratiometric Ca^{2+} imaging in freshly isolated split-opened connecting tubule and cortical collecting ducts, our group convincingly demonstrated that TRPV4 is functional on the apical plasma membrane. Blockade and genetic deletion of TRPV4 abolished elevations in $[Ca^{2+}]_i$ in both principal and intercalated cells in response to physiologically relevant levels of shear stress existing in an intact kidney (Berrout et al. 2012; Zaika et al. 2013; Mamenko et al. 2017). Consistently, siRNA-mediated silencing of TRPV4 expression disrupted flow-induced $[Ca^{2+}]_i$ responses in immortalized collecting duct M1 cells (Wu et al. 2007; Du et al. 2012). Of interest, TRPV4 might also serve as a flow sensor in epithelial cells other tubular segments. Pharmacological inhibition and shRNA knockdown of TRPV4 expression significantly decreased $[Ca^{2+}]_i$ responses to luminal flow in perfused thick ascending limbs (Cabral et al. 2015). As was discussed earlier, it is unknown whether increased flow activates TRPV4 directly or via a secondary pathway. It was shown that inhibition of PLA_2-signaling cascade blocks flow activation of TRPV4 in ciliated oviductal cells and cerebral arteries (Andrade et al. 2005; Marrelli et al. 2007). Furthermore, stimulation of protein kinase C with PMA potentiated TRPV4-dependent $[Ca^{2+}]_i$ elevations in response to high flow in split-opened collecting ducts, whereas protein kinase A facilitated TRPV4 translocation to the apical plasma membrane without affecting channel activity (Fig. 27.7) (Mamenko et al. 2013).

It is well described that alterations in tubular flow modulate electrolyte transport in the renal tubule (Satlin et al. 2006; Liu et al. 2007; Weinbaum et al. 2010). Distal nephron is the principal site of controlled K^+ secretion into the lumen, which is the major contributor to urinary K^+ excretion (Wang and Giebisch 2009; Giebisch and Wang 2010). Activity of the renal outer medullary K^+ (ROMK) channels underlies constitutive potassium secretion at baseline conditions (Wang et al. 1990; Hebert et al. 2005). Loop diuretics and high dietary K^+ intake significantly increase fluid delivery to the distal nephron, which, in turn, produces marked kaliuresis by stimulating flow-induced K^+ secretion (FIKS) (Kaufman and Hamburger 1996; Woda et al. 2001). FIKS blockade caused hyperkalemia and primary aldosteronism (Grimm et al. 2009). Activity of the Ca^{2+}-activated BK channel underlies FIKS (Taniguchi and Imai 1998; Woda et al. 2001; Liu et al. 2007; Taniguchi et al. 2007; Wu et al. 2007). It is reasonable to assume that flow-sensitive Ca^{2+}-permeable

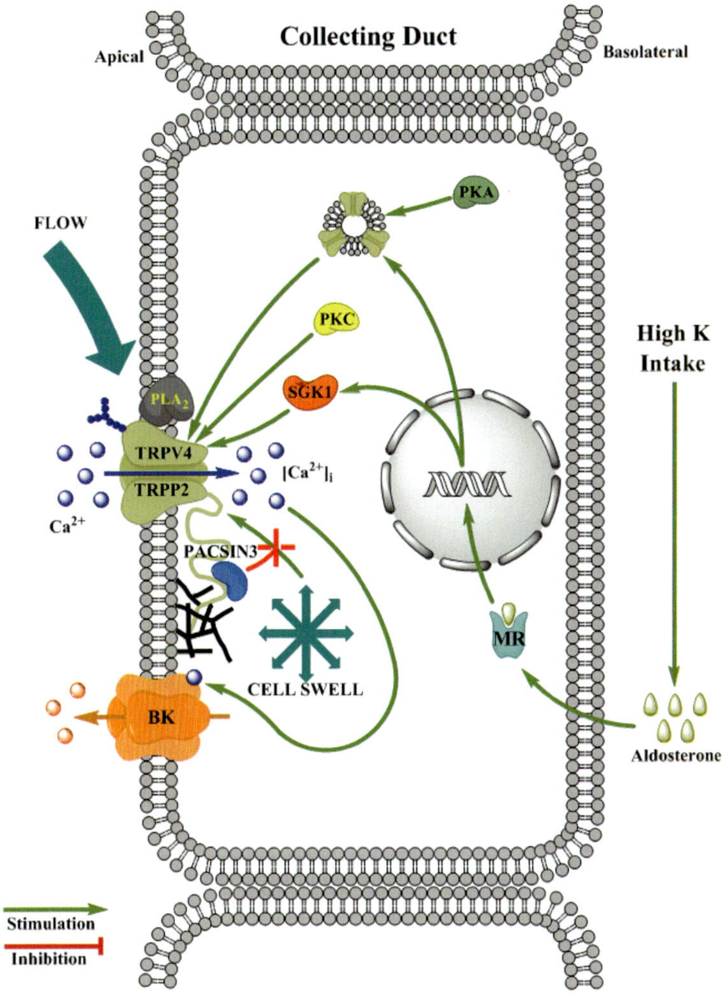

Fig. 27.7 Functional role and molecular mechanisms of TRPV4 channel regulation in collecting duct cells. Description of TRPV4 and TRPP2 represents functional heteromerization of both channels. *PLA*$_2$ phospholipase type A2, *PACSIN3* a protein kinase C and casein kinase substrate in neurons 3, *BK* Ca^{2+}-activated big conductance K^+ channel, *PKA* protein kinase A, *PKC* protein kinase C, *SGK1* serum glucocorticoid kinase 1, *MR* mineralocorticoid receptor

TRPV4 is perfectly suited to control BK activity and FIKS by influencing $[Ca^{2+}]_i$ levels. Indeed, stimulation of TRPV4 with luminal but not basolateral 4α-PDD increased K^+ secretion in perfused murine collecting ducts of WT but not TRPV4$^{-/-}$ mice (Taniguchi et al. 2007). Consistently, treatment with loop diuretic furosemide led to significantly higher urinary K^+ levels in WT than in TRPV4$^{-/-}$ mice (Taniguchi et al. 2007). In addition to BK channel having relatively low Ca^{2+}-binding affinity, small conductance KCa2.1 (KCNN1), KCa2.3 (KCNN3), and intermediate conductance KCa3.1 (KCNN4) with notably higher Ca^{2+}-binding affinity have been detected in native and cultured distal nephron cells (Berrout et al. 2014; Li et al. 2016). While it is not clear whether these K^+ channels contribute significantly to FIKS, it was proposed that flow-induced TRPV4-mediated Ca^{2+} influx first activates KCa2.1, KCa2.3, and KCa3.1. This hyperpolarizes the apical plasma membrane and further enhances Ca^{2+} entry necessary for activation of the low Ca^{2+}-binding affinity BK channel (Li et al. 2017b).

It appears that regulation of FIKS by TRPV4 is essential to maintain systemic K^+ balance during adaptation to increased dietary K^+ intake (Fig. 27.7) (Mamenko et al. 2017). Renal TRPV4 protein and mRNA levels were markedly increased in mice fed with 5% K^+ diet. This also led to TRPV4 translocation to the apical membrane and augmented flow-induced $[Ca^{2+}]_i$ elevations in split-opened murine collecting ducts. Furthermore, single-channel BK activity was drastically lower in collecting ducts from TRPV4$^{-/-}$ mice suggesting deficient FIKS. At the whole kidney level, TRPV4 knockouts lagged behind WT controls with respect to kaliuretic response to high-potassium diet developing hyperkalemia and hyperaldosteronism (Mamenko et al. 2017). Thus, TRPV4 dysfunction recapitulated the phenotype of BK channel deficiency leading to renal K^+ retention (Grimm et al. 2009; Mamenko et al. 2017). From the clinical perspective, TRPV4 polymorphism might indicate susceptibility to developing hyperkalemia, and stimulation of TRPV4 might be beneficial in correction of high plasma potassium. Blockade of mineralocorticoid receptors with spironolactone blunted upregulation of renal TRPV4 expression by dietary K^+, suggesting a critical contribution of aldosterone-signaling cascade (Mamenko et al. 2017). Of note, steroid hormones can also upregulate TRPV4 mRNA levels in extrarenal tissues (Boychuk et al. 2013; Kim et al. 2015). The aldosterone-sensitive serum glucocorticoid kinase 1 (SGK1) can directly phosphorylate TRPV4 at Ser^{824} residue leading to increased channel activity in overexpression systems (Shin et al. 2012). Consistently, SGK1$^{-/-}$ mice demonstrated impaired adaptation to dietary K^+ load (Huang et al. 2004). Further studies are required to accurately delineate the signaling mechanism of TRPV4 regulation by aldosterone-dependent and -independent pathways during variations of dietary K^+ intake.

Recent studies also suggested TRPV4 expression in the proximal tubule cells (Janas et al. 2016). However, it is currently not known whether TRPV4 also governs flow sensitivity at this site. Stimulation of TRPV4 with a selective agonist, GSK1016790A, enhanced albumin endocytosis in primary cultured proximal tubule cells from WT, but not TRPV4$^{-/-}$ mice (Janas et al. 2016). The relevance of this is unclear since proteinuria is not reported in TRPV4$^{-/-}$ mice. Preconditioning with stronger stimuli or particular conditions, such as high protein intake, might be necessary to uncover potential defects of reabsorption of albumin (or other solutes) in proximal tubule at the systemic level. It was also demonstrated that TRPV4 deficiency aggravated renal tubular damage after ischemia-reperfusion injury, one of the major clinically relevant problems (Mannaa et al. 2018). This was particularly attributed to enhanced tubular necrosis of AQP1-positive proximal tubule cells. It was speculated that hypoxia-induced Ca^{2+} influx via TRPV4 is beneficial for proximal cell survival via a currently unknown mechanism. It is also very likely, though, that the lack of TRPV4 in the renal vasculature also contributed to the observed necrosis in renal cortex of TRPV4$^{-/-}$ mice after ischemia-reperfusion injury.

An important aspect of TRPV4 channel function is its ability to form heteromers with other members of TRP family. Specifically, TRPV4 can oligomerize with polycystin 2 (PC2 or TRPP2) in a 2:2 ratio (Stewart et al. 2010) forming a functional Ca^{2+}-permeable channel with distinct biophysical properties and 23 pS conductance on the apical membrane of the collecting duct cells (Kottgen et al. 2008; Zhang et al. 2013b). Similarly, GSK1016790A-activated channel with estimated conductance ~20 pS was detected in primary cultures of human distal tubule cells consistent with TRPV4/TRPP2 heteromer properties (Tomilin et al. 2018). Of note, TRPV4/TRPP2 heteromer has augmented mechanosensitive Ca^{2+} influx compared to homomeric TRPV4 (Kottgen et al. 2008; Du et al. 2012). Mutations in PKD2 gene encoding TRPP2 caused ADPKD, one of the most common monogenic disorders, affecting 12.5 million people worldwide in all ethnic groups and is a major burden on public health (Harris and Torres 2009; Chapman et al. 2015). It is characterized by the development and enlargement of numerous renal cysts filled with fluid eventually progressing to the end-stage renal disease and decline in kidney function. The detailed roles of polycystins and their relation to polycystic kidney disease are covered in Chap. 29 of this volume. Relevant to the current topic, cystic cells have lower basal $[Ca^{2+}]_i$ levels and deficient $[Ca^{2+}]_i$ responses to mechanical stimuli (Siroky et al. 2006; Hovater et al. 2008; Xu et al. 2009). It was further demonstrated that disturbed $[Ca^{2+}]_i$ homeostasis drives de-differentiation and proliferation of cyst cells via cAMP-dependent B-Raf/ERK pathway (Harris and Torres 2009). Nonspecific stimulation of Ca^{2+} influx across the plasma membrane prevented the exaggerated proliferation of cyst cells, suggesting beneficial effects of this strategy in counteracting cystogenesis (Yamaguchi et al. 2006). While TRPV4 deletion abolished flow-dependent $[Ca^{2+}]_i$ elevations, TRPV4$^{-/-}$ mice and zebrafish do not develop renal cysts, at least in unstressed conditions (Liedtke and Friedman 2003; Kottgen et al. 2008). This suggests that the relation between cystogenesis and disruption of mechanosensitivity is more complex. On the other side, TRPV4 Ca^{2+} influx and single-channel activity were drastically reduced in freshly isolated monolayers of cystic cells in PCK 453 rats, a model of autosomal recessive PKD (Zaika et al. 2013), and in primary cells cultured from human ADPKD kidneys (Tomilin et al. 2018). There were no changes in TRPV4 abundance, but a drastic reduction in TRPV4 glycosylation in both cases (Zaika et al. 2013; Tomilin et al. 2018), suggesting the common mechanism of TRPV4 dysfunction during different forms of polycystic kidney disease. Previous reports demonstrated that TRPV4 is glycosylated by complex high mannose sugars at an N-linked glycosylation site (Asn^{651}) in the vicinity of S5–S6 pore domain (Xu et al. 2006; Wu et al. 2007). Importantly, TRPV4 mutations causing familial digital arthropathy-brachydactyly (FDAP) have deficient glycosylation. Mutant channels were unable to respond to mechanical stimuli and had impaired activation by GSK1016790A (Lamande et al. 2011). Similarly, blockade of TRPV4 glycosylation with tunicamycin markedly decreased channel activity and compromised mechanosensitive $[Ca^{2+}]_i$ signaling in primary cultures of human distal tubule cells (Tomilin et al. 2018), supporting a critical role of TRPV4 glycosylation in channel gating and mechanoactivation. It is quite possible that future development of specific molecular/genetic tools to augment TRPV4 glycosylation might be beneficial in treatment of ADPKD patients, especially considering the highest level of TRPV4 expression in the kidney compared to other organs/tissues (Liedtke et al. 2000). Consistently, systemic administration of TRPV4 activator, GSK1016790A partially restored mechanosensitive $[Ca^{2+}]_i$ responses in cyst cells and significantly diminished renal cystogenesis in PCK 453 rats (Zaika et al. 2013). Future studies are necessary to examine pharmacological potential and limitations of TRPV4-based strategy to counteract polycystic kidney disease in clinic.

27.7 TRPV5/TRPV6 and Renal Ca^{2+} Handling

TRPV5 (previously known as epithelial Ca^{2+} channel, ECaC or ECaC1) was originally cloned as a Ca^{2+}-permeable channel expressed in 1,25-dihydroxyvitamin D_3-responsive epithelia: proximal small intestine and distal segments of the renal nephron (Hoenderop et al. 1999b). Shortly after, calcium transport protein, TRPV6 (originally CaT1; 75% amino acid sequence identity with TRPV5), was cloned from rat duodenum with the proposed key role in the intestinal uptake of calcium (Peng et al. 1999). Both channels showed considerable homology with capsaicin-sensitive TRPV1 and were included into TRPV subfamily as fifth and sixth members, respectively. TRPV6 is more broadly expressed across the body, including esophagus, stomach, small intestine, colon, kidney, placenta, pancreas, prostate, uterus, salivary gland, and sweat gland (Hoenderop et al. 2005; Peng 2011). In contrast, TRPV5 is found predominantly in the kidney and small intestine (Hoenderop et al. 2005; Peng 2011). It is interesting that fish and birds contain only one TRPV6-like gene, whereas mammals have TRPV5 and TRPV6 genes indicating that duplication of TRPV6 gene during the evolution was important for the development of complex renal function (Peng 2011). The signature of both TRPV5 and TRPV6 is their remarkable selectivity for Ca^{2+} (permeability Ca^{2+} to Na^+ is more than 100:1) suggesting their specific roles in Ca^{2+} flux across the plasma membrane (Vennekens et al. 2000; Yue et al. 2001).

Ca^{2+} is the most abundant electrolyte in the body (approximately 1 kg versus 100 g of Na^+ and 140 g of K^+ in humans). Approximately 99% of Ca^{2+} is stored in skeletal bones and is interchangeable with Ca^{2+} in the extracellular compartment via processes of bone resorption/formation. Serum Ca^{2+} is maintained in the narrow physiological range of 2.2–2.7 mM/l by matching intestinal Ca^{2+} absorption with urinary Ca^{2+} excretion under tight control of several hormones with the major contribution of the parathyroid hormone (PTH) and the active form of vitamin D, 1,25-dihydroxyvitamin D_3 (Hoenderop et al. 2005; Peng 2011). The bulk of filtered ionized (i.e., not bound to serum proteins) Ca^{2+} is reabsorbed via the paracellular route in the proximal tubule and the thick ascending limb along with Mg^{2+}

secondary to the active Na^+ transport. The late distal convoluted tubule and the connecting tubule are the major sites of controlled transcellular Ca^{2+} reabsorption (Subramanya and Ellison 2014). In contrast to Mg^{2+} (see TRPM6 section), the early portion of distal convoluted tubule (DCT1) does not significantly reabsorb Ca^{2+}. The downstream segments of the collecting duct system can also participate in Ca^{2+} transport (see TRPC3 section), but their contribution is less appreciated.

As for the majority of TRPV channels, cryo-EM structures of full-length rabbit TRPV5 (Hughes et al. 2018) and rat TRPV6 have been recently solved (Saotome et al. 2016) directly confirming classical TRP channel tetrameric composition. The overall organization of ion permeation pore was similar to that of other TRPV channels having the outer selectivity gate and the inner lower gate with a large vestibule (see Fig. 27.2b). However, the position of four aspartate residues (Asp^{542} for TRPV5 and Asp^{541} for TRPV6) constitutes the narrow point of the selectivity filter for coordination of Ca^{2+} ions, which likely underlies high Ca^{2+}-selectivity of TRPV5 and TRPV6. In contrast, the selectivity filter of other TRPV channels is formed by nonpolar residues (such as G675 for TRPV4, see Fig. 27.2b). It has to be noted that there are apparent differences in the organization of conducting pore between TRPV5 and TRPV6. Specifically, unique spatial orientation of the respective Asp residues in the selectivity filter and several other residues could potentially contribute to discrete gating patterns of both channels. Moreover, the region connecting S5 and S6 (forming so-called vestibule) contains eight acidic residues, four of which face the ion conduction pathway to produce a highly electronegative "mouth" to the pore, which can further assist with coordinating of Ca^{2+} movement.

It is now well recognized that TRPV5 is expressed exclusively on the apical plasma membrane of the late distal convoluted tubule and early connecting tubule (see Fig. 27.3). Here, TRPV5 serves as the dominant pathway for Ca^{2+} reabsorption (Hoenderop et al. 1999b). TRPV6 mRNA can be also detected from the distal convoluted tubule to the medullary collecting ducts (Nijenhuis et al. 2003b). However, its expression is lower compared to TRPV5 (Van Cromphaut et al. 2001; Song et al. 2003). Overall, it is generally accepted that TRPV5 is essential for renal Ca^{2+} reabsorption, whereas it counter partner, TRPV6 governs transcellular Ca^{2+} absorption in intestinal epithelia (Peng et al. 1999). The transcellular Ca^{2+} flux requires the presence of Calbindin-D28k, which rapidly binds Ca^{2+} at the vicinity of TRPV5 and shuttles it to the basolateral membrane, where Ca^{2+} is exported through the Na^+/Ca^{2+} exchanger (NCX1) and the plasma membrane calcium-ATPase (PMCA1b) (Lambers et al. 2006a, b) (Fig. 27.8). TRPV5 is constitutively active in the presence of low $[Ca^{2+}]_i$ (less than 100 nM). Calbindin-D28K translocates toward the plasma membrane and directly associates with TRPV5 at low $[Ca^{2+}]_i$ levels to stimulate TRPV5-dependent Ca^{2+} influx (Lambers et al. 2006b). Mice lacking calbindin-D28K demonstrate deficient renal Ca^{2+} conservation (Lee et al. 2016). In contrast, TRPV5 is inactive when $[Ca^{2+}]_i$ raises above 100 nM (Hoenderop et al. 1999c), thereby precluding deleterious actions of sustained $[Ca^{2+}]_i$ elevation. It has been further demonstrated that inactivation of TRPV5 by high $[Ca^{2+}]_i$ levels occurs in a calmodulin-dependent manner. While an early study reported the inhibitory effect of calmodulin on TRPV6 but not on TRPV5 in HEK293 cells (Lambers et al. 2004), the

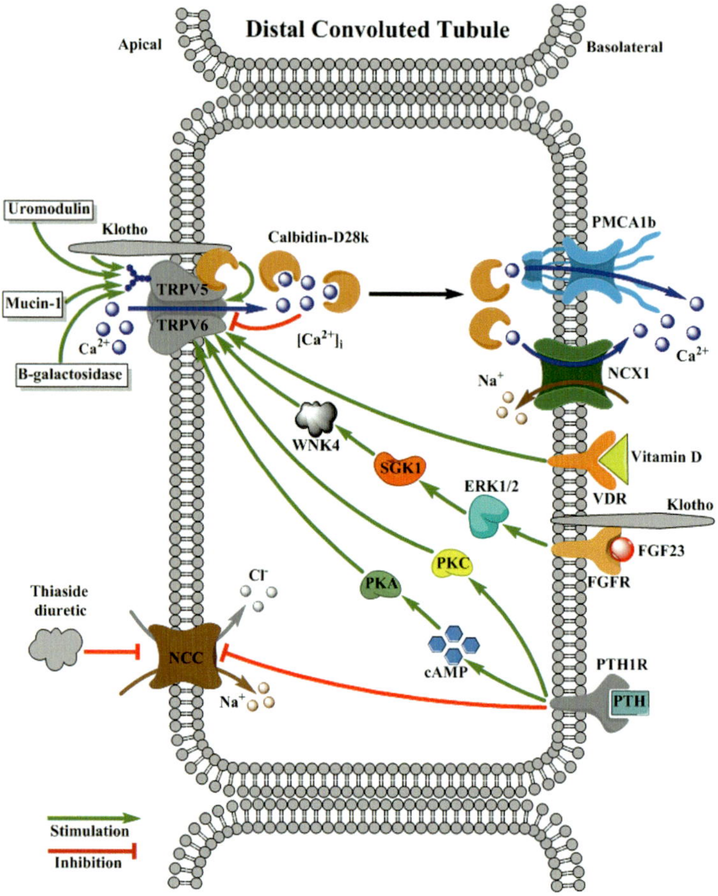

Fig. 27.8 Proposed role of TRPV5 channel in transcellular Ca^{2+} reabsorption in the distal convoluted tubule cells. Description of TRPV5 and TRPV6 represents functional heteromerization of both channels. *NCC* thiazide-sensitive sodium chloride cotransporter, SLC12A3, *WNK4* with-no-lysine(K) kinase 4, *SGK1* serum glucocorticoid kinase 1, *ERK1/2* extracellular-signal-regulated kinases, *PKC* protein kinase C, *PKA* protein kinase A, *cAMP* cyclic adenosine monophosphate, *PMCA1b* plasma membrane Ca^{2+} ATPase type 1b, *NCX1* Na^+/Ca^{2+} exchanger type 1, *VDR* vitamin D receptor, *FGF23* fibroblast growth factor 23, *FGFR* fibroblast growth factor receptor, *PTH* parathyroid hormone, *PTH1R* parathyroid hormone receptor type 1

following studies using high-resolution nuclear magnetic resonance (NMR) spectroscopy revealed a Ca^{2+}-dependent interaction between calmodulin and residues 696 to 729 of TRPV5 C-terminus (de Groot et al. 2011). TRPV5 mutants lacking calmodulin binding were much less sensitive to increased $[Ca^{2+}]_i$ levels (de Groot et al. 2011). Furthermore, recently determined crystal structure of TRPV5 C-terminus in complex with calmodulin revealed direct binding of the channel to calmodulin C-lobe at low $[Ca^{2+}]_i$ levels and more tight interaction with both C- and N-lobes of calmodulin at elevated Ca^{2+} levels (Bokhovchuk et al. 2018). A similar mechanism has been also demonstrated for TRPV6, where calmodulin forms a "bridge" between two TRPV6 monomers resulting in closure of the channel pore in the presence of high Ca^{2+} (Bate et al. 2018).

Basic and clinical evidence revealed an important contribution of TRPV5-dependent Ca^{2+} reabsorption on systemic Ca^{2+} balance and renal function. $TRPV5^{-/-}$ mice exhibited augmented urinary Ca^{2+} levels and demonstrated significant abnormalities of bone structure, such as reduced trabecular and cortical bone thickness (Hoenderop et al. 2003). Similarly, $TRPV6^{-/-}$ mice were also hypercalciuric and showed signs of reduced bone mineral density (Bianco et al. 2007). However, no mutations in TRPV5 and TRPV6 were reported to significantly affect serum Ca^{2+} levels in humans. This indicates that reduced TRPV5-dependent Ca^{2+} reabsorption by the kidney could be compensated by augmented TRPV6-mediated Ca^{2+} absorption in intestine and vice versa. It is also quite possible that concomitant mutations of both channels are necessary to compromise systemic Ca^{2+} balance. Luckily, the probability of this event is low. However, clinical studies found a significant correlation of TRPV5 polymorphisms (rs4236480, rs757494578) with the incidence of nephrolithiasis (kidney stone formation) in tested groups (Khaleel et al. 2015; Oddsson et al. 2015). L530R mutation in TRPV5 (which corresponded to the latter variant) abolished Ca^{2+} uptake upon channel overexpression in *Xenopus* oocytes (Wang et al. 2017). It is reasonable to assume that defective TRPV5-dependent Ca^{2+} reabsorption in the distal convoluted tubule will lead to increased Ca^{2+} delivery to the collecting duct system. This, in turn, increases the risk of Ca^{2+} precipitation in renal medulla particularly during antidiuresis. On the contrary, higher allele frequencies of TRPV5 and TRPV6 single-nucleotide polymorphisms associated with augmented channel activity in African population when compared to Caucasians might contribute to reduced urinary Ca^{2+} excretion, diminished risk of developing kidney stones, and lower incidence of osteoporosis-related fractures (reviewed in (Peng 2011)). Inhibition of Na^{+} reabsorption in the distal convoluted tubule via Na-Cl cotransporter (NCC, SLC12A3) by antihypertensive thiazide diuretics commonly leads to hypocalciuria (McCormick and Ellison 2015; Alexander and Dimke 2017). The mechanism is complex involving both increased passive Ca^{2+} reabsorption in the proximal tubule and stimulation of TRPV5-dependent Ca^{2+} transport in the distal convoluted tubule (Nijenhuis et al. 2003a, b; Lee et al. 2004; Jang et al. 2009). Similarly, Gitelman's syndrome due to loss-of-function mutations of SLC12A3 is characterized by hypocalciuria (Simon et al. 1996; Ellison 2000).

It is well accepted that decreases in extracellular Ca^{2+} levels stimulate secretion of PTH eliciting pleiotropic actions in kidneys, intestine, and bones to restore Ca^{2+} homeostasis (Blaine et al. 2015). PTH binds to its receptor PTH1R to stimulate TRPV5-dependent calcium reabsorption in the distal convoluted and connecting tubule segments. The molecular pathway involved activation of adenylyl cyclase-cAMP-PKA leading to TRPV5 phosphorylation at T709 residue (Fig. 27.8). This increased single-channel open probability and diminished channel binding to calmodulin (de Groot et al. 2009). It is interesting that this mechanism might not be operational in humans due to incomplete conservation of the PKA consensus motif (RR/QxT) of the channel. Thus, PTH stimulated rabbit TRPV5 overexpressed in

HEK293 cells in a PKA-dependent manner, but failed to increase activity of human TRPV5 (van Goor et al. 2017). PTH can also increase TRPV5 activity independent of PKA via channel phosphorylation by PKC (Friedman et al. 1996; Hoenderop et al. 1999a; Cha et al. 2008). This inhibited caveolae-mediated endocytosis of TRPV5 leading to accumulation of the channel levels on the plasma membrane (Cha et al. 2008). Moreover, PTH decreased NCC activity in cultured mDCT15 cells, which can indirectly augment TRPV5-dependent Ca^{2+} reabsorption (Hoover et al. 2016). Stimulation of β1-adrenergic receptors also led PKA-dependent TRPV5 phosphorylation at T709 residue and increased TRPV5-mediated Ca^{2+} uptake in HEK293 cells (van der Hagen et al. 2014) indicating that beta blockers commonly used in the treatment of congestive heart failure and other cardiovascular pathologies might potentially exert adverse calciotropic effects. Future clinical studies are needed to carefully evaluate this possibility.

The active form of vitamin D, 1,25-dihydroxyvitamin D_3, stimulates Ca^{2+} reabsorption in the distal convoluted and connecting tubule segments by binding to vitamin D receptor (VDR) to increase TRPV5 expression (Fig. 27.8) (Kumar et al. 1994; Hoenderop et al. 1999b; Lambers et al. 2006a; Woudenberg-Vrenken et al. 2009). Consistently, $VDR^{-/-}$ mice exhibited increased urinary Ca^{2+} excretion and reduced TRPV5 expression (Van Cromphaut et al. 2001). Low vitamin D levels and loss-of-function mutations in VDR lead to vitamin D-deficient rickets associated with compromised Ca^{2+} balance and hypocalcemia (Malloy and Feldman 1999; Durmaz et al. 2012). Ca^{2+} supplementation is an efficient strategy to treat rickets (Malloy and Feldman 1999), which potentially indicates that mechanisms causing TRPV5 activation would also be beneficial. Recent experimental evidence has suggested a critical role of a vitamin D-regulating hormone, fibroblast growth factor-23 (FGF23) in stimulation of TRPV5 activity in the distal convoluted tubule (Andrukhova et al. 2014b). FGF23 was able to regulate renal Ca^{2+} reabsorption in mice lacking VDR suggesting an independent signaling pathway (Andrukhova et al. 2014b). $FGF23^{-/-}$ mice were hypercalciuric and demonstrated reduced renal TRPV5 abundance. Double knockout of FGF23 and VDR further exacerbated urinary Ca^{2+} wasting (Andrukhova et al. 2014b). Binding FGF23 to FGF receptor (FGFR) activated a signaling cascade involving ERK1/2, SGK1, and with-no-lysine kinase 4 (WNK4) to promote TRPV5-dependent Ca^{2+} influx (Andrukhova et al. 2014b). It is interesting that FGF23 also stimulated NCC membrane in the distal convoluted tubule via the same pathway involving WNK4 phosphorylation (Andrukhova et al. 2014a). Injection of recombinant FGF23 induced distal tubule Na^+ retention, hypertension, and cardiac hypertrophy (Andrukhova et al. 2014a). The concomitant stimulation of Na^+ and Ca^{2+} reabsorption in the distal convoluted tubule by FGF23 is intriguing. Inhibition of NCC-mediated Na^+ reabsorption by thiazide diuretics and loss-of-function NCC during Gitelman's syndrome upregulate TRPV5-dependent Ca^{2+} reabsorption leading to hypocalciuria (Simon et al. 1996; Ellison 2000; Alexander and Dimke 2017). Furthermore, gain-of-function mutations in WNKs (WNK1 and WNK4) or in the components of ubiquitin system controlling WNK degradation (KLHL3 and CUL3), cause familial hyperkalemic hypertension in humans (FHHt, also known as pseudohypoladosteronism type II or Gordon syndrome) sensitive to thiazide diuretics (Wilson et al. 2001; Boyden et al. 2012; Shibata et al. 2013). FHHt patients usually exhibit hypercalciuria (Mayan et al. 2009; Mayan et al. 2015), indicative of diminished tubular Ca^{2+} reabsorption and TRPV5 activity. Future studies are needed to investigate how stimulation of WNK4-dependent signaling might differentially regulate Na^+ and Ca^{2+} transport in the distal tubule.

Originally discovered as an antiaging molecule, single transmembrane glycoprotein Klotho has also a notable role in regulation of renal calcium handling (Kuro-o et al. 1997; Torres et al. 2007). Klotho expression is the most abundant in the distal segments of renal tubule indicating its effect on TRPV5 activity (Fig. 27.8) (Chang et al. 2005). Indeed, Klotho$^{-/-}$ mice exhibited renal Ca^{2+} wasting due to a failure to absorb Ca^{2+} in the distal tubule via TRPV5 (Alexander et al. 2009). However, the mechanism of its action on TRPV5 is complex and is not completely understood. Intracellular actions of Klotho likely involved enhanced forward trafficking of the channel, whereas the extracellular action is due to inhibition of endocytosis by hydrolyzing extracellular sugar residues at N-glycosylation site Asn^{358} of TRPV5, thereby entrapping the channel in the plasma membrane (Chang et al. 2005; Wolf et al. 2014). Of interest, stimulation of TRPV5 activity via control of N-glycosylation has been recently demonstrated for β galactosidase (Leunissen et al. 2016), mucin-1 (Nie et al. 2016), and the most abundant protein in human urine, uromodulin (also commonly known as the Tamm-Horsfall protein) (Wolf et al. 2013). Mice lacking uromodulin develop hypercalciuria, renal calcium crystals, and calcifications (Liu et al. 2010). On the contrary, it was shown that Klotho functions as an obligatory coreceptor for ubiquitously expressed FGFR (Kurosu et al. 2006; Urakawa et al. 2006) to mediate specificity of renal anticalciuretic FGF23 actions to the distal renal tubule and does not operate as a calcium-regulating hormone (Andrukhova et al. 2014b).

27.8 Conclusions

It is becoming increasingly appreciated that TRP channels are functionally expressed in various segments of the renal tubule where they are critical for many physiological processes, including maintenance of glomerular barrier integrity, sensitivity/adaptation to external stimuli (alterations in tubular flow, ROS production, etc.), and electrolyte transport of Ca^{2+} and Mg^{2+}. Recent advances in basic studies utilizing high-resolution 3-D TRP channel structures, novel genetic animal models in combination with available clinical evidence suggest that TRP channels are important for the baseline kidney function and even more during adaptation to physiological stimuli, such as changes in dietary water and electrolyte intake, stimulation of endocrine axes, for example. Abnormal activity of TRP channels can lead to systemic pathologies associated with electrolyte imbalance (hypomagnesemia, hyperkalemia), and to structural damage of renal parenchyma (polycystic kidney disease, nephrolithiasis, loss of glomeruli, fibrosis). However, we are just at the beginning of a long and convoluted road to implement TRP-based pharmacological and potentially genetic strategies into the clinical realm at the maximal efficiency.

Acknowledgments We thank Drs. O. Palygin and A. Staruschenko (Medical College of Wisconsin) for helpful suggestions.

Research program in Pochynyuk's lab is supported by NIH-NIDDK DK095029 and DK119170 (to O. Pochynyuk), AHA 17GRNT33660488 (to O. Pochynyuk), and ASN Ben J. Lipps Research Fellowship (to V. N. Tomilin).

References

Alexander RT, Dimke H (2017) Effect of diuretics on renal tubular transport of calcium and magnesium. Am J Physiol Renal Physiol 312:F998–F1015. https://doi.org/10.1152/ajprenal.00032.2017

Alexander RT, Woudenberg-Vrenken TE, Buurman J, Dijkman H et al (2009) Klotho prevents renal calcium loss. J Am Soc Nephrol 20:2371–2379. https://doi.org/10.1681/ASN.2008121273

Alim I, Teves L, Li R, Mori Y et al (2013) Modulation of NMDAR subunit expression by TRPM2 channels regulates neuronal vulnerability to ischemic cell death. J Neurosci 33:17264–17277. https://doi.org/10.1523/JNEUROSCI.1729-13.2013

Ambrus L, Olah A, Olah T, Balla G et al (2015) Inhibition of TRPC6 by protein kinase C isoforms in cultured human podocytes. J Cell Mol Med 19:2771–2779. https://doi.org/10.1111/jcmm.12660

Anderson M, Kim EY, Hagmann H, Benzing T et al (2013) Opposing effects of podocin on the gating of podocyte TRPC6 channels evoked by membrane stretch or diacylglycerol. Am J Physiol Cell Physiol 305:C276–C289. https://doi.org/10.1152/ajpcell.00095.2013

Anderson M, Roshanravan H, Khine J, Dryer SE (2014) Angiotensin II activation of TRPC6 channels in rat podocytes requires generation of reactive oxygen species. J Cell Physiol 229:434–442. https://doi.org/10.1002/jcp.24461

Andrade YN, Fernandes J, Vazquez E, Fernandez-Fernandez JM et al (2005) TRPV4 channel is involved in the coupling of fluid viscosity changes to epithelial ciliary activity. J Cell Biol 168:869–874. https://doi.org/10.1083/jcb.200409070

Andrukhova O, Slavic S, Smorodchenko A, Zeitz U et al (2014a) FGF23 regulates renal sodium handling and blood pressure. EMBO Mol Med 6:744–759. https://doi.org/10.1002/emmm.201303716

Andrukhova O, Smorodchenko A, Egerbacher M, Streicher C et al (2014b) FGF23 promotes renal calcium reabsorption through the TRPV5 channel. EMBO J 33:229–246. https://doi.org/10.1002/embj.201284188

Astor MC, Lovas K, Wolff AS, Nedrebo B et al (2015) Hypomagnesemia and functional hypoparathyroidism due to novel mutations in the mg-channel TRPM6. Endocr Connect 4:215–222. https://doi.org/10.1530/EC-15-0066

Bate N, Caves RE, Skinner SP, Goult BT et al (2018) A novel mechanism for calmodulin-dependent inactivation of transient receptor potential vanilloid 6. Biochemistry 57:2611–2622. https://doi.org/10.1021/acs.biochem.7b01286

Bengele HH, Alexander EA, Lechene CP (1980) Calcium and magnesium transport along the inner medullary collecting duct of the rat. Am J Phys 239:F24–F29. https://doi.org/10.1152/ajprenal.1980.239.1.F24

Berrout J, Jin M, Mamenko M, Zaika O et al (2012) Function of TRPV4 as a mechanical transducer in flow-sensitive segments of the renal collecting duct system. J Biol Chem 287:8782–8791

Berrout J, Mamenko M, Zaika OL, Chen L et al (2014) Emerging role of the calcium-activated, small conductance, SK3 K^+ channel in distal tubule function: regulation by TRPV4. PLoS One 9:e95149. https://doi.org/10.1371/journal.pone.0095149
Bianco SD, Peng JB, Takanaga H, Suzuki Y et al (2007) Marked disturbance of calcium homeostasis in mice with targeted disruption of the Trpv6 calcium channel gene. J Bone Miner Res 22:274–285. https://doi.org/10.1359/jbmr.061110
Blaine J, Chonchol M, Levi M (2015) Renal control of calcium, phosphate, and magnesium homeostasis. Clin J Am Soc Nephrol 10:1257–1272. https://doi.org/10.2215/CJN.09750913
Blanchard MG, Kittikulsuth W, Nair AV, de Baaij JH et al (2016) Regulation of Mg^{2+} reabsorption and transient receptor potential melastatin type 6 activity by cAMP signaling. J Am Soc Nephrol 27:804–813. https://doi.org/10.1681/ASN.2014121228
Blumberg D, Bonetti A, Jacomella V, Capillo S et al (1998) Free circulating magnesium and renal magnesium handling during acute metabolic acidosis in humans. Am J Nephrol 18:233–236. doi:13342
Bockenhauer D, Feather S, Stanescu HC, Bandulik S et al (2009) Epilepsy, ataxia, sensorineural deafness, tubulopathy, and KCNJ10 mutations. N Engl J Med 360:1960–1970. https://doi.org/10.1056/NEJMoa0810276
Bokhovchuk FM, Bate N, Kovalevskaya NV, Goult BT et al (2018) The structural basis of calcium-dependent inactivation of the transient receptor potential vanilloid 5 channel. Biochemistry 57:2623–2635. https://doi.org/10.1021/acs.biochem.7b01287
Bouron A, Chauvet S, Dryer S, Rosado JA (2016) Second messenger-operated calcium entry through TRPC6. Adv Exp Med Biol 898:201–249. https://doi.org/10.1007/978-3-319-26974-0_10
Boychuk CR, Zsombok A, Tasker JG, Smith BN (2013) Rapid glucocorticoid-induced activation of TRP and CB1 receptors causes biphasic modulation of glutamate release in gastric-related hypothalamic preautonomic neurons. Front Neurosci 7:3. https://doi.org/10.3389/fnins.2013.00003
Boyden LM, Choi M, Choate KA, Nelson-Williams CJ et al (2012) Mutations in kelch-like 3 and cullin 3 cause hypertension and electrolyte abnormalities. Nature 482:98–102. https://doi.org/10.1038/nature10814
Buscher AK, Konrad M, Nagel M, Witzke O et al (2012) Mutations in podocyte genes are a rare cause of primary FSGS associated with ESRD in adult patients. Clin Nephrol 78:47–53
Cabral PD, Capurro C, Garvin JL (2015) TRPV4 mediates flow-induced increases in intracellular Ca in medullary thick ascending limbs. Acta Physiol (Oxf) 214:319–328. https://doi.org/10.1111/apha.12528
Cao G, van der Wijst J, van der Kemp A, van Zeeland F et al (2009) Regulation of the epithelial Mg^{2+} channel TRPM6 by estrogen and the associated repressor protein of estrogen receptor activity (REA). J Biol Chem 284:14788–14795. https://doi.org/10.1074/jbc.M808752200
Cao E, Liao M, Cheng Y, Julius D (2013) TRPV1 structures in distinct conformations reveal activation mechanisms. Nature 504:113–118. https://doi.org/10.1038/nature12823
Cha SK, Wu T, Huang CL (2008) Protein kinase C inhibits caveolae-mediated endocytosis of TRPV5. Am J Physiol Renal Physiol 294:F1212–F1221. https://doi.org/10.1152/ajprenal.00007.2008
Chang Q, Hoefs S, van der Kemp AW, Topala CN et al (2005) The beta-glucuronidase klotho hydrolyzes and activates the TRPV5 channel. Science 310:490–493. https://doi.org/10.1126/science.1114245
Chapman AB, Devuyst O, Eckardt KU, Gansevoort RT et al (2015) Autosomal-dominant polycystic kidney disease (ADPKD): executive summary from a kidney disease: improving global outcomes (KDIGO) controversies conference. Kidney Int 88:17–27. https://doi.org/10.1038/ki.2015.59
Chen L, Kassmann M, Sendeski M, Tsvetkov D et al (2015) Functional transient receptor potential vanilloid 1 and transient receptor potential vanilloid 4 channels along different segments of the renal vasculature. Acta Physiol (Oxf) 213:481–491. https://doi.org/10.1111/apha.12355

Chi X, Hu B, Yu SY, Yin L et al (2015) Losartan treating podocyte injury induced by Ang II via downregulation of TRPC6 in podocytes. J Renin-Angiotensin-Aldosterone Syst 16 (4):1118–1124. https://doi.org/10.1177/1470320315573682
Christensen AP, Corey DP (2007) TRP channels in mechanosensation: direct or indirect activation? Nat Rev Neurosci 8:510–521. https://doi.org/10.1038/nrn2149
Chubanov V, Gudermann T (2014) Trpm6. Handb Exp Pharmacol 222:503–520. https://doi.org/10.1007/978-3-642-54215-2_20
Chubanov V, Ferioli S, Wisnowsky A, Simmons DG et al (2016) Epithelial magnesium transport by TRPM6 is essential for prenatal development and adult survival. elife 5. https://doi.org/10.7554/eLife.20914
Ciura S, Liedtke W, Bourque CW (2011) Hypertonicity sensing in organum vasculosum lamina terminalis neurons: a mechanical process involving TRPV1 but not TRPV4. J Neurosci 31:14669–14676. https://doi.org/10.1523/JNEUROSCI.1420-11.2011
Corre T, Arjona FJ, Hayward C, Youhanna S et al (2018) Genome-wide meta-analysis unravels interactions between magnesium homeostasis and metabolic phenotypes. J Am Soc Nephrol 29:335–348. https://doi.org/10.1681/ASN.2017030267
Cosens DJ, Manning A (1969) Abnormal electroretinogram from a Drosophila mutant. Nature 224:285–287
Coulter M, Colvin C, Korf B, Messiaen L et al (2015) Hypomagnesemia due to two novel TRPM6 mutations. J Pediatr Endocrinol Metab 28:1373–1378. https://doi.org/10.1515/jpem-2014-0394
Cuajungco MP, Grimm C, Oshima K, D'Hoedt D et al (2006) PACSINs bind to the TRPV4 cation channel. PACSIN 3 modulates the subcellular localization of TRPV4. J Biol Chem 281:18753–18762. https://doi.org/10.1074/jbc.M602452200
Cuevas CA, Su XT, Wang MX, Terker AS et al (2017) Potassium sensing by renal distal tubules requires Kir4.1. J Am Soc Nephrol 28:1814–1825. https://doi.org/10.1681/ASN.2016090935
Curthoys NP, Moe OW (2014) Proximal tubule function and response to acidosis. Clin J Am Soc Nephrol 9:1627–1638. https://doi.org/10.2215/CJN.10391012
D'Hoedt D, Owsianik G, Prenen J, Cuajungco MP et al (2008) Stimulus-specific modulation of the cation channel TRPV4 by PACSIN 3. J Biol Chem 283:6272–6280. https://doi.org/10.1074/jbc.M706386200
Dai LJ, Ritchie G, Kerstan D, Kang HS et al (2001) Magnesium transport in the renal distal convoluted tubule. Physiol Rev 81:51–84. https://doi.org/10.1152/physrev.2001.81.1.51
de Baaij JH, Hoenderop JG, Bindels RJ (2015) Magnesium in man: implications for health and disease. Physiol Rev 95:1–46. https://doi.org/10.1152/physrev.00012.2014
de Baaij JH, Arjona FJ, van den Brand M, Lavrijsen M et al (2016) Identification of SLC41A3 as a novel player in magnesium homeostasis. Sci Rep 6:28565. https://doi.org/10.1038/srep28565
de Groot T, Lee K, Langeslag M, Xi Q et al (2009) Parathyroid hormone activates TRPV5 via PKA-dependent phosphorylation. J Am Soc Nephrol 20:1693–1704. https://doi.org/10.1681/ASN.2008080873
de Groot T, Kovalevskaya NV, Verkaart S, Schilderink N et al (2011) Molecular mechanisms of calmodulin action on TRPV5 and modulation by parathyroid hormone. Mol Cell Biol 31:2845–2853. https://doi.org/10.1128/MCB.01319-10
Delany NS, Hurle M, Facer P, Alnadaf T et al (2001) Identification and characterization of a novel human vanilloid receptor-like protein, VRL-2. Physiol Genomics 4:165–174
Deng Z, Paknejad N, Maksaev G, Sala-Rabanal M et al (2018) Cryo-EM and X-ray structures of TRPV4 reveal insight into ion permeation and gating mechanisms. Nat Struct Mol Biol 25:252–260. https://doi.org/10.1038/s41594-018-0037-5
Dietrich A, Mederos Y, Schnitzler M, Emmel J, Kalwa H et al (2003) N-linked protein glycosylation is a major determinant for basal TRPC3 and TRPC6 channel activity. J Biol Chem 278:47842–47852. https://doi.org/10.1074/jbc.M302983200
Dietrich A, Kalwa H, Rost BR, Gudermann T (2005) The diacylgylcerol-sensitive TRPC3/6/7 subfamily of cation channels: functional characterization and physiological relevance. Pflugers Arch 451:72–80. https://doi.org/10.1007/s00424-005-1460-0

Diez-Sampedro A, Lenz O, Fornoni A (2011) Podocytopathy in diabetes: a metabolic and endocrine disorder. Am J Kidney Dis 58:637–646. https://doi.org/10.1053/j.ajkd.2011.03.035
Dryer SE, Reiser J (2010) TRPC6 channels and their binding partners in podocytes: role in glomerular filtration and pathophysiology. Am J Physiol Renal Physiol 299:F689–F701. https://doi.org/10.1152/ajprenal.00298.2010
Du J, Xie J, Yue L (2009a) Intracellular calcium activates TRPM2 and its alternative spliced isoforms. Proc Natl Acad Sci U S A 106:7239–7244. https://doi.org/10.1073/pnas.0811725106
Du J, Xie J, Yue L (2009b) Modulation of TRPM2 by acidic pH and the underlying mechanisms for pH sensitivity. J Gen Physiol 134:471–488. https://doi.org/10.1085/jgp.200910254
Du J, Wong WY, Sun L, Huang Y et al (2012) Protein kinase G inhibits flow-induced Ca^{2+} entry into collecting duct cells. J Am Soc Nephrol 23:1172–1180
Duan J, Li Z, Li J, Santa-Cruz A et al (2018) Structure of full-length human TRPM4. Proc Natl Acad Sci U S A 115:2377–2382. https://doi.org/10.1073/pnas.1722038115
Durmaz E, Zou M, Al-Rijjal RA, Bircan I et al (2012) Clinical and genetic analysis of patients with vitamin D-dependent rickets type 1A. Clin Endocrinol 77:363–369. https://doi.org/10.1111/j.1365-2265.2012.04394.x
Durvasula RV, Shankland SJ (2008) Activation of a local renin angiotensin system in podocytes by glucose. Am J Physiol Renal Physiol 294:F830–F839. https://doi.org/10.1152/ajprenal.00266.2007
Ellison DH (2000) Divalent cation transport by the distal nephron: insights from Bartter's and Gitelman's syndromes. Am J Physiol Renal Physiol 279:F616–F625
Endlich N, Kress KR, Reiser J, Uttenweiler D et al (2001) Podocytes respond to mechanical stress in vitro. J Am Soc Nephrol 12:413–422
Estacion M, Sinkins WG, Jones SW, Applegate MA et al (2006) Human TRPC6 expressed in HEK 293 cells forms non-selective cation channels with limited Ca^{2+} permeability. J Physiol 572:359–377. https://doi.org/10.1113/jphysiol.2005.103143
Everaerts W, Nilius B, Owsianik G (2010a) The vanilloid transient receptor potential channel TRPV4: from structure to disease. Prog Biophys Mol Biol 103:2–17. https://doi.org/10.1016/j.pbiomolbio.2009.10.002
Everaerts W, Zhen X, Ghosh D, Vriens J et al (2010b) Inhibition of the cation channel TRPV4 improves bladder function in mice and rats with cyclophosphamide-induced cystitis. Proc Natl Acad Sci U S A 107:19084–19089. https://doi.org/10.1073/pnas.1005333107
Facemire CS, Mohler PJ, Arendshorst WJ (2004) Expression and relative abundance of short transient receptor potential channels in the rat renal microcirculation. Am J Physiol Renal Physiol 286:F546–F551. https://doi.org/10.1152/ajprenal.00338.2003
Fan C, Choi W, Sun W, Du J et al (2018) Structure of the human lipid-gated cation channel TRPC3. Elife 7:e36852. https://doi.org/10.7554/eLife.36852
Ferioli S, Zierler S, Zaisserer J, Schredelseker J et al (2017) TRPM6 and TRPM7 differentially contribute to the relief of heteromeric TRPM6/7 channels from inhibition by cytosolic mg^{2+} and Mg.ATP. Sci Rep 7:8806. https://doi.org/10.1038/s41598-017-08144-1
Ferre S, Hoenderop JG, Bindels RJ (2011) Insight into renal Mg^{2+} transporters. Curr Opin Nephrol Hypertens 20:169–176. https://doi.org/10.1097/MNH.0b013e3283435ee4
Fischer MJ, Edwardson JM (2014) V2A2lidating TRP channel heteromers. Temperature (Austin) 1:26–27. https://doi.org/10.4161/temp.29548
Friedman PA, Coutermarsh BA, Kennedy SM, Gesek FA (1996) Parathyroid hormone stimulation of calcium transport is mediated by dual signaling mechanisms involving protein kinase a and protein kinase C. Endocrinology 137:13–20. https://doi.org/10.1210/endo.137.1.8536604
Friedrich C, Endlich N, Kriz W, Endlich K (2006) Podocytes are sensitive to fluid shear stress in vitro. Am J Physiol Renal Physiol 291:F856–F865. https://doi.org/10.1152/ajprenal.00196.2005
Gao X, Wu L, O'Neil RG (2003) Temperature-modulated diversity of TRPV4 channel gating: activation by physical stresses and phorbol ester derivatives through protein kinase C-dependent and -independent pathways. J Biol Chem 278:27129–27137

Gao G, Wang W, Tadagavadi RK, Briley NE et al (2014) TRPM2 mediates ischemic kidney injury and oxidant stress through RAC1. J Clin Invest 124:4989–5001. https://doi.org/10.1172/JCI76042
Giebisch GH, Wang WH (2010) Potassium transport-an update. J Nephrol 23(Suppl 16):S97–S104
Gigante M, Caridi G, Montemurno E, Soccio M et al (2011) TRPC6 mutations in children with steroid-resistant nephrotic syndrome and atypical phenotype. Clin J Am Soc Nephrol 6:1626–1634. https://doi.org/10.2215/CJN.07830910
Glaudemans B, van der Wijst J, Scola RH, Lorenzoni PJ et al (2009) A missense mutation in the Kv1.1 voltage-gated potassium channel-encoding gene KCNA1 is linked to human autosomal dominant hypomagnesemia. J Clin Invest 119:936–942. https://doi.org/10.1172/JCI36948
Goel M, Schilling WP (2010) Role of TRPC3 channels in ATP-induced Ca^{2+} signaling in principal cells of the inner medullary collecting duct. Am J Physiol Renal Physiol 299:F225–F233. https://doi.org/10.1152/ajprenal.00670.2009
Goel M, Sinkins WG, Zuo CD, Estacion M et al (2006) Identification and localization of TRPC channels in the rat kidney. Am J Physiol Renal Physiol 290:F1241–F1252. https://doi.org/10.1152/ajprenal.00376.2005
Goel M, Sinkins WG, Zuo CD, Hopfer U et al (2007) Vasopressin-induced membrane trafficking of TRPC3 and AQP2 channels in cells of the rat renal collecting duct. Am J Physiol Renal Physiol 293:F1476–F1488. https://doi.org/10.1152/ajprenal.00186.2007
Goel M, Zuo CD, Schilling WP (2010) Role of cAMP/PKA signaling cascade in vasopressin-induced trafficking of TRPC3 channels in principal cells of the collecting duct. Am J Physiol Renal Physiol 298:F988–F996. https://doi.org/10.1152/ajprenal.00586.2009
Graham S, Ding M, Sours-Brothers S, Yorio T et al (2007) Downregulation of TRPC6 protein expression by high glucose, a possible mechanism for the impaired Ca^{2+} signaling in glomerular mesangial cells in diabetes. Am J Physiol Renal Physiol 293:F1381–F1390. https://doi.org/10.1152/ajprenal.00185.2007
Greka A, Mundel P (2011) Balancing calcium signals through TRPC5 and TRPC6 in podocytes. J Am Soc Nephrol 22:1969–1980. https://doi.org/10.1681/ASN.2011040370
Grimm PR, Irsik DL, Settles DC, Holtzclaw JD et al (2009) Hypertension of $Kcnmb1^{-/-}$ is linked to deficient K secretion and aldosteronism. Proc Natl Acad Sci U S A 106:11800–11805. https://doi.org/10.1073/pnas.0904635106
Groenestege WM, Hoenderop JG, van den Heuvel L, Knoers N et al (2006) The epithelial Mg^{2+} channel transient receptor potential melastatin 6 is regulated by dietary Mg^{2+} content and estrogens. J Am Soc Nephrol 17:1035–1043. https://doi.org/10.1681/ASN.2005070700
Guilak F, Leddy HA, Liedtke W (2010) Transient receptor potential vanilloid 4: the sixth sense of the musculoskeletal system? Ann N Y Acad Sci 1192:404–409. https://doi.org/10.1111/j.1749-6632.2010.05389.x
Guther T, Vormann J, Forster R (1984) Regulation of intracellular magnesium by Mg^{2+} efflux. Biochem Biophys Res Commun 119:124–131
Hardie RC, Minke B (1992) The trp gene is essential for a light-activated Ca^{2+} channel in Drosophila photoreceptors. Neuron 8:643–651
Harris PC, Torres VE (2009) Polycystic kidney disease. Annu Rev Med 60:321–337
Harteneck C, Gollasch M (2011) Pharmacological modulation of diacylglycerol-sensitive TRPC3/6/7 channels. Curr Pharm Biotechnol 12:35–41
Hebert SC, Desir G, Giebisch G, Wang W (2005) Molecular diversity and regulation of renal potassium channels. Physiol Rev 85:319–371. https://doi.org/10.1152/physrev.00051.2003
Heeringa SF, Moller CC, Du J, Yue L et al (2009) A novel TRPC6 mutation that causes childhood FSGS. PLoS One 4:e7771. https://doi.org/10.1371/journal.pone.0007771
Hiroi T, Wajima T, Negoro T, Ishii M et al (2013) Neutrophil TRPM2 channels are implicated in the exacerbation of myocardial ischaemia/reperfusion injury. Cardiovasc Res 97:271–281. https://doi.org/10.1093/cvr/cvs332

Hisatsune C, Kuroda Y, Nakamura K, Inoue T et al (2004) Regulation of TRPC6 channel activity by tyrosine phosphorylation. J Biol Chem 279:18887–18894. https://doi.org/10.1074/jbc.M311274200
Hoenderop JG, De Pont JJ, Bindels RJ, Willems PH (1999a) Hormone-stimulated Ca^{2+} reabsorption in rabbit kidney cortical collecting system is cAMP-independent and involves a phorbol ester-insensitive PKC isotype. Kidney Int 55:225–233. https://doi.org/10.1046/j.1523-1755.1999.00228.x
Hoenderop JG, van der Kemp AW, Hartog A, van de Graaf SF et al (1999b) Molecular identification of the apical Ca^{2+} channel in 1, 25-dihydroxyvitamin D3-responsive epithelia. J Biol Chem 274:8375–8378
Hoenderop JG, van der Kemp AW, Hartog A, van Os CH et al (1999c) The epithelial calcium channel, ECaC, is activated by hyperpolarization and regulated by cytosolic calcium. Biochem Biophys Res Commun 261:488–492. https://doi.org/10.1006/bbrc.1999.1059
Hoenderop JG, van Leeuwen JP, van der Eerden BC, Kersten FF et al (2003) Renal Ca^{2+} wasting, hyperabsorption, and reduced bone thickness in mice lacking TRPV5. J Clin Invest 112:1906–1914. https://doi.org/10.1172/JCI19826
Hoenderop JG, Nilius B, Bindels RJ (2005) Calcium absorption across epithelia. Physiol Rev 85:373–422. https://doi.org/10.1152/physrev.00003.2004
Hofmann T, Obukhov AG, Schaefer M, Harteneck C et al (1999) Direct activation of human TRPC6 and TRPC3 channels by diacylglycerol. Nature 397:259–263. https://doi.org/10.1038/16711
Hofmann T, Schaefer M, Schultz G, Gudermann T (2002) Subunit composition of mammalian transient receptor potential channels in living cells. Proc Natl Acad Sci U S A 99:7461–7466. https://doi.org/10.1073/pnas.102596199
Hofstra JM, Lainez S, van Kuijk WH, Schoots J et al (2013) New TRPC6 gain-of-function mutation in a non-consanguineous Dutch family with late-onset focal segmental glomerulosclerosis. Nephrol Dial Transplant 28:1830–1838. https://doi.org/10.1093/ndt/gfs572
Hoover RS, Tomilin V, Hanson L, Pochynyuk O et al (2016) PTH modulation of NCC activity regulates TRPV5 Ca^{2+} reabsorption. Am J Physiol Renal Physiol 310:F144–F151. https://doi.org/10.1152/ajprenal.00323.2015
Hovater MB, Olteanu D, Hanson EL, Cheng NL et al (2008) Loss of apical monocilia on collecting duct principal cells impairs ATP secretion across the apical cell surface and ATP-dependent and flow-induced calcium signals. Purinergic Signal 4:155–170. https://doi.org/10.1007/s11302-007-9072-0
Huang DY, Wulff P, Volkl H, Loffing J et al (2004) Impaired regulation of renal K^+ elimination in the sgk1-knockout mouse. J Am Soc Nephrol 15:885–891
Huang H, You Y, Lin X, Tang C et al (2017) Inhibition of TRPC6 signal pathway alleviates Podocyte injury induced by TGF-beta1. Cell Physiol Biochem 41:163–172. https://doi.org/10.1159/000455985
Huber TB, Kottgen M, Schilling B, Walz G et al (2001) Interaction with podocin facilitates nephrin signaling. J Biol Chem 276:41543–41546. https://doi.org/10.1074/jbc.C100452200
Huber TB, Schermer B, Muller RU, Hohne M et al (2006) Podocin and MEC-2 bind cholesterol to regulate the activity of associated ion channels. Proc Natl Acad Sci U S A 103:17079–17086. https://doi.org/10.1073/pnas.0607465103
Huber TB, Schermer B, Benzing T (2007) Podocin organizes ion channel-lipid supercomplexes: implications for mechanosensation at the slit diaphragm. Nephron Exp Nephrol 106:e27–e31. https://doi.org/10.1159/000101789
Hughes TET, Lodowski DT, Huynh KW, Yazici A et al (2018) Structural basis of TRPV5 channel inhibition by econazole revealed by cryo-EM. Nat Struct Mol Biol 25:53–60. https://doi.org/10.1038/s41594-017-0009-1
Hurd TW, Otto EA, Mishima E, Gee HY et al (2013) Mutation of the Mg^{2+} transporter SLC41A1 results in a nephronophthisis-like phenotype. J Am Soc Nephrol 24:967–977. https://doi.org/10.1681/ASN.2012101034

Hurst RS, Zhu X, Boulay G, Birnbaumer L et al (1998) Ionic currents underlying HTRP3 mediated agonist-dependent Ca^{2+} influx in stably transfected HEK293 cells. FEBS Lett 422:333–338
Huynh KW, Cohen MR, Jiang J, Samanta A et al (2016) Structure of the full-length TRPV2 channel by cryo-EM. Nat Commun 7:11130. https://doi.org/10.1038/ncomms11130
Ilatovskaya DV, Staruschenko A (2015) TRPC6 channel as an emerging determinant of the podocyte injury susceptibility in kidney diseases. Am J Physiol Renal Physiol 309:F393–F397. https://doi.org/10.1152/ajprenal.00186.2015
Ilatovskaya DV, Palygin O, Levchenko V, Staruschenko A (2013) Pharmacological characterization of the P2 receptors profile in the podocytes of the freshly isolated rat glomeruli. Am J Physiol Cell Physiol 305:C1050–C1059. https://doi.org/10.1152/ajpcell.00138.2013
Ilatovskaya DV, Palygin O, Chubinskiy-Nadezhdin V, Negulyaev YA et al (2014) Angiotensin II has acute effects on TRPC6 channels in podocytes of freshly isolated glomeruli. Kidney Int 86:506–514. https://doi.org/10.1038/ki.2014.71
Ilatovskaya DV, Levchenko V, Lowing A, Shuyskiy LS et al (2015) Podocyte injury in diabetic nephropathy: implications of angiotensin II-dependent activation of TRPC channels. Sci Rep 5:17637. https://doi.org/10.1038/srep17637
Ilatovskaya DV, Blass G, Palygin O, Levchenko V et al (2018) A NOX4/TRPC6 pathway in Podocyte calcium regulation and renal damage in diabetic kidney disease. J Am Soc Nephrol 29:1917–1927. https://doi.org/10.1681/ASN.2018030280
Janas S, Seghers F, Schakman O, Alsady M et al (2016) TRPV4 is associated with central rather than nephrogenic osmoregulation. Pflugers Arch 468:1595–1607. https://doi.org/10.1007/s00424-016-1850-5
Jang HR, Kim S, Heo NJ, Lee JH et al (2009) Effects of thiazide on the expression of TRPV5, calbindin-D28K, and sodium transporters in hypercalciuric rats. J Korean Med Sci 24(Suppl): S161–S169. https://doi.org/10.3346/jkms.2009.24.S1.S161
Jefferson JA, Shankland SJ, Pichler RH (2008) Proteinuria in diabetic kidney disease: a mechanistic viewpoint. Kidney Int 74:22–36. https://doi.org/10.1038/ki.2008.128
Jha JC, Banal C, Okabe J, Gray SP et al (2017) NADPH oxidase Nox5 accelerates renal injury in diabetic nephropathy. Diabetes 66:2691–2703. https://doi.org/10.2337/db16-1585
Jiang L, Ding J, Tsai H, Li L et al (2011) Over-expressing transient receptor potential cation channel 6 in podocytes induces cytoskeleton rearrangement through increases of intracellular Ca^{2+} and RhoA activation. Exp Biol Med (Maywood) 236:184–193. https://doi.org/10.1258/ebm.2010.010237
Jin J, Desai BN, Navarro B, Donovan A et al (2008) Deletion of Trpm7 disrupts embryonic development and thymopoiesis without altering Mg^{2+} homeostasis. Science 322:756–760. https://doi.org/10.1126/science.1163493
Kagami S (2012) Involvement of glomerular renin-angiotensin system (RAS) activation in the development and progression of glomerular injury. Clin Exp Nephrol 16:214–220. https://doi.org/10.1007/s10157-011-0568-0
Kanda S, Harita Y, Shibagaki Y, Sekine T et al (2011) Tyrosine phosphorylation-dependent activation of TRPC6 regulated by PLC-gamma1 and nephrin: effect of mutations associated with focal segmental glomerulosclerosis. Mol Biol Cell 22:1824–1835. https://doi.org/10.1091/mbc.E10-12-0929
Kaufman JS, Hamburger RJ (1996) Potassium transport in the connecting tubule. Miner Electrolyte Metab 22:242–247
Khaleel A, Wu MS, Wong HS, Hsu YW et al (2015) A single nucleotide polymorphism (rs4236480) in TRPV5 calcium channel gene is associated with stone multiplicity in calcium nephrolithiasis patients. Mediat Inflamm 2015:375427. https://doi.org/10.1155/2015/375427
Kim YH, Goyal M, Kurnit D, Wharram B et al (2001) Podocyte depletion and glomerulosclerosis have a direct relationship in the PAN-treated rat. Kidney Int 60:957–968. https://doi.org/10.1046/j.1523-1755.2001.060003957.x
Kim EY, Anderson M, Wilson C, Hagmann H et al (2013) NOX2 interacts with podocyte TRPC6 channels and contributes to their activation by diacylglycerol: essential role of podocin in

formation of this complex. Am J Physiol Cell Physiol 305:C960–C971. https://doi.org/10.1152/ajpcell.00191.2013
Kim JW, Ku SK, Han MH, Kim KY et al (2015) The administration of Fructus Schisandrae attenuates dexamethasone-induced muscle atrophy in mice. Int J Mol Med 36:29–42. https://doi.org/10.3892/ijmm.2015.2200
Kim JH, Xie J, Hwang KH, Wu YL et al (2017) Klotho may ameliorate proteinuria by targeting TRPC6 channels in podocytes. J Am Soc Nephrol 28:140–151. https://doi.org/10.1681/ASN.2015080888
Kim EY, Yazdizadeh Shotorbani P, Dryer SE (2018) Trpc6 inactivation confers protection in a model of severe nephrosis in rats. J Mol Med (Berl) 96:631–644. https://doi.org/10.1007/s00109-018-1648-3
Kiselyov K, Patterson RL (2009) The integrative function of TRPC channels. Front Biosci (Landmark Ed) 14:45–58
Kitajima N, Numaga-Tomita T, Watanabe M, Kuroda T et al (2016) TRPC3 positively regulates reactive oxygen species driving maladaptive cardiac remodeling. Sci Rep 6:37001. https://doi.org/10.1038/srep37001
Kobori H, Nangaku M, Navar LG, Nishiyama A (2007) The intrarenal renin-angiotensin system: from physiology to the pathobiology of hypertension and kidney disease. Pharmacol Rev 59:251–287. https://doi.org/10.1124/pr.59.3.3
Kolisek M, Nestler A, Vormann J, Schweigel-Rontgen M (2012) Human gene SLC41A1 encodes for the Na^+/Mg^{2+} exchanger. Am J Physiol Cell Physiol 302:C318–C326. https://doi.org/10.1152/ajpcell.00289.2011
Kottgen M, Buchholz B, Garcia-Gonzalez MA, Kotsis F et al (2008) TRPP2 and TRPV4 form a polymodal sensory channel complex. J Cell Biol 182:437–447. https://doi.org/10.1083/jcb.200805124
Krall P, Canales CP, Kairath P, Carmona-Mora P et al (2010) Podocyte-specific overexpression of wild type or mutant trpc6 in mice is sufficient to cause glomerular disease. PLoS One 5:e12859. https://doi.org/10.1371/journal.pone.0012859
Krapivinsky G, Krapivinsky L, Renthal NE, Santa-Cruz A et al (2017) Histone phosphorylation by TRPM6's cleaved kinase attenuates adjacent arginine methylation to regulate gene expression. Proc Natl Acad Sci U S A 114:E7092–E7100. https://doi.org/10.1073/pnas.1708427114
Kuang XY, Huang WY, Xu H, Shi Y et al (2013) 254C>G: a TRPC6 promoter variation associated with enhanced transcription and steroid-resistant nephrotic syndrome in Chinese children. Pediatr Res 74:511–516. https://doi.org/10.1038/pr.2013.144
Kumar R, Schaefer J, Grande JP, Roche PC (1994) Immunolocalization of calcitriol receptor, 24-hydroxylase cytochrome P-450, and calbindin D28k in human kidney. Am J Phys 266: F477–F485
Kuro-o M, Matsumura Y, Aizawa H, Kawaguchi H et al (1997) Mutation of the mouse klotho gene leads to a syndrome resembling ageing. Nature 390:45–51. https://doi.org/10.1038/36285
Kurosu H, Ogawa Y, Miyoshi M, Yamamoto M et al (2006) Regulation of fibroblast growth factor-23 signaling by klotho. J Biol Chem 281:6120–6123. https://doi.org/10.1074/jbc.C500457200
Lainez S, Schlingmann KP, van der Wijst J, Dworniczak B et al (2014) New TRPM6 missense mutations linked to hypomagnesemia with secondary hypocalcemia. Eur J Hum Genet 22:497–504. https://doi.org/10.1038/ejhg.2013.178
Lamande SR, Yuan Y, Gresshoff IL, Rowley L et al (2011) Mutations in TRPV4 cause an inherited arthropathy of hands and feet. Nat Genet 43:1142–1146. https://doi.org/10.1038/ng.945
Lambers TT, Weidema AF, Nilius B, Hoenderop JG et al (2004) Regulation of the mouse epithelial Ca^{2+} channel TRPV6 by the Ca^{2+}-sensor calmodulin. J Biol Chem 279:28855–28861. https://doi.org/10.1074/jbc.M313637200
Lambers TT, Bindels RJ, Hoenderop JG (2006a) Coordinated control of renal Ca^{2+} handling. Kidney Int 69:650–654. https://doi.org/10.1038/sj.ki.5000169

Lambers TT, Mahieu F, Oancea E, Hoofd L et al (2006b) Calbindin-D28K dynamically controls TRPV5-mediated Ca^{2+} transport. EMBO J 25:2978–2988. https://doi.org/10.1038/sj.emboj.7601186
Lameris AL, Nevalainen PI, Reijnen D, Simons E et al (2015) Segmental transport of Ca^{2+} and Mg^{2+} along the gastrointestinal tract. Am J Physiol Gastrointest Liver Physiol 308:G206–G216. https://doi.org/10.1152/ajpgi.00093.2014
Lechner SG, Markworth S, Poole K, Smith ES et al (2011) The molecular and cellular identity of peripheral osmoreceptors. Neuron 69:332–344. https://doi.org/10.1016/j.neuron.2010.12.028
Ledeganck KJ, Boulet GA, Bogers JJ, Verpooten GA et al (2013) The TRPM6/EGF pathway is downregulated in a rat model of cisplatin nephrotoxicity. PLoS One 8:e57016. https://doi.org/10.1371/journal.pone.0057016
Lee HS, Song CY (2010) Effects of TGF-beta on podocyte growth and disease progression in proliferative podocytopathies. Kidney Blood Press Res 33:24–29. https://doi.org/10.1159/000285844
Lee CT, Shang S, Lai LW, Yong KC et al (2004) Effect of thiazide on renal gene expression of apical calcium channels and calbindins. Am J Physiol Renal Physiol 287:F1164–F1170. https://doi.org/10.1152/ajprenal.00437.2003
Lee CT, Ng HY, Lee YT, Lai LW et al (2016) The role of calbindin-D28k on renal calcium and magnesium handling during treatment with loop and thiazide diuretics. Am J Physiol Renal Physiol 310:F230–F236. https://doi.org/10.1152/ajprenal.00057.2015
Lennon EJ, Piering WF (1970) A comparison of the effects of glucose ingestion and NH4Cl acidosis on urinary calcium and magnesium excretion in man. J Clin Invest 49:1458–1465. https://doi.org/10.1172/JCI106363
Letavernier E, Rodenas A, Guerrot D, Haymann JP (2012) Williams-Beuren syndrome hypercalcemia: is TRPC3 a novel mediator in calcium homeostasis? Pediatrics 129:e1626–e1630. https://doi.org/10.1542/peds.2011-2507
Leunissen EH, Blanchard MG, Sheedfar F, Lavrijsen M et al (2016) Urinary beta-galactosidase stimulates Ca^{2+} transport by stabilizing TRPV5 at the plasma membrane. Glycobiology 26:472–481. https://doi.org/10.1093/glycob/cwv172
Li M, Jiang J, Yue L (2006) Functional characterization of homo- and heteromeric channel kinases TRPM6 and TRPM7. J Gen Physiol 127:525–537. https://doi.org/10.1085/jgp.200609502
Li Y, Hu H, Butterworth MB, Tian JB et al (2016) Expression of a diverse Array of Ca^{2+}-activated K^+ channels (SK1/3, IK1, BK) that functionally couple to the Mechanosensitive TRPV4 channel in the collecting duct system of kidney. PLoS One 11:e0155006. https://doi.org/10.1371/journal.pone.0155006
Li W, Ding Y, Smedley C, Wang Y et al (2017a) Increased glomerular filtration rate and impaired contractile function of mesangial cells in TRPC6 knockout mice. Sci Rep 7:4145. https://doi.org/10.1038/s41598-017-04067-z
Li Y, Hu H, Tian JB, Zhu MX et al (2017b) Dynamic coupling between TRPV4 and Ca^{2+}-activated SK1/3 and IK1 K^+ channels plays a critical role in regulating the K^+-secretory BK channel in kidney collecting duct cells. Am J Physiol Renal Physiol 312:F1081–F1089. https://doi.org/10.1152/ajprenal.00037.2017
Liao M, Cao E, Julius D, Cheng Y (2013) Structure of the TRPV1 ion channel determined by electron cryo-microscopy. Nature 504:107–112. https://doi.org/10.1038/nature12822
Liao M, Cao E, Julius D, Cheng Y (2014) Single particle electron cryo-microscopy of a mammalian ion channel. Curr Opin Struct Biol 27:1–7. https://doi.org/10.1016/j.sbi.2014.02.005
Lichtenegger M, Groschner K (2014) TRPC3: a multifunctional signaling molecule. Handb Exp Pharmacol 222:67–84. https://doi.org/10.1007/978-3-642-54215-2_4
Liedtke W, Friedman JM (2003) Abnormal osmotic regulation in trpv4$^{-/-}$ mice. Proc Natl Acad Sci U S A 100:13698–13703
Liedtke W, Choe Y, Marti-Renom MA, Bell AM et al (2000) Vanilloid receptor-related osmotically activated channel (VR-OAC), a candidate vertebrate osmoreceptor. Cell 103:525–535

Lifton RP, Gharavi AG, Geller DS (2001) Molecular mechanisms of human hypertension. Cell 104:545–556

Lishko PV, Procko E, Jin X, Phelps CB et al (2007) The ankyrin repeats of TRPV1 bind multiple ligands and modulate channel sensitivity. Neuron 54:905–918. https://doi.org/10.1016/j.neuron.2007.05.027

Liu W, Xu S, Woda C, Kim P et al (2003) Effect of flow and stretch on the $[Ca^{2+}]_i$ response of principal and intercalated cells in cortical collecting duct. Am J Physiol Renal Physiol 285: F998–F1012. https://doi.org/10.1152/ajprenal.00067.2003

Liu W, Murcia NS, Duan Y, Weinbaum S et al (2005) Mechanoregulation of intracellular Ca^{2+} concentration is attenuated in collecting duct of monocilium-impaired orpk mice. Am J Physiol Renal Physiol 289:F978–F988

Liu W, Morimoto T, Woda C, Kleyman TR et al (2007) Ca^{2+} dependence of flow-stimulated K secretion in the mammalian cortical collecting duct. Am J Physiol Renal Physiol 293:F227–F235. https://doi.org/10.1152/ajprenal.00057.2007

Liu W, Wei Y, Sun P, Wang WH et al (2009) Mechanoregulation of BK channel activity in the mammalian cortical collecting duct: role of protein kinases A and C. Am J Physiol Renal Physiol 297:F904–F915. https://doi.org/10.1152/ajprenal.90685.2008

Liu Y, Mo L, Goldfarb DS, Evan AP et al (2010) Progressive renal papillary calcification and ureteral stone formation in mice deficient for Tamm-Horsfall protein. Am J Physiol Renal Physiol 299:F469–F478. https://doi.org/10.1152/ajprenal.00243.2010

Liu Y, Krueger K, Hovsepian A, Tepel M et al (2011) Calcium-dependent expression of transient receptor potential canonical type 3 channels in patients with chronic kidney disease. Arch Biochem Biophys 514:44–49. https://doi.org/10.1016/j.abb.2011.07.011

Lourdel S, Paulais M, Cluzeaud F, Bens M et al (2002) An inward rectifier K^+ channel at the basolateral membrane of the mouse distal convoluted tubule: similarities with $K_{ir}4.1$-$K_{ir}5.1$ heteromeric channels. J Physiol 538:391–404

Luan J, Li W, Han J, Zhang W et al (2012) Renal protection of in vivo administration of tempol in streptozotocin-induced diabetic rats. J Pharmacol Sci 119:167–176

Ma L, Fogo AB (2001) Role of angiotensin II in glomerular injury. Semin Nephrol 21:544–553

Magaldi AJ, van Baak AA, Rocha AS (1989) Calcium transport across rat inner medullary collecting duct perfused in vitro. Am J Phys 257:F738–F745. https://doi.org/10.1152/ajprenal.1989.257.5.F738

Malloy PJ, Feldman D (1999) Vitamin D resistance. Am J Med 106:355–370

Mamenko M, Zaika O, Jin M, O'Neil RG et al (2011) Purinergic activation of Ca-permeable TRPV4 channels is essential for mechano-sensitivity in the aldosterone-sensitive distal nephron. PLoS One 6:e22824

Mamenko M, Zaika OL, Boukelmoune N, Berrout J et al (2013) Discrete control of TRPV4 channel function in the distal nephron by protein kinases a and C. J Biol Chem 288:20306–20314. https://doi.org/10.1074/jbc.M113.466797

Mamenko M, Dhande I, Tomilin V, Zaika O et al (2016) Defective store-operated calcium entry causes partial nephrogenic diabetes insipidus. J Am Soc Nephrol 27:2035–2048. https://doi.org/10.1681/ASN.2014121200

Mamenko MV, Boukelmoune N, Tomilin VN, Zaika OL et al (2017) The renal TRPV4 channel is essential for adaptation to increased dietary potassium. Kidney Int 91:1398–1409. https://doi.org/10.1016/j.kint.2016.12.010

Mannaa M, Marko L, Balogh A, Vigolo E et al (2018) Transient receptor potential vanilloid 4 channel deficiency aggravates tubular damage after acute renal ischaemia reperfusion. Sci Rep 8:4878. https://doi.org/10.1038/s41598-018-23165-0

Marrelli SP, O'Neil RG, Brown RC, Bryan RM Jr (2007) PLA2 and TRPV4 channels regulate endothelial calcium in cerebral arteries. Am J Physiol Heart Circ Physiol 292:H1390–H1397. https://doi.org/10.1152/ajpheart.01006.2006

Mastrototaro L, Smorodchenko A, Aschenbach JR, Kolisek M et al (2016) Solute carrier 41A3 encodes for a mitochondrial Mg^{2+} efflux system. Sci Rep 6:27999. https://doi.org/10.1038/srep27999
Masuyama R, Vriens J, Voets T, Karashima Y et al (2008) TRPV4-mediated calcium influx regulates terminal differentiation of osteoclasts. Cell Metab 8:257–265. https://doi.org/10.1016/j.cmet.2008.08.002
Matsushita M, Kozak JA, Shimizu Y, McLachlin DT et al (2005) Channel function is dissociated from the intrinsic kinase activity and autophosphorylation of TRPM7/ChaK1. J Biol Chem 280:20793–20803. https://doi.org/10.1074/jbc.M413671200
Mayan H, Melnikov S, Novikov I, Holtzman EJ et al (2009) Familial hyperkalemia and hypertension: pathogenetic insights based on lithium clearance. J Clin Endocrinol Metab 94:3010–3016. https://doi.org/10.1210/jc.2008-2572
Mayan H, Carmon V, Oleinikov K, London S et al (2015) Hypercalciuria in familial hyperkalemia and hypertension with KLHL3 mutations. Nephron 130:59–65. https://doi.org/10.1159/000381563
McCormick JA, Ellison DH (2015) Distal convoluted tubule. Compr Physiol 5:45–98. https://doi.org/10.1002/cphy.c140002
McGoldrick LL, Singh AK, Saotome K, Yelshanskaya MV et al (2018) Opening of the human epithelial calcium channel TRPV6. Nature 553:233–237. https://doi.org/10.1038/nature25182
Meng K, Xu J, Zhang C, Zhang R et al (2014) Calcium sensing receptor modulates extracellular calcium entry and proliferation via TRPC3/6 channels in cultured human mesangial cells. PLoS One 9:e98777. https://doi.org/10.1371/journal.pone.0098777
Miller BA, Wang J, Hirschler-Laszkiewicz I, Gao E et al (2013) The second member of transient receptor potential-melastatin channel family protects hearts from ischemia-reperfusion injury. Am J Physiol Heart Circ Physiol 304:H1010–H1022. https://doi.org/10.1152/ajpheart.00906.2012
Mir S, Yavascan O, Berdeli A, Sozeri B (2012) TRPC6 gene variants in Turkish children with steroid-resistant nephrotic syndrome. Nephrol Dial Transplant 27:205–209. https://doi.org/10.1093/ndt/gfr202
Mizuno A, Matsumoto N, Imai M, Suzuki M (2003) Impaired osmotic sensation in mice lacking TRPV4. Am J Physiol Cell Physiol 285:C96–C101. https://doi.org/10.1152/ajpcell.00559.2002
Mochizuki T, Wu G, Hayashi T, Xenophontos SL et al (1996) PKD2, a gene for polycystic kidney disease that encodes an integral membrane protein. Science 272:1339–1342
Montell C (2003) Mg^{2+} homeostasis: the Mg^{2+}nificent TRPM chanzymes. Curr Biol 13:R799–R801
Montell C, Rubin GM (1989) Molecular characterization of the Drosophila trp locus: a putative integral membrane protein required for phototransduction. Neuron 2:1313–1323
Mount DB (2014) Thick ascending limb of the loop of Henle. Clin J Am Soc Nephrol 9:1974–1986. https://doi.org/10.2215/CJN.04480413
Mundel P, Reiser J (2010) Proteinuria: an enzymatic disease of the podocyte? Kidney Int 77:571–580. https://doi.org/10.1038/ki.2009.424
Muramatsu S, Wakabayashi M, Ohno T, Amano K et al (2007) Functional gene screening system identified TRPV4 as a regulator of chondrogenic differentiation. J Biol Chem 282:32158–32167. https://doi.org/10.1074/jbc.M706158200
Nair AV, Hocher B, Verkaart S, van Zeeland F et al (2012) Loss of insulin-induced activation of TRPM6 magnesium channels results in impaired glucose tolerance during pregnancy. Proc Natl Acad Sci U S A 109:11324–11329. https://doi.org/10.1073/pnas.1113811109
Navar LG (2014) Intrarenal renin-angiotensin system in regulation of glomerular function. Curr Opin Nephrol Hypertens 23:38–45. https://doi.org/10.1097/01.mnh.0000436544.86508.f1
Neal CR, Crook H, Bell E, Harper SJ et al (2005) Three-dimensional reconstruction of glomeruli by electron microscopy reveals a distinct restrictive urinary subpodocyte space. J Am Soc Nephrol 16:1223–1235. https://doi.org/10.1681/ASN.2004100822

Nie M, Bal MS, Yang Z, Liu J et al (2016) Mucin-1 increases renal TRPV5 activity in vitro, and urinary level associates with calcium nephrolithiasis in patients. J Am Soc Nephrol 27:3447–3458. https://doi.org/10.1681/ASN.2015101100
Nijenhuis T, Hoenderop JG, Loffing J, van der Kemp AW et al (2003a) Thiazide-induced hypocalciuria is accompanied by a decreased expression of Ca^{2+} transport proteins in kidney. Kidney Int 64:555–564. https://doi.org/10.1046/j.1523-1755.2003.00128.x
Nijenhuis T, Hoenderop JG, van der Kemp AW, Bindels RJ (2003b) Localization and regulation of the epithelial Ca^{2+} channel TRPV6 in the kidney. J Am Soc Nephrol 14:2731–2740
Nijenhuis T, Renkema KY, Hoenderop JG, Bindels RJ (2006) Acid-base status determines the renal expression of Ca^{2+} and Mg^{2+} transport proteins. J Am Soc Nephrol 17:617–626. https://doi.org/10.1681/ASN.2005070732
Nijenhuis T, Sloan AJ, Hoenderop JG, Flesche J et al (2011) Angiotensin II contributes to podocyte injury by increasing TRPC6 expression via an NFAT-mediated positive feedback signaling pathway. Am J Pathol 179:1719–1732. https://doi.org/10.1016/j.ajpath.2011.06.033
Nilius B, Honore E (2012) Sensing pressure with ion channels. Trends Neurosci 35:477–486. https://doi.org/10.1016/j.tins.2012.04.002
Nilius B, Owsianik G (2011) The transient receptor potential family of ion channels. Genome Biol 12:218. https://doi.org/10.1186/gb-2011-12-3-218
Nilius B, Voets T (2013) The puzzle of TRPV4 channelopathies. EMBO Rep 14:152–163. https://doi.org/10.1038/embor.2012.219
Nilius B, Vriens J, Prenen J, Droogmans G et al (2004) TRPV4 calcium entry channel: a paradigm for gating diversity. Am J Physiol Cell Physiol 286:C195–C205. https://doi.org/10.1152/ajpcell.00365.2003
O'Neil RG, Heller S (2005) The mechanosensitive nature of TRPV channels. Pflugers Arch 451:193–203. https://doi.org/10.1007/s00424-005-1424-4
Obeidova L, Reiterova J, Lnenicka P, Stekrova J et al (2012) TRPC6 gene variants in Czech adult patients with focal segmental glomerulosclerosis and minimal change disease. Folia Biol (Praha) 58:173–176
Oddsson A, Sulem P, Helgason H, Edvardsson VO et al (2015) Common and rare variants associated with kidney stones and biochemical traits. Nat Commun 6:7975. https://doi.org/10.1038/ncomms8975
Palygin O, Ilatovskaya DV, Levchenko V, Endres BT et al (2018) Nitric oxide production by glomerular podocytes. Nitric Oxide 72:24–31. https://doi.org/10.1016/j.niox.2017.11.005
Patel S, Rayanagoudar G, Gelding S (2016) Familial hypomagnesaemia with secondary hypocalcaemia. BMJ Case Rep 2016:bcr2016216870. https://doi.org/10.1136/bcr-2016-216870
Peng JB (2011) TRPV5 and TRPV6 in transcellular Ca^{2+} transport: regulation, gene duplication, and polymorphisms in African populations. Adv Exp Med Biol 704:239–275. https://doi.org/10.1007/978-94-007-0265-3_14
Peng JB, Chen XZ, Berger UV, Vassilev PM et al (1999) Molecular cloning and characterization of a channel-like transporter mediating intestinal calcium absorption. J Biol Chem 274:22739–22746
Perraud AL, Fleig A, Dunn CA, Bagley LA et al (2001) ADP-ribose gating of the calcium-permeable LTRPC2 channel revealed by Nudix motif homology. Nature 411:595–599. https://doi.org/10.1038/35079100
Perraud AL, Takanishi CL, Shen B, Kang S et al (2005) Accumulation of free ADP-ribose from mitochondria mediates oxidative stress-induced gating of TRPM2 cation channels. J Biol Chem 280:6138–6148. https://doi.org/10.1074/jbc.M411446200
Peti-Peterdi J, Sipos A (2010) A high-powered view of the filtration barrier. J Am Soc Nephrol 21:1835–1841. https://doi.org/10.1681/ASN.2010040378
Pollak MR, Quaggin SE, Hoenig MP, Dworkin LD (2014) The glomerulus: the sphere of influence. Clin J Am Soc Nephrol 9:1461–1469. https://doi.org/10.2215/CJN.09400913
Quamme GA (2008) Recent developments in intestinal magnesium absorption. Curr Opin Gastroenterol 24:230–235. https://doi.org/10.1097/MOG.0b013e3282f37b59

Reiser J, Polu KR, Moller CC, Kenlan P et al (2005) TRPC6 is a glomerular slit diaphragm-associated channel required for normal renal function. Nat Genet 37:739–744. https://doi.org/10.1038/ng1592
Riazanova LV, Pavur KS, Petrov AN, Dorovkov MV et al (2001) Novel type of signaling molecules: protein kinases covalently linked to ion channels. Mol Biol (Mosk) 35:321–332
Riehle M, Buscher AK, Gohlke BO, Kassmann M et al (2016) TRPC6 G757D loss-of-function mutation associates with FSGS. J Am Soc Nephrol 27:2771–2783. https://doi.org/10.1681/ASN.2015030318
Rohacs T (2014) Phosphoinositide regulation of TRP channels. Handb Exp Pharmacol 223:1143–1176. https://doi.org/10.1007/978-3-319-05161-1_18
Roselli S, Gribouval O, Boute N, Sich M et al (2002) Podocin localizes in the kidney to the slit diaphragm area. Am J Pathol 160:131–139. https://doi.org/10.1016/S0002-9440(10)64357-X
Roshanravan H, Dryer SE (2014) ATP acting through P2Y receptors causes activation of podocyte TRPC6 channels: role of podocin and reactive oxygen species. Am J Physiol Renal Physiol 306: F1088–F1097. https://doi.org/10.1152/ajprenal.00661.2013
Saliba Y, Karam R, Smayra V, Aftimos G et al (2015) Evidence of a role for fibroblast transient receptor potential canonical 3 Ca^{2+} channel in renal fibrosis. J Am Soc Nephrol 26:1855–1876. https://doi.org/10.1681/ASN.2014010065
Santin S, Ars E, Rossetti S, Salido E et al (2009) TRPC6 mutational analysis in a large cohort of patients with focal segmental glomerulosclerosis. Nephrol Dial Transplant 24:3089–3096. https://doi.org/10.1093/ndt/gfp229
Saotome K, Singh AK, Yelshanskaya MV, Sobolevsky AI (2016) Crystal structure of the epithelial calcium channel TRPV6. Nature 534:506–511. https://doi.org/10.1038/nature17975
Satlin LM, Carattino MD, Liu W, Kleyman TR (2006) Regulation of cation transport in the distal nephron by mechanical forces. Am J Physiol Renal Physiol 291:F923–F931. https://doi.org/10.1152/ajprenal.00192.2006
Schaefer M (2005) Homo- and heteromeric assembly of TRP channel subunits. Pflugers Arch 451:35–42. https://doi.org/10.1007/s00424-005-1467-6
Schaldecker T, Kim S, Tarabanis C, Tian D et al (2013) Inhibition of the TRPC5 ion channel protects the kidney filter. J Clin Invest 123:5298–5309. https://doi.org/10.1172/JCI71165
Schlingmann KP, Weber S, Peters M, Niemann Nejsum L et al (2002) Hypomagnesemia with secondary hypocalcemia is caused by mutations in TRPM6, a new member of the TRPM gene family. Nat Genet 31:166–170. https://doi.org/10.1038/ng889
Schlingmann KP, Waldegger S, Konrad M, Chubanov V et al (2007) TRPM6 and TRPM7-gatekeepers of human magnesium metabolism. Biochim Biophys Acta 1772:813–821. https://doi.org/10.1016/j.bbadis.2007.03.009
Schmitz C, Dorovkov MV, Zhao X, Davenport BJ et al (2005) The channel kinases TRPM6 and TRPM7 are functionally nonredundant. J Biol Chem 280:37763–37771. https://doi.org/10.1074/jbc.M509175200
Scholl UI, Choi M, Liu T, Ramaekers VT et al (2009) Seizures, sensorineural deafness, ataxia, mental retardation, and electrolyte imbalance (SeSAME syndrome) caused by mutations in KCNJ10. Proc Natl Acad Sci U S A 106:5842–5847. https://doi.org/10.1073/pnas.0901749106
Schrag D, Chung KY, Flombaum C, Saltz L (2005) Cetuximab therapy and symptomatic hypomagnesemia. J Natl Cancer Inst 97:1221–1224. https://doi.org/10.1093/jnci/dji242
Schweigel M, Martens H (2000) Magnesium transport in the gastrointestinal tract. Front Biosci 5: D666–D677
Seva PB, van der LN, Verdonk K, Roks AJ et al (2013) Key developments in renin-angiotensin-aldosterone system inhibition. Nat Rev Nephrol 9:26–36
Shi J, Mori E, Mori Y, Mori M et al (2004) Multiple regulation by calcium of murine homologues of transient receptor potential proteins TRPC6 and TRPC7 expressed in HEK293 cells. J Physiol 561:415–432. https://doi.org/10.1113/jphysiol.2004.075051

Shibata S, Zhang J, Puthumana J, Stone KL et al (2013) Kelch-like 3 and Cullin 3 regulate electrolyte homeostasis via ubiquitination and degradation of WNK4. Proc Natl Acad Sci U S A 110:7838–7843. https://doi.org/10.1073/pnas.1304592110

Shin SH, Lee EJ, Hyun S, Chun J et al (2012) Phosphorylation on the Ser 824 residue of TRPV4 prefers to bind with F-actin than with microtubules to expand the cell surface area. Cell Signal 24:641–651. https://doi.org/10.1016/j.cellsig.2011.11.002

Simon DB, Nelson-Williams C, Bia MJ, Ellison D et al (1996) Gitelman's variant of Bartter's syndrome, inherited hypokalaemic alkalosis, is caused by mutations in the thiazide-sensitive Na-Cl cotransporter. Nat Genet 12:24–30. https://doi.org/10.1038/ng0196-24

Singh AK, Saotome K, McGoldrick LL, Sobolevsky AI (2018) Structural bases of TRP channel TRPV6 allosteric modulation by 2-APB. Nat Commun 9:2465. https://doi.org/10.1038/s41467-018-04828-y

Siroky BJ, Ferguson WB, Fuson AL, Xie Y et al (2006) Loss of primary cilia results in deregulated and unabated apical calcium entry in ARPKD collecting duct cells. Am J Physiol Renal Physiol 290:F1320–F1328. https://doi.org/10.1152/ajprenal.00463.2005

Song Y, Peng X, Porta A, Takanaga H et al (2003) Calcium transporter 1 and epithelial calcium channel messenger ribonucleic acid are differentially regulated by 1,25 dihydroxyvitamin D3 in the intestine and kidney of mice. Endocrinology 144:3885–3894. https://doi.org/10.1210/en.2003-0314

Soni H, Peixoto-Neves D, Matthews AT, Adebiyi A (2017) TRPV4 channels contribute to renal myogenic autoregulation in neonatal pigs. Am J Physiol Renal Physiol 313:F1136–F1148. https://doi.org/10.1152/ajprenal.00300.2017

Sonneveld R, van der Vlag J, Baltissen MP, Verkaart SA et al (2014) Glucose specifically regulates TRPC6 expression in the podocyte in an AngII-dependent manner. Am J Pathol 184:1715–1726. https://doi.org/10.1016/j.ajpath.2014.02.008

Sonneveld R, Hoenderop JG, Isidori AM, Henique C et al (2017) Sildenafil prevents podocyte injury via PPAR-gamma-mediated TRPC6 inhibition. J Am Soc Nephrol 28:1491–1505. https://doi.org/10.1681/ASN.2015080885

Spassova MA, Hewavitharana T, Xu W, Soboloff J et al (2006) A common mechanism underlies stretch activation and receptor activation of TRPC6 channels. Proc Natl Acad Sci U S A 103:16586–16591. https://doi.org/10.1073/pnas.0606894103

Spires D, Ilatovskaya DV, Levchenko V, North PE et al (2018) The protective role of Trpc6 knockout in the progression of diabetic kidney disease. Am J Physiol Renal Physiol 315(4): F1091–F1097. https://doi.org/10.1152/ajprenal.00155.2018

Staruschenko A (2012) Regulation of transport in the connecting tubule and cortical collecting duct. Compr Physiol 2:1541–1584. https://doi.org/10.1002/cphy.c110052

Stewart AP, Smith GD, Sandford RN, Edwardson JM (2010) Atomic force microscopy reveals the alternating subunit arrangement of the TRPP2-TRPV4 heterotetramer. Biophys J 99:790–797. https://doi.org/10.1016/j.bpj.2010.05.012

Stritt S, Nurden P, Favier R, Favier M et al (2016) Defects in TRPM7 channel function deregulate thrombopoiesis through altered cellular Mg^{2+} homeostasis and cytoskeletal architecture. Nat Commun 7:11097. https://doi.org/10.1038/ncomms11097

Subramanya AR, Ellison DH (2014) Distal convoluted tubule. Clin J Am Soc Nephrol 9:2147–2163. https://doi.org/10.2215/CJN.05920613

Tang Q, Guo W, Zheng L, Wu JX et al (2018) Structure of the receptor-activated human TRPC6 and TRPC3 ion channels. Cell Res 28:746–755. https://doi.org/10.1038/s41422-018-0038-2

Taniguchi J, Imai M (1998) Flow-dependent activation of maxi K^+ channels in apical membrane of rabbit connecting tubule. J Membr Biol 164:35–45

Taniguchi J, Tsuruoka S, Mizuno A, Sato J et al (2007) TRPV4 as a flow sensor in flow-dependent K^+ secretion from the cortical collecting duct. Am J Physiol Renal Physiol 292:F667–F673

Tanner GA, Rippe C, Shao Y, Evan AP et al (2009) Glomerular permeability to macromolecules in the Necturus kidney. Am J Physiol Renal Physiol 296:F1269–F1278. https://doi.org/10.1152/ajprenal.00371.2007

Thebault S, Cao G, Venselaar H, Xi Q et al (2008) Role of the alpha-kinase domain in transient receptor potential melastatin 6 channel and regulation by intracellular ATP. J Biol Chem 283:19999–20007. https://doi.org/10.1074/jbc.M800167200

Thebault S, Alexander RT, Tiel Groenestege WM, Hoenderop JG et al (2009) EGF increases TRPM6 activity and surface expression. J Am Soc Nephrol 20:78–85. https://doi.org/10.1681/ASN.2008030327

Thilo F, Lee M, Xia S, Zakrzewicz A et al (2014) High glucose modifies transient receptor potential canonical type 6 channels via increased oxidative stress and syndecan-4 in human podocytes. Biochem Biophys Res Commun 450:312–317. https://doi.org/10.1016/j.bbrc.2014.05.116

Tian W, Salanova M, Xu H, Lindsley JN et al (2004) Renal expression of osmotically responsive cation channel TRPV4 is restricted to water-impermeant nephron segments. Am J Physiol Renal Physiol 287:F17–F24

Tomilin V, Reif GA, Zaika O, Wallace DP et al (2018) Deficient transient receptor potential vanilloid type 4 function contributes to compromised $[Ca^{2+}]_i$ homeostasis in human autosomal-dominant polycystic kidney disease cells. FASEB J 32:4612–4623. https://doi.org/10.1096/fj.201701535RR

Torres PU, Prie D, Molina-Bletry V, Beck L et al (2007) Klotho: an antiaging protein involved in mineral and vitamin D metabolism. Kidney Int 71:730–737. https://doi.org/10.1038/sj.ki.5002163

Toth B, Csanady L (2010) Identification of direct and indirect effectors of the transient receptor potential melastatin 2 (TRPM2) cation channel. J Biol Chem 285:30091–30102. https://doi.org/10.1074/jbc.M109.066464

Toth B, Csanady L (2012) Pore collapse underlies irreversible inactivation of TRPM2 cation channel currents. Proc Natl Acad Sci U S A 109:13440–13445. https://doi.org/10.1073/pnas.1204702109

Trebak M, St JBG, McKay RR, Birnbaumer L et al (2003a) Signaling mechanism for receptor-activated canonical transient receptor potential 3 (TRPC3) channels. J Biol Chem 278:16244–16252. https://doi.org/10.1074/jbc.M300544200

Trebak M, Vazquez G, Bird GS, Putney JW Jr (2003b) The TRPC3/6/7 subfamily of cation channels. Cell Calcium 33:451–461

Urakawa I, Yamazaki Y, Shimada T, Iijima K et al (2006) Klotho converts canonical FGF receptor into a specific receptor for FGF23. Nature 444:770–774. https://doi.org/10.1038/nature05315

Urban N, Neuser S, Hentschel A, Kohling S et al (2017) Pharmacological inhibition of focal segmental glomerulosclerosis-related, gain of function mutants of TRPC6 channels by semi-synthetic derivatives of larixol. Br J Pharmacol 174:4099–4122. https://doi.org/10.1111/bph.13977

Vallon V (2008) P2 receptors in the regulation of renal transport mechanisms. Am J Physiol Renal Physiol 294:F10–F27. https://doi.org/10.1152/ajprenal.00432.2007

Van Cromphaut SJ, Dewerchin M, Hoenderop JG, Stockmans I et al (2001) Duodenal calcium absorption in vitamin D receptor-knockout mice: functional and molecular aspects. Proc Natl Acad Sci U S A 98:13324–13329. https://doi.org/10.1073/pnas.231474698

van der Hagen EA, Tudpor K, Verkaart S, Lavrijsen M et al (2014) Beta1-adrenergic receptor signaling activates the epithelial calcium channel, transient receptor potential vanilloid type 5 (TRPV5), via the protein kinase a pathway. J Biol Chem 289:18489–18496. https://doi.org/10.1074/jbc.M113.491274

van der Wijst J, Glaudemans B, Venselaar H, Nair AV et al (2010) Functional analysis of the Kv1.1 N255D mutation associated with autosomal dominant hypomagnesemia. J Biol Chem 285:171–178. https://doi.org/10.1074/jbc.M109.041517

van Goor MK, Verkaart S, van Dam TJ, Huynen MA et al (2017) Interspecies differences in PTH-mediated PKA phosphorylation of the epithelial calcium channel TRPV5. Pflugers Arch 469:1301–1311. https://doi.org/10.1007/s00424-017-1996-9

Vazquez G, Wedel BJ, Trebak M, St John Bird G et al (2003) Expression level of the canonical transient receptor potential 3 (TRPC3) channel determines its mechanism of activation. J Biol Chem 278:21649–21654. https://doi.org/10.1074/jbc.M302162200
Venkatachalam K, Montell C (2007) TRP channels. Annu Rev Biochem 76:387–417. https://doi.org/10.1146/annurev.biochem.75.103004.142819
Vennekens R, Hoenderop JG, Prenen J, Stuiver M et al (2000) Permeation and gating properties of the novel epithelial Ca^{2+} channel. J Biol Chem 275:3963–3969
Voets T, Prenen J, Vriens J, Watanabe H et al (2002) Molecular determinants of permeation through the cation channel TRPV4. J Biol Chem 277:33704–33710. https://doi.org/10.1074/jbc.M204828200
Voets T, Nilius B, Hoefs S, van der Kemp AW et al (2004) TRPM6 forms the Mg^{2+} influx channel involved in intestinal and renal Mg^{2+} absorption. J Biol Chem 279:19–25. https://doi.org/10.1074/jbc.M311201200
Voolstra O, Huber A (2014) Post-translational modifications of TRP channels. Cell 3:258–287. https://doi.org/10.3390/cells3020258
Vriens J, Watanabe H, Janssens A, Droogmans G et al (2004) Cell swelling, heat, and chemical agonists use distinct pathways for the activation of the cation channel TRPV4. Proc Natl Acad Sci U S A 101:396–401. https://doi.org/10.1073/pnas.0303329101
Walder RY, Landau D, Meyer P, Shalev H et al (2002) Mutation of TRPM6 causes familial hypomagnesemia with secondary hypocalcemia. Nat Genet 31:171–174. https://doi.org/10.1038/ng901
Walder RY, Yang B, Stokes JB, Kirby PA et al (2009) Mice defective in Trpm6 show embryonic mortality and neural tube defects. Hum Mol Genet 18:4367–4375. https://doi.org/10.1093/hmg/ddp392
Wang WH, Giebisch G (2009) Regulation of potassium (K) handling in the renal collecting duct. Pflugers Arch 458:157–168. https://doi.org/10.1007/s00424-008-0593-3
Wang WH, Schwab A, Giebisch G (1990) Regulation of small-conductance K^{+} channel in apical membrane of rat cortical collecting tubule. Am J Phys 259:F494–F502
Wang L, Holmes RP, Peng JB (2017) The L530R variation associated with recurrent kidney stones impairs the structure and function of TRPV5. Biochem Biophys Res Commun 492:362–367. https://doi.org/10.1016/j.bbrc.2017.08.102
Wang X, Dande RR, Yu H, Samelko B et al (2018) TRPC5 does not cause or aggravate glomerular disease. J Am Soc Nephrol 29:409–415. https://doi.org/10.1681/ASN.2017060682
Watanabe H, Davis JB, Smart D, Jerman JC et al (2002) Activation of TRPV4 channels (hVRL-2/mTRP12) by phorbol derivatives. J Biol Chem 277:13569–13577. https://doi.org/10.1074/jbc.M200062200
Watanabe H, Vriens J, Janssens A, Wondergem R et al (2003a) Modulation of TRPV4 gating by intra- and extracellular Ca^{2+}. Cell Calcium 33:489–495
Watanabe H, Vriens J, Prenen J, Droogmans G et al (2003b) Anandamide and arachidonic acid use epoxyeicosatrienoic acids to activate TRPV4 channels. Nature 424:434–438. https://doi.org/10.1038/nature01807
Weinbaum S, Duan Y, Satlin LM, Wang T et al (2010) Mechanotransduction in the renal tubule. Am J Physiol Renal Physiol 299:F1220–F1236. https://doi.org/10.1152/ajprenal.00453.2010
Wilkes M, Madej MG, Kreuter L, Rhinow D et al (2017) Molecular insights into lipid-assisted Ca^{2+} regulation of the TRP channel Polycystin-2. Nat Struct Mol Biol 24:123–130. https://doi.org/10.1038/nsmb.3357
Wilson C, Dryer SE (2014) A mutation in TRPC6 channels abolishes their activation by hypoosmotic stretch but does not affect activation by diacylglycerol or G protein signaling cascades. Am J Physiol Renal Physiol 306:F1018–F1025. https://doi.org/10.1152/ajprenal.00662.2013
Wilson FH, Disse-Nicodeme S, Choate KA, Ishikawa K et al (2001) Human hypertension caused by mutations in WNK kinases. Science 293:1107–1112. https://doi.org/10.1126/science.1062844

Winn MP, Conlon PJ, Lynn KL, Farrington MK et al (2005) A mutation in the TRPC6 cation channel causes familial focal segmental glomerulosclerosis. Science 308:1801–1804. https://doi.org/10.1126/science.1106215

Wissenbach U, Bodding M, Freichel M, Flockerzi V (2000) Trp12, a novel Trp related protein from kidney. FEBS Lett 485:127–134

Woda CB, Bragin A, Kleyman TR, Satlin LM (2001) Flow-dependent K^+ secretion in the cortical collecting duct is mediated by a maxi-K channel. Am J Physiol Renal Physiol 280:F786–F793

Woda CB, Leite M Jr, Rohatgi R, Satlin LM (2002) Effects of luminal flow and nucleotides on $[Ca^{2+}]_i$ in rabbit cortical collecting duct. Am J Physiol Renal Physiol 283:F437–F446. https://doi.org/10.1152/ajprenal.00316.2001

Wolf MT, Wu XR, Huang CL (2013) Uromodulin upregulates TRPV5 by impairing caveolin-mediated endocytosis. Kidney Int 84:130–137. https://doi.org/10.1038/ki.2013.63

Wolf MT, An SW, Nie M, Bal MS et al (2014) Klotho up-regulates renal calcium channel transient receptor potential vanilloid 5 (TRPV5) by intra- and extracellular N-glycosylation-dependent mechanisms. J Biol Chem 289:35849–35857. https://doi.org/10.1074/jbc.M114.616649

Wong NL, Quamme GA, Dirks JH (1986) Effects of acid-base disturbances on renal handling of magnesium in the dog. Clin Sci (Lond) 70:277–284

Woroniecki RP, Kopp JB (2007) Genetics of focal segmental glomerulosclerosis. Pediatr Nephrol 22:638–644. https://doi.org/10.1007/s00467-007-0445-y

Woudenberg-Vrenken TE, Bindels RJ, Hoenderop JG (2009) The role of transient receptor potential channels in kidney disease. Nat Rev Nephrol 5:441–449. https://doi.org/10.1038/nrneph.2009.100

Woudenberg-Vrenken TE, Sukinta A, van der Kemp AW, Bindels RJ et al (2011) Transient receptor potential melastatin 6 knockout mice are lethal whereas heterozygous deletion results in mild hypomagnesemia. Nephron Physiol 117:p11–p19. https://doi.org/10.1159/000320580

Wu L, Gao X, Brown RC, Heller S et al (2007) Dual role of the TRPV4 channel as a sensor of flow and osmolality in renal epithelial cells. Am J Physiol Renal Physiol 293:F1699–F1713

Wu YL, Xie J, An SW, Oliver N et al (2017) Inhibition of TRPC6 channels ameliorates renal fibrosis and contributes to renal protection by soluble klotho. Kidney Int 91:830–841. https://doi.org/10.1016/j.kint.2016.09.039

Xu H, Fu Y, Tian W, Cohen DM (2006) Glycosylation of the osmoresponsive transient receptor potential channel TRPV4 on Asn-651 influences membrane trafficking. Am J Physiol Renal Physiol 290:F1103–F1109. https://doi.org/10.1152/ajprenal.00245.2005

Xu C, Shmukler BE, Nishimura K, Kaczmarek E et al (2009) Attenuated, flow-induced ATP release contributes to absence of flow-sensitive, purinergic Ca^{2+} signaling in human ADPKD cyst epithelial cells. Am J Physiol Renal Physiol 296:F1464–F1476. https://doi.org/10.1152/ajprenal.90542.2008

Yamaguchi T, Hempson SJ, Reif GA, Hedge AM et al (2006) Calcium restores a normal proliferation phenotype in human polycystic kidney disease epithelial cells. J Am Soc Nephrol 17:178–187. https://doi.org/10.1681/ASN.2005060645

Yu H, Kistler A, Faridi MH, Meyer JO et al (2016) Synaptopodin limits TRPC6 podocyte surface expression and attenuates proteinuria. J Am Soc Nephrol 27:3308–3319. https://doi.org/10.1681/ASN.2015080896

Yue L, Peng JB, Hediger MA, Clapham DE (2001) CaT1 manifests the pore properties of the calcium-release-activated calcium channel. Nature 410:705–709. https://doi.org/10.1038/35070596

Zaika O, Mamenko M, Berrout J, Boukelmoune N et al (2013) TRPV4 dysfunction promotes renal cystogenesis in autosomal recessive polycystic kidney disease. J Am Soc Nephrol 24:604–616. https://doi.org/10.1681/ASN.2012050442

Zaman MA, Oparil S, Calhoun DA (2002) Drugs targeting the renin-angiotensin-aldosterone system. Nat Rev Drug Discov 1:621–636. https://doi.org/10.1038/nrd873

Zhang H, Ding J, Fan Q, Liu S (2009) TRPC6 up-regulation in Ang II-induced podocyte apoptosis might result from ERK activation and NF-kappaB translocation. Exp Biol Med (Maywood) 234:1029–1036. https://doi.org/10.3181/0901-RM-11
Zhang Q, Ma J, Xie J, Wang Z et al (2013a) Screening of ACTN4 and TRPC6 mutations in a Chinese cohort of patients with adult-onset familial focal segmental glomerulosclerosis. Contrib Nephrol 181:91–100. https://doi.org/10.1159/000348471
Zhang ZR, Chu WF, Song B, Gooz M et al (2013b) TRPP2 and TRPV4 form an EGF-activated calcium permeable channel at the apical membrane of renal collecting duct cells. PLoS One 8: e73424. https://doi.org/10.1371/journal.pone.0073424
Zhang HT, Wang WW, Ren LH, Zhao XX et al (2016) The mTORC2/Akt/NFkappaB pathway-mediated activation of TRPC6 participates in adriamycin-induced podocyte apoptosis. Cell Physiol Biochem 40:1079–1093. https://doi.org/10.1159/000453163
Zhang S, Li N, Zeng W, Gao N et al (2017) Cryo-EM structures of the mammalian endo-lysosomal TRPML1 channel elucidate the combined regulation mechanism. Protein Cell 8:834–847. https://doi.org/10.1007/s13238-017-0476-5
Zhang Z, Toth B, Szollosi A, Chen J et al (2018) Structure of a TRPM2 channel in complex with Ca^{2+} explains unique gating regulation. elife 7:e36409. https://doi.org/10.7554/eLife.36409
Zhou Y, Castonguay P, Sidhom EH, Clark AR et al (2017) A small-molecule inhibitor of TRPC5 ion channels suppresses progressive kidney disease in animal models. Science 358:1332–1336. https://doi.org/10.1126/science.aal4178
Zhu X, Jiang M, Peyton M, Boulay G et al (1996) Trp, a novel mammalian gene family essential for agonist-activated capacitative Ca^{2+} entry. Cell 85:661–671
Zhu B, Chen N, Wang ZH, Pan XX et al (2009) Identification and functional analysis of a novel TRPC6 mutation associated with late onset familial focal segmental glomerulosclerosis in Chinese patients. Mutat Res 664:84–90. https://doi.org/10.1016/j.mrfmmm.2008.11.021

Chapter 28
P2X Receptors in Epithelia

Jens Leipziger

Abstract 2X receptors are ubiquitously expressed in all epithelial tissues but their physiological roles are not well understood. A comprehensive review is provided that assembles the current state of knowledge by focusing on functional effects of P2X receptor stimulation in secretory and in absorptive tissues. In glandular tissue, like the parotid gland or respiratory epithelia, P2X receptors stimulate secretion via an increase in cytosolic Ca^{2+}. In absorptive epithelia, like the renal tubule, P2X receptors mediate the inhibition of NaCl, Mg^{2+} and water transport as exemplified from the thick ascending limb, the distal convoluted tubule and the collecting duct, respectively. The underlying signalling pathways that inhibit epithelial absorption are currently not well understood. Epithelial P2X7 receptors show pronounced upregulation during various disease states highlighting a role of purinergic signalling in epithelial pathophysiology. Importantly, functional effects of epithelial P2X receptors cover numerous other aspects ranging from modulation of sound transmission, activation of apoptosis or production of oxygen radicals. Eventually, P2X receptors in epithelia are understudied but offer numerous novel and very attractive questions.

Keywords Epithelial transport · Exocrine secretion · Kidney · Renal tubule · Purinergic signalling · P2Y

28.1 Introduction

P2X receptors are non-selective, ligand-gated cation channels that specifically use ATP as their agonist (North 2002). Each P2X receptor is composed of three subunits, which all have intracellular N- and C-termini, two transmembrane spanning α-helices and a connecting acidic rich patch extracellular loop (Kawate et al. 2009). Three ATP

J. Leipziger (✉)
Department of Biomedicine – Physiology, Aarhus University, Aarhus, Denmark
e-mail: leip@biomed.au.dk

K. L. Hamilton, D. C. Devor (eds.), *Studies of Epithelial Transporters and Ion Channels*, Physiology in Health and Disease,
https://doi.org/10.1007/978-3-030-55454-5_28

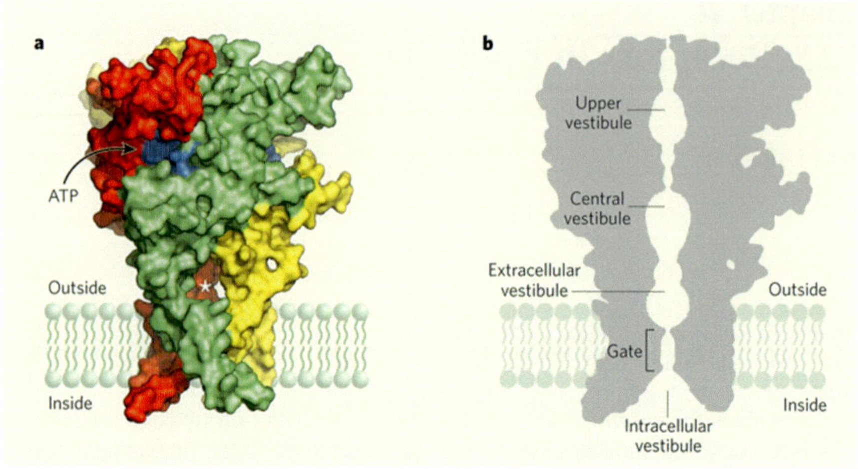

Fig. 28.1 Trimeric structure of the P2X receptor as taken from the original crystal structure description of the zebrafish P2X4 receptor (Kawate et al. 2009). This figure is taken with permission from Silberberg and Swartz (2009). (**a**) Surface representation of the P2X4 receptor with the three identical subunits coloured in green, red and yellow and the ATP-binding sites coloured in blue. The white asterisk marks one of three fenestrations between subunits where the extracellular vestibule opens to the exterior, providing a potential passage for ions to enter or exit from the extracellular side of the pore. (**b**) representation of the P2X receptor ion channel showing the four vestibules and a gate formed by a plug of hydrophobic amino-acid residues within the pore

binding pockets are localized in between the subunits and ATP binding likely conveys substantial spatial rearrangements of the receptor, eventually permitting its opening (Hattori and Gouaux 2012; Kawate et al. 2009). An image derived from the crystal structure of a closed P2X4 receptor (no ATP bound) is shown in Fig. 28.1 (Silberberg and Swartz 2009). Seven different P2X receptor genes have been identified in the mammalian genome and cells commonly express several different P2X receptor transcripts and proteins. Except for the P2X6 receptors, all P2X receptor subunits can form functional homomeric ATP-gated channels, but in the native situation heteromeric P2X receptors composed of two or three different subunits are believed to mediate the physiological response (North 2002). ATP binding to P2X receptors causes an increase in intracellular calcium ($[Ca^{2+}]_i$) and membrane depolarization via Na^+ entry and K^+ efflux. In excitable tissues like neurons or smooth muscle cells, P2X receptors stimulate the cells by virtue of membrane depolarization moving the membrane voltage towards the activating threshold. Purinergic P2X receptors are ubiquitously expressed in nearly all mammalian and non-mammalian cells including non-excitable cells. They serve a large variety of different cell and organ-specific functions. Also, in epithelia P2X receptors are widely expressed (Taylor et al. 1999). This chapter attempts to provide a comprehensive overview of P2X receptors in native mammalian epithelia and discusses their physiological roles in these tissues. As the main purpose of epithelia is to govern transport of solutes and water, the overview will mostly approach this theme, with the renal epithelium as major area of interest.

28.2 P2X Receptors and Excitation-Secretion Coupling

Exocrine secretion is one of the prime functions of epithelial cells. For example, the various glands of the gastro-intestinal tract serve to secrete the appropriate amount of digestive juices into the lumen of the gut to permit mixing, breakdown and eventually absorption of nutrients, electrolytes and water from the ingested food and beverages. Exocrine glands like the salivary, the gastric or the pancreatic glands are innervated from both autonomous and enteric nerve fibres and one of the prime agonists to activate secretion is the neurotransmitter acetylcholine (ACh). Excitation-secretion coupling involves the generation of inositol trisphosphate (IP_3) followed by an increase in $[Ca^{2+}]_i$ that activates basolateral Ca^{2+}-activated K^+ channels and in parallel apical Ca^{2+}-activated Cl^- secretion. The first functional description of an ATP-activated P2X receptor in epithelia was made in rat parotid acinar cells that displayed ATP-induced ionic inward currents and increases in $[Ca^{2+}]_i$ clearly independent of the generation of IP_3 (McMillian et al. 1988). These results were confirmed in lacrimal acinar cells and have led to a general extension of understanding how agonists stimulate exocrine secretion in some glands (Sasaki and Gallacher 1990) (Fig. 28.2). In addition to the "classical" IP_3/Ca^{2+}-dependent mechanism of secretion activation, also P2X receptors are secretion activating

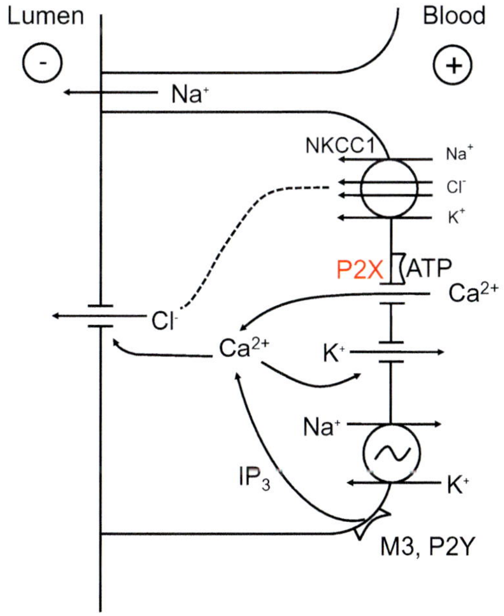

Fig. 28.2 Schematic cell model of the lacrimal and parotid exocrine acinar cell that highlights the central role of an increase in $[Ca^{2+}]_i$ for NaCl secretion activation. Note that a P2X receptor contributes to activate ion secretion via elevating $[Ca^{2+}]_i$

Table 28.1 P2X receptors in secretory epithelia

Epithelial tissue	Species	P2X receptor (subtype)	Intracellular signal	Regulated function	Reference
Parotid acinar cell	Rat	Undefined	$[Ca^{2+}]_i$	↑ ion secretion	McMillian et al. (1988)
Lacrimal acinar cell	Mouse	Undefined	$[Ca^{2+}]_i$	↑ ion secretion	Sasaki and Gallacher (1990)
Pancreatic duct	Rat	P2X1, P2X4, P2X7	$[Ca^{2+}]_i$ pH_i ↓	Suggested regulation of ductal ion transport	Hede et al. (1999), Henriksen and Novak (2003), Luo et al. (1999)
Epididymis	Mouse	P2X4 (and other P2X subunits	?	↑ H^+ secretion by the V-type ATPase	Belleannee et al. (2010) and unpublished
Collecting duct (α- and β-intercalated cell)	Mouse	P2X4	?	?	Chen et al. (2017)
Alveolar type II cells	Rat	P2X4	$[Ca^{2+}]_i$ (fusion activated Ca^{2+}entry)	Increase in exocytosis and surfactant secretion	Miklavc et al. (2011), Miklavc et al. (2013)
Alveolar type I and II cells	Mouse	P2X7 on alveolar type I cells	?	↑ surfactant secretion from alveolar type II cells (paracrine)	Mishra et al. (2011)
Airway secretory epithelia (goblet cells)	Human and mouse	P2X4	$[Ca^{2+}]_i$	Mucous secretion	Winkelmann et al. (2018)

effectors. P2X receptors that activate secretory epithelia are listed in Table 28.1. Noteworthy, not all exocrine acinar cells show unequivocal functional expression of P2X receptors. Apparent absence or low expression of P2X receptors was reported in rat pancreatic acinar cells (Novak et al. 2002) as well as in colonic crypts (Larsen and Leipziger 2013). It is worth remembering that ATP can also act via P2Y receptors to elevate $[Ca^{2+}]_i$ and ion secretion in several exocrine glands (Leipziger et al. 1997) (Fig. 28.2). P2X receptors are also involved in the regulation of cellular acid secretion. The P2X4 receptor was found in the apical membrane of the acid-secreting clear cells of the epididymis (Battistone et al. 2019). In addition, this tissue can secrete ATP into the ductal lumen and together with adenosine both purinergic agonists stimulate H^+ secretion via activation of the V-type ATPase in the ductal clear cells (Belleannee et al. 2010).

28.3 P2X Receptors in Absorptive Epithelia

Solute and water absorption are a key function of the gastrointestinal, the respiratory and the renal epithelium. Evidence is emerging that P2X receptors are abundantly expressed in most absorptive epithelia (Hede et al. 1999; Korngreen et al. 1998; Marques et al. 2012). A listing of P2X receptors in absorptive epithelia is shown in Table 28.2. In the kidney, the first observation of a P2X receptor-like function was in 1998. In the proximal tubule epithelial cell line LLC-PK_1 ATP activated a non-selective cation conductance and an increase in $[Ca^{2+}]_i$, which was fully dependent on the presence of extracellular Ca^{2+} (Filipovic et al. 1998). In the meantime, further work by many groups has identified specific mRNAs for many P2X receptor

Table 28.2 P2X receptors in absorptive epithelia

Epithelial tissue	Species	P2X receptor (subtype)	Intracellular signal	Regulated function	Reference
LLC-PK_1 Proximal tubular origin	Pig	P2X1	$[Ca^{2+}]_i$	?	Filipovic et al. (1998)
Distal convoluted tubule (MDCT cell line)	Mouse	P2X1-3 P2X4 P2X5	$[Ca^{2+}]_i$	↓cytosolic Mg^{2+} uptake = ↓ Mg^{2+} reabsorption	Dai et al. (2001)
Distal convoluted tubule (DCT)	Mouse	P2X4 P2X6	?	↓TRPM6	de Baaij et al. (2014)
Native medullary thick ascending limb (mTAL)	Mouse	P2X1 P2X4 P2X5	?	↓ NaCl reabsorption	Marques et al. (2012) #2259
Cortical collecting duct (CCD)	Rat	P2X1 P2X4 P2X6	PI3K	↓ and ↑ amiloride-sensitive currents	Wildman et al. (2008)
Entire renal tubule	Rat	P2X4 P2X6	?	Not studied	Turner et al. (2003)
Medullary collecting duct and S3 segment of PT	Rat	P2X5	?	Not studied	Turner et al. (2003)
A6, kidney epithelial cell line	*Xenopus laevis,* A6 cells	P2X4 like	?	↑ ENaC dependent Na^+ transport	Zhang et al. (2007)
IMCD-K2 Inner medullary collecting duct cell line	Mouse	P2X?	?	↓ NaCl reabsorption	McCoy et al. (1999)
Cochlear outer sulcus cells	Gerbil	P2X2?	?	Stimulation of cation absorption	Lee et al. (2001)

subtypes along the different renal epithelial tissues often supported by positive immunohistochemistry results (IHC) (de Baaij et al. 2014; Marques et al. 2012; Turner et al. 2003; Wildman et al. 2008). An integrative understanding of a physiological role of P2X receptors in renal epithelia, however, is still very limited. In 2001, a study using cultured mouse distal convoluted tubule (DCT) cells demonstrated that P2X receptor stimulation reduced cellular Mg^{2+} uptake (Dai et al. 2001). Because this segment is responsible for transcellular Mg^{2+} reabsorption, it was speculated that ATP could inhibit Mg^{2+} reabsorption. At that time, the apical ion channel for Mg^{2+} reabsorption in the distal convoluted tubule (DCT) had not been discovered yet, but is now identified as the TRPM6 channel (Schlingmann et al. 2002). Intriguingly, more recent results strongly support that P2X4 receptor stimulation indeed inhibits the TRPM6 channel (de Baaij et al. 2014). The P2X4 receptor is co-expressed with the TRPM6 channel in the DCT and in the distal colon, two important sites of epithelial Mg^{2+} reabsorption (de Baaij et al. 2014). Taken together, these results therefore strengthen the hypothesis that P2X receptor stimulation inhibits Mg^{2+} reabsorption. Direct evidence for this is currently pending and $P2X4^{-/-}$ mice on a normal diet show unchanged urinary Mg^{2+} excretion (de Baaij et al. 2014). In isolated perfused mouse medullary thick ascending limbs (mTAL), it was shown that basolateral P2X4 receptor stimulation triggers a fast and fully reversible (~25%) inhibition of NaCl reabsorption (Marques et al. 2012). This study was the first to identify that P2X receptor stimulation in native renal epithelium modulates renal ion transport. It is also noteworthy that infusion of P2X receptor agonists into rats caused increased urinary excretion of water and NaCl (Jankowski et al. 2011). Together, these results suggest that P2X receptor stimulation in renal epithelia inhibits reabsorption of NaCl, Mg^{2+} and water. Transcriptome data from the entire renal tubular system identified the P2X4 receptor as the most abundantly expressed P2X receptor isoform along nearly all tubular segments (Chen et al. 2017; Lee et al. 2015). The field of P2Y receptor research in renal epithelia has a much more developed research history (Praetorius and Leipziger 2010; Vallon 2008). Unequivocal results from many groups have identified that P2Y receptor stimulation inhibits tubular absorption in most tubule segments. Thus, local paracrine ATP signalling via both, P2Y and P2X receptors, inhibits renal tubular transport and thus opposes the pronounced absorption that is activated by a multi-hormone system involving, e.g. anti-diuretic hormone ADH and aldosterone (Fig. 28.3). These more recent discoveries addressing functional effects of P2X receptors in absorptive renal epithelia are in good agreement with the current understanding of intra-renal purinergic signalling: Local extracellular ATP signals likely present an endogenous local diuretic signalling system (Praetorius and Leipziger 2010). Eventually, the effect of ATP was tested on ENaC-mediated whole-cell currents in rat split open cortical collecting duct cells (CCD) (Wildman et al. 2008). Much in agreement with numerous other studies, a marked inhibition of the epithelial Na^+ channel (ENaC) was measured, which was mediated via the P2Y2 receptor. In addition, P2X receptor-mediated effects were reported. Interestingly, under high extracellular Na^+ (145 mM) P2X receptor activation inhibited the amiloride-sensitive currents but under low extracellular Na^+ (50 mM) P2X receptor activation apparently

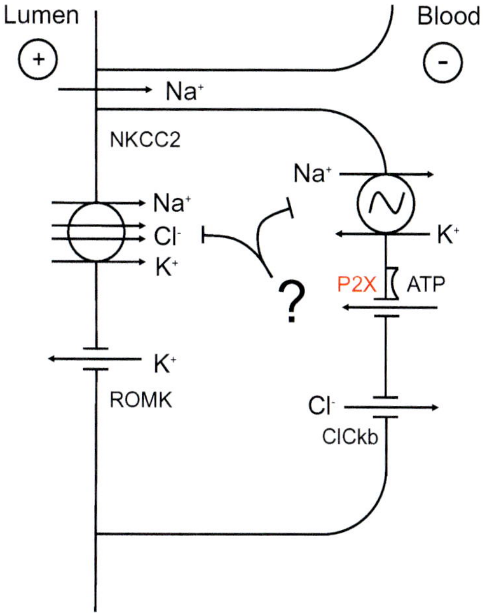

Fig. 28.3 P2X receptors are expressed in absorptive renal epithelia. Here, the example of the mouse medullary thick ascending limb (mTAL) is shown. Basolateral ATP inhibits NaCl absorption via a P2X4 receptor but the cellular mechanism of this inhibition is currently not understood

activated the amiloride-sensitive current in CCD cells. The authors therefore speculated that a P2X receptor could function as an extracellular Na^+ sensor (Wildman et al. 2008). These results will require further attention for a more comprehensive understanding. Nonetheless, the experimental evidence is interesting because an extracellular P2X receptor with Na^+ sensor features was previously suggested in rabbit airway ciliated cells. In ciliated cells, the application of ATP stimulated a P2X receptor-dependent non-selective cation current (Korngreen et al. 1998). Interestingly, this ATP-induced inward current was strongly inhibited by raising the extracellular Na^+ concentration above 10 mM (Ma et al. 1999, 2006). Again, this finding remains to be fully understood and integrated into a physiological concept. It is uncertain whether the Na^+ concentrations in the airway surface liquid layer can reach these very low values. Nonetheless, the research question of the role of purinergic signalling in pulmonary mucociliary clearance and airway surface liquid layer regulation is of greatest interest and clinical relevance (Boucher 2003). Luminal ATP activates ciliary beat frequency and ion secretion (Korngreen and Priel 1996). In addition, luminal nucleotides in bronchial epithelia also inhibit ENaC via P2Y2 receptors (Devor and Pilewski 1999). A specific role of epithelial P2X receptors in this physiology is not yet fully studied.

28.4 Identification of the Native P2X Receptor Responsible for a Physiological Response

Epithelia and most other cell types express multiple P2Y and P2X receptors (Praetorius and Leipziger 2010; Vallon 2008). By using specific P2Y receptor knock-out mice together with reasonably specific receptor agonists and antagonists, it was in many cases possible to identify which of the 8 different P2Y receptors is responsible for a certain nucleotide-induced effect. An example for this is the apical P2Y2 receptor in the collecting duct that markedly inhibits ENaC channels and thus Na^+ reabsorption (Pochynyuk et al. 2008).

The identification of P2X receptor-mediated cellular effects has also become reasonably straightforward and is based on a number of arguments that permit the safe conclusion that a nucleotide has activated a P2X receptor (1). P2X receptors are specifically activated by ATP and do not respond to pyrimidine nucleotides (North 2002), (2). The ATP-stimulated $[Ca^{2+}]_i$ increase is fully dependent on the influx of Ca^{2+} from the extracellular space (Jensen et al. 2007; Marques et al. 2012). This argument can, however, be confounded by co-expressed P2Y receptors in the same cell, leading to simultaneous release of Ca^{2+} from intracellular stores, when ionotropic and metabotropic receptors are stimulated in parallel. (3). A non-selective cation current is very rapidly (milliseconds) activated and deactivated after addition and removal of ATP (North 2002). (4). P2X receptors are inhibited by pre-treatment with oxidized ATP (irreversible) leaving P2Y receptor-mediated effects apparently unaffected (Murgia et al. 1993). (5). Finally, other subtype-specific pharmacological probes help to identify study P2X receptors (Burnstock 2007).

Three subunits assemble to form one P2X receptor, and theoretically, three different P2X receptor subunits can compose one functional P2X receptor. Indeed, it was shown that heterologously expressed and epitope-tagged P2X2, P2X4 and P2X6 receptors can form heterotrimeric channels (Antonio et al. 2014). An outlier in this structural theme is the P2X7 receptor, which is composed of three identical subunits (Pippel et al. 2017). For the other P2X receptors, the task to identify the molecular composition of a certain receptor can be a challenge. We have identified a P2X receptor in the basolateral membrane of mouse mTAL (Marques et al. 2012). On the level of mRNA, mouse mTAL expresses the P2X1, P2X4 and P2X5 receptor subunit (Marques et al. 2012). To unravel which P2X receptor subtype could mediate the physiological ATP effect, a knock-out strategy was applied. In mTALs from $P2X4^{-/-}$ mice, the ATP effect was blunted but still present. These results clearly support the involvement of the P2X4 receptor subunit. This raises the question which other P2X receptors could be activated in the $P2X4^{-/-}$. Apparently, P2X receptor subunits can interact promiscuously with numerous other P2X receptor subunits (North 2002). Most likely, the absence of one subunit in this example of the mTAL is compensated by the remaining isoforms, possibly leading to homomeric P2X5 or heteromeric P2X1/P2X5 receptors. A double knock-out of both receptor subunits, the P2X4 and the P2X5 could be attempted to resolve this question. This example illustrates the basic problem of identifying the exact subunit composition of P2X receptors in physiology. To the best of our knowledge, an unequivocal identification of the exact subunit composition of a native P2X receptor in epithelia has not been successfully achieved.

28.5 Intracellular Signalling Events Associated with Epithelial P2X Receptor Activation

In excitable cells, the activation of P2X receptors leads to a membrane voltage depolarization and therefore has a direct activating input. In excitation-secretion coupling of exocrine gland tissue, the receptor-dependent influx of Ca^{2+} increases $[Ca^{2+}]_i$ and activates relevant Ca^{2+}-sensitive ion channels and subsequently ion secretion directly (Fig. 28.2). Thus, in exocrine secretion, a main P2X receptor-associated intracellular transduction signal is Ca^{2+} as shown in lacrimal and parotid acinar cells (McMillian et al. 1988; Sasaki and Gallacher 1990). In renal absorptive epithelia, P2X receptors inhibit reabsorption but the associated signalling mechanisms are not deciphered. In all renal epithelial tissues, P2X receptor activation triggers the obligatory increase in $[Ca^{2+}]_i$ (Filipovic et al. 1998; Jensen et al. 2007). In rat TAL, P2X receptor stimulation triggers an intracellular increase in nitric oxide (NO) (Silva et al. 2006) and NO was suggested to inhibit NKCC2-dependent Cl^- reabsorption (Ortiz and Garvin 2002). More recent studies contradict these earlier ones and show that P2X receptor stimulation triggers NaCl transport inhibition independent of NO and also independent of $[Ca^{2+}]_i$ (Svendsen et al. 2017). Currently, the signalling pathway that leads to P2X4 receptor-mediated transport inhibition in the TAL is unknown. It may involve a direct effect on the basolateral Na^+/K^+ ATPase as suggested from P2X receptor effects in isolated proximal tubule cells (Jankowski et al. 2011). Possibly, the P2X receptor-stimulated epithelial NO production in the mTAL serves other functions, e.g. the regulation of renal medullary blood flow by modulating the vascular resistance of the vasa recta. In the study that had identified P2X4 receptor-stimulated inhibition of the TRPM6 Mg^{2+} channel, the authors also tested several protein kinase inhibitors (PKC, PKA, PI3K) but did not find any positive results (de Baaij et al. 2014). In summary, despite clear functional inhibitory effects on ion transport in renal epithelial cells, the underlying signalling events of P2X receptor stimulation remain not understood.

28.6 Apical and Basolateral P2X Receptors

Epithelial transport depends on polarized localization of ion pumps, transporters and channels in the apical or the basolateral membrane. Transcellular ion secretion follows the basic mechanism of "pump-leak," where an initial uptake of the ion

Table 28.3 Apical and basolateral localization of P2X receptor in epithelia

Epithelial tissue	Species	P2X receptor		Regulated function	Reference
		Apical	Basolateral		
Pancreatic duct	Rat	P2X7	P2X1 P2X4 P2X7	Ductal ion transport	Hede et al. (1999), Luo et al. (1999)
Medullary thick ascending limb	Mouse	None	P2X4 P2X5 P2X1	Inhibition of NaCl reabsorption	Marques et al. (2012)
Cortical collecting duct	Rat	P2X4 P2X6	P2X4 P2X6	Regulation of ENaC	Wildman et al. (2008)
Airway ciliated cells	Rabbit	P2X7? P2X4		? Activation of mucociliary clearance	Droguett et al. (2017), Ma et al. (2006)
Epididymis	Mouse	P2X4	?	Activation of H^+ secretion	Belleannee et al. (2010) and Battistone et al. (2019)
Retinal pigment epithelium	Human	P2X7	P2X7	Lysosomal alkalization	Guha et al. (2013)

over the basolateral membrane is followed by a passive leak over the apical membrane. In absorptive epithelial transport, this scheme operates in an "inverted" mode, were initially an apical membrane channel or transporter permits cytosolic uptake followed by a secondary removal step over the basolateral membrane. In addition, regulated paracellular movement of ions and waters through specialized "channel-like" junctional proteins like the claudins has been established as a major regulated pathway of transepithelial transport (Gunzel and Yu 2013). Regulation of these processes is highly complex and most often involves hormone, neurotransmitter and paracrine factor receptors in the basolateral membrane. It is also established that most epithelia can be regulated by agonist signals via membrane receptors in their apical membranes. This is especially well studied for the purinergic signalling system, where it was found that most epithelial cells express apical and basolateral P2Y receptors often in the same cell (Leipziger 2003). This has led to the appreciation that the luminal space, e.g. in the renal tubule, actually functions as an extracellular signalling compartment. The interested reader is referred to more comprehensive reviews on this topic (Leipziger 2003, 2011). Table 28.3 summarizes studies, which have addressed the apical and basolateral localization of epithelial P2X receptors. So far, the number of studies is small, but these studies clearly indicate that P2X receptors in native tissue are present in both the apical and basolateral membranes. One example is the P2X4 receptor, which can be found in the basolateral membrane of mouse mTAL (Marques et al. 2012) and apparently also in the apical membrane of rat CCD (Wildman et al. 2008), the rabbit ciliated respiratory epithelium (Ma et al. 2006) and the apical membrane of acid-secreting cells in the epididymis (Belleannee et al. 2010; Battistone et al. 2019). These results imply that some epithelia can pick-up the information of luminal ATP concentrations via apical P2X receptors.

28.7 Epithelial P2X Receptors in Pathophysiology

The number of studies that address the physiology of P2X receptors in epithelia is still small and an integrative understanding of their physiological roles is limited. Therefore, it is not surprising that only a few pathophysiological concepts involving epithelial P2X receptors have been suggested and these often remain speculative. The P2X4 receptor knock-out mouse is hypertensive, which was attributed to vascular and/or inflammatory effects (Yamamoto et al. 2006). In addition, we have recently identified a renal Na^{+}-sensitive component of the increased BP in P2X4 KO mice (Craigie et al. 2018). Single nucleotide polymorphisms in the P2X7 and the P2X4 receptor associate with slight elevation of BP in humans (Palomino-Doza et al. 2008). Much focus has been directed towards the P2X7 receptor because of its profound pro-inflammatory actions and is high expression in various immune cells. Interestingly, in normal renal epithelial cells including podocytes, P2X7 receptor expression apparently is low (Booth et al. 2012; Ilatovskaya et al. 2013; Marques et al. 2012; Vonend et al. 2004) but can be markedly up-regulated in diseased states. This was shown in a rat experimental diabetes mellitus model and in a rat hypertensive model with a distinctly increased P2X7 expression in podocytes (Vonend et al. 2004). P2X7 receptors also became significantly expressed in collecting ducts in a genetic model of recessive polycystic kidney disease and stimulation of in vitro cysts with BzATP, a potent P2X7 agonist, reduces cyst number (Hillman et al. 2004). Similarly, the P2X4 receptor is up-regulated in humans with diabetic nephropathy and this receptor is suggested to mediate pro-inflammatory effects (Chen et al. 2013). In a zebrafish model of cystogenesis by knock-down of PKD2 a role of the P2X7 receptor was identified, but here, P2X7 receptor inhibitors caused a reduction of cystogenesis (Chang et al. 2014). These results indicate that normally low P2X7 expressing "healthy" epithelial cells can change their phenotype during damaging influences and become targetable for high extracellular ATP. A more profound understanding of this phenotypical change remains to be developed. Similarly, also in salivary epithelial cells, the P2X7 receptor was recently described to trigger a pro-inflammatory response (Khalafalla et al. 2017; Woods et al. 2012). These results highlight a general and critical role of the P2X7 receptor and possibly, also the P2X4 receptor in epithelial disease processes.

Unilateral ureter obstruction (UUO) is a powerful model to induce chronic renal inflammation and interstitial fibrosis. In P2X7 receptor knock-out mice, the UUO-induced interstitial fibrosis was markedly reduced, consistent with its pro-inflammatory role (Goncalves et al. 2006). The same question was reinvestigated in $P2X4^{-/-}$ mice, and much to the authors surprise, they found the contrary, namely that $P2X4^{-/-}$ mice showed a significantly increased fibrosis (Kim et al. 2014). In normal renal epithelia, the P2X4 receptor is the most prominently expressed P2 receptor, and possibly in the kidney, it could mediate an anti-inflammatory effect, which is lost in the P2X4 KO mouse. In summary, epithelial P2X receptor physiology and pathophysiology is a steadily evolving field but requires much more basic work. Nevertheless, it has become clear that the P2X7 receptor is a major disease-causing element in a number of diseases challenged epithelia.

28.8 Other Functional Implications of Epithelial P2X Receptors

Beyond the effects of P2X receptors on the regulation of ion and water transport, other relevant epithelial functions have been identified which are modulated by P2X receptors and these are listed in Table 28.4. They cover a remarkably wide range of

Table 28.4 Other functions of epithelial P2X receptors

P2X receptor	Epithelial tissue	Function	References
P2X7	Human cervical epithelium, Alveolar type I cells, Intestinal epithelial cell	↑ apoptosis	Guo et al. (2014), Souza et al. (2012), Wang et al. (2004)
P2X4	Human cervical epithelium	↑ Paracellular resistance	Gorodeski (2002)
P2X7 and P2X4	Alveolar type I cells	↑ AQP5 expression ? Altered H_2O transport in the alveolus, Allergen-induced airway inflammation	Ebeling et al. (2014), Zech et al. (2016)
P2X7	Retinal pigment epithelium	Lysosomal alkalization	Guha et al. (2013)
P2X7/P2X4	Gingival epithelial cell	↑ reactive oxygen species	Hung et al. (2013)
P2X7	Corneal epithelium	↑ cell migration, ↑ proliferation, ↑ cell death	Mankus et al. (2011)
P2X2	Reissner membrane of the inner ear, supporting cells of inner ear	? Modulation of sound perception, ↓ vestibular function	King et al. (1998), Takimoto et al. (2018)
P2X4	Alveolar type II cells	FACE (= fusion activated Ca^{2+} entry) ↑ surfactant secretion	Miklavc et al. (2013)
P2X4	HeLa cells	Inhibition of *Chlamydia trachomatis* infection via epithelial ATP release	Pettengill et al. (2012)
P2X	Microvillous cells of the main olfactory epithelium	Odor transduction?	Fu et al. (2018)
P2X7	Pancreatic duct cancer cells	Growth stimulation, necrosis	Giannuzzo et al. (2015)
P2X4	Podocytes	Mechano-transduction	Forst et al. (2016)

different biological phenomena with no unifying theme. Several of the listed discoveries indicate a significant role of P2X receptors in the control of an epithelial innate immune/inflammatory response and tissue regeneration and repair. The listed functions of epithelial P2X receptors also point to a general role of purinergic signalling in the defence properties of the epithelial sheet. This research area continues to be underdeveloped and offers attractive perspectives worth pursuing.

28.9 Final Concluding Remarks

The topic of P2X receptor research is emerging and is beginning to extend our understanding of how barrier forming epithelial sheets function. Local paracrine ATP signals have been identified to modify epithelial transport function via both P2Y and P2X receptors. The physiology of P2X receptors is currently best understood in renal absorptive epithelia and here evidence from native mammalian tissue informs us that P2X receptor stimulation inhibits solute absorption thus producing a local diuretic event. Most intriguing is the rather dramatic upregulation of P2X receptors in processes that damage the epithelial tissue highlighting the possibility that P2X receptor biology is of substantial importance in tissue repair and regeneration.

Acknowledgements The author greatly thanks a very lively laboratory team, which is always willing to challenge the current working hypotheses and provides me with endless joy to go to work every day. These wonderful people are currently Mads Vaarby Sørensen, Peder Berg, Niklas Ayasse, Samuel Svendsen and Karen Sjødt Sørensen. I further thank the Danish Medical Research Council and the Lundbeck Foundation for many years of financial support for our research.

References

Antonio LS, Stewart AP, Varanda WA, Edwardson JM (2014) Identification of P2X2/P2X4/P2X6 heterotrimeric receptors using atomic force microscopy (AFM) imaging. FEBS Lett 588 (12):2125–2128. https://doi.org/10.1016/j.febslet.2014.04.048

Battistone MA, Merkulova M, Park YJ, Peralta MA, Gombar F, Brown D, Breton S (2019) Unravelling purinergic regulation in the epididymis: activation of V-ATPase-dependent acidification by luminal ATP and adenosine. J Physiol. https://doi.org/10.1113/JP277565

Belleannee C, Da SN, Shum WW, Brown D, Breton S (2010) Role of purinergic signaling pathways in V-ATPase recruitment to apical membrane of acidifying epididymal clear cells. Am J Physiol Cell Physiol 298:C817–C830

Booth JW, Tam FW, Unwin RJ (2012) P2 purinoceptors: renal pathophysiology and therapeutic potential. Clin Nephrol 78:154–163

Boucher RC (2003) Regulation of airway surface liquid volume by human airway epithelia. Pflügers Arch Eur J Physiol 445:495–498

Burnstock G (2007) Physiology and pathophysiology of purinergic neurotransmission. Physiol Rev 87:659–797

Chang M-Y, Lu J-K, Tian Y-C, Chen Y-C, Hung C-C, Huang Y-H, Chen Y-H, Wu M-S, Yang C-W, Cheng Y-C (2014) Inhibition of the P2X7 receptor reduces Cystogenesis in PKD. J Am Soc Nephrol 22:1696–1706

Chen K, Zhang J, Zhang W, Zhang J, Yang J, Li K, He Y (2013) ATP-P2X4 signaling mediates NLRP3 inflammasome activation: a novel pathway of diabetic nephropathy. Int. J Biochem. Cell Biol 45:932–943

Chen L, Lee JW, Chou CL, Nair AV, Battistone MA, Paunescu TG, Merkulova M, Breton S, Verlander JW, Wall SM, Brown D, Burg MB, Knepper MA (2017) Transcriptomes of major renal collecting duct cell types in mouse identified by single-cell RNA-seq. Proc Natl Acad Sci U S A 114:E9989–E9998

Craigie E, Menzies RI, Larsen CK, Jacquillet G, Carrel M, Wildman SS, Loffing J, Leipziger J, Shirley DG, Bailey MA, Unwin RJ (2018) The renal and blood pressure response to low sodium diet in P2X4 receptor knockout mice. Physiol Rep 6:e13899

Dai LJ, Kang HS, Kerstan D, Ritchie G, Quamme GA (2001) ATP inhibits Mg^{2+} uptake in mouse distal convoluted tubule cells via P2X purinoceptors. Am J Phys 281:F833–F840

de Baaij JH, Blanchard MG, Lavrijsen M, Leipziger J, Bindels RJ, Hoenderop JG (2014) P2X4 receptor regulation of transient receptor potential melastatin type 6 (TRPM6) mg channels. Pflugers Arch 466:1941–1952

Devor DC, Pilewski JM (1999) UTP inhibits Na^{+} absorption in wild-type and DeltaF508 CFTR-expressing human bronchial epithelia. Am J Phys 276:C827–C837

Droguett K, Rios M, Carreno DV, Navarrete C, Fuentes C, Villalon M, Barrera NP (2017) An autocrine ATP release mechanism regulates basal ciliary activity in airway epithelium. J Physiol 595:4755–4767

Ebeling G, Blasche R, Hofmann F, Augstein A, Kasper M, Barth K (2014) Effect of P2X7 receptor knockout on AQP-5 expression of type I alveolar epithelial cells. PLoS One 9:e100282

Filipovic DM, Adebanjo OA, Zaidi M, Reeves WB (1998) Functional and molecular evidence for P2X receptors in LLC-PK1 cells. Am J Phys 274:F1070–F1077

Forst AL, Olteanu VS, Mollet G, Wlodkowski T, Schaefer F, Dietrich A, Reiser J, Gudermann T, Mederos YS, Storch U (2016) Podocyte Purinergic P2X4 channels are Mechanotransducers that mediate cytoskeletal disorganization. J Am Soc Nephrol 27(3):846–862

Fu Z, Ogura T, Luo W, Lin W (2018) ATP and odor mixture activate TRPM5-expressing microvillous cells and potentially induce acetylcholine release to enhance supporting cell endocytosis in mouse Main olfactory epithelium. Front Cell Neurosci 12:71. https://doi.org/10.3389/fncel.2018.0007

Giannuzzo A, Pedersen SF, Novak I (2015) The P2X7 receptor regulates cell survival, migration and invasion of pancreatic ductal adenocarcinoma cells. Mol Cancer 14:203. https://doi.org/10.1186/s12943-015-0472-4

Goncalves RG, Gabrich L, Rosario A Jr, Takiya CM, Ferreira ML, Chiarini LB, Persechini PM, Coutinho-Silva R, Leite M Jr (2006) The role of purinergic P2X7 receptors in the inflammation and fibrosis of unilateral ureteral obstruction in mice. Kidney Int 70:1599–1606

Gorodeski GI (2002) Expression, regulation, and function of P2X(4) purinergic receptor in human cervical epithelial cells. Am J Physiol Cell Physiol 282:C84–C93

Guha S, Baltazar GC, Coffey EE, Tu LA, Lim JC, Beckel JM, Patel S, Eysteinsson T, Lu W, O'Brien-Jenkins A, Laties AM, Mitchell CH (2013) Lysosomal alkalinization, lipid oxidation, and reduced phagosome clearance triggered by activation of the P2X7 receptor. FASEB J 27:4500–4509

Gunzel D, Yu AS (2013) Claudins and the modulation of tight junction permeability. Physiol Rev 93:525–569

Guo Y, Mishra A, Weng T, Chintagari NR, Wang Y, Zhao C, Huang C, Liu L (2014) Wnt3a mitigates acute lung injury by reducing P2X7 receptor-mediated alveolar epithelial type I cell death. Cell Death Dis 5:e1286

Hattori M, Gouaux E (2012) Molecular mechanism of ATP binding and ion channel activation in P2X receptors. Nature 485:207–212

Hede SE, Amstrup J, Christoffersen BC, Novak I (1999) Purinoceptors evoke different electrophysiological responses in pancreatic ducts. P2Y inhibits K^+ conductance, and P2X stimulates cation conductance. J Biol Chem 274:31784–31791
Henriksen KL, Novak I (2003) Effect of ATP on intracellular pH in pancreatic ducts involves P2X7 receptors. Cell Physiol Biochem 13:93–102
Hillman KA, Woolf AS, Johnson TM, Wade A, Unwin RJ, Winyard PJ (2004) The P2X7 ATP receptor modulates renal cyst development in vitro. Biochem Biophys Res Commun 322:434–439
Hung SC, Choi CH, Said-Sadier N, Johnson L, Atanasova KR, Sellami H, Yilmaz O, Ojcius DM (2013) P2X4 assembles with P2X7 and pannexin-1 in gingival epithelial cells and modulates ATP-induced reactive oxygen species production and inflammasome activation. PLoS One 8: e70210
Ilatovskaya DV, Palygin O, Levchenko V, Staruschenko A (2013) Pharmacological characterization of the P2 receptors profile in the podocytes of the freshly isolated rat glomeruli. Am. J. Physiol Cell Physiol 305:C1050–C1059
Jankowski M, Szamocka E, Kowalski R, Angielski S, Szczepanska-Konkel M (2011) The effects of P2X receptor agonists on renal sodium and water excretion in anaesthetized rats. Acta Physiol (Oxf) 202:193–201
Jensen ME, Odgaard E, Christensen MH, Praetorius HA, Leipziger J (2007) Flow-induced $[Ca^{2+}]_i$ increase depends on nucleotide release and subsequent purinergic signaling in the intact nephron. J Am Soc Nephrol 18:2062–2070
Kawate T, Michel JC, Birdsong WT, Gouaux E (2009) Crystal structure of the ATP-gated $P2X_4$ ion channel in the closed state. Nature 460:592–598
Khalafalla MG, Woods LT, Camden JM, Khan AA, Limesand KH, Petris MJ, Erb L, Weisman GA (2017) P2X7 receptor antagonism prevents IL-1beta release from salivary epithelial cells and reduces inflammation in a mouse model of autoimmune exocrinopathy. J Biol Chem 292:16626–16637
Kim MJ, Turner CM, Hewitt R, Smith J, Bhangal G, Pusey CD, Unwin RJ, Tam FW (2014) Exaggerated renal fibrosis in P2X4 receptor-deficient mice following unilateral ureteric obstruction. Nephrol Dial Transplant 29:1350–1361
King M, Housley GD, Raybould NP, Greenwood D, Salih SG (1998) Expression of ATP-gated ion channels by Reissner's membrane epithelial cells. Neuroreport 9:2467–2474
Korngreen A, Priel Z (1996) Purinergic stimulation of rabbit ciliated airway epithelia: control by multiple calcium sources. J Physiol 497(Pt 1):53–66
Korngreen A, Ma W, Priel Z, Silberberg SD (1998) Extracellular ATP directly gates a cation-selective channel in rabbit airway ciliated epithelial cells. J Physiol 508(Pt 3):703–720
Larsen CK, Leipziger J (2013) P2Y receptors in the gastrointestinal epithelium. WIREs Membr Transp Signal 2:27–36
Lee JH, Chiba T, Marcus DC (2001) P2X2 receptor mediates stimulation of parasensory cation absorption by cochlear outer sulcus cells and vestibular transitional cells. J Neurosci 21:9168–9174
Lee JW, Chou CL, Knepper MA (2015) Deep sequencing in microdissected renal tubules identifies nephron segment-specific Transcriptomes. J Am Soc Nephrol 26:2669–2677
Leipziger J (2003) Control of epithelial transport via luminal P2 receptors. Am J Physiol Renal Physiol 284:F419–F432
Leipziger J (2011) Luminal nucleotides are tonic inhibitors of renal tubular transport. Curr Opin Nephrol Hypertens 20:518–522
Leipziger J, Kerstan D, Nitschke R, Greger R (1997) ATP increases $[Ca^{2+}]_i$ and ion secretion via a basolateral P2Y receptor in rat distal colonic mucosa. Pflügers Arch Eur J Physiol 434:77–83
Luo X, Zheng W, Yan M, Lee MG, Muallem S (1999) Multiple functional P2X and P2Y receptors in the luminal and basolateral membranes of pancreatic duct cells. Am J Phys 277:C205–C215
Ma W, Korngreen A, Uzlaner N, Priel Z, Silberberg SD (1999) Extracellular sodium regulates airway ciliary motility by inhibiting a P2X receptor. Nature 400:894–897

Ma W, Korngreen A, Weil S, Cohen EB, Priel A, Kuzin L, Silberberg SD (2006) Pore properties and pharmacological features of the P2X receptor channel in airway ciliated cells. J Physiol 571:503–517

Mankus C, Rich C, Minns M, Trinkaus-Randall V (2011) Corneal epithelium expresses a variant of P2X(7) receptor in health and disease. PLoS One 6:e28541

Marques RD, de Bruijn PI, Sorensen MV, Bleich M, Praetorius HA, Leipziger J (2012) Basolateral P2X receptors mediate inhibition of NaCl transport in mouse medullary thick ascending limb (mTAL). Am. J. Physiol Renal Physiol 302:F487–F494

McCoy DE, Taylor AL, Kudlow BA, Karlson K, Slattery MJ, Schwiebert LM, Schwiebert EM, Stanton BA (1999) Nucleotides regulate NaCl transport in mIMCD-K2 cells via P2X and P2Y purinergic receptors. Am J Phys 277:F552–F559

McMillian MK, Soltoff SP, Lechleiter JD, Cantley LC, Talamo BR (1988) Extracellular ATP increases free cytosolic calcium in rat parotid acinar cells. Biochem J 255:291–300

Miklavc P, Mair N, Wittekindt OH, Haller T, Dietl P, Felder E, Timmler M, Frick M (2011) Fusion-activated Ca^{2+} entry via vesicular P2X4 receptors promotes fusion pore opening and exocytotic content release in pneumocytes. Proc Natl Acad Sci U S A 108:14503–14508

Miklavc P, Thompson KE, Frick M (2013) A new role for P2X4 receptors as modulators of lung surfactant secretion. Front Cell Neurosci 7:171

Mishra A, Chintagari NR, Guo Y, Weng T, Su L, Liu L (2011) Purinergic P2X7 receptor regulates lung surfactant secretion in a paracrine manner. J Cell Sci 124:657–668

Murgia M, Hanau S, Pizzo P, Rippa M, Di VF (1993) Oxidized ATP. An irreversible inhibitor of the macrophage purinergic P2Z receptor. J Biol Chem 268:8199–8203

North RA (2002) Molecular physiology of P2X receptors. Physiol Rev 82:1013–1067

Novak I, Nitschke R, Amstrup J (2002) Purinergic receptors have different effects in rat exocrine pancreas. Calcium signals monitored by fura-2 using confocal microscopy. Cell Physiol Biochem 12:83–92

Ortiz PA, Garvin JL (2002) Role of nitric oxide in the regulation of nephron transport. Am J Physiol Renal Physiol 282:F777–F784

Palomino-Doza J, Rahman TJ, Avery PJ, Mayosi BM, Farrall M, Watkins H, Edwards CR, Keavney B (2008) Ambulatory blood pressure is associated with polymorphic variation in P2X receptor genes. Hypertension 52:980–985

Pettengill MA, Marques-da-Silva C, Avila ML, d'Arc dos Santos OS, Lam VW, Ollawa I, Abdul Sater AA, Coutinho-Silva R, Hacker G, Ojcius DM (2012) Reversible inhibition of chlamydia trachomatis infection in epithelial cells due to stimulation of $P2X_4$ receptors. Infect Immun 80:4232–4238

Pippel A, Stolz M, Woltersdorf R, Kless A, Schmalzing G, Markwardt F (2017) Localization of the gate and selectivity filter of the full-length P2X7 receptor. Proc Natl Acad Sci U S A 114: E2156–E2165

Pochynyuk O, Bugaj V, Rieg T, Insel PA, Mironova E, Vallon V, Stockand JD (2008) Paracrine regulation of the epithelial Na+ channel in the mammalian collecting duct by purinergic P2Y2 receptor tone. J Biol Chem 283(52):36599–36607. https://doi.org/10.1074/jbc.M807129200

Praetorius HA, Leipziger J (2010) Intrarenal purinergic signaling in the control of renal tubular transport. Annu Rev Physiol 72:377–393

Sasaki T, Gallacher DV (1990) Extracellular ATP activates receptor-operated cation channels in mouse lacrimal acinar cells to promote calcium influx in the absence of phosphoinositide metabolism. FEBS 264:130–134

Schlingmann KP, Weber S, Peters M, Niemann NL, Vitzthum H, Klingel K, Kratz M, Haddad E, Ristoff E, Dinour D, Syrrou M, Nielsen S, Sassen M, Waldegger S, Seyberth HW, Konrad M (2002) Hypomagnesemia with secondary hypocalcemia is caused by mutations in TRPM6, a new member of the TRPM gene family. Nat Genet 31:166–170

Silberberg SD, Swartz KJ (2009) Structural biology: Trimeric ion-channel design. Nature 460:580–581

Silva G, Beierwaltes WH, Garvin JL (2006) Extracellular ATP stimulates NO production in rat thick ascending limb. Hypertension 47:563–567

Souza CO, Santoro GF, Figliuolo VR, Nanini HF, de Souza HS, Castelo-Branco MT, Abalo AA, Paiva MM, Coutinho CM, Coutinho-Silva R (2012) Extracellular ATP induces cell death in human intestinal epithelial cells. Biochim Biophys Acta 1820:1867–1878

Svendsen SL, Isidor S, Praetorius HA, Leipziger J (2017) P2X receptors inhibit NaCl absorption in mTAL independently of nitric oxide. Front Physiol 8:18

Takimoto Y, Ishida Y, Kondo M, Imai T, Hanada Y, Ozono Y, Kamakura T, Inohara H, Shimada S (2018) P2X2 receptor deficiency in mouse vestibular end organs attenuates vestibular function. Neuroscience 386:41–50

Taylor AL, Schwiebert LM, Smith JJ, King C, Jones JR, Sorscher EJ, Schwiebert EM (1999) Epithelial P2X purinergic receptor channel expression and function. J Clin Invest 104:875–884

Turner CM, Vonend O, Chan C, Burnstock G, Unwin RJ (2003) The pattern of distribution of selected ATP-sensitive P2 receptor subtypes in normal rat kidney: an immunohistological study. Cells Tissues Organs 175:105–117

Vallon V (2008) P2 receptors in the regulation of renal transport mechanisms. Am J Physiol Renal Physiol 294:F10–F27

Vonend O, Turner CM, Chan CM, Loesch A, Dell'Anna GC, Srai KS, Burnstock G, Unwin RJ (2004) Glomerular expression of the ATP-sensitive P2X receptor in diabetic and hypertensive rat models. Kidney Int 66:157–166

Wang Q, Wang L, Feng YH, Li X, Zeng R, Gorodeski GI (2004) P2X7 receptor-mediated apoptosis of human cervical epithelial cells. Am J Physiol Cell Physiol 287:C1349–C1358

Wildman SS, Marks J, Turner CM, Yew-Booth L, Peppiatt-Wildman CM, King BF, Shirley DG, Wang W, Unwin RJ (2008) Sodium-dependent regulation of renal amiloride-sensitive currents by apical P2 receptors. J Am Soc Nephrol 19:731–742

Winkelmann VE, Thompson KE, Neuland K, Jaramillo AM, Fois G, Schmidt H, Wittekindt OH, Han W, Tuvim MJ, Dickey BF, Dietl P, Frick M (2018) Inflammation-induced upregulation of P2X4 expression augments mucin secretion in airway epithelia. Am J Physiol Lung Cell Mol Physiol 316(1):L58–L70

Woods LT, Camden JM, Batek JM, Petris MJ, Erb L, Weisman GA (2012) P2X7 receptor activation induces inflammatory responses in salivary gland epithelium. Am J Physiol Cell Physiol 303:C790–C801

Yamamoto K, Sokabe T, Matsumoto T, Yoshimura K, Shibata M, Ohura N, Fukuda T, Sato T, Sekine K, Kato S, Isshiki M, Fujita T, Kobayashi M, Kawamura K, Masuda H, Kamiya A, Ando J (2006) Impaired flow-dependent control of vascular tone and remodeling in $P2X_4$-deficient mice. Nat Med 12:133–137

Zech A, Wiesler B, Ayata CK, Schlaich T, Durk T, Hossfeld M, Ehrat N, Cicko S, Idzko M (2016) P2rx4 deficiency in mice alleviates allergen-induced airway inflammation. Oncotarget 7:80288–80297

Zhang Y, Sanchez D, Gorelik J, Klenerman D, Lab M, Edwards C, Korchev Y (2007) Basolateral P2X4-like receptors regulate the extracellular ATP-stimulated epithelial Na^+ channel activity in renal epithelia. Am J Physiol Renal Physiol 292:F1734–F1740

Chapter 29
The Polycystins and Polycystic Kidney Disease

Bonnie L. Blazer-Yost and Darren P. Wallace

Abstract The polycystins are transmembrane proteins that interact to form a nonselective, Ca^{2+} permeable, cation channel that is thought to be involved in the maintenance of intracellular Ca^{2+} homeostasis and epithelial differentiation of ductal structures. Mutations in *PKD1* or *PKD2,* genes that encode polycystin 1 (PC1) and polycystin 2 (PC2, TRPP2), respectively, are responsible for most cases of autosomal dominant polycystic kidney disease (ADPKD), a common disorder characterized by innumerous fluid-filled cysts in the kidneys and liver. PC1, a large receptor-like protein, and PC2, a transient receptor potential channel, colocalize to multiple cellular domains including primary cilia where they are thought to function in Ca^{2+} signaling in response to mechanosensation. Emerging data suggest that these proteins also function individually or in a complex to regulate other cellular functions in a variety of tissues. However, the mechanism by which the functional loss of the polycystins initiates cyst formation, their tissue-specific binding partners and ligands, and their normal physiological functions remain poorly understood.

Keywords Primary cilia · Renal epithelial cells · Autosomal dominant polycystic kidney disease · Transient receptor potential channels · cAMP · Tolvaptan

---

B. L. Blazer-Yost (✉)
Department of Biology, Indiana University, Purdue University, Indianapolis, IN, USA
e-mail: bblazer@iupui.edu

D. P. Wallace
Departments of Internal Medicine, and Molecular and Integrative Physiology, The Jared Grantham Kidney Institute, University of Kansas Medical Center, Kansas City, KS, USA
e-mail: dwallace@kumc.ed

K. L. Hamilton, D. C. Devor (eds.), *Studies of Epithelial Transporters and Ion Channels*, Physiology in Health and Disease,
https://doi.org/10.1007/978-3-030-55454-5_29

29.1 Discovery of the PKD Genes and the Polycystin Proteins

In 1994 and 1996, the causative genes for autosomal dominant polycystic kidney disease (ADPKD) were identified and termed *PKD1* (16p13.3) and *PKD2* (4q22), respectively (The European Polycystic Kidney Disease Consortium 1994; The American PKD1 Consortium 1995; The International Polycystic Kidney Disease Consortium 1995; Hughes et al. 1995; Mochizuki et al. 1996). The human *PKD1* gene contains 46 exons, spanning a genomic region of approximately 52 kb, and generates a 14-kb mRNA. There is a large genomic repeat encompassing two-thirds of the 5′ portion of the gene. This duplicon contains six highly homologous pseudogenes, sharing up to 99% sequence identity making the sequencing of *PKD1* particularly challenging and direct mutation analysis in the clinical setting more complex.

The human *PKD1* gene is unusual in that it contains two long polypyrimidine tracts in introns 21 and 22 (97% C + T), that form thermodynamically stable segments of triplex DNA (Van Raay et al. 1996; Blaszak et al. 1999). It has been proposed that these polypyrimidine tracts cause the DNA polymerase to stall, leading to frequent mutations in the gene. In addition, the polypyrimidine tracts also cause the RNA polymerase to stall leading to abnormal splicing events that decrease the number of full length *PKD1* gene transcripts. This suggests that humans are natural *PKD1* hypomorphs (Lea et al. 2018). Moreover, truncated proteins, produced from these abnormally spliced RNAs, may interfere with the function of the normal protein. It is possible that these unstable regions in the gene are responsible for renal cyst initiation in ADPKD patients who have a germline mutation in only one of the *PKD1* alleles. By contrast, mouse and nonprimate vertebrates lack the two polypyrimidine tracts and have infrequent renal cysts as *Pkd1* heterozygotes (Piontek and Germino 1998). The human *PKD1* gene is also unique in that intron 45 encodes a micro RNA (miRNA; miR-1225) with an unknown function (Rodova et al. 2003; Havens et al. 2012).

Polycystin-1 (PC1), the gene product of *PKD1*, is a large receptor-like protein with a large N-terminal extracellular region containing multiple domains thought to interact with extracellular matrix proteins and ligands (Hughes et al. 1995).

PKD2 contains at least 15 exons and encodes polycystin-2 (PC2), a transient receptor potential protein (TRPP2) that is thought to function as a Ca^{2+} permeable, nonselective cation channel (Hayashi et al. 1997; Giamarchi et al. 2006; Kottgen 2007). Hanaoka et al. was the first to show that PC1 and PC2 physically interact via their C-terminal domains to form a receptor-channel pair (Hanaoka et al. 2000).

29.2 The Functional Importance of the Polycystins: Autosomal Dominant Polycystic Kidney Disease

ADPKD is one of the most common monogenetic disorders of humans, and is characterized by the development of countless fluid-filled cysts in the kidneys. ADPKD has a greater frequency than cystic fibrosis, Huntington's disease, hemophilia, sickle cell anemia, Down's syndrome, and muscular dystrophy combined (Schrier et al. 2003). It is estimated that 600,000 Americans and as many as 12.5 million people worldwide have ADPKD, affecting both genders and all races, ethnicities, nationalities, and geographic locations equally. Renal cysts form in utero and progressively increase throughout life, leading to massively enlarged kidneys, interstitial fibrosis, and a decline in renal function in midlife (Fig. 29.1). Approximately, one half of the patients progress to end-stage renal disease (ESRD) by 60 years of age, accounting for 4–6% of all patients on renal replacement therapy. ADPKD is a systemic disorder that also affects multiple organs other than the kidney, including the liver, pancreas, and vascular system. Therefore, in addition to the renal complications, patients are at risk for extra-renal complications, especially from cardiovascular issues resulting in additional morbidity and mortality.

Mutations in *PKD1* account for 73–85% of ADPKD cases and lead to a more severe disease than mutations in *PKD2* (15–27% of cases) (Lanktree et al. 2018). The type of mutation is also important (Harris and Hopp 2013). Truncating or other large-scale mutations in *PKD1* (i.e., frameshift, nonsense, splice mutations, and large rearrangements) cause a more aggressive disease with a median age of 55 years for the onset of ESRD (Gall et al. 2013) whereas, most *PKD1* mutations with no truncating effects (i.e., small in-frame mutations and missense events) have on average a 12-year delay in the onset of ESRD (median age, 67 years). The median age of ESRD for *PKD2* patients is 79 years. Although the number of cysts and the rate of disease progression differ for *PKD1* and *PKD2* mutations, the presentation of

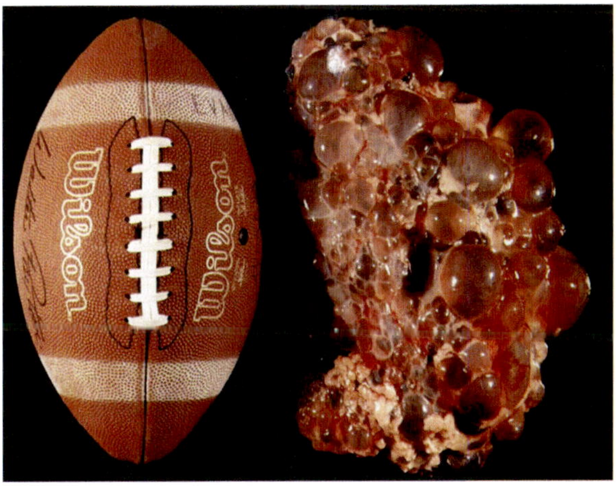

Fig. 29.1 Kidney of an adult male with polycystic kidney disease removed in preparation for renal transplantation with a normal kidney. Photograph courtesy of Dr. Andrew Evan, Department of Anatomy and Cell Biology, Indiana University School of Medicine

renal disease is remarkably similar supporting the hypothesis that PC1 and PC2 function together.

ADPKD is transmitted in a dominant fashion, requiring the inheritance of only one mutated allele. However, loss of the single PKD allele does not appear to be sufficient to induce global cyst formation throughout the entire kidney. In an ADPKD kidney, there are a few thousand detectable cysts, which would involve only a small percentage of the approximately one million nephrons (Grantham et al. 2012). While the mechanism for renal cyst initiation is not fully understood, it is generally thought that insufficient expression of the polycystins at a critical period of renal development or following injury induces a PKD cellular phenotype, leading to cyst formation. This "threshold effect" may involve either a second hit in the other PKD allele causing the loss of polycystin expression in specific cells (Qian et al. 1996) or a germline hypomorphic PKD mutation that reduces the expression of the protein close to a critical threshold level and sometimes below (Lea et al. 2018; Rossetti et al. 2009; Antignac et al. 2015).

Renal cyst growth involves aberrant proliferation of renal epithelial cells, accumulation of fluid in the cyst cavity, and remodeling of the extracellular matrix. Several signaling pathways have been implicated in PKD pathogenesis (Harris and Torres 2014); however, the 3′,5′-cyclic adenosine monophosphate (cAMP) pathway is central for cyst growth by stimulating both cell proliferation and transepithelial fluid secretion (Grantham 1993; Wallace 2011; Torres and Harris 2014). Converging evidence has supported the hypothesis that the polycystins form a nonselective cation channel that regulates intracellular Ca^{2+} (Hanaoka et al. 2000) and that functional loss of the polycystins induces a Ca^{2+}-sensitive phenotypic switch in the mitogenic responses to cAMP thus contributing to cyst formation and growth (Yamaguchi et al. 2004; Yamaguchi et al. 2006). Targeting pathways involved in renal cAMP production, including the arginine vasopressin V2 receptor, inhibit ADPKD cell proliferation and fluid secretion, and reduce the progression of disease (Belibi et al.2004; Reif et al. 2011; Gattone et al. 2003; Torres and Harris 2014). Tolvaptan, a V2 receptor antagonist, has been shown to slow kidney enlargement and the decline in estimated GFR in ADPKD patients (Torres et al. 2012), and has recently been approved by the FDA as the first therapy for the treatment of ADPKD (Torres et al. 2017).

29.3 Structure of the Polycystins

29.3.1 Polycystin 1

Polycystin 1 (PC1) is a large glycoprotein which contains 4302 amino acids and has a molecular weight of approximately 600 kDa when glycosylated (Harris et al. 1995). The N-terminal extracellular portion makes up approximately 75% of the protein, including two leucine-rich repeats, a C-type lectin receptor, LDL-A domain, 16 Ig-like (or PKD) domains, a domain homologous to the sea urchin sperm receptor

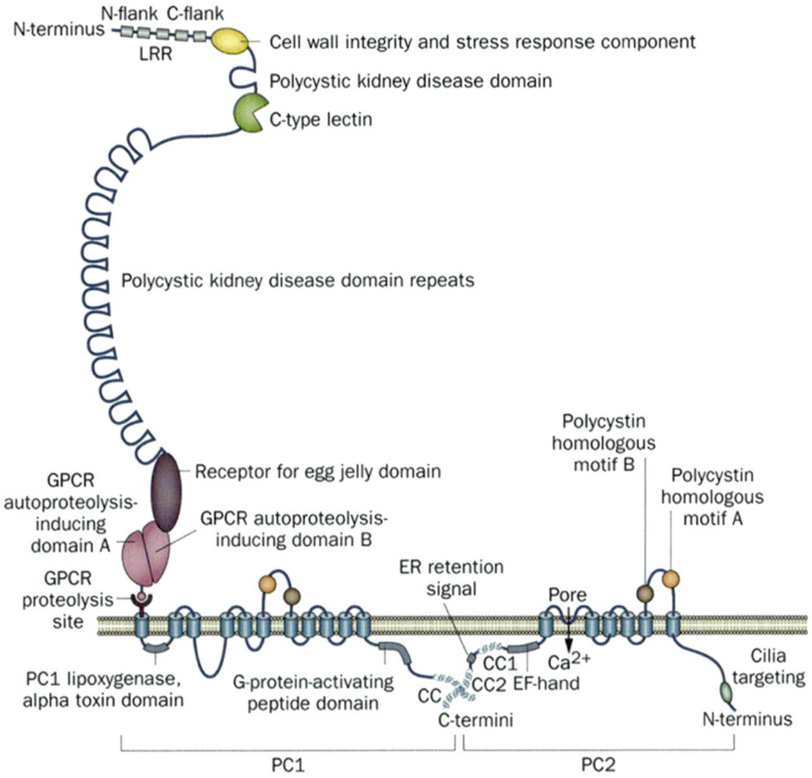

Fig. 29.2 Predicted structures of polycystin 1 (PC1) and polycystin 2 (PC2): PC1 is a receptor-like protein with a large ectodomain, 11 transmembrane domains, and a cytoplasmic tail consisting of 200 amino acids. The last six transmembrane domains of PC1 are homologous to the transmembrane region of PC2. PC2 is a transient receptor potential-like calcium channel that has an EF-hand motif and an endoplasmic reticulum (ER) retention signal in the carboxy (C) terminus and a proposed cilia targeting sequence in the amino (N) terminus. PC1 and PC2 physically interact through coiled-coil domains in the cytoplasmic tail of PC1 and in the carboxy-terminal tail of PC2. Reproduced from Chebib et al. ("Vasopressin and disruption of calcium signaling in polycystic kidney disease" Nat. Rev. Nephrol. 2015; 11:451–464) with permission of Nature Publishing Group. Abbreviations: GPCR, G protein-coupled receptor; LLR, leucine rich repeat

for egg jelly proteins (REJ), and a G protein-coupled receptor proteolytic site (GPS) (Hughes et al. 1995; Sanford et al. 1997; Ibraghimov-Beskrovnaya et al. 1997; Bycroft et al. 1999) (Fig.29.2). PC1 undergoes autoproteolytic cleavage at the conserved GPS located within the GAIN domain, which precedes the first transmembrane helix, a feature of adhesion-type G protein-coupled receptors (Ponting et al. 1999; Qian et al. 2002; Wei et al. 2007; Yu et al. 2007; Arac et al. 2012). The N-terminal fragment (NTF; ~325 kDa) and the C-terminal fragment (CTF; ~150 kDa) of PC1 remain noncovalently tethered to each other. While the role of the GPS cleavage remains unclear, mutations in this region that prevent cleavage lead to rapidly progressive cyst formation in mice, highlighting the importance of this region for proper PC1 function (Yu et al. 2007; Qian et al. 2002). However, mice with homozygous GPS cleavage mutations are born, indicating that this mutation does not cause a detrimental effect on kidney development per se. In contrast, mice with homozygous truncating mutations and other loss of function mutations in PC1 die in utero.

PC1 has 11 transmembrane domains and a short C-terminal cytoplasmic domain which contains a coiled-coil domain involved in protein interactions (Nims et al. 2011; Qian et al. 1997). The membrane proximal region of the cytosolic C-terminal tail of PC1 also contains a highly conserved region of approximately 70 amino acids that is likely to play a central role in PC1 function. This region of the protein has been described as containing a nuclear localization signal for a C-terminal PC1 cleavage product (Chauvet et al. 2004), a calmodulin-binding motif (Doerr et al. 2016), a protein kinase A phosphorylation site (Parnell et al. 1998), a protein phosphatase-1-binding motif (Parnell et al. 2012), and a G-protein binding and activation domain (Parnell et al. 1998).

In *Xenopus*, the G-protein binding domain (GBD) interacts with several classes of Gα subunits. PC1 constructs lacking the G-protein binding domain fail to rescue *Pkd1* zebrafish morphants, and disruption of Gα subunits induces a PKD phenotype (Zhang et al. 2018). In mice, the G-protein binding domain has also been shown to bind and sequester Gα12 heterotrimeric subunits, and genetic deletion of *Gna12* blocks cystogenesis (Yu et al. 2011; Wu et al. 2016). Numerous ADPKD-associated patient mutations affect residues in and around the G-protein activation domain (Fig. 29.2). A single amino acid deletion (L4132 Δ; ΔL) located immediately N-terminal to the activation domain disrupts G-protein signaling in vitro, disrupts PC1/PC2-dependent Ca^{2+}-channel activity in cells, and causes a severe PKD phenotype in mice suggesting that activation of G-protein signaling is an essential function of PC1 (Parnell et al. 2018). Collectively, these results demonstrate a key role for PC1 in the regulation of G-protein signaling, and further suggest that the membrane proximal region of the PC1 C-terminal tail may be a nucleation site for numerous signaling molecules critical for PC1 function.

29.3.2 Polycystin 2

Polycystin 2 (PC2; TRPP2) is a 968-amino acid polypeptide of approximately 110 kDa (Mochizuki et al. 1996; Hanaoka et al. 2000; Newby et al. 2002). The protein has six membrane spanning domains and both N- and C-termini are cytosolic (Mochizuki et al. 1996) (Fig.29.2). The C-terminus contains a coiled-coil domain, an ER retention signal, and EF-hand motif (Mochizuki et al. 1996; Tsiokas et al. 1999; Anyatonwu and Ehrlich 2004), and the N-terminus has a ciliary sorting motif. PC2 is a nonselective, Ca^{2+}-permeable cation channel that can be regulated by divalent cations, voltage, pH, phosphorylation, and interactions with other channels (Gonzalez-Perrett et al. 2001; Gonzalez-Perrett et al. 2002; Ramsey et al. 2006).

The structure of PC2 has been solved by electron cryomicroscopy at atomic resolution indicating a homotetrameric structure that is similar to other TRP channels (Grieben et al. 2017; Hulse et al. 2018). The structure of the PC2 channel at a 4.2 Å resolution showed a confirmation in which the external Ca^{2+} decreases monovalent cation currents by binding to an aspartate at the top of the pore filter (Grieben et al. 2017). The PC2 channel contains a tetragonal opening with "TOP domains" (also called "polycystin domains"), which forms from extensions of the voltage-sensor-like domain (VSLD). Extensive interactions among the TOP domain and the channel surface are important for the stability of the PC2 structure and may play a role in assembly into a tetrameric ring that covers the luminal surface of the VSLD and the pore.

PC1 also has a channel-like region homologous to that of PC2 (Mochizuki et al. 1996) and a TOP domain (Grieben et al. 2017), suggesting that the functional channel may consist of a 1:3 (PC1:PC2) heterotetramer (Yu et al. 2009; Zhu et al. 2011). More recently, Yu and associates generated a constitutively active gain-of-function mutant of PC2 (Pavel et al. 2016), enabling functional studies to prove that PC1 is a channel subunit and that PC1/PC2 transports Ca^{2+}. These top domains in the polycystin proteins are hot spots for pathogenic mutations for ADPKD, emphasizing the functional importance of this region.

While the PC1 family members have little overall structural similarity to transient receptor potential (TRP) channels, isolated domains have similarity to channel-forming domains of PC2, suggesting that these domains may participate in channel formation in specific cellular domains or under specific conditions. A potential breakthrough in the understanding of the importance of PC1-PC2 interactions occurred with the 3.2 Å resolution of a PC1-PC2 complex (Su et al. 2018). These investigators found a 1:3 (PC1:PC2) complex with PC1 forming an integral part of the pore. Among 11 transmembrane domains of PC1, six domains substitute as the fourth subunit of the heterotetramer. The PC1:PC2 complex likely has a different ion selectivity and gating structure than PC2 homotetrameric structures.

29.3.3 Other Polycystin Isoforms

Other members of the family of polycystin-related proteins include four PC1 homologues, PKD1L1, PKD1L2, and PKD1L3 (Yuasa et al. 2002, Yuasa et al. 2004; Li et al. 2003a), and PKD-REJ (receptor for egg jelly) (Hughes et al. 1999) and two analogues of PC2 named PKD2L1 and PKD2L2 (Nomura et al. 1998; Veldhuisen et al. 1999; Guo et al. 2000). Some of these additional members of the polycystin family, particularly the PC2-like proteins, have been shown to be cation channels. Because of the similarity in structure and function, the entire family of PC2 like proteins has been characterized as a branch of the TRP channels. PC2 is also known as TRPP2 and is the founding member of the TRPP subcategory; the other two members of this family are TRPP3 and TRPP5 (Delmas 2005; Giamarchi et al. 2006; Nilius and Owsianik 2011). PC2 (TRPP2) and PKD2L1 (TRPP3) localize to primary cilia. The homomeric form of TRPP3 was shown to be a calcium-modulated nonselective cation channel permeable to Na^+, K^+, and Ca^{2+}, suggesting that it may act as a transducer of Ca^{2+}-mediated signaling (Chen et al. 1999).

In addition to its expression in the cilium, TRPP3 has been localized to the centrosome, indicating a more direct involvement in cell proliferation (Bui-Xuan et al. 2006). The relevance of PC1-like and PC2-like proteins remains unclear since mutations in these genes have not been linked to disease in humans.

29.4 The Polycystin Channels

29.4.1 The Classic PC1-PC2 Channel

Deciphering the function of the polycystin channel complex is a major topic of research in the PKD field. There is strong historical evidence that PC1 and PC2 form a mechanosensitive cation channel in the plasma membrane and primary cilia (Somlo and Ehrlich 2001; Hanaoka et al. 2000; Yoder et al. 2002; Nauli et al. 2003). Mechanical bending of cilia and fluid flow over cilia have been shown to cause Ca^{2+} entry that is thought to communicate to signaling pathways within the cells (Praetorius and Spring 2001). Nauli et al. (2003) were the first to demonstrate that PC1 null renal epithelial cells failed to elicit a Ca^{2+} response to fluid flow and that PC2 blocking antibodies abolished the Ca^{2+} response to flow in normal cells. Direct electrophysiological evidence provided by Cantiello's group showed the presence of single channel current in cilia that was inhibited by an antibody to PC2 (Raychowdhury et al. 2005). These studies led to the hypothesis that PC1 is the mechanosensor and PC2 is the cation channel. Other studies suggested that the increased Ca^{2+} influx triggers the release of additional Ca^{2+} from IP3-dependent stores, a process that is amplified by nucleotide release (Praetorius and Leipziger 2009).

Using a unique single cell imaging technique with cells grown on microwires, Jin et al. (2014) demonstrated that fluid-shear stress can initiate Ca^{2+} signaling in the primary cilium via a PC2-dependent mechanism. These authors suggested that the Ca^{2+} signaling can spread to the cytoplasm or can be contained within the cilium depending on the nature of the stimulus. The fluid shear-induced increase in ciliary Ca^{2+} is amplified in the cytoplasm via the ryanodine receptor, whereas chemical stimuli do not result in a change in cytosolic Ca^{2+} (Jin et al. 2014). Consistent with role of PC1-PC2 regulation of intracellular Ca^{2+}, steady-state Ca^{2+} levels in cystic epithelial cells were found to be ~20 nM lower than cells cultured from noncystic tissues of human ADPKD kidneys or normal human kidneys (Yamaguchi et al. 2006).

29.4.2 *Alternate Polycystin Channels and Isoforms*

Recently, the idea that flow-induced ciliary Ca^{2+} signaling regulates cytosolic Ca^{2+} and cell signaling events has been challenged. Measurements of Ca^{2+} levels in cilia by using specialized cilia-restricted fluorescence indicators indicated that cilia-specific Ca^{2+} influx was not observed in physiological or even supraphysiological levels of fluid flow (Delling et al. 2016). Moreover, these investigators found that a heteromeric Ca^{2+} channel formed by PKD1L1 and PKD2L1 mediated an increase in ciliary Ca^{2+} (Delling et al. 2013). While the authors confirmed transcripts for PC1 and PC2 were present, only siRNA for PKD1L1 and PKD2L1 reduced the Ca^{2+} current. The channels were activated by external nucleotides but were not mechanosensitive (DeCaen et al. 2013). In contrast to previous studies, the Ca^{2+} wave within the cilium did not change the cytoplasmic Ca^{2+} levels. In addition, Ca^{2+} levels in cilia were maintained at a higher level than the cytoplasmic concentrations, indicating a separation between the two compartments and suggesting that the cilium is functionally distinct from the cell regarding ionic composition (Delling et al. 2013). The authors reasoned that the volume of the cilium is too small relative to the cell for enough Ca^{2+} ions to alter the cytosolic concentration (Delling et al.2013).

Polycystin channel conductance has become particularly controversial. The properties of the PC2 channel vary widely depending on experimental conditions (Kleene and Kleene 2017). Innovations in patch clamp methods and the ability to patch the cilia directly have led two groups to conclude that the PC2 channel preferentially conducts K^+ over Ca^{2+} (Kleene and Kleene 2017; Liu et al. 2018). The initial studies described a large conductance channel with permeability ratios of P_K:P_{Ca}:P_{Na} of 1:0.55:0.14 in a mouse inner medullary collecting duct cell line (mIMCD-3). The open probability was increased by membrane depolarization or increases in cytoplasmic Ca^{2+} (Kleene and Kleene 2017). A subsequent study in primary IMCD cells indicated somewhat different permeability ratios of P_K:P_{Ca}:P_{Na} of 1:0.4:0.025. These investigators also found that intraciliary Ca^{2+} enhances the open probability (Liu et al. 2018). Interestingly, the latter paper found that PC1 was not necessary for channel activity or trafficking to the membrane. The difference in permeability ratios between the two studies may be due to methodological differences, including where on the cilia, the patch pipette was placed. However, there is no discrepancy in the finding that the permeability of K^+ was higher than Ca^{2+}. Considering the extremely large gradient for Ca^{2+} across the membrane, it is likely that this channel plays a role in Ca^{2+} flux despite the lower permeability compared to K^+.

29.4.3 *Other Heteromeric Polycystin Channels*

The polycystins can form heteromultimers not only with each other but also with other members of TRP family of proteins, particularly those of the TRPC family involved in G-protein coupled receptor activation and the TRPV family. PC2 interacts with TRPC1 in transfected cells and in vitro via a C-terminal sequence as well as a domain located within the transmembrane segments S2 and S5 (Tsiokas et al. 1999). Heteromeric PC2/TRPC1 channels are found in the primary cilia where they exhibit single channel conductance, amiloride sensitivity, and ion permeability that are distinct from that shown by either homomeric channel (Bai et al. 2008). PC2 and TRPC1 have also been implicated in stretch-induced injury of the blood-brain barrier (Berrout et al. 2012). Similar to PC2 homomers, PC2-TRPC1 heteromers appear to be structural tetramers (Zhang et al. 2009).

TRPV4 plays a role in vertebrate mechanosensation and hypo-osmotic stress (Alessandri-Haber et al. 2003; Nilius et al. 2004; Wu et al. 2007). In Madin-Darby canine kidney (MDCK) cells, TRPV4/PC2 heteromeric channels are localized to the primary cilia where they form mechano-and thermo-sensitive channels (Kottgen et al. 2008). A PC2/TRPV4 heterotetramer has been found in the plasma membrane of renal collecting duct cells (Kottgen et al. 2008; Zhang et al. 2013; Berrout et al. 2018). Inhibitors as well as knock-down experiments were used to show that TRPV4 is an essential component of the ciliary flow sensor. Atomic force microscopy has indicated that both the TRPC1/PC2 and TRPV4/PC2 channels have a 2:2 stoichiometry with an alternating subunit arrangement (Kobori et al. 2009; Stewart et al. 2010). Prolonged stimulation with a TRPV4 agonist restored Ca^{2+} signaling in the cystic cells and reduced cyst growth in a polycystic rat model (Zaika et al. 2013). Unlike PC2 homotetramers, the combination channel is mechanosensitive (Kottgen et al. 2008).

Recently, it was demonstrated that TRPV4 was necessary for flow-dependent intracellular Ca^{2+} responses in primary cultures of normal human kidney epithelial cells. This response was diminished in primary ADPKD cyst epithelial cells where the authors noted a deficiency in TRPV4 glycosylation (Tomilin et al. 2018). These results substantiate previous studies showing similar results in rodent cells (Berrout et al.2012; Zaika et al. 2013; Mamenko et al. 2015; Tomilin et al.2018). While TRPV4 is also a nonselective cation channel, it exhibits a moderate Ca^{2+} selectivity (P_{Ca}/P_{Na} of 6–10) (Voets et al. 2002; Tomilin et al. 2018); therefore, when activated, it would result in substantial Ca^{2+} influx across the plasma membrane.

In summary, it remains unclear if and how the cilia mediate flow-dependent Ca^{2+} signaling in renal epithelial cells, if other polycystin family members are involved, and how defects in cilia induce PKD pathogenesis.

29.4.4 Other Potential Functions of the Polycystins

PC1 and PC2 localize to sites of cell-matrix interactions, cell-cell junctions, cell-substrate interactions, and the endoplasmic reticulum (ER). It is intriguing to speculate that the location and binding partners of the polycystins may be involved in force transmission and in sensing changes in mechanical stress on cells. How this type of signaling may contribute to cystogenesis remains speculative. While many functions of the polycystins via the cytoskeleton have been proposed, there is limited experimental evidence to support these disease-related hypotheses, particularly in vivo. As noted above, PC2 is thought to be a cation channel, but it may have additional functions. In proximal tubule cells, PC2 regulates the activity of Piezo-1-dependent stretch-activated channels that are important in the response to luminal pressure (Peyronnet et al. 2013). PC2 has also been shown to be an epidermal growth factor-activated channel on the plasma membrane of LLC-PK1 kidney cells that overexpress PC2 (Ma et al. 2005). Moreover, PC1 has been implicated in the cell cycle and growth and resistance to apoptosis (Boletta et al. 2000; Bhunia et al. 2002). Expression of PC1 in MDCK cells, which normally have low endogenous PC1, induced spontaneous tubulogenesis (Boletta et al. 2000). Thus, unlike many other pathological processes, localization of the polycystins does not explain their role in normal function or in disease progression and can only offer intriguing and often confusing possibilities.

29.5 PC1 and PC2 Localization and Binding Partners

29.5.1 Historical Studies

Analysis of the functional role of the polycystins has been complicated by their dynamic subcellular localization, which is dependent on the cell type, state of cellular differentiation, and cell-to-cell interactions. In early studies, the two proteins appeared to be expressed in very different compartments. PC2 was found almost exclusively in the ER while PC1 was predominantly on the plasma membrane and cilia. Moreover, low expression of endogenous PC1 made it difficult to characterize the subcellular localization of the native PC1-PC2 complex. As highly specific antibodies for the polycystin proteins became available, PC1 and PC2 were shown to interact to form a complex that localized to the lateral plasma membrane and primary cilia of renal cells (Newby et al. 2002). Subsequently, it was shown that the polycystins are expressed in several other subcellular locations, including centrosomes, exosome-like vesicles, and the ER. Recently, proteolytic products of PC1 were shown to translocate to the nucleus and the mitochondrial matrix (Lal et al. 2008; Padovano et al. 2017; Lin et al. 2018).

29.5.2 Polycystin 1 and 2 Expression in Primary Cilia

A substantial amount of research has been focused on the role of polycystins in the primary cilium, a microtubule structure that extends from the surface of most mammalian cells. PC1 and PC2 were first envisioned to interact by multiple domains, including the C-terminal coiled-coil domains of each protein (Qian et al. 1997), to form a flow-sensitive Ca^{2+} channel within the cilia. However, as noted above, more recent data suggest a heterotetrameric channel structure with both PC1 and PC2 participating as subunits (Yu et al. 2009). PC2 also colocalizes with fibrocystin, the protein mutated in autosomal recessive PKD (ARPKD) (Qian et al. 1997; Wu et al. 2006; Kim et al. 2008), suggesting that the polycystins and fibrocystin are components of the same signaling pathway.

There is growing evidence that primary cilia contain signaling molecules that orchestrate diverse cellular functions involved in tissue development and homeostasis (Choi et al. 2011; Ma et al. 2017; Viau et al. 2018). Mutations in genes that encode ciliary proteins are responsible for a group of diseases that are collectively referred to as ciliopathies. These include ADPKD, ARPKD, Bardet-Biedl syndrome (BBS), Meckel-Gruber syndrome, Jeune syndrome, and Joubert syndrome (Quinlan et al. 2008; Harris and Torres 2009). Ciliopathies present with a range of disease severity; however, all of these disorders are characterized by the presence of renal cystic disease.

Within the primary cilia, PC1 and PC2 have been historically characterized as critical binding partners with functional significance. Early studies confirmed that mutations that disrupt the PC1-PC2 interaction are associated with ADPKD pathogenesis, stressing the biological relevance of the complex. When overexpressed in Chinese hamster ovary cells, pathogenic *PKD1* mutation R4227X, which truncates the terminal 76 amino acids that contain the coiled-coil domain, reduced the capacity of PC1 to interact with PC2 and abolished the Ca^{2+} current generated by wild-type PC1 and PC2 (Hanaoka and Guggino 2000).

Interestingly, PC2 was found to be required for proper processing and trafficking of PC1 to cilia (Gainullin et al. 2015). The R742X mutation, which deletes the last 227 C-terminal amino-acids in PC2, prevented its interaction with PC1 and caused a loss of Ca^{2+} current (Hanaoka et al. 2000). Transport of PC1 into primary cilia, but not the plasma membrane per se, also requires binding to proteins of the BBSome, a complex named for Bardet-Biedl syndrome, a severe ciliopathy. Seven BBS proteins make up the BBSome, which localizes to the cilium and basal bodies and appears to gate protein trafficking into the cilium. PC1 binds BBS1, BBS4, BBS5, and BBS8; however, only the mutation or deletion of BBS1 impaired ciliary trafficking of PC1 in renal cells (Su et al. 2014).

Trafficking of PC2 to cilia does not appear to be dependent on PC1 since truncation of the C-terminus of PC1 did not prevent PC2 delivery to cilia (Geng et al. 2006). However, the N-terminal sequence motif R6V7xP8 of PC2 was found to be required for its ciliary localization (Geng et al. 2006). This localization motif was also enough to target other heterologous proteins to cilia.

PKD1L1 and PKD2L1 have been identified in primary cilia where they form a Ca^{2+} channel analogous to the channel formed by PC1-PC2 (DeCaen et al. 2013; Delling et al. 2013). PC2 also interacts with PKD1L1 and this interaction has been suggested to be important in establishing left-right asymmetry during development (Vogel et al. 2010; Field et al. 2011).

29.5.3 *PC1 in the Plasma Membrane and Cell-to-Cell Junctions*

PC1 was found to be expressed on the plasma membrane of renal epithelial cells, and to localize to lateral membranes at sites of cell-to-cell contact (Ibraghimov-Beskrovnaya et al. 1997; Bukanov et al. 2002). Overexpression of full-length PC1 in Madin-Darby canine kidney (MDCK) cells led to cell-cell junction localization (Boletta et al. 2001). Antibodies raised against the Ig-like (PKD) domains, a region of PC1 thought to mediate protein-protein interactions, disrupted cell-to-cell interactions. These PKD domains have strong homophilic interactions in vitro, suggesting that they are involved in cell-to-cell adhesion (Ibraghimov-Beskrovnaya et al. 2000).

PC1 was later found within desmosomes and localized with desmoplakins I and II (Scheffers et al. 2000), also consistent with a role in cell-to-cell contact (Scheffers et al. 2000; Streets et al. 2003). In ADPKD cells, the loss of colocalization of desmosomal proteins and PC1, led to a punctate staining of the desmosomal proteins within the cytoplasm (Russo et al. 2005). It has been suggested that the mislocalization of desmosomal proteins in cyst-lining epithelial cells makes these structures fragile and more sensitive to sheer stress (Silberberg et al. 2005).

The cytoplasmic tail of PC1 also binds the intermediate filament proteins, vimentin, cytokeratins 8 and 18, and desmin (Xu et al. 2001). Likewise, PC2 connects to the actin cytoskeleton by binding Hax-1 (Gallagher et al. 2000), tropomyosin-1 (Li et al. 2003b), filamin A (Sharif-Naeini et al. 2009), α-actinin (Li et al. 2005), and mDia1 (Rundle et al. 2004). The junctional complex connection between the actin cytoskeleton and the PC proteins supports the hypothesis that the PC1-PC2 complex is important for sensing pressure and flow, cell-cell adhesion, and cell morphogenesis and proliferation.

Yao et al. (2014) performed a yeast two-hybrid screen using PC1 as bait, and identified an interaction with protein kinase C and casein kinase substrate in neurons protein 2 (*PACSIN2*), a ubiquitously expressed member of a family of proteins that localize to sites of high actin turnover (Yao et al. 2014). The PC1 and the Pacsin 2 complex were also shown to contain neural Wiskott-Aldrich syndrome protein (N-Wasp) and this complex was necessary for N-Wasp interaction with Arp3, an actin filament nucleation protein. This study suggested that PC1 modulates cytoskeleton rearrangement and migration through the Pacsin2/N-Wasp/Arp2/3 complex and contributes to establishment and maintenance of the tubule architecture (Yao et al. 2014).

Adherens junctions connect epithelial cells at apical junctions and are important for the maintenance of cellular polarity, structure, and function. Adherens junctions are linked to the cortical actin network, connecting these structures to the cellular interior. PC1 was shown to colocalize with proteins involved in cell-cell adherens junctions (Huan and van Adelsberg 1999; Wilson et al. 1999; Geng et al. 2000) and is part of a complex containing the junctional proteins, E-cadherin, and α-, β-, and γ-catenins. In addition, PC1 binds annexin A5, a Ca^{2+}- and lipid-binding protein that is important for the initial recruitment of E-cadherin to the membrane (Markoff et al. 2007; Streets et al. 2009).

29.5.4 PC1 Interactions with the Extracellular Matrix (ECM)

The large external N-terminal domain of PC1 binds to ECM proteins including laminin, fibronectin, and collagens I, II, and IV (Weston et al. 2001; Malhas et al. 2002). Binding to extracellular substrates not only serves a structural role but also leads to intracellular changes and recruitment of both signaling intermediates and structural proteins. In aortic vascular smooth muscle cells, both PC1 and PC2 were found in dense plaques that are involved in cell-to-ECM interactions (Qian et al. 2003b). In human fetal collecting duct cells, a variety of techniques showed that PC1 associates with focal adhesion proteins, talin, vinculin, p130Cas, focal adhesion kinase pp125FAK, α-actinin, paxillin, and pp60c-src in subconfluent cultures (Wilson et al. 1999; Geng et al. 2000). The binding complex changed to include the cell-cell adherens junction proteins, E-cadherin, and β- and γ-catenins when the cells attained confluency. Some components of the multiprotein complex were altered by tyrosine phosphorylation and Ca^{2+} concentration (Geng et al. 2000). PC2 binds to CD2AP, an adapter protein that functions in the assembly of focal adhesion complexes (Lehtonen et al. 2000) via the C-terminal regions of the proteins, indicating that this interaction occurs within the intracellular compartment of the cell.

29.5.5 Polycystins in ER and Ca^{2+} Release Channels

PC2 is abundantly expressed in the ER (Cai et al. 1999), where it is thought to function as a Ca^{2+}-activated intracellular Ca^{2+} release channel and/or to modulate Ca^{2+} efflux through other ER-resident channels (Cai et al. 1999; Koulen et al. 2002; Li et al. 2005). PC2 interacts with the inositol 1,4,5-trisphosphate (IP3) receptor and the ryanodine receptor, two known ER Ca^{2+} channels (Li et al. 2005; Anyatonwu et al. 2007; Santoso et al. 2011; Mekahli et al. 2012). When stably expressed in LLC-PK1 kidney cells, PC2 behaved as an intracellular Ca^{2+} release channel augmenting IP3 receptor-mediated Ca^{2+} release from the ER (Koulen et al. 2002). In the oocyte expression system, exogenous PC2 and the IP3 receptor also interact and modulate intracellular Ca^{2+} signaling (Li et al. 2005). In a more detailed study, PC1 and PC2 were both shown to interact with the IP3 receptor. PC2- stimulated IP3-dependent increases in intracellular Ca^{2+}, whereas PC1 inhibited the process (Li et al. 2005). PC1 has been shown to bind stromal interaction molecule 1 (STIM1) which helps maintain its position in the ER membrane while inhibiting store-operated Ca^{2+} entry (SOCE) (Woodward et al. 2010). Based on inhibitor studies using MDCK cells with PC1 overexpression, it has been proposed that the PI3K/Akt pathway is at least partially responsible for modulating the competing effects of a PC1/PC2/STIM1/IP3 receptor complex (Santoso et al. 2011). In contrast, Booij et al. showed that inhibition of several targets, including aurora kinase, CDK, Chk, IGF-1R, Syk, and mTOR, but not PI3K, prevented forskolin-induced cyst expansion of *Pkd1*$^{-/-}$ kidney cells (Booij et al. 2017).

PC1 can stimulate the calcineurin/NFAT (nuclear factor of activated T-cells) signaling pathway through Gαq-mediated activation of phospholipase C (PLC). Inhibitors of PLC (U73122) and IP3 and ryanodine receptors (xestospongin and 2-aminophehylborate) diminished PC1-mediated NFAT activity. This indicates that PC1 signaling leads to sustained elevation of intracellular Ca^{2+} mediated by PC1 activation of Gαq, leading to increased activity of PLC, Ca^{2+} release from intracellular stores, and subsequently activation of calcineurin and NFAT (Puri et al. 2004.)

An ER-localized PC1/PC2 complex was confirmed in conditionally immortalized human proximal tubule epithelial cells where one or both of the polycystins were knocked down using miRNA (Mekahli et al. 2012). This study found that both polycystins were necessary for IP3-induced Ca^{2+} release under fixed Ca^{2+} concentrations, contradicting previous studies. Furthermore, neither polycystin by itself could modulate intracellular Ca^{2+} concentrations (Mekahli et al. 2012). PC2 has also been shown to interact with the ryanodine receptor of the sarcoplasmic reticulum and inhibit its Ca^{2+} channel activity in cardiac myocytes (Anyatonwu et al. 2007). Interestingly, an increased PC2-mediated Ca^{2+} influx was blocked by inhibitors of the ryanodine receptor suggesting a close interaction with influx and intracellular release. However, the flow-induced Ca^{2+} flux was not blocked by inhibitors of G-proteins, phospholipase C, or the IP3 receptor (Nauli et al. 2003).

In addition, a PC1 cleavage product may play a functional role in the ER membrane. A 100 kDa fragment, thought to be derived from cleavage in the third intracellular loop, contains the final six transmembrane domains and the C terminus (Li et al. 2009; Woodward et al. 2010). This protein fragment named P100 interacts with STIM1 (Hogan and Rao 2007). Under conditions of Ca^{2+}depletion, STIM1 interacts with and activates a plasma membrane protein, Orai, a subunit of the store-operated Ca^{2+}channels, thus increasing Ca^{2+} entry through store-operated Ca^{2+} release-activated Ca^{2+} (CRAC) channels (Prakriya et al. 2006). Binding of P100 to STIM1 causes an inhibition of the store-operated channels (Woodward et al. 2010). This C-terminal fragment also associates with the IP3 receptor to inhibit Ca^{2+} release for ER stores after IP3 stimulation (Li et al. 2009). Taken together, the results suggest that PC1 provides a brake for transient Ca^{2+} responses.

Recent evidence has demonstrated that PC2 trafficking through the ER and Golgi to the plasma membrane is controlled by phosphorylation and its interactions with specific proteins. PC1, glycogen synthase kinase-3 (GSK3) and the Golgi- and ER-associated protein-14 (PIGEA-14) have been shown to be important for PC2 transport in the secretory pathway (Hanaoka et al. 2000; Hidaka et al. 2004; Geng et al. 2006; Streets et al. 2006). By contrast, the phosphofurin acidic cluster proteins (PACS-1 and PASC-2) mediate PC2 phosphorylation, leading to its retention or retrieval in the ER and Golgi (Kottgen et al. 2005).

Taken together, these studies suggest a complex regulation of ER Ca^{2+} release with PC1 and PC2 playing opposing roles and other regulatory factors controlling the relative contribution of each of the PCs on a dynamic basis.

29.5.6 Other Potential Binding Partners

The studies demonstrate that PC1 and PC2 have numerous binding partners that may determine their function within specific cellular domains. Two global proteomic analyses have identified other potential interacting proteins. As many as 79 PC2-associated proteins were found in arterial myocytes (Sharif-Naeini et al. 2009) and a review by Retailleau and Duprat (2014) detailed 37 PC1 partners and 32 PC2 partners, most of which are characterized by the PC interacting domains, localization of the interaction, and proposed functional role.

Thus, it appears that both polycystins likely have tissue-specific binding partners that govern their functional activities. Additional research will be required to validate the role of these potential binding partners on polycystin trafficking, localization, and function as well as the exact nature of the channel subunits in each subcellular compartment.

29.6 Mechanisms Underlying Renal Cystogenesis: Further Clues to Polycystin Function

29.6.1 Intracellular Signaling Pathways

While the exact mechanism is a matter of debate, dysregulation of Ca^{2+} metabolism appears to be a proximal event for inducing a PKD cellular phenotype. Several studies have indicated that the changes in PC1/PC2 and Ca^{2+} homeostasis alter the level of intracellular cAMP and the proliferative response to cAMP. Regulation of cytoplasmic cAMP and the duration of a cAMP stimulatory response within specific cellular compartments are controlled by the expression and activity of adenylyl cyclases (ACs), enzymes that catalyze the formation of cAMP from ATP, and phosphodiesterases (PDEs), which degrade cAMP to AMP. Grantham and associates were the first to demonstrate increased cAMP levels in PKD kidneys (Yamaguchi et al. 1997) and this was later substantiated by other laboratories (Gattone et al. 2003; Smith et al. 2006; Starremans et al. 2008).

Arginine vasopressin (AVP), an antidiuretic hormone, binds the AVP V2 receptors (V2R) and increases intracellular cAMP levels in the distal nephron and collecting duct, major sites for cyst formation (Verani and Silva 1988; Wu et al. 1998; Thomson et al. 2003). A potential explanation for elevated renal cAMP in ADPKD is hyperactivation of the V2R in collecting duct-derived cystic cells.

Elevated expression of V2R in PKD kidneys (Gattone et al. 2003; Torres et al. 2004; Nagao et al. 2006) suggests that cystic cells, in situ, may be more responsive to AVP than normal collecting duct cells. Moreover, higher circulating AVP levels observed in ADPKD patients may be due to defects in the concentrating ability of the cystic kidneys (Danielsen et al. 1986; Michalski and Grzeszczak 1996).

Pinto et al. (2012) reported that steady-state cAMP levels were elevated in cultured ADPKD cells compared to normal human kidney cells, demonstrating that there are intrinsic differences in cAMP regulation. Torres proposed that reduced intracellular Ca^{2+}, secondary to mutations in the PKD genes, increases the activity of Ca^{2+}-inhibitable AC6, the primary adenylyl cyclase (AC) involved in V2R-mediated cAMP production, and reduces the activity of the Ca^{2+}/calmodulin-dependent PDE1, a key phosphodiesterase (Wang et al. 2010; Harris and Torres 2014). The combination of increased synthesis and decreased degradation of cAMP raises basal cAMP concentrations closer to the threshold for PKA activation, making cystic cells more sensitive to V2R stimulation.

Using single and double targeted collecting duct specific knockouts of PC1 and AC6, Rees et al. (2014) showed that ablation of AC6 mitigated cytogenesis and renal failure caused by the loss of PC1. Correspondingly, knockout of PDE1, a phosphodiesterase that appears to regulate the pool of cAMP important for controlling cell proliferation (Pinto et al. 2016), accelerated cystic disease in the $Pkd2^{WS25/-}$ mice (Ye et al. 2016). A kidney-specific knock-out of the main regulatory subunit of the cAMP activated PKA induced renal cystic disease (Ye et al. 2017). The complex intracellular signaling network in response to these changes has been the subject of multiple reviews (Harris and Torres 2014; Mangolini et al. 2016; Saigusa and Bell 2015; Chebib and Torres 2016; van Gastel and Torres 2017) and is illustrated in Fig.29.3. The changes in intracellular cAMP ultimately promote both cell proliferation and fluid secretion. These studies provided the basis for targeting renal cAMP by the V2R antagonist tolvaptan (Gattone et al. 1999, 2003; Reif et al. 2011; Torres et al. 2012, 2017) and suppressing plasma vasopressin by increased water intake (Nagao et al. 2006; Wang et al. 2013; Torres et al. 2009; Wong et al. 2018).

29.6.2 Cyst-Filling Fluid Secretion

Transepithelial fluid secretion by the cyst-lining cells is responsible for the accumulation of fluid within the cysts and the massive enlargement of ADPKD kidneys. Grantham and associates demonstrated that cyst-filling fluid secretion was driven by transcellular Cl^- secretion and stimulated by cAMP agonists including AVP (Grantham et al. 1989, 1995; Ye and Grantham 1993; Davidow et al. 1996; Mangoo-Karim et al. 1995; Wallace et al. 1996). Other groups confirmed the central role of Cl^--dependent fluid secretion in renal cyst growth (Hanaoka et al. 1996; Magenheimer et al. 2006; Reif et al. 2011; Terryn et al. 2011).

In ADPKD cells, Cl^- enters via a basolateral Na^+, K^+, $2Cl^-$ cotransporter, driven by the Na^+ gradient across the membrane (Fig. 29.3) (Wallace 2011). The movement

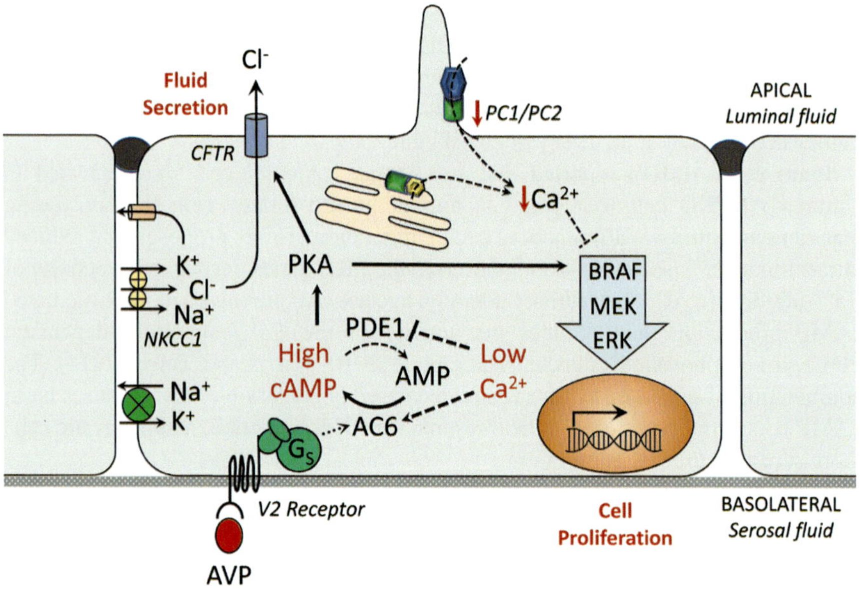

Fig. 29.3 Signaling pathways involved in cAMP-dependent cell proliferation and fluid secretion in ADPKD. Basal cAMP levels and the amplitude of the response to cAMP agonists are controlled by its synthesis by adenylyl cyclases (ACs), and its degradation to AMP by phosphodiesterases (PDEs). In normal renal cells, BRAF, a kinase involved in the activation of the MEK/ERK pathway, is repressed by the phosphorylation of an inhibitory site on BRAF by a Ca^{2+}-dependent mechanism. Increased cAMP due to arginine vasopressin (AVP) binding to the V2 receptor does not stimulate the proliferation of normal collecting duct cells. In ADPKD cystic cells, disruption of the PC1/PC2 complex reduces intracellular Ca^{2+} levels, leading to derepression of BRAF, and cAMP activation of the BRAF/MEK/ERK signaling pathway and cell proliferation. In addition, reduced Ca^{2+} levels in ADPKD cells increases the activity of AC6, a Ca^{2+}-inhibited AC, and decreases the activity of PDE1, a Ca^{2+}/calmodulin PDE isoform, causing elevated cAMP levels. Therefore, in ADPKD, reduced intracellular Ca^{2+} causes a phenotypic switch, such that cAMP is mitogenic, and amplifies the cAMP effect of AVP on ERK-mediated cell proliferation and CFTR-mediated Cl secretion, key components of cyst growth

of these ions into the cell raises Cl^- above its electrochemical gradient. The Na^+/K^+-ATPase pumps Na^+ out of the cell and K^+ leaves via basolateral K^+ channels. Factors that elevate intracellular cAMP, including cell-permeable cAMP, direct activators of adenylyl cyclase such as forskolin, phosphodiesterase inhibitors, and agonists for G_S-protein-coupled receptors, activate apical Cl^- channels allowing Cl^- efflux across the luminal surface of cell. The lumen-negative transepithelial potential favors the passive movement of Na^+ into the cyst via the paracellular pathway. Net accumulation of solute drives the osmotic accumulation of fluid into the cyst cavity. A similar mechanism was also shown to be responsible for hepatic cyst formation. Electrophysiological studies showed that cAMP promoted anion secretion by the cystic biliary epithelium obtained from freshly isolated liver cysts of a PKD mouse (Muchatuta et al. 2009).

The cystic fibrosis transmembrane conductance regulator (CFTR) is the major Cl^- channel responsible for anion secretion in both kidney and biliary cysts (Grantham et al. 1995; Davidow et al. 1996; Hanaoka et al. 1996; Magenheimer et al. 2006; Muchatuta et al. 2009; Reif et al. 2011). CFTR inhibition blocks anion and fluid secretion by human ADPKD cells and in animal models of PKD (Reif et al. 2011; Magenheimer et al. 2006; Mangoo-Karim et al. 1995) and is being considered as a potential therapeutic approach to inhibit cyst growth (Li et al. 2009; Yang et al. 2008). In addition, compounds that inhibit CFTR expression including PPARγ agonists are in preclinical and clinical studies (Blazer-Yost et al. 2010; Yoshihara et al. 2011).

29.6.3 Proliferation in Cyst Development

In ADPKD, cyst formation and growth involve aberrant cell proliferation, involving numerous cellular pathways (Chebib and Torres 2016). Central to the understanding of cyst expansion was the discovery that cAMP stimulates the proliferation of ADPKD cyst epithelial cells but inhibits the proliferation of normal tubule epithelial cells (Hanaoka et al. 2000; Yamaguchi et al. 2000). This phenotypic difference in the cAMP mitogenic response between normal and ADPKD cells has been linked to the differential regulation of the BRAF/MEK/ERK signaling pathway (Yamaguchi et al. 2003). The role of Ca^{2+} in the cAMP mitogenic response was demonstrated experimentally using normal human kidney cells and immortalized mouse collecting duct M-1 cells treated with Ca^{2+} channel blockers or EGTA, a Ca^{2+} chelator, to lower intracellular Ca^{2+} levels (Yamaguchi et al. 2004). In these experiments, Ca^{2+} restriction changed the normal cAMP growth-inhibited phenotype to a cAMP growth-stimulated phenotype, mimicking the cAMP response of ADPKD cells. This study showed that reduced cellular Ca^{2+} derepressed BRAF, allowing cAMP-dependent activation of the BRAF/MEK/ERK pathway and cell proliferation. In a reciprocal study, a graded increase in steady-state Ca^{2+} levels in ADPKD cells using either a Ca^{2+} channel activator or a very low concentration of a Ca^{2+} ionophore prevented BRAF activation and rescued the normal antimitogenic response to cAMP (Yamaguchi et al. 2006). Subsequent studies showed that PDE1 physically interacts with BRAF and A-kinase anchoring protein-79, a scaffolding protein for protein kinase A, and specifically controls the cAMP pool to regulate BRAF/MEK/ERK (Pinto et al. 2016).

Additional pathways downstream of the increased intracellular cAMP concentration have been identified in the pathogenesis of ADPKD, including mTOR, signaling transducer, and activator of transcription (STAT)3/6, tumor necrosis factor-α (TNF-α), and liver kinase B1 (LKB1)/AMP-activated protein kinase (AMPK) (Harris and Torres 2014). mTOR is a serine-threonine kinase and the core component for two distinct signaling complexes: complex 1 (mTORC1) and complex 2 (mTORC2) (Wullschleger et al. 2006). mTORC1 is an integration site for many pathways involved in cell cycle progression, protein translation, and cellular energy responses.

Downstream effectors of mTORC1, ribosomal S6 kinase (S6K), and eukaryotic initiation factor 4 binding proteins 1 and 2, are central for protein translation, cell growth, and proliferation. mTOR is activated by GTP-bound Rheb, a small GTPase that is regulated by guanine nucleotide exchange, and tuberin (tuberous sclerosis complex 2, TSC2), a GTPase-activating protein (GAP). TSC2 is phosphorylated by many kinases, including Akt and ERK, to inhibit its GAP activity resulting in upregulation of mTORC1 activity (Inoki et al. 2003); whereas other kinases, including AMPK, inhibit mTOR signaling. Components of the mTOR pathway (mTOR, S6K, and S6) were shown to be aberrantly phosphorylated in cyst epithelial cells of human ADPKD, autosomal recessive PKD, and PKD mouse models (Shillingford et al. 2006; Bonnet et al. 2009; Fischer et al. 2009).

The mTOR complex 1 signaling pathway may be overactive in the cyst epithelial cells for several reasons. *TSC2* and *PKD1* genes are in a tail-to-tail orientation on chromosome 16, suggesting that the regulation of the two proteins is linked. Mutations in *PKD1* may affect expression of *TSC2*. Approximately 2% of TSC patients have a contiguous gene syndrome in which both genes are affected (Brook-Carter et al. 1994). These patients have early onset of ADPKD and a more aggressive disease compared to mutations in the individual genes. PC1 regulates TSC2 by sequestering it at the membrane, thereby preventing Akt phosphorylation and repressing mTOR (Dere et al. 2010). The finding that the immunosuppressive drug rapamycin, which inhibits the mTOR pathway, decreases cyst growth in rodent models (Tao et al. 2005) has led to preclinical and clinical studies directed toward the use of this drug as a treatment option for ADPKD (Torres et al. 2010; Shillingford et al. 2010; Zafar et al. 2010; Serra et al. 2010; Walz et al. 2010). Unfortunately to date, the use of mTOR inhibitors has not been successfully translated to human patients (Watnick and Germino 2010; Pei 2010).

Cleavage of PC1 is required for its proper maturation and function, and multiple cleavage sites for PC1 have been described. The C-terminal tail of PC1 has been shown to be involved in the regulation of a variety of signaling cascades that are aberrantly regulated in ADPKD, including the Wnt signaling pathway (Kim et al. 1999a; Kim et al. 1999b), the AP-1 transcription factor complex (Le et al. 2004, 2005), NFAT (Puri et al. 2004; Calvet 2015), Janus kinase-signal transduction transducer and activator of transcription (JAK-STAT) pathway (Bhunia et al. 2002), and STAT6-dependent gene expression (Low et al. 2006). Regulation of the JAK-STAT pathway requires the autocatalytic cleavage of PC1 at a G protein-coupled receptor proteolytic site (GPS) that is located immediately following the REJ domain (Qian et al. 2002; Wei et al. 2007; Yu et al. 2007). After cleavage, most of the 370 kDa N-terminal fragment (NTF) remains noncovalently tethered to the 150 kDa C-terminal fragment (CTF) (Qian et al. 2002). The majority of the uncleaved PC1 resides in the ER. The cleaved, but associated, NTF/CTF and the dissociated NTF traffic through the secretory pathway to the plasma membrane (Kurbegovic et al. 2014).

A different cleavage reaction that is initiated in response to mechanical stimuli produces a C-terminal tail (CTT) of approximately 200 amino acids that enters the nucleus (Chauvet et al. 2004; Low et al. 2006). The CTT is released by the action of γ-secretase (Merrick et al. 2012). This cleavage product associates with β-catenin and disrupts canonical Wnt signaling (Lal et al. 2008), a pathway important for fate determination, proliferation, and planar cell polarity as well as renal development and morphology (Lancaster and Gleeson 2010). Therefore, the CTT cleavage product may contribute to aberrant Wnt signaling and abnormalities inherent in cyst formation.

Nuclear CTT of PC1 activates the signal transducer and activator of transcription-6 (STAT6) and STAT3 (Low et al. 2006; Talbot et al. 2011) as well as the common transcriptional coactivator p300 (Merrick et al. 2012). Genetic inactivation of STAT6 or treatment with a STAT6 inhibitor decreased cyst growth in PKD mice (Olsan et al. 2011); therefore, it is possible that nuclear CTT contributes to the regulation of cell proliferation and cyst growth.

29.6.4 Role of Metabolism in Cyst Development

PC1 mutations in mice resulted in enhanced aerobic glycolysis (Rowe and Boletta 2014), a pathway that generates substrates that are used as building blocks for cell growth. It has been suggested that polycystins may be modulated by cellular oxygen-sensing pathways and mitochondrial function (Padovano et al. 2017). PKD mouse and rat models, as well as cultured cells derived from human ADPKD cysts, have morphological and functional abnormalities in mitochondria, including increased superoxide production and abnormal fatty acid oxidation (Ishimoto et al. 2017; Lin et al. 2018; Menezes et al. 2016). However, at least some of the observed metabolic changes may arise as a direct effect from the polycystins. The CTT C-terminal cleavage product of PC1 was found to translocate to the mitochondrial matrix and overexpression of CTT in heterologous systems altered mitochondrial structure and function (Lin et al. 2018).

As the disease progresses, oxidative stress, mitochondrial dysfunction, and increased reactive oxygen species have been shown to upregulate another Cl^- channel, TMEM16a, in human and mouse kidneys as well as in cultured cell lines. This increased Cl^- secretion may further exacerbate cyst growth (Schreiber et al. 2019).

Within the cytoplasmic space, AMPK is a central energy sensor that serves as a metabolic checkpoint to regulate cell energy utilization and inhibit mTOR-mediated cell growth and proliferation during low energy periods (Hardie2011). While its role in kidney function is not well understood, AMPK appears to modulate solute transport in accordance with the energy status of the tubule cells (Hallows et al. 2010; Pastor-Soler and Hallows 2012). In ADPKD cells, elevated glucose metabolism and the metabolic switch to aerobic glycolysis, lead to reduced AMPK activity (Rowe et al. 2013). This reduced activity of AMPK may contribute to elevated mTOR activity, cell proliferation, and fluid secretion, as well as tissue inflammation and fibrosis. Moreover, persistent cAMP activation of the BRAF/MEK/ERK pathway may lead to the phosphorylation and inhibition of LKB1, the master kinase primarily responsible for AMPK activation. Metformin, a biguanide, inhibits mitochondrial complex I, resulting in an increased AMP:ATP ratio and AMPK activation (Viollet et al. 2012). Caplan and associates showed that metformin decreased mTOR activity, cell proliferation, and Cl^--dependent fluid secretion by MDCK cells, and delayed renal cyst growth in PKD mouse models (Takiar et al. 2011). Metformin is currently being considered for the treatment of ADPKD; however, the concentration of metformin necessary to increase AMPK activity in the kidneys may have side effects, such as lactic acidosis, particularly if circulating levels are not properly controlled as renal function declines. Recently, restriction of caloric intake, which would alter these metabolic pathways including increased AMPK activity and decreased mTOR signaling, was shown to decrease PKD progression in mice (Warner et al. 2016).

29.6.5 The Importance of Gene Dosage and Developmental Expression of the Polycystins

In ADPKD, every cell carries one germline-mutated allele; however, cysts form in a very small percentage of nephrons indicating that the loss of one PKD allele is not sufficient to induce cyst initiation. An early hypothesis for the focal nature of renal cyst formation in ADPKD was that a somatic mutation (second hit) during development or after birth causes the loss of the second allele. Homozygous deletion of *Pkd1* in mice causes massive cystic disease during embryogenesis, indicating that PC1 is critical for normal renal development and is consistent with the "two-hit" hypothesis. Inducible kidney-specific *Pkd1* knock-out models demonstrated that the timing of the loss of PC1 is critical for disease severity. Loss of PC1 before postnatal day 13 resulted in rapid onset and severe cystic disease (Piontek et al. 2007), and the mice died within ~30 days of renal failure. By contrast, loss of PC1 after day 14 resulted in a delayed onset of cystic disease that was slowly progressive (Piontek et al. 2007; Lantinga-van Leeuwen et al. 2007). These studies demonstrated that PC1 expression during a critical developmental period is essential for normal nephron maturation.

Although the hypothesis that cyst initiation is due to a recessive mechanism at the cellular level is attractive, recent data indicate that gene dosage is likely to be important too. Using a hypomorphic disease variant of *Pkd1* in mice ($Pkd1^{R3277C}$ mice), Hopp and colleagues found that PKD development and severity correlated with functional PC1 dosage (Hopp et al. 2012). Mice with two hypomorphic *Pkd1* alleles ($Pkd1^{RC/RC}$) had a slower development and milder disease than mice expressing one hypomorphic allele and one null allele ($Pkd1^{RC/-}$). It is easy to speculate that in human ADPKD kidneys, a somatic mutation may account for larger, but less common cysts, and insufficient expression of a normal allele, which may be influenced by factors in the microenvironment of the cystic kidney, accounts for the majority of smaller cysts (Leonard et al. 2015).

The observation that cyst formation was delayed several months when *Pkd1* was inactivated after renal development also suggests that a “third hit”, such as renal injury, may be important for cyst initiation. Evidence has demonstrated that several forms of renal injury initiate cystogenesis and lead to massive cystic disease in the adult onset PKD mice by triggering injury repair that promotes proliferation similar in rate to what occurs during renal development (Patel et al. 2008; Takakura et al. 2009; Happé et al. 2009).

PC1 appears to be important during early renal development, particularly in tubular morphogenesis (Ibraghimov-Beskrovnaya et al. 1997; Arnould et al. 1998). The effect of PC1 on renal tubular morphogenesis is mediated, in part, by GTP-binding proteins and PKCα resulting in the activation of the transcription factor AP-1 that modulates growth responses and apoptosis (Arnould et al. 1998). PC1 has been shown to associate with Par3, facilitating the formation of a Par3/aPKC complex that is important for cell polarity and oriented cell migration, both of which would establish the correct tubule diameter during development (Castelli et al. 2013).

PC1 expression is also required for the maturation and/or stability of vascular cells during angiogenesis (Kim et al. 2000). Knockout of *Pkd1* in mice resulted in embryonic death with edema, focal vascular leaks, and massive hemorrhage, demonstrating that PC1 is essential for structural integrity of blood vessels and lymphatic development (Outeda et al. 2014).

29.7 Role of Polycystin in Extra-Renal Pathology

29.7.1 Liver and Pancreatic Cysts

The most common extra-renal manifestation of ADPKD is liver cysts that arise from the aberrant proliferation of cholangiocytes lining the hepatic bile duct. Approximately 90% of ADPKD patients over the age of 35 years develop polycystic liver disease, during which the liver can become massively enlarged (Mikolajczyk et al. 2017). The cystic liver contributes to pain and discomfort; however, hepatic cysts rarely lead to organ failure in ADPKD patients. Symptoms associated with polycystic liver disease have become more common as the lifespan of these patients increases due to renal replacement therapy. These additional symptoms include dyspnea (shortness of breath), gastroesophageal reflux, and low-back pain due to the mass and weight of the enlarged liver. Hepatic cysts are more prevalent in women and the average cyst volume is higher in women compared to men. The involvement of liver cysts increases with age in women and is influenced by hormonal changes, particularly due to pregnancy or estrogen replacement therapy.

Liver cysts arise due to the excessive proliferation of cholangiocytes in the biliary duct, which are stimulated by estrogens and insulin-like growth factor. Like renal cysts, cystic cholangiocytes have reduced intracellular Ca^{2+} and elevated cAMP levels compared to normal cholangiocytes (Banales et al. 2009). As described for renal cysts, cAMP promotes the growth of liver cysts through activation of the BRAF/MEK/ERK pathway driving cell proliferation. Restoring Ca^{2+} levels in cystic cholangiocytes inhibited cAMP-dependent proliferation of the cystic epithelial cells. Also, as in the renal cysts, increased cAMP results in CFTR-mediated anion secretion (Muchatuta et al. 2009). Overall, it appears that many of the same pathways involved in renal cyst growth also promote the growth of liver cysts. However, since cholangiocytes do not express V2 receptor, arginine vasopressin and its antagonist tolvaptan do not affect cyst growth. Other approaches to modulate cAMP production in the liver include somatostatin analogues such as lanreotide and octreotide (Masyuk et al. 2007, 2013).

Cyst development in the pancreas is estimated to occur in ~10% of patients and is associated with increasing age, female sex, and *PKD1* mutation (Luciano and Dahl 2014). Pancreatic involvement is usually not life threatening; however, compression of the pancreatic duct can result in chronic pancreatitis.

29.7.2 Cardiovascular Abnormalities

ADPKD patients have increased risk of cardiovascular problems, including early-onset hypertension, aneurysms in coronary and cerebral blood vessels, and left ventricular hypertrophy (Schievink et al. 1992; Gabow 1993). Cardiac valve abnormalities (mitral valve prolapse) have been documented in ~25% of patients and intracranial aneurysms occur in ~6% of patients (Harris and Torres 2014; Mikolajczyk et al. 2017).

The demonstration of death from cardiovascular defects in PC1 and PC2 null mice suggest major roles for the polycystins in these tissues (Kim et al. 2000; Wu et al. 2000). In early studies, PC1 was detected in endothelial cells and elastic artery vascular smooth muscle cells from normal and PKD patients (Ibraghimov-Beskrovnaya et al. 1997; Griffin et al. 1997). Subsequently, the creation of a mouse expressing a targeted mutation in *Pkd1* which defines its expression using a lacZ reporter, showed that PC1 was involved in both vascular and skeletal development and was expressed at high levels in vascular smooth muscle, endothelial cells, cardiac valves, and myocytes (Boulter et al. 2001; Volk et al. 2003). In endothelial cells, PC1 localized predominately to primary cilia where it is thought to act as a putative mechanosensor facilitating the release of the vasodilator nitric oxide (Nauli et al. 2008). Likewise, mechanical injury of endothelial cells in microvessels of the brain induces an influx of Ca^{2+} and the synthesis of nitric oxide followed by actin rearrangement. TRPC1 and PC2 are crucial for this response which protects the blood-brain barrier (Berrout et al. 2012).

Haploinsufficiency resulting in a decrease in PC2 in vascular smooth muscle cells caused intracranial vascular abnormalities under hypertensive conditions in mice. These findings were coincident with lower intracellular Ca^{2+} and total sarcoplasmic reticulum (SR) Ca^{2+} stores due to a decreased activity of the store-activated Ca^{2+} channel (Qian et al. 2003a). Consistent with the regulation of intracellular Ca^{2+}

stores, PC2 localized to the SR membrane in vascular smooth muscle cells (Qian et al. 2003b). Decreases in smooth muscle Ca^{2+} have also been shown to increase cell proliferation and apoptosis (Kip et al. 2005).

In cardiac myocytes, PC2 binds to and inhibits the ryanodine receptor (RyR2) leading to decreased Ca^{2+} release from SR stores (Anyatonwu et al. 2007). In arterial myocytes, the PC1/PC2 ratio has been shown to be important for the cellular response to intraluminal pressure changes. PC2 alone inhibits the activity of stretch-activated channels, while the PC1/PC2 complex reverses this inhibition. Proteomic analysis revealed potential PC1/PC2 complex binding partners, filamin A, and actin, as potential intermediates in this effect (Sharif-Naeini et al. 2009).

29.7.3 *Extra-Tubular Renal Expression*

PC2 has also been found in nontubular renal tissue. In glomerular mesangial cells, angiotensin II stimulation results in an increased expression of membrane-bound PC2 and facilitates its interaction with TRPC1 and TRPC4 forming channel complexes that mediate the agonist-stimulated Ca^{2+} responses that are important for mesangial cell regulation of glomerular filtration (Du et al. 2008).

29.8 Closing Remarks

PC1 and PC2, the inaugural members of the polycystin family, have been intensely studied since they were first recognized as the proteins associated with ADPKD. The similarity in disease progression and severity when either PC1 or PC2 is mutated suggests a close relationship between at least some aspects of their function. There is little doubt that PC2 is a bona fide cation channel of the TRP channel family, capable to conducting monovalent cations and Ca^{2+}. However, due to the size and complexity of PC1, its role in normal cell physiology and how the loss of PC1 function leads to cyst initiation is less clear. PC1 may be a channel subunit that assembles with PC2 to form a functional tetramer, or an accessory protein required for PC2 targeting and activity. PC1 may have multiple other functions including the maintenance of epithelial characteristics of ductal structures through binding to the ECM and cytoskeleton, or modulating cell-cell adhesions.

The polycystin channel appears to have multiple functions depending on cell type and condition. Indeed, the propensity of TRP proteins to form heteromeric channels means that the polycystin channels may exhibit localization-specific differences in channel subunit partners and, consequently, channel properties.

The plethora of reported binding partners and proteolytic fragments and functions for the polycystins should be considered with a note of caution. Many of these studies rely on the overexpression of one or both of polycystin proteins, or complete knockout of key proteins thought to be involved in the pathways. These manipulations may not accurately reflect normal polycystin expression, binding to other proteins, or physiological function in the cell. To further complicate interpretation, in in vitro systems, the manipulations are often performed in immortalized cells that do not normally express detectable levels of the introduced proteins.

Even when cultured cells that normally express PC1/PC2 are used, they do not usually represent cells from the portions of the nephron that give rise to the cysts. Inexplicably, many studies attempting to link structure and function of the PC1/PC2 complex in ADPKD have been performed in primary or continuous cultures of the inner medullary collecting duct cells. Cysts are generally reported to initiate in the epithelia of the distal nephron and collecting duct (Verani and Silva 1988; Wu et al. 1998; Thomson et al. 2003). While the epithelial cells of these segments are similar, there are substantial variations in the expression of transport proteins (of note, urea transporters) and the cellular environments are notably different. This could lead to differential expression of other transporters and to the regulatory components that modulate them.

Another concern that haunts the PKD field is the lack of appropriate animal models. There are many orthologous and nonorthologous rodent models of polycystic kidney disease. In many ways, nonorthologous models more closely resemble human ADPKD, including the onset and progression of disease, and the contribution of fibrosis to the decline in renal function. There is no agreement in the field as to which models most accurately reflect the various disease components and should be required for preclinical drug testing. The next phase of research regarding the polycystins should, therefore, emphasize the role of physiologically relevant levels of proteins expressed in the cells, organs, and organisms, where they have been shown to have a functional role.

While the exact mechanisms causing cyst initiation are not firmly established, cystic cells have lowered intracellular Ca^{2+} altering the mitogenic and secretory responses to intracellular cAMP. The first successful therapy for the treatment of ADPKD, tolvaptan, targets the cAMP pathway to inhibit cell proliferation and CFTR-mediated fluid secretion. While this is an important first step for caring for ADPKD patients, more specific therapies to slow cyst growth are needed. Nevertheless, this milestone, having been reached, points to the central importance of the Ca^{2+}-regulated cAMP pathway in ADPKD by virtue of the fact that patients respond to tolvaptan therapy.

In summary, involvement of the polycystins in ADPKD has underscored the importance of these proteins in normal renal function. The fundamental research studies that followed the elucidation of the genes responsible for ADPKD have uncovered other potential functions of the polycystins individually and in a complex. Further studies are necessary to completely unravel all the complexities of the polycystin proteins.

Acknowledgments The authors are supported by grants from the National Institutes of Health (R01 DK081579 and P30 DK106912 to DPW) and (R01 FD004826 and the Polycystic Kidney

Disease Foundation to BLB-Y). We thank Dr. James Calvet for helpful suggestions during the preparation of the chapter.

References

Alessandri-Haber N, Yeh JJ, Boyd AE et al (2003) Hypotonicity induces TRPV4-mediated nociception in rat. Neuron 39:497–511

Antignac C, Calvet JP, Germino GG et al (2015) The future of polycystic kidney disease research--as seen by the 12 Kaplan awardees. J Am Soc Nephrol 26:2081–2095

Anyatonwu GI, Ehrlich BE (2004) Calcium signaling and polycystin-2. Biochem Biophys Res Commun 322:1364–1373

Anyatonwu GI, Estrada M, Tian X et al (2007) Regulation of ryanodine receptor-dependent calcium signaling by polycystin-2. Proc Natl Acad Sci U S A 104:6454–6459

Arac D, Boucard AA, Bollinger MF, Nguyen J, Soltis SM, Sudhof TC, Brunger AT (2012) A novel evolutionarily conserved domain of cell-adhesion GPCRs mediates autoproteolysis. EMBO J 31:1364–1378

Arnould T, Kim E, Tsiokas L et al (1998) The polycystic kidney disease 1 gene product mediates protein kinase C alpha-dependent and c-Jun N-terminal kinase-dependent activation of the transcription factor AP-1. J Biol Chem 273:6013–6018

Bai C-X, Giamarchi A, Rodat-Despoix L et al (2008) Formation of a new receptor-operated channel by heteromeric assembly of TRPP2 and TRPC1 subunits. EMBO Rep 9:472–479

Banales JM, Masyuk TV, Gradilone SA et al (2009) The cAMP effectors Epac and protein kinase a (PKA) are involved in the hepatic cystogenesis of an animal model of autosomal recessive polycystic kidney disease (ARPKD). Hepatology 49:160–174

Belibi FA, Reif G, Wallace DP et al (2004) Cyclic AMP promotes growth and secretion in human polycystic kidney epithelial cells. Kidney Intern 66:964–973

Berrout J, Jin M, O'Neil RG (2012) Critical role of TPPP2 and TRPC1 channels in stretch-induced injury of blood-brain barrier endothelial cells. Brain Res 1436:1–12

Berrout J, Jin M, Mamenko M et al (2018) Function of transient receptor potential cation channel subfamily V member 4 (TRPV4) as a mechanical transducer of flow-sensitive segments of renal collecting duct system. J Biol Chem 287:8782–8791

Bhunia AK, Piontek K, Boletta A et al (2002) PKD1 induces $p21^{waf1}$ and regulation of the cell cycle via direct activation of the JAK-STAT signaling pathway in a process requiring PKD2. Cell 109:157–168

Blaszak RT, Potaman V, Sinden RR et al (1999) DNA structural transitions within the PKD1 gene. Nuc Acids Res 27:2610–2617

Blazer-Yost BL, Haydon J, Eggleston-Gulyas T et al (2010) Pioglitazone attenuates cystic burden in the PCK rodent model of polycystic kidney disease. PPAR Res 2010:274376

Boletta A, Qian F, Onuchic LF et al (2000) Polycystin-1, the gene product of *PKD1*, induces resistance to apoptosis and spontaneous tubulogenesis in MDCK cells. Mol Cell 6:1267–1273

Boletta A, Qian F, Onuchic LF et al (2001) Biochemical characterization of bona fide polycystin-1 in vitro and in vivo. Am J Kidney Dis 38:1421–1429

Bonnet CS, Aldred M, von Ruhland C et al (2009) Defects in cell polarity underlie TSC and ADPKD-associated cystogenesis. Hum Mol Genet 18:2166–2176

Booij TH, Bang H, Leonhard WN et al (2017) High-throughput phenotypic screening of kinase inhibitors to identify drug targets for polycystic kidney disease. SLAS Discov 22:974–984

Boulter C, Mulroy S, Webb S et al (2001) Cardiovascular, skeletal, and renal defects in mice with a targeted disruption of the Pkd1 gene. Proc Natl Acad Sci U S A 98:12174–12179

Brook-Carter PT, Peral B, Ward CJ et al (1994) Deletion of the TSC2 and PKD1 genes associated with severe infantile polycystic kidney disease--a contiguous gene syndrome. Nat Genet 8:328–332
Bui-Xuan E-F, Li Q, Chen X-Z et al (2006) More than colocalizing with polycystin-1, polycystin-L is in the centrosome. Am J Physiol Renal Physiol 291:F395–F406
Bukanov NO, Husson H, Dackowski WR et al (2002) Functional polycystin-1 expression is developmentally regulated during epithelial morphogenesis in vitro: Downregulation and loss of membrane localization during cytogenesis. Hum Mol Genet 11:923–936
Bycroft M, Bateman A, Clarke J et al (1999) The structure of a PKD domain from polycystin-1: implications for polycystic kidney disease. EMBO J 18:297–305
Cai Y, Maeda Y, Cedzick A et al (1999) Identification and characterization of polycystin-1, the PKD2 gene product. J Biol Chem 274:28557–28565
Calvet JP (2015) The role of calcium and cyclic AMP in PKD. In: Li X (ed) Polycystic kidney disease. Codon Publications, Brisbane, Australia, pp 169–196
Castelli M, Boca M, Chiaravalli M et al (2013) Polycystin-1 binds Par3/aPKC and controls convergent extension during renal tubular morphogenesis. Nature Comm 4:2658
Chauvet V, Tian X, Husson H et al (2004) Mechanical stimuli induce cleavage and nuclear translocation of the polycystin-1 C terminus. J Clin Invest 114:1433–1443
Chebib FT, Torres VE (2016) Autosomal dominant polycystic kidney disease: Core curriculum 2016. Am J Kidney Dis 67:792–810
Chen XZ, Vassilev PM, Basora N et al (1999) Polycystin-L is a calcium-regulated cation channel permeable to calcium ions. Nature 401:383–386
Choi Y-H, Suzuki A, Hajarnis S et al (2011) Polycystin-2 and phosphodiesterase 4C are components of a ciliary A-kinase anchoring protein complex that is disrupted in cystic diseases. Proc Natl Acad Sci U S A 108:10679–10684
Danielsen H, Pedersen EB, Nielsen AH et al (1986) Expansion of extracellular volume in early polycystic kidney disease. Acta Med Scand 219:399–405
Davidow CJ, Maser RL, Rome LA et al (1996) The cystic fibrosis transmembrane conductance regulator mediates transepithelial fluid secretion by human autosomal dominant polycystic kidney disease epithelium in vitro. Kidney Int 50:208–218
DeCaen PG, Delling M, Vien TN et al (2013) Direct recording and molecular identification of the calcium channel of primary cilia. Nature 504:315–318
Delling M, DeCaen PG, Doerner JF et al (2013) Primary cilia are specialized calcium signaling organelles. Nature 504:311–314
Delling M, Indzhykulian AA, Liu X et al (2016) Primary cilia are not calcium-responsive mechanosensors. Nature 531:656–660
Delmas P (2005) Polycystins: polymodal receptor/ion-channel cellular sensors. Pflügers ArchEur J Physiol 451:264–276
Dere R, Wilson PD, Sanford RN et al (2010) Carboxy-terminal tail of polycystin-1 regulates localization of TSC2 to repress mTOR. PLoS One 5:e9239
Doerr N, Wang Y, Kipp KR et al (2016) Regulation of polycystin-1 function by calmodulin binding. PLoS One 11:e0161525
Du J, Ding M, Sours-Brothers S et al (2008) Mediation of angiotensin II-induced Ca^{2+} signaling by polycystin 2 in glomerular mesangial cells. Am J Physiol Renal Physiol 294:F909–F918
Field S, Riley K-L, Grimes DT et al (2011) Pkd1L1 establishes left-right asymmetry and physically interacts with Pkd2. Development 138:1131–1142
Fischer DC, Jacoby U, Pape L et al (2009) Activation of the AKT/mTOR pathway in autosomal recessive polycystic kidney disease (ARPKD). Nephrol Dial Transplant 24:1819–1827
Gabow PA (1993) Autosomal dominant polycystic kidney disease. New Engl J Med 329:332–342
Gainullin VG, Hopp K, Ward CJ et al (2015) Polycystin-1 maturation requires polycystin-2 in a dose-dependent manner. J Clin Invest 125:607–620
Gall C-LG, Andrezet M-P, Chen J-M et al (2013) Type of *PKD1* mutation influences renal outcome in ADPKD. J Am Soc Neprol 24:1006–1013

Gallagher AR, Cedzich A, Gretz N et al (2000) The polycystic kidney disease protein PKD2 interacts with Hax-1, a protein associated with the actin cytoskeleton. Proc Natl Acad Sci U S A 97:4017–4022
Gattone VH 2nd, Maser RL, Tian C et al (1999) Developmental expression of urine concentration-associated genes and their altered expression in murine infantile-type polycystic kidney disease. Dev Genet 24:309–318
Gattone VH 2nd, Wang X, Harris PC et al (2003) Inhibition of renal cystic disease development and progression by a vasopressin V2 receptor antagonist. Nat Med 9:1323–1326
Geng L, Burrow CR, Li H-P et al (2000) Modification of the composition of polycystin-1 multiprotein complexes by calcium and tyrosine phosphorylation. Biochim Biophys Acta 1533:21–35
Geng L, Okuhara D, Yu Z et al (2006) Polycystin-2 traffics to cilia independently of polycystin-1 by using an N-terminal RVxP motif. J Cell Sci 119:1383–1395
Giamarchi A, Padilla B, Coste B et al (2006) The versatile nature of the calcium-permeable cation channel TRPP2. EMBO Rep 7:787–793
Gonzalez-Perrett S, Kim K, Ibarra I et al (2001) Polycystin-2, the protein mutated in autosomal dominant polycystic kidney disease (ADPKD), is a Ca^{2+}-permeable nonselective cation channel. Proc Natl Acad Sci U S A 98:1182–1187
Gonzalez-Perrett S, Batelli M, Kim K et al (2002) Voltage dependence and pH regulation of human polycystin-2-mediated cation channel activity. J Biol Chem 277:24959–24966
Grantham JJ (1993) 1992 Homer Smith award: fluid secretion, cellular proliferation, and the pathogenesis of renal epithelial cysts. J Am Soc Nephrol 3:1841–1857
Grantham JJ, Mangoo-Karim R, Uchic ME et al (1989) Net fluid secretion by mammalian renal epithelial cells: stimulation by cAMP in polarized cultures derived from established renal cells and from normal and polycystic kidneys. Trans Assoc Am Phys 102:158–162
Grantham JJ, Ye M, Gattone VH 2nd et al (1995) In vitro fluid secretion by epithelium from polycystic kidneys. J Clin Invest 95:195–202
Grantham JJ, Mulamalla S, Grantham CJ et al (2012) Detected renal cysts are tips of the iceberg in adults with ADPKD. Clin J Am Soc Nephrol 7:1087–1093
Grieben M, Pike ACW, Shintre CA et al (2017) Structure of the polycystic kidney disease TRP channel Polycystin-2(PC2). Nat Struct Mol Biol 24:114–122
Griffin MD, Torres VE, Grande JP et al (1997) Vascular expression of polycystin. J Am Soc Nephrol 8:616–626
Guo L, Schreiber TH, Weremowicz S et al (2000) Identification and characterization of a novel polycystin family member, polycystin-L2, in mouse and human sequence expression, alternative splicing and chromosomal localization. Genomics 64:241–251
Hallows KR, Mount PF, Pastor-Soler NM et al (2010) Role of the energy sensor AMP-activated protein kinase in renal physiology and disease. Am J Physiol Renal Physiol 298:F1067–F1077
Hanaoka K, Guggino WB (2000) cAMP regulates cell proliferation and cyst formation in autosomal polycystic kidney disease cells. J Am Soc Nephrol 11:1179–1187
Hanaoka K, Devuyst O, Schwiebert EM et al (1996) A role for CFTR in human autosomal dominant polycystic kidney disease. Am J Physiol Cell Physiol 270:C389–C399
Hanaoka K, Qian F, Boletta A et al (2000) Co-assembly of polycystin-1 and -2 produces unique cation permeable currents. Nature 408:990–994
Happé H, Leonhard WN, van der Wal A et al (2009) Toxic tubular injury in kidneys from Pkd1-deletion mice accelerates cystogenesis accompanied by dysregulated planar cell polarity and canonical Wnt signaling pathways. Hum Mol Genet 18:2532–2542
Hardie DG (2011) AMP-activated protein kinase: an energy sensor that regulates all aspects of cell function. Genes Dev 25:1895–1908
Harris PC, Hopp K (2013) The mutation, a key determinant of phenotype in ADPKD. J Am Soc Nephrol 24:868–870
Harris PC, Torres VE (2009) Polycystic kidney disease. Annu Rev Med 60:321–337

Harris PC, Torres VE (2014) Genetic mechanisms and signaling pathways in autosomal dominant polycystic kidney disease. J Clin Invest 124:2315–2322

Harris PC, Ward CJ, Peral B et al (1995) Autosomal dominant polycystic kidney disease: molecular analysis. Hum Mol Genet 4:1745–1749

Havens MA, Reich AA, Duelli DM et al (2012) Biogenesis of mammalian microRNAs by a non-canonical processing pathway. Nucleic Acids Res 40:4626–4640

Hayashi T, Mochizuki T, Reynolds DM et al (1997) Characterization of the exon structure of the polycystic kidney disease 2 gene (PKD2). Genomics 44:131–136

Hidaka S, Konecke V, Osten L et al (2004) PIGEA-14, a novel coiled-coil protein affecting the intracellular distribution of polycystin-1. J Biol Chem 279:35009–35016

Hogan PG, Rao A (2007) Dissecting ICRAC, a store-operated calcium current. Trends Biochem Sci 32:235–245

Hopp K, Ward CJ, Hommerding CJ et al (2012) Functional polycystin-1 dosage governs autosomal dominant polycystic kidney disease severity. J Clin Invest 122:4257–4273

Huan Y, van Adelsberg J (1999) Polycystin-1, the PKD1 gene product, is a complex containing E-cadherin and the catenins. J Clin Invest 104:1459–1468

Hughes J, Ward C, Peral B et al (1995) The polycystic kidney disease 1 (PKD1) gene encodes a novel protein with multiple cell recognition domains. Nature Genet 10:151–160

Hughes J, Ward CJ, Aspinwall R et al (1999) Identification of a human homologue of the sea urchin receptor for egg jelly: a polycystic kidney disease-like protein. Hum Mol Genet 8:543–549

Hulse RE, Li Z, Huang RK et al (2018) Cryo-EM structure of the polycystin 2-l1 ion channel. elife 7:E36931

Ibraghimov-Beskrovnaya O, Dackowski WR, Foggensteiner L et al (1997) Polycystin: in vitro synthesis, in vivo tissue expression, and subcellular localization identifies a large membrane-associated protein. Proc Natl Acad Sci U S A 94:6397–6402

Ibraghimov-Beskrovnaya O, Bukanov NO, Donohue LC (2000) Strong hemophilic interactions of the Ig-like domains of polycystin-1, the protein product of an autosomal dominant polycystic kidney disease gene, PKD1. Human Mol Genetics 9:1641–1649

Inoki K, Zhu T, Guan KL (2003) TSC2 mediates cellular energy response to control cell growth and survival. Cell 115:577–590

Ishimoto Y, Inagi R, Yoshihara D et al (2017) Mitochondrial abnormality facilities cyst formation in autosomal dominant polycystic kidney disease. Mol Cell Bio 37:e00337–e00317

Jin X, Mohieldin AM, Muntean BS et al (2014) Cilioplasm is a cellular compartment for calcium signaling in response to mechanical and chemical stimuli. Cell Mol Life Sci 71:2165–2178

Kim E, Arnould T, Sellin L et al (1999a) Interaction between RGS7 and polycystin. Proc Natl Acad Sci U S A 96:6371–6376

Kim E, Arnould T, Sellin LK et al (1999b) The polycystic kidney disease I gene product modulates Wnt signaling. J Biol Chem 274:4947–4953

Kim K, Drummond I, Ibraghimov-Beskrovnaya O et al (2000) Polycystin 1 is required for the structural integrity of blood vessels. Proc Natl Acad Sci U S A 97:1731–1736

Kim I, Li C, Liang D et al (2008) Polycystin-2 expression is regulated by a PC2-binding domain in the intracellular portion of fibrocystin. J Biol Chem 283:31559–31566

Kip SN, Hunter LLW, Ren Q et al (2005) $[Ca^{2+}]_i$ reduction increases cellular proliferation and apoptosis in vascular smooth muscle cells: relevance to the ADPKD phenotype. Circ Res 96:873–880

Kleene SJ, Kleene NK (2017) The native TRPP2-dependent channel of murine renal primary cilia. Am J Physiol Renal Physiol 312:F96–F108

Kobori T, Smith GD, Sanford R et al (2009) The transient receptor potential (TRP) channels TRPP2 and TRPC1 form a heterotetramer with a 2:2 stoichiometry and an alternating subunit arrangement. J Biol Chem 284:33507–35513

Kottgen M (2007) TRPP2 and autosomal dominant polycystic kidney disease. Biochim Biophys Acta 1772:836–850

Kottgen M, Benzing T, Simmen T et al (2005) Trafficking of TRPP2 by PACS proteins represents a novel mechanism of ion channel regulation. EMBO J 24:705–716

Kottgen M, Buchholz B, Garcia-Gonzalez MA et al (2008) TRPP2 and TRPV4 form a polymodal sensory channel complex. J Cell Biol 182:437–447

Koulen P, Cai Y, Geng L et al (2002) Polycystin-2 is an intracellular calcium release channel. Nature Cell Biol 4:191–197

Kurbegovic A, Kim H, Xu H et al (2014) Novel functional complexity of polycystin-1 by GPS cleavage in vivo: role in polycystic kidney disease. Mol Cell Biol 34:3341–3353

Lal M, Song X, Pluznick JL et al (2008) Polycystin-1 C-terminal tail associates with β-catenin and inhibits canonical Wnt signaling. Human Mol Genetics 17:3105–3117

Lancaster MA, Gleeson JG (2010) Cystic kidney disease: the role of Wnt signaling. Trends Mol Med 6:349–360

Lanktree MB, Haghighi A, Guiard E et al (2018) Prevalence estimates of polycystic kidney and liver disease by population sequencing. J Am Soc Nephrol 29:2593–2600

Lantinga-van Leeuwen IS, Leonhard WN, van der Wal A et al (2007) Kidney-specific inactivation of the Pkd1 gene induces rapid cyst formation in developing kidneys and a slow onset of disease in adult animals. Human Mol Genetics 16:3188–3196

Le NH, van der Bent P, Huls G et al (2004) Aberrant polycystin-1 expression results in modifications of activator protein-1 activity, whereas Wnt signaling remains unaffected. J Biol Chem 279:27472–27481

Le NH, van der Wal A, van der Bent P et al (2005) Increased activity of activator protein-1 transcription factor components ATF2, cJun, and cFos in human and mouse autosomal dominant polycystic kidney disease. J Am Soc Nephrol 16:2724–2731

Lea WA, Parnell SC, Wallace DP et al (2018) Human-specific abnormal splicing of wild-type *PKD1* induces premature termination of polycystin-1. J Am Soc Nephrol 29:2482–2492

Lehtonen S, Ora A, Olkkonen VM et al (2000) In vivo interaction of the adapter protein CD2-associated protein with the type 2 polycystic kidney disease protein, polycystin-2. J Biol Chem 275:32888–32893

Leonard WN, Zandbergen M, Veraar K et al (2015) Scattered deletion of PKD1 in kidneys causes a cystic snowball effect and recapitulates polycystic kidney disease. J Am Soc Nephrol 26:1322–1333

Li A, Tian X, Sung SW et al (2003a) Identification of two novel polycystic kidney disease-1-like genes in human and mouse genomes. Genomics 81:596–608

Li Q, Dai Y, Guo L et al (2003b) Polycystin-2 associates with tropomyosin-1, an actin microfilament component. J Mol Biol 325:949–962

Li Y, Wright JM, Qian F et al (2005) Polycystin 2 interacts with type I inositol 1,4,5-trisphosphate receptor to modulate intracellular Ca^{2+} signaling. J Biol Chem 280:41298–41306

Li Y, Santoso NG, Yu S et al (2009) Polycystin-1 interacts with inositol 1,4,5-trisphosphate receptor to modulate intracellular Ca^{2+} signaling with implications for polycystic kidney disease. J Biol Chem 284:36431–36441

Lin C-C, Kurashige M, Terabayashi T et al (2018) A cleavage product of polycystin-1 is a mitochondrial matrix protein that affects mitochondria morphology and function when heterologously expressed. Sci Report 8:2742

Liu X, Vien T, Duan J, Sheu S-H et al (2018) Polycystin-2 is an essential ion channel subunit in the primary cilium of the renal collecting duct epithelium. elife 7:e33183

Low SH, Vasanth S, Larson CH et al (2006) Polycystin-1, STAT6 and P100 function in a pathway that transduces ciliary mechanosensation and is activated in polycystic kidney disease. Dev Cell 10:57–68

Luciano RL, Dahl NK (2014) Extra-renal manifestations of autosomal dominant polycystic kidney disease (ADPKD): considerations for routine screening and management. Nephrol Dial Transplant 29:247–254

Ma R, Li W-P, Rundle D et al (2005) PKD2 functions as an epidermal growth factor-activated plasma membrane channel. Mol Cell Biol 25:8285–8298

Ma M, Gallagher AR, Somlo S (2017) Ciliary mechanisms of cyst formation in polycystic kidney disease. Cold Spring Harb Perspect Biol 9:a028209

Magenheimer BS, St John PL, Isom KS et al (2006) Early embryonic renal tubules of wild-type and polycystic kidney disease kidneys respond to cAMP stimulation with cystic fibrosis transmembrane conductance regulator/Na^+, K^+, $2Cl^-$ Co-transporter-dependent cystic dilation. J Am Soc Nephrol 17:3424–3437

Malhas AN, Abuknesha RA, Price RG (2002) Interaction of the leucine-rich repeats of polycystin-1 with extracellular matrix proteins: possible role in cell proliferation. J Am Soc Nephrol 13:19–26

Mamenko M, Zaika O, Boukelmoune N et al (2015) Deciphering physiological role of the mechanosensitive TRPV4 channel in the distal nephron. Am JPhysio: Renal Physiol 308: F275–F286

Mangolini A, de Stephanis L, Aguiari G (2016) Role of calcium in polycystic kidney disease: from signaling to pathology. World J Nephrol 5:76–83

Mangoo-Karim R, Ye M, Wallace DP et al (1995) Anion secretion drives fluid secretion by monolayers of cultured human polycystic cells. Am J Phys 269:F381–F388

Markoff A, Bogdanova N, Knope M et al (2007) Annexin A5 interacts with polycystin-1 and interferes with the polycystin-1-stimulated recruitment of E-cadherin into adherens junctions. J Mol Biol 369:954–966

Masyuk TV, Masyuk AI, Torres VE et al (2007) Octreotide inhibits hepatic cystogenesis in a rodent model of polycystic liver disease by reducing cholangiocyte adenosine 3′,5′-cyclic monophosphate. Gastroenterology 132:1104–1116

Masyuk TV, Radtke BN, Stroope AJ et al (2013) Pasireotide is more effective than octreotide in reducing hepatorenal cystogenesis in rodents with polycystic kidney and liver diseases. Hepatology 58:409–421

Mekahli D, Sammels E, Luyten T et al (2012) Polycystin-1 and polycystin-2 are both required to amplify inositol-trisphosphate-induced Ca^{2+} release. Cell Calcium 51:452–458

Menezes LF, Lin CC, Zhou F et al (2016) Fatty acid oxidation is impaired in an orthologous mouse model of autosomal dominant polycystic kidney disease. EBioMedicine 5:183–192

Merrick D, Chapin H, Baggs JE et al (2012) The γ-secretase cleavage product of polycystin-1 regulates TCF and CHOP-mediated transcriptional activation through a p300-dependent mechanism. Develop Cell 22:197–210

Michalski A, Grzeszczak W (1996) The effect of hypervolemia on electrolyte level and level of volume regulating hormones in patients with autosomal dominant polycystic kidney disease. Pol Arch Med Wewn 96:329–343

Mikolajczyk AE, Te HS, Chapman AB (2017) Gastrointestinal manifestations of autosomal-dominant polycystic kidney disease. Clin Gastroenterol Hepatol 15:17–24

Mochizuki T, Wu G, Hayashi T et al (1996) PKD2, a gene for polycystic kidney disease that encodes an integral membrane protein. Science 272:1339–1342

Muchatuta M, Gattone V, Witzmann F et al (2009) Structural and functional analysis of liver cysts from the BALB/c-cpk mouse model of PKD. Exper Biol Med 234:17–27

Nagao S, Nishii K, Katsuyama M et al (2006) Increased water intake decreases progression of polycystic kidney disease in the PCK rat. J Am Soc Nephrol 17:2220–2227

Nauli SM, Alenghat FJ, Luo Y et al (2003) Polycystins 1 and 2 mediate mechanosensation in the primary cilium of kidney cells. Nat Genet 33:129–137

Nauli SM, Kawanabe Y, Kaminiski JJ et al (2008) Endothelial cilia are fluid-shear sensors that regulate calcium signaling and nitric oxide production through polycystin-1. Circulation 117:1161–1171

Newby LJ, Streets AJ, Zhao Y (2002) Identification, characterization, and localization of a novel kidney polycystin-1-polycystin-2 complex. J Biol Chem 277:20763–20773

Nilius B, Owsianik G (2011) The transient receptor potential family of ion channels. Genome Biol 12:218–229

Nilius B, Vriens J, Prenen J et al (2004) TRPV4 calcium entry channel: a paradigm for gating diversity. Am J Physiol Cell Physiol 286:C195–C205

Nims NM, Vassmer D, Maser RL (2011) Effect of PKD1 gene missense mutations on Polycystin-1 membrane topogenesis. Biochemist 50:349–355

Nomura H, Turco AE, Pei Y et al (1998) Identification of PKDL, a novel polycystic kidney disease 2-like gene whose murine homologue is deleted in mice with kidney and retinal defects. J Biol Chem 273:25967–25973

Olsan EE, Mukherjee S, Wulkersdorfer B et al (2011) Signal transducer and activator of transcription-6 (STAT6) inhibition suppresses renal cyst growth in polycystic kidney disease. Proc Natl Acad Sci U S A 108:18067–18072

Outeda P, Huso DL, Fisher SA et al (2014) Polycystin signaling is required for directed endothelial cell migration and lymphatic development. Cell Rep 7:633–644

Padovano V, Kuo IY, Stavola LK et al (2017) The polycystins are modulated by cellular oxygen-sensing pathways and regulate mitochondrial function. Mol Bio Cell 28:261–269

Parnell SC, Magenheimer BS, Maser RL et al (1998) The polycystic kidney disease-1 protein, polycystin-1, binds and activates heterotrimeric G-proteins in vitro. Biochem Biophys Res Comm 251:625–631

Parnell SC, Puri S, Wallace DP et al (2012) Protein phosphatase-1a interacts with and dephosphorylates polycystin-1. PLoS One 7:e36798

Parnell SC, Magenheimer BS, Maser RL et al (2018) A mutation affecting polycystin-1 mediated heterotrimeric G-protein signaling causes PKD. Hum Mol Genet 27:3313–3324

Pastor-Soler NM, Hallows KR (2012) AMP-activated protein kinase regulation of kidney tubular transport. Curr Opin Nephrol Hypertens 21:523–533

Patel V, Li L, Cobo-Stark P et al (2008) Acute kidney injury and aberrant planar cell polarity induce cyst formation in mice lacking renal cilia. Hum Mol Genet 17:1578–1590

Pavel M, Lv C, Ng C et al (2016) Function and regulation of TRPP2 ion channel revealed by a gain-of-function mutant. Proc Natl Acad Sci U S A 113:E2363–E2372

Pei Y (2010) Of mice and men: therapeutic mTOR inhibition in polycystic kidney disease. J Am Soc Nephrol 21:383–394

Peyronnet R, Martins JR, Duprat F et al (2013) Piezol-dependent stretch-activated channels are inhibited by polycystin-2 in renal tubular epithelial cells. EMBO Rep 14:1143–1148

Pinto CS, Raman A, Reif GA et al (2016) Phosphodiesterase isoform regulation of cell proliferation and fluid secretion in autosomal dominant polycystic kidney disease. J Am Soc Nephrol 27:1124–1134

Pinto CS, Reif GA, Nivens E, White C, Wallace DP (2012) Calmodulin-sensitive adenylyl cyclase mediates AVP-dependent cAMP production and Cl secretion by human autosomal dominant polycystic kidney cells. Am J Physiol Renal Physiol 303:F1412–F1424

Piontek KB, Germino GG (1998) Murine PHD1 introns 21 and 22 lack the extreme polypyrimidine bia present in human PKD1. Mamm Gen 10:194–196

Piontek K, Menezes LF, Garcia-Gonzalez MA et al (2007) A critical developmental switch defines the kinetic of kidney cyst formation after loss of Pkd1. Nature Med 13:1490–1495

Ponting CP, Hofmann K, Bork P (1999) A latrophilin/CL-1-like GPS domain in polycystin-1. Curr Biol 9:R585–R588

Praetorius HA, Leipziger J (2009) Released nucleotides amplify the cilium-dependent flow-induced $[Ca^{2+}]i$ response in MDCK cells. Acta Physiol 197:241–251

Praetorius HA, Spring KR (2001) Bending the MDCK cell primary cilium increases intracellular calcium. J Membr Biol 184:71–79

Prakriya M, Feske S, Gwack Y et al (2006) Orai is an essential pore subunit of the CRAC channel. Nature 443:230–233

Puri S, Magenheimer BS, Maser RL et al (2004) Polycystin-1 activates the calcineurin/NFAT (nuclear factor of activated T-cells) signaling pathway. J Biol Chem 279:55455–55464

Qian F, Watnick TJ, Onuchic LF et al (1996) The molecular basis of focal cyst formation in human autosomal dominant polycystic kidney disease type 1. Cell 87:979–987

Qian F, Germino FJ, Cai Y et al (1997) PKD1 interacts with PKD2 through a probable coiled-coil domain. Nat Genet 16:179–183
Qian F, Boletta A, Bhunia AK et al (2002) Cleavage of polycystin-1 requires the receptor for egg jelly domain and is disrupted by human autosomal-dominant polycystic kidney disease 1-associated mutations. Proc Natl Acad Sci U S A 99:16981–16986
Qian Q, Hunter LW, Li M et al (2003a) Pkd2 haploinsufficiency alters intracellular calcium regulation in vascular smooth cells. Human Mol Genetics 12:1875–1880
Qian Q, Li M, Cai Y et al (2003b) Analysis of the polycystins in aortic vascular smooth muscle cells. J Am Soc Nephrol 14:2280–2287
Quinlan RJ, Tobin JL, Beales PL (2008) Chapter 5 modeling ciliopathies: primary cilia in development and disease. Curr Top Dev Biol 84:249–310
Ramsey IS, Delling M, Clapham DE (2006) An introduction to TRP channels. Annu Rev Physiol 68:619–647
Raychowdhury MK, McLaughlin M, Ramos AJ et al (2005) Characterization of single channel currents from primary cilia of renal epithelial cells. J Biol Chem 280:34718–34722
Rees S, Kittikulsuth W, Roos K et al (2014) Adenylyl cyclase 6 deficiency ameliorates polycystic kidney disease. J Am Soc Nephrol 25:232–237
Reif GA, Yamaguchi T, Nivens E et al (2011) Tolvaptan inhibits ERK-dependent cell proliferation, Cl^- secretion, and in vitro cyst growth of human ADPKD cell stimulated by vasopressin. J Physiol Renal Physiol 301:F1005–F1013
Retailleau K, Duprat F (2014) Polycystins and partners: proposed role in mechanosensitivity. J Physiol 592:2453–2471
Rodova M, Islam MR, Peterson KR et al (2003) Remarkable sequence conservation of the last intron in the PKD1 gene. Mol Biol Evol 20:1669–1674
Rossetti S, Kubly VJ, Consugar MB et al (2009) Incompletely penetrant *PKD1* alleles suggest a role for gene dosage in cyst initiation in polycystic kidney disease. Kid Intern 75:848–855
Rowe I, Boletta A (2014) Defective metabolism in polycystic kidney disease: potential for therapy and open questions. Nephrol Dial Transplant 29:1480–1486
Rowe I, Chiaravalli M, Mannella V et al (2013) Defective glucose metabolism in polycystic kidney disease identifies a new therapeutic strategy. Nat Med 19:488–493
Rundle DR, Gorbsky G, Tsiokas L (2004) PKD2 interacts and co-localizes with mDia1 to mitotic spindles of dividing cells: role of nDia1 in PKD2 localization to mitotic spindles. J Biol Chem 279:29728–29739
Russo RJ, Husson H, Joly D et al (2005) Impaired formation of desmosomal junctions in ADPKD epithelia. Histochem Cell Biol 124:487–497
Saigusa T, Bell PD (2015) Molecular pathways and therapies in autosomal dominant polycystic kidney disease. Physiology (Bethesda) 30:195–207
Sanford R, Sgotto B, Aparicio S et al (1997) Comparative analysis of the polycystic kidney disease 1 (PKD1) gene reveals an integral membrane glycoprotein with multiple evolutionary conserved domains. Hum Mol Genet 9:1483–1489
Santoso NG, Cebotaru L, Guggino WB (2011) Polycystin-1, 2 and STIM1 interact with IP_3R to modulate ER Ca^{2+} release through the PI3K/Akt pathway. Cell Physiol Biochem 27:715–726
Scheffers MS, van der Bent P, Prins F et al (2000) Polycystin-1, the product of the polycystic kidney disease 1 gene, co-localizes with desmosomes in MDCK cells. Hum Mol Genet 9:2743–2750
Schievink WI, Torres VE, Piepgras DG et al (1992) Saccular intracranial aneurysms in autosomal dominant polycystic kidney disease. J Am Soc Nephrol 3:88–95
Schreiber R, Buchholz B, Kraus A et al (2019) Lipid peroxidation drives renal cyst growth in vitro through activation of TMEM16a. J Am Soc Nephrol 30:228–242
Schrier RW, McFann KK, Johnson AM (2003) Epidemiological study of kidney survival in autosomal dominant polycystic kidney disease. Kid Int 63:678–685
Serra AL, Poster D, Kistler AD et al (2010) Sirolimus and kidney growth in autosomal dominant polycystic kidney disease. N Engl J Med:3820–3829

Sharif-Naeini R, Folgering JH, Bichet D et al (2009) Polycystin-1 and -2 dosage regulates pressure sensing. Cell 139:587–596

Shillingford JM, Murcia NS, Larson CH et al (2006) The mTOR pathway is regulated by polycystin-1, and its inhibition reverses renal cystogenesis in polycystic kidney disease. Proc Natl Acad Sci U S A 103:5466–5471

Shillingford JM, Piontek KB, Germino GG et al (2010) Rapamycin ameliorates PKD resulting from conditional inactivation of Pkd1. J Am Soc Nephrol 21:489–497

Silberberg M, Charron AJ, Bacallao R et al (2005) Mispolarization of desmosomal proteins and altered intercellular adhesion in autosomal dominant polycystic kidney disease. Am J Physiol Renal Physiol 288:F1153–F1163

Smith LA, Bukanaov NO, Husson H et al (2006) Development of polycystic kidney disease in juvenile cystic kidney mice: insights into pathogenesis, ciliary abnormalities and common features with human disease. J Am Soc Nephrol 17:2821–2831

Somlo S, Ehrlich B (2001) Human disease: calcium signaling in polycystic kidney disease. Curr Biol 11:R356–R360

Starremans PG, Li X, Finnerty PE et al (2008) A mouse model for polycystic disease through a somatic in-frame deletion in the 5′ end of Pkd1. Kidney Int 73:1394–1405

Stewart AP, Smith GD, Sandford RN et al (2010) Atomic force microscopy reveals the alternating submit arrangement of the TRPP2-TRPV4 heterotetramer. Biophys J 99:790–797

Streets AJ, Newby LJ, O'Hare MJ et al (2003) Functional analysis of PKD1 transgenic lines reveals a direct role for polycystin-1 in mediating cell-cell adhesion. J Am Soc Nephrol 14:1804–1815

Streets AJ, Moon DJ, Kane ME et al (2006) Identification of an N-terminal glycogen synthase kinase 3 phosphorylation site which regulates the functional localization of polycystin-2 in vivo and in vitro. Hum Mol Genet 15:1465–1473

Streets AJ, Wagner BE, Harris PC et al (2009) Homophilic and heterophilic polycystin 1 interactions regulate E-cadherin recruitment and junction assembly in MDCK cells. J Cell Sci 122:1410–1417

Su X, Driscoll K, Yao G et al (2014) Bardet-Biedl syndrome proteins 1 and 3 regulate the ciliary trafficking of polycystic kidney disease 1 protein. Hum Mol Genet 23:5441–5451

Su Q, Hu F, Ge X et al (2018) Structure of the human PKD1/PKD2 complex. Science 361:eaat9819

Takakura A, Contrino L, Zhou X et al (2009) Renal injury is a third hit promoting rapid development of adult polycystic kidney disease. Hum Mol Genet 8:2523–2531

Takiar V, Nishio S, Seo-Mayer P et al (2011) Activating AMP-activated protein kinase (AMPK) slows renal cystogenesis. Proc Natl Acad Sci U S A 108:2462–2467

Talbot JJ, Shillingford JM, Vasanth S et al (2011) Polycystin-1 regulates STAT activity by a dual mechanism. Proc Natl Acad Sci U S A 108:7985–7990

Tao Y, Kin J, Schrier RW et al (2005) Rapamycin markedly slows disease progression in a rat model of polycystic kidney disease. J Am Soc Nephrol 16:46–51

Terryn S, Ho A, Beauwens R et al (2011) Fluid transport and cystogenesis in autosomal dominant polycystic kidney disease. Biochim Biophys Acta 1812:1314–1321

The American PKD1 consortium (1995) Analysis of the genomic sequence for the autosomal dominant polycystic kidney disease (PKD1) predicts the presence of a leucine rich repeat. Hum Mol Genet 4:575–582

The European Polycystic Kidney Disease Consortium (1994) The polycystic kidney disease gene encodes a 14 kb transcript and lies within a duplicated region on chromosome 16. Cell 77:881–894

The International Polycystic Kidney Disease Consortium (1995) Polycystic kidney disease: the complete structure of the PKD1 gene and its protein. Cell 81:89–298

Thomson RB, Mentone SA, Kim R et al (2003) Histopathological analysis of renal cystic epithelia in the $Pkd2^{WS25/-}$ mouse model of ADPKD. Am J Physiol Renal Physiol 285:F870–F880

Tomilin V, Reif GA, Zaika O et al (2018) Deficient transient receptor potential vanilloid type 4 contributes to compromised $[Ca^{2+}]_i$ homeostasis in human autosomal dominant polycystic kidney disease cells. FASEB J 32:4612–4623

Torres VE, Harris PC (2014) Strategies targeting cAMP signaling in the treatment of polycystic kidney disease. J Am Soc Nephrol 25:18–32
Torres VE, Wang X, Qian Q et al (2004) Effective treatment of an orthologous model of autosomal-dominant polycystic kidney disease. Nat Med 10:363–364
Torres VE, Bankir L, Grantham JJ (2009) A case for water in the treatment of polycystic kidney disease. Clin J Am Soc Nephrol 4:1140–1150
Torres VE, Boletta A, Chapman A et al (2010) Prospects for mTOR inhibitor use in patients with polycystic kidney disease and hamartomatous diseases. Clin J Am Soc Nephrol 5:1312–1329
Torres VE, Chapman AB, Devuyst O et al (2012) Tolvaptan in patients with autosomal dominant polycystic kidney disease. N Engl J Med 367:2407–2418
Torres VE, Chapman AB, Devuyst O et al (2017) Tolvaptan in later-stage autosomal dominant polycystic kidney disease. N Engl J Med 377:1930–1943
Tsiokas L, Arnould T, Zhu C et al (1999) Specific association of the gene product of PKD2 with the TRPC1 channel. Proc Natl Acad Sci U S A 96:3934–3939
Van Gastel MDA, Torres VE (2017) Polycystic kidney disease and the vasopressin pathway. Ann Nutr Metab 70:43–50
Van Raay TJ, Burn TC, Conners TD et al (1996) A 2.5 kb polypyrimidine tract in the PKD gene contains at least 23 H-DNA-forming sequences. Gen Sci Tech 4:317–327
Veldhuisen B, Spruit L, Dauwerse HG et al (1999) Genes homologous to the autosomal dominant polycystic kidney disease genes (PKD1 and PKD2). Euro J Hum Genet 7:860–872
Verani RR, Silva FG (1988) Histogenesis of the renal cysts in adult (autosomal dominant) polycystic kidney disease: a histochemical study. Mod Path 1:457–463
Viau A, Bienaime F, Lukas K et al (2018) Cilia-localized LKB1 regulates chemokine signaling, macrophage recruitment, and tissue homeostasis in the kidney. EMBO J 37:e98615
Viollet B, Guigas B, Sanz Garcia N et al (2012) Cellular and molecular mechanisms of metformin: an overview. Clin Sci (Lond) 122:253–270
Voets T, Prenen J, Vriens J et al (2002) Molecular determinants of permeation through the cation channel TRPV4. J Biol Chem 277:33704–33710
Vogel P, Read R, Hansen GM et al (2010) Situs inversus in Dpcd/poll$^{-/-}$, Nme7$^{-/-}$ and Pkd1l1$^{-/-}$ mice. Vet Path 47:120–131
Volk T, Schwoerer AP, Thiessen S et al (2003) A polycystin-2-like large conductance cation channel in rat left ventricular myocystes. Cardiovasc Res 58:76–88
Wallace DP (2011) Cyclic AMP-mediated cyst expansion. Biochim Biophys Acta 1812:1291–1300
Wallace DP, Grantham JJ, Sullivan LP (1996) Chloride and fluid secretion by cultured human polycystic kidney cells. Kidney Int 50:1327–1336
Walz G, Budde K, Mannaa M et al (2010) Everolimus in patients with autosomal dominant polycystic kidney disease. N Engl J Med 363:830–840
Wang X, Ward CJ, Harris PC et al (2010) Cyclic nucleotide signaling in polycystic kidney disease. Kidney Int 77:129–140
Wang CJ, Grantham JJ, Wetmore JB (2013) The medicinal use of water in renal disease. Kidney Int 84:45–53
Warner G, Hein KZ, Nin V et al (2016) Food restriction ameliorates the development of polycystic kidney disease. J Am Soc Nephrol 27:1437–1447
Watnick T, Germino GG (2010) mTOR inhibitors in polycystic kidney disease. New Engl J Med 363:879–881
Wei W, Hackmann K, Xu H et al (2007) Characterization of cis-autoproteolysis of polycystin-1, the product of human polycystic kidney disease 1 gene. J Biol Chem 282:21729–21737
Weston BS, Bagneris C, Price RG et al (2001) The polycystin-1 C type lectin domain binds carbohydrate in a calcium-dependent manner, and interacts with extracellular matrix proteins in vitro. Biochim Biophys Acta 1536:161–176
Wilson PD, Geng L, Li X et al (1999) The PKD1 gene product, "polycystin-1" is a tyrosine-phosphorylated protein that colocalizes with alpha2beta1-integrin in focal clusters in adherent renal epithelia. Lab Investig 79:1311–1323

Wong ATY, Mannix C, Grantham JJ et al (2018) Randomized controlled trial to determine the efficacy and safety of prescribed water intake to prevent kidney failure due to autosomal dominant polycystic kidney disease (PREVENT-ADPKD). BMJ Open 8:e018794

Woodward OM, Li Y, Yu S et al (2010) Identification of a polycystin-1 cleavage product, P100, that regulates store-operated Ca^{2+} entry through interactions with STIM1. PLoS One 5:e12305

Wu G, D'Agati V, Cai Y et al (1998) Somatic inactivation of *Pkd2* results in polycystic kidney disease. Cell 93:177–188

Wu G, Markowitz GS, Li L et al (2000) Cardiac defects and renal failure in mice with targeted mutations in Pkd2. Nature Genet 24:75–78

Wu Y, Dai X-Q, Li Q et al (2006) Kinesin-2 mediates physical and functional interactions between polycystin-2 and fibrocystin. Human Mol Genetics 15:3280–3292

Wu L, Gao X, Brown RC et al (2007) Dual role of the TRPV4 channel as a sensor of flow and osmolality in renal epithelial cells. Am J Physiol Renal Physiol 293:F1699–F1713

Wu Y, Xu JX, El-Jouni W et al (2016) Gα12 is required for renal cystogenesis induced by *Pkd1* inactivation. J Cell Sci 129:3675–3684

Wullschleger S, Loewith R, Hall MN (2006) TOR signaling in growth and metabolism. Cell 124:471–484

Xu GM, Sikaneta T, Sullivan BM (2001) Polycystin-1 interacts with intermediate filaments. J Biol Chem 276:46544–46552

Yamaguchi T, Nagao S, Kasahara M et al (1997) Renal accumulation and excretion of cyclic adenosine monophosphate in a murine model of slowly progressive polycystic kidney disease. Am J Kidney Dis 30:703–709

Yamaguchi T, Pelling JC, Ramaswamy NT et al (2000) cAMP stimulates the in vitro proliferation of renal cyst epithelial cells by activating the extracellular signal-regulated kinase pathway. Kidney Int 57:1460–1471

Yamaguchi T, Nagao S, Wallace DP (2003) cAMP activates B-Raf and ERK in cyst epithelial cells from autosomal dominant polycystic kidneys. Kidney Int 63:1983–1994

Yamaguchi T, Wallace DP, Magenheimer BS et al (2004) Calcium restriction allows cAMP activation of the B-Raf/ERK pathway, switching cells to a cAMP-dependent growth-stimulated phenotype. J Biol Chem 279:40419–40430

Yamaguchi T, Hempson SJ, Reif GA et al (2006) Calcium restores a normal proliferation phenotype in human polycystic kidney disease epithelial cells. J Am Soc Nephrol 17:178–187

Yang B, Sonawane ND, Zhao D et al (2008) Small-molecule CFTR inhibitors slow cyst growth in polycystic kidney disease. J Am Soc Nephrol 19:1300–1310

Yao G, Su X, Nguyen V et al (2014) Polycystin-1 regulates actin cytoskeleton organization and directional cell migration through a novel PC1-pacsin 2-N-wasp complex. Human Mol 23:2769–2779

Ye H, Wang X, Sussman CR et al (2016) Modulation of polycystic kidney disease severity by phosphodiesterase 1 and 3 subfamilies. J Am Soc Nephrol 27:1312–1320

Ye H, Wang X, Constans MM et al (2017) The regulatory 1α subunit of protein kinase a modulates renal cystogenesis. Am J Physiol Renal Physiol 313:F677–F686

Yoder BK, Hou X, Guay-Woodford LM (2002) The polycystic kidney disease proteins, polycystin-1, polycystin-2, polaris and cystin, are co-localized in renal cilia. J Am Soc Nephrol 13:2508–2516

Yoshihara D, Kurahashi H, Moriata M et al (2011) PPAR-γ agonist ameloriates kidney and liver disease in an orthologous rat model of human autosomal recessive polycystic kidney disease. Am J Physiol Renal Physiol 300:F465–F474

Yu S, Hackmann K, Gao J et al (2007) Essential role of cleavage of polycystin-1 at G protein-coupled receptor proteolytic site for kidney tubular structure. Proc Natl Acad Sci U S A 104:18688–18693

Yu Y, Ulbrich MH, Li MH et al (2009) Structural and molecular basis of the assembly of the TRPP2/PKD1 complex. Proc Natl Acad Sci U S A 106:1558–11563

Yu W, Ritchie BJ, Su X et al (2011) Identification of polycystin-1 and Galpha12 binding regions necessary for regulation of apoptosis. Cell Signal 23:213–221
Yuasa T, Venugopal B, Weremowicz S et al (2002) The sequence, expression and chromosomal localization of a novel polycystic kidney disease 1-like gene, PKD1L1, in human. Genomics 79:376–386
Yuasa T, Takakura A, Denker BM et al (2004) Polycystin-1L2 is a novel G-protein-binding protein. Genomics 84:126–138
Zafar I, Ravichandran K, Belibi FA et al (2010) Sirolimus attenuates disease progression in an orthologous mouse model of human autosomal dominant polycystic kidney disease. Kidney Int 78:754–761
Zaika O, Mamenko M, Berrout J et al (2013) TRPV4 dysfunction promotes renal cystogenesis in autosomal recessive polycystic kidney disease. J Am Soc Nephrol 24:604–616
Zhang P, Luo Y, Chasan B et al (2009) The multimeric structure of polycystin-2 (TRPP2): structural-functional correlates of homo- and hetero-multimers with TRPC1. Hum Mol Genet 18:1238–1251
Zhang ZR, Chu WF, Song B et al (2013) TRPP2 and TRPV4 form an EGF-activated calcium permeable channel at the apical membrane of renal collecting duct cells. PLoS One 8:e73424
Zhang B, Tran U, Wessely O (2018) Polycystin-1 loss-of-function is directly linked to an imbalance in G-protein signaling in the kidney. Development 145:dev158931
Zhu J, Yu Y, Ulbrich MH et al (2011) Structural model of the TRPP2/PKD1 C-terminal coiled-coil complex produced by a combined computational and experimental approach. Proc Natl Acad Sci U S A 108:10133–10138

Chapter 30
Renal Aquaporins in Health and Disease

Marleen L. A. Kortenoeven, Emma T. B. Olesen, and Robert A. Fenton

Abstract Aquaporins (AQPs) are a large family of membrane proteins that act as semiselective channels. The majority of AQPs are permeable to water, but a subset of the family can also transport glycerol, urea, and other small solutes. Currently, thirteen AQP homologues have been identified in mammals, termed AQP0–12. These aquaporins are highly abundant in epithelial and nonepithelial cells in various tissues, including the kidney, brain, liver, lungs, and salivary glands. In this chapter, we focus on AQPs expressed in kidney epithelial cells. We summarize the current knowledge with respect to their localization and function within the kidney tubule, and their critical role in mammalian water homeostasis. We describe a number of water balance disorders resulting from altered AQP function, and provide an overview of some of the treatment strategies for these disorders.

Keywords Urinary-Concentrating Mechanism · Nephrogenic Diabetes Insipidus · NDI · Hyponatremia · Vasopressin · Hypernatremia · Polyuria · Knockout Mouse

30.1 Introduction

Aquaporins (AQPs) are a large family of small membrane proteins (approximately 30 kDa) that act as semiselective channels. To date, thirteen aquaporins have been cloned in mammals, named AQP0–12, which are expressed in various tissues such as the kidney, brain, liver, lungs, and salivary glands (Ishibashi et al. 2009). Aquaporins have also been identified and cloned from amphibians, plants, yeast,

M. L. A. Kortenoeven · R. A. Fenton (✉)
Department of Biomedicine, Aarhus University, Aarhus, Denmark
e-mail: mkor@biomed.au.dk; rofe@biomed.au.dk

E. T. B. Olesen
Department of Biomedical Sciences, Faculty of Health and Medical Sciences, University of Copenhagen, Copenhagen, Denmark
e-mail: etbo@sund.ku.dk

K. L. Hamilton, D. C. Devor (eds.), *Studies of Epithelial Transporters and Ion Channels*, Physiology in Health and Disease,
https://doi.org/10.1007/978-3-030-55454-5_30

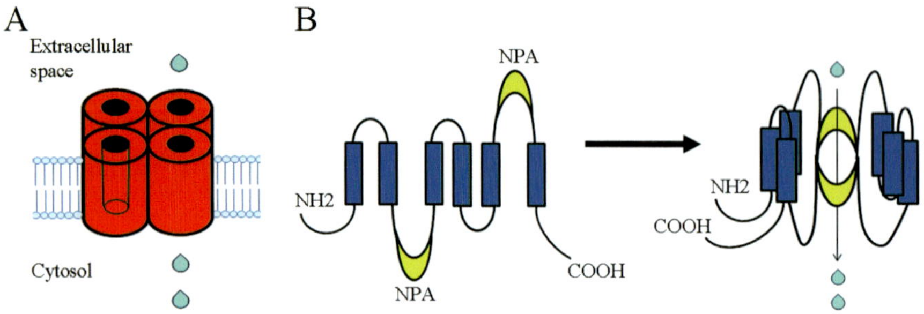

Fig. 30.1 Schematic representation of the structural organization of aquaporins. (**a**) Four AQP monomers assemble to form tetramers, which are the functional units in the membrane. Each AQP monomer has a central water-transporting pore. (**b**) A single monomer of an AQP channel consists of six membrane spanning alpha-helices and has short intracellular amino and carboxyl termini. Almost all AQPs have two highly conserved Asn-Pro-Ala (NPA) motifs in each monomer. The NPA sequences have been proposed to interact in the membrane to form the pathway for translocation of water across the plasma membrane. Modified from Moeller HB, Fenton RA (2012) Cell biology of vasopressin-regulated aquaporin-2 trafficking. Pflugers Arch 464:133–144

bacteria and various lower organisms, where they likely fulfill a variety of functions. This chapter specifically focuses on mammalian renal AQPs.

A single monomer of an AQP channel consists of six membrane spanning alpha-helices that have a central pore through which the majority of transport occurs (see Fig. 30.1) (Murata et al. 2000; Sui et al. 2001). AQPs have short intracellular amino and carboxyl termini, which contain major regulatory domains for channel function. Almost all AQPs have two highly conserved Asn-Pro-Ala (NPA) motifs in each monomer. Structural analysis of AQP1 has shown that these motifs reside on opposite sides of the AQP1 monomer and are important for water-selective pore formation (Murata et al. 2000). The two most recently identified mammalian AQPs, AQP11 and AQP12, have a conserved C-terminal NPA motif, but the N-terminal NPA motif is deviated to an Asn-Pro-Cys (NPC) motif for AQP11 and to an Asn-Pro-Thr (NPT) motif for AQP12 (Ishibashi et al. 2014). The function of AQP12 is still unclear, but AQP11 does have low water channel activity, despite the incompletely conserved NPA (Yakata et al. 2007, 2011). Four AQP monomers assemble to form tetramers, which are the functional units in the membrane. Precisely where these tetramers form is unknown, and most likely formation occurs differently for each AQP family member. For one of the mammalian AQPs, AQP4, these tetramers further assemble into supramolecular square arrays (Yang et al. 1996), which may further influence the activity and molecular regulation of AQP4 (Fenton et al. 2010; Silberstein et al. 2004).

Water, if given enough time, can cross the majority of biological lipid bilayers by simple diffusion. However, this process is relatively slow and energetically unfavorable. The water permeability of the lipid membrane is greatly enhanced by insertion of AQPs, which provide a semiselective pore through which molecules can rapidly translocate. Although termed AQPs based on their initial characterization, besides water, a subset of the mammalian AQPs, the aquaglyceroporins AQP3,

AQP7, AQP9, and AQP10, can transport glycerol and urea ((Ishibashi et al. 1994, 1997a, 2002; Ko et al. 1999) reviewed in (Rojek et al. 2008)), while both AQP7 and AQP9 are permeable for arsenite (Liu et al. 2002). In addition, AQP9 is permeable to a wide range of noncharged solutes, like mannitol, sorbitol, purines, and pyrimidines (Tsukaguchi et al. 1998). AQP6 may function primarily as an anion transporter (Yasui et al. 1999a; Ikeda et al. 2002). AQPs may also transport other small molecules and gases, including carbon dioxide, ammonia, nitric oxide, and hydrogen peroxide (Musa-Aziz et al. 2009; Miller et al. 2010; Wang and Tajkhorshid 2010). AQPs also play other roles in the cell, including a structural role in, e.g., cell-to-cell adhesion. For example, AQP0, a lens-specific AQP molecule, forms single- and double-layered crystals, each layer in an opposing membrane, suggesting that AQP0 has an adhesive role in the ocular lens fiber cells (Fotiadis et al. 2000). In addition, AQP2 interacts with integrins to promote renal epithelial cell migration, contributing to the structural and functional integrity of the mammalian kidney (Chen et al. 2012).

30.2 Renal Aquaporins

Nine of the known AQPs are expressed in the kidney, with various cellular and subcellular distributions (Fig. 30.2). AQP1 is constitutively localized in the plasma membranes of proximal tubule cells, cells of the descending thin limbs of long loops of Henle, and endothelial cells from the descending vasa recta (Nielsen et al. 1993b, 1995b; Zhai et al. 2007). AQP2, AQP3, and AQP4 are expressed in cells from the connecting tubule, collecting duct principal cells and inner medulla collecting duct cells. Whereas AQP2 is predominantly expressed in the apical plasma membrane and membranes of intracellular vesicles (it traffics between the two compartments, see later), AQP3 and AQP4 are constitutively localized to the basolateral plasma membrane (Terris et al. 1995; Fushimi et al. 1993; Nielsen et al. 1993a; Ecelbarger et al. 1995). AQP5 is expressed in the apical membrane of intercalated cells in the connecting tubule and the cortical part of the collecting duct (Procino et al. 2011b), cells that are predominantly involved in acid:base balance rather than water transport. AQP7 is localized in the apical plasma membrane of cells from the S3 segment of proximal tubules (Nejsum et al. 2000; Ishibashi et al. 2000). AQP6, AQP8, and AQP11 are not detected in the plasma membrane but are localized to intracellular membranes. AQP6 is localized in membrane-bound vesicles within intercalated cells of the connecting tubule and collecting duct (Yasui et al. 1999b; Ohshiro et al. 2001). Both AQP8 and AQP11 are localized intracellularly in proximal tubule cells, while AQP8 has also been detected in the collecting ducts (Elkjaer et al. 2001; Morishita et al. 2005).

In addition to removal of various waste substances from the body, maintaining acid:base balance and tight control of plasma electrolyte concentrations, the kidney plays a critical role in regulation of body water content. Body water balance is tightly controlled by regulating both water intake and urinary water excretion. In the kidney, 180 liters of plasma are filtered by the human glomeruli each day, of which less than

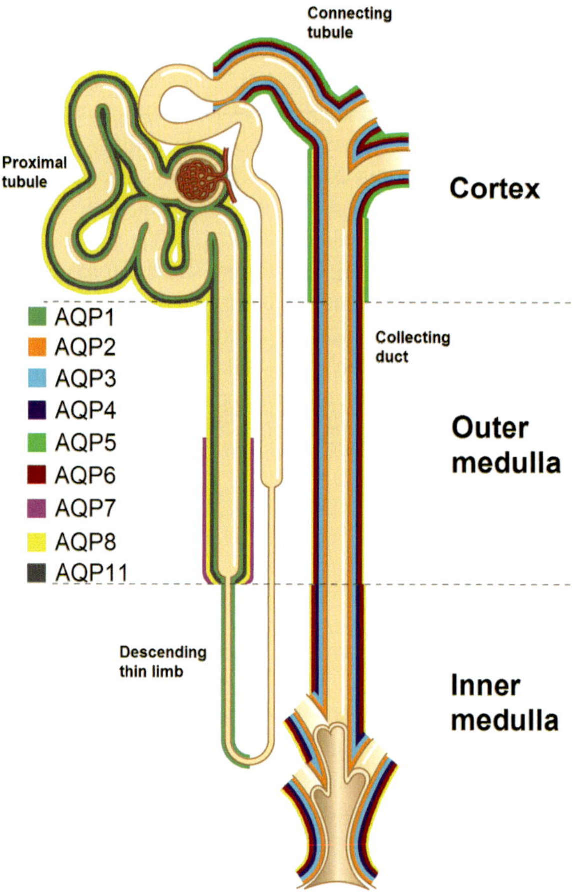

Fig. 30.2 Expression of aquaporins along the nephron. Blood is filtered at the glomerulus, and the filtrate is modified as it travels through the nephron to make the final urine. Most of the glomerular filtrate is reabsorbed through AQP1 in the proximal tubule and descending thin limbs of Henle, although AQP7 is also expressed in the S3 segment of the proximal tubule. AQP1 is also expressed in the descending vasa recta, facilitating the removal of water. In the connecting tubule and collecting duct, AQP2 is mainly expressed at the apical membrane and intracellular vesicles of principal cells, while AQP3 and AQP4 are present at the basolateral membrane, representing exit pathways for water reabsorbed via AQP2. AQP5 is expressed in the apical membrane of intercalated cells in the connecting tubule and the cortical collecting duct. In contrast to these AQPs, AQP6, AQP8 and AQP11 are localized in intracellular membranes only. AQP6 is localized to intercalated cells of the collecting duct and connecting tubule, AQP8 is expressed in proximal tubules and weakly in collecting ducts, while AQP11 is localized to proximal tubules. Modified from Kortenoeven ML, Fenton RA (2014) Renal aquaporins and water balance disorders. Biochim Biophys Acta 1840:1533–1549

1% is finally excreted in the urine (Stanton and Koeppen 1998). Approximately 67% of the filtered water is reabsorbed in the proximal tubule and 15% in the descending thin limb of Henle, which are both constitutive processes. Depending on the body's needs, the remaining fluid can be reabsorbed in the connecting tubule and collecting duct, defining the final urine concentration. The processes of water reabsorption are tightly regulated by a variety of hormones, and systemic and local factors. A major regulator of body water balance is the peptide hormone arginine vasopressin (AVP), the effects of which allow the body to adapt to periods of water load or water restriction (Stanton and Koeppen 1998). Of the nine AQPs expressed in the kidney, only AQP1–4 and possibly AQP7 have a role in the process of urine concentration and maintenance of body water balance. In the following section, we provide an overview of each of the renal AQPs, focusing on their role in renal epithelial cell and kidney function. Readers are directed toward an extensive catalogue of reviews that provide more details of individual AQP function and regulation (Rojek et al. 2008; Ishibashi et al. 2009, 2014; Verkman 2008; Badaut et al. 2014; Schey et al. 2014; Delporte 2014).

30.2.1 Aquaporin 1 (AQP1)

AQP1 is the first molecularly identified aquaporin and was originally cloned from erythrocytes. AQP1 is expressed in the apical and basolateral membrane of epithelial cells in the proximal tubule (Nielsen et al. 1993b), where the majority of fluid filtered by the glomerulus is reabsorbed by an active near-isomolar transport mechanism. AQP1 is also expressed in the descending thin limb of long-looped, but not short-looped, nephrons, and the descending vasa recta (Nielsen et al. 1995b; Zhai et al. 2007). In addition, AQP1 was also detected in the thick ascending limb of the loop of Henle (TAL) of mice and rats, where it is localized both intracellularly and on the basolateral plasma membrane (Cabral and Herrera 2012). This finding is controversial due to the impermeability of this segment to water. Although AQP1 expression is constitutively high, it can be modulated by hypertonicity and angiotensin II (Bouley et al. 2009). AQP1 expression is higher in male rats than in females, and AQP1 expression can be strongly increased by testosterone and moderately increased by progesterone (Herak-Kramberger et al. 2015). The antidiuretic hormone AVP does not regulate AQP1 expression in the proximal tubule, descending thin limb or vasa recta; however, the expression in the TAL was shown to be increased in rats by chronic infusion of the synthetic vasopressin analogue dDAVP (1-desamino-8-D-arginine vasopressin, Desmopressin), which could be either direct or secondary to an increased interstitial osmolality (Terris et al. 1996; Cabral and Herrera 2012). Surface expression of AQP1 is regulated on the short term in response to flow, with higher luminal perfusion rates in isolated perfused proximal tubules increasing AQP1 in both brush border apical and basolateral membranes (Pohl et al. 2015).

AQP1 knockout mice have a reduced urine osmolality and increased urine volume compared with wild-type mice, and urine osmolality is not increased in response to injection of dDAVP (Ma et al. 1998; Schnermann et al. 1998). In addition, AQP1 knockout mice do not increase urine osmolality in response to water deprivation, and become severely dehydrated and lethargic, altogether showing that AQP1 plays an essential role in maintaining body water balance. AQP1 deletion produces a 78% decrease in osmotic water permeability across the proximal tubule epithelium, indicating that the major pathway for osmotically driven water transport in the proximal tubule is transcellular and mediated by AQP1 water channels (Schnermann et al. 1998). In addition, the rate of proximal fluid reabsorption in AQP1 knockout mice is reduced by approximately 50% relative to control mice. Osmotic water permeabilities of microperfused long thin descending limbs of Henle and outer medullary descending vasa recta are reduced in AQP1 knockout mice (Chou et al. 1999; Pallone et al. 2000). However, micropuncture flow measurements of distal nephron segments demonstrated normal distal fluid delivery in AQP1 knockout mice, because of a decrease in the single-nephron glomerular filtration rate, probably caused by activation of the tubuloglomerular feedback mechanism (Schnermann et al. 1998). The increased urinary output, despite normal distal delivery, suggests that the urine-concentrating defect seen in AQP1 knockout mice results primarily from, ultimately, reduced fluid absorption in the collecting duct. The decreased water permeability in the thin limbs and vasa recta should reduce countercurrent multiplication and countercurrent exchange, respectively, which will prevent the formation of a hypertonic medullary interstitium and limit water transport across the collecting duct epithelium. This conclusion is supported by the absence of an increase in urine osmolality in AQP1 knockout mice after dDAVP stimulation of collecting duct water permeability (Ma et al. 1998). Consistent with the phenotype of AQP1 knockout mice, humans with a loss-of-function mutation in AQP1 are unable to maximally concentrate their urine when challenged by water deprivation (King et al. 2001).

AQP1 in the TAL is unlikely to be involved in the process of urine concentration, as the TAL is impermeable to water, indicating an absence of water channels at the apical membrane. However, TAL cells rapidly shrink and/or swell in response to acute changes in basolateral osmolality. AQP1 could therefore play an important role in the regulation of TAL cell volume subsequent to changes in interstitial osmolality, such as during a high-salt diet or water deprivation. This is supported by the 50% reduction in the rate of water flux across the basolateral side of TALs isolated from AQP1 knockout mice (Cabral and Herrera 2012).

30.2.2 Aquaporin 2 (AQP2)

AQP2 is abundantly expressed in the plasma membrane and intracellular membrane vesicles of the principal cells along the whole connecting tubule and collecting duct (Fushimi et al. 1993; Nielsen et al. 1993a). Although AQP2 is mainly associated with the apical plasma membrane, AQP2 is also found in the basolateral plasma membrane, especially in the connecting tubule and inner medulla collecting duct (Christensen et al. 2003). AQP2 localization and abundance are regulated by several hormones and signaling molecules, including AVP. During hypernatremia,

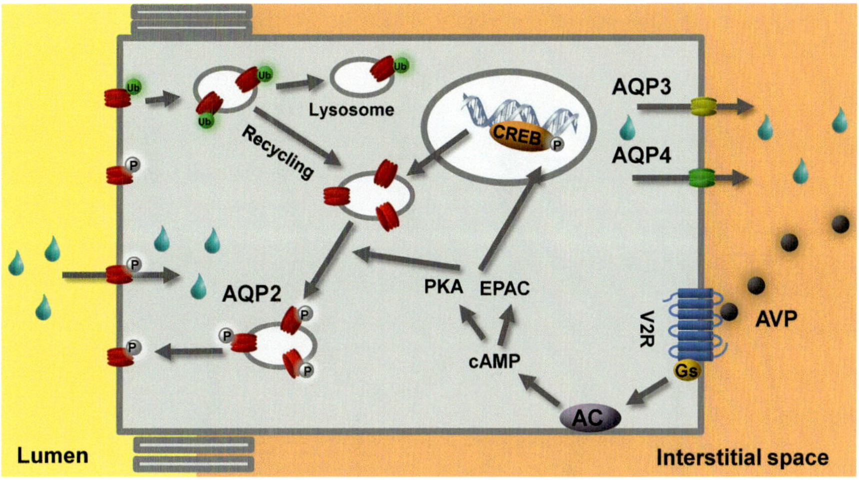

Fig. 30.3 Regulation of AQP2-mediated water reabsorption. AVP binds to the vasopressin type-2 receptor (V2R) at the basolateral membrane of collecting duct and collecting duct principal cells. This induces a signaling cascade, involving Gs protein mediated activation of adenylate cyclase (AC), a rise in intracellular cAMP, activation of protein kinase A (PKA) and possibly the Exchange factor directly activated by cAMP (EPAC), and subsequent phosphorylation of AQP2. This results in the redistribution of AQP2 from intracellular vesicles to the apical membrane. Driven by the transcellular osmotic gradient, water will then enter principal cells through AQP2 and leave the cell via AQP3 and AQP4, which are expressed in the basolateral membrane, resulting in concentrated urine. In the long term, vasopressin also increases AQP2 expression via phosphorylation of the cAMP responsive element binding protein (CREB), which stimulates transcription from the AQP2 promoter. Once the water balance is restored, AVP levels drop and AQP2 is internalized via ubiquitination. Internalized AQP2 can either be targeted to recycling pathways or to degradation via lysosomes. Modified from Kortenoeven ML, Fenton RA (2014) Renal aquaporins and water balance disorders. Biochim Biophys Acta 1840:1533–1549

hypovolemia, or hypotension, AVP is released from the posterior pituitary gland into the bloodstream (Baylis 1989; Bankir 2001; Voisin and Bourque 2002). AVP regulates the body's retention of water, increasing both the osmotic driving force for water reabsorption and the transcellular route for water transport, causing the kidneys to concentrate the urine. The transcellular route of regulation mainly occurs via modulating plasma membrane expression of AQP2. AVP interacts with the vasopressin type-2 receptor (V2R), present in the basolateral membrane of renal connecting tubule and collecting duct principal cells (Mutig et al. 2007; Fenton et al. 2007). This initiates a signal transduction cascade, involving Gs protein–mediated activation of adenylate cyclase, a rise in intracellular cAMP, activation of protein kinases, and subsequent phosphorylation of AQP2 water channels. This results in the redistribution of AQP2 from intracellular storage vesicles to the apical membrane (Kuwahara et al. 1995; Katsura et al. 1997; Nielsen et al. 1995a), greatly increasing the rate-limiting water permeability of the apical plasma membrane (see Fig. 30.3) (Strange and Spring 1987; Flamion and Spring 1990). Water will now enter the principal cells through AQP2 driven by the transcellular osmotic gradient of sodium

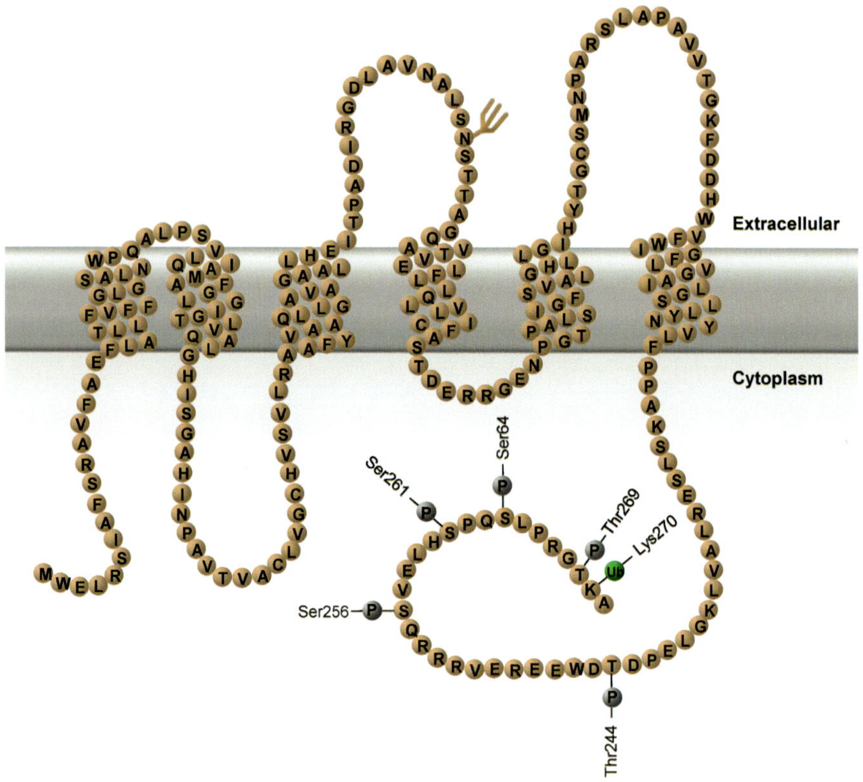

Fig. 30.4 Schematic representation of the human. AQP2 protein AQP2 is a 271 amino acid protein that is predicted to have six transmembrane domains with an intracellular N- and C-terminus. The known phosphorylation (P) and ubiquitination (Ub) sites are indicated. Modified from Kortenoeven ML, Fenton RA (2014) Renal aquaporins and water balance disorders. Biochim Biophys Acta 1840:1533–1549

and (in the medulla) urea, and will leave the cells through AQP3 and AQP4 in the basolateral membrane (Ecelbarger et al. 1995; Terris et al. 1995), leading to a concentration of the urine. Once water balance is restored, AVP levels return to normal and AQP2 is internalized, which returns the apical membrane water permeability to basal levels. AQP2 is retrieved to intracellular vesicles, resulting in either lysosomal degradation or intracellular storage and relocation to the plasma membrane upon restimulation with AVP (Kamsteeg et al. 2006; Katsura et al. 1996).

A number of post-translational modifications of AQP2 are required for the trafficking of AQP2 to/from the apical plasma membrane (for extensive review, see (Moeller et al. 2011)). AQP2 can be phosphorylated at five residues, Thr244, Ser256, Ser261, Ser264, and Ser269 (Thr269 in human AQP2, see Fig. 30.4) (Hoffert et al. 2006, 2012). The role of Thr244 phosphorylation is unknown. Phosphorylation of Ser256, Ser264, and Ser269 are increased in abundance upon dDAVP stimulation, while Ser261 phosphorylation is decreased (Hoffert et al. 2007,

2008; Fenton et al. 2008). Phosphorylated Ser261-AQP2 is mainly found in intracellular vesicles and seems to be absent from the plasma membrane (Hoffert et al. 2007). Ser261 phosphorylation is thought to stabilize AQP2 ubiquitination (Tamma et al. 2011). Phosphorylated Ser256-AQP2 and Ser264-AQP2 reside in intracellular vesicles and the plasma membrane, and upon stimulation with dDAVP there is a change in distribution (or increase in abundance) to the plasma membrane (Christensen et al. 2000; Fenton et al. 2008). The function of Ser264-AQP2 phosphorylation is currently unclear. Ser256 can be phosphorylated by protein kinase A (PKA) upon AVP stimulation and phosphorylation at this site is essential for AQP2 accumulation in the plasma membrane (Katsura et al. 1997; Fushimi et al. 1997; Van Balkom et al. 2002). A study using *Xenopus laevis* oocytes as a model system for exogenous AQP2 expression showed that Ser256 phosphorylation of at least three of four monomers of an AQP2 tetramer is needed for AQP2 to localize to the apical plasma membrane (Kamsteeg et al. 2000). Phosphorylated Ser269-AQP2 is solely found in the apical plasma membrane (Moeller et al. 2009). Ser269 phosphorylation is thought to be involved in apical plasma membrane retention of AQP2 (Hoffert et al. 2008) and interaction of AQP2 with proteins involved in endocytosis (Moeller et al. 2010, 2016). In addition to phosphorylation, AQP2 is also modified by ubiquitylation at Lys270, which is believed to mediate the endocytosis of AQP2 from the plasma membrane upon AVP removal (Kamsteeg et al. 2006). The ubiquitin ligase CHIP is involved in AQP2 ubiquitylation (Wu et al. 2018), and the ubiquitin-specific protease USP4 interacts with AQP2 to modulate its membrane accumulation (Murali et al. 2019). Although AQP2 Ser269 phosphorylation and Lys270 ubiquitylation can occur in parallel, the effects of phosphorylation can override the effect of ubiquitylation on AQP2 endocytosis, possibly by decreasing the interaction of AQP2 with proteins of the endocytic machinery (Moeller et al. 2014). AQP2 is subjected to S-glutathionylation in kidney and in HEK-293 cells stably expressing AQP2, which is modulated by changes in cellular content of reactive oxygen species (Tamma et al. 2014b). It appears that the cholesterol content of the basolateral plasma membrane is important for AQP2 post-translational modification and the regulated apical plasma membrane targeting of AQP2 (Moeller et al. 2018).

In addition to playing an acute role in AQP2 translocation, AVP-mediated increases in cAMP also increase AQP2 abundance via PKA-mediated phosphorylation of the cAMP responsive element–binding protein (CREB), which stimulates transcription from the AQP2 promoter (Hozawa et al. 1996; Yasui et al. 1997; Matsumura et al. 1997). The PKA-CREB pathway is involved in the initial increase in AQP2 abundance after AVP stimulation, but not in the long-term effect of AVP (Kortenoeven et al. 2012b). Instead, activation of the exchange protein directly activated by cAMP (EPAC) levels may be involved in the long-term regulation of AQP2 (Kortenoeven et al. 2012b). EPAC might also play a role in the regulation of AQP2 localization, as perfusion of mouse inner medullary collecting ducts with an EPAC activator resulted in intracellular Ca^{2+} mobilization and translocation of AQP2 to the apical membrane (Yip 2006).

In addition to cAMP, alternative signaling pathways that do not involve cAMP play a role in AVP-mediated AQP2 membrane targeting, although the precise nature of these alternative pathways has not currently been elucidated (Olesen and Fenton 2017; Olesen et al. 2016). Furthermore, several other factors besides AVP stimulate AQP2 expression or translocation, for example, extracellular tonicity, insulin, or long-term aldosterone stimulation (Hasler et al. 2003, 2005; Bustamante et al. 2005). Other factors decrease AVP-induced AQP2 abundance or trafficking, like extracellular purines, prostaglandin E2, and dopamine (Boone et al. 2011; Nejsum et al. 2005; Zelenina et al. 2000; Wildman et al. 2009) (for review see (Boone and Deen 2008)). Although prostaglandin E2 antagonizes AVP-induced water permeability, prostaglandin E2 or selective prostaglandin E2 receptor agonists increase the translocation and abundance of AQP2 in the absence of high levels of AVP (Olesen et al. 2011; Li et al. 2009). Studies on transgenic mice have demonstrated an essential role of AQP2 in water handling. Mice with total AQP2 gene deletion or with a Thr126Met mutation in the AQP2 gene, shown to result in a failure of delivery of mature AQP2 protein to the plasma membrane, die within the first few days of life (Rojek et al. 2006; Yang et al. 2000). Collecting duct-specific AQP2 knockout mice have polyuria and growth retardation, but are viable to adulthood, as are mice with an inducible AQP2 gene deletion resulting in >95% reduction of renal AQP2 (Rojek et al. 2006; Yang et al. 2006a). Both mouse models have severe polyuria, accompanied by a very low urine osmolality. In these mice, the urine osmolality is not increased after water deprivation. Mice with a specific deletion of AQP2 in the connecting tubules have defective renal water handling under basal conditions, with a 1.5-fold higher urine volume and reduced urine osmolality (Kortenoeven et al. 2013a), indicating that AQP2 in the connecting tubules also plays an important role in regulating body water balance. However, in contrast to the collecting duct-specific knockout, these mice are able to concentrate their urine under conditions of high circulating AVP, suggesting that the connecting tubule plays a minor, yet still significant contribution to water conservation, and the collecting duct has the major role in AVP-mediated antidiuresis. Alternatively, increased water reabsorption in the collecting duct can compensate for dysfunction of the connecting tubule. Although a loss-of-function mutation in the AQP2 gene is not lethal in humans, it does lead to a severe urinary-concentrating defect known as nephrogenic diabetes insipidus (NDI), confirming the importance of AQP2 in renal water handling (Robben et al. 2006a). Autoantibodies targeting AQP2 are also a cause of tubulointerstitial nephritis resulting in end-stage renal disease (Landegren et al. 2016).

30.2.3 Aquaporin 3 (AQP3)

AQP3 is constitutively expressed in the basolateral membrane of the connecting tubule and collecting duct principal cells (Ecelbarger et al. 1995; Ishibashi et al. 1997b; Coleman et al. 2000), where it functions as an exit pathway for water reabsorbed via AQP2. There is no evidence for short-term regulation of AQP3 by AVP-induced trafficking. However, water deprivation or AVP infusion, resulting in long-term increases in AVP levels, increase AQP3 protein and mRNA levels in cortex and medulla (Ishibashi et al. 1997b; Terris et al. 1996; Murillo-Carretero et al. 1999; Ecelbarger et al. 1995). In addition, hypoxia increases mRNA expression of AQP3 in the mouse kidney (Hoogewijs et al. 2016). AQP3 is acetylated at lysine 282 in rat inner medullary collecting ducts (Hyndman et al. 2018), although the functional significance of this modification is not yet known. AQP3 knockout mice have increased urine volume and a lower urine osmolality. In addition, osmotic water permeability of the cortical collecting duct basolateral membrane is more than three-fold reduced by AQP3 deletion, showing the importance of AQP3 for water transport in this segment (Ma et al. 2000). After water deprivation or dDAVP injection, AQP3 knockout mice are however able to concentrate their urine partially to around 30% of that in wild-type mice, which might be due to the remaining presence of basolateral AQP4 or AQP2 in the principal cells. In agreement with this, AQP3/AQP4 double-knockout mice have a greater impairment of urinary-concentrating ability than AQP3 single-knockout mice (Ma et al. 2000).

Besides water, AQP3 can transport glycerol, ammonia, and urea (Holm et al. 2005). The glycerol-transporting function of AQP3 plays an important role in skin hydration (Rojek et al. 2008), and AQP3 knockout mice develop hypotriglyceridemia, which may be related to the glycerol-transporting function of AQP3 in the kidney (Ma et al. 2000). The physiological role of renal AQP3-mediated glycerol, ammonia, or urea transport has however not been delineated in detail.

30.2.4 Aquaporin 4 (AQP4)

AQP4 is localized to the basolateral membrane of connecting tubule and collecting duct principal cells, with the highest levels in the inner medullary collecting duct, where it represents an additional pathway for water exit from the principal cells into the interstitium (Terris et al. 1995; Coleman et al. 2000; Mobasheri et al. 2007). AQP4 is expressed in the basolateral plasma membrane of proximal tubule S3 segments of mice also, which is likely to be species-specific, as no AQP4 expression was detected in the S3 segment of humans or rats (Van Hoek et al. 2000). Six alternatively spliced isoforms of AQP4 have been identified in rats. Three of these isoforms transport water to different extent: the short M23 consisting of 301 amino acids, M1 consisting of 323 amino acids, and the long Mz consisting of 364 amino acids (Moe et al. 2008; Fenton et al. 2010). The M23 isoform gives rise to formation of supramolecular square arrays (Furman et al. 2003). AQP4 is also phosphorylated on at least six different serine residues in the COOH terminus, but these do not appear to be required for proper plasma membrane localization of AQP4 or to act as a molecular switch to gate the water channel (Assentoft et al. 2014).

Long-term increased AVP levels do not increase AQP4 protein abundance in the renal inner medulla, where AQP4 is constitutively highly expressed (Terris et al. 1995, 1996). However, AQP4 mRNA is increased in the cortex, where unstimulated AQP4 expression is low, and this increase could partly be prevented by treatment with a V2R antagonist, suggesting that AVP regulates AQP4 expression (Murillo-Carretero et al. 1999). This was confirmed by a study in AVP-deficient Brattleboro rats, showing that chronic dDAVP administration increases AQP4 protein abundance in the connecting tubule, cortical collecting duct, and outer part of the inner medullary collecting duct, while decreasing AQP4 abundance in the middle and inner segments of the inner medullary collecting duct (Poulsen et al. 2013). In addition, long-term AVP infusion of Brattleboro rats increases the short AQP4 splice form M23 and decreases the long isoform M1 (Van Hoek et al. 2009).

Transepithelial osmotic water permeability in microperfused, AVP-stimulated, inner medullary collecting duct of AQP4 knockout mice is four-fold lower than in wild-type mice, indicating that AQP4 is the dominant basolateral membrane water channel in inner medullary collecting ducts (Chou et al. 1998). However, under basal conditions, AQP4 knockout mice do not have a difference in urine osmolality compared with wild-type mice (Ma et al. 1997). After water deprivation, urine osmolality is significantly reduced in AQP4 knockout mice compared to controls, and this is not further increased by injection of dDAVP, showing that AQP4 knockout mice have a mild urinary-concentrating defect. The large decrease in inner medullary water permeability and the absence of a more profound defect in urinary-concentrating ability in mice lacking AQP4 suggests that the majority of collecting duct water reabsorption occurs in segments proximal to the inner medullary collecting duct. This is in agreement with micropuncture studies showing that under antidiuretic conditions, the amount of water reabsorbed in the connecting tubule and initial collecting duct is much greater than that absorbed in the medullary collecting duct (Lassiter et al. 1961).

30.2.5 Aquaporin 5 (AQP5)

AQP5 is expressed in rat, mice, and human kidneys in the apical membrane of type-B intercalated cells in the connecting tubule and the cortical part of the collecting duct (Procino et al. 2011b). AQP5 was absent from intercalated cells in the medullary part of the collecting duct (Procino et al. 2011b). The absence of a detectable AQP in the basolateral membranes of type-B intercalated cells suggests that AQP5 does not mediate net transepithelial water reabsorption in this cell type. Instead, AQP5 has been speculated to be acting as an osmosensor in these cells, regulating the cell volume when exposed to the luminal hypotonicity produced by solute reabsorption (in the absence of water reabsorption) from the thick ascending limb and distal convoluted tubule. Chronic potassium depletion in mice, a treatment that elicits metabolic alkalosis, resulted in a strong reduction in both pendrin and AQP5 expression in type-B intercalated cells. Simultaneously, both proteins shifted distribution from the apical cell membrane to an intracellular compartment (Procino et al. 2013), suggesting that pendrin and AQP5 might be subject to a common regulatory pathway. AQP5-dependent water flux could be coupled to pendrin-mediated Cl^- influx. Supporting the localization and role of AQP5 in type-B intercalated cells, functional AQP5 plasma membrane expression can be induced by extracellular GTP in M1-CCD cells, a cell line that shows an intermediate phenotype between principal and intercalated cells (Mancinelli et al. 2014).

Additional evidence for a role of AQP5 in the kidney comes from studies of a renal-collecting duct/connecting tubule specific knockout model of Dot1l, a histone H3 K79 methyltransferase (Wu et al. 2013). In wild-type control mice, AQP5 was barely detectable. However, in the Dot1a knockout mice, AQP5 was robustly expressed in renal intercalated cells and principal cells. In the principal cells from knockout mice, AQP5 colocalized with AQP2. AQP5 could also be coimmunoprecipitated with AQP2, and AQP5 appears to impair the cell surface expression of AQP2. Despite this evidence for AQP5 expression in renal epithelial cells, the function of AQP5 in the kidney needs further investigation.

30.2.6 Aquaporin 6 (AQP6)

AQP6 is an unusual AQP, in that it possesses anion permeability that is activated by low pH or $HgCl_2$. When expressed in *Xenopus laevis* oocytes, AQP6 exhibits low basal water permeability; however, when exposed to Hg^{2+} or low pH, the water permeability of AQP6 expressing oocytes rises (Yasui et al. 1999a). In addition, AQP6 is permeated by anions in response to acidic pH or Hg^{2+} activation, with selectivity for nitrate (Ikeda et al. 2002; Yasui et al. 1999a). Each subunit of the AQP6 tetramer has an individual ion-conducting pore (Ikeda et al. 2002). AQP6 is expressed exclusively in the kidney (Ma et al. 1996; Ohshiro et al. 2001) where, in contrast to most other AQPs, it is localized only in intracellular membranes. Localization of AQP6 in these intracellular membranes is dependent on the N-terminus of AQP6 (Beitz et al. 2006). AQP6 has a relatively broad distribution in the kidney. It is detected in type-A intercalated cells of the renal connecting tubule and cortical- and medullary-collecting duct, where it colocalizes with the V-type H-ATPase in intracellular vesicles (Yasui et al. 1999a; Ohshiro et al. 2001). In addition, AQP6 has been detected in membrane vesicles within podocyte cell bodies and foot processes and within the subapical compartments of cells from segments 2 and 3 of the proximal tubule, although this localization is debated (Yasui et al. 1999b; Ohshiro et al. 2001). AQP6 expression is increased in a rat model of chronic alkalosis, but the cellular distribution of AQP6 is not altered (Promeneur et al. 2000). In addition, water-loading or lithium-induced nephrogenic diabetes insipidus increases AQP6 expression, while water deprivation decreases the expression of AQP6 (Promeneur et al. 2000; Ohshiro et al. 2001). Although the physiological role of AQP6 is still undefined, its location and anion permeability suggest a role in the acidification of intracellular vesicles. As nitrate can inhibit the V-type H-ATPase, AQP6 may be involved in the regulation of H-ATPase activity in acid-secreting type-A intercalated cells (Ikeda et al. 2002).

30.2.7 Aquaporin 7 (AQP7)

AQP7, a member of the aquaglyceroporins that can transport both water and glycerol, is expressed in the apical brush border of cells from the S3 segment of the proximal tubule (Nejsum et al. 2000; Ishibashi et al. 2000). Although AQP7 is a functional water channel, as emphasized by the reduced osmotic water permeabilities in apical membrane vesicles isolated from the proximal tubule of AQP7 knockout mice (Sohara et al. 2005), AQP7 knockout mice do not have impaired renal water handling or urinary-concentrating ability. This might be because in the S3 segment of the proximal tubule, AQP1 is abundant, making AQP7 water transport redundant. This hypothesis was confirmed in AQP1-AQP7 double knockout mice, which had a significantly greater urine volume accompanied by a proportional decrease in urine osmolality compared with AQP1 knockout mice, suggesting that the amount of water reabsorbed through AQP7 in the proximal straight tubules is physiologically substantial. In contrast to the absence of a defect in water handling, AQP7 knockout mice show significantly increased renal glycerol excretion (Sohara et al. 2005); thus, the main physiological role of AQP7 in the kidney might therefore be related to glycerol, rather than water, reabsorption.

30.2.8 Aquaporin 8 (AQP8)

AQP8 is permeable to water, but it can also transport ammonia (Liu et al. 2006; Holm et al. 2005). Ammonia is produced predominantly from the metabolism of glutamine in the mitochondria of the proximal tubule and is secreted into the urine by specialized transport processes in several nephron segments (Knepper et al. 1989). The regulation of renal ammonium excretion plays a central role in the maintenance of systemic acid-base balance (Knepper et al. 1989). AQP8 is expressed along the – length of the proximal tubule (Elkjaer et al. 2001), with weaker expression in the collecting ducts. In proximal tubule cells, AQP8 immunolabeling was confined to the cytoplasm, with no labeling of the apical brush border and an absence of, or very weak, labeling of basolateral plasma membrane domains. Similarly, in collecting duct cells, the weak labeling of AQP8 was associated with intracellular structures. Intracellularly, AQP8 is located in the inner mitochondrial membrane in the rat kidney, which is similar to its localization in rat hepatocytes (Lee et al. 2005). Despite its high water conductance, AQP8 does not contribute greatly to the facilitated transport of water across the mitochondrial membrane (Yang et al. 2006c; Calamita et al. 2006). Instead, AQP8 efficiently facilitates the transport of ammonia through mitochondrial membranes (Soria et al. 2010). Comparative studies demonstrated that the AQP8-mediated movement of ammonia across mitochondrial membranes was markedly higher than that of water (Soria et al. 2010), suggesting that ammonia diffusional transport is a major function for mitochondrial AQP8. In addition, AQP8 is expressed in the mitochondria of a human renal proximal tubule cell line, and knockdown of its expression decreased the rate of ammonia release into the culture medium, either from cells grown at neutral or acidic pH. AQP8 expression was also increased in this cell line in response to metabolic acidosis, together with a greater ammonia excretion rate (Molinas et al. 2012). In vivo, ammonia-loaded rats, which are a model of metabolic acidosis, have increased AQP8 expression in the kidney cortex (Molinas et al. 2012). Taken together, these data suggest that AQP8 could play an important role in the adaptive response of proximal tubule cells to acidosis, possibly by facilitating mitochondrial ammonia transport. However, a major physiological role of AQP8 in renal ammonia handling has been questioned, since AQP8 knockout mice show defects in renal ammonia excretion only in response to acidosis by chronic ammonium loading (Yang et al. 2006b). Since ammonia transport is critical to maintain systemic acid-base balance, it cannot be ruled out that other compensatory ammonia transport pathways may exist and compensate in the AQP8 knockout mice.

An additional role of AQP8 in renal water handling has been dismissed based on studies in knockout mice. Urine osmolalities were not different between wild-type and AQP8 knockout mice at baseline or after water deprivation (Yang et al. 2005). Furthermore, ruling out a compensatory role of AQP1, urine osmolality was not significantly different in AQP1 knockout mice relative to mice lacking AQP8 and AQP1 together (Yang et al. 2005). This confirms that there is no significant role of AQP8 in the urinary-concentrating ability in mice.

30.2.9 Aquaporin 11 (AQP11)

AQP11, in contrast to most other AQPs that have two highly conserved short "NPA" (asparagine-proline-alanine) sequences that form loops directed into the membrane constituting a pore, has unique "NPC" motifs (Morishita et al. 2005). Experiments using reconstituted AQP11 in proteoliposomes or Sf9 cell membrane vesicles revealed that AQP11, despite the incompletely conserved NPA motif and low homology with other AQP family proteins, has water channel activity, but this is relatively low (Yakata et al. 2007; Yakata et al. 2011). AQP11 is also permeable to glycerol (Madeira et al. 2014). Changing the "NPC" motif of AQP11 to "NPA" reduced the water permeability and oligomerization of AQP11 (Ikeda et al. 2011). In mice kidneys, AQP11 is expressed intracellularly in cells along the proximal tubule and is absent at the apical brush border membrane (Morishita et al. 2005; Inoue et al. 2014). Thus far, AQP11 antibodies that are suitable for subcellular localization studies, e.g., immunogold electron microscopy have not been developed, hampering studies to localize AQP11 to a specific organelle. However, in CHO-K1 cells and in vivo in mice expressing GFP or HA-tagged AQP11, AQP11 localizes to the endoplasmic reticulum (ER) (Morishita et al. 2005; Inoue et al. 2014; Ikeda et al. 2011).

The majority of AQP11 knockout mice die before weaning due to advanced renal failure from polycystic kidneys. The cysts in the AQP11 knockout mice originate mainly from the proximal tubules (Morishita et al. 2005; Inoue et al. 2014). Although these cysts are absent in the kidneys of newborn mice, they are apparent after 3 weeks when functional renal failure is evident. The cyst formation is preceded by epithelial cell swelling with intracellular vacuolization (Morishita et al. 2005). These vacuoles are limited to the proximal tubule, where AQP11 is expressed, suggesting that the dysfunction of AQP11 is the direct cause of the vacuole formation. As vacuolization precedes the cyst formation, the proximal tubule cysts may be caused by the vacuolization. The vacuoles strongly stain for an ER marker and ribosomes are attached to some of the vacuoles, suggesting that they originate from the ER (Morishita et al. 2005). A similar phenotype to the total AQP11 knockout mice is observed in another mouse model with a spontaneous amino acid substitution (Cys227Ser) in the predicted E-loop region of AQP11. These mice also present with proximal tubule injury, with formation of giant vacuoles in the renal cortex and die from renal failure at an early age (Tchekneva et al. 2008). Examining the total AQP11 knockout mouse model, it is unclear if the kidney failure is caused by functional insufficiency of the proximal tubule in the absence of AQP11, or by abnormal development of the nephron and structural malformation of the kidney. Recently, an inducible AQP11 knockout mouse model was developed, where AQP11 gene disruption can be temporally controlled by injection of tamoxifen (Rutzler et al. 2017). Proximal tubule cell vacuolization and apparent tubular cysts developed only in mice where *Aqp11* gene disruption was induced less than 8 days after birth. *Aqp11* gene deletion from 12 days onward did not result in a clear deficiency in renal development, proximal tubule injury, or cyst formation. In addition, intraperitoneal injection of biotinylated-dextran into adult mice resulted in endocytic dextran uptake in both cystic *Aqp11* knockout and control proximal tubule epithelium, suggesting that apparent cysts are not membrane-enclosed structures, but instead represent proximal tubule dilations (Rutzler et al. 2017). This suggests that the premature death of *Aqp11* knockout mice is caused by early abnormal kidney development, resulting in formation of dilated proximal tubules.

The kidneys from AQP11 knockout mice exhibit increased protein levels of polycystin-1 (PC-1) and decreased protein levels of polycystin-2 (PC-2). Despite the increased PC-1 levels, PC-1 is abnormally N-glycosylated and less is localized in the plasma membrane (Inoue et al. 2014). These differences in PC-1 and PC-2 could represent a key mechanism of cyst formation or dilations in the AQP11 knockout mice. PC-1 and PC-2 are usually located in the plasma membrane and primary cilia of renal epithelia, where they form part of a complex that acts as a flow sensor on the cilium. Mutations in the genes coding for PC-1 and PC-2 are the cause of autosomal dominant polycystic kidney disease (ADPKD), a disease characterized by cyst formation in the kidney that in many cases ultimately results in end-stage renal disease. Autosomal recessive polycystic kidney disease (ARPKD) can be caused by mutations in the gene coding for fibrocystin, which is in the same complex as PC-2 within the primary cilia (Harris and Torres 2009).

Although in PKD, lesions are localized to collecting ducts, murine models of ARPKD demonstrate an early phase of proximal tubular cystic involvement that disappears shortly after birth, but is followed by a phase of collecting duct cyst formation (Nakanishi et al. 2000). Similarly, in mice homozygous for Pkd1 mutations (the gene coding for PC-1), renal cyst formation begins in the proximal tubules, followed by cyst formation in the collecting duct (Lu et al. 1997). In humans, ARPKD has a transient phase of proximal tubular cyst formation during fetal development, with a gradual shift in the site of cystic nephron involvement, from proximal tubules to collecting ducts during the fetal period (Nakanishi et al. 2000). This suggests that AQP11 knockout mice and polycystic kidney disease share a common pathogenic mechanism for cyst formation.

Despite this association, the exact function of AQP11 in proximal tubule cells remains unclear. As proximal tubule cells exhibit ER vacuolization in AQP11 knockout mouse kidneys, and AQP11 knockout mice show increased ER stress response and oxidative stress, AQP11 may play an important role in the homeostasis of the ER (Atochina-Vasserman et al. 2013; Okada et al. 2008). In addition, glucose increases AQP11 expression in mouse kidney and in AQP11 insufficient proximal tubule cells, glucose potentiated the increase in reactive oxygen species, suggesting a role for AQP11 in preventing glucose-induced oxidative stress in proximal tubules (Atochina-Vasserman et al. 2013). A rs2276415 (G > A) single-nucleotide polymorphism in the human *AQP11* gene, resulting in a Gly102Ser substitution, has been shown to have a harmful effect on the graft survival after kidney transplantation (Park et al. 2015). In addition, the A allele of AQP11 increases the risk of chronic kidney disease (CKD) progression (Han et al. 2018). Interestingly, another study showed that the A allele of AQP11 is associated with increased risk of acute kidney injury and CKD in patients with diabetes mellitus, but not in nondiabetic individuals (Choma et al. 2016), further suggesting a role for glucose in modulating AQP11.

30.3 Renal Aquaporins in Disease

30.3.1 Water Balance Disorders Associated with Hyponatremia and Increased Aquaporin Levels

Disturbances of water balance are frequently occurring and are characterized by hyponatremia, hypernatremia, or polyuria. In individuals with maintained glomerular filtration, several of these pathologics result from too high or too low levels of renal AQPs, mainly AQP2. Hyponatremia with high AQP2 levels can be caused by either osmoregulation disorders or, most frequently, by diseases with low effective circulating blood volume. In these latter conditions, osmoregulation is overruled by the body's regulation of volume, leading to the development of hyponatremia as a consequence of the body's goal to maintain sufficient circulating volume (Rodriguez et al. 2019).

30.3.2 Congestive Heart Failure

Congestive heart failure (CHF) patients have high blood levels of AVP that contribute to hyponatremia and an increased extracellular volume (Schrier 2006). AQP2 excretion is increased in the urine of CHF patients (Pedersen et al. 2003) via secretion of apical exosomes rather than whole-cell shedding (Wen et al. 1999). Exosomes are formed from multivesicular bodies (MVBs) within the cell, which under certain conditions dock and fuse with the plasma membrane and empty their contents (exosomes) into the urinary space (Huebner et al. 2015). The urinary excretion of AQP2 in exosomes directly correlates with the amount of renal AQP2 at the apical membrane (Wen et al. 1999), thus can be used as a marker for the expression of AQP2 in the kidney. Rat models of CHF show increased renal AQP2 trafficking and total expression (Nielsen et al. 1997; Xu et al. 1997; Lin et al. 2011), whereas AQP1 and AQP3 levels are not increased, indicating that the effect on AQP2 in CHF is selective (Nielsen et al. 1997). The increased AQP2 levels are most likely the result of increased circulating levels of AVP. However, studies have shown that rats with CHF have increased protein levels of AQP2 even during conditions with unchanged circulating levels of AVP (Brond et al. 2013). In these rats, the V2 receptor had an increased ability to recycle to the plasma membrane, and, consequently, collecting ducts showed an increased accumulation of cAMP in response to AVP. This indicates increased AVP sensitivity in the collecting ducts, which might contribute to the increased AQP2 levels and hyponatremia seen in CHF (Brond et al. 2013). Administration of the V2R antagonist Mozavaptan to CHF rats inhibited the AQP2 upregulation and trafficking, and led to an increase in urine output and a rise in plasma osmolality, suggesting a major role for AQP2 in CHF-induced water retention (Xu et al. 1997). In agreement with these animal data, V2R antagonists increase urine volume and correct hyponatremia in CHF patients (Schrier et al. 2006; Aronson et al. 2011).

30.3.3 Hepatic Cirrhosis

Hepatic cirrhosis is a chronic disease associated with water retention, hyponatremia, and increased levels of AVP (Schrier 2006). The development of hyponatremia in cirrhosis patients is likely caused by systemic vasodilation, leading to increased secretion of AVP and consequently water retention and hyponatremia. The role of AQP2 in hepatic cirrhosis is not clear as, unlike for CHF, the changes in AQP2 levels vary considerably between different studies. In some rat models of hepatic cirrhosis induced by carbon tetrachloride, AQP2 protein and mRNA levels were increased

(Asahina et al. 1995; Fujita et al. 1995). The mRNA expression of renal AQP2 was furthermore found to correlate with the volume of ascites, suggesting that AQP2 plays a crucial role in abnormal water retention followed by the development of ascites in hepatic cirrhosis (Asahina et al. 1995). In another study, an increase in plasma membrane expression of AQP2 was observed, although total AQP2 abundance was not changed, while total expression of AQP1 and AQP3 was increased (Fernandez-Llama et al. 2000). In contrast, AQP2, AQP3, and AQP4 levels were decreased, and AQP1 levels unchanged, in a rat model of liver cirrhosis induced by chronic common bile duct ligation, suggesting that an increase in renal aquaporins is not essential for the development of excessive water retention with hyponatremia (Fernandez-Llama et al. 1999). However, there is increased excretion of AQP2 in the urine of cirrhosis patients compared with control subjects (Ivarsen et al. 2003; Chung et al. 2010; Nakanishi et al. 2016). AQP2 levels correlated with the clinical severity of the cirrhosis and were highest in patients with ascites, suggesting a critical role for AQP2 in cirrhosis-induced water retention. However, other studies have demonstrated no increase, or even a decrease, in urinary AQP2 in cirrhosis patients (Pedersen et al. 2003; Esteva-Font et al. 2006). An explanation for the differences between these studies is at present lacking.

30.3.4 Syndrome of Inappropriate Secretion of Antidiuretic Hormone (SIADH)

The syndrome of inappropriate secretion of antidiuretic hormone (SIADH) results from abnormally increased AVP levels, leading to excessive renal water reabsorption and compromised osmoregulation, which ultimately can result in life-threatening hyponatremia (Palmer 2003). The most common causes of SIADH are neoplasia (notably small-cell lung cancer), neurological diseases, lung diseases, and a wide variety of drugs, particularly psychoactive drugs and chemotherapy (Baylis 2003). AQP2 levels are markedly increased in SIADH rats, but can be reduced using V2R antagonists, a process that correlates closely with a marked diuresis and a normalization of serum sodium levels (Fujita et al. 1995). In agreement with these animal data, V2R antagonists induce diuresis and correct hyponatremia in SIADH patients (Wong et al. 2003; Schrier et al. 2006; Soupart et al. 2006). SIADH patients are normovolemic or slightly hypervolemic and often show excessive release of AVP that is not related to plasma osmolality, e.g., ectopic AVP production by neoplasms (Hoorn et al. 2014). Other patients continue to regulate water excretion normally, but do so around a lower plasma osmolality set-point compared to normal. The underlying mechanism of this is unclear (Baylis 2003).

30.3.5 Nephrogenic Syndrome of Inappropriate Antidiuresis (NSIAD)

Another cause of normovolemic hyponatremia is the Nephrogenic Syndrome of Inappropriate Antidiuresis (NSIAD), a rare disorder caused by a mutation in the *AVPR2 gene, encoding the* V2R. Although the clinical presentations of NSIAD resemble those of SIADH, i.e., hyponatremia and lack of urinary dilution, in NSAID patients the plasma AVP levels are undetectable or very low (Levtchenko and Monnens 2010). As the *AVPR2 is located on the X-chromosome,* NSIAD is an X-linked genetic disease affecting predominantly males. Although female carriers of the mutation are clinically asymptomatic, a proportion demonstrated an impaired ability to dilute urine during a water-load test (Ranchin et al. 2010; Decaux et al. 2007). The *AVPR2* mutation usually results in a change in amino acid 137 from an arginine into a cysteine or a leucine (Arg137Cys/Leu) (Feldman et al. 2005), but can also be caused by a Phe229Val, Ile130Asn, or Leu312Ser mutation (Carpentier et al. 2012; Tiulpakov et al. 2016; Erdelyi et al. 2015). The mutation of the *AVPR2* results in constitutive activation of the V2R, resulting in increased basal cAMP production compared to the wild-type V2R (Feldman et al. 2005; Tenenbaum et al. 2009; Carpentier et al. 2012). This constitutive activation of the V2R will result in enhanced AQP2 abundance and membrane targeting, which explains the inappropriate antidiuresis. While this mechanism is very similar for all known mutations, there are however also differences in how the mutation affects the receptor function. Although the Arg137 mutants constitutively recruit β-arrestin, this is not the case for the other 3 mutants. The V2R antagonist Tolvaptan does not reduce the constitutive cAMP production observed with the Arg137Cys/Leu mutant receptors, whereas it does for the three other mutant receptors (Erdelyi et al. 2015; Carpentier et al. 2012; Tiulpakov et al. 2016).

In contrast to the NSIAD-causing mutations Arg137Cys/Leu, mutation of amino acid 137 into a histidine or glycine results in the opposite disease phenotype. The V2R undergoes constitutive endocytosis, which causes impaired agonist binding, reduced AQP2 abundance and trafficking, and the kidney is no longer able to concentrate urine (Barak et al. 2001; Hinrichs et al. 2016). This indicates that the arginine residue at position 137 in the V2R receptor is pivotal for normal function.

30.4 Treatment of Pathologies Resulting from Increased Aquaporin Levels

Euvolemic or hypervolemic chronic hyponatremia, caused either by CHF, liver cirrhosis, SIADH or NSIAD is usually treated via simple fluid restriction (Elhassan and Schrier 2011). However, not all patients comply with this regime, and are subsequently supplemented with oral intake of urea (to induce an osmotic diuresis) or with loop diuretics, e.g., furosemide. In addition to this, newer more effective treatments are becoming more common, including the use of either demeclocycline or V2R antagonists.

30.4.1 Demeclocycline

Demeclocycline is a bacteriostatic antibiotic of the tetracycline group, use of which causes water diuresis and subsequently NDI (Singer and Rotenberg 1973; Forrest et al. 1974). Demeclocycline is one of the options currently used to treat sustained hyponatremia in patients with SIADH (Sherlock and Thompson 2010). In SIADH patients, demeclocycline restores sodium plasma concentration to normal levels, permitting unrestricted water intake in these patients (Forrest et al. 1978). Demeclocycline selectively inhibits water reabsorption in the distal part of the nephron (Wilson et al. 1973) and AVP-induced osmotic water flow in the toad urinary bladder (a model system of mammalian collecting duct) (Feldman and Singer 1974; Hirji and Mucklow 1991; Singer and Rotenberg 1973). In a mouse collecting duct cell line (mpkCCD), demeclocycline decreases dDAVP-induced cAMP generation and adenylate cyclase 3 and 5/6 abundances (Kortenoeven et al. 2013b), leading to reduced AQP2 gene transcription and subsequently reduced AQP2 protein abundance. In a rat model of SIADH, demeclocycline increases urine volume, decreases urine osmolality, and corrects the hyponatremia, a result of decreased AQP2 and adenylate cyclase 5/6 abundances in the renal inner medulla. Thus, demeclocycline limits hyponatremia by attenuating the AVP-induced signaling cascade and AQP2 abundance in the inner medullary collecting duct (Kortenoeven et al. 2013b).

30.4.2 V2R Antagonists

Vasopressin-receptor antagonists, or vaptans, have become promising drugs for the treatment of hyponatremia. Mozavaptan (OPC-31260) increased the urinary flow rate, lowered the urinary osmolality, and attenuated the hyponatremia in rat models of SIADH, CHF, and liver cirrhosis (Xu et al. 1997; Fujisawa et al. 1993; Tsuboi et al. 1994). Mozavaptan treatment also results in water diuresis and lessens the hyponatremia in SIADH patients (Saito et al. 1997a). Tolvaptan [Jynarque® (USA); Jinarc® (EU, Canada); Samsca® (Japan)] decreased urine osmolality and expression of AQP2 and AQP3 in rats, and blocked vasopressin's effects on phosphorylation of AQP2 in vitro (Miranda et al. 2014). In addition, Tolvaptan increases urine volume and corrects hyponatremia in patients with euvolemic or hypervolemic hyponatremia due to CHF, liver cirrhosis and SIADH (Schrier et al. 2006) as well as decreasing AQP2 in the urine of these groups of patients (Tamma et al. 2017; Nakanishi et al. 2016; Imamura et al. 2016). A high baseline urine AQP2 relative to plasma vasopressin levels predicts a response to Tolvaptan in CHF patients

(Imamura et al. 2014). In addition, Tolvaptan has recently been approved for the treatment of autosomal dominant polycystic kidney disease (ADPKD) (Blair 2019). Three other vaptans, the V2R antagonists Lixivaptan and Satavaptan, and the V2R/V1a antagonist Conivaptan also increase diuresis and correct hyponatremia in patients with euvolemic or hypervolemic hyponatremia due to CHF, liver cirrhosis, and SIADH (Wong et al. 2003; Ghali et al. 2006; Soupart et al. 2006; Gines et al. 2008; Aronson et al. 2011).

30.5 Water Balance Disorders Associated with Polyuria and Decreased Aquaporin Levels

Polyuria and polydipsia are symptoms of a range of diseases and disturbances to homeostasis, which can originate at extra-renal, intra-renal, or post-renal sites. Given the tight hormonal regulation of AQP2, it is not surprising that polyuria is often accompanied by altered levels of plasma AVP, which are either pathologically decreased, as seen in central and gestational diabetes insipidus, or compensatory increased in polyuria from other causes, such as diabetes mellitus.

30.5.1 Central Diabete Insipidus (CDI)

Diabetes Insipidus (DI) is characterized by impaired renal water reabsorption, leading to polyuria and hypotonic urine. DI patients can have hypernatremia, but most often have a normal plasma osmolality, provided the thirst mechanism is normal and there is adequate access to fluid. CDI, also known as neurohypophysial DI, is caused by impaired AVP release from the posterior pituitary gland. Under normal circumstances, AVP is produced by the magnocellular neurons of the supraoptic nucleus of the hypothalamus, which project axons into the posterior pituitary gland, from which they release AVP into the bloodstream upon activation (Voisin and Bourque 2002). CDI can be congenital, but is usually acquired as a result of local inflammatory or autoimmune disease, vascular disease, Langerhans cell histiocytosis, sarcoidosis, intracranial tumors, or trauma resulting from surgery or an accident (Di et al. 2012). It is treated by administration of the synthetic AVP analog dDAVP, which drastically decreases urine output in CDI patients. The Brattleboro rat is an animal model of central diabetes insipidus, which has undetectable levels of plasma AVP due to a random genetic mutation (Valtin et al. 1962) and displays decreased expression of AQP2, which can be reversed by chronic AVP infusion (Digiovanni et al. 1994). In agreement with this, central DI patients excrete decreased amounts of AQP2 in the urine compared with healthy controls and AQP2 excretion remains unchanged after hypertonic saline infusion, but increases after exogenous AVP administration (Saito et al. 1997b, 1999). AVP regulates a variety of proteins involved in urine concentration, such as AQP3 and AQP4, in connecting tubules and collecting ducts, urea transporters in inner medullary collecting ducts, and the loop of Henle and the Na-K-2Cl cotransporter 2 (NKCC2) in the thick ascending limb (Fenton and Knepper 2007), which may be hypothesized to also be affected in central DI patients.

The rarest form of CDI is familial neurohypophysial DI (FNDI), which is characterized by severe polyuria and polydipsia, and in almost all cases is caused by an inherited mutation in the AVP gene. The AVP gene encodes a 164-kD-long precursor peptide, consisting of a signal peptide, AVP, neurophysin II, and copeptin. This precursor is enzymatically cleaved to form active AVP (Land et al. 1982). At least 70 different mutations resulting in a defective precursor peptide and a deficiency of active AVP have been identified in FNDI. The majority of these mutations show an autosomal dominant pattern of inheritance (http://www.hgmd.org (Arima and Oiso 2010)), and most of the mutations are located in either the signal peptide or the neurophysin II moiety. The carriers display a normal phenotype at birth, but typically between 1 and 6 years of age, symptoms of compulsive drinking and polyuria appear, even though they carry one normal allele. FNDI is hypothesized to result from accumulation of mutant AVP precursor in the ER, resulting in an aberrant endoplasmic morphology and possible cell dysfunction and death (Ito and Jameson 1997; Russell et al. 2003; Arima and Oiso 2010). Mutant precursors form heterodimers with the wild-type AVP precursor, thereby impairing intracellular trafficking of both forms and reducing the bioavailability of active AVP by means of a dominant negative effect (Arima and Oiso 2010; Ito et al. 1999).

Patients with an autosomal recessive instead of dominant pattern of inheritance have a missense mutation in the region encoding the AVP domain resulting in reduced biological activity of the mutant AVP peptide (Willcutts et al. 1999; Abu et al. 2010). In contrast to other forms of central DI, circulating AVP levels in these patients are high. In one family with autosomal dominant FNDI, no mutations in the coding region, introns, or the promoter of the AVP gene were found (Ye et al. 2005). The gene(s) responsible for FNDI in this family are located on chromosome 20, implying mutations in other genes than the AVP gene can cause FNDI as well.

30.5.2 Gestational DI

A transient form of DI can occur during pregnancy. Gestational DI is due to an abnormal increase in an AVP-degrading enzyme produced by the placenta termed vasopressinase, leading to AVP deficiency (Ananthakrishnan 2009). This form of DI may be associated with pre-eclampsia and HELLP syndrome (consisting of Hemolysis, Elevated Liver enzymes, and Low Platelets), possibly because the decreased hepatic function in these conditions leads to a decrease in vasopressinase degradation (Katz and Bowes 1987; Hadi et al. 1985; Nasrat et al. 1997). dDAVP is resistant to degradation by vasopressinase and is an effective treatment for gestational DI (Ananthakrishnan 2009).

30.5.3 Nephrogenic DI

Nephrogenic DI (NDI) is a collective term for congenital or acquired polyuric conditions where the kidney is unable to normalize urine concentration in response to AVP stimulation.

30.5.3.1 Congenital NDI

Congenital NDI can be divided into X-linked, autosomal recessive, and autosomal dominant NDI. The urine-concentrating defect is present from birth, and symptoms usually arise during the first weeks of life. X-linked congenital NDI is the cause of disease in more than 90% of all congenital NDI patients and occurs mainly in males; however, due to skewed X-chromosome inactivation, some heterozygous females have variable degrees of clinical NDI symptoms (Arthus et al. 2000; Satoh et al. 2008). It is caused by loss-of-function mutations in the *AVPR2* gene, encoding the V2R (Lolait et al. 1992), and over 270 such gene mutations have been described to cause NDI (http://www.hgmd.org). Mutations in the *AVPR2* gene interfere with receptor signaling, thus making the principal cells of the collecting duct insensitive to AVP, resulting in a severe urinary concentration defect. The most common mechanism underlying this insensitivity is misfolding of the V2R protein caused by a missense mutation and consequently retention in the ER. Other mechanisms include mRNA instability, splicing errors, large deletions, frameshifts, or nonsense mutations resulting in truncation of the receptor, diminished binding of the Gs protein, reduced affinity for AVP, and misrouting of the V2R to different organelles in the cell (Robben et al. 2006a). Approximately 10% of patients diagnosed with NDI have mutations in the *AQP2* gene, which is an autosomal recessive disorder in more than 90% of cases. To date, sixty-three AQP2 mutations have been described, of which fifty-three are involved in recessive NDI (http://www.hgmd.org). The majority of these comprise missense mutations. Typically, mutations occur between the first and last transmembrane domains of AQP2, which form the channels water pore. Nearly all mutations cause misfolding of the protein, which is consequently trapped in the ER, and rapidly degraded by the proteasome (Robben et al. 2006a). AQP2 could not be detected in the urine of recessive NDI patients, presumably due to extensive degradation of the protein (Deen et al. 1996). Several AQP2 mutants are able to function as water channels when overexpressed in *Xenopus laevis* oocytes, where a fraction of the AQP2 mutants escape the cell's quality control system and reach the plasma membrane (Canfield et al. 1997; Marr et al. 2001, 2002; Mulders et al. 1997). Thus, in these cases, the phenotype is caused exclusively by ER retention of the protein. The healthy parents of patients with recessive NDI express both mutant and wild-type AQP2 proteins. However, most AQP2 mutants are not able to form heterotetramers with wild-type AQP2, leaving only the formation of the functional wild-type AQP2 homotetramers, which likely explains the autosomal recessive pattern of inheritance (Kamsteeg et al. 1999). In contrast, some AQP2 mutants can form heterotetramers with wild-type AQP2, which can functionally recover the mutant (El et al. 2016).

Currently, 10 different AQP2 mutations have been described to cause autosomal dominant NDI, which is the rarest form of the disease. The identified mutations comprise missense mutations and small nucleotide deletion or insertions. All mutations are found in the carboxyl-terminal tail of AQP2, which is essential for correct intracellular routing of the protein, and the mutations cause missorting of the protein to other cellular organelles than the apical plasma membrane, e.g., late endosomes/lysosomes and the basolateral plasma membrane (Robben et al. 2006a; Kamsteeg et al. 1999). Dominant AQP2 mutants are able to form heterotetramers with wild-type AQP2 causing misrouting of both forms of the protein and consequently severely compromised collecting duct water reabsorption. At least six of the mutations introduce a missorting signal, whereas two, Arg254Leu and Arg254Gln, interfere with Ser256 phosphorylation of AQP2 and prevent the translocation of AQP2 to the plasma membrane, leading to impaired water reabsorption (De Mattia et al. 2005; Savelkoul et al. 2009). Heterozygous dominant NDI patients display generally milder symptoms than recessive NDI patients, suggesting that some wild-type AQP2 homotetramers are formed that are able to reach the apical plasma membrane (Robben et al. 2006a).

Although mutations in the carboxyl tail of AQP2 usually lead to dominant NDI, one mutation, Pro262Leu, causes recessive NDI. Like dominant NDI mutants, Pro262Leu-AQP2 is missorted to intracellular vesicles and is able to form heterotetramers with wild-type AQP2 (De Mattia et al. 2004). However, when expressed together with wild-type AQP2, wild-type/Pro262Leu-AQP2 heterotetramers are targeted to the apical membrane of MDCK cells. This indicates that, in contrast to AQP2 mutants in dominant NDI, the apical sorting signal of wild-type AQP2 overrules the sorting signal in Pro262Leu-AQP2.

Few subjects have been identified with loss-of-function mutations in AQPs other than AQP2. Humans lacking functional AQP1 do not have polyuria and no obvious clinical phenotype under normal conditions. However, when challenged by water deprivation, they have an impaired ability to maximally concentrate their urine (King et al. 2001). Under nonstressed conditions, AQP3 null individuals do not suffer from any clear clinical abnormality (Roudier et al. 2002). Mutations in AQP5 in humans cause autosomal-dominant diffuse nonepidermolytic palmoplantar keratoderma (Blaydon et al. 2013). Three unrelated children with a homozygous Gly264Val mutation in the AQP7 gene, which previously has been shown in *Xenopus laevis* oocytes to result in a lack of both water and glycerol permeability, had psychomotor retardation of unknown etiology, a mild platelet secretion defect, and pronounced renal glycerol loss (Goubau et al. 2013; Kondo et al. 2002). However, the renal phenotype of individuals with defective AQP3, AQP5, or AQP7 has not been studied in detail. Human gene mutations abolishing the function of other renal AQPs have not been reported.

30.5.3.2 Acquired NDI

30.5.3.2.1 Electrolyte Disorders

Acquired forms of NDI are much more common than the rare hereditary disorders. Hypokalemia and hypercalcemia are common electrolyte disorders, which cause NDI by mechanisms that are not completely understood (Rubini 1961; Gill and Bartter 1961). In rat models of hypokalemia, the condition is associated with decreased AQP2 mRNA and protein expression and enhanced autophagic degradation of AQP2 (Marples et al. 1996; Jeon et al. 2007; Khositseth et al. 2015) alongside decreased expression of renal urea and sodium transporters in the distal nephron. This may contribute to the urinary-concentrating defect by reducing interstitial tonicity (Jung et al. 2003; Elkjaer et al. 2002; Jeon et al. 2007)

Hypercalcemia decreases AQP2 protein abundance and apical membrane accumulation in rats, but does not affect AQP2 mRNA levels (Earm et al. 1998). Enhanced autophagic degradation of AQP2 plays an important role in the initial mechanism of hypercalcemia-induced NDI (Khositseth et al. 2017). Moreover, hypercalcemic rats have decreased AQP1 and AQP3 levels, while AQP4 levels remain unchanged (Wang et al. 2002b). In addition, hypercalcemia decreases the expression of renal sodium transporters such as NKCC2, decreases NaCl reabsorption in the thick ascending limb, and reduces medullary tonicity, likely contributing to the decreased urine-concentrating ability (Wang et al. 2004; Levi et al. 1983; Peterson 1990; Wang et al. 2002a).

In children with enuresis, hypercalciuria is associated with a decrease in nocturnal AQP2 levels in the urine, and urine calcium excretion by means of a low-calcium diet is associated with an increase in urine AQP2 and decreased nocturnal urine volume in these children (Valenti et al. 2002; Valenti et al. 2000). In another study, a higher baseline urine AQP2 excretion was found in hypercalciuric subjects. In contrast to normocalciurics, dDAVP administration did not increase AQP2 excretion in these subjects, suggesting a disturbed AQP2 regulation (Procino et al. 2012). In addition, prolonged bed rest can induce bone demineralization and transient increases in urinary calcium followed by a transient decrease in AQP2 excretion, while the return of calciuria to baseline is followed by a recovery of AQP2 excretion, indicating that urinary calcium modulates AQP2 expression (Tamma et al. 2014a).

The effects of hypercalcemia on AQP2 have been attributed to activation of the apical calcium-sensing receptor (CaSR), which might be a defense mechanism to reduce the risk of calcium renal stone formation in states of high calcium in the tubular lumen. CaSR is a G-protein-coupled receptor, which is activated by interaction with calcium at the extracellular domain (Alfadda et al. 2014). It is thought to couple to Gq and activates intracellular signaling pathways, including phospholipase C and mitogen-activated protein kinases (Tfelt-Hansen et al. 2003). Sands et al. showed that the CaSR is expressed at the apical border of cells in the terminal inner medullary collecting duct and that purified subapical endosomes contain both CaSR and AQP2 (Sands et al. 1998). Activation of the CaSR by high extracellular calcium antagonizes forskolin-induced AQP2 trafficking to the apical plasma membrane and decreases AVP-induced AQP2 expression in cultured cells (Procino et al. 2004, 2012; Bustamante et al. 2008), suggesting that hypercalcemia-induced activation of the CaSR in vivo may have a similar effect resulting in subsequent urine concentration defects. Double-knockout mice lacking pendrin and the NaCl cotransporter (NCC) are hypercalciuric, but inhibition of the CaSR reversed the effects on AQP2 and AQP2 ubiquitylation. In addition, CaSR inhibition prevented the increase in the AQP2-targeting miRNA-137 in these mice, indicating that CaSR signaling reduces AQP2 abundance both via the AQP2-targeting miRNA-137 and increasing AQP2 ubiquitylation and degradation (Ranieri et al. 2018). Alternatively, high calcium may activate the protease calpain, which is expressed in the inner medullary collecting ducts, leading to degradation of AQP2 (Puliyanda et al. 2003).

30.5.3.2.2 Obstruction of the Urinary Tract

A relatively common condition associated with long-term impairment of urinary-concentrating ability is obstruction of the urinary tract, a serious clinical condition and a common cause of end-stage renal disease. In adults, obstruction is mostly caused by stones, enlargement of the prostate, or urinary tract neoplasms, whereas in children, it is usually due to congenital abnormalities (Iyasere et al. 2012). During the recovery phase after release of the urinary tract obstruction, the response of the collecting duct to AVP and water reabsorption in the collecting duct is decreased, resulting in polyuria (Klahr et al. 1988). Experimental bilateral or unilateral obstruction of the ureters is associated with markedly reduced expression of AQP1, AQP2, AQP3, and AQP4, as well as of renal sodium and urea transporters (Frokiaer et al. 1996; Li et al. 2001, 2003a, 2003b, 2003c, 2004). Both AQP2 protein and mRNA levels are decreased (Frokiaer et al. 1997; Stodkilde et al. 2011). After obstruction, AQP2 redistributes to early endosomes and lysosomes, suggesting degradation of AQP2 (Stodkilde et al. 2011). Following release of the obstruction, there is a marked polyuria during which period AQP1, AQP2, and AQP3 levels remain decreased, providing an explanation at the molecular level for postobstructive polyuria (Li et al. 2001).

Bilateral obstruction in rats is associated with increased renal cyclooxygenase-2 (COX-2) expression and increased prostaglandin E_2 (PGE_2) excretion in the urine (Cheng et al. 2004; Norregaard et al. 2005). Several prostaglandins are increased in the inner medulla, and to a lesser extent in the cortex, of rats with ureteral obstruction. In these conditions, COX-2 is the predominant isoform involved in prostaglandin synthesis, with minor, but significant, contributions from COX-1 (Norregaard et al. 2010). Ureteral obstruction did not significantly decrease AQP2 and AQP3 protein levels in cortex and outer medulla of COX-2 knockout mice, in contrast to wild-type mice, although a decrease in inner medullary AQP2 expression remained (Nilsson et al. 2012), suggesting that COX-2-derived prostaglandins may contribute in part to downregulation of AQPs in the collecting duct in ureteral obstruction. Treatment of rats with COX-2 inhibitors prevents the increase in PGE_2, whereas the downregulation of AQP2 and the development of polyuria are alleviated in some studies, (Cheng et al. 2004; Norregaard et al. 2005) but have no effect or are even augmented in others (Jensen et al. 2010). Thus, the effect of COX-2 inhibition in the model is not completely understood (for discussion see (Olesen and Fenton 2013)). Recently, it was shown that mitochondrial oxidative stress plays an important role in mediating AQP2 downregulation in the obstructed kidney, possibly via modulating COX-2 and PGE_2 (Liu et al. 2018).

30.5.3.2.3 Lithium-Induced NDI

The most common cause of NDI is lithium therapy. Lithium is a frequently prescribed drug used by 1 in 1000 of the western population (Timmer and Sands 1999; Manji et al. 2001) as a mood-stabilizing drug in such conditions as bipolar disorder, schizoaffective disorder, and recurrent depression. Unfortunately, more than 50% of patients undergoing lithium treatment develop impaired urinary-concentrating ability and in 20% of patients this leads to the development of clinical NDI (Boton et al. 1987). Plasma lithium levels must be maintained within a narrow therapeutic range and lithium-NDI patients are at risk for dehydration-induced lithium toxicity. Moreover, prolonged lithium treatment may result in development of renal microcysts and end-stage renal disease (Farres et al. 2003; Timmer and Sands 1999). However, since lithium is still regarded as the most powerful mood-stabilizing drug available, substitution of lithium therapy with one of the newer drugs is not always an option.

In the short term, lithium-NDI coincides with AQP2 and AQP3 downregulation and natriuresis in rats, without gross changes in renal morphology (Laursen et al. 2004; Marples et al. 1995; Mu et al. 1999). The lithium-induced natriuresis is suggested to be due to the reduced expression of salt-transporting proteins, including the NaCl-cotransporter (NCC), the β- and γ-subunits of ENaC, and the Na-K-ATPase (Nielsen et al. 2003; Kwon et al. 2000; Laursen et al. 2004). A study using connecting tubule-specific AQP2 knockout mice suggested that lithium does not severely affect AQP2 in the connecting tubule, in contrast to the collecting duct (Kortenoeven et al. 2013a). As early as 4 h after lithium exposure, urine output was increased and phosphorylation of both mitogen-activated protein kinases (MAPK) ERK1/2 and p38, as well as of Ser261-AQP2 were increased. Pretreatment with MAPK inhibitors reversed the increased Ser261-AQP2 phosphorylation, suggesting that ERK1/2 and p38 are early targets of lithium and may play a role in the onset of lithium-induced polyuria (Trepiccione et al. 2014).

In the collecting duct, chronic lithium treatment leads to a severe decrease in the fraction of principal cells and an increase in the fraction of intercalated cells, which are involved in acid/base balance regulation (Christensen et al. 2004, 2006), suggesting that part of the lithium-induced NDI might be explained by epithelial remodeling. Lithium treatment initiates proliferation of principal cells, but a significant percentage of these cells are arrested in the late G2 phase of the cell cycle, which might explain the reduced principal/intercalated cell ratio (de Groot et al.

2014). In agreement, a study in humans showed that 4-week therapy with lithium carbonate leads to a reduced urinary AQP2 excretion (Walker et al. 2005).

Lithium exerts its effects by entering principal cells through the epithelial sodium channel ENaC, which has a higher permeability for lithium than for sodium (Kellenberger et al. 1999). In the mouse collecting duct cell lines $mCCD_{c11}$ and mpkCCD, coincubation of lithium with the ENaC blockers amiloride or benzamil prevented the lithium-induced AQP2 downregulation and decreased the intracellular lithium concentration (Kortenoeven et al. 2009), showing that ENaC is the major cellular entry pathway for lithium. These cell studies were confirmed by experiments in rats, where simultaneous treatment with amiloride and lithium decreased urine output and increased urine osmolality relative to lithium alone (Kortenoeven et al. 2009). Dual treatment also attenuated AQP2 downregulation, and completely prevented the reduction of principal/intercalated cell ratios. Christensen et al. determined that collecting duct-specific αENaC knockout mice do not suffer from polyuria or a reduction in urine osmolality when treated with lithium, and AQP2 protein levels are only decreased in the inner medulla of these mice, confirming that ENaC is the entry pathway for lithium (Christensen et al. 2011). In agreement with these animal studies, blocking ENaC with amiloride in lithium-NDI patients significantly reduces urine volume and increases urine osmolality (Batlle et al. 1985; Kosten and Forrest 1986; Bedford et al. 2008).

After entry into the principal cell, lithium does not affect the pathway activated by AVP, as indicated by data in mpkCCD mouse collecting duct cells, showing that lithium does not affect AVP-induced cAMP generation, PKA-dependent AQP2 phosphorylation, or the phosphorylation of the AQP2 transcription factor CREB (Li et al. 2006). Furthermore, in AVP-deficient Brattleboro rats with clamped blood dDAVP levels, dDAVP-induced cAMP generation is not affected by lithium treatment (Li et al. 2006), and lithium treatment induces AQP2 downregulation and NDI development of a similar magnitude in adenylate cyclase 6 knockout mice and control mice (Poulsen et al. 2017).

Instead, lithium leads to inactivation of intracellular glycogen synthase kinase (Gsk) 3β, which is temporally related to increased COX-2 expression in the kidney and in cultured principal and interstitial cells (Rao et al. 2004, 2005; Kortenoeven et al. 2009, 2012a; Nielsen et al. 2008). Consistent with the role of COX-2 in prostaglandin production, lithium increased PGE_2 excretion in the urine of rats and mice (Rao et al. 2005; Kotnik et al. 2005). As PGE_2 reduces AVP-stimulated water reabsorption in the perfused collecting ducts (Nadler et al. 1992; Hebert et al. 1990), this suggests a possible role for PGE_2 in lithium-induced NDI development. This theory is supported by studies showing that blocking prostaglandin production by indomethacin reduces the urine output of lithium-treated rats (Kim et al. 2008) as well as in patients with lithium-induced NDI (Allen et al. 1989; Weinstock and Moses 1990). However, the specific COX-2 blocker parecoxib did not prevent lithium-NDI in rats (Kjaersgaard et al. 2014). As COX-1 expression is strongly increased by lithium, this suggests that the previously shown effect of indomethacin might be via blocking COX-1 instead of COX-2 (Kjaersgaard et al. 2014). In mpkCCD cells, prostaglandins do not affect AQP2 gene transcription, but increase lysosomal degradation of AQP2 (Kortenoeven et al. 2012a), suggesting that prostaglandins might have a role in lithium-induced NDI by decreasing AQP2 protein stability. In mpkCCD cells, lithium decreases AQP2 protein abundance as well as AQP2 transcription independent of endogenous prostaglandin production. The pathway leading to decreased AQP2 transcription is not known, but might involve one of the pathways driven by Gsk3, as not only lithium, but also other Gsk3 inhibitors lead to a downregulation of AQP2 expression in mpkCCD cells (Kortenoeven et al. 2009). In agreement with a Gsk3 regulation of AQP2, collecting duct-specific Gsk3β knockout mice are polyuric and have reduced AQP2 mRNA and protein levels after water deprivation compared with wild-type mice (Rao et al. 2010), while Gsk3α total knockout mice show reduced AQP2, polyuria, and increased urinary vasopressin levels at baseline, and a failure to concentrate their urine to the same level as wild-type littermates when water deprived.

30.5.3.2.4 Acute Tubulo-Interstitial Nephritis

Tubular dysfunction resulting in reduced water and salt reabsorption can occur by inflammatory processes in the kidney tubules and/or interstitium as a complication of a variety of pathological conditions, including rabdomyolysis, septicemia, leptospirosis, and renal vascular occlusion. This causes acute tubulo-interstitial nephritis (ATIN), leading to acute renal failure of intrarenal origin. Diuresis is frequently maintained despite reduced GFR, resulting in nonoliguric renal failure (Raghavan and Eknoyan 2014).

Leptospirosis-induced ATIN is typically nonoliguric and accompanied by a urinary-concentrating defect (Ooi et al. 1972; Seguro and Andrade 2013). Downregulation of AQP1 in proximal tubules coupled with upregulation of AQP2 has been observed in autopsies after lethal outcome of the disease (Araujo et al. 2010), but animal models have suggested that AQP2 is in fact downregulated in the acute phase. In one such study, AVP resistance in the inner medullary collecting duct was observed (Magaldi et al. 1992) and AQP2 downregulation occurs in response to injections of glycolipoprotein, which is thought to be the virulence factor of leptospirosis (Cesar et al. 2012).

Acute renal failure in septicemia can be pre- or intra-renal, and a urinary-concentrating defect is frequently observed in patients despite elevated plasma AVP levels (Bagshaw et al. 2006; Jochberger et al. 2009). Lipopolysaccharide (LPS) is a known virulence factor, which induces a urinary-concentrating defect alongside acute downregulation of AQP2 in experimental animal models (Olesen et al. 2009; Grinevich et al. 2004), and its effects have been ascribed in part to a direct molecular link between the proinflammatory mediator NFκB and the AQP2 promoter (Hasler et al. 2008).

Loss of urinary-concentrating ability in patients suffering from acute renal ischemia is common, and downregulation of renal AQP1, AQP2, and AQP3 has been observed in animal models of ischemia-reperfusion (Kwon et al. 1999). The decrease in renal aquaporins was prevented by the anti-inflammatory agent αMSH, suggesting that the concentrating defect is induced by an inflammatory process. Ischemia-reperfusion AQP3 knockout mice have impaired renal function, increased apoptosis, and an enhanced inflammatory response compared with controls. Overexpression of AQP3 in MDCK cells attenuated reduced cell viability following hypoxia-reoxygenation suggesting a novel role for AQP3 in protection against hypoxic injury (Lei et al. 2017).

Taken together, these studies suggest that inflammatory mediators cause acute, vasopressin-independent (dys)regulation of renal aquaporins and that this process plays a role in the pathophysiology of ATIN and explains maintained diuresis in nonoliguric acute renal failure.

30.5.3.2.5 Diabetes Mellitus

A major cause of polyuria is diabetes mellitus (DM). Type-1 DM often presents as frequent urination, and feared complications of the disease are ketoacidosis and the hyperglycemic hyperosmolar state (Maletkovic and Drexler 2013). As the name suggests, the latter is a state of excessive water loss and increased plasma osmolality. The regulation of AQPs in patients during this medical emergency has not been investigated. There is a dramatic upregulation of AQP2, AQP2 phosphorylated at Ser256, and AQP3 in streptozotocin-induced diabetes mellitus in rats (Nejsum et al. 2001). This occurs in the Brattleboro rat also and is thus independent from increases in plasma AVP levels (Kim et al. 2004a), but it is blunted in adrenelectomized rats (Klein et al. 2006), indicating a role for the renin-angiotensin-aldosterone system. However, candesartan, an angiotensin-II type-1 receptor antagonist, failed to prevent upregulation of renal AQPs (Blount et al. 2008). Thus, the polyuria in DM is likely caused by osmotic diuresis due to hyperglycemia and ketonuria and not by downregulation of renal AQPs. On the contrary, there is a compensatory upregulation of AQPs by an unknown mechanism, which is assumed to alleviate the acute hyperosmolar state in DM patients (Satake et al. 2010).

About one-third of diabetic patients develop diabetic nephropathy (DN). Wu et al. showed that AQP5 is upregulated in kidney biopsies from DN patients (Wu et al. 2013). Recently, it has been shown that urine excretion of both AQP2 and AQP5 is increased in patients with DN compared with diabetic patients without DN and healthy controls, and the levels of AQP5 and AQP2 correlated with the stage of DN, suggesting that these AQPs could be used as noninvasive biomarkers for diabetic nephropathy (Rossi et al. 2017; Lu et al. 2016).

30.6 Treatment of Polyuria

Administration of dDAVP can successfully treat central DI (Robinson 1976). It is also an effective treatment for gestational DI, since dDAVP is resistant to degradation by vasopressinase (Ananthakrishnan 2009). dDAVP can be administered either as an intranasal spray, tablet, or sublingual melting formulation. For the majority of NDI patients, treatment with dDAVP is however not effective. The main strategy for treating NDI is a sufficient supply of fluid to replace the urinary water losses, usually combined with a low-solute diet, which probably, by causing a slight hypovolemia and activation of the RAAS system, increases the proximal reabsorption of water and thereby decreases the urine volume (BEASER 1947; Blalock et al. 1977). Additionally, thiazide diuretics, alone or in combination with amiloride or indomethecin, can efficiently decrease the urine volume in NDI patients, although usually not to normal levels and not in all patients.

30.6.1 Diuretics

Currently, thiazide diuretics are used to reduce polyuria in NDI patients. Several studies have shown that in patients with congenital NDI, as well as lithium-induced NDI, thiazide diuretics effectively decrease urine output and increase urine osmolality (Forrest et al. 1974; Alon and Chan 1985; Jakobsson and Berg 1994; Crawford et al. 1960). The precise mechanism by which thiazide diuretics cause their paradoxical antidiuretic effect is largely unknown. The thiazide-sensitive NaCl-cotransporter, NCC, is expressed in the renal distal convoluted tubule, a segment which is water impermeable; therefore, the water-preserving effect of thiazides is unlikely related to a direct effect on this part of the nephron. However, the thiazide-mediated inhibition of NCC will cause an increased renal sodium excretion, which will induce a decrease in extracellular volume, leading to a decreased glomerular filtration rate (GFR), increased activation of the renin-angiotensin-aldosterone system, and a compensatory increase in proximal sodium and water reabsorption. Consequently, less water is delivered to the collecting duct, and less urine is produced. Additionally, thiazide blocks not only NCC, but also the Na^+-dependent Cl^-/HCO_3^- exchanger NDCBE in the intercalated cells of the collecting duct (Leviel et al. 2010), and carbonic anhydrases (Matsumoto et al. 1989), which may contribute to the mechanism. Treatment of NCC knockout mice with lithium and hydrochlorothiazide revealed a significantly reduced urine volume and increased cortical AQP2 abundance compared with NCC knockout mice treated with lithium alone (Sinke et al. 2014), indicating that part of the antidiuretic effect of thiazide in lithium-induced NDI is NCC independent. This may involve a reduction in GFR due to thiazide-mediated proximal tubular carbonic anhydrase inhibition or inhibition of NDCBE in the collecting duct (Sinke et al. 2014). However, thiazides also directly increase the water permeability of isolated collecting ducts (Cesar and Magaldi 1999), or increase the expression of AQP2 in rats with lithium-induced NDI or the mpkCCD mouse collecting duct cell line (Kim et al. 2004b; Sinke et al. 2014), suggesting a direct effect of thiazide on the collecting duct.

Combining treatment of hydrochlorothiazide with either a prostaglandin synthesis inhibitor, e.g., indomethacin, or with the ENaC-blocker amiloride is more effective in reducing urine volume than thiazide alone (Jakobsson and Berg 1994;

Monnens et al. 1984; Rascher et al. 1987; Alon et al. 1985; Konoshita et al. 2004; Kirchlechner et al. 1999). Besides this, amiloride also prevents hypokalemia and metabolic alkalosis, common side-effects of thiazide therapy (Konoshita et al. 2004; Alon et al. 1985).

30.6.2 *Future Therapies*

In recent years, many promising therapies for NDI have been identified, e.g., treatment with cell-permeable nonpeptide V2R antagonists. Most V2R mutations associated with X-linked NDI result in ER retention of the receptor due to misfolding, but the retained receptor may still be functional. A group of compounds known as "chemical chaperones," including glycerol and DMSO, can rescue the plasma membrane expression of this mutant V2R, probably by stabilization of the receptors conformation (Robben et al. 2006b). Nonpeptide V2R antagonists have been tested for their efficacy as chaperones as well, to increase specificity of the treatment and to reduce toxicity. Once the antagonist-bound V2R reaches the plasma membrane, the antagonist needs to be replaced by AVP or dDAVP to allow activation of the receptor (Fig. 30.5). Several cell-permeable V2R antagonists can stabilize ER-retained V2R mutants in cultured cells, inducing the receptor to escape from the ER and reach the plasma membrane (Morello et al. 2000; Robben et al. 2007; Mouillac and Mendre 2018). Subsequent AVP administration leads to the generation of cAMP in these cells, showing a functional rescue of the receptors (Robben et al. 2007). The V1a receptor antagonist SR49059 (Relcovaptan, with moderate affinity for V2R) can also rescue cell membrane expression of V2R mutants in cultured cells. Furthermore, SR49059 significantly reduces the urine output in X-linked NDI patients, proving in principle that nonpeptide antagonists can rescue V2R mutant activity in vivo (Bernier et al. 2006). However, because of hepatic toxicity, the development of SR49059 was discontinued during the clinical phase II of the study (Mouillac and Mendre 2018). A limitation of nonpeptide antagonists is that their efficiency may depend on the type and location of the specific V2R mutation, meaning that different mutations may require different antagonists to achieve V2R rescue and some mutants may be insensitive to all antagonists (Los et al. 2010).

Another possible treatment for X-linked NDI is the use of cell-permeable nonpeptide V2R agonists (Fig. 30.5). The nonpeptide V2R agonists OPC51803, VA999088, and VA999089 can induce a cAMP response in polarized renal cells, eventually leading to an increase in AQP2 trafficking to the apical plasma membrane in six out of seven tested dDAVP-insensitive V2R mutants (Robben et al. 2009). Interestingly, the nonpeptide agonists did not affect ER localization of the mutant receptors and failed to induce receptor maturation, indicating that the observed cAMP signal was derived from V2R still retained in the ER. Three structurally related nonpeptide agonists MCF14, MCF18, and MCF57 also increase cAMP concentrations of intracellularly retained V2R mutants. However, this coincided

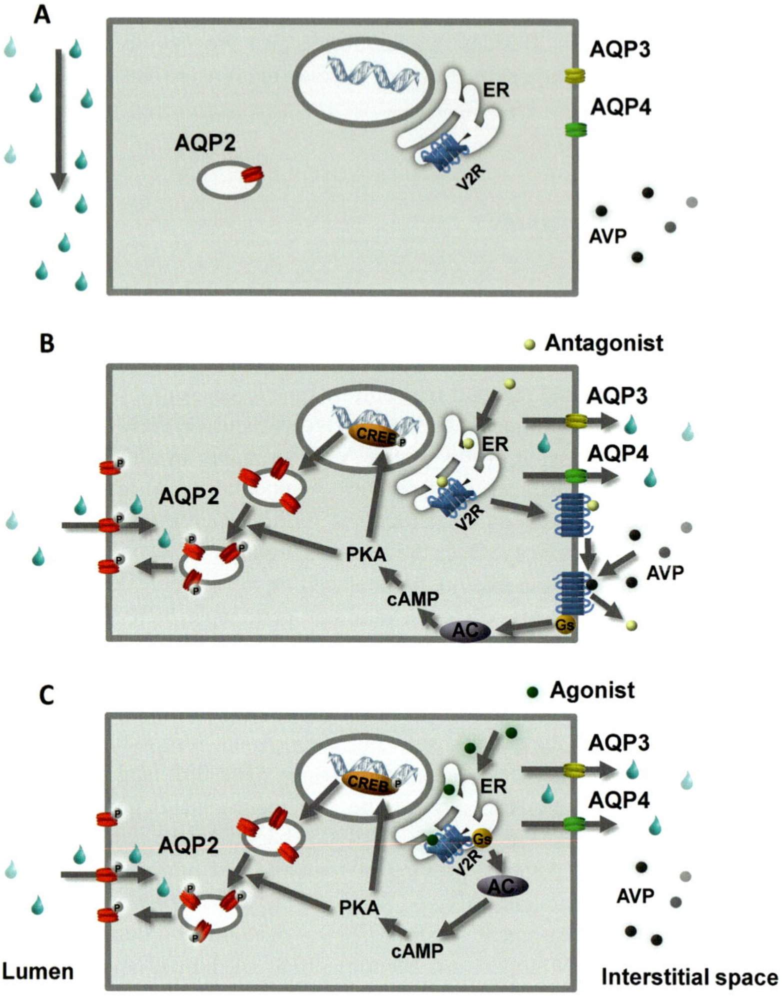

Fig. 30.5 Possible treatment strategy for X-linked NDI. (**a**) X-linked NDI is caused by mutations in the gene encoding the V2R. Most mutations are retained in the endoplasmic reticulum (ER) caused by misfolding of the protein. (**b**) Rescue of plasma membrane expression of the V2R by V2R antagonists. Antagonists bind to misfolded V2R mutants in the ER. This aids the correct folding of the V2R and allows the V2R to escape the ER and reach the plasma membrane. In the plasma membrane, the antagonist is displaced by AVP to allow activation of the receptor, leading to increased AQP2 at the plasma membrane. (**c**) Activation of misfolded V2R mutants by cell-permeable agonists. The agonists reach the misfolded V2R in the ER. This allows signaling to occur from ER-retained receptors, leading to increased AQP2 trafficking. Other cell-permeable agonists also aid proper folding of the V2R, leading to increased expression of V2R at the plasma membrane. Modified from Kortenoeven ML, Fenton RA (2014) Renal aquaporins and water balance disorders. Biochim Biophys Acta 1840:1533–1549

with an increased plasma membrane expression and increased maturation of the V2R mutants (Jean-Alphonse et al. 2009), indicating a different mechanism of rescue. Recently, using a high-throughput cell-based assay, 3 compounds were identified that could effectively reroute the mutant V2R receptor to the plasma membrane. These compounds, which were neither agonists nor antagonists, may be promising leads for further studies (Janovick et al. 2018).

Several alternative treatments might be useful for treatment of NDI, which bypass the V2R signaling cascade and increase trafficking or abundance of AQP2 via activation of another pathway. One option is to stimulate PGE_2 receptors. PGE_2 reduces AVP-induced water permeability; however, PGE_2 increases collecting duct water permeability in the absence of high levels of AVP, probably via activation of the PGE_2 receptors EP2 and/or EP4 (Nadler et al. 1992; Hebert et al. 1990; Sakairi et al. 1995). As mentioned earlier, inhibiting general prostaglandin synthesis by indomethacin can reduce urine volume in NDI patients. However, specific EP2 or EP4 agonists might be more effective, since no activation is expected of prostaglandin receptors involved in decreasing collecting duct water permeability. The EP2 agonist butaprost and the EP4 agonist CAY10580 increase AQP2 apical membrane accumulation in MDCK cells and butaprost increased AQP2 membrane accumulation in slices of kidney tissue (Olesen et al. 2011). The EP4 agonist ONO-AE-329 can decrease urine output, increase urine osmolality, and increase AQP2 abundance in a mouse model for X-linked NDI, while the EP2 agonist butaprost can reduce urine flow rate and increase urine osmolality in a rat model of NDI, suggesting that specific EP2 or EP4 agonists could be promising treatment strategies for NDI (Olesen et al. 2011; Li et al. 2009).

Another possible treatment option for NDI is the use of statins or HMG-CoA reductase inhibitors. Statins are used in the treatment of hypercholesterolemia, because they competitively inhibit the rate-limiting enzyme HMG-CoA in the biosynthesis of cholesterol. Fluvastatin, lovastatin, and simvastatin can increase the apical plasma membrane expression of AQP2 in cultured cells by inhibiting AQP2 endocytosis (Procino et al. 2010, 2011a; Li et al. 2011), which has been suggested to be due to a decrease in active Rho involved in regulation of the cytoskeleton and AQP2 translocation (Li et al. 2011). In vivo studies have shown that simvastatin, fluvastatin, and atorvastatin can decrease urine volume, increase urine osmolality, and increase AQP2 expression or membrane localization in different mice models (Li et al. 2011; Procino et al. 2011a, 2014; Danilovic et al. 2012). Simvastatin treatment increased urine osmolality in healthy individuals and reduced diuresis and increased urine osmolality and AQP2 in hypercholesterolemic patients (Procino et al. 2016; Bech et al. 2018). Cross-sectional data from lithium users using different statins showed that all events with a urine osmolality below 300 mOsm/kg occurred in statin nonusers suggesting a protective effect against lithium-induced NDI (Elie et al. 2015). Future studies have to determine if statins can indeed be used as a therapy for NDI.

An alternative method to bypass the V2R-pathway is increasing intracellular cGMP concentrations. An increase in cGMP, mediated by atrial natriuretic peptide or nitric oxide, can increase AQP2 abundance at the apical plasma membrane

(Boone et al. 2010; Wang et al. 2006; Bouley et al. 2000). Preventing cGMP degradation by the PDE5 inhibitor sildenafil results in an increased plasma membrane expression of AQP2 both in transfected cells and in AVP-deficient Brattleboro rats (Bouley et al. 2005). In addition, sildenafil can reduce urine volume and increase total membrane AQP2 levels in rats with lithium-induced NDI (Sanches et al. 2012), indicating that sildenafil could be a potential treatment strategy for NDI. Sildenafil treatment had no significant effect on urine osmolality in healthy individuals but increased urine osmolality and urinary AQP2 excretion in a boy with X-linked NDI (Bech et al. 2018; Assadi and Sharbaf 2015).

AG-490, an inhibitor of both the EGF receptor (EGFR) and JAK-2 kinase, increased AQP2 trafficking in vitro and decreased urine volume and increased urine osmolality in a rat model of central DI (Nomura et al. 2014). Erlotinib, a selective EGFR inhibitor, enhanced AQP2 apical membrane expression in collecting duct principal cells and reduced urine volume in mice with lithium-induced NDI, revealing a novel pathway that contributes to the regulation of AQP2-mediated water reabsorption and suggesting new potential therapeutic strategies for NDI treatment. (Cheung et al. 2016).

Metformin, a stimulator of the adenosine monophosphate kinase (AMPK) and currently registered for treatment of diabetes mellitus, increased AQP2 Ser256 phosphorylation and membrane accumulation in rat inner medullary collecting ducts (IMCDs) and increased water permeability in perfused rat terminal IMCDs (Klein et al. 2016). Stimulation of AMPK by 5-Aminoimidazole-4-carboxamide ribonucleotide (AICAR), inhibited cAMP-dependent apical accumulation and dDAVP-dependent phosphorylation of AQP2 (Al-Bataineh et al. 2016). The reason for this difference is unknown, but it might suggest that the effects of metformin are independent of activating AMPK. Metformin also increases urine osmolality and expression of AQP2 in tolvaptan-treated rats and in a mouse model of X-linked NDI (Efe et al. 2016). Metformin treatment had no significant effect on urine osmolality in healthy individuals (Bech et al. 2018).

Hydrochlorothiazide reduces lithium-induced NDI in mice lacking the thiazide-sensitive sodium–chloride cotransporter, which suggests that the mechanism of action may be mediated by carbonic anhydrase inhibition. The specific carbonic anhydrase-blocker acetazolamide was as effective as the combination of hydrochlorothiazide and amiloride in reducing polyuria in a mouse model of lithium-induced NDI but had fewer side effects (de Groot et al. 2016). Acetazolamide reduces the glomerular filtration rate by means of tubuloglomerular feedback and reduces renal prostaglandin secretion, which both may attenuate NDI-induced polyuria. In lithium-treated patients, acetazolamide did not reduce urine output, although these patients did not have overt lithium-induced polyuria (de Groot et al. 2017). Acetazolamide did reduce urine output in 2 patients with severe lithium-induced NDI (Macau et al. 2018; Gordon et al. 2016).

Another potential treatment strategy for lithium-induced NDI involves the ADP-activated P2Y12-receptor, which signals through Gi, ultimately reducing cellular cAMP levels. The P2Y12-R is expressed in principal cells and blockade of the P2Y12-R in primary cultures of rat inner medullary collecting duct cells potentiated the expression of AQP2 and AQP3 mRNA, and cAMP production induced by dDAVP. Clopidogrel, an inhibitor of the P2Y12-R and an FDA-approved antiplatelet drug, significantly increased urine concentration and AQP2 protein in the kidneys of normal rats but not in Brattleboro rats that lack AVP (Zhang et al. 2015). Clopidogrel or Prasugrel, another P2Y12-R inhibitor and FDA-approved antiplatelet drug, significantly ameliorated lithium-induced polyuria and increased AQP2 abundance (Zhang et al. 2017, 2019). In addition, deletion of P2Y12-R significantly ameliorates lithium-induced NDI (Zhang et al. 2019). Taken together, these data suggest that the pharmacological inhibition of P2Y12-R may potentially offer therapeutic benefits in lithium-induced NDI.

A variety of other approaches have been suggested to improve urinary-concentrating ability. For example, the renin inhibitor aliskiren increased AQP2 expression in mice, or mice with lithium-induced NDI or unilateral ureteral obstruction. In addition, aliskiren treatment improved the urinary-concentrating defect in lithium-treated mice (Lin et al. 2017; Wang et al. 2015). Another strategy concerns the selective estrogen receptor modulator tamoxifen, as estradiol has been shown to decrease AQP2 abundance in female mice, and tamoxifen-attenuated polyuria and downregulation of AQP2 protein expression in rats with lithium-induced NDI (Tingskov et al. 2018; Cheema et al. 2015). Inhibition of the nonreceptor tyrosine kinase Src using dasatinib also causes AQP2 trafficking, but its effects in vivo have not been tested (Cheung et al. 2019).

30.7 Conclusion

Nine AQPs are expressed in the kidney. Of these aquaporins, only AQP1–4 and possibly AQP7, have a role in the process of urine concentration, while others are involved in diverse functions such as acidification of intracellular vesicles and ammonia transport across mitochondrial membranes. Regulation of renal aquaporins is critical to osmoregulation and the maintenance of body water homeostasis. Considering the important role of aquaporins in water transport, it is not surprising that a number of aquaporins, especially AQP2, are involved in diseases associated with disturbed water homeostasis. In the last decade, our understanding of the molecular mechanisms of these disorders has increased enormously, which has opened up several possible treatment strategies. However, the exact molecular mechanisms that regulate the different aquaporins in different conditions are still largely unknown. Future basic and translational studies are needed to fully understand and treat various water balance disorders.

Acknowledgments Research in the authors' laboratory is supported by the Danish Medical Research Council, The Novo Nordisk Foundation, and the Carlsberg Foundation.

References

Abu LA, Levy-Khademi F, Abdulhadi-Atwan M, Bosin E, Korner M, White PC, Zangen DH (2010) Autosomal recessive familial neurohypophyseal diabetes insipidus: onset in early infancy. Eur J Endocrinol 162(2):221–226

Al-Bataineh MM, Li H, Ohmi K, Gong F, Marciszyn AL, Naveed S, Zhu X, Neumann D, Wu Q, Cheng L, Fenton RA, Pastor-Soler NM, Hallows KR (2016) Activation of the metabolic sensor AMP-activated protein kinase inhibits aquaporin-2 function in kidney principal cells. Am J Physiol Renal Physiol 311(5):F890–F900

Alfadda TI, Saleh AM, Houillier P, Geibel JP (2014) Calcium-sensing receptor 20 years later. Am J Physiol Cell Physiol 307(3):C221–C231

Allen HM, Jackson RL, Winchester MD, Deck LV, Allon M (1989) Indomethacin in the treatment of lithium-induced nephrogenic diabetes insipidus. Arch Intern Med 149(5):1123–1126

Alon U, Chan JC (1985) Hydrochlorothiazide-amiloride in the treatment of congenital nephrogenic diabetes insipidus. Am J Nephrol 5(1):9–13

Alon U, Wellons MD, Chan JC (1985) Hydrochlorothiazide, amiloride and tolmetin in the treatment of diabetes insipidus of Brattleboro rats. Med Biol 63(3):117–122

Ananthakrishnan S (2009) Diabetes insipidus in pregnancy: etiology, evaluation, and management. Endocr Pract 15(4):377–382

Araujo ER, Seguro AC, Spichler A, Magaldi AJ, Volpini RA, De BT (2010) Acute kidney injury in human leptospirosis: an immunohistochemical study with pathophysiological correlation. Virchows Arch 456(4):367–375

Arima H, Oiso Y (2010) Mechanisms underlying progressive polyuria in familial Neurohypophysial diabetes Insipidus. J Neuroendocrinol 22(7):574–577

Aronson D, Verbalis JG, Mueller M, Krum H (2011) Short- and long-term treatment of dilutional hyponatraemia with satavaptan, a selective arginine vasopressin V2-receptor antagonist: the DILIPO study. Eur J Heart Fail 13(3):327–336

Arthus MF, Lonergan M, Crumley MJ, Naumova AK, Morin D, De Marco LA, Kaplan BS, Robertson GL, Sasaki S, Morgan K, Bichet DG, Fujiwara TM (2000) Report of 33 novel AVPR2 mutations and analysis of 117 families with X- linked nephrogenic diabetes insipidus. J Am Soc Nephrol 11(6):1044–1054

Asahina Y, Izumi N, Enomoto N, Sasaki S, Fushimi K, Marumo F, Sato C (1995) Increased gene expression of water channel in cirrhotic rat kidneys. Hepatology 21(1):169–173

Assadi F, Sharbaf FG (2015) Sildenafil for the treatment of congenital Nephrogenic diabetes Insipidus. Am J Nephrol 42(1):65–69

Assentoft M, Larsen BR, Olesen ET, Fenton RA, MacAulay N (2014) AQP4 plasma membrane trafficking or channel gating is not significantly modulated by phosphorylation at COOH-terminal serine residues. Am J Physiol Cell Physiol 307(10):C957–C965

Atochina-Vasserman EN, Biktasova A, Abramova E, Cheng DS, Polosukhin VV, Tanjore H, Takahashi S, Sonoda H, Foye L, Venkov C, Ryzhov SV, Novitskiy S, Shlonimskaya N, Ikeda M, Blackwell TS, Lawson WE, Gow AJ, Harris RC, Dikov MM, Tchekneva EE (2013) Aquaporin 11 insufficiency modulates kidney susceptibility to oxidative stress. Am J Physiol Renal Physiol 304(10):F1295–F1307

Badaut J, Fukuda AM, Jullienne A, Petry KG (2014) Aquaporin and brain diseases. Biochim Biophys Acta 1840(5):1554–1565

Bagshaw SM, Langenberg C, Bellomo R (2006) Urinary biochemistry and microscopy in septic acute renal failure: a systematic review. Am J Kidney Dis 48(5):695–705

Bankir L (2001) Antidiuretic action of vasopressin: quantitative aspects and interaction between V1a and V2 receptor-mediated effects. Cardiovasc Res 51(3):372–390

Barak LS, Oakley RH, Laporte SA, Caron MG (2001) Constitutive arrestin-mediated desensitization of a human vasopressin receptor mutant associated with nephrogenic diabetes insipidus. Proc Natl Acad Sci U S A 98(1):93–98

Batlle DC, von Riotte AB, Gaviria M, Grupp M (1985) Amelioration of polyuria by amiloride in patients receiving long- term lithium therapy. N Engl J Med 312:408–414
Baylis PH (1989) Regulation of vasopressin secretion. Bailliere Clin Endocrinol Metab 3 (2):313–330
Baylis PH (2003) The syndrome of inappropriate antidiuretic hormone secretion. Int J Biochem Cell Biol 35(11):1495–1499
Beaser SB (1947) Renal excretory function and diet in diabetes insipidus. Am J Med Sci 213 (4):441–449
Bech AP, Wetzels JFM, Nijenhuis T (2018) Effects of sildenafil, metformin, and simvastatin on ADH-independent urine concentration in healthy volunteers. Physiol Rep 6(7):e13665
Bedford JJ, Weggery S, Ellis G, McDonald FJ, Joyce PR, Leader JP, Walker RJ (2008) Lithium-induced nephrogenic diabetes insipidus: renal effects of amiloride. Clin J Am Soc Nephrol 3 (5):1324–1331
Beitz E, Liu K, Ikeda M, Guggino WB, Agre P, Yasui M (2006) Determinants of AQP6 trafficking to intracellular sites versus the plasma membrane in transfected mammalian cells. Biol Cell 98 (2):101–109
Bernier V, Morello JP, Zarruk A, Debrand N, Salahpour A, Lonergan M, Arthus MF, Laperriere A, Brouard R, Bouvier M, Bichet DG (2006) Pharmacologic chaperones as a potential treatment for x-linked nephrogenic diabetes insipidus. J Am Soc Nephrol 17(1):232–243
Blair HA (2019) Tolvaptan: a review in autosomal dominant polycystic kidney disease. Drugs 79 (3):303–313
Blalock T, Gerron G, Quiter E, Rudman D (1977) Role of diet in the management of vasopressin-responsive and -resistant diabetes insipidus. Am J Clin Nutr 30(7):1070–1076
Blaydon DC, Lind LK, Plagnol V, Linton KJ, Smith FJ, Wilson NJ, McLean WH, Munro CS, South AP, Leigh IM, O'Toole EA, Lundstrom A, Kelsell DP (2013) Mutations in AQP5, encoding a water-channel protein, cause autosomal-dominant diffuse nonepidermolytic palmoplantar keratoderma. Am J Hum Genet 93(2):330–335
Blount MA, Sands JM, Kent KJ, Smith TD, Price SR, Klein JD (2008) Candesartan augments compensatory changes in medullary transport proteins in the diabetic rat kidney. Am J Physiol Renal Physiol 294(6):F1448–F1452
Boone M, Deen PM (2008) Physiology and pathophysiology of the vasopressin-regulated renal water reabsorption. Pflugers Arch 456(6):1005–1024
Boone M, Kortenoeven M, Robben JH, Deen PM (2010) Effect of the cGMP pathway on AQP2 expression and translocation: potential implications for nephrogenic diabetes insipidus. Nephrol Dial Transplant 25(1):48–54
Boone M, Kortenoeven ML, Robben JH, Tamma G, Deen PM (2011) Counteracting vasopressin-mediated water reabsorption by ATP, dopamine, and phorbol esters: mechanisms of action. Am J Physiol Renal Physiol 300(3):F761–F771
Boton R, Gaviria M, Batlle DC (1987) Prevalence, pathogenesis, and treatment of renal dysfunction associated with chronic lithium therapy. Am J Kidney Dis 10:329–345
Bouley R, Breton S, Sun T, McLaughlin M, Nsumu NN, Lin HY, Ausiello DA, Brown D (2000) Nitric oxide and atrial natriuretic factor stimulate cGMP-dependent membrane insertion of aquaporin 2 in renal epithelial cells. J Clin Invest 106(9):1115–1126
Bouley R, Pastor-Soler N, Cohen O, McLaughlin M, Breton S, Brown D (2005) Stimulation of AQP2 membrane insertion in renal epithelial cells in vitro and in vivo by the cGMP phosphodiesterase inhibitor sildenafil citrate (Viagra). Am J Physiol Renal Physiol 288(6):F1103–F1112
Bouley R, Palomino Z, Tang SS, Nunes P, Kobori H, Lu HA, Shum WW, Sabolic I, Brown D, Ingelfinger JR, Jung FF (2009) Angiotensin II and hypertonicity modulate proximal tubular aquaporin 1 expression. Am J Physiol Renal Physiol 297(6):F1575–F1586
Brond L, Mullertz KM, Torp M, Nielsen J, Graebe M, Hadrup N, Nielsen S, Christensen S, Jonassen TE (2013) Congestive heart failure in rats is associated with increased collecting duct vasopressin sensitivity and vasopressin type 2 receptor reexternalization. Am J Physiol Renal Physiol 305(11):F1547–F1554

Bustamante M, Hasler U, Kotova O, Chibalin AV, Mordasini D, Rousselot M, Vandewalle A, Martin PY, Feraille E (2005) Insulin potentiates AVP-induced AQP2 expression in cultured renal collecting duct principal cells. Am J Physiol Renal Physiol 288(2):F334–F344
Bustamante M, Hasler U, Leroy V, Seigneux de S, Dimitrov M, Mordasini D, Rousselot M, Martin PY, Feraille E (2008) Calcium-sensing receptor attenuates AVP-induced aquaporin-2 expression via a calmodulin-dependent mechanism. J Am Soc Nephrol 19(1):109–116
Cabral PD, Herrera M (2012) Membrane-associated aquaporin-1 facilitates osmotically driven water flux across the basolateral membrane of the thick ascending limb. Am J Physiol Renal Physiol 303(5):F621–F629
Calamita G, Gena P, Meleleo D, Ferri D, Svelto M (2006) Water permeability of rat liver mitochondria: a biophysical study. Biochim Biophys Acta 1758(8):1018–1024
Canfield MC, Tamarappoo BK, Moses AM, Verkman AS, Holtzman EJ (1997) Identification and characterization of aquaporin-2 water channel mutations causing nephrogenic diabetes insipidus with partial vasopressin response. Hum Mol Genet 6:1865–1871
Carpentier E, Greenbaum LA, Rochdi D, Abrol R, Goddard WA III, Bichet DG, Bouvier M (2012) Identification and characterization of an activating F229V substitution in the V2 vasopressin receptor in an infant with NSIAD. J Am Soc Nephrol 23(10):1635–1640
Cesar KR, Magaldi AJ (1999) Thiazide induces water absorption in the inner medullary collecting duct of normal and Brattleboro rats. Am J Phys 277(5 Pt 2):F756–F760
Cesar KR, Romero EC, de Braganca AC, Blanco RM, Abreu PA, Magaldi AJ (2012) Renal involvement in leptospirosis: the effect of glycolipoprotein on renal water absorption. PLoS One 7(6):e37625
Cheema MU, Irsik DL, Wang Y, Miller-Little W, Hyndman KA, Marks ES, Frokiaer J, Boesen EI, Norregaard R (2015) Estradiol regulates AQP2 expression in the collecting duct: a novel inhibitory role for estrogen receptor alpha. Am J Physiol Renal Physiol 309(4):F305–F317
Chen Y, Rice W, Gu Z, Li J, Huang J, Brenner MB, Van HA, Xiong J, Gundersen GG, Norman JC, Hsu VW, Fenton RA, Brown D, Lu HA (2012) Aquaporin 2 promotes cell migration and epithelial morphogenesis. J Am Soc Nephrol 23(9):1506–1517
Cheng X, Zhang H, Lee HL, Park JM (2004) Cyclooxygenase-2 inhibitor preserves medullary aquaporin-2 expression and prevents polyuria after ureteral obstruction. J Urol 172(6 Pt 1):2387–2390
Cheung PW, Nomura N, Nair AV, Pathomthongtaweechai N, Ueberdiek L, Lu HA, Brown D, Bouley R (2016) EGF receptor inhibition by Erlotinib increases aquaporin 2-mediated renal water reabsorption. J Am Soc Nephrol 27(10):3105–3116
Cheung PW, Terlouw A, Janssen SA, Brown D, Bouley R (2019) Inhibition of non-receptor tyrosine kinase Src induces phosphoserine 256-independent aquaporin-2 membrane accumulation. J Physiol 597(6):1627–1642
Choma DP, Vanacore R, Naylor H, Zimmerman IA, Pavlichenko A, Pavlichenko A, Foye L, Carbone DP, Harris RC, Dikov MM, Tchekneva EE (2016) Aquaporin 11 variant associates with kidney disease in type 2 diabetic patients. Am J Physiol Renal Physiol 310(5):F416–F425
Chou CL, Ma TH, Yang BX, Knepper MA, Verkman S (1998) Fourfold reduction of water permeability in inner medullary collecting duct of aquaporin-4 knockout mice. Am J Phys 43 (2):C549–C554
Chou CL, Knepper MA, Hoek AN, Brown D, Yang B, Ma T, Verkman AS (1999) Reduced water permeability and altered ultrastructure in thin descending limb of Henle in aquaporin-1 null mice. J Clin Invest 103(4):491–496
Christensen BM, Zelenina M, Aperia A, Nielsen S (2000) Localization and regulation of PKA-phosphorylated AQP2 in response to V(2)-receptor agonist/antagonist treatment. Am J Physiol Renal Physiol 278(1):F29–F42
Christensen BM, Wang W, Frokiaer J, Nielsen S (2003) Axial heterogeneity in basolateral AQP2 localization in rat kidney: effect of vasopressin. Am J Physiol Renal Physiol 284(4):F701–F717

Christensen BM, Marples D, Kim YH, Wang W, Frokiaer J, Nielsen S (2004) Changes in cellular composition of kidney collecting duct cells in rats with lithium-induced NDI. Am J Physiol Cell Physiol 286(4):C952–C964

Christensen BM, Kim YH, Kwon TH, Nielsen S (2006) Lithium treatment induces a marked proliferation of primarily principal cells in rat kidney inner medullary collecting duct. Am J Physiol Renal Physiol 291(1):F39–F48

Christensen BM, Zuber AM, Loffing J, Stehle JC, Deen PM, Rossier BC, Hummler E (2011) alphaENaC-mediated lithium absorption promotes nephrogenic diabetes insipidus. J Am Soc Nephrol 22(2):253–261

Chung SH, Jun DW, Kim KT, Chae JD, Park EK, Son BK, Kim SH, Jo YJ, Park YS (2010) Aquaporin-2 urinary excretion in cirrhosis: relationship to vasopressin and nitric oxide. Dig Dis Sci 55(4):1135–1141

Coleman RA, Wu DC, Liu J, Wade JB (2000) Expression of aquaporins in the renal connecting tubule. Am J Physiol Renal Physiol 279(5):F874–F883

Crawford JD, Kennedy GC, HILL LE (1960) Clinical results of treatment of diabetes insipidus with drugs of the chlorothiazide series. N Engl J Med 262:737–743

Danilovic A, Lopes RI, Sanches TR, Shimizu MH, Oshiro FM, Andrade L, Denes FT, Seguro AC (2012) Atorvastatin prevents the downregulation of aquaporin-2 receptor after bilateral ureteral obstruction and protects renal function in a rat model. Urology 80(2):485–420

de Groot T, Alsady M, Jaklofsky M, Otte-Holler I, Baumgarten R, Giles RH, Deen PM (2014) Lithium causes G2 arrest of renal principal cells. J Am Soc Nephrol 25(3):501–510

de Groot T, Sinke AP, Kortenoeven ML, Alsady M, Baumgarten R, Devuyst O, Loffing J, Wetzels JF, Deen PM (2016) Acetazolamide attenuates Lithium-induced Nephrogenic diabetes Insipidus. J Am Soc Nephrol 27(7):2082–2091

de Groot T, Doornebal J, Christensen BM, Cockx S, Sinke AP, Baumgarten R, Bedford JJ, Walker RJ, Wetzels JFM, Deen PMT (2017) Lithium-induced NDI: acetazolamide reduces polyuria but does not improve urine concentrating ability. Am J Physiol Renal Physiol 313(3):F669–F676

De Mattia F, Savelkoul PJ, Bichet DG, Kamsteeg EJ, Konings IB, Marr N, Arthus MF, Lonergan M, van Os CH, van der SP RG, Deen PM (2004) A novel mechanism in recessive nephrogenic diabetes insipidus: wild-type aquaporin-2 rescues the apical membrane expression of intracellularly retained AQP2-P262L. Hum Mol Genet 13(24):3045–3056

De Mattia F, Savelkoul PJ, Kamsteeg EJ, Konings IB, van der SP MR, Oksche A, Deen PM (2005) Lack of arginine vasopressin-induced phosphorylation of Aquaporin-2 mutant AQP2-R254L explains dominant Nephrogenic diabetes Insipidus. J Am Soc Nephrol 16(10):2872–2880

Decaux G, Vandergheynst F, Bouko Y, Parma J, Vassart G, Vilain C (2007) Nephrogenic syndrome of inappropriate antidiuresis in adults: high phenotypic variability in men and women from a large pedigree. J Am Soc Nephrol 18(2):606–612

Deen PMT, van Aubel RA, van Lieburg AF, van Os CH (1996) Urinary content of aquaporin 1 and 2 in nephrogenic diabetes insipidus. J Am Soc Nephrol 7:836–841

Delporte C (2014) Aquaporins in salivary glands and pancreas. Biochim Biophys Acta 1840 (5):1524–1532

Di IN, Napoli F, Allegri AE, Olivieri I, Bertelli E, Gallizia A, Rossi A, Maghnie M (2012) Diabetes insipidus--diagnosis and management. Horm Res Paediatr 77(2):69–84

Digiovanni SR, Nielsen S, Christensen EI, Knepper MA (1994) Regulation of collecting duct water channel expression by vasopressin in Brattleboro rat. Proc Natl Acad Sci U S A 91:8984–8988

Earm JH, Christensen BM, Frokiaer J, Marples D, Han JS, Knepper MA, Nielsen S (1998) Decreased aquaporin-2 expression and apical plasma membrane delivery in kidney collecting ducts of polyuric hypercalcemic rats. J Am Soc Nephrol 9(12):2181–2193

Ecelbarger CA, Terris J, Frindt G, Echevarria M, Marples D, Nielsen S, Knepper MA (1995) Aquaporin-3 water channel localization and regulation in rat kidney. Am J Phys 38:F663–F672

Efe O, Klein JD, LaRocque LM, Ren H, Sands JM (2016) Metformin improves urine concentration in rodents with nephrogenic diabetes insipidus. JCI Insight 1(11):e88409

El TA, Lussier Y, Da CS, Bissonnette P, Bichet DG (2016) Functional recovery of AQP2 recessive mutations through hetero-Oligomerization with wild-type counterpart. Sci Rep 6:33298

Elhassan EA, Schrier RW (2011) Hyponatremia: diagnosis, complications, and management including V2 receptor antagonists. Curr Opin Nephrol Hypertens 20(2):161–168

Elie D, Segal M, Low NC, Mucsi I, Holcroft C, Shulman K, Looper KJ, Rej S (2015) Statins in the prevention of lithium-associated diabetes insipidus: preliminary findings. Kidney Int 87(4):862

Elkjaer ML, Nejsum LN, Gresz V, Kwon TH, Jensen UB, Frokiaer J, Nielsen S (2001) Immunolocalization of aquaporin-8 in rat kidney, gastrointestinal tract, testis, and airways. Am J Physiol Renal Physiol 281(6):F1047–F1057

Elkjaer ML, Kwon TH, Wang W, Nielsen J, Knepper MA, Frokiaer J, Nielsen S (2002) Altered expression of renal NHE3, TSC, BSC-1, and ENaC subunits in potassium-depleted rats. Am J Physiol Renal Physiol 283(6):F1376–F1388

Erdelyi LS, Mann WA, Morris-Rosendahl DJ, Gross U, Nagel M, Varnai P, Balla A, Hunyady L (2015) Mutation in the V2 vasopressin receptor gene, AVPR2, causes nephrogenic syndrome of inappropriate diuresis. Kidney Int 88(5):1070–1078

Esteva-Font C, Baccaro ME, Fernandez-Llama P, Sans L, Guevara M, Ars E, Jimenez W, Arroyo V, Ballarin JA, Gines P (2006) Aquaporin-1 and aquaporin-2 urinary excretion in cirrhosis: relationship with ascites and hepatorenal syndrome. Hepatology 44(6):1555–1563

Farres MT, Ronco P, Saadoun D, Remy P, Vincent F, Khalil A, Le Blanche AF (2003) Chronic lithium nephropathy: MR imaging for diagnosis. Radiology 229(2):570–574

Feldman HA, Singer I (1974) Comparative effects of tetracyclines on water flow across toad urinary bladders. J Pharmacol Exp Ther 190(2):358–364

Feldman BJ, Rosenthal SM, Vargas GA, Fenwick RG, Huang EA, Matsuda-Abedini M, Lustig RH, Mathias RS, Portale AA, Miller WL, Gitelman SE (2005) Nephrogenic syndrome of inappropriate antidiuresis. N Engl J Med 352(18):1884–1890

Fenton RA, Knepper MA (2007) Mouse models and the urinary concentrating mechanism in the new millennium. Physiol Rev 87(4):1083–1112

Fenton RA, Brond L, Nielsen S, Praetorius J (2007) Cellular and subcellular distribution of the type-2 vasopressin receptor in the kidney. Am J Physiol Renal Physiol 293(3):F748–F760

Fenton RA, Moeller HB, Hoffert JD, Yu MJ, Nielsen S, Knepper MA (2008) Acute regulation of aquaporin-2 phosphorylation at Ser-264 by vasopressin. Proc Natl Acad Sci U S A 105 (8):3134–3139

Fenton RA, Moeller HB, Zelenina M, Snaebjornsson MT, Holen T, MacAulay N (2010) Differential water permeability and regulation of three aquaporin 4 isoforms. Cell Mol Life Sci 67 (5):829–840

Fernandez-Llama P, Turner R, Dibona G, Knepper MA (1999) Renal expression of aquaporins in liver cirrhosis induced by chronic common bile duct ligation in rats. J Am Soc Nephrol 10 (9):1950–1957

Fernandez-Llama P, Jimenez W, Bosch-Marce M, Arroyo V, Nielsen S, Knepper MA (2000) Dysregulation of renal aquaporins and Na-Cl cotransporter in CCl4- induced cirrhosis. Kidney Int 58(1):216–228

Flamion B, Spring KR (1990) Water permeability of apical and basolateral cell membranes of rat inner medullary collecting duct. Am J Phys 259(6 Pt 2):F986–F999

Forrest JN Jr, Cohen AD, Torretti J, Himmelhoch JM, Epstein FH (1974) On the mechanism of lithium-induced diabetes insipidus in man and the rat. J Clin Invest 53(4):1115–1123

Forrest JN Jr, Cox M, Hong C, Morrison G, Bia M, Singer I (1978) Superiority of demeclocycline over lithium in the treatment of chronic syndrome of inappropriate secretion of antidiuretic hormone. N Engl J Med 298(4):173–177

Fotiadis D, Hasler L, Muller DJ, Stahlberg H, Kistler J, Engel A (2000) Surface tongue-and-groove contours on lens MIP facilitate cell-to-cell adherence. J Mol Biol 300(4):779–789

Frokiaer J, Marples D, Knepper MA, Nielsen S (1996) Bilateral ureteral obstruction downregulates expression of vasopressin-sensitive AQP-2 water channel in rat kidney. Am J Phys 39:F657–F668

Frokiaer J, Christensen BM, Marples D, Djurhuus JC, Jensen UB, Knepper MA, Nielsen S (1997) Downregulation of aquaporin-2 parallels changes in renal water excretion in unilateral ureteral obstruction. Am J Phys 42:F213–F223
Fujisawa G, Ishikawa S, Tsuboi Y, Okada K, Saito T (1993) Therapeutic efficacy of non-peptide ADH antagonist OPC-31260 in SIADH rats. Kidney Int 44(1):19–23
Fujita N, Ishikawa S, Sasaki S, Fujisawa G, Fushimi K, Marumo F, Saito T (1995) Role of water channel AQP-CD in water retention in SIADH and cirrhotic rats. Am J Phys 38:F926–F931
Furman CS, Gorelick-Feldman DA, Davidson KG, Yasumura T, Neely JD, Agre P, Rash JE (2003) Aquaporin-4 square array assembly: opposing actions of M1 and M23 isoforms. Proc Natl Acad Sci U S A 100(23):13609–13614
Fushimi K, Uchida S, Hara Y, Hirata Y, Marumo F, Sasaki S (1993) Cloning and expression of apical membrane water channel of rat kidney collecting tubule. Nature 361:549–552
Fushimi K, Sasaki S, Marumo F (1997) Phosphorylation of serine 256 is required for cAMP-dependent regulatory exocytosis of the aquaporin-2 water channel. J Biol Chem 272:14800–14804
Ghali JK, Koren MJ, Taylor JR, Brooks-Asplund E, Fan K, Long WA, Smith N (2006) Efficacy and safety of oral conivaptan: a V1A/V2 vasopressin receptor antagonist, assessed in a randomized, placebo-controlled trial in patients with euvolemic or hypervolemic hyponatremia. J Clin Endocrinol Metab 91(6):2145–2152
Gill JR Jr, Bartter FC (1961) On the impairment of renal concentrating ability in prolonged hypercalcemia and hypercalciuria in man. J Clin Invest 40:716–722
Gines P, Wong F, Watson H, Milutinovic S, del Arbol LR, Olteanu D (2008) Effects of satavaptan, a selective vasopressin V(2) receptor antagonist, on ascites and serum sodium in cirrhosis with hyponatremia: a randomized trial. Hepatology 48(1):204–213
Gordon CE, Vantzelfde S, Francis JM (2016) Acetazolamide in Lithium-induced Nephrogenic diabetes Insipidus. N Engl J Med 375(20):2008–2009
Goubau C, Jaeken J, Levtchenko EN, Thys C, Di MM, Martens GA, Gerlo E, De VR, Buyse GM, Goemans N, Van GC, Freson K (2013) Homozygosity for aquaporin 7 G264V in three unrelated children with hyperglyceroluria and a mild platelet secretion defect. Genet Med 15(1):55–63
Grinevich V, Knepper MA, Verbalis J, Reyes I, Aguilera G (2004) Acute endotoxemia in rats induces down-regulation of V2 vasopressin receptors and aquaporin-2 content in the kidney medulla. Kidney Int 65(1):54–62
Hadi HA, Mashini IS, Devoe LD (1985) Diabetes insipidus during pregnancy complicated by preeclampsia. A case report. J Reprod Med 30(3):206–208
Han B, Wu X, Huang PP, Zhu FX, Liu S (2018) Aquaporin 11 rs2276415 variant and progression of chronic kidney disease. Nephrol Dial Transplant. https://doi.org/10.1093/ndt/gfy219
Harris PC, Torres VE (2009) Polycystic kidney disease. Annu Rev Med 60:321–337
Hasler U, Mordasini D, Bianchi M, Vandewalle A, Feraille E, Martin PY (2003) Dual influence of aldosterone on AQP2 expression in cultured renal collecting duct principal cells. J Biol Chem 278(24):21639–21648
Hasler U, Vinciguerra M, Vandewalle A, Martin PY, Feraille E (2005) Dual effects of hypertonicity on aquaporin-2 expression in cultured renal collecting duct principal cells. J Am Soc Nephrol 16 (6):1571–1582
Hasler U, Leroy V, Jeon US, Bouley R, Dimitrov M, Kim JA, Brown D, Kwon HM, Martin PY, Feraille E (2008) NF-kappaB modulates aquaporin-2 transcription in renal collecting duct principal cells. J Biol Chem 283(42):28095–28105
Hebert RL, Jacobson HR, Breyer MD (1990) PGE2 inhibits AVP-induced water flow in cortical collecting ducts by protein kinase C activation. Am J Phys 259:F318–F325
Herak-Kramberger CM, Breljak D, Ljubojevic M, Matokanovic M, Lovric M, Rogic D, Brzica H, Vrhovac I, Karaica D, Micek V, Dupor JI, Brown D, Sabolic I (2015) Sex-dependent expression of water channel AQP1 along the rat nephron. Am J Physiol Renal Physiol 308(8):F809–F821
Hinrichs GR, Hansen LH, Nielsen MR, Fagerberg C, Dieperink H, Rittig S, Jensen BL (2016) A novel mutation affecting the arginine-137 residue of AVPR2 in dizygous twins leads to

nephrogenic diabetes insipidus and attenuated urine exosome aquaporin-2. Physiol Rep 4(8): e12764
Hirji MR, Mucklow JC (1991) Transepithelial water movement in response to carbamazepine, chlorpropamide and demeclocycline in toad urinary bladder. Br J Pharmacol 104(2):550–553
Hoffert JD, Pisitkun T, Wang G, Shen RF, Knepper MA (2006) Quantitative phosphoproteomics of vasopressin-sensitive renal cells: regulation of aquaporin-2 phosphorylation at two sites. Proc Natl Acad Sci U S A 103(18):7159–7164
Hoffert JD, Nielsen J, Yu MJ, Pisitkun T, Schleicher SM, Nielsen S, Knepper MA (2007) Dynamics of aquaporin-2 serine-261 phosphorylation in response to short-term vasopressin treatment in collecting duct. Am J Physiol Renal Physiol 292(2):F691–F700
Hoffert JD, Fenton RA, Moeller HB, Simons B, Tchapyjnikov D, McDill BW, Yu MJ, Pisitkun T, Chen F, Knepper MA (2008) Vasopressin-stimulated increase in phosphorylation at Ser269 potentiates plasma membrane retention of aquaporin-2. J Biol Chem 283(36):24617–24627
Hoffert JD, Pisitkun T, Saeed F, Song JH, Chou CL, Knepper MA (2012) Dynamics of the G protein-coupled vasopressin V2 receptor signaling network revealed by quantitative phosphoproteomics. Mol Cell Proteomics 11(2):M111
Holm LM, Jahn TP, Moller AL, Schjoerring JK, Ferri D, Klaerke DA, Zeuthen T (2005) NH3 and NH4+ permeability in aquaporin-expressing Xenopus oocytes. Pflugers Arch 450(6):415–428
Hoogewijs D, Vogler M, Zwenger E, Krull S, Zieseniss A (2016) Oxygen-dependent regulation of aquaporin-3 expression. Hypoxia (Auckl) 4:91–97
Hoorn EJ, Monserez DA, Fenton RA, Overdevest I, Apperloo AJ, Zietse R, Hardillo JA (2014) Olfactory Neuroblastoma with Hyponatremia. J Clin Oncol 33(21):e88–e92
Hozawa S, Holtzman EJ, Ausiello DA (1996) cAMP motifs regulating transcription in the aquaporin 2 gene. Am J Phys 39:C1695–C1702
Huebner AR, Somparn P, Benjachat T, Leelahavanichkul A, Avihingsanon Y, Fenton RA, Pisitkun T (2015) Exosomes in urine biomarker discovery. Adv Exp Med Biol 845:43–58
Hyndman KA, Yang CR, Jung HJ, Umejiego EN, Chou CL, Knepper MA (2018) Proteomic determination of the lysine acetylome and phosphoproteome in the rat native inner medullary collecting duct. Physiol Genomics 50(9):669–679
Ikeda M, Beitz E, Kozono D, Guggino WB, Agre P, Yasui M (2002) Characterization of aquaporin-6 as a nitrate channel in mammalian cells: requirement of pore-lining residue threonine-63. J Biol Chem 277(42):39873–39879
Ikeda M, Andoo A, Shimono M, Takamatsu N, Taki A, Muta K, Matsushita W, Uechi T, Matsuzaki T, Kenmochi N, Takata K, Sasaki S, Ito K, Ishibashi K (2011) The NPC motif of aquaporin-11, unlike the NPA motif of known aquaporins, is essential for full expression of molecular function. J Biol Chem 286(5):3342–3350
Imamura T, Kinugawa K, Fujino T, Inaba T, Maki H, Hatano M, Yao A, Komuro I (2014) Increased urine aquaporin-2 relative to plasma arginine vasopressin is a novel marker of response to tolvaptan in patients with decompensated heart failure. Circ J 78(9):2240–2249
Imamura T, Kinugawa K, Komuro I (2016) Tolvaptan prolongs blockage of the vasopressin type II receptor over 24 hours in responders with stage D heart failure. Int Heart J 57(1):41–46
Inoue Y, Sohara E, Kobayashi K, Chiga M, Rai T, Ishibashi K, Horie S, Su X, Zhou J, Sasaki S, Uchida S (2014) Aberrant glycosylation and localization of Polycystin-1 cause polycystic kidney in an AQP11 knockout model. J Am Soc Nephrol 25(12):2789–2799
Ishibashi K, Sasaki S, Fushimi K, Uchida S, Kuwahara M, Saito H, Furukawa T, Nakajima K, Yamaguchi Y, Gojobori T (1994) Molecular cloning and expression of a member of the aquaporin family with permeability to glycerol and urea in addition to water expressed at the basolateral membrane of kidney collecting duct cells. Proc Natl Acad Sci U S A 91:6269–6273
Ishibashi K, Kuwahara M, Gu Y, Kageyama Y, Tohsaka A, Suzuki F, Marumo F, Sasaki S (1997a) Cloning and functional expression of a new water channel abundantly expressed in the testis permeable to water, glycerol, and urea. J Biol Chem 272(33):20782–20786

Ishibashi K, Sasaki S, Fushimi K, Yamamoto T, Kuwahara M, Marumo F (1997b) Immunolocalization and effect of dehydration on AQP3, a basolateral water channel of kidney collecting ducts. Am J Phys 41:F235–F241
Ishibashi K, Imai M, Sasaki S (2000) Cellular localization of aquaporin 7 in the rat kidney. Exp Nephrol 8(4–5):252–257
Ishibashi K, Morinaga T, Kuwahara M, Sasaki S, Imai M (2002) Cloning and identification of a new member of water channel (AQP10) as an aquaglyceroporin. Biochim Biophys Acta 1576 (3):335–340
Ishibashi K, Hara S, Kondo S (2009) Aquaporin water channels in mammals. Clin Exp Nephrol 13 (2):107–117
Ishibashi K, Tanaka Y, Morishita Y (2014) The role of mammalian superaquaporins inside the cell. Biochim Biophys Acta 1840(5):1507–1512
Ito M, Jameson JL (1997) Molecular basis of autosomal dominant neurohypophyseal diabetes insipidus - cellular toxicity caused by the accumulation of mutant vasopressin precursors within the endoplasmic reticulum. J Clin Invest 99:1897–1905
Ito M, Yu RN, Jameson JL (1999) Mutant vasopressin precursors that cause autosomal dominant neurohypophyseal diabetes insipidus retain dimerization and impair the secretion of wild-type proteins. J Biol Chem 274(13):9029–9037
Ivarsen P, Frokiaer J, Aagaard NK, Hansen EF, Bendtsen F, Nielsen S, Vilstrup H (2003) Increased urinary excretion of aquaporin 2 in patients with liver cirrhosis. Gut 52(8):1194–1199
Iyasere O, Xu G, Harris K (2012) Urinary tract obstruction. Br J Hosp Med (Lond) 73(12):696–700
Jakobsson B, Berg U (1994) Effect of hydrochlorothiazide and indomethacin treatment on renal function in nephrogenic diabetes insipidus. Acta Paediatr 83(5):522–525
Janovick JA, Spicer TP, Bannister TD, Smith E, Ganapathy V, Scampavia L (2018) Chemical validation and optimization of pharmacoperones targeting vasopressin type 2 receptor mutant. Biochem J 475(18):2941–2953
Jean-Alphonse F, Perkovska S, Frantz MC, Durroux T, Mejean C, Morin D, Loison S, Bonnet D, Hibert M, Mouillac B, Mendre C (2009) Biased agonist pharmacochaperones of the AVP V2 receptor may treat congenital nephrogenic diabetes insipidus. J Am Soc Nephrol 20 (10):2190–2203
Jensen AM, Bae EH, Norregaard R, Wang G, Nielsen S, Schweer H, Kim SW, Frokiaer J (2010) Cyclooxygenase 2 inhibition exacerbates AQP2 and pAQP2 downregulation independently of V2 receptor abundance in the postobstructed kidney. Am J Physiol Renal Physiol 298(4):F941–F950
Jeon US, Han KH, Park SH, Lee SD, Sheen MR, Jung JY, Kim WY, Sands JM, Kim J, Kwon HM (2007) Downregulation of renal TonEBP in hypokalemic rats. Am J Physiol Renal Physiol 293 (1):F408–F415
Jochberger S, Dorler J, Luckner G, Mayr VD, Wenzel V, Ulmer H, Morgenthaler NG, Hasibeder WR, Dunser MW (2009) The vasopressin and copeptin response to infection, severe sepsis, and septic shock. Crit Care Med 37(2):476–482
Jung JY, Madsen KM, Han KH, Yang CW, Knepper MA, Sands JM, Kim J (2003) Expression of urea transporters in potassium-depleted mouse kidney. Am J Physiol Renal Physiol 285(6): F1210–F1224
Kamsteeg EJ, Wormhoudt TA, Rijss JPL, van Os CH, Deen PMT (1999) An impaired routing of wild-type aquaporin-2 after tetramerization with an aquaporin-2 mutant explains dominant nephrogenic diabetes insipidus. EMBO J 18(9):2394–2400
Kamsteeg EJ, Heijnen I, van Os CH, Deen PMT (2000) The subcellular localization of an Aquaporin-2 tetramer depends on the stoichiometry of phosphorylated and nonphosphorylated monomers. J Cell Biol 151(4):919–930
Kamsteeg EJ, Hendriks G, Boone M, Konings IB, Oorschot V, van der SP KJ, Deen PM (2006) Short-chain ubiquitination mediates the regulated endocytosis of the aquaporin-2 water channel. Proc Natl Acad Sci U S A 103(48):18344–18349

Katsura T, Ausiello DA, Brown D (1996) Direct demonstration of aquaporin-2 water channel recycling in stably transfected LLC-PK1 epithelial cells. Am J Phys 39:F548–F553

Katsura T, Gustafson CE, Ausiello DA, Brown D (1997) Protein kinase a phosphorylation is involved in regulated exocytosis of aquaporin-2 in transfected LLC-PK1 cells. Am J Phys 41: F816–F822

Katz VL, Bowes WA Jr (1987) Transient diabetes insipidus and preeclampsia. South Med J 80 (4):524–525

Kellenberger S, Gautschi I, Schild L (1999) A single point mutation in the pore region of the epithelial Na+ channel changes ion selectivity by modifying molecular sieving. Proc Natl Acad Sci U S A 96(7):4170–4175

Khositseth S, Uawithya P, Somparn P, Charngkaew K, Thippamom N, Hoffert JD, Saeed F, Michael PD, Chen SH, Fenton RA, Pisitkun T (2015) Autophagic degradation of aquaporin-2 is an early event in hypokalemia-induced nephrogenic diabetes insipidus. Sci Rep 5:18311

Khositseth S, Charngkaew K, Boonkrai C, Somparn P, Uawithya P, Chomanee N, Payne DM, Fenton RA, Pisitkun T (2017) Hypercalcemia induces targeted autophagic degradation of aquaporin-2 at the onset of nephrogenic diabetes insipidus. Kidney Int 91(5):1070–1087

Kim D, Sands JM, Klein JD (2004a) Role of vasopressin in diabetes mellitus-induced changes in medullary transport proteins involved in urine concentration in Brattleboro rats. Am J Physiol Renal Physiol 286(4):F760–F766

Kim GH, Lee JW, Oh YK, Chang HR, Joo KW, Na KY, Earm JH, Knepper MA, Han JS (2004b) Antidiuretic effect of hydrochlorothiazide in lithium-induced nephrogenic diabetes insipidus is associated with upregulation of aquaporin-2, Na-Cl co-transporter, and epithelial sodium channel. J Am Soc Nephrol 15(11):2836–2843

Kim GH, Choi NW, Jung JY, Song JH, Lee CH, Kang CM, Knepper MA (2008) Treating lithium-induced nephrogenic diabetes insipidus with a COX-2 inhibitor improves polyuria via upregulation of AQP2 and NKCC2. Am J Physiol Renal Physiol 294(4):F702–F709

King LS, Choi M, Fernandez PC, Cartron JP, Agre P (2001) Defective urinary-concentrating ability due to a complete deficiency of aquaporin-1. N Engl J Med 345(3):175–179

Kirchlechner V, Koller DY, Seidl R, Waldhauser F (1999) Treatment of nephrogenic diabetes insipidus with hydrochlorothiazide and amiloride. Arch Dis Child 80(6):548–552

Kjaersgaard G, Madsen K, Marcussen N, Jensen BL (2014) Lithium induces microcysts and polyuria in adolescent rat kidney independent of cyclooxygenase-2. Physiol Rep 2(1):e00202

Klahr S, Harris K, Purkerson ML (1988) Effects of obstruction on renal functions. Pediatr Nephrol 2 (1):34–42

Klein JD, Kozlowski S, Antoun TA, Sands JM (2006) Adrenalectomy blocks the compensatory increases in UT-A1 and AQP2 in diabetic rat kidney. J Membr Biol 212(2):139–144

Klein JD, Wang Y, Blount MA, Molina PA, LaRocque LM, Ruiz JA, Sands JM (2016) Metformin, an AMPK activator, stimulates the phosphorylation of aquaporin 2 and urea transporter A1 in inner medullary collecting ducts. Am J Physiol Renal Physiol 310(10):F1008–F1012

Knepper MA, Packer R, Good DW (1989) Ammonium transport in the kidney. Physiol Rev 69 (1):179–249

Ko SB, Uchida S, Naruse S, Kuwahara M, Ishibashi K, Marumo F, Hayakawa T, Sasaki S (1999) Cloning and functional expression of rAOP9L a new member of aquaporin family from rat liver. Biochem Mol Biol Int 47(2):309–318

Kondo H, Shimomura I, Kishida K, Kuriyama H, Makino Y, Nishizawa H, Matsuda M, Maeda N, Nagaretani H, Kihara S, Kurachi Y, Nakamura T, Funahashi T, Matsuzawa Y (2002) Human aquaporin adipose (AQPap) gene. Genomic structure, promoter analysis and functional mutation. Eur J Biochem 269(7):1814–1826

Konoshita T, Kuroda M, Kawane T, Koni I, Miyamori I, Tofuku Y, Mabuchi H, Takeda R (2004) Treatment of congenital nephrogenic diabetes insipidus with hydrochlorothiazide and amiloride in an adult patient. Horm Res 61(2):63–67

Kortenoeven ML, Li Y, Shaw S, Gaeggeler HP, Rossier BC, Wetzels JF, Deen PM (2009) Amiloride blocks lithium entry through the sodium channel thereby attenuating the resultant nephrogenic diabetes insipidus. Kidney Int 76(1):44–53

Kortenoeven ML, Schweer H, Cox R, Wetzels JF, Deen PM (2012a) Lithium reduces aquaporin-2 transcription independent of prostaglandins. Am J Physiol Cell Physiol 302(1):C131–C140

Kortenoeven ML, Trimpert C, van den Brand M, Li Y, Wetzels JF, Deen PM (2012b) In mpkCCD cells, long-term regulation of aquaporin-2 by vasopressin occurs independent of protein kinase a and CREB but may involve Epac. Am J Physiol Renal Physiol 302(11):F1395–F1401

Kortenoeven ML, Pedersen NB, Miller RL, Rojek A, Fenton RA (2013a) Genetic ablation of aquaporin-2 in the mouse connecting tubules results in defective renal water handling. J Physiol 591(Pt 8):2205–2219

Kortenoeven ML, Sinke AP, Hadrup N, Trimpert C, Wetzels JF, Fenton RA, Deen PM (2013b) Demeclocycline attenuates hyponatremia by reducing aquaporin-2 expression in the renal inner medulla. Am J Physiol Renal Physiol 305(12):F1705–F1718

Kosten TR, Forrest JN (1986) Treatment of severe lithium-induced polyuria with amiloride. Am J Psychiatry 143(12):1563–1568

Kotnik P, Nielsen J, Kwon TH, Krzisnik C, Frokiaer J, Nielsen S (2005) Altered expression of COX-1, COX-2, and mPGES in rats with nephrogenic and central diabetes insipidus. Am J Physiol Renal Physiol 288(5):F1053–F1068

Kuwahara M, Fushimi K, Terada Y, Bai L, Marumo F, Sasaki S (1995) cAMP-dependent phosphorylation stimulates water permeability of aquaporin-collecting duct water channel protein expressed in Xenopus oocytes. J Biol Chem 270:10384–10387

Kwon TH, Fr F, Knepper MA, Nielsen S (1999) Reduced abundance of aquaporins in rats with bilateral ischemia-induced acute renal failure: prevention by alpha-MSH. Am J Phys 277(3 Pt 2):F413–F427

Kwon TH, Laursen UH, Marples D, Maunsbach AB, Knepper MA, Frokiaer J, Nielsen S (2000) Altered expression of renal AQPs and Na(+) transporters in rats with lithium-induced NDI. Am J Physiol Renal Physiol 279(3):F552–F564

Land H, Schutz G, Schmale H, Richter D (1982) Nucleotide sequence of cloned cDNA encoding bovine arginine vasopressin-neurophysin II precursor. Nature 295(5847):299–303

Landegren N, Pourmousa LM, Skov J, Hallgren A, Eriksson D, Lisberg Toft-Bertelsen T, MacAulay N, Hagforsen E, Raisanen-Sokolowski A, Saha H, Nilsson T, Nordmark G, Ohlsson S, Gustafsson J, Husebye ES, Larsson E, Anderson MS, Perheentupa J, Rorsman F, Fenton RA, Kampe O (2016) Autoantibodies targeting a collecting duct-specific Water Channel in Tubulointerstitial nephritis. J Am Soc Nephrol 27(10):3220–3228

Lassiter WE, Gottschalk CW, Mylle M (1961) Micropuncture study of net transtubular movement of water and urea in nondiuretic mammalian kidney. Am J Phys 200:1139–1147

Laursen UH, Pihakaski-Maunsbach K, Kwon TH, Ostergaard JE, Nielsen S, Maunsbach AB (2004) Changes of rat kidney AQP2 and Na,K-ATPase mRNA expression in lithium-induced nephrogenic diabetes insipidus. Nephron Exp Nephrol 97(1):e1–e16

Lee WK, Bork U, Gholamrezaei F, Thevenod F (2005) Cd^{2+}-induced cytochrome c release in apoptotic proximal tubule cells: role of mitochondrial permeability transition pore and Ca^{2+} uniporter. Am J Physiol Renal Physiol 288(1):F27–F39

Lei L, Wang W, Jia Y, Su L, Zhou H, Verkman AS, Yang B (2017) Aquaporin-3 deletion in mice results in renal collecting duct abnormalities and worsens ischemia-reperfusion injury. Biochim Biophys Acta Mol basis Dis 1863(6):1231–1241

Levi M, Peterson L, Berl T (1983) Mechanism of concentrating defect in hypercalcemia. Role of polydipsia and prostaglandins. Kidney Int 23(3):489–497

Leviel F, Hubner CA, Houillier P, Morla L, El MS, Brideau G, Hassan H, Parker MD, Kurth I, Kougioumtzes A, Sinning A, Pech V, Riemondy KA, Miller RL, Hummler E, Shull GE, Aronson PS, Doucet A, Wall SM, Chambrey R, Eladari D (2010) The Na+−dependent chloride-bicarbonate exchanger SLC4A8 mediates an electroneutral Na+ reabsorption process in the renal cortical collecting ducts of mice. J Clin Invest 120(5):1627–1635

Levtchenko EN, Monnens LA (2010) Nephrogenic syndrome of inappropriate antidiuresis. Nephrol Dial Transplant 25(9):2839–2843

Li C, Wang W, Kwon TH, Isikay L, Wen JG, Marples D, Djurhuus JC, Stockwell A, Knepper MA, Nielsen S, Frokiaer J (2001) Downregulation of AQP1, −2, and −3 after ureteral obstruction is associated with a long-term urine-concentrating defect. Am J Physiol Renal Physiol 281(1): F163–F171

Li C, Wang W, Knepper MA, Nielsen S, Frokiar J (2003a) Downregulation of renal aquaporins in response to unilateral ureteral obstruction. Am J Physiol Renal Physiol 284(5):F1066–F1079

Li C, Wang W, Kwon TH, Knepper MA, Nielsen S, Frokiaer J (2003b) Altered expression of major renal Na transporters in rats with bilateral ureteral obstruction and release of obstruction. Am J Physiol Renal Physiol 285(5):F889–F901

Li C, Wang W, Kwon TH, Knepper MA, Nielsen S, Frokiaer J (2003c) Altered expression of major renal Na transporters in rats with unilateral ureteral obstruction. Am J Physiol Renal Physiol 284 (1):F155–F166

Li C, Klein JD, Wang W, Knepper MA, Nielsen S, Sands JM, Frokiaer J (2004) Altered expression of urea transporters in response to ureteral obstruction. Am J Physiol Renal Physiol 286(6): F1154–F1162

Li Y, Shaw S, Kamsteeg EJ, Vandewalle A, Deen PM (2006) Development of lithium-induced nephrogenic diabetes insipidus is dissociated from adenylyl cyclase activity. J Am Soc Nephrol 17(4):1063–1072

Li JH, Chou CL, Li B, Gavrilova O, Eisner C, Schnermann J, Anderson SA, Deng CX, Knepper MA, Wess J (2009) A selective EP4 PGE2 receptor agonist alleviates disease in a new mouse model of X-linked nephrogenic diabetes insipidus. J Clin Invest 119(10):3115–3126

Li W, Zhang Y, Bouley R, Chen Y, Matsuzaki T, Nunes P, Hasler U, Brown D, Lu HA (2011) Simvastatin enhances aquaporin-2 surface expression and urinary concentration in vasopressin-deficient Brattleboro rats through modulation of rho GTPase. Am J Physiol Renal Physiol 301 (2):F309–F318

Lin QQ, Lin R, Ji QL, Zhang JY, Wang WR, Yang LN, Zhang KF (2011) Effect of exercise training on renal function and renal aquaporin-2 expression in rats with chronic heart failure. Clin Exp Pharmacol Physiol 38(3):179–185

Lin Y, Zhang T, Feng P, Qiu M, Liu Q, Li S, Zheng P, Kong Y, Levi M, Li C, Wang W (2017) Aliskiren increases aquaporin-2 expression and attenuates lithium-induced nephrogenic diabetes insipidus. Am J Physiol Renal Physiol 313(4):F914–F925

Liu Z, Shen J, Carbrey JM, Mukhopadhyay R, Agre P, Rosen BP (2002) Arsenite transport by mammalian aquaglyceroporins AQP7 and AQP9. Proc Natl Acad Sci U S A 99(9):6053–6058

Liu K, Nagase H, Huang CG, Calamita G, Agre P (2006) Purification and functional characterization of aquaporin-8. Biol Cell 98(3):153–161

Liu M, Sun Y, Xu M, Yu X, Zhang Y, Huang S, Ding G, Zhang A, Jia Z (2018) Role of mitochondrial oxidative stress in modulating the expressions of aquaporins in obstructive kidney disease. Am J Physiol Renal Physiol 314(4):F658–F666

Lolait SJ, O'Carroll AM, McBride OW, Konig M, Morel A, Brownstein MJ (1992) Cloning and characterization of a vasopressin V2 receptor and possible link to nephrogenic diabetes insipidus. Nature 357:336–339

Los EL, Deen PM, Robben JH (2010) Potential of nonpeptide (ant)agonists to rescue vasopressin V2 receptor mutants for the treatment of X-linked nephrogenic diabetes insipidus. J Neuroendocrinol 22(5):393–399

Lu W, Peissel B, Babakhanlou H, Pavlova A, Geng L, Fan X, Larson C, Brent G, Zhou J (1997) Perinatal lethality with kidney and pancreas defects in mice with a targetted Pkd1 mutation. Nat Genet 17(2):179–181

Lu Y, Chen L, Zhao B, Xiao Z, Meng T, Zhou Q, Zhang W (2016) Urine AQP5 is a potential novel biomarker of diabetic nephropathy. J Diabetes Complicat 30(5):819–825

Ma TH, Yang BX, Kuo WL, Verkman AS (1996) cDNA cloning and gene structure of a novel water channel expressed exclusively in human kidney: evidence for a gene cluster of aquaporins at chromosome locus 12q13. Genomics 35:543–550
Ma TH, Yang BX, Gillespie A, Carlson EJ, Epstein CJ, Verkman AS (1997) Generation and phenotype of a transgenic knockout mouse lacking the mercurial-insensitive water channel aquaporin-4. J Clin Invest 100:957–962
Ma TH, Yang BX, Gillespie A, Carlson EJ, Epstein CJ, Verkman AS (1998) Severely impaired urinary concentrating ability in transgenic mice lacking aquaporin-1 water channels. J Biol Chem 273(8):4296–4299
Ma T, Song Y, Yang B, Gillespie A, Carlson EJ, Epstein CJ, Verkman AS (2000) Nephrogenic diabetes insipidus in mice lacking aquaporin-3 water channels. Proc Natl Acad Sci U S A 97 (8):4386–4391
Macau RA, da Silva TN, Silva JR, Ferreira AG, Bravo P (2018) Use of acetazolamide in lithium-induced nephrogenic diabetes insipidus: a case report. Endocrinol Diabetes Metab Case Rep 2018:17–0154
Madeira A, Fernandez-Veledo S, Camps M, Zorzano A, Moura TF, Ceperuelo-Mallafre V, Vendrell J, Soveral G (2014) Human aquaporin-11 is a water and glycerol channel and localizes in the vicinity of lipid droplets in human adipocytes. Obesity (Silver Spring) 22(9):2010–2017
Magaldi AJ, Yasuda PN, Kudo LH, Seguro AC, Rocha AS (1992) Renal involvement in leptospirosis: a pathophysiologic study. Nephron 62(3):332–339
Maletkovic J, Drexler A (2013) Diabetic ketoacidosis and hyperglycemic hyperosmolar state. Endocrinol Metab Clin N Am 42(4):677–695
Mancinelli R, La Rovere RM, Fulle S, Miscia S, Marchisio M, Pierdomenico L, Lanuti P, Procino G, Barbieri C, Svelto M, Fano-Illic G, Pietrangelo T (2014) Extracellular GTP is a potent water-transport regulator via aquaporin 5 plasma-membrane insertion in M1-CCD epithelial cortical collecting duct cells. Cell Physiol Biochem 33(3):731–746
Manji HK, Moore GJ, Chen G (2001) Bipolar disorder: leads from the molecular and cellular mechanisms of action of mood stabilizers. Br J Psychiatry Suppl 41:s107–s119
Marples D, Christensen S, Christensen EI, Ottosen PD, Nielsen S (1995) Lithium-induced downregulation of aquaporin-2 water channel expression in rat kidney medulla. J Clin Invest 95:1838–1845
Marples D, Frokiaer J, Dorup J, Knepper MA, Nielsen S (1996) Hypokalemia-induced downregulation of aquaporin-2 water channel expression in rat kidney medulla and cortex. J Clin Invest 97:1960–1968
Marr N, Kamsteeg EJ, Van Raak M, van Os CH, Deen PMT (2001) Functionality of aquaporin-2 missense mutants in recessive nephrogenic diabetes insipidus. Pflugers Arch 442(1):73–77
Marr N, Bichet DG, Hoefs S, Savelkoul PJ, Konings IB, De Mattia F, Graat MP, Arthus MF, Lonergan M, Fujiwara TM, Knoers NVAM, Landau D, Balfe WJ, Oksche A, Rosenthal W, Muller D, van Os CH, Deen PMT (2002) Cell-biologic and functional analyses of five new Aquaporin-2 missense mutations that cause recessive Nephrogenic diabetes Insipidus. J Am Soc Nephrol 13(9):2267–2277
Matsumoto K, Miyazaki H, Fujii T, Hashimoto M (1989) Binding of sulfonamides to erythrocytes and their components. Chem Pharm Bull (Tokyo) 37(7):1913–1915
Matsumura Y, Uchida S, Rai T, Sasaki S, Marumo F (1997) Transcriptional regulation of aquaporin-2 water channel gene by cAMP. J Am Soc Nephrol 8:861–867
Miller EW, Dickinson BC, Chang CJ (2010) Aquaporin-3 mediates hydrogen peroxide uptake to regulate downstream intracellular signaling. Proc Natl Acad Sci U S A 107(36):15681–15686
Miranda CA, Lee JW, Chou CL, Knepper MA (2014) Tolvaptan as a tool in renal physiology. Am J Physiol Renal Physiol 306(3):F359–F366
Mobasheri A, Marples D, Young IS, Floyd RV, Moskaluk CA, Frigeri A (2007) Distribution of the AQP4 water channel in normal human tissues: protein and tissue microarrays reveal expression in several new anatomical locations, including the prostate gland and seminal vesicles. Channels (Austin) 1(1):29–38

Moe SE, Sorbo JG, Sogaard R, Zeuthen T, Petter OO, Holen T (2008) New isoforms of rat Aquaporin-4. Genomics 91(4):367–377
Moeller HB, Knepper MA, Fenton RA (2009) Serine 269 phosphorylated aquaporin-2 is targeted to the apical membrane of collecting duct principal cells. Kidney Int 75(3):295–303
Moeller HB, Praetorius J, Rutzler MR, Fenton RA (2010) Phosphorylation of aquaporin-2 regulates its endocytosis and protein-protein interactions. Proc Natl Acad Sci U S A 107(1):424–429
Moeller HB, Olesen ET, Fenton RA (2011) Regulation of the water channel aquaporin-2 by posttranslational modification. Am J Physiol Renal Physiol 300(5):F1062–F1073
Moeller HB, Aroankins TS, Slengerik-Hansen J, Pisitkun T, Fenton RA (2014) Phosphorylation and ubiquitylation are opposing processes that regulate endocytosis of the water channel aquaporin-2. J Cell Sci 127(Pt 14):3174–3183
Moeller HB, Slengerik-Hansen J, Aroankins T, Assentoft M, MacAulay N, Moestrup SK, Bhalla V, Fenton RA (2016) Regulation of the Water Channel Aquaporin-2 via 14-3-3theta and -zeta. J Biol Chem 291(5):2469–2484
Moeller HB, Fuglsang CH, Pedersen CN, Fenton RA (2018) Basolateral cholesterol depletion alters Aquaporin-2 post-translational modifications and disrupts apical plasma membrane targeting. Biochem Biophys Res Commun 495(1):157–162
Molinas SM, Trumper L, Marinelli RA (2012) Mitochondrial aquaporin-8 in renal proximal tubule cells: evidence for a role in the response to metabolic acidosis. Am J Physiol Renal Physiol 303 (3):F458–F466
Monnens LAH, Jonkman A, Thomas C (1984) Response to indomethacin and hydrochlorothiazide in nephrogenic diabetes insipidus. Clin Sci 66:709–715
Morello JP, Salahpour A, Laperriere A, Bernier V, Arthus MF, Lonergan M, Petaja-Repo U, Angers S, Morin D, Bichet DG, Bouvier M (2000) Pharmacological chaperones rescue cell-surface expression and function of misfolded V2 vasopressin receptor mutants [see comments]. J Clin Invest 105(7):887–895
Morishita Y, Matsuzaki T, Hara-chikuma M, Andoo A, Shimono M, Matsuki A, Kobayashi K, Ikeda M, Yamamoto T, Verkman A, Kusano E, Ookawara S, Takata K, Sasaki S, Ishibashi K (2005) Disruption of aquaporin-11 produces polycystic kidneys following vacuolization of the proximal tubule. Mol Cell Biol 25(17):7770–7779
Mouillac B, Mendre C (2018) Pharmacological chaperones as potential therapeutic strategies for misfolded mutant vasopressin receptors. Handb Exp Pharmacol 245:63–83
Mu J, Johansson M, Hansson GC, Lundgren O (1999) Lithium evokes a more pronounced natriuresis when administered orally than when given intravenously to salt-depleted rats. Pflugers Arch 438(2):159–164
Mulders SM, Knoers NVAM, van Lieburg AF, Monnens LAH, Leumann E, Wuhl E, Schober E, Rijss JPL, van Os CH, Deen PMT (1997) New mutations in the AQP2 gene in nephrogenic diabetes insipidus resulting in functional but misrouted water channels. J Am Soc Nephrol 8:242–248
Murali SK, Aroankins TS, Moeller HB, Fenton RA (2019) The Deubiquitylase USP4 interacts with the Water Channel AQP2 to modulate its apical membrane accumulation and cellular abundance. Cell 8(3):265
Murata K, Mitsuoka K, Hirai T, Walz T, Agre P, Heymann JB, Engel A, Fujiyoshi Y (2000) Structural determinants of water permeation through aquaporin-1. Nature 407(6804):599–605
Murillo-Carretero MI, Ilundain AA, Echevarria M (1999) Regulation of aquaporin mRNA expression in rat kidney by water intake [in process citation]. J Am Soc Nephrol 10(4):696–703
Musa-Aziz R, Chen LM, Pelletier MF, Boron WF (2009) Relative CO_2/NH_3 selectivities of AQP1, AQP4, AQP5, AmtB, and RhAG. Proc Natl Acad Sci U S A 106(13):5406–5411
Mutig K, Paliege A, Kahl T, Jons T, Muller-Esterl W, Bachmann S (2007) Vasopressin V2 receptor expression along rat, mouse, and human renal epithelia with focus on TAL. Am J Physiol Renal Physiol 293(4):F1166–F1177
Nadler SP, Zimpelmann JA, Hebert RL (1992) PGE2 inhibits water permeability at a post-cAMP site in rat terminal inner medullary collecting duct. Am J Phys 262(2 Pt 2):F229–F235

Nakanishi K, Sweeney WE Jr, Zerres K, Guay-Woodford LM, Avner ED (2000) Proximal tubular cysts in fetal human autosomal recessive polycystic kidney disease. J Am Soc Nephrol 11 (4):760–763

Nakanishi H, Kurosaki M, Hosokawa T, Takahashi Y, Itakura J, Suzuki S, Yasui Y, Tamaki N, Nakakuki N, Takada H, Higuchi M, Komiyama Y, Yoshida T, Takaura K, Hayashi T, Kuwabara K, Sasaki S, Izumi N (2016) Urinary excretion of the water channel aquaporin 2 correlated with the pharmacological effect of tolvaptan in cirrhotic patients with ascites. J Gastroenterol 51(6):620–627

Nasrat H, Wali FN, Warda A (1997) Diabetes insipidus, a rare complication of HELLP syndrome. Report of local experience and review of the literature. J Obstet Gynaecol 17(1):64–65

Nejsum LN, Elkjaer M, Hager H, Frokiaer J, Kwon TH, Nielsen S (2000) Localization of aquaporin-7 in rat and mouse kidney using RT-PCR, immunoblotting, and immunocytochemistry [in process citation]. Biochem Biophys Res Commun 277(1):164–170

Nejsum LN, Kwon TH, Marples D, Flyvbjerg A, Knepper MA, Frokiaer J, Nielsen S (2001) Compensatory increase in AQP2, p-AQP2, and AQP3 expression in rats with diabetes mellitus. Am J Physiol Renal Physiol 280(4):F715–F726

Nejsum LN, Zelenina M, Aperia A, Frokiaer J, Nielsen S (2005) Bidirectional regulation of AQP2 trafficking and recycling: involvement of AQP2-S256 phosphorylation. Am J Physiol Renal Physiol 288(5):F930–F938

Nielsen S, Digiovanni SR, Christensen EI, Knepper MA, Harris HW (1993a) Cellular and subcellular immunolocalization of vasopressin- regulated water channel in rat kidney. Proc Natl Acad Sci U S A 90:11663–11667

Nielsen S, Smith BL, Christensen EI, Knepper MA, Agre P (1993b) CHIP28 water channels are localized in constitutively water- permeable segments of the nephron. J Cell Biol 120:371–383

Nielsen S, Chou CL, Marples D, Christensen EI, Kishore BK, Knepper MA (1995a) Vasopressin increases water permeability of kidney collecting duct by inducing translocation of aquaporin-CD water channels to plasma membrane. Proc Natl Acad Sci U S A 92:1013–1017

Nielsen S, Pallone T, Smith BL, Christensen EI, Agre P, Maunsbach AB (1995b) Aquaporin-1 water channels in short and long loop descending thin limbs and in descending vasa recta in rat kidney. Am J Phys 268:F1023–F1037

Nielsen S, Terris J, Andersen D, Ecelbarger C, Frokiaer J, Jonassen T, Marples D, Knepper MA, Petersen JS (1997) Congestive heart failure in rats is associated with increased expression and targeting of aquaporin-2 water channel in collecting duct. Proc Natl Acad Sci U S A 94:5450–5455

Nielsen J, Kwon TH, Praetorius J, Kim YH, Frokiaer J, Knepper MA, Nielsen S (2003) Segment-specific ENaC downregulation in kidney of rats with lithium-induced NDI. Am J Physiol Renal Physiol 285(6):F1198–F1209

Nielsen J, Hoffert JD, Knepper MA, Agre P, Nielsen S, Fenton RA (2008) Proteomic analysis of lithium-induced nephrogenic diabetes insipidus: mechanisms for aquaporin 2 down-regulation and cellular proliferation. Proc Natl Acad Sci U S A 105(9):3634–3639

Nilsson L, Madsen K, Topcu SO, Jensen BL, Frokiaer J, Norregaard R (2012) Disruption of cyclooxygenase-2 prevents downregulation of cortical AQP2 and AQP3 in response to bilateral ureteral obstruction in the mouse. Am J Physiol Renal Physiol 302(11):F1430–F1439

Nomura N, Nunes P, Bouley R, Nair AV, Shaw S, Ueda E, Pathomthongtaweechai N, Lu HA, Brown D (2014) High-throughput chemical screening identifies AG-490 as a stimulator of aquaporin 2 membrane expression and urine concentration. Am J Physiol Cell Physiol 307(7): C597–C605

Norregaard R, Jensen BL, Li C, Wang W, Knepper MA, Nielsen S, Frokiaer J (2005) COX-2 inhibition prevents downregulation of key renal water and sodium transport proteins in response to bilateral ureteral obstruction. Am J Physiol Renal Physiol 289(2):F322–F333

Norregaard R, Jensen BL, Topcu SO, Wang G, Schweer H, Nielsen S, Frokiaer J (2010) Urinary tract obstruction induces transient accumulation of COX-2-derived prostanoids in kidney tissue. Am J Physiol Regul Integr Comp Physiol 298(4):R1017–R1025

Ohshiro K, Yaoita E, Yoshida Y, Fujinaka H, Matsuki A, Kamiie J, Kovalenko P, Yamamoto T (2001) Expression and immunolocalization of AQP6 in intercalated cells of the rat kidney collecting duct. Arch Histol Cytol 64(3):329–338

Okada S, Misaka T, Tanaka Y, Matsumoto I, Ishibashi K, Sasaki S, Abe K (2008) Aquaporin-11 knockout mice and polycystic kidney disease animals share a common mechanism of cyst formation. FASEB J 22(10):3672–3684

Olesen ET, Fenton RA (2013) Is there a role for PGE2 in urinary concentration? J Am Soc Nephrol 24(2):169–178

Olesen ET, Fenton RA (2017) Aquaporin-2 membrane targeting: still a conundrum. Am J Physiol Renal Physiol 312(4):F744–F747

Olesen ET, Seigneux de S, Wang G, Lutken SC, Frokiaer J, Kwon TH, Nielsen S (2009) Rapid and segmental specific dysregulation of AQP2, S256-pAQP2 and renal sodium transporters in rats with LPS-induced endotoxaemia. Nephrol Dial Transplant 24(8):2338–2349

Olesen ET, Rutzler MR, Moeller HB, Praetorius HA, Fenton RA (2011) Vasopressin-independent targeting of aquaporin-2 by selective E-prostanoid receptor agonists alleviates nephrogenic diabetes insipidus. Proc Natl Acad Sci U S A 108(31):12949–12954

Olesen ET, Moeller HB, Assentoft M, MacAulay N, Fenton RA (2016) The vasopressin type 2 receptor and prostaglandin receptors EP2 and EP4 can increase aquaporin-2 plasma membrane targeting through a cAMP-independent pathway. Am J Physiol Renal Physiol 311(5):F935–F944

Ooi BS, Chen BT, Tan KK, Khoo OT (1972) Human renal leptospirosis. Am J Trop Med Hyg 21 (3):336–341

Pallone TL, Edwards A, Ma T, Silldorff EP, Verkman AS (2000) Requirement of aquaporin-1 for NaCl-driven water transport across descending vasa recta. J Clin Invest 105(2):215–222

Palmer BF (2003) Hyponatremia in patients with central nervous system disease: SIADH versus CSW. Trends Endocrinol Metab 14(4):182–187

Park JI, Yang SH, Lee JP, Yoo SH, Kim YS (2015) Genetic predisposition of donors affects the allograft outcome in kidney transplantation: single-nucleotide polymorphism of aquaporin-11. Kidney Res Clin Pract 34(1):47–52

Pedersen RS, Bentzen H, Bech JN, Nyvad O, Pedersen EB (2003) Urinary aquaporin-2 in healthy humans and patients with liver cirrhosis and chronic heart failure during baseline conditions and after acute water load. Kidney Int 63(4):1417–1425

Peterson LN (1990) Vitamin D-induced chronic hypercalcemia inhibits thick ascending limb NaCl reabsorption in vivo. Am J Phys 259(1 Pt 2):F122–F129

Pohl M, Shan Q, Petsch T, Styp-Rekowska B, Matthey P, Bleich M, Bachmann S, Theilig F (2015) Short-term functional adaptation of aquaporin-1 surface expression in the proximal tubule, a component of glomerulotubular balance. J Am Soc Nephrol 26(6):1269–1278

Poulsen SB, Kim YH, Frokiaer J, Nielsen S, Christensen BM (2013) Long-term vasopressin-V2-receptor stimulation induces regulation of aquaporin 4 protein in renal inner medulla and cortex of Brattleboro rats. Nephrol Dial Transplant 28(8):2058–2065

Poulsen SB, Kristensen TB, Brooks HL, Kohan DE, Rieg T, Fenton RA (2017) Role of adenylyl cyclase 6 in the development of lithium-induced nephrogenic diabetes insipidus. JCI Insight 2 (7):e91042

Procino G, Carmosino M, Tamma G, Gouraud S, Laera A, Riccardi D, Svelto M, Valenti G (2004) Extracellular calcium antagonizes forskolin-induced aquaporin 2 trafficking in collecting duct cells. Kidney Int 66(6):2245–2255

Procino G, Barbieri C, Carmosino M, Rizzo F, Valenti G, Svelto M (2010) Lovastatin-induced cholesterol depletion affects both apical sorting and endocytosis of aquaporin-2 in renal cells. Am J Physiol Renal Physiol 298(2):F266–F278

Procino G, Barbieri C, Carmosino M, Tamma G, Milano S, De BL, Mola MG, Lazo-Fernandez Y, Valenti G, Svelto M (2011a) Fluvastatin modulates renal water reabsorption in vivo through increased AQP2 availability at the apical plasma membrane of collecting duct cells. Pflugers Arch 462(5):753–766

Procino G, Mastrofrancesco L, Sallustio F, Costantino V, Barbieri C, Pisani F, Schena FP, Svelto M, Valenti G (2011b) AQP5 is expressed in type-B intercalated cells in the collecting duct system of the rat, mouse and human kidney. Cell Physiol Biochem 28(4):683–692

Procino G, Mastrofrancesco L, Tamma G, Lasorsa DR, Ranieri M, Stringini G, Emma F, Svelto M, Valenti G (2012) Calcium-sensing receptor and aquaporin 2 interplay in hypercalciuria-associated renal concentrating defect in humans. An in vivo and in vitro study. PLoS One 7 (3):e33145

Procino G, Milano S, Tamma G, Dossena S, Barbieri C, Nicoletti MC, Ranieri M, Di MA, Nofziger C, Svelto M, Paulmichl M, Valenti G (2013) Co-regulated pendrin and aquaporin 5 expression and trafficking in type-B intercalated cells under potassium depletion. Cell Physiol Biochem 32(7):184–199

Procino G, Milano S, Carmosino M, Barbieri C, Nicoletti MC, Li JH, Wess J, Svelto M (2014) Combination of secretin and fluvastatin ameliorates the polyuria associated with X-linked nephrogenic diabetes insipidus in mice. Kidney Int 86(1):127–138

Procino G, Portincasa P, Mastrofrancesco L, Castorani L, Bonfrate L, Addabbo F, Carmosino M, Di CA, Svelto M (2016) Simvastatin increases AQP2 urinary excretion in hypercholesterolemic patients: a pleiotropic effect of interest for patients with impaired AQP2 trafficking. Clin Pharmacol Ther 99(5):528–537

Promeneur D, Kwon TH, Yasui M, Kim GH, Frokiaer J, Knepper MA, Agre P, Nielsen S (2000) Regulation of AQP6 mRNA and protein expression in rats in response to altered acid-base or water balance. Am J Physiol Renal Physiol 279(6):F1014–F1026

Puliyanda DP, Ward DT, Baum MA, Hammond TG, Harris HW Jr (2003) Calpain-mediated AQP2 proteolysis in inner medullary collecting duct. Biochem Biophys Res Commun 303(1):52–58

Raghavan R, Eknoyan G (2014) Acute interstitial nephritis - a reappraisal and update. Clin Nephrol 82(3):149–162

Ranchin B, Boury-Jamot M, Blanchard G, Dubourg L, Hadj-Aissa A, Morin D, Durroux T, Cochat P, Bricca G, Verbavatz JM, Geelen G (2010) Familial nephrogenic syndrome of inappropriate antidiuresis: dissociation between aquaporin-2 and vasopressin excretion. J Clin Endocrinol Metab 95(9):E37–E43

Ranieri M, Zahedi K, Tamma G, Centrone M, Di MA, Soleimani M, Valenti G (2018) CaSR signaling down-regulates AQP2 expression via a novel microRNA pathway in pendrin and NaCl cotransporter knockout mice. FASEB J 32(4):2148–2159

Rao R, Hao CM, Breyer MD (2004) Hypertonic stress activates glycogen synthase kinase 3beta-mediated apoptosis of renal medullary interstitial cells, suppressing an NFkappaB-driven cyclooxygenase-2-dependent survival pathway. J Biol Chem 279(6):3949–3955

Rao R, Zhang MZ, Zhao M, Cai H, Harris RC, Breyer MD, Hao CM (2005) Lithium treatment inhibits renal GSK-3 activity and promotes cyclooxygenase 2-dependent polyuria. Am J Physiol Renal Physiol 288(4):F642–F649

Rao R, Patel S, Hao C, Woodgett J, Harris R (2010) GSK3beta mediates renal response to vasopressin by modulating adenylate cyclase activity. J Am Soc Nephrol 21(3):428–437

Rascher W, Rosendahl W, Henrichs IA, Maier R, Seyberth HW (1987) Congenital nephrogenic diabetes insipidus-vasopressin and prostaglandins in response to treatment with hydrochlorothiazide and indomethacin. Pediatr Nephrol 1(3):485–490

Robben JH, Knoers NV, Deen PM (2006a) Cell biological aspects of the vasopressin type-2 receptor and aquaporin 2 water channel in nephrogenic diabetes insipidus. Am J Physiol Renal Physiol 291(2):F257–F270

Robben JH, Sze M, Knoers NV, Deen PM (2006b) Rescue of vasopressin v2 receptor mutants by chemical chaperones: specificity and mechanism. Mol Biol Cell 17(1):379–386

Robben JH, Sze M, Knoers NV, Deen PM (2007) Functional rescue of vasopressin V2 receptor mutants in MDCK cells by pharmacochaperones: relevance to therapy of nephrogenic diabetes insipidus. Am J Physiol Renal Physiol 292(1):F253–F260

Robben JH, Kortenoeven ML, Sze M, Yae C, Milligan G, Oorschot VM, Klumperman J, Knoers NV, Deen PM (2009) Intracellular activation of vasopressin V2 receptor mutants in nephrogenic diabetes insipidus by nonpeptide agonists. Proc Natl Acad Sci U S A 106(29):12195–12200

Robinson AG (1976) DDAVP in the treatment of central diabetes insipidus. N Engl J Med 294 (10):507–511

Rodriguez M, Hernandez M, Cheungpasitporn W, Kashani KB, Riaz I, Rangaswami J, Herzog E, Guglin M, Krittanawong C (2019) Hyponatremia in heart failure: pathogenesis and management. Curr Cardiol Rev 15(4):252–226

Rojek A, Fuchtbauer EM, Kwon TH, Frokiaer J, Nielsen S (2006) Severe urinary concentrating defect in renal collecting duct-selective AQP2 conditional-knockout mice. Proc Natl Acad Sci U S A 103(15):6037–6042

Rojek A, Praetorius J, Frokiaer J, Nielsen S, Fenton RA (2008) A current view of the mammalian aquaglyceroporins. Annu Rev Physiol 70:301–327

Rossi L, Nicoletti MC, Carmosino M, Mastrofrancesco L, Di FA, Indrio F, Lella R, Laviola L, Giorgino F, Svelto M, Gesualdo L, Procino G (2017) Urinary excretion of kidney Aquaporins as possible diagnostic biomarker of diabetic nephropathy. J Diabetes Res 2017:4360357

Roudier N, Ripoche P, Gane P, Le Pennec PY, Daniels G, Cartron JP, Bailly P (2002) AQP3 deficiency in humans and the molecular basis of a novel blood group system, GIL. J Biol Chem 277(48):45854–45859

Rubini ME (1961) Water excrtion in potassium-deficient man. J Clin Invest 40:2215–2224

Russell TA, Ito M, Ito M, Yu RN, Martinson FA, Weiss J, Jameson JL (2003) A murine model of autosomal dominant neurohypophyseal diabetes insipidus reveals progressive loss of vasopressin-producing neurons. J Clin Invest 112(11):1697–1706

Rutzler M, Rojek A, Damgaard MV, Andreasen A, Fenton RA, Nielsen S (2017) Temporal deletion of Aqp11 in mice is linked to the severity of cyst-like disease. Am J Physiol Renal Physiol 312 (2):F343–F351

Saito T, Ishikawa S, Abe K, Kamoi K, Yamada K, Shimizu K, Saruta T, Yoshida S (1997a) Acute aquaresis by the nonpeptide arginine vasopressin (AVP) antagonist OPC-31260 improves hyponatremia in patients with syndrome of inappropriate secretion of antidiuretic hormone (SIADH). J Clin Endocrinol Metab 82(4):1054–1057

Saito T, Ishikawa SE, Sasaki S, Nakamura T, Rokkaku K, Kawakami A, Honda K, Marumo F (1997b) Urinary excretion of aquaporin-2 in the diagnosis of central diabetes insipidus. J Clin Endocrinol Metab 82:1823–1827

Saito T, Ishikawa S, Ito T, Oda H, Ando F, Higashiyama M, Nagasaka S, Hieda M, Saito T (1999) Urinary excretion of aquaporin-2 water channel differentiates psychogenic polydipsia from central diabetes insipidus. J Clin Endocrinol Metab 84(6):2235–2237

Sakairi Y, Jacobson HR, Noland TD, Breyer MD (1995) Luminal prostaglandin E receptors regulate salt and water transport in rabbit cortical collecting duct. Am J Phys 269:F257–F265

Sanches TR, Volpini RA, Massola Shimizu MH, Braganca AC, Oshiro-Monreal F, Seguro AC, Andrade L (2012) Sildenafil reduces polyuria in rats with lithium-induced NDI. Am J Physiol Renal Physiol 302(1):F216–F225

Sands JM, Flores FX, Kato A, Baum MA, Brown EM, Ward DT, Hebert SC, Harris HW (1998) Vasopressin-elicited water and urea permeabilities are altered in IMCD in hypercalcemic rats. Am J Phys 274(5 Pt 2):F978–F985

Satake M, Ikarashi N, Ichikawa Y, Maniwa A, Toda T, Ito K, Ochiai W, Sugiyama K (2010) The role of renal aquaporin 2 in the alleviation of dehydration associated with diabetic polyuria in KKAy mice. Life Sci 87(15–16):475–480

Satoh M, Ogikubo S, Yoshizawa-Ogasawara A (2008) Correlation between clinical phenotypes and X-inactivation patterns in six female carriers with heterozygote vasopressin type 2 receptor gene mutations. Endocr J 55(2):277–284

Savelkoul PJ, De MF, Li Y, Kamsteeg EJ, Konings IB, van der Sluijs P, Deen PM (2009) P.R254Q mutation in the aquaporin-2 water channel causing dominant nephrogenic diabetes insipidus is due to a lack of arginine vasopressin-induced phosphorylation. Hum Mutat 30(10):E891–E903

Schey KL, Wang Z, Wenke L, Qi Y (2014) Aquaporins in the eye: expression, function, and roles in ocular disease. Biochim Biophys Acta 1840(5):1513–1523
Schnermann J, Chou CL, Ma T, Traynor T, Knepper MA, Verkman AS (1998) Defective proximal tubular fluid reabsorption in transgenic aquaporin-1 null mice. Proc Natl Acad Sci U S A 95 (16):9660–9664
Schrier RW (2006) Water and sodium retention in edematous disorders: role of vasopressin and aldosterone. Am J Med 119(7 Suppl 1):S47–S53
Schrier RW, Gross P, Gheorghiade M, Berl T, Verbalis JG, Czerwiec FS, Orlandi C (2006) Tolvaptan, a selective Oral vasopressin V2-receptor antagonist, for Hyponatremia. N Engl J Med 355(20):2099–2112
Seguro AC, Andrade L (2013) Pathophysiology of leptospirosis. Shock 39(Suppl 1):17–23
Sherlock M, Thompson CJ (2010) The syndrome of inappropriate antidiuretic hormone: current and future management options. Eur J Endocrinol 162(Suppl 1):S13–S18
Silberstein C, Bouley R, Huang Y, Fang P, Pastor-Soler N, Brown D, Van Hoek AN (2004) Membrane organization and function of M1 and M23 isoforms of aquaporin-4 in epithelial cells. Am J Physiol Renal Physiol 287(3):F501–F511
Singer I, Rotenberg D (1973) Demeclocycline-induced nephrogenic diabetes insipidus. In-vivo and in-vitro studies. Ann Intern Med 79(5):679–683
Sinke AP, Kortenoeven ML, de Groot T, Baumgarten R, Devuyst O, Wetzels JF, Loffing J, Deen PM (2014) Hydrochlorothiazide attenuates lithium-induced nephrogenic diabetes insipidus independently of the sodium-chloride cotransporter. Am J Physiol Renal Physiol 306(5): F525–F533
Sohara E, Rai T, Miyazaki J, Verkman AS, Sasaki S, Uchida S (2005) Defective water and glycerol transport in the proximal tubules of AQP7 knockout mice. Am J Physiol Renal Physiol 289(6): F1195–F1200
Soria LR, Fanelli E, Altamura N, Svelto M, Marinelli RA, Calamita G (2010) Aquaporin-8-facilitated mitochondrial ammonia transport. Biochem Biophys Res Commun 393(2):217–221
Soupart A, Gross P, Legros JJ, Alfoldi S, Annane D, Heshmati HM, Decaux G (2006) Successful long-term treatment of hyponatremia in syndrome of inappropriate antidiuretic hormone secretion with satavaptan (SR121463B), an orally active nonpeptide vasopressin V2-receptor antagonist. Clin J Am Soc Nephrol 1(6):1154–1160
Stanton BA, Koeppen BM (1998) Solute and water transport along the nephron: tubular function. In: Berne RM, Levy MN (eds) Physiology, 4th edn. Mosby, St. Louis, pp 699–714
Stodkilde L, Norregaard R, Fenton RA, Wang G, Knepper MA, Frokiaer J (2011) Bilateral ureteral obstruction induces early downregulation and redistribution of AQP2 and phosphorylated AQP2. Am J Physiol Renal Physiol 301(1):F226–F235
Strange K, Spring KR (1987) Cell membrane water permeability of rabbit cortical collecting duct. J Membr Biol 96(1):27–43
Sui H, Han BG, Lee JK, Walian P, Jap BK (2001) Structural basis of water-specific transport through the AQP1 water channel. Nature 414(6866):872–878
Tamma G, Robben JH, Trimpert C, Boone M, Deen PM (2011) Regulation of AQP2 localization by S256 and S261 phosphorylation and ubiquitination. Am J Physiol Cell Physiol 300(3):C636–C646
Tamma G, Di MA, Ranieri M, Svelto M, Pisot R, Bilancio G, Cavallo P, De Santo NG, Cirillo M, Valenti G (2014a) A decrease in aquaporin 2 excretion is associated with bed rest induced high calciuria. J Transl Med 12:133
Tamma G, Ranieri M, Di MA, Centrone M, Svelto M, Valenti G (2014b) Glutathionylation of the aquaporin-2 water channel: a novel post-translational modification modulated by the oxidative stress. J Biol Chem 289(40):27807–27813
Tamma G, Di MA, Ranieri M, Geller A, Tamma R, Zallone A, Valenti G (2017) The V2 receptor antagonist tolvaptan raises cytosolic calcium and prevents AQP2 trafficking and function: an in vitro and in vivo assessment. J Cell Mol Med 21(9):1767–1780

Tchekneva EE, Khuchua Z, Davis LS, Kadkina V, Dunn SR, Bachman S, Ishibashi K, Rinchik EM, Harris RC, Dikov MM, Breyer MD (2008) Single amino acid substitution in aquaporin 11 causes renal failure. J Am Soc Nephrol 19(10):1955–1964

Tenenbaum J, Ayoub MA, Perkovska S, Adra-Delenne AL, Mendre C, Ranchin B, Bricca G, Geelen G, Mouillac B, Durroux T, Morin D (2009) The constitutively active V2 receptor mutants conferring NSIAD are weakly sensitive to agonist and antagonist regulation. PLoS One 4(12):e8383

Terris J, Ecelbarger CA, Marples D, Knepper MA, Nielsen S (1995) Distribution of aquaporin-4 water channel expression within rat kidney. Am J Phys 38:F775–F785

Terris J, Ecelbarger CA, Nielsen S, Knepper MA (1996) Long-term regulation of four renal aquaporins in rats. Am J Phys 40:F414–F422

Tfelt-Hansen J, Schwarz P, Brown EM, Chattopadhyay N (2003) The calcium-sensing receptor in human disease. Front Biosci 8:s377–s390

Timmer RT, Sands JM (1999) Lithium intoxication. J Am Soc Nephrol 10(3):666–674

Tingskov SJ, Hu S, Frokiaer J, Kwon TH, Wang W, Norregaard R (2018) Tamoxifen attenuates development of lithium-induced nephrogenic diabetes insipidus in rats. Am J Physiol Renal Physiol 314(5):F1020–F1025

Tiulpakov A, White CW, Abhayawardana RS, See HB, Chan AS, Seeber RM, Heng JI, Dedov I, Pavlos NJ, Pfleger KD (2016) Mutations of vasopressin receptor 2 including novel L312S have differential effects on trafficking. Mol Endocrinol 30(8):889–904

Trepiccione F, Pisitkun T, Hoffert JD, Poulsen SB, Capasso G, Nielsen S, Knepper MA, Fenton RA, Christensen BM (2014) Early targets of lithium in rat kidney inner medullary collecting duct include p38 and ERK1/2. Kidney Int 86(4):757–767

Tsuboi Y, Ishikawa S, Fujisawa G, Okada K, Saito T (1994) Therapeutic efficacy of the non-peptide AVP antagonist OPC-31260 in cirrhotic rats. Kidney Int 46(1):237–244

Tsukaguchi H, Shayakul C, Berger UV, Mackenzie B, Devidas S, Guggino WB, Van Hoek AN, Hediger MA (1998) Molecular characterization of a broad selectivity neutral solute channel. J Biol Chem 273(38):24737–24743

Valenti G, Laera A, Pace G, Aceto G, Lospalluti ML, Penza R, Selvaggi FP, Chiozza ML, Svelto M (2000) Urinary aquaporin 2 and calciuria correlate with the severity of enuresis in children. J Am Soc Nephrol 11(10):1873–1881

Valenti G, Laera A, Gouraud S, Pace G, Aceto G, Penza R, Selvaggi FP, Svelto M (2002) Low-calcium diet in hypercalciuric enuretic children restores AQP2 excretion and improves clinical symptoms. Am J Physiol Renal Physiol 283(5):F895–F903

Valtin H, Schroeder HA, Benirschke K, Sokol HW (1962) Familial hypothalamic diabetes insipidus in rats. Nature 196:1109–1110

Van Balkom BWM, Savelkoul PJ, Markovich D, Hofman E, Nielsen S, van der Sluijs P, PMT D (2002) The role of putative phosphorylation sites in the targeting and shuttling of the aquaporin-2 water channel. J Biol Chem 277(44):41473–41479

Van Hoek AN, Ma T, Yang B, Verkman AS, Brown D (2000) Aquaporin-4 is expressed in basolateral membranes of proximal tubule S3 segments in mouse kidney. Am J Physiol Renal Physiol 278(2):F310–F316

Van Hoek AN, Bouley R, Lu Y, Silberstein C, Brown D, Wax MB, Patil RV (2009) Vasopressin-induced differential stimulation of AQP4 splice variants regulates the in-membrane assembly of orthogonal arrays. Am J Physiol Renal Physiol 296(6):F1396–F1404

Verkman AS (2008) Dissecting the roles of aquaporins in renal pathophysiology using transgenic mice. Semin Nephrol 28(3):217–226

Voisin DL, Bourque CW (2002) Integration of sodium and osmosensory signals in vasopressin neurons. Trends Neurosci 25(4):199–205

Walker RJ, Weggery S, Bedford JJ, McDonald FJ, Ellis G, Leader JP (2005) Lithium-induced reduction in urinary concentrating ability and urinary aquaporin 2 (AQP2) excretion in healthy volunteers. Kidney Int 67(1):291–294

Wang Y, Tajkhorshid E (2010) Nitric oxide conduction by the brain aquaporin AQP4. Proteins 78 (3):661–670
Wang W, Kwon TH, Li C, Frokiaer J, Knepper MA, Nielsen S (2002a) Reduced expression of Na-K-2Cl cotransporter in medullary TAL in vitamin D-induced hypercalcemia in rats. Am J Physiol Renal Physiol 282(1):F34–F44
Wang W, Li C, Kwon TH, Knepper MA, Frokiaer J, Nielsen S (2002b) AQP3, p-AQP2, and AQP2 expression is reduced in polyuric rats with hypercalcemia: prevention by cAMP-PDE inhibitors. Am J Physiol Renal Physiol 283(6):F1313–F1325
Wang W, Li C, Kwon TH, Miller RT, Knepper MA, Frokiaer J, Nielsen S (2004) Reduced expression of renal Na+ transporters in rats with PTH-induced hypercalcemia. Am J Physiol Renal Physiol 286(3):F534–F545
Wang W, Li C, Nejsum LN, Li H, Kim SW, Kwon TH, Jonassen TE, Knepper MA, Thomsen K, Frokiaer J, Nielsen S (2006) Biphasic effects of ANP infusion in conscious, euvolumic rats: roles of AQP2 and ENaC trafficking. Am J Physiol Renal Physiol 290(2):F530–F541
Wang W, Luo R, Lin Y, Wang F, Zheng P, Levi M, Yang T, Li C (2015) Aliskiren restores renal AQP2 expression during unilateral ureteral obstruction by inhibiting the inflammasome. Am J Physiol Renal Physiol 308(8):F910–F922
Weinstock RS, Moses AM (1990) Desmopressin and indomethacin therapy for nephrogenic diabetes insipidus in patients receiving lithium carbonate. South Med J 83(12):1475–1477
Wen H, Froklaer J, Kwon TH, Nielsen S (1999) Urinary excretion of aquaporin-2 in rat is mediated by a vasopressin-dependent apical pathway. J Am Soc Nephrol 10(7):1416–1429
Wildman SS, Boone M, Peppiatt-Wildman CM, Contreras-Sanz A, King BF, Shirley DG, Deen PM, Unwin RJ (2009) Nucleotides downregulate aquaporin 2 via activation of apical P2 receptors. J Am Soc Nephrol 20(7):1480–1490
Willcutts MD, Felner E, White PC (1999) Autosomal recessive familial neurohypophyseal diabetes insipidus with continued secretion of mutant weakly active vasopressin. Hum Mol Genet 8 (7):1303–1307
Wilson DM, Perry HO, Sams WM Jr, Dousa TP (1973) Selective inhibition of human distal tubular function by demeclocycline. Curr Ther Res Clin Exp 15(10):737–740
Wong F, Blei AT, Blendis LM, Thuluvath PJ (2003) A vasopressin receptor antagonist (VPA-985) improves serum sodium concentration in patients with hyponatremia: a multicenter, randomized, placebo-controlled trial. Hepatology 37(1):182–191
Wu H, Chen L, Zhang X, Zhou Q, Li JM, Berger S, Borok Z, Zhou B, Xiao Z, Yin H, Liu M, Wang Y, Jin J, Blackburn MR, Xia Y, Zhang W (2013) Aqp5 is a new transcriptional target of Dot1a and a regulator of Aqp2. PLoS One 8(1):e53342
Wu Q, Moeller HB, Stevens DA, Sanchez-Hodge R, Childers G, Kortenoeven MLA, Cheng L, Rosenbaek LL, Rubel C, Patterson C, Pisitkun T, Schisler JC, Fenton RA (2018) CHIP regulates Aquaporin-2 quality control and body water homeostasis. J Am Soc Nephrol 29(3):936–948
Xu DL, Martin PY, Ohara M, Stjohn J, Pattison T, Meng XZ, Morris K, Kim JK, Schrier RW (1997) Upregulation of aquaporin-2 water channel expression in chronic heart failure rat. J Clin Invest 99:1500–1505
Yakata K, Hiroaki Y, Ishibashi K, Sohara E, Sasaki S, Mitsuoka K, Fujiyoshi Y (2007) Aquaporin-11 containing a divergent NPA motif has normal water channel activity. Biochim Biophys Acta 1768(3):688–693
Yakata K, Tani K, Fujiyoshi Y (2011) Water permeability and characterization of aquaporin-11. J Struct Biol 174(2):315–320
Yang BX, Brown D, Verkman AS (1996) The mercurial insensitive water channel (AQP-4) forms orthogonal arrays in stably transfected Chinese hamster ovary cells. J Biol Chem 271:4577–4580
Yang B, Gillespie A, Carlson EJ, Epstein CJ, Verkman AS (2000) Neonatal mortality in an aquaporin-2 knock-in mouse model of recessive nephrogenic diabetes insipidus. J Biol Chem 276(4):2775–2779

Yang B, Song Y, Zhao D, Verkman AS (2005) Phenotype analysis of aquaporin-8 null mice. Am J Physiol Cell Physiol 288(5):C1161–C1170
Yang B, Zhao D, Qian L, Verkman AS (2006a) Mouse model of inducible nephrogenic diabetes insipidus produced by floxed aquaporin-2 gene deletion. Am J Physiol Renal Physiol 291(2): F465–F472
Yang B, Zhao D, Solenov E, Verkman AS (2006b) Evidence from knockout mice against physiologically significant aquaporin 8-facilitated ammonia transport. Am J Physiol Cell Physiol 291(3):C417–C423
Yang B, Zhao D, Verkman AS (2006c) Evidence against functionally significant aquaporin expression in mitochondria. J Biol Chem 281(24):16202–16206
Yasui M, Zelenin SM, Celsi G, Aperia A (1997) Adenylate cyclase-coupled vasopressin receptor activates AQP2 promoter via a dual effect on CRE and AP1 elements. Am J Phys 41:F443–F450
Yasui M, Hazama A, Kwon TH, Nielsen S, Guggino WB, Agre P (1999a) Rapid gating and anion permeability of an intracellular aquaporin. Nature 402(6758):184–187
Yasui M, Kwon TH, Knepper MA, Nielsen S, Agre P (1999b) Aquaporin-6: an intracellular vesicle water channel protein in renal epithelia. Proc Natl Acad Sci U S A 96(10):5808–5813
Ye L, Li X, Chen Y, Sun H, Wang W, Su T, Jiang L, Cui B, Ning G (2005) Autosomal dominant neurohypophyseal diabetes insipidus with linkage to chromosome 20p13 but without mutations in the AVP-NPII gene. J Clin Endocrinol Metab 90(7):4388–4393
Yip KP (2006) Epac-mediated Ca^{2+} mobilization and exocytosis in inner medullary collecting duct. Am J Physiol Renal Physiol 291(4):F882–F890
Zelenina M, Christensen BM, Palmer J, Nairn AC, Nielsen S, Aperia A (2000) Prostaglandin E (2) interaction with AVP: effects on AQP2 phosphorylation and distribution. Am J Physiol Renal Physiol 278(3):F388–F394
Zhai XY, Fenton RA, Andreasen A, Thomsen JS, Christensen EI (2007) Aquaporin-1 is not expressed in descending thin limbs of short-loop nephrons. J Am Soc Nephrol 18 (11):2937–2944
Zhang Y, Peti-Peterdi J, Muller CE, Carlson NG, Baqi Y, Strasburg DL, Heiney KM, Villanueva K, Kohan DE, Kishore BK (2015) P2Y12 receptor localizes in the renal collecting duct and its blockade augments arginine vasopressin action and alleviates Nephrogenic diabetes Insipidus. J Am Soc Nephrol 26(12):2978–2987
Zhang Y, Peti-Peterdi J, Brandes AU, Riquier-Brison A, Carlson NG, Muller CE, Ecelbarger CM, Kishore BK (2017) Prasugrel suppresses development of lithium-induced nephrogenic diabetes insipidus in mice. Purinergic Signal 13(2):239–248
Zhang Y, Hansson KM, Liu T, Magnell K, Huang Y, Carlson NG, Kishore BK (2019) Genetic deletion of ADP-activated P2Y12 receptor ameliorates lithium-induced nephrogenic diabetes insipidus in mice. Acta Physiol (Oxf) 225(2):e13191